HANDBOOK OF LATIN AMERICAN STUDIES: No. 55

A Selective and Annotated Guide to Recent Publications
in Anthropology, Economics, Geography, Government and Politics,
International Relations, Sociology, and Electronic Resources

VOLUME 56 WILL BE DEVOTED TO THE HUMANITIES:
ART, HISTORY, LITERATURE, MUSIC, AND PHILOSOPHY

EDITORIAL NOTE: Comments concerning the *Handbook of Latin American Studies* should be sent directly to the Editor, *Handbook of Latin American Studies*, Hispanic Division, Library of Congress, Washington, D.C. 20540.

ADVISORY BOARD

David Bushnell, *University of Florida*
Lambros Comitas, *Columbia University*
Frank N. Dauster, *Rutgers University, New Brunswick*
Roberto González Echevarría, *Yale University*
Peter T. Johnson, *Princeton University*
Betty J. Meggers, *Smithsonian Institution*
John V. Murra, *Cornell University*
Robert Potash, *University of Massachusetts*

ADMINISTRATIVE OFFICERS
LIBRARY OF CONGRESS

James H. Billington, *The Librarian of Congress*
Winston Tabb, *Associate Librarian for Library Services*
Carolyn T. Brown, *Assistant Librarian for Library Services*
Georgette M. Dorn, *Chief, Hispanic Division*

REPRESENTATIVE, UNIVERSITY OF TEXAS PRESS

Joanna Hitchcock, *Director*

HANDBOOK EDITORIAL STAFF

Katherine D. McCann, *Assistant to the Editor*
Ann R. Mulrane, *Editorial Assistant*
Amy S. Puryear, *Editorial Assistant*
Tracy North, *HLAS Webmaster*

HANDBOOK OF LATIN AMERICAN STUDIES: NO. 55
SOCIAL SCIENCES

*Prepared by a Number of Scholars
for the Hispanic Division of The Library of Congress*

DOLORES MOYANO MARTIN, *Editor*
P. SUE MUNDELL, *Assistant Editor*

1997

UNIVERSITY OF TEXAS PRESS   *Austin*

International Standard Book Number 0-292-75211-3
International Standard Serial Number 0072-9833
Library of Congress Catalog Card Number 36-32633
Copyright © 1997 by the University of Texas Press
All rights reserved
Printed in the United States of America

Requests for permission to reproduce material
from this work should be sent to
Permissions, University of Texas Press,
Box 7819, Austin, Texas 78713-7819.

First Edition, 1997

The paper used in this publication meets the minimum requirements of American National Standard for Information Sciences—Permanence of Paper for Printed Library Materials, ANSI Z39.48-1984. ∞

CONTRIBUTING EDITORS

SOCIAL SCIENCES

Juan M. del Aguila, *Emory University*, GOVERNMENT AND POLITICS
Benigno Aguirre-López, *Texas A&M University*, SOCIOLOGY
Amalia Alberti, *Independent Consultant, San Salvador*, SOCIOLOGY
G. Pope Atkins, *University of Texas at Austin*, INTERNATIONAL RELATIONS
Melissa H. Birch, *University of Virginia*, ECONOMICS
Eduardo Borensztein, *International Monetary Fund*, ECONOMICS
Jacqueline Braveboy-Wagner, *The City College-CUNY*, INTERNATIONAL RELATIONS
Roderic A. Camp, *Tulane University*, GOVERNMENT AND POLITICS
William L. Canak, *Middle Tennessee State University*, SOCIOLOGY
César Caviedes, *University of Florida*, GEOGRAPHY
Marc Chernick, *Georgetown University*, GOVERNMENT AND POLITICS
Harold Colson, *University of California, San Diego*, ELECTRONIC RESOURCES
Lambros Comitas, *Columbia University*, ANTHROPOLOGY
David W. Dent, *Towson State University*, GOVERNMENT AND POLITICS
Clinton R. Edwards, *University of Wisconsin-Milwaukee*, GEOGRAPHY
Gary S. Elbow, *Texas Tech University*, GEOGRAPHY
Malva Espinosa, *Ministerio del Trabajo, Santiago*, SOCIOLOGY
Gary Feinman, *University of Wisconsin-Madison*, ANTHROPOLOGY
Damián Fernández, *Florida International University*, INTERNATIONAL RELATIONS
Michael Fleet, *Marquette University*, GOVERNMENT AND POLITICS
Cornelia Butler Flora, *Iowa State University*, SOCIOLOGY
Jan L. Flora, *Iowa State University*, SOCIOLOGY
James W. Foley, *University of Miami*, ECONOMICS
Jeffrey Franks, *International Monetary Fund*, ECONOMICS
Daniel W. Gade, *University of Vermont*, GEOGRAPHY
Eduardo Gamarra, *Florida International University*, GOVERNMENT & POLITICS
José Zebedeo García, *New Mexico State University*, GOVERNMENT AND POLITICS
Manuel Antonio Garretón, *Universidad de Chile, Santiago*, SOCIOLOGY
Benjamín F. Hadis, *Montclair State University*, SOCIOLOGY
Kevin Healy, *Inter-American Foundation*, SOCIOLOGY
John Henderson, *Cornell University*, ANTHROPOLOGY
Jonathan Hill, *Southern Illinois University*, ANTHROPOLOGY
William Keegan, *Florida Museum of Natural History*, ANTHROPOLOGY
Roberto Patricio Korzeniewicz, *University of Maryland*, SOCIOLOGY
Danilo Levi, *Southeastern Louisiana University*, SOCIOLOGY
Paul Lewis, *Tulane University*, GOVERNMENT AND POLITICS
Robert E. Looney, *Naval Postgraduate School*, ECONOMICS
Peggy Lovell, *University of Pittsburgh*, SOCIOLOGY

Anthony Maingot, *Florida International University*, GOVERNMENT AND POLITICS
Markos J. Mamalakis, *University of Wisconsin-Milwaukee*, ECONOMICS
Tom L. Martinson, *Auburn University*, GEOGRAPHY
Nohra Rey de Marulanda, *Inter-American Development Bank*, ECONOMICS
Betty J. Meggers, *Smithsonian Institution*, ANTHROPOLOGY
Keith D. Muller, *Kent State University*, GEOGRAPHY
Robert Palacios, *The World Bank*, ECONOMICS
David Scott Palmer, *Boston University*, INTERNATIONAL RELATIONS
Ransford W. Palmer, *Howard University*, ECONOMICS
Jorge Pérez-López, *US Department of Labor*, ECONOMICS
Timothy J. Power, *Louisiana State University*, GOVERNMENT & POLITICS
Catalina Rabinovich, *Independent Consultant, Chevy Chase, Maryland*, ECONOMICS
Joanne Rappaport, *Georgetown University*, ANTHROPOLOGY
René Salgado, *Consultant, Inter-American Development Bank*, GOVERNMENT AND POLITICS
Claudio Sapelli, *The World Bank*, ECONOMICS
David W. Schodt, *St. Olaf's College*, ECONOMICS
Russell E. Smith, *Washburn University*, ECONOMICS
Dale Story, *University of Texas at Arlington*, INTERNATIONAL RELATIONS
Paul Sullivan, *Independent Consultant, Port Jefferson, New York*, ANTHROPOLOGY
Scott D. Tollefson, *Naval Postgraduate School*, INTERNATIONAL RELATIONS
Antonio Ugalde, *University of Texas at Austin*, SOCIOLOGY
Aldo C. Vacs, *Skidmore College*, INTERNATIONAL RELATIONS
Clarence Zuvekas, Jr., *Consulting Economist, Annandale, Virginia*, ECONOMICS

HUMANITIES

Edna Acosta-Belén, *State University of New York-Albany*, LITERATURE
Maureen Ahern, *The Ohio State University*, TRANSLATIONS
Severino João Albuquerque, *University of Wisconsin-Madison*, LITERATURE
Félix Angel, *Inter-American Development Bank*, ART
Uva de Aragón, *Florida International University*, LITERATURE
Barbara von Barghahn, *George Washington University*, ART
María Luisa Bastos, *Lehman College-CUNY*, LITERATURE
Alvaro Félix Bolaños, *Tulane University*, LITERATURE
Dain Borges, *University of California-San Diego*, HISTORY
John Britton, *Frances Marion University*, HISTORY
Francisco Cabanillas, *Bowling Green State University*, LITERATURE
Sara Castro-Klarén, *The Johns Hopkins University*, LITERATURE
Don M. Coerver, *Texas Christian University*, HISTORY
Edith B. Couturier, *National Endowment for the Humanities*, HISTORY
Edward Cox, *Rice University*, HISTORY
Joseph T. Criscenti, *Professor Emeritus, Boston College*, HISTORY
César Ferreira, *University of North Texas*, LITERATURE
Francisco Fonseca, *Princeton University*, ELECTRONIC RESOURCES
José Manuel García García, *State University of New Mexico*, LITERATURE
Magdalena García Pinto, *University of Missouri, Columbia*, LITERATURE

John Garrigus, *Jacksonville University*, HISTORY
Miguel Gomes, *University of Connecticut*, LITERATURE
Lance Grahn, *Marquette University*, HISTORY
María Cristina Guiñazú, *Lehman College-CUNY*, LITERATURE
Michael T. Hamerly, *University of Guam*, HISTORY
Robert Haskett, *University of Oregon*, HISTORY
José M. Hernández, *Professor Emeritus, Georgetown University*, HISTORY
Rosemarijn Hoefte, *Royal Institute of Linguistics and Anthropology, The Netherlands*, HISTORY
Joel Horowitz, *Saint Bonaventure University*, HISTORY
Regina Igel, *University of Maryland*, LITERATURE
Nils P. Jacobsen, *University of Illinois*, HISTORY
Peter T. Johnson, *Princeton University*, ELECTRONIC RESOURCES
Erick Langer, *Carnegie Mellon University*, HISTORY
Pedro Lastra, *State University of New York at Stony Brook*, LITERATURE
Asunción Lavrin, *Arizona State University*, HISTORY
Peter Linder, *University of Maine at Machias*, HISTORY
Maria Angélica Guimarães Lopes, *University of South Carolina*, LITERATURE
Carol Maier, *Kent State University*, TRANSLATIONS
Teresita Martínez-Vergne, *Macalester College*, HISTORY
David McCreery, *Georgia State University*, HISTORY
Joan Meznar, *Westmont College*, HISTORY
Molly Molloy, *New Mexico State University, Las Cruces, New Mexico*, ELECTRONIC RESOURCES
Elizabeth Monasterios, *State University of New York-Stony Brook*, LITERATURE
Naomi Hoki Moniz, *Georgetown University*, LITERATURE
José M. Neistein, *Brazilian-American Cultural Institute, Washington*, ART
José Miguel Oviedo, *University of Pennsylvania*, LITERATURE
Suzanne Pasztor, *Randolph-Macon College*, HISTORY
Daphne Patai, *University of Massachusetts-Amherst*, TRANSLATIONS
Anne Pérotin-Dumon, *Pontificia Universidad Católica de Chile*, HISTORY
Charles Perrone, *University of Florida*, LITERATURE
René Prieto, *Southern Methodist University*, LITERATURE
José Promis, *University of Arizona*, LITERATURE
Inés Quintero, *Universidad Central de Venezuela*, HISTORY
Susan Ramírez, *DePaul University*, HISTORY
Jane M. Rausch, *University of Massachusetts-Amherst*, HISTORY
Oscar Rivera-Rodas, *University of Tennessee, Knoxville*, LITERATURE
Humberto Rodríguez-Camilloni, *Virginia Polytechnic Institute*, ART
Mario A. Rojas, *Catholic University of America*, LITERATURE
Kathleen Ross, *Duke University*, TRANSLATIONS
William F. Sater, *California State University, Long Beach*, HISTORY
Jacobo Sefamí, *University of California-Irvine*, LITERATURE
Susan M. Socolow, *Emory University*, HISTORY
Robert Stevenson, *University of California, Los Angeles*, MUSIC
Barbara A. Tenenbaum, *Hispanic Division, The Library of Congress*, HISTORY
Juan Carlos Torchia Estrada, *Consultant, Hispanic Division, Library of Congress, Washington, D.C.*, PHILOSOPHY
Lilián Uribe, *Central Connecticut State University*, LITERATURE

Stephen Webre, *Louisiana Tech University*, HISTORY
Raymond Williams, *University of Colorado*, LITERATURE
Stephanie Wood, *University of Oregon*, HISTORY

Foreign Corresponding Editors

Teodoro Hampe-Martínez, *Universidad Pontificia Católica de Lima*, COLONIAL HISTORY MATERIALS IN GERMAN AND FRENCH LANGUAGES
Kotaro Horisaka, *Sophia University, Tokyo*, JAPANESE LANGUAGE
Mao Xianglin, *Chinese Academy of Social Sciences*, CHINESE LANGUAGE
Magnus Mörner, *Stockholm University, Sweden*, SCANDINAVIAN LANGUAGES
Inge Schjellerup, *Nationalmuseet, Denmark*, DANISH LANGUAGE

Special Contributing Editors

Marie Louise Bernal, *Library of Congress*, SCANDINAVIAN LANGUAGES
Christel Krause Converse, *Independent Consultant, College Park, Maryland*, GERMAN LANGUAGE
Barbara Dash, *Library of Congress*, RUSSIAN LANGUAGE
Georgette M. Dorn, *Library of Congress*, GERMAN AND HUNGARIAN LANGUAGES
Zbigniew Kantorosinski, *Library of Congress*, POLISH LANGUAGE
Vincent C. Peloso, *Howard University*, ITALIAN LANGUAGE
Juan Manuel Pérez, *Library of Congress*, GALICIAN LANGUAGE
Iêda Siqueira Wiarda, *Library of Congress*, SPECIAL MATERIAL IN PORTUGUESE LANGUAGE
Hasso von Winning, *Southwest Museum, Los Angeles*, GERMAN-LANGUAGE MATERIAL ON MESOAMERICAN ARCHAEOLOGY

CONTENTS

		PAGE
EDITOR'S NOTE		xv

ELECTRONIC RESOURCES　　　　　　*Harold Colson*　　　1

    Directories p. 6
    Surveys and Reviews p. 6
    Databases and Systems p. 7

    Journal Abbreviations: Electronic Resources　　　11

ANTHROPOLOGY

GENERAL　　　　　　　　　　　　　　　　　　　13

ARCHAEOLOGY

Mesoamerica　　　　　　　　*John S. Henderson and Gary Feinman*　　15

    General p. 16　Fieldwork and Artifacts p. 30　Native Sources and Epigraphy p. 55

Caribbean Area　　　　　　　　*William Keegan*　　61
    Lower Central America p. 62
    Caribbean Islands p. 64

South America　　　　　　　　*Betty J. Meggers*　　70
    General p. 71　Argentina p. 73
    Bolivia p. 75　Brazil p. 76
    Chile p. 78　Colombia p. 80
    Ecuador p. 82　The Guianas p. 84
    Peru p. 84　Uruguay p. 90
    Venezuela p. 90

ETHNOLOGY

Middle America　　　　　　　　*Paul Sullivan*　　91
West Indies　　　　　　　　　　*Lambros Comitas*　98
South America: Lowlands　　　　*Jonathan D. Hill*　110
    General p. 113　Brazil p. 115

        Colombia, Venezuela, and The
        Guianas p. 119   Peru and
        Ecuador p. 122   Paraguay,
        Argentina, and Bolivia p. 125

South America: Highlands	Joanne Rappaport	126

        General p. 128   Argentina p. 129
        Bolivia p. 130   Chile p. 133
        Colombia p. 134   Ecuador p. 137
        Peru p. 140

Journal Abbreviations: Anthropology	145

ECONOMICS

GENERAL	James W. Foley	151
MEXICO	Robert E. Looney	180
CENTRAL AMERICA	Clarence Zuvekas, Jr.	213

        General p. 215   Belize p. 219
        Costa Rica p. 219   El Salvador p. 221
        Guatemala p. 223   Honduras p. 224
        Nicaragua p. 225   Panama p. 227

THE CARIBBEAN AND THE GUIANAS (except Cuba and Puerto Rico)	Ransford W. Palmer	228

        General p. 228   Dominican
        Republic p. 231   Commonwealth
        Caribbean and Guyana p. 231
        French Caribbean and French
        Guiana p. 234

CUBA	Jorge F. Pérez-López	234
VENEZUELA	Robert Palacios	238
COLOMBIA	Nohra Rey de Marulanda	244
ECUADOR	David W. Schodt	249
CHILE	Markos Mamalakis	257
PERU	Catalina Rabinovich	265
BOLIVIA	Jeffrey Franks	272
PARAGUAY AND URUGUAY	Claudio Sapelli	277

        Paraguay p. 278   Uruguay p. 279

ARGENTINA	Eduardo Borensztein	283
BRAZIL	Melissa H. Birch and Russell E. Smith	290
Journal Abbreviations: Economics		306

GEOGRAPHY

GENERAL	*Clinton R. Edwards*	313
MIDDLE AMERICA	*Gary S. Elbow and Tom L. Martinson*	322

The Caribbean and The Guianas   323
Central America   326
    General p. 326   Belize p. 326
    Costa Rica p. 326
    El Salvador p. 328
    Guatemala p. 328   Honduras p. 329
    Nicaragua p. 329   Panama p. 329
Mexico   329

WESTERN SOUTH AMERICA *General p. 337* *Venezuela p. 339* *Colombia p. 340* *Ecuador p. 341 Peru p. 344* *Bolivia p. 347*	*Daniel W. Gade*	337
THE SOUTHERN CONE *General p. 349 Argentina p. 350* *Chile p. 353 Uruguay p. 358*	*César Caviedes*	349
BRAZIL	*Keith D. Muller*	359

Journal Abbreviations: Geography   369

GOVERNMENT AND POLITICS

GENERAL	*David W. Dent*	373
MEXICO	*Roderic A. Camp*	387
CENTRAL AMERICA *General p. 405 Belize p. 406* *Costa Rica p. 406* *El Salvador p. 407* *Guatemala p. 409* *Honduras p. 410* *Nicaragua p. 411* *Panama p. 413*	*José Z. García*	405
THE CARIBBEAN AND THE GUIANAS (except Cuba) *General p. 416* *Dominican Republic p. 418* *Haiti p. 420 Jamaica p. 420*	*Anthony Maingot*	415

    Lesser Antilles p. 421
    (British Commonwealth p. 421
    Dutch p. 421   French p. 422)
    Puerto Rico p. 422
    Suriname p. 423   Trinidad and
    Tobago p. 424

CUBA   *Juan del Aguila*   425

COLOMBIA AND ECUADOR   *Marc Chernick*   431
    Colombia p. 434
    Ecuador p. 443

VENEZUELA   *René Salgado*   448

BOLIVIA   *Eduardo Gamarra*   454

PERU   *David Scott Palmer*   467

CHILE   *Michael Fleet*   475

ARGENTINA, PARAGUAY, AND URUGUAY   *Paul H. Lewis*   491
    Argentina p. 493
    Paraguay p. 502   Uruguay p. 505

BRAZIL   *Timothy J. Power*   509

Journal Abbreviations: Government and Politics   525

INTERNATIONAL RELATIONS

GENERAL   *G. Pope Atkins*   531

MEXICO AND CENTRAL AMERICA   *Dale Story*   544
    Mexico p. 546   Central
    America p. 553

THE CARIBBEAN AND THE GUIANAS   *Damián Fernández and
Jacqueline Anne Braveboy-Wagner*   563
    General p. 565
    Hispanic Caribbean p. 568
    (Cuba p. 568   Dominican
    Republic p. 571   Puerto
    Rico p. 571)
    Haiti p. 572   Other
    Caribbean Islands p. 572
    The Guianas p. 574

SOUTH AMERICA (except Brazil)   *Aldo C. Vacs*   574
    General p. 576
    Argentina p. 581   Bolivia p. 586
    Chile p. 588   Colombia p. 592
    Ecuador p. 594   Paraguay p. 595
    Peru p. 596   Uruguay p. 600
    Venezuela p. 601

BRAZIL	*Scott Tollefson*	604
Journal Abbreviations: International Relations		610

SOCIOLOGY

GENERAL		615
MEXICO	*Antonio Ugalde*	624
CENTRAL AMERICA	*Jan L. Flora*	641
THE CARIBBEAN AND THE GUIANAS	*Benigno E. Aguirre-López*	660
COLOMBIA AND VENEZUELA	*William L. Canak and Danilo Levi*	678
Colombia p. 679		
Venezuela p. 682		
ECUADOR	*Amalia M. Alberti*	686
PERU	*Cornelia Flora*	689
BOLIVIA AND PARAGUAY	*Kevin J. Healy*	701
Bolivia p. 701		
Paraguay p. 706		
CHILE	*Manuel Antonio Garretón and Malva Espinosa*	707
ARGENTINA AND URUGUAY	*Roberto Patricio Korzeniewicz and Benjamín Hadis*	719
Argentina p. 722		
Uruguay p. 728		
BRAZIL	*Peggy A. Lovell*	729
Journal Abbreviations: Sociology		738

INDEXES

ABBREVIATIONS AND ACRONYMS	743
TITLE LIST OF JOURNALS INDEXED	753
ABBREVIATION LIST OF JOURNALS INDEXED	769
SUBJECT INDEX	787
AUTHOR INDEX	857

EDITOR'S NOTE

I. GENERAL AND REGIONAL TRENDS

ONE OF THE MORE INTERESTING TRENDS in the volume, as specifically noted by a sociologist but evident in many of the disciplines, is the increasing use of research to understand differences rather than form generalizations. There are now more empirical studies and fewer "reflections" on ideological theories or grandiose topics. These recent studies often employ creative combinations of qualitative and quantitative methodologies. Moreover, there is much less reductionism in viewing class struggle (p. 690). In Mesoamerican archaeology, on the other hand, emphases still "vary widely, from the largely material to the purely ideological" (p. 15).

A curious pattern, perhaps a result of increasing regional and economic integration, is that problems once thought to underscore one particular country's distinctiveness and exceptionalism have spread throughout the region. For example, drug-trafficking, organized crime, and social violence, problems long associated with Colombia, in recent years have proliferated throughout Mexico, Central America, and Brazil. The pioneering research on Colombia's difficulties serves scholars well as a model for studies on other Latin American nations (p. 432). Paulo Sérgio Pinheiro's contribution on the unacknowledged practice of extralegal police violence in Brazil is a most relevant work in light of recent tragedies there (item **3753**).

Also welcome are further studies of the "new social actors" who began to be examined in the last three social science volumes and who now finally command center stage. Among others, these groups comprise entrepreneurs, women's organizations, children and youth, rural and urban community organizations, local and municipal governments, and organized ethnic groups (p. 690).

This shift in focus from the large to the small and from the theoretical to the empirical is readily apparent in a surge of microeconomic studies, many of which use modern techniques of empirical analysis (p. 277). Or, as noted by another economist, "twenty-five years ago, much of the literature on Latin American economic problems was little more than lengthy, often emotional, diatribes intended to discredit opposing views, while supporting the author's own biases and ideology." In contrast, economic issues addressed in works annotated in this volume, "are debated using sophisticated mathematical and statistical techniques. Any resulting controversy is as likely to be concerned with the methodology employed as with the conclusions reached" (p. 152). Meanwhile, the formerly all-consuming interest in the debt crisis appears to be abating. Nevertheless, a study published in Argentina by Rudiger Dornbusch and Juan Carlos de Pablo deserves mention as it goes well beyond an examination of the debt crisis to serve as an excellent treatise on Argentina's macroeconomic issues (item **2060** and p. 283).

Another leading topic is the structural economic reform taking place throughout Latin America in varying degrees of intensity. These structural reforms have had a powerful impact on income distribution, employment, and growth (p. 151). Argentina leads the way in the breadth and depth of these economic changes, reform being the principal distinguishing element of that nation's economic policy in the 1990s (p. 284). Peru also instituted a major reform. The country achieved a successful turnaround by mid-decade, a truly remarkable accomplishment in the opinion of our political scientist who notes that Peru's inflation, which peaked in 1990 at the historically unprecedented rate of 7,650 percent, by 1994 had dropped to 15 percent. Moreover, "five successive years of negative net economic growth between 1988–92 were followed in 1993–94 by net positive growth, in 1994 the highest in the hemisphere (about four percent in 1993 and over nine percent in 1994)" (p. 467).

While structural economic reforms help calm raging inflation rates and improve economic growth rates, such reforms may have devasting social costs for both the middle and the marginalized classes. As one political scientist observed, the reform of Latin America's "antiquated State-dominated economic systems" is inevitable: "civilian politicians have discovered that there is no going back to the old populist economics." Nevertheless, the "bitter medicine of neoliberalism is causing widespread social discontent" (p. 491). A vivid example of the latter occurred in Venezuela where the introduction of neoliberal policies and their dreaded social consequences led to widespread protests (pp. 448–449).

Much of the social science literature examined in this volume considers the social and political consequences of structural reforms. Karen Remmer has written one of the best empirical examinations of the relationship between economics and political stability, a study that "breaks new ground in its interpretation of liberal democracy and regime change" (item **2793** and p. 373). A significant number of studies examine the relationship between the State and civil society, that is to say, between the government and the mass of the population, especially its poorer citizens, an increasingly problematic relationship in view of the persistent economic crises and adjustment policies (items **4838** and **4844** and p. 678). Or, according to another sociologist, these policies of adjustment have helped cause the "fragmentation of collective social action," a process that is debilitating the associative vitality of countries such as Peru (item **4911**). Moreover, the decline of the Peruvian State is especially evident in its "increasing inability to meet its traditional obligations to its citizens" (p. 689).

As noted above, this has been arguably an "unprecedented stage in Argentine economic history," involving the deepest structural reforms in its history, especially in areas such as international trade and payments, financial markets, and privatization of State enterprises (p. 283). The discontent engendered by such deep economic reforms in Argentina over the past several years has resulted in considerable research on citizen response to these painful adjustments. The response involves a fairly new constellation of forces, i.e., nongovernmental organizations, local communities, and State agencies that provide assistance at the local level. The most interesting findings of this research are emerging "from evaluations of the social and organizational dynamics taking place between and among the different agents involved in [these new types of] development projects" (item **5126** and p. 720).

Another subject that is attracting much interest is trade reform. The lowering of Latin American trade barriers and the formation of regional and extraregional trade blocks is perhaps the most visible outcome of a notable decline in Latin America's traditional economic nationalism and long-term prejudice against and resistance to free trade. Even in Ecuador, after years of protectionist policies, the nation has expanded its participation in regional free trade agreements (p. 250). The heightened levels of trade reform have been accompanied by increased scholarly attention to the topic; in Brazil, the literature on trade reform has temporarily displaced the previous focus on macroeconomic stabilization programs (p. 291). According to our sociologist for Mexico, the NAFTA Agreement is one of the most important events of recent years (p. 625); the preeminent concern of much of the current research on Mexican international relations also lies in the economic arena (p. 544). Publications on international economic relations now "indicate a long-term interest in redefining the direction and contents of trade and investment links as an essential component of the ongoing processes of economic opening and liberalization" (pp. 575–576). The North American Free Trade Agreement or NAFTA, signed in 1992, the South American Mercosur or Mercosul Treaty, established in 1991, and the possible future establishment of a Western Hemisphere Free Trade Area all point to the primary importance of such agreements for regional integration (p. 152). Finally, some of the most interesting publications concern "the new links between South America and Asia's newly industrializing countries" and between South America and the world's developed nations (p. 575).

Unfortunately, we are witnessing not only the expansion and success of free trade but also the more sinister expansion and success of international criminal cartels from Mexico and the Caribbean to South America. The political role of these cartels and the corruption they spawn are yet to be seriously analyzed, especially in the Caribbean. "What, for instance," asks one contributor, "is the political role of the vast amounts of money flowing through the Caribbean and seeking security in one of the many regional offshore tax havens?" (p. 416).

The corrupting profitability and destructive influence of the illegal drug trade continue to command much research attention. Among the more notable works on the topic are: Ivelaw Griffith's study of narcotics in the Caribbean (item **4148**); Adrián Bonilla's excellent and original book on drug trafficking in Ecuador (item **3214** and p. 434); Francisco Thoumi's "several publications that greatly enrich our understanding of the economics and politics of drug-trafficking in Colombia" (items **3198, 3199**, and **3200** and p. 433); and Mario De Franco and Ricardo Godoy's article on cocaine production in Bolivia (item **1976** and p. 273).

Traditional political corruption continues to stalk all Latin American countries, with the possible exception of Chile.[1] The most egregious example in recent years was the 1992 scandal involving Brazilian President Fernando Collor de Mello who "lost his bid to become the second civilian in Brazilian history to be democratically elected and serve a full term as president." For the most sophisticated analysis to date of Collor's Administration and eventual impeachment, see the study by Kurt Weyland (item **3779**). In his analysis of corruption during the Sarney Administration, José Carlos Bruzzi Castello reminds us that for Brazil corruption is not unique to the Collor Administration (item **3687** and p. 511). Another contributor looks to Argentina, emphasizing the "gross corruption permeating Menem's government" (items **3513** and **3585** and p. 492). And Francisco Leal and Andrés Dávila have written an outstanding study of clientelism in Colombia (item **3178**), which reveals how important the issue remains even after the tumultuous events of the 1980s and 1990s (p. 434).

"Building democracy in an era of controversial but unavoidable economic reform" is a very difficult endeavor (p. 492), primarily because of the complexity of political transformation and democratization, subjects that have dominated political science research on Latin America since *HLAS 51*. So far, "the struggle to understand political change in Latin America in the aftermath of the Cold War has not produced a consensus on the meaning and consequences of current democratic trends in the region" (p. 373). But in Brazil, the state of "permanent crisis" in which the nation has lived "has had the positive effect of sharpening the analytical rigor of Brazilian and Brazilianist political science. The sense of urgency has made the literature more diagnostic and politically relevant than ever before, as evinced by the burgeoning research on democratic institutions" (p. 510).

One of the more fascinating aspects of this gradual evolution towards democracy in Latin American countries has been the emergence of new political movements. Even in repressive Cuba, "a subculture of political opposition is emerging ... that fosters the activities of social movement organizations and quasi-organizations" (p. 661). These organizations growing throughout Latin America have challenged the traditional two-party rivalry in countries such as Uruguay and, especially, Paraguay. Benjamín Arditi describes the triumph of 'Asunción Para Todos' in the 1991 municipal elections in Paraguay's capital (item **3591**). Two years later, the "Encuentro Nacional" transformed Paraguay into a three-party system. For the story of the 1993 election, see Marcial A. Riquelme's work (item **3611**), "the best available description of the contemporary Paraguayan scene" (p. 493). In Bolivia, where there have been three national and five municipal elections since 1985, much attention is being paid to political parties and elections. On the national level, the most representative work can be found in Roberto Laserna's piece on the 1989 elections (item **3315** and p. 455). As in Bolivia, mayoral and gubernatorial elections in Venezuela are also fresh topics receiving rigorous treatment and much can be learned from works by José Vicente Carrasquero and Friedrich Welsch (item **3255**), Humberto Njaim (item **3270**), José Enrique Molina Vega (item **3267**), and Miriam Kornblith (item **3265** and p. 450). In Mexico, the volume of literature on elections and election data is overwhelming, but several works stand out for the quality of the data and analysis. This is especially true of the work by Guadalupe Pacheco, who explores "the relationship between urbanization, voting patterns, and political culture in the Federal District" (items **2900–2902** and p. 388). In one of the best contributions to the study of party underdevelopment, Scott Mainwaring argues that Brazil represents one of the most severe cases of this phenomenon (item **3729**). And, in a related paper, the same author "stresses the infelicitous combination of presidentialism and party fragmentation" that afflicts the country (item **3730** and p. 510).

Political studies about the Caribbean are less satisfactory. Most works on this area are characterized by their presentism, insularity, and the absence of all-inclusive or ontological paradigms. Moreover, "comparative studies of political cultures and the dynamics resulting from different social structures, or political studies of greater historical depth, continue to be few" (p. 415). Despite the Haitian crisis and the attention it commanded during these last few years, the partisan nature of much of the literature on this country is such that "students would do well to follow the sage advice of studying the historian before you study his or her history" (p. 416). For the Commonwealth Caribbean, exceptions to the above are found in works by the British Society for Caribbean Studies as exemplified by Paul Sutton's *Politics in the Commonwealth Caribbean* (item **3073**). Other publications that should be commended as comparative analyses of the region are Colin Clarke's *Society and politics in the Caribbean* (item **3043**) and Anthony Payne and Paul Sutton's *The contours of modern Caribbean politics* (item **3036**). And as usual, in terms of quantity if not of quality, for foreign policy studies Cuba continues to lead as "the most studied country in the region and perhaps in Latin America as a whole" (p. 564).

Two additional topics, the experience of "openness" and the role of the media, continue to attract much attention in this volume. A seminal study of the role of television and electoral politics in the process of democratization is Thomas Skidmore's *Television, politics and the transition to democracy in Latin America* (item **2803** and p. 374). For Mexico, where the role of the media "has been largely ignored in the 1980s," there is now a collection entitled *Así se calló el sistema: comunicación y elecciones en 1988*, edited by Pablo Arredondo Ramírez (item **2820**). This work is "a pathbreaking contribution, including excellent case studies of several urban centers" (p. 388). In Venezuela, despite the severe economic crisis afflicting the country, "three decades of uninterrupted democratic life have consolidated a national trend towards openness, one that may well be irreversible" (p. 450).

Parallel to the emergence of new social movements mentioned above, there is the continuing "rise of the indigenous movement as a major actor in national politics" (p. 432). Researchers are now focusing on symbolic, cultural, and ethnic aspects of these social movements (item **4922**), concentrating less on how they relate to the State and political parties, and more on how these movements connect with and relate to other social groups (item **4929** and pp. 689–690). In Ecuador, for example, "twice in the early 1990s, the indigenous movement organized nationwide protests that led to the reordering of the nation's political agenda to include indigenous concerns related to land, language, and national identity." For good analyses of this phenomenon, see Melina Selverston's study (item **3242** and p. 432). The emergence of indigenous social movements has coincided with the appearance of indigenous ethnographers and the creation of distinct Latin American anthropologies (p. 128). And in these new works there is more emphasis on ethnic and gender identity and less on class struggle. In addition, more and more works address the "internal tensions of movement formation and the need to understand differences among groups, thus departing from broad generalizations" (p. 490). For instance, in his well-researched and highly insightful comparison of Arawakan and Tukanoan religious movements of the 19th century, Robin Wright demonstrates "that such processes of cultural resistance follow distinct historical trajectories, depending upon underlying contrasts in indigenous political organization and patterns of interethnic relations with Western societies" (item **889** and p. 111). An ethnologist praises David Stoll's very timely work on the Ixil Triangle, an important study that "challenges many of our discipline's often taken-for-granted assertions about ethnicity, class, and political struggle in the Mesoamerican countryside" (item **765** and p. 91). The most sophisticated studies by native Andeans have been conducted, for the most part, in Bolivia, where the return of university-educated investigators to their communities has resulted in the production of in-depth diagnostics of local problems and the search for previously unheard voices (pp. 127–128). The most significant works are those sponsored by the Taller de Historia Oral Andina (e.g., items **946, 948, 954,** and **963**).

Black studies or research on Afro-Latin American populations is receiving less attention in this *Handbook* than in previous ones. One exception is found among the Andean nations, where the news of "greatest significance is the entry of Afro-Colombians into politics with their historic participation in the creation and implementation of a new multiethnic constitution." One notable contribution to the study of ethnicity and power in that country is Peter Wade's examination of racial attitudes, stereotypes, and metaphors in Colombia (item **1005** and p. 127). In Brazil, research on Brazilians of African descent has benefited from the work of sociologists studying how "gender identities and inequalities are shaped by class and race" (p. 729). Three excellent contributions on the changing dynamics of race relations are those by Thomas Skidmore (item **5202**), Howard Winant (item **5211**) and Nelson do Valle Silva and Carlos Alfredo Hasenbalg (item **5200**).

The literature reviewed in this *Handbook* includes a "fundamental reassessment of the nature of ethnic identity" (p. 91). The "enigma of ethnicity" is being addressed by more and more scholars such as Sampath on the creolisation of East Indian adolescent masculinity (item **829**), Daniel A. Segal on race and color in pre-independent Trinidad (item **834**), and Steven Vertovec on Hindu Trinidad (item **845**). In fact, "a great deal of the contemporary research on the Caribbean seems to be focused on questions related to ethnicity and identity" (p. 98). The same applies to Mesoamerica for which, emphasizes one ethnologist, the Chiapas rebellion of 1994 highlights the growing importance of ethnicity in Mesoamerican studies (p. 91).

Women as a focus of research continue to attract as much attention as in previous volumes of the *Handbook*. In Bolivia and Paraguay, for example, there is a plethora of studies on gender and women's topics. The new research attests to "an increase in the number of female researchers as well as of female-headed NGOs working with women's groups and analyzing the results of action programs" (items **4966**, **4983**, and **4978**). Moreover, some of this research is pathbreaking, such as the first comprehensive study of domestic workers in a Bolivian city (item **4959** and p. 701). For Peru, there is also an interesting study of domestic workers (item **4915**) as well as numerous studies of women and topics such as female-led organizations in the slums (items **4887** and **4921**), single mothers (item **4942**), fertility (item **4925**), and abortion (item **4886**). There are studies of prostitution in Paraguay and Uruguay (items **4991** and **5139**), a work on sexual violators in Paraguay (item **4993**), and an excellent study of female workers in the Uruguayan fishing industry (item **5135**). As mentioned above, in Brazil sociologists of gender have contributed a number of studies on the roles that class and race play in shaping gender identities and inequalities (p. 729). About Chile there is a pioneering historical study on female employment, *Siete décadas de registro del trabajo femenino, 1854–1920*, by Thelma Gálvez Pérez and Rosa Bravo Barja (item **1879** and p. 258). In Ecuador, María Mercedes Placencia and Eliana Franco use survey data to analyze women in the labor market during the 1980s, another topic that has received little attention to date (item **1826** and p. 250). Women are also included in a study of the changing socioeconomic context of Ecuadorian rural areas (item **4875** and p. 686). And finally, there is a most interesting study on a neglected topic: aging Ecuadorian women, older than 60 years (item **4871**). In Guyana, an interesting work examines the problem of violence against married women (item **4715** and p. 663); in Peru, there is a 10-year empirical study of family violence in Lima (item **4927**); in Argentina the proceedings of an interdisciplinary conference are devoted to violence against women (item **5096**) and a compelling monograph by Graciela Ferreira examines abused women and violent men (item **5099**).

Closely related to works on women is the emerging new field of children's studies, of which the most interesting contribution in this volume is Víctor Manuel Rodríguez's examination of the social and cultural adaptations abandoned street children (or *palomos*) use to survive in Santo Domingo (item **4775** and p. 662). There are also rigorous studies of Peruvian child and youth labor in Lima (items **4879** and **4892**) and a thorough analysis of the impact of poverty on Argentine children (item **5108**).

Violence continues as a dominant field in Colombia, the first country to develop *violentólogos* or "specialists in violence," as noted in previous "Editor's Notes" to the *Handbook*. Examples of the recent and notable Colombian contributions to this topic are: an English-language translation of Alonso Salazar's *No nacimos pa' semilla* (item **3195**); the Comisión Andina de Juristas' *Putumayo* (item **3193**); and Alejo Vargas Velásquez's *Colonización y conflicto armado: Magdalena Medio Santandereano* (item **3207** and p. 433).

Finally, the environment, a vitally important topic that crosses disciplinary and geographic boundaries, continues to dominate this volume as it did previous editions of the *Handbook*. This is especially true in the case of both the Amazon region and Brazil, and in disciplines such as economics, geography, and sociology. The trend is less evident in anthropology. For example, ecological studies about ethnology in the South American Lowlands have been in decline for nearly a decade and it is doubtful that they will ever again attain the prominence such works had in Amazonian ethnology during the 1970s and early 1980s (p. 112). The relation of economic growth to the environment—a topic virtually ignored until recently in works about the economics of Latin America as a whole—is now attracting increasing attention, as exemplified by a major volume by the Economic Commission on Latin America concerning *Planificación y gestión del desarrollo en áreas de expansión de la frontera agropecuaria en América Latina* (item **1123**) and proceedings from a conference entitled *Seminario Latinoamericano sobre Medio Ambiente y Desarrollo* (item **1270**). For Ecuador, Douglas DeWitt Southgate and Morris Whitaker show in their excellent study that "the process of environmental degradation is both broader and more longstanding than recent conflicts [among indigenous groups, the government, and foreign petroleum companies] would indicate" (item **1838** and p. 250). In Latin American geography as a whole, the environment is by far the topic most often treated (p. 313). For Mexico, Ezequiel Ezcurra's small historical monograph serves as a most important contribution by focusing on environmental problems in the Valley of Mexico (item **2377**). Costa Rica "emerges as a clear leader in ecological and environmental issues on Middle America this biennium." The number and quality of the nation's contributions "mirror Costa Rica's commitment to its own environment" (p. 322), a commitment we hope will be embraced more wholeheartedly by the rest of the region's nations.

II. CLOSING DATE FOR VOLUME 55

The closing date for works annotated in this volume was early 1995. Publications received and cataloged at the Library of Congress after that date will be annotated in the next social sciences volume, *HLAS 57*.

III. ELECTRONIC ACCESS TO THE *HANDBOOK*

The *HLAS* staff is proud to announce the availability of World Wide Web access to the *Handbook of Latin American Studies*. As of December 1996 all bibliographic records corresponding to *HLAS* volumes 1–58 are available at no charge to those with graphical browsers such as Netscape, WebExplorer, etc., as well as those using the non-graphical browser LYNX. The Uniform Resource Locator or URL for *HLAS Online* is **http://lcweb2.loc.gov/hlas/**. The project was generously supported by the family of Lewis U. Hanke (the *Handbook's* late founder), Plumsock Mesoamerican Studies, and the Institute for Civil Society. The electronic conversion of volumes 1–49 and the publication of the first 53 volumes on CD-ROM[2] was made possible by the Fundación MAPFRE América (now the Fundación Histórica Tavera) and its excellent, energetic team, with additional financial assistance from The Andrew W. Mellon Foundation.

The data from Volumes 50 and above were drawn from the computer system of the Library of Congress, a system designed by the Library's Information Technology Services (ITS). We would like to thank ITS for its earlier role in automating the *Handbook* and for its current assistance in mounting *HLAS Online* on the World Wide Web. In particular, we are grateful to Mary Ambrosio, Stan Lerner, and Dean Wilder for their unending patience, good humor, and esprit de corps. The Hispanic Division component of the *Handbook's* Web design team also deserves much credit for conceiving, designing, testing, and translating *HLAS Online*: Tracy North, Randy M. Wells, Miguel Valladares, María Lidia Buompadre, Katherine D. McCann, Victoria Funes, and William "Niko" Trentini.

IV. CHANGES IN VOLUME

Electronic Resources

A new chapter, Electronic Resources, inaugurated in *HLAS 54*, will now be included in each volume. This constantly changing and rapidly advancing field calls for a separate chapter written by a specialist in this area. Harold Colson of the University of California, San Diego, prepared this chapter for both *HLAS 54* and the present volume.

Bibliography and General Works

At the *Handbook* Advisory Board meeting of November, 1992, it was decided that Bibliography and General Works no longer should be covered as a separate chapter; rather, subject bibliographies and other similar works should be annotated in the relevant *HLAS* chapter by the appropriate contributor. It was also pointed out that the Seminar on the Acquisition of Latin American Library Materials (SALALM) issues an annual bibliographies publication. Thus, beginning with *HLAS 54*, the Bibliography and General Works chapter has been replaced by the above-mentioned Electronic Resources chapter.

Anthropology

Gary Feinman of the University of Wisconsin-Madison collaborated with John Henderson on preparation of the Mesoamerica chapter. William Keegan of the Florida Museum of Natural History annotated materials on the Caribbean.

Economics

Nohra Rey de Marulanda, Inter-American Development Bank, covered the literature on Colombia. The Peru chapter was prepared by Catalina Rabinovich; the Bolivia chapter by Jeffrey Franks of the International Monetary Fund. Paraguay and Uruguay were covered by Claudio Sapelli of The World Bank, and Argentina by Eduardo Borensztein, also of The World Bank. Melissa Birch, University of Virginia, and Russell Smith, Washburn University, assumed full responsibility for the chapter on Brazil.

Government and Politics

David Dent, Towson State University, annotated all general materials; René Salgado, Inter-American Development Bank, focused on Venezuela. Publications on Colombia and Ecuador were reviewed by Marc Chernick of Johns Hopkins University, and those on Bolivia by Eduardo Gamarra of Florida International University. The Brazil chapter was prepared by Timothy Power of Louisiana State University.

International Relations

Jacqueline Braveboy-Wagner of the City University of New York collaborated with Damián Fernández on the chapter focusing on the Caribbean and the Guianas.

Sociology

Benigno E. Aguirre-López of Texas A&M University completed the Caribbean and Guianas chapter. Publications on Peru were canvassed by Cornelia Flora of Iowa State University; Amalia Alberti focused exclusively on Ecuador materials. Roberto Korzeniewicz collaborated with Benjamín Hadis on the chapter covering Argentina and Uruguay.

Foreign Corresponding Editors

Inge Schjellerup of the Nationalmuseet of Denmark is our new corresponding editor for Danish-language materials.

Special Contributing Editors

Zbigniew Kantorosinski of the Library of Congress has agreed to canvass materials on Latin America in Polish.

Subject Index

The *Handbook* uses Library of Congress Subject Headings (LCSH) when they are consistent with usage among Latin Americanists. Differences in practice, however, make adaptation of LCSH headings necessary: 1) the *Handbook* index uses only two levels, while LC headings usually contain more; and 2) *Handbook* practice is to prefer a "subject-place" pattern, while LC practice generally uses a "place-subject" pattern. Automation of the *Handbook* has required that the subject index be compiled with two audiences in mind: users of the print edition and users of the on-line database. It also has demanded that index terms, once established, remain as stable as possible. Work has begun towards a complete thesaurus of *Handbook* subject index terms. In the meantime, cross references are included from subject terms used in *HLAS 48-HLAS 54* to new terms used in this volume. (A list of the subject headings used in indexing *HLAS 50-HLAS 58* is available from the *Handbook's* World Wide Web home page at **http://lcweb2.loc.gov/hlas/**.) Finally, since the *Handbook* is arranged by discipline, readers are encouraged to consult the table of contents for broad subject coverage.

V. ACKNOWLEDGMENTS

We would especially like to thank Claudia McNellis for her devotion to the *Handbook* over the past eight years. Not only was Claudia instrumental in guiding the *Handbook's* early automation, but she also played a leading role in its implementation. We are grateful to her for her willingness to serve in an extracurricular capacity as the *Handbook's* primary technical consultant across several very different stages of automation. In particular, she provided us with calming reassurances that we were ready and capable of launching into the realm of CD-ROM and World Wide Web publishing, brainstorming late into the night to discuss the best way to accomplish it.

We also are grateful to Terry Peet, Fehl Cannon, Basil Malish, Russell Marr, and the rest of the staff of the Library's Hispanic Acquisitions Section for ensuring the timely arrival of Latin American materials and working with the *Handbook's* contributing editors to fill the gaps in the Library's Latin American and Caribbean collection.

Finally, we are grateful to Library of Congress managers Winston Tabb, Carolyn T. Brown, and especially Hispanic Division Chief Georgette Dorn for their continuing support of the *Handbook's* print and electronic publications, especially during these past several years of budgetary constraints.

Dolores Moyano Martin, Editor
P. Sue Mundell, Assistant Editor

ENDNOTES

1. See, for example, "A Global Gauge of Greased Palms" in *The New York Times* (August 20, 1995, p. 3), in which Chile is listed as less corrupt than the US, France, and Japan, not to mention the rest of Latin America. The survey was compiled by Institute for Management Development (Lausanne), Political and Economic Risk Consultancy (Hong Kong), and Business International (New York).

2. The *HLAS/CD* is distributed by The University of Texas Press, P.O. Box 7819, Austin, TX, 78713.

ELECTRONIC RESOURCES

HAROLD COLSON, *Head of Public Services and Latin American Librarian, International Relations and Pacific Studies Library, University of California, San Diego*

THE ELECTRONIC SECTOR CONTINUES the promising and exhilarating transformations examined in *HLAS 54*, with news resources from and about Latin America posting especially dramatic gains. Despite the unfortunate demise of one premier Latin Americanist database, social science offerings are unquestionably stronger than they were just two years ago, due to numerous major products newly available on compact disc or via the Internet. Gladly, an increasing proportion of the electronic marketplace comes from Latin America itself, and present trends augur well for even better coverage of hemispheric affairs in the near term. At the same time, contrary to rising public expectations associated with the ongoing Internet boom, most of the jewels of the database sector remain outside the free portions of the worldwide net, although each day finds evermore sparkling bounty therein.

Numerous general and discipline-based files for bibliographic research in the social sciences were treated in the prior volume, and as a whole these databases still constitute the largest and most accessible corpus of electronic indexing and cataloging for the Latin Americanist researcher. Newcomers to computerized bibliographic research should consult *HLAS 54* (p. 3–22) and the earlier database survey by Colson and Stern (see *HLAS 54:10*) for descriptions of baseline products such as *WorldCat, Dissertation Abstracts Online, UNCOVER, Article1st, Sociological Abstracts, Social SciSearch, PAIS International, IntlEc, Economic Literature Index, Anthropological Literature,* and *HAPI Online.* Recent arrivals of note include the *International Bibliography of the Social Sciences* or *IBSS* (item **28**), with coverage of worldwide journal literature dating back to 1981, and *International Political Science Abstracts* (item **29**). The online *IPSA* fills a longstanding need for a comprehensive political science electronic resource, providing global coverage of articles drawn from over 2,000 journals, many of which are not treated by other databases. Moreover, its searchable article abstracts give *IPSA* a considerable topic-retrieval advantage over the various competing indexes that rely principally on title words as access points.

Both *IBSS* and *IPSA* are welcome additions to the bibliographic CD-ROM scene, but an even more powerful portable resource for Latin Americanists is *HLAS/CD* (item **26**), the long-awaited electronic version of the complete *Handbook of Latin American Studies,* vols. 1–53 (1936–1994). With some 250,000 searchable annotated citations drawn from books and journals spanning the social sciences and the humanities, *HLAS/CD* instantly ranks as a preeminent resource for bibliographic research in Latin American history and literature, outpacing *Historical Abstracts* and *MLA International Bibliography* in coverage intensity and retrospective depth. At the same time, it offers strong and often unique content in such social science fields as anthropology, economics, political science, and sociology, as well as in humanities disciplines such as art, music, and philosophy (in addition to the aforementioned strengths in history and literature). Standard indexing and abstracting databases in the social sciences tend to have light coverage of materials from Latin America, so *HAPI Online* (journal articles; see *HLAS 54: 46*) and now *HLAS/CD* (journal articles, books, and chapters) are two robust supplements on the market with much otherwise missing content. Considering its attractive, navigable search interface, its affordable list price ($150), and its unparalleled portfolio of key scholarship, *HLAS/CD* is the outstanding new compact disc product for 1995, a resource that should serve with distinction in academic libraries and faculty offices worldwide.

In the prior chapter on electronic resources (see *HLAS 54*, p. 6–9), the prospective *HLAS CD-ROM* product was characterized as the fourth of the Latin Americanist database "tigers," ready to join the established *INFO-SOUTH, Latin America Data Base (LADB)*, and *HAPI Online* files as a central resource for our field. As it turns out, there are now only three active database tigers, for less than three months after *HLAS/CD* was unveiled at a crowded LASA reception in the Library of Congress, the pioneering *INFO-SOUTH* current affairs database ceased updates. Despite its unchallenged role as the online source for article abstracts drawn from Latin American newspapers and news magazines, *INFO-SOUTH* never generated enough revenue during its six-year run as a commercial database to become wholly self-sufficient, and eventually fell victim to declining federal subsidies to its parent North-South Center.[1] At the time of its final update in Dec. 1995, *INFO-SOUTH* held approximately 94,000 newspaper, magazine, and journal entries dating back to 1988, a nonpareil cyber-chronicle of Latin American news literature offering coverage of diverse topics from cartels and contras through privatization, NAFTA, Mercosur, Zedillo, and beyond. Research conducted at subscribing government, academic, and corporate sites across the nation will now be marginally poorer for its loss, at what eventual cost to hemispheric understanding one can only speculate. The only positive note to this story is that negotiations are currently underway at the Univ. of Miami's North-South Center to allow use of the retrospective information already collected in the database.

The announcement of the *INFO-SOUTH* project seven years ago set the stage for the present boom in online Latin Americana, but the lack of future additions to this pacesetting bibliographic news file does not dim the scene completely. Enough timely full text sources emerged in 1995 to sustain and enhance the current affairs coverage formerly provided by the Miami product. Indeed, within a week of the North-South Center's decision to halt *INFO-SOUTH* operations, the federal government announced *World News Connection* (item **50**), an Internet World Wide Web (WWW) service with long-awaited online access to the complete texts of the worldwide *Daily Reports* from the US Foreign Broadcast Information Service (FBIS), key current affairs publications which for years have languished in paper- and microfiche-only format outside the federal intelligence community.[2] Like many key information commodities, the *World News Connection* is not free, but the reasonable subscription fees offer access to a searchable database of timely translations drawn from major press and broadcast sources across Latin America. With only a 48- to 72-hour delay from in-country radio, television, wire, or newspaper transmission to translated full-text articles on the Washington-area Internet site, *World News Connection* operates in a far more timely manner than did *INFO-SOUTH*. The service offers complete texts online rather than just abstracts, features that should position it to flourish in both the corporate and academic markets. Certain newswire files and newspaper homepages offer more rapid information from the region, but these scattered same-day sources lack the hemispheric breadth and multi-source scope of the powerful FBIS product.

The debut of *World News Connection* in Dec. 1995 was only one of many significant full-text news arrivals that year. Indeed, 1995 could be dubbed the year of the electronic newspaper, as many dailies from across Latin America inaugurated Internet editions, and two Mexican papers, *El Norte* and *Reforma*, became the first from the region to go online with major North American commercial database systems. *La Jornada* from Mexico started the newspaper boom in Feb. 1995 with a Web site sporting fresh loads of entire articles and even graphics, giving global Internet users access to much of the content enjoyed that day by Mexico City readers. For US research library denizens accustomed to long lags in the receipt of even airmailed papers from Latin America, the sight of *La Jornada*'s current front page and more on the computer screen was a delightful, almost stunning breakthrough.

Dozens of additional papers and news publications from Mexico, Costa Rica, Brazil, Argentina, Chile, Peru, and elsewhere across the hemisphere produced similar Web news sites during the year, giving wired Latin Americanists everywhere a strong and growing body of "hot" information and commentary direct from in-country press sources. Indeed, with searchable back issues and other research features now appearing on some sites, the overall power of free Latin American news on the Web is already quite compelling, and can rightfully be characterized as revolutionary considering the paucity of such online text resources just a few years ago. Although fee-based, commercial database systems still offer mammoth searchable libraries of full-text news information about Latin America, they have been slow to incorporate non-English-language newspapers, wires, magazines, and such from the region, and may be in danger of losing much emerging local information to postings on new Web sites. In the summer of 1995, the US-based vendors DataTimes and Dialog made a breakthrough in Latin American coverage by adding *Reforma* and *El Norte* to their massive repertoires of online publications. It remains to be seen if this welcome albeit overdue Spanish-language lead will fizzle out or spark a Latin American race in the commercial database sector. Certainly, the commercial sources provide better word-searching capability and greater historical depth (back to Apr. 18, 1995 on both systems) than the typical newspaper Web site, but the majority of wired Mexicanists should find the (presently) free Internet versions of the same titles more than adequate for casual use.

Not all full-text products are suitable or intended for access via the Internet, of course, and the CD-ROM format continued to garner some of the most notable and powerful database releases from Latin America. The Centro Nacional Editor de Discos Compactos at the Univ. de Colima (Mexico) enhanced its leading role as a CD-ROM producer with full-text output spanning journals and magazines, legislation, international treaties, and a biographical dictionary of the Mexican government. Long runs of *Comercio Exterior* (item **19**), *Investigación Económica* (item **30**), and *Revista Este País* (item **41**) are now fully portable and searchable thanks to the prolific Colima operation. In addition, CENEDIC's two-disc collection of treaty texts, *Tratados Internacionales Celebrados por México* (item **46**), makes an instant international law collection for many libraries. The important Mexican news magazine *Proceso* offers a free Web site with recent article texts, but its companion backfiles database on CD-ROM (item **40**) is the true research treasure, with over 13,000 complete articles drawn from the more than 300 issues published during the Salinas *sexenio* (1988–94). The forthcoming Colima CD-ROM of the Mexican magazine *Nexos* will cover the Salinas era and much more, reaching back with full texts from 1978–

94. Long resident online via LEXIS-NEXIS, *Latin American Weekly Report* and its kindred regional and specialty newsletters debuted on CD-ROM in 1995 (item **34**), giving users access to thousands of complete articles dating from Jan. 1990–June 1995. Like many other commercial publishers, the Latin American Newsletters firm maintains a Web site offering free access to only a few current articles, keeping the full complement of its electronic materials available via fee-based subscriptions.

Macroeconomic indicators, import-export figures, census enumerations, and other statistics underpin much current social science scholarship on Latin America. The database sector now offers several key resources for the numerically inclined investigator, although present data offerings lag behind the bibliographic and full-text markets in overall coverage breadth and access convenience. Many of the most cited statistical files are maintained by international organizations, but a number of rich data products have emerged from Latin America on CD-ROM or via Internet connection. The World Bank's new *World Data* compact disc (item **48**) stands out as an excellent value, providing some 700 annual time series dating back to the early 1960s for 200 countries on national accounts, balance of payments, trade, external debt and finance, social development, and natural resources, all for only $275. Likewise, the *World Trade Database* CD-ROM (item **51**) of United Nations data on global commodity-by-country trade flows (1980–93) is available from the Canadian government's statistical agency, Statistics Canada, at an affordably discounted price for academic institutions. (A free Internet source for much of this UN data is the *International Trade Information System* or *It-Is*, now available from the Center for the Study of Western Hemispheric Trade via The University of Texas at Austin's Latin American Network Information Center or UT-LANIC.) The *WTDB* is especially welcome because it is one of the most accessible sources for disaggregated trade flows not involving the US or the Organisation for Economic Co-operation and Development (OECD) countries. That is to say, there are many print and electronic products with worldwide or Latin American country import-export totals, and US and OECD sources add country-by-commodity breakdowns for Latin American trade partners, but before the production of the *WTDB* it was often difficult or expensive to obtain reasonably current statistics detailing, for example, Chile's trade with Thailand by sector. Similarly, the *Wistat* or *Women's Indicators and Statistics Database* CD-ROM (item **47**) from the United Nations provides a rich database of numbers for the study of women in Latin America and elsewhere, with over 1,600 country series covering population, education, households and fertility, health, economic activities, and more. Another new global resource with much useful Latin American content is *World Marketing Data and Statistics* on CD-ROM (item **49**), which contains over 450 tables with annual series (some dating back to 1977) covering demographic, economic, trade, consumer, business, and infrastructure indicators for 207 countries. Much of this market-related data can be difficult to find in standard government statistical products, although the popular and powerful *National Trade Data Bank* (see HLAS 54: 60) from the US Dept. of Commerce remains a great place to start. An excellent CD-ROM database for identifying Latin Americanist data and analyses reported—but often overlooked—in thousands of international government publications is *Statistical Masterfile* (item **44**), covering yearbooks, bulletins, special reports, working papers, journals, and books issued by the World Bank, the United Nations, the International Monetary Fund, the World Trade Organization, and dozens of other agencies.

Of special note to Mexicanists are the superior print and electronic statistical products of the OECD, which started reporting current and historical data for Mexico following its accession to membership in 1994. Respected OECD titles such as *Main Economic Indicators, Quarterly National Accounts Statistics, Monthly Statistics of Foreign Trade,* and *OECD Environmental Data* are now among the most accessible and thorough sources for many Mexican data series. Mexico itself remains a leader in census databases, with major INEGI releases on CD-ROM covering recent national agricultural, ejidal, and economic enumerations. Colima added to the bounty with a compact disc version of *La Economía Mexicana en Cifras* (item 22). NAFTA and the peso bailout helped bring an extensive collection of official Mexican economic and financial statistics to Internet users on Wall Street and beyond, courtesy of the free Web site maintained by Mexico's Consulate General in New York City. Basic country statistics from the Inter-American Development Bank, the Central Intelligence Agency, the US Agency for International Development, national governments, and other sources are also readily available on the Internet, although the timeliest and most comprehensive electronic sources of numerical data on Latin America are still generally reserved as revenue sources by their producers and allied redistributors who sell information commodities on tape, diskette, CD-ROM, or via online services rather than posting them freely on the networked "commons."

The Internet sector in early 1996 is substantially richer, faster, more accessible, and better organized for Latin Americanist work than it was at the close of the debut *HLAS* chapter on electronic resources some 24 months ago. Two years is practically an entire generation in the accelerating Internet realm, and the intervening time has brought a bevy of improved sites, browsers, search tools, and publications for the intrepid Web navigator, who can easily locate and retrieve free electronic treasures ranging from daily newspapers to country statistics to scholarly publications. The flagship UT-LANIC server (see *HLAS 54:73*) remains the best single point of entry for Latin Americanist forays across the Internet, offering convenient and well-maintained selections of new and established resource links organized by country and by subject area. Of course, the entire global Internet features a myriad of operational sites with Latin America-related content, and new Web homepages from universities, government agencies, political parties, research centers and institutes, publishers, opposition groups, and companies appear every week. The UT-LANIC helps organize these burgeoning resources with a point-and-click hierarchy, but a number of powerful new Internet search tools offer almost instant keyword retrieval from millions of texts across the globe. Lycos, Infoseek, Alta Vista, HotBot, Open Text Index, and other search engines can locate farflung and hitherto obscure electronic documents mentioning one or more specified terms, in much the same way DataTimes or LEXIS/NEXIS can bore through mountains of proprietary full-text newspapers, magazines, newsletters, and wires. Naturally, retrieval success will vary depending on the engine and the topic, but the collective power of the assorted Internet search tools is already quite compelling, and constitutes a major advance in unearthing the buried riches of the open Internet. Still, whether one approaches the Internet through an ordered gateway like UT-LANIC or a global sieve like Infoseek, it should be kept in mind that many of the core bibliographic, full-text, and statistical databases for Latin Americanist work are either protected by subscription-based password controls or sold offline on magnetic or optical storage media.

Notes

1 Ironically, North-South Center administrators decided to close the database just a few weeks after their own publicity newsletter ran an article headlined "INFO-SOUTH Becomes

Landmark on Information Superhighway." Press accounts of the INFO-SOUTH shutdown are available in *El Nuevo Herald* (Miami), Dec. 13, 1995, p. 1A, and *The Miami Herald*, Dec. 14, 1995, p. 2B.

2 Actually, the FBIS translations did enjoy a tantalizing albeit unauthorized run as a public database a couple of years ago, when an enterprising purveyor of online news converted portions of the print dailies to electronic form and briefly remarketed them on some major systems as *International Intelligence Reports*.

DIRECTORIES

1 ***Directory of United Nations Information Sources.*** 1994– . 5th edition. New York: Advisory Committee for the Co-ordination of Information Systems (ACCIS), United Nations.

Inventories over 1,100 information services, databases, and software packages produced by the UN, its specialized agencies, and related international organizations. More detailed and comprehensive than "outside" database directories listing UN resources.

2 **World databases in social sciences.**
Edited by C.J. Armstrong & R.R. Fenton. London; New Jersey: Bowker-Saur, 1996. 793 p.: indexes. (World databases series)

One of the first titles in a projected 23-volume series designed to cover every database distributed worldwide in any electronic format. This specialist-level directory thoroughly describes, compares, and evaluates files on demography, social services, women, children, urban studies, and employment, providing more detail than any other database guide. Later volumes in the series will treat global databases in management (including economics), government and politics (including international relations and defense), industry, environment, education, news, law, and company information.

SURVEYS AND REVIEWS

3 **Alonso Gamboa, Octavio** and **Rafael Reyna Espinosa.** Latin American databases: an analysis in the social sciences and humanities. (*Online CDROM Rev.*, 19:5, Oct. 1995, p. 247–254, bibl., graphs, tables)

Tabulates 476 databases from Latin America according to country of origin, producer, subject, access format, size, time span, geographical coverage, and other variables. Finds that most databases were produced for internal rather than commercial use and thus are little known and used outside local circles. Paper was first presented at the International Congress of Americanists (48th, Stockholm/Uppsala, Sweden, July 1994).

4 **Bates, Mary Ellen.** Online deskbook: Online magazine's essential desk reference for online and Internet searchers. Edited and with a foreword by Reva Basch. Wilton, CT: Pemberton Press, 1996. 256 p.: ill., index.

A great handbook for beginning and experienced searchers alike profiles consumer services (America Online, CompuServe, Prodigy, Microsoft Network), commercial vendors (DataStar, DataTimes, Dialog, LEXIS/NEXIS, Dow Jones, NewsNet), and the Internet/World Wide Web. To a greater or lesser degree, these systems offer Latin Americanist content, although with completely different search interfaces and at varying costs. Bates masterfully covers search basics, system strengths and weaknesses, types of information and key resources available, unique files, pricing, and more.

5 **Bogarín Navarro, Rodrigo.** Descubra el mundo de Internet. Cartago, Costa Rica: Editorial Tecnológica de Costa Rica, 1994. 171 p.: ill.

Perhaps the first book from Latin America to guide readers across the full spectrum of Internet functions as they existed before the Web explosion. Illustrated with many on-screen examples, it covers e-mail, file transfer, telnet, gopher, archie, newsgroups, WAIS, and early WWW. Now somewhat dated, but still a worthy Spanish-language introduction to Internet basics.

6 **Constance, Paul.** Latin America's info on-ramp. (*U.S./Latin Trade*, 2:9, Sept. 1994, p. 26–28, ill.)

An early profile of database, Internet, and e-mail use as aids to hemispheric commerce. Touches on progress of Internet hook-ups in Latin America, recounts some trader-oriented benefits of electronic bulletin boards and other resources, and interviews an executive with Innovative Telematics, producers of the ITINET online service.

7 Mader, Ron. Net gains: how to cut to the quick on Internet. *(Bus. Mex., 5:6, June 1995, p. 30–31, ill.)*

Focuses on news sources and listservs related to Mexico.

8 Norvell, Scott. Accessing @ mexico. com. *(Mex. Bus., 1:6, March 1995, p. 12–14, ill.)*

Surveys assortment of basic Mexico-related electronic sites and services of interest to traders and investors. Most of these resources are also treated in item **9.**

9 Norvell, Scott. Latin America on-line. *(U.S./Latin Trade, 3:7, July 1995, p. 74–78, ill.)*

An upbeat presentation of representative commercial and Internet resources, geared toward readers interested in doing business with Latin America. Covers consumer services such as CompuServe and America Online, mailing lists, newsgroups, WWW sites (UT-LANIC, NTDB), and commercial databases (including INFO-SOUTH, Dialog, LADB, NewsNet, Dow Jones, Reuters, Bloomberg, and DataTimes). To his credit, Norvell alerts readers that the most comprehensive news and information databases are generally fee-based and expensive, but it is hardly the case yet that the online "godsend" renders "trips to the library a thing of the past."

10 Oka, Christine K. *Latin American Studies: Volumes I and II.* (*CD-ROM Prof., 7:5, Sept./Oct. 1994, p. 149–152*)

Librarian's brief review of 2-disc collection of core Latin Americanist databases (see *HLAS 54:55*). Generally positive evaluation of the power and friendliness of ROMWright, the NISC-DISC search interface.

11 O'Leary, Mick. Online 100. Wilton, CT: Pemberton Press, 1995. 233 p.: index.

A respected and prolific database observer gives his always insightful and practical evaluations of 100 key multidisciplinary and subject files, which were selected by a panel of experts as the "most significant" (rather than, say, the most heavily used) products in the industry. Many fall into categories of general and business news, current events, law and government, social sciences and humanities, or general reference/multidisciplinary, and thus offer electronic support for Latin Americanist work.

12 Pagell, Ruth A. Electronic 'country' publications: electronic resources for information on specific countries. *(Database/Weston, 18:2, April 1995, p. 40–51, ill., map, tables)*

Expert review of various full-text products with economic, business, and political coverage by country. Online and CD-ROM resources from Economist Intelligence Unit (EIU), Political Risk Services, and other producers offer Latin American country profiles, analysis, news, statistics, and forecasts.

13 Resnick, Rosalind. Olé: Latin America's net presence is growing. *(Internet World, 6:4, April 1995, p. 86–90, ill.)*

Light survey of a dozen or so basic Internet sites on Latin America, drawn heavily from the work of Molly Molloy (see *HLAS 54:20*).

DATABASES AND SYSTEMS

14 *Anuario Estadístico de la República Argentina.* 1995– . Instituto Nacional de Estadística y Censos. Buenos Aires.

One of the first Argentine compact disc products contains over 650 tables of national and provincial data, maps, images, and video clips.

15 *Atlas de México.* Aguascalientes, Mexico: Instituto Nacional de Estadística, Geografía e Informática (INEGI), 1994. 1 computer laser optical disc.

Multimedia CD-ROM combines maps, texts, sound, photos, and video clips to present a geographical and statistical profile of Mexico. Includes coverage of the physical environment, society, infrastructure, productive sectors, and the states.

16 Censo nacional de población y vivienda. Buenos Aires: Instituto Nacional de Estadística y Censos, 1995. 1 computer laser optical disc.

Results from the 1991 census in Argentina provide basic population, housing, social, and urban indicators.

17   **Censos económicos 1994.** Aguascalientes, Mexico: Instituto Nacional de Estadística, Geografía e Informática (INEGI), 1994. 1 computer laser optical disc.

Collection of advance data from the Mexican economic census offers 94 series on productive activity in 243 urban localities. Includes census cartography.

18   **CIMA.** Aguascalientes, Mexico: Instituto Nacional de Estadística, Geografía e Informática (INEGI), 1994. 2 computer laser optical discs.

A compilation of statistics and cartography on Mexican municipalities presents 368 data series drawn from recent economic, agricultural, and general census enumerations.

19   *Comercio Exterior: 1973–1993.* Colima, Mexico: Centro Nacional Editor de Discos Compactos, Univ. de Colima; México: Bancomex, 1994? 1 computer laser optical disc.

A searchable full-text CD-ROM edition of an important Mexican economic publication, with articles, documents, statistics, and bibliography. The 3,663 articles treat a wide range of domestic and economic themes, including rural affairs, population and employment, debt and adjustment, economic integration, globalization, industrial development, and technology transfer.

20   **Cuenca del Pacífico.** Colima, Mexico: Centro Nacional Editor de Discos Compactos, Univ. de Colima; Red Nacional de Investigaciones sobre la Cuenca del Pacífico, 1994. 1 computer laser optical disc.

CD-ROM contains a bibliographic register of some 10,000 articles from Mexican and international journals on Pacific Rim affairs, statistical indicators for Mexico and Pacific nations, a directory of relevant Mexican researchers, and scanned images of articles from two Mexican journals on the Pacific.

21   **Diccionario biográfico del gobierno mexicano.** Colima, Mexico: Centro Nacional Editor de Discos Compactos, Univ. de Colima; México: Unidad de la Crónica Presidencial, Dirección General de Comunicación Social, Presidencia de la República, 1994. 1 computer laser optical disc.

Profiles 2,551 individuals from the executive, legislative, and judicial sectors. As a database, it offers many retrieval options not available through the counterpart print editions.

22   **La economía mexicana en cifras.** Colima, Mexico: Centro Nacional Editor de Discos Compactos, Univ. de Colima; México: Nacional Financiera, 1995. 1 computer laser optical disc: tables.

A key compendium of statistics covers evolution of the Mexican economy during the past decade or so. Includes many tables on the domestic and external sectors, along with some international comparisons.

23   **Estadísticas históricas de México.** Aguascalientes, Mexico: Instituto Nacional de Estadística, Geografía e Informática (INEGI), 1994. 1 computer laser optical disc.

A refreshingly inexpensive archive of key Mexican statistics on CD-ROM from the end of the colonial period to the 1990s. Coverage is heavy on demographic, social, and economic indicators. Also available in a 2-volume print edition.

24   **Estructura y dinámica poblacional.** Aguascalientes, Mexico: Instituto Nacional de Estadística, Geografía e Informática (INEGI), 1994. 1 computer laser optical disc.

A specialty CD-ROM with results from the 1992 Encuesta Nacional de la Dinámica Demográfica (ENADID) covers various sociodemographic data.

25   **GEMA: geomodelos de altimetría del territorio nacional.** Aguascalientes, Mexico: Instituto Nacional de Estadística, Geografía e Informática (INEGI), 1994. 1 computer laser optical disc.

CD-ROM contains 255 digital elevations of Mexico at 1:250,000 scale.

26   *Handbook of Latin American Studies CD-ROM: HLAS/CD, Vols. 1–53, 1936–1994.* 1995– . Madrid: Fundación MAPFRE América; Washington: Hispanic Division, Library of Congress; Austin: Distributed by Univ. of Texas Press.

The motherlode of Latin Americanist bibliography, now available in its entirety (v. 1–53, 1936–94) on searchable CD-ROM. A premier database for Latin American humanities and social sciences, it ties with World News Connection (item **50**) as the impact release of 1995, bringing an indispensable trove of information to scholars and libraries worldwide at an affordable price. Contains some 250,000 entries.

27   **Impex: Foreign Trade Statistics by Commodities = *Statistiques de commerce extérieur produits.*** 1993– . Paris: Organisation for Economic Co-operation and Development (OECD).

CD-ROM contains some 4.5 million statistical series covering the value of SITC imports and exports by OECD member countries (including Mexico) with their 200 worldwide partners, 1980–93. Compare with World Trade Database (item **51**) based on UN data.

28   ***International Bibliography of the Social Sciences, 1981– .*** London: British Library of Political and Economic Science, London School of Economics and Political Science.

Strong Latin American coverage is available in this massive new CD-ROM, which contains more than 600,000 records dating from 1981. Indexes journals and books in anthropology, economics, political science, and sociology. Treats core and specialty publications in 70 languages from more than 100 countries. Updated quarterly. Internet access also available.

29   ***International Political Science Abstracts, 1989– .*** Paris: International Political Science Association.

The top database of worldwide journal and yearbook literature in political science, public law, international law, and international relations includes many publications from and about the developing world. Offers more than 40,000 records. Unfortunately, coverage dates back only to 1989. Updated quarterly on CD-ROM and via Internet.

30   ***Investigación Económica.*** 1994– . Colima, Mexico: Centro Nacional Editor de Discos Compactos; México: Facultad de Economía, Univ. Nacional Autónoma de México.

Four CD-ROMs contain the complete texts of articles appearing in the first 204 issues of *Investigación Económica,* 1941–1993.

31   ***Jane's Sentinel.*** 1995– . Alexandria, Va.: Jane's Information Group.

Regional security assessments covering national military forces, threat conditions, economic indicators, political affairs, foreign investment opportunities, infrastructure, and more. CD-ROM sets for South America and Central America-Caribbean are available. Also online via LEXIS-NEXIS.

32   **LABORDOC.** Geneva, Switzerland: Central Library and Documentation Bureau, International Labour Office, s.d. 1 computer laser optical disc.

CD-ROM contains more than 200,000 citations to journals and reports on labor and employment issued worldwide since 1965. By far the most comprehensive bibliographic resource for these areas. Multilingual scope, with many entries on Latin American affairs. Online and Internet access also available.

33   ***Latin American Business Intelligence.*** 1995– . London: Economist Intelligence Unit.

Full-text versions of political, economic, and business reports on Latin America from the Economist Intelligence Unit (EIU). Updated every two months, this CD-ROM treats Argentina, Bolivia, Brazil, Chile, Colombia, Costa Rica, Ecuador, Mexico, Panama, Paraguay, Peru, Uruguay, and Venezuela. Content also available in separate CD-ROM editions and online.

34   ***Latin American Newsletters, 1990– .*** London: Latin American Newsletters.

Long available online via LEXIS-NEXIS, the respected London newsletters make their CD-ROM debut on this disk with complete page images from Jan. 1990–June 1995.

35   ***Latin American News.*** 1995– . Mountain View, Calif.: Knight-Ridder Information.

A new Dialog online file with full-text articles from Mexico's *El Norte* and *Reforma,* the Latin American Newletters family, and a marketing newsletter on Latin America.

36   **Latino 3.** Colima, Mexico: Centro Nacional Editor de Discos Compactos, 1995. 1 computer laser optical disc.

The third Colima collection of databases from Latin America and the Caribbean, following the now out-of-print first and second collections (see *HLAS 54:33* and *HLAS 54:34*). Offers 89 files dealing with economics, philosophy, literature, and demography. Publication of compact disc was supported by UNESCO.

37   *El Mercado de Valores:* la política económica de México, 1946–1994. Colima, Mexico: Centro Nacional Editor de Discos Compactos, Univ. de Colima; México: Nacional Financiera, 1995. 1 computer laser optical disc.

Full text of articles from this economic magazine.

38   **National Security Archive index.** Alexandria, Va.: Chadwyck-Healey; Washington: National Security Archive, 1994. 1 computer laser optical disc.

An index on CD-ROM to more than 180,000 pages of declassified US government policy documents available in the Archive's 12 "Making of US Policy" microfiche collections. The Latin Americanist sets cover Nicaragua (1978–90) and El Salvador (1977–84).

39   *OECD Statistical Compendium.* 1994– . Paris: OECD Electronic Publications.

A handy "best-of" compilation of Organisation for Economic Co-operation and Development (OECD) statistics spanning agriculture and food, general economic affairs, national accounts, financial and fiscal affairs, development and aid, economic indicators, energy, industry, science, and technology. Over 200,000 monthly, quarterly, and annual time series with current and retrospective coverage of new member Mexico. Updated twice a year.

40   *Proceso.* 1995– . México: Comunicación e Información.

A full-text chronicle on CD-ROM of Mexican affairs during the Salinas *sexenio* (1988–94) as reported in over 300 issues of this influential weekly. Includes images of all covers as well as 13,685 articles, interviews, chronicles, columns, and letters. Fully searchable.

41   *Revista Este País.* 1995– . Colima, Mexico: Centro Nacional Editor de Discos Compactos; México: Desarrollo de Opinión Pública.

This important Mexican magazine of trends and public opinion is reproduced in full text form on CD-ROM with 113 surveys, 12 debates, and 30 interviews in addition to the articles.

42   **SCINCE.** Aguascalientes, Mexico: Instituto Nacional de Estadística, Geografía e Informática (INEGI), 1994. 1 computer laser optical disc.

A companion to the *CODICE 90* census CD-ROM (see *HLAS 54:39*), this *Sistema para la consulta de informacion censal* presents 91 variables from the 1990 census of population and housing in tandem with census tract (AGEB) mapping. The SCINCE interface requires considerable practice to master.

43   **Sector agropecuario.** Aguascalientes, Mexico: Instituto Nacional de Estadística, Geografía e Informática (INEGI), 1994. 1 computer laser optical disc.

A data CD-ROM with final state and municipality results from the *VII Censo Agrícola-Ganadero* (1991) and the *VII Censo Ejidal* (1991), the former with 61 tables and the latter with 45. Includes 221 additional variables linked to census maps.

44   *Statistical Masterfile, 1974– .* Bethesda, Md.: Congressional Information Service.

The best CD-ROM database source for abstracts of statistical information published by the UN, World Bank, IMF, IDB, GATT/WTO, OECD, and other intergovernmental organizations. A real goldmine of leads to Latin Americanist data. Corresponds to *Index to International Statistics (IIS)*, *American Statistics Index (ASI)*, and *Statistical Reference Index (SRI)*. Microfiche archive of most publications is also available. International publications coverage dates from 1983.

45   **TRAINS.** Geneva: United Nations Conference on Trade and Development, 1994. 1 computer laser optical disc. (version 2.0)

This *Trade analysis and information system* CD-ROM contains databases on country tariff levels and other trade control measures, import statistics, trade classifications, and more, with substantial coverage of Latin America. Covers 1988–92, with some variations by country. One of the few accessible sources for country tariff details. Version 3.0 (1995) includes a Windows interface and data coverage from 1990–95.

46   **Tratados internacionales celebrados por México.** Colima, Mexico: Centro Nacional Editor de Discos Compactos; México: Secretaría de Relaciones Exteriores, 1993. 2 computer laser optical discs.

A set of CD-ROMs with a searchable index of Mexican treaties (1823–1993), linked to scanned images of the complete treaty texts. Corresponds to the multivolume print serial *Tratados Celebrados por México*.

47   **Wistat: women's indicators and statistics database.** Version 3. New York: UN Publications, 1995. 1 computer laser optical disc.

An affordable CD-ROM compendium of UN statistics on the situation of women worldwide, with 1,667 series for some 200 countries and areas. Topic areas: population composition and distribution; learning and educational services; economic activity; households, marital status, and fertility; housing conditions and human settlements; health and health services; public affairs and political participation; crime and criminal justice; national product and expenditure.

48   ***World Data.*** 1994–. Washington: International Bank for Reconstruction and Development, World Bank.

A superb portable database on CD-ROM of more than 700 World Bank annual data series on national accounts, balance of payments, trade, external debt and finance, social development, and natural resources. Covers over 200 countries, with some series dating from 1960 to the most recent estimates. Corresponds to *Social Indicators of Development*, *World Tables*, *World Debt Tables*, and *Trends in Developing Economies*. Updated annually.

49   **World marketing data and statistics.** 2nd. ed. London: Euromonitor, 1995. 1 computer laser optical disc.

More than 450 statistical tables on CD-ROM covering country demographic, economic, and marketing indicators, 1977–94. Many business-related series are difficult to find through other sources.

50   ***World News Connection.*** 1995–. Springfield, Va.: National Technical Information Service; Washington: Foreign Broadcast Information Service.

One of the best new releases in 1995, this internationalist news database on the Internet provides daily full-text translations of unclassified political, military, environmental, economic, and social information gathered by US government monitors. Sources include foreign political speeches, television programs, radio broadcasts, newspaper articles, periodicals, and books. Corresponds to the print *Daily Report: Latin America* and other regional titles from FBIS, which is phasing out hard copy production in favor of WNC. Coverage dates back to July 1994.

51   ***World Trade Database = Base de donées sur le commerce mondial, 1980–1992.*** Ottawa, Canada: International Trade Division, Statistics Canada.

A powerful CD-ROM product with country import and export flows (1980–92) disaggregated at the 4-digit SITC level. Reports UN data that has been cleaned up by Statistics Canada. One of the most accessible electronic sources for global country-by-commodity trade. Annual updates.

JOURNAL ABBREVIATIONS

Anu. Estad. Repúb. Argent.   Anuario Estadístico de la República Argentina. Instituto Nacional de Estadística y Censos. Buenos Aires.

Bus. Mex.   Business Mexico. American Chamber of Commerce of Mexico. México.

CD-ROM Prof.   CD-ROM Professional. Pemberton Press. Weston, Conn.

Database/Weston.   Database. Online, Inc., Weston, Conn.

Dir. U.N. Inf. Sources.   Directory of United Nations Information Sources. Advisory Committee for the Coordination of Information Systems (ACCIS), United Nations. New York.

HLAS/CD.   Handbook of Latin American Studies CD-ROM: HLAS/CD, Vols. 1–53. Fundación MAPFRE América, Madrid. Hispanic Division, Library of Congress, Washington. Distributed by Univ. of Texas Press, Austin .

Impex/OECD.   Impex: Foreign Trade Statistics by Commodities = Statistiques de commerce extérior produits. Organisation for Economic Co-operation and Development (OECD). Paris.

Int. Bibliogr. Soc. Sci.   International Bibliography of the Social Sciences. British Library of Political and Economic Science, London School of Economics and Political Science. London.

Int. Polit. Sci. Abstr.   International Political Science Abstracts. International Political Science Assn., Paris.

Internet World. Internet World. Meckler Corp. Westport, Conn.

Invest. Econ. Investigación Económica. Facultad de Economía, Univ. Nacional Autónoma de México. México.

Jane's Sentinel. Jane's Sentinel. Jane's Information Group. Alexandria, Va.

Lat. Am. Bus. Intell. Latin American Business Intelligence. Economist Intelligence Unit. London; New York.

Lat. Am. News/Knight-Ridder. Latin American News. Knight-Ridder Information. Mountain View, California.

Lat. Am. Newsl./London. Latin American Newsletters. London.

Mex. Bus. Mexico Business. Mexico Business Publishing Group. Houston, Tex.

OECD Stat. Compend. OECD Statistical Compendium. OECD Electronic Publications. Paris.

Online CDROM Rev. Online & CDROM Review. Learned Information. Oxford, England; Medford, N.J.

Proceso/México. Proceso. Comunicación e Información. México.

Rev. Este País. Revista Este País. Centro Nacional Editor de Discos Compactos. Colima, Mexico. Desarrollo de Opinión Pública. México.

Stat. Masterfile. Statistical Masterfile. Congressional Information Service. Bethesda, Md.

U.S./Latin Trade. U.S./Latin Trade. New World Communications. Miami.

World Data. World Data. World Bank. Washington.

World News Connect. World News Connection. National Technical Information Service. Springfield, Va.; Foreign Broadcast Information Service. Washington.

World Trade Database. World Trade Database. International Trade Division, Statistics Canada. Ottawa.

ANTHROPOLOGY

GENERAL

52 Alcina Franch, José. En torno al urbanismo precolombino de América: el marco teórico. (*Anu Estud. Am.*, 48, 1991, p. 3–47, tables)

Argues that urbanism has been confused with civilization and social organization, and identifies 10 characteristics to distinguish urban and non-urban settlements. [B. Meggers]

Baud, Michiel et al. Etniciteit als strategie in Latijns-Amerika en de Caraïben. See item **779**.

53 Cultural expression and grassroots development: cases from Latin America and the Caribbean. Edited by Charles D. Kleymeyer. Boulder, Colo.: L. Rienner Publishers, 1993. 292 p.: bibl., index.

An English-language edition of a previously-published anthology of case studies on cultural expression in grassroots development projects in Latin America, including cases from Bolivia, Chile, Colombia, Ecuador, and Peru. For comment on Spanish edition, see *HLAS 53:1182*. [J. Rappaport]

54 Fingerhut, Eugene R. Explorers of pre-Columbian America?: the diffusionist-inventionist controversy. Claremont, Calif.: Regina Books, 1994. 286 p.: bibl., index. (Guides to historical issues; 5)

Bipartisan review of principal claims for precolumbian transoceanic contacts is written in the form of a dialogue between protagonists with commentary by the author. [B. Meggers]

55 History of humanity. v. 1, Prehistory and the beginnings of civilization. Edited by Sigfried J. de Laet *et al.* London; New York: Routledge; Paris: UNESCO, 1994. 1 v: bibl., ill., index, maps. (Routledge reference)

Chapters on Mexico and Central America (J.L. Lorenzo), Caribbean and northern South America (M. Sanoja), non-Andean South America (O. Heredia), south Andean South America (L. Núñez), and central Andes (L.G. Lumbreras) synthesize information on cultural developments from initial peopling to ca. 5000 BP. [B. Meggers]

56 Ideology and pre-Columbian civilizations. Edited by Arthur Andrew Demarest and Geoffrey W. Conrad. Santa Fe, N.M.: School of American Research Press; Seattle: Univ. of Washington Press, 1992. 261 p.: bibl., ill., index, map. (School of American Research advanced seminar series)

Nine archaeologists draw on long-term fieldwork in Mesoamerica and central Andes to argue that ideology is the driving force behind sociopolitical complexity. Robert Carneiro defends priority of "the material conditions of existence." For Mesoamerican specialist's comment see item **120**. [B. Meggers]

Indianidad, etnocidio, indigenismo en América Latina. See item **936**.

57 Indigenous revolts in Chiapas and the Andean highlands. Edited by Kevin Gosner and Arij Ouweneel. Amsterdam: CEDLA, 1996. 296 p.: appendix, bibl., map. (CEDLA Latin American studies; 77)

Edited volume is divided in two parts: 1) six essays dealing with Chiapas rebellions, 1524–1994, and 2) five essays examining Andean rebellions, 1780–1980s. Work is result of one-day multidisciplinary seminar devoted to discussing and comparing indigenous revolts. One of the main questions addressed is whether these revolts can be labeled "indigenous." [R. Hoefte]

58 International Congress of Americanists, 47th, New Orleans, 1991. Five hundred years after Columbus: proceedings. Compiled by E. Wyllys Andrews and Elizabeth Oster Mozzillo. New Orleans, La.: Middle American Research Institute, Tulane Univ., 1993. 293 p.: bibl., index. (Publication; 63)

Contains abstracts of symposia papers presented at the 1991 Congress of Americanists. Arranged in 11 categories with subject and contributor index to facilitate use. [B. Meggers]

59 Langues et cultures en Amérique espagnole coloniale: colloque international, Université de la Sorbonne nouvelle-Paris III, 22–23 novembre 1991. Coordination de Marie-Cécile Bénassy-Berling, Jean-Pierre Clément et Alain Milhou. Paris?: Presses de la Sorbonne nouvelle, 1993. 327 p.: bibl.

Results of international colloquium (Univ. Paris III, Nov. 1991) in homage to Prof. André Saint-Lu, a well-known French Latin Americanist. Most papers focus on linguistics and culture in the colonial New World (e.g., political linkages; persistence of Quechua, Nahuatl, and African languages; and role played by linguistics in the evangelization process). Concluding remarks by Francisco de Solano (Madrid) emphasize the reality of a cultural *métissage* which prevailed over the desired Spanish linguistic imperialism. [T. Hampe-Martínez]

60 Latin American horizons: a symposium at Dumbarton Oaks, 11th and 12th October 1986. Edited by Don Stephen Rice. Washington: Dumbarton Oaks Research Library and Collection, 1993. 373 p.: bibl., ill., index.

Twelve contributions reassess validity of Horizon concept in precolumbian archaeology. Book follows a chronological organization that crosscuts Mesoamerica and South America, so that the Olmec horizon is juxtaposed with Chavín and Teotihuacán, the Maya with Huari and Tiahuanaco, and the Toltec and Aztec with the Inca. The majority of participants conclude that the Horizon concept has outlived its usefulness and requires reconsideration. [G. Feinman]

61 Marzal, Manuel María. Historia de la antropología indigenista: México y Perú. Barcelona: Anthropos Editorial del Hombre; Iztapalapa, Mexico: Univ. Autónoma Metropolitana, Unidad Iztapalapa, División de Ciencias Sociales y Humanidades, 1993. 543 p.: bibl., index. (Autores, textos y temas: Antropología; 29)

Critical review relates principal chronicles from initial contact to present. Writings surveyed reflect ancestry, social position, experiences, and perceptions of their authors. An informative introduction for non-specialists. [B. Meggers]

Milbrath, Susan. Representations of Caribbean and Latin American Indians in sixteenth-century European art. See item **510**.

62 Pitt-Rivers, Julian. La culture métisse: dynamique du statut ethnique. (*Homme/Paris*, 122/124, avril/déc. 1992, p. 133–148, bibl.)

Discusses cultural significance of *mestizaje* as expressed in ethnic status and its dynamic over time. Draws from both Mesoamerican and Andean societies, and emphasizes: 1) the existence of a wide-ranging terminology, but also the selection or elimination of certain terms depending on specific historical circumstances; and 2) relation of ethnic status to social position of an individual or a group. [A. Pérotin-Dumon]

63 Smith, Michael Ernest. Braudel's temporal rhythms and chronology theory in archaeology. (*in* Archaeology, Annales, and ethnohistory. Edited by A. Bernard Knapp. Cambridge, England: Cambridge Univ. Press, 1991, p. 23–34, bibl.)

Relates Braudel's model to current theoretical work on time and chronology in archaeology. Argues that Braudel's model encompasses both "ethnographic time" and "archaeological time," and that the different temporal rhythms associated with diverse socioeconomic processes are relevant when building methods for chronological refinement. [G. Feinman]

64 Las sociedades americanas y los orígenes de la producción de alimentos. (*Rev. Arqueol. Am.*, 6, 1992, p. 7–138, bibl., ill., maps)

Includes papers on: a "trilinear theory" of agricultural origins by R.S. MacNeish; early food production in lower Central America and Colombia by R. Cooke; diverse economic adaptations in Peru prior to 4500 BP by J.B. Richardson, III; the beginnings of food production in the Southern Cone by V. Castro and M.N. Tarragó; and the role of Zamia in the Dominican Republic by M. Veloz Maggiolo. [B. Meggers]

65 **Spores, Ronald.** Anthropology. (*in* Latin America and the Caribbean: a critical guide to research sources. Edited by Paula H. Covington. New York: Greenwood Press, 1992, p. 57–98, bibl.)

Review of trends in social anthropology, ethnohistory, archaeology, and anthrolinguistics since 1960s is followed by annotated bibliography of general sources (bibliographies, abstracts, indices, directories) for Latin America in general and by country, and of similar sources by topic (anthrolinguistics, archaeology, ethnohistory, social anthropology, periodicals). A useful introduction to existing literature. [B. Meggers]

Williams, Lynden S. Agricultural terrace evolution in Latin America. See item **2283**.

ARCHAEOLOGY
Mesoamerica

JOHN HENDERSON, *Professor of Anthropology, Cornell University*
GARY FEINMAN, *Professor of Anthropology, University of Wisconsin–Madison*

MESOAMERICAN ARCHAEOLOGISTS HAVE NEVER SHARED a single overarching theoretical perspective nor explicitly oriented their investigations in terms of common research themes and questions. Studies that deal with Mesoamerica as a whole or with large regions are most often synthetic rather than theoretically explanatory. The recent cycle of research covered in this overview continues to reflect this diversity of paradigms, perspectives, and interests.

Although it is not possible to characterize investigations in this extensive region in terms of a few tightly defined key themes, much current work can be seen as exploring broad issues of interrelationships both within societies—at the level of households, social groups, and communities—and between polities. Emphases vary widely, from the largely material to the purely ideological. Some studies focus on durable goods as indicators of economic systems, exchange networks, and the nature of craft production and specialization; some use settlement pattern data as a guide to political systems; others analyze domestic architecture and refuse as a reflection of household organization; still others mine hieroglyphic texts, monumental architecture, and art for clues to shared belief systems or to the specifics of political and social organization. While certain studies examine and emphasize the hierarchial relations within and between ancient Mesoamerican social groups, others stress the symbolic and conceptual systems that served to define and bind different populations and polities. Within this diversity of approaches, there is a common (but not always explicit) focus on interconnection and communication.

Studies of the "technology" of communication—iconography, writing, and other forms of notation—include analyses of inscriptions, painted documents, murals, and sculpture from all parts of Mesoamerica. Epigraphy and iconography continue to be well represented in the Maya literature. Teotihuacan art and notation are attracting increasing attention. Several studies focus on postclassic Mixtec painted codices and on late formative "pre-Maya" writing. A continuing controversy in this area revolves around the issue of whether these sources should be used as primary data to create detailed historical frameworks, or whether they are best treated as complementary to material data and the archaeological frameworks based on them (item **435**).

Research on late postclassic central Mexico includes a strong emphasis on the political structure of the Aztec state, on the nature of articulation among groups within it, on the composition of and linkages between households, and on the organization of craft production. On a larger scale, others explore the nature of the relations between the Aztec polity and other contemporaneous Mesoamerican states. In the same way, much current work on Teotihuacan focuses on the nature of the ties between the city and its hinterland and on the center's relationships with neighboring regions (item **73**). Analyses of burials, residential compounds, and urban barrios at Teotihuacan reflect a strong interest in internal social and economic organization within the city (item **182**). In central Mexico as well as elsewhere, characterization studies (especially of ceramics and obsidian) and lithic analyses (item **187**) represent key strategies for investigating inter-connections and craft production.

Large-scale excavation projects, focusing on individual cities and their hinterlands, and regional settlement survey programs continue in the Maya region (items **257** and **267**) as well as elsewhere in Mesoamerica. Within this framework, there is healthy and expanding emphasis on the residential contexts of ancient communities, including both aristocratic and non-elite segments of society, and on demography. In some cases, these house excavations have been buoyed by the application of new methods and techniques for the analysis of ancient living surfaces. Regional surveys and household archaeology are particularly well represented along the fringes of the Maya lowlands, especially in the southeast, which remains an intense focus of investigation (item **248**).

While Mesoamericanist scholars continue to collect a rich and diverse corpus of material and ideological insights on the similarities, differences, and changes that characterized the ancient Mesoamerican world, there seems to be little consensus concerning the specific frameworks that are necessary to synthesize and interpret this burgeoning empirical record. Nevertheless the interest noted above in how different social segments and groups expressed themselves is a productive trend, especially as the variability and change in their articulations are explored in more explicit fashion. From our perspective, the current overarching interest in the social, material, and ideological relations among individuals, households, kin groups, communities, regions, polities, and macroregions should foment healthy reevaluations and refinements of archaeological concepts and units, as well as the further development of useful theoretical bridges to other social, historical, ecological, and humanistic disciplines.

GENERAL

66 **Abrams, Elliot Marc.** How the Maya built their world: energetics and ancient architecture. Austin: Univ. of Texas Press, 1994. 176 p.: bibl., index., ill.

Presents a method for analyzing ancient architecture that converts building size and quality into quantified estimates of the labor involved in construction. This labor estimate can then be used as a measure of cost of construction in analysis of wealth and power differentials in ancient societies. The lowland Maya center of Copán is presented as a case study illustrating the method. [JSH]

67 **Agurcia Fasquelle, Ricardo.** Una síntesis de la arqueología de Honduras. (*Yaxkin/Tegucigalpa*, 12:1, 1989, p. 5–38, bibl., ill.)

Very brief overview of prehistory of Honduras gives particular attention to the Maya city of Copán. [JSH]

68 **Alcina Franch, José; Miguel León Portilla;** and **Eduardo Matos Moctezuma.** Azteca Mexica. Barcelona: Lunwerg Editores, 1992. 402 p.: bibl., ill. (some col.). (Col. Encuentros: Serie Catálogos)

Catalog presents materials from the ninth in a series of exhibitions of New World prehispanic materials organized by Comisión Nacional del Quinto Centenario del Descubrimiento de América. The exhibition was mounted by the Museo Arqueológico Nacional of Madrid, with the collaboration of Mexico's Instituto Nacional de Antropología e Historia. Chapters cover variety of topics including art and religion, calendrics, cosmology, markets and tribute, chinampas, the Templo Mayor, and life in Tenochtitlán. [GF]

69   Algaze, Guillermo. Expansionary dynamics of some early pristine states. (*Am. Anthropol.*, 95:2, June 1993, p. 304–333, bibl., maps)

Comparative examination of early civilizations, including Teotihuacán, provides empirical basis for recognizing presence of isolated outposts situated at key junctures with surrounding peripheries. Author proposes that these outposts served to regularize access to nonlocal resources; benefitted elites in the periphery; and served as collection points and distributional nodes. [GF]

70   The American southwest and Mesoamerica: systems of prehistoric exchange. Edited by Jonathon E. Ericson and Timothy G. Baugh. New York: Plenum Press, 1993. 302 p.: bibl., ill., index, maps. (Interdisciplinary contributions to archaeology)

Collection of papers explores patterns of exchange in these two cultural regions. Of the 10 contributions, four concentrate on prehispanic Mexico: R. Bradley discusses exchange of marine shells in northwest Mexico and southwestern US; P. Wiegand and G. Harbottle examine role of turquoise in ancient Mesoamerican trade; R. Stanley and C. Pool discuss prehispanic exchange relationships among Central Mexico, the Valley of Oaxaca, and the Gulf Coast; and P. McAnany looks at resources, specialization, and exchange in the Maya lowlands. [GF]

71   Andrews, Anthony P. Late postclassic lowland Maya archaeology. (*J. World Prehist.*, 7:1, 1993, p. 35–69, bibl., ill.)

Author synthesizes cultural history of lowland Maya area during last three centuries of precolumbian period. Summarizes history of investigation of late postclassic sites and identifies key issues in archaeology of the period. Emphasizes importance of understanding late lowland Maya societies on their own terms rather than dismissing them as impoverished descendants of more interesting classic-period predecessors. [JSH]

72   The archaeology of regions: a case for full-coverage survey. Edited by Suzanne K. Fish and Stephen A. Kowalewski. Washington; London: Smithsonian Institution Press, 1990. 277 p.: bibl., ill. (Smithsonian series in archaeological inquiry)

Includes two overviews of systematic regional surveys conducted in highland Mesoamerica. Jeffrey R. Parsons reflects on a decade of full-coverage regional survey in the Valley of Mexico, while Stephen A. Kowalewski stresses merits of full-coverage survey through examples from the Valley of Oaxaca. [GF]

73   Art, ideology, and the city of Teotihuacán: a symposium at Dumbarton Oaks, 8th and 9th October 1988. Edited by Janet Catherine Berlo. Washington: Dumbarton Oaks Research Library and Collection, 1992. 442 p.: bibl., ill., index, maps.

Important synthesis of 13 papers that collectively summarize state of knowledge about Teotihuacán. Includes chapters on mortuary practices, the city's foreign enclaves, art, iconography, murals, and economic organization. [GF]

74   Ashmore, Wendy. The theme *is* variation: recent publications on the archaeology of southern Mesoamerica. (*LARR*, 28:1, 1993, p. 128–140)

Review essay covers 14 recent works on archaeology and ethnohistory of southern Mesoamerica. Most of the publications are reports on archaeological excavations, and most deal with ancient Maya. Identifies heightened appreciation for variability in the archaeological record and increasing diversity in theoretical perspectives as major trends in the literature. [JSH]

75   Baker, Mary. Capuchin monkeys (*cebus capucinus*) and the ancient Maya. (*Anc. Mesoam.*, 3:2, 1992, p. 219–228, bibl., ill.)

Analysis of monkeys in Maya art and literature suggests that the monkey associated with the arts, and with scribes in particular, was the capuchin rather than the howler. [JSH]

76   Ball, Joseph W. Artifacts from the Cenote of Sacrifice, Chichén Itzá, Yucatán: textiles, basketry, stone, bone, shell, ceramics, wood, copal, rubber, other organic materials, and mammalian remains. Edited by Clemency Chase Coggins. Introduction by Gordon R. Willey. Cambridge, Mass.: Peabody Museum of Archaeology & Ethnology, Harvard Univ.; Harvard Univ. Press, 1992. 389 p.: bibl., ill. (Memoirs of the Peabody Museum of Archaeology & Ethnology, Harvard University; v. 10, no. 3)

Last volume in a series, work describes collections dredged from main cenote at Chichén Itzá by Edward H. Thompson in first decade of 20th century. Earlier volumes of this series dealt with the cenote in context of archaeology of Chichén Itzá, and with metal and jade artifacts. Present volume describes ceramics, ground and chipped stone artifacts, animal bone, shell, basketry, wood and other plant materials, leather, rubber, copal, and stucco. Coggins' introductory chapter on history of investigations in the cenote, the discussion in her chapter on wooden artifacts, and Ball's commentary in ceramics chapter provide useful interpretive assessments of nature and chronology of ritual activity at the cenote. [JSH]

77 Baquedano, Elizabeth and Michel Graulich. Decapitation among the Aztecs: mythology, agriculture and politics, and hunting. (*Estud. Cult. Náhuatl*, 23, 1993, p. 163–177, bibl., ill., photo)

Synthetic discussion of decapitation among the Aztecs traces roots of this widespread and ancient practice across Mesoamerica. Based on historical analysis and contextual consideration of Aztec decapitation events, the practice is inferred to have been closely linked to agricultural fertility. [GF]

78 Bauer, Arnold J. Millers and grinders: technology and household economy in Meso-America. (*Agric. Hist.*, 64:1, Winter 1990, p. 1–17)

Contrasts persistence of a laborious maize-grinding technology in Mesoamerica with animal- and water-powered milling technology that developed in early Egypt and Mediterranean Europe. Author outlines various factors that may account for this technological conservatism. For ethnohistorian's comment see *HLAS 54:432*. [GF]

79 Blanton, Richard E. *et al.* Ancient Mesoamerica: a comparison of change in three regions. 2nd ed. Cambridge, England; New York: Cambridge Univ. Press, 1993. 284 p.: bibl., ill., index, maps. (New studies in archaeology)

Substantially revised 2nd. ed. of an earlier work includes a new chapter on preceramic Mesoamerica. Offers comparative assessment of key similarities and differences in long-term prehispanic sequences of Central Mexico, the Valley of Oaxaca, and the eastern Maya lowlands. [GF]

80 Blanton, Richard E.; Stephen A. Kowalewski; and Gary M. Feinman. The Mesoamerican world system. (*Review/Braudel*, 15:3, Summer 1992, p. 419–426, bibl., map)

Briefly outlines two contrastive macroscale political economic strategies evident in long-term prehispanic social history of Mesoamerica. "Core-building" strategy leads to a hierarchical core-periphery structure, while "boundary" strategy blurred this hierarchy during times when dominant cores were weaker and regional boundaries more permeable. Through this contrast, authors endeavor to place macro-scale analysis in a more diachronic perspective. [GF]

81 Brumfiel, Elizabeth M. Weaving and cooking: women's production in Aztec Mexico. (*in* Engendering archaeology: women and prehistory. Edited by Joan M. Gero and Margaret W. Conkey. Oxford, England: Basil Blackwell Ltd., 1991, p. 224–251)

Synthethic analysis examines women's roles in late postclassic communities of Central Mexico. The most striking finding concerns greater-than-expected variability in women's work, which often has been described as limited to weaving and cooking. For ethnohistorian's comment, see *HLAS 54: 441*. [GF]

82 Burkhart, Louise M. Mujeres mexicas en "el frente" del hogar: trabajo doméstico y religión en el México azteca. (*Mesoamérica/Antigua*, 12:23, junio 1992, p. 23–54, facsims.)

Asserts that symbolic and social domain of Aztec women was distinct from, yet complementary to, that of men. Illustrates a relationship between women's work and male success in Aztec society. For ethnohistorian's comment, see *HLAS 54:442*. [GF]

83 **Cabrera Vargas, María del Refugio.** Una región histórica: la Sierra Norte de Puebla, época prehispánica. (*Bol. Antropol. Am.*, 22, dic. 1990, p. 113–122, ill., map, photos)

Uses concept of historical region to define and examine the northern mountains of Puebla. [GF]

84 **Carlson, John B.** Rise and fall of the City of the Gods. (*Archaeology/New York*, 46:6, Nov./Dec. 1993, p. 58–69, ill., map, photos)

Well-illustrated sketch discusses how warfare and blood sacrifice first sustained and then led to the downfall of Teotihuacán.

85 **Caso, Alfonso.** Alfonso Caso: de la arqueología a la antropología. México: Instituto de Investigaciones Antropológicas, UNAM, 1989. 215 p.: bibl., ill. (some col.). (Serie antropológica; 102: Etnología)

Anthology of articles written by Caso over his long and productive career. Pt. 1 is comprised of his writings on the prehispanic Mixtecs, and pt. 2 includes lectures he presented at El Colegio Nacional on various aspects of indigenous culture. [GF]

86 **Chiefdoms: power, economy, and ideology.** Edited by Timothy K. Earle. Cambridge, England; New York: Cambridge Univ. Press, 1991. 341 p.: bibl., ill., index. (School of American Research advanced seminar series)

Includes 10 case studies presented at a 1988 advanced seminar at the School of American Research. Authors address questions about origin, nature, and evolution of chiefdoms. Two papers discuss prehispanic Mesoamerica. Feinman considers demography, surplus, and inequality in a regional examination of early political formations in highland Mesoamerica, and Drennan compares and contrasts formative-period chiefdoms from three Mesoamerican regions with societies of comparable complexity in Central America and northern South America. [GF]

87 **Circumpacifica: Festschrift für Thomas S. Barthel.** Edited by Bruno Illius and Matthias Samuel Laubscher. Frankfurt; New York: P. Lang, 1990. 2 v.: bibl., ill.

Collection of essays honors a prominent specialist on Maya hieroglyphic writing. Includes papers on proposed decipherments of miscellaneous glyphs (Lyle Campbell, Michael Closs, Nikolai Grube, Jurij Knorozov, G. Jerschowa, and Berthold Riese), two unprovenanced Maya jade plaques (Dieter Dütting), towers in the Chenes region of Yucatán (Ursula Dyckerhoff and Hans Prem), Nicaraguan petroglyphs (Wolfgang Haberland), a gesture connoting sacrifice in Maya art (Kornelia Kurbjuhn), the syntax of Maya calendrical records (Floyd Lounsbury), mortuary iconography on two Maya polychrome vessels (Jacinto Quirarte), depictions of weapons in Maya art (Francis Robicsek), relationships between calendrical and astronomical cycles among the Quiche Maya (Barbara Tedlock), and Teotihuacán calendar signs (Jacques Soustelle). [JSH]

88 **Clendinnen, Inga.** Aztecs: an interpretation. Cambridge; New York: Cambridge Univ. Press, 1991. 398 p.: bibl., ill. (some col.), index, maps.

Speculative interpretation of Aztec society focuses on daily life on the eve of Spanish conquest. Author considers roles of victims, warriors, priests, merchants, and women. Reflections on ritual human sacrifice serve as a basis to unravel complexities and contradictions of Aztec life. For ethnohistorian's comment see *HLAS 54:452*. [GF]

89 **The *Codex Mendoza*.** Edited by Frances F. Berdan and Patricia Rieff Anawalt. Berkeley: Univ. of California Press, 1992. 4 v.: appendices, bibl., ill., indexes, maps.

Highly important pictorial manuscript prepared (ca. 1541) by Indian artist for Viceroy Mendoza was deposited in the Bodleian Library, Oxford. Consists of 71 folios depicting Aztec conquests, tribute lists, and ethnographic data. Impeccable scholarship and excellent drawings characterize this luxurious edition of a primary source for the study of prehispanic Central Mexico. See also **2410**. [H. von Winning]

90 **Coe, Michael D.** Breaking the Maya code. New York: Thames and Hudson, 1992. 304 p.: bibl., ill., index.

History of research on Maya hieroglyphic writing is enlivened with material based on Coe's own important contributions and his acquaintance with many of the other key figures in recent epigraphic research. Work is marred by attacks on those who disagree with Coe or epigraphers and iconographers of whom he does approve, especially with regard to ethical problems associated with unprovenanced objects. For ethnohistorian's comment see *HLAS 54:456*. [JSH]

91 Coe, Michael D. The Maya. 5th ed., fully rev. and expanded. New York: Thames and Hudson, 1993. 224 p.: bibl., ill., index. (Ancient peoples and places)

Updated ed. of general survey of lowland Maya archaeology. Revisions are most extensive in the addition of new data on preclassic societies and in new interpretations of classic-period Maya world based on recent epigraphic and iconographic work. A new final chapter deals with Maya societies since the Spanish invasion. [JSH]

92 Colegio de Michoacán. Centro de Estudios Antropológicos. Origen y desarrollo de la civilización en el Occidente de México: homenaje a Pedro Armillas y Angel Palerm. Coordinación de Brigitte Boehm de Lameiras y Phil C. Weigand. Zamora, Mexico: El Colegio de Michoacán, 1992. 382 p.: bibl., ill., maps. (Col. Memorias)

Compilation of 17 papers presented at the Centro de Estudios Antropológicos Round Table (4th, 1990) on western Mexico. Works focus on variety of themes including paleoenvironment, subsistence, agriculture, production activities, household units, settlement patterns, political territorial units, interaction networks, and iconography. [GF]

93 Coloquio en torno a la obra de un Mayista, *Mérida, Mexico, 1983.* Coloquio en torno a la obra de un Mayista: Sylvanus G. Morley, 1883–1948. Coordinación de Alfredo Barrera Rubio. México: Instituto Nacional de Antropología e Historia; Mérida, Mexico: Univ. Autónoma de Yucatán, 1992. 107 p., 16 leaves of plates: bibl., ill. (Col. científica. Serie Arqueología)

Nine essays deal with life and work of pioneering Maya archaeologist. Volume includes biographical data, assessments of Morley's ideas in context of Maya archaeology today, and a series of previously unpublished letters. [JSH]

94 Cowgill, George L. Comments on Andrew Sluyter: long-distance staple transport in western Mesoamerica; insights through quantitative modeling. (*Anc. Mesoam.*, 4:2, Fall 1993, p. 201–203, bibl.)

This constructive evaluation of Sluyter's model questions the argument that there is no limit to how far human bearers can transport staple foods and still have a net energy profit. [GF]

Davis, Wade and **Andrew T. Well.** Identity of a New World psychoactive toad. See *HLAS 54:459*.

95 Una definición de Mesoamérica. Edited by Jaime Litvak King. México: Instituto de Investigaciones Antropológicas, UNAM, 1992. 192 p.

Collection of articles focuses on definition of prehispanic Mesoamerica and Mesoamerican archaeology.

96 Dorweiler, Jane *et al. Teosinte* glume architecture 1: a genetic locus controlling a key step in maize evolution. (*Science/Washington*, 262, Oct. 8, 1993, p. 233–235, bibl., ill., photos)

Describes and maps genetic locus that appears to control major morphological change (from encased to exposed kernels) that distinguishes maize from its wild ancestor, teosinte. [GF]

97 Dunning, Nicholas P. *et al.* Classic Maya landscape archaeology. (*Anc. Mesoam.*, 5:1, 1994, p. 61–140, bibl., ill.)

Papers examine settlement patterns and political systems in Puuc region of western Yucatán; a water management system at Kinal in northeastern Guatemala; agricultural terracing in the upper drainage of the Belize River in western Belize; and use of tree crops in coastal southern Belize. [JSH]

98 Economic aspects of water management in the prehispanic New World. Edited by Vernon L. Scarborough and Barry L. Isaac. Greenwich, Conn.: JAI Press, 1993. 1 v. (Research in economic anthropology; 7)

Five of the eight articles in this volume consider water management in ancient Mesoamerica. V. Scarborough and P. Harrison each examine aspects of water management in southern Maya lowlands; D. Nichols and C. Frederick present recent research on irrigation canals and chinampas in northern Basin of Mexico; J. Angulo V. discusses water control and communal labor in Central Mexico during formative and classic periods; and P.

Weigand describes large-scale hydraulic works in prehistoric western Mesoamerica. [GF]

99   **Etnoarqueología: coloquio Bosch-Gimpera.** Edited by Yoko Sugiura Y. and Mari Carmen Serra P. México: Instituto de Investigaciones Antropológicas, UNAM, 1990. 594 p.

Ethnoarchaeology is the principal theme of this collection that covers a diverse range of topics and approaches. Papers were presented originally at a 1988 conference. [GF]

100   **Factional competition and political development in the New World.** Edited by Elizabeth M. Brumfiel and John W. Fox. Cambridge, England; New York: Cambridge Univ. Press, 1994. 234 p.: bibl., ill., index. (New directions in archaeology)

Series of articles (nine on Mesoamerica) examines how competing factions at both local and regional scales led to emergence of social inequality and development of institutional specialization in various prehispanic New World societies. Mesoamerican pieces examine this theme for coastal Chiapas, the Tarascans, the Aztecs, the Valley of Oaxaca, the Mixteca Alta, and the Maya. [GF]

101   **Feinman, Gary M.** and **Linda M. Nicholas.** Human-land relations from an archaeological perspective: the case of ancient Oaxaca. (*in* Understanding economic process. Edited by Sutti Ortiz and Susan Lees. Lanham, Md.: University Press of America, 1992, p. 155–178, bibl.)

Simulation of changing human-land relations in ancient Oaxaca finds that limiting factors on agricultural production were variable over the prehispanic sequence. A critique of the Boserupian approach. [GF]

102   **Feinman, Gary M.** and **Linda M. Nicholas.** New perpectives on prehispanic highland Mesoamerica: a macroregional approach. (*Comp. Civiliz. Rev.*, 24, Spring 1991, p. 13–33, bibl., table, map)

Authors propose that a multi-scale analytical approach that includes examination of macroregional relations is necessary to understand long-term change in Mesoamerica. Argument is illustrated through diachronic study of shifting prehispanic relationship between Valleys of Oaxaca and Ejutla and a consideration of late postclassic exchange networks. [GF]

103   **Florescano, Enrique.** Tiempo, espacio y memoria histórica entre los mayas. Tuxtla Gutiérrez, Mexico: Gobierno del Estado de Chiapas, Consejo Estatal de Fomento a la Investigación y Difusión de la Cultura, DIF-Chiapas; Instituto Chiapaneco de Cultura, 1992. 126 p.: bibl., ill. (Serie Antropología; 3)

Reconstruction of worldview and political symbolism of the ancient Maya draws on postclassic Maya ritual books, the *Popol Vuh* (the epic history of the highland Quiche Maya), and political symbolism of classic period lowland Maya cities (mainly following the interpretations of Schele and Freidel). [JSH]

104   **Ford, Anabel.** The ancient Maya of Belize: their society and sites. Santa Barbara, Calif.: CORI/Mesoamerican Research Center, 1994. 1 v.: bibl., ill.

Descriptions of 18 archaological sites, each with a simplified plan, comprise bulk of the text of this travellers' guide to archaeology of Belize. Sites are grouped geographically, and each description includes directions for reaching site. A very brief introductory section situates prehistory of Belize in the context of ancient Maya civilization. [JSH]

105   **The formation of complex society in southeastern Mesoamerica.** Edited by William R. Fowler. Boca Raton, Fla.: CRC Press, 1991. 286 p.: bibl., ill., index, maps.

Collection of essays considers archaeological evidence for beginnings of social complexity in formative period for Mesoamerica and Central America. Case studies focus on eastern Mesoamerica (Chiapas, coastal Guatemala, Honduras) and upper Central America. [JSH]

106   **Foro de Arqueología de Chiapas, *1st*, *Tuxtla Gutiérrez, Mexico*, *1990*.** Primer Foro de Arqueología de Chiapas: cazadores-recolectores-pescadores, agricultores tempranos. Tuxtla Gutiérrez, México: Gobierno del Estado de Chiapas, Consejo Estatal de Fomento a la Investigación y Difusión de la Cultura, DIF-Chiapas; Instituto Chiapaneco de Cultura, 1991. 145 p.: bibl., ill. (Serie Memorias; 4)

Collection of papers is based on presentations at a 1990 symposium on archaeology of Mexican state of Chiapas. Most of the essays (seven) focus on early farming societies in the Central Depression and along the Pacific Coast, and two deal with earlier foraging groups. [JSH]

107 **Fournier, Patricia** and **Andrea K.L. Freeman.** El razonamiento analógico en etnoarqueología: el caso de la tradición alfarera de Mata Ortiz, Chihuahua, Mexico. (*Bol. Antropol. Am.*, 23, julio 1991, p. 109–118, bibl., ill.)

Ethnoarchaeological study discusses traditional ceramic production in Mata Ortiz, Chihuahua. Addresses role of economic and social factors in stylistic change of pottery. [GF]

108 **Freidel, David A.; Linda Schele;** and **Joy Parker.** Maya cosmos: three thousand years on the shaman's path. Photographs by Justin Kerr and MacDuff Everton. New York: W. Morrow, 1993. 543 p.: bibl., ill. (some col.), index.

Highly interpretive treatment of Maya beliefs, especially about the structure of the universe, emphasizes elite belief systems as they are reflected in sculpture, painting, and hieroglyphic texts from lowland Maya cities of the classic period. Authors argue for close correspondence in belief among Maya societies, and accordingly use ethnohistorical sources, ethnography, and their own observations of modern Maya ritual activity to reconstruct ancient beliefs. Study also makes heavy use of unprovenanced objects in private collections. [JSH]

109 **Fritz, Gayle J.** Are the first American farmers getting younger? (*Curr. Anthropol.*, 35:3, June 1994, p. 305–309, bibl., table)

Summarizes recent accelerator dates made on early domesticated plant remains. Based on this review, proposes that transition to agriculture in highland Mesoamerica was much more recent than previously thought. Considers broader comparative, theoretical implications of this reconsideration. [GF]

110 **García Moll, Roberto; Felipe R. Solís Olguín;** and **Jaime Bali.** El tesoro de Moctezuma. México: Chrysler México, 1990. 205 p.: bibl., col. ill. (Col. Editorial de arte Chrysler)

Large-format book discusses Moctezuma, the last ruler of Aztec Tenochtitlán, and his meeting with Cortés. Includes lavish photographs of artifacts, stone sculpture, and codices. [GF]

111 **Gómez Serafín, Susana** and **Enrique Fernández Dávila.** Aprovechamiento de los moluscos durante el postclásico en Bahias de Huatulco. (*Cuad. Sur/Oaxaca*, 2:3, enero/abril 1993, p. 5–24, bibl., map, tables, ill.)

Studies the marine resources, particularly shellfish, available to past and present inhabitants of the Huatulco region on the Pacific Coast of Oaxaca. [GF]

112 **González Calderón, O.L.** Los señores de jade. Coatzacoalcos, Mexico: O.L. González Calderón, 1991. 135 p.: ill. (some col.), maps.

Extensively illustrated volume purports to identify reflections of Old World influence, especially from China, in the art of the Olmec, Mesoamerica's first complex society. [JSH]

113 **González Licón, Ernesto.** Zapotecas y mixtecas: tres mil años de civilización precolombina. Apéndice de María Luisa Franco Brizuela. Barcelona: Lunwerg, 1992. 239 p.: appendix, bibl., photos (some col.). (Corpus precolombino: Sección Las Civilizaciones mesoamericanas)

Lavishly illustrated book (with many full-page color photographs) on prehispanic Oaxaca focuses on well-known precolumbian sites in the Valley of Oaxaca. Author discusses history of research in the area and cultural history of the Zapotecs and Mixtecs. Includes appendix on Zapotec tomb at Huijazoo. [GF]

114 **Grove, David C.** The Olmec legacy: updating Olmec prehistory. (*Natl. Geogr. Res.*, 8:2, 1992, p. 148–165, bibl., ill.)

While Olmec legacy is still evident in later Mesoamerican civilizations, new data and fresh perspectives serve to challenge long-held views of widespread, direct Gulf Coast influence across Mesoamerica. [GF]

115 **Guillén, Ann Cyphers.** Women, rituals, and social dynamics at ancient Chalcatzingo. (*Lat. Am. Antiq.*, 4:3, Sept. 1993, p. 209–224, bibl., ill., map.)

Infers that preclassic (Cantera phase) female figurines had a role in female-focused ceremonies that created a range of social rights and obligations. These engendered societal relationships are proposed to have had a key role in processes by which particular households consolidated power and influence. [GF]

116 Harbottle, Garman and Phil C. Weigand. Turquoise in pre-Columbian America. (*Sci. Am.*, 266:2, Feb. 1992, p. 78–85, ill., map, photos)

Describes the prehispanic trade of turquoise between the American Southwest and Mesoamerica. The religious and economic importance of turquoise in Mesoamerica is proposed to have created extensive exchange networks emanating from source areas in northwestern Mexico and southwestern US. [GF]

117 Hassig, Ross. War and society in ancient Mesoamerica. Berkeley: Univ. of California Press, 1992. 337 p., 16 p. of plates: bibl., ill., index, maps.

Author proposes two key contrasts—aristocratic (fighting only by nobles) versus meritocratic (includes commoners) social organization; and territorial (directly administered) versus hegemonic (indirect) political control—to explain differences in militarism among Mesoamerican states. He emphasizes military technology and behavior rather than the broader significance of warfare. For ethnohistorian's comment see *HLAS 54:492.* [GF]

118 Hicks, Frederic. Gift and tribute: relations of dependency in Aztec Mexico. (*in* Early state economics. Edited by H.J.M. Claessen and Piet van de Velde. New Brunswick, N.J.: Transaction Publishers, 1991, p. 199–213)

Argues that Aztec empire was held together by two types of lord-commoner dependency relations, one based on gift exchange and the other on forcible extraction. [GF]

119 Historia general de Michoacán. v. 1, Escenario ecológico, época prehispánica. Coordinación general de Enrique Florescano. Morelia, Mexico: Gobierno del Estado de Michoacán, Instituto Michoacano de Cultura, 1989. 1 v.

As Vol. 1 of a set that presents history of Michoacán, this volume focuses on the physical environment, including flora and fauna, and on the prehispanic era. Topics discussed include ancient tombs found in the region, the Tarascans, and various geographic segments of Michoacán. For historian's comment on four-volume set see *HLAS 54:1134.* [GF]

120 Ideology and pre-Columbian civilizations. Edited by Arthur Andrew Demarest and Geoffrey W. Conrad. Santa Fe, N.M.: School of American Research Press; Seattle: Univ. of Washington Press, 1992. 261 p.: bibl., ill., index, map. (School of American Research advanced seminar series)

Contains 10 papers presented as an advanced seminar at the School of American Research. Seminar addressed role of ideology in evolution of precolumbian civilizations. The four papers on Mesoamerica include discussions of formative-period religion (D. Grove and S. Gillespie); Teotihuacán (G. Cowgill); and the classic Maya (D. Freidel and A. Demarest). For South America specialist's comment see item **56**. [GF]

121 Jones, Lindsay. The hermeneutics of sacred architecture: a reassessment of the similitude between Tula, Hidalgo and Chichén Itzá, Yucatán. Pt. 1–2. (*Hist. Relig.*, 32:3, Feb. 1993, p. 207–232 and 32:4, May 1993, p. 315–342)

In these two excellent articles author suggests model for reinterpreting the striking architectural similarities between Tula and Chichén Itzá. Asserts that theory of invasion and cultural imposition by the Toltecs ignores possibility that the apparently congruent architectural elements were used and interpreted differently in each region. Sees Chichén Itzá as center of a successful and aggressive Maya state whose rulers borrowed from Tula and other Central Mexican sites in a program termed "architectural allurement" geared to legitimize their hegemony. [R. Haskett and S. Wood]

122 El juego de pelota en Mesoamérica: raíces y supervivencia. Coordinación de María Teresa Uriarte. México: Siglo Veintiuno Editores, 1992. 413 p.: bibl., ill (some col.), maps. (Col. América nuestra; 39)

Extensively illustrated volume focuses on the ball game, one of the most distinctively Mesoamerican cultural complexes. Essays are based on presentations at a conference held in connection with the Festival Cultural Sinaloa that included performances by Mexican teams that still play a modern version of the game. Papers deal with all aspects of the precolumbian and post-conquest ball game: ball courts, ball game artifacts, representations of the game in precolumbian art, references to the game in ethnohistorical sources, and analyses of the symbolic dimensions of the game. [JSH]

123 Kelly, Joyce. An archaeological guide to Mexico's Yucatán Peninsula. Norman: Univ. of Oklahoma Press, 1993. 364 p., 16 p. of plates: ill. (some col.), maps.

Comprehensive guide to archaeological sites of Yucatán peninsula for travellers includes introduction to prehistory of Mexico with special attention to Maya civilization in Yucatán. [JSH]

124 Kobayashi, Munehiro. Tres estudios sobre el sistema tributario de los mexicas. México: Centro de Investigaciones y Estudios Superiores en Antropología Social; Kobe, Japan: Kobe City Univ. of Foreign Studies, 1993. 171 p.: bibl., ill., maps.

Three essays discuss tribute and its sociopolitical implications in the Basin of Mexico under imperial rule of the Mexica. The first focuses on food consumption in different social strata; the second examines Mexica intervention in Basin *señoríos*; and the third investigates role of prestige goods. [GF]

125 Kowalewski, Stephen A. The evolution of complexity in the Valley of Oaxaca. (*Annu. Rev. Anthropol.*, 19, 1990, p. 39–58, bibl.)

Quantitative, regional-scale analysis focuses on central places in the Valley of Oaxaca over a 3,000-year period (1500 BC–AD 1520). Eleven characteristics serve as basis for cross-phase comparisons. Concludes that most current explanations of complex society development are inadequate. [GF]

126 Kurtz, Donald V. and Mary Christopher-Nunley. Ideology and work at Teotihuacan: a hermeneutic interpretation. (*Man/London*, 28:4, Dec. 1993, p. 762–778, bibl.)

Reading physical remains of Teotihuacan as a "text open to interpretation," the authors argue that instead of coercion the creation of a shared, hegemonic ideology of work was the main sociocultural mechanism sustaining the population's will to support the economy, the State, and all of its enterprises. Highlights the significance of the nature and arrangement of the ceremonial precinct, the *Ciudadela*, wards and apartment compounds, murals, and plastic artifacts. [R. Haskett and S. Wood]

127 Lecturas históricas del estado de Oaxaca. v. 1, Época prehispánica. Recopilación de Marcus Winter y María de los Angeles Romero Frizzi. México: Instituto Nacional de Antropología e Historia, 1986. 1 v.: bibl., ill., maps. (Col. Regiones de México)

First in four-volume series covering Oaxacan history from prehispanic era through 1930. Vol. 1 includes 11 contributions that provide a general overview of prehispanic Oaxaca. Several chapters are included on the central Oaxacan highlands, the Mixteca Alta, and the Isthmus and other coastal regions. [GF]

128 León Portilla, Miguel. The Aztec image of self and society: an introduction to Nahua culture. Introduction by J. Jorge Klor de Alva. Salt Lake City: Univ. of Utah Press, 1992. 248 p.: bibl., ill., index.

Reviews how Aztecs and their neighbors viewed their past and preserved it through oral tradition and codices; describes mythical origins of various Nahuatl groups and the Mexica rise to power; discusses Mexica society and its search for legitimacy through creation of links to a legendary Toltec past; and presents Mexica worldview in the context of their religion. For ethnohistorian's comment see *HLAS 54:515*. [GF]

129 López Austin, Alfredo. Del origen de los mexicas: ¿nomadismo o migración? (*Hist. Mex.*, 39:3, enero/marzo 1990, p. 663–675, bibl.)

Presents series of arguments in support of Mexica migration. [GF]

130 Lowland Maya civilization in the eighth century A.D.: a symposium at Dumbarton Oaks, 7th and 8th October 1989. Edited by Jeremy A. Sabloff and John S. Henderson. Washington: Dumbarton Oaks Research Library and Collection, 1993. 482 p.: bibl., ill., index, maps.

Collection of essays based on presentations at a 1989 Dumbarton Oaks symposium focuses on classic Maya lowlands during its 8th-century heyday, on the eve of the transformations usually referred to as the "classic Maya collapse." Coverage includes economics, political organization, warfare, social organization, religion, environment, settlement patterns, architecture, art, inscriptions, pottery, and lithics. [JSH]

Malmström, Vincent H. Geographical diffusion and calendrics in pre-Columbian Mesoamerica. See item **2392**.

131 **Marcus, Joyce.** Political fluctuations in Mesoamerica: dynamic cycles of Mesoamerican states. (*Natl. Geogr. Res.*, 8:4, 1992, p. 392–441, bibl., ill., maps, photos)

Examination of four Mesoamerican regions (Central Mexico, Oaxaca, and northern and southern Maya lowlands) reveals repetitive patterns of state growth and decline. Most states consisted of a capital, a core region, and outlying provinces. In each case, maximum areal extent was reached early and was followed by contraction as outer provinces gained enough power to break away. [GF]

132 **Matos Moctezuma, Eduardo.** Vida y muerte en el Templo Mayor. México: Ediciones Océano, 1986. 143 p.: appendices, bibl., ill.

Incorporating historical documents with archaeological investigation, author discusses Nahuatl myth and cosmology, the foundation of Tenochtitlán, and significance and symbolism of the Templo Mayor. Book is written for the non-specialist. One appendix includes photographs of the Templo Mayor, while another recounts Nahuatl poetry.

133 **McAnany, Patricia A.** A theoretical perspective on elites and the economic transformation of classic period Maya households. (*in* Understanding economic process. Edited by Sutti Ortiz and Susan Lees. Lanham, Md.: Univ. Press of America, 1992, p. 85–103, bibl.)

Presents theoretical and synthetic consideration of classic Maya economics. Argues that most previous models are too simple to convincingly depict ancient Maya production and exchange. Concludes that Maya economy was pluralistic, and included sectors organized by real and fictive kin networks that coexisted, in conflict and accommodation, with dynastic structures of varying duration and power. [GF]

134 **McClung de Tapia, Emily.** The origins of agriculture in Mesoamerica and Central America. (*in* The origins of agriculture: an international perspective. Edited by C. Wesley Cowan and Patty Jo Watson. Washington: Smithsonian Institution Press, 1992, p. 143–171, bibl., ill., maps, tables)

This general overview describes the major Mesoamerican domesticated plants (with a table listing the site locations and temporal phases where a range of key species have been found archaeologically). Discussion of agricultural origins draws on findings from the Valleys of Tehuacán and Oaxaca, as well as the site of Zohapilco in the Basin of Mexico. Various models for the beginnings of plant domestication are assessed. [GF]

135 **Mesoamerican elites: an archaeological assessment.** Edited by Diane Z. Chase and Arlen Frank Chase. Norman: Univ. of Oklahoma Press, 1992. 375 p.: bibl., ill., index.

Collection of 19 articles illustrates difficulty of studying elites archaeologically, and diversity of viewpoints on how to approach this topic. Five contributions discuss general theoretical issues. Nine chapters consider Maya sites and regions including Caracol, Tikal, Altun Ha, Lamanai, Seibal, Sayil, Yaxuna, Santa Rita Corozal, Copán, the Quiché, and the southeastern zone. Five papers examine either Central Mexico or ancient Oaxaca. Collectively, papers draw on a wide assortment of empirical information. [GF]

136 **Miller, Mary Ellen** and **Karl A. Taube.** The gods and symbols of ancient Mexico and the Maya: an illustrated dictionary of Mesoamerican religion. New York: Thames and Hudson, 1993. 216 p.: bibl., ill., index, maps.

Presents encyclopedia-style overview of Mesoamerican religion, broadly defined to include symbolic dimensions of sociopolitical life and many specifics of iconography and writing systems. Includes brief introduction to Mesoamerican cultural tradition and a guide to the literature. [JSH]

137 **Morelos García, Noel.** Consideraciones teóricas sobre el proceso de urbanización en Mesoamérica. (*Bol. Antropol. Am.*, 23, julio 1991, p. 137–159, bibl., ill., maps.)

Comparative discussion examines classic period urbanization in Central Mexico, the Valley of Oaxaca, and the Maya region. Suggests that shifts in the nature of religion accompany rise of the state and urbanization, along with resultant amplification of class and rural divisions. [GF]

138 **New theories on the ancient Maya.** Edited by Elin C. Danien and Robert J. Sharer. Philadelphia: Univ. Museum, Univ. of Pennsylvania, 1992. 245 p.: bibl., ill., maps. (University Museum monograph; 77. University Museum symposium series; 3)

Collection of essays based on 1987 conference at Univ. Museum focuses on recent rapid changes in orthodox views of ancient Maya. Coverage includes new research in epigraphy and iconography, synthetic essays based on recent archaeological research, and new interpretations of ancient belief systems based on archeological and ethnohistoric data. [JSH]

139 **Novoa Magallanes, César.** Apuntes para el curso de desarrollo urbano en México correspondiente a la asignatura de análisis histórico del área de ciencias sociales. México: Univ. Nacional Autónoma de México, Facultad de Arquitectura, 1990. 216 p.: bibl., ill.

General and comparative discussion focuses on development of prehispanic cities during preclassic and classic periods, with special attention given to Tenochtitlán, Teotihuacán, and the Maya centers. [GF]

140 **O'Hara, Sarah L.; Alayne Street-Perrott; and Timothy P. Burt.** Accelerated soil erosion around a Mexican highland lake caused by prehispanic agriculture. (*Nature/London*, 362, March 1993, p. 48–51, bibl., ill., map, table)

Based on analysis of a series of sediment cores from Lake Pátzcuaro, Michoacán, authors define three prehispanic periods of accelerated erosion. They argue that prehispanic erosion rates were at least as high as those following contact, and therefore question utility of reviving indigenous agricultural practices. [GF]

141 **O'Mack, Scott.** Yacateuctli and Ehecatl-Quetzalcoatl: earth-divers in Aztec Central Mexico. (*Ethnohistory/Society*, 38:1, Winter 1991, p. 1–33, bibl., ill.)

Yacateuctli and Ehecatl-Quetzalcoatl, the two Aztec deities known as patrons of merchants, are proposed to have shared an association with diving waterfowl. The roots of this association are traced to a panamerican Earth-Diver tradition. Various ramifications of this implied relationship are advanced. [GF]

142 **Ortiz de Montellano, Bernard.** Aztec medicine, health, and nutrition. New Brunswick: Rutgers Univ. Press, 1990. 308 p.: appendices, bibl., ill., index.

Following introductory chapters that cover Aztec culture, demography, ecology, and nutrition, author focuses on illness and medicine, emphasizing Aztec methods of diagnosis and curing. Traces the Aztec roots of contemporary Mexican folk medicine. Includes appendices on nutritional value of Aztec foods and an examination of Aztec medicinal herbs. [GF]

143 **Pollard, Helen Perlstein.** The construction of ideology in the emergence of the prehispanic Tarascan state. (*Anc. Mesoam.*, 2:2, Fall 1991, p. 167–179, bibl., ill., maps, table)

Reconstructs how the Tarascan royal dynasty created a new worldview at the time when the Tarascan state (AD 1350–1520) was formed. [GF]

144 **Pollard, Helen Perlstein.** Taríacuri's legacy: the prehispanic Tarascan state. Introduction by Shirley Gorenstein. Norman: Univ. of Oklahoma Press, 1993. 1 v.: bibl., index. (The Civilization of the American Indian series; 209)

Combines ethnohistoric documents, archaeological research, and ecological data to provide comprehensive overview of late prehispanic Tarascan state. Concluding section draws stimulating contrasts with contemporaneous political domain centered at Tenochtitlán. [GF]

145 **Precolumbian jade: new geological and cultural interpretations.** Edited by Frederick W. Lange. Salt Lake City: Univ. of Utah Press, 1993. 378 p.: bibl., ill. (some col.), index.

Collection of essays is based on presentations at a 1987 conference. Eight essays deal with geological and mineralogical considerations; 13 focus on archaeological contexts and iconographic or stylistic analyses, mostly relating to Costa Rican, Maya, and Olmec jade. [JSH]

146 **Prehispanic domestic units in western Mesoamerica: studies of the household, compound, and residence.** Edited by Robert S. Santley and Kenneth G. Hirth. Boca Raton, Fla.: CRC Press, 1993. 296 p.: bibl., ill., index, maps.

Compilation of 14 articles focuses on domestic activities, house use, and community patterns in Central Mexico and South Gulf coast. Papers examine how archaeologists study and collect data on domestic space and areas of use. Contributors examine processes that create the archaeological record, the analytical means by which household artifactual and architectural features are investigated, use of residential space, and variation and change in household units in prehispanic Mesoamerica. [GF]

147 **Prem, Hanns J.; Ursula Dyckerhoff; and Helmut Feldweg.** Reconstructing Central Mexico's population. (*Mexicon/Berlin*, 15:3, Mai 1993, p. 50–57, bibl., ill., tables)

Reconstruction of population curve for early colonial Huejotzingo leads to a consideration of the size of the native population in Central Mexico in 1519. Estimates Central Mexican population was no higher than 15.1 million. [GF]

148 **Reents-Budet, Dorie et al.** Painting the Maya universe: royal ceramics of the classic period. Photographs by Justin Kerr. Durham, N.C.: Duke Univ. Press, 1994. 381 p.: bibl., ill. (some col.), index, maps.

Catalog of exhibition at Duke Univ. includes discussions of collecting and destruction of the archaeological record. Attempts to combine stylistic, iconographic, and epigraphic analysis of elaborate painted ceramics with data based on controlled archaeological excavation.

149 **Reinterpreting prehistory of Central America.** Edited by Mark Miller Graham. Niwot, Colo.: Univ. Press of Colorado, 1993. 336 p.: bibl., ill., index.

Collection of papers presented at 1990 "Central America and its Neighbors" conference re-assesses fields of archaeology and art history in Central America, on 50th anniversary of publication of *The Maya and their neighbors* (New York: 1940). Mark Miller Graham considers history of precolumbian art in Central America, especially in relation to Mesoamerican studies. Terence Grieder looks at Central America in context of the ancient Americas as a whole. Rosemary Joyce reconsiders the eastern frontier of Mesoamerica and the Ulúa polychrome style. Oscar Zamora looks at implications of art for belief in Late Period Costa Rica. Peter Briggs analyzes precolumbian Central American mortuary remains. Richard Cooke considers animal imagery in relation to ritual in ancient Panama. Mary Helms analyzes color symbolism in precolumbian ceramics from Panama. Whitney Davis provides a post-modernist perspective on the study of ancient Central American art from the vantage point of a specialist in ancient Egyptian art. Fred Lange provides an overview of conceptual structure of Central American studies, challenging the usual Mesoamerica-centric perspective. [JSH]

150 **Resources, power, and interregional interaction.** Edited by Edward M. Schortman and Patricia A. Urban. New York: Plenum Press, 1992. 259 p.: bibl., ill., index, maps. (The language of science. Interdisciplinary contributions to archaeology.)

Collection of papers presents a variety of perspectives on importance and nature of interregional contacts between ancient societies. The set includes two papers concerned primarily with prehispanic Mesoamerica. J. Whitecotton uses a world-systems perspective to look at exchange in postclassic Oaxaca, and G. Feinman and L. Nicholas take a diachronic perspective to examine the changing relationship between Valley of Oaxaca and adjacent Ejutla Valley. [GF]

151 **Sabloff, Jeremy A.** Interpreting the collapse of classic Maya civilization: a case study of changing archaeological perspectives. (*in* Metaarchaeology. Edited by Lester Embree. The Hague: Kluwer Academic Publishers, 1992, p. 99–119, bibl., ill.)

Examines development of archaeological perspectives on ancient Maya during last 50 years, with particular attention to the problem of explaining collapse of classic-period Maya societies. [JSH]

152 **Sanders, William T.** Ecology and cultural syncretism in 16th-century Mesoamerica. (*Antiquity/Cambridge*, 66, 1992, p. 172–190, bibl., ill., maps, photos, tables)

Uses an ecosystem model to reconstruct events and processes that contributed to 16th-century transformation of Mesoamerica. Asserts that the driving forces—the demands of a dominant foreign population on a declining indigenous labor force—led to degradation of both the physical and human components of the Mesoamerican ecosystem. [GF]

153 **Schele, Linda.** The founders of lineages at Copán and other Maya sites. (*Anc. Mesoam.*, 3:1, Spring 1992, p. 135–144, bibl., ill., table)

Discusses ways in which rulers were

listed in a numbered succession at Copán and other classic Maya sites. Argues that while in most cases descent was traced back to a founding historical male figure, it sometimes could be drawn from a mythological ancestor. [R. Haskett and S. Wood]

154 Seler, Eduard. Collected works in Mesoamerican linguistics and archaeology. v. 3. Edited by John Eric Sidney Thompson, Francis B. Richardson, and Frank E. Comparato. 2nd ed. Culver City, Calif.: Labyrinthos, 1992. 1 v.: bibl., ill.

Vol. 3 of translations from the German of Eduard Seler's contributions to Mesoamerican studies. Includes one paper on early excavations at Templo Mayor, nine on iconography of Aztec artifacts, and 20 translations of Aztec hymns to deities (from Sahagún's manuscript found in the Biblioteca del Palacio) with commentary and analysis. [JSH]

155 Simposio de Investigaciones Arqueológicas en Guatemala, 5th, Guatemala, 1991. Actas. Edición de Juan Pedro Laporte, Héctor L. Escobedo A. y Sandra Villagrán de Brady. Guatemala: Ministerio de Cultura y Deportes; Instituto de Antropología e Historia; Asociación Tikal, 1992. 392 p.: bibl., ill.

Collection of papers describing recent archaeological and historical investigations in Guatemala is based on presentations at a 1991 symposium held at the Museo Nacional de Arqueología y Etnología (Guatemala). Majority of the papers deal with Maya lowlands of northern Guatemala from the formative through the colonial period: 19 report on archaeological field investigations, three analyze classic period iconography and hieroglyphic texts, and one explores nature of classic Maya states more generally. Remainder focus on investigation of prehistoric and colonial sites in highland Maya regions (11) and non-Maya Pacific coast (2), and on drawing techniques for presenting excavation data (1). [JSH]

156 Sluyter, Andrew. Long-distance staple transport in western Mesoamerica: insights through quantitative modeling. (*Anc. Mesoam.*, 4:2, Fall 1993, p. 193–199, bibl., graph, maps, table)

Quantitive model for the transport of maize between Zempoala and Tenochtitlán is used to establish possibility of overland, long-distance transport of staple goods in Mesoamerica. For geographer's comment see item 2406. [GF]

157 Smith, Michael Ernest. New World complex societies: recent economic, social, and political studies. (*J. Archaeol. Res.*, 1:1, March 1993, p. 5–41, bibl.)

Synthetic overview of recent research concerning complex indigenous societies in Latin America focuses on the following topics: intensive agriculture, demography, exchange, households, urbanism, chiefdoms, and state-level polities. Long bibliography of recent work is appended to the article. [GF]

158 Smith, Michael Ernest. Rhythms of change in postclassic central Mexico: archeology, ethnohistory, and the Braudelian model. (*in* Archaeology, *Annales*, and ethnohistory. Edited by A. Bernard Knapp. Cambridge, England: Cambridge Univ. Press, 1991, p. 51–74, bibl., ill., maps, photos, tables)

Applies insights from Braudel's historical work to the problem of correlating native history and archaeological data in the reconstruction of long-term change in postclassic highland Central Mexico. Argues that different components of this episode of transition occurred at distinct temporal cycles. [GF]

159 Solís Olguín, Felipe R. Gloria y fama mexica. Edición de Mario de la Torre. México: Smurfit Cartón y Papel de México, 1991. 181 p.: bibl., col. ill.

Large-format book with many color photographs focuses on various prehispanic Central Mexican topics, especially Tenochtitlán. Text is written in both English and Spanish. [GF]

160 Sotelo Santos, Laura Elena and **María del Carmen Valverde.** Jaguar y Ocelo-xochitl: iconografía de un emblema de poder. (*Apunt. Arqueol./Guatemala*, 1:2, dic. 1991, p. 21–37, bibl., ill.)

Analyzes jaguar and flower motifs in classic and postclassic period Maya art as symbols of royal power and of the underworld. [JSH]

161 Spencer, Charles S. Human agency, biased transmission, and the cultural evolution of chiefly authority. (*J. Anthropol. Archaeol.*, 12:1, March 1993, p. 41–74, bibl., maps, tables)

Argues that perpetuation of centralized leadership involved a dual process in which expansion of internal authority beyond the aspiring leader's home village was coupled with an increase in regularized external contacts that were directed and mediated by this ascendant individual. Examines chiefly societies in Tehuacán (Mexico) and Barinas (Venezuela) to illustrate the theoretical argument. [GF]

162 **Šprajc, Ivan.** Venus and Temple 22 at Copán revisited. (*Archaeoastronomy/College Park*, 10, 1987–1988, p. 88–97, bibl., ill.)

Analyzes alignment of a window in a late classic palace in the Maya city of Copán in western Honduras. Concludes that the window is designed to relate appearances of the planet Venus to the beginning of the rainy season and the cultivation of maize. [JSH]

163 **Šprajc, Ivan.** The Venus-rain-maize complex in the Mesoamerican world view. pt. 2. (*Archaeoastronomy/England*, 18, 1993, p. s27-s53, bibl., ill., photo)

Explores origin and development of the so-called Venus-rain-maize complex. Draws on material from across Mesoamerica to consider historical change in the complex. [GF]

164 **Sten, María.** Ponte a bailar, tú que reinas: antropología de la danza prehispánica. México?: Editorial Joaquín Mortiz, 1990. 180 p.: bibl., ill. (some col.).

Compilation and symbolic interpretation of dances practiced in highlands of ancient Mesoamerica draws principally on late prehispanic Central Mexican sources. [GF]

165 **Stuart, Gene S.** and **George E. Stuart.** Lost kingdoms of the Maya. Washington: National Geographic Society, 1993. 248 p.: bibl., ill. (some col.), index, map.

Popular volume on ancient Maya is organized by topic: cities, social life, merchants, the arts, religion, hieroglyphs, the calendar, the Spanish conquest. Embellished with superb color photographs of sites and of modern Maya life. [JSH]

166 **Tedlock, Barbara.** Mayans and Mayan studies from 2000 B.C. to A.D. 1992. (*LARR*, 28:3, 1993, p. 153–73, bibl., ill.)

Reviews 18 very miscellaneous volumes on Maya archaeology, ethnohistory, and ethnography. Emphasizes cultural continuities within Maya cultural tradition. [JSH]

167 To change place: Aztec ceremonial landscapes. Edited by David Carrasco. Boulder: Univ. Press of Colorado, 1991. 254 p.: bibl., ill., index.

Collection of essays by archaeologists, ethnohistorians, art historians, and historians of religion focuses on spatial aspects of Aztec ritual and relates this to topographical landscape of the Basin of Mexico. Also includes discussions of recent archaeological discoveries at the Templo Mayor, Tlatelolco, and Mt. Tlaloc. For ethnohistorian's comment see *HLAS 54:577*. [GF]

168 **Townsend, Richard F.** The Aztecs. London: Thames and Hudson, 1992. 224 p.: bibl., index. (Ancient peoples and places; 107)

Portrait of Aztec civilization draws on data from recent excavations, studies of Aztec monuments, Spanish records, and illustrated codices. Topics covered include Spanish conquest, emergence and expansion of the Aztec State, and everyday life of farmers, artisans, traders, and elites in late prehispanic Central Mexico. [GF]

169 **Valdez, Fred.** La decadencia de la civilización maya clásica: evidencias de las tierras bajas centrales. (*Yaxkin/Tegucigalpa*, 12:2, julio/dic. 1989, p. 45–63, ill., map)

Considers processes involved in transformation of Maya societies at the end of the classic period, with special attention to evidence from Colha in northern Belize. [JSH]

170 **Viramontes, Carlos.** La producción tradicional de sal en un sitio de la Mixteca Baja, Oaxaca: un estudio comparativo. (*Cuad. Sur/Oaxaca*, 2:4, mayo-agosto 1993, p. 5–25, bibl., map, photos)

Diachronic study examines salt sources of Silcayoapán, Mixteca Baja, Oaxaca, where salt manufacture has been carried out for more than a half millennium. Principal focus is process of salt-making, which is amplified through comparisons with other highland sites where salt was produced. [GF]

171 **Weaver, Muriel Porter.** The Aztecs, Maya, and their predecessors: archaeology of Mesoamerica. 3rd ed. San Diego, Calif.: Academic Press, 1993. 567 p.: bibl., ill., index.

Extensive revisions found throughout this updated edition of standard synthesis of prehistory of Mexico and upper Central America reflect results of recent research. [JSH]

172 **Webster, David; William T. Sanders, and Peter van Rossum.** A simulation of Copán population history and its implications. (*Anc. Mesoam.*, 3:1, Spring 1992, p. 185–197, bibl., graphs, maps, tables)

Richly detailed work reconstructs Copán's demographic trajectory and its sociopolitical implications. Authors identify what they believe to have been local solutions to a population exceeding the carrying capacity of the core region for an extended period. [R. Haskett and S. Wood]

173 **Whitmore, Thomas M. and Billie Lee Turner.** Landscapes of cultivation in Mesoamerica on the eve of the conquest. (*Ann. Assoc. Am. Geogr.*, 82:3, Sept. 1992, p. 402–425, bibl., ill., maps)

Explores indigenous cultivated landscapes present in Mesoamerica when the Spanish arrived. Specifically examines three transects: 1) from the Gulf Coast to Central Mexico; 2) through the Yucatán Peninsula from north to south; and 3) from the Pacific coastal plain into highland Guatemala. Discusses suite of factors behind the environmental changes that occurred following the Spanish conquest. For geographer's comment see item **2416**. [GF]

174 **Wilkerson, S. Jeffrey K.** Escalante's *entrada:* the lost Aztec garrison of the Mar del Norte in New Spain. (*Res. Explor.*, 9:1, 1993, p. 12–31, bibl., ill., maps, photos)

Discusses first major Spanish defeat by the Aztec, at Mar del Norte near Zempoala in Veracruz, and its significant effect on the process of conquest. [GF]

175 **Willey, Gordon.** Horizonal integration and regional diversity: an alternating process in the rise of civilizations. (*Am. Antiq.*, 56:2, April 1991, 97–215, bibl.)

Precolumbian cultural sequences in Mesoamerica and Peru show alternating periods of Horizon-style unification and regional stylistic diversity. Author argues that this cyclical process is vital to the rise of complex civilizations. [GF]

176 **Zantwijk, Rudolf A.M. van.** El concepto del "Imperio Azteca" en las fuentes históricas indígenas. (*Estud. Cult. Náhuatl*, 20, 1990, p. 201–211, bibl.)

Outlines eight indigenous political concepts that author argues are necessary to understand the Aztec. Contrasts these views of power and social relations with political frame of reference held by 16th-century Europeans.[GF]

177 **Zeitlin, Robert N.** The energetics of trade and market in the early empires of Mesoamerica. (*Res. Econ. Anthropol.*, 13, 1991, p. 373–386, bibl.)

Presents critique of previously published models that have considered energy to be a reliable measure for long-distance procurement of bulk goods in prehispanic Mesoamerica. Argues that such energetics models are more appropriately conceived as provisional idealizations of procurement behavior rather than as accurate reflections of ancient economics. [GF]

FIELDWORK AND ARTIFACTS

178 **Abe, Masae.** Los sitios monumentales en la zona norte del Valle de Florida. (*Yaxkin/Tegucigalpa*, 11:2, 1988, p. 71–87, bibl., ill.)

Provides brief descriptions of three large centers with monumental architecture in La Florida valley of northwestern Honduras. [JSH]

179 **Acosta, Jorge R. and Javier Romero Molina.** Exploraciones en Monte Negro, Oaxaca: 1937–38, 1938–39 y 1939–40. Recopilación de José Luis Ramírez Ramírez. Coordinación de Lorena Mirambell Silva. México: Instituto Nacional de Antropología e Historia, 1992. 188 p.: bibl., ill., maps. (Antologías: Serie Arqueología)

Well-illustrated book presents findings from Acosta's archaeological investigations, completed more than 50 years ago at Monte Negro. This volume now represents the most complete published reference to this important formative period site in the Mixteca Alta. [GF]

180 **Adams, Richard E.W. and Hubert R. Robichaux.** Tombs of Río Azul, Guatemala. (*Natl. Geogr. Res.*, 8:4, 1992, p. 412–27, bibl., ill.)

Work surveys painted tombs from early classic city in northern Guatemala. Río Azul maintained close relationships with larger city of Tikal, which in turn shows indications of interaction with central Mexican city of Teotihuacán. [JSH]

181   Aldenderfer, Mark S. Defining lithics-using craft specialties in lowland Maya society through microwear analysis: conceptual problems and issues. (*in* The interpretative possibilities of microwear studies. Uppsala, Sweden: Societas Archaeologica Upsaliensis, 1990, p. 53–70, bibl., ill.)

Examines microwear traces on examples of three tool types common in Late Classic contexts in southern Maya lowlands, and explores nuances of drawing inferences from these data about the nature of household activities and craft production. [JSH]

182   **Anatomía de un conjunto residencial teotihuacano en Oztoyahualco: las excavaciones.** Edited by Linda Manzanilla. México: Instituto de Investigaciones Antropológicas, UNAM, 1993. 2 v.

Thorough, two-volume study examines prehispanic multi-room residential complex situated in northwestern sector of Teotihuacán. Describes architecture and activity areas that were defined through broad horizontal excavations. Various sections discuss ceramic artifacts and other categories of material remains; report on soil chemical analyses; and present an osteological analysis of recovered burials. [GF]

183   Andrews, Anthony P. and Gabriela Vail. Cronología de sitios prehispánicos costeros de la Península de Yucatán y Belice. (*Bol. Esc. Cienc. Antropol. Univ. Yucatán*, 18:104/105, 1990, p. 37–57, bibl., ill.)

Overview of occupation history of coastal Yucatán (including areas in the modern republics of Mexico and Belize) includes summary data on chronology of occupation at some 400 known sites, of which approximately 50 have had at least exploratory excavations. [JSH]

184   *Antropológicas.* Vol. 1, enero 1992– . México: Instituto de Investigaciones Antropológicas, Univ. Nacional Autónoma de México.

First issue of this new publication is comprised of 11 articles that briefly present history of archaeological research at various Mesoamerican sites and regions including Teotihuacán, Monte Albán, North Mexico, Michoacán, Oaxaca, Baja California, and the Maya area. [GF]

185   Aoyama, Kazuo. El estudio de la lítica en la región de La Entrada, Honduras. (*Yaxkin/Tegucigalpa*, 12:2, julio/dic. 1989, p. 65–99, ill., graphs, maps, tables)

Study of more than 11,000 chipped and ground stone artifacts from sites in La Entrada region of western Honduras includes preliminary reconstruction of patterns of production, distribution, and use mainly during middle preclassic and late classic periods. [JSH]

186   Aoyama, Kazuo. Observaciones preliminares sobre la lítica menor en el Valle de La Venta, Honduras. (*Yaxkin/Tegucigalpa*, 11:2, 1988, p. 45–69, bibl., ill.)

Summarizes morphological analysis of chipped stone from excavations at one middle preclassic and several late classic sites in La Florida valley of northwestern Honduras. Geologic sources of the obsidian artifacts are determined on the basis of visual comparison with source samples. [JSH]

187   Aoyama, Kazuo. Socioeconomic implications of chipped stone from the La Entrada region, western Honduras. (*J. Field Archaeol.*, 21:1, 1994, p. 133–45, bibl., ill.)

Analyzes changes in production and distribution of lithic artifacts in La Entrada region of western Honduras between middle preclassic and late classic periods. Specialized production of prismatic blades made of imported obsidian began in this region in late preclassic to early classic time range; by late classic period Copán was in control of distribution from Ixtepeque in this region and throughout southeastern Maya lowlands. [JSH]

188   Arana A., Raúl Martín. Proyecto Coatlán, área Tonatico-Pilcaya. México: Instituto Nacional de Antropología e Historia, 1990. 243 p.: bibl., ill., maps. (Col. científica; 200. Serie Arqueología)

Reports on 1980 surface surveys that recorded 36 archaeological sites in the terrain of Tonatico and Ixtapán de la Sal in southwestern part of the state of Mexico, and in Pilcaya and Teticpac in northern Guerrero. Includes site plans, photographs, and ceramic illustrations. [GF]

189   Arnold, Philip J. Dimensional standardization and production scale in Mesoamerican ceramics. (*Lat. Am. Antiq.*, 2:4, Dec. 1991, p. 363–370, bibl., tables)

Findings from the study of contemporary seasonal potters in southern Veracruz indicate that nonintensive ceramic producers can produce standardized products. Standardization is argued to be neither a sufficient nor necessary attribute of the goods made by craft specialists. [GF]

190   Arnold, Philip J. Domestic ceramic production and spatial organization: a Mexican case study in ethnoarchaeology. Cambridge, England; New York: Cambridge Univ. Press, 1991. 177 p.: bibl., ill., index. (New studies in archaeology)

Based on an ethnoarchaeological study of contemporary ceramic production and consumption at four communities in Los Tuxtlas region of Veracruz, author offers criteria to help archaeologists more consistently recognize prehispanic ceramic manufacture. Spatial concepts developed in the study are applied to recently documented ceramic production at Matacapan. [GF]

191   Arnold, Philip J., et al. Intensive ceramic production and classic-period political economy in the Sierra de los Tuxtlas, Veracruz, Mexico. (Anc. Mesoam., 4:2, Fall 1993, p. 175–191, bibl., ill., maps, photo)

Archaeological research at the Comoapán complex, an area on the southern edge of Matacapan, indicates that pottery manufacture was carried out intensively and on a large scale during the Classic period. Infers that the final ceramic products were distributed beyond the site and possibly outside the Tuxtlas. [GF]

192   Arnold, Philip J. The organization of refuse disposal and ceramic production within contemporary Mexican houselots. (Am. Anthropol., 92:4, Dec. 1990, p. 915–932, bibl., graphs, maps, photos, tables)

Ethnoarchaeological investigation of ceramic production serves as basis to illustrate how spatial constraints affect organization of craft production. [GF]

193   Arqueología. No. 2, julio/dic. 1989–. México: Dirección de Arqueología, Instituto Nacional de Antropología e Historia.

Includes seven articles on diverse topics: "La prehistoria en México y Centroamérica" (F. Rodriguez-Loubet); "Mirador-Plumajillo, Chiapas, y sus relaciones con cuatro sitios del horizonte olmeca en Veracruz, Chiapas y la Costa de Guatemala" (P. Agrinier); "Nuevo testimonio rupestre olmeca en el oriente de Guerrero" (S.L. Villela F.); "Una interpretación sobre el significado y función de la Estructura II de Hormiguero, Campeche" (L.A. Martos López); "Una genealogía zapoteca prehispánica" (J. Urcid and M. Winter); "Una etnografía arqueológica de la producción tradicional de sal en Nexquipayac, estado de México" (J.F. Parsons); and "La producción de sal en un sitio del postclásico tardío" (M. de J. Sánchez Vázquez). [GF]

194   Arqueología. No. 3, enero/junio 1990–. México: Dirección de Arqueología, Instituto Nacional de Antropología e Historia.

Contains 11 papers presented at a conference on the Olmec and their neighbors, held in 1989 as part of the 21st Mesa Redonda of the Sociedad Mexicana de Antropología. Includes papers on La Venta, Chalcatzingo, sites on the Pacific Coast of Guatemala and Chiapas, the colossal Olmec heads, and general discussions of the Olmec phenomenon. [GF]

195   Arqueología Mexicana. Vol. 1, No. 1, 1993–. México: Instituto Nacional de Antropología e Historia; Editorial Raíces.

New glossy journal has many color photographs and illustrations. Published by Mexico's Instituto Nacional de Antropología e Historia, periodical is designed to present the rich prehispanic history of Mexico to a wider audience. This first issue is devoted to Teotihuacán. [GF]

196   Arqueología Mexicana. Vol. 1, No. 3, 1993–. México: Instituto Nacional de Antropología e Historia; Editorial Raíces.

Entire issue of this lavish new journal focuses on Monte Albán and Oaxaca anthropology. Contributions address archaeological sequence for the region, origins of Monte Albán, art and architecture of Mitla, collected works of Ignacio Bernal, and diet of indigenous people in contemporary Oaxaca. [GF]

197   Arqueología Mexicana. Vol. 1, No. 4, 1993–. México: Instituto Nacional de Antropología e Historia; Editorial Raíces.

Elaborately illustrated issue focuses on Tenochtitlán and the Mexica. Articles discuss city's mythological past, Quetzalcoatl, warfare, the chinampas, and Mexica writing. [GF]

198   Arqueología Mexicana. Vol. 1, No. 5, 1994–. México: Instituto Nacional de Antropología e Historia; Editorial Raíces.

Archaeology of El Tajín and the Gulf Coast are the focus of this issue. Superbly illustrated pieces also examine Cempoala, la Mixtequilla Clásica, two ancient Veracruz dances (Volador and Huahua) that are still practiced today, the history and use of vanilla (*tlilxóchitl*), and an ethnoarchaeological consideration of ceramic production in the Sierra de los Tuxtlas. [GF]

199 *Arqueología Mexicana.* Vol. 1, No. 6, 1994– . México: Instituto Nacional de Antropología e Historia; Editorial Raíces.

Collection concentrates on Paquimé and North Mexico. General discussion of this desert frontier is followed by sections on: architecture of Paquimé (Casas Grandes); La Quemada site; prehispanic mining in Chalchihuites, Zacatecas; cave dwellings in Chihuahua; nomads of the coastal desert; and the life history of Gerónimo. [GF]

200 **Arroyo, Bárbara.** El formativo temprano en Chiapas, Guatemala y El Salvador. (*U tz'ib*, 1:1, nov. 1991, p. 7–14, bibl., maps)

Summarizes occupation history of Pacific coastal region of southern Mesoamerica, with emphasis on ceramic chronology, adaptation to the various environmental zones, and emergence of early complex societies. [JSH]

201 **Arroyo-Cabrales, Joaquín** and **Ticul Alvarez.** Restos óseos de murciélagos procedentes de las excavaciones en las grutas de Loltún. México: Instituto Nacional de Antropología e Historia, 1990. 103 p.: bibl., ill. (Col. científica; 194. Serie Prehistoria)

Reports on a study of bat remains from the caves of Loltún, Yucatán. In total, 24 bat species were identified, two of which had not been reported previously in the Yucatán Peninsula. [GF]

202 **Ashmore, Wendy.** Excavaciones en el sitio central de Gualjoquito, Santa Bárbara, Honduras, 1983–1985. (*Yaxkin/Tegucigalpa*, 10:2, julio/dic. 1987, p. 89–104, bibl., maps)

Summarizes 1983–85 excavations at Gualjoquito on the middle course of Río Ulúa in Honduras. Excavations focused on the site core and defined an occupation sequence extending from late preclassic through early postclassic. [JSH]

203 **Avila López, Raúl.** Chinampas de Iztapalapa, D.F. México: Instituto Nacional de Antropología e Historia, 1991. 183 p.: bibl., ill. (Col. científica; 225. Serie Arqueología)

Work examines Iztapalapa chinampas and associated prehispanic settlement. Modern development that is destroying these ancient agrarian features prompted initiation of the study in 1981. The Iztapalapa chinampas were built in the 13th century, with major growth occuring in the 15th century when area was dominated by Tenochtitlán. [GF]

204 **Avila López, Raúl** and **Ludwig Beutelspacher.** Investigaciones arqueológicas en Mexicaltzingo, D.F. México: Dirección de Salvamento Arqueológico, Instituto Nacional de Antropología e Historia, 1989. 104 p.: bibl., ill. (Cuaderno de trabajo; 7)

Reports on salvage excavations at a prehispanic ceremonial center situated below the colonial church of San Marcos Mexicaltzingo in southern Basin of Mexico. Discusses excavated prehispanic architecture and associated materials recovered during the excavations. [GF]

205 The Balberta Project: the terminal formative-early classic transition on the Pacific Coast of Guatemala = El Proyecto Balberta: la transición entre el formativo terminal y el clásico temprano en la Costa Pacífica de Guatemala. Edited by Frederick Joseph Bove *et al.* Translated by Jeffrey P. Blick, Ana María Boza-Arlotti, and Alvaro Higueras-Hare. Pittsburgh: Univ. of Pittsburgh, Dept. of Anthropology; Guatemala: Asociación Tikal, 1993. 201 p.: bibl., ill. (University of Pittsburgh memoirs in Latin American archaeology; 6)

Reports on investigations at the large early center on the Pacific coast of Guatemala. Includes essays on chronological and cultural context of the site, a description of excavations in the central sector, and analyses of ceramics, spindle whorls, obsidian, and burials. [JSH]

206 **Baños Ramos, Eneida.** Distribución de cerámicas prehispánicas en Tlatelolco-Tenochtitlán. (*Estud. Cult. Náhuatl*, 23, 1993, p. 221–249, bibl., ill., maps, photo, table)

Discusses 1984–87 salvage excavations at Tlatelolco. Main discussion concerns early and late postclassic ceramic assemblages that were recovered during the research. [GF]

207 **Barba Pingarrón, Luis Alberto** and **Agustín Ortiz.** Análisis químico de pisos de ocupación: un caso etnográfico en Tlaxcala, Mexico. (*Lat. Am. Antiq.*, 3:1, March 1992, p. 63–82, bibl., ill., map, photos)

Chemical analysis of modern habitational floors in San Vicente Xiloxochitla demonstrates utility of these procedures for identifying different use areas within the residential structure. Archaeological application of these techniques is stressed. [GF]

208 **Baudez, Claude-François.** Maya sculpture of Copán: the iconography. Norman: Univ. of Oklahoma Press, 1994. 300 p.: bibl., ill., index.

Detailed analysis examines iconography of classic period sculpture from the Maya city in northwestern Honduras. Includes detailed descriptions, photographs, and drawings of most of the major free-standing sculpture (stelae, "altars," etc.), and of much architectural sculpture. Iconographic interpretation reflects a structural approach. [JSH]

209 **Berdan, Frances.** Economic dimensions of precious metals, stones, and feathers: the Aztec state society. (*Estud. Cult. Náhuatl*, 22, 1992, p. 291–323, bibl., maps, tables)

Presents economic analysis of luxury goods exchanged in Aztec times. Discusses the different mechanisms—tribute, long-distance exchange, and markets—by which these items were distributed. Argues that in the Aztec economy tribute played a relatively minor role in the movement of precious goods. [GF]

210 *Boletín del Consejo de Arqueología.* 1990–. México: Instituto Nacional de Arqueología e Historia.

Reports on 63 recent and ongoing field projects in all regions of Mexico. Includes maps, site plans, and photographs. [GF]

211 *Boletín del Consejo de Arqueología.* 1991–. México: Instituto Nacional de Antropología e Historia.

Reports on 67 recent and ongoing field projects in all regions of Mexico. Includes maps, site plans, and photographs. [GF]

212 **Brady, James E.** and **George Veni.** Man-made and pseudo-karst caves: the implications of subsurface features within Maya centers. (*Geoarchaeology/New York*, 7:2, 1992, p. 149–167, bibl., ill.)

Describes caves, modified or entirely constructed by human labor, within archaeological sites in highland Guatemala, mainly of the postclassic period. Continuing modern ritual use of these caves, as well as their relationship to site layout, suggest their symbolic importance. [JSH]

213 **Braniff C., Beatriz.** La estratigrafía arqueológica de Villa de Reyes, San Luis Potosí. México: Instituto Nacional de Antropología e Historia, 1992. 182 p.: bibl., ill., maps. (Col. Científica; 265. Serie Arqueología)

Reports on excavations completed in 1966–67 at archaeological site of Villa de Reyes in the San Francisco Valley of San Luis Potosí, an area known archaeologically as Tunal Grande. Focus is on recovered ceramics and lithics, with much shorter discussions of the bone (both animal and human) and shell assemblages. [GF]

214 **Braniff C., Beatriz.** La frontera protohistórica Pima-Opata en Sonora, México: proposiciones arqueológicas preliminares. v. 1. México: Instituto Nacional de Antropología e Historia, 1992. 1 v.: bibl., maps, photos. (Col. Científica; 240–242. Serie Arqueología)

Uses ecological, archaeological, and documentary data to argue that the Río San Miguel in Sonora was a prehispanic cultural frontier between Mesoamerica and the cultural area of Northwest Mexico (which author sees as including the US Southwest). [GF]

215 **Braswell, Geoffrey E.** and **Michael D. Glascock.** A new obsidian source in the highlands of Guatemala. (*Anc. Mesoam.*, 3:1, 1992, p. 47–49, bibl., map, table)

Reports discovery of previously unknown obsidian source near Sansare in Guatemalan highlands, and provides trace element composition data based on neutron activation analysis. [JSH]

216 **Braswell, Geoffrey E.** Obsidian hydration dating, the Coner phase, and revisionist chronology at Copán, Honduras. (*Lat. Am. Antiq.*, 3:2, 1992, p. 130–147, bibl.)

Discusses sources of error in obsidian hydration dating, particularly the determination of rates and measurement of effective hydration temperature. Argues that recent interpretations of hydration dates from Copán that claim a continuation of basic patterns of late classic rural life into the postclassic period are premature because calculation of dates did not involve measurements of effective hydration temperature. [JSH]

217   Brown, Roy B. Arqueología y paleoecología del norcentro de México. México: Instituto Nacional de Antropología e Historia, 1992. 123 p.: bibl., ill., maps. (Col. Científica; 262. Serie Arqueología)

Describes research undertaken to test Pedro Armilla's hypothesis concerning role of environmental change in cultural transitions. Author summarizes archaeological and paleoecological data from north-central Mexico, and concludes that Armillas' hypothesis cannot be supported. [GF]

218   Brüggemann, Jürgen K.; Armando Pereyra Quinto; and Jaime Cortés Hernández. La cuenca del Actopan inferior: el análisis estadístico de un área. (*An. Antropol.*, 26, 1989, p. 15–97, bibl., maps, photos, tables)

Presents detailed report of prehispanic settlements recorded during a surface reconnaissance conducted along coastal strip of northern Veracruz, in an area that forms part of lower Actopan River Basin. Includes maps of topographic features and structures, as well as a discussion of recovered ceramic materials. [GF]

219   Brumfiel, Elizabeth M. Tribute and commerce in imperial cities: the case of Xaltocán, Mexico. (*in* Early state economics. Edited by H.J.M. Claessen and Piet van de Velde. New Brunswick, N.J.: Transaction Publishers, 1991, p. 177–198, bibl., maps, tables)

Interweaves archaeological and documentary findings to illustrate that the transformation of the late prehispanic center of Xaltocán from a tribute-receiving to a tribute-paying community was accompanied by a moderate decline in commercial activity. Argues that tribute and trade were mutually reinforcing in late postclassic Central Mexico, and that they should not be viewed as functional alternatives. [GF]

220   Cabrera Castro, Rubén; Saburo Sugiyama; and George L. Cowgill. The Templo de Quetzalcoatl project at Teotihuacán: a preliminary report. (*Anc. Mesoam.*, 2:1, Spring 1991, p. 77–92, bibl., ill., photos)

Describes project's principal findings, including recovery of 200 sacrificial victims and associated burial goods that were interred when the temple was constructed. Excavations have revealed that the temple itself was erected in a single episode during Miccaotli or Early Tlamimilolpa times. [GF]

221   Cabrera Guerrero, Martha Eugenia. Los pobladores prehispánicos de Acapulco: Proyecto Arqueológico Renacimiento. México: Instituto Nacional de Antropología e Historia, 1990. 245 p.: appendices, bibl., ill. (Col. científica; 211. Seria Arqueología)

Reports on salvage investigations in Acapulco, Guerrero. Research was conducted at six sites identified by the project as being in immediate danger of destruction. Recovered materials date from between TOM phase of middle formative to YAX phase of the classic. Two appendices present the petroglyphs recorded at two sites: R2 La Sabana and Palma Sola. [GF]

222   Cabrero García, María Teresa. Civilización en el norte de México. México: Univ. Nacional Autónoma de México, Instituto de Investigaciones Antropológicas, 1989. 360 p., 2 folded leaves of plates: bibl., ill. (Serie antropológica; 103. Arqueología)

Discusses archaeological survey and test excavations at 68 sites in La Cañada del Río Bolaños in Zacatecas and Jalisco. These sites date to period between 100 BC and AD 1100. Discussions focus on ceramic typology, chronology, settlement patterns, and social organization. Volume includes photos and drawings of the tested sites. [GF]

223   Cabrero García, María Teresa. Rescate arqueológico en Culiacán, Sinaloa. (*Antropológicas/México*, 3, 1989, p. 39–65, bibl., ill., maps, photos)

Reports on salvage excavations in a barrio of Culiacán. Five funerary urns, four direct primary burials, and a probable house floor were among the discoveries. [GF]

224   Cárdenas García, Efraín. Fases de ocupación prehispánica en la Cuenca de Pátzcuaro. (*An. Mus. Michoacano*, 2, 1990, p. 25–43, bibl., graphs, maps)

Presents seriation of ceramic materials surface-collected in 1983 from prehispanic settlements in Michoacán's Pátzcuaro Basin by the Pátzcuaro-Cuitzeo project. Proposed chronology suggests that more sites in the study region were inhabited during the postclassic than the classic period. [GF]

225   Carrasco, David and Eduardo Matos Moctezuma. Moctezuma's Mexico: visions of the Aztec world. Niwot, Colo.: Univ.

Press of Colorado, 1992. 188 p.: bibl., ill. (some col.), index.

Large, lavishly illustrated book is written for a general audience. Sculpture, architecture, and archaeological data provide basis for this discussion of Aztec history and cosmology. Also considers Aztec religion, archaeoastronomy, and indigenous concepts of land and community. [GF]

226   **Ceramic production and distribution: an integrated approach.** Edited by George J. Bey and Christopher A. Pool. Boulder, Colo.: Westview Press, 1992. 342 p.: bibl., ill., maps.

Edited collection contains three articles on Mesoamerica: 1) "Ceramic production in prehistoric La Mixtequilla, south-central Veracruz" (B. Stark); 2) "Middle classic pottery economics in the Tuxtla Mountains, southern Veracruz, Mexico" (C. Pool and R. Santley); and 3) "Ceramic production and distribution in late postclassic Oaxaca: stylistic and petrographic perspectives (G. Feinman *et al.*). [GF]

227   **Charlton, Cynthia L. Otis.** Obsidian as jewelry: lapidary production in Aztec Otumba, Mexico. (*Anc. Mesoam.*, 4:2, Fall 1993, p. 231–243, bibl., ill., maps, photos, tables)

Detailed presentation provides evidence for lapidary jewelry production at late postclassic city-state of Otumba. The careful study of this production site has enabled author to reconstruct techniques that were used to make ear spools, lip plugs, and beads from obsidian. [GF]

228   **Charlton, Thomas H.; Deborah L. Nichols; and Cynthia L. Otis Charlton.** Aztec craft production and specialization: archaeological evidence from the city-state of Otumba, Mexico. (*World Archaeol.*, 23:1, 1991, p. 99–114, bibl., maps)

Evidence on craft production at Otumba is used to evaluate two alternative models of city-state development in the Basin of Mexico subsequent to fall of Tula. Contrasts Otumba findings with earlier reconstructions based on previous research at Huexotla. [GF]

229   **Chinampas prehispánicas.** Recopilación de Carlos Javier González. México: Instituto Nacional de Antropología e Historia, 1992. 285 p.: bibl., ill. (Antologías: Serie Arqueología)

Collection of papers discusses various aspects of prehispanic chinampas in Central Mexico. The role of chinampas in sustaining large urban populations is stressed, as is importance of conserving them and promoting their use for future generations. [GF]

230   **Cobean, Robert H. *et al.*** High-precision trace-element characterization of major Mesoamerican obsidian sources and further analyses of artifacts from San Lorenzo Tenochtitlán, Mexico. (*Lat. Am. Antiq.*, 2:1, March 1991, p. 69–91, bibl., ill., maps, tables)

Reports on the instrumental neutron-activation analysis of 208 obsidian samples from 25 Mesoamerican obsidian sources. Compares these findings to 65 artifacts from the site of San Lorenzo that also were sourced. Although these latter artifacts could be matched to source areas, specific quarry sites or obsidian flows could not be definitively determined. [GF]

231   **Comalcalco.** Recopilación de Elizabeth Mejía Pérez Campos. Coordinación de Lorena Mirambell Silva. México: Instituto Nacional de Antropología e Historia, 1992. 335 p.: bibl., ill., maps. (Antologías: Serie Arqueología)

Provides useful compilation of material on westernmost Maya city of Comalcalco on Tabasco's Gulf coastal plain. Includes early travelers' accounts, a synthesis of regional archaeology, and overviews of conquest-period cultural and physical geography, as well as more focused investigations of Comalcalco itself. [JSH]

232   *Cuadernos de Arquitectura Mesoamericana.* No. 12, sept. 1991– . Jornadas de Arquitectura Prehispánica en Mesoamerica, I, pt. 1. México: Facultad de Arquitectura, UNAM.

Dedicated to Paul Gendrop, articles in this issue cover range of topics, with a focus on Río Bec, Uxmal, and other sites in the Yucatán Peninsula. One selection considers the carved stone heads at San Lorenzo Tenochtitlán. [GF]

233   *Cuadernos de Arquitectura Mesoamericana.* No. 13, oct. 1991– . Jornadas de Arquitectura Prehispánica en Mesoamérica, I, pt. 2. México: Facultad de Arquitectura, UNAM.

Contains articles dedicated to Paul Gendrop on various aspects of architecture (including domestic construction) at Teotihuacán, with several other contributions focused on prehispanic Chiapas. [GF]

234   *Cuadernos de Arquitectura Mesoamericana.* No. 15, dic. 1991– . Jornadas de Arquitectura Prehispánica en Mesoamérica, II. México: Facultad de Arquitectura, UNAM.

Series of articles dedicated to Ricardo de Robina covers range of general topics about Mesoamerican architecture, including discussions of Xochicalco, Teotihuacán, and the Maya area. [GF]

235   *Cuadernos de Arquitectura Mesoamericana.* No. 16, enero 1992– . Teoría e Historia del Urbanismo en México: Epoca Prehispánica, 1. México: Facultad de Arquitectura, UNAM.

Issue concentrates on the history of urbanism in Mexico, with articles on Teotihuacán and various sites in Guatemala. [GF]

236   *Cuadernos de Arquitectura Mesoamericana.* No. 18, marzo 1992– . Arquitectura de Oaxaca, 2. México: Facultad de Arquitectura, UNAM.

Focus of articles in this issue is on prehispanic architecture at Monte Albán and other sites in Oaxaca. [GF]

237   *Cuadernos de Arquitectura Mesoamericana.* No. 19, abril 1992– . Jornadas de Arquitectura Prehispánica en Mesoamérica, III. México: Facultad de Arquitectura, UNAM.

This volume, dedicated to Horst Hartung, compiles archaeoastronomical studies conducted at a series of prehispanic Mesoamerican sites. [GF]

238   *Cuadernos de Arquitectura Mesoamericana.* No. 23, enero 1993– . Teoría e Historia del Urbanismo en México: Epoca Prehispánica, 2. México: Facultad de Arquitectura, UNAM.

Theme of this issue is history of urbanism in Mexico. Several articles focus on Tenochtitlán; other contributions discuss research at Uaxactún and Paquimé (Casas Grandes). [GF]

239   *Cuadernos de Arquitectura Mesoamericana.* No. 24, feb. 1993– . Arquitectura del Altiplano, 2. México: Facultad de Arquitectura, UNAM.

This issue is focused on architecture in highland regions, with pieces on Xochicalco and other locations in Morelos. [GF]

240   *Cuadernos de Arquitectura Mesoamericana.* No. 25, marzo 1993– . Arquitectura de Centro y Occidente, 1. México: Facultad de Arquitectura, UNAM.

Theme of this issue is west and central Mexico, with articles covering a range of topics concerning sites in Guanajuato. [GF]

241   **Curet, Antonio.** Regional studies and ceramic production areas: an example from La Mixtequilla, Veracruz, Mexico. (*J. Field Archaeol.*, 20:4, 1993, p. 427–440, bibl., ill., maps, tables)

Author experimentally investigates factors that may account for general scarcity of direct evidence for specialized pottery production (e.g., overfired kilnwasters) recovered by surface survey projects. Quantitative findings from La Mixtequilla survey are then analyzed to identify comal production at several postclassic locations at the site. [GF]

242   **Darling, J. Andrew.** Notes on obsidian sources of the southern Sierra Madre Occidental. (*Anc. Mesoam.*, 4:2, Fall 1993, p. 245–253, bibl., maps, photos)

Describes a previously unknown obsidian source in the Huitzila-La Lobera area of southern Zacatecas/northern Jalisco. A comparison is drawn to the Llano Grande/Cerro Navajas source area in Durango. [GF]

243   **Day, Jane Stevenson.** Treasures from the Templo Mayor. (*Archaeology/New York*, 45:3, Sept./Oct. 1992, p. 42–47, photos)

Describes an exhibition at the Denver Museum of Natural History that focused on Aztec culture in 1518. [GF]

244   **Díaz Oyarzábal, Clara Luz.** Cerámica de sitios con influencia teotihuacana: catálogo de las colecciones arqueológicas del Museo Nacional de Antropología. México: Instituto Nacional de Antropología e Historia, 1991. 240 p.: ill.

Catalog describes Teotihuacán-style ceramics from various sites in central highlands of Mexico. Because of the large volume of Teotihuacán ceramics in Mexico's Museo Nacional de Antropología, this work features only those pieces that are in storage at the museum, and not on display. [GF]

245   **Díaz Oyarzábal, Clara Luz.** Colección de objetos de piedra, obsidiana, concha, metales y textiles del Estado de Guerrero: Museo Nacional de Antropología. México: Instituto Nacional de Antropología e Histo-

ria, 1990. 265 p.: bibl., ill., map. (Col. Catálogos de museos)

Catalog describes non-ceramic archaeological materials from Guerrero that are housed in the Museo Nacional de Antropología. Provides basic information, including a photograph and the catalog number, for each artifact. [GF]

246 **Dixon, Boyd W.** Estudio preliminar sobre el patrón de asentamiento del Valle de Comayagua: corredor cultural prehistórico. (*Yaxkin/Tegucigalpa,* 12:1, 1989, p. 40–76, bibl., ill.)

Summary of settlement history in the Comayagua valley in central Honduras includes comments on changing patterns of external relationships maintained by societies located astride a major natural route of communication. [JSH]

247 **Dixon, Boyd W. et al.** Formative-period architecture at the site of Yarumela, central Honduras. (*Lat. Am. Antiq.,* 5:1, 1994, p. 70–87, bibil., ill.)

Summarizes recent excavations at large formative center in Comayagua valley of central Honduras. Traces history of monumental architecture during middle and late formative periods in relation to other sociopolitical complexity indicators. [JSH]

248 **Dixon, Boyd W.** Prehistoric political change on the southeast Mesoamerican periphery. (*Anc. Mesoam.,* 3:1, 1992, p. 11–25, bibl., maps)

Analysis of changing settlement patterns in the Comayagua Valley of central Honduras suggests series of shifts in political organization: Yarumela's late preclassic centralized power gave way to political decentralization and regionalization in the classic period, which was followed by another episode of centralized political control by Las Vegas in the early postclassic, and then by a period of decentralization lasting until the Spanish invasion. [JSH]

249 **Dixon, Keith A.** La Cueva de la Pala Chica: a burial cave in the Guaymas region of coastal Sonora, Mexico. Nashville, Tenn.: Vanderbilt Univ., 1990. 97 p.: bibl., ill. (Vanderbilt University publications in anthropology; 38)

Provides descriptive account of late prehistoric/early historic burial cave. Includes skeletal analysis for the six interred individuals, as well as description of artifactual materials (including shell beads and fiber cordage). Suggests a Seri cultural affiliation. [GF]

250 **Dockall, John E. and Harry J. Shafer.** Testing the producer-consumer model for Santa Rita Corozal, Belize. (*Lat. Am. Antiq.,* 4:2, 1993, p. 158–79, bibl., ill.)

Analyzes production and consumption of stone tools in northern Belize during late preclassic period. Colha was a production center for artifacts made from locally available chert; Santa Rita Corozal, outside the chert-producing zone, represents a consumer community. [JSH]

251 **Douglas, John E.** Distant sources, local contexts: interpreting nonlocal ceramics at Paquimé, Casas Grandes, Chihuahua. (*J. Anthropol. Res.,* 48:1, Spring 1992, p. 1–24, bibl., ill., map)

Argues that nonlocal ceramics played a smaller role in interregional exchange at Paquimé than did other exotic goods. Discusses implications of this finding, and proposes a framework for understanding nonlocal goods acquisition that is not based on institutionalized networks. [GF]

252 **Dreiss, Meredith L. et al.** Expanding the role of trace-element studies: obsidian use in the late and terminal classic periods at the lowland Maya site of Colha, Belize. (*Anc. Mesoam.,* 4:2, 1993, p. 271–283, bibl., ill.)

Reports results of trace element analysis of 200 obsidian blades from late to terminal classic contexts at the Maya site in Belize. Explores variability in functional and social context as well as source area and time period. [JSH]

253 **Dunning, Nicholas P.** Lords of the hills: ancient Maya settlement in the Puuc region, Yucatán, Mexico. Madison, Wis.: Prehistory Press, 1992. 303 p.: bibl., ill. (Monographs in world archaeology, 1055–2316; 15)

Detailed study examines settlement patterns of Puuc hill region of western Yucatán in context of environmental conditions, especially soils and vegetation. Emphasis is on late and terminal classic population, settlement systems, and territorial organization. [JSH]

254 **Dunning, Nicholas P. and Timothy Beach.** Soil erosion, slope management, and ancient terracing in the Maya low-

lands. (*Lat. Am. Antiq.*, 5:1, 1994, p. 51–69, bibl., ill.)

Analyzes agricultural terracing in Petexbatun region of Guatemala and makes comparison with other regions. Most Maya terrace systems appear to have been constructed in late classic period, although there are indications of earlier origins in some cases. [JSH]

255   **Epstein, Jeremiah.** Cabeza de Vaca and the sixteenth-century copper trade in northern Mexico. (*Am. Antiq.*, 56:3, July 1991, p. 474–482, bibl., map)

Proposes that copper objects found among native peoples of northern Mexico by Cabeza de Vaca and others in 16th century were looted from abandoned Paquimé and were not obtained directly through trade with either Mesoamerica or west Mexico. [GF]

256   **Fähmel Beyer, Bernd Walter Federico.** La arquitectura de Monte Albán. México: Univ. Nacional Autónoma de México, Instituto de Investigaciones Antropológicas, 1991. 201 p.: bibl., map.

New analysis of excavated structures at this major prehispanic center provides descriptive detail on construction sequences for many of the monumental buildings on the Main Plaza. [GF]

257   **Fash, Barbara et al.** Investigations of a Classic Maya council house at Copán, Honduras. (*J. Field Archaeol.*, 19:4, 1992, p. 419–42, bibl., ill.)

Interprets a building at the Maya city in western Honduras as a reflection of emergence of a council of chiefs as a major political institution in 8th century, following Quirigua's supposed capture and sacrifice of the reigning king of Copán. [JSH]

258   **Fash, William L.** and **Ricardo Agurcia Fasquelle.** History carved in stone: a guide to the archaeological park of the ruins of Copán. Copán, Honduras: Asociación Copán, 1992. 1 v.: bibl., ill.

Guidebook for visitors to Maya center of Copán in western Honduras emphasizes interpretation of Copán's late classic history based on recent analysis of hieroglyphic texts and iconography of sculpture.

259   **Fauconnier, Françoise.** Projet Sierra del Nayar: résultats des travaux menés par la Mission archéologique belge au Mexique.

(*Mexicon/Berlin*, 14:2, März 1992, p. 24–30, bibl., maps, photos)

Brief report on investigations at Cerro de Huistle, Jalisco, shows that there were three phases of occupation at the site commencing around AD 900. [GF]

260   **La fauna en el Templo Mayor.** Coordinación de Oscar J. Polaco. México: Instituto Nacional de Antropología e Historia, Proyecto Templo Mayor; GV Editores; Asociación de Amigos del Templo Mayor, 1991. 263 p.: bibl., ill. (Col. Divulgación)

Volume includes six articles focusing on different aspects of faunal remains recovered from offerings at the Templo Mayor. General introduction is followed by more detailed discussions of fish bones, terrestrial vertebrates, Ofrenda H, and shell (mostly Atlantic species). Final chapter discusses references to aquafauna in the Florentine Codex. [GF]

261   **Fauvet-Berthelot, Marie France;** **Marie Charlotte Arnauld;** and **Alain Breton.** Tres cruces, tres épocas, tres enfoques. (*Apunt. Arqueol./Guatemala*, 1:2, dic. 1991, p. 39–56, bibl., maps, photos)

Preliminary interpretation of history of a small zone in Maya highlands of Guatemala is based on combined analysis of the archaeology of the postclassic period, ethnohistorical documents, and modern myths. [JSH]

262   **Feinman, Gary M.; Linda M. Nicholas;** and **William D. Middleton.** Craft activities at the prehispanic Ejutla site, Oaxaca, Mexico. (*Mexicon/Berlin*, 15:2, März 1993, p. 33–41, bibl., ill., maps, photos, table)

Evidence for a range of craft activities, including shell working, lapidary crafts, ceramic and figurine manufacture, and possibly cloth production, was found in stratigraphic association with a residential structure uncovered at the edge of the Ejutla site. [GF]

263   **Feinman, Gary M.** and **Linda M. Nicholas.** The Monte Albán state: a diachronic perspective on an ancient core and its periphery. (*in* Core/periphery relations in precapitalist worlds. Edited by Christopher Chase-Dunn and Thomas D. Hall. Boulder, Colo.: Westview Press, 1991, p. 240–276, bibl., ill., maps, table)

Draws on findings from two regional survey projects to consider nature of prehispanic relationship between Valley of Oaxaca and adjacent Ejutla Valley. Includes examination of prehispanic distribution of specialized craft activities in these regions. [GF]

264 **Feinman, Gary M.** and **Linda M. Nicholas.** Shell-ornament production in Ejutla: implications for highland-coastal interaction in ancient Oaxaca. (*Anc. Mesoam.*, 4:1, Spring 1993, p. 103–119, bibl., ill., maps, photos)

Recent investigations at the Ejutla site have documented classic-period manufacture of shell ornaments from Pacific marine species. Uses comparison with shell assemblages recovered elsewhere in Oaxaca to document shifts in shell-ornament manufacture and exchange during formative and classic periods. [GF]

265 **Feinman, Gary M.; Linda M. Nicholas;** and **Scott L. Fedick.** Shell working in prehispanic Ejutla, Oaxaca, Mexico: findings from an exploratory field season. (*Mexicon/Berlin*, 13:4, Aug. 1991, p. 67–77, bibl., ill., maps, photos, table)

Reports on first stage of test excavations at this highland site. Documents prehispanic shell-ornament manufacture in highland Oaxaca. A variety of shell ornaments (beads, pendants, mosaic plaques, disks, bracelets) were made from marine species native to the waters of the Pacific. [GF]

266 **Flannery, Kent V.** and **Joyce Marcus.** Early formative pottery of the Valley of Oaxaca. Technical ceramic analysis by William O. Payne. Ann Arbor: Museum of Anthropology, Univ. of Michigan, 1994. 399 p.: bibl., ill., map. (Memoirs of the Museum of Anthropology, University of Michigan; 27. Prehistory and human ecology of the Valley of Oaxaca; 10)

Detailed examination of three phases of early formative pottery is well illustrated, including a color frontispiece of key vessel varieties representative of each phase. Report considers ceramic variation at a series of different spatial scales, and concludes with a revised discussion of the Olmec and the Valley of Oaxaca. [GF]

267 **Ford, Anabel** and **Scott L. Fedick.** Prehistoric Maya settlement patterns in the Upper Belize River area: initial results of the Belize River Archaeological Settlement Survey. (*J. Field Archaeol.*, 19:1, Spring 1992, p. 35–49, bibl., maps)

Summarizes results of several seasons of survey and test excavation in the Upper Belize River drainage in western Belize. Discussion focuses on variation in settlements in upland, foothill, and valley bottom areas in relation to the distribution of key resources. [JSH]

268 **Ford, Anabel.** Problems with the evaluation of population from settlement data: examination of ancient Maya residential patterns in the Tikal-Yaxha intersite area. (*Estud. Cult. Maya*, 18, 1991, p. 157–186, bibl., ill.)

Analysis of late classic settlement patterns in central Petén region of Guatemala emphasizes factors which complicate population evaluation from surface survey data, especially difficulty of determining the function of buildings and establishing their precise contemporaneity. [JSH]

269 **Forsyth, Donald W.** The ceramic sequence at Nakbe, Guatemala. (*Anc. Mesoam.*, 4:1, 1993, p. 31–53, bibl., ill.)

Describes sequence of pottery types from Maya center in northern Guatemala spanning middle preclassic through late classic periods. [JSH]

270 **Fox, John G.** The ball court markers of Tenam Rosario, Chiapas, Mexico. (*Anc. Mesoam.*, 4:1, 1993, p. 55–64, bibl., ill.)

Analyzes iconography of ball court markers from a late to terminal classic lowland Maya center. Relates images of squatting figures on the markers to Central Mexican imagery and to themes of sacrifice, warfare, and regeneration. [JSH]

271 **Gallareta Negrón, Tomás; Anthony P. Andrews;** and **Rafael Cobos Palma.** Reconocimiento arqueológico de la península de Xkalak, Quintana Roo, Mexico. (*Bol. Esc. Cienc. Antropol. Univ. Yucatán*, 18:108/109, 1991, p. 48–74, bibl., ill.)

Presents results of an archaeological survey of a sector of the coast of eastern Yucatán and an analysis of the settlement patterns of the region, along with a brief account of recent history and cultural geography of the region. [JSH]

272 **Galván Villegas, Luis Javier.** Las tumbas de tiro del Valle de Atemajac, Jalisco. México: Instituto Nacional de Antropología e Historia, 1991. 331 p.: bibl., ill.,

maps, tables. (Col. Científica; 239. Serie Arqueología)

Report discusses Tabachines-phase shaft tombs excavated in Jalisco's Atemajac Valley. Provides descriptions of 23 tombs, including associated burial goods. [GF]

273 García Cruz, Florentino. Balamku: un sitio arqueológico maya en Campeche. (*Arqueología/México*, 4, 1990, p. 129–134, bibl., ill.)

Provides preliminary description of Maya site in western Yucatán with elaborate stucco reliefs probably dating from late formative-early classic periods. [JSH]

274 García Moll, Roberto et al. Catálogo de entierros de San Luis Tlatilco, México: temporada IV. Fotográfía de Ignacio Borja Aldana y Ramón Enríquez Rodríguez. México: Instituto Nacional de Antropología e Historia, 1991. 264 p.: bibl., photos. (Serie Antropología física-arqueología)

Detailed report examines burials uncovered at site of San Luis Tlatilco in western Basin of Mexico. Following brief introductory discussions, book describes 213 burials, including information on type of burial, sex and age of interred, and associated grave goods. Each burial is illustrated. Photographs of the artifacts comprise final section. [GF]

275 García Oropeza, Guillermo et al. Perros en las tumbas de Colima. Colima, Mexico: Univ. de Colima, 1991. 134 p.: bibl., photos.

Large format, heavily illustrated book includes full-page color photographs of the ceramic dogs found in the tombs of Colima. [GF]

276 **Gardens of prehistory: the archaeology of settlement agriculture in Greater Mesoamerica.** Edited by Thomas W. Killion. Tuscaloosa: Univ. of Alabama Press, 1992. 334 p.: bibl., ill., index.

Compilation of papers presents research on infield and residential gardening in a sample of New World precontact societies. Includes sections on highland Central Mexico and its northern peripheries, the humid tropical region of lowland Mesoamerica, and landscape modifications in El Salvador. Volume stresses diversity of indigenous gardening practices, the contribution of infield agriculture to subsistence, and relationships between these kinds of farming systems and spatial structure of prehispanic settlements. [GF]

277 Garza, Mercedes de la. Palenque. Tuxtla Gutiérrez, Mexico: Gobierno del Estado de Chiapas; México: M.A. Porrúa, 1992. 185 p.: bibl., ill. (some col.). (Chiapas eterno)

Extensively illustrated volume focuses on classic-period Maya city in Chiapas, Mexico. Includes summary of archaeological investigations at Palenque, descriptions of major buildings and sculptural monuments, and very brief summary of some aspects of dynastic history of the city recorded in its hieroglyphic texts. [JSH]

278 Gómez Serafín, Susana and Enrique Fernández Dávila. Costumbres funerarias de los años 800 a 1428 dne. en Tula, Hidalgo. (*Antropológicas/México*, 5, 1990, p. 29–40, bibl., ill., tables)

Presents analysis of over 100 burials recovered during salvage excavations at Tula. [GF]

279 González, Nancie L. and Charles D. Cheek. Patrón de asentamiento de los caribes negros a principios del siglo XIX en Honduras: la búsqueda de un modo de vida. (*Yaxkin/Tegucigalpa*, 11:2, 1988, p. 89–108, bibl., ill.)

Summarizes an investigation of the establishment and development of "Black Carib" (Garifuna) communities on the north coast of Honduras beginning in 19th century. Combines ethnographic, historical, and archaeological approaches. [JSH]

280 Graham, John A. and Larry Benson. Escultura olmeca y maya sobre canto en Abaj Takalik. (*Arqueología/México*, 3, 1990, p. 77–84, bibl., ill.)

Analyzes style and iconography of several monumental stone sculptures from Abaj Takalik at edge of the coastal plain in western Guatemala. Focuses on relationship of Abaj Takalik's sculpture to Olmec and early Maya styles. [JSH]

281 Graulich, Michel. On the so-called "Cuauhxicalli of Motecuhzoma Ilhuicamina," the Sánchez-Nava monolith. (*Mexicon/Berlin*, 14:1, Jan. 1992, p. 5–10, bibl., photos, tables)

Alternative interpretation attributes this monument, which is comparable to the

"Tizoc Stone," to Axayacatl instead of to the preceding King Motecuhzoma Ilhuicamina. [H. von Winning]

282 **Grove, David C. et al.** Five Olmec monuments from the Laguna de los Cerros hinterland. (*Mexicon/Berlin*, 15:5, Okt. 1993, p. 91–95, bibl., ill., map)

Describes five monuments at site of La Isla, 10 km northwest of Laguna de los Cerros. Not all Olmec monuments are found at Olmec settlements; some situated in isolated locations may have served as boundary markers, possibly marking sacred or political space. [GF]

283 **Guderjan, Thomas H.** Ancient Maya traders of Ambergris Caye. Benque Viejo del Carmen, Belize: Cubola Productions, 1993. 39 p.: bibl., ill. (some col.).

Briefly describes archaeological sites on Ambergris Caye, Belize. Many of these ancient coastal communities, especially those of later precolumbian period, were ports involved in active maritime commercial networks that ringed the Yucatán Peninsula. [JSH]

284 **Guillén, Ann Cyphers.** Estudio de cerámica y sociedad: Chalcatzingo, Morelos. México: Instituto de Investigaciones Antropólogicas, UNAM, 1992. 366 p. bibl., ill., maps.

Detailed report examines ceramic typology employed at this important preclassic site. [GF]

285 **Gussinyer i Alfonso, Jordi.** Notas sobre el patrón de asentamiento en las tierras bajas mayas. (*Bol. Am.*, 32:41, 1991, p. 203–259, bibl.)

Overview of basic issues in settlement archaeology of the Maya lowlands emphasizes houses and households. [JSH]

286 **Hasemann, George.** El patrón de asentamiento a lo largo del Río Sulaco durante el clásico tardío, Honduras. (*Yaxkin/Tegucigalpa*, 10:1, enero/junio 1987, p. 58–77, bibl., maps)

Summarizes results of settlement survey work done as part of a salvage program in advance of construction of the El Cajón Dam in central Honduras. Describes types of sites present in the region during late classic florescence and analyzes their form and distribution in terms of environmental parameters (especially agricultural potential) and economic networks. [JSH]

287 **Hatch, Marion Popenoe.** Kaminaljuyu: un resumen general hasta 1991. (*U tz'ib*, 1:1, nov. 1991, p. 2–6, bibl.)

Briefly summarizes occupation history of the major highland Maya center from early formative through late classic periods, emphasizing ceramic chronology and including results of recent and unpublished excavations. [JSH]

288 **Healan, Dan M.** Local versus non-local obsidian exchange at Tula and its implications for post-formative Mesoamerica. (*World Archaeol.*, 24:3, Feb. 1993, p. 449–466, bibl., ill., maps, table)

Examination of Tula's obsidian industry indicates that production at the site focused on prismatic blade segments for long-distance exchange. Argues that due to availability of obsidian sources in central Mexico, production for local exchange was limited. [GF]

289 **Healy, Paul F.** The ancient Maya ballcourt at Pacbitun, Belize. (*Anc. Mesoam.*, 3:2, 1992, p. 229–39, bibl., ill.)

Summarizes construction history of a ball court at the Maya center in western Belize, from late preclassic to early classic period. [JSH]

290 **Henderson, John S.** Investigaciones arqueológicas en el Valle de Sula. (*Yaxkin/Tegucigalpa*, 11:1, 1988, p. 5–30, bibl., ill.)

Summarizes long-term survey and excavation project in the Sula Valley of western Honduras. Presents chronological framework for formative period through the Spanish invasion, along with preliminary interpretations of changing settlement systems and patterns of interaction with the Maya world to the west. [JSH]

291 **Hendon, Julia A.** The interpretation of survey data: two case studies from the Maya area. (*Lat. Am. Antiq.*, 3:1, 1992, p. 22–42, bibl.)

Assesses adequacy of site typologies developed at Seibal and Copán by comparing them with results of excavations. Emphasizes importance of excavated data for understanding variation relevant to reconstructing site function and organization. [JSH]

292 **Hirth, Kenneth G.** La subsistencia y comercio prehispánicos en la región de El Cajón. (*Yaxkin/Tegucigalpa*, 10:1, enero/junio 1987, p. 39–50, bibl.)

Provides preliminary analysis of data recovered by salvage project in advance of construction of El Cajón Dam in central Honduras. Suggests a reconstruction of subsistence systems and commerce (especially obsidian exchange) in the region during its period of peak development in the late classic. [JSH]

293 **Hodge, Mary G.** Aztec market systems: the geographical structure of Aztec imperial-period market systems. (*Natl. Geogr. Res.*, 8:4, 1992, p. 428–445, bibl., ill., maps, photos, tables)

Study of spatial distribution of different stylistic varieties of Tenochtitlán-phase black-on-orange serving vessels indicates that certain exchanges in the core Aztec domain appear to have taken place within distinct market territories. The territories correspond to political confederations in existence before the empire formed. Market territories eventually became administrative units in the Aztec empire. [GF]

294 **Hodge, Mary G. et al.** Black-on-orange ceramic production in the Aztec empire's heartland. (*Lat. Am. Antiq.*, 4:2, June 1993, p. 130–157, bibl., ill., maps, tables)

Three different production areas in eastern and southern parts of the Basin of Mexico were distinguished on the basis of a neutron-activation analysis of 85 late Aztec black-on-orange ceramic samples. These results imply that this ceramic style was adopted by regional pottery production centers, and that manufacture of these decorated vessels was not restricted to a single locale. [GF]

295 **Hodge, Mary G. et al.** A compositional perspective on ceramic production in the Aztec empire. (*in* Chemical characterization of ceramic pastes in archaeology. Edited by Hector Neff. Madison, Wis.: Prehistory Press, 1992, p. 203–220, bibl., ill., maps, tables)

Neutron activation analyses of a large sample of late Aztec black-on-orange sherds indicate that there were at least four major production zones in the southern two-thirds of the Valley of Mexico during late Aztec times. [GF]

296 **Hohmann, Hasso.** Zur gestaltung des nördlichen durchganges in Tempel Str II von Copan. (*Arch. Völkerkd.*, 45, 1991, p. 69–72, bibl., ill., photo)

Proposes a reconstruction of the sculptural motifs flanking the doorway of the central room of Str. 11, a royal palace at the classic Maya city of Copán in western Honduras. [JSH]

297 **Hosler, Dorothy** and **Guy Stresser-Péan.** The Huastec region: a second locus for the production of bronze alloys in ancient Mesoamerica. (*Science/Washington*, 257, Aug. 1992, p. 1215–1220, bibl., ill., maps, photo, tables)

Chemical analyses of materials from two late postclassic sites were used to identify an area of bronze alloy production in the Huastec area of eastern Mesoamerica. Previously, bronze alloy production was thought to have taken place only in western Mesoamerica. [GF]

298 **Jackson, Lawrence J.** Lucky day at Tiger Mound. (*Archaeology/New York*, 47:1, 1994, p. 60–62, bibl., ill.)

Offers very brief summary of preliminary investigations at a small coastal settlement in southern Belize presumably involved in the exchange networks that ringed the Yucatán Peninsula. [JSH]

299 **Jiménez Valdez, Gloria Martha.** Poblaciones costeras de Tabasco y Campeche. (*An. Antropol.*, 26, 1989, p. 99–105, bibl.)

Brief overview describes prehispanic settlement patterns in coastal Tabasco and Campeche. [GF]

300 **Joyce, Arthur A.** Formative period social change in the lower Río Verde Valley, Oaxaca, Mexico. (*Lat. Am. Antiq.*, 2:2, June 1991, p. 126–150, bibl., maps, ill., tables)

Draws on archaeological and geomorphological studies to construct a sequence of human occupation for lower Río Verde region. Emphasizes role of environmental change and interregional interaction in the development of complex sociopolitical formations during the classic period. [GF]

301 **Joyce, Arthur A.** Interregional interaction and social development on the Oaxaca coast. (*Anc. Mesoam.*, 4:1, Spring 1993, p. 67–84, bibl., ill., maps, photos)

Data from the lower Río Verde valley and the southern isthmus indicate that elites attempting to symbolize and legitimize their special status dominated interregional interaction. In spite of contact with many powerful prehispanic political centers, the Oaxaca Coast is presumed to have remained politically autonomous for most of the prehispanic era. [GF]

302   Joyce, Arthur A. and **Raymond G. Mueller.** The social impact of anthropogenic landscape modification in the Río Verde drainage basin, Oaxaca, Mexico. (*Geoarchaeology/New York*, 7:6, 1992, p. 503–526, bibl., ill., maps, photos, table)

Argues that late formative population growth in valleys of highland Oaxaca led to increased erosion in the upper drainage of the Verde. Proposes that resulting flooding and alluviation along lower Río Verde increased the region's agricultural potential. The latter changes are suggested to account for population growth and social change in lower Verde region. [GF]

303   **Joyce, Rosemary A.** Innovation, communication, and the archaeological record: a reassessment of middle formative Honduras. (*J. Steward Anthropol. Soc.*, 20:1/2, 1992, p. 235–256, bibl., ill.)

Presents comparative analysis of Playa de los Muertos and contemporary early societies in Honduras. Identifies shared stylistic and iconographic features that reflect participation in broader patterns of interaction within Mesoamerica, but emphasizes distinctive features of each regional society. [JSH]

304   **Kennedy, Nedenia.** La cronología cerámica de Salitrón Viejo, región de El Cajón, Honduras. (*Yaxkin/Tegucigalpa*, 10:1, enero/junio 1987, p. 51–57, bibl.)

Briefly summarizes preliminary analysis of ceramics from excavations at Salitrón Viejo, the principal late classic center in El Cajón region of central Honduras. Tentatively identifies four ceramic complexes dating to late preclassic through late classic periods. [JSH]

305   **Lambert, Joseph B. et al.** Amber and jet from Tipu, Belize. (*Anc. Mesoam.*, 5:1, 1994, p. 55–60, bibl., ill.)

Magnetic resonance spectroscopy analysis of amber and jet beads from late postclassic and early colonial Maya town in western Belize indicates that the materials are of Old World origin. [JSH]

306   **Lara Pinto, Gloria** and **George Hasemann.** La sociedad indígena del noreste de Honduras en el siglo XVI: ¿son la etnohistoria y la arqueología contradictorias? (*Yaxkin/Tegucigalpa*, 11:2, 1988, p. 5–28, bibl., maps)

Analyzes ethnohistorical sources and newly discovered postclassic-period monumental archaeological sites from northeastern Honduras. Suggests that these sites may have been already abandoned by the time of the Spanish invasion, and that the peoples occupying the region during the early colonial period were not the descendants of the original inhabitants, but were instead later refugees from European rule. [JSH]

307   **Lentz, David.** Acromia mexicana: la palma de los antiguos mesoamericanos. (*Yaxkin/Tegucigalpa*, 12:1, 1989, p. 78–101, bibl., ill.)

Provides overview of utilization of the coyol palm in precolumbian Mesoamerica and Central America. [JSH]

308   **Leyenarr, Ted J.J.; Gerard W. van Bussel;** and **Gesine Weber.** Von Küste zu Küste, Prä-Kolumbische Skulpturen aus Mesoamerika = From coast to coast, pre-columbian sculptures from Mesoamerica. Kassel, Germany: Verlag Weber & Weidemeyer, 1992. 423 p.: bibl., ill., maps, plates.

Exhibition catalog has large illustrations of over 400 ceramic figures and vessels in two private collections in Germany and Holland. Good descriptions and introductory texts in German and English. [H. von Winning]

309   **Lira López, Yamile.** La cerámica de El Tajín, Norte de Veracruz, México. Münster, Germany: LIT, 1990? 282 p.: bibl., ill. (Beiträge zur Archäologie; 3)

Presents results of archaeometric study of 473 sherds from El Tajín and several other Gulf Coast sites. The ceramics were analyzed using x-ray diffraction and petrographic and mineralogical studies of thin sections. Concludes that plainware pottery was made locally at El Tajín, while fine wares were made elsewhere. [GF]

310   **Long, Austin et al.** First direct AMS dates on early maize from Tehuacán, Mexico. (*Radiocarbon/New Haven*, 31:3, 1989, p. 1035–1040, bibl., ill.)

AMS dates on corn specimens from Cueva San Marcos reveal that this oldest corn dates to ca. 3500 BC, 2500 years younger than previously thought. [GF]

311 López Luján, Leonardo. Las ofrendas del Templo Mayor de Tenochtitlán. México: Instituto Nacional de Antropología e Historia, 1993. 432 p.: bibl., ill.

Detailed report focuses on over 100 dedicatory offerings recovered in excavation of the Templo Mayor. Reconstructs aspects of ritual behavior in which offerings were entombed during episodes of architectural expansion. [GF]

312 López Luján, Leonardo and Noel Morelos García. Los petroglifos de Amecameca: un monumento dedicado a la elección de Motecuhzoma Xocoyotzin. (*An. Antropol.*, 26, 1989, p. 127–156, bibl., ill., map, photos)

Describes a large carved hemispherical stone monument near Amecameca. Suggests that monument was erected to commemorate election of Motecuhzoma Xocoyotzin. [GF]

313 Love, Michael W. La Blanca y el preclásico medio en la costa del Pacífico. (*Arqueología/México*, 3, 1990, p. 67–76, bibl., ill.)

Summarizes formative-period settlement patterns along western Pacific coast of Guatemala and excavations at La Blanca, the largest middle formative site in the region. [JSH]

314 Love, Michael W. Ceramic chronology and chronometric dating: stratigraphy and seriation at La Blanca, Guatemala. (*Anc. Mesoam.*, 4:1, 1993, p. 17–29, bibl., ill.)

Analysis of pottery from large center on western Pacific coast of Guatemala contributes to definition of fine chronological subdivisions within the middle preclassic period. [JSH]

315 Macias Goytia, Angelina. Huandacareo, lugar de juicios, tribunal. México: Instituto Nacional de Antropología e Historia, 1990. 222 p.: bibl., maps, photos. (Col. Científica; 222. Serie Arqueología)

Reports on archaeological investigations at Huandacareo in the Cuitzeo Valley of Michoacán, and emphasizes its mention in historical documents. Book is comprised largely of photographs and descriptions of the recovered materials. [GF]

316 Maldonado C., Rubén and Beatriz Repetto Tío. Los Tlalocs de Uxmal, Yucatán. (*Rev. Esp. Antropol. Am.*, 18, 1988, p. 9–19, bibl., photos, ill.)

Descriptive study compares the 10 reliefs of Tlaloc of Uxmal with representations of this deity elsewhere, primarily in Central Mexico. [R. Haskett and S. Wood]

317 Manzanilla, Linda *et al.* Caves and geophysics: an approximation to the underworld of Teotihuacán, Mexico. (*Archaeometry/Oxford*, 36:1, Feb. 1994, p. 141–157, bibl., ill., maps)

Describes results of a geophysical survey to locate a system of tunnels that runs beneath Teotihuacán. The system may have been dug originally to acquire construction materials, and later served as a representational model of the underworld. [GF]

318 Manzanilla, Linda. Los contextos de almacenamiento en los sitios arqueológicos y su estudio. (*An. Antropol.*, 25, 1988, p. 71–87, bibl.)

Presents analytical consideration of food storage and how it can be studied archaeologically. Proposes a multi-scale evaluation of storage facilities. [GF]

319 Manzanilla, Linda and Emilie A. Carreón Blaine. Un incensario teotihuacano en contexto doméstico: restauración e interpretación. (*Antropológicas/México*, 4, 1990, p. 5–18, bibl., photos)

Describes and interprets a "theater type" censer that was found in association with a burial excavated in the Oztoyohualco residential compound at Teotihuacán.

320 Márquez Morfín, Lourdes and Ernesto González L. La trepanación craneana entre los antiguos zapotecos de Monte Albán. (*Cuad. Sur/Oaxaca*, 1:1, mayo/agosto 1992, p. 25–50, bibl., ill., map, photos, tables)

Authors describe and discuss trephinated skulls found at Monte Albán. They consider antiquity of this practice, its geographical distribution, and techniques that were employed. [GF]

321 Martos López, Luis Alberto. Tres nuevos pendientes de jade del tipo "yelmo y babero." (*Arqueología/México*, 6, 1991, p. 121–126, bibl., ill.)

Reports discovery of three "bib and helmet" type jade pendants from site of Rancho INAH on east coast of Yucatán. [JSH]

322 Matos Moctezuma, Eduardo. Arqueología urbana en el centro de la Ciudad de México. (*Estud. Cult. Náhuatl*, 22, 1992, p. 133–141)

Discusses a project initiated in 1991 to study process of long-term urban development in the seven-block area at the heart of Mexico City. Study area once contained principal plaza of Aztec Tenochtitlán, and this account reviews the period from foundation of the prehispanic capital in the 14th century through this century. [GF]

323 **McCafferty, Sharisse D.** and **Geoffrey G. McCafferty.** Engendering Tomb 7 at Monte Albán: respinning an old yarn. (*Curr. Anthropol.*, 35:2, April 1994, p. 143–166, bibl., ill.)

Presents re-analysis of cultural materials from the tomb, including proposed spinning and weaving implements found with the burial. Suggests that tomb's principal individual was female, and that Tomb 7 may have served as a shrine to Lady 9 Grass. Includes *Current Anthropology* comments. [GF]

324 **Medrano, Sonia.** Culto al Dios Mundo de Santa Lucía Cotzumalguapa, Escuintla, Guatemala: un ritual realizado por indígenas emigrantes. (*Trace/México*, 21, junio 1992, p. 3–8, bibl., maps, photos)

Describes cult surrounding two classic-period sculptural monuments from archaeological site of El Baúl on Pacific slope of Guatemala. Highland Quiché Maya propitiate the monuments when they come to the coast for seasonal wage labor, and some highland ritual specialists make annual pilgrimages to visit them. [JSH]

325 **Mejía Pérez Campos, Elizabeth** and **Luis Alberto Barba Pingarrón.** El análisis de fosfatos en la arqueología: historia y perspectivas. (*An. Antropol.*, 25, 1988, p. 217–147, bibl.)

Reviews history of chemical analysis in archaeology and results from recent applications. Concludes that the most important application for this set of techniques is to identify and discriminate use areas on house floors. [GF]

326 **Mexico: splendors of thirty centuries.** Introduction by Octavio Paz. Translations by Edith Grossman *et al.* New York: Metropolitan Museum of Art; Boston: Little, Brown, 1990. 712 p.: bibl., ill. (some col.), index, maps.

Lavishly illustrated catalog presents more than 350 sculptures, paintings, and objects that were part of a major exhibition celebrating past 3,000 years of Mexican history. The prehispanic section examines eight major sites: La Venta, Izapa, Teotihuacán, Monte Albán, Palenque, El Tajín, Chichén Itzá, and Tenochtitlán. For art historian's comment see *HLAS 52:127.* [GF]

327 **Millet Cámara, Luis; Heber Ojeda M.;** and **Vicente Suárez A.** Tecoh, Izamal: nobleza indígena y conquista española. (*Lat. Am. Antiq.*, 4:1, 1993, p. 48–58, bibl., ill.)

Describes late prehispanic and early colonial architecture at this Maya town in northern Yucatán, emphasizing survival of the community after the Spanish invasion and adaptation of the Maya elite to colonial rule. [JSH]

328 **Moguel Cos, María Antonieta** and **Sergio Arturo Sánchez Correa.** Estudio comparativo entre una maqueta y elementos arquitectónicos de un sitio en Guanajuato. (*Antropológicas/México*, 5, 1990, p. 78–85, ill., maps, photos)

Proposes that petroglyph carved in rock at the site of El Cobre in southwestern Guanajuato may have been a miniature representation of the principal architectural complex at the site. [GF]

329 **Monte Albán: estudios recientes.** Coordinación de Marcus Winter. Oaxaca, Mexico: Proyecto Especial Monte Albán, 1994. 128 p.: bibl., ill., maps, photos, tables (Contribución; 2)

Collection of nine essays reports on recent work at Monte Albán and in surrounding regions. Includes several new radiocarbon dates from this major center in the Valley of Oaxaca. [GF]

330 **Murphy, Vincent.** La Cueva Pintada: un viaje al pasado. (*Yaxkin/Tegucigalpa*, 11:2, 1988, p.179–185, bibl., ill.)

Provides brief description of well-preserved prehispanic anthropomorphic, zoomorphic, and geometric paintings in a cave in central Honduras. [JSH]

331 **Nakamura, Seiichi.** Desarrollo y decaimiento en la periferia de Copán. (*Ann. Lat. Am. Stud.*, 14, 1994, p. 39–95, bibl., ill.)

Interpretive summary of cultural development from early formative through terminal classic periods in region of La Venta and La Florida Valleys in western Honduras is based on long-term program of survey and excavation. Emphasizes relationships between development of local centers and polities and changing political and economic fortunes of Copán during classic period. [JSH]

332 Nakamura, Seiichi. Frontera prehispánica en la encrucijada del sureste maya. (*Ann. Lat. Am. Stud.*, 12, 1992, p. 131–167, bibl., ill.)

Defines four patterns of material cultural remains in southeastern Mesoamerica based on settlement patterns, architecture, ceramics, and lithics. Interpretation focuses on relating this variability to eastern frontier of the Maya world during classic period. [JSH]

333 Nakamura, Seiichi. Proyecto arqueológico La Entrada, temporada de campo 1986–1987: resultados preliminares. (*Yaxkin/Tegucigalpa*, 11:2, 1988, p. 29–44, bibl., maps, tables)

Brief summary of settlement history of La Venta and La Florida Valleys in western Honduras is based on a long-term program of survey and test excavation sponsored by the Japanese government and Mitsubishi. [JSH]

334 Nakamura, Seiichi. Reconocimiento arqueológico en los Valles de La Venta y de Florida. (*Yaxkin/Tegucigalpa*, 10:1, enero/junio 1987, p. 1–38, bibl., maps)

Preliminary summary of settlement patterns of La Venta and La Florida Valleys in western Honduras is based on a long-term program of survey and test excavation. Interpretation emphasizes links with adjacent regions, particularly the classic Maya city of Copán. [JSH]

335 Nance, C. Roger. Guzmán Mound: a late preclassic salt works on the south coast of Guatemala. (*Anc. Mesoam.*, 3:1, 1992, p. 27–46, bibl., map)

Salvage excavations on Pacific coast of Guatemala document specialized late preclassic period salt-producing locality. Bulk of pottery consists of utility vessels that were apparently used to recover salt by evaporation over fires. [JSH]

336 Nárez, Jesús; Araceli Rivera Estrada; and José Luis Rojas Martínez. Materiales arqueológicos de Balcón de Montezuma, Tamaulipas. Presentación de Américo Villareal Guerra y Roberto García Moll. México: Instituto Nacional de Antropología e Historia, 1992. 261 p.: bibl., ill., photos.

Following brief discussion of site of Balcón de Montezuma, situated south of Ciudad Victoria in Tamaulipas, report presents a lengthy catalog containing descriptions and photographs of archaeological materials recovered at the site. [GF]

337 Nárez, Jesús. Materiales arqueológicos de Tlapacoya. México: Instituto Nacional de Antropología e Historia, 1990. 149 p.: bibl., ill., map. (Col. Científica; 204. Serie Arqueología)

Detailed descriptive report focuses on archaeological materials recovered from 1971–73 excavations at the largest (Tlapacoya XVIII) of 18 sites identified on edges of Cerro de Tlapacoya in southern Basin of Mexico. Excavated materials date to 8000–800 BC. [GF]

338 Nelson, Ben A.; J. Andrew Darling; and David A. Kice. Mortuary practices and the social order at La Quemada, Zacatecas, Mexico. (*Lat. Am. Antiq.*, 3:4, Dec. 1992, p. 298–315, bibl., ill., map, photo)

Structure in which epiclassic remains of 11–14 individuals were found is proposed to have been a charnel house that contained the bodies of revered community members. The continuity between this practice and earlier shaft tombs is discussed. [GF]

339 Neurath, Johannes. Xiuhuitzolli: Motecuhzoma's diadem of turquoise, fire, and time. (*Arch. Völkerkd.*, 46, 1992, p. 123–2148, bibl., ill.)

Discusses representations of the royal turquoise headband and its symbolic significance in relation to an Aztec cult of fire, tied to militarism and warfare. [H. von Winning]

340 Nichols, Deborah L.; Michael W. Spence; and Mark D. Borland. Watering the fields of Teotihuacán: early irrigation at the ancient city. (*Anc. Mesoam.*, 2:1, Spring 1991, p. 119–129, bibl., ill., maps, photo)

Recent excavations in the Oaxaca Barrio (Tlailotlacán) at Teotihuacán uncovered irrigation canals that are believed to date to the terminal formative period. [GF]

341 Nicholson, Henry B. Late pre-Hispanic Central Mexican ("Aztec") sacred architecture: the "Pyramid Temple." (*in* Circumpacifica: Festschrift für Thomas S. Barthel. Frankfurt am Main: P. Lang, 1990, p. 303–324, bibl., facsims., ill.)

Comprehensive survey of late prehispanic central Mexican sacred architecture draws on documentary and archaeological sources. Suggests that a relatively standard pyramid-temple pattern crystallized by the late postclassic period. This pattern is derived from earlier Mesoamerican sacred architectural traditions. [GF]

342 **Notas Mesoamericanas.** No. 13, 1991/ 1992– . Selecciones del Segundo Simposio de Cholula. Puebla, Mexico: Univ. de las Américas-Puebla.

Includes 20 papers from 1990 conference. Grouped under four themes, nine of the articles focus on archaeology and ethnohistory of Puebla and Tlaxcala, while seven papers consider either physical or social anthropology in the same region. The four remaining contributions discuss various topics relevant to archaeology of Oaxaca. [GF]

343 **Notas Mesoamericanas.** No. 14, 1992/ 1993– . Puebla, Mexico: Univ. de las Américas-Puebla.

Six of the eight articles in this issue focus on archaeological sites and data. John Paddock presents a retrospective on the site of Cholula; Andrés Noyola Cherpitel describes prehispanic houses excavated in Puebla; Damon E. Peeler and Marcus C. Winter discuss relationship of Mesoamerican site orientations to the 260-day ritual period; Terry Stocker considers Texacatlipoca at Tula to explore relationship between religion and economics; Sergio López Alonso and Carlos Serrano Sánchez present results of physical anthropological analyses from the Cholula Project; and Gabriela Uruñuela discusses preclassic burials from Colotzingo, Puebla. [GF]

344 **Orrego Corzo, Miguel.** Investigaciones arqueológicas en Abaj Takalik, El Asintal, Retalhuleu, año 1988. Reporte no. 1. Guatemala: Proyecto Nacional Abaj Takalik, Instituto de Antropología e Historia de Guatemala, Ministerio de Cultura y Deportes, 1990. 115 p.: bibl., ill., maps.

Presents preliminary synthesis of archaeology of Abaj Takalik on Pacific slope of Guatemala. Occupation there spans period from beginning of first milennium BC to first milennium AD; its sequence of stylistically-varied stone scupture should clarify the stylistic and iconographic relationships among Olmec, Izapa, and early Maya art. [JSH]

345 **Ortuño Cos, Francisco** and **Salvador Pulido Méndez.** Sitios arqueológicos en el área de la presa de Chilatán, Jalisco. (*An.*

Mus. Michoacano, 2, 1990, p. 9–23, bibl., ill., map, photos)

Presents findings from salvage work conducted near Chilatán, Jalisco. Three sites were recorded. Collectively, these occupations span from between 200 BC to AD 1200. [GF]

346 **Paillés Hernández, María de la Cruz** and **Rosalba Nieto Calleja.** Primeras expediciones a las ruinas de Palenque. (*Arqueología/México*, 4, 1990, p. 97–128, bibl., ill.)

Analyzes documents recording late 18th-century visits to Maya city of Palenque in Chiapas. Reproduces plans and drawings that record details of architecture and sculpture that did not survive into period of modern investigation. [JSH]

347 **Parsons, Jeffrey R.** Arqueología regional en la cuenca de México: una estrategia para la investigación futura. (*An. Antropol.*, 26, 1989, p. 157–257, bibl., maps, tables)

Synthesis of recent archaeological investigations in the Basin of Mexico provides a general overview for each prehispanic phase, as well as concise listing of key research problems that need to be addressed. [GF]

348 **Pastrana, Alejandro et al.** Querétaro prehispánico. Coordinación de Ana María Crespo y Rosa Brambila. México: Instituto Nacional de Antropología e Historia, 1991. 306 p.: bibl., ill. (Col. Científica; 238. Serie Arqueología)

Compilation of 11 papers discusses range of topics related to history of Querétaro. Collectively, the reported investigations highlight diversity of this Mesoamerican frontier zone, an area whose prehispanic inhabitants included both farming and hunting-gathering groups. [GF]

349 **Pendergast, David M.; Grant D. Jones;** and **Elizabeth Graham.** Locating Maya lowlands Spanish Colonial towns: a case study from Belize. (*Am. Antiq.*, 4:1, 1993, p. 59–73, bibl., ill.)

Analyzes archaeological and ethnohistorical data on early colonial towns in Belize. Explores implications of Belize cases for identifying remains of early postconquest communities in other regions. [JSH]

350 **Pendergast, David M.** Worlds in collision: the Maya/Spanish encounter in sixteenth and seventeenth century Belize. (*in*

The meeting of two worlds: Europe and the Americas, 1492–1650. Edited by Warwick Bray. Oxford: Oxford Univ. Press; The British Academy, 1993, p. 105–143, bibl., ill., map)

Evaluation of post-conquest archaeological data from two frontier sites in what is now Belize: Lamanai and Tipu. The evidence, though currently incomplete and impossible to support due to nonexistent ethnohistorical data, suggests a relatively weak Spanish cultural and religious impact on the area. Author also considers whether findings at these sites have a wider applicability in time and space. [R. Haskett and S. Wood]

351 **Piperno, Dolores R., and Deborah M. Pearsall.** Phytoliths in the reproductive structures of maize and teosinte: implications for the study of maize evolution. (*J. Archaeol. Sci.*, 20, 1993, p. 337–362, bibl., photos, tables)

Analysis of phytoliths from cobs of Argentine popcorn, a primitive race thought to resemble earliest archaeological maize at Tehuacán, shows them to have features distinct from other Zea phytoliths. Infers possiblity of differentiating ancient races of maize from teosinte in archaeological deposits. [GF]

352 **Pottery of prehistoric Honduras: regional classification and analysis.** Edited by John S. Henderson and Marilyn Beaudry-Corbett. Los Angeles: Institute of Archaeology, Univ. of California, Los Angeles, 1993. 312 p.: bibl., ill., index. (Monograph; 35)

Presents compact descriptions of prehistoric ceramics from regions of Honduras on which recent archaeological investigations have focused. Includes introductory essay on theoretical and methodological issues relating to ceramic classification and analysis. Presents an argument for consistency in basic descriptive analysis. [JSH]

353 **Proyecto Tajín.** v. 1–3. Coordinación de Jürgen K. Brüggemann. México: Dirección de Arqueología, Instituto Nacional de Antropología e Historia, 1991. 3 v.: bibl., ill., plates. (Cuaderno de trabajo; 8–10)

Three-volume set reports on results of 1984–85 Tajín Project. Excavations and surface surveys were conducted at Tajín and eight nearby sites. Includes history of work at Tajín, results of excavations and surface surveys, and description of restoration and conservation activities. [GF]

354 **Rattray, Evelyn Childs.** Nuevas interpretaciones en torno al Barrio de los Comerciantes. (*An. Antropol.*, 25, 1988, p. 165–180, bibl., ill., map, photo)

Recent excavations at the Barrio de los Comerciantes in Teotihuacán have revealed two major periods of occupation, from AD 400–550 and AD 550–650. Each of these occupation episodes is characterized by its own distinctive architecture, ceramic complex, and craft activities. [GF]

355 **Rattray, Evelyn Childs.** The Teotihuacán burials and offerings: a commentary and inventory. Nashville, Tenn.: Vanderbilt Univ., 1992. 236 p.: bibl., ill., maps. (Vanderbilt University publications in anthropology; 42)

Provides complete inventory of burials and offerings from apartment compounds (Tetitla, Zacuala, Yayahuala, and La Ventilla), the Tlajinga 33 compound, the Barrio de los Comerciantes, and the Oaxaca Barrio at Teotihuacán. [GF]

356 **Rivera Dorado, Miguel.** La religión maya en un solo lugar. (*Rev. Esp. Antropol. Am.*, 21, 1991, p. 53–76, photos)

Reconstruction of ceremonial activity and belief at the late classic Maya city of Oxkintok in northern Yucatán is based on recently excavated material. [JSH]

357 **Rivero Torres, Sonia E.** Laguna Miramar, Chiapas, México: una aproximación histórica-arqueológica de los lacandones desde el clásico temprano. Tuxtla Gutiérrez, Mexico: Gobierno del Estado de Chiapas, Consejo Estatal de Fomento a la Investigación y Difusión de la Cultura, DIF-Chiapas/Instituto Chiapaneco de Cultura; Instituto Nacional de Antropología e Historia, 1992. 196 p.: bibl., ill., maps. (Serie Antropología; 4)

Provides account of archaeological investigation of site of Lacam-Tún, Chiapas which was occupied from late formative period until after the Spanish invasion. Emphasis is on description of materials recovered, especially pottery and lithics, and on burials. [JSH]

358 **Rivero Torres, Sonia E.** Patrón de asentamiento rural en la región de San Gregorio, Chiapas, para el clásico tardío. México: Instituto Nacional de Antropología e Historia, 1990. 380 p.: bibl., ill. (Col. científica; 192. Serie Arqueología)

Detailed report provides analysis of

Late Classic settlement patterns in upper Rio Grijalva drainage in highlands of eastern Chiapas. Includes data on natural environment and analyses of ceramics and lithics, along with a spatial analysis of settlement distribution and interpretative discussion of socioeconomic organization in the region. [JSH]

359 Romero Rivera, María Eugenia. Aspectos de la navegación maya: la costa de Quintana Roo. *(Arqueología/México*, 5, 1991, p. 93–106, bibl., ill.)

Studies prehispanic Maya seafaring along east coast of the Yucatán peninsula. Combines ethnohistorical information on maritime activity in region with survey of coastal sites and information from modern navigation along the coast. [JSH]

360 Santamaría Estévez, Diana and **Joaquín García-Bárcena.** Puntas de proyectil, cuchillos y otras herramientas sencillas de Los Grifos. México: Subdirección de Servicios Académicos, Instituto Nacional de Antropología e Historia, 1989. 174 p.: bibl., ill. (Cuaderno de trabajo; 40)

Preliminary report contains classification and description of flaked tools, drills, projectile points, burins, and other stone tools from Los Grifos cave in Ocozocoautla, Chiapas. Includes discussions of general chronological placement of artifacts recovered in the cave. [GF]

361 Sarro, Patricia Joan. The role of architectural sculpture in ritual space at Teotihuacán, Mexico. *(Anc. Mesoam.*, 2:2, Fall 1991, p. 249–262, bibl., ill., photos)

Comparison of Pyramid of the Feathered Serpent (built in late 2nd century) and Palace of the Sacred Birds (AD 650–750) discusses how architectural form and decoration worked together to manipulate viewer's relationship to sacred space. [GF]

362 Schmidt Schoenberg, Paul. Arqueología de Xochipala, Guerrero. Apéndices de Luis Alberto Barba Pingarrón y Magalí Civera Cerecedo. México: Univ. Nacional Autónoma de México, Instituto de Investigaciones Antropológicas, 1990. 301 p.: appendices, bibl., ill., maps.

Reports on two episodes of fieldwork in Xochipala. Describes more than 90 sites that were recorded and the materials that were discovered during exploration and excavation in the area. Presents a sequence of settlement pattern changes, spanning the middle preclassic to the postclassic. [GF]

363 Schortman, Edward M. Archaeological investigations in the lower Motagua Valley, Izabal, Guatemala: a study in monumental site function and interaction. Philadelphia: Univ. Museum, Univ. of Pennsylvania, 1993. 292 p.: bibl., ill., index, map. (Quiriguá reports; 3. University Museum monograph; 80)

Reports on program of survey and excavation in lower Motagua Valley in eastern Guatemala. Emphasis is on work at three large late classic period sites in the region. Text provides detailed descriptions of excavations and analyses of ceramics, architecture, and site planning. Also interprets activities reflected within the sites, interactions among late classic communities in the lower Motagua Valley, and connections with adjacent regions. [JSH]

364 Schultz, Kevan C.; Jason J. Gonzalez; and **Norman Hammond.** Classic Maya ball courts at La Milpa, Belize. *(Anc. Mesoam.*, 5:1, 1994, p. 45–53, bibl., ill.)

Describes two terminal classic ball courts at the large Maya city in northwestern Belize. Authors suggest that, based on the contrasting form of the two contemporaneous ball courts, they were used simultaneously for different types of ball games. [JSH]

365 Scott, Sue A. Teotihuacán Mazapán figurines and the Xipe Totec statue: a link between the Basin of Mexico and the Valley of Oaxaca. Nashville, Tenn.: Vanderbilt Univ., 1993. 124 p.: bibl., ill., photos, plates. (Vanderbilt University publications in anthropology; 44)

Studies the roughly 2,000 figurines excavated from Xolalpán sector at Teotihuacán by Swedish archaeologist Sigvald Linné in 1932. Notes similarities between these Mazapán figurines and those found at Lambityeco in the Valley of Oaxaca. Includes b/w photographs. [GF]

366 Sempowski, Martha Lou and **Michael W. Spence.** Mortuary practices and skeletal remains at Teotihuacán. Addendum by Rebecca Storey. Salt Lake City: Univ. of Utah Press, 1994. 483 p., 17 p. of plates: bibl., ill., index, maps. (Urbanization at Teotihuacán, Mexico; 3)

Based on analysis of 373 burials recovered from pre-1970s excavations in Teotihuacán apartment compounds, Sempowski identifies varying levels of social status in Teotihuacán society. Spence's analysis of discrete traits of adult skeletons reveals greater homogeneity of males buried within compounds. Includes summary of the burial contents and addendum by R. Storey describing burials from Tlajinga 33. [GF]

367 **Sharer, Robert J.** Quiriguá: a classic Maya center & its sculptures. Durham, N.C.: Carolina Academic Press, 1990. 124 p.: bibl., ill., index, maps. (Centers of civilization)

Overview of ancient Maya city of Quiriguá in eastern Guatemala focuses on investigations by Univ. of Pennsylvania in 1970s. Also includes summary of earlier research, detailed description of the carved monuments with commentary on their iconography and texts, and a consideration of Quiriguá's place in the history of lowland Maya civilization. [JSH]

368 **Sheets, Payson D.** The Cerén Site: a prehistoric village buried by volcanic ash in Central America. Fort Worth, Tex.: Harcourt Brace Jovanovich College Publishers; 1992. 150 p.: bibl., ill., index, maps. (Case studies in archaeology series)

Cerén, in what is now El Salvador, was buried by volcanic ash about AD 600, preserving many aspects of household life that do not normally survive to become part of the archaeological record. Sheets, director of the excavations at Cerén, presents a non-technical summary of what the investigations reveal about life in the region in the first millenium AD. [JSH]

369 **Sheets, Payson D.** Tropical time capsule. (*Archaeology/New York*, 47:4, 1994, p. 30–33, bibl., ill.)

Briefly summarizes excavation of remarkably well-preserved remains of a village in western El Salvador buried under volcanic ash in late 6th century AD. [JSH]

370 **Sievert, April Kay.** Maya ceremonial specialization: lithic tools from the Sacred Cenote at Chichén Itzá, Yucatán. Madison, Wis.: Prehistory Press, 1992. 162 p.: bibl., ill. (Monographs in world archaeology; 1055–2316: 12)

Studies Maya ritual behavior, especially in the postclassic period in Yucatán, and its reflection in material remains recovered by archaeologists. Emphasis is on stone tools used in ceremonial activity, with lithics recovered from Sacred Cenote at Chichén Itzá providing primary data for developing a more general model of ritual use of lithics. [JSH]

371 **Sievert, April Kay.** Postclassic Maya ritual behavior: microwear analysis of stone tools from ceremonial contexts. (*in* The interpretative possibilities of microwear studies. Uppsala, Sweden: Societas Archaeologica Upsaliensis, 1990, p. 147–157, bibl., ill.)

Preliminary analysis of terminal classic and postclassic chipped stone artifacts from Sacred Cenote of Chichén Itzá suggests that ritual tools may be distinctive in terms of use and culturally deposited residues. [JSH]

372 **Sluyter, Andrew** and **Alfred H. Siemen.** Vestiges of prehispanic sloping-field terraces on the piedmont of central Veracruz, Mexico. (*Lat. Am. Antiq.*, 3:2, June 1992, p. 148–160, bibl., maps, photos)

Evidence of canals and maize planting platforms are noted in the wetlands south of Zempoala. Based on study of these agrarian features, authors argue that intensive farming was practiced in this area during the classic period. [GF]

373 **Smith, Michael Ernest et al.** Archaeological research at Aztec-period rural sites in Morelos, Mexico = Investigaciones arqueológicas en sitios rurales de la época azteca en Morelos. v. 1, Excavations and architecture. Translation by Ana María Boza-Arlotti. Pittsburgh: Univ. of Pittsburgh, Dept. of Anthropology, 1992. 1 v.: bibl., ill. (University of Pittsburgh memoirs in Latin American archaeology; 4)

Reports in detail on excavation of 50 houses and structures at three rural late postclassic sites. Architectural study considers function and significance of domestic constructions. Concludes that Aztec empire had little direct influence on these rural households. [GF]

374 **Smith, Michael Ernest** and **Frances Berdan.** Archaeology and the Aztec empire. (*World Archaeol.*, 23:3, Feb. 1992, p. 353–367, bibl., ill., photo, table)

Authors recognize that the bounds of the Aztec empire cannot be detected using archaeology alone. They present results of new archaeological and ethnohistoric research conducted outside the Basin of Mexico, and argue that both kinds of data are necessary to elucidate Aztec imperialism and its effects across postclassic Mesoamerica. [GF]

375   Smith, Michael Ernest and John F. Doershuk. Late postclassic chronology in western Morelos, Mexico. (*Lat. Am. Antiq.*, 2:4, Dec. 1991, p. 291–310, bibl., ill., map, tables)

A five-phase postclassic ceramic chronology, developed through integrated analysis of stratigraphy and quantitative seriation, is supported by archaeological fieldwork conducted at Capilco and Cuexcomate. [GF]

376   Smyth, Michael P. and Christopher D. Dore. Large-site archaeological methods at Sayil, Yucatán, Mexico: investigating community organization at a prehispanic Maya center. (*Lat. Am. Antiq.*, 3:1, 1992, p. 3–21, bibl., ill.)

Summarizes mapping and intensive surface collection at terminal classic Maya center in western Yucatán which was designed to explore structure of the entire community. Interpretation focuses on social variability, economic specialization, and spatial patterning of political and religious activity within the community. [JSH]

377   Sodi Miranda, Federica and Hugo Herrera Torres. Estudio de los objetos arqueológicos de la cultura matlatzinca: catálogo de las colecciones arqueológicas del Museo Nacional de Antropología, Instituto Nacional de Antropología e Historia. Proemio de Román Piña Chan. México: Instituto Nacional de Antropología e Historia, 1991. 175 p.: bibl., ill., map.

Catalogs ceramic vessels and other prehispanic artifacts from the Matlatzinca culture of Toluca. The materials are part of the Museo Nacional de Antropología collections. [GF]

378   Solís Olguín, Felipe R. and David Morales. Rescate de un rescate: colección de objetos arqueológicos de El Volador, Ciudad de México. Estudio histórico de la plaza y el mercado por José Guadalupe Victoria. México: Instituto Nacional de Antropología e Historia, 1991. 317 p.: bibl.

Catalogs ceramic vessels and carved stones in the Museo Nacional de Antropología that were recovered during 1936–37 salvage excavations in the Plaza de El Volador in Mexico City. Includes short post-conquest history of this famous plaza. [GF]

379   Sosa, Erasmo. Los petroglifos de Orealí, Municipio de Oropolí, Departamento de El Paraíso. (*Yaxkin/Tegucigalpa*, 12:1, 1989, p. 126–132, bibl., ill.)

Brief notice of discovery of precolumbian petroglyphs in southern Honduras. [JSH]

380   Sotelo Santos, Laura Elena. Yaxchilán. Presentación de José Patrocinio González B. Garrido. México: Gobierno del Estado de Chiapas, 1992. 190 p.: bibl., ill. (some col.).

Extensively illustrated volume describes late classic Maya city on Río Usumacinta in eastern Chiapas, Mexico. Includes summary of history of investigations at Yaxchilán, summary of city's dynastic history as recorded in its hieroglyphic texts, and discussions of the major architectural and sculptural monuments. [JSH]

381   Storey, Rebecca. Life and death in the ancient city of Teotihuacán: a modern paleodemographic synthesis. Tuscaloosa: Univ. of Alabama Press, 1992. 307 p.: bibl., ill., index.

Presents paleodemographic study of burial sample from Tlajinga area of Teotihuacán. Draws a comparison between prehispanic Teotihuacán and 17th-century London; both were cities with dense populations and poor sanitation. Although these cities were very different culturally and environmentally, the research suggests they were comparable in terms of health problems and death rates. [GF]

382   Storey, Rebecca. Residential compound organization and the evolution of the Teotihuacán state. (*Anc. Mesoam.*, 2:1, Spring 1991, p. 107–118, bibl., tables)

Presents and interprets mortuary findings from the lower-class Tlajinga 33 residential compound. Finds that this artisanal compound became poorer and had decreased autonomy in craft manufacture during its occupational sequence from early Tlamimilolpa to Metepec times (ca. AD 250–700). [GF]

383   Stresser-Péan, Guy and Dorothy Hosler. El cascabel de El Naranjo: uno de los más grandes y bellos de Mesoamérica. (*Trace/México*, 21, June 1992, p. 66–74, bibl., ill., map, photo)

Describes and chemically analyzes a large decorated copper bell from El Naranjo, Tamaulipas. Places the focal artifact in a broader artifactual context. [GF]

384   **Suárez Diez, Lourdes.** Conchas y caracoles: ese universo maravilloso—. México: Grupo Financiero Mexival, 1991. 192 p.: bibl., col. ill.

This lavishly illustrated book presents a range of topics on prehispanic use of shells. Sections focus on shell ornaments and how they were made, ritual significance of shells, and representations of shells in sculpture, murals, and the codices. [GF]

385   **Sugiura Y., Yoko** and **Emily McClung de Tapia.** Algunas consideraciones sobre el uso prehispánico de recursos vegetales en la Cuenca del Alto Lerma. (*An. Antropol.*, 25, 1988, p. 111–125, bibl., ill., map, table)

Provides preliminary analysis of macrobotanical remains recovered from six sites located in three distinct macrozones in the higher Lerma Basin of the Toluca Valley. [GF]

386   **Sugiyama, Saburo.** Worldview materialized in Teotihuacán, Mexico. (*Lat. Am. Antiq.*, 4:2, June 1993, p. 103–129, bibl., ill., map, photo)

Proposes that Teotihuacán was architecturally planned to express a specific worldview. According to this model, the Río San Juan divides city center into two sections, with watery underworld to the south represented by the Ciudadela. To the north, the Pyramid of the Sun represents the 260-day ritual calendar and a passageway from the underworld to the heavens. [GF]

387   **Tate, Carolyn Elaine.** Yaxchilán: the design of a Maya ceremonial city. Austin: Univ. of Texas Press, 1992. 306 p., 8 p. of plates: bibl., ill. (some col.), index, map.

Documents surviving architecture and sculpture at the late classic Maya city. Analysis focuses on iconography and spatial relationships of the monuments of late 7th- and 8th-century reigns of Shield Jaguar and Bird Jaguar IV as public art. [JSH]

388   **Tejada Bouscayrol, Mario.** El período preclásico en Chiapas: una síntesis. (*Anu. Inst. Chiapaneco Cult.*, 1990, p. 242–275, bibl., photos)

Provides brief descriptive synthesis of recent research in Chiapas, highlighting findings of New World Archaeological Foundation. Reviews changes during preclassic period, and includes supplemental bibliography of relevant literature. [GF]

389   **Teotihuacán: art from the City of the Gods.** Edited by Kathleen Berrin and Esther Pasztory. New York: Thames and Hudson; San Francisco: Fine Arts Museums of San Francisco, 1993. 288 p.: bibl., ill. (some col.), index, maps.

Lavishly illustrated book was published in conjunction with an exhibition of more than 200 pieces at the Fine Arts Museum, San Francisco. Volume consists of 11 papers covering a diversity of topics concerning Teotihuacán. These range from considerations of daily life and writing to human sacrifice and the ways that the later Mexica viewed the earlier city. [GF]

390   **Teotihuacán 1980–1982: nuevas interpretaciones.** Coordinación de Rubén Cabrera Castro, Ignacio Rodríguez García, y Noel Morelos García. México: Instituto Nacional de Antropología e Historia, 1991. 400 p.: bibl., ill., maps, plans (Serie Arqueología. Col. Científica; 227)

Collection of 19 articles that report on the 1980–82 investigations of the Teotihuacán Archaeology Project. Topics cover various aspects of Teotihuacán architecture (including the Ciudadela and *adoratorios* on the Street of the Dead), iconography (including ritual masks, painted murals, feline representations, and Teotihuacán figurines in Chengú, Hidalgo), the funerary system, and the role of hydraulic agriculture. [GF]

391   **Tesoros del Museo Regional de Oaxaca = Oaxaca Regional Museum treasures.** Coordinación de Ludwig Zellev. Descripciones de Marcus Winter. Traducción de Susana Wald. Fotografías de Cecilia Salcedo. Oaxaca, Mexico: Honorable Ayuntamiento de Oaxaca, 1994. 138 p.: bibl., col. ill., map.

Handsomely illustrated volume catalogs ceramic, stone, and bone artifacts in Museo Regional de Oaxaca. Accompanying text written by Marcus Winter. [GF]

392   **Tichy, Franz.** Der Fuβ zur Längenmessung in Mesoamerika, die Hand zur Winkelangabe? (*in* Circumpacifica: Festschrift für Thomas S. Barthel. Frankfurt am Main: P. Lang, 1990, v. 1, p. 437–454, bibl., ill., maps, tables)

Proposes that basic unit of measurement in architecture, city planning, and agriculture was the foot, standardized at 25 cm or its multiples. A sighting device, in Codex Seldon 14-IV, as well as a hand, may have served to determine angles. [H. von Winning]

393 **Torres Montes, Luis *et al.*** Informe del deterioro y propuesta para la conservación de la zona arqueológica de Malinalco, Estado de México. (*An. Antropol.*, 26, 1989, p. 107–126, bibl.)

Describes water percolation that is destroying the temples at Malinalco. Includes recommendations to enhance conservation. [GF]

394 **Tourtellot, Gair; Amanda Clarke; and Norman Hammond.** Mapping La Milpa, a Maya city in northwestern Belize. (*Antiquity/Cambridge*, 67:254, March 1993, p. 96–108, bibl., maps, photos)

Summarizes recent field investigations at large city of La Milpa in eastern Maya lowlands, occupied from late formative through early postclassic periods. Essay emphasizes mapping the architectural monuments in the city center and the smaller structures in the surrounding area, and includes brief descriptions of the carved monuments and hieroglyphic texts. [JSH]

395 **Trabajos arqueológicos en el centro de la Ciudad de México.** Coordinación de Eduardo Matos Moctezuma. 2. ed. México: Instituto Nacional de Antropología e Historia, 1990. 584 p.: appendix, bibl., ill. (Antologías: Serie Arqueología)

Compilation of 17 articles discusses various discoveries and investigations at the Templo Mayor. This historical overview includes many articles written at the turn of the century, or earlier (1792). Includes photographic record of stone sculpture and ceramic vessels found, and an appendix of key objects recovered during recent construction of the subway. [GF]

396 **Trombold, Charles D. *et al.*** Chemical characteristics of obsidian from archaeological sites in western Mexico and the Tequila source area: implications for regional and pan-regional interaction within the northern Mesoamerican periphery. (*Anc. Mesoam.*, 4:2, Fall 1993, p. 255–270, bibl., ill., maps, tables)

Obsidian samples from La Quemada, Totoate, Las Ventanas, and Laguna San Marcos were chemically sourced. Five sources were identified, with highest proportion having come from La Lobera source on Jalisco-Zacatecas border. A minute component came from the Tequila source, while the other groups have not yet been determined. [GF]

397 **Valadez Azúa, Raúl.** Informe preliminar acerca del material faunístico encontrado en el barrio oaxaqueño de Teotihuacán, 1987. (*Antropológicas/México*, 5, 1990, p. 67–78, bibl.)

Lists all faunal remains, including shell, recovered during 1987 excavation conducted by Michael Spence in the Oaxaca Barrio. [GF]

398 **Valdés, Juan Antonio and José Suasnávar.** La arquitectura maya y sus implicaciones: consideraciones iniciales sobre la zona de Petexbatún, Petén, Guatemala. (*Trace/México*, 21, June 1992, p. 9–21, bibl., ill., map, photo)

Analyzes settlement organization and architectural patterns in Petexbatún region of Maya lowlands in northern Guatemala. Focuses on comparison of patterns during 7th and early 8th centuries (when Dos Pilas was the dominant city in the region) with those of the period after decline of Dos Pilas in mid-8th century. [JSH]

399 **Valdés, Juan Antonio.** Proyecto arqueológico Petexbatún: desarrollo y datos históricos de los eventos sociopolíticos en el suroeste de Petén. (*Apunt. Arqueol./Guatemala*, 1:2, dic. 1991, p. 9–19, bibl., maps)

Overview examines recent archaeological investigations in Petexbatún region of Maya lowlands in northern Guatemala. Focuses on dynastic history of Dos Pilas and nearby cities, with emphasis on importance of warfare in the region at end of the classic period. [JSH]

400 **Véliz, Vito.** Copán en el mundo maya. Comayagüela, Honduras: Litográfica Comayagüela, 1991. 100 p.: bibl., ill., map.

Brief overview examines some of the recent work at ancient Maya city of Copán in western Honduras. [JSH]

401 **Von Winning, Hasso.** Altmexikanische Pyritspiegel mit reliefierter Rückseite. (*in* Circumpacifica: Festschrift für Thomas S. Barthel. Frankfurt am Main: P. Lang, 1990, v. 1, p. 455–481, bibl., facsims., ill.)

Provides iconographic interpretation of

reliefs on the backsides of four Mexican slate disks whose pyrite mosaic front served as mirrors. [H. von Winning]

402   Von Winning, Hasso. Malacates prehispánicos con figuras humanas en relieve. (*An. Inst. Invest. Estét.*, 64, 1993, p. 1–14, bibl., ill.)
   Studies of spindle whorls with Toltec portraits and of symbolic meaning of spindles (they are related to pregnancy), with reference to Tlazolteotl and lunations in Codex Borgia and Laud. [H. von Winning]

403   Webster, David; AnnCorinne Freter; and David Rue. The obsidian hydration dating project at Copán: a regional approach and why it works. (*Lat. Am. Antiq.*, 4:4, 1993, p. 303–24, bibl., ill)
   Reviews settlement data and extensive obsidian hydration samples from Maya city of Copán in western Honduras. Responds to an earlier critique (see item 216), defending methodology used and reiterating the conclusion that Copán experienced a prolonged population and political decline rather than a sudden collapse in 9th century. [JSH]

404   Wertime, Richard A. et al. Written in the stars. (*Archaeology/New York*, 46:4, July/Aug. 1993, p. 26–35, ill., photos)
   Popular summary focuses on some recent interpretations of Maya astronomy. Emphasizes interpretation of classic period texts by analogy with postclassic painted books and ethnohistoric and ethnographic information in order to reconstruct celestial aspects of ancient Maya belief. [JSH]

405   Widmer, Randolph J. Lapidary craft specialization at Teotihuacán: implications for community structure at 33:S3W1 and economic organization in the city. (*Anc. Mesoam.*, 2:1, Spring 1991, p. 131–147, bibl., ill., map, photo, tables)
   Excavations in this classic-period apartment compound unearthed two lapidary production loci. Micro-artifact analysis indicates that a number of different raw materials were worked, many imported. Volume of debitage recovered serves to infer that these manufacturing activities were not exclusively for local consumption. [GF]

406   Zárate Morán, Roberto. La Tumba 12 de Lambityeco. (*Cuad. Sur/Oaxaca*, 1:2, sept./dic. 1992, p. 5–22, bibl., ill., map)
   Discusses excavation of Tomb 12 at Lambityeco, Valley of Oaxaca. Includes description of objects found in this late classic tomb. [GF]

407   Zeitlin, Robert N. Pacific coastal Laguna Zope: a regional center in the terminal formative hinterlands of Monte Albán. (*Anc. Mesoam.*, 4:1, Spring 1993, p. 85–101, bibl., ill., maps, photos)
   Reports on excavations conducted in an elite-status area of terminal formative Laguna Zope. Considers nature of the political/economic relationship between the site and distant Monte Albán. [GF]

NATIVE SOURCES AND EPIGRAPHY

408   Anawalt, Patricia Rieff and Frances Berdan. The *Codex Mendoza*. (*Sci. Am.*, 266:6, June 1992, p. 70–79, ill., map)
   Discusses history and content of this codex that was prepared in 1541 at the behest of Mexico's first Spanish viceroy who wanted a firsthand account of Aztec life. For annotation of 4 vol. critical edition of the *Codex Mendoza* by the same authors, see item 89. [GF]

409   Anawalt, Patricia Rieff. Riddle of the emperor's cloak. (*Archaeology/New York*, 46:3, May/June 1993, p. 30–36, facsims., ill., map, photo)
   Suggests that the complex design motifs found on royal Aztec cloaks served to legitimize the ruler's claim of Toltec descent. [GF]

410   Barthel, Thomas S. Ein früher Schlüssel zur Indo-Mexikanistik. (*Indiana/Berlin*, 12, 1992, p. 123–145, bibl., ill.)
   Compares the nine hieroglyphic portraits on an early classic Maya bowl with the nine "Lords of the Night" as shown in the non-Maya codex Fejérváry-Mayer 1. Also discusses the "Matrix of Nine" by correlating the nine "Night Lords" in Codex Borgia 14 with Hindu patterns of planetary qualities, and postulates a syncretistic transmittal process from the southeastern donor cultures to the New World, beginning in 5th century. [H. von Winning]

Brotherston, Gordon and Ana Gallegos. El *Lienzo de Tlaxcala* y el *Manuscrito de Glasgow*: Hunter 242. See *HLAS 54:439*.

411 Couch, N.C. Christopher. The *Codex Ramírez*: copy or original? [*Estud. Cult. Náhuatl*, 21, 1991, p. 109–125, bibl., table]

Argues that the codex is more than just an abridgement of Durán's *Historia*. Instead, it is comprised of two separate works, the first of which condenses the first part of Durán's text, while the second is an original work containing an historically incorrect dynastic sequence. [GF]

412 Davoust, Michel. Nueva lectura de las inscripciones de Xcalumkin, Campeche, México. [*Mesoamérica/Antigua*, 12:22, dic. 1991, p. 249–276, ill., tables]

Studies hieroglyphic inscriptions of the classic-period Maya city of Xcalumkin in the Puuc hill region of western Yucatán. Emphasizes phonetic readings of glyphs, mainly names and titles, in Yucatec Maya. [JSH]

413 Demarest, Arthur Andrew et al. Arqueología, epigrafía y el descubrimiento de una tumba real en el centro ceremonial de Dos Pilas, Petén, Guatemala. [*U tz'ib*, 1:1, nov. 1991, p. 14–28, bibl., ill., map]

Describes discovery of a rich tomb within a large pyramid on main plaza of late classic Maya center of Dos Pilas in Pasión region of lowland Guatemala. Argues that occupant of the tomb can be identified with early 8th-century "Ruler 2," whose death and burial is the subject of the hieroglyphic text on a stela erected in front of the pyramid containing the tomb. [JSH]

414 Dibble, Charles E. The Boban Calendar Wheel. [*Estud. Cult. Náhuatl*, 20, 1990, p. 173–182, bibl., ill.]

Argues for an early date for the painting of this codex, ca. 1564. [GF]

415 Estrada Lugo, Erin Ingrid Jane. El *Códice Florentino*: su información etnobotánica. Montecillo, Mexico: Colegio de Postgraduados, Institución de Enseñanza e Investigación en Ciencias Agrícolas, 1989. 399 p.: bibl., ill. (some col.).

Detailed, computer-assisted realization of vast amount of ethnobotanical information contained in *Florentine Codex*. Most valuable for its numerous detailed tables that identify and arrange plants by scientific name, by taxonomy, by use, and by other important criteria. [R. Haskett and S. Wood]

416 Giesing, Cornelia. Die Grosse Göttin und der "Feuerphallus:" Tlazolteotl und ihre Partner im *Codex Borgia*. [*Indiana/Berlin*, 12, 1992, p. 147–205, bibl., ill.]

Extensive investigation examines multifarious aspects and functions of the Great Goddess Tlazolteotl and her associations with other deities—mainly her partner, the old fire god Huehuecoyotl. Giesing explains intricacies of triadic and other subsystems in *Codex Borgia* concerning variants of the lunar Tlazolteotl, the Great Spinner, and her historical/political connections. [H. von Winning]

417 Grigsby, Thomas L. and Carmen Cook de Leonard. Xilonen in Tepoztlán: a comparison of Tepoztecan and Aztec agrarian ritual schedules. [*Ethnohistory/Society*, 39:2, Spring 1992, p. 110–147, bibl., graph, photos, tables]

This comparison of the recent Tepoztecan agrarian ritual schedule with that of the Aztec in Book 2 of Sahagún's *Florentine Codex* reveals symbolic and functional similarities between the two. The key difference was that Aztec rituals were enacted 78 days earlier. A range of hypotheses are advanced to account for these findings and their implications. [GF]

418 Grube, Nikolai. Classic Maya dance: evidence from hieroglyphs and iconography. [*Anc. Mesoam.*, 3:2, 1992, p. 201–218, bibl., ill.]

Interprets scenes in classic Maya sculpture and painting as representations of ritual dances. Includes proposed decipherment of hieroglyphic phrase for dancing. Relies heavily on unprovenanced objects. [JSH]

419 Guil'liem Arroyo, Salvador. Descubrimiento de una pintura mural en Tlatelolco. [*Antropológicas/México*, 3, 1989, p. 145–150, bibl., ill.]

Describes the Templo Calendárico (de los Glifos) and its mural painting. [GF]

420 Harris, John Ferguson. New and recent Maya hieroglyph readings: a supplement to *Understanding Maya inscriptions*. Philadelphia: Univ. Museum of Archaeology and Anthropology, Univ. of Pennsylvania, 1993. 39 p.: bibl., ill.

Supplementing 1992 work (see item 421), author summarizes new interpretations of Maya hieroglyphic writing. Includes table of probable phonetic values of those glyphs that are thought to represent syllables, and a listing of recent readings of various glyphs and glyph groups—verbal phrases, titles, place names, personal names, deity names, and a variety of other signs (mostly nouns). [JSH]

421 **Harris, John Ferguson** and **Stephen K. Stearns.** Understanding Maya inscriptions: a hieroglyph handbook. Philadelphia: Univ. Museum of Archaeology and Anthropology, Univ. of Pennsylvania, 1992. 159 p.: bibl., ill., tables.

Introduction to Maya calendars and writing systems provides tables and step-by-step instructions for a variety of calendrical calculations, including converting Maya Long Counts to Christian dates. See also item **420**. [JSH]

422 **Hasel, Ulrike.** Die Grosse Göttin im *Codex Laud*: Studien zum altmexikanischen Polytheismus der Golfküste. Münster, Germany: Lit, 1993. 257 p.: bibl., ill., map. (Studien zur Altamerikanistik; 1)

In *Codex Laud*, Tlazolteotl occupies a dominant central position in the pantheon. Her diverse functions as Great Goddess, Great Mother, Mother of the Gods, Mother of Maize, Moon and Earth Goddess, and goddess of maize, cotton, spinning and weaving, childbirth, and death are elucidated. Study supports the Gulf Coast origin of this south-central Veracruz Codex, and links it with Highland traditions. [H. von Winning]

423 **Hasselkus, Hans.** Wooh: introducción al conocimiento de los códices mayas. Mexico: s.n., 1993. 1 v.: bibl., ill.

Non-technical essay focuses on Maya hieroglyphic writing, mainly as it is represented in the four surviving precolumbian codices. Emphasizes calendric and phonetic interpretation (often idiosyncratic) of various passages in the codices. Lacks scholarly apparatus; does not cite scholars who developed most of the interpretations presented. [JSH]

Hicks, Frederic. Subject states and tribute provinces: the Aztec empire in the northern Valley of Mexico. See *HLAS 54:496*.

424 **Hinz, Eike.** Kritische Rekonstruktion aztekischer Denk- und Handlungsstrukturen. (*Indiana/Berlin*, 12, 1992, p. 55–75, bibl.)

Structures of Aztec thought and action patterns are methodically investigated in four lengthy passages of Sahagún's writings. These structures are not openly discernible in translations, but require a reconstruction of theoretically autochthonous semantic components and their relationships for a better understanding of the thought processes in alien cultures. [H. von Winning]

425 **Houston, Stephen D.** Hieroglyphs and history at Dos Pilas: dynastic politics of the Classic Maya. Austin: Univ. of Texas Press, 1993. 181 p.: bibl., ill., index, maps.

Study of Maya city of Dos Pilas in Pasión region of Petén lowlands is based on author's 1987 Yale dissertation. Includes description of site, its environmental setting, architecture, and style and technique of relief sculpture, but focuses on iconography and especially on hieroglyphic texts on the carved monuments. Presents reconstruction of dynastic history of Dos Pilas during late classic period, along with general argument for importance of balancing epigraphic and other archaeological evidence in reconstructing Maya history. [JSH]

426 **Justeson, John S.** and **Terrence Kaufman.** A decipherment of epi-Olmec hieroglyphic writing. (*Science/Washington*, 259, 1993, p. 1703–1711, bibl., ill.)

Summarizes proposed decipherment of long hieroglyphic text on stela from La Mojarra, Veracruz which contains calendar dates placing it in 2nd century AD. Identifies language of the text as ancestral to the Zoquean languages. See also item **427**. [JSH]

427 **Kelley, David H.** The decipherment of the epi-Olmec script as Zoquean by Justeson and Kaufman. (*Rev. Archaeol./Salem*, 14:1, 1993, p. 29–32, bibl., ill.)

Assessment by one of the leading students of Maya hieroglyphic script of proposed decipherment of late formative writing system of the La Mojarra stela. Kelley places Justeson's and Kaufman's interpretation (see item **426**) in context of the history of investigation of Maya script and of early Mesoamerican writing and calendars generally. [JSH]

428 **Kristan-Graham, Cynthia.** The business of narrative at Tula: an analysis of the vestibule frieze, trade, and ritual. (*Lat. Am. Antiq.*, 4:1, March 1993, p. 3–21, bibl., ill.)

New contextual interpretation of 8-m frieze of the *caciques* at Tula identifies depicted figures as merchants engaged in ritual activities related to trade. Hypothesizes that merchants from Tula may have served as prototypes for Aztec *pochteca*. [GF]

429   **La Gamma, Alisa.** A visual sonata at Teotihuacán. (*Anc. Mesoam.*, 2:2, Fall 1991, p. 275–284, bibl., ill., photos)

Proposes that borders of mural paintings at Teotihuacán had a preeminent rather than peripheral role in these compositions. Suggests that borders allow for organization of diverse entities within a structured whole. [GF]

430   **Langley, James C.** The forms and usage of notation at Teotihuacán. (*Anc. Mesoam.*, 2:2, Fall 1991, p. 285–298, bibl., ill., photos)

Analyzes over 1,000 glyphic signs found on decorated pottery and in mural paintings. Suggests that ancient Teotihuacanos developed a notational system comprised of at least 120 signs. Discusses how this sign system supplemented pictorially conveyed information. [GF]

431   **Lidzinski, Silvia.** Beschreibung und Analyse der mit der Person "12 Wind, Rauchends Auge" verbundenen Ereignisse auf den Seiten 18, II-22 der sogenannten Vorderseite des *Codex Nuttall*. (*Indiana/Berlin*, 12, 1992, p. 207–224, bibl., ill.)

Author compares scenes in Codices Nuttall and Vindobonensis in which "12 Wind" represents, according to her analysis, a deified ancestor and probably a variant of the culture hero Quetzalcoatl. [H. von Winning]

432   **López Austin, Alfredo; Leonardo López Luján; and Saburo Sugiyama.** The Temple of Quetzalcoatl at Teotihuacán: its possible ideological significance. (*Anc. Mesoam.*, 2:1, Spring 1991, p. 93–105, bibl., ill.)

Based on iconographic studies and recent archaeological excavations, authors suggest that the temple was dedicated to the passage of time. [GF]

433   **Love, Bruce.** The Paris Codex: handbook for a Maya priest. Introduction by George E. Stuart. Austin: Univ. of Texas Press, 1994. 124 p.: bibl., ill., index, maps.

Detailed study analyzes one of the four surviving precolumbian Maya books, now in the Bibliothèque Nationale in Paris. Includes introductory material on Maya books and writing generally, a study of the history of the manuscript itself as well as of research on it, a detailed commentary on its contents, and a b/w photographic reproduction. [JSH]

434   **Marcus, Joyce and William J. Folan.** Una estela más del siglo V y nueva información sobre Pata de Jaguar, gobernante de Calakmul, Campeche, en el siglo VII. (*Gac. Univ./Campeche*, 4:15/16, 1994, p. 21–26, bibl., ill.)

Reports discovery of an early (Cycle 8) stela erected at the ancient Maya city in AD 431. Also notes discovery of a later stela with a text referring to Jaguar Paw, the powerful 7th-century king of Calakmul, and summarizes his dynastic career and his foreign alliances and conflicts. [JSH]

435   **Marcus, Joyce.** Mesoamerican writing systems: propaganda, myth, and history in four ancient civilizations. Princeton, N.J.: Princeton Univ. Press, 1992. 495 p.: bibl., ill., index, maps.

Study of four prehispanic hieroglyphic writing systems—Aztec, Zapotec, Mixtec, and Maya—highlights role of Mesoamerican writing as a political tool that enabled leaders to compete for power and prestige. Concludes that ancient Mesoamerican writing recorded a complicated web of propaganda, myth, and history, and thus contents of the documentary record should not be construed literally. [GF]

436   **Matos Moctezuma, Eduardo.** Las seis Coyolxauhqui: variaciones sobre un mismo tema. (*Estud. Cult. Náhuatl*, 21, 1991, 15–29, ill., map)

Describes six known sculptures of this deity, including two of the head only and four that portray the full body. All of these sculptures are found at or near the Templo Mayor. Specific attributes of these portrayals are constant, yet the six representations are shown to have important differences as well. [GF]

437   **Mayer, Karl Herbert.** Maya inscriptions from Dzibilnocac, Campeche, Mexico. (*Arch. Völkerkd.*, 46, 1992, p. 111–122, bibl., photos)

Brief work discusses hieroglyphic texts from classic-period Maya city of Dzibilnocac in the Chenes region of Yucatán. Includes brief descriptions of texts on one stela, three painted capstones, and five fragmentary inscribed stones. [JSH]

438 Máynez, Pilar. La fauna mexicana en la obra de Fray Bernardino de Sahagún. (*Estud. Cult. Náhuatl,* 21, 1991, p. 145–161, table)

Contains a list of the 459 Nahuatl terms for animal species that are described in the Florentine Codex. [GF]

439 McClung de Tapia, Emily. El *amamalócotl* en el arte teotihuacano. (*Antropológicas/México,* 3, 1989, p. 29–37, bibl., ill., photos)

This aquatic plant served as both a food and a medicine at Teotihuacán. Frequent representations of the plant on ceramic vessels and in murals suggest it also had ritual and symbolic importance. [GF]

440 Méluzin, Sylvia. An ancient Zapotec calendrical cosmogram. (*Archaeoastronomy/College Park,* 10, 1987/88, p. 139–147, bibl., ill., map, photos)

Argues that one of the inscriptions on Building J at Monte Albán reinforces interpretation that the building served an astronomical function. [GF]

441 Méluzin, Sylvia. The Tuxtla script: steps toward decipherment based on La Mojarrra Stela 1. (*Lat. Am. Antiq.,* 3:4, 1992, p. 283–297, bibl., ill.)

Analyzes late preclassic hieroglyphic texts on Stela 1 from La Mojarra, Veracruz, the Tuxtla statuette, and a few related objects. Proposes readings for calendrical glyphs based on similarities with homologous Yucatec Maya signs. [JSH]

442 Mohar Betancourt, Luz María. La escritura en el México antiguo. México: Plaza y Valdés; Casa Abierta al Tiempo, Unidad Xochimilco, 1990. 2 v.: bibl., ill. (some col.), index.

Based on thorough comparison of two pictorial documents, the *Matrícula de Tributos* and the *Codex Mendoza,* author argues that second part of the codex is not simply a copy of the former. Vol. 2 includes line drawings and color plates of both documents. [GF]

443 Ostrowitz, Judith. Second nature: concentric structures and gravity as represented in Teotihuacán art. (*Anc. Mesoam.,* 2:2, Fall 1991, p. 263–274, bibl., ill.)

Presents interpretation of landscapes painted by Teotihuacán muralists. Proposes that murals reproduced aspects of nature considered essential for structuring a metaphor for power. [GF]

444 Proskouriakoff, Tatiana. Maya history. Edited by Rosemary A. Joyce. Foreword by Gordon R. Willey. Biographical sketch by Ian Graham. Illustrations by Barbara C. Page. Austin: Univ. of Texas Press, 1993. 212 p.: bibl., ill., index.

Posthumous presentation analyzes historical information in Maya lowland monuments of Classic period, on which Proskouriakoff was working at the time of her death in 1985. Emphasis is on style and iconography of the monuments, with complementary interpretation of hieroglyphic texts. Joyce's sensitive editing preserves Proskouriakoff's insightful but fundamentally conservative approach to surviving evidence of classic Maya history. [JSH]

445 Sánchez Montañés, Emma. El cambio de función y la persistencia del culto a las estelas en el área maya: el caso de Cobá. (*Cuad. Prehispánicos,* 14, 1989/1990, p. 5–20, bibl., ill.)

Analysis of patterns of stela erection at classic city of Cobá in northeastern Yucatán shows strong similarities in this respect to southern lowland cities. Includes discussion of modern ritual activity involving prehispanic stelae. [JSH]

446 Saurwein, Anton. Ruler of Huexotla: glyphs and text sources. (*Mexicon/Berlin,* 13:1, Jan. 1991, p. 10–16, bibl., ill., map)

Examines two textual and glyphic sources from Huexotla. Warns that because Aztec glyphs were scarcely conventionalized, they have not been read consistently since early colonial times. [GF]

447 Smith, Mary Elizabeth and **Ross Parmenter.** The Codex Tulane. New Orleans, La.: Middle American Research Institute, Tulane Univ., 1991. 142 p.: bibl., ill., index, maps, plates. (Publication / Middle American Research Institute, Tulane Univ.; 61)

Presentation and discussion of *Codex Tulane,* an early colonial manuscript from Mixtec-speaking area of southern Mexico, includes 14 full-page color reproductions. [GF]

448 **Spores, Ronald.** Tututepec: a postclassic-period Mixtec conquest state. (*Anc. Mesoam.*, 4:1, Spring 1993, p. 167–174, bibl., tables)

Descriptive analysis examines political empire that formed on Oaxaca's Pacific Coast and rose to prominence around AD 1000 under leadership of Mixtec lord 8 Deer. Draws largely on documentary analysis to show that the Tututepec polity was linked to Mixtec states in the mountains to the north through both royal marital alliances and trade. For ethnohistorian's comment see *HLAS 54:574*. [GF]

449 **Stuart, David** and **Stephen D. Houston.** Classic Maya place names. Washington: Dumbarton Oaks Research Library and Collection, 1993. 102 p.: bibl., ill., index, maps. (Studies in pre-Columbian art & archaeology; 33)

Analyzes references to specific geographical locations (including places in mythical landscapes) in classic-period hieroglyphic inscriptions. Includes discussion differentiating these glyphs from Emblem Glyphs, which are royal titles and usually refer directly to polities and only indirectly to places. [JSH]

450 **Stuart, George E.** The carved stela from La Mojarra, Veracruz, Mexico. (*Science/Washington*, 259, 1993, p. 1700–1701, bibl., ill.)

Summary of circumstances of discovery of stela from La Mojarra includes very brief assessment of its significance. Style of the relief figure and form of the long hieroglyphic text are consistent with two calendrical dates (in a system closely related to the Maya Long Count) that place monument in 2nd century AD. The text therefore represents one of the earliest examples of complex writing in Mesoamerica. [JSH]

451 **Taube, Karl A.** The Bilimek pulque vessel: starlore, calendrics, and cosmology of late postclassic central Mexico. (*Anc. Mesoam.*, 4:1, Spring 1993, p. 1–15, bibl., ill., photo)

The Bilimek vessel contains a very complex set of images concerning starlore and cosmic battle that, along with other sources, reveal the integral part pulque played in Aztec concepts of warfare and cosmology. [GF]

452 **Taube, Karl A.** The Temple of Quetzalcoatl and the cult of sacred war at Teotihuacán. (*Res/Harvard*, 21, 1992, 53–87, bibl., ill., photo)

Identifies and distinguishes two serpent forms, Quetzalcoatl and the War Serpent, present at Teotihuacán. Compares these images to those found elsewhere in Mesoamerica, and infers that latter serpent may be associated with a war cult that had its genesis at Teotihuacán. [GF]

453 **Tedlock, Dennis.** On hieroglyphic literacy in ancient Mayaland: an alternative interpretation. (*Curr. Anthropol.*, 33:2, April 1992, p. 216–218, bibl.)

Discusses Maya linguistic terms related to books and writing, especially in Quichean sources of early colonial period. Argues that there was widespread literacy among Maya societies in precolumbian times. [JSH]

454 **Tedlock, Dennis.** Torture in the archives: Mayans meet Europeans. (*Am. Anthropol.*, 95:1, 1993, p. 139–152, bibl., ill.)

Essay considers 16th-century documents dealing with encounter between Europeans and Mayas. Highlights ways in which European scholars have privileged the texts produced by Europeans and proposes new readings and interpretations of several documents. [JSH]

455 **Tena, Rafael.** El calendario mexica y la cronografía. México: Instituto Nacional de Antropología e Historia, 1987. 129 p.: bibl., plates. (Col. científica; 161. Serie Historia)

Painstaking calendrical reconstruction offers a more precise coordination of Mexica and European calendars. Also aims to recover with more precision such things as exact sequencing of the "months" within the *xiuhpohualli*, or solar year, and of the solar years within the "century" of 52 years. [R. Haskett and S. Wood]

456 **Tichy, Franz.** Orientation calendar in Mesoamerica: hypothesis concerning their structure, use and distribution. (*Estud. Cult. Náhuatl*, 20, 1990, p. 183–199, bibl., ill., tables)

Comparative discussion examines building orientations at key Mesoamerican sites. Infers a relationship between calendric cycles and regional variation in these architectural orientations. [GF]

Tourtellot, Gair; Amanda Clarke; and **Norman Hammond.** Mapping La Milpa: a Maya city in northwestern Belize. See item **394**.

457 **Urcid, Javier.** The Pacific Coast of Oaxaca and Guerrero: the westernmost extent of Zapotec script. (*Anc. Mesoam.*, 4:1, Spring 1993, p. 141–165, bibl., ill., maps, photo)

Based on comparisons of carved monuments from coastal and highland Oaxaca, author concludes that coastal writing system was derived from central (Zapotec) Oaxaca. [GF]

458 **Vega Sosa, Constanza.** Relaciones intercalendáricas de los códices Azoyú 1, Humboldt Fragmento 1 y Azoyú 2. (*Estud. Cult. Náhuatl*, 21, 1991, p. 99–107, bibl., ill., map, photos, tables)

Describes and establishes calendric relationship between these three codices that relate tribute accounts and historical information about Tlapanecan, Mixtec, and Nahua groups who lived in kingdom of Tlachinollan in Guerrero from 14th-16th centuries. [GF]

459 **Whitecotton, Joseph W.** Zapotec elite ethnohistory: pictorial genealogies from eastern Oaxaca. Nashville, Tenn.: Vanderbilt Univ., 1990. 176 p.: bibl., ill., maps. (Vanderbilt University publications in anthropology; 39)

Presents comprehensive study of three 16th-century Zapotec documents: the Genealogy of Macuilxóchitl, the Etla Genealogy, and the Yale Zapotec Genealogy. Analysis of these pictorial genealogical histories provides clues for understanding late preconquest history of Oaxaca highlands. [GF]

460 **Winfield Capitaine, Fernando.** La Estela 1 de la Mojarra. México: Univ. Nacional Autónoma de México, Coordinación de Humanidades, Seminario de Estudios Prehispánicos para la Descolonización de México, 1990. 256 p.: bibl., ill., maps, plates.

Studies an early stela from Gulf Coast of Mexico with a long hieroglyphic text that includes dates in Maya Long Count system corresponding to 2nd century AD. Includes detailed description and comparative analysis of the hieroglyphic text. [JSH]

461 **Zantwijk, Rudolf A.M. van.** Chichimekische und Colhua-aztekische Gründungslegenden: Verschiedene Entstehungsgeschichten zweier unterschiedlicher mesoamerikanischer Bevölkerungsgruppen. (*Indiana/Berlin*, 12, 1992, p. 97–120, bibl.)

Scrutinizes various territorial foundation legends in ethnohistorical sources for comparisons of cultural elements among Chichimec groups vs. non-Chichimecs (Colhua-Azteca, Azteca-Mexicana). The former were led by military chiefs and had a simpler, less differentiated religion, while among the latter religion and the priesthood played a decisive role in establishment of settlements. [H. von Winning]

462 **Zeitlin, Judith Francis.** The politics of classic-period ritual interaction: iconography of the ballgame cult in coastal Oaxaca. (*Anc. Mesoam.*, 4:1, Spring 1993, p. 212–140, bibl., ill., maps, photos)

Presents examination and comparative analysis of stone sculptures from six classic-period sites located along Pacific coast of Oaxaca between Río Verde and Río de los Perros. Suggests that these monuments share iconographic elements and themes with contemporaneous work found in other coastal regions. Proposes that political networks crosscut these regions, but that no region was dominant. [GF]

Caribbean Area

WILLIAM KEEGAN, *Associate Curator, Florida Museum of Natural History, University of Florida*

THE CARIBBEAN CONTINUES TO BE a hotbed of activity, largely due to three trends. The recent publishing explosion can be attributed first to the Columbus Quincentenary which generated numerous critical responses (items **516** and **509**), along with synthetic volumes dealing with the prehistory of the region (items **519** and **500**). The fallout from the Quincentenary is expected to continue for several more years. Second, monographs reporting on the results of long-term research projects (e.g., item **495** on Barbados and item **485** on St. Eustatius), as well as a number of PhD dissertations (see below), have come to press recently. Finally, publication of the proceedings of the Congresses of the International Association for Caribbean Archaeology are finally back on track. The Congresses include papers by scholars and avocational archaeologists on method and theory; prehistoric technology; petroglyphs; interaction, adaptation and population movement; prehistoric, historical and underwater archaeology; and physical anthropology. Vol. 11 containing 49 papers was published in 1990; Vol. 12 with 23 papers was published in two parts in 1991 (item **503**); Vol. 13 with 65 papers was published in two parts in 1991; and Vol. 14 with 57 papers was published in 1993, although it carries a 1991 copyright date (item **504**). These volumes are available from the I.A.C.A. office in Martinique.

Archaeologists of Lower Central America continue to make strides in establishing that this geographic area was not just a marginalized periphery, but rather a center that held its own against pressures from both Mesoamerica and the Andes. Several recent works have greatly expanded the scope of analysis, such as in the area of exchange (item **466**). In addition, systematic regional surveys and excavations continue to expand the available body of knowledge (items **465** and **475**). Two works of major significance are Moscoso's study of the chiefdoms of Nicaragua (item **472**) and Fonseca's history of Costa Rica (item **467**).

RECENT DOCTORAL DISSERTATIONS

Carini, Stephen Peter. Compositional analysis of West Indian Saladoid ceramics and their relevance to Puerto Rican prehistory. University of Connecticut, 1991.

Curet, Luis Antonio. The development of chiefdoms in the Greater Antilles: a regional study of the Valley of Maunabo, Puerto Rico. Arizona State University, 1992.

Hardy, Ellen Teresa. The mortuary behavior of Guanacaste/Nicoya: an analysis of precolumbian social structure. University of California, Los Angeles, 1992.

Higuera-Gundy, Antonia. Antillean vegetational history and paleoclimate reconstructed from the paleolimnological record of Lake Miragoane, Haiti. University of Florida, 1991.

Newson, Lee Ann. Native West Indian plant use. University of Florida, 1993.

Siegel, Peter E. Ideology, power, and social complexity in prehistoric Puerto Rico. State University of New York at Binghamton, 1992.

LOWER CENTRAL AMERICA

463 **Aguilar F., Juan Manuel.** Museos y parques arqueológicos: breve síntesis histórica. Tegucigalpa: Instituto Hondureño de Antropología e Historia, 1991. 94 p.: bibl., ill., maps.

Useful guide to ten museums and the botanical garden in San Pedro Sula includes mention of planned museums and other places of interest.

464 **Ambrose, Stanley H.** and **Lynette Norr.** On stable isotopic data and prehistoric subsistence in the Soconusco Region. (*Curr. Anthropol.*, 33:4, 1992, p. 401–404, bibl.)

Authors point out that dietary reconstruction for Soconosco Region of Costa Rica and Panama proposed by Blake *et al.* (*Current Anthropology*, Vol. 33, No. 1, 1992, p. 83–84) suffers from a number of errors. As a result, the temporal and geographical distinctions they recognized in subsistence patterns are not supported by isotopic data.

Aoyama, Kazuo. Socioeconomic implications of chipped stone from the La Entrada region, western Honduras. See item **187**.

465   **Corrales Ulloa, Francisco.** Investigaciones arqueológicas en el Pacífico Central de Costa Rica. (*Vínculos/San José*, 16:1/2, 1990, p. 1–29, bibl., ill., maps, tables)

Paper presents overview of Proyecto Arqueológico Pacífico Central conducted from 1986–90. Chronology, settlement and subsistence patterns, ceramics, lithics, and relations with other areas are discussed with regard to the 66 sites investigated in the region. Work provides significant baseline for this region of Lower Central America.

466   **Corrales Ulloa, Francisco** and **Ifigenia Quintanilla Jiménez.** El Pacífico Central de Costa Rica y el intercambio regional. (*Vínculos/San José*, 16:1/2, 1990, p. 111–126, bibl., map, tables)

Focuses on exchange between Central Pacific archaeological region and Gulf of Nicoya area. Models of exchange are evaluated: direct access, reciprocal exchange in towns and on the frontier, down-the-line, and central distribution. Main item of exchange is polychrome pottery in seven main styles.

467   **Fonseca Z., Oscar M.** Historia antigua de Costa Rica: surgimiento y caracterización de la primera civilización costarricense. San José: Editorial de la Univ. de Costa Rica, 1992. 260 p.: bibl., ill. (some col.). (Col. Historia de Costa Rica)

Very important contribution to prehistory of Costa Rica provides extremely well written synthesis of Costa Rican cultural history from initial settlement of the country through climax of cacical societies at contact. Work is a comprehensive, well-illustrated, and thoroughly researched synthesis of Costa Rica's prehispanic past.

468   **González Cháves, Alfredo** and **Fernando González Vásquez.** Poblados amerindios de Costa Rica: antecedentes arqueológicos e históricos. San José: Biblioteca del Colegio de Licenciados y Profesores en Letras, Filosofía, Ciencias y Artes; Editorial de la Univ. de Costa Rica, 1992. 96 p.: bibl., ill. (Serie Ciencias sociales)

Whereas Fonseca Z. has written a comprehensive text (see item **467**), this work is a guidebook which presents a useful introduction to the Amerindians of Costa Rica from initial settlement through early colonial period.

469   **Guerrero M., Juan V.; Ricardo Vázquez L.;** and **Federico Solano B.** Entierros secundarios y restos orgánicos de ca. 500 A.C. preservados en una área de inundación marina: Golfo de Nicoya, Costa Rica. (*Vínculos/San José*, 17, 1992, p. 17–51, bibl., map)

Analyzes 16 secondary burials containing 28 individuals found in the tidal zone, discussing burial procedure, burial goods (including organic remains), osteology, and paleodemography. Shows that the burials were originally deposited in a coastal swamp, whose anaerobic mud helped preserve the organic wood and wrappings. Paper expands current knowledge of burial practices and of artifacts which usually are not preserved.

470   **Hoopes, John W.** The Trondora complex: early formative ceramics in northwestern Costa Rica. (*Lat. Am. Antiq.*, 5:1, 1994, p. 3–30)

Describes early to middle formative ceramics complex (ca. 2000 BC) which is associated with early horticulture and sedentism. Differences in ceramics in the region indicate a high degree of regionalism, despite participation in a broad interaction sphere. Because origins of this evolved ceramic complex are not identified, it is expected that even earlier ceramics remain to be discovered in the Central American isthmus.

471   **Kelly, Thomas C.** Preceramic projectile-point typology in Belize. (*Anc. Mesoam.*, 4:2, Fall 1993, p. 205–227, bibl., ill., maps, table)

Extremely detailed study of projectile points from Belize questions BAAR projectile-point typology, demonstrating instead that only two projectile point types are justified by the data. These are Lowe (ca. 2500–1900 BC) and the undated Sawmill type. A third provisional point, Allspice, is also discussed. This reevaluation of projectile point data has important implications for study of the preceramic period in Belize.

472 **Moscoso, Francisco.** Los cacicazgos de Nicaragua antigua. San Juan: Instituto de Estudios del Caribe, 1991. 129 p.: bibl. (Cuadernos de estudio; 10)

Moscoso is one of the most important Marxist scholars working with historical data in the region. Building on his earlier work *Tribu y clases en el Caribe antiguo* (San Pedro de Macorís, Dominican Republic: Univ. Central del Este, 1986) which investigated the Tainos in the Greater Antilles, he developes a preliminary, historical materialist interpretation of the transition from egalitarian to class society in Nicaragua. Use of historical data and refinement of Marxist concepts add substantially to the literature.

Nabel Pérez, Blas. Las culturas que encontró Colón. See item **511**.

473 **Odio Orozco, Eduardo.** La Pochota: un complejo cerámico temprano en las tierras bajas del Guanacaste, Costa Rica. (*Vínculos/San José*, 17, 1992, p. 1–16, bibl., map)

Author describes a new ceramic complex dated to middle formative (1500–500 BC), and questions chronological range for the Zoned Bichrome Period (500 BC–AD 500). The new complex is more similar to northern styles, perhaps due to similar emphasis on grain agriculture in contrast to horticulture in the south.

474 **Quijano, Edgardo** and **Ricardo Lindo.** Las estrellas y las piedras. San Salvador: Consejo Nacional para la Cultura y el Arte, Dirección General de Publicaciones e Impresos, 1992. 43 p., 32 leaves of plates: ill.

Major part of book is composed of 37 line drawings of petroglyphs from La Isla de Igualtepec, El Salvador. The very short accompanying text is not concerned with issues related to archaeology directly, but rather with situating the petroglyphs from Güija Lake "in the vast context of the History of Art."

475 **Quintanilla Jiménez, Ifigenia.** La Malla: un sitio arqueológico asociado al uso de recursos del Manglar de Tivives, Pacífico Central de Costa Rica. (*Vínculos/San José*, 16:1/2, 1990, p. 57–83, bibl., ill., maps, photos, table)

Registered sites located at Manglar de Tivives are described, as are details of excavations at La Malla undertaken in 1988–89.

Study focuses on ceramic production, salt production, mollusc collecting, hunting, and fishing.

476 **Reinterpreting prehistory of Central America.** Edited by Mark Miller Graham. Niwot, Colo.: Univ. Press of Colorado, 1993. 336 p.: bibl., ill., index.

Collection of papers continues trend of past several years in which scholars working in Central America have declared area's historic independence from negative hegemony of Mesoamerica. It is argued that during precolumbian times this region was not a marginalized periphery, but rather a center that held its own against pressures from both Mesoamerica and the Andes.

477 **Solís Alpízar, Olman E.** Jesús María: un sitio con actividad doméstica en el Pacífico Central, Costa Rica. (*Vínculos/San José*, 16:1/2, 1990, p. 31–56, bibl., ill., maps, photos)

The site of Jesús María in Central Pacific zone contains seven stone features which define remains of a small village or settlement. Three circular arrangements of stone reflect perimeters of house structures. Determination of activity areas within two of the structures was accomplished through functional analysis of pottery and lithics, chemical analysis of soils, and spatial correlation.

478 **Solís del Vecchio, Felipe** and **Anayensy Herrera Villalobos.** Lomas Entierros: un centro político prehispánico en la cuenca baja del Río Grande de Tárcoles. (*Vínculos/San José*, 16:1/2, 1990, p. 85–110, bibl., maps, photos, tables)

The Lomas Entierros site, originally defined as a late cemetery, is shown to be a political center which developed due to its strategic position on a hilltop adjacent to the Grande de Tárcoles River. Site played a role in transport of goods between the Central Valley and Nicoya Gulf between 800–1350 AD.

CARIBBEAN ISLANDS

479 **Agorsah, Emanuel Kofi.** Archaeology and resistance history in the Caribbean. (*Afr. Archaeol. Rev.*, 11, 1993, p. 175–195, bibl., maps)

Archaeological and ethnological evidence from Ghana is used to indicate African cultural heritage of people enslaved and exported to the Caribbean. Discusses heritage of Maroon communities in Jamaica, and their success at establishing independent communities which resisted colonial power. One of the first studies to draw extensively on archaeological research in both Africa and the Caribbean.

480 Agorsah, Emanuel Kofi. Archaeology and the Maroon heritage in Jamaica. (*Jam. J.*, 24:2, March 1992, p. 2–9, ill., maps)

Brief, clear, concise, and well-written overview surveys recent archaeological investigations, mostly performed by author, on Maroon settlements in Jamaica.

481 Agorsah, Emanuel Kofi. An objective chronological scheme for Caribbean history and archaeology. (*Soc. Econ. Stud.*, 21:1, 1993, p. 119–147, bibl., map)

In an important first step towards overhauling prevailing regional systematics, author questions chronological schemes developed for history and archaeology of the Caribbean, heavily cricitizing use of ceramic types to determine cultural traditions and time framework. Proposes a general framework that eliminates the inherent ambiguities and weaknesses of the former system.

482 Alexandrenkov, Eduardo and **Arístides Folgado.** El casabe. (*Anu. Etnol./Habana*, 1988, p. 36–49, ill.)

Discusses development of cassava production in Cuba since contact period, providing important record of how cassava is used today, and of the changes that have occured in its use since contact.

483 Amodio, Emanuele. Relaciones interétnicas en el Caribe indígena: una reconstrucción a partir de los primeros testimonios europeos. (*Rev. Indias*, 51:193, sept./dic. 1991, p. 571–606)

Author recognizes a number of significant problems with discussion of native Caribbean peoples in the *Handbook of South American Indians*. In an effort to resolve some of these problems, develops a theoretical model of interethnic relationships between indigenous societies based on concept of reciprocity and redistribution as reflected in the work of John Murra and Karl Polanyi. Author presents an extremely interesting discussion of these concepts in the context of contact period descriptions of indigenous societies of the region.

484 Antczak, Andrzej and **María Magdalena Antczak.** Análisis del sistema de los asentamientos prehistóricos en el archipiélago de Los Roques. (*Montalbán/Caracas*, 23, 1991, p. 335–386, bibl., maps)

Extremely important review examines prehistoric colonization and settlement of these small Venezuelan islands. Authors' investigations identified 26 prehistoric sites, of which four were excavated. Their findings suggest that two different groups settled on the islands, one from La Tortuga and the other from the Venezuelan coast.

485 The archaeology of St. Eustatius: the Golden Rock Site. Edited by Aad H. Versteeg and Kees Schinkel. St. Eustatius: St. Eustatius Historical Foundation; Amsterdam: Foundation for Scientific Research in the Caribbean Region, 1992. 284 p.: bibl., ill., maps. (Foundation for Scientific Research in the Caribbean Region; 131. St. Eustatius Historical Foundation; 2)

The Golden Rock Site on St. Eustatius is one of the most thoroughly investigated early Saladoid sites in the region, and the first at which extensive excavations of house structures has been undertaken. Contributors report on the physical environment, faunal remains, artifacts, house structures, wood charcoal, and physical anthropology of human remains. One of the most important recent contributions to prehistory of the region.

Between St. Eustatius and The Guianas: contributions to Caribbean archaeology. See item 650.

486 *Boletín del Museo del Hombre Dominicano.* No. 24, 1991– . Santo Domingo: Museo del Hombre Dominicano.

Contains three papers on Monoguayobo Region, including an overview of settlements, a report on excavations at the Santa Clara site, and a report on pollen analysis. A new aboriginal *duho* from Cuba is also descibed.

487 *Boletín del Museo del Hombre Dominicano.* No. 25, 1992– . Santo Domingo: Museo del Hombre Dominicano.

Includes papers on precolumbian art and culture, indigenous navigation, archaeology of Las Sardinas, ethnic identity of Macorix-Ciguayos, and an archaic (Guayabo Blanco) site at Playa de El Paraíso, Ovama, Santiago de Cuba.

488   *Bulletin du Bureau national d'ethnologie.* numéro spécial, 1987/1992– . Port-au-Prince: Bureau national d'ethnologie.

Special issue of *Bulletin* includes articles in French by Jean-Claude Selime (chronology), Clark Moore (lithic age), Irving Rouse (colonization), Kathleen Deagan (La Navidad), and Charles Ewen and Maurice Williams (Puerto Real). Volume is important as one of the few recent works on archaeology in Haiti.

489   **Burney, David A.** and **Lida Pigott Burney.** Holocene charcoal stratigraphy from Laguna Tortuguero, Puerto Rico, and the timing of human arrival on the island. (*J. Archaeol. Sci.*, 21, 1994, p. 273–281, bibl., map)

Historically documents 7,000 years of fire occurrence and sedimentation on north coast of Puerto Rico. An abrupt increase in charcoal is noted for around 5,300 cal-BP, which may reflect arrival of humans. A decline in fire occurrence to more moderate levels after 3,200 cal-BP was also noted.

490   **Chanlatte Baik, Luis A.** and **Yvonne M. Narganes Storde.** La nueva arqueología de Puerto Rico: su proyección en las Antillas. Santo Domingo: Taller, 1990. 49 p.: bibl., ill.

Research by authors into unique archaeological assemblages in Puerto Rico and Vieques Island, called Huecoid, have led them to propose an alternative model of cultural development in the West Indies. This clearly written overview of their model is important reading for archaeologists in the region.

491   Civilisations précolombiennnes [i.e., précolombiennes] de la Caraïbe: actes du colloque du Marin, août 1989. Conçu et réalisé par l'Office municipal de la culture du Marin, Martinique. Coordination de André Lucrèce. Saint-Joseph-de-la-Martinique: Presses universitaires créoles, Groupe d'études et de recherches en espace créolophone; Paris: L'Harmattan, 1991. 286 p.: bibl., ill., maps.

One of numerous recent publications spawned by the Quincentenary, volume is mixed bag for archaeologists. Chapters by Rivero de la Calle (pre-agricultural Cuba), the Antczaks (Venezuelan islands), and Morban Laucer (pictographs) are of interest to archaeologists.

492   **Davis, Dave D.** Archaic blade production on Antigua, West Indies. (*Am. Antiq.*, 58:44, 1993, p. 688–697, bibl., map)

Antigua is the major source of flint in the Lesser Antilles. In the first paper to address archaic flaked-stone technology in the Lesser Antilles, Davis' analysis of the largest excavated Archaic assemblage reveals that direct-percussion blades were the primary flaked-stone technology.

493   **Deagan, Kathleen** and **José María Cruxent.** From contact to criollos: the archaeology of Spanish colonization in Hispaniola. (*in* The meeting of two worlds: Europe and the Americas, 1492–1650. Edited by Warwick Bray. Oxford: Oxford Univ. Press; London: The British Academy, 1993, p. 67–103, bibl., ill., maps)

Drawing from three archaeological projects at very early Spanish colonial sites, examines processes and events of Spanish-American interaction during first decades of contact. Interactions of American Indians, Europeans, and Africans are considered. This seminal paper which draws on authors' decades of research and extensive database is essential reading for Spanish colonial archaeologists.

494   **Domínguez, Lourdes.** Arqueología del centro-sur de Cuba. La Habana: Editorial Academia, 1991. 102 p., 39 p. of plates: bibl., ill., maps.

Previous archaeological research in Cuba has tended to emphasize eastern and western extremes of the island. This major work describes research at 16 sites in south-central Cuba. Interpretation examines relations between development in this region and possible interaction with Archaic groups to the West and Taino groups to the East.

495   **Drewitt, Peter.** Prehistoric Barbados. London: Institute of Archaeology, Univ. College, London; Barbados: Barbados Museum and Historical Society, 1991. 196 p.: bibl., ill., maps.

From 1984–89 Drewitt and his team conducted a comprehensive survey of known sites, a systematic survey of coastal regions, and sample excavations at four sites. Book presents results of these studies in a landmark volume.

496 Dubelaar, C.N. Bibliography of South American and Antillean petroglyphs. Amsterdam: Foundation for Scientific Research in the Caribbean Region; Oranjestad, Aruba: Archaeological Museum Aruba, 1991. 134 p.: ill. (Archaeological Museum Aruba; 5. Foundation for Scientific Research in the Caribbean Region; 129)

Collection includes more than 1,600 references, with a handy key to locations on the right margin.

497 Elbow, Gary S. Migration or interaction?: reinterpreting pre-Columbian West Indian culture origins. (*J. Geogr.*, 91:5, Sept./Oct. 1992, p. 200–204, bibl., map)

Assimilates some of the new information and new ideas concerning the peopling of the West Indies; however, situation is not as simple as this author suggests. Paper provides important suggestions to teachers regarding how these data can be used.

498 England, Suzannah. An archaeological perspective on settlement patterns on British West Indian sugar estates. (*Caribena/Martinique*, 1, 1991, p. 105–122, bibl., map, tables)

Study looks at pattern of Negro housing on British West Indian sugar estates with intention of defining an Afro-Caribbean identity in the housing pattern. Statistical analyses failed to identify a general pattern, but inferences concerning control have been put forward. Results may also be useful for those looking for information on Negro housing on plantations for which plans are lacking.

499 Fariñas Gutiérrez, María Daisy. Paralelismo y transculturación en la religiosidad aborigen antes y después de la conquista. (*Rev. Cuba. Cienc. Soc.*, 9:27, enero/junio 1992, p. 109–122, bibl.)

Paper explores role of religion in Spanish/indigenous transculturation in Cuba. Parallelism in African, indigenous, and Spanish religions is described as mechanism through which syncretism was facilitated. Important contribution to emerging literature on contact-period archaeology, acculturation, and religious studies.

500 Guarch Delmonte, José M. Estructura para las comunidades aborígenes de Cuba. Holguín, Cuba: Ediciones Holguín, 1990. 78 p.: bibl. (Col. de la ciudad)

Presents chronological framework for prehispanic cultures of Cuba based on unilineal evolution of Morgan, Marx, and Engels. Describes four phases: 1) hunting; 2) fishing and shellfish collecting; 3) protoagriculture; and 4) agriculture in terms of grade of development, forces of production, technology, and superstructure. Of special note is author's identification of seven cultural variants at contact and his refinement of the portrait of protoagriculturalists in Cuba.

501 Guerrero, José and Marcio Veloz Maggiolo. Los inicios de la colonización en América: la arqueología como historia. San Pedro de Macorís, Dominican Republic: Univ. Central del Este, 1988. 117 p.: bibl., ill. (Univ. Central del Este; 67. Serie V Centenario; 1)

A significant trend has been the recognition that the term *Taino*, as used today, encompasses a number of different ethnic groups. Authors use pottery styles to distinguish Taino and Macorix ethnic groups in Hispaniola at contact. Their approach promotes perspective of archaeology as history.

502 Hulme, Peter. Making sense of the native Caribbean. (*Nieuwe West-Indische Gids*, 67:3/4, 1993, p. 189–220, bibl.)

Who were the native peoples of the West Indies at contact, and what should we call them? This critical essay reviews current nomenclature and historical sources of these names. Important contribution to current debate over degree of ethnic plurality at contact and accuracy of characteristics attributed to the various groups (e.g., Carib cannibalism).

503 International Congress for Caribbean Archaeology, *12th, Cayenne, 1987.* Proceedings. Edited by Linda Sickler Robinson. Martinique: A.I.A.C., 1991. 401 p.: bibl., ill., maps, plates.

Includes 23 papers from the Congress held in Cayenne, French Guiana, in 1987. Papers cover a wide range of subjects related to Caribbean archaeology and cultural history.

504 International Congress for Caribbean Archaeology, *14th, Barbados, 1991.* Proceedings. Edited by Alissandra Cummins and Philippa King. Martinique: A.I.A.C., 1991 [i.e., 1993]. 694 p.: bibl., ill., maps, plates.

Includes 57 papers covering wide range of archaeological subjects and Caribbean cultural history.

505 **Keegan, William F.; Morgan Maclachlan; and Bryan Byrne.** Los cimientos sociales de los caciques tainos. (*Eres/Tenerife*, 3:1, 1992, p. 7–16, bibl., map)

Social foundations of Taino chiefs are examined in terms of particulars of Taino social organization. The interpretation of Tainos as an avunculocal chiefdom is expanded and refined.

506 **Keegan, William F.** "Columbus murdered a continent:" present, past and future consequences. (*J. Bahamas Hist. Soc.*, 14:1, 1992, p. 2–8)

Aspects of initial encounter between Columbus and native peoples of the Bahamas are discussed, especially as these relate to estimating total population of the region at contact.

Keegan, William F. The people who discovered Columbus: the prehistory of the Bahamas. See *HLAS 54:1819*.

507 **Keegan, William F.** West Indian archaeology. pt. 1, Overview and foragers. (*J. Archaeol. Res.*, 2:3, 1994, p. 255–284, bibl., map)

Describes geographical characteristics of the five main West Indian archipelagos, with a focus on research conducted in the region during past five years. Recent investigations of the aceramic and protoagricultural groups are reviewed. First in an important series reviewing recent research in the region.

508 **Klinken, G.J. van.** Dating and dietary reconstruction by isotopic analysis of amino acids in fossil bone collagen—with special reference to the Caribbean. Amsterdam: Foundation for Scientific Research in the Caribbean Region, 1991. 113 p.: bibl., ill. (Publications; 128)

Describes stable isotope analysis of samples from prehistoric sites in the Greater Antilles and northern Lesser Antilles. Results suggest that a strong reliance on terrestrial food sources existed throughout Caribbean prehistory.

509 **Matos Moquete, Manuel.** Letrocentrismo en la versión de Ramón Pané sobre los aborígenes de Quisqueya. (*in* Descubrimiento, conquista y colonización de América: mito y realidad. Santo Domingo: Centro para la Investigación y Acción Social en el Caribe [CIASCA], 1992, p. 270–288)

Why do we not doubt the accuracy of the Spanish chronicles? Author addresses this question through a detailed evaluation of the work of Ramón Pané. Article provides a fascinating review of context of Pané's work and raises questions concerning use of contact-period documents.

510 **Milbrath, Susan.** Representations of Caribbean and Latin American Indians in sixteenth-century European art. (*Arch. Völkerkd.*, 45, 1991, p. 1–38, bibl., ill.)

The image of Native Americans in European art is discussed with regard to ethnographic content and use of visual imagery to promote colonial agendas. Although not focused on archaeology, work provides significant insights for evaluating contact-period imagery.

511 **Nabel Pérez, Blas.** Las culturas que encontró Colón. Quito: Abya-Yala; MLAL, 1992. 139 p.: bibl., ill., maps. (Col. 500 años; 52)

Provides brief description of cultures encountered by Columbus, including the Lucayans, the Tainos, the Ciguayos and Macorix, the Caribs, the Ciboneys, the Guanahatabeyes, the Arawaks, and the Chontals. Short, easy to read, introduction to these peoples.

512 **Ramos Gómez, Luis Javier.** Huellas de la relación mantenida por españoles e indios en La Isabela hasta la partida de Antonio de Torres, el 2 de febrero de 1494. (*Rev. Esp. Antropol. Am.*, 22, 1992, p. 75–88, bibl.)

Discusses in detail aspects of Spanish/Indian relations in Hispaniola before 1494 as reflected in the writings of the day. Demonstrates variability of such relations and their change through time. Major trend was from one of communication and informality to the more formal relationship that led to the *encomienda*.

513 **Reid, Basil.** Arawak archaeology in Jamaica: new approaches, new perspectives. (*Caribb. Q.*, 38:1/2, 1992, 17–20)

Reviews archaeological research in Jamaica and calls for a more systematic approach to archaeology on the island.

514 **Rodríguez, Miguel.** La colección arqueológica de Puerto Rico en el Museo Peabody de la Universidad de Yale. (*Rev. Cent. Estud. Av.*, 8, 1989, p. 27–41, bibl., map)

A museum collection has a tendency to become a black hole, especially when it is large and relevant publications are short and poorly illustrated. This article provides a comprehensive and well-illustrated overview of the Puerto Rican artifacts in Yale University's Peabody Museum.

515 **Siegel, Peter E.** and **K.P. Severin.** The first documented prehistoric goldcopper alloy artifact from the West Indies. (*J. Archaeol. Sci.*, 20, 1993, p. 67–79)

Describes a small fragment of metal from an early Saladoid deposit in the Maisabel site, Puerto Rico. Object is 55 percent copper, 5 percent silver, and 40 percent gold, and is the first archaeological example described in the literature. The early date for the find is also significant.

516 **Sued-Badillo, Jalil.** Facing up to Caribbean history. (*Am. Antiq.*, 57:4, 1992, p. 599–607, bibl.)

Sued-Badillo takes aim at colonialist character of Caribbean history and archaeology, and proposes a rereading of the early colonial period from an integral multidisciplinary perspective. Although some would dismiss this as postmodern ranting, it is food for thought for all who write the history of others.

517 **Veloz Maggiolo, Marcio.** La distribución de espacios y de asentamientos al momento de la conquista en las Antillas. (*Cienc. Soc./Santo Domingo*, 17:3, julio/sept. 1992, p. 293–308, bibl.)

Caribbean archaeology has historically focused on documenting changes in pottery styles through time. An important recent trend, reflected in this paper, is the study of settlement patterns and spatial distributions. Drawing largely from his work in Yuma Province, author describes changes in spatial arrangements at sites in the Dominican Republic.

518 **Veloz Maggiolo, Marcio.** Notas sobre la Zamia en la prehistoria del Caribe. (*Rev. Arqueol. Am.*, 6, julio/dic. 1992, p. 125–138, bibl.)

In general, very little research has been done on plant use among prehistoric West Indians. Author has been involved in long-term study of Zamia use, which he reviews in this paper.

519 **Veloz Maggiolo, Marcio.** Panorama histórico del Caribe precolombino. Santo Domingo: Banco Central de la República Dominicana, 1991. 262 p.: bibl., ill. (some col.), maps.

Drawing heavily from his own work over the past three decades, author presents overview of archaeology of the circum-Caribbean as it relates especially to cultural developments in the West Indies. Includes aspects of the pre-ceramic in coastal Central America and South America. Presented in chronological order from initial entry of people into the region through arrival of the Spanish. Describes the material cultures, lifeways, and social and political organizations of West Indian cultures. Extremely important perspective on Caribbean archaeology by leading Dominican scholar.

520 **Versteeg, Aad H.; J. Tacoma;** and **P. van de Velde.** Archaeological investigations on Aruba: the Malmok cemetery. Aruba: Archaeological Museum Aruba; Foundation for Scientific Research in the Caribbean Region, 1990. 83 p.: bibl., ill., maps. (Publication / Foundation for Scientific Research in the Caribbean Region; 126. Publication / Archaeological Museum Aruba; 2)

Given the fact that so few studies of physical anthropology have been published for prehistoric populations of the Caribbean, this work is an extremely valuable publication. Study is all the more important because it is the first recent comprehensive analysis of skeletons from an Archaic context. The work is presented in three parts: site descriptions (Versteeg), distribution of graves and gifts (van de Velde), and physical anthropology (Tacoma).

521 **Versteeg, Aad H.** and **Arminda C. Ruiz.** Reconstructing Brasilwood Island: the archaeology and landscape of Indian Aruba. Oranjestad, Aruba: Archaeological Museum Aruba, 1995. 116 p.: map, photos (Publications / Archaeological Museum Aruba; 6)

Presents results of survey of Aruban archaeological sites conducted in 1990. Purpose was to provide Archaeological Museum Aruba with an up-to-date site list and data on each site. Authors attempt to reconstruct precolumbian Aruba. They also look at island's Amerindian heritage in modern times. [R. Hoefte]

522 **Wagenaar Hummelinck, Pieter.** De rotstekeningen van Aruba = The prehistoric rock drawings of Aruba. Utrecht, The

Netherlands: Uitgeverij Presse-Papier, 1991. 228 p.: bibl., ill., maps.

Written by amateur archaeologist, work is nevertheless a comprehensive and well-illustrated guide to petroglyphs on the island of Aruba. In Dutch and English, work includes depictions from 19th-century surveys. Offers few interpretations, focusing instead on clear descriptions and depictions.

523 **Walter, Véronique.** Analyses pétrographiques et minéralogiques de céramiques précolombiennes de Martinique. (*Caribena/Martinique*, 1, 1991, p. 11–54, bibl., tables, photos)

Thirty-three prehistoric potsherds from various archaeological sites on Martinique were analyzed using optic microscopy of thin-sections and x-ray diffraction. This technique can distinguish pottery made in the northeast from pottery made in the south due to clear differences in mineral composition. Results provide a baseline for ceramic production in Martinique and demonstrate that the sherds were from vessels manufactured locally. Work is an extremely important study for the region because so many recent efforts have been devoted to examining exchange and inter- as well as intra-island relationships.

524 **Wild Majesty: encounters with Caribs from Columbus to the present day; an anthology.** Edited by Peter Hulme and Neil L. Whitehead. Oxford, England: Clarendon Press; New York: Oxford Univ. Press, 1992. 369 p.: bibl., ill., index, maps.

Anthology examines primarily historical writings about the Island Caribs of the Lesser Antilles. Includes recent writings which address survival of Caribs to present.

Editors provide brief introductions to each section and short overviews of each reading. Pt. 1, which contains six readings drawn from encounters in 15th and 16th centuries, and Pt. 2, containing seven 17th-century French accounts, are of relevance to archaeologists who are trying to grapple with ethnic identities and cultural practices of the peoples called Island Caribs.

525 **Wilson, Samuel M.** Structure and history: combining archaeology and ethnohistory in the contact period Caribbean. (*in* Ethnohistory and archaeology: approaches to postcontact change in the Americas. Edited by J. Daniel Rogers and Samuel M. Wilson. New York: Plenum Press, 1993, p. 19–30, bibl., map)

One product of the Quincentenary has been recognition that our use of ethnohistoric reports to reconstruct contact-period societies has been seriously flawed. Wilson discusses problem of "mixed epistemologies" which emerge from multidisciplinary research. He then uses his own research on Nevis to confront the problem of integrating archaeological and ethnohistorical research.

526 **Wing, Elizabeth S.** The realm between wild and domesticated. (*in* Skeletons in her cupboard: festschrift for Juliet Clutton-Brock. Oxford, England: Oxbow Books, 1993, p. 243–250, bibl., map)

Evaluates the possible domestication, management, and/or range extension of four kinds of rodents by precolumbian inhabitants of the West Indies, using faunal samples from 43 sites. A significant report from ongoing research into the question of animal domestication in the region.

South America

BETTY J. MEGGERS, *Research Associate, Department of Anthropology, Smithsonian Institution*

AFTER A SUBSTANTIAL INCREASE in the number of entries for *HLAS 53*, the total for this volume is the lowest in 30 years. This reduction is not a consequence of annotating symposia and proceedings of congresses as single works, since this policy was adopted several years ago. The number of entries per country remains proportionally the same, implying a general decline in productivity throughout South America. Peru, for example, accounts for 36 percent of the entries in *HLAS 53* and 34 percent here, although the total output is nearly 50 percent lower. For the first time, however, the number of Peruvian authors equals the number of foreign authors.

Another interesting development is a surge of regional and local syntheses authored by nationals for non-specialists. Two general works on Peru, one extending from its initial peopling to European contact (item **667**) and the other terminating at the end of the formative (item **680**), make explicit their desire to provide an Andean perspective on prehistoric cultural development. Similar motivation is reflected in a review of the prehispanic history of Ayacucho (item **676**) and the use of Inca sacred geography to interpret Inca site location and architecture (item **673**). A summary of Venezuelan prehistory was produced to counteract the general view that the precolumbian cultures are irrelevant to defining Venezuelan national identity (item **712**). Regional and local syntheses directed toward the general public have also appeared in Colombia (item **631**), Brazil (item **580**), Ecuador (item **638**), Chile (item **607**), and Argentina (item **4125**).

Growing interest in the past is evidenced by large university enrollments, especially in Peru, in spite of meager funds for fieldwork and poor job prospects. Suspension of traditional scientific series in Peru has stimulated efforts to develop new journals, notably the *Revista de Ciencias Sociales "Pacífico,"* which focuses on Chimbote, and the *Revista de Investigaciones*, edited by the Centro de Estudiantes de Arqueología, Univ. Nacional Mayor de San Marcos, which accepts articles on all aspects of archaeology. In Chile and Colombia, by contrast, high quality serials and books appear regularly. The three volumes of *Actas* from the 11th Congreso Nacional de Arqueología Chilena (items **529, 600,** and **601**) and the two volumes of *Actas* from its 12th congress are important general references (items **534** and **602**).

Among individual volumes deserving mention are the first major history of Brazilian archaeology (item **589**), a comprehensive analysis of the typological variations among metal plaques from northeastern Argentina as a basis for inferring their function and ideology (item **556**), and an exposé of the fallacious reasoning behind persistent rumors of "lost" Inca treasure (item **527**).

GENERAL

527 **Angles Vargas, Víctor.** El Paititi no existe. Cusco, Peru: Imprenta Amauta S.R. Ltda., 1992. 184 p.: bibl., ill., maps.

The persistent belief in hidden treasure among European colonists and their descendants reflects concepts and values alien to indigenous Americans and unsupported by any form of evidence.

528 *Arqueología Contemporánea.* No. 5, 1994– . Arqueología de Cazadores-Recolectores: Límites, Casos y Aperturas. Recopilación de José Luis Lanata y Luis Alberto Borrero. Buenos Aires: Programa de Estudios Prehistóricos.

Nine articles discuss aspects of theory, method, and technology of investigations in Argentina, Uruguay, and Chile; four focus on validity of conflicting theoretical positions.

529 **El arte y los símbolos como fuente de información arqueológica: simposio.** (*in* Congreso Nacional de Arqueología Chilena, *11th, Santiago, 1988*. Actas. Santiago: Museo Nacional de Historia Natural; Sociedad Chilena de Arqueología, 1991, v. 1, p. 29–94, bibl.)

Rock art from several regions in Bolivia and Chile is examined using ethnographic, archaeological, and statistical perspectives.

530 **Berberián, Eduardo E.** and **Rodolfo A. Raffino.** Culturas indígenas de los Andes Meridionales. Madrid: Alhambra, 1991. 223 p.: bibl., ill. (Estudios; 43)

Useful synthesis of cultural development from 11,000 BC to AD 1532 is divided into five general periods: hunter-gatherers, village farmers and herders, regional florescence, semi-urban societies, and the Inca Empire.

531 **Bouchard, Jean François.** El formativo final y el desarrollo regional en el litoral pacífico nor-ecuatorial. (*Gac. Arqueol. Andin.*, 6:22, 1992, p. 5–21, bibl., ill.)

Successful adaptation to mangrove and coastal swamp was disrupted ca. AD 300 when La Tolita and associated sites were abandoned, possibly as consequence of cessation of interaction with the Tumaco region.

532 **Bruhns, Karen Olsen.** Ancient South America. New York: Cambridge Univ. Press, 1994. 424 p.: bibl., ill., index, maps. (Cambridge world archaeology)

Misleading portrait of precolumbian cultural development.

533 **Cané, Ralph E.** Iconografía e interpretaciones en un marco shamánico: el caso de las culturas nasca (Perú), pascua (Chile) y tairona (Colombia). (*Bol. Lima*, 19:81, mayo 1992, p. 33–42, bibl., photos)

Similarities can be attributed to common ancestral heritage in Asia.

534 **Congreso Nacional de Arqueología Chilena, 12th, Temuco, Chile, 1991.** Actas. v. 1, Simposios. Edición de Hans Niemeyer Fernández. Santiago: Sociedad Chilena de Arqueología; Temuco, Chile: Dirección de Bibliotecas, Archivos y Museos, Museo Regional de la Araucanía, 1993. 1 v.: bibl. (Museo Regional de la Araucanía; 4)

Symposia deal with adaptive strategies in Fuego Patagonia, bioanthropology, and strategies of Inca domination in Kollasuyo.

535 **Cook, Anita Gwynn.** Wari y Tiwanaku: entre el estilo y la imagen. Lima: Pontificia Univ. Católica del Perú, Fondo Editorial, 1994. 344 p.: bibl., ill.

After extensive review of previous sequences and archaeological evidence, author aligns the Wari and Tiwanaku sequences, assesses their relationships, and concludes that they diverged ideologically from common roots in Pukara. Detailed analysis of iconographic and stylistic elements provides a new perspective on Wari-Tiwanaku interaction.

536 **Culturas indígenas de la Patagonia.** Edición de J. Roberto Bárcena. Madrid: Turner, 1990. 273 p.: bibl., ill. (Col. Encuentros: Serie Seminarios)

Chapters discuss initial colonization, rock art, and aspects of Selk'nam and Mapuche culture and adaptation.

537 **Dillehay, Tom D.** Archaeological trends in the Southern Cone of South America. (*J. Archaeol. Res.*, 1, 1993, p. 235–261, bibl.)

Discussion focuses on three themes: early hunter-gatherers, emergence of chiefdoms, and impact of Inca expansion.

538 **Domestic architecture, ethnicity, and complementarity in the south-central Andes.** Edited by Mark S. Aldenderfer. Iowa City: Univ. of Iowa Press, 1993. 178 p.: bibl., ill., index.

Utility of domestic architecture for distinguishing ethnic groups is evaluated by characteristics of residential units in nine locations along the southwestern margin of Lake Titicaca and in the Osmore (Moquegua) and Azapa Valleys draining toward the Pacific coast.

Dubelaar, C.N. Bibliography of South American and Antillean petroglyphs. See item **496.**

539 **Hocquenghem, Anne-Marie** *et al.*

Bases del intercambio entre las sociedades norperuanas y surecuatorianas: una zona de transición entre 1500 A.C. y 600 D.C. (*Bull. Inst. fr. étud. andin.*, 22, 1993, p. 443–466, bibl., ill.)

Suggests that a transition zone of less complex societies separated Peruvian and Ecuadorian loci of socioeconomic development.

540 **El Imperio inka: actualización y perspectivas por registros arqueológicos y etnohistóricos.** v. 2. Córdoba, Argentina: Comechingonia, 1991. 1 v: ill.

Eight articles deal with evidence of Inca occupation in Argentina, Chile, Bolivia, and Peru, with emphasis on ceramics as indices of impact.

541 **Langebaek, Carl Henrik.** Noticias de caciques muy mayores: origen y desarrollo de sociedades complejas en el noro-

riente de Colombia y norte de Venezuela. Bogotá: Univ. de los Andes, 1992. 256 p.: bibl., maps.

Principal factors used to account for increased sociocultural complexity—population growth, exchange, maize, technology, climate, environment—do not correlate with archaeological and ethnohistorical distributions of egalitarian and chiefdom societies. The peripheral locations of the former and central locations of the latter suggest a direction for further investigation.

542 Núñez Atencio, Lautaro. The western part of South America (southern Peru, Bolivia, Northwest Argentina, and Chile) during the Stone Age. (*in* History of humanity. London; New York: Routledge; Paris: UNESCO, 1994, v. 1, p. 348–362, bibl., ill.)

Leading specialist on southern Andean prehistory synthesizes temporal and spatial changes in cultural adaptation from 13,000 to 4000 BP.

543 Los primeros americanos. (*Rev. Arqueol. Am.*, 3, enero/junio 1991, p. 57–111, bibl.)

Includes a work on paleoecology and stratigraphy of preceramic sites in Colombia by Thomas van der Hammen; an article on Monte Verde in Chile by Tom Dillehay and Michael Collins; and a critique of Monte Verde by Thomas Lynch.

544 Las sociedades americanas del postpleistoceno temprano. (*Rev. Arqueol. Am.*, 4, 1991, p. 7–163)

Articles on: the rise of civilization in the central Andes by Thomas Patterson; Archaic adaptations in the Andean region by Karen Stothert and Jeffrey Quilter; cultural development 9000–4000 BP in tropical Brazil by Ondemar Dias; prehistoric groups and climatic change in southern Brazil by Arno Kern; and hunter-gatherers of Patagonia and adjacent regions by Francisco Mena.

545 Yoshio, Onuki; Yuji Seki; and Tsuyoshi Ushino. Tōkyō Daigaku Sōgō Kenkyū Shiryōkan shozō Minami Amerika tairiku senshi bijutsu kōgeihin katarogu = Catalog of South American prehistoric art objects in the Department of Cultural Anthropology, the University Museum, the University of Tokyo. pt. 1, Ceramics. Tōkyō: Tōkyō Daigaku Sōgō Kenkyū Shiryōkan, 1992. 92 p.: ill. (Tōkyō Daigaku Sōgō Kenkyū Shiryōkan hyōhon shiryō hōkoku, 0910-2566; dai 27-gō = The University Museum, the University of Tokyo material reports; 27)

Description and photos of prehistoric South American art objects in the University Museum (Tokyo). Most were collected in the Central Andean region of Peru between 1958–69 by a research team from the Univ. of Tokyo. Since 1988 the Dept. of Cultural Anthropology has been working to create a database of prehistoric South American art objects, especially ceramics and fabrics; this catalog is one of the results. [K. Horisaka]

ARGENTINA

546 Arqueología del Ambato. Córdoba, Argentina: Centro de Investigaciones de la Facultad de Filosofía y Humanidades, Univ. Nacional de Córdoba, 1992. 1 v.: bibl., ill. (Publicaciones del CIFFYH. Arqueología; 46)

Eight chapters describe excavations, ceramics, and other evidence identifying the Ambato Valley in Catamarca as the source of the Aguada culture.

547 Bárcena, J. Roberto and Alicia J. Román. Funcionalidad diferencial de las estructuras del tambo de Tambillos: resultados de la excavación de los recintos 1 y 2 de la unidad A del sector III. (*An. Arqueol. Etnol.*, 41/42, 1986/1987, p. 7–81, bibl., ill., map)

A pottery-making function is suggested by archaeological evidence and experimental use of tools and materials.

548 Borrero, Luis Alberto et al. Análisis espacial en la arqueología patagónica. Buenos Aires: Búsqueda de Ayllu S.R.L., 1992. 161 p.: bibl., ill. (Col. Estudios arqueológicos)

Approaches include systematic survey, occupational superposition, site catchment area, seasonality, and territorial boundaries.

549 Cardich, Augusto and Rafael S. Paunero. Arqueología de la Cueva 2 de Los Toldos, Santa Cruz, Argentina. (*An. Arqueol. Etnol.*, 46/47, 1991–92, p. 49–74, bibl., ill.)

Describes artifacts and stratigraphy, and compares with evidence from Los Toldos 1 and 3. Emphasis on Camelidae suggests incipient pastoralism.

Congreso Nacional de Arqueología Chilena, 11th, Santiago, 1988. Actas. v. 3, Comunicaciones Norte Chico, zonas central y austral de Chile y áreas vecinas. See item **601**.

Congreso Nacional de Arqueología Chilena, 12th, Temuco, Chile, 1991. Actas. v. 1, Simposios. See item 534.

550   Crivelli Montero, E.A.; D.E. Curzio; and M.J. Silveira. La estratigrafía de la cueva Traful I, Provincia de Neuquén. (*Prehistoria/Buenos Aires*, 1, 1993, p. 9–160, bibl., ill.)

Detailed description, tabulation, and illustration of stratigraphy and lithics defining four periods: initial occupation ca. 10,000 BP, Component I (Toldense) ca. 7800 BP, Component II (Complejo Norpatagoniense) ca. 6000–2200 BP, and final occupation associated with pottery.

Culturas indígenas de la Patagonia. See item 536.

551   Durán, Victor. Estudio tecno-tipológico de los raspadores del sitio El Verano, Cueva 1: Patagonia Centro Meridional, Santa Cruz, Argentina. (*An. Arqueol. Etnol.*, 41/42, 1986/1987, p. 129–163, maps, tables)

Detailed classification and description of 98 lithics from stratigraphic excavations; chronology extends from Toldense to Río Pinturas traditions.

552   Fernández, Jorge; Héctor O. Panarello; and Susana A. Valencio. Interrupciones seculares en el poblamiento prehistórico de la Cueva Haichol, Cordillera Andina del Neuquén, Argentina: aplicación de isótopos estables a la explicación de sus causas. (*Dédalo/São Paulo*, 28, 1990, p. 147–170, bibl., ill.)

Discontinuities in occupation implied by 36 C14 dates correlate with changes in isotopic values measured in molluscs and egg shells, implying periods of abandonment due to climatic deterioration during the past 4500 years.

553   Fernández Distel, Alicia A. Nuevos hallazgos de estólicas en el borde de la puna jujeña. (*Cuad. Prehispánicos*, 14, 1989/1990, p. 63–102, bibl., ill., maps, tables)

Detailed description of spear-thrower elements from excavations is supplemented by review of specimens in museums.

554   Gómez Otero, Julieta. The function of small rockshelters in the Magallanes IV-phase settlement system, South Patagonia. (*Lat. Am. Antiq.*, 4, 1993, p. 325–345, bibl., ill.)

Dimensions, artifacts, and faunal remains suggest small rockshelters served as occasional camps during emergencies, a strategy employed by contemporary Nunamiuts.

555   González, Alberto Rex. A cuatro décadas del comienzo de una etapa: apuntes marginales para la historia de la antropología argentina. (*Anu. IEHS*, 5, 1990, p. 13–28)

Analysis by Argentina's leading archaeologist examines the crisis in archaeological theory and method resulting from conflicting European and North American influences since 1950.

556   González, Alberto Rex. Las placas metálicas de los Andes del Sur: contribución al estudio de las religiones precolombinas. Mainz am Rhein, Germany: P. von Zabern, 1992. 311 p., 61 p. of plates: bibl., ill., maps. (Materialien zur allgemeinen und vergleichenden Archäologie, 0170-9518; 46)

Detailed descriptions of 409 metal plaques from museums and private collections in America and Europe provide the basis for examining typology, iconography, cultural context, use, function, evolution, relationships, origins, and religious significance. A monumental contribution to comprehension of the content and character of Andean religion.

557   Krapovickas, Pedro and Sergio Aleksandrowicz. Breve visión de la cultura de Yavi. (*An. Arqueol. Etnol.*, 41/42, 1986/1987, p. 83–127, bibl., ill.)

Details of vessel shape and decoration affiliate agricultural settlements with Yavi culture of the Bolivian Puna, implying expansion or vertical reciprocity.

558   Lagiglia, Humberto A. Pipas de fumar indígena de Mendoza y Neuquén, con un aporte al conocimiento de los narcotizantes y alucinógenos americanos. (*Rev. Mus. Hist. Nat. San Rafael*, 11, 1991, p. 15–41, 107–118, 157–166, 201–216)

Turtle effigy pipe provides basis for review of prehistoric use of tobacco in the Americas and its postcolumbian diffusion.

559   Núñez Regueiro, Víctor A. La metalurgia en Condorhuasi-Alamito, siglos III al V D.C. (*An. Arqueol. Etnol.*, 46/47, 1991/1992, p. 107–164, bibl., ill.)

Three sites in the Campo del Pucara

contain features compatible with metallurgical function, reinforcing hypothesis that this region was the progenitor of Aguada culture.

Orellana, Mario. Los antiguos pobladores de Chile: problemas e hipótesis. See item **609**.

560 **Ortiz-Troncoso, Omar R.** Desarrollo histórico de las investigaciones arqueológicas en Patagonia austral y Tierra del Fuego. (*An. Inst. Patagon./Soc.*, 20, 1991, p. 29–44, ill., photos, table)

Useful synthesis of archaeological investigations since AD 520 is divided into five periods.

561 **Pérez Gollán, José Antonio** and **Inés Gordillo.** Religión y alucinógenos en el antiguo noroeste argentino. (*Cienc. Hoy*, 4:22, enero/feb. 1993, p. 50–64, ill., maps)

Overview of historical and archaeological evidence for use of hallucinogenic plants beginning ca. 2000 BC.

562 **Raffino, Rodolfo A.** Inka, arqueología, historia y urbanismo del altiplano andino. Buenos Aires: Corregidor, 1993. 318 p.: bibl., ill.

Characteristics of architecture, burials, artifacts, fauna, and setting of La Huerta in the valley of Humahuaca provide basis for assessing degree of Inca impact and type of cultural transformations produced.

563 **Schiavini, Adrián.** Los lobos marinos como recurso para cazadores-recolectores marinos: el caso de Tierra del Fuego. (*Lat. Am. Antiq.*, 4, 1993, p. 346–366, bibl., ill.)

Analysis of relative energetic yield, human metabolic requirements, and trophic ecology of pinnipeds indicates their exploitation was critical for prehistoric human populations and was sustainable.

BOLIVIA

El arte y los símbolos como fuente de información arqueológica: simposio. See item **529**.

564 **Boero Rojo, Hugo.** Iskanwaya: la ciudadela que sólo vivía de noche. La Paz: Editorial Los Amigos del Libro, 1992. 192 p.: bibl., ill. (some col.). (Col. Bolivia mágica)

Travelogue most useful for color photos of landscape and architectural features (high walls, irrigation canals, burial cists).

565 **Conchupata: un panteón formativo temprano en el Valle de Mizque, Cochabamba, Bolivia.** Cochabamba, Bolivia: Univ. Mayor de San Simón, Instituto de Investigaciones Antropológicas y Museo Arqueológico, 1992. 43 p.: bibl., ill. (some col.), map. (Cuadernos de investigación: Serie Arqueología; 7. Biblioteca San Simón; Serie Arqueología)

Describes human remains and associated artifacts from a cemetery C14 dated 2890 ± 85 to 2625 ± 85 BP.

Cook, Anita Gwynn. Wari y Tiwanaku: entre el estilo y la imagen. See item **535**.

Goldstein, Paul. Tiwanaku temples and state expansion: a Tiwanaku sunken-court temple in Moquegua, Peru. See item **675**.

566 **Graffam, Gray.** Beyond state collapse: rural history, raised fields, and pastoralism in the South Andes. (*Am. Anthropol.*, 94:4, Dec. 1992, p. 882–904, bibl., maps, tables)

Argues that continued productivity of raised-field agriculture after collapse of the Tiwanaku state was essential for subsidizing extensive camelid herds.

567 **Janusek, John W.** Nuevos datos sobre el significado de la producción y uso de instrumentos musicales en el estado de Tiwanaku. (*Pumapunku/La Paz*, 2:4, enero 1993, p. 9–47, bibl., ill., tables)

Presence of 25 bone tubes of varying lengths in a structure on southwestern edge of Lukurmata implies specialized manufacture of pan pipes during Tiwanaku IV.

568 **Michel López, Marcos** and **Carlos Lémuz Aguirre.** Influencia barrancoide en el bajo Maniquí. (*Nuevos Aportes*, 1:1, 1992, p. 51–65, bibl., ill.)

Pottery decorated by incisions terminating in punctates from the Beni region supports hypothesis of dispersion of Arawak speakers from the central Amazonia (see *HLAS 33:873*.)

569 **Ortloff, Charles R.** and **Alan L. Kolata.** Climate and collapse: agro-ecological perspectives on the decline of the Tiwanaku state. (*J. Archaeol. Sci.*, 20, 1993, p. 195–221, bibl., ill.)

Archaeological evidence for progressive failure of drought-sensitive agricultural systems and abandonment of associated settlements correlates with significant decrease in precipitation documented in Quelccaya ice core, accounting for collapse of Tiwanaku state ca. AD 1000–1100.

570 Pantoja Andrade, Willy. Konchamarka: centro ceremonial de altura. (*Nuevos Aportes*, 1:1, 1992, p. 39–49, bibl., ill.)

Location at 4400 m and aspects of construction suggest a ceremonial function associated with Tiwanaku.

571 Ponce Sanginés, Carlos. La cerámica de la época I (aldeana) de Tiwanaku. (*Pumapunku/La Paz*, 2:4, enero 1993, p. 49–89, bibl., ill., tables)

Describes 35 vessels from three burials and a cist in Kalasasaya.

572 Ponce Sanginés, Carlos *et al.* Exploraciones arqueológicas subacuáticas en el Lago Titikaka: informe científico. Prólogo de Oswaldo Rivera Sundt. La Paz: Editorial La Palabra Producciones, 1992. 767 p.: bibl., ill., maps.

In spite of persistent beliefs in submerged cities, more than a dozen underwater explorations have retrieved only rare stone containers, metal objects, animal bones, pottery, and Spondylus shell. Extensive bibliography.

Raffino, Rodolfo A. Inka, arqueología, historia y urbanismo del altiplano andino. See item **562**.

BRAZIL

573 Arqueologia nos empreendimentos hidrelétricos da Eletronorte: resultados preliminares = Archeology in the hydroelectric projects of Eletronorte: preliminary results. Organização de Eurico T. Miller *et al.* Brasília: Centrais Elétricas do Norte do Brasil, 1992. 93 p.: bibl., ill. (mainly col.). (Arqueologia: ambiente/desenvolvimento = Archeology: environment/development)

Descriptions of environment, archaeological phases, and reconstructed historical sequences are based on survey and excavations in zones inundated by the Balbina (Rio Uatumã, Amazonas), Samuel (Rio Jamarí, Rondônia), and Tucuruí (Rio Tocantins, Pará) hydroelectric dams. Data support the adoption of subsistence-settlement behavior associated with surviving indigenous Amazonians by beginning of the Christian Era.

574 *Arquivos do Museu de História Natural*. Vol. 12, 1991–. Santana do Riacho, Tomo I. Coordenação do André Prous e I.M. Malta. Belo Horizonte, Brazil: Univ. Federal de Minas Gerais.

Exhaustive study describes stratigraphy, plant and animal remains, coprolites, lithics, pigments, textiles, and fibers from excavations in a rock shelter in Minas Gerais, including distribution by levels, composition, manufacture, use, and source. Occupations concentrated around 9000–8000 and 4000–2000 BP.

575 Barbosa, Altair Sales. Pre-história dos cerrados: período paleoíndio. Goiás, Brazil: Univ. Católica de Goiás, Instituto do Trópico Subúmido, 1993. 1 v. (Suma Arqueológica dos Cerrados; 5)

Nuclear areas within the Itaparica tradition, ca. 11,000-8000 BP, are described by present environment, number and type of site, site location, characteristics of lithics, and chronology.

576 Bryan, Alan L. and **Ruth Gruhn.** Archaeological research at six cave or rockshelter sites in interior Bahia, Brazil. Corvallis: Center for the Study of the First Americans, Oregon State Univ., 1993. 169 p.: bibl., ill. (Brazilian Studies)

Although available dates indicate occupations extending from at least 9000 BP, basic technology and material culture exhibit little change. Lithic tools are unshaped flakes modified by use; absence of stone projectile points contrasts with contemporary North American complexes.

577 Bryan, Alan L. The *sambaqui* at Forte Marechal Luz, state of Santa Catarina, Brazil. Corvallis: Center for the Study of the First Americans, Oregon State Univ., 1993. 114 p.: bibl., ill. (Brazilian Studies)

Descriptions of excavations, artifacts, and burials that distinguish seven occupation zones extending from ca. 2400 BC to AD 1350, each characterized by addition of new technological innovations to the pre-existing inventory.

578 Dias, Ondemar, Jr. Paulo Seda; and **Mónica T. Bello.** Escavações arqueológicas no norte de Minas Gerais: Varzelândia; o sítio do Zé Preto, MG-VG-27. (*Rev. Arqueol./ São Paulo*, 1, 1994, p. 75–89, bibl., ill.)

Describes stratigraphy, features, lithic

and faunal remains, and wall paintings from a rock shelter with C14 dates of 1785 ± 55 and 1096 ± 50 BP.

579 **Escavações arqueológicas do Pe. João Alfredo Rohr, S.J.: o sítio da Praia das Laranjeiras II; uma aldeia da tradição ceramista Itararé.** (*Pesquisas/São Leopoldo*, 49, 1993, bibl., ill.)

Detailed descriptions of lithics, ceramics, bone and shell artifacts, faunal remains, and 114 burials are accompanied by plans showing spatial distributions, providing comprehensive documentation of village life on coast of Santa Catarina between ca. AD 800–1300.

580 **Jacobus, André Luiz et al.** Arqueologia pré-histórica do Rio Grande do Sul. Edição de Arno Alvarez Kern. Porto Alegre, Brazil: Mercado Aberto, 1991. 356 p.: bibl., ill., maps. (Série Documenta; 26)

Overview of cultural development from ca. 12,000 BP to surviving groups reveals strong correlation between subsistence-settlement behavior and environment (coast, *planalto*, forest, savanna).

581 **Kneip, Lina Maria.** Cultura material e subsistência das populações pré-históricas de Saquarema, RJ. Rio de Janeiro: Univ. Federal do Rio de Janeiro; Museo Nacional, 1994. 120 p.: bibl., ill. (Documento de trabalho: Série Arqueologia; 2)

Describes, classifies, and tabulates lithic, bone, and shell artifacts, ceramics, and invertebrate and vertebrate remains from Beirada, Moa, and Pontinha shell middens.

582 **Kneip, Lina Maria** and **Lilia Cheuiche Machado.** Os ritos funerários das populações pré-históricas de Saquarema, RJ: Sambaquis da Beirada, Moa e Pontinha. Rio de Janeiro: Univ. Federal do Rio de Janeiro; Museo Nacional, 1993. 76 p.: bibl., ill., tables. (Documento de trabalho: Série Arqueologia; 1)

Detailed description of stratigraphic position of skeletal remains in three shell middens occupied sequentially between ca. 4500 and 1800 BP. Tables identify age, sex, position, orientation, associated ocher and artifacts, and type of preparation (primary or secondary direct burial or cremation).

583 **Meggers, Betty J.** Amazonia: real or counterfeit paradise? (*Rev. Archaeol./Salem*, 13, Fall 1992, p. 25–40, bibl.)

Summarizes evidence incompatible with existence of dense populations on Marajó and elsewhere in Amazonia prior to European contact.

584 **Meggers, Betty J.** Archaeological evidence for the impact of mega-Niño events in Amazonia during the past two millennia. (*Clim. Change*, 28, 1994, p. 1–18, bibl., ill.)

Transamazonian discontinuities in archaeological sequences are contemporary with mega-Niño events documented on the coast of Peru, implying the population suffered severe subsistence stress as a consequence of El Niño-related droughts ca. 1500, 1000, 700, and 400 BP.

585 **Paleopatologia e paleoepidemiologia: estudos multidisciplinares.** Coordenação de Adauto José Gonçalves de Araújo e Luiz Fernando Ferreira. Rio de Janeiro: Escola Nacional de Saúde Pública, Secretaria de Desenvolvimento Educacional, 1992. 234 p.: bibl., ill., maps. (Panorama ENSP)

Five articles discuss paleodemography, health, trauma, and diet inferred from skeletal remains from archaeological sites in Rio de Janeiro, Pernambuco, and Piauí.

586 **Pereira, Edithe.** Análise preliminar das pinturas rupestres de Monte Alegre, PA. (*Bol. Mus. Para. Goeldi*, 8, julho 1992, p. 5–24, bibl., ill.)

Analyzes anthropomorphic, zoomorphic, and geometric figures outlined in red, yellow, and black on walls and ceilings of six rockshelters.

587 **Perota, Celso** and **Valéria Soares de Assis.** O sítio Areal: influência da pressão ambiental sobre a população pré-histórica no litoral do Espírito Santo. (*UFES Rev. Cult.*, 48, 1993, p. 36–54, bibl., ill.)

Changes in sea level during past 4500 years are reflected in fluctuations from terrestrial to marine subsistence emphasis and associated lithic artifacts.

588 **A pré-história no século do descobrimento.** (*Rev. Arqueol./São Paulo*, 7, 1993, bibl.)

Eleven chapters provide archaeological perspectives on early European contact along the Brazilian coast.

589 **Prous, André.** Arqueologia brasileira. Brasília: Editora UnB, 1991. 605 p.: bibl. ill., index, maps.

Provides useful overview of history, methods, and results of archaeological investigations up to 1982, conducted principally by Brazilian archaeologists.

590 Revista de Arqueologia. No. 6, 1991–. São Paulo: Sociedade de Arqueologia Brasileira.

Seven articles draw on rock art, plant remains, animal bones, pottery, and human skeletal remains to reconstruct subsistence behavior among prehistoric populations of coastal Brazil.

591 Ribeiro, Pedro Augusto Mentz. Arqueologia do Vale do Rio Pardo, Rio Grande do Sul, Brasil. Pt. 1. (*Rev. CEPA*, 18: 21, dez. 1991, p. 1–184, bibl., ill.)

Systematic survey recorded 162 sites representing eight phases assigned to six traditions. Detailed examinations of the phases within the preceramic (Umbu, Humaitá) and early ceramic (Vieira and Taquara) traditions reveals regional variations are correlated with different environmental differences.

592 Scatamacchia, Maria Cristina Mineiro and **Francisco Moscoso.** Análise do padrão de estabelecimentos Tupi-Guarani: fontes etno-históricas e arqueológicas. (*Rev. Antropol./São Paulo*, 30/31/32, 1987/1988/1989, p. 37–53, bibl.)

Quotes descriptions of Tupinamba villages by 16th- and 17th-century observers relevant to interpreting archaeological sites.

593 Schmitz, Pedro Ignacio. Áreas arqueológicas no litoral e do planalto do Brasil. (*Rev. Mus. Arqueol. Etnol.*, 1, 1991, p. 3–20, bibl., ill.)

Spatial distributions of the principal lithic and ceramic traditions show general correlation with geobiological regions, implying adaptation to different sets of resources.

594 Sociedade de Arqueologia Brasileira, 6th, Rio de Janeiro, 1991. Anais. Rio de Janeiro: s.n., 1992. 2 v.: bibl., ill.

Eighty articles address all aspects of theory, interpretation, fieldwork, and analysis. Among major foci are paleobiology, ethnoarchaeology, historical archaeology, technology, and rock art.

595 Wust, Irmhild. A pesquisa arqueológica e etnoarqueológica na parte central do território Bororo, Mato Grosso: primeiros resultados. (*Rev. Antropol./São Paulo*, 30/31/32/, 1987/88/89, p. 21–35, bibl.)

Provides brief descriptions of the aspects of Bororo culture potentially identifiable archaeologically, and their distribution in known village sites. Survey also recorded 14 rock art locations, three lithic sites, and 41 pre-Bororo ceramic sites.

CHILE

596 Acha-2 y los orígenes del poblamiento humano en Arica. Arica, Chile: Ediciones Univ. de Tarapacá, 1993. 169 p.: bibl., ill.

Multidisciplinary approach, including environment, settlement pattern, subsistence remains, and physical and chemical analyses, permits reconstructing aspects of diet, health, economy, and society ca. 9000 BP on the north coast.

El arte y los símbolos como fuente de información arqueológica: simposio. See item **529**.

597 Benavente, María Antonieta. Determinación de especies de camélidos sudamericanos: un enfoque arqueozoológico. (*Rev. Chil. Antropol.*, 11, 1992, p. 41–59, bibl., tables)

Llama, guanaco and vicuña were identified in faunal assemblages from three sites in the middle Río Loa.

598 Borrero, Luis Alberto; José Luis Lanata; and **Pedro Cárdenas.** Reestudiando cuevas: nuevas excavaciones en Ultima Esperanza, Magallanes. (*An. Inst. Patagon./Soc.*, 20, 1991, p. 101–110, bibl., ill., maps, table)

Reexamination of stratigraphy and dating of Mylodon excrement support terminal Pleistocene extinction.

599 Cabeza, Angel et al. Desarrollo cultural y adaptación ambiental durante el período alfarero en la precordillera de Pirque, Chile Central. (*Rev. Chil. Antropol.*, 11, 1992, p. 61–86, bibl., ill.)

Survey followed by excavations in three rock shelters and one open site reveal differences in settlement and subsistence, indicating discontinuity between Early Ceramic Period hunter-gatherers and Late Ceramic Period agriculturalists.

600 Congreso Nacional de Arqueología Chilena, 11th, Santiago, 1988. Actas. v. 2, Comunicaciones Norte Grande de Chile y áreas vecinas. Edición de Hans Niemeyer Fernández. Santiago: Museo Nacional de Historia Natural; Sociedad Chilena de Arqueología, 1991. 210 p.: bibl., ill., maps (some folded).

Seventeen papers deal with hunter-gatherer adaptations, thermoluminescence dating, faunal resources, physical properties of lithic raw materials, ceramic styles, projectile point typology, and inter-regional communication.

601 **Congreso Nacional de Arqueología Chilena, 11th, Santiago, 1988.** Actas. v. 3, Comunicaciones Norte Chico, zonas central y austral de Chile y áreas vecinas. Edición de Hans Niemeyer Fernández. Santiago: Museo Nacional de Historia Natural; Sociedad Chilena de Arqueología, 1991. 241 p.: bibl., ill., maps (some folded).

Twenty-one papers discuss a variety of archaeological evidence, including the Inca road, sampling strategies, indications of complementarity, and sequences throughout the region.

Congreso Nacional de Arqueología Chilena, 12th, Temuco, Chile, 1991. Actas. v. 1, Simposios. See item **534**.

602 **Congreso Nacional de Arqueología Chilena, 12th, Temuco, Chile, 1991.** Actas. v. 2, Comunicaciones. Edición de Hans Niemeyer Fernández. Santiago: Sociedad Chilena de Arqueología; Temuco, Chile: Dirección de Bibliotecas, Archivos y Museos, Museo Regional de la Araucanía, 1993. 445 p: bibl., ill., maps. (Boletín / Museo Regional de la Araucanía; 4)

Twenty-nine articles deal with various aspects of Chilean prehistory from Archaic to Inca periods.

Culturas indígenas de la Patagonia. See item **536**.

603 **Dillehay, Tom D.** Humans and proboscideans at Monte Verde, Chile: analytical problems and explanatory scenarios. (*in* Proboscidean and paleoindian interactions. Waco, Tex.: Baylor Univ. Press, 1992, p. 191–210, bibl., ill.)

Reviews evidence for and against association between humans and fragmentary remains of ca. seven mastodons of varying ages.

604 **Grosjean, Martin** and **Lautaro Núñez Atencio.** Late glacial, early and middle holocene environments, human occupation, and resource use in the Atacama, northern Chile. (*Geoarchaeology/New York*, 9, 1994, p. 271–286, bibl., ill.)

Absence of evidence for human occupation between ca. 8500 and 5000 BP equates with change from moist to arid climate.

605 **Identidad y prestigio en los Andes: gorros, turbantes y diademas; exposición, noviembre 1993 a junio 1994.** Santiago: Museo Chileno de Arte Precolombino; Municipalidad de Santiago; Fundación Familia Larrain Echenique, 1993. 64 p.: bibl., ill. (some col.).

Well-illustrated chronological review examines headdresses and hair styles in northern Chile and includes suggestions concerning their sociopolitical implications.

606 **Massone M., Mauricio.** El estudio de las cenizas volcánicas y su implicancia en la interpretación de algunos registros arqueológicos de Chile Austral. (*An. Inst. Patagon./Soc.*, 20, 1991, p. 111–115, bibl.)

Stratigraphic relations between volcanic ash falls, Mylodon remains, and cultural evidence suggest human presence by 11,800 BP.

607 **Núñez Atencio, Lautaro.** Cultura y conflicto en los oasis de San Pedro de Atacama. Santiago: Editorial Universitaria, 1991. 273 p.: bibl., ill., maps. (Col. El Saber y la cultura)

Authoritative and readable overview surveys 11,000 years of human history in one of the driest environments on earth.

608 **Núñez Atencio, Lautaro et al.** Reconstrucción multidisciplinaria de la ocupación prehistórica de Quereo, centro de Chile. (*Lat. Am. Antiq.*, 5, 1994, p. 99–118, bibl., ill.)

Summarizes sedimentological, microfaunal, palynological, paleontological, and cultural evidence for four periods of occupation, the two earliest associated with extinct fauna.

609 **Orellana, Mario.** Los antiguos pobladores de Chile: problemas e hipótesis. (*Rev. Chil. Antropol.*, 11, 1992, p. 21–40, bibl.)

Absence of stone projectile points in the earliest Paleo-Indian sites (Quereo, Monte Verde, Los Toldos) suggests existence of a non-projectile point tradition.

610 **Orellana, Mario.** Reflexiones sobre el desarrollo de la arqueología en Chile. (*Rev. Chil. Antropol.*, 10, 1991, p. 11–23, bibl.)

In the absence of monumental architecture, rich burials, and beautiful artifacts, Chilean archaeologists have emphasized detailed description of fragmentary remains and employed varying theoretical perspectives.

Ortiz-Troncoso, Omar R. Desarrollo histórico de las investigaciones arqueológicas en Patagonia austral y Tierra del Fuego. See item **560.**

611 **Prieto, Alfredo.** Cazadores tempranos y tardíos en la Cueva 1 del Lago Sofía. (*An. Inst. Patagon./Soc.*, 20, 1991, p. 75–99, appendices, bibl., ill., graph, maps, photos, table)

Describes excavations, pollen sequence, faunal remains, and artifacts from a rock shelter with C14 dates of 11570 ± 60 and 12990 ± 490 BP.

612 **Rodman, Amy Oakland.** Textiles and ethnicity: Tiwanaku in San Pedro de Atacama, north Chile. (*Lat. Am. Antiq.*, 3, 1992, p. 316–3340, bibl., ill.)

Differences in tunic stripes and selvages and in associated headgear segregate 56 burials from the Coyo Oriental cemetery into two overlapping groups, one local and the other linked to Tiwanaku, implying coexistence of distinct ethnic populations from ca. AD 600–1000.

COLOMBIA

613 **Arte de la tierra: Sinú y Río Magdalena.** Bogotá: Fondo de Promoción de la Cultura, Banco Popular, 1992. 109 p.: bibl., ill. (some col.), maps. (Col. Tesoros precolombinos)

Articles on archaeological evidence for chiefdoms, symbolism of anthropomorphic burial urns, and the "burial urn horizon" supplement illustrations of 115 pottery objects.

Bouchard, Jean François. El formativo final y el desarrollo regional en el litoral pacífico nor-ecuatorial. See item **531.**

614 **Bray, Warwick.** ¿A donde han ido los bosques?: el hombre y el medio ambiente en la Colombia prehispánica. (*Boletín/Bogotá*, 30, enero/junio 1991, p. 42–65, bibl., ill.)

Patterns of climatic and cultural change in several Highland and coastal regions do not show consistent correlations indicative of causal relationship, except that deforestation occurred widely beginning with the Christian Era.

615 **Cadavid, Gilberto** and **Hernán Ordóñez.** Arqueología de salvamento en la vereda de Tajumbina, municipio de La Cruz, Nariño. Bogotá: Fundación de Investigaciones Arqueológicas Nacionales, Banco de la República; Instituto Colombiano de Antropología, 1992. 139 p.: bibl., ill., map.

Describes pottery, lithic, and metal artifacts from 24 shaft tombs and nine associated features assigned to the protohistoric Quillacinga population.

616 **Calima: diez mil años de historia en el suroccidente de Colombia.** Bogotá: Fundación Pro Calima, 1992. 188 p.: bibl., ill. (some col.), maps.

Illustrations of beautiful pottery and gold objects complement an overview of cultural development by the three periods: Ilama, Yotoco, and Sonso.

617 **Chaves Mendoza, Alvaro** and **Mauricio Puerta Restrepo.** Una nueva fecha para la cultura de Tierradentro. (*Universitas/Bogotá*, 20:33, enero/junio 1991, p. 31–37, bibl., ill., maps)

A C14 date of 2720 ± 240 BP, obtained from a bone sample, is significantly earlier than previous dates for Tierradentro.

618 **Cubillos, Julio César.** Informe de los trabajos de excavación y reconstrucción de las Tumbas No. 9 del Montículo No. 4 del Alto de las Piedras y No. 3 de la Meseta B del Alto de los Idolos, en San Agustín. (*Bol. Arqueol.*, 7:1, enero 1992, p. 3–36, ill.)

Well-illustrated work describes excavation and reconstruction of two stone-walled tombs and associated ceramics assigned to the formative period.

619 **Groot de Mahecha, Ana María.** Checua: una secuencia cultural entre 8500 y 3000 años antes del presente. Bogotá: Fundación de Investigaciones Arqueológicas Nacionales, Banco de la República, 1992. 100 p.: bibl., ill. (Publicaciones de la Fundación de Investigaciones Arqueológicas Nacionales; 54)

Differences in dwellings, lithics, bone artifacts, fauna, and other features distinguish four zones of occupation in an open site during transition from hunting/gathering to incipient plant domestication.

620 **Groot de Mahecha, Ana María** and **Eva María Hooykaas.** Intento de delimitación del territorio de los grupos étnicos pastos y quillacingas en el altiplano nariñense. Bogotá: Fundación de Investigaciones Arqueológicas Nacionales, Banco de la República, 1991. 166 p.: bibl., ill., maps. (Publicación de la Fundación de Investigaciones Arqueológicas Nacionales; 48)

Interesting effort to correlate territories of Pastos and Quillacingas defined by associating chronicles and place names with boundaries of archaeological phases. Detailed distributional maps make this an important reference.

Langebaek, Carl Henrik. Noticias de caciques muy mayores: origen y desarrollo de sociedades complejas en el nororiente de Colombia y norte de Venezuela. See item **541**.

621 **Llanos Vargas, Héctor.** La naturaleza del sur del Alto Magdalena como fundamento cultural prehispánico. (*Cespedesia/Cali*, 19:62/63, 1992, p. 199–221, bibl.)

Climatic changes ca. AD 100, 700, and 1400 correlate with florescence and decline of San Agustín culture and its replacement by Yalcomes culture.

622 **López Castaño, Carlos Eduardo.** Investigaciones arqueológicas en el Magdalena Medio: cuenca del Río Carare, Departamento de Santander. Bogotá: Fundación de Investigaciones Arqueológicas Nacionales, Banco de la República, 1991. 125 p.: bibl., ill. (Publicación de la Fundación de Investigaciones Arqueológicas Nacionales; 47)

Describes lithics and ceramics from 25 localities; four C_{14} dates extend from 1040 to 640 BP.

623 **Moreno González, Leonardo.** Arqueología de San Agustín: pautas de asentamiento agustinianas en el noroccidente de Saladoblanco, Huila. Bogotá: Fundación de Investigaciones Arqueológicas Nacionales, Banco de la República, 1991. 143 p.: bibl., ill., map.

Ceramics from stratigraphic excavations at El Monday define formative (1000 BC–AD 300) and recent (AD 800–1550) occupations on the periphery of San Agustín influence.

624 **Patiño, Diógenes.** Arqueología del Bajo Patía: fases y correlaciones en la costa pacífica de Colombia y Ecuador. (*Lat. Am. Antiq.*, 4, 1993, p. 180–199, bibl., ill.)

Two sequential phases are defined: Buena Vista, located on the alluvial plain, C_{14} dated ca. 200–500 AD; and Maina, located on hills surrounded by mangrove and dated ca. 800–1500 AD.

625 **Patiño, Diógenes** and **Cristóbal Gnecco.** Ocupación prehispánica del Alto Patía. (*Noved. Colomb.*, 5, 1992, p. 72–91, bibl., ill.)

Provides preliminary descriptions of ceramics and burials defining archaeological phases dating from ca. 500 BP to European contact.

626 **Patiño, Diógenes.** Sociedades Tumaco-La Tolita: costa pacífica de Colombia y Ecuador. (*Bol. Arqueol.*, 7:1, enero 1992, p. 37–58, bibl., ill.)

Review of nearly a millennium of successful domination of the mangrove environment raises question of why the population suffered a cultural decline and abandoned the region ca. AD 600.

627 **Peña León, Germán Alberto.** Exploraciones arqueológicas en la cuenca media del Río Bogotá. Bogotá: Fundación de Investigaciones Arqueológicas Nacionales, Banco de la República, 1991. 137 p.: bibl., ill.

Gives type descriptions of pottery from stratigraphic excavations assigned to Herrera, Muisca, and Pubenza periods, clearly differentiated in seriated ceramic sequences.

628 **Reichel-Dolmatoff, Gerardo** and **Alicia Reichel-Dolmatoff.** Arqueología del Bajo Magdalena: estudios de la cerámica de Zambrano. Bogotá: Banco Popular, Fondo de Promoción de la Cultura; Colcultura; Instituto de Investigaciones Culturales y Antropológicas, 1991. 385 p.: bibl., ill. (some col.), map. (Biblioteca Banco Popular: Col. Textos universitarios)

Describes pottery types defining five ceramic complexes based on excavations during late 1950s at Zambrano, Bucarelia, and Pacífico, and surface collections from 36 additional sites. Most useful for 140 b/w plates, 53 repeated in color.

629 **Rivera Escobar, Sergio.** Neusa: 9.000 años de presencia humana en el Páramo. Bogotá: Fundación de Investigaciones

Arqueológicas Nacionales, Banco de la República, 1992. 144 p.: bibl., ill. (Publicaciones de la Fundación de Investigaciones Arqueológicas Nacionales; 52)

Describes lithic and bone artifacts, pottery, and animal and plant remains from excavations at three rock shelters at 3,350 m elevation in the Cordillera Oriental.

630   **Rodríguez, Carlos Armando** and **David Michael Stemper.** Cambios medioambientales y culturales prehispánicos en el curso bajo del Río Bolo, municipio de Palmira, Valle del Cauca. (*Cespedesia/Cali*, 19: 62/63, 1992, p. 139–198, bibl., ill.)

Stratigraphic excavations at two sites northeast of Cali identify two cultural periods: Bolo, tentatively dated prior to AD 1000; and Quebrada Seca, extending to European contact.

631   **Rodríguez, Carlos Armando.** Tras las huellas del hombre prehispánico y su cultura en el Valle del Cauca. Colombia: Instituto Vallecaucano de Investigaciones Científicas; Fundación Hispanoamericano de Cali; Embajada de España en Colombia, 1992. 571 p.: bibl., ill., maps.

Well organized and informative synthesis of cultural development designed for secondary school teachers but useful for specialists. Describes physical and cultural features by period; illustrates typical ceramics; and lists $C14$ dates. Nearly half the text consists of extractions from 16th-century documents.

632   **Rodríguez Ramírez, Camilo.** Patrones de asentamiento de los agricultores prehispánicos en "El Limón," municipio de Chaparral, Tolima. Bogotá: Fundación de Investigaciones Arqueológicas Nacionales, Banco de la República, 1991. 108 p., 2 folded leaves of plates: bibl., ill.

Describes lithics from a preceramic occupation with $C14$ dates of 7370 ± 130 and 5600 ± 90 BP, and pottery from a ceramic occupation dated 1620 ± 70 BP.

633   **Salgado López, Héctor; Carlos Armando Rodríguez;** and **Vladimir Aleksandrovich Bashilov.** La vivienda prehispánica Calima. Cali, Colombia: Instituto Vallecaucano de Investigaciones Científicas (INCIVA), 1993. 176 p.: appendices, bibl., ill.

Extensive excavations at the Jiguales site, composed of 24 residential terraces in groups of two to five, permit reconstructing the shape and size of dwellings, which are compared with pottery representations. Six $C14$ dates extend from 1500 ± 60 to 400 ± 70 BP. Appendices provide detailed descriptions of pottery, lithics, and maize.

634   **Sotomayor Tribín, Hugo Armando.** Arqueomedicina de Colombia prehispánica. Bogotá: Caja de Compensación Familiar, 1992. 80 p.: bibl., ill. (some col.).

Identifies disease, injury, malformation, and other pathological conditions from ethnohistorical accounts and figurines, principally of the Tumaco culture.

635   **Sotomayor Tribín, Hugo Armando.** Enfermedades en el arte prehispánico colombiano. (*Boletín/Bogotá*, 29, oct./dic. 1990, p. 62–73, bibl., col. ill.)

Diagnosis of pathology from 19 figurines (16 from Tumaco) shows equal representation of genetic and acquired conditions.

636   **I tesori delle città perdute: oro della Colombia.** A cura di Stella Herrera Falcone. Genova, Italy: Edizioni Colombo, 1991. 173 p.: ill. (some col.).

Catalog of a 1991–92 exhibition at the Museo di Sant'Agostino (Genova) of a collection from the Museo del Oro in Bogotá. Contains representative examples of gold, pottery, and shell objects from the principal precolumbian cultures.

637   **Velandia Jagua, César Augusto.** San Agustín: arte, estructura y arqueología. Bogotá: Fondo de Promoción de la Cultura del Banco Popular, 1994. 152 p.: bibl., ill. (some col.). (Biblioteca Banco Popular. Col. Textos universitarios)

Subtitled "modelo para una semiótica de la iconografía precolombina," this work draws on ethnographic evidence to reconstruct the symbolic significance of form, color, and motifs.

ECUADOR

Bouchard, Jean François. El formativo final y el desarrollo regional en el litoral pacífico nor-ecuatorial. See item **531**.

638   **Bravomalo de Espinosa, Aurelia.** Ecuador ancestral. Quito: Artes Gráficas Señal, 1992. 280 p.: bibl., ill. (some col.), maps.

Descriptive summary of archaeological investigations by period, region, and cultural complex uses a standardized outline. Presentation is comprehensive and balanced; maps and illustrations are informative.

639 **Bray, Tamara.** Archaeological survey in northern highland Ecuador: Inca imperialism and the País Caranqui. (*World Archaeol.*, 24, 1992, p. 218–233, bibl., ill.)

Survey recorded 66 sites of all periods. Ceramic data suggest decline of long-distance exchange under Inca rule, implying political strategy to "divide and conquer."

640 **Guinea Bueno, Mercedes.** El desarrollo espacial del poblado de Atacames Esmeraldas, Ecuador. (*Rev. Esp. Antropol. Am.*, 24, 1993, p. 93–111, bibl., ill.)

Application of correspondence analysis to multiple surface collections indicates growth from a small dispersed hamlet to a large town during the Integration Period.

641 **Heras y Martínez, César M.** La cerámica de Integración de la costa nordecuatoriana: el caso esmeraldeño. (*Rev. Esp. Antropol. Am.*, 24, 1994, p. 113–138, bibl., ill.)

Provides relative frequencies of varieties of temper, texture, surface treatment, firing, surface color, hardness, and vessel shape, based on samples from excavations and surface collections at sites assigned to the Integration Period.

642 **Historia de la cerámica en el Ecuador.** Síntesis realizada por Segundo Moreno Yáñez y Jaime Peña N. Cuenca, Ecuador: Paul Rivet Fundación, 1992. 47 p.: bibl., ill.

Chapters summarize precolumbian pottery by regions and by colonial, folk, artistic, and industrial productions. Utility is diminished by absence of identification on illustrations.

Hocquenghem, Anne-Marie et al. Bases del intercambio entre las sociedades norperuanas y surecuatorianas: una zona de transición entre 1500 A.C. y 600 D.C. See item **539**.

643 **Meggers, Betty J.** Jōmon-Valdivia similarities: convergence or contact? (*NEARA J.*, 27, 1992, p. 23–32, bibl., ill.)

Refutes principal objections to a transpacific introduction of Valdivia ceramics from western Japan.

644 **Molina, Manuel Jesús.** Arqueología ecuatoriana: los cañaris. Rome: LAS; Quito: Ediciones Abya-Yala, 1992. 118 p.: bibl., ill.

Work is useful mainly for illustrations of stone carvings from collection of Padre Crespi, many acknowledged to be recent "folk art."

Patiño, Diógenes. Arqueología del Bajo Patía: fases y correlaciones en la costa pacífica de Colombia y Ecuador. See item **624**.

Patiño, Diógenes. Sociedades Tumaco-La Tolita: costa pacífica de Colombia y Ecuador. See item **626**.

645 **Pearsall, Deborah M.** Prehistoric subsistence and agricultural evolution in the Jama River Valley, Manabí Province, Ecuador. (*J. Steward Anthropol. Soc.*, 20, 1992, p. 181–207, bibl.)

Although maize phytoliths occur in terminal Valdivia flotation samples, charred kernels and cupules appear earliest in Chorrera (1500–500 BC).

646 **Regional archaeology in Northern Manabí, Ecuador = Arqueología regional del Norte de Manabí, Ecuador.** v. 1, Environment, cultural chronology, and prehistoric subsistence in the Jama River Valley. Edited by James A. Zeidler and Deborah M. Pearsall. Spanish translation by Ana María Boza-Arlotti and Alvaro Higueras-Hare. Pittsburgh: Univ. of Pittsburgh, Dept. of Anthropology, 1994. 247 p.: bibl., ill. (University of Pittsburgh memoirs in Latin American archaeology; 8)

Describes stratigraphic, ceramic, sediment, and botanical evidence supporting occupation from Late Valdivia to present. Tephra deposits cap Valdivia, Chorrera, and Jama-Coaque I occupations ca. 3620, 2845, and 1960 BP.

647 **Scott, David A. and E. Doehne.** La soldadura con aleaciones de oro en la América antigua: un análisis de dos pequeños adornos provenientes del Ecuador. (*Boletín/Bogotá*, 29, 1990, p. 52–61, bibl., ill.)

Analysis reveals differences in composition between the solder and the pieces joined.

648 **Stemper, David Michael.** The persistence of prehispanic chiefdoms on the Río Daule, Coastal Ecuador =La persistencia de los cacicazgos prehispánicos en el Río Daule, Costa del Ecuador. Spanish translation by Juana Camacho. Pittsburgh: Univ. of Pittsburgh, Dept. of Anthropology; Quito: Libri Mundi, 1993. 212 p.: bibl., ill. (Univ. of Pitts-

burgh memoirs in Latin American archaeology; 7)

Detailed report on results of survey and excavations in five regions totaling 10.6 sq. km, including ceramic chronology, raised field construction, metal analysis, and C14 dates. Ceramic seriation using 90 attribute associations defines two phases: Silencio, from ca. 400 BC–AD 250; and Yumes, from ca. AD 400–1600. Chiefdoms developed ca. 400–AD 600; theocratic factors best account for their persistence.

649 Stothert, Karen E. Un sitio de Guangala temprano en el suroeste del Ecuador. Guayaquil, Ecuador: Museo Antropológico; Banco Central, 1993. 1 v: bibl., ill.

Ceramics, burials, subsistence remains (including maize), and other artifacts from excavations in a habitation site adjacent to Valdivia suggest social stratification was still minimal.

THE GUIANAS

650 Between St. Eustatius and The Guianas: contributions to Caribbean archaeology. Edited by Aad H. Versteeg. St. Eustatius, Netherlands Antilles: St. Eustatius Historical Foundation, 1994. 300 p.: bibl., ill., map, photos. (Publications / St. Eustatius Historical Foundation; 3)

Includes three articles on background and first years of Golden Rock excavations (where Leiden Univ. excavated prehistoric sites), and an essay by S. Rostain on prehistory of the coastal zone of The Guianas. Versteeg's article includes 1923 diary of archaeologist J.P.B. de Josselin de Jong. Project supervisor Louwe Kooijmans presents methods and results. Paper by L. Delvoye is a reconnaissance of Corre Corre Bay. [R. Hoefte]

651 Tacoma, J. et al. On "amazonidi": precolumbian skeletal remains and associated archaeology from Suriname. Edited by L. J. van der Steen and Aad H. Versteeg. Amsterdam: Foundation for Scientific Research in the Caribbean Region, 1991. 105 p.: ill. (Publications / Foundation for Scientific Research in the Caribbean Region = Uitgaven / Natuurwetenschappelijke Studiekring voor het Caraïbisch Gebied; 127)

Gives age, sex, stature, deformation, measurements, and non-metric observations on remains of 23 individuals from Kwatta Tingiholo. C14 dates extend from ca. 1500 to 1000 BP.

652 Versteeg, Aad H. and F.C. Bubberman. Suriname before Columbus. (*Mededelingen/Paramaribo*, 49a, Dec. 1992, p. 3–64, bibl., maps, photos)

Concludes that the Suriname Amerindians lived in well-organized societies. Authors base this argument on complicated hydraulic systems used in permanent agriculture on raised fields and on the material culture. [R. Hoefte]

653 Williams, Denis. The forms of the shamanic sign in the prehistoric Guianas. (*Archaeol. Anthropol.*, 9, 1993, p. 3–21, bibl., ill.)

The similarities between Warao burial practices and burials in Archaic shell middens suggest the associated cosmology may also have been shared.

PERU

654 Anders, Martha B. et al. Estudios de arqueología peruana. Edición de Duccio Bonavia. Lima: FOMCIENCIAS, 1992. 417 p.: bibl., ill.

Includes reports of ongoing research by participants in a 1988 international coloquium. Coverage extends from Piura to Ica on the coast, from Chachapoyas to Ayacucho in the Highlands, and from preceramic to Inca periods.

655 Arsenault, Daniel. El personaje del pie amputado en la cultura mochica del Perú: un ensayo sobre la arqueología del poder. (*Lat. Am. Antiq.*, 4, 1993, p. 225–245, bibl., ill.)

Contexts depicting individuals with an artificial foot suggest they symbolized perpetuation of the power structure.

656 Bauer, Brian S. Ritual pathways of the Inca: an analysis of the Collasuyo *ceques* in Cuzco. (*Lat. Am. Antiq.*, 3, 1992, p. 183–205, bibl., ill.)

Comparing the courses of nine *ceques* described in ethnohistorical sources with archaeological evidence reveals numerous inconsistencies, implying their paths were more varied than literature suggests.

657 Bengtsson, Lisbeth and Curt Roslund. "Estaquería" revisited. (*Årstryck/Göteborg*, 1989/1990, p. 1–17, bibl., ill., photos, tables)

Two C14 dates place Estaquería in Nasca 8; similarity of the forked posts to wooden figurines suggests a ritual function.

658 Blasco Bosqued, María Concepción and Luis Javier Ramos Gómez. Catálogo de la cerámica Nazca del Museo de América. v. 2, Recipientes decorados con figuras humanas de carácter ordinario o con cabezas cortadas u otras partes del cuerpo humano. Madrid: Ministerio de Cultura, Dirección General de Bellas Artes, 1985. 1 v.

Detailed description and illustration of 288 vessels is divided into three categories: complete human figures, human heads, and painted decoration incorporating human parts.

659 Bonavia, Duccio and Claude Chauchat. Presencia del Paijanense en el desierto de Ica. (*Bull. Inst. fr. étud. andin.*, 19:2, 1990, p. 399–407, bibl., ill.)

Discovery of a small workshop site for biface production confirms previous reports of a local variety of Paiján culture.

660 Browne, David M.; Helaine Silverman; and Rubén García. A cache of 48 Nasca trophy heads from Cerro Carapo, Peru. (*Lat. Am. Antiq.*, 4, 1993, p. 274–294, bibl., ill.)

Preliminary observations indicate all skulls were deformed and represent robust males 20 to 45 years old.

661 Burger, Richard L. Chavín and the origins of Andean civilization. New York: Thames and Hudson, 1992. 248 p.: bibl., ill., index, maps.

Synthesis of cultural development from late preceramic through collapse of Chavín civilization is organized by period and region. Maps, plans, isometric reconstructions, illustrations of diagnostic artifacts and decorative styles, and list of C14 dates complement detailed descriptions and interpretations, making this a basic reference. See also item **680**.

662 Cabello Carro, Paz and Cruz Martínez. Catálogo de la colección arqueológica norperuana del Museo Casa de Colón del Cabildo Insular de La Gomera. Tenerife, Spain: Viceconsejería de Cultura y Deportes, Gobierno de Canarias, 1992. 58 p.: bibl., ill.

Illustrates 69 vessels of Chimu style.

663 Canziani Amico, José. Arquitectura y urbanismo del período Paracas en el Valle de Chincha. (*Gac. Arqueol. Andin.*, 6: 22, 1992, p. 87–117, bibl., ill.)

Evidence argues for independent development of both early urban settlements and methods of constructing monumental architecture.

664 Cárdenas Martín, Mercedes; Cirilo Huapaya Manco; and Jaime Deza Rivasplata. Arqueología del Macizo de Illescas, Sechura-Piura: excavaciones en Bayóvar, Nunura, Avic, Reventazón y Chorrillos. Lima: Pontificia Univ. Católica del Perú, Dirección Académica de Investigación, 1991. 248 p.: bibl., ill., maps.

Report on 1975–76 excavations includes description of paddle-marked pottery. Twenty-one C14 dates extend from 5590 BC to 1350 AD.

665 Clarkson, Persis and Ronald I. Dorn. Nuevos datos relativos a la antigüedad de los geoglifos y pukios de Nazca, Perú. (*Bol. Lima*, 13:78, nov. 1991, p. 33–47, bibl., maps, tables, photos)

C14 dates on organic material from beneath rock varnish provided by accelerator mass spectrometry (AMS) range from ca. 190 BC to 650 AD, placing construction during the Early Intermediate Period.

666 Coloquio sobre la Cultura Moche, 1st, Trujillo, Peru, 1993. Moche: propuestas y perspectivas. Edición de Santiago Uceda y Elias Mujica B. Trujillo, Peru: Univ. Nacional de la Libertad; Instituto Francés de Estudios Andinos; Asociación Peruana para el Fomento de las Ciencias Sociales, 1994. 549 p., 16 p. of plates: bibl., maps. (Travaux de l'Institut français d'études andines, 0768–424X; 79)

Chapters discuss recent investigations, new interpretations, iconography, ideology, and conservation. Bibliography (49 p.) of existing publications and brief overview of the history of Mochica investigations amplify the significance of this volume.

667 Compendio histórico del Perú. v. 1, Historia arqueológica del Perú: del paleolítico al Imperio Inca [de] Daniel Morales Chocano. Edición de Carlos Milla Batres. Lima: Editorial Milla Batres, 1993. 1 v.: bibl., ill., maps

Takes exception to domination of Peruvian archaeology by North Americans and presents comprehensive overview offering alternative interpretations compatible with Andean ecology, social organization, and worldview. Also important for citations of unpublished work by Peruvians.

668 Cook, Anita Gwynn. The stone ancestors: idioms of imperial attire and rank among Huari figurines. (*Lat. Am. Antiq.*, 3, 1992, p. 341–364, bibl., ill.)

Characteristics and similarities of 40 turquoise figurines from separate caches in the same structure at Pikillacta imply that historical individuals are represented, possibly the legendary 40 founding ancestors of the Huari polity.

Cook, Anita Gwynn. Wari y Tiwanaku: entre el estilo y la imagen. See item **535**.

669 Deza Rivasplata, Jaime. El apogeo de las lanzas: el paleolítico superior andino; la comunidad primitiva en la costa norte. Lima: Centro de Investigación de la Cultura Andina de la Asociación Peruana de Arqueología, 1991. 122 p.: bibl., ill. (some col.). (Serie Investigación A.P.A; 1)

Provides resumé of information on environmental context, settlement, lithic industry, and C14 dates of the Paiján tradition.

670 Diessel, Wilhelm G. Formativzeitliche beinschnitzereien aus Peru. (*Arch. Völkerkd.*, 45, 1991, p. 73–91, bibl., ill., photos)

Describes 16 bone tubes from the Chicama region, the majority with carved decoration in Chavín style, possibly used for inhaling hallucinogens.

671 Donnan, Christopher B. The blowgun in Moche art. (*in* Circumpacifica: Festschrift für Thomas S. Barthel. Frankfurt am Main: P. Lang, 1990, v. 1, p. 509–520, bibl., photos)

Five Moche IV depictions indicate blowguns were used for hunting birds.

672 Elera, Carlos; José Pinilla; and Víctor Vásquez. Bioindicadores zoológicos de eventos ENSO para el formativo medio y tardío de Puémape-Perú. (*Pachacamac/Lima*, 1:1, agosto 1992, p. 9–19, bibl., ill., maps, photos, tables)

Increased frequency of tropical molluscs during the Salinas Period in the Cupisnique Valley implies strong ENSO (El Niño and the Southern Oscillation) events; increased moisture permitted expansion of agriculture and altered settlement and social behavior.

673 Elorrieta Salazar, Fernando E. and **Edgar Elorrieta Salazar.** La gran pirámide de Pacaritanpu: entes y campos de poder en Los Andes. Cusco, Peru: Sociedad Pacaritanpu Hatha, 1992. 306 p.: bibl., ill. (some col.), map.

Argues that applying Andean cosmovision and sacred geography to Inca constructions reveals symbolic elements consistent with origin myths, discrediting interpretations by scholars trained in Western ideology.

674 Engel, Frédéric André. Un desierto en tiempos prehispánicos: Río Pisco, Paracas, Río Ica. Lima?: Didi de Arteta, 1991. 167 p., 1 folded leaf of plates: ill., maps.

Survey of the 160 km between the mouths of the Ica and Pisco Rivers documented 167 sites, indicating occupation of the arid environment from 10,000 BP to present. A summary of the sequence is supplemented by plans and descriptions, sources of water, and 46 C14 dates.

675 Goldstein, Paul. Tiwanaku temples and state expansion: a Tiwanaku sunken-court temple in Moquegua, Peru. (*Lat. Am. Antiq.*, 4, 1993, p.22–47, bibl., ill.)

Existence of a Tiwanaku sunken-court temple complex 300 km southwest of the type site indicates that Tiwanaku was an expansive state.

676 González Carré, Enrique. Historia prehispánica de Ayacucho. 2. ed. Ayacucho, Peru: Univ. Nacional San Cristóbal de Huamanga, Consejo General de Investigaciones, 1992. 130 p.: bibl., ill., maps.

Provides overview of cultural development from Paleo-Indian to Inca periods for nonspecialists. Extensive bibliography.

677 González Carré, Enrique. Los señorios chankas. Lima: Univ. Nacional de San Cristóbal de Huamanga; Instituto Andino de Estudios Arqueológicos, 1992. 153 p.: bibl., ill., maps.

Synthesis of archaeological and archival information is relevant for defining the Chanka nation and its relationship to Inca expansion.

678 Hastorf, Christine Ann. Agriculture and the onset of political inequality before the Inka. Cambridge, England; New York: Cambridge Univ. Press, 1993. 298 p.: bibl., ill., index. (New studies in archaeology)

Changes in agricultural production, population density, settlement size and location, and household goods imply increased social stratification in the Mantaro Valley between Wanka I and Wanka II (AD 199–1500).

Rejecting alternative explanations, author identifies principal mechanism of change as desire for power.

679 **Hastorf, Christine Ann** and **Sissel Johannessen.** Pre-hispanic political change and the role of maize in the Central Andes of Peru. (*Am. Anthropol.*, 95:1, March 1993, p. 115–138, bibl., graphs, ill., map)

Analysis of incrustations on pottery, C3/C4 ratio in skeletons, relative frequency of large jars, and frequency of grinding stones in Wanka I and Wanka II contexts suggest change in maize consumption from a basic food to a political tool for elite consolidation in the form of beer (chicha).

680 **Historia general del Perú.** v. 1, Los orígenes de la civilización andina: arqueología del Perú [de] Peter Kaulicke. Lima: Editorial Brasa, 1994. 606 p.: bibl., ill.

Well-illustratated authoritative work summarizes cultural development from initial colonization to the end of the formative period. Almost half the text discusses origins of the formative from perspectives of environment, subsistence, art, and religion. Use of same categories to examine the Chavín phenomenon highlights continuities. For a different perspective, see item **661.**

Hocquenghem, Anne-Marie *et al.* Bases del intercambio entre las sociedades norperuanas y surecuatorianas: una zona de transición entre 1500 A.C. y 600 D.C. See item **539.**

681 **Iconografía de Cajamarca.** v. 1. Dirección y producción de Alfredo Mires. Cajamarca, Peru: Cedepas Cajamarca; Asparderuc-P.E.C.; Asociación Editora Cajamarca, 1992. 1 v.: bibl., ill.

Representations on stone, pottery, and textiles document the artistic quality of precolumbian art.

682 **Los incas y el antiguo Perú: 3000 años de historia.** t.1, Texto. t.2, Catálogo. Madrid: Ayuntamiento de Madrid, Concejalía de Cultura; Centro Cultural de la Villa de Madrid; Quinto Centenario, 1991. 2 v.: bibl., ill. (some col.). (Col. Encuentros. Serie Catálogos)

Chapters by 32 leading Peruvianists provide context for exhibition of objects selected from European and American museums, creating an authoritative and spectacular insight into the quality and significance of Andean art.

683 **Julien, Daniel G.** Late pre-Inkaic ethnic groups in Highland Peru: an archaeological-ethnohistorical model of the political geography of the Cajamarca region. (*Lat. Am. Antiq.*, 4, 1993, p. 246–273, bibl., ill.)

Of the seven Inca administrative units in Cajamarca prov., five or six correspond to indigenous chiefdoms which exhibit incipient political integration enhanced by Inca administration.

684 **Kauffmann Doig, Federico.** Pinturas mágicas sobre placas de cerámica. (*Arqueológicas/Lima*, 21, 1992, p. 1–202, bibl., ill.)

Natural crevasses at Chucu, Chuquibamba region, southwest Peru, contain hundreds of rectanguloid to trapezoidal plaques produced from sherds of large vessels, painted on one surface with abstract patterns. Some 400 examples are illustrated and described.

685 **Kirkpatrick, Sidney.** Lords of Sipán: a tale of pre-Inca tombs, archaeology, and crime. New York: Morrow, 1992. 256 p.: bibl., col. ill., index, maps.

Popular account of the discovery and excavation of the New World's most spectacular royal tombs on the Peruvian north coast.

686 **Lavalle, José Antonio.** Oro del antiguo Perú. Lima: Banco de Crédito del Perú en la Cultura, 1992. 351 p.: bibl., col. ill. (Col. Arte y tesoros del Perú)

Color photographs of more than 300 spectacular objects document the skill and artistry of precolumbian artisans from Chavín to Inca periods. The informative accompanying text makes this a fundamental reference on the topic.

687 **Malpass, Michael A.** Ocupación precerámica del Valle de Casma, Perú. (*Bol. Lima*, 13:76, julio 1991, p. 79–95 bibl., ill., maps, tables)

The Paiján complex (9000–8000 BC) is followed by a previously undescribed complex, designated Mongoncillo (8000–5000 BC), associated with exploitation of the lomas.

688 **Matos Mendieta, Ramiro.** Pumpu: centro administrativo Inka de la puna de Junín. Lima: Editorial Horizonte; B.C.R. Fondo Editorial; Taraaxacum, 1994. 327 p.: bibl, ill. (Arqueología e historia; 10)

Archaeological, historical, ethnographic, and ecological data contribute to reconstructing the plan, architecture, internal organization, and function of residences, administrative structures, storage facilities, plazas, and other features of an Inca installation occupied during the last 70 years of indigenous rule.

689   **El mundo ceremonial andino.** Recopilación de Luis Millones y Yoshio Ōnuki. Lima: Editorial Horizonte, 1994. 299 p.: bibl., ill. (Etnología y antropología; 8)

Twelve chapters attempt to infer ideology and ceremonialism from artifacts, iconography, architecture, burials, chronicles, and contemporary ritual in various coastal and highland regions, with emphasis on the formative period.

690   **Paracas art & architecture: object and context in south coastal Peru.** Edited by Anne Paul. Iowa City: Univ. of Iowa Press, 1991. 445 p.: bibl., ill., index, maps.

Contributors discuss technical and iconographic analysis of painted textiles, headbands as system templates, physical and chemical analysis of fibers, inferences from embroidered images, social and political leadership inferred from habitation sites and ceramics, and evidence for ethnic differentiation among remains included within Paracas.

691   **Paternosto, César.** The stone and the thread: Andean roots of abstract art. Translated by Esther Allen. Austin: Univ. of Texas Press, 1996. 1 v.: bibl., ill., index.

Author, "a lifelong abstractionist," argues that European preference for realism and categorization of non-European products as "specimens" has prevented recognition of Inca masonry, sculptured outcrops, textile patterns, and other creations as sophisticated examples of abstract art. Numerous illustrations of elaborately carved rock, often incorporated into architectural complexes, support his thesis that Inca achievements cannot be excluded from mainstream art history.

692   **Pelegrin, Jacques** and **Claude Chauchat.** Tecnología y función de las puntas de Paiján: el aporte de la experimentación. (*Lat. Am. Antiq.*, 4, 1993, p. 367–382, bibl., ill.)

Elongated narrow tip of the projectile point and absence of large terrestrial fauna among subsistence remains suggest specialization for spearing large marine fish.

Ponce Sanginés, Carlos *et al.* Exploraciones arqueológicas subacuáticas en el Lago Titikaka: informe científico. See item **572**.

693   **Protzen, Jean-Pierre.** Inca architecture and construction at Ollantaytambo. Drawings by Robert Batson. New York: Oxford Univ. Press, 1993. 303 p.: bibl., ill., index.

An architect applies his expertise to interpret clues to Inca methods of cutting, transporting, handling, and fitting stones, planning and constructing buildings, and resolving physical and engineering problems. Detailed plans, photographs of structural details, and isometric reconstructions are illuminated by results of experiments to assess methods, manpower, and function. A basic reference on the subject.

694   **Provincial Inca: archaeological and ethnohistorical assessment of the impact of the Inca state.** Edited by Michael A. Malpass. Iowa City: Univ. of Iowa Press, 1993. 272 p., 4 p. of plates: bibl., index, maps.

Data from five regions demonstrate importance of combining ethnohistorical and archaeological evidence, and document flexibility of Inca administrative procedures for integrating diverse local groups.

695   **Ravines, Rogger.** Mates ornamentados del Perú: una tradición prehispánica. (*Bol. Lima*, 13:78, nov. 1991, p. 17–22, photos)

Provides glossary of colonial and current terms for gourds, and illustrations of decorated precolumbian examples.

Ravines, Rogger. Notas, testimonios, documentos. See *HLAS 54:2772*.

696   **Ravines, Rogger.** Un sello de cerámica de Huacaloma, Cajamarca. (*Bol. Lima*, 13:75, mayo 1991, p. 24–26, bibl.)

Describes a unique pottery stamp encountered during 1974 excavations, possibly used for facial decoration.

697   **Reiche, María.** Contribuciones a la geometría y astronomía en el antiguo Perú. Lima: Asociación María Reiche para las Líneas de Nasca; Epígrafe Editores S.A., 1993. 571 p.: ill. (some col.), maps.

Brief biographical sketch introduces 29 articles, 13 manuscripts, six scientific letters, six letters to newspapers, and 242 photographs and drawings of Nasca geoglyphs produced during a lifetime of tireless work. An essential reference on the subject.

698 **Reiche, María.** Conversaciones con María Reiche. Entrevista realizada por Clorinda Caller Iberico. Lima: Editorial Horizonte, 1992. 133 p.: ill. (Narrativa contemporánea; 14)

Thirteen "conversaciones" recorded by an intimate friend provide insight into the personality of the enigmatic investigator of the Nasca geoglyphs.

699 **Richardson, James B.** Early hunters, fishers, farmers and herders: diverse economic adaptations in Peru to 4500 BP. (*Rev. Arqueol. Am.*, 6, julio/dic. 1992, p. 71–90, bibl.)

Recognizes four general subsistence strategies: terrestrial and maritime hunting and gathering prior to ca. 6000 BP, followed by agriculture and camelid herding.

700 **Samaniego Román, Lorenzo.** Arte mural de Punkurí: aproximación. (*Pacífico/Chimbote*, 1:1, 1992, p. 11–37, bibl., ill.)

Iconography and architecture associate the Punkurí temple in Nepeña with the Sechín culture.

701 **Schuster, Angela M.H.** The Moche of Peru. (*Archaeology/New York*, 45:6, Nov./Dec. 1992, p. 30–45, ill., map, photos)

Well-illustrated popular account describes royal tombs of Sipán and their significance for reconstructing Moche religious practices.

702 **Shimada, Izumi.** Pampa Grande and the Mochica culture. Austin: Univ. of Texas Press, 1994. 323 p.: bibl., ill., index, maps.

Provides archaeologically based synthesis of origins, internal organization, economy, and environmental context of Mochica culture in general, along with a detailed case study of the emergence, composition, and demise of the urban center of Pampa Grande, ca. AD 500–750. Environmental, economic, social, strategic, and other factors are assessed as potential causal mechanisms for historical developments.

703 **Silva S., Jorge E.** Ocupaciones postformativas en el valle del Rímac: Huachipa-Jicamarca. (*Pachacamac/Lima*, 1:1, agosto 1992, p. 49–74, bibl., ill., maps)

Presents evidence for changes in settlement location and function, life style, and artistic expression, and examines their implications.

704 **Silverman, Helaine.** Cahuachi in the ancient Nasca world. Iowa City: Univ. of Iowa Press, 1993. 371 p.: bibl., ill., index, maps.

Provides detailed descriptions of excavations, burials, trophy heads, pottery, textiles, other artifacts, plant remains, and other evidence leading to the conclusion that Cahuachi was the focus of periodic mass pilgrimages and that Nasca society was a confederation of chiefdoms organized along Andean principles of moieties and ayllus.

705 **Silverman, Helaine.** Estudio de los patrones de asentamiento y reconstrucción de la antigua sociedad Nasca. (*Bol. Lima*, 14:82, julio 1992, p. 33–44, bibl., map, photo, table)

Based on 100 percent coverage of a 200 sq. km sector in the Valle del Ingenio, author concludes that Nasca social organization followed Andean principles of reciprocity between moieties, rather than the hierarchical state model.

706 **Stanish, Charles; Lee Steadman;** and **Matthew T. Seddon.** Archaeological research at Tumatumani, Juli, Peru. Chicago: Field Museum of Natural History, 1993. 1 v.: bibl., ill. (Fieldiana: Anthropology, new series; 23)

Survey along 40 km of southwestern Lake Titicaca shore permits defining Sillumocco and Tiwanaku settlement location and ceramic assemblages, the latter described in detail.

707 **Trabajos arqueológicos en Moquegua, Perú.** v. 1–3. Recopilación de Luis K. Watanabe, Michael Edward Moseley y Fernando Cabieses. Lima?: Programa Contisuyo del Museo Peruano de Ciencias de la Salud; Southern Peru Copper Corp., 1990. 3 v. : bibl., ill., maps.

Vol. 1 discusses general characteristics of the Archaic Period, settlement behavior, an open site (Asana), rock art, maritime adaptation (Ring Site), a pre-ceramic burial, pre-Inca ecology, Early Period ceramics, and prehispanic fortifications. Vol. 2 reports various aspects of the Wari and Tiwanaku periods, post-Tiwanaku economy, copper working, and agrarian collapse following Spanish occupation. Vol. 3 focuses on Late Period cemeteries, textiles, and excavations, followed by details on the wine industry during the colonial period.

708 **Valdez, Lidio M.** and **Cirilo Vivanco.** Arqueología de la Cuenca del Qaracha, Ayacucho, Perú. (*Lat. Am. Antiq.*, 5, 1994, p. 144–157, bibl., ill.)

Survey revealed differences in location of settlements during the Wari, post-Wari, and Inca periods, reflecting changes in political organization.

709 **Valencia Zegarra, Alfredo** and **Arminda Gibaja Oviedo.** Machu Picchu: la investigación y conservación del monumento arqueológico después de Hiram Bingham. Cusco, Peru: Municipalidad del Qosqo, 1992. 354 p.: bibl., ill.

Chronological resumé lists excavations by sector (including stratigraphy, architecture, artifacts, conservation, restoration), personnel, and evaluation of present needs.

URUGUAY

710 **Bracco Boksar, Roberto.** Desarrollo cultural y evolución ambiental en la región este del Uruguay. (*Ed. Quinto Centen.*, 1, 1992, p. 43–73, bibl.)

Differences between coastal and interior sites reflect adaptation to different subsistence resources; 15 C14 dates extend from 2450 BP to present.

711 **López Mazz, José M.** Aproximación a la génesis y desarrollo de los cerritos de la zona de San Miguel, Dpto. de Rocha. (*Ed. Quinto Centen.*, 1, 1992, p. 75–96, bibl., ill.)

Excavation in one mound produced occupation refuse and human burials.

VENEZUELA

712 **Alfaro Salazar, Luis** and **Antonio J. Vargas Ramírez.** Prehistoria de Venezuela: Venezuela antes de la llegada de los europeos. Caracas: Fondo Editorial Tropykos, 1992. 300 p.: bibl., ill., maps.

Educator and obstetrician attempt to overcome the "Eurocentric" view that precolumbian cultures are irrelevant for understanding Venezuelan national identity.

713 **Contribuciones a la arqueología regional venezolana.** Recopilación de Francisco Fernández y Rafael Gasson. Caracas: Fondo Editorial Acta Científica Venezolana, 1993. 223 p.: bibl., ill.

Eight chapters by Venezuelan archaeologists draw on recent fieldwork to modify accepted views of chronology, settlement pattern, population size, ethnicity communicated via ceramics, and petroglyph chronology.

714 **Durán, Reina.** Una aldea prehispánica en colinas de Queniquea. (*Bol. Inf./San Cristóbal*, 10:10, 1993, p. 1–63, bibl., ill.)

Reviews investigations of 30 occupational terraces in a 5 ha. area, the first reported occurrence of a well-known Colombian settlement pattern.

715 **Formas del inicio: la pintura rupestre en Venezuela.** Edited by Roberto Colantoni. Caracas: Consejo Nacional de la Cultura, Fundación Galería de Arte Nacional, 1992. 1 v. (unpaged): bibl., col. ill.

Plates illustrate variations in motif, color, and execution of paintings on rock.

Langebaek, Carl Henrik. Noticias de caciques muy mayores: origen y desarrollo de sociedades complejas en el nororiente de Colombia y norte de Venezuela. See item **541**.

716 **Spencer, Charles S.; Elsa M. Redmond; and Milagro Rinaldi.** Drained fields at La Tigra, Venezuelan llanos: a regional perspective. (*Lat. Am. Antiq.*, 5, 1994, p. 119–143, bibl., ill.)

Reconstruction of chronology, crops grown, productivity, labor requirements, and population size suggests that political/economic interests of an elite rather than subsistence demands of an increasing population account for construction of ridged fields between AD 550-1000.

Spencer, Charles S. Human agency, biased transmission, and the cultural evolution of chiefly authority. See item **161**.

717 **Wagner, Erika.** Diversidad cultural y ambiental en el occidente de Venezuela. (*in* Archaeology and environment in Latin America. Amsterdam: Institut voor Pre-en Protohistorische Archaeologie Albert Egges van Giffen, Univ. van Amsterdam, 1992, p. 207–221, bibl.)

Reviews diverse models for post-formative cultural development and inadequacy of evidence to evaluate their accuracy.

ETHNOLOGY
Middle America

PAUL SULLIVAN, *Independent Consultant, Port Jefferson, New York*

EVEN AS THE CHIAPAS REBELLION drew national and international attention to poverty and political conflict in at least one region of Mexico, the spectacle of masked Tzeltal rebels highlighted the growing importance of ethnicity in Mesoamerican studies. A quarter of the works cited in this section, and a great many lesser pieces not included here, explore one aspect or another of the complex interrelationship between ethnicity, class, political conduct, and religious change. Frans Schryer's very detailed, insightful and novel study of land invasions in the Huasteca is the most notable contribution to this strengthening trend in Mesoamerican anthropology and history (item **763**). Even as Schryer's data challenges any ready assumptions about ethnic identity, class position, and political affiliation, shorter works cited provoke a fundamental reassessment of the nature of ethnic identity and the factors which lead to its loss or, as we are ever surprised to discover, its resurgence. The rich, diverse anthology entitled *Zapotec struggles* (item **772**) assures that in any such reassessment the case of Zapotec revival and politicization in the Isthmus of Tehuantepec will receive deserved attention, while reports such as those of Warren (items **770** and **769**) alert us to the revival of ethnicity and the creation of a pan-Maya identity and nationalism in Guatemala with powerful potential to reshape ethnic realities beyond the borders of that nation. Concerning the still emerging consequences of civil war in Guatemala, special note should be taken of Stoll's very timely study of the Ixil Triangle (item **765**), an important study, which, like Schryer's mentioned above, challenges many of our discipline's often taken-for-granted assertions about ethnicity, class, and political struggle in the Mesoamerican countryside.

718 **Alvarez Santiago, Héctor.** El xochitlali en San Andrés Mixtla: ritual e intercambio ecológico entre los nahuas de Zongolica. Xalapa, Mexico: Comisión Estatal Conmemorativa del V Centenario del Encuentro de Dos Mundos, Gobierno del Estado de Veracruz, 1991. 166 p.: bibl., ill., maps. (Col. V centenario; 7)

Lengthy, detailed work describes related curing and agricultural rituals in a Nahua community of Veracruz.

Arnold, Philip J. Dimensional standardization and production scale in Mesoamerican ceramics. See item **189**.

Arnold, Philip J. Domestic ceramic production and spatial organization: a Mexican case study in ethnoarchaeology. See item **190**.

Arnold, Philip J. The organization of refuse disposal and ceramic production within contemporary Mexican houselots. See item **192**.

Arqueología Mexicana. Vol. 1, No. 3, 1993–. See item **196**.

Arqueología Mexicana. Vol. 1, No. 5, 1994–. See item **198**.

719 **Bartolomé, Miguel Alberto** and **Alicia Mabel Barabas.** La presa Cerro de Oro y el Ingeniero El Gran Dios: relocalización y etnocidio chinanteco en México. México: Dirección General de Publicaciones del Consejo Nacional para la Cultura y las Artes; Instituto Nacional Indigenista, 1990. 2 v.: bibl., ill., maps. (Col. Presencias; 19–20)

One of the most complete discussions of forced relocation available for Mexico, includes much ethnographic data on the Chinantec of Oaxaca and relates an interesting case of messianism in a crisis context.

720 **Binford, Leigh.** Peasants and petty capitalists in southern Oaxacan sugar cane production and processing, 1930–1980. (*J. Lat. Am. Stud.*, 24:1, Feb. 1992, p. 33–55)

Insightful study examines nature and decline of a type of petty commodity production—the cultivation of sugarcane and production of *panela*—with useful lessons for further studies of village-level production.

Brady, James E. and **George Veni.** Man-made and pseudo-karst caves: the implications of subsurface features within Maya centers. See item **212**.

721 **Brockmann, Andreas.** Santa Martha: Untersuchungen zur Ethnographie einer Tzotzilgemeinde in Mexiko. Münster, Germany: Lit, 1992. 241 p.: bibl., ill., maps. (Ethnologische Studien; 16)

Ethnographic dissertation on a Tzotzil Maya community of northern Chiapas is based on a year of fieldwork in late 1980s. Contains especially detailed descriptions of religious and civil authority structures, and major festivals associated with carnival and Easter. Perhaps most valuable for its description of small-scale indigenous coffee and sugar production and marketing.

Brysk, Alison and **Carol Wise.** Economic adjustment and ethnic conflict in Bolivia, Peru, and Mexico. See item **932**.

722 **Burns, Allan F.** Maya in exile: Guatemalans in Florida. Introduction by Jerónimo Camposeco. Philadelphia, Pa.: Temple Univ. Press, 1993. 255 p.: bibl., ill., index.

Highly original and important study examines community of Kanjobal Maya refugees and immigrants in southern Florida.

Caso, Alfonso. Alfonso Caso: de la arqueología a la antropología. See item **85**.

723 **Cohen, Jeffrey.** Danza de la pluma: symbols of submission and separation in a Mexican fiesta. (*Anthropol. Q.*, 66:3, 1993, p. 149–158)

Very stimulating discussion examines symbolism of a Zapotec dance of the conquest and its relevance to the social relations of production in an important weaving community of Oaxaca.

724 **Collier, George A.** The new politics of exclusion: antecedents to the rebellion in Mexico. (*Dialect. Anthropol.*, 19:1, May 1994, p. 1–44, maps)

Timely and fruitful study of religious and ethnic dimensions of class and political conflict in a Tzotzil region of highland Chiapas. Written by an anthropologist with over 20 years of experience in the region.

725 **Crafts in the world market: the impact of global exchange on Middle American artisans.** Edited by June C. Nash. Albany: State Univ. of New York Press, 1993. 264 p.: bibl., ill., index. (SUNY series in the anthropology of work)

Important and innovative collection of 10 short case studies examines impact of international markets, capital, and tourism on Mesoamerican artisan production and ethnicity.

726 **Dehouve, Danièle.** Compter l'argent: les indiens de Tlapa, Mexique. (*Annales/Paris*, 47:2, mars/avril 1992, p. 315–329, graph, photo, table)

Unusual piece on the obviously important and neglected question of how Indians count money. Insightfully examines the play of historical continuities and contemporary meanings in counting practices.

727 **Diechtl, Sigrid.** Cae una estrella: desarrollo y destrucción de la Selva Lacandona. México: Secretaría de Educación Pública; Programa Cultural de las Fronteras, 1988. 118 p.: bibl., ill. (Frontera)

Overview of various forms and phases of colonization of the Selva Lacandona from 1950s–80s. Particular attention given to colonist interactions with and effects upon the Lacandon people.

728 **Earle, Duncan.** Authority, social conflict and the rise of Protestantism: religious conversion in a Mayan village. (*Soc. Compass*, 39:3, Sept. 1992, p. 377–388, bibl.)

Discusses impact of conversion to Protestantism and proliferation of Protestant churches on social harmony and land tenure in a Cakchikel Maya community. Polemical, but raises anew issues of religious change and decline of traditional authority that merit study.

729 **Edelman, Marc.** Landlords and the devil: class, ethnic, and gender dimensions of Central American peasant narratives. (*Cult. Anthropol.*, 9:1, Feb. 1994, p. 58–93, bibl.)

Exemplary discussion of devil-pact narratives in rural Central America includes important observations concerning peasant ambivalence towards rich and successful neighbors and exploiters.

730 **Edelman, Marc.** The logic of the latifundio: the large estates of northwestern Costa Rica since the late nineteenth century. Stanford, Calif.: Stanford Univ. Press, 1992. 478 p.: bibl., ill., index.

Exceptional study of agrarian change in Costa Rica provides excellent discussion of divergent theoretical perspectives on nature of large landed estates and changing social relations of production. Among the best of such studies available for Central America. For historian's comment see *HLAS 54:1716*.

731 **Eweg, Erlijn M.** and **Elisabeth M. Joosten.** El valor del *Coyuchi:* aspectos culturales y agrícolas del algodón criollo mesoamericano. (*Am. Indíg.*, 53:3, julio/sept. 1993, p. 57–95, bibl., ill., tables)

Ethnographic study, with some recourse to sources on prehispanic beliefs and practices, of cotton cultivation. [R. Haskett]

732 **Franco Pellotier, Víctor Manuel.** Grupo doméstico y reproducción social: parentesco, economía e ideología en una comunidad otomí del Valle del Mezquital. México: CIESAS, 1992. 258 p.: bibl., ill., maps. (Col. Miguel Othón de Mendizábal)

Solid work examines kinship and households in an Otomi community of Hidalgo.

733 **Fry, Douglas P.** "Respect for the rights of others is peace:" learning aggression versus nonaggression among the Zapotec. (*Am. Anthropol.*, 94:3, Sept. 1992, p. 621–639, bibl., table)

Comparative study of real and play aggression among children in two Zapotec communities nicely highlights the relevance of social learning as a factor contributing to persistently high (or low) levels of intracommunity violence. A provocative study and valuable introduction to the relevant Mesoamerican literature.

Galarza, Joaquín. In amoxtli, in tlacatl = el libro, el hombre; códices y vivencias. See *HLAS 54:471*.

734 **Galinier, Jacques.** La mitad del mundo: cuerpo y cosmos en los rituales otomíes. Traducción de Angela Ochoa y Haydée Silva. México: Univ. Nacional Autónoma de México, Centro de Estudios Mexicanos y Centroamericanos, Instituto Nacional Indigenista, 1990. 746 p.: bibl., ill., index, maps.

Monumental work on contemporary Otomi ritual, religion, and body symbolism includes trenchant commentary on importance of and approaches to exploring the world view of an indigenous people.

735 **García Bresó, Javier.** Monimbó: una comunidad india de Nicaragua. Managua: Editorial Multiformas, 1992. 357 p.: bibl., ill., map.

Interesting ethnography of an urban Indian barrio in Nicaragua addresses in particular the loss and renewal of ethnic identity.

736 **García Valencia, Enrique Hugo.** San Miguel Aguazuelos: estrategias de residencia. Xalapa, Mexico: Comisión Estatal Conmemorativa del V Centenario del Encuentro de Dos Mundos, Gobierno del Estado de Veracruz, 1991. 190 p.: bibl., ill., maps. (Col. V Centenario; 3)

Useful work examines residence and kinship among lesser studied Totonacs of Veracruz.

737 **Gossen, Gary H.** La diáspora de San Juan Chamula: los indios en el proyecto nacional mexicano. (*in* De palabra y obra en el Nuevo Mundo. Madrid: Siglo Veintiuno Editores, 1992, v. 2, p. 429–456)

Provides additional data on colonization of and expulsions from Chamula, the major indigenous community of Chiapas. Uses a gender-informed interpretation of the survival of Chamula and its colonies in a world dominated by non-Indians.

738 **Gossen, Gary H.** Las variaciones del mal en una fiesta tzotzil. (*in* De palabra y obra en el Nuevo Mundo. Madrid: Siglo Veintiuno Editores, 1992, v. 1, p. 195–235, ill.)

Rich interpretation of Chamula (Maya) carnival celebration emphasizes symbolic representations of inter-ethnic relations. See also *HLAS 51:742*.

739 **Grigsby, Thomas L.** and **Carmen Cook de Leonard.** Xilonen in Tepoztlán: a comparison of Tepoztecan and Aztec agrarian ritual schedules. (*Ethnohistory/Society*, 39:2, Spring 1992, p. 110–147, appendix, bibl., graph, photos, tables)

Correlation between the Aztec calendar outlined in Book 2 of the Florentine Codex and the ritual agrarian one used in contemporary Tepoztlan, Morelos. Based on this comparison the authors posit a set of factors that would have determined the precise ordering of each system, and suggest a possible geographic point of origin for the pre-Hispanic system. An appendix presents texts describing the two systems in parallel columns. For archaeologist's comment, see item **417**. [R. Haskett]

740 Gutiérrez Estévez, Manuel. Alteridad étnica y conciencia moral: el juicio final de los mayas yucatecos. (*in* De palabra y obra en el nuevo mundo. Madrid: Siglo Veintiuno España Editores, 1992, v. 2, p. 295–322)

Presents interpretations of the nature and bases of ethnic identity on the Yucatán Peninsula, giving special emphasis to prophetic traditions of contemporary Yucatec Maya.

741 Gutiérrez Estévez, Manuel. Mayas y "mayeros:" los antepasados como otros. (*in* De palabra y obra en el Nuevo Mundo. Madrid: Siglo Veintiuno Editores, 1992, v. 1, p. 417–441)

Novel and provocative interpretation of the complicated and vexing question of ethnic identity on the Yucatan Peninsula focuses especially on historical consciousness and the meanings of pre-conquest ruins for contemporary Yucatec Mayas.

742 Hanks, William. Copresencia y alteridad en la práctica ritual maya. (*in* De palabra y obra en el nuevo mundo. Madrid: Siglo Veintiuno de España Editores, 1993, v. 3, p. 75–117)

Very detailed and carefully argued essay concerning forms and nature of Yucatec Maya shamanic curing is written by one of the foremost scholars of Mesoamerican shamanism.

743 Hard, Robert J. and **William L. Merrill.** Mobile agriculturalists and the emergence of sedentism: perspectives from northern Mexico. (*Am. Anthropol.*, 94:3, Sept. 1992, p. 601–620, bibl., graph, map, table)

Fine study of sedentism and residential mobility among Rarámuri (Tarahumara) agriculturalists of Chihuahua illustrates that an agricultural economy does not necessitate sedentism.

744 Hirabayashi, Lane Ryo. Cultural capital: mountain Zapotec migrant associations in Mexico City. Tucson: Univ. of Arizona Press, 1993. 157 p.: bibl., ill., index, map. (PROFMEX series)

Major study of rural-urban migration, migrant associations, and ethnicity in urban settings based on fieldwork from late 1970s. Provides an historical perspective on Zapotec migration to Oaxaca City and Mexico City beginning in 1940s, and enriches that perspective with ethnographic research in both rural and urban settings. For an earlier treatment of this data see *HLAS 49:810*.

745 Hostettler, Ueli. Staatliche Landpolitik und periphere Lage: Chance für das Überleben kultureller Eigenständigkeit: die Cruzoob-Maya in Quintana Roo, Mexiko. (*Bulletin/Geneva*, 53/54, 1989/90, p. 59–72, map, photos, tables)

Presents results of meticulous research on socioeconomic stratification in a Yucatec Maya village.

746 Jiménez Huerta, Fernando. ¿El vuelo del fénix?: Antorcha Campesina en Puebla. Puebla, Mexico: Benemérita Univ. Autónoma de Puebla, 1992. 198 p.: bibl. (Col. Crónicas y testimonios; 10)

Ethnography of agrarian and political conflict in rural Puebla is particularly interesting for light it sheds on structure and activities of Antorcha Campesina, a peasant organization.

747 Kane, Stephanie C. The phantom gringo boat: shamanic discourse and development in Panama. Washington: Smithsonian Institution Press, 1994. 266 p.: bibl., ill, index, map. (Smithsonian series in ethnographic inquiry)

First book-length ethnography of the Embera of Panama is based upon field research conducted in mid-1980s. Explores in dazzling and provocative prose diverse realms of Embera life—gardening, shamanism, gender, political organization, political economy, and more—giving particular attention to goals, forms, politics, and consequences of externally-planned and externally-promoted economic development (i.e., resource exploitation). Imaginative and effective.

748 Kasburg, Carola. Die Totonaken von El Tajín: Beharrung und Wandel über vier Jahrzehnte. Münster, Germany: Lit, 1992. 292 p.: bibl., ill. (Ethnologische Studien; 22)

Rich study of social and cultural change in a Totonac community is based on one year's field research and published and unpublished data collected at the same site in 1940s and 1960s.

749   **Köhler, Ulrich.** Schamanismus in Mesoamerika? (*in* Circumpacifica: Festschrift für Thomas S. Barthel. Frankfurt am Main: P. Lang, 1990, p. 257–275, bibl.)

Comparison of significant features of Siberian and Mesoamerican shamanism is based on studies of both cultures. Köhler rules out trans-Pacific influence. [C.K. Converse]

750   **Lancaster, Roger N.** Skin color, race, and racism in Nicaragua. (*Ethnology/Pittsburgh*, 30:4, Oct. 1991, p. 339–353, bibl.)

Masterful essay on contextualized use of racial/color terminology in Nicaragua argues that racism in Nicaragua is not structural, but an ubiquitous, ambiguous discursive practice of everyday life. Argument has potentially broad implications for study of race, ethnicity, and legacies of colonialism throughout Mesoamerica and Central America.

751   **Lok, Rossana.** Gifts to the dead and the living: forms of exchange in San Miguel Tzinacapan, Sierra Norte de Puebla, Mexico. Leiden, The Netherlands: Centre of Non-Western Studies, Leiden Univ., 1991. 115 p.: bibl., ill., index. (CNWS publications, 0925–3084; 4)

Fine, detailed discussion of mortuary practices among Nahua of Puebla includes observations concerning sacrifice, cults of the dead, and *compadrazgo* that merit exploration in other parts of Mexico.

752   **Manning, Roswitha.** Vrouwen met goddelijke kracht: godinnen en vroedvrouwen bij de Maya's [Women with divine power: Mayan goddesses and midwives]. Leiden, The Netherlands: s.n., 1993. 279 p.: bibl., ill.

Dissertation focuses on Mayan goddesses and midwives, past and present, of Yucatán Peninsula. Identifies goddesses that were portrayed in Dresden Codex. Using ethnographical and historical sources, studies participation of Mayan women in religious life. Also looks at traditional midwives as example of female religious specialists. Includes summaries in English and Spanish. [R. Hoefte]

753   **Marion Singer, Marie-Odile.** Los hombres de la selva: un estudio de tecnología cultural en medio selvático. México: Instituto Nacional de Antropología e Historia, 1991. 287 p.: bibl., ill. (Col. Regiones de México)

Solid, timely study of Lacandon focuses particularly on issues of production, technology, and ecology, and the myriad consequences of logging in Lacandon forest reserve.

Montejo, Victor. The bird who cleans the world: and other Mayan fables. See *HLAS 54: 4400*.

754   **Nachtigall, Horst.** West-Tarasken: Beiträge zur Archäologie, Ethnologie und Akkulturation eines westmexikanischen Volkes. Berlin: D. Reimer, 1992. 374 p.: bibl., ill. (some col.), index. (Marburger Studien zur Völkerkunde; 10)

Traditional, general ethnography of a Tarascan town in western Michoacán.

755   **Nutini, Hugo G.** and **John M. Roberts.** Bloodsucking witchcraft: an epistemological study of anthropomorphic supernaturalism in rural Tlaxcala. Tucson: Univ. of Arizona Press, 1993. 476 p.: bibl., index.

This monumental study of the social, psychological, and medical aspects of witchcraft beliefs in indigenous communities of Tlaxcala is the most thorough study of witchcraft ever written about Mesoamerica. Contains a wealth of historical, comparative, and ethnographic data.

756   **Paula de Teresa, Ana.** Crisis agrícola y economía campesina: el caso de los productores de henequén en Yucatán. México: Univ. Autónoma Metropolitana-Iztapalapa; M.A. Porrúa, Grupo Editorial, 1992. 305 p.: bibl., ill. (Col. Las Ciencias sociales)

Excellent ethnography of henequen producers in Yucatán explores paradoxical persistence of an unproductive enterprise by examining social relations of production at regional and national levels, and the history of household units of production in one village.

757   **Perera, Victor.** Unfinished conquest: the Guatemalan tragedy. Photographs by Daniel Chauche. Berkeley: Univ. of California Press, 1993. 382 p.: bibl., index, maps, photos.

Offers dark and highly informative

view of Guatemalan political, military, and social affairs in late 1980s and early 1990s. Interviews in various regions of Guatemala with people from all ranks of society, and author's long experience in his former country, provide important insights into recent history and the still volatile and uncertain circumstances of Guatemalan conflict and social change. For sociologist's comment see item **4650**.

758 Pérez Castro, Ana Bella. Entre montañas y cafetales: luchas agrarias en el norte de Chiapas. México: Univ. Nacional Autónoma de México, 1989. 235 p.: bibl., graphs, maps, tables. (Instituto de Investigaciones Antropológicas, Serie Antropológica; 85)

Important work describes agrarian history and recent conditions of a coffee-producing region of northern Chiapas inhabited by speakers of Tzotzil and Chol Maya.

Purcell, Trevor W. Banana fallout: class, color, and culture among West Indians in Costa Rica. See item **826**.

759 Reyes, Reynaldo and **Judith Kay Wilson.** Ráfaga: the life story of a Nicaraguan Miskito *comandante*. Edited by Tod Stratton Sloan. Norman: Univ. of Oklahoma Press, 1992. 224 p.: bibl., ill., index, map.

Fascinating life history providing unique perspective on indigenous leadership and resistance in Nicaragua. For political scientist's comment see item **3007**.

760 Royce, Anya Peterson. Music, dance, and fiesta: definitions of Isthmus Zapotec community. (*Lat. Am. Anthropol. Rev.*, 3:2, Winter 1991, p. 51–60, bibl., photos)

Fine short piece on Zapotec music and dance highlights continuing creation of "tradition" in the interest of ethnic assertion in Oaxaca.

761 Ruiz Lombardo, Andrés. Caficultura y economía en una comunidad totonaca. México: Dirección General de Publicaciones del Consejo Nacional para la Cultura y las Artes; Instituto Nacional Indigenista, 1991. 203 p.: bibl., ill., maps. (Col. Presencias; 40)

Two chapters of this book provide very useful information about indigenous small-scale coffee production and its impact on households otherwise devoted to subsistence corn farming.

762 Sault, Nicole L. The evil eye, both hot and dry: gender and generation among the Zapotec of Mexico. (*J. Lat. Am. Lore*, 16:1, Summer 1990, p. 69–89, bibl.)

Convincingly argues that to understand evil-eye beliefs in Mesoamerica one must examine symptoms, cures, preventions, and, importantly, social relations between "assailants" and victims.

763 Schryer, Frans J. Ethnicity and class conflict in rural Mexico. Princeton, N.J.: Princeton Univ. Press, 1990. 363 p.: bibl., ill., index.

Major exploration of complex relationship between class conflict and ethnicity in the Huasteca region is one of the few studies available from Mesoamerica of agrarian revolt in progress. This study of 1970s and 1980s land invasions by Nahua and mestizo residents of the Huasteca raises issues concerning class, ethnicity, and agrarianism that are of broad importance for Mesoamerican ethnology and history.

764 Sierra, María Teresa. Discurso, cultura y poder: el ejercicio de la autoridad en los pueblos hñähñús del Valle del Mazquital. México: Centro de Investigación y Estudios Superiores en Antropología Social; Pachuca, Mexico: Hidalgo, Gobierno del Estado, Archivo General del Estado, 1992. 281 p.: bibl.

Although devoted to sociolinguistic analysis of discursive practices in Otomi village assemblies and courts, work provides much useful ethnographic information about village stratification, formal and informal leadership, and biographical backgrounds of Otomi leaders.

765 Stoll, David. Between two armies in the Ixil towns of Guatemala. New York: Columbia Univ. Press, 1993. 383 p.: bibl., index, maps, photos, tables.

Major assessment of consequences of counterinsurgency and process of reconstruction in the Ixil Maya regions of Guatemala based on 1987–92 fieldwork. Documents changes in religious affiliation, land tenure, authority, demographics, and more. Provides a provocative, well-argued reconsideration of the nature of civil conflict in the region in 1980s and the current circumstances and interests of Mayas and Ladinos. Vitally important contribution to study of Guatemala, and to comparative research on conflict and its consequences in Mesoamerica.

766 **Stonich, Susan C.** "I am destroying the land!": the political ecology of poverty and environmental destruction in Honduras. Boulder, Colo.: Westview Press, 1993. 191 p.: bibl., ill. (Conflict and social change series)

Outstanding ethnography of southern Honduras focuses on political economy and human ecology of rural poverty and environmental degradation. Depth, lucidity, and sophisticated melding of diverse perspectives on poverty and environmental destruction make this book one of the best of its kind concerning Mexico and Central America. For sociologist's comment see item **4674**.

Suco Campos, Idalberto. La música en el complejo cultural del *walagallo* en Nicaragua. See *HLAS 54:5197*.

767 **Tyrtania, Leonardo.** Yagavila: un ensayo en ecología cultural. México: Univ. Autónoma Metropolitana, Unidad Iztapalapa, División de Ciencias Sociales y Humanidades, 1992. 332 p.: bibl., ill., maps. (Texto y contexto; 8)

Very well-written, solid cultural-ecological study of Zapotec community focuses on ecological disequilibrium occasioned by region's integration into national and international markets.

768 **Vogt, Evon Zartman.** Fieldwork among the Maya: reflections on the Harvard Chiapas Project. Albuquerque: Univ. of New Mexico Press, 1994. 451 p.: bibl., ill., index.

Memoir is first book-length history of Harvard Chiapas Project, one of the most important Mesoamerican Indian research projects undertaken by American anthropologists. Surveys project's changing goals and practices from 1957–80s, and provides a thorough orientation to project's principal personalities and publications. A major contribution to historiography of American anthropology, and an essential introduction to ethnography of Maya area.

769 **Warren, Kay B.** Interpreting *La Violencia* in Guatemala: shapes of Mayan silence and resistance. (*in* The violence within: cultural and political opposition in divided nations. Boulder, Colo.: Westview Press, 1993, p. 25–56, bibl.)

Explores how Mayas talk about the violence that engulfed them in 1970s and 1980s. Offers important and timely observations about cultural construction of terror, and raises issues concerning ethnicity, ethnic revivalism, and ethnic nationalism that are increasingly coming to the fore in Mesoamerican studies. See also items **770** and **771**.

770 **Warren, Kay B.** Transforming memories and histories: the meanings of ethnic resurgence for Mayan Indians. (*in* Americas: new interpretive essays. New York: Oxford Univ. Press, 1992, p. 189–219)

Short essay offers insightful commentary on the rise of Mayan ethnic nationalism in Guatemala. See also items **769** and **771**.

771 **Wilson, Richard.** Anchored communities: identity and history of the Maya-Q'eqchi'. (*Man/London*, 28:1, March 1993, p. 121–138, bibl.)

Explores revival of Mayan ethnicity in the wake of recent Guatemalan civil war. Important work sheds light on early stages of a seemingly more widespread emergence of ethnic revival and nationalism. See also items **769** and **770**.

772 **Zapotec struggles: histories, politics, and representations from Juchitán, Oaxaca.** Edited by Howard Campbell et al. Poetry translated by Nathaniel Tarn. Washington: Smithsonian Institution Press, 1993. 343 p.: bibl., ill., index. (Smithsonian series in ethnographic inquiry)

Unusual collection of approximately 40 articles, poems, songs, speeches, and anecdotes written by Zapotec people of Juchitán, Mexico, and by the anthropologists and historians who have studied them. Selections view from diverse angles the history, nature, successes, and failures of the Coalición Obrera Campesina Estudiantil del Istmo (COCEI), one of the most significant indigenous movements in Mesoamerica. Work is of considerable value for scholars and for use in undergraduate instruction.

773 **Zimmermann, Klaus.** Sprachkontakt, ethnische Identität und Identitätsbeschädigung: Aspekte der Assimilation der Otomí-Indianer an die hispanophone mexikanische Kultur. Frankfurt am Main, Germany: Vervuert, 1992. 500 p.: bibl., ill., maps. (Bibliotheca Ibero-Americana; 41)

Major contribution to the study of problems of linguistic acculturation (or assimilation) in modern Spanish America.

Based on terms of ethnic identity and language contact, Zimmermann's study focuses on the so-called "identity damnation" among the Otomí Indians of the Mezquital valley in central Mexico. Exemplary research notable for its theoretical rigor and methodological sophistication. [T. Hampe-Martínez]

West Indies

LAMBROS COMITAS, *Gardner Cowles Professor of Anthropology and Education, Teachers College, Columbia University and Director, Research Institute for the Study of Man*

THIS SECTION IS DESIGNED to include publications in sociocultural anthropology dealing with the Caribbean archipelago, the Guianas, Belize, and the several West Indian cultural enclaves located in other parts of the Caribbean mainland. In this issue, roughly two-thirds of the publications annotated deal with 20 countries or dependencies: Antigua, Barbados, Barbuda, Costa Rica, Cuba, Curaçao, Dominica, the Dominican Republic, French Guiana, Guadeloupe, Guyana, Haiti, Jamaica, Martinique, Nevis, Puerto Rico, St. Lucia, St. Vincent, Suriname, and Trinidad and Tobago. The remaining publications deal with the Caribbean in general or intra- or inter-regional comparisons of one sort or another. The territories or units receiving the most anthropological attention were, in order: the Caribbean in general, Trinidad, and French Guiana. For the reader's convenience, I list below, in several broad categories, those items annotated that deal with the most representative subjects or orientations.

I. DIACHRONIC AND HISTORICAL STUDIES

As I indicated in *HLAS 53*, Caribbean anthropology has long been ambivalent in its use of historical perspective. Nonetheless, as reflected in the number of publications with diachronic dimensions cited in that last issue, the interest in and value of history for anthropologists appeared to have grown considerably. This tendency has persisted during this report period. See, for example, Baker on the ethnohistory of Dominica (item **777**), Dreyfus on native political networks in western Guiana (see *HLAS 54:1901*), González on ethnic identity and inter-ethnic relations during and after the Carib War (item **796**), Lazarus-Black on law and society in Antigua and Barbuda (item **812**), and Olwig on the development of Nevisian cultural identity (item **821**). Anthropological contributions to the understanding of slavery also continue apace. For these materials, see Binder's collection on slavery in the Americas (item **836**), Hoogbergen on marronage and slave rebellions in Suriname (item **803**), Jamard's intra-regional comparison of slavery (item **807**), Mintz on food and eating habits of slaves (item **818**), Mörner on patterns of social stratification in the 18th and 19th centuries (item **819**), Palmie on ethnogenetic processes in Afro-American slave populations (item **824**), and the reedition of Rubin and Tuden's work on comparative perspectives of slavery in New World plantation societies (item **784**).

II. SYNCHRONIC STUDIES

a) Ethnicity and Identity. A great deal of the contemporary research on the Caribbean seems to be focused on questions related to ethnicity and identity. In addition to the studies of González and Olwig listed above, publications dealing with these themes from a more synchronic perspective include the following on the Dominican Republic: Davis on music and black ethnicity (item **789**), Douany on ethnicity, identity and the merengue (item **790**), and Nyberg-Sorensen on Creole culture and Dominican identity (item **839**). For publications with a Trinidadian focus, see Birth on coup, carnival, and calypso (item **781**), Eriksen on multiple traditions and cultural integration and on ethnicity and nationalism (items **793** and **794**), Gosine on the East Indian odyssey (item **791**), Houk on the Africanization of the Orisha tradition (item **804**), Khan on food pollution and hierarchy, and on what is a "Spanish" (items **808** and **809**), the Premdas collection on the enigma of ethnicity (item **792**), Sampath on the creolisation of East Indian adolescent masculinity (item **829**), Segal on race and color in pre-independent Trinidad (item **834**), Vertovec on Hindu Trinidad (item **845**), and Yelvington's edited work entitled *Trinidad ethnicity* (item **844**), together with his own two contributions to that volume (items **848** and **849**). For other contributions to these themes, see Lefever (item **813**) and Purcell (item **826**) on West Indians in Costa Rica, Spencer-Strachan on problems of self-identity among diasporic Africans (item **840**), M.G. Smith on theoretical aspects of race and ethnicity (item **837**), and Young on becoming a West Indian in St. Vincent (item **850**).

b) Maroon Culture and Society. Mainland research of this genre remains active. For instance, see Bilby *et al.* on vocabulary related to food and its usage among the Boni and Djuka (item **780**), Bruleaux on descriptions of native food resources in French Guiana (item **806**), Groot *et al.* on Aluku/Boni history (item **798**), Hurault on material culture and art styles of the Boni, Djuka, and Saramaka (item **805**), and Price and Price's diary of an ethnographic expedition to collect maroon artifacts (item **825**).

c) Gender Relations and Women's Studies. See Abraham on industrialization and female-headed households in Curaçao (item **774**), D'Amico on a reconsideration of female-headed households in Jamaica (item **788**), Greene on race, class, and gender in the future of the Caribbean (item **827**), Handwerker on empowerment and fertility transition in Antigua (item **800**), Schnepel on language and gender in the French Caribbean (item **833**), and Sobo on health, sickness, and gender relations among the Jamaican poor (item **838**).

d) Rural Studies/Peasantry. See Alvarado Ramos on rural settlement types in Cuba (item **776**), Crichlow on family land tenure (item **786**), Griffith *et al.* on proletarianization in Puerto Rican fisheries (item **797**), LeFranc on land tenure in St. Lucia, and on a small farming village in Dominica (items **814** and **815**), and Wylie's comparison of crises of glut in the Faroe Islands and Dominica (item **847**).

e) Language and Society. See Cooper on orality and gender in Jamaican popular culture (item **785**), Schieffelin and Doucet on Haitian Creole (item **830**), and Schnepel on the Creole movement and East Indians in Guadeloupe (items **831** and **832**).

f) Religion. Five books or collections make up this category: see Brandon on santería (item **783**), Murphy on working the spirit (item **820**), the Simposio Internacional on ancestor cults (item **835**), Glazier on African-derived religions in the Caribbean (item **841**), and Yelvington on *Traditional spirituality in the African diaspora* (item **843**).

g) Reviews and assessments. I include here Guanche Pérez and Campos Mitjans on Cuban cultural anthropology in the 20th century (item **799**), Kimber's geographical review of aboriginal and peasant cultures (item **810**), Olwig on Danish scholarship on the West Indies (item **822**), and Oriol's appreciation of the anthropological work on Haiti by Louis Price Mars (item **823**).

774 **Abraham, Eva.** Caught in the shift: the impact of industrialization on female-headed households in Curaçao, Netherlands Antilles. (*in* Where did all the men go? Female-headed/female-supported households in cross-cultural perspective. Edited by Joan P. Mencher and Anne Okongwu. Boulder, Colo.: Westview Press, 1993, p. 89–106, bibl., tables)

Changes in the social position of women (specifically as reflected in marriage rates and percentages of children born to unmarried mothers) are linked to major changes in the island's economy.

Agorsah, Emanuel Kofi. Archaeology and resistance history in the Caribbean. See item **479**.

775 **Allen, Rose Mary.** *Muzik di ingles tambe a bira di nos:* an overview of the Calypso on Curaçao in the period of its popularity. Curaçao: Archaeological-Anthropological Institute of the Netherlands Antilles, 1988. 60 p.: ill. (Report of the Institute of Archaeology and Anthropology of the Netherlands Antilles; 8)

Preliminary study of calypso on Curaçao by English-speaking West Indians. With lyrics in English Creole and Papiamentu, these calypsos, which had great success in the 1960s and 1970s, deal primarily with male-female relationships and difficult social situations. Local calypsonians are identified and calypso lyrics appended.

776 **Alvarado Ramos, Juan Antonio.** Algunos criterios para la clasificación etnográfica de los asentamientos rurales en la actualidad. (*Anu. Etnol./Habana*, 1988, p. 67–82, bibl.)

Basing study on classificatory scheme developed by Soviet ethnographer Vitov and utilizing data from 1981 Cuban census and field research, author delineates some aspects related to the definition of rural settlement types in present-day Cuba, an important theme in the organization of the *Ethnographic Atlas of Cuba*. Two major categories, the dispersed settlement and concentrated settlements, are considered with particular attention given to new types developed since 1959.

Amodio, Emanuele. Relaciones interétnicas en el Caribe indígena: una reconstrucción a partir de los primeros testimonios europeos. See item **483**.

777 **Baker, Patrick L.** Centring the periphery: chaos, order, and the ethnohistory of Dominica. Jamaica: The Press, Univ. of the West Indies, 1994. 251 p.: bibl., ill., index, maps.

Detailed account from precolumbian times to present utilizing the metaphor of center and periphery, "an attractor creating and re-creating order and chaos," as conceptual device for organizing a history that has not had a smooth linear progression. Of greater anthropological relevance are chapters on the peasantry, the mulatto elite, and capitalizing a subsistence economy.

778 **Basch, Linda G.; Nina Glick Schiller; and Cristina Szanton Blanc.** Nations unbound: transnational projects, postcolonial predicaments, and deterritorialized nation-states. Langhorne, Pa.: Gordon and Breach, 1993. 344 p.: bibl., index.

Concept of transnationalism ("the processes by which immigrants forge and sustain multi-stranded social relations that link together their societies of origin and settlement") is explored and considered. Three case studies of migration from St. Vincent, Grenada, and Haiti to the US, as well as a comparative study of transnational migration of Filipinos and Caribbean people to the US, provide ethnographic bases for the propositions put forth.

779 **Baud, Michiel et al.** Etniciteit als strategie in Latijns-Amerika en de Caraïben. Amsterdam: Amsterdam Univ. Press, 1994. 152 p.: bibl., index.

Argues that ethnicity is not simply a historical or social fact; rather, ethnicity often is a strategy, deliberately chosen by different groups to reach certain goals. Shows that in Latin American and Caribbean history the ideas about ethnicity and ethnic groups changed due to such major developments as European colonization, the building of the nation-state, and migration. [R. Hoefte]

780 **Bilby, Kenneth et al.** L'Alimentation des noirs marrons du Maroni: vocabulaire, pratiques, représentations. Cayenne: Institut français de recherche scientifique pour le développement en coopération, Centre ORSTOM de Cayenne, 1989. 393 leaves: bibl., ill.

Useful inventory of vocabulary related to food and food usage among the Aluku (Boni) and Djuka of French Guiana. Listed alphabetically, each item includes, as appropriate, linguistic, botanical, zoological, medicinal, and ritual data as well as notes on preparation.

781   **Birth, Kevin K.** Bakrnal: coup, carnival, and calypso in Trinidad. (*Ethnology/Pittsburgh*, 33:2, Spring 1994, p. 165–177, bibl.)

Young village males shared a cultural model of the attempted coup d'etat of July 1990, one that held the event to be a threat to their cultural construction of freedom. The coup became a dominant calypso theme during the Carnival that followed. Describes the creation of the villagers' cultural model and the way that Carnival participation, specifically the interaction between Carnival audience and performance, caused the model to change. The new model depicts Trinidad suffering periodic conflicts in which freedom and humor triumph over political repression and fear.

782   **Brana-Shute, Gary** An inside-out insurgency: the Tukuyana Amazones of Suriname. (*in* Size and survival: the politics of security in the Caribbean and the Pacific. Edited by Paul Sutton and Anthony Payne. London: Frank Cass, 1993, p. 54–69.)

Study of mobilization based on Carib ethnicity in guerrilla movement designed to secure aid, resources, territory, and recognition in a society that has ignored these people. Argues that Carib behavior can only be understood in relation to the situation of other ethnic groups in Suriname. [R. Hoefte]

783   **Brandon, George.** *Santería* from Africa to the New World: the dead sell memories. Bloomington: Indiana Univ. Press, 1993. 206 p.: bibl., ill., index, maps. (Blacks in the diaspora)

Guided by anthropological concepts and field work data from Ghana, Cuba, and New York, author relies heavily on historical sources and a processual framework for analyzing evolution of *santería*. Bracketed by chapters on African origins and contemporary patterns in the US are two chapters of specific interest to Caribbeanists: pre-*santería* and early *santería* in Cuba from 1492–1870 and latter-day *santería* in Cuba from 1870 to 1959.

784   **Comparative perspectives on slavery in New World plantation societies.** Edited by Vera Rubin and Arthur Tuden. New York: New York Academy of Sciences, 1993. 703 p.: bibl., ills., index.

This reprint of the 1977 publication, the New York Academy of Sciences' contribution to the Quincentenary, includes a new foreword by the anthropologist, Faye V. Harrison. Based on a 1976 multidisciplinary and multinational conference on slavery which brought together the leading scholars on the topic, this landmark publication includes substantial contributions from anthropology, not just from its two editors and conference organizers, but also from Richard Frucht, Angelina Pollak-Eltz, Luz María Martínez Montiel, Silvia de Groot, Richard Price, and Sidney Greenfield.

785   **Cooper, Carolyn.** Noises in the blood: orality, gender, and the "vulgar" body of Jamaican popular culture. Durham, N.C.: Duke University Press, 1995. 214p.: bibl., index.

Stimulating study of Jamaican as language and its contribution to Jamaican cultural life, of obvious value to anthropologists. Examines word-culture as exemplified in the work of local writers and performers, including that of Louise Bennett, Jean Binta Breeze, Mikey Smith, the Sistren Theatre Collective, Michael Thelwell, Bob Marley, as well as the "erotic play in the dancehall" of Jamaican DJs.

786   **Crichlow, Michaeline A.** An alternative approach to family land tenure in the Anglophone Caribbean: the case of St. Lucia. (*Nieuwe West-Indische Gids*, 64:1/2, 1994, p. 77–99, bibl., tables)

Argues against "more popular approach" taken by analysts treating family land as institutionally separated with definite and fixed characteristics. Considers family land part of the "small holder sector" which reflects the problems of that sector. Therefore, distinction between legal and supposedly non-legal forms of tenure often found in the literature need reconsideration. Finally, economic pressures have, in fact, led to the sale of family land with consequent declines in agricultural production and quality of life.

787   **Cross, Malcolm.** Ethnic pluralism and racial inequality: a comparison of colonial and industrial societies. Utrecht, The Netherlands: ISOR, 1994. 333 p.: appendix, bibl., graphs, tables.

First half of study focuses on ethnic relations in Guyana and Trinidad. Last half examines Caribbean migrants and ethnic relations in Dutch and British societies. Argues that ethnic differentiation and racial inequality have to be understood as historically grounded interactions which are heavily influenced by economic and political factors. Includes a critique of the plural society theory and revisits post-industrial theories. [R. Hoefte]

788 **D'Amico, Deborah.** A way out of no way: female-headed households in Jamaica reconsidered. (*in* Where did all the men go? Female-headed/female supported households in cross-cultural perspective. Edited by Joan P. Mencher and Anne Okongwu. Boulder, Colo.: Westview Press, 1993, p. 71–88, bibl.)

Reviews Caribbean family literature and discusses some of its "major tendencies." Concludes that female researchers "can begin to alter the destructive uses to which social science analyses of female-headed households have been party, by ensuring that our work contributes, in its process and product, to the empowerment of poor women and those with whom they share their lives."

789 **Davis, Martha Ellen.** Music and black ethnicity in the Dominican Republic. (*in* Music and black ethnicity: the Caribbean and South America. Edited by Gerard H. Béhague. New Brunswick, NJ: Transaction Publishers, 1994, p. 119–155, bibl.)

Excellent analysis/review of music and its relationship to Dominican identity and ethnicity. Includes informative sections on traditional Afro-Dominican musical culture, Dominican musical genres as symbols of ethnic, class, rural/urban, and pan-regional identity, and the merengue as ethnic marker.

790 **Douany, Jorge.** Ethnicity, identity, and music: an anthropological analysis of the Dominican merengue. (*in* Music and black ethnicity: the Caribbean and South America. Edited by Gerard H. Béhague. New Brunswick, NJ: Transaction Publishers, 1994, p. 65–90, bibl.)

Significance of the Afro-Caribbean merengue and its rise as the most popular music in the Dominican Republic is examined in light of ethnic relations and the emergence of Dominican identity. Argues that merengue "synthesizes" many features of this identity, and embodies Creole beliefs and customs in contrast to Haitian influences.

Dreyfus, Simone. Les Réseaux politiques indigènes en Guyane occidentale et leurs transformations aux XVIIIe siècles. See *HLAS 54: 1901.*

791 **The East Indian odyssey: dilemmas of a migrant people.** Edited by Mahine Gosine. New York: Windsor Press, 1994. 257 p.: bibl.

Collection of 32 papers, some of excellent quality, presented at the multidisciplinary Conference on East Indians (4th, New York, 1988), 19 of which (3 by anthropologists) focus on East Indian-related issues in the Caribbean region or on Caribbean East Indians in the US. For the contributions on Caribbean ethnology, see Schnepel (item **831**) and Mintz (item **842**), as well as Segal's related work (item **834**) in *Trinidad ethnicity.*

792 **The enigma of ethnicity: an analysis of race in the Caribbean and the world.** Edited by Ralph R. Premdas. Special essay on race and ethnicity by M.G. Smith. Foreword by Esmond Ramesar. St. Augustine, Trinidad and Tobago: Univ. of the West Indies, School of Continuing Studies, 1993. 378 p.: bibls.

Collection of 14 articles, seven on the Caribbean. For the contributions on Caribbean ethnology, see items **848, 837,** and **832.**

793 **Eriksen, Thomas Hylland.** Multiple traditions and the question of cultural integration. (*Ethnos*/Stockholm, 57:1/2, p. 5–29, bibl.)

Based on data drawn primarily from East Indians in Trinidad, essay explores an important issue on social anthropology—"the relationship between agency and structure or between holist and individualist orientations in social analysis." In this theoretical and epistemological context, argues that the identity of Indo-Trinidadians is created mainly through "abstract mediating structures," not through face-to-face relations.

794 **Eriksen, Thomas Hylland.** Us and them in modern societies: ethnicity and nationalism in Mauritius, Trinidad and beyond. Foreword by Bruce Kapferer. Oslo: Scandinavian Univ. Press, 1992. 208 p.; bibl., index.

Making use of considerable Trinidadian and Mauritanian ethnographic data, author explores, in interdisciplinary context, theoretical perspectives impinging on ethnicity, nationalism, and modernity. Difficulties of employing these concepts in modern situations are throughly discussed (for example, ethnicities and nations are seen as beset by a dual process of globalization and localization, by simultaneous cultural homogenization and differentiation).

795   **Gmelch, George.** Learning culture: the education of American students in Caribbean villages. (*Hum. Organ.*, 51:3, Fall 1992, p. 245–252, bibl.)

Discussion of the impact of ten weeks of field work in Barbadian rural villages on white undergraduates. Author/director contends that students gained a new awareness of race and social class, learned what it means to be a minority, experienced rural life, and gained knowledge of another culture, all of which provided them with "a more critical perspective on their own culture." In addition, author believes, without benefit of psychological data, that the field work experience had significant characterological benefits for student participants.

796   **González, Nancie L.** Identidad étnica y artificio en los encuentros interétnicos del Caribe. (*in* De palabra y obra en el Nuevo Mundo. Madrid: Siglo Veintiuno Editores, 1992, v. 2, p. 403–428, maps)

Discusses interracial and interethnic relations on St. Vincent in 1795–96 during Carib War and on Central American coast where defeated Caribs were sent in 1797. Focus is on encounter of Amerindians, Africans, and Europeans, the creation of the hybrid Black Carib, their ethnic and racial misidentification by English and French colonists, and the situational nature of Black Carib self-identity past and present.

797   **Griffith, David; Manuel Valdés Pizzini; and Jeffrey C. Johnson.** Injury and therapy: proletarianization in Puerto Rico's fisheries. (*Am. Ethnol.*, 19:1, Feb. 1992, p. 53–73, bibl.)

"Class" and "therapy" are two concepts "appropriated" by Puerto Rican "peasant" fishermen from the formal economy and "adapted" to the "politics and semantics" of their socioeconomic life. Focusing on these two remolded terms, authors explore the conceptual and political consequences of the conditions of fishermen who combine fishing with wage labor.

798   **Groot, Silvia W. de.; Wim S. M. Hoogbergen; and Kenneth Bilby.** Sur les traces de Boni: résumé des communications présentées le 22 avril 1989 à la Chambre de commerce et d'industrie de Cayenne. Cayenne: Conseil régional, 1989. 22 p.

Program for conference on Boni that includes extended abstracts of the three principal papers on Aluku/Boni history and on knowledge transmission.

799   **Guanche Pérez, Jesús and Gertrudis Campos Mitjans.** La antropología cultural en Cuba durante el presente siglo. (*Interciencia/Caracas*, 18:4, July/Aug. 1993, p. 176–183, bibl.)

Brief but informative Cuban perspective on history of Cuban cultural anthropology in the 20th century. Two major periods, each with sub-periods, are identified: "The Neocolonial Republic" (1902–1958) and "Following the triumph of the Cuban Revolution" (1959–1990). The anthropological activity (and methodological character) of the major institutions and representative individuals are detailed.

800   **Handwerker, W. Penn.** Empowerment and fertility transition on Antigua, WI: education, employment, and the moral economy of childbearing. (*Hum. Organ.*, 52:1, Spring 1993, p. 41–52, bibl.)

Report from a long-term, comparative study of gender relations on Barbados, Antigua, and St. Lucia. Presents a sophisticated argument based on the notion that "education by itself had almost no impact on Antigua's decline to replacement-level fertility, which is explained by the conjuction of new educational and employment opportunities for women."

801   **Ho, Christine G.T.** The internationalization of kinship and the feminization of Caribbean migration: the case of Afro-Trinidadian immigrants in Los Angeles. (*Hum. Organ.*, 52:1, Spring 1993, p. 32–40, bibl., ill.)

Argues that modern migration scholars confront the effects of a process of "globalization" (e.g., the emergence of "international families" linked to trends such as circular migration flows and the importance of kinship and primacy of women in these flows). To deal with these emerging phenomena, author favors network analyses as a more "fluid" method than household-based studies.

802 **Hoffmann, Léon-François.** Histoire, mythe et idéologie: la cérémonie du Bois-Caiman. (*Etud. créoles*, 13:1, 1990, p. 9–34, bibl.)

Rejects interpretation of Bois Caiman ceremony as historical event and argues that it is a myth whose origin is imputed to the malevolence of a "Français de Saint-Dominique." Traces development of myth and indicates its utility for illustrating cleavages in Haitian society.

803 **Hoogbergen, Wim S.M.** Marronage and slave rebellions in Surinam. (*in* Slavery in the Americas. Edited by Wolfgang Binder. Würzburg: Königshausen and Neumann, 1993, p. 165–195, bibl.)

Rebellion was an ineffective form of slavery resistance; marronage was far more successful method for escaping slavery. Nonetheless, close links existed between marronage and successful slave rebellions in Suriname; a slave uprising could hope to succeed only if Maroons were somehow involved. Several cases of revolts that had varying success are analyzed and a list of all plantation revolts in colonial Suriname is appended.

804 **Houk, James.** Afro-Trinidadian identity and the Africanisation of the Orisha religion. (*in* Trinidad ethnicity. Edited by Kevin Yelvington. Knoxville: Univ. of Tennessee Press, 1993, p. 161–179, bibl., ill.)

Orisha, an eclectic and syncretic Yoruba-derived religion, has over time taken on not only Catholic, Protestant, and Hindu elements, but also incorporated East Indian members. It is argued that this relatively recent influx of non-Africans has resulted in a concerted attempt to revitalize the religion by expurgating all non-African derived components.

805 **Hurault, Jean.** Africains de Guyane: la vie matérielle et l'art des noirs réfugiés de Guyane. Cayenne: Editions Guyane presse diffusion, 1989. 232 p.: bibl., ill.

Slightly modified second edition of a 1970 publication. This precisely detailed and beautifully illustrated volume, based on data from eight field expeditions to French Guiana (1948–65), deals with aspects of social structure, material culture, and art styles among the Boni, Djuka, and Saramaka of the Maroni.

806 **Images à croquer: l'alimentation guyanaise à travers l'iconographie ancienne.** Catalogue réalisé par Anne-Marie Bruleaux et Véronique Defrance *et al.* Cayenne: Archives départementales; Musée départemental, 1990. 113 p.: bibl., ill.

Exhibit catalog describing the native animals, fish, fruits, and vegetables underpinning French Guianese cuisine as well as traditional hunting and fishing techniques. Text is illustrated with 18th- and 19th-century drawings.

807 **Jamard, Jean-Luc.** Consommation d'esclaves et production de "races": l'expérience caraïbéenne. (*Homme/Paris*, 122/124, avril/déc. 1992, p. 209–234, bibl.)

Intra-regional comparison of types of slavery, changes to the system of slavery, abolition, and lasting effects, based on a model of slavery generated by author. Creates "transformist" analysis of the dynamics of social "races" and classes, and their consequences in Caribbean society.

Keegan, William F. The people who discovered Columbus: the prehistory of the Bahamas. See *HLAS* 54:1819.

808 **Khan, Aisha.** *Juthaa* in Trinidad: food, pollution, and hierarchy in a Caribbean diaspora community. (*Am. Ethnol.* 21:2, May 1994, p. 245–269, bibl.)

Examination of the concept of *juthaa* (food and drink "polluted" by being partially consumed), an element of the cultural repertoire of Trinidadian East Indians, sheds light on "the larger question of how indigenous traditions are invested with diverse meanings through which they gain significance and function under new conditions." This concept, although caste derived, has egalitarian connotations in Trinidad.

809 **Khan, Aisha.** What is "a Spanish?": ambiguity and "mixed" ethnicity in Trinidad. (*in* Trinidad ethnicity. Edited by Kevin A. Yelvington. Knoxville: Univ. of Tennessee Press, 1993, p. 180–207, bibl.)

Focusing on "Spanish," a specific manifestation of the Trinidadian category "mixed" (the latter considered by the author to be an overarching rubric for glossing ethnic or racial combinations), author deals with the importance and relevance of the existence of ambiguity for defining, maintaining, or resisting hierarchy in sharply stratified social systems. Includes historical as well as contemporary sketches of "Spanish" in Trinidad.

810   **Kimber, Clarissa.** Aboriginal and peasant cultures of the Caribbean. (*Yearbook/CLAG*, 17/18, 1990, p. 153–163, bibl.)

Geographer's terse review of post-1980 studies on aboriginal Caribbean cultures as well as contemporary Caribbean peasantries, and an anticipation of future research directions dealing with the latter.

811   **Kulakova, N.N.** Gaĭtiĭstsy: formirovaniė etnosa (kolonial'nasià epokha). Moskva: Rossiĭskasia akademisia nauk, In-t etnologii i antropologii im. N.N. Miklukho-Maklasia, 1993. 169 p.: bibl., map.

Writing for the Miklukho-Maklaia Institute of Ethnology and Anthropology of the Russian Academy of Sciences, the author traces the ethnogenesis of Haitians from the beginning of French colonization in the 1630s to independence in 1804. Introduces new historical ethnographic approaches, material based on colonial sources, and colonial period literature. Addresses social, economic, cultural, and religious developments. Includes tables and extensive references to Russian and Western sources. [B. Dash]

812   **Lazarus-Black, Mindie.** Legitimate acts and illegal encounters: law and society in Antigua and Barbuda. Washington: Smithsonian Institution Press, 1994. 357 p.: bibl., index. (Smithsonian series in ethnographic inquiry)

Provides informative and interesting exercise in historical anthropology, a diachronic exploration into kinship, class, and gender in colonial and post-colonial society as these "relate dialectically to systems of legalities and illegalities." Study provides framework for dealing with "class formation, family ideology and structure, and gender hierarchy within the wider contexts of slavery, post-emancipation society, and independence." Recommended reading.

813   **Lefever, Harry G.** Turtle Bogue: Afro-Caribbean life and culture in a Costa Rican village. Selinsgrove, Pa.: Susquehanna Univ. Press, 1992. 249 p.: bibl., ill., index, maps.

Study of Tortuguero, a "second step" African diaspora community whose residents trace ancestry to African slaves from the eastern Caribbean and their descendants who migrated to the western Caribbean and/or the east coast of Central America during the 19th and 20th centuries. Useful descriptions of the development and dynamics of Creole culture and social structure in northeastern Costa Rica, and of issues related to ethnic identity.

814   **LeFranc, Elsie.** Rural land tenure systems in St. Lucia. Mona, Jamaica: Univ. of the West Indies, Institute of Social and Economic Research, 1993. 92 p.: bibl., tables. (ISER working paper, 40)

Long delayed appearance of field work report completed in 1975 and submitted for publication in 1982. Four-month study of St. Lucian small-farm economy focuses on the relationships between farm family, land tenure, and production system. Introduction deals with the historical traditions which shaped the small-farm community of 1975 and the ties binding it to the urban sector.

815   **LeFranc, Elsie.** Status group formation in small communities: a case study of a Dominican small-farming village. Mona, Jamaica: Institute of Social and Economic Research, Univ. of the West Indies, 1993. 106 p.: bibl., ill. (Working paper; 39)

Another unfortunately delayed publication of field research. With issues revolving around the nature of Caribbean peasantry as context, traces village income, property, and kinship and how these relate to status group formation and persistence. Descriptions and analyses of social networks and property relations are particularly useful.

816   **L'Etang, Thierry.** Mythes et croyances de la mer. (*Caribena/Martinique*, 1, 1991, p. 83–104, bibl., photo, ill.)

Amerindian as well as current Antillean myths related to the sea. Particular attention is given to the distribution of the myth "Manman D" Lo and its origins, both Amerindian and African.

817   **Magaña, Edmundo.** El sacerdote caníbal: una visión kaliña de los misioneros. (*in* De palabra y obra en el Nuevo Mundo. Madrid: Siglo Veintiuno de España Editores, 1992, v. 1, p. 143–164, facsims.)

Study of the Kaliña representation of Europeans (hombres del mar). It is difficult to reconstruct this representation because the available information is fragmentary and Europeans rarely figure in myths. Europeans are generally represented as cannibals or are associated with water. Argues that Europeans have not radically altered the system of representation of "the other" among the peoples of the Guianas. [R. Hoefte]

818 **Mintz, Sidney W.** Tasting food, tasting freedom. (*in* Slavery in the Americas. Edited by Wolfgang Binder. Würzberg: Königshausen and Neumann, 1993, p. 257–275, bibl.)

Food and eating habits of slaves and their descendants in the Caribbean are described and analyzed in anthropological context with specific reference to origins, production, processing, and distribution of foods, as well as the emergence of cuisine.

819 **Mörner, Magnus.** Patterns of social stratification in the 18th- and 19th-century Caribbean: some comparative clarifications. (*Plant. Soc. Am.*, 3:2, 1993, p. 1–30, bibl., tables)

Of value not only to historians but to any Caribbeanist concerned with the development of contemporary Caribbean society, this article illuminates the methodological issues and difficulties inherent in the objective comparison of post-abolition stratification patterns in the region. Case material drawn from Saint-Domingue/Haiti, Martinique, Barbados, Jamaica, Cuba, and Puerto Rico.

820 **Murphy, Joseph M.** Working the spirit: ceremonies of the African diaspora. Boston: Beacon Press, 1994. 263 p.; bibl., index.

Eminently readable, well argued treatise on how religions of the African diaspora share a kindred sprituality drawn from an African past; how they "work the spirit" despite having evolved differently in the New World given differing historical circumstances. Separate chapters are devoted to Haitian voodoo, Brazilian Candomblé, Cuban and Cuban-American *santería*, Jamaican Revival Zion, and the Black church in the US.

821 **Olwig, Karen Fog.** Global culture, island identity: continuity and change in the Afro-Caribbean community of Nevis. Switzerland; Philadelphia, Pa.: Harwood Academic Publishers, 1993. 239 p.: bibl., ill., index. (Studies in anthropology and history, 1055–2464; 8)

Development of Nevisian cultural identity is viewed in historical anthropological perspective. Traces interplay of English conceptions of patriarchy and African conceptions of belonging during the formative period which enabled slave populations to assert a social presence in that colonial society; a succeeding period characterized by an English framework of respectability that offered the subordinated segments of the society opportunities for seeking social recognition before and even more so after Emancipation; and, finally, the unravelling of a territorially localized Nevisian society, massive emigration, the development of a transnational community, and the cultural implications of the process.

822 **Olwig, Karen Fog.** West Indian research in Denmark. (*Plant. Soc. Am.*, 3:2, 1993, p. 51–62, bibl.)

Review of Danish scholarship on the West Indies (mainly historical in nature and focused on their former colonies) identifies three phases: a national phase concerned with the role of Denmark as a colonial power; an international phase concerned with placing the research within a broader academic framework; and a West Indian phase, in which the Danish colonies are reexamined within a holistic Caribbean context. Anthropological perspectives during the latter phase are noted.

823 **Oriol, Jacques.** L'apport du Dr. Louis Price Mars a l'anthropologie socialeet culturelle haitienne. (*Bull. Bur. natl. ethnol.*, 1/2, 1986, p. 15–33, bibl.)

Discusses contributions of Louis Price-Mars (psychiatrist, ethnopsychiatrist, co-founder of the Haitian Institute of Ethnology, and son of celebrated Jean Price-Mars) to Haitian sociocultural anthropology. Bibliography of his work appended.

824 **Palmie, Stephan.** Ethnogenetic processes and cultural transfer in Afro-American slave populations. (*in* Slavery in the Americas. Edited by Wolfgang Binder. Würzburg: Königshausen and Neumann, 1993, p. 337–363, bibl.)

A model of transitional ethnogenetic processes for dealing with the development of

Afro-American cultures is proposed as a heuristic device complementing rather than contradicting the rapid early creolization model put forth by Mintz and Price.

825 **Price, Richard** and **Sally Price.** Equatoria. Sketches by Sally Price. New York: Routledge, 1992. 295 p.: bibl., ill., maps.

Combined diary of a one-month ethnographic expedition commissioned by French Guianese authorities to collect artifacts illustrative of Maroon life and material culture for a proposed Musée de l'Homme Guyanais. Interesting insights, forays, and asides by two well-known Maroon specialists on contemporary currents in anthropology, and on museology issues and approaches to the dissemination of knowledge.

826 **Purcell, Trevor W.** Banana fallout: class, color, and culture among West Indians in Costa Rica. Foreword by R.S. Bryce-Laporte. Los Angeles: Center for Afro-American Studies, Univ. of California, 1993. 197 p.: bibl., ill., index. (Afro-American culture and society, 0882–5297; 12)

Significant contribution to the understanding of the "adjustment of Afro-Costa Ricans, from their arrival as recruited migrant labor to their present position as an integral, but only partially accepted, ethnic minority." Provides theoretically sophisticated examination of the intertwined sociocultural factors that created inequality and dependency.

827 **Race, class & gender in the future of the Caribbean.** Edited by John Edward Greene. Mona, Jamaica: Institute of Social & Economic Research, Univ. of the West Indies, 1993. 138 p.: bibl., ill.

Collection of seven papers by non-anthropologists on topics of importance to anthropologists interested in the anglophone Caribbean. Four "state-of-the-art" reports provide theoretical context for empirical studies for a major research project on the Future of the Caribbean and are recommended reading: J. Edward Greene on race, class, and gender in the future of the Caribbean; J.G. LaGuerre on race and class; Rhoda Reddock on primacy of gender in race and class; and Hermione McKenzie on family, class, and ethnicity.

828 **Regards sur l'art boni aujourd'hui: Bureau du patrimoine ethnologique, Association Mi Wani Sabi, 22 avril – 13 mai 1989.** Exposition et catalogue réalisés par Marie-Paule Jean-Louis, avec le concours de Philippe Darcissac et Hugues Delorme. Cayenne: Conseil régional, 1989. 37 p.: ill. (some col.).

Catalogue of exposition on contemporary Boni art given in association with a conference on Boni history, society, and culture. See also Groot *et al.* (item *798*).

829 **Sampath, Niels M.** An evaluation of the "creolisation" of Trinidad East Indian adolescent masculinity. (*in* Trinidad ethnicity. Edited by Kevin Yelvington. Knoxville: Univ. of Tennessee Press, 1993, p. 235–253, bibl.)

Ethnographic study of how the complex interplay of village notions of creolization, adolescence, and masculinity lead to transformations of cultural identity at the local level.

830 **Schieffelin, Bambi B.** and **Rachelle Charlier Doucet.** The "real" Haitian Creole: ideology, metalinguistics, and orthographic choice. (*Am. Ethnol.*, 21:1, Feb. 1994, p. 176–200, bibl.)

"Competing representations of *kreyòl* and the symbolic importance of decisions taken in standardizing a *kreyòl* orthography" provides an interesting format for analysis of the role of language and the implications of orthographic debates in the forming of Haitian identity and in the vexed discourse about Haitianness.

831 **Schnepel, Ellen M.** The Creole movement and East Indians on the island of Guadeloupe, French West Indies. (*in* The East Indian odyssey: dilemmas of a migrant people. Edited by Mahin Gosine. New York: Windsor Press, 1994, p. 113–116)

Analysis of the movement to promote the Creole language and the relevance of this movement for the sociopolitical integration of East Indians in Guadeloupe in tandem with an examination of the political implications of the rise of *indianité* and the construction of an Indian identity on that island.

832 **Schnepel, Ellen M.** The Creole movement and its significance for the sociopolitical integration of East Indians on the island of Guadeloupe, French West Indies. (*in* The enigma of ethnicity: an analysis of race in the Caribbean and the wider world. Edited by Ralph R. Premdas. St. Augustine, Trinidad: Univ. of the West Indies, School of Continuing Studies, 1993, p. 197–220, bibl.)

The creole movement is defined as sociopolitical in nature with ethnocultural goals focused on the promotion, defense, and development of the Creole language. Article deals directly with East Indian reactions and relations to this movement. In this context, construction of an Indian identity (*indianité*) and the upswing in interest about India is seen as a response "to the politicalization and racialization of the cultural question in the quest to create an authentic Guadeloupean identity in the 1980s."

833 **Schnepel, Ellen M.** The other tongue, the other voice: language and gender in the French Caribbean. (*Ethnic Groups/New York*, 10:4, 1993, p. 243–268, bibl.)

Contributes empirical data from Martinique and Guadeloupe dealing with issues of linguistic duality, gender in language evaluation, orality and literacy in Creole, and the emerging Creole language movement in the region. Suggests areas for further research.

834 **Segal, Daniel A.** "Race" and "colour" in pre-independence Trinidad and Tobago. (*in* Trinidad ethnicity. Edited by Kevin A. Yelvington. Knoxville: Univ. of Tennessee Press, 1993, p. 81–115, bibl., tables)

Stimulating exploration of the semiotics of Trinidadian race and color terms used for the half-century before 1962 and useful suggestions for research on the social pragmatics of race and color during this period. Maintaining that racial categories and identities are socially constructed, or historically invented, argues that two quite different principles of subordination explain the range of socially intelligible actions which shaped qualitatively different patterns of social mobility for "East Indians" and "Africans" in that country. Comparison of a centenary celebration of East Indian achievements in Trinidad and the autobiography of Eric Williams help illustrate argument. A similarly titled article, somewhat differently organized and written, is published by the author in *The East Indian Odyssey* (item **791**).

835 **Simposio Internacional Cultos Religiosos a los Antepasados en el Caribe,** *Río Piedras, Puerto Rico, 1990.* Simposio Internacional Cultos Religiosos a los Antepasados en el Caribe. San Juan: Univ. de Puerto Rico, Recinto de Río Piedras, Depto. de Actividades Culturales y Recreativas, 1991. 50 p.: bibl., ill.

Program/calendar of a Nov. 1990 symposium held in Puerto Rico on Caribbean ancestor cults. Includes brief descriptions of the religious altars on exhibit and associated artifacts.

836 **Slavery in the Americas.** Edited by Wolfgang Binder. Würzburg: Königshausen & Neumann, 1993. 647 p.: bibl. (Studien zur Neuen Welt; 4)

Excellent collection of 32 papers of considerable quality given at an international, multidisciplinary conference held in Germany in 1989. For the contributions on Caribbean ethnology, see Hoogbergen (item **803**), Mintz (item **818**), and Palmie (item **824**).

837 **Smith, M.G.** Race and ethnicity. (*in* The enigma of ethnicity: an analysis of race in the Caribbean and the wider world. Edited by Ralph R. Premdas. St. Augustine, Trinidad: Univ. of the West Indies, School of Continuing Studies, 1993, p. 23–58, bibl.)

Published at the time of his death, this last essay of a distinguished Caribbeanist deals with concepts of race and ethnicity, emphatically rejecting current views in social science that tend "to assimilate racial and ethnic relations to one another in the undifferentiated category of 'intergroup relations.'" In tandem, the race concept in Western thought, and racism as cultural theory are examined.

838 **Sobo, Elisa Janine.** One blood: the Jamaican body. Albany: State Univ. of New York Press, 1993. 329 p.: bibl., ill., index. (SUNY series, the body in culture, history, and religion)

Conceptions about health and sickness held by poor, rural people living in a northeastern coastal district of Jamaica. Using social arena as an organizing principle, this descriptively rich study focuses on gender relations and ideas about kin and children. Contains 15 chapters that deal with notions about the Jamaican body, the ethnophysiology of reproduction, the social and moral order by which informants attempt to live, the relations between parents and children and between men and women, traditional health beliefs, and, "bad bellies" (e.g., menstrual taboos, binding "ties," abortion, "witchcraft babies").

839 **Sorensen, Ninna Nyberg.** Creole culture, Dominican identity. (*Folk/Copenhagen*, 35, 1993, p. 17–35, bibl., ill.)

Argues that cultural identity in the Dominican Republic has tended to be elastic and that the easy incorporation of immigrant groups into that society has been the rule except for people coming from neighboring Haiti. Utilizing data from the current transnational situation of Dominicans, rejects view that the nonincorporation of Haitians implies a Dominican rejection of an African heritage, questions the view of "Dominicanness" as a uniform identity, and challenges "traditional" concepts of culture.

840   Spencer-Strachan, Louise. Confronting the color crisis in the African diaspora: emphasis Jamaica. New York: Afrikan World Infosystems, 1992. 87 p.: bibl., index.

Jamaican-American anthropologist discusses race, class, and problems of self-identity among diasporic Africans.

841   Spiritual Baptists, shango, and others: African derived religions in the Caribbean. Edited by Stephen D. Glazier. (*Caribb. Q.*, 39:3/4, Sept./Dec. 1993, p. v-129)

Special double issue of *Caribbean Quarterly* contains ten articles, the majority by anthropologists: Stephen Glazier on funerals and mourning in the Spiritual Baptist and Shango traditions: Angelina Pollack-Eltz on the Shango cult and other African rituals in Trinidad, Grenada, and Carriacou; Father Ian A. Taylor on mourning rites in the Spiritual Baptist Church; James Houk on the role of the Kabbalah in the Trinidadian Afro-American religious complex; Roland Littlewood on appropriation and reinterpretation in Spiritual Baptist visions; Patrick J. Polk on African religion and Christianity in Grenada; Manfred Kremser on St. Lucian Djine in communion with their African kin; Donald J. Consention on Voudou Vatican or a prolegomenon for understanding authority in a syncretic religion; Maureen Warner-Lewis on African continuities in the Rastafari belief system and, Carole Yawney's comments on the Spiritual Baptist and Shango papers.

842   Stephanides, Stephanos. Victory over time in the Kali Puja and in Wilson Harris: *The Far Journey of Oudin*. (*in* The East Indian Odyssey: dilemmas of a migrant people. Edited by Mahin Gosine. New York: Windsor Press, 1994, p. 244–248)

" ... structural and symbolic parallels between the literary art of Wilson Harris, focusing on his 'East Indian novel' *The Far Journey of Oudin*, and the Madrassi tradition of Worship of the Mother Goddess in Guyana."

843   Traditional spirituality in the African diaspora. Edited by Patrick Bellegarde-Smith. (*J. Caribb. Hist.*, 9:1 & 2, Winter 1992/Spring 1993, 143 p., ill.)

Of value to anthropologists, this special issue on traditional spirituality includes McAlister's collective biography of seven voodoo priestesses in New York; Ocasio's essay on *santería* and contemporary Cuban literature; Aborampah on religious sanction and social order in traditional Akan communities in Ghana and Jamaica; Benson's observations on Islamic motifs in Haitian religious art; Desch on Capoeira as spiritual discipline; Gibson on the Guyanese Comfa dance; and Nodal on the concept of *Ebbo* as a healing mechanism in Santería.

844   Trinidad ethnicity. Edited by Kevin A. Yelvington. Knoxville, Tenn.: Univ. of Tennessee Press, 1993. 296 p.: bibls., ill., index.

Well-balanced, quite useful collection of 12 original articles on a persistently important topic. In addition to contributions by anthropologists, includes work on social conflict in the 19th century; the evolution of inequality; spatial patterns and social interaction; ethnic conflict; gender and ethnicity; ethnicity and social change in literature; and, ethnicity and calypso. For contributions on Caribbean ethnology, see Yelvington (item **849**), Segal (item **834**), Houk (item **804**), Khan (item **809**), and Sampath (item **829**).

845   Vertovec, Steven. Hindu Trinidad: religion, ethnicity, and socio-economic change. London: Macmillan Caribbean, 1992. 272 p: bibl., glossary, ill., index.

Sophisticated, comprehensive study of the development of Hindu society and culture in Trinidad. Key cultural transformations of the Indian population and their social, cultural, and economic impact are considered as are the diverse facilitating factors. Introductory section deals with the early Indian indentured diaspora and the differential modification of key elements of Indian culture (i.e., kinship and household, caste, and Hinduism) in foreign locales and provides excellent context for the sociocultural history of Trinidad Indians, the significance of Trinidadian socioeconomic development in that history, and the discussions of contemporaray Hinduism and its revitalization.

846 Wolves from the sea: readings in the anthropology of the native Caribbean. Edited by Neil L. Whitehead. Leiden: KITLV Press, 1995. 176 p.: bibl., index. (Caribbean series, 0921–9781; 14)

Collection of seven essays on the archaeology, linguistics, history, and sociocultural anthropology of native Caribbean groups, particularly that of the Island Carib. Central theme is the acknowledgement of the plurality of ethnic identities existing at the time of the European arrival. Rejects the way in which orthodox anthropology has blindly accepted colonial ethnological schema. [R. Hoefte]

847 Wylie, Jonathan. Too much of a good thing: crises of glut in the Faroe Islands and Dominica. (*Comp. Stud. Soc. Hist.*, 35:2, Apr. 1993, p. 352–389, bibl.)

The division of the spoils after a Faroean *grindadráp* (tumultuous and dangerous collective hunt and slaughter of pilot whale herds) is remarkably orderly, while the division after a Dominican *bonik* seining (considerably less difficult and dangerous collective hunt of shoals of skipjack tuna) is disorderly and chaotic. An exploration of this question builds some parameters for considering Faroean and Dominican cultural differences.

848 Yelvington, Kevin A. Ethnicity at work in Trinidad. (*in* The enigma of ethnicity: an analysis of race in the Caribbean and the wider world. Edited by Ralph R. Premdas. St. Augustine, Trinidad: Univ. of the West Indies, School of Continuing Studies, 1993, p. 99–122)

Utilizing ethnographic detail effectively, argues that "a paradoxical process characterizes the role of ethnicity at work in Trinidad." Occupational diversification has led to conditions that are changing ethnic groups and facilitating their incorporation into a "common but permutating class structure." However, ethnicity remains a "salient" factor in Trinidad's economic structure "because the process of class composition involves closure around a number of qualitative factors, including ethnicity."

849 Yelvington, Kevin A. Introduction: Trinidad ethnicity. (*in* Trinidad ethnicity. Edited by Kevin A. Yelvington. Knoxville: Univ. of Tennessee Press, 1994, p. 1–32, bibl., table)

This substantive introductory chapter to item **844**, discusses historical themes in Trinidadian ethnicity, as well as the "culture of ethnicity," ethnicity and politics, and competing theories of Trinidadian ethnic and cultural diversity.

850 Young, Virginia Heyer. Becoming West Indian: culture, self, and nation in St. Vincent. Washington: Smithsonian Institution Press, 1993. 229 p.: bibl., ill., index. (Smithsonian series in ethnographic inquiry)

Author "assesses the practices and symbols that constitute an adaptive way of life and a shared idea system" in St. Vincent. Explores the complex issue of culture and national identity and explains a Vincentian desire for regional cooperation. Three kinds of data were used: history for setting the context and large-scale frameworks; village-level ethnography for delineating regularized forms of behavior and thought; and analysis of the writings of the nation's intellegentsia as well as popular performances for providing local formulations of society and culture.

South America
Lowlands

JONATHAN D. HILL, *Professor of Anthropology, Southern Illinois University at Carbondale*

MAIN TRENDS IN THE FIELD OF AMAZONIAN ETHNOLOGY during the early 1990s have deepened and broadened the concern for studying indigenous social organization and religion from a variety of historical and interpretive perspectives.

Indigenous ritual and ceremonial practices, together with associated beliefs and mythic narratives, continue as a focal interest. However, it has become increasingly apparent that symbolic and other interpretive approaches to indigenous religion are most productive when integrated into broader issues of social organization and history.

One of the clearest examples of such integrated, interpretive, and historical approaches to social change is the collection of essays on *Cosmology, values, and inter-ethnic contact in South America* edited by Terence Turner (item **855**), a work in which the intricacies of indigenous cultural symbols and their meanings are analyzed as dynamic elements of political resistance to the loss of powers of self-determination. Through a well-researched and highly insightful comparison of Arawakan and Tukanoan religious movements of the 19th century, Robin Wright (item **889**) demonstrates that such processes of cultural resistance follow distinct historical trajectories, depending upon underlying contrasts in indigenous political organization and patterns of interethnic relations with Western societies. Oscar Agüero's fine study of religious movements among the Tupi-Cocama of Peru (item **908**) traces contemporary "ethno-dynamism" to the pan-Tupian complex of prophetism and the unique social history of the Upper Amazon region. In a valuable survey of religious movements in Lowland South America, Michael Brown (item **852**) argues that the specific cultural forms manifested in messianism must be studied in relation to divisions and contradictions internal to indigenous social orders as well as to external relations of resistance and accommodation to Western societies.

The ongoing creativity and historical longevity of indigenous religions are amply demonstrated in the collection of essays edited by E. Jean Langdon (item **862**). Through a number of case studies from various regions of Lowland South America, Langdon's collection explores the rich diversity of indigenous concepts of shamanistic power and the expressive styles through which such power is put into practice. Complementing Langdon's broadly comparative study of shamanism, Jonathan Hill provides an in-depth study (item **896**) of how the musical and verbal practices of Wakuenai shamans and chant-owners are used in the construction of an indigenous poetics of ritual power. Fusing the musical and visual dimensions of shamanic ritual into a seamless whole, Eduardo Luna and Pablo Amaringo interpret a magnificent chiaroscuro of indigenous paintings based on power songs and hallucinogenic visions induced by *ayahuasca*, or Banisteriopsis caapi (item **921**). Still another tribute to shamanic creativity is Jean-Pierre Chaumeil's survey (item **853**) of various mixtures between indigenous shamanism and Afro-and Euro-American religions in Latin America. Taken together, these studies imply that shamanic practices continue to develop in creative new ways and that they are in no way reducible to static relics of an archaic past.

A small number of very significant works employ discourse-centered approaches to the study of indigenous ritual performance and narrative discourse. Ellen Basso's attention to the language of Kalapalo history demonstrates the great potential for using discourse analysis as a tool for exploring indigenous histories (item **865**). Specifically, Basso's study reveals how the Kalapalo construct memories of the historical past through biographical stories about great warriors whose lives embodied the transition from a formative period of chaotic warfare to a more recent past characterized by relatively peaceful ties based on ritual and ceremonial exchanges among Upper Xingu peoples. Charles Briggs shows the value of discourse analysis for understanding power relations within Warao society through an exploration of women's ritual wailing as a musical and verbal means for challenging the authority of male political and religious leaders (item **891**). Laura Graham focuses discourse analysis on the political speeches of Xavante men's councils to show how these speeches dampen factionalism through performance practices that strengthen collective identities over and above individual speakers' voices (item **874**). Although the number of discourse-centered studies is somewhat lower than in previous years, the outstanding quality of these works and the success with which they use discourse analysis as a tool for illuminating broader issues of social organization, history, and religion is a sure sign that discourse analysis will continue to play an increasingly prominent role in Amazonian ethnology as it develops through the 1990s.

A number of studies focus on the material, political and economic processes of change that provide the broader context within which indigenous societies are enmeshed. Roberto Pineda Camacho offers a general survey of the economic roles of indigenous peoples as slave laborers and debt peons in the historical development of the Colombian Amazon between 1550 and 1945 (item **905**). R. Brian Ferguson marshals historical and ethnographic evidence to argue that the Yanomami war complex portrayed in recent ethnographies must be understood in terms of the long-term history of interethnic relations between the Yanomami and Western societies and the consequent disintegration of traditional practices of reciprocity (item **892**). Ferguson's study is particularly timely and significant in light of the recent massacres of Yanomami by Brazilian goldminers and the sterility of academic debates over the supposed causes of Yanomami warfare and violence. Another valuable work is the new edition of Expedito Arnaud's historical and sociological articles on the Galibi, Mundurucú, Cayapo, and other indigenous peoples of Brazil (item **864**). Arnaud's case studies span a 50-year period (1940–90) and bear tribute to an entire lifetime dedicated to indigenous advocacy informed by solid anthropological research and publication.

Given the centrality of interpretive and materialist approaches to indigenous histories in contemporary ethnology, it is not surprising to find that archaeologists are beginning to turn to these new historical studies for insights into pre- and postcontact processes of social change. In an interesting study of Wayu (Guajiro) historical origins, José Oliver uses colonial history, comparative linguistics, and archaeology to argue that the Wayu arrived in their present location at least 1,500 years ago, rather than more recently as a result of the breaking apart of the Caquetios during the colonial period (item **903**). In a similar manner, Alberta Zucchi argues that indigenous oral histories of the Piapoco and other Northern Arawakan groups outline patterns of past migrations that are consistent with archaeological evidence from the Llanos and adjacent forest areas to the south (item **907**). By taking advantage of the opportunities provided by recent ethnohistorical studies, these archaeological works could become a point of departure for productive collaborative research between archaeologists and ethnologists in the future.

As in past years, the number of works focusing on aspects of indigenous ecology have continued to diminish. Case studies in this area include an analysis of fluctuating game resources and the sexual division of labor among the Hiwi of Venezuela by Magdalena Hurtado and Kim Hill (item **898**), an intensive study of subsistence activities among the Siona-Secoya of Ecuador by William Vickers (item **926**), and a Spanish translation of Roland Bergman's ecological analysis of Shipibo subsistence economics (item **910**). The downward trend in ecological studies has been consistent for nearly a decade, and it is probable that such studies will never return to the prominence they had in Amazonian ethnology during the 1970s and early 1980s.

Studies of indigenous social organization have rebounded in the early 1990s to become one of the more productive research topics. Among the most important of these new studies are Irene Bellier's historical interpretation of the Mai Huna complex of patrilineal clans and uxorilocal residence (item **909**), Janet Chernela's monographic treatment of ranked social organization among the Wanano of the Brazilian Vaupés (item **867**), Stephen Fabian's analysis of ethnoastronomical knowledge as symbolic mediation of social processes among the Bororo of Central Brazil (item **870**), and Alcida Ramos' study of patrilineal sibs and accompanying ritual practices among the Sanuma (Yanomami) of Brazil (item **884**). Shorter works worth mentioning include a study of Cashinahua kinship as a historically dynamic process of adapting to Brazilian society by Cecilia McCallum (item **880**), a feminist interpretation of gender relations in the Colombian Vaupés region by Jean Jackson (item **900**), and an interesting study of Cayapo naming practices as evidence of matrilineal descent groups by Vanessa Lea (item **879**). The resurgence of interest in social organization indicates that contemporary ethnologists of Lowland South America have not lost interest in the major questions raised by earlier generations of ethnologists. On the contrary, today's ethnologists are developing imaginative theoretical approaches and methods, such as interpretive ethnohistory and discourse analysis, to explore old topics in fresh new ways.

GENERAL

851   **Boomert, Arie.** Gifts of the Amazons: "green stone" pendants and beads as items of ceremonial exchange in Amazonia and the Caribbean. (*Antropológica/Caracas*, 67, 1987, p. 33–54, bibl., maps, photos)

An archaeological and ethnohistorical examination of jade pendants worn by females of the society and used as the principal media in ceremonial exchanges. Examines the sociocultural context and underworld connotations of the pendants. Describes a series of exchange networks, since these pendants were significant as trade items throughout South America and the West Indies in Pre-Columbian and Proto-Columbian times. Also examines the origin myths and symbolism associated with the green stones. For archaeologist's comment see *HLAS 53:151*.

852   **Brown, Michael F.** Beyond resistance: a comparative study of utopian renewal in Amazonia. (*Ethnohistory/Society*, 38:4, Fall 1991, p. 388–413, bibl.)

Provides a valuable survey of millenarian movements in Lowland South America, including Ashaninca, Tukanoan and Arawakan groups of the northwest Amazon, Canela, Tupi-Cocama (Orden Cruzada), and Pemón (Hallelujah). Following Hélène Clastres' argument in *La terre sans mal* (1975), author suggests that millenarian movements go "beyond resistance" to colonial domination by reflecting internal processes of contradiction that are independent of relations with expanding European or national States.

853   **Chaumeil, Jean-Pierre.** Varieties of Amazonian shamanism. (*Diogenes/Philosophy*, 158, p. 101–113, bibl.)

Surveys the ways in which indigenous Amazonian shamanism has mixed with Afro- and Euro-American religions. Coverage includes mixed-blood shaman-healers and herbalists (*curanderos*), emergent shamanistic sects, and the spread of "Catholicized" shamanism from urban centers to rural indigenous communities. Although no specific case is studied in depth, provides a fascinating glimpse of the dynamic, creative potentialities of indigenous shamanism in Latin America.

854   **Chernela, Janet Marion.** The role of indigenous organizations in international policy development: the case of an Awa biosphere reserve in Colombia and Ecuador. (*in* International Congress of Ethnobiology, *1st, Belém, Brazil, 1988.* Ethnobiology: implications and applications. Belém, Brazil: Museu Paraense Emílio Goeldi, 1990, v. 2, p. 57–72, bibl.)

Describes events leading up to the foundation of a biosphere reserve for the Awa of Ecuador and Colombia. Although the Awa actively supported the early stages demarcating the reserve, they allied themselves with other indigenous groups in opposing the formal, binational meeting of Colombia and Ecuadorian officials. An important example of the difficulty of gaining formal recognition of indigenous lands in border regions without undermining indigenous ways of life.

855 Cosmology, values, and inter-ethnic contact in South America. Edited by Terence Turner. Bennington Vt.: Bennington College, 1993. (South American Indian Studies; 2)

Aims to re-integrate the concerns of two formerly divergent types of research: political studies of indigenous peoples' changing relations with national societies and symbolic analysis of sociocultural forms within indigenous societies. Coverage includes Kayapó (T. Turner), Guayanese Carib (K. Adams), Kagwahiv (W. Kracke), Wapisiana (N. Foster), Huaorani/Quichua (M.E. Reeve), Cashinahua (K. Kensinger), Upper Xingu (S. Schwartzman), Wakuénai (J. Hill), and three essays on Andean Highlands peoples. Demonstrates how indigenous Amazonian peoples are creatively adapting traditional sociocultural forms as an integral part of their political resistance to the loss of their lands, resources, and powers of self-determination. See also item **856**.

856 Discourses and the expression of personhood in South American inter-ethnic relations. Edited by Jonathan David Hill. Bennington, Vt.: Bennington College, 1993. (South American Indian Studies; 3)

Collection examines indigenous concepts of personhood in Lowland South America, their multiple transformations in inter-ethnic relations with Western societies, and the theoretical implications of such complex processes of change for ethnologists. Coverage includes Shuar (J. Hendricks), Kagwahiv (W. Kracke), Toba (E. Miller), Culina (D. Pollock), Vaupes (J. Jackson), and Cuiva (B. Arcand). Expands on earlier set of essays on cosmology, values, and inter-ethnic contact. (see item **855**).

857 Ferguson, R. Brian. Ecological consequences of Amazonian warfare. (*Ethnology/Pittsburgh*, 28:3, July 1989, p. 249–264, bibl.)

Shows that warfare alone does not produce a better distribution of population to available natural resources. Argues that despite negative consequences for population trends, warfare may not lead to a population/resource balance. Maintains that war creates a pattern of forced displacement in which populations shift to areas where population decline has occurred. This pattern can explain long-term stability in the Amazon. Concludes that warfare is not ecologically adaptive.

858 Ferguson, R. Brian. Game wars?: ecology and conflict in Amazonia. (*J. Anthropol. Res.*, 45:2, Summer 1989, p. 179–206, bibl.)

Argues that depletion of game can lead to increasing hostility and social conflict, scarcity alone tends to result in population movement, not warfare. Maintains that scarcity of game is a factor, but not the sole explanation for warfare.

859 Jaguar no sokuseki: Andesu, Amezon no shūkyo to girei [Jaguar's trace: religion and ritual in the Andes and the Amazon]. Edited by Hiroyasu Tomoeda and Ryozo Matsumoto. Tokyo: Tōkai Shuppankai, 1992. 186 p.: ills., maps

A two-year collaborative effort (Apr. 1986–March 1988) by 12 eminent Japanese ethnologists and prehistorians explores the "jaguar" belief—represented in myths, rituals, and personified icons—which flourished among major central Andean cultures such as the Chavin culture. Today, the jaguar continues to hold an important place among Lowland Amazonian indigenous societies. The seven articles (five based on analysis of the Andean region, one on the Amazonian region, and one on the Mayan civilization) explore unifying themes in an attempt to explain the existence of similar jaguar beliefs in two such distant regions. Conclude that jaguar symbols among indigenous societies share characteristics of "ambiguity," "connection," and "transference." [K. Horisaka]

860 Maybury-Lewis, David. Becoming Indian in Lowland South America. (*in* Nation-States and Indians in Latin America. Edited by Greg Urban and Joel Sherzer. Austin: Univ. of Texas Press, 1991, p. 207–235, bibl., tables)

Examines the treatment of Indian populations in Brazil, Argentina, and Chile, and the recent emergence of a distinct Indian consciousness. The overall national agenda of each country at the time it focused on the Indian question seems to have guided policy development. In each case, the Indian populations actively responded to protect their interests. Their struggles to organize against national Indian policies resulted in an emerging Indian "conciousness."

861 **Populações humanas e desenvolvimento amazônico.** Organização do Luis E. Aragón e Maria de Nazaré Oliveira Imbiriba. Belém, Brazil: Univ. Federal do Pará, Assessoria Especial de Relações Nacionais e Internacionais, Casa de Estudos Latinoamericanos, 1989. 351 p.: bibl., ill. maps. (Série Cooperação amazônica; 3)

Collection of papers from 1988 international symposium on Human Population and Amazonian Development. Presents detailed historical, demographic, social, and economic information on Amazonian territories in Brazil, Colombia, Ecuador, Peru, Guyana, and Venezuela. In addition to a wealth of useful, up-to-date information, concludes with proposals urging scholarly and governmental cooperation among countries sharing Amazonian territory.

862 **Portals of power: shamanism in South America.** Edited by E. Jean Matteson Langdon and Gerhard Baer. Albuquerque: Univ. of New Mexico Press, 1992. 350 p.: bibl., ill., index, map.

Major collection of case studies of South American shamanism by Latin American, European, and North American contributors who advocate symbolic approach to study of shamanism as a practice of mediating between human and extra-human worlds. Organized into four general topics: 1) native conceptions of shamans and power; 2) shamans and visionary experience; 3) expressive culture and shamanism; and 4) and shamanism and responses to change.

863 **Wahl Kleiser, Lissie et al.** La región del Madre de Dios: bibliografía anotada. Cusco, Peru: Centro de Estudios Regionales Andinos Bartolomé de Las Casas; Lima: Instituto Indigenista Peruano, 1991. 304 p.: indexes. (Debates andinos; 20)

Annotated bibliography of 1,052 entries citing monographs, periodical articles, government reports, and other publications about the Madre de Dios region. (Encompasses broader area than the department of the same name.) Each entry includes references to historical period, place or places, and subjects covered. Thoroughly indexed by author, subject, ethnic groups, geographic location, and date(s) of coverage. [A. Hartness]

BRAZIL

864 **Arnaud, Expedito.** O índio e a expansão nacional. Belém, Brazil: Edições CEJUP 1989. 485 p.: bibl., ill., maps.

Collection of Arnaud's major articles on indigenous peoples of Pará, Brazil, spanning a 50-year period (1940–90). Historical and sociological approach covers several indigenous groups: Galibi, Oyampik/Emerilon, Kayapó-Kararao, Mundurucú, Mirania, Gaviões, and Kayapó-Gorotire (see *HLAS 51: 862*, "A expansão dos Indios Kayapó-Gorotire e a Ocupação Nacional" by Arnaud). Also includes a valuable historical overview of indigenist activism in southern Pará between 1940–70. The author's status as an indigenous rights advocate adds stength to the essays.

865 **Basso, Ellen B.** Kalapalo biography: ideology and identity in a South American oral history. (*Hist. Relig.*, 29:1, Aug. 1989, p. 1–22)

Analyzes historical consciousness among the Kalapalo of Brazil through close attention to biographical stories about great warriors whose actions led to the emergence of peaceful alliances based on respect and reciprocity among Upper Xingu peoples. Author's discourse-centered approach to local history is a model for future ethnological research in Lowland South America.

866 **Burkhalter, Brian S.** and **Robert F. Murphy.** Tappers and sappers: rubber, gold and money among the Mundurucú. (*Am. Ethnol.*, 16:1, Feb. 1989, p. 101–116, bibl)

Presents an example of social change that occurred among the Mundurucú Indians from 1953–81. Demonstrates a change from a barter-credit economy to a cash economy, largely due to the region's gold boom. Economic transactions have become highly objectified, and have had an immense impact on social relations.

867 **Chernela, Janet Marion.** The Wanano Indians of the Brazilian Amazon: a sense of space. Austin: Univ. of Texas Press, 1993. 185 p.: bibl., ill.

Major ethnographic and historical study of the Guanano, an eastern Tukanoan group of the northwest Amazon. Examines history, social organization, ecology, and politics in order to understand the interplay between principles of hierarchy and equality. The work makes important contributions to the study of ranked societies and is a rich source of ethnographic information based on extensive fieldwork.

868   **Chibnik, Michael.** Quasi-ethnic groups in Amazonia. (*Ethnology/Pittsburgh*, 30:2, April 1991, p. 167–182, bibl., tables)

Looks at the determinants of ethnic boundaries among locally-born, non-tribal residents. Argues that local historic and demographic differences have resulted in the variations in defining group boundaries. Provides definitions or characteristics of each ethnic group from the three countries.

869   **Early, John D.** and **John F. Peters.** The population dynamics of the Mucajaí Yanomama. San Diego: Academic Press, 1990. 152 p.: bibl., ill., index, maps.

Important demographic study of the Mucajaí Yanoama, a Ninam-speaking subgroup of the Brazilian Yanoama. Includes detailed statistics on fertility and mortality rates from 1958–87, as well as ethnographic information on relevant cultural pratices. By demonstrating high rates of population growth in the early post-contact period, the study questions the widely accepted assumption of static population in small-scale foraging and horticultural societies.

870   **Fabian, Stephen Michael.** Space-time of the Bororo of Brazil. Gainesville: Univ. Press of Florida, 1992. 253 p.: bibl., ill., index, maps.

One of a growing number of ethnographies focusing on indigenous observations of solar, lunar, and other astronomical phenomena in Lowland South America. Argues that ethnoastronomical knowledge is not merely a metaphor for Bororo social processes, rather it is integral to their mediation between natural and social realms. Most important is demonstration of the layering of ethnoastronomical knowledge into complex webs of temporal devices, most of which are terrestrial cycles such as flowering of plants, rainfall, and animal activities.

Ferguson, R. Brian. A savage encounter: western contact and the Yanomani war complex. See item **892**.

871   **Fernandes, Florestan.** A organização social dos Tupinambá. São Paulo: Editora Hucitec; Brasília: Editora UnB, 1989. 325 p.: bibl., ill.

New edition of Fernandes' classic historical monograph (São Paulo, 1948) on the Tupinamba, the Tupi-speaking inhabitants of coastal Brazil during the early colonial period. Careful use of historical sources and ethnographic analogies makes this work valuable for all anthropologists; it remains a foundational text for studies of Tupi-speaking peoples. Especially significant is the analysis of Tupinamba migrations in terms of ecological determinants and culturally-induced search for a utopian "Land Without Evil." Sophisticated argumentation, consistent with contemporary research methods.

872   **Frechione, John; Darrell A. Posey;** and **Luiz Francelino da Silva.** The perception of ecological zones and natural resources in the Brazilian Amazon: an ethnoecology of Lake Coari. (Advances in economic botany; 7) (*in* Resource management in Amazonia: indigenous and folk strategies. Bronx, N.Y.: New York Botanical Garden, 1989, p. 260–282, bibl., ill., index, maps)

Provides an assessment of resources as perceived by an indigenous member of the society, the third author of the article. Gives detailed lists of natural resources available and shows how these resources are exploited in an integrated subsistence strategy. Knowledge of the interrelatedness between resources allows fine tuning of subsistence strategies to maximize gain and minimize time and effort. Good introduction to ethnoecology.

873   **Gallois, Dominique T.** O discurso Waiãpi sobre o ouro: um profetismo moderno. (*Rev. Antropol./São Paulo*, 30/31/32, 1987/88/89, p. 457–467, bibl.)

After showing how white people are classified as a residual, external category in Waiãpi cosmology, presents short myth texts that explain Waiãpi prophesy and the search for "The Land Without Evil" are presented. Demonstrates how Waiãpi leaders are using this complex of cataclysmic imagery to interpret gold mining and to resist land invasions by Brazilian miners and *garimpeiros*.

874 **Graham, Laura.** A public sphere in Amazonia?: the depersonalized collaborative construction of discourse in Xavante. (*Am. Ethnol.*, 20:4, Nov. 1993, p. 717–741, bibl., photo)

Transcription and analysis of public speaking in Xavante men's councils. Argues that accountability for public discourse is deliberately dissociated from individual speakers through polivocality. Most men speak continuously throughout evening meetings, resulting in a murmuring collage of voices that embody collective rather than individual identities. Xavante public speaking strengthens age- and gender-based hierarchies and dampens factionalism that could lead to village fission. An outstanding illustration of how the new discourse theory can be used to explore indigenous social organization.

875 **Hydroelectric dams on Brazil's Xingu River and indigenous peoples.** Edited by Leinad Ayer de O. Santos and Lúcia Mendonça Morato de Andrade. Translated by Robin Wright. Cambridge, Mass.: Cultural Survival; São Paulo: Pro-Indian Commission of São Paulo, 1990. 192 p.: bibl., ill., maps. (Cultural survival report; 30)

Compilation of articles giving precise information on effects of hydroelectric dam building in the Brazilian Amazon. Eduardo Viveiros de Castro provides a general survey and critical analysis of governmental Master Plans and Lucia M.M. de Andrade introduces specific articles on indigenous peoples and their environments along the lower and middle Xingu River, Brazilian energy policy, and the harmful effects of the planned Xingu hydroelectric proyect at Altamira.

876 **Indios de Roraima: Makuxí, Taurepang, Ingarikó, Wapixana.** Boa Vista, Brazil: Centro de Informação, Diocese de Roraima, 1989? 104 p.: bibl., ill., map. (Col. histórico-antropológica; 1)

Brief historical and cultural survey of Roraima and the Rio Branco basin of northern Brazil. Pt. I contains useful information on early colonial warfare, late colonial Portuguese domination, 19th-century missions, and the 20th-century founding of the Indian Protection Service (SPI) centers. Pt. 2 briefly sketches four indigenous cultures of Brazilian Roraima: Macuxi, Taurepang, Ingariko, and Wapixana. Ethnographic sources in bibliography are out-of-date, but historical sources are valuable.

877 **Kapfhammer, Wolfgang.** Der Yurupari-Komplex in Nordwest-Amazonien. München: Edition Anacon, 1992. 326 p., 7 p. of plates: bibl., ill., maps. (Münchener Amerikanistik Beiträge; 28)

Detailed description and analysis of two multi-faceted Yurupari rituals of the Tucano and Arhuaco tribes in the northwest Amazon. Emphasizes initiation and harvest rites, their paraphernalia, and significance for socialization and physical control of body orifices. [C. Converse]

878 **Kracke, Waud.** He who dreams: the nocturnal source of transforming power in Kagwahiv shamanism. (*in* Portals of power: shamanism in South America. Albuquerque: Univ. of New Mexico Press, 1992, p. 127–148)

Analyzes the close interrelationship between shamanistic power and dreaming among the Kagwahiv of Brazil. The indigenous term for shamans (*ipaji*) means "one possessed of power," and to a large extent Kagwahiv shamans exercised power in special "agentic" dreams. Whereas trance, song, and other ritual activities were used in curing ceremonies, shamanistic dreaming was focused on the conception of children and success in hunting. Analysis is based on memories of older informants, since the Kagwahiv have not had any active shamans for at least 25 years.

879 **Lea, Vanessa.** Mẽbengokre (Kayapó) onomastics: a facet of houses as total social facts in Central Brazil. (*Man/London*, 27:1, March 1992, p. 129–153, bibl., graphs)

Re-examines Mebengokre (Kayapó) naming practices in relation to social organization. Shows that "beautiful names" (*idzi mets* or *idzi Kati*) are a scarce symbolic resource through which uxorilocal extended family households, called "houses," maintain continuing identities through matrilineal descent. Concludes that Kayapó houses and matrilineal transmission of names form the basis of social reproduction and that Lowland specialists should re-open debate over unilineal descent groups.

880 **McCallum, Cecilia.** Language, kinship and politics in Amazonia. (*Man/London*, 25:3, Sept. 1990, p. 412–433, bibl., maps)

Analyzes kinship terms and political oratory among Cashinawa of western Brazil. Shows that Cashinawa social and political organization is both an historically dynamic means of embracing new cultural forms and a way of conserving indigenous identity. Author's claim to be developing a "novel pradigm" for understanding social organization in Lowland South America is doubtful, since numerous works by ethnologists in the 1980s have already demonstrated the historical dynamics of indigenous social organization in the region.

881 **Picchi, Debra.** The impact of an industrial agricultural project on the Bakairi Indians of Central Brazil. (*Hum. Organ.*, 50: 1, Spring 1991, p. 26–38, bibl., tables)

Examination of the negative impact of government-sponsored industrial agricultural projects on the traditional mode of production and economy of the Bakairi. Author discusses changes in the patterns of land use, labor patterns, and the economy, suggesting that development agencies must alter plans to avoid such negative changes, since these modifications have had a negative impact on the Bakairi instead of benefitting them. Shows that an ecological model can be applied successfully to the study of economic development and change.

882 **Pollock, Donald.** Conversion and "Community" in Amazonia. (*in* Conversion to Christianity: historical and anthropological perspectives on a great transformation. Berkeley: Univ. of California Press, 1993, p. 165–197)

Describes Catholic (CIMI) and Protestant (SIL) missionary activities among the Culina of Brazil and Peru. Also analyzes cultural reasons for failure of Catholic economic projects and concept of "community." Excellent historical and ethnographic survey of larger failure of Christian missionaries to induce conversions among indigenous Amazonians.

883 **Pollock, Donald.** Culina shamanism: gender, power, and knowledge. (*in* Portals of power: shamanism in South America. Albuquerque: Univ. of New Mexico Press, 1992, p. 25–40)

Describes shamanistic rituals as a gendered practice of defining masculinity and of opening a dialogue between men and women among the Culina of Brazil. Analyzes indigenous concept of *dori*, or ritual power, as a symbol of antisocial wildness that causes disease. Women play a critical role in Culina curing rituals by domesticating these unsocialized, dangerous powers.

884 **Ramos, Alcida Rita.** Memórias sanumá: espaço e tempo em uma sociedade yanomami. São Paulo: Editora Marco Zero; Brasília: Editora Univ. de Brasília, 1990. 343 p.: bibl., ill., maps.

Sociological and interpretive study of the Brazilian Sanuma (Yanomamo). Demonstrates that Sanuma organize themselves into patrilineal sibs, or named unilineal descent groups, and that these sibs are in turn loosely clustered into larger sets of groups that conceive of themselves as consanguines. Also important is the interpretation of ritual hunts for the names of newborn children as a process of linking infants to social networks beyond their patrilineal identities. Fills several major gaps in the existing literature on the Yanomamo.

885 **Ribeiro, Berta G.** Classificação dos solos e horticultura desâna. (*in* International Congress of Ethnobiology, *1st, Belém, Brazil, 1988*. Ethnobiology: implications and applications. Belém, Brazil: Museu Paraense Emílio Goeldi, 1990, v. 2, p. 27–49, appendices, bibl., photos)

Provides detailed information on Desana classification of soil types and their relation to indigenous horticultural and plant-gathering activities. Appendices include lists of useful and edible plant species by categories of soil and by their scientific classifications. Concludes that Desana horticulture and fishing serve as the basis for a widely spread "mode of production" among indigenous people living in nutrient-poor, highly acidic blackwater ecosystems.

886 **Silverwood-Cope, Peter L.** Os makú: povo caçador do noroeste da Amazônia. Brasília: Editora UnB, 1990. 205 p.: bibl., ill., maps. (Col. Pensamento antropológico)

The first major monograph about the Makú, a semi-nomadic hunting and fishing society living in interfluvial areas of the northwest Amazon region. Includes detailed descriptions of subsistence activities, social organization, and religious cosmology. Demonstrates that Makú have been misidentified as "hunter-gatherers," whereas they are more accurately understood as specialized hunter-fishermen who trade meat products for agricultural products grown by Tucanoan peoples. Like the latter, the Makú have patrilineal sibs, but these are dispersed rather than co-resident.

887 Viertler, Renate Brigitte. A refeição das almas: uma interpretação etnológica do funeral dos índios Bororo, Mato Grosso. São Paulo: Editora Hucitec: Editora da Univ. de São Paulo, 1991. 221 p.: bibl., ill., maps. (Ciências sociais; 27)

Detailed study of Bororo funeral rituals, including relevant aspects of cosmology and an analysis of the verses of a genre of sacred songs, called *roia*. Concludes that Bororo funeral rites are analogous to the famous *Kuarúp* ceremonies among Upper Xingu peoples insofar as both are syntheses of diverse rites of passage allowing people to explore the possibilities for social harmony and division.

888 Villas Bôas, Orlando and Claudio Villas Bôas. Xingu: los indios, sus mitos. Quito: ABYA-YALA; Roma: MLAL, 1991. 237 p.: ill. (Col. 500 años; 33)

Spanish translation of a classic work on Upper Xingu peoples and their myths. Introductory section provides overview of recent history in region and intercultural relations among indigenous peoples. A second part includes myth texts from the Kamayurá, Kuikuro, and Yuruna peoples. This new translation makes an important corpus of Amazonian myths accessible to scholars and students throughout Spanish-speaking Latin America.

889 Wright, Robin R. "Uma conspiração contra os civilizados:" história, política e ideologias dos movimentos milenaristas dos Arawak e Tukano do noroeste da Amazônia. (*Anu. Antropol.*, 89, 1992, p. 191–231, bibl.)

Historical and political analysis of millenarian movements among Guarequena (Arawakan) and Tukanoan peoples of the Northwest Amazon region in 1858–59. Comparison of these movements shows a basic contrast between Arawakan movements aimed at rejection of whites' domination and Tukanoan rebellions aimed at usurpation of whites' power. Also, Arawakan movements had long-term consequences, whereas Tukanoan incidents were more ephemeral. Traces this contrast to underlying differences in indigenous political organization and historical relations with non-indigenous peoples.

COLOMBIA, VENEZUELA, AND THE GUIANAS

890 Boven, Karen. De Wayana [The Wayana]. (*SWI Forum*, 9:1/2, okt. 1992, p. 145–161, bibl., ill., map)

Following an historic and economic introduction, describes the present situation of Wayana Amerindians in the village of Kawemhakan on the Lawa River. The Suriname Wayana have experienced many rapid changes during the last decade. Underlines the differences between Suriname Wayana, and Brazilian and French Guianan Wayana, who are less modern and "urbanized." [R. Hoefte]

891 Briggs, Charles L. "Since I am a woman, I will chastise my relatives:" gender, reported speech, and the (re)production of social relations in Warao ritual wailing. (*Am. Ethnol.*, 19:2, may 1992, p. 337–361, bibl.)

Describes and interprets ritual wailing performed by Warao women at funerals in eastern Venezuela as poetic discourses on power relations. Whereas public discourse and political leadership roles are exclusively male domains in Warao society, women's ritual wailing provides a musical and verbal means for challenging and constraining the authority of male shamans and political leaders. Very sophisticated and well-substantiated argument that builds on the growing number of discourse-oriented studies of Lowland South America.

892 Ferguson, R. Brian. A savage encounter: western contact and the Yanomani war complex. (*in* War in the tribal zone: expanding States and indigenous warfare. Santa Fe: School of American Research Press, 1992, p. 199–227)

Important historical analysis of violence and warfare among the Yanomamo of Venezuela and Brazil, demonstrating that the primary causes for indigenous warfare are rooted in interethnic relations with Western societies and the breakdown of traditional reciprocity practices. Chagnon's famous ethnography of "the Fierce People" is situated in this broader historical context of interethnic relations, resulting in a much more convincing explanation of the Yanomamo war complex.

893 **Frechione, John.** Supervillage formation in the Amazonian Terra Firme: the case of Asenona. (*Ethnology/Pittsburgh*, 29, 1990, p. 117–133)

A study of modifications to the socionatural environment and reactions to these modifications to allow the support of a large population in the Terra Firme zone. Shows that the Yecuana have found large, permanent settlements a beneficial strategy for protection and for access to desired goods. Demonstrates that human choice, or "adaptive dynamics," plays a key role in adaptive strategies. This strategy appears to be the result of the encroachment and presence of outside forces such as missionaries, colonists, and development organizations.

894 **Good, Kenneth** and **David Chanoff.** Into the heart: one man's pursuit of love and knowledge among the Yanomama. New York: Simon & Schuster, 1991. 349 p., 16 p. of plates: ill., index.

Autobiographical account of a North American anthropologist's life among the Yanomamo of Venezuela, based on almost ten years of fieldwork. At the request of a village elder, the author became married to a young Yanomamo woman. The resulting intimate view of Yanomamo social life and economic practices is unparalleled. Frankly relates many problems arising from his marriage, but the depth and sensitivity of the portrayal of his wife's culture outweighs the problems caused by his strong emotional attachments.

895 **Guss, David.** In the absence of gods: the Yekuana road to the sacred. (*Antropológica/Caracas*, 68, 1987, p. 49–58, bibl.)

Begins with a general introduction to the tradition of the "absent god" found among tribal peoples throughout Latin America. Demonstrates how the Yecuana have incorporated this concept into their oral and artistic traditions. Gives an account of the mythology surrounding the departure of the culture hero, Wanadi, as seen in four cycles of creation. In so doing, shows how the the absent god tradition has helped to determine the nature of the Yecuana narrative structure as well.

896 **Hill, Jonathan David.** Keepers of the sacred chants: the poetics of ritual power in an Amazonian society. Tucson: Univ. of Arizona Press, 1993. 245 p.: bibl., ill., index.

Analyzes complex genre of chanted and sung speech (*málikai*) performed in sacred rituals among the Arawakan Wakuénai of Venezuela. Demonstrates how musical and semantic transformations of everyday discourse integrate social relations and history into mythic processes of empowerment and the concept of primordial humanness as a musical naming power that opened up the world and created all natural species. Argues that indigenous practices of spirit-naming are keys for understanding how details of musicality and meaning are used to construct a poetics of ritual power.

897 **Hill, Jonathan David.** A musical aesthetic of ritual curing in the northwest Amazon. (*in* Portals of power: shamanism in South America. Edited by E. Jean Matteson Langdon and Gerhard Baer. Albuquerque: Univ. of New Mexico Press, 1992, p. 175–210)

Discusses shamanistic curing among the Arawakan Wakuénai of Venezuela. Focuses on sacred myths and the belief in the power of chanted and sung speech to mediate relations among mythic ancestors and their human descendants, both living and dead. Analyzes the activities of two kinds of specialists, shamans and chant-owners, as musical journeys through bodily, social, and cosmic space and time. Argues that the apparently opposite directions of these two journeys are grounded in an underlying musical aesthetic of ritual curing.

898 **Hurtado, A. Magdalena** and **Kim R. Hill.** Seasonality in foraging society: variation in diet, work effort, fertility, and sexual division of labor among Hiwi of Venezuela. (*J. Anthropol. Res.*, 46:3, fall 1990, p. 293–346, bibl., graphs, tables)

Describes seasonal variations in the diet, subsistence work effort, and sexual division of labor among the Hiwi. Maintains there is stability in game resources, but variation in vegetable foods. Fluctuations in subsistence work effort among males and females correspond to a sexual division of labor. Notes seasonal physical changes in the populations, as well as seasonal female fertility correlated with changes in energy intake. Provides a context for future studies to look at ecological determinants of sex differences in relation to subsistence strategies.

899 **Jackson, Jean E.** Being and becoming an Indian in the Vaupés. (*in* Nation-States and Indians in Latin America. Edited by Greg Urban and Joel Sherzer. Austin: Univ. of Texas Press, 1991, p. 131–155, bibl.)

Discusses recent and current changes among Eastern Tucanoan peoples of the Colombian Vaupés territory as they have become increasingly self-conscious of their indigenous cultures as valuable commodities in national society. Traces the history of this new ethnic consciousness to liberal elements of the Catholic Church, indigenous organizations, and Colombian anthropologists. Finds that new ideologies of indigenous culture of Vaupés region in many ways contradict how these cultures organize themselves internally.

900 **Jackson, Jean E.** Gender relations in the Central Northwest Amazon. (*Antropológica/Caracas*, 70, 1988, p. 17–38, bibl.)

An interpretation of gender relations among Tucanoan peoples of the northwest Amazon region. Argues that Tucanoan societies only partially fit Collier and Rosaldo's model of "brideservice" societies and that divergences from this model help to explain the different evaluations of women's status by ethnographers working in the region. Although Tucanoan women are rarely objects of physical violence, the author maintains that their exclusion from most sacred rituals is an indication of lower social status than men.

901 **Jackson, Jean E.** The meaning and message of symbolic sexual violence in Tukanoan ritual. (*Anthropol. Q.*, 65:1, Jan. 1992, p. 1–18)

Describes and analyzes ritual, mythic, and other symbolic forms of male violence directed at women among Eastern Tucanoan peoples of the Colombian Vaupés region. Concludes that all-male ritual practices, called *Yuruparí*, reflect the vulnerability of men's status in a patrilineal, patrilocal social order and the association of women with potentially dangerous relations with affines and other outsiders.

Kapfhammer, Wolfgang. Der Yurupari-Komplex in Nordwest-Amazonien. See item **877**.

902 **Mansutti Rodríguez, Alexander.** Pueblos, comunidades y fondos: los patrones de asentamiento Uwotjuja. (*Antropológica/Caracas*, 69, 1988, p. 3–35, bibl.)

Demonstrates the historical changes of Piaroa settlement patterns over the 250 years of sustained interethnic relations with Spanish and Venezuelan societies. From an original settlement pattern consisting of two or three extended families living in a single round-house in remote forest areas, the Piaroa have come to reside in larger, interdependent villages which are in turn culturally and economically connected to mestizo towns. Gives precise information on history of missionaries, rubber gatherers, and other intercultural agents.

903 **Oliver, José R.** Reflexiones sobre posibles orígenes del Wayu (Guajiro). (*in* La Guajira: de la memoria al porvenir, una visión antropológica. Edición de Gerardo Ardila. Bogotá: Univ. Nacional de Colombia, 1990, p. 81–138)

Study of early colonial history, comparative Arawakan linguistics, and archaeological styles in order to construct hypotheses about the cultural origins of Wayu people. Some historical sources support the notion that contemporary Wayu are descended from Caquetios, a large but extinct Arawakan people living in the Venezuelan llanos when Europeans arrived. However, linguistic and archaeological evidence leads the author to favor the hypothesis that Wayu ancestors arrived in the Guajira region at least 1500 years ago, with Caquetios entering the llanos in more recent times.

904 **Perrin, Michel.** Creaciones míticas y representacíon del mundo: el ganado en el pensamiento simbólico Guajiro. (*Antropológica/Caracas*, 67, 1987, p. 3–31, bibl., tables)

Collection of Goajiro stories about cattle provides an empirical basis for exploring what happens to indigenous "mythic thought" when confronted with novel experiences and social change. Although starting from a Levi-Straussian understanding of myths as a closed system, concludes that Goajiro search for mythic coherence always remains unfinished because of ever-increasing exterior influences on Goajiro society.

905 **Pineda C., Roberto.** Participación indígena en el desarrollo amazónico colombiano. (*Maguaré/Bogotá*, 7:8, 1992, p. 81–124, bibl., map, tables)

Surveys history of Colombian Amazon from 1500 to 1945, including overviews of missionary campaigns and indigenous slave trade during the colonial period. Provides a highly detailed account of the rise of rubber gathering and its impact on indigenous groups during the 19th and early 20th centuries. Argues that Amazonian Lowlands were integrated into the world economy as sources of slave labor and natural resources since the mid-16th century. For four centuries, enslavement, forced dependency, and extractive economics were the common themes in the history of interethnic relations between Western societies and indigenous Amazonian peoples.

906 **Schackt, Jon.** Hierarchical society: the Yukuna story. (*Ethnos/Stockholm*, 55:3/4, 1990, p. 200–213, bibl.)

An analysis and test of the theory of hierarchy and value developed by Dumont and his students. Shows that Yucuna society can easily fit into the Dumont model of hierarchical order, but criticizes the notion that such an order is rooted to an ultimate value. Clearly demonstrates that there is a ranking of descent groups as well as a Yucuna heritage maintained by the groups along the Miritiparaná River in Colombia. Further discusses the notion that no single factor created this order, but rather a multitude of factors could have contributed to its rise.

Scholtens, Ben; Gloria Wekker; Laddy van Putten; and **Stanley Diko.** Gaama duumi, buta gaama: overlijden en opvolging van Aboikoni, grootopperhoofd van de Saramaka Bosnegers [The Paramount Chief is asleep, installation of the Paramount Chief: death and succession of Aboikoni, Paramount Chief of the Saramaka Maroons]. See *HLAS 54:2217*.

907 **Zucchi M., Alberta.** Como ellos la cuentan: la ocupación de la Orinoquia según la historia oral de un grupo maipure del norte. (*in* Simposio Desarrollos Recientes en la Historia de los Llanos del Orinoco: Colombia y Venezuela, *New Orleans, 1991.* Café, caballo y hamaca: visión histórica del Llano. Quito: Ediciones Abya-Yala; Bogotá: Orinoquia Siglo XXI, 1992, p. 23–38, graphs, maps)

Summarizes recent works on oral histories and historical consciousness among the Maipuran, or Northern, Arawak-speaking peoples of the Upper Río Negro and middle Orinoco regions. Supports the view that Piapoco phratries of the Guaviare River are historically descended from Wakuénai phratries to the south. Argues that archaeologists need to become more aware of the value of indigenous oral histories rather than relying solely upon written historical sources.

PERU AND ECUADOR

908 **Aguero, Oscar Alfredo.** The Millenium among the Tupi-Cocama: a case of religious ethno-dynamism in the Peruvian Amazon. Uppsala, Sweden: Uppsala Research Reports in Cultural Anthropology, 1992.

Outstanding case study of messianic and prophetic movements initiated in the 1970s by Francisco da Cruz among Tupian Cocama of Peru and Brazil. Demonstrates how this movement engaged basic contradictions of region's interethnic relations by tapping into inherently dynamic Tupian beliefs in universal cataclysm *(mbae-megua)* and the Land without Evil *(yvy'mara'ey)*. Situates Tupi-Cocama complex of "ethno-dynamism" in social history of the region and in the rich literature about Tupi-Guarani religious movements.

909 **Bellier, Irene.** El temblor y la luna: ensayo sobre las relaciones entre las mujeres y los hombres mai huna. Lima: Instituto Francés de Estudios Andinos; Quito, Ecuador: Ediciones ABYA-YALA, 1991. 2 v.: bibl., ill., maps. (Travaux de l'Institut français d'études andines, 0768–424X; t. 58) (Col. 500 años; 44–45)

Two-volume ethnography of the Mai Huna, a Western Tukanoan people living on small rivers between the lower Napo and middle Putumayo rivers in northern Peru. Makes excellent use of historical sources to trace the emergence of contemporary Mai Huna from "Payaguas" of early colonial period. Focus is on complementarity of male-female relations in context of patrilineal clans and uxorilocal residence. Male dominance in myth and ritual is counterbalanced by female importance in food production and child rearing.

910 **Bergman, Roland W.** Economía amazónica: estrategias de subsistencia en las riberas del Ucayali en el Perú. Traducción por Martha Beingolea. Lima: Centro Amazónico de Antropología y Aplicación Práctica-CAAAP, 1990. 209 p.: bibl., ill.

Spanish translation of Bergman's 1980 book, *Amazon economics*, presenting quantitative data on gardening, fishing, and hunting among Shipibo of Lowland Peru. Data indicate that Shipibo subsistence activities are highly efficient and productive ways of adapting to humid, sub-Andean rain forest. Argues that Shipibo economics refute the concept of hunter-gatherers as "the original affluent societies" and support the idea that simple horticulture in combination with fishing and hunting can provide greater affluence for small-scale human populations.

911 **Brown, Michael F.** Ropes of sand: order and imagery in Aguaruna dreams. (*in* Dreaming: Anthropological and Psychological Interpretations. Edited by Barbara Tedlock. Cambridge: Cambridge Univ. Press, 1987, p. 154–170)

Looks at the roles of dreams and imagery among the Aguaruna. Shamans use visions to gain power, while warriors need vision to have power to fight. Sees the acquisition, interpretation, or control of dreams as important factors in demonstrating competence. Dreams are vehicles for the expression of authority. This is a preliminary account on the role dreams can play in bridging the gap between self and other.

912 **Brown, Michael F.** and **Eduardo Fernández.** Tribe and State in a frontier mosaic: the Asháninka of eastern Peru. (*in* War in the tribal zone: expanding States and indigenous warfare. Edited by R. Brian Ferguson and Neil L. Whitehead. Santa Fe: School of American Research Press, 1992, p. 175–197)

Surveys history of interethnic relations between the Ashaninka and expanding colonial and national States. Argues that level of violence in Ashaninka society has increased due to policies and activities of State and that indigenous group has consistently resisted, sometimes with violent force, attempts to force it into a lower status. Concludes that Ashaninka have fashioned a culture of survival through challenging State's claims over their lands.

913 **Cipolletti, María Susana.** El piri-piri y su significado en el shamanismo secoya. (*Amazonía Peru.*, 8:15, agosto 1988, p. 83–98, bibl., tables)

Describes the uses and meanings of semi-domesticated plants of the Ciperaceas family among the Secoya of eastern Ecuador. Although used in a variety of subsistence activities, these plants, called *nuni*, are primarily valued for their ritual uses in curing the sick and mediating between living and dead. Shows that ethnobotanical knowledge is inseparable from cosmology and indigenous ritual practices.

914 **Ehrenreich, Jeffrey.** Contacto y conflicto: el impacto de la aculturación entre los Coaiquer del Ecuador. Traducción de Héctor Dueñas. Quito: Ediciones ABYA-YALA; Otavalo, Ecuador: Instituto Otavaleño de Antropología, 1989. 303 p.: ill. (Col. 500 años; 16)

Brief history of the Awa people, or "Cuaiquer," living in northern coastal Lowlands of Ecuador. Cuaiquer have experienced major losses of land and autonomy due to Ecuadorian government's plans for development of their lands. Explains indigenous strategies for surviving these losses, including secrecy, dissimulation, and adopting Catholic religion. Appendix includes complete copies of government documents that outline land development plans. Concludes that externally imposed changes, regardless of their intent, lead to destructive processes of ethnocide.

915 **Gow, Peter.** Of mixed blood: kinship and history in Peruvian Amazonia. Oxford, England: Clarendon Press; New York: Oxford Univ. Press, 1991. 1 v.: bibl., index. (Oxford studies in social and cultural anthropology)

Ethnographic study of "mixed blood" peoples of the lower Urubamba River in eastern Peru. Demonstrates how contemporary peoples have developed an open system of interethnic marriages and how this marriage system is consistent with an emerging concept of Comunidad Nativa that negates cultural continuities with past ethnolinguistic regional groupings. Critique of "traditional" versus "acculturated" dichotomy is interesting but not as original as the author claims.

916 **Hendricks, Janet.** Symbolic counterhegemony among the Ecuadorian Shuar. (*in* Nation-States and Indians in Latin America. Edited by Greg Urban and Joel Sherzer. Austin: Univ. of Texas Press, 1991, p. 53–71, bibl.)

Shuar opposition to Ecuador's hegemonic expansion is seen as more than a fear of losing cultural identity and autonomy. Instead, author believes opposition stems from a belief that whites have failed to understand the impact of their actions and have thus fallen prey to destruction and greed which has culminated in a system of inequalities among people in Ecuadorian societes. Examines rise of Shuar federation in a sociopolitical context. Counterhegemony is achieved primarily through the use of correct rhetoric. The Shuar also emphasize visionary knowledge as the means to achieve correct rhetoric.

917 **Hern, Warren M.** Shipibo polygyny and patrilocality. (*Am. Ethnol.*, 19:3, Aug. 1992, p. 501–522, bibl., graphs, map, tables)

Gives detailed statistical information on population, marriage practices, and fertility rates in eight Shipibo villages of the Peruvian Amazon. Shows that Shipibo tradition of polygynous, patrilocal marriages is being abandoned in response to missionaries' teachings and that rapidly increasing fertility rates are statistically linked to this social change. Important argument that offers one of few studies in Lowland South America based on sophisticated demographic methods.

918 **Hurtado, A. Magdalena** and **Kim Hill.** Experimental studies of tool efficiency among Machiguenga women and implications for root-digging foragers. (*J. Anthropol. Res.*, 45:2, Summer 1989, p. 207–217, bibl., tables)

Observers found that when metal tools were used in place of traditional wooden tools, the time needed to gather and process manioc decreased. However, this change is not particularly significant because enough manioc to feed 25 adults can be procured in one hour using traditional wooden sticks. Thus, the total time allocated daily to subsistence tasks is not greatly affected. Metal tools may lead to a change in women's allocation of time to subsistence activities and dietary contributions if the rates of digging and peeling are significantly lower than those found among the Machiguenga.

919 **Junquera, Carlos.** The confirmation rite of female names among the Harakmbet of the Southwestern Peruvian Amazon Region. (*Bull. Int. Anthropol. Ethnol.*, 32/33, 1990/1991, p. 103–119, bibl., tables)

Briefly describes a collective female naming ritual performance and argues this is one of few contexts in which Harakmbet women have access to power, which is primarily wielded by male ritual specialists and political leaders. Shows how ritual practices remain important despite the transition from patrilineal, exogamous families to cognatic, nuclear family organization. Valuable ethnographic information.

920 **Langdon, E. Jean Matteson.** Dau: shamanic power in Siona religion and medicine. (*in* Portals of power: shamanism in South America. Edited by E. Jean Matteson Langdon and Gerhard Baer. Albuquerque: Univ. of New Mexico Press, 1992, p. 41–62)

Documents the shamanistic use of *yage* (B. caapi) among Sioni of Ecuador as a process of achieving different levels of power *(dau)*. Only a few men become master shamans with exceptional powers to cure and cause disease. Concludes that ritual use of hallucinogenic plants are essential for lending an aura of factuality to shamanistic power.

921 **Luna, Luis Eduardo** and **Pablo Amaringo.** Ayahuasca visions: the religious iconography of a Peruvian shaman. Berkeley, Calif.: North Atlantic Books, 1991. 160 p.: bibl., ill. (some col.), index, map.

Collection of paintings made and interpreted by Pablo Amaringo, a mestizo shaman from the Peruvian Amazon. Co-author Luna provides an excellent overview of the indigenous roots of ritual healing practices in Peruvian Lowlands, with primary attention to the hallucinogenic visions induced by drinking Banisteriposis caapi (or ayahuasca). Importance of power songs, called *icaros*, is noted, and several translations are included. The work makes outstanding comparative use of studies on ayahuasca use among Shipibo and other indigenous Amazonian peoples.

922 **Napolitano, Emanuela.** Shuar y anent: el canto sagrado en la historia de un pueblo. Quito: Ediciones ABYA-YALA, 1988. 200 p.: bibl., ill.

Description and interpretation of sacred songs, called *anent*, performed among the Shuar of eastern Ecuador. Coverage includes men's hunting songs, men's and women's social songs, and women's gardening songs, with primary attention given to the latter. Semiotic analysis of song texts is very similar to Michael Brown's earlier study (*Tsewa's Gift*, 1985) of *anen* songs among the Aguaruna of Peru. Also provides some analysis of formal musical structures in *anent*, Shuar musical instruments, and interrelations between textual meanings and musical sounds.

923   **Renard-Casevitz, France Marie.** Le banquet masqué: une mythologie de l'étranger chez les indiens Matsiguenga. Paris: Lierre & Coudrier: Distribution, Eadiff, 1991. 280 p., 4 p. of plates: bibl., ill. (some col.). (Col. Recherche)

Collection of myths from the Machiguenga, an Arawakan people of eastern Peru. Structural analysis focuses on theme of the stranger in Machiguenga mythology. Bibliography is weak, and like other French structuralist studies of myths, no real social or historical contexts are provided. Nevertheless, some of the analytical ideas could prove useful for ethnographers designing new field projects on indigenous narratives.

924   **Rojas Zolezzi, Enrique Carlos.** Concepciones sobre la relacion entre géneros: mito, ritual y organización del trabajo en la unidad doméstica Campa-Asháninka. (*Amazonía Perú.*, 22, oct. 1992, p. 175–220, bibl. tables)

Provides detailed information on the division of labor by sexes among the Ashaninka of Peru. Argues that myths and rituals legitimize the strict separation of gender roles. Although Ashaninka concepts of gender emphasize the complementarity of male and female roles in everyday production, a male-dominated gender hierarchy is constructed through rituals and myths that emphasize female pollution taboos.

Urteaga Cabrera, Luis. El universo sagrado: versión literaria de mitos y leyendas de la tradición oral shipibo-coniba. See *HLAS 54: 4025.*

925   **Vickers, William T.** Processes and problems of land demarcation for a native Amazonian community in Ecuador. (*Law Anthropol. Int. Jahrb. Rechtsanthropol.*, 3, p. 203–245)

Examines the struggles for land rights among Siona and Secoya peoples of Ecuadorian Amazon in contexts of land invasions by mestizo colonists, oil companies, and others. Demonstrates how land-granting laws are often poorly conceived due to lack of familiarity with local cultural and ecological conditions. Also shows how emergence of an Interinstitutional Commission partially succeeded in protecting Sioni and Secoya lands. Concludes that enforcement of land laws depends on indigenous peoples themselves.

926   **Vickers, William T.** Los sionas y secoyas: su adaptación al ambiente amazónico. Quito: Ediciones ABYA-YALA; Roma Italia: MLAL, 1989. 374 p., 10 leaves of plates: bibl., ill. (Col. 500 años; 9)

Intensive ecological study of horticulture, hunting, and other subsistence activities among the Sioni-Secoya, a Western Tukanoan people of the Ecuadorian Amazon. Finds that indigenous gardening activities are highly efficient (over 56-to-1 ratio of calories produced versus spent). Hunting and fishing are also very efficient in newly settled areas, and local groups split apart and relocate soon after hunting efficiency begins to decline. Argues that such movements explain the indigenous population's ability to manage fish and game resources without overexploitation of them.

927   **Whitten, Dorothea S.** and **Norman E. Whitten.** From myth to creation: art from Amazonian Ecuador. Urbana: Univ. of Illinois Press, 1988. 64 p.: bibl., ill. (some col.).

Interpretation of ceramic and other material arts of the Canelos Quichua of Lowland Ecuador. Includes numerous photographs of artifacts displayed at the Krannert Art Museum, Univ. of Illinois, April–May, 1988. Demonstrates that indigenous material arts form a major dimension of complex cognitive processes rooted in knowledge of rain forest ecology and undifferentiated power of mythic space-times. Canelos Quichua arts also embody indigenous ways of understanding historical changes. A powerful demonstration of indigenous creativity and resilience.

PARAGUAY, ARGENTINA, AND BOLIVIA

928   **Albó, Xavier.** La comunidad hoy. La Paz: Centro de Investigación y Promoción del Campesinado, 1990. 433 p.: bibl., ill.,

index, map. (Cuadernos de investigación; 32. Los Guaraní-chiriguano; 3)

Third volume in series of three works on Chiriguano of Bolivia. Examines recent history and contemporary social and economic relations in five Chiriguano communities, including migratory wage labor, employment on cattle ranches, and communal labor. Concludes by showing how Chiriguano have recently begun to form representative organization to deal with interethnic relations. Also describes a project of indigenous history in which the Chiriguano themselves are investigating the long-term processes that have led to their present condition of poverty and marginalization.

929 Meliá, Bartomeu. Ñande reko, nuestro modo de ser y bibliografía general comentada. La Paz: Centro de Investigación y Promoción del Campesinado, 1988. 222 p.: ill., index, maps.

First volume of three in series aimed at commemorating the Chiriguano of Bolivia a century after their military defeat in 1892. Author interprets Chiriguano history and religion in terms of the general patterns of Tupi-Guarani and concludes that the Chiriguano are representative of these overall patterns but that they have absorbed local cultural and ecological traits from their new location in the region between the Chaco and the Andean Highlands. Includes annotated bibliography on all historical and ethnological works about the Chiriguano.

930 Pifarré, Francisco. Historia de un pueblo. La Paz: Centro de Investigación y Promoción del Campesinado, 1989. 542 p., 3 folded leaves of plates: bibl., ill., index, maps. (Cuadernos de investigación; 31. Los Guaraní-chiriguano; 2)

Second volume in series of three works on Chiriguano history. In one of the most complete histories of an indigenous people of Lowland South America, author shows how Chiriguano identity emerged out of a complex matrix of interethnic relations during the early colonial period. Chiriguano resistance and adaptation to Jesuit and other missionaries during the colonial period was followed by military conquest and genocide during the 19th-century expansion of the Bolivian nation-state. Also examines 20th-century changes, such as the Agrarian Reform of 1953.

931 Verna, Maria Alejandra. Askiúna: une expression de la passion amoureuse dans la culture chorote. (*Schweiz. Amer. Ges.*, 52, 1988, p. 57–66, bibl.)

Analysis of the *askiúnata* among the Chorote of Gran Chaco, in which young men, dressed in special clothing and engaging in specific forms of dance, hunt young women and seduce them. The *askiúnata* is examined by reference to mythology and through a consideration of the techniques used for attracting young women. [J. Rappaport]

Weber, Jutta. Población indígena de las tierras bajas de Bolivia. See item **4980**.

Highlands

JOANNE RAPPAPORT, *Associate Professor of Spanish and Latin American Studies, Georgetown University*

DOES THE SOCIAL, ETHNIC, OR NATIONAL ORIGIN of the observer impact significantly upon what she or he has to tell us about Andean society? This is a question that anthropologists, both foreign and national, confront today as we enter indigenous communities, peasant villages, and urban shantytowns to conduct ethnographic research among populations that are increasingly aware of the benefits and shortcomings of being studied and are alert to how international events impact upon everyday life. The people we study are participants in (and frequently the architects of) new social movements nurtured by development projects, legislative innovations, and political activism, which link the village or the barrio to the wider world. Such transformations represent a challenge to ethnographic investigation. While authors in the north are increasingly answering the summons to innovate through the adoption of new themes, multisite ethnographies, and historically-oriented analysis, Andean investigators are defining distinctive national and ethnic anthropologies, demonstrating that the questions we ask and the sorts of answers we provide may well depend upon who is the one asking.

A number of important ethnographic monographs have appeared in English in this biennium, all of which examine how a variety of ethnic groups in the Andes have accommodated or resisted State institutions and other social arrangements of the dominant society. How do Quechuas and Aymaras navigate and negotiate the city? The issue of migration is addressed in novel ways by Lesley Gill, Sarah Skar, and Thomas Turino. Gill examines how ethnicity and class are represented through such symbolic means as clothing by female domestic workers and employers in the urban setting (item **953**). Skar suggests that Quechua urban life must be understood as part of a continuous movement from city to various rural settings, and that in each of these sites Andean people make sense of the surrounding reality by investing it with a unique metaphoric content (item **1061**). For Turino, music is a means to establishing a new identity in urban community centers, as well as a form of maintaining roots in the rural village, where urbanites participate in and sponsor festivals (item **1068**).

Hegemony and resistance are the themes of monographs by María Lagos, June Nash, Joanne Rappaport, Roberto Santana, and Harry Sanabria. Lagos investigates merchant-peasant relations in terms of class ideologies, hegemonic practices and ideologies, and peasant struggle (item **956**). Nash's life history of a tin miner explores proletarian resistance in the eyes of a single participant (item **964**). The creation of an anti-hegemonic ideology by native Colombians is scrutinized in Rappaport in her study of Cumbal history-making (item **999**). Santana suggests that the Indian movement in Ecuador can only be appreciated if we trace the development of an indigenous ideology in conjunction with similar developments in the broader mestizo society (item **1024**). A more hazardous response to inequality is the theme of Sanabria's inquiry into Bolivian peasant participation in the cocaine trade (item **966**).

The growing importance of Afro-Latin communities in national dialogues in the Andean countries is mirrored by several new book-length publications. Fernando Romero's monumental dictionary of Afro-Peruvianisms attests to the careful attention scholars are paying to the distinctiveness of African culture in the Andes (item **1053**). Among the Andean nations, of greatest significance is the entry of Afro-Colombians into politics with their historic participation in the creation and implementation of a new multiethnic constitution. Peter Wade's examination of racial attitudes, stereotypes, and metaphors in Colombia (item **1005**) mirrors similar excursions into ethnicity and power by students of indigenous Andean societies.

While foreign scholars' emphases tend to run toward historically-oriented investigations of the economic, political, and symbolic manifestations of interethnic relations in local settings, native Andean investigators are intent upon defining a distinctive brand of ethnographic discourse that concentrates, for the most part, upon the elaboration of new forms of cultural and historical description by providing a bird's-eye view of what it means to be Andean. The reedition of Manuel Quintín Lame's 1939 treatise on the place of Indians in Colombian society (item **995**) marks one of the earliest contributions to this growing movement. The most sophisticated studies by native Andeans have been conducted, for the most part, in Bolivia, where university-educated investigators have returned to their communities to produce in-depth diagnostics of local problems and to search out voices unheard before. Most significant is the work of the Taller de Historia Oral Andina and affiliated investigators (items **946, 948,** and **954**), although other institutions have also contributed to this body of indigenous-authored literature (item **961**). Indigenous organizations and individual authors in Colombia (items **985** and **986**) and in Ecuador (item **1014**), dissatisfied with the interpretations of outside observers, have also joined this dialogue. As a corollary to the rise of the indigenous ethnographers' movement, a number of significant collaborative studies between national or foreign investigators and their native colleagues have resulted in unique and novel transformations of the traditional ethnographer-informant relationship (items **947** and **1004**).

While native ethnographers find their voice, anthropologists belonging to the national cultures of Andean countries have also been engaged in the creation of distinct Latin American anthropologies. The most significant contributions to this new body of literature dwell upon pressing issues of social concern to the national societies. Teófilo Altamirano's continuing investigation into urban marginality in Lima, as well as other Peruvian cities (item **1026**), marks one direction in which Latin American anthropologists have moved in much greater numbers than their North American counterparts. A second avenue of research is exemplified by Ecuadorian Jorge León (item **1016**) and by the Colombian contributors to collections edited by François Correa (item **1004**) and by Esther Sánchez (item **988**), all of which inquire into the relationship of indigenous populations to the State. León's emphasis centers upon the diversity of indigenous political players in the recent Indian Uprising, while Correa and Sánchez's volumes both focus on the writing of ethnicity into Colombian legislation.

This biennium marks the loss of one of the fathers of Colombian anthropology, Gerardo Reichel-Dolmatoff, (1912–1991) whose classic monographs on the Desana and the Kogi have travelled well beyond the borders of Colombia and have survived the ravages of time. He will also be remembered as the founder of the Dept. of Anthropology at the Universidad de los Andes, a training ground for many of Colombia's most creative anthropological minds.

GENERAL

932 Brysk, Alison and **Carol Wise.** Economic adjustment and ethnic conflict in Bolivia, Peru, and Mexico. Washington: The Woodrow Wilson Center, 1995. 40 p. (Latin American Program, Working Papers; 216)

Examines impact of economic adjustment on ethnic conflict in Bolivia, Mexico, and Peru. Variations are traced to type of adjustment to economic crisis and availability of institutional channels for expressing ethnic demands. The most violent conflict is in Peru (delayed adjustment, no response to demands); the most peaceful patterns are in Bolivia (rapid adjustment, institutionalized incorporation).

933 Different places, different voices: gender and development in Africa, Asia, and Latin America. Edited by Janet Henshall Momsen and Vivian Kinnaird. London; New York: Routledge, 1993. 322 p.: bibl., ill., index, maps. (International studies of women and place)

The Latin America section of this volume concentrates on geographical treatments of gender and development, looking in particular at changing gender relations in the migration process to frontier zones in Colombia and Peru, and at women in the productive process in highland Bolivia.

934 Gentile Lafaille, Margarita E. Salud, dinero y amor: ensayo sobre amuletos andinos actuales. Prólogo de Luis Millones.

Buenos Aires: Casimiro Quiros Editor, 1989. 112 p.: bibl., ill., index.

Useful catalog of 30 amulets acquired in various Andean markets, accompanied by an analysis of their symbolism.

935 **Imashi! Imashi!: adivinanzas poéticas de los campesinos indígenas de la sierra andina ecuatoriana/peruana.** Recopilación de Charles D. Kleymeyer. Quito: Ediciones Abya-Yala, 1990. 129 p.

Quechua-English compendium of riddles collected in Ecuador, Peru, and Bolivia, organized by categories that range from types of plants and animals, to human material culture and the human body.

936 **Indianidad, etnocidio, indigenismo en América Latina.** Traducción de Ana Freyre de Zavala. 1a ed. en español. México: Instituto Indigenista Interamericano: Centre D'Études Mexicaines et Centraméricaines, 1988. 354 p.: bibl.

Collection of interdisciplinary reflections by French scholars on *indigenista* policies, multilingualism, and ethnic rights movements in Latin America. For the Andean region, the volume includes historical analyses of Indian policies in Peru (Bonilla, Favre, Vayssière), and Bolivia (Saignes), examinations of language policies in Colombia (Landaburu) and Peru (Escobar), and discussions of indigenous movements in Colombia (Gros), Bolivia (Le Bot) and Ecuador (Santana). Also note important contributions on indigenous movements of lowland Ecuador (Descola, Fauroux, Taylor) and Brazil (Albert, Menget).

Jaguar no sokuseki: Andesu, Amezon no shūkyo to girei [Jaguar's trace: religion and ritual in the Andes and the Amazon]. See item 859.

937 **Molinié Fioravanti, Antoinette.** Comparaisons transatlantiques. (*Homme/Paris*, 122/124, avril/dec. 1992, p. 165–183, bibl.)

Case-study of the "fat-taker," a symbolic figure found today in the Andean region, illustrates how cultural mestizaje operates. Draws both from historically-documented representations of 17th-century Spain and from ethnographies of precolumbian indigenous practices. [A. Pérotin-Dumon]

938 **Vigorización de la chacra andina.** Lima: PRATEC: PPEA (PNUMA), 1991. 236 p.: bibl., ill., maps.

Collection of case studies from Bolivia, Ecuador, Paraguay, and Peru examines indigenous agricultural knowledge. Also contains articles on indigenous political organizing and native education. Based on the assumption that Andean forms of knowledge can be contrasted with Western technologies and that this dichotomy must form the basis of any alternative development scheme.

ARGENTINA

939 **Arenas, Patricia.** Antropología en la Argentina: el aporte de los científicos de habla alemana. Buenos Aires: Institución Cultural Argentino-Germana; Museo Etnográfico J.B. Ambrosetti, Facultad de Filosofía y Letras de la U.B.A., 1991. 136 p.: bibl., ill.

A welcome addition to the small number of investigations into the history of Latin American anthropology, this brief volume provides concise sketches of the lives and the intellectual and institutional contributions of German ethnologists and archaeologists in Argentina.

940 **Barreda, Adriana.** Images of a forgotten nation. Buenos Aires: Gaglianone Establecimiento Gráfico, 1994. 168 p.: bibl., ill., map.

Exceptionally fine b/w photographs by Barreda of individuals, objects, dwellings, animals, etc., of Mapuches of Neuquén prov., Argentina. Project sponsored by UN, OAS, and several Argentine corporations. Includes bibliography on Neuquén's Mapuches and introductions by Carlos Valiente Noailles and Carlos Martínez Sarasola. [Ed.]

941 **Borrero, Luis Alberto.** Los selk'nam (onas): evolución cultural en la Isla Grande de Tierra del Fuego. Buenos Aires: Búsqueda-Yuchán, 1991. 128 p.: bibl., ill. (Col. Desde Sudamérica)

Investigates territorial strategies of the hunting and gathering Selk'nam as they faced European expansion at the end of the 19th century. Argues that excessive isolation in response to outside pressures led to a decrease in spatial movements and a consequent loss of adaptive flexibility.

942 **Chapman, Anne MacKaye.** El fin de un mundo: los selk'nam de Tierra del Fuego. Buenos Aires: Vázquez Mazzini Editores, 1989. 309 p.: bibl., ill.

Anthology of Chapman's publications on the Selk'nam includes anthropological essays reconstructing Selk'nam myth and ritual, song, social organization, and subsistence strategies, and a narrative of the destruction of autonomous Selk'nam society in the late 19th century. A film script and the life history of an Ona woman complete this collection, which is illustrated by numerous historical and contemporary photos.

943 Fischman, Gustavo and **Isabel Hernández.** La ley y la tierra: historia de un despojo en la tribu mapuche de Los Toldos. Buenos Aires: Centro de Estudios Avanzados, Univ. de Buenos Aires; Centro Editor de América Latina, 1990. 193 p.: bibl., ill. (Papeles políticos)

This meticulous history of land-loss in a Mapuche community during the late-19th and early-20th centuries includes genealogical reconstructions of the central actors, tabulations of data from court cases involving land, and an extensive documentary appendix.

944 Mychaszula, Sonia M.; Germán Pollitzer; and **Jorge L. Somoza.** Infant mortality in Junín de los Andes and in the Mapuche population in the south of Neuquén Province: studies carried out between 1984 and 1989. Buenos Aires: Fundación Cruzada Patagónica, 1991. 64 p.: bibl., ill.

Brief summary in Spanish and English of the results of four infant mortality studies among Mapuches, including a retrospective survey of rural socioeconomic conditions, a collection of data based on the previous-child method, a census of a small urban area, and a demographic survey. Results point to declining infant mortality in the 1980s, but underline the unreliability of estimating infant mortality from official registers.

945 Testimonios mapuches en Neuquén. Neuquén, Argentina: Fundación Banco Provincia del Neuquén, 1992. 364 p.: bibl., ill.

Collection of folktales, sayings, verses, personal reminiscences, drawings, and songs collected from a wide range of Mapuche narrators, identified by name, age, and residence, from Aluminé dept., Neuquén prov. Especially appealing are the reminiscences, which range from recipes and remedies to knowledge of natural and supernatural phenomena, and the personal memories, which include family histories, autobiographies, chronicles of local schools, and information on historical material culture.

BOLIVIA

946 Andean Oral History Workshop. The Indian Santos Marka T'ula, chief of the *ayllus* of Qallapa and general representative of the Indian communities of Bolivia. Translated by Emma Gawne-Cain. (*Hist. Workshop*, 34, Autumn 1992, p. 101–118)

Originally published as a mimeographed pamphlet, this history of a turn-of-the-century Aymara leader, Santos Marka T'ula, is one of the first modern examples of Aymara history-writing, produced by a collective of indigenous university students. A useful introduction by Olivia Harris.

947 Arnold, Denise; Domingo Jiménez Aruquipa; and **Juan de Dios Yapita.** Hacia un orden andino de las cosas: tres pistas de los Andes meridionales. Coordinación de Denise Y. Arnold. La Paz: Hisbol; ILCA, 1992. 274 p.: bibl., ill., map. (Biblioteca andina; 12)

A series of ethnographic texts written by Arnold in concert with two Aymara authors analyzes the common logic manifest in ritual language and song associated with domestic architecture, agricultural products, and wild animals. The care with which these Aymara texts are interpreted is unusual in Andean scholarship.

948 Ayllu Sartañäni. Pachamamax tipusiwa = La pachamama se enoja. I: Qhurqhi. La Paz: Aruwiyiri; Ayllu Sartañäni, 1992. 215 p.: bibl., ill., maps. (Serie Cuál desarrollo?; 1)

Ethnographic text written by a collective of young people from Ayllu Sartañäni, Qhurqhi, an Aymara-speaking community in Oruro dept. Analyzes the effects of an unequal integration of the national administrative system within the Andean system of *ayllus* and *parcialidades*. This results in an unequal balance of power fostering cultural subordination and the economic, social and cultural destruction of Aymara regional society.

949 Berg, Hans van den. Material bibliográfico para el estudio de los aymaras, callawayas, chipayas, urus. v. 5. Suplemento II.

Cochabamba, Bolivia: Univ. Católica Boliviana, Facultad de Filosofía y Ciencias Religiosas, 1988. 1 v.: index.

Exhaustive collection of recent English, French, German, and Spanish sources for the ethnohistory and ethnography of the Aymara, Callawaya, Chipaya and Uru supplements earlier bibliographies. Most of the items date from the 1970s and 1980s, and are organized by ethnic group, catalogued by author, and indexed by topic.

950 **Crandon-Malamud, Libbet.** Blessings of the Virgin in capitalist society: the transformation of a rural Bolivian fiesta. (*Am. Anthropol.*, 95:3, Sept. 1993, p. 574–596, bibl.)

Examines how *compadrazgo* ties, in the context of a fiesta in a highland town, are used as a funnel for resources by migrants to La Paz. From the town's vantage point, the fiesta can be seen as a failure of formal capitalism to develop in the periphery of the national economy; from the standpoint of La Paz, however, it represents a successful capitalist metamorphosis of an institution.

951 **Fortún, Julia Elena.** Festividad del gran poder. La Paz: Ediciones Casa de la Cultura, 1992. 60 p.: bibl., col. ill.

A detailed description of the history, participants, costumes, and dances in the celebration of Nuestro Señor del Gran Poder in La Paz.

952 **Gill, Lesley.** Precarious dependencies: gender, class, and domestic service in Bolivia. New York: Columbia Univ. Press, 1994. 175 p.: bibl., ill., index.

Examines gender, class, and ethnicity in urban Bolivia through ethnographic and historical study of domestic servants and their employers in La Paz. Discusses how changing power relationships sustain, alter, and recreate social, cultural, and ethnic distinctions between employers and domestic workers. For sociologist's comment see item **4959**.

953 **Gill, Lesley.** "Proper women" and city pleasures: gender, class and contested meanings in La Paz. (*Am. Ethnol.*, 20:1, Feb. 1993, p. 72–88, bibl.)

This outstanding article explores fashion as a cultural practice contested by women of various class and ethnic backgrounds in La Paz and argues that hegemonic notions of the ideal woman are not inclusive of all women and are constantly contested in the process of class, gender, and ethnic formation.

954 **Huanca L., Tomás.** El yatiri en la comunidad aymara. La Paz: Ediciones CADA, 1990. 244 p.: bibl.

Rigorous ethnolinguistic study of oral traditions associated with medical specialists (*yatiris*) among the Aymara. Includes extensive transcriptions of Aymara texts, Spanish translations of the same, and a morphological analysis of part of a text.

955 **Intipampa, Carlos.** Opresión y aculturación: la evangelización de los aymara. La Paz: HISBOL, 1991. 197 p.: bibl., ill. (Serie Religión y sociedad)

Poses and reflects upon the following theological question: can the conversion of the Aymara to Christianity be evaluated positively, or was the gospel manipulated in the interests of the Church? Uses theological, historical, and anthropological methodologies to conclude that Catholic and Protestant missionary activities have not brought the true message of the gospel to the Aymara.

956 **Lagos, Maria L.** Autonomy and power: the dynamics of class and culture in rural Bolivia. Philadelphia: Univ. of Pennsylvania Press, 1994. 206 p.: bibl., index, maps. (The Ethnohistory series)

Focusing on the ambiguities and paradoxes of daily life in Tiraque, a canton in Cochabamba, this ethnography analyzes the dynamics of domination and resistance through a study of the process of hegemony in social and cultural struggles. Examines the different forms hegemony takes in specific historical contexts and cultural arenas. For sociologist's comment see item **4962**.

957 **Lagos, Maria L.** "We have to learn to ask:" hegemony, diverse experiences, and antagonistic meanings in Bolivia. (*Am. Ethnol.*, 20:1, Feb. 1993, p. 52–71, bibl.)

Interprets the Cochabamba celebration of the Virgin of Urkupiña as a cultural arena of contestation in which hegemony is constituted. The cult has changed from a local indigenous fiesta to a multiclass national phenomenon in the course of a struggle over the meaning of shared concepts and symbols.

958 **Leons, Madeline Barbara.** Risk and opportunity in the coca/cocaine economy of the Bolivian Yungas. (*J. Lat. Am. Stud.*, 25: 1, Feb. 1993, p. 121–157, graph)

Explores impact of the cocaine trade on coca growers of the Yungas. Focuses on strategies for coping with the fluctuating fortunes of the coca economy and the economic and social differentiation created by the new technology. The cocaine trade must be investigated locally in order to understand the insertion of producers into the process.

959 **Mariño Ferro, Xosé Ramón.** Muerte, religión y símbolos en una comunidad quechua. Santiago de Compostela, Spain: Univ. de Santiago de Compostela, Servicio de Publicacións e Intercambio Científico, 1989. 198 p.: bibl. (Monografías da Universidade de Santiago de Compostela; 143)

A detailed descriptive ethnography of the rituals associated with death and the life cycle and with the calendar round of Chaquilla (Potosí dept.). Embellishes descriptions by using brief case studies of individuals and associated narratives. Similar treatment is given to a catalog of sacred beings.

960 **Platt, Tristan.** Writing, shamanism and identity, or voices from Abya-Yala. (*Hist. Workshop*, 34, Autumn 1992, p. 132–146)

The metaphor of literacy is commonly used in shamanic practice in the Potosí region. Explores the impact of literacy on native Andeans, ethnographically and historically, in this provocative article.

961 **Portugal M., Pedro.** La República Aymara de Laureano Machaka. (*Yachay/Cochabamba*, 9:16, 1992, p. 153–181, bibl., photo)

This Aymara-authored history of the 1956 establishment of an independent Aymara Republic by Laureano Machaka analyzes the event within the context of the 1952 Revolution, which fostered separatist mobilization in rural areas.

962 **Raíces de América: el mundo Aymara.** Recopilación de Xavier Albó. Madrid: Alianza Editorial, 1988. 607 p.: bibl., ill., index.

Superb collection of essays brings together some of the most distinguished Aymara specialists to reflect upon the history of Aymara culture. An anthology which will be consulted for years to come, the volume contains: John Murra's historical reflections upon the independent Aymara culture that predated Tawantinsuyu; Mauricio Mamani's detailed description of high-altitude agriculture; a portrait by Félix Palacios of camelid herding; Martha Hardman's excellent discussion of issues in contemporary Aymara linguistics; Olivia Harris and Thèrése Bouysse-Cassagne's philosophical analysis of the nature of the concept *pacha*; an excursion by Verónica Cereceda into the nature of Aymara aesthetics; a historical contemplation of Aymara political structure by Tristan Platt; William Carter and Xavier Albó's panorama of the forms of local organization among the contemporary Aymara; and Víctor Hugo Cárdenas' study of Aymara resistance from Inca times to the present.

963 **Rivera Cusicanqui, Silvia.** Pachakuti: los aymara de Bolivia frente a medio milenio de colonialismo. La Paz: Taller de Historia Oral Andina, 1991. 24 p.: bibl. (Serie Cuadernos de debate; 1)

Analyzes the contemporary Aymara movement in light of colonial resistance studies. Contemporary demands for an autonomous indigenous space built on the premise that autonomy can be achieved through coexistence of various ethnic groups in a heterogeneous society reflect ideas present in the struggles of colonial-era Aymaras.

964 **Rojas, Juan.** I spent my life in the mines: the story of Juan Rojas, Bolivian tin miner. Edited by June C. Nash. New York: Columbia Univ. Press, 1992. 390 p., 24 p. of plates: bibl., ill., index, map.

Juan Rojas, a tin miner, and his family first began to narrate their lives to June Nash in 1969. Completed in 1986, this story of a miner, his family, and his community documents personal hardship and family life against the backdrop of the growth of labor unions, the nationalization of the tin mines, and transformations in work-life over 60 years of recent Bolivian history. A welcome revision and English translation of a significant 1976 publication.

965 **Rojas Ramírez, Policarpio.** Historia de levantamientos indígenas en Bolivia, 1781–1985. Cochabamba, Bolivia: Editorial IDEAS UNIDAS, 1989. 76 p., 2 leaves of plates: bibl., ill.

Aymara author briefly describes major

moments in the history of Bolivian indigenous resistance. In addition to major uprisings and massacres, includes descriptions of the founding of ethnic, political and labor organizations.

966 **Sanabria, Harry.** The coca boom and rural social change in Bolivia. Ann Arbor: Univ. of Michigan Press, 1993. 277 p.: bibl., ill., index, maps. (Linking levels of analysis)

Ethnographic study of the incorporation of a rural community in Cochabamba dept. into the drug market. Argues that peasant participation in cocaine production provides a means of enhancing rural livelihoods in the face of impoverishing policies. Critically examines development policies that played a role in the emergence of the cocaine market and are today inextricably linked to the fate of the drug trade. For sociologist's comment see item **4972.**

967 **Wachtel, Nathan.** Gods & vampires: return to Chipaya. Translated by Carol Volk. Chicago: Univ. of Chicago Press, 1994. 153 p.: bibl., index, maps.

Translation from the original French (see *HLAS 53:1219*) of field research among the Uru of Chipaya. Focuses on sorcery, vampirism, and interethnic relations.

968 **Wachtel, Nathan.** Note sur le problème des identités collectives dans les Andes méridionales. (*Homme/Paris*, 122/124, avril/déc 1992, p. 39–52, bibl.)

Uses examples of the Yura, K'ulta and Chipaya to demonstrate that ethnicity can be comprehended through a combination of elements, including territoriality, dual organization, civil and religious offices, syncretic symbolic representations and practices. These combinations were crystallized into current identities during the 18th century.

CHILE

969 **Bengoa, José.** Quinquén: cien años de historia pehuenche. Santiago: Ediciones ChileAmérica, CESOC, 1992. 136 p.: maps.

Narrates history of the Pehuenches of Quinquén, concentrating on land ownership and property litigation. As a result of litigation, the Chilean government returned 30,000 hectares to the Pehuenches in 1992.

970 **Carrasco, Ana María.** Mujeres y participación social en la sociedad aymara contemporánea. (*in* Seminario Mujer y Antropología, Problematización y Perspectivas, *Santiago, 1992.* Huellas. Edición de Sonia Montecino y María Elena Boisier. Santiago: Centro de Estudio para el Desarrollo de la Mujer (CEDEM), 1993, p. 183–193, bibl.)

These brief descriptions of organizations and institutions catering to women in the Aymara area detail changes that have occurred due to Chilean democratization.

971 **Castro, Milka** and **Miguel Bahamondes.** Control de la tierra en la cabecera del Valle de Lluta. (*Rev. Chil. Antropol.*, 7, 1988, p. 99–113, bibl., maps, tables)

Aymara-speakers in Parinacota prov. maintain a complex network of relations for exchange of products extending to border towns in Peru and Bolivia, as well as to coastal valleys of Chile and Peru. Includes statistical data on landowners' places of origin, size of plots, and control of land.

972 **Castro, Milka *et al.*** Cultura, tecnología y uso del agua en un pueblo andino del norte de Chile. (*Rev. Chil. Antropol.*, 10, 1991, p. 45–69, bibl., ill., maps)

A study of technological, organizational, and ideological aspects of the use and maintenance of water resources in Parinacota prov. Includes details on rotational systems, irrigation technology, and land use.

973 **Dannemann, Manuel.** Las comunidades pehuenches y su relación con los proyectos hidroeléctricos del Alto Bío-Bío. (*Rev. Chil. Antropol.*, 10, 1991, p. 109–146, bibl., tables)

Hydroelectric projects on the Upper Bío-Bío are evaluated with regard to their possible impact on the Pehuenche population, an economically marginalized indigenous group transformed into poor peasants. Negative factors include loss of lands, influx of outsiders, and cultural change, while positive factors include infrastructural development and improved employment possibilities.

974 **Gundermann Kroll, Hans** and **Héctor González Cortez.** La cultura aymara: artesanías tradicionales del altiplano. Santiago: Depto. de Extensión Cultural del Ministerio de Educación; Museo Chileno de Arte Precolombino, 1989. 112 p.: bibl., ill. (some

col.), map. (Serie El Patrimonio cultural chileno. Col. Culturas aborígenes; 8)

Brief general descriptions of Aymara handicrafts, accompanied by excellent color photos, are followed by a catalog of the Museo Chileno de Arte Precolombino's collection of Aymara weavings, silverwork, ceramics, and other handicrafts.

975 **Laborde, Miguel.** La selva fría y sagrada. Santiago: Editorial Contrapunto, 1990. 143 p.: bibl., ill.

Explains Mapuche cosmology in literary language through the vehicle of letters from a father to his son.

976 **Marimán, José.** Cuestión mapuche: descentralización del Estado y autonomía regional. (*Caravelle/Toulouse*, 59, 1992, p. 189–205)

Reflections on Chilean indigenous policy under a new democratic government. The contemporary indigenous movement grew out of a Mapuche culture created by official national policies of pacification and ethnic isolation. Administrative decentralization must be achieved at the regional level, where ethnic differences can be taken into account, thus deepening the democratic process.

977 **Martínez, Gabriel.** Espacio y pensamiento: Andes meridionales. v. 1. La Paz: HISBOL, 1989. 1 v.: bibl., ill., maps. (Biblioteca andina; 6)

An important collection of Martínez's articles, some previously published and others appearing for the first time. Two deal with the organization of space in northern Chile: the sacred geography of Isluga and the irrigation space of Chiapa. The third examines a map of Lupaqa lands, developed from 16th and 17th century documents, and analyzes Lupaqa notions of space.

978 **Mege R., Pedro.** La imagen de las fuerzas: ensayo sobre un mito mapuche. (*Bol. Mus. Chil. Arte Precolomb.*, 5, 1991, p. 9–22, bibl., ill., map, photos)

Posits that a story's efficacy resides in its development of a mythic narrative that does not moralize, or make explicit its pedagogical intentions. Then explores the syntagmatic and paradigmatic axes of a Mapuche myth to determine its efficacy.

979 **Molina Otárola, Raúl.** El pueblo huilliche de Chiloé: elementos para su historia. Chiloé, Chile: Oficina Promotora del Desarrollo Chilote, 1987. 71 p.: bibl., ill., maps.

The product of collaborative research with the Consejo General Indígena Huilliche de Chilué, this brief mimeographed history narrates the loss of indigenous lands and the rise of a Huilliche organization at mid-century. Meant for use in the indigenous community.

980 **Morris von Bennewitz, Raúl.** Platería mapuche. Imágenes de Juan Carlos Gedda O. Santiago: Editorial Kactus, 1992? 95 p.: ill. (some col.)

Excellent collection of photos of Mapuche silver jewelry and implements, but lacks information identifying artists and motifs. Accompanied by archaeological comparisons.

981 **Seminario Desarrollo Andino y Cultura Aymara en el Norte de Chile, Iquique, Chile, 1989.** Desarrollo Andino y Cultura Aymara en el Norte de Chile. Edición de Héctor González Cortez y Bernardo Guerrero J. Iquique, Chile: Ediciones el Jote Errante, 1990. 156 p. (Serie Publicaciones ocasionales; 6)

The published minutes of a seminar on Aymara culture and development attended by Chilean scholars, development workers, educators, missionaries, and Aymara activists, as well as by foreign experts on the Aymara. Interesting interplay of opinions by representatives of different sectors.

982 **Soto-Heim, Patricia.** Isonymie et consanguinité chez les mapuches, Quetrahue, Chile. (*Bull. mém. Soc. anthropol. Paris*, 3:1/2, 1991, p. 97–111, bibl., map, tables)

Analyzes consanguinity through studies of isonymy and genealogy among the Mapuche of Quetrahue and reveals a high degree of consanguinity and an important degree of mobility of females, as evidenced by a higher heterogeneity of women's patronyms.

COLOMBIA

983 **Aptekar, Lewis.** Street children of Cali. Durham: Duke Univ. Press, 1988. 235 p.: bibl., ill., index.

Psycho-ethnography of *gamines* or street children of Cali provides vital information on their places of origins, their social and economic survival strategies, their language, and the social programs available to them. An important first step in understanding the 40 million children living on the streets in Latin America.

984 **Arnson, Cynthia** and **Robin Kirk.** State of war: political violence and counterinsurgency in Colombia. New York: Human Rights Watch, 1993. 149 p.: bibl.

An indispensable report for Colombianists, this study documents complicity between the Colombian military and paramilitary organizations in what the authors call a "total war" in the Colombian countryside. Also includes documentation of abuses by guerrilla organizations.

985 **Berichá.** Tengo los pies en la cabeza. Bogotá: Los Cuatro Elementos, 1992. 154 p., 16 p. of plates: ill. (some col.)

A mission-educated U'wa teacher, Berichá served as a linguistic informant before she became an ethnographer in her own right. Most of this book reconstructs oral traditions she learned from her mother, followed by a description of religious life, healing practices, productive life, and the family among the contemporary Tunebo. Illustrated with drawings by the author.

986 **Comunidad Camëntsá (Colombia).** Procesos de transformación y alternativas de autogestión indígena. Bogotá: Editorial ABC, 1989. 107 p.: ill., maps.

Ethnography prepared by a collective of Camëntsá authors covers the socioeconomic organization of their community, its political structure, problems within the community, and possible solutions. The community undertook this research because they felt that outside observers had not adequately captured the realities of Camëntsá life.

987 **Congreso de Antropología en Colombia, 6th, Bogotá, 1992.** Memorias: conflicto social & violencia; notas para una discusión. Bogotá: Sociedad Antropológia de Colombia; Institut Français d'Etudes Andines, 1993. 78 p.

Among the contributions to this timely anthology on anthropological approaches to violence are several fascinating studies: Camacho provides a sociological perspective on the relations between culture and the drug trade; García and Steiner explore social conflict in northern Antioquia, one of the most violent regions of Colombia; and Uribe reports on field research conducted in the emerald mines of Boyocá, where the violence of everyday life and political violence are intimately related.

988 **Congreso Nacional de Antropología, 6th, Bogotá, 1992.** Antropologia jurídica: normas, formas, costumbres legales. Recopilación de Esther Sánchez Botero. Bogotá: Sociedad Antropológica de Colombia, 1992. 163 p.:

One of the first contributions to a distinctly Colombian legal anthropology, this volume constitutes a dialogue between anthropologists and lawyers specializing in indigenous issues. Articles focus on local systems of social control, anthropologists as brokers in criminal cases, ecopolitics in indigenous areas, and the possibilities for ethnic self-definition opened by the 1991 Constitution. Especially interesting is François Correa's historical examination of indigenous policy since Independence.

989 **Congreso Nacional de Antropología, 6th, Bogotá, 1992.** La construcción de las Américas. Edición de Carlos A. Uribe. Bogotá: Univ. de los Andes, Facultad de Humanidades y Ciencias Sociales, Depto. de Antropología, 1993. 287 p.: bibl., ill.

Articles on ethnicity and identity by Colombian and foreign scholars written as the new Constitution ushered in a multiethnic era by enfranchising Afrocolombians and indigenous peoples. Of special interest are contributions on Afrocolombian identity and the State (Arocha, Friedemenn, Taussig, Wade) and indigenous ethnicity in a multiethnic nation (Jimeno, Henao).

990 **Digges, Diana** and **Joanne Rappaport.** Literacy, orality and ritual practice in Highland Colombia. (*in* The ethnography of reading. Edited by Jonathan Boyarin. Berkeley: Univ. of California Press, 1993, p. 139–155)

Argues that literacy and orality cannot be treated as distinct poles of expression. Using the Pasto community of Cumbal as an ethnographic example, illustrates the oral nature of literacy and the literate nature of orality in everyday and ritual life.

991 **Faust, Franz X.** Apuntes al sistema médico de los campesinos de la Sierra Nevada del Cocuy. (*Boletín/Bogotá*, 26, enero/marzo 1990, p. 43–63, bibl., map, photos)

Describes healing practices among mestizo peasants of the Sierra Nevada del Cocuy and includes information on concepts of calor/frío and medical specialistas. Contains some case studies.

992 **Field, Les W.** Harvesting the bitter juice: contradictions of Páez resistance in the changing Colombian nation-State. (*Identities/Yverdon*, 1:1, 1994, p. 89–108, bibl.)

Author's collaboration with Páez engaged in agricultural experiments was aimed at advocating cultural and territorial rights and challenging anthropology's relationship with the nation-State. This theoretically sophisticated article reflects on how the development project reproduced historical contradictions between resistance and nation-building.

993 **Gros, Christian.** Attention! un Indien peut cacher un autre: droits indigènes et nouvelle constitution en Colombie. (*Caravelle/Toulouse*, 59, 1992, p. 139–160)

Argues that the constitutional recognition of an indigenous population with its own specific rights does not represent the democratization of political life in Colombia. Instead, the constitution serves as a means for a formerly-weak State to ensure its presence and intervention in indigenous areas.

994 **Hombres de páramo y montaña: los Yanaconas del Macizo colombiano.** Bogotá: Instituto Colombiano de Antropología, Colcultura. 1993. 92 p.: bibl., maps, plates.

The Yanaconas of the southern Highlands have long been viewed as a de-indianized population. The contributors to this volume explore the realities of this invisible ethnic group, describing its history, oral tradition, subsistence strategies, and cosmology. Work of this sort marks an important shift to the study of hybrid communities in Colombia.

995 **Lame Chantre, Manuel Quintín.** Los pensamientos del indio que se educó dentro de las selvas colombianas. Bogotá: Organización Nacional Indígena de Colombia, 1987. 48 p.: ill.

Re-edition of Páez philosopher and activist Manuel Quintín Lame's classic treatise on the rights of indigenous peoples and their place in history. Published for use by indigenous communities, this new edition includes a prologue by the historian Juan Friede recounting his own experiences with Lame and a brief biography of the indigenous leader.

996 **Morales Gómez, Jorge.** Cuerpo humano y contexto cultural en el golfo de Morrosquillo. (*Rev. Colomb. Antropol.*, 29, 1992, p. 191–205, bibl.)

Fisherman on the Caribbean coast see similarities between the internal organs of humans and pigs. The author relates the model of the body as a machine for work with Bastien's model of the body as a system of fluids.

997 **Parra Rizo, Jaime Hernando** and **Claudia Afanador H.** Las guaguas de pan en San Pedro. (*in* Ecuador indígena: sincretismo e identidad en las culturas nativas de la Sierra Norte. Recopilación de Hernán Jaramillo Cisneros. Quito: Ediciones ABYA-YALA, 1991, p. 89–100, photos)

A brief description of the Fiesta de Guaguas de Pan, which takes place at the celebration of San Pedro, in Jongovito, Nariño, a mestizo town on the outskirts of Pasto. One of a small handful of ethnographic works on the descendants of the Quillacingas.

998 **Preuss, Konrad Theodor.** Visita a los indígenas Kágaba de la Sierra Nevada de Santa Marta: observaciones, recopilación de textos y estudios lingüísticos. Traducción de María Mercedes Ortiz. Bogotá: Instituto Colombiano de Antropología, 1993. 2 v.: ill.

Spanish translation of a classic study published in German in 1926, this two-volume ethnography of the Kagaba pays special attention to cosmology, mythology, and ritual. Includes an appreciation of Preuss' life and work by Carlos A. Uribe.

999 **Rappaport, Joanne.** Cumbe reborn: an Andean ethnography of history. Chicago: Univ. of Chicago Press, 1994. 245 p.: bibl., ill., index, map.

Myriad forms of historical knowledge, ranging from the mnemonics of space, through ritual and oral tradition, to the reading and production of legal papers, are explored as they relate to one another in what author terms an "ethnography of history" of a formerly Pasto community in the southern highlands.

1000 **Rodríguez, Jaime Arocha** and **Nina S. de Friedemann.** Marco de referencia histórico-cultural para la ley sobre los derechos étnicos de la comunidades negras en Colombia. (*Am. Negra*, 5, junio 1993, p. 155–172, bibl.)

In light of the legal recognition of the territorial rights of Afro-Colombian communities occupying public lands along the Pacific coast, this article provides lawmakers with an ethnographic and historical background to Afro-Colombian culture, highlighting the contributions of Black Colombians, the diversity of the Afro-Colombian experience, and central issues in Afro-Colombian history.

1001 **Salazar J., Alonso.** Born to die in Medellín. Translated by Nick Caistor. Introduction by Colin Harding. London: Latin America Bureau, 1992. 130 p.: maps.

This English-language edition of testimonies of young hired assassins in Medellín (for the Spanish original see *HLAS 53:1260*) provides a faithful translation from the original Spanish, but fails to capture the idiosyncracies of individual speakers and the flavor of their unusual argot.

1002 **Townsend, Janet G.** Gender and the life course on the frontiers of settlement in Colombia. (*in* Full circles: geographies of women over the life course. Edited by Cindi Katz and Janice Monk. New York: Routledge, 1993, p. 138–155, bibl., photos, tables)

Women's work and gender roles in three areas of colonization, and life histories of their inhabitants. Emphasizing the dangers of overgeneralization from variables as complex as age, gender, class, marital status, and personal history.

1003 **Vasco Uribe, Luis Guillermo; Abelino Dagua Hurtado;** and **Misael Aranda.** En el segundo día, la Gente Grande (Numisak) sembró la autoridad y las plantas y, con su jugo, bebió el sentido. (*in* Encrucijadas de Colombia Amerindia. Edited by François Correa R. Bogotá: Instituto Colombiano de Antropología/Colcultura, 1993, p. 9–48,)

A blending of a narrative of Guambiano history, collected by indigenous oral historians Dagua and Aranda, with ethnographic analysis directed by Colombian anthropologist Vasco. A successful model for future collaborative research in Andean settings.

1004 **Vasco Uribe, Luis Guillermo et al.** Encrucijadas de Colombia amerindia. Edición de François Correa R. Bogotá: Instituto Colombiano de Antropología, Colcultura, 1993. 334 p.: bibl., ill.

Articles by Colombian authors provide good examples of new Colombian perspectives in the study of indigenous populations. Indigenous groups are studied in relation to the nation-State (Findji, Uribe, Pineda, Correa, Romero, Alvarez, Carmona), in regional multiethnic contexts (Caycedo, Correa, Ramírez and Velásquez), and as participants in projects of resurgent ethnicity (Triana, Pardo). Of special interest is an oral history of Guambía, written as a dialogue between indigenous and national scholars (see item **1003**).

1005 **Wade, Peter.** Blackness and race mixture: the dynamics of racial identity in Colombia. Baltimore, Md.: Johns Hopkins Univ. Press, 1993. 415 p.: bibl., ill., index, maps. (The Johns Hopkins Studies in Atlantic history and culture)

Focusing on the Afrocolombian population of the Chocó, this ethnography uncovers what "blackness" means to Chocoanos and how it is reproduced and transformed in different social contexts. Deftly navigates the complexities of racial identity providing a sophisticated study of race, ethnicity and national ideology, reflecting upon the persistence of racism in a country that does not recognize itself as racist. For sociologist's comment see item **4830.**

1006 **Wade, Peter.** El movimiento negro en Colombia. (*Am. Negra*, 5, junio 1993, p. 173–191, bibl.)

Serious and reflective analysis of the context in which Afro-Colombian ethnic politics has arisen. With the current economic opening, the Pacific region has been opened up to capitalist exploitation, stimulating Black populations, who have found an appropriate vehicle in the political reforms leading to the 1991 Constitution, to defend their land rights.

1007 **Zuluaga Gómez, Víctor.** Dioses, demonios y brujos de la comunidad indígena chamí. Pereira, Colombia: Editorial Gráficas Olímpica, 1991. 239 p. (Col. de escritores de Risaralda; 6)

Myths and folktales in Spanish from Chamí shamans are embedded within the author's narrative of his own collaboration with land rights activists. Also includes a list of descriptions of 355 plants used by Chamí shamans.

ECUADOR

1008 **Antropología política en el Ecuador: perspectivas desde las culturas indígenas.** Recopilación de Jeffrey Ehrenreich. Quito: Ediciones ABYA-YALA, 1991. 294 p.: ill., maps.

An important contribution to the political anthropology of Ecuador by North American scholars. Of particular interest for Highland ethnography are two articles on the Coaiquer: 1) Ehrenreich's contribution on dissimulation and isolation as modes of cultural preservation; and 2) Kempf's study of curing practices as an informal politics promoting social cohesion and control where no formal system exists. Two studies of Otavalo examine disjunctures between community authorities and the State: Chávez studies weavers and Butler looks at the role of internal political authorities.

1009 Cifuentes, Mauro et al. Medicina andina: situaciones y respuestas. Quito: Centro Andino de Acción Popular (CAAP), 1992. 398 p.: ill. (Estudios y análisis)

This richly-documented series of ethnographic studies of folk medicine in Cotopaxi and Imbabura includes detailed information on diagnostic systems, the use of drugs by peasants, and the social context of medical knowledge. Accompanied by a valuable statistical annex on public health and health practices in Cotopaxi.

1010 Coba Andrade, Carlos Alberto G. et al. Ecuador indigena: sincretismo e identidad en las culturas nativas de la Sierra Norte. Recopilación de Hernán Jaramillo Cisneros. Quito: Ediciones ABYA-YALA, 1991. 217 p.: bibl., ill., map.

Uneven compilation of expressive culture primarily in Imbabura prov. Of greatest interest for their ethnographic detail are studies by Rohr of an Otavaleño re-reading of Mormonism, Jaramillo Cisneros on basketry in Imbabura, and a musicological analysis by Peter Banning of the *sanjuanito* in Otavalo. See also item **997**.

1011 La cultura popular en el Ecuador. v. 4, Esmeraldas. Coordinación de Marcelo Fernando Naranjo. Coordinación de la investigación de Juan Martínez Borrero. Cuenca, Ecuador: Centro Interamericano de Artesanías y Artes Populares, 1992. 1 v.: bibl.

Descriptions of popular medicine, popular religiosity, traditional music and dance, and oral tradition of the Afro-Ecuadorian and Chachi populations of Esmeraldas prov. are expertly cast within a more general ethnography and history of the region.

1012 La cultura popular en el Ecuador. v. 6, Cañar. Coordinación de Harald Einzmann y Napoleón Almeida. Coordinación de la investigación de Juan Martínez Borrero. Cuenca, Ecuador: Centro Interamericano de Artesanías y Artes Populares, 1991. 1 v.: bibl.

Provides brief information and uneven illustrations of the material culture and handicrafts, festivals, and oral traditions of mestizos and indigenous people of Cañar prov. Includes sporadic Spanish or Quichua texts of songs. More useful as a catalog or dictionary than as a detailed encyclopedia of popular culture.

1013 La cultura popular en el Ecuador. v. 7, Tungurahua. Coordinación de Marcelo Fernando Naranjo. Coordinación de la investigación de Juan Martínez Borrero. Cuenca, Ecuador: Centro Interamericano de Artesanías y Artes Populares, 1992. 1 v.: bibl.

A carefully-documented and meticulously-illustrated compendium of artisanal traditions and other popular representations of indigenous populations of Tungurahua prov. The collections of handicrafts, festival descriptions, oral and musical traditions, discussions of popular cuisine, architecture, medicine, and games are by no means exhaustive. Instead, the author has opted for an infinitely more valuable approach in which examples of popular tradition are employed as illustrations within an ethnography of popular sectors of the province.

1014 Indios, tierra y utopía: los mejores trabajos del 40 concurso de testimonios y 10 de dibujo y pintura indígena "A 500 años de resistencia, nuestros mayores cuentan su vida." Quito: CEDIS-CONAIE, 1992. 58 p.: ill. (some col.) (Serie Movimiento indígena en el Ecuador contemporáneo; 2)

Series of prize-winning personal testimonies collected by indigenous historians from Bolívar and Imbabura, accompanied by paintings by indigenous artists from Azuay, Bolívar, Cotopaxi, and Imbabura, all submitted to a contest sponsored by the Confederation of Indigenous Nationalities of Ecuador (CONAIE).

1015 Kowii, Ariruma. Mutsuctsurini. Quito: Llactamanta Quillcac Tantanacushca; Corporación Editora, 1988. 144 p.: ill.

Going far beyond the aspirations of other native authors, Kowii has embarked upon the project of creating a new Quichua-language literature. This slim volume contains 50 of his poems, written in the Imbabura variety of Quichua.

1016 **León Trujillo, Jorge.** De campesinos a ciudadanos diferentes: el levantamiento indígena. Quito: Centro de Investigación de los Movimientos Sociales del Ecuador-CEDIME, 1994. 235 p.

Analyzes the 1990 Indigenous Uprising in terms of its multiple social actors and its opponents, and their myriad actions and objectives. One of the most serious works on the *Levantamiento*, this volume illustrates the impact of such collective action on Ecuadorian social and political life. Includes commentaries by indigenous leaders and by scholars.

1017 **Manual de planeamiento andino comunitario: el PAC en el Ecuador.** Quito: COMUNIDEC, 1992. 127 p.: ill.

Workbook aimed at facilitating systematic reflection on community problems in order to encourage development planning by members of indigenous communities. Essentially an outline for indigenous peoples to produce their own goal-oriented ethnographies.

1018 **Maynard, Kent.** Protestant theories and anthropological knowledge: convergent models in the Ecuadorian Sierra. (*Cult. Anthropol.*, 8:2, May 1993, p. 246–267)

Maynard examines the contrastive process involved in evangelical identity as both church member and born-again Christian. He interprets the importance attributed to a self-conscious "decision for Christ" in order to understand his informants' theological conception of faith and its relationship to his anthropological interpretation of their social identities.

1019 **Miles, Ann.** Helping out at home: gender socialization, moral development, and devil stories in Cuenca, Ecuador. (*Ethos/Society*, 22:2, 1994, p. 132–157)

Devil stories told by migrant parents to children are considered as mechanisms for reinforcing moral lessons about gender behavior, reciprocity among kin, and family cooperation in the urban context of Cuenca. As in other areas of the Andes, these stories contain implicit critiques of capitalism and urbanization.

1020 **Moreno Yáñez, Segundo.** Antropología ecuatoriana: pasado y presente. Quito: EDIGUIAS C., 1992. 136 p.: bibl., ill. (Col. Primicias de la cultura de Quito; 1)

This general introduction to history of Ecuadorian anthropology covers research by both Ecuadorians and foreigners in archaeology, ethnohistory, and sociocultural anthropology. Concentrates on studies of indigenous peoples, but also briefly mentions work conducted among Afroecuadorians, mestizo peasants, and in urban areas.

1021 **Muratorio, Blanca et al.** Imágenes e imagineros. Edición de Blanca Muratorio. Quito: FLACSO-Sede Ecuador, 1994. 293 p.: bibl., ill. (Serie Estudios-Antropología)

Ecuadorian, North American, and European scholars examine contexts of the creation of images of indigenous people through studies of the language of visual representations, of travel narratives, of history, and of nationalistic discourse in the 19th and 20th centuries. Of special interest are Taylor's study of historical images of the Shuar and Guerrero's examination of political discourse surrounding indigenous people in the 19th century.

1022 **Rens, Marjan.** De helft van de wereld: vrouwen, identiteit en symboliek in Ecuador [Half of the world: women, identity, and symbolism in Ecuador.] Amsterdam: Het Spinhuis, 1993. 98 p.: bibl., ill.

Study of women living in indigenous community of Cochapamba focuses on two issues: 1) how women think of themselves and of their place in the community; and 2) the symbolic translation of gender relations. Based on five life stories collected during eight months of fieldwork. [R. Hoefte]

1023 **Santana, Roberto.** Actores y escenarios étnicos en Ecuador: el levantamiento de 1990. (*Caravelle/Toulouse*, 59, 1992, p. 161–188)

Santana argues that the June 1990 Indigenous Uprising represents a turning point in the notion of struggle espoused by CONAIE, insofar as it revealed the limits of indigenous peoples' bargaining power with the State. In addition, limitations to the possibilities of mass mobilization on the basis of ethnic demands were revealed at this juncture.

1024 **Santana, Roberto.** Les Indiens d'Equateur, citoyens dans l'ethnicité? Paris: Editions du Centre national de la recherche scientifique; Presses du CNRS, 1992. 240 p.: bibl., ill., index, map.

A thoughtful evaluation of the contemporary Indian movement. Argues that modern indigenous organizations cannot be understood simply as resistance to mestizo society, but rather as catalysts for a complete overhaul of the national political system, creating a multiethnic nation. Correspondingly, indigenous organizations need to create an autonomous ideology developed through internal political debate.

1025 **Weiss, Wendy A.** "Gringo . . . Gringita." (*Anthropol. Q.*, 66:4, 1993, p. 187–196, bibl.)

An analysis of the power dynamics of fieldwork using Bakhtin on dialogue and semiotic play. Shifts in the use of signifiers referencing the hegemonic hierarchy between nations, in particular shifts in the use of diminutives, mark transformations in the power relations of the field encounter.

PERU

1026 **Altamirano, Teófilo.** Cultura andina y pobreza urbana: aymaras en Lima metropolitana. Lima: Pontificia Univ. Católica del Perú, Fondo Eitorial, 1988. 209 p.: bibl., ill., maps.

Study of the survival strategies of the poorest Aymara migrants to Lima argues that it is impossible to consider urban poverty without accepting its intimate relation to rural poverty. Examines declining expectations of migrants as they adapt to urban poverty, and return migration to rural areas.

1027 **Altamirano, Teófilo.** De los Andes a las llanuras del oeste americano: el caso de los pastores. (*Rev. Peru. Poblac.*, 1, segundo semestre 1992, p. 107–128)

An ethnographic study of Quechua-speakers from the Mantaro Valley of Peru working as shepherds on ranches in Wyoming. Contrasts their experience with that of other immigrants, emphasizing their cultural isolation from North American society. One of the few extant publications on this group of immigrants.

1028 **Arguedas, José María.** Indios, mestizos y señores. 3a ed. Lima: Editorial Horizonte, 1989. 149 p.: bibl. (Etnología y antropología; 2)

Compilation of 44 articles published between 1939–44 in *La Prensa* of Buenos Aires on various themes, including mestizo expressive culture, indigenous rituals, and descriptions of various carnivals. Also includes a 1965 article on Peruvian *indigenismo*.

1029 **Béjar, Ana María.** Cultura, utopía y percepción social: los festivales por la vida y por la paz y la práctica musical juvenil en Sicuani. (*Allpanchis/Cusco*, 25:41, primer semestre 1993, p. 109–141, tables)

Based on the work of García-Canclini, Béjar explores the meaning of musical festivals for Sicuani youth in order to comprehend the meaning of these public events for the spectators, the role musical preference plays in group definition, and the utopic dreams of youth as evidenced in song lyrics.

1030 **Burga, Manuel.** Historia y memoria colectiva: violencia e identidad en el ritual andino. (*in* Congreso Nacional de Investigación Histórica, *1st, Lima, 1984*. Actas. Lima: Consejo Nacional de Ciencia y Tecnología (CONCYTEC), 1991, p. 201–226, tables)

Article weighs differing representations of Andeans and Europeans in rituals depicting the Spanish invasion. Reflects upon the process of defining a collective identity and a national Peruvian consciousness.

1031 **Cánepa, María Angela.** Recuerdos, olvidos y desencuentros: aproximaciones a la subjetividad de los jóvenes andinos. (*Allpanchis/Cusco*, 25:41, primer semestre 1993, p. 11–73, bibl.)

On the basis of the life histories of nine Quechua- and Aymara-speaking youths, this article explores the subjective worlds of the narrators from a psychological perspective, examining the meanings inherent in their discourse. Includes a highly critical commentary by Carlos Iván Degregori and more supportive comments by Andrés Gallego, Luis Herrera, Gonzalo Portocarrero.

1032 **Contreras, Carlos et al.** Comunidades campesinas y nativas: normatividad y desarrollo. Recopilación de Laureano del Castillo. Lima: Fundación Friedrich Naumann; Servicios Educativos Rurales, 1989. 182 p.: bibl. (Enfoques peruanos: Temas latinoamericanos; 12)

Valuable series of reflections on the legal nature of *comunidades campesinas* and *comunidades nativas* in Peru by legal scholars, development professionals, historians, and other scholars.

1033 **Eyzaguirre, Graciela.** Los escenarios de la guerra en la región Cáceres. (*Allpanchis/Cusco*, 23:39, primer semestre 1992, p. 155–180, tables)

Presents statistical data from 1989–91 profiling the participants in the violence, the most critical zones of conflict, and their relative regional importance in the area of Andrés Avelino Cáceres.

1034 **Fano, Hugo** and **Marisela Benavides.** Los cultivos andinos en perspectiva: producción y utilización en el Cusco. Cusco: Centro de Estudios Regionales Andinos Bartolomé de las Casas; Lima: Centro Internacional de la Papa, 1992. 86 p.: bibl., ill.

Analyzes production and consumption patterns of Andean crops in five rural communities and in Cusco. In Cusco consumption differs according to economic strata. Adaptation of these crops to the requirements of a growing urban population is a prerequisite to increasing their production while avoiding marginalization.

1035 **Irrigation at high altitudes: the social organization of water control systems in the Andes.** Edited by William P. Mitchell and David Guillet. Arlington, Va.: American Anthropological Association, 1994. 305 p.: bibl., ill. (Society for Latin American Anthropology publication series; 12)

A major contribution to our knowledge of the relationship between environment and social organization in the Andes, this volume examines irrigation in various parts of Highland Peru, concentrating on such issues as environmental variation, hydraulic technology, the political economy of irrigation, the impact of State/local polity relations on hydraulic systems, the place of ritual in irrigation, and the future of irrigation in the Andes. Unlike other edited volumes, this is not simply a collection of articles, but an integrated anthology in which contributors examine common issues.

1036 **Kapsoli, Wilfredo.** Los *pishtacos:* degolladores degollados. (*Bull. Inst. fr. étud. andin.*, 20:1, 1991, p. 61–77, bibl., tables)

Critical elements in the representation of the *pishtaco* are isolated through the comparison of twelve tales from the Central Sierra. Among the common elements are the triumph of bulls, dogs, and poor peasants over the *pishtaco* through the use of chili peppers and *ch'uño*.

1037 **Lindner, Bernardo.** Nadie en quien confiar: actitud política de los jóvenes campesinos del Altiplano puneño. (*Allpanchis/Cusco*, 25:41, primer semestre 1993, p. 77–108, bibl., table)

The product of collaborative research with Quechua-speaking youth of Puno, this article points out the ambiguities of political opinion among young peasants. While they feel they are victims of the political system, they also seek opportunities to participate politically and maintain, in the face of violence, ideals of peace and justice.

1038 **Malengreau, Jacques.** Espacios institucionales en los Andes. Lima: Instituto de Estudios Peruanos; Univ. Libre de Bruselas, 1992. 105 p.: bibl., ill. (Col. mínima, 1019–4479; 28)

Analyzes peasant land tenancy within the context of ever-broader relations, from the family to the region. Argues that land ownership is understandable only from the standpoint of these wider spheres. These higher levels of regional integration are studied from a political standpoint as strategic alliances.

1039 **Manrique, Nelson.** "El otro" de la modernidad: los pastores de puna. (*Pretextos/Lima*, 3/4, 1992, p. 103–125)

A critique of the thesis that migrants to urban areas transform and abandon tradition, entering the space of modernity. Using the case of herders in the *puna*, historical analysis and personal reminiscences demonstrate that periodic migration, coupled with exchange, has always been the rule. The rural=traditional and urban=modern equation is a gross oversimplification of a complex reality.

1040 **Mendizábal Losack, Emilio.** Continuidad cultural y textilaría en Pachitea andina. Lima: CONCYTEC, 1990. 246 p.: bibl., ill.

This posthumous publication of a richly-documented 1966 doctoral thesis concentrates on the technology of textile production in the Central Sierra, demonstrating cultural continuity from pre-Incaic times.

1041 Molinié Fioravanti, Antoinette. Sebo bueno, indio muerto: la estructura de una creencia andina. (*Bull. Inst. fr. étud. andin.*, 20:1, 1991, p. 79–92, bibl.)

Author analyzes the structural position of the *pishtaco* in Andean culture and history. The mythical fat collector cannot be considered a deity. Its indigenous meaning derives from the symbolic function of fat, while its colonial significance is tied to the practical properties of human fat in Europe.

1042 Morote Best, Efraín. Aldeas sumergidas: cultura popular y sociedad en los Andes. Revisado y corregido por Carlota Rosasco de Chacón. Cusco, Perú: Centro de Estudios Rurales Andinos Bartolomé de las Casas, 1988. 366 p.: bibl., ill., maps. (Biblioteca de la tradición oral andina; 9)

An assemblage of eleven articles by Efraín Morote Best, previously published in the late 1940s and the 1950s. Their publication in a single volume makes these classic studies in Peruvian folklore accessible to a broader public.

1043 Mossbrucker, Harald. Sharecropping: traditional economy, class relation, or social system?; towards a reevaluation. (*Anthropos/Switzerland*, 87:1/3, 1992, p. 49–61, bibl., tables)

Data from the community of Quinches (Lima dept.) indicate that an economic explanation of sharecropping is insufficient, insofar as it does not embrace the universe of social relations that support the institution. Generalizing statements, moreover, oversimplify the complexities of local situations, which must be examined on their own terms.

1044 Nolte Maldonado, Rosa María Josefa. Qellcay, arte y vida de Sarhua: comunidades campesinas andinas. Rome: Terra Nuova; Lima: Imagen Editores, 1991. 259 p.: bibl., ill. (some col.)

A stunning series of 196 color reproductions of painted *tablas* from Sarhua, Ayacucho and transcriptions of accompanying legends. The paintings, collected over a period of 15 years, are assembled in four thematic groups: paintings of love, of agriculture and festivals, cosmological representations, and relations with the broader world. Blurred reproductions and inadequate identification of artists and dates of paintings mar an otherwise impressive collection.

1045 Núñez Rebaza, Lucy. Los dansaq. Lima: Museo Nacional de la Cultura Peruana, 1990. 165 p.: bibl., ill.

Author examines the Danza de las Tijeras in the urban context as a vehicle for the maintenance of regional identity among Andean migrants in Lima and as an arena for the incorporation of metropolitan sociocultural forms into the migrants' expressive culture. Based on careful historical and ethnographic research, the history and content of the dance are contextualized within a study of the urban growth in Lima.

1046 Ossio A., Juan M. Parentesco, reciprocidad y jerarquía en los Andes: una aproximación a la organización social de la comunidad de Andamarca. Lima: Pontificia Univ. Catolica del Perú, Fondo Editorial, 1992. 407 p.: bibl., ill. (some col.), maps.

An adaptation of a 1978 doctoral dissertation, this ethnography of kinship in Ayacucho argues that society is unified through an asymmetrical hierarchical principle that involves individuals in collectivities. Social integration is prefigured by a symbolic unfolding of the community into two opposed, symmetrical, and complementary parts.

1047 Paerregaard, Karsten. Complementarity and duality: oppositions between agriculturists and herders in an Andean village. (*Ethnology/Pittsburgh*, 31:1, Jan. 1992, p. 15–26, bibl., tables)

Argues that the patterns of complementarity and duality that bind agriculturists and herders in the Colca valley can only be understood in the context of an aggregate analysis of three forms of exchange relations: barter, marriage exchanges, and moiety-based ritual life.

1048 Painter, Michael. Re-creating peasant economy in southern Peru. (*in* Golden ages, dark ages: imagining the past in anthropology and history. Edited by Jay O'Brien and William Roseberry. Berkeley: University of California Press, 1991, p. 81–106, bibl.)

Contemporary Andean forms of labor exchange, far from relics of the precolumbian past, are responses to widespread rural poverty that is the product of a particular form of capitalist expansion, minimizing cash expenditure to maintain agricultural production.

1049 Poole, Deborah A. and **Gerardo Rénique.** Peru: time of fear. London: Latin American Bureau; Monthly Review

Press, 1992. 212 p.: bibl., index, maps, photos.

An anthropologist and historian collaborate to interpret the rise of Sendero Luminoso and the impact of its violence on Peruvian society. Ranging from a discussion of *senderista* ideology, to the assumption of dictatorial powers by Fujimori, to a consideration of the coca trade, this is a significant analysis. For political scientist's comment see *HLAS 53:3935.*

1050 **Radcliffe, Sarah A.** "People have to rise up—like the great women fighters:" the State and peasant women in Peru. (*in* "Viva:" women and popular protest in Latin America. Edited by Sarah A. Radcliffe and Sallie Westwood. New York: Routledge, 1993, p. 197–218, bibl.)

Investigation of the negotiation of socially articulated peasant femininities and State-sponsored images of women during the past two decades. Concentrates on action of women in two major peasant unions, focusing in particular on Puno.

1051 **Rénique, José Luis.** Antropología e ideología: notas sobre un artículo controvertido. (*Debate Agrar.*, 15, oct./dic. 1992, p. 145–159)

Critique of Orin Starn's controversial 1991 article (see *HLAS 53:1381*). Historically situates one of Starn's two poles of comparison, the work of Antonio Díaz Martínez, arguing that it cannot be taken as a precursor of Sendero Luminoso, as Starn argues.

1052 **Rodríguez, Yolanda.** Los actores sociales y la violencia política en Puno. (*Allpanchis/Cusco*, 23:39, primer semestre 1992, p. 131–154)

Explores history of Sendero Luminoso's presence in the Puno region, focusing on the movement's strategies for gaining popular support among peasants and youth.

1053 **Romero, Fernando.** Quimba, fa, malambo, ñeque: afronegrismos en el Perú. Lima: Instituto de Estudios Peruanos, 1988. 311 p.: bibl. (Serie Lengua y sociedad; 9)

Exhaustively researched and meticulously annotated dictionary of Afro-Peruvian words and concepts by a leading student of African culture in Peru.

1054 **Sarkisyanz, Manuel.** Temblor en los Andes: profetas del resurgimiento indio en el Perú. Prólogo del Ignacio Sotelo. Quito: Ediciones ABYA-YALA; Roma: Movimiento Laicos para América Latina, 1992. 248 p.: bibl. (Col. 500 años; 49)

Views indigenismo in Peru as a series of mythic types: the myth of the indigenous past, the myth of Inca socialism, the myth of the ideal *ayllu*. Indigenismo, he argues, is best understood as a problem of the identity of intellectuals and of national integration. The history of the movement is studied from the 19th century to contemporary indigenous organizations.

1055 **Schaedel, Richard P.** La etnografía muchik en las fotografías de H. Brüning, 1886–1925. Lima: Ediciones COFIDE, 1989. 288 p., [1] folded leaf of plates: bibl., ill., map.

Based on work in the archives of German ethnologist Hans Heinrich Brüning, who lived on the north coast of Peru during the late 19th and early 20th centuries, Schaedel has created an impressive photographic ethnography built around data found in Brüning's diaries, field notes, and German-language publications. Also includes a catalog of Brüning's photographic archives.

1056 **Schmelz, Bernd.** Kontinuität und Wandel religiöser Feste im Departamento Lambayeque, Peru: eine historisch-ethnographische Analyse anhand dreier Fallbeispiele. Bonn: Holos Verlag, 1992. 358 p.: bibl., map. (Mundus Reihe Ethnologie; 57)

Uses historical-ethnographic data analysis to examine popular religion in three Peruvian case studies of origins and significance of religious festivities in Ciudad Eten (1649), Motupe (1868), and Olmos (1944). Concludes that, although still primarily religious, these local festivals serve as important socializing events, reaffirming solidarity of community, family, and friends and are even observed among far-away urban migrants. [C.K. Converse]

1057 **Seligmann, Linda J.** Between worlds of exchange: ethnicity among Peruvian market women. (*Cult. Anthropol.*, 8:2, May 1993, p. 187–213)

Through the interpretation of conversations between market women and other women, this fascinating article underlines the complexities and ambiguities of ethnic definition in Peru. Also focuses on forms of resistance used by market women in the marketplace.

1058 Seligmann, Linda J. The burden of visions amidst reform: peasant relations to law in the Peruvian Andes. (*Am. Ethnol.*, 20:1, Feb. 1993, p. 25–51, bibl.)

Examines the ways Quechua-speaking peoples interpret legal principles and use laws. Illustrates the complex ideological and socioeconomic relationships that exist between the nation-State and local communities, focusing on the efforts of rural inhabitants to transform the structure and content of legal principles themselves.

1059 Los seres del más acá: muestras sobrenaturales en la tradición oral cajamarquina. Recopilación de José Dammert Bellido y Alfredo Mires Ortiz. Selección y estudio de Alfredo Mires Ortiz. Chiclayo, Peru: Proyecto Enciclopedia Campesina, 1988. 225 p.: bibl., ill. (Biblioteca campesina: Serie Nosotros los cajamarquinos; 2)

A series of folktales and accompanying line drawings about supernatural beings, collected and drawn by peasants in the Cajamarca region.

1060 The Shining Path of Peru. Edited by David Scott Palmer. New York: St. Martin's Press, 1992. 270 p.: bibl., ill., index, maps.

Long-awaited collection of articles by anthropoligists, other social scientists, and journalists, some written for this anthology and others first published elsewhere. Includes analyses of the history, organization, and ideology of Sendero Luminoso (Degregori, De Wit and Gianotten, Gorriti, Marks, McClintock, Smith, Tarasona-Sevillano, Woy-Hazleton and Hazleton), as well as local case studies from Andahuaylas, Ayachucho, and the Huallaga Valley (Berg, Gonzales, Isbell).

1061 Skar, Sarah Lund. Lives together, worlds apart: Quechua colonization in jungle and city. Oslo: Scandinavian Univ. Press; New York: Oxford Univ. Press, 1994. 300 p., 8 p. of plates: bibl., index, maps. (Oslo studies in social anthropology)

This multi-site ethnography focuses on the movement of people between a sierra community, a colonization site in the Amazonian Lowlands, and Lima. Not an ethnography of migration or colonization *per se*, this richly-textured study examines the changing importance of objects, landscapes, boundaries, time, and personhood as Quechuas move back and forth between their Highland home and other residences.

1062 Skar, Sarah Lund. Marry the land, divorce the man: Quechua marriage the problem of individual autonomy. (*in* Carved flesh/cast selves: gendered symbols and social practices. Edited by Vigdis Broch-Due, Ingrid Rudie and Tone Bleie. Providence: Berg, 1994, p. 129–146, bibl.)

Quechua marriage rituals exemplify the reconstituting of the married person as an incomplete half of a paired unity. The ritual disrobes youths of their relative autonomy, which is made complete again in the merged complex of the paired couple. Contrasts completeness of Highland marriages with fragmentation of personal relationships in Lowland colonization zones.

1063 Skar, Sarah Lund. On the margin: letter exchange among Andean non-literates. (*in* Exploring the written: anthropology and the multiplicity of writing. Edited by Eduardo P. Archetti. Oslo: Scandinavian Univ. Press, 1994, p. 261–276, bibl.)

This suggestive article examines the relationship between the spoken word and the written text by studying the process of letter-writing among Andean migrants. Memory plays an essential role in the oral delivery of these messages, while their literate form legitimizes them and sometimes affords them supernatural power.

1064 Smith, Gavin. The production of culture in local rebellion. (*in* Golden ages, dark ages: imagining the past in anthropology and history. Edited by Jay O'Brien and William Roseberry. Berkeley: Univ. of California Press, 1991, p. 180–207, bibl.)

Rebellious peasants recognize the differences among themselves and produce an image of themselves as internally homogenous and externally distinctive. In the process of rebellion, components of culture are challenged and rearticulated, and history is reconstructed.

1065 Starn, Orin. Rethinking the politics of anthropology: the case of the Andes. (*Curr. Anthropol.*, 35:1, Feb. 1994, p. 13–38, bibl.)

The self-fashioned image of Andeanists as the "good outsiders" has glossed over colonial and postcolonial ambiguities in the enterprise. Based on his own experiences with *rondas campesinas*, Starn examines possibilities for contributions to social change by anthropologists. Followed by a series of critical commentaries by noted Andeanists.

1066 Stein, William W. El caso de los becerros hambrientos y otros ensayos de antropología económica. Lima: Mosca Azul Editores, 1991. 353 p.: bibl.

A compilation of translations into Spanish of various articles published since the late-1950s dealing with the anthropological study of peasant labor, the meaning of "Indian" in Peru, and the utility of notions of community vs. domestic unit for conducting socioeconomic research in rural Andean society.

1067 Steinhauf, Andreas. Diferenciación étnica y redes de larga distancia entre migrantes andinos: el caso de *sanka* y *colcha*. (*Bull. Inst. fr. étud. andin.*, 20:1, 1991, p. 93–114, bibl., ill.)

With the deepening national crisis in Peru and the consequent shrinking of the market in Lima, a significant long-distance spatial reorientation is emerging in urban migrant populations. Transregional networks based on kinship and locality have developed across time in different ways, depending upon ethnicity and region of origin.

1068 Turino, Thomas. Moving away from silence: music of the Peruvian Altiplano and the experience of urban migration. Chicago: Univ. of Chicago Press, 1993. 324 p.: bibl., ill., index. (Chicago studies in ethnomusicology)

A comparative musical ethnography of an indigenous rural Highland region and of migrants from that region in Lima and other cities. Illustrates how individuals and small groups interact dialectically with broader structural patterns and constraints in the shaping of social and musical life.

1069 Urrutia, Jaime. Comunidades campesinas y antropología: historia de un amor (casi) eterno. (*Debate Agrar.*, 14, junio/sept. 1992, p. 1–16)

This history of the scholarly treatment of peasant communities in Peru asks the following questions: 1) what historical junctures have given rise to studies of peasant communities?; 2) how have social scientists' perceptions of peasant communities changed over time?; and 3) what are peasant studies today?

1070 Valencia Espinoza, Abraham. La *Wak'a* de Suyu: deidad andina en plena función. Lima: Univ. Nacional Mayor de San Marcos, Seminario de Historia Rural Andina, 1987. 226 p.: bibl., ill.

Argues that the contemporary veneration of a pre-Incaic stela, called *Wak'a* by the faithful of Suyu, a town located near Sicuani (Cusco dept.), demonstrates the persistence of Andean cosmology. Includes considerable detail concerning the stela itself, its names, and the narratives that surround it.

JOURNAL ABBREVIATIONS

Afr. Archaeol. Rev. The African Archaeological Review. Cambridge Univ. Press. Cambridge, England.

Agric. Hist. Agricultural History. Agricultural History Society. Univ. of Calif. Press. Berkeley.

Allpanchis/Cusco. Allpanchis. Instituto de Pastoral Andina. Cusco, Peru.

Am. Anthropol. American Anthropologist. American Anthropological Assn., Washington.

Am. Antiq. American Antiquity. The Society for American Archaeology. Washington.

Am. Ethnol. American Ethnologist. American Ethnological Society. Washington.

Am. Indíg. América Indígena. Instituto Indigenista Interamericano. México.

Am. Negra. América Negra. Pontificia Univ. Javeriana. Bogotá.

Amazonía Peru. Amazonía Peruana. Centro Amazónico de Antropología y Aplicación Práctica, Depto. de Documentación y Publicaciones. Lima.

An. Antropol. Anales de Antropología. Univ. Nacional Autónoma de México, Instituto de Investigaciones Históricas. México.

An. Arqueol. Etnol. Anales de Arqueología y Etnología. Univ. Nacional de Cuyo, Facultad de Filosofía y Letras. Mendoza, Argentina.

An. Inst. Invest. Estét. Anales del Instituto de Investigaciones Estéticas. Univ. Nacional Autónoma de México. México.

An. Inst. Patagon./Soc. Anales del Instituto de la Patagonia: Serie Ciencias Sociales. Univ. de Magallanes. Punta Arenas, Chile.

An. Mus. Michoacano. Anales del Museo Michoacano. Centro Regional Michoacán del

INAH; Museo Regional Michoacano. Morelia, Mexico.

Anc. Mesoam. Ancient Mesoamerica. Cambridge Univ. Press. Cambridge, England.

Ann. Assoc. Am. Geogr. Annals of the Association of American Geographers. Lawrence, Kan.

Ann. Lat. Am. Stud. Annals of Latin American Studies. Nikon Raten Amerika Gakkai. Tokyo.

Annales/Paris. Annales. Centre national de la recherche scientifique de la VIe Section de l'Ecole pratique des hautes études. Paris.

Annu. Rev. Anthropol. Annual Review of Anthropology. Annual Reviews, Inc., Palo Alto, Calif.

Anthropol. Q. Anthropological Quarterly. Catholic Univ. of America, Catholic Anthropological Conference. Washington.

Anthropos/Switzerland. Anthropos. International Review of Ethnology and Linguistics. Anthropos-Institut. Freiburg, Switzerland.

Antiquity/Cambridge. Antiquity. A Quarterly Review of Archaeology. The Antiquity Trust. Cambridge, England.

Antropológica/Caracas. Antropológica. Fundación La Salle de Ciencias Naturales; Instituto Caribe de Antropología y Sociología. Caracas.

Antropológicas/México. Antropológicas. Instituto de Investigaciones Antropológicas, UNAM. México.

Anu. Antropol. Anuário Antropológico. Tempo Brasileiro. Rio de Janeiro.

Anu. Estud. Am. Anuario de Estudios Americanos. Consejo Superior de Investigaciones Científicas; Univ. de Sevilla, Escuela de Estudios Hispano-Americanos. Sevilla, Spain.

Anu. Etnol./Habana. Anuario de Etnología. Academia de Ciencias de Cuba; Editorial Academia. La Habana.

Anu. IEHS. Anuario IEHS. Univ. Nacional del Centro de la Provincia de Buenos Aires, Instituto de Estudios Histórico-Sociales. Tandil, Argentina.

Anu. Inst. Chiapaneco Cult. Anuario Instituto Chiapaneco de Cultura. Instituto Chiapaneco de Cultura. Tuxtla Gutiérrez, Mexico.

Apunt. Arqueol./Guatemala. Apuntes Arqueológicos. Universidad de San Carlos de Guatemala, Escuela de Historia. Guatemala.

Arch. Völkerkd. Archiv für Völkerkunde. Museum für Völkerkunde in Wien und von Verein Freunde der Völkerkunde. Vienna.

Archaeoastronomy/College Park. Archaeoastronomy. The Center for Archaeoastronomy, Univ. of Maryland. College Park, Md.

Archaeoastronomy/England. Archaeoastronomy. Science History Publications. Giles, England.

Archaeol. Anthropol. Archaeology and Anthropology. Ministry of Education and Cultural Development. Georgetown, Guyana.

Archaeology/New York. Archaeology. Archaeology Institute of America. New York.

Archaeometry/Oxford. Archaeometry. Oxford Univ., Oxford, England.

Arq. Mus. Hist. Nat. Arquivos do Museu de História Natural. Univ. Federal de Minas Gerais. Belo Horizonte, Brazil.

Arqueol. Contemp. Arqueología Contemporánea. Programa de Estudios Prehistóricos. Buenos Aires.

Arqueol. Mex. Arqueología Mexicana. Instituto Nacional de Antropología e Historia, Editorial Raíces. México.

Arqueología/México. Arqueología. Instituto Nacional de Antropología e Historia. México.

Arqueológicas/Lima. Arqueológicas. Museo Nacional de Antropología y Arqueología, Instituto Nacional de Cultura. Lima.

Arstryck/Göteborg. ° Arstryck. Etnografiska Museum. Göteborg, Sweden.

Bol. Am. Boletín Americanista. Univ. de Barcelona, Facultad de Geografía e Historia, Depto. de Historia de América. Barcelona.

Bol. Antropol. Am. Boletín de Antropología Americana. Instituto Panamericano de Geografía e Historia. México.

Bol. Arqueol. Boletín de Arqueología. Fundación de Investigaciones Arqueológicas Nacionales. Bogotá.

Bol. Cons. Arqueol. Boletín del Consejo de Arqueología. Instituto Nacional de Antropología e Historia. México.

Bol. Esc. Cienc. Antropol. Univ. Yucatán. Boletín de la Escuela de Ciencias Antropológicas de la Universidad de Yucatán. Mérida, Mexico.

Bol. Inf./San Cristóbal. Boletín Informativo. Museo de Táchira. San Cristóbal, Venezuela.

Bol. Lima. Boletín de Lima. Revista Cultural Científica. Lima.

Bol. Mus. Chil. Arte Precolomb. Boletín del Museo Chileno de Arte Precolombino. Santiago.

Bol. Mus. Hombre Domin. Boletín del Museo del Hombre Dominicano. Santo Domingo.

Bol. Mus. Para. Goeldi. Boletim do Museu Paraense Emílio Goeldi. Nova série: antropologia. Conselho Nacional de Desenvolvimento Científico e Tecnológico, Instituto Nacional de Pesquisas da Amazônia. Belém, Brazil.

Boletín/Bogotá. Boletín del Museo del Oro. Banco de la República. Bogotá.

Bull. Bur. natl. ethnol. Bulletin du Bureau national d'ethnologie. Bureau national d'ethnologie. Port-au-Prince, Haiti.

Bull. Inst. fr. étud. andin. Bulletin de l'Institut français d'études andines. Lima.

Bull. Int. Anthropol. Ethnol. Bulletin of the International Committee on Urgent Anthropological and Ethnological Research. International Union of Anthropological and Ethnological Sciences. Vienna.

Bull. mém. Soc. anthropol. Paris. Bulletins et mémoires de la Société d'anthropologie de Paris. Paris.

Bulletin/Geneva. Bulletin. Société suisse des américanistes; Musée et institut d'éthnographie. Geneva.

Caravelle/Toulouse. Caravelle. Cahiers du monde hispanique et luso-brésilien. Univ. de Toulouse, Institute d'études hispaniques, hispano-americaines et luso-brésiliennes. Toulouse, France.

Caribb. Q. Caribbean Quarterly. Univ. of the West Indies. Mona, Jamaica.

Caribena/Martinique. Caribena: cahiers d'études américanistes de la Caraïbe. Centre d'études et de recherches archéologiques (CERA). Martinique.

Cespedesia/Cali. Cespedesia. Depto. del Valle del Cauca. Cali, Colombia.

Cienc. Hoy. Ciencia Hoy. Asociación Ciencia Hoy; Morgan Antártica. Buenos Aires.

Cienc. Soc./Santo Domingo. Ciencia y Sociedad. Instituto Tecnológico de Santo Domingo.

Clim. Change. Climatic Change. Reidel Publishers. Boston.

Comp. Civiliz. Rev. Comparative Civilizations Review. Dept. of History, Dickinson College. Carlisle, Penn.

Comp. Stud. Soc. Hist. Comparative Studies in Society and History. Society for the Comparative Study of Society and History; Cambridge Univ. Press. London.

Cuad. Arquit. Mesoam. Cuadernos de Arquitectura Mesoamericana. Facultad de Arquitectura, Univ. Nacional Autónoma de México. México.

Cuad. Prehispánicos. Cuadernos Prehispánicos. Seminario de Historia de América, Univ. de Valladolid. Spain.

Cuad. Sur/Oaxaca. Cuadernos del Sur: Ciencias Sociales. Instituto de Investigaciones Sociológicas, Univ. Autónoma Benito Juárez de Oaxaca. Oaxaca, Mexico.

Cult. Anthropol. Cultural Anthropology: Journal of the Society for Cultural Anthropology. American Anthropological Assn.; Society for Cultural Anthropology. Washington.

Curr. Anthropol. Current Anthropology. Univ. of Chicago. Chicago, Ill.

Debate Agrar. Debate Agrario. Centro Peruano de Estudios Sociales (CEPES). Lima.

Dédalo/São Paulo. Dédalo. Univ. de São Paulo, Museu de Arqueologia e Etnologia. São Paulo.

Dialect. Anthropol. Dialectical Anthropology. M. Nijhoff. Dordrecht, The Netherlands.

Diogenes/Philosophy. Diogenes. International Council for Philosophy and Humanistic Studies (Paris); Berg Publishers. Oxford, England.

Ed. Quinto Centen. Ediciones del Quinto Centenario. Univ. de la República. Montevideo.

Eres/Tenerife. Eres. Museo Arqueológico y Etnográfico. Tenerife, Spain.

Estud. Cult. Maya. Estudios de Cultura Maya. Centro de Estudios Mayas, Univ. Nacional Autónoma de México. México.

Estud. Cult. Náhuatl. Estudios de Cultura Náhuatl. Instituto de Investigaciones Históricas, Univ. Nacional Autónoma de México. México.

Ethnic Groups/New York. Ethnic Groups. Gordon and Breach. New York.

Ethnohistory/Society. Ethnohistory. American Society for Ethnohistory. Duke Univ., Durham, N.C.

Ethnology/Pittsburgh. Ethnology. Univ. of Pittsburgh, Penn.

Ethnos/Stockholm. Ethnos. Statens Etnografiska Museum. Stockholm.

Ethos/Society. Ethos. Society for Psychological Anthropology; Univ. of California, Los Angeles.

Etud. créoles. Etudes créoles. Comité international des études créoles. Montréal.

Folk/Copenhagen. Folk: Journal of the Danish Ethnographic Society. Danish Ethnographic Society. Copenhagen.

Gac. Arqueol. Andin. Gaceta Arqueológica Andina. Instituto Andino de Estudios Arqueológicos. Lima.

Gac. Univ./Campeche. Gaceta Universitaria. Univ. Autónoma de Campeche, Mexico.

Geoarchaeology/New York. Geoarchaeology. John Wiley. New York.

Hist. Mex. Historia Mexicana. Colegio de México. México.

Hist. Relig. History of Religions. Univ. of Chicago. Chicago, Ill.

Hist. Workshop. History Workshop. Ruskin College, Oxford Univ., Oxford, England.

Homme/Paris. L'Homme. Laboratoire d'anthropologie, Collège de France. Paris.

Hum. Organ. Human Organization. Society for Applied Anthropology. New York.

Identities/Yverdon. Identities: Global Studies in Culture and Power. Gordon and Breach Publishers. Yverdon, Switzerland.

Indiana/Berlin. Indiana. Gebr. Mann., Berlin.

Interciencia/Caracas. Interciencia. Asociación Interciencia. Caracas.

J. Anthropol. Archaeol. Journal of Anthropological Archaeology. Academic Press. New York.

J. Anthropol. Res. Journal of Anthropological Research. Univ. of New Mexico. Albuquerque, N.M.

J. Archaeol. Res. Journal of Archaeological Research. Plenum Press. New York.

J. Archaeol. Sci. Journal of Archaeological Science. Academic Press. New York.

J. Bahamas Hist. Soc. Journal of the Bahamas Historical Society. Nassau, Bahamas?.

J. Caribb. Hist. The Journal of Caribbean History. Caribbean Univ. Press. St. Lawrence, Barbados.

J. Field Archaeol. Journal of Field Archaeology. Boston Univ., Boston, Mass.

J. Geogr. Journal of Geography. National Council of Geographic Education. Menasha, Wis.

J. Lat. Am. Lore. Journal of Latin American Lore. Univ. of California, Latin American Center. Los Angeles, Calif.

J. Lat. Am. Stud. Journal of Latin American Studies. Centers or Institutes of Latin American Studies at the Universities of Cambridge, Glasgow, Liverpool, London, and Oxford. Cambridge Univ. Press. London.

J. Steward Anthropol. Soc. Journal of the Steward Anthropological Society. Urbana, Ill.

J. World Prehist. Journal of World Prehistory. Plenum Press. New York.

Jam. J. Jamaica Journal. Institute of Jamaica. Kingston.

LARR. Latin American Research Review. Latin American Research Review Board. Univ. of New Mexico, Albuquerque, N.M.

Lat. Am. Anthropol. Rev. The Latin American Anthropology Review. Society for Latin American Anthropology. Fairfax, Va.

Lat. Am. Antiq. Latin American Antiquity. Society for American Archaeology. Washington.

Law Anthropol. Int. Jahrb. Rechtsanthropol. Law & Anthropology: Internationales Jahrbuch für Rechtsanthropologie. VWGö. Vienna; Klaus Renner Verlag. Hohenschäftlarn.

Maguaré/Bogotá. Maguaré. Depto. de Antropología, Univ. Nacional de Colombia. Bogotá.

Man/London. Man. The Royal Anthropological Institute. London.

Mededelingen/Paramaribo. Mededelingen. Stichting Surinaams Museum. Paramaribo.

Mesoamérica/Antigua. Mesoamérica. Centro de Investigaciones Regionales de Mesoamérica. Antigua, Guatemala.

Mexicon/Berlin. Mexicon. K.-F. von Flemming. Berlin, Germany.

Montalbán/Caracas. Montalbán. Univ. Católica Andrés Bello, Facultad de Humanidades y Educación, Institutos Humanísticos de Investigación. Caracas.

Natl. Geogr. Res. National Geographic Research. National Geographic Society. Washington.

Nature/London. Nature: International Weekly Journal of Science. Macmillan Magazines. London.

NEARA J. NEARA Journal. New England Antiquities Research Assn., Milford, N.H.

Nieuwe West-Indische Gids. Nieuwe West-Indische Gids. Martinus Nijhoff. The Hague.

Notas Mesoam. Notas Mesoamericanas. Univ. de las Américas-Puebla. Puebla, Mexico.

Noved. Colomb. Novedades Colombianas. Museo de Historia Natural, Univ. del Cauca. Papayán, Colombia.

Nuevos Aportes. Nuevos Aportes. Editorial San José S.R.L., La Paz.

Pachacamac/Lima. Pachacamac: Revista del Museo de la Nación. Museo de la Nación. Lima.

Pacífico/Chimbote. Pacífico: Revista de Ciencias Sociales. Chimbote, Peru.

Pesquisas/São Leopoldo. Pesquisas. Instituto Anchietano de Pesquisas. São Leopoldo, Brazil.

Plant. Soc. Am. Plantation Society in the Americas. Univ. of New Orleans.

Prehistoria/Buenos Aires. Prehistoria: Revista del Programa de Estudios Prehistóricos. Consejo Nacional de Investigaciones Científicas y Técnicas. Buenos Aires.

Pretextos/Lima. Pretextos. Centro de Estudios y Promoción del Desarrollo. Lima.

Pumapunku/La Paz. Pumapunku. Centro de Investigaciones Antropológicas Tiwanaku. La Paz.

Radiocarbon/New Haven. Radiocarbon. Supplement of the American Journal of Science. New Haven, Conn.

Res. Econ. Anthropol. Research in Economic Anthropology. JAI Press. Greenwich, Conn.

Res. Explor. Research & Exploration. National Geographic Society. Washington.

Res/Harvard. Res. Peabody Museum of Archaeology and Ethnology, Harvard Univ., Cambridge, Mass.

Rev. Antropol./São Paulo. Revista de Antropologia. Univ. de São Paulo, Faculdade de Filosofia, Letras e Ciências Humanas; Associação Brasileira de Antropologia. São Paulo.

Rev. Archaeol./Salem. The Review of Archaeology. Salem, Mass.

Rev. Arqueol. Am. Revista de Arqueología Americana. Instituto Panamericano de Geografía e Historia. México.

Rev. Arqueol./São Paulo. Revista de Arqueologia. Sociedade de Arqueologia Brasileira. São Paulo.

Rev. Cent. Estud. Av. La Revista del Centro de Estudios Avanzados de Puerto Rico y el Caribe. San Juan.

Rev. CEPA. Revista do CEPA. Centro de Ensino e Pesquisas Arqueológicas, Faculdades Integradas de Santa Cruz do Sul. Santa Cruz do Sul, Brazil.

Rev. Chil. Antropol. Revista Chilena de Antropología. Depto. de Antropología, Univ. de Chile. Santiago.

Rev. Colomb. Antropol. Revista Colombiana de Antropología. Ministerio de Educación Nacional, Instituto Colombiano de Antropología. Bogotá.

Rev. Cuba. Cienc. Soc. Revista Cubana de Ciencias Sociales. Centro de Estudios Filosóficos, Academia de Ciencias de Cuba. La Habana.

Rev. Esp. Antropol. Am. Revista Española de Antropología Americana. Facultad de Geografía e Historia. Univ. Complutense de Madrid.

Rev. Indias. Revista de Indias. Consejo Superior de Investigaciones Científicas, Instituto Gonzalo Fernández de Oviedo. Madrid.

Rev. Mus. Arqueol. Etnol. Revista do Museu de Arqueologia e Etnologia. Univ. de São Paulo.

Rev. Mus. Hist. Nat. San Rafael. Revista del Museo de Historia Natural de San Rafael. Mendoza, Argentina.

Rev. Peru. Poblac. Revista Peruana de Población. Asociación Multidisciplinaria de Investigación y Docencia en Población. Lima.

Review/Braudel. Review: Fernand Braudel Center. Fernand Braudel Center for the Study of Economics, Historical Systems, and Civilizations. Binghamton, New York.

Schweiz. Amer. Ges. Schweizerische Amerikanisten Gesellschaft. Société Suisse des Américanistes. Genève.

Sci. Am. Scientific American. Scientific American, Inc. New York.

Science/Washington. Science. American Assn. for the Advancement of Science. Washington.

Soc. Compass. Social Compass. The International Catholic Institute for Social-Ecclesiastical Research. The Hague.

Soc. Econ. Stud. Social and Economic Studies. Univ. of the West Indies, Institute of Social and Economic Research. Mona, Jamaica.

SWI Forum. SWI Forum voor Kunst, Kultuur en Wetenschop. De Stichting. Paramaribo, Suriname.

Trace/México. Trace. Centre d'études mexicaines et centraméricaines. México.

U tz'ib. U tz'ib. Asociación Tikal. Guatemala.

UFES Rev. Cult. UFES: Revista de Cultura. Univ. Federal do Espírito Santo, Brazil.

Universitas/Bogotá. Universitas. Pontificia Univ. Javeriana, Facultad de Derecho y Ciencias Socioeconómicas. Bogotá.

Vínculos/San José. Vínculos. Museo Nacional de Costa Rica. San José.

World Archaeol. World Archaeology. Routledge & Kegan Paul. London.

Yachay/Cochabamba. Yachay. Facultad de Filosofía y Ciencias Religiosas, Univ. Católica Boliviana. Cochabamba, Bolivia.

Yaxkin/Tegucigalpa. Yaxkin. Instituto Hondureño de Antropología e Historia. Tegucigalpa.

Yearbook/CLAG. Yearbook. Conference of Latin Americanist Geographers; Ball State Univ., Muncie, Ind.

ECONOMICS

GENERAL

JAMES W. FOLEY, *Associate Dean, School of Business, University of Miami, Coral Gables*

IT IS NOT SURPRISING THAT THE DEBT ISSUE and subsequent economic reforms continue to attract the attention of economists interested in Latin America. But, while these topics continue to dominate the literature, in recent years we have seen a significant change in the focus of inquiry. Earlier literature on the debt was concerned with its causes and immediate impact on growth. These subjects now seem to have been exhaustively documented, and, accordingly, Latin American specialists have shifted their attention to evaluating the timing, sequencing, and results of neoconservative policy reforms and market liberalization programs and the macroeconomic impact of these policies on income distribution, employment, and growth.

Within this broad classification, I particularly liked volumes edited by Gustafson (item **1143**), Ros (item **1147**), and Williamson (item **1240**). The latter is of particular interest since it analyzes the "political ingredients" of successful reform, a subject that is all too often ignored by theoretical economists. I also liked the neo-structuralist critique of recent policy reform by Ramírez (item **1247**).

Within the area of income distribution, I was impressed with an extensive work done by ECLAC which documents the extent of poverty and indigence in the region (item **1213**), as well as with a short article by Berry on "Distribution of Income and Poverty in Latin America" which describes changes in income distribution for individual countries during the 1980s (item **1096**). Also worth reading is a volume edited by Hausmann and Rigobón which empirically analyzes the impact of various types of government expenditures on income distribution (item **1181**).

In *HLAS 53* (p. 193) I commented that it was surprising that so little had been written evaluating the success, or failure, of privatization efforts. This deficiency has now, in large part, been corrected. ECLAC, for example, has published an extensive bibliography on this subject, *Documentos sobre privatización con énfasis en América Latina* (item **1139**). Readers will also want to examine Cardoso's article "La Privatización en América Latina" (item **1112**) and the 1993 spring volume of the *Columbia Journal of World Business* which devoted the entire issue to privatization programs in various parts of the world (item **1121**). I also found interesting an article by Glade on "Privatization in Rent-Seeking Societies" (item **1175**) describing the attempts made by special interest groups to influence government so as to profit from the privatization process. Also worthwhile is a volume by Martín del Campo and Winkler which examines not only privatization but also policy reforms to improve efficiency of State enterprises that have not been privatized (item **1218**).

In the last edition of the *Handbook*, my essay completely ignored regional integration since much of the literature I examined was little more than a rehash of previously discussed issues. During the last several years, however, this subject has assumed primary importance in the region. The recent formation of Mercosur and NAFTA, and the possible establishment of a Western Hemisphere Free Trade Area has rekindled scholarly interest in trade and regional integration. On these subjects, I particularly recommend volumes edited by Green (item **1154**), Bradford (item **1279**), and Salgado (item **1141**). Also worth reading is an article by Edwards which provides an overview of the region's past integration efforts (item **1148**) and a book published by SELA which examines more recent integration schemes and possible future integration scenarios (item **1231**).

Labor markets, in general, and the informal sector, in particular, continue to attract scholarly attention. For instance, Infante and Klein have provided an overview of recent trends in the region's labor markets (item **1194**). Those interested in the informal sector will want to examine an annotated bibliography, published by PREALC, on recent research on this topic (item **1097**).

Growth and the environment, a topic virtually ignored until recently, is attracting increasing attention. Noteworthy works include a volume by ECLAC on *Planificación y gestión del desarrollo en áreas de expansión de la frontera agropecuaria en América Latina* (item **1123**) and proceedings from a conference on development and the environment (item **1270**).

Perhaps the most striking characteristic of the recent economic literature is its increasing sophistication. Twenty-five years ago, much of the literature on Latin American economic problems was little more than lengthy, often emotional, diatribes intended to discredit opposing views, while supporting the author's own biases and ideology. Today, that type of literature is far less frequent. Instead, economic issues are debated using sophisticated mathematical and statistical techniques. Any resulting controversy is as likely to be concerned with the methodology employed as with the conclusions reached. Nevertheless, while this trend is laudable and certainly pleasing to professional economists, we should not forget that not all issues are subject to statistical verification, and that there is still a place for interdisciplinary work and subjective analysis.

1071 Abuhadba, Mario and Pilar Romaguera. Inter-industrial wage differentials: evidence for Latin American countries. (*J. Dev. Stud.*, 30:1, Oct. 1993, p. 190–205)

Using data from Brazil, Chile, Uruguay, and Venezuela, authors examine wage differentials between industries. Demonstrates that differentials are large and temporally stable, thereby casting doubt on competitive models of wage determination and supporting efficiency wage models.

1072 The administration of water resources in Latin America and the Caribbean. Santiago?: United Nations Economic Commission for Latin America and the Caribbean, 1991. 139 p.: bibl.

Describes regional water management policy. Focuses special attention on changes in water resource administration that have occurred subsequent to the UN Water Conference of 1977.

1073 Agrarnyĭ kapitalizm v Latinskoĭ Amerike: tendentsii 60–80-kh godov [Agrarian capitalism in Latin America: the tendencies of the 1960s–1980s.] Edited by I͡U.G. Onufriev and Igor' Konstantinovich Sheremet'ev. Moscow: Akademii͡a nauk SSSR, In-t Latinskoĭ Ameriki, 1990. 183 p.: bibl.

Twelve articles arising from a conference sponsored in late 1988 by the Economics Dept. of the Soviet Academy of Sciences' Institute of Latin American Studies examine the potential for modernization and the effect of US or capitalist influences on production in various Latin American countries. Most economic data are from non-Soviet sources. Includes a number of references to Lenin, as well as some tables. [B. Dash]

1074 La agricultura latinoamericana: crisis, transformaciones y perspectivas. Santiago: Grupo de Investigaciones Agrarias, Univ. Academia de Humanismo Cristiano;

CLACSO, Comisión de Estudios Rurales, 1991. 250 p.: bibl.

Series of papers by specialists on the current state of the region's agricultural sector and prospects for future change. Evaluates recent agricultural programs and policies in Argentina, Brazil, Chile, Ecuador, Peru, and Nicaragua. From a Sept. 1988 seminar in Punta de Tralca, Chile.

1075 Alam, Asad and **Sarath Rajapatirana.** Trade reform in Latin America and the Caribbean. (*Financ. Dev.*, 30:3, Sept. 1993, p. 44–47, graphs, ill., table)

Documents recent trade reforms in 16 countries. Trade reform sharply lowered average tariff rates, virtually eliminated quantitative import restrictions, and reduced overvalued exchange rates. Discusses political lessons to be learned from these reforms.

1076 Alba Vega, Carlos and **Dirk Kruijt.** The convenience of the minuscule: informality and microenterprise in Latin America. Amsterdam: Thela, 1994. 88 p.: bibl. (Latin American series; 3)

Five essays examine importance of small-scale production and heterogeneity of informal sector, with focus on structural dualization of society. Following introductory chapters on poverty, employment, and informal society, two essays deal with Mexico: micro-industry in Jalisco and role of small-scale enterprise in context of NAFTA. Final essay examines social policies designed to mitigate effects of poverty. [R. Hoefte]

1077 Albala-Bertrand, J. M. Natural disaster situations and growth: a macroeconomic model for sudden disaster impacts. (*World Dev.*, 21:9, Sept. 1993, p. 1417–1434, appendix, bibl., tables)

Author uses econometric techniques to estimate macroeconomic impact of major natural disasters in Ecuador, Dominican Republic, Guatemala, Honduras, Nicaragua, and Peru. Concludes that the resulting capital loss has only a minor impact on growth, and that relief funds are better spent to help victims than to replace lost capital.

1078 Allen, Chris; David Currie; T.G. Srinivasan; and **David Vines.** Policy interactions between the OECD countries and Latin America in the 1980s. (*Manch. Sch. Econ. Soc. Stud.*, 60:4, Dec. 1992, supplement, p. 1–20)

Presents econometric model which describes macroeconomic interaction between OECD countries and Latin America. Uses simulation techniques to estimate how global shocks will affect Latin America. Very sophisticated.

1079 América Latina: crítica del neoliberalismo. México: Centro de Estudios para un Proyecto Nacional Alternativo, 1992. 197 p.: bibl., ill.

Series of essays in honor of the late Mexican economist, Eduardo González Ramírez. Reviews the principle tenants of different schools of economic thought and describes their impact on Latin American political economy. Critically evaluates neoliberal reforms and proposes alternative policies.

1080 Armendariz de Aghion, Beatriz. On the pricing of LDC debt: an analysis based on historical evidence from Latin America. Paris: Organisation for Economic Co-operation and Development, 1991. 30 p.: bibl., ill. (Technical papers; 52)

Sophisticated presentation of mathematical debt valuation model, using Chilean and Colombian experiences during the 1930s as examples. Investor expectations about future levels of foreign exchange determine the value of old (defaulted) debt and new debt that is exchanged for the old at a reduced value.

1081 Arnade, C.A. Evaluación de dos modelos de comercio internacional para la agricultura latinoamericana. (*Invest. Agrar.*, 8:1, abril 1993, p. 13–27, bibl., tables)

Applies Heckscher-Olin and Markusen models of international trade to Latin American agriculture. Using econometric techniques, author concludes that Latin American trade is best explained by relative factor abundance, rather than different levels of technology.

1082 Arriagada, Irma. Latin American women and the crisis: impact in the work market. (*in* Alternatives. Edited by Neuma Aquiar and Thaís Corral. Rio de Janeiro: Editora Rosa dos Tempos, 1991, vol. 1, p. 67–95, bibl., tables)

Uses household survey data from five cities (Bogotá, Caracas, Panama City, San José, and São Paulo) to determine the impact of the debt crisis on the labor market. Concludes that the increase in unemployment and the decrease in wages was greater for women than for men.

1083 Arrigone, Jorge L. Low-income shelter strategies in Latin America. (*UNISA/Lat. Am. Rep.*, 7:2, 1991, p. 16–29, bibl., map, photos, tables)

Examines how government has responded to the squatter problem in Argentina (Mendoza), Brazil (Rio de Janeiro), and Chile (Santiago). Argues that successful upgrading requires squatter participation both in making decisions and supplying labor.

1084 Arrizau, Ricardo H. et al. Deuda interna y estabilidad financiera. v. 1, Aspectos analíticos. Edición de Carlos Massad y Roberto Zahler. Buenos Aires: Grupo Editor Latinoamericano; Emecé Editores, 1987. 1 v.: bibl. (Col. Estudios políticos y sociales)

Series of highly theoretical articles evaluate impact of external shocks on domestic financial sector stability. Discusses debt crisis and subsequent financial liberalization programs' effects on enterprise finances, private indebtedness, and various macroeconomic variables.

1085 Asociación Latinoamericana de Instituciones Financieras de Desarrollo. Del ajuste al desarrollo económico latinoamericano en los noventa: el papel de los bancos de fomento. (*Comer. Exter.*, 41:12, dic. 1991, p. 1124–1149, bibl.)

Examines changing role of development banks as governments adopt a market- and export-oriented approach to development. Provides short case studies for Argentina, Colombia, Ecuador, El Salvador, Mexico, Peru, and Uruguay.

1086 Atencio Bello, Heraclio. Deuda externa, inversión extranjera y transferencia de tecnología en América Latina. Caracas: Monte Avila Editores, 1986. 352 p.: bibl.

The text, which often reads like a business presentation, is mostly devoted to possible role of foreign direct investment by multinational companies. Sections on the debt crisis are entertaining reading with some useful information but the analysis is not profound by any means. Somewhat deceptive title since discussion of technology transfer is limited. [R. Palacios]

1087 Automatización flexible en la industria: difusión y producción de máquinas-herramienta de control numérico en América Latina. Recopilación de Gerard K. Boon y Alfonso Mercado. México: Technology Scientific Foundation; Noriega Editores; Editorial Limusa, 1990. 244 p.: bibl., ill.

Series of articles on automation and its potential impact includes case studies for Argentina, Brazil, Colombia, Mexico, and Peru. Contends that automation would be beneficial for Latin America in the apparel and footwear industries.

1088 Ayala Espino, José. Límites del mercado, límites del estado: ensayos sobre economía política del Estado. México: Instituto Nacional de Administración Pública, 1992. 202 p.: bibl.

General discussion of the role of the State in economic development. Balanced assessment of pros and cons of government intervention in the economy. Provides excellent bibliography. [R.E. Looney]

1089 Bacha, Edmar Lisboa et al. De espaldas a la prosperidad: América Latina y la economía internacional a fines de los ochenta. Edición de Roberto Bouzas. Buenos Aires: Grupo Editor Latinoamericano, 1989. 224 p.: bibl., ill. (Col. Estudios internacionales)

Series of well-written articles on Latin America's international economic relations emphasizes influence that the world economy has had on the region during the 1980s. Topics include: 1) macroeconomic policies of developed countries and their impact on the region; 2) trade trends; and 3) the Uruguay Round.

1090 Baer, Werner and Melissa Birch. La privatización y el rol cambiante del Estado en América Latina. (*Rev. Parag. Sociol.*, 29:85, sept./dic. 1992, p. 7–28, bibl.)

Describes changing role of private and public sectors as governments have replaced growth strategies based on import-substituting industrialization with those based on market liberalization. Analyzes costs and benefits of privatization and speculates as to the likely future role of the State.

1091 Baer, Werner. U.S.-Latin American trade relations: past, present and future. (*in* Conference on the Southwest Economy, *Dallas, Tex., 1991*. Free trade within North America: expanding trade for prosperity. Boston: Kluwer Academic Publishers, 1993, p. 53–71, bibl., tables)

Examines the region's shift from a growth strategy based upon import substituting industrialization to one focused on export growth and diversification. Discusses the likely implications of this shift on future Latin America-US trade relations.

1092 **Barham, Bradford L. et al.** Nontraditional agricultural exports in Latin America. (*LARR*, 27:2, 1992, p. 43–82, bibl., tables)

Analyzes impact of "nontraditional agricultural or natural-resource-based exports (NTAX)" on growth and income distribution. Uses data from Costa Rica, Chile, and Guatemala. Also contains extensive literature review.

1093 **Barkin, David; Rosemary L. Batt; and Billie R. DeWalt.** The substitution among grains in Latin America. (*in* Modernization and stagnation: Latin American agriculture into the 1990s. Edited by Michael J. Twomey and Ann Helwege. New York: Greenwood Press, 1991, p. 13–53, bibl., tables)

Examines production of agricultural commodities during 1961–86. Documents that a major shift has occurred away from traditional cereals (maize) consumed by the poor toward profitable grains (sorghum, wheat, and rice). Government policies that discriminate against small farmers in favor of commercial agriculture are responsible for this shift.

Bartholomew, Joy A. et al. Desarrollo sostenible y políticas económicas en América Latina. See item **1541**.

1094 **Baumann, Renato.** An appraisal of recent intra-industry trade for Latin America. (*CEPAL Rev.*, 48, Dec. 1992, p. 83–94, bibl., tables)

Describes pattern of intra-industry trade (i.e., exporting and importing similar products) for various Latin American countries (Argentina, Brazil, Chile, Colombia, Mexico, and Uruguay). During the 1970s–80s the absolute level and relative importance of intra-industry trade has increased.

1095 **Benavente, José Miguel.** Commodity exports and Latin American development. (*CEPAL Rev.*, 45, Dec. 1991, p. 41–60, bibl., tables)

Describes importance of non-oil primary commodities in Latin American exports. To maximize gains countries must engage in: 1) commodity processing; 2) more effective marketing strategies; and 3) technological innovation to lower production costs and find new uses for primary products.

1096 **Berry, Albert.** Distribution of income and poverty in Latin America: recent trends and challenges for the 1990s. (*in* Latin America to the year 2000: reactivating growth, improving equity, sustaining democracy. New York: Praeger, 1992, p. 67–80, bibl., tables)

Examines changes in income distribution in Latin American countries during the 1980s. Concludes that income distribution became less equal in Argentina and Chile, became more equal in Colombia, and was little changed in Brazil, Costa Rica, Peru, and Venezuela.

1097 **Bibliografía comentada sobre el sector informal urbano en América Latina, 1975–1987.** v. 1–3. Santiago: PREALC, 1989. 3 v. (1155 p.). (Documentos de trabajo; 332)

Annotated bibliography provides research published from 1975–87 on the region's urban informal sector. Countries covered include Bolivia, Brazil, Colombia, Dominican Republic, Ecuador, El Salvador, Guatemala, Honduras, Mexico, Panama, Peru, and Venezuela. Valuable contribution.

1098 **Birdsall, Nancy and David Wheeler.** Trade policy and industrial pollution in Latin America: where are the pollution havens? (*in* International trade and the environment. Edited by Patrick Low. Washington: World Bank, 1992, p. 159–171, tables)

Using econometric techniques, authors test the hypothesis that free trade will increase industrial pollution and environmental degradation. Using data for 25 Latin American countries, authors conclude that the opposite is true: the more open the economy, the cleaner industry is likely to be.

1099 **Bitar, Sergio et al.** América Latina en el mundo de mañana: ámbito internacional y regional. Coordinación de Gonzalo Martner. Caracas: Editorial Nueva Sociedad; Unitar/Profal, 1987. 367 p.: bibl. (Latinoamérica de fines de siglo)

Series of articles focuses on two themes: 1) Latin America's likely *future* role in the world economy; and 2) the likelihood and feasibility of greater economic and political cooperation between various countries of the region.

1100 Boisier, Sergio et al. La descentralización: el eslabón perdido de la cadena transformación productiva con equidad y sustentabilidad. Santiago: CEPAL-ILPES, Naciones Unidas, 1992. 79 p.: bibl. (Cuadernos del ILPES, 0020–4080; 36)

Calls for geographical decentralization of economic activity. Argues that the current centralization of economic activity at two or three locations within most Latin American countries has caused social inequities, and that geographical decentralization is necessary to alleviate this problem and achieve sustainable growth.

1101 Boltvinik, Julio. La medición de la pobreza en América Latina. (*Comer. Exter.*, 41:5, mayo 1991, p. 423–428, tables)

Reviews basic methods used to quantify the extent of poverty and unsatisfied basic needs. Proposes new technique of measurement. Includes data on poverty in Buenos Aires, Montevideo, and Peru.

1102 Bonturi, Marcos and **Montague J. Lord.** Latin America's trade in manufactures: an empirical study. (*in* Strategic options for Latin America in the 1990s. Edited by Colin I. Bradford, Jr. Paris: Development Centre of the Organisation for Economic Co-Operation and Development; Washington: Inter-American Development Bank, 1992, p. 21–99, bibl., graphs, tables)

Empirically describes dramatic growth in Latin American manufactured exports, from roughly 10 percent of total exports in 1970 to approximately one-third in 1987. Analyzes types of manufactured goods being exported, their factor intensity, and major trading partners. Includes 28 tables of data. Interesting.

1103 Borner, Silvio; Aymo Brunetti; and **Beatrice Weder.** Institutional obstacles to Latin American growth. San Francisco, Calif.: ICS Press, 1992. 47 p.: bibl. (Occasional papers; 24)

Authors contend that "institutional obstacles" play a role in explaining the region's stagnation. Specifically, uncertainty about future economic "rules of the game" has paralyzed private investment and innovation.

1104 Borzutzky, Silvia. Social security and health policies in Latin America: the changing role of the State and the private sector. (*LARR*, 28:2, 1993, p. 246–256)

Reviews eight books on health and social security issues and problems. Primary emphasis is placed on declining role of the State and its implications for health and social security.

1105 Bouzas, Roberto et al. Conversión de deuda externa y financiación del desarrollo en América Latina. Edición de Roberto Bouzas y Ricardo Ffrench-Davis. Buenos Aires: Grupo Editor Latinoamericano; EMECE Editores, 1990. 214 p.: bibl., ill. (Col. Estudios internacionales)

Describes various types of debt conversion mechanisms used by Latin American countries to reduce their external debt. Includes separate articles on Argentina, Brazil, Chile, Costa Rica, Mexico, and Uruguay.

1106 Bradford, Colin I. Options for Latin American reactivation in the 1990s. (*CEPAL Rev.*, 44, Aug. 1991, p. 97–103, bibl.)

Essentially a review of some recent economic literature relating to the likely performance of the region's economies during the 1990s. Stresses the need for improved competitiveness and the fact that *both* import substitution *and* export promotion are compatible with the region's growth objectives.

1107 Buitelaar, Rudolf and **Juan Alberto Fuentes.** The competitiveness of the small economies of the region. (*CEPAL Rev.*, 43, April 1991, p. 83–96, bibl., ill., tables)

Analyzes changes in export competitiveness from 1978–88 for Bolivia, Costa Rica, Dominican Republic, Ecuador, El Salvador, Guatemala, Honduras, Haiti, Jamaica, Paraguay, Trinidad, and Uruguay. Finds that competitiveness is primarily due to the availability of low-cost labor and unprocessed natural resources.

Bulavin, V.I. et al. Latinskaia Amerika: lesnye resursy i ikh ispol'zovanie [Latin America: forest resources and their use]. See item **2226**.

Burgueño Lomelí, Fausto. Economía en crisis: ensayos sobre México y América Latina. See item **1336**.

1108 **Calvo, Guillermo A.; Leonardo Leiderman; and Carmen M. Reinhart.** Capital inflows and real exchange rate appreciation in Latin America: the role of external factors. (*Staff Pap.*, 40:1, March 1993, p. 108–151, bibl., graphs, tables)

Authors use econometric techniques to analyze the determinants of the resurgence of capital inflows into Latin America during 1990–91. Argues that domestic reforms in Latin America only partially explain this change, and that the conditions of the US economy (recession, low interest rates, balance of payments) are also significant.

1109 **Cardoso, Eliana A. et al.** ¿Adonde va América Latina?: balance de las reformas económicas. Recopilación de Joaquín Vial. Santiago: CIEPLAN, 1992. 301 p.: bibl., ill.

Series of articles evaluate recent structural reforms (e.g., market liberalization, privatization, etc.). Separate articles on overall structural reform are provided for Bolivia, Brazil, Colombia, Mexico, and Peru, with articles focusing specifically on privatization in Brazil and Chile. Well worth reading. See also item **1112**.

1110 **Cardoso, Eliana A. and Ann Helwege.** Latin America's economy: diversity, trends, and conflicts. Cambridge, Mass.: MIT Press, 1992. 326 p.: bibl., ill., index, map.

Comprehensive text describes current state and historical evolution of the Latin American economy. Separate chapters cover import substitution and trade liberalization, the debt, inflation, stabilization, poverty, agrarian reform, and economic populism. Contains a great deal of data and an extensive bibliography. Valuable contribution.

1111 **Cardoso, Eliana A.** Private investment in Latin America. (*Econ. Dev. Cult. Change*, 41:4, July 1993, p. 833–848)

Uses econometric techniques to estimate determinants of private investment in six countries (Argentina, Brazil, Chile, Colombia, Mexico, and Venezuela) during 1970–85. Concludes that roughly 3/4 of private investment can be explained by increases in the level of public investment and GNP and improvements in the terms of trade.

1112 **Cardoso, Eliana A.** La privatización en América Latina. (*in* ¿Adonde va América Latina?: balance de las reformas económicas. Recopilación de Joaquín Vial. Santiago: CIEPLAN, 1992, p. 79–100, bibl.)

Reviews economic rationale for privatization in Latin America. Discusses different methods of implementation in Argentina, Brazil, Chile, and Mexico, as well as the extent to which this process has been pursued. Excellent and timely article with an extensive bibliography.

1113 **The Caribbean in the global political economy.** Edited by Hilbourne A. Watson. Boulder, Colo.: L. Rienner Publishers, 1994. 1 v.: bibl., index.

Thirty essays on various issues (e.g., regional integration, privatization, women's roles, international competitiveness, etc.) relevant to the region. Author contends that—like it or not—global capitalism is the dominant economic force in the world today and that the Caribbean must react in a positive manner to this reality. For political scientist's comment see item **4078**.

1114 **Carroll, Thomas F.** Intermediary NGOs: the supporting link in grassroots development. West Hartford, Conn.: Kumarian Press, 1992. 274 p.: bibl., index. (Kumarian Press library of management for development)

Comprehensive examination of nongovernmental organizations (NGOs) looks at their differing structures and functions and provides a methodology for measuring their performance. Includes case studies of NGOs in Chile, Costa Rica, and Peru. Significant contribution.

1115 **Cartas, José María.** Cambio estructural y costo social: algunas reflexiones. (*Contribuciones/Buenos Aires*, 9:2, abril/junio 1992, p. 61–75, bibl.)

Describes social cost of recently implemented structural adjustment policies. Focuses on challenges faced by democratic regimes as they simultaneously attempt to correct macroeconomic imbalances while meeting the social needs of the populace.

1116 **Castro Suárez, Pedro.** Teorías del desarrollo: crítica a la teoría de la CEPAL. Lima: Univ. Nacional Mayor de San Marcos, 1992. 247 p.: bibl., index.

Summary and critique of the various theories of economic development from earliest historical ones through modern theories. Highlights relevance of these theories for the region and compares them with those of the United Nations' Economic Commission for Latin America (CEPAL). [R.E. Looney]

Chudnovsky, Daniel. El futuro de la integración hemisférica: el Mercosur y la iniciativa para las Américas. See item **4166.**

1117 **Las ciudades latinoamericanas en la crisis: problemas y desafíos.** Recopilación de Martha Schteingart. México: Editorial Trillas, 1989. 286 p.: bibl., ill., indexes.

Series of articles by experts from various disciplines on the impact of the debt crisis and subsequent policy reforms (e.g., privatization, market liberalization, etc.) on Latin American citizens. Case studies of major cities focus on subjects such as urban growth, health, education, and miscellaneous public services.

1118 **Clavijo, Sergio.** Stabilization policies in Latin America: some lessons for the new decade. (*Estud. Econ./México*, 7:2, julio/dic. 1992, p. 209–224, bibl., graphs, tables)

Describes stabilization programs in Argentina, Bolivia, Brazil, and Colombia during 1984–85. Puts forth an alternative stabilization plan that combines orthodox and heterodox features.

1119 **Cohen, Benjamin J.** U.S. debt policy in Latin America: the melody lingers on. (*in* In the shadow of the debt: emerging issues in Latin America. New York: Twentieth Century Fund Press, 1992, p. 153–172)

Author puts forth thesis that US debt policy had as its primary goal ensuring the financial welfare of US banks, with little thought given to the burden this would impose on Latin America.

1120 **Cole, Julio Harold.** La falsa promesa del proteccionismo para América Latina. (*Contribuciones/Buenos Aires*, 9:2, abril/junio 1992, p. 77–95, bibl., tables)

Critical analysis of Prebisch's views on secular terms of trades and the Import Substituting Industrialization (ISI) model of development which resulted from these views. Argues that an incorrect interpretation of historical data led to a strategy (ISI) which wasted capital and contributed little to growth.

1121 *Columbia Journal of World Business.* No. 1, Spring 1993– . New York: Columbia Univ.

Entire issue devoted to privatization programs in various parts of the world. Includes four articles on privatization of the Argentine steel industry and telecommunication sector as well as privatization efforts in Chile and Mexico. Timely contribution.

1122 **Comisión Económica para América Latina y el Caribe.** Balance preliminar de la economía de América Latina y el Caribe. Santiago: Comisión Económica para América Latina y el Caribe, 1993. 65 p.: graphs, tables.

Brief review of economic evolution of Latin American countries during 1993. Provides preliminary data on output, employment, wages, prices, balance of payments, and foreign debt. Useful for individuals who "can't wait" for more complete, comprehensive, and accurate data to be published.

1123 **Comisión Económica para América Latina y el Caribe.** Planificación y gestión del desarrollo en áreas de expansión de la frontera agropecuaria en América Latina. Santiago: Naciones Unidas, Comisión Económica para América Latina y el Caribe, 1989. 113 p.: bibl., ill., maps.

Discusses methods that can be used to increase amount of land used for agricultural purposes, while simultaneously protecting the environment. Concludes that without government planning, these twin goals will not be achieved. Describes various appropriate planning methodologies. Includes short case studies for Argentina, Brazil, and Colombia.

1124 **Corbo, Vittorio.** Estrategias y políticas de desarrollo en América Latina: una perspectiva histórica. (*Economía/Lima*, 14:27, junio 1991, p. 9–55, bibl., tables)

Describes the Latin American development process from the beginning of the 20th century to the present. Identifies five distinct periods, with each being characterized by the particular growth strategy (import substitution, free market, etc.) employed. Includes excellent bibliography.

1125 **Crisis y crecimiento en América Latina: material para un diagnóstico.** Recopilación de Víctor Urquidi, Javier Villanueva y Carlos A. Cattaneo. Buenos Aires: Editorial Tesis; Fundación Raúl Prebisch, 1989. 457 p.: bibl., ill.

Interesting collection of articles focuses on the debt crisis, adjustment policies and problems, and the need for new development strategies.

1126 Cukierman, Alex; Miguel A. Kiguel; and Nissan Liviatan. How much to commit to an exchange rate rule?: balancing credibility and flexibility. (*Rev. Anál. Econ.*, 7:1, junio 1992, p. 73–89, bibl.)

Presents highly technical mathematic model that is used to explain why government policymakers are reluctant to make a strong commitment to a fixed exchange rate. Uses examples from various Latin American countries.

1127 Davis, Shelton and William Partridge. Promoting the development of indigenous people in Latin America. (*Financ. Dev.*, 31:1, March 1994, p. 38–40)

Indigenous people comprise eight percent of the region's population and are the poorest of the poor. Article examines recent role of the World Bank in improving the life of these people. Programs include providing technical assistance, access to credit, and help in obtaining legal rights to their land.

1128 De Janvry, Alain and Elisabeth Sadoulet. Market, State, and civil organization in Latin America beyond the debt crisis: the context for rural development. (*World Dev.*, 21:4, April 1993, p. 659–674)

Describes changing role of the market and the State in Latin America due to recent liberalization policies. Examines implications of these changes for the agricultural sector and for nongovernment organizations (NGOs) and grassroots organizations.

1129 De Melo, Jaime and Sumana Dhar. Lessons of trade liberalization in Latin America for economies in transition. Washington: Country Economics Dept., World Bank, 1992. 40 p.: bibl., ill. (Policy research working papers; WPS 1040)

Discusses recent trade liberalization programs in Argentina, Bolivia, Chile, Mexico, and Uruguay. Draws lessons from the experiences of these countries and the current liberalization attempts of East European and former Soviet Union countries. Concludes that successful liberalization depends on a stable real exchange rate.

1130 Decentralization of agricultural planning systems in Latin America. Rome: Food and Agriculture Organization of the United Nations, 1990. 62 p.: bibl. (FAO economic and social development paper; 92)

Four case studies (Brazil, Colombia, Mexico, and Peru) summarize approaches to and results of decentralization of agricultural systems.

1131 El desarrollo desde dentro: un enfoque neoestructuralista para la América Latina. Recopilación de Osvaldo Sunkel. México: Fondo de Cultura Económica, 1991. 507 p.: bibl. (Lecturas; 71)

Series of articles critically analyzes the region's development from 1950–90. From a neostructuralist perspective, authors describe the evolution of the structuralist approach, the rise of neoliberalism in the 1980s, and the current economic crisis.

1132 Desarrollo industrial y cambio tecnológico: políticas para América Latina y el Caribe en los noventa. Caracas: Editorial Nueva Sociedad; Sistema Económico Latinoamericano, 1991. 223 p.: bibl., ill.

Describes industrial development and technological change, both globally and within Latin America. Identifies and discusses reasons for Latin America's lagging performance. Argues that greater regional integration is necessary if Latin America is to have a dynamic industrial sector.

1133 Deuda externa, renegociación y ajuste en la América Latina. Edición de Stephany Griffith-Jones. México: Fondo de Cultura Económica, 1988. 436 p.: bibl., ill. (El Trimestre económico. Lecturas; 61)

Series of well-written articles on the debt crisis and subsequent adjustment process. Separate articles cover Brazil, Chile, Costa Rica, Mexico, Peru, and Venezuela.

1134 Development and democracy: aid policies in Latin America. Paris: OECD, 1992. 116 p.: bibl.

Position paper focuses on donor experiences and attitudes toward aid to Latin America. Places great emphasis on the need for economic and political pluralism, democracy, and human rights.

1135 Development from within: toward a neostructuralist approach for Latin America. Edited by Osvaldo Sunkel. Boulder, CO: L. Rienner, 1992. 1 v.

Consists of 13 articles on neostructuralist theory and policy as applied to problems of the region. Topics include the environment, technological change, labor markets, capital formation, agriculture, industry, trade, the appropriate role of the State, and income distribution. Well worth reading. For comment by the international relations specialist, see item **3806**.

1136 Devereux, John and **Michael Connolly.** Taxes, growth and the real exchange rate: theory and evidence for Latin America. (*JOICE*, 3:1, 1994, p. 1–20)

Econometric analysis of factors determining real exchange rates (i.e., relative national price levels) for 17 Latin American countries during 1960–85. Authors conclude that primary determinants are per capita income, external terms of trade, government spending, and openness of the economy.

1137 Devlin, Robert and **Martine Guerguil.** Latin America and the new finance and trade flows. (*CEPAL Rev.*, 43, April 1991, p. 23–50, bibl., tables, graphs)

Examines likely state of Latin American trade and international finance during the 1990s. Concludes that external factors will be unfavorable. Argues that the region must: 1) implement proactive industrial policies to increase the level of technological competitiveness, and 2) form regional blocs to improve its bargaining position.

1138 Diálogo con nuestro futuro común: perspectivas latinoamericanas del Informe Brundtland. Recopilación de Günther Maihold y Víctor L. Urquidi. México: Fundación Friedrich Ebert-México; Caracas: Ediciones Nueva Sociedad, 1990. 179 p.: bibl., ill.

Series of articles whose primary focus is "environmentally sustainable" growth and its relevance to Latin America. Includes separate articles on Argentina, Brazil, Chile, Colombia, and Costa Rica.

1139 Documentos sobre privatización con énfasis en América Latina. Santiago: Naciones Unidas, Comisión Económica para América Latina y el Caribe, Centro Latinoamericano de Documentación Económica y Social, 1991. 82 p. (Serie INFLOPLAN, temas especiales del desarrollo, 0259–0107; 7)

Extensive bibliography of Spanish and non-Spanish publications on the subject of privatization. Valuable contribution.

1140 Dornbusch, Rudiger and **Mario Henrique Simonsen.** Estabilización de la inflación con el apoyo de una política de ingresos: la experiencia de Argentina, Brasil e Israel. (*Trimest. Econ.*, 54:214, abril/junio 1987, p. 225–281, bibl., tables, graphs)

Examines various types of anti-inflation stabilization plans, focusing on those factors that determine their success or failure. Considers impact of inertial inflation, monetary illusion, and the behavior of different actors, using a game-theory perspective. Specifically analyzes stabilization plans in Argentina, Brazil, and Israel during 1985–86.

1141 Economía de la integración latinoamericana: lecturas seleccionadas. Recopilación de Germánico Salgado P. Buenos Aires: Banco Interamericano de Desarrollo, Instituto para la Integración de América Latina; Editorial Tesis, 1989. 2 v.: bibl., ill. (Publ.; 319)

Series of well-written articles on a broad range of topics relating to the potential economic integration of the region.

1142 *Economic and social progress in Latin America; annual report.* 1992– . Washington: Inter-American Development Bank.

Reviews economic policy and performance for each country during 1992. Includes a special section devoted to human resources. Discusses the "payoff" from investing in health, education, and nutrition. Appendix contains 75 tables of useful serial data for each country.

1143 Economic development under democratic regimes: neo-liberalism in Latin America. Edited by Lowell S. Gustafson. Westport, Conn.: Praeger, 1994. 264 p.: bibl., index.

Series of well-written articles on Latin America's experience with neoliberal economic policy reforms. Includes country-studies for Argentina, Bolivia, Colombia, Ecuador, and Venezuela. Also includes articles on subregional integration efforts such as NAFTA and Mercosur. Timely and important contribution.

1144 Economic reforms in Latin America: symposium held in November 1992 at the Georg-August-Universität Göttingen. Edited by Hermann Sautter. Frankfurt am Main:

Vervuert Verlag, 1993. 169 p.: bibl., ill. (Göttinger Studien zur Entwicklungsökonomik; 1)

Seven papers on current economic issues in the region. Topics include the debt, NAFTA, debt swaps, liberalization, foreign investment, and the current and changing role of the State.

1145 Economic strategies and policies in Latin America. Edited with an introduction by Jorge I. Domínguez. New York: Garland Pub., 1994. 388 p.: bibl., ill. (Essays on Mexico, Central and South America; 1)

Excellent series of articles by leading economists (Prebisch, Harberger, Hirschman, Dornbusch, Edwards, etc.) on the evolution of economic policy from the 1950s to the present. Discusses ascendency of import-substituting industrialization, subsequent disenchantment with it, and its replacement by free market, export-oriented policies.

1146 The economy of Latin America and the Caribbean: guidelines for a comprehensive approach to the foreign debt problem and possible action by the inter-American system; informative document. Organization of American States, Executive Secretariat for Economic and Social Affairs. Washington: General Secretariat, Organization of American States, 1986. 30 p.

Provides OAS perspective on the debt crisis. Argues that debtor nations need greater degree of "financial tranquility" if they are to deal with problems of growth, health, education, malnutrition, etc. Describes several approaches whose implementation would help achieve such tranquility.

1147 La edad de plomo del desarrollo latinoamericano. Recopilación de Jaime Ros. México: Instituto Latinoamericano de Estudios Transnacionales; Fondo de Cultura Económica, 1993. 325 p.: bibl., ill. (El Trimestre económico. Lecturas; 77)

Series of articles on the debt crisis and resulting stabilization programs implemented by Latin American nations. Analyzes impact of alternative types of stabilization programs on growth, trade, employment, wages, income distribution, etc. Contains interesting economic models and data.

1148 Edwards, Sebastian. Latin American integration: a new perspective on an old dream. (*World Econ.*, 16:3, May 1993, p. 317–338)

Readable review of past integration efforts: Latin American Free Trade Area (ALALC), Latin American Integration Association (ALADI), Central American Common Market (CACM), Caribbean Free Trade Association (CARIFTA), Caribbean Community (CARICOM), Andean Group, and South American Common Market (Mercosur). Unlike their predecessors, recent attempts are outward-oriented.

1149 El-Erian, Mohamed A. Restoration of access to voluntary capital market financing: the recent Latin American experience. (*Staff Pap.*, 39:1, March 1992, p. 175–194, bibl., graphs)

Documents the recent and gradual resumption of voluntary external financial flows (equity, loans, bonds) to certain Latin American countries. Provides detailed analysis of the various factors which contributed to this restoration (e.g., lower risk, decreased transaction costs, etc.).

1150 El-Erian, Mohamed A. Voluntary market financing. (*Financ. Dev.*, 29:1, March 1992, p. 28–41)

Describes resumption of voluntary capital flows to Chile, Mexico, and Venezuela, and to a lesser extent, Argentina and Brazil. Author attributes this to implementation of IMF-type policies, debt restructuring, and reduced transaction costs resulting from regulatory changes in the US and other developed countries.

1151 El-Saeed, Hala Helmy. The international banking system and the external debt of Latin American countries. Cairo: Nahdet Misr, 1992. 520 p.: bibl.

Impressive and comprehensive discussion of the debt crisis discusses its causes and structure, its management, and its impact on the international monetary system, as well as various proposals for alleviating the problem.

1152 Empresarios y estado en América Latina: crisis y transformaciones. Coordinación de Celso Garrido N. México: CIDE, 1988. 374 p.: bibl., ill.

Series of papers on the role of the State vis-à-vis entrepreneurs, with special emphasis on how the debt crisis and subsequent reforms have altered this relationship. Includes case studies for Argentina, Brazil, Chile, Mexico, Nicaragua, Peru, and Uruguay.

1153 **Engel, Eduardo** and **Patricio Meller.** Review of stabilization mechanisms for primary commodity exporters. (*in* External shocks and stabilization mechanisms. Edited by Eduardo Engel and Patricio Meller. Washington: Inter-American Development Bank, 1993, p. 1–23, bibl., graphs)

Examines advantages and disadvantages of two different types of stabilization mechanisms for short-run primary commodity export prices. These mechanisms are: 1) stabilization funds; and 2) financial instruments such as futures contracts. Focuses on the principle primary commodity exports of Bolivia, Chile, and Venezuela.

1154 **The Enterprise for the Americas Initiative: issues and prospects for a free trade agreement in the Western Hemisphere.** Edited by Roy E. Green. Westport, Conn.: Praeger, 1993. 209 p.: bibl., index.

Series of well-written technical essays that examine issues relating to western hemisphere economic integration. Within this framework articles are wide-ranging, with major focus on the Enterprise for the Americas and NAFTA.

1155 **Equidad y transformación productiva: un enfoque integrado.** Santiago: Naciones Unidas, Comisión Económica para América Latina y El Caribe, 1992. 254 p.: bibl., ill.

Economic Commission for Latin America and the Caribbean describes its "new philosophy" of economic growth and development. Argues that future growth must embody: 1) greater income equality; 2) environmental sustainability; 3) democracy; 4) technological innovation; and 5) international competitiveness.

1156 **Erzan, Refik** and **Alexander Yeats.** Free trade agreements with the United States: what's in it for Latin America? Washington: International Economics Dept., The World Bank, 1992. 66 p.: bibl. (Policy research working papers: WPS 827)

Using data for 11 major Latin American countries, authors use econometric techniques to analyze the impact (i.e., gains and losses) on each country of free trade with the US. Covers Argentina, Bolivia, Brazil, Chile, Colombia, Ecuador, Mexico, Paraguay, Peru, Uruguay, and Venezuela.

1157 **Estadísticas de intercambio comercial de los países latinoamericanos en la década de los ochenta.** Buenos Aires: BID-INTAL, 1991. 144 p.: bibl., ill. (Publ.; 369)

Provides extensive array of statistics on intra-country trade (exports and imports) during the 1980s. Especially useful for those who do empirical work on Latin American trade.

1158 **Estay Reyno, Jaime.** La concepción general y los análisis sobre la deuda externa de Raúl Prebisch. México: Siglo Veintiuno Editores, 1990. 133 p.: bibl. (Economía y demografía)

Summarizes Prebisch's writings on Latin American development and external debt published during the 1950s-60s. Compares these views on external debt with the actual experience of the region during the debt buildup and subsequent crisis. Interesting.

1159 **Fajnzylber, Fernando.** Unavoidable industrial restructuring in Latin America. Durham, N.C.: Duke Univ. Press, 1990. 207 p.: bibl., ill., index.

Calls for regional restructuring of industry. Goals should be to: 1) increase international competitiveness and trade; 2) reduce emphasis on non-renewable income (i.e., natural resources); and 3) reduce emphasis on manufacturing *per se,* instead placing more emphasis on sub-sectors that contribute to dissemination of new technologies.

1160 **Félix, David.** Privatizing and rolling back the Latin American State. (*CEPAL Rev.,* 46, April 1992, p. 31–46, bibl., tables)

Critical analysis of privatization process in Latin America. Author contends that due to a variety of factors, favorable results of privatization will be far less than anticipated. Well worth reading.

1161 **Fernández, Roque B.** *et al.* Testimonios sobre la actuación de la banca central. México: Centro de Estudios Monetarios Latinoamericanos, 1994. 380 p.: bibl.

Series of papers by central bank presidents and former central bankers outlining their philosophy towards monetary policy and economic stabilization. Countries included are Argentina, Brazil, Canada, Chile, Colombia, France, Jamaica, Japan, Mexico, Spain, Trinidad and Tobago, US, and Venezuela. [R.E. Looney]

1162 **Ffrench-Davis, Ricardo.** El desarrollo económico de América Latina desde 1950. (*Rev. Occident.*, 131, abril 1992, p. 51–62, table)

Describes development of Latin America since 1950, emphasizing role played by international markets. Analyzes three different periods: the years of growth (1950–80), the crisis of the 1980s, and the challenges of the 1990s.

1163 **Fichet, Gérard.** The competitiveness of Latin American industry. (*CEPAL Rev.*, 43, April 1991, p. 51–66, graphs, tables)

Compares industrial growth of Asia, the Mediterranean, and Latin America from 1970–85. Finds that Latin American industrial exports are far less than would be expected, given its industrial record.

1164 **Fishlow, Albert** and **Jorge Friedman.** Tax evasion, inflation and stabilization. (*J. Dev. Econ.*, 43:1, Feb. 1994, p. 105–123)

Using data from Argentina, Brazil, and Chile, authors test hypothesis that tax evasion increases when incomes decline and the rate of inflation increases. Results support this hypothesis.

1165 **Fondos y programas de compensación social: experiencias en América Latina y el Caribe.** Washington: Organización Panamericana de la Salud; Organización Mundial de la Salud, 1992. 286 p.: bibl., ill. (Serie Salud en el desarrollo)

Series of articles that review the Latin American and Caribbean experiences with Social Compensation Programs. These programs are intended to alleviate human-social costs associated with adjustment programs and reforms. Separate chapters are devoted to case studies for Bolivia, Costa Rica, Guatemala, Honduras, Jamaica, Nicaragua, and Peru.

1166 **Frank, André Gunder.** Latin American development theories revisited: a participant review. (*Lat. Am. Perspect.*, 19:2, Spring 1992, p. 125–139)

From a radical-heterodox perspective, author reviews five books on development theory. Interesting.

1167 **Frankman, Myron J.** Global income redistribution: an alternate perspective on the Latin American debt crisis. (*in* Latin America to the year 2000: reactivating growth, improving equity, sustaining democracy. New York: Praeger, 1992, p. 41–51, bibl., tables)

Documents that the debt crisis has resulted in a net transfer of resources from the South to the North (i.e., from Latin America to the US and other lender nations). Calls for advanced nations of the world to establish a formal system to reverse this trend.

1168 **Frediani, Ramón O.** Estabilización y reforma estructural en Latinoamérica durante la década del noventa. (*Contribuciones/Buenos Aires*, 4, oct./dic. 1991, p. 91–97, bibl.)

Summarizes Latin American stabilization efforts and economic reforms over the last two decades, emphasizing the shift from State regulation to free-market policies. Maintains that adoption of free market policies is not due to ideological conviction but rather to perceived failure of the State over the last 40 years.

1169 **Fuentes, Juan Alberto.** European investment in Latin America: an overview. (*CEPAL Rev.*, 48, Dec. 1992, p. 61–81, bibl., graphs, tables)

Empirically describes the pattern of European direct investment in Latin America in the 1980s. Indicates the geographic distribution by both investing and host countries, as well as the distribution of these investments by type of economic activity (e.g., mining, petroleum, metals, etc.)

1170 **Fujii, Gerardo.** Productivity: agriculture compared with the economy at large. (*CEPAL Rev.*, 44, Aug. 1991, p. 105–114, appendix, bibl., tables)

In Latin America, the ratio of labor productivity in agriculture to labor productivity in the rest of the economy is less than one-half. Author demonstrates that similar relationships prevailed in many developed countries (US, France, Germany) as recently as 1950 but have since narrowed.

1171 **The future development of maize and wheat in the Third World.** México: CIMMYT, 1987. 163 p.: bibl., col. ill.

Series of articles (with discussion) focuses on various technological and scientific innovations that have significantly increased the output of maize and wheat in Third World countries. Many examples are drawn from Latin America. A must read for the agricultural specialist.

1172 Garramón, C.J. et al. Ajuste macroeconómico y sector agropecuario en América Latina. 2da ed. corr. y aum. Buenos Aires: Legasa, 1991. 1 v.: bibl., ill.

Series of case studies on the impact of the debt crisis and subsequent economic reforms on the agricultural sector. Covers Argentina, Brazil, Central America, Chile, Colombia, Dominican Republic, Ecuador, Mexico, and Peru. Contains much interesting data. Especially useful for the agricultural specialist.

1173 Gestión para el desarrollo de cuencas de alta montaña en la zona andina. Santiago: Naciones Unidas, Comisión Económica para América Latina y el Caribe; New York: United Nations, 1988. 187 p.

Describes and analyzes policies and methods adopted by Bolivia, Colombia, Ecuador, Peru, and Venezuela to develop their highland areas. Focuses on appropriate strategies and management styles that should be used to exploit the economic potential of these regions without causing ecological degradation.

1174 Glade, William P. Economics: essay. (*in* Latin America and the Caribbean: a critical guide to research sources. Edited by Paula H. Covington. New York: Greenwood Press, 1992, p. 156–234)

Brief essay on the historical evolution of Latin American economic literature, followed by an annotated bibliography of 679 items, listed by country and topic. Emphasizes post-WWII empirical work.

1175 Glade, William P. Privatization in rent-seeking societies. (*World Dev.*, 17:5, 1989, p. 673–682, bibl.)

Interesting examination of recent privatization efforts in Argentina, Brazil, Chile, and Mexico. Focuses on attempt by special interest groups to influence government so as to affect a transfer of wealth and the impact that these activities have had on privatization.

1176 Glover, David and **Carlos Larrea Maldonado.** Changing comparative advantage, short term instability and long term change in the Latin American banana industry. (*Can. J. Lat. Am. Caribb. Stud.*, 16:32, 1991, p. 91–108, graphs, table)

Examines the changing structure of the Latin American banana industry over the past 100 years, emphasizing the role played by transnational corporations (TNCs) after WWII. Discusses response of TNCs to changing economic and political conditions.

1177 Glover, David. Trade and industrial policy in Latin America: issues for the 1990s. (*in* Latin America to the year 2000: reactivating growth, improving equity, sustaining democracy. New York: Praeger, 1992, p. 53–63, bibl.)

Author provides research agenda for Latin American trade and industrial policy. Covers issues such as the appropriate sequencing of liberalization policies, the relationship between exports and growth, etc. Well referenced.

1178 Goldin, Ian and **Dominique van der Mensbrugghe.** The forgotten story: agriculture and Latin American trade and growth. (*in* Strategic options for Latin America in the 1990s. Edited by Colin I. Bradford, Jr. Paris: Development Centre of the Organisation for Economic Co-Operation and Development; Washington: Inter-American Development Bank, 1992, p. 181–216, bibl., graphs, tables)

Uses simulation techniques to analyze the impact that various changes in world and domestic policies would have on Latin American agriculture. These policies include: 1) greater rural investment; 2) increase in net foreign transfers to the region; and 3) various degrees of world trade liberalization. Interesting!

1179 Gottret, M.V. et al. Investigación empírica sobre los efectos de las dificultades financieras en los flujos comerciales de los países del Caribe y América Latina. (*Invest. Agrar.*, 6:2, dic. 1991, p. 147–160, bibl., tables)

Uses econometric techniques to analyze the relationship between financial pressures (uses K. Bollen's Financial Pressure Index) and total imports. Estimates import elasticities for agriculture, non-agriculture, and total imports from the US and the rest of the world. Interesting.

1180 Gouvea Neto, Raúl de and **Geraldo M. Vasconcellos.** La diversificación de las exportaciones y la eficiencia de la cartera de exportación: estudio comparativo de los países del Sureste de Asia y de la América Latina. (*Trimest. Econ.*, 60:237, enero/marzo 1993, p. 29–52, appendix, bibl., graph, tables)

Uses econometric techniques to evaluate the success of export-led growth strategies. Concludes that countries that have successfully promoted export diversification (Brazil, Hong Kong, the Philippines, Singapore, and South Korea) enjoy greater export stability than countries that have not pushed diversification (Argentina, Mexico, and Thailand).

1181 **Government spending and income distribution in Latin America.** Edited by Ricardo Hausmann and Roberto Rigobón. Washington: Inter-American Development Bank, 1993. 221 p.: bibl., graphs, tables.

Five essays on government spending and taxation and their joint impact on income distribution in four countries (Chile, the Dominican Republic, Peru, and Venezuela). Concludes that far too many programs that are intended to redistribute income to the poor are in fact regressive.

1182 **Gregorio, José de.** El crecimiento económico en la América Latina. (*Trimest. Econ.*, 59, número especial, dic. 1992, p. 75–107, bibl., graph, tables)

Uses econometric techniques and cross-country data to quantify the determinants of income growth: domestic investment, foreign investment, government expenditures, inflation, etc. Results are as one would expect with one interesting exception: terms of trade are not a significant determinant of growth.

1183 **Grosh, Margaret E.** Administering targeted social programs in Latin America: from platitudes to practice. Washington: The World Bank, 1994. 174 p.: bibl., ill.

Analyzes costs and benefits of delivering social services to specific target groups. Draws conclusions from 30 programs (e.g., food stamps, school lunches, health care, health insurance, student loans, day care, etc.) in 11 countries. Provides methodology for assessing outcomes. Significant contribution.

1184 **Guerra-Borges, Alfredo.** La integración de América Latina y el Caribe: la práctica de la teoría. México: Instituto de Investigaciones Económicas, Univ. Nacional Autónoma de México, 1991. 253 p.: bibl. (México y América)

The first half of this book is devoted to various theories of and justifications for regional integration. The second half evaluates the success of various Latin American trading blocs in terms of their original goals.

1185 **Guidotti, Pablo E.** and **Carlos Alfredo Rodríguez.** Dollarization in Latin America: Gresham's Law in reverse. (*Staff Pap.*, 39:3, Sept. 1992, p. 518–544)

In countries with high rates of inflation (particularly Argentina, Bolivia, Peru, and Uruguay) the dollar has gradually replaced local currencies. Authors present mathematical model that explains this phenomenon. Concludes that this process will be very difficult to reverse. Very sophisticated.

1186 **Gutiérrez Santos, Luis E.** El análisis económico de proyectos de mejoramiento de carreteras. (*Invest. Econ.*, 47:183, enero/marzo 1988, p. 139–166, bibl., graphs)

Provides detailed methodology for estimating costs and benefits of road improvement projects.

1187 **Harberger, Arnold C.** Growth, industrialization, and economic structure: Latin America and East Asia compared. Reflections on social project evaluation. Taipei, Taiwan: Institute of Economics, Academia Sinica, 1988. 104 p.: bibl., ill. (some col.). (Chung-Hua series of lectures by invited eminent economists; 15)

The first of these two short essays compares economic structure and performance of Pacific Rim countries (Hong Kong, Korea, Singapore, etc.) with various Latin American countries, emphasizing their similarities and their differences. The second essay describes various techniques of cost-benefit analysis used in social project evaluation.

1188 **Hirschman, Albert O.** Journeys toward progress: studies in economic policy-making in Latin America. Boulder: Westview Press, 1993. 308 p.: bibl., ill., index, maps.

Reprint of 1963 classic (see *HLAS 25: 1436.*)

1189 **Hufbauer, Gary Clyde** and **Jeffrey J. Schott.** Western Hemisphere economic integration. Washington: Institute of International Economics, 1994. 1 v.: bibl., index.

Examines feasibility of forming a Western Hemisphere Free Trade Area (WHFTA). Develops "readiness indicators," rating each country on a scale of zero to five based on price stability, budget discipline, level of external debt, free-market orientation, emphasis on trade taxes, and level of democracy. Provocative.

1190 **Hutchinson, Gladstone A.** and **Ute Schumacher.** NAFTA's threat to Central America and Caribbean basin exports: a revealed comparative advantage approach. (*J. Interam. Stud. World Aff.*, 36:1, Spring 1994, p. 127–148)

Computes index of comparative advantage for the top 100 Caribbean Basin (CB) exports to the US. Concludes that competition from Mexico will have little impact on CB's overwhelmingly resource-based exports to the US, but will adversely affect its non-resource-based exports.

1191 **Iglesias, Enrique V.** Hacia una agenda económica para los años noventa. (*Estud. Int./Santiago*, 25:99, julio/sept. 1992, p. 322–340)

Head of the Inter-American Development Bank sets forth an agenda for the 1990s that will help Latin America achieve growth and greater social equity. Presents 10 policies (e.g., decreasing role of the State, strengthening democratization, integration, etc.) to promote these goals.

1192 **Iglesias, Enrique V.** The new Latin America and the Inter-American Development Bank. (*Wash. Q.*, 16:1, Winter 1993, p. 115–125)

President of the Inter-American Development Bank (IDB) states his views concerning reasons for the region's improved economic performance during the 1990s. States that the region and the IDB must do more to decrease poverty and increase employment and social integration.

Impex: Foreign Trade Statistics by Commodities = Statistiques de commerce extérieur produits. 1993- . See item 27.

1193 **La industria de bienes de capital en América Latina y el Caribe: su desarrollo en un marco de cooperación regional.** Santiago: Naciones Unidas, Comisión Económica para América Latina y el Caribe, 1991. 235 p.: bibl., ill. (Cuadernos e informes de la CEPAL, 0256-9795; 79)

Examines region's lagging capital goods industry. Argues that potential benefits from the industry are substantial, but because of limited market size in most countries, developing this industry will require regional cooperation and/or government subsidies to private or public producers. Discusses policies to encourage this critical sector.

1194 **Infante, Ricardo** and **Emilio Klein.** The Latin American labour market, 1950–1990. (*CEPAL Rev.*, 45, Dec. 1991, p. 121–135, bibl., graph, tables)

Compares Latin American labor markets during 1950–80 and 1980–90. Key patterns in the latter period include: 1) increase in unemployment; 2) significant increase in informal sector employment; 3) relative decline in public sector jobs; and 4) shift in private employment to lower-productivity (i.e., lower wage) sectors.

1195 **Inflación: economía, empresa y sociedad.** Recopilación de Antonio Francés y Lorenzo Dávalos. Caracas: IESA, 1991. 327 p.: bibl., ill. (Ediciones IESA)

Comparative study of growth and development in Scandinavian and Latin American countries, with explanations for their differences in growth performance. Covers Finland, the Netherlands, Norway, Sweden, Chile, Colombia, Ecuador, and Uruguay.

Información para el desarrollo. See *HLAS 54:51.*

1196 **Integración, deuda externa y relaciones económicas internacionales.** Conferencia Permanente de Partidos Políticos de América Latina (COPPPAL). Lima: Cambio y Desarrollo, Instituto de Investigaciones, 1992. 251 p.: bibl.

Presents the responses of various Latin American political parties to the Enterprise for the Americas Initiative and to the globalization of the world economy. Highly critical of the social costs of neoliberal reforms and privatization. Contends that development will require unification of Latin American countries.

1197 **The international common-carrier transportation industry and the competitiveness of the foreign trade of the countries of Latin America and the Caribbean.** Santiago: United Nations, Economic Commission for Latin America and the Caribbean,

1989. 116 p.: bibl. (Cuadernos de la CEPAL, 0252-2195; 64)

Examines use of containers for shipping and transportation. Due to removal of legal restrictions, "containerization" has integrated ocean and land transportation policy to increase Latin America's competitiveness and to maximize potential future gains.

IntlEc CD-ROM: the Index to International Economics, Development, and Finance. See *HLAS 54:52.*

Investigación Económica. 1994– . See item **30.**

1198 Izam, Miguel. Europe 92 and the Latin American economy. (*CEPAL Rev.*, 43, April 1991, p. 67–81, bibl.)

Analyzes possible impact of the European Economic Community (EEC). Latin American trade flows might increase, but could decrease if the EEC adopts a "fortress Europe" mentality. Concludes that potential negative effects would have a greater impact on Latin America than would potential positive effects—if either were to occur.

1199 Jornadas Bancarias de la República Argentina, 2nd, Buenos Aires?, 1991? La financiación del crecimiento. Argentina: Mercado, 1991. 192 p.

Proceedings of this high-quality meeting include papers touching on a variety of international financial issues and the experiences of several countries.

1200 Kagami, Mitsuhiro. The voice of East Asia: development implications for Latin America. Tokyo: Institute of Developing Economies, 1995. 145 p.: bibl., ill., index. (I.D.E. occasional papers series; 30)

Four comparative chapters on economic and business development in East Asia and Latin America provide a Japanese perspective on Latin America's industrial structure. Challenges Latin American policymakers to examine and adopt East Asian macro- and microeconomic approaches to industrialization. Based on two industrial surveys conducted just before the height of Brazilian and Mexican economic liberalization. [K. Horisaka]

1201 Kaimowitz, David. The role of nongovernmental organizations in agricultural research and technology transfer in Latin America. (*World Dev.*, 21:7, July 1993, p. 1139–1150, bibl.)

Explores possible role of nongovernmental organizations (NGOs) in increasing agricultural productivity via technological transfer. Concludes that NGOs have great potential but must markedly improve their technical competence. Proposes a partnership between research institutes and NGOs.

LABORDOC. See item **32.**

1202 Latin America in graphs: two decades of economic trends, 1971–1991. Washington: Inter-American Development Bank; Baltimore, MD: Johns Hopkins Univ. Press, 1992. 128 p.

Exhaustive graphical presentation of demographic and macroeconomic data for 26 Latin American countries for the period. Includes national income, balance of payments, debt, labor force, and population statistics.

Latin American Business Intelligence. 1995– . See item **33.**

Latin American Newletters. 1990– . See item **34.**

Latin American News. 1995– . See item **35.**

1203 Latinskaia Amerika: problemy i tendentsii razvitiia transporta [Latin America: problems and tendencies in the growth of transportation]. Edited by Konstantin Sergeevich Tarasov and A.IU. Teslenko. Moscow: ILA AN SSSR, 1991. 179 p.: bibl.

Seven authors from the then Soviet Institute of Latin American Studies contribute articles on Latin American transportation, including the trucking industry; the Brazilian automotive industry; railroad, water, and air transportation; the merchant marine; oil pipelines; public transportation; and technical and political problems relating to the Panama Canal. Many tables with data for individual countries and years; Russian and Western sources. [B. Dash]

Latinskaia Amerika: sobytiia i liudi; analiticheskii obzor [Latin America: events and people; an analytical survey]. See item **3836.**

1204 Lee, Terence and **Andrei Jouravlev.** Self-financing water supply and sanitation services. (*CEPAL Rev.*, 48, Dec. 1992, p. 117–128, bibl., tables)

Empirically analyzes the feasibility of pricing Latin American water and sanitary services so that such activities would be self-financing (i.e., revenues cover all costs of production). Concludes that this is indeed possible, but would likely require subsidies for the poor.

1205 **Leff, Nathaniel H.** and **Kazuo Sato.** Condiciones psicoculturales y desarrollo económico: comportamiento del ahorro y la inversión en el Asia Oriental y la América Latina. (*Trimest. Econ.*, 58:232, oct./dic. 1991, p. 707–728, appendix, bibl., tables)

Uses econometric techniques to test hypothesis that cultural factors explain differing rates of growth in eight Asian and 22 Latin American countries. Results are inconclusive and neither support nor refute this hypothesis.

1206 **León de Leal, Magdalena** and **Hobart A. Spalding.** An evaluation of SAREC's support to three women's research institutions in Latin America: Grupo de Estudios sobre la Condición de la Mujer Uruguaya in Montevideo (GRECMU), Centro de la Mujer Peruana Flora Tristán in Lima, Peru, and Centro de Investigación para la Acción Femenina in Santo Domingo, Dominican Republic (CIPAF). Stockholm: Swedish Agency for Research Cooperation with Developing Countries, 1992. 58 p. (Documentation)

Evaluates research productivity of three Latin American institutes that focus on women's and gender studies and are supported by the Swedish Agency for Research Cooperation with Developing Countries. Provides an interesting overview of general research topics undertaken by such institutions.

Levitt, Kari. Debt adjustment and development: looking to the 1990s. See item **1647.**

Lindenberg, Marc. The human development race: improving the quality of life in developing countries. See item **1553.**

1207 **Little, Marilyn.** Destabilization and debt in Latin America and Africa: comparative perspectives on "economic miracles." (*Yearbook/CLAG*, 17/18, 1990, p. 239–242, bibl.)

Author uses two variables (levels of social unrest and voluntary migration) to measure the "geography of well-being" (i.e., quality of life) in Latin America and Africa. Both levels are quite high. Author concludes that the future looks bleak.

1208 **Liu, Peter C.** Purchasing power parity in Latin America: a co-integration analysis. (*Weltwirtsch. Arch.*, 128:4, 1992, p. 662–880)

Uses econometric analysis to test hypothesis that the rate of exchange of two currencies is a function of relative prices in the two countries. Using data from Argentina, Bolivia, Brazil, Chile, Colombia, Mexico, Peru, Uruguay, and Venezuela, author concludes that this hypothesis is valid. Sophisticated.

1209 **Long-term trends in Latin American economic development.** Edited by Miguel Urrutia. Washington: Inter-American Development Bank; Johns Hopkins Univ. Press, 1991. 170 p.: bibl., ill., indexes.

Documents enormous degree of economic and social change in the region since 1913. Argues that the 1980s represent a cyclical downturn that should not be allowed to obscure positive long-term trends.

1210 **Machado, João Bosco M.** Integración económica y arancel aduanero común en el Cono Sur. (*Integr. Latinoam.*, 16:167, mayo 1991, p. 18–35, bibl., graphs)

Examines factors considered by Argentina, Brazil, and Uruguay in the formation and implementation of the South American Common Market (Mercosur). Provides data on the structure of customs duties in the three countries.

1211 **Machovec, Frank M.** The rise and fall of Prebischian economics. (*South East. Lat. Am.*, 36:1, Summer 1992, p. 9–21)

Highly critical evaluation of Prebisch's views, particularly those that called for a sharply enhanced role for State planning and a greatly reduced role for free markets.

1212 **Maggiora, Ferruccio.** Trade flows and economic integration in Latin America. (*Int. Spect./Rome*, 27:3, July/Sept. 1992, p. 87–100, bibl., tables)

Reviews the current status and performance of regional integration efforts. Covers the Andean Pact, Central American Common Market, Enterprise for the Americas, NAFTA, and the Southern Cone Common Market.

1213 **Magnitud de la pobreza en América Latina en los años ochenta.** Santiago: Naciones Unidas, Comisión Económica para América Latina y El Caribe; New York:

United Nations, 1991. 177 p.: bibl. (Estudios e informes de la CEPAL, 0256-9795; 81)

Documents extent of poverty and indigence (i.e., those unable to afford a minimally adequate diet) in 10 Latin American countries. Discusses various methodologies used to quantify these concepts. Estimates that 44 percent of the population was poor and 21 percent was indigent in 1989. Contains a wealth of data.

1214 Major changes and crisis: the impact on women in Latin America and the Caribbean. Santiago: Economic Commission for Latin America and the Caribbean, 1992. 279 p.: bibl. (Libros de la CEPAL; 27)

Describes major changes in Latin America over the past 30 years and their impact on women. Discusses rural and urban employment (both formal and informal), legal status, mortality and fertility rates, migration, and changing cultural attitudes and roles of women. Important contribution.

1215 Maldifassi, José O. and **Pier A. Abetti.** Defense industries in Latin American countries: Argentina, Brazil, and Chile. Westport, Conn.: Praeger, 1994. 260 p.: bibl., ill., index.

Analyzes economic impact (employment, foreign exchange generation, etc.) of defense industries in Argentina, Brazil, and Chile because they were net importers of military equipment in 1970, but net exporters by 1990. Finds that defense industries have a favorable impact on overall technological change and innovation. Significant contribution.

1216 Manzetti, Luigi. The political economy of Mercosur. (*J. Interam. Stud. World Aff.*, 35:4, Winter 1993/94, p. 101-141)

Describes basic features of the South American Common Market (Mercosur). Due to asymmetrical distribution of gains and losses resulting from integration and the inability to settle related disputes, author concludes that the four countries will not form a true common market, but rather a free trade area.

1217 Marks, Siegfried. Latin America's oil outlook. (*North-South/Miami*, 3:1, June/July 1993, p. 20-25)

Examines current state of region's petroleum industry. Debt crisis has caused governments to drastically reduce investments in oil exploration. Author estimates that investments in petroleum must double in order to maintain reserves. Analyzes production prospects in Argentina, Brazil, Colombia, Mexico, and Venezuela.

1218 Martín del Campo, Antonio and **Donald R. Winkler.** State-owned enterprise reform in Latin America. (*CEPAL Rev.*, 46, p. 47-67, bibl., graphs, tables)

Comprehensive examination of State-owned enterprise reforms in Argentina, Chile, and Mexico. These reforms include not just privatization but also measures to give management greater incentives and autonomy to improve efficiency in enterprises that have not been privatized. One result is significant improvements in labor productivity.

1219 McMahon, Gary. Hyperinflation in Latin America: the search for developmental solutions. (*in* Latin America to the year 2000: reactivating growth, improving equity, sustaining democracy. New York: Praeger, 1992, p. 23-39, bibl.)

Examines failure of heterodox anti-inflationary programs (price controls and incomes policy) in Argentina (1985), Brazil (1986), and Peru (1985), and the success of an orthodox (IMF-type) approach in Bolivia (1985). Concludes that heterodox programs can be successful, but were improperly implemented in the above examples.

1220 Meissner, Frank and **Nancy Morrison.** Seeds of change: stories of IDB innovation in Latin America. Washington: Inter-American Development Bank; Baltimore: Johns Hopkins Univ. Press, 1991. 110 p.: bibl., ill.

Consists of 28 "case studies" on the experience of the Inter-American Development Bank in projects designed to foster technological change in agriculture. Diverse range of topics (forestry, fisheries, dairy, livestock, coffee, honey, potatoes, marketing, irrigation, and credit).

1221 Mesa-Lago, Carmelo. Changing social security in Latin America: toward alleviating the social costs of economic reform. Boulder, Colo.: L. Rienner, 1994. 1 v.: bibl., index.

Economic adjustment and restructuring processes of the 1980s were implemented at the price of high social costs for a large portion of the population. This work describes the role that social security can play in ameliorating these conditions.

1222 Mesa-Lago, Carmelo. The current situation, limitations, and potential role of social security schemes for income maintenance and health care in Latin America and the Caribbean: focus on women. (*in* Coping with social change: programs that work. Washington: Women's Initiative, American Association of Retired Persons; International Federation on Aging, 1990, p. 7–14, bibl., tables)

Provides estimates of protection provided to women by the region's social security programs. Although not formally discriminated against, women are less protected than men due to women's lower participation rates and the fact that they often work as domestic servants or in the informal sector, areas not covered by social security.

1223 Mesa-Lago, Carmelo. El financiamento de la seguridad social en los países latinoamericanos. (*in* La seguridad social y el Estado moderno. México: Fondo de Cultura Económica, 1992, p. 221–248, bibl., tables)

Describes historical evolution of State-sponsored pension systems in Latin America. Discusses problems with the current systems and presents several reform proposals.

1224 Mesa-Lago, Carmelo and **Lothar Witte.** Regímenes pensionales en el Cono Sur y el área andina: problemas y propuestas. (*Nueva Soc.*, 122, nov./dic. 1992, p. 18–34, bibl.)

Describes history of State-operated pension systems, comparing those in Andean countries with Southern Cone countries. Outlines proposals to reform regional pension systems.

1225 Miranda R., Ernesto. Cobertura, eficiencias y equidad en el área de salud en América Latina: problemas y propuestas de solución. (*Estud. Públicos*, 46, otoño 1992, p. 163–248, bibl., tables)

Describes health conditions and characteristics of the health-care sector in various countries. Presents various proposals that when implemented would increase coverage, equity, and efficiency of the health care sector.

1226 Mobilising international investment for Latin America. Edited by Colin I. Bradford, Jr. Paris: OECD Publications, 1993. 252 p.: bibl., ill.

Twenty-one articles on capital flows to the region. Current inflows are often speculative and non-productive. Focuses on appropriate policies to implement in order to control financial inflows and ensure that they are productive.

1227 Money doctors, foreign debts, and economic reforms in Latin America from the 1890s to the present. Edited by Paul W. Drake. Wilmington, Del.: SR Books, 1994. xxxiii, 270 p.: bibl. (Jaguar books on Latin America; 3)

Series of well-written articles examines the role of "money doctors" in Latin America from 1890 to present. Money doctors are foreign economic advisors whose function is to advocate economic reforms that will ensure foreign debts will be repaid. Fascinating, unique, and important contribution.

1228 Mujica Miranda, Estela. ALIDE: más de dos décadas al servicio del financiamiento, el desarrollo y la integración de América Latina.

Describes creation, evolution, scope, function, and structure of Asociación Latinoamericana de Instituciones Financieras de Desarrollo (ALIDE).

1229 Nogués, Julio. La elección entre estrategias de liberalización comercial unilaterales y multilaterales. (*Estudios/Fundación Mediterránea*, 13:53, enero/marzo 1990, p. 21–28, bibl., graphs, tables)

Presents theoretical framework to analyze multilateral and unilateral trade liberalization strategies and uses Argentina as an example. Considers the costs and benefits of each strategy.

1230 Noyola Vázquez, Juan F. La evolución del pensamiento económico en el último cuarto de siglo y su influencia en América Latina. (*in* Desequilibrio externo e inflación. México: Facultad de Economía, UNAM, 1987, p. 49–65)

A critical analysis of the "Anglo-Saxon" school of economic thought. Contends that this method of analysis (classical and neoclassical) is dominant in Latin

America and has led to policies that have had a deleterious effect on the region's development.

1231 La nueva etapa de la integración regional. Secretaría Permanente del SELA. México: Sistema Económico Latinoamericano; Fondo de Cultura Económica, 1992. 164 p.: bibl.

Describes goals and accomplishments of recent Latin American and Caribbean integration efforts (Mercosur, Andean Pact, ALADI, etc.). Outlines possible future integration scenarios and assesses their probability of success.

1232 O'Brien, Philip J. Debt and sustainable development in Latin America. (*in* Environment and development in Latin America: the politics of sustainability. Edited by David Goodman and Michael Redclift. Manchester, England: Manchester Univ. Press; New York: St. Martin's Press, 1991, p. 24–47, bibl.)

Provides excellent overview of the response by the major players in the debt crisis (e.g., private creditor banks, international financial institutions such as the IMF, and debtor governments). Contends that the resulting large transfer of resources from Latin America to creditors contributed greatly to the region's environmental degradation.

OECD Statistical Compendium. 1994– . See item **39**.

1233 Ondarts, Guillermo. Exportaciones y crecimiento: una encuesta a firmas de Argentina, Brasil, Chile, Paraguay y Uruguay. (*Integr. Latinoam.*, 16:173, nov. 1991, p. 23–34, graphs, tables)

Presents results of a survey administered to industrial enterprises in Argentina, Brazil, Chile, Paraguay, and Uruguay. Purpose was to measure the extent to which industrial firms were reacting to new outward-oriented policies of their governments by undertaking investments that would promote exports.

1234 Palma, Norman. De las inversiones extranjeras en América Latina. (*in* América Latina: regiones en transición. Ciudad Real, Spain: Publicaciones de la Univ. de Castilla-La Mancha, 1991, p. 41–47)

Interesting article about the region's shift in policy, from emphasizing import substitution to stressing free markets and a smaller, more efficient State. Discusses resulting changes in economic structure and public social policies.

1235 *Panorama Económico de América Latina.* 1993– . Santiago: Comisión Económica para América Latina y el Caribe.

Series of tables of key economic variables for the region and certain countries (Argentina, Brazil, Chile, Colombia, Ecuador, Mexico, Peru, Uruguay, and Venezuela). Provides data for gross domestic product, interest rates, real wages, consumer prices, exports, imports, exchange rates, etc. Covers 1991 through the first half of 1993.

1236 Pastor, Manuel. Inversión privada y "efecto arrastre" de la deuda externa en la América Latina. (*Trimest. Econ.*, 49:233, enero/marzo 1992, p. 107–151, bibl., graphs, tables)

Analyzes impact of the debt crisis on private investment. Reviews general patterns of the 1970s (e.g., debt, stagnation, macroeconomic instability, investment deterioration). Develops interesting model, focusing on the primary determinants of private investment.

1237 Pastor, Manuel and **Eric Hilt.** Private investment and democracy in Latin America. (*World Dev.*, 21:4, April 1993, p. 489–507)

Using data for seven countries (Argentina, Brazil, Chile, Colombia, Mexico, Peru, and Venezuela) for the period 1973–86, authors use econometric techniques to test hypothesis that democracy and private investment are positively correlated. Results support this hypothesis.

1238 Pazos, Felipe. Medio siglo de política económica latinoamericana. v. 1–3. 2a. ed. Caracas: Academia Nacional de Ciencias Económicas, 1991. 3 v. (1313 p.): bibl., ill.

Three-vol. collection of the works of Felipe Pazos covers wide range of topics including development theories and models for open and closed economies, government planning, foreign investment, banking systems, inflation, income distribution, and the evolution of Latin American economic thought.

1239 Peres, Wilson. The internationalization of Latin American industrial firms. (*CEPAL Rev.*, 49, April 1993, p. 55–74, bibl., tables)

Examines direct foreign investment by Latin American industrial companies. Discusses its magnitude, concentration in certain industries and countries, and factors that encourage and inhibit it.

1240 The political economy of policy reform. Edited by John Williamson. Washington: Institute for International Economics, 1993. 601 p.: bibl., index, tables.

Series of high-quality papers that examine the political ingredients of successful economic reform. Includes papers on Australia, Indonesia, Portugal, Spain, former Soviet-bloc countries, as well as Brazil, Chile, Colombia, Mexico, and Peru. Also includes comments and discussion. Important contribution.

1241 Populismo económico: ortodoxia, desenvolvimentismo e populismo na América Latina. Organição de Luiz Carlos Bresser Pereira. São Paulo: Nobel, 1991. 249 p.: bibl., ill.

Vol. brings together nine essays on economic populism in Latin America, principally in the 1970s-80s. Although there are several country studies, the focus of most papers is regional. Several authors posit cycle of populism that alternates with periods of orthodox stabilization in the context of income inequality. [M.H. Birch and R.E. Smith]

1242 Portes, Alejandro and **Richard Schauffler.** The informal economy in Latin America: definition, measurement and policies. (*in* Work without protections: case studies of the informal sector in developing countries. Edited by Gregory K. Schoepfle and Jorge F. Pérez-López. Washington: US Dept. of Labor, Bureau of International Labor Affairs, 1993, p. 3–39, bibl., graphs, tables)

Overview of the region's informal sector discusses its origins, size, and characteristics. Provides excellent review of literature on this subject. Includes a short section on the impact that NAFTA will have on the Mexican informal sector.

1243 Psacharopoulos, George and **Harry A. Patrinos.** Indigenous people and poverty in Latin America. (*Financ. Dev.*, 31:1, March 1994, p. 41–43)

Summarizes World Bank study on indigenous people in Bolivia, Guatemala, Mexico, and Peru where 81 percent of the continent's indigenous people reside. Finds they are at the bottom in income, living conditions, health, and schooling. Concludes that more schooling, training, and health services are essential.

1244 Public finances in Latin America in the 1980s. Santiago: United Nations, Economic Commission for Latin America and the Caribbean, 1992. 96 p.: bibl., ill. (Cuadernos de la CEPAL, 0252-2195; 69)

Examines trends in public finance for 13 Latin American countries between 1977–79 and 1990. Variables include: aggregate public expenditures; distribution of public expenditures (social security, capital expenditure, etc.); interest on public, foreign, and domestic debt; aggregate level of taxes; relative importance of different types of taxes; fiscal capacity; etc.

1245 Rahnama-Moghadam, Mashaalah; Hedayeh Samavati; and **Lawrence J. Haber.** The determinants of debt rescheduling: the case of Latin America. (*South. Econ. J.*, 58:2, Oct. 1991, p. 510–517, bibl., tables)

Uses econometric techniques to identify financial determinants of debt rescheduling by Latin American countries. Finds that the greater the debt service ratio and the higher the proportion of debt in variable interest loans, the greater the probability of debt rescheduling. The level of international reserves is inversely related to rescheduling.

1246 Ramamurti, Ravi. The impact of privatization on the Latin American debt problem. (*J. Interam. Stud. World Aff.*, 34:2, Summer 1992, p. 93–125, bibl.)

Author contends that borrowing by State-owned enterprises was a major factor in creating the debt problem. Argues that privatization has the potential to substantially alleviate the problem.

1247 Ramírez, Miguel D. Stabilization and adjustment in Latin America: a neostructuralist approach. (*J. Econ. Issues*, 27:4, Dec. 1993, p. 1015–1040)

Discusses "devasting" impact of orthodox adjustment policies on welfare of majority of the region's population. Provides alternative set of neostructuralist policies.

1248 **Raúl Prebisch: un aporte al estudio de su pensamiento; las cinco etapas de su pensamiento sobre el desarrollo, su última intervención pública, bibliografía de su obra entre 1920 y 1986.** Santiago: Comisión Económica para América Latina y el Caribe, Naciones Unidas, 1987. 146 p.

Provides overview of the intellectual legacy of Raúl Prebisch. Focuses on his contributions to development theory and economic policy. Includes comprehensive bibliography of his work.

1249 **Reestructuración de la industria automotriz mundial y perspectivas para América Latina.** Santiago: Comisión Económica para América Latina y el Caribe, Naciones Unidas, 1987. 232 p.: bibl., ill. (Estudios e informes de la CEPAL, 0256-9795; 67)

Analyzes current state of Latin American automobile industry. Discusses government policies and their impact on the industry in Argentina, Brazil, Colombia, Peru, Venezuela, and Andean Group nations. Compares these policies with those in place in South Korea and Australia. Recommends new policies to strengthen the industry.

1250 **Reestructuración industrial y cambio tecnológico: consecuencias para América Latina.** Santiago: Naciones Unidas, Comisión Económica para América Latina y el Caribe, 1989. 105 p.: bibl. (Estudios e informes de la CEPAL, 0256-9795; 74)

Compares technological advances in developing and developed countries since WWII. Concludes that Latin America must transform and modernize its productive structure and must adopt technological change more efficiently than it has in the past. Proposes an agenda to achieve these goals.

1251 **Regional overview of food security in Latin America and the Caribbean with a focus on agricultural research, technology transfer and application.** Prepared for the World Food Council/United Nations Development Program Interregional Consultation on Meeting the Food Production Challenges of the 1990s and Beyond. San Jose: Inter-American Institute for Cooperation on Agriculture, 1991. 100 p.: bibl., ill. (Reports, results and recommendations from technical events series, 0253-4746; A1/SC-91-06)

Documents prevalence of food insecurity in the region. Describes types of technological transfer and agricultural research necessary for this problem to be solved.

1252 **Relación gobierno central-empresas públicas en América Latina.** Edición de Nuria Cunill y Juan Martín. Caracas: ILPES; CLAD, 1988. 335 p.: bibl.

Comprehensive description of the role of State enterprises. Separate chapters are devoted to Argentina, Bolivia, Brazil, Colombia, Costa Rica, Mexico, Nicaragua, Peru, Uruguay, and Venezuela. Discusses their problems and successes. Recommends various alternative management models that can be used to increase efficiency of these entities.

1253 **The restructuring of public-sector enterprises: the case of Latin American and Caribbean ports.** Santiago: United Nations, Economic Commission for Latin America and the Caribbean, 1992. 129 p.: bibl. (Cuadernos de la CEPAL, 0252-2195; 68)

Discusses potential private sector involvement in public ports for the purpose of increasing competitiveness of Latin American export-import sector. Examines various methods (e.g., full privatization, mixed public-private ventures, concessions, etc.) and legal and regulatory changes required to accomplish this goal.

1254 **Ríos, Roberto José.** Economic development and family size. (*Am. Econ.*, 35:2, Fall 1991, p. 81-85)

Uses regression analysis to explain why demographic transition in Europe (1850-1980) decreased family size, while in Latin America demographic transition (1905-85) increased family size. Concludes that higher incomes and increases in proportion of women attending high school will decrease Latin American fertility rates.

1255 **Rodríguez, Allen M.** All that glitters: Latin America's stock markets. (*North-South/Miami*, 3:5, Feb./March 1994, p. 43-47)

Anayzes reasons for recent growth of Latin American equities and their attractiveness to foreign investors. Reasons include desire to diversify risk, the need to finance privatization programs, and the more extensive use of American Depository Receipts (ADRs). Also discusses risks associated with such investments.

1256 Romero Pérez, Jorge Enrique. La crisis y la deuda externa en América Latina. San José: Editorial de la Univ. de Costa Rica, 1993. 231 p.: bibl.

Author, who is Law School Dean at Univ. of Costa Rica, provides legal perspective on the debt crisis. Focuses on legal aspects of neoliberal reforms and policies implemented to alleviate the crisis. Country case studies emphasize Costa Rica.

1257 Sachs, Jeffrey D. Conflicto social y políticas populistas en América Latina. (*Estud. Econ./México*, 5:2, julio/dic. 1990, p. 231–262, bibl., tables, graphs)

Analyzes populist experiences of Argentina (1946–49), Chile (1971–73), Brazil (1983–88), and Peru (1985–88). Argues that the highly unequal distribution of income created political pressures that forced populist regimes to impose policies to improve the welfare of the poor, despite such policies' adverse impact on growth.

1258 Sachs, Jeffrey D. Nuevo enfoques de la crisis de la deuda latinoamericana. (*Rev. Integr. Desarro. Centroam.*, 44, enero/junio 1989, p. 4–60, tables)

Comprehensive discussion of the region's economic situation as of the end of the 1980s. Reviews the evolution of the crisis, the Latin American policy response, the reaction of international banks, and the Baker Plan. Summarizes proposals that have been put forth to solve the debt problem.

1259 Salas Serrano, Julián. Contra el hambre de vivienda: soluciones tecnológicas latinoamericanas. Bogotá: Escala, 1992. 312 p.: bibl., ill., maps. (Tecnologías para viviendas de interés social)

Documents inadequacy of housing in Latin America. Proposes solutions for each country based on individual country characteristics. Provides interesting data and analysis of housing policies in different countries.

1260 Salazar-Carrillo, Jorge. Poder de compra y productividad en América Latina. México: Centro de Investigación y Docencia Económicas, 1991. 192 p.: bibl.

Calculates purchasing power parities for various Latin American countries. Uses these calculations to compute and compare gross domestic product in 41 countries (17 from Latin America). Concludes by analyzing Latin American trade potential based on its absolute and comparative advantage.

1261 Sampaio, Yoni. Ajuste económico y exportaciones agrícolas en Latinoamérica. (*Invest. Agrar.*, 8:1, abril 1993, p. 7–12, bibl., graphs, tables)

Uses econometric techniques to analyze impact of crisis-induced wage decreases and currency devaluation on Latin American agricultural exports. Both factors had a positive impact on agricultural exports, but wage decreases were of quantitatively greater importance.

1262 Sánchez Torres, Fabio. Rentabilidad de la deuda externa de la región andina durante el siglo XIX, 1840–1914: un análisis empírico. (*Cuad. Econ./Bogotá*, 6:16, segundo semestre 1991, p. 183–204, graphs, tables)

Historical study of Andean nations' (Colombia, Ecuador, Peru, and Venezuela) foreign debt with England from 1849–1914. Estimates England's rate of return on these loans.

1263 Sangmeister, Hartmut. Política de reformas en América Latina: oportunidades y riesgos del cambio en los paradigmas económicos. (*Contribuciones/Buenos Aires*, 4, oct./dic. 1991, p. 75–89)

Describes post-debt crisis evolution of the Latin American economy. Discusses increase in poverty, the rise of the informal economy, the redemocratization of the region, and the shift to a neoliberal model of development.

1264 Santos, Eduardo A. La internacionalización de la producción agroalimentaria y el comercio agrícola mundial: implicaciones para el desarrollo agrícola y rural de América Latina y el Caribe. Buenos Aires: Grupo Editor Latinoamericano; Emece Editores, 1992. 425 p.: bibl. (Col. Estudios internacionales)

Describes internationalization of agriculture and its impact on Latin America. Discusses agricultural trade policies of developed countries and implications for Latin America, assuming various alternative scenarios.

1265 Sarmiento Palacio, Eduardo. Growth and income distribution in countries at intermediate stages of development. (*CEPAL Rev.*, 48, Dec. 1992, p. 141–155, bibl., tables)

Notes that over the last four decades, Latin American countries have not been able to achieve *simultaneously* the objectives of growth and social equity. Examines reasons for this phenomenon, concluding that constraints are related to characteristics of countries that are in the intermediate stage of development.

1266 **Savastano, Miguel A.** The pattern of currency substitution in Latin America: an overview. (*Rev. Anál. Econ.*, 7:1, junio 1992, p. 29–72, bibl., graphs, tables)

Author uses econometric techniques to analyze the impact of currency substitution (i.e., dollarization) in Bolivia, Mexico, Peru, and Uruguay. Concludes that the use of foreign currency as part of the domestic financial system often increases inflationary pressures as well as adversely affecting the stability of an exchange rate regime. Excellent.

1267 **Scheetz, Thomas.** The evolution of public sector expenditures: changing political priorities in Argentina, Chile, Paraguay and Peru. (*J. Peace Res.*, 29:2, May 1992, p. 175–190, bibl., graphs, tables)

Examines government spending in four countries from 1969–87. Finds that military spending is understated in official sources, and is, in fact, the largest functional outlay of central government and is generally growing at a faster rate than social expenditures. Conclusion is contrary to prevailing views.

1268 **Schoepfle, Gregory K.** and **Jorge F. Pérez-López.** Work and protections in the informal sector. (*in* Work without protections: case studies of the informal sector in developing countries. Edited by Gregory K. Schoepfle and Jorge F. Pérez-López. Washington: US Dept. of Labor, Bureau of International Labor Affairs, 1993, p. 247–279, bibl., tables)

Documents lack of legal protections (right to unionize, work conditions, minimum wages, minimum age, etc.) provided workers in the informal sector. Argues that simply extending present laws to the informal sector will *not* work, since higher costs will force most informal firms to cease operations. Discusses alternative approaches.

1269 **Schott, Jeffrey J.** and **Gary C Hufbauer.** Free trade areas, the Enterprise for the Americas Initiative, and the multilateral trading system. (*in* Strategic options for Latin America in the 1990s. Edited by Colin I. Bradford, Jr. Paris: Development Centre of the Organisation for Economic Co-Operation and Development; Washington: Inter-American Development Bank, 1992, p. 249–277, bibl., tables)

Examines NAFTA and the Enterprise for the Americas Initiative. Focuses on their likely impact on various regional integration schemes (Andean Group, Central American Common Market, Caribbean Economic Community, and the Southern Cone Common Market).

1270 **Seminario Latinoamericano sobre Medio Ambiente y Desarrollo,** *San Carlos de Bariloche, Argentina, 1990.* Latinoamérica: medio ambiente y desarrollo; programa de medio ambiente. Buenos Aires: Fundación Jorge Esteban Roulet, 1990. 352 p.: bibl., ill., maps. (Col. Encuentros; 8)

Approximately 50 articles on Latin American development and its impact on the environment. Some articles are general, while others are quite specific. Specialists in the environmental area will want to examine this.

1271 **Serven, Luis** and **Andrés Solimano.** Debt crisis, adjustment policies and capital formation in developing countries: where do we stand? (*World Dev.*, 21:1, Jan. 1993, p. 127–140, bibl., graph, tables)

Examines the decrease in rates of investment that occurred in developing countries after 1982. Uses econometric techniques to estimate the impact of key variables (debt, macroeconomic instability, uncertainty) on investment. Covers 15 developing countries, nine of which are Latin American.

1272 **Sherman, Amy L.** Preferential option: a Christian and neoliberal strategy for Latin America's poor. Grand Rapids, Mich.: W.B. Eerdmans, 1992. 230 p.: bibl., index.

Analyzes role to be played by Christian service organizations in alleviating regional poverty. Argues that such organizations should support neoliberal, free-market capitalistic approaches to development.

1273 **Sistema financiero y asignación de recursos—experiencias latinoamericanas y del Caribe: Colombia, Costa Rica, Chile, República Dominicana, Venezuela.** Edición de Carlos Massad y Günther Held. Buenos Aires: Grupo Editor Latinoamericano; Emecé

Editores, 1990. 455 p.: bibl. (Col. Estudios políticos y sociales)

Evaluates solvency and efficiency of banking systems as well as banking deregulation in five Latin American countries.

1274 Skoczek, M. La colonización agrícola como factor formador de regiones nuevas en América Latina. (*in* América Latina: regiones en transición. Ciudad Real, Spain: Publicaciones de la Univ. de Castilla-La Mancha, 1991, p. 49–59, tables)

Describes features and characteristics of newly colonized agricultural areas in the region. Concepts from the field of economic geography are used to explain the pattern of population growth in these areas.

1275 Sosa Rodríguez, Raúl. Historia de las relaciones económicas internacionales de la América Latina. Caracas: Academia Nacional de Ciencias Económicas, 1992. 423 p.: bibl.

Describes Latin American economic relations with the European powers (England, France, Germany, and Italy) and the US. Identifies three periods: 1) independence to 1914 (European influence); 2) WWI to WWII (American influence); and 3) post-WWII. Emphasizes trade and contains interesting historical data.

1276 Soto, Hernando de. Caminando el otro sendero. Ed. restringida. Bogotá: FUNDES, 1990. 31 p.: ill. (Serie Diálogo; 1)

Meditations on author's *El otro sendero* (Lima: Editorial El Barranco, 1986; also published in English under the title *The other path*, New York: Harper & Row, 1989). Soto maintains that since the book appeared, the less developed countries have been reformulating their political debate. This validates the need to make institutional changes in order to eradicate misery. [C. Rabinovich]

Statistical Masterfile. 1974– . See item **44**.

1277 Sterner, Thomas. Nuclear energy and sustainability in Latin America. (*in* Environment and development in Latin America: the politics of sustainability. Edited by David Goodman and Michael Redclift. Manchester, England: Manchester Univ. Press; New York: St. Martin's Press, 1991, p. 97–115, bibl., tables)

Analyzes benefits, costs, and risks of nuclear power in Latin America. Describes experiences of Argentina, Brazil, Cuba, and Mexico with nuclear power. Concludes that due to its high cost and environmental and security risk, nuclear power is not appropriate for the region.

1278 Stone, Roger D. and Eve Hamilton. Global economics and the environment: toward sustainable rural development in the Third World. New York: Council of Foreign Relations Press, 1991. 1 v.: bibl.

Essay on the feasibility of sustainable rural growth without ecological degradation in Third World countries. Argues that such a goal is feasible but would require greater cooperation (trade, investment, etc.) by the developed countries of the world.

1279 Strategic options for Latin America in the 1990s. Edited by Colin I. Bradford, Jr. Paris: Development Centre of the Organisation for Economic Co-operation and Development; Washington: Inter-American Development Bank; OECD Publications and Information Centre, 1992. 287 p.: bibl., ill.

Articles examine potential future trade strategies for Latin American countries. Topics include trade in manufactures, trade in agricultural goods, Mexico and NAFTA, technological progress and competitiveness, and subregional integration efforts.

1280 Sunkel, Osvaldo. El desarrollo en los tiempos del cólera. (*Bol. Socioecon.*, 24/25, agosto/dic. 1992, p. 1–20, ill., graph, tables)

Historical examination of the Latin American development process with special emphasis on the State's role during this century. Author supports a strong, activist State and is highly critical of the political and social costs of neoliberal reforms. Uses data to support his views.

1281 Sunkel, Osvaldo. El futuro del desarrollo latinoamericano: algunos temas de reflexión. (*in* Neoliberalismo y políticas económicas alternativas. Quito: Corporación de Estudios para el Desarrollo, 1987, p. 19–58, graph, tables)

Critical appraisal of neoliberal policies adopted by Latin American countries as a reaction to the debt problem. Proposes alternative (heterodox) policies.

1282 Tomassini, Luciano. Desarrollo económico e inserción externa en América Latina: un proyecto elusivo. (*Estud. Int./Santiago*, 25:97, enero/marzo 1992, p. 73–116)

Historical description of Latin American growth and development. Discusses Spanish colonization, 19th-century liberalism, the inward-oriented development of the 20th century, the debt crisis of the 1980s, free-market policies, and redemocratization of the region.

1283 Towards a new development strategy for Latin America: pathways from Hirschman's thought. Edited by Simón Teitel. Washington: Inter-American Development Bank; Baltimore, MD: Johns Hopkins Univ. Press, 1992. 403 p.: bibl., ill., index.

Book includes two types of articles. One set describes economic growth and performance of Argentina, Brazil, Chile, Colombia, and Mexico from 1950–80. A second set examines Hirschman's contributions to development theory, with special emphasis on those concepts and theories useful for the region's future growth.

1284 Towards sustained development in Latin America and the Caribbean: restrictions and requisites. Santiago: Economic Commission for Latin America and the Caribbean, United Nations; New York: United Nations Publications, 1989. 93 p.: bibl., ill. (Cuadernos de la CEPAL, 0252-2195; 61)

Presents ECLAC's views on debt-induced negative transfer of financial resources from Latin America. Proposes policies that should be implemented to overcome this impediment to growth.

TRAINS. See item **45**.

1285 La transferencia de recursos externos de América Latina en la posguerra. Santiago: Naciones Unidas, Comisión Económica para América Latina y El Caribe, 1991. 92 p.: bibl., ill., tables. (Cuadernos de la CEPAL, 0252-2195; 67)

Predicts impact of the region's current indebtedness and negative transfer of resources on growth from 1990–95. Uses simulation techniques to describe alternative scenarios. Provides annual data by country and region on the net transfer of resources from 1950–89.

1286 Transformación ocupacional y crisis social en América Latina. Santiago: Naciones Unidas, Comisión Económica para América Latina y el Caribe, 1989. 243 p.: bibl., ill.

Analyzes social mobility and the changing structure of employment from 1960 to late 1980s. Discusses the decline of agricultural employment, the changing role of women, the emergence of the informal sector, etc. Provides extensive case studies for Bolivia, Brazil, Ecuador, Honduras, and Panama.

1287 Transnational banks and the international debt crisis. New York: United Nations, 1991. 148 p.: bibl., ill.

Examines role played by private transnational banks in extending loans to Latin America during the 1970s and early 1980s, as well as the role played by these banks in the debt restructuring process.

1288 Trayectorias divergentes: comparación de un siglo de desarrollo económico latinoamericano y escandinavo. Coordinación de Magnus Blomström y Patricio Meller. Santiago: CIEPLAN; Hachette, 1990. 283 p.: bibl., ill.

Comparative study of growth and development in Scandinavian and Latin American countries with explanations for their differences in growth performance. Covers Finland, The Netherlands, Norway, Sweden, Chile, Colombia, Ecuador, and Uruguay.

TSentral'naia Amerika i Kariby: nachalo 90-kh godov [Central America and the Caribbean: the beginning of the nineties]. See item **2804**.

1289 Tuller, Lawrence W. Doing business in Latin America and the Caribbean: including Mexico; the U.S. Virgin Islands and Puerto Rico; Central America; South America. New York: AMACOM, American Management Association, 1993. 348 p.: index.

Interesting and comprehensive presentation of the opportunities, advantages, risks, and methods of doing business with and in Latin America. Includes basic information on potential markets, tariffs, taxes, legal restrictions, business climate, as well as basic demographic and economic variables. Wealth of useful information.

1290 Twomey, Michael J. La crisis de la deuda y la agricultura latinoamericana. (*Economía/Lima*, 12:24, dic. 1989, p. 69–99, graphs, tables)

Describes impact of the debt crisis and subsequent macroeconomic policies on the agricultural sector.

1291 **Twomey, Michael J.** Términos de intercambio, afluencia extraordinaria de divisas y tipo de cambio real en América Latina. (*Economía/Lima*, 14:28, dic. 1991, p. 275-306, bibl., tables)

Uses econometric techniques to analyze the relationship between purchasing power parity and the terms of trade, as well as their sensitivity to macroeconomic changes.

1292 **Urban poverty alleviation in Latin America.** The Hague: Ministry of Foreign Affairs, Development Cooperation Information Dept., 1992? 92 p: bibl., map. (Poverty and development: analysis & policy; 3)

Examines extent and causes of urban poverty in Latin America. Discusses the informal sector, as well as the role of non-governmental organizations (NGOs) in alleviating poverty. Specifically focuses on urban poverty in Chile, Peru, Kingston (Jamaica), and San Pedro Sula (Honduras).

1293 **Urquidi, Víctor L.** The prospects for economic transformation in Latin America: opportunities and resistances. (*LASA Forum*, 22:3, Fall 1991, p. 1-8)

Documents the region's economic policy shift from one that is inward-oriented (import substitution) to one that is outward-focused (market liberalization and export promotion). While supportive of these changes, Urquidi concludes that unless the massive out-transfer of financial resources is sharply reduced, transformation and growth will not occur.

1294 **Valenzuela Feijóo, José.** Crítica del modelo neoliberal: el FMI y el cambio estructural. México: Facultad de Economía, Univ. Nacional Autónoma de México, 1991. 160 p.: bibl. (Col. América Latina)

Sophisticated critique of neoliberal economic policies argues that neoliberal theory is based on behavioral postulates and assumptions (e.g., competitive markets, resource mobility) that do not reflect Latin American realities. Accordingly, neoliberal policies will not succeed, thereby necessitating an activist role for the State.

1295 **Velasco, Andrés** and **Felipe Larraín.** The basic macroeconomics of debt swaps. (*Oxf. Econ. Pap.*, 45:2, April 1993, p. 207-228, graphs, tables)

Authors provide theoretical framework for analyzing debt-equity swaps, where the retired debt is public and swapped assets are private. Concludes that the macroeconomic impact will depend on how these transactions are financed.

1296 **Villamonte Blas, Ricardo N.** La conversión de deuda en inversión: una alternativa para el financiamento del desarrollo. (*Cienc. Econ./Lima*, 13:28, set./dic. 1992, p. 121-145, bibl., table)

Describes characteristics of Latin American secondary capital markets and the development by the region's governments of various mechanisms designed to convert foreign debt into investment. Provides examples for Argentina, Brazil, Chile, Ecuador, Mexico, and Venezuela.

1297 **Vuskovic, Pedro.** La crisis en América Latina: un desafío continental. México: Siglo Veintiuno Editores; Tokio: Editorial de la Univ. de las Naciones Unidas, 1990. 236 p.: bibl. (Biblioteca América Latina)

Marxist perspective on Latin America's economic problems argues that structural problems result from historical subordination to external powers (e.g., governments, business, and special interests).

1298 **Warner, Andrew M.** Did the debt crisis cause the investment crisis? (*Q.J. Econ.*, 107:4, Nov. 1992, p. 1161-1186)

Using data from 13 countries, 10 of which are Latin American, author tests hypothesis that the decrease in investment in highly indebted countries was due to the debt crisis, a view which is generally accepted. Results do not support this hypothesis.

1299 **Weintraub, Sidney.** Policy-based assistance: a historical perspective. (*in* The effects of receiving country policies on migration flows. Boulder, Colo.: Westview Press, 1991, p. 13-38, graphs)

Describes the changing role of US aid from post-WWII to present. Contends that US aid has always been policy-based; that is, aid is given in exchange for adoption of certain economic, political, or social policies by the recipient government.

1300 **Weisskoff, Richard.** Income distribution and the Enterprise for the Americas Initiative. (*J. Interam. Stud. World Aff.*, 33:4, Winter 1991, p. 111-132, bibl.)

Reviews changing views of academic economists on the relationship between growth and income distribution. Whereas in the 1960s economists believed that growth and inequality went hand-in-hand, now many economists believe that inequality causes political and economic instability which hinder growth. Discusses implications for Latin America.

1301 West, Peter J. Latin America's return to the private international capital market. (*CEPAL Rev.*, 44, Aug. 1991, p. 59–78, bibl., tables)

Explores feasibility of obtaining foreign financial resources from the private international credit market. Companies in three countries—Chile, Mexico, and Venezuela—have been successful in obtaining such funds by selling bonds and foreign portfolio investment. Notes the growing popularity of securities guaranteed by export income.

1302 Whalley, John. CUSTA and NAFTA: Can WHFTA be far behind? (*J. Common Market Stud.*, 30:2, June 1992, p. 125–141, bibl., tables)

Given the existence of NAFTA, the Canada-US Trade Agreement (CUSTA), and various Latin American integration organizations, author asks if we can expect the eventual formation of a Western Hemisphere Free Trade Area (WHFTA). Concludes that this is unlikely.

Wistat: women's indicators and statistics database. Version 3. See item **47**.

1303 Women in the Americas: bridging the gender gap. Washington: Inter-American Development Bank; distributed by The Johns Hopkins Univ. Press, 1995. 222 p.

Useful summary of economic and social policy issues with a focus on women's role in labor markets, politics, environment, and other social institutions is a product of a 1994 forum sponsored by IDB, ECLAC, and UNIFEM. Excellent bibliography and statistical appendix are complemented by a concluding "Agenda for Action" detailing specific policy objectives. [W. Canak]

1304 Won Choi, Dae. The Pacific Basin and Latin America. (*CEPAL Rev.*, 49, April 1993, p. 21–40, bibl., tables, graphs)

Examines economic relations between Latin America and various Southeast Asian countries (Hong Kong, Indonesia, Malaysia, Philippines, Republic of Korea, Singapore, Taiwan, and Thailand). Concludes that Latin America must adopt policies similar to those in place in these countries if it is to emulate their success.

World Data. 1994– . See item **48**.

World marketing data and statistics. See item **49**.

World Trade Database = Base de donées sur le commerce mondial. 1980/1992– . See item **51**.

1305 Wurgaft, José. Social investment funds and economic restructuring in Latin America. (*Int. Labour Rev.*, 131:1, 1992, p. 35–44, bibl.)

Examines growing prevalence of Social Investment Funds in the region. Such funds are intended to compensate the losers, usually labor, in the adjustment and restructuring process. Concludes that such funds, while desirable, are inadequate in scope.

1306 Yamaguchi, Yutaka. Change in development policies in Latin America: the new role of foreign aid and investment. (*Iberoam./Tokyo*, 15:1, primer semestre 1993, p. 31–48, table)

Explains why the structuralist (nonmarket) approach to development, dominant from World War II until the 1980s, was replaced by the neoclassical (market-oriented) approach. The debt crisis forced governments to adopt this new approach, as well as a more receptive attitude toward aid and direct foreign investment.

1307 Zulawska, Ursula. El papel económico del Estado latinoamericano en los ochenta. (*in* América Latina: regiones en transición. Ciudad Real, Spain: Publicaciones de la Univ. de Castilla-La Mancha, 1991, p. 31–39, tables)

Provides brief historical overview of the State's role in the development process from the "Great Depression" to the present. Special emphasis is placed on Argentina, Brazil, Chile, and Mexico. Describes how social problems exert pressures that affect State decision-making on economic policies.

MEXICO

ROBERT E. LOONEY, *Professor of National Security Affairs, Naval Postgraduate School*

MEXICO'S MOST RECENT ECONOMIC CRISIS took many in the international business community by surprise. In early Dec. 1994, the Blue Chip consensus forecast for Mexico's 1995 real Gross Domestic Product growth was 3.8 percent. A few weeks later, on Dec. 20, 1994, the devaluation of the Mexican peso rocked international financial markets. What first appeared to be a minor correction in Mexico's nominal exchange rate quickly developed into a broader financial crunch felt inside and outside Mexico. The Mexican government now expects the country's real GDP to fall about three percent in 1995; some private economists suggest an even greater decline. What caused Mexico's recent economic crisis, and how long will it take the country to recover? Were Mexico's economic reforms reality or illusion? Although it will be some time until in-depth analyses of the current crisis appear in published form, it is apparent that to assess Mexico's future, one must look at the country's past. In this regard, many of the works annotated in this chapter provide the framework and background necessary to understand the significance of the current emergency.

Unlike the period prior to Mexico's 1982 debt crisis, the recent trend in Mexico's economic policies has been toward greater economic integration in the world economy and less reliance on the government. Although Mexico may need several years to regain the investor confidence it lost during the recent economic crisis, the trend in Mexico's policies is more consistent with future low inflation and higher growth than the country's previous closed-market policies.

From the perspective of those previous policies, the country's recent achievements are impressive both in themselves and for the break with the past which they represent. Mexico, home of what can be regarded as the world's first socialist constitution in 1917, was one of the 1980s' most enthusiastic converts to the cause of economic liberalism. For one thing, it opened up large tracts of its economy to foreign competition. Since joining the General Agreement on Tariffs and Trade (GATT) in 1987, Mexico's average tariff on imports has dropped from 45 percent to nine percent. The North American Free Trade Agreement (NAFTA) recently signed with the US and Canada is the continuation of this process, not the start of it. In many industries Mexican businessmen already face the full brunt of foreign competition.

Mexico has also privatized large segments of its economy including the telephone company, the banks, and, more recently, agriculture. It has also liberated most aspects of commercial life, a sharp change from the old system of tight controls. Decision-making in both the private and public sectors is now largely dictated by economic rather than political factors. Deregulation is perhaps the biggest change of all to daily life and popular attitudes. Many of the writings annotated below attempt to assess the consequences of the country's recent policy shifts and dramatic improvements in economic performance in various fields (e.g., agriculture, energy, finance, foreign investment, income distribution, industry, labor, macroeconomics, population, regional development, services, social conditions, trade, and urbanization).

In contrast to the last several *HLAS* volumes, the economic literature on Mexico has shifted from a preoccupation with the country's debt crisis to that of assessing recent policy initiatives. Of recent developments, the signing of the North American Free Trade Agreement is by far the most important. While NAFTA does have its critics (items **1361, 1464,** and **1361**), most writers have concluded that both Mexico and the US should benefit from the new arrangement (e.g., items **1424, 1458, 1459**). First, the American economy is all important to Mexico. Even though Mexico is the US's third largest trade partner after Japan and Canada, bilateral trade has grown at a rate well over twice the average for both countries. Second, NAFTA provides the regulatory framework to encourage both Mexicans and foreign investors to believe that the economic reforms are irreversible. Third, NAFTA will help the country create more jobs thus slowing the flow of workers to the US. Job creation is especially important considering that the Mexican workforce is still expanding by three percent per annum (reflecting a higher population growth rate 15–20 years ago) even though population growth has slowed to just under two percent in recent years (see items **4545** and **1476**). Since more than a third of the population is under 15 years of age and more than 80 percent are under 40, Mexico must create more than one million jobs each year to provide employment for its working-age population. Fourth, NAFTA is vital for attracting foreign investment to Mexico (items **1350** and **1405**). Mexico's savings rate of around 19 percent of Gross Domestic Product (GDP) is simply not enough to finance the five to six percent sustained rate of growth that the country needs to reduce unemployment (item **1322**). NAFTA will create an established order in the new relationship with the US and Canada, with a specific timetable for the opening of most key industries.

On the other hand, the country is still nervous that foreign investment is not more diversified: the US accounts for two-thirds and the European Community for 25 percent, with only five percent coming from Japan. Furthermore, two-thirds of direct foreign investment is in manufacturing and 20 percent in tourism. On a brighter note, the border's famous *maquiladoras*—manufacturing plants which assemble components imported (beginning in 1965) tax-free for reexport—are no longer receiving a disproportionate share of foreign investment. Although total sales of *maquiladoras* reached about $15 billion in 1993, they accounted for only eight percent of total foreign investment. *Maquiladoras* tend to be labor-intensive operations where the level of capital investment is not large (items **1438, 1362,** and **1354**). In fact, because much of the *maquiladora* machinery is leased, it is not even registered as a Mexican asset. The chief attraction of the *maquiladoras* continues to be their cheap labor: in the automotive industry, for example, American workers still earn eight times more than their Mexican counterparts. The Mexican government now wants to encourage the domestic manufacture of components that are currently imported into the country for use in the *maquiladora* plants. With NAFTA, the flow of foreign money—into *maquiladoras* and elsewhere—should spread more rapidly away from the border areas (items **1437**).

The fate of the agricultural sector is more indirectly related to NAFTA (item **1514**). Here the consequences of dismantling the land tenure system created by the 1910 revolution are critical (item **1315**). At the onset of the revolution, 260 families owned 80 percent of Mexican territory, sparking cries for land reform. The result, Article 27 of the Mexican Constitution of 1917 (item **1341**), compelled the government to give land to any group of peasants who asked for it. To comply with these requests, the government was given the authority to expropriate land from private owners and form new ejidos or communal-ownership arrangements. The constitutional drafters were so anxious to avoid the reappearance of large land-holdings that the peasants who owned these ejidos could not sell the land nor rent it, nor even pledge it as collateral for loans. The core of the ejido system remained intact until 1991. The consequences of this attachment to collectivist dogma have become only too apparent (item **1530**). Agriculture accounted for only seven percent of GDP in the early 1990s, yet 23 percent of economically active Mexicans work in agriculture and 30 percent of the population live in rural areas. Out of the 40 million Mexicans considered living in poverty, 70 percent reside in the countryside. Agricultural GDP per head is now lower than it was in 1965 and the country is importing food, with the trade balance in food turning negative in 1989 for the first time in years. Mexico is now a net importer of maize and wheat and is the world's largest importer of milk powder.

In a dramatic departure from past policies, agriculture is to join the market economy as part of the reforms undertaken by President Salinas. Three new provisions will make this possible: 1) the constitutional right to be granted land by the State has been eliminated; 2) well-defined rights of private ownership of property have been reestablished; and 3) the ejido system has been reformed so that land once again can be rented, sold, or pledged for the purpose of borrowing money (though only to someone inside the ejido unless the majority of the community agrees). Yet the changes to agriculture are even bigger than those implied above. For instance, one must also consider the impact of NAFTA. Under the terms of the treaty, tariffs on the most sensitive agricultural goods such as maize, beans, and powdered milk will be reduced to zero over the next 15 years, and tariffs and other barriers for seasonal fruits and vegetables will be phased out. Mexico will also remove 82 percent of the tariffs on US agricultural goods, and the US will cut 95 percent of the tariffs on Mexican goods within ten years. The Mexican government knows that it cannot simply switch to international prices without wiping out, for example, more than 90 percent of the country's 2.4 million corn (maize) producers, whose families make up almost half of the rural population. The government is therefore committed to channeling welfare payments to these groups—but, as much as possible, in a way that will not encourage uneconomic production (item **1404**). Support programs, however, do not solve the problem of what will happen to most of the current subsistence farmers, though solidarity projects may provide work for some of them. The only viable long-term solution is more industrialization and the development of a larger services industry.

In sum, although NAFTA will cause Mexico considerable short-term pain it will be worth the cost for it represents the country's best hope for delivering a large number of its citizens out of extreme poverty. Politically, agricultural change will also compel the government to develop a more democratic system; electoral tampering is far easier in the countryside, and in the future more of the voters will be found in cities. Successful agricultural reform is the key to Mexico's modernization overall. To have nearly one third of the population living in rural areas is simply not sustainable if the country is to live up to its aspirations.

1308 Aboites, Jaime. Industrialización y desarrollo agrícola en México. México: Plaza y Valdés; Univ. Autónoma Metropolitana, Xochimilco, 1989. 201 p.: bibl., ill. (Col. Agricultura y economía)

Detailed examination of country's agricultural sector focuses on links to industry. Topics include sector's patterns of employment, investment, and external trade.

1309 El Acuerdo de Libre Comercio México-Estados Unidos. Centro de Investigación para el Desarrollo, A.C. México: Editorial Diana, 1991. 291 p.: bibl. (Alternativas para el futuro)

Examines implications for Mexico of the new international order. Interesting speculation as to the effects of the country's liberalization programs. Topics include future

US-Mexican trade, legal aspects of trade, and factors likely to affect competitiveness of Mexican imports.

1310 La adhesión de México al GATT: repercusiones internas e impacto sobre las relaciones México-Estados Unidos. Coordinación de Pamela S. Falk y Blanca Torres. México: Colegio de México, Centro de Estudios Internacionales, 1989. 423 p.: bibl., ill.

Series of essays on problems Mexico faces because of joining the General Agreement on Tariffs and Trade (GATT). Topics include the economic impact of GATT membership, agricultural problems, potential frictions with the US, and implications for the transport sector.

1311 Aggarwal, Raj *et al*. Problemas económicos de México: realidad y perspectivas. Recopilación de Francisco Carrada Bravo. México: Editorial Trillas, 1988. 354 p.: bibl., ill., index.

Leading authorities examine Mexico's recent (post-1982) economic crisis. Areas covered include debt, labor conditions, agriculture, petroleum, and the industrial sector. A number of constructive policies are proposed to deal with the identified problems.

1312 Ajuste estructural, mercados laborales y TLC. México: El Colegio de México, Centro de Estudios Sociológicos; Fundación Friedrich Ebert; El Colegio de la Frontera Norte, 1992. 400 p: bibl., ill.

Papers presented at seminar held at the Colegio de México (Oct. 23-26, 1991). Wide-ranging essays examine various aspects of labor conditions in the country, emphasizing impact of the goverment's stabilization program on working conditions, job security, and salaries.

Alba Vega, Carlos and **Dirk Kruijt.** The convenience of the minuscule: informality and microenterprise in Latin America. See item **1076.**

1313 Algunos enfoques sobre la restructuración económica en México. Recopilación de Silvia Tamez. México: Univ. Autónoma Metropolitana, Dirección de Difusión Cultural, Depto. Editorial, 1989. 107 p.: bibl., ill. (Col. Doble espiral; 2)

Assessment of impact of debt crisis and associated stabilization program on the country's pattern of employment and unemployment. Contains excellent discussion of the country's labor institutions and legislation.

Alvarez Mosso, Lucía and **María Luisa González Marín.** Industria y clase obrera en México, 1950-1980. See item **4460.**

1314 Alvarez Soberanis, Jaime. México: retos y oportunidades en el año 2,000. México: Editorial Jus, 1993. 232 p.

Examines implications for Mexican economic development of recent changes in the international system. In this regard the study examines GATT, the diffusion of technology and productivity, deregulation, and the role of multinational corporations in shaping the country's path to the year 2000.

1315 Análisis crítico de la nueva reforma agraria. Recopilación de Emilio López G. y Bernardino Mata G. Chapingo, Mexico: Univ. Autónoma Chapingo, Depto. de Sociología Rural, 1992. 266 p.: bibl.

Critique of Mexico's land tenure, land reforms, and relevant legislation focuses on identifying current contradictions and problems associated with the government's rural reforms.

1316 Antología del pensamiento económico de la Facultad de Economía, 1929-1989. Recopilación de Felipe Becerra Maldonado. México: Facultad de Economía, Univ. Nacional Autónoma de México, 1989. 2 v.: bibl., ill.

Compilation of classic papers by leading Mexican economists including Víctor Urquidi (*El concepto del multiplicador exterior*, 1941) and Miguel S. Wionczek (*La gran quiebra de la Bolsa de Valores de Nueva York en 1929*, 1956). Invaluable reference for those wishing to read these often-cited papers.

1317 La apertura comercial y la frontera México-Texas. Coordinación de Alejandro Dávila Flores. Saltillo, Mexico: Univ. Autónoma de Coahuila, 1993. 184 p.: bibl., ill.

Papers from the Conferencia Fronteriza México-Texas (2nd, Saltillo, Mexico, Oct. 19-20, 1990) cover various topics concerning economic integration and increased Mexican-US trade. Topics include conceptual issues involved in North American economic integration, trends in *maquiladora* investment, the changing nature of markets, and marketing strategies.

1318 **La apertura económica de México y la Cuenca del Pacífico: perspectivas de intercambio y cooperación.** Recopilación de Juan José Palacios Lara. Guadalajara, Mexico: Univ. de Guadalajara, 1992. 198 p.: bibl., ill. (Col. Jornadas académicas. Serie Coloquios.)

Papers examine trade and economic cooperation in the Pacific Basin region. Topics deal with US policy in the region, Mexican/Japan trade, and the relationship of the *maquiladora* industries to the newly industrialized countries of South-East Asia.

1319 **Appendini, Kirsten A. de et al.** Alternativas para el campo mexicano. Coordinación de José Luis Calva. México: Distribuciones Fontamara, PUAL-UNAM: Friedrich Ebert Stiftung, Fundación Friedrich Ebert, Representación en México, 1993. 2 v.: bibl., ill., maps. (Col. Economía y sociedad)

Essays address the range of problems currently facing the country's agricultural sector and rural economy. Topics include land reform, agricultural production and modernization, social conditions and problems of campesinos, resource development, ecological problems, and the development of new agricultural technologies.

1320 **Appendini, Kirsten A. de.** De la milpa a los tortibonos: la restructuración de la política alimentaria en México. México: Colegio de México, Centro de Estudios Económicos: Instituto de Investigaciones de las Naciones Unidas para el Desarrollo Social, 1992. 259 p.: bibl., ill., map.

Examines government policies towards the agricultural sector. Emphasizes pricing and government intervention in markets. Excellent assessment of the contribution of agriculture to the country's development. Also includes assessment of the role played by Conasupo (Compañia Nacional de Subsistencias Populares).

1321 **Argüelles, Antonio and José Antonio Gómez.** La desconcentración en el proceso de modernización económica de México: el caso de SECOFI. México: M.A. Porrúa, 1992. 166 p.: bibl., ill.

Thorough examination of the means by which public agencies are reorienting themselves to promote industry and geographical growth in the country.

1322 **Arrau, Patricio and Sweder van Wijnbergen.** Intertemporal substitution, risk aversion, and private savings in Mexico.

Washington: International Economics Dept., the World Bank, 1991. 28 p.: bibl., ill. (Policy, research, and external affairs working papers; WPS 682)

Examines factors involved in the decline in private savings in Mexico since 1982. Economic results suggest that this drop in savings can be linked to a substantial increase in public savings. This factor is more important than uncertainty or real interest rate developments.

1323 **Arriaga Conchas, Enrique.** Finanzas públicas de México. México: Instituto Politécnico Nacional, 1992. 233 p.: bibl.

Study provides excellent overview of the country's system of public finances. Detailed discussion of the main institutional structure of revenue collection covers national, regional, and local aspects of the country's tax system.

1324 **Arriola, Carlos et al.** Los empresarios y la modernización económica de México. Presentación de Jaime Serra Puche. Recopilación de Carlos Arriola. 2. ed. México: M.A. Porrúa, 1991. 308 p.

Examines role of entrepreneurship in modernizing Mexico's major industries. Leading authorities assess implications of government policy—especially stabilization programs of the 1980s—on exports and industrial structure.

1325 **Ascencio Franco, Gabriel.** Los mercaderes de la carne: causalidad estructural de la economía y relaciones personales en el mercado capitalista: el abasto de carne a Guadalajara. Zamora, Mexico: El Colegio de Michoacán, 1992. 192 p.: ill., maps. (Col. Investigaciones)

Examines development of markets in the Guadalajara region of Jalisco.

1326 **Aspe Armella, Pedro.** Economic transformation the Mexican way. Cambridge, Mass.: The MIT Press, 1993. 280 p.: bibl., ill., index. (The Lionel Robbins lectures)

Mexico's Finance Minister offers an informed, inside look at attempts to modernize Mexican economy through the 1970s–80s. Examines how Mexico has tried to stabilize its economy with measures such as economic deregulation, fiscal reform, privatization of State-owned enterprises, and realistic budget management. He argues that these changes have had not only profound economic effects, but social and political ones as well.

1327 Assessments of the North American Free Trade Agreement. Edited by Ambler H. Moss, Jr. Coral Gables, Fla.: North South Center, Univ. of Miami; New Brunswick, N.J.: Transaction Publishers, 1993. 109 p.: ill.

Five experts on Mexican-US economic relations assess the impact that the North American Free Trade Agreement will have on both countries. Each author presents a unique perspective on NAFTA's origins, and major features. Topics include international trade, the environment, political and security issues, economic theory, and public policy.

Atlas de México. See item **15**.

1328 Barajas Escamilla, Rocío. Reestructuración industrial: subcontratación internacional, cambio tecnológico y flexibilidad en la maquiladora. (*Estud. Front.*, 23, sept./dic. 1990, p. 33–54, bibl., tables)

Discusses theories which give support to the phenomenon of industrial restructuring in Mexico. Attempts to give explanation of determinants of subcontracting and their relation to technological change and flexibility of work systems. Studies export *maquiladora* industry in Mexico in order to explain the process. Concludes that as a result of the industrial restructuring a larger segmentation of the productive process occurs. This increases subcontracting activities in Mexico. Presently Mexican *maquiladoras* are incorporating new technologies as well as the latest organizational systems. The result is a technological duality.

1329 Barceló R., Víctor Manuel. La reforma agraria y la crisis. México: Centro de Estudios Históricos del Agrarismo en México, 1988. 268 p.: bibl.

Broad examination of country's progress at land reform. Contains detailed listing of relevant legislation, together with an assessment of problems remaining in rural areas.

1330 Berthélemy, Jean-Claude and Ann Vourc'h. Allégement de la dette et croissance: le cas mexicain. Paris: Organisation de coopération et de développement économiques, 1992. 49 p.: bibl., ill. (Documents techniques/Centre de développement de l'OCDE; 79)

Presents dynamic model simulating the Mexican economy focusing on effects of public indebtedness. Model is used to simulate effect on the Mexican economy of the Brady Plan, the creation of NAFTA, and the fall in international interest rates. Simulations suggest that complementarity exists between the Brady Plan and NAFTA.

1331 Betts, Dianne C.; Daniel J. Slottje; and Jesús Varga-García. Crisis on the Rio Grande: poverty, unemployment, and economic development on the Texas-Mexico border. Boulder: Westview Press, 1994. 195 p.: bibl., index, maps.

Study provides in-depth assessment of socioeconomic conditions along the Mexican-US border. Examines demographic changes that have transpired over the past decade and have contributed to problems that exist on both sides of the border. Also explores impact that a Free Trade Agreement could have on the region. Policy suggestions are made that might help alleviate current border problems and could lead the way in preventing future ones.

1332 Blanco Mendoza, Herminio. Las negociaciones comerciales de México con el mundo. México: Fondo de Cultura Económica, 1994. 281 p.: bibl. (Una Visión de la modernización de México)

Comprehensive examination of Mexico's recent negotiations to reduce trade barriers and become more integrated with the economies of the rest of North America and the Pacific. Study contains useful assessment of trade prospects for each of Mexico's major industries.

1333 Brailovsky, Vladimir; Roland Clarke; and Natán Warman. La política económica del desperdicio. México: Univ. Nacional Autónoma de México, Facultad de Economía, 1989. 500 p. (Economía de los 80)

Examines pursuit of stabilization policies following the 1982 crisis. Macroeconomic framework is used to project consequences of these policies into the early 1990s.

1334 Brown, Jonathan Charles. Oil and revolution in Mexico. Berkeley: Univ. of California Press, 1993. 453 p.: bibl., ill., index.

History of the country's oil industry focuses on frictions between Mexico and foreign oil companies. Argues that these frictions developed because every economic benefit that crossed Mexico's northern border also contained its own peril. As the experience of foreign oil companies makes clear, Mexicans sought a compromise between forces of economic change and modernity and a social heritage stronger than both. For historian's comment, see *HLAS 54:1461*.

1335 **Buffie, Edward F.** Economic policy and foreign debt in Mexico. (*in* Developing country debt and economic performance. Chicago: Univ. of Chicago Press, 1989, v. 2, p. 393–551, bibl., graphs, tables)

Sets out to critically examine macroeconomic policy in Mexico from 1958–86. Author develops a formal economic model to identify factors responsible for the dismal post-1982 record of falling per capita incomes, low real wages, high inflation, and widespread unemployment. Based on this model, links are made between capital accumulation, inflation, fiscal deficits, and financial innovation.

1336 **Burgueño Lomelí, Fausto.** Economía en crisis: ensayos sobre México y América Latina. México: Instituto de Investigaciones Económicas, Univ. Nacional Autónoma de México, 1991. 151 p.

Series of essays on the economic crisis facing Mexico and other Latin American countries provides general overview of causes of the debt crisis. Focuses on declining incomes, the role of the international system, and problems associated with external debt.

1337 **Calderón Salazar, Jorge A.** El Tratado de Libre Comercio y el desarrollo rural: impacto en la industria alimentaria y en la producción de granos básicos: propuestas alternativas. México: Centro de Estudios del Movimiento Obrero y Socialista, 1992. 126 p.: bibl.

Examines legal foundations of rural economic relationships in Mexico. Builds on this framework to assess problems currently affecting incomes and family life in Mexico's more underdeveloped areas.

1338 **Calva, José Luis.** The agrarian disaster in Mexico, 1982–89. (*in* Modernization and stagnation: Latin American agriculture into the 1990s. Edited by Michael J. Twomey and Ann Helwege. New York: Greenwood Press, 1991, p. 101–119, bibl.)

Argues that the true goal of adjustment policies imposed on Mexico by the International Monetary Fund during this period was to achieve monetary and financial equilibrium in the international economy rather than to strengthen the real economy (e.g., production, investment, employment, and social welfare) in Mexico. For this reason these policies, while serving the interests of international financial capital, leave in their wake a pattern of hunger and desolation.

1339 **Calva, José Luis** *et al.* La agricultura mexicana frente al Tratado Trilateral de Libre Comercio. 1. ed. en español. Chapingo, Mexico: CIESTAAM; México: J. Pablos Editor, 1992. 257 p.: bibl., ill.

Essays assess the likely impact that increased North American trade will have on the country's agricultural sector. Includes separate chapter for each major crop.

1340 **Calva, José Luis.** Crisis agrícola y alimentaria en México, 1982–1988. México: Distribuciones Fontamara, 1988. 230 p.: bibl. (Fontamara; 54)

Outlines problems confronting agriculture and food industry, especially during 1982–88. Study focuses on the impact of inflation, debt, and government stabilization policies on the country's agricultural development. Examines in detail impact of recent public policies on the sector's performance and prospects.

1341 **Calva, José Luis.** La disputa por la tierra: la reforma del Artículo 27 y la Nueva Ley Agraria. México: Distribuciones Fontamara, 1993. 244 p.: bibl.

Examines and compares Article 27 of the 1917 Constitution with the government's new approach to land reform introduced in 1992. Tries to anticipate the impact the new policy will have on rural life and the agricultural economy.

1342 **Calva, José Luis.** Probables efectos de un Tratado de Libre Comercio en el campo mexicano. México: Distribuciones Fontamara, 1991. 167 p.: bibl. (Fontamara; 134)

Assessment of problems created for Mexico's rural sector by the opening up of free trade, particularly that in the NAFTA region. Valuable interviews with leading participants.

1343 **Camarena L., Margarita.** La industria automotriz en México. México: Univ. Nacional Autónoma de México, Instituto de Investigaciones Sociales, 1981. 64 p.: bibl. (Cuaderno de investigación social; 6)

Provides convenient overview of the country's automotive sector. Contains numerous tables with useful information regarding production by firm, imports/exports, and capital formation in the industry.

1344 Cambiaso, Jorge E. and Carlos López.
México, la apuesta por el cambio: se advierte una transformación en todos los aspectos de la vida del país: y está surgiendo un México nuevo. México: Plaza y Valdés, 1992. 156 p.: ill.

Addresses increasing integration of Mexico into the world economy. Also covers Mexico's ability to compete in foreign markets and structure its industry and labor force to meet this challenge.

1345 Campos, Emma et al. La pobreza en México: causas y políticas para combatirla. Recopilación de Félix Vélez. México: Instituto Tecnológico Autónomo de México; Fondo de Cultura Económica, 1994. 302 p.: bibl., ill., map. (Lecturas; 78)

Papers by leading authorities explore factors responsible for persistent poverty in Mexico. Excellent discussion of changing patterns of income distribution and areas in which public policy might be able to play a productive role.

1346 Campos Alvarez Tostado, Ricardo.
Fondo Monetario Internacional, deuda externa mexiana [i.e., mexicana] y la administración pública. Toluca, México: Univ. Autónoma del Estado de México, 1991. 283 p.: bibl. (Col. Ciencias y técnicas; 22)

Beginning with the dollar devaluation in 1971, author assesses major developments in internal financial markets as they have affected Mexico. Emphasizes how the government has responded to these external challenges. Excellent critique of Mexican policymaking during this difficult period.

1347 Campos Alvarez Tostado, Ricardo. El Fondo Monetario Internacional y la deuda externa mexicana: crisis y estabilización. 2. ed. México: Plaza y Valdés Editores, 1993. 276 p.: ill.

Examines evolution of the International Monetary Fund and analyzes its approach to the Mexican debt crisis of 1982. Provides original critique of the IMF stabilization policies implemented in Mexico in the 1980s.

1348 Canabal Cristiani, Beatriz; Pablo Alberto Torres-Lima; and Gilberto Burela Rueda. La ciudad y sus chinampas: el caso de Xochimilco. México: Univ. Autónoma Metropolitana-Xochimilco, 1992. 177 p.: bibl., ill., maps. (Col. Ensayos)

Essays on various aspects of Xochimilco contain valuable data on the city's demographics and structure of production.

1349 Carney, Judith Ann. Triticale production in the central Mexican highlands: smallholders' experiences and lessons for research. México: CIMMYT, 1990. 48 p.: bibl., ill. (CIMMYT economics paper, 0188-2414; 2)

Triticale is a cross between wheat and rye that yields more than its parent species under two kinds of marginal conditions: highland areas where acid soils and foliar diseases are a problem and semiarid areas where drought frequently affects crop production. This paper examines triticale utilization patterns among smallholders in one tropical mountain region, the central Mexican highlands. Concludes that future research to meet smallholders' needs for forage and feed triticales may necessitate a reconsideration of the current emphasis on complete triticale types.

1350 Carrillo, Arturo J. The new Mexican revolution: economic reform and the 1989 regulations of the Law for the Promotion of Mexican Investment and the Regulation of Foreign Investment. (*J. Int. Law Econ.*, 24:3, 1991, p. 647–690)

Evaluates President Salinas' foreign investment reforms from a legal perspective and assesses the current configuration of Mexican foreign investment law. Emphasizes implications for future foreign investment in Mexico in light of the 1989 Regulation's legal and economic impact, modern Mexican economic policy and current events.

1351 Carrillo Dewar, Ivonne. Industria petrolera y desarrollo capitalista en el norte de Veracruz, 1900–1990. Xalapa, México: Univ. Veracruzana, 1993. 184 p.: bibl. (Biblioteca)

Essays examine historical evolution of the petroleum industry in the northern part of the state of Veracruz. Topics include the industry's impact on the local economy and its contribution to the national economy.

1352 Carrillo Huerta, Mario M. The impact of *maquiladoras* on migration in Mexico. (*in* The effects of receiving country policies on migration flows. Boulder, Colo.: Westview Press, 1991, p. 67–102, tables)

Study's main purpose is to assess the recent relationship between the *maquiladoras* and the migration of Mexicans to the northern border zone and the US. Evaluates importance of these plants as a magnet for migration. Predicts effect that a change in maquiladoras might have, not just on the internal and international distribution of the Mexican population, but also on employment and the standard of living in the regions where this industry operates.

1353 **Carrillo Viveros, Jorge; Alfredo Hualde;** and **Miguel Angel Ramírez.** Empresas maquiladoras y Tratado Trilateral de Libre Comercio: empleo, eslabonamientos y expectativas. Tijuana, Mexico: El Colegio de la Frontera Norte, 1992. 78 p.: bibl., ill. (Cuadernos; 4)

Essays examine several important aspects for the northern states stemming from increased regional trade. Topics include the automobile industry, migration to Tijuana, employment in the region's *maquiladora* firms, and labor markets in Tijuana, Ciudad Juárez, and Nuevo Laredo.

1354 **Carrillo Viveros, Jorge** and **Miguel Angel Ramírez.** Modernización tecnológica y cambios organizacionales en la industria maquiladora. (*Estud. Front.*, 22, sept./dic. 1990, p. 55–76, bibl., tables)

Analyzes technology in *maquila* plants, especially those implementing new forms of organization of work. Examines two hypotheses: 1) there exists a vast and heterogeneous diffusion of technology throughout the *maquila* plants, and alternatively, 2) there are significant differences in organizational structure between plants that use new organizational models and those that are more traditional. To carry out this research a probability inquiry was conducted in Ciudad Juárez, Tijuana, and Monterrey. Finds that the *maquiladora* industry is very important as a source of foreign exchange and employment at the national level, but also is important because of its impact at the regional level. At the present time the *maquiladora* industry is characterized by dynamism, heterogeneity, and restructuring. Hallmarks of restructuring include diffusion of automated processes, changes in employment, and growth of organizational flexibility.

1355 **Cartas, José María.** Orden político y orden económico en México: dilemas y alternativas. Buenos Aires: Centro Interdisciplinario de Estudios sobre el Desarrollo Latinoamericano, 1990. 127 p.: bibl., ill. (Serie Investigaciones externas)

General survey of economic development in Mexico since the 1910 Revolution is policy-oriented, with a good discussion of the structure of the economy. Examines likely consequences of alternative government programs.

1356 **Castillo, Víctor M.** Economía fronteriza y desarrollo regional. Mexicali, Mexico: Univ. Autónoma de Baja California, 1991. 126 p.: bibl., ill.

Overview of the economies of the northern Mexican states: Baja California, Sonora, Chihuahua, Coahuila, Nuevo León, and Tamaulipas. Provides analysis of factors affecting employment in the region. Assesses nature of links to the US. Contains numerous tables of hard-to-find data.

1357 **Catalán Valdés, Rafael.** Las nuevas políticas de vivienda. México: Fondo de Cultura Económica, 1993. 234 p.: bibl. (Una Visión de la modernización de México)

Historical assessment of Mexico's housing policies and programs. Includes an extensive discussion of the government's changing approach to the country's housing shortages and a number of innovative recommendations for improved policy design and implementation.

Censos económicos 1994. See item **17**.

1358 **Centeno, Miguel Angel.** "Still disputing after all these years:" recent work on the Mexican political economy. (*LARR*, 27:2, 1992, p. 167–179)

Review essay of six recent books examines long-standing debate between neoliberals committed to markets and nationalist-populists committed to the welfare State. Assesses books in terms of their ability to answer the question of why neoliberals have won the ideological battle in a country with one of the strongest nationalist traditions in politics and economics.

1359 **Chabat, Jorge.** Mexico: so close to the United States, so far from Latin America. (*Curr. Hist.*, 92:571, Feb. 1993, p. 55–58)

Examines Mexico's relations with the US and Latin America. Concludes that the country is still tightly linked to the US despite attempts by the Salinas government to diversify the economy through increased ties with other industrial countries. Instead of viewing Mexico as abandoning Latin America by joining NAFTA, it is more accurate to recognize that the country has never been part of the broader Latin American economic and political arena.

CIMA. See item 18.

1360 **COLEF (conference), *1st, Tijuana, Mexico, 1990.*** COLEF I. v. 1, Economía fronteriza y libre comercio. Tijuana, Mexico: Colegio de la Frontera Norte; Ciudad Juárez, Mexico: Univ. Autónoma de Ciudad Juárez, 1993. 1 v.: bibl., ill.

First vol. of proceedings of a conference held in Oct. 1990 at El Colegio de la Frontera Norte. Contains works by leading trade experts on various aspects of trade between Mexico, the US, and Canada. Emphasizes the border region, but also includes assessment of inflation, industrial restructuring, and Mexico's automotive industry.

Comercio Exterior: 1973–1993. See item 19.

1361 **Conroy, Michael E.** and **Amy K. Glasmeier.** Unprecedented disparities, unparalleled adjustment needs: winners and losers on the NAFTA "fast track." (*J. Interam. Stud. World Aff.*, 34:4, Winter 1992/93, p. 1–37, bibl., graphs, tables)

Explores potential distributive impacts of NAFTA. Who is most likely to gain in each country? Who is most likely to lose? What significance will the magnitude of the dissatisfied among the three partners have for the probable success of the endeavor? Are there comparable experiences from which to draw? Concludes that adjustment costs are high and that they will be even greater if they are borne without well-planned, carefully-created and fully-funded multilateral and national programs for job retraining, regional impact alleviation, subsidies to labor mobility, and other transitional programs.

1362 **Contreras Montellano, Oscar F.** *et al.* Condiciones de empleo y capacitación en las maquiladoras de exportación en México. Coordinación de Jorge Carrillo V. México: Secretaría del Trabajo y Previsión Social, Subsecretaría "B," Dirrección General de Empleo; Colegio de la Frontera Norte, Dirección General Académica, 1993. 286 p.: bibl., ill.

Series of papers examining the country's *maquiladoras* emphasizes patterns of employment, salaries, and working conditions. Several papers examine patterns of investment in these firms from the perspective of foreign companies.

1363 **Cordera Campos, Rolando** and **Enrique González Tiburcio.** Crisis and transition in the Mexican economy. (*in* Social responses to Mexico's economic crisis of the 1980s. Edited by Mercedes González de la Rocha and Agustín Escobar Latapí. San Diego, Calif.: Center for U.S.-Mexican Studies, Univ. of California, San Diego, 1991, p. 19–56, bibl., tables)

Attempts to define where Mexico stands in terms of its social, macroeconomic and productive capacities and to analyze some of the regional impacts of the 1980s economic crisis and policies meant to ameliorate it. Considers some of the quantitative implications of financial, material, and social welfare resources needed to satisfy the population's basic social needs by the year 2000. Finally, examines primary economic policy decisions that will appear on Mexico's national agenda in the near future.

1364 **Corona Rentería, Alfonso** and **Juan Sánchez Gleason.** Integración del norte de México a la economía nacional: perspectivas y oportunidades. México: SPP, Consejo Nacional para la Cultura y las Artes, 1989. 448 p.: tables.

Examines economies of Northern Mexico (Baja California Norte, Sonora, Chihuahua, Coahuila, Nuevo León, and Tamaulipas), providing detailed analysis of each state's demography, economic structure and trade patterns. Many useful tables.

1365 **Cruz Piñeiro, Rodolfo.** La fuerza de trabajo en los mercados urbanos de la frontera norte. Tijuana, Mexico: Colegio de la Frontera Norte, 1992. 75 p.: bibl., ill. (Cuadernos; 5)

Examines labor conditions along the US-Mexican border. Topics include labor disputes, labor conditions, employment migration, and dynamics of the region's demographics.

Cuenca del Pacífico. See item **20**.

1366 **Dávila, Enrique.** Ingresos y prestaciones del sector informal. México: Fundación Friedrich Ebert, 1989. 35 p.: bibl. (Documentos de trabajo; 20)

Assesses labor conditions in the country's informal sector. Topics include patterns of wages, employment, and relevant labor legislation. Makes series of recommendations for improving both employment and work conditions in the country.

1367 **Desarrollo urbano en Sinaloa, 1987–1992.** Culiacán: Gobierno del Estado de Sinaloa, Secretaría de Planeación y Desarrollo, 1992. 466 p.: bibl., ill., maps.

Essays on the state of Sinaloa emphasize city planning, urbanization, and the impact of the national development plan. Topics include the process of urbanization, demographics, and the effect of growth on the state's ecological situation.

1368 **Desarrollo y medio ambiente en Veracruz.** Coordinación de Eckart Boege e Hipólito Rodríguez. Jalapa, Mexico: Instituto de Ecología; México: Fundación Friedrich Ebert; Jalapa, Mexico: Centro de Investigaciones y Estudios Superiores en Antropología Social, 1992. 303 p.: bibl., ill., maps.

Examines environmental problems associated with economic growth in the Veracruz region. Contains basic overview of the economy with links made to the manner in which the ecology of the region has changed over time. Offers a number of policy proposals to achieve sustained growth in the region.

1369 **Díaz Serrano, Jorge.** La privatización del petróleo mexicano. México: Editorial Planeta Mexicano, 1992. 139 p.: bibl. (Col. México vivo)

Former head of PEMEX reflects on problems facing Mexico's oil sector. Provides arguments for privatizing the oil sector in light of the world's changing economic situation.

1370 **La disputa por los mercados: TLC y sector agropecuario.** Coordinación de Alejandro Encinas, Juan de la Fuente y Horacio Mackinlay. México: H. Cámara de Diputados, LV Legislatura: Editorial Diana, 1992. 388 p.: bibl., ill. (Territorios; 1)

Essays cover problems encountered by the agricultural sector as it gets increasingly integrated into world markets. Authors emphasize trade with the US. Contains valuable appendices on the relevant trade legislation.

1371 **La distribución del ingreso en México: encuesta sobre los ingresos y gastos de las familias, 1968.** México: Banco de México: Fondo de Cultura Económica, 1974. 173 p.: ill.

Official income distribution figures for Mexico are from 1968. Invaluable collections of tables presenting all facets of family income patterns throughout the country. Also contains data on the cost and standard of living.

1372 **La economía mexicana actual: pobreza y desarrollo incierto.** México: Univ. Autónoma Metropolitana, Unidad Iztapalapa, División de Ciencias Sociales y Humanidades, Depto. de Economía, 1991. 247 p.: bibl., ill. (Serie de investigación; 3)

Series of essays on the Mexican economy. Topics include income distribution, the agricultural sector, stabilization policy, economic integration with the US and Canada, inflation, and domestic finance.

La economía mexicana en cifras. See item **22**.

1373 **Economía sonorense más allá de los valles.** Coordinación de Miguel Angel Vázquez Ruiz. Hermosillo, Mexico: Univ. de Sonora, Depto. de Economía y Centro de Investigaciones Económicas y Sociales, 1991. 162 p.: bibl., tables.

Essays cover economy of the state of Sonora, including industrialization, agriculture, minerals and mining, exports, and *maquiladoras*. Many useful and unique tables of data on the state's economy.

1374 **Elizondo, Néstor.** Illegality in the urban informal sector of Mexico City. (*in* Beyond regulation: the informal economy in Latin America. Edited by Víctor E. Tokman. Boulder, Colo.: L. Rienner, 1992, p. 55–83, tables)

Analyzes functioning and economic performance of a number of informal production units that operate under varying conditions of legality and illegality in Mexico City. Finds that there are specific differences between companies according to the legal situation. Also finds that there is a clear correlation between the number of workers employed and the degree of legality. Enterprises employing up to two workers generally tend to be operating illegally.

1375 **Espinosa Villarreal, Oscar.** El impulso a la micro, pequeña y mediana empresa. México: Fondo de Cultura Económica, 1993. 152 p.: ill. (Una Visión de la modernización de México)

Examines small-business sector of the economy. Assesses major problems faced by these small firms and criticizes governmental policy for its failure to improve the environment in which these firms operate.

Estadísticas históricas de México. See item 23.

1376 **Estrategia, desarrollo y política económica.** Iztapalapa, Mexico: Univ. Autónoma Metropolitana, Unidad Iztapalapa, División de Ciencias Sociales y Humanidades, Depto. de Economía, 1989. 394 p.: bibl., ill. (Serie de investigación; 1)

Essays on international finance emphasize the 1982 Mexican debt crisis and the international capital markets. Topics include inflation, alternative sources of finance, foreign investment, and capital flight.

1377 **Evolución de la productividad total de los factores en la economía mexicana, 1970–1989.** México: Secretaría del Trabajo y Previsión Social, 1993. 178 p.: bibl., ill. (Cuadernos del trabajo; 1)

This volume is largely a compilation of statistics and analysis of production and productivity in Mexican industries. Covers production, productivity, and patterns of employment. Excellent discussion of alternative methods used to measure production and productivity.

1378 **Favret Tondato, Rita.** Tenencia de la tierra en el estado de Coahuila, 1880–1987. Buenavista, Mexico: Univ. Autónoma Agraria Antonio Narro, 1992. 247 p.: bibl., maps.

Study of land tenure arrangements and patterns takes historical approach to show the evolution of ownership and land use. Detailed maps of the region add to this excellent study.

1379 **Feliz, Raúl Aníbal** and **Fernando Atilio Torres.** Deuda y déficit público en México. (*Estud. Econ./México*, 6:1, enero/junio 1991, p. 91–109, tables, graphs)

Paper's objectives are to determine what restrictions the "sustainability" of the public debt (defined as the expectation of an intertemporal balanced public budget) impose on the observed behavior of the time series of the public debt and on the fiscal deficit. Author then contrasts these restrictions for Mexico during Jan. 1981–Dec. 1988.

1380 **Foreign direct investment and industrial restructuring in Mexico: government policy, corporate strategies and regional integration.** New York: United Nations Centre on Transnational Corporations, 1992. 114 p.: bibl., ill. (UNCTC current studies. Series A; 18)

Analyzes the largest enterprises with foreign capital in Mexico's manufacturing sector and their contribution to Mexico's industrial restructuring process in areas such as improving production efficiency, transferring competitive technology, expanding exports, and creating linkages with local supplier industries. Special consideration is given to an evaluation of the performance of firms in these respects and how such performance has been influenced by macroeconomic policies and regulations governing foreign direct investment. Study attempts to identify policy elements and competitive conditions that must prevail to induce transnational corporations to adopt corporate strategies that promote industrial restructuring in accordance with host country objectives.

1381 **Fox, Jonathan.** Agriculture and the politics of the North American trade debate: a report from the Trinational Exchange on Agriculture, the Environment and the Free Trade Agreement; Mexico City, November 14–17, 1991. (*LASA Forum*, 23:1, Spring 1992, p. 3–9)

Summarizes conference as one where: 1) farmers increasingly came to view their counterparts not as competitors but as people grappling with common issues, laying groundwork for further exchanges based on greater understanding of the traditional political and economic context in which they work the land; 2) participants highlighted the need for policies to take into account the survival of family farms and rural communities; and 3) participants agreed that downward harmonization (i.e. lowering) of environmental and food safety standards would be both unhealthy and undemocratic.

1382 **Frankman, Myron J.** North American economic cooperation: the wartime experience. (*Can. J. Lat. Am. Caribb. Stud.*, 16:32, 1991, p. 35–57, graphs, tables)

Paper examines extent to which Canadian and Mexican dependence on the US increased during World War II and the extent to which economic activity in Canada and Mexico either paralleled that in the US or was influenced by government decisions in Washington. Examples relating to foreign trade, import substituting industrialization, prices, and infrastructure development in Mexico and Canada are used to show Washington's expanded wartime role.

1383 **Gaceta informativa de la industria maquiladora.** Monterrey, Mexico?: Gobierno del Estado de Nuevo León, 1992? 1 v.: ill.

Contains basic data on the growth of *maquiladora* firms in Nuevo León. For major firms in the state, also includes useful section listing information such as firm's name, products, address, etc.

1384 **García, Rolando Víctor *et al*.** Modernización en el agro: ventajas comparativas, ¿para quién?: el caso de los cultivos comerciales en El Bajío. México: Cinvestav, 1988. 225 p.: bibl., ill. (IFIAS research series; 8)

Vol. 8 in the International Federation of Institutes for Advanced Study's *Research Series*. This is an association of 38 leading research institutes that collaborate to address major global problems of long-term importance in environment, economy, and science and technology. Volume examines environmental problems associated with commercial agriculture in Mexico. Includes excellent maps outlining the geography of the country.

1385 **García Bedoy, Humberto.** Neoliberalismo en México: características, límites y consecuencias. México: Centro de Reflexión y Acción Social; Centro de Reflexión Teológica; ITESO; 1992. 143 p.: bibl. (Serie Contextos y análisis; 2)

Critical examination of the neoliberal approach to the country's stabilization strategy. Focuses on problems associated with this approach to economic management.

1386 **García Montaño, Jorge.** Diagnóstico de largo plazo de la economía de Baja California, 1950–1980. Tijuana, México: Univ. Autónoma de Baja California, Facultad de Economía, 1988. 62 p. (Cuadernos de economía, 0187–6929; ser. 3, no. 4)

Brief overview of the economy of Baja California examines population, the pattern of production, trade, and economic links with the other northern states.

1387 **Gereffi, Gary.** The "old" and "new" *maquiladora* industries in Mexico: what is their contribution to national development and North American integration? (*Nuestra Econ.*, 2:8, mayo/agosto 1991, p. 39–63, bibl., tables)

Highlights some key issues concerning potential contributions of the *maquiladora* program to national development in Mexico and to the de facto process of North American integration. Several related arguments are developed in the paper. First, the stereotype of Mexico's *maquiladora* plants as being simple, female-dominated, labor-intensive operations is no longer accurate. The last decade has witnessed the emergence of new technology. Second, there are now capital-intensive *maquiladoras*, especially in automobile-related manufacturing and advanced electronics assembly. These plants employ much higher percentages of skilled workers. Third, Mexico's *maquiladora* program is different in a number of important ways from the East Asian export-processing zones with which it is often compared.

1388 **Gereffi, Gary** and **Lu-Lin Cheng.** The role of informality in East Asian development, with implications for Mexico. (*in* Work without protections: case studies of the informal sector in developing countries. Edited by Gregory K. Schoepfle and Jorge F. Pérez-López. Washington: US Dept. of Labor, Bureau of International Labor Affairs, 1993, p. 177–214, bibl., tables)

Examines significance and special features of informal sector in three East Asian NICs and discusses implications of the East Asian experience for Mexico. Based on their analysis the authors develop four alternative scenarios for the future: 1) export processing; 2) component-supply manufacturing; 3) specification contracting; and 4) domestic and overseas retailers of local brands of consumer goods. Argues that Mexico appears to be moving from Scenario 1 to Scenario 2, but still has a long way to go before it matches the success of the East Asian NICs which are now moving from Scenario 3 to Scenario 4.

1389 **Girón, Alicia.** Cincuenta años de deuda externa. México: Instituto de Investigaciones Económicas, Univ. Nacional Autónoma de México, 1991. 253 p.: bibl. (Col. La Estructura económica y social de México)

Survey of international financial system and Mexican economic policies that led up to the debt crisis of 1982, including a detailed assessment of government policy and developmental strategies in the 1970s. Examines stabilization programs followed by the government in the 1980s and assesses implications of the Brady Plan for Mexico.

1390 **Godínez Plascencia, Alberto.** El cambio tecnológico en la industria maquiladora de exportación en México: un enfoque metodológico. (*Estud. Front.*, 23, sept./dic. 1990, p. 9–31, bibl., tables)

Cricically reviews the most representative studies that analyze technological change of the *maquiladora* industry in Mexico. Objective is to assess the different elements studied in each work, with a special emphasis on the measuring instruments and databases. Argues that diversity of issues and conclusions are the result of heterogeneity of the *maquila* plants, but also arise from diversity of objectives, hypotheses, theories, variables indicators, measuring instruments, databases, etc., employed. Conclusions pose a new question: is there really a structural change in the technological level of Mexico's *maquiladora* industry?

1391 **Gómez Mont, Carmen.** El desafío de los nuevos medios de comunicación en México. México: AMIC; Diana, 1992. 180 p.: bibl., ill.

Surveys country's rapidly changing telecommunications system. Emphasizes changes over the last decade and introduction of new technologies. Contains excellent bibliography.

1392 **Gómez Sahagún, Lucila.** San Miguel Tlaixpan: cultivo tradicional de la flor. México: Univ. Iberoamericana, 1992. 124 p.: bibl., ill., maps. (Col. Tepetlaostoc; 1)

Historical examination of economic and social patterns in the San Miguel Tlaixpan region. Excellent discussion of how social conditions of the region have evolved over time.

1393 **Gómez Solórzano, Marco A.** *et al.* La reestructuración industrial en México: cinco aspectos fundamentales. Coordinación de Josefina Morales. México: Univ. Nacional Autónoma de México, Instituto de Investigaciones Económicas: Editorial Nuestro Tiempo, 1992. 206 p.: ill. (Col. La Estructura económica y social de México. Col. Desarrollo)

Essays on Mexican industry cover issues including modernization, structural conditions, finance, capital formation, and private investment.

1394 **González Vela, Gabriel.** Desarrollo económico y social del estado de Aguascalientes, 1986–1992. Aguascalientes, Mexico: Instituto Cultural de Aguascalientes: Instituto de Investigación Económica y Social Lucas Alamán, 1992. 199 p.: bibl., ill. (Contemporáneos)

Excellent general survey of social and economic conditions in Aguascalientes covers analysis of sectoral development, public finances, trade with other states, demographics, urban development, and infrastructural development. Contains invaluable compilations of data on the state.

1395 **Goulet, Denis** and **Kwan S. Kim.** Estrategias de desarrollo para el futuro de México. Guadalajara, México: Instituto Tecnológico y de Estudios Superiores de Occidente, 1989? 288 p: bibl.

Examines various problems confronting the economy: inflation, pollution, debt, rapid population growth, uncontrolled urbanization, and increased income inequality. Based on an analysis of these factors, authors from the Univ. of Notre Dame outline a development strategy for the country so that it will be capable of achieving sustainable growth.

1396 **Grammont, Hubert Carton de.** Empresarios agrícolas y el estado: Sinaloa, 1893–1984. México: Univ. Nacional Autónoma de México, Instituto de Investigaciones Sociales, 1990. 279 p.: bibl., map.

Broad examination of historical development in the state of Sinaloa. Examines development of agricultural finance and the sector's infrastructure. Structure of production is assessed as well as the regions exports and producer organizations.

1397 **Griffith, Kathleen Ann.** An interview with Dr. Jaime Serra Puche: Secretary of Trade and Industry of Mexico. (*Columbia J. World Bus.*, 27:1, Spring 1992, p. 52–58)

Interview covers Serra Puche's involvement in Mexican trade negoatiations and his impressions on what new treaties mean for the country. Some of the more thorny issues integral to NAFTA which he discusses are services, agricultural subsidies, and labor.

1398 Guadalajara en el umbral del siglo XXI. Recopilación de Jesús Arroyo Alejandre y Luis Arturo Velázquez. Guadalajara, Mexico: Editorial Univ. de Guadalajara: H. Ayuntamiento de Guadalajara, 1992. 407 p.: bibl., ill., maps. (Col. Aula magna)

Essays on all facets of the economy of Guadalajara and the surrounding region emphasize historical evolution of institutions and socioeconomic structures. Topics include demographics, migration, and industrialization.

1399 Gurría, José A. La política de la deuda externa. México: Fondo de Cultura Económica, 1993. xxvii, 274 p.: bibl., ill. (Una Visión de la modernización de México)

Assesses the country's external debt crisis during 1981–90. Focuses on restructuring of the debt and options open to the country to reduce the burden imposed by the government's excessive borrowing in international markets.

1400 Haber, Stephen H. Assessing the obstacles to industrialisation: the Mexican economy, 1830–1940. (*J. Lat. Am. Stud.*, 24:1, Feb. 1992, p. 1–32)

Paper seeks to understand the long lag in Latin American industrialization through an analysis of Mexico's experience from 1830–1940. The purpose is to look at the obstacles that prevented self-sustaining industrialization from taking place in Mexico, as well as to assess results of industrialization that did occur.

1401 Hacia una nueva política industrial. México: Editorial Diana, 1988. 175 p.: bibl. (Alternativas para el futuro)

Volume presents series of papers addressing the country's industrial policies. Topics include impact of the evolving world economy on Mexico's industry, industrial promotion, and policies to encourage regional diversification of industries. Interesting speculation about future industrial problems and policies needed to address these difficulties.

1402 Haigh, Robert W. Building a strategic alliance: the Hermosillo experience as a Ford-Mazda proving ground. (*Columbia J. World Bus.*, 27:1, Spring 1992, p. 60–74, tables)

Argues that rather than try to beat Japanese automobile makers, Ford Motor Co. decided to join Mazda Motor Corp. to make a better product. Since the late 1960s, Ford has been buying parts from Mazda for Ford cars sold in the Asia-Pacific region. In the 1980s, this distant Ford-Mazda alliance blossomed into a successful collaboration for the Ford/Mercury Tracer at the assembly and stamping plant in Hermosillo, Mexico. Concludes that this historic collaboration will pave the way for other strategic alliances and teach those who follow how to effectively collaborate cross-culturally.

1403 Heath, John Richard. Evaluating the impact of Mexico's land reform on agricultural productivity. (*World Dev.*, 20:5, May 1992, p. 695–711, bibl., tables)

In Mexico, land reform law and terms of credit suggest a priori that farmers in the land reform (ejido) sector should be less productive than private farmers. An analysis of studies comparing productivity of individual ejido parcels and small private holdings, however, reveals no significant difference between the two sectors. Between tenure categories there is a de jure but not de facto difference in small farmers' control over resources, partly because farmers circumvent restrictions implicit in the land reform. The case for formal privatization of the ejido sector may be more compelling than has been suggested.

1404 Heath, John Richard. Further analysis of the Mexican food crisis. (*LARR*, 27:3, 1992, p. 123–145, bibl., tables)

Examines hypothesis of Barkin and DeWalt that producers of basic foods in Mexico need more protection and more government support if food self-sufficiency is to be achieved. Barkin and DeWalt are in favor of preferential price supports for crops grown for direct human consumption. By inference the government should intervene in the market to redress the innate profitability of sorghum relative to maize. Author contends that the use of guaranteed price mechanisms in the past has fostered significant growth of basic food crops, but that the efficacy of these mechanisms in raising peasant incomes is highly questionable.

1405 Hecht, Laurence and **Peter Morici.** Managing risks in Mexico. (*Harv. Bus. Rev.*, 71:4, July/Aug. 1993, p. 32–40, graph)

Argues that despite NAFTA any company considering a move to Mexico must balance the risks and rewards based on long-term considerations, not the latest free trade rhetoric. Contends that the Free Trade Agreement will not cure the economy's problems. Even with NAFTA, Mexico will be plagued by commonplace corruption, environmental messes, and a history of excessive State involvement in the economy.

1406 Hernández Laos, Enrique. Crecimiento económico y pobreza en México: una agenda para la investigación. México: Univ. Nacional Autónoma de México, Centro de Investigaciones Interdisciplinarias en Humanidades, 1992. 268 p.: bibl. (Col. Alternativas)

Study provides good survey of Mexico's economic challenges mainly from 1963–88. Provides a number of useful recommendations for overcoming some of the social and economic difficulties facing policymakers.

1407 Hernández Laos, Enrique. Productividad y eficiencia en la industria mexicana del azúcar: un ensayo metodológico. México: Univ. Autónoma Metropolitana, Unidad Iztapalapa, División de Ciencias Sociales y Humanidades, 1992. 205 p.: bibl., ill. (Iztapalapa, texto y contexto; 12)

Study of factors underlying efficiency and productivity in the country's sugar industry provides detailed data on the industry, together with an assessment of the policies best designed to increase productivity and utilization of resources.

1408 Herrera Toledano, Salvador and **Javier Macedo Martínez.** El Tratado de Libre Comercio y la industria en el Estado de México: retos y perspectivas. Toluca, Mexico: Colegio Mexiquense, 1992. 129 p.: bibl.

Examines arguments for economic integration of the North American countries. Assesses ability of Mexican firms to compete in the new free trade environment, listing both advantages and disadvantages. Good discussion of the opportunities opened up for Mexico by the new free trade area.

1409 Historia y porvenir de México ante el Tratado de Libre Comercio. México: Instituto Politécnico Nacional, Dirección de Bibliotecas y Publicaciones, 1991. 158 p.

Essays trace evolution of Mexico's free trade policy. Contains numerous documents from both the US and Mexico outlining views of leading commentators and government officials.

1410 Hudgins, Edward Lee. Promoting prosperity through a U.S.-Mexico free trade area. Washington: Heritage Foundation, 1990. 8 p.: bibl. (The Heritage lectures; 272)

Study argues that the goal of US trade policy should be to open markets in the US and with all of our trading partners. Contends that Free Trade Areas (FTAs) are an important tool in any free trade strategy. Concludes that the US should offer to negotiate FTAs with any country that wishes to remove its own barriers to American imports in exchange for open US markets.

1411 Impactos regionales de la apertura comercial: perspectivas del tratado de libre comercio en Jalisco. Recopilación de Jesús Arroyo Alejandre y David Lorey. Guadalajara, Mexico: Univ. de Guadalajara; Los Angeles, Calif.: UCLA Program on Mexico, 1993. 372 p.: bibl., ill. (Serie Ciclos y tendencias en el México del siglo XX; 4)

Volume consists of papers focused on impact on the Jalisco economy stemming from changing international trade patterns and relationships. Several papers examine the likely effect NAFTA will have on the regional economy. In addition there is an excellent discussion of the trade impact on urbanization, migration, and the ecological conditions in the state.

1412 Indicadores de la modernización mexicana. Coordinación de Raúl H. Mora. México: Centro de Reflexión y Acción Social; Centro de Reflexión Teológica, 1992. 189 p.: bibl. (Serie Contextos y análisis; 1)

Essays focus on broad policy issues such as modernización, economic integration, theories of neoliberalism, and monetary policy.

1413 Industria, comercio y Estado: algunas experiencias en la Cuenca del Pacífico. Recopilación de Omar Martínez Legorreta. México: Colegio de México, Centro de Estudios de Asia y Africa, 1991. 419 p.: bibl., ill.

Papers presented at the seminar "Industria, Comercio y Estado" held Nov. 21-23, 1993 at the Colegio de México cover wide-ranging topics related to trade in the Pacific Basin. Focus is on commercial policy and economic cooperation in the region.

1414 **Industria y trabajo en México.** Coordinación de James W. Wilkie y Jesús Reyes Heroles González Garza. México: Univ. Autónoma Metropolitana Azcapotzalco, 1990. 332 p.: bibl., ill. (Serie Ciclos y tendencias en el México del siglo XX ; 1)

Leading authorities contribute chapters surveying facets of the country's industrialization experience. Discusses industrial exports, government policies toward the sector, and the impact of the 1982 debt crisis on the industry.

1415 **Instituto de Banca y Finanzas (Mexico). Centro de Investigación para el Desarrollo.** Tecnología e industria en el futuro de México. México: Editorial Diana, 1989. 231 p.: ill. (Alternativas para el futuro)

Examines problems of technological transfer to and indigenous technical development in Mexico. Proposes a number of policies that would speed the process of local technical development.

1416 **Inversión extranjera directa en México en la industria informática y automotriz.** Coordinación de Isaac Minian et al. México: Fundación Friedrich Ebert, 1988. 43 p.: bibl. (Documentos de trabajo; 13)

Examines foreign ownership in Mexico's automotive sector. Contains useful information on the sector's employment as well as investment and trade patterns.

Investigación Económica. 1994– . See item **30**.

1417 **Jeannot, Fernando; Raúl Conde Hernández; and Fernando Sancen Contreras.** Estudios sobre economía y Estado: identidad, regulación, integración y regímenes productivos. Introducción de Jorge Ruiz Dueñas. México: Secretaría de Energía, Minas e Industria Paraestatal: Univ. Autónoma Metropolitana; Fertilizantes Mexicanos; Fondo de Cultura Económica, 1988. 250 p.: bibl. (Industria paraestatal en México; 1)

Leading authorities examine various facts of government enterprises in Mexico, especially in the energy/mining industries. Issues stressed are capital formation, the legal environment, economic integration, and alternative means of organizing economic activity.

1418 **Keller, Tom.** Interview with Pedro-Pablo Kuczynski. (*Columbia J. World Bus.*, 26:2, Summer 1991, p. 108–114)

Interview with the Chairman of First Boston International and an Executive Director at Credit Suisse First Boston, London who argues that the recipe for Mexico's economic success involves fomenting more growth at lower-income levels, thus creating a mass market, the key to success in the advanced countries. Second area of importance is to keep the oil industry internationally competitive and expand its capacity. Finally, Kuczynski thinks the US should actively support Mexico's efforts by wholeheartedly endorsing the free trade area.

1419 **Kessel, Georgina** and **Ricardo Samaniego.** Apertura comercial, productividad y desarrollo tecnológico: el caso de México. Washington: Depto. de Desarrollo Económico y Social, Banco Interamericano de Desarrollo, 1992. 68 p.: bibl. (Serie de documentos de trabajo; 112)

Examines role of technology in affecting pattern and structure of Mexican trade. Focuses on the effects of productivity and the recent liberalization of Mexican trade restrictions in altering the country's pattern of imports and exports.

1420 **King, Jonathan.** Mexican perspectives on the North American Free Trade Area. San Germán, P.R.: Caribbean Institute and Study Center for Latin America (CISCLA), Inter-American Univ. of Puerto Rico, 1993. 36 p.: bibl., tables.

Presents critical discussion of Mexican literature on the Free Trade Agreement between Mexico, the US, and Canada. Finds number of contradictory opinions about the desirability of NAFTA from the Mexican perspective. Concludes that the ultimate success of NAFTA will depend on how the government deals with the country's social problems, and the leadership it provides in promoting private sector development.

KOMPASS México. See *HLAS 54:53*.

1421 Kras, Eva Simonsen. Modernizing Mexican management style: with insights for U.S. companies working in Mexico. Las Cruces, NM: Editts Publishing, 1994. 160 p.: bibl., index.

Revised translation of *La administración mexicana en transición,* published originally in Mexico. Book is concrete and practical handbook helping Mexican companies take the initial step in analyzing their present management style—including executives' attitudes and outlooks as well as organizational approaches. Outlines basic steps that many Mexican companies have successfully used in transition from traditional to modern management. Also identifies specific issues US companies need to consider when introducing modern management techniques in their Mexican operations.

1422 Labra, Armando et al. El sector social de la economía. Coordinación de Armando Labra. México: Siglo Veintiuno Editores; Centro de Investigaciones Interdisciplinarias en Humanidades, Univ. Nacional Autónoma de México, 1988. 270 p.: bibl. (Biblioteca México, actualidad y perspectivas)

Series of essays on wide-ranging topics concerning social conditions in the country, including modernization, rural life, *ejidos,* and working conditions.

Lee, Naeyoung and Jeffrey Cason. Automobile commodity chains in the NICs: a comparison of South Korea, Mexico, and Brazil. See item 3724.

1423 Levy, Santiago and Sweder van Wijnbergen. Maize and the Free Trade Agreement between Mexico and the United States. (*World Bank Econ. Rev.,* 6:3, Sept. 1992, p. 481–502, bibl., graphs, tables)

Argues that the government practice of setting the price of maize in rural Mexico above the world price is inefficient and likely to have negative distribution effects because many producers, and all landless workers, are net buyers. In fact, this policy screens out the relatively poor rather than the relatively rich. The policy objective, therefore, should be to move toward free trade. This would yield large gains in efficiency. Concludes that the Free Trade Agreement provides an ideal opportunity to pursue this objective. It will provide freer entrance into the US for other agricultural products as well as for a broad range of manufactured products. Insuring secure and sustained access for labor-intensive agricultural and manufactured products can help ease the impact on the labor market of transition away from subsistence maize cultivation.

1424 Levy, Santiago and Sweder van Wijnbergen. Mercados de trabajo, migración y bienestar: la agricultura en el Tratado de Libre Comercio entre México y los Estados Unidos. (*Trimest. Econ.,* 60:238, abril/junio 1993, p. 371–411, bibl., graphs, tables)

Argues that Mexico, like many other countries, provides its agricultural sector with substantial protection from international competition. Income distributional considerations and concerns about transitional problems once liberalization starts explain much of the persistence of such policies. Authors put forth four main points: 1) substantial efficiency gains can be expected from liberalization; 2) income distributional effects of protection are in fact regressive; 3) well-targeted adjustment programs that help rather than delay adjustment can be designed once a careful analysis of the precise distributional impact has been made; and 4) a compromise, two-sided liberalization by Mexico and its main export market (the US) would significantly reduce any adjustment problems. The latter point is made in the context of an assessment of the potential impact of the Free Trade Agreement on agriculture.

1425 Levy, Santiago and Sweder van Wijnbergen. Mexican agriculture in the free trade agreement: transition problems in economic reform. Paris: Organisation for Economic Co-operation and Development, 1992. 92 p.: bibl., ill. (Technical papers; 63)

Examines whether Mexican agriculture should be liberalized and if so the speed of liberalization and policies used to accompany the transition. Focuses on implications for policy design dealing with the absence of efficient capital markets; on the welfare costs of reforming only gradually; on incentive problems created by trade adjustments; and on the redistributive aspects of policy reform in the presence of realistic limits on available policy instruments. Key point is that adjustment should focus on increasing the value of the assets owned by the groups affected, in-

1426 **Levy, Santiago.** La pobreza extrema en México: una propuesta de política. (*Estud. Econ./México*, 6:1, enero/junio 1991, p. 47–89, tables, graphs)

Analyzes research on poverty indices, nutrition, fertility, incentives, and related topics for the purpose of designing poverty alleviation programs. Measures magnitude and regional composition of extreme poverty. Makes concrete proposal to deal with extreme poverty.

1427 **Levy, Santiago.** Poverty alleviation in Mexico. Washington: Country Dept. II, Latin America and the Caribbean Regional Office, World Bank, 1991. 94 p.: bibl., ill. (Policy, research, and external affairs working papers; WPS 679)

Presents evidence, analyzes economic determinants, discusses policy options, and assesses existing poverty programs. Surveys current government programs to alleviate poverty and offers some suggestions for improvement in concept and implementation.

1428 **Lifschitz, Edgardo.** El complejo automotor en México y América Latina. Azcapotzalco, Mexico: Univ. Autónoma Metropolitana, Unidad Azcapotzalco, División de Ciencias Sociales y Humanidades, 1985. 216 p.: bibl., ill., tables. (Serie Economía. Biblioteca de ciencias sociales y humanidades)

Examines development and expansion of the automotive sector in Mexico. Comparisons are made with Brazil and Argentina. Numerous original tables add to this excellent comparative study.

1429 **López de Alba, Federico.** Evaluación del impacto ambiental: instrumento para el desarrollo. México: IAPEM; ICATI, 1986? 66 p.: bibl., ill.

Develops economic efficiency model and framework for assessing environmental impact. Includes comprehensive discussion of Mexico's evolving environmental protection legislation.

1430 **López Esparza, Víctor Manuel.** Escenario del mercado bursátil mexicano. México: Nacional Financiera, S.N.C., 1992. 113 p.: bibl., ill. (Biblioteca NAFIN; 2)

Excellent survey of Mexico's financial system emphasizes historical evolution of the country's main financial markets. Contains comprehensive examination of the country's stock market.

1431 **López G., Julio.** Contractive adjustment in Mexico, 1982–1989. (*Banca Naz. Lavoro*, 178, p. 293–318, bibl., tables)

During the 1980s the IMF and other international agencies often prescribed an economic policy of austerity for Latin American countries. Since its emphasis lies in the reduction of domestic demand—coupled with drastic changes in relative prices—analysts have labelled the strategy "contractive adjustment." Paper's purpose is to assess the macroeconomic effects of this strategy. Concludes that while the Mexican experience has been less costly and in certain areas more successful than similar efforts in other Latin American countries, macroeconomic inefficiencies reflected in low degrees of utilization of capacities, underemployment, and persistent capital flight cannot be ignored. Also one cannot overlook the strong concentration of income which also took place. Author's main argument is that social costs of this policy were too high and that it is hard to believe that no alternative strategy existed.

1432 **Lorey, David E.** The emergence of the U.S.-Mexican border economy in the twentieth century: an overview of basic trends. (*Nuestra Econ.*, 2:8, mayo/agosto 1991, p. 111–147, appendix, tables)

Focuses on development of basic economic tendencies in the US-Mexican border region during the 20th century. Emphasizes the role that government policies and World War II played in creating a dynamic economy along the border.

1433 **Low, Patrick.** Trade measures and environmental quality: the implications for Mexico's exports. (*in* International trade and the environment. Edited by Patrick Low. Washington: World Bank, 1992, p. 105–120, tables)

Focuses on implications for Mexico's exports of the imposition by the US of a hypothetical "dirty" industry import tax that would equalize expenditures by US and Mexican industries on pollution abatement and control. From his analysis author concludes that an environmental tax of this type would amount to bad environmental policy and bad trade policy. In addition the tax would carry negative consequences for the GATT trading system.

1434 Lustig, Nora. Indices y ordenamientos de pobreza: una aplicación para Mexico. (*Estud. Econ./México*, 6:2, July/Dec. 1991, p. 271–283, bibl.)

Uses Atkinson's and Foster and Shorrosks' approach to generate poverty orderings for different population groups in Mexico. Poverty is measured using three poverty measures: the head-count ratio, the normalized poverty gap, and the Foster *et al.* P3 index. For comparisons undertaken here, the head-count ratio yields an unambiguous ordering.

1435 Lustig, Nora. Mexico, the remaking of an economy. Washington: The Brookings Institution, 1992. xviii, 186 p.: bibl., index.

Seeks to describe and interpret Mexico's economic experience during the 1980s. Particular emphasis is on the shift in development strategy from import-substituting, State-controlled economy to an open economy in which the State's intervention is limited by a new legal and institutional framework. Also of interest is an examination of the social costs of the economic crisis and adjustment during the 1980s. Speculates as to why the process of adjustment and reform in the 1980s may have been easier for Mexico than for other debt-ridden countries.

1436 Lustig, Nora. Mexico's integration strategy with North America. (*in* Strategic options for Latin America in the 1990s. Edited by Colin I. Bradford, Jr. Paris: Development Centre of the Organisation for Economic Co-Operation and Development; Washington: Inter-American Development Bank, 1992, p. 155–179, tables)

Examines policy choices open to Mexico. Concludes that: 1) dislocation costs of unilateral liberalization seem to have been far lower than predicted by the measure's opponents—there have been no massive bankruptcies or layoffs, and the open unemployment rate has been remarkably low and falling; 2) the decision to join GATT has given more credibility to the change in foreign trade regime, and has allowed Mexico to become an active participant in the process of defining new rules of the game which will guide the world trading system; and 3) the pursuit of a free trade agreement with the US and Canada has closed whatever credibility gap remained and, more importantly, it has provided Mexico with a platform to advertise its reforms and attract foreign investment.

1437 Las maquiladoras: ajuste estructural y desarrollo regional. Recopilación de Bernardo González-Aréchiga y Rocío Barajas Escamilla. Coordinación de Töns H. Hilker. Tijuana, Mexico: Colegio de la Frontera Norte; Fundación Friedrich Ebert, 1989. 339 p.: bibl., ill.

Essays on changes now taking place in the country's *maquiladora* industries cover following areas: economic policy, structural change, regional impact of the industry, patterns of *maquiladora* exports, the role of the *maquiladoras* under the General Agreement on Tariffs and Trade, labor conditions, and patterns of employment.

1438 Maquiladoras, su estructura y operación. Delegación Benito Juárez, Mexico: Instituto Mexicano de Ejecutivos de Finanzas; S.l.: Editorial PAC, 1986. 143 p.: ill. (Publicaciones IMEF)

Leading authorities discuss various aspects of Mexico's *maquiladora* system. Emphasizes export potential, imports from the US, and problems of finance.

1439 Marcadent, Philippe. Sociedades de pequeños productores de café: una alternativa de organización en zonas marginadas; dos experiencias, Las Tenerias y Tlapexcatl. México: Laboratorio de Investigación y Desarrollo Regional, 1992? 99 p.: bibl., map.

Survey of production in the region around Veracruz stresses the coffee industry and problems that sector currently faces.

1440 María y Campos, Mauricio de. Reestructuración y desarrollo de la industria automotriz mexicana en los años ochenta: evolución y perspectivas. Santiago: Naciones Unidas, Comisión Económica para América Latina y el Caribe, 1992. 191 p.: ill., maps. (Estudios e informes de la CEPAL, 0256–9795; 83)

Examines Mexican policy toward the automotive industry in the 1980s. Covers government involvement in the industry through its development plans. Assesses implications for the domestic auto industry of developments in other parts of the world.

1441 Martínez Almazán, Raúl. Las relaciones fiscales y financieras intergubernamentales en México: propuesta para un sistema nacional de asignación de recursos financieros a los estados y municipios. México: Instituto Nacional de Desarrollo Municipal, 1981. 308 p.: bibl., ill.

Examines patterns of inter-governmental finance in Mexico. Numerous comparisons are made with practices in other developing countries. From this analysis, author develops a number of recommendations for improving many of the country's procedures and methods of collection and disbursement.

1442 Marúm Espinosa, Elia. Empresa pública e intervencionismo estatal en México: un análisis alternativo. Guadalajara, México: Editorial Univ. de Guadalajara, 1992. 244 p.: bibl., ill. (Col. Fin de milenio)

Comprehensive examination of the evolution of public enterprises in the economy contains extensive data and quantitative estimates of major trends.

1443 Massolo, Alejandra et al. Procesos rurales y urbanos en el México actual. México: Univ. Autónoma Metropolitana, Unidad Iztapalapa, División de Ciencias Sociales y Humanidades, Depto. de Sociología, 1991. 219 p.: bibl.

Essays focus on various urban problems, examining urban links to the rural sector, as well as implementation of the government's urban policies.

1444 Memoria: energía, medio ambiente y desarrollo sustentable. México: ENEP Acatlán, UNAM; Friedrich Ebert Stiftung, 1992. 189 p.: bibl., ill.

Conference held in Mexico City on May 6–8, 1992. Proceedings cover all aspects of energy consumption, environmental policy, and policies needed to assure sustainable development in Mexico.

1445 Mendoza Fernández, María Teresa et al. Presencia y tendencia de la industria maquiladora en Yucatán. Mérida, México: Univ. Autónoma de Yucatán, Facultad de Contaduría y Administración, Unidad de Posgrado e Investigación, 1990. 128 p.: bibl., ill.

Traces historical development of the offshore assembly industry in the Yucatán. Detailed examination of the industry's structure, together with numerous tables on the production and employment patterns of the *maquiladora* industries.

El Mercado de Valores: **la política económica de Mexico, 1946–1994.** See item 37.

1446 Mesa-Lago, Carmelo. Social security and the informal sector in Latin America: the case of Mexico. (*in* Work without protections: case studies of the informal sector in developing countries. Edited by Gregory K. Schoepfle and Jorge F. Pérez-López. Washington: US Dept of Labor, Bureau of International Labor Affairs, 1993, p. 41–97, bibl., tables)

After defining both informality and social security and outlining her research methodology, author summarizes the magnitude and trends of the informal sector in Latin America and Caribbean countries and its coverage by social security. The rest of the chapter deals with Mexico: 1) the size, trends, and characteristics of its informal sector; 2) social security coverage of that sector and reasons for its low level of social protection; 3) the financial viability of the formal social security and public health systems to incorporate informal workers; 4) alternative non-conventional means of social protection; and 5) public policy recommendations for Mexico.

1447 The Mexican petroleum industry in the twentieth century. Edited by Jonathan C. Brown and Alan Knight. Austin: Univ. of Texas Press, 1992. 315 p.: bibl., ill., index, map. (Symposia on Latin America series)

Papers presented at a conference organized at the Univ. of Texas at Austin in Feb. 1988 mainly focus on history of the Mexican oil industry. Topics include origins of workers' unions, foreign investment, expropriation, growth of PEMEX, the oil industry, and Mexico's relations with industrial countries.

1448 The Mexican-U.S. border region and the Free Trade Agreement. Edited by Paul Ganster and Eugenio O. Valenciano. San Diego, CA: Institute for Regional Studies of the Californias, San Diego State Univ., 1992. 117 p.: bibl., ill.

Proceedings of a workshop on the impact of NAFTA. Basic premise of the conference organizers was that the free trade area will have very regional and sectoral impacts. In order to provide more concrete information and options to policymakers, the workshop sought to better clarify what impacts the free trade agreement will have on the various regions along the border.

1449 Mexico. Poder Ejecutivo Federal. Plan Nacional de Desarrollo, 1989–1994. Informe de ejecución. México: Secretaría de Programación y Presupuesto, 1991. 2 v.

Annual report on the performance under the National Plan between 1989–94. Outlines basic achievements during 1990 towards the Plan's goals. Lists official actions to deal with some of the country's current economic difficulties.

1450 Mexico. (*in* National human settlements: institutional arrangements; selected case studies. Nairobi: United Nations Centre for Human Settlements (Habitat), 1987, p. 188–261, bibl., tables)

Human Settlements Planning is a relatively recent phenomenon in Mexico. Strictly understood, it did not begin until 1976, although a number of government measures since the 1930s tried to affect the country's spatial structure and urban standard of living. This chapter's purpose is to list and briefly describe the major, somewhat scattered, actions that preceded the inauguration of a full-scale policy apparatus for dealing with the problems of human settlements.

1451 México: auge, crisis y ajuste. v. 2. Recopilación de Carlos Bazdresch et al. México: Fondo de Cultura Económica, 1993. 1 v.: bibl., ill. (Lecturas; 73)

Vol. 2 in series on Mexico's economic crisis, this outstanding volume is largely devoted to the debt crisis and the government's attempt to restore stability. First essay is by new president, Ernesto Zedillo Ponce de León. Zedillo examines balance of payments from 1973 to the debt crisis of 1982. Other essays are by former Minister of Finance Pedro Aspe, José Angel Gurria, and Clark Reynolds.

1452 México en la década de los ochenta: la modernización en cifras. Coordinación de Rosa Albina Garavito Elías y Augusto Bolívar Espinoza. Azcapotzalco, Mexico: Univ. Autónoma Metropolitana-Azcapotzalco, División de Ciencias Sociales y Humanidades, 1990. 466 p.: ill.

Focuses on major economic developments and trends from 1982–88. As the title suggests, this study is largely an extensive compilation of data on all facets of the economy. Very useful for anyone wishing to see a broad overview of developments in the 1980s.

1453 México-Estados Unidos: la interacción macroeconómica. Recopilación de Claudia Schatan, Cassio Luiselli, y Darryl McLeod. México: Centro de Investigación y Docencia Económica, 1989. 352 p.: bibl., ill.

Papers by leading authorities examine all facets of economic linkages between Mexico and the US. Particular attention is focused on employment patterns associated with trade, economic development of the borderlands, trade in manufactures, and the macroeconomic environment, and linkages in which trade between the two countries takes place.

1454 México: informe sobre la crisis, 1982–1986. Coordinación de Carlos Tello, Enrique González Tiburcio y Francisco Báez. México: Univ. Nacional Autónoma de México, 1989. 536 p.: bibl., ill. (Biblioteca México, actualidad y perspectivas)

Series of papers by leading authorities examines various facets of Mexico's economic crisis of 1982–86. Topics include employment problems, changes in income distribution, and the pattern of government expenditures.

1455 México: perspectivas de una economía abierta. México: Colegio Nacional de Economistas: M.A. Porrúa, 1993. 449 p.: bibl.

Series of essays on recent developments in the international economy.

1456 Mexico: the strategy to achieve sustained economic growth. Edited by Claudio Loser and Eliot Kalter. Washington: International Monetary Fund, 1992. 91 p.: bibl., ill. (Occasional paper, 0251–6365; 99)

Various papers prepared by International Monetary Fund staff members who have participated in the cooperation between Mexico and the IMF since the 1982 debt crisis began. Book does not seek to present a complete description of policies and developments. Rather, it reviews in detail some central aspects of the Mexican experience, particularly over the last several years. Includes discussions on macroeconomic policies; an analysis of the evolution of structural reforms in key areas such as in the financial system, trade and foreign investment; and the changing nature of external private market financing.

1457 México y el Tratado Trilateral de Libre Comercio: impacto sectorial. Recopilación de Eduardo Andere y Georgina Kessel. México: ITAM; McGraw-Hill, 1992. 384 p.: bibl.

Leading authorities examine wide-ranging issues associated with Mexico's increased integration with Canada and the US. Topics include the automotive market, macroeconomic impacts likely from free trade, the petroleum sector, financial services, regulation of industry, the transport sector, and lingering trade problems with the US.

1458 Morici, Peter. Grasping the benefits of NAFTA. (*Curr. Hist.*, 92:571, Feb. 1993, p. 49–54)

Examines implications of the new NAFTA Treaty and concludes that free trade in the region supports economic reform in Mexico, which in turn undermines the old corporatist system. Diffusion of economic decision-making loosens the PRI's grip on political control, adding pressures for genuine multiparty democracy. Overall, NAFTA will serve US global economic objectives as well as policy goals of supporting economic and political modernization in Mexico.

1459 Morton, Colleen Shores and Joseph A. Greenwald. An analysis of the initialed text of the North American Free Trade Agreement. Washington: U.S. Council of the Mexico-U.S. Business Committee, 1992. 68 p.

Analysis in this report was undertaken as a service to members of the US Council of the Mexico-US Business Committee. It is based on the Oct. 7 initialed text of the Agreement, official advisory committee reports, and conversations with negotiators and US Council members familiar with the negotiations. Concludes that the NAFTA agreement will strengthen the economies of the US, Mexico, and Canada and by drawing on the unique resources of each will ultimately strengthen the regional economy. In addition, the NAFTA negotiations complement the work underway in the GATT Uruguay Round.

1460 Mungaray Lagarda, Alejandro. Crisis, automatización y maquiladoras. Mexicali, Mexico: Univ. Autónoma de Baja California, 1990. 185 p.: bibl., ill.

Examines structural changes taking place in Mexico's *maquiladoras*. Focuses largely on technological change and restructuring. Draws a number of conclusions concerning impact on employment brought about by these changes.

1461 Los municipios de las fronteras de México. v. 1, Economía y trabajo [de] Margarita Nolasco. v. 2, El medio ambiente [de] María Luisa Acevedo. v. 3, Población, cultura y sociedad [de] Margarita Nolasco, Virginia Molina y Miguel A. Bravo. México: Centro de Ecodesarrollo; Centro Nacional de Desarrollo Municipal, 1990. 3 v.: bibl., ill., maps.

General examination of economic activity on Mexico's borders contains many useful tables depicting the main patterns and composition of activity along the US-Mexican border.

1462 Murcio, F. Javier. Toward free trade: how far south can we go? (*in* Conference on the Southwest Economy, *Dallas, Tex., 1991.* Free trade within North America: expanding trade for prosperity. Boston: Kluwer Academic Publishers, 1993, p. 73–101, appendices, graphs, tables)

Assesses how Mexico and Latin America fit into the worldwide scheme of emerging trading blocs. Although largely descriptive rather than analytical, work does include a number of informative graphs and charts.

1463 Narro R., José. La seguridad social mexicana en los albores del siglo XXI. México: Fondo de Cultura Económica, 1993. 158 p.: bibl., ill. (Una Visión de la modernización de México)

Examines progress towards improved quality of life in Mexico. Assesses factors responsible for lagging progress in this area and governmental policies to deal with the problem. Outlines a new approach towards the public provision of social services.

1464 North American free trade: assessing the impact. Edited by Nora Lustig, Barry P. Bosworth, and Robert Z. Lawrence. Washington: Brookings Institution, 1992. 274 p.: bibl., ill., index.

Papers presented at a Brookings Institution conference in April 1992. Studies focus on potential costs and benefits of NAFTA, with each treating a particular subject: labor markets, industry, agriculture, economy-wide modeling, the effect on the rest of the world, and non-trade issues. These papers share a consensus in that they anticipate the direct economic effects of NAFTA will be small for both Mexico and the US.

1465 North American Free Trade Agreement: U.S.-Mexican trade and investment data: report to the Honorable Richard A. Gephardt, Majority Leader, and to the Honorable Sander Levin, House of Representatives. Washington: US General Accounting Office; Gaithersburg, MD: US General Accounting Office, 1992. 119 p.: bibl.

This report to the US Congress provides information concerning: 1) Mexico's trade flows with the US, and its worldwide and selected bilateral trade flows for 1980–91; 2) foreign direct investment between Mexico and the US and selected countries' direct investment in Mexico between 1980–91; 3) *maquiladora* operations in Mexico between 1980–91; and 4) the top 50 companies, importers-exporters, and *maquiladoras* operating in Mexico in 1990. Finds that the US and Mexican government trade statistics differ, even after adjustment for *maquiladora* trade, because of variations in concepts and definitions used by the two countries. As a result of this report, the US and Mexico are currently engaged in a project to reconcile the differences in their bilateral trade data.

1466 La nueva era de la industria automotriz en México: cambio tecnológico, organizacional y en las estructuras de control. Coordinación de Jorge Carrillo V. Tijuana, Mexico: Colegio de la Frontera Norte, 1990. 364 p.: bibl., ill., tables.

Assesses expansion of the *maquiladoras*, particularly in the automotive sector. Topics include modernization of the sector, technology transfer, and the recent economic structure and performance of the sector. Many useful tables.

1467 El nuevo Estado mexicano. v. 1, Estado y economía. Coodinación de Jorge Alonso, Alberto Aziz Nassif, y Jaime Tamayo. México: Univ. de Guadalajara; Nueva Imagen, CIESAS, 1992. 1 v.: bibl., ill.

Vol. 1 of a major study examines Mexico's changing position in a world economy. Focuses on the international monetary system. Emphasizes evolution of Mexico's relationship with the US.

1468 Núñez de la Peña, Francisco J. La canasta de los números: información y análisis macroeconómicos. Guadalajara, Mexico: ITESO, 1992. 199 p.: bibl., ill.

Brief macroeconomic overview of the country. Critiques various macro measures of activity and discusses their strong and weak points. Includes considerable amount of data on the main macroeconomic aggregates.

OECD Statistical Compendium. 1994– . See item **39**.

1469 Orme, William A., Jr. The sunbelt moves south. (*NACLA*, 24:6, May 1991, p. 10–38, graph, photos)

Surveys arguments for and against NAFTA. Concludes in the long term even a flawed free-trade pact will set in motion economic and political forces that could potentially enrich both societies and set the stage for a fairer continental partnership.

1470 Orozco Alvarado, Javier et al. Economía, agroindustria y política agraria en Jalisco: cuatro ensayos. Zapopan, Mexico: Colegio de Jalisco, 1992. 172 p., 1 folded p. of plates: bibl., map.

Various authors examine development of agricultural and agrobusiness in the state of Jalisco. Numerous tables contain useful and hard-to-find information on the structure of the agroindustrial companies examined.

1471 Ortiz Cruz, Etelberto. Competencia y crisis en la economía mexicana. México: Univ. Autónoma Metropolitana-Xochimilco: Siglo Veintiuno Editores, 1994. 250 p.: bibl., ill. (Economía y demografía)

Marxist analysis of the country's major economic problems. Areas discussed include income distribution, labor conditions, economic crisis of the 1980s, industrial policy, and the country's integration into the world economy.

1472 Ortiz Martínez, Guillermo. La reforma financiera y la desincorporación bancaria. México: Fondo de Cultura Económica, 1994. 363 p.: bibl., ill. (Una visión de la modernización de México)

Excellent examination of the country's

financial institutions and the regulatory system that controls their operations. Focuses on programs proposed by President Salinas to modernize the nation's banking system. Provides good description of the government's new approach towards the financial sector and an assessment of impacts likely to occur if the reforms are completed. Pedro Aspe provides an introduction to the study.

1473 **Pacheco Méndez, Teresa** *et al.* Recursos y desarrollo de Chiapas hasta 1990. Tuxtla Gutiérrez, Mexico?: s.n., 1992. 169 p.: bibl.

General survey of the environment and resources of the state of Chiapas emphasizes not only the region's natural endowments, but also its man-made resources: infrastructure, educational system, financial sector, and other institutions aiding in the region's economic progress.

1474 **Pérez López, Emma Paulina; Orem Peralta Ramírez**, and **José María Martínez.** De mineros a ganaderos: un caso de incorporación campesina al desarrollo regional: La Colorada, Sonora, 1886–1984. Hermosillo, Mexico: Centro de Investigación en Alimentación y Desarrollo, 1986. 74 leaves: bibl., ill. (Cuaderno de trabajo; 3)

Essay covers economic growth and demographic evolution of the Colorada region in Sonora state. Numerous maps complement this excellent presentation of the state's history.

1475 **Perzabal, Carlos.** Acumulación de capital e industrialización compleja en México. México: CIDE; Siglo Veintiuno Editores, 1988. 168 p.: bibl. (Economía y demografía)

Develops highly sophisticated input-output model of the economy for the purposes of assessing sectoral linkages and the patterns of capital formation.

Piñeda Bañuelos, Gilberto J. and **Alfredo Madrigal Carmona.** Las maquiladoras en México y el proyecto maquilador en Baja California Sur. See item **4536.**

1476 **La población en el desarrollo contemporáneo de México.** Recopilación de Francisco Alba y Gustavo Cabrera. México: Colegio de México, Centro de Estudios Demográficos y de Desarrollo Urbano, 1994. 405 p.: bibl., ill.

Essays by leading authorities on Mexican demographics. Topics include patterns of birth and death rates, urbanization, internal migration, profiles of migrants to the US, structure of the workforce, educational attainment, and the government's demographic policies.

1477 **Productividad: distintas experiencias.** Coordinación de Enrique de la Garza y Carlos García. México: Univ. Autónoma Metropolitana, Unidad Iztapalapa: Fundación Friederich Ebert, 1993. 235 p.: bibl.

Leading authorities examine several important aspects of labor productivity in Mexico. Topics include productivity trends in private, autonomous public agencies and the government. Attention is given to the manner in which the regulative environment affects productivity. Most interesting are several case studies including that of the Volkswagen plant in Puebla.

1478 **Puga, Cristina.** México: empresarios y poder. México: Facultad de Ciencias Políticas y Sociales, UNAM; M.A. Porrúa, 1993. 207 p.: bibl. (Las Ciencias sociales)

Survey of the country's entrepreneurial class covers topics including family connections, philosophies, professional organization, links to politics, relationship to the Revolution and activities in industry. Speculates as to the future of entrepreneurship in the country.

1479 **Quintana Roo: los retos del fin de siglo.** Edición de Alfredo César Dachary, Daniel Navarro López y Stella M. Arnaiz Burne. Chetumal, Mexico: Centro de Investigaciones de Quintana Roo, 1992. 268 p.: bibl., maps.

Essays examine economic and environmental conditions in the state of Quintana Roo. Topics include environmental impact of tourism, industry, and agriculture, as well as the state's educational system.

1480 **Quiroga Garza, Julián; Francisco Blanco Celaya;** and **Irma Laura Murillo Lozoya.** La tasa natural de desempleo: conceptos, antecedentes y el caso del area metropolitana de Monterrey. Monterrey, Mexico: Univ. Autónoma de Nuevo León, Facultad de Economía, Centro de Investigaciones Económicas, 1990. 109 p.: bibl., ill.

Drawing on the concept of the natural rate of unemployment, this study assesses various facets of the unemployment problem in Monterrey. Contains a number of original sources of data on the city's demographics, labor markets, and pattern of employment.

1481 **Ramírez, Miguel D.** Stabilization and trade reform in Mexico, 1983–1989. (*J. Dev. Areas*, 27:2, Jan. 1993, p. 173–190, graph, tables)

Assesses nature and impact of Mexico's 1980s trade reform strategy in the context of severe macroeconomic stabilization. Shows that during those years, the Mexican government systematically opened the domestic market to world prices, especially after 1984 when import licensing was greatly reduced. Finds however that trade liberalization program may be seriously undermined by the severe stabilization measures implemented by the Mexican government. These deflationary policies have had a substantially negative effect on the country's rate of capital accumulation, particularly in infrastructure.

1482 **Ramírez Brun, J. Ricardo.** La política económica en México, 1982–1988: transición de la ortodoxia a la heterodoxia. México: Univ. Nacional Autónoma de México, 1989. 141 p.: bibl.

Focuses on economic policy-making during 1982–88. Provides an in-depth analysis of the change in the government's economic philosophy, and projects the likely direction of policy-making through 1994.

1483 **Ramírez García, Agustín.** Nueve décadas del comercio exterior de México: hacia el Tratado de Libre Comercio. México: Cárdenas Editor y Distribuidor, 1993. 484 p.: bibl., ill.

Overview of the evolution of the country's commercial trade. Period under examination begins with 1914–33 and ends with a discussion of the trade orientation of the Décimo Plan Sexenal (1988–94). Includes excellent assessment of implications for Mexico of the North American Free Trade Agreement.

1484 **Ramos Tercero, Raúl M.** Política monetaria óptima bajo tipo de cambio fijo: una evaluación empírica del caso mexicano. México: Centro de Estudios Monetarios Latinoamericanos, 1993. 45 p.: bibl., ill. (Serie Estudios)

Develops sophisticated macroeconomic model of the economy to assess the alternative policy options. Uses model to identify what might be termed an optimal monetary policy.

1485 **Relaciones industriales y productividad en el norte de México: tendencias y problemas.** Coordinación de Alejandro Covarrubias V. y Blanca Lara E. México: Fundación Friedrich Ebert, 1993. 280 p.: bibl., ill.

Leading authorities assess problems of manufacturing in the northern regions of Mexico. Topics include competitiveness, productivity, comparison of Mexican and US plants, structure of production, and technology transfer. There are two interesting essays on industrial development in Sonora and Tamaulipas.

1486 **Revenga, Ana; Michelle Riboud;** and **Hong Tan.** The impact of Mexico's retraining program on employment and wages. Washington: World Bank, 1992. 40 p.: bibl., ill. (Policy research working papers; WPS 1013)

Evaluates impact of labor participation in the PROBECAT (Programa de Becas de Capacitación para Trabajadores) retraining program initiated in 1984. Addresses four main question: 1) what is the impact of training on post-training employment of trainees?; 2) does training increase the speed with which workers move from being unemployed to employed?; 3) conditional on their finding employment, what effect does training have on monthly earnings, hours worked and hourly wages of trainees?; and 4) do the monetary benefits from program participation outweigh the costs of providing training?

1487 **Revenga, Ana** and **Michelle Riboud.** Unemployment in Mexico: its characteristics and determinants. Washington: World Bank, 1993. 29 p.: tables. (Policy Research Working Paper; 1230)

Study examines causes of unemployment in Mexico. Main issues examined are the characteristics of the unemployed. Finds that the overall structure of unemployment is broadly similar to that observed for other countries. Unemployment rates are highest for the young—particularly for those 16 to 25 years of age—and have been consistently higher among women than among men. Regarding education, the highest unemployment rates for males correspond to those with either incomplete or complete lower or secondary schooling (7-9 years of school). For females, the highest rates are found among those with either complete lower secondary or higher secondary schooling (9-12 years of school). This pattern of higher unemployment rates for secondary school graduates differs somewhat from that observed in other countries where unemployment appears to be more prevalent among the less educated.

1488 Reynolds, Clark Winton. Power, value, and distribution in the NAFTA. (*in* Political and economic liberalization in Mexico: at a critical juncture? Boulder, Colo.: L. Rienner Publishers, 1993, p. 69-92, graphs)

Argues that the present wave of interest in integration and policies on it differ in important ways from those of the past. In the 1950s-60s free trade areas and common markets were designed to extend market reserves and market shelters from individual countries to groups of countries. The process was analyzed in terms of trade creation and diversion, with regional integration treated as a second-best strategy in an environment not yet ready for global free trade. The new pattern of regional integration manifests a different approach, one heavily influenced by the expected benefits (over costs) of development in the direction of dynamic comparative advantage through trade and investment liberalization.

1489 Rich, Jan Gilbreath. Planning for the border's future: the Mexico-U.S. integrated border environmental plan. Austin, TX: U.S.-Mexican Policy Studies Program, LBJ School of Public Affairs, Univ. of Texas at Austin, 1992. 48 p.: bibl. (U.S.-Mexican occasional paper; 1)

Analysis begins with brief history of economic trends that led to creation of the Integrated Environmental Plan for the Mexico-US Border Area. Second section analyzes the Mexican and US border community response to the first publicly released draft of this plan in Aug. 1991. Final section discusses revised version of the plan and whether these modifications adequately address concerns of border communities affected by the plan. Concludes that the revised plan still has a number of weaknesses. In particular, author claims it lacks specific implementation plans for a number of environmental problems that border communities testified about in a Congressional hearing. In addition, work plans outlined in the areas of education, pollution monitoring, data gathering, geographic information surveys, technology transfers, and hazardous waste tracking are more ambitious than in the past, but lack sufficient financing to carry them out.

1490 Rivera Ríos, Miguel Angel. El nuevo capitalismo mexicano: el proceso de reestructuración, 1983-1989. México: Ediciones Era, 1992. 223 p.: bibl., ill. (Col. Problemas de México)

Examines economic changes that have taken place since the onset of the debt crisis of the early 1980s. Emphasizes structural changes in industry and commerce, together with the changing relationship between the State and the private sector.

1491 Roberts, Bryan R. The dynamics of informal employment in Mexico. (*in* Work without protections: case studies of the informal sector in developing countries. Edited by Gregory K. Schoepfle and Jorge F. Pérez-López. Washington: US Dept. of Labor, Bureau of International Labor Affairs, 1993, p. 101-125, bibl., tables)

Provides overview of current trends in the Mexican labor market, paying particular attention to labor conditions and the issue of the informal sector. Finds that the basic issue for Mexico is low pay, not labor-market dualism. Heterogeneity of the labor market is based on most households' need to supplement the income of the main breadwinner.

1492 Rogers, John H. The currency substitution hypothesis and relative money demand in Mexico and Canada. (*J. Money Credit Bank.*, 24:3, Aug. 1992, p. 300-318, bibl., graphs, tables)

Estimates models for the demand for US dollars relative to domestic currency for both Mexico and Canada. In the Mexican case the paper finds a negative and significant correlation between the ratio of Mexdollars to pesos and the expanded rate of depreciation of the peso. In the Canadian model, the relationship is positive and is inconsistent with the conventional currency substitution hypotheses. Considering the policies directed at the Mexdollar accounts by the Mexican government, the timing and magnitude of the capital flight from Mexico, and the absence of such shocks in Canada, concludes that holding Mexdollars is associated with convertibility risk. That is, when the central bank runs low on foreign reserves, all parties realize that it may not be possible to convert all Mexdollars into real dollars when desired.

1493 Romero, José. La teoría de la unión aduanera y su relevancia para México ante el Acuerdo de Libre Comercio con Estados Unidos y Canadá. (*Estud. Econ./México*, 6:2, julio/dic. 1991, p. 231–270, tables, graph)

Reviews aspects of Customs Unions theory that are relevant to Mexico given membership in the North American Free Trade Agreement. The following themes are discussed: effects of trade creation and diversion; production effects; terms of trade effects; common external tariffs; economies of scale; unilateral tariff reductions; foreign firms in the formation of customs unions; and economic development.

1494 Rosenzweig, Fernando. El desarrollo económico de México, 1800–1910. Zinacantepec, Mexico: Colegio Mexiquense; ITAM, 1989. 262 p.: bibl.

Chiefly a series of papers on the country's economic history published by the author between 1963–72. Topics include agriculture, mining, manufacturing, rural development, and trade.

1495 Rubio, Luis. Mexico: debt and reform. (*in* In the shadow of the debt: emerging issues in Latin America. New York: Twentieth Century Fund Press, 1992, p. 111–128, tables)

Argues that the success or failure of Mexico's transformation attempts will have implications far beyond its borders. Reforming a society frozen in time entails not only liberalization and deregulation but also direct hits against traditional vested interests. The transition from a closed society and economy into a modern, responsible one implies changes in structures, habits, traditions, and political culture. Success in transforming the society and polity along these lines will mean that Mexico has the ability not only to participate in international markets, but to become a modern democracy as well.

1496 Ruiz Durán, Clemente and **Carlos Zubirán Schadtler.** Cambios en la estructura industrial y el papel de las micro, pequeñas y medianas empresas en México. México: Nacional Financiera, 1992. 261 p.: ill., map. (Biblioteca de la micro, pequeña y mediana empresa; 2)

Largely a compilation of data on small industrial firms in Mexico. Much of this data—output, input, and the like—is original and not found in other sources.

1497 Salazar Sánchez, Héctor. Dinámica de crecimiento de ciudades intermedias en México: los casos de León, San Luis Potosí y Torreón, 1970–1980. México: El Colegio de México, Centro de Estudios Demográficos y de Desarrollo Urbano; Consejo Nacional de Población, 1984. 110 p.: bibl., ill.

Examines demography and patterns of population and employment in León, San Luis Potosí, and Torreón. Also assesses the ability of these municipalities to finance public services and other expenditures associated with rapid population growth.

1498 Salgado Vega, Jesús. Estado de México: evolución socioeconómica, 1989–1993. Toluca, Mexico: Univ. Autónoma del Estado de México, Facultad de Economía, 1993. 322 p.: bibl., ill.

Examines major economic and demographic development of the state of Mexico from 1989–93. Contains extensive and difficult-to-find data on the economy, population, and trade of the state.

1499 Sánchez, Manuel *et al.* The privatization process in Mexico: five case studies. (*in* Privatization in Latin America. Edited by Manuel Sánchez and Rossana Corona. Washington: Inter-American Development Bank, 1993, p. 101–199, bibl., graphs, tables)

Presents empirical analysis of five case studies that fully exemplify the privatization experience in Mexico, specific lessons drawn from them, and recommendations for future progress in the region. Concludes that there is a need for coordination between deregulation and privatization policies. In order to obtain the highest possible price, a sense of security must exist regarding various factors that affect the industry to which the company or companies to be privatized belong. In particular it is recommended that deregulation occur prior to initiating the sales process. On the other hand, when the privatization process involves a group of companies with different operating characteristics and levels of efficiency, it is possible to sell those companies in which there is no interest by creating balanced packages. In this way not only are subsidies to such companies substantially reduced but jobs are preserved as well.

1500 Schatan, Claudia. Out of the crisis: Mexico. (*in* The developing countries in world trade: policies and bargaining strategies. Edited by Diana Tussie and David Glover. Boulder, Colo.: Lynne Rienner Publishers, 1993, p. 79–98, tables)

Assesses Mexico's bilateral bargaining power and bargaining strategies as it attempted to improve its position in the US and world markets in the 1980s. Argues that the Mexican experience in trade bargaining in 1980–90 seems to demonstrate that bilateral negotatiations may be more efficient than multilateral ones for quick trade expansion and the removal of tariff barriers by the trading partner.

1501 Schteingart, Martha. The role of the informal sector in providing urban employment and housing in Mexico. (*in* Urban management: policies and innovations in developing countries. Edited by G. Shabbir Cheema with the assistance of Sandra E. Ward. Westport, Conn.: Praeger, 1993, p. 285–299, tables)

The employment situation has deteriorated in large urban centers in Latin America and in Mexico in particular, the unemployment rate has risen, the percentage of workers in the so-called informal sector has increased, the formal sector has lost its capacity to absorb workers, especially in the industrial sector, and the few advances made up to 1980 were nullified by the subsequent chronic economic crisis. A large sector of the population has been involved in illegal housing production, which has meant that the families have had to make great sacrifices and waste a tremendous amount of energy to obtain, at the end, a rather expensive and ramshackle home.

Sector agropecuario. See item **43**.

1502 La seguridad social y el estado moderno. Recopilación de José Narro Robles y Javier Moctezuma Barragán. México: Instituto Mexicano del Seguro Social; Fondo de Cultura Económica; Instituto de Seguridad y Servicios Sociales de los Trabajadores del Estado, 1992. 503 p.: bibl., ill. (Estructura económica y social de México—los noventa)

Essays compare Mexican social services with similar programs throughout the Latin American region. Also includes an extensive discussion of labor conditions and government legislation to improve the workplace environment.

1503 Seminario sobre el Acuerdo de Libre Comercio y su Impacto en la Agricultura, *Culiacán, Mexico, 1991.* Memoria. Coordinación técnica de José Luis López Duarte e Eduardo Zárate Márquez. Culiacán, Mexico?: Congreso Agrario Permanente de Sinaloa, 1991. 270 p.: ill.

Proceedings of conference held in Culiacán, Mexico on Feb. 26–28, 1991 attempt to anticipate effects of free trade on the agricultural sector. Divided into five sections: 1) views on NAFTA from the perspective of Mexico and Sinaloa; 2) NAFTA's agricultural provisions; 3) education and technological development within NAFTA framework; 4) export agriculture and workforce under NAFTA; and 5) NAFTA's impacts on the ejido system.

1504 Serrato, Marcela *et al.* Las relaciones económicas entre México y Japón: influencia del desarrollo petrolero mexicano. Coordinación de Miguel S. Wionczek y Miyohei Shinohara. México: Colegio de México, Programa de Energéticos, 1982. 246 p.: bibl., ill.

Examines impact of the petroleum sector on the economy. Focuses on the country's trade relations with Japan, especially Mexican export of hydrocarbons. Papers are written from both Mexican and Japanese perspectives.

1505 Snodgrass, B. Warren. Free trade and Mexican steel. (*Hemisphere/Miami*, 4:2, Winter/Spring 1992, p. 8–10, graphs)

Argues that Mexico's steel industry is ready to profit from growth in both foreign and national markets, and it is unlikely that a free trade agreement will hamper its ability to do so. There will certainly be competition from US producers, but not on a scale that spells doom for Mexico. Concludes that the future of regional free trade will be one in which the US and Mexican steel industries concentrate on their domestic clients while each exports to the other nation those goods in which they enjoy a comparative advantage.

1506 Soberanes Reyes, José Luis. La reforma urbana. México: Fondo de Cultura Económica, 1993. 398 p.: bibl. (Una Visión de la modernización de México)

Traces historical development of cities in Mexico. Assesses migration and the country's urbanization problems. Study contains large section assessing the country's efforts at urban planning. Makes a number of constructive recommendations to improve the country's urban planning methods and process.

1507 **Solís Soberón, Fernando.** Sobre la relevancia de quién paga el impuesto. (*Estud. Econ./México*, 6:1, enero/junio 1991, p. 111–123, bibl.)

Presents model of international trade with imperfect competition to show that in the absence of price discrimination it does indeed matter which economic agent pays the tariff. Analysis indicates that effects of trade protection (or liberalization) are not uniform across industries, and that for some industries the effects could in fact be negligible.

1508 **Sonora ante el Tratado de Libre Comercio.** Recopilación de Alejandro Covarrubias V. y José Luis Moreno V. México: El Colegio de Sonora, Fundación Friedrich Ebert, 1991. 179 p.: bibl., ill.

Papers presented at the Symposium "Sonora before the TLC," organized by the Fundación Friedrich Ebert and the Colegio de Sonora and held April 18–19, 1991. Authors examine implications of pending economic integration with the US and Canada (NAFTA). Topics include the impact of free trade on the *maquiladoras*, primary sector, and services.

1509 **Tabasco: realidad y perspectivas.** v. 2, Economía y desarrollo. México: Gobierno del Estado de Tabasco; M.A. Porrúa, 1993. 1 v.

Vol. 2 of three-volume work covering the state of Tabasco. Leading authorities contribute essays on the state's historical evolution, emphasizing sectoral development, urbanization, public finances, and trade.

1510 **Taddei Bringas, Cristina** and **Jesús Robles Parra.** La inversión japonesa en el norte de México: la industria maquiladora de exportación. Hermosillo, Mexico: Desarrollo Regional, Centro de Investigación en Alimentación y Desarrollo, 1992. 82 p.: bibl., ill., maps. (Cuaderno de trabajo; 5)

Examines a number of trade issues involving Japanese foreign investment in the country's *maquiladora* industries. Compiles an extensive set of original data summarizing Japanese investment patterns and the impact this capital formation has had on the country's composition and patterns of trade.

1511 **Téllez Kuenzler, Luis.** La modernización del sector agropecuario y forestal. México: Fondo de Cultura Económica, 1994. 307 p.: bibl., ill. (Una Visión de la modernización de México)

Study provides detailed examination of major problems confronting the agricultural and forestry industries. Focuses on agricultural production and distribution, trade problems with the US, the extent of rural poverty, technological change, and land reform.

1512 **Tello, Carlos.** Combatting poverty in Mexico. (*in* Social responses to Mexico's economic crisis of the 1980s. Edited by Mercedes González de la Rocha and Agustín Escobar Latapí. San Diego, Calif.: Center for U.S.-Mexican Studies, Univ. of California, San Diego, 1991, p. 57–65)

Argues that the Mexican government has failed to pull millions of Mexicans out of abject poverty, despite economic growth as well as policies, institutions, and social programs intended to benefit workers and peasants. Furthermore, it did not reduce inequalities in the distribution of income or lessen the unequal distribution by region of the benefits of progress. It also provoked a significant decline in Mexico's heretofore abundant natural resources, one of the basic ingredients of development.

1513 **Terríquez, Ernesto** *et al.* Colima al final del segundo milenio. Coordinación de Blanca Estela Gutiérrez Grageda. Colima, Mexico: Gobierno del Estado de Colima: Univ. de Colima, 1992. 318 p.: bibl., ill.

Excellent set of essays examines economic conditions, politics, and government of the state of Colima. Emphasizes the development of tourism, the agricultural sector, the state's *maquiladora* industries, urbanization, and patterns of migration.

1514 **Thompson, Gary D.** and **Philip L. Martin.** The potential effects of labor-intensive agriculture in Mexico on U.S.-Mexico migration. (*in* The effects of receiving country policies on migration flows. Boulder, Colo.: Westview Press, 1991, p. 103–136, appendix, graphs, tables)

Examines impact of trade liberalization in labor-intensive industries as a way to diminish undocumented migration that results from the pull of US jobs and the push of unemployment in Mexico. From a detailed analysis concludes that: 1) trade liberalization per se does not guarantee immediate increases in foreign production of commodities for export to the US; 2) trade-induced increases in labor requirements must be analyzed in the context of regional labor markets; and 3) the lack of jobs in countries of origin and an increased difficulty in finding jobs in the US are only two of the factors effecting levels of illegal immigration to the US. The primary economic incentive for illegal migration remains the gap between Mexican and US wages.

1515 **TLC: impactos en la frontera norte.** Coordinación de Alejandro Dávila Flores. México: Univ. Nacional Autónoma de México, Facultad de Economía, 1993. 309 p.: bibl., ill., maps. (Libros de Investigación económica)

Proceedings of first conference on "TLC: Impactos en la Frontera Norte," was held in Saltillo, Coahuila, May 1992. Most papers focus on some aspect of the impact NAFTA will have on the border states of Mexico. Topics include industrial development and growth, the future of the *maquiladoras*, industrial concentration in Nuevo León, migration, and urbanization.

1516 **Transborder data flows and Mexico: a technical paper.** New York: United Nations Centre on Transnational Corporations, 1991. 194 p.: bibl., ill., maps.

Assesses various aspects of Mexico's telecommunications services and infrastructure, main telecommunications suppliers, and government policies and regulations. Topics covered include the Mexican informatics market, the role of multinationals as suppliers of hardware and software, and public data service systems. Different public data-transmission and teleprocessing networks are also reviewed, as is the use of remote sensing.

1517 **Trejo, Guillermo et al. Contra la pobreza.** Coordinación de Guillermo Trejo y Claudio Jones. México: Cal y Arena, 1993. 309 p.: bibl.

Assesses problems of poverty in Mexico. Focuses on increase in poverty during the 1980s' stabilization period and outlines shifts in government policies and development strategies that might help to alleviate the problem.

1518 **U.S. exports to Mexico: a state-by-state overview, 1987–1991.** Washington: U.S. Dept. of Commerce, International Trade Administration; U.S. G.P.O., Supt. of Docs., 1992. 127 p.: bibl., ill.

Examines exports to Mexico, paying particular attention to sources and interpretations of US-Mexico trade. Statistical tables summarize exports to Mexico by US states and major regions (New England, Mid Atlantic, etc.). Final section contains two-page profile of 1987–91 exports to the Mexican market for each US state.

1519 **U.S.-Mexico trade: impact of liberalization in the agricultural sector: report to the Chairman, Committee on Agriculture, House of Representatives.** Washington: General Accounting Office, 1991. 46 p.

Comprehensive report: 1) examines current bilateral and unilateral efforts to remove impediments to agricultural trade between the US and Mexico; 2) explores benefits of increased agricultural trade and the nature of this trade between the two countries; 3) reviews trade barriers that will need to be addressed in free trade agreement negotiations; and 4) presents views of US producer groups regarding agricultural trade liberalization with Mexico. Concludes that increased bilateral agricultural trade during the 1980s benefitted the US and Mexico in different ways. Benefits of this trade to both countries is because much of the trade in agricultural products between the two countries is characterized by complementary production and comparative advantage.

1520 **U.S.-Mexico trade: extent to which Mexican horticultural exports complement U.S. production; briefing report to the Chairman, Committee on Agriculture, House of Representatives.** Washington: U.S. General Accounting Office, 1991. 37 p.: bibl., ill.

Presents data on harvest and marketing seasons for major horticultural commodities exported from Mexico that are also grown in the US. Analyzes how major Mexican horticultural exports complement, supplement, or compete with US domestic production. Finds that while there is a clear pattern of complementary production for some major Mexican horticultural exports to the US, there is significant overlap in harvest and marketing seasons for other commodities. However, US production of most of these horticultural commodities increased during the 1980s, despite rising Mexican exports and sizable seasonal overlap. US consumers benefited from increased fruit and vegetable imports in terms of greater supplies, greater variety, and in some markets, lower prices.

1521 Urbanización y desarrollo en Michoacán. Coordinación de Gustavo López Castro. Zamora, Mexico: Colegio de Michoacán; Gobierno del Estado de Michoacán, 1991. 337 p.: bibl., ill., maps.

Leading authorities contribute series of essays on urbanization and the economy of Michoacán. Topics include agricultural sector, income distribution, industrialization, and social conflicts associated with urbanization.

1522 Urbina Fuentes, Manuel and **Adolfo Sánchez Almanza.** Distribución de la población y desarrollo en México. (*Comer. Exter.*, 43:7, julio 1993, p. 652–661, graph, maps, tables)

Excellent overview of regional patterns of income and population in Mexico. Includes discussion of international migration patterns, urbanization, and changing regional structure of employment.

1523 Valenciano, Eugenio O. and **Paul Ganster.** El Acuerdo de Libre Comercio México-Estados Unidos y repercusiones en la frontera. Buenos Aires: Banco Interamericano de Desarrollo, Instituto para la Integración de América Latina, 1991. 50 p.: bibl., map. (Publ.; 374. Serie Economía regional)

Examines several issues associated with economic integration between the US and Mexico. Attempts to anticipate problems associated with increased trade flows. Emphasizes industrial development, migration, tourism, and infrastructural constraints.

1524 Vázquez Ruiz, Miguel Angel and **Guadalupe García de León P.** Modernización industrial en Sonora. Hermosillo, Mexico?: Gobierno del Estado de Sonora, Secretaría de Fomento Educativo y Cultura; Instituto Sonorense de Cultura, 1992. 196 p.: bibl.

Examines industrial development in Sonora, emphasizing structural transformation. Sections cover *maquiladoras*, the automotive industry, sources of financing, patterns of capital formation, and technological change.

1525 Verduzco Chávez, Basilio. Empleo y crecimiento urbano: aplicación del modelo de Czamanski al caso mexicano. (*Estud. Demogr. Urb.*, 6:2, mayo/agosto 1991, p. 261–283, bibl. graphs, tables)

Applies Stanislaw Czamanski's model ("A Model of Urban Growth," *Regional Science Association Papers*, 1964) to the Mexican situation. Finds that it explains most of the country's patterns of urbanization.

1526 Verduzco Igartúa, Gustavo. Una ciudad agrícola: Zamora del Porfiriato a la agricultura de exportación. México: El Colegio de México; Zamora, Michoacán; El Colegio de Michoacán, 1992. 282 p.: bibl., map.

Detailed examination of the rural economy in the Zamora region. Study provides excellent description of the process of city growth and market development in the area.

1527 Weiss, John. Trade policy reform and performance in manufacturing: Mexico, 1975–88. (*J. Dev. Stud.*, 29:1, Oct. 1992, p. 1–23, bibl., tables)

Examines impact of major trade liberalizations introduced in Mexico since the mid-1980s on efficiency in the manufacturing sector. Alternative indicators of efficiency are estimated and changes in these indicators are explained by regression analysis, including various measures of the degree of liberalization. Concludes that thus far liberalization has had a positive but relatively weak effect on the sector's performance.

1528 Whiting, Van R. The political economy of foreign investment in Mexico: nationalism, liberalism, and constraints on choice. Baltimore: Johns Hopkins Univ. Press, 1992. 313 p.: bibl., index.

Examines domestic and international forces that shape political choices made towards foreign investment in Mexico. Contends that neither dependency nor statism is sufficient to explain foreign investment policy in Mexico. Political preferences and political choices do matter, but domestic and international structural constraints bound the choices of policymakers. Globalization of capital and technology, for example, shapes policy options in a way that favors liberalization. In the first half of the book, Whiting examines Mexico's nationalist tradition and the limits of its foreign investment policies. The joint venture policy and the regulatory apparatus put in place in the 1970s did not succeed in replacing the capital, technology, and marketing capabilities of foreign firms. In the second half, he explains how the international industrial structure limited national policy and created greater opportunities for liberalization.

1529 Wilson, Fiona. Workshops as domestic domains: reflections on small-scale industry in Mexico. (*World Dev.*, 21:1, Jan. 1993, p. 67–80, bibl.)

Paper explores growth of workshop-based industry in contemporary Latin America using the theoretical approach of gender relations and the family. Attention is thus shifted away from the informal sector paradigm which has historically provided the principle guidelines for the study of urban poverty.

1530 Wilson, Paul N. and **Gary D. Thompson.** Common property and uncertainty: compensating coalitions by Mexico's pastoral *ejidatarios*. (*Econ. Dev. Cult. Change*, 41:2, Jan. 1993, p. 299–318, graph, tables)

Examines causes underlying breakdown of the *ejidos*. Argues that agricultural production in semiarid and arid zones requires resource mobility, particularly the freedom to graze livestock throughout a large, extensive land area. Common grazing lands therefore represent a hedge or insurance against uncertainty in rainfall patterns. Concludes that breakdown in *ejido* productivity on these extensive, livestock-herding areas is due to a deterioration in property management at the community level. Rights, duties, functions, and obligations of individual herders have not been clearly specified or enforced by *ejido* authorities. Yet this failure of group management has led to the formation of coalitions within smaller groups where cooperation is assured and benefits are enjoyed under very severe ecological conditions.

1531 Winters, Cecilia Ann. Unanticipated capital flows and the Mexican economy: an empirical investigation. (*J. Dev. Stud.*, 29:3, April 1993, p. 504–517, bibl., tables)

Explores differential impact of unanticipated and anticipated foreign capital flows on Mexico's economy from 1965–85. Assumes that if unanticipated flows cause appreciation of the real exchange rate and do not affect domestic expenditure, one can conclude that the country's foreign exchange constraint is not binding. Based on empirical evidence, this hypothesis can be rejected. Implies that Mexico's problems probably do not stem from over-borrowing. Anticipated capital flows do affect private spending, but a negative coefficient suggests that the private sector has borne the brunt of post-crisis adjustment. Results show that the Mexican government has dominated expenditure of foreign loans throughout the period.

1532 Wong, Rebecca and **Ruth E. Levine.** The effect of household structure on women's economic activity and fertility: evidence from recent mothers in urban Mexico. (*Econ. Dev. Cult. Change*, 41:1, Oct. 1992, p. 89–102, tables)

Attempts to isolate empirically the effect of a specific characteristic of household structure, the presence of a "mother substitute," as a determinant of pariticipation in economic activity, and family-formation behaviors. Finds that as a group, the availability of mother substitutes in the household increases likelihood of participating in the work force. Furthermore, it decreases the likelihood of having more children after controlling for other socioeconomic and demographic characteristics.

1533 Wong-González, Pablo. International integration and locational change in Mexico's car industry: regional concentration and deconcentration. (*in* Decentralization in Latin America: an evaluation. Edited by Arthur Morris and Stella Lowder. New York: Praeger, 1992, p. 161–178, bibl., graphs, map, tables)

Analyzes changing locational pattern of Mexico's motor industry. Finds important factors different from those traditionally considered in the field of industrial location. Frequently, nonspatial or sectoral factors have a deciding influence on the location decisions of productive facilities. In the case of Mexico's northern region, the peso-dollar exchange rate and aspects of free trade have had an enormous influence recently. Continuing devaluation of the peso has substantially reduced the cost of labor. Concludes that while government policy has had an impact on the geographic dispersion of the Mexican automotive sector, the government's role has not been vital.

1534 **Wood, Christopher.** Mexico: into the spotlight. (*Economist/London*, 326: 7798, Feb. 13-19, 1993, p. 56 (p. 1-22), graphs, photos)

Excellent survey of recent economic developments and governmental policies. Makes a number of forecasts about future growth and the country's political system.

1535 **Workshop El Acuerdo de Libre Comercio México-Estados Unidos y Repercusiones en la Frontera, Tijuana, Mexico, 1991.** La integración fronteriza en los acuerdos de libre comercio: memoria. Recopilación de Eugenio O. Valenciano y Paul Ganster. Buenos Aires: Banco Interamericano de Desarrollo, Instituto para la Integración de América Latina, 1992. 302 p.: bibl., ill. (Publicación; 393. Serie Economía regional)

Papers on topics including impact on the border region from increased US-Mexican trade, problems associated with migration to the region, changing industrial structure, and infrastructural constraints on further growth.

1536 **Young, Linda Wilcox.** Labour demand and agroindustrial development: the evidence from Mexico. (*J. Dev. Stud.*, 30:1, Oct. 1993, p. 169-189, bibl., graphs, tables)

Asserts that agro-industrialization has had a negative employment-generating effect on agriculture in El Bajío region, contrary to what Mexican policymakers have stated. Shows that impact of widespread crop substitution and the wholesale transfer of technology promoted by transnational processing plants have both reduced the total labor employed and increased the instability of employment through heightened seasonality of labor demand. Through a decomposition analysis, the author assesses role played by various determinants of labor demand in the alteration of regional agricultural employment resulting from agro-industrialization.

1537 **Yúnez-Naude, Antonio.** El Tratado de Libre Comercio y la agricultura mexicana: un enfoque de equilibrio general aplicado. (*Estud. Econ./México*, 7:2, julio/dic. 1992, p. 225-264, bibl., graphs, tables)

Paper presents main results obtained from an applied general equilibrium model of Mexico. The model tries to identify the impact on agricultural production and consumption of a North American Free Trade Agreement. Also discusses limits and potentialities of such models.

1538 **Zepeda Miramontes, Eduardo.** El gasto público en la frontera norte. (*Front. Norte*, 4:7, enero/junio 1992, p. 5-43, bibl., tables)

Describes pattern of public expenditures by state and local governments of the border states. Analyzes arguments for further decentralized public expenditures according to different functions they perform. Concludes that it is necessary to reverse the currently existing high degree of centralization in public expenditures, particularly in investment outlays; decentralization would be a better approach to government allocations.

CENTRAL AMERICA

CLARENCE ZUVEKAS, JR., *Consulting Economist, Annandale, Virginia*

THE CESSATION OF ARMED CONFLICTS in Central America (except in Guatemala), the break-up of the Soviet Union, and the end of the Cold War have dampened the ideological content and sharpened the objectivity of writings on the economies of the Central American isthmus. The gap in quality of analysis between Costa Rican economists and those in the rest of the region is narrowing, although studies by most Panamanian economists continue to be disappointingly weak.

Many works annotated for this volume deal directly or indirectly with issues of structural adjustment, which has been increasingly embraced as necessary for rapid and sustained economic growth. Miguel Angel Rodríguez's edited volume (item **1539**) and Juan Buttari's comparative study (item **1542**) are good starting points. Negative views of structural adjustment are best articulated in the volume edited by Pelupessy and Weeks (item **1548**). Trade issues for the region, including possible economic integration with North America, are clearly delineated in an essay by Saborío and Michalopoulos (item **1563**). The revival of intraregional trade and the lowering of trade barriers among the Central American countries since the mid-1980s are encouraging trends. At the same time, the relative priority given to intraregional trade has tended to decline.

Nontraditional agricultural exports (mainly to the US) have expanded rapidly over the past decade, but much skepticism remains about the long-term sustainability of this trend and its distributional and environmental consequences (e.g., see the collection edited by Mendizábal and Weller, item **1560**). Curiously, little attention is given to the even larger (and also rapidly growing) volume of nontraditional manufactured exports.

The trend toward more analytical writings on poverty alleviation and social development—themes that are certainly not new to the region—is especially welcome. Particularly valuable is the study by Marc Lindenberg, a former rector of the Instituto Centroamericano de Administración de Empresas or INCAE (item **1553**). Also of note is Carlos Briones' analysis of the 1988 and 1990 household survey data in El Salvador (item **1594**), although the weaknesses of these data need to be borne in mind.

The Coordinadora Regional de Investigaciones Económicas y Sociales (CRIES) sponsored a series of studies on the role of US economic assistance to individual Central American countries in the 1980s and early 1990s (not all of which were annotated for this volume). The best is Herman Rosa's excellent work on El Salvador (item **1603**).

Macroeconomic policy in Nicaragua, an analytically challenging topic, has been ably addressed by a number of scholars from outside the country. A good place to begin is Joseph Ricciardi's essay (item **1633**), which was written before the Chamorro government initiated a major reform program in March 1991. Another noteworthy work on Nicaragua is Brizio Biondi-Morra's analysis of food policy (item **1622**).

A controversial topic now receiving much attention throughout Latin America is labor-market policy. Although written in the mid-1980s, studies on Panama by Spinanger and by Butelman and Videla are especially valuable because they seek to quantify the negative employment and wage effects of allegedly "pro-labor" policies (the two essays were published together; see item **1637**). The functioning of labor markets in Costa Rica is skillfully analyzed in studies by Gindling (item **1575**) and Gindling and Berry (item **1574**). Edward Funkhouser's study of El Salvador shows how labor force participation is affected by external migration and remittances (item **1597**).

Other topics receiving attention include industrial policy, inflation, the determination of parallel-market exchange rates, environmental issues (item **1541**), agrarian reform, and the informal sector. On the latter, see in particular the Menjívar Larín and Pérez Sáinz volume's country studies (item **1559**) that provide gender-disaggregated data.

GENERAL

1539 Ajuste estructural y progreso social: la experiencia centroamericana. Recopilación de Miguel Angel Rodríguez. San José: Libro Libre, 1992. 277 p.: bibl. (Serie económica)

Three general essays and six country studies that provide succinct summaries of structural adjustment measures and generally sound policy analysis despite sometimes overdoing details. The editor, an able expositor of "free-market" economics, is joined by other contributors with similar viewpoints as well as those with more eclectic positions.

1540 Arias Sánchez, Oscar et al. Poverty, natural resources, and public policy in Central America. Edited by Sheldon Annis. New Brunswick, N.J.: Transaction Publishers, 1992. 199 p.: bibl., ill., maps. (U.S.-Third World policy perspectives; 17)

Generally well-balanced study oscillates between highlighting negative trends and providing grounds for optimism, although sometimes unduly negative about current trends. Editor's overview builds on themes of other contributors: international peace parks, citizen participation and reform of development assistance, conservation in Costa Rica, nontraditional agricultural exports, and land taxation. Policy recommendations include both the shopworn and the thought-provoking.

1541 Bartholomew, Joy A. et al. Desarrollo sostenible y políticas económicas en América Latina. Recopilación de Olman Segura Bonilla. San José: Maestría en Política Económica para Centroamérica y el Caribe: DEI, 1992. 311 p.: ill. (Col. Ecología-teología)

Proceedings of 1991 seminar with good balance of conceptual issues, policy considerations, and case studies. Most contributors are Central Americans and most adopt moderate positions, recognizing importance of economic growth but seeking to make it less environmentally destructive. Several stress importance of a long-term vision.

1542 Buttari, Juan J. Economic policy reform in four Central American countries: patterns and lessons learned. (*J. Interam. Stud. World Aff.*, 34:1, Spring 1992, p. 179–214, bibl., tables)

Valuable, comparative review of major economic policy reforms in Costa Rica (begining in 1983), Guatemala (1986), El Salvador (1989), and Honduras (1990). Examines liberal orientation of the reforms, timing, and implementation. Cautiously optimistic about prospects for success and ability of "fledgling democracies" (Guatemala, El Salvador, Honduras) to survive pain associated with reforms.

1543 Caballeros Otero, Rómulo. Reorientation of Central American integration. (*CEPAL Rev.*, 46, April 1992, p. 125–137, bibl., tables)

Argues that a revitalized integration movement in Central America must emphasize increased supply rather than return to a demand-oriented approach so that the regional economy can compete more effectively in world markets. Presents detailed, specific proposals for restructuring the institutional framework for regional integration.

1544 Cáceres, Luis René and Frederick Jiménez. La capacidad de absorción de inversiones de los países centroamericanos. (*Rev. Integr. Desarro. Centroam.*, 44/45, julio/dic. 1989 and enero/junio 1990, p. 65–94, graphs, tables)

Calculates absorptive capacity of investment, a concept of limited analytical utility. More constructively, uses simple two-gap model to determine external resource requirements for achieving a five percent GDP growth rate in each country. Regression analysis estimates effect of US prime rate on investment. Simplifying assumptions result in questionably large effects.

1545 Cáceres, Luis René. Notas sobre la fuga de capital en Centroamérica. Tegucigalpa: Banco Centroamericano de Integración Económica, Depto. de Planificación, 1990. 41 p.: bibl. (Cuadernos de economía y finanzas; 12)

Good review of alternative definitions of capital flight applied to Central America, which give widely divergent results. Uses econometric analysis to analyze determinants of capital flight (for which the explanatory power is low) and estimates its macroeconomic impact (incomplete, because the model employed is rather simple).

Crisis económica en Centroamérica y el Caribe. See item **1641**.

1546 Democracia sin pobreza: alternativa de desarrollo para el Istmo Centroamericano. Coordinación de Eduardo Stein y Salvador Arias Peñate. San José: Editorial Depto. Ecuménico de Investigaciones: Comité de Acción de Apoyo al Desarrollo Económico y Social de Centroamérica, 1992. 581 p.: bibl., ill. (Col. universitaria)

Dozen essays by various authors who generally accept Central America's need for greater linkages with world economy but criticize existing trade liberalization for neglecting interests of the poor. Their alternative strategy stresses agroindustrial development, a misguided focus on food self-sufficiency, greater regional integration, social development, and income and wealth redistribution.

1547 El desafío del desarrollo centroamericano. Edición de Ennio Rodríguez. San José: FEDEPRICAP; Editorial Univ. Estatal a Distancia, 1991. 271 p.: bibl.

Interesting, motley collection of papers on two themes: regional action and redefining boundaries between public and private sectors. First theme is misleading because topics include national-level actions and integration with North America as well as economic and financial cooperation within Central America.

1548 Economic maladjustment in Central America. Edited by Wim Pelupessy and John Weeks. Houndmills, England: Macmillan; New York: St Martin's Press, 1993. 198 p.: bibl., ill., index.

Eleven essays on a variety of regional and country-specific topics. Authors are generally critical of structural adjustment and relatively pessimistic or unenthusiastic about extraregional export prospects, emphasizing risks more than opportunities. They call for more attention to measures that reduce poverty and narrow income inequalities.

1549 Fleissig, Adrian and Thomas Grennes. The real exchange rate conundrum: the case of Central America. (*World Dev.*, 22:1, Jan. 1994, p. 115–128)

Uses econometric techniques to calculate alternative measures of the "real" exchange rate for Costa Rica, El Salvador, Guatemala, Honduras, Mexico, Nicaragua, and Panama during the 1962–88 period. Sophisticated. [J.W. Foley]

1550 Gordon C., Maribel. Centroamérica y Japón: presente y futuro. (*Rev. Centroam. Econ.*, 11:33, sept./dic. 1990, p. 9–30, tables)

Largely descriptive review of Japan's trade, investment, aid, and other economic relations with Central America (including Panama and Belize). Author suggests that Japan's involvement in the region is modest but growing, largely because of pressure from the US. Useful data but interpretation is uneven.

1551 El impacto económico y social de las migraciones en Centroamérica. Santiago: Naciones Unidas, Comisión Económica para América Latina y El Caribe; New York: United Nations, 1993. 78 p.: map. (Estudios e informes de la CEPAL, 0256–9795; 89)

Estimates numbers of Central Americans migrating in the 1980s across borders within the region (including Mexico and Belize as receiving countries), plus displaced persons within their home countries and repatriated persons. Total exceeds 10 percent of Central America's population. Provides labor force, production, and spending estimates, and discusses social effects.

1552 Jones, Jeffrey R. Colonization and environment: land settlement projects in Central America. Tokyo: United Nations Univ. Press, 1990. 155 p.: bibl., maps.

Useful review of directed and spontaneous colonization and its environmental effects. Controversial argument is that deforestation is less an economic strategy (beef production) than one of establishing claims to land by demonstrating active land use. Clearly, much deforestation is stimulated by institutional and policy deficiencies.

1553 Lindenberg, Marc. The human development race: improving the quality of life in developing countries. San Francisco, Calif.: ICS Press, 1993. 231 p.: bibl., ill., index.

Important book analyzes statistical relationship of human development to economic, social, and political variables in 90 developing countries. Second half examines these relationships in more detail for Central American countries (including Panama). Stable political environments and long-term political commitments are found to be strong explanatory variables.

1554 **Lizano Fait, Eduardo.** Factores económicos en la evolución sociopolítica de Centroamérica. (*Rev. Integr. Desarro. Centroam.*, 44/45, julio/dic. 1989 and enero/junio 1990, p. 212–246, graphs, table)

Insightful historical interpretation of relationship between economic structures and political behavior. Argues that social and political spheres have been influenced significantly by countries' different relative factor endowments. Elites responded to change by diversifying their economic interests and altering political tactics, seeking to maintain their dominance. Postponing solutions to social problems and reliance on repression produced armed conflict inside some countries.

1555 **López, José Roberto.** Market efficiency, purchasing power parity and cointegration in Central American black foreign exchange markets. (*Estud. Econ./México*, 8:1, Jan./June 1993, p. 111–153)

Applies theory of cointegration and sophisticated statistical analysis to test for market efficiency in the determination of black (parallel) market exchange rates in Costa Rica, El Salvador, and Guatemala. Results vary by country and time period. Rewarding for those with expertise in exchange-rate-determination models.

1556 **López, José Roberto.** La ruptura de la Cámara de Compensación Centroamericana: origen, desarrollo y perspectivas en el contexto del conflicto regional. (*Rev. Integr. Desarro. Centroam.*, 48, enero/junio 1991, p. 149–184, bibl., graphs, table)

Good review of operations of Central American payments clearinghouse from its inception in the early 1960s through 1987 emphasizes its breakdown during the 1980s. Draws heavily on interviews with key actors in the Cámara and related institutions.

1557 **Martínez, Daniel.** Les fonds d'investissement social en Amérique centrale et au Panama. (*Rev. Tiers-Monde*, 32:127, juillet/sept. 1991, p. 533–549, table)

Useful analysis of similarities to and differences between various (compensatory) social investment funds created in Central America since 1985 (plus Costa Rica's pre-existing program). Argues—questionably—that only Costa Rica's and Guatemala's programs are truly redistributive because they are financed entirely with national resources.

1558 **Melhado, Oscar.** Los impuestos y sus efectos en la integración económica centroamericana. San Salvador: Fundación Salvadoreña para el Desarrollo Económico y Social, Depto. de Estudios Económicos y Sociales, 1992. 48 p.: bibl., ill. (Documento de trabajo; 32)

Concludes that value added taxes (IVAs) in Central America generally have not affected price levels, and that differential IVA rates among countries have not significantly distorted regional trade. Also concludes that direct taxes do not significantly affect investment. Analytical techniques are rather crude, but conclusions seem reasonable.

1559 **Ni héroes ni villanas: género e informalidad urbana en Centroamérica.** Coordinación de Rafael Menjívar Larín y Juan Pablo Pérez Sáinz. San José: FLACSO, 1993. 580 p.: bibl.

Significant collection of gender-disaggregated studies is based on surveys and case studies conducted in informal sectors of metropolitan areas of the five Central American countries and Panama. Stresses heterogeneity of the urban informal sector, which is disaggregated into three categories. Includes data on household characteristics.

1560 **¿Promesa o espejismo?: exportaciones agrícolas no tradicionales; su análisis y evaluación en el Istmo Centroamericano.** Coordinación de Ana Beatriz Mendizábal P. y Jürgen Weller. Panamá: CADESCA, 1992. 418 p.: bibl., ill. (Temas de integración y desarrollo; 2)

Papers by economistas, sociologists, and others with different perspectives on costs and benefits of nontraditional agricultural exports. Issues discussed include distribution of benefits (mostly to large producers, including foreign firms, although small producers have benefited significantly in some circumstances); marketing and price-volatility problems; and pesticide poisoning.

1561 **The reconstruction of Central America: the role of the European Community.** Edited by Joaquín Roy. Coral Gables, Fla.: Univ. of Miami North-South Center, 1992. 414 p.: bibl.

Proceedings of 1991 conference with informative contributions from knowledgeable academic researchers and representatives of government agencies and international organizations in the Americas and Europe. Economic themes include role of the CACM, Central America, and the world economy, external assistance, stabilization and adjustment programs, and effects of the single European market on Central America.

1562 Revista de la Integración y Desarrollo de Centroamérica. No. 47, julio/dic. 1990– . Tegucigalpa: Banco Centroamericano de Integración Económica.

Valuable reference includes texts of the Declaration of Antigua and Plan de Acción Económica de Centroamérica (PAECA), discussion papers and proposals presented by various organizations, and studies prepared by CEPAL on industrial reconversion possibilities in a dozen or so specific industries.

1563 Saborío Alvarado, Sylvia and **Constantine Michalopoulos.** Centroamérica en una encrucijada. (*Rev. Integr. Desarro. Centroam.*, 49, julio 1991/dic. 1992, p. 151–185)

Clear, insightful analysis of changes in trade policies and patterns since 1980 and of likely effects of the Uruguay Round, 1992 reforms in the European Community, and NAFTA. Authors recommend three-pronged trade strategy including multilateral liberalization, greater intraregional integration, and bargaining for preferential agreements.

1564 Saborío Alvarado, Sylvia. U.S.-Central America free trade. (*in* The premise and the promise: free trade in the Americas. New Brunswick, N.J.: Transaction Publishers, 1992, p. 141–185)

Argues that Central America stands to lose by not joining the then-proposed North American Free Trade Agreement (NAFTA). Proposes additional temporary US preferences for Central America to offset lost preferences implied by Mexican membership in NAFTA. Demonstrates how increased intraregional cooperation can strengthen the benefits of broader trade liberalization.

1565 Uthoff B., Andras. Población y desarrollo en el Istmo Centroamericano. (*CEPAL Rev.*, 40, abril 1990, p. 139–158, bibl., graphs, tables)

Presents much useful data on demographic trends and patterns and their relationship to socioeconomic variables. Asserts that demographic factors, notably the inverse relationship between income and family size, are significant determinants of intergenerational transfers of poverty. Argues that population policy should include attention to nutrition and investment in education and training.

1566 Villasuso Estomba, Juan Manuel and **Marlon Yong Chacón.** La propiedad intelectual y el libre comercio en Centroamérica. San José: Federación de Entidades Privadas de Centroamérica y Panamá (FEDEPRICAP), 1993. 226 p.

Compares intellectual property rights legislation in Central America, Mexico, and the US. Recognizes need for legislative reform but warns against undue restrictions on technology transfer. Suggests how legislative changes might affect selected industries (pharmaceuticals, agro-chemicals, informatics, and biotechnology) and proposes more detailed studies of these industries.

1567 Willmore, Larry. Industrial policy in Central America. (*CEPAL Rev.*, 48, Dec. 1992, p. 95–105, bibl., tables)

Describes industrial policy as of mid-1992, focusing on foreign investment, registration of new investments, tariff protection, and incentives for extraregional exporting. Notes lack of uniformity in policies among Central American countries.

1568 Willmore, Larry and **Jorge Máttar.** Industrial restructuring, trade liberalization and the role of the State in Central America. (*CEPAL Rev.*, 44, Aug. 1991, p. 7–19, bibl., tables)

Based in part on interviews with 358 firms in seven manufacturing industry categories. Discusses obstacles in each industrial group and recommends public and private measures to improve efficiency and competitiveness in an environment of continuing trade liberalization.

1569 Zuvekas, Clarence. Central American debt. (*in* The Caribbean Basin: economic and security issues; study papers submitted to the Joint Economic Committee, Congress of the United States. Washington: U.S. Government Printing Office, 1993, p. 206–227, tables)

Thorough survey examines Central American debt during 1980s and assesses prospects for 1990s. Notes that debt problems in four of the countries rivaled or exceeded those of Mexico. Includes aggregate data for the region as well as individual country surveys. Valuable overview. [D.W. Schodt]

BELIZE

1570 Barham, Bradford L. Foreign direct investment in a strategically competitive environment: Coca-Cola, Belize, and the international citrus industry. (*World Dev.*, 20:6, June 1992, p. 841–857)

Argues that the (at least temporary) failure of a large proposed citrus investment by Coca-Cola Foods was due to its proponents' neglect of the investment's effects on strategic competition in an industry whose characteristics are described in detail. Presents simple counterfactual model suggesting effects of the investment.

COSTA RICA

1571 Badilla Portuguez, Marcos and **Nancy Montiel Masís.** Participación de los factores de producción en el comercio exterior de Costa Rica a la luz de la teoría de la proporción de factores. (*Cienc. Econ./San José*, 11:1/2, julio/dic. 1991, p. 181–220, bibl., graph, tables)

Data from comparable input-output models for 1957, 1968, 1980, and 1987 are used to test the Heckscher-Ohlin-Vanek (H-O-V) theorem regarding factor proportions in international trade. Costa Rica is found to export products which are intensive in the use of relatively abundant natural resources and low-skilled labor, thus supporting H-O-V.

1572 Bensión, Alberto *et al.* Costa Rica and Uruguay. Edited by Simon Rottenberg. New York: Oxford Univ. Press, 1993. 424 p.: bibl., ill., index. (A World Bank comparative study. The Political economy of poverty, equity, and growth)

Costa Rica chapters by Claudio González-Vega and Víctor Hugo Céspedes insightfully demonstrate how the growth-with-equity model, which was successful for a number of decades, contributed to the 1980s crisis. Editor's brief country chapters and comparative chapter argue that some redistributive policies adversely affected growth by distorting incentives.

Brenes, Lidiette. La nacionalización bancaria en Costa Rica: un juicio histórico. See *HLAS* 54:1701.

1573 Garnier, Leonardo *et al.* Costa Rica entre la ilusión y la desesperanza: una alternativa para el desarrollo. San José: Ediciones Guayacán, 1991. 191 p.: bibl., ill.

Four essays call for the Costa Rican State to be an active promoter of economic and social development. Overview essay on this theme is followed by essays on social policy and human resources, natural resources and sustainable development (the weakest of the group), and science and technology.

1574 Gindling, T.H. and **Albert Berry.** The performance of the labor market during recession and structural adjustment: Costa Rica in the 1980s. (*World Dev.*, 20:11, Nov. 1992, p. 1599–1616, appendix, bibl., tables)

Perceptively examines disaggregated labor force data (relatively good in Costa Rica) to show how labor-market institutions responded well, "in an institutional context rather distant from the free market model," to the economic crisis of the early 1980s. Output, employment, and real wages all recovered pre-crisis levels relatively quickly.

1575 Gindling, T.H. Women's wages and economic crisis in Costa Rica. (*Econ. Dev. Cult. Change*, 41:2, Jan. 1993, p. 277–297, appendix, tables)

Econometrically analyzes changes in male-female wage differentials during recession (1980–82), stabilization (1982–83), and recovery (1983–86). Widening differentials are explained during recession primarily by labor-force entry of less-educated women, and during stabilization mainly by increases in government employment, which is male-dominated. Male-female differentials narrowed during recovery.

1576 Hoffmaister, Alexander. The cost of export subsidies: evidence from Costa Rica. (*Staff Pap.*, 39:1, March 1992, p. 148–174, bibl., graphs, tables)

Statistically analyzes effects of export subsidies on export performance. During 1984–89 subsidies expanded exports by about 10 percent, but also increased intermediate imports. Each dollar of subsidy generated *net* exports of only 54 cents. Fiscal costs were high and rising. Additional currency devaluation is viewed as preferable to subsidies.

1577 Hoffmaister, Alexander. Costa Rica: trimestralización del PIB real y el ajuste dinámico ante shocks monetarios y de tipo de cambio. (*Cienc. Econ./San José*, 11:1/2, julio/dic. 1991, p. 145–179, appendix, bibl., graphs, tables)

Applies three alternative methodologies to estimate quarterly real GDP series (1963–69). Uses results in a model positing dynamic interrelationships among money supply, exchange rate, price level, and GDP. Inflation is explained largely by exchange-rate variations in the short run and mainly by money-supply changes in the long run.

1578 **Lizano Fait, Eduardo.** Programa de ajuste estructural en Costa Rica. San José: Academia de Centroamérica, 1990. 190 p.: bibl., ill. (Serie Estudios; 6)

Clear, well-argued exposition of rationale for structural adjustment by former president (1984–90) of Costa Rica's Central Bank, who played major role in facilitating his country's process of adjustment. Includes texts of policy statements and structural adjustment agreements with World Bank.

1579 **Raíces institucionales de la política económica costarricense.** Edición de Jorge Corrales Quesada. San José: Litografía e Imprenta LIL, 1993. 292 p.

Proceedings of 1992 conference applying public-choice theory to Costa Rica. Most authors and discussants are Costa Rican; US public-choice notables also contribute papers. Most participants advocate reducing size of the State and improving its efficiency. Generally sound analysis and stimulating debate. Ideological issues sometimes lurk in the background.

1580 **Ramírez Rodríguez, Edwin** and **Manuel Barahona Montero.** Costa Rica: zonas de mayor y menor desarrollo relativo. San José: Ministerio de Planificación Nacional y Política Económica; s.l.: Fondo de Población de las Naciones Unidas, 1991. 41 p.: ill., maps.

Calculates indices of social development based on simple averages of scaled scores for eight separate indicators covering education, health, social security, nutrition, and housing. Scales are based on indicators for all 420 of Costa Rica's districts. Indices are calculated only for the 169 districts outside the Central Region.

1581 **Reinert, Kenneth A.** Discriminatory export taxation in Costa Rica: a counterfactual history. (*J. Dev. Areas*, 28:1, Oct. 1993, p. 39–48)

Counterfactual experiments using six-sector general equilibrium model, with simplifying assumptions, alter traditional pattern of export taxes: relatively high for coffee and sugar and low or zero for bananas and rices, the main large-farm crops. Less discriminatory export tax structure would have maximized national welfare.

1582 **Rodríguez, Adrián G.** and **Stephen M. Smith.** A comparison of determinants of urban, rural and farm poverty in Costa Rica. (*World Dev.*, 22:3, March 1994, p. 381–397)

Applies logistic regression to identify factors affecting poverty, using 1988 household survey of urban and rural (farm and nonfarm) households. Sensitivity analysis shows that more education achieves greatest reduction in probability of being poor, followed by increased number of family members working, and type of job. Some policy recommendations are good, others questionable.

1583 **Rodríguez, Ennio.** The multiple tracks of a small open economy: Costa Rica. (*in* The developing countries in world trade: policies and bargaining strategies. Edited by Diana Tussie and David Glover. Boulder, Colo.: Lynne Rienner Publishers, 1993, p. 99–117, tables)

Well-written, well-informed analysis of significant changes in Costa Rica's trade strategies and policies since the early 1980s, with specific attention to GATT, CBI, EAI, NAFTA, and intraregional trade. Recommends that Costa Rica give more attention to bargaining as part of a coalition with its Central American neighbors.

Romero Pérez, Jorge Enrique. La crisis y la deuda externa en América Latina. See item **1256**.

1584 **Sáenz, O.,** and **J.R. Vargas.** El modelo macro: un instrumento para la prospección económica. (*Cienc. Econ./San José*, 12:2, dic. 1992, p. 69–101, bibl., graph)

Ambitious but flawed effort to develop macroeconomic model of Costa Rican economy. Problems include quirky behavioral equations, including questionable use of autoregressive variables, and suspiciously high R^2s. Authors report running successful policy simulations with their model but provide no examples.

1585 **Segura Bonilla, Olman.** Costa Rica y el GATT: los desafíos del nuevo orden del comercio mundial. San José: Editorial Porvenir: Maestría en Política Económica para Centroamérica y el Caribe, Univ. Nacional, 1991. 127 p.: bibl., ill. (Col. Debate)

Discusses pros and cons of Costa Rica's entry into GATT (completed in 1990), a process that generated surprisingly little debate. Reviews status of GATT negotiations as of early 1991, and examines Costa Rica's interests in specific outcomes. Cautiously optimistic that Costa Rica, on balance, can benefit from GATT membership.

1586 **Sojo, Carlos.** La utopía del estado mínimo: la influencia de AID en las transformaciones funcionales e institucionales del estado costarricense en los años ochenta. Managua: CRIES, 1991. 89 p.: bibl.

Provides detailed information on US economic assistance to Costa Rica during the 1980s and its conditionality, particularly regarding the role of the State. Despite significant privatization of State enterprises, concludes that the State's overall role was not significantly reduced. Analytically unexciting but informative.

1587 **Vries, Peter de.** Squatters becoming beneficiaries: the trajectory of an integrated rural development programme. (*Rev. Eur.*, 58, 1995, p. 45–70)

Studies relations between local development bureaucracy and settlers/beneficiaries by presenting a case study of State intervention following invasion of an hacienda by a Costa Rican peasant union. Argues in favor of an approach that takes into account the local history of relationships between State officials and farmers. [R. Hoefte]

1588 **Webb, Michael** and **James Fackler.** Learning and the time interdependence of Costa Rican exports. (*J. Dev. Econ.*, 40:2, April 1993, p. 311–329)

Uses dynamic economic model and Granger-causality tests to examine relationships between exports to Central American Common Market (CACM) and to other regions. The former apparently induced learning effects in some industries (machinery, intermediate manufactures, and leather) but not others. CACM probably inhibited exploitation of developed country markets for some products.

1589 **Zúñiga F., Norberto** and **Juan Enrique Muñoz G.** Costa Rica: determinantes de la tasa de interés en un contexto de liberalización financiera. (*Cienc. Econ./San José*, 8:1, primer semestre 1988, p. 23–37, bibl.)

Authors develop model to explain determination of interest rates over a 19-month period following their liberalization in 1985. An equation with expected devaluation, the interest rate on government bonds, and lagged money supply gave satisfactory results, but the number of observations is relatively low.

1590 **Zúñiga F., Norberto.** Un enfoque monetario del costo de la inflación en Costa Rica. (*Cienc. Econ./San José*, 12:1, julio 1992, p. 61–79, appendices, bibl., graphs, tables)

Estimates welfare costs of Costa Rican inflation during 1963–89 at inflation rates of 10, 20, 30, and 40 percent. Welfare costs tend to rise with the inflation rate. Equations are in linear, log-linear, and log-log form. Assumptions are restrictive, but results have some validity.

EL SALVADOR

1591 **Arriola Palomares, Joaquín.** Política industrial: ¿asignatura pendiente o asignatura imposible? (*ECA/San Salvador*, 46:515, sept. 1991, p. 815–827, photos, table)

Representative criticism of government's outward-oriented development strategy. Proposes highly State-directed industrial policy emphasizing domestic and Central American markets, and greater participation by labor and small business. Ignores poor worldwide track record of inward-looking strategies in improving workers' living standards.

1592 **Bitran, Ricardo A.** and **D. Keith McInnes.** The demand for health care in Latin America: lessons from the Dominican Republic and El Salvador. Washington: World Bank, 1993. 54 p.: ill. (EDI seminar paper; 46)

Statistically examines determinants of demand for curative ambulatory care from doctors in the two capitals. Finds that even the poorest people show strong preference for private health care, even when free public services are available. Public health care systems are found to be highly inefficient.

1593 **Briones, Carlos.** Ajuste estructural y desarrollo: ¿dónde estamos? (*ECA/San Salvador*, 47:527, sept. 1992, p. 741–755, bibl., photos, tables)

Constructive critique of the Cristiani government's structural adjustment policies. Recognizes their positive elements but argues that: 1) they confuse economic growth with sustainable development, which requires more attention to equity; and 2) they neglect human resource development and an active industrial policy, both of which are necessary to compete successfully internationally.

1594 Briones, Carlos. La pobreza urbana en El Salvador: características y diferencias de los hogares pobres, 1988–1990. San Salvador: UCA Editores, 1992. 148 p.: bibl. (Col. Estructuras y procesos: Serie mayor; 10)

Good, detailed analysis of poverty and characteristics of poor households, based on multipurpose household surveys. Combines poverty-line and unsatisfied-basic-needs methodologies to produce two-by-two matrix that defines three different types of poverty. Offers constructive policy suggestions for various targeted anti-poverty programs.

1595 Cáceres, Luis René and Oscar A. Núñez-Sandoval. La determinación del tipo de cambio en el mercado negro de El Salvador. (*Trimest. Econ.*, 58:230, abril/junio 1991, p. 249–262, bibl., graph, tables)

Reviews existing models of exchange-rate determination in a black market and constructs new model for El Salvador's black (actually legal, free) market. Money supply was a significant explanatory variable only when defined narrowly as currency. Dummy variable for seasonal influences was also significant.

1596 Cuevas, Carlos E.; Douglas H. Graham; and Julia A. Paxton. El sector financiero informal en El Salvador: informe final. Columbus, Ohio: Agricultural Finance Program, Ohio State Univ., Dept. of Agricultural Economics and Rural Sociology; La Libertad, El Salvador: Fundación Salvadoreña para el Desarrollo Económico y Social, 1991. 1 v.: bibl., ill., map. (Documentos de trabajo; 29)

Interviews with 2,000 borrowers and lenders provide quantitative estimates of size and scope of informal finance. Significant links were found between formal and non-formal financial markets, with nonfinancial firms as intermediaries. Expansion of sample yields surprisingly low estimate for remittances, casting doubt on reliability of remittance data or sample itself.

1597 Funkhouser, Edward. Mass emigration, remittances, and economic adjustment: the case of El Salvador in the 1980s. (*in* Immigration and the work force: economic consequences for the United States and source areas. Chicago: Univ. of Chicago Press, 1992, p. 135–175, appendices, bibl., graphs, tables)

Excellent analysis of external migration patterns and remittance flows, with detailed comments on data sources. Economic forces stimulate migration more than political forces. For non-migrants, migration lowers men's labor force participation, has insignificant effects on women's participation, raises wages, and reduces unemployment.

1598 Jiménez A., Arnoldo. Regulación de la industria azucarera en El Salvador: trabajo de investigación. San Salvador: Fundación Salvadoreña para el Desarrollo Económico y Social, 1992. 49 p.: bibl., ill., map. (Documento de trabajo/FUSADES; 27)

Detailed examination of cost structure of sugar industry and government regulation of production, processing, and marketing. Concludes that government policies promote inefficient resource use and damage the environment. Recommends gradual rather than abrupt liberalization to ease the burden of adjustment.

1599 Levy, Santiago and Roberto Rosales. Los tipos de cambio múltiples y el racionamiento de las divisas: la teoría y una aplicación al caso de El Salvador. (*Trimest. Econ.*, 58:231, julio/sept. 1991, p. 521–559, bibl., tables)

Develops CGE model of Salvadoran economy, with official and parallel exchange rates, and simulates effects of different foreign-exchange rationing mechanisms. Concludes that short-term benefits of multiple exchange rates are offset by penalties on traditional exports and negative long-run effects of unpredictable and unequal price and output shifts.

1600 Mallat, Gustavo. Economía y medio ambiente en El Salvador. San Salvador: Fundación Salvadoreña para el Desarrollo Económico y Social, Depto. de Estudios Económicos y Sociales, 1992. 86 p.: ill. (Documento de Trabajo/FUSADES; 30)

Balanced, well-written discussion of complex relationships between economy and environment. Explains how structural adjustment helps environment in short run by eliminating price distortions and in long run by reducing poverty. Recognizes that structural adjustment has some negative short-run effects but is essential for long-run environmental preservation.

1601 Pelupessy, Wim. Economic adjustment policies in El Salvador during the 1980s. Translated by John F. Uggen. (*Lat. Am. Perspect.*, 18:4, Fall 1991, p. 48–78, bibl., tables)

During 1980–88 two different economic models were seen as vying for supremacy: "structural reform," emphasizing expansion and broadening of domestic market, and "productive adjustment," based on export-led growth and reduction of role of the State. Insufficient attention given to world economic trends and domestic price distortions as determinants of economic performance.

1602 Rivera Campos, Roberto. La inflación en El Salvador. (*Real. Econ.-Soc.*, 1:1, enero/feb. 1988, p. 7–56, appendices, bibl., graphs, tables)

Develops cost-push model of inflationary process in El Salvador during 1971–86 and finds that it explains inflation better than several monetarist models, which are also tested econometrically. Specifications of cost-push model may be questioned, but study is still valuable.

1603 Rosa, Herman. AID y las transformaciones globales en El Salvador: el papel de la política de asistencia económica de los Estados Unidos desde 1980. Managua: CRIES, 1993. 133 p.: bibl.

Excellent, balanced description and analysis of trends in official US economic assistance to El Salvador since 1980. Rosa sees this assistance as an effort by the US government to find a solution to the Central American crisis favorable to its long-run interests by restructuring El Salvador's economy and society.

1604 Rubén, Raúl. El problema agrario en El Salvador: notas sobre una economía agraria polarizada. (*Cuad. Invest./San Salvador*, 2:7, abril 1991, p. 5–74, appendices, bibl., graphs, tables)

Attributes agricultural production and income problems to class structure and political power of oligarchic interests. Argues that structure of agricultural sector remains dualistic, even though agrarian reform in the 1980s reduced concentration of landholdings. Contains much useful data on agricultural structure and trends though 1987.

GUATEMALA

1605 Cáceres, Luis René and Oscar A. Núñez-Sandoval. Influencias internas y externas en la determinación del tipo de cambio en el mercado negro de Guatemala. (*Trimest. Econ.*, 49:234, abril/junio 1992, p. 297–310, bibl., graph, tables)

Granger-causality tests show that El Salvador's exchange rate is affected by Guatemala's. Model of exchange-rate determination shows that Guatemala's exchange rate is affected by El Salvador's and by Guatemala's money supply (M_1). For El Salvador, Guatemala's exchange rate has a significant effect but domestic money supply does not.

1606 Cole, Julio Harold. Inflación y masa en Guatemala: ¿cuál es el agregado monetario relevante? (*Banca Cent.*, 4:16, enero/marzo 1993, p. 40–53, bibl. tables)

With the quantity theory of money as a theoretical base, uses a relatively simple least squares regression model to determine whether inflation is better explained by M_1 (narrow money) or M_2 (broad money). M_1 is found to be the better explanatory variable.

1607 Immink, Maarten D.C. and Jorge A. Alarcón. Household income, food availability, and commercial crop production by smallholder farmers in the western highlands of Guatemala. (*Econ. Dev. Cult. Change*, 41:2, Jan. 1993, p. 319–342, tables)

Statistically analyzes data from 1987 survey of small farmers, separated into those growing predominantly maize, potatoes, wheat, and vegetables, respectively. Finds crop diversification related to access to credit, household labor-supply constraints, off-farm employment opportunities, and location, but not to farm size or household human capital. While diversification away from maize generally increased income, it aggravated economic vulnerability of smallest potato farmers. Rising incomes had little effect on nutrition.

1608 Morrison, Andrew R. Violence or economics: what drives internal migration in Guatemala? (*Econ. Dev. Cult. Change*, 41:4, July 1993, p. 817–830, tables)

Two alternative measures of violence are incorporated into standard economic model of internal migration. In three of four equations, violence is statistically significant determinant of migration. Still, a given percentage change in any of the independent economic variables produces more migration than the same percentage change in violence.

1609 Mujeres y empleo en Ciudad de Guatemala. Coordinación de Juan Pablo Pérez Sáinz y Eugenia Castellanos de Ponciano. Guatemala: FLACSO, 1991. 154 p.: bibl.

Analyzes socioeconomic data for women in Guatemala City from: 1) a 1989 employment survey whose methodology is not described; and 2) sample surveys of tortilla makers in 18 low-income neighborhoods (the most representative sample), employees in four *maquiladoras*, and domestic workers from a small universe defined by networking.

1610 Murray, Douglas L. and Polly Hoppin. Recurring contradictions in agrarian development: pesticide problems in Caribbean Basin nontraditional agriculture. (*World Dev.*, 20:4, April 1992, p. 597–608)

Discusses overuse and misuse of pesticides in horticulture production for export in the Dominican Republic and Guatemala, which have led to greater pest infestation, high pesticide residues resulting in rejection of products by US Food and Drug Administration, and health problems. Solutions to date have caused production to shift from small to large producers. Authors recommend that US assistance policy emphasize technologies more appropriate for small farmers.

1611 Samayoa Urrea, Otto Arturo. Factores determinantes en la tasa real de cambio: el caso de Guatemala. (*Banca Cent.*, 4:16, enero/marzo 1993, p. 27–39, bibl., graphs, tables)

Adapting an econometric model developed by Sebastian Edwards, author seeks to explain real exchange-rate movements in Guatemala. Monetary variables are found to be more important explanatory factors than real variables. Results are generally satisfactory, but monetary variables do not seem to be truly independent of each other.

1612 Schneider, Pablo R. *et al.* La economía informal en Guatemala: proyecto de investigación realizado para el programa INCAE-ROCAP 596-0147-0A-00-8621-00. Guatemala: Centro de Investigaciones Económicas (CIEN), 1992. 104 p.: bibl., ill.

Reviews different visions of informal economy and alternative analytical techniques for quantifying its size. Monetary approach using econometrics suggests it rose from under 10 percent of total output in early 1960s to about 35 percent by late 1980s. Other evidence corroborates the implied exaggeration in official statistics of declining per capita income in 1980s. Rewarding study.

1613 Schweigert, Thomas E. Commercial sector wages and subsistence sector labor productivity in Guatemalan agriculture. (*World Dev.*, 21:1, Jan. 1993, p. 81–91, bibl., tables)

Comparison of author's 1988 sample survey results on South Coast *parcelamientos* with similar 1965 data reveals that real wages were unchanged over time (and showed no significant seasonal variation), despite rapid agricultural growth. Real wages were equivalent to labor productivity in subsistence agriculture. These results support W.A. Lewis' classical labor-surplus model.

1614 Von Braun, Joachim and Maarten D.C. Immin123k. Cultivos de hortalizas (no tradicionales) para exportación por pequeños agricultores en Guatemala: impactos sobre su ingreso familiar y seguridad alimentaria. (*Cuad. Econ./Santiago*, 27:81, agosto 1990, p. 291–308, bibl., graph, tables)

Statistically analyzes data drawn from interviews at two points in time with randomly selected small-farmer households belonging to the "Cuatro Pinos" cooperative (N = 195) and non-member households (N = 204) in the same communities. Cultivation of nontraditional export crops resulted in significant increases in income but only modest improvements in food security.

HONDURAS

1615 Barham, Bradford L. and Malcolm Childress. Membership desertion as an adjustment process on Honduran agrarian reform enterprises. (*Econ. Dev. Cult. Change*, 40:3, April 1992, p. 587–613, tables)

Statistical analysis of panel data from 70 agrarian reform cooperatives effectively challenges the view that significant membership desertion reflects inefficiencies in the agrarian reform enterprises. Desertion instead resulted largely from rational adjustments to initial membership oversubscription in the face of lack of access to land and credit.

1616 **Baumeister, Eduardo.** El café en Honduras. (*Rev. Centroam. Econ.*, 11:33, sept./dic. 1990, p. 33–78, bibl., tables)

Detailed description and analysis of structure and evolution of coffee production in Honduras, where small and medium-size producers grow a large share of the crop. These growers sell much of their coffee to private intermediaries at prices said to be relatively low, with intermediaries capturing significant economic rents.

1617 **Cálix Suazo, Miguel** and **Zonia Vindel de Cálix.** Política económica antes y después de 1989. Tegucigalpa: Litografía López, 1991. 438 p.

Revised speeches by the senior author, a longtime Central Bank official, writing privately. Largely descriptive accounts of monetary and fiscal policy and macroeconomic trends, with much useful information. Authors stress importance of fiscal discipline and low inflation, but defend an exchange rate that in the late 1980s was clearly overvalued.

1618 **Honduras, el ajuste estructural y la reforma agraria.** Recopilación de Hugo Noé Pino y Andrew Thorpe. Tegucigalpa: Centro de Documentación de Honduras: Postgrado Centroamericano en Economía y Planificación del Desarrollo, 1992. 224 p.: bibl., ill.

Thirteen essays presented at a seminar, a document synthesizing the seminar, and a statement by Honduran College of Economics. Diverse viewpoints, but authors generally are critical of or ambivalent toward economic adjustment because of presumed (but undocumented) negative effects on the rural poor. Synthesis document states that economic adjustment is incomplete without agrarian reform.

1619 **Lewis, Maureen A.** User fees in public hospitals: comparison of three country case studies. (*Econ. Dev. Cult. Change*, 41:3, April 1993, p. 513–532, tables)

Comparative analysis of Honduras, the Dominican Republic, and Jamaica. Honduran public hospitals, which are financially accountable, have greatest flexibility in setting fees and deciding on their use. User fees account for a modest 13 percent of their nonpersonnel budgets, but still enough to permit improvements in quality of care.

1620 **Navarro, Jorge.** Poverty and adjustment: the case of Honduras. (*CEPAL Rev.*, 49, April 1993, p. 91–101, bibl., tables)

Although there is some indication that the incidence of poverty increased after stabilization and structural adjustment measures were introduced (little was done before 1990), author concludes correctly that data limitations preclude a conclusive determination that structural adjustment increased poverty, or that poverty would have been even worse without it.

NICARAGUA

1621 **Amador A., Freddy** and **Gerardo Ribbink.** Nicaragua: reforma agraria, propiedad y mercado de tierra. Managua: Univ. Nacional Autónoma de Nicaragua, Facultad de Ciencias Económicas, Escuela de Economía Agrícola, CIES-ESECA, 1992. 37 p.: bibl. (Serie CIES/ESECA; 92.1)

Good, informative summary of still-unsettled conditions in agrarian reform sector approximately two years after the Sandinistas' departure from government. Discusses legal insecurity and political instability, as well as economic pressures driving some agrarian reform beneficiaries to sell their land. Includes case studies.

1622 **Biondi-Morra, Brizio N.** Revolución y política alimentaria: un análisis crítico de Nicaragua. México: Siglo Veintiuno Editores, 1990. 342 p.: bibl., ill., map. (Economía y demografía)

Valuable, well-written, analytical study of food security policy failures during 1979–85. These failures are attributed less to external events than to: 1) ineffective linkages between macroeconomic policies (exchange rate, interest rate, wages, prices) and micro-level application of food policies; and 2) inadequate administrative mechanisms. Author interviewed over 200 State enterprise managers and administrators. Detailed data.

1623 Dijkstra, Geske. Industrialization in Sandinista Nicaragua: policy and practice in a mixed economy. Boulder, CO: Westview Press, 1992. 225 p.: bibl., index. (Series in political economy and economic development in Latin America)

Seeks to explain lack of success of industrialization policy given Sandinistas' commitment to a mixed economy. Based partly on interviews with managers of a (sometimes overly) stratified sample (N = 33) of all State and private manufacturing firms (N = 141) with 30 or more employees (excluding agroindustry). Analysis is generally sound but sometimes politically colored.

1624 Entre la agresión y la cooperación: la economía nicaragüense y la cooperación externa en el período 1979–1989. Managua?: Instituto de Investigación, Capacitación y Asesoría Económica, 1991. 189 p.

Provides valuable data on sources and uses of external assistance, which averaged $720 million annually during 1979–89. Discusses role of external assistance in mitigating economic crisis, attributed mainly to US aggression, although economic policy errors are also recognized. External resources were sometimes inefficiently used because of absorptive capacity problems.

1625 El "éxito" al renegociar la deuda: ¿cadena para el sandinismo? (*Envío/Managua*, 11:124, marzo 1992, p. 13–20, tables)

Provides useful details on Nicaragua's 1991 renegotiation with its Paris Club (official bilateral) creditors, while arguing that debt relief has provided less than meets the eye. Criticizes the government's acceptance of debt and other policies promoted by the international financial community, but provides no clear, viable alternative.

1626 Funkhouser, Edward. Migration from Nicaragua: some recent evidence. (*World Dev.*, 20:8, Aug. 1992, p. 1209–1218)

Data from two household surveys in 1989 show substantial emigration in the 1980s. Emigrants were disproportionately better educated and of working age. Remittance income in both studies was lower than previous estimates of about $80 million, but patterns differed. Statistical analysis shows that remittance income affects labor force participation negatively and the probability of self-employment positively.

1627 Gibson, Bill. Nicaragua. (*in* The rocky road to reform: adjustment, income distribution, and growth in the developing world. Cambridge, Mass.: MIT Press, 1993, p. 431–456, appendix, bibl., tables)

Good analytical review of economic policy and performance through 1991, focusing on both external constraints and internal policy decisions. A three-gap macroeconomic model projects very slow economic growth through the year 2000 (negative in per capita terms), even with relatively optimistic assumptions, because of balance-of-payments constraints.

1628 Godoy, Ricardo and Jeremy Hockenstein. Agricultural research bias in Nicaragua: the case of beans. (*World Dev.*, 20:11, Nov. 1992, p. 1685–1696, bibl., graphs, tables)

Relates declining bean yields since 1960 to limited and misguided research on food staples and concentration on export crops, even during the Sandinista years (when bean research favored urban consumers of more expensive red, rather than black, beans). Argues that appropriate investment in bean research would have high returns.

1629 Martínez, Philip R. Peasant policy within the Nicaraguan agrarian reform, 1979–89. (*World Dev.*, 21:3, March 1993, p. 475–487)

Argues that the Sandinista government significantly underestimated the importance of the peasantry as producers of both basic grains and exports. Documents how government policies remained biased against peasants despite belated recognition of importance of these producers and negative economic and political effects of discriminating against them.

1630 Mendoza Fletes, Orlando. Distribución del ingreso y seguridad alimentaria en un ambiente de ajustes cambiarios: los precios relativos en Nicaragua, 1986–1992. Managua: Univ. Nacional Autónoma de Nicaragua, Escuela de Economía Agrícola, Centro de Investigaciones Económicas y Sociales, 1992. 60 p.: bibl. (Serie CIES/ESECA; 92.2)

Argues that maxi-devaluations did not reallocate resources from nontradables to tradables, but rather produced a deterioration in terms of trade between campesino-produced crops and goods produced by domestic oligopolistic industries. Empirical evidence only partially supports this view.

1631 Neira Cuadra, Oscar and **Adolfo Acevedo V.** Nicaragua: hiperinflación y desestabilización. Managua: Coordinadora Regional de Investigaciones Económicas y Sociales (CRIES), 1992. 132 p.: appendices, bibl., graphs, ill., tables. (Cuadernos CRIES, Serie Ensayos; 21)

Argues that monetarist explanations of hyperinflation are invalid in Nicaragua, where supply shocks should be seen as the dominant causes. Orthodox adjustment measures are thus seen as inappropriate and harmful to the poor. Well-written exposition of this viewpoint, but no viable alternative package of remedies is proposed.

1632 Ocampo, José Antonio. Collapse and (incomplete) stablization of the Nicaraguan economy. With commentary by Ann Helwege (p. 361–365) and Arnold C. Harberger (p. 365–368). (*in* The macroeconomics of populism in Latin America. Chicago: Univ. of Chicago Press, 1991, p. 331–361)

Focusing primarily on the 1988–89 stabilization efforts, this essay by the head of the WIDER/SIDA economic policy mission to Nicaragua in the late 1980s views both external and internal obstacles and domestic policy shortcomings as important explanations of the country's economic travails during the 1980s.

1633 Ricciardi, Joseph. Economic policy. (*in* Revolution and counterrevolution in Nicaragua. Edited by Thomas W. Walker. Boulder, Colo.: Westview Press, 1991, p. 247–273, graphs)

Good summary of changes in macroeconomic policy and extreme price distortions under the Sandinista regime (1979–90). Central thesis is that internal tensions of managing a mixed economy contributed to hyperinflation in late 1980s, forced the adoption of stabilization measures antithetical to revolutionary objectives, and ultimately led to the Sandinistas' electoral demise.

1634 Stahler-Sholk, Richard. Stabilization, destabilization, and the popular classes in Nicaragua. (*LARR*, 25:3, 1990, p. 55–88)

Good analysis of Sandinistas' difficulties in reconciling economic and political objectives in a mixed-economy framework while coping with problems posed by the Contras. Illustrates how large relative price distortions, and changes therein, had major and often unintended distributional consequences. Relates economic policy changes to political objectives.

PANAMA

1635 Jované, Juan. Deuda externa: contradicciones y políticas alternativas. (*Tareas/Panamá*, 68, enero/abril 1988, p. 11–23)

The "contradictions" are between temporary growth stimulated by debt accumulation and the ultimate need to service the debt. Nationalist fervor leads author to recommend unrealistic policy prescriptions that fail to take into account Panama's small size and therefore its strong dependence on external trade and finance.

1636 Méndez, Roberto N. Innovación socioeconómica en Panamá, 1982–1986. (*in* ¿Hacia un nuevo orden estatal en América Latina? Los actores socio-económicos del ajuste estructural. Coordinación de Fernando Calderón y Mario R. dos Santos. Buenos Aires: CLACSO, 1988, v. 3, p. 413–435)

Clear, detailed review of (relatively moderate) economic liberalization measures undertaken by various administrations during 1982–86. Describes actions of organized labor, professional groups, industrialists, farmers, and political organizations opposed to these measures. Attributes their disappointing results to internal politics, unfavorable external conditions, and the measures themselves, without proposing viable alternatives.

1637 Spinanger, Dean; Andrea Butelman; and **Pedro Videla.** El mercado laboral en Panamá [por] Dean Spinanger. La política económica en Panamá y su efectos sobre salarios y empleo [por] Andrea Butelman y Pedro Videla. San Juan: Servicios Técnicos del Caribe, 1986. 196 p.: bibl., ill.

Two solid studies using, respectively, survey research and statistical analysis to quantify the effects of Panama's labor code on wages and employment. They demonstrate that ostensibly "pro-labor" policies had the perverse effect of reducing both employment and wages by sharply raising employers' total labor costs.

THE CARIBBEAN AND THE GUIANAS (Except Cuba and Puerto Rico)

RANSFORD W. PALMER, *Professor of Economics, Howard University*

MANY OF THE STUDIES ANNOTATED BELOW provide a comprehensive assessment of the economic performance of individual Caribbean countries over the past 30 years—a period during which most of them experienced relative economic decline (items **1674, 1679, 1676,** and **1640**). Much of the analysis in these assessments reflects a sense of despair about such intractable issues as the excessive debt burden, the external imbalance, declining currency values, and high unemployment rates (item **1659**). A dominant theme running through these studies is the social dislocation caused by structural adjustment programs required by the International Monetary Fund as a condition for receiving balance-of-payments assistance. The IMF is criticized for using a structural adjustment template which is unable to take into account national peculiarities (item **1675**). Many authors argue that the elimination of subsidies and the reduction of government spending have imposed a disproportionate burden on the poorest people (items **1651** and **1673**). It is also argued that the structural adjustment programs are geared more toward improving the ability of these debtor countries to pay their external debt than toward improving the standard of living of their populations (item **1647**).

While none of the studies lays out a clear vision of the future, many authors recognize that such a future must be one in which the Caribbean is able to compete in a global economy where traditional trade concessions will ultimately be replaced by competition. The foundation of Caribbean competitiveness is seen to rest on a more skilled labor force producing high value added products for export. Some see economic integration, especially within the Caribbean Community (CARICOM), as a necessary step toward this greater competitiveness (see items **1643, 1645, 1653,** and **1644**).

The assessments of the past three decades of Caribbean economic performance clearly suggest that the literature has arrived at some kind of watershed. As Caribbean countries restructure their economies to embrace a greater role for the private sector, the focus of research is expected to shift away from the role of the public sector and public institutions to the role of the private sector and private institutions. The impact of a more competitive Caribbean on the region's environment will also demand greater research attention than in the past. This means that the relevance of future economic research on the Caribbean will increasingly require a focus on those microeconomic issues affecting production efficiency and the environment.

GENERAL

1638 Blackman, Courtney N. The economic management of small island developing countries. (*Caribb. Aff.*, 4:4, Oct./Dec. 1991, p. 1–12)

Argues that extreme exposure of small Caribbean economies to risk requires economic managers to operate with a generous margin of error, and that the most potentially catastrophic outcome is loss of economic sovereignty to the IMF, the World Bank, or other creditors.

1639 Caribbean ecology and economics. Edited by Norman P. Girvan and David Simmons. St. Michael, Barbados: Caribbean Conservation Association, 1991. 260 p.: ill.

Selection of papers from the Caribbean Conference on Economics and the Environment focuses on the environmental degradation resulting from use and exploitation of the region's natural resources. The general conclusion is that governments of the region should carefully monitor structural adjustment, export-led growth, and development patterns that may jeopardize the very resource base on which their growth and development is predicated.

1640 **Célimène, Fred** and **Patrick Watson.** Economie politique caribéenne. Paris: Economica, 1991. 245 p.: bibl., ill. (Coll. Caraïbe-Amérique latine)

Claims to be first work written in French to examine Caribbean economic thought. Work is divided into three parts: 1) synthesis of Caribbean economic thought; 2) growth and structural transformation of the Caribbean economy; and 3) relationship between the Caribbean economy and the world economy.

1641 **Crisis económica en Centroamérica y el Caribe.** Edited by Mats Lundahl and Wim Pelupessy. San José: DEI, 1989. 297 p.: bibl. (Col. universitaria)

Papers discuss structure of individual Central American and Caribbean economies as well as problems of adjustment and agricultural production. Crisis in these economies is attributed largely to their vulnerability to world market conditions and their inability to absorb the labor force in productive employment.

1642 **Deere, Carmen Diana** and **Edwin Meléndez.** When export growth isn't enough: US trade policy and Caribbean Basin economic recovery. (*Caribb. Aff.*, 5:1, Jan./March 1992, p. 61–70, tables)

Addresses why export growth and diversification have contributed so little to economic growth. Highlights certain characteristics of non-traditional exports, principally the fast-growing assembly operations in free trade zones which have minimal linkages to the local economy.

1643 **Demas, William G.** Perspectives on the future of the Caribbean in the world economy. (*Caribb. Aff.*, 1:4, Oct./Dec. 1988, p. 6–26)

Emphasizes need for the emergence of a more active and autonomous Caribbean with its own identity, which will increasingly meet the material, psychological, and cultural needs of its people, while being integrated in a genuinely interdependent (rather than dependent) way into the international economy and community of nations. Argues that in this endeavor, it is the indigenous human capability of the Caribbean that really matters.

1644 **Farrell, Terrence W.** The political economy of Caribbean monetary and financial integration. (*Caribb. Aff.*, 4:4, Oct./Dec. 1991, p. 20–29)

Argues that monetary integration offers much—at least potentially—to a Caribbean intent on not merely surviving in tomorrow's world but also on being successful and competitive.

1645 **Hosten-Craig, Jennifer.** Regional disparity in the Caribbean: the case of the CARICOM LDCs. (*Bull. East. Caribb. Aff.*, 16:4/5, Sept./Dec. 1990, p. 24–36, bibl.)

Discusses the effort of Commonwealth Caribbean countries to integrate their economies to combat the effects of small size and geographic location. Argues that the mechanisms implemented to effect the Caribbean Common Market (CARICOM) have provided only limited developmental opportunities for member states, particularly the less developed countries (LDCs); thus, the hope for deepening the regional integration process has shifted from CARICOM to the LDCs that make up the Organization of Eastern Caribbean States (OECS).

1646 **Latorre, Eduardo.** Sobre azúcar. Santo Domingo: Instituto Tecnológico de Santo Domingo, 1988. 190 p.: bibl., ill.

An examination of the role of sugar in the development of Latin America and the Caribbean, and particularly in the Dominican Republic.

1647 **Levitt, Kari.** Debt adjustment and development: looking to the 1990s. (*Caribb. Aff.*, 3:3, July/Sept. 1990, p. 29–56)

Argues that adjustment is unavoidable, but the approach and prescription of the IMF and the World Bank are not necessarily in the best interest of countries which are increasingly in bondage to these agencies; that the leverage of the agencies may be reduced when an increasing number of countries, unable to service their multilateral debt, seek a moratorium on debt payments in exchange for a reduction in new balance of payments support.

1648 **Lewis, David E.** The North American Free Trade Agreement: its impact on Caribbean Basin economies. (*Caribb. Aff.*, 4:4, Oct./Dec. 1991, p. 56–67, table)

Argues that what the region needs to focus on is not so much threats and challenges posed by free trade in general, and NAFTA specifically, but on opportunities the current economic climate provides for economic reform and restructuring which will help the region to become more competitive internationally; and that these opportunities include not only the traditional preferential markets in North America and Europe but Latin American markets as well.

1649 **Lewis, Vaughan A.** Compulsions of integration. Black Rock, Barbados: West Indian Commission Secretariat, 1992. 26 p. (Occasional paper; 6)

Presents the position of St. Lucia on economic integration and other aspects of Caribbean cooperation. Includes commencement address by author at the Univ. of the West Indies in Barbados in which the significance of policy flexibility, continuing structural adaptation, political stability, and investment in human capital is underscored.

1650 **Mandle, Jay R.** Economic development in the Caribbean: facilitating technological innovation. (*Caribb. Aff.*, 3:3, July/Sept. 1990, p. 1–17, tables)

Argues that fundamental sources of development are found in Caribbean institutions and the role they play in either facilitating or limiting technical innovation; that the study of development should center on the impact of social institutions on the pace of change in production methods and the introduction of new products.

1651 **McAfee, Kathy.** Storm signals: structural adjustment and development alternatives in the Caribbean. Boston, MA: South End Press in association with Oxfam America, 1991. 259 p.: bibl., ill., index, map.

This work is based largely on interviews with private individuals and public officials in the Caribbean. Argues that the strategy of structural adjustment promoted by the US government, the IMF, and the World Bank is having a disastrous effect on the population of Caribbean countries. It is increasing poverty and imposing a great burden on future generations.

1652 **Stephens, Evelyne Huber** and **John D. Stephens.** Caribbean development options. (*Hemisphere/Miami*, 4:1, Fall 1991, p. 6–8)

Argues that a moderate form of democratic socialism holds more promise for Caribbean development. The State should promote special export markets, guide investment into areas with high export potential, establish linkages between agricultural and industrial development in both the use of local resources and the creation of a domestic market, and improve human resources.

1653 **Stone, Carl.** A unified regional European market and its implications challenges the Caribbean. (*Caribb. Aff.*, 3:3, July/Sept. 1990, p. 57–65)

Argues that this challenge for the Caribbean will demand joint regional policy strategies, regional thinking and collaboration by entrepreneurs, and a great deal of innovative and creative action by Caribbean business interests. This will enable them to struggle against the new trend in the global economy which has been progressively marginalizing Third World economies and preventing them from benefitting fully from the fruits of First World economic growth and development.

1654 **Tax reform in the Caribbean.** Edited by Karl Theodore. Mona, Jamaica: Regional Programme of Monetary Studies, Institute of Social and Economic Research, Univ. of West Indies, 1992. 209 p.: bibl., ill.

Collection of papers presented at a seminar at the Univ. of the West Indies, St. Augustine in 1987. Major issues addressed include direct versus indirect taxation, the role of the tax system in economic transformation, and the problems of tax administration and implementation.

1655 **West Indian Commission.** Overview of the Report of the West Indian Commission, *Time for action*. Black Rock, Barbados: West Indian Commission, 1992. 176 p.

Shortened version of the West Indian Commission report which examines prospects for greater Caribbean integration and the place of the region in the larger world of the 21st century.

1656 **Worrell, DeLisle.** A common currency for the Caribbean: a study. Black Rock, Barbados: West Indian Commission Secretariat, 1992. 32 p. (Occasional paper/West Indian Commission; 4)

Explores fiscal implications of a common currency for the Caribbean and its implications for mobility of the factors of production and for regional disparity. Discusses ways of ensuring that regional currency would be convertible and suggests alternative paths to a common currency.

DOMINICAN REPUBLIC

1657 **Camejo, Mary Jane.** Half measures: reform, forced labor and the Dominican sugar industry. New York: Americas Watch, 1991. 35 p.

Based on interviews with officials and workers, this report examines forced labor as well as poor living and working conditions in the sugar industry.

1658 **Carozzi, Federico.** El desarrollo industrial y la economía dominicana. Santo Domingo: Taller, 1991. 415 p.: bibl., ill. (Publicaciones de FUNDESIRE. Serie Ciencias, sociedad y desarrollo)

Detailed study of the structure of the industrial sector and its role in economic development between 1970–89. Also discusses technology transfer and the legal and institutional factors which affect it.

1659 **Guiliani Cury, Hugo.** Deuda externa: un proceso de renegociación. Santo Domingo: Centro de Estudios Monetarios y Bancarios, 1986. 169 p.: ill. (Serie general)

Examines evolution of the Dominican Republic's foreign debt and its renegotiation between 1985–86. Argues that the burden of foreign debt tranfers real resources abroad, thereby inhibiting growth and development.

Latorre, Eduardo. Sobre azúcar. See item **1646.**

Lewis, Maureen A. User fees in public hospitals: comparison of three country case studies. See item **1619.**

1660 **Martínez Moya, Arturo.** Inflación y estancamiento. Santo Domingo: Centro de Estudios Monetarios y Bancarios, 1989. 188 p. (Serie Estudio)

Papers presented by author at various seminars and conferences focus on monetary policy and structure of exchange rates between the early 1960s and the first half of the 1980s. Particular attention is paid to the multiple exchange rate system of the period.

1661 **Medina, Abraham.** 50 años de historia bancaria: Banco de Reservas. República Dominicana: Banco de Reservas de la República Dominicana, 1991. 546 p.: bibl., ill.

Detailed history of the Central Reserve Bank of the Dominican Republic since its founding in 1947 and its role as a protagonist in the country's development. A main objective in establishing the bank was to wrest control of economic and finanical decisions from foreign institutions.

Meyer, Carrie A. A step back as donors shift institution building from the public to the "private" sector. See item **1821.**

Murray, Douglas L. and **Polly Hoppin.** Recurring contradictions in agrarian development: pesticide problems in Caribbean Basin non-traditional agriculture. See item **1610.**

1662 **Sánchez Roa, Adriano.** FMI, agricultura y pobreza. Santo Domingo: Editora Corripio, 1991. 112 p.: bibl., ill.

Examines the IMF's role in the structural adjustment of the Dominican economy. Argues that a policy which channels the country's resources into export production in free trade zones and into tourism leads to high dependence on foreign consumption goods and neglect of such key sectors as agriculture.

1663 **Seminario ANJE, 12th, Santo Domingo?, 1990.** El Pacto de Solidaridad Económica y sus implicaciones económicas, sociales y políticas. Santo Domingo?: Asociación Nacional de Jóvenes Empresarios, 1990. 179 p.: ports.

Seminar's stated objective is to present ideas that will improve understanding of the origins, essence, and consequences of short-run economic adjustment in the Dominican Republic. The most important feature of the agreement between the government, the entrepreneurs, and the many other organizations and communities in El Pacto de Solidaridad Económica is considered to be customs tariff reform.

COMMONWEALTH CARIBBEAN AND GUYANA

1664 **Brown, Deryck R.** History of money and banking in Trinidad and Tobago from 1789 to 1989. Foreword by William G. Demas. Edited with an introduction by Terrence W. Farrell and Penelope Forde. Port of Spain?: Central Bank of Trinidad and Tobago, 1989. 284 p.: bibl., ill.

Traces establishment and early performance of banking institutions as well as the introduction of bank notes as the federal economy became more monetized. Describes a relatively well-developed financial sector, arguably the most advanced in the Commonwealth Caribbean, which includes indigenous financial institutions and a national currency issued by a national monetary authority.

1665 **Demas, William G.** Towards West Indian survival: an essay. Black Rock, Barbados: West Indian Commission Secretariat, 1990? xi, 74 p.: bibl. (Occasional paper; 1)

Argues that with political will and wise policies, the new major Commonwealth Caribbean countries (currently experiencing problems of stabilization, structural adjustment, and debt) can recover and be on the road to human development by the middle of this decade; that in spite of significant and fast changes in international economic, technical, and geopolitical environment, and the inherited deficiencies of the small West Indian economies, the region does have an important range of economic advantages and opportunities in today's and tomorrow's world; and that the region needs the four instruments of Training, Producing, Saving, and Uniting to achieve the four goals of Development, Identity, Self-respect, and Interdependence with the rest of the world in the 1990s and the 21st century.

1666 **Dijk, Meine Pieter van.** Guyana: economic recession and transition. (*Rev. Eur.*, 53, Dec. 1992, p. 95–110, bibl., tables)

Traces economic decline of Guyana since the 1970s, the consequences of this decline, and the attempt to execute a program of structural adjustment.

1667 **Farrell, Terrence W.** Central banking in a developing economy: a study of Trinidad and Tobago, 1964–1989. Mona, Jamaica: Regional Programme of Monetary Studies, Institute of Social and Economic Research, Univ. of the West Indies, 1990. 150 p.: bibl., index.

Examines difficult task of the Central Bank of Trinidad and Tobago as it seeks to define its functions, establish a degree of independence, and balance considerations of development and monetary stability. Argues that the Central Bank of Trinidad and Tobago has not been successful in balancing these complex roles and that it has failed to contribute adequately to the management of the petro-dollar boom.

1668 **French, Joan.** Hope and disillusion: the CBI in Jamaica: a case study. Kingston, Jamaica: Association of Development Agencies, 1990. x, 95 p.: bibl.

Describes grassroots perspectives on the impact of structural adjustment in rural and urban Jamaica. Argues that consultation with people should occur before, during, and after decisions are made about the country's development; that ways ought to be found to allow effective participation by a broad cross-section of people; and that the source of the Caribbean Basin Initiative's (CBI) success resides in the people and not in the marketplace.

1669 **Gayle, Dennis J.** Managing Commonwealth Caribbean tourism for development. (*Caribb. Aff.*, 3:4, Oct./Dec. 1990, p. 87–100, tables)

Argues that tourism is the only Caribbean Basin industry that has shown steady growth since 1973, but as the industry is integrated into the national and regional development strategy, it needs to achieve greater competitiveness.

1670 **Hope, Dianne L.** The traditional marketing system in Antigua: an economic analysis. Cave Hill, Barbados: Institute of Social and Economic Research (Eastern Caribbean), Univ. of the West Indies, 1991. vi, 37 p.: bibl. (Occasional paper; 22)

Examines economic efficiency of the traditional marketing system in Antigua and its ability to stimulate and/or respond to changes in output and consumer taste.

1671 **Jarvis, Leslie.** An econometric model of the Guyanese economy, 1957–1976. Lanham, Md.: Univ. Press of America, 1990. xv, 76 p.: bibl., index.

Exploratory econometric model of macroeconomic relationships which govern short-term fluctuations in the Guyanese economy.

1672 **Jones, E.** and **V. Subramaniam.** Jamaica: embracing privatization and seeking integration. (*Int. Rev. Adm. Sci.*, 59:4, Dec. 1993, p. 651–661)

Argues that formation of the middle class in the post-colonial Third World has been more bureaucracy-oriented than business-oriented; that the growth of public enterprises in Jamaica, the inability to manage them well, and the difficulty faced in privatizing these enterprises are an outgrowth of the peculiar way in which middle class bureaucracy was created during colonial times.

1673 **Levitt, Kari.** The origins and consequences of Jamaica's debt crisis, 1970–1990. Rev. and expanded. Mona, Jamaica: Consortium Graduate School of Social Sciences, 1991. 67 p.: bibl., ill.

Study, commissioned by the North-South Institute of Ottawa, Canada examines growth of Jamaica's external debt, the structural adjustment program imposed by the IMF and the World Bank, and social effects of structural adjustment. Author observes that "never in the post-war history of this country has there been a persistent trend of the enrichment of a relatively small element of the population accompanied by a sustained impoverishment of the majority of the population."

Lewis, Maureen A. User fees in public hospitals: comparison of three country case studies. See item **1619**.

1674 **McIntyre, Arnold M.** The economies of the Organisation of Eastern Caribbean States in the 1970s. Cave Hill, Barbados: Institute of Social and Economic Research (Eastern Caribbean), Univ. of the West Indies, 1986. xiv, 86 p.: bibl. (Occasional paper; 18)

Reviews macroeconomic performance of OECS countries against the background of the 1970s global recession.

1675 **Ramsaran, Ramesh.** The challenge of structural adjustment in the Commonwealth Caribbean. New York: Praeger, 1992. 206 p.: bibl., index.

Discusses some of the main issues involved in the structural adjustment of Commonwealth Caribbean economies, including privatization, public finance reform, and exchange rate adjustments. Author is critical of externally imposed adjustment programs because they "say little or nothing about the development of human resources. He feels that the universal structural adjustment template applied by the IMF is inappropriate for small Caribbean countries and argues that "adjustment requires a case by case approach in which national objectives and conditions cannot be ignored." (p. 176)

1676 **Ramsaran, Ramesh.** Growth, employment and the standard of living in selected Commonwealth Caribbean countries. (*Caribb. Stud.*, 25:1/2, Jan./July 1992, p. 103–122, tables)

Focuses on Barbados, Jamaica, and Trinidad and Tobago. Outlines recent growth experience of the region, the nature and size of the unemployment problem, factors affecting development, and recent trends in living standards.

1677 **Ryan, Selwyn D.** and **Lou Anne Barclay.** Sharks and sardines: blacks in business in Trinidad and Tobago. St. Augustine, Trinidad: I.S.E.R., The Univ. of the West Indies, 1992. 217 p.: bibl. (Culture and entrepreneurship in the Caribbean)

Examines experiences of black men and women who sought to become entrepreneurs after independence was achieved in 1962. On the basis of survey data, authors conclude that blacks in Trinidad and Tobago remain on the periphery of a business world which is dominated by whites and Indians.

1678 **Stone, Carl.** Putting enterprise to work: the Jamaican divestment experience. (*Caribb. Aff.*, 5:1, Jan./March 1992, p. 12–23, tables)

Argues that the success of Jamaican divestment experience has been unique in the Third World, involving small enterprises as well as large ones.

1679 **Worrell, DeLisle.** The economies of the English-speaking Caribbean since 1960. (*in* Democracy in the Caribbean: political, economic, and social perspectives. Baltimore: The Johns Hopkins Univ. Press, 1993, p. 189–211)

Survey of English-speaking Caribbean countries' economies between 1960–91 focuses on crises of 1971–75 and 1979–81. Author attributes the reason for the overall disappointing Caribbean performance to "adverse external circumstances, policy errors, and omissions and weaknesses in policy design."

FRENCH CARIBBEAN AND FRENCH GUIANA

1680 Eluther, Jean-Paul. La Guadeloupe ambitieuse. Paris: L'Harmattan, 1990. 91 p.

Argues that the development of Guadeloupe based on the departmental model, i.e., complete integration with metropolitan France, has created superficial industrialization, unemployment, monoculture, and a dependence on metropolitan transfers. Author believes that there ought to be a serious reassessment of this relationship.

1681 Lundahl, Mats. Politics or markets?: essays on Haitian underdevelopment. London; New York: Routledge, 1992. 519 p.: bibl., ill., index.

Argues that Haiti in 1992 is to a large extent beset by the same problems as 10 or 15 years ago. The main changes have occurred in the political area which has seen the fall of the Duvalier dynasty followed by a brief period of a democratically-elected president and his replacement by military rule. Essays focus on determinants of economic underdevelopment in Haiti, emphasizing the government's role in shaping some of the structures that keep Haiti in a state of poverty and underdevelopment.

CUBA

JORGE PEREZ-LOPEZ, *Bureau of International Labor Affairs, United States Department of Labor*

THAT CUBA IS FACING its most serious economic challenge in over 35 years of revolutionary rule is not an issue open to debate. Uncharacteristic of matters related to the island, there is consensus among government officials, analysts on the island, and scholars abroad that the economic crisis of the 1990s is extremely severe.

The tailspin of the economy has affected the quantity and thematic orientation of the professional economic literature. Shortages of paper and other problems have virtually eliminated Cuban publications: the statistical yearbook has been discontinued as of 1989 and most professional journals also have suspended publication. This near blackout on economic information from the island has been broken by an important article by Cuban economist Carranza Valdés (item **1683**), interviews granted by high-level official Carlos Lage (items **1694** and **1695**), and by publications abroad by some professional Cuban economists (items **1684, 1710,** and **1711**).

The breadth and depth of the economic crisis, coupled with the demise of socialism in the Soviet Union and Eastern Europe, have given rise to a literature that looks beyond the immediate predicaments and focuses on Cuba's post-socialist economy. Several books (items **1682, 1687, 1690,** and **1705**) and articles (items **1706** and **1707**) consider economic alternatives for Cuba—with most concluding that a market economy is the only viable one—or propose specific strategies to achieve a market economy on the island.

1682 Cardoso, Eliana A. and **Ann Helwege.** Cuba after communism. Cambridge, Mass.: MIT Press, 1992. xiii, 148 p.: bibl., ill., index, map.

Short book by experts on Latin American economics assesses Cuba's economic crisis of the 1990s and prospects. Authors see continued erosion of the economic base unless Cuba moves towards a more market-oriented system. They then lay out possible strategies for economic change.

1683 **Carranza Valdés, Julio.** Cuba: los retos de la economía. *(Cuad. Nuestra Am.,* 9:19, julio/dic. 1992, p. 131–158)

Extremely important contribution by Cuban economist associated with the Centro de Estudios de América. Provides analysis and heretofore unavailable statistics on the depth and breadth of the economic crisis of the 1990s. Essential reading for all researchers and students of the Cuban economy.

1684 **Carriazo Moreno, George.** Las relaciones económicas Cuba-Estados Unidos: una mirada al futuro. *(Estud. Int./Santiago,* 26:103, julio/sept. 1993, p. 480–499, table)

Orthodox analysis by Cuban economist of reasons for, and impact of, the US trade embargo. Argues that lifting the embargo could bring substantial benefits to certain US corporations.

1685 **Cuba at a crossroads: politics and economics after the Fourth Party Congress.** Edited by Jorge F. Pérez-López. Gainesville: Univ. Press of Florida, 1994. 282 p.: bibl., ill., index.

Eleven well written essays that examine the current state of the Cuban economy and prospects for the 1990s. Three essays deal with political issues, while eight focus upon the economy, covering such topics as food, labor, tourism, biotechnology, joint ventures, and possible economic reforms (e.g., marketization, market socialism, etc.). [J.W. Foley]

1686 **Cuba, handbook of trade statistics.** Directorate of Intelligence, Central Intelligence Agency. Washington: Directorate of Intelligence, Central Intelligence Agency; Springfield, VA: National Technical Information Service, 1993. vii, 65 p.

Second edition of very useful compendium of foreign trade statistics, compiled from official sources and from information reported by Cuba's foreign trade partners. Most data series run through 1992.

1687 **Cuba in transition.** Miami: Florida International Univ., 1992. 2 v.

Vol. 1 (1991) contains 12 papers presented at the Association for the Study of the Cuban Economy's first annual meeting, together with shorter papers, commentaries, and invited papers. Most contributions deal with issues related to Cuba's transition to a market economy. Vol. 2 (1992) contains 20 papers and commentaries.

1688 **Cuba's ties to a changing world.** Edited by Donna Rich Kaplowitz. Boulder, Co.: L. Rienner Publishers, 1993. 263 p.: bibl., ill., index, maps.

Collection of 14 essays dealing with Cuban relations with Asia, Africa, and the Middle East; Europe; Latin America and the Caribbean; and North America. Essays on relations with Japan (by Kanako Yamaoka), Britain (by Gareth Jenkins), and Brazil (by Luis Vasconcelos) have a focus on trade relations.

1689 **Deere, Carmen Diana** and **Mieke Meurs.** Markets, markets everywhere?: understanding the Cuban anomaly. *(World Dev.,* 20:6, June 1992, p. 825–839, tables)

Examines use of markets in Cuba's centrally-planned economy, focusing on farmers' free markets and government-operated parallel markets.

1690 **Desarrollo agrícola de Cuba: ciclo de estudios para el desarrollo agrícola de Cuba.** Coordinación de Arturo Pino Navarro *et al.* Selección y revisión de Darío Espina Pérez. Miami: Colegio de Ingenieros Agrónomos y Azucareros, 1992. vi, 315 p.: bibl., ill.

Collection of 20 essays on aspects of Cuban agriculture written by an impressive group of experts residing outside Cuba. Individual essays deal with topics such as land tenure, sugarcane cultivation and sugar production, cultivation of rice and coffee, animal husbandry, and fishing. Some essays make recommendations on how to restructure the agricultural sector in a market economy setting.

1691 **Díaz-Briquets, Sergio.** Collision course: labor force and educational trends in Cuba. *(Cuba. Stud.,* 23, 1993, p. 91–112)

High occupational expectations of well-educated Cubans meet the reality of an economy in crisis.

1692 **Espino, María Dolores.** Tourism in Cuba: a development strategy for the 1990s? *(Cuba. Stud.,* 23, 1993, p. 49–69, tables)

Tourism is an important source of hard currency revenue, but its contribution to national income and employment is limited.

1693 Figueras, Miguel Alejandro. Cambios estructurales en la economía cubana. (*Cuad. Nuestra Am.*, 7:15, julio/dic. 1990, p. 82–100, graphs)

Argues that socialist Cuba has successfully restructured its economy away from agriculture and from specialization on exports of a single commodity. Although conclusions are questionable, statistical comparisons are useful.

1694 Lage, Carlos. El desafío económico de Cuba. La Habana: Grupo Empresarial UFO, 1992. 73 p.

Transcript of two Cuban television programs, broadcast on Nov. 6, 1992 and Nov. 11, 1992, features high-level official Carlos Lage being interviewed by a group of national journalists. Thrust of the questioning was on problems facing the economy and strategies to address them. Contains information on foreign trade and other aspects of the economy not available elsewhere. (The two interviews are available in English as "Carlos Lage Comments on Economy," FBIS-LAT-92-219 (Nov. 12, 1992), p. 2–14 and "Carlos Lage Gives Interview on Economy 11 November," FBIS-LAT-92-224 (Nov. 19, 1992), p. 1–14). See also item **1695**.

1695 Lee, Susana. Entrevista a Carlos Lage. (*Granma/La Habana*, Oct. 30, 1993, p. 3–8)

Lengthy interview with Carlos Lage by journalist Susana Lee. Supplements public statements by Lage from a year earlier (see item **1694**) and updates dire economic situation. Also available in English as "Carlos Lage Interview on Economic Situation," FBIS-LAT-93-216-A (Nov. 10, 1993), p. 2–15.

1696 Marrero, Leví. Cuba: economía y sociedad. v. 15, Azúcar, Ilustración y conciencia, 1763–1868: part III. Madrid: Editorial Playor, 1992. 476 p.: bibl., ill., index, maps.

Final volume of monumental economic history of Cuba begun by author in 1972 (for reviews of previous volumes see *HLAS 51:2086* and *HLAS 53:2058*). Spans period between reestablishment of Spanish rule immediately after brief occupation by the British and the beginning of the Revolutionary War. Concentrates on education, political repression, and the emergence of national consciousness.

1697 Mesa-Lago, Carmelo. Cuba: un caso único de reforma anti-mercado; retrospectiva y perspectivas. (*Pensam. Iberoam.*, 2:22/23, 1992/93, p. 65–100)

Incisive analysis of Cuba's rectification policies and results.

1698 Mesa-Lago, Carmelo and Jorge F. Pérez-López. Cuban economic growth in current and constant prices, 1975–88: a puzzle on the foreign trade component of the material product system. (*in* Statistical abstract of Latin America. Edited by James W. Wilkie and Carlos Alberto Contreras. Los Angeles: UCLA Latin American Center Publications, 1991, v. 29, p. 599–615, bibl., graphs, tables)

Important contribution to debate over Cuban economic growth rates, article recalculates trade balance figures to produce a 10–50 percent lower rate of GDP growth than that reported in official data. Includes explanation of major components of Cuba's global social product (GSP). [D.W. Schodt]

1699 Mesa-Lago, Carmelo. Cuba's economic policies and strategies for confronting the crisis. (*in* Cuba after the Cold War. Edited by Carmelo Mesa-Lago. Pittsburgh, Penn.: Univ. of Pittsburgh Press, 1993, p. 197–257)

Description and analysis of strategies and policies implemented by Cuba to address the economic crisis of the 1990s.

1700 Mesa-Lago, Carmelo. The economic effects on Cuba of the downfall of socialism in the USSR and Eastern Europe. (*in* Cuba after the Cold War. Edited by Carmelo Mesa-Lago. Pittsburgh, Penn.: Univ. of Pittsburgh Press, 1993, p. 133–196)

Well-documented essay analyzes effects of the breakdown in trade with former socialist countries on specific sectors of the Cuban economy. Also published as "Efectos Económicos en Cuba del Derrumbe del Socialismo en la Unión Soviética y Europa Oriental" (in *Estudios Internacionales*, vol. 26 no. 103, julio/sept. 1993, p. 341–414).

1701 Meurs, Mieke. Popular participation and central planning in Cuban socialism: the experience of agriculture in the 1980s. (*World Dev.*, 20:2, Feb. 1992, p. 229–240, bibl., table)

Interesting case study of the implementation of the planning and management system, the Sistema de Dirección y Planificación de la Economía (SDPE), to the agricultural sector in the 1980s.

1702 Pérez-López, Jorge F. The Cuban economy: rectification in a changing world. (*Camb. J. Econ.*, 16:2, March 1992, p. 113–126, bibl., tables)

Documents Cuba's lackluster economic performance during the rectification period and the impact of the shocks brought about by the dissolution of the socialist economic community.

1703 Pérez-López, Jorge F. Economic reform in Cuba: lessons from Eastern Europe. (*in* Cuba at a crossroads: politics and economics after the Fourth Party Congress. Gainesville: Univ. Press of Florida, 1994, p. 238–263, bibl., tables)

Based on historical review of marketization efforts in Poland, Hungary, and Czechoslovakia, author argues that Cuba would need to address the following issues in creating a market economy: stabilization; creating an institutional framework for the market; privatizing; transforming government structures; and establishing a social safety net, especially for dislocated workers and the needy. [M. Mamalakis]

1704 Pérez-López, Jorge F. Islands of capitalism in an ocean of socialism: joint ventures in Cuba's development strategy. (*in* Cuba at a crossroads: politics and economics after the Fourth Party Congress. Gainesville: Univ. Press of Florida, 1994, p. 190–219, bibl., tables)

Thorough examination of the role of promoting foreign investment—particularly in the form of equity joint ventures—within Cuba's development strategy. Even though there has been an upturn since 1989 to about $500–600 million, author remains skeptical about the ability of foreign investment to solve Cuba's problems. [M. Mamalakis]

1705 Preeg, Ernest H. and **Jonathan D. Levine.** Cuba and the new Caribbean economic order. Washington: Center for Strategic & International Studies, 1993. 94 p. (Significant issues series, 0736-7136; v. 15, no. 2)

Thoughtful analysis of Cuba's economic situation and prospects in the context of other Caribbean nations. Assuming there is a fundamental market-oriented restructuring accompanied by lifting of the US embargo, author sees a strong, favorable reaction of the Cuban economy within a five-year period.

1706 Ritter, Archibald R.M. Cuba in the 1990s: economic reorientation and international reintegration. (*in* Cuba in transition: crisis and transformation. Edited by Sandor Halebsky and John M. Kirk. Boulder, Colo.: Westview Press, 1992, p. 115–135)

Cuba's challenge is to reorient its external economic relations toward new markets after the dissolution of the socialist trading bloc. Also published as "Cuba en los Noventa: Reorientación Económica y Reintegración Internacional" (in *Estudios Internacionales*, vol. 26 no. 103, julio-dic. 1993, p. 454–479).

1707 Ritter, Archibald R.M. Exploring Cuba's alternate economic futures. (*Cuba. Stud.*, 23, 1993, p. 3–31, tables)

Provocative essay. Author examines several possible economic futures for Cuba, arranged in the form of scenarios, and suggests outcome of each. Most likely—and desirable—outcome would be evolution toward a "market economy with a social face."

1708 Roca, Sergio G. Evolución del pensamiento cubano sobre Cuba y la economía mundial a través de las revistas económicas. (*Estud. Int./Santiago*, 26:103, julio/sept. 1993, p. 537–564, table)

Author traces views of professional economists in Cuba on international economic matters by analyzing their writings in leading journals. Concludes that the profession was ill-prepared to address challenges brought about by the collapse of economic relations with the Soviet Union and Eastern Europe.

1709 Rodríguez, José Luis. Los cambios en la política económica y los resultados de la economía cubana, 1986–1989. (*Cuad. Nuestra Am.*, 7:15, julio/dic. 1990, p. 63–81, tables)

Sympathetic description and evaluation of economic policies since 1986. Author was formerly with the Centro de Investigaciones de la Economía Mundial. Since 1993 he has been Minister-President of the Comité Estatal de Finanzas.

1710 **Rodríguez, José Luis.** Cuba en la economía internacional: nuevos mercados y desafíos de los años noventa. (*Estud. Int. / Santiago*, 26:103, julio/sept. 1993, p. 415–453, tables)

Describes Cuba's experience in diversifying its trade in the 1980s and intensification of efforts to do so in the 1990s.

1711 **Rodríguez, José Luis.** The Cuban economy in a changing international environment. (*Cuba. Stud.*, 23, 1993, p. 33–47)

Author analyzes adverse impact on the Cuban economy of changes in relations with former socialist countries and strategies being followed to overcome the economic crisis.

1712 **Rosenberg, Jonathan.** Cuba's free-market experiment: los mercados libres campesinos, 1980–1986. (*LARR*, 27:3, 1992, p. 51–89, bibl.)

Excellent article on historical development of farmers' free markets, their success, and consequent demise. Research relies on Cuban and US literature and on interviews with former Cuban officials and emigrés who participated in the short-lived free-market experiment.

1713 **Svejnar, Jan** and **Jorge F. Pérez-López.** A strategy for the economic transformation of Cuba based on the East European experience. (*in* Cuba after the Cold War. Edited by Carmelo Mesa-Lago. Pittsburgh, Penn.: Univ. of Pittsburgh Press, 1993, p. 323–351, tables)

Argues that because Cuba's economy shares many features with economies of most Central and Eastern European countries, many effective transition policies for Cuba can be derived from those countries' experiences. Also highlights important differences such as the relatively higher extent of State ownership and absence of a history of experimentation with reform in Cuba. Recommended. [D.W. Schodt]

1714 **Vega Vega, Juan.** Sobre inversiones extranjeras en Cuba: comentarios a la legislación cubana sobre asociaciones eco nómicas con empresarios extranjeros. (*Rev. Cuba. Derecho*, 5, enero/marzo 1992, p. 28–39)

Commentary to foreign joint venture law by prominent Cuban jurist.

1715 **Zimbalist, Andrew.** Teetering on the brink: Cuba's current economic and political crisis. (*J. Lat. Am. Stud.*, 24:2, May 1992, p. 407–418)

Sober analysis of the condition of the Cuban economy in the 1990s and of strategies being pursued to reverse the economic crisis.

VENEZUELA

ROBERT PALACIOS, *Economist, World Bank*

MANY OF THE WORKS CITED BELOW represent an initial reaction to the economic collapse of 1989. With "El Paquete," as the austerity measures have come to be known, President Pérez began to reverse many of the policies initiated by his own administration 15 years earlier. Less than two decades after massive nationalizations, airlines, hotels, the telephone company, and other State firms were put on the privatization block. After years of broad consumer subsidies, targetted social programs were espoused as more efficient in the fight against poverty. Furthermore, a new emphasis on primary education was a dramatic shift for a president who had concentrated resources on the expansion of the university system during his first term. The historic reversals were perhaps a fitting end to a period which was tragic in terms of opportunities lost since 1974.

These measures permanently altered the view held by many Venezuelans about their economic and political system and led to a rethinking of the role of the State in the economy. Some of the literature annotated in this chapter relates to emerging issues such as privatization (item **1741**) and fiscal decentralization (item **1728**). Other authors grapple with the impact of the first few years of austerity on different aspects of the economy. Cartaya and D'Elia (item **1721**), Fajardo Cortez (item **1727**), and Lovera (item **1735**) look at poverty, unemployment, and housing respectively, while Jongkind (item **1733**) provides a first glimpse at the effects of the 1989 economic policy changes on the country's industrial base.

Another major contrast between the first and second Pérez Administrations was the leader's relationship with multilateral development institutions. During his first administration, Pérez used massive oil revenues to increase his international stature and influence hemispheric politics. Venezuela became a net creditor to institutions like the World Bank and the Inter-American Development Bank. By 1989, with international reserves depleted and a huge external debt, the second Pérez Administration began to borrow from these institutions and to take some of their advice on structural adjustment measures. This new relationship helps explain the recent surge in the number of World Bank works on Venezuela covering topics such as health (item **1754**), education (item **1755**), and labor policy (item **1736**).

Many of the economists from the Instituto de Estudios Superiores en Administración (IESA) served as technocrats in the first years of the second Pérez Administration, which may explain the decline in published output from that institution as compared to previous years. Nevertheless, IESA authors continue to produce economic analysis of high quality, including the excellent study by the team of researchers directed by Márquez under the auspices of the Inter-American Development Bank (item **1737**) and the exhaustive study of the telecommunications privatization by Francés *et al.* (item **1730**). From Washington, IESA professor and former Pérez minister Moisés Naim adds to his earlier post-mortem analyses of the 1989–91 reforms (item **1739**). A somewhat contrasting view of the same period is found in the account by Bottome (item **1719**).

The remainder of the chapter is comprised of economic history (which, as in past writings by the same authors, is more history than economics) and several useful statistical publications. The new statistics from the Oficina Central de Estadística e Informática (OCEI) include 25 years of industrial survey data (item **1752**) and the detailed, state-level results of the housing census conducted as part of the 1990 population census (item **1747**).

1716 **Abreu Olivo, Edgar et al.** La agricultura: componente básico del sistema alimentario venezolano. Caracas: Fundación Polar, 1993. 432 p.: bibl., ill. (Sistema alimentario venezolano. Componentes)

Detailed analysis of Venezuela's agricultural sector. Looks at government subsidies, agricultural trade policy, and domestic production. Most data current through 1991.

1717 **Adriani, Alberto.** Labor venezolanista: Venezuela, las crisis y los cambios. Caracas: Fondo de Inversiones de Venezuela, 1987. 376 p.

Writings by Italian immigrant and prominent public official Alberto Adriani on the 50th anniversary of his passing (1931 essay by Adriani is also included in vol. 1 of *La economía contemporánea de Venezuela*, 1990, see *HLAS 53:2096*). Writings include sadly ironic discussions of devaluation, emergence of the social welfare State, and tax administration. The man who sought an equitable and efficient tax regime for Venezuela would have been greatly disappointed by the turn of events 50 years later.

1718 **Azpúrua, Pedro Pablo; Aurelio Useche K.; and Eloy Lares Monserratte.** Tres escenarios: la administración del Estado y la participación privada [de] Pedro Pablo Azpúrua. Proposición: la administración de los servicios públicos descentralizados y la participación del sector privado [de] Pedro Pablo Azpúrua, Aurelio Useche K. y Eloy Lares Monserratte. Caracas: Fundación Polar, 1988. 52 p.

Interesting proposals which reflect decay in public services experienced since the late 1970s. Recommendation to expand the private sector's role is not surprising given sponsorship of the study.

1719 Bottome, Robert. Venezuela: the struggle for reform. (*in* In the shadow of the debt: emerging issues in Latin America. New York: Twentieth Century Fund Press, 1992, p. 59–81, graphs, tables)

Another example of Venezuelan interaction with political and economic analysts in the US. Detailed account of reforms from 1989–91 and a good counterpoint to Naim's piece (item **1739**). Compares difficulties of reform in Venezuela to those in Hungary and Poland in terms of entrenched attitudes and economic culture. Author publishes the monthly economic newsletter, *VenEconomía Mensual*.

1720 Cartay Angulo, Rafael. Historia económica de Venezuela, 1830–1900. Valencia, Venezuela: Vadell Hnos., 1988. 331 p.: bibl., ill.

Extremely thorough description of the Venezuelan economy in the 19th century. Includes data on every aspect of life such as a detailed picture of industry and commerce, the number of head of cattle, and government spending on defense. Extensive bibliography.

1721 Cartaya, Vanessa and **Yolanda D'Elía.** Pobreza en Venezuela: realidad y políticas. Caracas: Cesap-Cisor, 1991. 243 p.: bibl., ill. (Enfoque social, 0798-2356; 1)

Very detailed account of poverty and Venezuelan social programs designed to alleviate it. Presents rich source of data including wages, income distribution, housing, social spending, and health through the severe recession of 1989 and, in some cases, through 1991. Good bibliography for each section.

1722 Coleman, Jonathan Roger and **Donald F. Larson.** Tariff-based commodity price stabilization schemes in Venezuela. Washington: International Economics Dept., World Bank, 1991. 46 p.: bibl. (Policy, research, and external affairs working papers; WPS 611)

Investigates possible price-stabilization schemes for Venezuela's agricultural sector in the context of early agricultural reforms during the second Pérez Administration. Concludes that the most appropriate policy for Venezuela would be a "price band" scheme based on a five-year moving average of monthly prices.

1723 Coronel, Gustavo. Venezuela: la agonía del subdesarrollo. Caracas: G. Coronel, 1990. 209 p.: bibl., index.

Subjective but lucid account of Venezuela's political economy and social history over the last few decades. Includes names and deeds of reformers and villains alike. Author, a geologist, does not use economic framework for the analysis.

1724 Cox, Donald and **George Psacharopoulos.** Female participation and earnings, Venezuela 1987. (*in* Case studies on women's employment and pay in Latin America. Washington: World Bank, 1992, p. 451–461, bibl., tables)

Using the 1987 *Encuesta de Hogares*, authors first estimate determinants of female labor force participation taking into account factors such as age, marital status, and education level attained. After verifying high rates of return to education for both males and females, male/female wage differentials are investigated. Authors conclude that "less than a third" of the 42 percent wage advantage to men can be explained by "observable" factors.

1725 Dunia A., Jonny E. Y . . . ¿qué pasó con la industrialización? Caracas: Univ. Central de Venezuela, Facultad de Ciencias Económicas y Sociales, 1991. 80 p.: bibl., graphs, tables.

Changes in the industrial sector between 1960–85 are outlined in painful detail and are related to policy measures undertaken. Abundance of sectoral data may be useful for specialists in this area, but it makes dense reading for the non-specialist or those interested in the current industrial situation.

1726 Duque Corredor, Román José *et al.* Política agraria y desarrollo. Recopilación de Oscar David Soto. Valencia, Venezuela: Univ. de Carabobo, Vicerrectorado Académico, Ediciones del Consejo de Desarrollo Científico y Humanístico, 1988. 274 p.

Series of short essays by many Venezuelan authors on topics related to agriculture ranging from agricultural credit to family planning in rural areas. Somewhat superficial and ideological and lacking in economic analysis. Includes some interesting historical data.

1727 **Farjado Cortez, Víctor.** Políticas económicas y paro forzoso: Venezuela 1989–1991. (*Cuad. CENDES*, 17/18, abril/dic. 1991, p. 17–65, bibl., tables)

Describes contrasting periods 1986–88 and 1989–91 with abundant descriptive statistics but shallow analysis. Critical assessment of the unemployment insurance program introduced in 1989–90 and known as the "Ley del Paro Forzoso" is useful. In general, however, lacks analytical clarity. Proposed solutions, which involve an exchange rate anchor and appreciation, are unconvincing.

1728 **Federalismo fiscal: el costo de la descentralización en Venezuela.** Coordinación de Rafael de la Cruz y Armando Barrios. Caracas: Comisión Presidencial para la Reforma del Estado (COPRE); Programa de las Naciones Unidas para el Desarrollo (PNUD); Editorial Nueva Sociedad, 1994. 303 p.: bibl., ill., map. (Serie Venezuela, la reforma del futuro)

Given the historical concentration of fiscal power in Caracas during this century, few studies have analyzed the role of subnational public spending in the development of the country. Study places the Venezuelan case of fiscal decentralization into the context of the phenomenon in Latin America. Provides extensive data and good analysis through the early 1990s when political and economic power began to devolve to state and local governments in Venezuela.

1729 **Ferrán, Lourdes de.** Participación económica de la mujer y la distribución del ingreso. Caracas: Banco Central de Venezuela, 1986. 246 p.: bibl., ill. (Col. de estudios económicos; 13)

Rich descriptive analysis of Venezuelan labor markets discusses the contribution of females. Makes interesting use of data such as calculating the value of housewives' labor using the *Encuesta de Hogares*. However, lacks some empirical and theoretical methods often used to analyze issues such as gender discrimination. See items **1724** and **1757**.

1730 **Francés, Antonio; Felipe Aguerrevere; Raquel Benbunan; and María Boza.** ¡Aló Venezuela!: apertura y privatización de las telecomunicaciones. Caracas: Ediciones IESA; Conatel, 1993. 427 p.: bibl., ill.

Thorough and well-documented account of privatization of CANTV along with projections for the future of Venezuela's telecommunications industry. Includes interesting details on everything from media coverage of the privatization process to the international experience with the cost of telephone service.

Fullerton, Thomas M. and **Ajay Kapur Jr.** Predicción de multiplicadores monetarios en Colombia, Ecuador y Venezuela. See item **1801**.

1731 **Historia de las finanzas públicas en Venezuela.** Siglo XX. v. 41. Planificación y ordenación por Tomás Enrique Carrillo Batalla. Caracas: Academia Nacional de la Historia, 1992. 1 v. (Biblioteca de la Academia Nacional de la Historia. Serie Economía y finanzas de Venezuela; 41)

Author and his team of researchers explore details of public finance in 1936–37 under President López Contreras. The continuing expansion of the government's role since the end of the Gómez era is documented in sometimes excruciating detail.

1732 **Izard, Miguel et al.** Política y economía en Venezuela, 1810–1991. Presentación por Alfredo Boulton. 2. ed. Caracas: Fundación John Boulton, 1992. 379 p.: bibl., ill.

Collaborative international effort to connect political and economic history during almost two centuries of Venezuelan history.

1733 **Jongkind, Fred.** Venezuelan industry under the new conditions of the 1989 economic policy. (*Rev. Eur.*, 54, June 1993, p. 65–93, bibl., tables)

Very basic account of the state of industry at the beginning of the second Pérez Administration. Includes usual indictment of past industrial policies.

1734 **Larios, F.; J. Caro; and L. Clemente.** La política crediticia en la agricultura venezolana. (*Debate Agrar.*, 16, enero/abril 1993, p. 127–141, tables)

Traces evolution of agricultural credit until 1989 when reforms were put into place. Good analysis of fairly clear reasons that subsidized credit policies failed to stimulate agricultural production in the 1974–88 period.

1735 Lovera, Alberto. Techos recortados: vivienda y ajuste económico en Venezuela. (*in* Políticas habitacionales y ajustes de las economías en los 80s. Guatemala: IDESAC; SIAP; San José: Consejo Superior Universitario Centroamericano, 1991, p. 96–116, graphs)

Looks at evolution of housing and construction during the 1980s. Criticizes government policies and emphasizes the *Ley de Política Habitacional*, which subsidized variable rate mortgages for middle and upper classes at great expense to the government. Includes data on the sector.

1736 Márquez, Gustavo. Cui bono?: regulation and outcomes in the labor market. Washington: World Bank, 1993. 124 p.: graphs, ill.

Best analysis of Venezuelan labor market since the 1989 crisis by leading authority on the issue in Venezuela. Data includes detailed sectoral breakdowns for formal and informal sectors. Covers all aspects of government intervention in labor markets ranging from wage controls and collective bargaining to the social insurance system.

1737 Márquez, Gustavo *et al.* Fiscal policy and income distribution in Venezuela. (*in* Government spending and income distribution in Latin America. Edited by Ricardo Hausmann and Roberto Rigobón. Washington: Inter-American Development Bank, 1993, p. 145–213, bibl., graphs, tables)

After a throrough discussion of increasing poverty rates during the 1980s, separate studies of fiscal incidence of public spending are presented for social insurance, electricity and gasoline subsidies, and social insurance programs. Good example of empirical economic research with strong policy implications, including the clear need to raise gasoline prices and reform the pension system.

1738 Maza Zavala, Domingo Felipe. Los procesos económicos y su perspectiva. Caracas: Academia Nacional de la Historia, 1990. 299 p.: bibl. (Biblioteca de la Academia Nacional de la Historia. Estudios, monografías y ensayos; 134)

Series of essays by one of Venezuela's "old guard" economists. Author's florid prose is in stark contrast to the new wave of economic writings emerging from the country.

1739 Naím, Moisés. The launching of radical policy changes, 1989–1991. (*in* Venezuela in the wake of radical reform. Edited by Joseph S. Tulchin with Gary Bland. Boulder, Colo.: Lynne Rienner Publishers, 1993, p. 39–94, graphs, tables)

Solid analysis focusing on post-1988 reforms. However, many of the interesting insights are already covered in the author's recent popular book *Paper tigers and minotaurs* (Washington: Carnegie Endowment for International Peace, 1993, 180 p.).

1740 Palma C., Pedro A. Contracción de los precios de exportación, deuda y crecimiento en economías en desarrollo: el caso de Venezuela. Caracas: Academia Nacional de Ciencias Económicas, 1989. 56 p.: bibl., ill. (Serie Cuadernos; 25)

Review of oil boom period is followed by an application of author's MODVEN economic forecasting model. Tests sensitivity of various indicators to different oil price scenarios through 1991. Like Hausmann, author sees a need to build international reserves to mitigate negative oil price movements.

1741 Recio Pinto, Alejandro. Privatización en Venezuela: primer caso, Banco Occidental de Descuento, B.O.D. Caracas: Librería Mundial, 1991. 206 p.: ill.

Author provides extremely detailed information on process of privatizing the Banco Occidental de Descuento, the first application of the *Ley de Mercado de Capitales*. Traces evolution of this early privatization including political opposition and media coverage. Also includes some illuminating discussion of Venezuelan private banks before the collapse of 1993.

1742 Sabino, Carlos. De cómo un estado rico nos llevó a la pobreza: hacia una nueva política social. Prólogo de Luis Pazos. Caracas: CEDICE, 1994. 143 p.

Well-written essay on basic problem of the overextended welfare State in Venezuela. It is surprising to read frequent citations of

Hayek by this professor from the Univ. Central, a bastion of socialist thinking. The antigovernment message—which is almost libertarian—is a direct attack on old ideas about the ability of the State to solve all problems.

1743 **Sachs, Jeffrey D.** Que se piensa en el exterior de la política económica de Venezuela. Caracas: Academia Nacional de Ciencias Económicas, 1989. 29 p.: ill.

Somewhat exaggerated title actually refers to a speech by Jeffrey Sachs at the National Academy of Economic Science during the first year of the second Pérez Administration. Sachs compares post-war Latin American economic policies with those in Southeast Asia and praises policies of the new Pérez government. Predictably, Sachs preaches continued orthodoxy and expresses optimism about Venezuela's future.

1744 **Salazar-Carrillo, Jorge** and **Roberto D. Cruz.** Oil and development in Venezuela during the twentieth century. Westport, Conn.: Praeger, 1994. 280 p.: bibl., index, tables.

Traces the role of oil revenues through the century. More of a historical account than economic analysis. Perhaps most useful for its detailed accounts of the less-documented 1910–57 period. Includes many tables of economic data.

1745 **Sánchez Guerrero, Gustavo.** La nacionalización del petróleo y sus consecuencias económicas. Caracas: Monte Avila Editores, 1990. 126 p.: bibl., ill. (Perspectiva actual)

Descriptive analysis of planning and organization of Venezuela's State-controlled energy sector focuses on relationship between PDVSA and central government. Finds system has not adjusted to post-nationalization realities. Criticizes inefficient decision-making and wasteful policies such as indiscriminant subsidies through domestic sales of petroleum derivatives below market prices.

1746 **Silva, Carlos Rafael.** Medio siglo del Banco Central de Venezuela. Caracas: Academia Nacional de Ciencias Económicas, 1990. 390 p.: bibl., ill.

Good, through description of Venezuela's economic and monetary history includes a review of laws and decrees that defined evolution of Venezuela's experience with a central monetary authority. Unfortunately, lacks analysis or critical thinking about the implications of the changes described.

1747 **Situación habitacional en Venezuela.** Caracas: Oficina Central de Estadística e Informática, 1994. 469 p.: tables.

Based on the 1990 Census, this housing survey includes important data disaggregated by states for the serious researcher. Useful for analyzing effect of the income declines of the 1980s from the perspective of housing quality and quantity.

1748 **Toro Hardy, José.** Venezuela: 55 años de política económica, 1936–1991. Caracas: Editorial PANAPO, 1992. 214 p.: bibl., ill.

Relying primarily on historical data drawn from Asdrúbal Baptista's work and Central Bank documents, author also makes use of graphs, tables, and extensive quotes from secondary material. Includes interesting discussion of contradictions of "el paquete" in Chap. 12.

1749 **Valecillos Toro, Héctor.** Acumulación del capital y desigualdades distributivas en la economía venezolana. Caracas: INAESIN, 1989. 251 p.: bibl., ill.

Highly descriptive analysis of patterns of public and private investment at the macroeconomic level and their possible relationship to the income distribution. Contains abundance of tables including historical series of income distribution numbers.

1750 **Valecillos Toro, Héctor.** Economía y política del trabajo en Venezuela. Caracas: Academia Nacional de Ciencias Económicas, 1990. 527 p.: bibl., ill.

Collection of analyses of the labor market. Includes reprints of interesting papers by important Venezuelan authors and some useful descriptive information. However, most of the empirical work is useful only for historical analysis and may frustrate the researcher expecting an explanation of policies and experiences of the 1980s.

1751 **Valecillos Toro, Héctor.** Proceso y crisis de la inversión privada en Venezuela. Venezuela: Univ. Central de Venezuela, Facultad de Ciencias Económicas y Sociales, 1990. 155 p.: bibl., ill.

Adequate description of investment

patterns since 1955. Author cites problems with orthodox adjustments such as real wage suppression. Repeats much of the analysis in his earlier 1989 text *Acumulación del capital y desigualdades distributivas en la economía venezolana* (see item **1749**).

1752 25 años, encuesta industrial. Caracas: Oficina Central de Estadística e Información (OCEI), 1989. 163 p.: bibl.

Description of definitions, methodology, and how these changed between 1961–86 is followed by more than 80 p. of historical data for Venezuela's industrial sector. Does not include any analysis of these detailed survey results.

1753 Velázquez, Efraín J. El déficit público y la política fiscal en Venezuela, 1980–1990. Caracas: Banco Central de Venezuela, 1991. 167 p.: bibl. (Col. de estudios económicos, 0798–3026; 14)

Interesting, although densely written, attempt to measure and analyze role of quasi-fiscal deficits to ascertain macroeconomic implications of fiscal deficits. Author finds that taking various subsidies into account—including large Central Bank losses resulting from foreign exchange subsidies to importers during the period of multiple exchange rates—results in larger fiscal deficits between 1983–89.

1754 Venezuela: health sector review. Washington: The World Bank; Population and Human Resources Division (Latin American and the Caribbean Regional Office); Country Dept. I, 1993. 2 v.: graphs, ill.

Compiles some of the best data available on the health sector, including causes of morbidity and mortality through 1988. Includes the World Bank's assessments of some targeted programs like the *Beca Alimentaria* implemented after 1989.

1755 Venezuela 2000: education for growth and social equity. Washington: The World Bank, Population and Human Resources Operations Division, Country Dept. I, 1993. 1 v.

Excellent comprehensive study of education sector at all levels. Best source of data available for US researchers on the topic of recent Venezuelan education policy. Most who know the subject, however, will not be surprised by findings nor policy recommendations except perhaps by the fact that education spending increased as a share of GDP between 1990–92.

1756 Vetencourt Guerra, Lola. Monopolios contra Venezuela, 1870–1914. Caracas: Ediciones FACES/UCV; Vadell Hnos., 1988. 252 p.: bibl.

Ideological account of role of foreign capital in the late 19th and early 20th centuries. Contains interesting tables on role foreign investment played in the infrastructure of Latin America and Venezuela in particular. Concludes that any development which took place only made the process of plundering the country's natural resources more efficient and did not help Venezuelans in general.

1757 Winter, Carolyn. Female earnings, labor force participation and discrimination in Venezuela, 1989. (*in* Case studies on women's employment and pay in Latin America. Washington: World Bank, 1992, p. 463–475, bibl., tables)

Essentially repeats Cox/Psachoropoulos study (item **1724**) using data from the 1989 *Encuesta de Hogares*. Results contrast with the first study in that male/female wage differentials are found to be much lower using the 1989 data. Nevertheless, quantitative analysis did not explain salary advantage for males over females of close to 20 percent.

COLOMBIA

NOHRA REY DE MARULANDA, *Manager of Integration and Regional Programs Department, Inter-American Development Bank*

ALTHOUGH CERTAIN TRENDS already highlighted in previous *HLAS* volumes continue (i.e., scarcity of theoretical works and emphasis on surveys which collect essays on key economic policy issues), it is encouraging to acknowledge the significant increase of quantitative and econometric research in Colombia, as reflected by a sizeable number of relatively short articles which appear in the main journals. The dominant subjects of these articles seem to be financial sector analysis, inflation, and monetary and fiscal policy. Among the journals, the largest number of contributions comes from the Central Bank's *Ensayos sobre Política Económica*.

In recent years, much of the policy debate among economists in Colombia and, indeed, in Latin America, has been related to the profound macroeconomic reforms which have taken place as a response to the deep crisis faced by the region in the decade of the 1980s, commonly referred to as "The Lost Decade." Not surprisingly, many of the works annotated below are related to research, conferences, and public debates on the nature of those reforms. In particular, the reform of the external sector, known in Colombia as "La Apertura," accounts for about one-third of the annotated works.

Two pieces deserve to be singled out: Londoño's examination of changes in income distribution from 1938–88 (item **1776**) and Urrutia's edited volume *40 años de desarrollo: su impacto social* (item **1767**). Both explore a subject which has not received much attention from mainstream economists, i.e., the analysis of the changes in the structure, level, and distribution of income, wealth, and welfare as a result of economic growth. Using different methodological approaches, both studies reach very positive conclusions on the progress achieved by Colombia in this area.

One should highlight also the contributions of Luis Jorge Garay (items **1773, 1771, 1772,** and **1778**) on Colombia's external sector. Both as an author and as an editor, his work is rigorous and valuable.

Finally, a theme emerging in the literature which deserves far more study and analysis is the impact of the drug trade on Colombia's economy and society. Interesting contributions to this new topic include two articles in a compilation of readings on capital inflows (item **1778**; see chapters by Urrutia and Pontón and by O'Byrne and Reina) and a work edited by C. Arrieta *et al.* on the impact of narcotrafficking in Colombia (item **1760**).

1758   **Aguilar P., José Alejandro *et al.*** El campesinado en Colombia hoy: diagnóstico y perspectivas. Edición de Edelmira Pérez C. Bogotá: Pontificia Univ. Javeriana, Facultad de Ciencias Económicas y Administrativas; ECOE Ediciones, 1991. 351 p.: bibl., ill., map. (Serie Investigacíon y desarrollo; 3)

Collection of papers presented at conference cover the rural sector and its people, both important for Colombia's development. Essays provide interesting information, including analyses of peasant society, organizations which represent them, and links to the urban context. Interesting points of view presented by NGOs working in rural development in Colombia.

1759   **Apertura económica y sistema financiero.** Edición de Florángela Gómez Ordóñez. Bogotá: Asociación Bancaria de Colombia, 1990. 439 p.: bibl., ill.

Conference proceedings organized by the National Association of Banks discuss the impact of trade liberalization and globalization of the economy on the financial sector. Contains articles by leading economists, high ranking policymakers, and high level representatives of the public and private banking systems who explore constraints and challenges that the new economic model poses to the banking sector. Examines issues like timing of reforms, upgrading of the sector, and need for more precise risk analysis.

1760   **Arrieta, Carlos Gustavo *et al.*** Narcotráfico en Colombia: dimensiones políticas, económicas, jurídicas e internacionales. Bogotá: Ediciones Uniandes; Tercer Mundo Editores, 1990. 374 p.: bibl., ill. (Sociología y política)

Very interesting book on narcotics, a topic which has not been well researched in Colombia. The problem is viewed from four angles: the economics of drugs; criminal justice; the connection between drugs and politics; and finally, the role this issue plays in

Colombia's bilateral relationship with the US. Four university professors and their associates contributed to this multidisciplinary work, which provides a useful framework for analyzing the origin and causes of this problem and inconsistencies in the way the State has handled it. For political scientist's comment see HLAS 53:3706.

1761 **Beltz Peralta, Hernán et al.** Apertura y modernización: las reformas de los noventa. Edición de Eduardo Lora. Bogotá: FEDESARROLLO; Tercer Mundo Editores, 1991. 298 p.: bibl., ill., maps. (Economía colombiana)

Collection of articles offers comprehensive account of nature and scope of massive economic reforms recently adopted in Colombia to stabilize public finances and globalize the economy. Introductory essay provides useful guideline of the magnitude of sectors covered by the recent reforms of trade, labor sector, exchange rate, tax, financial sector, rules regulating foreign investment, ports and transport sector, and institutional reforms. Other articles, many of which were prepared by members of the government's economic team, take a more in-depth look at individual sector reforms. Although not analytical, this is a very useful contribution towards understanding the magnitude of changes occurring in Colombia.

1762 **Botero, Libardo et al.** Neoliberalismo y subdesarrollo: un análisis crítico de la apertura económica. Bogotá: El Ancora Editores, 1992. 187 p.: bibl.

Academics and journalists strongly criticize Colombia's recent macroeconomic reforms. Criticism goes beyond Colombian policies and covers equity and efficiency problems of what authors perceive as economic neoliberalism in Latin America's new development trends.

1763 **Caballero, Carlos et al.** Apertura y crecimiento: el reto de los noventa. Edición de Eduardo Lora. Bogotá: FEDESARROLLO; Tercer Mundo Editores, 1991. 228 p.: bibl., ill. (Economía colombiana)

Collection of short essays by well-known Colombian economists on the most likely effects of macroeconomic and trade reforms on Colombia's growth pattern in the 1990s. Authors explore mechanisms by which structural reform will presumably enhance growth in years to come, as well as the most obvious risks and constraints of the model adopted.

Colmenares, Germán et al. Ensayos de historia económica de Colombia. See HLAS 54: 2627.

1764 **Colombia. Misión de Estudios del Sector Agropecuario.** El desarrollo agropecuario en Colombia: informe final. v. 1-2. Misión de Estudios del Sector Agropecuario. Bogotá: Ministerio de Agricultura; Depto. Nacional de Planeación, 1990. 2 v. (862 p.): bibl., ill., maps.

Very thorough and comprehensive report on characteristics, conditions, and changes which took place in the agricultural sector over the last 30 years. Although commissioned by the Government, study was carried out independently by a recognized group of economists and specialists directed by Albert Berry. The project's objective is to provide a broad framework within which to analyze the modernization process. Published in three volumes (see also item **1765**), first two volumes contain detailed subsector diagnosis on areas like the supply of agricultural goods, peasant economy, and cattle sector. Also covers major structural transformations such as demography, property and use of land, and regional dynamism. Also explores impact of macroeconomic policy on the sector.

1765 **Colombia. Misión de Estudios del Sector Agropecuario.** Estrategias y políticas para el desarrollo agropecuario: informe final de la Misión de Estudios del Sector Agropecuario. Bogotá: Ministerio de Agricultura; Depto. Nacional de Planeación, 1990. 241 p.: bibl.

Based on a comprehensive report on the agricultural sector in Colombia directed by Albert Berry (see item **1764**), this volume provides the medium and long-term policy recommendations for further development and modernization. Authors highlight advice on areas which they see as critical: the development of the agroindustrial sector, and the stimulus of exports of both basic and processed goods.

1766 **The Colombian economy: issues of trade and development.** Edited by Alvin Cohen and Frank R. Gunter. Foreword by Rodolfo Segovia S. Boulder, Colo.: Westview Press, 1992. 399 p.: bibl., ill., index.

Interesting collection of essays by Colombian and foreign economists and political scientists were presented at a conference on contemporary Colombian economics at Lehigh Univ. Provides very comprehensive overview of remarkable, very complex economic and political transformations in the last 50 years.

1767 **40 años de desarrollo: su impacto social.** Edición de Miguel Urrutia. Bogotá: Banco Popular, 1991. 207 p.: bibl., ill. (Biblioteca Banco Popular. Textos universitarios)

Interesting articles examine 40 years of major economic transformations in Colombia in light of their impact on families and individuals. Seeks to evaluate costs and benefits borne by individuals because of economic development. Innovative use of photography illustrates change during the period. Concludes with a positive view of the economic process by claiming that for Colombians, costs have been outweighed by benefits.

1768 **De Janvry, Alain** *et al.* Campesinos y desarrollo en América Latina. Bogotá: Fondo DRI; Tercer Mundo Editores, 1991. 259 p.: bibl., ill. (Sociología y política)

Detailed and comprehensive evaluation of the first 10 years of Colombia's Rural Development Program (DRI), performed by recognized national and foreign experts and researchers. Contains valuable statistical information on rural and regional development and provides programmatic and institutional policy recommendations.

1769 **Easterly, William Russell.** The macroeconomics of the public sector deficit: the case of Colombia. Washington: World Bank, Country Economics Dept., 1991. 96 p.: bibl., ill. (Policy, research, and external affairs working papers; WPS 626)

Interesting quantitative report on impact of fiscal deficits on macroeconomic outcomes. Reviews evolution of fiscal policy since the 1960s using econometric models. Highlights effect of fiscal deficits on the inflation rate, real interest rate, and real exchange rate.

1770 **Echeverri Correa, Fabio.** La industria: de los tiempos del proteccionismo a los de la apertura. Prólogo de Carlos Lleras Restrepo. Bogotá: Intermedio Editores, 1991. 239 p.

Written by former President (for 17 years) of the Association of Manufacturers in Colombia. Describes contemporary history of Colombian industrial development. Considers economic, political, and institutional factors and their role in the development process. Highlights need for an industrial policy and appropriate government intervention to develop the sector.

1771 **Estrategia industrial e inserción internacional.** Edición de Luis Jorge Garay. Bogotá: FESCOL, 1992. 444 p.: bibl., ill.

Collection of essays on effect of trade liberalization on industrial development and policy. After an interesting analysis of the limitations of conventional trade theory, presents current alternative theoretical developments in that field. Subsequent articles deal with industrial policy, technological policy, and trends in productivity.

Fullerton, Thomas M. and **Ajay Kapur Jr.** Predicción de multiplicadores monetarios en Colombia, Ecuador y Venezuela. See item **1801.**

1772 **Garay, Luis J.** Apertura y protección: evaluación de la política de importaciones. Bogotá: Tercer Mundo Editores; Facultad de Ciencias Económicas, Univ. Nacional de Colombia, 1991. 255 p.: bibl. (Economía colombiana)

Collection of essays written over the last 10 years on trade liberalization in Colombia. Covers analysis of the structure of imports, international financial markets, effective protection, and integration within the Andean Group. Also provides valuable information on pulp and paper. Consists of an interesting combination of author's vast and valuable experience in the formulation of trade policy and his solid theoretical background.

1773 **Garay, Luis J.** El manejo de la deuda externa de Colombia. Bogotá?: FESCOL; CEREC, 1991. 87 p.: bibl. (Serie Textos; 16)

Recognized economist and Colombia's chief external debt negotiator for many years analyzes the country's particular strategy in managing its external debt. Because of Colombia's exceptional status in the Latin American context in that Colombia did not renegotiate its debt, Colombia was able to establish a distinct and particular relationship with international financial institutions, all of which is analyzed in this book.

1774 Henao, Marta Luz and Oliva Sierra García. Pobreza urbana y distribución del ingreso en Colombia. Medellín, Colombia: Univ. de Antioquia, Centro de Investigaciones Económicas, 1991. 205 p.: bibl., ill.

Interesting empirical study on the incidence of poverty in four major cities from 1971–86, based on data obtained from household surveys. Presents characterization of the poor, examines determinants of income, and explores unemployment by levels of income and gender, among other things.

1775 Isaza, José Fernando and Luis Eduardo Salcedo S. Sucedió en la Costa Atlántica: los albores de la industria petrolera en Colombia. Bogotá: Ancora Editores, 1991. 223 p.: bibl., ill., map.

Historical account of origins of the oil sector written by well-known technocrats in the energy sector. Provides information which shows that investments in this field began in the late 1880s in the Atlantic Coast. Development of major fields in the center of the country began in the late 1920s. Discusses and questions popular and widespread belief that for less developed countries foreign direct investment has had overall negative effects.

Koonings, Kees. Industrialization, industrialists, and regional development in Brazil: Rio Grande do Sul in comparative perspective. See item **2162**.

1776 Londoño de la Cuesta, Juan Luis. Capital humano y distribución del ingreso: la experiencia colombiana. (*Rev. Planeac. Desarro.*, 23:2, sept. 1992, p. 5–70, bibl., graphs, tables)

Based on doctoral thesis presented at Harvard Univ. Examines change in income distribution from 1938–88. Rigorous quantitative treatment of a generally under-researched area. Provides set of inter-temporal indicators for a period characterized by profound distributive changes. Provides evidence that the Colombian experience deviates from the traditional Kuznet-curve explanation.

1777 Machado C., Absalón. El modelo de desarrollo agroindustrial de Colombia. Bogotá: CEGA; Siglo Veintiuno Editores de Colombia, 1991. 491 p.: bibl., ill.

Careful and comprehensive sectoral study on the main issues and conflicts faced by the agroindustrial sector from 1950–90. Articles seek to evaluate effects of import-substitution model on the economy. Looks into challenges and opportunities confronting the sector in the future because of the recent adoption of an open, export-led model of development. Provides ample statistical and institutional information on the subject.

1778 Macroeconomía de los flujos de capital en Colombia y América Latina. Recopilación de Mauricio Cárdenas Santa-María y Luis Jorge Garay S. Bogotá: Tercer Mundo Editores; FESCOL; Fedesarrollo, 1993. 335 p.: bibl., ill. (Economía colombiana)

Interesting essays on critical issue of recent massive capital inflows into the Colombian economy and their most likely causes. At the center of economic debate throughout the Latin American region, these inflows are analyzed rigorously by leading researchers. This book represents the state of the art on this debate in Colombia. Although all contributions are of a quantitative nature, authors use different perspectives and methodologies. Therefore, conclusions as to causes of the inflows differ. Two articles are of particular interest because of the novelty of the topic: Miguel Urrutia and Adriana Pontón questions the belief that capital inflows are greatly influenced by drug trade revenues while Andrés O'Bryne and Mauricio Reina reach the opposite conclusion.

1779 McCleary, William A. and Evamaría Uribe Tobón. Earmarking government revenues in Colombia. Washington: World Bank, 1990. 88 p.: bibl. (Policy, research, and external affairs working papers; WPS 425)

Detailed analysis of trends in the extent and structure of earmarking in Colombia's tax revenues since 1970. Contains valuable statistical information on the subject. Examines origins of the phenomenon, evaluates major examples of earmarking, and presents recommendations for change.

1780 Perry, Guillermo. Política petrolera: economía y medio ambiente. Bogotá: FESCOL; CEREC, 1992. 169 p.: bibl., ill. (Serie Textos/CEREC; 20)

Book analyzes important and growing influence of oil revenues in fiscal accounts and the economy, given recently discovered oil deposits. Explores policy issues which

1781 **Producción de café en Colombia.** v. 1. Coordinación de Roberto Junguito Bonnet y Diego Pizano Salazar. Bogotá: FEDESARROLLO; Fondo Cultural Cafetero, 1991. 1 v.: ill.

should be faced to ensure long-term export capacity in oil. Also briefly looks into institutional and legislative aspects of the impact of this productive activity on the environment.

Very comprehensive study on coffee sector's national and international trends compiled and written by two recognized economists in the field. Rigorous and quantitative analysis of productivity of crops under different technological options; costs and benefits of production under different technologies; and labor markets in different regions. Interesting analysis of results of policies designed to promote crop diversification in coffee regions. Authors conclude by examining Colombia's future prospects in coffee production and international market share.

1782 **Rodríguez, Oscar** *et al.* Estructura y crisis de la seguridad social en Colombia, 1946–1992. Bogotá: Centro de Investigaciones para el Desarrollo, 1992? 288 p.: bibl.

Collection of articles on important issue in Colombia's public debate: the crisis of the social security system. Provides wealth of interesting and comparative statistical information on different social security systems. Analyzes institutional structure of social security and looks into the system's financial problems and its need for structural reform.

Sarmiento, Luis Fernando and **Ciro Krauthausen.** Cocaína & Co.: un mercado ilegal por dentro. See item **3197.**

1783 **Sarmiento Palacio, Eduardo** *et al.* Cambios estructurales y crecimiento. Bogotá: Ediciones Uniandes; Tercer Mundo Editores, 1992. 288 p.: bibl., ill. (Economía colombiana)

Authors question accepted theoretical formulations on Colombian issues such as inertial inflation, endogenous character of fiscal and monetary policy, low response of savings to market solutions, weakness of models of comparative advantages, incapacity of structural reforms to promote high growth rates, and lack of a theory of income distribution.

Thoumi, Francisco E. The size of the illegal drugs industry in Colombia. See item **3199.**

1784 **Zerda Sarmiento, Alvaro.** Apertura, nuevas tecnologías y empleo. Bogotá: FESCOL, 1992. 194 p.: bibl., ill.

Analyzes impact of the adoption of new technologies on the structure of Colombia's manufacturing sector since the 1980s and examines likely future trends given recently adopted model of export-led growth. Concentrating heavily on labor and capital productivity, provides the reader with valuable empirical information on the effects of technology on overall employment levels and labor force structure.

ECUADOR

DAVID W. SCHODT, *Professor of Economics, St. Olaf College*

IN THE LATE 1980S AND EARLY 1990S, Ecuador experienced a variety of economic changes, some of which were new and some of which, although old, emerged with increased saliency in public discussions. From the point of view of the economic literature, one departure from the past was the appearance of more theoretically-grounded and empirically-based studies which sought to analyze these changes.

The article by Emanuel and Dahik (item **1789**) analyzes the preceding decade from the point of view of two of the architects of Febres Cordero's economic policy. Infante (item **1806**), Jaramillo (item **1809**), and Jácome (item **1808**) bring econometric tools to bear on what, for Ecuador, were historically high rates of inflation that accompanied the deterioration of the economy during the 1980s and early 1990s. The tension between policies of economic adjustment and social conditions is analyzed by De Janvry, Sadoulet, and Fargeix in their path-breaking study (item **1799**), while Calvache examines trends in social spending between 1988–90 (item **1795**). The series from ILDIS provides a discussion of government social policy (item **1800**). Placencia and Franco use survey data to analyze women in the labor market during the 1980s, a topic that has received little attention to date (item **1826**).

Privatization of State enterprises was an important component of the economic policies promoted by Sixto Durán Ballén's Administration. While sparking intense political debate, the subject has received little attention in the economics literature. A notable exception is Alberta Acosta's survey (item **1786**), which also contains a brief but interesting discussion of military enterprises. The *sucretización* program receives its most careful analysis to date in Younger's work (item **1845**).

After years of protectionist policies, Ecuador expanded its participation in regional free trade agreements. This important trend receives useful attention in a number of studies. Cáceres' empirical analysis of Ecuador's participation in the Andean pact is valuable (item **1792**), as are those studies that focus on the consequences of trade liberalization for small agricultural producers.

Environmental concerns have only recently become part of the national dialogue, when conflicts among indigenous groups in the Ecuadorian Amazon, the government, and foreign petroleum companies gained international attention. Nevertheless, as Southgate and Whitaker show in their excellent study (item **1838**), the process of environmental degradation is both broader and more longstanding than the recent conflicts would indicate. Non-governmental organizations, both Ecuadorian and international, have played an important role in bringing environmental problems to public attention. Meyer's works discuss Ecuadorian environmental NGOs, providing useful information on Fundación Natura, the most influential of these (items **1820** and **1821**). Finally, proceedings from a conference organized by the Escuela Politécnica Nacional provide some useful insights into Ecuadorian views in this area (item **1835**).

1785 **Abril-Ojeda, Galo** and **Rafael Urriola.** Incentivos de fomento industrial en el Ecuador, 1972–86. Quito: CEPLAES; CIID, 1990. 143 p.: bibl.

Analysis of effects of the Ley de Fomento Industrial on industrial performance (as measured by output growth, employment growth, and contributions to the balance of trade) in Ecuador from 1972–86. Uses input-output analysis to examine employment effects. Includes results of survey of 106 industrialists about role of industrial incentives in their decision-making. (For related work see item **1804**).

1786 **Acosta Espinosa, Alberto** and **Lautaro Ojeda Segovia.** Privatización. Quito: Centro de Educación Popular, 1993. 223 p.: bibl.

Survey of privatization in Ecuador. Somewhat journalistic, but contains useful information and sources are well documented. Includes one of few published discussions of military enterprises.

1787 **La acumulación del capital y los problemas de la macroeconomía ecuatoriana en el período de postguerra.** Ecuador: s.n., 1991? 83 p.: bibl., ill.

Report of a study financed by CONADE and the German Agencia de Cooperación Técnica on saving and investment behavior in Ecuador from 1965–89. Argues that the current economic crisis has structural roots in chronic under-accumulation. For a more systematic approach by Jácome and Arizaga, see "Determinantes Macroeconómicos del Ahorro en el Ecuador" (*HLAS* 53:2192).

1788 **Báez, René.** Ecuador: ¿genocidio económico o vía democrática? Quito: Corporación Editora Nacional, 1992. 100 p.: bibl. (Biblioteca de ciencias sociales; 40)

Sweeping critique of Latin American development models and proposal for an alternative for Ecuador which the author defines as "national and democratic." Similar to item **1824**.

1789 **Banco Central del Ecuador.** La afluencia extraordinaria de divisas y la política económica de Ecuador. (*in* Banca Central en América Latina: selección de textos. v. 3, Estabilización y ajuste. Recopilación de Jesús Silva-Herzog F. y Ramón Lecuona V. México: Centro de Estudios Monetarios Latinoamericanos, 1991?, p. 163–197, appendix, graphs, tables)

Carlos Emanuel and Alberto Dahik, two architects of Febres Cordero's economic policy, examine policies to neutralize the impact arising from a large increase in foreign exchange earnings. Includes theory and discussion of Ecuadorian economic policy during the petroleum boom of 1972–79.

1790 **Brea, Jorge A.** Migration and circulation in Ecuador. (*Tijdschr. Econ. Soc. Geogr.*, 82:3, 1991, p. 206–219, bibl., maps, tables)

Investigates labor mobility as function of socioeconomic structure of places and individuals. Focuses on role of agrarian change in altering patterns of migration and circulation at the canton level. Regression results point to increasing circulation of males who retain land ownership. Finds that a rural proletariat is not emerging. (See also item **1816**).

1791 **Cabrera, Yolanda; Judith Martínez;** and **Rolando Morales.** Medición de la pobreza en las áreas urbana y rural del Ecuador. Quito: Instituto Nacional de Empleo; UNICEF, 1993. 205 p.: bibl., ill.

Valuable results of 1990 study of poverty and income distribution undertaken by UNICEF and the Ecuadorian Instituto Nacional de Empleo (INEM). Based on rural and urban household surveys. (See also *HLAS 49:3545*.)

1792 **Cáceres, Luis René.** Ecuador y la integración andina: experiencias y perspectivas. (*Integr. Latinoam.*, 18:195, nov. 1993, p. 31–46, tables, graphs)

Analyzes benefits of Ecuador's participation in the Andean Pact. Includes empirical estimates of trade creation, economic growth, capital formation, economic stabilization, and industrialization. Finds limited benefits, but argues that much of this is caused by high transport costs, not tariff barriers. Valuable.

1793 **Calderón G., Fernando** *et al.* Guayaquil, realidades y desafíos: seminario realizado por la Corporación de Estudios para el Desarrollo, CORDES. Introducción de Osvaldo Hurtado Larrea. Quito: CORDES, 1991? 376 p.: bibl., ill., maps.

Proceedings of 1989 conference provide useful overview of contemporary problems in this city.

1794 **Calderón Viteri, Roberto.** El sistema aduanero en Ecuador: obstáculos y soluciones. (*in* Integración y burocracia: trabas no arancelarias. Coordinación de Cristina Barrera, Oswaldo Dávila, y Marc Meinardus. Caracas: Fundación Friedrich Ebert; Editorial Nueva Sociedad, 1991, p. 107–119, tables)

Overview of the current customs process for imports and proposal for its reform. Useful principally for its description of the actual non-tariff barriers facing imported products.

1795 **Calvache U., Gladys.** El gasto social del Presupuesto del Estado, 1988–1991. (*Cuest. Econ.*, 19, mayo 1992, p. 59–67, tables)

Analysis of social spending through the *Presupesto del Estado* (a significant component of public sector spending). Documents decline in the budget share going to social spending from 1988–90. Useful data.

1796 **Camacho, Patricio León.** Dos décadas de la producción, comercio exterior y reordenamiento del espacio. (*in* El Ecuador de la postguerra: estudios en homenaje a Guillermo Pérez Chiriboga. Quito: Banco Central del Ecuador, 1992, v. 1, p. 93–116)

Survey of economic changes that occurred in Ecuador from 1948–71, a relatively understudied period. Solid work by respected Central Bank economist.

1797 **Cornejo, Boris.** Ecuador: la integración andina y la apertura comercial. (*in* Las Nuevas Condiciones de la Economía Mundial y los Procesos de Investigación en América Latina, *Quito, 1991.* Integración latinoamericana: su última oportunidad. Bogotá: Corporación de Estudios para el Desarrollo, 1991, p. 187–261, tables)

General survey of costs and benefits of Ecuador's participation in the Andean Pact. Discusses legislative changes and includes statistics on changes in trade patterns.

1798 Cueva, Juan Martín and Mónica Sánchez. Bibliografía ecuatoriana sobre pequeña y mediana empresa. Quito: INSOTEC, Unidad de Investigaciones de Política Industrial, Centro de Documentación e Información, 1988? 134 p.: indexes.

Bibliography prepared by the Instituto de Investigaciones Socio-Económicas y Tecnológicas (INSOTEC) for the Federación Nacional de Cámaras de la Pequeña Industria. Contains material held in Ecuadorian libraries, principally those in Quito and Guayaquil. Indexed by author, subject, and geographic area.

1799 De Janvry, Alain; Elisabeth Sadoulet; and André Fargeix. Adjustment and equity in Ecuador. Paris: Development Centre of the Organisation for Economic Cooperation and Development; Washington: OECD Publications and Information Centre, 1991. 174 p.: bibl., ill. (Adjustment and equity in developing countries. Document Centre studies)

Innovative application of computable general equilibrium model to examine alternative approaches to stabilization in Ecuador. Provides detailed explanation of the model and results obtained. See also *HLAS 53:2179* for a summary of material presented here.

1800 Escobar, Santiago et al. Nuevas orientaciones de políticas sociales. Recopilación de Eduardo Palma y Jaime Ahumada. Santiago: ILPES-CEPAL; Quito: ILDIS, 1989? 159 p.: bibl. (Cuadernos de políticas sociales; 1)

Vol. 1 in a useful series of reports from a 1988–89 national dialogue program on social policy for Ecuador, sponsored by ILDIS and the Instituto Latinoamericano de Planificación Económica y Social (ILPES). Vol. 2 is *Políticas de desarrollo social y vivienda* (item **1831**) and vol. 4 is *Políticas de empleo* (item **1814**).

1801 Fullerton, Thomas M. and Ajay Kapur Jr. Predicción de multiplicadores monetarios en Colombia, Ecuador y Venezuela. (*Lect. Econ.*, 35, julio/dic. 1991, p. 53–86, appendices, bibl., graphs., tables)

Empirical estimation of money multipliers in Ecuador, Colombia, and Venezuela applies statistical techniques (univariate ARMA) previously employed to estimate multipliers in the industrialized nations.

1802 Guerrero Carrión, Trosky. Modernización agraria y pobreza rural en el Ecuador. Loja, Ecuador: Editorial Universitaria, 1992. 198 p.: bibl., ill.

Survey of changes in Ecuadorian agriculture since 1970 by Univ. of Loja agricultural economist. See *HLAS 53:2166* for a more in-depth survey.

1803 Hernández, Carmen and Rafael Urriola. Los pequeños productores agropecuarios y la apertura comercial. Quito: IICA; ILDIS, 1993. 116 p. (Serie Publicaciones misceláneos, 0534–5391)

Useful study of effects of trade liberalization on small agricultural producers. Compares Ecuador and Colombia and includes case studies of potato, rice, and coffee producers.

1804 Hidrobo Estrada, Jorge. Power and industrialization in Ecuador. Boulder, Colo.: Westview Press, 1992. 206 p.: bibl., index.

Study of Ecuadorian industrialists in the post-1972 period draws on survey research to examine both the ideology of industrialists and their ability to influence industrial policy. Based on personal interviews carried out during the early 1980s. First published in Spanish as *Industriales, Estado, industrialización en el Ecuador* (Quito: Centro de Promoción Universitaria, Univ. San Francisco de Quito, 1990). Complements Conaghan's *Restructuring domination* (see *HLAS 51:2172*) and David Hanson's "Political Decision-Making in Ecuador: The Influence of Business Groups" (unpublished dissertation, The Univ. of Florida, 1971).

1805 Hollihan, Mike. El medio circulante y la economía ecuatoriana en los años treinta. (*Rev. Ecuat. Hist. Econ.*, 3:6, 2nd semestre 1989, p. 15–101)

Argues that widespread lack of agreement on economic data for the 1920s–30s lends itself to numerous interpretations of the period. Author provides important contribution to monetary history of the period.

1806 Infante T.P., Sebastián et al. La inflación en Ecuador: interpretaciones y comentarios. Quito: Instituto Latinoamericano de Investigaciones Sociales, 1992. 83 p.: bibl., ill.

Two empirical studies of inflation in Ecuador in the period 1984–91, when the rate began to rise to historically high levels: 1) an econometric analysis of short-run price determination; and 2) a study focused on the distributional consequences of relative price changes during an inflationary period. Includes other material on inflation at this time.

1807 Instituto Nacional de Energía (Ecuador). Balances energéticos, 1980–1988. Quito: Instituto Nacional de Energía, 1990. 78 p.

Energy sector statistics and energy balances for Ecuador.

1808 Jácome, Luis Ignacio. Estabilización en el Ecuador: de la inflación crónica a la inflación moderada. S.l.: CORDES, 1994. 38 p. (Apunte técnico; 3)

Analyzes the macroeconomic stabilization program introduced in Sept. 1992 that reduced the rate of inflation to 31 percent from levels in excess of 55 percent prevailing during the previous five years. Useful data and analysis.

1809 Jaramillo Buendía, Fidel. Inflación, política fiscal y estabilización en el Ecuador: un análisis intertemporal. (*Cuest. Econ.*, 19, mayo 1992, p. 91–112, bibl., graphs, tables)

Reductions in the public sector deficit to reduce inflation appeared less successful in 1990–91 than in 1984–85. Article develops an intertemporal econometric model that is estimated for the two periods. Finds that inflationary cost of seignorage above given level is very high.

1810 Klak, Thomas. Recession, the State and working-class shelter: a comparison of Quito and Guayaquil during the 1980s. (*Tijdschr. Econ. Soc. Geogr.*, 83:2, 1992, bibl., graphs, tables)

Richly textured investigation of the well-being of the working class in Guayaquil and Quito uses shelter conditions as the primary indicator. Finds benefits of State policy helped offset effects of the recession in Quito, but not in Guayaquil.

1811 Larrea Stacey, Eduardo. Evolución de la política del Banco Central del Ecuador, 1927–1987. Quito: Centro de Investigación y Cultura, Banco Central del Ecuador, 1990. 157 p. (Biblioteca de historia económica; 3)

History of the Central Bank by former employee who worked there for many years, and who was its general manager from 1972–73.

1812 Lee, David; Andrés Guarderas; and Gregory Scott. La integración del análisis de la demanda de alimentos y la comercialización agrícola en la formulación de políticas alimentarias: estudio de un caso ecuatoriano. (*in* Mercadeo agrícola: metodologías de investigación. Lima: Centro Internacional de la Papa; San José: Instituto Interamericano de Cooperación para la Agricultura, 1991?, p. 164–182, tables)

Econometric model of demand for basic food using data from survey of consumption by urban households conducted in 1975–76. Identifies extreme diversity in food consumption patterns of lower income families. Argues that more such studies are needed to guide design of macro policies with implications for food prices. Useful.

1813 Maiguashca, Juan. Las clases subalternas en los años treinta. (*Rev. Ecuat. Hist. Econ.*, 3:6, 2nd semestre 1989, p. 165–197, tables)

Extension of author's previous work on the origins of "Velasquismo" in which he takes issue with both Rafael Quintero's and Agustín Cueva's theses. Argues that economic transformations in the 1920s–30s created important political openings for lower class political participation. Important.

1814 Márquez, Gustavo et al. Políticas de empleo. Recopilación de Santiago Escobar. Santiago: ILPES-CEPAL; Quito: ILDIS, 1989? 237 p.: bibl., ill. (Cuadernos de políticas sociales; 4)

For annotation, see item **1800**.

1815 Marshall Silva, Jorge. Ecuador: cuantificación, distribución y efectos del ingreso petrolero, 1973–1988. Santiago: J. Marshall, 1988. xxxii, 234 p.: bibl.

Expanded and updated version of a chapter written with Alan Gelb, "Ecuador: Windfalls of a New Exporter" (in Alan Gelb et al., *Oil windfalls: blessing or curse?*, New York: Oxford Univ. Press, 1988, p. 170–196). Analyzes distributional consequences of the two oil windfall periods (1974–78 and 1978–82) by sector. Points to failure to shift oil income to development of export manufacture sector. See also *HLAS 51:2168*. Valuable.

1816 Martínez Valle, Luciano. El empleo en economías campesinas productoras para el mercado interno: el caso de la sierra ecuatoriana. (*Rev. Eur.*, 53, Dec. 1992, p. 83–93, bibl., tables)

Study of rural employment that draws on the 1990 *Survey of Rural Households*. Finds evidence of increasing circulation of men and correspondingly greater participation of women in farm activities. Compare with Brea (item **1790**) for a more quantitative analysis.

1817 Martínez Valle, Luciano. El empleo rural en el Ecuador. Quito: INEM; ILDIS, 1992. 58 p.: bibl., ill. (Serie Documentos de investigación; 2)

Preliminary study of rural employment based on the Nov. 1990 survey of rural households conducted by INEM. Focuses on seasonal variation in employment and on regional differences between the coast and the Sierra. Notes important changes in rural employment, such as the diminished weight of agricultural employment and the significant role of women. Useful.

1818 Méndez de Herrera, Genoveva. El transporte urbano en Quito: análisis de su problemática y alternativas de mejoramiento. Quito: Instituto de Investigaciones Económicas, CONUEP, Facultad de Ciencias Económicas, Univ. Central, 1991. 330 p.: bibl., ill.

Study of urban transport in Quito by Central Univ. Professor of Economics contains a wealth of information, both qualitative and quantitative, about the history and current operations of the system. Valuable resource on this topic.

1819 Mesa-Lago, Carmelo. Instituto Ecuatoriano de Seguridad Social (IESS): evaluación económica y opciones para reforma. Quito: INCAE; PROGRESEC, 1993. 127 p.: bibl., tables.

Report prepared for USAID by acknowledged expert in the field provides systematic analysis of Ecuadorian social security system. See also *HLAS 53:2228* and item **1840**.

1820 Meyer, Carrie A. Environmental NGOs in Ecuador: an economic analysis of institutional change. (*J. Dev. Areas*, 27:2, Jan. 1993, p. 191–210, appendix)

Documents and analyzes emergence of 24 environmental NGOs operating in Ecuador since 1984. Useful discussion of Fundación Natura (more detailed than in item **1821**) contrasts it with more radical environmental NGOs.

1821 Meyer, Carrie A. A step back as donors shift institution building from the public to the "private" sector. (*World Dev.*, 20:8, Aug. 1992, p. 1115–1126, bibl.)

Draws on theory and case studies from Ecuador and the Dominican Republic to examine the shift in donor support from the public sector to the nonprofit sector (NGOs). Discusses experiences of USAID/Ecuador with Fundación Natura, IDEA, and FUNDAGRO. Raises concerns over expanding role of NGOs.

1822 Middleton, Alan. La dinámica del sector informal urbano en el Ecuador. Quito: Centro de Investigaciones de la Realidad Ecuatoriana, 1991. 201 p.: bibl. (Serie Punto crítico; 2)

Study of small informal-sector producers in Quito from 1974–90 is notable for its tracking of informal producers over time.

1823 Miras, Claude de; Gustavo Rodríguez; and Roberto Roggerio. Bibliografía comentada sobre el sector informal urbano. Quito: CEDIME; ORSTOM; ILDIS, 1992. 179 p.: forms, indexes. (Guayaquil futuro)

Bibliography of books and articles on the informal sector that are held in depositories in Quito, Guayaquil, and Cuenca. Organized according to those edited in Ecuador, and/or that deal with Ecuador; and those from other countries.

1824 Moncada Sánchez, José. Ecuador: ¿integración mundial o desintegración nacional? Quito: Corporación Editora Nacional, 1992. 126 p.: bibl. (Biblioteca de ciencias sociales; 39)

Analysis of Ecuador's economic crisis and critique of proposals for trade liberalization by one of the country's leading leftist critics is largely historical.

1825 **Oleas Montalvo, Julio et al.** Indice de tesis universitarias sobre temas económicos, 1900–1984. Quito: Banco Central del Ecuador, 1989. 409 p.: indexes. (Fuentes para la historia económica del Ecuador. Serie Indices de documentación; 2)

Second in the series *Fuentes para la Historia Económica del Ecuador*, this bibliography indexes theses written by graduates of Ecuadorian universities on economic topics. Indexed by author, subject, and province studied.

1826 **Placencia, María Mercedes** and **Eliana Franco.** Situación de la mujer en el mercado laboral ecuatoriano. (*in* Mujer y trabajo. Quito: CEPLAES, 1990, p. 39–68, bibl., graphs, tables)

Part of a larger study of Ecuadorian women undertaken by CEPLAES, this chapter examines labor market changes since 1974. Draws on data from the 1982 Census of Population and the 1987 Survey of Households. Useful introduction to a topic that has received little attention.

1827 **Ramos, Hugo H.** and **Mónica Acosta.** Impactos de la apertura comercial regional en el sector agropecuario ecuatoriano. (*Ecuad. Deb.*, 23, junio 1991, p. 59–71, graphs, tables)

Overview of potential gains to the Ecuadorian agricultural sector from trade liberalization. Some discussion of specific agricultural products. See also *HLAS 53:2166*.

1828 **Reforma estructural al sistema arancelario.** Quito: Ministerio de Finanzas y Crédito Público, 1990. 55 p.

Brief overview of tariff reforms proposed for implementation by the Borja Administration.

1829 **Rodas, Sonia** and **Jurgen Schuldt.** Impacto del proceso de ajuste económico sobre la reproducción social del Ecuador en los años ochenta. (*Ecuad. Deb.*, 27, dic. 1992, p. 49–62, bibl., graph)

Survey of changes in social conditions during the 1980s. Argues that the deterioration generated by structural forces resulted in a highly segmented internal market.

1830 **Rueda Novoa, Rocío.** El Obraje de San Jóseph de Peguchi. Quito: Ediciones ABYA-YALA; TEHIS, 1988. 159 p.: bibl.

Economic history of an *obraje* operating near Otavalo during the 17th century. Includes discussion of ecomomic linkages between Peru and Ecuador in this period.

1831 **Samán Mancero, Alfredo et al.** Políticas de desarrollo social y vivienda. Recopilación de Santiago Escobar. Santiago: ILPES-CEPAL; Quito: ILDIS, 1989? 222 p.: ill. (Cuadernos de políticas sociales; 2)

For annotation, see item **1800**.

1832 **Santacruz Guzmán, Fabián.** Economic growth and productivity in Ecuador. (*Anuario/Quito*, 1:1, julio 1990, p. 71–121, tables)

Examines productivity performance from 1973–86 using ICOR and neoclassical growth analysis. Results highly sensitive to choice of period. Includes discussion of industrial policy.

1833 **Seminario El Ecuador del Siglo XXI, Quito, 1991.** Memorias: perspectivas económicas del Ecuador en el siglo XXI. Textos de Alberto Cárdenas *et al.* Quito: IAEN; ILDIS, 1992. 110 p.

Transcripts from presentations made at 1991 conference provide general views of Ecuador's economic future. Notable principally for the insights provided by some of Ecuador's leading analysts, such as Osvaldo Hurtado and Walter Spurrier.

1834 **Seminario Internacional de Economía y Energía 1990–2000, Quito, 1990.** Memoria. Quito: Ministerio de Energía y Minas; Organización Latinoamericana de Energía; Federación Nacional de Economistas del Ecuador; Colegio de Economistas de Quito, 1990? 300 p.: bibl., ill., maps.

Interesting in that this conference brought together representatives from government, private petroleum companies, and environmental groups. Some data on petroleum sector.

1835 **Seminario Los Retos Tecnológicos del Ecodesarrollo, Quito, 1991.** Memoria. Quito: Escuela Politécnica Nacional, 1991. 262 p.: bibl. (Col. Tecnología y sociedad; 1)

Proceedings from conference organized by the Escuela Politécnica Nacional focus on role of technology in balancing economic development with environmental preservation. Largely confined to general observations but does provide insights into Ecuadorian concerns in this area.

1836 Sierra Valenzuela, Vitalia. Ecuador: inflación y respuestas; teoría, origen y características del proceso inflacionario y las respuestas de la política económica. Quito: EDIDAC, 1991. 276 p.: bibl., ill. (Serie Economía)

Overview of inflation from a structuralist perspective for 1970–90 period. Little theory but includes statistical information on costs and prices.

1837 Solano de la Sala Veintimilla, Germán. Indice de folletos sobre temas económicos y sociales del Ecuador. Quito: Banco Central del Ecuador, 1991. 299 p.: indexes. (Fuentes para la historia económica del Ecuador. Serie Indices de documentación; 3)

Third in series *Fuentes para la Historia Económica del Ecuador*, this bibliography indexes documents from regional libraries as well as from those in Quito and Guayaquil. Contains a wealth of difficult to locate material from the period 1817–1989. Indexed by author, subject, and date. Essential resource for economic historians.

1838 Southgate, Douglas DeWitt and Morris D. Whitaker. Economic progress and the environment: one developing country's policy crisis. New York: Oxford Univ. Press, 1994. 150 p.: bibl., ill., index, maps.

Overview of how economic policies have contributed to environmental deterioration in Ecuador, followed by case studies ranging from topics such as tropical deforestation to the impact of the shrimp industry on the coastal ecosystem. Includes recommendations for more environmentally-sound policies. Highly recommended.

1839 Thorp, Rosemary et al. Las crisis en el Ecuador: los treinta y ochenta. Quito: Corporación Editora Nacional, 1991. 305 p.: bibl. (Biblioteca de ciencias sociales; 33)

Papers from a project comparing economic and social crises of the 1930s and the 1980s. An uneven collection, with a number of articles reprinted from other sources. Chapter by Ana Lucía Armijos and Marco Flores provides useful overview of economic policy during the 1980s. Germánico Salgado provides a discussion of regional integration.

1840 Torres Rodríguez, Luis. IESS: una agonía en cifras. Quito: Fundación Ecuatoriana de Estudios Sociales, 1989. 103 p.: bibl., ill.

Descriptive statistics for the Ecuadorian Social Security System from 1928–88. See also *HLAS 53:2228*.

1841 Trabas no arancelarias al comercio andino en el Ecuador. Edición de Galo Chiriboga, Marc Meinardus y Rafael Urriola. Quito: Instituto Latinoamericano de Investigaciones Sociales; FAUS, 1991. 109 p.: bibl., ill.

Summary collection of empirical papers from 1991 seminar sponsored by ILDIS (originals available from that source). Based on interviews with 100 public and private firms. Remarks on complete lack of consensus in Ecuador with respect to regional integration.

1842 Vásquez de Labandera, Edwin. Pequeña industria: impactos de la crisis en el período 1986–1990. Quito: Instituto de Investigaciones Socio-Económicas y Tecnológicas, 1992. 26 leaves: ill. (Serie Estudios sectoriales)

Statistical study of the performance of small industry during the period 1986–90. Examines production, investment, employment, and productivity. Finds that small and medium industry performed much worse than large industry.

1843 Vázquez, Paciente and Iván González Aguirre. Empleo e ingreso en la construcción en Cuenca. Quito: Instituto Latoinamericano de Investigaciones Sociales, 1992. 107 p. (Cuadernos del austro; 5)

ILDIS-supported study of the construction industry in Cuenca during 1977–87. Documents important phenomenon of rapid growth, coupled with increasing rural-urban migration until 1983, followed by subsequent sharp slowdown in the employment and income in this sector. Useful for its regional focus. Numerous statistical tables.

1844 Vicuña Izquierdo, Leonardo. Ecuador: la política económica en la década de los años ochenta. Guayaquil, Ecuador: Depto. de Publicaciones, Facultad de Ciencias Económicas, Univ. de Guayaquil, 1992. 173 p.: bibl., graphs, ill., tables.

Study of economic stabilization policy in Ecuador. Limited analysis, but useful as a chronicle of economic policy-making by successive democratic governments from 1980–90. Some tables and graphs.

1845 Younger, Stephen D. The economic impact of a foreign debt bail-out for private firms in Ecuador. (*J. Dev. Stud.*, 29:3, April 1993, p. 484–503, bibl., graphs, tables)

Uses industry level data to analyze effects of the *sucretización* program on firms' financial viability and to determine whether transfers implicit in the program helped prevent a wave of bankruptcies in 1983–84. Concludes that transfers were large but unnecessary, and that few firms used transfers to strengthen their balance sheets. Important article that complements Nicklesburg (see *HLAS 53:2198*).

CHILE

MARKOS J. MAMALAKIS, *Professor of Economics, University of Wisconsin-Milwaukee*

THE DEMOCRATIC TRIUMPH OF CHILE, as exemplified by the smooth transition of the presidency from Patricio Aylwin to Eduardo Frei in 1994, reflects an improvement in the government's performance. The best publications annotated in this chapter share a high degree of consensus that the Pinochet presidency made a singular contribution to Chilean development through the formulation and implementation of liberal, nondiscriminatory, mesoeconomic, sectoral policies. Furthermore, it is increasingly recognized that without the unwavering support of Gen. Pinochet, the Liberal experiment could have failed. Thus, Pinochet is given as much credit for the Chilean success story as the Chicago Boys, if not more. On the other hand, it is universally recognized that the Pinochet presidency failed to create a politically and socially enlightened government.

Restoration of basic political and economic freedoms and human rights coincided with the 1989 election that brought Patricio Aylwin to power. Mesoeconomic policies in the areas of education, health, and welfare were redefined with increased emphasis placed on social justice and the welfare of the poor. The Aylwin presidency was characterized by phenomenal economic success based on a continuation and strengthening of the liberalization and privatization policies of the Pinochet era.

The Frei presidency began in 1994 with an aura of optimism. However, there was by then an increased awareness that unresolved problems of poverty, inequality, and often acute environmental degradation had to be addressed. Paradoxically, sustained prosperity undoubtedly is both a precondition for, as well as a consequence of, resolving these problems.

Major topics analyzed in the literature selected for this chapter include: foreign trade; income distribution and poverty; the mesoeconomic institutional reforms of the Pinochet and Aylwin presidencies; technology transfer; monetary policy; the nature of labor markets; the private pension scheme; agriculture; environmental decay; and so forth.

The vast majority of the scholarly work annotated below is characterized by objectivity, balance, and absence of extreme political bias. An impressive flow of truly outstanding publications emanated from the Departments and Institutes of Economics of the University of Chile, the Catholic University, the Center for Public Studies, The Central Bank of Chile, and the United Nations' Economic Commission for Latin America and the Caribbean (ECLAC). Among the many worthy publications, the following stand out in terms of their lasting contributions to Chilean economic historiography: 1) Juan Ricardo Couyoumdjian, René Millar, and Josefina Tocornal's *Historia de la Bolsa de Comercio de Santiago, 1893–1993: un siglo de mercado de valores en Chile*, an excellent history of the Santiago Stock Exchange (item **1866**); 2) ECLAC's "Una Estimación de la Magnitud de la Pobreza en Chile, 1987," which focuses on the profound, unresolved problem of persistent, large-scale, extreme poverty in Chile (item **1865**); 3) the historic document *Quo vadis, Chile: versión completa del 110 Encuentro Nacional de la Empresa,* which conveys the converging views of Chile's leaders regarding development and democracy (item **1876**); 4) Thelma Gálvez Pérez and Rosa Bravo Barja's pioneering historical study on female employment, *Siete décadas de registro del trabajo femenino, 1854–1920* (item **1879**); and 5) Aníbal Pinto's perceptive review of the Chilean neoliberal paradigm, "Las Raíces del Experimento Ortodoxo Chileno" (item **1907**).

1846 La agricultura chilena durante el gobierno de las Fuerzas Armadas y de Orden: base del futuro desarrollo. Santiago?: Ministerio de Agricultura, División de Estudios y Presupuesto, 1989? 522 p.: bibl.

Comprehensive examination of agriculture during the military dictatorship of Pinochet.

1847 Ahumada, Jorge. En vez de la miseria. 10. ed., rev. Santiago: Ediciones BAT, 1990. 168 p.

Classic on Chilean structuralism by the influential, late Jorge Ahumada. Major treatise in the history of Chilean and Latin American economic thought.

1848 Aldunate, Rafael. El mundo en Chile: la inversión extranjera. Santiago: Zig-Zag, 1990. 211 p. (Col. Temas de hoy)

Explores benefits to Chile of foreign investment.

1849 Aninat U., Eduardo et al. La empresa privada como factor esencial de desarrollo en una democracia estable. Concepción, Chile: Cámara de la Producción y del Comercio de Concepción, 1987? 103 p.: ill.

Numerous brief essays of this compendium examine manifold interaction mechanisms between private enterprise and stable democracy.

1850 Asociación de Administradoras de Fondos de Pensiones (Santiago). 10 años de historia del sistema de AFP: antecedentes estadísticos, 1981–1990. Preparado por Augusto Iglesias P., Rodrigo Acuña R. y Claudio Chamorro C. Santiago: AFP Habitat, 1991. 343 p.: bibl., tables.

Contains exhaustive description and analysis of Chile's truly remarkable experiment, the Pension Funds paradigm introduced in 1981.

1851 Balmaceda M., Felipe. Antigüedad en el empleo y los salarios en Chile. (*Estud. Econ./Santiago*, 19:1, junio 1992, p. 85–105, bibl., tables)

According to this study, seniority increases the wage level, whereas labor turnover decreases it. Because there exist natural incentives for both employees and employers to reduce job turnover, there is no need to regulate it. On the contrary, a highly flexible labor market is necessary to induce a high-rate labor absorption.

1852 Banco Central de Chile. Disposiciones sobre conversión de deuda externa. Ed. en castellano e inglés. Santiago: Banco Central de Chile, 1988. 57 p.

Invaluable document about converting external debt into domestic investment.

1853 Baraona Urzúa, Pablo. Chile en el último cuarto de siglo: visión de un economista liberal. (*Estud. Públicos,* 42, otoño 1991, p. 45–57)

Major advocate of economic liberalism largely attributes decline of socialism, autarchy, and corporatism as well as the rise of deregulation and market economies to the increasing role played in economic development by "human capital."

1854 Behrens, Roberto. Inversión extranjera y empresas transnacionales en la economía de Chile, 1974–1989. El papel del capital extranjero y la estrategia nacional de desarrollo. Santiago: Naciones Unidas, Comisión Económica para América Latina y el Caribe, 1992. 163 p.: bibl. (Estudios e informes de la CEPAL, 0256–9795; 86)

Excellent analysis of the nature and manifold forces affecting foreign investment and transnational enterprises in Chile during 1974–89.

1855 **Bengolea, Manuel** and **Luis Hernán Paúl Fresno.** Inversión directa de las empresas chilenas en el exterior: razones de una creciente necesidad. (*Estud. Públicos*, 43, invierno 1991, p. 229–244)

This fascinating article examines forces behind the rising trend of foreign investment by Chilean enterprises. Investment abroad, according to the author, may become as important to Chile as the 1973–90 export boom.

1856 **Benítez, Andrés.** Chile al ataque. Santiago: Zig-Zag, 1991. 124 p.: ill.

Vital signs and components of triumphant, dynamic "New Chile" are described in this insightful, journalistic, pleasant-to-read, larger-than-normal pamphlet.

Boisier, Sergio and **Verónica Silva.** Propiedad del capital y desarrollo regional endógeno en el marco de las transformaciones del capitalismo actual: reflexiones acerca de la región del Bío-Bío. See item **5006**.

1857 **Boisier, Sergio.** Los siete pecados capitales de la capital y el desarrollo de la región del Bío-Bío: un binomio para potenciar la calidad de vida. (*EURE/Santiago*, 15:46, oct. 1989, p. 7–15, bibl.)

Detailed review of serious problems affecting inhabitants of the Santiago megalopolis, especially environmental degradation including atmospheric contamination. Advocates regulation of size of the capital city both in physical and in economic terms.

1858 **Boloña & Büchi, estrategas del cambio: reflexiones para el desarrollo.** 2a ed. Lima: Agenda 2000 Editores, 1991. 280 p. (Serie Hacia la modernidad; 1)

Reviews Chilean experiment of structural liberalism.

1859 **Brock, Philip L.** External shocks and financial collapse: foreign-loan guarantees and intertemporal substitution of investment in Texas and Chile. (*Am. Econ. Rev.*, 82:2, May 1992, p. 168–173, graphs)

Examines relationship between external shocks, foreign-loan guarantees, intertemporal substitution of investment and financial collapse in Chile during 1981–85.

1860 **Büchi Buc, Hernán** *et al.* La transformación económica de Chile. Santiago: Univ. Nacional Andrés Bello, 1992. 225 p.: ill. (Cuadernos universitarios. Serie Debates; 1)

Compendium of essays containing excellent surveys of institutional reform during the Pinochet presidency.

1861 **Bustos Castillo, Raúl.** Analysis of a national private pension scheme: the case of Chile. (*Int. Labour Rev.*, 132:3, 1993, p. 407–416, tables)

Informative critique and response to Colin Gillion's and Alejandro Bonilla's article on Chilean national private pension scheme (see item **1881**). Important contribution to ongoing debate about optimum private pension scheme.

1862 **Chile: informe nacional a la Conferencia de las Naciones Unidas sobre Medio Ambiente y Desarrollo, Río de Janeiro, Brasil, junio de 1992.** Santiago: La Comisión Nacional del Medio Ambiente (CONAMA), 1992. 102 p.: ill., maps.

Comprehensive statistical compendium describes multifaceted impact of Chilean economic development on the environment, and environmental goals and policies of the government of President Patricio Aylwin.

1863 **Coeymans, Juan Eduardo** and **Yair Mundlak.** Sectoral growth in Chile, 1962–82. Washington: International Food Policy Research Institute, 1993. 152 p.: bibl., ill. (Research report; 95)

First-rate, econometric study of sectoral growth in Chile.

1864 **Coloma C., Fernando** and **Luis Oscar Herrera B.** Análisis institucional y económico del sector de telecomunicaciones en Chile. (*Cuad. Econ./Santiago*, 27:82, dic. 1990, p. 429–472, bibl., graphs)

Focuses on economic organization of Chilean telecommunications (telex, telegram, data transmission and telephone) sector and on the main problems it actually faces in terms of economic efficiency and competition.

1865 **Comisión Económica para América Latina y el Caribe.** Una estimación de la magnitud de la pobreza en Chile, 1987. (*Colecc. Estud, CIEPLAN*, 31, marzo 1991, p. 107–129, bibl., graphs, tables)

According to this study, the proportion

of households below the poverty line increased from 17 percent in 1970 to more than 38 percent in 1987. As a consequence, almost 5.5 million persons lived under conditions of poverty at the end of 1987.

1866 Couyoumdjian, Juan Ricardo; René Millar; and Josefina Tocornal. Historia de la Bolsa de Comercio de Santiago, 1893–1993: un siglo de mercado de valores en Chile. Santiago: Bolsa de Comercio de Santiago, 1993. 768 p.: index, tables.

This monumental study of the Santiago stock exchange from 1893–1993 makes a lasting contribution to Chilean economic historiography.

1867 Daher, Antonio. Debt conversion and territorial change. (*CEPAL Rev.*, 43, April 1991, p. 117–129, bibl., tables)

According to this solid study, conversion of Chile's external debt, which originated in various economic and social sectors and had a highly concentrated territorial distribution, involved a sectoral and regional reassignment of resources which has been reflected in marked territorial change.

1868 Daher, Antonio. Infraestructuras: regiones estatales y privadas en Chile. (*Estud. Públicos*, 49, verano 1993, p. 137–173, bibl., tables)

Examines regional distribution and concentration of public and private infrastructure investment. Traces implication for Chilean economic development.

Déniz Espinós, José A. La política económica neoliberal y sus efectos socioeconómicos: el caso de Chile. See item **5017**.

Los desafíos del Estado en los años 90. See item **3435**.

1869 El desarrollo regional desde el mundo social. Dirección y coordinación de Bernardo Castro Ramírez. Equipo de investigación de Bernardo Castro Ramírez et al. Concepción, Chile: Centro Itata; CETAL; SER; CASDE, 1992. 217 p.: bibl., ill., maps.

Thorough description and analysis of the imperfect economic, social, and political realities of the Bío-Bío Region.

1870 Después de las privatizaciones: hacia el Estado regulador. Edición de Oscar Muñoz G. Santiago: CIEPLAN, 1993. 359 p.: bibl., ill.

Ten studies examine impact and problems that have arisen in Chile due to privatization of public enterprises in the 1970s–80s.

1871 Durán, Esteban and Oscar Troncoso. El problema del hambre en Chile. Santiago: Grupo de Investigaciones Agrarias, 1989. 84 p.: bibl. (Cuadernillo de información agraria; 22)

Examines phenomenon of hunger in Chile during the military dictatorship of Gen. Augusto Pinochet Ugarte.

1872 *Economía y Trabajo en Chile*. Vol. 1, 1990/1991– . Santiago: Programa de Economía del Trabajo.

Solid examination of labor markets during 1990, the first year of the democratic government.

1873 Edwards, Sebastian and Alejandra Cox Edwards. Monetarism and liberalization: the Chilean experiment; with a new afterword. Chicago: Univ. of Chicago Press, 1991. 250 p.: bibl., ill., index.

Excellent study of the Chilean liberalization paradigm of the Pinochet presidency.

1874 "El Ladrillo:" bases de la política económica del gobierno militar chileno. Prólogo de Sergio de Castro. Santiago: Centro de Estudios Públicos, 1992. 193 p.: ill.

This classic liberal manifesto was prepared for Jorge Alessandri and became the economic policy blueprint for the Military Junta.

1875 Encuentro Nacional de la Empresa, 10th, Santiago?, 1988. La libre empresa y el futuro de Chile. Santiago: ICARE, 1989. 282 p.: ill.

Extremely valuable complete proceedings of the tenth National Meeting of Private Enterprises (ENADE).

1876 Encuentro Nacional de la Empresa, 11th, Santiago?, 1989. Quo vadis, Chile. Santiago: ICARE, 1990. 256 p.: ill.

Historic document reveals unique consensus and spirit of cooperation which prevailed in Chile as Gen. Pinochet was getting ready to pass the mantle of leadership to President-elect Aylwin. These proceedings include presentations by presidential candidates Patricio Aylwin, Hernán Büchi, and Francisco Javier Errázuriz as well as a soldierlike "farewell" address by one of Chile's most powerful, influential, and controversial presidents, Gen. Augusto Pinochet Ugarte.

1877 Ffrench-Davis, Ricardo. Trade liberalization in Chile: experiences and prospects. New York: United Nations Conference on Trade and Development, 1992. 125 p.: bibl., ill. (Trade policy series; 1)

Critical examination of Chilean trade liberalization policies.

1878 Figueroa B., Eugenio and **George Lever D.** Determinantes del precio de la vivienda en Santiago: una estimación hedónica. (*Estud. Econ./Santiago,* 19:1, junio 1992, p. 67–84, tables)

Careful examination of housing services market in Santiago, Chile.

1879 Gálvez Pérez, Thelma and **Rosa Bravo Barja.** Siete décadas de registro del trabajo femenino, 1854–1920. (*Estad. Econ.,* 5, dic. 1992, p. 1–52, graph, tables)

Excellent analysis of female employment in Chile.

1880 Garretón, Oscar Guillermo *et al.* Institucionalidad democrática y dinámica de la economía. Edición de Guillermo E. Martínez. Santiago: Corporación de Estudios Liberales, 1989. 296 p.: bibl., ill.

Proceedings of major conference on democracy and development.

1881 Gillion, Colin and **Alejandro Bonilla.** Analysis of national private pension scheme: the case of Chile. (*Int. Labour Rev.,* 131:2, p. 171–195, tables, graphs)

Comprehensive, balanced review of Chile's private pension scheme emphasizes its advantages and limitations.

Goldfrank, Walter L. Fresh demand: the consumption of Chilean produce in the United States. See item **5035.**

1882 Haindl Rondanelli, Erik; Ema Budinich Besoaín; and **Ignacio Irarrázaval Llona.** Gasto social efectivo: un instrumento para la superación definitiva de la pobreza crítica. Santiago: Presidencia de la República, Oficina de Planificación Nacional: Univ. de Chile, Facultad de Ciencias Económicas y Administrativas, 1989? 286 p.: bibl., ill.

Excellent study of impact of effective social expenditures on distribution of income, expenditure, and welfare in Chile, 1985–87.

1883 Hardy Raskovan, Clarisa and **Victoria Legassa.** La ciudad escindida: los problemas nacionales y la Región Metropolitana. Santiago?: Programa de Economía del Trabajo, 1989. 267 p.: bibl.

Thorough examination of segmented urbanization of the Metropolitan Region of Chile.

1884 Hernández T., Leonardo. Inflación y retorno bursátil: una investigación empírica; Chile, 1960–1988. (*Cuad. Econ./Santiago,* 27:82, dic. 1990, p. 382–406, bibl., table)

This careful empirical study finds no relationship between stock exchange real returns and inflation in two of three periods examined (1960s, 1970s, 1980s). It thus finds partial support for Fisher's generalized hypothesis.

1885 Hojman, David E. Non-govermental Organisations (NGOs) and the Chilean transition to democracy. (*Rev. Eur.,* 54, June 1993, p. 7–24, bibl.)

Thorough examination of role played by Nongovernmental Organizations in the Chilean transition to democracy.

1886 Huss, Torben. Transfer of technology: the case of the Chile Foundation. (*CEPAL Rev.,* 43, April 1991, p. 97–115, bibl., tables)

Fascinating study of the Chile Foundation, an innovative consulting and engineering design organization which demonstrates one manner of overcoming barriers for technology transfer. Contains excellent analysis of technology transfer involving establishment of salmon fisheries in Chile.

1887 Invirtiendo en Chile. Santiago: Banco Central de Chile, Comité de Inversiones Extranjeras, 1991. 256 p.

Valuable compendium of 54 papers presented at the 1990 International Seminar on Investment Opportunities in Chile. Includes presentations by President Patricio Aylwin, Minister of Finance Alejandro Foxley, Central Bank President Andrés Bianchi, and Minister of Economics Carlos Ominami.

1888 Jeftanovic, Pedro. El síndrome holandés: teoría, evidencia y aplicación al caso chileno, 1901–1940. (*Estud. Públicos,* 45, verano 1992, p. 299–331, tables)

Description and analysis of Dutch Disease (i.e. the negative impact of a primary sector export boom on other domestic activities) during the rise and fall of nitrate (1901–40). Emphasizes relevance of this historical experience to Chile's 1991 export boom.

1889 **Lagos M., Luis Felipe,** and **Alexander Galetovic P.** Los efectos de la indización cambiaria y salarial en el control de la inflación: el caso de Chile, 1975–1981. (*Cuad. Econ./Santiago,* 27:82, dic. 1990, p. 357–379, bibl., graphs, tables)

Reviews fiscal and monetary policy from a credibility point of view when indexation of wages and exchange rate to past inflation implied that inflation only gradually declined. Authors find inertia in the inflation process.

1890 **Larraín Arroyo, Luis et al.** Soluciones privadas a problemas públicos. Edición de Cristián Larroulet V. Santiago: Instituto Libertad y Desarrollo, 1991. 283 p.: bibl., ill.

Outstanding compendium of papers describes selected, generally successful, mesoeconomic constitutions and transformations implemented in Chile during the Pinochet presidency.

1891 **Leiva, Alicia Ximena.** El sector informal en Chile: análisis de sus componentes y mediciones posibles. (*Estad. Econ.,* 5, dic. 1992, p. 85–115, bibl., graphs, tables)

Careful examination of Chilean informal sector.

1892 **Mamalakis, Markos J.** Sectoral clashes, basic economic rights and redemocratization in Chile: a mesoeconomic approach. (*Ibero-Am./Stockholm,* 22:1, 1992, p. 31–44, bibl.)

Argues that mesoeconomic (i.e., sector specific) policies of the post-Allende period provided foundation for subsequent Chilean growth and development. Describes reforms and policies in the following sectors: government, trade, finance, industry, labor, and agriculture. [J.W. Foley]

1893 **Marshall R., Enrique.** El sistema financiero y el mercado de valores en Chile. México: Centro de Estudios Monetarios Latinoamericanos, 1991. 228 p. (Serie Estudios)

Detailed examination of Chile's financial system between 1973–89.

1894 **Marshall Silva, Jorge.** Banco Central: concepto, evolución y objetivos. Santiago: Univ. de Chile, Facultad de Ciencias Económicas y Administrativas, Editorial de Economía y Administración, 1991. 353 p.: bibl., ill.

Compendium of essays about the concept, evolution, and objectives of a central bank.

1895 **Marshall Silva, Jorge.** Políticas monetarias seguidas en Chile desde la creación del Banco Central. (*Cuad. Econ./Santiago,* 28:83, abril 1991, p. 29–54, graphs)

Leading specialist reviews major characteristics of monetary policies pursued in Chile since the establishment of the Central Bank in 1925.

1896 **McNelis, Paul D.** and **Klaus Schmidt-Hebbel.** Volatility reversal from interest rates to the real exchange rate: financial liberalization in Chile, 1975–82. Washington: Country Economics Dept., World Bank, 1991. 33 p.: bibl., ill. (Policy, research, and external affairs working papers; WPS 697)

Solid, technical analysis of financial liberalization in Chile during 1975–82.

1897 **Meller, Patricio.** Adjustment and equity in Chile. Paris: Development Centre of the Organisation for Economic Co-operation and Development; Washington: OECD Publications and Information Centre, 1992. 102 p.: bibl. (Adjustment and equity in developing countries. Development Centre studies)

Solid examination of Chilean development between 1980–87.

1898 **Meller, Patricio.** Adjustment and social costs in Chile during the 1980s. (*World Dev.,* 19:11, Nov. 1991, p. 1545–1561, bibl., tables)

Argues that adjustment measures of the 1980s were regressive even though the Chilean government was successful in targeting its expenditures toward the very poor during the period of fiscal retrenchment.

1899 **Messner, Dirk.** Shaping competitiveness in the Chilean wood-processing industry. (*CEPAL Rev.,* 49, April 1993, p. 117–137, bibl., table)

Examines international competitiveness in Chilean wood-processing industry from a dynamic and systemic perspective. Concludes that "it is very difficult to make the transition from the production and export of labor-intensive goods with a low level of processing to industrial products of greater added value, while at the same time ensuring and improving sustainable competitiveness."

1900 Mizala, Alejandra. International industrial linkages and export development: the case of Chile. (*CEPAL Rev.*, 46, April 1992, p. 151–175, bibl., graphs, tables)

Excellent analysis of role played by international industrial linkages in development of Chilean exports. Concludes that in an initial phase the export potential of national industries does not depend to a major extent on industrial linkages with enterprises in developed countries. These, however, facilitate penetration of protected markets in developed countries in later stages.

1901 Moyano Berrios, Eduardo. Trade opportunities in Chile. (*UNISA/Lat. Am. Rep.*, 7:2, 1991, p. 44–53, graphs, tables)

Contrasts pre-1973 protectionist policies followed in Chile with post-1973 "open" economy strategy of liberalization, privatization, and deregulation. Attention is drawn to transformation of the Chilean economy, its export growth, and many opportunities afforded by Chile for trade, investment, and partnership in numerous enterprises.

1902 Mujica, Patricio and Osvaldo Larrañaga. Social policies and income distribution in Chile. (*in* Government spending and income distribution in Latin America. Edited by Ricardo Hausmann and Roberto Rigobón. Washington: Inter-American Development Bank, 1993, p. 17–51, bibl., graphs, tables)

Critical examination of social policies and income distribution in Chile during 1970–89.

1903 Mujica, Rodrigo; Juan Ignacio Varas; and Rosa Marina Contesse. El impacto de las políticas de fomento a las exportaciones en la uva de mesa chilena. (*Cuad. Econ./Santiago*, 28:85, dic. 1991, p. 411–431, bibl. graphs, tables)

Careful examination of Chilean grape exports traces impact of external trade liberalization from 1974–79 on regional areas cultivated for production of grapes for export. Finds structural change in 1977.

1904 Muñoz, Oscar and Hugo Ortega. Chilean agriculture and economic policy, 1974–86. (*in* Modernization and stagnation: Latin American agriculture into the 1990s. Edited by Michael J. Twomey and Ann Helwege. New York: Greenwood Press, 1991, p. 161–188, bibl., tables)

Excellent review of the profound economic and institutional transformation of Chilean agriculture between 1974–86 highlights persistent heterogeneity and pluralism within agriculture and dynamism of export fruit production from Central Region.

Osobennosti razvitiia chiliĭskoĭ ėekonomiki i perspektivy rossiĭsko-chiliĭskogo delovogo sotrudnichestva. [Features of the development of the Chilean economy and perspectives on Russian-Chilean economic relations.] See item **4258**.

1905 Pardo V., Lucía and Ignacio Irarrázaval L. Características principales de las jefas de hogar en el Gran Santiago: algunos alcances de política. (*Cuad. Econ./Santiago*, 28:85, dic. 1991, p. 491–519, tables)

Seeks to determine implications of household authority in terms of employment, income, and other features as an aid to formulating policy measures to safeguard the family, with special emphasis on female authority. Based on information obtained from a June 1990 survey of Greater Santiago.

1906 Piñera Echenique, José. El cascabel al gato: la batalla por la reforma previsional. Santiago: Zig-Zag, 1991. 172 p. (Col. Testimonio y futuro)

Inside story of Chilean social security reform during the Pinochet presidency.

1907 Pinto, Aníbal. Las raíces del experimento ortodoxo chileno. (*Invest. Econ.*, 50:195, enero/marzo 1991, p. 9–19, table)

Master structuralist and one of Chile's most prolific and admired economists examines the post-1973 neoliberal, orthodox paradigm from an historical perspective. By tracing its historical roots, he demonstrates the high degree of continuity in the evolution of the Chilean economy.

1908 Provisions on conversion of external debt = Disposiciones sobre conversión de deuda externa. Santiago: Banco Central de Chile, 1988? 1 v.

Vital document describes rules of acquisition of foreign debt instruments.

1909 **Queisser, Monika.** Soziale Sicherung in der Krise: Die chilenische Rentenreform als Modell für Lateinamerika? (*Lat.am. Jahrb.*, 1993, p. 93–110, bibl.)

Concise and keen analysis of Chile's privatized, but government-regulated pension system, which author examines as a possible model for Latin America. Although the system performs well enough within Chile's current framework of favorable economic and political stability, only its technical aspects would be applicable elsewhere. [C.K. Converse]

Rabkin, Rhoda. How ideas become influential: ideological foundations of export-led growth in Chile, 1973–1990. See item **3480**.

1910 **Ritter, Archibald R.M.** Development strategy and structural adjustment in Chile: from the Unidad Popular to the *Concertación*, 1970–92. Ottawa: North-South Institute, 1992. 88 p.: bibl., ill., map.

Survey of economic policies reviews developmental strategies of Presidents Allende, Pinochet, and Aylwin.

1911 **Rivera Agüero, Rigoberto.** Organizaciones sociales para solucionar la pobreza: problemas y perspectivas. Santiago: Grupo de Investigaciones Agrarias, 1990? 109 p.: bibl. (Cuadernillo de información agraria; 23)

Examines nature of rural social organizations which aim to alleviate poverty.

1912 **Rodríguez, Daniel** and **Sylvia Venegas.** De praderas a parronales: un estudio sobre estructura agraria y mercado laboral en el Valle de Aconcagua. Santiago: Univ. Academia de Humanismo Cristiano, Grupo de Estudios Agro-Regionales, 1989. 261 p.: bibl., ill., map. (Serie "Abriendo caminos")

Thorough, historical-structural micro study of agrarian structure and labor market in the Aconcagua Valley.

1913 **Rosende R., Francisco.** Política monetaria: criterios para su diseño y evaluación. (*Estud. Públicos*, 42, otoño 1991, p. 59–90, bibl., graphs, tables)

Careful examination of traditional and new criteria used in evaluating efficiency of monetary policy. Rosende recommends that performance of monetary authorities be measured by using both types of criteria.

1914 **Rozas, Patricio** and **Gustavo Marín.** 1988, el "mapa de la extrema riqueza:" 10 años después. Santiago: Ediciones Chile-América CESOC, 1989. 1 v.: bibl.

Polemical examination of Chilean development from a Marxist-dependency perspective.

1915 **Rozas, Patricio.** Inversión extranjera y empresas transnacionales en la economía de Chile, 1974–1989: proyectos de inversión y estrategias de las empresas transnacionales. Santiago: Naciones Unidas, Comisión Económica para América Latina y el Caribe, 1992. 257 p.: bibl. (Estudios e informes de la CEPAL, 0256–9795; 85)

Detailed examination of foreign investment projects and strategies of transnational enterprises in Chile during 1974–89.

1916 **Rubinstein, Eugenia Muchnik de; Isabel Vial de Valdés; Andreas Strüver;** and **Bettina Harbart.** Oferta de trabajo femenino en Santiago. (*Cuad. Econ./Santiago*, 28: 85, dic. 1991, p. 463–489, bibl., tables)

Econometric study estimating female labor supply is based on primary data collected through an Oct. 1985 survey of over 3,000 households. Female labor supply is shown to be directly related to labor income and inversely related to number of preschool children and total household income (excluding the female labor income).

1917 **Santamaría, Marco.** Privatizing social security: the Chilean case. (*Columbia J. World Bus.*, 27:1, Spring 1992, p. 38–51, graphs, tables)

Examines role of Chile's private social security system within long-term process of economic development.

1918 **Schkolnik, Mariana.** The taxi market in Chile: regulation and liberalization policies, 1978–1987. (*in* Beyond regulation: the informal economy in Latin America. Edited by Víctor E. Tokman. Boulder, Colo.: L. Rienner, 1992, p. 207–245, graphs, tables)

Article examines principal changes that occurred in the taxi market following implementation of liberalization policies in Chile in 1978.

1919 **Sunkel, Osvaldo.** La consolidación de la democracia y del desarrollo en Chile: desafíos y tareas. (*Estud. Públicos*, 48, primavera 1992, p. 97–115)

Leading Chilean structuralist describes challenges being faced and goals that need to be achieved for Chile to enjoy both democracy and economic development.

1920 Trade and investment relations between Chile and the United States during 1989. Washington: Embassy of Chile, 1990. vi, 45 p.: ill.

Detailed survey of trade and investment relations between Chile and the US during 1989.

1921 Velis Meza, Héctor. Crónica personal del libro en Chile. Santiago: Ediciones Cerro Huelén, 1989. 124 p.: bibl.

Systematic review of Chilean book industry and trade (1976–87).

1922 Verlin, Evgueni. ¿Puede sernos útil la experiencia de Chile? (*Am. Lat. Enfoques Rusia*, 1, 1992, p. 27–30)

Russian economist who, along with four others, visited Chile and interviewed Gen. Pinochet describes central features of the Chilean liberal paradigm and the cardinal role played by Pinochet in its implementation. According to Kagalovski, a member of the team, "there are many dictators, but only one Pinochet." Author concludes by pointing out differences between Chile and present-day former Soviet Republics that limit the applicability of the Chilean Pinochet model for those republics.

PERU

CATALINA RABINOVICH, *Consultant, Hispanic Division, Library of Congress, Washington, D.C.*

CHANGE, CHANGE, AND MORE CHANGE has been Peru's script over the last three decades. Though the trend towards modernization started at the beginning of this century, it certainly has accelerated dramatically since the introduction of Gen. Velasco's reforms in 1968. The social turmoil generated both by his reforms and subsequent counterreforms peaked in 1990 with the political debut and overwhelming victory of independent presidential candidate Alberto Fujimori. Only two years later, this turbulence subsided with the Sept. 1992 capture of Shining Path leader Abimael Guzmán, which brought with it an atmosphere of peace not seen by Peruvians for almost 15 years.

Peru's social texture changed dramatically in the 1960s-90s period, so much so that it became possible for an unknown member of a small minority to become president of the most fragmented society in Latin America. Moreover, the effects of Fujimori's economic and institutional policies have accelerated the process of change and modernization to the point where now the "common man" in Peru feels that he or she could also occupy Lima's Presidential Palace. This new consciousness helps to explain the emergence of 27 candidates for president in 1995, the highest number ever in the history of Peru. Most of these individuals were unaffiliated with a party, as exemplified by the top two contenders: the victor, well-known President Fujimori, and former United Nations Secretary General, Javier Pérez de Cuéllar.

Among Fujimori's major achievements has been bringing under control the highest inflation in the country's history. In Aug. 1990, shortly after his election, Fujimori launched his anti-inflationary program which featured an initial shock-therapy approach. Afterwards, under Economics Minister Carlos Boloña the country regained the confidence of the international financial community and rapidly moved towards a true market economy, liberalizing the Peruvian economy and diminishing the size and scope of the State in accordance with world trends following the fall of the Berlin Wall. Simultaneously, the government fought against the Peruvian people's ideological distrust of private enterprise, an outlook stressed for decades by Peruvian politicians and intellectuals who demonized the notion of entrepreneurship at the individual level.

Although intellectuals have persisted in their anti-entrepreneurial mindsets, as of the 1980s an increasing number of the city poor were surviving and succeeding in the informal sector as either micro or small businessmen. The very rich, in turn, were reorganizing their businesses in more up-to-date and sophisticated ways. Even average Peruvians in rural areas were affected by this change in attitude. Fujimori, a savvy politician, has consistently capitalized on these developments. Although many problems still remain to be solved, it is clear that by now Peru has undergone an important attitudinal change in direction or a *cambio de rumbo*.

Most books and articles annotated below attest to how much is understood (as well as how much remains to be learned) about Peruvian events towards the end of the 20th century. Change, change, and more change is occurring within a democracy in progress, a process which has opened up the governing of the nation to almost anyone regardless of ethnic, social, or political origin. Previously, the nation's leaders belonged to either one of two categories: the white elite of Spanish descent, from which most democratically-elected former presidents traced their lineage, or the high-ranking mestizo members of the military who usually came to power by military coup.

Of particular interest is Boloña's appropriately titled book, *Cambio de rumbo*, which is highly recommended for those wishing to understand Fujimori's economic policies as well as future prospects for the country (item **1928**). Tulchin's *Peru in crisis* deals with more or less the same phenomena albeit from a different perspective (item **1959**). In *El nuevo capital financiero*, Alcorta analyzes how the very rich navigated through Peru's tumultuous changes (item **1925**). González de Olarte is among the first Peruvians to find *domestic* causes for the country's troubles in *El péndulo peruano* (item **1944**). Lastly, in *Empleo y pequeña empresa en el Perú*, Villarán places small industry in proper perspective, showing that it constitutes a pre-condition for a true market economy (item **1966**).

1923 Alarco, Germán and **Carmen Salas.**
Restructuración productiva en el Perú: propuestas nacionales. (*Social. Particip.*, 55, sept. 1991, p. 31–57, bibl.)

Discusses main arguments of experts such as Isaías Flit, Luis García-Núñez, and Fernando Villarán de la Puente on issues of science, technology, and restructuring of production being carried out in the country.

1924 Albuquerque Lloréns, Francisco *et al.* Nuevos rumbos para el desarrollo del Perú y América Latina. Recopilación de Efraín Gonzales de Olarte. Lima: Instituto de Estudios Peruanos, 1991. 206 p.: bibl., ill. (América problema; 15)

For its 25th anniversary, the Instituto de Estudios Peruanos convened national and international experts to address how to improve development policies in order to avoid any more consecutive economic failures. Includes variety of alternatives to the structural adjustments commonly used at the end of the 1980s, ranging from even greater State intervention all the way to its substantial elimination, using various combinations of heterodox and orthodox policies.

1925 Alcorta, Ludovico. El nuevo capital financiero: grupos financieros y ganancias sistémicas en el Perú. Lima: Fundación Friedrich Ebert, 1992. 411 p.: bibl., ill.

Alcorta asserts that financial groups are institutional systems created by the Peruvian families who control them with the specific purpose of exploiting systemic profits or synergies. Fundamental reasons why they are so different from Korean *chaebols* is their financial "aristocratic" conduct towards management and their unquestioned authority. As a result, planners lack both a strategic vision and technological training. If macroeconomic analysis had been used to question *why* things are as Alcorta describes, these would have been some powerful answers.

1926 Alvarado, Javier. Cajas rurales y fondos rotatorios: soluciones o mitos para el financiamiento rural en el Perú. (*Debate Agrar.*, 16, enero/abril 1993, p. 109–125, tables)

Peru's Banco Agrario traditionally financed rural areas. Once it was dissolved, the current government began encouraging the creation of rural savings banks as well as rotating funds. Author evaluates arguments, favorable and unfavorable, concerning these not-so-brand-new proposals, taking into account experience already available in the country.

Anaya Franco, Eduardo. Los grupos de poder económico: un análisis de la oligarquía financiera. See item **4883**.

Barrantes Lingán, Alfonso. La reinserción de Perú en el sistema financiero internacional en la perspectiva de la izquierda socialista. See item **4298**.

1927 Boada Rebata, Hugo. PRODERM, acciones de desarrollo y cambios en Anta. Cusco, Peru: PRODERM; Nederlands Economisch Instituut, 1991. 156 p.: bibl., col. ill., map.

Major changes have been occurring during the last decade in the rural area of Anta. Some are related to land ownership and others to technological changes. Ten years after the PRODERM Project got started, it was evaluated to determine if there had been any changes in the standard of living of Anta's population, wih the purpose of establishing whether such changes were a consequence of the project.

1928 Boloña Behr, Carlos. Cambio de rumbo: el programa económico para los '90. 2. ed. Lima?: Instituto de Economía de Libre Mercado, 1993. 253 p.: bibl.

Without a doubt this is *the* book to read to understand the present and future economic direction of the Fujimori Administration that began in July 1990. Book is valuable not so much for its theoretical aspects as for the changing intellectual environment it depicts exceptionally well. Shows how after decades of contempt for private enterprise, the Peruvians have begun to embrace it wholeheartedly.

Bustamante Belaunde, Luis. La reinserción del Perú en el sistema financiero internacional en la perspectiva del Frente Democrático. See item **4301**.

1929 Campodónico, Humberto. De poder a poder: grupos de poder, gremios empresariales y política macroeconómica; consorcio de Investigación Económica. Lima: DESCO, Centro de Estudios y Promoción del Desarrollo, 1993. 339 p.

Book's originality lies in its introduction and use of interest-group theory to shed light on relations between economic and political power. Groups that were powerful either economically or politically adopted new positions after the 1990s. These new attitudes attest to their capacities for adapting to complex political, economic, and social changes.

1930 Chávez, Eliana. Consequences of the legal and regulatory framework in Peru's taxi market. (*in* Beyond regulation: the informal economy in Latin America. Edited by Víctor E. Tokman. Boulder, Colo.: L. Rienner, 1992, p. 247–275, tables)

Analyzes changes in the workings of the taxicab market in Peru, incorporating macroeconomic variables that affect it. Concludes with hypothesis that Peru's free market, at least as far as taxis are concerned, does not provide even minimal security. This is indeed a refreshing article which uses innovative concepts. Chávez is one of the most knowledgeable experts on the informal sector.

1931 Chossudovsky, Michel. Ajuste económico: el Perú bajo el dominio del FMI. Lima: Mosca Azul Editores, 1992. 99 p.: bibl.

Harsh criticism of economic shock policy adopted by Fujimori in Aug. 1990. However, economic measures author proposes as an alternative program have been implemented already—at least partially— in Peru.

1932 Congreso Nacional de Gerencia, *8th, Lima, 1992.* Anales: gerencia—cambio de actitud, actitud de cambio. Lima: Instituto

Peruano de Administración de Empresas, 1992. 361 p.: ill.

Proceedings of annual congress reveal state of mind of entrepreneurs during dramatic period (July 1992). Includes thorough discussion of country's change in direction towards a much smaller State, a goal to be accomplished by Fujimori's government through dynamic privatization process. Also addresses extraordinary changes faced by the world since the 1989 fall of the Berlin Wall as well as technological innovations that accelerate the obsolescence of industries.

1933 Cornelissen, Wilhelmus Jan et al. Hiperinflación y crédito agrario. Cusco, Peru: PRODERM; CENES, 1991. 188 p.: ill.

Compilation of papers from the 4th Seminar-Workshop on Credit for Peasants (Cusco, Nov. 1989). Focuses, in the context of hyperinflation, on macroeconomic issues of credit policy, its influence on Andean peasants, and the attempts to decentralize and regionalize since late 1980s.

1934 Dammert, Manuel. El Perú: tarea pendiente; bases para un proyecto nacional descentralista. Lima: Centro Nacional de Estudios y Asesoría Popular, 1992. 101 p.: bibl.

Briefly retells Peruvian history since before the Spanish arrival, attributing all of its problems to its "colonial condition." Proposes decentralization to democratize the country.

1935 Deere, Carmen Diana. Familia y relaciones de clase: el campesinado y los terratenientes en la Sierra Norte del Perú, 1900–1980. Lima: Instituto de Estudios Peruanos, 1992. 414 p.: bibl., maps. (Serie Estudios históricos, 1019-4533; 13)

A jewel of a book portrays sprouting of capitalism in the Northern sierra through the eyes of peasant families who were part of the process. Incorporates gender as an analytical tool.

1936 Desarrollo sostenido de la selva: un manual técnico para promotores y extensionistas. Lima: INADE; APODESA, 1990. 319 p.: ill. (Serie Documentos técnicos; 25)

Even though the Peruvian Amazon Jungle comprises about two thirds of national territory and is regarded as "the promised land of Peru" it has been systematically ransacked since 1557; drug traffic only worsens the situation. Book is a unique contribution advocating a rational use of the forest's natural resources within the framework of ecological harmony. Study demystifies many "fantasies" about the unlimited wealth of jungle land, providing a realistic perspective regarding what to expect from the region.

1937 Las economías andinas: evolución y perspectivas. Edición de César A. Ferrari, Clark Winton Reynolds y Reinhart Wettmann. Lima: Fundación Friedrich Ebert del Perú, 1993. 255 p.: bibl.

Very informative book on development of economies of the Andean Group during the last two decades. Best chapters concern development of each country: Bolivia, Colombia, Ecuador, Peru, and Venezuela. Authors discuss economic future of these nations and where they stand on the issue of economic integration.

1938 Las empresas del Perú ante el mundo. Lima: Apoyo; Forum & Forum, 1992. 209 p.: ill. (some col.).

Beautiful book designed to serve as Peru's "business card" at the EXPO '92 in Sevilla, Spain. Provides overview of the Peruvian economy, emphasizing its 1,000 most important companies.

Espejo Ortega, Alberto Octavio. O plano de estabilização heterodoxo: a experiência comparada de Argentina, Brasil e Peru. See item **2137**.

1939 Estrategias para el desarrollo de la investigación agropecuaria en la costa central y sur del Perú: seminario taller; Arequipa, 25–26 de mayo 1989. Lima: Fundación para el Desarrollo del Agro (FUNDEAGRO), 1989. 216 p.: bibl., ill., maps.

Compilation of discussions held at seminar (Arequipa, May 1990). Researchers from both the public and private sectors met with farmers' representatives and international experts to address the question of organizing a National Agricultural Research System. They also examined the issue of regional products for both export and domestic markets.

1940 Ferrari, César A. Industrialización y desarrollo: políticas públicas y efectos económicos en el Perú. Lima: Fundación Friedrich Ebert, 1992. 386 p.: bibl., ill.

Ferrari maintains that the much maligned import-substitution strategy—to which most problems of economic development are ascribed—was never fully implemented in Peru, at least not in a systematic and coherent way. Analyzes possibilities for a new strategy as well as context and conditions in which it could be applied.

1941 Gómez G., Rosario and **Karen Weinberger V.** Problemática de la mujer peruana en el campo laboral: un ensayo bibliográfico. Lima: Univ. del Pacífico, Centro de Investigación, 1992. 200 p.: bibl.

Very systematic synthesis of the available bibliography on the condition of Peruvian women in the nation's labor market.

1942 Gonzales de Olarte, Efraín. Una economía bajo violencia: Perú, 1980–1990. Lima: Instituto de Estudios Peruanos, 1991. 29 p.: bibl., tables (Documentos de trabajo; 40)

Since 1980, first Shining Path and then MRTA terrorism as well as cocaine production and traffic have destabilized the Peruvian economy, making it highly risky and unpredictable. Fortunately democracy holds. Author deals with these issues from an economic perspective.

1943 Gonzales de Olarte, Efraín *et al.* La lenta modernización de la economía campesina: diversidad, cambio técnico y crédito en la agricultura andina. Lima: Instituto de Estudios Peruanos, 1987. 233 p.: bibl, ill. (Análisis económico; 12)

Demolishes myths about peasants being a homogeneous group refusing to face the future. Nonetheless, peasants' modernization does advance slowly, depending as it does on both their own motivation as well as the incentives of the national economy. Authors use three different elements to analyze how modernization develops: product diversity, technology, and credit. Since regular development policies have had little or no effect on peasants, author believes it is important to study how peasants are motivated in order to create policies geared towards them.

1944 Gonzales de Olarte, Efraín and **Lilian Samamé.** El péndulo peruano: políticas económicas, gobernabilidad y subdesarrollo, 1963–1990. Lima: Consorcio de Investigación Económica; Instituto de Estudios Peruanos, 1991. 129 p.: bibl., ill. (Serie Análisis económico; 14)

Authors seek reasons for Peru's underdevelopment in *domestic* economic and political instability rather than in external causes. Book's three main objectives are: 1) to analyze how the fluctuation in economic policies since the 1960s contributed to underdevelopment and the current economic crisis; 2) to explore how political instability feeds economic instability; and 3) to uncover the requirements and mechanisms for overcoming "ungovernability with underdevelopment."

1945 Gootenberg, Paul. Niveles de precios en Lima del siglo diecinueve: algunos datos e interpretaciones. (*Economía/Lima*, 12:24, dic. 1989, p. 137–205, tables)

Recreating price statistics for the 19th century when the State did not maintain them requires massive effort. In this pioneering work, author both determines these statistics and applies them to the social, economic, and political history of Peru. This outstanding paper is also scientifically impeccable.

1946 Hacia la estabilización y el crecimiento. Lima: Centro de Investigación, Univ. del Pacífico; Société générale de surveillance, 1990. 247 p.: bibl. (INTERCAMPUS)

Univ. del Pacífico convened this conference to address Peru's worst economic, political, and social crisis, which included hyperinflation and a 20-year sustained drop in income levels (1970s-80s). Distinguished local and foreign economists gathered to discuss how each proposed to stabilize the economy and create growth. Dornbusch singled out the fiscal deficit as his target and cautioned against over-reliance on income policies to resolve the crisis. Moreyra emphasized that domestic inflation should be lowered to levels similar to those of Peru's leading commercial partners. Salazar noted that a welfare program, designed within a framework of economic stabilization, "could not afford to fail."

1947 Hendriks, Jan. Promoción rural y proyectos de riego: la experiencia del Proyecto Rehabilitación del antiguo canal La Estrella-Mollepata, Cusco, Perú. Cusco, Peru: Centro Andino de Educación y Promoción José María Arguedas, 1988. 155 p., 8 p. of plates: ill. (some col.), maps.

Summarizes experience of rehabilitating an ancient irrigation channel in a rural

Cusco province, a collaborative effort by a rural-development NGO and peasants seeking improvement of their community.

1948 Iguíñiz, Javier; Rosario Basay; and Mónica Rubio. Los ajustes: Perú, 1975–1992. Lima: Fundación Friedrich Ebert, 1993. 276 p.: bibl., ill.

Authors analyze the success, failure, and social impact of each adjustment policy adopted during 1975–92, linking each with the individual responsible for it. Because of their uniqueness, it is difficult for any one of these adjustment policies to serve as a "prescription."

1949 Iguíñiz, Javier and Ismael Muñoz. Políticas de industrialización del Perú, 1980–1990. Lima: Consorcio de Investigación Económica; Centro de Estudios y Promoción del Desarrollo, 1992. 214 p.: bibl., ill. (Cuadernos Desco; 17)

Follow-up to 1984 work by Iguíñiz entitled *Política industrial peruana, 1970–1980: una síntesis* (see *HLAS 49:3656*). Authors review and analyze policies that directly or indirectly have had an effect on industry over the 1980s and show reasons why that sector has been on a steady decline despite investment, employment, and the generation of hard currency from exports.

1950 Iguíñiz, Javier and Noemí Montes. Proyecto nacional: empresarios y crisis, 1970–1987. Lima: Centro de Estudios y Promoción del Desarrollo, 1990. 230 p.: bibl. (Cuadernos Desco; 15)

Given the diversity of Peruvian society and the variety of economic structures, macroeconomic policies are not enough to promote development. Therefore, authors propose creation of intermediate policies—which they regard as mesoeconomic and pluri-institutional—which would encourage the interaction and cooperation of the State, microeconomic enterprises, and large entities or corporate bodies.

1951 La importancia económica de la minería en el Perú. Lima: IDEM, 1991. 175 p.: bibl., ill., col. maps (Instituto de Estudios Económicos Mineros; 5)

Useful guide to mining provides information on an industry that contributes roughly 50 percent of the country's export income.

1952 Jiménez, Félix. Acumulación y ciclos en la economía peruana: crisis de paradigmas y estrategia de desarrollo no-liberal. Lima: Centro de Estudios para el Desarrollo y la Participación, 1991. 146 p.: bibl. (Serie Realidad nacional)

Jiménez identifies ideas that underlie policies sponsored by ECLAC. Distinguishes between those that derive from neoclassical theory and those which do not. Finally, author proposes macroeconomic policy measures designed to democratize the nation.

Kisic, Drago. La reinserción internacional del Perú. See item **4308**.

1953 Mejía, José Manuel. Cooperativas azucareras: crisis y alternativas. Lima: Cambio y Desarrollo, Instituto de Investigaciones, 1992. 165 p.: bibl. (Col. Textos; 1)

Author describes reasons for the current crisis of Peru's sugar industry, which is among the few sugar industries in the world entirely owned by cooperatives. Mejía reviews State policies towards them during the last 22 years and speculates as to what to expect of this strategic national industry.

1954 Mujica, María Eugenia. ¿Por qué fracasó la heterodoxia?: análisis de la política de la primera etapa del gobierno de Alan García. (*Apuntes/Lima*, 27, segundo semestre, 1990, p. 31–43, graph)

Mujica acknowledges that there was a decline in President Alan García's credibility after he announced the expropriation of Peru's financial sector. Nevertheless, author emphasizes that the imbalances and gaps in the fiscal and monetary sectors began developing before July 1987.

1955 Niekerk, Nicolaas G.W. Desarrollo rural en los Andes: un estudio sobre los programas de desarrollo de organizaciones no-gubernamentales. Leiden, The Netherlands: Vakgroep CASNW, Rijksuniversiteit Leiden, 1994. 360 p.: bibl., map, tables. (Leiden development studies; 13)

Analyzes reasons why aid and development projects have limited impact in the rural Andes. First part discusses different approaches to development. Second part looks at four case studies in Peru and Bolivia. Author concludes that development interventions by NGOs contribute little to increasing family incomes. [R. Hoefte]

1956 Ortiz de Zevallos M., Felipe. Solving the Peruvian puzzle. (*in* In the shadow of the debt: emerging issues in Latin America. New York: Twentieth Century Fund Press, 1992, p. 37–57, tables)

In a few pages, author provides thorough depiction of changes implemented by President Fujimori since his surprising electoral victory in July 1990.

Pastor, Manuel. Inflation, stabilization, and debt: macroeconomic experiments in Peru and Bolivia. See item **1993**.

1957 Perspectivas de desarrollo de la Región Inka. Sicuani, Peru: Vicaría de Solidaridad, Prelatura de Sicuani; Cusco: Centro Bartolomé de Las Casas, 1990. 92 p.

Seminar on "Perspectives for the Development of the Inka Region" (Sicuani, March 1990) was held to discuss the future of the area encompassing the departments of Cusco, Apurímac, and Madre de Dios. Six essays deal with issues discussed at the conference, especially in regards to what actions to take in order for regional governments to further development among the local population.

1958 El Perú, el medio ambiente y el desarrollo. Edición de Eduardo Ferrero Costa. Lima: CEPEI, 1992. 238 p.: bibl., ill. (Serie Simposios internacionales, 1017–513X; 8)

Includes three particularly interesting environmental papers on Peru: 1) the sea and its meaning as an ecological Peruvian resource; 2) the Amazon—addressing topics beyond those usually dealt with; and 3) drug traffic and its impact on Peru's environment and development.

1959 Peru in crisis: dictatorship or democracy? Edited by Joseph S. Tulchin and Gary Bland. Boulder, Colo.: L. Rienner Publishers, 1994. 208 p.: bibl., ill., index. (Woodrow Wilson Center current studies on Latin America)

One of the most recent publications devoted to Peruvian current events. Analyzes the three most important developments in the Fujimori Administration to date: 1) the Fujishock or radical economic adjustment; 2) the *autogolpe* or self-coup; and 3) the capture of Shining Path's leader, Guzmán.

1960 Revilla C., Víctor. El proceso de liberalización comercial, 1978–1983: lecciones de una experiencia frustrada. Lima: Fundación Friedrich Ebert, 1990. 70 p.: bibl., ill. (Diagnóstico/debate; 47)

Notes that the 1982–83 crisis was not entirely due, as has been maintained, to the 1983 climatological phenomena, the 1982 debt problem, and commercial liberalization begun in 1979. Instead, all three developments aggravated an already existing crisis that author attributes to traditional causes (e.g., excessive government spending and fluctuations in global economic activity). Concludes with analysis, review, and suggestions as to how to avoid recreating these circumstances.

1961 Rojas-Suárez, Liliana. Currency substitution and inflation in Peru. (*Rev. Anál. Econ.*, 7:1, junio 1992, p. 153–176, bibl., graphs, tables)

Author demonstrates the existence and functioning of the long-term relationship between the expected rate of depreciation in black market exchange rates and the ratio of domestic-to-foreign currency in Peru.

1962 Scurrah, Martín J. and **Baltazar Caravedo.** Cambio tecnológico y comunidad campesina: módulos lecheros en los Andes del Perú. Con la colaboración de Eudosio Sifuentes y César Bedoya. Lima: Publicaciones SASE, 1991. 136 p.: bibl.

Although originally introduced by Spanish conquerors, five centuries later dairy cattle constitute an integral part of rural Andean life. At present, new technologies are being adapted to conditions in Highland peasant communities. Authors analyze impact of such technological innovations on these communities.

1963 Thijsen, Theo. Pequeñas empresas en comunidades campesinas. Cusco, Peru: T. Thijsen, 1991. 124 p., 6 leaves of plates: ill. (some col.), maps.

Describes recent experiences of PRODERM (Proyecto de Desarrollo Rural en Microregiones) in small rural industries in the Peruvian sierra. Examples not only prove that employment was generated among the poorest peasants, but also indicate how some difficult problems were solved (e.g., wheat grinding in very remote places). Final suggestions are of great interest: based on his rich knowledge of the subject, Thijsen highlights the need to define the goals of these particular small industries.

1964 Torres C., Víctor. El Perú frente a la Cuenca del Pacífico: flujo comercial con los países asiáticos. Lima: Instituto Peruano de Relaciones Internacionales; Programa de las Naciones Unidas para el Desarrollo, 1991. 245 p.: bibl., ill.

Author is well acquainted with present state of affairs in commercial relations between Peru and the US, Japan, "the four tigers" (South Korea, Taiwan, Hong Kong and Singapore), and the ASEAN countries (Malaysia, Philippines, Thailand, and Indonesia). Also discusses economic policies adopted by those Asian nations to attain their present level of industrial exports.

1965 Vega Centeno, Máximo. Desarrollo económico y desarrollo tecnológico. Lima: Consorcio de Investigación Económica; Pontificia Univ. Católica del Perú, Fondo Editorial, 1993. 232 p.: bibl., ill.

Important effort to incorporate technological developments within an economic development framework. Stresses need to define and adopt an explicit policy even in a market economy.

1966 Villarán de la Puente, Fernando. Empleo y pequeña empresa en el Perú. Lima: Fundación Friedrich Ebert; PEMTEC, 1993. 242 p.: bibl.

Author proposes solution for employment problems in Peru that not only takes into account the deficit in number of adequate jobs, but focuses on *productivity*. Maintains that promoting small modern enterprises would be entirely consistent with a market economy. The creation of numerous small enterprises that are modern and efficient is practically a precondition for a market economy.

1967 Villarán de la Puente, Fernando. El nuevo desarrollo: la pequeña industria en el Perú. Lima: ONUDI; PEMTEC, 1992. 182 p.: bibl., ill.

Describes small industrial enterprises in Peru as well as strategies and institutional framework required for their future development.

1968 Yepes, Ernesto. La modernización en el Perú del siglo XX: economía y política, ilusión y realidad. Lima: Mosca Azul Editores, 1992. 89 p.: bibl.

Addresses Peru's two modernization processes, from 1900s–30s and from 1950s to the present. Author notes that though neither one benefitted the entire population, their eventual impact was to produce the social turmoil that opened the door in 1990 to Fujimori's government.

BOLIVIA

JEFFREY FRANKS, *Economist, International Monetary Fund*

THE ECONOMIC LITERATURE ON BOLIVIA shows improvement in both quantity and quality, although it is still an understudied economy. This progress is due in part to the international attention Bolivia has received from its 1985 stabilization program and from the ongoing international cocaine crisis. Another contributing factor to the improvement has been a spate of recent high-quality works by a few academics, particularly economist Juan Antonio Morales and economic anthropologist Ricardo Godoy.

Papers on stabilization and structural adjustment continue to dominate the macroeconomic literature. With the New Economic Policy now a decade old, several studies explore the longer-term issue of why Bolivian growth has not lived up to the early post-1985 euphoric expectations. Particularly good publications in this area are Morales and Sachs' "Bolivia's Economic Crisis" (item **1987**), Morales, Espejo, and Chávez's "Temporary External Shocks and Stabilization Policies for Bolivia" (item **1991**), and Antezana's *Análisis de la nueva política económica* (item **1969**). Other recommended macroeconomic studies are Pastor's comparative study of Peru and Bolivia in the 1980s (item **1993**) and Clements and Schwartz's article on currency substitution (item **1974**).

There have been a number of recent works on the economics of coca and cocaine in Bolivia, many of them polemical and lacking in firm analysis. One exception is the recommended piece by De Franco and Godoy, "The Economic Consequences of Cocaine Production in Bolivia: Historical, Local, and Macroeconomic Perspectives" (item **1976**). For additional worthwhile studies on drugs and drug traffic in Bolivia, see the chapter on "Government and Politics: Bolivia" (pp. 454–467). Finally, there is the usual collection of sectoral studies of uneven quality, with mining, agriculture, and the informal sector receiving particular attention. Contreras and Pacheco's book on private sector commercial mining from 1939–89 provides needed information on the fastest growing sector of Bolivian mining (item **1975**).

1969 Antezana Malpartida, Oscar R. Análisis de la nueva política económica. La Paz: Editorial Los Amigos del Libro, 1988. 289 p.: bibl.

Useful analysis of the New Economic Policy (NEP) under Paz Estenssoro from orthodox economic perspective. Discusses economic problems during the 1970s and the collapse into hyperinflation in the early 1980s. Quantifies the NEP's results in terms of monetary, fiscal, external, and real effects, and examines prospects for future growth.

1970 Arrieta Abdalla, Mario et al. Agricultura en Santa Cruz: de la encomienda colonial a la empresa modernizada, 1559–1985. La Paz: ILDIS, 1990. 374 p.: bibl., ill., maps.

Comprehensive history includes agricultural economic history and discusses crucial role of government policies in agricultural development. Emphasizes respective roles of four crops: soya, cotton, sugar, and rice. Concludes that Santa Cruz suffers from excessive segmentation between poor, small-scale, technologically-backward farmers and wealthy, powerful, agroindustrialists.

1971 Arze Vargas, Carlos. Veleidades de la iniciativa privada. (*Rev. UNITAS*, 7, sept. 1992, p. 53–60, ill, tables)

Argues that preconditions for private investment have been fulfilled—low inflation, smaller government, freer trade, and liberalized capital markets—yet investment has responded anemically. An inefficient financial system and emphasis on short-term speculation are responsible and reveal Bolivia's subordination to international capital. Concise and well-written.

1972 Auty, Richard M. Bolivia: accelerating weakness despite external positive shocks. (*in* Sustaining development in mineral economics: the resource curse thesis. London: Routledge, 1993, p. 73–90, tables)

Looks at 1970s and 1980s as "resource curse decades"—in which resource booms led to macroeconomic policies that hurt the country. Argues that a commodity boom allowed the Banzer government to abort stabilization efforts and increase external indebtedness, setting the stage for the collapse in the 1980s. Interesting perspective.

1973 Baptista Gumucio, Fernando. La deuda externa de Bolivia. (*Síntesis/Madrid*, 14, mayo/agosto 1991, p. 325–345, tables)

Historical narrative on evolution of Bolivia's external debt by a former Minister of Finance. Interesting policy memoir, but short on analytical content and filtered through a partisan lens to show the Paz Estenssoro government in favorable light, while negatively viewing both previous and subsequent administrations.

1974 Clements, Benedict and **Gerd Schwartz.** Currency substitution: the recent experience of Bolivia. (*World Dev.*, 21:11, Nov. 1993, p. 1883–1893, bibl., tables, graphs)

Explores currency substitution (dollarization) after the 1985 stabilization. Paradoxically, dollarization continued rising despite low inflation and economic stability. Inflation, devaluation, and interest rate differential variables poorly explain the currency substitution trend; inertial factors dominate regressions. Evidence suggests that substitution increases with macroeconomic instability, but does not reverse when stability is achieved. Well-done, concise article.

1975 **Contreras C., Manuel E.** and **Mario Napoleón Pacheco T.** Medio siglo de minería mediana en Bolivia, 1939–1989. La Paz: Biblioteca Minera Boliviana, 1989. 164 p.: bibl., ill. (Biblioteca Minera Boliviana; 4)

Two papers on private "medium-sized" mining in Bolivia over the last 50 years contain useful analysis of this sector which now rivals COMIBOL in output and surpasses it in efficiency and innovation. Concludes that the future of Bolivian mining lies increasingly in these private firms.

1976 **De Franco, Mario A.** and **Ricardo Godoy.** The economic consequences of cocaine production in Bolivia: historical, local, and macroeconomic perspectives. (*J. Lat. Am. Stud.*, 24:2, May 1992, p. 375–406, appendices, graphs, tables)

Presents historical perspective on coca along with a Computable General Equilibrium model to estimate economic impact of the recent cocaine boom. Argues that coca has been important since colonial times and has contributed positively to the economy. Concludes that cocaine has not had highly negative effects on agriculture or on the overall economy. Good work on an extremely important subject.

1977 **De Franco, Mario A.** and **Ricardo Godoy.** Potato-led growth: the macroeconomic effects of technological innovations in Bolivian agriculture. (*J. Dev. Stud.*, 29:3, April 1993, p. 561–587, appendix, bibl., graph, tables)

Uses Computable General Equilibrium model to simulate results of agricultural innovation on the economy, which permits analysis of indirect effects not captured in a partial equilibrium framework. Results show that technological improvements in staple goods (particularly potatoes) have very positive macroeconomic effects: higher growth, declining unemployment, and rising real wages. Solid article using an innovative approach.

1978 **Desempeño y colapso de la minería nacionalizada en Bolivia: estudio técnico, económico, social y organizacional de la Corporación Minera de Bolivia.** La Paz: Centro de Estudios Minería y Desarrollo, 1990. xiv, 227 p.: bibl., ill.

Analysis of the State-owned mining company COMIBOL via a study of the Colquiri mine from 1952 nationalization through the radical restructuring of COMIBOL after 1985. Argues that COMIBOL maximized short-term production, ignoring its cost structure and neglecting investments necessary for longterm improvements in output and productivity. Despite radical cuts in COMIBOL, argues that the short-term mentality remains and problems will reemerge.

1979 **La economía campesina y el cultivo de la coca.** v. 2. La Paz: Instituto Latinoamericano de Investigaciones Sociales, 1988. 1 v.: bibl. (Debate agrario; 11)

Seminar proceedings on coca and the peasant economy. Lively, non-academic debate on Bolivian coca/cocaine policy involved both government and peasant representatives. Includes reproductions of government policy documents. Interesting, but lacks solid data and analysis.

1980 **Edwards, Sebastian.** Política cambiaria en Bolivia: avances recientes y perspectivas. (*Anál. Econ./La Paz*, 5, junio 1992, p. 7–49)

Study of behavior of real exchange rate after the 1985 stabilization program. Argues that the exchange rate has become overvalued due to structural shocks. Suggests that more rapid devaluation could cure the problem, but adjustments in fiscal and interest rate policy are also necessary. Very good empirical work.

1981 *Evaluación Económica.* No. 77, marzo 1993/No. 84, dic. 1993– . La Paz: Müller & Asociados.

Collection of monthly economic bulletins published by Müller & Asociados during 1993. Includes papers on the financial system, the electrical sector, social security, external debt, the Sánchez de Lozada government's economic programs and political reforms, and the overall macroeconomic performance. Timely and generally insightful.

1982 **Godoy, Ricardo** and **Mario A. De Franco.** High inflation and Bolivian agriculture. (*J. Lat. Am. Stud.*, 24:3, Oct. 1992, p. 617–637, graphs, tables)

Looks at interesting aspect of Bolivian hyperinflation and stabilization: interactions between agriculture, food prices, and hyperinflation. Argues that agriculture did well under hyperinflation since terms of trade favored that sector. Stabilization, on the other hand, has hurt agriculture due to reduced consumption.

1983 **Graham, Carol.** The politics of protecting the poor during adjustment: Bolivia's Emergency Social Fund. (*World Dev.*, 20:9, sept. 1992, p. 1233–1251, tables)

Examines Emergency Social Fund (ESF) organized to ameliorate social consequences of the 1985 stabilization policy. Concludes that the ESF provided significant real benefits to the poor, although it did not do well in reaching the very poorest. Well-done article, although focus is more political than economic.

1984 **Huibers, Bart** and **Peter de Kwaasteniet.** Handel op z'n Boliviaans: een actor-gericht onderzoek naar de graan- en uienhandel in Cochabamba [Trade the Bolivian way: an actor-oriented study of the grain and onion trade in Cochabamba]. Utrecht, The Netherlands: Faculteit Ruimtelijke Wetenschappen Univ. Utrecht, 1993. 133 p.: bibl., ill., maps, tables. (Diskussiestukken van de Vakgroep Sociale Geografie van Ontwikkelingslanden; 49)

Analyzes main characteristics of marketplace system and trade strategies used by individuals involved in the trade. Authors argue that two changes in Bolivian socioeconomic structure have had a definite impact on internal agricultural trade: the land reform act of 1953 and the economic adjustment process after 1985. Includes brief summary in English. [R. Hoefte]

1985 **Melvin, Michael** and **Kurt Fenske.** Dollarization and monetary reform: evidence from the Cochabamba region of Bolivia. (*Rev. Anál. Econ.*, 7:1, junio 1992, p. 125–138, table)

Uses data on informal loans in rural Cochabamba to examine determinants of dollarization. Finds that policy reform did not produce dedollarization: inflation, depreciation, and exchange-rate volatility all induce less dollarization. Lack of confidence in reform and coca dollar inflows may account for the surprising results. Interesting, but flawed model specification.

1986 **Morales, Juan Antonio et al.** Bolivia: ajuste estructural, equidad y crecimiento. La Paz: BAREMO; MILENIO, 1991. 375 p.: bibl.

Contributions from some of Bolivia's most important economists on issues of growth and poverty alleviation after the 1985 New Economic Policy stabilization effort. Includes analysis of the economic and social impact of reform and policies needed to help the poor. Papers are of differing quality, but several are quite good.

1987 **Morales, Juan Antonio** and **Jeffrey D. Sachs.** Bolivia's economic crisis. (*in* Developing country debt and economic performance. Chicago: Univ. of Chicago Press, 1989, v.2, p. 157–268, bibl., graphs, tables)

In-depth account of Bolivian economic problems from 1952–1980s. Authors argue that structural problems had been building up for 40 years, culminating in hyperinflation. Shock therapy achieved stabilization, but many of the structural problems have yet to be resolved. Bolivia must diversify exports, address social cleavages, and develop a competent State. First-rate article with compelling argument concerning Bolivia's development needs.

1988 **Morales, Juan Antonio.** Democracia y política económica en Bolivia. (*Síntesis/Madrid*, 14, mayo/agosto 1991, p. 311–323, tables)

Reviews New Economic Policy (NEP) and analyzes perspectives for growth. These policies produced a stable macroeconomic climate and liberalization of markets, but led to falling real wages. Future growth depends on stable policies, increased investment, and export orientation. Interesting and coherent, but democracy issue is not addressed adequately.

1989 **Morales, Juan Antonio.** Reformas estructurales y crecimiento económico en Bolivia. (*in* ¿Adonde va América Latina?: balance de las reformas económicas. Recopilación de Joaquín Vial. Santiago: CIEPLAN, 1992, p. 103–133, bibl., graphs, tables)

Focuses on structural reforms and stabilization policies of the New Economic

Policy (NEP). Recovery was slow because of depth of structural changes needed. Some important reforms were implemented simultaneously with stabilization, but other changes are needed if growth is to recover. Well-written. Mildly critical of the NEP.

1990 Morales, Juan Antonio; Justo Espejo; and Gonzalo Chávez. Shocks externos transitorios y políticas de estabilización para Bolivia. (*in* Shocks externos y mecanismos de estabilización. Edición de Eduardo Engel y Patricio Meller. Santiago: CIEPLAN; Banco Interamericano de Desarrollo, 1992, p. 185–230, appendices, bibl., graphs, tables)
See item **1991**.

1991 Morales, Juan Antonio; Justo Espejo; and Gonzalo Chávez. Temporary external shocks and stabilization policies for Bolivia. (*in* External shocks and stabilization mechanisms. Edited by Eduardo Engel and Patricio Meller. Washington: Inter-American Development Bank, 1993, p. 173–220, appendices, bibl., graphs, tables)

Interesting and sophisticated analysis of effects of commodity price changes on Bolivian GDP, government revenue, and exports. Examines price volatility using unit root tests and coefficients of variations. Presents suggestions on use of options markets and government policies to stabilize fluctuations, and explores establishment of macroeconomic stabilization fund to smooth out fluctuations. Also available in Spanish (see item **1990**.)

Niekerk, Nicolaas G.W. Desarrollo rural en los Andes: un estudio sobre los programas de desarrollo de organizaciones no-gubernamentales. See item **1955**.

1992 Palza Medina, Javier. La coca en la construcción nacional. La Paz: Editorial Signo A&G, 1991. 222 p.: bibl., col. ill.

In-depth narrative on the coca/cocaine problem. Discusses history of coca production, and recounts recent history of anti-drug policy in Bolivia. Argues that drug trafficking has nefarious impacts on morality and politics of Bolivia, and that it is generating increasing violence. Lays primary blame for cocaine problem on US policy and culture and advocates legalization of cocaine as the only solution to the problem. Provocative and polemical with interesting information, but not tightly argued.

1993 Pastor, Manuel. Inflation, stabilization, and debt: macroeconomic experiments in Peru and Bolivia. Boulder, Colo.: Westview Press, 1992. 176 p.: bibl., ill., index. (Westview special studies on Latin America and the Caribbean)

Provides comparative analysis of heterodox stabilization policy under García in Peru and the orthodox program in Bolivia. Although the orthodox policy has been more successful, there are ongoing difficulties in bringing sustained growth to Bolivia. In many ways the structuralist paradigm underlying heterodox policies better describes Latin American reality, but because it is more interventionist, it requires a more developed State which is lacking in most developing countries. Intelligent, well-written, and balanced book. Highly recommended.

1994 Prudencio B., Julio and Mónica Velasco L. La defensa del consumo: crisis de abastecimiento alimentario y estrategias de sobrevivencia. La Paz: Centro de Estudios de la Realidad Económica y Social, 1988. 271 p.: bibl., ill. (Seguridad alimentaria; 3)

Examines consumption of food and agricultural production during the pre-1985 economic crisis. Provides detailed information on consumption patterns, demonstrating inadequate calories and nutrition among many residents in poor neighborhoods. Argues that producers were also hurt by the crisis. Consumption study is fascinating, but production side is weaker.

1995 Seminario Nacional: Mujer, Género y Desarrollo, 1st, La Paz, 1992. Actas. La Paz: Coordinadora de la Mujer; Misión de Cooperación Técnica Holandesa; Plataforma de la Mujer, 1992. 88 p.: bibl.

Papers argue forcefully for incorporation of gender concerns into development projects. Participants advocate increasing participation of women in development projects and in politics. Also explores role of women in rural development and in coca-growing regions.

1996 Soria Galvarro, Carlos. La experiencia boliviana. (*in* Conversión de deuda en desarrollo: análisis de la situación de los países del área andina. Recopilación de Alfredo Angulo S. Bogotá: Cinep, 1993, p. 75–88, table)

Discusses commercial debt swaps. Bolivia pioneered debt-for-nature and debt-for-investment swaps and also conducted debt buybacks, but these programs had little impact. Commercial debt payments were already suspended, debt-for-nature affected little debt and did not improve the environment, and swaps were abused. Useful review, but narrow in coverage.

1997 Taller de Investigaciones Socio-Económicas, *12th, La Paz, 1991.* Programa de reforma estructural. Edición de Carlos F. Toranzo Roca. La Paz: ILDIS, 1992. 26 p.: ill.

Reports from seminar on structural reforms under Paz Zamora. Argues that economic growth must come from an increase in both efficiency and quantity of investment. Privatization, improvements in efficiency of government investment, and restructuring the Bolivian financial system are necessary to achieve this. Interesting perspective, though analysis lacks detail.

1998 Taller de Política Social, *4th, La Paz?, 1991.* Nueva política económica y sector informal urbano. Edición de Carlos F. Toranzo Roca. La Paz: ILDIS-BOLIVIA, 1992. 52 p.

Seminar on impact of the New Economic Policy (NEP) on the urban informal sector. Argues that the sector grew in size as household incomes suffered and indices of poverty rose. Calls for government action to support informal enterprises. Basically solid, but exaggerates negative effects of the NEP.

1999 Toranzo Roca, Carlos F. Bolivia: reproducción de capital y política. Santa Cruz, Bolivia: Editorial Universitaria, 1988? 306 p.: bibl. (Serie Estudios sociales, económicos, jurídicos)

Essays written between 1978–86 and originally published in journals explore economic and political issues ranging from policy analysis of the Banzer government to evaluation of the 1985 stabilization. Refreshingly well-written works in tradition of Marxist political economy school, with strong emphasis on class struggle and social injustice.

2000 Tovar Piérola, Raúl F. El Estado como empresario: un enfoque histórico. La Paz: Instituto Boliviano de Cultura, 1992. 222 p.: bibl., ill., maps.

Tour d'horizon of history and political economy of public enterprises and government involvement in the economy since the 19th century. Provides interesting narrative, but weak on statistical data and formal analysis.

2001 Zeballos Hurtado, Hernán. Agricultura y desarrollo económico. La Paz?: Asociación Boliviana de Iniciativas para el Desarrollo, 1988. 414 p.: bibl.

Comprehensive examination of agricultural sector with recommendations for increasing agricultural development. Focuses on economic analysis, but also includes examination of natural resources, geography, and politics of agriculture policy. Solid, wide-ranging reference work, but short on insightful analysis.

PARAGUAY AND URUGUAY

CLAUDIO SAPELLI, *Economist, World Bank*

ALTHOUGH IN THE MAKING FOR SOME TIME, there has been a noticeable change in the Uruguayan economic literature during this biennium: it has grown in volume, quality and coverage—both in the depth and breadth of topics addressed. For instance, there has been a surge of microeconomic studies and many of them use modern techniques of empirical analysis. And, although there is a wide variety of topics covered in this chapter, two themes stand out: Mercosur and human resources (from an analytic rather than a descriptive perspective). The Mercosur

Treaty—signed in Asunción, Paraguay on March 26, 1991 by Argentina, Brazil, Paraguay, and Uruguay—established a common market composed of all four countriesas of Jan. 1, 1995. As a result of this agreement, which opened Uruguay to competition from the efficient Brazilian industrial sector, much attention has been focused on the challenges of integration. The literature has centered on the analysis of: 1) costs and benefits of a complete opening to regional competition (items **2024** and **2011**); 2) the need for modernization of the industrial sector (item **2009**); and 3) the call to improve human resources and the operation of the labor market (item **2013**).

Among the considerable literature analyzing Uruguay's trade with Argentina and Brazil, many papers of analytical value have been excluded because of their narrow objectives. Unfortunately, many Uruguayan works are often not useful as references for further research since there appears to be a lack of discussion among Uruguayan authors who write on similar topics. Therefore, the literature is often repetitive and framed in response to foreign authors, rather than formulated on behalf of a more useful, healthy, and explicit domestic debate within Uruguay itself.

Writings on the economy of Paraguay are scant and are mostly descriptive as noted in previous *HLAS* volumes. Topics parallel those in Uruguay, e.g., the challenge of Mercosur (item **2006**) and the issue of human resources, especially the poor and, in particular, the rural poor. However, most studies that appear under standard economic titles are really sociological or philosophical discussions with very limited analysis of empirical evidence. Consequently, this chapter includes very few works.

PARAGUAY

2002 Borda, Dionisio. La estabilización de la economía y la privatización del Estado en el Paraguay, 1954–89. (*Estud. Parag.*, 17: 1/2, 1989/93, p. 37–89, bibl., graphs)

Interesting and provocative analysis of the Stroessner government. Weaves description of the evolution of economic policy during the period with that of the role of the State. Places particular emphasis on how the State apparatus was used for private gain and how deprivatizing the State is one of the main achievements of the new democratic regime.

2003 Ocampos, Lorraine. Paraguay: tipos de cambio y proceso de integración. Asunción: CEPPRO, 1993. 76 p.: appendix, bibl., tables. (Serie Estudios; 4)

In the framework set by Mercosur, this monograph analyzes possible consequences of lack of macroeconomic coordination. Reviews evolution of exchange rate policy in the four member countries, in particular Paraguay's bilateral real exchange rate with Brazil and Argentina. Shows that this exchange rate has fluctuated dramatically over the last decade and discusses its consequences. Concludes with a call for more macroeconomic coordination among member countries. Discusses need for presence of independent central banks in all four countries for this to be achieved.

2004 Palau, Tomás and **Carlos Verón.** Una contribución preliminar para el estudio de la frontera en el Paraguay y su impacto socioeconómico. (*in* Congreso Internacional sobre Fronteras en Iberoamérica Ayer y Hoy, *1st, Tijuana, Mexico, 1989.* Memoria. Edición de Alfredo Félix Buenrostro Ceballos. Mexicali, Mexico: Univ. Autónoma de Baja California, 1990, v. 1, p. 262–288, tables)

Excellent introductory study examines evolution of Paraguay's borders and frontier regions. Considers current economic significance of these regions in terms of transportation, tourism, energy production, legitimate trade and smuggling, and demographic features including population settlement and international migration. [A. Vacs]

2005 Schreiner, Jorge. El status del Banco Central en el sistema financiero nacional. Asunción: CEPPRO, 1993. 66 p.: bibl. (Serie Estudios; 3)

Brief historical survey of the evolution of the financial system and its regulation. Discusses need to reform the regulatory system and in particular the Central Bank's role in a new framework.

Stunnenberg, Peter and **Johan Kleinpenning.** The role of extractive industries in the process of colonization: the case of quebracho exploitation in the Gran Chaco. See item **2090.**

2006 Tappatá, Ricardo and **Jorge Vasconcelos.** El Paraguay y el Mercosur: situación actual, perspectivas y nuevas ventajas competitivas. Asunción: CEPPRO, 1994. 41 p.: tables. (Serie Estudios; 5)

Interesting analysis of the Paraguayan economy in the context of Mercosur. Starts with theoretical discussion of pros and cons of a common market, and ends with a discussion of the sectors that could have competitive advantages in the regional market.

URUGUAY

2007 Abuhadba, Mario. Models of wage determination and the industry wage structure in Uruguay. (*Rev. Anál. Econ.*, 6:1, junio 1991, p. 93–111, bibl., tables)

Based on a PhD dissertation, article examines wage structure in the manufacturing sector (1968–87). Tests competitive and efficiency models of wage determination. Results tend to support the efficiency-wage model.

2008 Alfie, Isaac. El mercado laboral uruguayo y la integración regional. (*Estudios/Fundación Mediterránea*, 14:59, julio/sept. 1991, p. 105–115, tables)

Paper analyzes costs and benefits of regional integration versus multilateral integration (i.e.: reducing tariffs to a region or to the world as a whole). Discusses whether Mercosur is a step towards opening markets to the world, an intermediate step to prepare productive sectors for world-wide competition, or a step to prevent world-wide competition. Argues in favor of worldwide economic integration rather than regional integration. Given the adoption of Mercosur however, describes policy measures to minimize the costs and maximize benefits of regional integration.

2009 Argenti, Gisela *et al.* Uruguay: el debate sobre la modernización posible. Edición de Gisela Argenti. Montevideo: Centro de Información y Estudios del Uruguay; Ediciones de la Banda Oriental, 1991. 275 p.: bibl., ill.

Good example of current Uruguayan microeconomic work. Includes six essays examining modernization of the industrial sector, a topic made more relevant by the challenges of Mercosur. Of the six, two are worth reading. The first, by Jean Ruffier, summarizes the literature on competitiveness and technology adoption in small countries and its application to Uruguay. The second, by Nelson Stratta, analyzes innovations in the industrial sector prompted by the increased openness to international competition during the last ten years and their implications for the modernization of Uruguayan industry.

2010 Arocena, José *et al.* El territorio, los hogares, la vivienda. Montevideo: Instituto Nacional del Libro, 1990. 163 p.: bibl. (La Pobreza en el Uruguay; 2)

Two essays on unrelated subjects. First provides an overview of the decentralization debate in Uruguay, analyzing the issue theoretical, historically, and politically. Second examines the poor in Uruguay and describes the process of their concentration in the "old city" downtown. Good analytical work.

Bensión, Alberto *et al.* Costa Rica and Uruguay. See item **1572.**

2011 Berretta, Nora; Fernando Lorenzo; and **Carlos Paolino.** En el umbral de la integración. Montevideo: Centro de Investigaciones Económicas; Ediciones de la Banda Oriental, 1991. 197 p.: bibl., ill. (Estudios CINVE; 14)

Contains four separate papers. Two describe and analyze Uruguayan trade relations and bilateral treaties with Brazil and Argentina before the 1991 adoption of Mercosur. The other two are more analytical. One quantifies regional intra-industry trade flows and analyzes their determinants; the second quantifies potential costs of regional bilateral agreements (the so-called "trade diversion"), concluding that costs were small. Book presents upbeat picture of Uruguay's benefits from past and present regional trade agreements, a conclusion supported by proficient quantification and adept use of econometric techniques.

2012 Buxedas, Martín. Oligopolios y dinámica industrial: el caso de Uruguay. Montevideo: CIEDUR, 1992. 160 p.: bibl., index.

One of the first attempts to apply industrial organization concepts to the manufacturing industry in Uruguay. Analyzes relationship of industrial structure to profits, dynamism of the sector, and wages. Good starting point to investigate possible effects of Mercosur on different sectors.

2013 **Capital humano y mercado de trabajo en el Uruguay.** Montevideo: Academia Nacional de Economía, 1989. 206 p.: bibl., ill.
Collection of ten papers presented during 1988 at the Academy includes a few worthwhile essays. Alberto Bension offers a brief survey of literature on human capital through 1988. Julio Preve provides a useful survey of literature on agricultural sector productivity. Both are worth reading.

2014 **Díaz, Ramón.** País pequeño debe ser país abierto: análisis de la estrategia de desarrollo óptimo para el Uruguay. (*Síntesis/Madrid*, 13, enero/abril 1991, p. 317–346, appendix, bibl., graphs, tables)
Absorbing analysis of Uruguay's trade policy since the last century and its relationship to growth. Asserts Uruguay's early wealth was due to an open trade policy and that closure of the economy in the late 19th century and again after the Great Depression is the main explanation for poor economic performance since the 1950s. Worth reading.

2015 **Diez de Medina, Rafael.** La estructura ocupacional y los jóvenes en Uruguay. Montevideo: Comisión Económica para América Latina y el Caribe, Oficina de Montevideo; Ministerio de Economía y Finanzas, 1993? 118 p.: bibl., ill.
Composed of two essays, work has broader coverage than title suggests. The first essay studies the occupational structure and behavior of young adults and the second examines overall income distribution, its evolution in 1984–88, and its relationship to the occupational structure. Essays contain excellent data analysis and helpful tables and graphs. Both include a wealth of data, and use econometric methods to analyze issues such as the labor supply of young adults. Very useful starting point for studying the Uruguayan labor market.

2016 **Diez de Medina, Rafael.** Los pasivos en el Uruguay: sus características sociales. Montevideo: Comisión Económica para América Latina y El Caribe, Oficina de Montevideo; Fundación de Cultura Universitaria, 1990. 36 p.: bibl. (Col. Temas nacionales; 24)
Short monograph uses the Household Survey to characterize population receiving pensions. Though mainly descriptive, an important addition to literature regarding social security policy. Shows that social security recipients live in households that are relatively better off than the population as a whole. In particular, households with social security recipients are underrepresented among poor households.

2017 **Diez de Medina, Rafael.** El sesgo de selección en la actividad de jóvenes y mujeres. (*Suma/Montevideo*, 7:13, oct. 1992, p. 69–85, bibl., tables)
Very good analytical article uses econometric methods to find determinants of young adults' and women's labor supply. Confirms results found elsewhere about behavior of the "secondary" labor force and importance of economic and demographic variables. One example of the domestic trend to use modern econometric methods to analyze microeconomic issues.

2018 **Favaro, Edgardo** and **Claudio Sapelli.** Promoción de exportaciones y crecimiento económico. San Francisco, Calif.: ICS Press, 1989. 188 p.
Authors carefully discuss and systematically analyze history of trade expansion and economic growth in Uruguay (1868–1986). They convincingly demonstrate that periods of economic stagnation have been clearly associated with trade restrictions and explain positive association between open trade systems and growth in terms of increased productivity of resources that generally results from export-oriented development strategies. Furthermore, authors demonstrate that partial trade expansion based on artificial incentives to protected sectors and exporting to protected markets does not necessarily raise productivity and can in fact be detrimental to economic growth. Executive summary of this work is available as *Export promotion and economic growth* (San Francisco, Calif: ICS Press, 1989). [M. Mamalakis]

2019 **Forteza, Alvaro.** Los convenios salariales de 1990 y la inflación. (*Suma/Montevideo*, 12, abril 1992, p. 7–36, bibl., tables)

Good analysis of the consequences of different indexation clauses in wage contracts, particularly regarding the possibility of inflation reduction.

2020 Haedo, Javier de and **Claudio Sapelli.** Simplificación y modernización del sistema tributario en Uruguay, 1972–1986. (*Rev. Econ./Uruguay*, 3:3, abril 1989, p. 79–114, graphs, tables)

In this careful study, authors demonstrate that fiscal reforms carried out in the 1970s-80s contributed to a more simplified, neutral, and equitable tax system. [M. Mamalakis]

2021 Laens, Silvia; Fernando Lorenzo; and **Rosa Osimani.** Macroeconomic conditions and trade liberalization: the case of Uruguay. (*in* Macroeconomic conditions and trade liberalization. Edited by Adolfo Canitrot and Silvia Junco. Washington: Inter-American Development Bank, 1993, p. 159–208, bibl., graphs, tables)

Worthwhile summary of evolution of trade policy and its interrelationship with macroeconomic policy (1970s–90s). Pays particular attention to domestic economy's reaction to trade liberalization (reaction of exports, imports, wages, and employment). Good survey of vast amount of literature from the last five to ten years. Concludes that most of the inward-looking industries have not yet been subject to external competition. Discusses the role of regional integration in the overall trade liberalization program.

2022 Longhi Zunino, Augusto and **Luis Stolovich.** La dinámica del mercado laboral uruguayo. Montevideo: CIEDUR, Depto. de Asesoramiento Técnico, Económico y Social, 1991. 96 p.: bibl., ill. (Cuadernos de información popular; 11)

Attempts to isolate the effects of increasing economic liberalization on the labor force, the labor market, and trade unions. Interesting perspective from trade union movement's point of view.

2023 Mercosur: claroscuro de una integración; ciclo de conferencias realizadas en la Facultad de Ciencias Económicas y Administración. v. 1–2 Montevideo: Editorial Fin de Siglo, 1991. 2 v. (Uruguay XXI)

Worthwhile and thorough work provides various perspectives on the issues and debates in Uruguay regarding Mercosur. Discusses historical precedents of the Treaty of Asunción, the customs union theory as it applies to the treaty, the contents of the treaty (the text is included in an appendix), its effects on agriculture, industry, workers, labor market, and the financial sector, and the role of the State. Includes presentations by many of Uruguay's leading figures. Useful introduction to Mercosur debate in Uruguay.

2024 Una nueva etapa en el proceso económico latinoamericano. Montevideo: Academia Nacional de Economía, 1992. 137 p.: bibl.

Papers presented at annual conference series organized by the Academia Nacional de Economía discuss regional integration, focusing on Mercosur. Contains valuable essay by Gustavo Magariños analyzing the content of the treaty that created Mercosur and comparing it to other trade agreements signed by Uruguay (ALADI, PEC, CAUCE); a good survey by Jorge Roldós examining recent literature on international trade and its relevance for the analysis of Mercosur; another good survey by Adolfo Díaz of the literature on financial markets and the impact of Mercosur on the Uruguayan capital market; and an interesting summary of the issues negotiated between 1991–92 (the common external tariff of the Custom Union, in particular) and the harmonization of macroeconomic policies (emphasizing fiscal policy) by a member of the Uruguayan negotiating team, Juan García Pelufo.

2025 Papadópulos, Jorge. Seguridad social y política en el Uruguay: orígenes, evolución y mediación de intereses en la restauración democrática. Montevideo: CIESU, 1992. 209 p.: bibl., ill.

Uruguay's social security expenditures are among the highest in the world, and reform attempts have been a matter of intense internal debate for over 15 years. Through a 1989 referendum, the Association of Pensioners succeeded in changing the Constitution to guarantee inflation indexation for pensions. Since then, social security reform has topped the political agenda. This work analyzes the political and historical dimensions of the problem from the client's (i.e.: pensioner's) point of view. Examines political and sociological reasons for the lack of reform characterizing the Uruguayan political system. Unlike other political scientists who attribute this resistance to change to the institutional organization of the system, this book investigates the endogenous forces accounting for such resistance. Provides good explanation of the success of the 1989 reform. However, analysis appears wanting regarding the reforms themselves, their economy-wide effects, and, in particular, the most obvious paradox implicit in the reform—why the Uruguayan society accepted a substantial redistribution of income to pensioners. Nonetheless, a useful addition to the literature.

2026 Rius, Andrés. El gobierno, la economía y el hombre de la calle. (*Suma/Montevideo*, 7:13, oct. 1992, p. 7–35, bibl., graph, tables)

Despite its title, this is a very good and extremely perceptive study of the influence of economic variables (in particular GDP growth and inflation) on opinion polls and voting patterns. Contains good review of relevant literature and concludes that a simple trend variable explains evolution of the popularity of presidents (popularity decreases steadily as the presidency unfolds). Reductions in inflation affect popularity positively and significantly (in the econometric sense), but the added explanatory power to a simple trend is small.

2027 Sapelli, Claudio. Desempleo, indexación y política económica. (*Rev. Econ./Uruguay*, 4:1, agosto 1989, p. 27–50, bibl., graphs, tables)

According to Sapelli, optimal indexation of labor contracts is determined by nominal or real nature of shocks received by the economy, parameters of the demand for and supply of labor, the permanent or transitory nature of shocks, and duration of labor contracts. Economic policy is effective only in changing the structure of nominal shocks. [M. Mamalakis]

2028 Sapelli, Claudio. La evolución del salario real en la década de los años 70. Montevideo: Centro de Estudios de la Realidad Económica y Social, 1989. 59 p.: bibl. (Temas Económico-Sociales; 2)

Empirical and theoretical study attempts to reconcile paradox of falling real wages with official statistics, and rising real welfare of workers with other indexes. Argues that the standard of living of workers between 1970–80 declined less than the level suggested by official statistics. [M. Mamalakis]

2029 Sapelli, Claudio. Tamaño del Estado, instituciones y crecimiento económico. Montevideo: Centro de Estudios de la Realidad Económico y Social; Centro Internacional para el Desarrollo Económico, 1992. 118 p.: appendices, bibl.

In this excellent study, author convincingly demonstrates that the "presence" of the State had at least as much impact on economic development in Uruguay as the size of its public sector. Through regulation and manipulation of institutions (e.g., rent controls, subsidized credit, and exchange controls providing selective protection to national industry), the power of the State was used to generate rents rather than promote development. Concludes that in Uruguay the level of State income has negatively affected the level of private income, that economic openness improved the efficiency of the State, and that regulation greatly affected welfare. [M. Mamalakis]

2030 Stolovich, Luis. Los empresarios, la apertura y los procesos de integración regional: contradicciones y estrategias; el caso de Uruguay en el Mercosur. (*Rev. Parag. Sociol.*, 29:84, mayo/agosto 1992, p. 53–90, bibl., tables)

Interesting analysis, from Marxist viewpoint, of political and economic motivations behind the regional integration movement and challenges ahead.

2031 Torello, Mariella. Las causas de una inversión insuficiente. (*Suma/Montevideo*, 7:13, oct. 1992, p. 37–67, bibl., graph, tables)

Interesting paper studies determinants of investment in Uruguay and investigates puzzle of recent high rates of GDP growth accompanied by low investment levels. Concludes that the small local market is the principal restriction to investment and that openness will be the most successful investment promotion policy.

2032 Trade policies and practices by sector. (*in* Trade policy review: Uruguay, 1992. Geneva: GATT Publication Services, 1992, v. 1, p. 153–213, graphs, tables)

Extremely useful analysis of trade policies by sector, with good sectoral descriptions and very good presentation of data. Main conclusion is that Uruguayan trade policy has shifted sequentially. The first shift, from quantitative quotas to tariffs, took place in the early 1960s. Nevertheless, since exchange controls remained in place, quotas continued to be enforced through them until 1974, when exchange controls were lifted. In the 1970s a gradual reduction of tariffs began but it was canceled in many ways by the concurrent increase in the use of gauge prices (which increased the base on which tariffs were applied). Gauge prices began to be eliminated only in the early 1990s.

ARGENTINA

EDUARDO BORENSZTEIN, *Chief, Developing Countries Studies Division, Research Department, International Monetary Fund*

SINCE THE MID-1980s, research on the Argentine economy has been particularly strong, coinciding with a time of deep economic change. Themes covered in the literature are: inflation, hyperinflation, stabilization plans of all persuasions, the international debt crisis, and, arguably, the deepest structural reforms in Argentine history in areas such as international trade and payments, financial markets, and privatization of State enterprises. Thus, books and articles annotated below have a special significance in that they provide an inside view of developments during this unprecedented stage in Argentine economic history. Among them are several publications that stand out for their originality and/or appeal. Two former finance ministers and the current one (1995) have authored books comprising a broad collection of lectures, speeches, newspaper articles, and political and economic works of great value for those interested in policy and history: Cavallo (item **2051**), Dagnino Pastore (item **2057**) and Martínez de Hoz (item **2074**). In addition, Vázquez Presedo's tracing of the trajectory of Argentine economic history since 1776 makes fascinating reading (item **2091**).

As expected, the inflationary process and stabilization attempts are among the leading research subjects. There is no question that a certain degree of consensus has been building to support the need to balance the nation's fiscal accounts as a crucial condition for stabilization. In fact, various authors have attributed the failure of many previous stabilization attempts to fiscal weaknesses; Kiguel (item **2073**), Rodríguez (item **2084**), and Chisari *et al.* (item **2052**) offer the most comprehensive and persuasive studies in this regard. An aspect which remains largely unexplored is the precise nature of the role played by Argentina's exchange-rate arrangement which was designed to support the "convertibility" plan (the heart of the 1991 price stabilization process). One should also note the pessimism shared by economists in Argentina regarding the possibility of price stabilization, at a time when the most successful stabilization plan was being launched.

Another major factor in Argentina's economic crisis at the time was the international debt problem. This is the subject of a highly recommended book by Dornbusch and De Pablo (item **2060**), a work that goes well beyond the debt crisis to serve as a treatise on Argentine macroeconomic issues. While the debt crisis appears to have receded as of 1995, the expected resilience of price stabilizations and balance of payments surpluses is yet to be determined, particularly if the early 1990s' relatively benign international environment worsens.

The effects of inflation and balance of payments adjustments on real wages and income distribution are also the subject of interesting research (items **2075** and **2044**). In addition, the literature on labor markets includes studies on the implications of informal markets and population trends (items **2080, 2085, 2086,** and **2069**). Informal markets and small-scale economic activities are not merely intriguing features of many Latin American economies, but a potential source of great economic dynamism as well.

The depth of structural economic reforms is the one element that distinguishes Argentine economic policy in the 1990s. Although the process is still incomplete, the transformation in areas such as deregulation, privatization, and the international opening of the economy has taken on historical significance. A number of works address the reform process in general, including examinations of the relation between reforms in different areas and their proper sequencing (items **2040, 2039, 2066,** and **2049**). Other works concentrate on aspects of specific reforms, such as international trade (items **2048** and **2078**), fiscal reform (item **2059**), and financial markets (item **2042**). Privatization was also the subject of much debate; work in this area covers micro- and macroeconomic implications (items **2067** and **2081**), political considerations (item **2061**), and case studies of different public enterprises (items **2033, 2071,** and **2054**). In addition, a number of volumes of conference proceedings present good contributions covering diverse topics such as international trade (item **2055**), international financial markets (item **1199**), and economic reforms (item **2034**).

2033 **Abdala, Manuel Angel.** Privatización y cambios en los costos sociales de la inflación: el caso de ENTEL Argentina. (*Desarro. Econ.*, 32:127, oct./dic. 1992, p. 358–380, bibl., graphs, tables)

Investigates implications of privatization of State enterprises on fiscal flows and social costs of inflation, with special reference to the case of ENTEL.

Acero, Liliana et al. Textile workers in Brazil and Argentina: a study of the interrelationships between work and households. See item **2094.**

2034 **Alemann, Roberto T. et al.** Argentina 2000: su futuro económico. Buenos Aires: Ediciones Macchi, 1988. 187 p.: bibl., ill.

Conference proceedings include articles proposing projections for different economic variables and essays on policy strategies to achieve long-term growth in Argentina.

2035 **Almansi, Aquiles A.** "Bipapelismo" e hiperinflación en Argentina. Buenos Aires: Centro de Estudios Macroeconómicos de Argentina (CEMA), 1990. 19 p.: bibl., graphs. (Documentos de trabajo; 73)

Presents analytical framework to study the monetary system which briefly functioned under "frozen deposits," a system in which two types of currency effectively coexisted.

2036 **Almansi, Aquiles A.** and **Carlos Alfredo Rodríguez.** Reforma monetaria y financiera en hiperinflación. Buenos Aires: Centro de Estudios Macroeconómicos de Argentina (CEMA), 1989. 35 p.: bibl., graphs, tables. (Documentos de trabajo; 67)

Studies the role of the "quasi-fiscal" deficit (resulting from Central Bank operations with financial institutions) in feeding the hyperinflationary process of 1989.

2037 **Anidjar, Leonardo et al.** La crisis argentina y los planes de estabilización. Buenos Aires: Fundación Omega Seguros, 1991. 40 p. (Col. Conferencias)

Series of lectures by economists and former policymakers on the inflationary problem and stabilization initiatives, including the "Convertibility Plan" of 1991.

Anuario Estadístico de la República Argentina. 1995– . See item **14.**

2038 **Argentina: from insolvency to growth.** Washington: World Bank, 1993. 332 p.: ill. (World Bank country study; 0253-2123)

Detailed study of all aspects of public finances with projections and recommendations for reform in various areas.

2039 **Argentina: hacia una economía de mercado.** Fundación de Investigaciones Económicas Latinoamericanas. Buenos Aires: Manantial, 1990. 174 p.: bibl., ill.

Summary of the experience of countries that accomplished a transition into market-oriented economic systems and some lessons applicable to the case of Argentina.

2040 **Argentina: la reforma económica, 1989–1991; balance y perspectivas.** Fundación de Investigaciones Económicas Latinoamericanas, Consejo Empresario Argentino. Buenos Aires: Manantial, 1991. 310 p.: bibl., ill.

This interesting book comprises analysis of economic policies of the first two years of the Alfonsín government and a set of prescriptions for reform in fiscal and monetary policy and economic regulations.

2041 **Argentina: tax policy for stabilization and economic recovery.** Washington: The World Bank, 1990. xx, 143 p.: bibl., ill. (A World Bank country study, 0253–2123)

Report from a World Bank mission that studied the tax system, tax administration, and intergovernmental fiscal relations.

2042 **Arnaudo, Aldo A.** Perspectivas del sistema financiero argentino. (*An. Acad. Nac. Cienc. Econ.*, 35, 1990, p. 109–145, tables, graphs)

Survey of the evolution of the financial sector since the 1960s and assessment of its possible future evolution, including consequent risks and challenges.

2043 **Artana, Daniel; Oscar Libonatti;** and **Carlos Rivas.** Algunas consideraciones sobre el endeudamiento y la solvencia del sector público argentino. (*Bol. Inf. Techint*, 266, abril/junio 1991, p. 53–71, graphs, tables)

Study of domestic debt in Argentina, including nontraditional forms such as compulsory bank deposits at the Central Bank, and its evolution through the successive inflation and adjustment periods since 1985.

2044 **Beccaria, Luis A.** Distribución del ingreso en la Argentina: explorando lo sucedido desde mediados de los setenta. (*Desarro. Econ.*, 31:123, oct./dic. 1991, graphs, tables)

Examines evolution of real wages and income distribution since the 1970s, and traces the relation to inflation and recession episodes.

2045 **Beckerman, Paul Ely.** Public sector "debt distress" in Argentina, 1988–89. Washington: World Bank, 1992. 36 p.: bibl. (Policy research working papers; WPS 902)

Study of two failed price stabilization attempts.

2046 **Berlinski, Julio** *et al.* Empleo, inflación y comercio internacional. Edición de Javier Villanueva. Buenos Aires: Editorial Tesis; Instituto Torcuato Di Tella, 1988. 306 p.: bibl., ill. (Serie Economía/Instituto Torcuato Di Tella)

Collection of seven mostly empirical articles by researchers of the Instituto Di Tella on labor markets, inflation, growth, and international trade. Somewhat technical but highly recommended.

2047 **Bour, Juan Luis; Jorge Sereno;** and **Nuria Susmel.** Incidencia de los impuestos indirectos en el gasto de las familias. Buenos Aires: Fundación de Investigaciones Económicas Latinoamericanas, 1989. 37 p.: bibl., graphs, tables. (Documentos de trabajo; 20)

Empirical study of indirect tax incidence both from geographical and income distribution perspectives.

2048 **Bouzas, Roberto.** Un acuerdo de libre comercio entre Estados Unidos/Mercosur: una evaluación preliminar. (*in* Liberalización comercial e integración regional: de NAFTA a Mercosur. Edición de Roberto Bouzas y Nora Lustig. Buenos Aires: FLACSO, 1992, p. 165–199, bibl., tables)

Reviews trade relations among Southern Cone countries and their relationship with the US to derive some implications for a possible free trade agreement between the US and Mercosur.

2049 **Bouzas, Roberto.** Beyond stabilization and reform: the Argentine economy in the 1990s. (*in* In the shadow of the debt: emerging issues in Latin America. New York: Twentieth Century Fund Press, 1992, p. 83–109, tables)

Considers structural reforms undergone by Argentina in international trade and capital movements, fiscal policy, regulation, etc. Evaluates challenges that the international and domestic environments could pose in the 1990s.

2050 Canitrot, Adolfo and **Silvia Junco.**
Macroeconomic conditions and trade liberalization: the case of Argentina. (*in* Macroeconomic conditions and trade liberalization. Edited by Adolfo Canitrot and Silvia Junco. Washington: Inter-American Development Bank, 1993, p. 31–80, appendices, bibl., graphs, tables)

Analysis of interactions between international trade and macroeconomic conditions, with an empirical review of Argentina since the 1970s.

2051 Cavallo, Domingo. Economía en tiempos de crisis. Buenos Aires: Editorial Sudamericana, 1989. 228 p.

Published weeks before the elections that brought President Menem to power (and eventually the author of this book to the Economic Ministry), this book contains a collection of newspaper articles, lectures, and debates from two hyperinflationary periods (1982–83 and 1987–88). Required reading not only for the light it sheds on the trascendental role played by the author as minister but also because of the good economics and sharp rebuttals that permeate the articles.

2052 Chisari, Omar; José M. Fanelli; Roberto Frenkel; and **Guillermo Rozenwurcel.** Argentina and the role of fiscal accounts. (*in* Savings and investment requirements for the resumption of growth in Latin America. Edited by Edmar Bacha. Washington: Inter-American Development Bank, 1993, p. 1–56, bibl., graphs, tables)

Studies role of fiscal policy in the Argentine economy during the last 20 years, including effects on the behavior of savings and investment and on price stability and efficiency.

2053 Cohen, Noemí. El servicio de empleo en Argentina. (*in* Modernización de los servicios públicos de empleo en América Latina. Lima: Organización Internacional del Trabajo, Centro Interamericano de Administración del Trabajo, 1991, p. 11–31, table)

Describes role and origins of the National Employment Service in Argentina. Agency's main task is to act as an intermediary between employers and employees. Analyzes agency's changing role in light of free-market reforms. [J.W. Foley]

2054 Coloma, Germán. Productividad global de los factores: teoría y aplicación al caso de una empresa pública argentina. (*Económica/La Plata*, 36:1/2, 1990, p. 21–51, bibl., graph, tables)

Applies a total factor productivity framework to evaluate the performance of a public utility in the energy sector in the province of Buenos Aires during the period 1983–88.

2055 El comercio exterior argentino en la década de 1990: agenda y cursos de acción, septiembre de 1991. Consejo Argentino para las Relaciones Internacionales. Recopilación de Felipe A.M. de la Balze. Buenos Aires: Ediciones Manantial; Fundación Banco República; Fundación Lloyds Bank, 1991. 525 p.: bibl., ill.

Interesting collection of 39 papers on different aspects of international trade, including trade and exchange rate policy instruments, multilateral and regional trade arrangements, and macroeconomic effects.

2056 Convención de Bancos Privados Nacionales, 8th, Buenos Aires, 1992.
Hacia una nueva organización del federalismo fiscal en Argentina: resumen. Buenos Aires?: Instituto Torcuato Di Tella, Centro de Investigaciones Económicas, 1992. 25 p.: graphs, tables.

Proposes designing a system to deal with a critical problem in order to foster macroeconomic stabilization and strengthen public sector efficiency.

2057 Dagnino Pastore, José María. Crónicas económicas: Argentina, 1969–1988. Buenos Aires: Crespillo, 1988. 313 p.: bibl., ill.

Broad collection of works includes public addresses as Finance Minister, papers presented at professional meetings, and "reserved circulation" consultant reports. An insider's view on aspects of economic policymaking.

2058 Damill, Mario et al. Las relaciones financieras en la economía argentina. Buenos Aires: Ediciones del IDES, 1988. 108 p.: bibl. (Ediciones del IDES, 0326–6133; 15. Col. economía y planificación)

This book focuses on the external debt problem and its repercussions on domestic financial markets and inflation, suggesting the use of "heterodox" tools for stabilization.

2059 **La desarticulación del pacto fiscal: una interpretación sobre la evolución del sector público argentino en las dos últimas décadas.** Buenos Aires: Comisión Económica para América Latina y el Caribe, 1990. 1 v.: ill. (Documento de trabajo; 36)

Detailed study of different aspects of public finances and of the structural and institutional reasons for the chronic fiscal imbalance that was at the core of persistent inflation and stabilization failures.

2060 **Dornbusch, Rudiger and Juan Carlos de Pablo.** Debt and macroeconomic instability in Argentina. (*in* Developing country debt and economic performance. Chicago: Univ. of Chicago Press, 1989, v. 2, p. 39–156, appendices, bibl., graphs, tables)

Authors present overview of Argentina's external and macroeconomic problems and provide an insightful consideration of short- and long-term policy options.

2061 **Empresas públicas y política de privatizaciones (22/11/88): ciclo de reuniones; debate sobre la coyuntura política argentina.** Buenos Aires: Centro Latinoamericano para el Análisis de la Democracia; Fundación Friedrich Ebert, 1988. 40 p.: bibl., ill. (Serie Relatorios)

Introductory article and proceedings of debate about privatization among economists associated with major political parties.

Espejo Ortega, Alberto Octavio. O plano de estabilização heterodoxo: a experiência comparada de Argentina, Brasil e Peru. See item **2137**.

2062 **Estudio sobre el desarrollo económico de la República Argentina: informe final.** Buenos Aires?: Agencia de Cooperación Internacional del Japón, 1987. 3 v. in 7: bibl., ill., maps.

Study conducted from 1985–86 on the design of a "Japanese-style" development strategy for Argentina. Comprises separate chapters dealing with the macroeconomic situation, agriculture, industry, transport, and exports, and a separate volume summarizing the Japanese experience.

2063 **Fernández, Roque B.** Exchange rate policy and hyperinflation. Buenos Aires: Centro de Estudios Macroeconómicos de Argentina, 1990. 18 p.: bibl., graphs. (Documentos de trabajo; 72)

Studies exchange rate policy under conditions of fiscal deficit and inflation, emphasizing portfolio and liquidity services of domestic and foreign currency.

2064 **Fernández, Roque B.** Real interest rate and the dynamics of hyperinflation: the case of Argentina. Buenos Aires: Centro de Estudios Macroeconómicos de Argentina (CEMA), 1990. 39 p.: bibl., graphs. (Documentos de trabajo; 69)

Presents model of inflation, with variable real interest rates that affect the budget (through public debt) and money demand, and an application to the Austral Plan.

2065 **Fuchs, Mariana.** Los programas de capitalización de la deuda externa argentina. Buenos Aires: Comisión Económica para América Latina y el Caribe, 1990. 39 leaves.

Detailed description of the different "swap" and "on-lending" programs for external debt.

2066 **García Ruiz, José Luis.** Sector financiero y apertura económica: una perspectiva comparada entre la Argentina y España. (*Rev. Ciclos*, 2:3, segundo semestre 1992, p. 95–112, graphs, tables)

Compares sequencing of financial reform, trade liberalization, and price stabilization in the case of Spain in the late 1950s to the reverse ordering adopted in Argentina since 1976.

2067 **Gerchunoff, Pablo and Germán Coloma.** Privatization in Argentina. (*in* Privatization in Latin America. Edited by Manuel Sánchez and Rossana Corona. Washington: Inter-American Development Bank, 1993, p. 251–299, bibl., tables)

Author analyzes largest privatization operations in 1990–91 and consider the effect of existing distortions in domestic product and labor markets as well as conflicts between immediate macroeconomic objectives and considerations of efficiency and long-term public finances.

2068 **Guerberoff, Simón L.** Crecimiento y restricción externa: el caso argentino en los años 90. Buenos Aires: Editorial Tesis; Instituto Torcuato di Tella, 1989. 111 p.: bibl., ill. (Serie Economía)

Applies traditional "two-gap model" ideas to the study of the foreign debt problems of Argentina, drawing what, in hindsight, look like overly pessimistic conclusions.

2069 Instituto de Estudios Contemporáneos (Buenos Aires). La economía informal en Argentina. *(in* El sector informal en América Latina: una selección de perspectivas analíticas. Recopilación de Jacobo Schatan, Dieter Paas, y Alvaro Orsatti. México: Centro de Investigación y Docencia Económicas; Fundación Friedrich Naumann, 1991, p. 345–355)

Attractive discussion of the reasons for existence of an informal economy in Argentina, its measurement, and its micro- and macroeconomic consequences, as well as political and social implications of the phenomenon.

2070 Jornadas Bancarias de la República Argentina, 1st?, Buenos Aires?, 1991. La banca en tiempos de ajuste. Buenos Aires: Asociación de Bancos de la República Argentina; Tesis, Grupo Editorial Norma, 1991. 405 p.: bibl., ill.

Conference proceedings centered around the then current (and failing) attempts at price stabilization in 1990, with some emphasis on effects on the financial sector.

2071 Juri, María E. and **Raúl Mercau.** Privatización en la Argentina: el caso de Bodegas y Viñedos Giol. *(Estudios/Fundación Mediterránea,* 13:53, enero/marzo 1990, p. 3–19, bibl., tables)

Detailed study of the long process of privatization of this large enterprise operating in the wine and fruit markets.

2072 Katz, Jorge M. and **Bernardo Kosacoff.** El proceso de industrialización en la Argentina: evolución, retroceso y prospectiva. Buenos Aires: Centro Editor de América Latina; CEPAL; 1989. 108 p.: bibl. (Bibliotecas universitarias. Economía)

Historical and analytical study of Argentina's industrial sector, including aspects related to international trade and direct foreign investment.

2073 Kiguel, Miguel A. Inflation in Argentina: stop and go since the Austral Plan. *(World Dev.,* 19:8, Aug. 1991, p. 969–986, bibl., graphs, tables)

Examines recurrent cycles of stabilization attempts and inflation explosions since the Austral Plan, tracing the root of repeated failures to structural fiscal problems.

Lavagna, Roberto. Cambios en la estructura productiva inducidos por el Mercosul. See item **2163.**

Lipietz, Alain *et al.* Instabilidade econômica: moeda e finanças. See item **2165.**

2074 Martínez de Hoz, José Alfredo. Quince años después. Buenos Aires: Emecé Editores, 1991. 305 p.: ill.

The former finance minister explains why he still believes his policies were right on target.

2075 Montuschi, Luisa. Algunas consecuencias de un proceso hiperinflacionario: la redistribución de los ingresos en el sector manufacturero argentino, 1989–1990. Buenos Aires: Centro de Estudios Macroeconómicos de Argentina (CEMA), 1991. 26 p.: bibl., graphs. (Documentos de trabajo; 78)

Presents analytical framework and empirical study of effects of hyperinflationary conditions on real wages in the manufacturing sector. Finds the wage-price spiral has an unambiguous negative effect on workers' income.

2076 Nicolini, José Luis. Un modelo de ajustes macroeconómicos con deuda externa. *(Desarro. Econ.,* 31:122, julio/sept. 1991, p. 235–250, appendix, bibl., graphs)

Model compares the limits of attempting to increase external debt-servicing through contraction of domestic absorption with the alternative of a debt reduction to induce higher growth and debt-repayment capacity over time.

2077 Nogués, Julio. El ajuste de la balanza de pagos de acuerdo al GATT: análisis económico y propuestas de modificación. *(Estudios/Fundación Mediterránea,* 14:60, oct./dic. 1991, p. 131–138, bibl., tables)

Analyzes and criticizes approach followed in the GATT provisions on balance of payments adjustment as heavily tilted towards quantitative restrictions, and not sufficiently reliant on domestic policy adjustment. Insightful paper on a timely issue.

2078 **Nogués, Julio.** Patents and pharmaceutical drugs: understanding the pressures on developing countries. (*J. World Trade*, 2:6, Nov. 1990, p. 81–104, bibl., tables)

Study of the reasons for international pressures on developing countries for the enforcement of patent protection for pharmaceutical drugs. Focuses on developments in industrial countries that prompted these pressures.

2079 **Ostiguy, Pierre.** Los capitanes de la industria: grandes empresarios, política y economía en la Argentina de los años 80. Buenos Aires: Legasa, 1990. 375 p.: bibl., chart, ill. (Ensayo crítico)

Book surveys the roster of the largest industrial conglomerates in Argentina, their founders, and leaders (the "capitanes de la industria"). Traces economic relations with the government and lobbying efforts of these industrial groups during the 1980s.

2080 **Panaia, Marta.** El trabajo negro en la Argentina. Buenos Aires: Univ. de Buenos Aires, Facultad de Ciencias Sociales, Instituto de Investigaciones, 1991. 81 p.: bibl., ill., map. (Cuadernos/Instituto de Investigaciones, Facultad de Ciencias Sociales; 4)

Study presents attempts to measure the "underground" or "informal" economy, and investigates the main determinants of the growth of this parallel economy.

2081 **Privatizaciones en Argentina = Privatizations in Argentina: conferencia, Bs. As., Argentina, Septiembre de 1991.** Buenos Aires: Bureau de Investigaciones Empresariales, 1991. 1 v.: ill.

Conference proceedings dealing with legal, tax, and regulatory aspects of privatization in different sectors of the economy.

2082 **Riveros, Luis A.** El enfoque de salarios de eficiencia y el ajuste económico en países en desarrollo. (*Desarro. Econ.*, 31:122, julio/sept. 1991, p. 189–208, bibl.)

Analysis of the "efficiency wage" theory of wage rigidity and its implications for adjustment policies, in particular, as well as for devaluations, reallocation of resources, and unemployment in developing countries.

2083 **Rodríguez, Carlos Alfredo.** The external effects of public sector deficits. Buenos Aires: Centro de Estudios Macroeconómicos de Argentina (CEMA), 1990. 22 p.: bibl. (Documentos de trabajo; 70)

Analyzes the effects of fiscal deficits and their financing on exchange rates, the trade account, the current account, and foreign debt.

2084 **Rodríguez, Carlos Alfredo.** The macroeconomic effects of public sector deficits: Argentina. Buenos Aires: Centro de Estudios Macroeconómicos de Argentina, 1991. 90 p.: bibl., ill. (Serie Documentos de trabajo; 76)

This extensive study focuses both on the measurement of fiscal deficits and the inflation tax in Argentina and on econometric estimates of the effects of fiscal deficits on domestic and international variables. Fairly technical but an excellent analysis.

2085 **Sánchez, Carlos E.** Población y subsidios sociales: su distribución regional en Argentina. (*Estudios/Fundación Mediterránea*, 13:54, abril/junio 1990, p. 39–46, tables)

Studies role of infrastructure investment and social subsidies in the concentration of resources and population in Buenos Aires and a few urban centers.

2086 **Schenone, Osvaldo H.** El "cuentapropismo," la legislación laboral y la acumulación de capital. Buenos Aires: Centro de Estudios Macroeconómicos de Argentina, 1988. 22 p.: bibl., graphs. (Documentos de trabajo; 64)

Studies effects of the social security system on capital accumulation and the implications of the existence of a vast sector of the economy with the ability to avoid (not evade) social security taxes.

2087 **Schwab y Etchebarne, Martín.** Exportar para crecer y ganar. Con un modelo económico realizado con Jorge L. Demaría. Prólogo de Domingo F. Cavallo. Prefacio de Oscar H. Saggese. Buenos Aires: Editorial Fraterna, 1989. 221 p.: bibl., ill.

Proposes export-oriented strategy to gain productive and State efficiency and achieve satisfactory investment and growth performance.

2088 **Setti, Eduardo Pablo.** La inflación estructural en la Argentina. Buenos Aires: Grupo Editor Latinoamericano; Emece

Editores, 1991. 213 p.: bibl. (Col. Estudios políticos y sociales)

Study of inflation in the 1976–89 period, a restatement of "nonmonetarist" policy explanation, and a prediction (in 1991) of failure of then just-initiated "Plan Cavallo."

2089 Sguiglia, Eduardo. El club de los poderosos: historia pública y secreta de los grandes *holdings* empresariales argentinos. 2. ed. Buenos Aires: Planeta, 1992. 155 p.: bibl., ill., index.

Another book designed to satisfy the endless fascination of Argentines as to who owns what among the nation's largest industrial conglomerates.

2090 Stunnenberg, Peter and Johan Kleinpenning. The role of extractive industries in the process of colonization: the case of quebracho exploitation in the Gran Chaco. (*Tijdschr. Econ. Soc. Geogr.*, 84:3, 1993, p. 220–229, bibl., map)

Case study of quebracho exploitation shows that this extractive industry has not initiated a process of economic growth and diversification. Attributes reasons to fact that area has no additional exploitable potential and that other parts of Argentina and Paraguay offer more promising prospects for colonization. [R. Hoefte]

2091 Vázquez-Presedo, Vicente. Auge y decadencia de la economía argentina desde 1776. Buenos Aires: Academia Nacional de Ciencias Económicas, Instituto de Economía Aplicada; Abeledo-Perrot, 1992. 208 p.: bibl.

Fascinating and very readable summary of Argentine economic history and international developments that had major impacts on the country's economy.

2092 Visintini, Alfredo. La dinámica del crecimiento económico en Argentina. (*Económica/La Plata*, 36:1/2, 1990, p. 97–143, tables, graphs)

Applies the general "two-gap" approach to analysis of external and internal restrictions on the rate of economic growth in Argentina since 1975.

BRAZIL

MELISSA H. BIRCH, *Graduate School of Business, University of Virginia*
RUSSELL E. SMITH, *School of Business, Washburn University*

WITH THE CREATION OF MERCOSUL, the South American Common Market, regional integration has become an area of significant research interest in the already large and increasingly complex Brazilian economic literature. Created on March 26, 1991 with the signing of the Treaty of Asunción by Argentina, Brazil, Paraguay, and Uruguay, Mercosul is initially a customs union which is intended eventually to become a fully integrated common market. The treaty provided for the reduction and elimination of most tariffs among the four countries over a several-year period, with other measures including a common external tariff, permanent institutions, and a dispute resolution procedure to be decided by Jan. 1, 1995. While the Mercosul project falls short of a political union with supranational institutions, it does envision the coordination and harmonization of macroeconomic policies.

The literature reviewed for this volume includes several books rich in information that establish the history and transitional mechanisms of Mercosul and place the Mercosul project in the context of other trade regimes and previous integration projects (items **2160, 2096,** and **2106**) or that deal specifically with Mercosul from a labor relations point of view (item **2101**). A fifth volume on Mercosul, Latin American integration, and links to Europe was sponsored jointly by Brazil's Central Unica dos Trabalhadores and Italy's Confederazione Generale Italiana del Lavoro (item

2110). The scope of the undertaking and the magnitude of the challenge is discussed in terms of the coordination of national policies (items **2139** and **2152**) and the impact of integration on one particular industry (item **2114**). The importance for Brazil of extra-regional trade is reflected in the literature on the patterns of Brazilian trade (item **2112**) and the implications of NAFTA (item **2169**) for Brazilian trade and investment.

The move toward regional integration took place in the context of unilateral neoliberal trade reforms in Brazil—with far-reaching implications for Brazilian industrial and technology policies (item **2145**)—and concurrently with significant modifications in world trade regimes represented by the Uruguay Round of the General Agreement on Tariffs and Trade. These trade liberalizing trends significantly impact Brazil's external sector, industrial and agricultural production, and international relations (item **2093**).

The literature on trade reform seemed to displace temporarily the previous focus on macroeconomic stabilization programs. There was relative peace on the macroeconomic front since the period of central focus for this essay, 1992–93, fell between the 1990 and 1991 Collor Plans and the 1994 Real Plan. Instead, the major contributions to the macroeconomic literature tended to come from award-winning masters' and doctoral theses which tried to make theoretical sense of the Brazilian inflation and stabilization experiences (items **2208, 2137, 2102,** and **2179**) and from books directed toward a broad non-professional audience with the goal of making the economic policy debate more accessible without sacrificing conceptual rigor (items **2126, 2185,** and **2150**). There were several high-quality additions to the well-established literatures on regional development such as those on Minas Gerais (item **2171**) and Rio Grande do Sul (item **2127**), on the Brazilian model of economic development (item **2148** and **2131**), and on Brazilian economic history (items **2168, 2129,** and **2212**).

Two conferences are of special note: the widely publicized 1992 United Nations Conference on Economic Development and the Environment and the less widely reported Encontro Nacional de Estudos do Trabalho (3rd, Rio de Janeiro, 1993) held by the Associação Brasileira de Estudos do Trabalho. The latter, the third biennial congress of a relatively new association of economists and sociologists founded in 1989, featured an ample and rich menu of sessions and papers, with topics ranging from teaching, data sources, and worker health and training, to labor relations, labor law, workplace flexibility and organization of production, race, and gender (item **2135**). The Rio environmental conference spawned a significant increase in the literature on economic development and environmental economics. While some had a distinctly regional focus (item **2207**), others dealt more broadly with conceptual issues (items **2128, 2172,** and **2119**) and applications of theory to Brazilian data (item **2156**). Still others had a more activist tone, examining ongoing social movements in the environmental context (item **2136**).

Several of the items annotated below are of particular interest for the massive amounts of raw data they contain. Notable in this regard are data from the national household surveys (Pesquisa Nacional por Amostra de Domicílios or PNAD), including national and regional results from 1981–87 (item **2184**), a critique of labor force participation based on the PNAD data (item **2130**), and social indicators given in the fourth volume of a multi-volume study on social policy (item **2187**). Three other data-rich works provide detailed information on the informatics industry (item **2157**), the infrastructure industry (item **2194**), and on foreign direct investment in Brazil (item **2215**).

2093 Abreu, Marcelo de Paiva. Trade policies and bargaining in a heavily indebted economy: Brazil. (*in* The developing countries in world trade: policies and bargaining strategies. Edited by Diana Tussie and David Glover. Boulder, Colo.: Lynne Rienner Publishers, 1993, p. 137–154, tables)

Considers formulation and implementation of trade policies in Brazil during the 1980s. Places particular emphasis on factors affecting the Brazilian stance in international trade negotiations, the relative development of bilateral and multilateral trade, and the increased leverage of multilateral financial institutions. Recent trade liberalization policies are analyzed in this context.

2094 Acero, Liliana et al. Textile workers in Brazil and Argentina: a study of the interrelationships between work and households. Tokyo: United Nations Univ. Press, 1991. 305 p.: bibl.

Two-country comparative study of the impact of industrial transformation and technical change on labor, and in particular on relationships within household groups. Based largely on systematic field surveys done from 1984–86, analyzes industry characteristics and workers' households, behaviors, life histories, and perceptions of both countries.

Alfie, Isaac. El mercado laboral uruguayo y la integración regional. See item **2008**.

2095 Almada, Vilma Paraíso Ferreira de. Estudos sobre estrutura agrária e cafeicultura no Espírito Santo. Vitória, Brazil: SPDC/UFES, 1993. 159 p.: bibl., ill., map. (Col. Cultura UFES; 17)

Study of coffee agriculture in the period following the abolition of slavery. Includes substantial analysis of available statistical and written primary sources and compares regions worked by slaves to those developed more recently by European immigrants.

2096 Almeida, Paulo Roberto de. O Mercosul no contexto regional e internacional. São Paulo: Edições Aduaneiras, 1993. 204 p.: bibl.

Useful general work on Mercosul written during the transitional phase. Discusses founding, basic structures, and provisions, as well as challenges faced. Places Mercosul in the context of the preceeding Argentina-Brazil Integration program and broader trade agreements, such as the General Agreement on Tariffs and Trade and the Latin American Free Trade Area.

2097 Amadeo, Edward J. et al. Distribuição de renda no Brasil. Organização de José Márcio Camargo e Fabio Giambiagi. Prefácio de Antônio B. Castro. São Paulo: Paz e Terra, Instituto dos Economistas do Rio de Janeiro, 1991. 237 p.: bibl., ill.

Ten distinct papers by seasoned researchers analyze various human and technical aspects of income distribution in Brazil. Some papers discuss broad and theoretical issues; others are sectorally focused, considering wage inequality and policy, regional inequality, the informal sector, agriculture, and the financial sector.

2098 Amadeo, Edward J. Restricciones institucionales a la política económica: negociación salarial y estabilización en el Brasil. (*Desarro. Econ.*, 33:129, abril/junio 1993, p. 29–47, bibl., graph, tables)

Discusses labor institutions in the 1980s and their relation to wage negotiation and economic stabilization. Notes increased centralization of the labor movement and its links to political parties and analyzes efficacy of political mechanisms in promoting economic stabilization.

2099 Amaral Filho, Jair do. A economia política do babaçu: um estudo da organização da extrato-indústria do babaçu no Maranhão e suas tendências. São Luís, Brazil: Serviço de Impr. e Obras Gráficas do Estado, 1989 [i.e. 1990]. 309 p.: bibl., ill., maps.

Detailed study of evolution and future prospects of the palm nut (*babassu*) industry of the state of Maranhão. Considers the industry's productive structure, including technological processes and labor utilization, as it moved from traditional harvesting to organization by commercial capital, and then to dominance by industrial processing.

2100 Anderson, Julie. Rural labor legislation and permanent agricultural employment in northeastern Brazil. (*World Dev.*, 21:5, May 1993, p. 705–719, bibl., tables)

Assesses impact of the 1963 Rural Labor Statute on permanent employment in the Zona da Mata in the 1960s and 1970s. Concludes that the law's severance pay provision reduced permanent employment by eliminating benefits to the employer of a subject work force disciplined by threat of termination.

2101 Andrade, Everaldo Gaspar Lopes de. O Mercosul e as relações de trabalho. São Paulo: Editora Ltr, 1993. 152 p.: bibl., ill.

Written by law professor and official of the labor judiciary, this work analyzes Mercosul from the point of view of labor law, international law, labor relations, and labor markets. Substantial and useful institutional information, especially with regard to Mercosul's Subgroup 11 on labor issues.

Argenti, Gisela et al. Uruguay: el debate sobre la modernización posible. See item **2009**.

2102 Aronovich, Selmo. Inflação, crescimento e decisões empresariais: uma abordagem neo-estruturalista para a economia brasileira. Rio de Janeiro: BNDES, Gabinete da Presidência, Depto. de Relações Institucionais, 1991. 119 p.: bibl., ill.

Prize-winning master's thesis examines process of inflation and dynamics of capital accumulation using a neo-structuralist model. On the basis of econometric testing, concludes that both inflation and capital accumulation were adversely affected by rising level of uncertainty associated with the deterioration of public finances.

2103 Arruda, Marcos et al. Privatizar é solução?: casos Mafersa, Acesita e Cobra. Rio de Janeiro: PACS/FASE, 1990. 104 p.: bibl., ill. (Série Economia popular; 2)

Non-polemical examination of privatization of Mafersa, Acesita, and Cobra. Includes some interesting data from the three firms.

2104 Assentamentos: a resposta econômica da reforma agrária. Organização de Sérgio Antônio Görgen e João Pedro Stédile. Petrópolis: Vozes, 1991. 184 p.: bibl., ill., maps.

Useful contribution to the debate on Brazilian land reform. Addresses the economic impact of land tenure and agrarian reform. Reports field research findings and attempts to demonstrate that land in the hands of those who work it (by virtue of limited reforms undertaken to date) is more productive than land in the hands of large land-owners, and that social problems of the landless are dramatically reduced by land reform.

2105 Barbieri, José Carlos and Walter Delazaro. Nova regulamentação da transferência de tecnologia no Brasil. (*Rev. Adm. Empres.*, 33:3, maio/junho 1993, p. 6–19, tables)

Discussion of changes in technology policy that will be required due to the transition to a more open model of economic development in Brazil. Whereas technology import-substitution policies required increasing bureaucratic regulation and control, the move toward international competitiveness will require a liberalization of technology policy. Reviews similar situations in other countries and identifies potential problems for Brazil during this transition.

2106 Barbosa, Rubens Antonio. Liberalização do comércio, integração regional e Mercado Comum do Sul: o papel do Brasil. (*Rev. Econ. Polít.*, 13:1, jan./março 1993, p. 64–81, ill.)

Useful overview from a historical or evolutionary perspective of regional integration in the Southern Cone, outlining role and achievements of ALADI, development of Argentine-Brazilian binational accords, and the emergence of Mercosur. Good discussion of conceptual as well as practical issues.

2107 Barros, Alexandre Rands. A periodization of the business cycles in the Brazilian economy, 1856–1985. (*Rev. Bras. Econ.*, 47:1, jan./março 1993, p. 53–82, appendix, bibl., tables, graphs)

Paper presents coherence analyses to demonstrate that Brazilian economic time series are reasonably correlated to business cycle frequencies. Provides an estimate of business cycle periodization for the period 1856–1985, including Kitchin, Juglar, and Kuznets cycles.

2108 Barros, Ricardo Paes de and Rosane Silva Pinto de Mendonça. Infância e adolescência no Brasil: as conseqüências da pobreza diferenciadas por gênero, faixa etária e região de residência. (*Pesqui. Planej. Econ.*, 21:2, agosto 1991, p. 355–376, bibl., graphs, tables)

Analyzes school non-attendance and labor market participation for children between seven and 17 in the Fortaleza, São Paulo, and Porto Alegre metropolitan areas in terms of household income. Finds that non-attendance and employment levels increase with age and decrease with household income, are higher for males, and higher in São Paulo than in Fortaleza.

2109 Barros, Souza. Velhos e novos problemas vinculados à economia de Pernambuco. Ed. especial. Recife, Brazil: Comissão de Desenvolvimento Econômico de Pernambuco, 1992. 71 p., 2 folded leaves: bibl., map. (Série Política econômica; 5)

Reissuance of a 1956 publication by the secretary general of the Comissão Estadual de Desenvolvimento de Pernambuco (CODEPE). Discusses various persistent themes of economic development in northeastern Brazil and the policy tools then being or beginning to be applied, including irrigation, agriculture, hydroelectricity, and fiscal incentives.

2110 Battaglini, Elena *et al*. Mercosul: integração na América Latina e relações com a Comunidade Européia. São Paulo: Editora Cajá; Projeto IRES/DESEP; Livraria e Editora, 1993. 307 p.: bibl.

Collection of eight thoughtful and informative papers on Mercosul, Latin American integration, and the possibility of strengthening relations with Europe. Produced as a joint project between the Central Unica dos Trabalhadores (Brazil) and the Confederazione Generale Italiana del Lavoro (Italy), papers cover full range of trade and economic integration topics, plus labor and social issues.

2111 Baumann, Renato. Befiex: efeitos internos de um incentivo à exportação. (*Rev. Bras. Econ.*, 44:2, abril/junho 1990, p. 167–189, bibl., tables)

Study of Befiex, a tax incentive program to encourage exports by subsidizing imports. Without disputing export benefits, raises questions about the social cost of Befiex as an industrial policy that subsidizes imports and consolidates the industrial structure.

2112 Baumann, Renato. A opção não regional: o Brasil e os blocos econômicos. (*Pesqui. Planej. Econ.*, 21:2, agosto 1991, p. 185–208, bibl., tables)

While Brazilian export promotion has been very successful, policy is not sustainable and changing patterns of production call for new trade arrangements. Traces Brazilian trade with various regions and groups of countries, suggesting that the role of trade with more developed countries outside the region may be particularly significant.

2113 Beauclair, Geraldo de. Raízes da indústria no Brasil. Rio de Janeiro: Studio F&S Editora, 1992. 204 p.: bibl.

Careful study of the origins of the industrial sector in Brazil based on what the author terms "case studies" of public policies and private sector initiatives, principally in manufacturing, in the first half of the last century.

2114 Bekerman, Marta. O setor petroquímico e a integração Argentina-Brasil. (*Pesqui. Planej. Econ.*, 22:2, agosto 1992, p. 369–398, bibl., tables)

Scholarly study of problems of Argentine-Brazilian economic integration focusing on a single rapidly growing and important industry. Concludes that comparative advantages in this sector can be improved only by future joint projects that take into account structural asymmetries, such as Brazil's market size and Argentina's natural gas resources.

2115 Bermudez, Jorge. Remédios, saúde ou indústria?: a produção de medicamentos no Brasil. Rio de Janeiro: Relume Dumará, 1992. 122 p.: bibl., ill.

Comprehensive yet concise review of the State's role in the pharmaceutical industry, from its controversial patent policy to its operation of state-owned laboratories and manufacturing facilities, includes relevant issues for the 1988 Constitution. Written by professional with administrative experience in a number of relevant government agencies.

Berretta, Nora; Fernando Lorenzo; and **Carlos Paolino.** En el umbral de la integración. See item **2011.**

2116 Bezerra, Jocildo Fernandes *et al*. Produção, emprego e distribuição de renda no Rio Grande do Norte. Coordenação de Maurício Costa Romão. Recife, Brazil: PROPESQ; Editora Universitária UFPE, 1990. 326 p.: bibl.

Comprehensive study of the economy of the northeastern state of Rio Grande do Norte analyzes behavior of the principal sectors in terms of productive structure, output, employment, and income. Set in the context of national economic policy and performance and specific local factors. Makes wide-ranging recommendations.

2117 **Bonelli, Regis; Gustavo Henrique Barroso Franco;** and **Winston Fritsch.** Macroeconomic conditions and trade liberalization in Brazil: lessons from the 1980s and 1990s. (*in* Macroeconomic conditions and trade liberalization. Edited by Adolfo Canitrot and Silvia Junco. Washington: Inter-American Development Bank, 1993, p. 31–80, appendices, bibl., graphs, tables)

Reviews Brazil's internal macroeconomic situation in the 1980s and its impact on external performance, trade liberalization, and competitiveness to assess how internal performance constrains external variables and policies. Finds that constraints are serious and that attention must be given to the internal disequilibria.

Borda, Dionisio. La estabilización de la economía y la privatización del Estado en el Paraguay, 1954–89. See item **2002.**

2118 **Borges, Fernando Tadeu de Miranda.**
Do extrativismo à pecuária: algumas observações sobre a história econômica de Mato Grosso, 1870 a 1930. Cuiabá, Brazil: Gráfica Genus, 1991. 198 p.: ill.

This master's thesis analyzes the economy of the state of Mato Grosso in terms of its export activities. Some extractive activity and large public imports characterized the 1870–90 period. Extractive exports dominated from 1890–1914, then declined as livestock activity expanded. Changes in transportation explain development and distribution of activities.

2119 **Braga, Carlos Alberto Primo.** Tropical forests and trade policy: the cases of Indonesia and Brazil. (*in* International trade and the environment. Edited by Patrick Low. Washington: World Bank, 1992, p. 173–196, graphs, tables)

Emphasizes role played by international trade in depletion of tropical forests, particularly the economic and environmental implications of tropical hardwood trade. Focuses on comparative experience of Indonesia and Brazil and includes discussion of role of GATT in regulating tropical hardwood trade.

2120 **Cacciamali, Maria Cristina.** Mudanças recentes no producto e no emprego: uma comparação entre os países industrializados e aqueles em desenvolvimento. (*Rev. Bras. Econ.*, 45:2, abril/junho 1991, p. 213–250, bibl., tables)

Analyzes changes in structures of production, employment, and occupation in the major industrialized and newly industrialized countries from 1970–87. Finds that the decline in urban productivity for Brazil in the 1980s was compensated for by gains in agriculture.

2121 **Câmara Neto, Alcino Ferreira et al.**
Aquarella do Brasil: ensaios políticos e econômicos sobre o governo Collor. Apresentação de Maria da Conceição Tavares. Rio de Janeiro: Rio Fundo Editora, 1990. 150 p.: bibl.

Collection of articles by UFRJ faculty on politics and economics of the Collor government's economic policies. Of special note is the series of the Collor government's impact on industrial policy and technology.

2122 **Cardoso, Eliana A.** Cyclical variations of earnings inequality in Brazil. (*Rev. Econ. Polít.*, 13:4, out./dez. 1993, p. 112–124, bibl., graphs, tables)

Analyzes cyclical behavior of income distribution. Finds that it can change dramatically in a short period of time, that the change can be explained macroeconomically rather than microeconomically, and that high inflation and unemployment increase inequality.

2123 **Cardoso, Eliana A.** and **Ann Helwege.**
Reforma agrária na América Latina: o que tem o Brasil a aprender com a Bolívia, o México e o Peru? (*Rev. Bras. Econ.*, 45:2, abril/junho 1991, p. 251–285, bibl., graphs, tables)

Analysis of agrarian reforms conducted in Bolivia, Mexico, and Peru with the aim of deriving lessons for reform in Brazil. Authors note high cost of effective reform and the need for adequate financing.

2124 **Carneiro, Dionísio** and **Rogério L.F. Werneck.** Obstacles to investment resumption in Brazil. (*in* Savings and investment requirements for the resumption of growth in Latin America. Edited by Edmar Bacha. Washington: Inter-American Development Bank, 1993, p. 57-102, appendices, bibl., graphs, tables)

This chapter on Brazil traces the evolution of investment in Brazil since the 1970s. Evaluates the problems of financing investment during this period, analyzes modest role of foreign savings and decline in public savings, and offers results of a simulation exercise which estimates the required recovery in public savings necesary under different scenarios.

2125 Castilhos, Clarisse Chiappini. O sistema brasileiro de inovação: uma proposta de configuração. (*Ensaios FEE*, 13:1, 1992, p. 88-114, bibl., ill.)

Contends that a "Brazilian system of innovation" exists and seeks to identify distinguishing characteristics of producing, assimilating, and diffusing technologies in Brazil. Discusses the impact of the 1980s economic crisis on innovation in Brazil, and contrasts the Brazilian situation with that of developed countries.

2126 Castro, Paulo Rabello de and Paulo Carlos de Brito. Brasil, este país tem jeito? Rio de Janeiro: Rio Fundo Editora, 1992. 120 p.: ill.

Written by a conservative economist and businessman and directed toward a broad audience, book examines causes of economic stagnation and political instability. Topics included are political organization, social debt, tax and financial reforms, real wages, and economic growth.

2127 Conceição, Octavio Augusto C. *et al.* A economia gaúcha e os anos 80: uma trajetória regional no contexto da crise brasileira. v. 1-3. Coordenação de Pedro Fernando Cunha de Almeida. Porto Alegre, Brazil: Fundação de Economia e Estatística Siegfried Emanuel Heuser, 1990. 3 v. (718 p.): bibl., ill., maps.

Essays in this three-volume collection analyze the economy of the state of Rio Grande do Sul in response to the economic crisis of the 1980s. Collection examines five themes: performance of the state's economy; regional distribution of growth and activities; performance by sector, including agriculture and livestock, industry, finance, and public administration; relations between capital and labor; and exports and Latin American integration.

2128 Contabilização econômica do meio ambiente: elementos metodológicos e ensaio de aplicação no Estado de São Paulo. São Paulo: Secretaria do Meio Ambiente, 1992. 111 p.: bibl., ill. (Série Seminários e debates, 0103-7722)

Series discusses alternative national income accounting methodologies that allow for environmental factors. Includes theoretical and applied essays.

2129 Costa, Francisco de Assis. Grande capital e agricultura na Amazônia: a experiência Ford no Tapajós. Belém, Brazil: Editora Universitária UFPA, 1993. 163 p.: bibl., ill., tables.

Outstanding case study of Ford Motor Company's attempt to establish a rubber plantation in Pará in the early 1920s. Originally written as a dissertation in 1981, study is very well documented and includes many tables.

2130 Costa, Letícia B. A força de trabalho paulista: análise crítica das fontes. (*Rev. Bras. Estud. Popul.*, 7:2, julho/dez. 1990, p. 125-161, bibl., graphs, tables)

Useful and detailed analysis of the Pesquisa Nacional Amostra de Domicílios (PNAD), a principal data source used to study labor force participation and the economically active population. Includes an analysis of labor supply trends in the state of São Paulo in the 1970s and 1980s.

2131 Costa Filho, Sydney M. da. Modelo de desenvolvimento brasileiro. Rio de Janeiro: Livraria Editora Cátedra, 1993. 95 p.: bibl.

Concise and serious economic history of Brazil from the end of the 19th century through the 1980s. Focuses on explanations for persistent problems of economic development that have plagued the Brazilian economy.

2132 Delfim Netto, Antônio. Moscou, Freiburg e Brasíla: ensaios. Rio de Janeiro: Topbooks, 1990? 173 p.

Volume brings together 51 newspaper columns written by a prominent economist and public official between 1988-90 on various public issues, including the implications of *perestroika*, the German social market economy, and Brazilian economic policy.

2133 **A economia brasileira em preto e branco.** Organização de Fabrício Augusto de Oliveira. São Paulo: Editora Hucitec; Campinas: Fundação Economia de Campinas, 1991. 212 p.: bibl., ill.

Collection of essays by the Campinas faculty examines economic policies of 1990, the first year of the Collor government, and evaluates their impact and results. Essays focus on level of economic activity and investment, behavior of employment and wages, monetary and fiscal policy, and the agricultural and external sectors of the Brazilian economy.

2134 *Empresas Siderúrgicas do Brasil.* Vol. 1, 1991- . Rio de Janeiro: IBS.

Prepared by the Instituto Brasileiro de Siderurgia, first annual volume reviews performance of the Brazilian steel industry and profiles individual companies in the industry. Rich source of data.

2135 **Encontro Nacional de Estudos do Trabalho, 3rd, *Rio de Janeiro*, 1993.** Anais. Apresentação do João Saboia. Rio de Janeiro: Associação Brasileira de Estudos do Trabalho, 1993. 2 vol.

Published proceedings include 54 papers from 22 sessions covering a range of labor market, labor relations, and workplace topics from economic, sociological, and legal perspectives, including pedagogy and research.

2136 **Environment and democracy.** Organized by Henri Acselrad. Rio de Janeiro: IBASE, 1992. 104 p.: bibl.

Published in English by IBASE in connection with the Rio Conference, this set of papers examines links between the Brazilian socio-environmental crisis and the country's chosen capitalist development model. Examines the role of civil society and democratic institutions in improving outcomes.

2137 **Espejo Ortega, Alberto Octavio.** O plano de estabilização heterodoxo: a experiência comparada de Argentina, Brasil e Peru. Rio de Janeiro: Depto. de Relações Institucionais, Gabinete da Presidência, 1989. 140 p.: bibl., ill.

Prize-winning master's thesis examines experiences of Argentina, Brazil, and Peru with heterodox stabilization policies. Using a macroeconomic model, discusses impact of heterodox policies on fiscal deficit, prices, and wages. Serious study with well-reasoned conclusions.

2138 **Ferreira, Muniz G.** Investimentos diretos japoneses na economia brasileira, 1951–85. (*Cont. Int.* 15:1, jan./junho 1993, p. 135–153, tables)

Based on a master's thesis, reviews evolution of Japanese investment in Brazil and identifies factors contributing to growth of investment in the 1970s and dramatic decline in the 1980s. Concludes with identification of obstacles to the resumption of Japanese investment.

2139 **Ferrer, Aldo** and **Roberto Lavagna.** Mercosul e a coordenação de políticas econômicas. (*Rev. Bras. Comér. Exter.*, 31, abril/junho 1992, p. 3–9, graphs)

Two well-known Argentine economists discuss the impact of Mercosur on the coordination of national policies, including the coordination of common stances in multilateral forums. The differing levels of technological and industrial development of member countries is highlighted as a potential obstacle to coordination.

2140 **Fórum Nacional Como Evitar uma Nova Década Perdida, *Rio de Janeiro*, 1991.** A ecologia e o novo padrão de desenvolvimento no Brasil. Organização de João Paulo dos Reis Velloso. São Paulo: Nobel, 1992. 184 p.: bibl.

This collection of essays on economic development and the environment in Brazil was published after the Rio Conference on the Environment. Includes a series focusing on various regions of the country.

2141 **Fórum Nacional Idéias para a Modernização do Brasil, *Rio de Janeiro*, 1988.** O Brasil e a nova economia mundial. Coordenação de João Paulo dos Reis Velloso. Rio de Janeiro: J. Olympio Editora, 1991. 160 p.: bibl.

Excellent collection of articles on insertion of Brazil into the world's industrialized economies. Authors contend that this is most effectively achieved with a well-defined industrial policy, adoption of new technologies, and exposure to international competition. Includes brief discussion of Brazilian investment abroad.

2142 Fórum Nacional Idéias para a Modernização do Brasil, Rio de Janeiro, 1988. A modernização do capitalismo brasileiro: reforma do mercado de capitais. Coordenação de João Paulo dos Reis Velloso. Rio de Janeiro: J. Olympio Editora, 1991. 148 p.: bibl., ill.

Another in the series on modernization of Brazil, this volume focuses on domestic capital markets. Provides an assessment of their immediate and structural problems, a proposal for reform designed to improve prospects for development of healthy capital markets, and an outlook for internationalization of the markets.

2143 Fox, M. Louise and Samuel A. Morley. Who paid the bill?: adjustment and poverty in Brazil, 1980–95. Washington: World Bank, 1991. 45 p.: bibl., ill. (Policy, research, and external affairs working papers; WPS 648)

World Bank study analyzes impact on poverty alleviation of expansionary fiscal policy followed by Brazil in the early 1980s which traded off immediate economic growth against inflation and larger external debt later in the decade. Study concludes that while wage policy was able to protect formal-sector incomes, failure to reduce consumption in the short-run resulted in reduced investment and income losses in later periods.

2144 Franco, Gustavo Henrique Barroso. A década republicana: o Brasil e a economia internacional, 1888–1900. Rio de Janeiro: IPEA, 1991. 111 p.: bibl., ill. (Série PNPE; 24)

Outstanding and original study of economic policy in Brazil in the 1890s. Begins with a thorough review of the workings of the international monetary system in the pre-1914 gold standard world, then reconstructs valuable international trade statistics for Brazil and reexamines the country's role in the international economic system.

2145 Fritsch, Winston and Gustavo Henrique Barroso Franco. Los avances de la reforma de la política comercial e industrial en Brasil. (*in* ¿Adónde va América Latina?: balance de las reformas económicas. Recopilación de Joaquín Vial. Santiago: CIEPLAN, 1992, p. 135–157, bibl., tables)

Systematic overview of trade policy reforms undertaken by the Collor government. Based on primary policy documents, identifies four key areas of policy focus (commercial policy, deregulation and internal competition, technology, and industrial targets) and discusses nature and implications of changes in these areas.

2146 Fritsch, Winston. Brazil's trade strategy for the 1990s. (*in* Strategic options for Latin America in the 1990s. Edited by Colin I. Bradford, Jr. Paris: Development Centre of the Organisation for Economic Co-Operation and Development; Washington: Inter-American Development Bank, 1992, p. 141–153, bibl., graphs, table)

Comprehensive overview of both domestic and international dimensions of key Brazilian trade policy issues, including trade liberalization, regional integration, and the Uruguay Round of the GATT. Describes previously existing policy, changes made in the 1990s, and speculates on future Brazilian strategy under various scenarios.

2147 Füchtner, Hans. Städtisches Massenelend in Brasilien: seine Entstehungsgeschichte, Ursachen und Absicherung durch politische Herrschaft und soziale Kontrolle. Mettingen, Germany: Institut für Brasilienkunde; Brasilienkunde-Verlag, 1991. 337 p.: bibl., map. (Aspekte der Brasilienkunde = Aspetos de Brasilologia; 12)

After years of research and development work in Brazil and using almost entirely Brazilian sources, author analyzes causes and aspects of urban poverty. In addition to providing statistical and economic data, author describes socio-psychological aspects of mass misery. Rejects theories of marginality and considers how the lower class plays intrinsic role within the scheme of Brazilian development. [C.K. Converse]

2148 Furtado, Celso. Brasil: a construção interrompida. São Paulo: Paz e Terra, 1992. 87 p.: bibl.

Examination of Brazil's recent economic history in a dramatically changed international context by one of Brazil's leading social scientists.

2149 O futuro do sindicalismo no Brasil: o diálogo social. Organização de Nelson Gomes Teixeira. São Paulo: Livraria Pioneira Editora, 1990. 205 p. (Col. Novos umbrais)

Volume compiles contributions of 23 speakers at the conference organized by the Fundação Instituto de Desenvolvimento Empresarial e Social. Evenly balanced between

European and Brazilian speakers, and employer and labor representatives, with some academics, the volume allows a comparison of the Brazilian and European experiences in unionism.

2150 Gonçalves, Antonio Carlos Porto et al. Plano Collor: avaliações e perspectivas. Organização de Clóvis de Faro. Rio de Janeiro: Livros Técnicos e Científicos Editora, 1990. 354 p.: bibl., ill.

Designed for a general audience, this collection examines the Collor Plan for economic stabilization in Brazil. Focuses on legal controversies caused by the Plan, its political strategy and macroeconomic consistency, monetary reform and monetary policy, public sector reform and fiscal policy, and finally, its impact on various sectors of the economy.

2151 Gonzalez, Manuel José Forero et al. O Brasil e o Banco Mundial: um diagnóstico das relações econômicas, 1949–1989. Brasília?: IPEA/IPLAN, 1990? 174 p.: bibl.

Part of a series of SEPLAN studies of Brazil's relations with multilateral financial organizations, this well-done study traces relations with the World Bank from 1949–89. Contains many useful tables.

2152 Guimarães, Eduardo Augusto. Sistemas e instrumentos de estímulos às exportações nos países do Mercosul. (*Rev. Bras. Comér. Exter.*, 9:35, abril/junio 1993, p. 26–42, bibl., tables)

Identifies principal asymmetries in export policy in the four Mercosul countries and proposes guidelines for harmonization of policies and a strategy for achieving common region-wide rules consistent with national traditions and the rules of GATT. Useful comparative data and sound recommendations.

2153 Haber, Stephen H. Lucratividade industrial e a grande depressão no Brasil: evidências da indústria têxtil de algodão. (*Estud. Econ./São Paulo*, 21:2, maio/agosto 1991, p. 241–270, tables)

Designed to examine more closely the effects of the Depression on manufacturing in Latin America by studying firm-level data from the textile industry in Brazil, article focuses on profitability of 10 large firms from 1925–87. Results sustain thesis of Fishlow and Stein regarding slow recovery of investment rates.

2154 História do Banco do Brasil. Brazil: Banco do Brasil, 1988. 259 p.: bibl., ill. (some col.), index.

Published by the bank itself, chronicles its evolution from 1808 through the important institutional changes of the mid-1980s. Contains helpful index but few tables.

2155 Humphrey, John. Japanese production management and labour relations in Brazil. (*J. Dev. Stud.*, 30:1, Oct. 1993, p. 92–114, bibl.)

Finds that Japanese-style production management is being implemented in Brazil and that labor's support is gained in exchange for improved employment and working conditions. Terms of exchange are explained by reference to union activities, labor market conditions, and internal controls.

2156 Impactos de grandes projetos hidroelétricos e nucleares: aspectos econômicos e tecnológicos, sociais e ambientais. Coordenação de Luiz Pinguelli Rosa, Lygia Sigaud e Otávio Mielnik. São Paulo: AIE/COPPE; Editora Marco Zero; Conselho Nacional de Desenvolvimento Científico e Tecnológico, 1988. 199 p.: ill.

Scholarly comparative study of two sources of electric energy focusing on economic, technological, environmental, and social aspects of large generation projects. Conducted by researchers from both the Graduate School of Engineering and the Graduate Program in Social Anthropology at the UFRJ.

2157 A indústria de informática: tendências e oportunidades. São Paulo: Editora Tama, 1989. 140 p.: ill.

Published by the Informatics Industry Association, this volume brings together a wealth of statistical information regarding development of the national industry and applications of technology in other sectors. Includes industry, company, and market information. A gold mine for anyone doing research on this important industry.

2158 Issler, João Victor. Testing exports underinvoicing under a dual exchange rate regime: evidence for Brazilian exports. (*Rev. Econ./Recife*, 12:1, abril 1992, p. 1–29, bibl., graph, tables)

Using econometric models, tests hypotheses regarding underinvoicing of exports in response to a dual exchange rate regime.

Finds underinvoicing by price, though not by quantity, large enough to imply capital flight sufficient to purchase a significant portion of Brazilian external debt at market prices.

2159 Jatobá, Jorge. Oferta de trabalho e flutuações econômicas: Brasil, 1979–1986. *(Rev. Bras. Estud. Popul.*, 7:2, julho/dez. 1990, p. 162–179, bibl., tables)

Analyzes the response of labor force participation to changes in macroeconomic activity for the period 1979–86. Found that participation slackened with economic downturn and effect was greater for the young and the old, and for lower income households. Distinguishes between northeastern and southeastern Brazil.

2160 Jesus, Avelino de. Mercosul: estrutura e funcionamento. São Paulo: Edições Aduaneiras, 1993. 166 p.: bibl., ill.

Useful compilation of information on Mercosul covering its founding, structures, and organizational activities through 1992. Includes chapters on binational firms, other western hemisphere regional integration programs, and statistical data for the Mercosul countries.

2161 Kon, Anita. A produção terciária: o caso paulista. São Paulo: Nobel, 1992. 140 p.: bibl.

Examines contribution of the service sector to economic growth from the 1950s-1980s with particular reference to São Paulo. Good mix of theory and empirical work.

2162 Koonings, Kees. Industrialization, industrialists, and regional development in Brazil: Rio Grande do Sul in comparative perspective. Amsterdam: Thela Publishers, 1994. 301 p.: bibl., tables (Thela Latin America series; 4)

Compilation of interrelated essays offers detailed look at role of industry in development of Rio Grande do Sul. Discusses history of industrialization and industrial bourgeoisie from 1880–1985. Final essay compares experiences of Rio Grande do Sul with industrialization in Antioquia, Colombia. Includes summaries in Portuguese and Dutch. [R. Hoefte]

2163 Lavagna, Roberto. Cambios en la estructura productiva inducidos por el Mercosul. *(Contribuciones/Buenos Aires*, 10:2, abril/junio 1993, p. 45–56, graphs, tables)

Thoughtful and somewhat technical examination by former Argentine policymaker of the potential impact of Mercosul on the industrial sectors of Argentina and Brazil. While his assessment is generally positive (Mercosul enhances region's readiness for NAFTA membership), some of the conclusions are quite provocative (further gains in efficiency will have high social costs).

Lee, Naeyoung and Jeffrey Cason. Automobile commodity chains in the NICs: a comparison of South Korea, Mexico, and Brazil. See item **3724**.

Lehman, Howard P. and Jennifer L. McCoy. The dynamics of the two-level bargaining game: the 1988 Brazilian debt negotiations. See item **3725**.

2164 Lima Netto, Roberto Procópio de. Volta por cima. Rio de Janeiro: Editora Record, 1993. 221 p.

The story of the turn-around of Compania Siderúgica Nacional (CSN) by the manager who did it. Important contribution to growing literature on management in Brazil.

2165 Lipietz, Alain et al. Instabilidade econômica: moeda e finanças. Organização de Mílton Santos. São Paulo: Editora Hucitec, 1993. 253 p.: bibl., ill.

Collection seeks to identify and explain monetary and financial causes of chronic instability suffered by Latin American economies beginning at the end of the 1970s. While Brazil is the focus, some discussion of Germany, Argentina, and Israel is also included.

2166 Longo, Carlos Alberto. Federal problems with VAT in Brazil. *(Rev. Bras. Econ.*, 48:1, jan./mar. 1994, p. 85–105, bibl., tables)

Paper describes tax system at federal and state levels and contrasts 1964–67 and 1988 tax reforms in Brazil. Identifies problems emanating from the system of centralized fiscal federalism. Principal recommendations include recourse to traditional tax bases, broadening of VAT tax base, adoption of destination principle, and minimization of intergovernmental transfers.

2167 Lopes, Francisco Lafaiete et al. Anais dos seminários temáticos. v. 1–5. Rio de Janeiro: Centrais Elétricas Brasileiras S.A.,

Eletrobrás, 1991. 5 v.: ill. (Cadernos do plano 2015)

Five-volume set of papers prepared in connection with the formulation of Eletrobras' Plano 2015. Volumes include discussion of outlook for the Brazilian economy and Brazilian industrial policy, environmental policy and the use of hydropower, selection and financing of investments, the use of thermal power, and the role of the electric energy sector in technological development.

2168 Lopes, Lucas. Memórias do desenvolvimento. Rio de Janeiro: Centro da Memória da Eletricidade no Brasil, Memória da Eletricidade, 1991. 346 p.: ill., maps.

In the tradition of *testemunhos*, this is a personal chronicle of a critical period of Brazilian economic development through the eyes of the former Minister of Finance (1958–59), one of the key actors in the development of Brazil's energy sector.

2169 Machado, João Bosco M. O NAFTA e os impactos sobre as exportações brasileiras. (*Rev. Bras. Comér. Exter.*, 9:36, julho/agôsto/set. 1993, p. 80–97, bibl., tables)

Interesting study evaluates impact of NAFTA on Brazilian exports to the US and examines its effects on the allocation of US foreign investment between Mexico and Brazil. Part of a larger study funded by Funcex, Dow Chemical, Odebrecht, and Petrobras.

2170 Magalhães, João Paulo de Almeida; Nelson Kuperman; and Roberto Crivano Machado. Proálcool: uma avaliação global. Rio de Janeiro: Assessores Técnicos, 1991. 194 p.: bibl., ill.

Scholarly evaluation of the government's National Alcohol Program (PROALCOOL) from 1975–86. One of a series of studies on the impact of government policy on employment and income distribution, it also discusses environmental and cultural impacts of the program.

2171 Mendes, Ana Gláucia et al. Contradições do desenvolvimento agrícola em Minas Gerais: uma perspectiva regional. Organização de Maria Regina Nabuco. Belo Horizonte, Brazil: Centro de Desenvolvimento e Planejamento Regional, Faculdade de Ciências Econômicas, Univ. Federal de Minas Gerais, 1990. 202 p.: bibl., ill., maps. (Ensaios econômicos CEDEPLAR; 4)

Study of regional differentiation in agriculture in the state of Minas Gerais. Some of the ten papers discuss differentiation and distribution of activities over the state as a whole, while others analyze agriculture, land, and labor and employment issues in specific regions.

2172 Mendes, Armando Dias et al. Para pensar o desenvolvimento sustentável. Organização de Marcel Bursztyn. São Paulo: Editora Brasiliense; Brasília?: Instituto Brasileiro de Meio Ambiente e Recursos Naturais Renováveis; Fundação Escola Nacional de Administração Pública, 1993. 161 p.: bibl., ill.

Collection of "think-pieces" on issues related to economic development, environmental preservation, and human progress from both a scientific and administrative point of view.

2173 Mercado de trabalho na Grande São Paulo. São Paulo: Secretaria de Economia e Planejamento; Fundação Sistema Estadual de Análise de Dados; Depto. Intersindical de Estatistica e Estudos Sócio-Econômicos; UNICAMP, 1989. 384 p.: ill., maps.

Dense and useful statistical collection reports the detailed results of mid-1980s survey administered in greater São Paulo. Presents approximately 20 sets of tables with narrative, reflecting diverse labor market concepts and issues, including residence, family status, unemployment, gender, race, age, migration, education, registered employment, and minimum wage.

Mercosur: claroscuro de una integración; ciclo de conferencias realizadas en la Facultad de Ciencias Económicas y Administración. v. 1–2 See item **2023**.

2174 Minella, Ary Cesar. Banqueiros: organização e poder político no Brasil. Rio de Janeiro: Espaço e Tempo; ANPOCS, 1988. 530 p.: bibl.

Well-documented sociological study of the role of banks and bankers in the Brazilian economy. Work covers the period 1960–80 and is based on interviews of industry leaders, including those of labor unions.

2175 Moldau, Juan Hersztajn et al. Instituições financeiras e desenvolvimento tecnológico autônomo: o Banco Nacional de Desenvolvimento Econômico e Social. Coordenação de Henrique Rattner. São Paulo:

IPE-USP; FIPE; FAPESP, 1991. 186 p.: bibl., ill. (Série Relatórios de pesquisa; rp-46)

Scholarly analysis of the role of BNDES and other Brazilian development institutions in promoting technological innovation. Includes brief overview of role of similar institutions in Europe and other developing countries.

2176 **Moreira, Marcílio Marques.** The Brazilian quandary revisited. (*in* In the shadow of the debt: emerging issues in Latin America. New York: Twentieth Century Fund Press, 1992, p. 129–152, table)

Written by Brazil's then ambassador to Washington and future economy minister under President Fernando Collor de Mello, paper argues that Brazil can anticipate an economic breakthrough. Assesses many challenges Brazil now faces and must face in the future.

2177 **Moura, Alexandrina Saldanha Sobreira de et al.** Política fundiária no Nordeste: caminhos e descaminhos. Coordenação de Dirceu Murilo Pessoa. Organização de Osmil Torres Galindo Filho e Alberto de Sousa Amorim Rosa. Recife, Brazil: Fundação Joaquim Nabuco, Editora Massangana, 1990. 384 p.: bibl., ill., map. (Série Estudos e pesquisas; 63)

Collection of essays on land tenure in the Northeast, the policy reforms of the New Republic, and a review of special topics of particular concern with regard to land tenure such as water rights and the role of women.

2178 **Moura, José Tupy Caldas de** and **Walter J.B. Graneiro.** Em busca da nova ordem financeira: uma contribuição às reformas dos sistemas financeiro e de seguros no Brasil. Rio de Janeiro: Rio Fundo Editora, 1992. 247 p.: bibl., ill.

Systematic study of Brazilian banking and insurance markets and institutions focuses on regulation and oversight. Written by two retired Central Bank officials with the aim of contributing to reform, it is more practical than theoretical and provides ample data.

2179 **Neri, Marcelo Cortes.** Inflação e consumo: modelos teóricos aplicados ao imediato pós-cruzado. Rio de Janeiro: BNDES, Depto. de Relações Institucionais, Gabinete da Presidência, 1990. 145 p.: bibl., ill.

Prize-winning master's thesis examines institutional and theoretical causes of the explosion of consumption in the aftermath of the Cruzado Plan, paying special attention to the availability of consumer finance.

2180 **Neto, Abdon Portela N. et al.** Causas e tendências do processo migratório piauiense. Equipe técnica de Olavo Ivanhoé de B. Bacellar. Coordenação de Gerson Portela Lima. Consultação de Hélio Augusto Moura. Teresina, Brazil: Governo do Estado do Piauí, Secretaria de Planejamento, Fundação Centro de Pesquisas Econômicas e Sociais do Piauí, Depto. de Estudos e Projetos, 1990. 300 p.: bibl., ill., maps. (Relatório de pesquisa; 12)

Thoughtful and thorough analysis of migration into and out of the state of Piauí. Considers structure of agricultural production and the magnitude and direction of migratory flows. Analyzes members of actual and potential migrant population and the causes of migration.

Una nueva etapa en el proceso económico latinoamericano. See item **2024**.

Ocampos, Lorraine. Paraguay: tipos de cambio y proceso de integración. See item **2003**.

2181 **Oliveira, Carlos Tavares de.** Comercio exterior e a questão portuária. São Paulo: Edições Aduaneiras, 1992. 326 p.

One of the founders of Cacex, a journalist and consultant on international trade, provides commentary on the Collor government's international trade policy, reviews conditions of international markets in areas of particular interest to Brazilian exporters, and discusses the state of Brazilian ports and the implications for Brazilian exports.

2182 **Pereira, Luiz Carlos Bresser.** A crise do Estado: ensaios sobre a economia brasileira. São Paulo: Nobel, 1992. 195 p.: bibl.

Written between 1987 (when author resigned as Minister of Finance) and 1992, this outstanding collection explains macroeconomic problems in terms of State failures and fiscal crisis. Provides policy recommendations to improve economic performance.

2183 **Pero, Valéria Lúcia.** A carteira de trabalho no mercado de trabalho metropolitano brasileiro. (*Pesqui. Planej. Econ.*, 22:2, agosto 1992, p. 305–342, bibl., graphs, tables)

Complex article assesses labor market formality versus informality in the 1980s in metropolitan Brazil using the signed or unsigned work card as the indicator of segmentation. Considers wage differences, degree of informality, and rate of unemployment.

2184 **Pesquisa nacional por amostra de domicílios.** Síntese de indicadores da pesquisa básica da PNAD de 1981–1989. Rio de Janeiro: Ministério da Economia, Fazenda e Planejamento, Fundação Instituto Brasileiro de Geografia e Estatística-IBGE, Diretoria de Pesquisas, Depto. de Emprego e Rendimento, 1990. 95 p.: map, tables.

Statistical reference book presents results of the Pesquisa Nacional por Amostra de Domicílios (PNAD) from 1981–89 for Brazil as a whole and for the five large census regions. Tables and text are provided for population, education, work, family characteristics, residence characteristics, and income.

2185 **Pinto, Nuno Renan de Figueiredo** *et al.* A economia da inflação. Organização de Nali de Jesus de Souza. Porto Alegre, Brazil.: Univ. Federal do Rio Grande do Sul, 1992. 188 p.: bibl., ill.

Collection of essays by UFRGS faculty attempts to explain Brazil's chronic inflation by examining economic, social, and political causes. Designed as a textbook for a wide audience, uses various theoretical models and simple language to discuss inflation and the nature of stabilization policies with special reference to the impact on employment and income distribution.

2186 **Prochnick, Víctor** and **Marcos Lisboa.** Brasil: perspectivas para el complejo textil. (*in* América Latina: inversión y equidad. Coordinación y edición de Alvaro García. Ginebra: Oficina Internacional del Trabajo; Programa Mundial del Empleo, 1990, p. 135–199, bibl., graphs, tables)

Detailed and thorough analysis of outlook for the Brazilian textile industry in the next decade is part of an International Labor Organisation series on investment and equity. Describes recent behavior (1980s) of the industry, forecasts future performance, and assesses government's textile policy.

2187 **Projeto: a política social em tempo de crise; articulação institucional e descentralização.** v. 4, Brasil: indicadores sociais selecionados. Brasília: Ministério da Previdência e Assistência Social (MPAS); Comisão Econômica para América Latina e Caribe (CEPAL), 1990. 1 v.: bibl., ill. (Economia e desenvolvimento; 7)

Statistical volume of a four-volume set on Brazilian social policy, this work provides a wealth of social statistics, including population, social expenditure, housing and sanitation, health, social security, and education. Statistics date from as early as 1940 or the 1970s to about 1985. Some regional disaggregation.

2188 **Quesada, Gustavo Martín; José Antônio Costa Beber;** and **Francisco Romualdo de Sousa Filho.** Cenários energéticos das principais culturas agrícolas do Cone Sul: ano 2000. (*Rev. Econ. Sociol. Rural*, 31:3, julho/set. 1993, p. 147–159, bibl., tables)

Paper attempts to forecast agricultural production of major Southern Cone crops to the year 2000. Concludes that in contrast to complementary regional economic activity of European and East Asian countries, the agricultural sectors of the Mercosul countries compete against one another.

2189 **Ramos, Lauro Roberto Albrecht.** A distribuição de rendimentos no Brasil, 1976/85. Rio de Janeiro: Instituto de Pesquisa Econômica Aplicada, 1993. 135 p.: bibl., ill. (IPEA; 141)

Econometric study of income distribution evolution in Brazil from 1976–85 using data from the Pesquisa Nacional por Amostra de Domicílios. Author finds no necessary contradiction between economic growth and distributive goals and attributes changes in distribution to the business cycle, economic structure, and growth strategy adopted.

2190 **Rangel, Inácio.** Do ponto de vista nacional. Rio de Janeiro: Bienal; BNDES, 1992. 139 p.

Collection of articles written by and published in honor of Inácio Rangel, former head of the Economics Department of the BNDES. The works originally appeared in *Ultima Hora* during the early 1960s.

2191 **Reis, Eustáquio J.** and **Rolando M. Guzmán.** Um modelo econométrico do desflorestamento da Amazônia. (*Pesqui. Planej. Econ.*, 23:1, abril 1993, p. 33–64, bibl., maps, tables)

Based on cross-section data at the municipal level, paper specifies, estimates, and simulates an econometric model of Amazonian deforestation and its contribution to carbon dioxide emissions. Outlines future research to refine methodology and improve results.

2192 Relatório Norte-Sul: o Brasil e o Nordeste. Fortaleza, Brazil: SEPLAN/IPLANCE, 1991. 183 p.: bibl.

Fine set of commissioned essays on various aspects of the development of the Northeast in the context of Brazil as a whole. The wide variety of themes considered includes the role of the State, poverty and regional inequality, the social debt, macroeconomic policies, agriculture, and industry.

2193 Repercussões sócio-econômicas do complexo industrial ALBRAS-ALUNORTE em sua área de influência imediata. Belém, Brazil: O Instituto do Desenvolvimento Econômico-Social do Pará, 1991. 337 p.: bibl., ill., maps. (Relatórios de pesquisa; 20)

Detailed study of the socioeconomic impact of two large export-oriented aluminum refineries in the state of Pará. Begins with the economic background, the decision to build, and physical characteristics of the projects. Analyzes impacted municipalities in terms of productive structure, public finance, and living conditions of the population.

2194 Rocha, Sonia et al. Os efeitos da infra-estrutura no crescimento da economia brasileira. Coordenação de Hamilton C. Tolosa. Consultação de Roberto C. Albuquerque. Brasília: Câmara Brasileira da Indústria da Construção, 1993? 115 p.: bibl., ill.

Produced with support of the relevant industry association, volume is rich source of data on construction and related industries in Brazil in the early 1980s.

2195 Ronci, Márcio Valério. Política econômica e investimento privado no Brasil, 1955-82. Rio de Janeiro: Editora da Fundação Getulio Vargas, 1991. 47 p.: bibl., ill. (Série Teses; 20)

Revision of doctoral dissertation monograph empirically tests the crowding-out hypothesis with regard to government spending and private investment in Brazil. Examines impact of government policy on private investment from the mid-1950s to the mid-1980s. Very well done with interesting and provocative conclusions.

2196 Rush, Howard and João Carlos Ferraz. Employment and skills in Brazil: the implications of new technologies and organizational techniques. (*Int. Labour Rev.*, 132:1, 1993, p. 75-93, bibl., tables)

Results of a survey of 132 leading firms in eight sectors in Brazilian industry regarding the employment and skill profile of their work forces and their training needs, especially relative to the adoption of microelectronically controlled equipment.

2197 Sampaio, Yoni. Impacto nutricional de los subsidios de alimentos para familias de bajos ingresos: el caso de PROAB en Brazil. (*Cuad. Econ./Santiago*, 27:81, agosto 1990, p. 219-239, bibl., tables)

Study of PROAB, an organization that provides food subsidies to low-income families, examines its impact in six large suburbs in Recife. Found little impact on the nutritional status of preschool children and birth weight due to product income inelasticities and a small transfer of income.

2198 Saul, Nestor. Análise de investimentos: critérios de decisão de desempenho nas maiores empresas do Brasil. Porto Alegre, Brazil: Ortiz, 1993. 238 p.: bibl., ill.

Based on a survey of 132 national and multinational firms, examines factors Brazilian firms use to make investment and budgeting decisions.

2199 Savedoff, William D. Os diferenciais regionais de salários no Brasil: segmentação versus dinamismo da demanda. (*Pesqui. Planej. Econ.*, 20:3, dez. 1990, p. 521-556, bibl., tables)

Analyzes regional wage differences in terms of conventional market and segmentation hypotheses. Concludes that differences derive primarily from differences in the composition of labor demand, rather than from skill or cost of living differences, or from segmentation and barriers to labor flows.

2200 Seminário Desenvolvimento Econômico, Investimento, Mercado de Trabalho e Distribuição da Renda, *Rio de Janeiro, 1992.* Anais. Rio de Janeiro: Sistema BNDES, Depto. de Relações Institucionais, Nações Unidas, 1993. 122 p.: bibl., ill.

Collection of four long conference papers written by seasoned scholars. Vol. considers four themes in labor market studies: well-being and inequality; the new human resources management strategies and quality; industrial relations and competitiveness; and flexibility, turnover, and productivity.

2201 Seminário sobre a Economia Mineira, 6th, Diamantina, Brazil, 1992. Anais: história econômica e demográfica; economia mineira; políticas sociais; previdência, saúde, trabalho, educação e meio ambiente. Belo Horizonte, Brazil: Centro de Desenvolvimento e Planejamento Regional, Faculdade de Ciências Econômicas, Univ. Federal de Minas Gerais, 1992. 398 p.: ill., maps.

Volume presents 18 conference papers from a regular conference on the economy of the state of Minas Gerais. Topics include economic history and demography, the economy of the state, and social policies regarding welfare, health, work, education, and environment.

2202 Shapiro, Helen. Engines of growth: the state and transnational auto companies in Brazil. Cambridge, England; New York: Cambridge Univ. Press, 1993. 267 p.: index, tables.

Examines the Brazilian automobile industry from 1957–90. Describes how the Brazilian government banned auto imports in 1956, thereby forcing foreign firms to produce locally. Discusses the bargaining process between the government and foreign producers, as well as the positive linkage effects. [J.W. Foley]

2203 Siqueira, Ethevaldo. Telecomunicações: privatização ou caos. Com depoimentos e artigos de José Antonio Alencastro e Silva *et al.* São Paulo: TelePress Editora, 1993. 176 p.: bibl., ill.

Interesting collection of articles by and interviews with important actors in the telecommunications industry of Brazil (including Gen. José Alencastro e Silva, and ex-Ministers Euclides Quandt de Oliveira and Ozires Silva) with a clear bias in favor of privatization.

2204 Solingen, Etel. Macropolitical consensus and lateral autonomy in industrial policy: the nuclear sector in Brazil and Argentina. (*Int. Organ.*, 47:2, Spring 1993, p. 263–298)

Comparative perspective on structure and development of industrial policy on nuclear industry in Brazil and Argentina. Identifies reason for emergence of certain sectoral arrangements as being rooted in domestic structural and institutional differences, in particular, the varying degrees of macropolitical consensus and the relative autonomy of the particular sectoral agency.

2205 Sowing the whirlwind: soya expansion and social change in southern Brazil. Edited by Geert Arent Banck and Cornelis Wilhelmus Maria den Boer. Amsterdam: CEDLA, 1991. 196 p.: bibl., ill. (Latin America studies; 61)

Collection of nine multidisciplinary articles on the soya boom in northern Rio Grande do Sul and southwestern Paraná. This crop is a metaphor for profound changes in structures of production and ensuing social unrest. The boom is analyzed in the wider context of Brazilian and international developments. [R. Hoefte]

2206 Sued, Ronaldo. O desenvolvimento da agroindústria da laranja no Brasil: o impacto das geadas na Flórida e da política econômica governamental. Rio de Janeiro: Editora da Fundação Getúlio Vargas, 1993. 213 p.: bibl., index. (Série Teses EPGE; 25)

Doctoral thesis develops model of international orange juice concentrate industry and estimates impact of freezes in Florida and government policies on the rapid development of this industry in Brazil in the last 25 years.

2207 Sustainable development of the Amazon: development strategy and investment alternatives. Rio de Janeiro: SUDAM; Secretaria do Desenvolvimento Regional, 1992. 39 leaves: ill.

Published by SUDAM in connection with the Rio Conference on Development and the Environment, this document seeks to initiate discussion with the business sector on investment and sustainable development of the Amazon region. Contains recommendations for appropriate development in five sectors: agribusiness, bio-technology, timber and forestry, fisheries, and tourism.

Tappatá, Ricardo and **Jorge Vasconcelos.** El Paraguay y el Mercosur: situación actual, perspectivas y nuevas ventajas competitivas. See item **2006**.

2208 Triches, Divanildo. Demanda por moeda no Brasil e a causalidade entre as variáveis monetárias e a taxa de inflação, 1972–87. Rio de Janeiro: BNDES, Gabinete da Presidência, Depto. de Relações Institucionais, 1992. 115 p.: bibl., ill.

Prize-winning master's thesis examines demand for money in Brazil, and the relation between and behavior of monetary variables and the rate of inflation. Using quarterly data, tests functional form and stability of the demand for money and direction of causality among a series of monetary variables.

2209 **Urani, André** and **Carlos D. Winograd.** Distributional effects of stabilization policies in a dual economy: the case of Brazil, 1981–88. (*Rev. Bras. Econ.*, 48:1, jan/mar. 1994, p. 71–84, bibl., graphs, tables)

Econometric model examining links between stabilization policies of 1981–88, the trade balance, economic growth, inflation, and relative incomes of workers in both formal and informal sector. Using data from household surveys, model demonstrates that despite stagnation, losses in the informal sector were smaller than those in the formal sector.

2210 **Vergara, Dulce Helena.** Diferenciais de salários entre os setores público e privado da economia brasileira. (*Ensaios FEE*, 12:1, 1991, p. 73–85, bibl., tables)

Analyzes the sources of wage differentials between the public and private sectors using the 1970 census, controlling for structure of employment, worker characteristics, etc. Finds differential in favor of the private sector, although it would favor the public sector if only State firms were considered.

2211 **Verhine, Robert Evan.** Educational alternatives and the determination of earnings in Brazilian industry. Frankfurt am Main; New York: P. Lang, 1993. 375 p.: bibl. (Empirische Schul- und Unterrichtsforschung, 0724–4460; Bd. 12)

Analyzes impact of educational alternatives on industrial earnings in an industrial park in Bahia in the 1970s. Both a case study and an application of human capital theory, the study adds extra-school education to the conventional equation and substantially improves the ability to predict earnings from industrial employment.

2212 **Vianna, Maria Lúcia Teixeira Werneck.** A administração do "milagre": o Conselho Monetário Nacional, 1964–1974. Petrópolis, Brazil: Vozes, 1987. 180 p.: bibl.

Well-documented case study of the important Conselho Monetário Nacional from its founding in 1964 to 1974, placed in the context of economic policy-making in a key period of capitalist expansion in Brazil. Focuses on agency's role as a mediator of class interests.

2213 **Vieira, Renata** and **Luiz Vieira.** Produtividade de P&D no setor de informática: uma análise comparativa entre Brasil e EUA na década de 80. (*Rev. Bras. Econ.*, 46:2, abril/junho 1992, p. 241–259, bibl., graphs, tables)

Comparative analysis of research and development spending in the informatics industry in Brazil and the US in the 1980s, and an assessment of the effect of government policy on the efficiency and productivity of investment in the sector. Study finds significant opportunities for competitive development of the industry in Brazil.

2214 **Willumsen, Maria José F.** and **Robert Cruz.** O impacto das exportações sobre a distribuição de renda no Brasil. (*Pesqui. Planej. Econ.*, 20:3, dez. 1990, p. 557–580, bibl., tables)

Using simulations from a computable general equilibrium model, authors find that while an increase in exports worsens income distribution, exports of non-primary products have a lesser negative impact than exports of agricultural products. Results are sensitive to changes in relative prices.

2215 **Zockun, Maria Helena Garcia Pallares.** A importância das empresas brasileiras de capital estrangeiro para o desenvolvimento nacional. 3a. ed. São Paulo: FIESP/CIESP, 1992. 103 p.: bibl.

Produced by the group of Brazilian Firms of Foreign Capital of FIESP, book examines role of foreign capital in the Brazilian economy in both quantitative and qualitative terms. Rich in statistical data.

JOURNAL ABBREVIATIONS

Am. Econ. The American Economist. Lubin Graduate School of Business, Pace Univ. New York.

Am. Econ. Rev. The American Economic Review. American Economic Assn., Evanston, Ill.

Am. Lat. Enfoques Rusia. América Latina: Enfoques desde Rusia. Academia de Ciencias de Rusia, Instituto de América Latina. Moscow.

An. Acad. Nac. Cienc. Econ. Anales de la Academia Nacional de Ciencias Económicas. Buenos Aires.

Anál. Econ./La Paz. Análisis Económico. Ediciones UDAPE (Unidad de Análisis de Políticas Económicas). La Paz.

Anuario/Quito. Anuario de la Fundación Los Andes de Estudios Sociales. Quito.

Apuntes/Lima. Apuntes. Univ. del Pacífico, Centro de Investigación. Lima.

Banca Cent. Banca Central. Banco de Guatemala. Guatemala.

Banca Naz. Lavoro. Banca Nazionale del Lavoro Quarterly Review. Banca Nazionale del Lavoro. Rome.

Bol. Inf. Techint. Boletín Informativo Techint. Organización Techint. Buenos Aires.

Bol. Socioecon. Boletín Socioeconómico. Centro de Investigaciones y Documentación Socioeconómico (CIDSE), Univ. del Valle. Cali, Colombia.

Bull. East. Caribb. Aff. Bulletin of Eastern Caribbean Affairs. Univ. of West Indies. Cave Hill, Barbados.

Camb. J. Econ. Cambridge Journal of Economics. Cambridge Political Economy Society. London; Academic Press. New York.

Can. J. Lat. Am. Caribb. Stud. Canadian Journal of Latin American and Caribbean Studies. Univ. of Ottawa. Ontario, Canada.

Caribb. Aff. Caribbean Affairs. Trinidad Express Newspapers Ltd.; Inprint Caribbean Ltd., Port of Spain, Trinidad and Tobago.

Caribb. Stud. Caribbean Studies. Univ. of Puerto Rico, Institute of Caribbean Studies. Río Piedras.

CEPAL Rev. CEPAL Review/Revista de la CEPAL. Naciones Unidas, Comisión Económica para América Latina. Santiago.

Cienc. Econ./Lima. Ciencia Económica. Facultad de Economía, Univ. de Lima.

Cienc. Econ./San José. Ciencias Económicas. Instituto de Investigaciones en Ciencias Económicas, Univ. de Costa Rica. San José.

Colecc. Estud. CIEPLAN. Colección Estudios CIEPLAN. Corporación de Investigaciones Económicas para Latinoamérica. Santiago.

Columbia J. World Bus. Columbia Journal of World Business. Columbia Univ., New York.

Comer. Exter. Comercio Exterior. Banco Nacional de Comercio Exterior. México.

Cont. Int. Contexto Internacional. Instituto de Relações Internacionais, Pontifícia Univ. Católica. Rio de Janeiro.

Contribuciones/Buenos Aires. Contribuciones. Estudios Interdisciplinarios sobre Desarrollo y Cooperación Internacional. Konrad-Adenauer-Stiftung; Centro Interdisciplinario de Estudios Sobre el Desarrollo Latinoamericano (CIEDLA). Buenos Aires.

Cuad. CENDES. Cuadernos del CENDES. Centro de Estudios del Desarrollo, Univ. Central de Venezuela. Caracas.

Cuad. Econ./Bogotá. Cuadernos de Economía. Univ. Nacional de Colombia. Bogotá.

Cuad. Econ./Santiago. Cuadernos de Economía. Pontificia Univ. Católica de Chile, Instituto de Economía. Santiago.

Cuad. Invest./San Salvador. Cuadernos de Investigación. Centro de Investigaciones Tecnológicas y Científicas. San Salvador.

Cuad. Nuestra Am. Cuadernos de Nuestra América. Centro de Estudios sobre América. La Habana.

Cuba. Stud. Cuban Studies. Univ. of Pittsburgh, Center for Latin American Studies. Pittsburgh, Penn.

Cuest. Econ. Cuestiones Económicas. Banco Central del Ecuador. Quito.

Curr. Hist. Current History. Philadelphia, Penn.

Debate Agrar. Debate Agrario. Centro Peruano de Estudios Sociales (CEPES). Lima.

Desarro. Econ. Desarrollo Económico. Instituto de Desarrollo Económico y Social. Buenos Aires.

ECA/San Salvador. Estudios Centro-Americanos: ECA. Univ. Centroamericana José Simeón Cañas. San Salvador.

Econ. Dev. Cult. Change. Economic Development and Cultural Change. Univ. of Chicago, Research Center in Economic Development and Cultural Change. Chicago, Ill.

Econ. Soc. Prog. Lat. Am. Economic and Social Progress in Latin America. Inter-American Development Bank. Washington.

Econ. Trab. Chile. Economía y Trabajo en Chile. Programa de Economía del Trabajo, Academia de Humanismo Cristiano. Santiago.

Economía/Lima. Economía. Depto. de Economía, Pontificia Univ. Católica del Perú. Lima.

Económica/La Plata. Económica. Univ. Nacional de La Plata, Facultad de Ciencias Económicas, Instituto de Investigaciones Económicas. La Plata, Argentina.

Economist/London. The Economist. London.

Ecuad. Deb. Ecuador Debate. Centro Andino de Acción Popular (CAAP). Quito.

Empres. Sider. Brasil. Empresas Siderúrgicas do Brasil. Instituto Brasileiro de Siderurgia. Rio de Janeiro.

Ensaios FEE. Ensaios FEE. Secretaria de Coordenacão e Planejamento. Fundação de Economia e Estatística. Porto Alegre, Brazil.

Envío/Managua. Envío. Univ. Centroamericana (UCA). Managua.

Estad. Econ. Estadística & Economía. Instituto Nacional de Estadísticas. Santiago.

Estud. Demogr. Urb. Estudios Demográficos y Urbanos. El Colegio de México. México.

Estud. Econ./México. Estudios Económicos. El Colegio de México. México.

Estud. Econ./Santiago. Estudios de Economía. Depto. de Economía, Univ. de Chile. Santiago.

Estud. Econ./São Paulo. Estudos Econômicos. Univ. de São Paulo, Instituto de Pesquisas Econômicas. São Paulo.

Estud. Front. Estudios Fronterizos. Instituto de Investigaciones Sociales, Univ. Autónoma de Baja California. Mexicali, Mexico.

Estud. Int./Santiago. Estudios Internacionales. Instituto de Estudios Internacionales, Univ. de Chile. Santiago.

Estud. Parag. Estudios Paraguayos. Univ. Católica Nuestra Señora de la Asunción. Asunción.

Estud. Públicos. Estudios Públicos. Centro de Estudios Públicos. Santiago.

Estudios/Fundación Mediterránea. Estudios. Instituto de Estudios Económicos sobre la Realidad Argentina y Latinoamericana; Fundación Mediterránea. Córdoba, Argentina.

EURE/Santiago. EURE: Revista Latinoamericana de Estudios Urbanos Regionales. Centro de Desarrollo Urbano y Regional, Univ. Católica de Chile. Santiago.

Eval. Econ. Evaluación Económica. Müller & Machicado Asociados. La Paz.

Financ. Dev. Finance and Development. International Monetary Fund; The World Bank. Washington.

Front. Norte. Frontera Norte. Colegio de la Frontera Norte. Tijuana, Mexico.

Granma/La Habana. Granma. La Habana.

Harv. Bus. Rev. Harvard Business Review. Graduate School of Business Administration, Harvard Univ., Boston.

Hemisphere/Miami. Hemisphere. Latin American and Caribbean Center, Florida International Univ., Miami, Fla.

Ibero-Am./Stockholm. Ibero-Americana: Nordic Journal of Latin American Studies. Institute of Latin American Studies, Univ. of Stockholm.

Iberoam./Tokyo. Iberoamericana. Univ. of Sofia. Tokyo.

Int. Labour Rev. International Labour Review. International Labour Office. Geneva.

Int. Organ. International Organization. World Peace Foundation; Univ. of Wisconsin Press. Madison.

Int. Rev. Adm. Sci. International Review of Administrative Sciences. International Institute of Administrative Sciences. Brussels.

Int. Spect./Rome. The International Spectator. Istituto Affari Internazionali. Rome.

Integr. Latinoam. Integración Latinoamericana. Instituto para la Integración de América Latina. Buenos Aires.

Invest. Agrar. Investigación Agraria: Economía. Instituto Nacional de Investigaciones Agrarias. Madrid.

Invest. Econ. Investigación Económica. Facultad de Economía, Univ. Nacional Autónoma de México. México.

J. Common Market Stud. Journal of Common Market Studies. Oxford, England.

J. Dev. Areas. The Journal of Developing Areas. Western Illinois Univ. Press. Macomb, Ill.

J. Dev. Econ. Journal of Development Economics. North-Holland Publishing Co., Amsterdam, The Netherlands.

J. Dev. Stud. The Journal of Development Studies. Frank Cass. London.

J. Econ. Issues. Journal of Economic Issues. California State Univ., Sacramento.

J. Int. Law Econ. Journal of International Law and Economics. George Washington Univ., The National Law Center. Washington.

J. Interam. Stud. World Aff. Journal of Interamerican Studies and World Affairs. Institute of Interamerican Studies, Univ. of Miami. Coral Gables, Fla.

J. Lat. Am. Stud. Journal of Latin American Studies. Centers or Institutes of Latin American Studies at the Universities of Cambridge, Glasgow, Liverpool, London, and Oxford. Cambridge Univ. Press. London.

J. Money Credit Bank. Journal of Money, Credit and Banking. Ohio State Univ. Press. Columbus.

J. Peace Res. Journal of Peace Research. International Peace Research Institute, Universitetforlaget. Oslo.

J. World Trade. Journal of World Trade: Law, Economics, Public Policy. Werner Publishing Company, Ltd., Geneva.

JOICE. Journal of International and Comparative Economics (JOICE). Physica-Verlag. Heidelberg, Germany.

LARR. Latin American Research Review. Latin American Research Review Board. Univ. of New Mexico, Albuquerque, N.M.

LASA Forum. LASA Forum. Latin American Studies Assn., Pittsburgh, Penn.

Lat.am. Jahrb. Lateinamerika Jahrbuch. Vervuert Verlag. Frankfurt.

Lat. Am. Perspect. Latin American Perspectives. Univ. of California. Newbury Park, Calif.

Lect. Econ. Lecturas de Economía. Univ. de Antioquia. Medellín, Colombia.

Manch. Sch. Econ. Soc. Stud. The Manchester School of Economic and Social Studies. Economics Dept., Manchester School. Manchester, England.

NACLA. NACLA: Report on the Americas. North American Congress on Latin America. New York.

North-South/Miami. North-South: The Magazine of the Americas. North-South Center, Univ. of Miami. Coral Gables, Fla.

Nuestra Econ. Nuestra Economía. Facultad de Economía de la Univ. Autónoma de Baja California. Tijuana, Mexico.

Nueva Soc. Nueva Sociedad. Caracas.

Oxf. Econ. Pap. Oxford Economic Papers. Oxford Univ. Press. London.

Panorama Econ. Am. Lat./ CEPAL. Panorama Económico de América Latina. División de Desarrollo Económico, Comisión Económica para América Latina y el Caribe, Naciones Unidas. Santiago.

Pensam. Iberoam. Pensamiento Iberoamericano. Instituto de Cooperación Iberoamericano (ICI) de España; Comisión Económica para América Latina y el Caribe (CEPAL). Madrid.

Pesqui. Planej. Econ. Pesquisa e Planejamento Econômico. Instituto de Planejamento Econômico e Social. Rio de Janeiro.

Q. J. Econ. The Quarterly Journal of Economics. Cambridge, Mass.

Real. Econ.-Soc. Realidad Económico-Social. Univ. Centroamericana José Simeón Cañas. San Salvador.

Rev. Adm. Empres. Revista de Administração de Empresas. Fundação Getúlio Vargas, Instituto de Documentação. São Paulo.

Rev. Anál. Econ. Revista de Análisis Económico. Programa de Postgrado en Economía, ILADES/Georgetown Univ., Santiago.

Rev. Bras. Comér. Exter. Revista Brasileira de Comércio Exterior. Fundação Centro de Estudos do Comércio Exterior. Rio de Janeiro.

Rev. Bras. Econ. Revista Brasileira de Economia. Fundação Getúlio Vargas, Instituto Brasileiro de Economia. Rio de Janeiro.

Rev. Bras. Estud. Popul. Revista Brasileira de Estudos de População. Associação Brasileira de Estudos Populacionais. São Paulo.

Rev. Centroam. Econ. Revista Centroamericana de Economía. Univ. Nacional Autónoma de Honduras, Programa de Postgrado Centroamericano en Economía y Planificación. Tegucigalpa.

Rev. Ciclos. Revista Ciclos en la Historia, Economía y la Sociedad. Fundación de Investigaciones Históricas, Económicas y Sociales, Facultad de Ciencias Económicas, Univ. de Buenos Aires. Buenos Aires.

Rev. Cuba. Derecho. Revista Cubana de Derecho. La Habana.

Rev. Econ. Polít. Revista de Economia Política. Centro de Economia Política. São Paulo.

Rev. Econ./Recife. Revista de Econometria. Sociedade Brasileira de Econometria; Univ. Federal de Pernambuco. Recife, Brazil.

Rev. Econ. Sociol. Rural. Revista de Economia e Sociologia Rural. Sociedade Brasileira de Economia e Sociologia Rural. Brasília.

Rev. Econ./Uruguay. Revista de Economía. Banco Central de Uruguay. Montevideo.

Rev. Ecuat. Hist. Econ. Revista Ecuatoriana de Historia Económica. Banco Central del Ecuador, Centro de Investigación y Cultura. Quito.

Rev. Eur. Revista Europea de Estudios Latinoamericanos y del Caribe = European Review of Latin American and Caribbean Studies. Center for Latin American Research and Documentation; Royal Institute of Linguistics and Anthropology. Amsterdam.

Rev. Integr. Desarro. Centroam. Revista de la Integración y el Desarrollo de Centroamérica. Banco Centroamericano de Integración Económica. Tegucigalpa.

Rev. Occident. Revista de Occidente. Madrid.

Rev. Parag. Sociol. Revista Paraguaya de Sociología. Centro Paraguayo de Estudios Sociológicos. Asunción.

Rev. Planeac. Desarro. Revista de Planeación y Desarrollo. Depto. Nacional de Planeación. Bogotá.

Rev. Tiers-Monde. Revue Tiers-Monde. Institut d'étude du développement économique et social, Univ. de Paris.

Rev. UNITAS. Revista UNITAS. UNITAS. La Paz.

Síntesis/Madrid. Síntesis. Asociación de Investigación y Especialización sobre Temas Latinoamericanos. Madrid.

Social. Particip. Socialismo y Participación. Ediciones Socialismo y Participación. Lima.

South East. Lat. Am. South Eastern Latin Americanist. South Eastern Council of Latin American Studies. Boone, N.C.

South. Econ. J. Southern Economic Journal. Chapel Hill, N.C.

Staff Pap. Staff Papers. International Monetary Fund. Washington.

Suma/Montevideo. Suma. Centro de Investigaciones Económicas. Montevideo.

Tareas/Panamá. Tareas. Centro de Estudios Latinoamericanos (CELA). Panamá.

Tijdschr. Econ. Soc. Geogr. Tijdschrift voor Economische en Sociale Geographie. Netherlands Journal of Economic and Social Geography. Rotterdam, The Netherlands.

Trimest. Econ.   El Trimestre Económico. Fondo de Cultura Económica. México.

UNISA/Lat. Am. Rep.   UNISA/Latin American Report. Univ. of South Africa. Pretoria.

Wash. Q.   The Washington Quarterly. Georgetown Univ., The Center for Strategic and International Studies. Washington.

Weltwirtsch. Arch.   Weltwirtschaftliches Archiv. Zeitschrift des Institut für Weltwirtschaft an der Christians-Albrechts-Univ. Kiel. Kiel, Germany.

World Bank Econ. Rev.   The World Bank Economic Review. World Bank. Washington.

World Dev.   World Development. Pergamon Press. Oxford, England.

World Econ.   The World Economy. Basil Blackwell. London.

Yearbook/CLAG.   Yearbook. Conference of Latin Americanist Geographers; Ball State Univ., Muncie, Ind.

GEOGRAPHY

GENERAL

CLINTON R. EDWARDS, *Professor of Geography, University of Wisconsin-Milwaukee*

SELECTIONS FOR THIS BIENNIUM include more than the usual number of commentaries on past and current geographical writings on Latin America, with extensive bibliographies. This is due in part to the several chapters in *Benchmark 1990* (Conference of Latin Americanist Geographers, Vol. 17/18, 1992) devoted to themes such as ecology, population, commerce, economic development, health, and historical and cultural geography. Various assessments are also available in a special issue of the *Annals of the Association of American Geographers* (Vol. 82, No. 3, Sept. 1992) which has a useful introduction by Karl W. Butzer (item **2227**) and summaries of topics of interest to students of precolumbian and conquest times. In a very thorough and useful summary, James J. Parsons traces geographical studies concerning Latin America and the Caribbean from early colonial times to the present (item **2264**). Briefer but also useful is William V. Davidson's review of literature on aboriginal and peasant communities (item **2235**).

Urban geography remains a strong focus, with two significant contributions from the late Jorge E. Hardoy providing a legacy of knowledge to the history and historical geography of Latin American cities (items **2248** and **2247**). Urban studies have undergone an interesting transformation over recent years. Earlier works emphasizing rural-urban migration and resultant economic and social problems have been replaced largely by commentary on the suburbanization of "barrios." Populations of the primate cities have burgeoned to the point that decentralization is being considered as just about the only solution to urban problems (item **2255**). As in several previous biennia, by far the most entries have to do with the environment. More and more, scholars and others are questioning the advisability of unplanned or inappropriately planned development. Adverse environmental consequences have become all too evident, ranging from the continued rapid disappearance of rainforests through air and water pollution to devastating effects on indigenous peoples.

According to César Caviedes, research by North Americans on physical geography and natural hazards has declined, but there has been an increase of studies by Latin American and European scholars (item **2229**). The recent behavior of "El Niño" and associated phenomena have come under more intense study, and one important result has been better predictability.

Historical geography continues as a strong interest, with contributions on perennial subjects such as Columbus (items **2244** and **2261**), the contributions of Alexander von Humboldt, and vexing questions about early mapping of the New World.

2216 **Alés, Catherine** and **Michel Pouyllau.** La conquête de l'inutile: les géographies imaginaires de l'Eldorado. (*Homme/Paris*, 122/124, avril/déc. 1992, p. 271–308, bibl., maps, photos)

Analyzes the El Dorado myth in terms of the history of ideas, cartography, and persistence of imaginary geographies in literature.

2217 **América Latina y el Caribe: el manejo de la escasez de agua.** Santiago: Naciones Unidas, Comisión Económica para América Latina y El Caribe, 1991. 148 p.: bibl., ill., maps. (Estudios e informes de la CEPAL, 0256-9795; 82)

Discusses regions of water shortage in most Latin American countries, with case studies of irrigation and problems of water use. Contains maps of arid and semi-arid regions.

2218 **Bähr, Jürgen** and **Günter Mertins.** Verstädterung in Lateinamerika. (*Geogr. Rundsch.*, 44:6, Juni 1992, p. 360–370, bibl., graphs, photos, tables)

Latin America's high level of urbanization is accompanied by serious problems of water, electricity, waste disposal, and air pollution. Recent trend is for inner city poor and new rural-urban migrants to concentrate in suburbs.

2219 **Bañas Llanos, Belén.** De la rima y el mangostán: un sueño frustrado de Carlos III. (*Rev. Esp. Pac.*, 1:1, julio/dic. 1991, p. 115–127, ill.)

Discusses abortive Spanish efforts in the 18th century to introduce tropical fruits from the South Sea Islands to the West Indies.

2220 **Becco, Horacio Jorge.** Crónicas de la naturaleza del Nuevo Mundo. Caracas: Lagoven, 1991. 147 p.: bibl., ill. (Cuadernos Lagoven: Serie Medio milenio)

Excerpts from works of many historians, naturalists, and others describing fauna and flora with many period illustrations.

2221 **Borsdorf, Axel.** The cultural context of urban morphology, with special emphasis on Latin America. (*Universitas/Stuttgart*, 31:1, Fall 1989, p. 37–46, bibl.)

Essay on original cultural traits shown by Latin American cities. Site and morphology of Indian cities were superimposed upon by Iberian conquerors and a hybrid city emerged. Afterwards, economic and cultural influences from North America contributed to the emergence of an urban picture in Latin America that expounds the aesthetic and economic values of North American culture, which is rather alien to Latin American culture.

2222 **Borsdorf, Axel.** El modelo y la realidad: la discusión alemana hacia un modelo de la ciudad latinoamericana. (*Rev. Interam. Planif.*, 22:87/88, julio/sept., oct./dic. 1989, p. 21–29, bibl., ill.)

German geographers have attempted for quite some time to design a morphological model of archtypical Latin American cities. Paper condenses views held on this issue by Jürgen Bähr, Axel Borsdorf, Günther Mertins, and E. Gormsen. Valuable insight into German efforts that have previously appeared mostly in German language. [C. Caviedes]

2223 **Borsdorf, Axel.** Räumliche Dimensionen der Krise Lateinamerikas. (*in* Lateinamerika, Krise ohne Ende?: Beiträge zu einer Ringvorlesung im Wintersemester 1993/94 an der Leopold-Franzens-Universität Innsbruck. Innsbruck, Austria: Institut für Geographie der Univ. Innsbruck, 1994, p. 27–42, bibl.)

Within general pessimistic tenor of lectures contained in the cycle "Lateinamerika, Krise ohne Ende," this contribution offers more insights into the reasons for underdevelopment and dependency in Latin America. The split personality of the continent between its Indian and European origin is regarded as main reason for the socioeconomic failure of Latin America. [C. Caviedes]

2224 **Borsdorf, Axel.** Wohin steuern die Städte Anglo- and Lateinamerikas? Eine Reflektion aus kulturgeographischer Sicht. (*in* Gefahren und Chancen des Wertewandels: Abhandlungen. Herausgegeben von Herbert Kessler. Mannheim, Germany: Verlag Humboldt-Gesellschaft, 1993, p. 289–316, bibl., ills.)

German vision of disturbing deterioration of urban life in both Americas. "Westernization" and "Americanization" of Latin American cities have contributed to their loss of originality, leading them to commit the same mistakes as their North American counterparts. Proposes a return to the Indian-

Hispanic city model to prevent further character deterioration. One wonders if this is a feasible solution.

2225 Brücher, Heinz. Difusión transamericana de vegetales útiles del neotrópico en la época pre-colombina. (*in* International Congress of Ethnobiology, *1st, Belém, Brazil, 1988*. Ethnobiology: implications and applications. Belém, Brazil: Museu Paraense Emílio Goeldi, 1990, v. 1, p. 265–283, bibl., table)

Selective addition to author's important work on neotropical ethnobotany, *Useful plants of neotropical origin and their wild relatives,* (Heidelberg: Springer, 1989). Plants are grouped under roots, tubers, and rhizomes; palms; oil plants; pulses and grains; stimulants and narcotics; and edible fruits. There is commentary on origins of domestication and distribution through human carriage for each type. Disagrees with former reconstructions by Vavilov, Bukasov, and others.

2226 Bulavin, V.I. et al. Latinskai͡a Amerika: lesnye resursy i ikh ispol'zovanie [Latin America: forest resources and their use]. Otv. redaktor K.S. Tarasov. Moskva: Rossii͡skai͡a akademii͡a nauk, In-t Latinskoĭ Ameriki, 1992. 151 p. bibl.

The seven authors of this monograph published by the Institute of Latin American Studies (of the newly renamed Russian Academy of Sciences) claim to provide the first complex study of the economics, ecology, and trade of Latin American forest resources that examines not only present circumstances, but also future growth. Includes 25 tables putting Latin American forest production, industry, and trade in worldwide perspective, as well as a bibliography of Russian and Western sources. [B. Dash]

2227 Butzer, Karl W. The Americas before and after 1492: an introduction to current geographical research. (*Ann. Assoc. Am. Geogr.,* 82:3, Sept. 1992, p. 345–368, bibl., tables)

Very useful introduction to articles by geographers in this special issue. Commentary on precolumbian agriculture, native maps, environmental degradation, depopulation, cultural diffusion, colonial landscapes, ecological myths, European perceptions, and the Columbus quincentenary.

2228 Butzer, Karl W. From Columbus to Acosta: science, geography, and the New World. (*Ann. Assoc. Am. Geogr.,* 82:3, Sept. 1992, p. 543–565, bibl., table)

Intellectual confrontation with natural history and ethnography in the Americas changed European methods of observing and recording. Examples are Christopher and Fernando Columbus, Fernández de Oviedo, Sahagún, Cieza de León, López de Velasco, and Acosta.

2229 Caviedes, César N. The study of the physical environment and natural hazards in Latin America: progress and challenges. (*Yearbook/CLAG,* 17/18, 1990, p. 19–30, bibl.)

Informative summary demonstrating decline of work in Latin America by North American physical geographers, but an increase in work by Latin Americans and Europeans during the 1980s. Major development has been increasing emphasis on environmental history and relationships between environmental stress and social, economic, and cultural history.

2230 Cerezo Martínez, Ricardo. La carta de Juan de la Cosa. (*Rev. Hist. Naval,* 10:39, 1992, p. 31–48, ill.)

New look at the origin and purpose of the map. Its political and operative significance must be taken into account or the map remains merely an historical curiosity, subject to analysis inappropriate to its purpose.

2231 Clawson, David L. Conservation of food crop genetic resources in Latin America. (*Yearbook/CLAG,* 17/18, 1990, p. 11–17, bibl., table)

Proponents of various conservation strategies must recognize that genetic pools of food crops are dependent on traditional agriculture for their preservation. Suggests strategies for preserving traditional agricultural systems. Urges accelerated collection of traditional food crop varieties and study of not only their genetic composition, but of niches they occupy within agricultural systems as well.

2232 Comisión Económica para América Latina y el Caribe. Medio ambiente y desarrollo en América Latina y el Caribe: bibliografía seleccionada. Santiago: Biblioteca, Comisión Económica para América Latina y el Caribe, 1992. 297 p.

Includes subject, geographic, author, and conference indices, and list of serial publications.

2233 Conservation of neotropical forests: working from traditional resource use. Edited by Kent Hubbard Redford and Christine Padoch. New York: Columbia Univ. Press, 1992. 475 p.: bibl., ill., index. (Biological resource management in the tropics)

Readings—mostly case studies—in Latin American context emphasize the Amazon Basin. Much is to be learned of forest conservation from indigenous forest dwellers. Sustainable development must include input from traditional societies. Time is of the essence in solving the problem of deforestation: "the costs of inaction are intolerable."

2234 Craig, Alan K. Depletion of natural resources and the status of conservation in Latin America. (*Yearbook/CLAG*, 17/18, 1990, p. 61–66, bibl., table)

Pessimistic views of continuing damage to natural resources, especially forests and fisheries. Non-renewable resources are declining because utilization is outstripping exploration for new reserves. People at poverty level are generally not concerned with conservation, and enforcement of regulations is poor.

2235 Davidson, William V. Commentary: the status of geographical research on the aboriginal and peasant communities of Latin America. (*Yearbook/CLAG*, 17/18, 1990, p. 189–190)

Brief but informative and inclusive summary of literature reviews in *Benchmark 1990, Conference of Latin American Geographers*, primarily an inventory of basic themes of food production. Studies are predominantly interdisciplinary, incorporating especially anthropology, history, and geography. Great diversity in time, scale, and purpose. Ethnogeography of rural and traditional "peasant" societies emerges as the principal focus of applied studies.

2236 Denevan, William M. The pristine myth: the landscape of the Americas in 1492. (*Ann. Assoc. Am. Geogr.*, 82:3, Sept. 1992, p. 369–385, bibl., map)

One of the most respected cultural/historical geographers working in Latin America attacks myth of a "pristine" New World at the time of the conquest. In fact, the Americas had been occupied by hunter-gatherers and farmers for thousands of years and was heavily humanized in 1492. Article should attract readers from many disciplines. [G. Elbow]

2237 Desarrollo y medio ambiente en América Latina y el Caribe: una visión evolutiva. Programa de las Naciones Unidas para el Medio Ambiente; Agencia Española de Cooperación Internacional; MOPU, Secretaría General de Medio Ambiente. Madrid: Ministerio de Obras Públicas y Urbanismo, 1990. 262 p.: bibl., ill., maps.

Brief descriptions of natural ecosystems and historical antecedents (e.g., encounter of two worlds, the colonial period), modern metamorphosis of environments since World War II, and the current crisis. Considers urban as well as rural problems.

2238 Ebanks, G. Edward. Las sociedades urbanizadas de América Latina y el Caribe: algunas dimensiones y observaciones. (*Notas Pobl.*, 21:57, junio 1993, p. 125–160, bibl., tables)

Latin America and the Caribbean are the most urbanized of the world's developing areas. Overpopulation and poverty in many large cities present critical social and environmental problems that demand urgent action. Decentralization is necessary if sustainable development, social equity, and environmental protection are to be achieved.

2239 Eyre, L. Alan. The tropical national parks of Latin America and the Caribbean: present problems and future potential. (*Yearbook/CLAG*, 16, 1990, p. 15–33, bibl., graph, maps, photos, table)

There are 168 national parks in 22 countries, but most of them are "paper parks" with administrative problems, lack of security, and inappropriate uses.

2240 Garzón Valdés, Ernesto. Verfassung und Stabilität in Lateinamerika. (*in* Lateinamerika, Krise onhe Ende?: Beiträge zu einer Ringvorlesung im Wintersemester 1993/94 an der Leopold-Franzens-Universität Innsbruck. Innsbruck, Austria: Institut für Geographie der Univ. Innsbruck, 1994, p. 43–60, bibl.)

Political instability in Latin American countries is viewed as the consequence of political constitutions that do not reflect the needed respect for the State and their institutions, and are thereby trespassed upon or ignored by power groups such as plutocracies and militaries. [C. Caviedes]

2241 Geisse, Guillermo et al. De Teotihuacán a Brasilia: estudios de historia urbana iberoamericana y filipina. Dirección y coordinación de Gabriel Alomar. Coordinación de la documentación gráfica de Javier Aguilera Rojas. Madrid: Instituto de Estudios de Administración Local, 1987. 487 p.: bibl., ill.

Eight of 11 chapters are readings on precolumbian and colonial urbanization. Subjects include Spanish legislation and regional treatment (Mexico, Central and South America, Brazil, and the Philippines). Nicely complements Lewis Mumford's *The City in History*.

Geobase. 1980–. See *HLAS 54:44*.

2242 Giddings, Lorrain. Zonas aparentes de la vegetación de Sud América vistas desde satélites meteorológicos. (*Interciencia/Caracas*, 17:4, julio/agosto 1992, p. 223–234, bibl., maps)

Data on vegetation types and distributions gathered by metereological satellites differ from those obtained by classical means. When refined, new techniques hold much promise for identifying vegetation types more accurately over large areas.

2243 González, Alfonso. Population geography of mainland Hispanic America: inventory of the 1980s. (*Yearbook/CLAG*, 17/18, 1990, p. 99–108, bibl.)

Classification and analysis of recent literature on population. Somewhat neglected is study of relationships between population and environment changes.

2244 Green, Duncan. Columbus, commodities and cocaine. (*Geographical/London*, 13:12, Dec. 1991, p. 14–16, photos)

Traces history of Latin American dependence on export of commodities. Modern economic development cannot be based exclusively on commodity-oriented systems. Industrialization is a significant partial answer, but many difficulties prevent easy conversion.

2245 Greenow, Linda. Geographic perspectives on Latin American women. (*Yearbook/CLAG*, 17/18, 1990, p. 231–237, bibl.)

Review of literature since 1975. Not all contributions reflect the "feminist perspective." List of subjects treated under aegis of women-oriented studies offers convincing evidence that women play important roles in social, economic, migrational, agricultural, labor, and other processes analyzed by geographers.

2246 Gutman, Margarita and Jorge Enrique Hardoy. Encarando los problemas ambientales. (*Medio Ambient. Urban.*, 9:38, marzo 1992, p. 3–19, table)

Suggests new approaches to preservation of valuable historic sites and buildings in urban areas.

2247 Hardoy, Jorge Enrique. Antiguas y nuevas capitales nacionales de América Latina. (*EURE/Santiago*, 17:52/53, oct./dic. 1991, p. 7–26, tables)

Latin American capital cities have grown explosively in recent decades. Traces history of their development, and suggests modern rationales for relocation.

2248 Hardoy, Jorge Enrique. Cartografía urbana colonial de América Latina y el Caribe. Buenos Aires: Instituto Internacional de Medio Ambiente y Desarrollo-IIED-América Latina; Grupo Editor Latinoamericano, 1991. 510 p. (Col. Estudios políticos y sociales)

City plans and illustrations have been largely ignored by map historians and in histories of cartography. Hardoy offers chronological arrangement of 220 examples of urban mapping from early colonial times to the end of 18th century. Useful appendices contain bibliography, archives and libraries consulted, biographical sketches, and a guide for studying the maps.

2249 Harley, J. Brian. Rereading the maps of the Columbian encounter. (*Ann. Assoc. Am. Geogr.*, 82:3, Sept. 1992, p. 522–542, bibl.)

Suggests alternative views of early European maps of the New World. Native American maps belong in the cartographic record of the European occupation.

2250 Kaufer, Erich. Lateinamerika: das verlorene Jahrzehnt. (*in* Lateinamerika, Krise ohne Ende?: Beiträge zu einer Ringvorlesung im Wintersemester 1993/94 an der Leopold-Franzens-Universität Innsbruck.

Innsbruck, Austria: Institut für Geographie der Univ. Innsbruck, 1994, p. 171–180)

Controversial view that regards the 1980s as the "lost decade" for Latin America, considering its growing indebtedness. It could be argued, however, that since the state of indebtedness subsided in many debtor countries before the decade ended, and because the 1980s saw an end to many military dictatorships, the decade was not really a lost one. [C. Caviedes]

2251 **Klak, Thomas.** Latin American urban development: review of the 1980s and prospects for the 1990s. (*Yearbook/CLAG*, 17/18, 1990, p. 283–292, bibl.)

During the 1980s there was a shift in the literature from the theoretical to the empirical, resulting in many detailed case studies. Interpretation of urban change must take into account global, social, political, and economic changes.

2252 **Lateinamerika, Krise ohne Ende?: Beiträge zu einer Ringvorlesung im Wintersemester 1993/94 an der Leopold-Franzens-Universität Innsbruck.** Herausgegeben von Axel Borsdorf. Innsbruck, Austria: Institut für Geographie der Univ. Innsbruck, 1994. 204 p.: bibl. (Innsbrucker geographische Studien; 21)

Collection of lectures given at the Univ. of Innsbruck between Oct. 1993 and Jan. 1994. Editor is noted Latin Americanist geographer and most contributors are social scientists from Austria and Germany who hold differing views on Latin America. Good compendium for gaining insight into perceptions on Latin America by contemporary Germanic Latin Americanists. [C. Caviedes]

2253 **Latin America and the Caribbean: inventory of water resources and their use.** v. 1–2. Santiago: United Nations, Economic Commission for Latin America and the Caribbean, 1990. 2 v.: bibl., ill., maps.

Summary of river basin planning information contains not only stream flow data but also climate, soils, maps, and population densities by large civil divisions. [T.L. Martinson]

Latinskasia Amerika: problemy i tendenstsii razvitisia transporta [Latin America: problems and tendencies in the growth of transportation]. See item 1203.

2254 **Lovell, W. George.** "Heavy shadows and black night:" disease and depopulation in colonial Spanish America. (*Ann. Assoc. Am. Geogr.*, 82:3, Sept. 1992, p. 426–443, bibl., tables)

Consensus attributes much of Native American depopulation to Old World diseases introduced from first contact to early 17th century. Examples include Hispaniola, central Mexico, northwest Mexico, Guatemala south of the Petén, and the central Andes.

2255 **Lowder, Stella.** Decentralization in Latin America: an evaluation of achievements. (*in* Decentralization in Latin America: an evaluation. Edited by Arthur Morris and Stella Lowder. New York: Praeger, 1992, p. 179–194, bibl.)

Decentralization takes many forms and has been subject to many theoretical approaches. Useful concept is that decentralization occurs as reaction to processes of centralization. Forms of decentralization discussed are largely those not amenable to geographical analysis.

2256 **Marray, Michael.** Natural forgiveness. (*Geographical/London*, 13:12, Dec. 1991, p. 18–22, graphs, photos, table)

"Debt for nature swaps"—reduction or cancellation of Third World debts to the US in return for attention to environmental problems and ecological preservation—have so far been minimal compared to the total debt. This will change as financial agencies work increasingly with non-governmental environmental organizations and local currencies become more available for conservation programs.

2257 **Meio ambiente: aspectos técnicos e econômicos.** Edição de Sérgio Margulis *et al.* Brasília: Instituto de Pesquisa Econômica Aplicada; Programa das Nações Unidas para el Desenvolvimento, 1990. 238 p.: bibl., ill.

Discusses environmental damage and cost factors. Topics include energy, water and air pollution, cost benefit analysis, natural resources, and ecology. General approach to topics; lacks case studies. [K.D. Muller]

2258 **Meyer-Arendt, Klaus J.** Geographic research on tourism in Latin America, 1980–1990. (*Yearbook/CLAG*, 17/18, 1990, p. 199–207, bibl., map, tables)

Review of literature during the 1980s includes the Caribbean. Geographical studies are prominent, emphasizing descriptive studies, development and historical geography. Also includes important contributions from other disciplines on the environmental, social and economic impacts of tourism. Not surprisingly, in view of the explosive development of tourism, there are "dense clusters" of studies on Mexico and certain Caribbean islands. However, other factors such as distance and language have also influenced distribution of studies; these explain the relative paucity of research in South America.

Mignolo, Walter. Putting the Americas on the map: geography and the colonization of space. See *HLAS 54:3617.*

2259 **Montecinos, Camila** and **Miguel Altieri.** Situación y tendencias en la conservación de recursos genéticos a nivel local en América Latina. (*Agroecol. Desarro.*, 2/3, julio 1992, p. 25–34, bibl. table, photos)

Lists and discusses factors that work against preservation of traditional crops and their genetic diversity. Some progress has been made by non-governmental organizations, with examples of local conservation of maize, beans, and potatoes.

2260 **Mower, Roland D.** Mines and minerals: a treasure house of Latin American research topics. (*Yearbook/CLAG*, 17/18, 1990, p. 209–213, bibl.)

Review of literature in English by North Americans on petroleum, mining, and mineral industries during the 1980s. Very few contributions by geographers were found, suggesting a wealth of research opportunities for geographers with the requisite talents.

2261 **Nunn, George E.** The geographical conceptions of Columbus: a critical consideration of four problems. With a new essay "The Test of Time" by Clinton R. Edwards. Expanded ed. Milwaukee: American Geographical Society Collection of the Golda Meir Library, Univ. of Wisconsin-Milwaukee; New York: American Geographical Society, 1992. 195 p.: bibl., ill., (some col.), index, maps. (American Geographical Society research series; 14)

Reprint of classic work originally published in 1924 with a new essay tracing course of Nunn's contribution in subsequent studies and evaluations by others. Although the four "problems" or questions continue to receive attention, none has been answered to the satisfaction of everyone concerned.

2262 **Paisajes Geográficos.** Vol. 13, no. 27, número especial 1993- . Quito: Centro Panamericano de Estudios e Investigaciones Geográficas.

Nine epistemological and research studies, indicative of a wide range of geographical concerns, authored by scholars from Ecuador, Brazil, Costa Rica, Chile, and the US. [D.W. Gade]

2263 **Palazón Ferrando, Salvador.** La emigración española a Latinoamérica, 1946–1990: reanudación y crisis de un flujo secular. (*Estud. Geogr./Madrid*, 54:20, enero/marzo 1993, p. 97–128, bibl., tables)

Two distinct stages of migration from Spain to Latin America occurred before and after the period that included the world depression of the 1930s, World War II, and restrictions during the early part of Franco's Administration. Analyzes volumes, causes, sources, destinations, and characteristics of migration during both periods. At present, emigration from Spain is minimal and the balance now favors immigration.

2264 **Parsons, James J.** Geography: essay. (*in* Latin America and the Caribbean: a critical guide to research sources. Edited by Paula H. Covington. New York: Greenwood Press, 1992, p. 267–290)

Very useful and thorough commentary on geographical work of Latin America from Columbus to the present by a foremost practitioner. Definition and description of uses of "geography" should be read by all geographers. Lengthy categorized regional bibliography of atlases, regional surveys, field guides, encyclopedias, and bibliographies by Tamara Brunnschweiler.

2265 **Petriella, Angel; Alberto Ford Hurtado;** and **Raúl Domingo Motta.** Prospectiva ecopolítica del cambio climático en América Latina y el Caribe. (*in* Sistemas políticos, poder y sociedad: estudios de casos en América Latina. Caracas: Asociación Latinoamericana de Sociología; Centro de Estudios sobre América; Editorial Nueva Sociedad, 1992, p. 193–205)

Social scientists and appropriate international agencies should be considering measures necessary to combat or alleviate possible consequences of future human-induced climatic changes.

2266 **Philander, S. George.** El Niño. (*Oceanus/Woods Hole*, 35:2, Summer 1992, p. 56–61, ill., map)

Reviews recent research on *El Niño* and discusses successful prediction of the phenomenon in 1991.

2267 **Poole, Peter.** Desarrollo de trabajo conjunto entre pueblos indígenas: conservacionistas y planificadores del uso de la tierra en América Latina. Turrialba, Costa Rica: Centro Agronómico Tropical de Investigación y Enseñanza, 1990. 103 p.: bibl., ill. (Política, planificación e investigación: Documento de trabajo/Banco Mundial; 245)

Advocates and describes instances of cooperation among all constituencies favoring rational land use.

2268 **Rees, Peter W.** Transportation in Latin America. (*Yearbook/CLAG*, 17/18, 1990, p. 191–197, bibl.)

Review of literature during the 1980s, a period dominated by historical studies of the role of transportation in economic and geographic change. Other studies concerned changes in ecology and settlement resulting from road-building. During this period, very few studies on transportation were written by geographers, makes recommendations for further geographical research.

2269 **Reilly, Charles A.** When do environmental problems become issues?: whose issues? And who manages them best? (*Centen. Rev.*, 35:2, Spring 1991, p. 423–437, bibl.)

Solution of environmental problems in Latin America lies in interchange of information between agencies like the Inter-American Foundation and local people. Resource management and cooperation of local people are inseparable. NGOs (nongovernmental grassroots environmental organizations) are the obvious and available vehicles for interchange.

2270 **Reseñas de documentos sobre desarrollo ambientalmente sustentable.** Santiago: Naciones Unidas, CEPAL-Comisión Económica para América Latina y El Caribe, Centro Latinoamericano de Documentación Económica y Social (CLADES), División de Medio Ambiente y Asentamientos Humanos, Unidad Conjunta CEPAL/PNUMA de Desarrollo y Medio Ambiente, 1992. 217 p. (Serie INFOPLAN, temas especiales del desarrollo, 0259–0107; 8)

Important for understanding current thought among Latin American agencies and individuals regarding relationships among conservation, environmental problems, poverty, and socioeconomic development.

2271 **Riaño, Yvonne** and **Rolf Wesche.** Changing informal settlements in Latin American cities. (*in* Latin America to the year 2000: reactivating growth, improving equity, sustaining democracy. New York: Praeger, 1992, p. 113–121, bibl.)

"Informal settlements," occupied in contravention of official planning regulations, contain from 25 to over 50 percent of inhabitants of large South American cities. Governments have accepted that they are permanent features, and must now reassess them in terms of changing economic and social situations and growing influence of low-income urban political groups.

2272 **Rodan, Bruce D.; Adrian C. Newton;** and **Adalberto Veríssimo.** Mahogany conservation: status and policy initiatives. (*Environ. Conserv.*, 19:4, Winter 1992, p. 331–338, bibl., ill.)

Selective logging of the best mahogany (*Swietenia* spp.) trees has resulted in genetic deterioration. Discussion of scientific and policy rationale for proposal by US and Costa Rica to regulate trade in the genus through Appendix II of Convention on International Trade in Endangered Species.

2273 **Serrera Contreras, Ramón María.** Tráfico terrestre y red vial en las indias españolas. Barcelona: Lunwerg, 1992. 336 p.: bibl., col. ill.

Depicts historical, technological, and social aspects of colonial transportation. Includes many original and modern maps showing routes as well as colonial and modern illustrations of different modes of transport. For historian's comment see *HLAS 54:1064*.

2274 **Sips, P.** Management of tropical secondary rain forests in Latin America: today's challenge, tomorrow's accomplished fact!? Wageningen, The Netherlands: IKC-

NLBF/Stichting BOS, 1993. 71 p.: bibl., ill., tables.

Defines tropical secondary rainforests and discusses their origin, extent, and rates of formation. Also examines management of these rainforests and systems used in management experiments. Concludes with recommendations to explore and use possibilities offered by tropical rain forests. [R. Hoefte]

2275 **Sojo, Ana.** La singularidad de las políticas de población en América Latina y el Caribe en las postrimerías del siglo XX. (*Notas Pobl.*, 21:57, junio 1993, p. 83–124, bibl., tables)

Analyzes and makes recommendations concerning population policy.

2276 **Southgate, Douglas DeWitt.** Policies contributing to agricultural colonization of Latin America's tropical forests. (*in* Managing the world's forests: looking for balance between conservation and development. Edited by Narendra P. Sharma. Dubuque, Iowa: Kendall/Hunt Pub. Co., 1992, p. 215–235, bibl.)

Despite rapid decline in opportunities for new land opening, government policies are still encouraging agricultural colonization. Most critical is agricultural pressure on tropical forests. Examples are from the Brazilian Amazon, Ecuador, and Guatemala.

2277 **Steger, Hanns-Albert.** Hat Lateinamerika noch eine Zukunft? (*in* Lateinamerika, Krise ohne Ende?: Beiträge zu einer Ringvorlesung im Wintersemester 1993/94 an der Leopold-Franzens-Universität Innsbruck. Innsbruck, Austria: Institut für Geographie der Univ. Innsbruck, 1994, p. 81–89, bibl.)

Interesting essay maintains that due to cultural traditions, expectations of change in Latin America are different from those of other continents, and therefore cannot be assessed according to European notions of progress. [C. Caviedes]

2278 **Toledo, Víctor M.** Utopía y naturaleza: el nuevo movimiento ecológico de los campesinos e indígenas de América Latina. (*Nueva Soc.*, 122, nov./dic. 1992, p. 72–85, bibl., ill., tables)

In the last decade there has been a rapid increase in the number of grassroots environmentalist organizations, but so far there has been little analysis of their impact or political significance.

2279 **Tulet, Jean-Christian.** Cafeiculteurs latino-americains: les vignerons du tropique. (*Caravelle/Toulouse*, 61, 1993, p. 7–25, bibl., tables)

French geographer provides overview of Latin America's 19 coffee-producing countries, which together account for two-thirds of world production. Analysis highlights: 1) coffee peasantry (characterized by medium and small-scale operations), family structure, and variety of growing and commercial conditions; 2) strong disparities in production performance; 3) notable organization and political influence; and 4) dynamism of these rural communities, with strong cultural identities and impressive capacities to adapt to a changing market. [A. Pérotin-Dumon]

United Nations Conference on Environment and Development, *Rio de Janeiro, 1992*. Earth summit = Sommet planète terre = Cumbre para la tierra. See *HLAS 54:72*.

2280 **Uribe, Maruja** and **Gladys Soche.** Fuentes de información sobre protección del medio ambiente y los recursos naturales en América Latina y el Caribe. Bogotá: El Colegio Verde de Villa de Leyva, 1992. 92 p.: bibl., ill.

Includes networks and information systems, reference works, specialized bibliographies, periodicals, and international institutions. Lists institutions by country.

2281 **Weil, Connie.** Medical geographical research in Latin America and the Caribbean in the 1980s and beyond. (*Yearbook/CLAG*, 17/18, 1990, p. 223–230, bibl., graphs)

Literature review finds basic geographical themes were the "human use of the environment, diffusion, migration, urbanization and urban structure, cultural pluralism and regional inequalities." Other major contributions were in the area of disease distribution and ecology, often on local scales but with country-wide studies as well.

2282 **Whiteford, Scott; David Wiley;** and **Kenneth Wylie.** In the name of development: transforming the environment in Africa and Latin America. (*Centen. Rev.*, 35:2, Spring 1991, p. 205–209, bibl.)

Despite hopes for economic growth and higher standards of living in the 1980s, there has been retrogression coupled with environmental deterioration.

2283 Williams, Lynden S. Agricultural terrace evolution in Latin America. [*Yearbook/CLAG*, 16, 1990, p. 82–93, bibl., graph, ill.]

Precolumbian terraces probably evolved through control of soil erosion, and were developed as a natural consequence of cultivation on slopes.

World marketing data and statistics. See item **49**.

MIDDLE AMERICA

TOM L. MARTINSON, *Professor of Geography, Auburn University, Alabama*
GARY S. ELBOW, *Professor of Geography, Texas Tech University*

COSTA RICA EMERGES AS A CLEAR LEADER in ecological and environmental issues on Middle America this biennium. The number and quality of the nation's contributions mirror Costa Rica's commitment to its own environment. The richness of materials on Costa Rica ranges from textbooks (item **2333**) and works for the general public (item **2327**) to analyses of protected natural areas (items **2335** and **2334**) and computer models that relate several aspects of the social and natural environment (item **2323**).

Other environmental contributions worthy of special mention are a bibliography on the tropical forest with over 3,100 entries (item **2318**), a water resource inventory (item **2294**) and bibliography (item **2310**), a database on the major ecosystems of the Caribbean (item **2302**), and another in a series of indispensable country environmental profiles (items **2313**).

The offerings on Mexico for *HLAS 55* follow a pattern similar to that of recent volumes with emphasis on high profile issues such as environmental pollution and urban sprawl in Mexico City, rationalizing the use of natural resources such as water and soil, and developments along the US-Mexico political boundary. As usual, historical geography is well represented: the literature is augmented this year with some excellent offerings from *The Americas before and after 1492: current geographical research,* a special Columbian Quincentennial issue of the *Annals of the Association of American Geographers* (Sept. 1992, see items **2362, 2398,** and **2416**).

The most important contribution to the literature on environmental problems is Ezcurra's small historical monograph on the Valley of Mexico (item **2377**). For the US-Mexico frontier area, Arreola and Curtis' work on Mexican border cities sets a standard for future comparative studies (item **2358**). The Colegio de la Frontera Norte's collection of papers on environmental problems along the Mexico-California border (item **2367**) also merits readers' attention.

West's short monograph on Sonora (item **2414**) dominates the work on historical geography. Also of interest are Arij Ouweneel's methodological work on reconstructing colonial demographic patterns based on tribute records (item **2395**) and Whitmore's work on disease and death in early colonial Mexico (item **2415**) which applies computer simulations to the problem of reconstructing past populations. Horst's study on San Juan Ostuncalco, Guatemala (item **2343**) adds still another approach, analysis of parish records, to the corpus of techniques for population estimation.

The literature on Guatemala is totally dominated by historical geography, a reflection, no doubt, of the weak development of the discipline of geography within Guatemala which tends to limit local studies, combined with social and political problems which discourage fieldwork by non-Guatemalans.

CARIBBEAN AND THE GUIANAS

2284 **Archivo Histórico Nacional (Spain). Sección de Ultramar.** Planos y mapas de Puerto Rico. Por María José Arranz Recio y María Angeles Ortega Benayas, bajo la dirección de María Teresa de la Peña Marazuela. Madrid: Ministerio de Cultura, Dirección General de Bellas Artes y Archivos, 1987. 156 p.: indexes. (Archivo Histórico Nacional: Sección de Ultramar; 6)

Listing of Puerto Rican maps in the Spanish Archivo Histórico Nacional provides only minimal information.

2285 **Ayisi, Eric O.** St. Eustatius, treasure island of the Caribbean. Trenton, N.J.: Africa World Press, 1992. 224 p.: bibl., ill., index, maps.

Ten years of intensive anthropological research yields insights into family and community structure.

2286 **Barbados. Town and Country Development Planning Office.** Barbados physical development plan, amended 1983. Prepared by the Town and Country Development Planning Office, Ministry of Finance and Planning, in technical co-operation with United Nations Centre for Human Settlements (HABITAT). Barbados: Govt. Print. Dept., 1989. 191 p.: col. maps.

Policy document details land use planning and physical development on Barbados.

2287 **Boomgaard, Peter.** The tropical rain forest of Suriname: exploitation and management, 1600–1975. (*Nieuwe West-Indische Gids,* 66:3/4, 1992, p. 207–235, bibl., tables)

Traces reason behind persistance of high proportion of forest cover in Suriname. Examines destruction of wooded areas through land-clearing, forest fires, and timber felling. Discusses role of government and Western enterprise in the exploitation of forests. [R. Hoefte]

2288 *Caribbean Perspectives.* Vol. 2, 1992– . Environment and Labor in the Caribbean. New Brunswick, N.J.: Transaction Publishers; University of the Virgin Islands

Vol. contains several studies, including one by Mills and Iniama that introduces statistical models predicting rainfall variability in context of urban water management, agriculture, and crop insurance.

2289 **Chantada, Amparo.** Medio ambiente, crisis y desarrollo: reflexión en torno a los ríos Ozama e Isabela. (*Estud. Soc./Santo Domingo,* 25:83, enero/marzo 1991, p. 5–36, map, tables)

Lists enviromental contaminants by industry type and location.

2290 **Chomereau-Lamotte, Marie.** A historical guide to Saint Pierre. Fort-de-France: Bureau du Patrimoine, Conseil régional de la Martinique, 1987. 96 p.: bibl., ill. (some col.).

Guide surveys vestiges of the city remaining after the eruption and efforts toward historical preservation.

2291 **Clement, David B.** An analysis of disaster: life after Gilbert. Kingston: Institute of Social and Economic Research, Univ. of the West Indies, 1990. viii, 57 p., 4 p. of plates: col. ill. (Working paper; 37)

Reviews damage to structures from the insurance industry's point of view.

2292 **Colchester, Marcus.** Forest politics in Suriname. Utrecht, The Netherlands: International Books, 1995. 96 p.: bibl., maps, photos.

Report written at request of Surinamese NGOs focuses on populations living in the rainforest and how these relate to coastal population and national government. Argues that uncontrolled logging may generate short-term profits but will undermine the peoples' lives and bring ruin to the forests. Recommends alternative path to sustainable development based on securing community rights to the forests. [R. Hoefte]

2293 **Congrès des forestiers de la Caraïbe, 3rd, Gosier, Guadeloupe, 1986.** L'aménagement récréatif dans les forêts des Petites Antilles: annales. Édité par Michel Vallance. Guadeloupe: Direction régionale de la Guadeloupe, Office national des forêts, 1988. 81 p.: bibl.

Brief summaries of conference presentations on various forest topics, from tourism to rare trees.

2294 Dominican Republic. Oficina Nacional de Planificación. Plan de desarrollo de la zona fronteriza. v. 1 República Dominicana: Secretaría General de la Organización de los Estados Americanos, Depto. de Desarrollo Regional, 1987. 1 v.: ill., maps.

Water resource development plan for the region bordering Haiti.

2295 Etude des relations entre la population et le développement régional en Haïti. Port-au-Prince: République d'Haïti, Ministére de l'économie et des finances, Institut haïtien de statistique et d'information (IHSI), Division d' analyse et de recherche démographique. 1989. 64 p.: ill.

Contains hard-to-find maps of transportation routes, social infrastructure, and population density in a context of economic development.

2296 Geoghegan, Tighe et al. Environmental guidelines for development in the Lesser Antilles. St. Croix, U.S. Virgin Islands: Eastern Caribbean Natural Area Management Programme, 1984. vi, 44 p.: bibl., ill. (some col.), col. maps. (Caribbean environment. Technical report; 3)

Handbook on protecting coastal marine resources as "development" occurs.

2297 Geoghegan, Tighe. Guidelines for integrated marine resource management in the Eastern Caribbean. Barbados?: Eastern Caribbean Natural Area Management Program, 1983. viii, 52 p.: bibl., ill. (Caribbean environment: Technical report; 2)

Discussion guidelines and checklist for resource management planning in the Eastern Caribbean.

2298 González, Geraldino. Ríos y arroyos de la República Dominicana: ¿quiénes los han matado? Santo Domingo: Editora Educativa Dominicana, C por A, 1992. 288 p.: bibl., ill.

Inventory of Dominican rivers affected by pollution includes list of offenders.

2299 Imbert, Daniel; François Bland; and François Russier. Les milieux humides du littoral guadeloupéen. Basse Terre: Office national des forêts, Direction régionale de la Guadeloupe, 1988. 61 p.: bibl., col. ill.

Well-illustrated introduction to the lowland tropical eocsystem—especially the mangroves—in Guadeloupe.

2300 Institutional analysis — 4 SKN: institutional analysis in the area of natural resource management; the case of St. Kitts-Nevis. Castries, St. Lucia, West Indies: General Secretariat of the Organization of American States and OECS, Natural Resource Management Unit, 1989. 81 p.: bibl.

Describes public organizations responsible for resource management.

2301 Janssen, René and Okke ten Hove. Historisch-geografisch woordenboek van Suriname: naar A.J. van der Aa, 1839–1851. Met medewerking van Wim S.M. Hoogbergen. Utrecht, The Netherlands: Vakgroep Culturele Antropologie, Univ. Utrecht, 1993. 176 p.: bibl., ill., maps. (Bronnen voor de studie van Afro-Surinaamse samenlevingen, 0922–3630; 14)

Collection of all entries on Suriname as published in Van der Aa's *Aadrijkskundig Woordenboek der Nederlanden* (1839–1851). Book is divided into five sections: 1) country and people; 2) plantations; 3) rivers, waterfalls, and mountains; 4) military posts and forts; and 5) Maroon and Amerindian villages. [R. Hoefte]

2302 Johnson, Timothy Hugh. Biodiversity and conservation in the Caribbean: profiles of selected islands. Cambridge, UK: International Council for Bird Preservation, 1988. xvii, 144 p.: bibl. (ICBP monograph; 1)

Written database contains general information, important flora and fauna, ecosystems, conservation infrastructure, conservation action activities, and references for selected major Caribbean islands.

2303 Noordegraaf, Wim and Marie-Annet van Grunsven. Suriname. Amsterdam: Koninklijk Instituut voor de Tropen; 's-Gravenhage: Novib; Brussels: NCOS, 1993. 75 p.: bibl., ill., maps. (Landenreeks, 0922–4939)

Concise introduction to Suriname concentrates on developments in last decade. Emphasizes social and cultural conditions and the economic situation. Ends with geographical, economic, and demographic statistics; historical chronology; and brief overview of relations with Belgium and the Netherlands. [R. Hoefte]

2304 **Núñez Jiménez, Antonio.** Cuba, la naturaleza y el hombre. v. 1, El archipiélago. v. 2, Bojeo. La Habana: Editorial Letras Cubanas, 1982–1984. 2 v. bibl., ill. (some col.).

Abundantly-illustrated introduction to shorelines of Cuba by an illustrious Cuban geographer.

2305 **Nweihed, Kaldone G.** El Caribe de la pesca: estudio acerca de las pesquerías del Caribe y áreas adyacentes; aspectos económico, social, político y jurídico. Caracas: Univ. Simón Bolívar, Instituto de Tecnología y Ciencias Marinas: Asociación de Universidades e Institutos de Investigación del Caribe, 1982. 2 v.: bibl., ill., maps.

Illustrates the role of fishing in Caribbean social life, a counterpoint to the many natural science studies on fish and fishing in the region.

2306 **Potter, Robert B.** A note concerning housing conditions in Grenada, St. Lucia and St. Vincent. (*Bull. East. Caribb. Aff.*, 16:4/5, Sept./Dec. 1990, p. 13–23, bibl., tables)

Presents and tests methodology for analysis of housing conditions in an effort to measure the extent to which these countries have met housing needs of their people.

2307 **Ragster, LaVerne E. and Tighe Geoghegan.** Resource management training in the Caribbean, a necessity for sustainable development: the consortium approach. (*Caribb. Perspect.*, 2, 1992, p. 1–17, bibl., tables)

Offers historical background as well as the current form and function of proposed multidisciplinary resource management degree program.

2308 **La région et l'environnement.** Martinique: Conseil régional de la Martinique, 1988. 35 p.: col. ill.

Sketches environmental protection plans for the island.

2309 **Ruff, Bernard.** La Guyane aujourd'hui. Paris: Editions J.A.; Guyane: Guyane Presse diffusion, 1989. 207 p.: bibl., col. ill., index.

Well-illustrated tourist guide to French Guiana (Guyane).

2310 **Salisbury, Lutishoor.** Water resources management in the Caribbean: a bibliography; prepared on the occasion of the Regional Workshop on Water Resources Management, May 7–9, 1990, St. Augustine, Trinidad and Tobago. St. Augustine, Trinidad and Tobago: Main Library, Univ. of the West Indies, 1990. 79 p.: bibl., indexes.

According to the introduction, "seeks to provide in a single source an extensive listing of material on water resources management in the Caribbean."

2311 **Sepúlveda-Rivera, Aníbal and Jorge Carbonell.** Cangrejos-Santurce: historia ilustrada de su desarrollo urbano, 1519–1950. 2. ed. San Juan: Centro de Investigaciones CARIMAR, Oficina Estatal de Preservación Histórica, 1988. 85 p.: bibl., ill., maps (some col.).

Invaluable for historic preservation activities because of its many maps, photos, and architectural drawings.

2312 **Sepúlveda-Rivera, Aníbal.** San Juan: historia ilustrada de su desarrollo urbano, 1508–1898. San Juan: CARIMAR, 1989. 335 p.: bibl., ill., maps (some col.).

Story of Latin American urbanization told in a sequence of maps and architectural drawings.

2313 **St. Kitts and Nevis country environmental profile.** St. Michael, Barbados: Caribbean Conservation Assn.; St. Thomas, U.S.V.I.: Island Resources Foundation, 1991. 277 p.: bibl., ill., maps.

Indispensible handbook for study of environmental issues.

2314 **Suriname: twintig jaar benarde onafhankelijkheid [Suriname: 20 years of perilous independence].** Edited by Jan van Mourik. (*Geografie/Utrecht*, 4:6, 1995, p. 3–28, map, photos, tables)

Collection of six essays discusses Suriname's economy (agriculture, gold mining, and ecotourism), social networks, and environmental planning. Also covers country's awkward relationship with The Netherlands. Slightly optimistic in tone. [R. Hoefte]

2315 **Taller Internacional sobre la Transformación del Medio Geográfico en Cuba, 1st, La Habana, 1988.** Tranformación del medio geográfico en Cuba. La Habana: Unidad de Producción No. 3, Empresa de Producción del Ministerio de Educación Superior, 1988? 239 p.: bibl., ill., maps.

Wide-ranging essays include works on economic and social development, water resources, recreation and tourism, and environmental protection.

CENTRAL AMERICA
General

2316 Hedström, Ingemar. Somos parte de un gran equilibrio: la crisis ecológica en Centroamérica. 2a ed., rev. y ampliada. San José: Editorial DEI, 1986. 149 p.: bibl., ill. (Col. Ecología-teología)

Reviews ecological change in Central America and introduces the concept of "ecoteleology" in support of current conservation efforts.

2317 Nations, James D. Terrestrial impacts in Mexico and Central America. (*in* Development or destruction: the conversion of tropical forest to pasture in Latin America. Boulder, Colo.: Westview Press, 1992, p. 191–203)

Documents rapid destruction of forests in region, including three-step process of deforestation: 1) road-building to extract forest resources; 2) spontaneous settlement of peasant farmers; and 3) following rapid decline in soil fertility, the arrival of export cattle ranchers. Solutions include land reform, an end to subsidized credit for forest conversion, and laws promoting forest conservation. [J. Flora]

2318 Plan de acción forestal tropical para América Central: bibliografía. Turrialba, Costa Rica: Centro Agronómico Tropical de Investigación y Enseñanza, Programa de Producción y Desarrollo Agropecuario Sostenido, Información y Documentación Forestal para América Tropical, Comisión Centroamericana de Ambiente y Desarrollo, 1991. 600 p.: bibl., indexes. (Serie Bibliotecología y documentación; 19)

Contains over 3,100 entries, some annotated. Includes author and subject indexes.

2319 Reiche C., Carlos *et al.* Costos del cultivo de árboles de uso múltiple en América Central. Turrialba, Costa Rica: Centro Agronómico Tropical de Investigación y Enseñanza, Programa de Producción y Desarrollo Agropecuario Sostenido, Area de Producción Forestal y Agroforestal, 1991. 70 p.: bibl., ill. (Serie técnica: Informe técnico; 182)

Basic data on costs of tree production should be invaluable for plantation investors.

2320 Sistema regional de áreas silvestres protegidas en América Central: plan de acción, 1989–2000. Edición de Róger Morales y Miguel Cifuentes. Turrialba, Costa Rica: Centro Agronómico Tropical de Investigación y Enseñanza, Programa Manejo Integrado de Recursos Naturales, 1989. 121 p.: maps (some col.).

Action plan is noted for its inventory of protected areas, including maps of their location.

Belize

2321 Ford, Robert E. Toponymic generics, environment, and culture history in pre-independence Belize. (*Names/Freeman*, 39:1, March 1991, p. 1–26)

Interesting analysis of Belizian place names that also sheds considerable light on the settlement history. Of particular interest for students of Belize, while scholars interested in the British West Indies may find useful comparisons with place terminology on the islands.

2322 Koop, Gerhard S. Pioneer years in Belize. Belize City: G.S. Koop, 1991. 133 p.: ill.

Personal recollections of the original Mennonite colonists.

Costa Rica

2323 Arcia, Gustavo *et al.* POMA: modelo interactivo de población y medio ambiente en Costa Rica 1990: análisis y proyecciones para el Valle Central. San José: Asociación Demográfica Costarricense, 1991. 217 p.: bibl., ill.

Presents computer model illustrating interaction between population growth and the environment, particularly air quality, solid waste management, rapid urbanization, and forest destruction.

2324 Berrangé, Jevan P. Gold from the Golfo Dulce Placer Province, Southern Costa Rica. (*Rev. Geol. Am. Central*, 14, oct. 1992, p. 13–37, bibl., maps, tables)

Detailed description of the Placer gold deposits, which are derived locally from the Nicoya Complex.

2325 Congreso Ambiental de Costa Rica, 1st, San José, 1985. El deterioro ambiental en Costa Rica: balance y perspectivas; memoria. Edición de José Gracia Bondía. San José: Editorial de la Univ. de Costa Rica; Fundación de Parques Nacionales; Fundación de Educación Ambiental; 1990. 305 p.: bibl., ill.

Papers at this first congress range from AIDS studies to economic development and environmental planning.

2326 Elbow, Gary S. Costa Rica. (*in* Latin American urbanization: historical profiles of major cities. Edited by Gerald Michael Greenfield. Westport, Conn.: Greenwood Press, 1994, p. 159–172, bibl., map, tables)

Sketch characterizes Costa Rica's urban growth, concentrating on San José, the capital city.

2327 Ellenberg, Ludwig. Geografía de Costa Rica en fotografías. San José: Lehmann Editores, 1989. 143 p.: bibl., ill. (some col.), maps.

Vivid color photographs accompany text that sketches contemporary life in Costa Rica.

2328 Enríquez Solano, Francisco J. and **Isabel Avendaño Flores.** El Cantón de Goicoechea: un reencuentro histórico-geográfico, 1891–1991. San José: Instituto de Fomento y Asesoría Municipal; Comisión Nacional de Conmemoraciones Históricas; Goicoechea: Municipalidad de Goicoechea, 1991. 144 p.: bibl., ill., maps.

Guide to growth of the suburb of Guadalupe serves as an excellent example and prototype for a needed urban study of San José.

2329 Fournier Origgi, Luis A. Desarrollo y perspectivas del movimiento conservacionista costarricense. San José: Editorial de la Univ. de Costa Rica, 1991. 113 p.: bibl.

Historical treatment of the conservation movement begins in pre-colonial times and extends through the formation of current governmental initiatives.

2330 Gaupp, Peter. Ecology and development in the tropics. (*Swiss Rev. World Aff.*, 42:6, Sept. 1992, p. 14–19, maps, photos)

Managing environmental protection and eco-tourism is difficult, as the experience of Costa Rica shows.

2331 Horn, Sally P. The Inter-American highway and human disturbance of páramo vegetation in Costa Rica. (*Yearbook/CLAG*, 15, 1989, p. 13–22, bibl., maps, photos)

Highway construction appears not to have produced change in the páramo of the Cordillera de Talamanca.

2332 Hurkmans, Dennis. Afhankelijke regional ontwikkeling: de ruimtelijke en sektorale integratie van een perifere agrarische economie, Huetar Norte, Costa Rica [Dependent regional development: the spatial and sectoral integration of a peripheral agrarian economy, Huetar Norte, Costa Rica]. Utrecht, The Netherlands: Faculteit Ruimtelijke Wetenschappen Univ. Utrecht, 1993. 115 p.: bibl., ill., maps, tables.

Analysis of influence of space and business economics on regional development. Emphasizes importance of relations between agriculture and other economic sectors, as well as relations between towns and their hinterlands. [R. Hoefte]

2333 Introducción a la problemática ambiental costarricense: principios básicos y posibles soluciones. Recopilación de Isabel María Chacón, Jaime E. García, y Estrella Guier Serrano. San José: Programa de Educación Ambiental, Univ. Estatal a Distancia: Editorial Univ. Estatal a Distancia, 1990. 217 p.: bibl., ill.

Textbook introduces causes, effects, and potential treatment of ecological problems in Costa Rica.

2334 Meza Ocampo, Tobías and **Alexander Bonilla.** Areas naturales protegidas de Costa Rica. Cartago, Costa Rica: Editorial Tecnológica de Costa Rica, 1990. 318 p.: bibl., ill. (some col.), index, map.

Basic ecological information on each of the country's protected areas.

2335 Meza Ocampo, Tobías. Areas silvestres de Costa Rica. San Pedro, Costa Rica: Alma Mater, 1988. 111 p., 11 leaves of plates: bibl., ill. (some col.).

Thumbnail sketches of wildlife refuges includes their general ecological description as well as details of their establishment.

2336 **Orozco Vílchez, Lorena.** Estudio ecológico y de estructura horizontal de seis comunidades boscosas en la Cordillera de Talamanca, Costa Rica. Turrialba, Costa Rica: Centro Agronómico Tropical de Investigación y Enseñanza, Programa de Producción y Desarrollo Agropecuario Sostenido, Area de Producción Forestal y Agroforestal, Proyecto Silvicultura de Bosques Naturales, 1991. viii, 34 p.: ill. (Serie técnica: Informe técnico; 176. Col. Silvicultura y manejo de bosques naturales; 2)

Six woodland communities in Costa Rica's Talamanca mountain range are studied to present their floristic and structural characteristics as a basis for possible further regeneration or management projects.

2337 **Pedersen, Art.** La Península de Osa & Parque Nacional Corcovado: guía turística = tourist guide. San José: Editorial Heliconia; Fundación Neotrópica, 1992. 68 p.: bibl., col. ill., col. maps.

Information on traveling and hiking in Corcovado, Costa Rica.

2338 **Ramírez Avendaño, Victoria** and **Juan Rafael Quesada Camacho.** Evolución histórica de los cantones Osa, Golfito, Corredores y Coto Brus. San José: Ministerio de Cultura, Juventud y Deportes; Organización de Estados Americanos, 1990. 70 p.: bibl., maps.

Background information on southwestern Costa Rica emphasizes role of multinational corporations in its economic development. There are many maps, but they are poorly reproduced.

2339 **Sistemas agroforestales: principios y aplicaciones en los trópicos.** San José: Organización para Estudios Tropicales: Centro Agronómico Tropical de Investigación y Enseñanza, 1986. 818 p.: bibl., ill., index.

Handbook on designing and conducting effective agro-forestry projects with case studies.

2340 **Wille Trejos, Alvaro.** Corcovado, meditaciones de un biólogo: un estudio ecológico. 2. ed. San José: Editorial Univ. Estatal a Distancia, 1987. 403 p., 9 p. of plates: bibl., ill. (some col.), index.

Second edition of this lauded volume describes author's lifetime work in one of the world's most complex ecosystems.

El Salvador

2341 **Elbow, Gary S.** El Salvador. (*in* Latin American urbanization: historical profiles of major cities. Edited by Gerald Michael Greenfield. Westport, Conn.: Greenwood Press, 1994, p. 252–272, bibl., map, tables)

A general survey of urbanization processes and urban population change in El Salvador from precolumbian times to the present. Contains separate profiles of San Salvador, Santa Ana, and San Miguel.

Guatemala

2342 **Elbow, Gary S.** Guatemala. (*in* Latin American urbanization: historical profiles of major cities. Edited by Gerald Michael Greenfield. Westport, Conn.: Greenwood Press, 1994, p. 273–293, bibl., map, tables)

A general survey of urbanization processes and urban population change in Guatemala from precolumbian times to the present. Contains separate profiles of Guatemala City, Antigua Guatemala, Quezaltenango, and Escuintla.

2343 **Horst, Oscar H.** La utilización de archivos eclesiásticos en la reconstrucción de la historia demográfica de San Juan Ostuncalco. (*Mesoamérica/Antigua*, 12:22, dic. 1991, p. 211–231, graphs, map, tables)

Church records are used to reconstruct colonial and early republican era population of the Parroquia de San Juan Ostuncalco (comprising nine contemporary municipios). Parish registers provide variety of data, such as births and deaths, causes of death and family names, that are not available from other sources and which may be useful in reconstructing past population. See also item **2395.**

2344 **Kramer, Wendy; W. George Lovell;** and **Christopher H. Lutz.** Encomienda and settlement: towards a historical geography of early colonial Guatemala. (*Yearbook/CLAG*, 16, 1990, p. 67–72, bibl., map, table)

The Guatemalan *encomienda* began at conquest, a point not often realized.

2345 **Lovell, W. George** and **Christopher H. Lutz.** Conquest and population: Maya demography in historical perspective. (*LARR*, 29:2, 1994, p. 133–142, bibl., table)

Analysis of Maya Indian population in

Guatemala. Authors conclude that the total Maya population has doubled since the conquest and risen by a factor of ten since independence, a record of survival unequaled elsewhere in the Americas.

2346 **Lovell, W. George.** Conquest and survival in colonial Guatemala: a historical geography of the Cuchumatán highlands, 1500–1821. Rev. ed. Montreal; Buffalo: McGill-Queen's Univ. Press, 1992. 279 p.: bibl., ill., indexes, maps.

Revised version of a landmark work in Guatemalan historical geography contains a new 20-page epilogue that includes new information and modification of earlier conclusions. See *HLAS 48:2238* for a review of the first edition.

2347 **Lovell, W. George** and **William R. Swezey.** Indian migration and community formation: an analysis of *congregación* in colonial Guatemala. (*in* Migration in colonial Spanish America. Edited by David J. Robinson. Cambridge: Cambridge Univ. Press, 1990, p. 18–40, maps, tables)

Describes process by which modern day *municipios* evolved from precolumbian and colonial period predecessors. Somewhat different processes yielded town nucleus and vacant-town types. Sacapulas is used as an example of Indian migration and settlement patterns.

Honduras

2348 **Herlihy, Peter H.** and **Andrew P. Leake.** Los sumus tawahkas: un delicado equilibrio dentro de la Mosquitia. (*Yaxkin*/Tegucigalpa, 11:1, enero/junio 1988, p. 109–121, bibl., photo)

Attempts to provide legal protection to the lands of this indigenous group have proved successful thus far.

2349 **Palacios, Sergio; Olga Maldonado;** and **Manuel Aguilar.** Guía histórica-turística de la ciudad de La Ceiba. Tegucigalpa: Instituto Hondureño de Antropología e Historia, 1991. 76 p.: bibl., ill., maps.

Brief introduction to growth and influence of La Ceiba on the north coast of Honduras.

2350 **Palmer-Moloney, Jean.** Honduras: studying the economic and political geography of a banana republic. (*J. Geogr.*, 90:3, May/June 1991, p. 121–128, bibl., graphs, maps, tables)

Despite factual errors, activity format may provide a good introduction for high school students.

2351 **Pastor Fasquelle, Rodolfo.** Biografía de San Pedro Sula, 1536–1954. San Pedro Sula, Honduras: Centro Editorial, 1990. 486 p.: bibl., ill., maps.

Popular history of this "capital" of northern Honduras concentrates on social change and economic growth.

Nicaragua

2352 **Solà, Roser.** Geografía y estructura económica de Nicaragua en el contexto centroamericano y de América Latina. Managua: Univ. Centroamericana, Facultad de Ciencias Económicas y Administrativas, 1989. 247 p.: bibl., ill., maps.

Introductory economic geography concentrates on a description of primary activities.

Panama

2353 **Cobos Morán, Jorge A.** Los recursos naturales renovables de Panamá. Panamá: Instituto Nacional de Recursos Naturales Renovables; Plan de Acción Forestal Tropical de Panamá, 1992. 25 p.: bibl., ill., maps.

Briefly reviews location and extent of protected forest lands in Panama.

2354 **Mendoza B., Rodolfo E.** and **José E. González.** Plantas acuáticas de Panamá—Tracheophyta. Panamá: Editorial Universitaria, 1991. 224 p.: bibl., col. ill., index. (Col. Ciencias naturales)

Botanical illustrations and descriptions.

MEXICO

2355 **Aguilar, Adrián Guillermo** and **Guillermo Olvera L.** El control de la expansión urbana en la ciudad de México: conjecturas de un falso planteamiento. (*Estud. Demogr. Urb.*, 6:1, enero/abril 1991, p. 89–115, bilb., map, tables)

Critique of the planning process in Mexico City. Inability to control illegal settlements has led to uncontrolled expansion of the urban area. Planning has failed to deal with causes of the problems.

2356 **Alemán Santillán, Trinidad** *et al.* El subdesarrollo agrícola en los altos de Chiapas. Coordinación de Manuel Roberto Parra Vázquez. Edición de Aurora González C. México: Univ. Autónoma Chapingo, Dirección de Difusión Cultural, Subdirección de Centros Regionales, 1989. 405 p.: bibl., ill., maps. (Col. Cuadernos universitarios: Serie Agronomía; 18)

Published before the Jan. 1994 uprising, this book now takes on more significance than it might otherwise command. Work contains a great deal of information on the indigenous economy of Chiapas but offers little analysis.

2357 **Arias Chávez, José** *et al.* El agua: recurso vital. Acatlima, Mexico: Univ. Tecnológica de la Mixteca, 1993. 147 p.: ill.

Collection of nine symposium papers on water and its problems. Topics include surface water, subterranean water, water problems in arid lands, water pollution, and integrated river basin management. Papers vary considerably in coverage of their topic.

2358 **Arreola, Daniel David** and **James R. Curtis.** The Mexican border cities: landscape anatomy and place personality. Tucson: Univ. of Arizona Press, 1993. 258 p.: bibl., ill., index, maps.

Comparison of 18 Mexican cities located on the US-Mexico border and paired with a US counterpart. Identifies characteristics of a distinct Mexican border city landscape but rejects the idea that the border and its cities comprise a distinct cultural area, neither totally Mexican nor US. They place border cities clearly within the Mexican cultural and urban tradition.

2359 **Arreola, Daniel David.** Mexican origins of south Texas Mexican Americans, 1930. (*J. Hist. Geogr.*, 19:1, Jan. 1993, p. 48–63)

Birth records of Mexican-born parents were used to identify Mexican towns that served as migrant sources. Migrants identified in the sample came overwhelmingly from centers near the border. Interesting case study increases understanding of pattern of Mexican migration to the US.

Atlas de México. See item 15.

2360 **Balance y perspectivas de los estudios regionales en México.** Coordinación de Carlos R. Martínez Assad. México: Centro de Investigaciones Interdisciplinarias en Humanidades, UNAM; M.A. Porrúa Grupo Editorial, 1990. 451 p.: bibl., ill., maps. (Col. México—actualidad y perspectivas)

Review of research on Mexican regions and regionalism from the perspective of different disciplines: history, geography, political science, anthropology, and communication. Chapters on Protestants, electoral politics, and mass media deal with timely themes.

2361 **Barral, Henri.** Bolsón de Mapimí, ayer y hoy. (*Trace/México*, 19, juin 1991, p. 53–58, bibl., map)

Brief, general review of physical resources and human activities of the Bolsón. Identifies two types of cattle culture, raising pure-bred stock for US markets and capture of semi-feral *ganado bronco* to serve local markets.

2362 **Butzer, Karl W.** and **Barbara J. Williams.** Addendum: three indigenous maps from New Spain dated ca. 1580. (*Ann. Assoc. Am. Geogr.*, 82:3, Sept. 1992, p. 536–542, facsims.)

Three maps from *relaciones geograficas* ca. 1580 are reproduced and interpreted. Maps are from Misantla, Zempoala, and Atenango-and-Mixquihuala. Valuable for historical geographers and ethnohistorians. Article is presented as an addendum to a longer article by late cartographic historian J.B. Harley which appears on p. 522–536.

2363 **Butzer, Karl W.** Ethno-agriculture and cultural ecology in Mexico: historical vistas and modern implications. (*Yearbook/CLAG*, 17/18, 1990, p. 139–152, bibl.)

Ambitious review of work on agroecology, primarily by geographers, from precolumbian times to the present is sprinkled with suggestions for further research. Essential article for scholars working on issues of human ecology and ethnic-agro systems in northern and central Mexico (Chiapas and Yucatán not included).

2364 **Cambrezy, Luc.** La movilidad de la población en el centro del estado de Veracruz: colonización agrícola y crisis de la tenencia de la tierra. (*Trace/México*, 19, juin 1991, p. 27–40, bibl., map, tables)

Innovative study uses population dispersion in centers of different sizes to reconstruct competition for land. Central Veracruz is characterized by small, rural population centers mainly occupied by small farmers.

2365 **Chias Becerril, Luis.** Articulación de las costas mexicanas. (*Rev. Mex. Sociol.*, 52:3, julio/sept. 1990, p. 69–84, bibl., maps)

Paper focuses on highway transportation. Poorly connected for most of Mexico's history, coastal areas have become better connected to the interior and with each other in recent years. Four stages in highway development on the coast are identified.

Chinampas prehispánicas. See item **229**.

CIMA. See item **18**.

2366 **Cochet, Hubert.** Agriculture sur brulis, élevage extensif et dégradation de l'environnement en Amérique Latine: un exemple en sierra Madre del Sur, au Mexique. (*Rev. Tiers-Monde*, 34:134, avril/juin 1993, p. 281–303, graphs, map)

Case study of development of cattle raising in a zone of traditional slash-and-burn farming. Systems have become interdependent and swidden cycles have been modified to accommodate pasture age. Argues that ecological relationships must be understood in social and historical context.

2367 **COLEF (conference), 1st, Tijuana, Mexico, 1990.** COLEF I. v. 5, Frontera y medio ambiente. Tijuana: Colegio de la Frontera Norte; Ciudad Juárez: Univ. Autónoma de Ciudad Juárez, 1992–1993. 8 v.: ill.

Collection of seven papers presented in journal style without introduction. All deal with environmental issues on the Mexico-California border, primarily having to do with water. Papers are of interest in part because they offer a Mexican perspective on border-area water problems.

2368 **Coll-Hurtado, Atlántida** and **María Teresa Sánchez-Salazar.** Pasado y presente de la minería mexicana: estructura y organización territorial a principios del decenio de los noventa. (*Estud. Geogr./Madrid*, 53:206, enero/abril 1992, p. 9–26, bibl. maps, tables)

Brief review of Mexican mining history—in which authors identify four distinct periods—is followed by description of contemporary mineral production. Useful article for up-to-date (1992) information on Mexican mining.

2369 **Collins, Charles O.** and **Steven L. Scott.** Air pollution in the Valley of Mexico. (*Geogr. Rev.*, 83:2, April 1993, p. 119–133)

Superficial review of published information that omits most of voluminous work in Spanish on the topic. Focuses on ozone and particulate matter pollution and government policy responses to the problem. Primarily of interest to general readers.

2370 **Contreras Sánchez, Alicia del C.** Historia de una tintórea olvidada: el proceso de explotación y circulación del palo de tinte, 1750–1807. Mérida, Mexico: Univ. Autónoma de Yucatán, 1990. 135 p.: bibl., 17 folded maps.

History of exploitation of the dye tree *Haematoxylum campechianum* in Campeche and Tabasco during 16th-19th centuries. Information on Belize is also included.

2371 **Curtis, James R.** Central business districts of the two Laredos. (*Geogr. Rev.*, 83:1, Jan. 1993, p. 54–65)

Economic and cultural differences account for marked contrast in the type of business establishments found in these two cities on either side of the border. Useful work for planners and others with an interest in urban morphology.

2372 **Delgado, Javier.** Centro y periferia en la estructura socioespacial de la Ciudad de México. (*in* Espacio y vivienda en la Ciudad de México. Coordinación de Martha Schteingart. México: El Colegio de México, Centro de Estudios Demográficos y de Desarrollo Urbano, 1991, p. 85–105, bibl., graphs, map, tables)

Mexico City's core is losing population and reducing population density while the periphery expands into surrounding small towns and farm areas. Situation is analyzed and proposals made to increase urban population densities and reduce the city's peripheral expansion.

2373 **Delgado, Javier.** Valle de México: el crecimiento por conurbaciones. (*in* América Latina: regiones en transición. Ciudad Real, Spain: Publicaciones de la Univ. de Castilla-La Mancha, 1991, p. 209–229, bibl., graphs, maps, tables)

Population growth causes the Mexico City metropolitan area to expand outward into surrounding agricultural communities. One way to avoid this is to restructure certain areas of the city to increase population density. The reality is the opposite: central city population densities are decreasing.

2374 **Delhoume, Jean-Pierre.** Una zona árida del norte de México: limitaciones para el desarrollo de la ganadería extensiva. (*Trace/México*, 19, juin 1991, p. 59–65, bibl., maps)

Experimental study of cattle raising practices in the Chihuahuan Desert identifies mis-match between water and forage availability. Author recommends pasturing cattle according to availability of certain forage plants and placing water retention dams as close to preferred pasture areas as possible.

2375 **Desarrollo y medio ambiente en México: diagnóstico, 1990.** México: Fundación Universo Veintiuno; Fundación Friedrich Ebert, Representación en México, 1990. 165 p.: bibl., ill., index, maps. (Col. Medio ambiente: Temas; 9)

Work summarizes contents of eight books on specific environmental systems (air, water, soils, etc.). Summary provides concise introduction to Mexico's physical environment and its problems. Lack of footnotes and an adequate bibliography reduces book's value for specialists.

2376 **Enríquez Hernández, Jorge.** Análisis geoeconómico del sistema regional de la Sierra Tarahumara. México: Univ. Nacional Autónoma de México, Facultad de Filosofía y Letras, Colegio de Geografía, 1988. 201 p.: bibl., maps. (Seminarios—investigación)

Detailed regional study will be of interest for those who want information on the spatial economy of this portion of the Sierra Madre Occidental.

Estructura y dinámica poblacional. See item **24.**

2377 **Ezcurra, Exequiel.** De las chinampas a la megalópolis: el medio ambiente en la Cuenca de México. 2. ed. México: SEP, 1991. 119 p.: bibl., ill., maps. (La Ciencia desde México; 91)

If one could read but one of the dozens of books and essays written during the past few years on environmental problems of Mexico City, this small monograph would be a good choice. Examines environmental deterioration from a historical perspective looking at environmental deterioration. Shorter version appeared in *The Earth as transformed by human action* by B.L. Turner et al. (Cambridge Univ. Press, 1989).

2378 **Frontera sur: historia y perspectiva.** Edición de Alfredo A. César Dachary y Stella Maris Arnáiz Burne. Chetumal, Mexico: Centro de Investigaciones de Quintana Roo, 1992? 284 p.: bibl., ill., maps.

Book focuses on the Mexico-Belize-Guatemala frontier and contains sections dedicated to history, the Mexico-Belize border, the Mexico-Belize-Guatemala border, and migration and refugees. Important work for those interested in Mexico's "other" frontier.

2379 **Fuentes Aguilar, Luis.** La industria electrónica en México. (*Bol. Inst. Geogr.*, 23, 1991, p. 71–87, bibl., map, tables)

Article describes characteristics of the Mexican electronics industry and focuses on *maquiladora* plants.

2380 **Garrocho Rangel, Carlos.** El sistema urbano de México: organización, crecimiento y estructura funcional. (*Estud. Territ.*, 38, enero/abril 1992, p. 115–137, bibl., maps, tables, graphs)

Straight-forward central place analysis using 1990 census and earlier data that will interest regional specialists. Compare this with similar study conducted by the Consejo Nacional de Población (item **2405**).

GEMA: geomodelos de altimetría del territorio nacional. See item **25.**

2381 **Gilbert, Alan G.** Self-help housing during recession: the Mexican experience. (*in* Social responses to Mexico's economic crisis of the 1980s. Edited by Mercedes González de la Rocha and Agustín Escobar Latapí. San Diego, Calif.: Center for U.S.-Mexican Studies, Univ. of California, San Diego, 1991, p. 221–242, bibl.)

Paper is divided into a theoretical section and a second part on the situation in Mexico. Conclusions are hard to draw because of many variables, local differences, and changing government policies. Finds that increasing poverty does not necessarily lead to more self-help housing.

2382 Hernández Cerdá, María Engracia. Delimitación espacial de las zonas áridas de México. (*Estud. Geogr./Madrid*, 53:206, enero/abril 1992, p. 27–43, bibl. maps, tables)

Attempt to reconcile widely divergent estimates of area and location of Mexico's arid lands. Several different indices of aridity are employed.

2383 Hoffman, Peter R. Tourism and language in Mexico's Los Cabos. (*J. Cult. Geogr.*, 12:2, Spring/Summer 1992, p. 77–92)

Short article offers somewhat more than its title suggests. Uses appearance of English as "the language of international tourism" as a foil to discuss emergence of Los Cabos as a resort and to highlight impacts of that process.

2384 Hoffmann, Odile. De los hacendados a los forestales: manejo del espacio, dominación y explotación del bosque en la Sierra Madre Oriental-Cofre de Perote. (*Palabra Hombre*, 70, abril/junio 1989, p. 87–116, maps, tables)

Paper recounts 100 years of exploitation—primarily of forest products—in an isolated sector of the state of Veracruz.

2385 Jáuregui O., Ernesto. Mexico City's urban heat island revisited. (*Erdkunde/Bonn*, 47:3, Okt. 1993, p. 185–195)

Long-term analysis of diurnal and seasonal characteristics of temperture in Mexico City. Overall nocturnal heat island effect has risen from 2 degrees Celsius in 1898 to 8–9 degrees Celsius in the 1980s.

2386 Jáuregui O., Ernesto and Elda Luyando. Patrones de flujo de aire superficial y su relación con el transporte de contaminantes en el Valle de México. (*Bol. Inst. Geogr.*, 24, 1992, p. 51–78, bibl., graph, maps, tables)

Data collected from network of 11 anemometers located around the Valle de México are used to determine diurnal and seasonal local wind patterns. These patterns are then related to atmospheric pollution problems in Mexico City.

2387 Jones, Gareth. The commercialisation of the land market?: land ownership patterns in the Mexican city of Puebla. (*Third World Plan. Rev.*, 13:2, May 1991, p. 129–153, graphs, maps, tables)

Interesting study highlights tension between private and *ejido* interests in an urban land market. Research was conducted before the law was changed to allow legal sale of *ejido* lands.

2388 Levi de Lopez, Silvana. Rural change and circular migration to the United States: a case study from Michoacán, México. (*Bol. Inst. Geogr.*, 23, 1991, p. 33–52, bibl., graphs, maps, tables)

Research was conducted in two contiguous *ejidos* of Yurecuaro municipio in which 79 percent of men over 15 years of age have worked in US. Focuses on impacts in a rural peasant community over a number of decades. Valuable addition to the literature on Mexican migration to the US.

2389 Lezama, José Luis. Mexico. (*in* Latin American urbanization: historical profiles of major cities. Edited by Gerald Michael Greenfield. Westport, Conn.: Greenwood Press, 1994, p. 350–395, bibl., map, tables)

A general survey of urbanization processes and urban population change in Mexico from precolumbian times to the present. Contains separate profiles of Guadalajara, Mérida, Mexico City, Monterrey, Oaxaca, Puebla, Veracruz, and Zacatecas.

2390 Liverman, Diana. The regional impact of global warming in Mexico: uncertainty, vulnerability and response. (*in* The regions and global warming: impacts and response strategies. Edited by Jurgen Schmandt. Oxford Univ. Press: New York, 1992, p. 44–68)

Global climate change models are applied to Mexico and the result used to predict impacts on water resources and agricultural productivity. Paper concludes with brief discussion of policy implications for the Mexican government.

2391 López Moreno R., Eduardo. La cuadrícula en el desarrollo de la ciudad hispanoamericana, Guadalajara, México: estudio de la evolución morfológica de la traza a partir de la ciudad fundacional. Guadalajara, Mexico: Editorial Univ. de Guadalajara, 1992. 275 p., 3 folded leaves of plates: bibl., ill., maps. (Col. Fin de milenio: Serie Arquitectura y urbanismo)

Sub-title better describes this work

than the title. Covers history of physical development of the city from its founding through the 1930s. A wealth of information for urban geographers, historians, and planners.

2392 Malmström, Vincent H. Geographical diffusion and calendrics in pre-Columbian Mesoamerica. (*Geogr. Rev.*, 82:2, April 1992, p. 113–127)

Computer programs are used to correlate solar and lunar eclipses with Mayan and Olmec dates inscribed on monuments. Results support J.E.S. Thompson's original 1927 correlation of Maya and Western calendars. Second part deals with Aztec-Toltec calendar correlations. Of interest mainly to students of calendrics.

2393 Musset, Alain. Congregaciones y reorganización del espacio: el caso del acueducto de Tenango, siglo XVI. (*in* Mundo rural, ciudades y población del Estado de México. Coordinación de Manuel Miño Grijalva. Toluca, México: El Colegio Mexiquense; Instituto Mexiquense de Cultura, 1990, p. 147–163, ill., map)

Colonial records and surviving aqueduct remains provide details of acquisition of water rights and construction of a 16th-century aqueduct.

Nations, James D. Terrestrial impacts in Mexico and Central America. See item **2317**.

O'Hara, Sarah L.; Alayne Street-Perrott; and **Timothy P. Burt.** Accelerated soil erosion around a Mexican highland lake caused by prehispanic agriculture. See item **140**.

2394 O'Hara, Sarah L. Historical evidence of fluctuation in the level of Lake Pátzcuaro, Michoacán, México over the last 600 years. (*Geogr. J.*, 159:1, March 1993, p. 51–62)

Lake-level changes reflect relatively short-term climate fluctuations in Central Mexico. Findings for Lake Pátzcuaro correlate with changes in lake level in the Basin of Mexico.

2395 Ouweneel, Arij. Growth, stagnation, and migration: an explorative analysis of the *tributario* series of Anáhuac, 1720–1800. (*HAHR*, 71:3, Aug. 1991, p. 531–577, graphs, maps, tables)

Tribute records are utilized to reconstruct local population for the valleys of Toluca, Mexico, and Puebla, along with adjacent mountain slopes. Identifies five distinct demographic patterns for the region based on relative rates of increase/decrease throughout the period. See also the article by Horst on Guatemala (item **2343**).

2396 Paisajes rurales en el norte de Michoacán. Coordinación de Dominique Michelet. México: Colegio de Michoacán; Centre d'études mexicaines et centraméricaines, 1991. 101 p.: bibl., ill., maps. (Col. Etudes mésoaméricaines, 0378–5726; II-11. Cuadernos de estudios michoacanos; 3)

Two studies of landscape change in the area of Zacapu. Cayetano Reyes focuses on impacts of an "industrial farm" established around 1900 and converted to *ejidos* in the aftermath of the Revolution. Olivier Gougeon analyzes landscapes in rural communities around Zacapu in the 20th century. Two interesting case studies of Mexican rural landscapes.

2397 Petróleo y ecodesarrollo en el sureste de México. Coordinación de Alejandro Toledo. México: Centro de Ecodesarrollo, 1982. 253 p.: bibl., ill., maps. (Serie sobre energía y sociedad; 2)

Summarizes results of a project that began in 1978. Partly funded by PEMEX, it examined environmental impacts related to development of petroleum and other activities in southeast Mexico. Covers very broad range of themes in ecology, politics, sociology, and other disciplines.

2398 Prem, Hanns J. Spanish colonization and Indian property in central Mexico, 1521–1620. (*Ann. Assoc. Am. Geogr.*, 82:3, Sept. 1992, p. 444–459, bibl., graphs, map, tables)

Review of process by which Spain acquired and allocated Indian lands during the first century of colonization in central Mexico. Provides useful summary of current knowledge of processes that led, ultimately, to the formation of Mexico's great haciendas.

2399 Reynoso, Víctor Manuel. Notas para una geografía electoral del estado de Sonora. (*Iztapalapa/México*, 11:23, julio/dic. 1991, p. 87–116, bibl., graphs, ill., maps, tables)

Elections for municipal president (1979–88) are used to identify regions based on voting behavior.

2400 Rodríguez, Victoria E. Mexico's decentralization in the 1980s: promises, promises, promises . . . (*in* Decentralization in Latin America: an evaluation. Edited by Arthur Morris and Stella Lowder. New York: Praeger, 1992, p. 127–143, bibl., map)

Primary result of the De la Madrid government's decentralization program was to diffuse federal control to the states. Municipal governments still lack revenue base necessary for true autonomy. Offers several explanations for this situation.

2401 Rodríguez Benitez, Leonel. La geografía en el Proyecto Nacional de México Independiente, 1824–1835: la fundación del Instituto Nacional de Geografía y Estadística. (*Interciencia/Caracas*, 17:3, May/June 1992, p. 155–160, bibl.)

State interests led to the creation of a scientific commission soon after independence, followed shortly by the establishment of the National Institute of Geography and Statistics and professionalization of the discipline in Mexico. Interesting and typical example of early development of geography in a Latin American country.

2402 Sánchez-Crispín, Alvaro. Las ciudades mineras de México: evolución de su población y de su población económicamente activa, 1950–1990. (*Estud. Geogr./Madrid*, 53:206, enero/abril 1992, p. 167–181, maps, tables)

Compares 13 mining centers with respect to economically active population. Total employment in mining has risen in some centers, but the percentage of economically active persons so employed has declined in all but two centers (Piedras Negras and Santa Barbara) since 1950, reflecting a decline in the relative importance of mining.

2403 Sánchez-Salazar, María Teresa. Algunas consideraciones sobre la agroindustria azucarera en México y sus principales problemas. (*Estud. Geogr./Madrid*, 53:206, enero/abril 1992, p. 183–197, bibl., graphs, maps)

Describes Mexican sugar industry following privatization of State-owned sugar mills. Contains useful data on the industry.

SCINCE. See item **42.**

Sector agropecuario. See item **43.**

2404 Servicios urbanos, gestión local y medio ambiente. Recopilación de Martha Schteingart y Luciano d'Andrea. México: El Colegio de México, Centro de Estudios Demográficos y de Desarrollo Urbano; Roma: C.E.R.FE., 1991. 479 p.: bibl., ill., maps.

Papers from a joint 1989 Mexico-Italy symposium on urban services and environment. Over half of the 24 papers focus on Mexico. Of special interest for geographers/planners are papers on the role of the State and citizen's groups in dealing with problems of urban services and environment. Also includes five case studies of large cities.

2405 Sistema de ciudades y distribución espacial de la población en México. México: Consejo Nacional de Población, 1991. 2 v. (418 p.): bibl., ill., maps.

Lengthy application of geographical theory to practical planning in Mexico. Study of the city system is accompanied by a set of detailed development project priorities. Based on 1980 census.

2406 Sluyter, Andrew. Long-distance staple transport in western Mesoamerica: insights through quantitative modeling. (*Anc. Mesoam.*, 4:2, Fall 1993, p. 193–199, bibl., graph, maps, table)

Challenge to Drennan's well-known hypothesis regarding precolumbian food staple transport. Argues that maize could have been transported profitably over long distances by supplying bearers en route with local food supplies. For archaeologist's comment see item **156.**

2407 Szasz Pianta, Ivonne. Regiones de atracción y expulsión de la población en el estado de México. (*in* Mundo rural, ciudades y población del estado de México. Coordinación de Manuel Miño Grijalva. Toluca, México: El Colegio Mexiquense; Instituto Mexiquense de Cultura, 1990, p. 483–506, bibl., graphs, maps, tables)

Study concludes that population growth rates of *municipios* in the state of Mexico are a function of Mexico's City urban expansion. Places near edges of urban expansion are growing, but places farther out in the state are losing population.

2408 Tamayo, Jesús and **Fernando Lozano Ascencio.** Las áreas expulsoras de mano de obra del estado de Zacatecas. (*Estud. Demogr. Urb.*, 6:2, mayo/agosto 1991, p. 347–379, maps, tables)

Synopsis of a report on migration to US prepared for the US Congress. Sending communities tend to be poor, rural, and traditional with respect to agricultural techniques. Much of the article deals with the state of Zacatecas.

2409 Toledo, Alejandro; Alfonso Vásquez Botello; and Mónica Herzig. El pantano: una riqueza que se destruye. México: Centro de Ecodesarrollo, 1987. 140 p.: bibl. (Serie Medio ambiente en Coatzacoalcos; 12)

Collection of essays on marshes of Gulf Coast of Mexico ranges from geochemical analysis to socioeconomic impacts.

2410 Tyrakowski, Konrad. Wallfahrer in Mexiko: eine form Traditionellen Fremdenverkehrs Zwischen "Kultischer Okonomie" und Tourismus. (*Erdkunde/Bonn*, 48:1, Marz 1994, p. 60–74)

Study of the practice of pilgrimage focuses on its origin, economic impact, and role in creating a Mexican national identity. Three sites were studied. Mexican pilgrims should not be classified with other tourists because they have a very different motivation for travel and have a different economic impact.

2411 Vachon, Michael. Onchocerciasis in Chiapas, Mexico. (*Geogr. Rev.*, 83:2, April 1993, p. 141–149)

Case study concludes that incidence of the disease—commonly known as "river blindness"—in southern Mexico is related to development of coffee farming in areas inhabited by the disease vector, a black fly. Unlike Africa, the disease seems to be restricted to a few limited areas in Latin America which are suitable habitat for the vector.

2412 Villers Ruiz, Lourdes. El impacto de los proyectos sectoriales sobre el ambiente en la República Mexicana. (*Estud. Geogr./Madrid*, 53:206, enero/abril 1992, p. 199–212, bibl. graphs, maps, tables)

Government-sponsored projects with the greatest environmental impact are large-scale tourism development, industrial ports, and basic industry. Fives areas of multiple impact are identified: Coatzacoalcos-Minatitlán, Altamira-Tampico, Lázaro Cárdenas, the Bajío, and the Mexico City metropolitan area.

2413 Watson, Rodney. Informal settlement and fugitive migration amongst the Indians of late colonial Chiapas, Mexico. (*in* Migration in colonial Spanish America. Edited by David J. Robinson. Cambridge: Cambridge Univ. Press, 1990, p. 238–278, graphs, tables)

Migration and living outside of direct Spanish control were common in Chiapas throughout the colonial period. These patterns were both a continuation of precolumbian practices and a response to colonial demands for Indian labor and tribute.

2414 West, Robert Cooper. Sonora: its geographical personality. Austin: Univ. of Texas Press, 1993. 191 p.: bibl., ill., index, maps.

Traditional historical geography by one of the masters of the genre. Book will probably be definitive work on the state for years to come.

2415 Whitmore, Thomas M. Disease and death in early colonial Mexico: simulating Amerindian depopulation. Boulder, Colo.: Westview Press, 1992. 261 p.: bibl., ill., index, map. (Dellplain Latin American studies; 28)

Computer simulation of population decline in the Valley of Mexico after the conquest fails to eliminate any of the wide-ranging estimates of population at conquest or subsequent decline, but it does assign highest probability to mid-range estimates. Technique has potential for use in other areas. For historian's comment, see *HLAS 54:1278*.

2416 Whitmore, Thomas M. and Billie Lee Turner. Landscapes of cultivation in Mesoamerica on the eve of the conquest. (*Ann. Assoc. Am. Geogr.*, 82:3, Sept. 1992, p. 402–425, bibl., maps)

Three transects (Veracruz-Mexico City, N-S through western Yucatan, and across the western highlands of Guatemala through Lake Atitlán) are used to illustrate the variety of Indian-cultivated landscapes in pre-conquest Mesoamerica. Complements Denevan's article in the same issue (item **2236**). For archaeologist's comment see item **173**.

2417 Wood, Stephanie. Gañanes y cuadrilleros formando pueblos: región de Toluca, época colonial. (*in* Mundo rural, ciudades y población del Estado de México. Coordinación de Manuel Miño Grijalva. Toluca, México: El Colegio Mexiquense; Instituto Mexiquense de Cultura, 1990, p. 93–143, bibl., maps, tables)

According to the laws of New Spain, a community of individuals without land could solicit up to 1.44 million square *varas* or more for their use. Hacienda and mine workers took advantage of this law to form *pueblos* and acquire land.

WESTERN SOUTH AMERICA

DANIEL W. GADE, *Professor of Geography, The University of Vermont*

PUBLISHED MATERIALS IN GEOGRAPHY on Western South America show little correlation with the size of each country's academic and scientific community. For example, Ecuador, less than a third the size of Colombia, receives more scholarly attention in this field. International interest in the Andes and adjoining Lowlands remains high; in fact, more than half of the monographs and articles annotated below are generated from outside the region. *Mountain Research and Development*, a quarterly produced in California but with a worldwide authorship, has emerged over the past decade as a major outlet for highland research, including that on the Andes.

Approximately one-third of the geographically-related publications on the Andean countries canvassed for this volume have been annotated. In addition to works by trained geographers, this chapter also includes research by scholars whose work reflects a basic geographical approach but who may identify primarily with other disciplines, since the pigeonholing of knowledge into the traditional disciplines is becoming increasingly arbitrary.

Major items to be highlighted here include two historical-geographical pieces on Colombia (items **2451** and **2461**), and the perception of the threat from the Cotopaxi volcano in Ecuador (item **2468**). The compelling interest which cultural geography holds for a serious understanding of Andean land and life, both past and present, is especially manifested in items **2461, 2426, 2507, 2519, 2519,** and **2424.**

GENERAL

2418 Baied, Carlos A. and **Jane C. Wheeler.**
Evolution of high Andean ecosystems: environment, climate, and culture change over the last 12,000 years in the Central Andes. (*Mt. Res. Dev.*, 13:2, May 1993, p. 145–156)

Changes in pollen profiles derived from a high Andean site at 4,000 m were used to infer timing of human arrival (from 10,000–9,000 BP) and the shift from food-gathering to food-producing economy (4,000–3,000 BP).

2419 Brawer, Moshe. Atlas of South America. Scales differ. New York; London: Simon & Schuster, 1991. 1 atlas (144 p.): bibl., ill. (some col.), indexes, col. maps.

Half illustrations (maps, charts, graphs) and half text, most statistics are from the mid-1980s. Disappointing cartography and a pedestrian text. Country and city maps lack detail that many atlas users seek.

2420 Clay, Jason W. Indigenous peoples and tropical forests: models of land use and management from Latin America. Cambridge, Mass.: Cultural Survival, 1988. 116 p.: bibl., ill. (Cultural survival report; 27)

Sets out research priorities to further understand indigenous peoples land use practices with the aim of preserving non-industrial ways of life of tropical America: gathering, hunting, fishing, swidden agriculture, and permanent farming.

2421 Comprendre l'agriculture paysanne dans les Andes centrales: Pérou et Bolivie. Édition de Pierre A. Morlon. Paris: Institut National de la Recherche Agronomique, 1992. 1 v.

Excellent collection on Andean peasant agriculture that features perspective and logic of the inhabitants rather than the researcher. Essays are organized into sections on: 1) *chaquitaclla* ("foot plow") and its persistence; 2) social organization and land use; 3) peasant yields; 4) production systems; and 5) peasant economy.

2422 Dollfus, Olivier. Territorios andinos: reto y memoria. Lima: Instituto Francés de Estudios Andinos; Instituto de Estudios Peruanos, 1991. 221 p.: bibl., ill., maps. (Historia andina; 18) (Travaux de l'Institut français d'études andines; 46)

Reflections on spatial, physical, and human character of Andean spaces from Colombia to Bolivia and how the four states have organized their respective territories. Largely a period piece, it ignores much new knowledge generated over the past two decades.

2423 Espinosa Goitizolo, Reinaldo. Atlas mínimo histórico, biográfico y militar Simón Bolívar. Guillermo Grau Guardarrama, compilador cartógrafo e investigador. Nelsy Babiel Guitiérrez, asesora literal. La Habana: Editorial Pueblo y Educación; Instituto Cubano de Geodesia y Cartografía, 1988. 63 p.: bibl., ill. (some col.), col. maps.

Attractively produced atlas on the life and military campaigns of Simón Bolívar. Chronologically organized text accompanies excellent maps.

2424 Gade, Daniel W. Landscape, system, and identity in the post-conquest Andes. (*Ann. Assoc. Am. Geogr.*, 82:3, Sept. 1992, p. 460–477, bibl., map, photos)

Since the 16th century, peasant livelihoods and material culture in Highland Peru, Ecuador, and Bolivia have been characterized by a duality of material traits as much southern European in origin as indigenous Andean.

2425 Gade, Daniel W. Leche y civilización andina: en torno a la ausencia del ordeño de la llama y alpaca. (*Yearbook/CLAG*, 19, 1993, p. 3–14)

Why Andean people never milked their two domesticated camelids is a cultural historical puzzle. Lack of an impetus to establish milk as a sacred product for the priestly caste might be the key in explaining this failure. In the prehistoric Near East cows began to be milked for religious and not economic reasons.

2426 Knapp, Gregory. Cultural and historical geography of the Andes. (*Yearbook/CLAG*, 17/18, 1990, p. 165–175, bibl.)

Positive assessment of Andeanist research in geography published during the 1980s based on a bibliography of 175 sources.

2427 Lancini V., Abdem Ramón. Alejandro de Humboldt: semblanza del viajero del Orinoco y de la Siberia. Venezuela?: Lagoven, S.A.—Filial de Petroleos de Venezuela, División de Oriente, 1989. 84 p.: bibl., photos.

Well-illustrated biography of the great German geographer (1769–1859) who made brilliant observations about Western South America and elsewhere.

2428 Lauer, Wilhelm. Human development and environment in the Andes: a geoecological overview. (*Mt. Res. Dev.*, 13:2, May 1993, p. 157–166)

Interpretive piece in the mold of Carl Troll on how Andean climate, vegetation, and land use determine and express verticality.

2429 Masson Meiss, Luis et al. De papa a patata: la difusión española del tubérculo andino. Edición científica e iconográfica de Javier López Linage. Barcelona: Lunwerg Editores, 1991? 213 p.: bibl., col. ill., col. maps.

Beautifully illustrated work on the potato largely in Andean culture, but the most original sections are on its history in Spain.

2430 Recursos naturales andinos. Edited by Shozo Masuda. Tokyo: Univ. de Tokio, 1988. 342 p.: bibl., ill., index.

Nine cultural historical papers on Western South America include coastal fishing, algae collection, potatoes, llamas, and desert resources.

2431 Las sabanas americanas: aspecto de su biogeografía, ecología y utilización. Simposio organizado en Guanare, Venezuela, bajo el auspicio del Programa Década de los Trópicos (IUBS, MAB-UNESCO). Recopilación de Guillermo Sarmiento. Mérida, Venezuela: Centro de Investigaciones Ecológicas de los Andes Tropicales, Facultad de Ciencias, Univ. de los Andes, 1990. 332 p.: bibl., ill., maps.

Eleven papers from a 1987 international ecological symposium, with most papers devoted to Brazil and the Llanos of Colombia and Venezuela.

2432 Seltzer, Geoffrey O. Late Quaternary glaciation as a proxy for climatic change in the Central Andes. (*Mt. Res. Dev.*, 13:2, May 1993, p. 129–138, ill.)

Radiocarbon-dated peat found in moraines was used to determine that snowlines in Andean glaciers were substantially raised between 12000 and 8000 BP. This deglaciation provides indirect evidence for a warming climate.

2433 **Stadel, Christoph.** Altitudinal belts in the tropical Andes: their ecology and human utilization. (*Yearbook/CLAG*, 17/18, 1990, p. 45–60, bibl., graphs, ill., maps, tables)

Verticality has defined Andean environmental knowledge since the time of Alexander von Humboldt. Author's remarks on the human utilization of these zones focus on Ecuador.

VENEZUELA

2434 **Amaya, Carlos Andrés.** Geografía urbana de una ciudad: el caso de Mérida. Mérida, Venezuela: Consejo de Publicaciones, Univ. de los Andes, 1989. 105 p.: bibl., ill., maps. (Textos de la Universidad de los Andes. Col. Tecnología. Serie Geografía)

Stages of growth, public services, and functional areas of an important western Venezuelan city.

2435 **Amend, Stephan.** Parque Nacional El Avila. Caracas: s.n., 1991. 186 p.: bibl., ill., maps (some col.). (Parques nacionales y conservación ambiental, 0798-2887; 2)

Founded in 1958, El Avila National Park covers 85,000 ha north and east of Caracas. Towns and farms that still lie within its boundaries create land-use conflicts.

2436 **Briceño Monzillo, Bernardo.** Para la historia hidrológica del Río Apure: la "inundación grande" de 1672. (*Rev. Geogr./Mérida*, 31, 1990, p. 81–94, bibl.)

The exceptional flood of the Apure River in June 1672 was the theme of colonial documents which give insight into the Llanos cattle trade of the period.

2437 **Cressa, Claudia *et al.*** Aspectos generales de la limnología en Venezuela. (*Interciencia/Caracas*, 18:5, sept./oct. 1993, p. 237–248, bibl., tables, maps)

Useful overview of lakes, rivers, and coastal lagoons with data on their water chemistry, biological productivity, pollution, and the effects of dam construction.

2438 **Escobar N., Marcos E.** and **Manuel Martínez S.** Los depósitos de carbón en Venezuela. (*Interciencia/Caracas*, 18:5, sept./oct. 1993, p. 224–229, bibl., graphs, maps, tables)

Venezuela swims in oil, but it also has lots of coal. Most of the production from abundant reserves in the states of Zulia and Táchira states is exported.

2439 **Ferrer Oropeza, Carlos *et al.*** Una visión geográfica del trayecto Mérida-Laguna de Mucubají, Estado Mérida, Venezuela. (*Rev. Geogr. Venez.*, 32:1, 1991, p. 117–148, bibl. maps)

Geographical synthesis of a 50-km transect in the Venezuelan Andes moving from the city of Mérida (1612 m) to Lake Mucubají (3560 m).

2440 **Lancini V., Abdem Ramón.** Alejandro de Humboldt: el viajero del Orinico, en el Oriente de Venezuela. Venezuela?: Lagoven, S.A.—Filial de Petróleos de Venezuela, División de Oriente, 1988. 44 p.: bibl., photos.

Richly illustrated booklet retraces Humboldt's journey to eastern Venezuela.

2441 **Meyer, Henry O.A.** and **Malcolm E. McCallum.** Diamonds and their sources in the Venezuelan portion of the Guyana Shield. (*Econ. Geol.*, 88:5, Aug. 1993, p. 989–998)

Diamond-bearing alluvial deposits are found in most streams south of the Orinoco River. Their geological origin may not relate to diamonds found in West Africa as had been reported.

2442 **Morales Mena, Faustino.** El rastrojo social en la depresión del Lago de Valencia: casos de Valencia y Maracay. Prólogo de Isbelia Sequera Segnini. Caracas: Academia Nacional de Ciencias Económicas, 1990. 439 p., 4 folded leaves of plates: bibl., ill., maps.

Land no longer used for agriculture but not yet urbanized is considered to be in "social fallow." That phenomenon is examined for the outskirts of Valencia and Maracay, two cities which have undergone rapid population increase and rampant land speculation.

2443 **Muller-Karger, Frank** and **Ramón J. Varela.** Influjo del Río Orinoco en el Mar Caribe: observaciones con el CZCS desde el espacio. (*Mem. Soc. Cienc. Nat.*, 49:131/132, enero/dic. 1989 and 50:133/134, enero/dic. 1990, p. 361–390, bibl., maps, photos, table)

Surface waters originating in the Orinoco and Amazon Rivers flow northward into the Caribbean Sea, but their trajectories change with the seasons.

2444 **Olivo Chacín, Beatriz.** Geografía de la región insular y del mar venezolano. Caracas: Editorial Ariel-Seix Barral Venezolana, 1989. 245 p.: bibl., ill. (Col. Geografía de Venezuela nueva)

Synthesis of Venezuela's offshore interests, part of a series of regional geographies of the country, ranges from geological origins to touristic impacts.

2445 **Perera, Miguel A.** Actividad cauchera e impacto ambiental en el Territorio Federal Amazonas, Venezuela. (*Rev. Esp. Antropol. Am.*, 20, 1990, p. 221–250, bibl., graphs, maps, tables)

Traces exploitation of the *Hevea brasiliensis* tree in Amazonas Territory from the 1860s–1960s. Demographic and environmental impacts of the rubber boom are discussed.

2446 **Price, Marie.** Hands for coffee: migrants and western Venezuela's coffee economy, 1870–1930. (*J. Hist. Geogr.*, 20:1, Jan. 1994, p. 62–80)

Archival marriage records are used to plot migrations of Venezuelan and Colombian peasants to the *tierra templada* to establish coffee farms. Sets the record straight on a misinterpreted aspect of Venezuela's *fiebre de café*.

2447 **Pulwarty, Roger S.; Roger G. Barry; and Herbert Riehl.** Annual and seasonal patterns of rainfall variability over Venezuela. (*Erdkunde/Bonn*, 46:3/4, 1992, p. 273–289, graphs, tables)

Using 1972–86 data from 60 Venezuelan weather stations, authors favor a more complex explanation than those previously proposed to account for some anomalous climatic patterns.

2448 **Sarmiento, L.; M. Monasterio; and M. Montilla.** Ecological bases, sustainability, and current trends in traditional agriculture in the Venezuelan High Andes. (*Mt. Res. Dev.*, 13:2, May 1993, p. 167–176, ill.)

Traditional fallowing at 3,000–4,000 m above sea level as observed at Gavidia creates mosaic landscape in which some fields are cropped. Introduction of chemical fertilizers threatens this ecologically adaptive farming system.

2449 **Schubert, Carlos.** The glaciers of the Sierra Nevada de Mérida, Venezuela: a photographic comparison of recent deglaciation. (*Erdkunde/Bonn*, 46:1, 1992, p. 58–64, ill.)

Over the past century, the snowline has risen from 4100 to more than 4700 m above sea level. At least three glaciers have disappeared since 1972. Raises possibility that atmospheric pollution in Mérida may be contributing to the accelerated glacier melt.

2450 **Veillon, Jean-Pierre.** Los bosques naturales de Venezuela. v. 1, El medio ambiente. Mérida, Venezuela: Instituto de Silvicultura, Univ. de los Andes, 1989. v. 1: bibl., ill. (some col.), maps.

Identifies character and distribution of natural vegetation as climatically controlled. However, that can only be part of the explanation.

COLOMBIA

2451 **Aprile Gniset, Jacques.** La ciudad colombiana v. 1, Prehispánica, de conquista e indiana. Bogotá: Banco Popular, Fondo de Promoción de la Cultura, 1991–1992. 2 v.: bibl., ill., maps. (Biblioteca Banco Popular. Colección Textos universitarios)

Rich feast of data and interpretation on form and function of human settlement in Colombia. Fresh perspectives on precolumbian settlement types, the plaza in the town fabric, and demographic changes of towns. Special attention is given to Cali, but the work contains a host of insights on many other Colombian cities.

2452 **Biota y ecosistemas de Gorgona.** Edición de Jaime Aguirre C. y Orlando Rangel Ch. Bogotá: Fondo FEN Colombia, 1990. 303 p.: ill. (some col.), maps (some col.).

Good natural history survey of the island of Gorgona, formerly a high-security prison and currently a national park since 1984.

2453 **Colombia. Ministerio de Gobierno.** Política del Gobierno Nacional para la defensa de los derechos indígenas y la conservación ecológica de la Cuenca Amazónica. Bogotá: Ministerio de Gobierno, 1990. 238 p.: appendices, bibl.

Covers policies of the Colombian government toward highland and lowland indigenous communities and how that relates to ecological conservation in the Amazon Basin.

Legislation in 1989 granted tribes formal land rights to a sizeable area of Amazonian rainforest.

2454 Colonización del bosque húmedo tropical. Bogotá?: Fondo de Promoción de la Cultura; Bogotá: COA, 1990? 303 p.: bibl., ill., maps.

Fourteen of 16 essays deal with colonization in the Amazon, Pacific, and Caribbean parts of Colombia.

2455 Correa, I.; B.S. Acosta; J. Sionneau; and J. Khobzi. Evaluación global del potencial para acuicultura de la franja costera del pacífico colombiano: sector Buenaventura-frontera con Ecuador. (*Rev. Univ. EAFIT*, 88, oct./nov./dic. 1992, p. 65–74, bibl., maps, table)

Using satellite imagery and other data sources, research team assesses natural conditions for intensive shrimp farming along the littoral from Buenaventura to the Ecuadorian border. Coastal areas with clay soil offer the best possibilities.

2456 Guhl, Ernesto. Las fronteras políticas y los límites naturales. Bogotá: Fondo FEN Colombia, 1991. 379 p.: bibl., ill., maps.

Wide-ranging study of political boundaries by Colombian geographer emphasizes those of Colombia.

2457 Londoño Paredes, Julio. La frontera terrestre colombo-venezolana: el proceso de la fijación de 1492–1941. Bogotá: Banco de la República, 1990. 498 p.: bibl., col. maps. (Historia colombiana) (Col. bibliográfica Banco de la República)

Analyzes land boundary between Venezuela and Colombia from a Colombian perspective.

2458 Marín Ramírez, Rodrigo. Estadísticas sobre el recurso agua en Colombia. 2. ed. Santa Fe de Bogotá: Ministerio de Agricultura, Instituto Colombiano de Hidrología, Meteorología y Adecuación de Tierras, 1992. 412 p.: bibl., col. ill., col. maps.

Useful assemblage of data on water in Colombia, ranging from kinds of water resources, data collection network, water use, problems associated with water, water management, and legal aspects of water resources.

2459 Mendoza Morales, Alberto. El ordenador: metodología del ordenamiento territorial. Bogotá: Corporación Univ. Piloto de Colombia, 1992. 187 p.: ill. (some col.).

Colombia's geography, more complicated than any other Latin American country, is made even more complex by jurisdictional and natural boundaries that often do not coincide. Abundance of maps in this work helps to define issues of space, territory, and region. The Sabana de Bogotá gets special attention.

2460 Montañez G., Gustavo; Oscar Arcila N.; and Juan Carlos Pacheco. Urbanización y conflicto en la Sabana de Bogotá. (*Coyunt. Soc.*, 3, nov. 1990, p. 131–151, bibl., graph, map, table)

Excellent study of causes and effects of urbanization in 26 *municipios* that comprise the Sabana de Bogotá. Outlying *municipios* are growing fast through immigration to work in intensive floriculture and as dormitory suburbs for the capital.

2461 Plazas, Clemencia; Ana Maria Falchetti; Thomas van der Hammen; and Pedro Botero. Cambios ambientales y desarrollo cultural en el bajo Río San Jorge. (*Boletín/Bogotá*, 20, enero/abril 1988, p. 55–88, bibl., ill., maps, photos, tables)

Brilliant piece on the ancient network of man-made canals draining 500,000 ha of the lower San Jorge River basin. Construction of this hydraulic system began around 800 BC when a great drought occurred. Work continued at least until the 10th century AD.

2462 Prahl, Henry von; Jaime R. Cantera; and Rafael Contreras. Manglares y hombres del Pacífico colombiano. Bogotá: COLCIENCIAS; FEN Colombia, 1990. 193 p., 5 leaves of plates: bibl., ill. (some col.).

Fetching presentation of botany and ecology of mangrove species, animal adaptations, and relationships of people and mangroves on the Pacific coast of Colombia.

ECUADOR

2463 Amazonía nuestra: una visión alternativa. Recopilación de Lucy Ruiz M. Quito: CEDIME; ABYA-YALA; ILDIS, 1991. 552 p.: bibl., maps.

Uneven yet useful papers from a symposium on the Amazonian environment, most dealing with Ecuador.

2464 Amazonía presente y – ?. Quito: Ed. Abya-Yala; Grupo Ecológico Tierra Viva; ILDIS, 1988? 420 p.: bibl.

Seventeen papers on the Ecuadorian Amazon cover ecology and development, 1987 earthquakes, environmental problems, and planning policies.

2465 Bendix, Jörg and **Wilhelm Lauer.** Die Niederschlagsjahrezeiten in Ecuador und ihre klimadynamische Interpretation. (*Erdkunde/Bonn*, 42:2, 1992, p. 118–134, graphs, tables)

Four main and two transitional types of rainfall regions in Ecuador are explained by means of a correlation analysis between the annual variation of rainfall, wind direction, and frequency of thunderstorms.

2466 Bock, Marie S. Guayaquil: arquitectura, espacio, y sociedad, 1900–1940. Quito: Corporación Editora Nacional, 1992. 130 p.: bibl., ill., maps. (Biblioteca de ciencias sociales; 38)

Historical study of urbanization in Ecuador's largest city since 1900. Major attention to definition of neighborhoods and building styles.

2467 Carrión, Fernando M. Quito: una política urbana alternativa. (*Medio Ambient. Urban.*, 9:38, marzo 1992, p. 55–69, maps)

Quito's historic core has become increasingly peripheral to the life of the city. Author gives insights to different plans for preventing its deterioriation so that its colonial character can be preserved.

2468 D'Ercole, Robert and **Juan Fernando Moncayo.** "Influents locaux" face à une situation d'urgence: une analyse selon l'hypothèse d'une éruption du volcan Cotopaxi (Équateur). (*Bull. Inst. fr. étud. andin.*, 20:1, 1991, p. 181–220, bibl., maps, tables)

Remarkable study of attitudes in the face of potential disaster. Extensive interviewing of town notables within 55 km of Cotopaxi volcano shows absence of "emergency thinking."

2469 Dugard, Jane. The influence of the Galápagos Islands on Charles Darwin and the subsequent development of his theory of evolution. (*UNISA/Lat. Am. Rep.*, 7:2, 1991, p. 4–15, bibl., ill., photos, port.)

Well-written, non-technical piece describes Darwin's fabulous insights into organic life on the Galápagos better than most papers.

2470 Fernández, María Augusta. Ecuador. (*in* Latin American urbanization: historical profiles of major cities. Edited by Gerald Michael Greenfield. Westport, Conn.: Greenwood Press, 1994, p. 215–251, bibl., map, tables)

Concise introduction to urbanization in Ecuador. Vignettes on Cuenca, Guayaquil, Machala, Quito, and Santo Domingo de los Colorados suggest diverse range of the reasons for and timing of urban growth.

2471 Flujos geográficos en el Ecuador: intercambios de bienes, personas e información. Coordinación de Juan León V., Alba Moya y Pierre Peltre. Quito: Corporación Editora Nacional; Colegio de Geógrafos del Ecuador, 1989. 111 p.: bibl., ill. (Estudios de geografía; 1)

Collection of essays on wheat imports, market flow of foodstuffs, movement of lumber, movement of cargo at ports, migration networks and movement of migrants, and international interchange.

2472 Harden, Carol P. Land use, soil erosion, and reservoir sedimentation in an Andean drainage basin in Ecuador. (*Mt. Res. Dev.*, 13:2, May 1993, p. 177–184, ill.)

Most slope erosion in the Paute drainage basin comes from abandoned farm lands. Displaced sediment washed from those slopes clogs the Amaluza reservoir whose dam supplies half of Ecuador's electric power.

2473 Jordan, Ekkehard. Die Mangrove Ecuadors. (*Geogr. Rundsch.*, 43:11, Nov. 1991, p. 664–671, graphs, maps, tables)

Excellent overview of mangroves which occupy 2,600 km^2 of the Ecuadorian coast in nine discrete areas, the largest of which is in the Gulf of Guayaquil. Expansion of aquaculture threatens their future.

2474 Klak, Thomas. Recession, the state and working-class shelter: a comparison of Quito and Guayaquil during the 1980s. (*Tijdschr. Econ. Soc. Geogr.*, 83:2, 1992, p. 120–137, bibl., graphs, tables)

Quito was able to withstand recessionary impacts of the 1980s through State oil revenues, whereas Guayaquil was largely excluded from central government funding. Overstatement of why working class housing in Guayaquil is more inadequate than in Quito.

2475 **Long, Brian.** Conflicting land-use schemes in the Ecuadorian Amazon: the case of Sumaco. (*Geography/London,* 77:4, Oct. 1992, p. 336–348, bibl., maps)

The Sumaco project in Napo Province was established in 1988 under Protestant Church auspices to resettle families displaced in the 1987 earthquakes. Emergency scheme has conflicted with a plan to preserve tropical forests.

2476 **Lowder, Stella.** The role of intermediate cities in decentralization: observations from Ecuador. (*in* Decentralization in Latin America: an evaluation. Edited by Arthur Morris and Stella Lowder. New York: Praeger, 1992, p. 59–75, bibl., map, table)

Provincial capitals of Cuenca, Loja, Machala, Ambato, and Santo Domingo de los Colorados are discussed to show how State allocation has, in concert with elite goals and production structures, created a range of economic outcomes in Ecuadorian cities.

2477 **Lozano Castro, Alfredo.** Quito, ciudad milenaria: forma y símbolo. Quito: Ediciones Abya-Yala; Ciudad, Centro de Investigaciones; Madrid: Centro de Investigación Urbana y Arquitectura Andina, 1991. 262 p.: bibl., ill., maps.

Imaginatively mining the early colonial chronicles, this book draws out the pre-conquest roots of Quito whose morphological configuration offers keys to understanding transculturation in Andean America beginning with the Spanish conquest.

2478 **Nieto C., Carlos.** The preservation of foods indigenous to the Ecuadorian Andes. (*Mt. Res. Dev.,* 13:2, May 1993, p. 185–188)

Program of the 1980s preserved germplasm of 12 Andean crops and cultivars in one of three ways: cold storage, field cultivation, and in vitro. Of these crops, quinoa is considered to offer the most future potential.

2479 **Nieuwolt, Simon.** Climatic uniformity and diversity in the Galapagos Islands and the effects on agriculture. (*Erdkunde/Bonn,* 45:2, 1991, p. 134–142)

Agriculture, the major source of income of many of the 6,000 inhabitants of these islands, is subject to severe limitations. Irrigation is essential in the lowlands, but even in the highlands, dry spells pose risk to crop growing.

2480 **El paisaje volcánico de la sierra ecuatoriana: geomorfología, fenómenos volcánicos y recursos asociados.** Coordinación de Patricia Mothes. Quito: Corporación Editora Nacional; Colegio de Geógrafos del Ecuador, 1991. 92 p.: bibl., ill., maps. (Estudios de geografía; 4)

Seven interesting articles on Ecuador's volcanic setting, mostly from a human perspective that involve both dangers of eruption and value of soils, rocks, and geothermal energy derived from vulcanism.

2481 **Peltre, Pierre.** Risque morphoclimatique urbain à Quito, Equateur, 1900–1988. (*Espace géogr.,* 21:2, 1992, p. 123–136)

In 88 years more than 550 disastrous episodes (floods, sidewalk cave-ins and mudflows) have occurred in the city. They are directly connected to the replacement of natural drainage by a sewer system incapable of carrying away surges of rainwater.

2482 **Southgate, Douglas DeWitt** and **Morris D. Whitaker.** Promoting resource degradation in Latin America: tropical deforestation, soil erosion, and coastal ecosystem disturbance in Ecuador. (*Econ. Dev. Cult. Change,* 40:4, July 1992, p. 787–807, bibl., map, tables)

Using examples from lowland Ecuador, author identifies what he considers to be reasons for resource degradation: inappropriate land tenure arrangements, government intervention in markets (including subsidies), and failure to invest in research.

2483 **Teltscher, Suzanne.** Gender differences in Ecuador's urban informal economy: survival to upward mobility. (*Yearbook/CLAG,* 19, 1993, p. 81–91)

Author interviewed vendors at the Calle Ipiales Market in Quito to determine how their economic situation is related to capital resources, kinds of sales units, and gender differences. More women than men are poor.

2484 **Wesche, Rolf.** Ecotourism and indigenous peoples in the resource frontier of the Ecuadorian Amazon. (*Yearbook/CLAG,* 19, 1993, p. 31–41)

Describes both lodge and backpacker types of ecotourism in Napo Province and focuses on lowland Quichua initiatives to control part of that activity. Ecotourism is viewed as a proactive strategy for native peoples who want to protect their forest environment.

PERU

2485 Barclay, Frederica. La colonia del Perené. Iquitos, Perú: Centro de Estudios Teológicos de la Amazonía, 1989. 258 p.: bibl., ill., maps. (Debate amazónico; 4)

In 1891 the Peruvian government ceded two million ha of forested land east of the Cordilleras to the British-held Peruvian Corporation. One part along the Perené River was used for agricultural colonization. Competent historical geography of early capitalist enterprise.

2486 Bedoya Garland, Eduardo. Las causas de la deforestación en la Amazonía peruana: un problema estructural. Lima: Centro de Investigación y Promoción Amazónica, 1991. 130 p.: bibl., ill. (Documento; 12)

Fine discussion of deforestation at different scales: Amazon drainage of Peru; the upper selva; and the Huallaga Valley where the forest has given way to coca plantings.

2487 Brisseau-Loaiza, Jeanine. Evolution des modèles de consommation et stratégies alimentaires au Pérou depuis les années 60. (*Ann. géogr./Paris*, 102:569, 1993 p. 53–76)

Peruvian government policies on food and agriculture during the García Administration failed miserably due to the application of an urban framework of consumption.

2488 Burga, Jorge and **Claire Delpech.** Villa El Salvador: la ciudad y su desarrollo; realidad y propuestas. San Isidro, Peru: Centro de Investigación, Educación y Desarrollo, 1988. 127 p.: bibl., ill. (some col.).

Fascinating study of changes in a satellite city of Lima founded as a shantytown in 1971. Emphasizes land use and zoning, improvements in the built environment, delivery of urban services, and a development plan designed to solidify the economic base and enhance liveability.

2489 Córdova Aguilar, Hildegardo. El desarrollo de la geografía cultural en el Perú. (*Espac. Desarro.*, 3, 1991, p. 57–69)

Peruvian geographer at the Univ. Católica discusses research themes of the past and proposes some for the future.

2490 Córdova Aguilar, Hildegardo. Firewood uses and the effect on the ecosystem: a case study of the Sierra de Piura, northwestern Peru. (*GeoJournal/Boston*, 26:3, 1991, p. 297–309)

Data on firewood use, quality of wood, time consumed in collection, and commercialization are provided for six peasant communities at different elevations. Strong need for firewood for cooking clashes with erosion created by cutting forests on steep slopes.

2491 Dávila-Flores, José and **Percy Jiménez-Milón.** El valle del Colca (Arequipa, Perú): zonas de vida natural, conservación y manejo. (*Bol. Lima*, 10:57, mayo 1988, p. 65–73, bibl., maps, tables)

Sketchy though it is, article lays out environmental diversity of the magnificent Colca canyon.

2492 Denevan, William M. The 1931 Shippee-Johnson aerial photography expedition to Peru. (*Geogr. Rev.*, 83:3, 1993, p. 238–251)

Account of a remarkable expedition in which thousands of photographs of settlement, land use, archaeological ruins, and physical landscape features were taken. Technically superb, these photos have taken on a new life as scholarly documents, as the author has shown in his own research.

2493 Dourojeanni, Marc J. Amazonia, ¿qué hacer? Iquitos, Perú: Centro de Estudios Teológicos de la Amazonia, 1990. 444 p., 58 p. of plates: bibl., ill. (some col.).

Major tome on past, present, and future of Amazonian natural resources by leading Peruvian scientist with impressive command of the literature. Believes that agriculture, forestry, and wildlife use may save—not destroy—the region if sensibly applied.

2494 Driant, Jean-Claude. Las barriadas de Lima: historia e interpretación. Traducción del francés de Zaida Lanning de Sánchez. 231 p.: bibl., ill., maps. 24 cm. Lima: IFEA; DESCO, (Travaux de l'Institut français d'études andines, 0768–424X; t. 60)

Describes evolution of Lima's shantytowns and analyzes government housing policies that encouraged them. Views *barriada* concept as a massive failure, yet claims that even if rural-to-city migration were to cease, more *barriadas* would have to be created to house the next generation of people who were born in them.

2495 **Echavarría, Fernando R.** Cuantificación de la deforestación en el Valle del Huallaga, Perú. (*Rev. Geogr./México*, 114, julio/dic. 1991, p. 37–53, bibl., graphs, maps, tables)

Comparison of vertical air photos for a 1,000 sq. km. sector between Tocache and Tingo María reveals that between 1963–76 deforested land rose from five to 30 percent. Expansion of coca growing was a leading cause of this change. Methodology of this study described in some detail.

2496 **Emperaire, L.** Végétation et action anthropique dans le département de Piura, Pérou. (*Bull. Inst. fr. étud. andin.*, 19:2, 1990, p. 335–349, bibl., graphs, map, tables)

Firewood and construction materials are derived from trees and shrubs in a range of plant formations from open woodland to savanna that correspond to an elevational gradient from zero to 500 m.

2497 **Goland, Carol.** Field scattering as agricultural risk management: a case study from Cuyo Cuyo, Department of Puno, Peru. (*Mt. Res. Dev.*, 13:4, Nov. 1993, p. 317–338)

Methodologically sophisticated study presents evidence to indicate that dispersion of fields is an effective response to the risks inherent in Andean farming.

2498 **Hocquenghem, Anne-Marie** and **Ortlieb, L.** Pizarre n'est pas arrivé au Pérou durant une année *El Niño*. (*Bull. Inst. fr. étud. andin.*, 19:2, p. 327–334, bibl., maps)

Refutation of 1987 article which asserted that Francisco Pizarro arrived on the north Peruvian coast during an *El Niño* event characterized by torrential rainfall where normally none occurs.

2499 **Johannessen, Sissel** and **Christine Ann Hastorf.** A history of fuel management (A.D. 500 to the present) in the Montaro Valley, Peru. (*J. Ethnobiol.*, 10:1, 1990, p. 61–92)

Demonstrates that even such a practice as fuel procurement in the tree-poor Andes has a continuity with the past.

2500 **Kent, Robert B.** Geographical dimensions of the Shining Path insurgency in Peru. (*Geogr. Rev.*, 83:4, Oct. 1993, p. 441–454)

Spatial and temporal spread of this fearsome organization from its original core in Ayacucho in 1980. Written in crystal clear prose, article serves as an excellent summary of the movement.

2501 **Kent, Robert B.** Peru. (*in* Latin American urbanization: historical profiles of major cities. Edited by Gerald Michael Greenfield. Westport, Conn.: Greenwood Press, 1994, p. 448–467, bibl., map, tables)

Concise summary, chronically organized, of the urban phenomenon in Peru. Good capsule profiles of Arequipa, Lima, and Trujillo.

2502 **Kuznar, Lawrence A.** Mutualism between Chenopodium, herd animals, and herders in the south central Andes. (*Mt. Res. Dev.*, 13:3, Aug. 1993, p. 257–265)

Brilliant deduction points to quinoa, whose wild ancestors have weedy tendencies, as having been domesticated in camelid corrals so rich in nutrients.

2503 **Loker, W.M. et al.** Identification of areas of land degradation in the Peruvian Amazon using a geographic information system. (*Interciencia/Caracas*, 18:3, May/June 1993, p. 133–141, bibl., graph, maps, table)

Demonstration of possibilities of GIS technology in handling data on crop distributions, agroecosystems, soils, and agricultural capability in the four easternmost departments of Peru.

2504 **López-Ocón Cabrera, Leoncio.** Medio siglo de actividades ciéntificas de la Sociedad Geográfica de Lima. (*Interciencia/Caracas*, 17:3, May/June 1992, p. 147–154, bibl.)

The Geographical Society of Lima, founded in 1888 by the positivist elite, has played major role over the last century in formulating and diffusing knowledge about Peru's natural resources and regions.

2505 **Maskrey, Andrew.** El manejo popular de los desastres naturales: estudios de vulnerabilidad y mitigación. Lima: Tecnología Intermedia, 1989. 208 p.: bibl., ill.

Three aspects of natural hazards are discussed: 1) Lima's vulnerability to earthquakes; 2) landslides and floods in the Rimac Valley; and 3) how such tragedies might be mitigated.

2506 **Novoa Goicochea, Daniel I.** La urbanización en el trópico húmedo de la región Inka. (*Espac. Desarro.*, 4, 1992, p. 43–72)

Cities in the departments of Cusco and Madre de Dios are a 20th-century phenomenon. Only Quillabamba and Puerto Maldonado, have surpassed 25,000 people.

2507 Orlove, Benjamin. The ethnography of maps: the cultural and social contexts of cartographic representation in Peru. (*Cartographica/Ontario*, 30:1, Spring 1993, p. 29–46)

Critical analysis of natural landscape and social groups on five maps of Lake Titicaca. Three are drawn by government officials and two by peasants. Shows how resources are perceived differently by various groups and how maps can become a vehicle for empowerment.

2508 Radcliffe, Sarah A. Spontaneous population decentralization in Peru. (*in* Decentralization in Latin America: an evaluation. Edited by Arthur Morris and Stella Lowder. New York: Praeger, 1992, p. 39–58, bibl., map, tables)

Shifts in migration flows have progressively increased percentage of urban and rural population in coastal and jungle regions. Population in many sierran cities has grown substantially, contrasting with stagnation and decline of rural areas in that region.

2509 Ravines, Tristán. Calles de Cajamarca: historia de su crecimiento urbano. (*Bol. Lima*, 14:82, julio 1992, p. 55–74, map, photos)

Historical-geographical study of Cajamarca's streets, focusing on name changes and notable buildings, institutions and residents.

2510 Ricalde, David G. Sobre el santuario histórico de Machu Picchu. (*Bol. Lima*, 10:60, nov. 1988, p. 61–66, bibl., map, photos)

Machu Picchu, Peru's prime tourist attraction, has environmental problems that threaten the integrity of the site, among them forest fires, slope erosion, uncontrolled growth of nearby settlements, and decline in its flora and fauna.

2511 Samanez Argumedo, Roberto. Ciudad de Cuzco. (*Medio Ambient. Urban.*, 9:38, marzo 1992, p. 99–108, maps, photos)

In 1950 Cusco occupied an area of 300 ha and had 50,000 inhabitants; by 1990 it covered 2,500 ha and contained 300,000 people. Such growth, along with earthquakes in 1950 and 1986 and subsequent reconstruction, have compromised its historic fabric.

2512 Santillana Cantella, Tomás Guillermo. Los viajes de Raimondi. Lima: Occidental Petroleum Company of Peru, 1989. 222 p., 13 leaves of plates: ill. (some col.), maps.

Severely edited log of journeys based on published versions of Raimondi's travels from 1851–69 in all three major regions of Peru. His original manuscripts were destroyed in the 1943 fire at Lima's National Library.

2513 Segura Altamirano, José. Problemática del agua en la Sierra de Salas y la búsqueda de alternativas. (*Alternativa/Chiclayo*, 16, nov. 1991, p. 9–50, bibl., ill., tables)

Well-crafted history of irrigation in Salas, a community on the lower west flank of the Andes. Greater prosperity depends on improvement of irrigation infrastructure to minimize water loss.

2514 Seminario-Taller Tecnología Apropiada para la Mitigación de Desastres, Moyobamba, Peru, 1990. Los desastres sí avisan: estudios de vulnerabilidad y mitigación II. Ponencias presentadas al Seminario-Taller "Tecnología Apropiada para la Mitigación de Desastres," realizado en Moyobamba del 27 al 29 de agosto de 1990. Edición de Juvenal Medina y Rocío Romero. Lima: Tecnología Intermedia, 1992. 172 p.: bibl., ill., maps.

Experience of natural hazards offers basis for understanding how best to respond to them. Papers on *El Niño* in Piura, floods in Arequipa, rockslide in Chosica, and earthquake in San Martín Dept.

Wahl Kleiser, Lissie et al. La región del Madre de Dios: bibliografía anotada. See item **863**.

2515 Watters, R.F. Poverty and peasantry in Peru's Southern Andes, 1963–90. Pittsburgh, PA: Univ. of Pittsburgh Press, 1993. 366 p.: bibl., index, photos, tables. (Pitt Latin American series)

Core of this work is an evaluation of agrarian reform in the community of Chilca on the Pampa de Anta, Cusco Dept., visited between 1964–90. Other half of book generalizes with not much originality about Andean peasantry, the Peruvian revolution, and economic development.

2516 **Young, Kenneth R.** National park protection in relation to the ecological zonation of a neighboring human community: an example from northern Peru. (*Mt. Res. Dev.*, 13:3, 1993, p. 267–280)

Though disjointed, this study discusses human impact on Río Abiseo National Park which was established in 1983. Points out how nature protection in the Andean communities needs to take indigenous land-use systems into account to minimize conflict.

2517 **Young, Kenneth R.** Tropical timberlines: changes in forest structure and regeneration between two Peruvian timberline margins. (*Arctic Alpine Res.*, 25:3, 1993, p. 167–174)

Describes upper (3450–3600 m) and lower (3200–3500 m) timberlines in Río Abiseo National Park in north central Peru. Spatial patterning of timberline forest varies with the differing reliefs of tropical mountains.

2518 **Zimmerer, Karl S.** Agricultura de barbecho sectorizada en las alturas de Paucartambo: luchas sobre la ecología del espacio productivo durante los siglos XVI Y XX. (*Allpanchis/Cusco*, 23:38, segundo semestre 1991, p. 189–225, bibl., map, tables)

Study of highland agriculture in Paucartambo Province (Cusco Dept.) disputes the "tragedy-of-the-common" explanation for the crisis in farming the hillside fields in fallow. Rather, privatization of plots and the decline of community institutions are seen as the cause.

2519 **Zimmerer, Karl S.** Agricultural biodiversity and peasant rights to subsistence in the central Andes during Inca rule. (*J. Hist. Geogr.*, 19:1, 1993, p. 15–32)

In the Inca period, crop surpluses for State redistribution were based on only a few crop species and varieties, whereas local family subsistence made use of the full range of Andean crop diversity. Original conceptualization puts past agricultural achievements in a more focused perspective.

2520 **Zimmerer, Karl S.** The loss and maintenance of native crops in mountain agriculture. (*GeoJournal/Boston*, 27:1, 1992, p. 61–72)

Fine paper based on field research in the Dept. of Cusco where four crops—potatoes, maize, ulluco, and quinoa—are studied to determine the patterns of *cultivar* (or variety) loss and maintenance. Each is different.

BOLIVIA

2521 **Bojanic, Alan.** Tenencia y uso de la tierra en Santa Cruz: evaluación de la estructura agraria en el área integrada de Santa Cruz. Ilustraciones de Carmelo Corzón M. La Paz: Centro de Estudios para el Desarrollo Laboral y Agrario, 1988. 213 p.: ill. (Talleres CEDLA; 4)

Five data-rich essays by various authors cover agrarian structure, seasonal labor, colonization schemes, indigenous groups, and land tenure of the agriculturally most dynamic dept. in Bolivia.

2522 **Eastwood, David A.** Planned colonisation in Bolivian and Ecuadorian Amazonia: the need for a re-assessment of successful planning policy. (*Rev. Eur.*, 50, June 1991, p. 115–134, bibl., maps)

Planned settlement east of the Andes so enthusiastically embraced in the 1960s–70s has been largely a failure. Compares in particular the Sushufindi Project in Ecuador and that of San Julián in Bolivia. Author still insists that planned colonization is better than the anarchy of spontaneous movement.

2523 **Evans, Brian.** Migration processes in Upper Peru in the seventeenth century. (*in* Migration in Colonial Spanish America. Edited by David J. Robinson. Cambridge: Cambridge Univ. Press, 1990, p. 62–85, graph, maps, tables)

Using colonial census data from the 1680s, author carefully documents demographic displacements in the Bolivian highlands a century after Viceroy Toledo created the *reducciones* in 1572. This migration changed native settlement beyond recognition.

2524 **Gierhake, Klaus.** Die Rolle der öffentlichen Investitionen im Prozess der Regionalentwicklung in Bolivien, 1987–1990. (*Geogr. Z.*, 79:4, 1991, p. 229–245, bibl., tables)

Assessment of public investments made in Bolivian departments, led by Santa Cruz, between 1987–90. Incomplete analysis because after 1985 the Bolivian government dramatically changed gears to make the private sector the central actor of economic development.

2525 **Haase, R.** Physical and chemical properties of savanna soils in northern Bolivia. (*Catena/Cremlingen*, 19:1, 1992, p. 119–134)

Better drained soils in the study area in the Beni are more likely to be degraded than those soils subject to flooding.

2526 **Historia natural de un valle en los Andes, La Paz.** Edición de Eduardo Forno y Mario Baudoin. La Paz: Instituto de Ecología, Carrera de Biología, Facultad de Ciencias Puras y Naturales, Univ. Mayor de San Andrés, 1991. xii, 559 p.: bibl., ill. (some col.), index, maps.

Guide to flora, vegetation, fauna, and aquatic environments with introduction to geology, climate, and human settlement. Another fine contribution from the Instituto de Ecología in La Paz.

2527 **Morales, Cecile B. de.** Bolivia: medio ambiente y ecología aplicada. La Paz: Instituto de Ecología, Univ. Mayor de San Andrés, 1990. iv, 318 p., 2 leaves of plates: bibl., ill.

Written with didactic intent, this well-illustrated volume provides fine exposition of Bolivia's physiography, soils, water resources, climate, pollution, vegetation, and fauna. Other chapters cover economic plants, forest resources, animal conservation, agricultural ecology, and human adaptations.

2528 **Muñoz Reyes, Jorge.** Geografía de Bolivia. 3. ed. La Paz: Librería Editorial Juventud, 1991. 521 p.: bibl., ill.

This latest edition of a standard work on geography of Bolivia will be welcomed for its wealth of useful information.

2529 **Solares Serrano, Humberto.** Historia, espacio y sociedad: Cochabamba, 1550–1950; formación, crisis y desarrollo de su proceso urbano. Cochabamba, Bolivia: Honorable Alcaldía Municipal de Cochabamba; Centro de Investigación y Desarrollo Regional; Instituto de Investigaciones de Arquitectura, 1990. v. 1– : bibl., ill., maps.

Historical geography of the city of Cochabamba, especially useful for changes in the built environment, urban transport and communications, public works, institutional life, and relationships of the city with its hinterland.

2530 **Stearman, Allyn MacLean** and **Kent Hubbard Redford.** Commercial hunting by subsistence hunters: Sirionó Indians and Paraguayan caiman in lowland Bolivia. (*Hum. Organ.*, 51:3, Fall 1992, p. 235–244, bibl., graphs, ill.)

Sirionó identity persists at the village of Ibiato in the Beni, but now these people are heavily involved in caiman hunting for sale. Includes abundance of data on hunting this creature.

2531 **Stearman, Allyn MacLean.** Making a living in the tropical forest: Yuqui foragers in the Bolivian Amazon. (*Hum. Ecol.*, 19:2, 1991, p. 245–260)

Analyzing the Yuqui people who historically were full-time foragers, author contradicts assertion that human foragers cannot survive in a tropical forest without some food provided by agriculture. Clever exploitation of open habitats, certain social adaptations, dietary breadth, and ability to deal with scarcity all enabled the Yuqui to survive as the author asserts.

2532 **Tichit, Muriel.** Los camélidos en Bolivia. La Paz: Fundación para Alternativas de Desarrollo, 1991. 154 p.: bibl., ill., maps.

Good source of information on llamas and alpacas in Bolivia, regions of production, traditional and modern husbandry practices, and the status of several camelid projects.

Weil, Jim and **Connie Weil.** Verde es la esperanza: colonización, comunidad y coca en la Amazonia. See item **4981.**

2533 **Zimmerer, Karl S.** Soil erosion and labor shortages in the Andes with special reference to Bolivia, 1953–1991: implication for conservation with development. (*World Dev.*, 21:10, 1993, p. 1659–1675)

Soil erosion in the Calicanto watershed near Tarata in the Dept. of Cochabamba accelerated after the Bolivian Revolution. Peasants used more of their time to work off the land. As a result, conservation measures were no longer applied. Weaves together three dimensions: environmental perception, peasant labor, and technological innovation.

THE SOUTHERN CONE

CESAR CAVIEDES, *Professor of Geography, University of Florida, Gainesville*

THE GEOGRAPHIC LITERATURE from and about Southern Cone countries is rather weak for the period covered by this chapter. Works on the environment and populations seem to have changed considerably over the last years as topics which were previously considered as "geographical" are now also the purview of urbanists, political economists, and/or urban sociologists. Rather than viewing this as a shortcoming on the part of geographers, it is mostly an indication of the increasingly eclectic character of a discipline which also encompasses related social sciences. Such cross-disciplinary incursions and expansions attest to the degree of intellectual maturity and sophistication among Argentine, Chilean, and Uruguayan social scientists.

Several of the recently published general works on Latin America should be consulted for valuable information on the Southern Cone. The volume edited by Axel Borsdorf entitled *Lateinamerika: Krise ohne Ende* (item **2252**) deserves special consideration as a reflection of perceptions shared by many German-language specialists concerning Latin America and its unending crises. Equally revealing are Borsdorf's articles on urban models proposed by German specialists (item **2222**), regionalization of Southern Cone countries (item **2534**), and the futures of Anglo-American and Latin American cities (item **2221**).

While in the recent past Argentina generated a large number of entries dealing with the Malvinas/Falkland War—thereby revealing not only an interest in geopolitical subjects but also concern for the perceived dismemberment of its national territory—books and papers on this theme have declined in recent years. The only article on this subject is a review essay on books analyzing the causes and consequences, as well as the future implications of the conflict (item **4199**). The bulk of Argentine geographical contributions is compiled in a special issue of the Madrid journal *Estudios Geográficos* (Vol. 53, No. 208, julio/sept. 1992), which contains collaborations by prominent Argentine geographers, such as Bolsi, Bruniard, Reboratti, and Roccatagliata. Moreover, one of the major additions to the geography of Argentina is the synthesis produced under Roccatagliata's direction (item **2538**), a work that reflects the state of the art in Argentine geographical research. Geographical production in Chile evinces a strong bias towards geopolitical subjects, a tendency that is exemplified by the relatively large number of articles published in the *Revista de Geopolítica* on the ongoing dispute with Bolivia pertaining to an opening into the Pacific Ocean (items **4241** and **2572**), the maritime vocation of the country (items **2571** and **2613**), and questions related to the meaning of boundaries in Chile (items **2590** and **2597**). Comparatively little has been issued by strictly geographical sources, since even the usually productive *Revista Geográfica Norte Grande* is represented below by only a few entries.

As in previous years, the geographic production of Uruguay and Paraguay is minimal compared with the published materials on Argentina and Chile.

GENERAL

2534 Borsdorf, Axel. La cuestión regional en el Cono Sur: concepciones y problemas de la regionalización en el sur de la América Latina. (*in* América Latina: regiones en transición. Ciudad Real, Spain: Publicaciones de la Univ. de Castilla-La Mancha, 1991, p. 61–81, bibl., ill., tables)

Regions established based on novel criteria such as colonization history, resource exploitation, and personal perceptions of inhabitants are proposed for Argentina and Chile.

2535 Caviedes, César N. Estudios sobre la variabilidad espacial del voto en América Latina. (*Rev. Interam. Bibliogr.*, 63:2, 1993, p. 257–267, bibl.)

Discussion of existing literature on electoral cartography in some South American countries, with emphasis on Argentina, Brazil, Chile, and Uruguay. This field is rather underdeveloped due to the fragile situation of democracy in the 1970s-80s. Also notes absence of knowledge about advanced techniques of mapping electoral results applied in Angloamerica and Europe.

2536 Dirié, Cristina. Movilidad funcional y geográfica de la fuerza de trabajo y libre circulación de trabajadores en el Cono Sur de América Latina. (*in* Integración latinoamericana: informe base. Buenos Aires: Consejo Federal de Inversiones, 1991, p. 91–99)

Article reviews legislation on workers' circulation within the Mercosur countries once this commercial alliance between Argentina and Brazil is implemented.

2537 Salomon, Jean-Noël. Le complexe touristico-industriel d'Iguaçu-Itaipu: Argentine-Brésil-Paraguay. (*Cah. Outre-Mer*, 177:45, jan./mars 1992, p. 5–19, bibl., ill.)

Author purports that at the corner where Argentina, Brazil, and Paraguay meet in the Iguazú Falls a new region of development is emerging. Tourism, commerce, and incipient industrialization are being fostered by the presence of the waterfalls and electrical energy produced at Itaipú.

ARGENTINA

2538 La Argentina: geografía general y los marcos regionales. Coordinación de Juan A. Roccatagliata. Buenos Aires: Grupo Editorial Planeta, c1988. 783 p.: bibl., ill.

Commendable effort to update geographical information on the country. Each chapter has been written by a prominent geographer displaying both originality and penetrating analysis. Considering the synthesizing character of this work, references are rather insufficient to convey a proper image of the abundance of existing research. Poor quality of illustrations and deficient typographic work (several pages are blank) diminish value of this informative volume.

2539 Bastida, Ricardo O. *et al.* Impacto ecológico y económico de las capturas alrededor de las Malvinas después de 1982: informe preparado en el Instituto Nacional de Investigacion y Desarrollo Pesquero (INIDEP). Dirección de Antonio E. Malaret. Mar del Plata, Argentina: INIDEP, 1986. 115 p.: bibl., ill., maps. (Serie contribuciones, 0325–6790; 513)

Study deals with fish stocks in conterminous waters of the Malvinas/Falkland Islands prior to the war of 1982. Banning Argentina from fishing in these waters resulted in depressed captures by Argentine fishing boats and a parallel increase in the captures by Polish and Russian boats during 1983.

2540 Bolsi, Alfredo S.C.; Marta Madariaga; and Ana E. Batista. Sociedad y naturaleza en el borde andino: el caso de Tafí del Valle. (*Estud. Geogr.*/Madrid, 53:208, julio/sept. 1992, p. 383–417, graph, maps, tables)

Describes present situation of the picturesque Tafí del Valle in the interior of Tucumán and offers an outline of its occupación from prehispanic times to present. Most estates in the valley breed horses, mules, cattle, and sheep.

2541 Bruniard, Enrique D. and Celia O. Moro. El ámbito subtropical en la República Argentina: climatología dinámica y límites climáticos. (*Estud. Geogr.*/Madrid, 53:208, julio/sept. 1992, p. 419–446, bibl., maps, tables)

Mobility of air masses and circulation patterns that explain weather changes in the subtropical belt of Argentina are analyzed with special emphasis on the limit between the temperate subtropical belt and the southern margin of the tropical zone lies.

2542 Brunstein, Fernando; Beatriz Cuenya; and Nora Clichevsky. Crisis y condiciones de vida en el Gran Buenos Aires. (*in* Las ciudades en conflicto: una perspectiva latinoamericana. Montevideo: Centro de Informaciones y Estudios del Uruguay; Ediciones de la Banda Oriental, 1989, p. 135–174, bibl., maps, tables)

Increased pauperization of Buenos Aires is attributed to capitalist penetration of Argentina. Survey of "unsatisfied basic needs" reveals that the poor populations are located in the periphery of the Federal District of Buenos Aires.

2543 **Bulgheroni, Raúl** and **Juan Carlos Agulla.** Argentina: imagen de un país. v. 2A, Summa andina. Buenos Aires: Bridas, 1987. v. 2A: appendices, ill. (some col.), maps (some col.).

Another volume in a series of well-illustrated regional monographs produced by this author. This time the basic topics of the Andean region of Argentina are presented in attractive vignettes.

Censo nacional de población y vivienda. See item **16**.

2544 **Clark, Ricardo.** Aves de Tierra del Fuego y Cabo de Hornos: guía de campo. Buenos Aires: Literature of Latin America, 1986. 294 p., 1 leaf of plates: bibl., ill., index, map.

Valuable guide to marine and terrestrial aviary of Tierra del Fuego and Patagonia. A must to take along when cruising in those scenic regions.

2545 **Dugini de De Candido, María Inés.** Las relaciones intra-americanas: William Wheelwright y el ferrocarril del Pacífico al Atlántico. (*Rev. Hist. Am. Argent.*, 15:29, 1989, p. 137–155, bibl., maps)

In 1845, the American entrepreneur/adventurer planned construction of a railway to connect the Atlantic with the Pacific coast from Catamarca to Caldera. Rising tensions between Argentina and Chile hindered the realization of this project. Good insights into the progress of these two nations in the early republican years.

2546 **Espizúa, Lydia E.** Quaternary glaciations in the Río Mendoza Valley, Argentine Andes. (*Quat. Res./New York*, 40:2, 1993, p. 150–162, bibl., maps, photos, tables)

Traits left by four glacial advances during the Pleistocene are thoroughly detailed and interpreted. Correlations with similar occurences on the Chilean side in the Aconcagua Valley are made. Excellent methodological work and enlightening findings.

2547 **Furlani de Civit, María E.** and **M.J. Gutiérrez de Manchón.** Aportación al acercamiento entre literatura y geografía: imágenes regionales de Mendoza, República Argentina. (*Rev. Geogr. Norte Gd.*, 17, 1990, p. 67–74, bibl.)

This analysis of literary depictions of Mendoza's landscapes and inhabitants proves that the constrast between pampa, mountains, and oases dominate geographically-inspired literature of the region.

2548 **Giddings, Lorrain.** Visión por satélite de las inundaciones extraordinarias en la cuenca del Río de la Plata. (*Interciencia/Caracas*, 18:1, Jan./Feb. 1993, p. 16–26, bibl., photos, tables, graphs)

Decimating floods in the Río de la Plata basin during 1983—caused by El Niño phenomenon—are mapped using satellite imagery (AVHRR) from NOAA satellites. Nearly 38,000 km^2 of riverine land were flooded.

2549 **Gutiérrez Colombres, Benjamín.** Toponimia histórica y geográfica de Tucumán. San Miguel de Tucumán, Argentina: Univ. Nacional de Tucumán, 1990. 197 p.: bibl. (Publicación; 1441)

Useful toponymic guide of the province of Tucumán, where Quechua and Diaguita linguistic influences are overridden by the Spanish dominance.

2550 **Iglesias, Alicia N.** Erosión eólica, desertificación y crisis de rentabilidad de la economía ganadera en Patagonia: el caso de la provincia de Santa Cruz. (*Estud. Geogr./Madrid*, 53:208, julio/sept. 1992, p. 447–479, bibl., graphs, maps, tables)

Desertification, excessive sheep grazing, and absence of improved pastures are regarded as reasons for the deterioration of environmental systems in a segment of the southern Patagonian plains.

2551 **Informe nacional a la Conferencia sobre Medio Ambiente y Desarrollo de las Naciones Unidas.** Secretaría General, Comisión Nacional de Política Ambiental. Buenos Aires: Argentina, Presidencia de la Nación, 1991. 582 p.: bibl., ill., maps.

General work that reports in a very pessimistic tone about environmental deterioration and production conditions across the country in the early 1990s.

2552 **Instituto de Estudios e Investigaciones sobre el Medio Ambiente (Argentina).** Política ambiental y gestión municipal: programa de medio ambiente; compilación de los seminarios realizados en 1989–1990. Buenos

Aires: Fundación Jorge Esteban Roulet, 1990. 115 p.: ill. (Col. Encuentros; 6)

Discussion led by Elva Roulet (see item **2560**) provides fair contribution to questions of urban environmental deterioration and public attitudes towards these problems in Argentina.

2553 Loret, John. The Río Deseado: a mosaic of marine life. (*Explor. J.*, 69:3, Fall 1991, p. 94–99, maps, photos)

Description of the coastal fauna along Río Deseado in Patagonia, such as penguins, whales, seals, and dolphins of cold Atlantic waters.

2554 Martínez, Elena. Ciudad de Salta, Argentina. (*Medio Ambient. Urban.*, 9:38, marzo 1992, p. 71–80, tables, maps)

Morphological and functional survey of Salta recalls the historical circumstances that accompanied development of the various boroughs. Generalized deterioration is noted in most quarters of this traditional city.

2555 Morris, Arthur. Neoconservative policies in Argentina and the decentralization of industry. (*in* Decentralization in Latin America: an evaluation. Edited by Arthur Morris and Stella Lowder. New York: Praeger, 1992, p. 77–91, bibl., maps, table)

Decentralization attempts in the industrial sector had different results: while they worked satisfactorily in Tucumán, they failed in Chubut and Tierra del Fuego.

2556 Muscar Benasayag, Eduardo F. and Teresa Franchini. Emplazamientos urbanos en zonas de riesgos naturales: el caso de Gran Resistencia en la Planicie Chaqueña. (*Estud. Geogr./Madrid*, 53:208, julio/sept. 1992, p. 481–502, bibl., maps)

Settlements in eastern Chaco are located in areas prone to flooding by the high waters of the Paraguay River. The city of Resistencia has experienced severe inundations in 1973, 1983, 1987, 1991 and 1992.

2557 Otero, Hernán. La inmigración francesa en Tandil: un aporte metodológico para el estudio de las migraciones en demografía histórica. (*Desarro. Econ.*, 32:125, abril/junio 1992, p. 79–106, bibl., graphs, maps, tables)

From 1850–1914 more than 1500 families from France, Navarre, and the Basque Provinces arrived in the famous region of Sierra del Tandil in the southern part of the Buenos Aires province. Author sees in this focused migration the result of a familial reference network that prompted the flow of migrants from the Pyrenees into Tandil.

2558 Reboratti, Carlos. Ambiente, producción y estructura agraria en el Umbral al Chaco. (*Estud. Geogr./Madrid*, 53:208, julio/sept. 1992, p. 503–522, bibl., maps)

The region of *Umbral al Chaco* or the fringe between the Chaco plain and the eastern border of the Andes was characterized by forest exploitation and large cattle ranches. In recent years beans and soybeans have been principal crops. Natural forests are being degraded by the expansion of a colonization front spearheaded by small entrepreneurs.

2559 Roccatagliata, Juan A. Relaciones entre políticas territoriales y políticas de transporte: el caso del transporte ferroviario. (*Estud. Geogr./Madrid*, 53:208, julio/sept. 1992, p. 523–542, bibl., map, table)

The railway in Argentina as a tool for development and spatial order is regarded as a failure: length of journeys, inefficiency, and obsolete equipment conspire against use of railways as a unifying force. Outlines recommendations for changes.

2560 Roulet, Elva. La nueva capital. Buenos Aires: Fundación Jorge Esteban Roulet, Centro de Participación Política, 1988. 144 p.: bibl.

Historical outline of events that led to the formulation of the project to convert Carmen de Patagones/Viedma into the national capital, produced by a then-governor of the province of Buenos Aires. Viewed in perspective, this ill-fated project reveals the extraordinary influence of Buenos Aires' local politics on national affairs.

2561 Sargent, Charles S. Argentina. (*in* Latin American urbanization: historical profiles of major cities. Edited by Gerald Michael Greenfield. Westport, Conn.: Greenwood Press, 1994, p. 1–37, bibl., mapg)

Urban history of Argentina has been determined by the predominance of Buenos Aires. In colonial times, however, it was rivaled by the economic significance of San Miguel de Tucumán and Córdoba. Modern urban development since the second half of the 19th century has centered on the emergence of

Greater Buenos Aires, Rosario, and La Plata, cities that mushroomed with the arrival of European immigrants and recently with the inflow of internal migrants.

2562 Sayago, José Manuel. El deterioro del ambiente en el noroeste argentino. (*Estud. Geogr./Madrid*, 53:208, julio/sept. 1992, p. 543–566, bibl., graphs, maps, table)

Environmental deterioration in northwestern Argentina is considered in terms of that region's aridity and occupation history. Careless cultivation and grazing practices over the course of the centuries have contributed to the present state of environmental degradation.

2563 Sejas, Lidia. Condicionantes territoriales en la integración fronteriza con los países limítrofes. (*in* Integración latinoamericana: informe base. Buenos Aires: Consejo Federal de Inversiones, 1991, p. 63–74, maps, tables)

Several "Boundary Committees" have been instituted to assure that land ownership along the boundaries of the country remains in Argentine hands. These committees also regulate the flow of people and merchandise in ports of entry. Explores future of these boundary crossings in an integrated Southern Cone community.

2564 Winograd, Alejandro. Areas naturales protegidas y desarrollo: perspectivas y restricciones para el manejo de parques y reservas en la Argentina. (*in* Seminar on the Acquisition of Latin American Library Materials (SALALM), 36th, *San Diego, Calif., 1991.* Latin American studies into the twenty-first century: new focus, new formats, new challenges. Albuquerque: SALALM Secretariat, General Library, Univ. of New Mexico, 1993, p. 105–122, bibl.)

Protected natural areas in Argentina are numerous, but, according to this author, public attitude and the State's interest in their preservation are far from satisfactory.

CHILE

2565 Abele, Gerhard. The interdependence of elevation, relief, and climate on the western slope of the central Andes. (*Zent.bl. Geol. Paläontologie*, 1:5/6, Nov. 1989, p. 1127–1139)

The rise of mountains during the Tertiary and Quaternary contributed to an increase of the aridity of the western slope of the Andes between latitudes 17 degrees and 24 degrees South.

2566 Abele, Gerhard. Landforms and climate on the western slope of the Andes. (*Z. Geomorphol. Suppl.bd.*, 84, Feb. 1992, p. 1–11, bibl., ill.)

It is argued that the western slope of the Andes in northern Chile was humid during the Tertiary but that it became dry during transit to the Quaternary. Thus, most reliefs display unchanged forms of Tertiary morphogenesis. Active landform development was possible only in the humid extra-tropical expanses of the Andes.

2567 Abele, Gerhard. Modelle zur Entwicklung des Hochgebirgsreliefs und ihre Anwendung am Beispiel der Anden. (*Mitt. Österr. Geogr. Ges.*, 135, 1993, p. 140–160, bibl., ill.)

Tectonic uplift of the Andes contributed to create particular conditions of rainfall across the mountains. The northern segment is dry, the eastern slope is characterized by summer rains, and the southern sector experiences frontal winter precipitation. Morphogenesis occurs in accordance with these rainfall regimes.

2568 Abele, Gerhard. Die Zertalung der nordchilenischen Anden in ihrer Abhängigkeit von Klima, Tektonik und Vulkanismus. (*in* Der Geograph im Hochgebirge: Beiträge zu Theorie und Praxis geographischer Forschung; Festschrift für Helmut Heuberger zum 70. Geburtstag von seinen Schülern gewidmet. Innsbruck, Austria: Selbstverlag des Institutes für Geographie der Univ. Innsbruck, 1993, p. 15–28, ill.)

Valley-formation in the dry northern Chilean cordillera is considered the result of a rapid rising of the Andes during periods of heightened precipitation (pluvial periods) of the Quaternary. Today the region is absolutely dry.

2569 Aravena Ricardi, Nancy. Un corredor territorial para Bolivia: propuestas y opciones. (*Rev. Chil. Geopolít.*, 6:1, dic. 1989, p. 17–28, maps)

Implications of a hypothetical plan granting Bolivia access to the Pacific Ocean are outlined. Author fears that such a corridor would allow Argentina and Brazil to use Bolivia for obtaining a "frontage" on the Pacific Ocean.

2570 **Aravena Ricardi, Nancy.** Un corredor territorial para Bolivia: ventajas y desventajas geopolíticas. (*Rev. Chil. Geopolít.*, 6:2, abril 1990, p. 19–38, tables)

Exposition of reasons given by Chile for rejecting Bolivia's claims for a corridor to the Pacific Ocean. Bolivia claims this solution will break its landlocked status and bring about overall development.

2571 **Arnello Romo, Mario.** Exigencias y orientaciones para construir el destino oceánico de Chile. (*Rev. Chil. Geopolít.*, 6:1, dic. 1989, p. 3–15, map)

Pedantic elaboration of the role to be played by Chile in the future of the Pacific Ocean, the "most important geopolitical space of the twenty-first century."

2572 **Bahamonde, Marcelo Alejandro.** Reflexiones geopolíticas sobre el mito de la mediterraneidad boliviana. (*Rev. Chil. Geopolít.*, 6:2, abril 1990, p. 49–59)

Member of the Instituto Geopolítico de Chile delivers caustic attack against Bolivian diplomats for demanding that Chile grant Bolivia "corridor to the sea." Author contends that Bolivia has been given ample access to Pacific and Atlantic oceans via railways and waterways through Argentina, Brazil, Peru, and Chile.

2573 **Berguño, Jorge.** El descubrimiento de las islas Shetland del Sur. (*Rev. Esp. Pac.*, 1:1, julio/dic. 1991, p. 129–158, appendices, tables)

Lengthy refutation of thesis that the South Shetland Islands were discovered by Flemish navigator Dirck Gherritz in 1600. Author believes that these pre-Antarctic Islands were first sighted by the Spanish Commodore Gabriel de Castilla in 1603.

2574 **Bernales Lillo, Mario.** Toponimia de Valdivia. Temuco, Chile: Ediciones Univ. de La Frontera, 1990. 141 p.: bibl., index, maps. (Serie Quinto centenario; 6)

Place names in the province of Valdivia reveal intentions of Indians, Spaniards, and non-Iberian colonists when founding settlements over more than five centuries of occupation.

2575 **Borsdorf, Axel.** Chile: Kunst- und Reiseführer mit Landeskunde und Exkursionsvorschlägen. Stuttgart, Germany: W. Kohlhammer, 1987. 338 p.: bibl., ill. (some col.), index, maps (some col.), plans. (Kohlhammer Kunst- und Reiseführer)

Attractive guide for the discriminating tourist. Contains pertinent scientific information about the country as well as valuable practical tips about Chilean mentality, customs, places of recreation, and places of scenic beauty.

2576 **Borsdorf, Axel.** Grenzen und Möglichkeiten der räumlichen Entwicklung in Westpatagonien am Beispiel der Region Aisén: natürliches Potential, Entwicklungshemmnisse und Regionalplanungsstrategien in einem lateinamerikanischen Peripherieraum. Stuttgart, Germany: F. Steiner Verlag Wiesbaden, 1987. 190 p.: bibl., ill., index, maps. (Acta Humboldtiana, 0400–4043; 11)

Development possibilities of the remote and scarcely populated region of Aisén in southern Chile are scrutinized after a careful survey of human and natural conditions. Author feels that regional development projects based on primary resource exploitation fostered by the State have been unsuccessful due to isolation and lack of infrastructures. Small-scale development projects based on apiculture, specialized agriculture, fine fish, and tourism enhancement could make possible the longed-for development of this region.

2577 **Calderón, Ernesto *et al.*** Santiago, dos ciudades: análisis de la estructura socio-económica espacial del Gran Santiago. Coordinación de Eduardo Dockendorff y Carlos Fuenzalida. Equipo de investigadores, María Bertrand *et al.* Santiago: Centro de Estudios del Desarrollo, 1990. 235 p.: bibl., ill., maps.

Thorough study of social fabric and spatial distribution of social classes in Santiago in the early 1980s. Basic primer for tackling questions of urban planning. Excellent documentation.

2578 **Castro, Milka** and **Miguel Bahamondes.** Cambios en la tenencia de la tierra en un pueblo de la precordillera del Norte de Chile: Socoroma. (*Rev. Chil. Antropol.*, 6, 1987, p. 35–57, graphs, maps, tables)

Land tenure patterns in a pastoral community of northernmost Chile have changed as population growth, though modest, has caused the fragmentation of property and an increase in minifundia.

2579 **Cooper, Marc.** Alerce dreams. (*Sierra/San Francisco*, 77:1, Jan./Feb. 1992, p. 122–129)

"Alerce," one of the largest conifers of South America, is rapidly declining due to illegal logging. Chile's natural forests are dwindling as wood chips are exported to Japan.

2580 **Daher, Antonio.** Ajuste económico y ajuste territorial en Chile. (*EURE/Santiago*, 18:54, abril 1992, p. 5–13)

Author make a connection between the expansion of the export economy in the 1980s and the emergence of new growth dynamics in different regions. Enlightening essay contrasts with often negative and unjustified criticisms of these developments.

2581 **Daskam, Thomas** and **Jürgen Rottman.** Aves de Chile. Santiago: Publicaciones Lo Castillo, 1984. 56 p.: ill. (some col.). (Col. Apuntes)

Beautiful—though brief—pictorial guide of Chilean birds, with passing references to habitat and customs.

2582 **Domínguez Díaz, Marta.** El medio ambiente en Chile hoy: informe y bibliografía. (*in* Seminar on the Acquisition of Latin American Library Materials (SALALM), 36th, San Diego, Calif., 1991. Latin American studies into the twenty-first century: new focus, new formats, new challenges. Albuquerque: SALALM Secretariat, General Library, Univ. of New Mexico, 1993, p. 130–156)

Rather rudimentary report on environmental deterioration in Chile accompanied by a bibliographic survey which is far from exhaustive or helpful.

2583 **Echenique, Marcial.** Ideas sobre el futuro de la ciudad de Santiago. (*Estud. Públicos*, 48, primavera 1992, p. 5–15, graphs, map)

General ideas about the future of Santiago as an *urbis*. Advances idea that the most feasible way of expansion is along major arteries of communication with the north, south, and west (Viña del Mar and Valparaíso).

2584 **Errázuriz K., Ana María** and **Mónica Gangas Geisse.** Atlas del desarrollo territorial de Chile: características demográficas. (*Rev. Chil. Geopolít.*, 6:3, agosto 1990, p. 29–42, bibl., maps)

Planned *Atlas of the territorial development of Chile* envisages addition of a section on "Demographic Characteristics." Article outlines subjects to be included.

2585 **Errázuriz K., Ana María** *et al.* Consideraciones geopolíticas acerca de la ocupación del territorio chileno. (*Rev. Chil. Geopolít.*, 6:3, agosto 1990, p. 43–51, graph, maps, tables)

Uneven spatial distribution of Chileans over national territory is considered to have negative influence on their overall effectiveness as a working force.

2586 **Ferrocarril de Arica a La Paz. Sección Chilena.** 75 años: revista aniversario. Arica, Chile: Ferrocarril de Arica a La Paz, Sección Chilena, 1988? 61 p.: ill. (some col.).

Revealing monograph on functions and background of the railway that connects Arica, Chile with La Paz, Bolivia.

2587 **Gangas Geisse, Mónica** and **Ana María Errázuriz K.** Características de la distribución espacial de la población: provincia de San Antonio, 1982. (*Rev. Geogr. Norte Gd.*, 17, 1990, p. 17–25, bibl., maps, tables)

Hierarchy and location analysis of settlements in the province of San Antonio display a conspicuous concentration along the coast and random dispersion in rural hinterland.

2588 **Gangas Geisse, Mónica.** Características demográficas de la población chilena entre 1952–1982. (*Rev. Chil. Geopolít.*, 6:3, agosto 1990, p. 53–80, bibl., graphs, maps, tables)

Tendencies of population growth in Chile since 1895 are analyzed. Interesting decrease in fertility is observed particularly since 1970, auguring modernization of Chile's demographic profile.

2589 **Gangas Geisse, Mónica.** Las migraciones de la provincia de San Antonio: características demográficas. (*Rev. Geogr. Norte Gd.*, 17, 1990, p. 27–36, bibl., tables, maps)

The port city of San Antonio, south of

Valparaíso, lost inhabitants—especially female population—between 1977–82. Outmigration is blamed on the decline of activities in the city and on the enormous attraction of Greater Santiago.

2590 Gangas Geisse, Mónica. El tema de las fronteras políticas en la segunda mitad del siglo XX. (*Rev. Chil. Geopolít.*, 6:2, abril 1990, p. 39–47, bibl., maps)

Political boundaries have been taking on a new connotation in this century: they are peripheral places of contested occupation. Notions of "maritime boundaries" which have elicited heated debates like those generated by the drafting of The Law of the Sea in the 1970s and 1980s have become particularly acute.

2591 Gonzales Leiva, José I. and **Félix Gajardo Maldonado.** Pensamiento geopolítico castellano e hispano a través de la cartografía colonial de Chile. (*Rev. Chil. Geopolít.*, 6:3, agosto 1990, p. 11–28, bibl., maps)

Colonial maps of South America reveal extension attributed by Spanish cartographers to the Kingdom of Chile. In many instances territories assigned to colonial Chile were larger than those claimed by the republic in 1826 in the Congress of Panama.

2592 Gross F., Patricio and **Orlando Acosta P.** Santiago de Chile: carácter patrimonial y rol funcional. (*Medio Ambient. Urban.*, 9:38, marzo 1992, p. 35–54, bibl., graphs, maps)

Quality of Santiago's residential life reveals that in general the situation is better than expected. Only "interior" slums or *conventillos* suffer from serious deterioration.

2593 Gwynne, Robert N. Non-traditional export growth and economic development: the Chilean forestry sector since 1974. (*Bull. Lat. Am. Res.*, 12:2, May 1993, p. 147–169, bibl., graphs, map)

Although recognizing impressive growth of the export-oriented forestry sector in Chile from 1985–91, author is still dissatisfied with development and complains that forestry has not created "a multiplier effect" in the Chilean economy and has not generated "local technology." So much for unrealistic expectations.

2594 Jornadas Territoriales, 2nd, Santiago, 1987? Chiloé y su influjo en la XI Región. Santiago: Instituto de Investigaciones del Patrimonio Territorial de Chile, Univ. de Santiago, 1988. 282 p., 1 folded leaf of plates: bibl., ill., indexes, maps. (Col. Terra nostra, 0716–5293; 12)

Another in a series of volumes dealing with specific regions of Chile. Combines talents of prominent historians, ecologists, and urbanists with expertise on the island of Chiloé. Contains useful list of references.

2595 Katz, Ricardo. Reflexiones sobre contaminación ambiental. (*Estud. Públicos*, 40, primavera 1990, p. 171–194)

Air pollution in Santiago is the consequence of the collision of private versus common interests, represented by the State. Some cases in Chile are mentioned where offenders were prosecuted by the State for violating environmental conservation measures. However, minor offendors are difficult to identify and even more difficult to prosecute.

2596 Larraín Navarro, Patricio and **Héctor Toledo Rivera.** Diferencias espaciales en los niveles de bienestar social en el Gran Santiago: implicancias conceptuales, metodológicas y políticas. (*EURE/Santiago*, 16:49, oct. 1990, p. 33–49, bibl., maps, tables)

Inequality in well-being and affluence are measured by multivariate analysis. Resulting map indicates that municipalities of west, north, and southeast Santiago are less favored than the northeastern municipalities, where high living standards are found.

2597 Latorre, Adolfo Paul. La frontera marítima austral: perspectiva de conflicto. (*Rev. Chil. Geopolít.*, 6:1, dic. 1989, p. 29–40, bibl., maps)

Conflict between Argentina and Chile is still expected by some Chilean geopoliticians because of misinterpretations and unprecise design of boundaries in the southernmost islands of Tierra del Fuego.

2598 Maino, Valeria. Las embarcaciones menores del Maule, 1860–1896. (*Universum/Talca*, 5, 1990, p. 51–62, ill., tables)

Barge building was one of the main specialities of communities on the Maule River in Central Chile during the 19th century. Barges of up to 48 tons sailed the Pacific Ocean from Central Chile to Antofagasta, Arica, and Callao.

2599 Mariangel Candia, Walter B. Aplicación al pie de Monte de la Comuna de la Reina, de un parámetro primordial para la expansión urbana: el riesgo físico. (*Rev. Geogr. Chile*, 32, 1990, p. 51–71, bibl., graphs, tables)

Cursory survey of landslides caused by the torrential rains of 1987–88 in La Reina borough, Greater Santiago.

2600 Martinic B., Mateo. La "Comisión Científica del Pacífico" en Magallanes, 1863. (*An. Inst. Patagon./Soc.*, 20, 1991, p. 7–18, photos)

In 1862 the King of Spain sent a scientific expedition to South America that became entangled in hostilities with Chile and Peru. The Spaniards lost two ships. No less auspicious were events which originated in Punta Arenas and were caused by the Spanish crew and the bad-tempered Commodore Luis Hernández Pinzón. There was not much scientific reporting from this mission.

2601 Mellén Blanco, Francisco. Estudio de nuevas copias de planos de la isla de Pascua (Rapa-Nui) de 1770. (*Rev. Esp. Pac.*, 1:1, julio dic. 1991, p. 33–46, ill., maps)

Two old maps from Easter Island surveyed in 1770 by the Spanish navigator Felipe González de Aedo reveal natural and human characteristics of that interesting island half a century after its discovery.

2602 Mingo M., Orlando; Miguel Contreras C.; Alicia Ross A. Proyecto Región Capital de Chile. (*EURE/Santiago*, 16:48, junio 1990, p. 7–24, map)

Concern is expressed over unbalanced growth of the capital of Chile. This is nothing new in Latin America where decentralization seems unfeasible.

2603 Misetic Yurac, Vladimir and Glenda Kapstein Lomboy. Circuitos turísticos en la II región de Antofagasta, Chile diagnóstico y análisis de requerimientos. (*Rev. Geogr. Chile*, 32, 1990, p. 25–49, bibl, graphs, tables)

Rather questionable analysis of tourist potential of the desert region of Antofagasta. Rudimentary infrastructures and high transportation costs conspire against potential attractiveness of places mentioned.

2604 Núñez Pinto, Jorge. Rodolfo Philippi y el Desierto de Atacama. (*Universum/Talca*, 5, 1990, p. 75–78, photos)

Rudolph A. Philippi, German physician and naturalist, was one of the initiators of German colonization in Chile and a pioneer in the exploration of arid northern Chile. His life and work are outlined in this paper.

2605 Parrochia Beguin, Juan. Semi urbano y semi humano. Recopilación e ilustraciones de María Isabel Pavez Reyes. Santiago: Univ. de Chile, Facultad de Arquitectura y Urbanismo, Depto. de Urbanismo, 1989. 242 p.: bibl., ill., maps.

Collection of miscellaneous essays by one of the most prolific of Chile's urbanists contains a strange mixture of plausible as well as bizarre and unrealistic ideas.

2606 Pinochet, Fernando. Los suelos forestales de la Región del Maule. (*Universum/Talca*, 6, 1991, p. 37–46, bibl., maps, graphs)

Soils supporting commercial forests in the region of Maule are described in detail with no reference to their advantages or limitations.

2607 Pizarro Tapia, Roberto and José Luis Saavedra Lucero. Predicción de caudales recesivos mediante modelos matemáticos. (*Universum/Talca*, 6, 1991, p. 31–36, bibl., graphs, photos, tables)

Generalities about runoff measurements and discharge estimations are illustrated very poorly with an example from the region of Maule.

2608 Quintanilla P., Víctor. Zonación de riesgos de montaña en base a la determinación de procesos de remoción en masa: estudio de casos en Los Andes de Chile Central, 32° latitud sur. (*Bol. Lima*, 14:84, nov. 1992, p. 33–43, bibl., graph, maps, photos)

Landslides and avalanches in upper reaches of the Aconcagua Valley are attributed to slopes in critical equilibrium, snow accumulation, and high water impregnation during winter thawing periods.

2609 Quiroz, Daniel; Patricio Poblete; and Juan C. Olivares. Los salineros en la costa de Chile Central. (*Rev. Chil. Antropol.*, 5, 1986, p. 103–120, bibl., ill., maps)

Salt collecting from the coastal lagoon of Cahuil and neighboring places continues to be a picturesque activity in Central Chile.

2610 Rioseco Hormazábal, Reinaldo. Indicadores socioeconómicos relativos a las condiciones de salud de la población en Chile. (*Rev. Geogr. Norte Gd.*, 14, 1987, p. 67–73, bibl., graphs, maps)

Discusses infant mortality and rates of contagious diseases of workers with and without health coverage. Results are obvious. Very elementary paper that does not acknowledge pioneer works by Scarpaci in Chile.

2611 Rojas Hoppe, Carlos. La terraza fluvial de "canchagua" en la ciudad de Valdivia: nuevos antecedentes estratigráficos y granulométricos. (*Rev. Geogr. Chile*, 32, 1990, p. 7–24, bibl., graphs, maps)

Sedimentological characteristics of "canchagua," an argillous sand of mixed glacial/volcanic origin, are investigated in the vicinity of the city of Valdivia.

2612 Santis Arenas, Hernán. La estructura del espacio político. (*Rev. Geogr. Norte Gd.*, 17, 1990, p. 53–65, bibl., graphs, tables, maps)

Using Chile as an example, article tests ideas of political space as expounded by Friedrich Ratzel, German geopolitician of the 19th century. As expected, Greater Santiago stands out as the primary core region in Chilean politics.

2613 Santis Arenas, Hernán. Papel de las telecomunicaciones como factor geopolítico de integración de un país marítimo y tricontinental. (*Rev. Chil. Geopolít.*, 8:1, dic. 1991, p. 47–63, bibl., maps)

Telecommunications in Chile are regarded by the author as a contributing factor towards furthering the hegemony of the centralizing State. Readers, however, are not told how such a process works.

2614 Santis Arenas, Hernán. Significado y contenido de Chile: país marítimo y tricontinental. (*Rev. Chil. Geopolít.*, 6:3, agosto 1990, p. 81–92, bibl., map)

Contends that Chile is a "tri-continental country" in that it holds territories in South America, the Antarctic, and the southern Pacific Ocean. Author feels that Chile's destiny is tied to the future of these three geographic entities.

2615 Scarpaci, Joseph L. Chile. (*in* Latin American urbanization: historical profiles of major cities. Edited by Gerald Michael Greenfield. Westport, Conn.: Greenwood Press, 1994, p. 106–133, bibl., map, tables)

Expansion of Chilean cities is regarded in the context of historical-economic development of the country. Short but pertinent profiles of the major cities of Santiago, Valparaíso, and Viña del Mar are also presented.

2616 Seminario sobre Administración de las Pesquerías Chilenas, *Santiago, 1989.* El desafío pesquero chileno: la explotación racional de nuestras riquezas marinas. Eduardo Bitrán y otros. Santiago: Hachette, 1989. 412 p.: bibl., ill. (Documento)

Several aspects of contemporary Chilean fisheries, an important pillar of the national economy, are discussed in this volume. Insightful remarks about legal problems concerning interference and overfishing by distant nations like Japan, Korea, and China.

URUGUAY

2617 Díaz de Guerra, María A. La zona de José Ignacio en el departamento de Maldonado y su incidencia en la evolución regional. (*Hoy Hist.*, 10:58, julio/agosto 1993, p. 27–53, ill.)

Using a small river as a unit of analysis, the historical geography of a segment of the Dept. of Maldonado is traced.

2618 Gans, Paul. Desarrollo económico y sector informal en América Latina: el ejemplo del comercio ambulante en Montevideo. (*Rev. Geogr./México*, 113, enero/junio 1991, p. 203–222, bibl., tables)

Tertiary commerce (the informal sector) on the streets of Montevideo is seen as an indicator of overall socioeconomic decay as it comprises 40 percent of the active labor force. Positive aspects of this sector, as underlined by De Soto, are not mentioned in this article.

2619 Kleinpenning, Jan M.G. Peopling the Purple Land: a historical geography of rural Uruguay, 1500–1915. Amsterdam: CEDLA, 1995. 355 p.: bibl., ill., index, maps. (Latin America studies; 73)

Historical geography of colonization and occupation of Uruguay is divided in two parts: colonial era and 1810–1915 period. Discusses development of cattle ranching, arable farming, and private landownership; foundation of settlements; construction of an infrastructure; socioeconomic problems; and population growth and immigration. Based on secondary sources, mainly from Britain and Montevideo. [R. Hoefte]

2620 Sargent, Charles S. Uruguay. (*in* Latin American urbanization: historical profiles of major cities. Edited by Gerald Michael Greenfield. Westport, Conn.: Greenwood Press, 1994, p. 468–485, bibl., map, table)

Uruguay is the South American country with the largest number of urban dwellers. Chapter outlines growth of the country through examination of its principle traits.

BRAZIL

KEITH D. MULLER, *Associate Professor of Geography, Kent State University*

AMAZONIA, ESPECIALLY ITS ENVIRONMENT, dominates the geographical literature on Brazil for this *HLAS* volume. Quality books with comprehensive views of Amazonia include Moran's study which emphasizes socioanthropological and ecological issues (item **2690**) and Goodman and Hall's collection of a wide range of Amazonian topics and authors (item **2665**). Noteworthy articles on Amazonia include: Fearnside's examination of sustainable extractive reserves (item **2660**); Hiraoka's thorough discussion of the viability of supplementing subsistence riverine agriculture with sustainable commercial economic systems (item **2673**); Browder and Godfrey's pertinent theoretical framework for future investigations on frontier urban networks in Amazonia (item **2638**); and Becker's (item **2632**) and Ab'saber's (item **2621**) appropriate insistence on the need to study and administer Amazonia in subregions.

Writings on the Northeast concentrate on rural and industrial development and are mostly case studies. For example, Caviedes and Muller question the role of large irrigation projects which do little to improve the situation of peasants (item **2642**) and Santos' report for SUDENE provides an overview of temporary rural laborers in Bahia (item **2703**).

The Instituto SPN (Sociedade, População e Natureza), has released 15 articles by five authors on a wide range of topics. The most significant include Mueller's description of how agricultural expansion primarily benefits the urban-industrial sector at the expense of the countryside—which must endure the subsequent negative environmental impacts (item **2691**)—and Mueller *et al.*'s overview of settlement patterns in the West Central region (item **2693**).

We hope that future studies address the recently created Mercosul, the economic union of Brazil, Argentina, Uruguay, and Paraguay.

2621 Ab'Saber, Aziz. Amazônia: proteção ecológica e desenvolvimento. (*São Paulo Perspect.*, 6:1/2, jan./junho 1992, p. 112–126, bibl.)
Essay calls for a greater number of ecological subdivisions with special attention to climatic and biogeographic differences to aid planning and administration.

2622 Ajara, Cesar et al. O Estado do Tocantins: reinterpretação de um espaço de fronteira. (*Rev. Bras. Geogr.*, 53:4, out./dez. 1991, p. 5–48, bibl., maps, tables)
Regional examination of settlement processes, urban forms, productive activities, and environmental impact.

2623 Allegretti, Mary Helena. Reservas extrativistas: uma proposta de desenvolvimento da floresta amazônica. (*Pará Desenvolv.*, 25, jan./dez. 1989, p. 3–29, bibl., map, tables)
Points to the need for test plots to aid implementation of extractive reserve programs.

2624 Almeida, Joaquim Anécio and Dino Magalhães Soares. Análise de variáveis sociais na questão do uso dos agrotóxicos: o caso da fumicultura. (*Ciênc. Amb.*, 3:4, jan./junho 1992, p. 85–104)
Detailed study of the effects of pesticides on the health of tobacco farm workers.

2625 Amazon: Nihonjin ni yoru 6onen no Ijūshi [Amazon: 60 years of Japanese immigration]. Edited by Associação Pan-Amazonia Nipo-Brasileira. Belém, Brazil: Associação Pan-Amazonia Nipo-Brasileira, 1994. 251 p.: photos.
Commemorates the 1929 arrival of the first Japanese immigrants to the Brazilian Amazon, a group of 43 families and 9 unmarried men. Chronology and description of major colonized regions in eight Amazon states details the history of Japanese settlements and their contributions to farming in the tropical rain forest. [K. Horisaka]

2626 Amazônia: a fronteira agrícola 20 anos depois. Organização de Philippe Léna e Adélia Engrácia de Oliveira. Belém, Brazil: Governo do Brasil, SCT/CNPq, Museu Paraense Emílio Goeldi; France: ORSTOM—França, 1991. 363 p.: bibl., ill., maps. (Col. Eduardo Galvão)
Results of seminar cover wide range of topics on frontier expansion and development. Includes indigenous peoples, the military, religious groups, agricultural colonization, land conflicts, deforestation and other environmental issues, economics, and Carajás.

2627 Anderson, Anthony B. Estratégias de uso da terra para reservas extrativistas da Amazônia. (*Pará Desenvolv.*, 25, jan./dez. 1989, p. 30–37, bibl., ill., tables)
Examines economic viability, social equality, and land-use strategies for extractive reserves.

2628 Andrade, Manuel Correia de Oliveira. A AGB e o pensamento geográfico no Brasil. (*Terra Livre*, 9, julho/dez. 1991, p. 143–152, bibl.)
Describes geographical thought of the Associação Brasileira de Geografia.

2629 Anjos, Rafael Sanzio Araújo dos. Configurações espaciais do crescimento urbano no Distrito Federal e seu entorno imediato, 1964–1990: leitura a partir de dados de sensoriamento remoto. (*Geosul/Florianópolis*, 6:11, 1991, p. 55–73, bibl., graph, maps)
Airphoto analysis denotes rapid peripheral growth of urban centers near Brasília.

2630 Aryeetey-Attoh, Samuel. Housing affordability ratios in Rio de Janeiro, Brazil. (*Yearbook/CLAG*, 15, 1989, p. 49–58, map, tables)
Detailed study of households of different income levels suggests that normal parameters of housing affordability are inadequate.

2631 Aumond, Juarês José; Edson Fortes; and Carlos Loch. Uso do sensoriamento remoto para análise do impacto ambiental resultante da atividade cerâmica no Vale do Rio Tijucas. (*Geosul/Florianópolis*, 6:11, 1991, p. 75–88, bibl., maps, photos)
Compares Landsat images and airphotos to evaluate environmental impact of extraction of clay minerals for the ceramic industry.

2632 Becker, Bertha K. Amazônia. São Paulo: Editora Atica, 1990. 112 p.: bibl., ill., maps. (Série Princípios; 192)
Comprehensive work emphasizes issues of land monopoly, colonization, land conflicts, migration, mineral exploitation, and formation of new subregions. Includes annotated bibliography.

2633 Becker, Bertha K. Fragmentação do espaço e formação de regiões na Amazônia—um poder territorial? (*Rev. Bras. Geogr.*, 52:4, oct./dez. 1990, p. 117–126, bibl., table)
Discusses emerging political significance of new subregions (see item **2632**).

2634 Beltrame, Angela da Veiga. A colonização do Vale do Itajaí-Mirim e os reflexos na degradação de seus recursos naturais renováveis. (*Geosul/Florianópolis*, 6:11, 1991, p. 91–100, bibl.)
Historical environmental impact study.

2635 Bicalho, Ana Maria de Souza Mello and Scott William Hoefle. Urban capital and pseudo-modernization of agriculture in the rural hinterland of Northeast Brazil. (*Yearbook/CLAG*, 15, 1989, p. 35–48, bibl., map, tables)
Suggests that cattle ranching in the

Agreste favors urban-based absentee landowners and results in displacement of peasants, unemployment, and social polarization.

2636 Brasil '92: perfil ambiental e estratégias. São Paulo: Secretaria do Meio Ambiente; Vitória, Brazil: Associação Brasileira das Entidades de Meio Ambiente, 1992. 218 p.: bibl.

Covers environmental proposals for regional policy implementation.

2637 Bressan, Delmar Antonio. A gestão racional dos ecossistemas. (*Ciênc. Amb.*, 3:4, jan./junho 1992, p. 33–53, table)

Dialectic discussion of microbasins and reforestation conservation systems.

2638 Browder, John O. and Brian J. Godfrey. Frontier urbanization in the Brazilian Amazon: a theoretical framework for urban transition. (*Yearbook/CLAG*, 16, 1990, p. 56–66, bibl., graph, map, table)

Useful historical-geographical model of urban transition that postulates settlement stages in Amazonia. Provides pertinent theoretical framework for future investigations of frontier urban networks as they function within the national economy and impact rural land use.

2639 Brown, I. Foster et al. Carbon storage and land-use in extractive reserves: Acre, Brazil. (*Environ. Conserv.*, 19:4, Winter 1992, p. 307–315, appendix, bibl., tables)

Technical study from the Chico Mendes Extractive Reserve concludes that sustainable forest extractive methods result in less release of carbon than does slash-and-burn agriculture. Suggests diversification and improved income to maintain rubber tappers in the region.

2640 Campos Ribeiro, Miguel Ângelo and Roberto Schmidt de Almeida. Análise da organização espacial da indústria nordestina através de uma tipologia de centros industriais. (*Rev. Bras. Geogr.*, 53:2, abril/junho 1991, p. 5–31, bibl., tables)

Classification of seven industrial centers based on value added, degree of diversification, and distribution.

2641 Cardoso, Maria Francisca Thereza. Organização e reorganização do espaço no Vale do Paraíba do Sul: uma análise geográfica até 1940. (*Rev. Bras. Geogr.*, 53:1, jan./março 1991, p. 81–135, bibl.)

Regional history.

2642 Caviedes, César N. and Keith D. Muller. Fruticulture and uneven development in Northeast Brazil. (*Geogr. Rev.*, 84:4, Oct. 1994, p. 380–393)

Irrigation project designed to improve economic condition of peasant cultivators. However, most peasants maintain traditional farming practices and are not incorporated into the thriving production system of international export enterprises.

2643 Cerrado: caracterização, ocupação e perspectivas. Organização por Maria Novaes Pinto. Brasília, Distrito Federal: Editora UnB; Governo do Distrito Federal, Secretaria do Meio Ambiente, Ciência e Tecnologia, 1990. 657 p.: bibl., ill., maps.

Majority of 20 authors discuss physical geography topics, while others cover human impact on the environment and conservation.

2644 Cleary, David. The "greening" of the Amazon. (*in* Environment and development in Latin America: the politics of sustainability. Edited by David Goodman and Michael Redclift. Manchester, England: Manchester Univ. Press; New York: St. Martin's Press, 1991, p. 116–140, bibl., maps, table)

A general narrative discussion.

2645 Clement, Charles R. Origin, domestication and genetic conservation of Amazonian fruit tree species. (*in* International Congress of Ethnobiology, *1st, Belém, Brazil, 1988.* Ethnobiology: implications and applications. Belém, Brazil: Museu Paraense Emílio Goeldi, 1990, v. 1, p. 249–263, bibl., graphs, maps, tables)

Calls for research and subsidies to preserve fruit tree species utilized by Amerindians that are threatened by development.

2646 Cordeiro, Helena Kohn and Denise Aparecida Bovo. A modernidade do espaço brasileiro através da rede nacional de Telex. (*Rev. Bras. Geogr.*, 52:1, jan./março 1990, p. 107–154, graphs, maps, tables)

Describes the Telex network.

2647 Costa, Nadja Maria Castilho da and Cláudia Rodrigues Segond. Plano de manejo ecológico da Reserva Particular de Bodoquena. (*Rev. Geogr./México*, 114, julio/dic. 1991, p. 91–100, bibl., maps, table)

Multidisciplinary descriptive essay

about a private reserve bordering the Pantanal with recommendations to revive the environment.

2648 Costa, Rogério Haesbaert da. RS—latifúndio e identidade regional. Porto Alegre, RS: Mercado Aberto, 1988. 98 p.: bibl., ill. (Documenta; 25. Geografia/história)

Emphasizes political regions as related to latifundia.

2649 Coy, Martin and Reinhold Lücker. Der brasilianische Mittelwesten: Wirtschafts- und sozialgeographischer Wandel eines peripheren Agrarraumes. Tübingen: Im Selbstverlag des Geographischen Instituts der Univ. Tübingen, 1993. 305 p.: bibl., ill., maps. (Tubinger Beiträge zur geographischen Lateinamerika-Forschung; Heft 9. Tübinger geographische Studien, 0932–1438; Heft 108)

Detailed monograph describes class structures, land occupation, and agrarian utilization of space in Mato Grosso and Mato Grosso do Sul. Reveals social inequality created by rapid inflow of landless colonists and land speculators with financial means, under the modernizing impact of export agriculture. Summarized version of this monograph is offered in item **2650**. [C. Caviedes]

2650 Coy, Martin. Cuiabá, Mato Grosso: wirtschafts- und sozialräumlicher Strukturwandel einer Regionalmetropole im brasilianischen Mittelwesten. (*Z. Wirtsch.geogr.*, 36:4, 1992, p. 193–209, bibl., maps, tables)

Assesses role played by the city of Ciuabá (401,461 inhabitants in 1991) in the economic expansion of west central Brazil. The city is growing fast and so are problems related to housing and environmental deterioration. [C. Caviedes]

2651 Coy, Martin and Reinhold Lücker. Mutations dans un espace périphérique en cours de modernisation: espaces sociaux dans le milieu rural du Centro-Oeste brésilien. (*Cah. Outre-Mer*, 46:182, avril/juin 1993, p. 153–174, bibl., graph, maps)

Conflicts of interest between upper classes engaged in extensive development projects of lumber, soybeans, maize, and lower classes involved in small-scale extractive industries characterize underdevelopment in most colonization fronts of centralwestern Brazil. See also item **2652**. [C. Caviedes]

2652 Coy, Martin. Regionalentwicklung und regionale Entwicklungsplanung an der Peripherie in Amazonien: Probleme und Interessenkonflikte bei der Erschliessung einer jungen Pionierfront am Beispiel des brasilianischen Bundesstaates Rondônia. Tübingen: Im Selbstverlag des Geographischen Instituts der Univ. Tübingen, 1988. 535 p.: bibl., ill., maps. (Tübinger Beiträge zur geographischen Lateinamerika-Forschung; Heft 5. Tübinger geographische Studien, 0932–1438; 97)

Comprehensive survey of colonization front development in western Brazil. Reviews status of colonization fronts in Rondônia including examinations of PoloNoroeste, a government-sponsored regional development plan, and PIC Ouro Preto, a migrant settlement subproject. [C. Caviedes]

2653 Davidovich, Fany. Brasil metropolitano e Brasil urbano não-metropolitano: algumas questões. (*Rev. Bras. Geogr.*, 53:2, abril/junho 1991, p. 127–133, bibl.)

Urban population is expected to incease from 75 to 80 percent from 1990–2000. Questions impact of urban growth and reflects on poverty, violence, land conflict, and the restructure of cities.

2654 Dawsey, Cyrus B., III. Migration in Brazil: research during the 1980s. (*Yearbook/CLAG*, 17/18, 1990, p. 109–116, bibl.)

Useful overview recognizes two dominant trends: 1) a rural-urban shift due to land consolidation, improved transportation, and agricultural capitalization; and 2) movement to Amazonia, principally to Rondônia where high densities are unlikely to be supported.

2655 Denevan, William M. and Mário Hiraoka. Geographic research on aboriginal and peasant cultures in Amazonia, 1980–1990. (*Yearbook/CLAG*, 17/18, 1990, p. 117–126, bibl.)

Relevant survey focuses on Indian subsistence, riverine peasants, and frontier settlement. Other themes include prehistoric agriculture and demography, protein scarcity, crop diversity, swidden-fallow management, floodplain cultivation, colonization, forest extraction, fishing, cattle ranching, deforestation, and soil decline.

2656 A desordem ecológica na Amazônia. Organizão por Luis E. Aragón. Belém, Brazil: Editora Universitária UFPA, 1991. 486

p.: bibl., ill., maps. (Série Cooperação amazônica; 7)

Commendable book from international conference on 25 environmental topics, many by prominent authors. Main ecological themes include development and conservation, man and his habitat, uses and abuses of Amazonia, and global implications of the above.

2657 **Dias, Sérgio da Fonseca.** Subsídios para a formulação de uma política de revitalização da produção de borracha natural para a Amazônia. (*Pará Desenvolv.*, 25, jan./dez. 1989, p. 72–87, bibl., graphs, maps, tables)

Cost analysis proposal for a hypothetical rubber plantation.

2658 **Encontro Nacional de Estudos sobre Meio Ambiente, 2nd, *Florianópolis, Brazil, 1989.*** Anais: 20. Encontro Nacional de Estudos sobre Meio Ambiente, Florianópolis, 24 a 29/09/89. v. 1–3. Florianópolis: Mestrado em Geografia, Coordenadoria de Pós-Graduação em Geografia, Depto. de Geociências, Centro de Ciências Humanas, Univ. Federal de Santa Catarina, 1989. 3 v.: bibl., ill., maps.

Consists of 75 articles on the environment ranging from five to 15 p. Topics include development, mining, coastal degradation, water resources, and policymaking. See also item **2659**.

2659 **Encontro Nacional de Estudos sobre Meio Ambiente, 3rd, *Londrina, Brazil, 1991.*** Anais. v. 1–2. Londrina, Brazil: Univ. Estadual de Londrina, Núcleo de Estudos do Meio Ambiente; Maringá: Univ. Estadual de Maringá, Departamento de Geografia; Presidente Prudente: UNESP de Presidente Prudente, Mestrado em Geografia, 1991–. v. 1–2: bibl., ill., index, maps.

Extensive coverage of environmental issues in more than 100 articles that range from five to 15 pages each. Degradation of drainage basins, urban and rural development, role of the university, and legislative policy are the main themes. See item **2658**.

2660 **Fearnside, Philip Martin.** Manejo florestal na Amazônia: necessidade de novos critérios na avaliação de opções de desenvolvimento. (*Pará Desenvolv.*, 25, jan./dez. 1989, p. 49–59, bibl., table)

Author believes that sustainable management of the Amazonian forest on a commercial scale is non-existent, that research is only in its infancy, and that more importance should be devoted to cost-benefit analysis.

2661 **Fearnside, Philip Martin.** A ocupação humana de Rondônia: impactos, limites e planejamento. Brasília: Programa Polonoroeste; SCT/PR; CNPq, 1989. 76 p.: bibl., ill., map. (Relatório de pesquisa/Programa Polonoroeste; 5)

Describes causes and distribution of the deforestation process, planning, limitations to settlement, and measures to halt deforestation.

2662 **Fox, David J.** Decentralization, debt, democracy, and the Amazonian frontierlands of Bolivia and Brazil. (*in* Decentralization in Latin America: an evaluation. Edited by Arthur Morris and Stella Lowder. New York: Praeger, 1992, p. 17–37, bibl., graphs, map)

Suggests that supranational development agencies foster cross-border cooperation.

2663 **Fundação Instituto Brasileiro de Geografia e Estatística. Diretoria de Geociências.** Geografia do Brasil. v. 3, Região norte. Rio de Janeiro: Diretoria de Geociências, 1991. 1 v.: bibl., ill. (some col.), maps.

Worthy IBGE regional volume that covers major physical and cultural subdivisions.

2664 **Fundação Instituto Brasileiro de Geografia e Estatística.** Diagnóstico geoambiental e sócio-econômico: área de influência da BR-364, trecho Porto Velho-Rio Branco. v. 1. Rio de Janeiro: IBGE, 1990–. 1 v.: bibl., ill., maps, tables.

Detailed report discusses impact of the BR-364 highway on migration, land use, agriculture, and levels of social conflict. Also presents future recommendations. Includes useful tables on Indian groups.

2665 **The future of Amazonia: destruction or sustainable development?** Edited by David Goodman and Anthony Hall. New York: St. Martin's Press, 1990. 419 p.: bibl., ill., index, maps.

Valuable collection by 15 authors focuses on current development strategies, frontier integration, environmental destruction, social conflict, and sustainable development.

2666 Gelpi, Adriana et al. O Rio Grande do Sul urbano. Organzação por Naia Oliveira e Tanya Barcellos. Porto Alegre, RS: Secretaria de Coordenação e Planejamento, Fundação de Economia e Estatística Siegfried Emanuel Heuser, 1990. 262 p.: bibl., ill., maps.

Several authors cover urban topics such as violence, housing, technology, planning, non-governmental organizations, transportation, and the need for interdisciplinary research.

2667 Góes-Filho, Luiz and Ricardo Forin Lisboa Braga. A vegetação do Brasil: desmatamento e queimadas. (*Rev. Bras. Geogr.*, 53:2, abril/junho 1991, p. 135–141, maps, table)

Analysis of interrelationship between deforestation and burning in Amazonia. Describes natural resource exploitation.

2668 Grabois, José and Mauro José da Silva. O Brejo de Natuba: estudo da organização de um espaço periférico. (*Rev. Bras. Geogr.*, 53:2, abril/junho 1991, p. 33–62, bibl., photos, table)

Describes land use transformation from coffee and oranges to bananas, a transition that results in decreasing demands for labor and a declining population. See item **2652**.

2669 Grabois, José; Maria Inez Medeiros Marques; and Mauro José da Silva. A organização do espaço no baixo Vale do Taperoá: uma ocupação extensiva em mudança. (*Rev. Bras. Geogr.*, 53:4, out/dez. 1991, p. 81–114, bibl., graphs, photos, tables)

Detailed case study examines cattle raising, small subsistence farms, and irrigation for garlic production that result in ecological degradation and concentration of income in the Northeast.

2670 Guimarães, Raúl Borges. A tecnificação da prática médica no Brasil: em busca de sua geografização. (*Terra Livre*, 9, julho/dez. 1991, p. 41–55, bibl. ill., map)

Attempts to explain spatial distribution and development of medical technology.

2671 Haffer, Jürgen. Ciclos de tempo e indicadores de tempos na história da Amazônia. (*Estud. Avançados*, 15, maio/junho 1992, p. 7–39, bibl., graphs, ill., map)

Astronomical cycles of Milankovitch are employed to explain climatic and vegetative fluctuations, which help to account for the richness of species.

2672 Hecht, Susanna B. and Darrell A. Posey. Indigenous soil management in the Latin American tropics: some implications for the Amazon Basin. (*in* International Congress of Ethnobiology, *1st, Belém, Brazil, 1988.* Ethnobiology: implications and applications. Belém, Brazil: Museu Paraense Emílio Goeldi, 1990, v. 2, p. 73–86, bibl., tables)

Native agriculture and modernized livestock production are compared. Argues for study of native systems such as that of the Cayapó who practice sustainable agriculture.

2673 Hiraoka, Mário. Mudanças no padrões econômicos de uma população ribeirinha do estuário do Amazonas. (*in* Povos das águas: realidade e perspectivas na Amazônia. Belém, Brazil: Museu Paraense Emílio Goeldi, 1993, p. 133–157)

Competent study examines viability of supplementing riverine agriforestry. Author values native systems, but also questions them as workable commercial models that simultaneously could lessen deforestation, recuperate degraded land, and increase the value of tropical forests. Suggests new products, sound planning, and government subsidies.

2674 Isenburg, Teresa. L'agricoltura nello stato di San Paolo, Brasile: agroindustria e scarsità alimentare. (*Riv. Geogr. Ital.*, 98:1, marzo 1991, p. 33–65, bibl., tables)

Spatial analysis of agriculture, rural life, culture, and the success of citrus fruit and sugar production. Impact of modern technology on land consolidation, pollution, and lack of food crops are also discussed.

2675 Kahn, Francis and Jean-Jacques de Granville. Palms in forest ecosystems of Amazonia. Berlin; New York: Springer-Verlag, 1992. 226 p.: bibl., ill., index, maps. (Ecological studies; 95)

Comprehensive biogeographical work emphasizes palm distribution as a function of soil drainage, topography, forest architecture and dynamics, and human activities.

2676 Kohlhepp, Gerd. Desenvolvimento regional adaptado: o caso da Amazônia brasileira. (*Estud. Avançados*, 6:16, set./dez. 1992, p. 81–102, bibl.)

Existing projects for sustaining the Brazilian rainforest are analyzed, particularly implications for biological preserves and the habitat of indigenous groups.

2677 **Kohlhepp, Gerd.** Mudanças estruturais na agropecuária e mobilidade da população rural no norte do Paraná, Brasil. (*Rev. Bras. Geogr.*, 53:2, abril/junho 1991, p. 79–94, bibl., maps, table)

Sound analysis of land use changes and impact of modernization of agriculture. Notes that ecologically and economically sound crop rotations and conservation practices combine to stabilize agriculture, but that many displaced small farmers are forced to migrate to pioneer zones or to urban areas.

2678 **Kohlhepp, Gerd.** The regional impact of major projects and possible spatial reorganization of the Amazon periphery. (*in* Ecological disorder and Amazonia. Paris: International Social Science Council; Rio de Janeiro: Editora Universitária Candido Mendes, 1991, p. 147–187, bibl.)

Author uses the Greater Carajás Development Project to assert that settlement and exploration of the Amazon Basin cannot succeed if ecological concerns are not respected and land speculation continues at its present pace. Pauperization of rural inhabitants will increase. [C. Caviedes]

2679 **Kohlhepp, Gerd.** Strukturwandlungen in der Landwirtschaft und Mobilität der ländlichen Bevölkerung in Nord-Paraná (Südbrasilien). (*Geogr. Z.*, 77:1, 1989, p. 42–62, bibl., maps)

Transition from coffee cultivation to stock raising and mechanized soy and wheat cultivation has created land tenure concentration and a mass of landless workers in the state of Paraná since the early 1970s. See item **2677** for Portuguese translation. [C. Caviedes]

2680 **Kohlhepp, Gerd.** Umweltpolitik zum Schutz tropischer Regenwälder in Brasilien. (*KAS Ausl.-Inf.*, 7:7, Juli 1990, p. 1–23, bibl.)

Institutional measures taken by the Brazilian government to curb rapid dwindling of its rainforests are critically evaluated. Also reviews efforts made by foreign agencies. Good summary of complex and politically-sensitive subject. [C. Caviedes]

2681 **Kvist, Lars Peter.** Hab for Amazonas. [Hope for Amazonas.] Arhus, Denmark: Nepenthes' Forlag, 1993. 104 p.: bibl., photos.

Provides ethnobotanical account of causes and consequences of deforestation of the Amazonian rainforest. Discusses the relationship between humans and nature at different levels. Outstanding photos. Highly recommended. [I. Schjellerup]

2682 **Laroche, Rose Claire.** Ecossistemas e impactos ambientais da modernização agrícola do Vale do São Francisco. (*Rev. Bras. Geogr.*, 53:2, abril/junho 1991, p. 63–77, bibl., ill.)

Integrated regional study examines environmental and socioeconomic problems in the context of modern agricultural technology and natural resource utilization.

2683 **Maimon, Dália.** Ensaios sobre economia do meio ambiente. Rio de Janeiro: APED Editora, 1992. 149 p.: bibl., ill.

Environmental essays discuss, in a general manner, topics such as education, present trends, policy, evaluation and quantification, industry, and debt conversion.

2684 **Mallet-Guy Guerra, Sinclair; Karenia Córdova Sáez;** and **Marco Antonio Bonn.** Perspectivas y estrategias para el gas natural en la América Latina: Brasil y Venezuela. (*Interciencia/Caracas*, 18:1, Jan./Feb. 1993, p. 24–28, maps, tables)

Examines natural gas potential.

2685 **Marini, Onildo J.** and **Emanuel T. Queiroz.** Main geologic-metallogenetic environments and mineral exploration in Brazil. (*Ciênc. Cult.*, 43:2, março/abril 1991, p. 153–161, maps, tables)

Overview of five geologic provinces and their distribution.

2686 **Martine, George.** Ciclos e destinos na migração para áreas de fronteira na era moderna: uma visão geral. Brasília: Instituto Sociedade, População e Natureza, 1992. 27 leaves: bibl., tables. (Documento de trabalho; 12)

Examines urban concentration on the agricultural frontier. The many unsuccessful farmers flee to the cities that lack infrastructure. See item **2687**.

2687 **Martine, George.** Processos recentes de concentração e desconcentração urbana no Brasil: determinantes e implicações. Brasília: Instituto Sociedade, População e Natureza, 1992. 29 leaves: bibl., graphs, tables. (Documento de trabalho; 11)

Due to economic failure on the agricultural frontier, cities are growing without the infrastructure to absorb unemployed farmers. See item **2686.**

2688 **Mello, João Baptista Ferreira de.** Geografia humanística: a perspectiva da experiência vivida e uma crítica radical ao positivismo. (*Rev. Bras. Geogr.*, 52:4, oct./dez. 1990, p. 91–115, bibl., map)

Dialectic overview of geographic thought.

2689 **Mello, Mauro Pereira de.** A questão de límites entre os estados do Acre, do Amazonas e de Rondônia: aspectos históricos e formação do território. (*Rev. Bras. Geogr.*, 52:4, oct./dez. 1990, p. 5–71, bibl., maps)

Lengthy historical attempt to clarify the Acre/Rondônia border dispute.

2690 **Moran, Emilio F.** A ecologia humana das populações da Amazônia. Petrópolis: Vozes, 1990. 367 p.: bibl., ill., maps. (Col. Ecologia & ecosofia)

Commendable socio-anthropologic work that considers ecological regional differences by underscoring climatic, pedalogic, and vegetative influences.

2691 **Mueller, Charles Curt.** Agriculture, urban bias development, and the environment: the case of Brazil. Brasília: Instituto Sociedade, População e Natureza, 1992. 10 leaves: bibl. (Documento de trabalho; 14)

Agricultural expansion mostly benefits the domestic urban-industrial sector and negatively impacts the environment.

2692 **Mueller, Charles Curt.** Colonization policies, land occupation, and deforestation in the Amazon countries. Brasília: Instituto Sociedade, População e Natureza, 1992. 22 leaves: bibl. (Documento de trabalho; 15)

Overview of development by Amazonian countries since World War II. Fear of foreign intervention has led to colonization projects for demographic expansion and economic integration, while Amerindians were mostly ignored.

2693 **Mueller, Charles Curt; Haroldo Torres;** and **George Martine.** Settlement and agriculture in Brazil's forest margins and savannah agrosystems. Brasília: Instituto Sociedade, População e Natureza, 1992. 79 leaves: bibl., graphs, tables. (Documento de trabalho; 10)

Worthy overview of the frontier since the 1950s that identifies settlement zones and agricultural systems.

2694 **Muller, Keith D.** The future of the long-lot in South Brazil: the case of West Paraná. (*Pap. Proc. Appl. Geogr. Conf.*, 17, 1994, p. 129–134)

Questions applicability of widespread and successful long-lot system of land division in which mechanized agriculture replaces traditional oxen-plow with hoe methods.

2695 **Muller, Keith D.** Pindorama: European agricultural settlement scheme, coastal Northeast Brazil. (*Pap. Proc. Appl. Geogr. Conf.*, 16, 1993, p. 1–7)

Peasant members have risen above subsistence in a region dominated by latifundia-sized sugarcane holdings. Project's success can be attributed to a cooperative that is involved in all levels of growing, processing, and marketing.

2696 **Nentwig Silva, Bárbara-Christine.** Análise comparativa da posição de Salvador e do Estado da Bahia no cenário nacional. (*Rev. Bras. Geogr.*, 53:4, out./dez. 1991, p. 49–79, bibl., graphs, tables)

Examines region's demographic growth with respect to economic potential of private corporations and public entities.

2697 **Neupert, Ricardo F.** La colonización brasileña en la frontera agrícola del Paraguay. (*Notas Pobl.*, 28/29:51/52, dic. 1990/abril 1991, p. 121–155, bibl., maps)

Examines lack of integration of Brazilian immigrants with modernized farms that contrast to small subsistence plots of Paraguayans.

2698 **Nugent, Stephen L.** The limitations of environment "management:" forest utilisation in the Lower Amazon. (*in* Environment and development in Latin America: the politics of sustainability. Edited by David Goodman and Michael Redclift. Manchester, England: Manchester Univ. Press; New York: St. Martin's Press, 1991, p. 141–154, bibl.)

Declares that without agrarian reform and recognition of Amazonia realities, rational forest extraction is not viable.

2699 **Oliveira, Elizabeth Homem et al.** Poder e participação política no campo. Coordenação por Eduardo Paes Machado. São Paulo: Cerifa; Editora Hucitec, 1987. 152 p.: bibl., ill.

Examines rural development and political participation of small farmers around Lake Sobradinho on the São Francisco River.

2700 **Oliveira, Márcio de.** A questão da industrialização no Rio de Janeiro: algumas reflexões. (*Terra Livre*, 9, julho/dez. 1991, p. 91–101, bibl.)

Historical study of the textile industry related to urban land use with a dialectic slant.

2701 **Pesquisa: trabalho temporário e migrações na agricultura baiana.** Salvador, Brazil: Governo do Estado da Bahia, Secretaria do Trabalho e Bem-Estar Social, Superintendência Baiana para o Trabalho, 1988. 209 p.: bibl., ill. (Série População e emprego; 22)

Overview of rural migrant laborers, which includes methodological discussion of surveying seasonal workers.

2702 **Pires, J.S.R.** and **Evlyn M.L.M. Novo.** Use of TM/Landsat data to identify silting areas in the Tucuruí reservoir. (*Ciênc. Cult.*, 43, Sept./Oct. 1991, p. 385–387, bibl., ill.)

Integrates three years of limnological and radiometric data.

2703 **O processo de urbanização no Oeste Baiano.** Coordenção por Milton Santos Filho. Recife: Ministério do Interior, Superintendência do Desenvolvimento do Nordeste, Diretoria de Planejamento Global, Depto. de Planejamento Sub-Regional e Urbano, Grupo de Desenvolvimento Urbano, 1989. 281 p.: bibl., maps. (Estudos urbanos; 1)

Overview of the *cerrado*, a recently opened pioneer area of modernized agriculture. Emphasizes need for urban infrastructure to support agriculture, as well as the importance of farm cooperatives.

2704 **Ramires, Julio Cesar de Lima.** As grandes corporações e a dinâmica socioespacial: a ação da Petrobrás em Macaé. (*Rev. Bras. Geogr.*, 53:4, out/dez. 1991, p. 115–151, bibl., maps, tables)

Addresses impact of petroleum and natural gas exploration and production in the state of Rio de Janeiro.

2705 **Ribeiro, Miguel Angelo Campos** and **Roberto Schmidt de Almeida.** Os pequenos e médios estabelecimentos industrias nordestinos: padrões de distribução e fatores condicionantes. (*Rev. Bras. Geogr.*, 53:1, jan./março 1991, p. 5–49, bibl., maps, tables)

Describes principal industries and their economic importance for rural communities. Products which include hammocks and fish nets, processed sugar and rum, flour, cigars, and leather goods.

2706 **Sampaio, Sílvia Selingardl** and **Vera Cristina Daltrini.** Da agro-indústria à produção de bens de capital: a evolução industrial do Município de Araraquara (SP). (*Geografia/Río Claro*, 15:1, abril 1990, p. 55–80, bibl., tables, map)

Focuses on historical expansion of agroindustry (especially oranges), describing structural changes and decentralization.

2707 **Saragoussi, Muriel; Jorge Hugo I. Martel;** and **Gilberto de Assis Ribeiro.** Comparação na composição de quintais de três localidades de terra firme do estado do Amazonas. (*in* International Congress of Ethnobiology, *1st, Belém, Brazil, 1988.* Ethnobiology: implications and applications. Belém, Brazil: Museu Paraense Emílio Goeldi, 1990, v. 1, p. 295–303, bibl., map, tables)

Detailed comparative study of subsistence gardens shows that recent settlers experiment with more crop species than do more established settlers.

2708 **Sawyer, Donald Rolfe.** Malaria and the environment. Brasília: Instituto Sociedade, População e Natureza, 1992. 37 leaves: bibl., tables. (Documento de trabalho; 13)

Widespread control of tropical diseases seems unrealistic and large-scale migration to Amazonia is unwise. Insecticides and antimalarial drugs are not practical because of cost and vastness of area.

2709 **Serra, Geraldo.** Urbanização e centralismo autoritário. São Paulo: EDUSP; Nobel, 1991. 172 p.: bibl., ill.

Discusses Brazilian urbanization in general terms, using as examples the cities of São Luiz, Natal, Rio Branco, Ribeirão Preto, Santa Maria, and Cuiabá.

2710 Silva, Ciléa Souza de; José Carlos Valim Rodrigues; and **Nelly Lamarão Câmara.** Saneamento básico e problemas ambientais na região metropolitana do Rio de Janeiro. (*Rev. Bras. Geogr.*, 52:1, jan./março 1990, p. 5–105, bibl., maps, graphs, tables)

Lengthy description of water and sewer systems and related environmental and health problems.

2711 Silva, Sylvio Bandeira de Mello e and **Jaimeval Caetano de Souza.** Análise da hierarquia urbana do estado da Bahia. (*Rev. Bras. Geogr.*, 53:1, jan./março 1991, p. 51–79, bibl., maps, tables)

Quantitative analysis of centrality based on urban functions.

2712 Sternberg, Rolf. Perspectivas geográficas nos sistemas hidroeléctricos. (*Rev. Bras. Geogr.*, 52:1, jan./março 1990, p. 157–186, graphs, tables)

Reviews literature and provides model for the geographical study of hydroelectric energy.

2713 Strohaecker, Tânia Marques and **Célia Ferraz de Souza.** A localização industrial intra-urbana: evolução e tendências. (*Rev. Bras. Geogr.*, 52:4, oct./dez. 1990, p. 73–89, bibl., ill.)

Review of industrial location literature related especially to intra-urban space, and to the analysis of dynamics, locational patterns, and trends in urban industrial activities.

2714 Struck, Ernst. Persistenz und Wandel des zentralörtlichen Gefüges im brasilianischen Bundesstaat Espírito Santo. (*Z. Wirtsch.geogr.*, 36:4, 1992, p. 229–237, bibl., ill.)

An investigation of urban centers in the state of Espírito Santo finds remarkable increases in population and numbers of private economic enterprises during the last 25 years. Reveals significant changes in spatial patterns of settlement hierarchy in central Brazil. [C. Caviedes]

2715 Torres, Haroldo and **George Martine.** Amazonian extractivism: prospects and pitfalls. Brasília: Instituto Sociedade, População e Natureza, 1991. 30 p.: bibl., tables. (Documento de trabalho; 5)

Suggests that the future of Amazonian rubber and nut production is limited by low absorptive capacity of extractive activities.

2716 Townroe, Peter. Decision making in decentralizing companies in a city region: the case of greater São Paulo. (*in* Decentralization in Latin America: an evaluation. Edited by Arthur Morris and Stella Lowder. New York: Praeger, 1992, p. 93–108, bibl., map, tables)

Considers industrial location decision-making from a sampling of over 500 private companies.

2717 Tricart, Jean L.F. Ce qu'apporte la télédétection à la connaissance du milieu et pour sa gestion, particulièrement en Amazonie: l'expérience de Radambrasil. (*Bull. Assoc. géogr., fr.*, 69:5, dec. 1992, p. 403–421, bibl.)

Calls for accurate ground verification and data registration to accompany the Radambrasil remote sensing program.

2718 Veríssimo, Adalberto *et al.* Impactos sociais, econômicos e ecológicos da exploração seletiva de madeiras numa região de fronteira na Amazônia oriental: o caso de Tailândia. (*Pará Desenvolv.*, 25, jan./dez. 1989, p. 95–116, bibl., graphs, maps, tables)

Provides cost efficiency details for small saw mills. Also calls for improved uses of discarded trees.

2719 Vieira, Eurípedes Falcão and **Susana Regina Salum Rangel.** Planície costeira do Rio Grande do Sul: geografia física, vegetação e dinâmica sócio-demográfica. Porto Alegre, RS: Sagra, 1988. 256 p.: bibl., ill., maps.

General descriptive geography.

2720 Vieira, Paulo Freire. A problemática ambiental e as ciências sociais no Brasil: 1980–1990. (*ANPOCS BIB*, 33, 1992, p. 3–32, bibl.)

Survey of methodological approaches to environmental problems in the social sciences, including human geography. Extensive bibliography.

2721 Wesche, Rolf and **Michael Small.** Brazilian Amazonia: from destruction to sustainable development. (*in* Latin American to the year 2000: reactivating growth, improving equity, sustaining democracy. New York: Praeger, 1992, p. 81–96, bibl., map)

Suggests that future investment will be spatially selective for mineral and hydropower development, and that the strong international connections of the Brazilian government will result in less deforestation.

2722 Willis, Edwin O. and **Yoshika Onki.**
Losses of São Paulo birds are worse in the interior than in Atlantic forests. (*Ciênc. Cult.*, 44:5, set./out. 1992, p. 326–328, tables)

Attempts to explain the disappearance of birds due to deforestation and human occupation. Useful biogeographical study.

JOURNAL ABBREVIATIONS

Agroecol. Desarro. Agroecología y Desarrollo. Consorcio Latinoamericano sobre Agroecología y Desarrollo (CLADES). Santiago.

Allpanchis/Cusco. Allpanchis. Instituto de Pastoral Andina. Cusco, Peru.

Alternativa/Chiclayo. Alternativa. Centro de Estudios Sociales Solidaridad. Chiclayo, Peru.

An. Inst. Patagon./Soc. Anales del Instituto de la Patagonia: Serie Ciencias Sociales. Univ. de Magallanes. Punta Arenas, Chile.

Anc. Mesoam. Ancient Mesoamerica. Cambridge Univ. Press. Cambridge, England.

Ann. Assoc. Am. Geogr. Annals of the Association of American Geographers. Lawrence, Kan.

Ann. géogr./Paris. Annales de géographie. Bulletin de la Société de géographie; Librairie Armand Colin. Paris.

ANPOCS BIB. Boletim Informativo e Bibliográfico de Ciências Sociais: BIB. Associação Nacional de Pós-Graduação e Pesquisa em Ciências Sociais. Rio de Janeiro.

Arctic Alpine Res. Arctic and Alpine Research. Institute of Arctic and Alpine Research, Univ. of Colorado. Boulder.

Bol. Inst. Geogr. Boletín del Instituto de Geografía. Univ. Nacional Autónoma de México. México.

Bol. Lima. Boletín de Lima. Revista Cultural Científica. Lima.

Boletín/Bogotá. Boletín del Museo del Oro. Banco de la República. Bogotá.

Bull. Assoc. géogr. fr. Bulletin de l'Association de géographes français. Paris.

Bull. East. Caribb. Aff. Bulletin of Eastern Caribbean Affairs. Univ. of West Indies. Cave Hill, Barbados.

Bull. Inst. fr. étud. andin. Bulletin de l'Institut français d'études andines. Lima.

Bull. Lat. Am. Res. Bulletin of Latin American Research. Society for Latin American Studies. Oxford, England.

Cah. Outre-Mer. Les Cahiers d'Outre-Mer. Institut de géographie de la Faculté des lettres de Bordeaux; Institut de la France d'Outre-Mer; Société de géographie de Bordeaux. Bordeaux, France.

Caravelle/Toulouse. Caravelle. Cahiers du monde hispanique et luso-brésilien. Univ. de Toulouse, Institute d'études hispaniques, hispano-americaines et luso-brésiliennes. Toulouse, France.

Caribb. Perspect. Caribbean Perspectives. Transaction Publishers. New Brunswick, New Jersey.

Cartographica/Ontario. Cartographica. Univ. of Toronto Press. Downsview, Ontario.

Catena/Cremlingen. Catena. Catena Verlag. Cremlingen, Wolfenbüttel, Germany.

Centen. Rev. The Centennial Review. Michigan State Univ., College of Science and Arts. East Lansing, Mich.

Ciênc. Amb. Ciência & Ambiente. Univ. Federal de Santa Maria. Santa Maria, Brazil.

Ciênc. Cult. Ciência e Cultura. Sociedade Brasileira para o Progresso da Ciência. São Paulo.

Coyunt. Soc. Coyuntura Social. Fundación para la Educación Superior y el Desarrollo (FEDESARROLLO); Instituto SER de Investigación. Bogotá.

Desarro. Econ. Desarrollo Económico. Instituto de Desarrollo Económico y Social. Buenos Aires.

Econ. Dev. Cult. Change. Economic Development and Cultural Change. Univ. of Chicago, Research Center in Economic Development and Cultural Change. Chicago, Ill.

Econ. Geol. Economic Geology. Economic Geology Publishing Co., Lancaster, Penn.

Environ. Conserv. Environmental Conservation. Foundation for Environmental Conservation; Elsevier Sequoia. Lausanne, Switzerland.

Erdkunde/Bonn. Erdkunde. Archiv für Wissenschaftliche Geographie. Univ. Bonn, Geographisches Institut. Bonn, Germany.

Espac. Desarro. Espacio y Desarrollo. Pontificia Univ. Católica del Perú, Depto. de Humanidades, Centro de Investigación en Geografía Aplicada. Lima.

Espace géogr. L'Espace géographique. Doin. Paris.

Estud. Avançados. Estudos Avançados. Univ. de São Paulo, Instituto de Estudos Avançados. São Paulo.

Estud. Demogr. Urb. Estudios Demográficos y Urbanos. El Colegio de México. México.

Estud. Geogr./Madrid. Estudios Geográficos. Instituto de Economía y Geografía Aplicadas, Consejo Superior de Investigaciones Científicas. Madrid.

Estud. Públicos. Estudios Públicos. Centro de Estudios Públicos. Santiago.

Estud. Soc./Santo Domingo. Estudios Sociales. Centro de Estudios Sociales Juan Montalvo, SJ. Santo Domingo.

Estud. Territ. Estudios Territoriales. Instituto del Territorio y Urbanismo; Ministerio de Obras Públicas y Transportes. Madrid.

EURE/Santiago. EURE: Revista Latinoamericana de Estudios Urbanos Regionales. Centro de Desarrollo Urbano y Regional, Univ. Católica de Chile. Santiago.

Explor. J. The Explorers Journal. New York.

Geogr. J. The Geographical Journal. The Royal Geographical Society. London.

Geogr. Rev. Geographical Review. American Geographical Society of New York.

Geogr. Rundsch. Geographische Rundschau. Zeitschrift für Schulgeographie. Georg Westermann Verlag. Braunschweig, Germany.

Geogr. Z. Geographische Zeitschrift. Franz Steiner Verlag. Wiesbaden, Germany.

Geografia/Rio Claro. Geografia. Associação de Geografia Teorética. Rio Claro, Brazil.

Geografie/Utrecht. Geografie. KNAG. Utrecht, The Netherlands.

Geographical/London. Geographical. Hyde Park Publications. London.

Geography/London. Geography. Geographical Assn., London.

GeoJournal/Boston. GeoJournal. D. Reidel Publishing Co., Boston, Mass.

Geosul/Florianópolis. Geosul. Depto. de Geociências, Univ. Federal de Santa Catarina. Florianópolis, Brazil.

HAHR. Hispanic American Historical Review. Conference on Latin American History of the American Historical Assn.; Duke Univ. Press. Durham, N.C.

Homme/Paris. L'Homme. Laboratoire d'anthropologie, Collège de France. Paris.

Hoy Hist. Hoy es Historia: Revista Bimestral de Historia Nacional e Iberoamericana. Editorial Raíces. Montevideo.

Hum. Ecol. Human Ecology. Plenum Publishing Corp., New York.

Hum. Organ. Human Organization. Society for Applied Anthropology. New York.

Interciencia/Caracas. Interciencia. Asociación Interciencia. Caracas.

Iztapalapa/México. Iztapalapa. Univ. Autónoma Metropolitana, División de Ciencias Sociales y Humanidades. México.

J. Cult. Geogr. Journal of Cultural Geography. Popular Culture Assn.; American Culture Assn.; Bowling Green State Univ., Bowling Green, Oh.

J. Ethnobiol. Journal of Ethnobiology. Center for Western Studies. Flagstaff, Ariz.

J. Geogr. Journal of Geography. National Council of Geographic Education. Menasha, Wis.

J. Hist. Geogr. Journal of Historical Geography. Academic Press. London; New York.

KAS Ausl.-Inf. KAS Auslands-Informationen. Konrad Adenauer-Stiftung. Bonn.

LARR. Latin American Research Review. Latin American Research Review Board. Univ. of New Mexico, Albuquerque, N.M.

Medio Ambient. Urban. Medio Ambiente y Urbanización. Comisión de Desarrollo Urbano y Regional, Consejo Latinoamericano de Ciencias Sociales. Buenos Aires.

Mem. Soc. Cienc. Nat. Memoria de la Sociedad de Ciencias Naturales La Salle. Caracas.

Mesoamérica/Antigua. Mesoamérica. Centro de Investigaciones Regionales de Mesoamérica. Antigua, Guatemala.

Mitt. Österr. Geogr. Ges. Mitteilungen der Österreichischen Geographischen Gesellschaft. Verleger, Herausgeber und Eigentümer. Vienna.

Mt. Res. Dev. Mountain Research and Development. International Mountain Society. Boulder, Colo.

Names/Freeman. Names: Journal of the American Name Society. Pine Hill Press. Freeman, S.D.

Nieuwe West-Indische Gids. Nieuwe West-Indische Gids. Martinus Nijhoff. The Hague.

Notas Pobl. . Notas de Población. Centro Latinoamericano de Demografía. Santiago.

Nueva Soc. Nueva Sociedad. Caracas.

Oceanus/Woods Hole. Oceanus. Oceanographic Institution. Woods Hole, Mass.

Paisajes Geogr. Paisajes Geográficos. Centro Panamericano de Estudios e Investigaciones Geográficas. Quito.

Palabra Hombre. La Palabra y el Hombre. Univ. Veracruzana. Xalapa, Mexico.

Pap. Proc. Appl. Geogr. Conf. Papers and Proceedings of the Applied Geography Conferences. Dept. of Geography, Kent State Univ.; State Univ. of New York (SUNY) at Binghamton.

Pará Desenvolv. Pará Desenvolvimento. Instituto de Desenvolvimento Econômico-Social do Pará. Belém, Brazil.

Quat. Res./New York. Quaternary Research. Academic Press. New York.

Rev. Bras. Geogr. Revista Brasileira de Geografia. Conselho Nacional de Geografia, Instituto Brasileiro de Geografia e Estatística. Rio de Janeiro.

Rev. Chil. Antropol. Revista Chilena de Antropología. Depto. de Antropología, Univ. de Chile. Santiago.

Rev. Chil. Geopolít. Revista Chilena de Geopolítica. Instituto Geopolítico de Chile. Santiago.

Rev. Esp. Antropol. Am. Revista Española de Antropología Americana. Facultad de Geografía e Historia. Univ. Complutense de Madrid.

Rev. Esp. Pac. Revista Española del Pacífico. Asociación Española de Estudios del Pacífico. Madrid.

Rev. Eur. Revista Europea de Estudios Latinoamericanos y del Caribe = European Review of Latin American and Caribbean Studies. Center for Latin American Research and Documentation; Royal Institute of Linguistics and Anthropology. Amsterdam.

Rev. Geogr. Chile. Revista Geográfica de Chile. Instituto Geográfico Militar. Santiago.

Rev. Geogr./Mérida. Revista Geográfica. Univ. de Los Andes. Mérida, Venezuela.

Rev. Geogr./México. Revista Geográfica. Instituto Panamericano de Geografía e Historia, Comisión de Geografía. México.

Rev. Geogr. Norte Gd. Revista de Geografía Norte Grande. Pontificia Univ. Católica de Chile. Santiago.

Rev. Geogr. Venez. Revista Geográfica Venezolana. Univ. de los Andes. Mérida, Venezuela.

Rev. Geol. Am. Central. Revista Geológica de América Central. Escuela Centroamericana de Geología, Univ. de Costa Rica. San José.

Rev. Hist. Am. Argent. Revista de Historia Americana y Argentina. Univ. Nacional de Cuyo, Instituto de Historia. Mendoza, Argentina.

Rev. Hist. Naval. Revista de Historia Naval. Instituto de Historia y Cultura Naval Armada Española. Madrid.

Rev. Interam. Bibliogr. Revista Interamericana de Bibliografía. Organization of American States. Washington.

Rev. Interam. Planif. Revista Interamericana de Planificación. Sociedad Interamericana de Planificación. Bogotá.

Rev. Mex. Sociol. Revista Mexicana de Sociología. Instituto de Investigaciones Sociales, Univ. Nacional Autónoma de México. México.

Rev. Tiers-Monde. Revue Tiers-Monde. Institut d'étude du développement économique et social, Univ. de Paris.

Rev. Univ. EAFIT. Revista Universidad EAFIT. Depto. de Comunicaciones, Univ. EAFIT. Medellín.

Riv. Geogr. Ital. Rivista Geografica Italiana. Societá di studi geografici di Firenze. Florence, Italy.

São Paulo Perspect. São Paulo em Perspectiva. Fundação SEADE. São Paulo.

Sierra/San Francisco. Sierra. Sierra Club. San Francisco.

Swiss Rev. World Aff. Swiss Review of World Affairs. Neue Zürcher Zeitung. Zürich, Switzerland.

Terra Livre. Terra Livre. Associação dos Geógrafos Brasileiros. São Paulo.

Third World Plan. Rev. Third World Planning Review. Liverpool Univ. Press. Liverpool, England.

Tijdschr. Econ. Soc. Geogr. Tijdschrift voor Economische en Sociale Geographie. Netherlands Journal of Economic and Social Geography. Rotterdam, The Netherlands.

Trace/México. Trace. Centre d'études mexicaines et centraméricaines. México.

UNISA/Lat. Am. Rep. UNISA/Latin American Report. Univ. of South Africa. Pretoria.

Universitas/Stuttgart. Universitas. Wissenschaftliche Verlagsgesellschaft. Stuttgart, Germany.

Universum/Talca. Universum. Univ. de Talca. Talca, Chile.

World Dev. World Development. Pergamon Press. Oxford, England.

Yaxkin/Tegucigalpa. Yaxkin. Instituto Hondureño de Antropología e Historia. Tegucigalpa.

Yearbook/CLAG. Yearbook. Conference of Latin Americanist Geographers; Ball State Univ., Muncie, Ind.

Z. Geomorphol. Suppl.bd. Zeitschrift für Geomorphologie Supplementband. Bebrüder Borntraeger. Berlin.

Z. Wirtsch.geogr. Zeitschrift für Wirtschaftgeographie. Hagen, Westfallen, Germany.

Zent.bl. Geol. Paläontologie. Zentralblatt für Geologie und Paläontologie. E. Schweizerbart'sche Verlagsbuchhandlung (Nägele u. Obermiller). Stuttgart, Germany.

GOVERNMENT AND POLITICS

GENERAL

DAVID DENT, *Professor of Political Science, Towson State University*

THE COMPLEXITY OF POLITICAL transformation and democratization are clearly the dominant subjects of political science research on Latin America, continuing a trend that started with *HLAS 51*. Yet, the struggle to understand political change in Latin America in the aftermath of the Cold War has not produced a consensus on the meaning and consequences of current democratic trends in the region. New and interesting topics of investigation are beginning to appear, adding more insights into the democratization debate and efforts to understand the government and politics of the region. The following topics and works illustrate major trends in the literature.

Democratic Theory. The theoretical dimensions of the democratization debate center on a number of conceptual problems confronting scholars. In his "On the State, Democratization and some Conceptual Problems" (item **2782**), O'Donnell argues that democratization cannot be understood unless current legal/constitutional dimensions of the State are revised. Loveman's *The constitution of tyranny* (item **2770**) is also cognizant of the need for fundamental changes in the constitutional foundations of Latin American governments, arguing that regimes of exception that allow for the suspension of constitutional protections will continue to impede democratic development. Karl and Schmitter claim that the mode of transition from military-authoritarian rule is the key to determining whether democracy will take hold in Latin America, particularly when transitions are by political pacts (item **2765**). In *The failure of presidential democracy,* Linz and Valenzuela (item **2750**) argue that presidential democracies are more unstable than parliamentarism. This observation is challenged by Nohlem in "Presidencialismo vs. Parlamentarismo en América Latina" (item **2779**).

Political Economy: Democratic and Authoritarian Outcomes. Studies that emphasize political economy continue to influence the debate on regime characteristics and political outcomes. In *Politician's dilemma* (item **2755**), Geddes explores some of the ways in which politicians resolve the painful dilemma of using State resources to foster economic development. Remmer's empirical examination of the relationship between economic conditions and electoral stability breaks new ground in its interpretation of liberal democracy and regime change (item **2793**). Moreover, in "The Political Economy of Elections in Latin America, 1980–1991" (item **2792**), Remmer provides an excellent quantitative analysis of the impact of elections on macroeconomic performance using eight Latin American cases.

Democratization. The issues and problems of democratization receive a great deal of attention in the literature on regime change. The best works on this subject include Mainwaring, O'Donnell, and Valenzuela, *Issues in democratic consolida-*

tion (item **2763**), and Remmer (item **2791**). Once again, Remmer makes an important point with her observation that the "current wave" of democratization is not the same as those studied in earlier periods. Skidmore's seminal work, *Television, politics, and the transition to democracy in Latin America* (item **2803**), examines the role of television and electoral politics in the process of democratization.

Military and Democracy. Although the military as a political actor in Latin America does not receive the research attention it once did, a few valuable studies have been issued recently. Nunn's treatment of professional militarism (1964–89) is excellent (item **2780**), as is Grace's *The coup* (item **2758**), an interesting examination of the essential ingredients of the coup process. By focusing on military autonomy, Pion-Berlin (item **2788**) refutes the common view that South America is experiencing another cycle in civil-military relations. Many analysts of the Latin American military are not ready to rule out the possibility of future coups in the region.

Ideological Trends and Political Groups. With the end of the Cold War, some scholars are beginning to ponder the impact that the world-wide collapse of communism has had on the Latin American left. The most valuable works in this category include Carr and Ellner, *The Latin American left* (item **2768**), a carefully crafted edited volume on the subject, and Castañeda, *Utopia unarmed* (item **2732**), an assessment of different segments of Latin America's left and its post-socialist agenda. The revival of populism, according to some Latin American scholars, is a sign that democratic consolidation has failed (item **2733**). In *Latin America in the time of cholera* (item **2787**), Petras and Morley offer a solid leftist critique of post-Cold War policies in the region, particularly the flaws in free markets and North-South cooperation.

Revolutions. The changing nature of Latin American politics has not eliminated the need to understand revolutionary change. Selbin's *Modern Latin American revolutions* (item **2798**) stands out as a valuable critique of structural theories of revolution stressing the importance of ideas, ideologies, and revolutionaries in the process of revolutionary consolidation. Colburn's *The vogue of revolution in poor countries* (item **2739**) is an interesting and useful refutation of Marxist and modernization paradigms as explanations for socialist revolutions. In *The state of revolution* (item **2742**), Crahan and Smith examine why revolutions succeed or fail, stressing origins, trajectories, and outcomes.

2723 **Adrianzén, Alberto** *et al.* Los partidos políticos en el inicio de los noventa: seis casos latinoamericanos. Coordinación de Manuel Antonio Garretón Merino. Santiago: Grupo de Trabajo de Partidos Políticos-CLACSO; Ediciones FLACSO, 1992. 85 p.: bibl.

Trabajos presentados en reunión del Grupo de Trabajo de Partidos Políticos de CLACSO, México 1991. En la introducción se hace un recuento de las situaciones de transición sobrevividas hacia fines de la década de los ochenta. Diversos autores tratan los casos de Brazil, México, Paraguay, Perú y Venezuela. [M.A. Garretón]

2724 **Alexander, Robert Jackson.** The ABC presidents: conversations and correspondence with the presidents of Argentina, Brazil, and Chile. Westport, Conn.: Praeger, 1992. 321 p.: bibl., index.

Conversations and correspondence between author and 20 presidents from Argentina, Brazil, and Chile. Attempts to provide a few insights into the careers of prominent politicians and their roles in the political history of their individual countries. Useful index.

2725 **Alexander, Robert Jackson.** The Bolivarian presidents: conversations and correspondence with presidents of Bolivia, Peru, Ecuador, Colombia, and Venezuela. Westport, Conn.: Praeger, 1994. 283 p.: bibl., index.

Interview transcripts and letters from many of the recent presidents of Bolivia, Ecuador, Colombia, and Venezuela. Interesting insights emerge from attitudes of key presidents from the 1950s through the 1980s.

2726 Americas: an anthology. Edited by Mark B. Rosenberg, A. Douglas Kincaid, and Kathleen Logan. New York: Oxford Univ. Press, 1992. 380 p.: ill., map.

General reader designed for students enrolled in the *Americas*, a 13-unit telecourse. Selections include both primary and secondary source materials—speeches, letters, government documents, cartoons, magazine articles, and essays—on historical, political, and social dynamics that have shaped the hemisphere.

2727 Apostolakis, Bobby E. Warfare-welfare expenditure substitutions in Latin America, 1953–87. (*J. Peace Res.*, 29:1, Feb. 1992, p. 85–98, bibl., tables)

Presents overarching view and empirical testing of the alleged trade-off between defense and other public needs (health, education, social security and public works) in 19 Latin American countries. Using sophisticated econometric method based on a time series approach, provides strong evidence that military expenses crowded out the potential allocation for social betterment. Concludes that the guns vs. butter trade-off is corroborated by data from the region.

2728 Aramouni, Alberto. Perspectiva de la política social en América Latina con relación al estilo de desarrollo. (*in* Encuentro Internacional sobre Política Social, *1st, Vitoria, Spain, 1990*. Actas. Vitoria, Spain: Servicio Central de Publicaciones del Gobierno Vasco, 1991, p. 103–118, bibl.)

Attacks consumerist capitalism for bringing about global political, social, and economic crises. Argues for an anti-bureaucratic and participationist socialism. [P.H. Lewis]

Aricó, José. 1917 y América Latina. See *HLAS 54:5313*.

2729 Barrios, Harald and **Petra Bendel.** Los sistemas políticos de América Latina: bibliografía comentada de obras recientes. (*Ibero-Am. Arch.*, 18:1/2, 1992, p. 291–310)

Series of bibliographic essays review recent works on Latin American politics and political systems, with emphasis on authoritarianism, redemocratization, institutional reform, political parties, and elections and electoral systems. Authors find an increase in cooperation between European and Latin American political analysts. Most studies cited are by German political scientists.

2730 Boersner, Demetrio *et al.* El estado periférico latinoamericano. Recopilación de Juan Carlos Rubinstein. Buenos Aires: Editorial Universitaria de Buenos Aires, 1988. 306 p.: bibl. (Manuales)

Collection of articles, varying in quality, about Latin America's social class structures and their relationship to the problems of underdevelopment and dependency. Basically, old-fashioned Marxism dressed up in social science jargon. [P.H. Lewis]

2731 Borzutzky, Silvia. Social security and health policies in Latin America: the changing roles of the States and the private sector. (*LARR*, 28:2, 1993, p. 246–256)

Review of eight books on economic and political dimensions of social security and health policies, emphasizing the role of the State, escalating costs, and the impact of democratization on these policies. On the basis of these recent studies, reviewer concludes that "despite the wave of privatization taking place today, the pressures for expanding and improving social security and health policies and services are likely to increase as democratization continues."

2732 Castañeda, Jorge G. Utopia unarmed: the Latin American left after the Cold War. New York: Knopf; Distributed by Random House, 1993. 498 p.: bibl., index.

Offers intriguing assessment of relevance of the left—populist, social-democratic, Castroist, and political-military—now that the Cold War is over and the socialist paradigm has disappeared. In describing past, present, and future of Latin American left, author argues that the left's "geopolitical" handicap has been lifted after years of Cold War struggles, thus affording it "the possibility of contending for power on a level playing field, freed from the handicaps that have infinitely weakened it over the past half century." However, the post-socialist agenda must rectify the left's past tendencies to make concessions regarding the absence of representative democracy, human rights violations, corruption, and non-democratic procedures when settling internal disputes. Vanguards and confrontational activism will have to be abandoned to meet future challenges. Important study of the role of ideology in bringing about economic growth with social equity in a region that has agonized over the problems and prospects of development.

2733 Castro Rea, Julián; Graciela Ducatenzeiler; and **Philippe Faucher.** Back to populism: Latin America's alternative to democracy. (*in* Latin America to the year 2000: reactivating growth, improving equity, sustaining democracy. New York: Praeger, 1992, p. 125–146, bibl.)

Critical assessment of recent Latin American efforts at democratic consolidation in which the author argues that the "populist temptation" is an illusory strategy designed to evade constraints to development and boost legitimacy of elitist domination. Resurgence of populism in Peru, Argentina, Brazil, and Mexico represents "a common denomination of regimes characterized by ineffectual political representation compensated by latent corporatism and flourishing clientelism." Thus, the populist revival is clearly a sign that democratic consolidation has failed.

2734 Cavarozzi, Marcelo. Beyond transitions to democracy in Latin America. (*J. Lat. Am. Stud.*, 24:3, Oct. 1992, p. 665–684, table)

Critique of the "transition paradigm's" ability to explain regime change. Focusing on recent trends in the Southern Cone, Mexico, and Brazil, author argues that long-term historic processes have a more decisive influence on politics of democratic consolidation than short-term prerequisites.

2735 Cerdas, Rodolfo; Juan Rial Roade; and **Daniel Zovatto.** Elecciones y democratización en América Latina: balance de una década. (*in* Una tarea inconclusa: elecciones y democracia en América Latina, 1988–1991. San José: Instituto Interamericano de Derechos Humanos, Centro de Asesoría y Promoción Electoral, 1992, p. 661–707, appendices, bibl., tables)

Detailed treatment of the role of elections and democratic governance in Latin America during the 1980s. The ability to get major players in the political game to accept elections as a means of transferring power is a factor in regime legitimation. Lengthy appendix offers detailed information on elections and electoral systems.

2736 Chaffee, Lyman G. Political protest and street art: popular tools for democratization in Hispanic countries. Westport, Conn.: Greenwood Press, 1993. 173 p.: bibl., ill., index. (Contributions to the study of mass media and communications, 0732–4456; 40)

Interesting historical and sociopolitical treatment of the tradition and culture of street art—posters, wall paintings, graffiti, and murals—as political expression in Spain and Latin America. Four Hispanic case studies—Spain, the Basque country, Argentina, and Brazil—are used to illustrate characteristics, motives, and effectiveness of street art as a form of political communication. Author notes that "as the democratic transition process stabilizes and works itself through, tensions lessen as does street art."

2737 Chaffee, Wilber A. The economics of violence in Latin America: a theory of political competition. New York: Praeger, 1992. 173 p.: bibl., index.

Study of political violence based on theory of political competition drawn from microeconomics and applied to four revolutions: Mexico, Bolivia, Cuba, and Nicaragua. By emphasizing positive political theory and comparative politics of Third World societies, author finds that collective goods "with an exclusionary mechanism tied to their distribution are more probably the motivation for political violence."

2738 Chilcote, Ronald H. Left political ideology and practice. (*in* The Latin American left: from the fall of Allende to perestroika. Boulder, Colo.: Westview Press, 1993, p. 171–186)

Survey of ideological trends in Latin America, focusing on traditional ideological currents and evolution of these ideas on Latin American politics since the overthrow of Allende in 1973. Author finds that part of the left is in retreat from Marxism (favoring a pluralism of interests), while other leftists continue to search for some workable form of Marxism free of the stifling model of centralist bureaucratic socialism.

2739 Colburn, Forrest D. The vogue of revolution in poor countries. Princeton, N.J.: Princeton Univ. Press, 1994. 135 p.: bibl., ill., index.

Inspiration for revolution in Asia, Africa, Latin America, and the Middle East is supplied by a fashionable political imagination rooted in European ideas of socialism. Although this political imagination inspired revolutions, it also contributed to basically flawed ideas of how to eliminate inequality and poverty. Valuable refutation of Marxist and modernization paradigms as explanations for outcomes of socialist revolutions. Logic of contemporary revolutions can be found in the intellectual culture: the values, expectations, phraseology, iconography, and rules employed by those in control of the government.

2740 **Collier, David** and **Deborah L. Norden.**
Strategic choice models of political change in Latin America. (*Comp. Polit.*, 24:2, Jan. 1992, p. 229–243, bibl., tables)

Review of three "strategic choice" models of political change—Hirschman, Przeworski, and O'Donnell—which emphasize actors and preference distributions, thresholds, subjective probabilities, costs, and signals and communication. In examining the overall utility of the models, authors find that strategic choice analysis should complement rather than supplant other research traditions currently in vogue.

2741 **Collier, Ruth Berins** and **David Collier.**
Shaping the political arena: critical junctures, the labor movement, and regime dynamics in Latin America. Princeton, N.J.: Princeton Univ. Press, 1991. 877 p.: bibl., index.

Excellent comparative-historical analysis of eight countries—Argentina, Brazil, Chile, Colombia, Mexico, Peru, Uruguay, and Venezuela—focuses on emergence of different forms of control and mobilization of the labor movement during initial incorporation period of development. By concentrating on alternative strategies of the State in shaping the labor movement, authors are able to explain different trajectories of national political change in countries with longest history of urban, commercial, and industrial development. While labor incorporation may not be easily understood by a single defining moment ("critical juncture"), or as a determinant in shaping subsequent economic, political, and social events, the insertion of labor movements in capitalist political systems deserves close research attention. Important and valuable work includes glossary of terms, extensive index (general and by country), and abbreviations of useful terms.

2742 **Crahan, Margaret E.** and **Peter H. Smith.** The state of revolution. (*in* Americas: new interpretive essays. New York: Oxford Univ. Press, 1992, p. 79–108, table)

Examines origins, trajectories, and outcomes of revolution in Latin America focusing on Mexico, Bolivia, Cuba, and Nicaragua. Valuable treatment of the revolutionary process and its key components, along with interesting insights into why revolutions succeed and fail. Authors argue that recent growth in social movements and decline in armed struggle reflects a rejection of a strong State, one of the main goals of classic revolutionary movements in Latin America.

2743 **Crespo Martínez, Ismael.** La problemática transicional y el desafío de la consolidación: Argentina, Uruguay y Chile. (*Rev. Estud. Polít.*, 74, oct./dic. 1991, p. 661–669, bibl.)

Bibliographical essay summarizes recent thinking of prominent social scientists of the region about the role of political parties, the military, and public attitudes about democracy in this period of transition. [P.H. Lewis]

2744 **Davis, Charles L.** and **John G. Speer.**
The psychological bases of regime support among urban workers in Venezuela and Mexico: instrumental or expressive? (*Comp. Polit. Stud.*, 24:3, Oct. 1991, p. 319–343, appendix, bibl., tables)

Empirical study of the relative influence of symbolic and normative orientations to politics and economic utility on the political attitudes of urban workers in Mexico and Venezuela. Authors find that in both countries economic advantage and material benefits have considerably less influence on regime support than do symbolic and normative orientations. Important implications for the study of political culture and modernization in Latin America.

2745 **Democracia 2000: los grandes desafíos en América Latina.** Bogotá: Fundación Simón Bolívar, Centro de Estudios Internacionales Foro Interamericano; Tercer Mundo Editores, 1991. 404 p.: bibl., ill.

Multi-authored work on the prospects for democracy based on conference papers (Bogotá, 1989). Essays are organized around several themes, including democracy and peace, security, economy, participation, and international political cooperation. Democratic rule is mostly taken as a given and observations tend to confirm what most Latin American elites think of democracy. Most participants are optimistic about democratic prospects for Latin America, except for parts of Central America.

2746 Dix, Robert H. Democratization and the institutionalization of Latin American political parties. (*Comp. Polit. Stud.*, 24:4, Jan. 1992, p. 488–511, bibl., tables)

By examining the institutionalization of political parties in the 1960s and 1990s using criteria developed by Huntington, author argues that prospects for consolidation of democracy have improved over the past 30 years.

2747 Domínguez, Jorge I. On understanding the present by analyzing the past in Latin America: a review essay. (*Polit. Sci. Q.*, 107:2, Summer 1992, p. 325–329)

Review essay of Collier and Collier, *Shaping the Political Arena* (item **2741**). Several criticisms are offered but the reviewer heaps praise on a work that "deserves to shape the intellectual arena for social scientists and historians for years to come."

2748 Dunkerley, James. Political suicide in Latin America and other essays. London; New York: Verso, 1992. 252 p.: bibl.

Collection of essays explores the meaning and explanation for political suicide in Latin America focusing on the political history of El Salvador, Guatemala, and Nicaragua during the 1980s. The conceptual analysis of political suicide, however, is quite loose and disjointed. Lack of an index and bibliography seriously weakens documentation offered in support of each essay.

2749 Ellner, Steve. Introduction: the changing status of the Latin American left in the recent past. (*in* The Latin American left: from the fall of Allende to perestroika. Boulder, Colo.: Westview Press, 1993, p. 1–21)

Introduction to edited volume on the transformation of the Latin American left. Author finds left's survival strategy is based on a newly adopted Gramscian ideology devoted to strengthening democratic institutions and practices. The Latin American left's survival and growth demonstrates that "it can no longer be understood solely on the basis of foreign influences, ideology (itself shaped from abroad), and performance in organized labor."

2750 The failure of presidential democracy. Edited by Juan J. Linz and Arturo Valenzuela. Baltimore: Johns Hopkins Univ. Press, 1994. 436 p.: bibl., index.

Essays devoted to long-term viability of democratic governments emphasize the presidential experience in Uruguay, Brazil, Colombia, Ecuador, Peru, and Venezuela. Analysis focuses on some structural problems inherent in presidential democracies. Linz argues that "presidentialism seems to involve greater risk for stable democratic politics than contemporary parliamentarism." For comments on individual articles see items **3490, 3175, 3218, 3390,** and **3257.**

2751 Fernández Baeza, Mario. Las políticas sociales en el Cono Sur, 1975–1985: análisis de sus determinantes políticas y socioeconómicas. Santiago: Instituto Latinoamericano y del Caribe de Planificación Económica y Social, 1989. 139 p.: bibl., ill. (Cuadernos del ILPES: 0020–4080; 34)

Scholarly, comparative study of economic and social effects of recent military rule in Argentina, Chile, and Uruguay. For each country, work examines changes in income distribution, occupational structure, and investment and consumption patterns. [P.H. Lewis]

Franco, Jean. Gender, death, and resistance: facing the ethical vacuum. See *HLAS 54: 3470.*

2752 Garretón Merino, Manuel Antonio. Human rights in processes of democratisation. (*J. Lat. Am. Stud.*, 26, 1994, p. 221–234)

Brief examination of human rights problems during political transformation from military to democratic regimes in the Southern Cone countries. Stresses need for reformulation of the human rights to include the "right to life" in post-authoritarian societies. Valuable assessment of human rights investigations and subsequent publication of information on past abuses by several post-military governments.

Garretón Merino, Manuel Antonio. New State-society relations in Latin America. See item **4415**.

2753 **Gaupp, Peter.** Latin America's flexible drug trade. (*Swiss Rev. World Aff.*, 41:9, Dec. 1991, p. 20–21, tables)

Brief assessment of the ability of Latin America's cocaine industry to react and survive despite formulation and enaction of antinarcotics policies.

2754 **Geddes, Barbara.** A game theoretic model of reform in Latin American democracies. (*Am. Polit. Sci. Rev.*, 85:2, June 1991, p. 371–392, bibl., tables)

Using a game-theoretic model to explain political reform in Latin American democracies, author concludes that electoral rules which result in proliferation of candidate lists reduce the probability of reform.

2755 **Geddes, Barbara.** Politician's dilemma: building State capacity in Latin America. Berkeley: Univ. of California Press, 1994. 246 p.: bibl., index. (California series on social choice and political economy; 25)

Explores how politicians resolve painful dilemma of using State resources to foster economic development thereby complicating their own struggle for political survival. Using the contradictory pressures of the "politician's dilemma" as a theoretical basis, work discusses circumstances under which politicians support or oppose reforms that increase State capacity. Author finds that presidents who need not fear military intervention and can rely on legislative allies are more likely to resolve the dilemma by hiring experts and building State capacity.

2756 **Gillespie, Richard.** Guerrilla warfare in the 1980s. (*in* The Latin American left: from the fall of Allende to perestroika. Boulder, Colo.: Westview Press, 1993, p. 187–203)

Assesses the evolution of guerrilla movements in Latin America, emphasizing responses to their defeat in South America and the success of armed struggle in Nicaragua after 1979. Changes that ushered in the decade of the 1990s—the crisis of communism, the electoral defeat of the Sandinistas, and the quickly fading Cuban regime—will require serious efforts at adaptation by guerrilla movements if they are to remain influential. Author also suggests that marginal sectors, traditionally ignored by the left, could become a source of political violence in the 1990s.

2757 **González Encinar, José Juan et al.** El proceso constituyente: enseñanzas a partir de cuatro casos recientes; España, Portugal, Brasil y Chile. (*Ibero-Am. Arch.*, 18: 1/2, 1992, p. 151–179)

Article offers overview and comparison of constitution-making in four countries in which democratic governments have come to power after lengthy periods of authoritarian rule. Noting that in three cases (Spain, Portugal, and Brazil) constitutions were drafted precisely to facilitate democratic transition already underway, while the fourth (Chile) involved an attempt to institutionalize the authoritarian regime permanently, authors analyze the constitutional experiences of each country in terms of context, political function, procedures, and actual constitutional contents. They arrive at the less than startling conclusion that the terms and impacts of these processes are likely to be determined by a country's historical-political context, and by the political and related (e.g., economic) objectives of leading political forces. [M. Fleet]

2758 **Grace, Alexander M.** The coup: tactics in the seizure of power. Westport, Conn.: Praeger, 1994. 222 p.: bibl., index.

Interesting examination of the causes, stages, and tactics of a number of well-known *golpes de estado*. Case studies of both failed and successful coups are offered to illustrate a set of general principles and essential ingredients of the coup process. A final chapter provides governments with guidelines for countering the coup d'état. Grace is one of the few authors to predict "that we will see another wave of coups and coup attempts throughout Latin America in the coming years."

2759 **Gritti, Roberto.** Evoluzione e caratteristiche dei partiti politici in Argentina, Brasile, Uruguay e Cile. (*in* Democrazia e partiti in America Latina. Milano, Italy: FrancoAngeli, 1991, p. 249–303)

Summarizes general political trends in the four countries and the impact of the return to democracy on their political parties. Reviews the volume's other essays on each country and adds summaries of the major parties in each of them, a task the analytical essays did not undertake. Intended as a supplement to preceding essays, this piece ably summarizes leadership, platform, and recent legislative and electoral history of each party until 1989. [V. Peloso]

2760 **Harris, Richard Legé.** Marxism, socialism, and democracy in Latin America. Boulder, Colo.: Westview Press, 1992. 234 p.: bibl., index. (Latin American perspectives series; 8)

Comparative analysis of the relevance of Marxism and socialism for Latin America and the Caribbean focuses on recent revolutionary regimes and their attempts at socialist transformation. Argues that democratic consolidation and solutions to Latin America's economic and social problems will ultimately require some form of democratic socialism. Generalizations or "lessons" proffered in the last chapter stress that the basic political-economic concepts developed by Marx and Engels are still applicable, despite the collapse of communism elsewhere.

2761 **Hellman, Judith Adler.** The study of new social movements in Latin America and the question of autonomy. (*in* The making of social movements in Latin America: identity, strategy, and democracy. Edited by Arturo Escobar and Sonia E. Álvarez. Boulder, Colo.: Westview Press, 1992, p. 52–61)

Brief treatment of similarities and differences of new social movements in Latin America and Western Europe.

2762 **The Human Rights Watch global report on prisons.** New York: Human Rights Watch, 1993. xxxvii, 303 p.: bibl.

World-wide human rights report on prison conditions in 20 countries including Brazil, Cuba, Mexico, Jamaica, and Peru. Interesting report, valuable for its cross-national implications.

International Political Science Abstracts. 1989– . See item **29**.

2763 **Issues in democratic consolidation: the new South American democracies in comparative perspective.** Edited by Scott Mainwaring, Guillermo A. O'Donnell, and Julio Samuel Valenzuela. Notre Dame, Ind.: Helen Kellogg Institute for International Studies; Univ. of Notre Dame Press, 1992. 357 p.: bibl., ill., index.

Focusing on the new South American democracies, authors examine democratization as a process of political transformation, from the breakdown of military-authoritarian rule to the establishment of a consolidated democratic regime. Emphasis on the latter phase of democratic consolidation reveals that key issues pertaining to the "second" transition include: 1) the difficulty of subordinating the military to civilian government control; 2) the weakness of interest group connections to parties and policymaking; 3) the difficulties of political accountability and conduits for political representation; and 4) the inadequacy of transitional political institutions. In his examination of the logic of transitions to democracy, Przeworski detects an interesting paradox: "For democracy to be established, it must protect to some degree the interests of the forces capable of subverting it, above all capitalists and the armed forces." Excellent treatment of problems confronting new democracies.

2764 **Kaláshnikov, Nikolái.** Velar en común por proteger la naturaleza. (*Am. Lat./Moscow*, 12:168, dic. 1991, p. 30–36)

A Russian expresses concern about ecological destruction in Latin America, especially the Amazon rain forest. [P.H. Lewis]

2765 **Karl, Terry Lynn** and **Philippe C. Schmitter.** Modes of transition in Latin America, Southern and Eastern Europe. (*Age Dem.*, 128, May 1991, p. 269–284, photos, tables)

Brief exploration of hypothetical notion that the mode of transition from autocratic rule is the key to determining whether democracy will take hold. Transitions "by pact" are the most likely to contribute to political democracy, rather than those in which elites are completely destroyed.

2766 **Korzeniewicz, Roberto P.** Contested arenas: recent studies on politics and labor. (*LARR*, 28:2, 1993, p. 206–220)

Systematic review of five recent studies of the politics of labor emphasizes tension between control and autonomy as key factors for understanding political change and regime dynamics. Lengthy examination of Collier and Collier's *Shaping the Political Arena* item **2741** is valuable for its insights and criticisms.

LABORDOC. See item **32**.

2767 **Latin America faces the twenty-first century: reconstructing a social justice agenda.** Edited by Susanne Jonas and Edward J. McCaughan. Boulder, Colo.: Westview Press, 1994. 222 p.: bibl., index.

Eclectic compilation of essays devoted to a "massive social crisis" attributed to a failure of capitalist development models since World War II. With their new vision and projects for a more just and equitable social transformation throughout the region, authors challenge neoliberal formulas.

2768 **The Latin American left: from the fall of Allende to Perestroika.** Edited by Barry Carr and Steve Ellner. Boulder, Colo.: Westview Press; London: Latin American Bureau, 1993. 256 p.: bibl., index. (Latin American perspectives series; 11)

Important study of the transformation of the Latin American left since the early 1970s argues that despite the collapse of communism and the fading virtues of Guevaraism, the left is alive and well. Based on eight country studies, each contributor examines lessons drawn from the failure of guerrilla strategies in the 1960s, the current emphasis on democratic reforms over socioeconomic change, and affects of the changing international environment—the erosion of US influence in the region and the opposition to IMF-imposed structural adjustment policies—on the future role of leftist parties. Valuable for understanding the cross-currents of leftist ideologies and guerrilla movements on Latin American politics.

Latin American Newletters. 1990– . See item **34**.

Latin American News. 1995– . See item **35**.

2769 **Latinskai͡a Amerika: demokratii͡a i vlast' [Latin America: democracy and power.** Otvetstvennyĭ redaktor B.M. Merin. Moskva: Akademii͡a nauk SSSR, In-t Latinskoĭ Ameriki, 1991. 107 p.

This collection of eight papers was prepared under the auspices of the Institute of Latin American Studies of the then Soviet Academy of Sciences for the 15th World Congress of the International Political Science Association held in Buenos Aires in 1991. Examines democratization in the USSR and Latin America, the scientific-technical revolution, and new social movements in Latin America. One paper outlines the difficulties of studying Russian and other immigrant groups in Latin America and identifies contributions of the Russian diaspora in Paraguay. Table of contents in Spanish and Russian. [B. Dash]

Latinskai͡a Amerika: sobytii͡a i li͡udi; analiticheskiĭ obzor [Latin America: events and people; an analytical survey]. See item **3836**.

2770 **Loveman, Brian.** The constitution of tyranny: regimes of exception in Spanish America. Pittsburgh: Univ. of Pittsburgh Press, 1993. 481 p.: bibl., index. (Pitt Latin American series)

Comprehensive study of constitutional foundations of dictatorships and political repression in Spanish America that have impeded democratization and facilitated military rule. Unless there is fundamental change in the constitutional framework of Latin American politics, regimes of exception, which allow the suspension of constitutional protections, the nullification of rights and liberties, and the expansion of governmental authority, will flourish.

2771 **Lynch, Edward A.** Latin America's Christian democratic parties: a political economy. Westport, Conn.: Praeger, 1993. 197 p.: bibl., index.

Examines how and why Christian Democracy gained electorally in the 1960s but diminished in the 1980s. Author suggests that Christian Democracy failed because it lost sight of its ideological roots and while in power became too anxious to involve the State in economic policy-making. To compete with the neoliberal politics of the 1990s, Christian Democrats need to rediscover their Catholic social roots which offer the advantage of coherence, consistency, and compassion along with friendliness toward the free market.

2772 **Mansilla, H.C.F.** Posibilidades y dilemas de los procesos de democratización en América Latina. La Paz: Centro Boliviano de Estudios Multidisciplinarios, 1991. 73 p.: bibl. (Cuadernos del CEBEM; 3)

Brief collection of essays on democratization process in Latin America. Author surveys legacy of patrimonialism and authoritarianism. Concludes that in addition to political culture, these factors constitute a serious obstacle to consolidation of democracy in the region. Final chapter applies Mansilla's conclusions about Latin America to Bolivian case study. Overall, a useful interpretation of the democratization process, although author's cultural determinism is excessive at times. [E. Gamarra]

2773 **Manzo, Kate.** Modernist discourse and the crisis of development theory. (*Stud. Comp. Int. Dev.*, 26:2, Summer 1991, p. 3–36, bibl.)

Current crisis of development theory—developmentalism versus dependency—is examined within a three-fold framework designed to refine the meaning and interpretation of dependency theory. Author argues that: 1) the major achievements of dependency theory have been unrecognized or misrepresented; 2) dependency theory remains trapped within a distorted modernist discourse; and 3) development theory should be grounded in knowledge of local histories and experiences rather than on general conceptual categories and Western assumptions. Valuable for those interested in comparative international development.

2774 **Massenmedien in Lateinamerika.** v. 2, Chile, Costa Rica, Ecuador, Paraguay. Jürgen Wilke (Hrsg.). Frankfurt am Main: Vervuert, 1994. 1 v.: bibl. (Americana Eystettensia. Serie B, Monographien, Studien, Essays; 5 = Publikationen des Zentralinstituts für Lateinamerika-Studien. Serie B, Monographien, Studien, Essays; 5)

Vol. 2 concerns mass media in four countries noted above. Provides in-depth synopses for all of them with special emphasis on political and social transitions affecting their development. Brief histories discuss development of laws governing the press, radio, and television. Also represents legal aspects, ownership, financing, and programming as well as special perspectives and problems in the mass media of each country. [C. Converse]

2775 **Morlino, Leonardo** and **Alberto Spreafico.** Introduzione. (*in* Democrazia e partiti in America Latina. Milano, Italy: FrancoAngeli, 1991, p. 9–38, bibl.)

Introduction notes that the book's essays examine the relationship between a political system founded on choice and the organization of power of an informed electorate. Editors raise issues later examined in essays on selected Latin American countries: why parties are founded, their central organization, their objectives, the formation of their rank-and-file membership, their meaning for systems of power, their institutional assets and weaknesses, and their prospects for success. Authors address dynamics of institutions and power in Argentina, Brazil, Chile, and Uruguay. Closing essay provides historical synopsis of major political parties in each country. Some essays reach back to party origins in the 19th century. [V. Peloso]

2776 **Munck, Gerardo L.** Between theory and history and beyond traditional area studies: a new comparative perspective on Latin America. (*Comp. Polit.*, 25:4, July 1993, p. 475–498, bibl.)

Review article of Collier and Collier's *Shaping the Political Arena* item **2741** and Touraine's *Palavra e Sangue: Política e Sociedade na América Latina* (São Paulo: Editorial UNICAMP, 1989) focusing on their conceptual frameworks, use of comparative methodology, and substantive findings dealing with the dynamics of Latin American politics. Reviewer praises the Colliers' "critical-juncture model" for its theoretical sophistication and Touraine's "national-populism model" for its "looseness" in use of data within a broad theoretical framework.

2777 **Murillo Castaño, Gabriel** and **Juan Carlos Ruiz Vásquez.** Elecciones, partidos políticos y democracia en los países andinos. (*in* Una tarea inconclusa: elecciones y democracia en América Latina, 1988–1991. San José: Instituto Interamericano de Derechos Humanos, Centro de Asesoría y Promoción Electoral, 1992, p. 405–454)

Comparative analysis of recent elections and aspects of democratic rule in Bolivia, Colombia, Ecuador, Peru, and Venezuela during the 1980s. Descriptive treatment of parties and elections is more useful than information on economic development and prospects for the future of the Andean democracies.

2778 **Nef, Jorge** and **Remonda Bensabat.** "Governability" and the receiver State in Latin America: analysis and prospects. (*in* Latin America to the year 2000: reactivating growth, improving equity, sustaining democracy. New York: Praeger, 1992, p. 161–176, bibl.)

Critical assessment of redemocratization trends in the 1990s based on four fallacious assumptions of political transition and pluralistic political theories. Authors argue that "As the current *fórmulas de recambio* run their course and the present crisis deepens, it is likely that the weak and limited civilian regimes of today will be replaced, once again, by equally weak, yet violently repressive, military regimes."

2779 **Nohlen, Dieter.** Presidencialismo vs. parlamentarismo en América Latina: notas sobre el debate actual desde una perspectiva comparada. (*Rev. Estud. Polít.*, 74, oct./dic. 1991, p. 43–54, bibl.)

Interesting article challenges thesis popularized by J. Linz, according to which Latin American nations would be more stable if their political systems were parliamentary instead of presidential. Argues that such a stance overlooks other important political dimensions, notably, political culture, social consensus, and international factors.

2780 **Nunn, Frederick M.** The time of the generals: Latin American professional militarism in world perspective. Lincoln: Univ. of Nebraska Press, 1992. 349 p.: bibl., index.

In-depth assessment of Latin American professional militarism from 1964–89 emphasizes what army officers thought and how they perceived their roles in society. Eight chapters cover principal ingredients of professional militarism in Argentina, Brazil, Chile, and Peru. Author argues that the line between professionalism and militarism in Latin America rested on the perceived degree to which democracy, unlike communism, needed to be defended from its own weaknesses. Valuable treatment of the anatomy of professional militarism. For historian's comment see *HLAS 54:1117*.

2781 **O'Donnell, Guillermo A.** Acerca del Estado, la democratización y algunos problemas conceptuales: una perspectiva latinoamericana con referencias a países poscomunistas. (*Desarro. Econ.*, 33:130, julio/sept. 1993, p. 163–184, bibl.)

Current conceptions of the State must now be revised to achieve a better understanding of democratization. By contrasting different legal forms of the new democracies—representative, consolidated, and polyarchical—author offers new concepts he feels are essential to better understand what is going on politically in most newly democratized countries in Eastern Europe and Latin America. For English version see item **2782**.

2782 **O'Donnell, Guillermo A.** On the State, democratization and some conceptual problems: a Latin American view with glances at some postcommunist countries. (*World Dev.*, 21:8, Aug. 1993, p. 1355–1369, bibl.)

See item **2781**.

2783 **O'Kane, Rosemary H.T.** The revolutionary reign of terror: the role of violence in political change. Aldershot, England; Brookfield, Vt.: E. Elgar, 1991. 304 p.: bibl., index.

General study of revolutionary reigns of terror using eight cases (including Cuba and Nicaragua) designed to achieve a better understanding of the role of the State in bringing about revolutionary change.

2784 **Parker, Dick.** Trade union struggle and the left in Latin America, 1973–1990. (*in* The Latin American left: from the fall of Allende to perestroika. Boulder, Colo.: Westview Press, 1993, p. 205–223)

General discussion of the left and unions focuses on how economic and political factors have shaped the organizational capacity of unions and influenced the priorities of the left. Author argues that "the weakening of unions, the dramatic increase in the size of the informal economy, and the emergence of new social movements have raised doubts as to whether the unions can be expected to continue to play an important role in popular mobilizations or in left strategy as they have up until now."

2785 **Los partidos y la transformación política de América Latina.** Edición de Manuel Antonio Garretón Merino. Santiago: Ediciones FLACSO-Chile, 1993. 103 p.: bibl.

Series of short essays devoted to impact of recent political-institutional change on political parties in Bolivia, Brazil, Venezuela, Peru, Argentina, Chile, and Mexico. For sociologist's comment see item **4437**.

2786 **Periodismo y medio ambiente: memoria del seminario realizado en Quito, entre el 28 de noviembre y el 1 de diciembre de 1990.** Quito: Centro Internacional de Estudios Superiores de Comunicación para América Latina; Education Development Center; Servicio de Cultura e Información de los Estados Unidos; Waste Management International, 1991. 341 p.: bibl. (Encuentros; 2)

Papers presented at conference hosted by the Centro Internacional de Estudios Superiores de Comunicación para América Latina. In this factual and informative piece, Latin American journalists and environmentalists present their recommendations on the educational role of the press regarding environmental issues.

2787 Petras, James F. and Morris H. Morley. Latin America in the time of cholera: electoral politics, market economics, and permanent crisis. New York: Routledge, 1992. 208 p.: bibl., index.

Leftist critique of post–Cold War policies in Latin America emphasizes free markets, democratization and north-south cooperation for development. Socialism is the only way to "confront the irrational tide of market madness that threatens to defend general misery with unrestrained violence."

2788 Pion-Berlin, David. Military autonomy and emerging democracies in South America. (*Comp. Polit.*, 25:1, Oct. 1992, p. 83–102, tables)

Interesting attempt to evaluate the behavior and motives of the armed forces within emerging democracies of South America by focusing on military autonomy or its decision-making authority in Argentina, Brazil, Uruguay, Peru, and Chile. Author compares these five countries examining twelve issues related to military autonomy. Concludes that there are too many discontinuities with the past to "support the view that South America is simply experiencing another 'turn in the political cycle.'"

2789 Politics and social change in Latin America: still a distinct tradition? Edited by Howard J. Wiarda. 3rd ed., fully rev. and updated. Boulder, Colo.: Westview Press, 1992. 354 p.: bibl.

Collection of 16 lively essays expanding and conceptually updating well-known 1974 edition. Answer to subtitle's question is "yes:" understanding region's unique sociopolitical heritage is indeed critical. [G.P. Atkins]

El principio del pez gordo: estrategias para combatir la corrupción. See item **3192.**

2790 Relaciones laborales y modelos de acción sindical: experiencias europeas y latinoamericanas. Edición de Jaime Ensignia, Malva Espinosa y Luise Rürup. Santiago: Fundación Friedrich-Ebert, 1994. 105 p.

Comparative analysis of labor relations and forms of union activity in Brazil, Argentina and Chile, countries that have enacted policies of economic adjustment and democratic consolidation since 1989. These short essays stem from a discussion group in Santiago in late 1993. For sociologist's comment see item **4445.**

2791 Remmer, Karen L. Democratization in Latin America. (*in* Global tranformation and the third world. Edited by Robert O. Slater, Barry M. Schutz, and Steven R. Dorr. Boulder, Colo.: L. Rienner Publishers, 1993, p. 91–111)

Important essay argues that the recent process of democratization differs from earlier periods and that the "current wave" poses "major theoretical challenges to established interpretations of Latin American politics."

2792 Remmer, Karen L. The political economy of elections in Latin America, 1980–1991. (*Am. Polit. Sci. Rev.*, 87:2, June 1993, p. 393–407, bibl., tables)

Quantitative analysis of the impact of elections on macroeconomic performance in eight Latin American nations during the 1980s. Findings indicate that competitive elections have enhanced rather than undermined the capacity of political leaders to face problems of macroeconomic management. Author concludes that the relationship between democracy and economics is captured more adequately by a "political capital model" than by the traditional "political business cycle model."

2793 Remmer, Karen L. The political impact of economic crisis in Latin America in the 1980s. (*Am. Polit. Sci. Rev.*, 85:3, Sept. 1991, p. 777–800, bibl., graphs, tables)

Empirical examination of the relationship between economic conditions (crises) and electoral instability suggests that previous theoretical frameworks predicting political breakdowns are associated more with the mediation by party system structure than democratic longevity. Important implications implications for the study of liberal democracy and regime change.

2794 **Rial Roade, Juan.** Las fuerzas armadas en los años 90: una agenda de discusión. Montevideo: Peitho, Sociedad de Análisis Político, 1990. 73 p.: bibl.

Collection of previously published articles and conference papers about the problems being faced by the Southern Cone's newly-established democracies in trying to subordinate their armies to civilian rule. [P.H. Lewis]

2795 **Roniger, Luis.** Public trust and the consolidation of Latin American democracies. (*in* Latin America to the year 2000: reactivating growth, improving equity, sustaining democracy. New York: Praeger, 1992, p. 147–160, bibl.)

Brief analysis of internal sociopolitical factors that affect prospects for democratic consolidation and the extension and institutionalization of public trust. Interestingly, author suggests several ways in which institutional trust can be generated but is not sanguine about the elimination of democratic breakdowns, given the difficulties of consolidating and enhancing trust in the public realm.

2796 **Rosenberg, Tina.** Latin America's magical liberalism. (*Wilson Q.*, 16:4, Autumn 1992, p. 58–74, photos)

A warning to Latin American leaders that resurgent faith in constitutional democracy and free markets is not likely to take hold in Latin America because "an economically and politically liberal State demands not fewer rules but new rules—rules that apply to all, sustained not through violence but through the shared conviction that it is law, not power, that governs the new world."

2797 **Salinas, Maximiliano.** The Church in the Southern Cone: Chile, Argentina, Paraguay and Uruguay. (*in* The Church in Latin America, 1492–1992. Edited by Enrique Dussel. Maryknoll, N.Y.: Orbis Books, 1992, p. 295–309, bibl., map)

Liberation theology from 1492–1992, in twelve easy-to-read pages. [P.H. Lewis]

2798 **Selbin, Eric.** Modern Latin American revolutions. Boulder, Colo.: Westview Press, 1993. 244 p.: bibl., index.

Critical analysis of structural theories of revolution highlights importance of ideas, ideologies, and revolutionaries to explain the different paths taken by postrevolutionary regimes in Bolivia, Cuba, Nicaragua, and Grenada. Emphasizes successful social revolutions are found "only in those cases where there has been a significant degree of institutionalization and consolidation." Stresses that social scientists need to refocus discussion of revolutionary change on the power and possibility of individuals to control their own destiny.

Sistemas eleitorais e processos políticos comparados: a promessa de democracia na América Latina e Caribe. See item **3771**.

2799 **Smith, William C.** Reestructuración neoliberal, pugnas distributivas y escenarios de consolidación democrática en América Latina. (*Real. Econ./Buenos Aires*, 110, 16 agosto/30 sept. 1992, p. 105–123, bibl.)

Author is concerned about social ramifications of the free market paradigm that is now dominant in Latin America. In particular, it threatens to widen inequalities of income and therefore clashes with democracy's promise of greater popular influence over government. Author sees little hope of reversing this, however, because national governments are increasingly dependent on the world economy. [P.H. Lewis]

2800 **Snyder, Richard.** Explaining transitions from neopatrimonial dictatorships. (*Comp. Polit.*, 24:4, July 1992, p. 379–399, graphs)

Using the concept of "neopatrimonial dictatorship," author examines a number of variables that account for different paths to political development for neopatrimonial regimes. Three variables—institutional autonomy of the military, strategies and organizational power of moderate groups opposed to the dictator, and organizational strength of opposition revolutionary groups—play a key role in accounting for alternative transitions in Cuba, Nicaragua, Haiti, Paraguay, Panama, and Venezuela. These factors are influenced in turn by the structure of the patrimonial state including the military, domestic actors, and foreign powers.

2801 **Social democracy in Latin America: prospects for change.** Edited by Menno Vellinga. Boulder: Westview Press, 1993. 327 p.: bibl., index.

Latin American and European scholars examine the political crisis by focusing on whether the European experience with social democracy offers a model of development that suits Latin America. Authors exhibit varying degrees of optimism and pessimism regarding the possibility of a social democratic solution to Latin America's current ills.

2802 Souza, Ayda Connia de et al. Democracia, partidos e cultura política na América Latina. Porto Alegre, Brazil: NUPESAL; Editora KUARUP, 1989. 224 p.: bibl.

Four essays review literature on parties and political culture. Except for Portuguese translation of Scott Mainwaring's essay on political parties in Brazil and the Southern Cone (see *HLAS 51:3366*), papers are generally not of high quality. [T. Power]

Statistical Masterfile. 1974- . See item **44**.

Stoll, David. Is Latin America turning Protestant?: the politics of Evangelical growth. See *HLAS 53:1302*.

2803 Television, politics, and the transition to democracy in Latin America. Edited by Thomas E. Skidmore. Washington: Woodrow Wilson Center Press; Baltimore: Johns Hopkins Univ. Press, 1993. 188 p.: bibl., index, map.

Valuable work devoted to role of television and electoral politics in four Latin American countries involved in redemocratization: Chile (1988), Mexico (1988), Brazil (1989), and Argentina (1989). Nine media analysts—both US and Latin American—examine the power and limitations of television in the new democratic era, emphasizing the twin functions of dissemination of information and political mobilization of voters.

2804 TSentral'nasia Amerika i Kariby: nachalo 90-kh godov [Central America and the Caribbean: the beginning of the nineties]. Edited by Anatoliĭ Danilovich Bekarevich. Moscow: Akademia nauk SSSR, Institut Latinskoĭ Ameriki, 1991. 192 p.: bibl.

This late Soviet collection of articles from the Institute of Latin American Studies continues a series on the economy and politics of Central America and the Caribbean. Drawing on US and Latin American sources of economic data, the editor and 10 authors give a last Soviet view of Central American problems and the balance of US, West European, and Soviet power in the region. Individual articles focus on foreign economic relations, debt, tourism, and other issues, with particular attention to Mexico, Haiti, Nicaragua, Costa Rica, Guatemala, and Panama. One contributor explores British and West European relations with the Caribbean community. [B. Dash]

2805 Vilas, Carlos M. Latin American populism: a structural approach. (*Sci. Soc.*, 56:4, Winter 1992/1993, p. 389–420, bibl.)

Brief article designed to formulate some general propositions on the economic and political basis of populism. As a product of a specific level of development of peripheral capitalism, populism contains a number of inherent contradictions, particularly the popular mobilization that angers the Latin American bourgeoisie even though this is necessary to promote its economic interests.

2806 Whitehead, Laurence. The alternatives to "liberal democracy:" a Latin American perspective. (*Polit. Stud.*, 40, special issue, 1992, p. 146–159)

Critical assessment of viability of liberal democratic regimes argues that consolidated democracies will tend to develop a range of features at variance with the neoliberal model touted by many non-Latin American analysts.

2807 Wickham-Crowley, Timothy P. Exploring revolution: essays on Latin American insurgency and revolutionary theory. Armonk, N.Y.: M.E. Sharpe, 1991. 241 p.: bibl., ill., index.

Theories of revolution and social movements focusing on guerrilla-peasant relationships. Aim is "adapting revolutionary theories to Latin American realities" in order to explain peasant participation in insurrectionary movements. [G.P. Atkins]

2808 Wickham-Crowley, Timothy P. Guerrillas and revolution in Latin America: a comparative study of insurgents and regimes since 1956. Princeton, N.J.: Princeton Univ. Press, 1992. 424 p.: bibl., index.

Rigorous, data-based comparative analysis examines longstanding debate about causes of revolution. Assesses reasons for Latin American guerrilla successes (in Cuba and Nicaragua) and failures (in Colombia, Bolivia, Guatemala, Peru, and Venezuela) while probing social and political characteristics of guerrillas, their movements, and target regimes. [G.P. Atkins]

World News Connection. 1995 . See item **50**.

2809 **Wynia, Gary W.** Politics: essay. (*in* Latin America and the Caribbean: a critical guide to research sources. Edited by Paula H. Covington. New York: Greenwood Press, 1992, p. 627–681)

Excellent introduction to research trends in political science literature on Latin America since the 1960s is followed by a bibliography of 500 sources organized into useful reference categories. Author finds that political science research has been characterized by intellectual diversity, not consensus, and despite an enormous increase in publications there is still a great deal that needs to be studied.

MEXICO

RODERIC A. CAMP, *Professor of Political Science, Tulane University*

THERE IS MUCH TO BE PLEASED ABOUT in recent scholarship on Mexican politics and government. As political liberalization increasingly dominates Mexican policy, the issue of elections, electoral reform, and voting behavior exerts strong influence in the scholarly literature. Nevertheless, some traditionally neglected topics have received more attention, including the Catholic Church, the media, opposition parties, and non-governmental organizations.

Mexican and American scholars alike have been taken with the notion of where Mexico is going politically, and how the process of political liberalization, i.e., democratization, occurs. One of the most interesting overviews of this topic is Steven Morris' "Political Reformism in Mexico: Salinas at the Brink" (item **2893**), which argues that crisis management has become the PRI's present mode of operation and will continue to be so in the future. One of Mexico's most thoughtful political analysts, a historian by training, is Lorenzo Meyer, whose classic article on authoritarianism is brought up-to-date in his latest exploration of the interrelationships between civil society, the political arena, and the State, and their consequences for democratic transformation (item **2890**).

Perhaps the most compelling theoretical issue raised in the literature of the early 1990s is the interrelationship between economic and political liberalization. The most comprehensive work on this topic, Riordan Roett's edited book, *Political and economic liberalization in Mexico: at a critical juncture?* (item **2907**), combines Mexican and American research and thoroughly explores possible linkages between economic and political development. To date, few works on this topic have been written from an explicitly comparative perspective. A notable exception is the very useful case study of Poland and Mexico by Judith Gentleman and Voytek Zubek who provide valuable insights about the East European and Mexican experiences (item **2858**). Jaime Sánchez Susarrey, who frequently writes about democratization, presents an optimistic picture of political liberalization within the context of economic and political interactions (item **2920**). One of the few essays to explore the actual consequences of economic liberalization on policy is Stephen Mumme's analysis of the contradictions in Salinas' environmental policies (item **2895**).

Political liberalization has also encouraged the expansion of social movements and non-governmental organizations. A broad overview of such movements, and changes taking place in the Mexican corporate structure, is offered by Susan Street (item **2925**). Moving to a more basic grassroots level, Jonathan Fox and Luis Hernández provide numerous insights into the role of NGOs in Mexico (item **2856**), and how their behavior differs from that of NGOs elsewhere in Latin America.

Another by-product of political liberalization is a change in the media, and in media-State relations. The role of the media in Mexican politics has been largely ignored in the 1980s, therefore the collection *Así se calló el sistema: comunicación y elecciones en 1988*, edited by Pablo Arredondo Ramírez, is a pathbreaking contribution, including excellent case studies of several urban centers (item **2820**). Ilya Adler also contributes an important essay on the media in the 1988 presidential elections (item **2811**).

Although the volume of literature on elections and election data is overwhelming, several items stand out for the quality of the data and analysis. This is particularly true of the work by Guadalupe Pacheco, who explores the relationship between urbanization, voting, and political culture in the Federal District (items **2900, 2901,** and **2902**). Other authors, who have long contributed to this topic, provide several outstanding collections. The first of these, *Insurgencia democrática: las elecciones locales* (item **2869**), compiled by Jorge Alonso and Silvia Gómez Tagle, focuses on highly disputed state elections, providing a much needed work on local patterns. Also, Alberto Aziz Nassif and Jacqueline Peschard's work on the PRI's comeback in the 1991 elections provides extensive statistical data and analysis (item **2853**). The broadest overview of the topic, including in-depth data by district on party representation in the Chamber of Deputies, is Juan Molinar's *El tiempo de la legitimidad* which covers the 1940s through 1988 (item **2892**).

The attention given to elections has also increased interest in other political parties, particularly the National Action Party (PAN). Among the most useful, fresh analyses of PAN is one exploring its institutional role, by Leticia Barraza and Ilán Bizberg (item **2823**), and a 1991 electoral analysis assessing the consequences of the PRI-PAN electoral reform alliance, by Mario Alejandro Carrillo (item **2839**). Although the PRD has not received as much serious attention, interviews contained in Eduardo de Castillo's *20 años de búsqueda* (item **2841**) provide an excellent historical memory of the left's development. Finally, although opposition parties have long controlled local governments, and more recently state administrations, Victoria Rodríguez and Peter Ward are among the first scholars to explore the consequences of opposition control (item **2914**).

Although in past *HLAS* volumes it was counted among the understudied interest groups and institutions, the Catholic Church finally has begun to attract the attention it deserves. Roberto Blancarte contributes an excellent analysis of the Church's role during the Salinas years (item **2830**). A useful summary of background information is also available in a clear overview by Allan Metz (item **2887**). Charles L. Davis (item **2848**), who has been interested in religion and partisanship for several decades, provides an excellent empirical study of working-class voters in 1979–80 and their relationship to political parties. The private sector, another major interest group, has been largely ignored in the last several years, with the exception of Blanca Heredia's insightful analysis of business-State relations in the 1980s (item **2864**).

Government leadership and structures have received some attention, and Miguel Angel Centeno and Sylvia Maxfield's excellent work explores both policy implications and elite composition (item **2843**). Rogelio Hernández Rodríguez, who is best known for his analysis of the Mexican leadership (item **2867**), offers one of the few works of political biography, the most balanced account of PRI reformer Carlos A. Madrazo yet published (item **2866**). One of the few studies incorporating actual budget expenditures and their policy implications is that of Judith Teichman, which explores the status of State-owned enterprises in the 1980s (item **2927**). The most neglected of governmental institutions, the Chamber of Deputies, is the subject of an interesting analysis of labor representation by Juan Reyes del Campillo (item **2912**).

More attention needs to be focused on State-military and State-labor relations, and especially on the changes in the corporatist structures. Although the recent work on elections is commendable, much of it is repetitive, and more attention could be paid to the three leading parties. What is needed is careful analyses of changes in the party structures, ideology, and leadership, and their relation to the State and society. Again, the most difficult topics remain largely untouched, including work on decision-making, political biography, and civil-military relations, the latter especially in light of what occurred in Chiapas in Jan. 1994. For discussion of the Chiapas events from an anthroplogical perspective, see p. 91.

2810 Ackroyd, William S. Military professionalism, education, and political behavior in Mexico. (*Armed Forces Soc.*, 18:1, Fall 1991, p. 81–96)

Based on a PhD dissertation, provides one of the few detailed examinations of the education of army officers in Latin America generally, and Mexico specifically. Reveals important socialization features that affect the behavior of the officer corps and its relationship with civil authorities.

2811 Adler, Ilya. The Mexican case: the media in the 1988 presidential election. (*in* Television, politics, and the transition to democracy in Latin America. Washington: Woodrow Wilson Center Press, 1993, p. 145–173, tables)

A very useful analysis of an important and neglected topic of Mexican political behavior. Argues that the 1988 elections brought a new openness to media coverage, but, in general, the media did not respond to societal changes and conducted its reporting along traditional lines. Concludes that a democratic Mexico requires changing journalistic practices.

2812 Aguayo, Sergio *et al.* Las elecciones federales en México según los noticieros 24 Horas de Televisa y Hechos de Televisión Azteca, 30 de mayo a 30 de junio de 1994. México: Academia Mexicana de Derechos Humanos; Alianza Cívica/Observación 94; 1994. 6 p.: graphs.

Spanish-language version of item **2885**.

2813 Aguila M., Marcos Tonatiuh. La casaca de Don Plutarco. (*Memoria/CEMOS*, 45, agosto 1992, p. 5–9, plates)

Thoughtful exploration of the issue of no reelection. Examines its historical impact on the political development of elite leadership in Mexico and compares the period of Plutarco Elías Calles (1924–28) to that of Carlos Salinas de Gortari. Although their differences are as strong as their similarities, much can be learned from the contrast.

2814 Alonso, Jorge and **Manuel Rodríguez Lapuente.** La cultura política y el poder en México. (*in* Cultura y política en América Latina. Coordinación de Hugo Zemelman. México: Siglo Veintiuno Editores, 1990, p. 343–378, bibl.)

A serious attempt to reexamine the political culture of various institutions and groups to determine changes in Mexican attitudes. Concludes that Mexican population is experiencing numerous developments in attitudes which translate into popular behavior at the grassroots level, providing the potential for altering the political status quo.

2815 Alonso, Jorge *et al.* El nuevo Estado mexicano. Coordinación de Jorge Alonso, Alberto Aziz Nassif y Jaime Tamayo. México: Univ. de Guadalajara; Nueva Imagen, CIESAS, 1992. 4 v.: bibl., ill.

Alonso, author of previous works on the Mexican State, brings together a superior collection of essays by leading Mexican social scientists on changes affecting the State, elections, parties, and regional groups. In particular, the chapters on the military and the State, human rights, the 1988 elections, and neo-Cardenism offer many fresh insights and useful material.

2816 Alvarado, Arturo and **Nelson Minello.** Política y elecciones en Tamaulipas: la relación entre lo local y lo nacional. (*Estud. Sociol./México*, 10:30, sept./dic. 1992, p. 619–650, graphs)

Interesting case-study analysis of elections in the northern border state of Tamaulipas. Examines the 1988, 1989, and 1991 elections within the context of a decade of elections (1979–91), in particular their impact on State and federal relations. Among other conclusions, argues that rural Tamaulipas has remained largely unchanged, and that no other parties except PRI have penetrated the area.

2817 Alvarez, Luis H. Las ideas de los adversarios: luchar, gobernar, dialogar. *(Ideas Polít., 1:1, mayo/junio 1992, p. 211–223)*

The president of PAN outlines his campaign rhetoric and policy goals in a speech to party faithful at a national meeting in Sept. 1991.

2818 Alvarez, Luis H. *et al.* Los partidos políticos mexicanos en 1991. Recopilación de Federico Reyes Heroles. México: Fondo de Cultura Económica, 1991. 446 p. (Col. popular; 440)

Comparative sourcebook of party platforms (National Action Party, Institutional Revolutionary Party, Popular Socialist Party, Democratic Revolutionary Party, and Authentic Party of the Mexican Revolution) enables readers to compare the positions of prominent party leaders in 1991.

2819 Anguiano, Arturo. El eclipse de la izquierda en México. *(in* El socialismo en el umbral del siglo XXI. Coordinación de Arturo Anguiano. México: Univ. Autónoma Metropolitana, Unidad Azcapotzalco y Unidad Xochimilco, 1991, p. 355–390)

As the title suggests, presents a pessimistic future for the left in Mexico, including the Party of the Democratic Revolution (PRD), claiming they do not have even minimal credibility. Events after March 1991, when this was written, suggest stronger possibilities for the left, in elections and otherwise, than those presented in this analysis.

2820 Arredondo Ramírez, Pablo; Gilberto Fregoso Peralta; and **Raúl Trejo Delarbre.** Así se calló el sistema: comunicación y elecciones en 1988. Guadalajara, Mexico: Univ. de Guadalajara, 1991. 268 p.: bibl., ill. (Col. Jornadas académicas: Serie Compilaciones)

The role of the media within the larger context of Mexican politics has been woefully ignored in the last decade by North American political scientists, a gap filled well by several excellent essays in this volume, focusing on media coverage of the 1988 presidential elections. In addition to several studies of the national media, this collection includes two fine chapters on Guadalajara and Mexico City, and a comprehensive overview by Raúl Trejo Delarbre.

2821 Arroyo Alejandre, Jesús and **Stephen D. Morris.** The electoral recovery of the PRI in Guadalajara, Mexico, 1988–1992. *(Bull. Lat. Am. Res., 12:1, Jan. 1993, p. 91–102, bibl., tables)*

Using survey data from elections in Guadalajara, one of Mexico's most important urban centers and a locus of left and right opposition strength, the authors focus on voters' electoral confidence. They argue that the initial transition toward democracy has actually given new life to Mexican authoritarianism, rather than to political liberalization.

2822 Arvide, Isabel. 23 diálogos con gobernadores. México: Encuadernación Progreso, 1990. 319 p.

Arvide, a political journalist, interviews 23 governors, including some of the most controversial figures during the Salinas Administration. Most of the answers are unrevealing; however, when the interviewer achieves a rapport with the interviewees, interesting responses are recorded. The comments of Beatriz Paredes, one of the few female governors in Mexico, suggest how women cope in the male-dominated world of Mexican politics.

2823 Barraza, Leticia and **Ilán Bizberg.** El Partido Acción Nacional y el régimen político mexicano. *(Foro Int., 31:3, enero/marzo 1991, p. 418–445)*

Most of the work which has been done on the National Action Party since that of Mabry and Von Sauer in the early 1970s has been polemical or non-scholarly, but this is an exception. Well-researched and thoughtful, going far beyond the traditional election analysis. This article examines the potential future of PAN vis-à-vis the State, the PRD, and political liberalization.

2824 Barrera Zapata, Rolando and **María del Pilar Conzuelo Ferreyra.** Descentralización y administración pública en los estados federados: ensayo analítico-metodológico. Toluca, Mexico: Instituto de Administración Pública del Estado de México; Univ. Autónoma del Estado de México, 1989. 209 p.: bibl., ill.

Among the studies of Mexican politics, there exists much up-to-date information on the relationship between state and national institutions. This work, while providing a serious examination of the theoretical aspects of some of the issues of decentralization, does not provide detailed material or case studies of the Mexican situation.

2825 **Becerra Chávez, Pablo Javier.** El COFIPE y las elecciones federales de 1991. (*Iztapalapa/México*, 11:23, julio/dic. 1991, p. 49–64, bibl., ill.)

Analysis of the role that the Mexican Federal Electoral Institute might play in elections, specifically the off-year congressional election of 1991. Argues that this institution is not the one to bring about the transition to clean and fair elections. The Institute has undergone repeated and numerous alterations since publication of this article.

2826 **Bellinghausen, Hermann.** Algo ocurre en la esquina de esta nación. (*Ojarasca/México*, 29, feb. 1994, p. 43–60, photos)

Captures some background details about the cause and origins of the Zapatista National Liberation Army in Chiapas as well as the flavor of the communities and participants through perceptive photography.

2827 **Bensusan Areous, Graciela Irma** *et al.* Política y gobierno en la transición mexicana. Coordinación de Manuel Canto Chac y Víctor Manuel Durand Ponte. México: Univ. Autónoma Metropolitana, 1990. 245 p.: bibl., ill.

Diverse collection of essays on recent changes in Mexican politics. The most interesting, by Pedro Moreno Salazar, examines recent policy perspectives on social expenditures at the federal level, providing a useful overview and empirical economic data on government spending trends from 1980–88.

2828 **Bermejillo, Eugenio.** Hasta donde llegan las armas. (*Ojarasca/México*, 29, feb. 1994, p. 22–37)

Description of events during the Chiapas uprising, including a quite useful and detailed chronology of the crucial month of January, which assesses all the pertinent actors, and a more general chronology showing events from 1966 to Dec. 1993.

2829 **Biblioteca de las entidades federativas.** v. 1, Aguascalientes. v. 6, Chihuahua. v. 8, Colima. v. 12, Hidalgo. v. 14, Estado de México. v. 15, Michoacán. v. 17, Nayarit. v. 22, Quintana Roo. v. 25, Sonora. v. 26, Tlaxcala. v. 29, Veracruz. v. 31, Zacatecas. Coordinación de Pablo González Casanova y Jorge Cadena Roa. Edición de Francisco José Paoli Bolio y Guadalupe Valencia García. México: Centro de Investigaciones Interdisciplinarias en Humanidades, Coordinación de Humanidades, Univ. Nacional Autónoma de México, 1988–1992 12 v.: bibl., ill., maps.

This multivolume series attempts to provide politically relevant information on the culture, society, and economy of the Mexican states, including many of the less politically and economically powerful states. Although short on analysis, the series provides difficult to obtain statistical information and offers an excellent place to start exploring general features of the political and economic landscape.

2830 **Blancarte, Roberto.** El poder salinismo e Iglesia Católica: ¿una nueva convivencia? México: Grijalbo, 1991. 318 p.: bibl., ill. (Política mexicana)

One of the most important books written about the Catholic Church and the Mexican government in many years. Blancarte, who has studied the Church extensively and published a larger historical work, raises questions about Church-State relations, and the consequences of recent reforms for Church involvement in electoral processes and human rights.

2831 **Blancarte, Roberto.** Religion and constitutional change in Mexico, 1988–1992. (*Soc. Compass*, 40:4, Dec. 1993, p. 555–569)

Mexico's leading scholar of Church-State relations provides a superior and insightful assessment of the impact of constitutional changes under Salinas, believing that the changes enacted, and the source of those altered laws, may provoke more, rather than fewer, conflicts in the next few years.

2832 **Borrego Estrada, Genaro.** El debate de las ideas: las ideas del partido de la Revolución Mexicana. (*Ideas Polít.*, 1:1, mayo/junio 1992, p. 185–209)

Borrego, briefly the president of the

National Executive Committee of PRI, outlines his goals for the party at his acceptance speech on May 14, 1992.

2833 Brachet-Márquez, Viviane. Explaining sociopolitical change in Latin America: the case of Mexico. (*LARR*, 27:3, 1992, p. 91–122, bibl.)

Broad examination of four perspectives on political permanence and change in Mexico: clientelism, pluralism, authoritarianism-corporatism, and class perspective. The author concludes that individual models provide only partial views, and that to understand change fully, elements from each perspective must be linked.

2834 Burgoa, Ignacio et al. La participación política del clero en México. Coordinación de Luis J. Molina Piñeiro. México: Univ. Nacional Autónoma de México, Facultad de Derecho, 1990. 238 p.: bibl.

Collection of studies from a 1990 conference, one of the few works on the involvement of the Catholic Church in Mexican politics. The work, while very uneven in scholarly quality, contains commentary from the conference, including presentations from politicians and the clergy.

2835 Cadena Roa, Jorge. Fuentes para el estudio del Estado mexicano: 1968–1988. México: Coordinación de Humanidades, Univ. Nacional Autónoma de México, 1989. 154 p. (Cuadernos del CIIH: Serie Fuentes; 3)

This excellent bibliography, broadly conceived, of studies on the Mexican State and politics from 1968–88 is indexed by author and subject.

2836 Camp, Roderic Ai. Mexican political biographies, 1935–1993. 3rd ed. Austin: Univ. of Texas Press, 1995. 985 p.: bibl. (ILAS special publication)

Revised and updated version of earlier (1982) volume of biographies of prominent public figures in Mexican political life from 1935 to mid-1993. By including both biographical information and lists of the most important elective, appointive, and party positions in Mexico, the book can serve as "a selective version of a government organizational manual" for contemporary Mexico. Excellent reference work for mapping trends in leadership recruitment and socialization. [D. Dent]

2837 Camp, Roderic Ai. Political liberalization: the last key to economic modernization in Mexico? (*in* Political and economic liberalization in Mexico: at a critical juncture? Boulder, Colo.: L. Rienner Publishers, 1993, p. 17–33)

Examines theories of economic and political liberalization in Mexico in which the author argues that political reform is not essential to immediate economic modernization but is nevertheless an integral part of long-term political development. Structural, cultural, and psychological variables are examined for their explanatory capacity in the Mexican case. [D. Dent]

2838 Cancino, César and Víctor Alarcón Olguín. La relación gobierno-partido en un régimen semicompetitivo: el caso de México. (*Rev. Mex. Cienc. Polít. Soc.*, 38:151, enero/marzo 1993, p. 9–33)

Thoroughly researched exploration of the peculiarities of the relationship between the Mexican government and political parties. While set within the larger Latin American context, variables unique to Mexico are identified. Suggests that despite changes to the electoral process and within the PRI, Mexico has progressed little toward establishing a democratic, competitive party system.

2839 Carrillo, Mario Alejandro. Tres años, el largo trecho: el PAN ante las elecciones federales de 1991. (*Cotidiano/México*, 7:44, nov./dic. 1991, p. 27–34, photos)

Excellent analysis of PAN's internal problems, and the electoral consequences of the party's strategies and ideological preferences in the 1991 federal elections. Concludes that by accepting the electoral changes PRI proposed, PAN lost its role as an incisive, destabilizing electoral force, a position the author claims brought PAN success in the 1980s.

2840 Casillas Hernández, Roberto. Fuerzas de presión en la estructura política del estado. 3. ed. Aguascalientes, Mexico: Instituto Cultural de Aguascalientes, 1991. 292 p. (Contemporáneos)

Casillas, who served as José López Portillo's private secretary, would be in a position to provide many insights into how interest groups influence Mexican decision-making. Instead, like so much of this literature in Mexico, book provides a general, theoretical discussion which does not examine the Mexican case. For example, the author says nothing about the role of the Mexican military.

2841 Castillo, Eduardo del. 20 años de búsqueda. México: Palabra en Vuelo Ediciones; Ediciones de Cultura Popular; Claves Latinoamericanas, 1991. 320 p.

Will prove invaluable to researchers seeking the motivations behind the founding of the Party of the Democratic Revolution (PRD) in 1989. Contains interviews with the most prominent leaders of the PRD, in which each is asked similar questions. It is particularly valuable to have recent political developments examined in the context of 1968, since many of those interviewed were actors or close witnesses to the events of that year.

2842 Castro y Castro, Fernando. El acontecer de un funcionario público: etapas previas 1947–1981 y efemérides 1981–1988. México: Editorial Diana, 1992. 600 p.: ill.

Author, who was prominent in public life in the 1960s, 1970s, and 1980s, records his daily thoughts in a diary from 1947–81. Although not a major figure, Castro y Castro alludes to being a generational representative of Mexicans born in the mid-1920s, and consequently his remarks offer some insight into the life and times of a political cohort.

2843 Centeno, Miguel Angel and Sylvia Maxfield. The marriage of finance and order: changes in the Mexican political elite. (*J. Lat. Am. Stud.*, 24:1, Feb. 1992, p. 57–85)

Largely based on Centeño's excellent PhD dissertation, authors offer some helpful and provocative interpretations of trends in Mexican political leadership. Discusses ramifications for the role of the political-economic elite and the implications of bureaucratic in-fighting among specific agencies, as well as possibilities for democratizing the electoral system.

2844 Cornelius, Wayne A. and Ann L. Craig. The Mexican political system in transition. La Jolla, CA: Center for U.S.-Mexican Studies, Univ. of California, San Diego, 1991. 124 p.: bibl., ill., maps. (Monograph series; 35)

An excellent general, introductory survey of Mexican politics by two leading Mexicanists.

2845 Cornelius, Wayne A. The politics and economics of reforming the *ejido* sector in Mexico: an overview and research agenda. (*LASA Forum*, 23:3, Fall 1992, p. 3–10)

A leading scholar of Mexican immigration to the United States provides an extremely useful, straightforward outline of the ejido reforms, including the rationale for advocating the reforms, and a list of real and potential consequences of reforms once enacted. An excellent starting point for researchers of this policy issue.

2846 Craske, Nikki. Women's political participation in *Colonias populares* in Guadalajara, Mexico. (*in* "Viva:" women and popular protest in Latin America. Edited by Sarah A. Radcliffe and Sallie Westwood. London; New York: Routledge, 1993, p. 112–135)

An important analysis of the role and impact of women in popular, political organizations. Concludes that women are fundamental to existence of such organizations, providing the bulk of the rank and file, but rarely filling leadership roles.

2847 Cuéllar, Alfredo. El triunfo de un partido de oposición y la calidad educativa. (*Rev. Cienc. Soc./Río Piedras*, 29:1/2, enero/junio 1990, p. 159–186, bibl., tables)

Traces the development of the National Action Party (PAN) platform on improving the quality of Mexican public education, and discusses the party's proposal for implementing its ideology in Baja California. Considers educational policy more than the political consequences of policy implementation.

Davis, Charles L. and John G. Speer. The psychological bases of regime support among urban workers in Venezuela and Mexico: instrumental or expressive? See item **2744**.

2848 Davis, Charles L. Religion and partisan loyalty: the case of Catholic workers in Mexico. (*West. Polit. Q.*, 45:1, March 1992, p. 275–297, bibl., tables)

One of the few analysts who has examined the impact of Mexican religious beliefs on partisanship over time explores a sample of working class voters from 1979–80, discovering that, contrary to popular assumptions, Catholic workers tend to vote more consistently for the establishment party, the PRI, rather than for the opposition PAN.

2849 Descentralización y democracia en México. Recopilación de Blanca Torres. México: El Colegio de México, 1986. 280 p.: bibl.

The Colegio de México conducted a conference in 1985 on one of the seven principle themes of the Miguel de la Madrid campaign: decentralization. This work combines academic and public sector presentations which tend to focus on administrative and economic decentralization between the Federal District and the provinces, although a contribution by Francisco Gil Villegas does analyze the issue of democracy and decentralization theoretically, without reference to the Mexican case.

2850 **Díaz, María del Carmen.** Lo político y el sistema partidario en México. (*Iztapalapa/México*, 11:23, julio/dic. 1991, p. 25–48, ill.)

Historical overview of the relationship of the PRI to the State focuses on liberal reforms within the party. Argues that political space exists for certain groups to express their views, but that there is potential for mobilization and confrontation from rural groups, a forecast prescient of the 1994 rebellion in Chiapas.

Diccionario biográfico del gobierno mexicano. See item **21**.

2851 **Díez Duarte, Raúl et al.** Libertad religiosa y autoridad civil en México: elementos para el análisis de las relaciones Iglesia-Estado; simposio universitario, marzo 14–16 de 1989. México: Univ. Pontificia de Mexico, 1989. 273 p.: bibl. (Selecta UPM; 3)

One of the most important issues of the Salinas Administration—Church-State relations—is debated in this collection of essays, unfortunately, only from an essentially legal standpoint. Nevertheless, because some of the scholars are among the best on this topic, the essays provide useful historical insights.

2852 **Las elecciones federales de 1988 en México.** Edición de Juan Felipe Leal, Jacqueline Peschard y Concepción Rivera. México: Facultad de Ciencias Políticas y Sociales, Univ. Nacional Autónoma de México, 1988. 492 p.: bibl., ill. (Col. Procesos electorales; 4)

Although this is one of the best collections of articles on the 1988 presidential elections, drawing as it does on the work of many leading Mexican social scientists, many of the articles are brief transcriptions of oral presentations rather than serious, researched explorations of political events. Nevertheless, includes important contributions on the electoral code, the candidate selection process, campaigns, political parties, the media, and balloting.

2853 **Las elecciones federales de 1991.** Coordinación de Alberto Aziz Nassif y Jacqueline Peschard. México: Centro de Investigaciones Interdisciplinarias en Humanidades, UNAM; M.A. Porrúa Grupo Editor, 1992. 245 p.: bibl., ill., maps. (Democracia en México)

An outstanding collection of essays on the 1991 elections by some of Mexico's leading scholars of the electoral process. These essays make extensive use of statistical data and analysis to examine controversial questions surrounding the 'comeback' election for PRI. Provides a detailed exploration of the vote increases for PRI in various regions and districts, as well as an assessment of the credential and vote-counting processes.

2854 **The evolution of the Mexican political system.** Edited by Jaime E. Rodríguez O. Wilmington, Del.: SR Books, 1993. 322 p.: bibl., ill., index. (Latin American silhouettes)

Based on a series of presentations organized by the editor at the Univ. of California, Irvine in 1990, these essays analyze political developments in Mexico from 1810 to the present, focusing on the 19th and early 20th centuries. The concluding essays evaluate provocative questions that arose during the conference and comment on the themes raised in the individual contributions.

2855 **Foweraker, Joe.** Popular mobilization in Mexico: the teachers' movement, 1977–87. Cambridge, England; New York: Cambridge Univ. Press, 1993. 204 p.: bibl, index.

A contribution to the body of work on popular movements in Mexico, focusing on the National Union of Education (SNTE). Goes beyond a case study to develop a general discussion of the newly important political role of Mexican popular movements.

2856 **Fox, Jonathan** and **Luis Hernández.** Mexico's difficult democracy: grassroots movements, NGOs, and local government. (*Alternatives/Boulder*, 17:2, Spring 1992, p. 165–208)

No topic deserves more attention than the role of NGOs in Mexico. The authors perceptively examine Mexican NGOs, and more importantly, offer a comparison to NGO experiences elsewhere in Latin America. They conclude that in Mexico, NGOs are not, for the most part, leading the social and political transition.

2857 **Frontera norte: una década de política electoral.** Coordinación de Tonatiuh Guillén López. México: El Colegio de México; Tijuana, Mexico: El Colegio de la Frontera Norte, 1992. 281 p.: bibl., ill.

One of Mexico's most careful scholars of state and local electoral politics brings together six articles covering the northern border states during the 1980s. This excellent comparative study of Tamaulipas, Chihuahua, Coahuila, Baja California, Nuevo León, and Sonora is especially valuable since two of these states were the first to elect opposition governors.

2858 **Gentleman, Judith** and **Voytek Zubek.** International integration and democratic development: the cases of Poland and Mexico. (*J. Interam. Stud. World Aff.*, 34:1, Spring 1992, p. 59–109, bibl.)

Despite the international patterns of global democratization, surprisingly few analysts have compared Mexico with other countries, making this work an outstanding and perceptive exploration of the differences and similarities between the Polish and Mexican cases. Greatly expands knowledge of the issues and consequences of a democratic transition.

Ghiringhelli, Robertino and **Rafael Pérez Miranda.** Lo stato degli studi e delle ricerche sulla classe politica in Messico. See item **4493.**

2859 **Goldrich, Daniel** and **David V. Carruthers.** Sustainable development in Mexico?: the international politics of crisis or opportunity. (*Lat. Am. Perspect.*, 19:1, Winter 1992, p. 97–122, bibl.)

An imaginative essay in which the authors, in addition to offering the usual criticisms of the Salinas Administration, propose an alternative development strategy. Essentially, they believe small-scale production and labor-intensive environmental reconstruction will revive decaying rural economies and absorb unemployed and underemployed labor currently headed northward.

2860 **González Casanova, Pablo** *et al.* México, el 6 de julio de 1988: segundo informe sobre la democracia. México: Siglo Vientiuno Editores; Centro de Investigaciones Interdisciplinarias en Humanidades, Universidad Nacional Autónoma de México, 1990. 185 p.: bibl. (Biblioteca México)

Two articles stand out among these analyses of the disputed 1988 presidential elections. The first by Silvia Gómez Tagle provides a superb overview, data, and analysis of contested electoral districts in Mexico. The second, a joint effort by Alberto Aziz Nassif and Juan Molinar, uses statistical probabilities from previous election data to examine the likely accuracy of official vote tallies.

2861 **González Hinojosa, Manuel** *et al.* Las bases de la modernidad: 1970–1987. México: PAN; EPESSA, 1991. 218 p.: ill. (Col. Informes de los presidentes de Acción Nacional; 3)

The third volume in a series of works presenting the 'state of union' addresses of National Action Party presidents to their national councils, covering the period 1969–87.

2862 **Grindle, Merilee S.** The response to austerity: political and economic strategies of Mexico's rural poor. (*in* Social responses to Mexico's economic crisis of the 1980s. Edited by Mercedes González de la Rocha and Agustín Escobar Latapí. San Diego, Calif.: Center for U.S.-Mexican Studies, Univ. of California, San Diego, 1991, p. 131–153, bibl., tables)

Grindle, who has published some of the most perceptive work on Mexican decision-making and interest groups, focuses on the behavior of the rural poor under conditions of economic crisis. Concludes that peasants pursue complex household strategies linked to external economic institutions, which, in turn, affect local political behavior.

2863 **Guerrero, Omar.** El Estado y la administración pública en México: una investigación sobre la actividad del estado mexicano en retrospección y prospectiva. México: Instituto Nacional de Administración Pública, 1989. 812 p.: bibl.

A detailed legal analysis of the evolution of the State within the context of public administration in Mexico. Ignores the informal practices which directly affect the State's impact on decision-making and domestic politics.

2864 **Heredia, Blanca.** Profits, politics, and size: the political transformation of Mexican business. (*in* The right and democracy in Latin America. Edited by Douglas A. Chalmers, María do Carmo Campello de Souza, and Atilio A. Boron. New York: Praeger; Institute of Latin American and Iberian Studies, Columbia Univ., 1992, p. 277–302, bibl., tables)

An excellent analysis of business-State relations in the late 1980s documenting the changing strategies of the private sector and its increasingly activist political posture vis-à-vis the government and the National Action Party (PAN).

2865 **Hernández Madrid, Miguel J.** Una lectura política sobre la calidad de vida urbana en Zamora, Michoacán: la década de los ochenta. (*Relaciones/Zamora*, 12:49, invierno 1992, p. 35–61, bibl., graphs, tables)

An analysis of 142 interviews conducted in 1991 in the important provincial city of Zamora, Michoacán, the heart of the Democratic Revolutionary Party. Focuses on perceptions of quality of life and the resulting consequences for grassroots political behavior and attitudes.

2866 **Hernández Rodríguez, Rogelio.** La formación del político mexicano: el caso de Carlos A. Madrazo. México: Colegio de México, Centro de Estudios Sociológicos, 1991. 207 p.: bibl., ill.

The author, who has written some of the best work on Mexican leadership, offers the most balanced, comprehensive biography yet published of PRI reformer Carlos A. Madrazo. Provides many interesting insights into the evolution of his ideas, and clarifies his personal role as a reformer and party politician.

2867 **Hernández Rodríguez, Rogelio.** La reforma interna y los conflictos en el PRI. (*Foro Int.*, 32:2, oct./dic. 1991, p. 222–249)

A thoughtful and objective interpretation of the divisions within PRI leadership and the sources of conflict within the party. Among the interesting issues the author raises is the disagreement over territorial versus organizational structure, including the transformation of the National Federation of Popular Organizations (CNOP), one of the major traditional sectors of the party. One of the best analyses of internal party reforms.

2868 **Hurtado, Flor de María.** Documents and publications of the government of President Miguel de la Madrid. (*in* Seminar on the Acquisition of Latin American Library Materials (SALALM). *33rd, Berkeley, California, 1988.* Latin American frontiers, borders, and hinterlands: research needs and resources. Alburquerque: SALALM Secretariat, General Library, Univ. of New Mexico, 1990, p. 364–366)

A concise outline of the documents available at the *Centro de Documentación de la Gestión Gubernamental* of the *Unidad de la Crónica Presidencial*, the same agency publishing the extremely useful biographical directories of government officials. A highly valuable source of material for the Miguel de la Madrid Administration (1982–88).

2869 **Insurgencia democrática: las elecciones locales.** Compilación de Jorge Alonso y Silvia Gómez Tagle. Guadalajara, Mexico: Univ. de Guadalajara, 1991. 214 p.: bibl., map. (Col. Jornadas académicas: Serie Compilaciones)

The editors, two of Mexico's leading scholars of electoral processes, have compiled an excellent collection on local elections, generally covering the significant transition period from 1984–89. The case studies focus on the more disputed locales where opposition strength from the National Action Party and the Democratic Revolutionary Party has been strongest, including Chihuahua, Michoacán, Baja California, and Oaxaca, as well as states less frequently in the electoral limelight, such as Hidalgo and Nayarit.

2870 **Intermediación social y procesos políticos en Michoacán.** Coordinación de Jesús Tapia Santamaría. Zamora, Mexico: El Colegio de Michoacán, 1992. 470 p.: bibl.

Essays on local Michoacán politics presented at a conference at the Colegio de Michoacán in Nov. 1987, plus commentaries made by participants following each presentation. Selections contain vaulable analyses of rural political conditions, focusing on mobilization, *caciquismo*, political brokers or intermediaries, and indigenous movements. Some of the latter may offer interesting comparisons to uprisings in Chiapas.

2871 **Kaller, Martina.** Bauren zwischen produktion und subsistenz: hintergründe zur rebellion in Chiapas aus historischer sicht. (*Z. Lat.am. Wien*, 46/47, 1994, p. 27–37, bibl.)

Provocative analysis captures spirit of Zapatista *indigenismo* as it applies to recent local revolutionary activities. Examines antimodern perspective of Chiapas' peasants in their attempt to defend their traditions vis-à-vis a destructive modern State and economy. [C. Converse]

Latin American News. 1995- . See item 35.

2872 **Lau, Rubén; Vicente Jaime; and Víctor Orozco.** Sistema político y democrácia en Chihuahua. México: Instituto de Investigaciones Sociales de la UNAM; Ciudad Juárez, México: Univ. Autónoma de Ciudad Juárez, 1986. 135 p.: bibl. (Estudios regionales; 1)

Of the three essays in this short collection on politics in the northern state of Chihuahua, the only one which offers some new and helpful perspectives is that by Rubén Lau, who analyzes pressure groups in Ciudad Juárez, one of the most important centers of urban opposition groups in Mexico.

2873 **Laurell, Ana Cristina.** Democracy in Mexico: will the first be the last? (*New Left Rev.*, 194, July/Aug. 1992, p. 33–53, table)

A useful overview of the political and economic events leading to the 1988 presidential election and the crisis of legitimacy in Mexico. The analysis of the 1988–92 period is insightful, primarily for recording a series of human rights abuses in various sectors.

2874 **Legorreta, Jorge.** Expansión urbana, mercado del suelo y estructura de poder en la Ciudad de México. (*Rev. Mex. Cienc. Polít. Soc.*, 36:145, julio/sept. 1991, p. 45–76)

Useful only insofar as article identifies many leading local, informal political bosses in the Federal District.

2875 **Linares Zapata, Luis.** La conveniencia y un voto que mira al norte. (*Cotidiano/México*, 7:44, nov./dic. 1991, p. 3–13, photos, tables)

A brief but interesting analysis of the 1991 elections which compares national electorate's concerns (primarily economic and ecological), with those of Mexico City residents (economic, ecological, and safety). Concludes that voters, given the problems of the 1980s, looked to the economic model of the north, and possible linkages with the US, as a solution to their problems, thus voting for PRI.

2876 **Loaeza, Soledad.** Los partidos y el cambio político en México. (*Rev. Estud. Polít.*, 74, oct./dic. 1991, p. 389–403)

One of Mexico's leading political scientists concludes that political parties have not assisted the democratic transformation of the Mexican system. Further, the electoral process, rather than enhancing political liberalization in the first three years of the Salinas Administration, demonstrated its weakness as a tool of political transformation.

2877 **Loaeza, Soledad.** The role of the right in political change in Mexico, 1982–1988. (*in* The right and democracy in Latin America. Edited by Douglas A. Chalmers, María do Carmo Campello de Souza, and Atilio A. Boron. New York: Praeger; Institute of Latin American and Iberian Studies, Columbia Univ., 1992, p. 129–141)

A brief but helpful overview of the influence exercised by the National Action Party (PAN) on the electoral structure in Mexico under Miguel de la Madrid. Basically generalizations without extensive research.

2878 **Loret de Mola, Rafael.** Las entrañas del poder: secretos de campaña. México: Grijalbo, 1991. 226 p. ; (Política mexicana)

The author, son of the late journalist and PRI politician, became a candidate of PARM (Authentic Party of the Mexican Revolution) for mayor of Mérida, the capital city of Yucatán. His work, which basically relies on recollected conversations with other politicians, provides a picture of the 1989 electoral process, and is useful background for the 1993 disputed gubernatorial race.

2879 **Lutz, Ellen L.** Human rights in Mexico: cause for continuing concern. (*Curr. Hist.*, 92:571, Feb. 1993, p. 78–82)

Specialist in Mexican human rights provides a brief overview of efforts under the Salinas Administration to respond to accusations of human rights abuses. Highly critical conclusions paint an unfavorable picture of the administration's efforts to curb abuses, and indicate government involvement in the perpetration of abuses.

2880 **MacEwan, Arthur.** Banishing the Mexican Revolution. (*Mon. Rev.*, 43:6, Nov. 1991, p. 16–27)

Basically a review of two books: James Cypher's *State and capital in Mexico* and David Barkin's *Distorted development*, which argues that a free-trade agreement is likely to exacerbate the worst aspects of Mexican development and US-Mexican relations.

2881 **La magnitud y la integración de la administración pública en el Estado de México: ángulo de interpretación cuantificable.** México: INAP México; IAPEM, 1989. 139 p.: bibl., ill.

A detailed, statistical description of expenditures and categories of public spending in the state of México.

2882 **Maldonado Bautista, Samuel.** Orígenes del Partido de la Revolución Democrática. Morelia, Mexico: S. Maldonado Bautista, 1989. 303 p.

A description of the evolution of the Democratic Revolutionary Party. Contains interesting nuggets of information for the reseacher attempting a more serious analysis of the party's origins and evolution.

2883 **Martínez Rodríguez, Antonia** and **Mauricio Merino Huerta.** México: en busca de la democracia. (*Rev. Estud. Polít.*, 74, oct./dic. 1991, p. 405–430)

Authors comment on elements affecting the evolution of political liberalization in Mexico, in particular placing attempted internal PRI reforms within the context of larger societal changes and tensions. They also note the possibility for internal party democracy within PRI, despite a continuing authoritarian posture toward other parties.

2884 **Maxfield, Sylvia.** The international political economy of bank nationalization: Mexico in comparative perspective. (*LARR*, 27:1, 1992, p. 75–103, bibl.)

The author, who has examined Mexico's international finance in her work *Governing capital* (1991), here offers a comparative study of three recent bank nationalizations. Concludes that government actors in each case perceived their countries as losing control over monetary affairs and economic development policy, and that international economic theories do have utility in Third World countries, including Mexico.

2885 **The media and the 1994 federal elections in Mexico: a content analysis of television news coverage of the political parties and presidential candidates.** Washington: Washington Office on Latin America; México: Academia Mexicana de Derechos Humanos; Alianza Cívica/1994 Observación, 1994. 35 p: graphs, tables.

An outstanding, clear, methodologically balanced evaluation of radio and television coverage of Mexican political parties. Clearly demonstrates quantitative and qualitative advantage of the PRI in media coverage.

2886 **Medios, democracia y fines.** México: Dirección General de Apoyo y Servicios a la Comunidad; Fundación Friedrich Naumann; Agencia Mexicana de Noticias; Dirección General de Fomento Editorial de la UNAM, 1990. 336 p.: ill.

A collection of essays, of varying quality, from a conference on the media held at the National University in Sept. 1989. The papers are divided into 5 themes: democracy, elections, political culture, education, and popular communication. The media is an important topic in Mexico, and the essay by Enrique Sánchez Ruiz on education, the media, and democracy is particularly excellent.

2887 **Metz, Allan.** Mexican Church-State relations under President Carlos Salinas de Gortari. (*J. Church State*, 34:1, Winter 1992, p. 111–130)

Provides an excellent, up-to-date overview of the changed relationship between the government and the Catholic Church leading up to, but not including, the approval of constitutional amendments and reestablishment of relations with the Vatican. More useful for background information than for analysis of the relationship.

2888 **Mexico: dilemmas of transition.** Edited by Neil Harvey. London; Institute of Latin American Studies, Univ. of London and British Academic Press; New York: Distributed by St. Martin's Press, 1993. 381 p.: bibl., ill., index.

Based on a series of presentations given during a seminar on contemporary Mexico at the Institute of Latin American Studies, London, from 1989–90, these articles cover a broad range of topics, focusing on recent political and economic change. Valuable case studies on the State, State-social relations, and the Mexican political crisis during the 1980s.

2889 **Mexico: torture with impunity.** London: Amnesty International, 1991. 54 p.: ill., map.

This short work documents human rights abuses in Mexico, primarily since the Salinas Administration. While noting federal efforts to improve human rights monitoring, Amnesty International gives Mexico poor overall marks in protecting human rights, indicating that abuse is part of the structure of the criminal justice system.

2890 **Meyer, Lorenzo.** La prolongada transición mexicana: ¿del autoritarismo hacia dónde? (*Rev. Estud. Polít.* 74, oct./dic. 1991, p. 363–387)

Lorenzo Meyer, who has provided some of the most imaginative and insightful analyses of Mexican authoritarianism, examines the interrelationships between civil society, the political arena, and the State, and the resulting consequences for democratic transformation. Concludes that the political leadership offers no real commitment to political reform, instead, they have chosen economic liberalism within political authoritarianism.

2891 **Molinar Horcasitas, Juan and Jeffrey Weldon.** Elecciones de 1988 en México: crisis del autoritarismo. (*Rev. Mex. Sociol.*, 52:4, oct./dic. 1990, p. 229–262, graphs, tables)

An empirical examination of voting trends for the Institutional Revolutionary Party (PRI), the National Action Party (PAN), and the National Democratic Front (FDN) from 1979–88, testing the importance of independent variables such as geography, urbanization, immigration, level of education, and industrialization. Identifies the variables that affected voter outcomes for each party.

2892 **Molinar Horcasitas, Juan.** El tiempo de la legitimidad. México: Cal y Arena, 1991. 265 p.: bibl., ill.

Molinar, one of the best electoral analysts among Mexican social scientists, provides a comprehensive empirical study of the evolution of the electoral appeal of PRI and the opposition, from the mid-1940s through 1988. The book provides a wealth of relevant data on voting patterns and party representation in the districts, parties, and chamber of deputies. A necessity for Mexican electoral research.

2893 **Morris, Stephen D.** Political reformism in Mexico: Salinas at the brink. (*J. Interam. Stud. World Aff.*, 34:1, Spring 1992, p. 27–57, bibl.)

The first author to analyze in detail political corruption in Mexico offers here a broad interpretation of on-going political change under Salinas, arguing that crisis management is likely to characterize the PRI governments through the end of the century.

2894 **Morris, Stephen D.** Political reformism in Mexico: past and present. (*LARR*, 28:2, 1993, p. 191–205)

A review of five Mexican works on political and electoral reform which argues that present patterns duplicate those of the past and that the presidency is the cornerstone of any such reforms.

2895 **Mumme, Stephen P.** System maintenance and environmental reform in Mexico: Salinas's preemptive strategy. (*Lat. Am. Perspect.*, 19:1, Winter 1992, p. 123–143, bibl.)

One of the few analyses available that examines the policy implications, broadly conceived, of environmental policy and economic liberalization in Mexico. Focuses on the contradictions between Salinas' short-term economic liberalization strategy and sound environmental concerns.

2896 **Negociación y conflicto laboral en México.** Coordinación de Graciela Irma Bensusan Areous y Samuel León. México: FLACSO, Sede México; Friedrich Ebert Stiftung, 1990. 278 p.: bibl.

Case studies, primarily from the 1980s, examining union movements and conflicts with management, presented in 1990 at FLASCO in Mexico City. Some excellent essays on topics ranging from major unions, such as the National Teachers Organization and the National Petroleum Workers Union, to the important Cananea strike and subsequent military occupation under the Salinas Administration.

2897 **Nuestra palabra: el fraude electoral de 1991 y la participación ciudadana en la lucha por la democracia.** México: Convergencia de Organismos Civiles por la Democracia, 1992? 387 p.: bibl.

Despite some short and undeveloped essays, this is an extremely useful collection, providing detailed analyses and case studies of fraud during the widely-touted PRI comeback elections of 1991. An extensively documented study of Cuidad Juárez, Chihuahua is one of the best articles available on how fraud actually occurs. These case studies from all regions in Mexico will be essential to analyzing electoral fraud.

2898 Núñez, Arturo. El nuevo sistema electoral mexicano. México: Fondo de Cultura Económica, 1991. 345 p.: bibl., ill. (Col. popular; 451)

The director of the Federal Electoral Institute provides a comprehensive legal overview of the electoral system in Mexico. Although rapid changes in electoral laws will date this presentation, it provides a clear theoretical understanding of the process.

2899 Ouweneel, Arij. Alweer die Indianen: ... de jaguar en het konijn in Chiapas ... [Those Ameridians again: .... the jauguar and the rabbit in Chiapas ...] Amsterdam: Thela Publishers, 1994. 245 p.: bibl., graphs, ill., map, photos. (Latijas Amerika Bibliotheek)

Following a rather disjointed historical introduction, author explains the Ejército Zapatista de Liberación Nacional (EZLN) revolt in Chiapas. Shows that in media accounts negative as well as positive stereotypes dominate the discussion. Argues that the revolt was essentially political rather than ethnic. The EZLN presented itself as a revolutionary socialist movement. The fact that Amerindians partook in this revolt didn't make it an Amerindian movement. [R. Hoefte]

2900 Pacheco Méndez, Guadalupe. Geografía distrital del voto por el FDN: una pluralización regionalizada. (*Relaciones/Coyoacán*, 3, 1990, p. 39–53, graphs, maps, tables)

Despite reservations about the quality of the official election data from the 1988 presidential contest, the author attempts a careful statistical analysis of the regional strengths of the parties forming the Frente Democrático Nacional led by Cuauhtémoc Cárdenas. Demonstrates regional differences among the parties in the electoral alliance, as well as their weaknesses compared to those of the official party.

2901 Pacheco Méndez, Guadalupe. El PRI en los procesos electorales de 1961 a 1985. México: Univ. Autónoma Metropolitana, Unidad Xochimilco, División de Ciencias Sociales y Humanidades, 1988. 145 p.: ill. (Breviarios de la investigación; 5)

An excellent empirical study making statistical correlations between social factors and voting levels for PRI from 1960–85. Author provides very sophisticated statistical breakdowns for such variables as urban-rural residency and literary and PRI support. For example, that there is a high correlation in communities of fewer than 5,000 people, while PRI support varies widely in middle-sized communities and states.

2902 Pacheco Méndez, Guadalupe. Urbanización, elecciones y cultura política: el Distrito Federal de 1985 a 1988. (*Estud. Sociol./México*, 10:28, enero/abril 1992, p. 177–218, bibl. maps, tables)

Very useful empirical analysis of data on the relationship between urbanization, voting patterns, and political culture in the Federal District. Compares opinions of PRI supporters versus Democratic Front for National Reconstruction (FDN) sympathizers from the 1985 and 1988 elections, uncovering the pessimism with which the marginalized urban population views Mexico's economic development.

2903 Partido Acción Nacional (Mexico). El México de la oposición. v. 1, 7 plata formas presidenciales. México: Comisión Editorial del Partido Acción Nacional, 1986–1990. 1 v.

Collection of seven conservative national platforms from the National Action Party from 1946–88. Useful reference tool.

2904 El partido en el poder: seis ensayos. México: PRI, IEPES, 1990. 443 p.: bibl., ill. (El día en libros; 43: Sección Ciencias políticas)

Six essays on the evolution of the PRI and the problems it faces in changing from a dominant to a competitive political organization. Useful because it collects in one place the proposed 1990 resolutions for changing the internal structure of the party. Valuable for its reflection of intentions rather than descriptions of actual, structural changes.

2905 Peschard, Jacqueline. México: los partidos políticos en la coyuntura electoral. (*Secuencia/México*, 17, mayo/agosto 1990, p. 11–20, ill.)

A thought piece about electoral change and political parties. Argues that although

Salinas successfully revived the legitimacy of the presidency, if completed alterations are not accompanied by a strengthening of party institutions, the changes will enhance an authoritarian, rather than a democratic transformation.

2906 **Pimentel González, Nuri** and **J. Francisco Rueda Castillo.** Las elecciones del 18 de agosto: ¿avance democrático o estancamiento autoritario? (*Cotidiano/México*, 7:44, nov./dic. 1991, p. 35–42, photos, tables)

An analysis of the recovery of the Institutional Revolutionary Party (PRI) in the congressional elections of Aug. 1991. Concludes that some of the reasons for PRI's success were: State controlled fraud; National Solidarity funds in the right locations; stabilization of macroeconomic trends; errors and tactical limitations of the opposition parties, especially the PRD; and the restoration of a strong presidential system.

2907 **Political and economic liberalization in Mexico: at a critical juncture?** Edited by Riordan Roett. Boulder, Colo.: L. Rienner, 1993. 1 v.: bibl., index.

Explores the timely issue of the interrelationship between economic and political liberalization from both Mexican and US points-of-view, offering many interesting speculations about possible linkages and their consequences in the Mexican case. Also focuses on prospects for political change and contains a section by political party leaders, including Luis Donaldo Colosio's interesting interpretation of the PRI's victory in the 1991 elections. Available in Spanish from Siglo XXI.

2908 **Ponce G., Dolores** and **Antonio Alonso Concheiro.** México hacia el año 2010: política interna. México: Centro de Estudios Prospectivos de la Fundación Javier Barros Sierra; Noriega Editores, Editorial Limusa, 1989. 400 p.: bibl., ill., maps. (Foro México 2010)

Applying a political-risk analysis approach to Mexican political and economic development, and extrapolating from statistical data, the study predicts scenarios for Mexico through the year 2010. Interesting, provocative work examining topics relevant to internal growth.

2909 **Ramos, Alejandro**; **José Martínez;** and **Carlos Ramírez.** Salinas de Gortari: candidato de la crisis. 2a ed. México: Plaza y Valdés Editores, 1988. 391 p.

Three authors, all journalists, combine their efforts to analyze the process of presidential succesion in 1987, which resulted in the designation of Carlos Salinas. They provide a useful overview of the process, including references to many commentaries from the period that suggest other possible leading pre-candidates, giving some of the reasons why Salinas emerged as the nominee.

2910 **Reding, Andrew.** Mexico: the crumbling of the "perfect dictatorship." (*World Policy J.*, 8:2, Spring 1991, p. 255–284)

Reding, a frequent op-ed contributor to the *Wall Street Journal* and *New York Times* on Mexico, offers highly critical interpretations of Mexico's political weaknesses. Provides a detailed list of anti-democratic conditions, calling for Congress to establish pre-approved political conditions before accepting a trade pact with Mexico and Canada.

2911 **Relaciones Iglesia-Estado: cambios necesarios; tésis del Partido Acción Nacional.** Compilación y síntesis de María Elena Alvarez de Vicencio. México: Epessa, 1990. 186 p.

Examination of the views of the National Action Party (PAN) regarding Church-State relations is based on party documents and articles and interviews with leading Panistas. Also includes the party's position on the 1992 constitutional reform proposals.

Revista Este País. 1995- . See item **41.**

2912 **Reyes del Campillo, Juan.** El movimiento obrero en la Cámara de Diputados, 1979–1988. (*Rev. Mex. Sociol.*, 52:3, julio/sept. 1990, p. 139–160, tables)

The Chamber of Deputies, an institution frequently neglected in the literature on Mexican politics, is analyzed in this essay examining the labor movement's representation in seats held by the PRI from 1979–88. The author concludes that the type of union, and its institutional resources, is an important determinant of labor success within the official party.

2913 **Reyes Heroles, Federico.** El poder: la democracia difícil. México: Grijalbo, 1991. 249 p.: bibl.

A collection of popular essays written from 1987–90 by one of Mexico's leading intellectuals, an articulate, perceptive commentator on political affairs. In these articles Reyes Heroles, editor of the important monthly *Este País,* covers a wide range of topics tied to democratization, providing stimulating opinions from a Mexican perspective.

Reynoso, Víctor Manuel. Notas para una geografía electoral del estado de Sonora. See item **2399.**

2914 **Rodríguez, Victoria E.** and **Peter M. Ward.** Opposition politics, power, and public administration in urban Mexico. (*Bull. Lat. Am. Res.,* 10:1, 1991, p. 23–36, bibl.)

These are the first authors in many years to address the consequences of opposition parties administering towns and states within a political system dominated by a single party. This article raises a number of important theoretical issues related to intergovermental relations and public administration within such an environment.

2915 **Rodríguez, Victoria E.** The politics of decentralisation in Mexico: from *municipio libre* to *solidaridad.* (*Bull. Lat. Am. Res.,* 12:2, May 1993, p. 133–145, bibl.)

This expert on Mexican decentralization explores the movement toward the distribution of power and decision-making at local levels. She is guardedly optimistic about future decentralization, but argues that electoral reforms and democratization have done more than administrative changes to increase the possibilities of distribution of power.

2916 **Rodríguez Prats, Juan José.** El poder presidencial: Adolfo Ruiz Cortines. 2a. ed. corr. México: Miguel Angel Porrúa Grupo Editorial, 1992. 318 p.: bibl.

This is not a scholarly biography, but Adolfo Ruiz Cortines, one of Mexico's most astute politicians and a respected president, deserves much more attention in the political historiography. Provides many interesting anecdotes from important figures who knew and worked with the president. Will be a useful source for completing a more comprehensive and analytical portrait of this 1950s' public figure.

2917 **Rubio, Luis.** Hacia un nuevo sistema político. (*Vuelta/México,* 16:183, feb. 1992, p. 57–61)

Rubio, a contributor of op-ed pieces to numerous Mexican publications and a member of an independent think tank in Mexico, offers a broad sketch of political reform, arguing that not since the 1920s have the problems in Mexico been as clear. Believes the priority is finding a balance between the demands of reality and the old, inflexible political structures.

2918 **Salyano Rodríguez, Raúl.** Tendencias del proceso electoral en el Estado de México, noviembre de 1990. Toluca, Mexico: Univ. Autónoma del Estado de México, 1990. 316 p.: bibl., ill. (Biblioteca científica; 2)

The Autonomous University of the State of Mexico sponsored this research during Summer 1990 as part of its graduate political science program. The results of surveying approximately 10,000 voters are presented here without significant analysis, in more than 200 pages of tables.

2919 **Sánchez González, Agustín.** Fidel: una historia de poder. México: Planeta, 1991. 271 p.: bibl., ill., index. (Espejo de México; 6)

Although not a scholarly approach, potentially helpful in providing background information on Mexico's leading labor figure, Fidel Velásquez, head of the Mexican Federation of Labor for many decades.

2920 **Sánchez Susarrey, Jaime.** Antecedentes y perspectivas de la reforma del Estado en México. (*Cienc. Polít.,* 27, 1992, p. 87–100)

The author, who has written several excellent essays on democracy and political liberalization, develops some interesting arguments about the interaction of political and economic reforms since 1970, particularly in terms of the consequences for the Mexican State. Concludes optimistically that the future of ongoing political liberalization is assured.

2921 **Santiago Quijada, Guadalupe** *et al.* Como se hizo el fraude en las pasadas elecciones: el caso de Ciudad Juárez. Coordinación de Hugo Almada Mireles. Ciudad Juárez, Mexico: Centro de Estudios Regionales y Comunicación Alternativa, 1991. 47 p.: bibl., ill.

Brief, but detailed study of voting fraud in Ciudad Juárez in 1991 makes specific suggestions for improving the voting process in 1992. Contains many useful, pragmatic insights and statistics.

2922 **Schmidt, Samuel.** Elite lore in politics: humor versus Mexico's presidents. (*J. Lat. Am. Lore,* 16:1, Summer 1990, p. 91–108, bibl.)

Argues that humor plays an important role in the political process. Suggests, for example, that humor can provide a release valve for political frustration and violence. More importantly, author believes, although he does not provide evidence, that Mexican presidents originate humor about themselves to diffuse potential opposition.

2923 **Seminario sobre el Sistema Político Mexicano, *Cuernavaca, Mexico, 1988.*** Sociedad, desarrollo y sistema político en México. Coordinación de Franciso López Cámara. México: Univ. Nacional Autónoma de México, Centro Regional de Investigaciones Multidisciplinarias, 1989. 90 p.

Collection of essays from a fall 1988 conference examining the contemporary political situation. The contributors, some of them party leaders, and others, prominent Mexican scholars, focused primarily on the role of the PRI, the relationship between economic policy and political development, and the implications of the 1988 presidential elections.

2924 **El sistema presidencial mexicano: algunas reflexiones.** México: Instituto de Investigaciones Jurídicas, Univ. Nacional Autónoma de México, 1988. 465 p.: bibl. (Serie G—Estudios doctrinales; 116)

While including the typical legal-administrative essays on the Mexican State's evolution, the collection also offers several helpful articles on the relationships between the executive and judicial branches, the legislative and executive branches, and interest groups and the State. Although in-depth analysis is lacking, the articles provide insight and useful citations to relationships rarely examined in the literature.

2925 **Street, Susan.** Movimientos sociales y el análisis del cambio sociopolítico en México. (*Rev. Mex. Sociol.,* 53:2, abril/junio 1991, p. 141–158, bibl.)

An excellent review of the literature on social change and social movements which argues government institutions and the bureacracy are failing to respond to the formation of new social organizations and the transformation of corporatism in Mexico. Based on a paper presented at the Latin American Studies Association (Miami, 1989).

2926 **Suárez Farías, Francisco José.** Familias y dinastías políticas de los presidentes del PNR-PRM-PRI. (*Rev. Mex. Cienc. Polít. Soc.,* 38:151, enero/marzo 1993, p. 51–79, bibl.)

One of Mexico's leading scholars of political elites identifies extended family ties of the 30 presidents of the PRI National Executive Committee who served from 1929–92. Does not examine analytically the consequences or significance of such extended kinship linkages, however.

2927 **Teichman, Judith.** The State and economic crisis in Mexico: restructuring the parastate sector. (*in* Latin America to the year 2000: reactivating growth, improving equity, sustaining democracy. New York: Praeger, 1992, p. 221–234, bibl., tables)

Teichman, one of the few analysts to examine Mexican decision-making in recent years, explores parastatal budget changes during the De la Madrid and early Salinas years. Suggests that the changes had negative consequences within the labor movement, but received positive political responses from the private sector. Concludes that economic policy has contributed to greater political conflict.

2928 **Transición a la democracia y reforma del Estado en México.** Compilación de José Luis Barros Horcasitas, Javier Hurtado y Germán Pérez Fernández del Castillo. Guadalajara, Mexico: Univ. de Guadalajara; México: Miguel Angel Porrúa; FLACSO, Sede México, 1991. 374 p.: bibl. (Col. Las Ciencias sociales)

Essays by some of Mexico's leading intellectuals, social scientists, and political figures who attended a colloquium sponsored by the Univ. of Guadalajara on the consequences of economic tranformation in Mexico. Includes provocative discussions on the political consequences for the State, for the electoral system, for a new political culture, and for politics in general. Covers many relevant political liberalization themes.

2929 **Trejo Delarbre, Raúl.** Crónica del sindicalismo en México, 1976–1988. México: Siglo Veintiuno Editores, 1990. 420 p.: bibl., index. (Historia)

Unlike most studies of labor organizations in Mexico which are generally collaborative, this study was written by a single author who examines new corporatism in the relationship between unions and the State from the administration of José López Portillo through that of Miguel de la Madrid. The author specifically examines various labor sectors, including automotive, teaching, pharmaceutical, petroleum, and publishing unions, among others.

2930 Trueba Lara, José Luis. Crónica de una venta anunciada. (*in* Simposio de Historia y Antropología de Sonora, *15th, Hermosillo, 1992.* Memoria. Hermosillo, Mexico: Instituto de Investigaciones Históricas, Univ. de Sonora, 1991, v. 2, p. 47–94, tables)

A detailed chronology of the political and economic conflicts involved in the sale of the Cananea mining enterprise which resulted in President Salinas sending troops to protect the plant. This useful basic reference also contains an excellent bibliography of primary sources.

2931 Valderrábano, Azucena. Historias del poder: el caso de Baja California. 2. ed. México: Grijalbo, 1990. 235 p.: bibl., ill., index. (Política mexicana)

Baja California, one of the states most electorally disputed between PRI and the opposition, is a fertile locale for analysis. Provides a brief description of opposition experiences, but focuses primarily on the campaign between PAN and PRI in 1989, leading to the election of the first opposition governor in PRI's history, Panista Ernesto Ruffo Appel. More descriptive than analytical.

2932 Valdés Zurita, Leonardo. Elecciones y partidos en México: 1988–1990. (*Secuencia/México,* 17, mayo/agosto 1990, p. 21–38, ill., tables)

An empirical assessment of the level of competition among political parties in Mexico during the 1988–90 electoral period. Analyzes vote tallies, concluding that independent or uncommitted voters are the single most important variable in the electoral outcome.

2933 Vargas González, Pablo E. Hidalgo: dos elecciones locales después de 1988. (*Iztapalapa/México,* 11:23, julio/dic. 1991, p. 117–130, ill., tables)

A case study which examines the 1990 local elections in Hidalgo, concluding that the opposition has achieved some electoral successes, but confronts, not surprisingly, a resistant authoritarian culture. Nonetheless, opposition forces were encouraged by societal pressures emerging from the 1988 national elections.

2934 Velázquez Hernández, Emilia. Política, ganadería y recursos naturales en el trópico húmedo veracruzano: el caso del municipio de Mecayapan. (*Relaciones/Zamora,* 12:50, primavera 1992, p. 23–63, bibl., map)

Anthropological case study of the allocation of natural resources in a small rural community in the Veracruz sierra is part of a larger study. Concludes, unsurprisingly, that the ejido comisariat played the most significant role in political-economic conflicts, and that purely political disputes typically involved conflicts with the mayor.

2935 Vicencio Acevedo, Gustavo. Memorias del PAN. v.4 2. ed. México: Editorial Jus, 1992. 1 v.: bibl., ill., indexes.

Fourth volume in a series of *memorias* of the National Action Party (PAN) provides indispensible information on state- and national-level party candidates during the 1950s, information which would otherwise be difficult to obtain, particularly in a single work.

2936 Vicencio Tovar, Abel. La reforma del Estado. México: PAN, 1992. 205 p.: bibl.

A collection of articles which the author, a leading figure in the National Action Party (PAN), contributed to journals, seminars, and conferences between 1980–92. The primary focus is political and constitutional reforms.

2937 Villar, Luis del. Los que mandan. México: Editorial Quehacer Político, 1990. 577 p.: ill.

This uncritical, descriptive work provides background information otherwise difficult to obtain about President Carlos Salinas and members of his cabinet, including Luis Donaldo Colosio, the PRI candidate for president in 1993. The book also includes interesting family photographs and copies of some personal documentation, including Salinas' birth certificate.

CENTRAL AMERICA

JOSE Z. GARCIA, *Associate Professor, New Mexico State University*

DURING THE EARLY 1990s some of the best writings on government and politics in Central America were historical studies that shed light on relatively unexplored subjects. Among the most impressive of these were Knut Walter's *The regime of Anastasio Somoza, 1936–1956* (item **3015**), a major work on the Somoza regime's support structure and strategic imperatives; Harry Vanden and Gary Prevost's *Democracy and socialism in Sandinista Nicaragua* (item **3013**), an unusually nonpolemical description of Sandinista democratic political theory as it evolved in power; Víctor Valle's *Siembra de vientos* (item **2971**), about formative experiences of future guerrillas in El Salvador during the 1960s; José Luis Chea's *Guatemala: la cruz fragmentada* (item **2976**), the first detailed analysis of the political role of the Catholic Church in recent Guatemalan history; Graciela García's *Porque quiero seguir viviendo* (item **2991**), a personal accounting of several decades of the labor movement in Honduras; and Manuel Solís' *Costa Rica: ¿reformismo socialdemócrata o liberal?* (item **2957**), a new thesis about the driving forces behind the ideological evolution of the Partido de Liberación Nacional (PLN).

Otherwise, reporting on human rights has improved (see especially item **2980** on Guatemala), as have country-specific works on parties, elections, the prospects for democracy as in works by Salazar Mora (item **2955**); Córdova Macías, (item **2961**); Monica Toussaint (item **2944**); Guillermo Castro (item **3018**); Oscar René Vargas (item **3014**); and Juan Arancibia Córdova (item **2987**). Unfortunately the domestic politics of Guatemala, Belize, and Panama remain seriously understudied. Finally, in spite of many attempts, there is little examination and even less analysis of how Central American domestic and international politics have been affected by the same transnational forces that are affecting countries everywhere: new patterns and politics in migration, new forms of organized crime, new manifestations of de-territoriality and re-territoriality, changes in regional/local identities, new political discourses and intellectual agendas, reductions in—and in other places consolidations of—sovereignty, and so on. With few exceptions, the selection of most topics and the use of conceptual tools in works annotated below belong to the out-of-date perspectives of the 1980s.

GENERAL

2938 **Barry, Tom.** Central America inside out: the essential guide to its societies, politics, and economies. New York: Grove Weidenfeld, 1991. 501 p.: bibl., index.

Argues that in order to understand the persistence of crisis in Central America, it is necessary to look deep inside the region, i.e., to look at Central America from the inside out. This means studying the region's internal dynamics as well as its international relations. [R. Palmer]

2939 **Brockett, Charles D.** Measuring political violence and land inequality in Central America. (*Am. Polit. Sci. Rev.*, 86:1, March 1992, p. 169–176, bibl.)

Compelling discussion of weaknesses in political violence data (in the *World Handbook of Political and Social Indicators*) and problems with conceptualizations of land inequality (the gini coefficient and other measures) for Central America, which make comparative analysis impossible.

2940 **Caldera T., Hilda** and **Benjamín Santos M.** La democracia cristiana en Centroamérica. Tegucigalpa?: INCEP, 1987? 432 p.: bibl., ill.

Brief histories of Christian Democratic parties in each Central American country—including Panama—based on interviews more than on primary or secondary sources. Information is more complete for Honduras. No comparative analysis.

2941 **Centroamérica, escenario de "desaparición forzada."** San José: CODEHUCA, 1992? 78 p.: bibl., ill. (Serie Cuadernos centroamericanos de derechos humanos; 3)

Brief comparative analysis of "disappeared" persons in Central America, using limited sources.

Les Forces politiques en Amérique centrale. See *HLAS 54:1723.*

2942 **Political parties and democracy in Central America.** Edited by Louis W. Goodman, William M. LeoGrande, and Johanna Mendelson Forman. Boulder, Colo.: Westview Press, 1992. 407 p.: bibl., index.

Anthology of 20 articles about political parties in Central America in the 1980s.

2943 **Weinberg, Bill.** War on the land: ecology and politics in Central America. London; Atlantic Highlands, N.J.: Zed Books, 1991. 203 p.: bibl., ill., index.

Ecological interpretation of the Central American crisis of the 1980s. This unusual volume, written by a journalist active in the environmental movement, provides examples of recent environmental degradation in each country, often—but not always—explaining these in conventional leftist terms.

BELIZE

2944 **Toussaint Ribot, Mónica.** Las elecciones en Belice: del espejismo bipartidista a la realidad neocolonialista. (*Secuencia/México*, 18, sept./dic. 1990, p. 5–16, ill.)

Superb class-oriented introduction to Belizean politics. Criticizes political parties as being clientilistic and personalistic, without clear-cut class bases. Argues that the threat from Guatemala has unified Belizeans but has also distracted them from political development issues.

COSTA RICA

2945 **Anderson, Leslie.** Bendiciones mezcladas: disrupción y organización entre uniones campesinas en Costa Rica. (*Rev. Hist./Heredia*, 25, enero/junio 1992, p. 97–156)

For sociologist's comment (on English version) see item **4572.**

2946 **Araya Monge, Rolando.** Vino nuevo en odres nuevos. San José: Editorial Univ. Estatal a Distancia, 1991. 425 p.

Taking into account recent global changes, young Partido de Liberación Nacional (PLN) intellectual and politician tries to chart a new (neoliberal) course for his party, which has dominated Costa Rican politics since 1948.

2947 **Arias Sánchez, Oscar.** La semilla de la paz: selección de discursos. San José: Presidencia de la República, 1990. 391 p.: ill.

Presidential speeches given by the Nobel Peace Prize-winning statesman.

2948 **Coto Molina, Walter.** Vamos hacia la reforma del partido. San José?: Partido Liberación Nacional, 1990. 74 p.

National PLN party chairman offers program of neoliberal reform for his party. Candid, insightful glimpse of the internal organization of the party during the late 1980s.

2949 **El Estado y la financiación de los partidos políticos en Costa Rica.** San José: Centro de Estudios para la Justicia Social con Libertad, 1989. 125 p.

Seminar funded by Friedrich Naumann Foundation discussed history and current status of public financing of parties, which was at a rate of two percent of the average national budget for the three years prior to elections. Debates by representatives of the major parties.

2950 **Gutiérrez Gutiérrez, Carlos José.** Neutralidad y democracia combativa. Heredia, Costa Rica: Fundación Friedrich Ebert, República Federal de Alemania, Centro de Estudios Democráticos de América Latina, 1987. 215 p.: bibl.

Speeches and memoirs of Costa Rica's Foreign Minister from 1984–86 as he struggled to maintain a neutral policy toward Nicaragua in the face of considerable domestic and international pressure to do otherwise.

2951 Gutiérrez Saxe, Miguel and **Jorge Vargas Cullell.** Costa Rica es el nombre del juego. San José: Instituto Costarricense de Estudios Sociales, 1986. 140 p.

Case study of a multifaceted 1984 international power struggle to allow private banks to obtain credit from foreign banks. Author believes US interests in Central America coincided with private-sector domestic interests to produce a political crisis and pressure the government to acquiesce.

2952 Hernández Valle, Rubén. Democracia y participación política. San José: Editorial Juricentro, 1991. 192 p.: bibl.

Eight essays on Costa Rican democracy, two of which provide lengthy discussions about evolution of constitutional law in Costa Rica.

2953 Nichaus, Bernd. El narcotráfico en Costa Rica. (*in* Democracia 2000: los grandes desafíos en América Latina. Bogotá: Fundación Simón Bolívar; Tercer Mundo Editores, 1991, p. 89–125)

Summary and analysis of Costa Rican legislative hearings.

2954 Reforma del Estado en Costa Rica. San José?: Comisión de Reforma del Estado Costarricense (COREC), 1990. xxviii, 249 p.

World Bank-financed study by a blue-ribbon, bipartisan committee in 1989–90 analyzes history of the State, participation, centralization, regionalism, income distribution, human resources, and environment. Suggests program of reform and continual formal review of the functioning of the State. Highly critical in spite of in-house character of the review. Superb work.

2955 Salazar Mora, Orlando and **Jorge Mario Salazar Mora.** Los partidos políticos en Costa Rica. San José: Editorial Univ. Estatal a Distancia, 1991. 194 p.: bibl.

Analysis of appearance, evolution, and interaction of political parties as reflections of evolution of elite composition and economic and sectoral change in Costa Rica from 1889 to the present. Major contribution, excellent bibliography.

2956 Silva Hernández, Ana Margarita. Municipalidad, educación y democracia en Costa Rica. San José: Instituto de Fomento y Asesoría Municipal, 1986. 55 p.: bibl. (Serie Documentos históricos municipales; 04)

Brief history of municipal education in Costa Rica.

2957 Solís, Manuel Antonio. Costa Rica: ¿reformismo socialdemócrata o liberal? San José: FLACSO, 1992. 434 p.: bibl.

Origins and evolution of PLN ideology. Argues that the party consists of both a reformist-conservative wing dating to the turn of the century and a post-WWII international wing advocating economic development through State intervention. Continuity and change in PLN doctrine is traceable in part to tensions between these two factions, in part to exogenous changes. New interpretation. Outstanding work about one of the world's most successful social democratic parties.

2958 Trejos, Gerardo. La oposición democrática. Prólogo de Isaac Felipe Azofeifa. San José: Editorial Juricentro, 1990. 167 p.: bibl.

Proposals for political and electoral reforms written by an advocate of the Partido de Progreso, a minor party. Opposes the near-monopoly held by the two major parties in the legislature. Well-argued, insightful sections on the system of electoral representation and campaign financing.

2959 Verschoor, Gerard. Intervenors intervened: farmers, multinationals and the State in the Atlantic zone of Costa Rica. (*Rev. Eur.*, 57, Dec. 1994, p. 69–87, bibl., map)

Case study demonstrates how a specific form of planned intervention by State agencies does not automatically lead to an expected outcome but takes concrete form *en route*. Local actors—farmers, middlemen, extensionists, bureaucrats, and multinationals—all have influence in the shaping of the project. Illustrates the complex relationship between knowledge and power. [R. Hoefte]

EL SALVADOR

2960 Arnson, Cynthia and **David Holiday.** El Salvador and human rights: the challenge of reform. New York: Human Rights Watch, 1991. 90 p.: bibl. (An Americas Watch report)

Americas Watch report on human rights violations by both sides near the end of the war includes information on Jesuit murders.

2961 Córdova Macías, Ricardo. Procesos electorales y sistema de partidos en El Salvador, 1982–1989. (*Foro Int.*, 32:4, abril/sept. 1992, p. 519–559, graphs, tables)

Electoral study modifying theories of Giovanni Sartori speculates on the evolution of the party system. Argues El Salvador's party system may be evolving from "bipolar, polarized pluralism" to "polarized pluralism."

2962 Los derechos humanos en El Salvador en 1989. San Salvador: Instituto de Derechos Humanos, Univ. Centroamericana José Simeón Cañas (IDHUCA), 1990. 382 p., 1 leaf of plates: bibl., ill. (Col. Informes anuales; fasc. 7)

Comprehensive, day-by-day, case-by-case descriptions of human rights violations in 1989, a year of extraordinary political violence in El Salvador, including the murders of Jesuit priests. Includes overall analyses of violence against targeted groups. Invaluable reference and resource.

Documento de la Prensa Gráfica: el conflicto en El Salvador. See *HLAS 54:1713*.

2963 Ellacuría, Ignacio. Veinte años de historia en El Salvador, 1969–1989: escritos políticos. v. 1–3. San Salvador: UCA Editores, 1991. 3 v.: tables. (Col. Estructuras y procesos; 9)

Collected writings of one of the slain Jesuit priests, rector of Univ. Centroamericana Simeón Cañas (UCA) at the time of his death in 1989. Most of these writings were originally published in the superb journal, *Estudios Centroamericanos (ECA)*, occasionally under pseudonym, and provided commentaries on current events.

2964 El estado democrático de derecho en El Salvador. San Salvador: Instituto de Estudios Jurídicos de El Salvador, 1990. 192 p.

Unusually frank presentations at an event funded by Friedrich Ebert review various facets of the often dubious state of law in El Salvador in 1990. Presenters, including judges and lawyers, reveal frustration, fear, and doubt.

2965 Guidos Béjar, Rafael. La crisis del socialismo en El Salvador. (*in* El socialismo en el umbral del siglo XXI. Coordinación de Arturo Anguiano. México: Univ. Autónoma Metropolitana, Unidad Azcapotzalco y Unidad Xochimilco, 1991, p. 324–336, table)

Summary of Frente Farabundo Martí de Liberación Nacional's ideological responses to the global retreat of socialism after 1987, written before the FMLN constituted itself as a party.

Harnecker, Marta. Con la mirada en alto: historia de las Fuerzas Populares de Liberación Farabundo Martí a través de entrevistas con sus dirigentes. See *HLAS 54:1737*.

2966 Lungo Uclés, Mario. El Salvador en los años 80: contrainsurgencia y revolución. La Habana: Casa de las Américas, 1991. 222 p.: bibl. (Ensayo)

One of the best analysts of the Salvadoran civil war describes changes—in the military situation, in the economy and society, and in guerrilla thinking—that occurred from 1979, when a right-wing government collapsed, to 1988, when a new right-wing government took power through elections.

2967 Rojas U., Javier. Conversaciones con el comandante Miguel Castellanos. San Salvador: Editorial UNSSA, 1988. 184 p.: ill.

Interviews with the late FMLN commander who quit the FPL guerrillas in 1985 after ten years on the inside, and who was killed by the FMLN a few years later. Invaluable source for understanding the early organization and motivation of the Fuerzas Populares de Liberación (FPL), one of the organizations that comprised the FMLN.

2968 Romero, Oscar Arnulfo. Mons. Oscar Arnulfo Romero: su diario desde el 31 de marzo de 1978 hasta jueves 20 de marzo de 1980. San Salvador: Arzobispado de San Salvador, 1990. 471 p.

During the turbulent months before his death, Mons. Romero was consulted by conspirators prior to the October 15th coup, and met with leaders of virtually all political sectors in El Salvador. His diary reveals both a growing awareness of politics during a moment of rapid political change, and increasing frustration as human rights abuses worsened.

2969 Seminario El Movimiento Sindical en la Asamblea Legislativa, *El Salvador, 1991.* Actas. San Salvador: CENITEC-DISE, 1991? 55 p. (Col. de seminarios; 10)

Four labor leaders (for STRABIF, FESINCONTRANS, UNTS, AND ONOC) discuss the labor situation, elections, and peace after the Alianza Renovadora Nacionalista (ARENA) took control of the legislative and executive branches in 1989.

2970 **Stanley, William.** Risking failure: the problems and promise of the new civilian police force in El Salvador. Cambridge, Mass.: Hemisphere Initiatives, Inc.; Washington: Washington Office on Latin America, 1993. 28 p.

Summary update on the creation of a new civilian police force discusses problems and policy recommendations.

2971 **Valle, Víctor Manuel.** Siembra de vientos: El Salvador 1960–69. San Salvador: CINAS, 1993. 554 p.

Current leader of the Movimiento Nacional Revolucionario (MNR) discusses student organizing and politics during the 1960s. Valle was deeply involved during that decade, as were many other 1980s rebel leaders. Important contribution for understanding this crucial but still ignored decade. Includes documents.

2972 **Walter, Knut** and **Philip J. Williams.** The military and democratization in El Salvador. (*J. Interam. Stud. World Aff.*, 35:1, 1993, p. 39–88, bibl., tables)

History of military involvement in politics, with section on the armed forces after the peace accords.

GUATEMALA

2973 **Aguilera Peralta, Gabriel Edgardo** *et al.* Los problemas de la democracia. Guatemala: FLACSO, 1993? 142 p.: bibl.

Somewhat abstract essays about the need for interethnic participation, a negotiated peace, and institutional support for democracy.

2974 **Amaro, Nelson.** Guatemala: historia despierta. Prólogo de Juan José Arévalo. Epílogo de Marco Vinicio Cerezo. Guatemala: IDESAC; Miami: Distribuidora Universal, 1992. 284 p.: bibl., ill.

Analysis of instability argues that Guatemalan history has a cyclical pattern of mobilization of traditional groups followed by violent demobilization. Focuses on period from 1944 to the present. Skillfully written, important contribution.

2975 **Arias, Arturo.** La democracia en Guatemala: actualidad y perspectivas. (*in* La democracia en América Latina: actualidad y perspectivas. Coordinación de Pablo González Casanova y Marcos Roitman Rosenmann. Madrid: Univ. Complutense de Madrid; México: Centro de Investigaciones Interdisciplinarias en Humanidades, UNAM, 1992, p. 477–507, bibl.)

Prospects for democracy within theoretical context of formation of the State. Gives optimistic and pessimistic scenarios.

2976 **Chea, José Luis.** Guatemala: la cruz fragmentada. San José: DEI; FLACSO, 1988. 356 p.: bibl., maps. (Col. Sociología de la religión)

History of the conservative political role of the Catholic Church under Archbishops Rosell and Casariego (1939–64 and 1964–83). Includes organizational structure of the Church, survey research of priests, discussion of internal conflicts, responses to Medellín, and excellent bibliography. Major contribution.

2977 **Delli Sante, Angela.** Nightmare or reality: Guatemala in the 1980s. Amsterdam: Thela, 1996. 439 p.: bibl., maps. (Thela Latin America series; 5)

Studies refugees fleeing war and repression in Guatemala. Focuses on nature of the repression and causal factors that made it possible. Also considers effects of repression on the population in general and on Mayan people in particular. Includes 296 p. of text and 98 p. of footnotes. [R. Hoefte]

2978 **Gleijeses, Piero.** The agrarian reform of Jacobo Arbenz. (*J. Lat. Am. Stud.*, 21:1, Feb. 1989, p. 453–480, bibl.)

Argues agrarian reform was successful economically and politically, yet met with US opposition nonetheless. Concludes US opposition, which ended in the overthrow of Arbenz in 1954, was based on fear of communists surrounding Arbenz, not US economic interests.

2979 **Guatemala: Amnesty International's current human rights concerns.** New York: Amnesty International U.S.A., 1991. 11 p.: bibl., ill.

Discusses human rights abuses in 1990.

2980 **Guatemala, getting away with murder: an Americas Watch and Physicians for Human Rights report.** New York: Americas Watch; Somerville, MA: Physicians for Human Rights, 1991. 89 p.: bibl., ill.

Unusual, penetrating analysis of Guatemala's system of death investigation, based on two fact-finding missions. Forensic specialists examined six medicolegal investigations into human rights abuses, exhumed clandestine graves to verify evidence of military human rights abuse, presented summary findings, and offered recommendations. Solid contribution.

Jonas, Susanne. The battle for Guatemala: rebels, death squads, and U.S. power. See *HLAS 54:1741*.

2981 *The Mayan People and Human Rights.* 1992- . Guatemala: Pro Justice & Peace Committee of Guatemala.

Discussion of human rights problems perpetrated by the government on the Mayan people in Guatemala. Refers to a multitude of Mayan indigenous organizations seeking redress.

2982 **Rosada-Granados, Héctor.** Elecciones y democracia en América Latina: Guatemala, 1990-1991. (*in* Una tarea inconclusa: elecciones y democracia en América Latina, 1988-1991. San José: Instituto Interamericano de Derechos Humanos, Centro de Asesoría y Promoción Electoral, 1992, p. 67-94, tables)

Critical analysis of the role of elections in promoting democracy in Guatemala. Excellent background, with superb statistical summaries of elections since 1944.

2983 **Schirmer, Jennifer.** Guatemala: los militares y la tesis de estabilidad nacional. (*in* Militares y Sociedad Civil en América Latina, *1st, San José, 1991*. América Latina: militares y sociedad. San José: FLASCO, 1991, v. 1, p. 183-219, bibl.)

Rare article on the Guatemalan military informed by a reading of contemporary theory on the subject and interviews with officers. Not surprisingly, concludes the armed forces are not yet subordinate to civilian power but have modified their doctrine.

2984 **Seminario sobre el Rol de los Partidos Políticos, *4th, Guatemala?, 1988*** Los partidos políticos y la función parlamentaria. Guatemala: Asociación de Investigación y Estudios Sociales, 1989. 296 p.

Proceedings of symposium with participation of Guatemalan politicians. Strong theme is the value of reducing power of the president vis-à-vis the legislature through a stronger party system.

2985 **Seminario sobre el Rol de los Partidos Políticos, *7th, Guatemala, 1991*.** En el proceso de elecciones, 1990-91. Guatemala: Asociación de Investigación y Estudios Sociales, 1991. 128 p.

Seminar on parties discusses elections, with some data on the party system.

HONDURAS

2986 **Amnesty International.** Honduras: civilian authority, military power; human rights violations in the 1980s. London: Amnesty International Publications, 1988. 54 p.: ill.

Account of the relatively few violations during the early 1980s. Does not compare Honduras with her more violent neighbors.

2987 **Arancibia Córdova, Juan.** Las perspectivas de la democracia en Honduras. (*in* La democracia en América Latina: actualidad y perspectivas. Coordinación de Pablo González Casanova y Marcos Roitman Rosenmann. Madrid: Univ. Complutense de Madrid; México: Centro de Investigaciones Interdisciplinarias en Humanidades, UNAM, 1992, p. 427-446, bibl.)

Places elections and democracy in historical perspective. Extensive use of previous books on Honduran politics.

2988 **Becerra, Longino.** Cuando las tarántulas atacan. Tegucigalpa: Baktún Editorial, 1987. 266 p.

In 1982 a gifted university student leader was kidnapped, tortured, and killed by a death squad apparently organized by Gen. Gustavo Alvarez. The victim's father, author of this biography, was a longstanding leftist leader. Anecdotes about Gen. Alvarez and the political milieu at the time.

2989 **Cáceres Lara, Víctor.** Gobernantes de Honduras en el siglo 20: de Terencio Sierra a Vicente Tosta. Tegucigalpa: Banco Central de Honduras, 1992. 344 p.: bibl.

Useful reference summarizes presidential succession, legislative functioning and major events from 1900–25. Four military men and five civilians (all but one deposed by armed force) ruled during this turbulent period in which the budget quintupled and the National Party, through which General Tiburcio Carias would later rule, was organized.

2990 Delgado Fiallos, Aníbal. Lecturas de política. San Pedro Sula, Honduras: Centro de Investigación y Acción para el Desarrollo, 1993. 180 p.: bibl.

Textbook contains excellent chapter providing periodization and summary of national political history.

2991 García, Graciela A. and Rina Villars. Porque quiero seguir viviendo—. Tegucigalpa: Editorial Guaymuras, 1991. 359 p.: bibl., ill. (Col. Talanquera. Documentos y testimonios)

Important, carefully crafted autobiography of an extraordinary feminist and tireless labor organizer from 1920s–40s who knew most major figures on the left throughout Central America. She also managed to write an important book in 1948 and several during the 1970s. Extremely insightful book based on two years of interviews by Rina Villars.

2992 Molina Chocano, Guillermo. Elecciones y consolidación democrática en Honduras en la última década. (*in* Una tarea inconclusa: elecciones y democracia en América Latina, 1988–1991. San José: Instituto Interamericano de Derechos Humanos, Centro de Asesoría y Promoción Electoral, 1992, p. 95–115, tables)

Competent summary of elections of 1980, 1981, 1985, and 1989, including regional analysis.

2993 Partidos = pleitos = politiquería. (*Envío/Managua*, 11:132, nov. 1992, p. 39–55)

Highly informative, insightful discussion of the popular base, internal organization, and competition within Honduran parties as they began to prepare for presidential and municipal elections held in 1993.

Posas, Mario. El proceso de democratización en Honduras. See item **4654**.

2994 Salomón, Leticia. Política y militares en Honduras. Tegucigalpa: Centro de Documentación de Honduras, 1992. 157 p.

Five excellent, balanced essays discussing different aspects of the role of the armed forces during the 1980s: civil-military crises, changes in military structure, the military's relationship with parties, demilitarization of society, and new roles for the armed forces.

2995 Schulz, Donald E. Cómo Honduras evitó la violencia revolucionaria. 1. ed. en español. Tegucigalpa: Centro de Documentación de Honduras, 1993. 48 p.: bibl.

Army War College analyst discusses important topic, stressing success of agrarian reform during the 1970s, the relative strength of the two-party system, inclusiveness of the political system, and a collective decision-making tradition.

NICARAGUA

2996 Arnaiz Quintana, Angel. Christianity and revolution: Nicaragua, 1979–90. (*in* The Church in Latin America, 1492–1992. Edited by Enrique Dussel. Maryknoll, N.Y.: Orbis Books, 1992, p. 425–434, bibl., table)

Four-part periodization of activist Christian involvement in politics during the Sandinista years, including intense polarization. Includes ideologically committed but informed introduction.

2997 Asociación Nicaragüense Pro-Derechos Humanos. Documentos: 23 octubre 1989 al 2 de marzo 1990. San José: Asociación Nicaragüense Pro-Derechos Humanos (ANPDH), 1990? 1 v.

Document analyzes climate of violence in Northern Nicaragua, where most Contras repatriated after Violeta Chamorro took office. They found the Sandinista army still in place, arming paramilitary sympathizers. The ANPDH, funded by the US government, published many works about Nicaragua during the late 1980s and early 1990s.

Borge, Tomás. The patient impatience: from boyhood to guerilla: a personal narrative of Nicaragua's struggle for liberation. See *HLAS* 54:1700.

Cardenal, Gloria et al. La guerra en Nicaragua. See *HLAS* 54:1707.

2998 Los evangélicos son cada vez más políticos. (*Envío/Managua*, 11:132, nov. 1992, p. 22–34, tables)

Good work uses survey data to analyze the growing political impact of Protestant evangelical groups.

2999 Gutiérrez Mayorga, Alejandro. Municipalidades y revolución. Managua?: CINASE; Fundación Friedrich-Ebert Stiftung, 1988. 98 p.: bibl.

Study of the organization, tax structure, and financing of Nicaragua's municipalities during the first seven years of the Sandinista government. Documents the rise and fall of municipalities under Sandinista rule (after rising, funds were severely cut in 1985). Adjusts for inflation, and compares data with previous periods.

3000 Guzmán, Luis Humberto. Políticos en uniforme: un balance del poder del EPS. Ciudad de Guatemala: Instituto Centroamericano de Estudios Políticos, 1992. 132 p.: bibl. (Panorama centroamericano: Ensayos socialcristianos; 1)

Discusses Sandinista armed forces in the post-Sandinista period.

3001 Jiménez Ruvalcaba, María del Carmen. El estado de emergencia y el periodismo en Nicaragua. Guadalajara, México: Univ. de Guadalajara, 1987. 102 p.

Content analysis comparing coverage of three major newspapers (one opposed to the regime—La Prensa—and two in favor) in Managua before and during the state of emergency decreed by the Sandinista government in March 1982. Interesting data on media in Sandinista Nicaragua. Pro-Sandinista author justifies censorship.

3002 Kuant, Elia María and **Trish O'Kane.** Nicaragua: partidos políticos y elecciones 1990. Managua: Coordinadora Regional de Investigaciones Económicas y Sociales, 1991? 40 p.: bibl. (Cuadernos de trabajo = Working paper)

Excellent briefing on all political parties—origins, ideological orientation, support structure—just prior to presidential elections.

3003 Martínez Cuenca, Alejandro and **Roberto Pizarro.** Nicaragua, una década de retos. Entrevista de Roberto Pizarro a Alejandro Martínez Cuenca. Cronología de María Roza Renzi. Prólogo de Sergio Ramírez. Managua: Editorial Nueva Nicaragua; FIDEG, 1990. 276 p.

US-trained economist who was Nicaraguan minister of commerce and then of planning during the 1980s is interviewed about Sandinista economic policies. Argues Violeta Chamorro won primarily because of years of neglect of the economy. Includes useful chronology of economic and other events.

3004 Nicaragua: a country study. Edited by Tim Merrill. 3rd ed. Washington: Federal Research Division, Library of Congress; Supt. of Docs. US GPO, 1994. xxxvii, 300 p.: bibl., ill., index, maps. (Area handbook series, 1057–5294. DA pam; 550–88)

Reliable introduction to contemporary Nicaragua, including chapters on history, social structure, the economy, government and politics, and national security. This edition is especially strong in covering various dimensions of the transition from Sandinista to Chamorro rule.

3005 Obando y Bravo, Miguel. Agonía en el bunker. Managua: Comisión de Promoción Social Arquidiocesana, 1990. 203 p.: ill., map.

Nicaragua's famous prelate relates his participation (often as mediator) in several confrontational episodes between the Sandinista Front and the Somoza regime from 1974 until the collapse of the National Guard on July 19, 1979. At one point the Guard was willing to allow Obando to become president.

Ramírez, Sergio. Confesión de amor. See *HLAS 54:3791.*

3006 Reimann, Elisabeth and **Fernando Rivas Sánchez.** Los tigres vencidos. Buenos Aires: Ediciones Reunir, 1988. 256 p.: bibl.

Eight Somoza supporters jailed by the Sandinista government for human rights violations were interviewed twice by Reimann. The first interviews (1980) are accompanied by documentation on offenses committed. Seven years later, six were still jailed and one was an officer in the Sandinista army.

3007 Reyes, Reynaldo and **Judith Kay Wilson.** Ráfaga: the life story of a Nicaraguan Miskito Comandante. Edited by Tod Stratton Sloan. Norman: Univ. of Oklahoma Press, 1992. 192 p.: bibl., ill., index, map.

Autobiography transcribed from tape recordings of the MISURA commander who led hundreds of villagers on a trek (called Operation Alpha One) across the Río Coco to Honduras to protest Sandinista treatment of the Miskitos. Reyes was ambivalent about MISURA leader Steadman Fagoth and CIA support. For ethnologist's comment see item 759.

3008 **Rivera, Carlos Alá Santiago.** Labor relations during the Sandinista government. (*Caribb. Stud.*, 24:3/4, 1991, p. 241–266)

Description of various major labor organizations and their growth and evolution during Sandinista rule. Excellent summary.

3009 **Rodríguez Gil, Adolfo.** Centralismo, municipio, regionalización y descentralización en Nicaragua. Managua: Fundación Friedrich Ebert, 1992. 163 p.: bibl. (Serie Descentralización y desarrollo municipal; 2)

Concise history of municipal reform in Nicaragua during the 1980s by an accomplished scholar of municipalities in Nicaragua. Argues strongly in favor of decentralization. Excellent bibliography.

3010 **Schneider, Robin.** Rama and the Sandinist revolution. Translated by Carol Baerg. Photographs by Klaudine Oland. Berlin: Reimer, 1989. 178 p.: bibl., ill., index. (Speaking with the tiger; 1)

Anthropologist who conducted field research from 1980–81 on Rama Cay (an island inhabited by Creole-English and Rama-speaking indigenous people, descendants of the Rama tribe) discusses tensions between islanders and the new Sandinista government at the time.

3011 **Schwartz, Stephen.** A strange silence: the emergence of democracy in Nicaragua. San Francisco, Calif.: ICS Press; Lanham, Md.: National Book Network, 1992. 156 p.: bibl.

Unusual, insightful work about what author believes was hypocrisy of some US leftists who supported Sandinista Marxist-Leninism long after the left elsewhere grew disenchanted. Author, an intellectual with a leftist past, criticizes the news media and sectors of the US intelligence. Excellent bibliography.

3012 **Sollis, Peter.** The Atlantic Coast of Nicaragua: development and autonomy. (*J. Lat. Am. Stud.*, 21:1, Feb. 1989, p. 481–520, bibl.)

Historical analysis from colonial times through Sandinista rule provides excellent background to the drive for more autonomy in the region.

3013 **Vanden, Harry E.** and **Gary Prevost.** Democracy and socialism in Sandinista Nicaragua. Boulder, Colo.: L. Rienner, 1993. 172 p.: bibl., ill., index.

Superb primer on origins and evolution of Sandinista thought and practice regarding democracy, mass organizations, and elections. Rare non-polemical view, informed by an excellent discussion of democracy in general.

3014 **Vargas, Oscar-René.** Nicaragua: los partidos políticos y la búsqueda de un nuevo modelo. Managua: Centro de Investigación y Desarrollo ECOTEXTURA; Comunicaciones Nicaragüenses, 1990. 203 p.

Excellent discussion of 21 political parties that contested the 1990 elections, written by one of the founding members of the Frente Sandinista de Liberacion Nacional (FSLN).

Vilas, Carlos M. La contribución de la política económica y la negociación internacional a la caída del gobierno sandinista. See item **4685.**

3015 **Walter, Knut.** The regime of Anastasio Somoza, 1936–1956. Chapel Hill: Univ. of North Carolina Press, 1993. 303 p.: bibl., ill., index, map.

Analysis of foundations of the dynastic regime that collapsed in 1979. Argues that a key to Somoza's success was his ability to co-opt the conservative opposition behind an export-oriented government and neutralize powerful regional caudillos by creating a broad coalition of national interest groups. First rate scholarship, major contribution.

PANAMA

3016 **Adames Mayorga, Enoch.** Política social e invasión: las opciones del estado panameño. Panama: IDEN, 1990. 49 p.: ill. (Col. Panamá '90; 4)

Documents deterioration of the Panamanian economy from 1987–90, arguing that it represented not only the political failure of the Noriega regime but also the end product of 20 years of populist rule.

3017 Alemancia, Jesús. Panamá: crisis política e identidad nacional. (*Rev. Panameña Sociol.*, 7, 1991, p. 369–384, photo)

Argues that the growing political role of the armed forces after the Canal treaties forced Panamanian citizens to choose between two negative national identities: authoritarian nationalism defended by the armed forces and a pro-US right-wing traditional elite.

3018 Castro H., Guillermo. Panamá, 1970–1990: transitismo, nación y democracia. (*in* La democracia en América Latina: actualidad y perspectivas. Coordinación de Pablo González Casanova y Marcos Roitman Rosenmann. Madrid: Univ. Complutense de Madrid; México: Centro de Investigaciones Interdisciplinarias en Humanidades, UNAM, 1992, p. 339–358)

Analysis of Panamanian democracy in the context of changes in State formation.

3019 Ernesto de la Guardia Navarro: el hombre, el político y el estadista: foro de conferencias presentadas en el acto celebrado el día 12 de enero de 1989. Panama: Instituto Latinoamericano de Estudios Avanzados (ILDEA), 1989? 38 p.

Four brief, friendly, and informative essays about Ernesto de la Guardia, elected president for the 1956–60 period. De la Guardia continued the "officialist" coalition known as the Coalición Patriótica Nacional that came to power with military assistance after elections in 1952.

Fuchs, Jochen. Costa Rica: von der Conquista bis zur "Revolution;" historische, ökonomische und soziale Determinanten eines konsensualistisch-neutralistischen Modells in Zentralamerika. See *HLAS 54:1628*.

3020 Gandásegui, Marco A. Policía militar o ejército nacional. (*Rev. Panameña Sociol.*, 7, 1991, p. 305–320, bibl.)

Argues a need for the creation of a national military force rather than just a police force.

3021 Human rights in post-invasion Panama: justice delayed is justice denied. New York: Americas Watch, 1991. 13 p.

Asserts human rights conditions improved after the invasion of Panama in 1989, but complains about overcrowded prisons, prolonged detentions without trial, and politicization of courts.

3022 Menéndez D'Avila, Lionel and **Alvaro Uribe.** Tercera fuerza y opción política. (*Rev. Panameña Sociol.*, 7, 1991, p. 225–240, tables)

Polls indicate a majority of Panamanians feel unsatisfied with traditional political parties. Using census data, article tries to sort out sectoral, class, and geographic origins of this marginalized majority, suggesting room for another party.

3023 Moral, Octavio del. Diagnóstico crítico de la democracia en Panamá: hacia una Asamblea Nacional Constituyente. Panamá: Pro Ley, 1992. 41 p.: bibl.

Group of lawyers argues for a constitutional convention.

3024 Pérez, Fernando; Ricardo Villalba; and **Carlos Francisco Changmarín.** Panamá: la coordenada militar. Panama?: Ediciones Unidad, 1989. 100 p.

The Partido del Pueblo, a communist party, published this set of essays by party members just prior to the US invasion. Essays written in 1968, 1973, 1986, and 1989 wrestle with the relationship between the armed forces after 1968, the US government, and the "revolutionary vanguard."

3025 Ropp, Steve C. Things fall apart: Panama after Noriega. (*Curr. Hist.*, 92:572, March 1993, p. 102–105)

Deep-seated legitimacy problems plagued the Endara government prior to the 1994 elections. Compares this period to the one just prior to the Torrijos coup of 1968.

3026 Rueda, Guillermina G. de and **Elia G. de Ferro.** Problemas políticos y socio económicos de Panamá. 23. ed. Panamá: s.n., 1992. 225 p., 1 folded leaf of plates: bibl., ill., maps.

High school textbook includes history and statistical summaries, with keen interest in the concept of "nation."

3027 Smith W., David A. Elecciones y democracia en América Latina, 1988–1990: el caso de Panamá. (*in* Una tarea inconclusa: elecciones y democracia en América Latina, 1988–1991. San José: Instituto Interamericano de Derechos Humanos, Centro de Asesoría y Promoción Electoral, 1992, p. 151–173, appendices, tables)

Summary of electoral reforms, parties, candidates, and elections since 1983.

3028 Soler, Giancarlo; Andrés Bolaños Herrera; and José Eulogio Torres Abrego. Panamá, fuerzas armadas y cuestión nacional. Panamá?: Taller de Estudios Laborales y Sociales, 1988? 105 p.: bibl., ill.

Three essays by Giancarlo Soler, Andrés Bolaños, and José E. Torres Abrego provide periodization and characterization of the role of the Panamanian armed forces during the national period. Does not provide much analysis for the Noriega period.

3029 Sterling Arango, Rolando. La batalla de San Miguelito. Panamá: CELA, 1991. 57 p.: ill.

Sympathizer discusses organization of the "Dignity Batallions" that fought US forces during Operation Just Cause. Criticizes Panamanian Defense Force leaders for poor training of troops and offers account of one battle. Includes an excerpt by Marta Iturralde and Juan Carlos Espinar from "Así se Organizaron los Batallones de la Dignidad" (1991).

THE CARIBBEAN AND THE GUIANAS (Except Cuba)

ANTHONY MAINGOT, *Professor of Sociology, Florida International University*

THE DOMINANT FEATURES of the literature on Caribbean government and politics during this period were its presentism, its insularity, and the absence of all-inclusive or ontological paradigms. This last feature is new; it reflects the passing of ideology evident elsewhere in the world. The first two features, presentism and insularity, are not. As in the past, a small part of the literature reflects the contemporary political concerns of subregions of the area, while the other, much larger, part is concerned with the contemporary politics of individual islands. Judging from recent scholarship, the first generation since independence—which the late Gordon K. Lewis calls the "third seminal period" after Discovery and Emancipation—has continued the scholarly insularism so often attributed to the colonial atomization of the Caribbean. And this, let it be said, at a time when governments are engaging in constant talks and efforts at various forms of regional association. Whether it be widening and deepening CARICOM, joining NAFTA, or creating the Association of Caribbean States, Caribbean leaders have been actively searching for ways to overcome the economic disadvantages of smallness in an age of regional geoeconomics. This political activity has been several paces ahead of the scholarship.

In addition to the insular bias in the analysis of politics, the preference for formal institutional analysis continues. Comparative studies of political cultures and the dynamics resulting from different social structures, or political studies of greater historical depth, continue to be few. Exceptions to the latter assertion are works by members of the British Society for Caribbean Studies, such as Paul Sutton's *Politics in the Commonwealth Caribbean* (item **3073**) which reflects the high quality of their scholarship, and by the Fundación Cultural Dominicana and its president, Bernardo Vega. The Fundación specializes in primary research in Dominican Republic-US and Dominican Republic-Haiti topics and has recently been publishing key documents in US-Caribbean diplomatic history. This still leaves major research voids in critical areas of Caribbean political life. Absent, for instance, are studies of the political influence of the private sectors, including those formally so described, such as commerce and agriculture, as well as the all-important informal sector. The nature and role of this informal sector, handled mostly by marketing women who have a

Caribbean-wide reach, have yet to be studied. Also awaiting analysis is the political role of international criminal cartels and the corruption they spawn. What, for instance, is the political role of the vast amounts of money flowing through the Caribbean and seeking security in one of the many regional offshore tax havens? The fact is that the Caribbean has entered into a new phase of development which emphasizes the private sector, the service sector, and export-driven growth, with very few academic studies to provide the policymakers with intellectual support and guidance.

The closest thing to a comparative analysis of the region's politics continues to be the edited book, by now a hallowed Caribbean genre. Colin Clarke's *Society and politics in the Caribbean* (item **3043**) and Anthony Payne and Paul Sutton's *The contours of modern Caribbean politics* (item **3036**) are outstanding examples of this approach: bring together a fine group of scholars, acknowledged experts in their particular areas; assign them the study of specific subregions of the area; and provide an encompassing summary introduction to pull the various strands into an intelligible whole. Because Clarke, Payne, and Sutton have done all this so well, their books will have an enduring shelf life. And yet, as valuable as these readers are, their division of the region by type of political system reflects the persistent bias towards institutional political analysis rather than the examination of political cultures.

Less valuable but also noteworthy in this genre is *Caribbean visions* (item **3030**), a collection of 10 presidential speeches of the Caribbean Studies Association, the premier gathering of scholars working on the Caribbean. The volume's value is reduced by the absence of a good analytical introduction or conclusion.

Evidence that official political enquiry in the Caribbean has outpaced formal scholarship is the *Report of the Stone Committee* (item **3070**), produced by the late Carl Stone of Jamaica. As chairperson of a government-appointed committee with relatively narrow terms of reference—to study the performace of Jamaica's parliament—Stone led his committee into a comparison of parliamentary and presidential executive systems. Given the variety of systems existent in the region, and given the scholarly trend to associate political with economic changes, this certainly is an area waiting for some serious comparative analysis.

Given this presentist and insular concern, it is perhaps predictable that Haiti would receive the greatest attention from both Haitian and foreign writers. The partisan nature of this literature is such, however, that students would do well to follow the sage advice of studying the historian before you study his or her history.

GENERAL

3030 Caribbean visions: ten presidential addresses of ten presidents of the Caribbean Studies Association. Edited by Simon B. Jones-Hendrickson. Frederiksted, V.I.: Eastern Caribbean Institute, 1990. 257 p.: bibl., index, ports.

Includes 10 addresses to the Caribbean Studies Association, the premier gathering of scholars working on the Caribbean, by presidents of the organization. Covers 1979–89 and reflects range of ideological and disciplinary positions in the field.

3031 Caribbean Workers Conference, 5th, St. John's, Antigua and Barbuda, 1985. The Caribbean workers struggle for real democracy: main documents of the 5th Caribbean Workers Conference, convened by CWC and CLAT. Caracas: DECOS; CLAT; CARISFORM, 1987. 180 p.: ill.

Speeches cover situation in the British, French, and Dutch Caribbean. Of interest because at time of conference, Central Latinoamericana de Trabajadores (CLAT) was a significant opponent of the various dictatorships in the area (Cuba, Haiti, Guyana, Suriname).

3032 **Dew, Edward M.** Caribbean paths in the dark. (*LARR*, 28:1, 1993, p. 162-173)

Masterful review examines 11 books on political economy and international relations of the Caribbean. Author draws sobering but not totally pessimistic conclusions from the many studies reviewed.

3033 **Domínguez, Jorge I.** The Caribbean question: why has liberal democracy (surprisingly) flourished? (*in* Democracy in the Caribbean: political, economic, and social perspectives. Baltimore: The Johns Hopkins Univ. Press, 1993, p. 1-25)

"The Caribbean is the only part of the world (with the partial exception of Rwanda) where the descendants of slaves govern sovereign countries." Surprisingly, says the author, they are successful democracies.

3034 **From Tortola to Kingstown: a report on eastern Caribbean political union.** Saint Lucia?: Voice Press, 1991. 49 p.: ill.

Contains collection of speeches surrounding Jan. 14, 1991 meeting at which at least 80 persons from islands of Dominica, Grenada, Saint Lucia, Saint Vincent, and the Grenadines came together to speak about forming a regional union. While there has been no significant movement towards a true political union, speeches reflect mood of major sectors at the time.

3035 **Islands of the Commonwealth Caribbean: a regional study.** Edited by Sandra W. Meditz and Dennis M. Hanratty. Washington: Library of Congress, Federal Research Division, 1989. 771 p.: bibl., ill., index, maps. (DA pam.; 550-33. Area handbook series)

Provides most useful treatment of social and political conditions for each of the English-speaking islands of the Caribbean. Maps, charts, and economic statistical tables enhance clearly written analyses. Individual chapters on regional security system and economic integration also contribute to volume's general utility.

3036 **Payne, Anthony** and **Paul K. Sutton.** The contours of modern Caribbean politics. (*in* Modern Caribbean politics. Baltimore, Md.: The Johns Hopkins Univ. Press, 1993, p. 1-27)

Very informative introduction lays the basis for the four issues dealt with in the book: 1) the imperial experience; 2) beginning of modern Caribbean politics; 3) a more specific look at "crisis" years of 1970s; and 4) key themes of politics of 1980s, dominated by the Reagan agenda.

3037 **Phillips, Fred.** Caribbean life and culture: a citizen reflects. Kingston: Heinemann Publishers (Caribbean) Limited, 1991. 252 p.: bibl., ill., index.

Despite title, work is a rather chatty autobiography. A distinguished West Indian constitutionalist and educator traces his life from his youth in Saint Vincent, through World War II, the West Indies Federation, and governorship of Saint Christopher-Nevis.

3038 **Reaching for the future: a timely trilogy.** Barbados: West Indian Commission Secretariat, 1991? 31 p. (West Indian Commission occasional paper; 2)

Speeches by Prime Ministers of Barbados, Jamaica, and Saint Vincent deal with need to integrate given uncertainties created by a changed global situation.

3039 **Regional Constituent Assembly of the Windward Islands. Meeting, *1st, Kingstown, Saint Vincent, and the Grenadines, 1991.*** Report. S.l.: s.n., 1991? 1 v.

In addition to verbatim transcripts, contains summary of contributions made by delegates, observers, and members of the public on need for political union in the Windward Islands.

3040 **Sealy, Theodore.** Sealy's Caribbean leaders. Kingston: Eagle Mechant Bank of Jamaica; Kingston Publishers, 1991. 207 p.: ill., map.

Popularly written and undocumented biographical sketches of 11 West Indian political leaders are mostly of anecdotal interest.

3041 **El servico civil en el Caribe = The civil service in the Caribbean.** Edited by Jorge Morales Yordán. Santo Domingo: PUCMM, 1990. 593 p.: bibl. (Col. Documentos; 149)

Contains papers (in Spanish and English) presented to a seminar on the civil service in the Caribbean held in the Dominican

Republic in 1989, attended by public administration scholars and senior civil servants from various Caribbean countries. Functioning of civil service system in each of the countries represented is explained.

3042 **Sir Arthur Lewis: an economic and political portrait.** Edited by Ralph R. Premdas and Eric St. Cyr. Mona, Jamaica: Regional Programme of Monetary Studies, Institute of Social and Economic Research, Univ. of the West Indies, 1991. 125 p.: bibl., port.

Contains papers presented at conference of the Caribbean Studies Association *14th, Barbados, 1989.* Arthur Lewis, a native of St. Lucia and recipient of the Nobel Prize in Economics, was in many ways the architect of "industrialization by invitation," the economic development model adopted by Caribbean countries (including Puerto Rico) from 1950s on.

3043 **Society and politics in the Caribbean.** Edited by Colin G. Clarke. New York: St. Martin's Press, 1991. 295 p.: bibl., ill., index, maps.

A group of the more recognized Caribbeanists divides topic into three "types" of politics and society: "independent parliamentary democracies" (Jamaica, Trinidad and Tobago, Belize, Eastern Caribbean states, Grenada, Dominican Republic, Venezuela); "independent authoritarian regimes" (Haiti, Cuba); and "territories under metropolitan control" (French Antilles, Puerto Rico).

DOMINICAN REPUBLIC

3044 **Bosch, Juan.** El PLD, un partido nuevo en América. Santo Domingo: Editora Alfa & Omega, 1989. 178 p.: ill.

Despite title, the most interesting essays in this compilation of articles deal with founding of the Partido Revolucionario Dominicano, precursor to the Partido de la Liberación Dominicana (PLD), and its anti-Trujillo activities in exile. Articles were previously published in *Vanguardia del Pueblo,* organ of the PLD.

3045 **Bosch, Juan.** 33 artículos de temas políticos. Santo Domingo: Editora Alfa & Omega, 1988. 265 p.

Essays were written in journal *Política: Teoría y Acción,* established in 1980 by the Partido de la Liberación Dominicana. Many throw new light on Bosch's actions during his long exile from Dominican Republic.

3046 **Cassá, Roberto.** Los doce años: contrarrevolución y desarrollismo. v. 1. Santo Domingo: Editora Alfa & Omega, 1986. 1 v.: bibl.

Offers straightforward Marxist interpretation of political events after assassination of Trujillo in 1961. Section dealing with political and ideological foundations of Joaquín Balaguer's thought provides important analysis. Considers Balaguer the ideological architect of the Trujillato and a classical "Bonapartist" leader.

3047 **Collado, Faustino.** Balaguer y el futuro del PRSC. Santo Domingo?: Ediciones del Centro para la Democracia, 1990. 202 p.: bibl. (El Partidismo político dominicano; 1)

Analyzes role of Joaquín Balaguer in ideology and activities of the Partido Reformista Social Cristiano. Given the fact that in 1994 Balaguer was 88 years old, it is important to ask whether "balaguerismo" will survive passing of its caudillo.

3048 **Collado, Faustino.** La democracia electoral: propuestas para las reformas políticas. Santo Domingo?: Ediciones del Centro para la Democracia, 1992. 240 p.

Very complete and informative analysis of Dominican electoral system contains useful statistical descriptions and computations of votes cast.

3049 **La corrupción gubernamental.** Selección de textos y supervisión general de Víctor M. Abreu Páez. Santo Domingo?: ICPARD, 1987. 196 p.: ill.

Major political leaders provide a somber assessment of pervasive role of corruption in both private and public spheres of the Dominican Republic.

3050 **Despradel, Fidelio.** Operación verdad: de héroes y traidores. Santo Domingo: Editora Alfa y Omega, 1990. 220 p., 10 p. of plates: ill.

Part of ongoing debate over origins, motivations, and interests involved in 1973 landing, from Cuba, of Col. Francisco Caamaño Deñó. Written by an insider from the left with intimate knowledge of both betrayals and heroics of intrigue and Cuban involvements, this work is not easily deciphered by non-specialists. See also item **3055.**

3051 **Dilla Alfonso, Haroldo** and **Félix Calvo.** Crisis y evolución del sistema político dominicano, 1982–85. La Habana: Centro de Estudios sobre América, 1986. 113 p.: bibl. (Avances de investigación; 23)

Two prominent Cuban social scientists use classical Marxist categories to analyze conflict-ridden years of 1982–85. Despite supercharged ideological language, this is an insightful study of divisions in the bourgeoisie and in leftist political parties.

3052 **Dominika imin wa kimin datta: sengo nikkei imin no kiseki [Abandoned people: Japanese immigrants in the Dominican Republic].** Edited by Toshihiko Imano and Takahashi Yukihiro. Tokyo: Akashi Shoten, 1993. 278 p.

In-depth analysis, based on immigrants' testimony, of post-WWII emigration from Japan to the Dominican Republic. According to the editors, immigration was a government strategy to relieve high unemployment resulting from overpopulation. Without regard for difficult conditions, the Japanese government sent 249 families to the Dominican Republic between 1956–58. By 1962, 133 families had returned to Japan, 70 other families had been forced to move to South America, and the remaining immigrants had been denied reentry into Japan for various reasons. In 1987, those remaining in the Dominican Republic took their "human rights violation" case to court in Tokyo. The editors attempt to explain the human rights violations suffered by both the immigrants who remained and those forced to leave the land they had settled. [K. Horisaka]

3053 **Espinal, Rosario.** Procesos electorales en la República Dominicana, 1978–1990. (*Caribb. Stud.*, 24:3/4, 1991, p. 43–58, tables)

Reviews what is known about stability of political identification of voters and relative institutionalization of political parties in the Dominican Republic. Also reviews results of 1978, 1982, 1986, and 1990 presidential elections and findings from political polls. [B. Aguirre-López]

3054 **González Canahuate, L. Almanzor.** El juicio del siglo: proceso Jorge Blanco-Cuervo Gómez. Santo Domingo: L.A. Gónzalez Canahuate, 1992. 978 p.: ill., index.

Contains transcripts of criminal trial of ex-president Salvador Jorge Blanco and ex-secretary of the armed forces, Major Gen. Manuel Antonio Cuervo Gómez. Trial lasted 27 months and was broadcast live on TV.

3055 **Hermann, Hamlet** and **Ramiro Matos González.** El guerrillero y el general. Santo Domingo: Editora Alfa y Omega, 1989. 341 p.: ill., index, maps.

Relates dialogue between a general (Ramiro Matos González) and a guerrilla (Hamlet Hermann) about campaign against 1973 invasion from Cuba by Col. Francisco Caamaño Deñó, an event which now goes by the name "Playa Caracoles." See also item **3050.**

3056 **Lozano, Wilfredo.** Las elecciones dominicanas de 1990: del reacomodo político a la crisis de legitimidad de los populismos reales. (*Rev. Mex. Sociol.*, 52:4, oct./dic. 1990, p. 391–420, tables)

1990 elections gave Joaquín Balaguer his sixth presidential term. This careful and balanced analysis throws light on many irregularities surrounding this election, thereby laying a framework for study of other elections in the Dominican Republic.

3057 **Peña Sánchez, Julián.** Debilidad del movimiento sindical en la República Dominicana. Santo Domingo: Ediciones de Taller, 1986. 104 p.: bibl. (Biblioteca Taller; 196)

Self-described "orthodox Leninist" believes that trade unionism is "authentic" only when it advances the class struggle.

3058 **Tavares, Froilán J.R. et al.** Elecciones '90: discursos pronunciados ante la Cámara Americana de Comercio. Santo Domingo: American Chamber of Commerce of the Dominican Republic; Báez Guerrero & Asociados, 1990. 137 p.: ports.

Contains speeches by major candidates competing in 1990 elections. Reprinting of question-and-answer sessions following each speech makes for valuable reading.

3059 **Vega, Bernardo.** En la década perdida. Santo Domingo: Fundación Cultural Dominicana, 1991. 403 p.: bibl., ill., index.

Set of 61 articles on Dominican Republic written between 1984–90 covers wide range of topics (political and economic conditions, relations with Haiti, migration, and historical themes).

3060 **Velázquez-Mainardi, Miguel Angel.** El narcotráfico y el lavado de dólares en República Dominicana. Prólogo de Marino Vinicio Castillo. Santo Domingo: Editora Corripio, 1992. 236 p.

Includes series of articles on corruption, narcotrafficking, and money laundering in the Dominican Republic. Claims that the CIA introduced drugs into Dominican Republic to weaken the rebels in 1965 uprising.

HAITI

3061 **Aristide, Jean-Bertrand.** La vérité! En vérité!: dossier de défense présenté à la Sacrée Congrégation pour les religieux et les instituts séculiers. Port-au-Prince: Impr. Le Natal, 1989. 125 p. .

Aristide uses theology of liberation arguments to defend himself against charges by Catholic Church hierarchy that his radical politics were unbecoming of a priest. Letters are reprinted in their entirety and there is a very valuable chronology of confrontation with Salesian order.

3062 **Barthélémy, Gérard.** Les Duvaliéristes après Duvalier. Paris: L'Harmattan, 1992. 143 p.: bibl.

Catalogs, rather than studies, Duvalierist thought from 1989–90. Analyzes philosophical bases of Duvalierist thought on culture, nationalism, color, and political institutions.

3063 **Ferguson, James.** The Duvalier dictatorship and its legacy of crisis in Haiti. (*in* Modern Caribbean politics. Baltimore, Md.: The Johns Hopkins Univ. Press, 1993, p. 73–97)

Duvalierism, in its original sense, had changed significantly before 1986 collapse. Jean-Claude Duvalier was "a corrupt and inefficient" version of the original dictatorship.

3064 **Marrero, Rosita.** Testigo de cargo: entrevista a fondo con el presidente de Haití Jean-Bertrand Aristide. Río Piedras, P.R.: NCM Editores, 1992. 107 p.: ill. (Col. Blanco y negro)

Although flawed by many misspelled names and incorrect dates, this long interview with Aristide conducted just after his 1991 overthrow is of great value. Aristide believed that a tightened embargo would return him to power. Marrero describes Aristide's method of analysis as "socialist."

3065 **Moïse, Claude et al.** Les espaces du politique: politique intérieure; la transition démocratique difficile. (*in* La République Haïtienne: état des lieux et perspectives. Sous la direction de Gérard Barthélemy et Christian Girault. Paris: Karthala, 1993, p. 153–232)

Careful analysis examines 1987 Constitution and its possible effect on political processses in Haiti which heretofore have been essentially despotic.

3066 **Packer, George.** Choke hold on Haiti. (*Dissent/New York*, Summer 1993, p. 297–308, photo)

Offers interesting interpretation of role of corruption in Haitian society and politics.

3067 **Richman, Karen.** "A *Lavalas* at Home / a *Lavalas* for Home": inflections of transnationalism in the discourse of Haitian President Aristide. (*Ann. N.Y. Acad. Sci.*, 645, 1992, p. 189–200, bibl.)

Good analysis examines role of *Dizyèm Depatman an* ("Tenth Department" or Haitian diaspora), in Aristide's rhetoric and political actions.

3068 **Weinstein, Brian** and **Aaron Segal.** Haiti: the failure of politics. New York: Praeger, 1992. 203 p.: bibl., index, map.

Authors outline various possible scenarios for Haiti's political future. They conclude that since Haiti has "a pervasive tendency towards authoritarianism," a "limited dictatorship" is likely. They believed (in 1992) that an external military intervention was "highly unlikely."

JAMAICA

3069 **Bakan, Abigail Bess.** Ideology and class conflict in Jamaica: the politics of rebellion. Montreal; Buffalo, N.Y.: McGill-Queen's Univ. Press, 1990. 183 p.: bibl., index.

Attempts to make parallels between slave resistance, 19th-century peasant rebellions, and 20th-century trade union action. This account of the legacy of small-scale, private agricultural cultivation and landholding from slavery days forward is very compelling. The peasant experience described had a significant impact on Jamaican working class ideology.

3070 **Jamaica. Committee Appointed to Advise the Jamaican Government on the Performance, Accountability, and Responsibilities of Elected Parliamentarians.** Report. Kingston: Bustamante Institute of Public and International Affairs, 1991. 60 p.

The late professor Carl Stone chaired a committee on the performance of members of parliament and ministers of government. Committee branched out into a comparison of parliamentary and executive systems.

Keith, Nelson W. and **Novella Zett Keith.** The social origins of democratic socialism in Jamaica. See item **4750**.

LESSER ANTILLES
British Commonwealth

3071 **Fergus, Howard A.** Rule Britannia: politics in British Montserrat. Montserrat: Univ. Centre, Univ. of the West Indies, 1985. 115 p.: bibl.

Author describes Montserrat as an "essentially conservative and anglophile colony." He nevertheless expects independence to arrive some day and offers this analysis of "the characteristics of the Montserrat political market place . . ."

3072 **Lewis, Gordon.** The challenge of independence in the British Caribbean. (*in* Caribbean freedom: society and economy from emancipation to the present. Kingston: Randle, 1993, p. 511–518)

Perhaps the last published piece by the acknowledged *doyen* of Caribbean scholarship, work is written in an asperous style and projects a dark future for the region.

3073 **Sutton, Paul K.** Politics in the Commonwealth Caribbean: the post-colonial experience. (*Rev. Eur.*, 51, Dec. 1991, p. 51–66)

Analyzes unique combination of historical and socioeconomic factors which explain both permanence of democracy and fragmentation of the region. Emphasis is on metropolitan tradition: local acceptance of and resistance to it.

3074 **Yearwood, Gladstone Lloyd** and **Mike Richards.** Broadcasting in Barbados: the cultural impact of the Caribbean Broadcasting Corporation. Bridgetown, Barbados: Lighthouse Communications, 1989. 55 p.: ill.

Brief work examines history of broadcasting in Barbados, with focus on cultural impact of Caribbean Broadcasting Corporation (established in 1964) on the island. Role of the media is measured in terms of its ability to foster "cultural autonomy," and through that, to contribute to larger process of national development.

Dutch

3075 **Castilo, O.A.** Nos gobernashon: Antilliaanse staatsinrichting in een politicologisch perspecktief [Our government: the Antillian Constitution from a political science perspective]. Curaçao: s.n., 1992. 99 p.: bibl., graphs.

Student guide to constitution and administrative structure of Netherlands Antilles emphasizes existing political reality rather than legal aspects. First chapters discuss general topics including democracy and basic rights, while second part focuses on government and administration of Netherlands Antilles. [R. Hoefte]

3076 **Giacalone, Rita.** Aruba y el status aparte, 1969–1989. (*in* Curazao y Aruba: entre la autonomía y la independencia. Mérida, Venezuela: Univ. de Los Andes, Consejo de Desarrollo Científico, Humanístico y Tecnológico, 1990, p. 125–167)

A bit rambling but still interesting work describes evolution of modern Aruban politics. Explains why Aruba does not favor independence, and, even less, integration into a Netherlands Antilles dominated by its island neighbor, Curaçao.

Giacalone, Rita. Cambios políticos y sociales en Curazao, 1969–1989. See item **4730**.

3077 **Munneke, Harold F.** Ambtsuitoefening en onafhankelijke controle in de Nederlandse Antillen en Aruba: juridische en beheersmatige controle als waarborg voor deugdelijk bestuur [The exercise of official duties and independent control in the Netherlands Antilles and Aruba: juridical and administrative control as a guarantee for good governance]. Nijmegen, The Netherlands: Ars Aequi Libri, 1994. 247 p.: bibl., ill.

Looks at social and independent control of politicians and civil servants in autonomous Netherlands Antilles and Aruba. Premise is that in small societies dependency may diminish objectivity of the administration. Includes number of recommendations for improving control and administration of the islands. Also contains short summaries in English and Papiamentu. [R. Hoefte]

3078 **Oostindie, Gert.** The Dutch Caribbean in the 1990s: decolonization or recolonization? (*Caribb. Aff.*, 5:1, Jan./March 1992, p. 103–119)

Given the fact that 14 percent of all Caribbean people still live—by their own choice—in territories which are not fully independent, it is important to know what motivates these people to remain semi-colonial. Both local and metropolitan Dutch views are clearly described in this analysis. See also item **3079**.

3079 **Oostindie, Gert.** Knellende Koninkrijksbanden: Nederland en zijn Caraïbische partners [The galling bonds of the Kingdom: the Netherlands and its Caribbean partners]. (*Int. Spect./Hague*, 47:2, feb. 1993, p. 102–106)

Author argues that is a definite trend towards terminating previous policy of decolonization and thinks it unlikely that transatlantic Kingdom of the Netherlands will be dismantled. Expects that the *metropole* will hold an increasingly firmer grip on the overseas administration. See also item **3078**. [R. Hoefte]

3080 **Reinders, Alex.** Politieke geschiedenis van de Nederlandse Antillen en Aruba, 1950–1993 [Political history of the Netherlands Antilles and Aruba, 1950–1993]. Zutphen, The Netherlands: Walburg Pers, 1993. 429 p.: bibl., ill.

Useful chronicle is divided into 21 chapters. Each discusses one administration, providing overview of previous administration, campaign and election results, formation of cabinet, legislation, and political and constitutional developments. Provides brief analysis at end of each chapter, but overall analysis is sorely missing. Includes short biographies of some 100 politicians. [R. Hoefte]

French

3081 **Guadeloupe. Conseil régional.** Débat sur le devenir institutionnel de la Guadeloupe face aux échéances européennes. Guadeloupe: Conseil Régional, 1990? 200 p.: ill. (some col.).

Debate over details of political-administrative impact on Guadeloupe (a *département d'Outre-Mer*) of new processes unfolding in Europe. Of interest to the specialist in French administrative law.

3082 **Marie Jean Robert, Joseph.** Ma vie à travers la Révolution Nationale. Fort-de-France?: Organisation martiniquaise pour le développement des arts et de la culture, 1988. 95 p.: ill., ports.

Interesting account examines rarely studied phenomenon and period: the "époque de l'Amiral Robert," the pro-Vichy supreme commander of the French Antilles, based in Martinique.

PUERTO RICO

3083 **Bothwell, Reece B.** Orígenes y desarrollo de los partidos políticos de Puerto Rico, 1869–1980. San Juan?: Editorial EDIL, 1987. 255 p.: bibl., indexes.

Given the stretch of time covered, most subjects are analyzed only superficially. Valuable as a chronology of political developments in Puerto Rico, however.

Frambes-Buxeda, Aline. Sociología política puertorriqueña. v. 1. See item **4724**.

3084 **Mattos Cintrón, Wilfredo.** The struggle for independence: the long march to the twenty-first century. (*in* Colonial dilemma: critical perspectives on contemporary Puerto Rico. Edited by Edwin Meléndez and Edgardo Meléndez. Boston: South End Press, 1993, p. 201–214)

Author asks: "Why is independence not a stronger movement today?" Part of the reason is that there never was a strong independence movement in Puerto Rico, even under Spanish rule. Another is that US's "illustrated colonialism" made statehood attractive. Interestingly enough, this pro-statehood movement was led by the working class, what Mattos Cintrón calls "a partial bourgeois-democratic revolution." Once this was merged with US hegemony, the fate of the pro-independence forces was sealed for years to come.

3085 **Quintero Rivera, Angel G.** Soberanía y cotidianidad: las contradictorias perspectivas de las prácticas democráticas en Puerto Rico. (*in* La democracia en América

Latina: actualidad y perspectivas. Coordinación de Pablo González Casanova y Marcos Roitman Rosenmann. Madrid: Univ. Complutense de Madrid; México: Centro de Investigaciones Interdisciplinarias en Humanidades, UNAM, 1992, p. 603–625, bibl.)

Reflects maturing of the leftist critique of US presence in Puerto Rico. Quintero analyzes "social movements" which reflect new collective cultural identities and social action, from informal economic activities to music. Author views this trend as a "cultural democratization."

SURINAME

3086 Boerboom, Harmen and **Joost Oranje.** De 8-decembermoorden: slagschaduw over Suriname [The December 8th murders: casting a shadow over Suriname]. The Hague: BZZToh, 1992. 160 p.: index.

Provides good journalistic account of planning and execution of murder of 15 prominent opponents of Bouterse military regime. Authors argue that the military, under the influence of alcohol and drugs, eliminated the opposition without the knowledge of (leftist) civil politicians. [R. Hoefte]

Brana-Shute, Gary An inside-out insurgency: the Tukuyana Amazones of Suriname. See item **782.**

3087 Brana-Shute, Gary. Suriname: a military and its auxiliaries. (*Caribb. Aff.*, 7:2, 1994, p. 79–93)

Military and political history of Suriname emphasizes events of last two decades. Argues that there are no military coups in sight in Suriname, but that Bouterse and his men will not go away. Expects that violence by small ethnic groups, some military officers, and some politicians will be part and parcel of Suriname society for a long time to come. [R. Hoefte]

3088 Dew, Edward M. Suriname: transcending ethnic politics the hard way. (*in* Resistance and rebellion in Suriname: old and new. Edited by Gary Brana-Shute. Williamsburg, Va.: Dept. of Anthropology, College of William and Mary, 1990, p. 189–212, tables)

Dew notes that despite tenets of "consociational" theory, plural societies find it easier to affirm political power-sharing than to apply it. Ethnic outbidding and ideological extremism are two clear threats to democracy in culturally plural societies.

3089 Dew, Edward M. The trouble in Suriname, 1975–1993. Westport, Conn.: Praeger, 1994. 243 p.: bibl., index, map.

Study based largely on Dutch newspaper accounts presents pessimistic, not wholly satisfying account of political developments in Suriname. Work is too detailed and lacks analysis. Major issues such as growing presence of a group of businessmen with ties to the military are barely touched upon. [R. Hoefte]

3090 Jansen van Galen, John. Kapotte plantage: Suriname, een Hollandse erfenis [Broken plantation: Suriname, a Dutch legacy]. Amsterdam: Balans, 1995. 85 p.

Fascinating collection of articles previously published in Dutch weeklies is supplemented with new research on Dutch policy regarding Suriname nationalism and independence. Most articles are based on visits to Suriname spanning the 1970–95 period. Failure to provide information on dates of these materials and to identify the previously published articles is problematic. Also lacks footnotes. [R. Hoefte]

3091 Meel, Peter. The march of militarization in Suriname. (*in* Modern Caribbean politics. Baltimore, Md.: Johns Hopkins Univ. Press, 1993, p. 125–146)

Fine analysis of Suriname politics in 1980s distinguishes four phases: social democracy, a socialist people's republic, a military-led democracy, and a parliamentary democracy. None of these models functioned adequately since they lacked popular support and merely served to hide rampant militarization. [R. Hoefte]

3092 Meel, Peter. Verbroederingspolitiek en nationalisme: het dekolonisatievraagstuk in de Surinaamse politiek [Fraternization policy and nationalism: the question of decolonization in Suriname politics]. (*Bijdr. Meded. Betreffende Geschied. Nederlanden*, 109:4, 1994, p. 638–659)

Important article analyzes how respective governments have tackled decolonization and independence. Focuses on Emanuels Administration (1958–61), which tried to expand Suriname's autonomy. Meel argues that economic, social, and political stability in 1960 should have led to constitutional changes; however, ethnic rivalry caused increasing polarization. When independence came in 1975, economic and political conditions for such a step were no longer favorable. [R. Hoefte]

3093 Sedoc-Dahlberg, Betty. Suriname: the politics of transition from authoritarianism to democracy, 1988–1992. (*in* Democracy in the Caribbean: myths and realities. Westport, Conn.: Praeger, 1994, p. 131–145)

Analyzes process of transformation from authoritarianism to democracy. Concludes that civilian government failed because of its inability to explain to the electorate the processes and mechanisms that led to democracy, and because of continued military control of the State. Civilian regime's survival strategies did not work because of government's lack of access to State power and unsuccessful negotiations with key local and Dutch actors. [R. Hoefte]

3094 Suriname in het jaar 2000 [Suriname in the year 2000]. Edited by A.J. Brahim *et al.* Baarn, The Netherlands: Bosch en Keuning, 1994. 404 p.: graphs, tables.

Collection of 34 rather uneven essays on future of Suriname is divided into four parts: 1) transition to a market economy; 2) restructuring of the administrative apparatus; 3) international relations; and 4) Suriname's position in the international market. Includes brief summaries of each paper. [R. Hoefte]

TRINIDAD AND TOBAGO

3095 Anthony, Michael. A better and brighter day. Port-of-Spain: Circle Press, 1987. 180 p.: bibl., index.

Well-known Trinidadian novelist relates post-WWII political history of Trinidad up to independence (1962). Although based nearly exclusively on newspaper accounts and hardly a critical history in the traditional sense, the easy writing style and imaginative selection for a chronology of events make this a useful study.

3096 John, George. 50 years of the ballot: a political history of Trinidad and Tobago. Port of Spain: Trinidad Express Newspapers Ltd., 1991 80 p.: photos. (Express books; 6)

Contains short biographies of some of the major political leaders. Of interest for the many photographs not readily available elsewhere.

3097 Panday, Basdeo. An enigma answered: a first volume of speeches. Foreword by John Gaffar La Guerre. Trinidad and Tobago: Chakra Publishing House, 1991. 487 p., 64 p. of plates: ill.

Collected speeches of then-official leader of opposition in Trinidad and Tobago covers period 1966–91, i.e., from radical trade union days to various political alliances of 1980s.

3098 A party politics for Trinidad and Tobago: the flowering of an idea. Edited by Lloyd Best and Allan Harris. Port of Spain: Trinidad and Tobago Institute of the West Indies, 1991. 91 p.

Contains selection of writings by Trinidad intellectuals who founded the New World Group in 1968 and, later that year, established the Tapia House Group and its political arm, the Tapia House Movement. Mostly of historical value since most recent piece was written in 1988.

3099 Ryan, Selwyn D. Good innings, 1901–1990: the life & times of Ray Edwin Dieffenthaller. Port-of-Spain: McEnearney Alstons; Paria Pub. Co., 1990. 82 p., 72 p. of plates: ill.

Dieffenthaller was a pioneer in development of the hardware and oilfield equipment business in Trinidad. More than a biography, work traces history of some of the more important business firms on the island. An all too rare contribution to the study of the private sector, especially the black and colored members of that sector.

3100 Ryan, Selwyn D. The Muslimeen grab for power: race, religion, and revolution in Trinidad and Tobago. Port of Spain: Inprint Caribbean, 1991. 345 p.: bibl., ill.

Provides most authoritative account thus far of coup attempt of Abu Bakr and his Jamaat al Muslimeen followers on July 27, 1990. Presents debatable thesis that uprising was "in the tradition of rebellion and revolution that was born when the first slaves were brought to Trinidad and Tobago." Ryan himself concludes that Bakr "soiled Trinidad and Tobago's and the region's record as a place where constitutional democracy prevailed..."

CUBA

JUAN M. DEL AGUILA, *Associate Professor of Political Science, Emory University*

THE LITERATURE ON CUBA in the mid 1990s focuses on the maintenance of political control in the midst of stressful socioeconomic circumstances. Several authors point to the Cuban regime's resilience and its will to overcome mounting internal difficulties and external pressures. Other works like Andrew Zimbalist's "Teetering on the Brink: Cuba's Current Economic and Political Crisis" (item **3151**) assess an incipient process of macroeconomic reforms characterized by economic decentralization, a partial opening to foreign capital in some sectors like tourism, and the legalization of some forms of self-employment. Sources of internal crisis with a destabilizing potential effect are identified in *Cuba at a crossroads: politics and economics after the Fourth Party Congress* (item **1685**), edited by Jorge Pérez-López, and in *Conflict and change in Cuba* (item **3116**), edited by Enrique Baloyra and James Morris.

Cuba's record on human rights comes in for much academic and political criticism. For instance, the Inter-American Commission on Human Rights' report on the situation in Cuba (item **3133**) documents past and recurring abuses while criticizing the Cuban Communist Party's domination of State and society. Individual works probe this subject from various perspectives. In *Fidel Castro, el fin de un mito* (item **3135**), César Leante examines the revolution's crumbling cultural myths and the disaffection of the domestic intelligentsia and foreign "literatti."

Speeches by President Castro (items **3111** and **3113**) and Minister of Defense Raúl Castro (item **3114**) serve as primary sources for anyone interested in how Cuba's top officials explain or rationalize policies, failures, or foreign crises. Lengthy interviews with President Castro by sympathetic journalists like Gianni Mina (item **3110**) and others offer unedited views of Castro on democracy, socialism, human rights, revolution and the US. In sum, though hagiographic literature is still being produced, critical thinking is evident as a new and dominant trend in the 1990s, especially among scholarly works, non-academic contributions and testimonials.

3101 Aguila, Juan M. del. Cuba, dilemmas of a revolution. 3rd ed. Boulder, Colo.: Westview Press, 1994. 222 p.: bibl., ill., index, map. (Nations of contemporary Latin America)

Third edition of perceptive, useful, and balanced overview of and introduction to socialist Cuba (for first edition, 1984, see *HLAS 47:6217*; for second edition, 1988, see *HLAS 50:1729*, and *HLAS 53:3566*). [Ed.]

3102 Aguila, Juan M. del. The Party, the Fourth Congress, and the process of counter-reform. (*Cuba. Stud.*, 23. 1993, p. 71–93)

Critically examines Fourth Congress of the Partido Comunista de Cuba (1991), the first party congress held following the collapse of the Soviet Union. Argues that this congress focused on "counter-reforms" designed to perfect the notion of democratic centralism, to refuel the process of ideological invigoration, and to make intraparty affairs more "democratic." Despite some changes in the ruling Political Bureau, the Fourth Congress merely reaffirmed primacy of existing leadership, suggesting that "its stranglehold on the [political] system is fundamentally unaffected by internal difficulties and external crises." Most of author's conclusions do not augur well for either "liberal reforms" or "radical departures." Same article also available in *Cuba at a crossroads* (see item **1685**). [D.W. Dent]

3103 Andrade Terán, Ramiro. Fidel, acorralado y sin salida. Bogotá: Intermedio Editores, 1991. 199 p.: bibl.

Short pieces and newspaper columns written by former Colombian ambassador to Cuba. Filled with personal anecdotes and diplomatic experiences. Neither academic nor impressive in its scope and depth. Useful in part because it challenges some dogmas of the Latin American left.

3104 **Arguedes, Sol.** Will changes in Eastern Europe reach Cuba? (*Voices Mex.*, 15, Oct./Nov./Dec. 1990, p. 12–18, ill., photos)

Shallow and emotional analysis of conditions in Cuba, argues that since "socialism and independence" are indisssolubly linked in most Cubans' hearts, the society will find its way out of the current dead-end. Wonders whether Fidel Castro "will see his dream of offering the continent a socialist world crumble" (p. 12) but surprisingly concludes that this is not so. Good example of academic propaganda.

3105 **Aroca, Santiago.** Fidel Castro: el final del camino. Barcelona, Spain: Planeta, 1992. 324 p.: bibl., ill., index. (Col. Documento; 314)

Very revealing account of the sordid and hypocritical practices that surround Cuba's maximum leader. Stunning interviews with the widow of Gen. Ochoa, with the parents of the de la Guardia brothers, and with Alina Fernández. All attest to the degeneration of the regime into little more than a dingy and dirty police state. Aroca, a Spanish journalist, was subsequently expelled from Cuba.

3106 **Asiaín, Aurelio.** Nostalgia habanera. (*Vuelta/México*, 16:184, marzo 1992, p. 58–62)

Position paper by Mexican tourist in La Habana. Describes deterioration of sociopolitical and economic conditions in Cuba during the *período especial*, and explores the contradictions inherent in Castro's policies. [E. Sacerio-Garí]

3107 **Benemelis, Juan F.** El último comunista. San Juan: Publicaciones Puertorriqueñas, 1992. 395 p.: bibl., ill.

Very solid analysis of the structure of power in Cuba, emphasizing the interlocking relationships among the leader, the party, the State, and ancillary organizations and institutions. Also examines Castro's personality, his megalomania, and his deep-seated revolutionary convictions.

3108 **Blanco, Juan Antonio.** Cuba: utopía y realidad, treinta años después. (*Cuad. Nuestra Am.*, 7:15, julio/dic. 1990, p. 10–26)

Cuban scholar argues that praxis over 30 years "legitimates the historical option assumed by the people" to build socialism (p. 11). Sees revolutionary development as a historical need, especially as a bulwark against US influences. Raises questions regarding what model of socialism is appropriate, but concludes categorically that "capitalism offers no solutions for the problems confronting Cuba" (p. 25). In a flight of fancy, Blanco maintains that "the Cuban case constitutes a historical paradigm" due to its effort to achieve national independence and stable development (p. 26), a judgement that is both as foolish as unwarranted given the mounting evidence of a social catastrophe on the island.

3109 **Cardoso, Eliana A.** and **Ann Helwege.** Cuba after communism. Cambridge, Mass.: MIT Press, 1992. 148 p.: bibl., ill., index, map.

Somewhat disjointed analysis by two economists of post-communist Cuba. Suggests measures that Cuba can adopt in order to stave off economic collapse, but argues that the regime's fall is not imminent. Includes good data, economic indicators, and useful analysis of how communism's collapse affected Cuba.

3110 **Castro, Fidel.** Fidel. Intervista di Gianni Minà. Prefazione di Jorge Amado. Traduzione di Gianni Minà, Paolo Tufano, e Elena Grechi. Milano: Sperling & Kupfer, 1991. xxxii, 229 p. (Saggi; 89)

Lengthy interview with Castro by adoring Italian journalist who covers events, crisis, and issues of the late 1980s. Useful reference to trace contradictions and inconsistencies in Castro's discourse over the years.

3111 **Castro, Fidel.** Fidel Castro: ideología, conciencia y trabajo político, 1959–1986. La Habana: Editora Política, 1986. 427 p.: bibl.

Pronouncements by Castro on politics, ideology, social consciousness, and socialism and "the new society." Somewhat incoherent, because it moves from topic to topic without much context or perspective.

3112 **Castro, Fidel.** Un grano de maíz: conversación con Fidel Castro. Entrevista de Tomás Borge. México: Fondo de Cultura Económica, 1992. 273 p. (Tierra firme)

Interview with Fidel Castro by former Sandinista comandante and Minister of Interior, Tomás Borge. His sentimental introduction asserts that "utopia is achievable," and maintains that he was very impressed with Castro's "persuasive abilities," and his "explanations of how human rights are protected in Cuba." Text covers trends and events following the demise of the Soviet Union, and includes lengthy philosophical discussions by Castro.

3113 **Castro, Fidel.** Por el camino correcto: compilación de textos, 1986–1989. 3. ed. La Habana: Editora Política, 1989. 333 p.: bibl. (Olivo colección)

Third in a series of speeches and reports by President Castro in various fora. Speeches cover everything under the sun, from urgings to "be more intransigent than ever," to evaluations of the rectification campaign. Very useful as a primary source.

3114 **Castro Ruz, Raúl.** Selección de discursos y artículos. La Habana: Editora Política, 1988. v. 2: ill.

Selection of articles and speeches by Raúl Castro (1976–86). Includes the younger Castro's views on Soviet-Cuban military collaboration, the role of the Communist Party in the military, civic-military education, and relations among "fraternal" countries and armies. Events have made this collection obsolete.

3115 **Causa 1/89: fin de la conexión cubana.** La Habana: Editorial José Martí, 1989. 481 p.: ill.

Official record and interpretation of the Ochoa scandal. Includes valuable documents, text covering the interrogation of Ochoa and his collaborators, and Castro's demand that the Council of State sanction the executions.

3116 **Conflict and change in Cuba.** Edited by Enrique A. Baloyra and James A. Morris. Albuquerque: Univ. of New Mexico Press, 1993. 347 p.: bibl., index, map.

Series of essays examine the Cuban crisis from various perspectives. Most of them focus on internal sources of conflict, economic difficulties, reasons for the regime's resilience, and on the prospects for change. Very solid, balanced book.

3117 **Cuba 1990: realidad y futuro.** Edición de Carlos Robles Piquer. Santiago de Compostela, Spain: Fundación Alfredo Brañas, 1991. 161 p.: bibl. (Col. América; 2/1991)

Essays by experts and critics cover Cuba's economy, society, relations with Caribbean and Central American nations, and human rights.

3118 **Cuba, a different America.** Edited by Wilber A. Chaffee, Jr. and Gary Prevost. Totowa, N.J.: Rowman & Littlefield, 1989. 181 p.: bibl., ill., index.

Informative essays on various aspects of Cuban life, society, economy, and polity. Strong piece on political control, with useful biographical data on Central Committee members (ca. 1986). Authors tend to be sympathetic to regime and revolution.

3119 **The Cuban Revolution into the 1990s: Cuban perspectives.** Edited by Centro de Estudios sobre América. Boulder, Colo.: Westview Press, 1992. 197 p.: bibl., ill. (Latin American perspectives series; 10)

Assessment of the revolution by Cuban scholars, edited by one of Havana's think tanks, the Center for the Studies of America (CEA). Partisan and ideological in substance, tone and content, many essays examine various topics from explicitly Marxist perspectives. Some include heavy rhetoric; others are less influenced by political imperatives. Useful, but to be read with skepticism.

3120 **Díaz-Briquets, Sergio** and **Jorge F. Pérez-López.** Cuba's labor adjustment policies during the special period. (*in* Cuba at a crossroads: politics and economics after the Fourth Party Congress. Gainesville: Univ. Press of Florida, 1994, p. 118–146, bibl., tables)

Analysis of how the Cuban government is adjusting to its loss of markets and subsidies from former communist countries. Examines new strategies for allocating resources, maintaining employment and compensation levels, and on how workers (labor) are affected by difficult economic circumstances. Conclusion is that economy has yet to bottom out, so dislocations and "economic pain" continue.

3121 **Domínguez, Jorge I.** Leadership strategies and mass support: Cuban politics before and after the 1991 Communist Party Congress. (*in* Cuba at a crossroads: politics and economics after the Fourth Party Congress. Gainesville: Univ. Press of Florida, 1994, p. 1–18, bibl.)

Analysis of some of the personnel changes and "policy" debates carried out during the Fourth Congress of the Communist Party. Explains why Cuba rejected wholesale reforms, but introduced limited changes in the economy and society. Interestingly, author finds that "civil society" is slowly reemerging, but that liberalization or democratization are not part of the reform package.

3122 **Erisman, H. Michael.** The odyssey of revolution in Cuba. (*in* Modern Caribbean politics. Baltimore, Md.: The Johns Hopkins Univ. Press, 1993, p. 212–237)

Balanced overview of development through the early 1990s, emphasizing internal development, the international context, and future challenges. Argues that adjustments to a new world order will force *fidelistas* to be more prudent, and put their efforts into domestic priorities.

3123 **Espín de Castro, Vilma.** La mujer en Cuba. La Habana: Editora Política, 1990. 179 p. (Col. Olivo)

Speeches and interviews by Vilma Espín, president of the Cuban Women's Federation, on the role the organization plays in the revolution. Topics cover changing roles for women, participation, and social militancy.

3124 **Feinsilver, Julie M.** Will Cuba's wonder drugs lead to political and economic wonders?: capitalizing on biotechnology and medical exports. (*Cuba. Stud.*, 22, 1992, p. 79–111)

Very useful contribution places development of biotechnology in the wider effort to replace markets lost in the former socialist community. Still, long-term success is not predicted due to technological insufficiencies and lack of international market competitiveness.

3125 **Fernández Ríos, Olga.** Formación y desarrollo del Estado socialista en Cuba. La Habana: Editorial de Ciencias Sociales, 1988. 256 p.: bibl. (Filosofía)

Conventional, Marxist interpretation of how and why a "popular" revolution turned into "the dictatorship of the proletariat." Keys to the consolidation of a revolutionary State were the destruction of the old class system, distributional measures, and the convictions of the "revolutionary vanguard." Not much that is new but good as a reference aid.

3126 **Fogel, Jean-François** and **Bertrand Rosenthal.** Fin de siècle à La Havane: les secrets du pouvoir cubain. Paris: Editions du Seuil, 1993. 600 p.: bibl., index, maps. (L'Histoire immédiate)

Extremely useful, yet critical account of "the secrets of power in Cuba" by noted European writers. Well documented and insightful, it strips away the romanticism of a revolution and shows it turning upon itself.

3127 **Frayde, Martha.** El estado de los derechos humanos en Cuba. (*Vuelta/México*, 15:179, sept. 1991, p. 47–54)

Stinging critique of the Cuban government's human rights practices, including brief narrative on the rise of political dissidence. Maintains that the Cuban State denies all rights listed in the UN's Declaration of Human Rights, and that "the vigilant state" (p. 48) intrudes into every aspect of life.

Fung Riverón, Thalia and **Pablo Guadarrama González.** El desarrollo del pensamiento filosófico en Cuba. See *HLAS 54:5407*.

3128 **Geldof, Lynn.** Cubans: voices of change. 1st U.S. ed. New York: St. Martin's Press, 1992. 358 p.: bibl., map.

Interesting, largely personal accounts of life in Cuba in the late 1980s. Those interviewed include ordinary Cubans, exiles, and three academics. Contrast is remarkable, partly because many of the predictions regarding the future of socialism and the Soviet Union turned out to be entirely wrongheaded.

3129 **Griffin, Clifford E.** Cuba, the domino that refuses to fall: can Castro survive the "special period?" (*Caribb. Aff.*, 5:1, Jan./March 1992, p. 24–42, tables)

Good analysis of how and why poor economic conditions affect Cuba's regime. Concludes that its fall is not iminent.

3130 **Gunn, Gillian.** Cuba's search for alternatives. (*Curr. Hist.*, 91:562, Feb. 1992, p. 59–64)

Brief but solid review of options available to the Cuban government in economic affairs. Partly based on Gunn's interview in 1991 with President Castro, article analyzes development strategy now in place. At the time, Gunn noted that "the strategy will probably produce significant improvement only in the mid-1990s," as wrongheaded a prediction as one can think of in 1994 (p. 61). Warns overthrow of *fidelista* regime not a foregone conclusion (p. 64), and repeats usual criticisms of US policy.

3131 **Guy, James J.** "No man is an island:" Fidel Castro and the future of the Cuban revolution. (*Rev. Interam. Bibliogr.*, 41:3, 1991, p. 405–421, tables)

Lengthy analysis of how Castro and Cuba fare in times of crisis. As others have done, author probes Castro's personality and psychology, and how these have shaped the revolution. Isolated and on the defensive, Castro still aims to sustain "his international following to magnify his revolutionary image abroad" (p. 417) in order to maintain legitimacy at home. Author believes socialism in some form will remain even after Castro is gone, due to deep cultural, social and psychological changes wrought by the revolution.

3132 **Harnecker, Marta.** Crisis del socialismo: ¿qué pasa en Cuba? (*Tareas/Panamá*, 76, sept./dic. 1990, p. 41–73)

Thoughtful review of how crisis of Marxist ideology and socialism have affected Cuba. Critical of single-party model, but argues that international (hostile) context must be factored into the equation. Maintains that most Cubans support the revolution, and that youth in particular remains committed. Does not anticipate the regime falling, but exaggerates Cuba's potential for economic recovery. In fact, the situation has definitely worsened since Harnecker's article appeared.

3133 **Inter-American Commission on Human Rights.** The situation of human rights in Cuba: seventh report. Washington: General Secretariat, Organization of American States, 1983. 183 p.: bibl. ([OAS official records]; OEA/Ser.L/V/II.61. doc 29 rev. 1)

Documents past and recurring abuses while criticizing the Cuban Communist Party's domination of State and society.

3134 **Kirk, John M.** (Still) waiting for John Paul II: the Church in Cuba. (*in* Conflict and competition: the Latin American Church in a changing environment. Edited by Edward L. Cleary and Hannah Stewart-Gambino. Boulder, Colo.: Lynne Rienner Publishers, 1992, p. 147–165)

Good balanced overview of where Church and State stand in the early 1990s. Argues that the Church "demands to be taken seriously by the government" (p. 149), but finds resistance from orthodox communists and Marxist ideologues. Anticipates the Church becoming more critical of regime, expanding its international contacts, and increasing the number of believers. Laments that John Paul II has not been to Cuba, but does not really explore the political reasons for the lack of a visit.

3135 **Leante, César.** Fidel Castro: el fin de un mito. Madrid: Editorial Pliegos, 1991. 150 p. (Col. Aquí y ahora)

Series of short pieces on how literary and cultural myths regarding socialism and the revolution are slowly falling apart. Author once worked for Prensa Latina and for Cuba's Ministry of Culture. Worth pondering are his views on how and why Gabriel García Márquez remains as the lone holdout and apologist for Castro and Cuba.

Lie, Nadia. Casa de las Américas y el discurso sobre el intelectual, 1960–1971. See *HLAS 54:3866.*

3136 **Marín, Raúl.** ¿La hora de Cuba? Madrid: Editorial Revolución, 1991. 167 p.

Interesting look at problems Cuba faces from a friendly but not uncritical perspective. Questions past policies, especially internationalism, and emphasizes the problem of how a revolutionary generation can satisfy economic demands and cope with political disloyalty.

3137 **Martínez Heredia, Fernando.** Cuba: problemas de la liberación, el socialismo, la democracia. (*Cuad. Nuestra Am.*, 8:17, julio/dic. 1991, p. 124–148)

Attempt to rationalize official policies by restating the commitment to socialism under difficult economic conditions. Maintains that "democracy will be deepened" in the 1990s even as Cuba struggles through crisis. Castro is said to "have the most revolutionary ideas inside the system" (p. 146), and hold immense moral authority. Concludes that the revolution continues, as does "a socialist project" (p. 148). Major flaw is author's refusal to consider the exhaustion of the very system he struggles to rehabilitate.

3138 **Mesa-Lago, Carmelo** and **Horst Fabian.** Analogies between East European socialist regimes and Cuba: scenarios for the future. (*in* Cuba after the Cold War. Edited by Carmelo Mesa-Lago. Pittsburgh, Penn.: Univ. of Pittsburgh Press, 1993, p. 353–380)

Excellent analysis of how socialism's collapse has devastated the Cuban economy. Also examines Cuba's reactions to that crisis, and draws different scenarios for the 1990s. Concludes that the "current politico-economic model is not viable and will soon be replaced" (p. 377), but not without democratization.

3139 **Mesa-Lago, Carmelo.** Introduction: Cuba, the last communist warrior. (*in* Cuba after the Cold War. Edited by Carmelo Mesa-Lago. Pittsburgh, Penn.: Univ. of Pittsburgh Press, 1993, p. 3–16)

Introduction by prominent Cuban-American economist to a book on Cuba after the Cold War. Mesa-Lago briefly summarizes the impact on Cuba of socialism's collapse, and looks at why Cuba rejects macroeconomic and political reforms.

3140 **Mesa-Lago, Carmelo.** El proceso de rectificación en Cuba: causas políticas y efectos económicos. (*Rev. Estud. Polít.*, 74, oct./dic. 1991, p. 497–530)

Comprehensive assessment of origins, impact, and failure of the Rectification Campaign. Most useful for professional economists, it provides a wealth of data and information on the workings of the Cuban economy. Subjective factors contribute to economic decline, but Castro maintains that no major reforms are needed.

3141 **Orozco, Román.** Cuba roja: cómo viven los cubanos con Fidel Castro. Prólogo de Manuel Leguineche. Madrid: Información y Revistas S.A. Cambio 16, 1993. 912 p.: appendix, bibl.

Journalist draws interesting portraits, at the personal level, of well-known exiled Cubans as well as dissidents and Cuban government officials. No final judgment is rendered and no conclusions drawn. Provides excellent appendix with political, economic and social data.

3142 **Pensar al Ché.** La Habana: Centro de Estudios sobre América; Buenos Aires: Ediciones Dialéctica, 1989. 2 v.: bibl.

Lengthy essays on Che Guevara's thought, but offers little more than an outdated hagiography. Views Guevara as a great humanitarian, an outstanding thinker, and committed revolutionary who sacrificed for humanity. His "communist passion" and anti-imperialism were the essence of his life and being. Makes no attempt to critically assess Guevara's disastrous legacy, nor examines why his internationalist "projects" failed miserably as well.

3143 **Pérez-López, Jorge F.** Cuban politics and economics in the 1990s. (*in* Cuba at a crossroads: politics and economics after the Fourth Party Congress. Gainesville: Univ. Press of Florida, 1994, p. ix-xviii)

Introduction to series of chapters on Cuba at a crossroads (1993). Offers brief summary of each chapter and sets the context for economic difficulties and political challenges in the 1990s. Pérez writes that "these essays provide readers with a foundation for understanding" changes in regime and society that will surely develop (p. xviii).

Pérez-Stable, Marifeli. The Cuban Revolution: origins, course, and legacy. See item **4769.**

3144 **Pino, Rafael del.** Proa a la libertad. México: Planeta, 1991. 436 p.: ill. (Col. Fábula (México)

Gen. Rafael del Pino's account of his life as a revolutionary, and of his defection in 1987. Good information on the relationship between Castro and the Cuban armed forces, on the war in Angola, and on how top-level military officers were forced to agree to the execution of Gen. Ochoa in 1989. Del Pino often justifies his commitment to a cause in which he no longer believes.

3145 **Un plebiscito a Fidel Castro.** Recopilación de Reinaldo Arenas y Jorge Camacho. Madrid: Editorial Betania, 1990. 150 p.: ill. (Col. Documentos)

Open letter to Fidel Castro calling for a plebiscite in Cuba regarding Castro's continued rule. Signed by prominent artists, writers, intellectuals, scientists and political figures from Europe, Latin America, and the Cuban community in exile. Documents include official Havana's crass reaction to the letter, and its view that "Cubans celebrated a plebiscite 31 years ago" when Castro came to power.

3146 Preston, Julia. El juicio de Arnaldo Ochoa. (*Cuad. Marcha,* tercera época, nov. 1991, p. 7–18)

Extremely detailed and objective look at Gen. Ochoa's trial, exploring the reason for his execution. Perceives trial as a political exercise, and raises questions about the Castro brothers' prior knowledge and approval of what Ochoa was accused of doing. Points out how members of the political and military elites reacted to Ochoa's execution, but concludes that through purges and personnel shuffles, the Castros reestablished control.

3147 Rabkin, Rhoda. Cuban socialism: ideological responses to the era of socialist crisis. (*Cuba. Stud.,* 22, 1992, p. 27–50, bibl.)

Critical look at how Cuba's ideological shifts shape a discourse that is increasingly unfocused and incoherent. Rabkin sees Marxism-Leninism as being downplayed while nationalism and anti-imperialism are reasserted. Neither symbolically nor factually is Cuba able to rally the masses any more, and its ideology is seen as little more than "utopian social protest" (p. 47). Very good piece.

3148 Ritter, Archibald R.M. Prospects for economic and political change in Cuba in the 1990s. (*in* Latin America to the year 2000: reactivating growth, improving equity, sustaining democracy. New York: Praeger, 1992, p. 235–252, bibl., table)

Outlines prominent economic scenarios that might develop in Cuba in this decade. These go from US-induced liberalization to "Romanian-style collapse." Data provided is useful and conclusions sound.

Rodríguez-Menier, Juan Antonio. Cuba por dentro: el MININT. See item **4776**.

3149 Rosenberg, Jonathan. Cuba's free-market experiment: los mercados libres campesinos, 1980–1986. (*LARR,* 27:3, 1992, p. 51–89, bibl.)

Very good analysis of origins and abolition of free peasant markets (1980–86). Places policy in its proper context, and explanations emphasize political and subjective criteria over strictly economic factors. Rise and fall of free peasant markets "best understood as the outcomes of political conflict" (p. 81), with Castro the ultimate arbiter. Speculates that if food shortages become critical, the State could be pitted against consumers.

3150 Schulz, Donald E. Can Castro survive? (*J. Interam. Stud. World Aff.,* 35:1, 1993, p. 89–117, bibl.)

Good analysis of why regime survives in the midst of deepening economic difficulties and international isolation. Views politics of exiles as helping the regime raise nationalist threat. Criticizes US policy and concludes that if his health holds up, "Castro can be around for some time" (p. 116). Good piece, but should have paid more attention to indigenous reasons for the regime's endurance.

3151 Zimbalist, Andrew. Teetering on the brink: Cuba's current economic and political crisis. (*J. Lat. Am. Stud.,* 24:2, May 1992, p. 407–418)

Excellent analysis of Cuba's economic and political crisis. Based on premise that "in 1992, Cuba is in the worst crisis of the revolution" (p. 407), and that government's response is inadequate. Reviews macroeconomic changes, looks at what resources can be expanded, and urges more internal privatization and "market mechanisms" (p. 416). Concludes that the "deep demoralization of the Cuban people" (p. 417) demands radical political and economic change.

COLOMBIA AND ECUADOR

MARC CHERNICK, *Director, Andean and Amazonian Studies Project, Center for Latin American Studies, Georgetown University*

POLITICAL STUDIES OF COLOMBIA AND ECUADOR now have a higher profile outside of the Andean region than they have had in the past. One of the reasons is the increase in the quantity and quality of scholarship being produced in the region. Another is the consolidation of important research and teaching centers in both countries. Finally, it seems that by the early 1990s, political processes in each of these Andean countries began to pique the interest of political scientists. In the past, Colombia and Ecuador were often neglected in political anthologies, largely because they did not fit the broader paradigms generated by political events in the Southern Cone, Brazil, or Mexico that have dominated much of the theoretical literature. In the case of Colombia, political scientists were often compelled to explain why certain events, such as military intervention and regime breakdown, did not occur, or why more traditional political patterns, such as entrenched elite-dominated multiclass parties, have endured.

Much has changed. For one, there are few paradigms left. Further, many aspects of political life in Colombia and Ecuador in the 1990s are proving to be in the forefront of contemporary research in the region. In Colombia, the issues of drug-trafficking, organized crime, and social violence are generating much scholarship. At one time, these issues were thought to underscore Colombia's distinctiveness and exceptionalism. Today this work is pioneering the analysis of political and social issues that increasingly define politics in other regions, such as Mexico, Central America, and Brazil.

In Ecuador, although there is still a lag between politics and published scholarship, one of the key themes stimulating scholarly interest and research is the rise of the indigenous movement as a major actor in national politics. Twice in the early 1990s, the indigenous movement organized nationwide protests that led to the reordering of the nation's political agenda to include indigenous concerns related to land, language, and national identity. Selverston provides a good account of this phenomenon (item **3242**); we can also expect several doctoral dissertations on the subject over the next few years.

At the same time, both Colombia and Ecuador have followed the regional patterns of neoliberal economic reforms. In Ecuador, these have been pursued by a conservative government; in Colombia, they have been implemented by administrations associated with the more reformist wing of the Liberal Party. Foreign and national scholars are beginning to address the political consequences of these economic policies from a comparative perspective.

Collectively, all the changes outlined above make the countries more accessible to comparative research. Having recently spent two years in the region as a visiting profesor at the National University of Colombia and a regular lecturer at FLACSO-Ecuador (made possible through an individual research grant from the Harry Frank Guggenheim Foundation), I was able to witness closely the expanded activities of the social science communities in both countries as well as the upsurge in interest by foreign scholars. The increase in entries during this biennium reflects both the greater quantity of good scholarship, and also my extended stay in the region.

COLOMBIA

Violence and drug-trafficking continue to shape the political debate and scholarly production in and on Colombia. However, whereas two years ago scholars, researchers and journalists were mostly absorbed in analyzing the nature of the crisis, today there has emerged an extensive literature on the response of the government, especially during the administration of César Gaviria from 1990–94. On the peace process, see García Durán's *De la Uribe a Tlaxcala* (item **3170**); *La paz: más allá de la* *guerra* (item **3188**), which contains an interesting article by Jesús Antonio Bejarano; the very important study by the Comisión de Superación de la Violencia, *Pacificar la paz* (item **3187**); Vélez de Piedrahita's *El diálogo y la paz* (item **3208**); and *El Tolima: una respuesta pacífica* (item **3201**).

On the drug trade, Francisco Thoumi has several publications that greatly enrich our understanding of the economics and politics of drug-trafficking in Colombia (items **3198, 3199,** and **3200**). Additionally two excellent chapters on the rise of the Colombian mafias and the relationship of drugs and violence can be found in *Violencia en la región andina: el caso colombiano* (item **3172**). Alvaro Camacho also has a very interesting review essay (item **3156.**)

Perhaps the most significant response to political crisis was the Constitutional Assembly that met from Feb. to July 1991. This latter produced a new Constitution which replaced the 1886 one. Manuel José Cepeda has written three important books on the subject (items **3159, 3158,** and **3161**). Dugas (item **3163**) and Lleras Restrepo (item **3181**) also make important contributions to this topic.

At the same time, there is a growing concern about the deterioration of human rights. In the past few years, several Colombian human rights organizations have been created. Their voice is increasingly being heard, and their work is now an essential source for primary data and analysis. Most notable of the non-governmental organizations is the Comisión Andina de Juristas, Seccional Colombiana. Two important works by Colombian scholars and human rights activists are *Guerra y constituyente* (item **3174**), and *Justicia, derechos humanos e impunidad*, published by the Consejería Presidencial para los Derechos Humanos (item **3176**). For other excellent human rights work, turn to Valencia Villa's *La justicia de las armas* (item **3205**), *Desplazamiento, derechos humanos y conflicto armado* (item **3165**), and *La verdad del '93* (item **3210**).

Studies of urban and regional violence have also proliferated. See, for instance, an English-language translation of Alonso Salazar's *No nacimos pa' semilla* (item **3195**); *Putumayo*, produced by the Comisión Andina de Juristas (item **3193**); and Alejo Vargas Velásquez's *Colonización y conflicto armado: Magdalena Medio Santandereano* (item **3207**).

Civil-military relations continue to be the subject of research and debate, especially as the country, beginning in 1991, returned to the tradition of a civilian Minister of Defense. Francisco Leal's *El oficio de la guerra* (item **3179**) is an important analysis of the evolution of military doctrine in Colombia. Elsa Blair Trujillo's *Las fuerzas armadas: una mirada civil* (item **3154**) provides a useful interpretive history of the military's role in politics. Two books address this theme by revisiting the 1985 takeover of the Palace of Justice: Ana Carrigan's *The Palace of Justice: a Colombian tragedy* (item **3157**), a first-rate piece of investigative reporting as well as a gripping narrative, and *Militares, guerrilleros y autoridad civil* (item **3209**).

Although the continued escalation of political and social violence shaped much of the research, as well as the politics, of the country, there were some good studies reflecting other aspects of the political regime and society. Pilar Gaitán *et al.* produced a thorough study on decentralization and the direct election of mayors (item **3169**). Hartlyn addressed the issue of presidentialism, once again adroitly applying larger theoretical concerns in the discipline of political science to the study of Colombia (item **3175**). Oscar Fresneda *et al.* researched and published under the auspices of the United Nations an extensive and well-documented study on the relationship between poverty and violence (item **3168**). Chernick and Jiménez, a political scientist and an historian, teamed up to write a history of the Colombian left which challenges some of the traditional interpretations (item **3162**). Torres Carrillo investigated popular urban struggles in *La Ciudad en la sombra: barrios y luchas populares en Bogotá* (item **3202**). Findji provides a good analysis on the politics of the indigenous movements (item **3167**) and government functionaries and politicians address corruption (item **3192**).

Revisiting a theme which used to be central to the study of Colombian politics, Francisco Leal and Andres Dávila have written an outstanding study of clientelism (item **3178**). This detailed case study, an important contribution to the literature on Colombia, demonstrates that the issue remains important even after the tumultuous events of the 1980s and 1990s.

ECUADOR

Over 60 years after he was first elected president, and more than 20 years after his fifth and last government was overthrown by the military coup of 1972, populist leader José María Velasco Ibarra is still attracting scholarly attention and debate over his role in shaping Ecuadorian politics. Among the best of these new works is Carlos de la Torre Espinosa's, *La seducción velasquista* (item **3243**). Also insightful is Agustín Cueva's volume, *El proceso de dominación política en el Ecuador.* For other works, see *La agonía del populismo* (item **3226**) and *Populismo* (item **3236**).

Other central themes in recent scholarship are the military and democracy, not surprising in a country where the military has retained much of its popularity, and where in the past 15 years the electorate has lurched from right to left, and back to right in a frustrated attempt to realize the promise of the democratic transition of 1979. A major contribution to the literature on democratic transition in Ecuador is Anita Isaac's *Military rule and transition in Ecuador 1972–92* (item **3230**), an example of first-rate social science research applied to area and country studies and an achievement in its own right.

Concerning the military, see *El Estado y las F.F.A.A.* (item **3227**); for a discussion and analysis of Ecuadorian democracy in the 1990s, see Menéndez-Carrión et al. (item **3212**). "La Hora de las Elecciones en Ecuador" by the late Ecuadorian sociologist, Agustín Cueva (item **3222**) and Ninfa León's interview with César Verduga (item **3245**) are also important contributions.

Bonilla has written an excellent and original book on drug-trafficking in Ecuador which should be required reading (item **3214**) and Bustamante has written a comprehensive article on Ecuadorian foreign policy (item **3215**). Conaghan turns to the issue of presidentialism, skillfully using her broad knowledge of Ecuadorian politics to insert this central Andean country into the wider debate on presidentialism vs. parliamentarism (item **3218**). Conaghan and Malloy have written a major book on neoliberalism in the Central Andes, ably comparing Ecuador with Peru and Bolivia (item **3219**). This book demonstrates first-rate comparative scholarship. Lind has written a very good study on women's organizations which serves as both a history and an analysis of the shaping of collective identities (item **3233**).

As mentioned above, Selverston has begun to address the issue of indigenous politics in a pioneering effort that helps lead the way for future research (item **3242**).

COLOMBIA

3152 Alape, Arturo. Tirofijo: las vidas de Pedro Antonio Marín, Manuel Marulanda Vélez. Bogotá: Planeta, 1989. 399 p.: bibl., ill. (Col. Documento)

Biography of Manuel Marulanda Vélez, leader of the Fuerzas Armadas Revolucionarias de Colombia, Colombia's oldest guerrilla group formed in 1966 and still active in the 1990s. The first of a scheduled two volumes, this book covers Marulanda's formative years focusing on events that first led him to take up arms during La Violencia in the 1940s as a Liberal guerrilla. It then describes Marulanda's rise as a leader of the group of armed peasants that would emerge in the 1960s as a communist rebel movement, the Fuerzas Armadas Revolucionarias de Colombia (FARC).

3153 **Arenas, Jacobo.** Correspondencia secreta del proceso de paz. Recopilación, notas y comentarios de Jacobo Arenas. Bogotá?: La Abeja Negra, 1989. 295 p.: ill., ports.

Fascinating first-person account of author's correspondence with major actors in peace negotiations between the guerrilla movement and the government. Arenas was a leader of the Fuerzas Armadas Revolucionarias de Colombia (FARC). Well-written and thoughtful, letters offer personal insight and important inside information about the mid-1980s peace process.

3154 **Blair Trujillo, Elsa.** Las fuerzas armadas: una mirada civil. Bogotá, Colombia: Cinep, 1993. 200 p.: bibl. (Col. Sociedad y conflicto)

Chronological sketch of 20th-century history of the armed forces provides a valuable overview of the institution. Author discusses the military policies of civilian leaders, the evolution of the armed forces' autonomy, and the corruption of the institution through illegal activities and abuses.

3155 **Braun, Herbert.** Our guerrillas, our sidewalks: a journey into the violence of Colombia. Niwot, Colo.: Univ. Press of Colorado, 1994. 239 p.: bibl., ill.

Fascinating account of the kidnapping of Jake Gambini, a US oilman working in Colombia. This first-person narrative is by his brother-in-law, historian Herbert Braun, who has written extensively on Colombia. He documents the negotiation process and day-to-day activities of the guerrilla captors. This rather engaging book explains many of the guerrillas' justifications for their activities, while providing insight into their relationship with Colombian society and government. Gripping, informative, and well-written work.

3156 **Camacho Guizado, Alvaro.** Narcotráfico y sociedad en Colombia: contribución a un estudio sobre el estado del arte. (*Bol. Socioecon.*, 24/25, agosto/dic. 1992, p. 76–96)

Review of literature about narcotics trafficking that has emerged from Colombia since the late 1970s. Also discusses impact of drug exports on the national economy, and places the problem in an international (principally US/Colombian) context. Concludes with a warning that any measures taken to eradicate drug-related violence should not overstep the boundaries of the Colombian juridical framework, as such a move would ultimately undermine State legitimacy.

3157 **Carrigan, Ana.** The Palace of Justice: a Colombian tragedy. New York: Four Walls Eight Windows, 1993. 303 p.: ill.

Fascinating narrative account of the Nov. 1985 seizure of the Palace of Justice by the M-19 Revolutionary Movement, which resulted in the deaths of over 100 people, including 11 Supreme Court Justices. While largely descriptive rather than analytical, book provides vivid portrait of actors involved in this horrific event. Important and graphic depiction of the Colombian guerrilla movement and the armed forces' responses to its activities.

3158 **Cepeda, Manuel José.** La constituyente por dentro. Santa Fe de Bogotá: Presidencia de la República, Consejería para el Desarrollo de la Constitución, 1993. 331 p.

One of three well-researched and comprehensive books on the 1991 Constitutional Assembly by the Special Advisor to the President for the drafting of the new Constitution (see also items **3159** and **3160**). All these are required reading for scholars of human rights, international law, and the nation's political system. Explains both the Constituent process and rights guaranteed by the new Constitution, these works are accessible, well organized, and written from the vantage of an insider who chronicled every move from his seat in the presidential palace.

3159 **Cepeda, Manuel José.** Los derechos fundamentales en la Constitución de 1991. Bogotá: Consejería Presidencial para el Desarrollo de la Constitución; Editorial Temis, 1992. 350 p.

See item **3158**.

3160 **Cepeda, Manuel José.** Introducción a la Constitución de 1991: hacia un nuevo constitucionalismo. Bogotá: Presidencia de la República, Consejería para el Desarrollo de la Constitución, 1993. 449 p.

See item **3159**.

3161 Cepeda, Manuel José. La tutela: materiales y reflexiones sobre su significado. Bogotá: Presidencia de la República, Consejería para el Desarrollo de la Constitución, 1992. 171 p.

With an introduction by former President Gaviria, this report documents background and purpose of the *tutela*, and the debate surrounding it in the Constituent Assembly. Important record of this unprecedented legislation which has brought the guarantee of a legal hearing to every Colombian citizen.

3162 Chernick, Marc W. and Michael F. Jiménez. Popular liberalism, radical democracy, and Marxism: leftist politics in contemporary Colombia, 1974–1991. (*in* The Latin American left: from the fall of Allende to perestroika. Boulder, Colo.: Westview Press, 1993, p. 61–81)

Chapter describes roots and ideological bases of leftist organizations from the National Front period to the present. The left is viewed in the context of long-standing opposition to elite rule in the face of major changes in Colombian politics and economics. Argues that many analysts have overstated the influence of the Cuban revolution and Marxist revolutionary thought on Colombian insurgent movements while underestimating the long tradition of rebellion in the Colombian countryside. Conclusion looks at differences among leftist groups as they entered the 1990s and offers insights into the continuing difficulty of incorporating a leftist opposition into the legal political system.

3163 La Constitución de 1991: ¿un pacto político viable? Recopilación de John Dugas. Bogotá: Dept. de Ciencia Política, Univ. de los Andes, 1993. 225 p.: bibl.

Eight essays and a prologue analyzing the 1991 Constitution describe the process that led to its creation. Informative works cover development of the Constituent Assembly, economic reforms in the Constitution, and other important social and political changes encompassed by the 1991 document. Good basic overview of the Constitutional reform.

3164 Daza Molina, Romelio Elías and Jorge Arturo Salazar Manrique. Democracia y prensa en Colombia. Bogotá: Univ. Javeriana, 1991. 153 p.: bibl. (Serie Monografías; 1)

Starting with the premise that the press is an ideological apparatus of the State, book analyzes the four major Colombian newspapers (*La Prensa, El Espectador, El Tiempo,* and *El Siglo*) to judge whether they promote established social, economic, and political norms. By citing numerous articles and editorials, authors reinforce their original contention but offer little additional analysis.

3165 Desplazamiento, derechos humanos y conflicto armado. Recopilación de Jorge Enrique Rojas. Bogotá: CODHES, 1993. 206 p.: bibl., ill.

Report prepared by the Consultoría para los Derechos Humanos y el Desplazamiento (CODHES) addresses the human rights of populations displaced by armed conflict. Topics include historical examination of internal migration, the role of the UN and international NGOs in resolving the problem, etc., with special attention to the effect of forced displacement on women and children.

3166 Fals Borda, Orlando. La accidentada marcha hacia la democracia participativa en Colombia. (*in* La democracia en América Latina: actualidad y perspectivas. Coordinación de Pablo González Casanova y Marcos Roitman Rosenmann. Madrid: Univ. Complutense de Madrid; México: Centro de Investigaciones Interdisciplinarias en Humanidades, UNAM, 1992, p. 319–335, bibl.)

Rather optimistic piece on prospects for Colombia's political future analyzes historical and political events leading to the 1991 Constitution. Author, one of the writers of the new Constitution, considers it a nonpartisan document that should serve a durable social pact, aiding in the nation's democratic participation and development.

3167 Findji, María Teresa. From resistance to social movement: the indigenous authorities movement in Colombia. (*in* The making of social movements in Latin America: identity, strategy, and democracy. Edited by Arturo Escobar and Sonia E. Alvarez. Boulder, Colo.: Westview Press, 1992, p. 112–133)

Interesting analysis of the indigenous movement in Colombia since the early 1970s. Within the constitutional reform process, the indigenous community asserted its political identity and became a visible actor in national politics. Author raises important questions relating to the significance of this social movement and its meaning for the future of Colombian democracy.

3168 Fresneda, Oscar et al. Pobreza, violencia y desigualdad: retos para la nueva Colombia. Bogotá: Programa de las Naciones Unidas para el Desarrollo, 1991. 537 p.

Important work published by a UN agency examines extent of poverty and the government's efforts to eradicate it from 1986–90. Offers useful statistics on income distribution, urban versus rural poverty and disaggregates the situation by individual cities. Final chapters address the National Plan of Rehabilitation, an effort undertaken by President Gaviria, and his last speech (1991) before the National Assembly.

3169 Gaitán de Pombo, Pilar et al. Comunidad, alcaldes y recursos fiscales. Bogotá: Fundación Friedrich Ebert de Colombia, 1991. 222 p.: bibl., ill.

This edited volume provides careful analysis of the impact of the direct election of mayors, a practice which was first instituted in 1988. Volume is based on interviews, polls, and careful evaluation of the first three years of this constitutional reform that was designed to increase democratic participation and accountability. Although authors find many deficiencies in decentralization laws that were also implemented, on balance they are supportive of the laws.

Gaitán Pavía, Pilar and **Carlos Moreno Ospina.** Poder local: realidad y utopía de la descentralización en Colombia. See item **4815**.

3170 García Durán, Mauricio. De la Uribe a Tlaxcala: procesos de paz. Bogotá: CINEP, 1992. 321 p.: bibl., ill. (Col. Sociedad y conflicto)

Well-documented, detailed account of seven years of negotiations between the government and guerrilla groups (1984–91). Traces historical development of insurgent organizations and gives excellent data on the armed conflict, as well as statistics on victims of violence. Excellent work not only accurately describes the Colombian guerrilla movement, but also precisely analyzes the delicate, high-stakes negotiation process between the government and the guerrillas.

3171 Gobernabilidad en Colombia: retos y desafíos. Edición de Elisabeth Ungar Bleier. Bogotá: Depto. de Ciencia Política, Univ. de los Andes, 1993. 234 p.: appendix, bibl. (Col. Departamento de Ciencia Política 25 años)

History of Colombia's democratic "governability," beginning with the National Front and ending with the second year of the Gaviria Administration. Five chapters and an appendix cover public administration, political economy, decentralization, and government policies during the peace processes. The appendix provides insight into issues confronting the national government as of the early 1990s, outlining the greatest challenges to the Gaviria presidency.

3172 González, Fernán E. et al. Violencia en la región andina: el caso Colombia. Bogotá: Centro de Investigación y Educación Popular (CINEP); Asociación Peruana de Estudios e Investigaciones para la Paz (APEP), 1993. 357 p.: bibl., ill.

Important book is part of a larger project initiated by Peruvian scholar and activist Father Felipe MacGregor, S.J. to study violence throughout the Andes. The Colombian research team was centered at CINEP. Book provides thorough treatment of Colombian violence, examining various issues from cultural identity to everyday violence and crime. Two articles on drug trafficking are particularly illuminating: Jackeline Barragán and Ricardo Vargas, "Economía y violencia del narcotráfico en Colombia: 1981–1991" and Martha Luz García and Dario Betancourt, "Narcotráfico e historia de la mafia colombiana."

3173 González Arias, José Jairo. Espacios de exclusión: el estigma de las Repúblicas Independientes 1955–1965. Bogotá: Centro de Investigación y Educación Popular, 1992. 195 p.: bibl., ill., maps. (Col. Sociedad y conflicto)

Carefully researched text describes the rise and role of Colombia's "Independent Republics," administrative areas organized by the Communist Party in regions where there was little State presence or control. The Republics were eventually destroyed by the government, causing mobile guerrilla movements to take up operations in the affected zones. The book, full of facts, maps, and statistics, offers little analysis but is still one of the few good studies on the Independent Republics.

3174 Guerra y constituyente. Recopilación de Gustavo Gallón Giraldo. Bogotá: Comisión Andina de Juristas Seccional Colombiana, 1991. 173 p.: bibl.

Essays on the human rights situation in Colombia from 1988–90 includes important recommendations for policymakers and draft version of the proposal for the creation of the Commission to Overcome Violence (Comisión de Superación de la Violencia). Report presents organized, useful data on human rights violations and the Colombian State's policies, reactions, and involvement in violence. By analyzing the process under which the new Constitution was written, book also examines the legal framework for upholding human rights in Colombia.

3175 **Hartlyn, Jonathan.** Presidentialism and Colombian politics. (*in* The failure of presidential democracy. Edited by Juan J. Linz and Arturo Valenzuela. Baltimore: Johns Hopkins Univ. Press, 1994, p. 294–327, bibl., tables)

In the context of the debate between presidential and parliamentary systems, author explores pros and cons as well as the complexity of presidentialism in Colombia. Concluding discussion of changes resulting from the 1991 Constitution is useful, as are historical descriptions of the evolution of presidential powers in the institutionalized two-party system.

3176 **Justicia, derechos humanos e impunidad.** Bogotá: Consejería Presidencial para los Derechos Humanos, 1991. 314 p.

Collection of 15 essays written by Colombian and international politicians, human rights workers, and policymakers. Rates of impunity continue to hover at nearly 100 percent, and the numbers of homicides and other human rights violations continue to increase; the introduction reports that the official number of murders in Colombia from 1985–90 was 378,488. Offers useful insights into the judicial system and constructive analysis of the roots of and solutions for the problem of impunity.

3177 **Landazábal Reyes, Fernando.** El equilibrio del poder. Bogotá: Plaza & Janés, 1993. 206 p.

Interesting essay on the balance between military and political power written by a former Minister of Defense.

3178 **Leal Buitrago, Francisco** and **Andrés Dávila L.** Clientelismo: el sistema político y su expresión regional. Bogotá: Instituto de Estudios Políticos y Relaciones Internacionales; Tercer Mundo Editores, 1990. 382 p.: bibl., ill., maps. (Sociología y política)

Excellent, comprehensive analysis of the role of clientelism in Colombian politics. First section offers overview of the phenomenon in its relationship to the Colombian State. The remaining two-thirds of the work analyzes dynamics at the municipal level. Case studied is the city of Rionegro, in the department of Santander. Conclusions explain corrupting role of clientelism in the two-party system and call for fundamental reform of the Colombian political structure.

3179 **Leal Buitrago, Francisco.** El oficio de la guerra: la seguridad nacional en Colombia. Bogotá: IEPRI; TM Editores, 1994. 298 p.: bibl.

Excellent detailed history of Colombian national security looks at the evolution of its principal actors: the military, the police, and the government. Book begins by placing the Colombian case in a larger Latin American context, and concludes with a redefinition of national security given two important factors: the tremendous wave of internal violence Colombia is currently combatting, and new international challenges of the post–Cold War period.

3180 **El libro blanco de la tutela: Vicepresidencia de la República.** Bogotá: Consejeria Presidencial para el Desarrollo Institucional, 1995. 234 p.

Brief volume provides statistics and discusses the background of the *tutela*, the new legal guarantee that allows Colombians to bring a legal action to court in any instance in which they feel that their constitutional rights have been violated. This important development in the country's jurisprudence is one of the greatest direct benefits of the new Constitution. This book accurately reflects how, when, and by whom it is being utilized.

3181 **Lleras Restrepo, Carlos.** ¿Constituyente o congreso? Seleccíon y prólogo de Carlos Gutiérrez-Cuevas. Bogotá: Nueva frontera; Ancora Editores, 1990. 310 p.

Ex-president of Colombia reflects on constitutional reforms and political processes of the government.

3182 **López Vigil, María.** Camilo camina en Colombia. México: Editorial Nuestro Tiempo, 1989. 272 p.: maps. (Col. La Lucha por el poder)

Sympathetic profile of Father Camilo Torres and his successors in the Ejército de Liberación Nacional (ELN). Author interviews the ELN's two highest-ranking members, political leader Father Manuel Pérez and military leader Nicolás Rodríguez. Discussing topics ranging from drugs to Christianity to various groups in the armed struggle, book is a positive and informative portrayal of the history and convictions of those active in Colombia's long-standing guerrilla movement.

3183 **Martz, John D.** Party elites and leadership in Colombia and Venezuela. (*J. Lat. Am. Stud.*, 24:1, Feb. 1992, p. 87–121)

Comparative study of leaders and party elites in Venezuelan and Colombian political parties, tracing their evolution and organizational structure. Author concludes that Venezuelan party elites have been more responsive to participatory demands and needs, and have demonstrated somewhat greater "flexibility to survive" than their Colombian counterparts.

3184 **Medina Gallego, Carlos** and **Mireya Téllez Ardila.** La violencia parainstitucional, paramilitar y parapolicial en Colombia. Bogotá: Rodríguez Quito Editores, 1994. 254 p.: bibl., maps.

Thorough analysis of structure and types of paramilitary activities operating in Colombia, which the authors contend are directly tied to the State security apparatus. Text provides historical background and explains "justification" for paramilitary violations under the aegis of the National Security Doctrine. In addition to numerous individual and regional case studies, authors outline relationship between the armed forces and paramilitary groups, and ask how these activities continue to be conducted with impunity. Concludes by reviewing several recent reports on human rights in Colombia written by international NGOs. Hard-hitting, well-documented analysis that clearly presents the most serious problem faced by contemporary Colombia.

3185 **Orozco Abad, Iván** and **Alejandro David Aponte.** Combatientes, rebeldes y terroristas: guerra y derecho en Colombia. Bogotá: Instituto de Estudios Políticos y Relaciones Internacionales, Univ. Nacional; Temis, 1992. 327 p.: bibl., indexes.

Three-part essay on political violence and drug trafficking. The first part describes theoretical aspects of political crimes and legal structures. The second part focuses on the history of violence in the nation since independence, and the final third describes principal actors and themes involved in Colombia's "internal war." The concluding discussion of the new Constitution and its effect on the legal system is especially useful.

3186 **Pabón Tarantino, Elvyra Elena.** Colombia y su revolución pacífica: la nueva Constitución del 5 de julio de 1991; inicio de un marco institucional dentro de un contexto político pluralista. (*Rev. Estud. Polít.*, 79, enero/marzo 1993, p. 161–208, bibl., graph, tables)

Detailed explanation of reforms of the 1991 Constitution and institutions and legislation enacted. In addition to comparing the 1991 reforms to previous constitutional changes, author asserts that the new Constitution will enhance the federal system and aid Colombia in economic development. Also traces background of the 1991 process. Conclusions are highly optimistic about the positive effects of the Constitution on Colombian society.

3187 **Pacificar la paz: lo que no se ha negociado en los acuerdos de paz.** Dirección de Alejandro Reyes Posada *et al.* Edición de Hernán Darío Correa. Bogotá: Comisión de Superación de la Violencia, 1992. 301 p.: bibl.

Concise treatment of perpetrators and victims of violence in the context of the peace processes of 1990 and 1991. This official report analyzes problems of demobilizing and reincorporating several guerrilla movements after they signed cease-fire and peace agreements with the government. Its four chapters report specifically on levels of violence in different regions, describe the actors (paramilitary groups, State actors, narcotraffickers, and guerrilla fighters), give special attention to violence against indigenous peoples, and call for strengthening civil society to extend the peace process to other guerrilla movements.

3188 **La paz: más allá de la guerra.** Bogotá: Centro de Investigación y Educación Popular, 1993. 116 p. (Documentos ocasionales; 68)

Collection of essays by analysts and

negotiators of the peace process is essential text for anyone interested in the subject. Includes comparison of Salvadoran and Colombian peace processes by Jesús Antonio Bejarano and a study of the re-insertion of the EPL (Ejército Popular de Liberación) after they signed agreements in 1990.

3189 Pearce, Jenny. Colombia, inside the labyrinth. London: Latin America Bureau (Research and Action); New York: Monthly Review Press, 1990. 311 p.: bibl., ill., index, maps.

Overview of history and politics is well-written and comprehensive. Examines mobilization from above and below, focusing on popular responses to the government. Conclusion examines the counteroffensive against guerrillas and suggests options for ending Colombian violence.

3190 Pizarro, Juan Antonio. Carlos Pizarro. Bogotá?: Editorial Printer Latinoamericana, 1991. 291 p.: bibl., ill.

Detailed biography of Carlos Pizarro, M-19 leader and presidential candidate who was assassinated on April 26, 1990. Written by a family member, account consists of first-person interviews with close friends, relatives, and political acquaintances.

3191 Pizarro Leongómez, Eduardo. Las FARC, 1949–1966: de la autodefensa a la combinación de todas las formas de lucha. Con la colaboración de Ricardo Peñaranda. Prólogo de Pierre Gilhodes. Bogotá: UN, Instituto de Estudios Políticos y Relaciones Internacionales; Tercer Mundo Editores, 1991. 245 p., 24 p. of plates: ill., maps. (Sociología y política)

Historical account of the Fuerzas Armadas Revolucionarias de Colombia (FARC), one of the strongest and oldest guerrilla movements in Colombia. Tracing the movement's roots back to the period of La Violencia, author describes how the biparty system's exclusion of leftist politics inevitably led to the emergence of an armed opposition. Also describes the relationship of Colombia's Communist Party to the FARC.

3192 El principio del pez gordo: estrategias para combatir la corrupción. Bogotá: Planeta, 1993. 179 p. (Col. Documento)

Collection of essays written primarily by politicians and other government functionaries on corruption, with an introduction by the Unión de Partidos Latinoamericanos (UPLA). One-half of the book consists of essays on corruption in Colombia; the remaining chapters are essays from Spain and nine Latin American nations. Conclusion offers general recommendations on how countries should fight corruption.

3193 Putumayo. Bogotá: Comisión Andina de Juristas, Seccional Colombiana, 1993. 178 p.: bibl., maps. (Serie Informes regionales de derechos humanos)

Detailed and thoroughly-documented profile of human rights violations in Putumayo, a poor southwestern province of Colombia. Describes complex interactions of drug traffickers, guerrilla forces, military and paramilitary actors, and the victims, who are frequently civilians displaced by violent confrontation or caught in the cross-fire. Conclusion offers sound, unbiased recommendations for the resolution of the region's human rights problems and suggestions for paths to improve overall quality of life for Putumayo's inhabitants.

3194 Restrepo Moreno, Luis Alberto. Síntesis '93: anuario social, político y económico de Colombia. Bogotá: Tercer Mundo Editores; Instituto de Estudios Políticos y Relaciones Internacionales de la Univ. Nacional de Colombia, 1993. 242 p.

Work neatly reviews important political and economic events of 1992. Nine social scientists cover topics such as drug trafficking, social spending, violence, and income distribution. The second half provides a month-by-month account of major events, statistics, composition of the Congress, the national budget, and other economic indicators. First volume of what is expected to be an annual summary of political events during the previous year.

3195 Salazar J., Alonso. Born to die in Medellín. Translated by Nick Caistor. Introduction by Colin Harding. London: Latin America Bureau, 1992. 130 p.: maps.

Interviews with teenage *sicarios*, contract killers who are hired to commit acts of violence in Medellín. Interesting and shocking look at urban life in Colombia reflects high levels of street crime surrounding the narcotics trade, and also the cultural context in which the gang members operate. For Spanish-language original, see *HLAS 53: 1260*.

3196 **Sánchez David, Rubén** and **Patricia Pinzón de Lewin.** Elecciones y democracia en Colombia. (*in* Una tarea inconclusa: elecciones y democracia en América Latina, 1988–1991. San José: Instituto Interamericano de Derechos Humanos, Centro de Asesoría y Promoción Electoral, 1992, p. 287–309, tables)

Analyzes 1990 electoral results and effects of the 1991 Constitution on the electoral process.

3197 **Sarmiento, Luis Fernando** and **Ciro Krauthausen.** Cocaína & Co.: un mercado ilegal por dentro. Bogotá: Univ. Nacional de Colombia, Instituto de Estudios Políticos y Relaciones Internacionales; Tercer Mundo Editores, 1991. 239 p.: bibl. (Sociología y política)

Good description of the Colombian drug industry provides details of its structure, size, international links, and individuals and sectors involved. Approaches narcotics trafficking as a business, examining its underground networks and competitiveness, and concludes that the industry operates much like most legal multinationals, but faces different obstacles to success due to its illegality.

3198 **Thoumi, Francisco E.** Economía política y narcotráfico. Traducción de Pedro Valenzuela. Bogotá: TM Editores, 1994. 339 p.: bibl., index.

Exhaustive study of the drug industry and its economic and political implications for Colombia. Conclusions offer alternatives for formation of new policies to solve the drug problem, and acknowledge the need for US and other international collaboration. This book is the definitive work on the political economy of the drug business in Colombia.

3199 **Thoumi, Francisco E.** The size of the illegal drugs industry in Colombia. Coral Gables, Florida: North-South Center, Univ. of Miami, 1993. 20 p.: bibl., tables. (The North-South agenda papers; 3)

Paper examines difficulties in determining impact of illegal psychoactive drug industry on the Colombian economy and discrepancies in reporting by various studies conducted for over a decade. In addition to critiquing authors and agencies who estimate the size of the drug trade, Thoumi analyzes numerous national and international actors, production and transportation costs, capital flows, and interests involved in the Colombian drug industry. This interesting examination of data and methodology employed in assessing the industry's scope effectively challenges previous reporting on the subject. Thoumi concludes by citing proof of the drug trade's impact in some areas of the Colombian economy and by claiming that accurate, comprehensive estimates are impossible to obtain.

3200 **Thoumi, Francisco E.** Why the illegal psychoactive drugs industry grew in Colombia. (*J. Interam. Stud. World Aff.*, 34:3, Fall 1992, p. 37–63, bibl.)

Article explores traditional political, economic, and geographic reasons for development of Colombia's advantage in the drug trade. History of violence, tradition of contraband, and weak State vigilance and institutions are all considered factors in the industry's success. Conclusion provides several policy alternatives for eliminating cocaine production and trafficking, but eventually concedes that the Colombian government can do little by itself to combat the problem. Thoumi's assessment of the industry and his conclusions are realistic and accurate.

3201 **El Tolima: una respuesta pacífica.** Ibagué, Colombia: Consejo Municipal de Ibagué, 1989. 196 p. (Col. Documentos históricos)

Collection of documents and interviews which emerged from the Feb. 1989 Encuentro Nacional por la Paz at El Tolima, Ibagué. Includes letters, messages, and speeches from political leaders, guerrilla leaders, and religious representatives from all sides of the political spectrum. Concludes with declarations of meetings which were instrumental in forging successful peace agreements with several guerrilla movements.

3202 **Torres C., Alfonso.** La ciudad en la sombra: barrios y luchas populares en Bogotá, 1950–1977. Bogotá: Centro de Investigación y Educación Popular (CINEP), 1993. 222 p.: bibl., map.

Good report on rural to urban migration and the growth of Colombia's cities during a period of rapid expansion. Discussion centers around problems facing residents of the new barrios, including access to public services, legal barriers, and lack of economic opportunities. Methods of land invasion and incidences of public protest are also documented. Conclusion calls for public attention to this under-studied topic and for construction of a viable, autonomous identity for the subaltern social class.

3203 Uprimny, Rodrigo et al. Marginalidad política y económica: lectura social y teológica. Bogotá: Asociación de Teólogos de Colombia Koinonía, 1992. 131 p.: bibl. (Documentos koinonía; 8)

Series of five essays on human rights in Colombia. The first half provides statistical and historical analysis of State-sanctioned violence, while the second half is dedicated to a theological analysis of human rights violations. Also critiques neoliberal and privatization policies using the Catholic Church's teachings as a basis for examination.

3204 Valencia Villa, Hernando. Conflicto armado y éxodo interno en Colombia. San José: Instituto Interamericano de Derechos Humanos, 1991. 21 p. (Exodus en América Latina; 3)

Writer identifies internal migration due to armed conflict as a significant human rights issue in Colombia. Article calls for judicial reforms, investigation by the UN High Commission on Refugees, and attention by both domestic and international bodies to end displacement of civilians. Important article succinctly discusses under-studied aspect of the human rights situation in Colombia.

3205 Valencia Villa, Hernando. La justicia de las armas. Prólogo de Fernando Savater. Bogotá: Tercer Mundo Editores, 1993. 144 p.

Book constitutes series of thoughtful essays by prominent human rights lawyer and scholar on the question of internal war and the possible solutions to the decades' old conflict. Author argues for establishment of certain international norms for the conduct of war which would be applicable to both sides. Given the failure of several attempts at negotiating an end to the war, book concludes that the application of international humanitarian law would open up the possibility for an eventual peace.

3206 Vargas, Mauricio. Memorias secretas del revolcón: la historia íntima del polémico gobierno de César Gaviria, revelada por uno de sus protagonistas. Bogotá: T-M Editores, 1993. 304 p.: bibl. (Temas de actualidad)

Inside look at the Gaviria Administration, skillfully recounted by his media advisor. Intriguing first-person account offers rare snapshot of top-level Colombian policymakers. Well worth reading.

3207 Vargas Velásquez, Alejo. Colonización y conflicto armado: Magdalena Medio Santandereano. Bogotá: CINEP, 1992. 359 p.: bibl., maps. (Col. Sociedad y conflicto)

Well-written description of Magdalena Medio, one of Colombia's most conflictive zones. This chronological account traces effect of political, demographic, and economic changes on the area's landscape, with special focus on the role of the biparty system as one of the roots of conflict. Also details different armed political actors operating in the zone. Good case study of the effects of long-term violence on one poor, rural area of Colombia.

3208 Vélez de Piedrahita, Rocío. El diálogo y la paz: mi perspectiva. Bogotá: Tercer Mundo Editores, 1988. 229 p.: bibl., ill. (Periodismo)

Well-written journalistic account of 1980s Colombian peace process by a member of the Comisión de Negociación, Diálogo y Verificación. Book is divided into nine clear sections including an explanation of the peace process, points of departure, dialogues, protagonists on both sides, spectators, and an evaluation. Final chapters, offering an evaluation and recommendations, are insightful and useful.

3209 Vélez R., Humberto and **Adolfo L. Atehortúa Cruz.** Militares, guerrilleros y autoridad civil: el caso del Palacio de Justicia. Cali, Colombia: Univ. del Valle; Univ. Javeriana, 1993. 300 p.: bibl.

Detailed analysis of the M-19 guerrilla movement's 1985 attack on the Palace of Justice, and the response of the armed forces. By comparing military strategies of both sides of the conflict, authors conclude that the army acted in a manner consistent with its behavior for the past 30 years. However, their response in the case of the Palace was the first occasion in which all of Colombia witnessed their actions and brought the conflict under close scrutiny.

3210 La verdad del '93: paz, derechos humanos y violencia. Bogotá: Centro de Investigación y Educación Popular (CINEP), 1994. 237 p.: bibl., ill.

Report on human rights in Colombia, documented by the Centro de Investigación y

Educación Popular. Provides useful statistics on violence and human rights violations, describes and critiques current government policies and procedures, and sets objectives for achieving peace.

3211 Vergara, Rafael. Colombia: ¿dónde encontrar la necesitada paz?; ¿En las elecciones? (*Secuencia/México*, 18, sept./dic. 1990, p. 95–104, graph, ill.)

Author juxtaposes Colombia's reputation as a long-standing democracy with its history of violence. This piece asserts that results of the 1988 elections reflect a positive change in the two-party status quo, signalling municipal successes of the Unión Patriótica as evidence of an opening in the system. The M-19's electoral victory of 1990 is offered as further proof of the hypothesis. However, the ongoing assassinations of politicians leads the chapter to a more pessimistic conclusion.

ECUADOR

3212 Acosta Espinosa, Alberto et al. Ecuador, la democracia esquiva. Introducción de Amparo Menéndez-Carrión. Quito: ILDIS, 1991. 205 p.: bibl.

Well-documented analysis of challenges to restoring Ecuadorian democracy during the 1980s. Series of five essays compiled by the Instituto Latinoamericano de Investigaciones Sociales accurately describes the nation's political economy in the context of the debt crisis, addressing topics of social mobilization, the role of the judicial system, governance, and effects of structural adjustment policies on democratic consolidation. Introduction and tables of economic indicators are especially useful.

3213 Ayala Mora, Enrique. Los partidos políticos en el Ecuador: síntesis histórica. 2. ed. actualizada. Quito: Ediciones La Tierra, 1989. 94 p. (Ediciones La Tierra; 3)

Brief, updated edition of a basic reference book on Ecuadorian political parties from 1947–88 includes descriptions of all legally recognized parties plus electoral results from 1948–88.

3214 Bonilla, Adrián. Las sorprendentes virtudes de lo perverso: Ecuador y narcotráfico en los 90. Quito: FLACSO-Sede Ecuador, 1993. 103 p.: bibl. (Serie Ciencias políticas)

Two-part volume analyzes drug trafficking in Ecuador and throughout the Andes by placing the problem against the backdrop of US relations with the region. Effectively argues that Ecuador is neither a major cocaine producer, center of money laundering, nor transit point for drugs. However, the country has adopted the US discourse on the war on drugs and has altered its penal system, legislation, and financial regulations accordingly, rather than creating a more appropriate and autonomous response to the situation. Given the fragile financial and political condition of Ecuador, author fears that fragmentation caused by the "creation" of a drug war could prove extremely detrimental to the nation's institutions. Highly convincing approach to a complex issue sheds light on US policies, their consequences, and the Ecuadorian response.

3215 Bustamante, Fernando. Ecuador: putting an end to ghosts of the past? (*J. Interam. Stud. World Aff.*, 34:4, Winter 1992/93, p. 195–224, bibl.)

Clearly written article by FLACSO researcher outlines major Ecuadorian foreign policy issues under the Borja regime, among them multilateralism, regional peacekeeping, democracy and human rights, peace in Central America, environment, drug trafficking, and integration with the other Andean countries. Analysis is set in the context of policy changes made since the last administration. Author is clearly supportive of the Social Democratic approach to foreign relations. The second section, discussing the Ecuador-Peru conflict, analyzes history and current status of the conflict from the Ecuadorian perspective. Also defines points of convergence and disagreement between US and Ecuadorian foreign policies during the Borja years. Thoughtful, succinct article provides accessible overview of the issues.

3216 Cárdenas Reyes, María Cristina. Velasco Ibarra: ideología, poder y democracia. Quito: Corporación Editora Nacional, 1991. 110 p.: bibl. (Biblioteca de ciencias sociales; 32)

Interesting assessment of origins and content of political ideas of Ecuadorian populist José María Velasco Ibarra. Author briefly reviews the Latin American context of populism and then attempts to analyze the relationship between Velasquismo and Ecuador's unarticulated class structure. Also discusses

Velasco's ambiguous and conflictive relationship with Ecuador's parties and successive Constitutions.

3217 **Centro Andino de Acción Popular (Quito).** Los derechos humanos en el Ecuador: una aproximación cuantitativa. (*Ecuad. Deb.*, 28, abril 1993, p. 67–78, ill.)

Refutes notion of Ecuador as an "Island of Peace" by examining human rights record in the context of the Universal Declaration of Human Rights. Article systematically reviews rights guaranteed under the Declaration and offers socioeconomic statistics demonstrating that, in addition to an increase in cases of torture and other mistreatment, the country fails to provide basic rights such as housing, employment, women's and children's rights, and access to free education to large sectors of the population. This informative piece asserts that human rights conditions are steadily deteriorating in Ecuador; it is a succinct and powerful presentation of this argument.

3218 **Conaghan, Catherine M.** Loose parties, "floating" politicians, and institutional stress: presidentialism in Ecuador, 1979–1988. (*in* The failure of presidential democracy. Edited by Juan J. Linz and Arturo Valenzuela. Baltimore: Johns Hopkins University Press, 1994, p. 328–359, tables)

Conaghan describes obstacles to presidentialism in Ecuador as having roots in historical executive-legislative conflicts, traditional oligarchic domination of the system, and a "loose" multiparty structure. Characterizes the latter as a weak, densely populated party system in which politicians' alliances "float" among numerous parties and political fronts. Traces trajectory of presidentialism both before and after the most recent military regime, ending with considerations of a parliamentary option (which she rejects), and institutional changes that might help resolve the executive-legislative impasse and lead Ecuador to establish stronger democratic structures. Useful historical look at the multiparty system, constitutional reform affecting the presidency, and challenges to consolidation of civilian rule in Ecuador.

3219 **Conaghan, Catherine M. and James M. Malloy.** Unsettling statecraft: democracy and neoliberalism in the central Andes. Pittsburgh, Penn.: Univ. of Pittsburgh Press, 1994. 303 p.: bibl., ill., index. (Pitt Latin American series)

Well-researched and carefully argued book compares neoliberal experiments in Ecuador, Peru, and Bolivia. Analysis begins in the early 1980s when each country experienced a democratic transition bringing to power a conservative government that implemented neoliberal policies. In Ecuador and Peru this first phase of neoliberal policy collapsed. Yet in both cases, the neoliberal project was resurrected in the 1990s. Authors argue that although outcomes were different in each country, the political processes were similar. Moreover, they conclude that the neoliberal program is fundamentally anti-democratic and weakens the democratic project in each country. First-rate study that should be widely read.

3220 **Consuelo Benavides: las razones y los sueños; biografía, testimonios, documentos.** Guayaquil, Ecuador: Editorial Univ. de Guayaquil, 1990. 87 p.: ill.

Thin volume of testimony, biography, and documents concerning the life of a young woman who was accused of being a terrorist during the Febres Cordero government, and then was assassinated. Of interest to those concerned with human rights violations in Ecuador during the 1980s.

3221 **La cuestión regional y el poder.** Edición de Rafael Quintero. Quito: Corporación Editora Nacional, 1991. 304 p.: bibl., ill. (Biblioteca de ciencias sociales; 29)

Regional differences have traditionally divided Ecuator politically and shaped the course of its governmental trajectory. Book offers the most comprehensive historical analysis of the debate available. Essays cover topics such as voting patterns, the rise of populism on the Coast, and the political economy of the State's evolution. Conclusion discusses the regionalism problem by placing it in the framework of Ecuador's political culture.

3222 **Cueva, Agustín.** La hora de las elecciones en Ecuador. (*Secuencia/México*, 18, sept./dic. 1990, p. 105–112, ill.)

Transcript of talk given by Ecuadorian sociologist, Agustin Cueva, is an engaging piece that provides anecdotal inside view of the country's presidents and political system. Colloquially delivered, short chapter offers a frank, well-informed opinion of the modern Ecuadorian political scene that is well worth consideration.

3223 **Cueva, Agustín.** El proceso de dominación política en el Ecuador. Ed. corregida y actualizada. Quito: Planeta, 1988. 188 p.: bibl. (Col. País de la mitad; 8)

Class-based examination of the evolution of Ecuador's contemporary political landscape. Volume includes excellent analysis of the rise and fall of Velasquismo, and concludes with responses to critiques of his earlier edition made by other Ecuadorian political scientists. A "must read" by one of Ecuador's leading contemporary social scientists.

3224 **Una década de opinión en el Ecuador: la página editorial de *Hoy* de 1982 a 1992.** Estudio introductorio y selección de Samuel Guerra Bravo y Susana Hidalgo Saa. Quito: Hoy, 1992. 318 p.: ill.

Clever compilation of editorials and cartoons from the newspaper *Hoy* satirizes Ecuadorian political and social conditions in the 1980s.

3225 **El Ecuador frente al siglo XXI: seguridad y geopolítica.** Recopilación de Oswaldo Jarrín R. Quito: Ministerio de Defensa Nacional, Dirección de Protocolo, Prensa y Relaciones Públicas, 1992. 178 p.: bibl., maps.

Published by the Ministry of National Defense, book is an interesting collection of essays, written by senior officers plus a few invited foreign scholars reflecting on internal military debates as well as on the role of the Ecuadorian armed forces in the post–Cold-War world. Essays are of varied quality and somewhat oriented toward refuting the notion that there is no role for the armed forces within the new world order. Good contribution by Jack Child classifies current conflicts in Latin America.

3226 **Fernández Espinosa, Iván and Gonzalo Ortiz Crespo.** ¿La agonía del populismo?: informe urgente sobre las elecciones presidenciales de 1988. Quito?: Editorial Plaza Grande, 1988. 168 p.

Written following the 1988 elections, volume explores the electoral defeat of Abdalá Bucaram and whether his loss signifies the end of populism in Ecuador. Provides rather simplistic description of populism, but summarizes major populist movement in Ecuador: Velasquismo and "Cefepeismo" of the Concentración de Fuerzas Populares (CFP). Work focuses mainly on campaign dynamics and political players of modern Ecuador.

3227 **García Gallegos, Bertha.** El Estado y las F.F.A.A. (*Ecuad. Deb.*, 24, dic. 1991, p. 65–77, bibl.)

Discussion of the evolution of the Ecuadorian State describes barriers to development, including geographical separations, low levels of political participation, and historically weak institutional structures, along with the role of the armed forces in economic development. Military bureaucracy is identified as the catalyst to modernization and state-building. By juxtaposing economic progress and governmental organization of the military government to that which succeeded it, García Gallegos provokes reflection on the efficacy of different regime types in Ecuador.

3228 **Guerrero, Andrés.** De sujetos indios a ciudadanos étnicos: de la manifestación de 1961 al levantamiento indígena de 1990. (*in* Democracia, etnicidad y violencia política en los países andinos. Lima: Instituto de Estudios Peruanos; Instituto Francés de Estudios Andinos, 1993, p. 83–101, table)

Good analysis and narrative of emergence of Ecuador's strong indigenous confederation CONAIE (Confederación de Nacionalidades Indígenas de Ecuador) and of the national uprising that they led in 1990. Article argues that the indigenous population emerged as a major social and political actor by taking advantage of the weakening position of the Ecuadorian State, particularly at the local level, as well as the weakening of other political movements such as traditional political parties and labor.

3229 **Hurtado, Osvaldo.** La dictadura civil. Quito: Fundación Ecuatoriana de Estudios Sociales, 1988. 525 p.: bibl.

Collection of interviews, speeches, and press conferences given by former President Osvaldo Hurtado, comparing his administration to that of his successor, León Febres Cordero. Politically-charged text mainly consists of harsh condemnations of the Febres Cordero Administration and praise for the accomplishments of the Hurtado period.

3230 **Isaacs, Anita.** Military rule and transition in Ecuador, 1972–92. Pittsburgh, Pa.: Univ. of Pittsburgh Press, 1993. 178 p.: bibl., index. (Pitt Latin American series)

Excellent study of the Ecuadorian military and its relationship to civil society. Following a very insightful review of the wider literature on the military in Latin American politics, author skillfully uses the Ecuadorian case to rethink some major theories on the military and democratic transition. In so doing, she provides detailed discussion of the Ecuadorian military's internal dynamics, structure of civilian opposition, and characteristics of the military regime that precluded a smooth transition to a civilian government. Concludes that Ecuadorian democracy remains fragile and unconsolidated and leaves open the possibility of a return to authoritarian rule.

3231 **Isaacs, Anita.** Problems of democratic consolidation in Ecuador. (*Bull. Lat. Am. Res.*, 10:2, 1991, p. 221–238, tables)

Critical assessment of democratic consolidation in Ecuador. Explores obstacles to democratic stability and factors involved in sustaining a fragile civilian political order. Paradoxically, author finds that the unconsolidated nature of Ecuadorian democracy lies in the "success" of recent military governments. Because of the military's sense of responsibility for national security and development, a return to military rule should not be ruled out either. [D. Dent]

3232 **León Velasco, Juan Bernardo.** Las elecciones en el Ecuador: concejales cantonales, 1978 y 1990. Quito: CIESA, 1992. 166 p.: ill., maps.

Study of the "electoral geography" of Ecuador in the context of the proliferation of both political parties and elections since 1978. Analysis focuses on electoral behavior in city council elections in different political regions. Study attempts to understand the spatial distribution of political identities in country's varied socioeconomic regions—coast, sierra, Amazonian, as well as urban and rural. Good microanalysis of electoral data. Broader conclusions are weak.

3233 **Lind, Amy.** Power, gender, and development: popular women's organizations and the politics of needs in Ecuador. (*in* The making of social movements in Latin America: identity, strategy, and democracy. Edited by Arturo Escobar and Sonia E. Alvarez. Boulder, Colo.: Westview Press, 1992, p. 134–149)

Chapter traces roots and motivations for Ecuador's 80 women's rights organizations. Also discusses relationship of the women's movement to the State, and offers conclusions about the organizations' role in creating a "collective identity" and, ultimately, empowering Ecuadorian women. One of the only detailed studies on Ecuadorian women's organizations available in English.

Maiguashca, Juan. Las clases subalternas en los años treinta. See item **1813**.

3234 **Moreano, Alejandro** *et al.* Gobierno y política en el Ecuador contemporáneo: selección de textos. Edición de Luis Verdesoto. Quito: ILDIS, 1991. 558 p.: bibl.

Selection of essays from some of the leading Ecuadorian social scientists, including Agustín Cueva, Osvaldo Hurtado, and Amparo Menéndez-Carrión. Essays focus on the reinterpretation of Velasquismo, contemporary issues such as the military, social movements, and political parties, and an analysis of voting behavior. Good basic reader on standard political themes.

Moreno Yáñez, Segundo and **José Figueroa.** El levantamiento indígena del Inti Raymi de 1990. See *HLAS 53:1292*.

3235 **Muñoz Vicuña, Elías.** El movimiento obrero ecuatoriano: sus primeros pasos. Guayaquil: Editorial de la Univ. de Guayaquil, 1990. 46 p. (Col. Movimiento ecuatoriano; 13)

Transcript of a speech on Ecuadorian labor history, by a noted Ecuadorian labor historian.

3236 **Peñaherrera Padilla, Blasco** *et al.* Populismo. Prólogo de Juan J. Paz y Miño. Quito: ILDIS; El Duende; Ediciones Abya-Yala, 1992. 213 p.: bibl. (Realidad nacional)

Extensive look at populism in the Ecuadorian context asserts that it is the predominant force in the nation's modern political history. Interesting discussion of Abdalá Bucaram and José Maria Velasco Ibarra (the latter was president of Ecuador for five non-consecutive terms). Also analyzes "over-analysis" of populism in academic research. Concluding chapter by FLACSO director Amparo Menéndez-Carrión urges study beyond populism and into new realms of political science.

Periodismo y medio ambiente: memoria del seminario realizado en Quito, entre el 28 de noviembre y el 1 de diciembre de 1990. See item **2786**.

3237 **Los políticos y los indígenas: diez entrevistas a candidatos presidenciales y máximos representantes de partidos políticos del Ecuador sobre la cuestión indígena.** Recopilación de Erwin H. Frank, Ninfa Patiño y Marta Rodríguez. Quito: Ediciones Abya-Yala; ILDIS, 1992. 160 p.

Interviews were made in the pre-electoral period leading up to the presidential elections of 1992 and in the wake of the Indian uprising of 1990. Good indicator of the abyss between politicians and the indigenous movement.

3238 **Rivera, Ramiro.** Ecuador: elecciones y democracia, 1978–1990. (*in* Una tarea inconclusa: elecciones y democracia en América Latina, 1988–1991. San José: Instituto Interamericano de Derechos Humanos, Centro de Asesoría y Promoción Electoral, 1992, p. 311–343, tables)

Detailed description of post-1978 Ecuadorian elections provides statistics on votes received by the major political parties in presidential and congressional elections, including break-downs by province and abstention rates. Thoughtful analysis also traces trends in political orientation and a brief examination of legislation regarding political parties. Introductory section on the history of political power, originating in the nation's haciendas examines succession of military governments and explores the transition to civilian rule, successfully explaining the nature and roots of Ecuador's contemporary political system.

3239 **Robalino Bolle, Isabel.** El sindicalismo en el Ecuador. 2. ed. Quito: INEDES; CONUEP; PUCE, Ediciones de la Pontificia Univ. Católica del Ecuador, 1992. 273 p.: bibl.

Updated second edition covers historical antecedents beginning in the precolumbian period of Ecuadorian labor movements and detailed descriptions of different contemporary labor organizations. Author argues that organizing predates the industrial revolution and counts as unions the different labor and agricultural cooperatives and communes, as well as certain popular organizations such as the indigenous confederations. From a detailed statistical breakdown, by union and by region, that comprises the second part of the book she derives the relatively high statistic that 35.79 percent of the economically active population is organized, a figure comparable with the most industrialized countries of Europe.

3240 **Rodas, Raquel.** Nosotras que del amor hicimos—. Fotografías de Rolf Blomberg. Quito: R. Rodas, 1992. 111 p.: bibl., ill.

Biographical tribute to Luisa (Lucha) Gómez de la Torre, a teacher and political activist who worked closely with Ecuador's indigenous movement in the 1940s. One of the only works available about a woman's participation in Ecuador's political arena. Contains good documentary photography.

3241 **Sánchez Parga, José.** Ecuador en el engranaje neoliberal. (*Nueva Soc.*, 123, enero/feb. 1993, p. 12–17, table)

Article assesses social and political costs of Sixto Durán Ballén's economic policies, which the author fears may lead to "neoliberal totalitarianism." Asserts that socialist totalitarianism, which eliminates the private sector, and neoliberal totalitarianism, in which the private sector dominates the public sector, achieve the same result: liquidation of civil society as the setting for public/private relations. Views drastic reduction of the Ecuadorian State dictated by neoliberal policies as a challenge to its fragile democracy. Although the rhetoric seems overstated, article's position is clear and with merit.

3242 **Selverston, Melina.** The politics of culture: indigenous peoples and the State in Ecuador. New York: Institue of Latin American and Iberian Studies, Columbia Univ., 1993. 25 p. (Papers on Latin America)

Covers politics of indigenous movements in Ecuador in the 1990s. Author succinctly describes different indigenous federations and their regional bases. She then delineates recent government policies and the emergence of a national federation that began to challenge the government successfully and win major concessions. Interesting story and good analysis.

Solano de la Sala Veintimilla, Germán. Indice de folletos sobre temas económicos y sociales del Ecuador. See item **1837**.

3243 **Torre, Carlos de la.** La seducción velasquista. Quito: Ediciones Libri Mundi, Enrique Grosse-Luemern; FLACSO-Sede Ecuador, 1993. 261 p.: bibl., ill.

Detailed analysis of the particular brand of populist politics which brought José

María Velasco Ibarra to power five times as president of Ecuador. Excellent and original use of the social and economic data to explain the conditions that facilitated Velasco's rise to power. Important contribution to the general literature on Latin American populism and on one of the most important leaders in 20th-century Ecuadorian politics.

3244 Velarde, Patricio. Entre la encrucijada estatal y el clientelismo político: la gestión municipal en ciudades intermedias ecuatorianas. (*in* Viejo escenario, nuevos actores: problemas y posibilidades de la gestión municipal en cuidades intermedias en América Latina. Quito: CIUDAD, 1991, p. 87–102, bibl.)

Discussion of political and economic conditions in Ecuador's medium-sized cities focuses on their relationship to the State. Low levels of participation in local politics as well as the municipalities' clientelistic dependence on the central government are deemed to be the fundamental hindrances to the development of cities in this category. Conclusion offers several constructive options for enhancing local governments' economic and political authority and increasing citizens' democratic participation.

3245 Verduga, César. El arte de diferenciar: diálogos con César Verduga. Entrevistas de Ninfa León. Quito: Fundación Grupo Esquel-Ecuador; ILDIS, 1992. 140 p.

Extensive interview with César Verduga, Ecuadorian politician, academician, former Minister of Labor and Interior, and human rights worker. Covers topics ranging from governance to the political party structure, the military, and the nation's external debt. Frank critique of the political, economic, and social state of Ecuador, with insightful analysis and a left-leaning perspective on possible options for change.

3246 Ycaza Cortez, Patricio. Lucha sindical y popular en un período de transición, 1948–1970. (*in* El Ecuador de la postguerra: estudios en homenaje a Guillermo Pérez Chiriboga. Quito: Banco Central del Ecuador, 1992, v. 2, p. 543–569, bibl.)

Essay traces development of worker organization and mobilization during the period in which Ecuador moved from an agro-export economy to incipient industrial development. Author characterizes this period as one in which foreign capital and the petroleum bourgeoisie transformed the economy, forcing corresponding changes in labor activism.

VENEZUELA

RENE SALGADO, *Consultant, Inter-American Development Bank*

VENEZUELA'S POLITICAL developments have become more and more puzzling and unexpected since 1988 when Carlos A. Pérez won the presidential elections. Pérez became the first ex-President to be reelected after a mandatory two-term hiatus, despite his having been censured by his party's Ethical Commission and nearly condemned by Congress for malfeasance. Up to the 1988 elections however, his popular support remained high to the point that Venezuelans, nostalgic for the petro-bonanza Pérez presided over from 1974–79, failed for the first time in 20 years to vote out the party in power. Nonetheless, by the end of his turbulent term in office his prestige had sunk so low that he was even attacked for alleged immoral behavior.

Pérez's decline in popular support was to a large extent triggered by his introduction of neoliberal policies. Apparently many people felt betrayed by Pérez's failure to implement his populist campaign pledges. For instance, as soon as Pérez took office he cut State subsidies and initiated a privatization program that included the publicly-owned telephone company and airline, a course of action that negated in every sense the nationalization policies that had transformed him into a national hero back in the mid-1970s.

Throughout the late 1980s, the decline in presidential popularity continued unabated. Pérez's initial announcements resulted in price increases in public transportation which, in turn, ignited violent popular rioting in late Feb. 1989. Food-price liberalizations stirred protests from the middle classes who demonstrated by banging pots and pans in several cities on different occasions, especially during the early years of his presidency.

Pérez's neoliberal program deeply divided his own Acción Democrática (AD) Party, and contributed to the fracture of the well-established party system traditionally dominated by his party and the Christian Democratic COPEI. Austerity measures such as the elimination of price controls on basic products, a proposed sales tax, and mass layoffs in the public sector without programs to facilitate reemployment alienated Pérez from AD's labor sector and fueled disrupting opposition from COPEI. He never regained popularity with AD labor leaders despite resorting several times to the tactic of buying time by announcing temporary halts to his measures. On several other occasions he turned to blatant cooptation, designating party rivals to critical cabinet posts and even bringing in two cabinet ministers from COPEI.

None of these strategies, however, were enough to help him succeed, and three years after his inauguration it became increasingly clear that his reform package was in serious trouble with most of his reform ministers and associates forced to resign because of widespread opposition. Additional plans for balancing the budget, increasing privatization, lowering import tariffs, and introducing a value-added tax never materialized.

Finally, the military relinquished its characteristic neutrality of the past three decades: Pérez was the victim of two unsuccessful military coup attempts by midranking officers in Feb. and Nov. 1992. At one point an armored car bombardment of the front gate of Miraflores—the presidential office complex in the center of the city—forced the President to slip through a secret tunnel into the white, hilltop presidential palace.

Pérez survived both coup attempts, but by then the political class had become increasingly convinced that his stay in power would seriously harm democratic stability. After several months of demonstrations aimed at forcing him out of office, the Supreme Court decided in mid-1993 to indict Pérez on corruption charges, and next, in an unprecedented move, the Senate voted unanimously to suspend him from office—the first such suspension of an elected Venezuelan president. Congress then selected independent Senator Ramón J. Velásquez as interim Head of State and Government to complete his term, to end in Dec. 1993.

Pérez was sent to jail. In the meantime cleavages within the center-right COPEI party led, in mid-1993, to the virtual expulsion of elder statesman Rafael Caldera, along with a handful of other prominent national leaders. Several years before, Caldera, who had cast himself as a defender of the poor, had become the foremost critic of Pérez's economic package, clashing with the business sector over the 1991 Labor Law. Through an unusual political maneuver, Caldera became the 1993 presidential candidate representing the Movimiento al Socialismo and other leftist parties, winning the election on a strong anti-neoliberal platform. He reversed Pérez's trends immediately after taking office, imposing, among other measures, price and foreign-exchange controls.

Despite these dramatic political developments in Venezuela, there are no empirical academic works on these themes, probably because rigorous analysis takes time to prepare and publish. Nevertheless, the nation's contemporary political bibliography is, to an extent, still relevant and interesting. One of the outcomes of this political turmoil is the proliferation of relevant books and articles written by politicians, labor leaders, and even military officers, all of which attest to the fact that three decades of uninterrupted democratic life have consolidated a national trend towards openness, one that may well be irreversible. Several other recent books and articles are historical or normative in nature, while a few, the minority, provide more elaborate systematic discussions of some contemporary political developments, notably local and gubernatorial elections, as well as electoral reform.

Among works devoted to past political leaders there is much to learn about ex-President Rómulo Betancourt's leadership role and political ideology in *Antología política* (item **3251**) and *La segunda independencia de Venezuela: compilación de la columna "Economía y Finanzas" del diario "Ahora," 1937–1939* (item **3252**). *Alberto Carnevali: pasión de libertad* (item **3254**) is useful for a better understanding of Carnevali's political personality, while *Jovito Villalba en la historia política de Venezuela* (item **3258**) is a useful account of the evolution of Villalba's political ideas.

On formal aspects of the political system and its political implications, we should note Kornblith's cogent account of the two most recent Venezuelan constitutions (item **3264**) and Molina's examination of current electoral law which maintains a good balance between documentary and empirical analysis (item **3268**).

The outcomes of mayoral and gubernatorial elections are fresh topics receiving rigorous treatment. Readers will learn much from Carrasquero and Welsch (item **3255**), Njaim (item **3270**), Molina (item **3267**), and Kornblith (item **3265**).

The regulatory capabilities of police institutions are the subject of Navarro and Pérez Perdomo (item **3273**), Muller Rojas (item **3269**), and Hernández (item **3263**), all of whom look into institutional and legal factors that affect the government's ability to handle effectively socially dysfunctional behavior.

Reflections on current political difficulties and future prospects of Venezuelan democracy may be found in *Situación y perspectivas de la democracia venezolana* (item **3248**), *Tiempo de Páez, social democracia y régimen de coaliciones* (item **3266**), *Liderazgo e ideología* (item **3247**), *¿Cuándo se jodió Venezuela?* (item **3276**), and *El reto ideológico de los partidos políticos venezolanos* (item **3256**), all of which, with the exception of the last work, constitute collections of articles, some written by politicians and some by political and policy analysts.

Other topics addressed among works annotated below include civil-military relations and State reform, the latter a salient political issue during the 1980s. The first is the subject of *Todos los golpes a la democracia venezolana* (item **3253**) and *Militares y democracia* (item **3260**), while the second is the focus of *Antecendentes de la reforma del Estado* (item **3250**) and *Venezuela: centralización y descentralización del Estado* (item **3274**).

3247 Aguilar, Pedro Pablo *et al.* Liderazgo e ideología. Presentación de Isidro Morales Paúl. Dirección de Manuel Vicente Magallanes. Caracas: Consejo Supremo Electoral, 1991. 318 p.: bibl. (Col. del cincuentenario; 11)

Reflections on Venezuelan leadership and other hot political topics by prominent party leaders and analysts. Insightful essay on Rómulo Betancourt by COPEI's leader Pedro P. Aguilar describes challenges of party's democratization. Article by J.E. Molina offers useful account of 1989 local election results.

3248 Aguilar, Pedro Pablo *et al.* Situación y perspectivas de la democracia venezolana. Introducción de Aníbal Romero. Caracas: Editorial Fundación Rómulo Betancourt, 1991. 120 p.: bibl. (Col. Tiempo vigente; 2)

Venezuelan politicians wrote four of the five articles that make up this collection. Teodoro Petkoff, leader of the Movimiento al

Socialismo, contributed a short but persuasive appraisal of Venezuelan governance enlivened by pungent observations about political corruption. Fun reading.

3249 Alvarez D., Angel E. Noticias de sucesos, versiones y manipulaciones: qué hizo, quién y porqué el 27 de febrero según la prensa. (*Politeia/Caracas*, 13, 1989, p. 155–185, tables)

Examines content, style, and format of articles, and relative importance granted to popular demonstrations of late Feb. 1989 in top five Caracas newspapers. Includes a particularly interesting discussion of the different underlying messages about the cause of these events. Well organized and clearly written.

3250 Antecedentes de la reforma del Estado. Caracas: Comisión Presidencial para la Reforma del Estado, 1990. 154 p.: bibl. (Col. Reforma del estado; 2. Documentos para la discusión nacional)

Examines the scope of different State reform attempts since 1958, looking at the units responsible for reform proposals and their accomplishments. Uses a combination of documentary analysis and interviews.

3251 Betancourt, Rómulo. Antología política. v. 1. Selección, estudio preliminar y notas de Aníbal Romero, Elizabeth Tinoco y María Teresa Romero. Caracas: Editorial Fundación Rómulo Betancourt, 1990. 1 v.: bibl., index, port.

Anthology of Rómulo Betancourt's writings from 1928–35 edited by Aníbal Romero and several Venezuelan political scientists. Contains four chronologically-arranged sections roughly corresponding to the evolution of Betancourt's political ideology and activities: 1928–29 (anti-dictatorship writings); 1930 (anti-imperialist and nationalistic writings); 1931–32 (shaping of political strategy); 1933–35 (radicalization).

3252 Betancourt, Rómulo. La segunda independencia de Venezuela: compilación de la columna "Economía y Finanzas" del diario *Ahora*, 1937–1939. v. 1–3. Estudio introductorio de Arturo Sosa Abscal. Caracas: Fundación Rómulo Betancourt, 1992. 3 v.: bibl. (Col. Tiempo vigente; 3–5)

Fundamental reading for anyone interested in the evolution of Betancourt's political economic theory. Includes more than 600 articles he published between 1937–39 in the newspaper *Ahora*. The early economic and social policy preferences of Acción Democrática were largely derived from Betancourt's writings.

3253 Capriles Ayala, Carlos and **Rafael del Naranco.** Todos los golpes a la democracia venezolana. Prólogo de Ruth Capriles Méndez. Caracas: Consorcio de Ediciones Capriles, 1992. 239 p.: bibl., ill.

Descriptive accounts of failed coup attempts since 1958. Contains information about coup leaders and violent incidents surrounding events.

3254 Carnevali, Alberto. Alberto Carnevali: pasión de libertad; escritos. v. 1–3. Compilación de Ramón Rivas Aguilar. Mérida, Venezuela: Acción Democrática, Univer. Popular Alberto Carnevali, 1989. 3 v.: bibl., ill.

A chronology of the political life of Alberto Carnevali, one of the founders of Acción Democrática. Contains his newspaper articles published between 1935–53 as well as comments and information about activities in Venezuela and exile.

3255 Carrasquero, José Vicente and **Friedrich Welsch.** Las elecciones regionales y municipales de 1989 en Venezuela. (*Cuad. CENDES*, 12, sept./dic. 1989, p. 9–29, graphs, ill., tables)

Shows that abstention was widespread in Venezuela's local elections of Dec. 1989. Discusses political alliances and coalitions at the local level while arguing that Acción Democrática's electoral strategies were politically appropriate, despite unfavourable outcomes.

3256 Combelas, Ricardo. El reto ideológico de los partidos políticos venezolanos. (*Contribuciones/Buenos Aires*, 2, abril/junio 1991, p. 9–18)

Regrets the abandonment of ideological principles by political parties. Asserts that to regain legitimacy, party leaders should foster ideological dialogue and debate within their local units and the larger community.

3257 Coppedge, Michael. Venezuela: democratic despite presidentialism. (*in* The failure of presidential democracy. Edited by Juan J. Linz and Arturo Valenzuela. Baltimore: Johns Hopkins Univ. Press, 1994, p. 396–421, tables)

Examines the presidential system as the cause of executive-legislative stalemates. Fails to consider that deep disagreements and political deadlock are often the product of presidential policies, rather than a fatalistic outcome of institutional arrangements.

3258 Croce, Arturo. Jóvito Villalba en la historia política de Venezuela. Caracas: Edición Homenaje, 1990. 483 p.: ill.

Biography of Villalba, a founding father of contemporary Venezuelan democracy, based on newspaper reports and informal interviews. Organized around major events in Venezuelan history, the work begins with the final years of the Gómez regime. Chapter on the Unión Republicana Democrática, a party founded by Villalba, offers a useful explanation for its relatively short life and weak popular support.

3259 Dabaguián, Emil. Carlos Andrés Pérez: hombre, patriota, internacionalista. (*Am. Lat./Moscow*, 9, 1991, p. 52–64, photos)

Historian of the Academy of Science of former USSR praises ex-president Pérez. Reads like apologia to Soviet dictators.

3260 Daniels Hernández, Elías Rafael. Militares y democracia: papel de la Institución Armada de Venezuela en la consolidación de la democracia. Edición de José Agustín Catalá. Caracas: Centauro, 1992. 276 p.: bibl., ill.

Written by an Army vice-admiral who provides a useful account of the evolution and organization of the armed forces. Emphasizes the military's loyalty to civilian governments and democracy.

Davis, Charles L. and **John G. Speer.** The psychological bases of regime support among urban workers in Venezuela and Mexico: instrumental or expressive? See item **2744**.

3261 Ellner, Steve. The deepening of democracy in a crisis setting: political reform and the electoral process in Venezuela. (*J. Interam. Stud. World Aff.*, 35:4, Winter 1993/94., p. 1–42, bibl.)

Useful summary of the political reform process—electoral reform, party management and organization, decentralization of power, and strengthening of civil society. Notes that although Venezuela appears to be moving toward a more democratic and participative society, the process entails risks. If the incorporation of formerly inactive groups is stalemated and their rising political expectations are not met, the danger of anarchy or military dictatorship exists.

3262 Ellner, Steve. The Venezuela left: from years of prosperity to economic crisis. (*in* The Latin American left: from the fall of Allende to perestroika. Boulder, Colo.: Westview Press, 1993, p. 139–154)

Argues that the left's emphasis on electoral politics has alienated the poor. Notes that leftist parties neglected grassroots political work and therefore played no role in popular demonstrations during recent economic difficulties. Although an interesting and valid criticism of the left, it should be noted that this new strategy enabled the left to catapult Caldera to the presidency.

Gómez Calcaño, Luis and **Margarita López Maya.** El tejido de Penélope: la reforma del estado en Venezuela, 1984–1988. See item **4838**.

3263 Hernández, Tosca. "Extraordinary" police operations in Venezuela. (*in* Vigilantism and the State in modern Latin America: essays on extralegal violence. New York: Praeger Publishers, 1991, p. 157–165)

Examines the legal foundations, unfairness, and political impact of the so-called "extraordinary" police operations which often result in arrests without cause, particularly among the marginal social sectors. Argues that these ill-conceived operations exacerbate crime.

3264 Kornblith, Miriam. Proceso constitucional y consolidación de la democracia en Venezuela: las constituciones de 1947 y 1961. (*Politeia/Caracas*, 13, 1989, p. 283–329, bibl.)

Thoughtful and well-articulated discussion of political processes and developments leading to the enactment of the 1947 and 1961 (still in force) Constitutions. Argues that degree of political consensus is a critical factor in explaining the relative longevity of the 1961 Constitution.

3265 Kornblith, Miriam. Sistema de partidos y reforma electoral en Venezuela. (*in* Los partidos políticos en el inicio de los noventa: seis casos latinoamericanos. Santiago: Ediciones FLACSO, 1992, p. 27–48)

Argues that 1989 electoral reforms

sanctioning democratic elections of governors and mayors will positively affect political development. However, the current erosion of the public image and legitimacy of political parties, combined with deteriorating living conditions, will adversely impact long-term democratic stability.

3266 **Magallanes, Manuel Vicente et al.**
Tiempo de Páez; social democracia y régimen de coaliciones. Dirección de Manuel Vicente Magallanes. Caracas: Consejo Supremo Electoral, 1990. 251 p.; bibl. (Col. del cincuentenario; 10)

Eight short essays by prominent Venezuelan political scientists. With the exception of the first (devoted to the political ideas of J.A. Páez, one of Bolívar's generals) and the last (examining Bolívar's ideology), the essays focus on Venezuela's current political system. Particularly interesting are articles on electoral tendencies during the last 30 years by A. Torres and C. Codetta, on elections of state governors by H. Njaim, and on civil society by I. Quintero.

Martz, John D. Party elites and leadership in Colombia and Venezuela. See item **3183**.

3267 **Molina Vega, José Enrique.** Elecciones y democracia en Venezuela, 1988–1990. (*in* Una tarea inconclusa: elecciones y democracia en América Latina, 1988–1991. San José: Instituto Interamericano de Derechos Humanos, Centro de Asesoría y Promoción Electoral, 1992, p. 388–404, bibl., tables)

Useful discussion of 1988 general election and the local elections of 1989 in the context of two public demands: the deepening of political democracy and the improvement of social conditions. Concludes that progress has already been made on the first, but not on the second front.

3268 **Molina Vega, José Enrique.** El sistema electoral venezolano y sus consecuencias políticas. Valencia, Venezuela: Vadell Hermanos Editores; San José, Costa Rica: Instituto Interamericano de Derechos Humanos, Centro de Asesoría y Promoción Electoral, 1991. 243 p.: bibl.

Asserts that Venezuela's electoral system is democratic, pluralistic, trustworthy, and participatory. Uses a legal and political approach to analyze electoral legislation. Useful.

3269 **Müller Rojas, Alberto A.** Las fuerzas del orden en la crisis de febrero. (*Politeia*/Caracas, 13, 1989, p. 115–154, bibl.)

Written by a retired general of the Venezuelan armed forces. Interesting, especially for commentary on the evolution and organization of contemporary police institutions in the country, some of which he believes are dysfunctional despite their adequate performance.

3270 **Njaim, Humberto.** La época de los gobernadores designados, 1959–1989: regularidades observadas e hipótesis explicativas.

From 1959–89, governors were designated by presidents. Presidents often treated governors as second-class cabinet ministers and, as is statistically demonstrated, gubernatorial rotation was very high. The electoral law enacted in 1989 established gubernatorial races and a fixed three-year term. Notes that state government stability will be one of the positive outcomes. Interesting and well written.

3271 **Olivo, Francisco.** Francisco Olivo: conciencia del sindicalismo venezolano. Caracas: Ediciones del CEN de Acción Democrática, 1991. 165 p.: ill.

A collection of writings by Olivo, the Acción Democrática labor secretary from 1959–74.

3272 **Oropeza, Luis José et al.** La democracia venezolana: hipótesis de un plebiscito—los derechos humanos y la paz. Presentación de Alberto López Oliver. Dirección de Manuel Vicente Magallanes. Caracas: Consejo Supremo Electoral, 1988. 276 p.: bibl. (Col. del cincuentenario; 6)

Collection of eight essays by Venezuelan analysts offering broad, at times ill-articulated, discussions on Venezuela's political system. Article by Felice Castillo describing organizational and legal aspects of electoral administration is very useful.

3273 **Seguridad personal: un asalto al tema.**
Compilación de Juan Carlos Navarro, Rogelio Pérez Perdoma, con la colaboración de Tibisay Lucena. Caracas: Ediciones IESA, 1991. 285 p.: bibl., ill.

Includes five major papers and briefer contributions presented at a conference on personal safety at the Instituto de Estudios Superiores de Administración. The general theme is that weak policy design, poor implementation capabilities of Venezuelan police and correctional institutions, and the courts' inappropriate law enforcement are critical factors in explaining the high degree of criminal behavior.

3274 **Urdaneta, Alberto; Leopoldo Martínez Olavarría;** and **Margarita López Maya.** Venezuela: centralización y descentralización del Estado. Caracas: CENDES, 1990. 88 p.: bibl., map. (Col. Luis Lander; 1)

Explores centralization and decentralization trends during the 1980s. Examines three broad themes: political centralization, proposals of the Comisión para la Reforma del Estado, and decentralization models.

3275 **Velásquez, Andrés.** Tres entrevistas con Andrés Velásquez: 1986, 1990, 1991. Entrevistado por Farruco. Caracas: Ediciones del Agua Mansa, 1992. 256 p.: ill. (Col. Radical)

Like Fujimori in Peru and Perot in the US, the Radical Cause (Causa R) ascribes all the nation's ills to politicians, whom it derisively refers to as "the political class." This compilation of interviews provides insights into the political ideology of Andrés Velásquez, the leader of the movement who won a congressional seat in 1988.

3276 **Velásquez, Ramón J.** *et al.* Cuándo se jodió Venezuela. Prólogo de Ruth Capriles Méndez. Caracas: Consorcio de Ediciones Capriles, 1992. 238 p.; bibl.

Well-known public figures, including interim-President Ramón J. Velásquez, discuss the causes of contemporary economic and social difficulties. Contrasting viewpoints offer interesting reading.

BOLIVIA

EDUARDO GAMARRA, *Associate Professor of Political Science, and Acting Director, Latin American and Caribbean Center, Florida International University, University Park Campus, Miami*

SINCE THE TRANSITION to democracy in the early 1980s, a boom in publications has occurred in Bolivia. Most noteworthy are the numerous essays, articles, and books on democratization written by Bolivian social scientists who have conducted systematic research on the major political trends of the past decade. René Antonio Mayorga (items **3325, 3327,** and **3326**), Roberto Laserna (items **3316, 3313, 3315,** and **3314**), and Fernando Calderón (item **3290**) are undoubtedly the best of the crop. Their essays provide insightful, theoretical, and practical interpretations of the unfolding democratization process and the challenges that lie ahead. Foreign scholars have paid less attention to Bolivia and some provide superficial analyses that contribute little beyond the work of the aforementioned Bolivian analysts. One notable exception continues to be the work produced by James M. Malloy on the problems of democratic governance (item **3322**). Malloy's work on Bolivia spans three decades and his knowledge and analysis sets a standard other analysts have had difficulty matching.

Within the democratization literature, one of the most significant bodies of work engages the issue of institutional reform and administrative decentralization. As one of the most pressing issues affecting democratizing Bolivia, authors have devoted a great deal of their efforts to the study of this topic. Again, the works of Laserna (items **3316, 3313, 3315,** and **3314**) and Calderón (item **3290**) are worthy of note. Authors studying democratization have focused much of their attention on political parties and elections. Since 1985 Bolivia has held three national and five municipal electoral rounds; as a result, ample material has been generated concerning those processes. On national elections, the most representative work can be found in Roberto Laserna's piece on the 1989 elections (item **3315**). Works on municipal elections are incipient, and journal articles represented in this *HLAS* volume are mainly descriptive.

Another major trend in Bolivian publications has to do with the cultivation of coca leaves and narcotics trafficking. This issue has become the single most important topic affecting that nation's international relations (see also p. 575). Numerous articles and books in both Spanish and English have appeared in recent years; the most outstanding authors are Jorge Malamud Gotti and Raúl Barrios Morón (items **3321** and **3283**). The bulk of the literature produced on Bolivia's drug industry, however, remains largely superficial and anecdotal, and provides few insights about the coca and cocaine industry. Most resort to the usual litany of complaints about current drug policy and few provide any alternatives.

This *Handbook* chapter also includes a set of pamphlets and other political-party documents. Bolivian parties have always produced a fascinating array of documents ranging from the speeches of their leaders to party platforms and histories. These documents are especially useful for specialists on elections, political parties, and democracy in general. Other annotations concern transcripts of labor union congresses that provide a written record of the activities of Bolivian organized labor in the last decade.

3277 Aguilar Gómez, Aníbal *et al.* Debate regional: coca por desarrollo y militarización. La Paz: ILDIS; CERES, 1991. 1 v.

Edited booklet version of paper by Aníbal Aguilar Gómez, one of Bolivia's leading experts on coca-production and cocaine-trafficking and an adviser to the current government. Provides good critical summary of the evolution of counternarcotics policy and also warns about dangers of militarizing the drug war in the Andean region. Two commentaries on Aguilar's paper are appended.

3278 "Ama sua, ama llulla, ama quella:" entrevista con el General Hugo Banzer Suárez, líder de la ADN y ex-Presidente de Bolivia, y Guillermo Fortún Suárez, principal ideólogo del partido. (*Am. Lat./Moscow*, 11, nov. 1990, p. 40–44)

Magazine interview published in former Soviet Union with former dictator Hugo Banzer Suárez and his party chief, Guillermo Fortún Suárez. Worth noting chiefly because of Banzer's claims that his party, Acción Democrática y Nacionalista, is centrist; that he respects Fidel Castro; and that the Bolivian government (1989–93) was doing its utmost to fight drug trafficking.

3279 Añez Ribera, Lucio. Antecedentes históricos de la legislación de la coca: 1827–1983. La Paz: Diagrama, 1991. 158 p.

Good compilation of the most significant pieces of legislation approved by Bolivian governments since 1827 to regulate coca and cocaine. Includes legislation signed by former Presidents Banzer, Siles Zuazo, and Paz Estenssoro. Author, retired general, is former head of Bolivia's Fuerza Especial de Lucha Contra el Narcotráfico (FELCN).

3280 Aquino Huerta, Armando. Legislación y procedimiento en narcotráfico. v. 1. Prologado por Edgar Oblitas Fernández. La Paz: EDVIL, 1990- 1 v.: bibl., ill.

Includes compilation of legislation pertaining to drug-related offenses. Surveys Law 1008 and several significant decrees regulating Bolivia's counternarcotics efforts.

3281 Araníbar, Antonio; Chacho González; and **Pablo Ramos Sánchez.** Encuentro sobre la problemática nacional. La Paz: Imprenta de la Univ. Mayor de San Andrés, 1987. 109 p.

Transcripts of 1987 seminar sponsored by the Movimiento Bolivia Libre to analyze Bolivia's economic, social and political situation two years after the launching of neoliberal austerity measures that ended a record-breaking hyperinflation rate. Seminar participants are drawn mainly from left-of-center political parties who engage in an interesting dialogue regarding the crisis of the

Bolivian left. Nevertheless, conclusions are predictable: combat neoliberalism; renege on foreign debt payments; and construct a viable socialist state.

3282 Area promoción de la mujer: partidos políticos; ¿espacios de participación democrática para la mujer? La Paz: Fundación San Gabriel; Asociación de Centros de Madres de la Zona Este, 1992. 93 p.: bibl.

Record of seminar organized to commemorate International Women's Day. Representatives of a significant number of political parties analyze the role of women in their respective organizations. Although consisting mostly of prepared party documents, this is a useful publication about the growing significance of women in Bolivian politics.

3283 Barrios Morón, Raúl. Bolivia & Estados Unidos: democracia, derechos humanos y narcotráfico, 1980–1982. La Paz: HISBOL; FLACSO, 1989. 203 p.: bibl.

Bolivia's leading foreign policy analyst provides the most comprehensive and well-researched book on the early 1980s drug war. Includes very useful background up to July 17, 1980 and the international context of the military takeover, followed by an analysis of the responses of the Carter and Reagan Administrations to the military government. In one of the most lucid accounts of the period, the author discusses impact of international isolation stemming from US policies on the domestic political arena. Finally, Barrios provides a very interesting account of the role of US Ambassador Edwin Corr during the transition to democracy in 1982.

3284 Bedregal Gutiérrez, Guillermo. Declaración programática del Movimiento Nacionalista Revolucionario: presentada a la XV Convención del Partido. La Paz: Tall. Gráf. de Empresa Editora Urquizo, 1988. 46 p.

Pamphlet provides Movimiento Nacionalista Revolucionario platform presented to the XV National Party Convention in 1988. Useful statement of the evolution of the MNR from the party that led the 1952 revolution to the party that stabilized the economy and ended hyperinflation by implementing the 1985 New Economic Policy. Pamphlet provides interesting set of statements regarding the state of organized labor, drug trafficking, and administrative decentralization.

Bedregal Gutiérrez, Guillermo and **Ruddy Viscarra Pando.** La lucha boliviana contra la agresión del narcotráfico. See item **4229.**

3285 Bedregal Gutiérrez, Guillermo and **Ruddy Viscarra Pando.** La lucha boliviana contra la agresión del narcotráfico. La Paz: Editorial Los Amigos del Libro, 1989. 614 p.: bibl. (Col. Texto y documento)

Valuable book provides very detailed account of Bolivia's view of negotiations with the US (1985–89). Bedregal and Viscarra begin with standard chapter on the coca leaf as an integral part of Bolivian society and traditions. They also provide good account of Bolivia's efforts at a UN Convention (Vienna, 1988). Appendices are also useful references.

3286 Bedregal Gutiérrez, Guillermo. Nación y democracia. La Paz: Editorial Abril, 1990. 63 p.

Collection of short essays on wide variety of topics ranging from a critique of capitalism to the first 100 days of the Paz Zamora Administration (1989–93). Bedregal, one of Bolivia's most versatile politicians and the author of numerous books and articles, does not provide any systematic analytical framework nor does he conduct research to substantiate his arguments. Chiefly a collection of author's personal opinions on numerous subjects.

3287 Bedregal Gutiérrez, Guillermo. Semillas de liberación: hacia una economía social para la democracia. La Paz: Librería Editorial Juventud, 1991. 109 p.: bibl.

Another collection of author's personal opinions on topics ranging from economy to politics. These essays, however, are connected by an underlying theme. Author argues that Bolivia must pursue a social market economy that combines the State's positive role with a market-oriented economy in order to address key social issues.

3288 Bolivia. Congreso Nacional. Cámara de Diputados. La H. Cámara de Diputados discute los antecedentes, el presente y el futuro del narcotráfico en Bolivia. La Paz: s.n., 1987. 1 v.

Provides written transcript of seminar sponsored by Bolivia's Chamber of Deputies (June 10–13, 1987) to debate drug trafficking and coca production. Seminar was organized in response to increasing US pressure resulting from Operation Blast Furnace and confrontations between police and peasant groups in the Chapare region.

3289 **Bolivia. Congreso Nacional. Cámara de Diputados.** Proyecto de Ley de sustancias controladas y discutido en el Seminario-Debate de Cochabamba. La Paz: Honorable Cámara de Diputados, 1988? 1 v.

Copy of bill submitted to Congress which, with some modifications, became known as Law 1008 following its implementation in Dec. 1988. Law 1008 is Bolivia's version of the US zero tolerance ordinances and has become the hallmark of all Bolivian counternarcotics legislation.

3290 **Calderón G., Fernando.** Cuestionados por la sociedad: los partidos en Bolivia. (*in* Los sistemas políticos en América Latina. Coordinación de Lorenzo Meyer y José Luis Reyna. México: Siglo Veintiuno Editores; Tokyo: Univ. de las Naciones Unidas, 1989, p. 197–213)

Brief history of political parties in Bolivia (mid-1940s to mid-1980s) traces evolution of those such as Movimiento Nacionalista Revolucionario (MNR) and discusses emergence of newer ones such as Movimiento Revolucionario Túpac Katari de Liberación (MRTKL). Concludes that parties in Bolivia have become increasingly disconnected from civil society and that the latter has bypassed parties through its own organizations. Calderón highlights private sector and civic committees, which at one time rejected political party affiliation.

3291 **Calla Ortega, Ricardo.** Nueva derecha/vieja casta: algunas hipótesis sobre la modernización política en Bolivia. (*Estad. Soc.*, 6:7, 1990, p. 93–103)

Argues that a "new right" has emerged in Bolivia with characteristics that resemble similar social groups in other Latin American countries. Includes useful sociological description of the new right noting that its members belong mainly to traditional political parties. Its members are also primarily technocrats who were trained in US and European universities, espouse neoliberal policies, and are determined to rule Bolivia.

3292 **Camacho Romero, Juan Carlos.** Luis Sandoval Morón: patriota, revolucionario y combatiente. Santa Cruz, Bolivia: s.n., 1989. 139 p.

Short biography of Luis Sandoval Morón, significant member of the Movimiento Nacionalista Revolucionario who presumably ruled Santa Cruz's eastern dept. during the 1950s. Written by a friend and fellow party member, book glorifies Sandoval Morón and attempts to clear him of alleged wrongdoing. Nevertheless, work does include interesting discussion of topics and issues that outlived Sandoval, such as administrative decentralization.

3293 **Coca-cronología: 100 documentos sobre la problemática de la coca y la lucha contra las drogas, Bolivia, 1986–1992.** La Paz: ILDIS; Cochabamba; CEDIB, 1992. 732 p.: ill.

One of the most useful books published in Bolivia in recent years dealing with the coca problematique and with US-Bolivian relations. No other single source provides all the documents related to US-Bolivian bilateral counternarcotics policies from 1986–92, a period of escalation in the US-led drug war in Bolivia. Includes all bilateral accords, beginning with the 1987 framework agreement that established the structure for all current accords. Additionally, provides a valuable set of Bolivian documents. Highly recommended source for scholars interested in the conduct of Bolivia's drug war.

Conaghan, Catherine M. and **James M. Malloy.** Unsettling statecraft: democracy and neoliberalism in the central Andes. See item **3219.**

3294 **Cortéz Hurtado, Roger.** La guerra de la coca: una sombra sobre los Andes. La Paz: Centro de Información para el Desarrollo; Facultad Latinoamericana de Ciencias Sociales, 1992. 242 p.: bibl.

Primarily articles and speeches delivered by Cortéz, a well-known Socialist politician and former member of the Bolivian Congress who became one of the principal critics of the drug war in Bolivia. Includes valuable history of the evolution of Bolivia's drug war but presents a rather slanted interpretation of US policy. As with other books written by Bolivians, author provides no analysis of US documentation.

3295 **Cortez Romero, Enrique.** Marcelo Quiroga Santa Cruz: pensamiento político; Estado dependiente, FF.AA. y golpe militar en

Bolivia. Santa Cruz, Bolivia: Univ. Autónoma Gabriel René Moreno, 1992. 231 p.: bibl.

Interpretation of the political philosophy of Marcelo Quiroga Santa Cruz, prominent socialist leader brutally assassinated in 1980 by paramilitary squads under direct orders from the Bolivian military. Author provides good synthesis of the evolution of Quiroga's political thought on the 1952 revolution, the military period, and democracy.

3296 De Cartagena a San Antonio, Texas: entre el desarrollo y la interdicción; a propósito de la Segunda Cumbre Presidencial Antidrogas, 26–27 de febrero de 1992. Cochabamba, Bolivia: Centro de Documentación, Información y Biblioteca (CEDIB), 1992. 96 p.: ill.

Good compilation of principal counternarcotics agreements between US and Bolivia. Compiled by CEDIB in 1992, days before the San Antonio Summit, this was the only publication that reproduced a confidential US proposal to the Andean presidents. Among key documents found in this compilation are: the Cartagena Declaration; Andean Initiative; Andean Commission of Jurists Report (evaluation of Cartagena after one year); preliminary US proposal to the Presidents of Bolivia, Colombia, Ecuador, Peru, and Mexico prior to the San Antonio Summit; and the final report of the Second Annual Inter-Parliamentary Meeting.

3297 Democracia y descentralización en Bolivia. Edición de Carlos F. Toranzo Roca. La Paz?: FLACSO; ILDIS, 1988. 199 p.: bibl.

Transcripts of conference on decentralization sponsored by FLACSO and the Instituto Latinoamericano de Investigaciones Sociales (ILDIS) in July 1987. The most important speakers provide thorough introduction to one of the most pressing issues affecting Bolivian democracy. Useful collection that offers short and succinct analysis of critical legal, economic, and political issues affecting administrative decentralization.

3298 Desafíos para la izquierda. Edición de Carlos F. Toranzo Roca. La Paz: ILDIS, 1991. 157 p.

Transcripts of 1991 seminar sponsored by the Instituto Latinoamericano de Investigaciones Sociales (ILDIS) on the state of the Bolivian left in the context of neoliberal stabilization, a profound crisis among leftist political parties, the collapse of organized labor, and the emergence of drug trafficking as the principal ingredient of US-Bolivian relations. Because transcripts are unedited versions of unprepared texts, quality of articles is poor. Nevertheless, the compilation presents some interesting predictions of the left's future by a few of Bolivia's leading social scientists.

3299 Desarrollo alternativo: utopías y realidades. Coordinación de Mario Arrieta Abdalla. La Paz: ILDIS, 1993. 177 p.: ill.

Summary of an ILDIS-sponsored conference entitled Primer Seminario Regional Plan de Desarrollo del Trópico (La Paz, July 1992) which brought together five coca-grower federations (i.e, federaciones de trabajadores del trópico de Cochabamba). Includes good overview of federation campesinos' position.

3300 Diplomacia de la hoja de coca: documentos de información. La Paz: COCAYAPU, 1993. 1 v.

Collection of four documents which Bolivia's government used to launch its worldwide campaign to decriminalize coca leaf. Written by COCAYAPU, an organization staffed by Bolivian and foreign anthropologists and sociologists, these documents provide cultural historical, environmental, economic, and political perspectives on the coca leaf. Along with a coca-leaf lapel pin, these documents were distributed by President Paz Zamora and his entourage during the July 1992 meeting of Iberoamerican Presidents in Spain. This strategy became known as "coca no es cocaína."

3301 Dux, César. Contrarrevolución y resistencia popular en Bolivia. v. 1. La Paz: Ediciones Nueva Cultura, 1988. 1 v.: bibl. (Col. La Realidad boliviana contemporánea; 1)

Critical analysis of the military government headed by Gen. Hugo Banzer Suárez that ruled Bolivia from 1971–78. Reviews period's main events and provides critical opinions of the military's style of governance. Unfortunately, book is ultimately a series of short essays rather than a systematic attempt at explaining the period.

3302 Espejos: el cementerio de los rehabilitados. La Paz?: Honorable Cámara de Diputados, Comisión de Derechos Humanos, República de Bolivia, 1992? 473 p.: ill.

Transcripts of congressional hearings on one of the most significant cases of human rights abuse in recent Bolivian history. Presents judiciary proceedings, forensic reports, and other judicial documents pertaining to the execution of common prisoners by guards at the Espejos Rehabilitation Farm over several years and under democratic rule. Not merely an indictment of government-sponsored repression, but rather a sad report on Bolivia's judicial system.

3303 **Estrategia nacional de desarrollo alternativo 1990.** La Paz: Presidencia de la República, 1990. 1 v.

Official document consists of Paz Zamora's counternarcotics strategy. Developed by team of economists at Unidad de Análisis de Políticas Económicas (UDAPE), a government think tank, this document became the backbone of the official so-called "Estrategia Nacional de Lucha contra el Tráfico Ilícito de Drogas." Government argues that to end the drug war in Bolivia the entire coca-cocaine economy will have to be replaced. This thesis, also known as "coca for development," became the basis for alternative development programs developed by the US and Bolivia following the Feb. 1991 Cartagena Presidential Summit. Includes several useful statistical appendices.

3304 **Federación Sindical Unica de Trabajadores Campesinos de Cochabamba. Congreso Ordinario, 6th, Aiquile, Bolivia, 1993.** Actas. Cochabamba, Bolivia : s.n., 1993. 104 p.: ill.

Complete transcript of the VI Congress of the Federation of Peasant Workers of Cochabamba. Important document for labor historians and others interested in Bolivian peasant unions. Proceedings reveal union's rejection of political parties and government economic policies, as well as internal disputes within the federation.

3305 **Frente Revolucionario de Unidad Campesina (Bolivia).** Documentos políticos del FRUC. La Paz: Ediciones Wiphala, 1991. 90 p.

Brief pamphlet collects several FRUC (Frente Revolucionario de Unidad Campesina) speeches and documents (1989–91). Documents range from speeches on party's platform to anti-electoral declarations and calls for an increase in agrarian reform gains.

3306 **Gamarra, Eduardo.** Entre la droga y la democracia: la cooperación entre Estados Unidos-Bolivia y la lucha contra el narcotráfico. La Paz?: ILDIS, 1994. 218 p.: bibl.

Brief history of US-Bolivian antinarcotic efforts over the past three decades. Regardless of regime type, Bolivian drug policy decisions have been closed to the public, antidemocratic, and dictated by US policymakers and international financial lending institutions. A solid account of the major actors in the war against drugs in Bolivia. [D. Dent]

3307 **Gamarra Zorrilla, José.** Muerte blanca: fiebre de la coca y la cocaína. La Paz: ENLACE, 1991. 142 p.: ill., maps (some col.).

Presents general history of coca and cocaine in Bolivia. Despite 1991 publication date, book relies on dated information and presents a superficial analysis of the drug trade. For example, author neglects to discuss contacts between García Meza's government (1980–81), the armed forces, and the cocaine industry. However, author does provide good descriptions of the Chaparé region and the commercialization of coca.

Graham, Carol. The politics of protecting the poor during adjustment: Bolivia's Emergency Social Fund. See item **1983.**

3308 **Guevara Arze, Walter.** Bases para replantear la revolución nacional: con el Manifiesto de Ayopaya. La Paz: Libreria Editorial Juventud, 1988. 234 p.

Former President Walter Guevara Arze attempts to analyze Bolivian politics in the context of the end of military authoritarianism, transition to democracy, and imposition of neoliberal austerity by the Movimiento Nacionalista Revolucionario (MNR), the party that led the great revolution of the 1950s. While some of the analysis is interesting, book is chiefly a political statement seeking reunification of the MNR after nearly three decades of internal splits under the premise of rethinking the gains of the 1952 National Revolution.

3309 **Hargreaves, Clare.** Snowfields: the war on cocaine in the Andes. New York: Holmes & Meier, 1992. 202 p.: bibl., index, maps.

Describes the drug trade in the Andean region, specifically Bolivian and US involvement. British journalist, who conducted extensive research in Bolivia during 1991, offers useful snapshot of the "war on drugs" two years after its launching as result of former President George Bush's Andean strategy. Covers events through the spring of 1992 and offers the best available published discussion of the linkages between Bolivian politicians and the drug industry. Author's view of Bolivia's political players suggests that linkages are indeed widespread but not as pervasive as in other places. Nevertheless, despite author's access to many politicians, she relies primarily on US Embassy and DEA charges against government officials.

3310 **Irusta Medrano, Gerardo.** Narcotráfico, hablan los arrepentidos: personajes y hechos reales. La Paz: G. Irusta Medrano, 1992. 230 p.: ill.

Account of capture and expulsion from Bolivia of Luis Arce Gómez, Bolivia's nefarious Minister of Interior who protected drug-traffickers in 1980. Also summarizes role played by trafficker Herlan Echavarría who struck a deal with the DEA in which lenient treatment was exchanged for his testimony against Arce Gómez. Also provides very useful summary of declarations of Bolivia's so-called *narco-arrepentidos*, seven major traffickers who turned themselves over to authorities (July-Nov. 1992) in exchange for leniency. The drug lords' declarations shed some light on the nature and structure of Bolivian drug-trafficking organizations.

3311 **Lanza, Gregorio.** Del arrepentimiento a la militarización: los vaivenes de la política antidroga. Entrevistas de Ricardo Zelaya. La Paz: CEEPSA Ediciones, 1990. 26 p.

Document consists of two interviews conducted by journalist Ricardo Zelaya with Gregorio Lanza, a deputy in the lower chamber of Congress. Lanza indicts US policy and criticizes the Paz Zamora Administration for entering into agreements with Washington that would militarize the drug war in Bolivia.

3312 **Lanza, Gregorio.** Relaciones Bolivia-EE.UU.: militarización y anexos firmados en Washington. La Paz: s.n., 1990. 1 v.

Transcript of speech delivered by Lanza on the floor of the Bolivian Chamber of Deputies (June 1990) when the entire cabinet was summoned to explain agreements signed by President Paz Zamora in Washington. Lanza's speech consists of a defense of the alternative development strategy and indicts the "militarization" of the drug war in Bolivia.

3313 **Laserna, Roberto.** La acción social en la coyuntura democrática. (*Síntesis/Madrid*, 14, mayo/agosto 1991, p. 213–262, graph, tables)

One of the most thorough analyses of the behavior of social groups during the democratic period. Laserna provides a comprehensive discussion of the role played by labor, the private sector, and newly emerging social groups, such as regional and neighborhood associations.

3314 **Laserna, Roberto.** La democracia en Bolivia: problemas y perspectivas. (*in* La democracia en América Latina: actualidad y perspectivas. Coordinación de Pablo González Casanova y Marcos Roitman Rosenmann. Madrid: Univ. Complutense de Madrid; México: Centro de Investigaciones Interdisciplinarias en Humanidades, UNAM, 1992, p. 203–229)

Interpretation of Bolivia's democratic process describes principal components and outlines future difficulties and challenges. Well-known sociologist Laserna provides historically grounded interpretation of the authoritarian period, transition to democracy, and current phase of democratization. Overview of convoluted process is especially useful for those unfamiliar with Bolivian politics. Provides provocative analysis of problems that may hinder consolidation of democracy in Bolivia. Argues for a more representative political system capable of effectively incorporating key social groups, and finally notes that the most serious threats to democracy are patronage politics, the persistence of *caudillismo*, and drug trafficking. Nevertheless, warns Laserna, misguided US counterdrug policies are as much a threat to democracy as drug traffickers.

3315 **Laserna, Roberto.** 1989: elecciones y democracia en Bolivia. (*Rev. Mex. Sociol.*, 52:4, oct./dic. 1990, p. 205–226, map, tables)

Superb account of the 1989 elections includes good, brief political history of Bolivia. Discusses mechanisms that govern elections, examines principal candidates, and provides blow-by-blow account of the electoral process. Also provides excellent interpretation of the results, noting the significance of the national distribution of votes. Finally, Laserna describes the congressional election of a new president following a three-way tie at the polls and notes role of coalitions in bringing Jaime Paz Zamora to office. Also includes discussion of emergence of populist parties.

3316 **Laserna, Roberto.** Productores de democracia: actores sociales y procesos políticos en Bolivia, 1971–1991. Cochabamba, Bolivia: CERES; FACES, 1992. 280 p.: bibl., ill.

Collection of author's previously published works on authoritarianism, democratization, and social movements in Bolivia. Among the best collections available on Bolivian political history of the last three decades.

3317 **Lazarte Rojas, Jorge.** Ideas para un nuevo sistema electoral en Bolivia. (Estad. Soc., 6:7, primer semestre 1990, p. 105–111)

Proposes six changes to the Bolivian electoral system. Sociologist Lazarte, a member of Bolivia's National Electoral Court, argues that these reforms are necessary to strengthen democracy in the country. A seventh proposal is aimed at political parties with the expectation that someday they will offer better representation of an increasingly plural and heterogeneous society.

3318 **Lazarte Rojas, Jorge.** Movimiento sindical y transformaciones del sistema político boliviano. (in Movimientos sociales y política: el desafío de la democracia en América Latina. Santiago: CES Ediciones; Consejo Latinoamericano de Ciencias Sociales, 1990, p. 77–95)

Attempts to explain the crisis facing Bolivia's labor movement. Faced with uncertainty after the imposition of austerity measures in 1985, labor has failed to react to profound transformations of the political system. Thus, Lazarte argues, organized labor needs to transform itself, to adjust to new political realities of the country, and above all, to find a formula that can transform its demands for social justice into concrete political action.

3319 **Lora, Guillermo.** Lucha revolucionaria de los 60. La Paz?: Ediciones Muela del Diablo, 1991. 35 p.

A 1960s pamphlet by Guillermo Lora, Bolivia's famous Trotskyist leader, discusses the military establishment. Written in the 1960s when the country was still under military rule, this mimeograph provides a critique of the armed forces and proposes a framework for a popular uprising led by the left. Author updated a few footnotes in 1991 to contextualize his analysis of the 1960s.

3320 **Lora, Guillermo.** Rol de la UCS y de Fernández. La Paz: Ediciones La Colmena; Distribuye, Mi Kiosc, 1991. 79 p.

Short essay provides the Partido Obrero Revolucionario's explanation about the emergence of Max Fernández and his party, the Unidad Cívica Solidaridad, in the late 1980s. Lora, the guru of Bolivian Trotskyism, critiques Bolivian electoral politics and explains the emergence of UCS as yet another example of bourgeois posturing to attract popular sector support. Classic example of Lora's political analysis. Heavy on ideology, poor on research, and useful mainly as a reflection of the thought of the old Trotskyist titan of Bolivian politics.

3321 **Malamud Goti, Jaime E.** Smoke and mirrors: the paradox of the drug wars. Boulder, Colo.: Westview Press, 1992. 117 p.: bibl., ill.

Presents a good analysis, although at times superficial, of the "War on Drugs" in Bolivia. Relies principally on anecdotal material from visits to Bolivia as an Argentine diplomat in mid 1980s and follow-up visits in early 1990s. Much of the narrative is based on interviews Malamud held with members of Bolivian counternarcotics forces, the DEA, and Argentine Embassy personnel. On the positive side, conclusions regarding the cocaine war in Bolivia are insightful, and valid criticism is presented concerning policies aimed at escalating repression and interdiction activities. Malamud demonstrates how bureaucratic turf battles between and within US and Bolivian counternarcotics agencies have undermined the drug war's efforts.

3322 **Malloy, James M.** Democracy, economic crisis and the problem of governance: the case of Bolivia. (Stud. Comp. Int. Dev., 26:2, Summer 1991, p. 37–57, bibl.)

Presents broad political history of Bolivia and analyzes context under which democratically elected governments attempted in the mid-1980s to stabilize an economy in extreme crisis. Malloy, the most noteworthy US-based political scientist to study Bolivia, concludes that apart from purely structural factors, the leadership skills of former president Víctor Paz Estenssoro were the key to successfully stabilizing the economy in 1985. Article also raises interesting questions about other Latin American countries and the conflict between the consolidation of democracy and attempts to push through rigid austerity policies.

3323 **Mansilla, H.C.F.** Economía informal e ilegitimidad estatal en Bolivia. (*Nueva Soc.*, 119, mayo/junio 1992, p. 36–44, photos)

Argues convincingly that Bolivia's public bureaucracy has lost the capacity to resolve social problems due to corrupt practices of varying types including drug related offenses. Author notes that political parties, the State, labor, and the private sector have maintained an ambiguous posture toward corruption. As a result, the Bolivian State has no legitimacy and civil society has developed an aversion toward the "political class."

Mansilla, H.C.F. Posibilidades y dilemas de los procesos de democratización en América Latina. See item **2772**.

3324 **Mayorga, René Antonio.** Bolivia: ¿democracia como gobernabilidad? (*in* Estrategias para el desarrollo de la democracia en Perú y América Latina. Recopilación de Julio Cotler. Lima: Instituto de Estudios Peruanos (IEP), 1990, p. 159–193)

Argues as in other articles that in order for democratic consolidation to occur in Bolivia, a modern institutional framework must be developed. Essay provides analysis of governability of problems facing Bolivia in the 1980s in the aftermath of a profound economic crisis that shook the country. Interprets results of the 1989 elections as a fundamental step towards consolidating democracy in Bolivia.

3325 **Mayorga, René Antonio.** ¿De la anomia política al orden democrático?: democracia, Estado y movimiento sindical en Bolivia. La Paz: Centro Boliviano de Estudios Multidisciplinarios, 1991. 309 p.: bibl.

Comprehensive and theoretical examination of the transition to democracy in Bolivia. Surveys dilemmas of transition, politics of economic reform under democracy, and problems of mass participation. Although basically an updated collection of author's previously published essays, book makes an excellent addition to the literature on Bolivian democratization.

3326 **Mayorga, René Antonio.** La democracia en Bolivia: el rol de las elecciones en las fases de transición y consolidación. (*in* Una tarea inconclusa: elecciones y democracia en América Latina, 1988–1991. San José: Instituto Interamericano de Derechos Humanos, Centro de Asesoría y Promoción Electoral, 1992, p. 245–286, tables)

Succinct yet detailed comparative analysis of four general elections in Bolivia. Provides excellent interpretation of electoral politics during the last 15 years. In addition to diagnosing Bolivia's major institutional and constitutional problems, Mayorga offers some suggestions towards the consolidation of democracy therein. Argues that Bolivian democracy has a very weak institutional framework, a fact which only recently drew the attention of Bolivian policymakers.

3327 **Mayorga, René Antonio.** Tendencias y problemas de la consolidación de la democracia en Bolivia. (*Síntesis/Madrid*, 14, mayo/agosto 1991, p. 155–170)

Published version of 1989 Latin American Studies Association paper. Presents excellent analysis of problems confronting Bolivian democracy in the aftermath of the 1989 general elections. Describes changes in Bolivian politics and enumerates institutional problems that must be addressed if democracy is to be consolidated.

3328 **Méndez Valenzuela, Juan** and **Alvaro Pinedo Antezana.** El Condor de los Andes en subasta: el galardón al magnate Stanley Ho. La Paz: Ediciones Nogales, 1992. 187 p.: ill.

First-hand account of how Hong Kong magnate, Stanley Ho, was allegedly tricked by former Bolivian government officials into becoming honorary consul in Macao; donating $500,000 to a Bolivian university that never received the money; and purchasing 70,000 hectares of land he never received. Book's central argument, however, should be weighed against the fact that Méndez was deeply involved in this affair.

3329 **Mesa Gisbert, Carlos D.** Bolivia: Municipales 91; neoliberalismo vs. populismo. (*Nueva Soc.*, 117, enero/feb. 1992, p. 15–19, photos)

Brief essay analyzes the results of 1991 municipal elections. Identifies the principal contestants, analyzes their strengths and weaknesses, and provides a few perceptive insights into the nature of municipal elections in Bolivia. Meza notes significance of two populist parties that emerged in the late 1980s and challenged traditional parties in local elections. More importantly, article stresses the tension between the neoliberal adjustment programs pushed by traditional parties and the greater social spending backed by newly emerging populist parties.

3330 **Montenegro, Carlos; Luis Antezana Ergueta; and Guillermo Bedregal Gutiérrez.** Origen, fundación y futuro del M.N.R. La Paz: Ediciones Abril, 1992. 57 p.

Pamphlet presents speeches and essays by three prominent figures of the Movimiento Nacionalista Revolucionario (MNR), the party that led the revolution in 1952. Montenegro's essay is a classic MNR rebuttal to a 1950 Trotskyist critique of the party. Antezana Ergueta's essay is a 1970s historical interpretation of the party and revolution. Also includes speech delivered to the MNR National Convention by Guillermo Bedregal, current party subchief. These snapshots of the MNR over a 40-year period help explain shifts in party's ideology.

3331 **Movimiento Bolivia Libre. Congreso Nacional, *1st, La Paz, 1991.*** ¡Por una Bolivia libre!: documentos. La Paz: Movimiento Bolivia Libre, 1991. 189 p.

Transcripts of MBL's first party convention (1991). Important because they provide a record of the MBL's platform for the 1992 municipal and 1993 general elections. After 1993, the MBL joined the MNR in the government and has since abandoned its traditional leftist position.

Palza Medina, Javier. La coca en la construcción nacional. See item **1992.**

3332 **Paz Zamora, Jaime.** Diálogo con el pueblo. La Paz?: Movimiento de la Izquierda Revolucionaria, Secretaría Nacional Obrera, 1990. 38 p.

Transcripts of speech delivered by then President Jaime Paz Zamora (Jan. 11, 1990). Pamphlet includes four executive decrees signed by Paz Zamora in which he delineates his government's economic policy.

3333 **Peña Cazas, Waldo.** El lenguaje político en Bolivia: guía para entender al oficialismo y a la oposición. Cochabamba, Bolivia: Ediciones Runa, 1991. 128 p. (Col. Esta América; 6)

Attempts to explain uses and misuses of language in Bolivian politics. Sociolinguist ridicules Bolivian "double speak" and the lack of originality of political clichés and slogans. Some passages are informative and provide insight into how politicians can manipulate language.

3334 **Pozzo Medina, Julio.** Geopolítica y descentralización: zonificación integrada del Estado boliviano. La Paz: Librerías Don Bosco, Gisbert y El Ateneo, 1992. 162 p.: bibl., col. maps.

Bolivian army general proposes administrative decentralization in Bolivia based on a "geopolitical integrating triangle" that would divide the country into three federal administrative zones. Proposal never went beyond book's publication and never became part of national debate over decentralization that occurred in Bolivia in the past decade.

3335 **Programa del Partido Obrero Revolucionario: aprobado en el 32 Congreso.** La Paz: Partido Obrero Revolucionario, 1991. 87 p.

Pamphlet consists of platform of the Partido Obrero Revolucionario presented at its 32nd Congress (July 1991). Interesting political statement from the most resilient of the few remaining Trotskyist parties in Bolivia and Latin America. Platform includes usual calls for a worker-peasant insurrection and the establishment of a United Socialist States of Latin America. Pamphlet's significance is mainly historical as POR's membership dwindles and as Guillermo Lora, its octogenarian leader, approaches the twilight of his life.

3336 **Queiser Morales, Waltrud.** Bolivia: land of struggle. Boulder, Colo.: Westview Press, 1992. 234 p.: ill., map.

Broad overview of Bolivian history based mainly on secondary material covers colonial era and 120 years of republican rule. Paradoxically, while book's historical approach is its strong point, it also accounts for a major flaw. Because the author is a political scientist, one would expect a contemporary section adding to the current body of knowledge about Bolivian politics. However book's ambitious historical review leaves little room for any indepth examination of the democratization period.

3337 **Quintana Condarco, Raúl de la.** Breves reseñas sobre periódicos bolivianos. La Paz: Sindicato de Trabajadores de la Prensa de La Paz, Centro de Estudios de la Información y la Comunicación, 1992. 42 p.: bibl. (Cuadernos de estudio; 3)

Short book provides brief history of Bolivian newspaper journalism (mid 1800s–1920s). Demonstrates that—at least in La Paz—there was an abundant and often lively press. Useful document for press historians and students of Latin American journalism.

3338 **Quiroga T., José Antonio.** Coca/cocaína: una visión boliviana. La Paz: Asociación de Instituciones de Promoción y Educación, Programa de Contención Migratoria; Centro de Estudios del Desarrollo Laboral y Agrario, Centro de Información para el Desarrollo, 1990. 129 p.: bibl.

Introduction claims that the study's objective is to prepare a basic document to guide future research on coca and cocaine in Bolivia. Two chapters deal exclusively with the production and consumption of the coca-leaf and cocaine. Includes description of Bolivia's coca growing regions. Final two chapters assess impact of both coca and cocaine on Bolivian society. Concludes with weak analysis of Bolivian government policies.

3339 **Rivera Pizarro, Alberto.** ¿Qué sabemos del Chaparé? Cochabamba, Bolivia: CERES-CLACSO, 1991. 1 v.

Excellent, comprehensive description of the demography and land tenure patterns of Chaparé, Bolivia's principle coca-producing region. Also includes good section on the structure of Chaparé's coca grower's unions. Funded by the Institute for the Anthropology of Development in Binghamton, New York, this is one of the best available studies on Chaparé.

3340 **Roddick, Jacqueline and N.C.M. van Niekerk.** Bolivia. (*in* The State, industrial relations and the labour movement in Latin America. Edited by Jean Carrière, Nigel Haworth and Jacqueline Roddick. New York: St. Martin's Press, 1989, v. 1, p. 128–177, bibl., tables)

Detailed history of Bolivian organized labor traces its evolution through the MNR period, years of military rule, and the first democratic government of the early 1980s. Article provides good account of period although it relies heavily on some outdated material.

3341 **Rolón Anaya, Mario.** Política y partidos en Bolivia. 2. ed. La Paz: Librería Editorial Juventud, 1987. 739 p.: bibl.

Updated edition of 1967 book traces history of Bolivia's political party system. Useful work for first year political science students of Bolivia and for Latin American specialists interested in political institutions. Main contribution is description of political parties in Bolivia since the 19th century. Writing however is not conducive to easy reading, so general historical information should be sought elsewhere. Groups parties by ideology and provides very good summary of party history and party platforms. At least 70 new parties emerged in Bolivia since the mid 1960s, complicating the author's job. Work is now somewhat dated as several political parties have been established since book's publication.

3342 **Romero Pittari, Salvador.** Actores y estrategias en la reforma del Estado boliviano. La Paz: FLACSO, Programa Bolivia, 1988. 26 p. (Documento de Trabajo; 25)

Paper presented at conference on State reform (Bogotá, 1988). Provides interpretation of the crisis of Bolivian states and concomitant emergence of new social actors. Author notes that with the 1982 return to democracy, a new set of relations developed between the Bolivian State and society. Although author's discussion of electoral cycles is somewhat confusing, it is still valuable as one of the few works to address the significance of municipal elections during the democratization process. Also notes types of alliances fostered between political parties and "civic committees." Romero's article is especially good at addressing the role of the civic committees and the importance of regionalism in modern Bolivian politics.

3343 **Romero Pittari, Salvador.** El nuevo regionalismo. La Paz: FLACSO, Programa Bolivia, 1989. 31 p.: tables. (Documento de Trabajo; 32)

Brief history of Bolivian regionalism. Discusses factors that led to its emergence and its sustainment over the past 170 years. Romero notes changes that occurred in the regional movement after the 1952 Revolution and during the military dictatorships (1960s–70s). This "new regionalism" consolidated in the 1980s under democratic rule is characterized mainly by the presence of strong regional "civic committees." Regional groups have demanded administrative decentralization and the strengthening of local government.

3344 Rueda Peña, Mario. Un debate entre gitanos: crónica de una entrevista memorable. Entrevista de Carlos D. Mesa Gisbert. La Paz: Banco Boliviano Americano; Periodistas Asociados Televisión, 1991? 255 p.: bibl., ill.

Compilation of essays about one of the most noteworthy television interviews in recent Bolivian history. Carlos D. Mesa Gisbert, Bolivia's best known television anchor, interviewed Mario Rueda Peña, minister of information, during a critical stage of the Paz Zamora Administration. Includes interview transcript, press articles, and commissioned essays about the interview.

3345 Salazar Paredes, Fernando. El Parlamento Boliviano frente a la problemática coca-cocaína. La Paz: Ediciones CERID, 1992. 61 p.: bibl., facsim.

Very superficial description of the role of Bolivia's National Congress in legislating and overseeing coca and cocaine-related matters. Concludes that Congress follows the Executive's initiative on these issues, chiefly providing oversight. Book, not based on any substantive research, consists primarily of author's opinions on the role of Congress.

3346 Seminario de Periodismo, 1st, Cochabamba, Bolivia, 1990. Cochabamba 2000. Cochabamba, Bolivia: Oficialía Mayor de Cultura, Honorable Municipalidad de Cochabamba, 1990. 223 p.: ill.

Transcripts of conference on the state of Bolivian journalism. Some of the country's best-known journalists were joined by other outstanding Latin American journalists in a discussion of the media's role in an era of democratization and market-oriented reforms. Includes a few excellent presentations, such as those by Mesa Gisbert, Claure, and Sabat.

3347 Sheinin, David. The 1989 Bolivian national election: the bankruptcy of the center-left as the revolution's legacy. (*MACLAS Lat. Am. Essays*, 4, 1991?, p. 217–229, table)

Overview of 1989 elections and their outcome. Author provides general political history and chronological account of the electoral process. Also includes superficial—and at times biased—account of the electoral process and its aftermath. Essay relies on outdated view of Bolivian politics and misses the nuances found in other essays about the same process (e.g., see Laserna, item **3315**, and Mayorga, item **3327**).

3348 Sindicalismo latinoamericano: el desafío del cambio. La Paz: ILDIS, 1991. 72 p.: bibl.

Transcripts of a seminar on the state of organized labor in Latin America and Bolivia. Julio Godio provides a useful interpretation examining Latin America in its entirety and providing perspectives of organized labor in the 1990s. Jorge Lazarte, Bolivia's best analyst of labor issues, examines the situation therein. As is the case with most transcripts, quality of the essays varies.

3349 Soruco, Juan Cristóbal. Bolivia: ocaso de un ciclo histórico. (*Nueva Soc.*, 94, marzo/abril 1988, p. 13–20, photo)

Analysis of 1987 municipal elections in Bolivia. Provides interpretation of results and their impact on regions, political parties, and the labor movement.

3350 Tapia Montaño, Rafael. Conflicto de baja intensidad: análisis y propuesta; operaciones en perspectiva. La Paz: Editorial Julio Méndez, 1992. 143 p.: bibl., ill.

Author, a Bolivian general, describes book, more or less accurately, as a military-technical manual. Aim of work is to update all national security matters, "not as part of a military-police effort, but to prevent conflictual situations." First part provides confusing discussion of low-intensity conflict theory and adds little to existing literature. Rest of book includes excellent compilation of national security legislation and bilateral accords on counternarcotics matters signed by the Bolivian government with several nations.

3351 Toranzo Roca, Carlos F. Democracia y política cconómica en Bolivia. (*in* Escenarios y caminos para América Latina. Recopilación de Mauricio Betancourt y Orlando Gutiérrez. Bogotá?: Fondad-Comité Colombia, 1993, p. 107–130)

Reflections on process of market-oriented reforms and democratization in Bolivia. Evaluates positive and negative aspects of the 1985 Nueva Política Económica that transformed Bolivia's political economy. Following a brief analysis of the role of traditional actors—such as the private entrepreneurs' confederation—Toranzo explores the role of new sociopolitical actors in strengthening Bolivian democracy. Calls for a more democratic process of economic decision-making and for a broadening of popular participation.

3352 **Torrico Flores, Gonzalo.** Un desafío para el siglo XXI. La Paz: Editorial Los Amigos del Libro, 1993. 377 p.

Author served as undersecretary for social defense in the Ministry of Interior during the Paz-Zamora Administration. As a relatively well-known militant of Gen. Banzer's Acción Democrática y Nacionalista, Torrico made every effort to uphold his party's image and to demonstrate its commitment to counter the narcotics trade. Torrico's account of the drug war during the Paz Zamora Administration (1989–93) provides a good history of the period and explains the administration's official motivations of the Paz Zamora government for entering the drug war on US terms. Torrico lauds Paz Zamora's antidrug efforts but overpraises the DEA and other US government agencies. In his rush to defend the policy, he accuses anyone opposed to the US-Bolivia efforts as useful fools for drug traffickers.

3353 **Torrico V., Erick R.** Comunicación, política y emisión ideológica. La Paz: Sindicato de Trabajadores de la Prensa de La Paz, Centro de Estudios de la Información y la Comunicación, 1992. 127 p.: bibl.

Author argues that Bolivia's press has always played a role in legitimizing the ruling classes and the governments they control. Since the 1985 launching of the Nueva Política Económica which halted hyperinflation, Torrico claims that the press' role has intensified in this regard by propagating the notion that freedom of expression is possible only under a free-market economy. Worth reading chiefly as self-criticism of the role of Bolivia's press.

3354 **Verdesoto, Luis** and **Gloria Ardaya Salinas.** Entre la presión y el consenso: escenarios y previsiones para la relación Bolivia—Estados Unidos. La Paz: UDAPEX, Min.RR.EE.; ILDIS, 1993. 269 p.: appendix, bibl.

Among the first attempts at understanding US-Bolivian relations through a broad framework that examines the international context, the agenda for bilateral relations, US assistance programs, and a proposal for an alternative policy. Generally worthwhile interpretation of bilateral relations although a cumbersome writing style detracts from insightful remarks. Based largely on interviews with key actors in both US and Bolivian governments. Some important interviews are reproduced in the appendix but, unfortunately, not those with US actors. Also includes very useful chronology of US-Bolivian relations (1986–93).

3355 **Violencias encubiertas en Bolivia.** v. 1, Cultura y política [de] Silvia Rivera Cusicanqui y Raúl Barrios Morón. Coordinación de Xavier Albó y Raúl Barrios Morón. La Paz: CIPCA; Aruwiyiri, 1993. 1 v.: bibl. (CIPCA; 38)

In the chapter entitled, "La elusiva paz de la democracia boliviana," Barrios presents excellent interpretation of Bolivian politics while providing a broader framework for understanding the emergence of drug-trafficking and violence related to law enforcement. Unlike other works on the subject this chapter relies on solid social science analysis. For sociologist's comment see item **4979.**

3356 **Zapata Pericón, Hugo.** El origen de la corrupción: tecnoburocracía y poder en las instituciones del Estado. La Paz: Univ. Mayor de San Andrés, 1992. 235 p., 1 folded leaf: bibl.

Master's thesis in development sciences argues that a technobureaucratic class controls the Bolivian State and uses political parties and instruments of repression to serve its own class interests. Concludes that acts of corruption go unpunished as a result. Unfortunately author's argument is not persuasive and adds little to existing explanations for the prevalence of corruption in Bolivian politics.

3357 **Zavaleta Mercado, René.** La formación de la conciencia nacional. Cochabamba, Bolivia: Editorial Los Amigos del Libro, 1990. 171 p. (Col. Obras completas)

One of Zavaleta Mercado's classic works on Bolivian political economy. Zavaleta, the late dean of Bolivian social sciences, provides an exhaustive interpretation of Bolivian politics since the 1952 Revolution. Unfortunately, volume is jargon-laden and reads with difficulty. Nevertheless, for those interested in the writings of the most influential sociologist of mid- to late-20th-century Bolivia, this book is required reading.

PERU

DAVID SCOTT PALMER, *Professor of International Relations and Political Science, and Founding Director, Latin American Studies Program, Boston University*

FOR A COUNTRY with so many problems in the late 1980s and early 1990s as to appear on the verge of collapse, Peru's successful turnaround by mid-decade is truly remarkable. A few numbers tell a good part of the story: Inflation peaked in 1990 at the historically unprecedented rate of 7,650 percent (following two years of equally disastrous quadruple digit rates of 1,600 and 2,700 percent), but then dropped steadily, to 139 percent in 1991, 57 percent in 1992, 40 percent in 1993, and 15 percent in 1994. Five successive years of negative net economic growth between 1988–92 were followed in 1993–94 by net positive growth, in 1994 the highest in the hemisphere (about four percent in 1993 and over nine percent in 1994).

Political violence in terms of incidents (I) and deaths (D) rebounded in the late 1980s and continued at unacceptably high levels through 1992 (1988: I = 2,792, D = 1,509; 1989: I = 2,113, D = 2,877; 1990: I = 2,779, D = 3,745; 1991: I = 2,144, D = 3,044; 1992: I = 1,956, D = 2,633). However, after careful police intelligence work brought about the capture of Shining Path leader Abimael Guzmán Reynoso in Sept. 1992, which led in turn to the subsequent roundup of hundreds of other guerrilla cadres, incidents and violence dropped dramatically (1993: I = 1,021, D = 1,187; projection for 1994, based on Jan. through May figures: I = 650, D = 550).

President Alberto Fujimori's *autogolpe* of April 1992 suspended formal democracy and the 1979 Constitution, a move that appeared to play directly into the hands of the insurgents. These individuals, in turn, sought to polarize Peruvian politics as part of their strategy to undermine the institutional order so as to take power themselves. However, a combination of guerrilla leadership hubris and adroit political maneuvering by Fujimori forced Shining Path onto the defensive with the resulting decimation of their cadre, a development which enabled the government to legitimate a new set of rules for the political game. The Nov. 1992 elections placed a majority of Fujimori supporters in the unicameral congress/constitutional assembly, which wrote a new constitution, narrowly approved (53 percent to 47 percent) in a national referendum held in Oct. 1993. Besides reconcentrating power in the central government and imposing the death penalty for terrorism, the 1993 Constitution provided the legal framework for the 1995 elections in which, for the first time in Peru's history, a sitting president could run for immediate reelection to a second term. While many critics decried what they viewed as President Fujimori's authoritarian appropriation of the levers of political power, most of the population (and much of the international community as well) appeared to support the changes as necessary for the restoration of stability and order in Peru.

Unfortunately, given the lag in research and publication, most works annotated below focus overwhelmingly on the multiple problems Peru was facing in the early 1990s rather than on the solutions that began to emerge as mid-decade approached.

3358 **Abugattás A., Juan et al.** Estado y sociedad: relaciones peligrosas. Lima: Centro de Estudios y Promoción del Desarrollo, 1990. 205 p.: bibl.

Leading social-democratic scholars provide path-breaking analysis of the Peruvian state's profound crisis circa 1990. Traces the cause to a centralizing government apparatus that distributed capital rather than income, thereby enriching elites at the expense of the general public and sowing seeds of its own destruction.

3359 **Alegría, Claribel** and **Darwin J. Flakoll.** Fuga de Canto Grande. San Salvador: UCA Editores, 1992. 223 p. (Col. Testigos de la historia; 6)

The gripping "truth is stranger than fiction" tale of Movimiento Revolucionario Túpac Amaru (MRTA) prisoners who carefully planned their July 1990 escape from Peru's new maximum security prison via a tunnel dug by one of the principals from a nearby safe house.

3360 **Avila P., Diana et al.** Perú hoy, en el oscuro sendero de la guerra. Lima: Instituto de Defensa Legal, 1992. 314 p.: ill., index, maps, tables.

IDL's third annual study of political violence in Peru (1991), complete with tables, department maps with major incidents listed by province, and comprehensive analysis of Sendero Luminoso (SL), Movimiento Revolucionario Túpac Amaru (MRTA), counterinsurgency, judicial framework, human rights, drug trafficking, and case studies of violence by region. Richly detailed, valuable contribution.

3361 **Balaguer, Alejandro** and **Verónica Saenz Porras.** Rostros de la guerra = The faces of war. Fotografía de Alejandro Balaguer. Textos de Verónica Saenz Porras. Lima: Peru Reporting, 1993. 135 p.: chiefly ill.

Arresting photographs by the Argentine-born *Caretas* photographer chronicling human tragedies of Peru's insurgency, with accompanying narrative. Includes sections on Ayacucho, the Huallaga Valley, refugees, Ashaninkas, Huancayo, children, and Lima.

3362 **Balbi, Carmen Rosa.** Del golpe del 5 de abril al CCD: los problemas de la transición a la democracia. (*Pretextos/Lima*, 3/4, 1992, p. 41–61, tables)

Compelling, sophisticated, data-rich study of Peruvian politics in 1992. Argues that Fujimori's authoritarian approach had wide popular support because of: 1) his identification with democratic values rather than procedures; 2) his ethnicity; and 3) the efficiency of government counter-terrorism efforts. Author indicates strong doubts for future political stability. Outstanding analysis.

3363 **Bernales B., Enrique et al.** Actual escenario internacional y la defensa nacional. Edición de Alejandro Deustua Caravedo. Lima: CEPEI, 1992. 235 p.: bibl. (Serie Seminarios, mesas redondas y conferencias, 1017–5121; 12)

Overview of the country's difficult security situation as of early 1992 and challenges posed for civil-military relations, written by several of Peru's most important military analysts, including retired generals, civilian scholars, and a senator.

3364 **Calderón Cockburn, Julio** and **Rocío Valdeavellano.** Izquierda y democracia: entre la utopía y la realidad: tres municipios en Lima. Lima: Instituto de Desarrollo Urbano, CENCA, 1991. 224 p.: bibl.

Comprehensive analysis of local-level activities of leftist organizations in three Lima neighborhoods: San Martín de Porras (1981–89), Comas (1981–89), and El Agustino (1988–91). While critical of representative democracy at the national level, finds that the left made valuable contributions at the grassroots level.

3365 **El camino de la revolución peruana: documentos del II C.C. del MRTA.** Lima: Empresa Editora, Prensa Peruana, 1988. 83 p.: ill.

Full text of documents approved by the second Movimiento Revolucionario Túpac Amaru (MRTA) central committee meeting of Aug. 1988. Includes MRTA's interpretations of Peruvian class struggle, history, society, the revolution, and the political situation, along with the party's program, statutes, and political-military plan.

3366 **Castro Contreras, Jaime.** Violencia política y subversión en el Perú, 1924–1965. Lima: Editorial San Marcos, 1992? 130 p.: bibl.

Advances and documents argument that the Sendero Luminoso and Movimiento Revolucionario Túpac Amaru (MRTA) insurgencies of the 1980s represent the most recent manifestations of socialist armed struggles that have been present in Peru since the 1920s as political alternatives to elections or *golpes.*

Conaghan, Catherine M. and **James M. Malloy.** Unsettling statecraft: democracy and neoliberalism in the central Andes. See item **3219.**

3367 **Construir la paz: la violencia en la vida de los pobladores del Cono Norte de Lima.** Lima: Democracia y Socialismo, Centro de Investigación, Publicaciones y Educación Popular, 1989. 80 p. (Serie Estrategia integral de paz)

Based on surveys and workshops, views of 250 community leaders regarding political violence and local-level coping strategies as of Nov. 1988. For 40 percent, the most significant problem was the economic crisis; for 15 percent, it was violence. Includes presentations by Izquierda Unida (IU) mayors (Medina, Solis, Moreno) and senators (Bernales, Ames).

3368 **Cotler, Julio.** Descomposición política y autoritarismo en el Perú. Lima: Instituto de Estudios Peruanos, 1993. 34 p.: bibl. (Serie Sociología y política; 7. Documento de trabajo; 51)

Incisive overview of Peru's inability to develop democratic practices after 1980. Attributes failure to "pre-modern" institutional development and personalistic/clientelistic practices of key leaders. Views Fujimori's anti-institutionalist and neo-patrimonialist stance as both the president's and Peru's Achilles' heel, with a military coup possible.

3369 **Degregori, Carlos Iván.** Jóvenes y campesinos ante la violencia política: Ayacucho, 1980–1983. (*in* Coloquio Internacional del Grupo de Trabajo "Historia y Antropología Andinas," 2nd, Quito, 1990. Poder y violencia en los Andes. Edición de Henrique Urbano y Mirko Lauer. Cusco, Peru: Centro de Estudios Regionales Andinos Bartolomé de Las Casas, 1991, p. 395–417, appendix, bibl., tables)

Valuable analysis of Shining Path's early years. Presents hypothesis that it had only limited presence and appeal in the Ayacucho countryside before 1980. Its transformation into a significant political force occurred only after declaring the "People's War," and was largely due to gross government ineptness.

3370 **Degregori, Carlos Iván** and **Carlos Rivera.** Peru 1980–1993: fuerzas armadas, subversión, y democracia, 1980–1993. Lima: Instituto de Estudios Peruanos, 1993. 28 p.: bibl. (Documentos de Política; 5. Documento de Trabajo; 53)

Definitive, detailed, and theoretically-grounded analysis of Peru's civil-military dynamic convincingly demonstrating the progressive regression of the principle of civilian control of the military under formal democracy as well as after the *autogolpe.* Result is growing polarization of the military and militarization of society.

3371 **Durand, Francisco.** Business and politics in Peru: the State and the national bourgeoisie. Boulder, Colo.: Westview Press, 1994. 227 p.: bibl., index.

Important study of the political role and political evolution of Peru's national bourgeoisie (1960–90), based largely on interviews and business associations' materials. Identifies distinctive segments of the bourgeosie and traces growing political influence of the *grupos* in governing coalitions after the return to democracy in 1980.

3372 **Eyzaguirre, Graciela.** Los escenarios de la guerra en la región Cáceres. (*Allpanchis/Cusco*, 23:39, primer semestre 1992, p. 155–180, tables)

Detailed analysis of political violence and its historical context in Junín, Pasco, and Huánuco from 1989–91. These areas accounted for about 25 percent of all Sendero activity nationwide during the period studied. Also studies the shift of insurgents to the defensive in 1991 in response to the military's counter-offensive and widespread peasant opposition.

3373 **Gagnon, Mariano; William Hoffer;** and **Marilyn Mona Hoffer.** Warriors in Eden. New York: William Morrow and Co., 1993. 319 p.: ill., index, maps.

Compelling autobiography of a New Hampshire-born, Peru-educated Franciscan priest in the Cutivirene jungle mission of Ashaninka Indians, from his arrival in 1969 to his 1990 evacuation with surviving parishioners after savage Sendero attacks. Gagnon's struggles with the bureaucracies of Peru, the US, and the Catholic Church are equally poignant.

3374 **García Pérez, Alan.** La revolución regional. Lima: Editorial e Impr. Desa, 1990. 341 p.: maps.

President García lays out the rationale (partly ideological justification), key legislation, and physical characteristics for each administrative region. Largely reversed by 1993 Constitution.

3375 **Gaupp, Peter.** The sorry fate of Peruvian peasants—in pictures. (*Swiss Rev. World Aff.*, 9, Sept. 1993, p. 16–21, photos)

Using photos and accompanying text, reviews the Sendero insurgency through the folk art of displaced artisans from the peasant community of Sarhua (Víctor Fajardo prov., Ayacucho dept.).

3376 **González, Raúl.** Las heridas del ejército peruano: una entrevista con el General Alberto Arciniega. (*Quehacer/Lima*, 81, enero/feb. 1993, p. 8–15, photos)

Dramatic interview with one of the Peruvian Army's most successful officers in the struggle against Sendero. Arciniega points out the military's internal divisions as well as military leaders' decision to let the army's political role take precedence over fight against the guerrillas.

3377 **Grompone, Romeo.** Perú: la vertiginosa irrupción de Fujimori: buscando las razones de un sorprendente resultado electoral. (*Rev. Mex. Sociol.*, 52:4, oct./dic. 1990, p. 177–203, bibl., graphs)

Comprehensive and sophisticated discussion of Peru's 1990 presidential election, expanding on Rospigliosi (item **3408**) to include effect of informal sector vote and impact of communications through interpersonal networks.

3378 **Grompone, Romeo.** La representación política en la transición democrática peruana. (*in* ¿Qué queda de la representación política? Caracas: Consejo Latinoamericano de Ciencias Sociales; Editorial Nueva Sociedad, 1992, p. 61–92)

Detailed analysis of tension between political organization and true representative democracy in Peru, focusing on Partido Aprista Peruano (APRA) and the left. Concludes that Peru's mass public has not lost its faith in democracy even though parties and other popular organizations have failed to firmly establish democratic procedures and practices.

3379 **Guzmán, Abimael.** "Exclusive" comments by Abimael Guzmán. (*World Aff.*, 156:1, Summer 1993, p. 52–57)

Excerpts from interviews with Shining Path leader taped by Peruvian intelligence personnel shortly after his Sept. 1992 capture. Provides valuable information about his background and views.

3380 **Hinojosa, Iván.** Entre el poder y la ilusión: Pol Pot, Sendero y las utopías campesinas. (*Debate Agrar.*, 15, oct./dic. 1992, p. 69–93)

Systematic comparison of Shining Path and the Communist Party of Kampuchea, emphasizing differences based on geography, settlement patterns, history, and circumstances of their creation. Some similarities appear, mainly in party organization, Maoist and Stalinist ideology, and selective terrorism to gain power.

3381 **Hinojosa, Iván.** Sendero y el espejo camboyano. (*Quehacer/Lima*, 86, nov./dic. 1993, p. 34–39, photos)

Useful summary comparison of Shining Path and the Khmer Rouge, with details of Pol Pot's tactics and career. Writing after Guzmán's capture, author sees possibilities for Sendero's continuation in the form of armed bands (like the Khmer Rouge) with access to illegal resources, occupying territory where government's presence is weak or absent.

3382 **Jaime Barreto, Wilson.** Marketing político: elecciones 1990. Lima: Univ. del Pacífico, Centro de Investigación, 1991. 200 p.: bibl.

Novel application of business marketing principles to Peruvian politics in general and the 1990 presidential campaign in particular. Reviews the "electoral market" (1978–85), "products" (i.e. candidates) to be marketed in 1990, "sales" results, and the future of political marketing in Peru.

3383 **Klaiber, Jeffrey.** The Church in Peru: between terrorism and conservative restraints. (*in* Conflict and competition: the Latin American Church in a changing environment. Edited by Edward L. Cleary and Hannah Stewart-Gambino. Boulder, Colo.: Lynne Rienner Publishers, 1992, p. 87–103)

Seminal description and interpretation of the changing Catholic Church in Peru. Documents growing strength of conservative clergy and the challenge posed by guerrillas. Concludes that progressive priests retain sufficient influence within the Church to play a vital role in Peru's struggle against Sendero.

3384 **Kruijt, Dirk.** Perú: relaciones entre civiles y militares, 1950–1990. (*in* América Latina: militares y sociedad. Coordinación de Dirk Kruijt y Edelberto Torres-Rivas. San José: FLACSO, 1991, vol. 2, p. 29–142, bibl.)

Dutch scholar with extensive experience in Peru provides detailed, quite definitive analysis of the Peruvian military as political actor, with ample data and multiple interviews of key actors.

3385 **Kruijt, Dirk.** La revolución por decreto: Perú durante el gobierno militar. Traducción de R.B. Smith, revisada y autorizada por el autor. 2. ed. en español. San José: FLACSO, Secretaría General; Lima: Mosca Azul Editores, 1991. 344 p.: bibl.

Spanish translation of *Revolutie per decreet* (Amsterdam, 1989, see *HLAS 51: 3787*). This important case study of reforms from above focuses on the social revolution and its effects under regime of Gen. Velasco. Based on interviews, diaries, and secondary social science literature. [R. Hoefte]

3386 **López, Sinesio.** El dios mortal: Estado, sociedad y política en el Perú del siglo XX. Lima: Instituto Democracia y Socialismo, 1991. 263 p.: bibl., ill.

Powerful, articulate critique—backed by extensive data—of the view that Peru's economic, social, and political crises can be largely explained by big, inefficient government. Thesis is that Peru's fundamental problem is that it is too private (i.e. oriented to elite interests) rather than too public.

Luna Vegas, Ricardo. Contribución a la verdadera historia del APRA, 1923–1988. See *HLAS 54:2752*.

3387 **Manitzas, Elena S.** All the minister's men: paramilitary activity in Peru. (*in* Vigilantism and the State in modern Latin America: essays on extralegal violence. New York: Praeger Publishers, 1991, p. 85–103)

Description and analysis of the 1988–90 activities of the Rodrigo Franco Democratic Command (CRF) and its ties to the government, along with those of other, local paramilitary groups. Core document reviewed is the Minority Report of the Congressional Investigative Commission.

3388 **Mauceri, Philip.** Military politics and counter-insurgency in Peru. (*J. Interam. Stud. World Aff.*, 33:4, Winter 1991, p. 83–109, bibl.)

Useful analysis of effect of Shining Path and Movimiento Revolucionario Túpac Amaru (MRTA) insurgencies on the military institution and its changing strategies to combat them. Traces the roots of military's internal divisions over counterinsurgency approaches to its *docenio* in power (1968–80).

3389 **McClintock, Cynthia.** Peru's Fujimori: a caudillo derails democracy. (*Curr. Hist.*, 92:572, March 1993, p. 112–119)

Cogent, detailed presentation and interpretation of Peru's dramatic developments in 1992, including the *autogolpe*, Guzmán's capture, and national constituent assembly/legislative elections. Concludes on the lugubrious note that "democracy will be harder than ever to build in Peru."

3390 **McClintock, Cynthia.** Presidents, messiah, and constitutional breakdowns in Peru. (*in* The failure of presidential democracy. Edited by Juan J. Linz and Arturo Valenzuela. Baltimore: Johns Hopkins Univ. Press, 1994, p. 360–395, bibl., tables)

Discussion of severe limitations of executive-dominated democracy in Peru in the 1980s with reflections on why a parliamentary system would probably not be a satisfactory alternative. Based on local interviews.

3391 **McCormick, Gordon H.** From the sierra to the cities: the urban campaign of the Shining Path. Prepared for the United States Under-Secretary of Defense for Policy. Santa Monica, CA: RAND, 1992. 78 p.: bibl., ill., map. (RAND; R-4150-USDP)

Tight analysis of Sendero's increasing involvement since 1985 in building urban support, particularly in Lima, and the problems encountered. Sees support slowly growing as of early 1992 and concludes that Sendero's capacity in the countryside combined with government incapacity, will be determinate in the revolution's victory. Inaccurate on some of the particulars regarding Peru, but a penetrating analysis of guerrilla strategies and tactics.

3392 **Mejía Navarrete, Julio Víctor.** Estado y municipio en el Perú. Lima: Consejo Nacional de Ciencia y Tecnología, 1990. 224 p.: bibl., ill.

Sober, data-rich comprehensive study of local government in Peru (1892–1987) and its relationship to central government is buttressed by dozens of valuable tables and organizational charts. Author is a strong advocate of increasing local government powers, resources, and responsibilities, all fundamental prerequisites for any true expansion of democracy.

3393 **Melgar, Ricardo.** Los fantasmas del poder: coca, guerrillas y elecciones en Perú. (*Secuencia/México*, 18, sept./dic. 1990, p. 113–134, ill.)

Valuable overview of Peru's political panorama just before the 1990 presidential election. Details dynamics of campaign against the backdrop of US counter-drug and counter-insurgency policies. Includes postscript explaining why no analyst could have predicted the last-minute surge of Fujimori.

3394 **La mujer y la ciudad: propuestas de políticas a los gobiernos locales.** Lima: Servicios Urbanos y Mujeres de Bajos Ingresos (SUMBI), 1989. 80 p.: bibl. (Documento de trabajo)

Analysis of adverse effects of Peru's economic crisis on poor urban women in a Ford Foundation-sponsored study, with proposals for addressing their problems. Includes sections on neighborhood public dining rooms, emergency expansion of the Vaso de Leche program, local government role in child care, and health programs.

3395 **Murakami, Yusuke.** Peru no 1995nen senkyo ni kansuru ichikosatsu = Un análisis de las elecciones generales del Perú en 1995. (*Iberoam./Tokyo*, 17:1, primer semestre 1995, p. 17–34, graphs, tables)

In-depth analysis of Peruvian general elections of Apr. 1995. Examines significance of event as the first election held under the 1993 Constitution after the April 1992 *autogolpe*, and also as an opportunity for the Peruvian people to determine the legitimacy of the Fujimori government. Discusses the Oct. 1993 national referendum; its effect on election strategies for both President Fujimori and opposition leaders; and concludes with the difficulties faced by the re-elected Fujimori government and prospects for Peruvian democratization. [K. Horisaka]

3396 **Palmer, David Scott.** Perú 1992: la sorpresa de abril de Fujimori. (*Estud. Int./Santiago*, 25:99, julio/sept. 1992, p. 378–384)

Brief analysis of Fujimori's *autogolpe* of 1992 and its internal and external political consequences. Author emphasizes careful political planning that preceded the coup and the president's belief that an "institutional coup" would assist in dealing with the economy and Sendero Luminoso. [D. Dent]

3397 **Palmer, David Scott.** Peru, the drug business and Shining Path: between Scylla and Charybdis. (*J. Interam. Stud. World Aff.*, 34:3, Fall 1992, p. 65–89, bibl., table)

Records Peru's depressing development problems, especially with drugs, debt, and insurgency, and provides concrete recommendations for reestablishing some semblance of political order and economic progress. [D. Dent]

3398 **Palmer, David Scott.** Peru's persistent problems. (*Curr. Hist.*, 89:543, Jan. 1990, p. 5–8)

An attempt to understand Peru's ability to maintain formal democratic practices and procedures despite severe economic difficulties, a truculent guerrilla movement, and human rights abuses and violence. [D. Dent]

3399 **Pásara, Luis.** El rol del parlamento: Argentina y Perú. (*Desarro. Econ.*, 32:128, enero/marzo 1993, p. 603–624, bibl.)

Comparative analysis, based on interviews of parallel failures of Peruvian and Argentinean legislatures to fulfill their roles either as decision makers or as reflecting the interplay of the countrys' different social sectors. Views successful performance of those roles as crucial to strengthening democracy in both countries.

3400 **Peñaflor G., Giovanna.** Y el ganador es... (*Quehacer/Lima*, 80, nov./dic. 1992, p. 30–37, photos, tables)

Preliminary analysis of the Nov. 1992 constituent assembly/legislative elections which gave Fujimori supporters a working majority and helped re-legitimate his regime after the *autogolpe*. Key conclusion is that established parties have yet to appreciate "the gulf between their interests and those of most of the population."

3401 El Perú de los 90. Edición de Alfredo Barnechea. Lima: Instituto del Sur para la Cooperación Democrática, 1990. 429 p.: bibl., ill.

Comprehensive, data-rich overview of Peru's domestic and international economic situation as of 1990 and proposals for change by 20 important figures in Peruvian finance, commerce, and industry.

Peru in crisis: dictatorship or democracy? See item **1959**.

Pion-Berlin, David. The ideology of State terror: economic doctrine and political repression in Argentina and Peru. See item **3568**.

3402 Proceso de retorno a la institucionalidad democrática en el Perú. Edición de Eduardo Ferrero Costa. Lima: CEPEI, 1992. 276 p. (Serie Seminarios, mesas redondas y conferencias, 1017–5121; 14)

Valuable collected interpretations by leading Peruvian scholars and statesmen at a July 1992 CEPEI conference reflecting on the *autogolpe* and possibilities for democratic restoration. Includes key government documents and statements from April 5–Sept. 1, 1992, along with responses of interest groups and political parties.

3403 Prospects for democracy & peace in Peru. Washington: Washington Office on Latin America, 1993. 45 p.: appendices. (Conference briefing series)

Conference co-sponsored by the Washington Office on Latin America (WOLA) and George Washington Univ. on April 28, 1993. Short but incisive analysis as of early 1993 of Peru's politics, economy, post-Guzmán Shining Path, and political violence by important Peruvian scholars-practitioners (Rolando Ames, Felipe Ortiz de Zevallos, retired Col. José Bailetti, and Enrique Bernales).

3404 Rénique, José Luis. La batalla por Puno: violencia política en la Sierra del Perú. (*Hist. Soc./Río Piedras*, 4, 1991, p. 107–137, ill.)

Definitive case study of the political economy of Puno (1960–90), rich in details of the military government's agrarian reform, NGO presence, and the rise of an elected left alternative Partido Unificado Mariateguista (PUM) to Shining Path and its "people's war."

3405 Reyna, Carlos et al. Sendero: ¿el principio del fin? (*Quehacer/Lima*, 79, sept./oct. 1992, p. 30–53, graph, ill., photos)

Series of short but incisive articles by several leading Peruvian scholars who provide their initial analyses of the impact of Guzmán's capture on Shining Path. Particularly valuable are two analyses of the significant role of women in Sendero.

3406 Rocha V., Alberto. El redescubrimiento de la democracia en Perú: aproximación general al debate en la década de los años ochenta. (*Estud. Cult. Contemp.*, 5:15, 1993, p. 105–138, bibl.)

Helpful presentation and analysis of the Lima-based intellectual wellsprings of Peruvian democracy in the 1980s. Provides political map of Lima's think-tanks, with names, numbers, and contributions of the significant players.

3407 Rodríguez, Yolanda. Los actores sociales y la violencia política en Puno. (*Allpanchis/Cusco*, 23:39, primer semestre 1992, p. 131–154)

Details Sendero's presence in Puno dept. during the 1980s and its 1989–91 expansion from its original center in the provinces of Melgar and Azángaro.

3408 Rospigliosi, Fernando. Las elecciones peruanas de 1990. (*in* Una tarea inconclusa: elecciones y democracia en América Latina, 1988–1991. San José: Instituto Interamericano de Derechos Humanos, Centro de Asesoría y Promoción Electoral, 1992, p. 345–388, appendices, graphs, tables)

Careful analysis by leading Peruvian journalist of the 1990 presidential elections. Explains Fujimori's victory as due to loss of respect for traditional parties, social and ethnic polarization, and political polarization. Includes national-level data and departmental-level interpretation.

3409 Smith, Michael L. Entre dos fuegos: ONG, desarrollo rural y violencia política. Lima: Instituto de Estudios Peruanos, 1992. 151 p.: bibl. (Col. mínima; 26)

Analysis of the role of grass roots support organizations in the 1980s as they tried to respond to rural Peruvians' needs in a context of growing political violence and declining State capacity. Includes case studies of rural development centers at Allpachaka (Ayacucho) and Waqrani (Puno) and their destruction by Sendero.

3410 Spier, Fred. Religious regimes in Peru: religion and State development in a long-term perspective and the effects in the Andean village of Zurite. Amsterdam: Amsterdam Univ. Press, 1994. 328 p.: bibl., ill., index, maps.

Important, innovative, inter-disciplinary study of religion and politics from Peru's earliest known history (8000 BC) to 1991. Zurite, a village near Cusco, provides setting for analysis of State formation and development, relations between Church and State, and the role of various religious groups. [R. Hoefte]

3411 Starn, Orin. Sendero, soldados y ronderos en el Mantaro. (*Quehacer*/Lima, 74, nov./dic. 1991, p. 60–68, bibl., map, photos)

Valuable case study of the military's successful counterinsurgency strategy in this key central Highlands region. By maintaining a presence, organizing defense committees in willing peasant communities, and providing arms and rudimentary training, army units helped anti-Sendero peasants force a significant insurgent retreat from Peru's "bread basket" in 1990–91.

Thornberry, Guillermo. Un caso de internacionalización de conflictos regionales: narcotráfico, guerrilla y conservación ecológica en la Amazonia peruana. See item **4316**.

3412 Valdés Palacio, Arturo. Una revolución itinerante. Lima: INPET, 1989. 403 p.: appendix, bibl., ill.

First-person narrative of the most important nationalization initiatives of Peru's military government from 1968–77, particularly the IPC case, by the reformist lawyer-general who was an original member of Velasco's COAP, secretary of the cabinet, and attorney-general. Includes appendix of pertinent documents.

3413 Valencia Cárdenas, Alberto. Los crímenes de Sendero Luminoso en Ayacucho. Lima: Editorial Impacto, 1992. 129 p.: ill., maps.

Report presented to the United Nations Commission on Human Rights before the *autogolpe* by Ayacucho congressman Partido Aprista Peruano (APRA) provides detailed information on Sendero human rights abuses in Ayacucho, 1981–91. Criticizes human rights NGOs for their scanty coverage of guerrilla abuses compared with those committed by the government.

3414 Vargas Llosa, Alvaro. La contenta barbarie. Barcelona, España: Planeta, 1993. 249 p.: bibl., index. (Col. Documento)

Sub-title provides the core themes: "El fin de la democracia en el Perú y la futura revolución liberal como esperanza de la América Latina." Detailed critical analysis of factors leading to the 1992 *autogolpe* and projection of Peru's problems with democratic governance to the rest of Latin America.

Vargas Llosa, Mario. El pez en el agua. See *HLAS 54:4030*.

3415 Vega-Centeno B., Imelda. Aprismo popular: cultura, religión y política. Lima: CISEPA-PUC; TAREA, 1991. 599 p.: bibl.

Path-breaking study of Partido Aprista Peruano (APRA) supporters and the party's impact on them, based on extensive, open-ended interviews of a national representative sample of up to 168 and using a sophisticated analytical framework to determine core popular values and orientations. Presents data and interpretations in 60 tables and figures. Major contribution.

3416 Vegas Torres, José Martín. Fuerzas policiales, sociedad y constitución. Lima: Instituto de Defensa Legal, 1990. 135 p.: bibl.

Useful overview of the history of police in Peru, its perceived image, legal attributes and functions, and mechanisms for controlling activities. While author sees no necessary incompatibility between police and respect for human rights, he believes major restructuring is necessary for proper police functioning in a democratic society.

3417 Violencia y pacificación en 1991. Lima: Senado de la República, Comisión Especial de Investigación y Estudio sobre la Violencia y Alternativas de Pacificación, 1992. 209 p.: bibl., ill., maps (some col.).

Comprehensive, detailed presentation and analysis of political violence in Peru in

1991 in what turned out to be the fourth and last annual volume published by the Special Senate Commission (headed by Senator Enrique Bernales). The most complete and least subjective coverage available, with many tables and maps.

CHILE

MICHAEL FLEET, *Associate Professor of Political Science, Marquette University*

THIS BIENNIUM'S MATERIALS on Chilean politics continue to focus on the transition to and consolidation of democracy, but they also include a significant number of studies of pre-1973 politics and of the military years. Both Chilean and foreign scholars, it seems, believe that aspects of the country's past will play a role in determining its future.

Several studies of the pre-1973 period are worth singling out. The most impressive and instructive is Moulian's reflections on the Chilean party system from 1932–73 (item **3470**). Also extremely useful is Alan Angell's depiction of the Alessandri, Frei, and Allende governments (item **3420**), each driven by a utopian vision impossible for others to embrace. Paul Sigmund's review of the US government's Chilean policy over the last 30 years offers a similar overview of political forces and outcomes (item **3484**). Scully and Valenzuela, on the other hand, point to the striking (although not entirely unexpected) similarity between the 1988 plebiscite and the 1989 general election results and those of the 1969 and 1973 congressional and the 1970 presidential elections (item **3483**). Also worthy of note is the first of four projected volumes of reminiscences of veteran communist Orlando Millas, covering the period 1932–47 (item **3468**).

Studies of the Pinochet years offer interesting reflections on Pinochet himself and on the structural changes undertaken during his years in power. Both, of course, are likely to condition Chilean economics and politics for years to come. Chilean journalists Raquel Correa and Elizabeth Subercaseaux's 1989 interview of the Captain General succeeds in pressing him on several points and is surprisingly revealing (item **3476**). So too are the responses that Díaz and Devés elicit from their interviewees (item **3431**), and Marras' interviews of four prominent Chilean generals (Baeza, Medina Lois, Toro, and Danús) who worked closely with Pinochet (item **3465**). Codevilla's *Foreign Affairs* article (item **3432**) and Spooner's book (item **3487**) offer more explicit and engaging discussions of Pinochet's policies, although they may leave academic readers somewhat dissatisfied. More useful to the latter are Angell and Pollack's edited book (item **3460**), and Carol Graham's landmark study of approaches to poverty under Pinochet and Aylwin (item **3452**). Huneeus' analysis of 1986 survey data was a standard reference for assessing the regime's strength and the possibilities for mobilizing popular support against it (item **3456**). And, finally, Kay (item **3458**) and Kay and Silva (item **3436**) offer extremely useful assessments of agricultural policy under Pinochet and, to a lesser extent, Aylwin.

Suggestive, perhaps, of a decline in intensity of interest in establishing accountability in matters of human rights violations, this period's materials included only three pieces dealing explicitly with this subject. Luz Arce's account of her years in the DINA (item **3422**) and González and Contreras' account of a working-class

Air Force recruit who worked for 10 years with the Joint Military Intelligence Command (item **3451**) are moving descriptions, extremely engaging because they involve "confessions" of two people who participated directly in torture and other abusive practices for extended periods of time. Also of note is René García's discussion of cases which he tried to pursue before being forced to abandon them and to resign as the judge of Santiago's 20th Criminal Court (item **3446**).

This chapter also contains a number of valuable journalistic portrayals of the transition to democracy. This is the case, for example, with the Spaniard Luis Ignacio López's treatment of the period between the 1988 plebiscite and the 1989 elections (item **3463**) which includes separate chapters on such personalities as Ricardo Lagos, Andrés Zaldívar, Clodomiro Almeyda, Orlando Sáenz, and Manuel Feliú; the Spaniard José Antonio Gurriarán's interviews of Gen. Leigh and Gen. Díaz Estrada, former Minister Madariaga, and former Press Secretary Willoughby, and his discussion of Pinochet's plans for a "soft coup" (item **3454**); and the rambling account of the transition by a former Ministry of Interior Legal Adviser under Allende (item **3492**). In addition, Jeffrey Puryear has written a very informative account of the role of intellectuals (and of the foreign foundations that supported them) in the restoration of democratic rule (item **3478**). Abraham Santibáñez (item **3481**) and *La Campaña del NO vista por sus creadores* (item **3427**) offer fascinating and informative details about the opposition's campaign to defeat Pinochet in the plebiscite, an event which marked the arrival in Chile of modern campaign planning and propaganda techniques. Eduardo Silva analyzes how Pinochet's adoption of "pragmatic" (as opposed to "radical") neoliberalism shored up his relations with Chilean entrepreneurs, forced the opposition to accept the terms of his transition, and then to commit themselves to maintaining those very same policies in the post-military period (item **3485**).

Thoughtful, though divergent perspectives on the Aylwin government and on efforts to deepen and consolidate Chile's newly restored "democratic" institutions also characterize this biennium's materials. The principal controversy concerns Aylwin's decision not to pursue more redistributive economic policies or to push more aggressively to remove the remaining anti-democratic features of the 1980 constitution and to reassert civilian superiority over the military. *Los desafíos del Estado en los años 90* contains a number of essays (in particular Sunkel's) written before Aylwin took office which stress the need to go beyond Pinochet's pragmatic neoliberal policies (item **3435**). Garretón, writing in mid-1991, argues that much more aggressive efforts to limit the military's power are both possible and necessary (item **3449**). Finally, and somewhat ironically, Jilberto Fernández blames the left (and its internationally-supported social science advisors like Garretón) for embracing a weak, social-democratic "third way" that includes neoliberal economics policies (item **3443**).

On the other side of the divide are Aylwin himself (item **3424**), and Valenzuela's relatively favorable mid-1991 assessment (item **3491**), in which he points out that Pinochet's shadow has actually helped the Concertación remain united. In addition, Oscar Godoy brings together a number of observers to reflect on the merits of presidential and parliamentary forms of government (item **3423**); a young Socialist intellectual, Jaime Lizama, argues persuasively that in the post-military period politics will be dominated by forces adept at and financially capable of packaging and projecting images in the mass media rather than those building "bases" in neighborhoods, factories, or universities (item **3462**); and Cristián Bofill has written a revealing account of the 1992 bugging scandal that undermined the presidential candidacies of right-wing politicians Sebastián Piñera and Evelyn Matthei (item **3425**).

3418 **Aggio, Alberto.** Democracia e socialismo: a experiência chilena. São Paulo: Editora UNESP, Fundação para o Desenvolvimento da UNESP, 1993. 170 p.: bibl. (Prismas)

Brief Gramscian analysis of the Allende years by Brazilian social scientist. More interesting, perhaps, as a reflection of directions being taken by younger generation of left-oriented social scientists in Brazil than for its revelations regarding the Allende years. Argues persuasively that the UP's basic strategy was the most "advanced" of any undertaken to that point, but was saddled with limitations of leftist political culture of the period.

Alario, Margarita. Environmental policy enactment under the military: some generalities between Brazil and Chile. See item **4999**.

3419 **Allende Gossens, Salvador.** Obras escogidas: período 1939–1973. Presentación de Víctor Pey C. Prólogo de Joan E. Garcés. Recopilación de Gonzalo Martner. Santiago: Ediciones del Centro de Estudios Políticos Latinoamericanos Simón Bolívar; Madrid: Fundación Presidente Allende, 1992. 671 p.: bibl., ill. (Col. Chile en el siglo veinte; serie I-2)

Selection of Allende's writings and speeches, some of which are published for the first time. Volume begins with speech given when Allende was Minister of Public Health in the Popular Front government of Pedro Aguirre Cerda, includes most of his important speeches and correspondence while president, and ends with his final radio broadcast the day of the coup, just prior to his suicide. Materials included reveal greater consistency than might be expected of public pronouncements of an able and eminently practical political figure such as Allende. The impression of Allende one comes away with, however, is probably less clearly and less reliably defined than those available from the accounts of those who knew him or worked closely with him.

3420 **Angell, Alan.** Chile de Alessandri a Pinochet: en busca de la utopía. Santiago: Editorial Andrés Bello, 1993. 172 p.: bibl.

Thoughtful, well-written analysis of Chilean politics from 1958–90 by widely respected British expert on the Chilean labor movement. Author ably exploits advantages of hindsight to look again at the Alessandri, Frei, and Allende governments. His treatment of the Pinochet regime, whose "utopianism" he juxtaposes to that of the Frei and Allende governments, is less impressive. It goes no further than most Chilean accounts, but is generally competent.

3421 **Angell, Alan.** The transition to democracy in Chile: a model or an exceptional case? (*Parliam. Aff.*, 46:4, Oct. 1993, p. 563–578)

Angell analyzes Chilean democracy in early 1993. Notes that preparations for elections were proceeding without fear for the stability of the system as a whole, but asks how long the thus-far patient poor will wait for more equitable share of the fruits of the country's growth, and whether legitimacy of future governments might not require a more aggressive approach to vestiges of military power. Angell concedes that Chile's economic solvency makes its transition relatively unique, but he argues that all transitional regimes would do well to imitate the honesty of the country's politicians and the responsiveness of its political parties.

3422 **Arce, Luz.** El infierno. Santiago: Planeta, 1993. 397 p.: index.

Extraordinary memoirs of one-time member of President Allende's personal security force (the so-called GAP) who was arrested and tortured by the Dirección de Inteligencia Nacional (DINA), began collaborating and later (late 1970s) became one of its leading operatives, resigned and went into exile in 1985, and returned in 1991 to testify before the Rettig Commission. Her testimony and subsequent declarations were detailed and explicit in terms of the people participating in torture sessions and the methods which they employed. She also offers interesting portraits of ex-DINA head Manuel Contreras and other high-ranking officials.

3423 **Arriagada Herrera, Genaro et al.** Cambio de régimen político: encuentros. Edición de Oscar Godoy Arcaya. Santiago: Ediciones Univ. Católica de Chile, 1992. 375 p.: bibl.

Collection of papers on Chilean presidentialism and possibilities of its replacement by one or another form of parliamentary government. Includes thoughtful introductory essay by Godoy (who sees Chilean presidentialism as inherently polarizing and destructive of consensus-building), an essay by Giovanni Sartori (who opposes pure presidentialist and parliamentary systems with equal intensity), essays on the Chilean presidential system (by Genaro Arriagada, José Luis Cea, and Tomás Moulian), on the mixed French system (by Enrique Barros), and on problems of building coalitions in presidential and parliamentary systems (by Manuel Antonio Garretón and Angel Flisfisch).

3424 Aylwin Azócar, Patricio. La transición chilena: discursos escogidos, marzo 1990-1992. Edición de la Secretaría de Comunicación y Cultura, Ministerio Secretaría General de Gobierno. Santiago: Editorial Andrés Bello, 1992. 500 p.

Collection of President Aylwin's principal speeches during the first two years of his presidency. Although plagued with the limitations of most public presentations of policy-related materials (i.e., their one-sidedness and their inadequate attention to political context), this collection is more useful than most thanks largely to its thematic organization [e.g., it includes sections on accountability for human rights violations, democratization of institutions established by the military regime, economic policy (i.e., "growth with equity"), the environment, and foreign relations].

3425 Bofill, Cristián. Los muchachos impacientes. Prólogo de Roberto Pulido E. Santiago: Editorial Copesa, 1992. 224 p.: ill.

Fast-paced and revealing account of the bugging scandal that undermined the rival presidential candidacies of two of the right-wing Renovación Nacional party's leading younger lights, Sebastián Piñera and Evelyn Matthei. Fascinating glimpse of the political infighting plaguing the Chilean right and of the high stakes and powerful temptations of national politics in the post-military period.

Brito, Alexandra Barahona de. Truth and justice in the consolidation of democracy in Chile and Uruguay. See item **3622.**

3426 Brown, Cynthia G. Human rights and the "politics of agreements:" Chile during President Aylwin's first year. New York: Americas Watch, 1991. 100 p.: bibl. (An Americas Watch report)

Brief assessment of the Aylwin government's handling of human rights issues through mid-1991. Contains useful summary of the Rettig report, its impact, and the military's reaction to it, and the Aylwin government's handling of the politically delicate issue of terrorism.

3427 La campaña del NO vista por sus creadores. Santiago: Ediciones Melquiades, 1989. 188 p.: ill. (Serie Ensayo)

Proceedings from a conference held shortly after the 1988 plebiscite in which opposition leaders, advisers, and technical staffers candidly discuss strategies and tactics that helped them reject Pinochet's bid for an extension of his term, thereby setting the transition to democracy in motion. Of particular interest is Eugenio Tironi's account of the efforts of opposition and social scientists to identify fears and concerns of the potential electorate and to devise a campaign using television and other media that played to people's long repressed desires for social reconciliation and cohesiveness.

3428 Cavallo, Ascanio; Manuel Salazar Salvo; and Oscar Sepúlveda Pacheco. Chile, 1973-1988: la historia oculta del régimen militar. Rev. ed. Santiago: Editorial Antártica, 1989. 608 p.: bibl., ill.

This chronicle of the years of military rule offered Chileans their first real look at the regime's inner workings. It took the country by storm, selling out its first (1988) edition of more than 10,000 copies in a matter of weeks, and winning that year's Interamerican Press Society's Human Rights prize. Written by three of the country's leading opposition journalists (Cavallo wrote for *La Prensa*, the anti-regime daily that the Christian Democratic party helped to launch in 1987), it quickly became a standard reference for subsequent accounts and analyses of the Pinochet regime. Although some of its depictions of crucial conversations and events, including Pinochet's alleged efforts to void the results of the Oct. 1988 plebiscite, have been challenged in accounts by protagonists and others, most have held up remarkably well. And after seven years of free press and civilian rule it remains the most comprehensive, and one of the most measured and reliable, accounts of these years.

3429 Chaparro N., Patricio. La Iglesia cree en la democracia: elementos para una educación política democrática; la perspectiva de los obispos chilenos, 1970–1988. Santiago: Editorial Patris, 1989. 45 p.: bibl. (Serie Nueva evangelización)

Brief essay argues persuasively that the Chilean bishops believe that democracy is the political system that most effectively promotes human rights and citizen participation, and that they are helping to strengthen and perfect it as Chile returns to civilian rule. Chaparro, a layman who teaches at the Jesuit-run Latin American Institute for Doctrine and Social Studies (ILADES), stresses the Church's support for dialogue, a rejection of violence, the search for censensus, political pluralism, and other values favoring democracy. Unfortunately, he does not consider the effects of the bishops' opposition to tolerance or pluralism on moral questions (divorce, birth control, abortion, etc.) that become political issues.

3430 **Chile. Comisión Nacional de Verdad y Reconciliación.** Informe de la Comisión Nacional de Verdad y Reconciliación. v. 1–3. Santiago: Secretaría de Comunicación y Cultura, Ministerio Secretaría General de Gobierno, 1991. 3 v.

Contains detailed accounts of circumstances leading to the deaths of more than 2,200 people as the result of excessive or unwarranted use of force by security forces during the years of military rule. Offers additional information dealing with cases in which compelling or definitive evidence is not available, and includes moving interviews in which relatives of victims express opinions on a variety of subjects. Released in March 1991, its potential impact on public opinion and civil-military relations was greatly reduced by the assassination in early April of Pinochet loyalist Jaime Guzmán and the not implausible fear that pressing matters at that point would possibly provoke a military coup and yet more repression.

3431 **100 chilenos y Pinochet.** Recopilación de Jorge Díaz Saenger y Eduardo Devés V. Santiago: Zig-Zag, 1989. 191 p.: index.

Responses of one hundred Chileans to the question: "What do you think of Pinochet?" Those interviewed included intellectuals, public figures, and collaborators who had access to Pinochet and could claim to know him, and others who did not but may be taken as representative of some segment or current of public opinion, both pro and con. The image of Pinochet that emerges is a mixture of good and bad, of virtue and defect, that is probably closer to the "real" character than most of the portrayals offered by individual authors to this point.

3432 **Codevilla, Angelo.** Is Pinochet the model? (*Foreign Aff.*, 72:5, Nov./Dec. 1993, p. 127–140, photo)

A Hoover Institution Fellow praises Pinochet and his Chicago boy advisors for their successful assault on the Chilean economy's statist structures and practices. He views political costs paid along the way as both excessive and unnecessary, however, and questions whether countries that have not been subjected to the full logic of statism (as was Chile under Allende) will be capable of making an abrupt about-face. More nuanced observers of Chilean and Latin American politics are likely to find this piece insufficiently attentive to the economic costs (e.g., high-level unemployment, disastrously rigid exchange-rate policies, etc.) of radical liberalization schemes, and the continuing need for both stategic and tactical intervention by State authorities.

3433 **Correa, Jorge et al.** Justicia y libertad en Chile. Edición de Guillermo E. Martínez. Santiago: Corporación Libertas, 1992. 104 p.: bibl.

Collection of dispassionately written and relatively accessible essays on the Chilean judicial system and its potential reform. Judicial reform is a highly sensitive political issue because of the passive, subservient role played by the judiciary during the years of military rule, and the demands of many for an accounting of those responsible for violations of human rights. Several of the authors attempt to place reform in a broader historical and geographical context. Unfortunately, they do not help the reader understand the range of positions in the debate or political considerations likely to influence its outcome.

3434 **Cultura, autoritarismo y redemocratización en Chile.** Edición de Manuel Antonio Garretón, Saúl Sosnowski, and Bernardo Subercaseaux. México: Fondo de Cultura Económica, 1993. 303 p.: photos

Stimulating collection of essays assessing impact of cultural activities in Chile (and among Chileans in exile during the years of military rule) on both democratization and the post-military period (under real and imagined threats of a return to authoritarian rule). Contributing authors include social scientists, historians, cultural activists, and artists writing on different cultural phenomena. All treat cultural activities as having their own integrity and not simply as reflections of realities in other spheres (e.g., class, social, political, etc.) or for their impact on these spheres.

3435 **Los desafíos del Estado en los años 90.** Edición de Matías Tagle D. Santiago: Corporación de Promoción Universitaria, 1991. 171 p.: bibl., ill.

Papers presented at a conference in late 1990 attended by representatives from CEPAL (the United Nation's Economic Commission for Latin America and the Caribbean), FLACSO, CIEPLAN, and the Catholic Univ. of Chile's Institute of Political Science. Includes a *tour de force* essay by economist Osvaldo Sunkel attacking neoliberal models and proposing in their stead CEPAL's "neo-structuralist" accommodation of market forces and mechanisms, as well as pieces by Norbert Lechner, Roberto Durán, Oscar Godoy, Dagmar Raczynski, and Oscar Muñoz. Very useful articulation of the logic that led many of the country's leading social scientists (most of the Christian Democrats and Socialists) to a qualified embrace of market-based development strategies. As Muñoz puts it, "the empirical evidence has shown that in the medium run (and perhaps earlier) a subjectivist redistribution that was not based on objective conditions of growth of productivity and employment would end up self-destructing and making matters even worse."

3436 **Development and social change in the Chilean countryside: from the pre-land reform period to the democratic transition.** Edited by Cristóbal Kay and Patricio Silva. Amsterdam: Centre for Latin American Research and Documentation, 1992. 326 p.: bibl. (Latin America studies; 62)

Collection of essays by leading British, US, and Chilean experts (Kay and Silva, David Lehman, Brian Loveman, and Maurice Zeitlin, among others) on Chilean agriculture in the Pre-Frei, Frei, Allende, and Pinochet periods. In addition to offering a wealth of relevant empirical data in a single volume, collection juxtaposes differing views as to the importance of agriculture in the larger economic and political arenas.

3437 **Díaz Corvalán, Eugenio.** Nuevo sindicalismo, viejos problemas: la concertación en Chile. (*Nueva Soc.*, 124, marzo/abril 1993, p. 114–121, ill.)

Very interesting assessment of the first two and a half years of the Aylwin government by a moderate, Christian Left trade union analyst and strategist. Díaz acknowledges that the labor movement must find ways of defending legitimate worker interests without jeopardizing a precarious and vulnerable transition. Author also notes that it will take considerable work and resources to get rank-and-filers to understand changes in basic notions of labor activism required by economic changes sweeping the world of late. He worries that once the initial social pact had been reached, Chilean entrepenuers no longer felt the need to negotiate with the labor movement, and argues that the latter will have to develop greater organizational clout (even at the risk of frightening potential adversaries) and begin winning more tangible and significant benefits for workers. It is interesting to note that this essay was not published in Chile but in Venezuela.

3438 **Drago, Tito.** Chile: un doble secuestro. Madrid: Editorial Complutense, 1993. 241 p.: bibl. (Col. Andreía)

Detailed account of the Allende years by a veteran Argentine journalist who defends Allende as a thoughtful political leader whose social democratic strategy was the most appropriate and potentially viable of the several from which he was forced to choose. Provides valuable coverage of the Unidad Popular's divisions and of crucial moments (e.g., Allende's negotiations with the Partido Demócrata-Cristiana (PDC) and his conversations with General Prats) along the way to his seemingly inexorable end. Like most journalistic narratives, book suffers from insufficient attention to the structures and dynamics by which elite relations and negotiations took place.

3439 **Enríquez Frödden, Edgardo.** Edgardo Enríquez Frödden: testimonio de un destierro. Transcripciones por Jorge Gilbert. Santiago: Mosquito Editores, 1992. 201 p. (Biblioteca setenta & 3)

The memoirs of a former rector of the Univ. de Concepción and Minister of Education under Allende. Interesting primarily for their account of the university reform process under President Frei and for details regarding

Enríquez' sons, Edgardo Jr. and Miguel, who were prominent Movimiento de Izquierda Revolucionaria (MIR) activists with whom he shared "ends but not means."

3440 **Escobar Cerda, Luis.** Mi testimonio. Santiago: Editorial Ver, 1991. 242 p.: bibl.

Unapologetic memoirs of a formerly independent rightist economist, now a self-described Social Democrat, who served Pinochet briefly as his treasury minister.

3441 **Farreras Sanz, Leonor.** Operación "Swaps:" el caso Bardón; ¿escándalos en el Banco del Estado? Santiago?: Editorial La Noria, 1993. 165 p.: index.

Interesting portrayal of the life, times, and excesses of Alvaro Bardón, one of the more colorful "Chicago boys" who moved in and out of a number of important economic policy-making positions during the years of military rule.

3442 **Fernández Huidobro, Eleuterio** and **Graciela Jorge.** Chile roto: uruguayos en Chile, 11/9/1973. Montevideo: Tae, 1993. 279 p.: bibl., ill.

Interesting and surprisingly self-critical account of Allende's fall and the ensuing years of military rule written by Tupamaro militants who were in Chile in semi-exile during Allende's presidency. Offers a radical-left critique of the Unidad Popular (UP) and valuable insights into the organization and internal dynamics of the Chilean Movimiento de Izquierda Revolucionaria (MIR), with which many Uruguayans worked closely.

3443 **Fernández Jilberto, Alex E.** Internationalization and social democratization of politics in Chile. (*in* Social democracy in Latin America: prospects for change. Edited by Menno Vellinga. Boulder, Colo.: Westview Press, 1993, p. 163–185)

Class analysis of the renovation of the Chilean left, and of the Socialist Party in particular, emphasizes structural changes (e.g., increased importance of the export sector, the shrinking and weakening of industrial and rural labor forces) effected during the years of military rule. Author maintains a generally evenhanded tone, but clearly blames much of the left's willingness to embrace a "social democratic" third way position (which somehow embraces or is consistent with neoliberal policies) on FLACSO and other social scientists in the pay of European NGOs, and yet fails to consider seriously whether other policies are either economically or politically viable.

3444 **Frei Montalva, Eduardo.** Eduardo Frei Montalva, 1911–1982: obras escogidas (período 1931–1982). Selección y prólogo de Oscar Pinochet de la Barra. Santiago?: Ediciones del Centro de Estudios Políticos Latinoamericanos Simón Bolívar; Fundación Eduardo Frei Montalva, 1993. 663 p.: port., facsim. (Col. Chile en el siglo veinte; serie I-3)

Collection of the former Christian Democratic president's lectures, essays and speeches from his student days in the early 1930s until his unexpected death in 1982, roughly half-way through the period of military rule. It brings together previously published materials (chapters or excerpts taken from his many books), private correspondence, and unpublished essays and speeches. Of particular interest is the full text of his 1973 letter to Italian Christian Democratic leader Mariano Rumor, in which he explains the Partido Demócrata-Cristiana's condemnation of Allende's government and its willingness to collaborate with the new military government. On this point, some observers (see item **3463**) insist that over the next four or five years Frei reluctantly concluded that the military's "cure" was much worse than the disease it was designed to treat.

3445 **Frías F., Patricio.** Construcción del sindicalismo chileno como actor nacional. v. 1–2. Santiago: Central Unitaria de Trabajadores, CUT; Programa de Economía del Trabajo, PET, 1993. 2 v.: bibl., ill.

Very useful study of the Chilean labor movement, its collapse and partial recovery under the hostile and repressive military government, its role in the transition to democracy, and new dilemmas it faces in the post-military period. Author provides very clear picture of the movement's changing composition (in line with changes in the structure of the Chilean economy) and the tensions between traditional and newer approaches to relations with its entrepreneurial partners. He also offers thoughtful discussion of and suggestions for improving management-labor and State-labor relations within economic and political constraints of civilian rule in the early and mid-1990s.

García Márquez, Gabriel. La aventura de Miguel Littín, clandestino en Chile: un reportaje. See *HLAS 54:3952.*

3446 García Villegas, René. Soy testigo: dictadura, tortura, injusticia. Santiago: Amerinda, 1990. 264 p.

The personal testament of a conscientious judge who served on 20th Criminal Court in Santiago during the late-1980s, and was ultimately removed by Pinochet because of his willingness to actually pursue human rights abuse cases. Most cases that came before him were turned over (on appeal) to military tribunals and were never followed up. In the specific cases that he does review, little is added to what is already known, but his testimony generally confirms accounts kept and released by the *Vicaría de Solidaridad*. A valuable insider's look at the operation of the country's judicial system toward the end of the military period.

3447 Garretón Merino, Manuel Antonio; Marta Lagos; and Roberto N. Méndez. Los chilenos y la democracia: la opinión pública, 1991–1994; informe 1991. Santiago: Ediciones Participa, s.d. 95 p.: appendix, bibl., graphs, tables.

First of a four-part study of public attitudes toward recently restored but still only partially democractic institutions. Based on a 1,500-person national sample (six major urban centers), survey includes questions on the meaning of democracy, the performance of democratic institutions to date and their impact on various aspects of national life, as well as asking whether any changes are needed. Attitudes and concerns are broken down by age, political and ideological position, and socioeconomic level. Most Chileans appear aware that their institutions are not yet fully democratic. Citing evidence of "civic moderation," respect for democratic procedures, absence of high levels of economic expectation, attitudes critical of authoritarianism and its remaining vestiges, and increasing "modern" attitudes, however, authors conclude that Chilean democracy is "well on its way to consolidation." Longstanding tradition of survey research in Chile spearheaded by Eduardo Hamuy in the 1950s is carried on by one of his former associates, Marta Lagos, who lends further value and credibility to the study.

3448 Garretón Merino, Manuel Antonio; Marta Lagos; and Roberto N. Méndez. Los chilenos y la democracia: la opinión pública, 1991–1994; informe 1992. Santiago: Ediciones Participa, s.d. 117 p.: appendix, tables.

Most interesting finding of the second part of a four-part public opinion survey is that fewer Chileans are satisfied (*conforme*) with the performance of their democracy in 1992 (57 percent) than the previous year (59.6 percent). More people thought it weaker than the previous year and in greater need of modification, and the number of people thinking that conflicts between rich and poor, the government and armed forces, and workers and employers were serious (*gran conflictos*, as opposed to *conflictos menores*) increased as well. Prospects of continued and relatively stable civilian rule remain good, however, as most people are prepared to be patient with respect to full democratization.

3449 Garretón Merino, Manuel Antonio. Discutir la "transición:" estrategias y escenarios de la democratización política chilena. Santiago: FLACSO, 1991. 19 p. (Estudios Políticos; 15)

A year and a half into the Aylwin government, Chile's most prolific social scientist indirectly charges it with failing to pursue either the completion of the transition to democracy or the initiation of its consolidation (as he sees it, describing the Aylwin government as a "transition" government assumes something that is not yet clear; that it will successfully carry out the institutional changes necessary for full democracy). According to Garretón, pushing democratization only as far or as fast as extreme right-wing and military elements are willing to accept gives *them* control of the process, and may well preclude reforms and advances that he believes feasible over the long run. While persuasively formulated, his argument for further democratization failed to carry the day; Chilean democracy remains far from full and a long way from effective consolidation.

3450 Garretón Merino, Manuel Antonio. Ni tanto ni tan poco: cambio y continuidad en la política chilena. Santiago: FLACSO, 1992. 19 p. (Estudios Políticos; 18)

Author rejects Klugman's views (item **3459**) that Chileans are losing interest in politics; that traditional options of left, center, and right are no longer valid; and that contemporary politics are dominated by themes that span the spectrum of classical political options. Garretón makes good use of both his knowledge of post-1990 survey data and the reasons why people respond as they do in a given survey. He concludes, however, that while classical options remain valid (as alternative perspectives on questions of freedom and equality), the content of politics and political positions with respect to them are undergoing redefinition.

3451 **González, Mónica** and **Héctor Contreras.** Los secretos del Comando Conjunto. Santiago: Ediciones del Ornitorrinco, 1991. 299 p.: ports.

Account of operations of the Joint Military Intelligence Command written by a *Vicaría de Solidaridad* lawyer and a Santiago journalist. Features extensive testimony of a conscious-stricken Air Force recruit who was a member of the unit for more than ten years (1974–84) and could confirm fates and the cover-ups employed in the most publicized cases.

González Encinar, José Juan *et al.* El proceso constituyente: enseñanzas a partir de cuatro casos recientes; España, Portugal, Brasil y Chile. See item **2757**.

3452 **Graham, Carol.** From emergency employment to social investment: politics, adjustment, and poverty in Chile. (*in* Safety nets, politics, and the poor: transitions to market economies. Washington: Brookings Institution, 1994, p. 21–53)

Slightly expanded version of author's landmark piece on policies affecting Chile's poor under the Pinochet and Aylwin governments. Using Chilean government, UN, and other data, she concludes that almost all other socioeconomic strata (e.g., the merely poor, working classes, and peasantry) lost ground both relatively and absolutely, but that Pinochet's emergency employment and entitlement programs helped to shield the extremely poor (roughly 15 percent of the population) from the otherwise fatal effects of his structural economic reforms. As for the Aylwin government, she praises its more constructive and empowering approach (its FOCIS program, for example), but adds that its success in raising living standards (within a continuing commitment to structural reforms undertaken earlier) is in part attributable to foundations laid and resources freed under Pinochet.

3453 **Guastavino, Luis.** Caen las catedrales. Santiago: Hachette, 1990. 215 p.

Correspondence and interviews featuring a veteran and more recently dissident Communist leader between 1987–90, a period in which he grew increasingly critical of Marxist-Leninist dogma, and policies and practices of the Communist party's Political Commission.

3454 **Gurriarán, José Antonio.** Chile: el ocaso del General. Madrid: El País/Aguilar, 1989. 283 p.: ill.

Leading Spanish journalist's superbly written account of Chile's transition from military to civilian rule. While less detailed and more narrowly focused than Cavallo's *La historia oculta del régimen militar* (item **3428**), it probes more deeply and argues more persuasively in dealing with the mid-1988 and early-1989 period and the principal players therein. This is particularly true, for example, of the so-called soft-coup through which Pinochet and his allies purportedly sought to set aside the Oct. 1988 plebiscite that went against him. Of particular interest are Gurriarán's interviews with Generals Leigh and Díaz Estrada, former Minister Mónica Madariaga, and former Press Secretary Federico Willoughby, all of whom speak candidly and critically of Pinochet. Also instructive are parallels and comparisons that the author draws between the Chilean and Spanish transitions.

3455 **Hola A., Eugenia** and **Gabriela Pischedda.** Mujeres, poder y política: nuevas tensiones para viejas estructuras. Santiago: Ediciones CEM, 1993? 258 p.: bibl., ill.

Lengthy essay analyzes gender relations followed by interviews of nine politically active women (city councilwomen, deputies, senators, and other high-ranking government officials) representing right, center, and left. Authors review recent literature on women's social and political roles, distinguishing culturally-constructed gender roles from biologically-determined sex status, and assess extent of women's social and political participation. Although they advocate various strategies for addressing the multiple discriminations that Chilean women face, they readily concede the need to overcome fragmentation into isolated groups working at different levels.

3456 **Huneeus, Carlos.** Los chilenos y la política: cambio y continuidad bajo el autoritarismo. Santiago?: Centro de Estudios de la Realidad Contemporánea, Academia de Humanismo Cristiano; Instituto Chileno de Estudios Humanísticos, 1987. 241 p.: bibl., ill.

This study, based on a comprehensive survey of public opinion in greater Santiago completed in 1986 by the *Academia de Humanismo Cristiano's* Center for the Study of Contemporary Reality, was the first of its kind carried out in Chile since Eduardo Hamuy's last survey in 1973. It became the standard reference for assessing the regime's strength and the likelihood of enlisting popular support for one or another of the competing anti-regime strategies of the day.

3457 **Huneeus, Carlos.** Il sistema dei partiti in Cile: continuita e cambiamento. (*in* Democrazia e partiti in America Latina. Milano, Italy: FrancoAngeli, 1991, p. 205–248, tables)

Pointing out that parties that emerged to guide the redemocratization process in Chile were precisely those that had existed before 1973, author seeks the sources of strength that kept them alive, especially during the Pinochet dictatorship and its "guided succession" policy embedded in the 1980 Constitution. Contains tables on voter party allegiance in parliamentary elections from 1912–73, party allegiance of detained/disappeared persons by age as known in 1986, ideological distribution of the population (1958–86), and other unusual tables. Brief postscript on the election of Patricio Aylwin in 1990. [V. Peloso]

3458 **Kay, Cristóbal.** Agrarian policy and democratic transition in Chile: continuity or change? The Hague: Institute of Social Studies, 1991. 33 p.: bibl. (Working paper series; 101)

Brief, but useful assessment of agricultural policies under the Pinochet and Aylwin governments. Following the lead of Gómez and Echeñique, Kay stresses positive as well as negative elements of reforms undertaken during the military years. Also finds that apart from the greater emphasis given to peasant farmers and non-export sectors things have not changed much under Aylwin. Acknowledging difficulties involved, he advocates greater public investment and involvement in assisting peasant producers (who farm almost 40 percent of the land).

3459 **Klugmann, Mark.** La paradoja de la mayoría electoral: ¿dónde está el centro? (*Estud. Públicos*, 42, otoño 1991, p. 135–153, graphs, tables)

This speechwriter for Presidents Reagan and Bush who has worked as a political consultant in Chile since the late 1980s, thinks the traditional left-center-right spectrum needs to be modified. He argues that in most electorates the greatest number of votes are *not* in the "center" (i.e., equidistant between left and right polar extremes), and that while Chilean parties and their leaders still think in the left-center-right terms, fewer and fewer voters do (more than a third refuse to think this way, and many that do clearly misuse these labels). He concludes that contemporary politics are too complex or multidimensional to be captured on a single axis (measuring greater or lesser quantities of single values), and that in Chile the "center" (where the greatest number of votes are) probably lies at the intersection of as many as five distinct axis.

3460 **The legacy of dictatorship: political, economic and social change in Pinochet's Chile.** Edited by Alan Angell and Benny Pollack. Liverpool, U.K.: Institute of Latin American Studies, Univ. of Liverpool, 1993. 225 p.: bibl., ill. (Monograph series; 17)

Very useful collection of essays on various aspects of policy under Pinochet. Editors' introductory essay and Carol Graham's careful analysis of the regime's approach to the alleviation of poverty (as compared to the Aylwin government's) stand out among pieces dealing with social and economic issues. Chileans Cristián Gazmuri (on the military), Sofía Correa (on the right), Ignacio Walker (on the Christian Democrats), Eduardo Ortiz (on the Socialists), and Patricio Silva (on intellectuals and technocrats) contribute brief, but informative pieces on political parties and other developments.

3461 **Legassa, Victoria.** Gobierno local y políticas sociales en el Gran Santiago. Santiago: Programa de Economía de Trabajo, 1993. 129 p.: bibl.

Timely analysis of social and economic

programs and projects carried out at the local level through the collaboration of local governments (*Municipios*), non-governmental organizations, and local self-help organizations. Describes structure of local government agencies and procedures, NGO efforts in the areas of job training, technical and financial assistance to microenterprises, and housing and health services. Despite substantial progress to date, problems that need to be dealt with include excessive bureaucracy, scant experience of program coordinators and beneficiaries, and disparity between need and expectation on the one hand, and available resources on the other.

3462 **Lizama, Jaime.** Los nuevos espacios de la política. Santiago: Ediciones Documentos, 1991. 84 p. (Documentas/Estudio)

Short, but extremely interesting essay by young Socialist intellectual who argues that ideological parties (including his own) will find it increasingly difficult to compete effectively in the modern political arena. While conceding that social movements played a role in bringing an end to military rule in Chile, argues that the "natural" spaces in which such movements operated (neighborhoods, factories, universities, etc.) have been superseded by a new political arena in which parties and other groups compete for influence by packaging and projecting "mediated" images, in effect reconstructing reality for an increasingly ill-informed and disengaged mass public.

3463 **López, Luis Ignacio.** La derrota de las armas. Buenos Aires: Grupo Editorial Zeta, 1989. 334 p. (Serie Reporter; 40)

Account of the period between the Oct. 1988 plebiscite and the 1989 presidential and congressional elections by a journalist who returned to Chile after 15 years of "exile" in his native Spain. Having grown up in Chile, and having been active in leftist circles in the 1960s and 1970s, López deftly allows political figures themselves to describe changes in their own thinking and the emergence of a broad consensus in support of liberal democratic institutions. Among the more interesting chapters are those dealing with the Church; young people; political figures such as Ricardo Lagos, Andrés Zaldívar, Clodomiro Almeyda, and Andrés Allamand; and entrepreneurial leaders such as Orlando Sáenz and Manuel Feliú.

3464 **Márquez de la Plata Yrarrázaval, Alfonso.** El salto al futuro. Santiago: Zig-Zag, 1992. 173 p. (Col. Testimonio y futuro)

Vigorous defense of the Pinochet regime by a wealthy landowner and former minister of labor who attacks its "politicized" critics for "twisting the facts" and failing to acknowledge its achievements.

3465 **Marras, Sergio.** Palabra de soldado: entrevistas. Santiago: Ornitorrinco, 1989. 179 p.

Progressive Chilean journalist interviews at length Generals Ernesto Baeza, Alejandro Medina Lois, Horacio Toro, and Luis Danús, each of whom held important positions in the military regime. Medina Lois and Danús are both supportive of the initial decision to intervene, the ensuing years of military rule, the military's autonomy vis-à-vis civilian authorities and its broadly defined political role. Baeza and Toro are more critical on these issues. Toro draws parallels between the Ibáñez (1927–31) and Pinochet regimes, characterizing the latter as "neo-fascist," and stressing importance of a truly independent judiciary and military acceptance of civilian supremacy. Although Medina and Danús are doubtless more representative of army officers as a whole (both older and younger generation officers), Toro's remarks offer a ray of hope for the democratic conversion of the Chilean military.

3466 **Méndez, Roberto N.** Nuevas dimensiones en la política chilena. (*Estud. Públicos*, 45, verano 1992, p. 229–271, tables)

Using a 1991 CEP-Adimark survey which he directed, Méndez performs factor analysis on responses to eight dichotomous questions involving political issues. He identifies four "new" dimensions afoot in Chilean politics: statism, conservatism/innovation, participation/individualism, and authoritarianism. He then looks at how peoples' ideological (center-left-right) and partisan political affiliations affect their views toward each. People identifying with the left, center, and right hold generally predictable views on three of the four dimensions, but not very consistently. Range of views with respect to partisan affiliation is even broader. Turning matters around, Méndez warns that only future surveys will determine whether individual's views with respect to these dichotomies will be better predictors of political preferences than self-described left-center-right identification. Lucía Santa Cruz, Manuel Antonio Garretón, and Mark Klugman comment on Méndez's work, with Santa Cruz raising the interesting question of whether the number of questions (eight) was used to develop the new dimensions.

3467 Meneses C., Aldo. El poder del discurso: la Iglesia Católica chilena y el gobierno militar, 1973–1984. Santiago?: ILADES; CISOC, 1989. 227 p.: bibl., ill.

Scholarly comparison of formal pronouncements of Catholic Church and military officials during the first 11 years of military rule. Written as a PhD dissertation in sociology at Louvain Univ. in Belgium, it is more concerned with statements of general purpose and/or commitment (value meaning, interpretation, reconciliation, etc.) than with concrete actions or applications. By failing to examine the evolving socioeconomic and political contexts, and the diverse ideological and ecclesial tendencies represented among bishops and other Catholic groups, author sheds relatively little light on the Church's role in the transition to democracy. For sociologist's comment see item **5051**.

3468 Millas, Orlando. La alborada democrática en Chile: memorias. v. 1, 1932–1947: en tiempos del Frente Popular. Santiago: CESOC Ediciones, 1993. 1 v.: ill.

First of four volumes of reminiscences of the veteran Communist intellectual and political figure who died in 1992. Highly readable and informative reflections on the formative years of the Chilean left, during which Communists and Socialists learned to work with one another, on the opportunistic Radicals, and on the political right.

3469 Modellfall Chile?: ein Jahr nach dem demokratischen Neuanfang. Edited by Jaime Ensignia, Detlef Nolte. Münster: Lit, 1992. 208 p.: bibl. (Schriftenreihe des Instituts für Iberoamerika-Kunde; 34)

Assessment of the success of Chilean democracy after the first year of the Aylwin Administration by 12 German and Chilean experts who discuss positive and negative aspects including the economy, social disparities, human rights, and political consensus. Authors ask whether Chile is a regional model for democratization and development. [C.K. Converse]

3470 Moulian, Tomás. La forja de ilusiones: el sistema de partidos, 1932–1973. Santiago: Univ. ARCIS; Facultad Latinoamericana de Ciencias Sociales (FLACSO), 1993. 307 p.: bibl.

Superb collection of essays on post-1932 Chile written over the last six or seven years by one of the country's leading political historians. Blends thorough distillation of existing Chilean and international scholarship with his own unrivaled grasp of political forces and trends. Moulian brings to life the remarkably stable and accommodating politics of the 1930s and 1940s, and then traces the ideological radicalization and polarization to which they gave way and under whose weight they collapsed. Particularly enlightening discussion of the cultural underpinnings and historical influences of the various political atmospheres and stakes since 1932.

3471 Munck, Gerardo L. Authoritarianism, modernization, and democracy in Chile. (*LARR*, 29:2, 1994, p. 188–211)

Very useful review of recent books on Chile's economy and politics, around which Munck crafts an imaginative and stimulating assessment of the process of democratization to date and the tasks that lie ahead. Among the books reviewed are Scully (*HLAS 53: 4021*), Arriagada (*HLAS 53:3791*), Tironi (*HLAS 53:5480*), Brunner et al. (*HLAS 53: 5392*), and Kay and Silva (item **3436**). Arriagada's book is more concerned with how Pinochet amassed power and less with the regime's long-term effects on Chilean politics, therefore it fits less readily into this scheme. Munck, however, makes excellent use of the other materials (ten books in all). An invaluable resource for the reader who wants to catch up in a hurry.

3472 Ortega Frei, Eugenio. Historia de una alianza política: el Partido Socialista de Chile y el Partido Demócrata Cristiano, 1973–1988. Santiago: CED; CESOC, 1992. 376 p.: bibl.

Undergraduate thesis by a grandson of former Christian Democratic President Eduardo Frei examines relations between Chilean Christian Democrats and Socialists during the years of military rule. Thanks to author's unlimited access to party documents and materials, the study is an exhaustively detailed account by a young historian more interested in setting down information (along the lines of Carvallo *et al.*'s *La Historia Oculta*) (item **3428**) on which others might then reflect, than in interpreting it himself.

3473 Parrini Roces, Vicente. Matar al minotauro: ¿Chile, crisis moral o moral en crisis? Conversaciones con Antonio Bentué *et al.* Santiago: Planeta, 1993. 250 p. (Col. Debates de la transición)

Series of reflections offered by a conservative priest, a relatively liberal theologian (Antonio Bentué), a philosopher (Humberto Gianini), a young agnostic (Martin Hopenhayn), and a leading secular writer (Dimela Eltit). They respond to a 1991 pastoral letter issued by the Chilean Catholic bishops that took the country's younger generation to task for its permissiveness with respect to sex and other moral matters. They respond as much to one another as they do to the bishops, and in general give the reader a fairly good idea of the new and divergent cultural foundations on which Chile's newly restored democratic institutions are being built.

3474 Pastor, Aníbal *et al.* De Lonquén a los Andes. Santiago: Ediciones Rehue, 1993. 205 p.: ill.

Collection of essays and chronological details that form the backdrop for interviews of Chilean Catholic bishops, theologians, priests, and lay activists who helped to define the Church's opposition to the Pinochet regime. Very useful portrayal of the people themselves (Mons. Sergio Contreras, Mons. Alfonso Baeza, Fernando Castillo, Mons. Cristián Precht, María Luisa Sepúlveda, P. José Aldunate, P. Ronaldo Muñoz, and Luisa Riveros), of the Church's unity and diversity, and of its likely role in the post-military period.

3475 Pinochet Ugarte, Augusto. Camino recorrido. v. 3, pt. 1, Memorias de un soldado. Santiago: Tall. Gráf. del Instituto Geográfico Militar de Chile, 1993. 1 v.: ill.

Another volume in Pinochet's seemingly endless reconstruction of his years in power, this time covering the period 1981–1986. Beginning with a warning of the dangers of democratic ideas and sentiments, it includes newspaper and other materials designed to "prove" author's contention that as late as the early-and mid-1980s his government faced an ongoing "informal war," and that its harsh security measures were thus fully justified.

3476 Pinochet Ugarte, Augusto. Ego sum Pinochet. Entrevistas realizadas por Raquel Correa y Elizabeth Subercaseaux. Santiago: Zig-Zag, 1989. 157 p.: ill.

Extended 1989 interview of Pinochet by two of Chile's leading journalists. Format employed (imposed?) offers interviewee ample opportunity to project the image he wants and avoid questions or subjects with which he would be less comfortable. On occasion, however, as with the subject of human rights violations in the early years of military rule, Correa and Subercaseaux succeed in pressing him more effectively and he comes off much less favorably.

3477 Pinochet Ugarte, Augusto. Transición y consolidación democrática, 1984–1989. Santiago: Centro de Estudios Sociopolíticos, 1989. 266 p.: col. ill.

Collection of Pinochet's speeches and reflections on various aspects of the transition to civilian rule. Useful as a compendium or record of Pinochet's efforts to portray the process in the terms he hoped it would be perceived. Thematic sections deal with topics such as the place of the military in the new order, the role of women, the role of "intermediate-level" social organizations, and "disruptive developments" in the political arena.

3478 Puryear, Jeffrey. Thinking politics: intellectuals and democracy in Chile, 1973–1988. Baltimore: Johns Hopkins Univ. Press, 1994. 206 p.: bibl., index.

Puryear, a former Ford Foundation representative for Chile and Peru, examines contribution of Chilean social scientists (mostly Christian Democrats or members of other Popular Unity parties) to the restoration of democratic institutions after 17 years of military rule. According to Puryear, a handful of private, externally funded research centers (e.g., FLACSO, the Academia de Humanismo Cristiano, CED, SUR, and others) provided sanctuaries for dissident thought, helped to produce both a more democratic left and a more tolerant and flexible center, and ultimately enabled the coalition for the "No" to mount and carry out a successful 1988 plebiscite campaign. Intellectuals, he warns, are not always influential with potentially dominant political elites. In this case, however, they were aided by their generally high standing in Chilean political culture, by the trauma and exclusion to which Chilean political elites were subjected for much of the military period, and by external institutions' considerable investment in Chilean social sciences since the 1960s.

3479 **Rabkin, Rhoda.** The Aylwin government and "tutelary" democracy: a concept in search of a case? (*J. Interam. Stud. World Aff.*, 34:4, Winter 1992/93, p. 119–194, appendix, bibl., table)

Lengthy article—57 p. of text and 19 p. of footnotes and bibliography—covers the first two years of the Aylwin government. Author defends its cautious, go-slow approach to the elimination of military prerogatives and other restrictions of democratic rule, arguing that in any regime the amount of actual power a government exercises is a function of both formal and informal political forces (hence her preference for Chalmers' notion of "politicized" democracy); that the Chilean military "constrains" but does not "control" political outcomes; that even with its restrictions, Chilean "democracy" is more complete than many others (e.g., Brazil under Sarney) that have not been labelled "tutelary;" and that while the military may have had a lot to do with the shape of Chile's political order, the more significant restraints that the current government faces may stem from changes in ideological currents, political leadership, and the social forces that drive its institutions on a daily basis.

3480 **Rabkin, Rhoda.** How ideas become influential: ideological foundations of export-led growth in Chile, 1973–1990. (*World Aff.*, 156:1, Summer 1993, p. 3–25, tables)

In this very interesting article, Rabkin makes a case for the impact of ideas and changes in ideas on the political arena. In Chile, ideas in question are traditional capitalist and the change involved is their re-legitimation despite bitter opposition from a once powerful labor movement and political left. She attributes consolidation of Chile's export-oriented free-market economy to a thoroughgoing reassessment of cultural and political attitudes in the years that followed the traumatic tenure of Allende's Popular Unity government, years during which Chileans grew ever more distrustful of mobilization politics, more disposed to entrepreneurial initiative, and less expectant regarding egalitarian social and economic change. While other Latin American nations have as traumatic an experience on which to draw, Rabkin thinks that the Chilean economy's dynamism since 1986 will lure other countries in the region into following its example.

3481 **Santibáñez, Abraham.** El plebiscito de Pinochet: cazado en su propia trampa. Santiago: Editorial Atena, 1988. 171 p.: bibl.

Engagingly written account of the period 1980–88 is a more narrowly focused, concisely written version than Cavallo *et al.*'s *La historia oculta* (item **3428**). Less notable for its new revelations than for its anticipation of later accounts of the same period. Contains interesting insights into Pinochet's character, military politics, and the opposition's successful campaign for the 1988 plebiscite.

3482 **Schneider, Cathy.** Radical opposition parties and squatters movements in Pinochet's Chile. (*in* The making of social movements in Latin America: identity, strategy, and democracy. Edited by Arturo Escobar and Sonia E. Alvarez. Boulder, Colo.: Westview Press, 1992, p. 260–275)

Extremely interesting analysis of urban protest movements of 1983–86. Schneider rejects explanations of scholars such as Arriagada who attribute them to rising levels of deprivation, social dissolution, and anger in the wake of the 1982 economic crisis and those such as Leiva and Petras who emphasize development of autonomous local neighborhood organizations. She argues that protests were most successful (mobilized many more people) in neighborhoods organized by the Communist party and with long-standing traditions of militancy. She further contends that Catholic priests and nuns tended to respond to demands and requests of already mobilized groups rather than initiate things themselves. Unfortunately, her interview data illustrates but does not substantiate either of these conclusions nor her contention that local Communist success was a function of national party leadership and/or resources.

3483 **Scully, Timothy R.** and **Julio Samuel Valenzuela.** De la democracia a la democracia: continuidad y variaciones en las preferencias del electorado y en el sistema de partidos en Chile. (*Estud. Públicos*, 51, invierno 1993, p. 195–228, tables)

Article compares results of the 1988 plebiscite and the 1989 general elections with the 1970 presidential and the 1969 and 1973 congressional elections. They conclude, not unexpectedly for the first several elections following a long hiatus, that the pre-coup pattern of three thirds (right, center, and left) remained in effect, although they do note a broader consensus (which they describe as a centripetal tendency) among various parties on political and economic policy questions. They are aware, of course, that two such elections does not a party system make, and that the real tests will be the extent to which political consensus endures as well as the electorate's response in future elections.

3484 Sigmund, Paul E. The United States and democracy in Chile. Baltimore: Johns Hopkins Univ. Press, 1993. 254 p.: bibl., index.

Useful review of US policy in Chile over the last 30 years. To his credit, author concedes that the imposition of an invisible blockade during the Allende years predated nationalization of US-owned copper mines. But for the most part, he continues to press points that he has argued steadfastly for many years (most of which stand up remarkably well in the light of ensuing events and a more complete public record). Unfortunately, the book is largely a descriptive narrative designed to set the Chilean record straight, and not to reflect on broader, theoretical questions such as the economic and political impact of economic leverage, the interplay between external pressures and internal political dynamics, or the dilemmas of reconstituting full and unrestricted democracy and not just restoring civilian rule.

3485 Silva, Eduardo. Capitalist coalitions, the State, and neoliberal economic restructuring: Chile, 1973–88. (*World Polit.*, 45:4, July 1993, p. 526–559)

Silva uses the work of Haggard, Gourevitch, and Frieden to analyze the influence of capitalist forces on policy-making and policy shifts during the Pinochet regime. As in his piece in Drake and Jaksic (*HLAS 53:4024*), he distinguishes among the initial gradualist, subsequently radical neoliberal, and ultimately pragmatic neoliberal phases under Pinochet, relating shifts of position to changes in the relative leverage of different groups at different junctures. He argues that adoption of pragmatic neoliberal policies following the 1981–83 crisis underscores the regime's susceptibility to external developments and to shifts in the correlation of forces among domestic entrepreneurial groups.

3486 Silva, Eduardo. Capitalist regime loyalties and redemocratization in Chile. (*J. Interam. Stud. World Aff.*, 34:4, Winter 1992/93, p. 77–117, bibl.)

Interesting article by a political scientist specializing in economic policy during the years of military rule. Silva stresses importance of Pinochet's abandonment of radical neoliberalism in the early 1980s, and its replacement by a pragmatic version that the regime worked out with leading entrepreneurial elements. Argues that this was a crucial step in repairing the breach between the two, and that it forced the opposition first to accept terms of Pinochet's transition (with all its limitations), and then to commit itself to maintaining the very same pragmatic neoliberal policies in the post-military period.

3487 Spooner, Mary Helen. Soldiers in a narrow land: the Pinochet regime in Chile. Berkeley: Univ. of California Press, 1994. 304 p.: bibl., index.

Readable, narrative account of the Pinochet regime written by a US journalist who lived in Chile from 1980–89. Based on interviews of former regime officials, military officers, and both pro- and anti-regime Chileans from all walks of life, it basically reaffirms earlier journalistic accounts such as Cavallo et al. *La historia oculta* (item **3428**), Gurriarán's *El ocaso* (item **3454**), and López's *La derrota de las armas* (item **3463**), which were all published in 1989. While not as revealing of Pinochet himself as Correa and Subercaseaux's *Ego Sum Pinochet* (item **3476**), nor as analytical as Drake and Jaksic (*HLAS 53: 4024*), Arriagada (*HLAS 53:3971*), and Varas' *Los militares en el poder* (*HLAS 51:4945*) (1987), it is nonetheless a valuable and accessible book for the lay English-language reader.

3488 Tomic, Radomiro. Tomic testimonios. Recopilación de Jorge Donoso Pacheco. Prólogo de Jaime Castillo Velasco. Santiago: Centro Latinoamericano Simón Bolivar; Editorial Emisión; Ediciones Copygraph, 1988. 549 p.

Collection of speeches, lectures, interviews, and published essays of the late Radomiro Tomic (d. 1992), one of the Chilean

Christian Democratic party's founders and leading figures for more than 40 years. Includes interesting material from the party's early Falange period (1930s and 1940s), from Tomic's 1970 presidential campaign (run on a platform quite similar to Allende's), from the period of Allende's presidency in which the PDC grew increasingly more hostile, and from shortly before and after the coup itself, in which Christian Democrats found themselves at odds over what to do and how to respond. Of particular interest, perhaps, is Tomic's letter to Gen. Carlos Prats the day after he ceded his position as Army Commander-in-Chief to Gen. Pinochet.

3489 Urzúa Valenzuela, Germán. Historia política de Chile y su evolución electoral: desde 1810 a 1992. Santiago: Editorial Jurídica de Chile, 1992. 784 p.: bibl., maps.

The twelfth and most recent work of an unrepentently old-fashioned Chilean lawyer/political scientist who pieces together a wholly descriptive narrative using public records almost exclusively. Offers a wealth of detailed information on the emergence and demise of parties, winning and losing candidates for executive and legislative offices, and the composition of virtually all cabinets over the almost two full centuries it covers. It is less helpful in terms of posing or helping to shed light on fundamental questions regarding the extent or fullness of Chile's democratic politics at any particular point and/or the cultural, socioeconomic, and other foundations on which it rested, or might have rested. It is hard to imagine that the author reads other political scientists. The more searching questions and critiques of Chilean politics that have occupied them since 1973 do not appear to interest or engage him in the least.

3490 Valenzuela, Arturo. Party politics and the crisis of presidentialism in Chile: a proposal for a parliamentary form of government. (*in* The failure of presidential democracy. Edited by Juan J. Linz and Arturo Valenzuela. Baltimore: Johns Hopkins Univ. Press, 1994, p. 165–224, tables)

Valenzuela offers refined and updated version of the case he has been making since 1985 for the adoption of parliamentary government. Argues that a parliamentary system, and not the current presidential regime, is most likely to bridge the centrifugal realities of Chilean politics and to promote consensus on policies and the political rules of the game among its apparently perennial left-, center-, and right-wing referents. He terms the Aylwin government the most successful in contemporary Chilean history precisely because it has maintained its unity of purpose throughout its term (thanks to the obviously substantial power position of its opponents and the related disciplined support of both Concertación parties and labor and neighborhood organizations). But he does not think that such unity is likely to survive Aylwin's government and therefore urges adoption of mechanisms requiring executive accomodation of shifts in parliamentary majorities.

3491 Valenzuela, Arturo. Political and economic challenges for Chile's transition to democracy. (*in* In the shadow of the debt: emerging issues in Latin America. New York: Twentieth Century Fund Press, 1992, p. 13–35, table)

This essay, written in mid-1991, assesses progress of Chile's transition to democracy and identifies challenges that governments following Patricio Aylwin's are likely to face. Author is optimistic about the country's prospects given its deeply rooted democratic tradition, the current strength of its economy, and the sense of purpose and perspective shared by its centrist Partido Demócrata-Cristiana (PDC) and leftist Partido Socialista (PS) and Partido por la Democracia (PPD) parties. Noting that restrictions on democratic rule and the persistent shadow of Pinochet have helped to keep the Concertación united in support of moderate, consensual economic and political objectives, he does worry that this could make the achievement and consolidation of full democracy more difficult, and that, in any case, additional decisions and challenges await future governments.

3492 Vega, Luis. Estado militar y transición democrática en Chile. Madrid: Prensa y Ediciones Iberoamericanas, 1991. 277 p.: bibl. (El Dorado; 9)

Rambling, and somewhat histrionic account of the military regime and the transition to democracy by a former legal adviser (in the area of state security) to the Ministry of Interior in Valparaíso, and the author of *La caída de Allende*. Its most valuable contributions may be its revelation of how much the

UP government's security forces knew about coup-planning in the months before the coup itself, and its discussion of security operations of each of the four services during the years of military rule.

3493 Viera-Gallo, José Antonio. Chile, un nuevo camino. Chile: CESOC, Ediciones ChileAmérica, 1989. 363 p.: bibl.

This collection brings together original and previously published essays and speeches of one of Chile's most versatile and respected politicians/intellectuals, José Antonio Viera-Gallo. A member of the Socialist Party and the president of the Chamber of Deputies, author offers persuasive defense of his party's embrace of moderation: "Politics ought to begin with an analysis of the country's real problems, and not general models or schemes. We should retrain ourselves again in the habits of analysis, dialogue, compromise, and mediation of divergent and even conflicting interests in a country (like ours) immersed in a world in rapid flux." He also writes movingly of the experience of exile, analyzes the role of the Catholic Church in democratization and in the post-military period, and stresses importance of accountability for human rights violations and judicial reform.

3494 Viera-Gallo, José Antonio. La fuerza de las ideas. Santiago: Ediciones ChileAmérica CESOC, 1993. 398 p.

Collection includes essays and speeches on many issues that arose during the first four years of the post-military period. In several devoted to political and economic models, author defends liberal democracy as the "best political system of government." He concedes that capitalism is the most efficient economic regime, but argues that the free market can and should be regulated so as to overcome basic inequalities. Other speeches and essays deal with the military's past and potential political roles, and Chilean political culture. Essays are well written and generally persuasive, revealing a wide-ranging and politically astute mind.

3495 Yáñez Rojas, Eugenio. La Iglesia chilena y el gobierno militar: itinerario de una difícil relación, 1973–1988. Santiago: Editorial Andante, 1989. 135 p.: bibl. (Libros para la democracia)

Useful account of Church-State relations during the Pinochet regime offers a nuanced discussion of the different tendencies among the Chilean bishops. While conceding divisions among the bishops in terms of their political ideas, author argues that they remain cohesive on ecclesial and theological issues. He also argues that Pope John Paul II's appointment of more conservative bishops during the 1980s reflected a desire to prepare the Church to defend "Catholic" values in the post-military period, not to disarm it politically or to pursue more cordial relations with the regime.

ARGENTINA, PARAGUAY, AND URUGUAY

PAUL H. LEWIS, *Professor of Political Science, Tulane University*

THE THREE LA PLATA COUNTRIES are gradually turning their attention from the horrors of military dictatorship to the challenges of building stable democracies and reforming their antiquated State-dominated economic systems. In all three countries civilian politicians have discovered that there is no going back to the old populist economics, but the bitter medicine of neoliberalism is causing widespread social discontent.

Argentina's former military regime still elicits studies, most of them polemical and written from the viewpoint of the left, such as Andersen (item **3499**) and Hodges (item **3541**). Curiously, Argentine writers seem better able to analyze the "Dirty War" with more objectivity than their US counterparts. Alicia García's compilation of documents on the national security doctrine (item **3520**), Acuña and

Smulovitz's monograph on the trial of the ex-Junta leaders (item **3496**) and Terán's history of the rise of Argentina's "New Left" are important contributions to understanding that period (item **3582**).

Building democracy in an era of controversial but unavoidable economic reform is an even trickier subject. Argentine politics have been characterized for decades by a gridlock among intolerant, intransigent interest groups: see, for example, Lewis (item **3549**), Manzetti (items **3553** and **3554**) and Smith (item **3578**). The Alfonsín Administration fell foul of this same gridlock and gave up on reforming the economy after a half-hearted attempt called the Austral Plan. By 1989 inflation and shortages were so bad that his party was defeated in the elections and he was forced to turn over the presidency early to his successor, Carlos Menem. Alfonsín's failure fostered a spate of books and articles, most of them blaming the Argentine upper classes and international capitalism (items **3552, 3566, 3581, 3567,** and **3580**).

Menem has so far avoided Alfonsín's fate by actually privatizing the economy and opening it to foreign capital, all of which contradicts his peronist past and leads to bitter debate. Bertín (item **3505**) and Repetto (item **3571**) support this program on the grounds that the new world economy allows no other choice: a position that Smith (item **2799**) also adopts, but more reluctantly. Manzetti (item **3555**) concentrates on the technical problems involved in privatization. The left is unforgiving, however, and the gross corruption permeating Menem's government presents them with a fat target (items **3513** and **3585**).

Uruguay's recent return to democracy has prompted some interesting studies of how the military was forced to hand over power. Charles Gillespie's work (item **3633**) is a model of scholarly research, while Luis Eduardo González (item **3636**) makes the interesting point that Uruguay's democratic traditions actually made the military surrender the government, once they realized how truly unpopular they were.

The most fascinating recent development in Uruguay occurred when the 1989 elections gave the traditionalist Blanco Party control of the national government and the leftist Frente Amplio control of the capital city of Montevideo, where a third of Uruguay's population lives. This has given rise to speculation that Uruguay's traditional two-party competition between the Blancos and Colorados may be breaking down and that a whole new party system might emerge (items **3642** and **3647**). There are a couple of really good articles on the Frente Amplio's ambitious attempts at improving Montevideo (items **3648** and **3651**) which should be read in conjunction with Arana and Giordano's work on the city's decay under the military government (item **3620**).

Like Argentina, Uruguay is struggling with the challenge of neoliberal economic reform (items **3626** and **3643**), but so far the Uruguayans have resisted biting the bullet. President Lacalle's referendum on privatizing certain State corporations was defeated in Dec. 1992 (item **3665**). Strong labor unions form the backbone of popular resistance, but it appears likely that the relentless logic of the world economy will force changes (item **3626**).

As usual, the literature on Paraguay is relatively sparse. Most of it concentrates on the struggle to establish (not reestablish) democracy, although there are still some backward glances at the *stronato*. Mella Latorre and Ortellado (items **3602** and **3606**) both spent time in Stroessner's prisons and have written about their experiences. Carter's article on Catholic Church politics (item **3594**) and Segura Covalón's on the peasantry (item **3613**) start in the Stroessner period but carry their development forward to the present.

The really significant drama, however, revolves around the gradual evolution toward a democratic regime. This is a process that has its ups and downs, and is still incomplete. One of its more fascinating aspects is the emergence of new political movements which, as in Uruguay, challenge the traditional two-party rivalry of the Colorados and Liberals. Arditi (item **3591**) describes the triumph of "Asunción Para Todos" in the 1991 municipal elections in Paraguay's capital. Two years later, the "Encuentro Nacional" transformed Paraguay into a three-party system. Riquelme's work (item **3611**) relates the story of that campaign and stands as the best available description of the contemporary Paraguayan scene.

ARGENTINA

3496 Acuña, Carlos H. and **Catalina Smulovitz.** Ni olvido ni perdón: derechos humanos y tensiones cívico-militares en la transición argentina. Buenos Aires: CEDES, 1991. 56 p. (Documento CEDES; 69)

Long, detailed, and objective work describes struggles among Alfonsín, the military, and human rights groups about the extent to which soldiers accused of torture and killing under the Proceso should be prosecuted. Essential reading for anyone interested in this subject.

3497 Adrogué, Gerardo *et al.* Reforma institucional y cambio político. Recopilación de Dieter Nohlen y Liliana de Riz. Buenos Aires: Editorial Legasa;: CEDES, 1991. 358 p.: bibl., ill. (Ensayo crítico)

Collection of essays discusses relative advantages and disadvantages of political decentralization, presidentialism, and the two-party system. Last section includes commentaries by notable party leaders.

3498 Alvarez Guerrero, Osvaldo. Las razones de la libertad: las plataformas de la U.C.R., 1937–1989. Buenos Aires: Lugar Editorial, 1990. 309 p.

Contains original texts of all Unión Cívica Radical electoral platforms since 1937. Includes essays in which author attempts (not always successfully) to formulate some common themes that would constitute a coherent ideology.

3499 Andersen, Martin Edwin. *Dossier secreto:* Argentina's *desaparecidos* and the myth of the "Dirty War." Boulder, Colo.: Westview Press, 1993. 412 p.: bibl., index.

Based on "secret sources within the CIA," book purports to show that Argentine military and US Government exaggerated guerrilla threat in order to justify military rule. Thus, the "Dirty War" was only a ruse to break the power of labor unions and other "progressive" elements so as to impose a neo-liberal economic order. Military brutalities are described in considerable detail; guerrilla attacks are often dismissed as having been staged by the armed forces themselves.

Anuario Estadístico de la República Argentina. 1995– . See item **14**.

3500 Aramouni, Alberto and **Ariel H. Colombo.** Críticas al liberal-menemismo. Buenos Aires: Fundación Proyectos para el Cambio, 1992. 316 p.

As title suggests, work is a polemic against Menem's privatizations and his attempts to replace statism with a free market economy.

3501 Aznar, Luis. Discontinuidad y cultura política en la Argentina: ¿hacia una consolidación democrática débil? (*in* Le ombre del passato: dimensioni culturali e psicosociali di un processo di democratizzazione: Argentina e i suoi fantasmi. Torino, Italy: G. Giappichelli, 1992, p. 225–254, graph, tables)

Democracy is still fragile in Argentina because authoritarian attitudes are widespread. Interesting combination of survey research and historical methods.

3502 Azpiazu, Daniel; Eduardo M. Basualdo; and **Hugo Nochteff.** La revolución tecnológica y las políticas hegemónicas: el complejo electrónico en la Argentina. Buenos Aires: Editorial Legasa, 1988. 273 p.: bibl. (Ensayo crítico)

The electronics revolution has exacerbated the weakness of peripheral nations by making them more technologically dependent. Third World countries must do more to protect their industries and their own independent technological development.

3503 **Baizán, Mario** and **Silvia Mercado.** Oscar Smith: el sindicalismo peronista ante sus límites. Buenos Aires: Puntosur Editores, 1987. 226 p.

Oscar Smith was a leader of the Buenos Aires Luz y Fuerza union who "disappeared" during a strike held under the military government. Besides investigating his kidnapping, authors explore ambiguous position of trade unions in Argentina, as they are both linked to, yet antagonistic to, the State.

3504 **Barkey, Henri J.** Politics and the failure of the Austral Plan. (*MACLAS Lat. Am. Essays*, 5, 1992, p. 176–209, graphs)

Work is essentially an autopsy of the Alfonsín Administration's 1985 Austral Plan. Although author puts most of the blame on organized pressure groups' resistance to reform, he recognizes that Alfonsín and his Unión Cívica Radical were not really devoted to long-term restructuring of the Argentine economy.

3505 **Bertín, Hugo D.** La transición política en Argentina, 1983–1991. (*Cienc. Polít.*, 30, 1993, p. 109–122)

Transition that started in 1983 was not just a change toward civilian rule, but the beginning of a new economic trend that would make corporatism obsolete.

3506 **Borón, Atilio.** Memorias del capitalismo salvaje: Argentina de Alfonsín a Menem. Buenos Aires: Ediciones Imago Mundi, 1991. 195 p.: bibl.

Work is largely a polemic against Menem's neoliberal policies.

3507 **Botana, Natalio R.** La libertad política y su historia. Buenos Aires: Editorial Sudamericana, 1991. 232 p.: bibl. (Col. Historia y sociedad)

Contains essays about birth of the Argentine Republic and its formative values as perceived by two important statesmen, Bartolomé Mitre and Vicente Fidel López. These are followed by essays on French and American Revolutions and on De Tocqueville's and Sarmiento's ideas about liberty. Unifying purpose is to clarify the meaning of liberty and liberalism.

3508 **Bouvard, Marguerite Guzman.** Revolutionizing motherhood: the mothers of the Plaza de Mayo. Wilmington, Del.: Scholarly Resources, Inc., 1994. 278 p.: bibl., ill., index. (Latin American silhouettes)

Admiring study of the Madres de Plaza de Mayo is written by a feminist. Author, who has even participated in their marches, takes their side in all controversies affecting them, although she does express some concern about Communist infiltration of their ranks.

3509 **Buchrucker, Cristián et al.** Racionalidad del peronismo. Edición de José Enrique Miguens y Frederick C. Turner. Buenos Aires: Grupo Editorial Planeta, 1988. 254 p.: bibl., ill.

Several scholars contribute essays which, taken together, constitute a history of peronism as well as an explanation for its continuing popularity.

3510 **Carlson, Marifran.** A tragedy and a miracle: Leonor Alonso and the human cost of State terrorism in Argentina. (*in* Surviving beyond fear: women, children, and human rights in Latin America. Fredonia, N.Y.: White Pine Press, 1993, p. 71–85)

Relates story of a baby girl who was born in a torture center where her mother was killed. The child was subsequently raised by foster parents. Located years later by human rights activists, she became the subject of a controversial trial that ended with her being turned over to her biological grandmother.

3511 **Castro Madero, Carlos** and **Esteban A. Takacs.** Política nuclear argentina: avance o retroceso?. Buenos Aires: El Ateneo, 1991. 249 p.: bibl., col. ill. (Serie Sociología y ciencias políticas)

Principal author is a vice-admiral, and book argues for Argentina to continue developing its own nuclear energy.

3512 **Cavarozzi, Marcelo** and **Maria Grossi.** Radicalismo e peronismo in Argentina. (*in* Democrazia e partiti in America Latina. Milano, Italy: FrancoAngeli, 1991, p. 39–74, bibl.)

Beginning with Alfonsín regime in 1983, authors survey political issues of national importance, especially the economic crisis and human rights during the military government. They review weaknesses of the Austral Plan and failure of political system to overcome the claims of peronism and radicalism. They particularly focus on tensions internal to the Unión Cívica Radical that crippled party's ability to live up to its promises of reform; yet they assert that emergence of Alfonsín and his program represented a nationalization of the party for the first time. Authors suggest how party spread its influence beyond Buenos Aires region, and then analyze its efforts to deal with national problems, especially inflation. They conclude by explaining rise of Menem in context of collapse of the peronist response to democratic politics. [V. Peloso]

3513 **Cerruti, Gabriela** and **Sergio Ciancaglini.** El Octavo Círculo: crónicas y entretelones del poder menemista. Buenos Aires: Planeta, 1991. 287 p. (Espejo de la Argentina)

Popular journalistic exposé focuses on corruption inside Menem's Administration.

Chaffee, Lyman G. Political protest and street art: popular tools for democratization in Hispanic countries. See item **2736.**

3514 **Chelala, César A.** Women of valor: an interview with the mothers of Plaza de Mayo. (*in* Surviving beyond fear: women, children, and human rights in Latin America. Fredonia, N.Y.: White Pine Press, 1993, p. 58–69)

Relates the "Dirty War" as seen by Hebe de Bonafini, president of the Madres de la Plaza.

3515 **Chumbita, Hugo.** Los carapintada: historia de un malentendido argentino. Buenos Aires: Planeta, 1990. 287 p.: bibl.

Offers sympathetic treatment of Colonels Aldo Rico and Mohamed Alí Seineldín, who led military rebellions against Alfonsín's government between 1987 and 1989.

3516 **Ciria, Alberto.** Argentina: an underdeveloping country? (*in* Latin America to the year 2000: reactivating growth, improving equity, sustaining democracy. New York: Praeger, 1992, p. 195–207, bibl.)

Short survey examines Argentina's evolution in 20th century.

3517 **Ciria, Alberto.** Party and unions in peronism during the democratic transition in Argentina. (*in* Forging identities and patterns of development in Latin America and the Caribbean. Edited by Harry P. Díaz, Joanna W.A. Rummens, and Patrick D.M. Taylor. Toronto: Canadian Scholars' Press, 1991, p. 143–156)

Once-over-lightly history of the Partido Peronista and its trade unions notes that both are structurally weak.

3518 **Claves del periodismo argentino actual.** Recopilación de Jorge B. Rivera y Eduardo Romano. Argentina: Ediciones Tarso, 1987. 303 p.

Collection of essays and interviews aims to present a history of journalism in post-World War II Argentina. Slightly polemical, but full of worthwhile information.

3519 **Deiner, John T.** Durability of the transition to civilian rule in Argentina. (*MACLAS Lat. Am. Essays*, 5, 1992, p. 210–217)

The peronist movement has never been fully committed to democracy. Therefore, even though civilian rule now seems well established, real test will come "when and if the peronists leave office through an electoral defeat."

3520 **La doctrina de la seguridad nacional, 1958–1983.** v. 1–2. Edición de Alicia S. García. Buenos Aires: Centro Editor de América Latina, 1991. 2 v.: bibl. (Biblioteca Política argentina; 333–334)

Two-volume collection of articles and speeches by top officers in Argentina's armed forces (1958–83) aims primarily to show how national security doctrine evolved there. In the introduction, author argues that roots of that doctrine can actually be traced to beginning of the century, and that Perón was an important contributor to it. Important work demonstrates a logical consistency to ideas contained in national security doctrine, and treats these ideas as constituting an ideology in defense of traditional interests.

3521 **Domínguez, Roberto F.** Participación y responsabilidad de la dirigencia civil en los golpes de estado. (*Rev. Arg. Estud. Estrateg.*, 8:14, enero/dic., p. 13–37)

Argues that civilians have always used the military to settle their partisan disputes and that coups are usually initiated by the former.

3522 **Echegaray, Fabián.** Adiós al bipartidismo imperfecto?: elecciones y partidos provinciales en la Argentina. (*Nueva Soc.*, 124, marzo/abril 1993, p. 46–52, table)

Argues that provincial parties in Argentina are winning an ever larger percentage of the vote and thereby threaten present two-party system.

3523 Erro, Davide G. Resolving the Argentine paradox: politics and development, 1966–1992. Boulder, Colo.: Lynne Rienner Publishers, 1993. 254 p.: bibl., index.

Onganía, Perón, Videla, and Alfonsín all failed as presidents because they could not overcome entrenched corporatist entities that dominated the system and prevented rational economic policymaking. Menem's success was made possible by the military's disgrace, which eliminated that institution as a powerbroker. Also, the old corporatist order finally collapsed under economic crisis of 1989, allowing for a new, neoliberal paradigm to become dominant.

3524 Escudé, Carlos. Cultura política y contenidos educativos: el caso de Argentina. (*in* Le ombre del passato: dimensioni culturali e psicosociali di un proceso di democratizzazione: Argentina e i suoi fantasmi. Torino, Italy: G. Giappichelli, 1992, p. 111–154, table)

The patriotic hysteria that swept Argentina during Falkland/Malvinas War moved author to study sources of the phenomenon, which he located in curricula and textbooks of Argentina's primary and secondary schools. Study goes back in time and shows that xenophobia has been increasingly imparted in the schools.

3525 Fraga, Rosendo. Argentina en las urnas: 1931–1991. Buenos Aires: Editorial Centro de Estudios Unión para la Nueva Mayoría, 1992. 225 p. (Col. Análisis político; 8)

Extremely useful compendium of election statistics, broken down to provincial level, covers presidential, congressional, and gubernatorial races. Also includes brief nuggets of information that help to clarify various local party orientations.

3526 Fraga, Rosendo and **María Gabriela Malacrica.** El centro-derecha: de Alfonsín a Menem. Buenos Aires: Editorial Centro de Estudios Unión para la Nueva Mayoría, 1990. 215 p.: ill. (Col. Análisis político; 5)

Serious, empirical work studies Argentina's conservative and provincial parties during 1980s: their strength at national and provincial levels and their dealings with the two major parties.

3527 Fraga, Rosendo and **Valeria Leslie.** La cuestión militar, 1987–1989. Buenos Aires: Editorial Centro de Estudios Unión para la Nueva Mayoría, 1989. 191 p.: ill. (Col. Análisis político; 3)

Surveys Argentine military's situation in relation to society as a whole. Also discusses Argentine military's place in interAmerican system and its strength in comparison to military forces of neighboring countries. The Ley de Obediencia Debida, the *carapintada* revolts, and resurgence of leftwing terrorists at La Tablada get particular attention.

3528 Fraga, Rosendo and **Valeria Leslie.** La cuestión sindical. Buenos Aires: Editorial Centro de Estudios Unión para la Nueva Mayoría, 1991. 176 p.: bibl. (Col. Análisis político; 6)

Concludes that trade unions are badly divided and weakened by a shaky economy, and have an even worse image in the public's eye than does the military. Good empirical study.

3529 Fraga, Rosendo. Historia y análisis político. Buenos Aires: Editorial Centro de Estudios Unión para la Nueva Mayoría, 1992. 77 p. (Col. Estudios; 11)

Collection of articles that appeared in pro-business daily *Ambito Financiero* deals mostly with current domestic politics. Author is a well-trained political analyst with good connections to elite groups.

3530 Fraga, Rosendo and **Eduardo Ovalles.** Menem y la cuestión militar. Buenos Aires: Editorial Centro de Estudios Unión para la Nueva Mayoría, 1991. 207 p.: map. (Col. Análisis político; 7)

Careful, descriptive study examines Menem's relations with the military during 1990 and early 1991. Looks at ramifications of his basic policy of close alignment with US foreign policy, and includes description of Argentina's participation in the Gulf War.

3531 Garay, Alfredo. Buenos Aires en los noventa: gobierno y ciudad. (*in* Ciudades y gobiernos locales en la América Latina de los noventa. México: FLACSO, Sede México, 1991, p. 23–38)

Like many other big cities, Buenos Aires is losing industry and suffering a deterioration in its housing, transportation, and public services. Causes are largely beyond the control of the city government, although high taxes and powerful municipal trade unions contribute to the problem.

3532 Garzón Valdés, Ernesto. La democracia argentina actual: problemas ético-políticos de la transición. (*Rev. Cienc. Soc./ Valparaíso*, 34/35, 1989/90, p. 339–360, bibl.)

Criticizes the Ley de Obediencia Debida, which allowed many junior officers to escape trial for human rights violations.

3533 Gilbert, Isidoro. El largo verano del 91: de la ilusión menemista a la realidad todmaniana. Buenos Aires: Editorial Legasa, 1991. 217 p.: bibl., index. (Nueva información)

An old-line peronist takes President Menem to task for turning against nationalism and casting his lot with Yankee imperialism. US Ambassador Terence Todman plays the villain in this journalistic account of factional fighting inside the Casa Rosada.

3534 Goldar, Ernesto. ¿Qué hacer con Perón muerto?: los mitos de la izquierda peronista. Buenos Aires: Textos de Utopías del Sur, 1990. 181 p. (Pensamiento y creación. Col. de ensayo Círculo Vicioso)

In several short essays a peronist attacks movement's radical left, the Montoneros and their sympathizers. Argues that workers will never embrace Marxism or Guevarism, and therefore the left has no future in the peronist movement.

3535 Gómez, Alejandro. Radicalismo y petróleo. Buenos Aires: Editorial Plus Ultra, 1991. 240 p.: bibl., ill.

Author was once Arturo Frondizi's vice president, but was forced to resign for plotting. Argues that real reason was his patriotic opposition to leasing out Argentina's oilfields to foreigners.

3536 González, Jorge and **Vilma Osella.** El otro Menem: relatos. Buenos Aires: Ediciones Tu Llave, 1989. 152 p.: ill.

Adulatory biography of Argentina's current president.

3537 Grillo, Oscar Jorge. Articulación entre sectores urbanos populares y el estado local: el caso del barrio de La Boca. Buenos Aires: Centro Editor de América Latina, 1988. 141 p.: bibl. (Biblioteca Política argentina; 234)

Based on both observation and interviews, this study of the Boca area of Buenos Aires shows how ordinary citizens in today's democratic Argentina try to influence local government bureaucracy. Packed with information, work presents case studies about issues like housing and clash of cultures between old residents and new arrivals.

3538 Guelar, Diego R. El pueblo nunca se equivoca: los dirigentes a veces sí. Buenos Aires: Editorial Sudamericana, 1988. 259 p.: ill. (Crónicas de la transición; 2)

Critical description of Alfonsín Administration is written by a peronist legislator who wanted Argentina to repudiate its foreign debt.

3539 Hernández, Pablo José. La Tablada: el regreso de los que no se fueron. Buenos Aires: Editorial Fortaleza, 1989. 267 p.: ill.

Written from right-wing perspective, work uses Trotskyite attack on La Tablada barracks in Jan. 1989 to denounce Alfonsín for being soft on the left.

3540 Historia del P.R.T.: 25 años en la vida política argentina. Buenos Aires?: Editorial 19 de Julio, 1990. 78 p.: bibl.

Brief, sympathetic history relates origins, development, and destruction of Trotskyite party that was political arm of Roberto Santucho's Ejército Revolucionario del Pueblo guerrillas.

3541 Hodges, Donald Clark. Argentina's "Dirty War": an intellectual biography. Austin: Univ. of Texas Press, 1991. 387 p.: appendix, bibl., index.

In a polemic aimed at justifying the leftist guerrillas of 1970s, author also admits that military's use of counterterrorism was the only effective way of dealing with them. He also admits that the Madres de Plaza de Mayo are political soulmates of the Montoneros. Most interesting part of the book is author's interview with the imprisoned Mario Firmenich, contained in the appendix.

3542 Hodges, Donald Clark. The Argentine left since Perón. (*in* The Latin American left: from the fall of Allende to perestroika. Boulder, Colo.: Westview Press, 1993, p. 155–170)

Describes Argentina's Marxist parties and their frequent, unsuccessful attempts to find unity. Most useful section deals with Movimiento Todos por la Patria and its attempt to revive guerrilla warfare in 1989.

3543 **Horowicz, Alejandro.** Los cuatro peronismos. Buenos Aires: Planeta, 1991. 360 p.: bibl. (Espejo de la Argentina)

Work is a reissue of a long, popular essay published originally in 1985 (see *HLAS 51:3937*). A new prologue fits President Carlos Menem into the schema.

3544 **Jackisch, Carlota.** Los partidos políticos en América Latina: desarrollo, estructura y fundamentos programáticos; el caso argentino. Buenos Aires: Centro Interdisciplinario de Estudios sobre el Desarrollo Latinoamericano, 1990. 130 p.: bibl., ill.

Descriptive, legal-institutional approach to understanding Argentina's parties includes hard-to-find information about their internal organization.

3545 **Jofré Barroso, Haydée M.** La política de los argentinos: reportaje a los aciertos, errores, dudas, certezas, intuiciones e interrogantes políticos de los argentinos. Buenos Aires: Editorial Galerna, 1990. 351 p.: bibl.

About two-thirds of book consists of essays about Argentine culture and its effect on politics. Remaining one-third presents interviews with various kinds of notables. Interesting insights.

3546 **Kóssov, Igor** and **Igor' Konstantinovich Sheremet'ev.** Los peronistas y el sector estatal en la Argentina. (*Am. Lat./Moscow*, 7:163, julio 1991, p. 59–69, ill., photos)

Points out that Menem's austerity and privatization policies go directly counter to traditional peronist support for State enterprise and regulation.

3547 **Kvaternik, Eugenio.** Crisis sin salvataje: la crisis político-militar de 1962–63. Buenos Aires: IDES, 1987. 149 p.: bibl. (Ediciones del IDES, 0326-6133; 12; Col. América Latina)

Interesting, detailed study examines civil-military relations during provisional government of José María Guido, who was Argentina's nominal head from fall of Frondizi to inauguration of Illia.

3548 **Lewis, Daniel K.** A false turning point in Argentina's recent political history: Carlos Menem, *Los Renovados* and the unions; 1987–1989 regionalism. (*MACLAS Lat. Am. Essays*, 4, 1991?, p. 183–189)

In an unduly pessimistic view of Menem's government written after government was only a few months in power, author concludes that nothing is likely to change in Argentina.

3549 **Lewis, Paul.** The right and military, 1955–1983. (*in* The Argentine right: its history and intellectual origins, 1910 to the present. Wilmington, Del.: Scholarly Resources Inc., 1993, p. 147–180)

Examines inherent weaknesses of the ability of the Argentine armed forces to rule effectively. Using Finer's thesis as a point of departure, author argues that recruitment of civilian personnel from the political right to provide technical expertise and a legitimizing ideology ultimately fails to sustain the military in power. Focus on Nacionalistas and right-liberal *técnicos* provides important information on civil-military relations from 1955–83, particularly the reasons for the demise of the military and the eventual return to civilian rule. [D. Dent]

3550 **López Echagüe, Hernán.** El enigma del General Bussi: de la Operación Independencia a la Operación Retorno. Buenos Aires: Editorial Sudamericana, 1991. 239 p.: bibl.

The enigma refers to the fact that Gen. Bussi, involved in counterterrorism in Tucumán and later made governor of the province by the Junta, became a powerful vote-getter there after democracy was restored.

3551 **Majul, Luis.** Los dueños de la Argentina. v. 1, La cara oculta de los negocios. Buenos Aires: Editorial Sudamericana, 1992. 1 v.: bibl.

Muckraking portraits of Argentina's richest magnates reveal "scandalous" secrets about how they made their fortunes and how they abuse their power.

3552 **Majul, Luis.** Por qué cayó Alfonsín: el nuevo terrorismo económico; los personajes, las conexiones, las claves secretas. Buenos Aires: Editorial Sudamericana, 1990. 292 p.

Contends that Alfonsín was victim of corrupt businessmen and financial speculators.

3553 **Manzetti, Luigi.** The evolution of agricultural interest groups in Argentina. (*J. Lat. Am. Stud.*, 24:3, Oct. 1992, p. 585–616)

Rather than viewing development of pressure groups in agricultural sector as growing out of Argentina's peculiar conditions, author prefers to interpret their actions as conforming to a rational choice model of collective action.

3554 Manzetti, Luigi. Institutions, parties, and coalitions in Argentine politics. Pittsburgh, Pa.: Univ. of Pittsburgh Press, 1993. 382 p.: bibl., index. (Pitt Latin American series)

Interest groups are, appropriately, the center of attention in this study. Author sees their all-or-nothing approach to politics as chief cause of Argentina's instability and stagnation. Above all, they prevent democratic institutions from taking root, and Menem, for all his economic reforms, has done little to strengthen legislative or judicial branches of government. This makes author somewhat skeptical about Argentina's future.

3555 Manzetti, Luigi. The political economy of privatization through divestiture in lesser developed economies. (*Comp. Polit.*, 25:4, July 1993, p. 429–454, bibl., tables)

Article focuses on Menem's difficulties in privatizing the State telephone company and Argentine airlines.

3556 Massuh, Héctor Daniel. El mal argentino. Buenos Aires: Planeta, 1991. 124 p.

Written by an Argentine industrialist, work is essentially a Keynesian plea for government to "pump-prime" the economy.

3557 Mattini, Luis. Hombres y mujeres del PRT-ERP: la pasión militante. Buenos Aires: Editorial Contrapunto, 1990. 525 p.: bibl. (Col. Pensamiento crítico)

Forget the tiresome Marxist jargon and tendency to dwell on minor points of difference among Trotskyites; this work is a valuable source for understanding a guerrilla movement from the viewpoint of its practitioners.

3558 McGuire, James W. Union political tactics and democratic consolidation in Alfonsín's Argentina, 1983–1989. (*LARR*, 27:1, 1992, p. 37–74, appendix, bibl., tables)

Article is extremely useful for understanding confusing welter of trade union factions and their links to various sectors of the Partido Peronista during 1980s. Author does a good job of delving below personalistic quarrels to show a certain ideological consistency in these factional disputes.

3559 Mignone, Emilio Fermín. Derechos humanos y sociedad: el caso argentino. Buenos Aires: Centro de Estudios Legales y Sociales; Ediciones del Pensamiento Nacional, 1991. 175 p.: bibl.

Useful information about human rights groups in Argentina is provided by one of their leaders. Includes a critique of President Menem's first years in office.

3560 Minoliti, Claudia and Pedro Pírez. La gestión municipal en ciudades intermedias en Argentina. (*in* Viejo escenario, nuevos actores: problemas y posibilidades de la gestión municipal en cuidades intermedias en América Latina. Quito: CIUDAD, 1991, p. 37–56, bibl.)

Medium-sized cities of La Rioja, Resistencia, and Zarate are taken as examples of how municipal government, with few resources, tries to provide services and help its neediest citizens. Article is a summary of a much longer study.

3561 Morales Solá, Joaquín. Asalto a la ilusión. Buenos Aires: Planeta, 1990. 348 p. (Espejo de la Argentina)

Journalist for newspaper *Clarín* claims to give us a look at power politics behind the scenes in Alfonsín and Menem Administrations.

3562 Mustapic, Ana M.; Matteo Goretti; and Alejandro Corbacho. Los diputados frente al acuerdo: las experiencias comparadas de Argentina e Italia. (*in* Le ombre del passato: dimensioni culturali e psicosociali di un processo di democratizzazione: Argentina e i suoi fantasmi. Torino, Italy: G. Giappichelli, 1992, p. 201–218, tables)

Argentine legislators feel more independent of party discipline than do their Italian counterparts. And although less inclined to form coalition governments, they are willing to cooperate when it is necessary to keep democratic institutions functioning.

3563 Mustapic, Ana M. and Matteo Goretti. Gobierno y oposición en el Congreso: la práctica de la cohabitación durante la presidencia de Alfonsín, 1983–1989. (*Desarro. Econ.*, 32:126, julio/sept. 1992, p. 251–269, tables)

Under Alfonsín the Radicals failed to control a majority in Congress. Yet much of their legislation passed. What kinds of coalitions were most frequent, and on what sorts of issues was it easier to find a majority? These and other questions are answered by studying the recorded votes.

3564 Neilson, James. El fin de la quimera: auge y ocaso de la Argentina populista. Buenos Aires: Emecé Editores, 1991. 277 p.

Intelligent and perceptive essay examines modern Argentina and the myths that have driven it. Author accuses everyone, from left to right, of distorting reality during 1970s and 1980s. Argentina's chimera is the belief, held by nearly all groups in society, that only some anti-national conspiracy keeps the country from being a world power.

3565 Nun, José. Crisis económica y despidos en masa: dos estudios de casos. Buenos Aires: Editorial Legasa, 1989. 164 p.: bibl. (Nueva información)

What happens to factory workers who lose their jobs to privatization, "downsizing," technological change, or factory closings? Survey research reveals some surprising consequences of cutbacks in Argentina's automotive industry in 1970s and 1980s.

3566 Nun, José and Mario J. Lattuada. El gobierno de Alfonsín y las corporaciones agrarias. Buenos Aires: Manantial, 1991. 205 p.: bibl., ill. (Col. "Sentido común y política")

From Perón forward, almost every government has tried regulatory boards, taxes, and threats of land reform to break the power of the *estancieros*. Alfonsín, like his predecessors, ran into solid resistance and saw his plans fall apart. Three case studies show how agrarian elites defend their interests.

3567 Olmos, Alejandro. Todo lo que usted quiso saber sobre la deuda externa y siempre se lo ocultaron. Buenos Aires: Editorial de los Argentinos, 1990. 240 p. (Col. Los Problemas nacionales)

An old peronist militant demonstrates sinister and occult machinations of the alliance between foreign bankers and the oligarchy against the long-suffering *pueblo*.

Pásara, Luis. El rol del parlamento: Argentina y Perú. See item **3399**.

3568 Pion-Berlin, David. The ideology of State terror: economic doctrine and political repression in Argentina and Peru. Boulder, Colo.: L. Rienner Publishers, 1989. 227 p.: bibl., index.

The assault on human rights by military governments in 1970s was not really in response to guerrilla violence. The *real* underlying motive was to eliminate all progressive social forces that might oppose application of monetarist (i.e., free market) economic policies. Behind those army thugs and torturerers stands the ever-sinister IMF, which unfortunately the Menem government seems to embrace also.

3569 Puyau, Hermes A. Estabilidad política en la Argentina: una aplicación de la teoría de catástrofes. (*Signos Univ.*, 10:20, julio/dic. 1991, p. 53–64, ill., map)

Attempts to apply scientists' recent "catastrophe theory" to Argentine politics between 1930 and 1962.

3570 Quiroga, Hugo. Autoritarismo y reforma del Estado. Buenos Aires: Centro Editor de América Latina, 1989. 110 p.: bibl. (Biblioteca Política argentina; 276)

Avoiding free markets vs. regulatory State debate, author calls for direct democracy to create a decentralized government "committed to social distribution."

3571 Repetto, Fabián. La construcción de un nuevo orden: o el final de una época. (*Real. Econ./Buenos Aires*, 16 nov./31 dic. 1993, p. 18–40, ill., tables)

Since 1976 on, old statist and populist system has been obsolete. Relentless pressure by global capitalism has made the egalitarian ideal give way to a more hierarchical order in which some people enjoy enormous opportunities while others face a radical decline in their living standards. What can be done about this?

3572 Riz, Liliana de. El debate sobre la reforma electoral en la Argentina. (*Desarro. Econ.*, 32:126, julio/sept. 1992, p. 163–184, bibl., tables)

Thorough examination of recent proposals for reforming Argentina's electoral laws. Presents different views about minority representation and effect of electoral changes on political stability. Surveys electoral reform experiences of several provinces.

3573 Riz, Liliana de and Eduardo Feldman. Guía del parlamento argentino: poder legislativo; conformación, naturaleza y funciones. Ed. rev. y actualizada al 10/12/1989. Buenos Aires: Fundación Friedrich Ebert, 1990. 107 p.: appendices, bibl., ill.

Contains compendium of useful facts about congressional procedures and organization. Appendices have interesting data about political, professional, and personal characteristics of current deputies and senators.

3574 Riz, Liliana de. Régimen de gobierno y gobernabilidad: ¿parlamentarismo en Argentina? (*in* Reforma política y consolidación democrática: Europa y América Latina. Caracas: Editorial Nueva Sociedad, 1988, p. 273–285, tables)

Written while Alfonsín was president and peronists were disunited, essay seems dated at first. However, a 1994 constitutional convention will consider the same kind of proposal for a dual executive that failed to gain approval under Alfonsín.

3575 Sagastizábal, Leandro de *et al.* Argentina 1880–1943: Estado, economía y sociedad, aproximaciones a su estudio. Buenos Aires: Editorial Biblos, 1990. 177 p.: bibl., ill. (Col. Ciencias sociales; 12)

Between 1880 and 1943 Argentina went through three social transformations. First was an agrarian revolution that started the process of modernization and brought the economy into the world system. Second was the introduction of universal suffrage and rise of mass politics. The third, starting with the Great Depression, saw the army become the ultimate political arbiter. At the same time, the old upper class initiated industrialization as official policy and made the State a major actor in the economy.

3576 Samoilovich, Daniel. ¿Cuánta agua bajo qué puentes?: reforma política en Argentina desde 1983. (*in* Reforma política y consolidación democrática: Europa y América Latina. Caracas: Editorial Nueva Sociedad, 1988, p. 159–176, table)

Are Argentine politics really different today? Sobering experience of the *proceso* may have made Argentines more willing to strive toward making democracy work, but their patience will be sorely tested by current economic crisis for which there are no short-term solutions.

Sikkink, Kathryn. Las capacidades y la autonomía del Estado en Brasil y la Argentina: un enfoque neoinstitucionalista. See item **3764**.

3577 Simeoni, Héctor Rubén and Eduardo Allegri. Línea de fuego: historia oculta de una frustración. Buenos Aires: Editorial Sudamericana, 1991. 369 p.

Focuses on internal politics of Argentine army during Alfonsín's Administration. Authors identify various factions, explain their motives, and try to illustrate how political decisions are made regarding the military. Excellent analysis and essential reading.

3578 Smith, William C. Restructuring Argentina. (*Hemisphere/Miami*, 4:1, Fall 1991, p. 22–27, graphs)

Menem's free market reforms threaten old corporatist coalition of labor and small businessmen producing for domestic market. These reforms also run counter to the traditional peronist emphasis on economic equality, while favoring the very groups, such as export agriculture and big business, that Perón used to revile.

3579 Smith, William C. State, market and neoliberalism in post-transition Argentina: the Menem experiment. (*J. Interam. Stud. World Aff.*, 33:4, Winter 1991, p. 45–82, bibl., tables)

Excellent study describes macroeconomic policies pursued during Menem's first two years in office. Author fears that neoliberalism's success will be achieved at expense of poorer classes, but he sees no likelihood of returning to now-discredited statist policies of the past.

3580 Sobrino, Raúl Augusto. La crisis moral argentina, 1955–1991. Buenos Aires: Grupo Editor Latinoamericano; Emecé Editores, 1992. 301 p. (Col. Estudios políticos y sociales)

Nationalist blames Argentina's problems squarely on the oligarchy and its liberal philosophy. He thinks Argentina needs another Perón to guide it.

3581 Strubbia, Mario. ¿Por qué fracasó Alfonsín? Rosario, Argentina: Editorial Fundación Ross, 1989. 261 p.

Another autopsy of the Alfonsín government is written by a very hostile lawyer for the trade unions.

3582 Terán, Oscar. Nuestros años sesentas: la formación de la nueva izquierda intelectual en la Argentina, 1956–1966. Buenos Aires: Puntosur Editores, 1991. 193 p.: bibl.

Analyzes rise and fall of Argentina's "New Left" which culminated in guerrilla movements of 1970s. Good intellectual history shows how young, educated elites shift ideologically from one generation to the next.

3583 Vega, Abel U. de la. Espejismos y realidades en torno a la contraguerra revolucionaria argentina. Prólogo de Arturo Frondizi. Buenos Aires: Distribuidora y Editora Theoría, 1989. 157 p. (Biblioteca de ensayistas contemporáneos)

Impassioned defense of Junta leaders during the "Dirty War" is written by a military officer. Author bitterly attacks Alfonsín for putting Junta leaders on trial.

3584 Verbitsky, Horacio. La educación presidencial: de la derrota del '70 al desguace del Estado. Buenos Aires: Editora/12; Puntosur, 1990. 298 p.

According to this popular journalist and former Montonero, Argentina has been under the heel of foreign capital since 1976. For Argentina, a country laden with debt, industries destroyed or taken over, and threatened with more capital flight, democracy is a sham. Presidents like Alfonsín, who try to be independent, quickly learn how true this is.

3585 Verbitsky, Horacio. Robo para la corona: los frutos prohibidos del árbol de la corrupción. Buenos Aires: Planeta, 1991. 396 p.: bibl., index. (Espejo de la Argentina)

Argentine "best-seller" describes corruption in Menem's government.

3586 Wynia, Gary W. The peronists triumph in Argentina. (*Curr. Hist.*, 89:543, Jan. 1990, p. 13–16)

Clear, concise work describes Menem's 1989 electoral victory and his first months in office.

3587 Zavala, Juan Ovidio. Racionalización para el desarrollo. Prólogo de Arturo Frondizi. Buenos Aires: Ediciones Depalma, 1991. 294 p.: bibl., ill.

Favors privatization as surest way to reduce corruption in government and achieve economic growth. With economic growth it will be possible to reform public administration and make it efficient.

3588 Zuleta-Puceiro, Enrique. The Argentine case: television in the 1989 presidential campaign. (*in* Television, politics and the transition to democracy in Latin America. Washington: Woodrow Wilson Center Press, 1993, p. 55–81, tables)

Informative article examines decentralized character of Argentina's television industry and its relatively low impact on voters during electoral campaigns. So far, candidates using media "blitzes" have wasted their time and money, although this may be due to unsophisticated techniques.

PARAGUAY

3589 Acuña, Edith et al. El precio de la paz. Coordinación de José M. Blanch. Asunción: Centro de Estudios Paraguayos Antonio Guash, 1991. 574 p.: bibl., ill.

Catholic human rights group's documentation of atrocities committed by Stroessner regime is Paraguayan equivalent of *Nunca más*.

3590 Arbo, Higinio. Política paraguaya. Asunción: Archivo del Liberalismo, 1991. 70 p.: bibl. (Cuadernos históricos: 1015–2415; 22)

Reprint of 1946 essay by a Partido Liberal leader of that era (see *HLAS 13:1101*).

3591 Arditi, Benjamín. Elecciones municipales y democratización en el Paraguay. (*Nueva Soc.*, 117, enero/feb. 1992, p. 48–57, photos)

Article is especially interesting for its description of how Paraguay's candidates are incorporating modern campaign techniques to shake up the old political structures.

3592 Arditi, Benjamín. Elecciones y partidos en el Paraguay de la transición. (*Rev. Mex. Sociol.*, 52:4, oct./dic. 1990, p. 83–98, appendix)

In Paraguay "liberalization" is a cautious process, directed from the top. It does not mean a return to democracy, since that has never existed in Paraguay, but rather a gradual loosening of the authoritarian tradition. The process is fraught with pitfalls, but there are hopeful signs that democracy will eventually take root.

3593 Bouvier, Virginia M. Decline of the dictator: Paraguay at a crossroads. Washington: Washington Office on Latin America, 1988. 72 p.: photos, tables

Very brief summary of history of the *stronato* emphasizes 1980s when dictatorship was in decline.

3594 Carter, Miguel. La iglesia católica paraguaya: antes y después del golpe. (*Rev. Parag. Sociol.*, 28:81, mayo/agosto 1991, p. 177–207, bibl.)

Provides splendid overview of political tendencies inside Paraguay's Catholic Church and of Church-State relations over past 50 years. Changes in clerical hierarchy and current trends toward democratizing the State have reduced Church's political role.

3595 Carter, Miguel. El papel de la Iglesia en la caída de Stroessner. Asunción: RP Ediciones, 1991. 168 p.: bibl., ill.

Promises more than it delivers, since only about two and a half chapters actually deal with Paraguayan Catholic Church's struggles with Stroessner.

3596 Dictadura, corrupción y transición: compilación preliminar de los delitos económicos en el sector público durante los últimos años en el Paraguay. Recopilación de Tomás Palau, Félix Lugo y Gloria Estragó. Asunción: BASE, Investigaciones Sociales, 1990. 396 p. (Documento de trabajo; 24)

Very lengthy and thorough compendium lists all forms of corruption practiced by Stroessner government, classified by policy area and agencies of government. Includes names of those responsible.

3597 Franco, Rafael. Memorias militares. v. 2. Asunción?: Nueva Edición, 1990. 1 v.

Vol. 2 of memoirs of the Chaco War by one of the most dashing Paraguayan officers who fought in it. For annotation of Vol. 1 see *HLAS 52:2816*.

3598 García Lupo, Rogelio. Paraguay de Stroessner. Barcelona: Ediciones B; Buenos Aires: Grupo Zeta, 1989. 241 p.: bibl., ill., index. (Serie Reporter)

Provides "breathtaking, now-it-can-be-told" revelations about Stroessner's connections to international networks of crime and neo-Nazism.

3599 Iákovlev, Petr. El caleidoscopio paraguayo. (*Am. Lat./Moscow*, 12:168, dic. 1991, p. 49–61)

Provides general overview of Paraguay's struggle to transform itself from an authoritatian to a democratic State.

3600 Martini, Carlos. Del golpe militar a las elecciones municipales: el final de monopolio político del Partido Colorado en Paraguay. (*in* Los partidos políticos en el inicio de los noventa: seis casos latinoamericanos. Santiago: Ediciones FLACSO, 1992, p. 77–85)

Good think-piece written after triumph of Asunción para Todos in Asunción's 1991 municipal elections. Unfortunately, work has been dated by subsequent events.

3601 Martini, Carlos and Carlos María Lezcano. Las izquierdas y la transición en el Paraguay. (*Contribuciones/Asunción*, 8, junio 1991, p. 5–22, bibl.)

Provides capsule descriptions of the various parties of the left in Paraguay. With exception of "Asunción para Todos," none of them is really significant.

3602 Mella Latorre, Alejandro. Somoza y yo: crónica de un calvario en Paraguay. Asunción: Ediciones Ñandutí Vive; Intercontinental Editora, 1990. 209 p.

Author, a Chilean journalist, was picked up in a Paraguayan police dragnet after dramatic assassination of Anastasio Somoza in 1980. He was jailed for six years and claims he was victimized because of a quarrel over his hotel bill.

3603 Miranda, Carlos R. The Stroessner era: authoritarian rule in Paraguay. Boulder, Colo.: Westview Press, 1990. 177 p.: bibl., index, map.

Analysis of *stronato* places it within Paraguay's cultural traditions. Interesting thesis. For historian's comment see *HLAS 52: 2826*.

3604 Morínigo, José Nicolás; Ilde Silvero; and María Susana Villagra. Coyuntura electoral y liderazgos políticos en el Paraguay: resultados de una encuesta de opinión. Asunción: Editorial Histórica; Fundación Friedrich Naumann: Univ. Católica Nuestra Señora de la Asunción, 1988. 486 p.: bibl., ill.

Published before 1989 coup, these opinion surveys may seem somewhat dated, but the high degree of skepticism expressed toward traditional political parties already pointed to emergence of new political movements like Asunción Para Todos and Encuentro Nacional.

3605 **Neild, Rachel.** Paraguay: una transición en busca de la democracia. Washington: Washington Office on Latin America; Asunción: Ediciones Ñandutí Vive, Intercontinental Editora, 1991. 90 p.: bibl.

Sponsored by Washington Office on Latin America, work provides overview of how Rodríguez government has moved back, forth, and sideways in face of demands for a faster transition to truly representative democracy.

3606 **Ortellado Jiménez, Hilario.** Memorias de un oficial paraguayo: el caso Ortigoza; el Ejército Paraguayo doblegado y humillado por el stronismo. Asunción: RP Ediciones, 1990. 220 p.: ill., ports.

Arrested in 1962 for being involved in a plot against Stroessner, author spent eight years in one of Paraguay's grim prisons.

3607 **Partido Liberal Radical Auténtico (Paraguay).** Programa de gobierno: democracia auténtica sin corrupción. Asunción: Partido Liberal Radical Auténtico, 1989. 103 p.: ill. (some col.).

Platform of Domingo Laíno, the twice-unsuccessful presidential candidate of the Partido Liberal Radical Auténtico.

3608 **Prieto, Justo José.** Constitución y régimen político en el Paraguay. Prólogos de Rodrigo Campos Cervera. Asunción: El Lector, 1987. 348 p.: appendix, bibl. (Col. Realidad nacional; 4)

First part of book consists of several essays on constitutional law, from a liberal perspective. Second part describes a course of study on constitutional law that author developed at Asunción's Univ. Católica.

3609 **Radio Ñandutí (Paraguay)** Memoria escrita. Asunción: Radio Ñandutí, s.d. unpaged.

Pamphlet contains newspaper articles and brief notes that give a history of the Stroessner government's harassment of Radio Ñandutí, a private broadcasting station.

3610 **Riart, Gustavo Adolfo.** El Partido Liberal y el ejército. Asunción: Archivo del Liberalismo, 1990. 113 p.: bibl. (Cuadernos históricos: 1015–2415; 16)

Brief history of development and professionalization of Paraguay's armed forces is written by a leader of the Partido Liberal. Much of the book is devoted to defending the Liberals against the criticism that, while in the government, they failed to prepare for the Chaco War.

3611 **Riquelme, Marcial A.** *et al.* Negotiating democratic corridors in Paraguay: the report of the Latin American Studies Association delegation to observe the 1993 Paraguayan national elections. Pittsburgh, Pa.: Latin American Studies Assn., Univ. of Pittsburgh, 1994. 117 p.: appendices, bibl., tables.

Based on reports by LASA's 1993 team of election observers, work offers best available overview of Paraguay's gradual transition from dictatorship to democracy. Describes parties, candidates, issues, and recent elections.

3612 **Roett, Riordan** and **Richard Scott Sacks.** Paraguay: the personalist legacy. Boulder, Colo.: Westview Press, 1991. 188 p.: bibl., ill., index, maps. (Westview profiles: Nations of contemporary Latin America)

Offers good, brief introduction to Paraguay's history, economy, and politics.

3613 **Segura Covalón, Jerónimo F.** Cultura campesina y poder político en Paraguay. (*in* Cultura y política en América Latina. Coordinación de Hugo Zemelman. México: Siglo Veintiuno Editores, 1990, p. 177–206, bibl., map)

Provides brief history of emergence of today's autonomous and militant peasant organizations, such as Movimiento Campesino Paraguayo and Coordinación Nacional de Productores Agrícolas, which grew out of former Ligas Agrarias sponsored by the Catholic Church in the previous decade.

3614 **Simón G., José Luis.** Drug addiction and trafficking in Paraguay: an approach to the problem during the transition. (*J. Interam. Stud. World Aff.*, 34:3, Fall 1992, p. 155–200, bibl.)

Well-researched article briefly examines drug situation and policies under Stroessner Administration and discusses changes in this area resulting from transition to democracy. Particularly interesting is examination of new drug legislation enacted after 1989, creation of new agencies to deal with the problem, and initiatives taken at the international level for cooperation with other countries in the suppression of narcotraffic. [A. Vacs]

3615 **Sondrol, Paul C.** The emerging new politics of liberalizing Paraguay: sustained civil-military control without democracy. (*J. Interam. Stud. World Aff.*, 34:2, Summer 1992, p. 127–163, bibl.)

Despite recent moves toward democracy, Paraguay remains essentially a liberalized autocracy, somewhat like Mexico's one-party system.

3616 **Sondrol, Paul C.** The Paraguayan military in transition and the revolution of civil-military relations. (*Armed Forces Soc.*, 19:1, Fall 1992, p. 105–122)

Once-over-lightly review focuses on military's role under the *stronato*, in overthrow of Stroessner, and in the process of transition.

3617 **Testimonio de la represión política en Paraguay, 1954–1974.** Edición de José Luis Simón G. Asunción: Comité de Iglesias, 1991. 473 p.: ill. (Serie Nunca más; 2)

Interviews with victims of government persecution during Stroessner era are of interest for those concerned with human rights issues.

URUGUAY

3618 **Alsina, Andres.** Diferencias de las FF. AA. con Lacalle. (*Cuad. Marcha*, 8:83, mayo 1993, p. 40–46)

The generals find themselves at odds with current conservative, free-market government. President Lacalle wants to pay off Uruguay's foreign debt by reducing military budget.

3619 **Alsina, Andres.** Uruguay: moonlight sonata. (*Index Censorsh.*, 22:5/6, May/June 1993, p. 40–41, ill.)

Informal censorship exists in "democratic" Uruguay because of newspapers' dependence on government advertising and also because of their close ties to the political parties.

3620 **Arana, Mariano** and **Fernando Giordano.** Montevideo: between participation and authoritarianism. (*in* Rethinking the Latin American city. Edited by Richard M. Morse and Jorge E. Hardoy. Washington: Woodrow Wilson Center Press, 1993, p. 149–161, bibl.)

Written before the Frente Amplio won control of Montevideo, article condemns military and Sanguinetti for the city's deterioration.

3621 **Bradley, C. Paul.** The 1989 election in Uruguay: a report. Flint, Mich.?: C.P. Bradley, 1990. 42 leaves: bibl.

Describes parties, personalities, and issues involved in 1989 campaign. Brief overview of Uruguay's recent history sets election in its proper context. Not much analysis, but offers clear, useful introduction to the polity and its problems.

3622 **Brito, Alexandra Barahona de.** Truth and justice in the consolidation of democracy in Chile and Uruguay. (*Parliam. Aff.*, 46:4, Oct. 1993, p. 579–593)

Complains that not enough has been done to bring human rights violators to justice.

3623 **Caetano, Gerardo** and **José P. Rilla.** Raíces y permanencias de la partidocracia uruguaya. (*Secuencia/México*, 22, enero/abril 1992, p. 143–172, bibl., facsims.)

Interpretive essay examines, from a classist perspective, formation of Uruguay's two traditional political parties.

3624 **Cagnoni, José Aníbal** *et al.* Debate: reflexiones sobre la reforma política. (*Cuad. Marcha*, 8:82, abril 1993, p. 26–43)

Various political commentators offer their opinions about whether Uruguay should change its constitution to replace presidential system of government with a parliamentary system.

3625 **La caída de la democracia: las bases del deterioro institucional, 1966–1973.** Montevideo: Ediciones de la Banda Oriental; Estocolmo: Instituto de Estudios Latinoamericanos, 1987. 123 p.: bibl. (Monografía; 14)

Five essays judged to be the best in a competition sponsored by Univ. of Stockholm's Institute of Latin American Studies focus on theme of how Uruguay's democratic institutions deteriorated from 1966–73.

3626 **Cocchi, Angel M.** Las asignaturas pendientes de la redemocratización uruguaya. (*in* Reforma política y consolidación democrática: Europa y América Latina. Caracas: Editorial Nueva Sociedad, 1988, p. 187–197)

Think-piece highlights main dilemmas confronting Uruguay's newly-restored democracy. Although welfare state is bankrupt, public is not ready for insecurities of laissez-faire. Nor do political parties seem up to the task of finding acceptable alternatives. Yet, spectre of a return to military rule forces all civilian players to keep current system going as if it were viable.

3627 Congresos y documentos del Partido Comunista de Uruguay. Montevideo: Comisión Nacional de Propaganda del Partido Comunista de Uruguay, 1988. 365 p.: bibl.

Contains compendium of reports from PCU's Central Committee, written by Rodney Arismendi, party's chief ideologue. Covers period 1958–85.

3628 Crespo Martínez, Ismael; Pablo Mieres; and **Romeo Pérez Antón.** Uruguay: de la quiebra institucional a la presidencia de Lacalle, 1971–1991. (*Rev. Estud. Polít.*, 74, oct./dic. 1991, p. 297–321, tables)

Good interpretive overview focuses on recent political history and includes discussion of some problems facing four main political parties.

3629 Ecología, medio ambiente y sindicatos. Montevideo: Instituto de Formación e Investigación Sindical; Centro Uruguayo Independiente; Instituto de Investigación y Desarrollo, 1990. 71 p.

Conference papers focus on how to reconcile need for economic development and industrial employment with equally pressing need to protect environment. Participants include union representatives, scientists, and politicians from all the La Plata countries, but especially from Uruguay.

3630 Fernández Faingold, Hugo and **José Miguel Busquets.** Políticas públicas sociales: un examen no exhaustivo de algunas opciones en el debate. (*Cuad. CLAEH*, 17:62, set. 1992, p. 19–26)

Painful dilemmas confront modern welfare states such as Uruguay, where challenge of international competition requires controlling costs and expenditures. Where should government cut its budget, and how do you get powerful organized interests and public opinion to accept sacrifices? Authors pose these questions but offer no clear answers.

3631 Gatto, Hebert. ¿Que reforma política es posible y necesaria? (*Cuad. Marcha*, 8:76, oct. 1992, p. 1–8)

Presents arguments for and against retaining current presidential system of government in Uruguay.

3632 Gestión del estado y desburocratización. Recopilación de Rubén Correa Freitas y Rolando Franco. Montevideo: Libro Libre, 1989. 286 p.: bibl. (Col. Estudios políticos y sociales)

As title suggests, work deals with reducing bureaucracy and increasing role of private sector. Separate section of book focuses on Uruguay, and given all the government featherbedding there, the subject is important for that country. Examples from France, Brazil, Mexico, and Costa Rica point out difficulties of reform, however.

3633 Gillespie, Charles Guy. Negotiating democracy: politicians and generals in Uruguay. Cambridge, England; New York: Cambridge Univ. Press, 1991. 264 p.: bibl., index. (Cambridge Latin American studies; 72)

First-rate study examines breakdown and revival of Uruguay's political parties under military rule. Rich in detail and astute in analysis, work is essential reading for an understanding of the period.

3634 González, Luis Eduardo. Las perspectivas de la democracia en el Uruguay. (*Cuad. Marcha*, 7:74, agosto 1992 p. 30–33, ill.)

Brief essay argues for a parliamentary system of government, based on proportional representation. Under present presidential system, no chief executive has been able to command a majority in the legislature nor form a true coalition cabinet. Author hopes that a parliamentary system might end the institutional deadlock.

3635 González, Luis Eduardo. Political structures and democracy in Uruguay. Notre Dame, Ind.: Univ. of Notre Dame Press, 1991. 201 p.: bibl.

Scholarly analysis examines Uruguay's political parties and elite attitudes. Democracy broke down there because of uncontrolled party fragmentation, and history seems about to repeat itself.

3636 González, Luis Eduardo. Uruguay: una apertura inesperada. (*Síntesis/Madrid*, 13, enero/abril 1991, p. 167–185)

Uses concept of political culture to explain why Uruguay's military voluntarily gave up power after losing 1980 referendum.

3637 **González, Rodolfo.** Elecciones nacionales de 1989. Montevideo: CELADU, 1990. 134 p.: bibl.
Based on election statistics supplemented by opinion surveys, work provides important analysis of electoral trends in Uruguay.

3638 **Gutiérrez, Marcos.** Referéndum y conflictos. (*Cuad. Marcha*, 7:78, dic. 1992, p. 29–31, bibl.)
Dec. 13, 1992 referendum on privatizing certain State corporations was attended by series of labor conflicts including strikes by policemen, transport workers, hospital personnel, and teachers.

3639 **Gutiérrez, Marcos.** Somos los abogados del cambio responsable. Entrevista al Luis Alberto Lacalle Herrera. (*Cuad. Marcha*, 8:77, nov. 1992, p. 16–18, photo)
President Lacalle is interviewed about forthcoming referendum on issue of privatizing certain State enterprises. Lacalle is pledged to reduce scope of State activity.

3640 **Introducción al Uruguay de los 90.**
Montevideo: Centro de Investigaciones Económicas; Ediciones de la Banda Oriental, 1990. 149 p.: bibl.
Provides concise, readable introduction to Uruguay's society, economy, and politics.

3641 **León, Eduardo de.** Paradojas a propósito del 13 de diciembre. (*Cuad. Marcha*, 7:78, dic. 1992, p.32–36)
Using debate over referendum about privatizing State corporations as his foil, author attacks neoliberalism. Since he admits that import-substituting industrialization has failed, it is not clear just what he advocates.

3642 **Mieres, Pablo.** Elecciones de 1989 en Uruguay: una interpretación del cambio del sistema de partidos. (*Rev. Mex. Sociol.*, 52:4, oct./dic. 1990, p. 25–47, bibl., tables)
Good analytical study examines steady rise of support for non-traditional parties in Uruguay. Author notes that clientelistic politics have been on the wane as welfare state goes bankrupt and patronage becomes scarce.

3643 **Panizza, Francisco E.** Uruguay, batllismo y después: Pacheco, militares y tupamaros en la crisis del Uruguay batllista. Montevideo: Ediciones de la Banda Oriental, 1990. 204 p.: bibl. (Temas del siglo XX; 41)

Using a Gramscian interpretation of recent history, author argues that *batllista* liberalism was a spent force long before 1973 coup and that it is surprising that Uruguay's democracy didn't collapse sooner. Provocative and challenging essay avoids usual economic explanations of Uruguay's crisis in order to focus on cultural and ideological factors.

3644 **Pareja, Carlos.** Entre las falacias presidencialistas y el parlamentarismo furtivo de los uruguayos. (*Cuad. Marcha*, 7:78, dic. 1992, p. 37–40, bibl.)
Claims that Uruguay should not abandon its political traditions by adopting a presidential form of government.

3645 **Peña Hasbún, Paula.** Antecedentes ideológicos del Movimiento de Liberación Nacional (Tupamaros). Montevideo: Mim. Pesce Impresos, 1990. 62 p.: bibl.
Somewhat superficial study attempts to show influence of Lenin, Debray, Guevara, the Tricontinental, and various Uruguayan nationalist intellectuals on Tupamaros' political and strategic thought.

3646 **Pérez Antón, Romeo.** El parlamentarismo: la reforma necesaria para Uruguay. (*in* Reforma política y consolidación democrática: Europa y América Latina. Caracas: Editorial Nueva Sociedad, 1988, p. 287–300)
Presidentialism, with its separation of powers and checks and balances, is inefficient; moreover, it encourages authoritarianism.

3647 **Pérez Antón, Romeo.** Los partidos en el Uruguay moderno. (*Síntesis/Madrid*, 13, enero/abril 1991, p. 187–204)
Thoughtful overview examines role of political parties in Uruguay's political culture. Traditional two-party system is being challenged by rise of more ideological parties. That trend may cause a radical restructuring of the system, in which old parties may have to merge in order to confront the left.

3648 **Pérez Piera, Adolfo.** La descentralización en Montevideo: un itinerario innovador. (*Cuad. CLAEH*, 17:62, set. 1992, p. 93–107)
The Frente Amplio that controls Montevideo's city government has instituted a process of administrative decentralization that puts more responsibility for providing services in the hands of 18 neighborhood cen-

ters. Aim is not only to eliminate layers of bureaucracy, but to encourage direct citizen participation in policymaking. Interesting experiment, and worthwhile article.

3649 Pérez Santarcieri, María Emilia. Partidos políticos en el Uruguay: síntesis histórica de su origen y evolución. Montevideo: Impr. Valgraf, 1989. 80 p.

Offers chatty thumbnail histories of Uruguay's main political parties.

3650 Pessoa, Ricardo. Pablo Millor: entre el "Estado Baqueano" y el "Estado Desertor." (*Cuad. Marcha*, 7:73, julio 1992, p. 25–29, ill.)

Interview with a Partido Colorado politician of the old *batllista* school.

3651 Portillo, Alvaro. Montevideo: la primera experiencia del Frente Amplio. (*in* Ciudades y gobiernos locales en la América Latina de los noventa. México: FLACSO, Sede México, 1991, p. 53–66)

Interesting article by a member of Montevideo's municipal government relates Frente Amplio's attempts to revive a decaying city in the face of general economic deterioration and a hostile national government.

3652 Rama, Germán W. La democracia en el Uruguay: una perspectiva de interpretación. Montevideo: Arca, 1989. 238 p.: bibl.

Work of historical interpretation starts with "innovating project" of nation's founding fathers and shows how their vision inspired and integrated the society. The crisis that led to military rule in 1970s resulted from original "innovating project" having played itself out. Essay does not go past 1984.

3653 Ramírez, Gabriel. La cuestión militar: ¿democracia tutelada o democracia asociativa?; el caso uruguayo. Montevideo: Arca, 1989. 275 p.: bibl., ill.

Lengthy discussion examines military's role in recent transition to democracy. Writer believes that Uruguay's armed forces are in the pockets of local oligarchy and American imperialism.

3654 Real de Azúa, Carlos. Uruguay: ¿una sociedad amortiguadora? (*Síntesis/Madrid*, 13, enero/abril 1991, p. 109–145)

In interpretive essay examining Uruguay's political development in 20th century, author identifies certain constant tendencies. Thoughtful and stimulating.

3655 Reyes, Ana María and **Blanca Chiesa.** El Congreso del Pueblo y su significación en el proceso de lucha, movilización y unificación sindical, años 1950–1966. (*Hoy Hist.*, 10:56, marzo/abril 1993, p. 40–60, bibl.)

In 1965 the Convención Nacional de Trabajadores, the Communist-controlled labor front, called a "Congreso del Pueblo" to which other leftist groups were invited. Aim was to draft a program to solve deepening economic crisis, but instead it further polarized Uruguayan society and helped bring on collapse of democracy.

3656 Rial Roade, Juan. Competizione partitica e democrazia in Uruguay. (*in* Democrazia e partiti in America Latina. Milano, Italy: FrancoAngeli, 1991, p. 137–248, tables)

Examines breakdown of Uruguayan consensus system of governance from 1968 onward. Author finds problem rooted in government concerns about "grave internal disorder," linked largely to Tupamaro guerrilla movement. Government reliance on armed forces led military to seize power and thus brought military into politics. Article assesses return of parliamentarism and rebuilding of political parties, and weighs ability of parties to transcend social class differences in Uruguay. Emphasizes accomplishments since Naval Club Accord of 1984 and rise of Frente Amplio with military consent. [V. Peloso]

3657 Rial Roade, Juan. Los militares en tanto "partido político sustituto" frente a la redemocratización en Uruguay. (*Síntesis/Madrid*, 13, enero/abril 1991, p. 229–257, bibl.)

Splendid essay examines professionalization of Uruguay's armed forces. Shows that creation of a professional ethic is no guarantee against coups. This position challenges Samuel Huntington's long-established thesis.

3658 Rodríguez López, Jorge and **Claudio Trobo.** Construcción: historia de un sindicato. Montevideo: Proyección, 1989. 178 p.

Written in part by a professor of architecture who is also an advisor to the Sindicato Unico de la Construcción y Afines (SUNCA), work is obviously a partisan history. Still, study is useful for facts about union's organizational development, factional battles, and strike actions, and for examination of labor legislation.

3659 **La ruptura del Frente Amplio: antecedentes y documentos.** Montevideo: Cencadec, 1989. 43 p.: ill.

Always a coalition of potentially antagonistic groups, Frente Amplio saw one of its components, the Partido Demócrata Cristiano, break away at end of 1988. This is rebels' version of why that happened.

3660 **Sendic, Raúl.** Cartas desde la prisión. La Habana: Editorial Gente Nueva, 1988. 230 p.: indexes.

Contains letters from Tupamaro leader to his children and niece while in prison. First published in Cuba, they cover a wide variety of ostensibly non-political topics, dealing usually with history or character-building.

3661 **Sondrol, Paul C.** 1984 revisited?: a reexamination of Uruguay's military dictatorship. (*Bull. Lat. Am. Res.*, 11:2, May 1992, p. 187–203)

Applies analytical concepts of "totalitarianism" and "authoritarianism" to former military regime in Uruguay, and concludes that regime fell into latter category.

3662 **Stolovich, Luis** and **Juan Manuel Rodríguez.** Poder económico y diálogo social en la coyuntura uruguaya actual, 1990. Montevideo: Centro Uruguay Independiente, 1990. 125 p.: bibl. (Serie Los poderosos; 4)

Two long essays discuss struggle between Uruguay's neoliberal national government and trade unions. Struggle also involves Montevideo's Frente Amplio-controlled municipal government, which backs the unions—as do both of these authors.

3663 **Trías, Vivián.** Banca y neoliberalismo en el Uruguay. Selección y prólogo de Alberto Couriel. Montevideo: Ediciones de la Banda Oriental, 1990. 309 p.: bibl., port. (Selección de obras de Vivián Trías. Serie Patria chica; 9)

Essays and parliamentary speeches by Socialist writer and politician who died in 1980 constitute mainly diatribes against machinations of Uruguay's upper class, the IMF, and US imperialism.

3664 **Venturini, Angel R.** Estadísticas electorales, 1917–1989 y temas electorales. Montevideo: Ediciones de la Banda Oriental, 1989. 154 p.: ill.

Brief but valuable compendium of electoral statistics is broken down mostly to departmental (i.e., provincial) level, but sometimes to even smaller units.

3665 **Waksman, Guillermo.** Uruguay: la gran derrota de Lacalle. (*Nueva Soc.*, 124, marzo/abril 1993, p. 17–21)

Analyzes Dec. 1992 defeat of a referendum that would have allowed President Lacalle to privatize various State enterprises.

BRAZIL

TIMOTHY J. POWER, *Assistant Professor of Political Science, Louisiana State University*

THE CURRENT LITERATURE ON BRAZILIAN POLITICS AND GOVERNMENT reflects the perception that Brazil is a country in permanent crisis. When 21 years of military-authoritarian rule ended in 1985, many scholars and political practitioners were optimistic about the prospects for democratization and development. In retrospect, this optimism may have been justified in light of the heroic struggle against the military dictatorship, which coincided with a rebirth of Brazilian civil society in which social scientists themselves played no small part. But in recent years, Brazilians have had little to celebrate. Repeated bouts with hyperinflation, the impeachment and subsequent resignation of President Fernando Collor de Mello on corruption charges in 1992, and most importantly, the persistence and even aggravation of scandalous social inequalities have robbed the post-1985 democracy of its promise. Whereas in the early 1980s both Brazilians and Brazilianists wrote glowingly about the *abertura* in the "country of the future," a decade later they were more likely to meet at cheerless conferences and lament the economic crisis, the dysfunctional political system, and the generalized sense of ungovernability.

If what is bad for society is often good for social science, then perhaps it is no surprise that the crisis has had the positive effect of sharpening the analytical rigor of Brazilian and Brazilianist political science. The sense of urgency has made the literature more diagnostic and politically relevant than ever before, as evinced by the burgeoning research on democratic institutions. This trend has coincided with generational change in the scholarly community, both North and South. The laboratory and the frame of reference for younger, emerging researchers is not the military regime of 1964–85, but rather the democratic New Republic. In Brazil, the generation of political scientists who trained abroad in the 1970s and returned to build prestigious research institutions (such as IUPERJ, CEBRAP, IDESP, and CEDEC) has now shaped a new cohort of productive young PhDs. On the US side, a similar phenomenon is occurring. In the works reviewed here, those senior North American scholars who were first tagged as "Brazilianists" in the 1960s and 1970s are virtually absent. Rather, one notes the strong presence of their doctoral students. Many dissertations on Brazilian politics were completed in the late 1980s and early 1990s, and a stream of related books and articles has followed. Illustrating this trend, political science was prominent at the founding meetings of the new Brazilian Studies Association (BRASA) in March 1994.

Because of the widespread perception that the consolidation of Brazilian democracy is incomplete, general appraisals of the New Republic continue to appear. The ills of democracy are discussed in Jaguaribe (item **3714**) and Silva (item **3770**). Houaiss and Coutto provide a sobering report on the failure of the Brazilian model of development (item **3711**), while Almeida addresses the question of why a social democratic solution did not succeed in the early years of democracy (item **3672**). Sosnovskii (item **3774**) and Jerez (item **3715**) offer general treatments of the New Republic written for Russian and Spanish audiences, respectively, with the latter serving as a useful literature review on the new democracy. The symposium edited by Reis Velloso is a serious overview of public policy issues facing the nation in the 1990s (item **3751**), while Kinzo's edited volume is a rare collection of Brazilian analyses published in English (item **3684**).

The multiparty system of the New Republic continues to be fertile ground for studies of parties and elections. An excellent starting point is the bibliographic essay by Lima Jr. *et al.* which provides comprehensive coverage of Brazilian works published between 1978–92 (item **3727**). A compendium of basic information on parties is provided by Monteiro and Oliveira (item **3737**). General works on the party system include those by Lamounier (item **3722**) and Souza (item **3775**), both stressing the weakness of parties. The best contribution on party underdevelopment is by Mainwaring (item **3729**), who in a related paper stresses the infelicitous combination of presidentialism and party fragmentation (item **3730**). Balbachevsky offers an unconventional contribution to the study of party identification in Brazil (item **3678**). Topical studies on voting include the dense essay by Reis and Castro on region and class (item **3758**) and the short paper by Berquo and Alencastro on race (item **3681**). All of these works concern the post-1985 democracy, but the multiparty system of 1946–1964 is revisited in the works by Neves (item **3742**) and D'Araujo (item **3675**).

The study of electoral law, formerly considered esoteric, has flourished recently due to a growing dissatisfaction with Brazil's unique variant of proportional representation (PR). Mainwaring's highly original paper on politicians' shaping of electoral rules has been widely read and cited in Brazil, and is the best introduction to the debate (item **3731**). Two other competent studies of Brazilian PR are those by Silva (item **3768**) and Nicolau (item **3743**). The essays in Pedone's edited volume place the Brazilian electoral system in comparative perspective (item **3771**).

The presidential election of 1989, the first direct popular election of a chief executive since 1960, has understandably provoked a flurry of books and articles. Campaign memoirs were written by Pomar (item **3754**) and Figueiredo (item **3700**), advisers to the two finalists in the presidential runoff, Luis Inácio Lula da Silva of the Workers' Party (PT) and the eventual winner, Fernando Collor de Mello. Sophisticated studies of voting in the 1989 election are provided by Ames (item **3674**) and Kinzo (item **3717**). José Alvaro Moisés used the occasion of the election to conduct some fascinating surveys on Brazilian political culture (item **3735**). Likewise, Silva (item **3765**) and Straubhaar *et al.* (item **3776**) seized the moment to produce the first empirically-grounded studies of the impact of television on voting.

Another election—the plebiscite of April 21, 1993, in which Brazilians rejected a proposed change to parliamentary rule and chose instead to maintain the current presidential system of government—was responsible for another wave of studies, many of which showed political scientists in their advocacy role. Lamounier was the most visible proponent of the parliamentary option (item **3721**). In this he was joined by Almeida (item **3671**), whose essay earned a presidentialist rejoinder from Alencastro (item **3669**). The ongoing debate on constitutional reform led the union lobby DIAP to conduct a valuable survey of federal legislators (item **3685**).

The corruption scandal of 1992, in which Fernando Collor lost his bid to become the second civilian in Brazilian history to be democratically elected and serve a full term as president, inspired several works, including those by Flynn (item **3702**), Silva (item **3766**), and Sives (item **3772**). To date, the most sophisticated analysis of Collor's Administration and eventual impeachment is that of Weyland (item **3779**). A reminder that corruption was not unique to the Collor government is provided by Castello (item **3687**).

Studies of the working class and especially the Workers' Party (PT) remain numerous, in part reflecting the left's dominance in Brazilian academia. Union politics are discussed in Duarte (item **3697**), Silva (item **3691**), and Delgado (item **3695**). Werneck Vianna (item **3778**) and Abramo and Karepovs (item **3741**) study the former Communist Party (PCB). Rodrigues' collection of essays is a sometimes provocative look at the relationship between trade unions and party politics (item **3760**). The PT is examined in many works, reflecting not only its anomalous status as a truly ideological and cohesive party, but also its growing electoral strength. The most useful of these are the compendium by Gadotti (item **3673**) and the book-length analysis of the PT's early years by Meneguello (item **3733**). The party's capture of the São Paulo city government in 1988 led to two analyses of the administration of Mayor Luiza Erundia (items **3713** and item **3718**). Bolaffi analyzes why the party was thrown out of city hall in 1992 (item **3683**), and Novaes warns against the increasing bureaucratization of the PT (item **3744**).

The growing influence of neoliberalism in Brazil has focused new attention on conservative groups. Nylen's article (item **3745**) and Payne's book (item **3749**) are useful studies of business elites in the democratic transition. Pierucci (item **3752**) and Mariano and Pierucci (item **3732**) examine right-wing electoral strategies. This new focus on civilians is complemented by the enduring tradition of research on the most important rightist actor, the armed forces. Hunter provides an excellent discussion of the military's changing mission in democratic Brazil (item **3712**). Conca provides a competent study of military involvement in the defense and high-technology industries (item **3692**), while Wood and Schmink (item **3780**) and Mendonça Barreto (item **3680**) examine military influence in the Amazon region. Revisiting the authoritarian regime of 1964–85, Baffa's exposé on the intelligence community in the 1970s is a chilling reminder of the ongoing importance of civil-military relations (item **3677**). Pinheiro's contributions on extralegal police violence is timely in light of recent tragedies in Brazil (item **3753**). Without a doubt, the political role of the police merits further research in Brazil and in Latin America as a whole.

Among other prominent political actors, the Catholic Church is the subject of five contributions, two of which point to internal divisions within the institution (items **3696** and **3667**). State elites are examined in Hochman's paper on the early social security administration (item **3710**), while Raichelis studies the foot soldiers of social policy in the 1980s (item **3756**). Other miscellaneous contributions worthy of mention are a study of the crisis of municipal governments (item **3698**), an entertaining journalistic account of Brazil's computer industry in the 1980s (item **3694**) and a new biography of Juscelino Kubitschek (item **3670**).

Summarizing the foregoing, if there is an underlying theme linking most of the works on the New Republic, it is the issue of governability. Brazilians and Brazilianists are seeking to discover how the promise of democratization might be reclaimed by effecting social, economic, and political reforms. The quality of many of the works here illustrate the breadth and methodological rigor of Brazil's social-science community, which is arguably the best in the Third World. If Brazil's political practitioners showed the same sophistication and dedication as its political analysts, the crisis of the New Republic would have solved itself long ago.

3666 **Abbaszadeh, Babak.** Urban popular movements, the State and the political system: the case of São Paulo within the Brazilian political structure. (*in* Forging identities and patterns of development in Latin America and the Caribbean. Edited by Harry P. Díaz, Joanna W.A. Rummens, and Patrick D.M. Taylor. Toronto: Canadian Scholars' Press, 1991, p. 185–196)

Provides short introduction to urban social movements in São Paulo, distinguishing them from other poor peoples' movements. Contains little data or analysis.

3667 **Adriance, Madeleine.** Agents of change: the roles of priests, sisters, and lay workers in the grassroots Catholic Church in Brazil. (*J. Sci. Stud. Relig.*, 30:3, 1991, p. 292–305)

This study of nine parishes in Maranhão claims that local organizing is a central variable in the vitality of base communities, thereby calling into question the assertion that episcopal support is the determining factor. Thus current conservative stance of the Vatican may not result in destruction of the grassroots communities.

3668 **Affonso, Almino.** Raízes do golpe: da crise da legalidade ao parlamentarismo. São Paulo: Marco Zero, 1988. 147 p.: appendix, bibl., ill. (Col. Nossos dias)

A seasoned São Paulo politician discusses crisis of 1961–64. Author, a former minister and Goularts floor leader in Congress, was driven into exile by military coup of 1964. Appendix contains relevant speeches and correspondence.

Aggio, Alberto. Democracia e socialismo: a experiência chilena. See item **3418**.

Alario, Margarita. Environmental policy enactment under the military: some generalities between Brazil and Chile. See item **4999**.

3669 **Alencastro, Luiz Felipe de.** Cultura democrática e presidencialismo no Brasil. (*Novos Estud. CEBRAP*, 35, março 1993, p. 21–30)

Subtle and careful argument against eliminating a directly elected president. Brazilian culture is traditionally presidentialist, and this tendency was exacerbated by the plebiscitarian elections under military rule. Eliminating the presidency would frustrate the millions of new voters who fought for democracy. Subnational politics will always cause politicians (governors, mayors) to harbor presidential ambitions.

3670 Alexander, Robert Jackson. Juscelino Kubitschek and the development of Brazil. Athens, Ohio: Ohio Univ. Center for International Studies, 1991. 429 p.: bibl., map. (Monographs in international studies. Latin America series; 16)

In contrast to previous studies of Kubitschek's Administration, this admiring biography emphasizes Kubitschek the man. Contains the predictable treatments of industrialization and of Brasília, but also includes unconventional chapters on foreign policy, the "cultural renaissance," and Kubitschek the ex-president. Highly descriptive: everything you ever wanted to know about Juscelino.

3671 Almeida, Maria Hermínia Tavares de. Em defesa da mudança. (*Novos Estud. CEBRAP*, 35, março 1993, p. 15–20)

Provides most of the standard arguments for the abandonment of presidentialism and the adoption of a parliamentary government in Brazil. Argues that party and electoral system reform would be more likely under parliamentarism.

3672 Almeida, Maria Hermínia Tavares de. Reformismo democrático em tempos de crise. (*Lua Nova*, 22, dez. 1990, p. 189–205, bibl.)

Short, provocative essay on failure of social democracy in Brazil asks why there are so many so-called reformers but so little reform. Transition to democracy was a long-awaited opportunity for social democrats, but democratization coincided with the crisis of political institutions and the broadening appeal of neoliberal agendas.

3673 Alves, Maria Helena Moreira. Something old, something new: Brazil's Partido dos Trabalhadores. (*in* The Latin American left: from the fall of Allende to perestroika. Boulder, Colo.: Westview Press, 1993, p. 225–242)

Brief, sympathetic introduction to the Partido dos Trabalhadores focuses on party's organizational innovation and on its performance in 1989 presidential election. Author is reluctant to discuss the many problems that plague the party.

3674 Ames, Barry. The reverse coattails effect: local party organization in the 1989 Brazilian presidential election. (*Am. Polit. Sci. Rev.*, 88:1, March 1994, p. 95–111)

Sophisticated study shows how election was fought on the ground and demonstrates ongoing importance of political parties. Presidential candidates did significantly better in municipalities where their party controlled town hall. Author also examines how spatial factors affect strategies of local politicians, particularly their endorsements and alliances. Not for the statistically squeamish. Recommended.

3675 Araújo, Maria Celina Soares d'. O PTB na cidade do Rio de Janeiro, 1945–55. (*Rev. Bras. Estud. Polít.*, 74/75, jan./junho 1992, p. 182–231, tables)

Lengthy analysis examines first ten years of the old Partido Trabalhista Brasileiro (PTB), emphasizing development of its political machine in the capital city of Rio de Janeiro and tendency to 'bossism' within the party structure.

3676 Assies, Willem. Urban social movements and local democracy in Brazil. (*Rev. Eur.*, 55, 1993, p. 39–58, bibl.)

Discusses experiments in democratic administration at the municipal level. Emphasizes 1989–92 period, when number of experiments increased in the wake of 1988 municipal elections and implementation of the new Brazilian constitution affected municipal administration. [R. Hoefte]

3677 Baffa, Ayrton. Nos porões do SNI: o retrato do monstro de cabeça ôca. Rio de Janeiro: Editora Objetiva, 1989. 171 p.

Prize-winning exposé on the intelligence agency of former military regime has been compiled by a tenacious journalist. Reprints hundreds of reports, documents, and wiretaps illustrating the "antisubversive" domestic surveillance of the Médici and Geisel governments. Little analysis, but documents stand on their own as eloquent testimony to the victims of military repression in 1970s. Highly recommended as primary source material.

3678 Balbachevsky, Elizabeth. Identidade partidária e instituições políticas no Brasil. (*Lua Nova*, 26, 1992, p. 133–165, bibl., tables)

Basing her argument on survey research, author claims that level of party identification is not greatly different from that observed in other countries. Concludes that while Brazilian parties are indeed weak, reasons for this situation are not to be found in characteristics of the electorate. Provocative essay with interesting data.

3679 **Banck, Geert Arent.** Cultura política brasileira: que tradição é esta? (*Rev. Bras. Estud. Polít.*, 76, jan. 1993, p. 41–53, bibl.)

Writing from perspective of cultural anthropology, author takes impressionistic look at recent work on political culture by Brazilian political scientists. Emphasizes culture rather than politics, but more can be gained by reading the original works.

3680 **Barreto, Kátia Marly Mendonça.** Forças armadas e Amazônia na conjuntura política atual. (*Cadernos/Belém*, 22, out./dez. 1990, p. 51–77, bibl.)

Despite democratization, armed forces have not abandoned but rather updated their national security ideology. Anticommunism has been replaced by struggle to prevent alleged "internationalization" of Amazônia, and the "internal enemies" are now non-governmental organizations (NGOs) active in the region. Revealing analysis of internal documents from the Escola Superior de Guerra.

Barretto, Vicente. Evolução do pensamento político brasileiro. See *HLAS 54:5476*.

3681 **Berquó, Elza** and **Luiz Felipe de Alencastro.** A emergência do voto negro. (*Novos Estud. CEBRAP*, 33, julho 1992, p. 77–88, tables)

In this rare study of the impact of race on Brazilian elections, authors find that Afro-Brazilians generally do not support idea of voting for candidates based on their race, and conclude that black movement has underestimated its own electoral clout. Title of the article and optimistic conclusions are not supported by authors' own data, but unconventional topic of this study makes it worthwhile reading.

3682 **Betto, *Frei.*** Lula: biografia política de um operário. 2a. ed. São Paulo: Estação Liberdade, 1989. 79 p., 16 p. of plates: photos.

Short deification of the labor leader and presidential candidate is written by a liberation theologian close to the Partido dos Trabalhadores. Covers mostly the early years of Lula's political career. Marketed for campaign of 1989, book is rich in photographs.

3683 **Bolaffi, Gabriel.** A campanha eleitoral de Eduardo Suplicy. (*Novos Estud. CEBRAP*, 35, março 1993, p. 238–245)

Presents critical analysis of defeat of Partido dos Trabalhadores (PT) in the 1992 São Paulo election. The PT candidate unwisely tried to dissociate himself from the incumbent PT mayor. The argument is "let the PT be the PT." Interesting for understanding internal debates within the PT, which is the only party in Brazil that bothers to have them.

3684 **Brazil, the challenges of the 1990s.** Edited by Maria D'Alva Gil Kinzo. London: Institute of Latin American Studies, Univ. of London; British Academic Press, 1993. 214 p: bibl., index.

Collection of useful essays examines economic policy, social policy, and political and electoral issues. The contributors are all recognized authorities in their fields, and papers provide broad overview of the performance of democracy in the New Republic. Recommended.

3685 **A cabeça do Congresso: quem é quem na revisão constitucional.** São Paulo: Oboré, 1993. 383 p.: ill., index.

This most comprehensive survey of federal legislators ever undertaken in Brazil was conducted in advance of the constitutional revision process scheduled for 1993–94. Although the special assembly went nowhere, this survey by the Departamento Intersindical de Assessoria Parlamentar remains a valuable register of elite opinion. Presents short biographies of all 570 legislators who served from 1991–95, as well as the opinions of 418 respondents to the wide-ranging 1993 survey on political and economic reforms. A valuable research tool.

Câmara Neto, Alcino Ferreira *et al.* Aquarella do Brasil: ensaios políticos e econômicos sobre o governo Collor. See item **2121**.

3686 **Campos, Roberto de Oliveira.** O século esquisito: ensaios. Rio de Janeiro: Topbooks, 1990. 287 p.

Contains collection of op-ed columns by the high priest of Brazilian conservatism, published from 1988–90. Having principles and a consistent ideology, author is a true *rara avis* on the Brazilian right. His incisive, provocative, and often infuriating columns argue relentlessly for his radical brand of laissez-faire capitalism, and in the process cover everything from Collor to the Amazon, from Dukakis to the Berlin Wall.

3687 Castello, José Carlos Bruzzi. Os crimes do presidente: 117 dias na CPI da Corrupcão. Porto Alegre, Brazil: L&PM Editores, 1989. 153 p.

Title may bring Fernando Collor to mind, but this book examines corruption scandals which plagued the Sarney Administration of 1985–90. Provides in-depth coverage of investigation by a special congressional committee, whose damaging 1988 report was swept under the rug and later forgotten due to Collargate.

Chaffee, Lyman G. Political protest and street art: popular tools for democratization in Hispanic countries. See item **2736**.

3688 Chico Mendes: the defence of life. (*J. Peasant Stud.*, 20:1. Oct. 1992, p. 160–176)

This interview with the famed organizer of forest peoples was conducted in September 1988, three months before he was killed on orders from local landowers. Mendes discusses extensively the social, environmental, and political problems of the Amazon region.

3689 Cleary, David. After the frontier: problems with political economy in the modern Brazilian Amazon. (*J. Lat. Am. Stud.*, 25:2, May 1993, p. 331–349)

Offers revisionist political economy of Amazonian development. Most interpretations rely on the concept of frontier as a theoretical construct. Classical frontier theory is increasingly irrelevant, as Amazonia has moved into the post-frontier era. Provides good review of critical social science thinking on the region.

3690 Comin, Alvaro A. and **Carlos Alberto Marques Novaes.** Alguém tem que perder. (*Novos Estud. CEBRAP*, 36, julho 1993, p. 87–107)

In wide-ranging interview, Tasso Jereissati, the respected former governor of Ceará, speaks in his capacity as president of the Partido da Social Democracia Brasileira. Interesting discussion of problems of poverty, development, and redistribution of income in a democratic context.

3691 Comin, Alvaro A. Contra a maré: entrevista de Vicente Paulo da Silva a Alvaro A. Comin. (*Novos Estud. CEBRAP*, 33, julho 1992, p. 129–146)

Relates interview with "Vicentinho," a leader of the Central Unica dos Trabalhadores (CUT), the labor confederation affiliated with the Partido dos Trabalhadores. Interviewee discusses internal politics and organization of the CUT, providing insight into how the radical trade union movement viewed neoliberal economic reforms undertaken by Collor government in early 1990s.

3692 Conca, Ken. Technology, the military, and democracy in Brazil. (*J. Interam. Stud. World Aff.*, 34:1, Spring 1992, p. 141–177, bibl.)

Disputing contention that military industrialization has a professionalizing effect on the armed forces, author argues that persistent military control over the defense sector has impeded the advance of democratization in Brazil. Special attention is given to military's role in the defense, nuclear, and aerospace industries.

3693 Contreras, Mario *et al.* Perfil del Brasil contemporáneo. México: Univ. Nacional Autónoma de México, Coordinación de Humanidades, 1987. 207 p.: bibl., ill. (Nuestra América; 16)

Uneven collection of essays looks at various aspects of Brazilian politics. Chapters by Bolívar Lamounier on urban voting and Paul Israel Singer on economic history are worthwhile.

3694 Dantas, Vera. Guerrilha tecnológica: a verdadeira história da política nacional de informática. Rio de Janeiro: Livros Técnicos e Científicos, 1988. 302 p.: bibl., ill.

Well-written journalistic account examines Brazil's attempt to build a national computer industry under the "market reserve" policy in 1980s. Contains wealth of information on the intrigue surrounding sector bureaucrats, the intelligence community, and the nascent informatics firms. Recommended.

3695 **Delgado, Lucília de Almeida Neves.**
Entre o velho e o novo: a CGT em discussão. (*Análise Conjunt.*, 6:1, jan./abril 1991, p. 97–115, bibl.)

Descriptive study examines rebirth of Brazilian unionism after 1979, focusing on the Confederação Geral dos Trabalhadores (known as Conferência das Classes Trabalhadoras until 1986). A valuable corrective to the literature, which has tended to emphasize mostly the Partido dos Trabalhadores-affiliated Central Unica dos Trabalhadores (CUT). Excellent for understanding ideological and tactical controversies within organized labor as a whole.

3696 **Drogus, Carol Ann.** Popular movements and the limits of political mobilization at the grassroots in Brazil. (*in* Conflict and competition: the Latin American Church in a changing environment. Edited by Edward L. Cleary and Hannah Stewart-Gambino. Boulder, Colo.: Lynne Rienner Publishers, 1992, p. 87–103)

Presents case study of women's participation in Catholic grassroots group in São Paulo. Author finds that women activists in São Paulo Christian Base Communities are alienated by conservative turn of the archdiocese in recent years. The Church created its own dilemma by awakening women to lay activism but simultaneously limiting their role within the Church.

3697 **Duarte, Ozeas.** Os mercadores de ilusões: uma análise crítica do "sindicalismo de resultado." São Paulo: Brasil Debates Editora, 1988. 62 p.: bibl.

In a brief tract, a supporter of the Partido dos Trabalhadores-affiliated Central Unica dos Trabalhadores (CUT) attacks political strategies of moderate labor leaders, especially the current associated with Antônio Rogério Magri. Useful for understanding ideological struggle within different sectors of the labor movement.

3698 **Fernandes, Edésio.** Juridico-political aspects of metropolitan administration in Brazil. (*Third World Plan. Rev.*, 14:3, Aug. 1992, p. 227–243, tables)

City planner from Belo Horizonte discusses basic problems of urban planning in Brazil. Argues that the Constitution of 1988, although permitting some decentralization of power, continues to prioritize the traditional (legal) definition of "municipalities" over the emerging reality of "metropolitan areas." Metropolitan areas are increasingly the relevant units for policy action.

3699 **Figueira, Ricardo Rezende.** Violência no campo: depoimento na Comissão Parlamentar de Inquérito da Assembléia Legislativa do Estado do Pará. (*Cad. CEAS*, 134, julho/agôsto 1991, p. 13–26, map, table)

In his testimony before the Pará state legislature, Father Ricardo Rezende Figueira, a leader of the struggle for land reform in the municipality of Rio Maria, denounces the ongoing violence and death threats against rural organizers.

3700 **Figueirêdo, Ney Lima** and **José Rubens de Lima Figueiredo Júnior.** Como ganhar uma eleição: lições de campanha e marketing político. São Paulo: Cultura Editores Associados, 1990. 217 p.: bibl., ill.

This felicitous collaboration between a political scientist and a political consultant who advised the 1989 Collor campaign illustrates how imported campaign techniques have changed the nature of Brazilian elections, and makes for excellent reading. Essential documentation for study of the 1989 election.

3701 **Fiorin, José Luiz.** O regime de 1964: discurso e ideologia. São Paulo: Atual Editora, 1988. 158 p.: bibl. (Série Lendo)

This impenetrable attempt to deconstruct the discourse of the post-1964 military regime is strictly for connoisseurs of postmodern linguistics.

3702 **Flynn, Peter.** Collor, corruption and crisis: time for reflection. (*J. Lat. Am. Stud.*, 25:2, May 1993, p. 351–371)

Written in aftermath of Fernando Collor's impeachment and the inauguration of Itamar Franco, article is mostly a retrospective study of the Collor years. Author is optimistic about democratic forces unleashed during the impeachment process and about prospects for Franco government.

3703 **Uma foice longe da terra: a repressão aos sem-terra nas ruas de Porto Alegre.** Organização de Sérgio Antônio Görgen. Petrópolis, Brazil: Vozes, 1991. 176 p.: ill.

Documents, interviews, and analysis concern the bloody clash between police and landless peasants in the streets of Porto Alegre on Aug. 8, 1990, presenting protesters' version of events. Title is a macabre reference to the killing of a policeman. A one-sided account of the incident, but also illustrates pressing need for agrarian reform.

3704 **Gadotti, Moacir** and **Otaviano Pereira.** Pra que PT: origem, projeto e consolidação do Partido dos Trabalhadores. Prefácio de José Dirceu de Oliveira e Silva. Posfácio de José Genoíno Neto. São Paulo: Cortez Editora, 1989. 370 p.: bibl., ill.

Edited by intellectuals close to the Partido dos Trabalhadores (PT), this thick collection of party documents and short analyses amounts to a historical handbook of the PT. Purpose of book is to "sell" the PT, but the compendium is a valuable source for study of the party.

3705 **Galvan, Cesare Giuseppe.** Ciência, tecnologia e programas nucleares brasileiros: os militares. (*Geosul/Florianópolis*, 6:11, 1991, p. 7–34, bibl.)

Retrospective essay examines nuclear energy program begun under the military regime. The emphasis on nuclear energy is explained as an outcome of technological catch-up and as part of the legitimation strategy of the military government.

3706 **German, Cristiano.** Igreja versus Governo: opções políticas na transição democrática brasileira. (*Hist. Quest. Debates*, 10:18/19, junho/dez. 1989, p. 235–259, bibl., tables)

Analyzes political impact of the Catholic Church from 1985–88. Reviewing Church lobbying during Constituent Assembly of 1987–88 on several important issue-areas, author concludes that the new constitution represents a compromise between the progressive Church and the conservative Sarney government.

González Encinar, José Juan et al. El proceso constituyente: enseñanzas a partir de cuatro casos recientes; España, Portugal, Brasil y Chile. See item **2757.**

3707 **Grupos de pressão.** Organização de João Paulo Machado Peixoto e Walter Costa Porto. Brasília: Instituto Tancredo Neves; Fundação Friedrich Naumann, 1988. 122 p.

Contains transcript of a seminar on group politics held in the Brazilian Congress. Because most of the participants are politicians, presentations are delivered in diplomatese with little analytical content.

3708 **Gurgel, Cláudio.** Estrelas e borboletas: PT, origens e questões de um partido a caminho do poder. Rio de Janeiro: Editora Papagaio, 1989. 149 p.: bibl., ill.

Written in anticipation of 1989 presidential elections, book addresses question of whether the Partido dos Trabalhadores (PT) is ready for power. For the author, a longtime party militant, the answer is obvious. However, contains some useful descriptive material about founding of the PT.

3709 **Hewitt, W.E.** The Roman Catholic Church and environmental politics in Brazil. (*J. Dev. Areas*, 26, Jan. 1992, p. 239–258)

Reviews Church's policies and programs relating to the environment. In its support of agrarian reform and protection of indigenous peoples, Church's environmental activism is indirect. But Vatican's undermining of the progressive Church may force a downplaying of social justice strategies, once again making the environment a non-issue.

3710 **Hochman, Gilberto.** Os cardeais da Previdência Social: gênese e consolidação de uma elite burocrática. (*Dados/Rio de Janeiro*, 35:3, 1992, p. 371–401)

Well-researched study examines early years of the Instituto de Aposentadoria e Pensões dos Industriários (IAPI), the industrial workers' pension fund created by Vargas in 1937, and the bureaucratic elites who dominated the social security system until the 1960s. Predominance of these elites is explained as an outcome of administrative reform and of Vargas' desire to effect a controlled political incorporation of workers.

3711 **Houaiss, Antônio** and **Pedro de Coutto.** Brasil: o fracaso do conservadorismo. São Paulo: Editora Atica, 1989. 79 p. (Série Temas; 13. Estudos políticos)

Long essay mounts devastating attack on social costs of the contemporary Brazilian model of development. Effectively points out contradiction between discourse of reformers (as in *tudo pelo social*) and the fact that nothing ever changes. Depressing but necessary to digest.

3712 **Hunter, Wendy.** The Brazilian military after the Cold War: in search of a mission. (*Stud. Comp. Int. Dev.*, 28:4, Winter 1994, p. 31–49)

Provides excellent overview of identity crisis facing armed forces in 1990s. With external threats gone, military is reluctantly beginning to assume some domestic noncombatant functions. The way that civilian leaders handle this transformation of mission is critical to success of democratization. Refreshing for its balanced, non-deterministic treatment of military's changing role in democratic Brazil.

3713 **Jacobi, Pedro.** Gestión municipal y conflicto: el municipio de *São Paulo.* (*in* Ciudades y gobiernos locales en la América Latina de los noventa. México: FLACSO, Sede México, 1991, p. 95–108)

Sympathetic overview of Partido dos Trabalhadores government in São Paulo under Mayor Luiza Erundina de Souza (1989–93) is written by an academic close to the party. Discusses tradeoffs between ensuring popular participation in the decision-making process and producing immediate administrative results.

3714 **Jaguaribe, Hélio.** O caso brasileiro: o sistema público brasileiro. (*in* Sociedade, Estado e partidos na atualidade brasileira. São Paulo: Paz e Terra, 1992, p. 199–235)

Comprehensive study examines both the State apparatus and the entire political system. Argues that reform of the State and of representative institutions must be simultaneous and total or Brazil will continue on its drift toward ungovernability. Cogently presents the current crisis and its historical antecedents.

3715 **Jerez, Ariel.** Política en Brasil. (*Rev. Estud. Polít.*, 74, oct./dic. 1991, p. 645–656, bibl.)

Published in Spain for a European audience, this useful review essay covers eight books written between 1988–90 on the Brazilian transition to democracy. Analyzes works written or edited by Alfred Stepan, Guillermo O'Donnell, and Bolívar Lamounier, among others.

3716 **Kinzo, Maria D'Alva Gil.** La cuestión partidaria en Brasil. (*in* Los partidos políticos en el inicio de los noventa: seis casos latinoamericanos. Santiago: Ediciones FLACSO, 1992, p. 7–15, bibl.)

In transcript of conference remarks author discusses weakness of political parties in Brazil, but singles out the experience of the Partido dos Trabalhadores as an anomaly which holds promise for future change.

3717 **Kinzo, Maria D'Alva Gil.** The 1989 presidential election: electoral behaviour in a Brazilian city. (*J. Lat. Am. Stud.*, 25: 2, May 1993, p. 313–330, tables)

Analyzes 1989 presidential election in municipality of Presidente Prudente in São Paulo state. Profiles supporters of the four candidates who performed well in the city, giving special attention to the contrasting social bases of Collor and Lula. Survey research measures politicization and prevalence of democratic attitudes within the electorate.

3718 **Kowarick, Lúcio** and **André Singer.** A experiência do Partido dos Trabalhadores na prefeitura de São Paulo. (*Novos Estud. CEBRAP*, 35, março 1993, p. 195–216, tables)

Excellent study examines 1989–93 administration of Partido dos Trabalhadores (PT) mayor Luiza Erundina de Souza. The PT became increasingly pragmatic, shifting from a *basista* style of government to one in which the city administration itself became the engine for progressive change. Incisive analysis of both problems and undeniable achievements of this remarkable experiment in democracy. Recommended.

3719 **Krischke, Paulo J.** Church base communities and democratic change in Brazilian society. (*Comp. Polit. Stud.*, 24:2, July 1991, p. 186–210)

Theoretically well-grounded study, influenced by Habermas' theory of communicative action, examines role of Christian Base Communities (CBCs) in transition to democracy. CBCs emphasize motivational capacity rather than direct politicization, and thus reduce submissiveness to authoritarian traditions. These changes are examined in urban São Paulo and in two rural areas.

3720 **Lamounier, Bolívar.** El modelo institucional de los años treinta y la presente crisis brasileña. (*Desarro. Econ.*, 32:126, julio/sept. 1992, p. 185–198, tables)

Interprets model of presidentialism and the strong State which emerged in Brazil in the 1930s and continues to this day. Argues that the model has exhausted itself, and makes the case for sweeping reforms in the form of government, the party system, and the electoral system.

3721 Lamounier, Bolívar. Parlamentarismo, sistema eleitoral e governabilidade. (*Nova Econ.*, 2:2, nov. 1991, p. 9–25, bibl., table)

Straightforward comparison of presidentialism and parliamentarism is made by a prominent political scientist who advocates the latter. Emphasis is on the electoral systems that are compatible with these two forms of government. Writing to reach nonspecialists, author argues that the solution for Brazil is a combination of parliamentarism with proportional representation.

3722 Lamounier, Bolívar. Partidos e utopias: o Brasil no limiar dos anos 90. São Paulo: Edições Loyola, 1989. 150 p.: bibl., ill. (Col. Temas brasileiros; 7)

Five interesting essays written during 1987–88 focus on the party system and governability. Topics include Constituent Assembly of 1987–88 and 1988 municipal elections. Contains wealth of data on elections and congressional behavior of parties. Recommended.

3723 Lamounier, Bolívar and Rachel Meneguello. Partiti e consolidamento democratico: il caso brasiliano. (*in* Democrazia e partiti in America Latina. Milano, Italy: FrancoAngeli, 1991, p. 75–136, bibl.)

Brief history of Brazilian political parties demonstrates that, in contrast to surrounding countries, none of present-day parties existed before World War II, a factor causing a "permanent structural problem" in Brazilian society. Authors then undertake a comparative analysis of redemocratization and democratic consolidation in Brazil in that context. They explore persistent fragility of Brazilian parties, inability of the constitution of 1946 to stop the process of party disappearance, the present viability of parties, and possibilities for reinforcing party stability. In light of their findings, authors conclude by calling for more intensive study of the era following the Estado Novo. [V. Peloso]

3724 Lee, Naeyoung and Jeffrey Cason. Automobile commodity chains in the NICs: a comparison of South Korea, Mexico, and Brazil. (*in* Commodity chains and global capitalism. Edited by Gary Gereffi and Miguel Korzeniewicz. Westport, Conn.: Praeger, 1994, p. 223–243, bibl., graphs, tables)

Authors compare the three countries in terms of their different strategies of integration into the global auto market, focusing on parts supply networks, assembly networks, and marketing networks. Corrects common misconception that only in East Asia have newly industrializing countries made headway in exporting automobiles.

3725 Lehman, Howard P. and Jennifer L. McCoy. The dynamics of the two-level bargaining game: the 1988 Brazilian debt negotiations. (*World Polit.*, 44:4, July 1992, p. 600–644)

Debt negotiations are modeled as a two-level game in which each of the parties (the Brazilian government and international creditors) must somehow achieve an international accord while simultaneously satisfying their domestic constituencies. Paradoxically, the weakness of the Brazilian State during the democratic transition may have strengthened its bargaining position vis-à-vis foreign creditors.

3726 Lehman, Howard P. Strategic bargaining in Brazil's debt negotiations. (*Polit. Sci. Q.*, 108:1, Spring 1993, p. 133–155)

Presents game-theoretic analysis of negotiations to reschedule Brazil's foreign debt held in late 1980s. Author demonstrates how Brazilian government and international creditors learned to respect one another's interests, eventually producing a mutually acceptable second-best agreement.

3727 Lima Júnior, Olavo Brasil de; Rogério Augusto Schmitt; and **Jairo César Marconi Nicolau.** A produção brasileira recente sobre partidos, eleições e comportamento político: balanço bibliográfico. (*ANPOCS BIB*, 34, segundo semestre 1992, p. 3–66, tables)

Lengthy annotated bibliography lists works on Brazilian parties and elections, updating similar 1978 work by Bolívar Lamounier and Maria D'Alva Gil Kinzo (see *HLAS 42:48*). Reviews 54 books, 50 M.A. and Ph.D. theses, and 121 scholarly articles published in Brazil from 1978–92. An indispensable guide to the literature. Recommended.

3728 MacRae, Edward. Homosexual identities in transitional Brazilian politics. (*in* The making of social movements in Latin America: identity, strategy, and democracy.

Edited by Arturo Escobar and Sonia E. Alvarez. Boulder, Colo.: Westview Press, 1992, p. 185–203)

Examines emergence and development of gay and lesbian activism from late 1970s–early 1990s, placing homosexual groups in general context of the rebirth of civil society and the transition to democracy.

3729 **Mainwaring, Scott.** Brazilian party underdevelopment in comparative perspective. (*Polit. Sci. Q.*, 107:4, Winter 1992/1993, p. 677–707, tables)

Argues that Brazil presents one of the most severe cases of party underdevelopment among contemporary democracies. Explains party weakness as a result of four factors: characteristics of the electorate, the strength of the State, the combination of presidentialism and multipartyism, and the behavior of freewheeling politicians acting under permissive electoral laws. Recommended.

3730 **Mainwaring, Scott.** Democracia presidencialista multipartidária: o caso do Brasil. (*Lua Nova*, 28/29, 1993, p. 21–74, bibl., tables)

Examines problems of combining presidential rule with fragmented multiparty systems. Emphasis is on executive-legislative relations under minority presidents. Presidents cannot depend on party support in the legislature and thus must seek support elsewhere, with deleterious consequences for democratic stability. Brazil is placed in comparative perspective.

3731 **Mainwaring, Scott.** Politicians, parties, and electoral systems: Brazil in comparative perspective. (*Comp. Polit.*, 24:1, Oct. 1991, p. 21–43, tables)

Excellent analysis explains why Brazilian democracy is weakened by permissive electoral laws. Brazilian politicians have consistently shaped electoral legislation to encourage clientelism rather than representation of clearly defined social groups. An innovative approach that does not take electoral laws as a given, but rather explains them as the conscious choices of rational politicians. Recommended.

3732 **Mariano, Ricardo** and **Antônio Flávio Pierucci.** O envolvimento dos pentecostais na eleição de Collor. (*Novos Estud. CEBRAP*, 34, nov. 1992, p. 92–105)

Argues that Pentecostals supported Collor in 1989 presidential election because of traditional anticommunism and because of fears that a Partido dos Trabalhadores victory would bring the progressive Catholic Church to power in a Lula government. Contains useful information, but analysis is colored by authors' politics. Pentecostal ministers are portrayed as ignorant reactionaries.

3733 **Meneguello, Rachel.** PT: a formação de um partido, 1979–1982. São Paulo: Paz e Terra, 1989. 228 p: bibl.

The most important book-length study of Partido dos Trabalhadores (PT) yet to appear in Brazil, work is well researched and documented, and informed by relevant political science literature on party formation. Handicapped by its temporal restriction, but one of the best sources on how the PT was born of labor activism in the late 1970s.

3734 **Miranda, Edson.** Chapéu de palha: o segundo governo Arraes. São Paulo: Editora Alfa-Omega, 1991. 129 p.: ill. (Biblioteca Alfa-Omega de cultura universal. Série 1a., Col. Esta América; 27)

Memoir of second administration of Miguel Arraes in Pernambuco (1987–90) is written by his chief assistant for agrarian reform and rural relief. Interesting study of Northeast politics.

3735 **Moisés, José Alvaro.** Elections, political parties and political culture in Brazil: changes and continuities. (*J. Lat. Am. Stud.*, 25, Oct. 1993, p. 575–611)

Well-crafted, provocative study of political culture and democracy in Brazil relies on surveys conducted during 1989 presidential election. Plebiscitarian voting patterns shaped under military rule continue under democracy. Without the military to vote against, the electorate now seeks to punish corruption and patrimonial style of politicians. A gold mine of data on political opinion. Recommended.

3736 **Monclaire, Stéphane** and **Clóvis de Barros Filho.** A política da constituinte: de fevereiro/87 a março/88. Brasília: Instituto Tancredo Neves, 1988. 80 p.

In an unusual book, a graduate student interviews a São Paulo jurist and a Sorbonne Brazilianist during the constitutional convention. A play-by-play analysis of the politics of the assembly.

3737 **Monteiro, Brandão** and **Carlos Alberto Pereira Oliveira.** Os partidos políticos. São Paulo: Global Editora, 1989. 190 p.: bibl.

Descriptive review of contemporary party system is followed by reprints of party manifestos and platforms. Worthwhile only for convenience of having all the party documents collected in one place.

3738 **Montoro, André Franco.** El futuro de la democracia en América Latina. (*in* Democracia 2000: los grandes desafíos en América Latina. Bogotá: Fundación Simón Bolívar; Tercer Mundo Editores, 1991, p. 345–373)

Address by former governor of São Paulo to a conference on democracy in Latin America consists mostly of platitudes about necessity of consolidating democracy and moving toward regional integration.

3739 **Moraes, João Quartim de.** A esquerda militar no Brasil. v. 1, Da conspiração republicana à guerrilha dos tenentes. São Paulo: Edições Siciliano, 1991. 1 v: bibl.

Vol. 1 of a valuable history of leftist tendencies within the military is restricted to Old Republic. Useful for understanding early years of the republican army.

3740 **Movimentos sociais no campo.** Prefácio de Octavio Ianni. Curitiba, Brazil: Criar Edições; Scientia et Labor-Ed. da Univ. Federal do Paraná, 1987. 146 p.: bibl., maps.

Collection of essays by sociologists and anthropologists focuses on rural social movements and struggle for land reform in Paraná state. Essays are generally well-researched and theoretically informed.

3741 **Na contracorrente da história: documentos da Liga Comunista Internacionalista, 1930–1933.** Organização de Fulvio Abramo e Dainis Karepovs. Prefácio de Pierre Broué. São Paulo: Brasiliense, 1987. 179 p.: port.

Collection of documents is useful for studying early years of Partido Comunista Brasileira.

3742 **Neves, Maria Manuela Renha de Novis.** Elites políticas: competição e dinâmica partidário-eleitoral; o caso de Mato Grosso. Rio de Janeiro: Instituto Universitário de Pesquisas do Rio de Janeiro; São Paulo: Vértice, 1988. 228 p.: bibl. (Formação do Brasil; 7)

Author's revised master's thesis examines party competition among Mato Grosso elites from 1945–65. Well-researched study of factionalism illustrates that all politics is local. Important addition to literature on pre-1964 multiparty system, which has traditionally focused on national-level competition.

3743 **Nicolau, Jairo César Marconi.** A representação política e a questão da desproporcionalidade no Brasil. (*Novos Estud. CEBRAP*, 33, julho 1992, tables)

Competent study of proportional representation in modern Brazil focuses on disproportionality in the representation of federal units in the Câmara dos Deputados. Argues for creation of a supplementary, nation-wide electoral district which would give parliamentary access to parties underrepresented by the current system.

3744 **Novaes, Carlos Alberto Marques.** PT: dilemas da burocratização. (*Novos Estud. CEBRAP*, 35, março 1993, p. 217–237, tables)

Valuable study examines internal structure of the Partido dos Trabalhadores (PT). PT is too dependent on resources generated by electoral success. It must reduce this dependency or risk the bureaucratization which weakened European socialist parties. Author's status as party militant does not compromise his analysis, which is self-critical and based on excellent empirical research.

3745 **Nylen, William R.** Selling neoliberalism: Brazil's Instituto Liberal. (*J. Lat. Am. Stud.*, 25:2, May 1993, p. 301–311)

Interesting study relates how business elites organized to propagate neoliberal ideology in 1980s. Compared to their counterparts in early 1960s, the neoliberals are democratic insofar as they oppose authoritarian State intervention in the economy. By 1989 these groups had seen their ideas adopted by virtually every sector except the unreconstructed left.

3746 **Oliveira, Ariovaldo Umbelino de.** A geografia das lutas no campo. São Paulo: Editora Contexto; Editora da Univ. de São Paulo, 1988. 101 p.: ill. (Repensando a geografia)

Social geography of political flashpoints in Brazilian interior focuses on struggle for agrarian reform and ongoing threats to indigenous peoples. Politically charged, but a solid guide to the issues.

3747 **Oliveira, Odete Maria de.** A questão nuclear brasileira: um jogo de mandos e desmandos. Florianópolis, Brazil: Editora da UFSC, 1989. 201 p.: bibl.

Adequate history of Brazil's nuclear program written in a dry, factual style, but with useful information on many aspects of the program. Brazil's foray into nuclear energy is criticized as a history of bureaucratic errors and poorly informed decision making.

3748 **Passarinho, Jarbas Gonçalves.** Na planície. Belém, Brazil: Cultural CEJUP, 1990. 175 p.: bibl.

In the first installment of his planned multivolume autobiography, this luminary of the military regime covers the period from his military commission to his appointment as a minister in the Costa e Silva government. A defense of the authoritarian regime by one of its key players, and an essential memoir of the period.

3749 **Payne, Leigh A.** Brazilian industrialists and democratic change. Baltimore, Md.: Johns Hopkins Univ. Press, 1993. 216 p.: bibl., index.

Excellent study examines politics of business elites during transition to democracy. The conditions that caused industrialists to throw their support behind authoritarianism in 1964 will not necessarily obtain in the New Republic. Business elites desire competent, legitimate, and stable governments, and may now be ready to coexist with a truly social democratic government. Recommended.

3750 **Pellizzari, Deoni.** A grande farsa da tributação e da sonegação. Petrópolis, Brazil: Vozes, 1990. 132 p.: bibl.

Overview of tax evasion in Brazil is written in an appropriately ironic tone by a State official. In Brazil taxes are not paid by corporations, but by consumers. The latter rarely complain, while former whine loudly, break the law, or both. Not for an academic audience, but contributes to an understanding of the fiscal crisis of the Brazilian State.

3751 **As perspectivas do Brasil e o novo governo.** Coordenação de João Paulo dos Reis Velloso. São Paulo: Nobel, 1990. 304 p.

Papers from the Jan. 1990 meeting of the Fórum Nacional during the presidential transition from Sarney to Collor include wide-ranging overview of presentations by notables from all important sectors of politics and society. Some essays are serious analyses of economic and social policy. A useful dialogue on challenges facing Brazil.

3752 **Pierucci, Antônio Flávio** and **Marcelo Coutinho de Lima.** São Paulo 92, a vitória da direita. (*Novos Estud. CEBRAP*, 35, março 1993, p. 94–99, graphs)

Interesting work analyzes victory of rightist candidate Paulo Maluf in 1992 São Paulo mayoral election. Growth of the Partido dos Trabalhadores in recent years has generated increasing support for Maluf in city's wealthier neighborhoods. Distribution of Maluf votes in 1992 was reminiscent of the Aliança Renovadora Nacional/Partido Democrático Social vote under military rule.

3753 **Pinheiro, Paulo Sérgio.** Police and political crisis: the case of the military police. (*in* Vigilantism and the State in modern Latin America: essays on extralegal violence. New York: Praeger Publishers, 1991, p. 167–188)

Historical overview of extralegal police violence in Brazil is written by a political scientist known for courageous denunciations of official abuses. Author advocates separation of military and police powers. Useful introduction to police repression, which in mid-1990s is arguably the most pressing human rights issue in Brazil.

3754 **Pomar, Wladimir.** Quase lá: Lula, o susto das elites. São Paulo: Editora Brasil Urgente, 1990. 125 p.

Diary of 1989 presidential election by Lula's campaign manager is an essential source on the election.

3755 **A proposta social-democrata: a social-democracia na atualidade européia, hispano-americana e brasileira.** Organização de Hélio Jaguaribe. Rio de Janeiro: José Olympio Editora, 1989. 302 p.: bibl. (Col. Documentos brasileiros; 206)

Contains speeches given at a 1987 Rio de Janeiro conference on social democracy featuring a large number of speakers on Europe and Latin America. Only a handful of presentations deal with Brazil, and these are unremarkable.

3756 **Raichelis, Raquel.** Legitimidade popular e poder público. São Paulo: Cortez Editora, 1988. 211 p.: bibl.

Presents Marxist interpretation of social service in Brazil. Traditionally the State maintained the dependency of the poor through State benefits. In recent years, however, a cleavage has developed between social workers who practice the traditional *assistencialismo* and those who side with the popular movements. Theoretically dogmatic, but a rare study of the politics of social workers.

3757 Ramos, Roberto. Manipulação e controle da opinião pública: a grande impresa e o Plano Cruzado. Rio de Janeiro: Espaço e Tempo, 1987. 130 p.

Acerbic unmasking of 1986 Plano Cruzado is written by a journalist. Illustrates electoral manipulation of economic policy by Sarney government and denounces connivance of the mainstream press.

3758 Reis, Fábio Wanderley and **Mônica Mata Machado de Castro.** Regiões, classe e ideologia no processo eleitoral brasileiro. (*Lua Nova*, 26, 1992, p. 81–131, tables)

Sophisticated analysis of class and issue voting is based on survey research conducted in 1982 in rural areas and in seven state capitals. Authors find evidence for surprising cross-cutting cleavages. While an important contribution on Brazilian voting, article is handicapped by excessive length and inaccessible style.

3759 Resende, Maria Efigênia Lage de. Às vésperas de 37: o novo/velho discurso da ordem conservadora. (*Rev. Bras. Estud. Polít.*, 73, julho 1991, p. 7–51, bibl.)

Focusing on Minas Gerais, author discusses changing discourse of conservative elites on eve of the Estado Novo in 1936–37. Gives heavy emphasis to ideological convergence of state elites and the traditional Church.

3760 Rodrigues, Leôncio Martins. Partidos e sindicatos: escritos de sociologia política. São Paulo: Editora Atica, 1990. 151 p.: bibl. (Série Temas; 17: Sociologia e política)

Five essays are written by leading sociologist of labor in Brazil. Notable for controversial paper on leadership of Partido dos Trabalhadores (PT), which demonstrates growing influence of intellectuals, liberal professionals, and white collar unionists in the PT. Also contains a study of automobile workers. Essential for labor studies.

3761 Roett, Riordan and **Scott D. Tollefson.** The year of elections in Brazil. (*Curr. Hist.*, 89:543, Jan. 1990, p. 25–29)

Written between first and second rounds of 1989 presidential election, article describes economic and political crises facing Brazil at turn of the decade, with some attention to foreign policy.

3762 Sadek, Maria Tereza Aina. Poder local: perspectivas da nova ordem constitucional. (*São Paulo Perspect.*, 5:2, abril/junho 1991, p. 9–15, bibl.)

Provides overview of long-running debates over government decentralization and municipal autonomy in Brazil. Weakness of municipal governments has traditionally contributed to clientelism and patronage politics in Brazil. Constitution of 1988 strengthened municipal governments, thus holding out promise of reform.

3763 Santos, Guarino Fernandes dos. Nos bastidores da luta sindical. São Paulo: Icone Editora, 1987. 132 p.;

Memoirs of a former organizer of rail workers' union in São Paulo covers four decades. Useful for understanding labor movement at mid-century.

3764 Sikkink, Kathryn. Las capacidades y la autonomía del Estado en Brasil y la Argentina: un enfoque neoinstitucionalista. (*Desarro. Econ.*, 32:128, enero/marzo 1993, p. 543–574, graphs, table)

Compares State-building and State capacity in Brazil and Argentina during developmentalist period. Kubitschek's relative success when compared to Frondizi is not explained simply by the size of Brazilian public sector, but rather by intersection of ideas, innovation, and recruitment of intellectual and technical talent into embryonic State agencies.

3765 Silva, Carlos Eduardo Lins da. The Brazilian case: manipulation by the media? (*in* Television, politics, and the transition to democracy in Latin America. Washington: Woodrow Wilson Center Press, 1993, p. 137–144)

Analyzes influence of television in 1989 presidential election. Argues that TV is only one factor among many in explaining voter preferences, and claims that role of Globo network in Collor campaign was greatly exaggerated. Collor's victory is better explained by his anti-corruption discourse and "unadulterated demagoguery."

3766 **Silva, Carlos Eduardo Lins da.** Brazil's struggle with democracy. (*Curr. Hist.*, 92:572, March 1993, p. 126–129)

Short introduction to events surrounding impeachment of Fernando Collor de Mello in 1992 is written by one of Brazil's top political journalists.

3767 **Silva, Hélio.** A vez e a voz dos vencidos. Petrópolis, Brazil: Vozes, 1988. 367 p.: bibl. (Col. Memória dos vencidos; 1)

Study examines the purging of more than 7,500 members of the three branches of the armed forces following the coup of 1964. Contains a wealth of information including documents and court decisions relating to the struggle for amnesty.

3768 **Silva, José Afonso da.** Los efectos corporativos de la representación proporcional en el Brasil. San José: Centro Interamericano de Asesoría y Promoción Electoral, Instituto Interamericano de Derechos Humanos, 1988. 59 p.: bibl. (Serie Cuadernos de CAPEL; 25)

Solid study of proportional representation shows how Brazil's open-list system lends itself to "corporativist" candidacies (by lawyers, teachers, farmers, Protestants, ethnic groups, animists, etc., on behalf of their constituencies), rather than to the more common types based on partisan or territorial affiliation. Valuable for presentation of complex details of how proportional representation functions in Brazil.

3769 **Silva, José Wilson da.** O reacionarismo militar na terra de Santa Cruz. Porto Alegre, Brazil: Editora Sulina, 1989. 287 p.: bibl., ill.

Self-serving account of military politics in 1960s is written by a *gaúcho* captain purged from the army after coup of 1964. Work is so dependent on direct quotations from secondary sources that it verges on plagiarism, and is interesting only for its surprisingly unreconstructed politics. Author attempts to resuscitate the leftist-nationalist-Brizolist current within the military.

3770 **Silva, Paulo Napoleão Nogueira da.** Democracia e realidade brasileira. São Paulo: Editora Alfa-Omega, 1989. 176 p.: bibl. (Biblioteca Alfa-Omega de cultura universal. Série 1a., Col. Esta América; 25)

Essayistic overview of ills of democracy in the New Republic focuses on impunity and absence of republican practices on the part of elites. Concludes with a review of possibilities for institutional reform. Written for a general audience.

3771 **Sistemas eleitorais e processos políticos comparados: a promessa de democracia na América Latina e Caribe.** Redação de Luiz Pedone. Washington: Organization of American States; Brasília: Conselho Nacional de Desenvolvimento Tecnológico; Univ. de Brasília, 1993. 1 v.

Collection of papers presented at a Dec. 1992 conference on party and electoral systems in Latin America contains many interesting contributions covering region as a whole. The five Brazilian papers, including those on legislative elections (Ames), campaign finance (Fleischer), and the role of television (Lima) are all worthwhile.

3772 **Sives, Amanda.** Elites behaviour and corruption in the consolidation of democracy in Brazil. (*Parliam. Aff.*, 46:4, Oct. 1993, p. 549–562)

General overview of problems facing Brazilian democracy after 1989 was written during "Collorgate" scandal. Highly pessimistic appraisal of possibilities for progressive change assigns blame to elite indifference, a lack of democratic accountability, and widespread corruption. Diagnosis is not new, and is better sustained by more empirically oriented studies.

3773 **Soares, Gláucio Ary Dillon** and **Maria Celina Soares d'Araújo.** A imprensa, os mitos, e os votos nas eleições de 1990. (*Rev. Bras. Estud. Polít.*, 76, jan. 1993, p. 163–189, graphs, tables)

Content analysis of major Brazilian newspapers shows that media coverage of 1990 elections was seriously distorted. Myths concerning decline of the left, preference for old political warhorses, and contention that parties mean nothing in Brazilian elections are all disputed by means of a dispassionate analysis of election returns.

3774 **Sosnovskiĭ, Anatoliĭ Aleksandrovich.** Brasil rumo à democracia. São Paulo: Editora Alfa-Omega, 1989. 112 p. (Biblioteca Alfa-Omega de cultura universal. Série 1a., Col. Esta América; 24)

Broad overview of transition to democracy is written by a Brazilianist at the

Academy of Sciences of former USSR. Written in 1988 at height of Gorbachev's power, book is informed by the winds of change blowing in Moscow at the time. Result is a warm, encouraging treatment of prospects for democracy and development in the New Republic. More noteworthy as example of effects of *glasnost* on Soviet Latin American studies than as a study of Brazilian politics.

3775 **Souza, Amaury de.** O caso brasileiro: o sistema político-partidário. (*in* Sociedade, Estado e partidos na atualidade brasileira. São Paulo: Paz e Terra, 1992, p. 157–198, graphs, tables)

Valuable descriptive overview of development of Brazilian party system focuses mostly on period since 1985. Argues that institutional reform is necessary to gain legitimacy for the parties and other representative institutions. Useful tabular data and graphs.

3776 **Straubhaar, Joseph; Organ Olsen;** and **Maria Cavaliari Nunes.** The Brazilian case: influencing the voter. (*in* Television, politics, and the transition to democracy in Latin America. Washington: Woodrow Wilson Center Press, 1993, p. 118–136, tables)

Long-overdue study examines influence of television in Brazilian politics. Rather than demonize TV Globo, authors employ survey research to illustrate impact of free political advertising in 1989 presidential election. They conclude that free TV time succeeded in increasing awareness among voters, but caution that television influences candidate choice in conjunction with established interpersonal networks. Recommended.

3777 **Veiga, Sandra Mayrink** and **Isaque Fonseca.** Volta Redonda, entre o aço e as armas. Petrópolis, Brazil: Vozes, 1990. 222 p., 16 p. of plates: bibl., ill.

Study of army attack on striking workers of Volta Redonda steel mill in Nov. 1988 is prepared by an historian and by one of the union activists involved. Contains testimony and photographs that are essential primary source material, but account is highly partial.

3778 **Vianna, Luiz Werneck.** A transição: da Constituinte à sucessão presidencial. Rio de Janeiro: Editora Revan, 1989. 174 p.: bibl. (Série Pensamento brasileiro)

Series of essays on the political transition focuses on neoliberalism and crisis within the Marxist left. Written from 1987–89, essays illustrate well the transformation of theory and praxis in the Partido Comunista Brasileiro.

3779 **Weyland, Kurt.** The rise and fall of President Collor and its impact on Brazilian democracy. (*J. Interam. Stud. World Aff.*, 35:1, 1993, p. 1–37, bibl.)

Penetrating analysis examines Collor, corruption, and politics of presidential impeachment in 1992. Collor sought to maximize his personal autonomy by governing "above" parties. This increased the incentive for corruption and left the president without a base of legislative support when the "Collorgate" scandal broke. Essential for understanding Collor's demise.

Wolkmer, Antônio Carlos. Uma interpretação das idéias políticas no Brasil. See *HLAS 54: 5491.*

3780 **Wood, Charles H.** and **Marianne Schmink.** The military and the environment in the Brazilian Amazon. (*J. Polit. Mil. Sociol.*, 21:1, Summer 1993, p. 81–105, bibl., map)

Empirically rich and theoretically informed analysis examines impact of military rule on frontier expansion in Amazonia. The military's aggressive developmentalist policies from 1964–85 set into motion many of the social, environmental, and political crises which beset the region today. Despite democratization, the military continue to view the region through the lens of national security ideology. For sociologist's comment see item **5213.**

JOURNAL ABBREVIATIONS

Age Dem. The Age of Democracy. Basil Blackwood; UNESCO.

Allpanchis/Cusco. Allpanchis. Instituto de Pastoral Andina. Cusco, Peru.

Alternatives/Boulder. Alternatives. Lynne Rienner Publishers. Boulder, Colorado.

Am. Lat./Moscow. América Latina. Academia de Ciencias de la Unión de Repúblicas Soviéticas Socialistas. Moscow.

Am. Polit. Sci. Rev. American Political Science Review. American Political Science Assn.; Ohio State Univ., Columbus, Ohio.

Análise Conjunt. Análise & Conjuntura. Fundação João Pinheiro. Belo Horizonte, Brazil.

Ann. N.Y. Acad. Sci. Annals of the New York Academy of Sciences. New York.

ANPOCS BIB. Boletim Informativo e Bibliográfico de Ciências Sociais: BIB. Associação Nacional de Pós-Graduação e Pesquisa em Ciências Sociais. Rio de Janeiro.

Armed Forces Soc. Armed Forces and Society. Inter-Univ. Seminar on Armed Forces & Society. Univ. of Chicago. Chicago, Ill.

Bijdr. Meded. Betreffende Geschied. Nederlanden. Bijdragen en Mededelingen Betreffende de Geschiedenis der Nederlanden. Nederlands Historisch Genootschap. Utrecht, The Netherlands.

Bol. Socioecon. Boletín Socioeconómico. Centro de Investigaciones y Documentación Socioeconómico (CIDSE), Univ. del Valle. Cali, Colombia.

Bull. Lat. Am. Res. Bulletin of Latin American Research. Society for Latin American Studies. Oxford, England.

Cad. CEAS. Cadernos do Centro de Estudos e Ação Social (CEAS). Salvador, Brazil.

Cadernos/Belém. Cadernos. Centro de Filosofia e Ciências Humanas, Univ. Federal do Pará. Belém, Brazil.

Caribb. Aff. Caribbean Affairs. Trinidad Express Newspapers Ltd.; Inprint Caribbean Ltd., Port of Spain, Trinidad and Tobago.

Caribb. Stud. Caribbean Studies. Univ. of Puerto Rico, Institute of Caribbean Studies. Río Piedras.

Cienc. Polít. Ciencia Política. Instituto de Ciencia Política de Bogotá; Tierra Firme Editores. Bogotá.

Comp. Polit. Comparative Politics. The City Univ. of New York, Political Science Program. New York.

Comp. Polit. Stud. Comparative Political Studies. Sage Publications, Thousand Oaks, Calif.

Contribuciones/Asunción. Contribuciones. Centro de Documentación y Estudios (CDE). Asunción.

Contribuciones/Buenos Aires. Contribuciones. Estudios Interdisciplinarios sobre Desarrollo y Cooperación Internacional. Konrad-Adenauer-Stiftung; Centro Interdisciplinario de Estudios Sobre el Desarrollo Latinoamericano (CIEDLA). Buenos Aires.

Cotidiano/México. El Cotidiano: Revista de la Realidad Mexicana Actual. Univ. Autónoma Metropolitana, Unidad Azcapotzalco. México.

Cuad. CENDES. Cuadernos del CENDES. Centro de Estudios del Desarrollo, Univ. Central de Venezuela. Caracas.

Cuad. CLAEH. Cuadernos del CLAEH. Centro Latinoamericano de Economía Humana. Montevideo.

Cuad. Marcha. Cuadernos de Marcha. Eon Editores. Montevideo.

Cuad. Nuestra Am. Cuadernos de Nuestra América. Centro de Estudios sobre América. La Habana.

Cuba. Stud. Cuban Studies. Univ. of Pittsburgh, Center for Latin American Studies. Pittsburgh, Penn.

Curr. Hist. Current History. Philadelphia, Penn.

Dados/Rio de Janeiro. Dados. Instituto Univ. de Pesquisas. Rio de Janeiro.

Debate Agrar. Debate Agrario. Centro Peruano de Estudios Sociales (CEPES). Lima.

Desarro. Econ. Desarrollo Económico. Instituto de Desarrollo Económico y Social. Buenos Aires.

Dissent/New York. Dissent. Dissent Publishing Assn., New York.

Ecuad. Deb. Ecuador Debate. Centro Andino de Acción Popular (CAAP). Quito.

Envío/Managua. Envío. Univ. Centroamericana (UCA). Managua.

Estad. Soc. Estado & Sociedad: Revista Boliviana de Ciencias Sociales. Facultad Latinoamericana de Ciencias Sociales (FLACSO). La Paz.

Estud. Cult. Contemp. Estudios sobre las Culturas Contemporáneas. Centro Universitario de Investigaciones Sociales, Univ. de Colima. México.

Estud. Int./Santiago. Estudios Internacionales. Instituto de Estudios Internacionales, Univ. de Chile. Santiago.

Estud. Públicos. Estudios Públicos. Centro de Estudios Públicos. Santiago.

Estud. Sociol./México. Estudios Sociológicos. Centro de Estudios Sociológicos de El Colegio de México. México.

Foreign Aff. Foreign Affairs. Council on Foreign Relations, Inc. New York.

Foro Int. Foro Internacional. El Colegio de México. México.

Geosul/Florianópolis. Geosul. Depto. de Geociências, Univ. Federal de Santa Catarina. Florianópolis, Brazil.

Hemisphere/Miami. Hemisphere. Latin American and Caribbean Center, Florida International Univ., Miami, Fla.

Hist. Quest. Debates. História, Questões e Debates. Associação Paranaense de História. Curitiba, Brazil.

Hist. Soc./Río Piedras. Historia y Sociedad. Depto. de Historia, Univ. de Puerto Rico. Río Piedras.

Hoy Hist. Hoy es Historia: Revista Bimestral de Historia Nacional e Iberoamericana. Editorial Raíces. Montevideo.

Ibero-Am. Arch. Ibero-Amerikanisches Archiv. Ibero-Amerikanisches Institut. Berlin.

Iberoam./Tokyo. Iberoamericana. Univ. of Sofia. Tokyo.

Ideas Polít. Ideas Políticas: Revista de Análisis y Debate. Cambio XXI Fundación Mexicana. México.

Index Censorsh. Index on Censorship. Writers & Scholars International. London.

Int. Spect./Hague. Internationale Spectator. Nederlandsch Genootschap voor Internationale Zaken. The Hague.

Iztapalapa/México. Iztapalapa. Univ. Autónoma Metropolitana, División de Ciencias Sociales y Humanidades. México.

J. Church State. Journal of Church and State. J.M. Dawson Studies in Church and State, Baylor Univ., Waco, Tex.

J. Dev. Areas. The Journal of Developing Areas. Western Illinois Univ. Press. Macomb, Ill.

J. Interam. Stud. World Aff. Journal of Interamerican Studies and World Affairs. Institute of Interamerican Studies, Univ. of Miami. Coral Gables, Fla.

J. Lat. Am. Lore. Journal of Latin American Lore. Univ. of California, Latin American Center. Los Angeles, Calif.

J. Lat. Am. Stud. Journal of Latin American Studies. Centers or Institutes of Latin American Studies at the Universities of Cambridge, Glasgow, Liverpool, London, and Oxford. Cambridge Univ. Press. London.

J. Peace Res. Journal of Peace Research. International Peace Research Institute, Universitetforlaget. Oslo.

J. Peasant Stud. The Journal of Peasant Studies. Frank Cass & Co., London.

J. Polit. Mil. Sociol. Journal of Political & Military Sociology. Northern Illinois Univ., Dept. of Sociology. DeKalb, Ill.

J. Sci. Stud. Relig. Journal for the Scientific Study of Religion. Society for the Scientific Study of Religion. Storrs, Conn.

LARR. Latin American Research Review. Latin American Research Review Board. Univ. of New Mexico, Albuquerque, N.M.

LASA Forum. LASA Forum. Latin American Studies Assn., Pittsburgh, Penn.

Lat. Am. Perspect. Latin American Perspectives. Univ. of California. Newbury Park, Calif.

Lua Nova. Lua Nova. Editora Brasiliense. São Paulo.

MACLAS Lat. Am. Essays. MACLAS Latin American Essays. Middle Atlantic Council of Latin American Studies. New Brunswick, N.J.

Mayan People Hum. Rights. The Mayan People and Human Rights. Pro Justice and Peace Committee of Guatemala. Guatemala.

Memoria/CEMOS. Memoria: Boletín de CEMOS. Centro de Estudios del Movimiento Obrero y Socialista. México.

Mon. Rev. Monthly Review. New York.

New Left Rev. New Left Review. New Left Review, Ltd., London.

Nova Econ. Nova Economia. Depto. de Ciências Econômicas, Univ. Federal de Minas Gerais. Belo Horizonte, Brazil.

Novos Estud. CEBRAP. Novos Estudos CEBRAP. Centro Brasileiro de Análise e Planejamento. São Paulo.

Nueva Soc. Nueva Sociedad. Caracas.

Ojarasca/México. Ojarasca. Pro-México Indígena. México.

Parliam. Aff. Parliamentary Affairs. Oxford Univ. Press. London.

Polit. Sci. Q. Political Science Quarterly. Columbia Univ., The Academy of Political Science. New York.

Polit. Stud. Political Studies. Political Studies Assn. of the United Kingdom; Clarendon Press. Oxford, England.

Politeia/Caracas. Politeia. Instituto de Estudios Políticos, Univ. Central de Venezuela. Caracas.

Pretextos/Lima. Pretextos. Centro de Estudios y Promoción del Desarrollo. Lima.

Quehacer/Lima. Quehacer. Centro de Estudios y Promoción del Desarrollo (DESCO). Lima.

Real. Econ./Buenos Aires. Realidad Económica. Instituto Argentino para el Desarrollo Económico (IADE). Buenos Aires.

Relaciones/Coyoacán. Relaciones. Univ. Autónoma Metropolitana, Unidad Xochimilco. Coyoacán, México.

Relaciones/Zamora. Relaciones. El Colegio de Michoacán. Zamora, Mexico.

Rev. Arg. Estud. Estrateg. Revista Argentina de Estudios Estratégicos. Olcese Editores. Buenos Aires.

Rev. Bras. Estud. Polít. Revista Brasileira de Estudos Políticos. Univ. de Minas Gerais. Belo Horizonte, Brazil.

Rev. Cienc. Soc./Río Piedras. Revista de Ciencias Sociales. Univ. de Puerto Rico, Colegio de Ciencias Sociales. Río Piedras.

Rev. Cienc. Soc./Valparaíso. Revista de Ciencias Sociales. Facultad de Derecho y Ciencias Sociales, Univ. de Valparaíso. Chile.

Rev. Estud. Polít. Revista de Estudios Políticos. Centro de Estudios Constitucionales. Madrid.

Rev. Eur. Revista Europea de Estudios Latinoamericanos y del Caribe = European Review of Latin American and Caribbean Studies. Center for Latin American Research and Documentation; Royal Institute of Linguistics and Anthropology. Amsterdam.

Rev. Hist./Heredia. Revista de Historia. Univ. Nacional de Costa Rica, Escuela de Historia. Heredia, Costa Rica.

Rev. Interam. Bibliogr. Revista Interamericana de Bibliografía. Organization of American States. Washington.

Rev. Mex. Cienc. Polít. Soc. Revista Mexicana de Ciencias Políticas y Sociales. Facultad de Ciencias Políticas y Sociales, Univ. Nacional Autónoma de México. México.

Rev. Mex. Sociol. Revista Mexicana de Sociología. Instituto de Investigaciones Sociales, Univ. Nacional Autónoma de México. México.

Rev. Panameña Sociol. Revista Panameña de Sociología. Univ. de Panamá, Depto. de Sociología. Panamá.

Rev. Parag. Sociol. Revista Paraguaya de Sociología. Centro Paraguayo de Estudios Sociológicos. Asunción.

São Paulo Perspect. São Paulo em Perspectiva. Fundação SEADE. São Paulo.

Sci. Soc. Science and Society. New York.

Secuencia/México. Secuencia. Instituto Mora. México.

Signos Univ. Signos Universitarios: Revista de la Universidad del Salvador. Univ. del Salvador. Buenos Aires.

Síntesis/Madrid. Síntesis. Asociación de Investigación y Especialización sobre Temas Latinoamericanos. Madrid.

Soc. Compass. Social Compass. The International Catholic Institute for Social-Ecclesiastical Research. The Hague.

Stud. Comp. Int. Dev. Studies in Comparative International Development. Transaction Periodicals Consortium, Rutgers Univ., New Brunswick, N.J.

Swiss Rev. World Aff. Swiss Review of World Affairs. Neue Zürcher Zeitung. Zürich, Switzerland.

Tareas/Panamá. Tareas. Centro de Estudios Latinoamericanos (CELA). Panamá.

Third World Plan. Rev. Third World Planning Review. Liverpool Univ. Press. Liverpool, England.

Voices Mex. Voices of Mexico. Univ. Nacional Autónoma de México. México.

Vuelta/México. Vuelta. México.

West. Polit. Q. Western Political Quarterly. Western Political Science Assn.; Pacific Northwest Political Science Assn.; Southern California Political Science Assn.; Univ. of Utah, Institute of Government. Salt Lake City.

Wilson Q. The Wilson Quarterly. Woodrow Wilson International Center for Scholars. Washington.

World Aff. World Affairs. The American Peace Society. Washington.

World Dev. World Development. Pergamon Press. Oxford, England.

World Policy J. World Policy Journal. World Policy Institute. New York.

World Polit. World Politics. Princeton Univ., Center of International Studies. Princeton, N.J.

Z. Lat.am. Wien. Zeitschrift für Lateinamerika Wien. Österreichisches Lateinamerika-Institut. Vienna.

INTERNATIONAL RELATIONS

GENERAL

G. POPE ATKINS, *Research Fellow, Institute of Latin American Studies, University of Texas at Austin, and Professor Emeritus of Political Science, United States Naval Academy*

THE SCHOLARLY LITERATURE on the general international relations of Latin America during the period surveyed was cast largely in terms of the ensuing developments consequent to the end of the Cold War and the political and economic changes in the Latin American region. The impact of these changes is reflected on all levels of Latin America's international relations—the global system and the regional subsystem, the foreign policies of the individual state and nonstate actors involved, and the substantive issues giving rise to foreign policy decisions and international interactions.

While analysis has caught up with the international and regional transformations and the new realities underlying the treatments of specific topics, few works were devoted to general theoretical considerations or the systemic levels as such. Those that did followed established approaches to the general study of international relations, and recognized the revived importance of regionalism and the increased linkages between domestic and international politics for most states. Particularly useful among them is the book by Luciano Tomassini *et al.* (item **3869**), which places Latin America firmly, in theoretical and substantive terms, in the current multidimensional international system. A number of studies addressed dependency theories, which have been in decline among analysts for some two decades and all but abandoned as the bases for Latin American policies in favor of economic neoliberalism. The book authored by Robert Packenham (item **3848**) and the collection edited by Osvaldo Sunkel (item **3806**) present directly opposing viewpoints, the former worrying about the long-term negative influence of those theories on scholars and scholarship and the latter seeking a continuing role for structuralism in the current era. The contributing authors to Howard Wiarda's new edition of an established treatise (item **2789**) underscore Latin America's awkward fit into a global Third World conceptualization or structure.

A paucity of general foreign policy analyses is observed on the Latin American State actor level, following several steady years of solid contributions. Those cited see domestic influences on foreign policy decisions and actions as paramount, in particular the idea that the transformation from military regimes and closed economies to constitutional neoliberal systems has been completed, although with exceptions and many unknowns and risks in both instances. The book edited by Barry Levine (item **3874**) is an important critique of State-run economies in defense of neoliberalism. Jeffrey Frieden (item **3818**) places emphasis on social groups in political-economic change in consonance with his definition of "modern political economy." Analysts also continue to emphasize *concertación*—the concept of Latin American foreign policies "acting in concert" in the sense of "harmonization."

Studies of the United States' Latin American foreign policy comprise the single largest number of items cited. Following the general trend noted in HLAS 53, they address the difficulties faced by analysts and policy makers in dealing with the new set of high priority inter-American issues in a transformed post–Cold War security environment. Robert Pastor authored a book (item **3850**) that is helpful in sorting out US priorities, alternatives, and issues in the unclear future. An original endeavor by Martha Cottam (item **3801**) studies images held by decision makers and their influence on interventionist policies. Frederick Pike (item **3852**) adds another comprehensive volume to his distinguished list of studies on intellectual and international history with a systematic exploration of certain aspects of perceptions and policy. In the arena of US-Latin American relations, Chilean and US contributors to the book edited by Jonathan Hartlyn, Lars Schoultz, and Augusto Varas (item **3870**) see a period when a commonality of problems and interests, long favored in diplomatic rhetoric, is real and abiding. Abraham Lowenthal and Gregory Treverton edited a volume (item **3833**) in which a distinguished group addresses the substantive post–Cold War inter-American issues.

The quantity and quality of work on international institutions were mixed. A large number of items relating to Latin American economic integration and hemispheric free trade were surveyed but few are cited. Among the latter, a special issue of the *Annals of the American Academy of Political and Social Science* (item **3783**), organized and edited by Sidney Weintraub, and a book edited by Sylvia Saborío (item **3846**) are highlighted here. In a reversal of the trend noted in HLAS 53, the institutions of the Inter-American System received surprisingly little substantive treatment, even though they continued to undergo something of a resurgence. Among them, however, the book by Carlos Stoetzer (item **3866**) provides a thorough reference work on the Organization of American States. A significant new trend in the literature noted in HLAS 53 continues: an accelerated regard for Canada's new role in inter-American relations, the result of Canadian decisions taken in the late 1980s, especially its commitment to act through the Inter-American System and UN conflict-resolution mechanisms. Two books are particularly informative, one edited by Jerry Haar and Edgar Dosman (item **3810**) and another authored by James Rochlin (item **3855**).

European policies received substantial scholarly notice, more often than not treated in the context of the European Community (which has evolved into the European Union). Most of the works were written before the Maastricht Treaty provisions were changed or their implementation delayed, but still offer useful information and insights. Nonetheless, the quantity of writing on the subject seems to have dropped since early 1993.

The former Soviet Union, Russia, and Eastern Europe received attention, generally as a matter of updating and analyzing recent events and recognizing Moscow's virtual withdrawal (sometimes with the qualifier "temporary") from the region. An interesting line of analytic speculation was the usefulness of the Latin American experience with redemocratization and economic opening as a model for the former Soviet and East European states. The fact that no significant book-length works appeared may reflect the Sovietologists' general dilemma in that their field has lost much of its current and immediate future analytic and policy relevance.

A modicum of attention was devoted to specific aspects regarding other actors.

Cited below are works on Japan, China, Sweden, and Spain, as well as on the Ibero-American summit of heads of states from Spain, Portugal, and Latin America. In the case of Japan, the book edited by Barbara Stallings and Gabriel Székely (item **3828**) implicitly underscores the small role played by the US in the "trilateral relationship." Non-nation-state actors cited include the Holy See, the Socialist International, and insurgent groups. On the latter, the book by Timothy Wickham-Crowley (item **2808**) is especially recommended.

Works are cited on all of the issues currently at the top of state policy and academic research agendas (none of which are new): economic questions, democracy and human rights, narcotraffic, immigration and refugees, and environmental problems. The debt problem (after a temporary lull noted in *HLAS 53*) and narcotraffic received the most attention on a roughly equal level. In the case of the external debt, attention is called to the books edited by Robert Grosse (item **3854**) and by Alfred McCoy and Alan Block (item **3875**). Christopher Mitchell edited a good collection of essays on hemispheric immigration (item **3807**).

3781 **Almeida, Paulo Roberto de.** Propiedade intelectual: os novos desafios para a América Latina. (*Estud. Avançados*, 5:12, maio/agosto 1991, p. 187–203, bibl., photos)

Critical tirade against intellectual property protection provided by the Uruguay Round of GATT, which creates a system of "technological apartheid," benefiting the North. Recommends that the South seek technological cooperation on a regional level. For Brazil, Mercosul offers the best framework for such cooperation. [S. Tollefson]

3782 **Alvarez, José E.** Promoting the "Rule of Law" in Latin America: problems and prospects. (*J. Int. Law Econ.*, 25:2, 1991, p. 281–331)

Examines origins and development of the US government's Administration of Justice Program (AOJ Program) within the context of 1960s "law and development" projects. [D. Dent]

3783 *The Annals of the American Academy of Political and Social Science.* Vol. 526, special edition, March 1993– . Free Trade in the Western Hemisphere. Edited by Sidney Weintraub. Philadelphia: American Academy of Political and Social Science.

Thorough multiauthored analysis of new emphasis on hemispheric—including bilateral and subregional—free trade, with due attention to historical background and conceptual and institutional contexts.

3784 **Bagley, Bruce Michael.** After San Antonio. (*J. Interam. Stud. World Aff.*, 34:3, Fall 1992, p. 1–12)

Updates and analyzes developments of "war on drugs" by Bush Administration centering on seven-country anti-drug presidential summit in Feb. 1992.

3785 **Ballester, Horacio P.** Proyecciones geopolíticas hacia el tercer milenio: el dramático futuro latino americano caribeño. Prólogo de Liber Seregni. Argentina: Ediciones Fin de Siglo, 1993. 297 p.: appendices.

Interesting set of critical essays written by one of the founders and directors of the Centro de Militares para la Democracia Argentina discusses, from a democratic nationalistic perspective, the geopolitical situation of the Latin American countries. Useful source for those interested in ideas and interpretations of the relatively small group of military who define themselves as anti-imperialistic, democratic, and in favor of regional integration. [A. Vacs]

3786 **Berger, Mark T.** Civilising the South: the U.S. rise to hegemony in the Americas and the roots of 'Latin American Studies' 1898–1945. (*Bull. Lat. Am. Res.*, 12:1, Jan. 1993, p. 1–48)

Imaginative detailed interpretation of pre-Cold War evolution of US historiography on Latin America, seeing academic activity as a complement to, not distinct from, US policy.

3787 **Blasier, Cole.** Latin America without the USSR. Coral Gables, Fla.: North-South Center, Univ. of Miami, 1993. 5 p. (North-South Issues; 2:4)

This article and following one (see item **3788**) provide authoritative post–Cold

War update on process and substance of Russian decision making in recalculation of policies toward Latin America, especially Cuba.

3788 Blasier, Cole. Moscow's retreat from Cuba. (*Probl. Communism,* 40, Nov./Dec. 1991, p. 91–99, photos)

For annotation, see item **3787**.

3789 Brigagão, Clóvis et al. Ecologia e política mundial. Organização de Héctor Leis. Rio de Janeiro: Vozes; FASE; AIRI/PUC-Rio, 1991. 183 p.: bibl., ill., map. (Col. Ecologia & ecosofia)

Studies by three Brazilian and two Argentine political scientists on ecology, the international system, and Latin America (with special reference to Brazil). Excellent chapter by Clóvis Brigagão on the Amazon, Antarctic, and "ecological security." [S. Tollefson]

3790 Bustamante, Fernando. La política de Estados Unidos contra el narcotráfico y su impacto en América Latina. (*Estud. Int./Santiago,* 23:90, abril/junio 1990, p. 240–271)

Examines critically US international drug policies in 1980s and their impact on Latin America. Concludes that policies fail in part because of basic contradiction between US "modern" approach that considers drug use a criminal law and health problem and Latin American "non-modern" perspective that views drug production and consumption in cultural, religious, and economic terms. Moreover, US policies contribute to debilitating the already weak Latin American states while encouraging consolidation and mobilization of a network of drug-related interests and groups. [A. Vacs]

Cardona, Diego et al. Colombia y la integración americana. See item **4268**.

3791 Caro, Isaac. América Latina y el Caribe en el mundo militar. Santiago: Facultad Latinoamericana de Ciencias Sociales, 1988. 245 p.: bibl.

Overview of Latin America's military relations (cooperation, arms trade, military production, nuclear policy, other military programs) with every region of the world. [D. Fernández]

3792 The challenge of integration: Europe and the Americas. Edited by Peter H. Smith. Miami: North-South Center, Univ. of Miami; New Brunswick, N.J.: Transaction Publishers, 1993. 1 v.: bibl.

Thorough, useful collection of essays with due attention to both Latin American and European points of view.

3793 Child, Jack. El pensamiento geopolítico. (*in* Los militares y la democracia: el futuro de las relaciones cívico-militares en América Latina. Recopilación de Louis W. Goodman, Johanna S.R. Mendelson y Juan Rial. Montevideo: PEITHO, 1990, p. 213–233, tables)

Updated and well-documented analysis examines evolution of Latin American geopolitical thinking in last decades. Lists most important publications; examines topics most frequently discussed in geopolitical literature; and discusses possibility of emergence of a less nationalistic and aggressive geopolitical thought in the new democratic context. [A. Vacs]

3794 Chomsky, Noam. Year 501: the conquest continues. Boston: South End Press, 1993. 331 p.: bibl., index.

According to Chomsky, "the great work of subjugation and conquest" in the world that began with Columbus' voyages continues to be carried on prominently by the US in Latin America.

3795 Cline, William R. International debt reexamined. Washington: Institute for International Economics, 1995. 535 p.: bibl., ill., index.

Expert on Latin America includes substantial analysis of the region in terms of new economic and political conditions.

3796 Cochrane, James D. Latin America in the international arena. (*LARR,* 26:3, 1991, p. 213–225)

Together these two valuable comparative review articles evaluate 16 books published between 1985–91 (see also item **3797**).

3797 Cochrane, James D. The troubled and misunderstood relationship: the United States and Latin America. (*LARR,* 28:2, 1993, p. 232–245)

For annotation, see item **3796**.

3798 Cole, Daniel H. Debt-equity conversions, debt-for-nature swaps, and the continuing world debt crisis. (*Columbia J. Transnatl. Law,* 30:1, 1992, p. 57–88, bibl., appendix)

Substantial article clearly and thoroughly explains complexities of debt-for-equity and debt-for-nature swaps.

3799 **Conflict and competition: the Latin American church in a changing environment.** Edited by Edward L. Cleary and Hannah W. Stewart-Gambino. Boulder, Colo.: Lynne Rienner Publishers, 1992. 234 p.: bibl., index.

Solid study of the Latin American Church includes aspects of the Holy See's role and relations.

3800 **Connelly, Marisela** and **Romer Cornejo Bustamante.** China-América Latina: génesis y desarrollo de sus relaciones. México: Colegio de México, Centro de Estudios de Asia y Africa, 1992. 196 p.: bibl., index.

Chronicle of Chinese-Latin American relations begins with first contacts and emphasizes Chinese foreign policy and Latin American role in commercial calculations.

3801 **Cottam, Martha L.** Images and intervention: U.S. policies in Latin America. Pittsburgh: Univ. of Pittsburgh Press, 1994. 1 v.: bibl., index. (Pitt Latin American series)

Original analysis of old subject applies concepts related to formation of images by US policymakers toward Latin America (especially Circum-Caribbean). Interventions will not cease in the post-Cold War era if decisions continue to be shaped by paternalistic views of Latin American cultures and people.

3802 **Croan, Melvin.** Is Latin America the future of Eastern Europe? (*Probl. Communism,* 41:3, May/June 1992, p. 44–57)

Interesting adaptation of five-member academic panel discussion regarding lessons that post-Communist Eastern Europe might learn from Latin American experience with social transformation.

3803 **Cumbre Iberoamericana,** *1st, Guadalajara, Mexico, 1991.* Primera Cumbre Iberoamericana, Guadalajara, México, 1991: discursos, Declaración de Guadalajara y documentos. 2. ed. México: Fondo de Cultura Económica, 1992. 364 p.: bibl. (Col. popular; 467)

Following brief introduction by Leopoldo Zea about summit's purposes and themes, reproduces speeches by participating heads of state, Declaration of Guadalajara, statistical analyses, and other documents. Similar in content to item **3853.**

3804 **Demange, Nilson Joseph; Lili Katsuco Kawamura;** and **Akihito Ito.** Internacionalização no Japão e a América Latina. Nagoya, Japan: Univ. de Nanzan, Centro de Estudios de América Latina, 1994. 113 p. (Working paper: Monografía; 12)

Optimistic article by Demange on the state of Latin American studies in Japan discusses increasing level of communication and interchange between Japanese and Latin American scholars, with special reference to the Iberoamerican Institute and the Luso-Brazilian Center at Sophia University in Tokyo, the Center for Latin American Studies at Nanzan University in Nagoya, and the Institute for Developing Economics (IDE). As part of a larger research project funded by UNICAMP and Fundação de Amparo à Pesquisa do Estado de São Paulo on Brazilian workers in Japan, Kawamura studies companies in Aichi, Shizuoka, Gunma, and Tokyo, focusing particularly on Brazilian workers' qualifications within the technologically advanced country of Japan. Ito offers a unique analysis of public information services for Brazilian residents, "novos cidadãos," of Toyota, a city with a considerable Brazilian population. Also examines Brazilian-Japanese relations facilitated by the International Association in Toyota, part of a "consciousness of internationalization" initiative undertaken by the municipal government. [K. Horisaka]

3805 **Desch, Michael Charles.** When the Third World matters: Latin America and United States grand strategy. Baltimore: Johns Hopkins Univ. Press, 1993. 1 v.: bibl., index.

Deals with problem of future US "grand strategy" toward Latin America in post–Cold War environment.

3806 **Development from within: toward a neostructuralist approach for Latin America.** Edited by Osvaldo Sunkel. Boulder, Colo.: L. Rienner, 1993. 441 p.: bibl., ill., index.

Written in response to enthusiasm for economic neoliberalism throughout hemisphere and abandonment of structural dependency theory and prescriptions, important figure in latter movement guides multiauthored effort to justify dependency theory in current political-economic environment.

3807 **Domínguez, Jorge I. et al.** Western Hemisphere immigration and United States foreign policy. Edited by Christopher Mitchell. University Park, Pa.: Pennsylvania State Univ. Press, 1992. 314 p.: bibl., ill., index.

Expert on the subject orchestrates analyses of this increasingly important subject, combining scholarly objectivity with humane sensibilities.

3808 **Dosman, Edgar J.** Canada and Latin America: the new look. (*Int. J./Toronto*, 47:3, Summer 1992, p. 529–554)

Canadian discusses his country's policy reorientations toward Latin America since 1989. Should be read in conjunction with item **3810**.

3809 **Duncan, W. Raymond.** Russian-American cooperation in Latin America since Gorbachev. (*in* The end of superpower conflict in the Third World. Edited by Melvin A. Goodman. Boulder, Colo.: Westview Press, 1992, p. 49–74)

Explores move from conflict to cooperation between Moscow and the US in Latin America. Policy transformation seen as a result of Gorbachev's "new thinking," changing international and domestic settings, US pressures, and the idea of "dividing issues into negotiable parts."

3810 **A dynamic partnership: Canada's changing role in the Hemisphere.** Edited by Jerry Haar and Edgar J. Dosman. New Brunswick, N.J.: Transaction Publishers; Miami: Univ. of Miami North-South Center, 1993. 192 p.: bibl.

Excellent explanation of Canada's activist and increasingly influential role in inter-American relations following its apparently irrevocable policy reorientation in late 1980s.

3811 **Economic development and social change: United States–Latin American relations in the 1990s.** Edited by Antonio Jorge. New Brunswick, N.J.: Transaction Publishers, 1992. 142 p.: bibl.

Contributors emphasize compelling saliency of economic and social challenges for inter-American relations in new global and Latin American environments.

3812 **Estay Reyno, Jaime** and **Héctor Sotomayor.** El desarrollo de la Comunidad Europea y sus relaciones con América Latina.

Puebla, Mexico: Benemérita Univ. Autónoma de Puebla; México: Instituto de Investigaciones Económicas de la UNAM, 1992. 142 p.: bibl.

Straightforward treatment of evolution of European Community and its Latin American policies, emphasizing recent developments, commercial relations, and bilateral problems.

3813 **Fifer, J. Valerie.** United States perceptions of Latin America, 1850–1930: a 'New West' south of Capricorn? Manchester; Manchester Univ. Press; New York: St. Martin's Press, 1991. 203 p.: bibl., index, maps.

Using Frederick Jackson Turner's frontier thesis about western development and optimism as point of departure, argues that experience and self-image influenced US perceptions of and entrepreneurial activities in Argentina and Chile.

3814 **El fin del fantasma: las relaciones interamericanas después de la Guerra Fría.** Recopilación de Heraldo Muñoz. Santiago: Hachette, 1992. 177 p.: bibl.

Based on a 1990 academic conference, 20 analysts discuss numerous aspects of inter-American relations in post-Cold War context: security, US policy, crisis situations, drug traffic, ecology, and human migration.

3815 **Fitch, J. Samuel.** The decline of US military influence in Latin America. (*J. Interam. Stud. World Aff.*, 35:2, Summer 1993, p. 1–49, bibl., graphs, tables)

Comprehensive assessment of post-Cold War trends in US-Latin American military relations in which the primary instruments of US influence in the region are changing with an increased emphasis on economics. Concludes that the decline in US military influence has been most acute in the major nations of South America, the US–Latin American military alliance against "communism" is dead, and the threat of "narco-terrorism" provides a weak basis for resurrecting the alliance. Prior habits of policymakers need to change. [D. Dent]

Flor Belaunde, Pablo de la. Japón en la escena internacional: sus relaciones con América Latina y el Perú. See item **4305**.

3816 **La formación de la imagen de América Latina en España, 1898–1989.** Coordinación de Montserrat Huguet Santos, Antonio Niño Rodríguez y Pedro Pérez Herrero. Madrid: Organización de Estados Iberoamericanos para la Educación, la Ciencia y la Cultura (OEI), 1992. 437 p.: bibl., ill. (Cuadernos de cultura iberoamericana)

Narrowly focused contributions add up to original analysis of image formation and perception, and its impact on Spanish ideology and policy within government entities.

3817 **Freres, Christian L.; Alberto van Klaveren; and Guadalupe Ruiz-Giménez.** Europa y América Latina: la búsqueda de nuevas formas de cooperación. (*Síntesis/Madrid,* 18, sept./dic. 1992, p. 91–178, bibl., tables)

Lengthy article on policies of EC and certain member states toward Latin America, analyzing principal accomplishments and challenges in light of profound changes in international system.

3818 **Frieden, Jeffry A.** Debt, development, and democracy: modern political economy and Latin America, 1965–1985. Princeton, N.J.: Princeton Univ. Press, 1991. 280 p.: bibl., ill., index.

Comparative study of Argentina, Brazil, Chile, Mexico, and Venezuela from 1968–85. Argues that economic and class interests of social groups explain variations in economic responses to and political effects from their external environments.

3819 **Goodman, Louis Wolf and Johanna S.R. Mendelson.** La amenaza de las nuevas misiones: las fuerzas armadas lationamericanas y la guerra contra la droga. (*in* Los militares y la democracia: el futuro de las relaciones cívico-militares en América Latina. Compilación de Louis W. Goodman, Johanna S.R. Mendelson and Juan Rial. Montevideo: PEITHO, 1990, p. 259–266)

Critical discussion looks at dangers associated with a Latin American military role in the "war against drugs." Although recognizing dimension of problem and lack of easy solutions, authors favor renewed police activity, border controls, and international cooperation as alternative to military action, which should be considered only as a last resort. [A. Vacs]

3820 **Grabendorff, Wolf.** European integration: implications for Latin America. (*in* Strategic options for Latin America in the 1990s. Edited by Colin I. Bradford, Jr. Paris: Development Centre of the Organisation for Economic Co-Operation and Development; Washington: Inter-American Development Bank, 1992, p. 217–247, bibl., tables)

Argues probable impact of European integration resulting from adoption of Single European Market will be "mainly an unplanned side effect" and "relatively less potent for Latin America than for other regions." For Spanish versions see *Integración Latinoamericana,* Vol. 17, No. 180, julio 1992, p. 16–42 and *Síntesis,* No. 18, sept./dic. 1992, p. 15–52.

3821 **Gutiérrez Bermedo, Hernán and Manfred Wilhelmy von Wolff.** Concepciones latinoamericanas y asiáticas sobre cooperación regional. (*Estud. Int./Santiago,* 24:96, oct./dic. 1991, p. 472–517)

Predicts historical divergences between Latin American and Asian conceptions of regional cooperation will disappear as Asian countries continue to adopt Latin American approach to changing international system, emphasizing regional integration and consultation.

3822 **Hakim, Peter.** Clinton and Latin America: facing an unfinished agenda. (*Curr. Hist.* 92:572, March 1993, p. 97–101)

"Most Latin American governments seemed to prefer the continuity of the Bush Administration to the 'change' promised by the Clinton candidacy;" they especially liked Bush's hemispheric trade initiatives. "The task of the Clinton Administration is not to break new ground but to consolidate and build on what has been accomplished."

3823 **Hakim, Peter.** President Bush's southern strategy: the Enterprise for the Americas Initiative. (*Wash. Q.,* 15:2, Spring 1992, p. 93–106)

Summary and analysis of the Enterprise for the Americas Initiative for free trade, debt relief, and investment, written in early stage of implementation.

3824 **Hakim, Peter.** The United States and Latin America: good neighbors again? (*Curr. Hist.,* 91:562, Feb. 1992, p. 49–53)

Positive evaluation of inter-American developments and US roles in negotiating or contemplating free trade agreements, OAS efforts to restore to power deposed Haitian president Aristide, and the Central American peace process. Nonetheless, conditions US–

Latin American future cooperation on the existence of mutual interests and on Latin American democratic and economic success.

3825 Iglesias, Enrique V. Reflections on economic development: toward a new Latin American consensus. Washington: Inter-American Development Bank, 1992. 158 p.

The president of the Inter-American Development Bank reflects optimistically on positive attitudes of Latin American nations toward each other and outside world. These improved relations have helped to revive political and economic reform within Latin America.

3826 El impacto del capital financiero del narcotráfico en el desarrollo de América Latina: simposio internacional. La Paz: Ediciones CERID; Fondo Fiduciario Manuel Pérez-Guerrero para la Cooperación Económica y Técnica entre Países en Desarrollo, 1991. 475 p.: bibl.

Based on conference report of proceedings, offers broad-ranging analyses of financial impact of narcotraffic in the region.

Impex: Foreign Trade Statistics by Commodities = Statistiques de commerce extérieur produits. 1993– . See item **27**.

3827 Inter-American Dialogue. Convergence and community: the Americas in 1993; a report of the Inter-American Dialogue. Washington: Aspen Institute, 1992. xvi, 74 p.: ill.

Periodic report addresses potential Western Hemispheric free trade area, collective defense of democracy and human rights as matter of international security, and problems of Latin American poverty and social inequality.

International Political Science Abstracts. 1989– . See item **29**.

Jane's Sentinel. 1995– . See item **31**.

3828 Japan, the United States, and Latin America: toward a trilateral relationship in the Western Hemisphere. Edited by Barbara Stallings and Gabriel Székely. Baltimore: Johns Hopkins Univ. Press, 1993. 240 p.: bibl., ill., index.

Strongly emphasizes Japanese role. Only introductory chapter interpreting "new trilateralism" includes discussion of US role, with remainder devoted to "Perspectives from Japan" and Japanese relations with Brazil, Mexico, Peru, Chile, and Panama.

3829 Kaplan, Marcos. La internacionalización del narcotráfico latinoamericano y Estados Unidos. (*Relac. Int./Mexico*, 57, enero/marzo 1993, p. 75–86)

Drawing heavily on perspectives and historical-structural context in item **3830**, author focuses on heavy US role in international aspects of Latin America's *narcoeconomía, narcosociedad, narcopolítica,* and danger of a *narcoEstado*.

3830 Kaplan, Marcos. El narcotráfico latinoamericano en una perspectiva transdisciplinaria. (*Foro Polít.*, 5, agosto 1992, 69–106)

Inasmuch as drug addiction and narcotraffic are conditioned by numerous factors—economic, social, cultural, ideological, political, juridical, and military—analysis and policy should adopt interdisciplinary approaches.

3831 Klaveren, Alberto van. Entendiendo las políticas exteriores latinoamericanas: modelo para armar. (*Estud. Int./Santiago*, 25:98, abril/junio 1992, p. 169–216, bibl., ill.)

In recent years theoretical investigations of Latin American foreign policy have advanced and been consolidated but still have limitations; offers suggestions for synthesizing integrated analytic framework based on comparative analysis of foreign policy decision-making processes.

3832 Laredo, Iris Mabel. Definición y redefinición de los objetivos del proceso de integración latinoamericana en las tres últimas décadas, 1960–1990. (*Integr. Latinoam.*, 16:171/172, sept./oct. 1991, p. 3–25)

Traces development of changing concepts underlying Latin American integration, and argues need for Latin Americans to achieve a more favorable regional position at a time when North-North connections have transcended North-South relations.

3833 Latin America in a New World. Edited by Abraham F. Lowenthal and Gregory F. Treverton. Boulder, Colo.: Westview Press, 1994. 265 p.: bibl., index. (An Inter-American Dialogue book)

Examination of broad range of subjects concerning evolving international role of

Latin America. Covers post–Cold War developments and both domestic and international political and economic changes in the region.

3834 **The Latin American debt.** Edited by Antonio Jorge and Jorge Salazar-Carrillo. New York: St. Martin's Press, 1992. 210 p.: bibl., ill., index.

Two experienced political economists again collaborate for fresh analysis of the decade-old debt problem as it became subsumed under broader questions of inter-American trade and investment.

3835 **Latin America's international relations and their domestic consequences: war and peace, dependency and autonomy, integration and disintegration.** Edited with an introduction by Jorge I. Domínguez. New York: Garland Publishing, 1994. 451 p.: bibl. (Essays on Mexico, Central and South America; 6)

Reproduces 16 previously published journal articles (between 1955–91) by 18 authors with particular (but not exclusive) focus on international economics.

3836 **Latinskaia Amerika: sobytiia i liudi; analiticheskiĭ obzor [Latin America: events and people; an analytical survey].** Otvetstvennyĭ redaktor Anatoliĭ Nikolaevich Glinkin. Moskva: Rossiĭskaia akademiia nauk, In-t Latinskoĭ Ameriki, 1993. 119 p. bibl.

Staff of the Institute of Latin American Studies (Russian Academy of Sciences) focus on events in Latin America following the visit there of Russian Vice President Aleksandr Rutskoi in 1992. Special attention is given to possibilities for the development of Russian–Latin American relations; market reforms in Argentina, Chile, and Peru; consequences of the impeachment of Brazilian President Collor de Mello; Cuban-US relations; and President Clinton's policy toward Latin America. A professor at the Catholic University of Rio de Janeiro contributed one article, appearing in Russian translation, on the Brazilian ex-president. Some Russian, but mostly Western sources. Includes a 1992 chronology of Latin American political events of interest to Russian researchers. [B. Dash]

3837 **Lavados, Iván.** Tendencias e impactos de la cooperación internacional en América Latina. (*Estud. Soc.*/Santiago, 57, 1988, p. 49–70, tables)

Brief and informative analysis examines evolution of international cooperation activities of multilateral agencies, official sources, and foundations. Argues that Latin America was particularly favored in the first and second stages when these institutions' activities were oriented toward promotion of economic growth and then toward scientific and technological modernization. However, in the current stage, with attention shifting to elimination of extreme poverty, resources and activities have been focused on Africa and Asia. Also discusses impact of international cooperation on creation of developmental institutions, shaping of public policies and priorities, and formation and mobilization of human and financial resources in Latin America. [A. Vacs]

3838 **Li, He.** Sino-Latin American economic relations. New York: Praeger, 1991. 179 p.: bibl., index.

Tightly written political-economic analysis traces chronological development of relations through four stages from 1949–90 and analyzes specific processes, problems, issues, and prospects.

3839 **Lowenthal, Abraham F.** Latin America: ready for partnership? (*Foreign Aff.*, 72:1, 1992/93, p. 74–92)

Examines US-Latin American relations early in the Clinton Administration. Analyzes and proposes new US stance especially in terms of economic relations, Latin American political-economic transformations, and US domestic politics and economic revitalization.

3840 **Lowenthal, Abraham F.** U.S. policy toward Latin America. (*in* U.S. foreign policy: the search for a new role. New York: Macmillan Publishing Company, 1993, p. 358–382)

Summarizes transformations in US post–Cold War policy, reshaped by combination of global changes; revised agenda of inter-American issues; Latin American social, political, and economic developments; and shifting policy processes within the US ("the single biggest factor").

3841 **Mace, Gordon; Louis Bélanger; and Jean Philippe Thérien.** Regionalism in the Americas and the hierarchy of power. (*J. Interam. Stud. World Aff.*, 35:2, Summer 1993, p. 115–157, bibl., graphs, tables)

Reviews analytic trends in recent scholarly rediscovery of regionalism, especially notion of world economy restructuring around major trading blocs. In the Americas regional cooperation is likely to grow around US and Brazilian poles of power.

3842 Malamud Goti, Jaime E. Smoke and mirrors: the paradox of the drug wars. Boulder, Colo.: Westview Press, 1992. xxi, 117 p.: bibl., ill.

Emphasizes futility of current antinarcotics policies because they are countervailed by bureaucratic and personal interests. Interesting observations using techniques of journalism; virtually no mention of alternative strategies.

3843 McGurk, Russell and **Claudia Nierenberg.** U.S. economic policy and sustainable growth in Latin America. New York: Council on Foreign Press, 1992. 41 p.

Description and critique of developmental problems and policies, given changing policy-making and systemic environments.

3844 Medina, Manuel et al. Unidad y heterogeneidad en Iberamérica [i.e., Iberoamérica]. Huelva, Spain: Univ. Hispanoamericana Santa María de la Rábida, 1992. 275 p.: bibl.

Spanish academics and officials explore numerous aspects of Ibero-America's international relations with special reference to European and Iberian connections.

3845 Moreno, Rafael. El Nuevo Orden Internacional y América Latina. (*Universum/Talca*, 6, 1991, p. 15–26)

Examines main political, economic, and strategic transformations that typify emergence of a new world order, and considers ways in which Latin America inserts itself into this scenario. Lists emergence of market economies, democratization, and trend toward regional integration as crucial Latin American developments in recent past, and emphasizes that main problems facing the region are poverty, crime, population growth, and mismanagement of natural resources. Concludes that first priority is to reach political and economic agreement among the Latin American countries in order to negotiate as a block with the developed countries for more favorable conditions for insertion into the world order. [A. Vacs]

3846 Morici, Peter et al. The premise and the promise: free trade in the Americas. Edited by Sylvia Saborío Alvarado. New Brunswick, N.J.: Transaction Publishers, 1992. 282 p.: bibl., ill. (U.S.–Third World policy perspectives; 18)

Contributors (mostly Canadian and Latin American) generally agree that President Bush's Enterprise for the Americas Initiative, if implemented, will be a policy of historic proportions.

O'Donnell, Mario. El descubrimiento de Europa: la Comunidad Europea y sus nuevas relaciones con la Argentina y el resto de Latinoamérica. See item **4215**.

3847 Oliveri, Ernest J. Latin American debt and the politics of international finance. Westport, Conn.: Praeger, 1992. 235 p.: bibl., ill., index.

Forceful analysis focuses on the actors—the major creditor and debtor states and the IMF—and their shifting and not always credible political-economic strategies.

3848 Packenham, Robert A. The dependency movement: scholarship and politics in development studies. Cambridge, Mass.: Harvard Univ. Press, 1992. 362 p.: bibl., index.

Acknowledges accuracy of some dependency propositions but sees fundamental errors in its advocates' faulty policy prescriptions and mistaken attempts to combine scholarship with political activism.

3849 Pastor, Robert A. Estados Unidos y América Latina en los noventa. (*Cienc. Polít.*, 30, primer trimestre 1993, p. 71–83)

US-Latin American relations will assume higher priority on US security agenda with formation of regional trading blocs, increased global commerce, and increase of Latin American emigration to the north (among other factors).

3850 Pastor, Robert A. Whirlpool: U.S. foreign policy toward Latin America and the Caribbean. Princeton, N.J.: Princeton Univ. Press, 1992. 338 p.: bibl., index. (Princeton studies in international history and politics)

In the past the US overemphasized problems created by small states or drifted into policy of neglect, a pattern that will not necessarily be broken with the end of the Cold War. Prescribes new orientation based on multilateralism emphasizing promotion of democracy and free trade. An important book.

3851 Pensamiento Iberoamericano. No. Extraordinario, 1991- . La Nueva Europa y el Futuro de América Latina. Madrid: Instituto de Cooperación Iberoamericano (ICI); Comisión Económica para América Latina y el Caribe (CEPAL).

Substantial (437 p.) special journal edition based on conference proceedings contains large number of presentations and commentaries offering broad range of views.

3852 Pike, Fredrick B. The United States and Latin America: myths and stereotypes of civilization and nature. Austin: Univ. of Texas Press, 1992. 442 p.: bibl., ill., index.

North American and Latin American stereotypes of each other, dating from first colonial-period contacts and still impinging on relations, derive from former's view of latter's civilization as primitive because it accepts rather than dominates nature.

3853 Primera cumbre iberoamericana: memoria. Guadalajara, Mexico: Dirección General del Acervo Histórico Diplomático de la Secretaría de Relaciones Exteriores, 1991. 1 v.

Comprehensive *memoria* in Spanish and Portuguese includes speeches by 21 participating heads of state or government (from Spain, Portugal, and Latin America), a document regarding Cuba, Declaration of Guadalajara, and six issue-oriented "working documents." Similar in content to item **3803**.

3854 Private sector solutions to the Latin American debt problem. Edited by Robert E. Grosse. New Brunswick, N.J.: Transaction Publishers, 1992. 188 p.: bibl., ill.

Analysts offer prescriptions to the debt problem largely assuming that the "debt crisis" has ended.

Ramírez León, José Luis *et al.* Colombia y América Latina frente a Europa 1992. See item **4277**.

3855 Rochlin, James. Canada as a hemisphere actor. Toronto: McGraw Hill-Ryerson, 1992. 1 v.

Survey emphasizes revitalization and extension of Canada's inter-American activities since the late 1980s, providing ample insight.

3856 Rossiia i Latinskaia Amerika: k novomu partnerstvu [Russia and Latin America: to a new partnership]. Edited by Viktor Vatslavovich Vol'skiĭ. Moskva: Rossiĭskaia akademiia nauk, Institut Latinskoĭ Ameriki, 1992. 74 p.

Seven researchers writing for the Institute of Latin American Studies of the Russian Academy of Sciences identify parallels between the "new Russia" and Latin America in the early 1990s, opportunities for economic cooperation between the two, and examples in recent Latin American political and economic experience from which Russian leaders might learn. Discussions include new Russian policy in Latin America, including Cuba; a comparison of new Russian economic market models and the Latin American experience; foreign debt; and regional conflict. [B. Dash]

3857 Russian views of Russian–Latin American relations in the post-Cold War world. Edited by Richard Downes. Miami, Fla.: North-South Center, Univ. of Miami, 1993. 15 p. (The North-South agenda papers; 5)

Interesting commentary by five Russian analysts who pay special attention to Inter-American System, US policy, and problem of Cuba, arguing among other things that Russia can learn from Latin America's recent political and economic reform.

3858 Sanderson, Steven E. The politics of trade in Latin American development. Stanford, Calif.: Stanford Univ. Press, 1992. 292 p.: bibl., index.

Eschewing increasingly accepted notion of interdependence, offers political-economic explanations of linkages between international trade and Latin American development in tradition of dependency theory. For political scientist's comment, see *HLAS 53: 3268*.

3859 Saragossi, Maggy. Persuasion et séduction: le discours politico-commercial du Canada sur l'Amérique latine, 1982–1985. Candiac, Canada: Editions Balzac, 1991. 455 p.: bibl., index. (Col. L'Univers des discours)

Detailed and imaginative political-economic analysis of a three-year period in Canada's relations with Latin America.

3860 **Serbín, Andrés et al.** El Grupo de los tr3s [i.e., tres]: políticas de integración. Bogotá: FESCOL, 1992. 319 p.: bibl.

Edited volume on dynamics of creating regional spheres of power within developing regions—in this case Mexico, Venezuela, and Colombia in Central America and the Caribbean. [D. Story]

3861 **Sharbach, Sarah E.** Stereotypes of Latin America, press images, and U.S. foreign policy, 1920–1933. New York: Garland Pub., 1993. 233 p.: bibl., ill., index. (Modern American history)

Historical treatment of US views of Latin American race and culture. Discusses uninspiring views of diplomats, journalists, and academics in conceptual and diplomatic contexts.

3862 **Sheinin, David.** Making democracy safe for the world: the neo-liberal agenda and the new isolationism in relations with Latin America. (*Int. J./Toronto,* 48:1, Winter 1992/93, p. 100–123)

Argues that analyses of US post–Cold War foreign policy in terms of neo-isolationism are based on anachronistic concepts with little relevance for Latin American relations.

3863 **Shelton, Dinah L.** The inter-American human rights system. (*in* Guide to international human rights practice. Edited by Hurst Hannum. Philadelphia: Univ. of Pennsylvania Press, 1992, p. 119–132)

Serviceable description of structures and processes for observation and protection of human rights in the Organization of American States.

3864 **Shugart, Matthew Soberg.** Guerillas and elections: an institutionalist perspective on the costs of conflicts and competition. (*Int. Stud. Q.,* 36:2, June 1992, p. 121–151, bibl.)

Well-done theoretical analysis, including international implications, encompasses Venezuelan, Colombian, Nicaraguan, and Salvadoran cases.

3865 **Social democracy in Latin America: prospects for change.** Edited by Menno Vellinga. Boulder, Colo.: Westview Press, 1993. 327 p.: bibl., index.

Includes good chapters on German, Spanish, British, and Swedish social democracies' relations with and influences on Latin America.

Statistical Masterfile. 1974– . See item **44**.

3866 **Stoetzer, O. Carlos.** The Organization of American States. 2nd ed. Westport, Conn.: Praeger, 1993. 443 p.: bibl., index.

Thorough chronological description of institutional development of OAS from 19th-century antecedents through 1992.

3867 **Suárez, Carlos O. et. al.** La estrategia neocolonial del imperio para los años '90. Buenos Aires: Ediciones de GenteSur, 1990. 160 p.: bibl. (Los Libros de GenteSur: Ensayos)

Set of four articles written by members of Argentine chapter of the Tribunal Anti-imperialista de Nuestra América denounces US foreign policy approach to Latin America and particularly use of the war against drugs to interfere in internal affairs of countries in the region. Useful as compendium of main imputations against US foreign policy advanced by nationalistic and leftist critics in 1990s. [A. Vacs]

3868 **Suecia-Latinoamérica: relaciones y cooperación.** Edición de Weine Karlsson, Åke Magnusson y Carlos Vidales. Stockholm: Latinamerika-institutet, 1993. 250 p.: bibl., ill. (some col.). (Monografía, 0284–6675; 24)

Comprehensive treatment of relatively little-known subject, with chapters devoted to historical background from 1600, political contacts, cooperation, NGOs, and cultural and economic relations.

3869 **Tomassini, Luciano; Carlos Juan Moneta; and Augusto Varas.** La política internacional en un mundo postmoderno. Buenos Aires: Programa de Estudios Conjuntos sobre las Relaciones Internacionales de América Latina; Grupo Editor Latinoamericano, 1991. 302 p.: bibl., ill. (Col. Estudios internacionales)

Sophisticated exploration of "postmodernity" as concept and reality; evolution of global system in terms of "regional spaces," strategic relations, transformations of power and interdependence, and changes in the State and its actions and interactions. Last chapter has insightful discussion of research on Latin America's international relations.

TRAINS. See item **45**.

T︠S︡entral'na︠i︡a Amerika i Kariby: nachalo 90-kh godov [Central America and the Caribbean: the beginning of the nineties]. See item **2804**.

3870 **The United States and Latin America in the 1990s: beyond the Cold War.** Edited by Jonathan Hartlyn, Lars Schoultz, and Augusto Varas. Chapel Hill: Univ. of North Carolina Press, 1992. 328 p.: bibl., index, map.

Important attempt by 14 expert analysts to study early post-Cold War era and beyond in terms of changing international system, US interests, Latin American societies and politics, and inter-American issues.

3871 **Valdés Paz, Juan.** Notas sobre el nuevo sistema internacional, el Tercer Mundo y América Latina. (*Cuad. Nuestra Am.*, 8:17, julio/dic. 1991, p. 67–91, tables)

Some original insights on evolution of the international system since the 1940s from a Cuban expert. [D.J. Fernández]

3872 **Valle, Carlos del.** La deuda externa de América Latina: relaciones norte-sur, perspectiva ética. Prólogo de Paulo Evaristo Arns. Estella, Spain: Editorial Verbo Divino, 1992. 706 p.: bibl., ill. (Misión sin fronteras)

Detailed two-part treatment of Latin American external debt, in context of: 1) North-South economic tensions, North's concepts of Third World poverty and Latin American development and underdevelopment, and alternative developmental theories; and 2) international systemic elements.

3873 **Varas, Augusto.** De la coerción a la asociación: ¿hacia un nuevo paradigma de cooperación hemisférica? (*Estud. Int./IRIPAZ*, 2:3, enero/junio 1991, p. 23–35)

Analyzes principal effects of global changes from polarized paradigms of hemispheric control in 1970s and 1980s to new one of hemispheric cooperation. Posits different scenarios of specific security regimes.

3874 **Vargas Llosa, Mario *et al.*** El desafío neoliberal: el fin del tercermundismo en América Latina. Recopilación de Barry B. Levine. Barcelona, Spain: Grupo Editorial Norma, 1992. 518 p.: bibl., index. (Literatura y ensayo)

Important book by 22 distinguished contributors who emphasize negative consequences of state-run economies and approve of free market reforms.

3875 **War on drugs: studies in the failure of U.S. narcotics policy.** Edited by Alfred W. McCoy and Alan A. Block. Boulder, Colo.: Westview Press, 1992. 359 p.: bibl.

Good collection of studies on numerous aspects of counternarcotics actions, especially US actions. Subtitle sums up tenor of analysis.

3876 **Washington Office on Latin America.** Elusive justice: the U.S. Administration of Justice Program in Latin America. Washington: Washington Office on Latin America, 1990. 52 p.

Report on the Administration of Justice Program is result of 1989 workshop organized by WOLA and the School of International Service of American Univ., attended by Latin American and US officials, scholars, congressional aides, and human rights representatives. Includes general overview of history, goals, structure, and funding of the program, and case studies of program operations in El Salvador, Guatemala, and Colombia. Conclusions criticize subordination of program's long-term goal (improvement of justice administration) to short-term objectives (control of political violence, anti-drug enforcement), and point out lack of accepted criteria for evaluating program's effectiveness. [A. Vacs]

3877 **Washington Office on Latin America.** U.S. electoral assistance and democratic development: Chile, Nicaragua and Panama. Washington: Washington Office on Latin America, 1990. 33 p.

Summarizes debate and conclusions of a conference on the topic sponsored by WOLA in January 1990. Valuable critical analysis of role played by US in these three cases. [A. Vacs]

3878 **Wiarda, Howard J.** American foreign policy toward Latin America in the 80s and 90s: issues and controversies from Reagan to Bush. New York: New York Univ. Press, 1992. 363 p.: bibl., index, map.

Collection of Wiarda's papers grouped under categories of US policy in Latin America, policy issues, theoretical perspectives, country and regional studies, and "toward the future."

3879 **Wilhelmy von Wolff, Manfred.** Los objetivos en la política exterior latinoamericana. (*Estud. Int./Santiago*, 24:94, abril/junio 1991, p. 176–193)

As public policies, Latin American foreign policies should be better adapted, analytically and operationally, to changes in international scene. Policies should also be made in more integrated form.

3880 **Wilson, Suzanne** and **Marta Zambrano.** Cocaine, commodity chains, and drug politics: a transnational approach. (*in* Commodity chains and global capitalism. Edited by Gary Gereffi and Miguel Korzeniewicz. Westport, Conn.: Praeger, 1994, p. 297–315, bibl., table)

Interesting analysis, from the world system perspective, considers cocaine as final product of a transnational commodity chain. Argues that linkages to legal productive and financial sectors of global economy and concentration of profits in core countries contribute to explaining selective nature of US antidrug policy, which focuses on illegal and informal components of the chain. [A. Vacs]

World News Connection. 1995- . See item **50.**

3881 **Wright, Thomas C.** Latin America in the era of the Cuban Revolution. New York: Praeger, 1991. 236 p.: bibl., ill., index, maps.

Considers impact of the Cuban Revolution on Latin American and US policy with chapters on rural and urban guerrilla warfare, military regimes, Chile, Nicaragua, and Peru.

3882 **Yopo H., Boris.** The Rio Group: decline or consolidation of the Latin American *Concertación* policy? (*J. Interam. Stud. World Aff.*, 3:4, Winter 1991, p. 27–44, bibl.)

Traces evolution of Contadora Group into the Rio Group: "an association of democratic countries" with much broader agenda seeking foreign policy coordination and harmonization (*concertación*).

MEXICO AND CENTRAL AMERICA

DALE STORY, *Professor and Chair, Department of Political Science, University of Texas at Arlington*

AS REVOLUTION IN CENTRAL AMERICA is replaced in the headlines by the North American Free Trade Agreement (NAFTA), scholarly literature on Mexico's international relations has gained considerably in output relative to that on Central America. Some of the most compelling work on Mexico places that country's foreign policy in the context of broader changes in the international arena. As one example, the noted Mexican publishing house Fondo de Cultura Económica has published *La política exterior de México en el nuevo orden mundial* (item **3939**), a very extensive collection of essays stressing the impact of the end of the Cold War on the dynamics of Mexico's international policies. In addition to its theoretical contributions, this volume provides one of the most comprehensive surveys of the full spectrum of bilateral issues facing post-Cold War Mexico. This theme of both the limitations and opportunities afforded Mexico by the changing international context is also reflected in Aguayo Quezada and Bagley's compilation on national security (in Spanish, item **3909**; in English, item **3934**), in which various authors revisit the concept of national security in the Mexican experience and address a number of specific manifestations, such as in the areas of refugees and narcotics.

While important contributions continue to be made in the traditional areas of balance of power, military relations, diplomatic initiatives, and the like, undoubtedly the preeminent concern of much of the current research on Mexican international relations lies in the economic arena. And the major initiative here has been the North American Free Trade Agreement. What was almost unthinkable in Mexico of the 1980s became a reality in the 1990s. Thus, an important component to the research agenda on NAFTA should be the political dynamics that made its ratification not only acceptable but surprisingly popular. Sadly, too much of the literature ignores the political dimensions of this crucial economic reform. One exception is the excellent article by Weintraub and Baer (item **3960**). In tackling the complexities of the linkages between economic and political reform, they begin by suggesting that the successes of the economic liberalization eventually will make possible the democratic reform of the political system. They also posit that the economic changes can be (and have been in Mexico) much more dramatic than the political openings. Other arguments to be explored here include the necessity for political concessions given the exhaustion of the postwar economic model as well as the external pressures for reform in both the economic and political arenas. In a somewhat more analytical piece, Peter Smith presents four different propositions regarding the impact of NAFTA on domestic politics in Mexico (item **3952**). Rather than reaching any definite conclusions, Smith urges students of Mexican politics to examine these contending theories in the context of empirical analysis.

Turning more specifically to the treaty itself, three noteworthy contributions should be highlighted. Robert Pastor has provided a very readable book concentrating on the negotiations on NAFTA through its signing by Bush, Salinas, and Mulroney in Dec. 1992 (item **3938**). The Mexican Secretary of Commerce has edited a surprisingly objective volume that focuses on the Mexican economic background to the push for NAFTA (item **3937**). This perspective is important for understanding the political and economic changes occurring during the Salinas Administration. Jaime Ros explains the growing popularity of economic integration in North America in terms of a "silent" integration that had already been drawing Mexico and the US closer, a new ideological climate, and the necessities of competing in the international economic marketplace of the 1990s (item **3946**).

A number of excellent publications have concentrated on the environmental components of NAFTA. Roberto Sánchez Rodríguez utilizes two case studies—drainage in Tijuana and San Diego and air pollution from smelting plants in Arizona and Sonora—to explore the bilateral negotiations on the environment (item **3948**). Kelly and her co-authors argue the salience of environmental issues that will arise predictably with the implementation of NAFTA (item **3917**). Finally, Vargas and Bauer contribute an edited book with the most extensive coverage on the full range of environmental issues arising from NAFTA (item **3933**).

Two final areas of bilateral concern along the US-Mexico border warrant recognition: immigration and narcotics. In a study commissioned by the US government (item **3957**), the authors focus on the "push" (or supply) side of the equation in arguing for the need for economic development in Mexico. Focusing on the changing characteristics of Mexican immigrants into the US, Wayne Cornelius provides a wealth of information on the demographics of the migrant workers and concludes that the United States' 1986 Immigration Reform Act has had little impact (item **3900**). On the difficult issue of narcotics, Reuter and Ronfeldt argue that results from law enforcement efforts are not as important as the "integrity" of the effort (item **3945**).

Sadly, Central America seems to receive considerable scholarly attention only when a case of US intervention has occurred; typically, this literature tends to be highly polemical. With the short-term success of peaceful negotiations to end the military conflicts in Nicaragua and El Salvador, the most recent wave of writings has concentrated on Panama. The two most commendable contributions are relatively short treatises published by the Center for Strategic and International Studies: Falcoff and Millett predict a tenuous future for Panama as it strives for democratization (item **4002**) and contributors to a monograph edited by Eva Loser posit a negative impact on Panamanian political development from US ties (item **3980**). Utilizing the same theme of democratization but in a different national context, Joseph

Tulchin has edited a book that cautiously suggests some of the key factors in any potential transition to a democratic regime in El Salvador (item **4022**).

Returning to the post-revolutionary regimes in Nicaragua, two books are noteworthy. William Robinson has written an exceptionally well-researched volume on US intervention in the 1990 elections (item **4057**). Exploring a somewhat earlier time period, Morris Morley takes a broader and somewhat more analytical perspective and reaches some provocative conclusions about continuity of US policy toward Nicaragua (item **4036**).

Finally, three books provide useful overviews of Central America. Jan Adams looks at international diplomacy toward Central America from the Soviet perspective during the Gorbachev years (item **3965**). In another commendable volume by Jack Child, the former US Army Latin American specialist stresses issues of verification and confidence-building in the ongoing peace process (item **3985**). And finally, John Coatsworth has integrated some of the best research on Central American history and written what may become the ultimate source on 20th-century Central American development (item **3988**).

MEXICO

3883 **Abella, Gloria.** La política exterior de México en el gobierno de Carlos Salinas de Gortari: ¿una nueva concepción? (*Rev. Mex. Cienc. Polít. Soc.*, 37:148, abril/junio 1992, p. 63–86, bibl.)

Carlos Salinas has been widely depicted as bringing enormous changes to Mexican domestic and international trade policies. Also argues that profound initiatives have occurred in foreign policy.

3884 **Alvarez Soberanis, Jaime.** Necesidad de fortalecer el derecho internacional: la posición de México frente a la decisión número 91–712 de la Suprema Corte de Justicia de los Estados Unidos de América. (*Rev. Mex. Polít. Exter.*, 39, verano 1993, p. 24–40)

Written by an official of the Mexican Foreign Ministry, this article presents the Mexican perspective on the legal challenge in the US Supreme Court to the "kidnapping" of Humberto Alvarez Machain in April 1990 in Guadalajara.

3885 ***Anuario Mexicano de Relaciones Internacionales.*** Parte 1, 1987– . México: Univ. Nacional Autónoma de Mexico, Escuela Nacional de Estudios Profesionales Acatlán.

Encyclopedic reference work that contains numerous essays on Mexican international relations as well as documents and chronologies for 1987.

3886 **La apertura de México al Pacífico.** Coordinación, compilación y edición de Instituto Matías Romero de Estudios Diplomáticos. México: Secretaría de Relaciones Exteriores, 1990. 161 p.: bibl.

Publication of the Mexican Foreign Ministry focuses on Mexico's potential for stronger role in the Pacific Basin.

3887 **Bendesky, León; Fernando Chávez;** and **Jordy Micheli T.** México-Estados Unidos: vecinos y socios; un análisis por sectores y regiones económicas. Coordinación de Jorge Alcocer. México: Centro de Estudios para un Proyecto Nacional, Nuevo Horizonte Editores, 1993. 276 p.: bibl., ill.

Economic analysis of trade relations between Mexico and the US in the 1980s and 1990s. Disaggregated analysis of various US states is useful, as are many of the statistical tables.

3888 **Brown, George E., Jr.; J. William Goold;** and **John Cavanagh.** Making trade fair. (*World Policy J.*, 9:2, Spring 1992, p. 309–327)

Essentially a perspective piece written in the midst of the NAFTA debate. Authors, including one member of Congress, argue that the US negotiators should set highest priorities on protecting jobs, wages, and the environment.

3889 **Brutality unchecked: human rights abuses along the U.S. border with Mexico.** New York: Americas Watch, 1992. 81 p.

Noted human rights group reviews abuses by the US Immigration and Naturalization Service along the Mexican border. Findings are appropriately described as "appalling"—a reminder that violations occur on both sides of the border.

3890 **Bustamante, Jorge A.** Frontera México-Estados Unidos: reflexiones para un marco teórico. (*Estud. Cult. Contemp.*, 4:11, 1991, p. 11–35)

Very theoretical piece by one of the most prominent scholars on the US-Mexico border. Argues need to focus on the topic of social interactions of the region as well as the asymmetry of power relations.

Bustamante, Jorge A. Interdependence, undocumented migration, and national security. See item **4467**.

3891 **The California-Mexico connection.** Edited by Abraham F. Lowenthal and Katrina Burgess. Stanford, Calif.: Stanford Univ. Press, 1993. 364 p.: bibl., ill., index.

Compilation of over a dozen contributions from a 1991 symposium. From a commendable, theoretical introduction by James Rosenau to various specific issues, book provides excellent perspective that goes beyond that of a single state.

3892 **Calzada Falcón, Fernando; Abelardo Aníbal Gutiérrez Lara; and José Manuel Herrera Núñez.** Un tratado en marcha. México: El Nacional, 1992. 144 p.

As a compilation of academic and journalistic perspectives, this monograph provides useful depiction of both events as well as arguments surrounding negotiations between Mexico and the US.

3893 **Cárdenas, Héctor and Evgeni Dik.** Historia de las relaciones entre México y Rusia. México: Secretaría de Relaciones Exteriores; Fondo de Cultura Económica, 1993. 282 p.: bibl. (Sección de obras de historia)

An unprecedented description of bilateral relations between Mexico and Russia from the late 16th century to the present. Particular focus is an evaluation of Mexican domestic politics in this context. Includes excellent chronology.

3894 **Casteñeda, Jorge E. and Carlos Heredia.** After NAFTA: what a good agreement should offer. (*World Policy J.*, 9:4, Fall/Winter 1992, p. 673–685)

A progressive or nationalist alternative to the NAFTA negotiated by Bush and Salinas.

3895 **Castro Rea, Julián.** Canadá, ¿aliado o adversario?: un punto de vista mexicano. (*Rev. Mex. Polít. Exter.*, 38, primavera 1993, p. 42–62, tables)

Though the significance of Canada for Mexico pales in comparison to that of the US, work does contribute to understanding the increasing relations between Canada and Mexico.

3896 **Chabat, Jorge.** Mexico's foreign policy in 1990: electoral sovereignty and integration with the United States. (*J. Interam. Stud. World Aff.*, 33:4, Winter 1991, p. 1–25)

Discusses various themes reflected in Mexican foreign policy in the 1980's: international economic reforms, narcotrafficking, and foreign pressures for democratization.

3897 **Conchello, José Angel.** El TLC: un callejón sin salida. 2. ed. México: Grijalbo, 1992. 289 p.: bibl. (Política mexicana)

Mexican perspective that argues against joining NAFTA. Concludes that Mexico should not sacrifice so much to gain so little.

3898 **Conference on Regional Impacts of Mexico-U.S. Economic Relations, 1st, Guanajuato, Mexico, 1991.** Primer Encuentro sobre Impactos Regionales de las Relaciones Económicas, México-Estados Unidos = The First Conference on Regional Impacts of United States-Mexico Economic Relations: 8 al 11 de julio de 1981 en la ciudad de Guanajuato, México. México: EL Encuentro, 1982. 3 v.: bibl., ill., maps.

Dated summary of bilateral conference on the effect along the border of economic relations between the US and Mexico. Papers are too disjointed to make a valuable compilation.

3899 **Congreso Internacional sobre Fronteras en Iberoamérica Ayer y Hoy, 1st, Tijuana, Mexico, 1989.** Memoria del I Congreso Internacional sobre Fronteras en Iberoamérica Ayer y Hoy. Edición de Alfredo Félix Buenrostro Ceballos. Mexicali, Mexico: Univ. Autónoma de Baja California, 1990. 2 v.: bibl.

Very mixed collection with broad historical and geographical ranges. Ultimately fails to integrate essays in a common theme.

3900 **Cornelius, Wayne A.** *Los Migrantes de la Crisis:* the changing profile of Mexican migration to the United States. (*in* Social responses to Mexico's economic crisis of the 1980s. Edited by Mercedes González de la Rocha and Agustín Escobar Latapí. San Diego, Calif.: Center for U.S.-Mexican Studies, Univ.

of California, San Diego, 1991, p. 155–193, bibl., tables)

Fairly exhaustive review of available data on the number and other demographic characteristics of Mexican migrants. Concludes that the 1986 Immigration Reform Act has not reduced the total number of migrants.

3901 Cornelius, Wayne A. and **Philip L. Martin.** The uncertain connection: free trade and rural Mexican migration to the United States. (*Int. Migr. Rev.*, 27:103, Fall 1993, p. 484–512, bibl., graphs, tables)

Argues that even though NAFTA may create significant displacement of rural workers in Mexico, the treaty will not necessarily lead to increased migration to the US.

Cuenca del Pacífico. See item **20**.

3902 Dávila Pérez, Consuelo. La política exterior de México y el movimiento de los países No Alineados, 1961–1991. (*Relac. Int./Mexico*, 14:53, enero/abril 1992, p. 65–71)

Posits this analysis of Mexico's foreign policy vis-à-vis the Non-Aligned Movement as a demarcation point for understanding Mexico's independence in its diplomatic relations.

3903 Dávila Pérez, Consuelo. Las relaciones de México con la Asociación Latinoamericana de Integración (ALADI): evolución y perspectivas. (*Relac. Int./Mexico*, 13, mayo/agosto 1991, p. 25–34, tables)

While most attention is focused on North America, this article examines Mexican commitment to economic integration in Latin America.

3904 Davis, Diane E. Mexico's new politics: changing perspectives on free trade. (*World Policy J.*, 9:4, Fall/Winter 1992, p. 655–671)

Briefly summarizes Mexican political transition from protectionism to dominance of free trade.

3905 Una década de refugio en México: los refugiados guatemaltecos y los derechos humanos. Recopilación de Graciela Freyermuth Enciso y Rosalva Aída Hernández Castillo. México: Centro de Investigaciones y Estudios Superiores en Antropología Social; Tuxtla Gutiérrez, Chiapas: Instituto Chiapaneco de Cultura; México: Academia Mexicana de Derechos Humanos, 1992. 409 p.: bibl., ill.

Very lengthy edited volume provides comprehensive overview of all aspects of the exodus of Guatemalan refugees to Mexico.

3906 Delpar, Helen. The enormous vogue of things Mexican: cultural relations between the United States and Mexico, 1920–1935. Tuscaloosa: Univ. of Alabama Press, 1992. 274 p.: bibl., ill., index.

Reminds us that US fascination with Mexican culture did not begin with recent focus of attention on Frida Kahlo. Analyzes the years 1920–35, depicting evolution of cultural relations between the US and Mexico as reaching a zenith in this period.

3907 Doyle, Kate. The militarization of the drug war in Mexico. (*Curr. Hist.*, 92:571, Feb. 1993, p. 83–88, tables)

Very readable essay heaps considerable criticism on both Mexican and US policymakers for continuing bankrupt approaches to countering narcotics trade.

3908 Durán, Esperanza. Guerra y revolución: las grandes potencias y México, 1914–1918. México: Colegio de México, Centro de Estudios Internacionales, 1985. 277 p.: bibl.

A well-researched historical analysis juxtaposes international repercussions of World War I with the Mexican Revolution.

3909 En busca de la seguridad perdida: aproximaciones a la seguridad nacional mexicana. Recopilación de Sergio Aguayo y Bruce Michael Bagley. México: Siglo Veintiuno Editores, 1990. 416 p.: bibl., ill., index. (Sociología y política)

Highly recommendable essays on Mexican national security in the era of the new world political order. Explores conceptual and theoretical components as well as applied examples.

3910 Entre la guerra y la estabilidad política: el México de los 40. Coordinación de Rafael Loyola Díaz. 1. ed. en la colección Los Noventa. México: Grijalbo; Consejo Nacional para la Cultura y las Artes, 1990. xi, 396 p.: bibl. (Los Noventa; 9)

Some 13 essays examine a wide range of topics relevant to Mexico in the 1940s. The recurrent theme is the effect of World War II on Mexican domestic and foreign policies.

3911 **Estados Unidos y el occidente de México: estudios sobre su interacción.** Recopilación de Adrián De León Arias. Guadalajara, Mexico: Univ. de Guadalajara, 1992. 172 p.: bibl., ill. (Col. Jornadas académicas. Serie Compilaciones)

Human rights report on violence against Mexican migrant workers on both sides of the border.

3912 **García y Griego, Manuel** and **Mónica Verea.** México y Estados Unidos frente a la migración de indocumentados. México: Coordinación de Humanidades; M.A. Porrúa, 1988. 175 p.: bibl. (Col. Las Ciencias sociales)

Though somewhat dated, provides Mexican perspective and critique of US immigration reform, specifically the Simpson-Rodino bill.

3913 **Gómez Chiñas, Carlos.** La política de negociaciones comerciales de México y el Acuerdo de Libre Comercio México-Estados Unidos-Canadá. (*Anál. Econ./México*, 9:17, mayo/agosto 1991, p. 75–86, bibl.)

Overview of objectives (both stated and proposed) of the Mexican government in international trade negotiations, particularly GATT and NAFTA.

3914 **González-Souza, Luis.** México en la estrategia de Estados Unidos: enfoques a la luz del TLC y la democracia. México: Siglo Veintiuno Editores, 1993. 320 p.: bibl.

Noted Mexican scholar analyzes debate over the Free Trade Agreement in the US. Examines critical societal and state actors involved in the policy process.

Harrison, Benjamin T. Dollar diplomat: Chandler Anderson and American diplomacy in Mexico and Nicaragua, 1913–1928. See *HLAS 54:1739.*

3915 **Huchim, Eduardo.** TLC: hacia un país distinto. México: Nueva Imagen, 1992. 265 p.: bibl.

Readable description of negotiations that produced the North American Free Trade Ageement.

3916 **Interdependencia: ¿un enfoque útil para el análisis de las relaciones México–Estados Unidos?** Coordinación de Blanca Torres Ramírez. México: El Colegio de México, Centro de Estudios Internacionales, 1990. 310 p.: bibl., ill.

Interesting compilation of essays by US and Mexican scholars stressing theoretical components of Mexico-US relations. A few case studies are mixed with more theoretical analysis.

3917 **Kelly, Mary E. *et al.*** México–Estados Unidos: negociaciones sobre libre comercio y medio ambiente; análisis de temas. (*Integr. Latinoam.*, 17:181/182, agosto/sept. 1992, p. 55–72)

Argues that environmental issues and problems between Mexico and the US will only increase with free trade and that such issues should become a priority for both nations.

3918 **Kondracke, Morton.** Mexico and the politics of free trade. (*Natl. Int.*, 25, Fall 1991, p. 36–43)

Non-academic essay on political motivations behind NAFTA. Credits Salinas with sparking negotiations, then devotes attention to the debate in the US.

3919 **Loaeza, Soledad.** La iglesia católica mexicana y las relaciones internacionales del Vaticano. (*Foro Int.*, 32:2, oct./dic. 1991, p. 199–221)

Examines roles of the Mexican state, the Mexican Catholic Church, and the Vatican in the "normalization" of relations that occurred in 1991.

3920 **Lorey, David E.** El surgimiento de la región fronteriza entre Estados Unidos y México en el siglo XX. (*Rev. Mex. Sociol.*, 53:3, julio/sept. 1991, p. 305–347, bibl., tables)

Quantitative historical depiction of evolution of the "frontier" region (the ten states of Mexico and the US along the border) covers issues from population to economy to religion. Contains many useful tables.

3921 **Lustig, Nora.** Mexico's integration strategy with North America. (*in* Strategic options for Latin America in the 1990s. Edited by Colin I. Bradford Jr. Paris: Development Centre of the Organisation for Economic Co-operation and Development; Washington: Inter-American Development Bank, 1992, p. 155–179, tables)

Commendable overview of reasons Mexico entered negotiations in 1980 leading to NAFTA. Focuses on the need to move beyond the inward-looking industrialization stategy.

3922 **Lutz, Ellen L.** State-sponsored abductions: the human rights ramifications of Alvarez Machain. (*World Policy J.*, 9:4, Fall/Winter 1992, p. 687–703)

Human rights perspective on abduction of Humberto Alvarez Machain. Commendable survey of events that began with the 1985 kidnapping of Enrique Camarena Salazar.

3923 **Los mares nos unen: México y la Cuenca del Pacífico.** Realización y diseño de Beatrice Trueblood. México?: Banca Serfín, Dirección de Relaciones Públicas, 1989. 204 p.: bibl., ill. (some col.), index.

Beautifully illustrated and expensively designed, this "trade" book extolls virtues of Mexico looking westward across the Pacific for broadened relations.

3924 **Margain, Hugo B.** The war on drugs: a Mexican perspective. (*Voices Mex.*, 15, Oct./Nov./Dec. 1990, p. 3–8, photos)

Noted Mexican provides brief and readable perspective on need for more respect from US policy-makers for Mexican efforts in fighting narcotic flows.

3925 **Méndez Asensio, Luis.** Caro Quintero al trasluz: más allá de la *Mexican connection*. México: Plaza & Janés, 1985. 178 p., 40 p. of plates: ill., ports.

One of the relatively few publications on narcotics trafficking in Mexico. Provides probably the most detailed description of the imprisonment of Rafael Caro Quintero.

3926 **Mentz, Brígida von et al.** Los empresarios alemanes, el Tercer Reich y la oposición de derecha a Cárdenas. México: Centro de Investigaciones y Estudios Superiores en Antroplogía Social, Ediciones de la Casa Chata, 1988. 2 v.: bibl., ill. (Col. Miguel Othón de Mendizábal; 11–12)

Enlightening history of German-Mexican relations from 1870 through the 1940s focuses on German industrialists in Mexico and the *Sexenio* of Lázaro Cárdenas.

3927 **Messina, Ernesto G.** Unión México–Estados Unidos de América: revolución mundial. México: EDAMEX, 1988. 330 p.: bibl.

Analyzing Mexican history from the perspective of lost opportunities for democratic development, reaches controversial conclusion advocating political union with the US.

3928 **The Mexican-U.S. border region and the Free Trade Agreement.** Edited by Paul Ganster and Eugenio O. Valenciano. San Diego, Calif.: Institute for Regional Studies of the Californias, San Diego State Univ., 1992. vii, 117 p.: bibl., ill.

Summarizes proceedings of a binational conference in 1991 on the border impact of the free trade agreement. Topics include industrialization, migration, environment, infrastructure, and education.

3929 **Mexico. Secretaría de Relaciones Exteriores.** Las relaciones franco-mexicanas. v. 4, 1884–1911. Recopilación de Jorge Silva Castillo. México: Secretariá de Relaciones Exteriores, 1987. 1 v.: bibl., indexes. (Archivo histórico diplomático mexicano. Guías para la historia diplomática de México; 6)

Another volume in a set of fairly esoteric reference books that could, however, serve as a crucial tool for those interested in France-Mexico relations during this time period.

3930 **México en Centroamérica: expediente de documentos fundamentales, 1979–1986.** Recopilación de Raúl Benítez Manaut y Ricardo Córdova Macías. México: Univ. Nacional Autónoma de México, 1989. 387 p.: bibl. (Serie Antologías)

Mexican compilation of most significant documents relating to Mexican foreign policy toward Central America in the 1980s.

3931 **México en el comercio internacional.** México: Secretaría de Comercio y Fomento Industrial, 1990. 90 p.: ill.

Official overview from the Ministry of Commerce and Industrial Development of Mexico's policies in terms of international trade. Though the US receives primary attention, the book includes other trading partners and regions as well.

3932 **México en las Naciones Unidas.** México: Secretaría de Relaciones Exteriores, 1986. 424 p.: bibl.

This publication of the Mexican government includes 18 chapters by noted scholars on Mexico's role in a wide variety of international issues debated in the UN.

3933 **México-Estados Unidos: energía y medio ambiente.** Edición de Rosío Vargas y Mariano Bauer. México: Univ. Nacional Autónoma de México, 1993. 259 p.: bibl.

Compilation of over 20 essays on environmental issues relevant to the North American Free Trade Agreement.

3934 **Mexico: in search of security.** Edited by Bruce Michael Bagley and Sergio Aguayo. Coral Gables, Fla.: North-South Center, Univ. of Miami; New Brunswick, N.J.: Transaction Publishers, 1993. 367 p.: bibl.

Mexican and US scholars examine evolving role of the Mexican military in the context of economic crisis, drug trafficking, immigration, and even environmental concerns.

3935 **Meyer, Lorenzo.** México, Estados Unidos y un Tratado de Libre Comercio: el gran viraje de una relación. (*Rev. Occident.*, 131, abril 1992, p. 130–150, bibl.)

Noted Mexican historian reflects on the journey in US-Mexico relations from "distant neighbors" to "a marriage of convenience."

3936 **Morici, Peter.** Free trade with Mexico. (*Foreign Policy*, 87, Summer 1992, p. 88–104)

Surveys major issues of NAFTA for a general audience but contributes no new analysis.

3937 **Ojeda, Mario** *et al.* Hacia un tratado de libre comercio en América del Norte. Presentación de Jaime Serra Puche. México: M.A. Porrúa, 1991. 324 p.: bibl.

Impressive book with an obvious perspective edited by the Mexican secretary of commerce. Nevertheless, many chapters are written objectively by academics. Volume represents a significant contribution.

3938 **Pastor, Robert A.** Integration with Mexico: options for U.S. policy. New York: Twentieth Century Fund Press, 1993. 133 p.: bibl., ill., index.

A very readable book intended for a general audience. Provides historical overview of US/Mexico relations, the chronology of NAFTA negotiations, and potential impact of the trade agreement.

3939 **La política exterior de México en el nuevo orden mundial: antología de principios y tesis.** Prólogo de Juan María Alponte. México: Fondo de Cultura Económica, 1993. 428 p.: bibl. (Sección de obras de política y derecho)

In this grandiose project, collaborators of the Fondo de Cultura Económica depict evolution of Mexican foreign policy against the backdrop of the changing international arena of the 1980s.

Proceso. 1995– . See item **40**.

3940 **Raat, William Dirk.** Mexico and the United States: ambivalent vistas. Athens: Univ. of Georgia Press, 1992. 277 p.: bibl., ill., index, maps. (The United States and the Americas)

Unsuccessful attempt to combine numerous disciplinary and theoretical approaches in a study of Mexico's history. Blending diplomatic history, world-systems theories, comparative analysis of cultures, and macro-economic analysis proves an impossible task.

3941 **Ramírez López, Berenice Patricia.** Las relaciones económicas de México con América Latina, 1970–1990. México: Instituto de Investigaciones Económicas, Universidad Nacional Autónoma de México, 1991. 181 p.: bibl. (Col. La Estructura económica y social de México)

Analysis of economic relations and their political precursors between Mexico and four South American nations.

3942 **Ramos, José María.** Estados Unidos–México: entre un conflicto y la cooperación gubernamental, 1981–1990. (*Rev. Mex. Polít. Exter.*, 30, primavera 1991, p. 25–61)

Fairly lengthy essay attempts to categorize evolution of US-Mexico relations in the 1980s.

3943 **Ramos, José María.** El narcotráfico en el marco de una nueva fase gubernamental entre México y Estados Unidos. (*Can. J. Lat. Am. Caribb. Stud.*, 16:32, 1991, p. 77–90)

Examines US policy toward Mexico in the late 1980s regarding drug trafficking. Focuses on two major issues: decertification and militarization.

3944 **Reuter, Peter** and **David F. Ronfeldt.** Quest for integrity: the Mexican-US drug issue in the 1980s. (*J. Interam. Stud. World Aff.*, 34:3, Fall 1992, p. 89–153, bibl., tables)

Unusually lengthy article on Mexican-US relations in the area of narcotics trafficking. Concludes that actually stemming the flow is not as significant to the US as is a "good-faith" effort by Mexico (the so-called "quest for integrity").

3945 Reuter, Peter and David F. Ronfeldt. Quest for integrity: the Mexican-U.S. drug issue in the 1980s. Santa Monica, CA: Rand, 1992. xiii, 62 p.: bibl. (Rand note; N-3266-USDP)

Though prepared for the US secretary of defense, this concise publication provides relevant introduction to bilateral issues related to narcotics flow across the US-Mexican frontier.

3946 Ros, Jaime. Free trade area or common capital market?: notes on Mexico-US economic integration and current NAFTA negotiations. (*J. Interam. Stud. World Aff.*, 34:2, Summer 1992, p. 53–91, bibl., tables)

In the midst of NAFTA negotiations, posits that such a treaty is a logical outcome of recent Mexican liberalization policies, as well as a reflection of greater capital mobility and increasing bilateralism.

3947 Sáenz Carrete, Erasmo. Política del gobierno de México frente a los refugiados. (*Rev. Mex. Polít. Exter.*, 36/37, otoño/invierno 1992, p. 63–68)

As the US faces the presence of migrants from its southern neighbor, Mexico also must address the challenge of Guatemalan refugees in its southern territories. Brief article surveys issues created by 44,000 Guatemalans in three Mexican states.

3948 Sánchez Rodríguez, Roberto. El medio ambiente como fuente de conflicto en la relación binacional México–Estados Unidos. Tijuana, México: Colegio de la Frontera Norte, 1990. 134 p.: bibl., ill.

Very good analysis of bilateral negotiations between Mexico and the US, including a conceptual focus and two case studies. Important contribution to an area of vital concern.

3949 Serra Puche, Jaime. Principios para negociar el tratado de libre comercio de América del Norte. (*Comer. Exter.*, 41:7, julio 1991, p. 653–660, appendices)

Useful as an official statement of Mexican policy-makers in the midst of the NAFTA negotiations.

3950 El Servicio Exterior Mexicano. Prólogo de Bernardo Sepúlveda Amor. Coordinación y compilación del Instituto Matías Romero de Estudios Diplomáticos. México: Secretaría de Relaciones Exteriores, 1987. 167 p.: bibl., ill. (Archivo histórico diplomático mexicano. Cuarta época; 30)

Somewhat self-serving, this government publication compiles various articles describing and at times extolling the Mexican professional diplomatic corps.

3951 Silva-Herzog Flores, Jesús. Beyond the crisis: Mexico and the Americas in transition. Stanford, CA: Americas Program, Stanford Univ., 1987. 75 p. (Americas program visiting lecturer series)

Focusing on the debt crisis, this noted Mexican scholar and politician presents long-term analysis of the post-crisis era and consequences for Mexico and the rest of Latin America.

3952 Smith, Peter H. The political impact of free trade on Mexico. (*J. Interam. Stud. World Aff.*, 34:1, Spring 1992, p 1–25, bibl., table)

As a corollary to the hypothesis linking economic and political reform, this article examines four contending propositions regarding the potential impact of NAFTA on political outcomes in Mexico.

3953 Thorup, Cathryn L. The politics of free trade and the dynamics of cross-border coalitions in U.S.-Mexican relations. (*Columbia J. World Bus.*, 26:2, Summer 1991, p. 13–26)

Written before the passage and implementation of NAFTA, presents political issues debated on both sides of the border.

3954 El tratado de libre comercio — y usted. Recopilación de Ignacio Rodríguez Castro. Villahermosa, Mexico: Univ. Juarez Autónoma de Tabasco, Dirección de Difusión Cultural, 1992. 219 p.: bibl., ill.

Essays and documents covering all aspects of the free trade agreements then being negotiated. Discussion of constitutional issues in Mexico concerning the trade agreement is intriguing, albeit too brief.

Tratados internacionales celebrados por México. See item **46**.

3955 Trava Manzanilla, José Luis. et al. Manejo ambientalmente adecuado del agua en la frontera México–Estados Unidos: situación actual y perspectivas. Compilación de José Luis Trava Manzanilla, Jesús Román

Calleros, y Francisco A. Bernal Rodríguez. Tijuana, Mexico: El Colegio de la Frontera Norte, 1991. 265 p.

A precursor to concerns raised by NAFTA, volume analyzes issues related to ground water contamination along the border.

3956 U.S.-Mexico relations: labor market interdependence. Edited by Jorge A. Bustamante, Clark Winton Reynolds, and Raúl A. Hinojosa Ojeda. Stanford, Calif.: Stanford Univ. Press, 1992. 495 p.: bibl., index.

Very lengthy volume that brings together a large number of US and Mexican scholars focusing on models of interdependence between Mexico and the US in the area of labor migration. Much of the analysis begins with the 1986 Immigration and Control Act, providing some perspectives on the impact of this legislation.

3957 United States. Commission for the Study of International Migration and Cooperative Economic Development. Unauthorized migration: an economic development response. Washington: Commission for the Study of International Migration and Cooperative Economic Development, 1990. 1 v. (various pagings): bibl., col. ill., col. maps.

Focuses admirably on "push" factors of illegal immigration into the US. One of the major conclusions is the need for expanded trade between the US and the "sending" countries.

3958 Vázquez, Josefina Zoraida. La influencia de Estados Unidos en México. (*Secuencia/México*, 19, enero/abril 1991, p. 33–42, ill.)

Historical and cultural perspective on US influence in Mexico. Appealing to a broad audience.

3959 Vázquez Colmenares, Pedro. Las relaciones entre México y Guatemala. (*Rev. Mex. Polít. Exter.*, 36/37, otoño/invierno 1992, p. 16–25)

Too often recognition of Mexican bilateral relations with a bordering nation-state focuses on the US. This article is a significant reminder of the salience of Mexico's relations with its southern neighbor: Guatemala.

3960 Weintraub, Sidney and M. Delal Baer. The interplay between economic and political openings: the sequence in Mexico. (*Wash. Q.*, 15:2, Spring 1992, p. 187–201)

One of the few studies examining the potentially critical link between economic liberalization and political reform. Unfortunately for scholars, authors are writing for a more general audience.

3961 Weintraub, Sidney. US-Mexico free trade: implications for the United States. (*J. Interam. Stud. World Aff.*, 34:2, Summer 1992, p. 29–52, bibl.)

Useful overview written in the midst of US debate over "pro" and "con" arguments of NAFTA's impact on the US.

3962 Zepeda Miramontes, Eduardo. La frontera norte y el Tratado de Libre Comercio: efecto y desarrollo. (*Rev. Mex. Sociol.*, 53:3, julio/sept. 1991, p. 185–200, bibl.)

While admitting that the northern frontier is not the "big picture" for Mexico, nonetheless attempts to detail NAFTA's impact specifically on border regions of Mexico.

3963 Zermeño, Sergio. Desidentidad y desorden: México en la economía global y en el libre comercio. (*Rev. Mex. Sociol.*, 53:3, julio/sept. 1991, p. 15–64, bibl.)

Analyzes tensions among economic growth, societal development, and neo-liberal economic policies of Mexico.

3964 Zoghbi Pérez, Jorge Alberto; Lilia Jiménez Mejía; and Alfredo Rojas Díaz Durán. Deuda externa y seguridad nacional: geopolítica del endeudamiento externo mexicano. México: Instituto Mexicano de Estudios Internacionales de Deuda Externa, 1990. 193 p.: bibl., ill.

Despite its title, offers more description of the Mexican external debt and lacks sufficient focus on the "geopolitical" components. Still, a useful summary of many recent policies, particularly the Brady Plan.

CENTRAL AMERICA

3965 Adams, Jan S. A foreign policy in transition: Moscow's retreat from Central America and the Caribbean, 1985–1992. Durham, N.C.: Duke Univ. Press, 1992. 248 p.: bibl., index.

Very erudite volume on changing Soviet policy toward Central America in the Gorbachev years. Useful for scholars of both Central America and the former Soviet Union.

3966 Aguayo, Sergio. Del anonimato al protagonismo: los organismos no gubernamentales y el éxodo centroamericano. (*Foro Int.*, 32:3, enero/marzo 1992, p. 323–341)

Though lacking detail and in-depth analysis, an intriguing exploration of nongovernmental actors in the Central American crisis, particularly regarding the refugee issue.

3967 Alas de Franco, Carolina. Los desafíos de una zona de libre comercio entre México y Centroamérica. (*ECA/San Salvador*, 47:523/524, mayo/junio 1992, p. 461–479, photos, table)

While Mexico and the US enter into a free trade accord, author speculates on advantages of a similar accord between a different set of unequal nation-states: Mexico and Central America.

3968 Alvarez Icaza, Pablo. Belice: la crisis, el neocolonialismo y las relaciones con México, 1978–1986. Coordinación de Adolfo Aguilar Zinser y Rodrigo Jauberth Rojas. México: Programa de Estudios de Centroamérica, Centro de Investigación y Docencia Económicas, 1987. 137 p. bibl., ill. (Relaciones Centroamérica-México)

One of a series by a Mexican think tank. Examines Mexico's relations with the emerging nation of Belize.

3969 Arancibia Córdova, Juan. Honduras: en busca del encuentro, 1978–1986. Coordinación de Adolfo Aguilar Zinser y Rodrigo Jauberth Rojas. México: Programa de Estudios de Centroamérica, 1987. 156 p.: bibl. (Relaciones Centroamérica-México)

Another book in a laudable series on Central America by a Mexican think tank. Examines Honduran economic development, relations with Mexico, and viewpoints of Honduran entrepreneurs.

3970 Araúz, Virgilio. ¿A quién beneficia el canal?: porque lucharon los mártires. Panamá?: s.n., 1989? 54 p.: map.

A brief primer on the Panamanian perspective that the "Colossus of the North" continues to exploit Panama.

3971 Arias Peñate, Salvador. Seguridad o inseguridad alimentaria: un reto para la región centroamericana: perspectivas para el año 2000. San Salvador: Uca Editores, 1989. 184 p.: bibl. (Col. Estructuras y procesos. Serie mayor; 6)

Essentially a study of agricultural policy in Central America, positing that Western aid serves only to increase dependency of the region.

3972 Barry, Tom. El conflicto de baja intensidad: un nuevo campo de batalla en Centroamérica. 1a ed. en español. Tegucigalpa: Centro de Documentación de Honduras, 1988. 59 p.: bibl., ill.

Brief volume written by US scholar who is critical of US policy in Central America in the 1980s.

3973 Baumeister, Eduardo et al. Centroamérica: los desafíos, los intereses, las realidades. Recopilación de Eduardo Gitli. México: Univ. Autónoma Metropolitana-Azcapotzalco; México:Gernika, 1989. 288 p.: bibl. (Ensayos; 29)

Somewhat disjointed, edited volume of essays on issues ranging from a Central American Common Market to the role of the Catholic Church.

3974 Belice y Centroamérica: una nueva etapa. Coordinación de Gabriel Edgardo Aguilera Peralta. Guatemala: FLACSO; Fundación Friedrich Ebert; Society for the Promotion of Education and Research (Belize), 1992. 150 p.: bibl., maps.

Addressing the role of a much neglected nation, provides a series of essays analyzing unique situation of Belize as a country with both Central American and Caribbean identities.

3975 Beluche, Olmedo. La verdad sobre la invasión. Panamá: CELA, 1990. 143 p.: ill.

One of the most critical volumes written about the 1989 US invasion of Panama. Includes photographs and several personal accounts of the impact of the invasion.

3976 Benítez Manaut, Raúl. La ONU y el proceso de paz en El Salvador, 1990–1992. (*Rev. Mex. Polít. Exter.*, 34, primavera 1992, p. 35–52, appendices)

Aside from examining the role of the UN, author does an excellent job analyzing the military stalemate in the early-1990s in El Salvador in terms of four major elements.

3977 Brunner, Markus; Wolfgang Dietrich; and Martina Kaller. Projekt Guatemala: Vorder- und Hintergründe der österreichischen Wahrnehmung eines zentramerikanischen Landes. Frankfurt a.M.: Brandes & Apsel, 1993. 305 p.: bibl., ill. (Wissen & Praxis; 45)

Analyzes official and non-governmental relations between a peripheral Central American state and a minor industrialized state. Provides valuable information regarding Austrian activities, trade, aid and immigration in Guatemala, although authors' criticism of development aid is somewhat didactic.

3978 Bulmer-Thomas, V. et al. Central America: crisis and possibilities. Edición de Rigoberto García G. Stockholm: Latinamerika-institutet, 1988. 263 p.: bibl. (Monograph/Institute of Latin American Studies, 0284-6675; 16)

Series of articles resulting from seminars organized in Stockholm add little to existing literature on the Central American "crisis" of the 1980s.

3979 Bulmer-Thomas, V. et al. Central American integration: report for the Commission of the European Community. Miami: North-South Center, Univ. of Miami; European Community Research Institute, 1992. 115 p.: ill.

While recognizing potential benefits of Central American integration, authors delimit series of obstacles facing such efforts. In English and Spanish.

Cardenal, Gloria et al. La guerra en Nicaragua. See *HLAS 54:1707*.

3980 Carothers, Thomas et al. Conflict resolution and democratization in Panama: implications for U.S. policy. Edited by Eva Loser. Foreword by Georges Fauriol. Washington: Center for Strategic and International Studies, 1992. x, 95 p.: bibl. (Significant issues series; v. 14, no. 2)

Includes contributions from six leading scholars on the precursors to the US invasion of Panama and prospects for democratization. Concise essays provide critical view of US policy.

3981 Casillas Ramírez, Rodolfo. Migratory policy in Mexico regarding Central American migratory flows in the current context. (*Estud. Int./IRIPAZ*, 3:6, julio/dic. 1992. p. 70–80, bibl.)

Broad-based analysis of Central American patterns of migration into and through Mexico. Attempts to cover too much territory in such a short article.

3982 Castañeda Sandoval, Gilberto. Guatemala. Coordinación de Adolfo Aguilar Zinser y Rodrigo Jauberth Rojas. México: Programa de Estudios de Centroamérica, Centro de Investigación y Docencia Económicas, 1987. 133 p.: bibl. (Relaciones Centroamérica-México)

First examines model of capital accumulation in Guatemala and then analyzes bilateral relations with Mexico.

3983 Castillero Pimentel, Ernesto. Panamá y los Estados Unidos, 1903–1953. Panamá: s.n., 1988. 336, cxli, 92 p.: ill.

The fifth printing of important resource on bilateral relations between Panama and the US in the first half of this century.

3984 Central America: braving the new world. London: Catholic Institute for International Relations, 1992. 27 p.: bibl., map. (Comment)

English-language "primer" to the peace process in Central America after the electoral defeat of the Sandinistas.

3985 Child, Jack. The Central American peace process, 1983–1991: sheathing swords, building confidence. Boulder, Colo.: L. Rienner Publishers, 1992. 200 p.: bibl., ill., index, map.

Provides empirically sound and detailed description of the Central American peace process in the 1980s, in addition to insightful analysis and theoretical conclusions. Focus on issues of conflict resolution (such as "peace verification") sets this book apart. Also recognizes indigenous elements unique to the region.

3986 Child, Jack. Verification of the Central American peace process. (*MACLAS Lat. Am. Essays*, 5, 1992, p. 245–271)

"Peace verification" in Central America is explored in light of conflict resolution theories and their recent application in the Middle East.

3987 Clark, Paul Coe. The United States and Somoza, 1933–1956: a revisionist look. New York: Praeger, 1992. 239 p.: bibl., index.

Provides historical account of the years Anastasio Somoza Garcia was in power. The "revisionist book" criticizes arguments that he was nothing more than a tool of US policy.

3988 Coatsworth, John H. Central America and the United States: the clients and the colossus. New York: Twayne Publishers; Toronto: Maxwell Macmillan Canada; New York: Maxwell Macmillan International, 1994. 277 p.: bibl., index. (Twayne's international history series; 12)

Though relying predominantly on secondary sources, author has provided impressive volume that is likely to become required reading for any student of Central American history.

3989 Cottam, Martha L. The Carter administration's policy toward Nicaragua: images, goals, and tactics. (*Polit. Sci. Q.*, 107:1, Spring 1992, p. 123–146)

In an unusual tack, author looks back to the Carter Administration to examine US policy in Nicaragua. Fails to provide any generalizable conclusions.

3990 Crónica de una guerra no imaginaria: cronología de las relaciones Estados Unidos-Nicaragua, 1979–1984. Managua: Editorial de Ciencias Sociales-INIES; Instituto Nicaragüense de Investigaciones Económicas y Sociales, 1986. 60 p.: bibl., ill. (Serie Cronología)

Useful as annotated chronology of events from the Sandinista perspective.

3991 Cronologías de los procesos de paz: Guatemala y El Salvador. Guatemala: IRIPAZ, 1991. 189 p. (Serie Cooperación y paz; vols. 1–2)

Encyclopedic chronology of events and peace negotiations in two Central American nations during the latter 1980s.

3992 Cronologías de los procesos de paz: Guatemala y El Salvador. Guatemala: IRIPAZ, 1991. 189 p. (Serie Cooperación y paz; 1–2)

Essentially a resource volume, contains detailed, annotated chronology of events related to peace negotiations in Guatemala and El Salvador from 1987–91.

3993 Dabène, Olivier. La dimensión interna de las políticas exteriores en un contexto de crisis en Centroamérica. (*Trace/México*, 18, déc. 1990, p. 36–52)

Interesting examination of the thesis that nation-states utilize foreign policy to serve interests of domestic politics. While examples of Mexico, Cuba, and Argentina are well known, author explores the case of Central America.

3994 Darling, Jonathan. The 1969 war between El Salvador and Honduras: a case study in Central American peace-keeping. (*MACLAS Lat. Am. Essays*, 5, 1992, p. 272–302, bibl.)

Overly descriptive look at the misnamed "Soccer War." Concludes that internal problems (such as land distribution) were the principal causes of the war.

3995 De Guttry, Andrea. Hacia la integración política de Centro América: perfiles institucionales y problemas jurídicos. (*Panorama Centroam. Pensam.*, 29, enero/marzo 1993, p. 5–34)

Lengthy article details historical and current efforts at political integration in Central America.

3996 Dickson, Sandra H. Press and U.S. policy toward Nicaragua, 1983–1987: a study of the New York Times and Washington Post. (*Journal Q.*, 69:3, Autumn 1992, p. 562–571, tables)

Content analysis of these newspapers concludes that their coverage legitimized US policy in Nicaragua.

3997 Dodd, Thomas J. Managing democracy in Central America: a case study; United States election supervision in Nicaragua, 1927–1933. Coral Gables, FL: North-South Center, Univ. of Miami; New Brunswick, N.J.: Transaction Publishers, 1992. 161 p.: bibl., ill.

Details the history of US electoral supervision in Nicaragua in the 1920s and 1930s. The depiction of the early US intervention in Central America provides a contrast to images of exclusively military intervention.

3998 Drago, Tito. Centroamérica, una paz posible. Madrid: El País/Aguilar, 1988. 278 p.: ports.

Detailed historical analysis of 20th-century US influence in the six nations of Central America.

3999 Eguizábal, Cristina. De Contadora a Esquipulas: Washington y Centroamérica en un mundo cambiante. (*Anu. Estud. Centroam.*, 18:1, 1992, p. 5–15)

Argues that despite enormous efforts,

the US cannot profoundly alter events in Central American politics due to internal dynamics of Central American nations.

4000 Elton, Charlotte. ¿Rivales o aliados?: Japón y Estados Unidos en Panamá. Prólogo de Xavier Gorostiaga. Panamá: Centro de Estudios y Acción Social Panameño, 1990. xvii, 142 p.: bibl. (Serie Panamá hoy; 4)

Scholarly look at multi-polar interests of the US and Japan in Panama.

4001 Escoto, Jorge and Manfredo Marroquín. La AID en Guatemala: poder y sector empresarial. Managua: CRIES; AVANCSO, 1992. 166 p.: bibl.

Analyzes evolution of US financial assistance to Guatemala from the focus on national security to the concern with economic development, international markets, and entrepreneurship.

4002 Falcoff, Mark and Richard Millett. Searching for Panama: the U.S.-Panama relationship and democratization. Foreword by Georges A. Fauriol. Washington: Center for Strategic and International Studies, 1993. xi, 42 p.: bibl. (Significant issues series; v. 15, no. 6)

Two of the most respected scholars on Central America provide a brief and very readable primer on Panamanian politics and US-Panama relations that focuses on propects for democratic change.

4003 Flanagan, Edward M. Battle for Panama: inside Operation Just Cause. Washington: Brassey's, Inc., 1993. 251 p.: bibl., index. (An AUSA book)

Dedicated to "the Troops," this US Army publication is an effort to tell its story of a "well done" mission: the Panama invasion of 1989.

4004 The foreign policies of Caribbean and Central American countries: selections from PROSPEL'S 1986 Anuario *Las políticas exteriores de América Latina y el Caribe—continuidad en la crisis*, edited by Heraldo Muñoz. Edited by Jane G. Marchi. Foreword by Jaime Suchlicki. Miami?: Institute of Interamerican Studies at the Graduate School of International Studies, Univ. of Miami, 1990. 282 p.: bibl.

Diverse views on foreign policy issues from the perspectives of individual countries. Fails to provide any unifying themes.

4005 Francis, Anselm. The current phase of the Belize/Guatemala territorial dispute. (*Caribb. Aff.*, 5:1, Jan./March 1992, p. 71–85)

As was displayed in the Falklands/Malvinas dispute, Latin America is replete with underlying border disputes. One of the most recent examples is examined here.

4006 Gándara Gallegos, Mauricio. Panamá: la internacionalización del canal. Quito: Fundación Ascencio Gándara, 1990. 340 p.: bibl., ill. (Ensayo geopolítico)

Ecuadorean author examines history of Panama from its establishment through the US capture of Noriega.

4007 Gandásegui, Marco A. Democracia, intervención y elecciones: Panamá 1989. (*Rev. Mex. Sociol.*, 52:4, oct./dic. 1990, p. 371–389, tables)

Rather than focus on the Dec. 1989 invasion as an isolated event, this commendable work examines democratization and the election of 1989.

Gleijeses, Piero. Shattered hope: the Guatemalan Revolution and the United States, 1944–1954. See *HLAS 54:1730*.

4008 González-Camino, Fernando. Alta es la noche: Centroamérica ayer, hoy y mañana. Madrid: Instituto de Cooperación Iberoamericana, Ediciones de Cultura Hispánica, 1990. 348 p.: bibl. (Ensayo)

Interesting reflections of a Spanish diplomat on issues facing Central America in this century.

4009 González Mejía, Hernán. Centroamérica en crisis. Heredia, Costa Rica: Editorial de la Univ. Nacional, 1992. 131 p.: bibl., ill. (Col. Guayabo; 5)

Relatively brief book attempts to analyze Central America's limited economic development and unequal distribution of wealth through an examination of the economic and geopolitical history of the region.

4010 Grupo de Trabajo sobre la Cuenca del Canal de Panamá. Informe del Grupo de Trabajo sobre la Cuenca del Canal de Panamá: sumario ejecutivo. Pamaná: Impretex, 1986. 44 p.: ill. (some col.).

In this brief pamphlet, an official Panamanian working group argues that environmental protection of the Canal area is vital to Panama.

4011 Gurdián Guerra, Reymundo. La invasión militar y los desafíos de la política exterior panameña. Panamá: Imprenta Universitaria, 1990. 43 p.: bibl.

Volume contributes to the documentation on the US invasion of Panama through its discussion of Panamanian foreign policy.

Harrison, Benjamin T. Dollar diplomat: Chandler Anderson and American diplomacy in Mexico and Nicaragua, 1913–1928. See *HLAS 54:1739*.

4012 Henkin, Louis. The invasion of Panama under international law: a gross violation. (*Columbia J. Transnatl. Law*, 29:2, 1991, p. 293–317)

Aruges that the invasion of Panama was unjustified from a purely legal standpoint. Fails to recognize political realities of international relations.

4013 Hernández Milián, Jairo. Las negociaciones Costa Rica–Estados Unidos en las administraciones Arias y Bush. (*Rev. Cienc. Soc./San José*, 51/52, marzo/junio 1991, p. 23–33)

Interesting look at bilateral relations between the US and Costa Rica during the initiation of the Arias Plan.

4014 Hernández Ortiz, Evelyn. La adhesión de Centroamérica al Gatt. (*Rev. Cienc. Soc./San José*, 51/52, marzo/junio 1991, p. 47–55, bibl.)

Much has been made of Mexico's entry into the GATT. Here the author addresses significance of the agreement for Central America.

4015 Herrera, René. Relaciones internacionales y poder político en Nicaragua. México: Colegio de México, 1991. 157 p.: bibl., ill.

Long-term perspective on the evolution of political power in Nicaragua.

4016 Hey, Jeanne A.K. and **Lynn M. Kuzma.** Anti-U.S. foreign policy of dependent states: Mexican and Costa Rican participation in Central American peace plans. (*Comp. Polit. Stud.*, 26:1, April 1993, p. 30–62, bibl.)

Explores apparent contradiction of nations heavily dependent on the US (specifically Mexico and Costa Rica) pursuing foreign policies in Central America that are "notably anti-United States."

4017 Holden, Robert H. The real diplomacy of violence: United States military power in Central America, 1950–1990. (*Int. Hist. Rev.*, 15:2, May 1993, p. 283–322, graphs)

Very detailed look at full range of US military penetration in Central America during the postwar era. Includes appendix of comparative data and graphs on military transfers.

4018 Holiday, David and **William Stanley.** La construcción de la paz: las lecciones preliminares de El Salvador. (*ECA/San Salvador*, 48:531/532, enero/feb. 1993, p. 39–59, photos)

Focuses on potential role for the UN in maintaining and verifying peace in El Salvador following the 1992 Peace Agreement.

4019 Hurwitz, Jon; Mark Peffley; and **Mitchell A. Seligson.** Foreign policy belief systems in comparative perspective: the United States and Costa Rica. (*Int. Stud. Q.*, 37:3, Sept. 1993, p. 245–269, appendix, bibl., tables)

Authors test the universality of the hierarchial model of foreign policy belief systems. Originally developed in the context of US public opinion, the model is applied to Costa Rica.

4020 Impacto de la agresión del gobierno de los Estados Unidos de América sobre la economía y la sociedad panameña. Panamá: Ministerio de Planificación y Política Económica, Dirección de Planificación Económica y Social, 1989. 7 leaves.

Extremely brief Panamanian government publication criticizing US economic sanctions against Panama in the late 1980s.

4021 Informe blanco sobre los avances logrados en el proceso de cumplimiento del acuerdo de paz para Centroamérica "Esquipulas II" a los noventa días de haberse firmado. San José: Facultad Latinoamericana de Ciencias Sociales; Consejo Superior Universitario Centroamericano; Univ. para la Paz, 1988? 138 p.: ill.

Useful reference source for events and chronologies of the Central American peace process in the 1980s.

4022 Is there a transition to democracy in El Salvador? Edited by Joseph S. Tulchin and Gary Bland. Boulder, Colo.: L. Rienner Publishers, 1992. 213 p.: bibl., index. (Woodrow Wilson Center current studies on Latin America)

Commendable collection of essays from notable scholars recognizing recent positive moves toward democracy in El Salvador as well as historical obstacles to existing democratic institutions. While many articles provide excellent analysis (including a concluding assessment of democratization), book lacks integrating thesis.

4023 Johns, Christina Jacqueline and **P. Ward Johnson.** State crime, the media, and the invasion of Panama. Westport, Conn.: Praeger, 1994. 157 p.: bibl., index. (Praeger series in criminology and crime control policy, 1060–3212)

Explores the three "lies" used to justify the US invasion of Panama: restoration of democracy, protection of US lives, and termination of drug traffic. Detailed in much of its research, but overly polemical.

Jonas, Susanne. The battle for Guatemala: rebels, death squads, and U.S. power. See *HLAS 54:1741*.

4024 Lamboglia, Ramón. Panamá, de la narcodictadura a colonia yanqui: resumen del juicio de Noriega. San José?: R. Lamboglia, 1992. 164 p., 7 leaves of plates: ill.

Highly critical treatise of Noriega written by a long-time political opponent.

4025 Laudy, Marion. Nicaragua ante la Corte Internacional de Justicia de la Haya. México: Siglo Veintiuno Editores, 1988. 236 p.: bibl. (Sociología y política)

Interesting study of Nicaraguan case against US intervention presented before the International Court of Justice in the mid-1980s.

4026 Leis Romero, Raúl Alberto. Comando Sur, poder hostil. Panamá: Centro de Estudios y Acción Social Panameño, 1985. 124 p.: bibl., ill. (Serie Panamá hoy; 1)

Argues that US Southern Command in Panama is an ideological and military threat to that nation.

4027 Lungo Uclés, Mario. Centroamérica: la unidad regional solo es posible con la participación popular. (*Estud. Soc. Centroam.*, 54, sept./dic. 1990, p. 47–57)

Historical review of plans for political integration of Central America, indicating relevance of more recent civil unrest in the region.

4028 Martínez, José de Jesús. La invasión de Panamá. Bogotá: Causadías Editores, 1991. 152 p. (Col. Testimonio)

Reminder that the US perspective of the "successful" invasion of Panama does not necessarily reflect the reality of that event in Panama.

4029 Martínez, Milton. La crisis y la invasión: una periodización para su entendimiento. (*Rev. Panameña Sociol.*, 7, 1991, p. 149–167, ill.)

In contrast to many other essays on the invasion of Panama, article attempts to present the precursors in terms of domestic crisis evolving in Panama in the 1980s.

4030 Martínez Blanco, Gerardo. Enfoque histórico y jurídico de la controversia limítrofe entre Honduras y El Salvador. Tegucigalpa: Univ. Nacional Autónoma de Honduras, Editorial Universitaria, 1991. 594 p. (Col. Realidad nacional; 35)

Juridical analysis and useful reference on territorial disputes between Honduras and El Salvador.

4031 Martínez Velasco, Germán. Migración y poblamiento guatemalteco en Chiapas. (*Mesoamérica/Antigua*, 14:25, junio 1993, p. 73–100, tables, ill.)

Historical overview of Guatemalan migration into the Mexican state of Chiapas.

4032 Méndez Dávila, Lionel. Invasión USA a Panamá: modelo para no olvidar (y cinco presagios estructurales). Panamá: Fundación Omar Torrijos, 1991. 281 p.: bibl., ill. (Col. Panamá)

One of the lengthiest treatises on the 1989 US invasion of Panama written in light of one century of US intervention.

4033 Monge, Luis Alberto; Román Arrieta Villalobos; and **Armando Vargas Araya.** Neutralidad o guerra. San José: Secretaría de Información y Comunicación, 1986. 52 p.

Brief volume is essentially official restatement of Costa Rican neutrality in the midst of Central American revolutionary upheavals in the 1980s.

4034 Morales Carazo, Jaime. La Contra. México: Planeta, 1989. 452 p.: bibl., ill., index, maps. (Col. Documento)

Well-documented and informative account of ten years of Sandinista rule and Contra opposition. Written by a Nicaraguan entrepreneur and journalist (colleague of Pedro Joaquín Chamorro) who lived for years in the mountains with the Contras.

4035 Moreno, Dario. The struggle for peace in Central America. Gainesville: Univ. Press of Florida, 1994. 251 p.: bibl., index.

Overly optimistic view of the peace process in Central America. Praises the Arias plan for linking domestic and international dynamics.

4036 Morley, Morris H. Washington, Somoza, and the Sandinistas: state and regime in U.S. policy toward Nicaragua, 1969–1981. Cambridge, England; New York: Cambridge Univ. Press, 1994. 343 p.: bibl., index.

Utilizing a distinction between the State and the regime (essentially strategic versus tactical motivation), argues that long-term interests of the State are more salient to US policy in Central America.

4037 Musicant, Ivan. The banana wars: a history of United States military intervention in Latin America from the Spanish-American War to the invasion of Panama. New York: Macmillan, 1990. 470 p., 8 p. of plates: bibl., ill., index, maps.

Very readable, historical account of nine cases of US military intervention in the region since 1885. Recommended reading for any student of diplomatic or military history.

National Security Archive index. See item **38**.

4038 Navarrete Talavera, Ela. Panamá, ¿invasión o revolución? México: Planeta, 1990. 356 p.: ill. (Col. Documento)

Well-written, journalistic account of the Noriega years and the US invasion of Panama.

4039 Notario Castro, Nelson. Desde Panamá reportamos. Habana: Editorial José Martí, 1990. 129 p.: ill.

Depiction of the 1989 US invasion of Panama from the perspective of a Cuban reporter.

4040 Ochoa García, Carlos and **Nat Holmes.** The role of the refugees in the actual Guatemalan peace process. (*Estud. Int./ IRIPAZ*, 3:6, julio/dic. 1992, p. 59–69, bibl.)

Very useful overview of the issue of Guatemalan refugees in Mexico. Includes summaries and tables of major chronological events, principal camps, and international actions.

4041 Orozco, Concepción and **Jorge Castillo.** Cronología de la crisis panameña. (*Rev. Invest. Econ.*, 2, 1991, p. 41–55, photos)

Reference piece provides an almost daily chronology of events from 1987 to the January 3, 1990 surrender of Noriega.

4042 Panamá: autodeterminación contra intervención de Estados Unidos. Caracas: Ediciones Centauro; México: Programa de Estudios Centroamericanos del Centro de Investigación y Docencia Económicas, 1989. 218 p.: bibl., maps.

One of the more objective critiques of US policy in Panama.

La paz: más allá de la guerra. See item **3188**.

4043 Paz Barnica, Edgardo. Democracia e integración en América Latina. Buenos Aires: Grupo Editor Latinoamericano, 1991. 353 p.: bibl. (Colección Estudios internacionales)

Unsuccessful effort to combine themes of European contribution to Central American democracy, Nicaraguan-Honduran relations, and regional integration.

4044 Paz Barnica, Edgardo. Honduras y las cumbres presidenciales. Tegucigalpa: s.n., 1990. 175 p.

Noted Honduran diplomat and legal scholar examines the Honduran participation in Central American diplomacy in the late 1980s. Fails to include the role of the US.

4045 Perspectives on war and peace in Central America. Edited by Sŭng-ho Kim and Thomas W. Walker. Athens: Ohio Univ. Center for International Studies, 1992. 146 p.: bibl. (Monographs in international studies. Latin America series; 19)

Proceedings of 1988 seminar at Ohio

University that attempted to counterpose liberal and conservative views of the peace process then taking place in Central America.

4046 Pizzurno Gelós, Patricia. Harmodio Arias Madrid y las relaciones internacionales. Panamá: s.n., 1991. 48 p.: bibl.

Brief booklet provides interesting overview of contributions of this important Panamanian "nationalist."

4047 Pizzurno Gelós, Patricia. Informe sobre las elecciones presidenciales en Panamá en 1912 y la supervisión norteamericana. Panamá: s.n., 1991. 39 p.: bibl.

Working paper examines not only the 1912 Panamanian elections, but also origins of the republic and US involvement in the country.

4048 La política exterior norteamericana hacia Centroamérica: reflexiones y perspectivas. Coordinación de Mónica Verea y José Luis Barros Horcasitas. México: Centro de Investigaciones sobre Estados Unidos de América, Univ. Nacional Autónoma de México; M.A. Porrúa; FLACSO, Sede México, 1991. 442 p.: bibl. (Las Ciencias sociales)

Recognizing the unequal role of Mexico in Central America, the National Univ. of Mexico (UNAM) organized this publication of essays from US, Mexican, and Central American authors. Though the quality varies, many articles further our understanding of post-Cold War international relations in the region.

4049 Política exterior y estabilidad estatal. Guatemala: Asociación para el Avance de las Ciencias Sociales en Guatemala, 1989. 138 p.: bibl., ill. (Cuadernos de investigación; 5)

Scholarly examination links Guatemalan foreign policy to the "stabilization" project of the Guatemalan military.

4050 Prieto González, Alfredo and Julio Carranza Valdés. Las elecciones nicaragüenses y la prensa norteamericana. (*in* Sistemas políticos, poder y sociedad: estudios de casos en América Latina. Caracas: Asociación Latinoamericana de Sociología; Centro de Estudios sobre América; Editorial Nueva Sociedad, 1992, p. 295–326, bibl.)

Objective analysis of US press coverage of the 1990 Nicaraguan elections.

4051 Proceso de paz en Centro América: compendio. Guatemala: INFORPRESS Centroamericana, 1988? 1 v. (unpaged).

Anthology of documents, letters, and newspaper articles on the initial impact of the "Plan Arias" of 1988.

4052 El proceso de paz en El Salvador. El Salvador: Presidencia de la República de El Salvador, Secretaría Nacional de Comunicaciones, 1992? 1 v. (various pagings).

Book praises peace efforts of the Cristiani regime.

4053 Proceso de paz en El Salvador: la solución política negociada. Managua: CRIES; San Salvador: IDESES, 1992. 99 p.: appendix, ill. (Cuadernos CRIES. Serie documentos; 6)

Commendable, brief overview of peace process in El Salvador with a very useful appendix of documents.

4054 Quijano, Carlos. Nicaragua: ensayo sobre el imperialismo de los Estados Unidos. Prólogo, selección de documentos en anexo, notas y cronología por Roberto Cajina. Managua: Editorial Vanguardia, 1988. 331 p. (Rescate; 1)

Originally published in 1928, book critically analyzes financial relations between Nicaragua and the US in the first three decades of this century.

4055 Ramírez Ocampo, Augusto. Contadora: pedagogía para la paz y la democracia; la diplomacia de la verdad. Bogotá: Fondo Rotatorio del Ministerio de Relaciones Exteriores, 1986. 289 p.: ill. (some col.).

Motives for and justification of Colombia's participation in the Contadora process.

4056 Las relaciones entre España y América Central, 1976–1989. Presentación de Jordi Solé Tura. Barcelona: CIDOB; AIETI, 1989. 174 p.: bibl.

Essays examine relations between post-Franco Spain and Central America during the turbulent years surrounding the Nicaraguan Revolution.

4057 Robinson, William I. A faustian bargain: U.S. intervention in the Nicaraguan elections and American foreign policy in the post-Cold War era. Afterwords by Alejandro Bendaña and Robert A. Pastor. Boulder,

Colo.: Westview Press, 1992. 310 p.: bibl., ill., index.

Important though controversial book argues that the US intervened in and thus influenced the 1990 Nicaraguan elections.

4058 **Rodríguez, Mario Augusto.** La Operación "Just Cause" en Panamá: ensayo de reportaje periodístico sobre la invasión armada y la ocupación militar de Panamá a partir de 1989. Panamá: Fundación Omar Torrijos, 1991. 410 p.: bibl.

Detailed, journalistic account of the "Just Cause" invasion of Panama.

4059 **Rojas Aravena, Francisco** and **Luis Guillermo Solís Rivera.** ¿Súbditos o aliados?: la política exterior de Estados Unidos y Centroamérica. San José: Editorial Porvenir; FLACSO, 1988. 159 p.: bibl.

Analyzes US policy toward Central America in the 1980s from the perspective of the primacy of political relations and the significance of historical precursors.

4060 **Rosenberg, Robin L.** Spain and Central America: democracy and foreign policy. New York: Greenwood Press, 1992. 266 p.: bibl., index. (Contributions in political science, 0147–1066; 288)

While recognizing that Central America is not a major focus of Spanish foreign policy, author examines degree to which Spain uses the example of its own democratic transition to influence Central American nations.

4061 **Saborío Alvarado, Sylvia.** Centroamérica en los años noventa: el reto de la apertura. (*in* Liberalización comercial e integración regional: de NAFTA a Mercosur. Edición de Roberto Bouzas y Nora Lustig. Buenos Aires: FLACSO, 1992, p. 89–112, bibl., tables)

Vigorous and analytical look at Central American development in the 1990s focuses on potential for economic integration and trade liberalization.

4062 **Santamaría Gómez, Arturo.** La izquierda norteamericana y los trabajadores indocumentados. México: Ediciones de Cultura Popular; UAS Editorial, 1988. 245 p.: bibl.

Interesting look at undocumented workers in the US. Emphasizes the relationship of these workers with US and leftist groups and their labor organizations. Also discusses the historical antagonism of traditional organized labor toward undocumented workers.

4063 **Schulz, Donald E.** and **Deborah Sundloff Schulz.** The United States, Honduras, and the crisis in Central America. Boulder, Colo.: Westview Press, 1994. 368 p.: bibl., index, map. (Thematic studies in Latin America)

Very readable work that describes how Honduras became a critical base for US operations in the 1980s. To its credit, book also analyzes how Honduras avoided much of the upheaval that plagued the region.

4064 **Scranton, Margaret E.** Panama. (*in* Intervention into the 1990s: U.S. foreign policy in the Third World. Boulder, Colo.: L. Rienner Publishers, 1992, p. 343–360)

Provides brief historical background to US involvement in Panama, then explores linkages with the Noriega era leading up to the invasion.

4065 **Segura, Jorge Rhenán.** La Sociedad de las Naciones y la política centroamericana: 1919–1939. San José: Euroamericana de Ediciones, 1993. 327 p.: bibl., ill.

Provides not only a brief introduction to the often-forgotten League Of Nations, but also an insightful description of the League's role in Latin America.

4066 **Seligson, Mitchell A.** Cuba and the Central American connection. (*in* Cuba after the Cold War. Edited by Carmelo Mesa-Lago. Pittsburgh, Penn.: Univ. of Pittsburgh Press, 1993, p. 259–289, graphs)

Utilizes survey data to demonstrate the effects of the Cuban experience—both positive and negative—on Central America.

4067 **Sklar, Holly.** Washington's war on Nicaragua. Boston, MA: South End Press, 1988. 472 p., 2 p. of plates: bibl., index, maps.

Detailed journalistic account of US policy toward Nicaragua from the Carter Administration through Reagan Administration. Title aptly describes orientaton of this book.

4068 **Sojo, Ana** *et al.* Estado y crisis en Centroamérica en la década de los 80. San José: Instituto Centroamericano de Documentación e Investigación Social, 1986. 115 p. (Proyecto crisis y alternativas en Centroamérica—documentos de trabajo)

Mid-1980s publication offers fairly novel focus on economic intervention in Central America, while also exploring political and military factors.

4069 Sojo, Carlos. Costa Rica: política exterior y sandinismo. San José: FLACSO, 1991. 237 p.: bibl.

Explores Costa Rican foreign policy in the face of the revolution in Nicaragua. Commendably focuses on the impact of domestic social dynamics on Costa Rican international relations.

4070 Soler, Giancarlo. La "pequeña guerra" de Bush: la invasión a Panamá y la génesis del nuevo orden mundial. (*Tareas/Panamá*, 82, sept./dic. 1992, p. 3–27)

One of the better essays on the US invasion. Places action in the context of the "new world order" as well as domestic political dynamics in the US.

4071 Soto, Willy. Costa Rica y la Federación Centroamericana: fundamentos históricos del aislacionismo. (*Anu. Estud. Centroam.*, 17:2, 1991, p. 15–30, bibl., tables)

Historical article examining "isolationism" of Costa Rica in the context of the Central American region. Attempts to trace this phenomenon to the colonial period.

4072 Torrijos Herrera, Omar et al. 75 años de relaciones entre Panamá y Estados Unidos. Panamá?: Frente de Profesionales del P.R.D., 1989. 251 p.: ill.

Overview of Panama's relations with the US from the perspective of a Panamanian "nationalist struggle."

4073 Villarreal P., Alonso. La crisis: el componente político internacional. (*Rev. Panameña Sociol.*, 6, 1990, p. 83–98)

One of the more detailed analyses of US actions toward Panama in 1989. Gives considerable attention to events other than the invasion.

4074 Vuskovič Céspedes, Pedro. Centroamérica: fisonomía de una región. Coordinación de Adolfo Aguilar Zinser y Rodrigo Jauberth Rojas. México: Programa de Estudios de Centroamérica, 1986. 107 p.: bibl., ill. (Relaciones Centroamérica-México)

Though now dated, a fine introduction to the region and its problems of instability, war, intervention, and development in the 1980s.

4075 Yariv, Danielle and Cynthia Curtis. Después de la guerra: una mirada preliminar al papel de la ayuda militar de EE.UU. en la reconstrucción pos-guerra en El Salvador. (*ECA/San Salvador*, 48:531/532, enero/feb. 1993, p. 61–74, photos)

Prescribes new orientation for US military aid to reconstruct El Salvador after years of civil war.

THE CARIBBEAN AND THE GUIANAS

DAMIAN FERNANDEZ, *Associate Professor of International Relations, Florida International University*
JACQUELINE ANNE BRAVEBOY-WAGNER, *Professor of Political Science, The City College and The Graduate School & University Center of the City University of New York*

DUE TO THE INCREASE IN RESEARCH and publications on the Caribbean area, this volume introduces a new contributing editor to share responsibilities for this chapter. Damián Fernández will continue to cover the general works on the area and scholarship on Cuba, the Dominican Republic, and Puerto Rico, while Jacqueline Braveboy-Wagner will examine publications on the English-speaking islands as well as Haiti, The Guianas, and the smaller island nations.

GENERAL AND HISPANIC CARIBBEAN

The increased number of works on the region as a whole as well as on the Hispanic Caribbean attests to the growing emphasis on this field of study in the post-Cold War era. The extent of professional interest in this region may surprise some, but is indicative of the area's importance outside of traditional geostrategic considerations as well as proof of the large number of experts working on the Caribbean. Still, greater quantity does not necessarily mean a corresponding improvement in quality.

We included and annotated works that met one or more of the following criteria: 1) value for future research (i.e., primary documents); 2) explorations of neglected topics and/or subjects about which there is a dearth of information, even in cases where the analysis is mostly routine (e.g., studies of Haitian and Dominican foreign relations); and 3) the quality of the scholarship, in terms of sound and/or original methodology and new light shed on a topic. Unfortunately, exemplary works of scholarship are few. One notable exception is Harold Dana Sims' article on US policy (item **4130**) toward the pre-Castro leftist labor movement in Cuba.

It would seem that three topics command the most attention from researchers: 1) the international political economy (e.g., the Caribbean Basin Initiative and the impact of North American integration on the region); 2) foreign policy studies, with Cuba leading the way as the most studied country in the region and perhaps in Latin America as a whole; and 3) the role of the US. Overall, however, one theme underlies most scholarship on the region and that is the interconnections between domestic and international politics.

Such a perspective, however, has not led scholars to apply postmodern approaches to the study of the international politics of the region. While the postmodern point of view is leaving its mark on the field of international relations, in our region the discipline is still dominated by traditional paradigms. In that sense we continue to lag behind now, as in the past. An emerging and welcome trend is the avoidance, on the part of scholars and writers, of the most polemical and overtly ideological approaches—both the simplistic neomarxist and new-right perspectives—that have dominated much of the literature since the 1950s. [DJF]

NON-HISPANIC CARIBBEAN

That the literature on the international relations of the non-Hispanic Caribbean has grown quantitatively is a healthy development. Unfortunately, the dearth of theoretical studies continues and, if anything, theory seems to have been abandoned altogether in favor of descriptive and, in particular, prescriptive approaches. One notable exception in this regard is Michael Erisman's work (item **4080**). To some extent, this is the inevitable consequence of the changes in the international system, given the fact that the 1990s are a time of strategic reassessment for both policy-makers and scholars. But the result of this prescriptive preoccupation is a spate of studies that are very similar in tone and content and that contribute little to the long-term evolution of the field.

On the other hand, the end of the Cold War and the rise of international economic and social preoccupations have given impetus to research in areas other than the geopolitical sphere that so preoccupied scholars in the 1970s–80s. Although security studies are still in plentiful supply, even the most traditional now include some discussion of broader issues, for example narcotics trafficking or environmental issues, exemplified by studies such as Ivelaw Griffith's work (items **4148**) and Andrés Serbín's compilation (item **4088**).

Not surprisingly, the early 1990s are also witnessing a spurt of studies on economic trends and policies, especially various aspects of integration and free trade. While most studies are rather general, a few are notable for their detail and specialized analysis as evident in Guillermo Hillcoat and Carlos Quenan's publication (item **4082**) and the compilation edited by Hilbourne Watson (item **4078**).

A number of historical retrospectives have been written, resulting in some insightful backward glances at the role of the US in the Caribbean (Fraser, item **4147**); the breakup of the West Indies Federation (Wallace, item **4158**); and—of particular interest—the role of the Jamaican left in the 1950s (Munroe, item **4154**).

Finally, while the general thematic literature is expanding, there are still few studies that deal either specifically or inclusively with Suriname (only one work annotated below, item **4097**). Haiti, with only five works included below, is also underrepresented, despite recent crises experienced by that country, events that usually trigger an outpouring of publications. Also noticeable is the dearth of in-depth studies of Guyana.

To conclude, the publication of informative studies on new themes in international relations is a highly positive development. However, the disregard for theory and the neglect of certain areas remain as serious problems. Without theory, Caribbean international relations is less of a scholarly pursuit than a sort of coincidental categorization of a region. [JAB-W]

GENERAL

4076 **Alzugaray Treto, Carlos.** Seguridad nacional en la cuenca del Caribe. (*in* Sistemas políticos, poder y sociedad: estudios de casos en América Latina. Caracas: Asociación Latinoamericana de Sociología; Centro de Estudios sobre América; Editorial Nueva Sociedad, 1992, p. 21–44, bibl.)

Although this article has a distinct anti-US bias leading to some questionable assertions about past Soviet and future US policies, it offers some thought-provoking suggestions regarding post–Cold War strategic policy. Recommendations include Caribbean commitment to non-use of force, self-determination for Puerto Rico, and normalization of relations with Cuba. [J. Braveboy-Wagner]

4077 **Braveboy-Wagner, Jacqueline Anne et al.** The Caribbean in the Pacific century: prospects for Caribbean-Pacific cooperation. Boulder: L. Rienner Publishers, 1993. 217 p.: bibl., ill., index, map.

Examines political economic context and linkages of Asian-Caribbean relations.

4078 **The Caribbean in the global political economy.** Edited by Hilbourne A. Watson. Boulder, Colo.: L. Rienner Publishers, 1994. 261 p.: bibl., ill., index.

Specialized text whose basic theme is the need in an era of technological globalization for Caribbean adaptation through improved technological competitiveness. Offers trenchant critiques of past and current economic policies, especially, restructuring. General perspectives on Cuba, Puerto Rico, and Haiti are included. Very useful addition to the political economic literature. For economist's comment see item **1113**. [J. Braveboy-Wagner]

4079 **DeMar, Margaretta.** Constraints on constrainers: limits on external economic policy affecting nutritional vulnerability in the Caribbean. (*Lat. Am. Perspect.*, 20:2, Spring 1993, p. 54–73, bibl.)

Discusses impact of the perspectives of the International Monetary Fund, World Bank, U.S. Agency for International Development, Inter-American Development Bank, Caribbean Development Bank, and Caribbean Community on national policy with regard to human development in general, not simply "nutrition" as the title suggests. Makes interesting connection between external and domestic priorities. [J. Braveboy-Wagner]

Dew, Edward M. Caribbean paths in the dark. See item **3032**.

4080 **Erisman, H. Michael.** Pursuing post-dependency politics: South-South relations in the Caribbean. Boulder, Colo.: L. Rienner Publishers, 1992. 164 p.: bibl., ill., index.

Focuses on topic neglected in the literature: cooperation among developing countries. Work analyzes South-South cooperation as a Caribbean alternative to clientalism. Exceptional in being one of the few theoretically oriented works in international relations. [J. Braveboy-Wagner]

4081 Griffith, Winston H. Una zona hemisférica de libre comercio y los países del Caribe. (*Integr. Latinoam.*, 17:180, julio 1992, p. 43–55, bibl., table)

Hypothesizes that a free trade zone in the Americas would not lead to economic benefits, at least in the short run.

4082 Hillcoat, Guillermo and Carlos Quenan. L'intégration régionale dans les Caraïbes: antécédents et perspectives. (*Cah. Am. lat.*, 12, 1991, p. 139–164, graphs, tables)

Concise summary and analysis of economic aspects of Caribbean integration, including discussion of Caribbean Basin Initiative, North American Free Trade Area, and Lomé Convention issues. The latter is a preferential trade and aid agreement between the European Community and 70 countries in Africa, the Caribbean, and the Pacific. [J. Braveboy-Wagner]

4083 Hillcoat, Guillermo and Carlos Quenan. Reestructuración internacional y reespecialización productiva en el Caribe. (*in* El Caribe hacia el 2000: desafíos y opciones. Caracas: Editorial Nueva Sociedad, 1991, p. 59–101, graphs, tables)

Within a macro-international, political-economy perspective, authors focus on free industrial export zones and provide assessment of their benefits and limitations.

4084 Jiménez, Dolores. Estado actual y perspectiva de las relaciones del Japón con el Caribe. (*Caribe Contemp.*, 23, julio/dic. 1991, p. 55–64)

Noteworthy for dealing with a relatively new topic, not for its in-depth analysis. No bibliography.

4085 Maingot, Anthony P. The internationalization of corruption and violence: threats to the Caribbean in the post–Cold War world. (*in* Democracy in the Caribbean: political, economic, and social perspectives. Baltimore: The Johns Hopkins Univ. Press, 1993, p. 42–56)

Argues that "the greatest menace to the security of the Caribbean Basin stems from a new level of corruption that should best be called the internationalization of corruption and violence." Important, if brief, contribution to recent study of post–Cold War security.

4086 Maingot, Anthony P. The United States and the Caribbean: challenges of an asymetrical [i.e., asymmetrical] relationship. Boulder, Colo.: Westview Press, 1994. 260 p.: bibl., index, maps.

Intelligent, original, and rich analysis of US/Caribbean relations from an interdisciplinary perspective. Beginning with the early 19th century, traces patterns of the interaction between the US and Caribbean nations and highlights cases and themes neglected in other texts (i.e., the internationalization of corruption and intellectual underpinnings of US policy). Central argument that transnational relations between both actors can best be described as one of complex interdependence is a valuable contribution to the literature. Recognizes measure of autonomy among Caribbean states and, at the same time, the capability and limits of the US.

4087 McCoy, Terry L. U.S. policy and the Caribbean Basin sugar industry: implications for migration. (*in* The effects of receiving country policies on migration flows. Boulder, Colo.: Westview Press, 1991, p. 39–65, tables)

The paper "examines the sugar industry and US policy as factors contributing to migration" from 12 Caribbean countries. Argues that "US policies inadvertently effect Caribbean Basin migration by impacting development in the region." Well-crafted, provocative, and policy relevant.

4088 Medio ambiente, seguridad y cooperación regional en el Caribe. Coordinación de Andrés Serbín. Caracas: Instituto Venezolano de Estudios Sociales y Políticos; Centro de Investigaciones de Quintana Roo; Editorial Nueva Sociedad, 1992. 146 p.: bibl., maps.

One of the few integrated and extensive discussions of environmental concerns and policies affecting the Caribbean. Contributors point out important distinctions between sustainable development and ecodevelopment, inventory a wide spectrum of environmental problems, and detail existing bilateral, regional, and international agreements in this area. Important work. [J. Braveboy-Wagner]

4089 Pastor, Robert A. and **Richard D. Fletcher.** El Caribe en el siglo XXI. (*Cienc. Polít.*, 25, cuarto trimestre 1991, p. 105–119)

Prescriptions for Caribbean economic policy in the 1990s. Suggests liberalization but with government intervention for social development, closer links with the US, incentives for the return of trained nationals, and establishment of discussion group of North Americans interested in the Caribbean basin. [J. Braveboy-Wagner]

4090 Pastor, Robert A. and **Richard D. Fletcher.** Twenty-first century challenges for the Caribbean and the United States: toward a new horizon. (*in* Democracy in the Caribbean: political, economic, and social perspectives. Baltimore: The Johns Hopkins Univ. Press, 1993, p. 255–276)

Useful overview of international and domestic political and economic challenges facing the region. Authors draw from the area's past to project tasks and policy choices for the next century. Includes discussion on security, immigration, and US policy, among other issues.

4091 Peace, development, and security in the Caribbean: perspectives to the year 2000. Edited by Anthony T. Bryan, John Edward Greene, and Timothy M. Shaw. New York: St. Martin's Press, 1990. 332 p.: bibl., index, map.

Analysis of small-state security problems (prior to the end of the Cold War) from geopolitical, administrative, economic, demographic, societal, and external perspectives. Part three (External Influences) is especially informative. Although quality of the contributions is uneven, in general thoughtful insights are provided. [J. Braveboy-Wagner]

4092 Quick, Stephen A. The international economy and the Caribbean: the 1990s and beyond. (*in* Democracy in the Caribbean: political, economic, and social perspectives. Baltimore: The Johns Hopkins Univ. Press, 1993, p. 212–228)

Argues that the Caribbean's reorientation of economic policy toward export-led growth, necessitated by the changes in the global economy, will produce both economic benefits and political costs. Encapsulates emerging consensus in the literature from a balanced perspective.

4093 Ramsaran, Ramesh. Domestic policy, the external environment, and the economic crisis in the Caribbean. (*in* Modern Caribbean politics. Baltimore, Md.: The Johns Hopkins Univ. Press, 1993, p. 238–258)

Provides useful overview of interaction between domestic policies and external factors in the Commonwealth Caribbean in the 1980s, emphasizing structural adjustment, the IMF/World Bank, and the US.

4094 Richardson, Bonham C. The Caribbean in the wider world, 1492–1992: a regional geography. Cambridge, England; New York: Cambridge Univ. Press, 1992. 235 p.: bibl., ill., index, maps. (Geography of the world-economy)

Overview of the Caribbean from geographical, historical, anthropological, political, and economic perspectives. Very good historical analysis, but less effective in dealing with events in the 1970s and 1980s. [J. Braveboy-Wagner]

4095 Sanders, Ronald. The drug problem: social and economic effects; policy options for the Caribbean. (*Caribb. Aff.*, 3:3, July/Sept. 1990, p 18–28)

Examines consequences of drug traffic on Caribbean nations: corruption; drug abuse; rise in crime; diversion of foreign exchange; and threat to sovereignty.

4096 Serbín, Andrés. Peace in the Caribbean: is it an achievable utopia in a world full of threats? (*Caribb. Aff.*, 1:4, Oct./Dec. 1988, p. 148–167)

Addresses important, yet largely neglected, topic.

4097 Size and survival: the politics of security in the Caribbean and the Pacific. Edited by Paul K. Sutton and Anthony Payne. London; Portland, Or.: F. Cass, 1993. 200 p.: bibl.

General overview of security concerns of both the Caribbean and the Pacific small states, interspersed with specific case studies of 1991 coup in Trinidad, Amerindian insurgency in Suriname, and events in Papua New Guinea and Micronesia. Useful addition to security literature. See also items **4099, 4148,** and **4100.** [J. Braveboy-Wagner]

4098 **Smith, Robert Freeman.** The Caribbean world and the United States: mixing rum and Coca-Cola. New York: Twayne Publishers; Toronto: Maxwell Macmillan Canada; New York: Maxwell Macmillan International, 1994. xv, 120 p.: bibl., ill., index, maps.

Brief, basic historical description of US-Caribbean relations. Includes relatively new primary material on the Cuban Missile Crisis and some useful details on other issues. The tone is unapologetically pro-US and the overall discussion is rather superficial. [J. Braveboy-Wagner]

4099 **Sutton, Paul K.** The politics of small state security in the Caribbean. (*J. Commonw. Comp. Polit.*, 31:2, July 1993, p. 1–32)

Useful summary of security problems of Caribbean nations in the 1990s, including regime instability, economic difficulties, environmental hazards, drug abuse, and extraterritorial jurisdiction and secession. [J. Braveboy-Wagner]

4100 **Sutton, Paul K.** and **Anthony Payne.** Towards a security policy for small island and enclave developing states. (*J. Commonw. Comp. Polit.*, 31:2, July 1993, p. 193–201)

Summarizes a range of policies that small-state governments are pursuing and should pursue in the security area, including regional cooperation, extra-regional association, and diplomatic coordination. Unlike most other works, this one offers a British perspective, downplaying relations with the US somewhat in favor of association with Britain and Canada. [J. Braveboy-Wagner]

4101 **USA, USSR, and the Caribbean:** the new realities; a symposium held by the Bustamante Institute of Public & International Affairs, in association with the Press Association of Jamaica, July 19 and 20, 1990 at the Wyndham Hotel, Kingston, Jamaica. Kingston?: Bustamante Institute of Public & International Affairs, 1990. 90 p.: ill.

Speeches of workshop participants from the US, former Soviet Union, and Jamaica. Consensus is that the Caribbean risks marginalization in the post-Cold War era and should move toward more self-reliance and regional integration. [J. Braveboy-Wagner]

4102 **Watson, Hilbourne A.** Coalition security development: military industrial restructuring in the United States and military electronics production in the Caribbean. (*Caribb. Stud.*, 24:1/2, Jan./June 1991, p. 223–247, bibl.)

Sectoral analysis of the US defense industry and its relation to production of military goods in the Caribbean through the Caribbean Basin Initiative (CBI). Argues that although US policy meshes regional economic development with US national security interests, prospects for the CBI area to develop sustainable competitiveness in defense production "are few if any."

HISPANIC CARIBBEAN

Cuba

Aguirre, Benigno E. Cuban mass migration and the social construction of deviants. See item **4691**.

4103 **Alvarez, Alberto.** Cuba-América Latina: el caso de las relaciones interestatales con Colombia. (*Cuad. Nuestra Am.*, 7:15, julio/dic. 1990, p. 170–195, tables)

Solid contribution to the study of bilateral relations between Cuba and Colombia from 1959 to the late 1980s.

4104 **Angulo Rivas, Alfredo.** Relaciones Venezuela-Cuba: un caso de conflicto diplomático. (*Bol. Acad. Nac. Hist./Caracas*, 74:296, oct./dic. 1991, p.89–106)

Valuable diplomatic history of Cuban-Venezuelan relations from the 1940s to the 1960s, largely based on foreign ministry sources and testimonials.

4105 **Blasier, Cole.** El fin de la asociación soviético-cubana. (*Estud. Int./Santiago*, 26:103, julio/sept. 1993, p. 296–340, tables)

Rich analysis of the end of the special relationship between Cuba and the USSR in 1991. Focuses on the connection between changes in Soviet domestic politics and the breakdown of bilateral relations

4106 **Blight, James G.**; **Bruce J. Allyn**; and **David A. Welch.** Cuba on the brink: Castro, the missile crisis, and the Soviet collapse. Foreword by Jorge I. Domínguez. New

York: Pantheon Books, 1993. 509 p.: bibl., index.

Invaluable addition to the bibliography on the Cuban Missile Crisis. Based on an international conference held in Havana on Jan. 9–12, 1992 which brought together key policy-makers from the US, the USSR, and Cuba, book includes meeting transcripts and introductory analytical essays. Offers new information from key participants in Oct. 1962 events, including Fidel Castro, Robert McNamara, Olef Darusendov, and Aleksander Alekseev.

4107 Brenner, Philip. Kennedy and Khrushchev on Cuba: two stages, three parties. (*Probl. Communism*, 41, special issue, Spring 1992, p. 24–27, photo)

Summarizes new key lessons of the Cuban Missile Crisis after the 1992 tripartite conference in Havana.

4108 Brugioni, Dino A. Eyeball to eyeball: the inside story of the Cuban missile crisis. Edited by Robert F. McCort. New York: Random House, 1991. 622 p.: bibl., ill., index.

An insider's valuable contribution to the recent bibliography on the Cuban Missile Crisis. Author, former CIA expert on satellite imagery and aerial photography, interpreted reconnaissance photography of missile sites during the crisis. Most innovative aspects of the book cover the technology of security.

4109 Carotenuto, Gennaro. La Gran Bretagna e Cuba rivoluzionaria. (*Latinoamerica/Rome*, 13:48, ott./dic. 1992, p. 75–86)

Highly specialized, short, and sketchy account of Great Britain's policy towards the Cuban revolutionary government in 1959. Useful for scholars working on in-depth study of a topic which has been ignored in the literature.

4110 Castro, Fidel. Mensaje de Fidel Castro a la Primera Cumbre Iberoamericana. (*Caribe Contemp.*, 24, enero/junio 1992, p. 131–158)

Long speech covering issues of integration, human rights, and economic decline of Latin America in the changed international context. Defense of Cuba's political and social record.

Cotman, John Walton. The gorrión tree: Cuba and the Grenada Revolution. See item **4145**.

4111 Cuba after the Cold War. Edited by Carmelo Mesa-Lago. Pittsburgh, Penn.: Univ. of Pittsburgh Press, 1993. 383 p.: bibl., ill. (Pitt Latin American series)

Solid contribution by top scholars on impact of end of the Cold War on Cuba's domestic politics and international relations. Chapters include analogies between Cuba and Eastern Europe, discussions of economic effects of socialism's breakup in the USSR and Eastern Europe, political impact of the breakdown of communist regimes, perspectives of Central American elites on Cuba, and the repercussions of the Soviet collapse and the Cuban crisis on the Left in South America.

4112 Cuba y Estados Unidos: dos enfoques. Edición y compilación de Juan Tokatlian. Bogotá: CEREC; Grupo Editor Latinoamericano, 1984. 247 p.: bibl.

Very useful source on Cuban foreign policy in the 1980s from the perspective of US and Cuban specialists. One of the best of its kind.

4113 Cuba's ties to a changing world. Edited by Donna Rich Kaplowitz. Boulder, Co.: L. Rienner Publishers, 1993. 263 p.: bibl., ill., index, maps.

Case studies of how Cuban foreign policy has adapted to the changed global situation since the late 1980s. Includes chapters on the US, China, Japan, Africa, Middle East, Britain, the European Community, selected Latin American countries, and Canada.

4114 Díaz Vilches, Patricia. Los servicios: cooperación Cuba-América Latina; el caso de la rama pecuaria. Havana: Instituto Superior de Relaciones Internacionales Raúl Roa García, 1990. 132 p.: bibl., ill. (Trabajo de diploma; 28)

Original study of the role of the service sector (specifically in the fishing industry) in Cuba's relations with Latin America and the possibility for cooperation.

4115 Dobrynin, Anatoly. The Caribbean crisis: an eyewitness account. (*Int. Aff./Moscow*, 8, Aug. 1992, p. 47–60)

Important document for students of the Cuban Missile Crisis by the Soviet Ambassador to the US at the time. His recollections are available in other sources as well.

4116 Faya, Ana Julia and Estervino Montesino. La reincorporación de Cuba a la OEA: ¿voluntad latinoamericana o rechazo norteamericano? (*Cuad. Nuestra Am.*, 9:18, enero/junio 1992, p. 111–125)

Documents resurgence of the Cuban issue within the Organization of American States in the 1980s. Argues for Cuba's reincorporation in the organization, without addressing the Declaration of Santiago as a possible obstacle.

4117 Fernández, Damián J. Continuity and change in Cuba's international relations in the 1990s. (*in* Cuba at a crossroads: politics and economics after the Fourth Party Congress. Gainesville: Univ. Press of Florida, 1994, p. 41–66, bibl., ill.)

Well-written general assessment of Cuba's external policies in the 1990s within the framework of selective adjustments in the "operational code" of Cuba's decision makers. [J. Braveboy-Wagner]

4118 Fernández, Damián J. Opening the blackest of black boxes: theory and practice of decision making in Cuba's foreign policy. (*Cuba. Stud.*, 22, 1992, p. 53–78, bibl., tables)

Solid, imaginative study applies crisis decision theory to mysteries of Cuba's foreign policy formation regarding Angola and finds it "strikingly similar to that of most other states." [G.P. Atkins]

4119 Hernández, José M. Cuba and the United States: intervention and militarism, 1868–1933. Austin: Univ. of Texas Press, 1993. 266 p.: bibl., index.

Noteworthy diplomatic history (based on primary sources) on the impact of US interventions in Cuba (1898–1902 and 1906–09), focusing on elites and the army.

4120 Hernández, Rafael. ¿Había pedido Fidel Castro que se asestara un golpe preventivo contra EE.UU. con cohetes nucleares? (*Am. Lat./Moscow*, 7:163, julio 1991, p. 38–58, ill. photos)

Presents Cuba's role in and perspective of the Missile Crisis. Argues that the island's government was an important although often neglected actor, and that the island's policy offers significant lessons for small countries.

4121 Hernández, Rafael. El ruido y las nueces II: el ciclo en la política de los Estados Unidos hacia Cuba. (*Cuad. Nuestra Am.*, 9:18, enero/junio 1992, p. 4–21)

Multidimensional analysis of US-Cuban relations from four levels: 1) US domestic politics; 2) US policy towards Latin America and the Third World; 3) US policy towards the Soviet Union; and 4) bilateral relations. Includes discussion of Cuban national security and argues for a change in Washington's policy vis-à-vis the island.

4122 Hernández, Rafael. Sobre las relaciones con la comunidad cubana en los Estados Unidos. (*Cuad. Nuestra Am.*, 8:17, julio/dic. 1991, p. 149–160)

Briefly examines aspects of social, economic and ideological dynamics of Cuban-American immigration in the context of US-Cuban relations.

4123 Khazanov, Anatoly. The war in Angola, postscript. (*Int. Aff./Moscow*, 11, 1991, p. 49–60, bibl.)

Russian scholar argues that external factors were the main causes of the Angolan War and offers revisionist perspective by claiming that "the Cuban intervention had extremely negative consequences." Weak bibliography.

4124 Miyar Bolio, María Teresa. La política de Cuba hacia la comunidad cubana en el contexto de las relaciones Cuba–Estados Unidos, 1959–1980. (*Caribe Contemp.*, 24, enero/junio 1992, p. 91–107)

One of the few serious studies on dialogue between segments of the Cuban-American community and Castro's government.

4125 Monreal González, Pedro M. Cuba y América Latina y el Caribe: apuntes sobre un caso de inserción económica. (*Estud. Int./Santiago*, 26:103, julio/sept. 1993, p. 500–536, tables)

Serious analysis by a Cuban economist of opportunities and challenges for Cuba's economic reinsertion in Latin America and the Caribbean.

4126 Monreal González, Pedro M. Cuba y la nueva economía mundial: el reto de la inserción en América Latina y el Caribe. (*Cuad. Nuestra Am.*, 8:16, enero/junio 1991, p. 36–68, tables)

Provides overview of the Cuban economic crisis since the mid-1980s and its relationship to the international economy. Also pinpoints economic challenges and opportunities facing the island's attempt to reinsert itself into the Latin American and Caribbean regions.

4127 La paz de Cuito Cuanavale: documentos de un proceso. La Habana: Editora Política, 1989. xi, 97 p.: ill.

Post-1987 documents on negotiations between Cuba, Angola, and South Africa.

4128 El regreso de Fidel a Caracas, 1989. Moderación de Julio García Luis. Caracas: Ediciones de la Biblioteca de la Univ. Central de Venezuela; CEA, 1989. 55 p. (Col. Rectorado. Rueda de prensa)

Transcript of the Feb. 1989 interview of Fidel Castro in Venezuela by 16 Latin American reporters.

4129 Schulz, Donald E. The United States and Cuba: from a strategy of conflict to constructive engagement. (*J. Interam. Stud. World Aff.*, 35:2, Summer 1993, p. 81–102, bibl.)

Insightful discussion on wisdom of charting a new course in US-Cuban relations identified as "constructive engagement."

4130 Sims, Harold Dana. Collapse of the house of labor: ideological divisions in the Cuban labor movement and the U.S. role, 1944–1949. (*Cuba. Stud.*, 21, 1991, p. 123–147)

First-rate scholarship uncovers efforts of Cuban anti-communist labor leaders to secure Washington's support in the 1940s and Washington's hesitation to align itself with them. Shatters prevalent view that the US and the American Federation of Labor consistently undermined communist labor movements and sided with non-leftist organizations throughout Latin America.

4131 Suchlicki, Jaime. Myths and realities in US-Cuban relations. (*J. Interam. Stud. World Aff.*, 35:2, Summer 1993, p. 103–113, bibl.)

Characteristic of the posture of the traditional right, responds to Schulz (see item **4129**) that US-Cuba rapproachment will be elusive as long as Fidel Castro remains in power.

Dominican Republic

4132 Arias, Luis. La política exterior en la era de Trujillo. Santiago: PUCMM, 1991. 245 p.: bibl. (Col. Estudios; 156)

Covers US-Dominican relations, Haitian-Dominican relations, conflicts in the Caribbean, and external factors which contributed to Trujillo's fall.

Betances, Emelio. The formation of the Dominican capitalist State and the United States military occupation of 1916–1924. See item **4699**.

Dominika imin wa kimin datta: sengo nikkei imin no kiseki [Abandoned people: Japanese immigrants in the Dominican Republic]. See item **3052**.

Price-Mars, Jean. Mémoire sur les relations haitiano-dominicaines: adressé à S.E. Monsieur Dumarsais Estimé, Président de la République. See item **4142**.

4133 Puig, Max. Haití y República Dominicana: un esquema de relaciones puesto en entredicho. (*Caribe Contemp.*, 24, enero/junio 1992, p. 109–127)

Traces the "new" (post-1980s) issues in Haitian-Dominican relations, with special attention on labor issues.

4134 Vega, Bernardo. Eisenhower y Trujillo. Santo Domingo: Fundación Cultural Dominicana, 1991. 286 p.: bibl., ill., index.

Detailed diplomatic history from an anti-US perspective.

4135 Vega, Bernardo. Kennedy y los Trujillo. Santo Domingo: Fundación Cultural Dominicana, 1991. 423 p.: bibl., ill., index.

Important source for study of domestic politics and international relations of the Dominican Republic in the 1960s.

Zaglul, Jesús M. Una identificación nacional "defensiva": el antihaitianismo nacionalista de Joaquín Balaguer. See item **4802**.

Puerto Rico

4136 Heine, Jorge and **Juan M. García-Passalacqua.** Political economy and foreign policy in Puerto Rico. (*in* Modern Caribbean politics. Baltimore, Md.: The Johns Hopkins Univ. Press, 1993, p. 198–211)

Valuable analysis of Puerto Rico's foreign policy initiative since 1985 focuses on the Caribbean Basin Initiative Section 936, and the Twin Plants Strategy. Also provides new periodization of the island's foreign relations from the Cold War to the 1980s.

4137 **Libby, Justin H.** A wedding of contradictions: Caribbean/U.S. relations. (*J. Caribb. Stud.*, 8:3, Winter 1991/Spring 1992, p. 197–218)

Somewhat specialized discussion of three historical-constitutional cases that helped define the status of Puerto Rico as an incorporated territory. [J. Braveboy-Wagner]

HAITI

4138 **Cénatus, Bérard et al.** Les espaces du politique: politique extérieure (la scèna internationale). (*in* La République Haïtienne: état des lieux et perspectives. Sous la direction de Gérard Barthélemy et Christian Girault. Paris: Karthala, 1993, p. 233–262)

Series of short exposés on Haiti's international relations in the 1980s. Useful to anyone working on an in-depth study of this period, particularly regarding the role of France, Venezuela, and the French Caribbean.

4139 **The Haitian challenge: U.S. policy considerations.** Edited by Georges A. Fauriol. Foreword by representative James L. Oberstar. Washington: Center for Strategic & International Studies, 1993. xiii, 80 p.: bibl., map.

Collection of essays by scholars, policy-makers, and advocates on multiple challenges confronting the US as it crafts a policy towards Haiti in the early to mid-1990s. Includes discussion of the 1990 election results, immigration, multilateralism, and the role of the OAS.

4140 **Lawless, Robert.** Haiti's bad press. Rochester, Vt.: Schenkman Books, 1992. xxvii, 261 p.: bibl., ill., index.

Aims to establish relationship between the world political order and the message conveyed to the public about that order. Analyzes various international biases toward Haiti which apparently originate in a preference for anthropological "folk models" rather than "analytic models." While the basic premise is interesting, the analysis is rather overwrought. [J. Braveboy-Wagner]

4141 **McCalla, Jocelyn.** The Haitian refugee crisis: origins, causes and responses. San José: Instituto Interamericano de Derechos Humanos, 1991. 35 p.: table. (Serie Exodos en América Latina; 2)

Report submitted at a 1990 meeting held to discuss the refugee crisis in Central America, the Caribbean, and the Andean region. Useful if brief summary is critical of treatment of refugees by US, Bahamian, and Dominican Republic authorities. [J. Braveboy-Wagner]

4142 **Price-Mars, Jean.** Mémoire sur les relations haitiano-dominicaines: adressé à S.E. Monsieur Dumarsais Estimé, Président de la République. (*Rev. Soc. haïti.*, 48:172, mars 1992, p. 15–32)

Of some historical relevance for the study of Haitian-Dominican relations in the 1940s.

Puig, Max. Haití y República Dominicana: un esquema de relaciones puesto en entredicho. See item **4133**.

Zaglul, Jesús M. Una identificación nacional "defensiva": el antihaitianismo nacionalista de Joaquín Balaguer. See item **4802**.

OTHER CARIBBEAN ISLANDS

4143 **Compton, John G.M.** From the labours of Sisyphus to the light of a new dawn in U.S.-Caribbean understanding. (*Caribb. Aff.*, 1:4, Oct./Dec. 1988, p. 132–135)

Address of Prime Minister of St. Lucia at the 11th Annual Miami Conference on the Caribbean. Focuses on pitfalls of US policy, specifically the Caribbean Basin Initiative.

4144 **Coram, Robert.** Caribbean time bomb: the United States' complicity in the corruption of Antigua. New York: Morrow, 1993. 278 p.: bibl., ill., index.

Journalist's detailed account of various corruption schemes involving the Antiguan government. Also offers some insights into the US relationship with Antigua. Fascinating if controversial look at the less-than-pristine activities of the Bird political dynasty. [J. Braveboy-Wagner]

4145 **Cotman, John Walton.** The gorrión tree: Cuba and the Grenada Revolution. New York: P. Lang, 1993. 272 p.: bibl. (American university studies. Series X, Political science, 0740–0470; 38)

More useful for information it provides on Cuban-Grenadian relations (1979 and 1983) than for its analysis.

4146 **Cumberbatch, Janice A.** and **Neville C. Duncan.** Illegal drugs, USA policies and Caribbean responses: the road to disaster. (*Caribb. Aff.*, 3:4, Oct./Dec. 1990, p. 150–181)

Detailed study on responses of Caribbean nations (specifically Barbados, Jamaica, and Montserrat) to US drug policies. Focuses on legal reforms and criticizes unidimensional drug policies of the Caribbean state and the US.

4147 **Fraser, Cary.** Ambivalent anti-colonialism: the United States and the genesis of West Indian independence, 1940–1964. Westport, Conn.: Greenwood Press, 1994. 233 p.: bibl., index. (Contributions in Latin American studies, 1054–6790; 3)

Historical analysis of US role in the West Indies vis-à-vis Britain up to the 1960s. A valuable contribution to diplomatic history, this analysis offers new insights into societal influences on US officialdom, and uses primary material innovatively in analyzing US perspectives on the W.I. federation and British Guiana crisis. [J. Braveboy-Wagner]

4148 **Griffith, Ivelaw L.** Drugs and security in the Commonwealth Caribbean. (*J. Commonw. Comp. Polit.*, 31:2, July 1993, p. 70–102, tables)

Comprehensive analysis of narcotics issues, including drug production, transshipment, money laundering, and security implications of the drug trade (corruption, crime, economic costs, resource limitations, and regional initiatives). [J. Braveboy-Wagner]

4149 **Griffith, Ivelaw L.** The quest for security in the Caribbean: problems and promises in subordinate states. Armonk, N.Y.: M.E. Sharpe, 1993. 320 p.: bibl., ill., maps.

Thorough examination of security stances and interests of the English-speaking Caribbean nations. Probably the most detailed work of its kind published in English. Contains valuable background information on the military in these small countries. [J. Braveboy-Wagner]

4150 **Huntley, Earl.** The Union of East Caribbean States: thoughts on a form. St. Lucia?: Voice Pub. Co., 1988? 33 p.

Booklet offers suggestions on constitutional options open to Eastern Caribbean countries wishing to form a political union. Although written in 1988, suggestions are still relevant since the proposed union is far from realization. [J. Braveboy-Wagner]

4151 **Lasserre, Guy** and **Albert Mabileau.** The French Antilles and their status as overseas departments. (*in* Caribbean freedom: society and economy from emancipation to the present. Kingston: Randle, 1993, p. 444–454, tables)

Discussion of status of French Overseas Departments (DOM), possibly intended to be a historical essay but presented as a curiously-dated current analysis. Nevertheless, some aspects of the discussion are still relevant and useful. [J. Braveboy-Wagner]

4152 **McDougall, Derek.** The French Caribbean during the Mitterrand Era. (*J. Commonw. Comp. Polit.*, 31:3, Nov. 1993, p. 92–110)

Offers some useful perspectives on the relationship between French territories and France in the Mitterand/Chirac era within the context of the prospective merits of "decolonization through integration" and "decolonization through independence." [J. Braveboy-Wagner]

4153 **Mills, Don.** The new Europe: the new world order, Jamaica and the Caribbean. Kingston: Grace, Kennedy Foundation, 1991. xii, 88 p.: bibl. (Grace, Kennedy Foundation lecture; 1991)

Overview of world events and trends affecting Jamaica in 1991. [J. Braveboy-Wagner]

4154 **Munroe, Trevor.** The Cold War and the Jamaican left, 1950–1955: reopening the files. Kingston: Kingston Publishers, 1992. 242 p.: bibl.

Essential reading for Jamaica specialists, book is significant as one of the first published reassessments by the Caribbean left in the post-Cold War period. In this historical analysis of leftist activities in Jamaica, author provides balanced view of nationalist aspects of Jamaican Marxism, and their relevance today. [J. Braveboy-Wagner]

4155 **Payne, Anthony** and **Paul K. Sutton.** The Commonwealth Caribbean in the New World Order: between Europe and North

America? (*J. Interam. Stud. World Aff.*, 34:4, Winter 1992/93, p. 39-75, bibl.)

Summary of options facing the Caribbean with respect to the European Community and the North American Free Trade Area. Concludes with reason that the Caribbean is relocating itself geoeconomically within the Americas. [J. Braveboy-Wagner]

4156 **Ramphal, Shridath.** Time to act. (*Caribb. Q.*, 39:1, March 1993, p. 1-17)

Head of important West Indian task force established to assess future of Caribbean regionalism speaks about task force findings and proposes establishment of a Caribbean Community Commission and a broad-based Association of Caribbean States. [J. Braveboy-Wagner]

4157 **¿Vecinos indiferentes?: el Caribe de habla inglesa y América Latina.** Recopilación de Andrés Serbín y Anthony T. Bryan. Caracas: Instituto Venezolano de Estudios Sociales y Políticos; Editorial Nueva Sociedad, 1990. 250 p.: bibl., ill.

One of the most extensive and useful analyses of relations between Latin America and the Anglophone Caribbean. Includes discussions of cultural and literary issues as well as the more common political and economic perspectives. [J. Braveboy-Wagner]

4158 **Wallace, Elizabeth.** The break-up of the British West Indies Federation. (*in* Caribbean freedom: society and economy from emancipation to the present. Kingston: Randle, 1993, p. 455-475)

Well-written analysis of much-discussed historical topic with cogent commentary on reasons for the break-up of the federation.

THE GUIANAS

Consalvi, Simón Alberto. Grover Cleveland y la controversia Venezuela-Gran Bretaña: la historia secreta. See item **4332.**

Fraser, Cary. Ambivalent anti-colonialism: the United States and the genesis of West Indian independence, 1940-1964. See item **4147.**

4159 **Giacalone, Rita.** Guyana y los poderes regionales caribeños frente al conflicto del Esequibo, 1970-1984. (*in* El Caribe, objeto de investigación. Recopilación de José Moreno Colmenares. Caracas: Univ. Central de Venezuela, Consejo de Desarrollo Científico y Humanístico; Fondo Editorial Acta Científica Venezolana, 1988, p. 259-289)

Analyzes policies of three middle-range powers (i.e. Brazil, Cuba, and Venezuela) vis-à-vis Guyana and the Essequibo dispute. Traces Guyanese foreign policy from 1966-84, highlighting changes from an independent position to one increasingly responsive to Brazilian and Venezuelan aspirations.

SOUTH AMERICA (Except Brazil)

ALDO C. VACS, *Associate Professor of Government, Skidmore College*

DURING THIS BIENNIUM there has been an intensification of interest in the new context and contents of foreign policies formulated by South American nations, particularly in aspects such as their security, economy, politics, and diplomacy. We can attribute this interest to many factors, among them, the end of the Cold War, the consolidation of economic liberalism, the acceleration of the globalization and transnationalization process, and finally, the expansion of democratic trends throughout most of South America.

As a result, there has been a notable increase in the number of publications devoted to issues such as: 1) the new security challenges; 2) the reinsertion of South American countries into the world economy; and 3) the effect of international developments on the strengthening of democracy. As in the past, the majority of these works focus again on the largest countries of the region—Argentina, Chile, Colombia, Peru and Venezuela—and are written by specialists in those nations. However, the production of international relations studies in smaller countries such as Bolivia, Ecuador, and Uruguay continues to grow, while the recent democratization of Paraguay has been accompanied by the opening of a fruitful debate on the features and prospects of its foreign policies.

There is an apparent decline in writing on traditional subjects such as territorial confrontations and geopolitical issues as well as in the nationalistic biases often associated with such studies. Despite this decline, there continues to be a significant number of studies of territorial disputes (e.g., Argentine-British, Bolivian-Chilean, Ecuadorian-Peruvian, Colombian-Venezuelan). Other topics that have gained prominence in recent years include the impact of the end of the Cold War on regional and national security, particularly in reference to the transformation of the inter-American defense system and the problem of drug trafficking; the features and prospects of neoliberal foreign economic policies, especially as regards the integration processes, relations with Asian countries, and restructuring the links to developed economies; and the relationship between democratic regimes and international relations, as illustrated by the formulation and implementation of foreign policies by elected governments.

Among studies analyzing the overall impact of the end of the Cold War on regional security, we should note works sponsored by the Comisión Sudamericana para la Paz (items **4169,** and **4184**) and Gamba-Stonehouse's article (item **4173**). These publications offer the best appraisal of the new situation and advance reasonable policy recommendations regarding hemispheric and regional defense systems. Concerning the impact of the New World Order on foreign policies of particular countries there are some excellent works including volumes edited by Russell on Argentina (item **4219**), Varas on Chile (item **4251**), and Ferrero Costa on Peru (item **4315**), as well as the works by Tokatlian and Cardona on Colombia (item **4279**), and Abreu and Fillol on Uruguay (item **4320**). The security dimension of the drug problem, particularly in reference to the role played by the US, is critically assessed in articles by Andreas *et al.* (item **4161**), Bustamante (item **3790**), Goodman and Mendelson (item **3819**) and Perl (item **4182**), while important aspects of the Bolivian, Colombian, and Peruvian security dilemmas and options are adequately examined in studies by Bedregal and Viscarra (item **4229**), Botero (item **4267**), and Thornberry (item **4316**) respectively.

In the realm of international economic relations, the most interesting contributions are those related to the study of integration processes and new links between South America and Asia's newly industrializing countries and between South America and the developed nations. Useful examinations of economic integration in the La Plata basin are found in Alimonda (item **4321**), Chudnovsky (item **4166**), Hirst (items **4176** and **4177**), and Sánchez-Gijón (item **4183**) while some of the integration initiatives implemented in the Andean countries are discussed by Deustra (item **4170**), Ondarza Linares (item **4235**), Cardona *et al.* (item **4268**), and Carrera de la Torre (item **4283**). The economic role and prospects for the region in the Pacific basin are competently examined in a work edited by Armanet (item **4160**) while the particular cases of Chile and Peru are adequately evaluated by Valdivieso *et al.* (item **4245**) and Torres (item **4317**). The significant number of valuable publications on the current features and prospects of economic relations established by the region as well as by several individual countries with the US, Western Europe, and Japan makes it impossible to enumerate all of them but taken as a whole they indicate a long-term interest in redefining the direction and contents of trade and investment links as an essential component of the ongoing processes of economic opening and liberalization.

Interest in analyzing and explaining the relationship between a democratic regime and its foreign policies was already noted in this chapter in the previous volume (see *HLAS 53*, p. 630–632). This recent increase in interest has resulted in several interesting theoretical and empirical studies. Although most authors acknowledge that the correlation between democracy and specific types of foreign policies is weak, the majority agree in pointing out: 1) the more cooperative nature of civilian administrations' initiatives vis-à-vis those of non-democratic regimes; and 2) the importance of international factors in strengthening—or weakening—these democratic regimes. Among the most valuable contributions on this topic are works by Paradiso on Argentina (item **4217**), Barrios Morón on Bolivia (items **4228** and **4227**), the edition by Muñoz on Chile (item **4246**), Pardo and Tokatlian on Colombia (item **4275**), Simón on Paraguay (item **4295**), Kisic on Peru (item **4308**), and Luján on Uruguay (item **4327**).

The literature on territorial problems and geopolitical issues, though still much lacking in objectivity and dominated by nationalistic prejudices, includes some commendable exceptions. Among studies of territorial disputes which transcend such biases are: Borón and Faúndez's edited work (item **4220**) and Cardoso *et al.'s* updated edition (item **4198**), both on the Falklands/Malvinas conflict; Montenegro's analysis of the Bolivian-Chilean negotiations (item **4234**); Orrego Vicuña's edited volume on Chile-Argentina (item **4247**); Barrera's collection on Colombia's border relations with Ecuador and Venezuela (item **4269**); Valencia's study of the Ecuadorian-Peruvian dispute (item **4289**); Palau's analysis of Paraguay's border regions (item **2004**); and Granda's work on Peru's border situation (item **4306**). Valuable geopolitical analyses that rise above the conventional nationalistic mold are those written by Ballester (item **3785**) and Mercado Jarrín (item **4179**) as well as Child's study of the evolution and prospects of geopolitical thought in the region (item **3793**).

GENERAL

4160 América Latina en la Cuenca del Pacífico: perspectivas y dimensiones de la cooperación. Edición de Pilar Armanet Armanet. Santiago: Instituto de Estudios Internacionales, Univ. de Chile, 1987. 153 p.: ill. (Col. Estudios internacionales)

Competent set of articles written by Chilean experts examines seldom studied aspect of South America's international relations. Well-documented work analyzes common and conflicting interests held by Pacific Basin countries, and role and prospects of Latin America in commercial, security, maritime resource, and cultural aspects of intrabasin relations.

4161 Andreas, Peter R. *et al.* **La guerra contra la droga: callejón sin salida.** (*Cienc. Polít.*, 27, 1992, p. 121–139)

Excellent article written by noted specialists criticizes US supply-side approach to war on drugs. Reviews Bush Administration's anti-drug strategy in Andean region, attributing its failure to lack of political will of South American governments, corruption of security forces, lack of incentives for producers to replace coca with other crops, and, ultimately, US inability to create local capacity and lack of will to pursue this strategy. Concludes that what is necessary is demand-side strategy aimed at reducing drug consumption and curbing drug-related violence through treatment, education and urban development in the US.

Aravena Ricardi, Nancy. Un corredor territorial para Bolivia: propuestas u opciones. See item **4241**.

Aravena Ricardi, Nancy. Un corredor territorial para Bolivia: ventajas y desventajas geopolíticas. See item **4242**.

4162 Azambuja, Marcos Castrioto de. Nuclear non-proliferation and confidence-building in the Southern Cone. [*Disarmament/UN*, 16:2, 1993, p. 123–133, bibl.]

Brief analysis written by Brazilian ambassador to Argentina examines factors that led to bilateral nuclear agreements between Argentina and Brazil and acceptance by both countries of the Tlatelolco Treaty.

4163 Baquedano Muñoz, Manuel. La seguridad ecológica en América del Sur. Santiago: Comisión Sudamericana de Paz, 1989. 84 p.: bibl. (Documento de estudio; 3)

Brief but valuable study examines notion of ecological security; analyzes main ecological problems affecting South America; and advances several policy recommendations aimed at making ecological concerns and international environmental cooperation central components of a more adequate approach to collective security in the region.

4164 Bizzozero, Lincoln J. La relación entre el Mercosur y la comunidad Europea: ¿un nuevo parámetro de vinculación? [*Estud. Int./Santiago*, 26:101, enero/marzo 1993, p. 37–56, bibl.]

Analyzes relations between Mercosur and European Union. Examines development of institutional cooperation links, pointing out the crucial role played by Spain. Also discusses difficulties in reaching a commercial agreement, underscoring importance of European non-tariff barriers. Concludes that the cooperation agreement, changes in EU agricultural policies, possibilities of industrial cooperation, and convergence in foreign policies of Mercosur countries facilitate the establishment of closer collaborative links.

Bustamante, Fernando. La política de Estados Unidos contra el narcotráfico y su impacto en América Latina. See item **3790.**

4165 Campbell, R. Keith. The maritime balance in the South Atlantic. [*UNISA/Lat. Am. Rep.*, 6:2, Sept. 1990, p. 11–22, bibl., maps]

Informative examination of balance of naval forces in the South Atlantic is written by a South African specialist. Concerning South America, author assesses Argentine, Brazilian, Chilean, and Uruguayan capabilities and, after emphasizing Brazilian superiority, concludes that balance of power in region favors the Latin American navies.

Child, Jack. El pensamiento geopolítico. See item **3793.**

4166 Chudnovsky, Daniel. El futuro de la integración hemisférica: el Mercosur y la iniciativa para las Américas. [*Desarro. Econ.*, 32:128, enero/marzo 1993, p. 483–511, bibl., tables]

Excellent economic analysis of evolution of regional integration in context of Mercosur. Assesses main gains and shortcomings before presenting possible scenarios for future development of integration process. Last section examines potential impact of Initiative for the Americas on the region, and discusses advantages and disadvantages of an eventual incorporation of Argentina and Brazil into a continental free trade zone.

4167 Comisión Sudamericana de Paz. Sesión Plenaria, *2nd, Montevideo, 1988.* Seguridad democrática regional. Santiago: Comisión Sudamericana de Paz, 1988? 67 p.

Concise and valuable report was issued in 1988 by a commission of influential South American policymakers, politicians, businessmen, and religious and cultural figures. After reviewing main security problems facing the region and discussing decline of traditional notions of security, authors advocate a democratic concept of security and advance a number of proposals aimed at fostering economic development with social justice, strengthening democratic institutions, and creating a zone of peace and cooperation in South America.

4168 Desarme y desarrollo: condiciones internacionales y perspectivas. Buenos Aires: Fundación Arturo Illia para la Democracia y la Paz; Grupo Editor Latinoamericano, 1989. 245 p.: bibl., ill. (Col. Estudios políticos y sociales)

Contains articles delivered by Argentine, Brazilian, Chinese, and Swedish specialists at a 1988 seminar on disarmament, development, and peace sponsored by Fundación Arturo Illia para la Democracia y la Paz, a foundation of the Unión Cívica Radical. Particularly interesting are articles by Mônica Hirst on Latin American disarmament initiatives and Roberto Russell on Argentine positions on disarmament and non-proliferation.

4169 Después de la Guerra Fría: los desafíos a la seguridad de América del Sur. Coordinación de Carlos Contreras Quina.

Santiago: Comisión Sudamericana de Paz; Caracas: Editorial Nueva Sociedad, 1990. 153 p.: bibl.

Collection of four articles debates impact of post-Cold War international circumstances on the security of the region. Emphasis is on possible impact of new European situation of 1990s on regional security and on prospects for European-South American cooperation.

4170 Deustua Caravedo, Alejandro. Situación actual y perspectivas de la integración andina en el nuevo contexto internacional. (*in* Integración latinoamericana: su última oportunidad. Bogotá: Corporación de Estudios para el Desarrollo (CORDES), 1991, p. 149–186)

Valuable study assesses impact of new global and regional trends on Andean integration process. After examining centrifugal and centripetal forces affecting Andean countries, concludes that current international context provides a new opportunity to more adequately link the local economies with those of developed nations.

4171 Du Plessis, Anton. Contemporary strategic factors and the issue of co-operation in the South Atlantic area. (*UNISA/Lat. Am. Rep.*, 6:2, Sept. 1990, p. 4–10, bibl., map)

Briefly assesses possibilities for cooperation among South American and African countries of South Atlantic region in light of recent global strategic developments. After reviewing integrative factors such as geographic circumstances and desire to promote economic development and democratization, concludes that disintegrative factors such as distance, oceanic fragmentation, and diverse national interests, cultures, languages, religions, political systems, trade patterns, etc., are stronger and hinder emergence of an integrated regional sub-system in foreseeable future.

4172 Encuentro Latinoamericano por la Democracia y la Integración, *Bogotá, 1990.* Actas. Bogotá: Fundación Luis Carlos Galán Sarmiento, 1991. 234 p., 16 p. of plates: ill.

Includes addresses on regional integration and democracy by influential Latin American politicians, policymakers, and experts delivered at a conference in homage of Luis Carlos Galán. Useful presentation of positions on Latin American integration held by governmental representatives and well-known specialists.

4173 Gamba-Stonehouse, Virginia. Alternativas para el logro de una seguridad colectiva en sudamérica. (*SER/Buenos Aires*, 2, sept. 1992, p. 50–69, ill.)

Compact, thoughtful essay examines different perceptions of external threats in South and North America by end of 1980s. Focuses on South American security concerns and military role including civil-military relations, new professionalism, and possibilities of multilateral cooperation. Advances some sensible proposals for creation of a new regional security system.

4174 Gamba-Stonehouse, Virginia. Strategy in the southern oceans: a South American view. New York: St. Martin's Press, 1989. 155 p.: bibl., index, maps. (Studies in contemporary maritime policy and strategy)

Examines changing perceptions of and values attributed to the sea by South American countries, focusing on two case studies: Bolivia's claims for an outlet to the Pacific, and Argentina's and Brazil's postures concerning western South Atlantic. Concludes that perception of value of the southern oceans is changing from a traditional view of the sea as a barrier to interactions, to one that perceives the sea as facilitating international interactions while compelling regional maritime cooperation and integration to advance South American national aspirations.

4175 García Lupo, Rogelio; Newton Carlos; and Juan Jorge Faúndes Merino. El arsenal sudamericano de Saddam Hussein. Buenos Aires: Grupo Editorial Zeta, 1991. 273 p.: ill. (Serie Reporter)

Informative but highly speculative set of articles written by Argentine, Brazilian, and Chilean investigative journalists examines role played by military industries of their respective countries in supplying weapons to Saddam Hussein. García Lupo's article focuses on controversies surrounding development of the Condor II missile and the connection with Iraq; Carlos discusses Brazilian military exports to Iraq and role played by Brazilian military in training Hussein's forces; while Faúndes deals with role of Chile's Cardoen armaments industries and the mysterious death of an inquisitive British journalist.

Goodman, Louis Wolf and **Johanna S.R. Mendelson.** La amenaza de las nuevas misiones: las fuerzas armadas lationamericanas y la guerra contra la droga. See item **3819.**

4176 Hirst, Mônica. Continuidad y cambio del programa de integración Argentina-Brasil. Buenos Aires: Facultad Latinoamericana de Ciencias Sociales (FLACSO), 1990. 87 p.: bibl., tables (Documentos e informes de investigación; 108)

Updates and enlarges analysis presented in author's earlier study of the integration program (see item **4177**). Discusses results attained in economic, technological, transportation, and other areas, and examines projection of the program in Uruguay, Chile, and Paraguay.

4177 Hirst, Mônica. El programa de integración y cooperación Argentina-Brasil: los nuevos horizontes de vinculación económica y complementación industrial. Buenos Aires: Facultad Latinoamericano de Ciencias Sociales (FLACSO), 1990. 159 p.: appendix, bibl. (Documentos e Informes de Investigación; 81)

Excellent study examines domestic and international political and economic factors that led to formulation of the bilateral integration and cooperation program. Discusses characteristics of program's implementation in its first two years—with special reference to industrial cooperation—and briefly analyzes Uruguay's insertion into the project. Includes extensive and very useful statistical appendix. For update of this article see item **4176.**

4178 Jarpa, Sergio G. The defense of shipping off South America. (*Nav. War Coll. Rev.*, 43:3, Summer 1990, p. 62–76, maps, tables)

Although outdated in its references to potential threat posed by USSR to sea lanes of communication off the Americas, article—written by a Chilean rear admiral and chief of naval operations—rejects feasibility of a unified continental defense command. Discusses possibility of dividing responsibility for defense of shipping into six areas made up of ncighboring countries with similar interests while creating a general coordination committee composed of representatives of each area. For a US naval officer's viewpoint, see item **4185.**

4179 Mercado Jarrín, Edgardo. Un sistema de seguridad y defensa sudamericano. Lima: Centro Peruano de Estudios Internacionales, 1989. 240 p.: bibl. (Serie Investigaciones; 6)

Study of South American geopolitical situation in late 1980s with recommendations concerning creation of a regional security system is written by well-known Peruvian retired general, former commander-general of the army and prime minister during the Velasco Alvarado Administration. Offers critical analysis of decline of inter-American security system, and advances series of ambitious proposals—such as creation of a South American zone of peace and a regional security organization, military cooperation, establishment of a regional joint chiefs of staff group, non-proliferation, and the signing of a regional defense treaty—aimed at promoting peace, democracy, and economic development in a more independent context.

4180 Morris, Michael A. South American Antarctic policies. (*in* Ocean Yearbook 7. Edited by Elisabeth Mann Borgese, Norton Ginsburg, and Joseph R. Morgan. Chicago: Univ. of Chicago Press, 1988, p. 356–371, maps)

Brief and informative study relates South American positions concerning Antarctica. Analyzes central position Antarctica occupies in Argentine and Chilean ocean policies, discussing their similarities and differences and emphasizing that, although presenting potential for conflict, the countries' respective interests have been asserted pragmatically. Argues that Antarctica is likely to remain a secondary concern for other South American countries.

4181 Obando Arbulú, Enrique. Adquisición de armamentos y dependencia en América del Sur. Lima: Centro Peruano de Estudios Internacionales, 1990. 115 p.: bibl. (Documentos de trabajo; 13)

Brief and competent analysis, written from dependency perspective, examines importation of weapons by South American countries, importance of these purchases as part of military budgets, and domestic and external reasons for importation of these armaments.

4182 **Perl, Raphael F.** United States Andean drug policy: background and issues for decisionmakers. (*J. Interam. Stud. World Aff.*, 34:3, Fall 1992, p. 13–35, appendix, bibl., tables)

Competent work examines contents of President Bush's "Andean Initiative" to reduce drug activities in Bolivia, Colombia, and Peru. Provides balanced assessments, up to 1992, of soundness of this strategy, effectiveness of US leadership, degree of cooperation of the South American countries, and dangers associated with a single-issue foreign policy approach.

4183 **Sánchez-Gijón, Antonio.** La integración en la Cuenca del Plata. Madrid: Instituto de Cooperación Iberoamericana; Ediciones de Cultura Hispánica, 1990. 278 p.: bibl., index. (Sociología y política)

Excellent account and analysis of evolution of integration process in the La Plata basin examines historical and geopolitical roots of rivalry and cooperation between countries in the area. Also discusses the region's physical and human resources, the legal and political framework of the La Plata system, and features and impact of recent integration agreements between Argentina, Brazil, and Uruguay.

4184 **Seguridad democrática regional: una concepción alternativa.** Recopilación de Juan Somavía y José Miguel Insulza. Santiago: Comisión Sudamericana de Paz; Editorial Nueva Sociedad, 1990. 348 p.: appendix, bibl.

Knowledgeable collection of articles on different aspects of South America's regional security was assembled and published by the Comisión Sudamericana de Paz in 1990. First section deals with international insertion of South America and discusses political/strategic interests of the US, the former USSR, Europe, China, and Japan in the region. Second part discusses security interests shared by the South American countries. Third section analyzes civilian-military relations and role of the armed forces in preserving democratic regional security. Fourth section focuses on institutional aspects of regional security, particularly notion of a "peace zone" and importance of reciprocal trust measures. Appendix of this well-rounded and enlightening book includes proposals of the Comisión Sudamericana de Paz for strengthening democracy and formulating a more cooperative notion of regional security.

4185 **Stetson, Jeane H.** Defense of shipping in the Western Hemisphere: a second look. (*Nav. War Coll. Rev.*, 45:2, Spring 1992, p. 108–110)

Written by a commander in the US Naval Reserve, article agrees with some of the recommendations offered by Chilean Admiral Jarpa (see item **4178**), but advocates instead a more active use of existing cooperation mechanisms such as the plans for defense of inter-American maritime traffic, UNITAS, and joint naval games.

4186 **United States. Congress. House. Committee on Government Operations.** United States anti-narcotics activities in the Andean region: thirty-eighth report. Washington: U.S. G.P.O., 1990. 102 p.: bibl. (House report: 101st Congress, 2d session; 101–991)

Valuable report by joint delegation of members and staff of House sub-committees who traveled to Andean cocaine-producing countries (Bolivia, Colombia, and Peru) reviews impact of US anti-narcotics policy in the region and assesses prospects of President Bush's Andean Initiative to increase US activities in the area. After analyzing administration's anti-drug strategy in general and in each of the three countries, majority of delegation criticizes exclusive emphasis on military and law-enforcement responses to the problem and the meager results obtained. Recommends a counter-narcotic strategy focused on reducing economic dependency on coca in the region, promoting sustained and balanced economic growth, and strengthening democratic governments.

4187 **Vinhosa, Maria Celina Arraes.** Países de la Cuenca del Plata: una evaluación de la reciente relación comercial. (*Integr. Latinoam.*, 16:165, marzo 1991, p. 34–43, bibl., tables)

Informative analysis of evolution of commercial exchanges between La Plata Basin countries in 1980s. Documents existence of regional commercial disequilibria, pointing out the persistent Brazilian trade surpluses, the small volume of regional trade of the three smaller countries—Bolivia, Paraguay and Uruguay, and trend toward bilateral trade over regional integration. To solve some of these problems, author recommends implementing non-traditional financial policies such as a regional payments system, export financing measures, and increased aid to less developed countries in the Basin.

ARGENTINA

4188 Aja Espil, Jorge A. El mundo en la década del 80: testimonios de la vida internacional. Buenos Aires: Consejo Argentino para las Relaciones Internacionales, 1991. 480 p.: index.

Contains compilation of articles written between 1981–90 by former Argentine ambassador to US, and published in various newspapers. Useful for presenting viewpoint of noted member of Argentine diplomatic establishment on a number of issues including US, Soviet, and European politics, Latin American developments in 1980s, and Argentina's foreign policies since 1982.

4189 Antártica al iniciarse la década de 1990: contribución al 30 aniversario de la entrada en vigencia del Tratado Antártico. Recopilación de Calixto A. Armas Barea y Juan Carlos M. Beltramino. Buenos Aires: Ediciones Manantial, 1992. 314 p.

Contains useful compilation of essays delivered by Argentine specialists at a seminar on Antarctica in the 1990s organized by Consejo Argentino para las Relaciones Internacionales to commemorate 30th anniversary of Antarctic Treaty. Articles examine features and prospects of the Antarctic Treaty System as well as its impact on development of scientific, economic, touristic, ecological, and strategic activities on the continent. Brief but remarkably objective reference to Argentine role.

4190 Argentina en el mundo: 1973–1987. Edición de Rubén M. Perina y Roberto Russell. Buenos Aires: Grupo Editor Latinoamericano; Distribuidor exclusivo, Emecé Editores, 1988. 301 p.: bibl. (Col. Estudios internacionales)

Comprehensive description of Argentina's contemporary foreign policies and international relations illustrates analytical approaches favored by Argentine scholars. Articles examine study of international relations and policies in Argentina; evolution of country's foreign policies since independence; nature and contents of Peronist foreign policies from 1973–76; foreign policies pursued from 1976–83 by authoritarian regime; characteristics of foreign policies pursued by democratic regime since 1983; evolution of Argentine-Soviet relations; Argentina's approach to Latin American cooperation; Argentine nuclear policies and attitude toward non-proliferation; and nature and impact on foreign policies of Argentine territorial nationalism. Includes adequate bibliography on Argentine foreign policies.

4191 Arnaud, Vicente Guillermo. Política internacional argentina de protección del medio ambiente. (*Rev. Arg. Estud. Estrateg.*, 8:14, enero/dic. 1991, p. 83–92)

Brief but useful presentation examines main international initiatives taken by Argentina on environmental issues and Argentina's participation in relevant agreements.

4192 Barcelona, Eduardo and Julio Villalonga. Relaciones carnales: la verdadera historia de la construcción y destrucción del misil Cóndor II. Buenos Aires: Planeta, 1992. 252 p.: ill. (some col.), index. (Espejo de la Argentina)

Contains most complete journalistic account of development of missile industry in Argentina, its international connections, and negotiations that led to its termination under the Menem Administration.

4193 Bologna, Alfredo Bruno. Los derechos de la República Argentina sobre las Islas Malvinas, Georgias del Sur—San Pedro y Sandwich del Sur. Buenos Aires: Ediar, 1988. 293 p.: bibl.

Attempts to substantiate from legal, historical, political, and economic perspectives Argentina's sovereign rights over Falklands/Malvinas and dependencies. While no more convincing than similar efforts by other Argentine authors, work is valuable as a succinct review of historical and documentary evidence tending to support Argentine claims.

4194 Bologna, Alfredo Bruno. Dos modelos de inserción de Argentina en el mundo: las presidencias de Alfonsín y Menem. Rosario, Argentina: Centro de Estudios de Relaciones Internacionales de Rosario, 1991. 95 p.: bibl. (Cuadernos de política exterior argentina; 0326–7806: Informes sobre proyectos de investigación; 2)

Brief comparative analysis of Radical and Peronist foreign policy approaches examines their historical antecedents under Irigoyen and Perón. Highlights differences between Alfonsín's "globalistic" approach and Menem's emphasis on establishment of a "privileged relationship with the United States."

4195 Bouzas, Roberto and **Saúl Keifman.** Deuda externa y negociaciones financieras en la década de los ochenta: una evaluación de la experiencia argentina. Buenos Aires: FLACSO, 1990. 83 p.: bibl., tables (Documentos e informes de investigación; 98)

Concise and enlightening study of evolution of Argentina's foreign debt negotiations examines creation of the debt, successive attempts to manage the crisis, and roles played by international financial institutions, developed countries' governments, and commercial banks.

4196 Carasales, Julio César. El desarme de los desarmados: Argentina y el Tratado de No Proliferación de Armas Nucleares. Buenos Aires: Editorial Pleamar, 1987. 360 p.: bibl.

Analysis of origins and contents of Treaty on the Non-Proliferation of Nuclear Weapons is written by head of Argentina's delegation to Conference on Disarmament in Geneva (1981–85). Most of book is devoted to general analysis of the treaty, but last chapter briefly examines evolution of Argentine position in this regard.

4197 Carasales, Julio César. National security concepts of states: Argentina. New York: United Nations, 1992. 131 p.: bibl.

Excellent monograph describes and analyzes Argentina's basic security concepts. Examines legislation and issues, security policy decision-making actors and processes, and main features of actual policies implemented at regional and global levels. Special emphasis given to Argentine positions on disarmament.

4198 Cardoso, Oscar Raúl; Ricardo Kirschbaum; and **Eduardo van der Kooy.** Malvinas, la trama secreta. Nueva ed. ampliada, corr. y definitiva. Buenos Aires: Planeta, 1992. 526 p.: bibl., ill. (Espejo de la Argentina)

Updated edition of most complete journalistic account produced in Argentina of Falklands/Malvinas crisis and its outcome (see *HLAS 47:6508*). Excellent example of competent investigative journalism, this new edition adds section examining policies and initiatives implemented during last few months of military government and throughout Alfonsín and Menem Administrations up to reestablishment of diplomatic relations between Argentina and UK in February 1990.

4199 Caviedes, César N. Conflict over the Falkland Islands: a never-ending story? (*LARR*, 29:2, 1994, p. 172–187)

Review essay of major books published in Argentina, Great Britain, and the US reveals that controversy still rages on the interpretation of the Malvinas War outcome. Future sovereignty over the islands is still pending. While British hold that their ownership is unquestionable, Argentines maintain that they would continue to claim the islands, and that the adverse result of the war was only a temporary setback.

4200 Diamint, Rut Clara. Cambios en la política de seguridad: Argentina en busca de un perfil no conflictivo. (*Fuerzas Arm. Soc.*, 7:1, enero/marzo 1992, p. 1–16)

Informative work examines changes effected in Argentina's national security policies under Menem, including military reform and relations with civilian authorities as well as nuclear, aerospace, drug, and territorial policies. Underscores shift toward cooperative notions of security and alignment with US, but doubts that this would result in the amount of foreign aid and training expected by Argentine officials.

4201 Documentos sobre el conflicto argentino-chileno en la zona austral. Buenos Aires: Congreso de la Nación, Dirección de Información Parlamentaria, 1984. 608 p.: bibl., maps. (Serie Estudios e investigaciones; 1)

Valuable collection of documents relating to Argentine-Chilean border dispute over Beagle Channel includes chronology. Reproduces bilateral instruments approved by Argentina's Congress as well as those not ratified. Reproductions of arbitral decision of 1977, declarations issued and notes exchanged by Argentine and Chilean governments after 1977, and peace treaty signed in 1984 ending the dispute are also included.

4202 Ferrer Vieyra, Enrique. Segunda cronología legal anotada sobre las Islas Malvinas/Falkland Islands. Prólogo de Enrique de Gandía. Córdoba, Argentina Lerner, 1992. 643 p.: bibl.

Valuable annotated chronology and compilation of diplomatic documents found in Spanish, French, and British archives covers evolution of legal aspects of the Falklands/Malvinas dispute from first papal decrees of late 1400s to Argentine-British agreement of 1990. For historian's comment see *HLAS 54:2959.*

4203 García, Miguel V. Argentina en el Golfo. Buenos Aires: Editorial Pleamar, 1992. 294 p.: bibl., ill. (some col.). (Estrategia y política)

Chronicle of Argentine participation in Persian Gulf crisis of 1990–91 focuses mainly on role of Argentine Navy in blockade of Iraq. Includes brief account of domestic and international political repercussions of Menem Administration decision to join US-led coalition.

4204 Gough, Barry M. The Falkland Islands/Malvinas: the contest for empire in the South Atlantic. London; Atlantic Highlands, N.J.: Athlone Press, 1992. 212 p.: bibl., index, maps.

Readable and informative historical account examines origins of Falklands/Malvinas conflict between Great Britain and emerging Argentine Republic. Interesting references to US role in this early phase of the dispute.

4205 Granovsky, Martín. Misión cumplida: la presión norteamericana sobre la Argentina, de Braden a Todman. Buenos Aires: Planeta, 1992. 368 p.: bibl., ill., index. (Espejo de la Argentina)

Journalistic account of relations between Argentina and US is highly critical of Argentine shift toward pro-US alignment and cooperation. Particularly informative concerning relations established between US Ambassador Terence Todman and Alfonsín and Menem Administrations in late 1980s and early 1990s.

4206 Guadagni, Alieto Aldo. La reforma del Estado y la política exterior argentina. (*Integr. Latinoam.*, 16:173, nov. 1991, p. 3–13, table, graph)

The Secretario de Relaciones Económicas Internacionales of Argentina examines country's economic evolution in 20th century, pointing out that post-1930 decline was associated with an autarchic, protectionist, and State-led model that led to crisis of 1980s. Argues that market-oriented reforms implemented by Menem Administration were inevitable and will lead to important foreign policy consequences including trade liberalization, debt relief, regional and border integration, guarantees for foreign investment, and new credits.

4207 Klich, Ignacio. Perón, Braden y el antisemitismo: opinión pública e imagen internacional. (*Rev. Ciclos*, 2:2, 1992, p. 5–38)

Well-researched article assesses validity of accusations of antisemitism cast against Perón in 1940s by US Ambassador Spruille Braden. After examining evidence, concludes—without denying existence of antisemitic groups among Perón's followers—that accusations were unwarranted and resulted from Braden's distorted attacks on Perón.

4208 Landaburu, Carlos Augusto. La Guerra de las Malvinas. Buenos Aires: Círculo Militar, 1989. 676 p.: bibl. ill. (Biblioteca del oficial; 739)

Detailed military analysis examines Falklands/Malvinas conflict from the Argentine perspective. Useful mainly for readers interested in Argentine Army's official version of military operations on the islands and the "technical lessons" learned by army officers.

4209 Mack, Carlos. Der Falkland (Malvinas)-Konflikt: eine Konstellationsanalyse des britisch-argentinischen Konfliktes unter besonderer Berücksichtigung der argentinischen Entscheidung zur Invasion. Frankfurt am Main; New York: P. Lang, 1992. 296 p.: bibl. (Europäische Hochschulschriften. Reihe XXXI, Politikwissenschaft: 0721–3654; 191 = Publications universitaires européennes. Série XXXI, Sciences politiques; 191 = European university studies. Series XXXI, Political science; 191)

Despite complex analytical presentation of foreign and domestic agendas of Argentina, Great Britain, and the US, work provides a clear and interesting evaluation of the Falklands/Malvinas conflict. [C.K. Converse]

4210 Margheritis, Ana and **Laura Tedesco.** Malvinas: los motivos económicos de un conflicto. Buenos Aires: Centro Editor de América Latina, 1991. 145 p.: bibl., ill. (Biblioteca Política argentina; 342)

Notwithstanding title, work does not demonstrate existence of significant economic motivations leading to conflict over Falklands/Malvinas, at least until 1982 crisis. Nevertheless, offers useful appraisal of non-renewable and renewable resources found in the area, such as oil and fish, and emphasizes importance of Anglo-Argentine cooperation for efficient resource exploitation.

4211 **Menem, Carlos Saúl.** Estados Unidos, Argentina y Carlos Menem. San Isidro, Argentina: Editorial CEYNE, 1990. 205 p.

Comprehensive work presents Menem's ideas concerning relationships to be established between Argentina and US, and between Argentina and other western countries. Important for understanding reasoning behind shift from "third position" to western alignment in Peronist foreign policies.

4212 **Menem, Carlos Saúl.** Integración americana. San Isidro, Argentina: Editorial CEYNE, 1991. 239 p.: ill.

Collection of essays briefly introduces Menem Administration's foreign policy approach to the Western Hemisphere, particularly concerning topics such as Latin American integration, Mercosur, the Initiative for the Americas, and drug traffic.

4213 **Musacchio, Andrés.** La Alemania nazi y la Argentina en los años '30: crisis económica, bilateralismo y grupos de interés. (*Rev. Ciclos*, 2:2, 1992, p. 39–67)

Well-researched study examines little-known but important aspect of Argentina's international economic relations during 1930s. Analyzes features of bilateral trade and payments agreement of 1934 and discusses its impact on commercial relations. Shows how German interest in purchasing grains gave Argentina negotiating leverage necessary for increasing beef exports to German market, thereby solving some of the problems faced by local producers.

4214 **Muschietti, Ulises M.** *et al.* Conflictos en el Atlántico Sur: siglos XVII–XIX. Buenos Aires: Círculo Militar, 1988. 314 p.: bibl. (Biblioteca del oficial; 736)

Interesting collection of historical essays deals mostly with British and US interventions and hegemonic initiatives during Argentina's colonial and early independence period. Analyses are impaired by authors' legalistic and nationalistic viewpoints, but some articles offer interesting information on early aspects of Falklands/Malvinas controversy, British invasions of 1806–07, and British role in creation of Uruguay.

4215 **O'Donnell, Mario.** El descubrimiento de Europa: la Comunidad Europea y sus nuevas relaciones con la Argentina y el resto de Latinoamérica. Buenos Aires: Editorial Sudamericana, 1992. 203 p.

Brief but informative work examines European Union and current importance and prospects of its relations with Argentina and other Latin American countries. Contains excerpts of cooperation agreements signed between Argentina and EU, Spain, and Italy in 1980s.

4216 **Oliveri López, Angel M.** Malvinas: la clave del enigma. Buenos Aires: Grupo Editor Latinoamericano; Distribuidor exclusivo, Emece Editores, 1992. 256 p.: bibl., maps. (Col. Controversia)

Author of this interesting contribution to analysis of Falklands/Malvinas dispute is a former Argentine representative at UN, director of the Antarctica and Malvinas desk at the Argentine foreign relations ministry, and participant in negotiations with British until 1981. Examines British sources to validate thesis about weakness of Great Britain's rights over islands and to support Argentine claims. Recommends cooperative solution based on combination of partial Argentine control, British long-term lease-back arrangements, and joint binational administration on different islands and maritime zones.

4217 **Paradiso, José.** Debates y trayectoria de la política exterior argentina. Buenos Aires: Grupo Editor Latinoamericano, 1993. 212 p.: bibl. (Col. Estudios internacionales)

Insightful study by well-known international relations expert examines evolution of Argentina's foreign policy from 1860–1993. Excellent periodization and critical examination of Argentina's foreign policies underscores links between them and phases of country's political economic development, and relates influence of domestic actors in shaping policies. Welcome addition to the limited number of analytical studies of history of Argentina's international relations.

4218 **Pertusio, Roberto L.** Una marina de guerra: para hacer qué? Buenos Aires: Centro Naval, Instituto de Publicaciones Navales, 1989. 255 p.: bibl., ill. (Ediciones del Instituto de Publicaciones Navales; 85. Col. Ciencia y técnica; 18)

Retired rear admiral offers post-Malvinas assessment of role to be played by Argentine Navy. Useful for those interested in Argentine military perspective on importance of naval power and country's maritime interests. Includes detailed examination of Navy's resources. Interesting call for cooperation between Argentine, Brazilian, and Chilean navies in South Atlantic Ocean and Antarctic Sea.

4219 **La política exterior argentina en el nuevo orden mundial.** Edición de Roberto Russell. Buenos Aires: Facultad Latinoamericana de Ciencias Sociales; Grupo Editor Latinoamericano, 1992. 274 p.: bibl., ill. (Col. Estudios internacionales)

Excellent collection of articles and comments by noted Argentine and US specialists analyzes Argentina's foreign policies in post-Cold War period. Based on papers delivered at a FLACSO seminar in March 1992, work includes, among its high-quality contents, examinations of external context of Argentina's foreign policies (Roberto Russell); the new international situation and Argentine foreign policy prospects (Juan A. Lanús, Carlos Pérez Llana, and Atilio Borón); relations between changes in Argentina's political culture and its foreign policies (Carlos Escudé); and foreign policy and Argentine public opinion (Manuel Mora y Araujo *et al.*). Indispensable reference for anyone interested in evolution and prospects of Argentina's foreign policies in 1980s and 1990s.

4220 **Puig, Juan Carlos** *et al.* Malvinas hoy: herencia de un conflicto. Recopilación de Atilio Borón y Julio Faúndez. Buenos Aires: Puntosur Editores, 1989. 471 p.: bibl., maps.

Articles in this excellent collection were written by Argentine and British authors for 1988 seminar on Anglo-Argentine relations after South Atlantic conflict. Examines diplomatic, economic, strategic, legal, political, and cultural repercussions of the Falklands/Malvinas crisis on bilateral relations. Indispensable source for anyone interested in analyzing legacy of the conflict. For geographer's comment see *HLAS 53:2952.*

4221 **Rapoport, Mario** and **Claudio Spiguel.** Crisis económica y negociaciones con los Estados Unidos en el primer peronismo, 1949–1950: ¿un caso de pragmatismo? (*Rev. Ciclos,* 1:1, 1991, p. 65–116, graphs, tables)

Interesting and well-documented historical study of inconclusive Argentine-US negotiations of 1949–50 examines reasons they took place and failed. Highlights extent to which negotiations foretold bilateral rapprochement of 1952.

4222 **Rein, Raanan.** The Franco-Perón alliance: relations between Spain and Argentina, 1946–1955. Translated by Martha Grenzeback. Pittsburgh, Pa.: Univ. of Pittsburgh Press, 1993. 329 p.: bibl., index. (Pitt Latin American series)

Instructive study examines well-known but insufficiently analyzed ties between Perón's Argentina and Franco's Spain. Underscores importance of Argentina's political, diplomatic, and economic aid in strengthening Franco during his period of isolation between end of World War II and early 1950s, and highlights impact of domestic and international factors in rise and deterioration of bilateral cooperation. For historian's comment see *HLAS 54:3046.*

4223 **Thompson, Andrew.** Informal empire?: an exploration in the history of Anglo-Argentine relations, 1810–1914. (*J. Lat. Am. Stud.,* 24:2, May 1992, p. 419–436)

Provocative essay discusses application of notion of "informal empire" to relationship established between Great Britain and Argentina between 1810–1914. After examining extent of British indirect political hegemony and economic domination of Argentina, author concludes that "Britain's 'informal empire' in Argentina is in essence a myth."

4224 **Van Der Karr, Jane.** Perón y los Estados Unidos. Edición de Saad Chedid y Eduardo Machicote. Traducción de Gerardo Víctor Isler. Buenos Aires: Editorial Vinciguerra, 1990. 315 p., 8 p. of plates: bibl., ill.

Regrettably author died before completing her analysis of Argentina/US relations during Perón's first government. Nevertheless, book contains useful compilation of documents from the US Department of State dealing with bilateral relations from 1945–55. Particularly interesting are reports sent to Washington by officials at US Embassy in Buenos Aires.

4225 **Vázquez Ocampo, José María.** Política exterior argentina, 1973–1983: de los intentos autonómicos a la dependencia. Buenos Aires: Centro Editor de América Latina, 1989. 2 v.: bibl. (Biblioteca Política argentina; 261–262)

This Spanish translation of sections on Argentina from *Autoritarismo e Democracia na Argentina e Brasil: uma década de política exterior* (see *HLAS 53:4808*) provides well-written and thoughtful analysis of foreign policies of Peronist and military governments.

BOLIVIA

Aravena Ricardi, Nancy. Un corredor territorial para Bolivia: propuestas u opciones. See item **4241**.

Aravena Ricardi, Nancy. Un corredor territorial para Bolivia: ventajas y desventajas geopolíticas. See item **4242**.

4226 Balderrama G., Adalid. Mar boliviano. La Paz: Empresa Editora Urquizo, 1991. 250 p.: bibl., ill.

Historical and legal examination, from a Bolivian nationalistic perspective, of the secular problem of Bolivia's landlocked status. Includes a number of policy recommendations aimed at recovering Pacific territories lost to Chile in the late 19th century.

4227 Barrios Morón, Raúl. Bolivia: retórica y realidad de la diplomacia en linea directa, 1989–1990. (*Síntesis/Madrid*, 14, mayo/agosto 1991, p. 347–367)

Brief but informative analysis examines evolution of Bolivia's foreign policy up to late 1990. Argues that President Paz Zamora's personal diplomacy had mixed results: positive in economic terms, leading to favorable commercial and financial agreements; but negative in political/strategic terms as it failed to solve border problems and led to acceptance of agreements on drug traffic that promoted militarization and limited Bolivia's sovereignty.

4228 Barrios Morón, Raúl. ¿Buscando protagonismo? la política exterior boliviana en 1989. (*Estad. Soc.*, 6:7, 1990, p. 119–136)

Good review examines Bolivia's foreign policies during last months of Paz Estenssoro Administration and initial period of Paz Zamora's government. Emphasizes transition from a bureaucratic-diplomatic approach to a style characterized by presidential protagonism in elaboration and implementation of foreign policies.

4229 Bedregal Gutiérrez, Guillermo and **Ruddy Viscarra Pando.** La lucha boliviana contra la agresión del narcotráfico. La Paz: Editorial Los Amigos del Libro, 1989. 614 p.: bibl. (Col. Texto y documento)

Comprehensive account and exegesis of Bolivian foreign and domestic policies concerning drug production and traffic was written by the then minister of foreign relations and the director of ministry's drug office. Valuable examination from official perspective of legal, political, economic, and diplomatic initiatives taken in late 1980s by Bolivian government to deal with drug problem, as well as useful reproduction of governmental and diplomatic documents related to these issues.

Bedregal Gutiérrez, Guillermo and **Ruddy Viscarra Pando.** La lucha boliviana contra la agresión del narcotráfico. See item **3285**.

4230 Blanes J., José. Estado actual de la problemática fronteriza en Bolivia: elementos para su replanteo. (*in* Congreso Internacional sobre Fronteras en Iberoamérica Ayer y Hoy, *1st, Tijuana, Mexico, 1989.* Memoria. Edición de Alfredo Félix Buenrostro Ceballos. Mexicali: Univ. Autónoma de Baja California, 1990, v. 1, p. 207–261, bibl.)

Sound essay analyzes from political, economic, and social perspectives evolution of Bolivia's policies of territorial occupation with particular reference to the border areas. Evaluates main problems and makes some policy recommendations aimed at promoting Bolivian development. Persuasively argues that one of the main problems is centralistic nature of Bolivian development model. Advocates a decentralized model that acknowledges regional socioeconomic and political diversity.

4231 Brackelaire, Vincent. Coca, développement et coopération internationale en Bolivie. (*Rev. Tiers-Monde*, 33:131, juillet/ sept. 1992, p. 673–719, bibl.)

Interesting European analysis of attempts by Bolivian government, with international support, to eradicate coca production and replace it with alternative crops in specific regions. Points out failure of these narrow attempts, criticizing particularly militarization of production areas promoted by US. Recommends implementation of multilateral cooperation projects integrated into a broader program for Bolivian national socioeconomic development and strengthening of democracy.

4232 **Cajías, Lupe.** Las relaciones de Bolivia, Chile y Perú: el problema marítimo boliviano: guía mínima bibliográfica. La Paz: Ediciones CERID, 1992. 124 p.: index.

Useful annotated bibliography indicates contents and location of publications related to Bolivia's maritime problem available in La Paz's libraries.

Deustua Caravedo, Alejandro. El altiplano peruano-boliviano y el Lago Titicaca: proyección y alternativas internacionales. See item **4302.**

4233 **Guevara Arze, Walter.** Radiografía de la negociación del gobierno de las fuerzas armadas con Chile. La Paz: Librería Editorial Juventud, 1988. 285 p.

Second edition, with minor additions, of a text originally published in 1978 (see *HLAS 43:7422*). Analyzes failed 1975–77 negotiations between Banzer and Pinochet Administrations on possibility of granting to Bolivia an exit to the sea.

Hargreaves, Clare. Snowfields: the war on cocaine in the Andes. See item **3309.**

Lanza, Gregorio. Relaciones Bolivia–EE.UU.: militarización y anexos firmados en Washington. See item **3312.**

4234 **Montenegro, Walter.** Oportunidades perdidas: Bolivia y el mar. La Paz: Editorial Los Amigos del Libro, 1987. 243 p.: bibl., ill. (Col. Texto y documento)

Refreshing attempt by a Bolivian diplomat to study perennial issue of Bolivia's outlet to the sea. Self-critical perspective aims at examining—and preventing repetition—of political and diplomatic mistakes that have made it impossible for Bolivia to secure a satisfactory solution. Makes well-founded criticism of Bolivia's negotiating shortcomings (impatience and break of diplomatic relations, emphasis on sovereignty, lack of flexibility and improvisation), and offers reasonable recommendations aimed at long-term solution (existence of a stable and popular Bolivian government, confidential preparatory contacts with Chile, adequate assessment and guidance of local public opinion, search for Peruvian consent, emphasis on economic integration, etc.).

Morelli Pando, Jorge. Los acuerdos de Ilo en el marco de las relaciones del Perú con Bolivia. See item **4309.**

4235 **Ondarza Linares, Franz.** Bolivia y la integración: problemas de ayer y de hoy. La Paz: Graficolor, 1989. 493 p.: bibl.

Compilation of informative and well thought-out articles, speeches, and papers examines challenges and opportunities that economic integration in LAFTA, Andean Group, and bilateral level agreements between neighboring countries pose for Bolivia. Author—a prominent Bolivian economist, diplomat, and Christian Democratic politician—stresses failure of State-centered and protectionist integration processes. Recommends that new attempts consider current regional prevalence of market-oriented and liberal economic policies.

4236 *Opiniones y Análisis.* No. 17, julio 1993- . Seminario Internacional "Bolivia y sus Relaciones con los Países Limítrofes." La Paz: Fundación Boliviana para la Capacitación Democrática y la Investigación.

Contains useful compilation of lectures delivered in 1993 by Bolivian, Argentine, Brazilian, Chilean, and Peruvian politicians and diplomats at a seminar organized by the Comisión de Política Internacional of the Cámara de Diputados of Bolivia's Congreso Nacional. Essays examine economic and diplomatic bilateral relations from different national perspectives. Appendix reproduces Bolivia's foreign service law promulgated in 1993.

4237 **Paz Zamora, Jaime.** La diplomacia de la coca. (*Nueva Soc.*, 124, marzo/abril 1993, p. 168–172)

Brief essay by Bolivian president discusses problems associated with coca production in Bolivia, making clear distinction between coca and cocaine. Using economic, political, and cultural arguments, recommends international diplomatic negotiations aimed at legalizing production and commercialization of the leaf while curbing elaboration and traffic of cocaine.

Pinochet de la Barra, Oscar. Puerto para Bolivia?: centenaria negociación. See item **4259.**

4238 *Política exterior de Bolivia.* La Paz: Ministerio de Relaciones Exteriores y Culto, 1989. 96 p.: ports.

Brief but comprehensive presentation examines main features of Bolivia's foreign policy during Paz Estenssoro Administration. Analyzes contents of bilateral relations with numerous countries, Bolivian positions in different international organizations, and country's approach to a number of issues such as narcotraffic, trade, environment, and foreign debt.

4239 Salazar Paredes, Fernando. Tierra de hombres libres: temas de política internacional boliviana. La Paz: Ediciones CERID, 1988. 489 p., 24 p. of plates: ill., indexes.

Includes speeches, statements, and lectures delivered by author during his tenure as Bolivian ambassador to OAS appointed by Siles Zuazo Administration. Useful presentation and explanation of Bolivian positions concerning topics such as inter-American system, exit to the sea, drug traffic, economic relations, and democratization in period 1982–85.

4240 Vázquez Machicado, Humberto. Para una historia de los límites entre Bolivia y el Brasil. La Paz: Librería Editorial Juventud, 1990. 528 p.: bibl., ill., maps.

Historical study of origins and evolution of border dispute between Bolivia and Brazil focuses on early phases of the controversy (1825), negotiations of 1860s, and treaty of 1867. Also examines problems related to demarcation of the border in Rio Verde region in which author participated as chief of the Bolivian delegation to joint Brazilian-Bolivian boundary commission. Although it does not examine conflict of 1899–1903, work is important due to scarcity of Bolivian sources on this topic.

Verdesoto, Luis and **Gloria Ardaya Salinas.** Entre la presión y el consenso: escenarios y previsiones para la relación Bolivia–Estados Unidos. See item **3354**.

CHILE

4241 Aravena Ricardi, Nancy. Un corredor territorial para Bolivia: propuestas u opciones. (*Rev. Chil. Geopolít.*, 6:1, dic. 1989, p. 17–28, maps)

Briefly analyzes, from Chilean perspective, successive proposals and negotiations aimed at granting Bolivia an exit to the Pacific. Argues that these attempts failed due to contradictory Bolivian attitudes and Peruvian refusal to make concessions. Recommends establishment of trilateral joint commission to study the problem and design a mutually acceptable formula. Although biased in its judgment, offers useful summary of Chilean approach to this perennial problem. See also item **4242**.

4242 Aravena Ricardi, Nancy. Un corredor territorial para Bolivia: ventajas y desventajas geopolíticas. (*Rev. Chil. Geopolít.*, 6:2, abril 1990, p. 19–38, tables)

Supplements author's previous work (see item **4241**) and assesses in regional context the national advantages and disadvantages associated with concession to Bolivia of an exit to the sea through Chilean territory. After examining positions and proposals advanced by Argentina, Brazil, Bolivia, Peru, and Chile, argues that regional integration plans supported by Argentina and Brazil use Bolivia to conceal their expansionistic and hegemonic intentions toward the Pacific nations, and that for Chile and Peru—as well as for Bolivia—it would be more advantageous to reach a tripartite agreement on a territorial corridor.

4243 Armanet Armanet, Pilar. Política de Chile en la Cuenca del Pacífico: perspectivas para la decada del noventa. (*Estud. Int.*/Santiago, 25:97, enero/marzo 1992, p. 41–72, tables)

Brief but comprehensive account of Chile's economic and diplomatic relations with Pacific Basin countries and regional organizations. Includes examination of prospects for commercial and political developments in light of democratization process, and assesses domestic obstacles to be overcome in order to promote a more beneficial association.

4244 Balderrama G., Adalid. El litoral boliviano y los postulados geopolíticos de Augusto Pinochet. La Paz: Empresa Editora Urquizo, 1991. 183 p.: bibl., ill., maps.

Forceful and polemical work refutes Gen. Pinochet's geopolitical views disclosed in his book *Geopolítica* (see *HLAS 39:8877*), particularly his observations on Bolivia's lack of rights to territories in the Pacific.

Benavides Correa, Alfonso. Una difícil vecindad: los irrenunciables derechos del Perú en Arica y los recusables acuerdos peruano-chilenos de 1985. See item **4299**.

4245 Chile en la Cuenca del Pacífico: experiencias y perspectivas comerciales en Asia y Oceanía. Edición de Sergio Valdivieso Eguiguren y Eduardo Gálvez Carvallo. Santiago: Editorial Andrés Bello; Editorial Jurídica de Chile, 1989. 289 p.: bibl., ill.

Valuable and comprehensive collection of articles presented at a 1988 seminar sponsored by the Chilean Ministerio de Relaciones Exteriores and the Oficina de Planificación Nacional (ODEPLAN). Well-known economists and international relations experts from Chile and abroad analyze role of Pacific Basin in the international economic order, development strategies pursued by countries in the area, commercial experiences and exchanges between Chile and the Asian countries and Oceania, and investment and financial relations and opportunities in the Basin.

4246 Chile: política exterior para la democracia. Edición de Heraldo Muñoz. Santiago: Pehuén, 1989. 262 p.: bibl. (Ensayo)

Essays by respected Chilean social scientists examine country's position in the international arena and advance a number of recommendations aimed at designing a new foreign policy for the nascent democratic regime. Particularly valuable for understanding international challenges faced by the democratic government, and for well thought-out proposals for reinsertion of Chile at global and regional levels. Discusses foreign policy decision-making process, security matters, and economic interactions.

4247 Chile y Argentina: nuevos enfoques para una relación constructiva. Dirección de Francisco Orrego Vicuña. Santiago: Pehuén, 1989. 112 p. (Ensayo)

Collection of essays explores prospects for more cooperative diplomatic, political, and economic relations between Chile and Argentina. Optimistic assessment of opportunities for collaboration and integration is based on endorsement of the Tratado de Paz y Amistad of 1984 and existence of democratic regimes in both countries.

Documentos sobre el conflicto argentino-chileno en la zona austral. See item **4201**.

4248 Durán, Roberto. A política exterior chilena no primeiro ano da transição democrática: balanço e perpectivas imediatas. (*Cont. Int.*, 13:1, jan./junho 1991, p. 81–93)

Analysis of features of Chile's foreign policy in 1990 highlights development of closer ties with US, the EC, and Latin American countries, as well as Chile's more active participation in international organizations. Argues that Aylwin Administration intended to revive pragmatic foreign policy tradition followed until 1973 and that this task will require the professionalization and modernization of the country's foreign service.

4249 Fermandois Huerta, Joaquín. De una inserción a otra: política exterior de Chile, 1966–1991. (*Estud. Int./Santiago*, 24:96, oct./dic. 1991, p. 433–455)

Analysis of evolution of Chile's foreign policies argues that after 17 years of political isolation under Pinochet the new democratic government is promoting country's international reinsertion by resuming some of its pre-1973 features—domestic consensus, international pragmatism, regional rapprochement, and cordial relations with US—while preserving some post-1973 components, particularly regarding economic foreign policies and country's low diplomatic profile.

4250 González Abuter, Tulio. Negociaciones chileno-argentinas de límites, 1871–1881: historia de una década. Santiago: Instituto de Investigaciones del Patrimonio Territorial de Chile, Univ. de Santiago, 1988. 168 p.: bibl., ill., index. (Col. Terra nostra; 13)

Polemical but not completely convincing work attempts to demonstrate that, due to Chilean elites' narrow focus on northern border conflicts and lack of interest in southern lands disputed with Argentina, Chile's valid legal claims to those territories were not adequately pursued by late 19th-century Chilean governments.

Guevara Arze, Walter. Radiografía de la negociación del gobierno de las fuerzas armadas con Chile. See item **4233**.

4251 Hacia el siglo XXI: la proyección estratégica de Chile, Edición de Augusto Varas. Santiago?: Facultad Latinoamericana de Ciencias Sociales, 1989. 275 p.;

Valuable collection of articles by well-known Chilean civilian and military specialists examines country's strategic position and options in changing international environment of late 20th century. First part focuses on Chile's strategic relations with US, Europe, and Southern Hemisphere. Second part deals with prospects for specific aspects of Chile's strategic policies including national objectives, commercial relations, Easter Island and the Pacific, Antarctica, and South American military coordination.

4252 Latorre, Adolfo Paul. La frontera marítima austral: perspectiva de conflicto. (*Rev. Chil. Geopolít.*, 6:1, dic. 1989, p. 29–40, bibl., maps)

Polemical essay that, after probing what author considers Argentine tradition of territorial expansionism, argues that bilateral Peace and Friendship Treaty signed in 1984 does not guarantee elimination of tensions between Argentina and Chile. Useful review, from Chilean perspective, of potential for conflictive situations on the southern maritime boundary as a result of ambiguities contained in the treaty.

4253 Maldonado Prieto, Carlos. "La Prusia de América del Sur:" acerca de las relaciones militares chileno-germanas, 1927–1945. (*Estud. Soc./Santiago*, 73: trimestre 3, 1992, p. 75–102)

Interesting study examines links established between Chilean and German military from late 1920s to mid-1940s, and resulting spread of Fascist ideology among a significant group of Chilean officers. Shows that bilateral relations were only temporarily suspended as a consequence of World War I, and discusses role played by Gen. Carlos Ibañez in promoting resumption of military visits, training, and sales. Also explores close ties between Chilean military-led nationalistic groups and German Nazi officers that facilitated propagation of fascism within Chile's armed forces.

4254 Matta, Javier Eduardo. Chile y la República Popular China, 1970–1990. (*Estud. Int./Santiago*, 24:95, julio/sept. 1991, p. 347–367)

Informative article examines an often-mentioned but seldom-researched aspect of Chile's international relations. Analyzes evolution of bilateral ties throughout the Allende and Pinochet governments. Particularly interesting is examination of the close diplomatic relations between Beijing and Santiago after 1973 based on common hostility toward USSR and growth in commercial exchanges resulting from the pragmatic search for economic gains in the Pacific Basin.

4255 Meneses Ciuffardi, Emilio. El factor naval en las relaciones entre Chile y los Estados Unidos, 1881–1951. Santiago: Ediciones Pedagógicas Chilenas; Librería Francesa, 1989. 229 p.: bibl., ill. (Col. Histo-Hachette)

Historical study examines evolution of Chilean and US naval policies and their impact on relations between both countries. Advances hypothesis that a direct correlation existed between Chilean resistance to US naval hegemony and degree of bilateral hostility in the political-diplomatic sphere between 1880s and 1929. Argues that hostility was replaced by cooperation only as US naval supremacy was established and Chilean Navy was relegated to a dependent position, culminating in the signing of the Military Aid Pact of 1952.

4256 Morandé Lavín, José. Chile y los Estados Unidos: distanciamientos y aproximaciones. (*Estud. Int./Santiago*, 25:97, enero/marzo 1992, p. 3–22, table)

Essay on evolution of Chilean/US relations highlights primacy of conflictive aspects throughout their history as a consequence of political, economic, and cultural factors. Argues nevertheless that new conditions generated by Chile's democratization, the end of the Cold War and the US Initiative for the Americas have facilitated rise of a realistic approach that may lead to long-term improvement in the bilateral relationship.

4257 Orrego Vicuña, Francisco. Decisión de la Comisión para la Solución de Controversias entre Chile y Estados Unidos sobre el Caso Letelier Moffit, 11 de enero de 1992. (*Estud. Int./Santiago*, 25:97, enero/marzo 1992, p. 137–153)

Includes text of the majority decision together with Orrego Vicuña's concurring opinion concerning agreement between Chile and US for resolution of the dispute resulting from 1976 assassination of Letelier and Moffit by Chilean government agents. Although refusing to accept formal responsibility for the deaths, the Aylwin Administration representatives agreed to pay compensation to the victims' relatives and to the survivor of the attack in order to facilitate normalization of bilateral relations.

**4258 Osobennosti razviti︠a︡ chiliĭskoĭ ėkonomiki i perspektivy rossiĭsko-chiliĭskogo delovogo sotrudnichestva. [Features of the development of the Chilean economy and perspectives on Russian-Chilean economic relations.] Redaktsionnaia kollegiia, Nikolaĭ Grigor'evich Zaĭtsev, Pavel Nikolaevich Boĭko, Lev L'vovich Klochkovskiĭ. Moscow: Rossiĭskaia akademiia nauk, In-t Latinskoĭ Ameriki, 1992. 72 p.: tables.

Produced by researchers at the Institute of Latin American Studies of the newly renamed Russian Academy of Sciences, this monograph begins with an overview of the history of Russian-Chilean relations since the early 19th century. Follows with an examination of virtually all areas of the Chilean economy and opportunities for economic cooperation between the new Russian Federation and Chile, including in the Asian-Pacific region. Tables on Chilean industrial production, imports, exports, foreign investment. [B. Dash]

4259 Pinochet de la Barra, Oscar. Puerto para Bolivia?: centenaria negociación. Santiago: Editorial Salesiana, 1987. 112 p.: bibl., maps.

Brief and serviceable historical review of evolution of Chilean-Bolivian conflictive relations since the War of the Pacific focuses on unsuccessful negotiations aimed at giving Bolivia an outlet to the Pacific. More objective than other analyses of this topic offered by Chilean and Bolivian authors.

4260 Prieto Vial, Daniel. Defensa Chile 2000: una política de defensa para Chile. Santiago: FLACSO, 1990. 290 p.: bibl., ill. (Serie Libros FLACSO Chile)

Comprehensive set of proposals for modernization of Chilean armed forces departs from assumption that a South American coordinated defense policy would enhance national and collective security and strengthen democratic regimes in the region.

4261 Relacje Polska-Chile: doświadczenia i stan obecny. [Polish-Chilean relations: past and present.] Warszawa: Centrum Studiów Latynoamerykańskich, Uniwersytet Warszawski, 1992. 11 p.: bibl., map (Dokumenty robocze; 6)

Working paper prepared by Chilean and Polish scholars is divided into four sections: 1) Stocks provides current information on Chile's geography, population, and economy; 2) Smolana outlines Polish-Chilean relations from 1838 arrival of Polish engineer Domeyho through 1992; covers Polish immigrant organizations in Chile dating from 19th century; and examines diplomatic relations between the two countries, providing lists of representatives of both countries beginning with 1920s (when Poland regained its independence after 123 years of foreign occupation); 3) Kinast discusses politico-economic relations since 1920s; and 4) Norambuena reviews economic reforms undertaken in Chile beginning in 1950s. Work includes select bibliography of works on Chile published in Poland. [Z. Kantorosinski]

4262 Robledo, Marcos. Las relaciones Chile-Estados Unidos en el ámbito militar. (*Fuerzas Arm. Soc.*, 6:2, abril/junio 1991, p. 5–25)

Good analysis of recent military relations between Chile and US, particularly since reestablishment of democracy. Examines new features of security policies implemented by both countries in 1990s; discusses current agenda; and ponders prospects for relations concerning issues such as defense of democracy, military equipment and training, joint operations, drug traffic, regional conflicts, Inter-American security, and nuclear proliferation.

4263 Santis Arenas, Hernán. La naturaleza de la solución del conflicto entre Chile y Argentina por la delimitación de la frontera marítima, 1984. (*Rev. Chil. Geopolít.*, 6:1, dic. 1989, p. 41–54, bibl., maps)

Informative account and analysis examines Argentine-Chilean maritime boundary conflict in Beagle Channel area, including historical background, evolution of negotiations mediated by the Vatican, and contents of the agreement reached in 1984. To his credit, author, a Chilean expert, avoids nationalistic bias characteristic of most studies of the topic by authors from disputing countries.

4264 Wilkinson, Michael D. The Chile Solidarity Campaign and British government policy towards Chile. (*Rev. Eur.*, 52, June 1992, p. 57–74)

Studies role played by the lobbying group Chile Solidarity Campaign under friendly Labour governments (1974–79) and antagonistic Conservative administrations (1979–90) in shaping British policy towards Pinochet Administration. Concludes that group's limited success in isolating Chile resulted from not only Conservative opposition, but also Labour's refusal to cancel contracts as well as Campaign's organizational and financial weaknesses and political inexperience.

COLOMBIA

4265 **Ardila Ardila, Martha.** ¿Cambio de norte?: momentos críticos de la política exterior colombiana. Bogotá: Univ. Nacional de Colombia, Instituto de Estudios Políticos y Relaciones Internacionales; Tercer Mundo Editores, 1991. 252 p.: bibl.

Informative study relates historical evolution of Colombia's foreign policies from interdependency perspective. After discussing general characteristics and periods, work focuses on analysis of international strategies pursued by administrations of Suárez (1918–21), López Pumarejo (1942–45), Valencia (1962–66), and Betancourt (1982–86). Argues that following Suárez Administration, Colombia's foreign policy adopted an "active subordination" configuration in which general alignment with US has been partially counteracted by some autonomous initiatives.

4266 **Bagley, Bruce Michael.** Dateline Drug Wars: Colombia—the wrong strategy. (*Foreign Policy*, 77, Winter 1989/90, p. 154–171)

Analysis of failure of Colombian and US initiatives aimed at eradicating drug traffic was written in aftermath of assassination of presidential candidate Luis Galán. Criticizes overwhelming emphasis by US on military repressive actions aimed at suppliers. Recommends devoting more resources to demand-reduction programs in US and economic and institution-building assistance for Colombia.

4267 **Botero, Ana Mercedes.** Un aporte interpretativo en torno a las drogas y su papel en las relaciones entre Colombia y Estados Unidos. Bogotá: Centro de Estudios Internacionales de la Univ. de los Andes, 1989. 29 p.: bibl. (Serie Documentos ocasionales; 12)

Brief essay analyzes Colombia's foreign policy concerning drug problem, particularly in relation to US. After discussing Colombia's foreign policy decision-making process and evolution of its anti-drug policies, concludes that one of the main problems is lack of a long-term, balanced, consistent, and well-informed governmental policy.

4268 **Cardona, Diego et al.** Colombia y la integración americana. Bogotá: CLADEI; FESCOL, 1992. 135 p.: bibl.

Collection of articles is devoted to analysis of current role and future prospects of Colombia in process of hemispheric integration. Particularly interesting are articles by Diego Cardona on harmonization of interests and activities of the Grupo de Los Tres (Colombia, Mexico, and Venezuela) and by Alfredo Fuentes on commerce and integration in the Grupo.

4269 **Crisis y fronteras: relaciones fronterizas binacionales de Colombia con Venezuela y Ecuador.** Recopilación de Cristina Barrera. Bogotá: Ediciones Uniandes, 1989. 439 p.: bibl., ill. (Serie Historia contemporánea y realidad nacional; 24)

Excellent collection of academic articles analyzes border interactions between Colombia and Ecuador and Venezuela from the perspective of promoting integration and development of the frontier regions. Contains well-researched examinations of impact of national macroeconomic policies on the frontier areas; Colombian frontier development policies; territorial, migration, and guerrilla problems; and characteristics and regional consequences of commercial flows.

4270 **Fernández de Soto, Guillermo.** Reflexiones sobre política internacional. Bogotá: Editorial Kimpres, 1992. 348 p.

Articles and lectures written by a Colombian diplomat, professor, and journalist during 1980s and early 1990s examine different aspects of international politics. Particularly valuable are sections devoted to Colombia's insertion in the international arena as well as those focused on human rights policies, Contadora, and relations between Colombia and Venezuela, issues in which author was personally involved as a member of Colombia's foreign service.

4271 **Grosse, Robert E.** The economic impact of Andean cocaine traffic on Florida. (*J. Interam. Stud. World Aff.*, 32:4, Winter 1990, p. 137–159, bibl., maps, tables)

Well-documented study examines economic costs and benefits to state of Florida resulting from Colombian cocaine traffic. Estimates costs associated with this activity in terms of crime protection/law enforcement, medical expenses, and productive loss, as well as economic gains resulting from drug-related money inflows. Concludes that net economic impact is quite negative—at least $1.5 billion in 1988.

4272 Jaramillo Correa, Luis Fernando. La política exterior colombiana. (*Rev. Cancillería San Carlos*, 9, agosto, p. 3–31)

Useful review by Colombia's foreign relations minister surveys international policies pursued by Gaviria Administration in its first year. Examines relations with other Latin American countries, the European Community, the US, the Vatican, and several international organizations. Also discusses drug enforcement, environmental, and human rights initiatives.

4273 Murillo Castaño, Gabriel and **María Victoria Llorente Sardi.** Las relaciones colombo-venezolanas contemporáneas. (*in* Congreso Internacional sobre Fronteras en Iberoamérica Ayer y Hoy, *1st, Tijuana, Mexico, 1989*. Memoria. Edición de Alfredo Félix Buenrostro Ceballos. Mexicali: Univ. Autónoma de Baja California, 1990, v. 1, p. 289–319, bibl.)

Useful and balanced study examines main issues affecting Colombian/Venezuelan relations in recent years, including boundary dispute in the Gulf of Venezuela and development of economic relations and integration initiatives. Underscores negative influence of political/territorial considerations on economic relationship, and suggests suitability of adopting a pragmatic position concerning the former while depoliticizing the latter.

4274 Obregón T., Liliana and **Carlo Nasi L.** Colombia-Venezuela: conflicto o integración. Bogotá: Fundación Friedrich Ebert de Colombia; C.E.I., Uniandes, 1990. 134 p.: bibl., maps.

Concise study analyzes Colombian and Venezuelan territorial conflicts and steps taken toward cooperation and integration since late 1980s with creation of bilateral commissions to deal with frontier issues of common interest.

4275 Pardo, Rodrigo and **Juan Tokatlian.** Política exterior colombiana: ¿de la subordinación a la autonomía? Bogotá: Ediciones Uniandes; Tercer Mundo Editores, 1988. 237 p.: bibl. (Col. Uniandes 40 años. Relaciones internacionales)

Enlightening essays by two eminent Colombian specialists examine evolution and features of Colombia's foreign policies. Following initial theoretical chapter, work presents detailed annotated bibliography on Colombia's foreign policies and briefly examines its historical evolution from 1914–87. Also discusses interaction between external and domestic factors under Turbay, Betancourt, and Barco Administrations; reviews Colombia's role in Central American crisis (1978–86); and analyzes in detail foreign policies of the Barco Administration. Indispensable for anyone interested in Colombia's past and current foreign policies.

4276 Pedraza Gallardo, Consuelo. Bibliografía del Ministerio de Relaciones Exteriores de Colombia y antecedentes de su organización y función. v. 1–2. Bogotá: Ministerio de Relaciones Exteriores, 1991. 2 v.

Vol. 1 of this official publication details functions, evolution, and legislative history of the Ministerio de Relaciones Exteriores, legislative history of the foreign service, and a historical bibliography of memoirs sent to the Congreso Nacional. Vol. 2 contains a bibliography—organized chronologically and by titles and themes—of all materials published by the Ministerio and its predecessors from 1828–1990.

4277 Ramírez León, José Luis et al. Colombia y América Latina frente a Europa 1992. Bogotá: CLADEI; CEI, UNIANDES; FESCOL, 1991. 235 p.: bibl.

Contains four articles analyzing Colombian and Latin American relations with European Union at the threshold of 1993 European unification. Particularly useful are articles by Ramírez León examining evolution of Colombia's political/diplomatic and economic relations with the EU and by Marta Osorio analyzing in detail the features and prospects of bilateral commercial relations.

4278 Ramírez Ocampo, Augusto. Colombia y su política internacional. Bogotá: Fotolito 24 Horas, 1985. 115 p.: col. ill.

Collection of speeches delivered by Colombia's former minister of foreign relations at different international forums including UN General Assembly, Conference of the Nonaligned Movement, OAS, and the Grupo Andino. Informative presentation of Colombia's positions concerning border issues, the Central American crisis, and reorganization of the OAS.

4279 Tokatlian, Juan and **Diego Cardona.** Desafios e dilemas: segurança, cenário internacional e política externa colombiana. (*Cont. Int.*, 13:1, jan./junho, p. 95–110)

Analysis of features and prospects of Colombia's international relations in post–Cold War international order highlights modernization of Colombia's foreign service, economic policies, and regional integration efforts. Also examines influence of domestic conditions such as drug traffic, dialogue with guerrilla forces, and development of frontier areas on country's foreign policies.

4280 Vázquez Cobo, Alfredo. Pro patria: la expedición militar al Amazonas en el Conflicto de Leticia. Bogotá: Banco de la República, Depto. Editorial, 1985. 455 p.: bibl., ill., index. (Col. Banco de la República)

Interesting memoir is written by Colombian ambassador to France (1927–34) and chief of military expedition to recover the "Leticia Trapeze" in 1932–33. Clarifies origins and evolution of Colombian/Peruvian territorial dispute in Amazonian region, while throwing light on Colombia's domestic political circumstances that facilitated a negotiated solution to the conflict.

ECUADOR

4281 A 50 años del protocolo de Rio de Janeiro: opiniones de actualidad. Coordinación de Nelson Gómez E. y Juan Paz y Miño C. Quito: Sección de Historia y Geografía de la Casa de la Cultura, 1991. 117 p.: bibl., maps. (Estudios de la Sección de Historia y Geografía de la Casa de la Cultura; 1)

Essays by Ecuadorian academics published on 50th anniversary of the Protocol of Rio examine the dispute with Peru. Although contributors agree on considering the Protocol null and void, they tend to emphasize need for a peaceful resolution of the dispute in the framework of bilateral diplomatic negotiations and multilateral integration programs.

4282 Altamirano Escobar, Hernán Alonso. El por qué del ávido expansionismo del Perú. Quito: Instituto Geográfico Militar, 1991. 397 p.: bibl., maps (some col.).

Author, an Ecuadorian army captain, engages in a violent denunciation of Peru's expansionistic aims, attributing Peru's attacks on Ecuador to a mix of imperialistic ambitions and frustration resulting from military defeats at the hands of other Latin American countries. Highly prejudiced version of the bilateral confrontation but interesting source as it appears to have endorsement of senior military officers and Ministry of Education; thus, work reflects a widely-held Ecuadorian hostility toward Peru.

Bustamante, Fernando. Ecuador: putting an end to ghosts of the past? See item **3215.**

4283 Carrera de la Torre, Luis. El proyecto binacional Puyango-Tumbes. Quito: AFESE; ILDIS, 1990. 242 p.: ill., maps.

Analyzes from historical, diplomatic, hydrographic, and legal perspectives the agreement signed in 1971 between Ecuador and Peru for utilization of binational basin of the Puyango-Tumbes rivers. Author, an Ecuadorian participant in the bilateral negotiations, defends the agreement, emphasizing that it does not affect territorial rights, but rather creates conditions for cooperation, integration, and peaceful resolution of the territorial dispute in the area.

4284 Ecuador y Perú, vecinos distantes: seminario organizado por la Corporación de Estudios para el Desarrollo (CORDES), del 7 al 10 de diciembre de 1992, Quito, Ecuador. Quito: CORDES, 1993. 451 p.: bibl.

Papers delivered at a seminar on Ecuadorian-Peruvian relations intentionally leave aside discussion of the territorial dispute, and focus instead on analysis of themes such as prospects for collective security, frontier integration, public opinion, economic cooperation, historical and sociocultural commonalities and foreign policy convergence. As a result, volume not only fills a vacuum in a literature fixated on territorial problems, but offers a realistic appraisal of prospects for political, economic, and security cooperation.

4285 Guzmán Polanco, Manuel de. Ecuador en lo internacional: un cuarto de siglo, 1945–1970. (*in* El Ecuador de la postguerra: estudios en homenaje a Guillermo Pérez Chiriboga. Quito: Banco Central del Ecuador, 1992, v. 2, p. 411–457)

Historical review of Ecuador's foreign policies from end of World War II to 1970 focuses mostly on relations with Latin American countries and Ecuadorian role in interamerican organizations and the UN. Although analysis seemingly overstates impact and correctness of Ecuador's initiatives, article is useful as a brief survey of the most important foreign policy developments during this period.

4286 **Krestianinov, Vladimir.** Pedro Antonio Saad—"Las perspectivas sí, hay, y son muy prometedoras:" entrevista con el Embajador Extraordinario y Plenipotenciario de la República del Ecuador en la URSS. (*Am. Lat. / Moscow*, 12:168, dic. 1991, p. 4–11, ill.)

In informative interview, Ecuadorian ambassador to Moscow discusses features and prospects of bilateral relations and each country's domestic situation.

4287 **Rengel, Jorge Hugo.** La cuestión Puyango-Tumbes. Loja, Ecuador: Editorial La Emancipada, 1991. 149 p.

Critically assesses 1971 Ecuadorian-Peruvian agreement for utilization and development of the Puyango-Tumbes river basin. Considers that, particularly following bilateral exchange of reversal notes in 1985, Ecuadorian administration not only was resigning its sovereign rights over territory disputed with Peru, but also was promoting a project that would have negative economic consequences for the local population.

4288 **Seminario Internacional Relaciones Ecuador-Comunidad Europea 1992,** *Guayaquil, Ecuador, 1990.* Realidades y perspectivas. Edición y elaboración de Galo A. Chiriboga Zambrano y Rafael Urriola. Quito: Instituto Latinoamericano de Investigaciones Sociales, 1991? 97 p.: bibl.

Brief, useful report of conclusions of a seminar on relations between Ecuador and the EEC in which Ecuadorian and European specialists participated. Discusses features and prospects of commercial, industrial, scientific and technological, and financial bilateral relations in the 1990s.

4289 **Valencia Rodríguez, Luis.** El conflicto territorial ecuatoriano-peruano: programa de cooperación política regional—material de investigación. Quito: Casa de la Cultura Ecuatoriana Benjamín Carrión, 1988. 137 p., 4 leaves of plates: bibl., maps. (Serie Identificación y objetivación de tensiones y conflictos territoriales)

Brief account relates evolution of territorial dispute with Peru. Although presenting an interpretation favorable to Ecuador, rejects intransigent positions and advocates peaceful solution based on moderate demands and bilateral cooperation.

4290 **Villacrés Moscoso, Jorge W.** Historia de las relaciones culturales de la República del Ecuador. Guayaquil, Ecuador: Univ. de Guayaquil, Comisión de Defensa del Patrimonio Nacional, 1991. 442 p.

Volume, arranged by regions and countries, details diverse aspects of Ecuador's cultural relations and contacts with the rest of the world. Useful reference handbook.

4291 **Villacrés Moscoso, Jorge W.** Las invasiones y desmembraciones efectuadas por los estados limítrofes vecinos al territorio ecuatoriano. Guayaquil: Talleres Gráficos de Editorial Justicia y Paz, 1990. 112 p.: ill., map.

Denounces seizure of Ecuadorian territories by Brazil, Colombia, and Peru, and condemns diplomatic complicity of successive Ecuadorian administrations in these misappropriations. Embittered reminder that territorial disputes and irredentism remain a source of international and domestic tensions in several South American countries.

PARAGUAY

4292 **EE.UU. y el régimen militar paraguayo, 1954–1958: documentos de fuentes norteamericanas.** Recopilación de Anibal Miranda. Asunción: El Lector, 1987. 215 p. (Col. Realidad nacional; 7)

Compendium of documents produced by US Embassy in Asunción and State Department officials concerning Paraguayan political and economic situations and bilateral relations during first four years of Stroessner dictatorship. Critical comments by author clarify historical context of most documents and interpret impact of US actions on Paraguay's domestic political and economic situations and foreign policies. Documents reveal that, although well informed about corrupt and repressive characteristics of the Paraguayan regime, the Eisenhower Administration preferred for strategic and economic reasons to maintain a friendly but low-profile bilateral relationship.

4293 Fernández Estigarribia, José Félix. Perspectivas de cambio de la política exterior paraguaya. (*Síntesis/Madrid,* 10, enero/abril 1990, p. 325–334)

Examines initial changes and outlook of Paraguay's foreign policy following overthrow of Stroessner. Analyzes domestic situation and evolution of relations with Argentina, Brazil, and the US. Concludes that the new foreign policy has been characterized by a conservative pragmatism that precludes any significant changes in the near future.

4294 González, Luis J. Paraguay, prisionero geo-político. 2. ed. Asunción: Instituto Paraguayo de Estudios Geopolíticos e Internacionales, 1990. 284 p.: bibl.

Paraguayan edition of a book originally published in Bolivia in 1946 by an exiled military opponent of Stroessner traces history of Paraguay's relations with its neighbors, the US, and the rest of the Hemisphere. Emphasizes dangers of the country's isolation, and recommends a foreign policy guided by principles of neutrality, peaceful resolution of conflicts, inter-American cooperation, and regional integration.

Palau, Tomás and **Carlos Verón.** Una contribución preliminar para el estudio de la frontera en el Paraguay y su impacto socioeconómico. See item **2004.**

4295 Simón G., José Luis. La política exterior paraguaya en 1991: modernización insuficiente, carencia de una visión global y condicionamientos de un estado prebendario en crisis. Asunción: Centro Paraguayo de Estudios Sociológicos, 1991. 56 p.: bibl. (Cuadernos de discusión)

Insightful work evaluates Paraguay's foreign policies in the three years since fall of Stroessner. Analyzes roles played by new president Gen. Andrés Rodríguez, the ministry of foreign relations, and parliament, as well as relations established with Latin America, the US, Europe, Japan, and international organizations. Concludes that, although a democratic Paraguay has been able to reinsert itself in the world under more favorable conditions, much remains to be done at the domestic level to eliminate legacy of authoritarianism and generate a more modern, democratic, and efficient foreign policy organization and strategy.

4296 Yopo H., Mladen. Paraguay, Stroessner: la política exterior del régimen autoritario, 1954–1989. Santiago: Programa de Seguimiento de las Políticas Exteriores Latinoamericanas, 1991. 112 p.: bibl.

Concise and comprehensive study examines Paraguay's foreign policies under Stroessner. Excellent analysis of impact of authoritarian tradition on country's foreign policies and of its other domestic and external determinants. Gives balanced examination of international relations with Latin American countries, the US, the Vatican and the rest of the world. Concludes with insightful appraisal of Paraguay's foreign policies in more democratic context created by fall of Stroessner.

PERU

A 50 años del protocolo de Rio de Janeiro: opiniones de actualidad. See item **4281.**

4297 Abugattás A., Juan. El Perú y los retos del entorno mundial. (*in* Desde el límite: Perú, reflexiones en el umbral de una nueva época. Lima: IDS, 1992, p. 11–108)

Discusses from philosophical, political, economic, and social perspectives the emergence of the post-Cold War global order and its likely impact on Peru's international insertion. Although pessimistic, argues that Peru's recovery from its current decay requires an elite able to mobilize the majority of the population in pursuit of development, to build up a legitimate State, and to reinsert the country into the region and the world on a more cooperative basis.

Altamirano Escobar, Hernán Alonso. El por qué del ávido expansionismo del Perú. See item **4282.**

4298 Barrantes Lingán, Alfonso. La reinserción de Perú en el sistema financiero internacional en la perspectiva de la izquierda socialista. (*in* La reinserción del Perú en el sistema financiero internacional. Edición de Eduardo Ferrero Costa. Lima: Centro Peruano de Estudios Internacionales (CEPEI), 1990, p. 173–182)

Lecture by presidential candidate of

the Movimiento Izquierda Socialista during 1990 electoral campaign explains his position concerning Peru's international financial crisis. Rejects radical nationalistic positions and extreme liberal economic strategies. Recommends an approach aimed at introducing structural domestic transformations while negotiating with creditors for reductions in principal and interest and transformation of the debt into long-term public bonds.

4299 Benavides Correa, Alfonso. Una difícil vecindad: los irrenunciables derechos del Perú en Arica y los recusables acuerdos peruano-chilenos de 1985. Lima: Editorial de la Univ. Nacional Mayor de San Marcos, 1988. 391 p.: bibl., ill.

Polemical volume analyzes evolution of Chilean-Peruvian territorial dispute since late 19th century. Denounces bilateral agreement of 1985 as an invalid act relinquishing Peru's sovereign rights on the disputed territory. Work is helpful for understanding complicated mix of nationalism and political, legal, and historical factors that continue to generate tensions in the area and hinder chances of Bolivia's obtaining an exit to the sea.

4300 Berrios, Rubén and **Cole Blasier.** Peru and the Soviet Union, 1969–1989: distant partners. (*J. Lat. Am. Stud.*, 23:2, May 1991, p. 365–384, tables)

Comprehensive study and balanced assessment of evolution of Soviet-Peruvian relations in 1970s-80s includes examination of commercial, diplomatic, and military aspects, and of role of Partido Comunista del Perú. Even though writing prior to collapse of the USSR, authors nevertheless conclude that the close relations had already run their course and that existence of limited mutual interests anticipates normal but not intense economic and political relations in the future.

4301 Bustamante Belaunde, Luis. La reinserción del Perú en el sistema financiero internacional en la perspectiva del Frente Democrático. (*in* La reinserción del Perú en el sistema financiero internacional. Edición de Eduardo Ferrero Costa. Lima: Centro Peruano de Estudios Internacionales (CEPEI), 1990, p. 161–172)

Critical analysis of Peru's financial crisis and prospects for recovery is written from perspective of Vargas Llosa's supporters during 1990 presidential campaign. Denounces Alan García's economic policies and recommends free market and trade policies as well as renewed cooperation with international financial institutions and private banks as the only way to solve the debt problem.

Carrera de la Torre, Luis. El proyecto binacional Puyango-Tumbes. See item **4283**.

4302 Deustua Caravedo, Alejandro. El altiplano peruano-boliviano y el Lago Titicaca: proyección y alternativas internacionales. Lima: Centro Peruano de Estudios Internacionales, 1989. 187 p.: bibl. (Serie Investigaciones; 5)

Comprehensive study examines characteristics of highland territory and lake shared by Peru and Bolivia. From international relations viewpoint, most interesting chapters are those devoted to past and current joint initiatives for development of the area including utilization of natural resources, transportation, environmental protection, and commercial relations.

4303 Deustua Caravedo, Alejandro. Tres aproximaciones a la seguridad externa del Perú. Lima: Centro Peruano de Estudios Internacionales, 1990. 134 p.: bibl. (Serie Investigaciones: 1017–5113; 7)

Includes three essays on national and regional security delivered by author at different seminars in 1989. First essay offers updated and comprehensive analysis of border conflicts involving Peru. Other two address, in general terms, the evolution and current situation of Latin American, particularly South American, collective and regional security, highlighting importance of security threats such as drug traffic and terrorism and calling for a specifically South American security system.

Ecuador y Perú, vecinos distantes: seminario organizado por la Corporación de Estudios para el Desarrollo (CORDES), del 7 al 10 de diciembre de 1992, Quito, Ecuador. See item **4284**.

4304 Ferrero Costa, Eduardo. Las relaciones del Perú con los Estados Unidos. (*Anal. Int.*, 1, enero/marzo 1993, p. 5–22)

Study of Peru/US relations focuses on their evolution during Fujimori Administration until 1992. Distinguishes between initial period of Fujimori government in which relations improved and revolved around drug and economic issues, and period following presidential coup of April 1992 during which relations worsened as US focused on democracy and human rights.

4305 Flor Belaunde, Pablo de la. Japón en la escena internacional: sus relaciones con América Latina y el Perú. Lima: Centro Peruano de Estudios Internacionales; Cotecna Inspection; Omic International, 1991. 287 p.: bibl., ill. (Serie Investigaciones: 1017–5113; 9)

Useful study examines facet of Latin American and particularly Peruvian international relations that has not been given the necessary attention. Analyzes Japanese migration to Latin America, origins of bilateral relations, and Japan's new role in the global system. Making use of considerable supporting data, also discusses evolution of economic relations in areas such as trade, investment, finance, and development cooperation.

4306 Granda, Juan Velit. Situación de las fronteras del Perú. (*in* Congreso Internacional sobre Fronteras en Iberoamérica Ayer y Hoy, *1st, Tijuana, Mexico, 1989*. Memoria. Edición de Alfredo Félix Buenrostro Ceballos. Mexicali, Mexico: Univ. Autónoma de Baja California, 1990, v. 1, p. 328–342)

Survey of Peruvian border situation in relation to each of its neighbors offers brief but comprehensive account of past and current problems and of attempts to overcome tensions and disputes.

4307 Hacía una política marítima nacional. Edición de Alberto Indacochea Queirolo. Lima: Centro Peruano de Estudios Internacionales; Instituto de Estudios Histórico-Marítimos del Perú, 1990. 149 p.: ill. (Serie Seminarios, mesas redondas y conferencias; 8)

Works presented by civilian and military specialists at a 1990 round table on Peru's maritime policies. Offers thorough examination of economic, commercial, security, and environmental aspects of Peruvian maritime policies. Includes policy recommendations aimed at securing defense of maritime interests.

4308 Kisic, Drago. La reinserción internacional del Perú. (*Anal. Int.*, 1, enero/marzo 1993, p. 33–44, tables)

Examines Fujimori Administration's attempts since August 1990 to reinsert Peru into the international financial system. Discusses stabilization and adjustment programs and negotiations with international financial organizations, private banks, and foreign governments. Points out that positive trend toward reinsertion can continue only if processes of redemocratization and privatization are completed in the near future.

4309 Morelli Pando, Jorge. Los acuerdos de Ilo en el marco de las relaciones del Perú con Bolivia. (*Anal. Int.*, 1, enero/marzo 1993, p. 63–77)

Provides legal and political analysis of agreements on transportation, road communication, and industrial development signed by Peru and Bolivia in 1992. Agreements granted Bolivia use of free customs and tax-exempt areas in port of Ilo, and are seen by author as a positive development facilitating Bolivian trade expansion and contributing to development of southern Peruvian region in a context of regional cooperation.

4310 Obando Arbulú, Enrique. La subversión: situación interna y consecuencias internacionales. (*Anal. Int.*, 1, enero/marzo 1993, p. 45–62, graph)

Analyzes domestic and international Peruvian situation after capture of Abimael Guzmán Reynoso. Examines Sendero Luminoso's strategic errors that led to its decline, as well as international consequences of the subversive activity, particularly its impact on US policy toward Peru in relation to security issues, drug traffic, human rights, and democracy.

4311 Relaciones del Perú con Brasil, Colombia y Ecuador. Edición de Ramón Bahamonde Bachet. Lima: Centro Peruano de Estudios Internacionales, 1990. 235 p.: bibl. (Serie Seminarios, mesas redondas y conferencias; 7)

Includes essays by Peruvian civilian and military specialists delivered at third seminar on Peru's national defense and international relations, which focused on the situation and perspectives of Peru's relations with Brazil, Colombia and Ecuador. Consistently well-researched articles analyze traditional issues such as border disputes with Ecuador, relations in the Amazon Basin, economic relations with Brazil, and new developments such as ecological problems in the Amazonian region, military industries and limitation of military expenditures, drug traffic as a regional problem, and prospects for integration in the area. See also items **4313, 4315,** and **4312.**

4312 Relaciones del Perú con Chile y Bolivia. Edición de Eduardo Ferrero Costa. Lima: Centro Peruano de Estudios Internacionales, 1989. 217 p.: bibl., ill. (Serie Seminarios, mesas redondas y conferencias; 5)

Includes essays by Peruvian specialists delivered at second seminar on Peru's national defense and international relations which focused on relations with Chile and Bolivia. Particularly interesting are articles on Peru's maritime interests in relation to Chile, limitation of military expenditures, and commercial relations between the three countries. See also items **4313, 4315,** and **4311.**

4313 Relaciones del Perú con los países vecinos. Edición de Eduardo Ferrero Costa. Lima: Centro Peruano de Estudios Internacionales, 1988. 213 p.: bibl. (Serie Seminarios, mesas redondas y conferencias; 3)

Includes articles on evolution of Peru's relations with neighboring countries delivered by Peruvian experts at first seminar on national defense and international relations organized by Centro Peruano de Estudios Internacionales and Centro de Altos Estudios Militares. Particularly useful for their concise but comprehensive review of historical and contemporary issues are articles by Juan Miguel Bákula on relations with Colombia and Gino F. Costa on Brazilian-Peruvian relations. See also items **4312, 4315,** and **4311.**

4314 Relaciones económicas del Perú con la Comunidad Europea. Edición de Bruno Podestà, Harald Klein, y Eduardo Ferrero Costa. Lima: GREDES; CEPEI; Fundación Friedrich Naumann, 1991. 235 p.: bibl.

Excellent collection of essays delivered in 1990 by Peruvian and European specialists at a seminar sponsored by Fundación Friedrich Naumann examines Peru's relations with the European Union. Articles consider features of and prospects for evolution of bilateral relations after 1992, addressing policymaking institutions and mechanisms involved in the relationship; commercial, financial, and investment links; and possibilities for development cooperation.

Rengel, Jorge Hugo. La cuestión Puyango-Tumbes. See item **4287.**

4315 La seguridad del Perú frente al nuevo contexto internacional. Edición de Eduardo Ferrero Costa. Lima: Centro Peruano de Estudios Internacionales, 1991. 178 p.: bibl. (Serie Seminarios, mesas redondas y conferencias: 1017–5121; 9)

Includes articles delivered by Peruvian and foreign experts at fourth seminar on national defense and international relations. Participants analyze impact of end of the Cold War and other international transformations on strategic situation and security issues faced by Peru, and examine Peru's positions on the inter-American defense system, disarmament, subversion, drug traffic, and environmental problems. In general, contributors agree that in the new circumstances the traditional security doctrine should be replaced by another that emphasizes integral and interdependent nature of security (not only military but also economic, social, political, and ecological) and relies on cooperation rather than on confrontation. See also items **4313, 4312,** and **4311.**

4316 Thornberry, Guillermo. Un caso de internacionalización de conflictos regionales: narcotráfico, guerrilla y conservación ecológica en la Amazonia peruana. (*in* Los militares y la democracia: el futuro de las relaciones cívico-militares en América Latina. Compilación de Louis W. Goodman, Johanna S.R. Mendelson y Juan Rial. Montevideo: PEITHO, 1990, p. 177–190)

Interesting study examines intersection and clashes of international and domestic pressures in dealing with drug traffic, guerrilla subversion, and ecological problems in the Peruvian Amazon. Concludes that any adequate response to these three problems requires reassertion of State presence in the area, implementation of a program for socioeconomic development, and a less interventionist and more cooperative approach on the part of foreign states and international organizations.

4317 Torres, Victor. El Perú y la Cuenca del Pacífico. (*Polit. Int.*, 28, abril/junio 1992, p. 19–37, graphs, tables)

Informative article briefly describes economic features and development strategies of the Asian Pacific Basin countries and examines Peru's commercial ties with these countries.

Torres C., Víctor. El Perú frente a la Cuenca del Pacífico: flujo comercial con los países asiáticos. See item **1964**.

4318 Valdez Pérez del Castillo, Eduardo. Experiencias diplomáticas. Peru?: Talleres Gráficos de Grafía, 1992. 217 p.

In these interesting recollections a Peruvian diplomat recalls his experiences as representative in Chile, Panama, Nicaragua, Bolivia, and China.

Valencia Rodríguez, Luis. El conflicto territorial ecuatoriano-peruano: programa de cooperación política regional—material de investigación. See item **4289**.

4319 Youngers, Coletta. After the *autogolpe:* human rights in Peru and the U.S. response. Washington: Washington Office on Latin America, 1994. 63 p.

Critical study examines human rights situation in Peru following Fujimori's 1992 presidential coup and policies implemented afterwards by Bush and Clinton Administrations. After acknowledging temporary decline in human rights abuses following capture of Sendero Luminoso leader Abimael Guzmán Reynoso, report condemns lack of judicial guarantees under anti-terrorist legislation enacted after the coup, as well as impunity with which security personnel continue to engage in human rights violations, particularly in Huallaga region. Also criticizes focus on drug traffic and relative lack of concern with human rights abuses shown by Bush and Clinton Administrations, and recommends use of US diplomatic and economic pressure on Peru to promote democracy and respect for human rights.

URUGUAY

4320 Abreu, Sergio and **Alejandro Pastori Fillol.** Uruguay y el nuevo orden mundial. Montevideo: Fundación de Cultura Universitaria, 1992. 146 p.: bibl.

Assesses position of Uruguay in the new political, economic, and strategic circumstances of 1990s with special reference to economic and trade issues and regional integration. Discusses challenges faced by a small country in this new world order, and recommends a foreign policy aimed at promoting the strengthening of international law, economic opening, regional integration, and pragmatic flexibility.

4321 Alimonda, Héctor. Una agenda democrática frente al Mercosur. (*Cuad. Marcha,* 7:79, enero 1992, p. 18–24)

Critical Uruguayan assessment of likely impact of Mercosur warns against consequences of a liberal economic, export-oriented strategy on weakest nations and on socioeconomic groups involved in the integration process. Recommends democratic debate and political cooperation among these sectors to ensure that restructuring would not inordinately affect workers, small producers, and less-developed areas.

4322 Arteaga Sáenz, Juan José. Uruguay y Santa Sede: sus relaciones. Montevideo: República Oriental del Uruguay, Presidencia de la República, 1987. 60 p.: bibl., ill., ports.

Official publication issued before visit of John Paul II to Uruguay in 1987 describes some aspects of the history of bilateral relations but ignores periods of tension.

4323 Bizzozero, Lincoln J. La política exterior del Uruguay en una perspectiva histórica. (*Síntesis/Madrid,* 13, enero/abril 1991, p. 347–358, bibl.)

Brief interpretive work surveys main sources on Uruguay's international relations and synthesizes debates concerning Uruguay's territorial sovereignty, the internationalization of its domestic politics, and relations between internal and international developments.

4324 Bizzozero, Lincoln J. Las relaciones Uruguay-Unión Soviética y el tema de la pesca. (*Rev. Parag. Sociol.,* 26:76, sept.–dic. 1989, p. 79–89, bibl.)

Examines protracted negotiations that preceded signing of 1988 fishing agreement between Uruguay and USSR. Analyzes Soviet proposal and domestic Uruguayan opposition to some of its contents. Opposition finally led to an agreement that limited Soviet annual catches, precluded Soviet vessels from using Uruguayan port facilities, and prevented Aeroflot from operating at Carrasco Airport.

4325 Camou, María Magdalena. Los vaivenes de la política exterior uruguaya ante la pugna de las potencias: las relaciones con el Tercer Reich, 1933–1942. Montevideo: Fundación de Cultura Universitaria, 1990. 73 p.: bibl. (Cuadernos de investigación y docencia. Cuadernos de interguerras; 3)

Historical account of Uruguayan foreign policy toward Third Reich from 1933–42. Includes interesting information concerning Nazi influence on German-descent groups, conservative politicians and publications, and the military.

4326 Jerozolimski, José. Uruguay e Israel: fraternales relaciones. Montevideo: República Oriental del Uruguay, Presidencia de la República, 1989. 99 p.

Official publication issued in 1989 to commemorate visit of Israeli president to Uruguay. Contains historical account of Uruguayan role in the UN supporting creation of Israel, and discusses evolution of bilateral diplomatic and economic relations. Reproduces several agreements concluded between the two countries.

4327 Luján, Carlos. Redemocratización y política exterior en el Uruguay. (*Síntesis*/Madrid, 13, enero/abril 1991, p. 359–377, bibl.)

Examines impact of regime change on Uruguay's foreign policies. After analyzing the domestic changes associated with restoration of democracy and the features of the country's foreign relations since 1985, concludes that in practice the connection was weak and changes in foreign policy minimal.

Muschietti, Ulises M. et al. Conflictos en el Atlántico Sur: siglos XVII-XIX. See item **4214**.

4328 Uruguay-URSS: 60 años de relaciones diplomáticas, 1926–1986 — documentos y materiales. Montevideo: Ministerio de Relaciones Exteriores del Uruguay; Moscow: Ministerio de Relaciones Exteriores de la URSS, 1989. 221 p.: bibl., ill.

Useful compendium contains official Soviet and Uruguayan documents concerning establishment and evolution of bilateral relations, including texts of communiques, letters, notes, protocols, and agreements. Not surprisingly (but raising doubts about the objectivity of the compilation), volume does not reproduce notes and declarations exchanged and issued by both sides between Dec. 1935,

when relations were suspended and harsh accusations exchanged, and Jan. 1943, when relations were restored.

VENEZUELA

4329 Arenas, Nelly M. La denuncia del Tratado de Reciprocidad Comercial entre Venezuela y los Estados Unidos. Caracas: Centro Venezolano Americano, 1990. 125 p.: bibl.

Well-researched monograph looks at 1971 termination by Venezuela of reciprocal trade agreement signed with US in 1939 and modified in 1952. Points out motivations behind the Venezuelan decision—agreement's incompatibility with the Andean Pact, progress in the process of import-substitution industrialization, and US discrimination against Venezuelan oil exports, as well as the reasons for the calm acceptance of decision by the US.

4330 Cardozo de Da Silva, Elsa. Continuidad y consistencia en quince años de política exterior venezolana, 1969–1984. Caracas: Univ. Central de Venezuela, Consejo de Desarrollo Científico y Humanístico, 1992. 239 p.: bibl. (Col. Estudios)

Valuable analysis of Venezuela's foreign policies during Caldera, Pérez, and Herrera Administrations. Applying content analysis techniques to a database containing information on Venezuelan foreign policies toward Western Hemisphere countries, particularly Colombia, author seems to validate her hypothesis concerning fundamental policy continuity and consistency by showing that foreign policies of successive administrations pursued similar goals: domestic and international economic restructuring, strengthening of the democratic regime, diversification of Venezuela's external relations, defense of natural resources, and emphasis on peace and negotiated solutions. See also item **4346**.

4331 Cardozo de Da Silva, Elsa. La política exterior de Venezuela, 1984–1989: entre las vulnerabilidades económicas y los compromisos políticos. (*Econ. Cienc. Soc.*, 28, enero/dic. 1989, p. 127–165, bibl.)

Study of Venezuela's international relations during Lusinchi Administration underscores restrictive impact of the economic crisis on the country's foreign policies. Crisis led to a retreat from independent and nonaligned economic and diplomatic initiatives, particularly in terms of financial negotiations, oil policies, and Central American conflicts.

4332 Consalvi, Simón Alberto. Grover Cleveland y la controversia Venezuela–Gran Bretaña: la historia secreta. Washington: Tierra de Gracia Editores, 1992. 223 p.: bibl. (Historias de papel; 3)

Interesting analysis, written by former Venezuelan foreign relations minister and ambassador to US, relates role played by President Cleveland in forcing arbitration to resolve Venezuelan-British boundary conflict in Essequibo region. Reproduces Cleveland's historical memoir on the controversy, the notes exchanged between US Secretary of State Richard Olney and British Prime Minister Lord Salisbury, and presidential message sent by Cleveland to Congress in 1895. Useful for those interested not only in the border dispute but also in evolution and application of the Monroe Doctrine by US governments.

4333 Consalvi, Simón Alberto. Una política exterior democrática en tiempos de crisis. Caracas: Pomaire, 1988. 161 p.

Collection of lectures and essays delivered in 1986–87 by then minister of foreign relations of Venezuela. Discusses challenges for Venezuela's foreign policy in times of domestic political economic crisis and international transformation. Offers insider's view on topics such as formation and features of Venezuelan foreign policies, Contadora efforts, relations with US and other developed countries, and creation and importance of the Rio Group.

4334 Lara Peña, Pedro José. Las tesis excluyentes de soberanía colombiana en el Golfo de Venezuela. Caracas: Editorial Ex Libris, 1988. 689 p., 6 folded leaves of plates: bibl., ill. (some col.).

Massive compendium of historical, cartographic, legal, political, and diplomatic materials is aimed at sustaining Venezuela's exclusive sovereign rights to the Gulf of Venezuela waters disputed with Colombia.

4335 Morales Paúl, Isidro. Política exterior y relaciones internacionales. Caracas: Academia de Ciencias Políticas y Sociales, 1989. 332 p.: bibl., ill., map. (Biblioteca de la Academia de Ciencias Políticas y Sociales. Serie Estudios; 36)

Introduction to the study of foreign policy and international relations is valuable for its analysis of Venezuela. Includes historical discussion of ideas of Andrés Bello and general features of Venezuelan foreign policy, particularly the emphasis on negotiated solutions.

Murillo Castaño, Gabriel and **María Victoria Llorente Sardi.** Las relaciones colombo-venezolanas contemporáneas. See item **4273.**

Obregón T., Liliana and **Carlo Nasi L.** Colombia-Venezuela: conflicto o integración. See item **4274.**

4336 Olavarría, Jorge. El Golfo de Venezuela: es de Venezuela. Caracas?: E. Armitano, 1988. 190 p.: ill. (some col.), maps.

Includes transcripts of six shows presented on Venezuelan television by author in 1987 after a naval incident rekindled Colombian-Venezuelan dispute on territorial delimitation of Gulf waters. Although interpretation is, as expected, biased in favor of Venezuela, volume merits examination for its brief but comprehensive presentation of Venezuelan side of the dispute and excellent maps illustrating the arguments.

4337 Parra Pérez, Caracciolo. Caracciolo Parra Pérez, canciller de Venezuela, 1941–1945. Prólogo de Rafael Armando Rojas. Caracas: Fundación Biblioteca de Política Exterior, Ministerio de Relaciones Exteriores, 1989. 663 p. (Biblioteca de política exterior; 7)

Useful collection of official diplomatic documents signed by Parra Pérez during his tenure as Venezuela's minister of foreign relations. Good source for those interested in Venezuelan foreign policies during World War II, particularly in relation to US, Latin America, and the Axis powers.

4338 Pérez, Carlos Andrés. Hacia nuevos horizontes: Venezuela y América Latina en un mundo en transformación. Caracas: Ediciones Altair, 1988. 103 p.

Concise presentation of Pérez's ideas and policy recommendations concerning Venezuela's international relations was published before the beginning of his ill-fated second presidency. Examines, from Latin American and Venezuelan perspectives, prospects for integration, relations with US, the Central American crisis, disarmament, North-South relations, foreign debt, and role of social-democratic movements.

4339 **Pérez Luciani, Ramiro.** Con Colombia ya basta! Caracas: s.n., 1988. 497 p., 29 folded leaves of plates: bibl., ill. (some col.).

Lengthy and, as title conveys, strongly anti-Colombian volume on Maracaibo Gulf dispute is written by retired Venezuelan vice admiral. Useful for historical references and transcriptions of documents and conversations between Colombian and Venezuelan negotiators. Also may have value as an indication of unyielding nationalistic position of Venezuelan Navy and other military service personnel.

4340 **Romero, Carlos A.** *et al.* Los orígenes del pensamiento internacional de Acción Democrática, 1928–1945. (*Cuad. INVESP*, 1, 1990, p. 5–55)

Collection of articles analyzes some historical antecedents of the international ideas held by Acción Democrática (AD). Includes discussion of the impact of the German-Soviet pact on Venezuelan marxists and on Betancourt (Carlos Romero), the Americanist ideology embraced by Betancourt during World War II (María Teresa Romero), and evolution of AD's nationalism in the context of Roosevelt's "Good Neighbor Policy" (Hernández Arvelo).

4341 **Romero, Carlos A.** Las relaciones entre Venezuela y la URSS: diplomacia o revolución. Caracas: Univ. Central de Venezuela, Consejo de Desarrollo Científico y Humanístico, 1992. 203 p.: bibl. (Col. Estudios)

Comprehensive and insightful study examines evolution of Venezuelan-Soviet relations from 1960s-1990. Discusses lack of importance of economic factors while underscoring significance of political considerations in conditioning the hostile or friendly character of the bilateral relationship. Interesting examination of role played by changing tactics of the Partido Comunista de Venezuela and Castro's Cuba in shaping nature of the bilateral links.

4342 **Romero, María Teresa.** La administración Lusinchi y su política exterior hacia el Caribe anglófono. (*Caribe Contemp.*, 19, julio/dic. 1989, p. 23–37)

Analyzes evolution of Venezuela's policies toward English-speaking Caribbean under Lusinchi Administration. Argues that Venezuela pursued an active foreign policy strategy motivated by security interest in preserving stability in the region, but was forced to maintain policy at a low profile due to Venezuelan economic crisis.

4343 **Sequera Tamayo, Isbelia** *et al.* Guayana Esequiba: espacio geopolítico. Caracas: Academia Nacional de Ciencias Económicas, 1992. 190 p.: appendix, bibl., maps.

Collection of articles by Venezuelan authors examines aspects of territory disputed with Guyana. First and last sections, by Sequera Tamayo, analyze evolution of dispute, bilateral negotiations, and prospects for resolution. Remainder of book is devoted to physical analysis of the territory. Appendix includes bilateral agreements and protocols ratified by Venezuela.

4344 **Sureda Delgado, Rafael Angel.** La Guayana Esequiba: dos etapas en la aplicación del Acuerdo de Ginebra. Caracas: Academia Nacional de la Historia, 1990. 672 p.: bibl., ill., index, maps, plates. (Biblioteca de la Academia Nacional de la Historia. Estudios, monografías y ensayos; 129)

Well-researched and profusely documented study examines, from Venezuelan perspective, the history of the Essequibo Region territorial dispute with Guyana. Covers period from Geneva Agreement of 1966 negotiated by Great Britain, Guyana, and Venezuela for peaceful resolution of the controversy to 1990. Analyzes dispute under successive Venezuelan and Guyanese administrations, highlighting domestic and international factors that blocked a solution favorable to Venezuela's territorial demands.

4345 **Toro Hardy, Alfredo.** El desafío venezolano: como influir las decisiones políticas estadounidenses. Prólogo de Miguel Angel Burelli Rivas. Caracas: Instituto de Altos Estudios de América Latina de la Univ. Simón Bolívar, 1988. 286 p.: bibl.

Interesting manual explains characteristics of US decision-making process. Offers practical recommendations for development of a relatively well-organized Venezuelan lobby able to influence US decisions, particularly concerning issues such as oil imports, commercial relations, and foreign debt negotiations.

4346 Toro Hardy, Alfredo. La maldición de Sísifo: quince años de política externa venezolana. Caracas: Editorial Panapo, 1991. 140 p.: bibl., ill., maps.

Focusing on 1974–89, argues—in direct opposition to Cardozo de Da Silva's thesis (see item **4330**)—that Venezuela's foreign policy changes course every five years. Analysis of Pérez, Herrera, and Lusinchi Administrations' foreign policies concerning boundary disputes, economic issues, and relations with Caribbean Basin, Latin America, and the US shows policy inconsistencies and some contradictions; however, these do not demonstrate absolute lack of continuity regarding major foreign policy objectives.

4347 The United States and Venezuela: new opportunities in an established relationship; the report of the CSIS-CAUSA Working Group on U.S.-Venezuelan Relations. Foreword by David M. Abshire and Hans Neumann. Preface by Georges Fauriol and José Guillermo Castillo. Washington: Center for Strategic & International Studies; Caracas: Center for the Analysis of the U.S.A., 1991. 32 p.

Informative report on development of and prospects for Venezuelan-US relationship summarizes findings of a binational group of specialists, policymakers, and representatives of the private sector. Concise historical analysis and examination of contemporary situation is followed by thoughtful policy recommendations aimed at fostering mutually beneficial cooperation in energy, diplomatic, commercial, and law enforcement areas.

4348 Urdaneta, Alberto and **Ramón León.** Relaciones fronterizas entre Venezuela y Colombia: desde la perspectiva venezolana. Caracas: CENDES, 1991. 62 p. (Col. Luis Lander; 4)

Brief but informed study analyzes background, contents, and expected consequences of 1989 Declaración de Ureña signed by Colombian and Venezuelan presidents. According to authors, this cooperative declaration established political framework necessary for ending a period of hostility generated by territorial disputes, and created conditions for mutually beneficial integration on frontier areas.

BRAZIL

SCOTT TOLLEFSON, *Assistant Professor, Department of National Security Affairs, Naval Postgraduate School*

THERE ARE SEVEN MAJOR THEMES that continue to dominate the literature on Brazil's international relations: diplomatic history, the nation's position within the international system, bilateral relations (especially those with the US and Argentina), ecology, integration, external security, and international law. These themes overlap considerably, and are supplemented by others, such as foreign economic relations and the process of formulating foreign policy.

Diplomatic history is the most traditional field, and receives the greatest attention. With the untimely death of Gerson Moura, the field has unfortunately lost one of its best and most prolific young writers (for an example of his excellent work, see item **4376**).

The changing international system has led many analysts to consider Brazil's position within that system. While most such studies contribute little to our understanding of Brazil's international relations, some are insightful (items **4382** and **4368**).

In terms of Brazil's bilateral relations, the book by Weiss on Brazilian-US relations is especially good (item **4392**). It is complemented by the publication of various studies on Brazil's relations with Argentina, Chile, Cuba, Guyana, and Italy.

Brazil's ecological policies are receiving increased scrutiny. Much of the literature on this theme is mediocre, but a notable exception is the article by Barbosa (item **4352**), which utilizes a broad theoretical framework.

In terms of other themes, Bresser Pereira and Thorstensen accurately assess some of the difficult choices facing Brazil in integration (item **4378**). Studies on Brazil's external security examine confidence-building measures with Argentina (item **4359**) and the state of Brazil's armaments industry. Caubet's book on the Itaipú Dam is a rich and thorough study (item **4356**), from an international law perspective.

The literature on Brazil's international relations continues to suffer from a lack of theoretical and methodological rigor, but the studies highlighted above are among the prominent exceptions. Mello e Silva astutely observes that there is an excessive emphasis on external factors as explanatory variables of Brazil's foreign policy, "underestimating, or even excluding, the analysis of internal variables" (item **4372**). In that vein, the literature is becoming gradually more sophisticated. Silva's study of the relationship between regime change and foreign policy in Brazil and Argentina is an example of such erudition (item **4386**), as is Almeida's analysis of the role played by Brazil's political parties in the formulation of foreign policy (item **4349**).

Indeed, as Brazil continues to consolidate its democracy internally, and to face a changing international environment externally, its international relations will become more complex, requiring analysis that is even more refined, focusing on the interaction between internal and external variables. For example, the role of Brazil's Congress and even the Ministry of Foreign Relations in the formulation of foreign policy is poorly understood. With time, these and other gaps hopefully will be addressed.

4349 Almeida, Paulo Roberto de. Os partidos políticos nas relações internacionais do Brasil, 1930–90. (*Cont. Int.*, 14:2, julho/dez. 1992, p. 161–208)

Excellent chronological examination of the nearly marginal role played by Brazil's political parties in foreign affairs. Notes, however, that with the 1988 Constitution, the Congress and political parties began to participate more fully in Brazil's international relations, culminating in the appointment of a party leader (Fernando Henrique Cardoso) to head the Ministry of Foreign Relations.

4350 Almeida, Paulo Roberto de. Relações internacionais do Brasil: introdução metodológica a um estudo global. (*Cont. Int.*, 13:2, julho/dez. 1991, p. 161–185)

Overview of various methodologies utilized in studying Brazil in the international system. Historical in orientation, with many European sources. Ambitious effort, marred by lack of attention to domestic factors in explaining Brazil's international relations and excessive reliance on systemic variables.

4351 Azambuja, Marcos Castrioto de *et al.* O Brasil e o Plano Bush: oportunidades e riscos numa futura integração das Américas. Organização de João Paulo dos Reis Velloso. São Paulo: Nobel; Fórum Nacional, 1991. 172 p.: bibl., ill., maps.

Brazilian political scientists, economists, and diplomats discuss implications of the Bush Plan for Brazil. A common theme is Brazil's reticence in allowing the US to take the lead on regional integration. Good contribution to the broader literature on integration.

4352 Barbosa, Luiz C. The "greening" of the ecopolitics of the world-system: Amazonia and changes in the ecopolitics of Brazil. (*J. Polit. Mil. Sociol.*, 21:1, Summer 1993, p. 107–134, bibl.)

Excellent application of world-system theory (Wallerstein, Bunker) to ecopolitics and Amazonia. Concludes that "Brazil's links with the capitalist world-economy fueled the destruction of Amazonia." Ironically, "the dependency links of Brazil with the capitalist world-economy can help preserve the Amazon rain forest, if the ecopolitics of the world-system continues to green ... "

4353 Barboza, Mario Gibson. Na diplomacia, o traço todo da vida. Rio de Janeiro: Editora Record, 1992. 330 p.: ill., index.

Brazil's minister of foreign relations from 1969–74 traces his diplomatic career. Insightful and engaging.

Bartelt, Dawid. Fünfte Kolonne'ohne Plan: Die Auslandsorganisation der NSDAP in Brasilien, 1931–1939. See *HLAS 54:3239.*

4354 Batista Júnior, Paulo Nogueira. Perspectivas da rodada Uruguai: implicações para o Brasil. (*Estud. Avançados*, 6:16, set./dez. 1992, p. 103–116)

Speech by Brazilian diplomat on the General Agreement on Tariffs and Trade (GATT), assessing its impact on Brazil. Defensive of Brazilian interests, but offers no data to support arguments.

4355 Bergsten, Fred *et al.* O Brasil e a nova ordem internacional. Rio de Janeiro: Fundação Getúlio Vargas, Comitê de Cooperação Empresarial; Centro de Economia Mundial; Expressão e Cultura, 1991. 152 p.

Collection of speeches and essays sponsored by the Fundação Getúlio Vargas that relate loosely to the topic of Brazil's position in the new international order. Contributors include Fred Bergsten, Mário Henrique Simonsen, Francis Fukuyama, John Moberly, Raymond Barre, Nigel Lawson, Michael A. Walker, Armeane M. Choksi, and Carlos Geraldo Langoni.

4356 Caubet, Christian. As grandes manobras de Itaipú: energia, diplomacia e direito na Bacia do Prata. São Paulo: Editora Acadêmica, 1991. 385 p.: bibl., ill., maps.

Brilliant and exhaustive account of negotiations between Brazil, Argentina, and Paraguay concerning the Itaipú dam constructed on the Paraná river between Brazil and Paraguay. Focuses on international law.

4357 Cervo, Amado Luiz and **Clodoaldo Bueno.** História da política exterior do Brasil. São Paulo: Editora Atica, 1992. 432 p.: bibl. (Série Fundamentos; 81)

Broad historical overview of Brazil's foreign policy from 1822 to 1980s. Based almost exclusively on documents from the Brazilian Ministry of Foreign Relations. Solid but uninspired.

4358 Cervo, Amado Luiz. As relações históricas entre o Brasil e a Itália: o papel da diplomacia. Apresentação de Giovanni Agnelli. Sao Paulo: Istituto Italiano di Cultura; Brasília, DF: Editora UnB, 1991. 261 p.: bibl.

Insightful study regarding role of diplomacy in bilateral relations (case study: Brazil and Italy from 1861 to 1980s). Author concludes that societies of each country have been much more dynamic than their States, and argues that a vigorous foreign policy can compensate for such an imbalance.

4359 Costa, Thomaz Guedes da. A idéia de medidas de confiança mútua (CBMs) em uma visão Brasileira. (*Cont. Int.*, 14:2, julho/dez. 1992, p. 297–307)

Leading expert on Brazilian security affairs notes that although Brazil has not adopted explicitly the notion of confidence-building measures in its foreign policy discourse, that in practice CBMs have been adopted in the improved relations with Argentina, the process of regional economic integration, and various accords on nuclear and missile technology.

4360 Cruz Junior, Ademar Seabra; Antonio Ricardo F. Cavalcante; and **Luiz Pedone.** Brazil's foreign policy under Collor. (*J. Interam. Stud. World Aff.*, 35:1, 1993, p. 119–144, bibl.)

Broad overview of Brazil's foreign policy from 1990–91, based on the theory of hegemonic stability (Charles Kindleberger and Robert Keohane).

4361 Frota, Luciara Silveira de Aragão e. Brasil-Argentina: divergências e convergências. Brasília: Centro Gráfico do Senado Federal, 1991. 207 p.: bibl.

Brief history of Brazilian-Argentine relations within a global context. Although the author is Brazilian, the book is based almost exclusively on Argentine sources, including two dozen interviews with civilian and military officials in Buenos Aires. Lacks major argument and suffers from an imbalance in the sources.

4362 Hirst, Mônica and **Magdalena Segre.** La política exterior de Brasil en 1988: los avances posibles. (*Estud. Int./Santiago*, 22:88, oct./dic. 1989, p. 463–488)

Examines Brazil's relations in 1988 with other Latin American nations, the US, the USSR, and China.

4363 Hirst, Mônica. O pragmatismo impossível: a política externa do segundo Governo Vargas, 1951–1954. Rio de Janeiro: Fundação Getúlio Vargas, Centro de Pesquisa e Documentação de História Contemporânea do Brasil, 1990. 60 leaves: bibl. (Textos CPDOC)

Based on extensive primary documentation, concludes that during Vargas' second presidency, he was no longer able to extract major concessions from the US because of the latter's "total disinterest" in Latin America.

4364 Kirton, Mark. Towards greater inter-regional collaboration: new directions in Guyana-Brazil relations. (*J. Caribb. Stud.*, 8:1/2, Winter 1990/Summer 1991, p. 13–22)

Argues that Brazil's relations with Guyana have improved and intensified as Brazil has moved from a sub-imperialist role (1960s and early 1970s) to an independent and autonomous foreign policy (mid-1970s to 1990).

4365 Lafer, Celso. Perspectivas e possibilidades da inserção internacional do Brasil. (*Polít. Extern.*, 1:3, dez./jan./fev. 1992/93, p. 100–121)

Leading expert on Brazil's foreign policy examines Brazil's options in a rapidly changing external environment.

4366 Lamaziere, Georges and **Roberto Jaguaribe.** Beyond confidence-building: Brazilian-Argentine nuclear cooperation. (*Disarmament/UN*, 15:3, 1992, p. 102–117, bibl.)

Conceptual and chronological overview of Brazil's nuclear cooperation with Argentina by the two Brazilian diplomats most involved in those relations.

4367 Lyra, Heitor. A diplomacia brasileira na Primeira República, 1889–1930; e outros ensaios. Rio de Janeiro: Instituto Histórico e Geográfico Brasileiro, 1992. 63 p.: ill. (Col. Pedro Calmon; 1)

Lyra (1892–1973) was a Brazilian diplomat and diplomatic historian. These three short essays, published decades after his death, examine Brazil's diplomacy during the First Republic, focusing on leading diplomats (Oliveira Lima, Manuel de Araújo Porto-Alegre).

4368 Martins, Luciano. A nova ordem internacional e o Brasil. (*Polít. Extern.*, 1:3, dez./jan./fev. 1992/93, p. 172–176)

A short essay arguing that Brazil can increase its bargaining power in the new international order, but that its technological and educational deficiencies may leave the country on the margins of the "third industrial revolution."

4369 Mattos, Carlos de Meira. O Brasil no mundo em transição: seu poder e suas potencialidades. (*Defesa Nac.*, 756, abril/junho 1992, p. 18–28, bibl.)

Noted geopolitical theorist and retired army general, Meira Mattos considers Brazil's options in the new world order. Predictable, with no significant insights.

4370 Mello, Celso Albuquerque. O Brasil e o direito internacional da Nova Ordem Mundial. (*Rev. Bras. Estud. Polít.*, 76, jan. 1993, p. 7–26)

Author criticizes Brazil's lack of a well-defined foreign policy and argues that greater attention should be given to international law, "the weapon" of the weak.

4371 Mello, José Octávio de Arruda. A República no Brasil: ideologia, partidos e relações exteriores. Prefácio de Nelson Saldanha. Nota complementar de Gonzaga Rodrigues. João Pessoa, Brasil: Fundação Casa de José Américo, 1990. 103 p.: bibl.

Final chapter provides historiographical synthesis of Brazil's international relations, with a useful bibliography.

4372 Mello e Silva, Alexandra de. Desenvolvimento e multilateralismo: um estudo sobre a operação Pan-Americana no contexto da política externa de JK. (*Cont. Int.*, 14:2, julho/dez. 1992, p. 209–239)

Well-written analysis of President Juscelino Kubitschek's "Operation Pan America," proposed in 1958 to improve Latin American living standards. Concludes that the initiative was driven by political and economic concerns, as well as a desire for multilateralism.

4373 Miceli, Sergio. A desilusão americana: relações acadêmicas entre Brasil e Estados Unidos. São Paulo: IDESP; Editora Sumaré; Conselho Nacional de Desenvolvimento Científico e Tecnológico, 1990. 80 p.: bibl., ill.

Self-pitying yet fascinating assessment of academic relations between Brazil and the US. Includes case study of the Ford Foundation, which author claims is an instrument of US cultural domination.

4374 **Motoyama, Shozo.** Brasil e Japão: ocidente e oriente na cultura universal. (*Iberoam./Tokyo*, 17:2, 1996, p. 1–15)

Discusses rarely-examined relationship between Brazil and Japan in the natural science fields (astronomy, meteorology, biology, physics, etc.) before and after WWII. Includes episodes such as the 1931 foundation of the Instituto Kurihara de Ciência Natural Brasileira in Mirandopolis, São Paulo, and Japanese-Brazilians' post-WWII support for the financially stressed Japanese Society for Basic Physics. [K. Horisaka]

4375 **Moura, Gerson.** O alinhamento sem recompensa: a política externa do Governo Dutra. Rio de Janeiro: Fundação Getúlio Vargas, Centro de Pesquisa e Documentação de História Contemporânea do Brasil, 1990. 113 p: bibl. (Textos CPDOC)

Concludes that under President Dutra (1946–50) the policy of "automatic alignment" with the US failed to benefit Brazil. Essential reading for understanding Brazil's foreign policy during this period.

4376 **Moura, Gerson.** Sucessos e ilusões: relações internacionais do Brasil durante e após a Segunda Guerra Mundial. Rio de Janeiro: FGV, Editora da Fundação Getúlio Vargas, 1991. 116 p.: bibl.

Splendid book by the late Gerson Moura addresses Brazil's foreign relations during the 1940s. Focuses on Brazil's relations with the US. Notes pendular swing in Brazil's foreign policy between pompous rhetoric and servile action, both of which author considered inappropriate.

4377 **Nippon Brazil Kōryūshi = História das relações nipo-Brasileiras.** Edited by Commissão Organizadora do Centenário do Tratado de Amizade Japão-Brasil. Tokyo: Nippon Brazil Chūō Kyōkai (Associação Central Nipo-Brasileira), 1995. 448 p.: chronology, photos, tables.

Commemorating the 100th anniversary of the signing of the Treaty of Friendship, Commerce, and Navigation by Japan and Brazil, work examines the history of Brazilian-Japanese relations. Contains contributions by 23 Brazilian (including *nisei*) and Japanese scholars and journalists, and former Japanese officials involved with diplomatic and economic relations, immigration, economic and cultural cooperation, and so forth. Useful chronology of the bilateral relationship. [K. Horisaka]

4378 **Pereira, Luiz Carlos Bresser** and **Vera Thorstensen.** Do Mercosul à integração americana. (*Polít. Extern.*, 1:3, dez./jan./fev. 1992/93, p. 122–145, appendix, bibl., tables)

Influential economists boldly call for Brazil to abandon its multilateral strategy (favored by the Ministry of Foreign Relations) and adopt a strategy of American integration (even if under US leadership) as an "insurance policy" against uncertainties created by economic blocs.

4379 **Prieto, Jaire Brito** *et al.* Uma interpretação da conjuntura internacional e perspectivas para a indústria bélica brasileira. (*Defesa Nac.*, 756, abril/junho 1992, p. 53–60)

Military officers assess and bemoan the impact of changes in the international system on Brazil's arms industry. In dated and self-interested fashion, they call for continued government support for the industry.

4380 **Raiol, Osvaldino da Silva.** A utopia da terra na fronteira da Amazônia: a geopolítica e o conflito pela posse da terra no Amapá. Macapá, Brasil: Editora Gráfica o Dia Ltda., 1992. 240 p.: bibl., ill., maps.

Examination of geopolitical conflicts in the Brazilian Amazon focuses on the territory of Amapá. Laments that the Brazilian Amazonian border has been "militarized" with the Calha Norte program. Claims that such militarization has assisted "large national and international capitalist enterprises." Impassioned style detracts from argument.

4381 **Ricupero, Rubens.** Crônica de uma negociação na conferência do meio-ambiente e desenvolvimento, Rio/92. (*Lua Nova*, 28/29, 1993, p. 265–282)

One of Brazil's leading diplomats provides a personal and frank account of negotiations concerning "Agenda 21," one of the major documents of the 1992 environmental conference in Rio de Janeiro.

4382 **Ricupero, Rubens et al.** O futuro do Brasil: a América Latina e o fim da guerra fria. Organização de José Alvaro Moisés. São Paulo: Paz e Terra; USP Política Internacional & Comparada, 1992. 191 p.: bibl.

Examines Brazil's future within "the new international order." First-rate contributors include diplomats, scholars, and entrepreneurs: Rubens Ricupero, Ronaldo Motta Sardenberg, Marcílio Marques Moreira, José Augusto Guilhon Albuquerque, John Chipman, Geraldo de Figueiredo Forbes, Kotaro Horisaka, and Robert Keohane.

4383 **Saraiva, José Flávio Sombra.** Brazil's African policy: historical dimension. (*UNISA/Lat. Am. Rep.*, 9:2, 1993, p. 26–30, bibl., photo)

Brief synthesis of author's PhD dissertation (Univ. of Birmingham, UK), which argues that Brazil's close ties with Africa since 1961 were driven as much by diplomatic and economic Realpolitik as by cultural considerations.

4384 **Sarmento, Walney Moraes.** Subdesenvolvimento e política externa: o exemplo do Brasil. Salvador, Bahia: Ianamá, 1991. 96 p.: bibl.

Neo-Marxist explanation of the relationship between Brazil's "underdevelopment" and its foreign policy. Weak analysis is further clouded by facile interpretations of dependency theory.

4385 **Silva, José Luiz Foresti Werneck da.** As duas faces da moeda: a política externa do Brasil monárquico, 1831–1876. Rio de Janeiro: Univ. Aberta, 1990. 94 p.: bibl. (Uma história geral do Brasil; livreto 23)

Diplomatic historian juxtaposes two "faces" of Brazil's foreign policy during the Empire: a structural dependence on Great Britain and autonomous relationships with Uruguay, Paraguay, and Argentina, in which Brazil sought to maintain the status quo (namely, an independent Uruguay). Concise and well-written.

4386 **Silva, Patricio.** Democratization and foreign policy: the cases of Argentina and Brazil. (*in* Democratization and the State in the Southern Cone: essays on South American politics. Edited by Benno Galjart and Patricio Silva. Amsterdam: CEDLA, 1989, p. 83–102, bibl., tables)

Insightful analysis that compares the impact of regime change on foreign policy in Argentina ("profound reorientation") and Brazil ("remarkable degree of continuity"). Attributes the difference to Brazil's strong diplomatic bureaucracy, which achieved "significant autonomy from the executive in the formulation of foreign policy."

4387 **Temas de política externa brasileira.** v. 1. Organização de Gelson Fonseca Júnior y Valdemar Carneiro Leão. Brasília: Fundação Alexandre de Gusmão, Instituto de Pesquisa de Relações Internacionais; Editora Atica, 1989. 288 p.: bibl. (Col. Relações internacionais)

Essays by 11 Brazilian diplomats and one scholar (Vicente Marotta Rangel) on factors that affect Brazil's foreign policy. The final chapter by Gelson Fonseca Jr. examines recent studies on Brazil's foreign policy and is especially good.

4388 **Vasconcelos, Luiz L.** Um repasse sobre as relações Brasil-Cuba. (*Cont. Int.*, 13:2, julho/dez. 1991, p. 187–203, tables)

Succinct and even-handed examination of Brazil's relations with Cuba since 1964, when Brazil broke diplomatic relations with the island. Those relations were reestablished in June 1986, under President Sarney.

Vázquez Machicado, Humberto. Para una historia de los límites entre Bolivia y el Brasil. See item **4240**.

4389 **Veiga, Pedro da Motta; Paulo Guilherme Correa;** and **João Bosco Machado M.** Efeitos do AAP3 sobre as relações comerciais Brasil-Chile. (*Rev. Bras. Comér. Exter.*, 9:34, jan./março 1993, p. 11–26, bibl., tables)

Rich in data, article examines Brazil's trade with Chile from 1970–90, focusing on the (allegedly) positive effects of the Acordo de Alcance Parcial No. 3, signed in 1983 between the two countries.

4390 **Vidigal, Armando Amorim Ferreira.** O Brasil e a nova ordem mundial. Rio de Janeiro: Serviço de Documentação Geral da Marinha, 1991. 67 p.: bibl., col. maps.

Idealistic essay concerning Brazil's options in the new world order. Lacks theoretical orientation. Concludes with a set of recommendations for Brazil's foreign policy.

4391 Vinholes, Luiz Carlos. Intercâmbio cultural e artístico nas relações Brasil-Japão. (*Iberoam./Tokyo*, 17:2, segundo semestre 1995, p. 17–37, bibl.)

Discusses cultural and artistic exchanges between Brazil and Japan during their past 100 years of diplomatic relations. Covers literature and poetry (including the influence of haiku, a 17-syllable Japanese verse, on Brazilian poetry), music, theater, plastic arts, architecture, mass media, religious activities, and sports.

4392 Weis, W. Michael. Cold warriors & coups d'etat: Brazilian-American relations, 1945–1964. Albuquerque: Univ. of New Mexico, 1993. 262 p.: bibl., index.

Important contribution to our understanding of the changing Brazilian-US relationship in the post-World War II era. Argues that the "unwritten alliance" between Brazil and the US gave way to growing hostility, especially under Juscelino Kubitschek's *Operação Pan Americana*, which inspired Brazil's independent foreign policy.

JOURNAL ABBREVIATIONS

Am. Lat./Moscow. América Latina. Academia de Ciencias de la Unión de Repúblicas Soviéticas Socialistas. Moscow.

AMRI. Anuario Mexicano de Relaciones Internacionales. UNAM, ENEP Acatlán, Escuela Nacional de Estudios Profesionales. México.

Anál. Econ./México. Análisis Económico. Unidad Azcapotzalco, Univ. Nacional Autónoma de México. México.

Anál. Int. Análisis Internacional. Centro Peruano de Estudios Internacionales. Lima.

Ann. Am. Acad. Polit. Soc. Sci. The Annals of the American Academy of Political and Social Science. Philadelphia, Penn.

Anu. Estud. Centroam. Anuario de Estudios Centroamericanos. Univ. de Costa Rica. San José.

Bol. Acad. Nac. Hist./Caracas. Boletín de la Academia Nacional de la Historia. Caracas.

Bull. Lat. Am. Res. Bulletin of Latin American Research. Society for Latin American Studies. Oxford, England.

Cah. Am. lat. Cahiers des Amériques latines. Paris.

Can. J. Lat. Am. Caribb. Stud. Canadian Journal of Latin American and Caribbean Studies. Univ. of Ottawa. Ontario, Canada.

Caribb. Aff. Caribbean Affairs. Trinidad Express Newspapers Ltd.; Inprint Caribbean Ltd., Port of Spain, Trinidad and Tobago.

Caribb. Q. Caribbean Quarterly. Univ. of the West Indies. Mona, Jamaica.

Caribb. Stud. Caribbean Studies. Univ. of Puerto Rico, Institute of Caribbean Studies. Río Piedras.

Caribe Contemp. El Caribe Contemporáneo. Univ. Nacional Autónoma de México. México.

Cienc. Polít. Ciencia Política. Instituto de Ciencia Política de Bogotá; Tierra Firme Editores. Bogotá.

Columbia J. Transnatl. Law. Columbia Journal of Transnational Law. Columbia Univ. School of Law. New York.

Columbia J. World Bus. Columbia Journal of World Business. Columbia Univ., New York.

Comer. Exter. Comercio Exterior. Banco Nacional de Comercio Exterior. México.

Comp. Polit. Stud. Comparative Political Studies. Sage Publications, Thousand Oaks, Calif.

Cont. Int. Contexto Internacional. Instituto de Relações Internacionais, Pontifícia Univ. Católica. Rio de Janeiro.

Cuad. INVESP. Cuadernos de INVESP. Instituto Venezolano de Estudios Sociales y Políticos. Caracas.

Cuad. Marcha. Cuadernos de Marcha. Eon Editores. Montevideo.

Cuad. Nuestra Am. Cuadernos de Nuestra América. Centro de Estudios sobre América. La Habana.

Cuba. Stud. Cuban Studies. Univ. of Pittsburgh, Center for Latin American Studies. Pittsburgh, Penn.

Curr. Hist. Current History. Philadelphia, Penn.

Defesa Nac. A Defesa Nacional: Revista de Assuntos Militares e Estudo de Problemas Brasileiros. Rio de Janeiro.

Desarro. Econ. Desarrollo Económico. Instituto de Desarrollo Económico y Social. Buenos Aires.

Disarmament/UN. Disarmament. United Nations. New York.

ECA/San Salvador. Estudios Centro-Americanos: ECA. Univ. Centroamericana José Simeón Cañas. San Salvador.

Econ. Cienc. Soc. Economía y Ciencias Sociales. Facultad de Ciencias Económicas y Sociales, Univ. Central de Venezuela. Caracas.

Estad. Soc. Estado & Sociedad: Revista Boliviana de Ciencias Sociales. Facultad Latinoamericana de Ciencias Sociales (FLACSO). La Paz.

Estud. Avançados. Estudos Avançados. Univ. de São Paulo, Instituto de Estudos Avançados. São Paulo.

Estud. Cult. Contemp. Estudios sobre las Culturas Contemporáneas. Centro Universitario de Investigaciones Sociales, Univ. de Colima. México.

Estud. Int./IRIPAZ. Estudios Internacionales: Revista del IRIPAZ. Instituto de Relaciones Internacionales y de Investigaciones para la Paz. Guatemala.

Estud. Int./Santiago. Estudios Internacionales. Instituto de Estudios Internacionales, Univ. de Chile. Santiago.

Estud. Soc. Centroam. Estudios Sociales Centroamericanos. Programa Centroamericano de Ciencias Sociales. San José.

Estud. Soc./Santiago. Estudios Sociales. Corporación de Promoción Universitaria. Santiago.

Foreign Aff. Foreign Affairs. Council on Foreign Relations, Inc. New York.

Foreign Policy. Foreign Policy. National Affairs Inc.; Carnegie Endowment for International Peace. New York.

Foro Int. Foro Internacional. El Colegio de México. México.

Foro Polít. Foro Político. Instituto de Ciencias Políticas, Univ. del Museo Social Argentino. Buenos Aires.

Fuerzas Arm. Soc. Fuerzas Armadas y Sociedad. Centro Latinoamericano de Defensa y Desarme; FLACSO. Santiago.

Iberoam./Tokyo. Iberoamericana. Univ. of Sofia. Tokyo.

Int. Aff./Moscow. International Affairs. Moscow.

Int. Hist. Rev. The International History Review. Univ. of Toronto Press. Downsview, Ontario, Canada.

Int. J./Toronto. International Journal. Canadian Institute of International Affairs. Toronto, Canada.

Int. Migr. Rev. International Migration Review. Center for Migration Studies. New York.

Int. Stud. Q. International Studies Quarterly. Wayne State Univ., Detroit, Mich.

Integr. Latinoam. Integración Latinoamericana. Instituto para la Integración de América Latina. Buenos Aires.

J. Caribb. Stud. Journal of Caribbean Studies. Assn. of Caribbean Studies. Coral Gables, Fla.

J. Commonw. Comp. Polit. The Journal of Commonwealth & Comparative Politics. Univ. of London, Institute of Commonwealth Studies. London.

J. Int. Law Econ. Journal of International Law and Economics. George Washington Univ., The National Law Center. Washington.

J. Interam. Stud. World Aff. Journal of Interamerican Studies and World Affairs. Institute of Interamerican Studies, Univ. of Miami. Coral Gables, Fla.

J. Lat. Am. Stud. Journal of Latin American Studies. Centers or Institutes of Latin American Studies at the Universities of Cambridge, Glasgow, Liverpool, London, and Oxford. Cambridge Univ. Press. London.

J. Polit. Mil. Sociol. Journal of Political & Military Sociology. Northern Illinois Univ., Dept. of Sociology. DeKalb, Ill.

Journal. Q. Journalism Quarterly. Assn. for Education in Journalism; American Assn. of Schools and Depts. of Journalism; Kappa Tau Alpha Society.; Univ. of Minnesota. Minneapolis, Minn.

LARR. Latin American Research Review. Latin American Research Review Board. Univ. of New Mexico, Albuquerque, N.M.

Lat. Am. Perspect. Latin American Perspectives. Univ. of California. Newbury Park, Calif.

Latinoamerica/Rome. Latinoamerica. Edizioni Associate. Rome.

Lua Nova. Lua Nova. Editora Brasiliense. São Paulo.

MACLAS Lat. Am. Essays. MACLAS Latin American Essays. Middle Atlantic Council of Latin American Studies. New Brunswick, N.J.

Mesoamérica/Antigua. Mesoamérica. Centro de Investigaciones Regionales de Mesoamérica. Antigua, Guatemala.

Natl. Int. The National Interest. National Affairs. Washington.

Nav. War Coll. Rev. Naval War College Review. Newport, R.I.

Nueva Soc. Nueva Sociedad. Caracas.

Opin. Anál. Opiniones y Análisis. Fundación Boliviana para la Capacitación Democrática y la Investigación. La Paz.

Panorama Centroam. Pensam. Panorama Centroamericano: Pensamiento y Acción. Instituto Centroamericano de Estudios Políticos (INCEP). Guatemala.

Pensam. Iberoam. Pensamiento Iberoamericano. Instituto de Cooperación Iberoamericano (ICI) de España; Comisión Económica para América Latina y el Caribe (CEPAL). Madrid.

Polít. Extern. Política Externa. Paz e Terra. São Paulo.

Polít. Int. Política Internacional. Revista de la Academia Diplomática del Peru. Lima.

Polit. Sci. Q. Political Science Quarterly. Columbia Univ., The Academy of Political Science. New York.

Probl. Communism. Problems of Communism. United States Information Agency. Washington.

Relac. Int./Mexico. Relaciones Internacionales. Centro de Relaciones Internacionales. Facultad de Ciencias Políticas y Sociales, Univ. Nacional Autónoma de México. México.

Rev. Arg. Estud. Estrateg. Revista Argentina de Estudios Estratégicos. Olcese Editores. Buenos Aires.

Rev. Bras. Comér. Exter. Revista Brasileira de Comércio Exterior. Fundação Centro de Estudos do Comércio Exterior. Rio de Janeiro.

Rev. Bras. Estud. Polít. Revista Brasileira de Estudos Políticos. Univ. de Minas Gerais. Belo Horizonte, Brazil.

Rev. Cancillería San Carlos. Revista Cancillería de San Carlos. Ministerio de Relaciones Exteriores. Bogotá.

Rev. Chil. Geopolít. Revista Chilena de Geopolítica. Instituto Geopolítico de Chile. Santiago.

Rev. Ciclos. Revista Ciclos en la Historia, Economía y la Sociedad. Fundación de Investigaciones Históricas, Económicas y Sociales, Facultad de Ciencias Económicas, Univ. de Buenos Aires. Buenos Aires.

Rev. Cienc. Soc./San José. Revista de Ciencias Sociales. Univ. de Costa Rica. San José.

Rev. Eur. Revista Europea de Estudios Latinoamericanos y del Caribe = European Review of Latin American and Caribbean Studies. Center for Latin American Research and Documentation; Royal Institute of Linguistics and Anthropology. Amsterdam.

Rev. Invest. Econ. Revista de Investigaciones Económicas. Centro de Investigación, Facultad de Economía, Univ. de Panamá.

Rev. Mex. Cienc. Polít. Soc. Revista Mexicana de Ciencias Políticas y Sociales. Facultad de Ciencias Políticas y Sociales, Univ. Nacional Autónoma de México. México.

Rev. Mex. Polít. Exter.   Revista Mexicana de Política Exterior. Secretaría de Relaciones Exteriores; Instituto Matías Romero de Estudios Diplomáticos (IMPED). México.

Rev. Mex. Sociol.   Revista Mexicana de Sociología. Instituto de Investigaciones Sociales, Univ. Nacional Autónoma de México. México.

Rev. Occident.   Revista de Occidente. Madrid.

Rev. Panameña Sociol.   Revista Panameña de Sociología. Univ. de Panamá, Depto. de Sociología. Panamá.

Rev. Parag. Sociol.   Revista Paraguaya de Sociología. Centro Paraguayo de Estudios Sociológicos. Asunción.

Rev. Soc. haïti.   Revue de la Société haïtienne d'histoire et géographie. Port-au-Prince.

Rev. Tiers-Monde.   Revue Tiers-Monde. Institut d'étude du développement économique et social, Univ. de Paris.

Secuencia/México.   Secuencia. Instituto Mora. México.

SER/Buenos Aires.   Seguridad, Estrategia Regional en el 2000. Impresión Zona Gráfica. Buenos Aires.

Síntesis/Madrid.   Síntesis. Asociación de Investigación y Especialización sobre Temas Latinoamericanos. Madrid.

Tareas/Panamá.   Tareas. Centro de Estudios Latinoamericanos (CELA). Panamá.

Trace/México.   Trace. Centre d'études mexicaines et centraméricaines. México.

UNISA/Lat. Am. Rep.   UNISA/Latin American Report. Univ. of South Africa. Pretoria.

Universum/Talca.   Universum. Univ. de Talca. Talca, Chile.

Voices Mex.   Voices of Mexico. Univ. Nacional Autónoma de México. México.

Wash. Q.   The Washington Quarterly. Georgetown Univ., The Center for Strategic and International Studies. Washington.

World Policy J.   World Policy Journal. World Policy Institute. New York.

SOCIOLOGY

GENERAL

4393 Acosta, Maruja and **Roberto Briceño-León.** Ciudad y capitalismo. Caracas: Univ. Central de Venezuela, Ediciones de la Biblioteca, 1987. 286 p.: bibl., ill. (Col. Ciencias económicas y sociales; 33)

Provides interesting analysis of various aspects of urban sociology. Pt. 1 reviews ecological, social/psychological, developmentalist, and dependency theories. Pt. 2 focuses on ideology, exploring role of dominant cultures in the spatial organization and semiotic deconstruction of the city as a set of symbols. Pt. 3 compares impact of social class on urban growth and renovation in Venezuela, Cuba, and Nicaragua. [D. Levi]

4394 Albornoz, Orlando. Sociología y tercer mundo. Caracas: Univ. Central de Venezuela, Ediciones de la Biblioteca, 1991. 182 p.: bibl. (Col. Ciencias económicas y sociales; 34)

Contains stimulating and challenging essays from the World Congress of Sociology (12th, Madrid, 1990) and other forums on fundamental theoretical and practical problems posed by internationalization/globalization of society and sociology. Rejects "provincialist" and "chauvinist" tendencies that would "relegate sociology to intellectual oblivion" in favor of a "singular sociology" that "approximates society as a global, and at the same time, historically and culturally concrete, fact." Incisive discussions on "the social responsibility of the profession," "the indigenization of sociology," "the education-development relationship," "the State/civil society relationship in Latin America and the Caribbean," "orthodox-scientific" vs. "militant" research, the "Third World" concept, etc. [D. Levi]

4395 Alternative Lateinamerika: das deutsche Exil in der Zeit des Nationalsozialismus. Edited by Karl Kohut and Patrik von zur Mühlen. Frankfurt am Main: Vervuert, 1994. 257 p.: bibl. (Bibliotheca Ibero-Americana; 51)

Very useful collection includes 17 research papers presented by international panel of scholars at international symposium on German exiles in Spain, Portugal, and Latin America. Covers broad range of topics including countries of transit, immigration policies in Latin America, Jewish immigrants, Marxist refugees, cooperation with allies, exile literature, and assimilation. All papers but one include bibliographies. [C. Converse]

4396 Ambiente, Estado y sociedad: crisis y conflictos socio-ambientales en América Latina y Venezuela. Coordinación de María-Pilar García Guadilla. Caracas: Univ. Simón Bolívar, Centro de Estudios del Desarrollo, 1991. 408 p.: bibl., ill.

Argues that ecological concerns should be integral part of development strategies in Latin America; however, civil society and the State have failed to understand that "the environmental question synthesizes all the factors that affect the quality of life." Scientific/technical revolution promotes ecological awareness, restructuring of the State, and democratization. Includes comparative studies of Venezuela, Mexico, and Colombia. [D. Levi]

4397 Anderson, Jeanine. Estrategias de sobrevivencia revisitadas. (*in* Las mujeres y la vida de las ciudades. Recopilación de María del Carmen Feijoó y Hilda María Herzer. Buenos Aires: Grupo Editor Latinoamericano; IIED-América Latina, 1991, p. 33–62, bibl.)

Critical analysis examines concept of and research on survival strategies. Notes changes in such strategies over the course of economic crisis in Peru, as well as response of dominant classes to survival networks established by the poor, the dominant classes often having managed to either control or redirect these networks. Anderson argues for need to look at emerging macro as well as micro structures of survival strategies. [C.B. Flora]

4398 Barsky, Osvaldo. Políticas agrarias en América Latina. Buenos Aires?: Ediciones Imago Mundi; Grupo Esquel, 1990. 134 p.: bibl. (Col. América debate)

Very good study examines growing rural differentiation among production units (regardless of relative size). Suggests that rural development strategies should now be aimed principally at viable small producers (and perhaps landless rural workers). Taking into account heterogeneity of productive units, these strategies should enhance local access to credit, technology, markets, services, and ownership. [R.P. Korzeniewicz]

4399 Bastian, Jean-Pierre. Le protestantisme en Amérique latine: une approche socio-historique. Geneva, Switzerland: Editions Labor et Fides, 1994. 324 p.: bibl., tables. (Histoire et société; 27)

Professor of sociology and reputed expert on Protestantism in Mexican history, Bastian presents general survey on the diffusion of Protestant doctrines in Latin America since early colonial times. Mostly devoted to the development of such ideals in the aftermath of independence and in connection with liberal movements, religious tolerance, anti-Catholic radicalism, struggles for democracy, and populist regimes. According to Bastian, in recent times (1960-present) Protestantism has undergone a curious "mutation" on the continent, as evidenced by its enormous expansion, authoritarian style, and domination mechanisms. Appendices include bibliographical essay and comprehensive list of publications. [T. Hampe-Martínez]

4400 Beneker, Tine. Stedengroei en binnenlandse migratie in Centraal-Amerika met speciale aandacht voor de kleine steden [Urban growth and internal migration in Central America with special emphasis on small towns]. Utrecht, The Netherlands: Faculteit Ruimtelijke Wetenschappen Univ. Utrecht, 1993. 42 p.: bibl., ill., maps, tables. (Diskussiestukken van de Vakgroep Social Geografie van Ontwikkelingslanden; 50)

Study of postwar urbanization in Central America is based on secondary statistical data. Author looks at population density, population growth, settlement patterns, and urban growth. She also analyzes areas that attract migrants, attempting to link internal migration and urban growth. [R. Hoefte]

4401 Bifani, Patricia. Disponibilidad, derecho y gestión del espacio vital. (*Nueva Soc.*, 123, enero/feb. 1993, p. 84–93, ill.)

Utilizando estas nociones como herramientas analíticas, la autora hace una rápida panorámica de las condiciones medioambientales y de pobreza del Tercer Mundo, señalando algunos patrones profundos de evolución y apuntando la necesidad de reexaminar la relación entre ambos términos. [M.A. Garretón]

4402 Calderón G., Fernando and **Patricia Provoste Fernández.** Autonomía, estabilidad y renovación: los desafíos de las ciencias sociales en América Latina. 2. ed. Buenos Aires: Consejo Latinoamericano de Ciencias Sociales, 1992. 275 p.: bibl., ill.

Based on an exhaustive study, book provides extensive data on disparate characteristics and uneven development of over 100 social science centers, showing increasing differentiation of research from teaching activities (to the detriment of the latter) as well as new academic hierarchies between and among countries, centers, and scholars. Very thorough regional overview of social science research and teaching institutions. [R.P. Korzeniewicz]

4403 Calderón G., Fernando and **Mario R. dos Santos.** Hacia un nuevo orden estatal en América Latina: veinte tesis sociopolíticas y un corolario. Buenos Aires: CLACSO; Santiago: Fondo de Cultura Económica, 1991. 166 p. (Sección de obras de sociología)

Síntesis en torno a las situaciones y perspectivas sociopolíticas de la región referidas a las transformaciones del Estado. Contiene las 20 tesis resultantes, un conjunto de notas con ejemplos y referencias empíricas divididas en una área sociopolítica y una socioeconómica, y cuatro comentarios globales de personalidades. [M.A. Garretón]

4404 Calderón G., Fernando; Alejandro Piscitelli; and **José Luis Reyna.** Social movements: actors, theories, expectations. (*in* The making of social movements in Latin America: identity, strategy, and democracy. Edited by Arturo Escobar and Sonia E. Alvarez. Boulder, Colo.: Westview Press, 1992, p. 19–36)

Authors examine changes in characteristics of contemporary social movements and their implications for social sciences, presenting current landscape of social movements in Latin America and outlining main reasons for their appearance. Significant changes in nature and scope of collective action are identified. Also explores relationship between new manifestations of collective action and social science models through which they are studied, suggesting necessary changes for Latin American social sciences. [B. Aguirre-López]

4405 Carrasco Reyes, Ella. La formación de los trabajadores sociales en América Latina, 1987–1989. Lima: CELATS, 1991. 177 p.: bibl. (Nuevos cuadernos; 18)

Growth and professionalism of social work as a career in the Latin American context is presented in terms of social work pedagogy of 40 academic units in 17 countries of Latin America. Characteristics of students, their subsequent employment, and content of curricula are analyzed. The best pedagogy involves theory and practice, has an integrated focus, and builds a strong professional identity. As a result, a number of innovative social work practices have developed in response to changing Latin American context. Research is increasingly part of social work curriculum. [C.B. Flora]

4406 Chinchilla, Norma Stoltz. Marxism, feminism, and the struggle for democracy in Latin America. (*in* The making of social movements in Latin America: identity, strategy, and democracy. Edited by Arturo Escobar and Sonia E. Alvarez. Boulder, Colo.: Westview Press, 1992, p. 37–51)

Discusses relationship of feminism and Marxism to ideas about democracy and socialism, and relationship of Marxist and feminist perspectives to each other in context of contemporary social movements in Latin America. Argues that issues that were once considered irreconcilable are no longer seen as such, and that new conceptions of the relationship of class to gender and of daily life to the struggle for democracy and socialism are emerging. Contains brief review of evolution of contemporary feminist and New Marxist movements in Latin America. [B. Aguirre-López]

4407 Confronting the crisis in Latin America: women organizing for change. Santiago: Isis International; Development Alternatives with Women for a New Era (DAWN), 1988. 112 p.: bibl., ill. (Book series; 1988/2)

Series of articles prepared by women researchers organized by Development Alternatives with Women for a New Era (DAWN) examines effects of economic, social, and political crises during 1980s on lives of women in Latin America; policies enacted by governments; and methods used by women to create new forms of participation in subsistence economy and social movement organizations. [B. Aguirre-López]

Cultural expression and grassroots development: cases from Latin America and the Caribbean. See item **53**.

4408 Cumbre Iberoamericana, *1st, Guadalajara, Mexico, 1991*. Nota sobre el desarrollo social en América Latina. Santiago: Naciones Unidas, Comisión Económica para América Latina y El Caribe, 1992? 51 p.: bibl., ill.

Concise report on developmental trends from 1945–1980s emphasizes social-economic connection. Growth and disarticulation up to 1970s were followed in 1980s by "profound crisis" and regressive "adjustment costs" that undermined democratization. Education and training are essential to meet the "social challenge" in future decades. Examines population, employment, education, health, income, social inequality, environment, etc. [D. Levi]

Demange, Nilson Joseph; Lili Katsuco Kawamura; and Akihito Ito. Internacionalização no Japão e a América Latina. See item **3804**.

4409 Direitos reprodutivos. Coordenação de Sandra Azeredo e Verena Stolcke. São Paulo: Fundação Carlos Chagas, Concurso de Pesquisa sobre Direitos Reprodutivos, PRODIR, 1991. 186 p: bibl.

Includes papers and debates presented at a conference on reproductive rights organized by PRODIR, a Latin American program of training and research on reproductive rights. Topics include theory, methodology, population growth, fertility change, reproductive rights, and race issues. [P. Lovell]

4410 Elizalde, Antonio. Desarrollo a escala humana, economía social y microproyectos: desafíos y alternativas. (*in* Encuentro Internacional sobre Política Social, *1st, Vito-*

ria, Spain, 1990. Actas. Vitoria, Spain: Servicio Central de Publicaciones del Gobierno Vasco, 1991, p. 279–299)

Argumentación en torno a la idea de un modelo de desarrollo "autodependiente" que, partiendo de las condiciones reales a nivel local, sustente un proyecto de desarrollo ajustado a sus características específicas. [M.A. Garretón]

Encuentro de Historia y Realidad Económica y Social del Ecuador, 6th, Cuenca, Ecuador, 1989. Los campesinos en el proceso latinoamericano de los años ochenta y sus perspectivas. v. 1–2. See item **4859**.

Encuentro Nacional de Centros de Prevención de la Violencia Doméstica y Asistencia a la Mujer Golpeada, 1st, Chapadmalal, Argentina, 1988. Mujer golpeada. See item **5096**.

Ensayos críticos para el estudio de las organizaciones en México. See item **4488**.

4411 Era de nieblas: derechos humanos, terrorismo de Estado y salud psicosocial en América Latina. Edición de Horacio Riquelme U. Caracas: Editorial Nueva Sociedad, 1990. 190 p.: bibl.

Catorce ensayos de profesionales europeos y latinoamericanos del área psicosocial, articulados en torno a dos ejes centrales: uno enfocado a la teoría y práctica psicoterapéutica frente a la violencia organizada, y otro dedicado a la esfera cultural bajo el terrorismo del Estado. [M.A. Garretón]

4412 Errázuriz, Margarita M. El gobierno local como espacio para la acción con mujeres: promesa que requiere reflexiones. (*in* Políticas sociales, mujeres y gobierno local. Santiago: Corporación de Investigaciones Económicas para Latinoamérica, 1992, p. 31–49, bibl.)

El documento plantea que, pese a las potencialidades de los gobiernos locales para generar acciones en torno a las necesidades prácticas de género, está en discusión la capacidad estratégica de una contribución efectiva a la modificación de la condición de la mujer. [M.A. Garretón]

4413 Escobar, Arturo. Culture, economics, and politics in Latin American social movements: theory and research. (*in* The making of social movements in Latin America: identity, strategy, and democracy. Edited by Arturo Escobar and Sonia E. Alvarez. Boulder, Colo.: Westview Press, 1992, p. 62–85)

Focuses on epistemological and political context within which contemporary theory and research on social movements are being produced, especially in Latin America. Argues that crisis of development which generates social movements cannot be conceptualized in economic and political terms alone; the crisis in the cultural discourse of modernity is also a factor. Presents critical view of "epistemo-political" context of social movements and offers framework and methodology for considering cultural aspects of contemporary collective action. [B. Aguirre-López]

4414 García, Jesús Alberto. Afroamericano soy. Caracas: La Espada Rota: TIDCAV, 1987. 105 p.: bibl., ill.

Examination of African experience in the Americas covers slave trade, *quilombo* as "cultural refuge," European colonists' deportation of "freed" slaves from Jamaica and US back to Africa, and political ideas of Garvey, DuBois, etc. Focuses especially on African influences on music (merengue, rumba, etc.) and religion (*santería*, Rastafarianism, etc.). [D. Levi]

4415 Garretón Merino, Manuel Antonio. New State-society relations in Latin America. (*in* Redefining the State in Latin America. Paris: Organisation for Economic Co-operation and Development, 1994, p. 239–249)

Latin America's new sociohistorical context creates a need to redefine relations between the State and civil society and to undertake reform of the State in a manner that will promote autonomy and development for the State, politics, economics, and society. [M.A. Garretón]

4416 Garretón Merino, Manuel Antonio and **Malva Espinosa.** ¿Reforma del Estado o cambio en la matriz sociopolítica? (*Perf. Latinoam.*, 1:1, dic. 1992, p. 133–170)

Se propone una perspectiva analítica sobre la reforma del Estado que incluya las diferentes dimensiones de éste en sus relaciones con el sistema política y la sociedad civil bajo el concepto de matriz sociopolítica. Se describe, bajo esa perspectiva, las transformaciones del Estado chileno en los últimos decenios. [M.A. Garretón]

4417 **Gomes, Angela Maria de Castro** *et al.* Estado, corporativismo y acción social en Brasil, Argentina y Uruguay. Buenos Aires: Editorial Biblos; Fundación Simón Rodríguez, 1992. 111 p.: bibl. (Col. Cuadernos Simón Rodríguez; 22)

Three case studies (on Brazil, Uruguay, and Argentina) from a 1991 conference examine new role of the State after the Great Depression. Detailed analysis of Fundación Eva Peron during first Peronist Administration is the most useful contribution. [R.P. Korzeniewicz]

4418 **Gomezjara, Francisco A.** and **Herminia C. Foo.** La sociología frente a la nueva derecha y la posmodernidad. (*in* Sistemas políticos, poder y sociedad: estudios de casos en América Latina. Caracas: Asociación Latinoamericana de Sociología; Centro de Estudios sobre América; Editorial Nueva Sociedad, 1992, p. 129–148, bibl.)

Offers provocative critique of sociology. In Latin America sociology is a "colonized science" whose shifting predilections (e.g., modernizationist, developmentalist, neoliberal, Marxist, postmodernist, etc.) reflect its essential function as legitimator of national and international order under hegemonic capital. Provides guidelines for a reflexive, autonomous, and interventionist sociology (*sociología autogestiva*). [D. Levi]

4419 **González Cervera, Alfonso S.** and **Rosario Cárdenas Elizalde.** La medición de la mortalidad infantil: los problemas y las alternativas. México: Univ. Autónoma Metropolitana, Unidad Xochimilco, 1992. 76 p.: appendix, bibl., ill. (Col. Ensayos)

Short introductory text presents and discusses direct (single round, multiround, and dual record surveys) and indirect methods (Brass, Sullivan) used in measuring infant mortality. Annex includes 1946–80 infant mortality rates of a number of Latin American countries and industrialized nations. [A. Ugalde]

4420 **Huggins, Martha K.** U.S.-supported State terror: a history of police training in Latin America. (*in* Vigilantism and the State in modern Latin America: essays on extralegal violence. New York: Praeger Publishers, 1991, p. 119–242)

Offers brief review of US training of Latin American police from late 19th century through presidencies of Roosevelt, Truman, Eisenhower, Kennedy, Johnson, Nixon, and Reagan. Traces various police training projects in context of changing US foreign policy agendas, and examines consequences of such training for human rights conditions in Latin American countries. [B. Aguirre-López]

4421 **Huggins, Martha K.** Vigilantism and the State: a look south and north. (*in* Vigilantism and the State in modern Latin America: essays on extralegal violence. New York: Praeger Publishers, 1991, p. 1–18)

Discussion of Latin American vigilantism includes recent citizen violence against authority, violence among citizens (lynchings), citizen quasi-official violence against citizens (*justiciero* violence), covert State violence against citizens (death squads and paramilitary/parapolice violence), and on-duty official police violence against alleged criminals and subversives. Discusses commonalities and differences among these types of vigilantism, motivations for vigilantism, and variables that influence vigilante acts. Explores relationship between vigilantism and Latin American State formation in context of foreign economic and political influences. Seeks to understand uniqueness of Latin American vigilantism through comparative studies of US vigilantism. [B. Aguirre-López]

4422 **The legacy of the disinherited: popular culture in Latin America; modernity, globalization, hybridity and authenticity.** Edited by Ton Salman. Amsterdam: CEDLA, 1996. 278 p.: map, tables.

Collection of 13 essays discusses popular culture from colonial times to present. Contributors review popular culture's emancipatory or countervailing potential. Work is result of international interdisciplinary conference held in 1994. Participants were from US, Brazil, and The Netherlands. [R. Hoefte]

4423 **Levine, Daniel H.** Popular voices in Latin American Catholicism. Princeton, N.J.: Princeton Univ. Press, 1992. 403 p.: bibl., ill., index. (Studies in church and state)

Analyzes base communities and other popular groups in Latin American Catholic Church. First part develops theoretical framework used by author to examine link between liberation theology movement and emergence of popular groups, and relates experience of popular groups in Colombia and Venezuela. Second part has chapters on practice of religion, needs and aspirations of the people, reformulation of both the priesthood and other pastoral agents brought about by these changes, and life histories of peasants, women, and a lay pastoral worker. Third and concluding section returns to theoretical considerations of linkages between lives of the people and structures of their societies. [B. Aguirre-López]

4424 Lindert, Paul van and **Otto Verkoren.** Het stedelijk woningvraagstuk in Latijns Amerika [The urban housing problem in Latin America]. Utrecht, The Netherlands: Geografisch Instituut, Rijksuniversiteit Utrecht, 1992. 87 p.: bibl., ill., maps. (Diskussiestukken van de Vakgroep Sociale Geografie van Ontwikkelingslanden; 47)

Authors show that the State plays important role in development of real estate and housing markets in major cities. State intervention, or lack thereof, determines whether low-income groups will be able to improve their housing situation. In general, State policy seems to reproduce social inequality. [R. Hoefte]

4425 Logan, Kathleen. Women in public office. (*Hemisphere/Miami*, 4:1, Fall 1991, p. 10–11)

Examines experiences of Caribbean and Latin American women in government work, focusing on their gains in that area and the difficulties they face in other sectors of the economy. [B. Aguirre-López]

4426 Medina Cano, Federico. La telenovela: una historia verosímil. (*Rev. Univ. Pontif. Boliv.*, 40:132, julio 1991, p. 39–45, bibl., photos)

Offers analysis of the characteristics of the *telenovela* and its impact on the culture and life of viewers. [B. Aguirre-López]

4427 Melo M., Cándida. Contaminación ambiental: su impacto en la cotidianidad de la mujer. (*in* Mujer y medio ambiente en América Latina y el Caribe. Quito: Fundación Natura; CEPLAES, 1991, p. 107–114, bibl.)

Examines relationship between population growth and environmental pollution produced by industrialization in Latin America and the Caribbean. Offers special focus on the situation in Santo Domingo, relating impact of environmental pollution on women's daily lives there. [B. Aguirre-López]

Montecino Aguirre, Sonia. Madres y huachos: alegorías del mestizaje chileno. See item **5052.**

Montecino Aguirre, Sonia. Símbolo mariano y constitución de la identidad femenina en Chile. See item **5053.**

4428 Mujeres latinoamericanas en cifras: avances de investigación. v. 3, Trabajo (Empleo). Coordinación de Teresa Valdés y Enrique Gomáriz. Santiago: FLACSO, 1992. 81 p.: tables. (Estudios sociales; 22)

Parte de una compilación estadística sobre la situación de la mujer. Este volumen se ocupa de su participación económica, mostrando variables como distribución del empleo, sindicalización, discriminación, desempleo y subempleo. Se ofrecen datos comparables desde 1976 hasta 1989. Ver también item **4429.** [M.A. Garretón]

4429 Mujeres latinoamericanas en cifras: avances de investigación. v. 7, Participación sociopolítica. Coordinación de Teresa Valdés y Enrique Gomáriz. Santiago: FLACSO, 1992. 81 p.: tables. (Estudios Sociales; 19)

Describe la participación femenina en organizaciones mixtas. Incluye una completa exposición de antecedentes históricos, y una compilación exhaustiva de la escasa información disponible sobre el tema en relación a los poderes del Estado, partidos políticos, organizaciones sociales, religión, televisión, y opiniones sobre cuestiones de género. Ver también item **4428.** [M.A. Garreton]

4430 Muller, Frits; Emma Rubín de Celis T.; and **Alfredo Rurizo Callejas.** Pobreza, participación y salud: casos latinoamericanos. Medellín, Colombia: Editorial Univ. de Antioquia, 1991. 325 p.: ill.

Focuses on unequal access to health care in Latin America, with case studies on Peru, Colombia, and Guatemala. Peruvian case study examines health care efforts for the poor (community, Ministerio de Salud, UNICEF, NGOs, cooperatives, Caja Nacional de Seguro Social). Best results are achieved when efforts are preventative and part of community-based participatory development, although little impact on health status is discernible. [C.B. Flora]

4431 Mundigo, Axel. Los programas de planificación familiar y su función en la transición de la fecundidad en América Latina. (*Notas Pobl.*, 20:55, junio 1992, p. 11–41, bibl., graphs, tables)

Summarizes what is known about origins of Latin American family planning programs and their effectiveness in reducing population growth in the region. Examines demographic transition in Latin America and in this context identifies important changes in values, political will, and medical personnel's awareness of the problems of population expansion. [B. Aguirre-López]

4432 Muñoz Dálbora, Adriana. Fuerza de trabajo femenina: evolución y tendencias. (*in* Género, clase y raza en América Latina: algunas aportaciones. Recopilación de Lola G. Luna. Barcelona: Univ. de Barcelona, 1991, p. 63–130, bibl., tables)

Análisis de las tendencias que ha seguido la fuerza de trabajo femenina en los últimos 30 años. Se discute el contexto teórico y la diversidad de aproximaciones al tema. Se estudian estadísticamente las relaciones desarrollo económico/fuerza de trabajo, y género/mercado de trabajo. [M.A. Garretón]

4433 Una nueva lectura: género en el desarrollo. v. 1. Recopilación de Virginia Guzmán, Patricia Portocarrero y Virginia Vargas. Lima: Ediciones Entre Mujeres, 1991. 1 v: bibl.

Outstanding collection of essays underscores importance of taking gender into account to better achieve goals of development in both developed and less developed countries. Topics covered include access to and training for new technology including agricultural technology, literacy and its relation to poverty, social costs of violence against women, and a rethinking of strategies especially those for use by NGOs. Underlying theme is cooperation for betterment of all. [A.M. Alberti]

4434 Oliveira, Orlandina de and **Bryan R. Roberts.** Los antecedentes de la crisis urbana: urbanización y transformación ocupacional en América Latina, 1940–1980. (*in* Las ciudades en conflicto: una perspectiva latinoamericana. Montevideo: Centro de Informaciones y Estudios del Uruguay; Ediciones de la Banda Oriental, 1989, p. 23–80, bibl., tables)

Traces changes in structure of urban areas in six Latin American countries (Argentina, Brazil, Colombia, Chile, Mexico, Peru) over three historical periods. Singles out for study the effects of rapid urbanization, industrialization, and growth of service sector, particularly growth of women's employment. [B. Aguirre-López]

4435 Osorio, Jaime. La democracia ordenada: análisis crítico de la nueva sociología del Cono Sur latinoamericano. (*Estud. Sociol./México*, 11:31, enero/abril 1993, p. 111–132, bibl.)

Busca caracterizar la "nueva sociología" post-autoritarismos, revisando algunos de los factores que definen su contexto, y analizando críticamente sus orientaciones temáticas y políticas. Argumenta en torno a los elementos de ruptura con la sociología anterior, y sus actuales limitaciones. [M.A. Garretón]

4436 Panorama social de América Latina. Santiago: Naciones Unidas, Comisión Económica para América Latina y el Caribe, 1991. 75 p.: ill.

Describes worsening patterns of social inequality in Latin America that marked 1980s. Documents changes in distribution of income, increases in poverty, unemployment, underemployment, and educational dropout rates, as well as continuing discrimination against women in the labor force. [B. Aguirre-López]

4437 Los partidos y la transformación política de América Latina. Edición de Manuel Antonio Garretón Merino. Santiago: Grupo de Trabajo Partidos Políticos-CLACSO, 1993. 103 p.

Trabajos presentados en la reunión del Grupo de Trabajo de Partidos Políticos de CLACSO en Córdoba, Argentina, en 1992. Se hace una caracterización de las transformaciones sociopolíticas del continente y su incidencia en el sistema de partidos, y se analizan los casos de Argentina, Chile, Perú y México. For political scientist's comment see item **2785**. [M.A. Garretón]

4438 Pedrazzini, Yves and **Magaly Sánchez.** Malandros, bandas y niños de la calle: cultura de urgencia en las metrópolis latinoamericanas. Caracas: Vadell Hermanos Editores, 1992. 247 p., 16 p. of plates: bibl., ill., tables.

Authors argue that "social marginality" and the violence and crime it engenders have become "normative" in poorer sectors of Latin American metropolitan centers. Clash between this survivalist "culture of urgency" and the culture of elites has led to a "democratic model of repression" in which police brutality and police crime are ignored, and "armed arbitrary repression" of the poor is not only tolerated but legitimated. [D. Levi]

4439 **The popular use of popular religion in Latin America.** Edited by Susannna Rostas and André Droogers. Amsterdam: CEDLA, 1993. 233 p.: bibl., ill., index. (Latin America studies; 70)

Collection of 13 essays explores how popular religions are reformulated by their users in a process of continuous reinterpretation or invention. Authors emphasize currently existing religious heterogeneity from which people can make choices according to their needs, requirements, or ambitions. [R. Hoefte]

4440 **Portes, Alejandro.** La urbanización de América Latina en los años de crisis. (*in* Las ciudades en conflicto: una perspectiva latinoamericana. Montevideo: Centro de Informaciones y Estudios del Uruguay; Ediciones de la Banda Oriental, 1989, p. 81–134, bibl., maps, tables)

Offers detailed analysis of socioeconomic conditions and of urban and population growth in the major cities of Latin America from 1970–90. [B. Aguirre-López]

4441 **Raczynski, Dagmar** and **Claudia Serrano.** Abriendo el debate: descentralización del Estado, mujeres y políticas sociales. (*in* Políticas sociales, mujeres y gobierno local. Santiago: Corporación de Investigaciones Económicas para Latinoamérica, 1992, p. 11–30, bibl.)

La introducción examina la situación de la mujer y su rol en la estrategia familiar de vida, los cambios ocurridos en el Estado y las políticas sociales y los desafíos de la incorporación de la variable género en la política social local. [M.A. Garretón]

4442 **Ramdas, Anil.** In mijn vaders huis [In my father's house]. Amsterdam: Jan Mets, 1993. 128 p.

Contains written version of four TV interviews on treacherous presence of Western civilization in the Third World. If accepted, one loses one's own culture; and if rejected, one has no chance against Western rationality and drive. Includes, inter alia, interviews with Anthony Appiah (Ghana) and Stuart Hall (Jamaica) on the diaspora and racism. [R. Hoefte]

4443 **Redclift, Michael** and **David Goodman.** The machinery of hunger: the crisis of Latin American food systems. (*in* Environment and development in Latin America: the politics of sustainability. Edited by David Goodman and Michael Redclift. Manchester, England: Manchester Univ. Press; New York: St. Martin's Press, 1991, p. 48–78, bibl., tables)

Reviews problems in Latin American food systems. Examines relationship between Latin American agriculture and international food system, problems of unequal land distribution and landlessness, land conversion and land use, and problems in system of food production and consumption in the context of population growth and changing diet patterns. Author argues that problems in the food systems concern the very sustainability of Latin American development. [B. Aguirre-López]

4444 **Refugee and displaced women in Latin America and the Caribbean.** Santiago: United Nations, 1990. 27 p.: bibl. (Serie Mujer y desarrollo; 4)

Presents information on refugee and displaced women in Latin America and the Caribbean. Contains sections on population movements in Latin America and the Caribbean, demographic studies of victims, and activities undertaken to address the problems. [B. Aguirre-López]

4445 **Relaciones laborales y modelos de acción sindical: experiencias europeas y latinoamericanas.** Edición de Jaime Ensignia, Malva Espinosa y Luise Rürup. Santiago: Friedrich Ebert Stiftung, 1994. 105 p.

Análisis comparativo de Europa y América Latina fue realizado por expertos laborales y sindicalistas. Cubre tres casos de sindicalismos fuertes e institucionalizados (Alemania, España y Italia) y tres casos de sindicalismos en transición (Brasil, Argentina y Chile). Se destacan similitudes, diferencias y posibles líneas de acción sindical. [For political scientist's comment see item **2790**.] [M.A. Garretón]

4446 **Roggenbuck, Stefan.** Strassenkinder in Lateinamerika: sozialwissenschaftliche Vergleichsstudie: Bogotá, Kolumbien;

São Paulo, Brasilien; und Lima, Peru. Frankfurt am Main; New York: P. Lang, 1993. 320 p.: bibl., ill. (Bochumer Schriften zur Entwicklungsforschung und Entwicklungspolitik: 0572–6654; 32)

Provides fresh perspective on study of Latin America's street children phenomenon. Based on field research, author establishes framework covering multi-faceted aspects of social structures (sociopolitical, socioeconomic, environmental), and emphasizes nuances of children's individual characteristics. Concludes that there are no typical street children and that multi-causal problems require multi-causal solutions. [C.K. Converse]

Sgambatti, Sonia. La mujer: ciudadano de segundo orden. See item **4852**.

4447 Smoking and health in the Americas: a 1992 report of the Surgeon General, in collaboration with the Pan American Health Organization. Atlanta, Ga.: CDC, 1992. 213 p.: bibl., index. (DHHS publication; CDC 92–8419)

Reflects concern for broader problems posed by tobacco consumption. Describes historical, social, economic, and regulatory aspects of smoking and tobacco control activities in the countries of the Americas. [B. Aguirre-Lopez]

4448 Sternbach, Nancy Saporta et al. Feminisms in Latin America: from Bogotá to San Bernardo. (*Signs/Chicago*, 17:2, Winter 1992, p. 393–434)

Sketches general picture of political trajectory of Latin American feminism during 1970s and 1980s. Traces growth of Latin American feminism in last decade, and also dispels myth that Latin American women do not define themselves as feminists. [B. Aguirre-López]

4449 Tecnología y modernidad en Latinoamérica: ética, política y cultura. Recopilación de Eduardo Sabrovsky J. Santiago: ILET-CORFO; Hachette, 1992. 194 p.: bibl., ill.

En este diálogo interdisciplinario, 14 autores abordan el tema de la tecnología y sus implicaciones en tres unidades temáticas: 1) innovación tecnológica y cultura latinoamericana; 2) la técnica en las utopías y antiutopías contemporáneas; y 3) el impacto y posibilidades sociales de la tecnología. [M.A. Garretón]

4450 Urban poverty alleviation in Latin America. The Hague: Ministry of Foreign Affairs, Development Cooperation Information Dept., 1992? 92 p: bibl., map. (Poverty and development: analysis & policy; 3)

Presents results of 1991 workshop on urban poverty alleviation chaired by Dutch Minister of Development Cooperation. Aim of workshop was to provide Ministry of Foreign Affairs with up-to-date expert reviews on latest policy developments in this field. Includes contributions by social scientists, economists (including Hernando de Soto), and the mayor of San Pedro Sula, Honduras. [R. Hoefte]

4451 Vásquez, Ana and Ana María Araújo. La maldición de Ulises: repercusiones psicológicas del exilio. Santiago: Editorial Sudamericana, 1990. 251 p.: bibl.

Apoyada en estudios de caso, testimonios y la experiencia acumulada de doce años de investigación, las autoras analizan el exilio como proceso de transculturación, extrayendo conclusiones sobre las etapas, implicaciones y consecuencias profundas de esta experiencia sobre las diversas dimensiones de la vida personal. [M.A. Garretón]

4452 Violence in the Andean region. Edited by Felipe E. Mac Gregor. Assen, The Netherlands: Van Gorcum, 1994. 139 p.: bibl.

Collection of seven essays analyzes how violence occurs from day to day (culturally, within the State, within the media), and how violence is used to generate money through drug trafficking. Work is product of larger project to conduct interdisciplinary research into the violence affecting Andean countries. [R. Hoefte]

4453 La vulnerabilidad de los hogares con jefatura femenina: preguntas y opciones de política para América Latina y el Caribe. Santiago: Naciones Unidas, 1991. 43 p. (Serie Mujer y Desarrollo)

Studies chronic vulnerability of female-headed households in Latin America and Caribbean region. Includes chapters on usefulness of concept of female-headed households for setting social policy, economic situation of such households, and sociopsychological effects on such women and their children. [B. Aguirre-López]

4454 Weinberger, Mary Beth. Cambios en la combinación de métodos anticonceptivos durante la transición de la fecundidad:

América Latina y El Caribe. (*Notas Pobl.*, 20: 55, junio 1992, p. 41–79, bibl., graphs, tables)

Summarizes what is known about contraceptive use and other means of controlling natality in Latin America and the Caribbean. Includes sections on prevalence of contraceptive use and methods employed, changes over time in contraceptive use in a number of countries in the region, and use of coitus interruptus, the Ogino method, and voluntary sterilization. [B. Aguirre-López]

Wistat: women's indicators and statistics database. Version 3. See item **47**.

MEXICO

ANTONIO UGALDE, *Professor of Sociology, University of Texas, Austin*

THE INCREASE IN THE NUMBER of articles and books on Mexican sociology maintains the momentum reported in *HLAS 53* (p. 659–660) and consequently publications annotated below represent only a fraction of published materials. It should be added that every biennium the percentage of publications on Mexican sociology authored by foreigners decreases. A second characteristic of Mexican publications noted in this chapter is that many of them have not been published by commercial presses but by government agencies (federal and state), universities, and research centers. The severe economic crisis has had a devastating impact on commercial book sales, forcing a large number of book stores (some observers claim as many as half) to go out of business. It is not clear how many of the books published by noncommercial houses follow a rigorous review process rather than simply responding to pressures by researchers to see their works published.

As in the past, the vast majority of sociologists studying Mexico continue to examine various aspects of political sociology; for this volume, more than half of the entries are in this speciality. Within political sociology, writers are concerned with party organization, voting, electoral returns, political participation, and cooptation. Other aspects include labor unions and organization, social movements, and the development processes. Theoretically, the traditional Marxist-Leninist paradigm and the neo-Marxist approaches—which were until recently very predominant among Mexican social scientists—have disappeared, and the influence of French scholars such as Bourdie and Touraine is on the increase.

Because of the time lag between research and publication, and the *HLAS* review process, we are now beginning to examine studies about two events that occurred a few years ago: the economic and political crisis of the 1980s and the electoral and municipal reforms approved in the late 1970s and early 1980s (see, for example, items **4564, 4565**, and **4500**). As could be expected, the causes and consequences of the economic crisis caught the interest of Mexican sociologists, both in terms of the political system and the crisis' impact on the daily life of the poor. In addition, many writers have interpreted the elections of 1988 and the appearance of the new PRD party as a break-away from the PRI to be the consequences of the two above-mentioned events.

With few exceptions, political sociologists raise questions about the stability of the present Mexican political system. Most authors are in agreement that unless profound changes and reforms are implemented, the future of the PRI will be troublesome. Some of the most insightful analyses of the PRI's predicament are by Núñez González (item **4528**), Rubio (item **4550**), Alonso *et al.* (item **4527**), Zermeño and Cuevas Díaz (item **4522**), and, at the state level, Gutiérrez (item **4498**). In a way, it could be said that studies such as these predicted the political crisis of 1995. The PRI's use of violence at the national, state, and local levels continues to be well documented by the authors reviewed in this volume (see, for example, items **4525** and **4551**). Studies such as these make it easier to understand the saga of political murders that unfolded in 1994–95, and even lend credibility to the alleged involvement of PRI leaders in the crimes. On another political topic, for a thorough sociocultural examination of the Chiapas rebellion, see the *Handbook's* ethnology chapter written by anthropologist Paul Sullivan (p. 91-98).

A second area of major concern is international migration, an area that, by and large, examines migration from Mexico to the US, but has begun to include a few significant studies of Guatemalan migrants to Mexico as well (items **4520, 4473,** and **4542**). It is not surprising that Mexican scholars' concerns—and at times their interpretation of migration issues—do not coincide with those of their US colleagues. For instance, Mexicans tend to emphasize the asymmetry of power from a framework of dependency theory. Among works by US scholars, the most thorough and comprehensive study is by Wayne Cornelius (item **4476**). From the Mexican side of the border, the work carried out by scholars at the Colegio de la Frontera Norte in Tijuana is notable. Various studies examine the magnitude of the remittances, which are an important source of foreign currency in Mexico and, according to some, explain the lack of interest of the Mexican government to control the exodus towards the North (item **4557**). If this line of reasoning is correct, it would suggest that the profound economic crisis of 1995 will reduce further the government's interest in and efforts towards curbing international migration. Nonetheless, according to many Mexican scholars the primary beneficiary of the cheap labor offered by migrants is the US. In addition, Mexican researchers also present compelling data showing that migration to the US has very negative consequences for Mexico such as consumerism, social stratification, and cultural dysfunctionality (item **4529**), while US scholars tend to emphasize the positive consequences (items **4501** and **4494**). A number of US writers have focused on the impact of the 1986 Immigration Reform and Control Act on reducing migration; there is consensus that the Reform had no major impact in this regard (item **4512**). The shifting characteristics of the migration—such as the increasing number of women crossing the border and the growth of migrants' employment in the industrial and service sectors—are the topic of analysis of several articles.

The North American Free Trade Agreement is one of the most important events which has taken place in this biennium. Social scientists are beginning to analyze the potential consequences of NAFTA (see, for example, the volume compiled by Bensusan Areous, item **4546**), but we will have to wait until the next *HLAS* volume for a more thorough analysis.

The *maquila* industry along the US-Mexican border has traditionally been negatively viewed by sociologists. Researchers alleged that the border's industrial plants exploit the workers, that labor conditions are detrimental to their health, and that employers impose unfavorable contracts in order to undercut unionized companies on the other side of the border. The research reviewed in this volume shows the beginning of a shift in the appraisal of *maquiladoras;* Carrillo Huerta is one of the first to present a contrasting view (item **4472**).

To conclude, we should note a new journal, *Acta Sociológica,* which appeared for the first time in 1990 and is published three times a year by the Facultad de Ciencias Políticas y Sociales of the Univ. Nacional Autónoma de México.

4455 Alarcón, Rafael. *Norteñización:* self-perpetuating migration from a Mexican town. (*in* U.S.-Mexico relations: labor market interdependence. Stanford, Calif.: Stanford Univ. Press, 1992, p. 302–318, tables)

Findings from fieldwork in a small community near Zamora, Michoacán are used to describe community's economic dependence on migration and the cultural impact of migration on individuals and local institutions. Migration has become a rite of passage for male adults. Over the years, social institutions that make migration possible and necessary have emerged.

4456 Alcayá Moya, Graciela. Migrantes, pescadores y mujeres en Puerto Madero, Chiapas, México. (*Mesoamérica/Antigua*, 14:25, junio 1993, p. 101–114)

Ethnographic report studies fishing activities and role of undocumented Guatemalan migrants in development of the fishing industry. Examines reasons for hiring foreign workers, pragmatic approach of Mexican authorities towards presence of these migrants, and role of Guatemalan and Mexican women in assisting migrants to settle in Mexico.

4457 Aldrete-Haas, José Antonio. La deconstrucción del Estado mexicano: políticas de vivienda, 1917–1988. México: Alianza Editorial, 1991. 163 p.: bibl. (Estudios)

After identifying needs and demands of actors (bureaucracies, labor unions, private sector, and the poor), second chapter critiques the housing policies of each presidency. In last two chapters, author studies two institutions created to satisfy housing needs of the poor: Fondo Nacional para la Vivienda de los Trabajadores (INFONAVIT) and Fondo Nacional para las Habitaciones Populares (FONHAPO). Evidence presented confirms failure of Mexican corporatist model to resolve housing deficiencies.

4458 Alduncin Abitia, Enrique. Los valores de los mexicanos. v. 1, México, entre la tradición y la modernidad. México: Fomento Cultural Banamex, 1986. 1 v.: bibl., ill.

First battery of questions from 1981 national survey of 3,800 persons gathered information on attitudes towards poverty, the future, deliquency, success, and social well-being. Following Ackoff-Emery typology, author classifies Mexicans into four psychosocial types. Remainder of monograph presents responses to large number of questions regarding life objectives, family, sex, and work, cross-tabulated by basic sociodemographic variables.

4459 Alduncin Abitia, Enrique. Los valores de los mexicanos. v. 2, México en tiempos de cambio. México: Fomento Cultural Banamex, 1991. 1 v.: bibl., ill.

Results of survey on Mexican attitudes towards the economy, politics, and social conditions. Well organized presentation of the main findings. Period covered is largely 1981–87. [R.E. Looney]

Allub, Leopoldo. Impactos sociales de las grandes obras públicas: diagnóstico, evaluación y gestión. See item **5087**.

4460 Alvarez Mosso, Lucía and **María Luisa González Marín.** Industria y clase obrera en México, 1950–1980. México: Ediciones Quinto Sol, 1987. 183 p.: bibl.

Framed in traditional Marxist-Leninist paradigm and terminology, these two essays criticize well-known contradictions of capitalism, the social problems it creates, and widening gap between rich and poor. Unfortunately, volume adds no new insights to Marxist theory or to our understanding of Mexico's political economy.

4461 Anguiano, María Eugenia. Migración y derechos humanos: el caso de los mixtecos. (*Estud. Front.*, 26, sept./dic. 1991, p. 55–69, bibl., tables)

Discusses violations of human rights of Mixtec migrants from Oaxaca by Mexican police. Poverty pushes migrants to find employment in other parts of Mexico, mostly in the Northwest, and in the US. Because Mixtecs frequently do not carry documents, they are discriminated against and robbed by police both while living in Mexico and when returning from US. In 1989 Mexican government launched the *programa paisano* to reduce police abuses, but program's effects were not lasting.

4462 Baños Ramírez, Othón. Yucatán: ejidos sin campesinos. Prólogo de Esteban Krotz. Mérida, Mexico: Ediciones de la Univ. Autónoma de Yucatán, 1989. 336 p.: bibl., ill.

Presents innovative view of maguey ejidatarios. Creation of the ejidos after 1917 did not transform indebted peons into free *campesinos*. Ejidatarios did not own the land and were not free to choose crops or to make agricultural decisions. Through historical records, participant observation, life histories, and surveys, author studies evolution of ejidos and continuous exploitation of the ejidatarios.

4463 Barragán López, Esteban. Más allá de los caminos: los rancheros del Potrero de Herrera. Zamora, Mexico: Colegio de Michoacán, 1990. 208 p., 24 p. of plates: bibl., ill., maps.

Except for first two chapters which describe the region and its socioeconomic structure, monograph can best be characterized as a social history of the municipality of Tocumbo (Michoacán). Materials are organized by historical periods (1550–1945, 1945–65, and 1965–85). Explains in some detail the features and evolution of cattle ranching, the region's economic base.

4464 Barrera Bassols, Dalia. Condiciones de trabajo en las maquiladoras de Ciudad Juárez: el punto de vista obrero. México: Instituto Nacional de Antropología e Historia, 1990. 94 p.: bibl. (Serie Antropología social. Col. científica; 209)

Study is based on 1981–82 survey of 162 female workers and on unspecified number of open interviews with Instituto de Seguro Social personnel, public oficials, and labor leaders. Labor conditions and union membership are main focus of analysis. Includes very interesting materials on interlabor union conflicts.

4465 Behar, Ruth. Translated woman: crossing the border with Esperanza's story. Boston: Beacon Press, 1993. 372 p.: bibl., ill.

Relates life history of an elderly woman from a village near San Luis Potosí. First part focuses on village history as narrated by the woman, while in second part author's and woman's voices are interwoven in an effort to build a meta-history. These parts are followed by conversations between the biographer and the woman. In last part, author offers series of insightful reflections on how a woman's life can be read from a feminist and historical perspective.

4466 Bennett, Vivienne. The evolution of urban popular movements in Mexico between 1968–1988. (*in* The making of social movements in Latin America: identity, strategy, and democracy. Edited by Arturo Escobar and Sonia E. Alvarez. Boulder, Colo.: Westview Press, 1992, p. 240–259, table)

Well-documented history of urban movements explains strategies used by the groups and government responses. Behavioral and organizational diversity is explained by differences in groups' relations with local and federal governments, and regional context variations. Organizational role of students and women and their participation are also analyzed.

4467 Bustamante, Jorge A. Interdependence, undocumented migration, and national security. (*in* U.S.-Mexico relations: labor market interdependence. Stanford, Calif.: Stanford Univ. Press, p. 21–41)

Undocumented Mexican migration to US responds to political and economic interdependence between the two countries that is characterized by an asymmetry of power. From this premise author discusses four possible scenarios of migration and its impact on bilateral relations based on a four-cell matrix with two variables: stability-instability of the Mexican economy, and expansion-recession of US economy.

4468 Bustamante, Jorge A. Undocumented migration from Mexico to the United States: preliminary findings of the Zapata Canyon Project. (*in* Undocumented migration to the United States: IRCA and the experience of the 1980s. Santa Monica, Calif.: Rand Corporation; Washington: The Urban Institute, 1990, p. 211–226, bibl., graphs)

Questionnaires administered daily in several border cities to a sample of border crossers, along with photographs taken daily at peak crossing hours at Tijuana border, show that by 1988, two years after passage of the US Immigration Reform and Control Act (IRCA), the total number of undocumented migrants had decreased. Paper cautions about establishing cause-effect links between the decrease and the IRCA.

4469 Calderón Rodríguez, José María. Inflación y descentralización como estrategias capitalistas y su impacto sobre la fuerza de trabajo: México 1982–1988. (*in* Organización y luchas del movimiento obrero latinoamericano, 1978–1987. Coordinación de Mario A. Trujillo Bolio. México: Siglo Veintiuno, 1988, p. 257–297)

According to author, inflation and administrative/productive decentralization are two main economic policies used in 1980s by capital and the State to break down organization of labor and exploit the worker. Unfortunately, lengthy article is very short on qualitative or quantitative data, and reader can easily question logic of the argumentation.

4470 Camposortega Cruz, Sergio. Análisis demográfico de la mortalidad en México, 1940–1980. México: Centro de Estudios Demográficos y de Desarrollo Urbano, Colegio de México, 1992. 441 p.: bibl., ill.

Demographic monograph is technical, comprehensive, and qualitative. Includes information on quality of sources, estimates using different techniques, and cross tabs of mortality by age, sex, education, and other sociodemographic variables. Also includes time series.

4471 Campuzano Montoya, Irma. El impacto de la crisis en la CTM. (*Rev. Mex. Sociol.*, 52:3, julio/sept. 1990, p. 161–190, tables)

Well-written and informative account examines history of Confederación de Trabajadores Mexicanos (CTM). Presents its origins and subordination to the PRI, relations with other labor federations, and internal conflicts by presidential term. Last part is critical analysis of tensions between CTM and government caused by 1982 economic crisis.

4472 Carrillo Huerta, Mario M. The impact of *maquiladoras* on migration in Mexico. (*in* The effects of receiving-country policies on migration flows. Boulder, Colo.: Westview Press, 1991, p. 67–102, tables)

Findings from random survey of 1,200 workers, 40 *maquiladora* executives, and 40 opinion leaders in Tijuana, Ciudad Juárez, Nuevo Laredo, and Guadalajara indicate that *maquiladoras* have positive impact on living conditions of workers and on the region. According to author, *maquiladoras* do not promote Mexican migration to US.

4473 Castillo G., Manuel Angel. Población y migración internacional en la frontera sur de México: evolución y cambios. (*Rev. Mex. Sociol.*, 52:1, enero/marzo 1990, p. 169–184, bibl.)

After brief review of push and pull factors, author examines contemporary changes taking place in Guatemalan migration flow to coffee plantations in Chiapas. Interviews with laborers suggest that recent migration is more a response to deterioration of living conditions in Guatemala than to labor needs of the plantations.

4474 Castro Martignoni, Jorge. México: estimación de la migración internacional en el período 1960–1980. (*Estud. Front.*, 26, sept./dic. 1991, p. 71–122, bibl., graphs, tables)

Technical demographic study considers net migration flows in Mexico by sex and age group. Emigration was measured using Warren and Passel's estimation based on US population census and data from CELADE's Latin American international migration research; immigration figures were calculated from the Mexican population census. According to findings, return migration increased beginning around 1975.

4475 Chávez, Leo R. Paradise at a cost: the incorporation of undocumented Mexican immigrants into a local-level labor market. (*in* U.S.-Mexico relations: labor market interdependence. Stanford, Calif.: Stanford Univ. Press, p. 271–301, tables)

Reviews and summarizes previous publications on labor supply and demand, on impact of migration on local health and education services, and on manner in which type of labor demand influences migration. Useful summary of well-known information.

4476 Cornelius, Wayne A. From sojourners to settlers: the changing profile of Mexican immigration to the United States. (*in* U.S.-Mexico relations: labor market interdependence. Stanford, Calif.: Stanford Univ. Press, p. 155–195, tables)

Very insightful and thorough paper by leading scholar examines Mexican migrants' shift from rural to urban employment, shift from individual to family migration, increasing number of female migrants, and growing number of permanent settlers. Contains excellent critical analysis of consequences of the US Immigration Reform and Control Act on migration and migrants.

4477 Cornelius, Wayne A. Impacts of the 1986 U.S. immigration law on emigration from rural Mexican sending communities. (*in* Undocumented migration to the United States: IRCA and the experience of the 1980s. Santa Monica, Calif: Rand Corporation; Washington: The Urban Institute, 1990, p. 227–249, bibl., tables)

Very solid analysis of impact of US Immigration Reform and Control Act (IRCA) is based on 954 sample survey interviews conducted in 1988–89 in three rural communities in west-central Mexico. Respondents were very familiar with IRCA provisions. Act seems to have increased the average length of stay in US; has produced no massive return of undocumented migrants; and has resulted in slightly decreased remittances.

4478 Cortés, Fernando and **Rosa María Rubalcava.** Algunas determinantes de la inserción laboral en la industria maquiladora de exportación de Matamoros. (*Estud. Sociol./México*, 11:31, enero/abril 1993, p. 59–91, bibl., tables)

Identifies demographic and social determinants of employment in *maquiladora* industry. Studies in other border towns have shown that employers prefer young female workers, and this is also the case in Matamoros. Relatives and informal networks facilitate finding employment. Employment probability increases for those having family members already working in *maquiladoras*.

4479 Couffignal, Georges. La gran debilidad del sindicalismo mexicano. (*Rev. Mex. Sociol.*, 52:3, julio/sept. 1990, p. 191–210, graphs, tables)

First part explains history of labor organization and movements from 1910, with particular emphasis on labor's interaction with government. According to author, 1982 economic crisis signaled end of corporativist model and raised questions about future of traditional labor federations once their leaders had lost power and legitimacy.

4480 Covarrubias V., Alejandro. La flexibilidad laboral en Sonora: un análisis comparativo de la flexibilidad de los contratos colectivos de trabajo de la industria en Sonora, en la década de los ochenta. Hermosillo, Mexico: Colegio de Sonora; México: Fundación Friedrich Ebert, 1992. 229 p.: bibl., ill.

Globalization of the economy and the 1980s economic crisis have demanded more flexibility in conditions under which labor works. Using variables such as work schedules and hours, work organization, salary and fringe benefits, mobility, employment security, and restrictions on labor union activities, author classifies different industries (mining, *maquiladoras*, cement, automotive, and food and beverage) along four levels of flexibility.

4481 Crisis, conflicto y sobrevivencia: estudios sobre la sociedad urbana en México. Guadalajara, Mexico: Univ. de Guadalajara; México: CIESAS, 1990. 474 p.: bibl. (Col. Jornadas académicas. Serie Coloquios)

Thirty-one articles are grouped under seven headings that give idea of their content: urban migration, urban labor force, informal sector, urban health problems, urban health services, household organization and survival strategies, and social movements. Quality of contributions is uneven; several are revisions of previously published materials.

4482 Cuevas Seba, Teresa and **Miguel Mañana Plasencio.** Los libaneses de Yucatán. Mérida, Mexico: Impresiones Profesionales, 1990. 125 p.: bibl., ill., map.

Offers light account of migration and assimilation process of Lebanese in Yucatán. Despite descriptive nature, volume could be of interest to students of migration and those concerned with ethnic relations.

4483 Los días eran nuestros: vida y trabajo entre los obreros textiles de Atlixco. Coordinación de Luis Felipe Crespo Oviedo. Recopilación de Francisco Javier Gómez Carpinteiro. 2. ed. México: SEP, Subsecretaría de Cultura, Dirección General de Culturas Populares, 1988. 188 p. (Serie Testimonios)

Presents oral history of textile industries in Atlixco (Puebla), many of which did not survive technological innovations introduced in 1960s. Thirteen unedited narrations by former workers and other employees provide glimpses of labor struggles, labor union in-fights, survival strategies, and family life.

4484 Díaz Montes, Fausto. Los municipios: la disputa por el poder local en Oaxaca. Oaxaca, Mexico; Comunicación Social, Difusión Institucional, 1992. 168 p.: bibl., ill. (Col. Del barro nuestro; 4)

Evaluates impact of 1977 electoral and 1983 municipal reforms on clientism, cooptation, political party conflicts, and role of the PRI in the 570 municipalities of Oaxaca. Levels of socioeconomic development and modernization, rurality, ethnicity, and population growth are also taken into consideration. Includes case study of town of Tlacolula.

4485 Donato, Katherine M. Current trends and patterns of female migration: evidence from Mexico. (*Int. Migr. Rev.*, 27:104, Winter 1993, p. 748–771, bibl., tables)

Data were collected in 10 urban and rural communities in states of Jalisco, Michoacán, Guanajuato, and Nayarit (n=between 150 and 200), and in US (n=100), between 1987–89. Findings from regression analysis suggest probability that undocumented female migrants may have decreased in late 1980s, while authorized migrants may have increased. Study also examines influence of other dependent variables on female migration flow.

4486 Durand, Carlos A. Algunas consideraciones acerca de la etnia de Oaxaca, República Mexicana. (*Rev. Geogr./Mérida*, 30, 1989, p. 37–60, bibl., tables, maps)

The Trique are a little-known tribal group. Following a short historical and ethnographic presentation, paper illustrates social problems faced by Trique such as exploitation by caciques, land expropriation, unemployment, and lack of adequate educational and health services. Failure of public programs is also discussed.

4487 Encuentro sobre Investigaciones en Ciencias Sociales en Yucatán, 2nd, Mérida, Mexico, 1988. Memorias. Edición de Luis A. Várguez Pasos. Mérida, Mexico: Univ. Autónoma de Yucatán, 1990. 161 p.: bibl.

Twelve articles written by historians, anthropologists, and sociologists were presented at a 1988 symposium. Contributions summarize methodologies and findings on topics as diverse as Spanish marriages in colonial days, family organization in a poor neighborhood of Mérida, Maya ethnic and regional society, popular religion and traditions in Yucatán, and political conflict in a village.

4488 Ensayos críticos para el estudio de las organizaciones en México. Recopilación de Eduardo Ibarra Colado y Luis Montaño Hirose. México: Univ. Autónoma Metropolitana, Unidad Iztapalapa; Miguel Angel Porrúa, 1991. 244 p.: bibl., ill. (Col. Las Ciencias sociales)

The seven articles in this collection make a valuable contribution to sociology of complex organizations, a speciality understudied in Mexico. Three articles review theoretical issues and their applicability to the Latin American context. Other articles discuss issues such as labor motivation, social marketing, and various aspects of industrial organization.

4489 Estancamiento económico y crisis social en México, 1983–1988. v. 2, Sociedad y política. Recopilación de Jesús Lechuga Montenegro y Fernando Chávez. México: Univ. Autónoma Metropolitana-Azcapotzalco, 1989. 1 v.: bibl. (Serie Economía)

Two-part volume includes: 1) articles on political sociology, labor union organization, and party organizations of the right and the left; and 2) articles on various public policies such as housing, labor, ecology, public health, education, human rights, drug control, and political reform. Useful introduction to these topics.

4490 El estudio de los movimientos sociales: teoría y método. Recopilación de Víctor Gabriel Muro y Manuel Canto Chac. Zamora, Mexico: Colegio de Michoacán; Coyoacán, Mexico: Univ. Autónoma Metropolitana, Unidad Xochimilco, 1991. 194 p.: bibl.

Nine articles present case studies of diverse social movements (rural, urban, religious, and regional). Authors theorize and raise questions regarding mechanisms of group identity, nature of solidarity links, and group dynamics. Papers were presented at a working group meeting, El Colegio de Michoacán, 1989.

4491 Flores Lúa, Graciela et al. El Estado, los cañeros y la industria azucarera, 1940–1980. Coordinación de Luisa Paré. México: Instituto de Investigaciones Sociales, UNAM; Univ. Autónoma Metropolitana, 1987. 295 p.: bibl., ill. (Serie Sociología. Biblioteca de ciencias sociales y humanidades)

Provides carefully reconstructed history of organization of sugarcane workers. According to authors, exploitation of workers by the State and the estates has contributed to capital accumulation and industrialization of country. Case studies from Veracruz, Oaxaca, and Sinaloa are used to illustrate internal conflicts in the Confederación Nacional de Campesinos and mechanisms of political control and mobilization.

4492 Fuentes Navarro, Raúl. La investigación de comunicación en México: sistematización documental, 1956–1986. México: Ediciones de Comunicación, 1988. 656 p.: indexes.

Lists 877 annotated books and articles and 1,225 theses and dissertations. Appendices classify contents by discipline (communi-

4493 Ghiringhelli, Robertino and **Rafael Pérez Miranda.** Lo stato degli studi e delle ricerche sulla classe politica in Messico. (*in* Studi e ricerche sulla classe politica in Italia, Argentina, e Messico: materiali per un'analisi comparata. Padova, Italy: Facoltá di Scienze Politiche Dell'Univ. di Padova; Casa Editrice Dott. Antonio Milani, 1991, p. 301–326)

Provides brief review of political science studies in Mexico from 1970–90. Describes theoretical approaches of research centers and departments of political science. Second part presents selection of authors and their contribution to a variety of subdisciplines such as elite studies, social movements, political parties, and international relations.

4494 Goldring, Luin. Development and migration: a comparative analysis of two Mexican migrant circuits. (*in* The effects of receiving country policies on migration flows. Boulder, Colo.: Westview Press, 1991, p. 137–174, appendix, tables)

Compares migration patterns of two small rural communities in Zacatecas and Michoacán. Describes sociodemographic and employment characteristics of migrants and non-migrants, types of migration, length of stay, and employment experiences in US. US-generated wealth has distinct impact on the two villages; variations are attributed to local sociocultural differences.

4495 González Block, Miguel Angel. Economic crisis and the decentralization of health services in Mexico. (*in* Social responses to Mexico's economic crisis of the 1980s. Edited by Mercedes González de la Rocha and Agustín Escobar Latapí. San Diego, Calif.: Center for U.S.-Mexican Studies, Univ. of California, San Diego, 1991, p. 67–88, bibl., graph, table)

Scholarly discussion examines political dimensions of decentralizing services and the institutional in-fighting between Instituto Mexicano del Seguro Social and the Secretaría de Salud. Compares financial and health consequences of decentralization by comparing a non-centralized state (Oaxaca) and a decentralized one (Guerrero). According to author, equity suffered with decentralization.

4496 González Chávez, Humberto. Los empresarios en la agricultura de exportación en México: un estudio de caso. (*Rev. Eur.*, 50, June 1991, p. 87–114, bibl.)

Biography of a successful agricultural entrepreneur in coastal region of Jalisco is used to demonstrate that national actors can compete with multinational corporations in food production, access to national and international markets, and raising capital, as well as in adopting foreign technologies and in developing their own technological innovations.

4497 González Sánchez, Jorge A. Sociología de las culturas subalternas. Mexicali, México: Univ. Autónoma de Baja California, 1990. 176 p.: bibl., ill.

First half of the book is discussion/review of the literature on popular culture theory (Tylor, Gramsci, Bourdieu). Following a historical, sociodemographic, and economic description of the region (a municipality in the Atlacáyotl sierra, Veracruz), author analyzes briefly belief system of its inhabitants.

4498 Gutiérrez, Irma Eugenia. Hidalgo: sociedad, economía, política y cultura. México: Centro de Investigaciones Interdisciplinarias en Humanidades, Coordinación de Humanidades, Univ. Nacional Autónoma de México, 1990. 117 p.: bibl. (Biblioteca de las Entidades Federativas; 12)

Short introduction to the sociodemographic and economic characteristics of Hidalgo is followed by excellent analysis of state's political system. Political authors, interest groups, social and political mobilization of labor and peasantry, political cliques, and electoral context are critically discussed. Presents perceptive and balanced view of PRI's local behavior.

4499 Hernández Bringas, Héctor H. Las muertes violentas en México. México: Univ. Nacional Autónoma de México, Centro Regional de Investigaciones Multidisciplinarias; Asociación Mexicana de Población, 1989. 168 p.: bibl., ill., tables.

Primarily descriptive book on deaths caused by violence includes all types of accidental deaths (vehicle, accidental poisoning, drowning, other accidents, homicides, and suicides). The 86 pages of text are followed by as many pages of tables in which types of violent deaths are cross-tabulated by state, age, and sex for 1972–82.

Hinds, Harold E. and **Charles M. Tatum.** Not just for children: the Mexican comic book in the late 1960s and 1970s. See *HLAS 54:1514*.

4500 **Hoffmann, Odile.** Renovación de los actores sociales en el campo: un ejemplo en el sector cafetalero en Veracruz. (*Estud. Sociol./México*, 10:30, sept./dic. 1992, p. 523–554, bibl., ill., table)

Carefully documented case study examines political and economic organization of coffee growers and its marketing in Veracruz during crisis of 1980s. Discusses political patronage, causes of inefficiency and corruption of public institutions, conflicts between private and public sectors and among different actors of private sector, and market deficiencies.

4501 **Jones, Richard C.** Movilidad económica alternativa: migración internacional en Zacatecas rural. (*Iztapalapa/México*, 9:17, enero/junio 1989, p. 123–135, tables)

Jones claims that contemporary research on impact on the sending community of migrants returning to Mexico from US is influenced by dependency theory. Dependency affirms that return migration reduces investments in traditional economic activities, increases consumerism and social stratification, and produces cultural dysfunctions. His findings from a survey of 300 households with return migrants in a rural municipality led to opposite conclusions.

4502 **Kaiser, Lucia L.** and **Kathryn G. Dewey.** Migration, cash cropping and subsistence agriculture: relationships to household food expenditures in rural Mexico. (*Soc. Sci. Med.*, 33:10, 1991, p. 1113–1126)

Examines relationship of household resource allocation to source of income (wages, cash cropping, migrant remittances, etc.), to women's contribution to income, and to subsistence production level. According to findings, resource allocation patterns are influenced by income level and by household strategies through which income is generated. Based on cross-sectional survey of 178 households in three rural communities.

4503 **Langer, Ana; Rafael Lozano;** and **José Luis Bobadilla.** Effects of Mexico's economic crisis on the health of women and children. (*in* Social responses to Mexico's economic crisis of the 1980s. Edited by Mercedes González de la Rocha and Agustín Escobar Latapí. San Diego, Calif.: Center for U.S.-Mexican Studies, Univ. of California, San Diego, 1991, p. 195–219, bibl., graphs, tables)

Infant mortality rates continued to decline in Mexico during 1982–88 economic crisis, but neonatal mortality increased and rate of decline in preschool mortality slowed down. Among low-income groups and unemployed, uneducated mothers, and in small population centers and rural areas, reduction of infant mortality rates was smaller. A similar pattern was found for maternal mortality.

4504 **León, Arturo** and **Margarita Flores de la Vega.** Desarrollo rural: un proceso en permanente construcción. México: Univ. Autónoma Metropolitana, Unidad Xochimilco, 1991. 204 p.: bibl.

Examines transformation of mode of production brought about by exploitation of coffee for export among the Mazateca in Oaxaca and the Chol in Chiapas. Participant observation, action research, surveys, and in-depth interviews were used in a well-designed project aimed at critically determining the role of external agents, particularly the decentralized Instituto Mexicano del Cafe, in the organization and "modernization" of the indigenous populations.

4505 **Lezama, José Luis.** Ciudad y conflicto: usos del suelo y comercio ambulante en la Ciudad de México. (*in* Espacio y vivienda en la Ciudad de México. Coordinación de Martha Schteingart. México: El Colegio de México, Centro de Estudios Demográficos y de Desarrollo Urbano, 1991, p. 121–135, bibl.)

The 1980s economic crisis has increased the number of street vendors. Author describes different types of street vendors, from marginal poor trying to make a living to organized distributors of illegally imported goods. Analysis includes insightful discussion of conflict between established merchants and street vendors, role of the PRI, and inconsistencies between city ordinances and actual use of public spaces.

4506 **Liguori, Ana Luisa.** De campesina a obrera: el caso de la Unidad Industrial Ernesto Peralta. México: Instituto Nacional de Antropología e Historia, 1991. 62 p.: bibl. (Col. científica; 233. Serie Antropología social)

Provides valuable contribution to understanding of social transformation which occurs when peasant women are incorporated into industrial labor force. Discusses political, social, cultural, and sexual behavior changes. Also explores exploitation of the peasantry and advantages found by capital in employing persons who continue to work in agriculture.

4507 Lima Barrios, Francisca G. Familia popular, sus prácticas y la conformación de una cultura. México: Instituto Nacional de Antropología e Historia, 1992. 103 p.: bibl., ill. (Col. científica; 254. Serie Antropología social)

Survey of 45 families and in-depth studies of a handful of families form database for an ethnographical study of a low-income neighborhood in the Distrito Federal. Family organization, living conditions, work, and recreational activities are the main topics.

4508 López Cámara, Francisco. Apogeo y extinción de la clase media mexicana. Cuernavaca, Mexico: Univ. Nacional Autónoma de México, Centro Regional de Investigaciones Multidisciplinarias, 1990. 183 p.: bibl.

Three distinct essays are brought together in this volume. In first essay, author, who is sympathetic towards President Echeverría (1970–76), presents a general discussion of political role of middle classes from 1910 Revolution to 1988. Second essay discusses a 1975 PRI-organized conference on the middle classes, in which author seems to have played a role. Third essay is a journalistic assessment of evolution of the middle classes from 1970 to present.

4509 Marroquín, Enrique. La cruz mesiánica: una aproximación al sincretismo católico indígena. Oaxaca, Mexico: Univ. Autónoma Benito Juárez de Oaxaca; México: Palabra Ediciones, 1989. 246 p.: bibl. (Col. V centenario de la conquista de América)

A Catholic priest and anthropologist presents scholarly and provocative interpretation of religious syncretism among the 16 ethnic groups living in Oaxaca. From historical records, tales, and artifacts, author explains transformation and adaptation of concepts such as the supernatural, God, good and evil spirits, and of symbols such as the cross. Meanings and uses of rituals and sacraments are also examined.

4510 Martin, JoAnn. When the people were strong and united: stories of the past and the transformation of politics in a Mexican community. (*in* The paths to domination, resistance, and terror. Berkeley: Univ. of California Press, 1992, p. 177–189, bibl.)

The 1988 electoral victory of Cárdenas' Partido de la Revolución Democrática in a Morelos village is interpreted as village's culture of resistance to corruption. According to author, storytelling by elders reminded villagers of past revolutionary heroic efforts, while rampant rumors of PRI's corruption reminded them of need to fight in the spirit of their ancestors.

4511 Martínez Borrego, Estela. Organización de productores y movimiento campesino. México: Siglo Vientiuno Editores; Univ. Nacional Autónoma de México, 1991. 253 p.: bibl., ill., maps. (Sociología y política)

Presents case study of an independent agricultural cooperative organized in late 1970s in northern highlands of Puebla. Aspects discussed in some detail include: interactions of cooperative with bureaucracy and government programs such as Sistema Alimentario Mexicano, leadership patterns, system of production and marketing, and impact of political system on cooperative's internal organization.

4512 Massey, Douglas S.; Katherine M. Donato; and **Zai Liang.** Effects of the Immigration Reform and Control Act of 1986: preliminary data from Mexico. (*in* Undocumented migration to the United States: IRCA and the experience of the 1980s. Santa Monica, Calif: Rand Corporation; Washington: The Urban Institute, 1990, p. 183–210, bibl., tables)

To measure the impact of the US Immigration Reform and Control Act (IRCA) in reducing migration, in 1987–88 two surveys (total n=400) were conducted in state of Guanajuato, a primary source of migration. One survey took place in a small community of 70,000 and the other in León whose population numbers more than one million. Age-period analyses of the likelihood of illegal migration revealed no discernable impact of IRCA.

4513 **Mejía Barquera, Fernando.** La industria de la radio y la televisión y la política del estado mexicano. v. 1. México: Fundación Manuel Buendía, 1989. 1 v.: bibl.

Explores reasons why private radio and TV in Mexico support political system's ideology. Monograph is devoted primarily to analysis of legislation instrumental in coopting communication elites, and of non-written rules that produce mutually satisfactory support between PRI and leading stations. Period covered: 1920–60.

4514 **Mexico** (*in* Tobacco or health: status in the Americas. Washington: PAHO, 1992, p. 250–261, bibl., tables)

Informative and useful study examines economics of tobacco industry including its advertising practices and profits. Smoking behavior and changes over time by sex and age groups are presented, and health effects are estimated. Article closes with examination of government policies and efforts to combat tobacco use.

4515 **Meyers de Ortiz, Carol.** Pequeño comercio de alimentos en colonias populares de Ciudad Nezahualcóyotl: análisis de su papel en la estructura socioeconómica urbana. Guadalajara, Mexico: Editorial Univ. de Guadalajara, 1990. 115 p.: bibl. (Col. Fin de milenio)

Following a Marxist theoretical framework, author studies small food distributors in public markets. Conclusions are based on one year of participant observations in a large metropolitan area (population: over two million) with more than 15,000 small vendors. According to findings, small vendors are part of circulation of capital, even though they cannot be considered capitalists; rather, they are self-employed. Their main function is to facilitate accumulation of capital for the bourgeoisie.

4516 **Mier y Terán, Marta.** Descenso de la fecundidad y participación laboral femenina en México. (*Notas Pobl.*, 20:56, dic. 1992, p. 143–171, bibl., tables)

Establishes association between fertility decline and female labor force participation. Among findings: 1) first birth cohort to experience lower fertility rate (1942–47) also had a higher rate of labor force participation—these women began to work after birth of their last child; and 2) younger cohorts increasingly entered the labor force while raising the family. Based on secondary analysis of 1976 Encuesta Mexicana de Fecundidad and 1987 Encuesta Nacional de Fecundidad y Salud.

4517 **Módena, María Eugenia.** Madres, médicos y curanderos: diferencia cultural e identidad ideológica. México: Centro de Investigaciones y Estudios Superiores en Antropología Social, 1990. 229 p.: bibl. (Ediciones de la Casa Chata; 37)

Anthropologist studies health conditions (nutrition, water availability, feces disposal, and immunization levels) in a small town in southern Veracruz. Utilization patterns of different types of health services (private and public modern care, and traditional), provider-user interactions, and health attitudes are also analyzed in this informative monograph.

4518 **Montiel H., Yolanda.** Proceso de trabajo, acción sindical y nuevas tecnologías en Volkswagen de México. México: Centro de Investigaciones y Estudios Superiores en Antropología Social; Ediciones de la Casa Chata, 1991. 263 p.: bibl., ill. (Col. Miguel Othón de Mendizábal)

Exhaustive study examines industrial-labor relations from workers' perspective. The result of seven years of fieldwork, book makes valuable contribution to our understanding of labor unions and labor organization. Also exposes reader to multinational corporate behavior and tactics. Based on in-depth interviews and archival materials.

4519 **Moreno, Lorenzo.** The linkage between population and economic growth in Mexico: a new policy proposal? (*LARR*, 26:3, 1991, p. 159–170, bibl., tables)

Because past economic rates of growth cannot be sustained in 1990s, government decided to reduce population growth rates and, as a result, increase per capita GNP. After reviewing fertility and mortality changes, article predicts a lower per capita GNP than officially forecast.

4520 **Mosquera Aguilar, Antonio.** Trabajadores guatemaltecos en México: consideraciones sobre la corriente migratoria de trabajadores guatemaltecos estacionales a Chiapas, México. Guatemala: Tiempos Modernos, 1990. 160 p.: bibl., ill., maps.

Short monograph examines migration characteristics of temporary Guatemalan workers on coffee, sugar, and banana plantations in Chiapas. Migration routes, employment characteristics, control of migrants, salaries, and working and living conditions are examined critically. According to author, amidst indifference of Guatemalan government and abuses by Mexican government, exploitation of laborers is extreme.

4521 Movimientos sociales en México, 1968–1987. Recopilación de Elke Köppen. México: Coordinación de Humanidades, Univ. Nacional Autónoma de México, 1989. 136 p.: indexes. (Cuadernos del CIIH. Serie Fuentes; 4)

A total of 1,268 published and unpublished studies on social movements are listed and classified by topics. The term "social movement" is understood in a very broad sense. Includes studies of social classes, labor unions, voting behavior, and political sociology, as well as general books on the peasantry.

4522 Movimientos sociales en México durante la década de los 80. Coordinación de Sergio Zermeño y Jesús Aurelio Cuevas Diáz. México: Univ. Nacional Autónoma de México, Centro de Investigaciones Interdisciplinarias en Humanidades, Coordinación de Humanidades, 1990. 252 p.: bibl., ill., map. (Serie Antologías)

Solid collection includes nine studies from Hidalgo, Veracruz, Oaxaca, Chihuahua, Puebla, Nayarit, Chiapas, Durango, and the Distrito Federal. Topics are equally varied: land demands, urban squatter fights, business opposition to public sector and PRI, women's participation in labor movements, electoral conflicts, and student uprisings. Provides insightful view of fragility of the Mexican political system.

4523 La mujer jalisciense: clase, género y generación. Recopilación de Lucía Mantilla. Guadalajara, Mexico: Univ. de Guadalajara, 1991. 442 p.: bibl., ill. (Col. Jornadas académicas. Serie Coloquios)

The 23 papers in this volume were presented at a 1988 symposium organized by the Centro de Investigación Educativa of the Univ. de Guadalajara. Thematic classification provides idea of contents: women in history, women in the labor force, professional women, women in literature and in movies, biological and social differences, and ideology and power. Quality of contributions is uneven, and sociologists will be interested mostly in articles which discuss female labor force.

4524 Mujeres y ciudades: participación social, vivienda y vida cotidiana. Recopilación de Alejandra Massolo. México: Colegio de México, Programa Interdisciplinario de Estudios de la Mujer, 1992. 297 p.: bibl.

First of 10 articles in volume describes the Primer Encuentro Nacional de Mujeres del Movimiento Urbano Popular, held in 1983. Other articles discuss how women learn to deal with bureaucracies; changing role of women's participation in urban affairs; conflicts between men and women within urban movements; and health, family relations, and fertility of women who build their own houses.

4525 Nava Hernández, Eduardo. Lucha política y movilizaciones sociales en Michoacán, 1988–1989. (*Iztapalapa/México*, 10:21, 1990, p. 123–144, ill.)

Examination of PRI's debacle in 1988 and 1989 elections in Michoacán, the cradle of Cárdenas' opposition party, provides wealth of information at the state level on PRI-PRD conflict, PRI's use of violence, and electoral fraud. Also analyzes basis of Cárdenas' political support and strengths and limitations of the new nationalist and populist Partido de la Revolución Democrática (PRD).

4526 Navarrete, Emma Liliana and Marta G. Vera Bolaños. Diagnóstico de la evolución demográfica en el Estado de México, 1990. México: Gobierno del Estado de México, Consejo Estatal de Población; Colegio Mexiquense, 1992. 61 p.: bibl., ill., maps.

Short monograph estimates demographic changes and forecasts total population of state of Mexico for 1993 by sex and age. Sources include Censos Nacionales de Población for 1970, 1980, and 1990; Encuesta de Migración Interna for 1989; and official vital statistics from state of Mexico (1985–88).

4527 El nuevo Estado mexicano. v. 3, Estado, actores y movimientos sociales. Coordinación de Jorge Alonso, Alberto Aziz Nassif y Jaime Tamayo. Guadalajara, Mexico: Univ. de Guadalajara; México: Nueva Imagen, CIESAS, 1992. 1 v.

Economic crisis of 1980s brought profound transformations to Mexican political corporatist system. Nine articles written by leading social scientists discuss consequences of breakdown of Mexican model on labor unions, peasantry, civil servants, business elite, intellectuals, Christian grassroots movements, and indigenous populations.

4528 Núñez González, Oscar. Innovaciones democrático-culturales del movimiento urbano popular: ¿hacia nuevas culturas locales? México: Univ. Autónoma Metropolitana, Unidad Azcapotzalco, División de Ciencias Sociales y Humanidades, 1990. 295 p.: bibl. (Serie Sociología)

Solid monograph is grounded in European theorists such as Bourdie, Althusser, Rossanvallon, and Gramsci. Fully discusses democratic potential of the movements and PRI's attempts to control them. Internal and external conflicts, leadership patterns, and organizational issues are analyzed. Covers 1970–85, and includes leftist movements and Christian Base Communities organized in Mexico City and its surroundings.

4529 Orozco, Juan Luis. El negocio de los ilegales: ¿ganancias para quién? Guadalajara, Mexico: Instituto Libre de Filosofía, 1992. 601 p.: bibl.

Three small communities (of between 50 and 100 inhabitants) in three municipalities of Altos de Jalisco are the subject of in-depth economic analysis. Author, an economist, assesses socioeconomic factors that promote international migration and discusses, citing field data, the positive and negative impact of migration on the communities.

4530 Ortiz Pérez, Irene and **Ruth Joffre Lazarini.** Así es pues: trabajadoras domésticas de Cuernavaca. México: Colectivo Atabal, 1991. 214 p.: ill.

Findings from 1984 survey of 100 domestic workers form basis for data presented. Limitations of sample are explained in introduction. Sociodemographic and family characteristics of workers, place of origin, and working, economic, and living conditions are some of variables tabulated. Also includes seven vignettes of workers.

4531 Palma Cabrera, Yolanda; Juan Guillermo Figueroa Perea; and **Alejandro Cervantes Carson.** Dinámica del uso de métodos anticonceptivos en México. (*Rev. Mex. Sociol.*, 52:1, enero/marzo 1990, p. 51–81, tables)

Useful analysis of birth control practices is based on data from 1987 Encuesta Nacional sobre Fecundidad y Salud de México. Sociodemographic variables such as age, place of residence (rural-urban), education, and parity are cross-tabulated with levels of utilization and birth control methods. Levels of utilization and methods are also cross-tabulated with institutions providing medical care (Secretaría de Salud and Instituto Mexicano del Seguro Social).

4532 Palma Mora, María Dolores Mónica. Veteranos de guerra norteamericanos en Guadalajara. México: Instituto Nacional de Antropología e Historia; Guadalajara, Mexico: Gobierno del Estado de Jalisco, 1990. 151 p.: bibl. (Col. Regiones de México)

Two-thirds of book is dedicated to background information. Only in last two chapters does author address organization and living conditions of US war veterans residing in Guadalajara. Weak design and difficulties encountered by author in eliciting information limit contribution of the research.

4533 Paré, Luisa. Los pescadores de Chapala y la defensa de su lago. Guadalajara, Mexico: Instituto Tecnológico y de Estudios Superiores de Occidente (ITESO), 1989. 144 p.: bibl., ill.

Interesting book describes ecological destruction of the lake caused by chemicals, sewage, uncontrolled fishing, and a general lack of planning. Study identifies social actors (fishermen, agricultural users of water, tourist industry, ecological groups, and government agencies) that could benefit by collective action in defense of the lake.

4534 Pick, James B. and **Edgar W. Butler.** The Mexico handbook: economic and demographic maps and statistics. Boulder, Colo.: Westview Press, 1994. 422 p.: bibl., ill., index, maps.

Useful collection of data provides late 1980s–90 information on demographics, migration, and housing; industrial and agricultural output and other basic economic information; employment and labor force characteristics by economic sector; and electoral results. Generally, the information is disaggregated by state. Historical trends are also presented.

4535 Pick, James B.; Edgar W. Butler; and Raúl S. González Ramírez. Projection of the Mexican National Labor Force, 1980–2005. *(Soc. Biol.,* 40:3/4, fall/winter 1993, p. 161–190, bibl., tables, graphs)

Various rates of vital statistics, economic activity, and international migration are used to project size of labor force. Authors estimate that labor force will grow at average annual rate of about one million workers, will age slightly, and will have a much higher proportion of females. Implications for migration to US are also discussed.

4536 Piñeda Bañuelos, Gilberto J. and Alfredo Madrigal Carmona. Las maquiladoras en México y el proyecto maquilador en Baja California Sur. La Paz, Mexico: Centro Autónomo de Investigaciones Sociales en el Estado de Baja California Sur, 1988. 63 p.: bibl.

Provides short introduction to *maquiladora* industry in Baja California Sur, a state where presence of *maquiladoras* is relatively new (only 17 by 1988). Authors review advantages that Baja California Sur offers to the industry and use case studies to discuss industry-labor relations, labor conditions, productivity, and salaries.

4537 Población y sociedad en México. Coordinación de Humberto Muñoz García. México: Coordinación de Humanidades, Univ. Nacional Autónoma de México;: M.A. Porrúa, Grupo Editorial, 1992. 331 p.: bibl., ill. (Las Ciencias sociales)

By and large, the 11 articles in this volume make a solid contribution to various fields of demography. Topics covered include: fertility and mortality changes, internal and international migration (surprisingly, migration to US is given little attention), transformation of labor force, and population policies. Rodolfo Corona Vázquez's examination of internal migration surveys is a useful contribution.

4538 Población y trabajo en contextos regionales. Edición de Gail Mummert. Zamora, Mexico: Colegio de Michoacán, 1990. 214 p.: bibl., ill., maps.

Collection of six articles focuses on very different topics. Two study the shift from agricultural to industrial labor, two others examine characteristics of labor force on US-Mexico border, one looks at household adjustments during labor market changes, and last explores social stratification among workers.

4539 Poblamiento: desarrollo agrícola y regional. Recopilación de Carolina Martínez y Susana Lerner. México: Sociedad Mexicana de Demografía, 1992. 177 p.: bibl., ill., maps.

The eight essays in this volume are the result of a working group meeting on population and rural development, and revolve around themes as diverse as international and internal migration, availability of food and health services, impact of foreign investment on the labor force, and the encroachment of Mexico City into neighboring communities.

4540 Política y región: Los Altos de Jalisco. Coordinación de Jorge Alonso y Juan García de Quevedo. México: SEP, 1990. 306 p.: bibl., ill., maps. (Cuadernos de la Casa Chata; 171)

General introduction to the region is followed by five articles. One focuses on socioeconomic transformation (1940–80), a second on the apparel industry, and another on tequila agribusiness. Longest article (100 p.) is a study of nine municipalities which includes analysis of social movements, social classes, and party politics. Last piece is on local culture. Comprehensive and informative volume.

4541 Potreros, vegas y mahuechis: sociedad y ganadería en la sierra sonorense. Coordinación de Ernesto Camou Healy. Hermosillo, Mexico: Gobierno del Estado de Sonora, Secretaría de Fomento Educativo y Cultura, Instituto Sonorense de Cultura, 1991. 472 p.: bibl., ill., maps. (Sonora; 35)

Several of the 11 articles will be of interest to sociologists who study social change and North-South relations. Sonora's proximity to US has caused a chronic economic dependence. Transformation and modernization of agriculture—imposed in part by increasing exports of cattle to US—has had major impact, not always favorable, on inhabitants' living conditions. Data sources include historical and official records.

4542 Los procesos migratorios centroamericanos y sus efectos regionales. Recopilación de Rodolfo Casillas Ramírez. México: Sede Académica de México, Facultad Latinoamericana de Ciencias Sociales, 1992. 127 p.: bibl., maps. (Cuadernos de FLACSO; 1)

The armed conflict in Central America has caused displacement of hundreds of thousands of persons within and outside the region. Volume compiles six papers which analyze Central American migration to Mexico, Mexican migration policies towards Central American refugees, forced internal migration within Central America, effects of remittances, and migration movements on Mexico-Belize border.

4543 Quilodrán, Julieta. Niveles de fecundidad y patrones de nupcialidad en México. México: El Colegio de México, Centro de Estudios Demográficos y de Desarrollo Urbano, 1991. 244 p.: bibl., ill.

Mexico was one of the countries that participated in 1976 World Fertility Study. Using data from Mexican survey, author cross-tabulates fertility rates by rural/urban and age groups, types of marital status and unions, geographical regions, educational levels, occupational status of head of household and spouse, and generational cohorts.

4544 Quintero Ramírez, Cirila. La sindicalización en las maquiladoras tijuanenses, 1970–1988. México: Consejo Nacional para la Cultura y las Artes, Dirección General de Publicaciones, 1990. 246 p.: bibl. (Regiones)

The *maquila* industry has been able to generate a specific type of labor union, which author labels as a subordinate union. Monograph describes characteristics of these unions such as membership, attitudes of unions towards management, and management's mechanisms of labor control. Data are from in-depth interviews of labor leaders and management, examination of labor contracts, and small survey of workers.

4545 Ramírez Rodríguez, Juan Carlos. Tres interpretaciones sobre el fenómeno reproductivo: el caso del Consejo Nacional de Población, el Instituto Mexicano de Seguro Social y la Universidad de Guadalajara. Guadalajara, Mexico: Editorial Univ. de Guadalajara, 1991. 75 p.: bibl. (Cuadernos de divulgación. Segunda época; 40)

Examines various problems associated with reproduction in Mexico. Contains listing of governmental programs and laws to deal with the situation. [R.E. Looney]

4546 Las relaciones laborales y el Tratado de Libre Comercio. Coordinación de Graciela Irma Bensusan Areous. México: Friedrich Ebert Stiftung; Univ. Autónoma Metropolitana, Unidad Xochomilco; FLACSO, Sede México; Grupo Editorial Miguel Angel Porrúa, 1992. 269 p.: bibl. (Las Ciencias sociales)

Collection of 11 articles examines challenges faced by Mexican organized labor as country enters the North American Free Trade Agreement. How a labor force based on authoritarian corporatist model and labor bossism will be able to modernize and become more flexible without compromising welfare of the workers is leading question. Case studies of PEMEX and of automotive and *maquiladora* industries are presented to show possibilities and limits of change.

4547 Reyes Heroles González Garza, Jesús. La reforma del Estado en México. (*Perf. Latinoam.*, 1:1, dic. 1992, p. 171–194)

In recent years Mexico has experienced major political and economic transformations. Author describes them, and interjects his own views of what should be changed and how changes should take place. Decentralization, reduction of economic role of the State, strategies to improve wealth distribution, agricultural transformation, and educational reforms are a few of the topics discussed.

4548 Reyna, José Luis. Hacia la utopía: tenemos que ser menos desiguales. (*in* Los años noventa: ¿desarrollo con equidad? Coordinación de Adolfo Gurrieri y Edelberto Torres-Rivas. San José: FLACSO, 1990, p. 329–354, table)

Reviews 1940–80 socioeconomic transformation of Mexico from a small rural/agricultural to a large industrial/urban society. For decades the poor were willing to postpone sharing in economic benefits for the sake of accelerated growth. The 1982 economic crisis changed that, pushing the impoverished middle and lower classes to the point where they were no longer willing to wait for their share to trickle down. If peaceful social class coexistence is to continue, Mexican system will need to redistribute wealth and become democratic.

4549 Rubin, Rebecca B.; Carlos Fernández Collado; and Roberto Hernández-Sampieri. A cross-cultural examination of interpersonal communication motives in Mexico and the United States. (*Int. J. Intercult. Relat.*, 16:2, Spring 1992, p. 145–157, bibl., tables)

Mexican scores were not significantly higher than US scores on interpersonal control, relaxation, and escape motives, but were significantly lower on interpersonal affection, pleasure, and inclusion motives. Significant negative relationships between interpersonal communication motives and age were found in US but not in Mexico. Based on sample of Mexican students (n=225) and US samples of 504 persons and 477 Kent State students.

4550 Rubio, Luis. Tres años de reforma del Estado en México. (*Perf. Latinoam.*, 1:1, dic. 1992, p. 195–217)

Well-balanced and insightful review focuses on profound economic transformations that have taken place in Mexico as a result of economic crises of 1980s and their political ramifications. Author raises questions regarding future of the PRI and dilemmas it will have to face.

4551 Ruvalcaba Mercado, Jesús. Sociedad y violencia: extracción y concentración de excedentes en la Huasteca. México: SEP, Centro de Investigaciones y Estudios Superiores en Antropología Social, 1991. 155 p.: bibl. (Cuadernos de la Casa Chata)

Historical review considers agrarian violence from colonization to present. According to author, violence in the Huasteca has been consistently caused by dominant landlords and their supporters such as religious leaders and law enforcement authorities. Includes detailed chronograph of all violent events from 1975–85.

4552 Shao, Lixin. Reflections on democratic revolutions in Mexico and China. (*in* Modernization and revolution in Mexico: a comparative approach. Edited by Omar Martínez Legorreta. Tokyo: The United Nations Univ., 1989, p. 151–159)

Briefly compares similarities and differences of Mexico's 1821 and 1910 revolutions and Cárdenas' land reform to 1911 and 1949 Chinese revolutions. Absence of theoretical paradigm reduces value of the study.

4553 Solís de Alba, Ana Alicia and Alba Martínez Olivé. Trabajadoras mexicanas. México: Univ. Autónoma Metropolitana/Iztapalapa, 1990. 161 p.: bibl. (Cuadernos universitarios / División de Ciencias Sociales y Humanidades; 56)

Book is divided into three distinct sections: 1) analysis of documents—mostly letters—from archives of President Adolfo Ruiz Martínez (1952–58) by women about women's issues; 2) two studies of labor conditions, one based on a sample of 36 female industrial workers and the other on information from 16 women's focus groups; and 3) annotated bibliography of books and listing of articles about Latin American women.

4554 Stavenhagen, Rodolfo. Democracia: modernización y cambio social en México. (*Nueva Soc.*, 124, marzo/abril 1993, p. 27–45)

Leading social scientist decries new policies based on neoliberal economics. According to author, globalization of the economy, entry of transnational corporations without controls, dismantling of the ejido system, elimination of price controls for basic foods, reduction of the public sector role, and weakening of labor unions will increase social stratification and the suffering of the poor, and will cause extermination of indigenous minorities, a massive internal and international migration, and enrichment of US at the expense of Mexico.

4555 Suárez Farías, Francisco José. Elite politica e tecnocrazia in Messico. (*in* Studi e ricerche sulla classe politica in Italia, Argentina, e Messico: materiali per un'analisi comparata. Padova, Italy: Facoltá di Scienze Politiche Dell'Univ. di Padova; Casa Editrice Dott. Antonio Milani, 1991, p. 356–365)

Secondary analysis examines elite transformation that took place between 1934–88 (end of De la Madrid presidency). Author relies heavily on Roderic Camp's work, and contrasts roles and functions of traditional political elites (who were more frequently recruited in early years) with those of political technocrats of more recent times.

4556 Tamayo, Jesús and Fernando Lozano Ascencio. The economic and social development of high emigration areas in the state of Zacatecas: antecedents and policy alternatives. (*in* Regional and sectoral development in Mexico as alternatives to migration. Edited by Sergio Díaz-Briquets and Sidney Weintraub. Boulder, Colo.: Westview Press, 1991, p. 15–46, tables)

Suggests policies that could be carried out to develop economy of Zacatecas state, thereby decreasing flow of migration to US. Authors discuss: policy actions to maximize economic use of remittances; specific agricultural and water programs; mining strategies; and public services.

4557 **Tamayo, Jesús** and **Fernando Lozano Ascencio.** Mexican perceptions on rural development and migration of workers to the United States and actions taken, 1970–1988. (*in* Regional and sectoral development in Mexico as alternatives to migration. Edited by Sergio Díaz-Briquets and Sidney Weintraub. Boulder, Colo.: Westview Press, 1991, p. 363–387, tables)

Article examines often contrasting views held by four Mexican presidents on causes of migration, and reasons for limited impact of socioeconomic development and job-creating policies in out-migration regions. According to authors, economic benefits produced by remittances explain Mexican government's failure to formulate effective policies during the period.

4558 **Textos y pre-textos: once estudios sobre la mujer.** Coordinación de Vania Salles y Elsie Mc Phail. México: Colegio de México, 1991. 502 p.: bibl., ill.

These 11 articles are the result of the first research efforts of the Programa Interdisciplinario de Estudios de la Mujer of the Colegio de México. Articles are grouped under four thematic titles: political participation, women and family, female industrial labor (textile industry) at home and in the factory, and women's condition and life cycle. Some articles are theoretical; others are secondary analyses; while still others are findings from field studies.

4559 **Trejo Delarbre, Raúl.** La sociedad ausente. México: Cal y Arena, 1992. 244 p.: bibl.

The 12 essays collected in this volume were previously published in newspapers and magazines. They explore various aspects of influence of the mass media (radio, TV, and press) on Mexican political system and daily life. Approach is critical, and author questions freedom of mass media in Mexico.

4560 **Valenzuela Arce, José Manuel.** Empapados de sereno: reconstrucción testimonial del movimiento urbano popular en Baja California, 1928–1988. Tijuana, Mexico: Colegio de la Frontera Norte, 1991. 223 p.: bibl., ill.

Baja California's cities have experienced a phenomenal population growth. In the face of government's unwillingness to provide basic urban services, inhabitants organized movements to demand accessibility to housing. Using materials primarily from participant interviews, author reconstructs urbanization process of Baja. Analysis of urban movements is partially successful.

4561 **Várguez Pasos, Luis A.** Cultura obrera en crisis: el caso de los cordeleros de Yucatán. (*Estud. Sociol./México*, 11:31, enero/abril 1993, p. 93–110, bibl.)

Brief and interesting historical account examines labor and social organization of henequen workers from beginning of the industry in mid-19th century to present. Organizational capabilities of workers changed throughout the years, from participation in an aggressive socialist union to a PRI-controlled one. Demise of the union in 1991 followed collapse of henequen prices and privatization of the industry. Following Touraine, author suggests that union failure was due to workers' low level of concern for political action and exclusive interest in economic issues.

4562 **Vidrio, Martha.** Estudio descriptivo del abuso sexual en Guadalajara: violación, incesto, atentado al pudor y estupro. Guadalajara, Mexico: Editorial Univ. de Guadalajara, 1991. 130 p.: bibl., ill. (Col. Fin de milenio. Serie Psicología social)

Sexual crime is a topic little studied in Mexico. Informative and well-written monograph examines reported cases of rape, incest, and child abuse in Guadalajara from 1983–86. Cases are classified by types and cross-tabulated by sex, age, and education of victim and offender, and by place of crime.

4563 **Villavicencio, Judith.** Vivienda compartida y arrimados en la zona metropolitana de la ciudad de México. (*Secuencia/México*, 25, enero/abril 1993, p. 31–40, bibl., ill.)

Studies a rapidly expanding living arrangement modality: sharing of house/lot among close kin. An adjacent structure or second floor is added to original one. Most respondents do not consider overcrowding to be a problem. Based on a small sample (n=80) of households in two poor neighborhoods that were settled by land invasions about 20 years ago, which now have access to most urban services.

4564 Woldenberg, José. El proceso electoral en México en 1988 y su secuela. (*in* Una tarea inconclusa: elecciones y democracia en América Latina, 1988–1991. San José: Instituto Interamericano de Derechos Humanos, Centro de Asesoría y Promoción Electoral, 1992, p. 117–134, bibl., tables)

Informative and cogent discussion focuses on political and constitutional reforms which began in 1977. Impact of these changes on party realignments and on electoral results at the national, state, and municipal levels is analyzed in light of electoral returns. Author examines voters' participation and preferences in 1988 presidential elections, in 1989 gubernatorial, congressional, and municipal elections, and in 1990 state elections.

4565 Zamora, Gerardo. La política laboral del Estado Mexicano, 1982–1988. (*Rev. Mex. Sociol.*, 52:3, julio/sept. 1990, p. 111–138, tables)

Excellent review examines salary and employment policies during economic crisis of 1980s and labor responses. According to author, President De la Madrid favored industry at expense of workers, and was able to control traditionally subservient labor leadership by creating divisions among labor federations, making false promises, and exploiting fears of mass unemployment.

4566 Zapata Novoa, Juan. El mercado de las conciencias: sectas y cultos en Monterrey. Monterrey, Mexico: Ediciones Castillo, 1990. 252 p.: bibl.

Book lacks clear theoretical paradigm and adds little to our knowledge of sociology of religion. Nevertheless, because of scarcity of studies on the topic, those interested in new religions in Mexico could profit from some of the information presented.

4567 Zermeño, Sergio. Los intelectuales y el Estado en la década perdida. (*Rev. Mex. Sociol.*, 52:3, julio/sept. 1990, p. 213–235)

Offers provocative interpretation of Mexican intellectuals' shift to the right. According to author, successful cooptation of intellectuals by Cárdenas' Partido de la Revolución Democrática reduced number of intellectuals who were concerned with empowering the poor. Essay also discusses limitations of development models imposed on Mexico and other Third World countries.

CENTRAL AMERICA

JAN FLORA, *Professor of Sociology, Iowa State University*

THE CENTRAL AMERICAN SOCIOLOGICAL LITERATURE of the early 1990s focuses on: 1) changes resulting from the end of guerrilla insurgencies on the isthmus (although in Guatemala, negotiations dragged on) and the return of formal democracy to all countries of the region; 2) the effects of structural adjustment and lack of resources for reconstruction, including problems in provision of social services (especially in Nicaragua and El Salvador); 3) the reduction in the role and efficacy of the State, amd the subsequent growth of social and/or protest movements, evangelical and charismatic movements, and the so-called "new" social movements, with the latter by-passing the State altogether, since they view it as irrelevant; 4) the massive displacement and immigration occurring in the 1980s, and the repatriation of refugees which accelerated in the 1990s; and 5) deterioration of the environment, a major—and growing—problem for the area.

In matters related to the transition to democracy, Torres Rivas (items **4679** and **4678**) argues that the shift from military to civilian rule in Guatemala, El Salvador, and Honduras in the first half of the 1980s and the Central American peace accords near the end of that decade afford the popular classes political legitimacy and at least a limited opportunity to organize their own space (see also items **4654** and **4660** for more pessimistic views). Numerous discussions of agrarian policies, repression, and shifting relations between the military and the oligarchy in those countries and Panama appeared. The defeat of the Sandinistas in the 1990 elections caused a spate of studies. Authors offered various reasons: overzealous expansion of the "social wage" in the early part of Sandinista rule (item **4673**); industrial pricing and accumulation policy (item **1623**); mass organizations lacking in autonomy and linked to a State which became less and less able to deliver economically (items **4627** and **4598**); contradictions between electoral pluralism and a Leninist vanguard party (item **4665**); and the failure of the basic grains program to reduce food costs for the urban population (items **4604** and **4682**). A particularly important and well-documented study is that of Enríquez and Llanes (item **4608**) who conclude from fieldwork in the countryside that peasants given land in the latter half of the Sandinista government in order to reduce the political base of the Contras were transformed into peasant capitalists who came to define their interests as fundamentally distinct from those of the FSLN.

Structural adjustments, the new global economy, and social movements had different impacts in various countries. Costa Rica, the only welfare state in the region, was the first to feel the strong effects of structural adjustment (items **4589, 4671,** and **4626**). Protest movements developed in that nation to petition the State for redress (items **4639, 4572,** and **4607**). In Panama, female-headed households were particularly hard hit by the debt crisis and structural adjustment (item **4625**). There, non-political Urban Neighborhood Organizations (OBUs)—one example of a "new" social movement—sprang up to deal with deteriorating services (item **4583**), while indigenous groups established their own geographic spaces and linkages with an international indigenous movement (item **4592**). Neighborhood self-help groups also developed in Tegucigalpa, Honduras (item **4587**). Regarding social policy during Sandinista rule, which included an FSLN-initiated "structural adjustment" in the late 1980s, a number of useful works assess the pluses and minuses in education (item **4575**), health care (item **4615**), housing and urban land policy (item **4632** and **4640**), and the status of children (item **4669**) and women (item **4586**).

The countries with insurgencies were somewhat insulated from the changes in the global economy occurring throughout the 1980s. Movements in those countries were focused on changing power configurations, rather than simply affecting the distribution of goods and services. Peasant organizations in Guatemala such as the Comité de Unidad Campesina (item **4613**) and the land movement of Father Girón (item **4588**) engaged the State directly. On the other hand, the global economy manifested itself in Guatemala through the growth of the *maquila* industry. See *La maquila en Guatemala* (item **4629**) for an excellent critique of the limited impacts this form of multinational corporate involvement has on a nation's development.

From the many articles and books on evangelical protestantism, especially Pentecostalism, it is clear that there is concern among Central American intellectuals and others about this growing movement. Several of the treatises assess evangelical groups from a doctrinal perspective, concluding that their impact is conservative. Another suggests a good deal of diversity among evangelical groups (item **4657**). The most useful paper on the subject is by Aguilar *et al.*, since it contains hard data (item **4569**). Analyzing surveys conducted in 1988–89 by Father Martín-Baró, the authors conclude that Protestants in El Salvador do not differ markedly from their

Catholic counterparts in either level of political activism or degree of conservatism. In contrast, it appears that in Guatemala there is a small but influential, politically-engaged group of upper-class charismatic Pentecostals with a right-wing agenda (item **4590**). There continues to be a good deal of analysis of the Catholic Church—oriented toward understanding the shift away from the Church of the Poor and the consolidation of an institutionalized pastoral model.

The other movement about which much was written in the early 1990s was that of women in organizations, particularly in Nicaragua. Conclusions included the following: 1) AMNLAE, the Sandinista women's organization, was hampered in its goal of articulating a women's agenda by the need to defend the FSLN (item **4595**); 2) competition between reproductive and community roles created conflicts with husbands (item **4651**); 3) although the Sandinista period opened spaces for women, it did not transform patriarchal structures (item **4646**); and 4) as the war wound down, there was a resurgence of traditional values (item **4594**). Furthermore, the extent of the Chamorro government's commitment of resources to implement existing legal equality is questioned (item **4642**). The most significant contribution on this topic is a multi-regional study of Nicaraguan peasant women's subordination that documents widespread physical abuse by husbands, suggesting that the Sandinista Revolution did not make much headway in transforming peasant men's gender consciousness (item **4612**). On the other hand, case studies of politically involved Salvadoran and Guatemalan women indicated that through their political work they also became aware of gender issues (items **4662** and **4630**).

There are a number of works on migration, population, and ecology that should be noted. According to Hamilton and Stoltz Chinchilla, the massive population movements of the 1980s within Central America and to Mexico and the US were due to global economic changes, economic crises, and political conflict (item **4619**), a phenomenon reflected in numerous articles on migration and refugees. Topics include: adjustment in the US (item **4687**); patterns of rural to urban migration (items **4688** and **4653**); problems and successes in repatriation (items **4648, 4574,** and **4570**); informality and undocumented workers (items **4623, 4652,** and **4675**); and, in Belize, increasing ethnic tensions resulting from the in-migration of Spanish speakers (item **4633**). Finally, population pressures and expansion of capitalist agriculture have contributed to environmental degradation (items **4667, 4609, 2317,** and **4674**).

In addition to works explicitly mentioned above as seminal, one should note two additional publications worthy of special attention. Efforts by the Jesuits in El Salvador to bring about a "third force"—which actually came to fruition after the murder of Father Ellacuría and five others—is documented in a collection of their writings (item **4576**). In another important document, also written by a Jesuit, Ricardo Falla carefully documents the Ixcan massacres under Gen. Ríos Montt in Guatemala through interviews with survivors in camps in Chiapas (item **4610**).

4568 **Adames Mayorga, Enoch.** Crisis y reconstitución del Estado panameño: el giro hacia las políticas radicales. (*Rev. Panameña Sociol.*, 7, 1991, p. 47–60, ill., tables)

Examines crisis of Panama's social policy. Under Torrijos the inequalities of "international services platform" development were ameliorated by expansion in education and health services, which then declined sharply in 1980s. The US invasion returned an anti-statist bourgcois coalition to power. Living standards continue to decline and malnutrition grows.

4569 **Aguilar, Edwin Eloy *et al.*** Protestantism in El Salvador: conventional wisdom versus survey evidence. (*LARR*, 28:2, 1993, p. 119–140, bibl., tables)

Based on two surveys conducted in 1988 and 1989 by late Father Martín-Baró's survey center, study concludes that Protestants are recruited from poorest sectors, favor changing social structure by changing people, vote at a somewhat higher rate than do the religiously unaffiliated poor, vote a little more conservatively than Catholics, and do not represent an overt political challenge.

4570 Aguilar Zinser, Adolfo. CIREFCA: the promises and reality of the International Conference on Central American Refugees. Washington: Hemispheric Migration Project, Center for Immigration Policy and Refugee Assistance, Georgetown Univ., 1991. 73 p.: bibl.

Presents independent evaluation of International Conference on Central American Refugees, Returnees and Displaced Persons (CIREFCA), a regional assistance effort resulting from Central American peace process. While CIREFCA secured better refugee treatment in Mexico and Belize, the organization has not presented donors with many well-conceived projects for repatriation and improvement of refugee conditions. Restructuring recommendations are made.

4571 Aguilera Peralta, Gabriel Edgardo. Ejército y transición en Guatemala. (*Trace*/México, 18, déc. 1990, p. 53–60)

Argues that 1982 coup initiated transition from repressive authoritarian military rule to democratic civilian government. Gen. Ríos Montt sought to legitimate transition through extreme measures against the insurgency and through institutional changes to prepare for civilian control. Transition remains incomplete because military maintains a veto and peace remains elusive.

4572 Anderson, Leslie. Mixed blessings: disruption and organization among peasant unions in Costa Rica. (*LARR*, 26:1, 1991, p. 111–143)

Using Piven's and Cloward's theory about poor people's movements, examines three regional peasant unions which emerged from crisis of 1970s-80s. Concludes that unions have benefitted members, although membership may have immobilized some landless members. Documents unions' histories and cooptative and discrediting mechanisms used by State. For Spanish version, see item **2945**.

4573 Aproximación diagnóstica a las violaciones de mujeres en los distritos de Panamá y San Miguelito. Coordinación de Amelia Márquez de Pérez. Panamá: CEDEM, 1991. 111 p.: bibl., ill. (Col. Denuncia de la violencia contra la mujer)

Study is based on 1988–89 survey of juvenile court and police records, which found that majority of perpetrators were acquainted with their victims and between 19–25 years of age. Few cases showed extenuating circumstances (alcohol or drug abuse, previous sexual offenses, or mental illness). Most cases were brought by mothers of victims. Incest perpetrators received lighter sentences than other rapists.

4574 Ardittis, Solon. Targeted reintegration of expatriate brains into developing countries of origin: the EEC-IOM experience in Central America. (*Int. Migr.*, 29:3, Sept. 1991, p. 371–388, graphs)

Examines five-year project to reintegrate 80 young expatriate professionals into economic sectors of Costa Rica, Honduras, Nicaragua, Panama, and the Dominican Republic. Surveys of returned professionals and their bosses determined: 1) project's contribution to local development; 2) patterns of reintegration; and 3) cost-benefit ratio of project. Project had positive results, although limitations were noted.

4575 Arnove, Robert F. and **Anthony Dewees.** Education and revolutionary transformation in Nicaragua, 1979–1990. (*Comp. Educ. Rev.*, 35:1, Feb. 1991, p. 92–109)

Documents Sandinista efforts at transforming education: Literacy Crusade followed by nonformal popular education collectives and later by decentralization of adult education. Chronicles partly successful efforts to expand and improve quality of formal education. Discusses shortage of resources, the war's impact, and contradictions between popular participation and a vanguard party.

4576 Arroyo Lasa, Jesús *et al.* Universidad y cambio social: los jesuitas en El Salvador. México: Magna Terra, 1990. 207 p.: bibl. (Col. Universidades en América; 1)

Collection of essays by Jesuits on social change in El Salvador's universities reprinted from 1972–87 issues of *Estudios Centroamericanos*, the journal of Univ. Centroamericana José Simeón Cañas. Salvadoran journalist's introduction summarizes political history of the university and Rector Ignacio Ellacuría's philosophy of the third force, and ends with description of Salvadoran military's 1989 assassination of Ellacuría and five fellow Jesuits.

4577 **Asentamientos precarios y pobladores en Guatemala.** México: Ciencia y Tecnología para Guatemala, 1991. 85 p.: bibl., ill. (Formación y capacitación; 4)

Examines extent and problems of marginal and squatter settlements and government policies toward them during Cerezo Administration. Chronicles settlers' organizations, land invasions, and other movements, and assesses hypotheses of why no generalized movement has developed.

4578 **Bachmann, Sybille.** Acerca de la problemática de las comunidades eclesiales de base (CEB) en Centroamérica. (*Islas/Santa Clara*, 92, enero/abril 1989, p. 104–116)

Provides interpretation of origins, principles, and leftist political links of grassroots Christian movements. Author identifies three historical stages in development of the Christian Base Communities: 1) inception to creation of self-help movements (mid 1960s–early 1970s); 2) development of political linkages (mid- to late-1970s); and 3) final state of political commitment and action. [A. Ugalde]

4579 **Barre, Marie-Chantal.** La presencia indígena en los procesos sociopolíticos contemporáneos de Centroamérica. (*Cuad. Am.*, 6:18, nov./dic. 1989, p. 120–143)

Useful summary relates size and organization of indigenous groups and their relations with government for each of the seven Central American countries. For example, summarizes complex relationship between Sandinista government and indigenous groups in Nicaragua, Guatemalan government repression of indigenous peoples and their participation in guerrilla groups, and significance of the Coordinadora Regional de Pueblos Indios de Centroamérica.

4580 **Baumeister, Eduardo.** Productores y políticas estatales: experencias centroamericanas. (*Debate Agrar.*, 13, enero/mayo 1992, p. 313–327, table)

Comparative study examines Nicaragua's Unión Nacional de Agricultores y Ganaderos (UNAG, the Sandinista-linked organization of medium and small farmers) and expansion of small coffee growers in Honduras. Discusses independence of UNAG from Sandinistas, and dilemmas under Chamorro government. Question is whether clientelism of poor farmers with governments and Church will be replaced by clientelistic relations with medium producers and nongovernmental organizations in 1990s free market situation.

4581 **Beluche, Olmedo.** El movimiento obrero y popular panameño frente a la ocupación norteamericana. (*Rev. Panameña Sociol.*, 7, 1991, p. 249–264, bibl., photo)

Argues that Panamanian labor movement seriously damaged its position with Panamanian masses through alliance with inheritors of Torrijismo, particularly with Noriega whose blatant anti-democratic actions alienated a broad cross-section of Panamanian society. Suggests ways of strengthening labor movement.

4582 **Blanco, Gustavo** and **Jaime Valverde.** Honduras: Iglesia y cambio social. San José: Depto. Ecuménico de Investigaciones, 1987. 228 p.: bibl. (Col. Sociología de la religión)

Examines three social change modes for Catholic Church: hierarchical, developmentalist, and prophetic. The hierarchical Church was consolidated in 1957–67 and again after 1975 Olancho massacre. Developmentalism was salient from 1967–75. The prophetic Church (1970–75) grew from developmentalism. Concludes that prophetic Church fostered only limited social transformation because of weak organization of Honduras' oppressed classes.

4583 **Bolaños, Vicky.** Organizaciones barriales urbanas (OBUs): estudio exploratorio para su caracterización: distritos de Panamá, San Miguelito, Colón y Arraiján. (*Rev. Panameña Sociol.*, 7, 1991, p. 179–214, bibl., tables)

From survey of 100 *organizaciones barriales urbanas* (OBUs) in Panama's major cities, author concludes that they are quite diverse socioeconomically. A majority of OBUs are concerned with deteriorating or inadequate services, while a significant minority, generally Church-related, are concerned about growing social problems. Many choose apolitical orientation. Most have a formal democratic structure.

4584 **Bolaños de Aguilera, Aura Azucena.** Situación actual de la mujer guatemalteca. (*in* Jornadas de Investigación Interdisciplinaria sobre la Mujer, *9th, Madrid, 1992.* La mujer latinoamericana ante el reto del siglo XXI. Madrid: Ediciones de la Univ. Autónoma de Madrid, 1993, p. 205–224)

Surveys present legal, educational, health, economic, political, and social situation of Guatemalan women. Briefly examines gamut of women's organizations and State policies regarding women. Concludes with call for strengthening women's organizations and opportunities in order to increase their influence at national level.

4585 Booth, John A. Socioeconomic and political roots of national revolts in Central America. *(LARR,* 26:1, 1991, p. 33–73)

Explains origins of national revolt in Nicaragua, El Salvador, and Guatemala and reasons for absence of revolt in Honduras and Costa Rica. Examines differences in economic growth and in income and wealth distribution. Finds that rapid post-WWII growth of industrialization and export agriculture reduced living standards of working class. Modest reforms in Costa Rica and Honduras purchased stability; repression strengthened resistance in the other countries.

Bourgois, Philippe. West Indian immigration to Central America and the origins of the banana industry. See item **4702.**

4586 Brenes Peña, Ada Julia *et al.* La mujer nicaragüense en los años 80. Managua: Ediciones Nicarao, 1991. 305 p.: bibl.

Presents following gender studies: 1) survey of women and men in five Managua neighborhoods regarding informal sector survival strategies; 2) examination of effects of 1988 structural adjustment reforms by gender, age, and salaried/non-salaried status; 3) essay on women's movements within the FSLN, within the opposition, within religious organizations, and on the Atlantic Coast; and 4) interviews with eight women leaders. Concludes with directory of women's organizations in Nicaragua.

4587 Caballero Zeitun, Elsa Lily. El trabajo comunitario en la consolidación de los asentamientos populares, instalación de servicios básicos y el sistema de procedimientos legales. *(Rev. Univ./Tegucigalpa,* 6:25, agosto 1989, p. 27–39, bibl., photos, tables)

Examines post-WWII growth of poor Tegucigalpa neighborhoods and self-help efforts led by women and children. Collaborative community initiative is curtailed when completed facility is ceded to city government. Interpersonal and intergroup linkages are replaced by bureaucratic regulations and official corruption. Recommends rewriting regulations to allow for continued community-city collaboration.

4588 Cambranes, J.C. Agrarismo en Guatemala. Guatemala: Serviprensa Centroamericana, 1986. 255 p.: bibl., ill. (Monografía / Centro de Estudios Rurales Centroamericanos; 1)

This examination of agrarian movement in Guatemala and ideas of Father Andrés Girón consists of: 1) Girón's "sermon" to his parishioners requesting that President Cerezo require landlords to sell idle land to peasants; 2) Girón's life and thought; 3) article based on interviews with Girón's followers; and 4) reflections on Girón's pro-land movement.

4589 Campos V., Mariana. Del intervencionismo a la privatización: un balance del estado costarricense en su desarrollo reciente. *(Herencia/San José,* 3:1/2, 1991, p. 160–167)

Presents useful summary of economic and social program of Costa Rican State in each of three periods: 1) 1948–70, when State played a redistributive role through establishment of major social programs; 2) 1970s, when State intervention increased, including creation of public production enterprises; and 3) 1980s, when State is reduced and "rationalized," with decline in social programs and quality of life.

4590 Cantón Delgado, Manuela. Protestantismo pentecostal en Guatemala: discurso religioso y conciencia política. *(in* Encuentros Debate América Latina Ayer y Hoy, *3rd, Barcelona, 1991.* Conquista y resistencia en la historia de América = Conquista i resistència en la història d'Amèrica. Coordinación de Pilar García Jordán y Miquel Izard. Barcelona: Publicaciones Univ. de Barcelona, 1992, p. 353–364, bibl.)

Work is based on interviews with Guatemalan leaders of interdenominational Pentecostal and neo-Pentecostal organizations such as Women Aglow and International Businessmen's Full Gospel Fraternity that link politics and religion inextricably. These groups provided support for the candidacy of President Jorge Serrano Elías, an evangelical Protestant.

Castillo G., Manuel Angel. Población y migración internacional en la frontera sur de México: evolución y cambios. See item **4473**.

4591 Castillo G., Manuel Angel. Procesos de pacificación y reestructuración económica: impactos sobre la migración y refugio centroamericano. (*Estud. Int./IRIPAZ*, 3:6, julio/dic. 1992, p. 41–58, bibl.)

Prior to 1970s migration patterns were related to agrarian capitalist development. Crisis that began in 1970s and ensuing violence vastly increased number of refugees and internally displaced persons. Pacification cannot yet be declared a success, and economic restructuring has not caused people to return home.

4592 Castro, Carlos. Estados y movilización étnica en Panamá. (*Estud. Soc. Centroam.*, 48, sept/dic. 1988, p. 115–124, bibl.)

Since 1970s ethnicity and national identity have been approached in two ways: 1) through nationalist-populist Torrijista movement; and 2) through ethnically-identified geographic spaces, both urban and rural. These latter multi-class movements—a new form of integrated political, economic, and cultural organization—invalidate concept of petty bourgeoisie.

4593 Castro Orellana, José Rodolfo and Deborah Barry. La guerra de baja intensidad y la militarización de Centroamérica. (*Iztapalapa/México*, 10:20, julio/dic. 1990, p. 13–32, photos)

Well-documented examination focuses on low-intensity warfare as modified from Vietnam era by Reagan Administration and implemented by its Central American allies. Discusses cases of Nicaragua (where US supported insurgent forces against the government) and El Salvador (where US supported government against leftist insurgents). Mentions contradictions inherent in approach.

4594 Chamorro, Amalia. La mujer: logros y límites en 10 años de revolución. (*Cuad. Sociol.*, 9/10, enero/agosto 1989, p. 117–143, bibl., photo)

Assesses Nicaraguan women's situation, emphasizing central role of Asociación de Mujeres Nicaragüenses Luisa Amanda Espinosa (AMNLAE), and evolution of Sandinistas' programs affecting women's freedoms. Examines three periods: 1) 1979–82, in which social programs improved women's condition; 2) 1983–87, in which resources were diverted to Contra war, but spaces opened for women's participation; and 3) 1988–89, in which AMNLAE consolidated its organization, but traditional values reemerged as war wound down.

4595 Chinchilla, Norma Stoltz. Feminism, revolution, and democratic transitions in Nicaragua. (*in* The women's movement in Latin America: participation and democracy. Edited by Jane S. Jaquette. Boulder, Colo.: Westview Press, 1994, p. 177–197)

Assesses women's organizations during Sandinista period and into Chamorro presidency. Argues that among all Sandinista mass organizations, Asociación de Mujeres Nicaragüenses Luisa Amanda Espinosa (AMNLAE) had greatest difficulty articulating its own agenda. Women trade unionists progressed in linking women's practical and strategic needs. Political space opened by Nicaragua's revolution has given Nicaraguan women leadership in Central American women's movement.

4596 El Chorrillo: situación y alternativas. Panama: IDEN, 1990. 110 p.: bibl., ill., maps. (Cuadernos nacionales; 5)

The 1989 US invasion destroyed the poor neighborhood of El Chorrillo, next door to an installation of the Fuerzas de Defensa de Panamá. Interdisciplinary study examines neighborhood's history and urban function, problems in housing the victims, plans for rebuilding, psychosocial effects, and patterns in press reporting.

4597 Cleary, Edward L. Evangelicals and competition in Guatemala. (*in* Conflict and competition: the Latin American Church in a changing environment. Edited by Edward L. Cleary and Hannah Stewart-Gambino. Boulder, Colo.: Lynne Rienner Publishers, 1992, p. 167–195, table)

Evangelical Protestantism offers converts an explanation for suffering; provides emotional release; and presents belief system similar to traditional religious practices. For well-to-do, it is a politically conservative alternative to Guatemalan Catholicism. Suggests Protestant growth is leveling off following spiritual revival in Guatemalan Catholic Church, but Catholic Church no longer is the exclusive religious interpreter of politics.

4598 Cochran, Augustus B., III and **Catherine V. Scott.** Class, State and popular organizations in Mozambique and Nicaragua. (*Lat. Am. Perspect.*, 19:2, Spring 1992, p. 105–124)

Examines extent of democratization of State power in two countries where popular movements seized power from dictatorships after armed struggles. Emphasizing women's organizations, concludes that both governments tolerated organizations independent of party-affiliated mass organizations, which expanded pressure on the government and perhaps increased responsiveness of mass organizations.

4599 Córdova Macías, Ricardo and **Raúl Benítez Manaut.** Reflexiones en torno al Estado en Centroamérica. (*in* El Estado en América Latina: teoría y práctica. Coordinación de Pablo González Casanova. México: Siglo Veintiuno Editores, 1990, p. 505–541, tables)

Characterizes State power from colonial times to 1980s. Liberal reforms of late 1800s paved the way for export-oriented oligarchy. Great Depression led to militarized oligarchic domination (except in Costa Rica). With 1980s crisis, popular organization forced transition from dictatorial regimes to elected governments—either "storefront democracies" or governments in which popular classes became legitimate actors.

4600 Coutin, Susan Bibler. The culture of protest: religious activism and the U.S. sanctuary movement. Boulder, Colo.: Westview Press, 1993. 250 p.: bibl., index. (Conflict and social change series)

Examines role of religious activists in assisting Central American refugees during late 1980s. After explaining origins and functioning of the movement, author examines resistance tactics employed by the movement against abusive behavior of immigration officers and their questionable implementation of immigration laws. Methodology used for this excellent case study is participant observation and study of legal cases. [A. Ugalde]

4601 Daudelin, Jean. Political dependence and religious policy: Protestants and the State in pre-revolutionary Nicaragua, 1937–1979. (*J. Church State*, 34:2, Spring 1992, p. 229–258)

Useful history of Nicaraguan Protestantism discusses legal framework in which Protestant institutions operate. Examines specific groups' relationships with the State: Baptists, Moravians, Assemblies of God, and CEPAD, the Protestant relief organization established after 1972 earthquake. Finds little evidence linking Protestantism, now led by Nicaraguan nationals, to US imperialism.

4602 Dennis, Philip A.; Gary S. Elbow; and **Peter L. Heller.** Development under fire: the Playa Grande colonization project in Guatemala. (*Hum. Organ.*, 47:1, Spring 1988, p. 69–76, bibl., map.)

Based on their 1984 visit to area, authors examine history of an official colonization project initiated in 1979 in Ixcán lowlands using dual optics of rural development and insurgency/repression. During intense military repression in 1982, three project villages were obliterated; however, remaining villages received substantial government assistance which improved living standards.

4603 Disciplina social para la sociedad guatemalteca desde tres ópticas pentecostales. México: Ciencia y Tecnología para Guatemala, A.C., 1990. 77 p.: bibl. (Cuadernos; 18)

Through study of documents, interviews, and limited participant observation, examines doctrines and practices of two major Pentecostal groups and Jehovah's Witnesses. Doctrine of Second Coming leads to feverish evangelization but not extreme asceticism. Other-worldliness allows social-issue involvement, but acceptance of existing powers is unconditional. Participation allows for social approbation and mobility within group.

4604 Dore, Elizabeth. La respuesta campesina a las políticas agrarias y comerciales en Nicaragua, 1979–1988. (*Estud. Soc. Centroam.*, 49, enero/abril 1989, p. 25–43, tables)

Why did the Sandinista government have limited success in expansion of marketed basic grains? In 1979–85, policies favored cheap food for urban population. Policy shifted to expanded peasant control over land and policy, but government's inability to guarantee timely inputs encouraged bartering and subsistence activities.

4605 Dore, Elizabeth and **John Weeks.** "Up from feudalism." (*NACLA*, 26:3, Dec. 1992, p. 38–44)

Argues that because oligarchies squelched reformist movements, Marxism was banner under which modernization in Central America was carried forward. Reactionaries' alliance with US government limited success of Marxist-led reforms. In Sandinista Nicaragua pre-capitalist agrarian forms were eliminated and land was widely distributed. Ultimate fate of land reform in El Salvador's FMLN-controlled areas remains unclear.

4606 **Edelman, Marc** and **Rodolfo Monge Oviedo.** Costa Rica: the non-market roots of market success. (*NACLA*, 26:4, Feb. 1992, p. 22–29)

Chronicles relation of IMF accords and World Bank structural adjustment loans to government's economic policy in 1980s and early 1990s. Greatest success was shift to 50 percent non-traditional exports. Concludes that in view of the high external debt, decline in wages, deterioration of social services, and increasing poverty, stabilization and adjustment have been less than successful.

4607 **Edelman, Marc.** La cultura política de una protesta campesina contra el ajuste estructural económico en Guanacaste, Costa Rica, 1988. (*Rev. Hist./San José*, 23, enero/junio 1991, p. 145–190, bibl.)

Examines a farmer organization and its road blockade resulting from failure of government to pay indemnization for drought losses. Suggests protest movements are better predicted by linking economic disruptions with local political culture, rather than from increased inequality per se. Notes declining legitimacy of benevolent Costa Rican State as it enforces structural adjustment policies.

4608 **Enríquez, Laura J.** and **Marlen I. Llanes.** Back to the land: the political dilemmas of agrarian reform in Nicaragua. (*Soc. Probl.*, 40:2, May 1993, p. 250–265, bibl., tables)

Sandinistas implemented agrarian reform to generate peasant support for socialist transformation. Through interviews with policymakers and beneficiaries of Los Patios reform project and analysis of 1990 election results, authors conclude that reform improved beneficiaries' economic situation but transformed them into peasant capitalists seeking capitalist-oriented policies.

4609 **Faber, Daniel.** Imperialism, revolution and the ecological crisis of Central America. (*Lat. Am. Perspect.*, 19:1, Winter 1992, p. 17–44)

Argues that ecological degradation results directly from expansion of capitalist export agriculture (cotton and cattle are examples) and indirectly through semi-proletarianization of peasantry who must exploit soil, water, trees, and wildlife to survive. Highlights Nicaraguan effort at developing integrated socialist resource policy as alternative to disarticulated development.

4610 **Falla, Ricardo.** Masacres de la selva: Ixcán, Guatemala, 1975–1982. Guatemala: Editorial Universitaria, 1992. 253 p.: bibl., ill., maps. (Col. 500 años; 1)

Careful documentation of massacres committed by Guatemalan military, aimed at destroying insurgency in this indigenous region, is written by noted Guatemalan anthropologist. Data gathered from survivors in Chiapas in 1983–84 cover two phases: selective repression (1975–81) and scorched earth (1982).

4611 **Fandino, Mario.** Land titling and peasant differentiation in Honduras. (*Lat. Am. Perspect.*, 20:2, Spring 1993, p. 45–55, bibl.)

Argues that USAID-funded titling will benefit only a small elite among recipients. Program fails to realize that these small farmers are semi-proletarians; prohibition against selling, renting, or leasing titled land limits their economic alternatives. Inability of most small-farmer recipients to obtain bank loans prevents them from following implied "farmer" road.

4612 **Fauné, Angélica** *et al.* Cooperación y subordinación en las familias campesinas. Managua: Centro para la Promoción, Investigación y el Desarrollo Rural y Social, 1990. 292 p.

Consists of two field studies: first examines peasant family and community social relations in a northern Nicaragua community to determine compatibility with cooperative development and peasant organization; second study uses interviews in eight regions to assess women's subordination in different peasant strata by examining women's time budgets, their testimonies of abuse by their husbands, and family planning perspectives.

4613 Fernández Fernández, José Manuel. El Comité de Unidad Campesina: origen y desarrollo. Guatemala: Centro de Estudios Rurales Centroamericanos (CERCA), 1988. 64 p. (Cuadernos; 2)

Examines origins of this important peasant movement and its organizational appearance in 1978. Discusses strikes in cotton, sugarcane, and coffee, and the difficulties in establishing alliances between *ladino* workers on southern coast and proletarianizing peasants from western highlands, many of whom were brought to the coast to work.

4614 García Quesada, Ana Isabel. La participación de las mujeres en la toma de decisiones sobre paz en Costa Rica, 1978–1990: un asunto de poder. (*in* Jornadas de Investigación Interdisciplinaria sobre la Mujer, 9th, *Madrid, 1992.* La mujer latinoamericana ante el reto del siglo XXI. Madrid: Ediciones de la Univ. Autónoma de Madrid, 1993. p. 73–93, bibl.)

Documents that although Costa Rican women have had legal equality for 40 years their participation in national public life is quite limited. While number of women at vice ministerial or ministerial level has increased notably from 1978–90, their participation in the four ministries related to the Central American peace process was almost nil. Only the President's wife played a visible, official role.

4615 Garfield, Richard S. and Glen Williams. Health care in Nicaragua: primary care under changing regimes. New York: Oxford Univ. Press, 1992. 240 p.: bibl., ill., index.

Divides health system under Sandinistas into two periods: expansion (1979–84) and crisis management (1984–90). Result was a reasonably efficient primary care program, decentralized and participatory. Limitations include gradual appearance of a two-tiered system—urban and rural—and a continued curative orientation in medical schools and hospitals.

4616 Goldin, Liliana R. Work and ideology in the Maya highlands of Guatemala: economic beliefs in the context of occupational change. (*Econ. Dev. Cult. Change*, 41: 1, Oct. 1992, p. 103–123, tables)

Random sample of villagers of San Pedro Almolonga examines relationships between occupational, religious, and belief changes. Occupation was a much stronger independent contributor to belief changes than was conversion to Protestantism. Suggests that market development alone does not change one's values, but becoming a professional trader may.

4617 González, Nancie L. The Christian Palestinians of Honduras: an uneasy accommodation. (*in* Conflict, migration, and the expression of ethnicity. Edited by Nancie L. González and Carolyn S. McCommon. Boulder, Colo.: Westview Press, 1989, p. 75–90)

Christian Palestinians dominate commerce and industry in San Pedro Sula. Their incomplete acceptance by larger Honduran society is explained by Palestinians' strong ethnic identification due in part to their original migration strategies, nature of their settlement in Honduras, and the protracted conflict in Palestine. Predicts growing interethnic conflict.

4618 González Davison, Fernando. Democratización del Estado guatemalteco, 1984–1987. (*in* ¿Hacia un nuevo orden estatal en América Latina?: democratización/modernización y actores socio-políticos. Buenos Aires: CLACSO, 1988, v. 1, p. 317–385, bibl., tables)

Assesses formal democracy as outgrowth of the "Seguridad y Desarrollo" plan promulgated after 1982 coup. Dicussses election for the Asamblea Nacional Constituyente (1984), the new constitution itself, and presidential election of 1985. Provides year-by-year documentation of first three years of Cerezo Administration: economic and social policy, administrative decentralization, tax reform, administration of justice.

4619 Hamilton, Nora and Norma Stoltz Chinchilla. Central American migration: a framework for analysis. (*LARR*, 26:1, 1991, p. 75–110, bibl.)

Massive Central American population movements of 1980s, both within region and to Mexico and US, are due to: 1) region's economic crisis; 2) each country's different historic incorporation of changes in capitalist world economy (export crop expansion, in particular); and 3) political conflict arising from contradictions between modernization and backward socioeconomic structures maintained by repressive States.

4620 **Handy, Jim.** Anxiety and dread: State and community in modern Guatemala. (*Can. J. Hist.*, 26, April 1991, p. 43–65)

Reform governments of 1944–54 fostered State policies centered on agrarian reform, which began integrating corporate communities into national life. In reaction, succeeding coercive governments excluded highland communities. By 1970s, unable to disengage from national society, indigenous community members embraced external organizations seeking fundamental State alteration. This created another, more brutal, clash between community and the State.

4621 **Honduras. Secretaría de Planificación, Coordinación y Presupuesto.** Política nacional para la mujer. Tegucigalpa: Secretaría de Planificación, Coordinación y Presupuesto, 1989. 145 p.: bibl.

Collaborative effort of the Secretaría de Planificación, Coordinación y Presupuesto, Honduran women's organizations, and two UN agencies, publication assesses women's situation by sector and makes recommendations for achieving equality of opportunity and participation. Provides gender-disaggregated statistics on education, health, employment, land and housing, nutrition, and the media. Proposes institutional structure for a national policy for women.

4622 **Houtart, François and Geneviève Lemercinier.** El campesino como actor: sociología de una comarca de Nicaragua, El Comején. Managua: Ediciones Nicarao, 1992. 185 p.: bibl., ill., maps.

Study of rural district of El Comején, Masaya attempts to measure cultural change based on survey of individual residents. Four levels of analysis: individual, family, cooperative, and district. Concludes that traditional culture persisted in face of Sandinista revolution; cooperatives culturally affected a minority of residents.

4623 **Instituto Interamericano de Derechos Humanos (Costa Rica).** Undocumented and illegal in Central America. Washington: Hemispheric Migration Project, Center for Immigration Policy and Refugee Assistance, Georgetown Univ., 1991. 43 p.: bibl., maps.

In 1980s the number of undocumented migrants from political repression and violence expanded tremendously. Migration followed traditional labor migrant routes. Report details location and characteristics of illegal migrants, their socioeconomic conditions, and their access to health services, education, and employment. Describes migrants' impact on receiving areas, and argues for establishing internationally-recognized human rights.

4624 **Jiménez, Helga.** "Los Estudios de la Mujer" y su inserción en la educación superior en Centroamérica. (*Estud. Soc. Centroam.*, 55, enero/abril 1991, p. 15–24)

Describes two-year project conducted by the public universities' Consejo Superior Universitario Centroamericano to initiate women's studies in each member university. Support of *rectores* was first obtained. Some 125 female professors were trained, in graduate-level modules, in gender studies and the situation of Central American women. Women's studies commissions, centers, or programs were institutionalized in all but one university.

4625 **León de Bernal, Aracelly de.** Mujer, deuda y pobreza. (*Tareas/Panamá*, 81, mayo/agosto 1992, p. 65–75)

Panamanian debt crisis and structural adjustment brought median income of female-headed households in 1990 to 75 percent of that of male-headed households; female unemployment was 50 percent higher than male. Women's work day was extended, requiring daughters to leave school earlier than sons. Feminization of poverty contributes to extremely high child poverty rates and high fertility for less-educated daughters.

4626 **Li Kam, Sui Moy.** Costa Rica ante la internacionalización de la agricultura. (*Rev. Cienc. Soc./San José*, 57, sept. 1992, p. 87–96, bibl.)

Describes Costa Rica's recent neoliberal policies and their effects on agriculture. Explores contradictions between food security and export promotion. Suggests polarization between those benefitting from and those disadvantaged by internationalization of Costa Rica's agriculture, and growth of protest movements among the latter.

4627 **Luciak, Ilya A.** Democracy in the Nicaraguan countryside: a comparative analysis of Sandinista grassroots movements. (*Lat. Am. Perspect.*, 17:3, Summer 1990, p. 55–75)

Analyzes worker-peasant alliance

under Sandinista rule through Asociación de Trabajadores del Campo (ATC) and Unión Nacional de Agricultores y Ganaderos (UNAG). ATC sought popular hegemony; UNAG promoted national unity. UNAG was the most independent of all mass organizations; ATC suffered from over-identification with government, its members' employer.

4628 Madrigal Pana, Johnny and Jacobo Shifter. Primera encuesta nacional sobre SIDA: informe de resultados. San José: Asociación Demográfica Costarricense, 1990. 223 p.: bibl., ill., map.

Survey of males and females in Costa Rica between 15–49 years of age provides information regarding not only attitudes and behavior related to AIDS but also sexuality in general. Concludes that male and female behavior and attitudes differ substantially, and that sexual behavior is not easily changed.

4629 La maquila en Guatemala. México: Ciencia y Tecnología para Guatemala, 1991. 72 p.: bibl. (Cuadernos; 21)

Discusses development of *maquilas* in Guatemala, their role in the development process, working conditions, and efforts to form labor unions. Concludes that while foreign exchange earnings are significant, there are few backward and no forward linkages. Wages are low; unions are difficult to establish; little technology is transferred; and firms have short time horizons.

4630 Martínez Portilla, Isabel María. Lucha y resistencia desde el refugio: mujeres guatemaltecas en el sur de México. (*in* Encuentros Debate América Latina Ayer y Hoy, 3rd, Barcelona, 1991. Conquista y resistencia en la historia de América = Conquesta i resistència en la història d'Amèrica. Coordinación de Pilar García Jordán y Miquel Izard. Barcelona: Publicacions Univ. de Barcelona, 1992, p. 375–386)

Recounts how a women's organization focusing on family health needs in one refugee camp grew to encompass female Guatemalan refugees in three Mexican states. Illustrates how women's mutual aid efforts can be transformed into a consciousness-raising movement which includes among its goals "to defend the right of women to organize and educate themselves and to participate in equality with men."

4631 Martínez Rocha, Abelino. Hegemonía popular y catolicismo popular en Centroamérica. (*Estud. Soc. Centroam.*, 51, sept./dic., 1989, p. 125–142)

Theoretical study examines popular Catholicism's contribution to empowerment of the majority. Tentative conclusions: 1) popular Catholicism is historically central to identity of the popular classes; 2) it has been a cultural bulwark against the Central American crisis; and 3) popular religion is best understood not from a political perspective but as part of civil society.

4632 Mathéy, Kosta. An appraisal of Sandinista housing policies. (*Lat. Am. Perspect.*, 17:3(66), Summer 1990, p. 76–99, bibl.)

Presents good summary of FSLN housing policies and problems. Because of rural-urban migration exacerbated by the war, urban housing deficit grew throughout 1980s, and population decentralization efforts were frustrated. Expropriation and transfer of unused urban land to poor people somewhat ameliorated housing problems.

4633 McCommon, Carolyn S. Refugees in Belize: a cauldron of ethnic tensions. (*in* Conflict, migration, and the expression of ethnicity. Edited by Nancie L. González and Carolyn S. McCommon. Boulder, Colo.: Westview Press, 1989, p. 91–102)

Following overview of historical ethnic diversity, focuses on sociopolitical impacts of refugees from Central American conflicts. Discusses 1984 amnesty for illegal aliens, and ethnic politics in subsequent electoral loss by long-reigning People's United Party. Concludes that Latinization threatens nation's Caribbean identity and intensifies latent racial and ethnic conflict.

4634 Medina, Laurie Kroshus. Contest for continuity or change: a local level view of the organization of power relations in the Belizean citrus industry. (*Caribb. Stud.*, 23: 3/4, 1990, p. 51–67, bibl, maps)

Within a dependency framework, provides overview of interactive processes through which small-scale citrus growers, processing companies, and organized labor have historically structured relations among themselves. The State, through Citrus Ordinance, contributed to strengthening power of the producers; processing plant unions were not so favored. See also item **4637**.

4635 **Meléndez, Guillermo.** The Catholic Church in Central America: into the 1990s. (*Soc. Compass*, 39:4, Dec. 1992, p. 553–570)

Relates struggles of the *Iglesia Popular* in Central America and Church's efforts to counteract growth of evangelical movements in the region. Includes thumbnail sketches of Church's stance in each of five Central American countries. See also item **4636**

4636 **Meléndez, Guillermo.** Iglesias y sociedad en la actual coyuntura centroamericana. (*Perf. Latinoam.*, 2:2, junio 1993, p. 7–50)

Examines relationship of the Vatican's conservative Catholic restoration offensive to the *Nuevos Movimientos Religiosos* (NMRs)—especially Pentecostalism—and to the *Iglesia Popular* (IP) within Catholicism. Discusses reasons for growth of NMRs, Vatican's containment of the IP, and recognition by Central American Bishops Conference that IP helps counter NMRs. See also item **4635.**

4637 **Moberg, Mark.** Citrus and the State: factions and class formation in rural Belize. (*Am. Ethnol.*, 18:2, May 1991, p. 215–233, tables)

Examines local politics and effect on class formation in two Belizean villages. Charlestown and More Hope exhibit differing degrees of patronage-based factionalism and polarization because of differing State involvement in local politics. In context of expanding export citrus economy, Charlestown has played off national parties against one another to gain resources. See also item **4634.**

4638 **Montes, Segundo.** Levantamientos campesinos en El Salvador. (*Real. Econ.-Soc.*, 1:1, enero/feb. 1988, p. 79–100, bibl.)

Contrasts peasant uprisings of 1833, 1932, and 1980s. Unlike the 1980s uprisings, which developed alliances with other poor classes and embraced a more universal ideology, the two earlier uprisings lacked a defined ideology and revolutionary theory, thus isolating them geographically and socially. In recent years, US intervention has prevented peasant victory, but the movement has not been defeated.

4639 **Mora, Jorge A.** Movimientos campesinos en Costa Rica. San José: Facultad Latinoamericana de Ciencias Sociales, 1992. 63 p.: bibl., tables. (Cuadernos de Ciencias Sociales; 53)

Argues for both agency and structure: political, economic, organizational, cultural, and moral factors are important for understanding peasant mobilization. Divides last four decades into three periods: 1) 1950–78, a time of capitalist expansion and squatter movements; 2) 1979–82, a period of transition to an open economy; and 3) 1983–90, a period of structural adjustment, strengthening exports, and strong oppositional peasant mobilization.

4640 **Morales Ortega, Ninette** and **Mario Lungo Uclés.** La gestión de la tierra urbana pública en Managua durante el Gobierno Sandinista. (*Estud. Soc. Centroam.*, 55, enero/abril 1991, p. 109–125, tables)

Well-documented study examines Sandinista urban land policies. Through expropriation of mostly idle land held for specuation, the State redirected one-third of Managua area to popular sectors. Argues that government did not anticipate continued existence of land markets, thus ignoring impact of macroeconomic and financial policies on housing policy.

4641 **Morán de Ferrer, Rhina.** Las niñas trabajadoras: una estrategia de supervivencia. (*UTEC/San Salvador*, 2:3, marzo/abril/mayo 1991, p. 20–23, photos)

Survey of girls working as vendors in the streets of San Salvador found that 68 percent began working by age 10; only 20 percent currently attended school; and 83 percent were not in an intact household. Of those in intact families, the father was often identified as a sexual abuser. Over 90 percent desired to at least learn to read and write.

4642 **Morgan, Martha I.** The *Mother Law*: Nicaraguan women and democratization in the 1990s. (*SECOLAS Ann.*, 23, March 1992, p. 138–151)

Discusses stance of new Chamorro government in context of gender equality provisions of 1987 constitution. Examines three concerns: equality, dignity (freedom from domestic abuse), and reproductive freedom. Concludes that Nicaraguan women have won considerable rights, but the question is whether resources will be marshalled to implement them.

4643 La mortalidad en la niñez: Centroamérica, Panamá, y Belice. San José: Instituto de Nutrición de Centro América y Panamá, Fondo de las Naciones Unidas para la Infancia, Centro Latinoamericano de Demografía, 1990. 7 v. in 1: bibl., ill. (Serie OI; 1007)

Separate publications for each country of Central America (including Belize and Panama) provide comparable data on mortality rates for children under five years by geographic locality and by cause.

Mosquera Aguilar, Antonio. Trabajadores guatemaltecos en México: consideraciones sobre la corriente migratoria de trabajadores guatemaltecos estacionales a Chiapas, México. See item **4520**.

4644 Nieto, Elba María. El delito de la violación sexual en Honduras. Tegucigalpa?: Comité Hondureño de Mujeres por la Paz Visitación Padilla, 1987. 58 p.: bibl., tables. (Cuadernos Visitación Padilla; 2)

Work is based on accusations and testimonies of sexual assault made to criminal courts of Tegucigalpa during 1988. Tables show age and sex of victims, characteristics of the accused, and location of crimes. Data are presented for male, female, child, adolescent, and incest victims. Urges programs for prevention and for psychological and legal support.

4645 El nuevo rostro de Costa Rica. Edición de Juan Manuel Villasuso Estomba. Heredia, Costa Rica: Centro de Estudios Democráticos de América Latina, 1992. 547 p.: bibl., ill.

Articles written especially for this volume by diverse group of intellectuals assess changes in Costa Rica during 1980s. Major topics include values as reflected in social institutions, social organizations and movements, political institutions, State reform, and structural economic changes.

4646 Olivera, Mercedes; Malena de Montis; and Mark A. Meassick. Nicaragua: el poder de las mujeres. Managua: Cenzontle, 1992. 248 p.: bibl. (Col. Realidades; 2)

Through in-depth interviews with 14 women in public positions, assesses gender problems associated with participating in patriarchal political institutions. Describes the dual demands placed on women holding public office who continue to be single-handedly responsibile for the care of the home as well. From a random sample of voting-age women from Pacific region, authors determine level and type of women's organizational involvement. Concludes that Sandinista period was positive for women, but did not transform patriarchal structures.

4647 Opazo Bernales, Andrés. Costa Rica: la Iglesia Católica y el orden social: entre el Dios de la polis y el Dios de los pobres. San José: Depto. Ecuménico de Investigaciones, 1987. 217 p.: bibl. (Col. Sociología de la religión. Iglesia y pueblo)

Examines social role of Catholic Church (1960–84), which has involved blessing Costa Rica's welfare state. In 1970s welfare state entered into crisis, and poverty grew. Describes two urban and two rural efforts to respond as Church of the Poor. Discusses possibilities for reconciling two conceptions of the Church.

4648 Ortega, Marvin. Reintegration of Nicaraguan refugees and internally displaced persons. Washington: Hemispheric Migration Project, Center for Immigration Policy and Refugee Assistance, Georgetown Univ., 1991. 41 p.: bibl. (HMP policy brief; 3)

Compares reintegration of Miskitos with that of mestizo peasants. Miskito reintegration was relatively conflict-free because of ethnic cohesion, non-politicized traditional leadership, and attachment to ancestral lands. Individualistic mestizos were influenced by organized opposition to Sandinistas in refugee camps. Failure of Chamorro government to provide land promised to them has exacerbated conflicts.

4649 Pedroni, Guillermo. Territorialidad kekchi: una aproximación al acceso a la tierra: la migración y la titulación. Guatemala?: Facultad Latinoamericana de Ciencias Sociales (FLACSO)-Guatemala, 1991. 51 p.: bibl., ill. (Debate; 8)

Examines Kekchi's relation to the land in traditional and colonization areas. Based on a survey and case studies, assesses land scarcity, migration, and tenure patterns for selected communities in Polochic watershed (Panzos and El Estor municipalities). Emphasizes urgency of land titling. Predicts increased landlord-peasant conflict.

4650 Perera, Victor. Unfinished conquest: the Guatemalan tragedy. Photographs by Daniel Chauche. Berkeley: Univ. of California Press, 1993. 382 p.: bibl., index, map, photos.

Guatemalan-born journalist documents effects of military counterinsurgency on villages of Ixil Triangle, placing situation in larger historical and social context. Field research carried out during Vinicio Cerezo's presidency (1985–90). Very readable. For ethnologist's comment see item **757**.

4651 Pérez Alemán, Paola. Women's movement, crisis, and food: the case of Nicaragua. (*in* Alternatives. Edited by Neuma Aquiar and Thaís Corral. Rio de Janeiro: Editora Rosa dos Tempos, 1991, v. 2, p. 79–111, bibl., table)

As part of a three-country (Nicaragua, Peru, Chile) study of women's nutrition organizations, author attended meetings of two neighborhood groups in Managua and interviewed 17 members plus leaders. Found that conflicts between reproductive and community roles generated conflicts with their husbands. Recommends that Asociación de Mujeres Nicaragüenses Luisa Amanda Espinosa (AMNLAE) assist such groups to address strategic needs.

4652 Pérez Sáinz, Juan Pablo. Informalidad e identidades sociales en Area Metropolitana de Guatemala. (*Rev. Eur.*, 52, June 1992, p. 7–31, bibl., tables)

Based on case studies, concludes that identity of persons in metropolitan informal sector is rooted not only in their location within production system, as Alejandro Portes suggests, but also in their domestic situation. Finds that equivalence of home and workplace, gender and status in household, mobility within informal sector, and multiple occupations require contextualization of identities.

4653 Pérez Sáinz, Juan Pablo; Manuela Camus; and Santiago Bastos. Todito, todito es trabajo: indígenas y empleo en Ciudad de Guatemala. Guatemala: FLACSO, 1992. 137 p.: bibl.

In-depth study of 85 indigenous workers was drawn from a larger study of urban workers in Guatemala City. For indigenous labor, household is appropriate unit of analysis. Indigenous workers receive lower incomes and fewer benefits than ladinos, obligating more family members to work. Women perform both productive and reproductive work, and children forego education.

4654 Posas, Mario. El proceso de democratización en Honduras. (*Síntesis/Madrid*, 8, mayo/agosto 1989, p. 183–203)

Defines democratization as a movement to promote certain groups' interests. In 1970s, oligarchy promoted representative democracy against Gen. López Arrellano's socially progressive regime. Democratization in 1980s required oligarchy to forge alliances with the military and the US. Both traditional parties developed internal democracy, but because parties excluded urban middle sectors and the poor, democracy did not serve interests of those two groups.

Los procesos migratorios centroamericanos y sus efectos regionales. See item **4542**.

4655 Protestantismos y procesos sociales en Centroamérica. Recopilación de Luis Samandú. San José: Editorial Universitaria Centroamericana, 1990. 302 p.: bibl. (Serie investigaciones / Programa Centroamericano de Investigaciones; 4)

Presents results of collaborative research project coordinated by Consejo Superior Universitario Centroamericano. Topics include quantitative geographic distribution of Protestants, the Protestant challenge, Protestantism and society, and evangelical strategies among indigenous Guatemalans, as well as individual chapters on El Salvador, Honduras, Nicaragua, and Costa Rica.

4656 500 años de lucha por la tierra: estudios sobre propiedad rural y refoma [i.e., reforma] agraria en Guatemala. Edición de J.C. Cambranes. Guatemala: FLACSO, 1992. 2 v.: bibl., ill.

Vol. 1 examines land tenure and social class in colonial period, development of 19th-century capitalist agriculture, and land reform and counter-reform of 1950s. Vol. 2 assesses contemporary agrarian conflicts, beginning with survey of post-WWII period. Important reference by very knowledgeable contributors.

4657 Recio Adrados, Juan Luis. Incidencia política de las sectas religiosas: el caso de Centroamérica. (*ECA/San Salvador*, 48: 531/532, enero/feb. 1993, p. 75–91, photos)

In El Salvador most Protestant groups are supportive of political authoritarianism, while in Nicaragua Protestantism has played a more socially progressive role. Notes Protestant shift from separateness to greater political involvement in El Salvador and Guatemala. Discusses conservative Salvadoran charismatic sect which seeks to recruit political elites.

4658 **Rodríguez Solera, Carlos Rafael.** Tierra de labriegos: los campesinos en Costa Rica desde 1950. San José: FLACSO, 1993. 240 p.: bibl., ill., maps.

Using census information at cantonal level, examines de-peasantization and re-peasantization, and resulting social and demographic impacts, for period 1950–84. Country experienced rapid de-peasantization until 1973. Then process slowed—and in certain regions reversed—due to international economic and capitalist agricultural crises. Examines related State policies, especially agrarian reform.

4659 **Ruiz, María Teresa.** Racismo, algo más que discriminación. San José: Departamento Ecuménico de Investigaciones, 1988. 181 p.: bibl., ill. (Col. Análisis)

A verbal and visual survey was carried out among children in their last year of primary school in urban and rural parts of Costa Rica and Panama. Children were shown pictures of faces of persons with different social and racial characteristics. Anti-black prejudice and stereotyping were quite high, particularly among higher-class white and indigenous respondents. Blacks evidenced low levels of self esteem and low anti-white prejudice. Patterns held in both Costa Rica and Panama.

4660 **Salomón, Leticia.** Honduras: la transición de la seguridad a la democracia. (in América Latina: militares y sociedad. Coordinación de Dirk Kruijt y Edelberto Torres-Rivas. San José: FLACSO, 1991, v. 1, p. 93–117, bibl.)

Military security doctrine dominated in Honduras in 1980s. Politically legitimate claims of domestic groups were interpreted as subversive. Security doctrine was discredited as Central American conflicts were resolved and US troops withdrawn. Need for institutional and cultural changes remains to ensure enduring civilian control of the military.

4661 **Samandú, Luis; Hans Siebers; and Oscar Sierra.** Guatemala: retos de la Iglesia Católica en una sociedad en crisis. San José: Depto. Ecuménico de Investigaciones, 1990. 183 p.: bibl., maps.

From a sociology of religion perspective, authors examine 1970s grassroots movements (*Iglesia de los Pobres*) in Guatemalan Catholic Church, and transformation and institutionalization of pastoral functions during 1980s, with violence of early 1980s as the dividing point. Case studies in three parishes enrich the analysis.

4662 **Schirmer, Jennifer.** The seeking of truth and the gendering of consciousness: the CoMadres of El Salvador and the CONAVIGUA widows of Guatemala. (in "Viva:" women and popular protest in Latin America. Edited by Sarah A. Radcliffe and Sallie Westwood. London; New York: Routledge, 1993, p. 30–64, facsim., photo)

Based on in-depth interviews with widows and mothers of disappeared persons, study concludes that, in questioning first the "*official* truth" and then the "claimers of truth," the women see their ethnicity and class through the gender optic. As seekers of the "real truth," the majority were abused by agents of the State. What the women initially saw as individual acts of rape became perceived as part of a culture of gender abuse. History and programs of two groups are linked to members' growing gender awareness.

4663 **Schmölz-Häberlein, Michaela.** Die Grenzen des Caudillismo: die Modernisierung des guatemaltekischen Staates unter Jorge Ubico, 1931–1944: eine regionalgeschichtliche Studie am Beispiel der Alta Verapaz. Frankfurt am Main; New York: Lang, 1993. 241 p.: bibl., maps. (Europäische Hochschulschriften. Reihe III, Geschichte und ihre Hilfswissenschaften, 0531–7320; 567)

Uses Alta Verapaz, an important coffee-producing region and major German settlement in Guatemala, to analyze sociohistorical consequences of modernization and reforms on culture of indigenous peasant population. Concludes that the most consequential example of vertical social mobility was German-Indian symbiosis which, through gifts of land and houses, propelled many Indian women and their offspring into ladino society. [C.K. Converse]

4664 **Schrading, Roger.** El movimiento de repoblación en El Salvador. San José: Instituto Interamericano de Derechos Humanos, Area de Promoción y Asistencia a ONG, Programa para Refugiados, Repatriados y Desplazados, 1991. 103 p.: bibl. (Exodos en América Latina; 4)

Traces patterns of displacement of rural Salvadorans, and remedies attempted. Response organized by displaced persons on their own behalf has been most effective. Examines two early instances of repopulation. Repopulated villages formed national organizations which, because of their effective organization and domestic and international alliances, became significant players on the national scene.

4665 **Serra, Luis H.** Democracy in times of war and socialist crisis: reflections stemming from the Sandinista revolution. (*Lat. Am. Perspect.*, 20:2, Spring 1993, p. 21–44, bibl.)

Argues that Sandinistas combined elements from two political models: electoral pluralism and a vanguard Leninist party. FSLN directed a development-oriented bureaucratic State, from which popular organizations received resources in exchange for political loyalty, limiting organizations' political participation. Contradictions led to FSLN's 1990 electoral defeat. Speculates whether FSLN can adapt structurally to a multiparty system.

4666 **Serra, Luis H.** Movimiento cooperativo campesino: su participación política durante la revolución sandinista, 1979–1989. Managua: Impr. Univ. Centroamericana, 1991. 297 p.: bibl.

Seeks to determine conditions which help or hinder peasantry in becoming political protagonists during social upheaval. Presents historical antecedents of peasant movement (1950–79) and efforts to build participatory economic democracy through agricultural cooperatives (1979–89). In-depth individual and group interviews and surveys of members, women, youth, and ex-members in 11 cooperatives throughout the country were conducted in 1989.

4667 **Sierra Mejía, Marcio E.** Habitat y desarrollo urbano en el distrito central. (*Rev. Univ./Tegucigalpa,* 6:25, agosto 1989, p. 5–26, photos, tables)

Documents rapid urbanization of Tegucigalpa. Concludes that growth has been helter skelter, and that poverty and violence have lowered quality of life for all, with negative environmental effects. Argues for population and human settlements policy to strengthen intermediate cities and generate agroindustrial and industrial employment.

4668 **Simposio sobre la Reforma Agraria Nicaragüense,** *Amsterdam,* **1988.** El debate sobre la reforma agraria en Nicaragua: transformación agraria y atención al campesinado en nueve años de reforma agraria, 1979–1988. Coordinación de Raúl Rubén y Jan P. de Groot. Managua: Editorial Ciencias Sociales, 1989. 407 p.: bibl.

Contains 13 papers on agrarian reform's first nine years, presented at 1988 International Congress of Americanists. Examines planning context and accumulation model; reform structure and social classes; peasant organization, including cooperatives and women's role in agrarian reform; markets, prices, and labor policy; and land reform on the Atlantic Coast. Impressive group of contributors.

4669 **Situación de la niñez nicaragüense.** Coordinación de Ivonne Siu y Carlos Hernández. Managua?: UNICEF, 1990. 48 p.: bibl., col. ill., map.

Summarizes children's status under Sandinista rule. Economic crisis and civil war placed much of the population, including children, at risk. However, institutional commitment to child welfare, popular participation, and incipient administrative decentralization improved, or slowed deterioration of, specific conditions. Presents statistical trends in health, education, and demographics.

4670 **Smith, Carol A.** The militarization of civil society in Guatemala: economic reorganization as a continuation of war. (*Lat. Am. Perspect.*, 17:4, Fall 1990, p. 8–41, bibl.)

Argues that economic control has replaced military coercion as instrument for neutralizing resistance of western Guatemalan indigenous communities, which must rely on the State for economic survival. While dominance of oligarchy has been challenged by the military, mutual interest in suppressing insurgency contributes to coordination among government, military, and oligarchy.

4671 Sojo, Carlos. La utopía del estado mínimo: la influencia de AID en las transformaciones funcionales e institucionales del Estado costarricense en los años ochenta. Managua: CRIES, 1991. 89 p.: bibl.

Discusses relationship of Costa Rica's strategic plan to USAID agreements. Efforts to transform public sector, including reduction in number of public employees and transformation of public housing programs, had limited success (due partly to public sector unionism). AID then concentrated on privatization of State enterprises and establishing parallel activities in the private sector.

4672 Solà, Roser and María Pau Trayner. Ser madre en Nicaragua: testimonios de una historia no escrita. Managua: Editorial Nueva Nicaragua; Barcelona: Icaria, 1988. 255 p., 16 p. of plates: ill. (Totum revolutium; 33)

In-depth interviews were conducted with 42 members of the Comité de Madres de Héroes y Mártires de Matagalpa. Includes women whose children or husbands died during FSLN's clandestine period, during open insurrection (1978-79), and during Contra war. Examines impact on women's religious convictions, and their view of themselves as women, as mothers, and as revolutionaries.

4673 Stahler-Sholk, Richard. Stabilization, destabilization, and the popular classes in Nicaragua, 1979-1988. (*LARR*, 25:3, 1990, p. 55-88, bibl., tables)

Sandinista mixed economy and multiclass polity created demands incompatible with defense and long-term development. "Social wage," expanded through volunteer labor and mobilization, only partially substituted for money wages. State-led, capital-intensive agricultural investment drove peasants into parallel economy. In response, government de-emphasized collective sectors to gain wartime allegiance of the peasantry.

4674 Stonich, Susan C. "I am destroying the land!": the political ecology of poverty and environmental destruction in Honduras. Boulder, Colo.: Westview Press, 1993. 191 p.: bibl., ill. (Conflict and social change series)

Using a political ecology and case study approach, links agricultural development, demographic change, impoverishment, and environmental decline in southern Honduras. Concludes that pluri-activity contributes to environmental degradation as poorest farmers adopt less labor-intensive agricultural practices. Moderate-sized farmers (over five hectares) are most amenable to environmentally sound practices. For ethnologist's comment see item **766**.

4675 Stonich, Susan C. Rural families and income from migration: Honduran households in the world economy. (*J. Lat. Am. Stud.*, 23:1, Feb. 1991, p. 131-161, tables)

Unprecedented semi-proletarianization has been occurring in southern Honduras since 1950. Migration of selected members is one aspect of household diversification strategies, requiring linkages with rural communities, urban and rural labor markets, small-scale commodity production, and the informal and export agriculture sectors. Concludes that peasantization and de-peasantization are occurring simultaneously.

4676 Thomson, Marilyn. Las organizaciones de mujeres en El Salvador. (*Estud. Soc. Centroam.*, 54, sept./dic. 1990, p. 119-135)

Includes thumbnail sketches of 1980s women's organizations of the right and left: Liga Feminina Salvadoreña, Mujeres de ARENA (Alianza Republicana Nacionalista), Sección de Mujeres del Comité de Presos Políticos de El Salvador (COPPES), Comités de Madres y Familiares de los Desaparecidos, Asesinados y Presos Políticos (COMADRES), and Asociación de Mujeres de El Salvador (AMES).

4677 Tierra, café y sociedad: ensayos sobre la historia agraria centroamericana. Recopilación de Héctor Pérez Brignoli y Mario Samper K. San José: Programa Costa Rica, FLACSO, 1994. 589 p.: bibl., ill.

Examines coffee industry in Central America from mid-19th to mid-20th century. Includes local and country-specific monographs, and comparative and synthetic studies. Based on 1990 symposium organized by the Escuela de Historia, Universidad Nacional (Costa Rica), work represents important scholarship on societal impact of Central American coffee cultures.

4678 Torres-Rivas, Edelberto. La recomposición del orden: elecciones en Centroamérica. (*Rev. Esp. Invest. Sociol.*, 50, abril/junio 1990, p. 111-121)

Ten years of civil war and economic

crisis have ended with elections whose results were accepted by all major forces as legitimate. This has opened limited political space for continued partisan competition and popular organization. While experiences in each country differ, relations are redefined among dominant groups, the military, and society.

4679 Torres-Rivas, Edelberto. El sistema político y la transición a la democracia en Centroamérica. San José: Secretaría General, Facultad Latinoamericana de Ciencias Sociales, 1990. 99 p.: bibl., ill. (Cuadernos de ciencias sociales; 30)

Contains four essays on long-term transition to full democracy after era of armed conflict. Economic reactivation is necessary for present "low-intensity" (authoritarian) democracy to survive. *Full* democratization requires strong popular organizations which are separate from the State. While political parties can act as mediating forces to resolve conflicting interests among elites, they are less effective vehicles for popular classes.

4680 Una tragedia campesina: testimonios de la resistencia. Recopilación de Alejandro Bendaña. Managua: Editora de Arte, 1991. 271 p.: bibl., ill. (Col. Perspectiva)

The editor, spokesperson for the FSLN Foreign Ministry, seeks to understand political culture of middle class and rich peasants—and Sandinista policy errors—which allowed the Resistencia Nacional (the Contras) to recruit them. In collaboration with two Nicaraguan research organizations, Bendaña presents eight interviews with disarmed Contras, allowing them to speak for themselves. Editor's introduction provides international and domestic context.

4681 Turner, Jorge. La nueva etapa obrerista en Panamá: década de los ochenta. (*in* Organización y luchas del movimiento obrero latinoamericano, 1978–1987. Coordinación de Mario A. Trujillo Bolio. México: Siglo Veintiuno, 1988, p. 240–256)

Contends that organized labor in Panama entered new era with institutionalized coordination of labor federations, and with combined worker, peasant, student, and professional protest against three 1986 laws on labor, agriculture, and industry. These laws, it is argued, were imposed on President Ardito Barletta by the World Bank and IMF.

4682 Utting, Peter. The peasant question and development policy in Nicaragua. Geneva: URISD, 1988. 35 p.: bibl. (Discussion paper / United Nations Research Institute for Social Development, 1012–6511; 2)

Analyzes changes in food and development policy from 1984–88 as they bear on peasant question. Deals with four central issues: growth of overall food production; extraction of surplus; increased standard of living for rural poor; and integration of peasantry into dominant class alliance.

4683 Valverde, Jaime. Las sectas en Costa Rica: pentecostalismo y conflicto social. San José: Editorial Depto. Ecuménico de Investigaciones, 1990. 95 p.: bibl. (Col. Sociología de la religión)

Documents accelerating growth of Pentecostalism in Costa Rica. Concludes that, as with Christian Base Communities (CBCs), adherence to Pentecostalism is a response to oppression, is life-affirming, and promotes fraternal action. However, unlike CBCs, Pentecostalism's other-worldly orientation supports status quo. Includes case studies in communities of Guápiles and Hatillo.

4684 Vega Carballo, José Luis. Partidos, desarrollo político y conflicto social en Honduras y Costa Rica: un análisis comparativo. (*Síntesis/Madrid*, 8, mayo/agosto 1989, p. 363–383, bibl., tables)

Argues that Costa Rica's 1948 revolution successfully realigned oligarchy and a "new class" of reformers. Subsequently, differing interests were negotiated within the parties or the State. In Honduras, however, whenever conflicts between party insiders and outside groups became too great, the military shut down the parties, preserving an unequal class system.

4685 Vilas, Carlos M. La contribución de la política económica y la negociación internacional a la caída del gobierno sandinista. (*Rev. Mex. Sociol.*, 52:4, oct./dic. 1990, p. 329–351, bibl., table)

Sandinista government, seeking to attract middle groups in 1990 elections, made concessions to domestic right and bowed to international pressures. Nicaraguan voters, enduring structural adjustment policies and Sandinistas' inability to end Contra war, concluded that risking the unknown was preferable to sticking with the known.

4686 Vilas, Carlos M. Family affairs: class, lineage and politics in contemporary Nicaragua. (*J. Lat. Am. Stud.*, 24:2, May 1992, p. 309–341)

Examines transformation of lineage, centered on León-Granada liberal-conservative division, in recent Nicaraguan history. Argues that Somocismo marginalized traditional families from government and challenged them economically, more than did Sandinismo. Importance of conservative families from Granada in Sandinista government was a reaction to competition from Somocista newcomers.

4687 Vlach, Norita. The quetzal in flight: Guatemalan refugee families in the United States. Westport, Conn.: Praeger, 1992. 175 p.: bibl., ill., index.

Rich ethnographic study by community mental health worker examines migration motivations of six Guatemalan families and their coping strategies in US. Relates initial stress event, family resources, and cultural perceptions to coping process. Links post-migration Church support systems (Catholic and Pentecostal) to differing family patterns.

4688 Vonós a la capital: estudio sobre la emigración rural reciente en Guatemala. Guatemala: Asociación para el Avance de las Ciencias Sociales en Guatemala, 1991. 93 p.: bibl., ill. (Cuadernos de investigación; 7)

Presents results of 1989 survey of households in selected rural regions which focused on living conditions and out-migration patterns of youth from those households. The youth who resided in Guatemala City and could be located were interviewed in-depth on reasons for migrating and their situation in the capital. Historical census information also presented.

4689 Weaver, Frederick Stirton. Inside the volcano: the history and political economy of Central America. Boulder, Colo.: Westview Press, 1994. 276 p.: bibl., ill., index. (Series in political economy and economic development in Latin America)

Presents crisp historical political economy interpretation of recent Central American crises. Includes chapters on different societies resulting from Spanish conquest, colonial reforms and independence movements, rise of coffee and banana export production, the Great Depression and WWII, postwar agricultural modernization and development of Common Market, and struggle toward democracy in past dozen years.

THE CARIBBEAN AND THE GUIANAS

BENIGNO E. AGUIRRE-LOPEZ, *Professor, Department of Sociology, Texas A&M University*

REFLECTING THE TREND FOR LATIN AMERICA as a whole, the literature on the Caribbean region is dominated by the study of rapid social change processes such as the Cuban Revolution, activities of social movement organizations, and instances of collective behavior, especially mass migrations. Recent UN-sponsored research on refugees and displaced women is particularly noteworthy (item **4444**), as is research documenting the declining quality of life in Puerto Rico (item **4724**). The US Department of Health and Human Services and the Pan American Health Organization have sponsored research on the impact of tobacco smoking (items **4447** and **4766**). There are also new publications on health status and health service utilization (item **4751**) and education (item **4720**) in the Commonwealth Caribbean.

An important theme in recent scholarship is the demography of the Caribbean, with most interest focused on the dynamics of regional migratory patterns. Considerable research attention has also focused on the social and cultural effects of the economic transformations taking place in the region, particularly their impacts on individual economic well-being. Dominant themes in this body of research include the effect of the economic crisis on the region's agricultural systems, families, and women.

As in previous years, Cuba continues to attract significant attention from specialists. Especially noteworthy areas of research are the family as well as the sexuality and ideology of Cuban youth. The latter segment of the population appears particularly recalcitrant to the message of the Cuban government, as evidenced by their disproportionate participation in recent mass emigrations from Cuba and their victimization by the repressive system (item **4776**). Aguirre (item **4691**) argues that it is useful to conceptualize the periodic mass migrations from Cuba to the US—especially those during the Mariel crisis—as oppositional serial surges in which collective deviance labeling often occurs.

A number of works examine aspects of Cuban culture. For example, research on the subculture of art, especially painting and sculpture, is available (item **4692**). Other research focuses on Afro-Cuban cultural traits, such as religious funeral ceremonies, Haitian influences on beliefs, and Cuban material culture. Important research has also been conducted on fertility. New publications include bibliographies of Cuban and Caribbean mass media (items **4701** and **4700**) as well as detailed examinations of Cuban labor laws and practices (item **4725**) and the State's repression of the Catholic Church (item **4784**).

The Cuban Catholic Church is experiencing a rebirth in the aftermath of the bishops' Sept. 1993 "Love Is All Powerful" pastoral letter, one of the most revealing documents on the condition of contemporary life in Cuba (item **4692**). This post-Vatican II pastoral letter asks why so many Cubans from various social spheres want to leave their homeland; it calls for a plurality of perspectives from which to consider national problems; it reminds the authorities of the material needs which continue to be unmet despite the many resources available on the island; it reviews the effect of the crisis on the family and society, including increases in delinquency, theft, prostitution, violence, alcoholism, and suicide; and it condemns human rights violations. The letter calls for national dialogue and reconciliation based on mutual respect and democratic procedures. Espousing a Catholic perspective, Cuba's Movimiento Cristiano de Liberación has requested that the Pope take a more active role in Cuba. The organization openly supported the bishops' pastoral letter and requested the bishops to become more involved in guiding the nation's transition to democracy and capitalism. It also proposed a referendum to determine whether a majority of the Cuban people want to begin a peaceful transition. In a pioneering work which sheds light on a central institution long neglected by students of revolutionary Cuba, Evenson studied the Cuban legal system (item **4723**). The emergence of both the Church and the legal system as institutions relatively independent of the State is indicative of an important transformation of Cuban society facilitated by the current crisis.

In fact, a subculture of political opposition is emerging in Cuba that fosters the activities of social movement organizations and quasi-organizations, as well as collective and mass-protest behavior against the government and the socialist State. The origins of this subculture are several: the failures of the official institutions and the resulting "shadow" institutions of Cuban society; the discrepancy between the ideology and the practices of the regime; the work of religious believers, intellectuals, and artists; the continuation of a cultural tradition of political criticism through *chistes* or jokes; the paradoxical and unintended unifying effect of the repressive activities of the government security systems; the radicalizing effects of the generalized institutional implosion and the concomitant emergence of new institutions; and the transformation of systems of mass communication available in Cuba.

The government can no longer control the information the Cuban people receive about events in Cuba and elsewhere. This decline in State control of the media was caused by both the economic crisis and the new electronic means of mass communication (e.g., Radio Martí, the availability of black market electronic disks to tap satellite transmissions from major hotels in Havana, the introduction of electronic mail, and improved telephone service). An important effect of these changes is that Cubans receive information about Cuba from news sources outside the island. Increasingly, anti-hegemonic political participation in Cuba is not a national, but an international process. Collective action is often planned to coincide with the political sensitivities and agendas of foreign agencies and organizations. For instance, Cubans' political participation, spurred by events in Cuba, often is carried out in the US and elsewhere via cybernetic means.

With few exceptions, the social movement organizations (SMOs) that are emerging in Cuba sponsor only pacifist ideological opposition. So far, their dissident strategy has been to appeal to the Cuban authorities for peaceful change rather than to advocate widespread civil disobedience. Moreover, very few SMOs have developed an explicit ideology rejecting the ideals of the Cuban Revolution and the interpretation of pre-1959 Cuban society advanced by Fidel Castro and his followers. Instead, there is continuity in the use of revolutionary cultural symbols, or what Staniski, in the Polish case, refers to as the "inner structure" of communist culture. Nevertheless, the Cuban government's unwillingness to negotiate with the SMOs, combined with the country's growing social, political, and economic crises, may prod these movements to abandon their pacifist strategies in the future.

Many of the new systems of governmental repression in Cuba, such as the rapid action brigades, are ambulatory, rather than stationary (e.g., the earlier Committees for the Defense of the Revolution, or CDR). The Cuban government is no longer able to command total control of Cuba. The new and old systems are not necessarily complementary; they are based on mutually conflicting logics. In a number of documented cases, the people who manned the stationary systems such as the CDRs have reacted against the presence of brigade participants in their neighborhood. This is not surprising, for the old system was based on loyalty, belonging, and a sense of responsibility to a place; now these neighborhoods have been invaded by strangers who are charged with "defending" the Revolution, often at the expense of the individuals living in the neighborhood.

Paradoxically, the operational freedom of the government's repressive system and its crudeness is an important, albeit unwitting, mechanism in creating anti-hegemonic collective participation. In Cuba, as elsewhere, this usually results from breakdowns in the internal organization of bureaucracies, lack of supervision, and jealousy of lower-echelon officials. While probably effective as short-term social control measures, the use of rapid action brigades, mass arrests, and other tactics of intimidation helps to foster a culture of popular resistance to the government. Sharing similar experiences undoubtedly creates networks of social relations among the victims and their families which in turn facilitate discussion of their shared problems and the construction of new collective and individual identities. It creates popular consciousness about the need for social change outside the boundaries of the State's hegemonic vision of right.

Turning to the Dominican Republic, some of the most significant scholarship examines its declining standard of living. Rodríguez analyzes the problem of abandoned children, *palomos*, in the capital city, along with the social and cultural adaptations they use to survive (item **4775**). Others look at the health status of the Dominican nation and the problem of illegal drugs. The pioneering monograph edited by Rondón (item **4745**) presents a scientific effort to understand violence against women in the Dominican Republic and documents the limitations of the present judicial system in providing protection to the victims. As in previous years, the demography of the Dominican nation continues to be a topic of scientific investigation, particularly the means of population control. Scholars continue to examine relations between Haiti and the Dominican Republic, Haitian immigration to coffee growing regions, the presence of *Vodu* among Dominicans, and the continuing legacy of racism and discrimination against Haitians. In regard to the latter, Zaglul has analyzed the influence of Joaquín Balaguer on Dominican culture and politics and his presumed *antihaitianismo* (item **4802**). Pérez and Artiles' monograph on Dominican social movement organizations and their protest activities during 1984–90 is a significant contribution to the scholarship on this topic (item **4768**); it traces the evolution of these protests and places them in the broad context of national and international politics.

Researchers' attention has also focused on the economic and social bases of the current political crisis in Haiti. Guerrier provides background information on the crisis (item **4738**) and Souffrant studies famine in Haiti (item **4787**). Reflecting the important presence of Haitian immigrants in the US, Charles discusses their experiences in New York City and their evolving ethnic identities and group loyalties (item **4711**).

Scholarship on the Rastafari Movement in Jamaica continues to thrive, as does UN-sponsored research on the condition of children, women, and the elderly. Very noteworthy is Keith and Keith's recently published, comprehensive work on the political system. Examining the composition and interests of social classes in Jamaica to gain an understanding of Michael Manley's government and the ideology and resulting practices of democratic socialism, the authors reinterpret Manley's control of the State, seeing this as a program for social change based on an ideology of "national populism" (item **4750**).

For the French West Indies and French Guiana, a recent monograph examines demographic changes from 1982–90, including mortality, fertility, and migration patterns (item **4708**). As in other countries, recent scholarship for Guyana examines the problem of violence against married women (item **4715**).

One of the most important themes in contemporary scholarship is the problem of Puerto Rican cultural autonomy (items **4706** and **4759**). As in many parts of the mainland, the politics of language (item **4759**) and the conflicts surrounding bilingualism (item **4753**) reveal many of the key points in the cultural autonomy debate. Similarly, authors writing on the African roots of the Puerto Rican culture argue, at times implicitly, for the maintenance of a Puerto Rican national identity. Another important body of research examines Puerto Rican politics: Frambes-Buxeda's monograph on the elections of 1980 and 1984 (item **4724**) is particularly noteworthy, as is González's examination of the political actions of popular groups on behalf of the preservation of Puerto Rican culture (item **4733**). For Puerto Rico, as well as other countries in the region, scholarly attention has focused on health issues once again, especially the threat of AIDS. An important collection of articles on this topic was published by the University of Puerto Rico (item **4785**).

4690 Acevedo Marrero, Carlos Aníbal. Cristianismo y homosexualidad: una perspectiva puertorriqueña. Vega Alta, P.R.: C.A. Acevedo Marrero, 1992. 112 p.: bibl.

Presents historical study of homosexuality in Puerto Rico in context of Episcopal

Church's theology and social practice. Includes chapters on experience of this church in Puerto Rico, the controversy surrounding homosexuality, and emerging understanding of the problem by members of the church.

4691 **Aguirre, Benigno E.** Cuban mass migration and the social construction of deviants. (*Bull. Lat. Am. Res.*, 13:2, 1994, p. 155–183)

Studies 1980 Mariel mass migration from Cuba to US as an oppositional serial surge in which collective deviant labeling occurred. Includes sections on historical background of the crisis, and on creation of collective deviance through three distinct labeling processes: 1) moral passage, brought about by US and Cuban governments; 2) deviance amplification through offical acts; and 3) media support for transformation of moral identity of the Marielitos.

4692 **Alfonso, Pablo M.** Cuba: el diálogo ignorado. Miami: Ediciones Cambio, 1993. 109 p.

Reprints the Cuban Catholic bishops' pastoral letter (8 Sept. 1993) requesting changes from the government. Includes chapters on the Cuban bishops and their motivations, the pastoral letter itself ("El Amor Todo lo Espera"), and official and public reaction to it.

4693 **Anderson, Patricia Y.** Pérdida de potencial humano y adecuación de empleo en los trabajadores calificados de Jamaica, 1976–1985. (*in* Fronteras permeables: migración laboral y movimientos de refugiados en América. Buenos Aires: Planeta, 1991, p. 115–149, bibl., tables)

Studies emigration from Jamaica of professionals and skilled workers in search of work (1976–85). Examines their emigration as a function of political decisions of the Jamaican government. Includes sections on emigrants' labor force participation, emigration and demographic change, importance of US as a country of destination, motives of the emigrants, and relation of emigration to the national economy.

Ardittis, Solon. Targeted reintegration of expatriate brains into developing countries of origin: the EEC-IOM experience in Central America. See item **4574.**

4694 **Báez, Clara.** Mujeres: fuerza laboral y sector informal. (*Estud. Soc./Santo Domingo*, 25:88, abril/junio 1992, p. 99–116, bibl., tables)

Focuses on increased participation by women in the Dominican labor force (1980–90). Principal themes are growth in women's participation in the service sector, and effects of increased number of working women on Dominican industry and economy.

4695 **Bailey, Adrian J.** and **Mark Ellis.** Going home: the migration of Puerto Rican–born women from the United States to Puerto Rico. (*Prof. Geogr.*, 45:2, May 1993, p. 148–158, bibl., tables)

Analyzes migration of Puerto Rican–born women from US to Puerto Rico. Study indicates that length of sojourn in US and decision to return to Puerto Rico are functions of wage trends and community characteristics on the mainland as well as of education, marriage, and childbearing status.

4696 **Bakker, Eveline** *et al.* Geschiedenis van Suriname: van stam tot staat [History of Suriname: from tribe to state]. Zutphen, The Netherlands: De Walburg Pres, 1993. 176 p.: bibl., ill. (some col.), index, maps.

Despite title, this beautifully produced book containing more than 200 color and b/w illustrations is a useful introduction to Suriname. Authors stress cooperation rather than conflict, which may account for their lack of critical analysis of economic and political problems of present-day Suriname. [R. Hoefte]

4697 **Bansart, Andrés.** Cultura, ambiente, desarrollo: el caso del Caribe insular. Caracas: Univ. Simón Bolivar, Instituto de Altos Estudios de América Latina, 1991? 274 p.: bibl.

Uses Caribbean countries to illustrate applicability of a model of the relationship of culture to environment. Modeled relationship between these complex concepts is set in a context of rapid social change produced by economic development.

4698 **Barclay, Lou Anne.** The Syrian/Lebanese community in Trinidad/Tobago: a case study of a commercial ethnic minority. (*Caribb. Aff.*, 5:1, Jan./March 1992, p. 129–146, tables)

Explores relationship of Syrian/Lebanese community with other ethnic groups in the society. Focus is on entrepreneurial activities of this "middle-man" community.

4699 Betances, Emelio. The formation of the Dominican capitalist State and the United States military occupation of 1916–1924. (*MACLAS Lat. Am. Essays*, 4, 1991, p. 231–253, bibl.)

Offers historico-sociological analysis of US policy on Dominican Republic and formation of Dominican capitalist State. Focuses on establishment of the military government, the US-supported "occupation regime," and US conflicts with nationalist forces and local political elites.

4700 Bibliographic guide to Caribbean mass communication. Compiled by John A. Lent. Westport, Conn.: Greenwood Press, 1992. 301 p.: indexes. (Bibliographies and indexes in mass media and communications, 1041–8350; 5)

Includes chapters on Commonwealth Caribbean, Dominican Republic, French Caribbean, Haiti, Netherlands Caribbean, and US Caribbean, plus one chapter listing comparative and regional studies. Covers advertising, broadcasting (radio and television), development-focused communication, film, freedom of the press, history of the media, journalism education, news agencies, popular culture, print media, and telecommunications. [Ed.]

4701 Bibliography of Cuban mass communications. Compilation by John A. Lent. Westport, Conn.: Greenwood Press, 1992. 357 p.: indexes. (Bibliographies and indexes in mass media and communications, 1041–8350; 6)

Compilation of articles and other material on Cuban mass media includes chapters on history of the mass media covering colonial, republican, and post-1959 periods, and on sources consulted by the author. Bibliography has three parts: resources, contemporary perspectives, and historical perspectives. First part deals with anthologies, bibliographies, catalogs, collections, dictionaries, indexes, and other research materials concerning various forms of mass communications. Second has subdivisions on broadcasting, comic and graphic arts, film, freedom of press, news agencies, popular culture, print media, Radio and TV Martí, training and education, and women and media. Third part pulls together items of historical significance.

4702 Bourgois, Philippe. West Indian immigration to Central America and the origins of the banana industry. (*Cimarrón/New York*, 11:1/2, Spring/Summer 1989, p. 58–86, bibl.)

Detailed study examines development of labor movements in Costa Rica and Panama that ensued from increased banana production in Bocas and Limón divisions of United Fruit Company. Claims that employers encouraged additional West Indian immigration to offset labor union activism among those already residing in these areas. Includes information on health conditions and treatment of workers (1800–50s).

4703 Brandon, George. African religious influences in Cuba, Puerto Rico and Hispaniola. (*J. Caribb. Stud.*, 7:2/3, Winter 1989/Spring 1990, p. 201–231, bibl.)

Documents historical and contemporary presence of African religious influences in Spanish Caribbean islands of Cuba, Puerto Rico, and Hispaniola. Special attention is devoted to Cuban situation and Afro-Cuban *santería*. Concludes with brief discussion of ways in which emigration to US has brought about changes in religious practices and ethnic identity of Cubans and Puerto Ricans.

4704 Burgos Ortiz, Nilsa M. Los movimientos sociales como alternativa de organización para la mujer: el caso de Puerto Rico. (*in* Jornadas de Investigación Interdisciplinaria sobre la Mujer, *9th, Madrid, 1992.* La mujer latinoamericana ante el reto del siglo XXI. Madrid: Ediciones de la Univ. Autónoma de Madrid, 1993, p. 193–204, bibl.)

Reviews various women's social movement organizations in Puerto Rico, including the labor and voting rights movements, the more recent communal organizations, the Federación de Mujeres Puertorriqueñas, and the Comisión para el Mejoramiento de los Derechos de la Mujer.

4705 Camnitzer, Luis. New art of Cuba. Austin: Univ. of Texas Press, 1994. 400 p.: bibl., ill., index.

Social history of world of art concentrates mainly on post-1959 Cuba. Includes chapters on impact of State ideology on artists, organization of art education, and relation of Cuban art to postmodernism, as well as description of "generations" of artists. For art historian's comment see *HLAS 54:306*.

4706 Carrión, Juan Manuel. The national question in Puerto Rico. *(in* Colonial dilemma: critical perspectives on contemporary Puerto Rico. Edited by Edwin Meléndez and Edgardo Meléndez. Boston: South End Press, 1993, p. 67–75)

Contemporary study of Puerto Ricans' political struggle for national independence from US. Considers impact of social class divisions on development of a nationalist movement and ideology.

4707 Casimir, Jean. Cultura y poder en el Caribe. *(in* Cultura y política en América Latina. Coordinación de Hugo Zemelman. México: Siglo Veintiuno Editores, 1990, p. 207–227)

Offers historical context of Caribbean plantations and their domination by the State. Also examines political role of the *nuevos libres* (freemen or free blacks).

4708 Cazenave, J. and **Gérard Gautier.** Mouvement démographique, 1982–1990: Antilles-Guyane. Pointe-à-Pitre, Guadeloupe: INSEE, Direction interrégionale Antilles-Guyane, 1991. 68 p.: bibl. (Les Dossiers Antilles-Guyane; 17)

Study of demographic evolution of French West Indies and French Guiana from 1982–90 includes information on growth of population, fertility, mortality, migration, marriage and divorce. Also describes methodology used for the study.

4709 Ceballos, Zenón. República Dominicana: la esterilización femenina en los últimos años. Santo Domingo: Consejo Nacional de Población y Familia, 1987. 44 p.: bibl., ill. .

Study of Dominican women who have undergone voluntary sterilization includes chapters on their socioeconomic characteristics, formal education, geographic distribution, and fertility history.

4710 Cela, Jorge. Escuchar el clamor de los pobres. *(Estud. Soc./Santo Domingo,* 25:89/90, julio/dic. 1992, p. 61–83, graph, photo, tables)

Survey of poverty in Dominican Republic in 1980s includes sections on identity of the poor, types of poverty (e.g., accidental and structural), new forms of poverty, and contemporary mobilization activities of the poor.

4711 Charles, Carolle. Transnationalism in the construct of Haitian migrants' racial categories of identity in New York City. *(Ann. N.Y. Acad. Sci.,* 645, 1992, p. 101–123, bibl.)

Discusses construction of Haitian migrants' ethnic identity in New York City. Argues that multiple ethnic identities which Haitian migrants display stem from: 1) rejection of US racial categories used to construct racial inequality; 2) reconstruction of meanings of blackness used in their home society; and 3) perception and meanings given to their immigration experiences.

4712 Clarke, Colin G. Spatial pattern and social interaction among Creoles and Indians in Trinidad and Tobago. *(in* Trinidad ethnicity. Edited by Kevin A. Yelvington. Knoxville: Univ. of Tennessee Press, 1993, p. 116–135, bibl., maps, tables)

Examines changing racial proportions and patterns of residence among Creoles and East Indians in Trinidad and Tobago from 1960–80. Then focuses on changes in one key aspect of social interaction—intermarriage—among Creoles and East Indians during same period. Relationship between racial segmentation and segregation at the national level is also considered. Critically evaluates Beshers' hypothesis concerning link between social group, residence, and endogamy.

4713 Cotto, Liliana. The *rescate* movement: an alternative way of doing politics. *(in* Colonial dilemma: critical perspectives on contemporary Puerto Rico. Edited by Edwin Meléndez and Edgardo Meléndez. Boston: South End Press, 1993, p. 119–129)

Addresses relationship between the State and urban popular sectors through analysis of Puerto Rican government's response to land invasions, or *rescates*. Criteria examined are: type of organization, form of negotiation, use of violence, and role of external political organizations.

4714 Dann, Graham M.S. Family reunification as an emigration factor in the Eastern Caribbean. *(in* Comparative sociology of family, health and education: a volume in memory of Ferran Valls i Taberner. Málaga, Spain: Cátedra de Historia de Derecho y de las Instituciones, Facultad de Derecho de la Univ. de Málaga, 1991, p. 6161–6178, bibl., tables)

Analyzes motivations underpinning Eastern Caribbean emigration, information field of prospective migrants, types of migrants, and effects of emigration on the family.

4715 Danns, George K. and **Basmat Shiw Parsad.** Domestic violence in the Caribbean: a Guyana case study. Georgetown?: Women's Studies Unit, Univ. of Guyana, 1989. 147 p: bibl.

Examination of violence within marital relationships in Guyana focuses primarily on use of violence against women by conjugal partners. Other topics considered are conjugal violence among East Indians, marital violence within black households, analysis and meaning of marital violence, and conjugal violence and public policy. Includes chapters on history and emergence of conjugal marital relations in Guyana, conjugal violence among East Indians and blacks, and public policy needed to ameliorate the problem of domestic violence.

4716 Deere, Carmen Diana; Mieke Meurs; and **Niurka Pérez.** Toward a periodization of the Cuban collectivization process: changing incentives and peasant responses. (*Cuba. Stud.*, 22, 1992, p. 115–149, bibl., tables)

Focuses on Cuba's agricultural collectivization programs, the response of the peasants, and effectiveness of the cooperatives.

4717 Díaz-Briquets, Sergio. Cuba. (*in* Latin American urbanization: historical profiles of major cities. Edited by Gerald Michael Greenfield. Westport, Conn.: Greenwood Press, 1994, p. 173–187, bibl., map, tables)

Offers demographic and economic urban history of Cuba. Profiles major cities: Camagüey, Havana, and Santiago de Cuba.

4718 Dietz, James L. and **Emilio Pantojas-García.** Puerto Rico's new role in the Caribbean: the high-finance/*maquiladora* strategy. (*in* Colonial dilemma: critical perspectives on contemporary Puerto Rico. Edited by Edwin Meléndez and Edgardo Meléndez. Boston: South End Press, 1993, p. 103–115)

Focuses on evolution of *maquiladora* industry in Puerto Rico, 1970s–90s. Describes public and private policies influencing *maquiladora* strategy and post-1980 emergence of an international division of labor.

4719 Dijk, Frank Jan van. Sociological means: colonial reactions to the radicalization of Rastafari in Jamaica. (*Nieuwe West-Indische Gids*, 69:1/2, 1995, p. 67–101, bibl.)

One of the first works on the subject, article examines formative stage in development of the Rastafarian movement. Concerns period of radicalization and heightened expectations of an imminent return to Africa. Focuses on colonial government's reactions to Rastafari and on social unrest movement created. Recently released files from the Colonial Office reveal that authorities tried to suppress movement by both repression and manipulation. [R. Hoefte]

4720 Education and society in the Commonwealth Caribbean. Edited by Errol Miller. Mona, Jamaica: Institute of Social and Economic Research, UWI, 1991. 271 p.: bibl.

Contains research papers gathered from 1989 seminar sponsored by Research Institute for the Study of Man and Univ. of the West Indies' Faculty of Education. Focuses on educational reform in Saint Christopher-Nevis; impact of revolution on education in Grenada; education and research in Barbados and Trinidad and Tobago; and effects of educational access, achievement, and socialization on Jamaica.

4721 Enchautegui, María E. The value of U.S. labor market experience in the home country: the case of Puerto Rican return migrants. (*Econ. Dev. Cult. Change*, 42:1, Oct. 1993, p. 169–191, graphs, tables)

Based on data from Puerto Rico's 1980 Population Census for male Puerto Rican migrants who returned from US between 1970–80, analyzes transferability of human capital accumulated abroad by return migrants and economic assimilation process of returnees. Investigates whether return migrants capitalize at home on the years of labor market experience accumulated abroad.

4722 Encuesta nacional de fecundidad, 1987, Cuba. La Habana: Editorial Estadística, 1991. 415 p.: bibl., ill., map.

Discusses the results of the 1987 Encuesta Nacional de Fecundidad for Cuba. Includes information on population growth, fertility, mortality, migration, marriage and divorce, and status of women.

4723 Evenson, Debra. Revolution in the balance: law and society in contemporary Cuba. Boulder, Colo.: Westview Press, 1994. 1 v.: bibl., index. (Latin American perspectives series; 14)

Studies legal system of socialist Cuba. Includes chapters on the early transformation of the legal system, the Cuban Constitution and individual rights, evolution of the legal profession and its practice, court system, and family, criminal, economic, and private property law.

4724 Frambes-Buxeda, Aline. Sociología política puertorriqueña. v. 1. San Juan: Editorial Tortuga Verde, 1990. 1 v: bibl., ill. (Col. Libros Tortuga Verde)

Examines political dynamics that characterized 1980 and 1984 elections in Puerto Rico. Includes chapters on Puerto Rican political parties and on the economic situation and its links to international and national political behaviors and interests. Final section is devoted to quality of life, urbanism, and other important topics.

4725 Fuller, Linda. Work and democracy in socialist Cuba. Philadelphia, Pa.: Temple Univ. Press, 1992. 274 p.: bibl., index. (Labor and social change)

Considers factors that facilitate and impede democratization of the workplace in Cuba. Includes chapters on labor unions and their functions, activities of the Partido Comunista de Cuba, relative autonomy of workers and their ability to affect production plans and practices, management and the resolution of labor conflicts, and changes in the control of production.

4726 Funkhouser, Edward and Fernando A. Ramos. The choice of migration destination: Dominican and Cuban immigrants to the mainland United States and Puerto Rico. (*Int. Migr. Rev.*, 27:103, Fall 1993, p. 537–556, bibl., tables)

Uses 1980 US population census to examine relative importance of earnings and culture for Dominican and Cuban immigrants in determining their choice of destination, either US or Puerto Rico.

4727 García Tamayo, Eduardo and José Ramón Rodríguez. La situación rural dominicana. (*Estud. Soc./Santo Domingo*, 25:89/90, julio/dic. 1992, p. 85–102, bibl., photo)

Discusses crisis of rural society in the Dominican Republic. Points out growing urban-rural inequalities and transformations in land ownership, agricultural production, and rural demographics, especially as related to peasant families and poverty distribution.

4728 Garrido de Colasante, Henderglaist. El movimiento rastafari: un estudio sobre sus significaciones e implicaciones. (*in* El Caribe, objeto de investigación. Recopilación de José Moreno Colmenares. Caracas: Univ. Central de Venezuela, Consejo de Desarrollo Científico y Humanístico; Fondo Editorial Acta Científica Venezolana, 1988, p. 67–80, bibl.)

Reviews history of Rastafarian religious movement in Jamaica and its origins as a movement of resistance against European cultural dominance. Describes evolution of the goals of the movement, its theology, symbols, and rituals, and importance of reggae music as a means of expression and diffusion of movement's ideas.

4729 Georges, Eugenia. Gender, class, and migration in the Dominican Republic: women's experiences in a transnational community. (*Ann. N.Y. Acad. Sci.*, 645, 1992, p. 81–99, bibl.)

Examines experiences of women in the village of Los Pinos, Dominican Republic as these relate to transnational networks of social relations that now link the village to New York City, the destination of the great majority of Dominican migrants. First part is a synopsis of historical patterns of Los Pinos' linkages to changing conditions of global capitalism. Second part examines interplay between the Pinero cultural construction of gender and economic choices available to both women and men as a result of migration. Third part discusses ways in which various women have experienced transnationalization of their society and their cultural reponses to it.

4730 Giacalone, Rita. Cambios políticos y sociales en Curazao, 1969–1989. (*in* Curazao y Aruba: entre la autonomía y independencia. Mérida, Venezuela: Univ. de Los

Andes, Consejo de Desarrollo Científico, Humanístico y Tecnológico, 1990, p. 85–124)

Presents alternative interpretation of political dynamics in Curaçao after tumultuous events of May 30, 1969. Reviews creation of political parties, activities of political leaders, importance of selected mass communication outlets, voters' behavior, and changes in the political constitution and in relationship of Curaçao with Holland.

4731 **Giacalone, Rita.** La mujer en los procesos políticos y socioeconómicos del Caribe. (*in* El Caribe hacia el 2000: desafíos y opciones. Caracas: Editorial Nueva Sociedad, 1991, p. 201–222)

Analyzes women's contribution to process of socioeconomic change, and women's political participation in the region. Concludes with list of medium-range options for improving status of women in the Caribbean region.

4732 **Gjerset, Heidi.** First generation Rastafari in St. Eustatius: a case study in the Netherlands Antilles. (*Caribb. Q.*, 40:1, 1994, p. 64–77, bibl.)

The first Rastas appeared on St. Eustatius in early 1970s. Most of these Statian Rastas had spent time in other Caribbean islands, Suriname, or The Netherlands. Author compares evolution of Statian Rastafarian ideology, social life, and customs with developments in Jamaica. Looks at settlement and subsistence patterns, social structure, ritual smoking, philosophy, Rasta speech, gatherings, and music. [R. Hoefte]

4733 **González García, Lydia Milagros.** Cultura y grupos populares en la historia viva de Puerto Rico hoy. (*in* ¿Hacia un nuevo orden estatal en América Latina?: innovación cultural y actores socio-culturales. Coordinación de Fernando Calderón y Mario R. dos Santos. Buenos Aires: CLACSO, 1990, v. 8, p. 323–342)

Examines Puerto Rico's cultural problems and island's dependency on US. Considers development of culture and the State during 1970s. Also discusses progressive social movements' role in formation of a more equitable and democratic political system for the territory to preserve Puerto Rican culture.

4734 **González Kirby, Diana.** A survey of the literature on the Cuban immigration to the U.S. before and since the Mariel Boatlift. (*Rev. Interam. Bibliogr.*, 41:3, 1991, p. 504–517, bibl.)

Review divides literature into pre-Mariel, Mariel, and post-Mariel time periods. Includes section on Marielitos' demographics and adaptation, the causes of the mass migration, and the public and official reactions in US.

4735 **Guanche Pérez, Jesús** and **Dennis Moreno.** Caidije. Santiago de Cuba: Editorial Oriente, 1988. 139 p.: bibl., ill., map.

Analyzes influence of Haitian immigrants in Caidije, a *batey* in municipality of Minas, Cuba. Includes chapters on history of Haitian immigration to the area, background and practices of material and spiritual culture of Caidije, and community's familial practices and popular celebrations.

4736 **Guengant, Jean-Pierre.** Whither the Caribbean exodus?: prospects for the 1990s. (*Int. J./Toronto*, 48:2, Spring 1993, p. 335–354, map, tables)

Considers impact of post-WWII industrial transformations of large and small countries in Caribbean region on patterns of emigration, especially to US and Canada.

4737 **Guerrero, Natividad et al.** Algunas regularidades del desarrollo de la personalidad en la población juvenil cubana. La Habana: Editora Abril, 1987. 127 p.: bibl.

Examines sexual behavior of young people in Cuba. Also offers a diagnostic of their moral development as well as chapters on their use of free time, political development, and knowledge of Cuban law.

4738 **Guerrier, L. Arnault.** Haïti: comment arrêter sa chute libre: dimension économique et sociale de la crise politique actuelle. Delmas, Haiti: Impr. Grafos, 1993. 29 p.

Offers historical examination of social and economic dimensions of present-day political crisis, as well as of failures of the traditional Haitian State.

4739 **Hamm, Lyta.** Archipelago lessons: AIDS in the islands; a comparative study of Cuba, Haiti and Hawaii. (*Interciencia/Caracas*, 18:4, julio/agosto 1993, p. 184–189, bibl.)

Analyzes AIDS rates and efficiencies of health programs for prevention of AIDS in Cuba, Hawaii, and Haiti. Also considers na-

4740 Hidalgo, Ariel. Disidencia: ¿segunda revolución cubana? Miami: Ediciones Universal, 1994. 411 p.: bibl., ill., index, ports. (Col. Cuba y sus jueces)

History of human rights movement in Cuba is written by one of its founders. Contains thorough and detailed information on origins of the movement, importance of international support for movement activists, important human rights organizations active in Cuba, various repressive activities of State security systems, and experiences of political prisioners.

4741 Hill, Nancy P. and Stephen A. Bender. Jamaica. (*in* Latin American urbanization: historical profiles of major cities. Edited by Gerald Michael Greenfield. Westport, Conn.: Greenwood Press, 1994, p. 331–349, bibl., maps, tables)

Offers demographic and economic urban history of Jamaica, followed by profiles of major cities: Kingston, Montego Bay, Port Royal, Port Antonio, and Spanish Town (Villa de la Vega).

4742 Houk, James. Afro-Trinidadian identity and the Africanisation of the Orisha religion. (*in* Trinidad ethnicity. Edited by Kevin A. Yelvington. Knoxville: Univ. of Tennessee Press, 1993, p. 161–179, bibl., graph)

Examines historical transformation of Orisha religion in Trinidad in relation to Afro-Trinidadian identity. Briefly reviews history of the religion in Trinidad, and examines its social structures and rituals. Argues that the religion has been open to influences from island's East Indian population and that recent movements towards Africanization of the religion represent passive resistance to these syncretic tendencies.

4743 Hulst, Hans van and Jeanette Bos. Pan i rèspèt: criminaliteit van geimmigreerde Curaçaose jongeren: aard en omvang van de criminaliteit van geimmigreerde Curaçaose jongeren van 12–24 jaar in drie politieregios in de periode 1989-1991 [Bread and respect: criminality among migrant Curaçaoan youth: character and scope of the criminality of migrant Curaçaoan youth of from 12–14 years of age in three police districts in the period 1989–1991]. Utrecht, The Netherlands: OKU, 1994. 226 p.: appendix, bibl., map, tables.

Analyzes crime among Curaçaoan youth in The Netherlands during period 1989–91. Devotes much attention to socioeconomic, cultural, educational, and ideological background of these (male) youth in Curaçao. Also discusses motives for migration; in most cases it is perceived as the only escape hatch. [R. Hoefte]

4744 Hurbon, Laënnec. La Caraïbe à l'épreuve: le racisme et la migration haïtienne en République Dominicaine. (*Chemins crit.*, 2:2, sept. 1991, p. 5–15)

Reviews effects on Haitian society of Dominican Government's recent decision to repatriate thousands of Haitians residing in Santo Domingo. Reviews various stages of Haitian immigration and importance of Haitian workers to Dominican economy. Presents expulsion as result of avowedly racist ideology of President Balaguer.

4745 Incriminación a la violencia contra la mujer. Edición de Melania Rondón. Santo Domingo: Producciones Centro de Servicios Legales para la Mujer, 1991. 169 p.: bibl., ill.

Uses Santo Domingo judicial registers to study violence against women. Documents social and demographic characteristics of the victims and forms of aggression used against them, as well as operation of the legal system. Final part presents synthesis of what is known about the assailants.

4746 Irish, J.A. George. Visions of liberation in the Caribbean. Plymouth, Montserrat: JAGPI Productions; New York: Caribbean Research Center, Medgar Evers College, City Univ. of New York, 1992. 92 p.: bibl.

Historico-literary study examines pan-Caribbean experience. Discusses emancipation process in Spanish Caribbean, the Afro-Haitian experience, and broader themes of freedom and identity. Chapter titles are: 1) The Emancipation Process in the Spanish Caribbean; 2) An Interpretation of the Haitian Revolution; 3) Pedro Mir: National Poet of the Dominican Republic; 4) Magical Realism: a Search for Caribbean and Latin American Roots; 5) The Unexplored Dimension: Notes on the Caribbean Perspective of Luis Palés

Matos; 6) Nationalist Currents in Spanish American Literature; and 7) A Framework for Latin American Studies at the University of West Indies.

4747 James, Joel. Aproximación al carnaval de Santiago de Cuba. (*J. Caribb. Stud.*, 7:2/3, Winter 1989/Spring 1990, p. 151–171)

Traces historical development of *carnaval* as a cultural form or object in Cuba. Considers carnival's origins in Spain, similarities with and differences from European traditions, and importance of the practice for the Catholic Church. Advances hypothesis that carnival procession (*comparsa*) is derived from the religious procession, and then traces this evolution in actions of the *cabildos de nación* or associations of Africans.

4748 James, Winston. Migration, racism and identity: the Caribbean experience in Britain. (*New Left Rev.*, 193, May/June 1992, p. 15–55, bibl.)

Focuses on experience of black migrants in Britain. First section examines taxonomies of color and class, pride of black ethnicity, and burden of a negative self-image. Second, on the British experience, considers explosion of mystique of whiteness, emergence of ethnic consciousness, and ambivalent experience of exile. Final section deals with the second generation and beyond, pan-Caribbeanization and integration, and symbolic ethnicity. Author concludes with examination of boundaries of the imagined black community, lasting antipathies, and black-brown-Indian ethnic relations.

4749 Juteram, Dolores. Sangre Grande: the role of women in its development. Port of Spain?: D. Juteram, 1991. 67 p.: bibl., ill., maps.

Attempts to record involvement and contribution of women in development of Sangre Grande, a small market town in the Republic of Trinidad and Tobago. Includes sections on women's role in area's economic development and their contributions to stability of the family, voluntary services, nutrition education, child care, religion, sports, and politics.

4750 Keith, Nelson W. and Novella Zett Keith. The social origins of democratic socialism in Jamaica. Philadelphia, Pa.: Temple Univ. Press, 1992. 320 p.: bibl., index.

Offers reinterpretation of Manley's democratic socialist government platform. Includes chapters on historical roots of political crisis of 1970s, politics and organizational practices of the People's National Party, social classes, emergence of national populism, and relative effectiveness of State-directed social change programs.

4751 LeFranc, Elsie. Health status and health services utilization in the English-speaking Caribbean. Mona, Jamaica: Institute of Social and Economic Research, Univ. of the West Indies, 1990. 123 p.: bibl., ill.

First section uses aggregate data to analyze regional morbidity and mortality trends. Reviews general trends in health status and health services utilization in CARICOM region from 1978–86. Second section uses data from a community-based study and from two surveys conducted in St. Vincent, Grenada, and Trinidad and Tobago to investigate in greater detail some of the conclusions suggested in first section. Examines, at individual and household levels, current patterns in health status and health services utilization as well as relationships between social factors and health status.

4752 Lewis, David E. El sector informal y los nuevos actores sociales en el desarrollo del Caribe. (*in* El Caribe hacia el 2000: desafíos y opciones. Caracas: Editorial Nueva Sociedad, 1991, p. 223–239)

Examines economic crisis and its impact on Caribbean societies. Discusses unemployment, limitations of tourism industry, women's economic role, informal sector, and new development strategies.

Logan, Kathleen. Women in public office. See item **4425.**

4753 López Laguerre, María M. El bilingüismo en Puerto Rico: actitudes socio-lingüísticas del maestro. Río Piedras, P. R.: M.M. López Laguerre, 1989. 279 p.: bibl., ill.

Linguistic study focuses on attitudes of teachers towards language and bilingual education in Puerto Rico. Includes chapters on history of language use, and on different viewpoints on controversy surrounding language and bilingualism. Second part presents results of a linguistic field investigation which included psychometric tests.

4754   **Lozano, Wilfredo** and **Franc Báez Evertsz.** Migración internacional y economía cafetalera: estudio sobre la migración estacional de trabajadores haitianos a la cosecha cafetalera en la República Dominicana. 2. ed. Santo Domingo: Centro de Planificación y Acción Ecuménica (CEPAE), 1992. 254 p.: bibl., ill., maps.

Examines seasonal migration of Haitians to coffee-growing regions in Dominican Republic. Includes chapters on economic characteristics of Dominican coffee industry, its dependence on seasonal migrants, proletarianization of migrant labor force, primary methods of recruitment, and control of workers through labor contracts and modalities of payment.

4755   **Maingot, Anthony P.** Race, color, and class in the Caribbean. (*in* Americas: new interpretive essays. New York: Oxford Univ. Press, 1992, p. 220–247, tables)

Studies relationships between State and nation in Haiti, Dominican Republic, Martinique, Curaçao, and Trinidad. Traces influence of regional patterns of race relations and nationalistic ideals on the formation of states.

4756   **Martínez, Regino.** La lucha por la tierra: fe, cultura y solidaridad: la experiencia de Sanché. (*Estud. Soc./Santo Domingo*, 24:86, oct./dic. 1991, p. 58–84)

Examines peasant struggles led by the Unión Campesina Autónoma to recover land in Dajabón province, Dominican Republic. Presents detailed description of non-violent campaign in 1989–90 to recover the Sanché lands.

4757   **McCoy, Terry L.** U.S. policy and the Caribbean Basin sugar industry: implications for migration. (*in* The effects of receiving country policies on migration flows. Boulder, Colo.: Westview Press, 1991, p. 39–65, tables)

Considers US sugar policy and its implications for 12 Caribbean Basin sugar-producing countries currently assigned US import quotas. Argues that current US sugar policy indirectly encourages Caribbean Basin migration to US by contributing to unemployment, reducing rate of economic growth, and diminishing foreign exchange earnings.

4758   **McPherson, Everton S.P.** Rastafari and politics: sixty years of a developing cultural ideology: a sociology of development perspective. Clarendon, Jamaica: Black International Iyahbinghi Press, 1991. 408 p.: bibl. (A H.I.M. centenary celebration work; 1)

Offers historical examination of Rastafarian movement and its impact on Jamaica's politics. Includes chapters on ideology of the movement and its Ethiopian roots, political activism of the believers, movement's shift from a millenarian to an emancipatory focus, and a hermeneutic analysis of the Rastafaris as a religious movement.

Melo M., Cándida. Contaminación ambiental: su impacto en la cotidianidad de la mujer. See item **4427**.

4759   **Méndez, José Luis.** Puerto Rico, ¿español o inglés?: un debate sobre su identidad. (*Cuad. Am.*, 7:40, julio/agosto 1993, p. 84–96)

Considers function of the Spanish language in preserving Hispanic identity of Puerto Rican people. In context of language politics, work presents an historical account of Puerto Rico's struggle for cultural autonomy from North America and of action and ideology of nationalist political parties and electoral campaigns.

4760   **Mergal, Margarita.** Puerto Rican feminism at a crossroad: challenges at the turn of the century. (*in* Colonial dilemma: critical perspectives on contemporary Puerto Rico. Edited by Edwin Meléndez and Edgardo Meléndez. Boston: South End Press, 1993, p. 131–141)

Looks at women's movements in Puerto Rican society. Reviews their contemporary participation in the macro-political process and future expectations of the feminist movement in Puerto Rico.

4761   **Mills, Frank L.** Determinantes y consecuencias de la cultura migratoria de St. Kitts-Nevis. (*in* Fronteras permeables: migración laboral y movimientos de refugiados en América. Buenos Aires: Planeta, 1991, p. 55–88, bibl., tables)

Based in part on results of a survey of households in Saint Christopher-Nevis, work examines cultural practices that facilitate emigration from the islands, characteristics of emigration experience, reasons for emigrating, demographic characteristics of emigrants, their places of destination, and their information fields. Concludes with test of hypotheses regarding emigration.

4762 Miranda Francisco, Olivia. La autoconciencia nacional cubana en sus orígenes: reflexiones en el 500 aniversario. (*Rev. Cuba. Cienc. Soc.*, 9:27, enero/junio 1992, p. 78–92, bibl.)

Examines African and Hispano-European origins of the Cuban people, and post-1959 processes of syncretism and transculturation.

4763 Molendijk, Mathilde. Inheemsen in Paramaribo, 1964–1991 [Natives in Paramaribo, 1964–1991]. (*SWI Forum*, 9:1/2, okt. 1992, p. 128–144, bibl., graphs, map, tables)

Results of a 1991 study of the social and economic position of Amerindians living in Paramaribo are compared to 1964 census data. Concludes that socioeconomic position of Amerindians shows improvement vis-à-vis that of other ethnic groups. [R. Hoefte]

4764 Murray, Gerald F. Campesino, árbol e Iglesia: un proyecto de reforestación intermediado por la Iglesia. (*Estud. Soc./Santo Domingo*, 24:84, abril/junio 1991, p. 73–85, bibl.)

Compares results of 1980s reforestation projects carried out in Haiti, one directed by the government and the other by the Church.

4765 Nelen, J.M. and J.J.A. Essers. Veel voorkomende criminaliteit op de Nederlandse Antillen [Frequent criminality in the Netherlands Antilles]. Arnhem, The Netherlands: Gouda Quint, 1993. 75 p.: bibl., ill., map (Onderzoek en beleid; 122)

Part of investigation into alleged misconduct of police in Netherlands Antilles, report measures rate of criminality in Curaçao, Bonaire, and St. Maarten. Crime worries substantially more people now that it did a decade ago. In absolute terms, crime rate is highest in Curaçao, but the greatest increase is in Bonaire where rates have doubled during the last 11 years. Includes summaries in French, English, and Papiamentu. [R. Hoefte]

4766 Pan American Health Organization. Caribbean Area. (*in* Tobacco or health: status in the Americas. Washington: PAHO, 1992, p. 163–348, bibl., tables)

Documents tobacco industry, tobacco use, tobacco use and health, and tobacco use prevention and control activities in Cuba, Guyana, Haiti, Jamaica, and six member-countries of the Organization of Eastern Caribbean States.

4767 Pedraza, Silvia. Cubans in exile, 1959–1989: the state of the research. (*in* Cuban studies since the Revolution. Gainesville: Univ. of Florida, 1992, p. 235–257)

Review essay covers extensive literature on Cubans in exile. Includes sections on Cuban "success story" and experiences of Cuban women. Also identifies needed research.

4768 Pérez, César and Leopoldo Artiles. Movimientos sociales dominicanos: identidad y dilemas. Santo Domingo: Instituto Tecnológico de Santo Domingo, 1992. 158 p.: bibl., ill. (Serie Investigaciones; 10)

Surveys contemporary social movements and forms of collective behavior in the Dominican Republic. First part situates movements and protests in political context; second considers collective identities formed in the struggles.

4769 Pérez-Stable, Marifeli. The Cuban Revolution: origins, course, and legacy. New York: Oxford Univ. Press, 1993. 236 p.: bibl., index.

Study of Cuban politics in 20th century includes useful reviews of known historical events and characteristics of pre-revolutionary society, and of effects of legacy of mediated sovereignty arising from the Spanish-American War. Offers well-reasoned, albeit highly controversial, account of factors that facilitated Fidel Castro's triumph in 1959. Analysis of Cuban Government's "rectification" campaign in 1986, and of subsequent dilemmas posed by 1989 disappearance of USSR and Eastern European allies, is a major contribution. Includes chapters on politics and society during 1902–58, 1959–61 (early years of revolution), 1961–70, and 1971–86, as well as one on post-1989 situation.

Ramdas, Anil. In mijn vaders huis [In my father's house]. See item **4442**.

4770 Ramírez Calzadilla, Jorge. Libertad de conciencia y religión en Cuba. (*Rev. Cuba. Cienc. Soc.*, 9:25, enero/junio 1991, p. 133–156, bibl.)

Offers historical analysis of Cuban society from colonial times to the present, focusing on importance of religion in Cuba. Also examines current situation of religion in Cuba.

4771 **Ramos, Fernando A.** Out-migration and return migration of Puerto Ricans. (*in* Immigration and the work force: economic consequences for the United States and source areas. Chicago: Univ. of Chicago Press, 1992, p. 49–66, bibl., graph, tables)

Uses US Population Census to investigate Puerto Rico/US migration and return migration decisions of Puerto Ricans born there and in US. These data are used to test Borjas' income maximization hypothesis. Skill composition of Puerto Rican migration flows is consistent with predictions of the model. Migrants in US have less advantageous observable socioeconomic characteristics (such as education). Return migrants to Puerto Rico tend to be more skilled than Puerto Ricans who remain in US. Furthermore, US-born Puerto Ricans moving to Puerto Rico also have more human capital than US-born Puerto Ricans who choose to remain in US.

4772 **Reca Moreira, Inés et al.** Análisis de las investigaciones sobre la familia cubana, 1970–1987. La Habana: Editorial de Ciencias Sociales, 1990. 233 p.: bibl., ill. (Sociología)

Reviews what is known about family and marriage in Cuba. Includes most recent information on marriage, divorce, household composition, use of time, division of labor, gender relations, family values, socialization of youth, and patterns of family communication.

4773 **República Dominicana: encuesta demográfica y de salud, 1991.** Santo Domingo: Instituto de Estudios de Población y Desarrollo; Oficina Nacional de Planificación; Columbia, Md.: IRD/Macro International, 1992. 284 p.: ill., maps.

Report on health and demography of Dominican people is part of international research program sponsored by USAID. Includes chapters on geography and morphology of Dominican Republic, general household characteristics, use of contraceptives, other determinants of fecundity, infant mortality and health, and availability of family health and family planning services.

4774 **Ríos, Palmira N.** Export-oriented industrialization and the demand for female labor: Puerto Rican women in the manufacturing sector, 1952–1980. (*in* Colonial dilemma: critical perspectives on contemporary Puerto Rico. Edited by Edwin Meléndez and Edgardo Meléndez. Boston: South End Press, 1993, p. 89–101, tables)

Analyzes patterns of gender-typing in Puerto Rico's export-oriented industrialization program and the high proportion of women in Puerto Rican manufacturing firms.

4775 **Rodríguez, Víctor Manuel.** Sub-cultura del palomo: un estudio sobre el menor de la calle de Santo Domingo. Santo Domingo: Publicaciones Graphics Art, 1990. 164 p.: bibl., ill.

Considers problem of street children in Dominican Republic. Includes chapters on their subculture (practices, language, and beliefs that maximize their survival), demographic and social characteristics, typical risks associated with the experience, and failures of existing institutions. Calls for new vision of education to respond to their needs.

4776 **Rodríguez-Menier, Juan Antonio.** Cuba por dentro: el MININT. Miami: Ediciones Universal, 1994. 1 v.

Account of Cuba's Ministerio del Interior (MININT) is written by a former top official, now exiled. Includes chapters on Ministry's history, normative and operational structure, operations, relations with other mass and State organizations in Cuba and abroad, and typical career path for agents. Second part presents detailed outline and analysis of organizational structure and functions of the Ministry and its various departments and directorates.

4777 **Rolfes, Irene.** Women in the Caribbean: a bibliography, 1986–1990. pt. 3, 1986–1990. Leiden, The Netherlands: Dept. of Caribbean Studies, Royal Institute of Linguistics and Anthropology, 1992. 263 p.: indexes.

Lists more than 1,550 items on women in the Caribbean and on Caribbean women in North America and Europe. Includes monographs, articles, and unpublished works such as theses and conference papers. Also includes periodicals list, abbreviations list, and subject and geographical indexes, as well as co-authors, editors, and organizations. [R. Hoefte]

4778 **Rosario Quiles, Luis Antonio.** Comunicación y sociedad puertorriqueña. San Juan: Iberoamericana de Ediciones, 1991. 300 p.: bibl., ill.

First part reviews communications theory and nature of mass media. Second part, focusing on Puerto Rico, includes chapters on mass media and culture, the legal system, social impact of mass media on the island, and regional public policy, as well as separate chapters on the press, advertising, television, radio, and cinema.

4779 **Rummens, Joanna W.A.** Identity and perception: the politicalization of identity in St. Martin. (*in* Forging identities and patterns of development in Latin America and the Caribbean. Edited by Harry P. Díaz, Joanna W.A. Rummens, and Patrick D.M. Taylor. Toronto: Canadian Scholars' Press, 1991, p. 265–278)

Discusses identity construction and intergroup relations in St. Martin, in context of decolonization, growing nationalism, and various economic development dilemmas.

4780 **Sampath, Niels M.** An evaluation of "creolisation" of Trinidad East Indian adolescent masculinity. (*in* Trinidad ethnicity. Edited by Kevin A. Yelvington. Knoxville: Univ. of Tennessee Press, 1993, p. 235–253, bibl.)

Based on analysis of fieldwork carried out in Indian Wood, a southern Trinidad farming community, author examines clash between modern and traditional gender roles and gender identities. Also discusses continuation of traditional patriarchal power structures in post-colonial period.

4781 **San Miguel, Pedro L.** El estado y el campesinado en la República Dominicana: el Valle del Cibao, 1900–1960. (*Hist. Soc./Río Piedras*, 4, 1991, p. 42–74, ill.)

Case study of Valle del Cibao, Dominican Republic, is centered on its rural society and ways in which it has adjusted to external pressures and demands. Work singles out problem of land tenure, marketing of agricultural products, financing, and action of the State, particularly during Trujillo dictatorship.

4782 **Sánchez, Antulio.** El pueblo de Cuba, entre el drama y la esperanza. (*Hoy Hist.*, 10:57, abril/mayo 1993, p. 49–53)

Examines "Opción Cero," or the well-known economic crisis afflicting the Cuban people. Discusses the problem in several of its most important manifestations such as discontent among youth, prostitution, and negative impact on tourism.

4783 **Sansone, Livio.** Hangen boven de oceaan: het gewone "overleven" van Creoolse jongeren in Paramaribo [Hanging above the ocean: the ordinary "survival" of Creole youth in Paramaribo]. (*SWI Forum*, 10: 1, 1993, p. 5–30, bibl.)

Portrays life of Creole youth in Suriname. These lower-class boys and girls survive by engaging in all sorts of (illegal) activities. They dream of emigrating to the US or to The Netherlands. Even if they have never been abroad, they already feel like "hanging above the ocean." Also published in book form (Amsterdam, 1992). [R. Hoefte]

4784 **Short, Margaret I.** Law and religion in Marxist Cuba: a human rights inquiry. Coral Gables, Fla.: North-South Center, Univ. of Miami; New Brunswick, N.J.: Transaction Publishers, 1993. 209 p.: bibl.

Presents account of the human rights situation of religious communities, especially the Catholic Church, in Cuba. Includes chapters on constitutional law, State ideology, and religious persecution on the island.

4785 **El SIDA en Puerto Rico: acercamientos multidisciplinarios.** Edición de Ineke Cunningham, Carlos Gil Ramos-Bellido y Reinaldo Ortiz-Colón. Río Piedras: Univ. de Puerto Rico, 1991. 314 p.: bibl., ill.

Contains collection of articles on AIDS epidemic in Puerto Rico written by faculty members of Univ. de Puerto Rico. Includes material on epidemiology of the disease, treatment modalities, myths, public perceptions, economic impact, effects on heterosexual relations, therapeutic interventions and health ideologies, risk-taking behaviors, homosexuality, and legal and ethical considerations.

4786 **Situation analysis of the status of children and women in Jamaica.** Kingston: United Nations Children's Fund; Planning Institute of Jamaica, 1991. 148 p.: bibl., index.

Undertakes assessment of children's and women's status in Jamaica at end of 1980s, following 15 years of economic contraction or slow growth, and six years of decreases in social services expenditures mandated by Structural Adjustment Program. Book begins by reviewing structural causes that undermine life chances of Jamaican children and women, the skewed distribution of income, and the severe imbalance in resource allocation between urban and rural areas. Analysis reviews a wide range of studies, reports, and official data in order to establish: 1) the situation of children; 2) the situation of the women caring for the children; and 3) services available for child development. Identifies 10 critical problem areas and makes recommendations to address them.

4787 Souffrant, Claude. La société haitienne: une société de la faim. (*Bull. Bur. natl. ethnol.*, 1/2, 1986, p. 35–51, bibl.)

Points out that chronic famine in Haiti is not an agricultural but a political and social problem. Includes sections on disorganization of Haiti's agricultural sector, and on cultural practices such as inadequate rural education and small land holdings that create and sustain problem of famine in the country.

4788 Stewart, Thelma. The elderly. Kingston: T. Stewart, 1991? 71 p.: bibl., ill.

Examines composition and growth of elderly in Jamaica. Reviews facts on aging and gives advice on coping with it. Contains sections on concepts, myths, and realities of aging; sensory impairments of older people; and problems encountered by families. Offers advice for the elderly and for caregivers, with emphasis on importance of emotional and financial preparation for retirement.

4789 Stoffle, Richard W. Caribbean fishermen farmers: a social assessment of Smithsonian king crab mariculture. Ann Arbor: Survey Research Center, Institute for Social Research, Univ. of Michigan, 1986. 141 p.: bibl., ill. (Research report series / Institute for Social Research)

Assesses social and cultural factors that influence transfer of Caribbean king crab mariculture developed in two West Indian project sites. Includes sections on history of the mariculture project and on positive and negative effects of the project in Antigua and Dominican Republic.

4790 Stone, Carl. Hard drug use in a black island society: a survey of drug use in Jamaica. (*Caribb. Stud.*, 24:3/4, 1991, p. 267–288, tables)

Analysis of drug use in Jamaica is based on national survey carried out by author between Aug. and Oct. 1990. Includes sections on awareness of drug problem, consumption of legal and illegal drugs, and drug distribution.

4791 Subervi-Velez, Federico A.; Nitza M. Hernández López; and Aline Frambes-Buxeda. Los medios de comunicación masiva en Puerto Rico. (*Hómines/San Juan*, 15:2/16:1, oct. 1991/dic. 1992, p. 39–60)

Authors assert that mass media in Puerto Rico are best characterized as "commercial and liberal" despite being controlled by US capital, subject to federal regulation, and influenced by the colonial relationship. Article reviews newspapers and magazines, radio and television, and their public consumption. Authors propose thorough research on cultural impact of new technologies and their content. [T. Martínez-Vergne]

4792 Suriname jaarboek 1995: radioprogramma Zorg en Hoop [Suriname yearbook 1995: radio program Zorg en Hoop]. Edited by Roy Khemradj. The Hague: Amrit, 1996. 752 p.: indexes, photos.

Worthwhile experiment publishes full texts of all radio broadcasts of weekly Dutch-Suriname program "Zorg en Hoop," i.e., everything from holiday greetings by callers to interviews on cultural, political, and socioeconomic subjects. Includes very useful indexes on topics and personal names. [R. Hoefte]

4793 Taylor, Patrick. Ethnicity and social change in Trinidadian literature. (in Trinidad ethnicity. Edited by Kevin A. Yelvington. Knoxville: Univ. of Tennessee Press, 1993, p. 254–274, bibl.)

Surveys Trinidadian literature and the relationship between ethnicity and sociopolitical processes transforming 20th-century Trinidad. Includes review of pluralist and nationalist themes in Trinidadian literature, but focuses mainly on writings of Sam Selvon and Earl Lovelace.

4794 La tercera raíz: presencia africana en Puerto Rico. San Juan?: Centro de Estudios de la Realidad Puertorriqueña; Instituto de Cultura Puertorriqueña, 1992. 162 p.: bibl., ill. (some col.).

Beautifully-crafted catalog accompanying traveling exhibit on African presence in Puerto Rican history, music, art, literature, and religion includes original essays on these subjects by key academic figures. Photographs, scholarly books and articles, and oral interviews appear both as source material and as text. [T. Martínez-Vergne]

4795 Valdés, Yrmino. Ceremonias fúnebres de la santería afrocubana: Ituto y honras de Egún. San Juan: Sociedad de Autores Libres, 1991. 67 p.: ill.

Presents a compilation of funeral rituals and practices of the *orishas*, or Cuban *santería*, as recorded in the written accounts or "libretas" of one of the most distinguished practitioners of the religion, the *oriate* Yrmino Valdés.

4796 Valdez, José et al. Drogas y sociedad: investigación diagnóstica sobre la realidad de la drogadicción en República Dominicana. Santiago de los Caballeros, Dominican Republic: Pontificia Univ. Católica Madre y Maestra, Federación Internacional de Universidades Católicas, 1991. 194 p.: bibl., ill. (Col. Documentos)

Reviews various therapies for treatment of drug addiction and results of previous studies of use of illegal drugs in Dominican Republic. Second part includes diagnostic plans based on population of known addicts undergoing treatment, and summarizes what is known about population of patients in treatment facilities throughout the country. Concludes with prognostic statements about treatment modalities and goals.

4797 Villamán P., Marcos. Modernidad, crisis y constitución de los sujetos políticos: diversidad en temporalidades y factor religioso: Santo Domingo, 1961–1990. (*Estud. Soc./Santo Domingo*, 25:87, enero/marzo 1992, p. 67–85, bibl.)

Focuses on some features of political and economic modernization in the Dominican Republic and their impact on the present political crisis. Also considers importance of the Church in the formation of popular movements.

4798 Villamán P., Marcos. Perfil religioso en el Caribe hispano-parlante: el caso de la República Dominicana. (*Perf. Latinoam.*, 2:2, junio 1993, p. 51–83, bibl.)

Reviews relevant religious practices of Catholics and Protestants in the Dominican Republic. Contextualizes these practices in a society in crisis and prone to violence and in the social movement activities that have punctuated country's recent past.

4799 Wekker, Gloria. Ik ben een gouden munt, ik ga door vele handen, maar verlies mijn waarde niet [I am gold money, I pass through all hands, but I don't lose my value]. Amsterdam: Vita, 1994. 207 p.: bibl., tables.

Offers portrait of lower-class African-Surinamese women. Their daily life and ideas are based on Creole popular culture, which in turn is based on West African traditions. Author highlights free sexual culture, including relations with both women and men. Based on fieldwork and interviews conducted in Paramaribo in 1990–91. [R. Hoefte]

4800 Wiltshire, Rosina. Implications of transnational migration for nationalism: the Caribbean example. (*Ann. N.Y. Acad. Sci.*, 645, 1992, p. 175–187, bibl.)

Focuses on Commonwealth Caribbean. Research conducted on impact of regional and international migration examines movements of Grenadians and Vincentians to Trinidad (the major regional host) and to US (the major international host). Discusses implications of transnational migration and problems of multiple loyalties by looking at economic, gender, and socialization dimensions of migrant flows and linkages.

4801 Wooding, Charles J. Afrosurinamese ethnopsychiatry: a transcultural approach. Rijswijk, The Netherlands: C.J. Wooding, 1988? 48 p.: bibl., ill.

Presents transcultural approach to treatment of mental and emotional disorders. Afrosurinamese ethnomedical system is analyzed to show that folk healers function as ethnopsychiatrists or ethnopsychotherapists. A transcultural approach is proposed which argues that: 1) it is not imperative that a healer be taken into possession by a guardian spirit in order to heal; and 2) other types of diagnosis and therapy can be added to those of the popular healers.

4802 Zaglul, Jesús M. Una identificación nacional "defensiva": el antihaitianismo nacionalista de Joaquín Balaguer. (*Estud. Soc./Santo Domingo*, 25:87, enero/marzo 1992, p. 29–65, appendix, tables)

Historical study examines sociopolitical and cultural relations between Haitians and Dominicans. Summarizes influence of Joaquín Balaguer in the political history of the Dominican Republic and Balaguer's preoccupation with the Haitian problem.

COLOMBIA AND VENEZUELA

WILLIAM L. CANAK, *Associate Professor of Sociology, Middle Tennessee State University*
DANILO LEVI, *Assistant Professor of Sociology, Southeastern Louisiana University*

COLOMBIA

COLOMBIAN SOCIETY IN THE EARLY 1990s epitomizes the bundle of seemingly contradictory processes evident throughout Latin America. Expanding trade, urbanization, demographic maturation and progressive social action are married to violence, strong illegal and informal economies, impoverishment, and repression. Within this turbulent mosaic, Colombian sociologists continue their crucial role in providing diverse and highly competent profiles of Colombian society.

The well established and highly professional community of sociologists working in Colombian universities, government agencies and private institutes continues to mark new directions for the 1990s. The US- and European-trained generation of the 1960s-70s, most based in a few universities and private institutes supported by international monies, is still producing a prodigious volume of excellent works. But the next generation, often trained in Colombia, is also making substantial contributions across the board. In addition, sociologists based both at provincial universities and outside Colombia are producing works on diverse topics not always well represented in the past. Thus, cultural studies of race and ethnic relations (items **4814, 4813,** and **4830**), industrial and labor relations (items **4828** and **4818**), and public health (item **4807**) bode well for the continuing intellectual growth and vitality of Colombian research and its relevance to sociologists elsewhere. Meanwhile, traditional sub-fields of Colombian sociology such as political economy, women's studies, and rural and urban studies are still represented by important works, with implications for sociologists studying other nations.

Violence continues as a dominant or underlying theme throughout the field of Colombian sociology. Whether based in the politicized economic relations of regional economies or the narco-traffic industry, violence permeates every niche of Colombian society. Thus the sociologists in Colombia are a particularly hardy breed whose research necessarily makes them vulnerable, even as they build an institutional base for maintaining their community in the face of local threats and declining international resources. Sociology journals, books, conferences, and well-established departments at many universities document the vitality and strength of Colombian sociology in the 1990s. [WLC]

VENEZUELA

POLITICAL ECONOMY continues to be of salient concern to Venezuelan sociology. The impact of the petroleum economy remains a common theme in a number of important works focusing on a variety of social processes (items **4834, 4850,** and **4846**).

Another major theme is the relationship between the State and civil society, an increasingly problematic one in view of the persistent economic crises and "adjustment" policies (items **4838** and **4844**). The State-society relationship is also central in the burgeoning work on urban sociology, which notes the importance of community and environmental movements as the common ground in which diverse collective and institutional actors interact dialectically to further pluralism and democratization (items **4841, 4842,** and **4393**). The confluence of ecological and democratization movements suggests a potentially valuable resource that Latin American societies can deploy in their political and economic relations with core States.

Broader theoretical considerations of these processes and their linkage to major concerns seem necessary to promote the regional solidarity and coordination that could bring such potential to fruition. The literature published during the period under review here, however, does not apply a regional focus, instead favoring more limited empirical studies. Future efforts should attempt syntheses of empirical case studies and more inclusive theoretical generalizations.

A Latin American tradition of deepening concern with the adaptive strategies of social groups continues to enrich the scope of cultural and ethnic studies which range from urban survival tactics (items **4835, 4438,** and **4840**) to cultural diffusion, pluralism, and assimilation (items **4414, 4831, 4836, 4839, 4848,** and **4853**).

Finally, it is important to note the stimulating contributions of Venezuelans to the theoretical and practical development of sociology in general, with works that address crucial questions about the meaning and mission of the discipline and its practitioners, from a decidedly Latin American perspective (items **4394** and **4418**). [DL]

COLOMBIA

4803 Arango, Luz Gabriela. Mujer, religión e industria: Fabricato, 1923–1982. Medellín, Colombia: Editorial Univ. de Antioquia; Univ. Externado de Colombia, 1989. 339 p.: bibl., ill. (Col. Clío de historia colombiana; 2)

Theoretically and methodologically sophisticated longitudinal case study of women textile workers links forms of work culture, religion, and patrimonial authority with family survival strategies. Major contribution to international history of labor.

4804 Arango, Luz Gabriela. Mujeres ejecutivas en Colombia: ¿una nueva generación? (*Rev. ANDI*, 120, enero/feb. 1993, p. 51–72, bibl., photos, tables)

Useful benchmark report is based on 1989–91 survey of 553 largest Colombian business enterprises in Bogotá, Medellín, Bucaramanga, and Ibagué, and of 987 executive women. Also reports data from Colombia's Departamento Administrativo Nacional de Estadística on women's roles at various levels of responsibility.

4805 Barrera Restrepo, Efrén. Los círculos del poder en Colombia. Bogotá: Escuela Superior de Administración Pública; Medellín, Colombia: Comité de Desarrollo de la Investigación, Univ. de Antioquia, 1991. 290 p.: bibl. (Documentos ESAP)

Unique Colombian example of national-level "power structure" research traces genealogical, social, and economic links of political figures, bureaucrats, and business leaders.

4806 Bonilla Castro, Elssy and **Penélope Rodríguez Sehk.** Fuera del cerco: mujeres, estructura y cambio social en Colombia. Bogotá: Agencia Canadiense de Desarrollo Internacional, 1992. 258 p.: bibl., ill.

Major contribution to Colombian women's studies uses empirical data from diverse sources to profile history and current status of Colombian women. Reviews demographic, legal, familial, labor market, and political issues. Notable for integrating women's studies within political economy of Colombian development, this insightful work is a key reference.

4807 Bonilla Castro, Elssy *et al.* Salud y desarrollo: aspectos socioeconómicos de la malaria en Colombia. Bogotá: Univ. de los Andes, Facultad de Economía, Centro de Estudios sobre Desarrollo Económico; Plaza & Janés, 1991. 262 p.: bibl., ill., maps.

Exemplary biostatistical and epidemiological study of malaria funded by World Health Organization is based on 1980s field data from two communities. Documents variable risk factors and impact of public health policy and organizations.

4808 Botero Herrera, Fernando. Urabá: colonización, violencia y crisis del Estado. Medellín, Colombia: Editorial Univ. de

Antioquia, 1990. 200 p.: bibl., ill., maps. (Col. Clío de historia colombiana)

Longitudinal analysis of regional rural development in Antioquian zone of banana export production considers role of colonization, violence, and economic development trends in absence of State regulation. Concludes that without State presence, violence becomes an integral feature of economic relations.

4809 El campesino contemporáneo: cambios recientes en los países andinos. Edición de Fernando Bernal C. Bogotá: CEREC; Tercer Mundo Editores; Fundación Friedrich Ebert de Colombia, Estudios Rurales Latinoamericanos, 1990. 583 p.: bibl. (Sociología y política)

Pt. 3 contains 11 chapters on Colombian agriculture that focus on recent trends, regional peasant associations, sharecropping, and technology. This solid contribution by experienced researchers on central issues of rural development is required reading for understanding the current situation in rural Colombia.

4810 La cultura popular en el Caribe. v. 1, Ciénaga. Bogotá: Organización de Estados Iberoamericanos para la Educación, la Ciencia y la Cultura, Sede Bogotá, 1990. 1 v.: photos. (La Cultura popular en Colombia)

Part of a series of publications on the culture of the Caribe, Oriental, Altiplano, Cundiboyacense, Pacífica, and Antioqueña regions of the country, this first vol. in the Caribe subseries includes information on Ciénaga's music, festivals, literature, painting, and history. [B. Aguirre-López]

4811 Flórez Nieto, Carmen Elisa; Rafael Echeverri Perico; and Elssy Bonilla Castro. La transición demográfica en Colombia: efectos en la formación de la familia. Tokyo: Univ. de las Naciones Unidas; Bogotá: Ediciones Uniandes, Univ. de los Andes, 1990. 242 p.: bibl., ill.

Important benchmark data from UN-funded study compares two Colombian female cohorts before and after demographic transition. Part of multinational regional program, work includes quantitative and qualitative studies of attitudes, perceptions, and behavior in rural and urban areas.

4812 Forero de Saade, María Teresa; Leonardo Cañón Orgetón; and Javier Armando Pineda Duque. Mujer trabajadora: nuevo compromiso social. Bogotá: Instituto de Estudios Sociales Juan Pablo II, 1991. 221 p.: bibl. (Col. Horizontes de solidaridad; 2)

Broad-ranging review of public policy on Colombian women's social roles and economic participation includes descriptive, demographic, education, and labor market data. Chapters review legislation and social programs that target women.

4813 Foro Nacional Para, Con, Por, Sobre, De Cultura, *Bogotá, 1990*. Imágenes y reflexiones de la cultura en Colombia: regiones, ciudades y violencia. Bogotá: Colcultura, 1991. 446 p.: bibl., ill.

Proceedings of 1990 forum sponsored by Instituto Colombiano de Cultura during Colombia's Año Nacional de Cultura includes over two dozen papers organized into four themes: 1) perspectives on culture; 2) regional cultures; 3) impact of violence on culture; and 4) urban cultures. Volume reveals range of current concerns and insights from Colombia's intellectual elite.

4814 Friedemann, Nina S. de. Criele criele son, del Pacífico negro: arte, religión y cultura en el litoral Pacífico. Bogotá: Planeta, 1989. 200 p.: bibl., ill., index, map. (Espejo de Colombia)

Rich ethnographic history focuses on trajectories of cultural resistance against missionaries and harsh forces of nature. This empirical study constructs a vision of regional social stratification, brutal conflict, and a complex ideological tapestry, enabling one to understand meaning of cultural objects for reproducing a rich Afro-Colombian culture.

4815 Gaitán Pavía, Pilar and Carlos Moreno Ospina. Poder local: realidad y utopía de la descentralización en Colombia. Bogotá: Instituto de Estudios Políticos y Relaciones Internacionales; Tercer Mundo Editores, 1992. 315 p.: bibl., ill. (Sociología y política)

Multifaceted review of Colombian local politics analyzes decentralization policies, mayoral elections, and mayoral administrations in light of federal administrative and fiscal policies.

4816 Guerrero Barón, Javier. Los años del olvido: Boyacá y los orígenes de la violencia. Bogotá: Instituto de Estudios Políticos y Relaciones Internacionales; Tercer Mundo Editores, 1991. 268 p.: bibl., ill., maps. (Sociología y política)

Important contribution to scholarship on La Violencia focuses on Santander, Norte de Santander, and Boyacá in the 1930s. Complex historical analysis locates underlying causes of 1945 onset of La Violencia in the dynamics of State and party-led liberalization of conservative culture, class interests, and resulting resistance that produced a culture of violence. Roles of Roman Catholic Church, international fascism, and electoral campaigns are central to the explanation.

4817 León de Leal, Magdalena. Estrategias para entender y transformar las relaciones entre trabajo doméstico y servicio doméstico. (*in* Género, clase y raza en América Latina: algunas aportaciones. Recopilación de Lola G. Luna. Barcelona: Univ. de Barcelona, 1991, p. 25–61, tables)

Leading researcher on women's labor presents a detailed and theoretically sophisticated review of a five-city "action research" project (1981–83) to study and improve social and labor relations affecting domestic servants.

4818 Mayor Mora, Alberto. Institucionalización y perspectivas del Taylorismo en Colombia: conflictos y subculturas del trabajo entre ingenieros, supervisores y obreros en torno a la productividad, 1959–1990. (*Bol. Socioecon.*, 24/25, agosto/dic. 1992, p. 202–242, tables)

Presents important overview of industrial engineering's growth and influence on Colombian industrial shopfloor relations. Reveals trend from rigid Taylorism and conflict to emphasis on improved human relations and participatory management. Reviews union and worker responses.

4819 Molano, Alfredo. Aguas arriba: entre la coca y el oro. Bogotá: El Ancora Editores, 1990. 177 p.: map.

Focused study examines violent social relations dominating economic and social development in southeastern frontier region of Colombia where State regulation and law enforcement is minimal.

Morgan, María de la Luz et al. Sistematización, propuesta metodológica y dos experiencias: Perú y Colombia. See item **4919**.

4820 Muñoz V., Cecilia and **Ximena Pachón C.** Mortalidad infantil, crecimiento demográfico y control de la natalidad: una lucha por la supervivencia de la infancia bogotana, 1900–1989. (*Maguaré/Bogotá*, 6:6/7, 1988/91, p. 101–152)

Useful historical review of 20th-century infant mortality in Bogotá traces levels, causes, social policies, and secular trends.

4821 Ortiz Sarmiento, Carlos Miguel. Los estudios sobre La Violencia en las tres últimas décadas. (*Bol. Socioecon.*, 24/25, agosto/dic. 1992, p. 47–76)

Useful introductory review article provides coverage of diverse perspectives on La Violencia in Colombia.

4822 Rojas H., Fernando. El Estado colombiano: desde la dictadura de Rojas Pinilla hasta el gobierno de Betancur, 1948–1983. (*in* El Estado en América Latina: teoría y práctica. Coordinación de Pablo González Casanova. México: Siglo Veintiuno Editores, 1990, p. 442–481, bibl., table)

Densely argued and theoretically sophisticated chronological work examines State/society relations in successive administrations. Integrates structural interpretations of emerging forms of capitalist organization with conjunctural analysis of social struggles and State-centered imperatives.

4823 Sarmiento, Luis Fernando and **Ciro Krauthausen.** Cocaína & Co.: un mercado ilegal por dentro. Bogotá: Univ. Nacional de Colombia, Instituto de Estudios Políticos y Relaciones Internacionales; Tercer Mundo Editores, 1991. 239 p.: bibl. (Sociología y política)

Detailed analysis examines narcotics industry, including its social organization, means, resources, and commercial networks. Concludes that it closely resembles legitimate industries, varying only in regards to the distinct rules operating for illegal markets.

4824 Silva Téllez, Armando. Imaginarios urbanos, Bogotá y São Paulo: cultura y comunicación urbana en América Latina. Bogotá: Tercer Mundo Editores, 1992. 293 p.: bibl., ill. (Comunicación social)

Presents comparative analysis of urban dwellers' cognitive maps of their city and process by which collective understanding of physical and social demarcations is constructed. Atypical multidisciplinary focus echoes similar studies in North America and Europe.

4825 Tecnologías urbanas socialmente apropiadas: experiencias colombianas: Red Colombiana de Tecnología Apropiada. v. 1–2. Edición de Jean-Jacques Guibbert. Bogotá: ENDA-América Latina, 1987. 2 v.: bibl., ill. (Documentos Tercer Mundo; 38-39-40, 47-48-49)

Papers from 1986 conference on appropriate technology and community organization include diverse and important set of contributions to the debate over relationship of private micro initiatives to public services.

Ulloa, Alejandro. La Salsa en Cali. See *HLAS 54:5215.*

4826 Ungar Bleier, Elisabeth and **Helena Useche Aldaña.** Impacto de la recesión venezolana en la migración de regreso a Colombia: el caso de las principales áreas urbanas emisoras. (*in* Fronteras permeables: migración laboral y movimientos de refugiados en América. Buenos Aires: Planeta, 1991, p. 89–113, bibl.)

Studies variable impact of return migration on the five principal cities from which migrants originated. Variations are mostly a direct result of regional recession, but reduced remittances have lowered family living standards and expanded informal sector activity.

4827 Uribe Alarcón, María Victoria. Limpiar la tierra: guerra y poder entre esmeralderos. Bogotá: CINEP, 1992. 150 p.: bibl., ill., maps. (Col. Sociedad y conflicto)

Insightful case study examines violent local conflicts rooted in historical political economy of emerald production during period of La Violencia. Excellent study of local power dynamics operating in situations removed from Church and national government influences, wherein the culture and justice of blood feuds prevail.

4828 Urrea, Fernando. Nuevas tecnologías y relaciones industriales. (*Bol. Socioecon.*, 23, enero 1992, p. 131–154, bibl., ill.)

Study based on 1980s macro- and micro-level industry data analyzes impact of new technology. The study's findings—greater labor market segmentation, widening gap between formal and informal firms, deskilling jobs, worker participation schemes, and an anti-union drift—reveal trends parallel to those in advanced capitalist nations.

4829 Villarreal Méndez, Norma. Género y clase: la participación política de la mujer de los sectores populares en Colombia, 1930–1991. (*in* Jornadas de Investigación Interdisciplinaria sobre la Mujer, 9th, Madrid, 1992. La mujer latinoamericana ante el reto del siglo XXI. Madrid: Ediciones de la Univ. Autónoma de Madrid, 1993 p. 127–161, bibl.)

Narrative interpretation of Colombian women's suffrage campaign, voting behavior, organizational participation, and social campaigns (violent and peaceful) links women's political activities to legislative change and broad range of social issues.

4830 Wade, Peter. Blackness and race mixture: the dynamics of racial identity in Colombia. Baltimore, Md.: Johns Hopkins Univ. Press, 1993. 415 p.: bibl., ill., index, maps. (Johns Hopkins studies in Atlantic history and culture)

Carefully researched study examines racial and ethnic identity in the Chocó (Colombia's Pacific region), and considers their relationship to the ideology of Colombian national and regional identity. For anthropologist's comment see item **1005**.

Whitten, Norman E. Pioneros negros: la cultura afro-latinoamericana del Ecuador y de Colombia. See item **4877**.

VENEZUELA

4831 Ajuria, Peru and **Koldo San Sebastián.** El exilio vasco en Venezuela. Vitoria, Spain: Servicio Central de Publicaciones, Gobierno Vasco, 1992. 215 p.: bibl., ill. (some col.). (America y los vascos = Amerika eta euskaldunak; 7)

Traces Basque immigration and assimilation in Venezuela from Basque shipping companies of 1880s. The Spanish Civil War (1936–39) prompted immigration accords. Book includes sections on adaptation of early immigrants, Basque politics, social institutions, cultural assimilation, etc.

Ambiente, Estado y sociedad: crisis y conflictos socio-ambientales en América Latina y Venezuela. See item **4396**.

4832 Aranda, Sergio. Las clases sociales y el Estado: el caso Venezuela. Caracas: Pomaire; Fuentes, 1992. 235 p.: bibl.

Theoretical discussions on classes and

the State are complemented by empirical analyses of Venezuelan sociopolitical development from 1950s-80s.

4833 Bengoa y Lecanda, José María. Sanare hace 50 años: medicina social en el medio rural venezolano. 3. ed. Caracas: Ediciones Cavendes, 1992. 278 p.: bibl., ill., map. (Col. Temas y autores sanareños; 5)

New edition of work first published in 1940 (see *HLAS 6:1261*) updates application of Sanare approach to medicine in rural Venezuela. The organicist, functional, and causal models of medicine must be complemented by consideration of "the social multicausality of infectious and degenerative diseases." Includes sections on rural life, nutrition, housing, income, infant mortality, tuberculosis, sexual and other diseases, health co-ops, social security, etc.

4834 Briceño-León, Roberto. Los efectos perversos del petróleo. Prólogo de Germán Carrera Damas. Caracas: Fondo Editorial Acta Científica Venezolana; Consorcio de Ediciones Capriles, 1990. 230 p.: bibl., ill.

The town of Tinaquillo serves as "metaphor" for perverse impact of petroleum economy. Up to 1970s, region is largely excluded from petroleum bonanza, its traditional agro-export and subsistence sectors ravaged by international depression and emigration. Incorporation into petroleum economy brings rampant land speculation and immigration, agricultural deterioration, and dependent development, while pursuit of profit replaces traditional values as the social norm. Work was awarded first prize by Consejo Nacional de Investigaciones Científicas y Tecnológicas for research in the social sciences.

4835 Briceño-León, Roberto. Venezuela: clases sociales e individuos: un enfoque pluriparadigmático. Prólogo de Ruth Capriles Méndez. Caracas: Fondo Editorial Acta Científica Venezolana; Consorcio de Ediciones Capriles, 1992. 235 p.: bibl., ill.

Explores nexus between social class and the individual by effecting theoretical and empirical comparisons of class-as-mode-of-production (clase-categoría) and class-as-lifestyle (clase-situación) in Venezuela. Impact of these class dimensions is explored through analyses of "life-stories" of individuals in different class positions.

4836 Carciente, Jacob. La comunidad judía de Venezuela: síntesis cronológica, 1610–1990, y referencias bibliográficas para su estudio: crónicas sefardíes. Caracas: Asociación Israelita de Venezuela; Centro de Estudios Sefardíes de Caracas, 1991. 247 p.: bibl., ill. (Biblioteca popular sefardí: 0798–1953; 10)

Chronological outline of history of Sephardic Jewish immigrants in Venezuela is complemented by essays on contributions to contemporary culture, civic associations, synagogues, genealogy, etc.

4837 Clarac de Briceño, Jacqueline. La enfermedad como lenguaje en Venezuela. Mérida, Venezuela: Univ. de Los Andes, Consejo de Desarrollo Científico, Humanístico y Tecnológico; Consejo de Publicaciones, 1992. 514 p.: bibl., ill. (Textos de la Univ. de Los Andes. Col. Actual. Serie Temas de interés nacional y regional)

Examines definitions of illness and health, and of life and death, as historical-cultural constructions. The author argues that this process reflects the clash of (multi)-ethnicity with Western medicine, as well as the "violent sociocultural change and generalized insecurity" that characterizes Venezuelan society.

4838 Gómez Calcaño, Luis and **Margarita López Maya.** El tejido de Penélope: la reforma del estado en Venezuela, 1984–1988. Caracas: CENDES; APUCV-IPP, 1990. 215 p.: bibl. (Col. José Agustín Silva Michelena; 2)

Developmental trends reflect interactions among important national and transnational actors. Focuses on efforts by Venezuela's Comisión Presidencial para la Reforma del Estado (COPRE). These reflect broader Latin American democratization and local "conjunctural crises:" exhaustion of the "developmental model" and weakening of the State and "hegemonic actors."

4839 González Ordosgoitti, Enrique Alí. Diez ensayos de cultura venezolana. Caracas: Fondo Editorial Tropykos; Asociación de Profesores de la U.C.V., 1991. 173 p.: bibl.

Essays on various aspects of Venezuelan culture examine contemporary indigenous culture, the Iberian cultural heritage of "Spanish" settlers (especially religious tolerance and intolerance among Christian, Muslim, and Jewish peoples of Iberia throughout the centuries), contemporary Venezuelan culture, regional cultures, education and national identity, transculturation and cultural aggression, urban culture in Caracas, concept of culture in anthropology, etc.

4840 González Ordosgoitti, Enrique Alí.
Ensayos sobre la cultura urbana caraqueña. Caracas: Fundarte; Centro de Investigaciones Socioculturales de Venezuela, 1992. 190 p.: bibl. (Col. Rescate. Caracas toma Caracas; 7)

Focuses on "popular urban culture" of Caracas—the interactive means through which individuals attempt to satisfy community needs. Spatial intersection of class, ethnicity, modernity, and traditionalism generates axes of social solidarity that identify various residential subcultures. While most physical history of the city has been destroyed by modernization and development, traditional lore survives as part of an emergent, and dynamic, popular urban culture.

4841 Guerra T., Helena. Las asociaciones de vecinos como espacio de la política social del estado venezolano. Caracas: Univ. Central de Venezuela, Facultad de Ciencias Económicas y Sociales, 1990. 200 p.: bibl.

Community associations in marginal neighborhoods provide a nexus where social policies (codified in the Fifth and Sixth National Plans) of the capitalist, oil-dependent State interact dialectically with popular demands for housing, electricity, schools, health centers, etc. Marxist analysis is complemented by "other critical theories."

4842 Hurtado Salazar, Samuel. Dinámicas comunales y procesos de articulación social: las organizaciones populares. Caracas: Fondo Editorial Tropykos; Asociación de Profesores de la U.C.V., 1991. 185 p.: bibl., ill., maps.

Neighborhood organizations are emergent foci of popular solidarity where society-as-community is "articulated" with the State. Traces emergence, mobilization, decline, etc. of various community organizations. Reviews theoretical conceptions of "social marginality," the State-city relationship, social movements' influence on the "Populist State," etc.

4843 Justicia y pobreza en Venezuela. Recopilación de Rogelio Pérez Perdomo. Caracas: Monte Avila Editores, 1987. 257 p.: bibl., ill. (Tiempo de Venezuela)

Sponsored in part by the Ministerio de Justicia, this collection of studies focuses on various aspects of legal representation of the poor in Venezuela. Theoretical and historical-comparative essays are complemented by investigation of penal codes, labor and family law, legal assistance, social security, etc.

4844 Maingón, Thais and Heinz R. Sonntag.
Las elecciones en Venezuela en 1988 y 1989: del ejercicio del rito democrático a la protesta silenciosa. (*Rev. Mex. Sociol.*, 52:4, oct./dic. 1990, p. 127–154, bibl., graph, tables)

Authors argue that since the 1980s Latin America has been experiencing two contradictory processes: 1) exhaustion of the "developmental model," derived in part from the debt crisis and leading to regressive "adjustment" policies; and 2) "redemocratization" of society after long periods of authoritarian rule. This paradox is manifested in ritualistic electoral politics accompanied by increasing rates of voting abstention. The latter represents a "silent protest," signaling a "crisis of democracy."

4845 Méndez de Pérez, Betty. Análisis nutricional antropométrico: una encuesta de salud en tres grupos de la Amazonia Venezolana. Caracas: Instituto de Investigaciones Económicas y Sociales, Facultad de Ciencias Económicas y Sociales, Univ. Central de Venezuela, 1989. 90 leaves: tables.

Survey of nutritional and health data on three indigenous groups (Guajibo, Piapoco, Piaroa) from Territorio Federal de Amazonas covers children between 7 and 18 years old, enrolled in 1984–85 school year. Discusses methodological concerns and findings, including inter-group variations and deviations from expected phenotypical norms (height, weight, fat, etc.) owing to malnutrition and other deficiencies. Contains statistical tables.

4846 Muñoz L., Carlos A. El Estado venezolano y su política regional. Mérida, Venezuela: Univ. de los Andes, Consejo Editorial; Facultad de Ciencias Forestales, 1990. 88 p.: bibl. (Textos de la Univ. de los Andes. Col. Tecnología. Serie Geografía)

Regional policies of the State are important, but not exclusive, determinants of sociopolitical and spatial development patterns. Venezuelan development is characterized by tensions between an agro-export sector made up of peripheral "regional bourgeosies" dominant during 19th century, and a petroleum-based sector comprised of a "bourgeoisie of the center" favoring a more centralized model of industralization that became dominant during 20th century.

4847 Olinto Camacho, Oscar and **Ariana Tarhan.** Alquiler y propiedad en barrios de Caracas. Prólogo de Alan Gilbert. Ottawa: International Development Research Centre; Caracas: Univ. Central de Venezuela, Facultad de Arquitectura y Urbanismo, Centro de Estudios Urbanos, 1991. 197 p.: bibl., ill., maps.

Commercialization of and speculation on "peripheral lands"—fostered by intensifying municipal controls, escalating costs of living in peripheral areas (e.g., rising transportation costs), and declining family incomes linked to economic crises—make home ownership increasingly difficult. Study complements much research on growth of marginal neighborhoods by focusing on patterns of renting rather than ownership.

4848 Pollak-Eltz, Angelina. Umbanda en Venezuela. Caracas: Fondo Editorial Acta Científica Venezolana, 1993. 178 p.: bibl. (Cuadernos de investigación social; 2)

Examines history, rituals, beliefs, magic, initiation, sacraments, organization, etc. of Umbanda religion in Venezuela. Originating in Brazil from Yoruba traditions in early part of 20th century, this Afro-American religion has undergone significant adaptations as it spread into more modern urban centers in Venezuela.

4849 Reyna de Roche, Carmen Luisa. Patria potestad y matricentrismo en Venezuela: estudio de una disfuncionalidad. Caracas: Univ. Central de Venezuela, Facultad de Ciencias Jurídicas y Políticas, 1991. 241 p.: bibl., ill.

Explores family dysfunctions engendered by laws that "consecrate legal male supremacy" in a largely "matricentric" culture.

4850 Ríos de Hernández, Josefina and **Gastón Carvallo.** Análisis histórico de la organización del espacio en Venezuela. Caracas: Univ. Central de Venezuela, Consejo de Desarrollo Científico y Humanístico, 1990. 238 p.: bibl., maps. (Col. Estudios)

Contemporary environmental problems reflect spatial organization as a historical process (1500–1960) whereby social relations influence: appropriation of natural resources; economic production, circulation, and distribution; and sociopolitical dynamics. Analyzes three periods of spatial organization: pre-colonial (*indígena*), agro-export, and petroleum economy.

4851 Seguridad personal: un asalto al tema. Recopilación de Juan Carlos Navarro, Rogelio Pérez Perdoma y Tibisay Lucena. Caracas: Ediciones IESA, 1991. 285 p.: bibl., ill.

Multidisciplinary essays examine personal crime and victimization in Venezuela. Police action and State policies tend to reflect public perception more than actual criminal patterns. Congestion, corruption, negligence, and brutality plague criminal justice system. Rising crime rates may reflect fiscal constraints on the State, which gives relatively low priority to crime prevention and law enforcement.

4852 Sgambatti, Sonia. La mujer: ciudadano de segundo orden. 2a ed. Caracas: Ediciones del Congreso de la República, 1988. 357 p.

Author finds basis for widespread discrimination against women in Latin American law. Focuses expecially on uxoricide, the killing of an adulterous wife by her husband, legally accepted in various Latin American countries. Chronicles nine years of challenges to such discriminatory laws in Venezuela.

4853 Vargas, César. Estudio etnográfico del comportamiento mágico religioso en la Venezuela contemporánea. Maracaibo, Venezuela: Ediciones Astro Data, 1987. 181 p.: bibl.

Explores socio-psychological foundations and applications of magical-religious beliefs in daily life. Vitality of these beliefs rests on two axes: 1) economic interests of the *brujos* (producers of and traffickers in esoteric objects); and 2) the physical, mental, and emotional problems; economic insecurity; and lack of scientific knowledge on the part of individuals.

ECUADOR

AMALIA M. ALBERTI, *Independent Consultant, c/o United States Agency for International Development, San Salvador*

IN ECUADOR, SEVERAL SUBJECT AREAS dominate the social science literature canvassed and annotated below. First, a number of investigations focused on the changing sociocultural context of rural Ecuador from the perspective of indigenous households and communities (items **4865, 4867,** and **4856**), while others highlighted the changing socioeconomic context of rural Ecuador, particularly from the perspective of farmers and laborers, including women (items **4862, 4861,** and **4875**). Peasant uprisings were documented (items **4859** and **4858**), with the work of Uggen (item **4876**) providing an especially thought-provoking analysis. However, works on environmental awareness and related changes, on community and individual responses to environmental issues, and on the continuing evangelization of the countryside and its impact were subjects not covered in publications canvassed for this chapter.

The second dominant theme for this biennium is social welfare, including works on population, health, and care of the elderly. Much of the emphasis is on promoting preventative measures and providing alternative health care models (items **4873, 4869,** and **4866**). Studies that disaggregate target groups by gender or age or both are also much in evidence. Within this latter subject area, the works prepared under the aegis of the Centro de Población y Paternidad Responsable (CEPAR) deserve special mention. These provincial-level sociodemographic profiles (item **4872**) provide a wealth of information and data compiled from various secondary sources and organized in a meaningful way. Portions of the data presented appear specific enough to serve as a baseline for future endeavors. It is hoped that authors will persevere in their efforts and prepare a companion document for the country as a whole.

4854 Burgwal, Gerrit. Collective clientelism in contemporary Quito: some comments on its structure and culture. (*Anthropol. Verkenn.*, 11:2, 1992, p. 23–47, bibl., graphs)

Author distinguishes collective clientelism from dyadic clientelism. He sees a direct relationship between a patron clique, the (Christian Democratic) municipal government, and collective clients—the inhabitants of Quito's low-income settlements. He describes clientelism as a political machine. [R. Hoefte]

4855 Burgwal, Gerrit. Struggle of the poor: neighborhood organization and clientelist practice in a Quito squatter settlement. Amsterdam: CEDLA, 1995. 260 p.: bibl., maps, photos. (Latin American studies; 74)

Collective biography reconstructs history of Lucha de los Pobres settlement in Quito. Traces several lines of clientelism. Argues that clientelism and brokerage do not necessarily obstruct neighborhood improvement or urban social change. Based on 1990–91 fieldwork. [R. Hoefte]

4856 Castelnuovo, Allan and **Germán Creamer.** La desarticulación del mundo andino: dos estudios sobre educación y salud. Quito: Depto. de Sociología, PUCE-Q; Ediciones Abya-Yala, 1987. 260 p.: bibl., ill.

Two carefully-conducted, thought-provoking studies were developed with an interdisciplinary approach. One focuses on role of the school in promoting cultural change, with ensuing conflict with ethnic identity. Other work explores relationship between migration and health.

4857 Correa León, Sandra. Mujer, situación social y jurídico-laboral en el Ecuador. Quito: Fundación Friedrich Naumann; Ministerio de Bienestar Social; Comité Ecuatoriano de Cooperación con la Comisión Interamericana de Mujeres, 1990. 283 p.: appendices, bibl.

Work's main contribution is identification of changes in civil and labor codes that have occurred over past 25 years. Appendices give revised wording.

4858 Dubly, Alain and **Alicia Granda.** Desalojos y despojos: los conflictos agrarios en Ecuador, 1983–1990. Quito: El Conejo; CEDHU, 1991. 226 p.: bibl., ill., maps.

Investigation of select conflicts in the Sierra, coastal, and Amazon regions from 1983–90 differentiates among three types of land takeovers: 1) physical displacement, with destruction of homes and resources; 2) "official" takeovers of traditional landholdings; and 3) extraction of resources in a destructive manner and without compensation. Documents occurrence of each type, and includes brief analysis.

4859 Encuentro de Historia y Realidad Económica y Social del Ecuador, 6th, *Cuenca, Ecuador, 1989.* Los campesinos en el proceso latinoamericano de los años ochenta y sus perspectivas. v. 1–2. Coordinación de Iván González Aguirre. Cuenca, Ecuador: Univ. de Cuenca, Instituto de Investigaciones Sociales, 1991. 2 v.: bibl., ill., maps.

Nearly half of papers included in this two-volume work focus on Ecuador. Themes of vol. 1 include social reproduction of the peasantry and modernization. Vol. 2 looks at peasant conflicts and peasant movements.

4860 Entre los límites y las rupturas: las mujeres ecuatorianas en la década del 80. Quito: ACDI-CEPLAES, 1992. 408 p.: bibl., ill.

Collection of carefully investigated and well-documented essays provides diagnosis of Ecuadorian women's situation in 1980s. Topics covered include governmental policy, politics and the law, women's participation in the labor force and in the political process, and women's access to health and to formal and non-formal education.

4861 Espinosa, Roque. Parentesco y reproducción en Manabí: el caso de Membrillal. Quito: Fundación Ecológica Ecuatoriana; Ediciones Abya-Yala, 1990. 221 p.: bibl., ill.

Examines agricultural production and kinship in a rural agricultural coastal community. Excellent in-depth investigation emphasizes importance of social relations, especially direct familial and marital ties, in all aspects of daily life including access to labor, production, and commerce.

4862 Field, Leonard. Sistemas agrícolas campesinos en la Sierra Norte. Quito: Centro Andino de Acción Popular, 1991. 192 p., 1 folded leaf of plates: bibl., ill., maps. (Ciencia y tecnológia)

Presents results of study conducted over three-year period with approximately 50 farm households. Pursues question of what provokes change in traditional agricultural production. Concludes that modernization of agriculture is a necessary but insufficient cause, and looks rather to changing local conditions and demands affecting agricultural production. Provides good production data.

4863 Fine, Kathleen Sue. Cotocollao: ideología, historia y acción en un barrio de Quito. Traducción de V. Dueñas. Quito: Ediciones Abya-Yala, 1991. 278 p.: bibl.

Interdisciplinary study examines social organization and social action in the Cotocollao sector of Quito. Combining historical data with analysis of activities of neighborhood organizations, and following in-depth interviews with three local residents, investigator concludes that the State and the community are both participants in the process of social change, with the community exercising more influence on the State than is normally acknowledged by either side.

4864 García, Mauricio and **Carmen Hernández.** ¿Tiempo de jugar?: niños y adolescentes trabajadores de las familias populares urbanas. Quito: CEPLAES, 1992. 101 p.: bibl.

Descriptive study of contributions of youth to productive and domestic activities of their households is based on secondary data analysis. Discussion covers formal and informal sectors, and remunerated and unremunerated labor. Much of the data is gender disaggregated.

4865 Guerrero, Andrés. De la economía a las mentalidades. Quito: Editorial El Conejo, 1991. 198 p.: bibl., ill., map. (Col. Ecuador hoy)

Four essays analyze changes in composition, social relations, obligations, and expectations of labor force on select haciendas in Otavalo region from 1950s–80s. Author provides excellent details including kinship ties, family composition, and variations in monetary and non-monetary exchanges between labor force and specific haciendas.

4866 **Instrucción de la mujer y fecundidad.** Quito: Centro de Población y Paternidad Responsable, 1991. 39 p.: bibl., ill.

Compares findings from surveys conducted in 1979 and 1989 about fertility and socioeconomic differences. Investigators conclude that educational level is variable most associated with differences in fertility, but add that levels of attainment within the educational system tend to be associated with levels of social and economic well-being within the society-at-large.

4867 **Lentz, Carola.** Buscando la vida: trabajadores temporales en una plantación de azúcar. Quito: Ediciones ABYA-YALA, 1991. 204 p.: bibl., ill., maps.

Traces changing role of, and demand for, seasonal workers in sugarcane production, particularly from 1950s–1983. Describes erosion of kinship and social relations as factors in contacting and contracting the overwhelmingly indigenous core of migrant laborers from the Sierra, and discusses increasingly depersonalized conditions and increased demands encountered. Includes photos with migrant workers' commentary.

4868 **El levantamiento indígena y la cuestión nacional.** Quito: Ediciones ABYA-YALA; CDDH, Comisión por la Defensa de los Derechos Humanos, 1990. 114 p.: bibl.

Includes three essays directly addressing a 1990 indigenous uprising that sought to have Ecuador declared a multinational state. Also presents text of five documents issued over last 40 years from various international sources that describe and/or clarify what are deemed to be basic rights of indigenous communities.

Maiguashca, Juan. Las clases subalternas en los años treinta. See item **1813.**

4869 **Mauro, Amalia.** La vida es larga y nos importa mucho: la salud en las familias de sectores populares urbanos. Quito: CEPLAES, 1992. 114 p.: bibl.

One of several essays based on case studies of 23 families and a survey of over 400 families in urban areas conducted by Centro de Planificación y Estudios Sociales between 1990–91. Volume gives overview of health practices, use of formal and informal healing services including maternity care, and rationale for choices made.

4870 **Moreno Yáñez, Segundo et al.** Ecuador multinacional: conciencia y cultura. Quito: Ediciones ABYA-YALA; Centro Ecuatoriano de Desarrollo de la Comunidad, 1989. 272 p.: bibl., ill.

Collection of essays highlights aspects of ethnocultural pluralism in Ecuador. Papers were used to inform two 1988 seminars that attempted to differentiate the psychological make-up of persons of mestizo and indigenous heritage.

4871 **La mujer de la tercera edad en el Ecuador.** Quito: Ediciones CEDATOS, 1991. 114 p.: ill.

Study based on 560 inverviews—half in Quito and half in Guayaquil—is unique in that its subjects are women 60 years of age and over. Although methodologically weak, with tables poorly organized, work provides data in a variety of areas often overlooked regarding subjects who are all too often neglected.

4872 **Perfil socio-demográfico provincial.** v. 1–7. Quito: Centro de Estudios de Población y Paternidad Responsable, 1992. 7 v.: ill.

Each volume presents sociodemographic data and discussion specific to one province of Ecuador. Researchers appear to have spared no effort to disaggregate data by gender, age, and location (urban, rural). Discussion often compares recent and earlier findings, thus providing added perspective.

4873 **Ribadeneira, Juan Carlos.** Pobreza urbana: enfermedad y comportamiento popular. Quito: Centro Andino de Acción Popular, 1991. 208 p.: bibl., ill. (Estudios y análisis)

Presents case study of a barrio clinic near Quito, founded about 15 years before data were collected. Reviews information on one-third of cases treated over three-year period, associating types of illnesses and remedies (prescribed and household) with select socioeconomic characteristics. Investigators are seeking an alternative approach to health care that takes account of sociocultural as well as medical factors.

4874 **La situación de la mujer en la economía informal: caso ecuatoriano.** Quito: Instituto Ecuatoriano de Investigaciones y Capacitación de la Mujer, 1991. 61 p.: bibl., ill.

Findings are based on analysis of data obtained from weighted sample of women interviewed in four separate areas in vicinity of Quito. Criteria used by investigators in identifying micro-enterprises and determining their status as part of the informal economy are unclear.

Solano de la Sala Veintimilla, Germán. Indice de folletos sobre temas económicos y sociales del Ecuador. See item **1837**.

4875 **Stølen, Kristi Anne.** A media voz: ser mujer campesina en la sierra ecuatoriana. Quito: CEPLAES, 1987. 173 p.: bibl., tables.

Analysis of gender relations in Ecuadorian highlands combines theory with historical accounts of development of the area. Women who span several generations comment extensively on a range of subjects including division of labor, kinship and familial relations, abuse of women, and hopes and expectations. Excellent presentation of how these women perceive themselves and their world.

4876 **Uggen, John F.** Tenencia de la tierra y movilizaciones campesinas: zona de Milagro. Quito: ACLAS, 1993. 126 p.: bibl., maps. (Ecuador; 1)

In-depth study of peasant movements along coastal zone of Milagro is based on premise that peasant uprisings are most likely to occur when traditional landed elite becomes destabilized. Success of peasant efforts is then constrained to the extent that competing commercial and industrial interests exist. The role of the military and government agencies is also examined.

4877 **Whitten, Norman E.** Pioneros negros: la cultura afro-latinoamericana del Ecuador y de Colombia. Traducción de Cindy Lepeley y Oscar Lepeley. Quito: Centro Cultural Afro-Ecuatoriano, 1992. 252 p.: bibl., ill., maps.

Originally written in 1974 (see *HLAS 37:1555*) and revised in 1986, this is work's first translation into Spanish. Based on his own extensive field work and on secondary sources, author traces many phases of the process of cultural and environmental adaptation of first Africans brought to Ecuador and Colombia as slaves in the beginning of 16th century. Topics explored include historical background, adaptive strategies, secular and religious rituals, and ethnic considerations.

PERU

CORNELIA FLORA, *Professor of Sociology, North Central Regional Center for Rural Development, Iowa State University*

THE COMPLEXITY AND BREADTH of sociological studies of Peru continue to increase. Analyses in the 1980s emphasized circulation and consumption, while those of the 1970s emphasized the social relations of production. Both of these two previous decades of research focused on the Peruvian State as a principal protagonist. In the 1990s, however, the responsibility of ensuring social reproduction has shifted from the State to society as a consequence of the decline in the power and relevance of the State, including the loss of the State's monopoly on the use of arms, increasing State corruption, and its increasing inability to meet its traditional obligations to its citizens.

Current research no longer uses Marxist structural determinism to assess the Peruvian situation. Increasing attention is being paid to the concept of agency (particularly collective agency) and its ability to influence structure. For example, researchers now focus on symbolic, cultural, and ethnic aspects of social movements and social change (item **4922**). There is less concentration on the movements' relations with the State and political parties, and more on their relations with other social groups (item **4929**). There is also less emphasis on the political and economic aspects of movements and more on their internal dynamics, forms of leadership, internal conflicts, and differentiation (items **4934** and **4900**). Research is less to justify and legitimate the movements than to understand them holistically. Indigenous knowledge is increasingly recognized as a means to build identity and local action (items **4939, 4941,** and **4928**).

In the absence of an effective State, non-governmental organizations are becoming much more important in addressing social issues, including the delivery of basic social services (items **4940, 4904, 4919, 4905,** and **4917**). While this model is accepted and even applauded by most authors annotated below, others suggest that the increasing influence of such organizations decreases the possibility of democratic participation and increases international control over their activities (items **4891** and **4926**).

Research analyzes new social actors, rather than just examining repressive structures. New social actors include entrepreneurs (items **4883** and **4916**), women's groups (items **4890, 4893, 4909,** and **4921**), youth (items **4879** and **4892**), and community organizations (items **4900** and **4907**) in both rural and urban areas (item **4908**). And new space for social activity, particularly within local communities and municipal governments, receives more attention than in the past (items **4896, 4913,** and **4924**). Neither is structure forgotten. Many researchers draw connections between policies of adjustment and the fragmentation of collective social action, a process that is debilitating the associative vitality of Peru (item **4911**).

The research approaches for this biennium are influenced by Gramsci, Etienne Henry, and the polemic between Haya and Mariátegui. As the role of human agency and identity, particularly ethnic and gender identity, receive more emphasis than in the past (item **4930**), reductionism in viewing class struggle is less prevalent. This new trend extends to demographic analyses of particular ethnic groups as well as to studies of Peruvians abroad (items **4912, 4882,** and **3370**). More research addresses the internal tensions of movement formation and the need to understand differences among groups, thus departing from broad generalizations (item **4885**). In fact, research is increasingly designed to understand differences, instead of to generalize. There are more empirical studies (in contrast to "reflections"), and these studies often are creative combinations of qualitative and quantitative methodologies.

4878 Abarca Begazo, Noel *et al.* Desarrollo urbano y vivienda popular, Arequipa. Arequipa, Peru: Centro de Estudios para el Desarrollo Regional, 1989. 230 p., 34 folded leaves of plates: bibl., ill.

Urban housing markets and implications of speculation in land and housing have given rise to an important alternative movement: owner-constructed housing, particularly in the *pueblos jóvenes.* Traces urban housing policy from late 1940s when increased rural-urban migration created new political alliances. Neighborhood organizations that develop housing, facilitate migration, and influence policy are analyzed in terms of residential segregation and local empowerment. A typology of functional spatial organization and popular housing is derived.

4879 Alarcón Glasinovich, Walter. Entre calles y plazas: el trabajo de los niños en Lima. Lima: Asociación Laboral para el Desarrollo; Instituto de Estudios Peruanos; Fondo de las Naciones Unidas para la Infancia, 1991. 179 p.: bibl. (Serie Urbanización, migraciones y cambios en la sociedad peruana; 11)

With 26 percent of children between the ages of eight and 13 working in the summer of 1988, child labor has become an integral although minimal part of household economy of many poor families in Lima. Deteriorating economic conditions and the shock of 1990 increased child labor by 113 percent. Child workers are perdominantly male, live with their families, and work in the informal sector. A substantial proportion works for their own family, which decreases their maltreatment. Poverty alone does not explain child labor. Parents who value education for their children keep them from the labor force, while those families whose culture values work seek employment for them. The majority of child workers attend school, with work having less effect than socioeconomic status on their scholastic progress. Policy should not prohibit child labor (which often pays for school costs, particularly since 1990), but rather ensure that conditions of work are not exploitative or harmful. Educational system should adapt to children's work schedule and conditions.

4880 Alber, Erdmute. Flexibilität contra tradition: zur attraktivität der *comunidad campesina* in Perú. (*Sociologus/Berlin*, 43:2, 1993, p. 97–117, bibl.)

Using her previously published research on Comunidad San Agustín-Huayopampa, author proposes that the flexibility inherent in legal status of indigenous communities, rather than Andean tradition, is basis for their stability, modernization, and proliferation. [C.K. Converse]

4881 Altamirano, Teófilo. Exodo: peruanos en el exterior. Lima: Pontificia Univ. Católica del Perú, Fondo Editorial, 1992. 224 p.: bibl.

Peruvian out-migration has increased dramatically in the 1970s–80s for all social classes and areas of the country. High rates of migration to US are augmented by migration rates to new receiving areas, including Japan (as Peruvians of Japanese ancestry have special migration privileges and serve as a "Japanese" low-skilled labor force), Spain, Italy, and Canada. Remittances form an increasingly important part of the Peruvian economy, and return migration has increased.

4882 Altamirano, Teófilo. Los que se fueron: peruanos en Estados Unidos. Lima: Pontificia Univ. Católica del Perú, Fondo Editorial, 1990. 194 p.: bibl., maps.

A model of Peruvian migration to US is presented in terms of push-pull factor in Peru and pull factors of US. Contradictions of assimilation and cultural survival in US are examined, based on a case study of Peruvian communities in New York and Paterson, N.J. Social clubs are important for maintaining Peruvian identity, as is regular return migration. Finds that social divisions present in Peru are reproduced in US.

4883 Anaya Franco, Eduardo. Los grupos de poder económico: un análisis de la oligarquía financiera. Lima: Editorial Horizonte, 1990. 177 p.: bibl. (Realidad peruana; 17)

Peru's financial oligarchy is analyzed through examination of financial data and network analysis of 4,500 companies. Describes the manner in which banks, the nucleus of the various financial groups, control their dependent enterprises. Network analysis reveals complexity of the interlinkages, internal hierarchy of each group, and interrelationships among powerful families, suggesting the national base of their financial power.

Anderson, Jeanine. Estrategias de sobrevivencia revisitadas. See item **4397**.

4884 Ansión, Juan. Desde el rincón de los muertos: el pensamiento mítico en Ayacucho. Lima: GREDES, 1987. 244 p.: bibl., ill. (Investigaciones/ensayos/documentos; 6)

Presents sociology of knowledge approach to myths in Ayacucho. Clash of cultures and syncretism are embodied in evolving regional myths, which contrast Biblical myths to indigenous ones. Myths' present-day embodiment as witches and other forms link spiritual and physical realms for both good and evil. Myths are evolving ways of knowing and of situating self in terms of nature and society.

4885 Balbi, Carmen Rosa. Identidad clasista en el sindicalismo: su impacto en las fábricas. Lima: Centro de Estudios y Promoción del Desarrollo, 1989. 226 p.: bibl., ill.

Two tendencies in Peruvian labor unions, clientalist vs. classist, are analyzed through intensive interviews with labor leaders about events from 1968–85. Unions became more class-based and emerged as vital new social actors during late 1970s. Workers not only demanded better wages, they demanded more information and participation in all aspects of production; during that period they greatly increased their political power.

4886 Barrig, Maruja et al. Aproximaciones al aborto. Lima: SUMBI & The Population Council, 1993. 113 p.: bibl., ill.

Abortion rates in Peru have increased in response to economic crisis, lack of availability of contraceptives, and women's lack of power relative to men. Through case studies and systematic empirical analysis, high social and economic costs of abortion are analyzed. Abortion represents a public health problem, which is made worse through legal restriction of abortion. Availability of health care and education about sexuality and reproduction are more effective alternatives.

4887 Barrig, Maruja. Quejas y contentamientos: historia de una política social, los municipios y la organización femenina en la ciudad de Lima. (*in* Políticas sociales, mujeres y gobierno local. Santiago: Corporación de Investigaciones Económicas para Latinoamérica, 1992, p. 51–71, bibl.)

Analysis of the "Glass of Milk" program in Lima reveals decline of federal government's legitimacy, increasing importance of municipal in contrast to federal government, and conversion of basic services programs run by women into the most visible social service programs in 1980s–90s.

4888 Bebbington, Anthony. Farmer knowledge, institutional resources and sustainable agricultural strategies: a case study from the eastern slopes of the Peruvian Andes. (*Bull. Lat. Am. Res.*, 9:2, 1990, p. 203–228, bibl., tables)

Cultivation of the potato, a staple of the highlands, was introduced and modified by colonists on slopes on the edge of the Amazon jungle as they experimented to develop production systems adapted to their physical and socioeconomic situation. As market integration increased, so did instability. Structural disadvantages mediated against adaptation of a more sustainable system using indigenous research methodologies previously employed effectively.

4889 Béjar, Héctor. La organización popular y el rol de las ONGDs: una aproximación. (*Social. Particip.*, 63, nov. 1993, p. 39–57)

Analyzing popular movements that have arisen in a wide variety of cultural settings, Béjar documents increasing importance of nongovernmental developmental organizations in supporting these movements in absence of government activity. Organizations' challenge is to give popular movements greater autonomy in effective and productive use of their resources.

4890 Blondet M., Cecilia. Las mujeres y el poder: una historia de Villa El Salvador. Lima: Instituto de Estudios Peruanos, 1991. 196 p.: bibl. (Serie Urbanización, migraciones y cambios en la sociedad peruana; 10)

Relates history and structure of a strong women's organization in a slum neighborhood of Lima that developed from a land invasion. Women's difficulties in legitimizing their coming together, to themselves and others, are overcome by their efforts to deal with economic crisis, beginning with assistance projects such as soup kitchens. Evolution and problems of Federación Popular de Mujeres de Villa El Salvador are presented through organizational analysis and case histories of members.

4891 Boggio, Ana et al. La organización de la mujer en torno al problema alimentario: una aproximación socio-analítica sobre los comedores populares de Lima Metropolitana, década del '80. Lima: CELATS, 1990. 104 p.: bibl.

Based on 1986 survey, analyzes Lima's *comedores populares*, which numbered 2,300 by 1990. Focuses on origin, organization, meaning for participants, and resulting social relations and networks. Due to family obligations, crises, and different degrees of commitment, participation is irregular and groups difficult to build. Their major function is to channel subsidized international food from NGOs.

4892 Carrión, Julio. La juventud popular en el Perú. Lima: IEP Ediciones, 1991. 140 p.: bibl., ill. (Mínima IEP; 21)

Youth are a growing proportion of Peruvian population. With increasing urbanization, levels of education are increasing. As a result of staying in school longer, percent of employed youth declined from 50 percent to 40 percent during 1961–81. There is a large group of student workers, particularly in Lima. Looking at working-class youth, an increasing proportion were born in Lima and achieved secondary education. The higher the education, the greater the unemployment rate. Youth of all educational levels are employed in temporary jobs in the service sector. Migrant female youth are doubly disadvantaged. Youth were more likely to be unemployed and poorly paid in 1984 compared to 1975.

4893 Casos Huamán, Victoria. La mujer campesina en la familia y la comunidad. Lima: Ediciones Flora Tristán, 1991. 73 p.: bibl. (Avances de investigación)

Analysis of productive and reproductive roles of rural women stresses women's importance for peasant production, highlighting their function of guaranteeing social relations of production and community cohesion. Analyzes women's organizations, women's importance in community organizations, and dependence of family survival on such organizations. Study was conducted in three peasant communities in 1987.

4894 Centro de Estudios Cristianos y Capacitación Popular (Peru). Organizaciones de sobrevivencia y problemática alimentaria: reflexiones y propuestas. Arequipa, Peru: Centro de Estudios Cristianos y Capacitación Popular, 1991. 174 p.: ill.

"Survival organizations" have responded most effectively to increasing malnutrition resulting from economic crisis and structural adjustment policies. The organizations are now looking at strategies involving regional nutritional development to supplant soup kitchens. Those strategies link food production and consumption through alliances of grassroots groups and NGOs.

4895 Cosamalón, Ana Lucía. El lado oculto de lo cholo: presencia de rasgos culturales y afirmación de una identidad. (*Allpanchis*/Cusco, 25:41, 1993, p. 211–226)

Intensive interviews of members of youth organizations in *pueblos jóvenes* in Lima explored meaning of a *cholo* identity. This identity served to link the modern and urban with the traditional and rural, giving a sense of solidarity among poor youths of indigenous origin and physiognomy, with poor Spanish, who otherwise felt excluded. Work was rejected by some as negative.

4896 Cuba García, Herberth. Salud, una experiencia dentro de la crisis, Lima: H. Cuba García, 1992. 141 p.: ill., index.

Health care is presented as a unique service, requiring strong delivery system and State support. Despite fiscal difficulties of Peruvian State, participatory models of health delivery have been created that include local councils and health delivery systems linked to hospitals and infrastructure.

4897 Degregori, Carlos Iván et al. Tiempos de ira y amor: nuevos actores para viejos problemas. Lima: DESCO, Centro de Estudios y Promoción del Desarrollo, 1990. 246 p.: bibl.

Using literary analysis of love, including respect, dignity, and manifest attraction, addresses basic contemporary Peruvian cultural themes of marginality and violence.

4898 Delgado Súmar, Hugo E. Pervivencia ritual vs. plustrabajo. Ayacucho, Peru: Univ. Nacional San Cristóbal de Huamanga, Facultad de Ciencias Sociales, 1989. 41 p.: bibl. (Serie Cuadernos de investigación; 10)

Examines prevalence and meaning of rituals surrounding house construction in the face of modernization and migration to urban areas (Ayacucho sample, 1986). Rituals maintain identity for some and social connections for others in urban areas, and generate cash for the contractor. Collective cooperation and mutual aid are no longer present. Form is separated from its traditional content.

4899 Delpino, Nena. Saliendo a flote: la jefa de familia popular. Lima: Fundación Friedrich Naumann; Taller de Capacitación e Investigación Familiar, 1990. 135 p.: bibl.

Life histories and participant observation (1980–85) of nine female household heads and three women from male-female Lima slum households are presented. Discusses situations, survival strategies, and potential as actors for social change through new women's organizations. Female household heads have or are moving toward a progressive social ideology which they are transmitting to their children.

4900 Escamilo Cárdenas, Simón. Producción e intercambio en el valle de Chusgón: estrategias de la economía campesina en la sierra de La Libertad. Lima: Univ. Nacional Mayor de San Marcos, Facultad de Ciencias Sociales, 1989. 150 p.: bibl., ill., maps.

Peasant communities such as those in Sánchez Carrión province linked systems of production with changing forms of social organization, relating organization of work and capital to land distribution. Collective work and reciprocity remain, but in forms very different from the traditional *minga;* such systems are now employed by the powerful to exploit the labor of other community members. Religious festivals are designed in part to overcome internal inequalities in these communities.

4901 Espinosa, Cristina. Implicancias del género en el proceso de cambio técnico en sistemas de producción andinos. (*Rev. Peru. Cienc. Soc.*, 3:1, 1992, p. 67–97, bibl., tables)

Division of labor by gender and age has created complex, complementary mixed-crop and livestock systems in the Peruvian Andes. Detailed quantitative analysis by agricultural enterprise of who does what suggests need to involve women more fully in generation and transfer of agricultural technologies.

4902 Ferradas, Pedro. Quirio, prevención de desastres, tradición y organización popular en Chosica. Lima: PREDES, 1992. 93 p.: bibl., ill., maps (some col.).

Vulnerability to natural disasters both limits and facilitates community development in a slum suburb of Lima. Multidisciplinary professional study examines earthquake risk in the area, as well as urban and socioeconomic factors which effect vulnerability, including community organization, cultural tradition, and level of awareness of potential for disaster. Proposes an educational and technical response.

4903 Gavagnin Taffarel, Osvaldo. Estilos de desarrollo y realidad nacional, siglos XIX y XX: ensayos sociológicos. Lima: Arius, 1989. 199 p.: bibl., ill.

Describes a socioeconomic model linking underdevelopment to export of primary products and uses it to track major historical periods, beginning with guano export. Through an examination of the evolving forms of productive organization, traces power groups who benefitted from underdevelopment and who collaborated with foreigners to maintain the system. Relation of Peruvian industrialization to the world system is analyzed over time and is correlated to different forms of worker organization and to relations with the State.

4904 Gestión popular del hábitat: 7 experiencias en el Perú. Lima: Comisión Hábitat—ALTERNATIVA, CENCA, CIDAP, CIPUR, DESCO, IDEAS, IPADEL, SEA, 1991? 229 p.: bibl., ill., maps.

Popular participation in urban housing development is examined through seven case studies throughout urban Peru. In response to a failure of the State, strong grassroots neighborhood movements with city-wide influence arise. Their effectiveness requires establishing new, horizontal means of working with government bodies and among neighborhood associations, thereby creating effective new social actors despite dispersed organizations and decreased income due to economic crisis and ensuing restructuring.

4905 Gestión popular en salud: organizaciones no gubernamentales de desarrollo y políticas sociales; algunas experiencias peruanas. Edición de Carolina Carlessi. Lima: Alternativa; CESIP, 1990. 178 p.

Describes role of NGOs in development of health policy. Based on popular participation, these organizations have redefined dialogue between politics and civil society, creating important space for grassroots organizations. This approach has resulted in changing roles for professionals, as fiscal crisis of the State forces them to redesign health care delivery in response to local organizations. Much of the innovation comes from the organizations, as illustrated by case studies.

Gonzales de Olarte, Efraín *et al.* La lenta modernización de la economía campesina: diversidad, cambio técnico y crédito en la agricultura andina. See item **1943.**

4906 González, José Luis. La religión popular en el Perú: informe y diagnóstico. Cusco, Peru: Instituto de Pastoral Andina, 1987. 397 p.: bibl.

Massive 1981 nationwide study of religion is based on 2,219 survey responses, 107 in-depth interviews, 25 interviews with pastors, and 33 life histories. Examines beliefs (God, Christ, the Virgin, saints) and practices (sin, social action, sacraments). Popular religiosity is more powerful than established church authority in everyday life. Focuses on symbols and identity of popular religiosity and its importance for community.

4907 Grandón, Alicia. Discriminación y sobrevivencia. Lima: Pontificia Univ. Católica del Perú; Fundación Friedrich Naumann, 1990. 182 p.: bibl.

Domestic and extra-domestic work of poor women is analyzed through empirical study of El Agustino, a working-class neighborhood. Women's reproductive role has been made much more difficult by economic crisis; in such circumstances traditional division of labor by gender has intensified domestic oppression. Occupational discrimination keeps women in abusive situations. Women's organizations have grown markedly in El Agustino; by addressing problems related to gender and class they have reduced oppression for their members.

4908 Grompone, Romeo. El velero en el viento: política y sociedad en Lima. Lima: Instituto de Estudios Peruanos, 1991. 219 p.: bibl. (Serie Urbanización, migraciones y cambios en la sociedad peruana; 12)

Organizations of poor youth, informal sector workers, and women are analyzed and contrasted to political parties and their impact on civic culture. Traditional political parties have failed to articulate interests of these groups. Using secondary data, analyzes new social actors represented by these organizations. Concludes that new space is opening, although it is not institutionalized or well-defined.

4909 Guillén Velarde, Rosa and **Verónica de Kwant.** Ganarse la vida y el respeto: proyectos productivos y mujer rural; módulo de capacitación. Lima: Red Nacional Mujer Rural, C.M.P. Flora Tristán, 1991. 60 p.: bibl., ill. (Módulos de capacitación)

Different production project strategies for rural women, based on women's demands for income and economic integration, are related to women's strategic and practical needs. Projects are contextualized and practical social and economic aspects are considered, with case examples. Focuses on participation, respecting cultural differences, generating income and not welfare projects, and creating autonomous organizations.

4910 Gutiérrez Neyra, Javier. Los que llegaron después—: estudio del impacto cultural de las denominaciones religiosas no católicas en Iquitos. Iquitos, Peru: Centro de Estudios Teológicos de la Amazonía (CETA), 1992. 381 p.: bibl., ill.

Impact of non-Catholic religious groups in Iquitos is examined through direct observation, inventory of religious groups, analysis of documents, interviews, surveys, and participant observation. New religious groups including mainline Protestants, other Christian groups (Jehovah's Witnesses, Mormons, Adventists), cults (Mahikari, Seicho No, Mission Rama, Black and White Brotherhoods), and messianic-millenarian movements are compared to the Catholic Church in the Amazon region. Analyzes conditions for growth of these groups and their general impact.

4911 Huamantinco Cisneros, Francisco. Los refugiados internos en el Perú: un estudio de aproximación en dos asentamientos humanos de Lima. Lima: Ediciones y Publicidad de Comercio Exterior, 1990. 69 p.: bibl.

Beginning in 1980, deteriorating social and economic conditions in Ayacucho, particularly the rise of Sendero Luminoso and subsequent military repression, created a large group of refugees that migrated to Lima and formed their own communities there. Suffering continued in Lima, marked by little material progress and increasing economic uncertainty. Neither the State nor leftist parties provided appropriate responses.

4912 INTERCAMPUS. Meeting, *29th, Lima, 1990.* Población: presente y futuro del Perú. Lima: Univ. del Pacífico, Centro de Investigación, 1991. 279 p.: bibl., ill.

Discusses characteristics of Peruvian population in light of late 1980s crisis. Particular focus is on women and young people in the face of failed development models. Policy alternatives focus on interaction between employment, development, and population distribution, while examining efficacy of models based on family planning, health, and education. Extrapolation shows that model based on education is most effective in reducing population, which suggests that greatest attention should be given to youth, particularly young women.

4913 Larrea, José Enrique. Poblaciones urbanas precarias: el derecho y el revés; el caso Ancieta Alta. Lima: Servicios Educativos El Agustino Ediciones, 1989. 104 p.: bibl., ill., map. (Serie Creación; 1)

Empirical study examines old neighborhood in Lima. Housing and spatial arrangements typical of slum neighborhoods represent social bonds and identity. Land is a source of social cohesion for some residents, and of capital speculation for others. The neighborhood organization serves internally to preserve and reproduce traditional power, and externally to defend the neighborhood against State claims on its terrain. *Criollo* leadership's links to indigenous base are weak, reducing organizational effectiveness.

4914 Lenten, Roelie. Cooking under the volcanoes: communal kitchens in the southern Peruvian city of Arequipa. Amsterdam: CEDLA, 1993. 216 p.: bibl., ill. (Latin America studies; 68)

Focuses on degree of autonomy of the *comedores* (communal kitchens), where women take turns preparing the daily meal for their families. Also examines internal organization, and women's motives for participating in a *comedor*. Includes short summaries in Spanish and Dutch. [R. Hoefte]

4915 Loza, Martha et al. Así, ando, ando como empleada. Lima: Instituto de Publicaciones, Educación y Comunicaciones José Cardijn; Asociación de Capacitación y Servicio de Trabajadoras del Hogar, 1990. 232 p.: bibl., ill.

Examines domestic service, including recruitment and working conditions, using historical analysis, census data, special surveys, and case studies. Presents the need for organization and blockages women face when attempting to organize. Formation of organizations of domestic employees is analyzed through case studies of those organizations and life histories of their leaders. Concrete methods of how to organize are presented.

4916 Mastro V., Marco del. Los hilos de la modernización: empresarios agrarios en Chincha. Lima: Centro de Estudios y Promoción del Desarrollo, 1991. 165 p.: bibl., map.

Businesspersons in Chincha, a coastal area of high migration, are analyzed as new social actors. Initially Italian immigrants and large landowners, they lost property in the agrarian reform. Organizing and protesting through new, powerful organizations, they redefined their economic and social activity away from agriculture (a process begun prior to the agrarian reform).

4917 Mendoza, Iván. La promoción y el desarrollo rural desde las ONGD: enfoques y experiencias en el Perú, 1975–1990. (*Debate Agrar.*, 13, enero/mayo 1992, p. 221–244)

Almost all grassroots movements in rural Peru now have close ties to non-governmental development organizations (NGDOs). Analyzes movements created by NGDOs by period of activity, noting their evolution by 1990s to centers of development, services, and technical assistance. Options for such centers, with their implications, are discussed.

4918 Mora, Hernán. Autogestión y capacitación en el Perú: las cooperativas agrarias de producción en el Valle Sagrado de los Incas. San José: EUNED, 1991. 151 p.: bibl.

Addresses self-management (*autogestión*) theoretically in Peruvian context, based on training needed to implement the system. Using the Cooperativa Agraria de Producción (CAP) of the Valle Sagrado de los Incas as a case, analyzes the training offered to the peasants there, and uses the institutions and course content to design relevant training.

4919 Morgan, María de la Luz et al. Sistematización, propuesta metodológica y dos experiencias: Perú y Colombia. Lima: CELATS, 1991. 165 p.: bibl., ill. (Nuevos cuadernos, 0258–2678; 17)

Comparative case studies of community-based social action in Peru and Colombia are based on need for social change agents to document their processes. Peruvian case is popular health education, while Colombian case involves housing. Addressing integrated health issues, rather than single health problems, proved most effective in involving community organizations in health education.

4920 Morimoto, Amelia. Población de origen japonés en el Perú: perfil actual. Lima: Comisión Conmemorativa del 90 Aniversario de la Inmigración Japonesa al Perú, 1991. 217 p.: bibl., ill., maps.

Description of Peruvian population of Japanese descent was published in celebration of initial Japanese migration there. There were 45,644 Japanese-Peruvians in 1989, mostly second or third generation. Each generation is analyzed separately. Japanese-Peruvians are concentrated mainly around Lima. An increase in educational level and a decrease in fertility were noted for each succeeding generation. They tend to be self-employed professionals, or employed in business or education, and are primarily Catholic (92%). One-half speaks Japanese at home, and one-quarter appreciates Japanese music. There has been considerable acculturation.

4921 El movimiento popular de mujeres como respuesta a la crisis. Selección y edición de Helen Orvig. Recopilación de Gladys Cámere y Rosario Bustamante. Lima:

Centro de Documentación sobre la Mujer, 1992? 1 v.: bibl., ill. (Paquete informativo; 2)

Women in poor neighborhoods responded to economic crisis and accompanying terror and violence by organizing in groups. Groups focused first on family survival, moved to community support, and then began to influence urban politics, eventually becoming a new political force. Soup kitchens were organized without outside support in the neighborhoods, although later they were supported by NGOs. The *vaso de leche* governmental program was implemented—and taken over—by groups of neighborhood women. These groups were not homogeneous: migrant women formed their own groups and were excluded from others. Participation in these groups often resulted in intrafamilial conflicts, which led the women to realize the cost of becoming an agent of change—and the need to change women's position as well.

4922 Movimientos sociales: elementos para una relectura. Lima: Centro de Estudios y Promoción de Desarrollo, 1990. 255 p.: bibl., ill.

Movements of four sets of social actors—entrepreneurs, slum dwellers, workers, and poor women's neighborhood associations—are reinterpreted in light of redemocratization. Analyzes their complexity as well as their relationships with the State and other social actors. Focus is on dialectic of structure and collective agency. Emphasizes internal structure and symbolic, cultural, and ethnic aspects of movements.

4923 Mujer en el desarrollo: balance y propuestas. Edición de Patricia Portocarrero. Lima: Innovación y Redes para el Desarrollo; Flora Tristán Centro de la Mujer Peruana, 1990. 291 p.: bibl., ill.

Four essays on women in development, written by feminists. Using secondary data, authors present current situation of women in Peru, along with different ideological frameworks for improving their conditions. Government policy and activities of NGOs are analyzed in terms of approach: 1) welfare, and 2) individual and collective income gencration projects. Programs which built women's organizations often became multifaceted and a vehicle for organizing to promote a wide variety of social changes.

Muller, Frits; Emma Rubín de Celis T.; and **Alfredo Rurizo Callejas.** Pobreza, participación y salud: casos latinoamericanos. See item **4430.**

4924 Olivera Cárdenas, Luis; María del Carmen Piazza; and **Ricardo Vergara.** Municipios: desarrollo local y participación. Lima: DESCO, 1991. 164 p.: bibl., maps. (Cuadernos Desco; 16)

Rural *municipios* in Huaylas are increasing in importance as a result of agrarian reform and decentralization. Grassroots participation in municipal affairs began as a specific survival strategy oriented to health and food distribution. Because of weak municipal governments, such organizations took on important political as well as social service roles, suggesting new models of development with new social actors.

4925 Ortiz, Jorge and **Elsa Alcántara.** Cambios en la fecundidad peruana. Lima: Centro de Investigación en Población; Cusco, Peru: Univ. Nacional San Antonio Abad del Cusco, 1988. 103 p.: bibl., ill.

Based on national sample taken in 1969–70, demographic models of fertility were tested and used to predict changes in fertility between 1969–78. Changes in intermediate variables, including nuptuality, contraceptive use, and breast-feeding, explained most of the variance. Fertility decline has not been uniform however, and is influenced by region, degree of urbanization, and women's education.

4926 Pásara, Luis *et al.* La otra cara de la luna: nuevos actores sociales en el Perú. Buenos Aires: Manantial, 1991. 208 p.: bibl. (Col. Sentido común y política)

Describes new social actors: poor women who have organized community soup kitchens; microindustrials working in the informal sector; and neoindigenous peasant self-defense patrols which have replaced the State. Their linkages to burgeoning number of nongovernmental development organizations (NGDOs) are presented and discussed as a replacement for the State, but these ties do not provide democratic accountability.

4927 Pimentel Sevilla, Carmen. Familia y violencia en la barriada. Lima: Talleres Infantiles Proyectados a la Comunidad, 1988. 185 p.: bibl., ill.

A 10-year empirical study of family life in a poor area of Lima reveals ubiquitousness and complexity of family violence resulting from interplay of class and patriarchy. Analyzes interfamily as well as intrafamily violence. Conditions of authoritarianism and its physical consequences are presented in terms of clinical work aimed at improving family relations.

4928 Ploeg, Jan Douwe van der. Sistemas de conocimiento, metáfora y campo de interacción: el caso del cultivo de la patata en el altiplano peruano. (*Agric. Soc.*, 56, julio/sept. 1990, p. 143–166, bibl., tables)

While some similarities exist, indigenous and scientific knowledge systems vary in key ways. Interaction between knowledge systems is influenced by complex power relations and different political and economic interests. For potato cultivation in the Altiplano, indigenously-derived technology often works better in practice than does technology derived from scientific knowledge.

4929 Pobreza urbana: interrelaciones económicas y marginalidad religiosa. Edición de Marcel Valcárcel C. Lima: Pontificia Univ. Católica del Perú, Facultad de Ciencias Sociales, 1990. 201 p., 4 folded leaves of plates: bibl., ill.

Examines growth of Lima's informal sector, its composition (based on 1987 survey), links to formal sector industries, and relation of informal sector participation vis-à-vis new religious groups. Microindustries, begun in response to unemployment, purchase inputs from formal sector and sell to very poor. Lack of market prevents expansion of such industries. Millennial movement and attraction to cults on the part of the most marginalized are related to same conditions, constituting an alternative to economic responses.

4930 Portocarrero Maisch, Gonzalo; Isidro Valentín; and Soraya Irigoyen. Sacaojos: crisis social y fantasmas coloniales. Lima: TAREA, 1991. 263 p.: bibl., ill.

Turbulent times that created insecurity and fear provided the setting for urban myth of gringo doctors who use black bullies to kidnap children and take out their eyes. This myth created panic in Lima's *pueblos jóvenes*, and reached Ayacucho. Narrative uses intensive interviews to reconstruct the panic and to reveal ethnic-racial consciousness.

4931 Radcliffe, Sarah A. Multiple identities and negotiation over gender: female peasant union leaders in Peru. (*Bull. Lat. Am. Res.*, 9:2, 1990, p. 229–246, bibl.)

Gender relations can create gender-related interests expressed in political activity. Ethnic and peasant identities give women power within Peruvian peasant confederations. Women in peasant unions utilize these identities, stressing gender complementarity and need to deal with women's agendas to reinforce, not replace, peasant and ethnic identity.

4932 Radcliffe, Sarah A. The role of gender in peasant migration: conceptual issues from the Peruvian Andes. (*in* Different places, different voices: gender and development in Africa, Asia and Latin America. Edited by Janet H. Momsen and Vivian Kinnaird. New York: Routledge, 1993, p. 278–287)

Uses gender as a major conceptual tool in understanding organization of peasant migration due to gender-based criteria in allocation of household labor. In southern Peruvian peasant communities, men and women have different destinations, undertake different work, and maintain different ties with rural households. Presents central lines of inquiry to determine difference in the way migration is gendered.

4933 Raíces y bosques: San Martín, modelo para armar. Edición de Andrew Maskrey, Josefa Rojas y Teócrito Pinedo. Lima?: Tecnología Intermedia, 1991. 235 p.: bibl., ill., maps.

Describes development and change in San Martín department (encompassing high and low jungle) from precolumbian times to present. A major coca-growing area, its development was spurred by financial crisis of 1980s. Proposes a plan for urban development that takes into account coca and the violence associated with it. Growth of grassroots organizations culminated in a strike in 1990 which focused on creating an autonomous region and seeking solutions to agrarian crisis.

4934 La reforma agraria peruana, 20 años después. Edición de Angel Fernández de la Gala y Alberto Gonzales-Zúñiga Guzmán. Chiclayo, Peru: Centro de Estudios Sociales Solidaridad; CONCYTEC, 1990. 350 p.: bibl.

Examines current (1989) status of agrarian reform in context of rural violence and economic restructuring. Analyzes shift from State control to decentralization through peasant mobilization, including conflicts among popular groups in sugar industry. Despite little improvement in living conditions, reform was a partial success, increasing participation of peasantry through producer and community oragnizations.

4935 **Revesz, Bruno.** Agro y campesinado: coyuntura [i.e., coyunturas] nacionales y perspectiva regional. Piura, Peru: Centro de Investigación y Promoción del Campesinado, 1989. 237 p.: bibl., ill.

Presents historical examination of peasantry of Piura in light of regional agricultural development. Peasant organization and agrarian crisis are linked. Also discusses agrarian reform and its implications for peasantry. Few peasants were reform beneficiaries, and emphasis on export agriculture increased class divisions. These factors, combined with rapid population growth, led to deterioration in living conditions.

4936 **Rivera Agüero, Rigoberto.** Reforma agraria y transformaciones sociales y económicas del campesinado andino. (*in* El campesino contemporáneo: cambios recientes en los países andinos. Edición de Fernando Bernal C. Bogotá: Tercer Mundo Editores, 1990, p. 234–278, bibl.)

Analysis of changes in rural Peru through regional and national social and economic relations of the peasantry demonstrates their active roles in effecting change, particularly through building support networks. Division of labor within the household by age and gender—but also including migrants and temporary migrants—is critical to this process, which increases household diversification and resilience.

4937 **Schibotto, Giangi.** Niños trabajadores: construyendo una identidad. Lima: Movimiento de Adolescentes y Niños Trabajadores Hijos de Obreros Cristianos; Instituto de Publicaciones, Educación y Comunicación, 1990. 436 p.: bibl., ill. (Publicaciones MANTHOC; 5)

Informal sector work for pay is increasingly part of childhood in Lima, and critical for household survival. Presents gender differences, attempts to organize child workers, and organizations that offer them support. *Niños y adolescentes trabajadores* (NATs) are theorized to be integral to dependent-model development in Peru, requiring a reconceptualization of both childhood and intersection of class and age.

4938 **Segura Altamirano, José; Sara de Jesús D.;** and **Lindaura Rodríguez T.** Diagnóstico: comunidad campesina "San Francisco de Asís" de Salas. Chiclayo, Peru: Centro de Estudios Sociales Solidaridad, 1990. 129 p.: ill. (Documentos de trabajo)

Peasant community of San Francisco de Asís began to organize with the help of Centro de Estudios Sociales Solidaridad. A participatory diagnosis was carried out, leading to a number of community development projects related to agriculture and livestock. All were designed to bring about wide participation and decentralization, and to maintain natural resources and increase income and community well-being.

4939 **Seminario-Taller sobre Agricultura Andina y Proyecto Campesino, 1st, Ayacucho, Peru, 1987.** Manchay tiempo: proyectos de desarrollo en tiempos de temor en Ayacucho. Lima: PRATEC, 1989. 317 p.: bibl., ill. (Serie Eventos de técnicos)

Despite a climate of great fear in Ayacucho during 1980s, a number of NGOs were able to carry out development projects working with peasant communities. Using a holistic systems approach to agricultural development, participatory research based on indigenous knowledge helped foment agricultural change.

4940 **Los servicios del Estado en los pueblos jóvenes y la participación ciudadana: situación actual y perspectivas.** Lima: Instituto Nacional de Administración Publica, Centro Superior de Investigación en Desarrollo Administrativo Integral, Dirección Ejecutiva de Investigación Político Social, 1988. 81 p.

Based on interviews, assesses quality of education, health, housing, and basic services delivered by public sector to Lima slums. Services of all kinds were felt to be inadequate. Determined possibilities for mobilizing grassroots organizations. Building service improvement on already existing neighborhood organizations has greatest potential for increasing service quality.

4941 Solorio P., Fortunata and **Esther Revilla C.** Enfoques sobre alimentación andina. Puno, Peru: Centro de Proyectos Integrales Andinos, 1992. 147 p.: bibl., ill. .

Analyzes food resources in a peasant community in Juliaca-Puno in 1990–91. Women's organizations and their use of food-for-work projects are related to income generation, marketing agricultural products, and revival of indigenous technologies. Although peasant food production is inadequate to supply families year-round, peasants do produce food for a variety of reasons: for consumption, to improve nutritional status, and to uphold culture surrounding food, which in turn serves to maintain cultural identity and to insure food security.

4942 Tocón Armas, Carmen and **Armando Mendiburu Mendocilla.** Madres solteras, madres abandonadas: problemática y alternativas. Chimbote, Peru: La Casa de la Mujer, 1990. 146 p.: bibl., ill.

Single and abandoned mothers constitute a large and increasing portion of the population—over 10 percent in late 1980s—of slums and urban zone of Chimbote. Examines their situation and possible solutions, using a household census, survey data, and in-depth interviews with both mothers and leaders of grassroots organizations that have developed to meet their needs. Their major need is better employment to increase their income, which would be facilitated by provision of day care. More ways to organize in their own self-interests and access to contraceptives are also needed.

4943 Trahtemberg Siederer, León. Demografía judía del Perú: un estudio demográfico vocacional y de actitudes hacia lo judaico de la comunidad judía de Lima, Perú. Lima: Unión Mundial ORT, 1988. 114 p.: bibl.

Demographic study of Peruvian Jewish community was completed to predict enrollment at the Lima Jewish school. Decreasing fertility and out-migration due to problems of insecurity and the economy imply decreasing enrollments despite strong maintenance of Jewish identity and relatively high socioeconomic status of the Jewish community.

4944 Varillas Montenegro, Alberto and **Patricia Mostajo de Muente.** La situación poblacional peruana: balance y perspectivas. Lima: Instituto Andino de Estudios en Población y Desarrollo, 1990. 524 p.: bibl., ill.

History and projections of Peruvian population's change in size, distribution, and characteristics are systematically analyzed through surveys, historical documents, census, and vital statistics, and are related to political context. Rapid population growth was ignored by government, political parties, and the Church until 1960s. Policies and programs are now in place, and desired family size has decreased. Yet demand for family planning services has not been met, and problem of abortion has not been confronted.

4945 Velarde, Julio. Desarrollo rural en la sierra del Perú: análisis y perspectivas. Lima: Asociación Acción y Pensamiento Democrático, Instituto José Faustino Sánchez Carrión, 1988. 132 p.: bibl., ill. (Realidad nacional; 3)

Describes general situation of rural Andean region, using secondary data. Focuses particularly on different forms of organizing production (peasant communities, reformed collective enterprises, and small, medium, and large holders). Within the setting—which includes discussion of technology, crops, and credit—presents family and community characteristics. Stresses importance of community as source of identity even for migrants, and as an intermediary for social and economic activity.

4946 Vildoso Chirinos, Carmen. Sindicalismo clasista: certezas e incertidumbres. Lima: Edaprospo, c1992. 189 p.: bibl.

Labor unions in Peru became increasingly "classist" during 1970s, replacing clientelist unions. As a result, their struggles are more political and involve more than workers in a particular firm. New communication networks were established between *criollos* and *serranos*, union leaders and intellectuals, young people in and out of unions, and male and female workers, thereby creating a new, classist identity and set of mobilizing symbols. New leaders emerged in 1980s, favoring wide participation and negotiation with the company to defend both the union and the enterprise.

4947 **Vries, Jaap de.** Planificación urbana y participación popular: el caso del Cusco. Cusco, Peru: Instituto de Investigación UNSAAC-NUFFIC; Servicio Holandés de Cooperación Técnica, 1991. 186 p.: bibl., ill. (some col.), maps.

Examines potential of popular participation to influence urban planning, using Cusco as case study. Despite good mobilization of traditional Andean organizations and newer grassroots organizations, lack of efficient planning structures and unstable political situation made grand results impossible. Modest planning and implementation was carried out, and participatory process begun.

4948 **Zurita, Dante** and **Víctor Caballero Martín.** Puno: tierra y alternativa comunal; experiencias y propuestas de política agraria. Lima: Tall. Gráf. de Asociación Gráf. Educativa Tarea, 1991. 157 p.: bibl., ill.

Analyzes economy of Puno, a very poor mixed agricultural and livestock area of the Altiplano. Alternative development plans are presented in light of insertion of the local economy into the regional one. The area's economic potential, risk and uncertainty, technological change, extension and training, credit, and agroindustry are examined in relation to agrarian reform and new forms of peasant organization.

BOLIVIA AND PARAGUAY

KEVIN J. HEALY, *Foundation Representative, Inter-American Foundation, Rosslyn, Virginia*

PUBLISHED WORKS ANNOTATED FOR this section attest to the plethora of studies on gender and women's topics in both Bolivia and Paraguay. The new research reflects an increase in the number of female researchers as well as of female-headed NGOs working with women's groups and analyzing the results of action programs (items **4966, 4983,** and **4978**). Some of the research is pathbreaking, such as the first comprehensive study of domestic workers in a Bolivian city (item **4959**) and studies of prostitution (item **4991**) and sexual violence (item **4993**) in Paraguay. Others reflect the priority research required to satisfy programming needs of international and national developmental and population institutions in both urban as well rural areas (items **4960** and **4967**).

Several works from Bolivia point to the growing attention devoted to issues about native peoples, such as Amazonian nationalism (item **4956**), conditions of internal colonialism (item **4979**), and native knowledge and crops (items **4970** and **4961**). Another important book shows the reverse of this cultural coin by demonstrating how traditional Andean cultural practices can be used for class domination and exploitation in Highland communities such as Cochabamba (item **4962**). Two new books on coca and cocaine from the Chapare area of Bolivia are based upon fieldwork from the late 1970s (item **4981**) and early 1980s (item **4972**), presenting an interesting overview of social, economic, and political changes for over a decade. In contrast to previous years, when Paraguay's rural political economy literature was dominated by agrarian studies, for this biennium two of the most important books on Paraguay examine rural power structures (item **4987**) and peasant organizational development in recent decades (item **4984**).

BOLIVIA

4949 **Análisis de la realidad migratoria en Bolivia.** La Paz: Secretariado Nacional de Pastoral Social; Conferencia Episcopal Boliviana, 1991. 78 p.: bibl., ill.

Rather cursory and very sketchy treatment of rural-urban migration trends and underlying factors, written by a group of researchers from the Bolivian Bishops Conference.

4950 Antezana J., Luis H. La diversidad social en Zavaleta Mercado. La Paz: CEBEM, Centro Boliviano de Estudios Multidisciplinarios, 1991. 166 p.: bibl.

Analyzes some of the writings of distinguished Bolivian intellectual Zavaleta Mercado on revolution and popular participation in capitalist democracies.

4951 Antezana Villegas, Mauricio. El Alto desde El Alto II. La Paz: UNITAS, 1993. 356 p.: ill. (some col.).

Compendium on city of El Alto, which broke off from capital city of La Paz to form its own municipal government, contains maps, list of barrios, and demographic, economic, health, and migration data derived from surveys and official and other secondary sources. Useful and essential guide for a quick overview of many important characteristics of this rapidly growing Altiplano city. See also item **4954.**

4952 Archondo, Rafael. Compadres al micrófono: la resurrección metropolitana del ayllu. La Paz: Hisbol, 1991. 254 p.: bibl., ill. (Serie Movimientos sociales; 6)

Provides descriptive analysis of phenomenon of *palenquismo*, referring to Carlos Palenque's cultural and political trajectory from media celebrity to politician and presidential candidate. Author places his topic in broader context of Andean studies literature on social movements.

4953 Ayllón, Virginia and **Fernando Machicado.** De tanto haber andado yo ya soy otra: bibliografía de la mujer boliviana, 1986–1991. La Paz: Centro Documental de la Mujer Adela Zamudio, 1991. 110 p.: ill., index.

Current annotated bibliography lists publications by Bolivians relating to women in their country. Thematic subdivisions for individual chapters include health, politics, donated food aid, communications, law, and oral history.

4954 Beijnum, Paul van and **Ronald H. Kranenburg.** De zone 16 de Julio na het HAM-BIRF Project: de effecten van buurtverbetering in El Alto, Bolivia [The 16 de Julio area after the HAM-BIRF Project: the effects of neighborhood improvement in El Alto, Bolivia]. Utrecht, The Netherlands: Faculteit Ruimtelijke Wetenschappen Univ. Utrecht, 1993. 189 p.: bibl., ill., maps, tables. (Diskussiestukken van de Vakgroep Sociale Geografie van Ontwikkelingslanden; 48)

Looks at effects of upgrading in general, and on mobility of households in particular, during period 1984–89. Does improving the infrastructure attract more affluent citizens while pushing out poorer individuals? Authors analyze social and economic status of stayers and movers. Includes summary in Spanish. See also item **4951.** [R. Hoefte]

4955 Criales, Lucila. El amor a piedra: relaciones de subordinación en la pareja aymara urbana. La Paz: Centro de Promoción de la Mujer Gregoria Apaza, 1994. 117 p.: bibl. (Mujer y cultura)

Short study and set of reflections considers situation of Aymara migrant women within a family in a barrio of El Alto. Depicts multiple forms of subordinación y discriminación, both within the family and in public sphere of this new milieu.

4956 Debate Regional, 5th, Santa Cruz, Bolivia, 1991. La situación de los indígenas en la Amazonia boliviana. La Paz: ILDIS; UAGRM; C.C.C., 1991. 59 p.: bibl.

Short compilation of remarks by Bolivian intellectuals considers some of the economic and political issues of indigenous peoples in the Amazon.

4957 Debate Regional, 6th, Santa Cruz, Bolivia, 1992. La situación de la mujer en la ciudad de Santa Cruz. La Paz: ILDIS; UAGRM; C.C.C., 1992. 79 p.: bibl., ill.

Edited transcript of seminar held in Santa Cruz discusses basic demographic, economic, educational, and income data. See also item **4967.**

4958 Dibbits, Ineke. Lo que puede el sentimiento: la temática de la salud a partir de una experiencia de trabajo con mujeres de El Alto Sur. La Paz: TAHIPAMU, 1994. 194 p.: bibl.

Discussion of experiences in a community health program with Aymara indigenous women in city of El Alto is inserted within a wider literature of similar experiences elsewhere. Contains discussion of approaches to health care using Andean health concepts in addition to Western approaches. Also discusses family planning programs and reproductive rights in Bolivia. See also item **4966.**

4959 Gill, Lesley. Precarious dependencies: gender, class, and domestic service in Bolivia. New York: Columbia Univ. Press, 1994. 175 p.: bibl., ill., index.

Pathbreaking book examines an understudied social group in Bolivian social science literature: domestic workers in the capital city of La Paz. Study has historic depth and is placed within broader context of social class and ethnic relations in Bolivia and its capital city. Contains many rich insights and findings. For ethnologist's comment see item **952**.

4960 Gondrie, Peter; Cristina Mejía' and N.C.M. van Niekerk. Políticas sociales y ajuste estructural: Bolivia, 1985–93. La Paz: CID; COTESU; MCTH, 1993. 94 p.: bibl., ill.

Three articles address state of social development in Bolivia. Topics considered are: 1) social impact of structural adjustment policies on low-income sectors; 2) performance of public health sector, including comparison with other Latin American countries; 3) new educational reform law promulgated in 1993.

4961 Kietz, Renate. Compendio del amaranto: rescate y revitalización en Bolivia. La Paz: ILDIS, 1992. 175 p.: bibl., ill., maps.

German author provides rich body of information on native amaranth plant and its revival as an important cereal by peasant farmers and nongovernmental development organizations in Bolivia. Offers a view of plant's history and nutritional aspects, and argues for its suitability as an alternative to food donated by international aid programs in Bolivia.

4962 Lagos, Maria L. Autonomy and power: the dynamics of class and culture in rural Bolivia. Philadelphia: Univ. of Pennsylvania Press, 1994. 206 p.: bibl., index, maps. (The Ethnohistory series)

Important study examines peasant landowning history, agrarian structure, and politics in Tiraque canton of highland Cochabamba. Author shows how, from post-agrarian reform through 1980s, cultural practices such as reciprocity and barter, community and ritual kin, relations of exploitation, and domination of the poor by wealthier families are manipulated. For ethnologist's comment see item **956**.

4963 Lindert, Paul van. Huisvestingsstrategieën van lage-inkomensgroepen in La Paz [Shelter strategies of low-income groups in La Paz]. Utrecht, The Netherlands: Koninklijk Nederlands Aardrijkskundig Genootschap; Faculteit Ruimtelijke Wetenschappen Rijksuniversiteit Utrecht, 1991. 313 p.: bibl., ill., maps. (Nederlandse geografische studies: 0169–4839; 136)

Study of access to housing, land, and collective services for migrants in La Paz is drawn from base-line questionnaire surveys carried out in low-income residential areas in 1980s. Macro focus is complemented by analysis of primary census data. Includes summary in English. [R. Hoefte]

4964 El menor trabajador asalariado en la ciudad de Cochabamba. Cochabamba, Bolivia?: Defensa de los Niños—Internacional, 1989. 77 p.

Survey of 1,300 street children in four main zones of Cochabamba examines work conditions, social characteristics, aspirations, etc. Offers series of interesting and thoughtful conclusions about this workforce.

4965 Montaño García, Jaime. El proceso de urbanización en Bolivia, 1976–1992. La Paz: Ministerio de Desarrollo Humano, Unidad de Política de Población, 1994? 78 p.: bibl., ill.

Modest short comparison of a few demographic change indicators is based on 1976 and 1992 censuses.

4966 Mujer, género y desarrollo local urbano. Edición de Carmen Beatriz Ruiz. La Paz: Centro de Promoción de la Mujer Gregoria Apaza, 1993. 157 p.: bibl.

Series of articles examines women's participation in civil society and municipal government in El Alto, a city adjacent to La Paz. Major focus is on work of the Centro de Promoción de la Mujer Gregoria Apaza in relation to socioecomomic development and on role of Aymara women in urban barrios of El Alto. See also item **4958**.

4967 Mujer, trabajo y reproducción humana en tres contextos urbanos de Bolivia, 1986–1987. La Paz: Consejo Nacional de Población, Ministerio de Planeamiento y Coordinación; S.I.: Pathfinder Fund, 1989. 260 p.: bibl., ill.

Analyzes survey of 2,600 low-income

women in cities of La Paz, Santa Cruz, and Cochabamba. Central focus is on women's participation in urban economy and relationship to childbearing practices. See also item **4957**.

4968 Paredes Candia, Antonio. La chola boliviana. La Paz: Ediciones ISLA, 1992. 644 p.: bibl., ill. (some col.).

Rich, detailed study describes the *chola* (mestizo woman) in Bolivian urban culture. Focuses primarily on 20th century, although also considers earlier historical material. Discusses regional differences in costume styles as shown in use of bowler hats, skirts, shawls, and jewelry. Contains abundant early 20th-century photographs of *cholas* and other graphic descriptions of *chicheras*, cuisine, and humor.

4969 Pobreza y salud en Bolivia. La Paz: ILDIS, 1994. 254 p.: bibl., ill.

Wide-ranging discussion of inequitable access to health care in Bolivia, written by noted authorities on the subject. Analyzes both health conditions and strategies promising improvements in the future.

4970 Raqaypampa: los complejos caminos de una comunidad andina; estrategias campesinas, mercado, revolución verde. Edición de Pablo Regalsky. Cochabamba, Bolivia: CENDA, 1994. 234 p.: bibl., photos, tables.

Very detailed study examines peasant economy in Raqaypampa zone of Cochabamba Mizque region. Team conducted research from 1986–94, collecting an abundance of empirical data on different agricultural technologies, peasant resource management strategies, potato production, and local and regional marketing. Authors make a critique of conventional modernization approaches and argue for alternative strategies based upon Andean cultural resources and environmental awareness. Book is amply documented through tables and photos.

4971 Realidade dos seringueiros brasileiros na Bolívia: pesquisa. Pando, Bolivia: Vicariato de Pando; Rio Branco, Brazil: Diocese de Rio Branco; Ji-Paraná, Brazil: CEPAMI, 1991. 63 p.: maps.

Jointly sponsored research ethnography looks at religious orientation and conditions of Brazilian rubber tappers in Bolivian territory. Also contains maps documenting locations of communities found.

4972 Sanabria, Harry. The coca boom and rural social change in Bolivia. Ann Arbor: Univ. of Michigan Press, 1993. 277 p.: bibl., ill., index, maps. (Linking levels of analysis)

Detailed study examines economic strategies of peasant households tied to coca-cocaine economy in eastern part of Cochabamba. Author shows relationship between falling potato production and increased coca leaf cultivation among those families migrating between the Highlands and Lowlands. Study is framed within broader changes in Bolivian economy and US/Bolivian counternarcotics policy. Author argues with considerable evidence that US/Bolivian drug policies are failing and should be abandoned. For ethnologist's comment see item **966**.

4973 Sandoval Z., Godofredo and **Virginia Ayllón.** La memoria de las ciudades: bibliografía urbana de Bolivia, 1952–1991. La Paz: CEP-ILDIS, 1992. 474 p, 10 p. of plates: ill., maps.

Presents some basic social data, cultural aspects, and historical summaries comparing Bolivian cities. Also includes cartographic maps and an extensive bibliography on urban areas for the period 1952–91.

Seminario Nacional: Mujer, Género y Desarrollo, *1st, La Paz, 1992.* Actas. See item **1995**.

4974 Simposio Recursos Documentales sobre la Mujer en Bolivia, *La Paz, 1990.* Simposio Recursos Documentales sobre la Mujer en Bolivia. La Paz: Centro de Información y Desarrollo de la Mujer, Centro Documental de la Mujer Adela Zamudio, 1991. 82 leaves, 4 leaves of plates: ill.

Contains short papers and commentary by distinguished representatives of various nongovernmental and public research institutions about documentary sources for research on women.

4975 Situación alimentaria y nutricional de Bolivia, 1992. La Paz: INAN, 1993. 119 p.: bibl., ill.

Based on data used by public health programs to monitor height and weight throughout the country, work analyzes nutritional status of Bolivian population, focusing especially on low-income women and children. Makes regional comparisons and describes and analyzes basic facts about industrialization and food production and distribution. Attributes low nutritional levels to macroeconomic policies and changes; and also shows, through a rather conventional analysis, how vicious circle of rural poverty and low productivity contributes to undernutrition.

4976 **Subirats Ferreres, José** and **Ivonne Nogales Taborga.** Maestros, escuelas, crisis educativa: condiciones del trabajo docente en Bolivia. Santiago: UNESCO/OREALC, 1989. 132 p.: bibl.

Provides brief introduction to critical problems facing public primary education in Bolivia. Principal focus is on public school teachers' working and social conditions and educational skills and background.

4977 **Tenencia actual de la tierra en Bolivia.** Recopilación de Mario Arrieta Abdalla. La Paz: ILDIS, 1993. 192 p.

Transcripts of papers and comments from a workshop analyzing current land tenure system in Bolivia. Contributors include Bolivian intellectuals, government officials, World Bank and FAO representatives, and a peasant leader. Provides updated profile of a dualistic agrarian structure, and includes discussion of issues of indigenous territorial rights and forestry laws.

4978 **Tuijtelaars de Quitón, Christiane** and **María Rodríguez Amurrio.** La campesina de Carrasco, 1983–1993. Bolivia: Centro de Investigaciones para Planificación y Desarrollo, 1994. 204 p.: bibl., ill., map.

Documents organization and 10-year evolution of Central de Mujeres de Carrasco, a peasant organization in Carrasco region of Cochabamba. Analyzes both problems internal to the organization and the difficult economic and physical context. Also documents how development institutions fail to recognize important role of women in livestock management and to include women in their training programs for producers. Offers recommendations for working with women in Bolivian rural development programs.

4979 **Violencias encubiertas en Bolivia.** v. 1, Cultura y política [de] Silvia Rivera Cusicanqui y Raúl Barrios Morón. Coordinación de Xavier Albó y Raúl Barrios Morón. La Paz: CIPCA; Aruwiyiri, 1993. 1 v.: bibl. (CIPCA; 38–39)

Several of Bolivia's most lucid social science writers discuss important facets of internal colonialism in relation to various mechanisms of ethnic oppression. Examines formation of a mestizo identity and difficulties and discrimination faced by excluded rural sector when channeling its demands within national political system. Also considers other factors limiting Bolivia's democracy such as role of the military in modern political history and US-inspired "drug war." For political scientist's comment, see item **3355**.

4980 **Weber, Jutta.** Población indígena de las tierras bajas de Bolivia. Santa Cruz, Bolivia: Apoyo para el Campesino-Indígena del Oriente Boliviano, 1994. 53 p.: bibl., ill., maps.

Gives short, introductory profile of the 36 indigenous groups in Lowlands of eastern Bolivia. Briefly discusses local economies, natural resources, social organization, territorial rights, and new indigenous federations pushing for national political and economic reforms to benefit their peoples. Book barely scratches the surface of indigenous issues of eastern Lowlands.

4981 **Weil, Jim** and **Connie Weil.** Verde es la esperanza: colonización, comunidad y coca en la Amazonia. Cochabamba, Bolivia: Editorial Los Amigos del Libro, 1993. 222 p.: bibl., ill., maps. (Col. Texto y documento; NA 1150)

Based on authors' combined field research in the coca-leaf-growing Chaparé region in mid-1970s. Provides detailed analysis of peasant cropping systems, social organization, health strategies, etc. Authors place their study in wider literature on colonization and environmental decline in humid tropical areas of Latin America, especially the Amazon region.

4982 **Zalles Cueto, Alberto Augusto.** Balseros, horticultores itinerantes y barranquilleros: lecos, quechuas y aymaras en tierras de transición; ensayos etnográficos. La Paz: Editorial Ceja del Alto, 1993. 121 p.: bibl., maps.

Rather superficial series of ethnographic and historical essays looks at several ethnic groups in tropical Larecaja region of La Paz.

PARAGUAY

4983 Bareiro, Line and Celsa Vega. Campesinas frente a la pobreza: condiciones de vida de las familias organizadas de la Cordillera. Asunción: Centro de Documentación y Estudios, 1994. 229 p.: bibl., ill., map.

Study examines organization of peasant women and effects of foreign aid programs on their socioeconomic status.

4984 Campos Ruiz Díaz, Daniel and Dionisio Borda. Organizaciones campesinas en la década de los '80: sus respuestas ante la crisis. Asunción: Comité de Iglesias para Ayudas de Emergencia, 1992. 324 p.: bibl., ill.

Analyzes Paraguay's most important peasant organizations in terms of organizing strategies, strengths and weaknesses, and structural barriers to social change which they face. Emphasizes role of NGOs in revitalization of peasant organizations during 1980s. This important work mixes excellent analysis of agrarian structures with changes and developments in peasant organizations of the country.

4985 Fogel, Ramón B. La ciencia y la tecnología en Paraguay: su impacto socioambiental. Asunción: Centro de Estudios Rurales Interdisciplinarios, 1994. 341 p.: bibl., ill., maps.

Critique reveals how adoption of modern technologies in agriculture and hydroelectric power, and in reforestation in the rural areas, has contributed to environmental degradation and social inequities.

4986 Fogel, Ramón B. and Edith A. Pantelides. Determinantes principales de la fecundidad en áreas rurales del Paraguay: el caso de Itapúa. Asunción?: FNUAP; CERI, 1994. 160 p.: bibl.

Socioeconomic survey of 588 households in Itapúa region of rural Paraguay contains interesting data on marriage and reproductive patterns, birth rates, and female out-migration for domestic employment in Argentina.

4987 Galeano, Luis and Fátima Myriam Yore. Poder local y campesinos. Asunción: Centro Paraguayo de Estudios Sociológicos, 1994. 193 p.: bibl.

Examines changes and continuities of local power structures in rural Paraguay resulting from democratic changes underway since fall of military dictatorship in 1989. Analyzes changing forms of dependency. Shows that while the Partido Colorado lost its almost monopolistic grip over the rural population, forms of *caudillismo*—tied to State and economic structures and a traditional political culture of clientelism—persist.

4988 Gómez, Miguel and Susana Sottoli. En paños menores: un estudio basado en los datos de la Campaña por los Derechos del Niño y otras fuentes secundarias. Asunción: Centro de Defensa del Menor; Rädda barnen, 1991. 215 p.: bibl., ill.

Interesting, pioneering national study examines social condition of children in Paraguay. Book's framework is influenced by UN convention on rights of children. Using both primary and secondary data, study covers aspects of status, access to health services, mortality, family life, trafficking of children, parental abuse, youth organizations, and defense of children. Offers recommendations for enhancing children's legal rights and improving treatment by the criminal justice system. Also contains annotated bibliography of other published works on Paraguayan children's issues.

4989 Guanes de Laíno, Rafaela. Familias sin tierra en Paraguay. Asunción: Ñandutí Vive; Intercontinental Editora, 1993. 145 p.: bibl.

Traces history of land ownership patterns and dispossession from precolonial era, through colonial period, and up through 1982. Shows how State's favoritism of latifundios over poor peasant majority in land disputes is a thread running through Paraguayan history. Offers analysis of historical process that created conditions of landlessness among rural population.

4990 Hacia una presencia diferente: mujeres, organización y feminismo. Asunción: CDE, Area Mujer; Solidaridad Internacional; Instituto de la Mujer, 1992. 121 p.: bibl. (Serie Feminismo)

Provides overview of different expressions of feminism in Paraguayan capital through presentation of women's organizations, their public demands, and relations with international organizations.

4991 Haciendo la prostitución: menores prostituídos en la terminal de ómnibus de Asunción. Asunción: CEDEM; GCS, 1992. 52 p.: bibl.

Short book discusses prostitution among minors in the capital city. Contains a sociological profile and personal testimonies of young prostitutes.

4992 Schiavoni, Lidia. Pesadas cargas, frágiles pasos: transacciones comerciales en un mercado de frontera. Asunción: CPES; Posadas, Argentina: Editorial Universitaria, Univ. Nacional de Misiones, 1993. 117 p.: bibl. (Los Tesistas)

Ethnographic study examines economic behavior (survival and employment strategies) of female petty traders in Encarnación (Paraguay)/Posadas (Argentina) border area, on both sides of the Río Paraná. Access to foreign products enables these traders to maintain a relatively higher standard of living than other groups operating in Paraguay's informal commercial sector.

4993 La violación sexual en el Paraguay: aspectos psicológico, social y jurídico. Asunción: Centro Interdisciplinario de Derecho Social y Economía Política, Univ. Católica, 1993. 464 p.: bibl. (Serie Investigaciones; 23)

Broad study of sexual offenses in Paraguay's capital is based on police and first-aid records. Legal and social science scholars as well as pyschotherapists from Asunción's Univ. Católica were involved in the study. Includes interviews with offenders and makes recommendations for changes in public policy and legal codes.

CHILE

MANUEL ANTONIO GARRETON, *Professor, Department of Sociology, Universidad de Chile*
MALVA ESPINOSA, *Researcher, Dirección del Trabajo, Ministerio del Trabajo, Chile*

TWO PRINCIPAL TENDENCIES UNDERSCORE recent developments in Chilean sociology. First, there has been a shift away from global and interpretive studies or essays on Chilean society towards monographic and sectoral empirical studies, with a special emphasis on the methodological and technical aspects of data collection and analysis. Second, global and interpretive studies, in turn, tend to refer to four different processes. These individual processes no longer are encompassed in a single overarching process, but rather each is considered to have its own dynamics or to be related to the others in a non-deterministic way. The first of these processes is the construction of political democracy; the discussion in these studies now centers on the quality and relevance of democracy, rather than the establishment or consolidation of democratic institutions (item **5004**). The second process is social democratization, defined in these works as overcoming social inequalities and extreme poverty (item **5080**). The third concerns the effects of structural economic adjustments and the transition to a new development model (item **5056**). Finally, the fourth concerns the model of modernity, that is, the relationship between globalization, and national and particular identities (item **5052**).

The trends evident in current works are related to the changing role of sociology in Chile; increasingly sociology is regarded as a profession, rather than an intellectual discipline. Various factors have influenced this shift towards professionalization. The independent academic institutions where sociology flourished in the 1980s have weakened during the last decade, while, at the same time, universities continue to suffer ill-effects from the attacks that occurred under the military regime. On a more positive note, a significant number of new schools of sociology have opened in private universities, teaching in the public universities has been reformed, and the number of sociology students has increased. Still lacking, however, is a parallel effort to augment resources for research. With competition from other professions increasing, the further professionalization of the discipline appears likely.

The principal studies in the field (items **5014, 5016,** and **5018**) tend to be compilations, readings, or collected works, rather than monographic works by a single scholar. As a result, very good specialized studies exist providing a solid base of data. These works elaborate or evaluate public policies in several social spheres. Some topics studied in the past are being reconsidered, such as rural transformations (item **5018**). Other relatively neglected areas of study, such as women and gender (items **4429, and 4428**), decentralization and local life (items **5043 and 5066**), and environmental problems and policies (items **5057**) are now receiving greater attention.

Interpretative analysis and debate about the future of society and the social sciences are not part of the mainstream of current Chilean sociology. There are several important critical analyses that examine the social impact of development and modernization. These studies are particularly concerned with social inequalities, discriminations, and repressive behavior patterns. The general cultural climate, however, is one of social and political smugness; the tendency is to avoid discussions of the past and ignore problems of the future. Chilean society and current intellectual thought demonstrate more concern with narrow, short-term issues and pragmatic solutions than with long-term questions and a search for alternative social models. There is an absence of both social debate and effective social actors. Thus, social analyses tend to study sectoral situations, social categories, and public opinion, not the contradictions and conflicts of classic and new actors, and their alternative social models and projects.

4994 Acuña, Eduardo and **Gabriel Valdivieso.** Factores asociados al desempeño de la mujer como dirigente sindical. (*Estud. Soc./Santiago,* 57, trimestre 3, 1988, p. 71–92, bibl., tables)

Estudio de caso que analiza los diferentes factores que inhiben o estimulan la participación sindical de la mujer. Se estudia el impacto de variables como estado civil, maternidad y relaciones familiares, estableciéndose correlaciones tentativas.

4995 Adriance, Madeleine. The paradox of institutionalization: the Roman Catholic Church in Chile and Brazil. (*Sociol. Anal.,* 50, supplement, 1992, p. S51-S62, bibl.)

Using both fieldwork and published sources, author analyzes how the Catholic Church in both countries institutionalizes innovations. Argues that although both allow room for the expression of progressive tendencies, changes nevertheless are derived through conservative means.

4996 Agar, Lorenzo. Los habitantes de Santiago poniente: quiénes son y que piensan en la perspectiva de un proceso de renovación urbana. (*EURE/Santiago,* 17:52/53, oct./dic. 1991, p. 127–142, bibl., tables)

Análisis de los resultados de una encuesta sobre la disposición organizacional y psicosocial de los habitantes de este sector ante las transformaciones derivadas de un proceso de renovación urbana, en la perspectiva de considerar su participación en los niveles de decisión y acción.

4997 Aguilar, Renato. Medición de la discriminación en el mercado laboral. (*Estad. Econ.,* 5, dic. 1992, p. 53–64, bibl.)

Concisa revisión de los principales enfoques metodológico-estadísticos disponibles para el análisis de la discriminación por sexo, raza u origen, entre otras, que basados en las funciones "mincerianas y de frente" apuntan a detectar y cuantificar la presencia de discriminación bajo hipótesis estadísticas y económicas generales.

4998 Alaminos, Antonio. Chile: transición política y sociedad. Madrid: Centro de Investigaciones Sociológicas; Siglo Veintiuno de España Editores, 1991. 170 p.: bibl., ill. (Col. Monografías; 115)

Con apoyo de fuentes documentales como encuestas y publicaciones emitidas por diversos actores, el libro hace un análisis en profundidad de la sociedad chilena en la transición a la democracia, tomando como punto de partida el proyecto de sociedad autoritaria intentado por Pinochet.

4999 Alario, Margarita. Environmental policy enactment under the military: some generalities between Brazil and Chile. (*Int. J. Comp. Sociol.*, 34:3/4, Sept./Dec. 1993, p. 222–230, bibl., table)

Brief critical review of the environmental impact of economic policies of authoritarian regimes in Chile and Brazil. Hypothesizes that the supposed macroeconomic advances of these regimes are based on a pillaging of natural resources and the environment.

5000 Alvarez, Jorge. Los hijos de la erradicación. Santiago: PREALC, 1988. 234 p.: bibl., ill.

Once testimonios de adolescentes configuran un cuadro narrativo claro del proceso de erradicación y sus efectos sociales, desde la perspectiva de los propios afectados. Inserción laboral, socialización, delincuencia, marginalización son algunos de los problemas y contradicciones posteriormente recogidos en comentarios y análisis de expertos.

5001 Arnold, Marcelo; María Teresa Prado; and Vivian Saidel. Análisis del contenido de cuatro estudios de religiosidad popular en Chile. (*Rev. Chil. Antropol.*, 5, 1986, p. 73–101, bibl., table)

Reseña sistemática de los cuatro textos sobre religiosidad popular más consultados por los investigadores del tema, mediante la aplicación de un procedimiento para el control de la calidad de los datos recolectados, que proporcione las bases para una codificación con fines comparativos.

5002 Aylwin Azócar, Arturo et al. Toma de decisiones en el Estado contemporáneo. Edición de Guillermo E. Martínez. Santiago: Corporación Libertas, 1990? 138 p.

Textos y comentarios presentados en un seminario sobre el funcionamiento del Estado en 1990. En relación a procesos de decisión, se cubren temas como: funciones estatales, administración, gestión pública, coordinación y funcionamiento interno y asesoría política.

5003 Baño, Rodrigo and Angel Flisfisch. El colapso de la Unidad Popular y de la democracia chilena. (*in* Reforma política y consolidación democrática: Europa y América Latina. Caracas: Editorial Nueva Sociedad, 1988, p. 41–61)

Ensayo sobre los factores institucionales de la instauración del autoritarismo, que concluye que más que las reglas y rutinas del proceso político, el factor principal en el colapso democrático fue la impermeabilidad de instituciones como la alta burocracia, las Fuerzas Armadas y el Poder Judicial.

5004 Baño, Rodrigo. De Augustus a Patricios: la última (do)cena política. Santiago: Editorial Amerinda, 1992. 180 p.: ill.

Perceptivo análisis de 12 años de política chilena (1980–92), que da cuenta de los procesos que condicionaron la transición a la democracia y definen la actual coyuntura política. El tono irreverente y humorístico no le resta calidad ni seriedad al análisis.

5005 Benavides, Marisela. La mujer campesina en la práctica política de una comunidad agrícola: el caso de Huentelauquén IV Región. (*in* Seminario Mujer y Antropología, Problematización y Perspectivas, *Santiago, 1992*. Huellas. Edición de Sonia Montecino y María Elena Boisier. Santiago: Centro de Estudios para el Desarrollo de la Mujer (CEDEM), 1993, p. 205–216, bibl.)

Mediante observación participante y entrevistas, la autora hace una aguda observación de las prácticas y dinámicas de acción política al interior de la comunidad, enfatizando los distintos aspectos de las relaciones de género en el ámbito público de la vida comunitaria.

5006 Boisier, Sergio and Verónica Silva. Propiedad del capital y desarrollo regional endógeno en el marco de las transformaciones del capitalismo actual: reflexiones acerca de la región del Bío-Bío. (*EURE/Santiago*, 16:47, dic. 1989, p. 91–124, bibl.)

En el marco de una exploración de los sentidos, componentes y posibilidades del desarrollo regional endógeno, el trabajo avanza hacia una definición funcional del concepto de empresa regional, analizando a nivel empírico la propiedad del capital en el sector industrial manufacturero en la región.

5007 Brunner, José Joaquín. Chile: entre la cultura autoritaria y la cultura democrática. (*in* Cultura y política en América Latina. Coordinación de Hugo Zemelman. México: Siglo Veintiuno Editores, 1990, p. 85–98, bibl.)

Interpretación del golpe militar como reacción defensiva de un cierto orden hegemónico amenazado por el intento transformador de la Unidad Popular. El período autoritario es visto como refundación del modelo cultural, lo que lleva a interrogarse respecto a su impacto sobre la identidad social.

5008 Castañeda, Tarsicio. Para combatir la pobreza: política social y descentralización en Chile durante los '80. Santiago: Centro de Estudios Públicos, 1990. 310 p.: bibl., ill.

Examen de los cambios en políticas sociales durante los 80, desde la perspectiva de sus aspectos positivos y de continuidad. Se detallan los métodos de ejecución y logros, con recomendación de correcciones y de elementos de focalización y subsidio por parte del Estado.

5009 Castillo, Fernando et al. Teología de la liberación y realidad chilena. Santiago: Ediciones Rehue, 1989. 96 p.: bibl. (Serie de estudios del CEDM; 3)

Aborda las transformaciones en la sociedad chilena y su significado para la teología de la liberación. Los bloques temáticos, seguidos de comentarios teológicos, son: el debate sobre la teoría de la dependencia; los cambios en la estructura social; los movimientos sociales; y la cultura e identidad populares.

5010 Centro de Estudios de la Mujer (Chile). La condición de la mujer rural en Chile. (*in* Mujeres campesinas: América Latina. Santiago: Ediciones Isis Internacional de las Mujeres, 1987, p. 49–80, bibl., ill., photos, tables)

Estudio profundo, con testimonios, de las consecuencias de las transformaciones agrarias para la mujer rural desde la década del 30. Se describen diversos procesos, sus condiciones de trabajo, la diversificación de sus estrategias de subsistencia, y las iniciativas organizacionales y cooperativas en la actualidad.

5011 Chile. Servicio Nacional de Menores. El Servicio Nacional de Menores de Chile y su acción en torno a los niños de la calle. (*in* Seminário de Políticas Públicas para Crianças de Rua na América Latina, *1st, São Paulo, 1989.* Seminário de políticas públicas para crianças de rua na América Latina. São Paulo: Secretaria do Menor do Estado de São Paulo, 1990, v. 2, p. 35–40, tables)

Documento oficial que describe las principales líneas de asistencia desarrolladas por el Servicio, su modo de operación y el grado de cobertura que alcanza respecto al problema señalado.

5012 Chile: proyecciones y estimaciones de población, por sexo y edad; comunas 1980–1995. Santiago: Ministerio de Economía, Fomento y Reconstrucción, Instituto Nacional de Estadísticas, División Estad. Demog. y Laborales Continuas, Subdivisión Estad. Demográficas, 1989. 5 v.: bibl. (Fascículo F/CHI; 5a-5e)

Se utilizó el método de "relación de cohortes," descrito en el fascículo, para estimaciones del período 1980–2000. Los resultados 1980–85 se consideran plausibles dada la susceptibilidad de la población comunal a los cambios demográficos en el tiempo.

5013 Chile: proyecciones y estimaciones de población, por sexo y edad; total país y regiones, 1980–2000; urbano-rural. Santiago: Ministerio de Economía, Fomento y Reconstrucción, Instituto Nacional de Estadísticas, División Estad. Demog. y Laborales Continuas, Subdivisión Estad. Demográficas; Centro Latinoamericano de Demografía, 1989. 180 p.: bibl., tables. (Fascículo F/CHI; 6)

La proyección a nivel nacional fue elaborada con el método demográfico "de los componentes," que considera la evolución de la fecundidad, mortalidad y migración. A nivel regional, se usó el método de "tabla cuadrada."

5014 Congreso Chileno de Sociología, 4th, Santiago? 1992. Modernización, democracia y descentralización. Santiago: Comisión Organizadora, Congreso Chileno de Sociología, 1993. 300 p.: bibl., ill.

La más importante compilación reciente de textos sociológicos incluye discursos y una selección de 16 ponencias del congreso. Los títulos de las secciones son: "Estado y Procesos Políticos;" "Decentralización y

Poder Local;" "Modernización Productiva y Actores Sociales"; "Pobreza y Política Social;" "Opinión Pública y Comunicación;" "Educación y Juventud."

5015 Constructores de ciudad: nueve historias del Primer Concurso "Historia de las Poblaciones." Selección y edición de Alfredo Rodríguez, Alex Rosenfeld y Paulina Matta. Santiago: Ediciones SUR, 1989. 158 p.: ill.

Nueve pobladores escriben la historia de sus poblaciones en un esfuerzo de recuperación de su memoria colectiva. Entrecruzando lo individual y lo colectivo, los relatos expresan desde su perspectiva los conflictos personales y sociales que atraviesan la lucha por una vivienda.

5016 Cultura, autoritarismo y redemocratización en Chile. Edición de Manuel Antonio Garretón Merino, Saúl Sosnowski y Bernardo Subercaseaux. Santiago: Fondo de Cultura Económica, 1993. 303 p.

Libro con 21 trabajos presentados a un seminario realizado por la Univ. de Maryland en 1991. Constituye una reflexión sistemática sobre los cambios acaecidos en diferentes dimensiones de la cultura en Chile durante las últimas décadas, desde ópticas diversas que incluyen a creadores, analistas, formuladores de políticas culturales.

5017 Déniz Espinós, José A. La política económica neoliberal y sus efectos socioeconómicos: el caso de Chile. (*Cuad. Am.*, 32:2, marzo/abril 1992, p. 77–87, bibl., tables)

Con Chile como caso paradigmático de un "neoliberalismo real," el ensayo intenta brevemente extraer conclusiones generales sobre sus efectos en dos planos: 1) el productivo y macroeconómico, donde estarían sus mayores logros; y 2) el distributivo, donde se registrarían sus mayores deficiencias.

5018 Development and social change in the Chilean countryside: from the pre-land reform period to the democratic transition. Edited by Cristóbal Kay and Patricio Silva. Amsterdam: Centre for Latin American Research and Documentation, 1992. 326 p.: bibl. (Latin America studies; 62)

Volume presents a unified vision of the historical development of the Chilean rural sector and scholarship in the field. Carefully selected and edited essays cover three periods of historical development: the hacienda (1850–1964); agrarian reform (1964–73); and the neoliberal period (1973–90).

5019 Doggenweiler S., René. Hacia una ecología social: porque los pobres no pueden esperar. Santiago: Ediciones Paulinas-ILADES; Librería San Pablo, 1990. 69 p.: bibl.

Basado en una experiencia de participación laboral en la gestión empresarial, el autor desarrolla, desde una perspectiva cristiana sobre empresa, pobreza y desigualdad, algunas ideas en torno a los conflictos entre capital y trabajo y las instancias posibles de mediación y diálogo.

5020 Dubet, François and Eugenio Tironi Barrios. Pobladores: luttes sociales et démocratie au Chili. Paris: Harmattan, 1989. 190 p.: bibl. (Col. Logiques sociales)

Libro basado en una investigación realizada en 1985–86 tiene inspiración en el método de la intervención sociológica desarrollado por Alain Touraine. Estudia al actor poblacional en sus orientaciones, elecciones, y conflictos con otros actores. También discute las teorías sobre la marginalidad en la sociología latinoamericana.

5021 Espinosa, Malva. De la cultura del poder al poder de la cultura. (*in* Cuadernos del Foro 90. Santiago: CIEPLAN; CINDE; CPU; FLACSO; ICP/UC, 1992, p. 15–34)

Ensayo de interpretación sobre los cambios culturales de la sociedad chilena bajo el autoritarismo, relevando las potencialidades y vacíos para promover desde el ámbito político una cultura democrática en la transición.

5022 Espinosa, Malva. Los empresarios en la transición democrática. (*in* Cuadernos del Foro 90. Santiago: CIEPLAN; CINDE; CPU; FLACSO; ICP/UC, 1992, p. 55–89, bibl.)

Síntesis de las principales conclusiones de una investigación sobre el posicionamiento público de las agrupaciones gremiales y líderes empresariales durante los dos primeros años del gobierno de Aylwin. Analiza las posturas frente al Acuerdo Marco, la reforma tributaria y las reformas laborales del gobierno democrático.

5023 Espinosa, Malva; Hugo Yanes; and Antonio Frey. Trabajadores y empresarios de la PYMI frente a la modernización productiva. Santiago: Servicio de Cooperación Técnica (SERCOTEC), 1994. 105 p.: bibl., table (Serie Documentos de trabajo; 30)

Resultados de encuestas a empresarios y trabajadores realizadas en 299 empresas. Se concluye que en las empresas de menor tamaño relativo hay actitudes y procesos de modernización relacionados más con la renovación tecnológica y menos con la gestión de recursos humanos, lo que precariza las relaciones laborales.

5024 **Estudio sobre violencia doméstica en mujeres pobladoras chilenas.** Organización de Cecilia Moltedo Castaña. Santiago?: s.n., 1989? 48 p.

Estudio exploratorio basado en un cuestionario autoaplicado a 222 mujeres, que describe situaciones y formas de violencia doméstica en mujeres de estratos de bajos recursos. Muestra estadísticamente no representativa, pero con datos de varias regiones. Interesante dada la escasez de investigaciones empíricas sobre el tema.

5025 **Fontaine Talavera, Arturo** and **Harold Beyer.** Retrato del movimiento evangélico a la luz de las encuestas de opinión pública. (*Estud. Públicos*, 44, primavera 1991, p. 63–124, graphs, tables)

A través del análisis cuantitativo de dos encuestas realizadas en 1990 y 1991 a nivel nacional, los autores presentan una acabada caracterización del movimiento evangélico, aportando importantes datos en relación a sus dimensiones, conductas y valores así como sugiriendo posibles líneas de investigación a futuro.

5026 **Formación cívico-política de la juventud: desafío para la democracia.** Recopilación de Cristián Parker G. y Pablo Salvat. Santiago: Producciones del Ornitorrinco, 1992. 163 p.: bibl.

Intervenciones en un seminario (dic. 1990), que recoge los principales temas para diversas instituciones sociales y políticas en relación a los jóvenes y la transición democrática. El punto de partida de las visiones expuestas fue la insuficiente formación cívico-política del sector.

5027 **Fruhling, Hugo.** Resistance to fear in Chile: the experience of the Vicaría de la Solidaridad. (*in* Fear at the edge: State terror and resistance in Latin America. Berkeley: Univ. of California Press, 1992, p. 121–141)

Important study on the role of the Vicaría de la Solidaridad in the weakening or transformation of the military regime. Describes its evolution, analyzes its contribution to the formation of resistance groups, and considers its impact on the power of the regime and its policies of social control.

5028 **Gangas Geisse, Mónica.** La natalidad en la provincia de San Antonio y la influencia de las políticas de población. (*Rev. Chil. Geopolít.*, 8:1, dic. 1991, p. 65–75, bibl., tables)

Luego de una breve revisión de la evolución de las políticas de población, se analiza el ejemplo de San Antonio. Interpretando indicadores y tasas (1960–88), se indaga en la relación entre descenso de la natalidad y políticas públicas.

5029 **Garretón Merino, Manuel Antonio;** **Marta Lagos;** and **Roberto N. Méndez.** Los chilenos y la democracia: la opinión pública, 1991–1994. Santiago: Ediciones Participa, 1993. 3 v.

Cada volumen presenta resultados de una encuesta nacional anual destinada a una evaluación de las tendencias de la opinión pública sobre la modernización, la democracia, las instituciones sociales, la política. Las series se presentan por género, edad, nivel socioeconómico y posición política.

5030 **Garretón Merino, Manuel Antonio.** La faz sumergida del *iceberg*: estudios sobre la transformación cultural. Santiago: LOM; CESOC Ediciones, 1993. 233 p.

Analiza desde diversas perspectivas los cambios culturales acaecidos en la sociedad chilena en los últimos 20 años. Se aborda la cultura política, el comportamiento de algunos actores e instituciones y la evolución de las ciencias sociales, y se intenta una redefinición del debate sobre modernidad y modernización.

5031 **Garretón Merino, Manuel Antonio.** The political dimension of processes of transformation in Chile. (*in* Democracy, markets and structural reform in contemporary Latin America: Argentina, Bolivia, Brazil, Chile, and Mexico. New Brunswick, N.J.: Transaction Publishers, 1993, p. 217–235)

Proposes a sociopolitical perspective on the economic transformations that occurred under the military regime and, above all, their evolution under the first democratic government. Economic successes are contrasted with the weaknesses of a democracy limited by authoritarian enclaves and the absence of social actors.

Garretón Merino, Manuel Antonio and **Malva Espinosa.** ¿Reforma del Estado o cambio en la matriz sociopolítica? See item **4416**.

5032 **Garretón Merino, Manuel Antonio** and **Tomás Moulian.** La Unidad Popular e el conflicto político en Chile. Nueva edición. Santiago: CESOC-LOM, 1993. 227 p.

Análisis de los procesos y actores políticos del período 1970–73, editado bajo el régimen militar (ver *HLAS 51:4894*), es reeditado con ocasión de los 20 años del término del gobierno de la Unidad Popular. Los autores concluyen que todo proyecto político transformador exige sólidas mayorías dentro de las reglas del juego democrático.

5033 **Giaconi, Juan.** Readecuación del sistema de salud chileno. (*Adm. Econ. UC*, primavera 1991, p. 16–19, graphs)

Breve exposición de los principales puntos de las propuestas de readecuación del sector salud hechas por el Ministerio de Salud y la Asociación de ISAPRES, a partir del diagnóstico oficial hecho en 1991.

5034 **Gissi Bustos, Jorge.** Psicoantropología de la pobreza: Oscar Lewis y la realidad chilena. Prólogo de Alfredo Moffat. Santiago?: Psicoamérica Ediciones, 1990. 127 p.: bibl.

Preocupado de la "cultura de la pobreza" en sus implicaciones teóricas y empírico-metodológicas, el autor revisa crítica y comparativamente las ideas de Lewis y las investigaciones chilenas al respecto. Busca una visión interdisciplinaria del problema para una conceptualización y diagnósticos más apropiados y eficaces.

5035 **Goldfrank, Walter L.** Fresh demand: the consumption of Chilean produce in the United States. (*in* Commodity chains and global capitalism. Edited by Gary Gereffi and Miguel Korzeniewicz. Westport, Conn.: Praeger, 1994, p. 267–279, bibl., table)

Within the context of the wave of international trade of fresh produce and the growth of the Chilean fruit export industry, author briefly examines produce consumption in the US from the perspective of marketing, distribution networks, and transformations in dietary patterns.

5036 **Gonzáles, Raúl.** Organismos no-gubernamentales, políticas sociales y mujer. (*in* Políticas sociales, mujeres y gobierno local. Santiago: Corporación de Investigaciones Económicas para Latinoamérica, 1992, p. 217–247, bibl.)

Trazado histórico de estas organizaciones a partir de 1973, con menciones paralelas a las particularidades en el caso de aquellas directamente relacionadas con la mujer como sujeto/objeto de trabajo. El estudio se aborda desde una preocupación por su actual significación en la formulación de políticas sociales.

5037 **Harriet, María Inés** and **Gabriel Valdivieso.** Actitudes, valores y opiniones de alumnos de cuartos medios de colegios católicos. (*Estud. Soc./Santiago*, 64, 1990, p. 59–84, graph, tables)

Presentación de los resultados de una encuesta que, sin extraer conclusiones, ofrece información empírica sobre varios temas como relaciones familiares, creencias religiosas, comportamiento sexual, y percepción de la Iglesia y la sociedad en general.

5038 **Hirmes, María Eugenia.** The Chilean case: television in the 1988 plebiscite. (*in* Television, politics, and the transition to democracy in Latin America. Washington: Woodrow Wilson Center Press, 1993, p. 82–96)

Describes very well the principal aspects of the strategy and presentation of the "Sí" and "No" campaigns, situating them in the context of Chilean television of the period. Conclusion briefly questions the real significance of the television campaign on the election results.

5039 **Hola A., Eugenia** and **Rosalba Todaro C.** Los mecanismos del poder: hombres y mujeres en la empresa moderna. Chile: Centro de Estudios de la Mujer, 1992. 221 p.: bibl.

Abordando las relaciones de género como relaciones de poder en el ámbito laboral, el libro examina las estructuras y prácticas organizacionales empresariales. A propósito del sector financiero chileno, se hace una reflexión en torno a la cultura empresarial como lugar de reproducción de patrones discriminatorios.

5040 Huneeus, Carlos. La imagén pública de las Fuerzas Armadas en Chile. *(Fuerzas Arm. Soc.*, 1, marzo 1990, p. 49–56, tables)

Breve examen de la opinión civil sobre las Fuerzas Armadas y otras instituciones políticas y sociales en el marco de una estrategia de transición a la democracia vía reforma, sobre la base de algunas encuestas de opinión pública de 1988–89.

5041 Imbusch, Peter. Unternehmer und Politik in Chile: eine Studie zum politischen Verhalten der Unternehmer und ihrer Verbände. Frankfurt am Main: Vervuert, 1995. 493 p.: ill.

Well-written, researched, and documented analysis of Chilean entrepreneurs as a social category. Concentrates on their political behaviors, roles, and influence of their associations during the 1970s–80s. Examines their relationship to the State and commitment to democracy. Concludes that the entrepreneurs and their interest groups are less concerned with the type of government than with its policies towards business in general. [C.K. Converse]

5042 Jansana, Loreto. El pan nuestro: las organizaciones populares para el consumo. Santiago: Programa de Economía del Trabajo, 1989. 181 p.: bibl. (Col. Experiencias populares; 6)

Se recoge la experiencia de investigación participativa durante 1987 con organizaciones de subsistencia para el consumo alimentario de Santiago. Selección de estudios de caso, con información estadística y testimonios de algunas organizaciones representativas, describiendo su origen, modos de operación, trabajo cotidiano y cobertura.

5043 Jofré, Manuel Alcides. La cultura local. Santiago: Documentas; CENECA, 1991. 256 p.: bibl.

Estudio empírico y tipológico de las agrupaciones juveniles poblacionales dedicadas a la creación artística y cultural. Se examina el funcionamiento, estructura, objetivos, actividades y demandas de más de un centenar de centros y grupos de acción cultural.

5044 Kessel, Juan van. Los aymaras bajo el régimen militar de Pinochet, 1973–1990. Iquique, Chile: Centro de Investigación de la Realidad del Norte; Ediciones El Jote Errante, 1990. 75 p.: maps. (Cuaderno de investigación social; 29)

Síntesis y balance de las transformaciones introducidas en el mundo aymara y sus reacciones al acelerado proceso de "chileanización" e integración. Se analizan la planificación del gobierno militar, la economía andina, el conflicto religioso, y la acción y reacción sociales.

5045 Le Goff, Jean-Luis. Les technologies de la qualité au Chili: nouvel enjeu de développement? *(Sociol. trav.*, 34:2, 1992, p. 171–191)

Mediante el estudio de una empresa chilena del sector metal-mecánico, el autor quiere ilustrar una nueva tendencia en la concepción de la transferencia tecnológica, centrada en las nociones de calidad y control de calidad implicadas en la organización productiva de las empresas.

5046 Letelier L., Marta. Diagnóstico social de la Población "Puertas Negras" de Playa Ancha, Valparaíso, 1988. *(Rev. Cienc. Soc./Valparaíso*, 32/33, 1989, p. 313–343, tables)

Breve diagnóstico operacional, producto de un estudio censal de la población efectuado en 1988, para implementar un proyecto de desarrollo social. Incluye un resumen descriptivo y el delineamiento de sus principales necesidades.

5047 MacDonald, Joan. Mujer, vivienda y desarrollo local. (*in* Seminario Internacional Mujer y Municipio, *Quito, 1991.* Una nueva presencia comunitaria en el desarrollo local de América Latina. Quito: IULA/CELCADEL; RHUDO/SA—AID, 1991, p. 38–45)

La autora, autoridad de gobierno, revisa el sistema habitacional y su relación con el aparato de gestión local y la comunidad, además de los rasgos esenciales de la participación de la mujer particularmente en el área de la vivienda, indicando las acciones puestas en marcha para mejorarla.

5048 Marinović Pino, Milan. Fuerzas armadas y sociedad: marco teórico para un debate político. *(Fuerzas Arm. Soc.*, 1, marzo 1990, p. 11–28, bibl., table)

Se propone analizar las relaciones cívico-militares desde la perspectiva del estructuralismo genético desarrollada por George

Luckas. Plantea temas para el debate en torno a las fuerzas armadas en la arena política de la transición.

5049 Martínez, Javier. Fear of the state, fear of society: on the opposition protest in Chile. (*in* Fear at the edge: State terror and resistance in Latin America. Berkeley: Univ. of California Press, 1992, p. 142–160)

Considering fear as a political factor, studies civil protests from 1983–84. Analyzes the conditions that made them possible, their role in overcoming fear of the State, and their decline, attributed to protests' inability to define a social alternative to authoritarian order.

5050 Medel R., Julia; Soledad Olivos M.; and Verónica Riquelme G. Las temporeras y su visión del trabajo: condiciones de trabajo y participación social. Santiago: Programa Mujer Campesina y Asalariada Agrícola, Centro de Estudios de la Mujer, 1989. 146 p.: bibl., ill.

Valorizando la experiencia personal y lo cotidiano como espacio de intervención, presenta y analiza un completo cuadro del discurso, motivaciones, percepciones y conflictos de las temporeras en relación a sus condiciones de vida y trabajo, recogidos mediante grupos de discusión.

5051 Meneses C., Aldo. El poder del discurso: la Iglesia Católica chilena y el gobierno militar, 1973–1984. Santiago?: ILADES; CISOC, 1989. 227 p.: bibl., ill.

Constatando similitudes en las doctrinas de la Iglesia y el Gobierno Militar a pesar de su antagonismo, el autor analiza discursivamente ambos actores, estableciendo su estructuración, sistemas simbólicos y de legitimación y las diferencias en los significados conferidos a conceptos claves para la convivencia social. Ver también la reseña del politólogo, item **3467**.

5052 Montecino Aguirre, Sonia. Madres y huachos: alegorías del mestizaje chileno. Santiago: Editorial Cuarto Propio-CEDEM, 1991. 162 p.: bibl. (Serie Ensayo)

Importantísimo ensayo contiene una lectura aguda de algunos núcleos simbólicos de la cultura chilena, relacionados a las cuestiones cruciales de la identidad latinoamericana, nacional y de género. Tematiza las nociones de mestizaje, marianismo y el peso de lo feminino y lo materno recurriendo a fuentes históricas, literarias y orales.

5053 Montecino Aguirre, Sonia. Símbolo mariano y constitución de la identidad femenina en Chile. (*Estud. Públicos*, 39, invierno 1990, p. 283–290)

El ensayo señala lúcidamente el predominio de la figura materna en la configuración de nuestro *ethos* cultural, y la pervivencia del modelo mariano como núcleo simbólico de nuestra identidad, de donde surgirían interrogantes para repensar la cuestión de la identidad femenina en el contexto cultural latinoamericano.

5054 Morales, Eduardo. La crisis urbana en el Cono Sur: paradigmas y enfoques; la ciudad de Santiago de Chile. (*in* Las ciudades en conflicto: una perspectiva latinoamericana. Montevideo: Centro de Informaciones y Estudios del Uruguay; Ediciones de la Banda Oriental, 1989, p. 223–238, maps)

El autor traza una breve historia de la localización de sectores urbanos-marginales desde 1964, reseñando los procesos de asentamiento y posterior relocalización. Se revisa sus comportamientos políticos y se plantea la existencia de un patrón de crecimiento segregatorio y la necesidad de su transformación.

5055 Muñoz Dálbora, Adriana. The women's movement in Chile: a desired reality. (*in* Alternatives. Edited by Neuma Aquiar and Thaís Corral. Rio de Janeiro: Editora Rosa dos Tempos, 1991, v. 2, p. 113–134, bibl.)

Beginning with observations on the emergence and consolidation of diverse women's organizations during the authoritarian regime, author evaluates the current situation of these organizations and reflects on their advances in relation to social movements in general.

5056 Muñoz Gomá, Oscar. La reforma del Estado y la economía chilena: trayectoria de una relación difícil. (*in* Los nuevos límites del Estado. Quito: Corporación de Estudios para el Desarrollo (CORDES), 1990, p. 161–208, bibl., tables)

Analiza las transformaciones recientes en el rol del Estado bajo orientación neoliberal, tendientes a sustituir centralización e intervencionismo por una mayor autonomía empresarial. Sugiere la definición de un nuevo papel de "Estado Concertador," ya que éste aún mantiene un importante rol en el desarrollo.

5057 **Muñoz V., Mario.** La contaminación atmosférica en Santiago: impacto sobre la salud de la población. [*Estud. Públicos,* 45, verano 1992, p. 175–228, bibl., graphs, tables]

Detallado estudio de la contaminación atmosférica, tanto en sus fuentes, factores y componentes como en sus efectos. También se hace una revisión crítica de las principales medidas para contrarrestarla y se hacen sugerencias en torno a una política global para combatirla.

5058 **Necochea, Andrés** and **Ana María Icaza.** Una estrategia democrática de renovación urbana residencial: el caso de la comuna de Santiago. [*EURE/Santiago,* 16:48, junio 1990, p. 37–65, table]

Analiza resultados de una encuesta acerca de la motivación de los diferentes grupos que habitan las áreas centrales de la ciudad. Considerando la heterogeneidad como valor urbano democrático, se proponen programas de consolidación residencial para distintos estratos socioeconómicos con énfasis en la gestión municipal.

5059 **Opazo Bernales, Andrés.** Escuchando a la juventud poblacional. Edición de Miguel Díaz S. Santiago: Centro de Estudios del Desarrollo, 1991. 88 p.

El documento expone los resultados de una investigación del campo con entrevistas en profundidad a jóvenes pobladores. Se desglosan cuidadosamente los principales temas y preocupaciones detectados, analizándose algunas percepciones y vivencias claves para comprender el fenómeno, y se sugieren posibles líneas de acción.

5060 **Petras, James F.; Fernando Ignacio Leiva;** and **Henry Veltmeyer.** Democracy and poverty in Chile: the limits to electoral politics. Boulder, Colo.: Westview Press, 1994. 215 p.: bibl., index. (Series in political economy and economic development in Latin America)

Criticizing the transition to democracy in Chile, authors contrast legal and political changes under democracy with the continuation of policies perpetuating inequality that were inherited from the military regime.

5061 **Piñuel Raigada, J.L.** La cultura política del ciudadano y la comunicación política en TV, en la transición política del plebiscito chileno, octubre 1988. [*Rev. Esp. Invest. Sociol.,* 50, abril/junio 1990, p. 125–237, graphs, tables]

Acabado estudio de la campaña plebiscitaria como mediación comunicacional. Además de un complejo análisis de contenido de las franjas televisivas, se analizan datos recogidos a través de una encuesta nacional aplicada en 1988, en torno a las representaciones sociales y políticas vehiculadas y sus modos de recepción.

5062 **Pozo, José del.** Los militantes de base de la izquierda chilena: orígenes sociales, motivaciones y experiencias en la época de la Unidad Popular y en los años anteriores. [*Rev. Eur.,* 52, June 1992, p. 31–55, bibl., tables]

Parte de una investigación sobre experiencias políticas de izquierda entre 1970–73, con análisis de entrevistas a militantes a fin de lograr una mejor percepción de la vida de los partidos. Los testimonios van desde 1930–73, pero el grueso corresponde a los 60s.

5063 **Puente, Patricio de la; Emilio Torres Rojas;** and **Patricia Muñoz Salazar.** Satisfacción residencial en soluciones habitacionales de radicación y erradicación para sectores pobres de Santiago. [*EURE/Santiago,* 16:49, oct. 1990, p. 7–22, bibl., tables]

Estudio empírico referido a la satisfacción en sectores beneficiarios de programas habitacionales entre 1980–87. En la perspectiva del concepto de Hábitat Residencial Urbano, se analiza la incidencia en los niveles de insatisfacción de las dimensiones vinculadas a la vivienda, el equipamiento y a su entorno.

5064 **Ramón, Armando de.** La población informal: poblamiento de la periferia de Santiago de Chile, 1920–1970. [*EURE/Santiago,* 16:50, dic. 1990, p. 5–17, bibl., tables]

Se revisan numerosos artículos de los últimos 20 años sobre la evolución de la habitación popular en Santiago, analizando el impacto de la ocupación del espacio por los más pobres sobre la conformación de la ciudad y estableciéndose diversas etapas según formas de ocupación.

5065 **Reyes, Paulina de los.** The rural poor: agrarian changes and survival strategies in Chile, 1973–1989. Uppsala, Sweden:

Dept. of Economic History, Univ. of Uppsala; Stockholm: Almqvist & Wiksell International, 1992. 196 p.: bibl., ill. (Acta Universitatis Upsaliensis. Uppsala studies in economic history: 0346–6493; 34)

Examines evolution of farming and the household unit in relation to the factors involved in the reproduction and permanence of poverty. Postulates that in addition to macroeconomic conditions, the dynamics established at the micro level are also factors.

5066 Sabatini, Francisco. Participación de pobladores en organizaciones de barrio. (*EURE*/Santiago, 15:46, oct. 1989, p. 47–68, bibl., tables)

Estudio de caso en torno al nivel de participación en organizaciones de barrio, que tendería a confirmar la hipótesis de que ésta es obstaculizada por diversos factores subjetivos y de situación que neutralizan el impulso participativo originado en conciencia crítica e inseguridad.

5067 Sabatini, Francisco. Precios del suelo y edificacíon de viviendas: 4 conclusiones sobre Santiago relevantes para políticas urbanas. (*EURE*/Santiago, 16:49, oct. 1990, p. 63–72, bibl.)

El artículo resume las principales conclusiones de investigaciones del autor durante la década del 80 sobre el mercado inmobiliario, centrándose en la función económica que juegan los precios del suelo. Ejemplifica brevemente algunas alternativas habitacionales antisegregativas desarrollables por el Estado.

5068 Saiz, José L. et al. Medición del prejuicio negativo hacia el mapuche: una investigación metodológica. (*Estud. Soc.*/Santiago, 57, 1988, p. 111–126, bibl.)

Estudio de las propiedades métricas de la escala establecida por Gajardo (1983) para medir el prejuicio hacia la etnia mapuche, según criterios estadísticos.

5069 Salman, Ton. De verlegen beweging: desintegratie, inventiviteit en verzet van de Chileense pobladores, 1973–1990 [The shy movement: disintegration, inventiveness and resistance of the Chilean *pobladores*, 1973–1990]. Amsterdam: CEDLA, 1993. 436 p.: bibl., map, photos. (Latin America studies; 71)

Discusses informal organization in the *poblaciones* and cycle of protest days. Author argues that collective action should be understood primarily in light of specific circumstances in which it occurs. Whether or not criteria of what constitutes a "social movement" are met is of lesser importance. Includes brief summaries in Spanish and English. [R. Hoefte]

5070 Sepúlveda, Narciso. Breve síntesis histórica del movimiento pentecostal en Chile. (*in* Pentecostalismo y liberación: una experiencia latinoamericano. Edición de Carmelo Alvarez. San José: Editorial Depto. Ecuménico de Investigaciones (DEI), 1992, p. 37–45)

Abarca el origen y evolución de las iglesias pentecostales originadas en Chile a comienzos de siglo, describiendo sus cismas y apuntando a la necesidad de unidad ecuménica.

5071 Serrano, Claudia. Economic crisis and women of the urban popular sector in Santiago de Chile. (*in* Alternatives. Edited by Neuma Aquiar and Thaís Corral. Rio de Janeiro: Editora Rosa dos Tempos, 1991, v. 1, p. 109–123, bibl.)

Studies women's behavior in light of the socioeconomic conditions generated by the authoritarian regime. Analyzes behavior at three levels: domestic organization and subsistence, neighborhood organizations, and the network of governmental assistance.

5072 Serrano, Claudia. Estado, mujer y política social en Chile. (*in* Políticas sociales, mujeres y gobierno local. Santiago: Corporación de Investigaciones Económicas para Latinoamérica, 1992, p 195–216, bibl.)

Revisa los nexos entre participación de la mujer y las políticas públicas sociales desde el Estado benefactor, pasando por el modelo neoliberal militar hasta el período democrático.

5073 Silva, Patricio. Autoritarismo, neoliberalismo y sindicalismo agrario en Chile. (*Estud. Rural. Latinoam.*, 12:2, mayo/agosto 1989, p. 143–171, bibl., table)

Análisis del sindicalismo agrario durante el régimen autoritario en relación a los cambios a nivel estatal. Se intenta una periodización en el contexto nacional, bajo el supuesto de homogeneización de las condiciones del sindicalismo urbano y rural.

5074 Smith, Sherry. The political, social, and artistic power of the Chilean *arpilleras*: popular art movements under authoritarianism. (*MACLAS Lat. Am. Essays*, 4, 1991?, p. 167–181, bibl.)

Defines the creation of *arpilleras* as a popular art movement. Traces movement's history, concentrating on the growing politicization of its themes within the context of authoritarianism, praising its cultural resistance and alternative symbolic production roles.

5075 Stewart-Gambino, Hannah W. Redefining the changes and politics in Chile. (*in* Conflict and competition: the Latin American Church in a changing environment. Edited by Edward L. Cleary and Hannah Stewart-Gambino. Boulder, Colo.: Lynne Rienner Publishers, 1992, p. 21–44)

Speculates on possible future role of the Catholic Church in the context of redemocratization and Church-State relations, tentatively examining Church's situation before and during the authoritarian regime and its contribution to the transition to democracy.

5076 Tironi Barrios, Eugenio. Pobladores en Chile: protesta y organización. (*in* El sector informal en América Latina: una selección de perspectivas analíticas. Recopilación de Jacobo Schatan, Dieter Paas, y Alvaro Orsatti. México: Centro de Investigación y Docencia Económicas; Fundación Friedrich Naumann, 1991, p. 143–165, bibl.)

Los pobladores y su movilización durante los 80 son considerados una expresión del proceso desintegrador de la sociedad chilena. Se estudia la brecha entre su autopercepción y los modelos analíticos con que han sido definidos, por un lado, y la distancia entre movilización y organización por otro.

5077 Tramas para un nuevo destino: propuestas de la Concertación de Mujeres por la Democracia. Edición de Sonia Montecino Aguirre y Josefina Rossetti. Santiago?: Arancibia Hnos., 1990. 216 p.: bibl.

Reúne las propuestas programáticas de la Concertación de Mujeres por la Democracia que sirvieron de base para la elaboración de las políticas del Servicio Nacional de la Mujer. Cada trabajo consta de un diagnóstico y políticas sobre las diversas áreas del quehacer nacional.

5078 Valdés, Teresa and **Marisa Weinstein C.** Organizaciones de pobladoras y construcción democrática en Chile: notas para un debate. (*in* Las mujeres y la vida de las ciudades. Recopilación de María del Carmen Feijoó y Hilda María Herzer. Buenos Aires: Grupo Editor Latinoamericano; IIED-América Latina, 1991, p. 111–140, bibl.)

El artículo resume las condiciones en que se han gestado y desarrollado las organizaciones femininas bajo el regimen autoritario, sus características y la relación que han establecido con el Estado y la política hasta ahora.

5079 Valenzuela, María Helena. The military regime, women and dictatorship in Chile. (*in* Alternatives. Edited by Neuma Aquiar and Thaís Corral. Rio de Janeiro: Editora Rosa dos Tempos, 1991, v. 1, p. 97–108, bibl., table)

Briefly reviews consequences of authoritarianism and militarization on the condition of women, their political participation, and the formation of their identity.

5080 Vergara, Pilar. Market economy, social welfare and democratic consolidation in Chile. (*in* Democracy, markets, and structural reform in contemporary Latin America: Argentina, Bolivia, Brazil, Chile, and Mexico. New Brunswick, N.J.: Transaction Publishers, 1993, p. 237–261)

Rigorous and enlightening work describes economic reforms and social policies under the military regime and the first democratic government, establishing breaks and continuities with the past. Warns of a possible social bifurcation resulting from the policies.

5081 Vicaría de la Solidaridad: historia de su trabajo social. Santiago: Ediciones Paulinas; Librería San Pablo, 1991. 167 p.: bibl. (Col. Testigos del Evangelio; 4)

Historia testimonial de la Vicaría de la Solidaridad desde dentro. Se hace la crónica de su gestación y desarrollo, de su configuración estructural y organizacional, de sus relaciones con el resto de la Iglesia Católica y de su trabajo concreto en diferentes áreas.

5082 Weinstein, José. Los jóvenes pobladores en las protestas nacionales, 1983–1984: una visión socio-política. Santiago: Centro de Investigación y Desarrollo de la Educación, 1989. 169 p.: bibl.

Ensayo exploratorio apunta a sugerir elementos de interpretación de la relación entre juventud, subproletariado y comportamientos sociopolíticos. Una panorámica del sector en los 80 y consideraciones generales de orden teórico sirven de base para formular un conjunto de hipótesis sobre el fenómeno.

5083 Ya nunca me verás como me vieras. Entrevistas por Mili Rodríguez Villouta. Santiago: Ediciones del Ornitorrinco, 1990. 278 p.

Constituido por 12 entrevistas a chilenos, el libro entrega una visión múltiple y subjetiva de la doble experiencia del exilio y el retorno.

5084 Young, Frank. Concentración urbana y agrupación de los asentamientos rurales en Chile. (*EURE/Santiago*, 17:51, junio 1991, p. 23–31, tables)

Estudio empírico con análisis de regresión de cifras del Censo de 1982 busca describir y explicar esta doble proceso de nuclearización, tal como se ha desarrollado en el Valle Central de Chile.

5085 Zemelman, Hugo. Chile: el régimen militar, la burguesía y el Estado; panorama de problemas y situaciones, 1974–1987. (*in* El Estado en América Latina: teoría y práctica. Coordinación de Pablo González Casanova. México: Siglo Veintiuno Editores, 1990, p. 291–322)

Ensayo interpretativo de la coyuntura política pre-plebiscito critica la participación vía la institucionalidad establecida en la Constitución del 1980 y aboga por la configuración de un sistema político alternativo.

ARGENTINA AND URUGUAY

ROBERTO PATRICIO KORZENIEWICZ, *Assistant Professor, Department of Sociology, University of Maryland, College Park*
BENJAMIN HADIS, *Associate Professor of Sociology, Montclair State University*

ARGENTINA

WHEREAS THE 1970S AND THE EARLY 1980S were accompanied by increased funding of the social sciences and a phenomenal growth in the number of research centers servicing these disciplines, the late 1980s and early 1990s witnessed a reversal of these trends. Financial resources are increasingly scarce due to the adoption of new funding priorities by international agencies, the implementation of fiscal readjustment programs under the Menem Administration, and the impact of unfavorable exchange rates on grants denominated in foreign currency. Facing this dramatic decline in available resources, many independent research centers have been forced to reduce substantially their activities and personnel.

Scholars and research centers face limited options as they seek to overcome their financial constraints. Some institutions are creating or expanding instructional programs, a strategy designed to tap into the high-tuition market created by a rapidly growing demand for professional certificates. Other centers have sought greater institutional support by tailoring their research projects more closely to the policy needs of international organizations or State agencies. Many institutions have had to cut expenses severely, in some instances by closing down altogether. (Most deeply affected were those centers that had either limited sources of funding, or a narrow area of study and/or methodological approach.) Despite some exceptions among the most prestigious scholars and centers, the scope and depth of academic research in private and independent research centers have been restricted, at a time when the public university system continues to suffer from a lack of the most basic resources (item **4402**).

Stringent financial conditions have had a noticeable impact on recent publications in the social sciences. Due to budget cuts, independent research centers have drastically curtailed their own publications. Moreover, book prices continue to be very high, restricting the willingness of commercial presses to publish academic monographs attractive to only a limited market. As a result, scholars have been moving away from the traditional book format as the primary mechanism for publication. Instead, they rely heavily on occasional, and often informal, monographic series (e.g., research papers which are simply photocopied and sold at cost).

Thus, the social science monographs that do get published by commercial presses are frequently written for popular consumption. Generally characterized by less rigorous attention to methodology and research design, these books tend to adopt the form of essays and/or personal observations on issues of public concern (such as the persistence of authoritarianism or gender inequalities). For example, the study of gender continues to generate many contributions, but many of these publications are intended as general introductions to women's studies or to topics of public interest. Few monographs on gender provide new data or significant insights and high quality monographs tend to be scarce. Nevertheless, a number of publications annotated below do make important contributions to the field.

Since the early 1990s, the analysis of the social impact of economic restructuring has become a central area of concern. Studies focus on several dimensions of change in the distribution of wealth. Some monographs explore how economic restructuring has led to a greater concentration of income among elites. Other works similarly interpret the redistribution of wealth, but concentrate on the growing prevalence of poverty and unemployment among urban households. Several studies differentiate among types of poverty. The term "structurally poor" is used to refer to low-income households that fail to meet basic standards in housing and health, while "pauperized" refers to those meeting the latter basic needs, but with incomes below the poverty line (item **5114**). These studies suggest that one of the most pernicious effects of hyperinflation in the 1980s was to increase drastically the number of "pauperized" middle-class households.

Related to economic restructuring, but more closely tied to policy concerns, are several significant contributions assessing the social and economic impact of development projects in areas such as welfare and public health, with particularly interesting research on programs that target drug abuse. Some of these studies analyze the institutional trajectory of specific projects, which are generally initiated by non-governmental organizations in conjunction with local communities, although with considerable direct involvement of State agencies in some fields. These monographs also evaluate the appropriateness of State responses to these social programs. The most interesting findings, however, emerge from evaluations of the social and organizational dynamics taking place between and among the different agents involved in development projects (item **5126**).

More traditional studies of class structure and social change have also appeared (items **5095** and **5100**). Several areas that received considerable attention in the past are now attracting less research interest, however. For example, there are fewer recent studies on labor, either those examining trade union organization or workplace and labor market arrangements. Likewise, there are fewer sociological inquiries into migration and urbanization, although there is a heightened interest in the social construction of ethnic and cultural identity among the urban poor. (Advancing innovative interpretations, this latter set of studies follows a long tradition of urban research within Argentine sociology.) Important recent phenomena such as the rapid growth of medium-sized cities in the interior provinces and the impact of deindustrialization and impoverishment on working-class and middle-class neighborhoods generated relatively little new work.

Key areas of sociological inquiry continue to be characterized by an uneven balance of macro and micro studies, with most monographs adopting a broad and general approach. In the study of enterprises, for example, there is a continuing concern with the impact of business organizations on politics and the corporate arrangements through which large economic groups might influence policies. In addition, the negotiations involved in the privatization of public enterprises are examined. In works on political institutions, inquiries into the military as a political actor have continued to decline. Nevertheless, there are several general analyses of the persistence of authoritarian attitudes within civil society, a number of which seek to explain these attitudes as a cultural legacy of early Spanish colonization in the Americas.

Fewer sociological studies present a micro or detailed analysis of these spheres. In regard to enterprises, for example, there are virtually no monographs that provide in-depth perspectives on the organizational structure of business firms, the operation of internal labor markets, or the micro-level impact of economic restructuring on production and marketing. Likewise, there are few studies of the internal structure of State agencies or organizations of political representation, particularly in relation to recent changes.

The disparity between macro and micro studies is less pronounced in the study of households, an area that has received considerable attention. The study of household dynamics fits well with research on the social impact of economic restructuring, as well as with the social construction of identities (particularly in studies on gender and youth).

Regardless of the relative scarcity of detailed studies, salient monographs in the study of enterprises and the State have appeared. For example, scholars focus on the growing heterogeneity of enterprises, particularly in rural areas (item **5101**). In the area of political sociology, there is an excellent contribution seeking to reconstruct the evolution of political discourse (item **5122**). Studies in this area also provide important insights into the historical evolution of social programs, the relationship between politics and marketing, and the impact of corruption in shaping State agencies and policies.

URUGUAY

Many of the trends noted above for Argentina also apply to Uruguay. Fiscal constraints have had a detrimental impact on academic production, and this is reflected in the volume and quality of research publications. A noteworthy exception can be found in several publications of the Centro de Informaciones y Estudios del Uruguay, particularly regarding a long-term project focusing on social movements and forms of collective action (items **5134** and **5136**). There are also a significant number of monographs dealing with gender studies. As in Argentina, many of these publications are oriented towards the general public and provide little original research. However, there are some notable exceptions, such as an intelligent case study of female labor participation in the fish industry (item **5135**).

ARGENTINA

5086 Agulla, Juan Carlos. Estructura ocupacional de la Argentina. Buenos Aires: Instituto de Investigaciones Jurídicas y Sociales Ambrosio L. Gioja, Facultad de Derecho y Ciencias Sociales, U.B.A., 1989. 53 p.: bibl. (Cuadernos de investigaciones; 16)

Using data on economically active population (1960, 1970, 1980), brief monograph suggests that occupational characteristics continue to combine with other variables (caste, stratum, or class) in shaping social status in Argentina. Findings are reasonable, but no evidence is offered to substantiate claims that relative weight of education varies across the nation. [RPK]

5087 Allub, Leopoldo. Impactos sociales de las grandes obras públicas: diagnóstico, evaluación y gestión. San Juan, Argentina: Univ. Nacional de San Juan; Consejo Nacional de Investigaciones Científicas y Técnicas, 1990. 164 p., 1 leaf of plates: bibl., ill., maps.

Introductory chapters on how to study social impact of public infrastructural work are followed by three reports about author's work in Mexico and Argentina (San Juan prov.). Latter is based on a survey of farmers' perceptions about causes of soil erosion. [BH]

5088 Aufgang, Lidia G. Las puebladas: dos casos de protesta social, Cipolletti y Casilda. Introducción de Beba Balvé. Buenos Aires: Centro Editor de América Latina, 1989. 127 p.: bibl. (Biblioteca Política argentina; 252)

The Cordobazo and Rosariazo did not exhaust massive urban social protest during the 1966-72 "Revolución Argentina." This work does a remarkable comparison of two additional movements. Casilda, with its class structure and class antagonism, fits the pattern of unrestrained violence involved in the *azo* model. Cipolletti, an exclusively bourgeois town, maintains a corporative, restrained antagonism towards outsiders. [BH]

5089 Basualdo, Eduardo M. and **Miguel Khavisse.** El nuevo poder terrateniente: investigación sobre los nuevos y viejos propietarios de tierras de la provincia de Buenos Aires. Buenos Aires: Planeta, 1993. 374 p.: bibl., map. (Espejo de la Argentina)

Book for specialists uses original sources to analyze complex changes in land ownership such as growing importance of land subdivision (used by landowners to reduce their tax burden) and emergence of new forms of property. Book is extremely detailed and repetitive, but somewhat superficial in its analysis of general trends. [RPK]

5090 Castells, Julia. Ficción y realidad de la política social para la ancianidad. v. 1-2. Buenos Aires: Centro Editor de América Latina, 1992. 2 v. (Biblioteca Política argentina; 382-383)

Work evalutes services available to senior citizens in six provinces of Argentina (Córdoba, La Rioja, Mendoza, Río Negro, Santa Fe, Tucumán). Weighs relative importance of public and private sectors, and identifies severe shortages of resources and qualified personnel. This useful study includes a detailed survey of retirement homes and clients in the six areas. [RPK]

5091 Castiglione, Marta. La militarización del Estado en la Argentina, 1976/1981. Buenos Aires: Centro Editor de América Latina, 1992. 94 p.: bibl., ill. (Biblioteca Política argentina; 350)

Attempting to analyze relationship between armed forces and political power, monograph is at its best in showing that institutional scope of military interventions grew consistently over time, together with administrative and organizational capabilities. Most important was development of formal and informal negotiating mechanisms to organize administrative appointments and policy design/implementation through the military hierarchy. [RPK]

5092 Centro de Investigaciones sobre Pobreza y Políticas Sociales en la Argentina. El país de los excluídos: crecimiento y heterogeneidad de la pobreza en el conurbano bonaerense. Buenos Aires: Centro de Investigaciones sobre Pobreza y Políticas Sociales en la Argentina, 1991. 112 p.: ill.

Important contribution suggests a growth in pauperized (below poverty line but with basic needs satisfied) and structurally poor (below poverty line and with basic needs unsatisfied) populations. Latter population is particularly affected by health and housing problems, and finds it more difficult to gain social mobility through education. Public policies should consider variables such as age, gender, ethnicity, length of residence, property rights. [RPK]

5093 **Chapp, María Ester et al.** Religiosidad popular en la Argentina. Buenos Aires: Centro Editor de América Latina, 1991. 96 p.: bibl. (Biblioteca Política argentina; 332)

Theoretical introductory chapter on study of religion and cults is followed by short but competent accounts of a series of popular cults in provinces of Argentina. Case studies provide adequate exploration of hypotheses on symbolic representation raised in the introduction. [RPK]

5094 **Conurbano bonaerense: aproximación a la determinación de hogares y población en riesgo sanitario a través de la Encuesta Permanente de Hogares.** Buenos Aires: República Argentina, Presidencia de la Nación, Secretaría de Planificación, Instituto Nacional de Estadística y Censos, 1991. 21 p.: maps.

Reports results of 1988 survey on housing conducted to determine proportion of population experiencing deficiencies that might increase health risks (particularly in face of cholera epidemic). Proportion of households at risk was calculated at 49 percent, although with considerable variation among relevant districts. [RPK]

5095 **Después de Germani: exploraciones sobre *Estructura social de la Argentina*.** Recopilación de Jorge Raúl Jorrat y Ruth Sautu. Buenos Aires: Paidos, 1992. 278 p.: bibl., ill., maps. (Estado y sociedad; 9)

Heterogenous but interesting collection of essays focuses on Gino Germani and his title work (see *HLAS 19:6092*), and on social relations in Argentina. Most useful are eight monographs that use new data to analyze changes in social structure of urban and rural areas over past decades. Marta Panaia explains a decline in trade union membership as the product of changes in the labor force (growing unemployment, informal-sector activity, and female participation). [RPK]

5096 **Encuentro Nacional de Centros de Prevención de la Violencia Doméstica y Asistencia a la Mujer Golpeada, 1st, Chapadmalal, Argentina, 1988.** Mujer golpeada. Coordinación de Leonor Vain. Buenos Aires: Editorial Besana, 1989. 282 p. (Col. Ensayos)

Proceedings of an interdisciplinary conference focus on violence against women in Argentina. Two of the panelists, Pedro David and Silvia Chejter, are sociologists. Chejter focuses on Argentina and specifically on rape, pointing out how abstract notion of violence is one of many factors that serve to hush up rape. Analysis of texts of judicial processes shows how police and the courts muzzle the violence against women implicated in rape. Typical of a patriarchal production system, the public and private spheres are separated, and rape is considered a private crime. Indeed, rape is the only crime that requires the victim to press charges in order to put prosecution in motion, a factor which further silences it. [BH]

5097 **Estructura social de la Argentina: indicadores de la estratificación social y de las condiciones de vida de la población en base al *Censo de población y vivenda de 1980*.** v. 1–27. Buenos Aires: Consejo Federal de Inversiones; Comisión Económica para América Latina y el Caribe de las Naciones Unidas, 1988–1991. 27 v.: bibl., ill. (Estructura socioeconómica argentina)

Volumes contain useful but somewhat dated synthesis of household survey data on variables such as economically active population, living conditions, educational attendance, and demography. Vol. 1 presents data for nation as a whole; others deal in greater detail with each province and Buenos Aires metropolitan area. [RPK]

5098 **Feijoó, María del Carmen.** Alquimistas en la crisis: experiencias de mujeres en el Gran Buenos Aires. Buenos Aires: UNICEF-Argentina; Siglo Veintiuno de España Editores; Catálogos S.R.L., 1991. 99 p., 6 p. of plates: bibl., ill.

Presents useful overview of benefits and problems of two programs of social development implemented in poor neighborhoods. A child care program was innovative in that it sought to train a group of mothers in the area to provide (under the supervision of professional teachers) part of the care for children at the center. Second experience involved creation of very small production and marketing cooperatives among women. [RPK]

5099 **Ferreira, Graciela B.** Hombres violentos, mujeres maltratadas: aportes a la investigación y tratamiento de un problema social. Buenos Aires: Editorial Sudamericana, 1992. 430 p.: bibl., ill.

Provides anecdotal evidence of family violence in Argentina; selectively reviews existing literature on the topic; and evaluates public policies on the issue. Monograph fulfills an important service in bringing this compelling issue to the attention of the general public. [RPK]

5100 **Forni, Floreal H.; Roberto Benencia; and Guillermo Neiman.** Empleo, estrategias de vida y reproducción: hogares rurales en Santiago del Estero. Buenos Aires: Centro Editor de América Latina; Centro de Estudios e Investigaciones Laborales del Consejo Nacional de Investigaciones Científicas y Técnicas, 1991. 178 p.: bibl., ill., map. (Bibliotecas universitarias. Trabajo y sociedad)

Provides excellent analysis of rural household strategies. After giving statistical overview of productive and occupational structures in the province, authors make careful use of qualitative data (gathered through on-site interviews) to depict employment, migratory, and demographic strategies adopted by households and their members to enhance their individual and collective resources. [RPK]

5101 **Giarracca, Norma and Susana Aparicio.** Los campesinos cañeros: multiocupación y organización. Buenos Aires: Instituto de Investigaciones, Facultad de Ciencias Sociales, 1991. 115 p.: bibl. (Cuadernos; 3)

Impressive quantitative and qualitative study reveals changes and growing heterogeneity of sugar production over last 20 years. Rise and subsequent evolution of small sugarcane farmers prior to 1970s was tied to labor shortages and State policies that have regulated the industry since 1920s. Small- and medium-sized family farms have retained an important role in production through multiplicity of strategies designed to increase household resources. [RPK]

5102 **Gibaja, Regina Elena.** Imágenes de la condición femenina. Buenos Aires: Editorial Universitaria de Buenos Aires, 1990. 213 p.: bibl. (Col. Temas)

Very well written "qualitative, descriptive and exploratory" study is based on intensive, structured interviews with 29 middle and upper-middle class women (homemakers, professionals, school teachers and university staff) in a middle-sized Argentine city. Inquiry focuses on interviewees' perceptions of real women's conditions and their understanding of dominant—as well as their own—opinions about gender roles. [BH]

5103 **Godio, Julio.** El movimiento obrero argentino, 1910–1930: socialismo, sindicalismo y comunismo. Buenos Aires: Editorial Legasa, 1988. 442 p.: bibl. (Omnibus)

With copious, convincing evidence, author advances thesis that Argentine unions and worker party leaders' focus on European concerns and models made them uninterested in and blasé about progressive role that Yrigoyen and the Radicals were playing against conservative forces in Argentina. According to author, this historical error prevented the working-class movement from acquiring the strength and autonomy it should have achieved and deterred the nation from following an agroindustrial path of development similar to Australia's. [BH]

5104 **González Bombal, Inés.** Los vecinazos: las protestas barriales en el Gran Buenos Aires, 1982–83. Buenos Aires: Ediciones del IDES; Editorial Humanitas, 1988. 108 p.: bibl. (Ediciones del IDES, 0326–6133; 14. Col. Hombre y sociedad)

Adequate but largely descriptive study examines neighborhood-based protest movements that emerged towards the end of the military regime. Traces political unrest to military's failure to create effective mechanisms of political mediation at the community level. Combined with the crisis, this failure provided space for emergence of nontraditional political movements. [RPK]

5105 **Grassi, Estela.** La mujer y la profesión de asistente social: el control de la vida cotidiana. Buenos Aires: Editorial Humanitas, 1989. 246 p.: bibl. (Col. Desarrollo social)

Appealing inquiry examines role of social workers in Argentina. Combines historical approach to the profession and its predominantly female practitioners and students with a look at the ideological mechanisms that control daily life of the poor and reproduce class society. Of particular interest for women and those pursuing gender and ideology studies. [BH]

5106 **Gravano, Ariel and Rosana Gúber.** Barrio sí, villa también: dos estudios de antropología urbana sobre producción ideológica de la vida cotidiana. Prólogo de Hugo Ratier. Buenos Aires: Centro Editor de América Latina, 1991. 108 p.: bibl. (Biblioteca Política argentina; 320)

Includes two anthropological studies

which, although brief, provide important insights into social construction of urban identity. Gúber uses general data and field research to analyze how perceived identity of urban poor (*villeros*) differs among scholars, other urban resdents, and urban poor themselves. Gravano uses oral histories to analyze how neighborhood identity mediates class-, gender-, and age-related tensions. [RPK]

5107 **Hintze, Susana; Estela Grassi;** and **Mabel Grimberg.** Trabajos y condiciones de vida en sectores populares urbanos. Buenos Aires: Centro Editor de América Latina, 1991. 119 p.: bibl. (Biblioteca Política argentina; 327)

Essays present original data. Hintze argues that participation in informal labor markets represents a necessity rather than a choice for the poor. Grassi argues that *villas miserias* constitute an alternative housing strategy for workers rather than for unemployed poor. Grimberg indicates that health patterns among printing workers are related to occupational hazards and characteristics. [RPK]

5108 **Infancia y pobreza en la Argentina.** Dirección de Jorge Carpio, Alberto Minujin Z. y Pablo Vinocur. Coordinación de Susana Checa. Buenos Aires: UNICEF; INDEC; Siglo Veintiuno de España Editores; Catálogos S.R.L., 1990. 210 p.: bibl., ill.

Thorough analysis examines impact of poverty on Argentine children's health, education, and labor. Authors' adept methodology distinguishes between structurally poor children and those impoverished by the 1980s crisis. Universalism of childhood social policies is portrayed as benefiting children of non-poor households more than poor children, the intended beneficiaries. [BH]

5109 **Kalinsky, Beatriz; Wille Arrúe;** and **Diana Rossi.** La salud y los caminos de la participación social: marcas institucionales e históricas. Buenos Aires: Centro Editor de América Latina, 1993. 108 p.: bibl. (Biblioteca Política argentina; 405)

Provides interesting but anecdotal review of two expcricnces in the field of community health: specific groups (alcoholics, elderly, prisoners) participating in programs in Neuquén, and a health program in the periphery of the city of Buenos Aires. These programs sought active community involvement, but were seriously undermined by lack of resources and institutional support. [RPK]

5110 **Kornblit, Ana Lía; Ana M. Mendes Diz;** and **Azucena Bilyk.** Sociedad y droga. Buenos Aires Centro Editor de América Latina, 1992. 154 p.: bibl. (Aquí, Argentina)

Collection of essays provides interesting insights into drug consumption in Argentina. Interesting fourth chapter analyzes attitudes towards AIDS among drug users, and provides policy suggestions. Essays are well informed by international literature on the topic, but tend to be somewhat narrow in their field of study, making volume most useful for specialists.

5111 **Lattuada, Mario J.** El Estado argentino y los intereses industriales, 1983–1989. Buenos Aires: Centro Latinoamericano para el Análisis de la Democracia, 1990. 83 p.: bibl., ill. (Serie Informes de investigación; 1)

Useful study of political representation of manufacturing sector emphasizes organizational heterogeneity that characterizes both these interests and State agencies that cater to them. Agreements and compromises tend to be fragile due to difficulties encountered in trying to encapsulate diverse and often contradictory interests under a single set of demands or mode of representation. [RPK]

5112 **Lucchini, Cristina.** Apoyo empresarial en los orígenes del peronismo. Prólogo de Torcuato S. Di Tella. Buenos Aires: Centro Editor de América Latina, 1990. 128 p.: bibl. (Biblioteca Política argentina; 292)

In this analysis of speeches, statements, and editorials, author first traces Argentine industrialists' political and economic motivation for supporting military's project in 1930s and 1940s. Second part focuses on convergence of military's and industrialists' concerns and their proposed solutions. However, no evidence is presented to support thesis that industrialists actually backed Peronism. [BH]

5113 **Merklen, Denis.** Asentamientos en La Matanza: la terquedad de lo nuestro. Buenos Aires: Catálogos Editora, 1991. 205 p.: bibl., ill., maps.

Preliminary study examines land takeovers. Relying on selected oral histories, and loosely framing his discussion within broader literature on social movements, author reconstructs history of three communities to identify networks of leadership, solidarity, conflict, and organization. Existence of alternative models of urban development and linkages to existing political party system are both treated as important variables shaping the relative success of these communities. [RPK]

5114 Minujin Z., Alberto et al. Cuesto abajo: los nuevos pobres: efectos de la crisis en la sociedad argentina. Buenos Aires: UNICEF; LOSADA, 1992. 300 p.: bibl., ill.

While number of structurally poor declined slightly during 1980s, income inequalities have become more pronounced, affecting in particular middle sectors who have become impoverished. Pt. 3 provides a qualitative and quantitative assessment of strategies used by households to deal with financial constraints. Together, these monographs provide a substantial contribution to evaluation of income distribution and poverty in Argentina. [RPK]

5115 Moreno Ocampo, Luis Gabriel. En defensa propia: cómo salir de la corrupción. Buenos Aires: Editorial Sudamericana, 1993. 397 p.: bibl., ill.

Lawyer actively involved during 1980s in prosecuting military for human rights violations and rebellions offers interesting observations on corruption. Critical and somewhat biographical reflection on design and implementation of public policy, work is one of the few examples of its kind. Suggests alternative administrative practices to challenge public and private corruption. [RPK]

5116 Las mujeres en la imaginación colectiva: una historia de discriminación y resistencias. Recopilación de Ana María Fernández. Buenos Aires: PAIDOS, 1992. 363 p.: bibl.

Collection of essays provides good overview of current status of women's studies in Argentina. Many of the essays are of a theoretical orientation, ranging from such issues as rape to intersections between psychoanalysis and feminism. Some of the essays provide important insights on current social problems (such as teenage pregnancies). [RPK]

5117 Muraro, Heriberto. Poder y comunicación: la irrupción del *marketing* y de la publicidad en la política. Prólogo de Oscar Landi. Buenos Aires: Ediciones Letra Buena, 1991. 143 p.: bibl. (Col. Pensamiento científico)

This monograph is at its best in providing revealing insights into how traditional politics has been transformed by the new interplay among marketing, publicity, and political discourse. The empirical report on public opinion is more dated. Regrettably, book suffers from some major typographical and printing errors (such as deletion of a section on pp. 66–67). [RPK]

5118 Nino, Carlos Santiago. Un país al margen de la ley: estudio de la anomia como componente del subdesarrollo argentino. Buenos Aires: Emecé Editores, 1992. 273 p.: bibl., ill.

Author has a long and recognized trajectory in the legal field. To explain inability of Argentina to sustain democracy and high rates of economic growth, book emphasizes importance of institutional anomie (a lack of belief in and/or respect for law). Author explores how this way of thinking is reflected in a multiplicity of fields (politics, transit patterns, economic operations, academic promotions), and suggests that adequate norms could provide a better path for growth and democracy. [RPK]

5119 Recalde, Héctor. Beneficencia, asistencialismo estatal y previsión social. v. 1–2. Buenos Aires: Centro Editor de América Latina, 1991. 2 v.: appendix, bibl. (Biblioteca Política argentina; 335, 339)

Traces origin of social welfare programs to action of Catholics, doctors, and mutual aid societies. Growth of labor unrest and social movements at turn of the century significantly shaped these programs. Appendix includes several speeches representing approach of Catholic militants around 1910. Good use of historical evidence and careful organization make this a useful study of the topic. [RPK]

5120 Redondo, Nélida. Ancianidad y pobreza: una investigación en sectores populares urbanos. Buenos Aires: Centro de Promoción y Estudio de la Vejez; Editorial Humanitas, 1990. 276 p.: bibl.

Based on qualitative life course histories, this splendid book is part exploratory analysis of how an aged population copes with poverty and declining health, and part oral history of a 0.25 square mile pocket of the Boca neighborhood of Buenos Aires where 85.5 percent of the residents are over 85. [BH]

5121 **Reitano, Emir.** Manuel A. Fresco, antecedente del gremialismo peronista. Buenos Aires: Centro Editor de América Latina, 1992. 93 p.: appendix, bibl. (Biblioteca Política argentina; 385)

Joining other revisionist studies of the period, book argues that 1930s witnessed a concerted effort by certain State agencies (in this case, provincial authorities of Buenos Aires led by Governor Fresco) to channel labor movements towards formal institutional arrangements. These arrangements, as envisioned by Fresco, were influenced by the Italian Fascist experience. Includes appendix of speeches and writings by Fresco. [RPK]

5122 **Sidicaro, Ricardo.** La política mirada desde arriba: las ideas del diario La Nación, 1909–1989. Buenos Aires: Editorial Sudamericana, 1993. 545 p.: bibl. (Col. Historia y cultura)

Excellent study of political discourse analyzes newspaper editorials to provide comprehensive analysis of major debates and political transformations that engaged elites through most of 20th century. Considering its scope and depth, book constitutes a major contribution to the field. [RPK]

5123 **Sigal, Silvia.** Intelectuales y poder en la década del sesenta. Buenos Aires: Puntosur Editores, 1991. 259 p.: bibl. (La Ideología argentina)

Excellent, thorough discourse analysis covers leading intellectual streams in 1960s Argentina. Author addresses issue of separation between politics and society as evidenced by intellectuals' reluctance to participate in politics as well as by the State's misgivings about intellectuals' potential contributions. [BH]

5124 **Toer, Mario.** Como son los estudiantes: perfil socioeconómico y cultural de los estudiantes de la UBA. Buenos Aires?: Catálogos Editora; Ediciones Culturales Argentinas, 1990. 174 p.

Very interesting report covers three surveys of Univ. de Buenos Aires students (1,323 in 1985, 2,510 in 1986, and 2,315 in 1988). Data touch on wide spectrum of variables, including opinions on politics, religion, and sexuality. Despite their negative evaluations of their teachers, sociology students seem to be the most interesting students to teach. [BH]

5125 **Torrado, Susana.** Estructura social de la Argentina, 1945–1983. Buenos Aires: Ediciones de la Flor, 1992. 556 p.: bibl.

Detailed analysis examines census data on such variables as occupations, living conditions, and labor force participation in Argentina. Lacks analytical depth of Germani's classic studies (see item **5095**), but provides useful descriptive information on general social trends. Some government officials have recently challenged author's emphasis on deepening social inequalities. Statistics are by now somewhat dated. [RPK]

5126 **La trama solidaria: pobreza y microproyectos de desarrollo social.** Recopilación de Roberto Martínez Nogueira. Buenos Aires: Ediciones Imago Mundi; GADIS, 1991. 177 p.: bibl. (Col. Argentina debate)

Interesting essays evaluate experiences of small-scale development projects. Besides reviewing Argentina's recent trajectory in this area, essays are particularly good at identifying sources of tension affecting internal dynamics of grassroot organizations and their relationship to other forces involved (e.g., funding agencies, the community at large, State agencies). [RPK]

5127 **Vapnarsky, César A.** and **Néstor Gorojovsky.** El crecimiento urbano en la Argentina. Buenos Aires: Grupo Editor Latinoamericano; IIED-América Latina; Emecé Editores, 1990. 159 p.: bibl., ill., maps. (Col. Estudios políticos y sociales)

Magnificent analysis of census materials focuses on growth of *aglomeraciones de tamaño intermedio* (population clusters ranging from 50,000 to 999,999 inhabitants) for 1950–80. Rather than showing the continued primacy of Buenos Aires, period studied reveals an expansion of the smallest *aglomeraciones* (up to 200,000). This urban growth would be ideal for a decentralized industrial expansion catering to the domestic market. [BH]

5128 **Veeris, Milena.** Dierbare herinneringen die rondspoken: het verleden in een Duitse gemeenschap in Argentinië [Fond memories wandering about: the past in a German community in Argentina]. (*Etnofoor/Netherlands*, 6:1, 1993, p. 45–81, bibl.)

Report covers five-month period of fieldwork (1989–90) in Villa General Belgrano, Córdoba prov. Concludes that despite appearance of order and prosperity, German population has not yet come to grips with its traumatic past. Community's collective fantasy of order hides guilt and shame about World War II. [R. Hoefte]

5129 Wortman, Ana. Jóvenes desde la periferia. Buenos Aires: Centro Editor de América Latina, 1991. 93 p.: bibl. (Biblioteca Política argentina; 324)

Interesting study of social construction of youth combines interpretation of secondary materials with oral histories. Provides insights into major differences between males and females in their experience and definition of youth, and suggests that youth culture has been shaped by a persistent economic crisis. [RPK]

URUGUAY

5130 Bayce, Rafael. Drogas, prensa escrita y opinión pública. Montevideo: Fundación de Cultura Universitaria, 1990. 198 p.: bibl., facsims.

Combining systems theory with an interactionist approach, sociological study critically evaluates (through content analysis) coverage of drug issue in the mass media. Author claims that the mass media has itself made drugs an issue of public concern rather than the public's concern about drugs having generated the coverage. Study provides some interesting insights on the problem. [RPK]

5131 Boulet, Michel. Uruguay: ¿país en transición?; estudios semióticos. Montevideo: Ediciones de J. Darién, 1992. 115 p.: bibl., ill.

Uneven collection of semiotic essays covering a variety of themes is of limited interest for some specialists. [RPK]

5132 Errandonea, Alfredo and Marcos Supervielle. Las cooperativas en el Uruguay: análisis sociológico del primer relevamiento nacional de entidades cooperativas. Montevideo: Fundación de Cultura Universitaria, 1992. 115 p.: bibl., ill., maps.

One out of three to four Uruguayans participates in the cooperative system in some manner. Monograph uses 1989 national survey to provide in-depth study of this system. Cooperative experience appears to be associated with urbanization and some minimal living standards. Detailed data are provided on this phenomenon. Work is descriptive rather than analytical. [RPK]

5133 Genta Dorado, Gustavo. La colectividad japonesa en Uruguay. Montevideo: Ediciones de la Crítica, 1993. 89 p.: bibl., ill.

Drawing on interviews, author provides adequate portrayal of Japanese community in Uruguay. Study indicates that organizational ties within this community are relatively weak (as compared to those among Japanese communities in other Latin American countries such as Brazil or Argentina). Author attributes this weakness to small absolute size of the community (600 members, or .02 percent of the population). [RPK]

5134 González, Mariana. Las redes invisibles de la ciudad: las comisiones vecinales de Montevideo, 1985–1988. Montevideo: CIESU, 1992. 192 p.: bibl., ill.

Introductory chapter by Carlos Filgueira places this empirical study within broader literature on social organization and collective action. Survey data are used to show that neighborhood organizations tend to be informal, weak, linked to traditional political structures, and centered around very specific demands. They are most prevalent in lower-middle and middle class areas, and more recent organizations tend to have younger and more women participants. Excellent contribution to the field.

5135 López, Luz et al. Un mar de mujeres: trabajadoras en la industria de la pesca. Montevideo: Ediciones Trilce; Grecmu, 1992. 143 p.: bibl., ill., map.

Excellent study of female labor force participating in fishing industry of Uruguay. Provides both statistical overview and qualitative data based on interviews with workers, supervisors, and employers on the use, status, and trade union participation of female workers. [RPK]

5136 Midaglia, Carmen. Las formas de acción colectiva en Uruguay: movimientos de derechos humanos y el cooperativismo de vivienda por ayuda mutua. Montevideo: CIESU, 1992. 135 p.: bibl., ill.

Useful study uses secondary materials and interviews to reconstruct patterns of action of human rights organizations. Argues that specificity of their demands, and participation in these demands by political parties (generally from the left), weakened these organizations. Similar dynamics gradually came to characterize housing organizations. Results of study lead author to conclude that traditional forms of political action continue to prevail. [RPK]

5137 Rama, Claudio and **Gustavo Delgado.** El Estado y la cultura en Uruguay: análisis de las relaciones entre el Estado y la actividad privada en la producción de bienes y servicios culturales. Montevideo: Fundación de Cultura Universitaria; Ministerio de Educación y Cultura, 1992. 140 p.: bibl.

Adequate descriptive overview examines past organization and current characteristics of each of the major artistic spheres in the country. [RPK]

5138 Risso, Marta R. and **J. Yolanda Boronat.** La vivienda de interés social en el Uruguay: 1970–1983. Montevideo: Fundación de Cultura Universitaria; Instituto de Historia de la Arquitectura, Facultad de Arquitectura, Univ. de la República, 1992. 235 p.: bibl., ill., maps, photos.

Traces in detail recent history of low-cost housing in Uruguay, emphasizing legal framework and specific projects carried out over time. Includes great number of photographs of these projects, but little by way of analytical evaluation of their efforts. [RPK]

5139 Urruzola, María. El huevo de la serpiente: tráfico de mujeres Montevideo-Milán; ¿el nacimiento de una mafia? Montevideo: Ediciones de la Pluma, 1992. 174 p.

Journalistic account of Uruguayan prostitution rings in Milan argues that prostitutes were controlled by men who were not merely engaging in criminal association, but represented an actual mafia. Highlights lack of action by police in Uruguay. [RPK]

5140 Uruguayos en Argentina y Brasil: movimientos de población entre los países del Plata. v. 1. Montevideo?: Organización Internacional para las Migraciones; Comisión Económica para América Latina y el Caribe, 1991. 1 v.

Descriptive articles provide detailed assessment of both Uruguayan migration and Uruguayan communities in Argentina and Brazil. Although studies present useful evaluation, most of statistics are rather outdated (from 1970s–early 1980s). [RPK]

BRAZIL

PEGGY A. LOVELL, *Associate Professor of Sociology, University of Pittsburgh*

THE SOCIOLOGICAL LITERATURE on Brazil reviewed in this section includes articles and books published between 1990–94. Confronted by deepening social inequalities during this period, the tendency among Brazilian and Brazilianist scholars is toward studies that examine the ways in which gender, racial, religious, class, and regional differences are interrelated. Sociologists of gender, for example, have contributed a number of studies on how gender identities and inequalities are shaped by class and race (items **5175** and **5201**). Research on Brazilians of African descent also reflects this trend. Three excellent contributions on the changing dynamics of race relations are those by Thomas Skidmore (item **5202**), Howard Winant (item **5211**) and Nelson do Valle Silva and Carlos Alfredo Hasenbalg (item **5200**).

With the consolidation of civilian rule in the early 1990s, a central theme among political sociologists is the role of social movements in the political transition. Studies of labor movements investigate the role of unions (item **5195**), and Ruth Corrêa Leite Cardoso analyzes the negotiation process among social movements, political parties, and the State (item **5152**). Other studies examine the role of the Catholic Church and Christian Base Communities (CBCs). Works include studies of Catholicism and political action (item **5148**), the influence of the Church and the feminist movement on women (item **5176**), and empirical studies of CBCs (items **5173** and **5149**). Studies by sociologists of religion also analyze the growing religious diversity among the poor (item **5179**).

In the Editor's Note to *HLAS 52* and *HLAS 53*, Dolores Moyano Martin noted that the study of violence has become a recognized specialty in Latin America. In Brazil, scholars who study violence often examine the effects of violence on women and children. Notable examples of such writing are Marlise Vinagre Silva's *Violência contra a mulher: quem mete a colher?* which analyzes the sociopolitical significance of police delegations specialized in attending women (item **5198**) and Gilberto Dimenstein's investigation of forced child prostitution in the Brazilian Amazon (item **5165**).

Environmental issues in Brazil continue to command attention. Research includes Emilio F. Moran's analysis of the range of human and ecological diversity present in the Amazon Basin (item **5181**), Ana Luiza Ozorio de Almeida's study of directed colonization in the 1970s (item **5143**), and Wood and Schmink's article on the role and changing strategies of the military in developing the Amazon (item **5213**). Set against the backdrop of the current international debate on population growth and development, several publications discuss the changes in fertility in Brazil and the reproductive rights of women (items **5168, 5156**, and **4409**). An emerging area of sociological inquiry is popular culture, especially the media (items **5163** and **5192**).

5141 **Abreu, Alice R. de Paiva.** Mudança tecnológica e gênero no Brasil. (*Novos Estud. CEBRAP*, 35, março 1993, p. 121–132)
Draws on recent theoretical approaches within sociology of labor to discuss impact of new microelectronics technology on sexual division and organization of labor.

Adriance, Madeleine. The paradox of institutionalization: the Roman Catholic Church in Chile and Brazil. See item **4995**.

5142 **African-American reflections on Brazil's racial paradise.** Edited by David J. Hellwig. Philadelphia: Temple Univ. Press, 1992. 258 p.: bibl.
Unique collection of commentaries examines race relations from century's first decade through 1980s. Written by African-American scholars, journalists, and educators, documents provide wide range of insights into black American social thought on racial situation in both Brazil and US. Includes writings from W.E.B. du Bois and E. Franklin Frazier.

Alario, Margarita. Environmental policy enactment under the military: some generalities between Brazil and Chile. See item **4999**.

5143 **Almeida, Anna Luiza Ozorio de.** The colonization of the Amazon. Austin: Univ. of Texas Press, 1992. 371 p.: bibl., ill., index, maps. (Translations from Latin America series)
Excellent comprehensive study examines directed colonization of the Amazon during 1970s. Provocative results based on detailed primary and secondary data demonstrate success of small farmers. Concludes that land distribution can be an effective policy of rural income redistribution provided that correct pricing and other economic conditions are met.

5144 **Alves, Maria Helena Moreira.** Cultures of fear, cultures of resistance: the new labor movement in Brazil. (*in* Fear at the edge: State terror and resistance in Latin America. Berkeley: Univ. of California Press, 1992, p. 184–211)
Traces the experience of metal workers of the ABC region of São Paulo. Focuses on how base-level democratic politics influenced the organization, growth of solidarity, and success of the movement.

5145 **Barbosa, Lívia.** O jeitinho brasileiro: a arte de ser mais igual que os outros. Prefácio de Roberto da Matta. Rio de Janeiro: Editora Campus, 1992. 153 p.: bibl.
Socioanthropological investigation examines a social practice known and legitimized by all segments of society. Author locates analysis in a discussion of equality and social hierarchy.

5146 **Barbosa, Luiz C.** The "greening" of the ecopolitics of the world-system: Amazonia and changes in the ecopolitics of Brazil. (*J. Polit. Mil. Sociol.*, 21:1, Summer 1993, p. 107–134, bibl.)

Examines role of NGOs, grassroots movements, and media coverage in bringing about change in Brazilian government's ecopolitics. Argues that Brazil's economic dependence on international loans to stabilize its economy and implement new development projects added to political leverage of environmentalists.

5147 **Beozzo, José Oscar.** Brasil: 500 anos de migrações: povos indígenas, escravos africanos e brasileiros, imigrantes europeus e asiáticos. São Paulo: Edições Paulinas; Centro de Estudos Migratórios, 1992. 110 p.: bibl. (Conscientizar)

Chronicles 19th-century migration of indigenous peoples, African slaves, and European immigrants.

5148 **Bruneau, Thomas C.** and **W.E. Hewitt.** Catholicism and political action in Brazil: limitations and prospects. (*in* Conflict and competition: the Latin American Church in a changing environment. Edited by Edward L. Cleary and Hannah Stewart-Gambino. Boulder, Colo.: Lynne Rienner Publishers, 1992, p. 45–62)

Examines domestic and international secular and religious factors that have influenced the politicization of the Catholic Church. Argues that during military dictatorship the Brazilian Church redefined its role as defender of the poor and oppressed. Loss of Vatican support, the threat of competing value movements, and severe political and economic stress led to Catholic Church's decreased activism in political sphere.

5149 **Burdick, John.** Looking for God in Brazil: the progressive Catholic Church in urban Brazil's religious arena. Berkeley: Univ. of California Press, 1993. 280 p.: bibl., ill., index.

Provocative study of Christian Base Communities describes range of religious and ideological alternatives available to different sectors of the population. Focuses on context in which certain groups (married women, youth, Afro-Brazilians) make religious choices. Develops a methodology applicable to study of social movements.

5150 **Burdick, John.** Rethinking the study of social movements: the case of Christian Base Communities in urban Brazil. (*in* The making of social movements in Latin America: identity, strategy, and democracy. Edited by Arturo Escobar and Sonia E. Alvarez. Boulder, Colo.: Westview Press, 1992, p. 171–184)

Drawing from comparative study of a Christian Base Community and a Pentecostal church, Burdick proposes a new unit of analysis in the study of social movements. Argues that rather than focusing only on participants or movements themselves, a more comprehensive approach would include study of non-participants and the broader arenas in which each social movement exists.

5151 **Cândido, Antônio.** A Faculdade no centenário da abolição. (*Novos Estud. CEBRAP*, 34, Nov. 1992, p. 21–30)

Analyzes historical contribution of Univ. de São Paulo's Faculdade de Filosofia, Letras e Ciências Humanas to changes in the study of Brazilian society. Focusing on Afro-Brazilian studies, author considers contributions of the Faculdade to current debate on racial inequality.

5152 **Cardoso, Ruth Corrêa Leite.** Popular movements in the context of the consolidation of democracy in Brazil. (*in* The making of social movements in Latin America: identity, strategy, and democracy. Edited by Arturo Escobar and Sonia E. Alvarez. Boulder, Colo.: Westview Press, 1992, p. 291–302)

Analyzes process of negotiation among social movements, political parties, and the State in order to better understand how new sociopolitical identities are forged. Concludes with analysis of contribution of popular movements to Brazil's democratization process.

5153 **Carelli, Mario.** Cultures croisées: histoire des échanges culturels entre la France et le Brésil, de la découverte aux temps modernes. Préface de Gilbert Durand. Paris: Editions Nathan, 1993. 250 p.: bibl., index. (Col. Essais & recherches)

French-Brazilian researcher Carrelli, an extremely productive writer, contributes an extant (though light) overview of reciprocal cultural transfer between France and Brazil since the 16th century. Book may also be read as a compilation of short essays dealing with a series of privileged witnesses from both nations. French testimonies (the majority) include those by Thevet, Léry, Montaigne, La Condamine, Debret, Ternaux Compans, Claudel, Bernanos, Caillois, and Bastide. Useful perspective on the rich, transatlantic confluence of French and Brazilian nature and culture. [T. Hampe-Martínez]

5154 **Carta-tema: a assistência social no Brasil, 1983–1990.** Coordenação de Aldaíza de Oliveira Sposati. São Paulo: Cortez Editora, 1989. 94 p.: appendices, bibl.

Documents debates and conferences held to discuss inclusion of social security in 1988 constitution. Appendices provide associated declarations and a copy of the bill included in the constitution.

5155 **Carvalho, José Alberto Magno de** and **Claudio Caetano Machado.** Quesitos sobre migrações no Censo Demográfico de 1991. (*Rev. Bras. Estud. Popul.*, 9:1. jan./julho 1992, p. 22–34, bibl., table)

Discusses innovations introduced in 1991 Brazilian Censo Demográfico to measure migration. Examines analytical potential as well as comparability of these data with last two censuses.

5156 **Carvalho, José Alberto Magno de** and **Laura Rodríguez Wong.** La transición de la fecundidad en Brasil: causas y consecuencias. (*Notas Pobl.*, 20:56, dic. 1992, p. 107–141, bibl., graph, tables)

Documents decline in fertility across all social classes in Brazil from 1940–85. Reduction was most pronounced over two periods, 1970–75 and 1980–85, periods for which, paradoxically, indicators pointed to a substantial deterioration in the quality of life for Brazil's poorest.

5157 **Castro, Nadya** and **Antônio Sérgio Alfredo Guimarães.** Trabalhadores afluentes, indústrias recentes: revisitando a tese da aristocracia operária. (*Dados/Rio de Janeiro*, 35:2, 1992, p. 173–191)

Discusses formation of class identity among skilled, affluent petrochemical workers in Bahia. Presents and criticizes classic arguments that reject possibility of labor classes emerging in markets characterized by high salaries, stable employment, infusion of State capital, and developmentalist ideologies.

5158 **Collins, Jane L.** Gender, contracts and wage work: agricultural restructuring in Brazil's São Francisco Valley. (*Dev. Change*, 24:1, Jan. 1993, p. 53–82, bibl.)

Examines development of highly segmented agricultural labor market in domestic and global fruit and vegetable production in Brazilian Northeast. Based on interviews with agribusiness firms, Collins found that, depending on quality and timing of the crop, either seasonal migrants or local women and children are recruited. Following patterns of industrial and agricultural production in other parts of the world, these firms are establishing labor arrangements that tap the most vulnerable segments of the workforce.

5159 **Corcoran-Nantes, Yvonne.** Female consciousness or feminist consciousness?: women's consciousness raising in community-based struggles in Brazil. (*in* "Viva:" women and popular protest in Latin America. Edited by Sarah A. Radcliffe and Sallie Westwood. London; New York: Routledge, 1993, p. 136–155)

Examines development of political consciousness and solidarity among women of the popular urban movements in São Paulo. By focusing on forms of consciousness raising, self-help groups, oral history, and the struggle for literacy, author demonstrates complexity of women's political consciousness and interrelationships among gender, class, and social protest.

5160 **Costa Sobrinho, Pedro Vicente.** Chico Mendes: a trajetória de uma liderança. (*São Paulo Perspect.*, 6:1/2, jan./junho 1992, p. 175–186)

Excerpts from previously unpublished interview with rubber tapper leader Chico Mendes describe his rise to international prominence, his political activism, the defense of the Amazon rainforest, and original goals of the Xapuri rubber tappers' movement. Prefaced by analysis of rubber tappers and their unionization in Acre.

5161 **Curtis, Siân L.; Ian Diamond;** and **John W. McDonald.** Birth interval and family effects on postneonatal mortality in Brazil. (*Demography/Washington*, 30:1, Feb. 1993, p. 33–43, bibl., tables)

Findings from Brazil's 1986 Demographic and Health Survey show significant variation among families' risk of experiencing postneonatal deaths, even after controlling for socioeconomic characteristics. Suggests that increased use of contraception to space births could make an important contribution to reducing risks of infant death in the less-developed Northeast region where short birth intervals are common.

5162 **Dassin, Joan.** Testimonial literature and the armed struggle in Brazil. (*in* Fear at the edge: State terror and resistance in Latin America. Berkeley: Univ. of California Press, 1992, p. 161–183)

Dassin reviews and categorizes a variety of testimonial literature written about the 1969–73 Brazilian guerrilla movement. Analyzes relationship between works published from 1979–84 and the politics of *abertura*, focusing on questions of human rights, free expression, and social dynamics of fear.

5163 **De Lima, Venicio A.** Brazilian television in the 1989 presidential election: constructing a president. (*in* Television, politics, and the transition to democracy in Latin America. Washington: Woodrow Wilson Center Press, 1993, p. 97–117)

Examines media's role, especially the free electoral broadcast time (*horário gratuito*), in the 1989 presidential elections. Discusses theoretical, legal, and political implications for conducting research on relationship of media and politics.

5164 **De Onis, Juan.** The green cathedral: sustainable development of Amazonia. New York: Oxford Univ. Press, 1992. 280 p.: bibl., index, maps.

Author, a former correspondent for *The New York Times*, discusses often conflicting perspectives of First World environmentalists and Amazon residents on ways to approach the goal of sustainable development. Drawing on visits to the Brazilian Amazon over a 20-year period, he calls for establishment of a macro-ecological framework through which micro-ecological actions become the elements of sustainable economic growth and a reasonable ecological balance.

5165 **Dimenstein, Gilberto.** Meninas da noite: a prostituição de meninas-escravas no Brasil. 2a. ed. São Paulo: Editora Atica, 1992. 161 p.: ill., maps.

Internationally-recognized journalist reports results of his investigation of forced child prostitution in Brazilian Amazon. Vividly illustrated with photographs and life stories, work documents enslavement of young girls on Amazonian frontier.

5166 **Faria, Vilmar.** A conjuntura social brasileira: dilemas e perspectivas. (*Novos Estud. CEBRAP*, 33, julho 1992, p. 103–114, tables)

Evaluates social developments in Brazil within a comparative framework, based on welfare indicators used by UN Development Programme. Special attention is given to 1980s and to current challenges to development.

5167 **Figueira, Ricardo Rezende.** Rio Maria: song of the earth. Translated and edited by Madeleine Adriance. Maryknoll, N.Y.: Orbis; London: CIIR; Dublin: Trócaire, 1994. 146 p.

Vignettes portray poor farmers and indigenous peoples in their daily struggles against poverty and violence in the state of Pará.

5168 **Frias, Luiz Armando de Medeiros** and **Juarez de Castro Oliveira.** Níveis, tendências e diferenciais de fecundidade no Brasil a partir da década de 30. (*Rev. Bras. Estud. Popul.*, 8:1/2, jan./dez. 1991, p. 72–111, bibl., graphs, tables)

Offers historical profile of level and age pattern of fertility for entire country and for major geographic regions. Compares and evaluates several methodologies.

5169 **Gomes, Cândido Alberto da Costa.** O jovem e o desafio do trabalho. São Paulo: Editora Pedagógica e Universitária, 1990. 125 p.: bibl., ill.

Based on empirical results, book examines child labor in contemporary Brazil. Analyzes choices often made between children's schooling and their employment opportunities in tertiary sector. Compares Brazilian case with experiences of other countries in order to propose alternatives to children's work.

5170 **Goza, Franklin.** Causes and consequences of migration in the Jequitinhonha Valley of Minas Gerais. (*Sociol. Inq.*, 62:2, Spring 1992, p. 147–168, bibl., tables)

Based on migration histories of households in four communities, study documents temporary inter-regional labor migration in central-west Brazil. Explanations for and consequences of this migration are analyzed in light of Brazil's economic crisis.

5171 Grün, Roberto. Sindicalismo & antisindicalismo e a gênese das novas classes médias brasileiras. (*Dados/Rio de Janeiro*, 35:3, 1992, p. 435–471)

Examines the emergence of the modern middle class in Brazil, transformations in the economy, and the resulting anti-union sentiment of "neoliberal" ideology.

5172 Heinz, W.S. De mensenrechten in de nieuwe Braziliaanse democratie [Human rights in the new Brazilian democracy]. (*Anthropol. Verkenn.*, 11:4, 1992, p. 23–32)

Despite transition to democracy, severe human rights violations continue to occur in Brazil. The political context, however, differs from that of the previous dictatorship. Nowadays, local landowners and urban businessmen, rather than the armed forces, order killings. Author calls for a broad dialogue involving large sectors of society to discuss the violence. [R. Hoefte]

5173 Hewitt, W.E. The evolution of lay leadership within Brazilian *Comunidades Eclesiais de Base* (CEBs). (*in* Forging identities and patterns of development in Latin America and the Caribbean. Edited by Harry P. Díaz, Joanna W.A. Rummens, and Patrick D.M. Taylor. Toronto: Canadian Scholars' Press, 1991, p. 197–213, tables)

Based on a four-year study (1984–88) in São Paulo, article examines internal dynamics of 22 Christian Base Communities (CBCs). Results suggest a potentially debilitating crisis of leadership and an enduring tendency to both elitism and stagnation. Concludes that these problems may be resolved as the Church shifts support away from CBCs, thereby forcing them to assume greater responsibility for their own affairs.

5174 Ianni, Octávio. Ensaios de sociologia da cultura. Rio de Janeiro: Civilização Brasileira, 1991. 212 p.: bibl.

Collection of essays, some written during the military dictatorship, is divided into three parts: "O Romance da História;" "A Imaginação da Sociedade;" and "Cultura e Hegemonia." Ianni reflects on various aspects of Brazilian culture including abuse of power, consumption, cultural expressions of minorities, and industrial culture.

5175 Lovell, Peggy A. Race, gender and development in Brazil. (*LARR*, 29:3, 1994, p. 7–35, bibl., tables)

Using sample data from 1960 and 1980 demographic censuses, study examines differential gains made by Afro-Brazilian and white women and men in urban workforce. Results show that in over two decades of economic change, women and Afro-Brazilians benefited in absolute terms (in education, occupational distribution, and wages); yet they continued to suffer relative disadvantages.

5176 Machado, Leda Maria Vieira. "We learned to think politically:" the influence of the Catholic Church and the feminist movement on the emergence of the health movement of the Jardim Nordeste area in São Paulo, Brazil. (*in* "Viva:" women and popular protest in Latin America. Edited by Sarah A. Radcliffe and Sallie Westwood. London; New York: Routledge, 1993, p. 1–29, map)

Traces influence of Christian Base Communities and feminist movement on mobilization of low-income women during the military dictatorship. Chronicles women's increasing politicization, changing attitudes, and perceptions of themselves.

MacRae, Edward. Homosexual identities in transitional Brazilian politics. See item **3728**.

5177 Maia, Sylvia M. dos Reis. Market dependency as subsistence strategy: the small producers in Sapeaçu, Bahia. (*Bull. Lat. Am. Res.*, 10:2, 1991, 193–219, bibl., tables)

Ethnographic study surveys Northeastern community producing subsistence crops and tobacco for the market. Analyzes constraints of underdevelopment, relates household survival strategies (e.g., migration) used to cope with such constraints, and describes effects of these strategies on women.

5178 Margolis, Maxine L. Little Brazil: an ethnography of Brazilian immigrants in New York City. Princeton, N.J.: Princeton Univ. Press, 1994. 329 p.: bibl., ill., index.

This rich and wonderfully written ethnography—illustrated with vignettes—chronicles immigrant experiences of Brazilians in New York City.

5179 Mariz, Cecília Loreto. Coping with poverty: Pentecostals and Christian base communities in Brazil. Philadelphia, Pa.: Temple Univ. Press, 1994. 195 p.: bibl., index.

Examines growing religious diversity among the poor. Focuses on growth of Pentecostalism and Afro-Spiritism in light of broader social and economic changes, as well as on role of each religion in everyday life.

5180 **Morales, Anamaria.** Blocos negros em Salvador: reelaboração cultural e símbolos de baianidade. (*Cad. CRH*, suplemento 1991, p. 72–93, bibl.)

Examines "re-Africanization" of carnival in Bahia in terms of both cultural identity and political activism.

5181 **Moran, Emilio F.** Through Amazonian eyes: the human ecology of Amazonian populations. Iowa City: Univ. of Iowa Press, 1993. 230 p.: bibl., ill., index.

Analyzes range of human and ecological diversity present in the Amazon Basin. Examines a number of ecosystems, their ecological characterization, and some workable human strategies of resource use. Recommended.

5182 **Mulher e políticas públicas.** Rio de Janeiro: IBAM/UNICEF, 1991. 227 p.: bibl., ill.

Important collection of interdisciplinary essays is divided into three parts: "Trabalhando com Mulheres," "Trabalhando para Mulheres," "Mulheres Trabalhando." Includes writings on relationship between gender and class, methodology, family planning, social programs, and health.

5183 **Na corda bamba: doze estudos sobre a cultura da inflação.** Organização de José Ribas Vieira et al. Rio de Janeiro: Relume Dumará, 1993. 196 p.: bibl., ill.

Unique collection of interdisciplinary essays examines inflation as a social phenomenon and considers its implications for society and the individual. Includes a chapter by Roberto da Matta.

5184 **Neri, Anita Liberalesso.** Envelhecer num país de jovens: significados de velho e velhice segundo brasileiros não idosos. Campinas, Brazil: Editora da UNICAMP, 1991. 155 p.: bibl., ill. (Teses)

Empirical and methodological contribution to study of aged in Brazil is based on survey of attitudes of non-elderly toward aging.

5185 **Nugent, Stephen L.** Amazonian *caboclo* society: an essay on invisibility and peasant economy. Providence, R.I.: Berg, 1993. 278 p.: bibl., ill., index, maps. (Explorations in anthropology)

Ethnohistory of Amazonian peasant communities in Santarém and Combú (state of Pará) examines way in which ecological, modernization, and (more recently) sustainable development perspectives have treated *caboclo* societies.

5186 **Oliven, Ruben George.** A parte e o todo: a diversidade cultural no Brasil-nação. Petrópolis, Brazil: Vozes, 1992. 143 p.: bibl.

Examines question of cultural diversity, giving special attention to state of Rio Grande do Sul.

5187 **Ortiz, Renato; Silvia Helena Simões Borelli; and José Mário Ortiz Ramos.** Telenovela: história e produção. São Paulo: Editora Brasiliense, 1989. 197 p.: bibl.

Documents history, evolution, and production of Brazilian soap operas, based on empirical data from TV Globo and TV Manchete.

5188 **Pan American Health Organization.** Brazil. (*in* Tobacco or health: status in the Americas. Washington: PAHO, 1992, p. 64–80, bibl., tables)

Documents mortality effects of population's increased exposure to smoking as revealed in the increasing rates of lung cancer among both men and women. In Brazil at least 32,400 deaths per year are attributed to smoking. Discusses national and local efforts to control tobacco use.

5189 **Papma, Frans.** Contesting the household estate: Southern Brazilian peasants and modern agriculture. Amsterdam: Centre for Latin American Research and Documentation, 1992. 267 p.: bibl., ill., index, map. (Latin America studies; 67)

Case study of peasant community of São Judas Tadeu shows that household estate is contested between parents and children. Struggle leads to various types of production relations between parents and married children. Work also discusses peasant movement and its proposals to foster peasant agriculture. [R. Hoefte]

5190 **Paula, Sergio Goes de.** Morrendo à toa: causas da mortalidade no Brasil. São Paulo: Editora Ática, 1991. 160 p.: bibl. (Ensaio; 134)

Begins with well-written discussion of limits of demographic theory. Provides empirical analysis of rates and causes of death in Brazil from 1940-80, with a regional breakdown.

5191 A questão social no Brasil. Organização de João Paulo dos Reis Velloso. São Paulo: Nobel, 1991. 269 p.: bibl., ill.

Collection of papers, primarily by economists of the Instituto de Pesquisa Econômica Aplicada, documents increase in absolute poverty in Brazil during 1980s. Divided into three parts, work addresses social question and its demographic base, education and housing, and debate about social inequality.

5192 Rede imaginária: televisão e democracia. Organização de Adauto Novaes. São Paulo: Secretaria Municipal de Cultura; Companhia das Letras, 1991. 315 p.: bibl.

Includes collection of papers presented at a conference ("Rede Imaginária—Televisão e Democracia") held in São Paulo in 1990. Divided in two parts, work discusses social construction and consequences of television images.

5193 Rizzini, Irene. Infância, adolescência e pobreza na década de 1980: a situação da menina no Brasil. (*Humanidades/Brasília*, 8:1, 1992, p. 28-36, bibl., photos, table)

Examines interrelationship of poverty, prostitution, and violence in the exploitation of young girls.

5194 Rodrigues, Gilda de Castro. Planejamento familiar. São Paulo: Editora Atica, 1990. 94 p.: bibl. (Série Princípios; 197)

Examines family planning debate in light of both political interests expressed at the national level and women's concerns expressed at individual level. Provides limited statistics on abortion, contraceptive use, infanticide, abandonment, and family planning.

5195 Sandoval, Salvador A.M. Social change and labor unrest in Brazil since 1945. Boulder, Colo.: Westview Press, 1993. 245 p.: bibl., ill., index.

Examines relationship between labor and the State in Brazil from end of World War II to 1989. Drawing from a variety of sources, Sandoval chronicles changes in form, frequency, and content of strike activity. By analyzing importance of political, economic, and organizational processes in transformation of labor conflict, author contributes a new perspective to debate over economic development and democratization.

5196 Seidman, Gay W. Manufacturing militance: workers' movements in Brazil and South Africa, 1970-1985. Berkeley: Univ. of California Press, 1994. 361 p.: bibl., index.

Traces emergence of labor movements, and movements' similarities, in two different political, legal, and racial contexts. Challenges prevailing theories of development, of labor movement dynamics, and of transition from authoritarian rule, while in the process developing a broad analysis of how industrialization strategies shape the opportunities for labor.

5197 Silva, Maria A. Moraes. A nova divisão sexual do trabalho na agricultura. (*São Paulo Perspect.*, 4:3/4, julho/dez. 1990, p. 20-31, tables)

Based on census and survey data, study examines changes in agricultural production in São Paulo state from 1970-84. Over that time period women increased their labor participation, yet remained in the lowest-paying categories and were twice as likely as men to work without pay.

5198 Silva, Marlise Vinagre. Violência contra a mulher: quem mete a colher? Prefácio (ou Post-fácio?) de Heleieth I.B. Saffioti. São Paulo: Cortez Editora, 1992. 180 p., 2 leaves of plates: bibl., ill.

Pioneering theoretical and empirical study examines sociopolitical significance of police delegations specialized in attending women who are victims of violence. Considers relationships between public and private exercise and abuse of power, and articulates a theoretical and practical debate regarding class, race, and gender relations. Discusses possibility for a new role opening for the Serviço Social to include attention to women's civil rights. Foreword by Heleieth I. B. Saffioti.

5199 Silva, Nelson do Valle. O caso brasileiro: a sociedade. (*in* Sociedade, Estado e partidos na atualidade brasileira. São Paulo: Paz e Terra, 1992, p. 65-115, tables)

Provides thorough summary of sociodemographic change (e.g., changing levels in income, schooling, fertility) occurring since 1950. Based on census and survey data, documents rapid and uneven character of social change in Brazilian society.

5200 Silva, Nelson do Valle and **Carlos Alfredo Hasenbalg.** Relações raciais no Brasil contemporâneo. Rio de Janeiro: Rio Fundo Editora, 1992. 173 p.: bibl., ill.

Two of Brazil's leading scholars on race relations analyze racial stratification in Brazilian society during 1980s. Examines interracial marriage, miscegenation, racial inequality in educational and occupational levels, poverty, and the black movement. This important volume—based on the most recent statistical data available—demonstrates persistent racial inequality in contemporary Brazilian society.

Silva Téllez, Armando. Imaginarios urbanos, Bogotá y São Paulo: cultura y comunicación urbana en América Latina. See item **4824**.

5201 Simpson, Amelia S. Xuxa: the mega-marketing of gender, race, and modernity. Philadelphia, Pa.: Temple Univ. Press, 1993. 238 p.: bibl., ill., index.

Fascinating book explores international rise to fame of Brazil's blond megastar of children's television. Simpson argues that Xuxa's representation of femininity, her privileging of a white ideal of beauty, and her encouragement of consumerism perpetuate social inequality.

5202 Skidmore, Thomas E. Bi-racial U.S.A. vs. multi-racial Brazil: is the contrast still valid? (*J. Lat. Am. Stud.*, 25:2, May 1993, p. 373–386)

Provocative commentary suggests that during 20th century Brazil and US have come closer together on racial classification. Skidmore offers evidence indicating that color gradation is more important in the US than conventional wisdom suggests, whereas in Brazil it appears that color distinctions are becoming less important. Concludes with suggestions for further research.

5203 Souza-Lobo, Elisabeth. A classe operária tem dois sexos: trabalho, dominação e resistência. Apresentação de Helena Hirata. São Paulo: Secretaria Municipal de Cultura, Prefeitura do Município de São Paulo; Editora Brasiliense, 1991. 285 p.: bibl.

Collection of theoretical writings was published in memory of the author, a feminist labor scholar and political activist. Organized in three parts: 1) "Práticas e Discursos das Operárias, Processos de Trabalho e Lutas Sindicais no Brasil: os Anos 1970 e 1980;" 2) "O Gênero no Trabalho: Perspectivas Teóricas e Metodológicas;" and 3) "Movimentos Sociais de Mulheres: Igualdade e Diferença."

5204 Subervi-Velez, Federico A. and **Omar Souki Oliveira.** Negros (e outras etnias) em comerciais da televisão brasileira: uma investigação exploratória. (*Comun. Soc.*, 10:17, agôsto 1991, p. 79–101, bibl., tables)

Qualitative study focuses on limited representation of individuals of African descent in Brazilian commercial advertising. Observing nearly 60 hours of prime-time television and 1,500 commercials, authors found that Afro-Brazilians appeared in only 47 advertisements. Even more striking in a country where nearly one-half of the population is non-white is the fact that in only nine of those commercials did Afro-Brazilians have speaking roles.

5205 Sydenstricker, John M. and **Haroldo Torres.** Mobilidade de migrantes: autonomia ou subordinação na Amazônia legal? (*Rev. Bras. Estud. Popul.*, 8:1/2, jan./dez. 1991, p. 33–54, bibl., graphs, tables)

Focuses on mobility of the population settled in Machadinho, a rural colonization project in the state of Rondônia. Based on two years of household surveys, social mobility is analyzed according to general household indicators (age and sex composition, migratory experience, etc.). Differential mobility patterns are examined in context of recent debates on household survival strategies and on the peasantry in Brazilian frontier.

5206 Teixeira, Faustino Luiz Couto. Base church communities in Brazil. (*in* The Church in Latin America, 1492–1992. Edited by Enrique Dussel. Kent, England: Burns & Oats; Maryknoll, N.Y.: Orbis Books, 1992, p. 403–418, bibl.)

Thorough review examines emergence of Christian Base Communities within broader social, political, and cultural context of Brazilian society. Special attention is given to domestic and international ecclesial developments.

5207 Telles, Edward E. Industrialization and racial inequality in employment: the Brazilian example. (*Am. Sociol. Rev.*, 59:1, Feb. 1994, p. 46–63, bibl., tables)

Based on data from 1980, examines how racial inequality in occupations varies with level of industrialization across 74 Brazilian metropolitan areas. Finds that industrialized areas have lower occupational inequality overall; yet at higher occupational levels, racial inequality is either greater or unaffected by industrialization.

5208 Vainsencher, Semira Adler. O projeto de vida do menor institucionalizado. Recife, Brazil: Fundação Joaquim Nabuco, Instituto de Pesquisas Sociais, Depto. de Educação, 1989. 140 p.: bibl.

Presents results of research on socioeconomic background and daily life of abandoned children residing in a state institution in Recife. Provides detailed data on the children, including their family, education, exposure to violence, and use of drugs. Also includes results of interviews with the children on their opinions of the institution and their future goals.

5209 Velho, Otávio. Preventing or criticizing the process of modernization?: the case of Brazil. (*Ciênc. Cult.*, 44:1, jan./fev. 1992, p. 16–19, bibl.)

Argues that people sometimes resist modernization through development of alternatives normally associated with tradition. Velho exemplifies his argument with the recent debate on parliamentary monarchy.

5210 Vidas em risco: assassinatos de crianças e adolescentes no Brasil. Rio de Janeiro: Movimento Nacional de Meninos e Meninas de Rua; Instituto Brasileiro de Análises Sociais e Econômicas; São Paulo: Núcleo de Estudos da Violência, Univ. de São Paulo, 1991. 111 p.: bibl., ill.

Based on newspaper accounts from Recife, São Paulo, and Rio de Janeiro, book documents assassinations of children in 1989. Includes bibliographical references and synopsis of a report by Amnesty International.

5211 Winant, Howard. Racial conditions: politics, theory, comparisons. Minneapolis: Univ. of Minnesota Press, 1994. 199 p.: bibl., index.

Winant offers provocative discussion of theory, politics, and meaning of race at end of 20th century. Much of analysis is devoted to contrast between US and Brazil.

5212 Women in Brazil. London: Latin American Bureau; New York: Monthly Review Press, 1993. 139 p.: bibl.

Collection of articles, poems, and interviews assembled by Brazilian women's organization Caipora explores conditions of sexism, racism, and exploitation faced by poor women. Describes how women have engaged in social protest through organizing in shantytown neighborhood groups, Christian Base Communities, peasant associations, trade unions, and the women's movement.

5213 Wood, Charles H. and **Marianne Schmink.** The military and the environment in the Brazilian Amazon. (*J. Polit. Mil. Sociol.*, 21:1, Summer 1993, p. 81–105, bibl., map)

Analyzes role and changing strategies of the military from 1964–85 in developing Brazilian Amazon. Authors argue that in unsuccessful attempt to maintain its political legitimacy, regime adopted a form of military populism. Patterns of land settlement, deforestation, and resource use on the frontier are shown to be the outcome of a series of economic, social, and political contingencies. For political scientist's comment see item **3780**.

JOURNAL ABBREVIATIONS

Adm. Econ. UC. Administración y Economía UC. Revista de la Facultad de Ciencias Económicas y Administrativas, Pontificia Univ. Católica de Chile. Santiago.

Agric. Soc. Agricultura y Sociedad. Ministerio de Agricultura, Pesca, y Alimentación. Madrid.

Allpanchis/Cusco. Allpanchis. Instituto de Pastoral Andina. Cusco, Peru.

Am. Ethnol. American Ethnologist. American Ethnological Society. Washington.

Am. Sociol. Rev. American Sociological Review. American Sociological Assn., Washington.

Ann. N.Y. Acad. Sci. Annals of the New York Academy of Sciences. New York.

Anthropol. Verkenn. Antropologische Verkenningen. Coutinho. Muiderberg, The Netherlands.

Bol. Socioecon. Boletín Socioeconómico. Centro de Investigaciones y Documentación Socioeconómico (CIDSE), Univ. del Valle. Cali, Colombia.

Bull. Bur. natl. ethnol. Bulletin du Bureau national d'ethnologie. Bureau national d'ethnologie. Port-au-Prince, Haiti.

Bull. Lat. Am. Res. Bulletin of Latin American Research. Society for Latin American Studies. Oxford, England.

Cad. CRH. Caderno CRH. Centro de Recursos Humanos. Salvador, Brazil.

Can. J. Hist. Canadian Journal of History. Univ. of Saskatchewan. Saskatoon, Canada.

Caribb. Aff. Caribbean Affairs. Trinidad Express Newspapers Ltd.; Inprint Caribbean Ltd., Port of Spain, Trinidad and Tobago.

Caribb. Q. Caribbean Quarterly. Univ. of the West Indies. Mona, Jamaica.

Caribb. Stud. Caribbean Studies. Univ. of Puerto Rico, Institute of Caribbean Studies. Río Piedras.

Chemins crit. Chemins critiques: revue haïtiano-caraïbéenne. Société sciences-arts-littérature. Port-au-Prince, Haiti.

Ciênc. Cult. Ciência e Cultura. Sociedade Brasileira para o Progresso da Ciência. São Paulo.

Cimarrón/New York. Cimarrón. City Univ. of New York, Assn. of Caribbean Studies. New York.

Comp. Educ. Rev. Comparative Education Review. Comparative Education Society. New York.

Comun. Soc. Comunicação e Sociedade. Instituto Metodista do Ensino Superior. São Paulo.

Cuad. Am. Cuadernos Americanos. Editorial Cultura. México.

Cuad. Sociol. Cuadernos de Sociología. Univ. Centroamericana. Managua.

Cuba. Stud. Cuban Studies. Univ. of Pittsburgh, Center for Latin American Studies. Pittsburgh, Penn.

Dados/Rio de Janeiro. Dados. Instituto Univ. de Pesquisas. Rio de Janeiro.

Debate Agrar. Debate Agrario. Centro Peruano de Estudios Sociales (CEPES). Lima.

Demography/Washington. Demography. Population Assn. of America. Washington.

Dev. Change. Development and Change. Mouton. The Hague.

ECA/San Salvador. Estudios Centro-americanos: ECA. Univ. Centroamericana José Simeón Cañas. San Salvador.

Econ. Dev. Cult. Change. Economic Development and Cultural Change. Univ. of Chicago, Research Center in Economic Development and Cultural Change. Chicago, Ill.

Estad. Econ. Estadística & Economía. Instituto Nacional de Estadísticas. Santiago.

Estud. Front. Estudios Fronterizos. Instituto de Investigaciones Sociales, Univ. Autónoma de Baja California. Mexicali, Mexico.

Estud. Int./IRIPAZ. Estudios Internacionales: Revista del IRIPAZ. Instituto de Relaciones Internacionales y de Investigaciones para la Paz. Guatemala.

Estud. Públicos. Estudios Públicos. Centro de Estudios Públicos. Santiago.

Estud. Rural. Latinoam. Estudios Rurales Latinoamericanos. Consejo Latinoamericano de Ciencias Sociales. Bogotá.

Estud. Soc. Centroam. Estudios Sociales Centroamericanos. Programa Centroamericano de Ciencias Sociales. San José.

Estud. Soc./Santiago. Estudios Sociales. Corporación de Promoción Universitaria. Santiago.

Estud. Soc./Santo Domingo. Estudios Sociales. Centro de Estudios Sociales Juan Montalvo, SJ. Santo Domingo.

Estud. Sociol./México. Estudios Sociológicos. Centro de Estudios Sociológicos de El Colegio de México. México.

Etnofoor/Netherlands. Etnofoor. Antropologisch-Sociologisch Centrum. Amsterdam, The Netherlands.

EURE/Santiago. EURE: Revista Latinoamericana de Estudios Urbanos Regionales. Centro de Desarrollo Urbano y Regional, Univ. Católica de Chile. Santiago.

Fuerzas Arm. Soc. Fuerzas Armadas y Sociedad. Centro Latinoamericano de Defensa y Desarme; FLACSO. Santiago.

Hemisphere/Miami. Hemisphere. Latin American and Caribbean Center, Florida International Univ., Miami, Fla.

Herencia/San José. Herencia. Programa de Rescate y Revitalización del Patrimonio Cultural. San José.

Hist. Soc./Río Piedras. Historia y Sociedad. Depto. de Historia, Univ. de Puerto Rico. Río Piedras.

Hómines/San Juan. Hómines. Univ. Interamericana de Puerto Rico. San Juan.

Hoy Hist. Hoy es Historia: Revista Bimestral de Historia Nacional e Iberoamericana. Editorial Raíces. Montevideo.

Hum. Organ. Human Organization. Society for Applied Anthropology. New York.

Humanidades/Brasília. Humanidades. Editora Univ. de Brasília.

Int. J. Comp. Sociol. International Journal of Comparative Sociology. York Univ., Dept. of Sociology and Anthropology. Toronto, Canada.

Int. J. Intercult. Relat. International Journal of Intercultural Relations. Society for Intercultural Education, Training, and Research; Pergamon Press. New York.

Int. J./Toronto. International Journal. Canadian Institute of International Affairs. Toronto, Canada.

Int. Migr. International Migration = Migrations Internationales = Migraciones Internacionales. Intergovernmental Committee for European Migration; Research Group for European Migration Problems; International Organization for Migration. The Hague, Netherlands; Geneva, Switzerland.

Int. Migr. Rev. International Migration Review. Center for Migration Studies. New York.

Interciencia/Caracas. Interciencia. Asociación Interciencia. Caracas.

Islas/Santa Clara. Islas. Univ. Central de Las Villas. Santa Clara, Cuba.

Iztapalapa/México. Iztapalapa. Univ. Autónoma Metropolitana, División de Ciencias Sociales y Humanidades. México.

J. Caribb. Stud. Journal of Caribbean Studies. Assn. of Caribbean Studies. Coral Gables, Fla.

J. Church State. Journal of Church and State. J.M. Dawson Studies in Church and State, Baylor Univ., Waco, Tex.

J. Lat. Am. Stud. Journal of Latin American Studies. Centers or Institutes of Latin American Studies at the Universities of Cambridge, Glasgow, Liverpool, London, and Oxford. Cambridge Univ. Press. London.

J. Polit. Mil. Sociol. Journal of Political & Military Sociology. Northern Illinois Univ., Dept. of Sociology. DeKalb, Ill.

LARR. Latin American Research Review. Latin American Research Review Board. Univ. of New Mexico, Albuquerque, N.M.

Lat. Am. Perspect. Latin American Perspectives. Univ. of California. Newbury Park, Calif.

MACLAS Lat. Am. Essays. MACLAS Latin American Essays. Middle Atlantic Council of Latin American Studies. New Brunswick, N.J.

Maguaré/Bogotá. Maguaré. Depto. de Antropología, Univ. Nacional de Colombia. Bogotá.

Mesoamérica/Antigua. Mesoamérica. Centro de Investigaciones Regionales de Mesoamérica. Antigua, Guatemala.

NACLA. NACLA: Report on the Americas. North American Congress on Latin America. New York.

New Left Rev. New Left Review. New Left Review, Ltd., London.

Nieuwe West-Indische Gids. Nieuwe West-Indische Gids. Martinus Nijhoff. The Hague.

Notas Pobl. Notas de Población. Centro Latinoamericano de Demografía. Santiago.

Novos Estud. CEBRAP. Novos Estudos CEBRAP. Centro Brasileiro de Análise e Planejamento. São Paulo.

Nueva Soc. Nueva Sociedad. Caracas.

Perf. Latinoam. Perfiles Latinoamericanos. Facultad Latinoamericana de Ciencias Sociales. México.

Prof. Geogr. The Professional Geographer. Assn. of American Geographers. Washington.

Real. Econ.-Soc. Realidad Económico-Social. Univ. Centroamericana José Simeón Cañas. San Salvador.

Rev. ANDI. Revista ANDI. Asociación Nacional de Industriales (ANDI). Medellín, Colombia.

Rev. Bras. Estud. Popul. Revista Brasileira de Estudos de População. Associação Brasileira de Estudos Populacionais. São Paulo.

Rev. Chil. Antropol. Revista Chilena de Antropología. Depto. de Antropología, Univ. de Chile. Santiago.

Rev. Chil. Geopolít. Revista Chilena de Geopolítica. Instituto Geopolítico de Chile. Santiago.

Rev. Cienc. Soc./San José. Revista de Ciencias Sociales. Univ. de Costa Rica. San José.

Rev. Cienc. Soc./Valparaíso. Revista de Ciencias Sociales. Facultad de Derecho y Ciencias Sociales, Univ. de Valparaíso. Chile.

Rev. Cuba. Cienc. Soc. Revista Cubana de Ciencias Sociales. Centro de Estudios Filosóficos, Academia de Ciencias de Cuba. La Habana.

Rev. Esp. Invest. Sociol. Revista Española de Investigaciones Sociológicas. Centro de Investigaciones Sociológicas. Madrid.

Rev. Eur. Revista Europea de Estudios Latinoamericanos y del Caribe = European Review of Latin American and Caribbean Studies. Center for Latin American Research and Documentation; Royal Institute of Linguistics and Anthropology. Amsterdam.

Rev. Geogr./Mérida. Revista Geográfica. Univ. de Los Andes. Mérida, Venezuela.

Rev. Hist./San José. Revista de Historia. Centro de Investigaciones Históricas, Univ. de Costa Rica. San José.

Rev. Interam. Bibliogr. Revista Interamericana de Bibliografía. Organization of American States. Washington.

Rev. Mex. Sociol. Revista Mexicana de Sociología. Instituto de Investigaciones Sociales, Univ. Nacional Autónoma de México. México.

Rev. Panameña Sociol. Revista Panameña de Sociología. Univ. de Panamá, Depto. de Sociología. Panamá.

Rev. Peru. Cienc. Soc. Revista Peruana de Ciencias Sociales. Asociación Peruana para el Fomento de las Ciencias Sociales (FOMCIENCIAS). Lima.

Rev. Univ. Pontif. Boliv. Revista Universidad Pontificia Bolivariana. Medellín, Colombia.

Rev. Univ./Tegucigalpa. Revista de la Universidad. Univ. Nacional Autónoma de Honduras. Tegucigalpa.

São Paulo Perspect. São Paulo em Perspectiva. Fundação SEADE. São Paulo.

SECOLAS Ann. SECOLAS Annals. Southeastern Conference on Latin American Studies; West Georgia College. Carrollton, Ga.

Secuencia/México. Secuencia. Instituto Mora. México.

Signs/Chicago. Signs. The Univ. of Chicago Press. Chicago, Ill.

Síntesis/Madrid. Síntesis. Asociación de Investigación y Especialización sobre Temas Latinoamericanos. Madrid.

Soc. Biol. Social Biology. Society for the Study of Social Biology. Port Angeles, Wash.

Soc. Compass. Social Compass. The International Catholic Institute for Social-Ecclesiastical Research. The Hague.

Soc. Probl. Social Problems. Society for the Study of Social Problems; American and International Sociological Associations. Kalamazoo, Mich.

Soc. Sci. Med. Social Science and Medicine. Pergamon Press. New York.

Social. Particip. Socialismo y Participación. Ediciones Socialismo y Participación. Lima.

Sociol. Anal. Sociological Analysis. American Catholic Sociological Society. Worcester, Mass.

Sociol. Inq. Sociological Inquiry. National Sociology Honor Society; Dept. of Sociology, Univ. of Omaha. Omaha, Neb.

Sociol. trav.   Sociologie du travail. Association pour le développement de la sociologie du travail. Paris.

Sociologus/Berlin.   Sociologus. Berlin.

SWI Forum.   SWI Forum voor Kunst, Kultuur en Wetenschop. De Stichting. Paramaribo, Suriname.

Tareas/Panamá.   Tareas. Centro de Estudios Latinoamericanos (CELA). Panamá.

Trace/México.   Trace. Centre d'études mexicaines et centraméricaines. México.

UTEC/San Salvador.   UTEC: Revista de la Universidad Tecnológica. Univ. Tecnológica. San Salvador.

ABBREVIATIONS AND ACRONYMS

Except for journal abbreviations which are listed: 1) at the end of each major disciplinary section (e.g., Anthropology, Economics, Geography, etc.); 2) after each journal title in the *Title List of Journals Indexed* (p. 753); and 3) in the *Abbreviation List of Journals Indexed* (p. 769).

ALADI	Asociación Latinoamericana de Integración
a.	annual
ABC	Argentina, Brazil, Chile
A.C.	antes de Cristo
ACAR	Associação de Crédito e Assistência Rural, Brazil
AD	Anno Domini
A.D.	Acción Democrática, Venezuela
ADESG	Associação dos Diplomados de Escola Superior de Guerra, Brazil
AGI	Archivo General de Indias, Sevilla
AGN	Archivo General de la Nación
AID	Agency for International Development
a.k.a.	also known as
Ala.	Alabama
ALALC	Asociación Latinoamericana de Libre Comercio
ALEC	*Atlas lingüístico etnográfico de Colombia*
ANAPO	Alianza Nacional Popular, Colombia
ANCARSE	Associação Nordestina de Crédito e Assistência Rural de Sergipe, Brazil
ANCOM	Andean Common Market
ANDI	Asociación Nacional de Industriales, Colombia
ANPOCS	Associação Nacional de Pós-Graduação e Pesquisa em Ciências Sociais, São Paulo
ANUC	Asociación Nacional de Usuarios Campesinos, Colombia
ANUIES	Asociación Nacional de Universidades e Institutos de Enseñanza Superior, Mexico
AP	Acción Popular
APRA	Alianza Popular Revolucionaria Americana, Peru
ARENA	Aliança Renovadora Nacional, Brazil
Ariz.	Arizona
Ark.	Arkansas
ASA	Association of Social Anthropologists of the Commonwealth, London
ASSEPLAN	Assessoria de Planejamento e Acompanhamento, Recife
Assn.	Association
Aufl.	Auflage (edition, edición)
AUFS	American Universities Field Staff Reports, Hanover, N.H.
Aug.	August, Augustan
aum.	aumentada
b.	born (nació)
B.A.R.	British Archaeological Reports
BBE	Bibliografia Brasileira de Educação
b.c.	indicates dates obtained by radiocarbon methods

BC	Before Christ
bibl(s).	bibliography(ies)
BID	Banco Interamericano de Desarrollo
BNDE	Banco Nacional de Desenvolvimento Econômico, Brazil
BNH	Banco Nacional de Habitação, Brazil
BP	before present
b/w	black and white
C14	Carbon 14
ca.	*circa* (about)
CACM	Central American Common Market
CADE	Conferencia Anual de Ejecutivos de Empresas, Peru
CAEM	Centro de Altos Estudios Militares, Peru
Calif.	California
Cap.	Capítulo
CARC	Centro de Arte y Comunicación, Buenos Aires
CARICOM	Caribbean Common Market
CARIFTA	Caribbean Free Trade Association
CBC	Christian base communities
CBD	central business district
CBI	Caribbean Basin Initiative
CD	Christian Democrats, Chile
CDI	Conselho de Desenvolvimento Industrial, Brasília
CEB	comunidades eclesiásticas de base
CEBRAP	Centro Brasileiro de Análise e Planejamento, São Paulo
CECORA	Centro de Cooperativas de la Reforma Agraria, Colombia
CEDAL	Centro de Estudios Democráticos de América Latina, Costa Rica
CEDE	Centro de Estudios sobre Desarrollo Económico, Univ. de los Andes, Bogotá
CEDEPLAR	Centro de Desenvolvimento e Planejamento Regional, Belo Horizonte
CEDES	Centro de Estudios de Estado y Sociedad, Buenos Aires; Centro de Estudos de Educação e Sociedade, São Paulo
CEDI	Centro Ecumênico de Documentos e Informação, São Paulo
CEDLA	Centro de Estudios y Documentación Latinoamericanos, Amsterdam
CEESTEM	Centro de Estudios Económicos y Sociales del Tercer Mundo, México
CELADE	Centro Latinoamericano de Demografía
CELADEC	Comisión Evangélica Latinoamericana de Educación Cristiana
CELAM	Consejo Episcopal Latinoamericano
CEMLA	Centro de Estudios Monetarios Latinoamericanos, Mexico
CENDES	Centro de Estudios del Desarrollo, Venezuela
CENIDIM	Centro Nacional de Información, Documentación e Investigación Musicales, Mexico
CENIET	Centro Nacional de Información y Estadísticas del Trabajo, Mexico
CEPADE	Centro Paraguayo de Estudios de Desarrollo Económico y Social
CEPA-SE	Comissão Estadual de Planejamento Agrícola, Sergipe
CEPAL	Comisión Económica para América Latina y el Caribe
CEPLAES	Centro de Planificación y Estudios Sociales, Quito
CERES	Centro de Estudios de la Realidad Económica y Social, Bolivia
CES	constant elasticity of substitution
cf.	compare
CFI	Consejo Federal de Inversiones, Buenos Aires
CGE	Confederación General Económica, Argentina
CGTP	Confederación General de Trabajadores del Perú
chap(s).	chapter(s)
CHEAR	Council on Higher Education in the American Republics
Cía.	Compañía
CIA	Central Intelligence Agency

CIDA	Comité Interamericano de Desarrollo Agrícola
CIDE	Centro de Investigación y Desarrollo de la Educación, Chile; Centro de Investigación y Docencias Económicas, Mexico
CIE	Centro de Investigaciones Económicas, Buenos Aires
CIEDLA	Centro Interdisciplinario de Estudios sobre el Desarrollo Latinoamericano, Buenos Aires
CIEDUR	Centro Interdisciplinario de Estudios sobre el Desarrollo Uruguay, Montevideo
CIEPLAN	Corporación de Investigaciones Económicas para América Latina, Santiago
CIESE	Centro de Investigaciones y Estudios Socioeconómicos, Quito
CIMI	Conselho Indigenista Missionário, Brazil
CINTERFOR	Centro Interamericano de Investigación y Documentación sobre Formación Profesional
CINVE	Centro de Investigaciones Económicas, Montevideo
CIP	Conselho Interministerial de Preços, Brazil
CIPCA	Centro de Investigación y Promoción del Campesinado, Bolivia
CIPEC	Consejo Intergubernamental de Países Exportadores de Cobre, Santiago
CLACSO	Consejo Latinoamericano de Ciencias Sociales, Secretaría Ejecutiva, Buenos Aires
CLASC	Confederación Latinoamericana Sindical Cristiana
CLE	Comunidad Latinoamericana de Escritores, Mexico
cm	centimeter
CNI	Confederação Nacional da Indústria, Brazil
CNPq	Conselho Nacional de Pesquisas, Brazil
Co.	Company
COB	Central Obrera Boliviana
COBAL	Companhia Brasileira de Alimentos
Col.	Collection, Colección, Coleção
col.	colored, coloured
Colo.	Colorado
COMCORDE	Comisión Coordinadora para el Desarrollo Económico, Uruguay
comp(s).	compiler(s), compilador(es)
CONCLAT	Congresso Nacional das Classes Trabalhadoras, Brazil
CONDESE	Conselho de Desenvolvimento Econômico de Sergipe
Conn.	Connecticut
COPEI	Comité Organizador Pro-Elecciones Independientes, Venezuela
CORFO	Corporación de Fomento de la Producción, Chile
CORP	Corporación para el Fomento de Investigaciones Económicas, Colombia
Corp.	Corporation, Corporación
corr.	corrected, corregida
CP	Communist Party
CPDOC	Centro de Pesquisa e Documentação, Brazil
CRIC	Consejo Regional Indígena del Cauca, Colombia
CSUTCB	Confederación Sindical Unica de Trabajadores Campesinos de Bolivia
CTM	Confederación de Trabajadores de México
CUNY	City University of New York
CUT	Central Unica de Trabajadores (Mexico); Central Unica dos Trabalhadores (Brazil); Central Unitaria de Trabajadores (Chile; Colombia); Confederación Unitaria de Trabajadores (Costa Rica)
CVG	Corporación Venezolana de Guayana
d.	died (murió)
DANE	Departamento Nacional de Estadística, Colombia
DC	developed country; Demócratas Cristianos, Chile
d.C.	después de Cristo
Dec./déc.	December, décembre
Del.	Delaware

dept.	department
depto.	departamento
DESCO	Centro de Estudios y Promoción del Desarrollo, Lima
Dez./dez.	Dezember, dezembro
dic.	diciembre, dicembre
disc.	discography
DNOCS	Departamento Nacional de Obras Contra as Secas, Brazil
doc.	document, documento
Dr.	Doctor
Dra.	Doctora
DRAE	*Diccionario de la Real Academia Española*
ECLAC	UN Economic Commision for Latin America and the Caribbean, New York and Santiago
ECOSOC	UN Economic and Social Council
ed./éd.(s)	edition(s), édition(s), edición(es), editor(s), redactor(es), director(es)
EDEME	Editora Emprendimentos Educacionais, Florianópolis
Edo.	Estado
EEC	European Economic Community
EE.UU.	Estados Unidos de América
EFTA	European Free Trade Association
e.g.	*exempio gratia* (for example, por ejemplo)
ELN	Ejército de Liberación Nacional, Colombia
ENDEF	Estudo Nacional da Despesa Familiar, Brazil
ESG	Escola Superior de Guerra, Brazil
estr.	estrenado
et al.	*et alia* (and others)
ETENE	Escritório Técnico de Estudos Econômicos do Nordeste, Brazil
ETEPE	Escritório Técnico de Planejamento, Brazil
EUDEBA	Editorial Universitaria de Buenos Aires
EWG	Europaische Wirtschaftsgemeinschaft. *See* EEC.
facsim(s).	facsimile(s)
FAO	Food and Agriculture Organization of the United Nations
FDR	Frente Democrático Revolucionario, El Salvador
FEB	Força Expedicionária Brasileira
Feb./feb.	February, Februar, febrero, febbraio
FEDECAFE	Federación Nacional de Cafeteros, Colombia
fev./fév.	fevereiro, février
ff.	following
FGTS	Fundo de Garantia do Tempo de Serviço, Brazil
FGV	Fundação Getúlio Vargas
FIEL	Fundación de Investigaciones Económicas Latinoamericanas, Argentina
film.	filmography
fl.	flourished
Fla.	Florida
FLACSO	Facultad Latinoamericana de Ciencias Sociales
FMI	Fondo Monetario Internacional
FMLN	Frente Farabundo Martí de Liberación Nacional, El Salvador
fold.	folded
fol(s).	folio(s)
FRG	Federal Republic of Germany
FSLN	Frente Sandinista de Liberación Nacional, Nicaragua
ft.	foot, feet
FUAR	Frente Unido de Acción Revolucionaria, Colombia
FUCVAM	Federación Unificadora de Cooperativas de Vivienda por Ayuda Mutua, Uruguay

FUNAI	Fundação Nacional do Indio, Brazil
FUNARTE	Fundação Nacional de Arte, Brazil
FURN	Fundação Universidade Regional do Nordeste
Ga.	Georgia
GAO	General Accounting Office, Wahington
GATT	General Agreement on Tariffs and Trade
GDP	gross domestic product
GDR	German Democratic Republic
GEIDA	Grupo Executivo de Irrigação para o Desenvolvimento Agrícola, Brazil
gen.	gennaio
Gen.	General
GMT	Greenwich Mean Time
GPA	grade point average
GPO	Government Printing Office, Washington
h.	hijo
ha.	hectares, hectáreas
HLAS	*Handbook of Latin American Studies*
HMAI	*Handbook of Middle American Indians*
Hnos.	hermanos
HRAF	Human Relations Area Files, Human Relations Area Files, Inc., New Haven, Conn.
IBBD	Instituto Brasileiro de Bibliografia e Documentação
IBGE	Instituto Brasileiro de Geografia e Estatística, Rio de Janeiro
IBRD	International Bank for Reconstruction and Development (World Bank)
ICA	Instituto Colombiano Agropecuario
ICAIC	Instituto Cubano de Arte e Industria Cinematográfica
ICCE	Instituto Colombiano de Construcción Escolar
ICE	International Cultural Exchange
ICSS	Instituto Colombiano de Seguridad Social
ICT	Instituto de Crédito Territorial, Colombia
id.	*idem* (the same as previously mentioned or given)
IDB	Inter-American Development Bank
i.e.	*id est* (that is, o sea)
IEL	Instituto Euvaldo Lodi, Brazil
IEP	Instituto de Estudios Peruanos
IERAC	Instituto Ecuatoriano de Reforma Agraria y Colonización
IFAD	International Fund for Agricultural Development
IICA	Instituto Interamericano de Ciencias Agrícolas, San José
III	Instituto Indigenista Interamericana, Mexico
IIN	Instituto Indigenista Nacional, Guatemala
ILDIS	Instituto Latinoamericano de Investigaciones Sociales
ill.	illustration(s)
Ill.	Illinois
ILO	International Labour Organization, Geneva
IMES	Instituto Mexicano de Estudios Sociales
IMF	International Monetary Fund
Impr.	Imprenta, Imprimérie
in.	inches
INAH	Instituto Nacional de Antropología e Historia, Mexico
INBA	Instituto Nacional de Bellas Artes, Mexico
Inc.	Incorporated
INCORA	Instituto Colombiano de Reforma Agraria
Ind.	Indiana
INEP	Instituto Nacional de Estudios Pedagógicos, Brazil
INI	Instituto Nacional Indigenista, Mexico

INIT	Instituto Nacional de Industria Turística, Cuba
INPES/IPEA	Instituto de Planejamento Econômico e Social, Brazil
INTAL	Instituto para la Integración de América Latina
IPA	Instituto de Pastoral Andina, Univ. de San Antonio de Abad, Seminario de Antropología, Cusco, Peru
IPEA	Instituto de Pesquisa Econômica Aplicada, Brazil
IPES/GB	Instituto de Pesquisas e Estudos Sociais, Guanabara, Brazil
IPHAN	Instituto de Patrimônio Histórico e Artístico Nacional, Brazil
ir.	irregular
IS	Internacional Socialista
ITT	International Telephone and Telegraph
Jan./jan.	January, Januar, janeiro, janvier
JLP	Jamaican Labour Party
Jr.	Junior, Júnior
JUC	Juventude Universitária Católica, Brazil
JUCEPLAN	Junta Central de Planificación, Cuba
Kan.	Kansas
km	kilometers, kilómetros
Ky.	Kentucky
La.	Louisiana
LASA	Latin American Studies Association
LDC	less developed country(ies)
LP	long-playing record
Ltd(a).	Limited, Limitada
m	meters, metros
m.	murió (died)
M	mille, mil, thousand
M.A.	Master of Arts
MACLAS	Middle Atlantic Council of Latin American Studies
MAPU	Movimiento de Acción Popular Unitario, Chile
MARI	Middle American Research Institute, Tulane University, New Orleans
MAS	Movimiento al Socialismo, Venezuela
Mass.	Massachusetts
MCC	Mercado Común Centro-Americano
Md.	Maryland
MDB	Movimiento Democrático Brasileiro
MDC	more developed countries
Me.	Maine
MEC	Ministério de Educação e Cultura, Brazil
Mich.	Michigan
mimeo	mimeographed, mimeografiado
min.	minutes, minutos
Minn.	Minnesota
MIR	Movimiento de Izquierda Revolucionaria, Chile and Venezuela
Miss.	Mississippi
MIT	Massachusetts Institute of Technology
ml	milliliter
MLN	Movimiento de Liberación Nacional
mm.	millimeter
MNC	multinational corporation
MNI	minimum number of individuals
MNR	Movimiento Nacionalista Revolucionario, Bolivia
Mo.	Missouri
MOBRAL	Movimento Brasileiro de Alfabetização
MOIR	Movimiento Obrero Independiente y Revolucionario, Colombia

Mont.	Montana
MRL	Movimiento Revolucionario Liberal, Colombia
ms.	manuscript
M.S.	Master of Science
msl	mean sea level
n.	nació (born)
NBER	National Bureau of Economic Research, Cambridge, Massachusetts
N.C.	North Carolina
N.D.	North Dakota
NE	Northeast
Neb.	Nebraska
neubearb.	neubearbeitet (revised, corregida)
Nev.	Nevada
n.f.	neue Folge (new series)
NGO	nongovernmental organization
NGDO	nongovernmental development organization
N.H.	New Hampshire
NIEO	New International Economic Order
NIH	National Institutes of Health, Washington
N.J.	New Jersey
NJM	New Jewel Movement, Grenada
N.M.	New Mexico
no(s).	number(s), número(s)
NOEI	Nuevo Orden Económico Internacional
NOSALF	Scandinavian Committee for Research in Latin America
Nov./nov.	November, noviembre, novembre, novembro
NSF	National Science Foundation
NW	Northwest
N.Y.	New York
OAB	Ordem dos Advogados do Brasil
OAS	Organization of American States
Oct./oct.	October, octubre, octobre
ODEPLAN	Oficina de Planificación Nacional, Chile
OEA	Organización de los Estados Americanos
OIT	Organización Internacional del Trabajo
Okla.	Oklahoma
Okt.	Oktober
op.	opus
OPANAL	Organismo para la Proscripción de las Armas Nucleares en América Latina
OPEC	Organization of Petroleum Exporting Countries
OPEP	Organización de Países Exportadores de Petróleo
OPIC	Overseas Private Investment Corporation, Washington
Or.	Oregon
OREALC	Oficina Regional de Educación para América Latina y el Caribe
ORIT	Organización Regional Interamericana del Trabajo
ORSTOM	Office de la recherche scientifique et technique outre-mer (France)
ott.	ottobre
out.	outubro
p.	page(s)
Pa.	Pennsylvania
PAN	Partido Acción Nacional, Mexico
PC	Partido Comunista
PCCLAS	Pacific Coast Council on Latin American Studies
PCN	Partido de Conciliación Nacional, El Salvador
PCP	Partido Comunista del Perú

PCR	Partido Comunista Revolucionario, Chile and Argentina
PCV	Partido Comunista de Venezuela
PD	Partido Democrático
PDC	Partido Demócrata Cristiano, Chile
PDS	Partido Democrático Social, Brazil
PDT	Partido Democrático Trabalhista, Brazil
PDVSA	Petróleos de Venezuela S.A.
PEMEX	Petróleos Mexicanos
PETROBRAS	Petróleo Brasileiro
PIMES	Programa Integrado de Mestrado em Economia e Sociologia, Brazil
PIP	Partido Independiente de Puerto Rico
PLN	Partido Liberación Nacional, Costa Rica
PMDB	Partido do Movimento Democrático Brasileiro
PNAD	Pesquisa Nacional por Amostra Domiciliar, Brazil
PNC	People's National Congress, Guyana
PNM	People's National Movement, Trinidad and Tobago
PNP	People's National Party, Jamaica
pop.	population
port(s).	portrait(s)
PPP	purchasing power parities; People's Progressive Party of Guyana
PRD	Partido Revolucionario Dominicano
PREALC	Programa Regional del Empleo para América Latina y el Caribe, Organización Internacional del Trabajo, Santiago
PRI	Partido Revolucionario Institucional, Mexico
Prof.	Professor, Profesor(a)
PRONAPA	Programa Nacional de Pesquisas Arqueológicas, Brazil
prov.	province, provincia
PS	Partido Socialista, Chile
PSD	Partido Social Democrático, Brazil
pseud.	pseudonym, pseudónimo
PT	Partido dos Trabalhadores, Brazil
pt(s).	part(s), parte(s)
PTB	Partido Trabalhista Brasileiro
pub.	published, publisher
PUC	Pontifícia Universidade Católica
PURSC	Partido Unido de la Revolución Socialista de Cuba
q.	quarterly
rev.	revisada, revista, revised
R.I.	Rhode Island
s.a.	semiannual
SALALM	Seminar on the Acquisition of Latin American Library Materials
SATB	soprano, alto, tenor, bass
sd.	sound
s.d.	*sine datum* (no date, sin fecha)
S.D.	South Dakota
SDR	special drawing rights
SE	Southeast
SELA	Sistema Económico Latinoamericano
SENAC	Serviço Nacional de Aprendizagem Comercial, Rio de Janeiro
SENAI	Serviço Nacional de Aprendizagem Industrial, São Paulo
SEP	Secretaría de Educación Pública, Mexico
SEPLA	Seminario Permanente sobre Latinoamérica, Mexico
Sept./sept.	September, septiembre, septembre
SES	socioeconomic status
SESI	Serviço Social da Indústria, Brazil

set.	setembro, settembre
SI	Socialist International
SIECA	Secretaría Permanente del Tratado General de Integración Económica Centroamericana
SIL	Summer Institute of Linguistics (Instituto Lingüístico de Verano)
SINAMOS	Sistema Nacional de Apoyo a la Movilización Social, Peru
S.J.	Society of Jesus
s.l.	*sine loco* (place of publication unknown)
s.n.	*sine nomine* (publisher unknown)
SNA	Sociedad Nacional de Agricultura, Chile
SPP	Secretaría de Programación y Presupuesto, Mexico
SPVEA	Superintendência do Plano de Valorização Econômica da Amazônia, Brazil
sq.	square
SSRC	Social Sciences Research Council, New York
SUDAM	Superintendência de Desenvolvimento da Amazônia, Brazil
SUDENE	Superintendência de Desenvolvimento do Nordeste, Brazil
SUFRAMA	Superintendência da Zona Franca de Manaus, Brazil
SUNY	State University of New York
SW	Southwest
t.	tomo(s), tome(s)
TAT	Thematic Apperception Test
TB	tuberculosis
Tenn.	Tennessee
Tex.	Texas
TG	transformational generative
TL	Thermoluminescent
TNE	Transnational enterprise
TNP	Tratado de No Proliferación
trans.	translator
UABC	Universidad Autónoma de Baja California
UCA	Universidad Centroamericana José Simeón Cañas, San Salvador
UCLA	University of California, Los Angeles
UDN	União Democrática Nacional, Brazil
UFG	Universidade Federal de Goiás
UFPb	Universidade Federal de Paraíba
UFSC	Universidade Federal de Santa Catarina
UK	United Kingdom
UN	United Nations
UNAM	Universidad Nacional Autónoma de México
UNCTAD	United Nations Conference on Trade and Development
UNDP	United Nations Development Programme
UNEAC	Unión de Escritores y Artistas de Cuba
UNESCO	United Nations Educational, Scientific and Cultural Organization
UNI/UNIND	União das Nações Indígenas
UNICEF	United Nations International Children's Emergency Fund
Univ(s).	university(ies), universidad(es), universidade(s), université(s), universität(s), universitá(s)
uniw.	uniwersytet (university)
Unltd.	Unlimited
UP	Unidad Popular, Chile
URD	Unidad Revolucionaria Democrática
URSS	Unión de Repúblicas Soviéticas Socialistas
US	United States
USAID	*See* AID.
USIA	United States Information Agency

USSR	Union of Soviet Socialist Republics
UTM	Universal Transverse Mercator
UWI	Univ. of the West Indies
v.	volume(s), volumen (volúmenes)
Va.	Virginia
V.I.	Virgin Islands
viz.	*videlicet* (that is, namely)
vol(s).	volume(s), volumen (volúmenes)
vs.	versus
Vt.	Vermont
W.Va.	West Virginia
Wash.	Washington
Wis.	Wisconsin
WPA	Working People's Alliance, Guyana
WWI	World War I
WWII	World War II
Wyo.	Wyoming
yr(s).	year(s)

TITLE LIST OF JOURNALS INDEXED

For journal titles listed by abbreviation, see *Abbreviation List of Journals Indexed*, p. 769.

Administración y Economía UC. Revista de la Facultad de Ciencias Económicas y Administrativas, Pontificia Univ. Católica de Chile. Santiago. (Adm. Econ. UC)

The African Archaeological Review. Cambridge Univ. Press. Cambridge, England. (Afr. Archaeol. Rev.)

The Age of Democracy. Basil Blackwood; UNESCO. (Age Dem.)

Agricultura y Sociedad. Ministerio de Agricultura, Pesca, y Alimentación. Madrid. (Agric. Soc.)

Agricultural History. Agricultural History Society. Univ. of Calif. Press. Berkeley. (Agric. Hist.)

Agroecología y Desarrollo. Consorcio Latinoamericano sobre Agroecología y Desarrollo (CLADES). Santiago. (Agroecol. Desarro.)

Allpanchis. Instituto de Pastoral Andina. Cusco, Peru. (Allpanchis/Cusco)

Alternativa. Centro de Estudios Sociales Solidaridad. Chiclayo, Peru. (Alternativa/Chiclayo)

Alternatives. Lynne Rienner Publishers. Boulder, Colorado. (Alternatives/Boulder)

Amazonía Peruana. Centro Amazónico de Antropología y Aplicación Práctica, Depto. de Documentación y Publicaciones. Lima. (Amazonía Peru.)

América Indígena. Instituto Indigenista Interamericano. México. (Am. Indíg.)

América Latina. Academia de Ciencias de la Unión de Repúblicas Soviéticas Socialistas. Moscow. (Am. Lat./Moscow)

América Latina: Enfoques desde Rusia. Academia de Ciencias de Rusia, Instituto de América Latina. Moscow. (Am. Lat. Enfoques Rusia)

América Negra. Pontificia Univ. Javeriana. Bogotá. (Am. Negra)

American Anthropologist. American Anthropological Assn., Washington. (Am. Anthropol.)

American Antiquity. The Society for American Archaeology. Washington. (Am. Antiq.)

The American Economic Review. American Economic Assn., Evanston, Ill. (Am. Econ. Rev.)

The American Economist. Lubin Graduate School of Business, Pace Univ. New York. (Am. Econ.)

American Ethnologist. American Ethnological Society. Washington. (Am. Ethnol.)

American Political Science Review. American Political Science Assn.; Ohio State Univ., Columbus, Ohio. (Am. Polit. Sci. Rev.)

American Sociological Review. American Sociological Assn., Washington. (Am. Sociol. Rev.)

Anales de Antropología. Univ. Nacional Autónoma de México, Instituto de Investigaciones Históricas. México. (An. Antropol.)

Anales de Arqueología y Etnología. Univ. Nacional de Cuyo, Facultad de Filosofía y Letras. Mendoza, Argentina. (An. Arqueol. Etnol.)

Anales de la Academia Nacional de Ciencias Económicas. Buenos Aires. (An. Acad. Nac. Cienc. Econ.)

Anales del Instituto de Investigaciones Estéticas. Univ. Nacional Autónoma de México. México. (An. Inst. Invest. Estét.)

Anales del Instituto de la Patagonia: Serie Ciencias Sociales. Univ. de Magallanes. Punta Arenas, Chile. (An. Inst. Patagon./Soc.)

Anales del Museo Michoacano. Centro Regional Michoacán del INAH; Museo Regional Michoacano. Morelia, Mexico. (An. Mus. Michoacano)

Análise & Conjuntura. Fundação João Pinheiro. Belo Horizonte, Brazil. (Análise Conjunt.)

Análisis Económico. Unidad Azcapotzalco, Univ. Nacional Autónoma de México. México. (Anál. Econ./México)

Análisis Económico. Ediciones UDAPE (Unidad de Análisis de Políticas Económicas). La Paz. (Anál. Econ./La Paz)

Análisis Internacional. Centro Peruano de Estudios Internacionales. Lima. (Anál. Int.)

Ancient Mesoamerica. Cambridge Univ. Press. Cambridge, England. (Anc. Mesoam.)

Annales. Centre national de la recherche scientifique de la VIe Section de l'Ecole pratique des hautes études. Paris. (Annales/Paris)

Annales de géographie. Bulletin de la Société de géographie; Librairie Armand Colin. Paris. (Ann. géogr./Paris)

Annals of Latin American Studies. Nikon Raten Amerika Gakkai. Tokyo. (Ann. Lat. Am. Stud.)

The Annals of the American Academy of Political and Social Science. Philadelphia, Penn. (Ann. Am. Acad. Polit. Soc. Sci.)

Annals of the Association of American Geographers. Lawrence, Kan. (Ann. Assoc. Am. Geogr.)

Annals of the New York Academy of Sciences. New York. (Ann. N.Y. Acad. Sci.)

Annual Review of Anthropology. Annual Reviews, Inc., Palo Alto, Calif. (Annu. Rev. Anthropol.)

Anthropological Quarterly. Catholic Univ. of America, Catholic Anthropological Conference. Washington. (Anthropol. Q.)

Anthropos. International Review of Ethnology and Linguistics. Anthropos-Institut. Freiburg, Switzerland. (Anthropos/Switzerland)

Antiquity. A Quarterly Review of Archaeology. The Antiquity Trust. Cambridge, England. (Antiquity/Cambridge)

Antropológica. Fundación La Salle de Ciencias Naturales; Instituto Caribe de Antropología y Sociología. Caracas. (Antropológica/Caracas)

Antropológicas. Instituto de Investigaciones Antropológicas, UNAM. México. (Antropológicas/México)

Antropologische Verkenningen. Coutinho. Muiderberg, The Netherlands. (Anthropol. Verkenn.)

Anuário Antropológico. Tempo Brasileiro. Rio de Janeiro. (Anu. Antropol.)

Anuario de Estudios Americanos. Consejo Superior de Investigaciones Científicas; Univ. de Sevilla, Escuela de Estudios Hispano-Americanos. Sevilla, Spain. (Anu. Estud. Am.)

Anuario de Estudios Centroamericanos. Univ. de Costa Rica. San José. (Anu. Estud. Centroam.)

Anuario de Etnología. Academia de Ciencias de Cuba; Editorial Academia. La Habana. (Anu. Etnol./Habana)

Anuario de la Fundación Los Andes de Estudios Sociales. Quito. (Anuario/Quito)

Anuario Estadístico de la República Argentina. Instituto Nacional de Estadística y Censos. Buenos Aires. (Anu. Estad. Repúb. Argent.)

Anuario IEHS. Univ. Nacional del Centro de la Provincia de Buenos Aires, Instituto de Estudios Histórico-Sociales. Tandil, Argentina. (Anu. IEHS)

Anuario Instituto Chiapaneco de Cultura. Instituto Chiapaneco de Cultura. Tuxtla Gutiérrez, Mexico. (Anu. Inst. Chiapaneco Cult.)

Anuario Mexicano de Relaciones Internacionales. UNAM, ENEP Acatlán, Escuela Nacional de Estudios Profesionales. México. (AMRI)

Apuntes. Univ. del Pacífico, Centro de Investigación. Lima. (Apuntes/Lima)

Apuntes Arqueológicos. Universidad de San Carlos de Guatemala, Escuela de Historia. Guatemala. (Apunt. Arqueol./Guatemala)

Archaeoastronomy. The Center for Archaeoastronomy, Univ. of Maryland. College Park, Md. (Archaeoastronomy/College Park)

Archaeoastronomy. Science History Publications. Giles, England. (Archaeoastronomy/England)

Archaeology. Archaeology Institute of America. New York. (Archaeology/New York)

Archaeology and Anthropology. Ministry of Education and Cultural Development. Georgetown, Guyana. (Archaeol. Anthropol.)

Archaeometry. Oxford Univ., Oxford, England. (Archaeometry/Oxford)

Archiv für Völkerkunde. Museum für Völkerkunde in Wien und von Verein Freunde der Völkerkunde. Vienna. (Arch. Völkerkd.)

Arctic and Alpine Research. Institute of Arctic and Alpine Research, Univ. of Colorado. Boulder. (Arctic Alpine Res.)

Armed Forces and Society. Inter-Univ. Seminar on Armed Forces & Society. Univ. of Chicago. Chicago, Ill. (Armed Forces Soc.)

Arqueología. Instituto Nacional de Antropología e Historia. México. (Arqueología/México)

Arqueología Contemporánea. Programa de Estudios Prehistóricos. Buenos Aires. (Arqueol. Contemp.)

Arqueología Mexicana. Instituto Nacional de Antropología e Historia, Editorial Raíces. México. (Arqueol. Mex.)

Arqueológicas. Museo Nacional de Antropología y Arqueología, Instituto Nacional de Cultura. Lima. (Arqueológicas/Lima)

Arquivos do Museu de História Natural. Univ. Federal de Minas Gerais. Belo Horizonte, Brazil. (Arq. Mus. Hist. Nat.)

Årstryck. Etnografiska Museum. Göteborg, Sweden. (Årstryck/Göteborg)

Banca Central. Banco de Guatemala. Guatemala. (Banca Cent.)

Banca Nazionale del Lavoro Quarterly Review. Banca Nazionale del Lavoro. Rome. (Banca Naz. Lavoro)

Bijdragen en Mededelingen Betreffende de Geschiedenis der Nederlanden. Nederlands Historisch Genootschap. Utrecht, The Netherlands. (Bijdr. Meded. Betreffende Geschied. Nederlanden)

Boletim do Museu Paraense Emílio Goeldi. Nova série: antropologia. Conselho Nacional de Desenvolvimento Científico e Tecnológico, Instituto Nacional de Pesquisas da Amazônia. Belém, Brazil. (Bol. Mus. Para. Goeldi)

Boletim Informativo e Bibliográfico de Ciências Sociais: BIB. Associação Nacional de Pós-Graduação e Pesquisa em Ciências Sociais. Rio de Janeiro. (ANPOCS BIB)

Boletín Americanista. Univ. de Barcelona, Facultad de Geografía e Historia, Depto. de Historia de América. Barcelona. (Bol. Am.)

Boletín de Antropología Americana. Instituto Panamericano de Geografía e Historia. México. (Bol. Antropol. Am.)

Boletín de Arqueología. Fundación de Investigaciones Arqueológicas Nacionales. Bogotá. (Bol. Arqueol.)

Boletín de la Academia Nacional de la Historia. Caracas. (Bol. Acad. Nac. Hist./Caracas)

Boletín de la Escuela de Ciencias Antropológicas de la Universidad de Yucatán. Mérida, Mexico. (Bol. Esc. Cienc. Antropol. Univ. Yucatán)

Boletín de Lima. Revista Cultural Científica. Lima. (Bol. Lima)

Boletín del Consejo de Arqueología. Instituto Nacional de Antropología e Historia. México. (Bol. Cons. Arqueol.)

Boletín del Instituto de Geografía. Univ. Nacional Autónoma de México. México. (Bol. Inst. Geogr.)

Boletín del Museo Chileno de Arte Precolombino. Santiago. (Bol. Mus. Chil. Arte Precolomb.)

Boletín del Museo del Hombre Dominicano. Santo Domingo. (Bol. Mus. Hombre Domin.)

Boletín del Museo del Oro. Banco de la República. Bogotá. (Boletín/Bogotá)

Boletín Informativo. Museo de Táchira. San Cristóbal, Venezuela. (Bol. Inf./San Cristóbal)

Boletín Informativo Techint. Organización Techint. Buenos Aires. (Bol. Inf. Techint)

Boletín Socioeconómico. Centro de Investigaciones y Documentación Socioeconómico (CIDSE), Univ. del Valle. Cali, Colombia. (Bol. Socioecon.)

Bulletin. Société suisse des américanistes; Musée et institut d'éthnographie. Geneva. (Bulletin/Geneva)

Bulletin de l'Association de géographes français. Paris. (Bull. Assoc. géogr. fr.)

Bulletin de l'Institut français d'études andines. Lima. (Bull. Inst. fr. étud. andin.)

Bulletin du Bureau national d'ethnologie. Bureau national d'ethnologie. Port-au-Prince, Haiti. (Bull. Bur. natl. ethnol.)

Bulletin of Eastern Caribbean Affairs. Univ. of West Indies. Cave Hill, Barbados. (Bull. East. Caribb. Aff.)

Bulletin of Latin American Research. Society for Latin American Studies. Oxford, England. (Bull. Lat. Am. Res.)

Bulletin of the International Committee on Urgent Anthropological and Ethnological Research. International Union of Anthropological and Ethnological Sciences. Vienna. (Bull. Int. Anthropol. Ethnol.)

Bulletins et mémoires de la Société d'anthropologie de Paris. Paris. (Bull. mém. Soc. anthropol. Paris)

Business Mexico. American Chamber of Commerce of Mexico. México. (Bus. Mex.)

Caderno CRH. Centro de Recursos Humanos. Salvador, Brazil. (Cad. CRH)

Cadernos. Centro de Filosofia e Ciências Humanas, Univ. Federal do Pará. Belém, Brazil. (Cadernos/Belém)

Cadernos do Centro de Estudos e Ação Social (CEAS). Salvador, Brazil. (Cad. CEAS)

Cahiers des Amériques latines. Paris. (Cah. Am. lat.)

Les Cahiers d'Outre-Mer. Institut de géographie de la Faculté des lettres de Bordeaux; Institut de la France d'Outre-Mer; Société de géographie de Bordeaux. Bordeaux, France. (Cah. Outre-Mer)

Cambridge Journal of Economics. Cambridge Political Economy Society. London; Academic Press. New York. (Camb. J. Econ.)

Canadian Journal of History. Univ. of Saskatchewan. Saskatoon, Canada. (Can. J. Hist.)

Canadian Journal of Latin American and Caribbean Studies. Univ. of Ottawa. Ontario, Canada. (Can. J. Lat. Am. Caribb. Stud.)

Caravelle. Cahiers du monde hispanique et luso-brésilien. Univ. de Toulouse, Institute d'études hispaniques, hispano-americaines et luso-brésiliennes. Toulouse, France. (Caravelle/Toulouse)

Caribbean Affairs. Trinidad Express Newspapers Ltd.; Inprint Caribbean Ltd., Port of Spain, Trinidad and Tobago. (Caribb. Aff.)

Caribbean Perspectives. Transaction Publishers. New Brunswick, New Jersey. (Caribb. Perspect.)

Caribbean Quarterly. Univ. of the West Indies. Mona, Jamaica. (Caribb. Q.)

Caribbean Studies. Univ. of Puerto Rico, Institute of Caribbean Studies. Río Piedras. (Caribb. Stud.)

El Caribe Contemporáneo. Univ. Nacional Autónoma de México. México. (Caribe Contemp.)

Caribena: cahiers d'études américanistes de la Caraïbe. Centre d'études et de recherches archéologiques (CERA). Martinique. (Caribena/Martinique)

Cartographica. Univ. of Toronto Press. Downsview, Ontario. (Cartographica/Ontario)

Catena. Catena Verlag. Cremlingen, Wolfenbüttel, Germany. (Catena/Cremlingen)

CD-ROM Professional. Pemberton Press. Weston, Conn. (CD-ROM Prof.)

The Centennial Review. Michigan State Univ., College of Science and Arts. East Lansing, Mich. (Centen. Rev.)

CEPAL Review/Revista de la CEPAL. Naciones Unidas, Comisión Económica para América Latina. Santiago. (CEPAL Rev.)

Cespedesia. Depto. del Valle del Cauca. Cali, Colombia. (Cespedesia/Cali)

Chemins critiques: revue haïtiano-caraïbéenne. Société sciences-arts-littérature. Port-au-Prince, Haiti. (Chemins crit.)

Ciência & Ambiente. Univ. Federal de Santa Maria. Santa Maria, Brazil. (Ciênc. Amb.)

Ciência e Cultura. Sociedade Brasileira para o Progresso da Ciência. São Paulo. (Ciênc. Cult.)

Ciencia Económica. Facultad de Economía, Univ. de Lima. (Cienc. Econ./Lima)

Ciencia Hoy. Asociación Ciencia Hoy; Morgan Antártica. Buenos Aires. (Cienc. Hoy)

Ciencia Política. Instituto de Ciencia Política de Bogotá; Tierra Firme Editores. Bogotá. (Cienc. Polít.)

Ciencia y Sociedad. Instituto Tecnológico de Santo Domingo. (Cienc. Soc./Santo Domingo)

Ciencias Económicas. Instituto de Investigaciones en Ciencias Ecónomicas, Univ. de Costa Rica. San José. (Cienc. Econ./San José)

Cimarrón. City Univ. of New York, Assn. of Caribbean Studies. New York. (Cimarrón/New York)

Climatic Change. Reidel Publishers. Boston. (Clim. Change)

Colección Estudios CIEPLAN. Corporación de Investigaciones Económicas para Latinoamérica. Santiago. (Colecc. Estud. CIEPLAN)

Columbia Journal of Transnational Law. Columbia Univ. School of Law. New York. (Columbia J. Transnatl. Law)

Columbia Journal of World Business. Columbia Univ., New York. (Columbia J. World Bus.)

Comercio Exterior. Banco Nacional de Comercio Exterior. México. (Comer. Exter.)

Comparative Civilizations Review. Dept. of History, Dickinson College. Carlisle, Penn. (Comp. Civiliz. Rev.)

Comparative Education Review. Comparative Education Society. New York. (Comp. Educ. Rev.)

Comparative Political Studies. Sage Publications, Thousand Oaks, Calif. (Comp. Polit. Stud.)

Comparative Politics. The City Univ. of New York, Political Science Program. New York. (Comp. Polit.)

Comparative Studies in Society and History. Society for the Comparative Study of Society and History; Cambridge Univ. Press. London. (Comp. Stud. Soc. Hist.)

Comunicação e Sociedade. Instituto Metodista do Ensino Superior. São Paulo. (Comun. Soc.)

Contexto Internacional. Instituto de Relações Internacionais, Pontifícia Univ. Católica. Rio de Janeiro. (Cont. Int.)

Contribuciones. Centro de Documentación y Estudios (CDE). Asunción. (Contribuciones/Asunción)

Contribuciones. Estudios Interdisciplinarios sobre Desarrollo y Cooperación Internacional. Konrad-Adenauer-Stiftung; Centro Interdisciplinario de Estudios Sobre el Desarrollo Latinoamericano (CIEDLA). Buenos Aires. (Contribuciones/Buenos Aires)

El Cotidiano: Revista de la Realidad Mexicana Actual. Univ. Autónoma Metropolitana, Unidad Azcapotzalco. México. (Cotidiano/México)

Coyuntura Social. Fundación para la Educación Superior y el Desarrollo (FEDESARROLLO); Instituto SER de Investigación. Bogotá. (Coyunt. Soc.)

Cuadernos Americanos. Editorial Cultura. México. (Cuad. Am.)

Cuadernos de Arquitectura Mesoamericana. Facultad de Arquitectura, Univ. Nacional Autónoma de México. México. (Cuad. Arquit. Mesoam.)

Cuadernos de Economía. Pontificia Univ. Católica de Chile, Instituto de Economía. Santiago. (Cuad. Econ./Santiago)

Cuadernos de Economía. Univ. Nacional de Colombia. Bogotá. (Cuad. Econ./Bogotá)

Cuadernos de INVESP. Instituto Venezolano de Estudios Sociales y Políticos. Caracas. (Cuad. INVESP)

Cuadernos de Investigación. Centro de Investigaciones Tecnológicas y Científicas. San Salvador. (Cuad. Invest./San Salvador)

Cuadernos de Marcha. Eon Editores. Montevideo. (Cuad. Marcha)

Cuadernos de Nuestra América. Centro de Estudios sobre América. La Habana. (Cuad. Nuestra Am.)

Cuadernos de Sociología. Univ. Centroamericana. Managua. (Cuad. Sociol.)

Cuadernos del CENDES. Centro de Estudios del Desarrollo, Univ. Central de Venezuela. Caracas. (Cuad. CENDES)

Cuadernos del CLAEH. Centro Latinoamericano de Economía Humana. Montevideo. (Cuad. CLAEH)

Cuadernos del Sur: Ciencias Sociales. Instituto de Investigaciones Sociológicas, Univ. Autónoma Benito Juárez de Oaxaca. Oaxaca, Mexico. (Cuad. Sur/Oaxaca)

Cuadernos Prehispánicos. Seminario de Historia de América, Univ. de Valladolid. Spain. (Cuad. Prehispánicos)

Cuban Studies. Univ. of Pittsburgh, Center for Latin American Studies. Pittsburgh, Penn. (Cuba. Stud.)

Cuestiones Económicas. Banco Central del Ecuador. Quito. (Cuest. Econ.)

Cultural Anthropology: Journal of the Society for Cultural Anthropology. American Anthropological Assn.; Society for Cultural Anthropology. Washington. (Cult. Anthropol.)

Current Anthropology. Univ. of Chicago. Chicago, Ill. (Curr. Anthropol.)

Current History. Philadelphia, Penn. (Curr. Hist.)

Dados. Instituto Univ. de Pesquisas. Rio de Janeiro. (Dados/Rio de Janeiro)

Database. Online, Inc., Weston, Conn. (Database/Weston)

Debate Agrario. Centro Peruano de Estudios Sociales (CEPES). Lima. (Debate Agrar.)

Dédalo. Univ. de São Paulo, Museu de Arqueologia e Etnologia. São Paulo. (Dédalo/São Paulo)

A Defesa Nacional: Revista de Assuntos Militares e Estudo de Problemas Brasileiros. Rio de Janeiro. (Defesa Nac.)

Demography. Population Assn. of America. Washington. (Demography/Washington)

Desarrollo Económico. Instituto de Desarrollo Económico y Social. Buenos Aires. (Desarro. Econ.)

Development and Change. Mouton. The Hague. (Dev. Change)

Dialectical Anthropology. M. Nijhoff. Dordrecht, The Netherlands. (Dialect. Anthropol.)

Diogenes. International Council for Philosophy and Humanistic Studies (Paris); Berg Publishers. Oxford, England. (Diogenes/Philosophy)

Directory of United Nations Information Sources. Advisory Committee for the Coordination of Information Systems (ACCIS), United Nations. New York. (Dir. U.N. Inf. Sources)

Disarmament. United Nations. New York. (Disarmament/UN)

Dissent. Dissent Publishing Assn., New York. (Dissent/New York)

Economía. Depto. de Economía, Pontificia Univ. Católica del Perú. Lima. (Economía/Lima)

Economía y Ciencias Sociales. Facultad de Ciencias Económicas y Sociales, Univ. Central de Venezuela. Caracas. (Econ. Cienc. Soc.)

Economía y Trabajo en Chile. Programa de Economía del Trabajo, Academia de Humanismo Cristiano. Santiago. (Econ. Trab. Chile)

Economic and Social Progress in Latin America. Inter-American Development Bank. Washington. (Econ. Soc. Prog. Lat. Am.)

Economic Development and Cultural Change. Univ. of Chicago, Research Center in Economic Development and Cultural Change. Chicago, Ill. (Econ. Dev. Cult. Change)

Economic Geology. Economic Geology Publishing Co., Lancaster, Penn. (Econ. Geol.)

Económica. Univ. Nacional de La Plata, Facultad de Ciencias Económicas, Instituto de Investigaciones Económicas. La Plata, Argentina. (Económica/La Plata)

The Economist. London. (Economist/London)

Ecuador Debate. Centro Andino de Acción Popular (CAAP). Quito. (Ecuad. Deb.)

Ediciones del Quinto Centenario. Univ. de la República. Montevideo. (Ed. Quinto Centen.)

Empresas Siderúrgicas do Brasil. Instituto Brasileiro de Siderurgia. Rio de Janeiro. (Empres. Sider. Brasil)

Ensaios FEE. Secretaria de Coordenacão e Planejamento. Fundação de Economia e Estatística. Porto Alegre, Brazil. (Ensaios FEE)

Envío. Univ. Centroamericana (UCA). Managua. (Envío/Managua)

Environmental Conservation. Foundation for Environmental Conservation; Elsevier Sequoia. Lausanne, Switzerland. (Environ. Conserv.)

Erdkunde. Archiv für Wissenschaftliche Geographie. Univ. Bonn, Geographisches Institut. Bonn, Germany. (Erdkunde/Bonn)

Eres. Museo Arqueológico y Etnográfico. Tenerife, Spain. (Eres/Tenerife)

L'Espace géographique. Doin. Paris. (Espace géogr.)

Espacio y Desarrollo. Pontificia Univ. Católica del Perú, Depto. de Humanidades, Centro de Investigación en Geografía Aplicada. Lima. (Espac. Desarro.)

Estadística & Economía. Instituto Nacional de Estadísticas. Santiago. (Estad. Econ.)

Estado & Sociedad: Revista Boliviana de Ciencias Sociales. Facultad Latinoamericana de Ciencias Sociales (FLACSO). La Paz. (Estad. Soc.)

Estudios. Instituto de Estudios Económicos sobre la Realidad Argentina y Latinoamericana; Fundación Mediterránea. Córdoba, Argentina. (Estudios/Fundación Mediterránea)

Estudios Centro-Americanos: ECA. Univ. Centroamericana José Simeón Cañas. San Salvador. (ECA/San Salvador)

Estudios de Cultura Maya. Centro de Estudios Mayas, Univ. Nacional Autónoma de México. México. (Estud. Cult. Maya)

Estudios de Cultura Náhuatl. Instituto de Investigaciones Históricas, Univ. Nacional Autónoma de México. México. (Estud. Cult. Náhuatl)

Estudios de Economía. Depto. de Economía, Univ. de Chile. Santiago. (Estud. Econ./Santiago)

Estudios Demográficos y Urbanos. El Colegio de México. México. (Estud. Demogr. Urb.)

Estudios Económicos. El Colegio de México. México. (Estud. Econ./México)

Estudios Fronterizos. Instituto de Investigaciones Sociales, Univ. Autónoma de Baja California. Mexicali, Mexico. (Estud. Front.)

Estudios Geográficos. Instituto de Economía y Geografía Aplicadas, Consejo Superior de Investigaciones Científicas. Madrid. (Estud. Geogr./Madrid)

Estudios Internacionales. Instituto de Estudios Internacionales, Univ. de Chile. Santiago. (Estud. Int./Santiago)

Estudios Internacionales: Revista del IRIPAZ. Instituto de Relaciones Internacionales y de Investigaciones para la Paz. Guatemala. (Estud. Int./IRIPAZ)

Estudios Paraguayos. Univ. Católica Nuestra Señora de la Asunción. Asunción. (Estud. Parag.)

Estudios Públicos. Centro de Estudios Públicos. Santiago. (Estud. Públicos)

Estudios Rurales Latinoamericanos. Consejo Latinoamericano de Ciencias Sociales. Bogotá. (Estud. Rural. Latinoam.)

Estudios sobre las Culturas Contemporáneas. Centro Universitario de Investigaciones Sociales, Univ. de Colima. México. (Estud. Cult. Contemp.)

Estudios Sociales. Corporación de Promoción Universitaria. Santiago. (Estud. Soc./Santiago)

Estudios Sociales. Centro de Estudios Sociales Juan Montalvo, SJ. Santo Domingo. (Estud. Soc./Santo Domingo)

Estudios Sociales Centroamericanos. Programa Centroamericano de Ciencias Sociales. San José. (Estud. Soc. Centroam.)

Estudios Sociológicos. Centro de Estudios Sociológicos de El Colegio de México. México. (Estud. Sociol./México)

Estudios Territoriales. Instituto del Territorio y Urbanismo; Ministerio de Obras Públicas y Transportes. Madrid. (Estud. Territ.)

Estudos Avançados. Univ. de São Paulo, Instituto de Estudos Avançados. São Paulo. (Estud. Avançados)

Estudos Econômicos. Univ. de São Paulo, Instituto de Pesquisas Econômicas. São Paulo. (Estud. Econ./São Paulo)

Ethnic Groups. Gordon and Breach. New York. (Ethnic Groups/New York)

Ethnohistory. American Society for Ethnohistory. Duke Univ., Durham, N.C. (Ethnohistory/Society)

Ethnology. Univ. of Pittsburgh, Penn. (Ethnology/Pittsburgh)

Ethnos. Statens Etnografiska Museum. Stockholm. (Ethnos/Stockholm)

Ethos. Society for Psychological Anthropology; Univ. of California, Los Angeles. (Ethos/Society)

Etnofoor. Antropologisch-Sociologisch Centrum. Amsterdam, The Netherlands. (Etnofoor/Netherlands)

Etudes créoles. Comité international des études créoles. Montréal. (Etud. créoles)

EURE: Revista Latinoamericana de Estudios Urbanos Regionales. Centro de Desarrollo Urbano y Regional, Univ. Católica de Chile. Santiago. (EURE/Santiago)

Evaluación Económica. Müller & Machicado Asociados. La Paz. (Eval. Econ.)

The Explorers Journal. New York. (Explor. J.)

Finance and Development. International Monetary Fund; The World Bank. Washington. (Financ. Dev.)

Folk: Journal of the Danish Ethnographic Society. Danish Ethnographic Society. Copenhagen. (Folk/Copenhagen)

Foreign Affairs. Council on Foreign Relations, Inc. New York. (Foreign Aff.)

Foreign Policy. National Affairs Inc.; Carnegie Endowment for International Peace. New York. (Foreign Policy)

Foro Internacional. El Colegio de México. México. (Foro Int.)

Foro Político. Instituto de Ciencias Políticas, Univ. del Museo Social Argentino. Buenos Aires. (Foro Polít.)

Frontera Norte. Colegio de la Frontera Norte. Tijuana, Mexico. (Front. Norte)

Fuerzas Armadas y Sociedad. Centro Latinoamericano de Defensa y Desarme; FLACSO. Santiago. (Fuerzas Arm. Soc.)

Gaceta Arqueológica Andina. Instituto Andino de Estudios Arqueológicos. Lima. (Gac. Arqueol. Andin.)

Gaceta Universitaria. Univ. Autónoma de Campeche, Mexico. (Gac. Univ./Campeche)

Geoarchaeology. John Wiley. New York. (Geoarchaeology/New York)

Geografia. Associação de Geografia Teorética. Rio Claro, Brazil. (Geografia/Rio Claro)

Geografie. KNAG. Utrecht, The Netherlands. (Geografie/Utrecht)

Geographical. Hyde Park Publications. London. (Geographical/London)

The Geographical Journal. The Royal Geographical Society. London. (Geogr. J.)

Geographical Review. American Geographical Society of New York. (Geogr. Rev.)

Geographische Rundschau. Zeitschrift für Schulgeographie. Georg Westermann Verlag. Braunschweig, Germany. (Geogr. Rundsch.)

Geographische Zeitschrift. Franz Steiner Verlag. Wiesbaden, Germany. (Geogr. Z.)

Geography. Geographical Assn., London. (Geography/London)

GeoJournal. D. Reidel Publishing Co., Boston, Mass. (GeoJournal/Boston)

Geosul. Depto. de Geociências, Univ. Federal de Santa Catarina. Florianópolis, Brazil. (Geosul/Florianópolis)

Granma. La Habana. (Granma/La Habana)

Handbook of Latin American Studies CD-ROM: HLAS/CD, Vols. 1–53. Fundación MAPFRE América, Madrid. Hispanic Division, Library of Congress, Washington. Distributed by Univ. of Texas Press, Austin. (HLAS/CD)

Harvard Business Review. Graduate School of Business Administration, Harvard Univ., Boston. (Harv. Bus. Rev.)

Hemisphere. Latin American and Caribbean

Center, Florida International Univ., Miami, Fla. (Hemisphere/Miami)

Herencia. Programa de Rescate y Revitalización del Patrimonio Cultural. San José. (Herencia/San José)

Hispanic American Historical Review. Conference on Latin American History of the American Historical Assn.; Duke Univ. Press. Durham, N.C. (HAHR)

Historia Mexicana. Colegio de México. México. (Hist. Mex.)

História, Questões e Debates. Associação Paranaense de História. Curitiba, Brazil. (Hist. Quest. Debates)

Historia y Sociedad. Depto. de Historia, Univ. de Puerto Rico. Río Piedras. (Hist. Soc./Río Piedras)

History of Religions. Univ. of Chicago. Chicago, Ill. (Hist. Relig.)

History Workshop. Ruskin College, Oxford Univ., Oxford, England. (Hist. Workshop)

Hómines. Univ. Interamericana de Puerto Rico. San Juan. (Hómines/San Juan)

L'Homme. Laboratoire d'anthropologie, Collège de France. Paris. (Homme/Paris)

Hoy es Historia: Revista Bimestral de Historia Nacional e Iberoamericana. Editorial Raíces. Montevideo. (Hoy Hist.)

Human Ecology. Plenum Publishing Corp., New York. (Hum. Ecol.)

Human Organization. Society for Applied Anthropology. New York. (Hum. Organ.)

Humanidades. Editora Univ. de Brasília. (Humanidades/Brasília)

Ibero-Americana: Nordic Journal of Latin American Studies. Institute of Latin American Studies, Univ. of Stockholm. (Ibero-Am./Stockholm)

Ibero-Amerikanisches Archiv. Ibero-Amerikanisches Institut. Berlin. (Ibero-Am. Arch.)

Iberoamericana. Univ. of Sofia. Tokyo. (Iberoam./Tokyo)

Ideas Políticas: Revista de Análisis y Debate. Cambio XXI Fundación Mexicana. México. (Ideas Polít.)

Identities: Global Studies in Culture and Power. Gordon and Breach Publishers. Yverdon, Switzerland. (Identities/Yverdon)

Impex: Foreign Trade Statistics by Commodities = Statistiques de commerce extérior produits. Organisation for Economic Cooperation and Development (OECD). Paris. (Impex/OECD)

Index on Censorship. Writers & Scholars International. London. (Index Censorsh.)

Indiana. Gebr. Mann., Berlin. (Indiana/Berlin)

Integración Latinoamericana. Instituto para la Integración de América Latina. Buenos Aires. (Integr. Latinoam.)

Interciencia. Asociación Interciencia. Caracas. (Interciencia/Caracas)

International Affairs. Moscow. (Int. Aff./Moscow)

International Bibliography of the Social Sciences. British Library of Political and Economic Science, London School of Economics and Political Science. London. (Int. Bibliogr. Soc. Sci.)

The International History Review. Univ. of Toronto Press. Downsview, Ontario, Canada. (Int. Hist. Rev.)

International Journal. Canadian Institute of International Affairs. Toronto, Canada. (Int. J./Toronto)

International Journal of Comparative Sociology. York Univ., Dept. of Sociology and Anthropology. Toronto, Canada. (Int. J. Comp. Sociol.)

International Journal of Intercultural Relations. Society for Intercultural Education, Training, and Research; Pergamon Press. New York. (Int. J. Intercult. Relat.)

International Labour Review. International Labour Office. Geneva. (Int. Labour Rev.)

International Migration = Migrations Internationales = Migraciones Internacionales. Intergovernmental Committee for European Migration; Research Group for European Migration Problems; International Organization for Migration. The Hague, Netherlands; Geneva, Switzerland. (Int. Migr.)

International Migration Review. Center for Migration Studies. New York. (Int. Migr. Rev.)

International Organization. World Peace Foundation; Univ. of Wisconsin Press. Madison. (Int. Organ.)

International Political Science Abstracts. International Political Science Assn., Paris. (Int. Polit. Sci. Abstr.)

International Review of Administrative Sciences. International Institute of Administrative Sciences. Brussels. (Int. Rev. Adm. Sci.)

The International Spectator. Istituto Affari Internazionali. Rome. (Int. Spect./Rome)

International Studies Quarterly. Wayne State Univ., Detroit, Mich. (Int. Stud. Q.)
Internationale Spectator. Nederlandsch Genootschap voor Internationale Zaken. The Hague. (Int. Spect./Hague)
Internet World. Meckler Corp. Westport, Conn. (Internet World)
Investigación Agraria: Economía. Instituto Nacional de Investigaciones Agrarias. Madrid. (Invest. Agrar.)
Investigación Económica. Facultad de Economía, Univ. Nacional Autónoma de México. México. (Invest. Econ.)
Islas. Univ. Central de Las Villas. Santa Clara, Cuba. (Islas/Santa Clara)
Iztapalapa. Univ. Autónoma Metropolitana, División de Ciencias Sociales y Humanidades. México. (Iztapalapa/México)
Jamaica Journal. Institute of Jamaica. Kingston. (Jam. J.)
Jane's Sentinel. Jane's Information Group. Alexandria, Va. (Jane's Sentinel)
Journal for the Scientific Study of Religion. Society for the Scientific Study of Religion. Storrs, Conn. (J. Sci. Stud. Relig.)
Journal of Anthropological Archaeology. Academic Press. New York. (J. Anthropol. Archaeol.)
Journal of Anthropological Research. Univ. of New Mexico. Albuquerque, N.M. (J. Anthropol. Res.)
Journal of Archaeological Research. Plenum Press. New York. (J. Archaeol. Res.)
Journal of Archaeological Science. Academic Press. New York. (J. Archaeol. Sci.)
The Journal of Caribbean History. Caribbean Univ. Press. St. Lawrence, Barbados. (J. Caribb. Hist.)
Journal of Caribbean Studies. Assn. of Caribbean Studies. Coral Gables, Fla. (J. Caribb. Stud.)
Journal of Church and State. J.M. Dawson Studies in Church and State, Baylor Univ., Waco, Tex. (J. Church State)
Journal of Common Market Studies. Oxford, England. (J. Common Market Stud.)
The Journal of Commonwealth & Comparative Politics. Univ. of London, Institute of Commonwealth Studies. London. (J. Commonw. Comp. Polit.)
Journal of Cultural Geography. Popular Culture Assn.; American Culture Assn.; Bowling Green State Univ., Bowling Green, Oh. (J. Cult. Geogr.)
The Journal of Developing Areas. Western Illinois Univ. Press. Macomb, Ill. (J. Dev. Areas)
Journal of Development Economics. North-Holland Publishing Co., Amsterdam, The Netherlands. (J. Dev. Econ.)
The Journal of Development Studies. Frank Cass. London. (J. Dev. Stud.)
Journal of Economic Issues. California State Univ., Sacramento. (J. Econ. Issues)
Journal of Ethnobiology. Center for Western Studies. Flagstaff, Ariz. (J. Ethnobiol.)
Journal of Field Archaeology. Boston Univ., Boston, Mass. (J. Field Archaeol.)
Journal of Geography. National Council of Geographic Education. Menasha, Wis. (J. Geogr.)
Journal of Historical Geography. Academic Press. London; New York. (J. Hist. Geogr.)
Journal of Interamerican Studies and World Affairs. Institute of Interamerican Studies, Univ. of Miami. Coral Gables, Fla. (J. Interam. Stud. World Aff.)
Journal of International and Comparative Economics (JOICE). Physica-Verlag. Heidelberg, Germany. (JOICE)
Journal of International Law and Economics. George Washington Univ., The National Law Center. Washington. (J. Int. Law Econ.)
Journal of Latin American Lore. Univ. of California, Latin American Center. Los Angeles, Calif. (J. Lat. Am. Lore)
Journal of Latin American Studies. Centers or Institutes of Latin American Studies at the Universities of Cambridge, Glasgow, Liverpool, London, and Oxford. Cambridge Univ. Press. London. (J. Lat. Am. Stud.)
Journal of Money, Credit and Banking. Ohio State Univ. Press. Columbus. (J. Money Credit Bank.)
Journal of Peace Research. International Peace Research Institute, Universitetforlaget. Oslo. (J. Peace Res.)
The Journal of Peasant Studies. Frank Cass & Co., London. (J. Peasant Stud.)
Journal of Political & Military Sociology. Northern Illinois Univ., Dept. of Sociology. DeKalb, Ill. (J. Polit. Mil. Sociol.)
Journal of the Bahamas Historical Society. Nassau, Bahamas?. (J. Bahamas Hist. Soc.)
Journal of the Steward Anthropological Society. Urbana, Ill. (J. Steward Anthropol. Soc.)
Journal of World Prehistory. Plenum Press. New York. (J. World Prehist.)
Journal of World Trade: Law, Economics, Pub-

lic Policy. Werner Publishing Company, Ltd., Geneva. (J. World Trade)

Journalism Quarterly. Assn. for Education in Journalism; American Assn. of Schools and Depts. of Journalism; Kappa Tau Alpha Society.; Univ. of Minnesota. Minneapolis, Minn. (Journal. Q.)

KAS Auslands-Informationen. Konrad Adenauer-Stiftung. Bonn. (KAS Ausl.-Inf.)

LASA Forum. Latin American Studies Assn., Pittsburgh, Penn. (LASA Forum)

Lateinamerika Jahrbuch. Vervuert Verlag. Frankfurt. (Lat.am. Jahrb.)

The Latin American Anthropology Review. Society for Latin American Anthropology. Fairfax, Va. (Lat. Am. Anthropol. Rev.)

Latin American Antiquity. Society for American Archaeology. Washington. (Lat. Am. Antiq.)

Latin American Business Intelligence. Economist Intelligence Unit. London; New York. (Lat. Am. Bus. Intell.)

Latin American News. Knight-Ridder Information. Mountain View, California. (Lat. Am. News/Knight-Ridder)

Latin American Newsletters. London. (Lat. Am. Newsl./London)

Latin American Perspectives. Univ. of California. Newbury Park, Calif. (Lat. Am. Perspect.)

Latin American Research Review. Latin American Research Review Board. Univ. of New Mexico, Albuquerque, N.M. (LARR)

Latinoamerica. Edizioni Associate. Rome. (Latinoamerica/Rome)

Law & Anthropology: Internationales Jahrbuch für Rechtsanthropologie. VWGö. Vienna; Klaus Renner Verlag. Hohenschäftlarn. (Law Anthropol. Int. Jahrb. Rechtsanthropol.)

Lecturas de Economía. Univ. de Antioquia. Medellín, Colombia. (Lect. Econ.)

Lua Nova. Editora Brasiliense. São Paulo. (Lua Nova)

MACLAS Latin American Essays. Middle Atlantic Council of Latin American Studies. New Brunswick, N.J. (MACLAS Lat. Am. Essays)

Maguaré. Depto. de Antropología, Univ. Nacional de Colombia. Bogotá. (Maguaré/Bogotá)

Man. The Royal Anthropological Institute. London. (Man/London)

The Manchester School of Economic and Social Studies. Economics Dept., Manchester School. Manchester, England. (Manch. Sch. Econ. Soc. Stud.)

The Mayan People and Human Rights. Pro Justice and Peace Committee of Guatemala. Guatemala. (Mayan People Hum. Rights)

Mededelingen. Stichting Surinaams Museum. Paramaribo. (Mededelingen/Paramaribo)

Medio Ambiente y Urbanización. Comisión de Desarrollo Urbano y Regional, Consejo Latinoamericano de Ciencias Sociales. Buenos Aires. (Medio Ambient. Urban.)

Memoria: Boletín de CEMOS. Centro de Estudios del Movimiento Obrero y Socialista. México. (Memoria/CEMOS)

Memoria de la Sociedad de Ciencias Naturales La Salle. Caracas. (Mem. Soc. Cienc. Nat.)

Mesoamérica. Centro de Investigaciones Regionales de Mesoamérica. Antigua, Guatemala. (Mesoamérica/Antigua)

Mexico Business. Mexico Business Publishing Group. Houston, Tex. (Mex. Bus.)

Mexicon. K.-F. von Flemming. Berlin, Germany. (Mexicon/Berlin)

Mitteilungen der Österreichischen Geographischen Gesellschaft. Verleger, Herausgeber und Eigentümer. Vienna. (Mitt. Österr. Geogr. Ges.)

Montalbán. Univ. Católica Andrés Bello, Facultad de Humanidades y Educación, Institutos Humanísticos de Investigación. Caracas. (Montalbán/Caracas)

Monthly Review. New York. (Mon. Rev.)

Mountain Research and Development. International Mountain Society. Boulder, Colo. (Mt. Res. Dev.)

NACLA: Report on the Americas. North American Congress on Latin America. New York. (NACLA)

Names: Journal of the American Name Society. Pine Hill Press. Freeman, S.D. (Names/Freeman)

National Geographic Research. National Geographic Society. Washington. (Natl. Geogr. Res.)

The National Interest. National Affairs. Washington. (Natl. Int.)

Nature: International Weekly Journal of Science. Macmillan Magazines. London. (Nature/London)

Naval War College Review. Newport, R.I. (Nav. War Coll. Rev.)

NEARA Journal. New England Antiquities Research Assn., Milford, N.H. (NEARA J.)

New Left Review. New Left Review, Ltd., London. (New Left Rev.)
Nieuwe West-Indische Gids. Martinus Nijhoff. The Hague. (Nieuwe West-Indische Gids)
North-South:The Magazine of the Americas. North-South Center, Univ. of Miami. Coral Gables, Fla. (North-South/Miami)
Notas de Población. Centro Latinoamericano de Demografía. Santiago. (Notas Pobl.)
Notas Mesoamericanas. Univ. de las Américas-Puebla. Puebla, Mexico. (Notas Mesoam.)
Nova Economia. Depto. de Ciências Econômicas, Univ. Federal de Minas Gerais. Belo Horizonte, Brazil. (Nova Econ.)
Novedades Colombianas. Museo de Historia Natural, Univ. del Cauca. Papayán, Colombia. (Noved. Colomb.)
Novos Estudos CEBRAP. Centro Brasileiro de Análise e Planejamento. São Paulo. (Novos Estud. CEBRAP)
Nuestra Economía. Facultad de Economía de la Univ. Autónoma de Baja California. Tijuana, Mexico. (Nuestra Econ.)
Nueva Sociedad. Caracas. (Nueva Soc.)
Nuevos Aportes. Editorial San José S.R.L., La Paz. (Nuevos Aportes)
Oceanus. Oceanographic Institution. Woods Hole, Mass. (Oceanus/Woods Hole)
OECD Statistical Compendium. OECD Electronic Publications. Paris. (OECD Stat. Compend.)
Ojarasca. Pro-México Indígena. México. (Ojarasca/México)
Online & CDROM Review. Learned Information. Oxford, England; Medford, N.J. (Online CDROM Rev.)
Opiniones y Análisis. Fundación Boliviana para la Capacitación Democrática y la Investigación. La Paz. (Opin. Anál.)
Oxford Economic Papers. Oxford Univ. Press. London. (Oxf. Econ. Pap.)
Pachacamac: Revista del Museo de la Nación. Museo de la Nación. Lima. (Pachacamac/Lima)
Pacífico: Revista de Ciencias Sociales. Chimbote, Peru. (Pacífico/Chimbote)
Paisajes Geográficos. Centro Panamericano de Estudios e Investigaciones Geográficas. Quito. (Paisajes Geogr.)
La Palabra y el Hombre. Univ. Veracruzana. Xalapa, Mexico. (Palabra Hombre)
Panorama Centroamericano: Pensamiento y Acción. Instituto Centroamericano de Estudios Políticos (INCEP). Guatemala. (Panorama Centroam. Pensam.)
Panorama Económico de América Latina. División de Desarrollo Económico, Comisión Económica para América Latina y el Caribe, Naciones Unidas. Santiago. (Panorama Econ. Am. Lat./CEPAL)
Papers and Proceedings of the Applied Geography Conferences. Dept. of Geography, Kent State Univ.; State Univ. of New York (SUNY) at Binghamton. (Pap. Proc. Appl. Geogr. Conf.)
Pará Desenvolvimento. Instituto de Desenvolvimento Econômico-Social do Pará. Belém, Brazil. (Pará Desenvolv.)
Parliamentary Affairs. Oxford Univ. Press. London. (Parliam. Aff.)
Pensamiento Iberoamericano. Instituto de Cooperación Iberoamericano (ICI) de España; Comisión Económica para América Latina y el Caribe (CEPAL). Madrid. (Pensam. Iberoam.)
Perfiles Latinoamericanos. Facultad Latinoamericana de Ciencias Sociales. México. (Perf. Latinoam.)
Pesquisa e Planejamento Econômico. Instituto de Planejamento Econômico e Social. Rio de Janeiro. (Pesqui. Planej. Econ.)
Pesquisas. Instituto Anchietano de Pesquisas. São Leopoldo, Brazil. (Pesquisas/São Leopoldo)
Plantation Society in the Americas. Univ. of New Orleans. (Plant. Soc. Am.)
Politeia. Instituto de Estudios Políticos, Univ. Central de Venezuela. Caracas. (Politeia/Caracas)
Política Externa. Paz e Terra. São Paulo. (Polít. Extern.)
Política Internacional. Revista de la Academia Diplomática del Peru. Lima. (Polít. Int.)
Political Science Quarterly. Columbia Univ., The Academy of Political Science. New York. (Polit. Sci. Q.)
Political Studies. Political Studies Assn. of the United Kingdom; Clarendon Press. Oxford, England. (Polit. Stud.)
Prehistoria: Revista del Programa de Estudios Prehistóricos. Consejo Nacional de Investigaciones Científicas y Técnicas. Buenos Aires. (Prehistoria/Buenos Aires)
Pretextos. Centro de Estudios y Promoción del Desarrollo. Lima. (Pretextos/Lima)
Problems of Communism. United States In-

formation Agency. Washington. (Probl. Communism)

Proceso. Comunicación e Información. México. (Proceso/México)

The Professional Geographer. Assn. of American Geographers. Washington. (Prof. Geogr.)

Pumapunku. Centro de Investigaciones Antropológicas Tiwanaku. La Paz. (Pumapunku/La Paz)

The Quarterly Journal of Economics. Cambridge, Mass. (Q. J. Econ.)

Quaternary Research. Academic Press. New York. (Quat. Res./New York)

Quehacer. Centro de Estudios y Promoción del Desarrollo (DESCO). Lima. (Quehacer/Lima)

Radiocarbon. Supplement of the American Journal of Science. New Haven, Conn. (Radiocarbon/New Haven)

Realidad Económica. Instituto Argentino para el Desarrollo Económico (IADE). Buenos Aires. (Real. Econ./Buenos Aires)

Realidad Económico-Social. Univ. Centroamericana José Simeón Cañas. San Salvador. (Real. Econ.-Soc.)

Relaciones. El Colegio de Michoacán. Zamora, Mexico. (Relaciones/Zamora)

Relaciones. Univ. Autónoma Metropolitana, Unidad Xochimilco. Coyoacán, México. (Relaciones/Coyoacán)

Relaciones Internacionales. Centro de Relaciones Internacionales. Facultad de Ciencias Políticas y Sociales, Univ. Nacional Autónoma de México. México. (Relac. Int./Mexico)

Res. Peabody Museum of Archaeology and Ethnology, Harvard Univ., Cambridge, Mass. (Res/Harvard)

Research & Exploration. National Geographic Society. Washington. (Res. Explor.)

Research in Economic Anthropology. JAI Press. Greenwich, Conn. (Res. Econ. Anthropol.)

Review: Fernand Braudel Center. Fernand Braudel Center for the Study of Economics, Historical Systems, and Civilizations. Binghamton, New York. (Review/Braudel)

The Review of Archaeology. Salem, Mass. (Rev. Archaeol./Salem)

Revista ANDI. Asociación Nacional de Industriales (ANDI). Medellín, Colombia. (Rev. ANDI)

Revista Argentina de Estudios Estratégicos. Olcese Editores. Buenos Aires. (Rev. Arg. Estud. Estrateg.)

Revista Brasileira de Comércio Exterior. Fundação Centro de Estudos do Comércio Exterior. Rio de Janeiro. (Rev. Bras. Comér. Exter.)

Revista Brasileira de Economia. Fundação Getúlio Vargas, Instituto Brasileiro de Economia. Rio de Janeiro. (Rev. Bras. Econ.)

Revista Brasileira de Estudos de População. Associação Brasileira de Estudos Populacionais. São Paulo. (Rev. Bras. Estud. Popul.)

Revista Brasileira de Estudos Políticos. Univ. de Minas Gerais. Belo Horizonte, Brazil. (Rev. Bras. Estud. Polít.)

Revista Brasileira de Geografia. Conselho Nacional de Geografia, Instituto Brasileiro de Geografia e Estatística. Rio de Janeiro. (Rev. Bras. Geogr.)

Revista Cancillería de San Carlos. Ministerio de Relaciones Exteriores. Bogotá. (Rev. Cancillería San Carlos)

Revista Centroamericana de Economía. Univ. Nacional Autónoma de Honduras, Programa de Postgrado Centroamericano en Economía y Planificación. Tegucigalpa. (Rev. Centroam. Econ.)

Revista Chilena de Antropología. Depto. de Antropología, Univ. de Chile. Santiago. (Rev. Chil. Antropol.)

Revista Chilena de Geopolítica. Instituto Geopolítico de Chile. Santiago. (Rev. Chil. Geopolít.)

Revista Ciclos en la Historia, Economía y la Sociedad. Fundación de Investigaciones Históricas, Económicas y Sociales, Facultad de Ciencias Económicas, Univ. de Buenos Aires. Buenos Aires. (Rev. Ciclos)

Revista Colombiana de Antropología. Ministerio de Educación Nacional, Instituto Colombiano de Antropología. Bogotá. (Rev. Colomb. Antropol.)

Revista Cubana de Ciencias Sociales. Centro de Estudios Filosóficos, Academia de Ciencias de Cuba. La Habana. (Rev. Cuba. Cienc. Soc.)

Revista Cubana de Derecho. La Habana. (Rev. Cuba. Derecho)

Revista de Administração de Empresas. Fundação Getúlio Vargas, Instituto de Documentação. São Paulo. (Rev. Adm. Empres.)

Revista de Análisis Económico. Programa de Postgrado en Economía, ILADES/Georgetown Univ., Santiago. (Rev. Anál. Econ.)

Revista de Antropologia. Univ. de São Paulo, Faculdade de Filosofia, Letras e Ciências Humanas; Associação Brasileira de Antropologia. São Paulo. (Rev. Antropol./São Paulo)

Revista de Arqueologia. Sociedade de Arqueologia Brasileira. São Paulo. (Rev. Arqueol./São Paulo)

Revista de Arqueología Americana. Instituto Panamericano de Geografía e Historia. México. (Rev. Arqueol. Am.)

Revista de Ciencias Sociales. Facultad de Derecho y Ciencias Sociales, Univ. de Valparaíso. Chile. (Rev. Cienc. Soc./Valparaíso)

Revista de Ciencias Sociales. Univ. de Puerto Rico, Colegio de Ciencias Sociales. Río Piedras. (Rev. Cienc. Soc./Río Piedras)

Revista de Ciencias Sociales. Univ. de Costa Rica. San José. (Rev. Cienc. Soc./San José)

Revista de Econometria. Sociedade Brasileira de Econometria; Univ. Federal de Pernambuco. Recife, Brazil. (Rev. Econ./Recife)

Revista de Economía. Banco Central de Uruguay. Montevideo. (Rev. Econ./Uruguay)

Revista de Economia e Sociologia Rural. Sociedade Brasileira de Economia e Sociologia Rural. Brasília. (Rev. Econ. Sociol. Rural)

Revista de Economia Política. Centro de Economia Política. São Paulo. (Rev. Econ. Polít.)

Revista de Estudios Políticos. Centro de Estudios Constitucionales. Madrid. (Rev. Estud. Polít.)

Revista de Geografía Norte Grande. Pontificia Univ. Católica de Chile. Santiago. (Rev. Geogr. Norte Gd.)

Revista de Historia. Centro de Investigaciones Históricas, Univ. de Costa Rica. San José. (Rev. Hist./San José)

Revista de Historia. Univ. Nacional de Costa Rica, Escuela de Historia. Heredia, Costa Rica. (Rev. Hist./Heredia)

Revista de Historia Americana y Argentina. Univ. Nacional de Cuyo, Instituto de Historia. Mendoza, Argentina. (Rev. Hist. Am. Argent.)

Revista de Historia Naval. Instituto de Historia y Cultura Naval Armada Española. Madrid. (Rev. Hist. Naval)

Revista de Indias. Consejo Superior de Investigaciones Científicas, Instituto Gonzalo Fernández de Oviedo. Madrid. (Rev. Indias)

Revista de Investigaciones Económicas. Centro de Investigación, Facultad de Economía, Univ. de Panamá. (Rev. Invest. Econ.)

Revista de la Integración y el Desarrollo de Centroamérica. Banco Centroamericano de Integración Económica. Tegucigalpa. (Rev. Integr. Desarro. Centroam.)

Revista de la Universidad. Univ. Nacional Autónoma de Honduras. Tegucigalpa. (Rev. Univ./Tegucigalpa)

Revista de Occidente. Madrid. (Rev. Occident.)

Revista de Planeación y Desarrollo. Depto. Nacional de Planeación. Bogotá. (Rev. Planeac. Desarro.)

La Revista del Centro de Estudios Avanzados de Puerto Rico y el Caribe. San Juan. (Rev. Cent. Estud. Av.)

Revista del Museo de Historia Natural de San Rafael. Mendoza, Argentina. (Rev. Mus. Hist. Nat. San Rafael)

Revista do CEPA. Centro de Ensino e Pesquisas Arqueológicas, Faculdades Integradas de Santa Cruz do Sul. Santa Cruz do Sul, Brazil. (Rev. CEPA)

Revista do Museu de Arqueologia e Etnologia. Univ. de São Paulo. (Rev. Mus. Arqueol. Etnol.)

Revista Ecuatoriana de Historia Económica. Banco Central del Ecuador, Centro de Investigación y Cultura. Quito. (Rev. Ecuat. Hist. Econ.)

Revista Española de Antropología Americana. Facultad de Geografía e Historia. Univ. Complutense de Madrid. (Rev. Esp. Antropol. Am.)

Revista Española de Investigaciones Sociológicas. Centro de Investigaciones Sociológicas. Madrid. (Rev. Esp. Invest. Sociol.)

Revista Española del Pacífico. Asociación Española de Estudios del Pacífico. Madrid. (Rev. Esp. Pac.)

Revista Este País. Centro Nacional Editor de Discos Compactos. Colima, Mexico. Desarrollo de Opinión Pública. México. (Rev. Este País)

Revista Europea de Estudios Latinoamericanos y del Caribe = European Review of Latin American and Caribbean Studies. Center for Latin American Research and Documentation; Royal Institute of Linguistics and Anthropology. Amsterdam. (Rev. Eur.)

Revista Geográfica. Instituto Panamericano

de Geografía e Historia, Comisión de Geografía. México. (Rev. Geogr./México)
Revista Geográfica. Univ. de Los Andes. Mérida, Venezuela. (Rev. Geogr./Mérida)
Revista Geográfica de Chile. Instituto Geográfico Militar. Santiago. (Rev. Geogr. Chile)
Revista Geográfica Venezolana. Univ. de los Andes. Mérida, Venezuela. (Rev. Geogr. Venez.)
Revista Geológica de América Central. Escuela Centroamericana de Geología, Univ. de Costa Rica. San José. (Rev. Geol. Am. Central)
Revista Interamericana de Bibliografía. Organization of American States. Washington. (Rev. Interam. Bibliogr.)
Revista Interamericana de Planificación. Sociedad Interamericana de Planificación. Bogotá. (Rev. Interam. Planif.)
Revista Mexicana de Ciencias Políticas y Sociales. Facultad de Ciencias Políticas y Sociales, Univ. Nacional Autónoma de México. México. (Rev. Mex. Cienc. Polít. Soc.)
Revista Mexicana de Política Exterior. Secretaría de Relaciones Exteriores; Instituto Matías Romero de Estudios Diplomáticos (IMPED). México. (Rev. Mex. Polít. Exter.)
Revista Mexicana de Sociología. Instituto de Investigaciones Sociales, Univ. Nacional Autónoma de México. México. (Rev. Mex. Sociol.)
Revista Panameña de Sociología. Univ. de Panamá, Depto. de Sociología. Panamá. (Rev. Panameña Sociol.)
Revista Paraguaya de Sociología. Centro Paraguayo de Estudios Sociológicos. Asunción. (Rev. Parag. Sociol.)
Revista Peruana de Ciencias Sociales. Asociación Peruana para el Fomento de las Ciencias Sociales (FOMCIENCIAS). Lima. (Rev. Peru. Cienc. Soc.)
Revista Peruana de Población. Asociación Multidisciplinaria de Investigación y Docencia en Población. Lima. (Rev. Peru. Poblac.)
Revista UNITAS. UNITAS. La Paz. (Rev. UNITAS)
Revista Universidad EAFIT. Depto. de Comunicaciones, Univ. EAFIT. Medellín. (Rev. Univ. EAFIT)
Revista Universidad Pontificia Bolivariana. Medellín, Colombia. (Rev. Univ. Pontif. Boliv.)

Revue de la Société haïtienne d'histoire et géographie. Port-au-Prince. (Rev. Soc. haïti.)
Revue Tiers-Monde. Institut d'étude du développement économique et social, Univ. de Paris. (Rev. Tiers-Monde)
Rivista Geografica Italiana. Societá di studi geografici di Firenze. Florence, Italy. (Riv. Geogr. Ital.)
São Paulo em Perspectiva. Fundação SEADE. São Paulo. (São Paulo Perspect.)
Schweizerische Amerikanisten Gesellschaft. Société Suisse des Américanistes. Genève. (Schweiz. Amer. Ges.)
Science. American Assn. for the Advancement of Science. Washington. (Science/Washington)
Science and Society. New York. (Sci. Soc.)
Scientific American. Scientific American, Inc. New York. (Sci. Am.)
SECOLAS Annals. Southeastern Conference on Latin American Studies; West Georgia College. Carrollton, Ga. (SECOLAS Ann.)
Secuencia. Instituto Mora. México. (Secuencia/México)
Seguridad, Estrategia Regional en el 2000. Impresión Zona Gráfica. Buenos Aires. (SER/Buenos Aires)
Sierra. Sierra Club. San Francisco. (Sierra/San Francisco)
Signos Universitarios: Revista de la Universidad del Salvador. Univ. del Salvador. Buenos Aires. (Signos Univ.)
Signs. The Univ. of Chicago Press. Chicago, Ill. (Signs/Chicago)
Síntesis. Asociación de Investigación y Especialización sobre Temas Latinoamericanos. Madrid. (Síntesis/Madrid)
Social and Economic Studies. Univ. of the West Indies, Institute of Social and Economic Research. Mona, Jamaica. (Soc. Econ. Stud.)
Social Biology. Society for the Study of Social Biology. Port Angeles, Wash. (Soc. Biol.)
Social Compass. The International Catholic Institute for Social-Ecclesiastical Research. The Hague. (Soc. Compass)
Social Problems. Society for the Study of Social Problems; American and International Sociological Associations. Kalamazoo, Mich. (Soc. Probl.)
Social Science and Medicine. Pergamon Press. New York. (Soc. Sci. Med.)
Socialismo y Participación. Ediciones Socialismo y Participación. Lima. (Social. Particip.)

Sociological Analysis. American Catholic Sociological Society. Worcester, Mass. (Sociol. Anal.)
Sociological Inquiry. National Sociology Honor Society; Dept. of Sociology, Univ. of Omaha. Omaha, Neb. (Sociol. Inq.)
Sociologie du travail. Association pour le développement de la sociologie du travail. Paris. (Sociol. trav.)
Sociologus. Berlin. (Sociologus/Berlin)
South Eastern Latin Americanist. South Eastern Council of Latin American Studies. Boone, N.C. (South East. Lat. Am.)
Southern Economic Journal. Chapel Hill, N.C. (South. Econ. J.)
Staff Papers. International Monetary Fund. Washington. (Staff Pap.)
Statistical Masterfile. Congressional Information Service. Bethesda, Md. (Stat. Masterfile)
Studies in Comparative International Development. Transaction Periodicals Consortium, Rutgers Univ., New Brunswick, N.J. (Stud. Comp. Int. Dev.)
Suma. Centro de Investigaciones Económicas. Montevideo. (Suma/Montevideo)
SWI Forum voor Kunst, Kultuur en Wetenschop. De Stichting. Paramaribo, Suriname. (SWI Forum)
Swiss Review of World Affairs. Neue Zürcher Zeitung. Zürich, Switzerland. (Swiss Rev. World Aff.)
Tareas. Centro de Estudios Latinoamericanos (CELA). Panamá. (Tareas/Panamá)
Terra Livre. Associação dos Geógrafos Brasileiros. São Paulo. (Terra Livre)
Third World Planning Review. Liverpool Univ. Press. Liverpool, England. (Third World Plan. Rev.)
Tijdschrift voor Economische en Sociale Geographie. Netherlands Journal of Economic and Social Geography. Rotterdam, The Netherlands. (Tijdschr. Econ. Soc. Geogr.)
Trace. Centre d'études mexicaines et centraméricaines. México. (Trace/México)
El Trimestre Económico. Fondo de Cultura Económica. México. (Trimest. Econ.)
U.S./Latin Trade. New World Communications. Miami. (U.S./Latin Trade)
U tz'ib. Asociación Tikal. Guatemala. (U tz'ib)
UFES: Revista de Cultura. Univ. Federal do Espírito Santo, Brazil. (UFES Rev. Cult.)
UNISA/Latin American Report. Univ. of South Africa. Pretoria. (UNISA/Lat. Am. Rep.)

Universitas. Wissenschaftliche Verlagsgesellschaft. Stuttgart, Germany. (Universitas/Stuttgart)
Universitas. Pontificia Univ. Javeriana, Facultad de Derecho y Ciencias Socioeconómicas. Bogotá. (Universitas/Bogotá)
Universum. Univ. de Talca. Talca, Chile. (Universum/Talca)
UTEC: Revista de la Universidad Tecnológica. Univ. Tecnológica. San Salvador. (UTEC/San Salvador)
Vínculos. Museo Nacional de Costa Rica. San José. (Vínculos/San José)
Voices of Mexico. Univ. Nacional Autónoma de México. México. (Voices Mex.)
Vuelta. México. (Vuelta/México)
The Washington Quarterly. Georgetown Univ., The Center for Strategic and International Studies. Washington. (Wash. Q.)
Weltwirtschaftliches Archiv. Zeitschrift des Institut für Weltwirtschaft an der Christians-Albrechts-Univ. Kiel. Kiel, Germany. (Weltwirtsch. Arch.)
Western Political Quarterly. Western Political Science Assn.; Pacific Northwest Political Science Assn.; Southern California Political Science Assn.; Univ. of Utah, Institute of Government. Salt Lake City. (West. Polit. Q.)
The Wilson Quarterly. Woodrow Wilson International Center for Scholars. Washington. (Wilson Q.)
World Affairs. The American Peace Society. Washington. (World Aff.)
World Archaeology. Routledge & Kegan Paul. London. (World Archaeol.)
The World Bank Economic Review. World Bank. Washington. (World Bank Econ. Rev.)
World Data. World Bank. Washington. (World Data)
World Development. Pergamon Press. Oxford, England. (World Dev.)
The World Economy. Basil Blackwell. London. (World Econ.)
World News Connection. National Technical Information Service. Springfield, Va.; Foreign Broadcast Information Service. Washington. (World News Connect.)
World Policy Journal. World Policy Institute. New York. (World Policy J.)
World Politics. Princeton Univ., Center of International Studies. Princeton, N.J. (World Polit.)
World Trade Database. International Trade

Division, Statistics Canada. Ottawa. (World Trade Database)

Yachay. Facultad de Filosofía y Ciencias Religiosas, Univ. Católica Boliviana. Cochabamba, Bolivia. (Yachay/Cochabamba)

Yaxkin. Instituto Hondureño de Antropología e Historia. Tegucigalpa. (Yaxkin/Tegucigalpa)

Yearbook. Conference of Latin Americanist Geographers; Ball State Univ., Muncie, Ind. (Yearbook/CLAG)

Zeitschrift für Geomorphologie Supplementband. Bebrüder Borntraeger. Berlin. (Z. Geomorphol. Suppl.bd.)

Zeitschrift für Lateinamerika Wien. Österreichisches Lateinamerika-Institut. Vienna. (Z. Lat.am. Wien)

Zeitschrift für Wirtschaftgeographie. Hagen, Westfallen, Germany. (Z. Wirtsch.geogr.)

Zentralblatt für Geologie und Paläontologie. E. Schweizerbart'sche Verlagsbuchhandlung (Nägele u. Obermiller). Stuttgart, Germany. (Zent.bl. Geol. Paläontologie)

ABBREVIATION LIST OF JOURNALS INDEXED

For journal titles listed by full title, see *Title List of Journals Indexed*, p. 753.

Adm. Econ. UC. Administración y Economía UC. Revista de la Facultad de Ciencias Económicas y Administrativas, Pontificia Univ. Católica de Chile. Santiago.

Afr. Archaeol. Rev. The African Archaeological Review. Cambridge Univ. Press. Cambridge, England.

Age Dem. The Age of Democracy. Basil Blackwood; UNESCO.

Agric. Hist. Agricultural History. Agricultural History Society. Univ. of Calif. Press. Berkeley.

Agric. Soc. Agricultura y Sociedad. Ministerio de Agricultura, Pesca, y Alimentación. Madrid.

Agroecol. Desarro. Agroecología y Desarrollo. Consorcio Latinoamericano sobre Agroecología y Desarrollo (CLADES). Santiago.

Allpanchis/Cusco. Allpanchis. Instituto de Pastoral Andina. Cusco, Peru.

Alternativa/Chiclayo. Alternativa. Centro de Estudios Sociales Solidaridad. Chiclayo, Peru.

Alternatives/Boulder. Alternatives. Lynne Rienner Publishers. Boulder, Colorado.

Am. Anthropol. American Anthropologist. American Anthropological Assn., Washington.

Am. Antiq. American Antiquity. The Society for American Archaeology. Washington.

Am. Econ. The American Economist. Lubin Graduate School of Business, Pace Univ. New York.

Am. Econ. Rev. The American Economic Review. American Economic Assn., Evanston, Ill.

Am. Ethnol. American Ethnologist. American Ethnological Society. Washington.

Am. Indíg. América Indígena. Instituto Indigenista Interamericano. México.

Am. Lat. Enfoques Rusia. América Latina: Enfoques desde Rusia. Academia de Ciencias de Rusia, Instituto de América Latina. Moscow.

Am. Lat./Moscow. América Latina. Academia de Ciencias de la Unión de Repúblicas Soviéticas Socialistas. Moscow.

Am. Negra. América Negra. Pontificia Univ. Javeriana. Bogotá.

Am. Polit. Sci. Rev. American Political Science Review. American Political Science Assn.; Ohio State Univ., Columbus, Ohio.

Am. Sociol. Rev. American Sociological Review. American Sociological Assn., Washington.

Amazonía Peru. Amazonía Peruana. Centro Amazónico de Antropología y Aplicación Práctica, Depto. de Documentación y Publicaciones. Lima.

AMRI. Anuario Mexicano de Relaciones Internacionales. UNAM, ENEP Acatlán, Escuela Nacional de Estudios Profesionales. México.

An. Acad. Nac. Cienc. Econ. Anales de la Academia Nacional de Ciencias Económicas. Buenos Aires.

An. Antropol. Anales de Antropología. Univ. Nacional Autónoma de México, Instituto de Investigaciones Históricas. México.

An. Arqueol. Etnol. Anales de Arqueología y Etnología. Univ. Nacional de Cuyo, Facultad de Filosofía y Letras. Mendoza, Argentina.

An. Inst. Invest. Estét. Anales del Instituto de Investigaciones Estéticas. Univ. Nacional Autónoma de México. México.

An. Inst. Patagon./Soc. Anales del Instituto de la Patagonia: Serie Ciencias Sociales. Univ. de Magallanes. Punta Arenas, Chile.

An. Mus. Michoacano. Anales del Museo Michoacano. Centro Regional Michoacán del INAH; Museo Regional Michoacano. Morelia, Mexico.

Anál. Econ./La Paz. Análisis Económico. Ediciones UDAPE (Unidad de Análisis de Políticas Económicas). La Paz.

Anál. Econ./México. Análisis Económico. Unidad Azcapotzalco, Univ. Nacional Autónoma de México. México.

Anál. Int. Análisis Internacional. Centro Peruano de Estudios Internacionales. Lima.

Análise Conjunt. Análise & Conjuntura. Fundação João Pinheiro. Belo Horizonte, Brazil.

Anc. Mesoam. Ancient Mesoamerica. Cambridge Univ. Press. Cambridge, England.

Ann. Am. Acad. Polit. Soc. Sci. The Annals of the American Academy of Political and Social Science. Philadelphia, Penn.

Ann. Assoc. Am. Geogr. Annals of the Association of American Geographers. Lawrence, Kan.

Ann. géogr./Paris. Annales de géographie. Bulletin de la Société de géographie; Librairie Armand Colin. Paris.

Ann. Lat. Am. Stud. Annals of Latin American Studies. Nikon Raten Amerika Gakkai. Tokyo.

Ann. N.Y. Acad. Sci. Annals of the New York Academy of Sciences. New York.

Annales/Paris. Annales. Centre national de la recherche scientifique de la VIe Section de l'Ecole pratique des hautes études. Paris.

Annu. Rev. Anthropol. Annual Review of Anthropology. Annual Reviews, Inc., Palo Alto, Calif.

ANPOCS BIB. Boletim Informativo e Bibliográfico de Ciências Sociais: BIB. Associação Nacional de Pós-Graduação e Pesquisa em Ciências Sociais. Rio de Janeiro.

Anthropol. Q. Anthropological Quarterly. Catholic Univ. of America, Catholic Anthropological Conference. Washington.

Anthropol. Verkenn. Antropologische Verkenningen. Coutinho. Muiderberg, The Netherlands.

Anthropos/Switzerland. Anthropos. International Review of Ethnology and Linguistics. Anthropos-Institut. Freiburg, Switzerland.

Antiquity/Cambridge. Antiquity. A Quarterly Review of Archaeology. The Antiquity Trust. Cambridge, England.

Antropológica/Caracas. Antropológica. Fundación La Salle de Ciencias Naturales; Instituto Caribe de Antropología y Sociología. Caracas.

Antropológicas/México. Antropológicas. Instituto de Investigaciones Antropológicas, UNAM. México.

Anu. Antropol. Anuário Antropológico. Tempo Brasileiro. Rio de Janeiro.

Anu. Estad. Repúb. Argent. Anuario Estadístico de la República Argentina. Instituto Nacional de Estadística y Censos. Buenos Aires.

Anu. Estud. Am. Anuario de Estudios Americanos. Consejo Superior de Investigaciones Científicas; Univ. de Sevilla, Escuela de Estudios Hispano-Americanos. Sevilla, Spain.

Anu. Estud. Centroam. Anuario de Estudios Centroamericanos. Univ. de Costa Rica. San José.

Anu. Etnol./Habana. Anuario de Etnología. Academia de Ciencias de Cuba; Editorial Academia. La Habana.

Anu. IEHS. Anuario IEHS. Univ. Nacional del Centro de la Provincia de Buenos Aires, Instituto de Estudios Histórico-Sociales. Tandil, Argentina.

Abbreviation List of Journals Indexed / 771

Anu. Inst. Chiapaneco Cult.   Anuario Instituto Chiapaneco de Cultura. Instituto Chiapaneco de Cultura. Tuxtla Gutiérrez, Mexico.

Anuario/Quito.   Anuario de la Fundación Los Andes de Estudios Sociales. Quito.

Apunt. Arqueol./Guatemala.   Apuntes Arqueológicos. Universidad de San Carlos de Guatemala, Escuela de Historia. Guatemala.

Apuntes/Lima.   Apuntes. Univ. del Pacífico, Centro de Investigación. Lima.

Arch. Völkerkd.   Archiv für Völkerkunde. Museum für Völkerkunde in Wien und von Verein Freunde der Völkerkunde. Vienna.

Archaeoastronomy/College Park.   Archaeoastronomy. The Center for Archaeoastronomy, Univ. of Maryland. College Park, Md.

Archaeoastronomy/England.   Archaeoastronomy. Science History Publications. Giles, England.

Archaeol. Anthropol.   Archaeology and Anthropology. Ministry of Education and Cultural Development. Georgetown, Guyana.

Archaeology/New York.   Archaeology. Archaeology Institute of America. New York.

Archaeometry/Oxford.   Archaeometry. Oxford Univ., Oxford, England.

Arctic Alpine Res.   Arctic and Alpine Research. Institute of Arctic and Alpine Research, Univ. of Colorado. Boulder.

Armed Forces Soc.   Armed Forces and Society. Inter-Univ. Seminar on Armed Forces & Society. Univ. of Chicago. Chicago, Ill.

Arq. Mus. Hist. Nat.   Arquivos do Museu de História Natural. Univ. Federal de Minas Gerais. Belo Horizonte, Brazil.

Arqueol. Contemp.   Arqueología Contemporánea. Programa de Estudios Prehistóricos. Buenos Aires.

Arqueol. Mex.   Arqueología Mexicana. Instituto Nacional de Antropología e Historia, Editorial Raíces. México.

Arqueología/México.   Arqueología. Instituto Nacional de Antropología e Historia. México.

Arqueológicas/Lima.   Arqueológicas. Museo Nacional de Antropología y Arqueología, Instituto Nacional de Cultura. Lima.

Årstryck/Göteborg.   Årstryck. Etnografiska Museum. Göteborg, Sweden.

Banca Cent.   Banca Central. Banco de Guatemala. Guatemala.

Banca Naz. Lavoro.   Banca Nazionale del Lavoro Quarterly Review. Banca Nazionale del Lavoro. Rome.

Bijdr. Meded. Betreffende Geschied. Nederlanden.   Bijdragen en Mededelingen Betreffende de Geschiedenis der Nederlanden. Nederlands Historisch Genootschap. Utrecht, The Netherlands.

Bol. Acad. Nac. Hist./Caracas.   Boletín de la Academia Nacional de la Historia. Caracas.

Bol. Am.   Boletín Americanista. Univ. de Barcelona, Facultad de Geografía e Historia, Depto. de Historia de América. Barcelona.

Bol. Antropol. Am.   Boletín de Antropología Americana. Instituto Panamericano de Geografía e Historia. México.

Bol. Arqueol.   Boletín de Arqueología. Fundación de Investigaciones Arqueológicas Nacionales. Bogotá.

Bol. Cons. Arqueol.   Boletín del Consejo de Arqueología. Instituto Nacional de Antropología e Historia. México.

Bol. Esc. Cienc. Antropol. Univ. Yucatán.   Boletín de la Escuela de Ciencias Antropológicas de la Universidad de Yucatán. Mérida, Mexico.

Bol. Inf./San Cristóbal.   Boletín Informativo. Museo de Táchira. San Cristóbal, Venezuela.

Bol. Inf. Techint.   Boletín Informativo Techint. Organización Techint. Buenos Aires.

Bol. Inst. Geogr.   Boletín del Instituto de Geografía. Univ. Nacional Autónoma de México. México.

Bol. Lima.   Boletín de Lima. Revista Cultural Científica. Lima.

Bol. Mus. Chil. Arte Precolomb.   Boletín del Museo Chileno de Arte Precolombino. Santiago.

Bol. Mus. Hombre Domin.   Boletín del Museo del Hombre Dominicano. Santo Domingo.

Bol. Mus. Para. Goeldi.   Bolctim do Museu Paraense Emílio Goeldi. Nova série:

antropologia. Conselho Nacional de Desenvolvimento Científico e Tecnológico, Instituto Nacional de Pesquisas da Amazônia. Belém, Brazil.

Bol. Socioecon. Boletín Socioeconómico. Centro de Investigaciones y Documentación Socioeconómico (CIDSE), Univ. del Valle. Cali, Colombia.

Boletín/Bogotá. Boletín del Museo del Oro. Banco de la República. Bogotá.

Bull. Assoc. géogr. fr. Bulletin de l'Association de géographes français. Paris.

Bull. Bur. natl. ethnol. Bulletin du Bureau national d'ethnologie. Bureau national d'ethnologie. Port-au-Prince, Haiti.

Bull. East. Caribb. Aff. Bulletin of Eastern Caribbean Affairs. Univ. of West Indies. Cave Hill, Barbados.

Bull. Inst. fr. étud. andin. Bulletin de l'Institut français d'études andines. Lima.

Bull. Int. Anthropol. Ethnol. Bulletin of the International Committee on Urgent Anthropological and Ethnological Research. International Union of Anthropological and Ethnological Sciences. Vienna.

Bull. Lat. Am. Res. Bulletin of Latin American Research. Society for Latin American Studies. Oxford, England.

Bull. mém. Soc. anthropol. Paris. Bulletins et mémoires de la Société d'anthropologie de Paris. Paris.

Bulletin/Geneva. Bulletin. Société suisse des américanistes; Musée et institut d'éthnographie. Geneva.

Bus. Mex. Business Mexico. American Chamber of Commerce of Mexico. México.

Cad. CEAS. Cadernos do Centro de Estudos e Ação Social (CEAS). Salvador, Brazil.

Cad. CRH. Caderno CRH. Centro de Recursos Humanos. Salvador, Brazil.

Cadernos/Belém. Cadernos. Centro de Filosofia e Ciências Humanas, Univ. Federal do Pará. Belém, Brazil.

Cah. Am. lat. Cahiers des Amériques latines. Paris.

Cah. Outre-Mer. Les Cahiers d'Outre-Mer. Institut de géographie de la Faculté des lettres de Bordeaux; Institut de la France d'Outre-Mer; Société de géographie de Bordeaux. Bordeaux, France.

Camb. J. Econ. Cambridge Journal of Economics. Cambridge Political Economy Society. London; Academic Press. New York.

Can. J. Hist. Canadian Journal of History. Univ. of Saskatchewan. Saskatoon, Canada.

Can. J. Lat. Am. Caribb. Stud. Canadian Journal of Latin American and Caribbean Studies. Univ. of Ottawa. Ontario, Canada.

Caravelle/Toulouse. Caravelle. Cahiers du monde hispanique et luso-brésilien. Univ. de Toulouse, Institute d'études hispaniques, hispano-americaines et luso-brésiliennes. Toulouse, France.

Caribb. Aff. Caribbean Affairs. Trinidad Express Newspapers Ltd.; Inprint Caribbean Ltd., Port of Spain, Trinidad and Tobago.

Caribb. Perspect. Caribbean Perspectives. Transaction Publishers. New Brunswick, New Jersey.

Caribb. Q. Caribbean Quarterly. Univ. of the West Indies. Mona, Jamaica.

Caribb. Stud. Caribbean Studies. Univ. of Puerto Rico, Institute of Caribbean Studies. Río Piedras.

Caribe Contemp. El Caribe Contemporáneo. Univ. Nacional Autónoma de México. México.

Caribena/Martinique. Caribena: cahiers d'études américanistes de la Caraïbe. Centre d'études et de recherches archéologiques (CERA). Martinique.

Cartographica/Ontario. Cartographica. Univ. of Toronto Press. Downsview, Ontario.

Catena/Cremlingen. Catena. Catena Verlag. Cremlingen, Wolfenbüttel, Germany.

CD-ROM Prof. CD-ROM Professional. Pemberton Press. Weston, Conn.

Centen. Rev. The Centennial Review. Michigan State Univ., College of Science and Arts. East Lansing, Mich.

CEPAL Rev. CEPAL Review/Revista de la CEPAL. Naciones Unidas, Comisión Económica para América Latina. Santiago.

Cespedesia/Cali. Cespedesia. Depto. del Valle del Cauca. Cali, Colombia.

Chemins crit. Chemins critiques: revue haïtiano-caraïbéenne. Société sciences-arts-littérature. Port-au-Prince, Haiti.

Ciênc. Amb. Ciência & Ambiente. Univ. Federal de Santa Maria. Santa Maria, Brazil.

Ciênc. Cult. Ciência e Cultura. Sociedade Brasileira para o Progresso da Ciência. São Paulo.

Cienc. Econ./Lima. Ciencia Económica. Facultad de Economía, Univ. de Lima.

Cienc. Econ./San José. Ciencias Económicas. Instituto de Investigaciones en Ciencias Ecónomicas, Univ. de Costa Rica. San José.

Cienc. Hoy. Ciencia Hoy. Asociación Ciencia Hoy; Morgan Antártica. Buenos Aires.

Cienc. Polít. Ciencia Política. Instituto de Ciencia Política de Bogotá; Tierra Firme Editores. Bogotá.

Cienc. Soc./Santo Domingo. Ciencia y Sociedad. Instituto Tecnológico de Santo Domingo.

Cimarrón/New York. Cimarrón. City Univ. of New York, Assn. of Caribbean Studies. New York.

Clim. Change. Climatic Change. Reidel Publishers. Boston.

Colecc. Estud. CIEPLAN. Colección Estudios CIEPLAN. Corporación de Investigaciones Económicas para Latinoamérica. Santiago.

Columbia J. Transnatl. Law. Columbia Journal of Transnational Law. Columbia Univ. School of Law. New York.

Columbia J. World Bus. Columbia Journal of World Business. Columbia Univ., New York.

Comer. Exter. Comercio Exterior. Banco Nacional de Comercio Exterior. México.

Comp. Civiliz. Rev. Comparative Civilizations Review. Dept. of History, Dickinson College. Carlisle, Penn.

Comp. Educ. Rev. Comparative Education Review. Comparative Education Society. New York.

Comp. Polit. Comparative Politics. The City Univ. of New York, Political Science Program. New York.

Comp. Polit. Stud. Comparative Political Studies. Sage Publications, Thousand Oaks, Calif.

Comp. Stud. Soc. Hist. Comparative Studies in Society and History. Society for the Comparative Study of Society and History; Cambridge Univ. Press. London.

Comun. Soc. Comunicação e Sociedade. Instituto Metodista do Ensino Superior. São Paulo.

Cont. Int. Contexto Internacional. Instituto de Relações Internacionais, Pontifícia Univ. Católica. Rio de Janeiro.

Contribuciones/Asunción. Contribuciones. Centro de Documentación y Estudios (CDE). Asunción.

Contribuciones/Buenos Aires. Contribuciones. Estudios Interdisciplinarios sobre Desarrollo y Cooperación Internacional. Konrad-Adenauer-Stiftung; Centro Interdisciplinario de Estudios Sobre el Desarrollo Latinoamericano (CIEDLA). Buenos Aires.

Cotidiano/México. El Cotidiano: Revista de la Realidad Mexicana Actual. Univ. Autónoma Metropolitana, Unidad Azcapotzalco. México.

Coyunt. Soc. Coyuntura Social. Fundación para la Educación Superior y el Desarrollo (FEDESARROLLO); Instituto SER de Investigación. Bogotá.

Cuad. Am. Cuadernos Americanos. Editorial Cultura. México.

Cuad. Arquit. Mesoam. Cuadernos de Arquitectura Mesoamericana. Facultad de Arquitectura, Univ. Nacional Autónoma de México. México.

Cuad. CENDES. Cuadernos del CENDES. Centro de Estudios del Desarrollo, Univ. Central de Venezuela. Caracas.

Cuad. CLAEH. Cuadernos del CLAEH. Centro Latinoamericano de Economía Humana. Montevideo.

Cuad. Econ./Bogotá. Cuadernos de Economía. Univ. Nacional de Colombia. Bogotá.

Cuad. Econ./Santiago. Cuadernos de Economía. Pontificia Univ. Católica de Chile, Instituto de Economía. Santiago.

Cuad. INVESP. Cuadernos de INVESP. Instituto Venezolano de Estudios Sociales y Políticos. Caracas.

Cuad. Invest./San Salvador. Cuadernos de Investigación. Centro de Investigaciones Tecnológicas y Científicas. San Salvador.

Cuad. Marcha. Cuadernos de Marcha. Eon Editores. Montevideo.

Cuad. Nuestra Am. Cuadernos de Nuestra América. Centro de Estudios sobre América. La Habana.

Cuad. Prehispánicos. Cuadernos Prehispánicos. Seminario de Historia de América, Univ. de Valladolid. Spain.

Cuad. Sociol. Cuadernos de Sociología. Univ. Centroamericana. Managua.

Cuad. Sur/Oaxaca. Cuadernos del Sur: Ciencias Sociales. Instituto de Investigaciones Sociológicas, Univ. Autónoma Benito Juárez de Oaxaca. Oaxaca, Mexico.

Cuba. Stud. Cuban Studies. Univ. of Pittsburgh, Center for Latin American Studies. Pittsburgh, Penn.

Cuest. Econ. Cuestiones Económicas. Banco Central del Ecuador. Quito.

Cult. Anthropol. Cultural Anthropology: Journal of the Society for Cultural Anthropology. American Anthropological Assn.; Society for Cultural Anthropology. Washington.

Curr. Anthropol. Current Anthropology. Univ. of Chicago. Chicago, Ill.

Curr. Hist. Current History. Philadelphia, Penn.

Dados/Rio de Janeiro. Dados. Instituto Univ. de Pesquisas. Rio de Janeiro.

Database/Weston. Database. Online, Inc., Weston, Conn.

Debate Agrar. Debate Agrario. Centro Peruano de Estudios Sociales (CEPES). Lima.

Dédalo/São Paulo. Dédalo. Univ. de São Paulo, Museu de Arqueologia e Etnologia. São Paulo.

Defesa Nac. A Defesa Nacional: Revista de Assuntos Militares e Estudo de Problemas Brasileiros. Rio de Janeiro.

Demography/Washington. Demography. Population Assn. of America. Washington.

Desarro. Econ. Desarrollo Económico. Instituto de Desarrollo Económico y Social. Buenos Aires.

Dev. Change. Development and Change. Mouton. The Hague.

Dialect. Anthropol. Dialectical Anthropology. M. Nijhoff. Dordrecht, The Netherlands.

Diogenes/Philosophy. Diogenes. International Council for Philosophy and Humanistic Studies (Paris); Berg Publishers. Oxford, England.

Dir. U.N. Inf. Sources. Directory of United Nations Information Sources. Advisory Committee for the Coordination of Information Systems (ACCIS), United Nations. New York.

Disarmament/UN. Disarmament. United Nations. New York.

Dissent/New York. Dissent. Dissent Publishing Assn., New York.

ECA/San Salvador. Estudios Centroamericanos: ECA. Univ. Centroamericana José Simeón Cañas. San Salvador.

Econ. Cienc. Soc. Economía y Ciencias Sociales. Facultad de Ciencias Económicas y Sociales, Univ. Central de Venezuela. Caracas.

Econ. Dev. Cult. Change. Economic Development and Cultural Change. Univ. of Chicago, Research Center in Economic Development and Cultural Change. Chicago, Ill.

Econ. Geol. Economic Geology. Economic Geology Publishing Co., Lancaster, Penn.

Econ. Soc. Prog. Lat. Am. Economic and Social Progress in Latin America. Inter-American Development Bank. Washington.

Econ. Trab. Chile. Economía y Trabajo en Chile. Programa de Economía del Trabajo, Academia de Humanismo Cristiano. Santiago.

Economía/Lima. Economía. Depto. de Economía, Pontificia Univ. Católica del Perú. Lima.

Económica/La Plata. Económica. Univ. Nacional de La Plata, Facultad de Ciencias Económicas, Instituto de Investigaciones Económicas. La Plata, Argentina.

Economist/London. The Economist. London.

Ecuad. Deb. Ecuador Debate. Centro Andino de Acción Popular (CAAP). Quito.

Ed. Quinto Centen. Ediciones del Quinto Centenario. Univ. de la República. Montevideo.

Empres. Sider. Brasil. Empresas Siderúrgicas do Brasil. Instituto Brasileiro de Siderurgia. Rio de Janeiro.

Ensaios FEE. Ensaios FEE. Secretaria de Coordenacão e Planejamento. Fundação de Economia e Estatística. Porto Alegre, Brazil.

Envío/Managua. Envío. Univ. Centroamericana (UCA). Managua.

Environ. Conserv. Environmental Conservation. Foundation for Environmental Conservation; Elsevier Sequoia. Lausanne, Switzerland.

Erdkunde/Bonn. Erdkunde. Archiv für Wissenschaftliche Geographie. Univ. Bonn, Geographisches Institut. Bonn, Germany.

Eres/Tenerife. Eres. Museo Arqueológico y Etnográfico. Tenerife, Spain.

Espac. Desarro. Espacio y Desarrollo. Pontificia Univ. Católica del Perú, Depto. de Humanidades, Centro de Investigación en Geografía Aplicada. Lima.

Espace géogr. L'Espace géographique. Doin. Paris.

Estad. Econ. Estadística & Economía. Instituto Nacional de Estadísticas. Santiago.

Estad. Soc. Estado & Sociedad: Revista Boliviana de Ciencias Sociales. Facultad Latinoamericana de Ciencias Sociales (FLACSO). La Paz.

Estud. Avançados. Estudos Avançados. Univ. de São Paulo, Instituto dc Estudos Avançados. São Paulo.

Estud. Cult. Contemp. Estudios sobre las Culturas Contemporáneas. Centro Universitario de Investigaciones Sociales, Univ. de Colima. México.

Estud. Cult. Maya. Estudios de Cultura Maya. Centro de Estudios Mayas, Univ. Nacional Autónoma de México. México.

Estud. Cult. Náhuatl. Estudios de Cultura Náhuatl. Instituto de Investigaciones Históricas, Univ. Nacional Autónoma de México. México.

Estud. Demogr. Urb. Estudios Demográficos y Urbanos. El Colegio de México. México.

Estud. Econ./México. Estudios Económicos. El Colegio de México. México.

Estud. Econ./Santiago. Estudios de Economía. Depto. de Economía, Univ. de Chile. Santiago.

Estud. Econ./São Paulo. Estudos Econômicos. Univ. de São Paulo, Instituto de Pesquisas Econômicas. São Paulo.

Estud. Front. Estudios Fronterizos. Instituto de Investigaciones Sociales, Univ. Autónoma de Baja California. Mexicali, Mexico.

Estud. Geogr./Madrid. Estudios Geográficos. Instituto de Economía y Geografía Aplicadas, Consejo Superior de Investigaciones Científicas. Madrid.

Estud. Int./IRIPAZ. Estudios Internacionales: Revista del IRIPAZ. Instituto de Relaciones Internacionales y de Investigaciones para la Paz. Guatemala.

Estud. Int./Santiago. Estudios Internacionales. Instituto de Estudios Internacionales, Univ. de Chile. Santiago.

Estud. Parag. Estudios Paraguayos. Univ. Católica Nuestra Señora de la Asunción. Asunción.

Estud. Públicos. Estudios Públicos. Centro de Estudios Públicos. Santiago.

Estud. Rural. Latinoam. Estudios Rurales Latinoamericanos. Consejo Latinoamericano de Ciencias Sociales. Bogotá.

Estud. Soc. Centroam. Estudios Sociales Centroamericanos. Programa Centroamericano de Ciencias Sociales. San José.

Estud. Soc./Santiago. Estudios Sociales. Corporación de Promoción Universitaria. Santiago.

Estud. Soc./Santo Domingo. Estudios Sociales. Centro de Estudios Sociales Juan Montalvo, SJ. Santo Domingo.

Estud. Sociol./México. Estudios Sociológicos. Centro de Estudios Sociológicos de El Colegio de México. México.

Estud. Territ. Estudios Territoriales. Instituto del Territorio y Urbanismo; Ministerio de Obras Públicas y Transportes. Madrid.

Estudios/Fundación Mediterránea. Estudios. Instituto de Estudios Económicos sobre la Realidad Argentina y Latinoamericana; Fundación Mediterránea. Córdoba, Argentina.

Ethnic Groups/New York. Ethnic Groups. Gordon and Breach. New York.

Ethnohistory/Society. Ethnohistory. American Society for Ethnohistory. Duke Univ., Durham, N.C.

Ethnology/Pittsburgh. Ethnology. Univ. of Pittsburgh, Penn.

Ethnos/Stockholm. Ethnos. Statens Etnografiska Museum. Stockholm.

Ethos/Society. Ethos. Society for Psychological Anthropology; Univ. of California, Los Angeles.

Etnofoor/Netherlands. Etnofoor. Antropologisch-Sociologisch Centrum. Amsterdam, The Netherlands.

Etud. créoles. Etudes créoles. Comité international des études créoles. Montréal.

EURE/Santiago. EURE: Revista Latinoamericana de Estudios Urbanos Regionales. Centro de Desarrollo Urbano y Regional, Univ. Católica de Chile. Santiago.

Eval. Econ. Evaluación Económica. Müller & Machicado Asociados. La Paz.

Explor. J. The Explorers Journal. New York.

Financ. Dev. Finance and Development. International Monetary Fund; The World Bank. Washington.

Folk/Copenhagen. Folk: Journal of the Danish Ethnographic Society. Danish Ethnographic Society. Copenhagen.

Foreign Aff. Foreign Affairs. Council on Foreign Relations, Inc. New York.

Foreign Policy. Foreign Policy. National Affairs Inc.; Carnegie Endowment for International Peace. New York.

Foro Int. Foro Internacional. El Colegio de México. México.

Foro Polít. Foro Político. Instituto de Ciencias Políticas, Univ. del Museo Social Argentino. Buenos Aires.

Front. Norte. Frontera Norte. Colegio de la Frontera Norte. Tijuana, Mexico.

Fuerzas Arm. Soc. Fuerzas Armadas y Sociedad. Centro Latinoamericano de Defensa y Desarme; FLACSO. Santiago.

Gac. Arqueol. Andin. Gaceta Arqueológica Andina. Instituto Andino de Estudios Arqueológicos. Lima.

Gac. Univ./Campeche. Gaceta Universitaria. Univ. Autónoma de Campeche, Mexico.

Geoarchaeology/New York. Geoarchaeology. John Wiley. New York.

Geogr. J. The Geographical Journal. The Royal Geographical Society. London.

Geogr. Rev. Geographical Review. American Geographical Society of New York.

Geogr. Rundsch. Geographische Rundschau. Zeitschrift für Schulgeographie. Georg Westermann Verlag. Braunschweig, Germany.

Geogr. Z. Geographische Zeitschrift. Franz Steiner Verlag. Wiesbaden, Germany.

Geografia/Rio Claro. Geografia. Associação de Geografia Teorética. Rio Claro, Brazil.

Geografie/Utrecht. Geografie. KNAG. Utrecht, The Netherlands.

Geographical/London. Geographical. Hyde Park Publications. London.

Geography/London. Geography. Geographical Assn., London.

GeoJournal/Boston. GeoJournal. D. Reidel Publishing Co., Boston, Mass.

Geosul/Florianópolis. Geosul. Depto. de Geociências, Univ. Federal de Santa Catarina. Florianópolis, Brazil.

Granma/La Habana. Granma. La Habana.

HAHR. Hispanic American Historical Review. Conference on Latin American History of the American Historical Assn.; Duke Univ. Press. Durham, N.C.

Harv. Bus. Rev. Harvard Business Review. Graduate School of Business Administration, Harvard Univ., Boston.

Hemisphere/Miami. Hemisphere. Latin American and Caribbean Center, Florida International Univ., Miami, Fla.

Herencia/San José. Herencia. Programa de Rescate y Revitalización del Patrimonio Cultural. San José.

Hist. Mex. Historia Mexicana. Colegio de México. México.

Hist. Quest. Debates. História, Questões e Debates. Associação Paranaense de História. Curitiba, Brazil.

Hist. Relig. History of Religions. Univ. of Chicago. Chicago, Ill.

Hist. Soc./Río Piedras. Historia y Sociedad. Depto. de Historia, Univ. de Puerto Rico. Río Piedras.

Hist. Workshop. History Workshop. Ruskin College, Oxford Univ., Oxford, England.

HLAS/CD. Handbook of Latin American Studies CD-ROM: HLAS/CD, Vols. 1–53. Fundación MAPFRE América, Madrid. Hispanic Division, Library of Congress, Washington. Distributed by Univ. of Texas Press, Austin.

Hómines/San Juan. Hómines. Univ. Interamericana de Puerto Rico. San Juan.

Homme/Paris. L'Homme. Laboratoire d'anthropologie, Collège de France. Paris.

Hoy Hist. Hoy es Historia: Revista Bimestral de Historia Nacional e Iberoamericana. Editorial Raíces. Montevideo.

Hum. Ecol. Human Ecology. Plenum Publishing Corp., New York.

Hum. Organ. Human Organization. Society for Applied Anthropology. New York.

Humanidades/Brasília. Humanidades. Editora Univ. de Brasília.

Ibero-Am. Arch. Ibero-Amerikanisches Archiv. Ibero-Amerikanisches Institut. Berlin.

Ibero-Am./Stockholm. Ibero-Americana: Nordic Journal of Latin American Studies. Institute of Latin American Studies, Univ. of Stockholm.

Iberoam./Tokyo. Iberoamericana. Univ. of Sofia. Tokyo.

Ideas Polít. Ideas Políticas: Revista de Análisis y Debate. Cambio XXI Fundación Mexicana. México.

Identities/Yverdon. Identities: Global Studies in Culture and Power. Gordon and Breach Publishers. Yverdon, Switzerland.

Impex/OECD. Impex: Foreign Trade Statistics by Commodities = Statistiques de commerce extérior produits. Organisation for Economic Co-operation and Development (OECD). Paris.

Index Censorsh. Index on Censorship. Writers & Scholars International. London.

Indiana/Berlin. Indiana. Gebr. Mann., Berlin.

Int. Aff./Moscow. International Affairs. Moscow.

Int. Bibliogr. Soc. Sci. International Bibliography of the Social Sciences. British Library of Political and Economic Science, London School of Economics and Political Science. London.

Int. Hist. Rev. The International History Review. Univ. of Toronto Press. Downsview, Ontario, Canada.

Int. J. Comp. Sociol. International Journal of Comparative Sociology. York Univ., Dept. of Sociology and Anthropology. Toronto, Canada.

Int. J. Intercult. Relat. International Journal of Intercultural Relations. Society for Intercultural Education, Training, and Research; Pergamon Press. New York.

Int. J./Toronto. International Journal. Canadian Institute of International Affairs. Toronto, Canada.

Int. Labour Rev. International Labour Review. International Labour Office. Geneva.

Int. Migr. International Migration = Migrations Internationales = Migraciones Internacionales. Intergovernmental Committee for European Migration; Research Group for European Migration Problems; International Organization for Migration. The Hague, Netherlands; Geneva, Switzerland.

Int. Migr. Rev. International Migration Review. Center for Migration Studies. New York.

Int. Organ. International Organization. World Peace Foundation; Univ. of Wisconsin Press. Madison.

Int. Polit. Sci. Abstr. International Political Science Abstracts. International Political Science Assn., Paris.

Int. Rev. Adm. Sci. International Review of Administrative Sciences. International Institute of Administrative Sciences. Brussels.

Int. Spect./Hague. Internationale Spectator. Nederlandsch Genootschap voor Internationale Zaken. The Hague.

Int. Spect./Rome. The International Spectator. Istituto Affari Internazionali. Rome.

Int. Stud. Q. International Studies Quarterly. Wayne State Univ., Detroit, Mich.

Integr. Latinoam. Integración Latinoamericana. Instituto para la Integración de América Latina. Buenos Aires.

Interciencia/Caracas. Interciencia. Asociación Interciencia. Caracas.

Internet World. Internet World. Meckler Corp. Westport, Conn.

Invest. Agrar. Investigación Agraria: Economía. Instituto Nacional de Investigaciones Agrarias. Madrid.

Invest. Econ. Investigación Económica. Facultad de Economía, Univ. Nacional Autónoma de México. México.

Islas/Santa Clara. Islas. Univ. Central de Las Villas. Santa Clara, Cuba.

Iztapalapa/México. Iztapalapa. Univ. Autónoma Metropolitana, División de Ciencias Sociales y Humanidades. México.

J. Anthropol. Archaeol. Journal of Anthropological Archaeology. Academic Press. New York.

J. Anthropol. Res. Journal of Anthropological Research. Univ. of New Mexico. Albuquerque, N.M.

J. Archaeol. Res. Journal of Archaeological Research. Plenum Press. New York.

J. Archaeol. Sci. Journal of Archaeological Science. Academic Press. New York.

J. Bahamas Hist. Soc. Journal of the Bahamas Historical Society. Nassau, Bahamas?.

J. Caribb. Hist. The Journal of Caribbean History. Caribbean Univ. Press. St. Lawrence, Barbados.

J. Caribb. Stud. Journal of Caribbean Studies. Assn. of Caribbean Studies. Coral Gables, Fla.

J. Church State. Journal of Church and State. J.M. Dawson Studies in Church and State, Baylor Univ., Waco, Tex.

J. Common Market Stud. Journal of Common Market Studies. Oxford, England.

J. Commonw. Comp. Polit. The Journal of Commonwealth & Comparative Politics. Univ. of London, Institute of Commonwealth Studies. London.

J. Cult. Geogr. Journal of Cultural Geography. Popular Culture Assn.; American Culture Assn.; Bowling Green State Univ., Bowling Green, Oh.

J. Dev. Areas. The Journal of Developing Areas. Western Illinois Univ. Press. Macomb, Ill.

J. Dev. Econ. Journal of Development Economics. North-Holland Publishing Co., Amsterdam, The Netherlands.

J. Dev. Stud. The Journal of Development Studies. Frank Cass. London.

J. Econ. Issues. Journal of Economic Issues. California State Univ., Sacramento.

J. Ethnobiol. Journal of Ethnobiology. Center for Western Studies. Flagstaff, Ariz.

J. Field Archaeol. Journal of Field Archaeology. Boston Univ., Boston, Mass.

J. Geogr.   Journal of Geography. National Council of Geographic Education. Menasha, Wis.

J. Hist. Geogr.   Journal of Historical Geography. Academic Press. London; New York.

J. Int. Law Econ.   Journal of International Law and Economics. George Washington Univ., The National Law Center. Washington.

J. Interam. Stud. World Aff.   Journal of Interamerican Studies and World Affairs. Institute of Interamerican Studies, Univ. of Miami. Coral Gables, Fla.

J. Lat. Am. Lore.   Journal of Latin American Lore. Univ. of California, Latin American Center. Los Angeles, Calif.

J. Lat. Am. Stud.   Journal of Latin American Studies. Centers or Institutes of Latin American Studies at the Universities of Cambridge, Glasgow, Liverpool, London, and Oxford. Cambridge Univ. Press. London.

J. Money Credit Bank.   Journal of Money, Credit and Banking. Ohio State Univ. Press. Columbus.

J. Peace Res.   Journal of Peace Research. International Peace Research Institute, Universitetforlaget. Oslo.

J. Peasant Stud.   The Journal of Peasant Studies. Frank Cass & Co., London.

J. Polit. Mil. Sociol.   Journal of Political & Military Sociology. Northern Illinois Univ., Dept. of Sociology. DeKalb, Ill.

J. Sci. Stud. Relig.   Journal for the Scientific Study of Religion. Society for the Scientific Study of Religion. Storrs, Conn.

J. Steward Anthropol. Soc.   Journal of the Steward Anthropological Society. Urbana, Ill.

J. World Prehist.   Journal of World Prehistory. Plenum Press. New York.

J. World Trade.   Journal of World Trade: Law, Economics, Public Policy. Werner Publishing Company, Ltd., Geneva.

Jam. J.   Jamaica Journal. Institute of Jamaica. Kingston.

Jane's Sentinel.   Jane's Sentinel. Jane's Information Group. Alexandria, Va.

JOICE.   Journal of International and Comparative Economics (JOICE). Physica-Verlag. Heidelberg, Germany.

Journal. Q.   Journalism Quarterly. Assn. for Education in Journalism; American Assn. of Schools and Depts. of Journalism; Kappa Tau Alpha Society.; Univ. of Minnesota. Minneapolis, Minn.

KAS Ausl.-Inf.   KAS Auslands-Informationen. Konrad Adenauer-Stiftung. Bonn.

LARR.   Latin American Research Review. Latin American Research Review Board. Univ. of New Mexico, Albuquerque, N.M.

LASA Forum.   LASA Forum. Latin American Studies Assn., Pittsburgh, Penn.

Lat. Am. Anthropol. Rev.   The Latin American Anthropology Review. Society for Latin American Anthropology. Fairfax, Va.

Lat. Am. Antiq.   Latin American Antiquity. Society for American Archaeology. Washington.

Lat. Am. Bus. Intell.   Latin American Business Intelligence. Economist Intelligence Unit. London; New York.

Lat.am. Jahrb.   Lateinamerika Jahrbuch. Vervuert Verlag. Frankfurt.

Lat. Am. News/Knight-Ridder.   Latin American News. Knight-Ridder Information. Mountain View, California.

Lat. Am. Newsl./London.   Latin American Newsletters. London.

Lat. Am. Perspect.   Latin American Perspectives. Univ. of California. Newbury Park, Calif.

Latinoamerica/Rome.   Latinoamerica. Edizioni Associate. Rome.

Law Anthropol. Int. Jahrb. Rechtsanthropol.   Law & Anthropology: Internationales Jahrbuch für Rechtsanthropologie. VWGö. Vienna; Klaus Renner Verlag. Hohenschäftlarn.

Lect. Econ.   Lecturas de Economía. Univ. de Antioquia. Medellín, Colombia.

Lua Nova.   Lua Nova. Editora Brasiliense. São Paulo.

MACLAS Lat. Am. Essays. MACLAS Latin American Essays. Middle Atlantic Council of Latin American Studies. New Brunswick, N.J.

Maguaré/Bogotá. Maguaré. Depto. de Antropología, Univ. Nacional de Colombia. Bogotá.

Man/London. Man. The Royal Anthropological Institute. London.

Manch. Sch. Econ. Soc. Stud. The Manchester School of Economic and Social Studies. Economics Dept., Manchester School. Manchester, England.

Mayan People Hum. Rights. The Mayan People and Human Rights. Pro Justice and Peace Committee of Guatemala. Guatemala.

Mededelingen/Paramaribo. Mededelingen. Stichting Surinaams Museum. Paramaribo.

Medio Ambient. Urban. Medio Ambiente y Urbanización. Comisión de Desarrollo Urbano y Regional, Consejo Latinoamericano de Ciencias Sociales. Buenos Aires.

Mem. Soc. Cienc. Nat. Memoria de la Sociedad de Ciencias Naturales La Salle. Caracas.

Memoria/CEMOS. Memoria: Boletín de CEMOS. Centro de Estudios del Movimiento Obrero y Socialista. México.

Mesoamérica/Antigua. Mesoamérica. Centro de Investigaciones Regionales de Mesoamérica. Antigua, Guatemala.

Mex. Bus. Mexico Business. Mexico Business Publishing Group. Houston, Tex.

Mexicon/Berlin. Mexicon. K.-F. von Flemming. Berlin, Germany.

Mitt. Österr. Geogr. Ges. Mitteilungen der Österreichischen Geographischen Gesellschaft. Verleger, Herausgeber und Eigentümer. Vienna.

Mon. Rev. Monthly Review. New York.

Montalbán/Caracas. Montalbán. Univ. Católica Andrés Bello, Facultad de Humanidades y Educación, Institutos Humanísticos de Investigación. Caracas.

Mt. Res. Dev. Mountain Research and Development. International Mountain Society. Boulder, Colo.

NACLA. NACLA: Report on the Americas. North American Congress on Latin America. New York.

Names/Freeman. Names: Journal of the American Name Society. Pine Hill Press. Freeman, S.D.

Natl. Geogr. Res. National Geographic Research. National Geographic Society. Washington.

Natl. Int. The National Interest. National Affairs. Washington.

Nature/London. Nature: International Weekly Journal of Science. Macmillan Magazines. London.

Nav. War Coll. Rev. Naval War College Review. Newport, R.I.

NEARA J. NEARA Journal. New England Antiquities Research Assn., Milford, N.H.

New Left Rev. New Left Review. New Left Review, Ltd., London.

Nieuwe West-Indische Gids. Nieuwe West-Indische Gids. Martinus Nijhoff. The Hague.

North-South/Miami. North-South: The Magazine of the Americas. North-South Center, Univ. of Miami. Coral Gables, Fla.

Notas Mesoam. Notas Mesoamericanas. Univ. de las Américas-Puebla. Puebla, Mexico.

Notas Pobl. Notas de Población. Centro Latinoamericano de Demografía. Santiago.

Nova Econ. Nova Economia. Depto. de Ciências Econômicas, Univ. Federal de Minas Gerais. Belo Horizonte, Brazil.

Noved. Colomb. Novedades Colombianas. Museo de Historia Natural, Univ. del Cauca. Papayán, Colombia.

Novos Estud. CEBRAP. Novos Estudos CEBRAP. Centro Brasileiro de Análise e Planejamento. São Paulo.

Nuestra Econ. Nuestra Economía. Facultad de Economía de la Univ. Autónoma de Baja California. Tijuana, Mexico.

Nueva Soc. Nueva Sociedad. Caracas.

Nuevos Aportes. Nuevos Aportes. Editorial San José S.R.L., La Paz.

Oceanus/Woods Hole. Oceanus. Oceanographic Institution. Woods Hole, Mass.

OECD Stat. Compend. OECD Statistical Compendium. OECD Electronic Publications. Paris.

Ojarasca/México. Ojarasca. Pro-México Indígena. México.

Online CDROM Rev. Online & CDROM Review. Learned Information. Oxford, England; Medford, N.J.

Opin. Anál. Opiniones y Análisis. Fundación Boliviana para la Capacitación Democrática y la Investigación. La Paz.

Oxf. Econ. Pap. Oxford Economic Papers. Oxford Univ. Press. London.

Pachacamac/Lima. Pachacamac: Revista del Museo de la Nación. Museo de la Nación. Lima.

Pacífico/Chimbote. Pacífico: Revista de Ciencias Sociales. Chimbote, Peru.

Paisajes Geogr. Paisajes Geográficos. Centro Panamericano de Estudios e Investigaciones Geográficas. Quito.

Palabra Hombre. La Palabra y el Hombre. Univ. Veracruzana. Xalapa, Mexico.

Panorama Centroam. Pensam. Panorama Centroamericano: Pensamiento y Acción. Instituto Centroamericano de Estudios Políticos (INCEP). Guatemala.

Panorama Econ. Am. Lat./CEPAL. Panorama Económico de América Latina. División de Desarrollo Económico, Comisión Económica para América Latina y el Caribe, Naciones Unidas. Santiago.

Pap. Proc. Appl. Geogr. Conf. Papers and Proceedings of the Applied Geography Conferences. Dept. of Geography, Kent State Univ.; State Univ. of New York (SUNY) at Binghamton.

Pará Desenvolv. Pará Desenvolvimento. Instituto de Desenvolvimento Econômico-Social do Pará. Belém, Brazil.

Parliam. Aff. Parliamentary Affairs. Oxford Univ. Press. London.

Pensam. Iberoam. Pensamiento Iberoamericano. Instituto de Cooperación Iberoamericano (ICI) de España; Comisión Económica para América Latina y el Caribe (CEPAL). Madrid.

Perf. Latinoam. Perfiles Latinoamericanos. Facultad Latinoamericana de Ciencias Sociales. México.

Pesqui. Planej. Econ. Pesquisa e Planejamento Econômico. Instituto de Planejamento Econômico e Social. Rio de Janeiro.

Pesquisas/São Leopoldo. Pesquisas. Instituto Anchietano de Pesquisas. São Leopoldo, Brazil.

Plant. Soc. Am. Plantation Society in the Americas. Univ. of New Orleans.

Polít. Extern. Política Externa. Paz e Terra. São Paulo.

Polít. Int. Política Internacional. Revista de la Academia Diplomática del Peru. Lima.

Polit. Sci. Q. Political Science Quarterly. Columbia Univ., The Academy of Political Science. New York.

Polit. Stud. Political Studies. Political Studies Assn. of the United Kingdom; Clarendon Press. Oxford, England.

Politeia/Caracas. Politeia. Instituto de Estudios Políticos, Univ. Central de Venezuela. Caracas.

Prehistoria/Buenos Aires. Prehistoria: Revista del Programa de Estudios Prehistóricos. Consejo Nacional de Investigaciones Científicas y Técnicas. Buenos Aires.

Pretextos/Lima. Pretextos. Centro de Estudios y Promoción del Desarrollo. Lima.

Probl. Communism. Problems of Communism. United States Information Agency. Washington.

Proceso/México. Proceso. Comunicación e Información. México.

Prof. Geogr. The Professional Geographer. Assn. of American Geographers. Washington.

Pumapunku/La Paz. Pumapunku. Centro de Investigaciones Antropológicas Tiwanaku. La Paz.

Q. J. Econ. The Quarterly Journal of Economics. Cambridge, Mass.

Quat. Res./New York. Quaternary Research. Academic Press. New York.

Quehacer/Lima. Quehacer. Centro de Estudios y Promoción del Desarrollo (DESCO). Lima.

Radiocarbon/New Haven. Radiocarbon. Supplement of the American Journal of Science. New Haven, Conn.

Real. Econ./Buenos Aires. Realidad Económica. Instituto Argentino para el Desarrollo Económico (IADE). Buenos Aires.

Real. Econ.-Soc. Realidad Económico-Social. Univ. Centroamericana José Simeón Cañas. San Salvador.

Relac. Int./Mexico. Relaciones Internacionales. Centro de Relaciones Internacionales. Facultad de Ciencias Políticas y Sociales, Univ. Nacional Autónoma de México. México.

Relaciones/Coyoacán. Relaciones. Univ. Autónoma Metropolitana, Unidad Xochimilco. Coyoacán, México.

Relaciones/Zamora. Relaciones. El Colegio de Michoacán. Zamora, Mexico.

Res. Econ. Anthropol. Research in Economic Anthropology. JAI Press. Greenwich, Conn.

Res. Explor. Research & Exploration. National Geographic Society. Washington.

Res/Harvard. Res. Peabody Museum of Archaeology and Ethnology, Harvard Univ., Cambridge, Mass.

Rev. Adm. Empres. Revista de Administração de Empresas. Fundação Getúlio Vargas, Instituto de Documentação. São Paulo.

Rev. Anál. Econ. Revista de Análisis Económico. Programa de Postgrado en Economía, ILADES/Georgetown Univ., Santiago.

Rev. ANDI. Revista ANDI. Asociación Nacional de Industriales (ANDI). Medellín, Colombia.

Rev. Antropol./São Paulo. Revista de Antropologia. Univ. de São Paulo, Faculdade de Filosofia, Letras e Ciências Humanas; Associação Brasileira de Antropologia. São Paulo.

Rev. Archaeol./Salem. The Review of Archaeology. Salem, Mass.

Rev. Arg. Estud. Estrateg. Revista Argentina de Estudios Estratégicos. Olcese Editores. Buenos Aires.

Rev. Arqueol. Am. Revista de Arqueología Americana. Instituto Panamericano de Geografía e Historia. México.

Rev. Arqueol./São Paulo. Revista de Arqueologia. Sociedade de Arqueologia Brasileira. São Paulo.

Rev. Bras. Comér. Exter. Revista Brasileira de Comércio Exterior. Fundação Centro de Estudos do Comércio Exterior. Rio de Janeiro.

Rev. Bras. Econ. Revista Brasileira de Economia. Fundação Getúlio Vargas, Instituto Brasileiro de Economia. Rio de Janeiro.

Rev. Bras. Estud. Polít. Revista Brasileira de Estudos Políticos. Univ. de Minas Gerais. Belo Horizonte, Brazil.

Rev. Bras. Estud. Popul. Revista Brasileira de Estudos de População. Associação Brasileira de Estudos Populacionais. São Paulo.

Rev. Bras. Geogr. Revista Brasileira de Geografia. Conselho Nacional de Geografia, Instituto Brasileiro de Geografia e Estatística. Rio de Janeiro.

Rev. Cancillería San Carlos. Revista Cancillería de San Carlos. Ministerio de Relaciones Exteriores. Bogotá.

Rev. Cent. Estud. Av. La Revista del Centro de Estudios Avanzados de Puerto Rico y el Caribe. San Juan.

Rev. Centroam. Econ. Revista Centroamericana de Economía. Univ. Nacional Autónoma de Honduras, Programa de Postgrado Centroamericano en Economía y Planificación. Tegucigalpa.

Rev. CEPA. Revista do CEPA. Centro de Ensino e Pesquisas Arqueológicas, Faculdades Integradas de Santa Cruz do Sul. Santa Cruz do Sul, Brazil.

Rev. Chil. Antropol. Revista Chilena de Antropología. Depto. de Antropología, Univ. de Chile. Santiago.

Rev. Chil. Geopolít. Revista Chilena de Geopolítica. Instituto Geopolítico de Chile. Santiago.

Rev. Ciclos. Revista Ciclos en la Historia, Economía y la Sociedad. Fundación de Investigaciones Históricas, Económicas y Sociales, Facultad de Ciencias Económicas, Univ. de Buenos Aires. Buenos Aires.

Rev. Cienc. Soc./Río Piedras. Revista de Ciencias Sociales. Univ. de Puerto Rico, Colegio de Ciencias Sociales. Río Piedras.

Rev. Cienc. Soc./San José. Revista de Ciencias Sociales. Univ. de Costa Rica. San José.

Rev. Cienc. Soc./Valparaíso. Revista de Ciencias Sociales. Facultad de Derecho y Ciencias Sociales, Univ. de Valparaíso. Chile.

Rev. Colomb. Antropol. Revista Colombiana de Antropología. Ministerio de Educación Nacional, Instituto Colombiano de Antropología. Bogotá.

Rev. Cuba. Cienc. Soc. Revista Cubana de Ciencias Sociales. Centro de Estudios Filosóficos, Academia de Ciencias de Cuba. La Habana.

Rev. Cuba. Derecho. Revista Cubana de Derecho. La Habana.

Rev. Econ. Polít. Revista de Economia Política. Centro de Economia Política. São Paulo.

Rev. Econ./Recife. Revista de Econometria. Sociedade Brasileira de Econometria; Univ. Federal de Pernambuco. Recife, Brazil.

Rev. Econ. Sociol. Rural. Revista de Economia e Sociologia Rural. Sociedade Brasileira de Economia e Sociologia Rural. Brasília.

Rev. Econ./Uruguay. Revista de Economía. Banco Central de Uruguay. Montevideo.

Rev. Ecuat. Hist. Econ. Revista Ecuatoriana de Historia Económica. Banco Central del Ecuador, Centro de Investigación y Cultura. Quito.

Rev. Esp. Antropol. Am. Revista Española de Antropología Americana. Facultad de Geografía e Historia. Univ. Complutense de Madrid.

Rev. Esp. Invest. Sociol. Revista Española de Investigaciones Sociológicas. Centro de Investigaciones Sociológicas. Madrid.

Rev. Esp. Pac. Revista Española del Pacífico. Asociación Española de Estudios del Pacífico. Madrid.

Rev. Este País. Revista Este País. Centro Nacional Editor de Discos Compactos. Colima, Mexico. Desarrollo de Opinión Pública. México.

Rev. Estud. Polít. Revista de Estudios Políticos. Centro de Estudios Constitucionales. Madrid.

Rev. Eur. Revista Europea de Estudios Latinoamericanos y del Caribe = European Review of Latin American and Caribbean Studies. Center for Latin American Research and Documentation; Royal Institute of Linguistics and Anthropology. Amsterdam.

Rev. Geogr. Chile. Revista Geográfica de Chile. Instituto Geográfico Militar. Santiago.

Rev. Geogr./Mérida. Revista Geográfica. Univ. de Los Andes. Mérida, Venezuela.

Rev. Geogr./México. Revista Geográfica. Instituto Panamericano de Geografía e Historia, Comisión de Geografía. México.

Rev. Geogr. Norte Gd. Revista de Geografía Norte Grande. Pontificia Univ. Católica de Chile. Santiago.

Rev. Geogr. Venez. Revista Geográfica Venezolana. Univ. de los Andes. Mérida, Venezuela.

Rev. Geol. Am. Central. Revista Geológica de América Central. Escuela Centroamericana de Geología, Univ. de Costa Rica. San José.

Rev. Hist. Am. Argent. Revista de Historia Americana y Argentina. Univ. Nacional de Cuyo, Instituto de Historia. Mendoza, Argentina.

Rev. Hist./Heredia. Revista de Historia. Univ. Nacional de Costa Rica, Escuela de Historia. Heredia, Costa Rica.

Rev. Hist. Naval. Revista de Historia Naval. Instituto de Historia y Cultura Naval Armada Española. Madrid.

Rev. Hist./San José. Revista de Historia. Centro de Investigaciones Históricas, Univ. de Costa Rica. San José.

Rev. Indias. Revista de Indias. Consejo Superior de Investigaciones Científicas, Instituto Gonzalo Fernández de Oviedo. Madrid.

Rev. Integr. Desarro. Centroam. Revista de la Integración y el Desarrollo de Centroamérica. Banco Centroamericano de Integración Económica. Tegucigalpa.

Rev. Interam. Bibliogr. Revista Interamericana de Bibliografía. Organization of American States. Washington.

Rev. Interam. Planif. Revista Interamericana de Planificación. Sociedad Interamericana de Planificación. Bogotá.

Rev. Invest. Econ. Revista de Investigaciones Económicas. Centro de Investigación, Facultad de Economía, Univ. de Panamá.

Rev. Mex. Cienc. Polít. Soc. Revista Mexicana de Ciencias Políticas y Sociales. Facultad de Ciencias Políticas y Sociales, Univ. Nacional Autónoma de México. México.

Rev. Mex. Polít. Exter. Revista Mexicana de Política Exterior. Secretaría de Relaciones Exteriores; Instituto Matías Romero de Estudios Diplomáticos (IMPED). México.

Rev. Mex. Sociol. Revista Mexicana de Sociología. Instituto de Investigaciones Sociales, Univ. Nacional Autónoma de México. México.

Rev. Mus. Arqueol. Etnol. Revista do Museu de Arqueologia e Etnologia. Univ. de São Paulo.

Rev. Mus. Hist. Nat. San Rafael. Revista del Museo de Historia Natural de San Rafael. Mendoza, Argentina.

Rev. Occident. Revista de Occidente. Madrid.

Rev. Panameña Sociol. Revista Panameña de Sociología. Univ. de Panamá, Depto. de Sociología. Panamá.

Rev. Parag. Sociol. Revista Paraguaya de Sociología. Centro Paraguayo de Estudios Sociológicos. Asunción.

Rev. Peru. Cienc. Soc. Revista Peruana de Ciencias Sociales. Asociación Peruana para el Fomento de las Ciencias Sociales (FOMCIENCIAS). Lima.

Rev. Peru. Poblac. Revista Peruana de Población. Asociación Multidisciplinaria de Investigación y Docencia en Población. Lima.

Rev. Planeac. Desarro. Revista de Planeación y Desarrollo. Depto. Nacional de Planeación. Bogotá.

Rev. Soc. haïti. Revue de la Société haïtienne d'histoire et géographie. Port-au-Prince.

Rev. Tiers-Monde. Revue Tiers-Monde. Institut d'étude du développement économique et social, Univ. de Paris.

Rev. UNITAS. Revista UNITAS. UNITAS. La Paz.

Rev. Univ. EAFIT. Revista Universidad EAFIT. Depto. de Comunicaciones, Univ. EAFIT. Medellín.

Rev. Univ. Pontif. Boliv. Revista Universidad Pontificia Bolivariana. Medellín, Colombia.

Rev. Univ./Tegucigalpa. Revista de la Universidad. Univ. Nacional Autónoma de Honduras. Tegucigalpa.

Review/Braudel. Review: Fernand Braudel Center. Fernand Braudel Center for the Study of Economics, Historical Systems, and Civilizations. Binghamton, New York.

Riv. Geogr. Ital. Rivista Geografica Italiana. Societá di studi geografici di Firenze. Florence, Italy.

São Paulo Perspect. São Paulo em Perspectiva. Fundação SEADE. São Paulo.

Schweiz. Amer. Ges. Schweizerische Amerikanisten Gesellschaft. Société Suisse des Américanistes. Genève.

Sci. Am. Scientific American. Scientific American, Inc. New York.

Sci. Soc. Science and Society. New York.

Science/Washington. Science. American Assn. for the Advancement of Science. Washington.

SECOLAS Ann. SECOLAS Annals. Southeastern Conference on Latin American Studies; West Georgia College. Carrollton, Ga.

Secuencia/México. Secuencia. Instituto Mora. México.

SER/Buenos Aires. Seguridad, Estrategia Regional en el 2000. Impresión Zona Gráfica. Buenos Aires.

Sierra/San Francisco. Sierra. Sierra Club. San Francisco.

Signos Univ. Signos Universitarios: Revista de la Universidad del Salvador. Univ. del Salvador. Buenos Aires.

Signs/Chicago. Signs. The Univ. of Chicago Press. Chicago, Ill.

Síntesis/Madrid. Síntesis. Asociación de Investigación y Especialización sobre Temas Latinoamericanos. Madrid.

Soc. Biol. Social Biology. Society for the Study of Social Biology. Port Angeles, Wash.

Soc. Compass. Social Compass. The International Catholic Institute for Social-Ecclesiastical Research. The Hague.

Soc. Econ. Stud. Social and Economic Studies. Univ. of the West Indies, Institute of Social and Economic Research. Mona, Jamaica.

Soc. Probl. Social Problems. Society for the Study of Social Problems; American and International Sociological Associations. Kalamazoo, Mich.

Soc. Sci. Med. Social Science and Medicine. Pergamon Press. New York.

Social. Particip. Socialismo y Participación. Ediciones Socialismo y Participación. Lima.

Sociol. Anal. Sociological Analysis. American Catholic Sociological Society. Worcester, Mass.

Sociol. Inq. Sociological Inquiry. National Sociology Honor Society; Dept. of Sociology, Univ. of Omaha. Omaha, Neb.

Sociol. trav. Sociologie du travail. Association pour le développement de la sociologie du travail. Paris.

Sociologus/Berlin. Sociologus. Berlin.

South East. Lat. Am. South Eastern Latin Americanist. South Eastern Council of Latin American Studies. Boone, N.C.

South. Econ. J. Southern Economic Journal. Chapel Hill, N.C.

Staff Pap. Staff Papers. International Monetary Fund. Washington.

Stat. Masterfile. Statistical Masterfile. Congressional Information Service. Bethesda, Md.

Stud. Comp. Int. Dev. Studies in Comparative International Development. Transaction Periodicals Consortium, Rutgers Univ., New Brunswick, N.J.

Suma/Montevideo. Suma. Centro de Investigaciones Económicas. Montevideo.

SWI Forum. SWI Forum voor Kunst, Kultuur en Wetenschop. De Stichting. Paramaribo, Suriname.

Swiss Rev. World Aff. Swiss Review of World Affairs. Neue Zürcher Zeitung. Zürich, Switzerland.

Tareas/Panamá. Tareas. Centro de Estudios Latinoamericanos (CELA). Panamá.

Terra Livre. Terra Livre. Associação dos Geógrafos Brasileiros. São Paulo.

Third World Plan. Rev. Third World Planning Review. Liverpool Univ. Press. Liverpool, England.

Tijdschr. Econ. Soc. Geogr. Tijdschrift voor Economische en Sociale Geographie. Netherlands Journal of Economic and Social Geography. Rotterdam, The Netherlands.

Trace/México. Trace. Centre d'études mexicaines et centraméricaines. México.

Trimest. Econ. El Trimestre Económico. Fondo de Cultura Económica. México.

U.S./Latin Trade. U.S./Latin Trade. New World Communications. Miami.

U tz'ib. U tz'ib. Asociación Tikal. Guatemala.

UFES Rev. Cult. UFES: Revista de Cultura. Univ. Federal do Espírito Santo, Brazil.

UNISA/Lat. Am. Rep. UNISA/Latin American Report. Univ. of South Africa. Pretoria.

Universitas/Bogotá. Universitas. Pontificia Univ. Javeriana, Facultad de Derecho y Ciencias Socioeconómicas. Bogotá.

Universitas/Stuttgart. Universitas. Wissenschaftliche Verlagsgesellschaft. Stuttgart, Germany.

Universum/Talca. Universum. Univ. de Talca. Talca, Chile.

UTEC/San Salvador. UTEC: Revista de la Universidad Tecnológica. Univ. Tecnológica. San Salvador.

Vínculos/San José. Vínculos. Museo Nacional de Costa Rica. San José.

Voices Mex. Voices of Mexico. Univ. Nacional Autónoma de México. México.

Vuelta/México. Vuelta. México.

Wash. Q. The Washington Quarterly. Georgetown Univ., The Center for Strategic and International Studies. Washington.

Weltwirtsch. Arch. Weltwirtschaftliches Archiv. Zeitschrift des Institut für Weltwirtschaft an der Christians-Albrechts-Univ. Kiel. Kiel, Germany.

West. Polit. Q. Western Political Quarterly. Western Political Science Assn.; Pacific Northwest Political Science Assn.; Southern California Political Science Assn.; Univ. of Utah, Institute of Government. Salt Lake City.

Wilson Q. The Wilson Quarterly. Woodrow Wilson International Center for Scholars. Washington.

World Aff. World Affairs. The American Peace Society. Washington.

World Archaeol. World Archaeology. Routledge & Kegan Paul. London.

World Bank Econ. Rev. The World Bank Economic Review. World Bank. Washington.

World Data. World Data. World Bank. Washington.

World Dev. World Development. Pergamon Press. Oxford, England.

World Econ. The World Economy. Basil Blackwell. London.

World News Connect. World News Connection. National Technical Information Service. Springfield, Va.; Foreign Broadcast Information Service. Washington.

World Policy J. World Policy Journal. World Policy Institute. New York.

World Polit. World Politics. Princeton Univ., Center of International Studies. Princeton, N.J.

World Trade Database. World Trade Database. International Trade Division, Statistics Canada. Ottawa.

Yachay/Cochabamba. Yachay. Facultad de Filosofía y Ciencias Religiosas, Univ. Católica Boliviana. Cochabamba, Bolivia.

Yaxkin/Tegucigalpa. Yaxkin. Instituto Hondureño de Antropología e Historia. Tegucigalpa.

Yearbook/CLAG. Yearbook. Conference of Latin Americanist Geographers; Ball State Univ., Muncie, Ind.

Z. Geomorphol. Suppl.bd. Zeitschrift für Geomorphologie Supplementband. Bebrüder Borntraeger. Berlin.

Z. Lat.am. Wien. Zeitschrift für Lateinamerika Wien. Österreichisches Lateinamerika-Institut. Vienna.

Z. Wirtsch.geogr. Zeitschrift für Wirtschaftgeographie. Hagen, Westfallen, Germany.

Zent.bl. Geol. Paläontologie. Zentralblatt für Geologie und Paläontologie. E. Schweizerbart'sche Verlagsbuchhandlung (Nägele u. Obermiller). Stuttgart, Germany.

SUBJECT INDEX

Aasassinations. Guatemala, 2980.
Abaj Takalik Site (Guatemala), 280, 344.
Abandoned Children, 4446. Brazil, 5208.
Abolition (slavery). Caribbean Area, 807, 4746.
Abortion. Brazil, 5194. Peru, 4886.
Abstract Art. Incas, 691.
Abu Bakr, Yasin, 3100.
Abused Children. *See* Child Abuse.
Acapulco, Mexico (city). Excavations, 221. Precolumbian Civilizations, 221. Rock Art, 221. Salvage Archaeology, 221.
Acción Democrática (Venezuela), 3252, 3254–3255, 3271, 4340.
Acción Democrática y Nacionalista (Bolivia), 3278.
Acción Nacional (Mexico), 2903.
Acculturation, 779. Aymara, 955. Blacks, 4877. Caribbean Area, 817, 846. Colombia, 994. Cuaiquer, 914. Cuba, 499, 4762. Indigenous Peoples, 915. Mayas, 327. Mexico, 4482. Otomi, 773. Peru, 1045, 4920. Piaroa, 902. Precolumbian Civilizations, 536. Quechua, 1027. Rama, 3010. Sipibo, 917. Tucanoan, 899.
Aché. *See* Guayaqui.
Acordo de Alcance Parcial. No. 3, 4389.
Acosta, Jorge R., 179.
Acre, Brazil (state). Boundary Disputes, 2689. Forests and Forest Industry, 2639. Land Use, 2639. Rubber Industry and Trade, 2639.
Adolescents. *See* Teenagers; Youth.
Adriani, Alberto, 1717.
Advertising. Blacks, 5204. Brazil, 5204. Political Culture, 5117.
Affonso, Almino, 3668.
African Influences, 783, 4414. Brazil, 5179–5180. Caribbean Area, 479, 841, 843, 4703, 4746. Colombia, 1000, 4877. Costa Rica, 813. Cuba, 4703, 4747, 4762, 4795. Dominican Republic, 789. Ecuador, 1011, 4877. Hispaniola, 493. Jamaica, 4719, 4728, 4758. Netherlands Antilles, 816. Peru, 1053.
Popular Music, 4414. Popular Religion, 841. Puerto Rico, 4794. Religion, 820. Spanish Language, 1053. Suriname, 4799. Trinidad and Tobago, 4742. Venezuela, 4848.
Afro-Americans. *See* Africans; Blacks.
Aged. Argentina, 5090, 5120. Brazil, 5184. Ecuador, 4871. Jamaica, 4788. Poverty, 5120. Women, 4871.
Agrarian Reform. *See* Land Reform.
Agribusiness. Brazil, 5158.
Agricultural Colonization, 2276. Amazon Basin, 2625, 5143. Brazil, 2625, 5143. Mexico, 2364. Peru, 2485. Precolumbian Civilizations, 109.
Agricultural Credit. Peru, 1933, 1943.
Agricultural Development, 2283. Bolivia, 4978. Brazil, 2099, 2171, 2205, 2635, 2677, 2694. Caribbean Area, 1652. Costa Rica, 4658. Dominican Republic, 1610. Ecuador, 4862. Guatemala, 1610, 4656. Peru, 1939, 2520, 4918, 4939, 4948. Women, 4978.
Agricultural Development Projects. *See* Development Projects.
Agricultural Ecology. Mexico, 2363.
Agricultural Geography, 2231, 2259. Andean Region, 2519. Databases, 43. Ecuador, 2479. Galápagos Islands, 2479. Mesoamerica, 134, 173. Mexico, 372. Peru, 2497. Precolumbian Civilizations, 173, 302.
Agricultural Industries. Bolivia, 3338. Brazil, 2129, 2206, 2706. Colombia, 1765, 1777. Mexico, 1470, 1536, 2403.
Agricultural Labor. Bolivia, 3304. Brazil, 2100, 5158, 5170, 5197. Chile, 1912, 5050, 5073. Colombia, 1764. Dominican Republic, 4754. Ecuador, 4865, 4867. Guatemala, 4613. Indigenous Peoples, 4867. Labor Movement, 5073. Mexico, 1536, 4506. Migration, 5170. Nicaragua, 4627. Women, 5050, 5197.
Agricultural Policy, 1074, 1093, 1123, 1264, 1290, 4398. Argentina, 3566. Brazil, 2206, 2691. Central America, 3971, 4609. Chile, 3436, 3458. Costa Rica, 2959, 4626, 4639,

4658. Cuba, 1690, 1693, 1701, 1712, 4716. Ecuador, 1827. Environmental Pollution, 4609. Haiti, 4787. Mexico, 1320, 1423–1425, 1514. Nicaragua, 1622, 1629, 4604, 4609, 4668, 4673, 4682. Peru, 2487, 4934–4935. Sandinistas, 4604, 4673, 4682. Venezuela, 1722, 1734.

Agricultural Productivity, 1073, 1093, 1170–1172, 1201, 4398, 4443. Andean Region, 2519. Bolivia, 4975. Brazil, 2120, 2188. Coffee Industry and Trade, 2279. Dominican Republic, 4727. Ecuador, 2479, 4861. El Salvador, 1604. Galápagos Islands, 2479. Guatemala, 1607, 1613–1614. Haiti, 4787. Mesoamerica, 101. Mexico, 1319, 1403, 1511, 1530, 1537, 2390, 4496. Nicaragua, 1628. Precolumbian Civilizations, 101. Saint Lucia, 786. Uruguay, 2013. Venezuela, 1734.

Agricultural Subsidies. Mexico, 1397.

Agricultural Systems, 2231, 4443. Andean Region, 938. Aztecs, 197, 203. Bolivia, 4970, 4981. Brazil, 2104. Costa Rica, 730. Cuba, 4716. Ecuador, 4862. Honduras, 766, 4674. Indigenous Peoples, 2363, 4970. Jamaica, 3069. Mayas, 254. Mesoamerica, 134, 254, 276. Mexico, 229, 372, 743, 2356, 2363, 2366, 4496. Peru, 1043, 2497, 4888, 4901, 4918. Precolumbian Civilizations, 140, 229, 276, 340, 372, 390, 652, 716. Venezuela, 716.

Agricultural Technology. Bolivia, 4970. Brazil, 2099, 2682. Colombia, 4809. Indigenous Influences, 4928. Mesoamerica, 78, 97. Mexico, 1319, 4496. Peru, 2503, 4888, 4901, 4928. Precolumbian Civilizations, 78, 2283.

Agriculture, 1081, 1092, 1130, 1178–1179, 1251, 1261, 1264, 1274, 1768. Amazon Basin, 2672. Andean Region, 1035, 1803. Argentina, 3553. Bolivia, 1982, 3299. Brazil, 2095, 2214, 2672, 2686–2687, 2693, 2701, 2703. Capitalism, 1073. Caribbean Area, 1251. Central America, 1540. Chile, 1846, 1863, 1904, 3458. Colombia, 1765, 1777. Costa Rica, 2332. Databases, 43. Development Projects, 1220. Dominican Republic, 1662. Ecuador, 1802. Indigenous Peoples, 938. Mayas, 162. Mexico, 18, 1308, 1319, 1329, 1339–1340, 1342, 1348, 1372, 1384, 1396, 1422, 1470, 1503, 1511, 1513, 1526, 2414, 2416. Modernization, 1073. North America, 1381. Peru, 1034, 1939, 1943. Precolumbian Civilizations, 64, 2416. Saint Lucia, 814. Venezuela, 1716, 1726, 2448. Virgin Islands, 2288.

Agroindustry. *See* Agricultural Industries.

Aguaruna (indigenous group). Dreams, 911. Social Life and Customs, 911.

Aguascalientes, Mexico (state). Economic Conditions, 1394. Social Conditions, 1394.

AIDS. Argentina, 5110. Costa Rica, 4628. Cuba, 4739. Haiti, 4739. Puerto Rico, 4785.

Aisén del General Carlos Ibáñez del Campo, Chile (region), 2576, 2594.

Akwẽ Shavante. *See* Xavante.

ALADI. *See* Asociación Latinoamericana de Integración.

Alakaluf. *See* Alacaluf.

Alcohol Fuel Industry. Brazil, 2170.

Alfonsín, Raúl, 3504, 3512, 3538, 3552, 3561, 3581, 4194.

Alianza Popular Revolucionaria Americana. *See* APRA (Peru).

Allende Gossens, Salvador, 3438.

Almeida, Laura, 3240.

Alonso, Leonor, 3510.

Alta Verapaz, Guatemala (dept.). Modernization, 4663.

Aluminum Industry and Trade. Brazil, 2193.

Amapá, Brazil (state). Boundaries, 4380. Land Tenure, 4380.

Amarakaeri. *See* Mashco.

Amazon Basin. Agricultural Colonization, 5143. Agriculture, 2672. Cattle Raising and Trade, 2672. Colonization, 2522, 2692. Conservation, 2645. Deforestation, 2191, 2475, 2673, 2681, 2692, 2721. Demography, 2692. Development, 861, 3689. Droughts, 584. Ecology, 4311, 5164, 5181. Economic Development, 2692, 5164. Economic Integration, 2692. Economic Policy, 2207, 2493. Environmental Degradation, 2665. Environmental Policy, 2493, 3780, 5213. Environmental Protection, 2678, 2764, 5160, 5164. Environmental Protection Groups, 3688. Forests and Forest Industry, 1936. Fruit Trade, 2645. Geography, 2233, 2644. Human Adaptation, 573. Human Ecology, 893, 5181. Human Geography, 858. Indigenous/Non-Indigenous Relations, 893, 905. Indigenous Peoples, 855. Jade, 851. Japanese, 2625. Land Settlement, 861, 2522, 2665, 2678, 5143. Millennialism, 852. Missionaries, 882. Natural Resources, 857. Poverty, 2678. Precolumbian Land Settlement Patterns, 583. Precolumbian Trade, 851. Prostitution, 5165. Protestants, 2475. Rainforests, 2681. Regional Government, 2633. Religious Life and Customs, 853. Rubber Industry and Trade, 2129, 5160. Rural Development, 1936. Shamanism, 853. Social

Policy, 2493. Sustainable Development, 2207, 2665, 5164. Warfare, 857–858, 2207.

Amazonas, Brazil (state). Astronomy, 2671. Climatology, 2671. Land Settlement, 2707. Time, 2671. Peasants, 2655.

Ambergris Cay (Belize). Precolumbian Trade, 283.

Amecameca, Mexico (town). Monuments, 312.

American Federation of Labor, 4130.

Americans (US). Mexico, 4532. Nicaragua, 3011.

Amulets. Andean Region, 934.

Ancash, Peru (dept.). Economic Anthropology, 1066.

Andamarca, Peru (city). Social Structure, 1046.

Andean Pact. Economic Development, 1937. Ecuador, 1792, 1797.

Andean Region. Agricultural Geography, 2519. Agricultural Productivity, 2519. Agricultural Systems, 938. Agriculture, 1035. Amulets, 934. Anthropologists, 1065, 1069. Anthropology, 1065. Cattle Raising and Trade, 2425, 2502. Climatology, 2428, 2566, 2568. Commerce, 1841. Cosmology, 1070. Cultural Development, 530. Cultural Geography, 2426. Cultural History, 2425. Cultural Identity, 968, 1029. Democracy, 2777. Development Projects, 1955. Drug Enforcement, 3296, 4161, 4182, 4186. Drug Traffic, 4161, 4452. Ecology, 2418. Economic Conditions, 1048. Economic Development, 2515. Economic Integration, 1803, 4170. Economic Policy, 1173. Elections, 2777. Ethnoarchaeology, 689. Food Supply, 2418, 2478. Geographical History, 2565–2566, 2568. Geology, 2480. Geomorphology, 2565–2567. Historical Geography, 2418, 2424, 2426. Human Ecology, 1035. Human Geography, 2418, 2422, 2425, 2428. Indigenous Policy, 936. Insurrections, 57. Irrigation, 1035, 2513. Land Reform, 2515. Land Tenure, 1038. Land Use, 2428. Law and Legislation, 1058. Literacy and Illiteracy, 1063. Mestizos and Mestizaje, 937. Myths and Mythology, 1041. Peasants, 1038, 1069, 2424, 2515, 2519. Political Parties, 2777. Popular Culture, 1042. Precolumbian Civilizations, 544, 661. Precolumbian Land Settlement Patterns, 539. Presidents, 2725. Rain and Rainfall, 2567. Regional Integration, 4170. Rites and Ceremonies, 689. Sex Roles, 933. Social Life and Customs, 1063. Social Structure, 1047. Textiles and Textile Industry, 1040. Violence, 1033, 4452. Volcanoes, 2480. Water Rights, 1035. Water Supply, 1173, 2513. Women, 1050. Youth, 1031.

Andes, Chile. See Los Andes, Chile (city).

Andoque. See Andaqui.

Animal Remains. Caribbean Area, 526. Chile, 597–598, 603, 611. Mexico, 201, 260. Teotihuacán, 397.

Animals. Aztecs, 438. Caribbean Area, 2302. Cuba, 1690.

Anta, Peru (prov.). Cost and Standard of Living, 1927.

Antarctic Treaty System, 4189.

Antarctica, 4180. Argentina, 4189.

Anthrolinguistics. Aztecs, 424.

Anthropologists. Andean Region, 1065, 1069. Argentina, 939. Biography, 768. Fieldwork, 1025. Germany, 939. Haiti, 823. Historiography, 61. Indigenous Peoples, 61. Mexico, 768. Peru, 1051. Social Change, 1065. US, 768.

Anthropology. Bibliography, 65. Cuba, 799. Ecuador, 1020. Law and Legislation, 988. Research, 65. Social Change, 992.

Anthropometry. Mexico, 366.

Anti-Positivism. See Positivism.

Antisemitism. Argentina, 4207.

Antofagasta, Chile (prov.). Tourism, 2603.

Antorcha Campesina (peasant organization, Mexico), 746.

Apalakiri (indigenous group). Brazil, 865. Cultural Identity, 865.

Apinayé. See Apinagé.

Apostolado Positivista do Brasil. See Igreja Positivista do Brasil.

Appropriate Technology. Colombia, 4825. Community Development, 4825.

APRA (Peru), 3378, 3413, 3415.

Aqueducts. Mexico, 2393.

Araucanian. See Mapuche.

Araucano. See Mapuche.

Arawak (indigenous group). Archaeology, 513. Millennialism, 889. Oral History, 907. Precolumbian Pottery, 568.

Arbenz Guzmán, Jacobo, 2978.

Arbo, Higinio, 3590.

Arce, Luz, 3422.

Archaeoastronomy. Aztecs, 225. Mayas, 404, 433. Mesoamerica, 162–163, 237, 404. Mexico, 440. Peru, 697.

Archaeological Dating, 464, 481, 565, 598, 600, 665. Argentina, 552. Caribbean Area, 508. Colombia, 617. Corn, 310. Guatemala, 314. Honduras, 216. Mesoamerica, 79, 109,

158, 183, 200. Mexico, 196, 310, 329, 375. Peru, 674. Puerto Rico, 489. Theory, 63, 403.

Archaeological Geography. Belize, 104, 305. Brazil, 591, 593. Caribbean Area, 507. Chile, 601. Costa Rica, 464, 478. Honduras, 306. Mesoamerica, 95, 131. Mexico, 123, 214, 231. Panama, 464. Peru, 674. Precolumbian Civilizations, 358.

Archaeological Parks. Conservation and Restoration, 393. Honduras, 463.

Archaeological Surveys. Argentina, 548. Aruba, 521. Barbados, 495. Belize, 267, 298, 349, 394. Brazil, 573, 575–577, 589, 591, 595. Caribbean Area, 507. Colombia, 622, 624. Costa Rica, 465, 475. Cuba, 494. Dutch Caribbean, 521, 650. Ecuador, 624, 638–639, 646, 648. Guatemala, 155, 344, 363, 399. Hispaniola, 493. Honduras, 67, 178, 286, 290, 306, 331, 333–334, 400. Jamaica, 480, 513. Maroons, 480. Mayas, 71, 91. Mesoamerica, 72, 113, 146, 171, 183–184, 188, 291, 315. Mexico, 188, 210–211, 218, 222, 224, 271, 273, 336–337, 342–343, 345–348, 353, 362, 376. Patagonia, 548, 560. Peru, 674, 706–709. Southern Cone, 528. Suriname, 650. Theory, 291. Tierra del Fuego, 560. Venezuela, 713–714.

Archaeologists. Americans (US), 667. Argentina, 939. Central America, 476. Cultural Identity, 667, 673. Mesoamerica, 93. Mexico, 179. Peru, 667, 698. South America, 528.

Archaeology. Argentina, 555. Caribbean Area, 491, 503–504. Central America, 476. Chemistry, 325. Chile, 610. Congresses, 60. Costa Rica, 467. Dominican Republic, 487. Haiti, 488. Mesoamerica, 193–194. Methodology, 226, 555, 608, 610, 694. Peru, 654. Puerto Rico, 490, 514. Research, 171, 325, 347, 349, 374, 388, 654.

Archives. Guatemala, 2343.

Archives, US. Databases, 38. El Salvador, 38. Nicaragua, 38.

Archivo Histórico Nacional (Spain). Sección de Ultramar, 2284.

Arciniega, Alberto, 3376.

Arenas, Jacobo, 3153.

Arequipa, Peru (city). Communal Kitchens, 4914. Housing, 4878. Neighborhood Associations, 4878.

Arequipa, Peru (dept.). Floods, 2514.

Argentina. Congreso de la Nación, 3563, 3573.

Argentina. Congreso de la Nación. Cámara de Diputados de la Nación, 3562.

Arhuaco (indigenous group). Rites and Ceremonies, 877.

Arias Madrid, Harmodio, 4046.

Arias Peace Plan (1987), 3998, 4013, 4021, 4051.

Arias Sánchez, Oscar, 2947, 4013.

Arica, Chile (city). Archaeology, 596.

Arica, Chile (dept.). Boundary Disputes, 4299.

Aristide, Jean-Bertrand, 3061, 3064, 3067.

Armaments. South America, 4181.

Armed Forces. *See* Military.

Armillas, Pedro, 217.

Arms Control. Argentina, 4168, 4197. South America, 4168.

Arms Industry and Trade. Argentina, 4175. Brazil, 4175. Chile, 4175. South America, 4181.

Arraes, Miguel, 3734.

Art Catalogs, 545. Ecuador, 927. Indigenous Peoples, 927. Suriname, 828.

Art History. Cuba, 4705. Mexico, 326.

Art Schools. Cuba, 4705.

Artifacts. Argentina, 549. Aztecs, 395. Belize, 471. Bolivia, 565. Brazil, 577, 581–582. Chile, 611. Colombia, 615, 629, 633. Huastec, 336. Jade, 321. Matlatzinca, 377. Mayas, 305. Mesoamerica, 405. Mexico, 182, 245, 319, 360, 383. Mixtec, 391. Peru, 670. Puerto Rico, 514–515. South America, 545. Underwater Archaeology, 572. Zapotec, 391.

Artisanry. Chile, 974. Mesoamerica, 725. Peru, 1044. Precolumbian Civilizations, 262–263, 265, 405.

Artisans. Peru, 3375. Precolumbian Civilizations, 686.

Artists. Brazil, 4391. Cuba, 4705. Political Ideology, 4705.

Ashaninca (indigenous group). Indigenous/Non-Indigenous Relations, 912. Missions, 3373. Myths and Mythology, 924. Sex Roles, 924. Social Conditions, 3373. Violence, 912.

Asian Influences. Mayas, 410. Precolumbian Civilizations, 533.

Asociación de Administradoras de Fondos de Pensiones (Santiago), 1850.

Asociación de Mujeres Nicaragüenses Luisa Amanda Espinosa (AMNLAE), 4594–4595.

Asociación de Trabajadores del Campo (Nicaragua), 4627.

Asociación Latinoamericana de Instituciones

Financieras de Desarrollo (ALIDE), 1085, 1228.
Asociación Latinoamericana de Integración (ALADI), 1148.
Assassinations. Bolivia, 3295. Brazil, 5210. Ecuador, 3220. Street Children, 5210.
Assimilation. *See* Acculturation.
Astronomy. Bororo, 870. Brazil, 2671.
Asunción, Paraguay (city). Crime and Criminals, 4993. Local Elections, 3600. Prostitution, 4991. Rape, 4993.
Asunción para Todos (Paraguay), 3600.
Atacama Desert (Chile). Colonization, 2604. Germans, 2604. Human Ecology, 607.
Atacameño (indigenous group). Human Ecology, 607.
Atlantic Coast (Nicaragua). *See* Mosquitia (Nicaragua and Honduras).
Atlases. Chile, 2584. Databases, 15. Mexico, 15.
Atlixco, Mexico (town). Textiles and Textile Industry, 4483.
Austerity Measures. Argentina, 3546. Bolivia, 3318, 3322, 3329. Chile, 3432, 3452. Mexico, 2862.
Authoritarianism, 2772. Argentina, 3501, 3570. Bolivia, 2772, 3308, 3316. Brazil, 3748. Chile, 2757, 3428, 3456, 3471, 3477, 4998, 5003, 5007, 5016, 5021, 5027, 5049, 5071, 5079. Ecuador, 3230, 3241. Haiti, 3068. Mexico, 2821, 2833, 2890. Paraguay, 3592, 3598–3599, 3603, 4296. Peru, 3362, 3368. Southern Cone, 2752. Uruguay, 3620, 3661. Women, 5079.
Autobiography. Caribbean Area, 3037. Mexico, 2842.
Automobile Industry and Trade, 1249, 1428, 2202. Argentina, 3565. Brazil, 3724, 3760. Mexico, 1343, 1402, 1416, 1428, 1440, 1466, 1524, 1533, 3724, 4518.
Autonomy, 3835. Aymara, 961. Dutch Caribbean, 3077. Ecuador, 3214. Mapuche, 976. Nicaragua, 3012.
Awa. *See* Cuaiquer.
Ayacucho, Peru (dept.). Agricultural Development, 4939. Archaeological Surveys, 708. Archaeology, 676. Artisanry, 1044. Development Projects, 4939. Human Rights, 3413. Indigenous Peoples, 4884. Myths and Mythology, 4884. Painting, 1044. Precolumbian Civilizations, 676. Precolumbian Land Settlement Patterns, 708. Revolutions and Revolutionary Movements, 3369. Terrorism, 3413. Violence, 3413.

Aylwin Azócar, Patricio, 3424, 3426, 3449, 3479, 4248.
Aymara (indigenous group), 5044. Acculturation, 955. Autonomy, 961. Bibliography, 949. Commerce, 971. Congresses, 981. Cosmology, 977. Cultural Destruction, 948. Cultural History, 962. Discourse Analysis, 1031. Ethnography, 948. Family and Family Relations, 4955. Government Relations, 963. Human Ecology, 4982. Insurrections, 965. Land Tenure, 971. Material Culture, 974. Migration, 1026. Oral History, 946. Religion, 955. Shamanism, 954. Social Life and Customs, 947. Social Movements, 961, 963. Traditional Medicine, 954, 4958. Urbanization, 4955. Women, 970, 4955, 4958, 4966. Youth, 1031.
Ayoreo. *See* Moro.
Aysén, Chile. *See* Aisén, Chile (prov.).
Aztec Influences. Mesoamerica, 374.
Aztecs. Agricultural Systems, 197, 203. Animals, 438. Archaeoastronomy, 225. Artifacts, 110, 378, 395. Calendrics, 414, 419, 451, 455, 2392. Center/Periphery Relations, 373. City Planning, 139. Clothing and Dress, 409. Codices, 89, 110. Consumption (economics), 124. Cosmology, 225, 416, 451. Costume and Adornment, 227, 409. Cults, 339. Cultural History, 88, 128. Cultural Identity, 424. Dance, 164. Deities, 141, 154, 416, 431, 436. Epigraphy, 89, 197, 446. Ethnoarchaeology, 374. Excavations, 204. Exhibitions, 68. Historiography, 395, 408. Iconography, 154. Imperialism, 374. Kings and Rulers, 110, 281, 339, 409. Language and Languages, 154, 438. Legends, 461. Linguistics, 424. Luxuries, 209. Manuscripts, 89. Material Culture, 209, 243. Migration, 129. Militarism, 339. Monuments, 281, 312. Myths and Mythology, 77, 128, 141, 197. Nutrition, 142. Obsidian, 227. Political Culture, 176. Political Development, 168, 228, 293. Precolumbian Architecture, 220, 225, 341, 419. Precolumbian Art, 159, 419. Precolumbian Land Settlement Patterns, 203. Precolumbian Pottery, 293–295. Precolumbian Sculpture, 110, 159, 225, 436. Precolumbian Trade, 209, 293. Production (economics), 228. Pulque, 451. Religious Life and Customs, 88, 225, 461. Research, 171. Rites and Ceremonies, 88, 167, 260, 311, 417, 428, 739. Rock Art, 312. Sacred Space, 167, 220. Sacrifice, 77, 88, 220. Sex Roles, 82. Social Life and Customs, 88,

118, 124, 168, 225. Social Structure, 118, 124. Spanish Conquest, 168, 174. Symbolism, 82. Taxation, 118, 124, 209, 442. Tombs, 220. Traditional Medicine, 142. Urbanization, 139. Warfare, 197, 451. Women, 81–82. Writing, 442, 446.

Bahia, Brazil (state). Agriculture, 2701, 2703. Archaeological Surveys, 576. Carnival, 5180. Demography, 2696. Migrant Labor, 2701. Occupational Training, 2211. Population Growth, 2696. Private Enterprises, 2696. Regional Planning, 2703. Urban Areas, 2711. Urbanization, 2703, 2711. Wages, 2211.

Baja California, Mexico (state). Commerce, 1386. Demography, 1386. Economic Conditions, 1386. Elections, 2931. Social Movements, 4560. Urban Sociology, 4560. Urbanization, 4560.

Baja California Sur, Mexico (state). Commerce, 1386. Demography, 1386. Economic Conditions, 1386. Maquiladoras, 4536. Social Movements, 4560. Urban Sociology, 4560. Urbanization, 4560.

Bajío, Mexico (region), 1384.

Bakairi (indigenous group). Development Projects, 881.

Balaguer, Joaquín, 3046–3047, 4802.

Balamku Site (Mexico). Archaeological Surveys, 273.

Balance of Payments, 1122. Argentina, 2077. Caribbean Area, 1647. Nicaragua, 1627.

Balberta Site (Guatemala), 205.

Balcón de Montezuma Site (Mexico), 336.

Ball Games. Belize, 289. Cults, 462. Mayas, 270, 364. Mesoamerica, 122. Precolumbian Architecture, 289. Precolumbian Civilizations, 122.

Banana Trade, 1176. Central America, 4702. Colombia, 4808.

Banco Central del Ecuador, 1811.

Banco del Estado de Chile, 3441.

Banco do Brasil, 2154.

Banco Nacional de Desenvolvimento Econômico e Social (Brazil), 2175.

Banco Occidental de Descuento (Zulia, Venezuela), 1741.

Banking and Financial Institutions, 1085, 1151, 1161, 1232, 1273, 1287. Argentina, 2036, 2042–2043, 2070, 3567, 4195. Brazil, 2154, 2174–2175, 2178. Canada, 1492. Chile, 1894–1895, 3441. Costa Rica, 2951. Dominican Republic, 1661. Ecuador, 1811. Mexico, 1472, 1492, 2884. Paraguay, 2005.

Peru, 1926, 4883. Trinidad and Tobago, 1664, 1667. Uruguay, 3663. US, 1119, 3847. Venezuela, 1741, 1746.

Bankruptcy. Ecuador, 1845.

Banzer Suárez, Hugo, 3278, 3301.

Baptista, Asdrúbal, 1748.

Baptists. Caribbean Area, 841.

Barboza, Mario Gibson, 4353.

Barca, Mexico. *See* La Barca, Mexico (town).

Barco, Virgilio, 4275.

Bardón M., Alvaro, 3441.

Bari. *See* Motilon.

Barrientos Ortuño, René, 3319.

Barthel, Thomas S., 87.

Basic Christian Communities. *See* Christian Base Communities.

Basques. Venezuela, 4831.

Bats. Mexico, 201.

Beagle Channel (Argentina and Chile), 4201.

Becerra Lanza, José Eduardo, 2988.

Beliefs and Customs. *See* Religious Life and Customs; Social Life and Customs.

Bello, Andrés, 4335.

Benavides Cevallos, Elisa Consuelo, 3220.

Beni, Bolivia. *See* El Beni, Bolivia (dept.).

Berichá, 985.

Bernal, Ignacio, 196.

Betancourt, Rómulo, 3251–3252, 4340.

Betancur, Belisario, 3153.

Bibliography. Anthropology, 65. Bolivia, 4232, 4953, 4973. Brazil, 3727. Caribbean Area, 4700, 4777. Chile, 4261. Chiriguano, 929. Colombia, 4275–4276. Cuba, 4701. Economic History, 1174. Environmental Policy, 2232. Ethnography, 949. Forests and Forest Industry, 2318. Geography, 2264. Geopolitics, 3793. Maritime Law, 4232. Mass Media, 4492, 4700–4701. Mexico, 2835. Pacific Area, 20. Peru, 863. Political Conditions, 2729. Political Culture, 3727. Political Science, 2809. Political Systems, 2729. Precolumbian Civilizations, 157. Social Movements, 4521. Social Sciences, 28. Women, 4553, 4953.

Bilingual Education. Puerto Rico, 4753.

Bilingualism. Puerto Rico, 4753.

Bioarchaeology. Patagonia, 534.

Biobío, Chile (region). Economic Development, 1857. Regional Development, 5006.

Biogeography. Amazon Basin, 2675. Brazil, 2621, 2722.

Biography. British Caribbean, 3040. Colombia, 3152, 3190. Mexico, 21. Mosquito, 3007.

Biological Diversity. Caribbean Area, 2302.
Bird, Vere Cornwall, 4144.
Birds. Argentina, 2544. Brazil, 2722. Chile, 2544, 2581.
Birth Control, 4431, 4454. Bolivia, 4958. Brazil, 5161, 5194. Colombia, 4811, 4820. Congresses, 4409. Dominican Republic, 4709, 4773. Mexico, 4531.
Black Carib (indigenous group). Cultural Contact, 796. Cultural Development, 279. Cultural Identity, 796. Land Settlement, 279. Sources, 524. Views of, 524.
Black Indians. *See* Rikbaktsa.
Black Market. Central America, 1555. El Salvador, 1595, 1605. Guatemala, 1605. Peru, 1961.
Blacks, 4414. Acculturation, 4877. Advertising, 5204. Brazil, 3681, 5142, 5175, 5180, 5200, 5204. British Caribbean, 498. Caribbean Area, 801, 824, 4707. Colombia, 989, 1000, 1005–1006, 4814, 4830, 4877. Costa Rica, 813, 826, 4659. Cultural Identity, 5180. Dominican Republic, 789–790. Ecuador, 1011, 4877. French Guiana, 805. Panama, 4659. Political Participation, 3681. Public Opinion, 5142. Religious Life and Customs, 820. Saint Kitts and Nevis, 821. Suriname, 803. Trinidad and Tobago, 804, 1677. US, 5142.
Blood Pressure, High. *See* Hypertension.
Boban, Eugène, 414.
Bogotá, Colombia (city). Demography, 4820. Infant Mortality, 4820. Poor, 3202. Slums, 3202. Squatter Settlements, 3202. Street Children, 4446. Urban Sociology, 4824. Violence, 3157.
Bogotá River Valley (Colombia). Precolumbian Pottery, 627.
Bolaños River Valley (Mexico). Archaeological Surveys, 222.
Bolívar, Simón, 2423.
Bonafina, Hebe de, 3514.
Boni (Surinamese people). Art Catalogs, 828. Congresses, 798. Dictionaries, 780. Social Life and Customs, 805.
Books. Chile, 1921.
Borja, Rodrigo, 3215.
Bororo (indigenous group). Archaeological Surveys, 595. Astronomy, 870. Cosmology, 870, 887. Mortuary Customs, 887. Religious Life and Customs, 870. Social Life and Customs, 870.
Borrego Estrada, Genaro, 2832.
Bosch, Juan, 3045.

Botany, 2220, 2225. Colombia, 2462. Corn, 96. Costa Rica, 2334. Panama, 2354.
Boundaries, 2456, 3899. Argentina, 2563, 2597. Belize, 2378. Bolivia, 4232–4234, 4241, 4302. Brazil, 4380. Chile, 2590, 2597. Colombia, 2456–2457. Guatemala, 2378. Mexico, 1448, 2378. Paraguay, 2004. Peru, 4302. US, 1448. Venezuela, 2457.
Boundary Disputes. Argentina/Chile, 4201, 4250, 4252, 4263. Belize/Guatemala, 4005. Bolivia/Brazil, 4240. Bolivia/Chile, 4226, 4241–4242, 4244, 4259. Brazil, 2689. Brazil/Ecuador, 4291. Central America, 4030. Chile/Peru, 4299. Colombia/Ecuador, 4291. Colombia/Peru, 4280. Colombia/Venezuela, 2457, 4273–4274, 4348. Ecuador/Peru, 3215, 4281, 4283, 4287, 4289, 4291, 4311. El Salvador/Honduras, 4030. Guyana/Venezuela, 4332, 4343–4344. Peru, 4303, 4306.
Braden, Spruille, 4207.
Brasília, Brazil (city). Urban Areas, 2629.
Braudel, Fernand, 63, 158.
Brazil. Congresso Nacional, 3685, 3687, 3722, 3743.
Brazil. Escola Superior de Guerra, 3680.
Brazil. Serviço Nacional de Informações, 3677.
Brazilian Influences. France, 5153. Paraguay, 2697.
Brazilianists. US, 4373.
Brazilians. Bolivia, 4971. Paraguay, 2697. US, 5178.
British. Peru, 2485.
British Caribbean. Biography, 3040. Democracy, 3073. Economic Conditions, 3035. Economic Integration, 3038. Political Conditions, 3035, 3072. Political Culture, 3073. Politicians, 3040. Social Conditions, 3035. Sovereignty, 3072–3073.
Bronze. Precolumbian Civilizations, 297.
Brüning, Hans Heinrich, 1055.
Bucaram Ortiz, Abdalá, 3226, 3236.
Budget. Argentina, 2064. Ecuador, 1795. Mexico, 1379.
Buenos Aires, Argentina (city). Aged, 5120. Capitalism, 2542. Community Development, 5098. Economic Conditions, 3531. Housing, 5094. Municipal Government, 3531, 3537. Political Participation, 3537. Poor, 2542, 5092. Protests, 5104. Public Health, 5094. Public Works, 3531. Social Movements, 3537, 5104. Urban Sociology, 3537, 5106.

Buenos Aires, Argentina (prov.). Energy Supply, 2054. Historical Demography, 2557. Land Invasions, 5113. Migration, 2557.

Bureaucracy, 2755, 2772. Bolivia, 2772, 3323. Congresses, 3632. Jamaica, 1672. Peru, 3373. Uruguay, 3632.

Burials. *See* Cemeteries; Mortuary Customs; Tombs.

Business Administration, 1200. Mexico, 4488. Women, 4804.

Bussi, Antonio Domingo, 3550.

Caamaño Deñó, Francisco Alberto, 3050, 3055.

Cabinet Officers. Brazil, 2168. Mexico, 2937. Nicaragua, 3003.

Caboclos. *See* Mestizos and Mestizaje.

CACM. *See* Mercado Común Centroamericano.

Cahuachi Site (Peru), 704.

Cajamarca, Peru (city). Folklore, 1059. Historical Geography, 2509. Maps and Cartography, 2509.

Cajamarca, Peru (dept.). Haciendas, 1935. Households, 1935. Iconography, 681. Inca Influences, 683. Peasants, 1935. Precolumbian Art, 681. Precolumbian Civilizations, 683.

Cajón Reservoir Region (Honduras). Salvage Archaeology, 286, 292.

Cakchikel (indigenous group). Protestantism, 728. Social Change, 728.

Cakchiquel. *See* Cakchikel.

Calakmul Site (Mexico), 434.

Calendrics. Aztecs, 414, 419, 451, 455. Mayas, 87, 421, 423. Mesoamerica, 414, 426–427, 450–451, 455–456, 458. Mexico, 386, 440–441, 2392. Olmecs, 426–427. Precolumbian Architecture, 456.

Cali, Colombia (city). Street Children, 983.

California, US. Water Supply, 2367.

Calima Culture, 616.

Calima River Basin (Colombia). Archaeology, 633.

Callahuaya (indigenous group). Bibliography, 949.

Calypso. Curaçao, 775. Trinidad and Tobago, 781.

Camagüey, Cuba (city). History, 4717.

Campaign Funds. Costa Rica, 2949, 2958.

Campaña por los Derechos del Niño (Paraguay), 4988.

Campeche, Mexico (state). Archaeological Surveys, 273. Precolumbian Land Settlement Patterns, 299.

Campos, Roberto de Oliveira, 3686.

Camsa (indigenous group). Social Life and Customs, 986.

Canals. Colombia, 2461.

Cananea Consolidated Copper Co. (Mexico), 2930.

Cañar, Ecuador (prov.). Popular Culture, 1012.

Cañari (indigenous group). Precolumbian Sculpture, 644.

Candomblé (cult). Brazil, 820.

Canelo (indigenous group). Art, 927. Exhibitions, 927. Myths and Mythology, 927.

Canelos Quichua. *See* Canelo.

Canoeiro. *See* Rikbaktsa.

Canto Grande (prison, Peru), 3359.

Capilco Site (Mexico). Archaeological Dating, 375.

Capital, 1108, 1149, 1271, 1285, 1296, 1301, 1778. Argentina, 2086. Brazil, 2142, 2198. Colombia, 1778. Drug Traffic, 3826. Mexico, 1335, 1393, 1481, 1531. Regional Development, 5006.

Capital Flight, 1376. Brazil, 2158. Central America, 1545. Mexico, 1431.

Capital Goods, 1193.

Capital Movements, 1226. Mexico, 1531. South America, 1150.

Capitalism, 1417, 2728. Agriculture, 1073. Argentina, 2542, 3506, 3571, 3584. Bolivia, 950. Chile, 3480, 3485–3486, 3494. Cities and Towns, 4393. Colombia, 4822. Cuba, 1704. Guatemala, 3982. Mexico, 1490, 4460. Peru, 1935.

Caracas, Venezuela (city). Cost and Standard of Living, 4847. Housing, 4847. Popular Culture, 4840. Slums, 4847. Social Life and Customs, 4840. Urban Sociology, 4840, 4847.

Carajás Project. *See* Programa Grande Carajás.

Carare River Valley (Colombia). Material Culture, 622.

Cárdenas Solórzano, Cuauhtémoc, 2900, 4525.

Carib (indigenous group). Myths and Mythology, 817. Sources, 524. Views of, 524.

Caribbean Area. Abolition (slavery), 4746. African Influences, 479, 4703, 4746. Animal Remains, 526. Archaeological Dating, 508. Archaeological Geography, 507. Archaeological Surveys, 507. Archaeology, 491, 503–504. Autobiography, 3037. Bibliography, 4700, 4777. Blacks, 4707. Civil Service, 3041. Cultural Development, 4697.

Cultural History, 503–504. Democracy, 3033. Demography, 4751. Economic Conditions, 2804, 4752. Economic Development, 3042, 4752. Economic Models, 3042. Educational Sociology, 4720. Family and Family Relations, 4714. Fish and Fishing, 2305, 4789. Fisheries, 4789. Historiography, 481, 516, 525. Human Ecology, 4697. Indigenous Peoples, 483, 511. Industry and Industrialization, 4736. Informal Sector, 4752. Inter-Tribal Relations, 483. International Relations, 3032. Jade, 851. Labor and Laboring Classes, 3031. Labor Movement, 3031. Latin American Area Studies, 4746. Mass Media, 4700. Migration, 4714, 4736, 4800. Mortuary Customs, 520. Nationalism, 4746, 4755. Paleobotany, 518. Physical Anthropology, 520. Plantations, 4707. Political Conditions, 2804. Political Culture, 3043. Political Development, 3036. Political Economy, 3032. Political History, 3036. Political Integration, 3034. Precolumbian Civilizations, 490–491, 497, 502, 507–508, 511, 519, 526. Precolumbian Land Settlement Patterns, 517. Precolumbian Trade, 851. Public Health, 4751. Race and Race Relations, 792, 827, 3043, 4755. Regional Integration, 3034. Rock Art, 496. Slaves and Slavery, 479. Social Conditions, 2804. State, The, 4707, 4755. Stone Implements, 492. Sugar Industry and Trade, 4757. Tobacco Industry and Trade, 4766. Tobacco Use, 4766. Women, 4425, 4731, 4777.

Caribbean Area Studies, 3030.

Caribbean Basin Initiative (CBI), 1583, 1668, 4082, 4102, 4136, 4143.

Caribbean Broadcasting Corporation (Barbados), 3074.

Caribbean Community (CARICOM), 1148, 1645, 1655, 3038, 4080.

Caribbean Studies Association, 3030.

Caribbean Workers Council, 3031.

Caribbeanists. Denmark, 822.

Caribbeans. Costa Rica, 813. Great Britain, 787. Holland, 787.

CARICOM. *See* Caribbean Community.

Carnevali, Alberto, 3254.

Carnival. Brazil, 5180. Cuba, 4747. Mayas, 738. Trinidad and Tobago, 781.

Caro Quintero, Rafael, 3925.

Cartagena Agreement. *See* Andean Pact.

Cartography. *See* Maps and Cartography.

Cartoons. Ecuador, 3224.

Casas Grandes Site (Mexico), 251, 255. Urban History, 238. Precolumbian Architecture, 199.

Cashinawa (indigenous group). Cultural Identity, 880. Kinship, 880.

Casilda, Argentina (town). Social Conflict, 5088.

Caso, Alfonso, 85.

Cassava. Cuba, 482.

Castellanos, Miguel, 2967.

Castro, Fidel, 3103, 3105, 3107, 3110, 3112–3113, 3131, 3135, 3145, 4106, 4124, 4128, 4131.

Castro y Castro, Fernando, 2842.

Catamarca, Argentina (prov.). Precolumbian Civilizations, 546.

Catholic Church, 3799, 4423. Argentina, 5119. Brazil, 3667, 3696, 3706, 3709, 3759, 4995, 5148–5149, 5173, 5176, 5206. Central America, 3973, 4635–4636. Chile, 3429, 3467, 3473–3474, 3482, 3493, 3495, 4995, 5027, 5037, 5051, 5075, 5081. Colombia, 3203, 4423. Costa Rica, 4647. Cuba, 3134, 4692, 4784. Democratization, 5075. Dominican Republic, 4797. El Salvador, 2963, 2968. Environmental Protection, 3709. Guatemala, 2976, 4597, 4661. Haiti, 3061, 4764. Honduras, 4582. Land Reform, 4588. Mexico, 2830, 2834, 2851, 2887, 2911, 3919, 4509. Military Government, 3467, 5051. Paraguay, 3594–3595, 3613. Peasant Movements, 3613. Peru, 3373, 3383, 4906. Secondary Education, 5037. Sex and Sexual Relations, 3473. Social Change, 4582. Social Movements, 4797. Social Welfare, 5119. Southern Cone, 2797. Uruguay, 4322. Venezuela, 4423. Women, 3696.

Catholicism, 4423. Central America, 4631. Dominican Republic, 4798. Indigenous Peoples, 4509. Mexico, 2848. Peru, 4906. Syncretism, 4509.

Cattle Raising and Trade. Amazon Basin, 2672. Andean Region, 2425, 2502. Argentina, 2540, 2558. Brazil, 2635, 2669. Cayapo, 2672. Mexico, 1474, 2361, 2366, 2374, 4463, 4541. Peru, 2502.

Causa R (political party, Venezuela), 3275.

Cavallo, Domingo, 2051.

Caves. Geophysics, 317. Guatemala, 212. Mayas, 212. Mexico, 317, 360. Myths and Mythology, 317. Precolumbian Art, 330. Precolumbian Civilizations, 199. Rites and Ceremonies, 212.

Cayambe, Ecuador (town). Sex Roles, 1022. Women, 1022.

Cayapo (indigenous group). Agriculture, 2672. Cattle Raising and Trade, 2672. Households, 879. Social Structure, 879.
CBI. *See* Caribbean Basin Initiative.
CD-ROMs, 26. Reviews, 10, 12.
Censorship. Nicaragua, 3001. Paraguay, 3609. Uruguay, 3619.
Censuses. Argentina, 14, 16, 5097, 5125, 5127. Bolivia, 2523. Brazil, 5155. Chile, 5084. Databases, 14, 42–43. Ecuador, 1826. Mexico, 17–18, 24, 42–43. Venezuela, 1747.
Center/Periphery Relations. Aztecs, 373. Precolumbian Civilizations, 331, 407, 448.
Central American Common Market. *See* Mercado Común Centroamericano.
Central American Monetary Council. *See* Consejo Monetario Centroamericano.
Central Bank of Trinidad and Tobago, 1667.
Central de Mujeres de Carrasco (Bolivia), 4978.
Central Intelligence Agency. *See* United States. Central Intelligence Agency.
Central Latinoamericana de Trabajadores, 3031.
Central-Local Government Relations. Ecuador, 3221, 3244. Mexico, 2914. Peru, 3392, 4887. Uruguay, 3651.
Central Unica dos Trabalhadores (Brazil), 2110, 3691, 3695.
Central Valley (Costa Rica). *See* Valle Central, Costa Rica (region).
Centro de Estudios Sociales Solidaridad (Peru), 4938.
Centro de Promoción de la Mujer Gregoria Apaza (Bolivia), 4966.
CEPAL. *See* Comisión Económica para América Latina y el Caribe.
Ceramics. Mesoamerica, 189–190. Mexico, 192.
Ceremonies. *See* Rites and Ceremonies.
Cerén Site (El Salvador), 368.
Cerezo Arévalo, Vinicio, 4618.
Cerro de Huistle Site (Mexico). Excavations, 259.
Chaco, Paraguay (region). Indigenous Peoples, 931.
Chaco War (1932–1935), 3597, 3610.
Chalcatzingo Site (Mexico), 284. Rites and Ceremonies, 115. Sex Roles, 115. Social Life and Customs, 115.
Chamí (indigenous group). Ethnobotany, 1007. Folklore, 1007. Land Tenure, 1007.
Chamula. *See* Tzotzil.
Chanca (indigenous group). History, 677.

Chaparé, Bolivia (region). Agricultural Systems, 4981. Coca, 3288, 3307, 3321, 3339, 3842, 4981. Cocaine, 3307. Demography, 3339. Drug Enforcement, 3288, 3307. Drug Traffic, 3288, 3307, 3321, 3842. Environmental Pollution, 4981. Land Tenure, 3339. Peasants, 4981.
Chaquilla, Bolivia (town). Religious Life and Customs, 959.
Charcoal. Puerto Rico, 489.
Chavín (indigenous group). Artifacts, 661. Cultural Development, 661, 680.
Checua Site (Colombia). Precolumbian Land Settlement Patterns, 619.
Chemistry. Archaeology, 325.
Chiapas, Mexico (state). Agricultural Systems, 2356. Archaeological Geography, 358. Archaeology, 106, 388. Artifacts, 360. Chronology, 2828. Coffee Industry and Trade, 758, 2411. Diseases, 2411. Environmental Policy, 1473. Fish and Fisheries, 4456. Forced Labor, 2413. Forced Migration, 737. Guatemalans, 4473, 4520. Guerrillas, 2826, 2828. Historical Demography, 2413. Indigenismo and Indianidad, 2871. Indigenous Peoples, 57, 737, 758, 2356. Indigenous Policy, 724, 737. Insurrections, 57, 2899. Mayas, 768. Migrant Labor, 4473, 4520. Migration, 2413, 4031. Paleoecology, 358. Peasant Movements, 2871. Peasant Uprisings, 2871. Precolumbian Architecture, 233, 380. Precolumbian Civilizations, 106. Precolumbian Land Settlement Patterns, 358. Revolutions and Revolutionary Movements, 2871. Rural Sociology, 758. Social Conflict, 724. Social Marginality, 724. Spanish Conquest, 2413. Stone Implements, 360. Underdevelopment, 2356.
Chicama River Valley (Peru). Artifacts, 670.
Chicanos. *See* Mexican Americans; Mexicans.
Chicha. Bolivia, 4968. Precolumbian Civilizations, 679.
Chichén Itzá Site (Mexico), 76, 121, 370–371.
Chichimecs (indigenous group). Legends, 461. Religious Life and Customs, 461.
Chiefdoms. Ecuador, 648. Mayas, 257. Mesoamerica, 86. Mexico, 161. Nazca, 704. Peru, 683. Precolumbian Civilizations, 86. Venezuela, 161.
Chihuahua, Mexico (state). Agricultural Systems, 743. Caves, 199. Political Systems, 2872. Precolumbian Pottery, 107. Water Supply, 2374.
Child Abuse. Mexico, 4562.

Child Development. Argentina, 5098. Paraguay, 4988.
Childbirth. Brazil, 5161. Caribbean Area, 800.
Children. Argentina, 5108. Brazil, 5169. Central America, 4643. El Salvador, 4641. Employment, 4641, 4879, 4937, 4964, 5169. Jamaica, 4786. Nicaragua, 4669. Paraguay, 4988, 4991. Peru, 4879, 4937. Prostitution, 4991, 5165. Social Policy, 5108.
Chile. Dirección de Inteligencia Nacional, 3422.
Chile. Ministerio de Relaciones Exteriores, 4248.
Chile. Servicio Nacional de la Mujer, 5077.
Chile. Servicio Nacional de Menores, 5011.
Chile Solidarity Campaign (Great Britain), 4264.
Chileans. Paraguay, 3602.
Chiloé, Chile (prov.), 2594.
Chimbote, Peru (city). Single Mothers, 4942.
Chimu (indigenous group). Precolumbian Pottery, 662.
Chinampas. Mexico, 203, 229.
Chinantec (indigenous group). Forced Migration, 719. Messianism, 719.
Chincha Valley (Peru). Precolumbian Architecture, 663.
Chinese Influences. Olmecs, 112. Precolumbian Sculpture, 112.
Chipaya (indigenous group). Bibliography, 949. Ethnic Groups and Ethnicity, 968.
Chiriguano (indigenous group). Bibliography, 929. Cultural Destruction, 930. Cultural Identity, 930. Economic Conditions, 928. History, 929–930. Religion, 929. Social Marginality, 928.
Chocó, Colombia (dept.). Race and Race Relations, 1005, 4830.
Chocó (indigenous group, Brazil). See Shocó.
Chol (indigenous group). Agrarian Reform, 758.
Cholula, Mexico (town). Archaeology, 343.
Chontal (indigenous group). Archaeological Geography, 231.
Choroti (indigenous group). Social Life and Customs, 931.
Christian Base Communities, 4423. Brazil, 3667, 3696, 3719, 5149–5150, 5173, 5176, 5179, 5206. Central America, 4578. Costa Rica, 4683. Mexico, 4528. Women, 3696.
Christian Democracy, 2771. Central America, 2940.
Christianity. Revolutions and Revolutionary Movements, 2996.
Chronology, 481. Central America, 3991. El Salvador, 3991. Falkland Islands, 4202. Guatemala, 3991. Mexico, 2828, 2930. Peru, 2512. Trinidad and Tobago, 3095.
Church History, 3799. Chile, 3474. Southern Cone, 2797.
Church Records. Guatemala, 2343.
Church-State Relations. Brazil, 3706. Chile, 3467, 3474, 3495, 5051, 5075. Costa Rica, 4647. Cuba, 3134, 4692, 4770, 4784. Guatemala, 2976. Mexico, 2830–2831, 2834, 2851, 2887, 2911. Nicaragua, 4601. Paraguay, 3594–3595. Peru, 3410. Uruguay, 4322.
CIA. See United States. Central Intelligence Agency.
Ciénaga, Colombia (town). History, 4810. Popular Culture, 4810.
Cimientos Site. See Los Cimientos Site (Mexico).
Cipoletti, Argentina (town). Social Conflict, 5088.
Cities and Towns, 1117, 2221–2222, 2224, 2238, 2247, 2271, 4440. Andean Region, 2477. Argentina, 3531, 3560. Belize, 349. Bolivia, 4973. Brazil, 2666, 2687. Capitalism, 4393. Chile, 2583. Colombia, 2451. Colonial History, 349, 2248. Decentralization, 2476. Ecuador, 2476, 3244. History, 2241. Mexico, 1497, 1521, 2358, 2405, 2417, 4481. Peru, 2506. Political Culture, 4393. Venezuela, 2434.
City Planning, 2224, 2271. Brazil, 2709, 3698. Chile, 2577, 2605, 4996, 5054, 5058. Colombia, 2451, 3168. Ecuador, 2467, 2476. History, 2241. Honduras, 4667. Mesoamerica, 139. Mexico, 1367, 1506, 2355, 2371–2373, 2391, 2405. Peru, 4947. Puerto Rico, 2311–2312.
Ciudad Juárez, Mexico (city). Election Fraud, 2897. Labor and Laboring Classes, 4464. Maquiladoras, 4464. Pressure Groups, 2872.
Civil-Military Relations, 2763, 2770, 2780, 2800. Argentina, 3496, 3515, 3521, 3527, 3530, 3547, 3549. Bolivia, 3295. Brazil, 3769. Chile, 3428, 3454, 3465, 3477, 3479, 5048. Colombia, 3154, 3177, 3209. Ecuador, 3218, 3230. El Salvador, 2970, 2972. Guatemala, 2983, 4670. Honduras, 2994, 4660, 4684. Mexico, 2810. Nicaragua, 3000. Panama, 3028. Paraguay, 3615–3616. Peru, 1959, 3363, 3370, 3384–3385. South America, 2788, 4173, 4184. Southern Cone, 2794. Suriname, 3093. Uruguay, 3618, 3633, 3653, 3656–3657. Venezuela, 3260.

Civil Rights, 3876. Argentina, 3559. Brazil, 3767. Chile, 3426, 3430. Colombia, 995, 3161, 3180. Cuba, 3133, 4723. El Salvador, 2962. Honduras, 2986. Indigenous Peoples, 995. Mayas, 2981. Paraguay, 3617. Women, 5198.

Civil Service, 2755. Caribbean Area, 3041. Women, 4425.

Civil War. Angola, 4127. Colombia, 3185. El Salvador, 2966, 4075, 4593. Guatemala, 765. Nicaragua, 2997, 4593.

Civilization, 59, 2236, 3852. Brazil, 2648, 5153. France, 5153. Haiti, 4140. Mexico, 2360. Suriname, 2301.

Class Conflict. *See* Social Classes; Social Conflict.

Clergy. El Salvador, 2968. Peru, 3383.

Clientelism, 2733. Brazil, 3731. Colombia, 3178. Ecuador, 3244, 4854–4855. Mexico, 2833. Paraguay, 4987. Peru, 3368. Uruguay, 3642.

Climatology, 2265–2266. Andean Region, 2428, 2566. Argentina, 2541. Brazil, 2671. Chile, 2568. Ecuador, 2465, 2479. Galápagos Islands, 2479. Mexico, 2385–2386, 2390, 2394. Venezuela, 2447.

Clothing and Dress. Bolivia, 4968. Precolumbian Civilizations, 612.

Clothing Industry, 1087.

Coahuila, Mexico (state). Land Tenure, 1378.

Coaiquer. *See* Cuaiquer.

Coal Mining. *See* Minerals and Mining Industry.

Coalición Obrera Campesina Estudiantil del Istmo/COCEI (Mexico), 772.

Coalitions. Argentina, 3562–3563. Uruguay, 3659. Venezuela, 3255.

Coastal Areas. Argentina, 2553. Cuba, 2304. Guadeloupe, 2299.

Cobá Site (Mexico). Stelae, 445.

Coca. Bolivia, 958, 966, 1976, 1979, 1985, 3277, 3279, 3285, 3288, 3293–3294, 3299–3300, 3303, 3307, 3310, 3321, 3338–3339, 3352, 3842, 4231, 4237, 4972, 4981. Colombia, 4819. Cultural History, 3300. Peru, 1049, 1942, 2486, 2495, 3393.

Cocaine, 2753, 3880. Bolivia, 958, 1976, 1979, 1992, 3285, 3293–3294, 3300, 3303, 3307, 3309–3310, 3338, 3345, 3352, 4229, 4237. Colombia, 3197, 3200, 4271, 4823. Peru, 3373. US, 4271.

Cocama (indigenous group). Messianism, 908.

Cocamilla. *See* Cocama.

Cochabamba, Bolivia (city). Commerce, 1984.

Cultural Identity, 957. Markets, 1984. Street Children, 4964.

Cochabamba, Bolivia (dept.). Agriculture, 3299. Coca, 966, 3299, 4972. Historical Geography, 2529. Peasant Movements, 3304. Rural Development, 3299. Social Change, 4972. Social Structure, 956. Trade Unions, 3304.

CODESA. *See* Corporación Costarricense de Desarrollo.

Codices. Azoyú 1, 458. Azoyú 2, 458. Aztecs, 110. Borgia, 410, 416. Dresden, 752. Fejérváry-Mayer 1, 410. Florentine, 260, 415, 417, 438, 739. Humboldt, Fragmento 1, 458. Laud, 422. Mayas, 423, 433. Mendoza, 89, 408, 442. Nuttall, 431. Paris, 433. Ramírez, 411. Tulane, 447. Vindobonensis, 431.

Código federal de instituciones y procedimientos electorales (Mexico), 2825.

Coffee Industry and Trade, 2279. Brazil, 2095, 2668, 2679. Central America, 4677. Colombia, 1781. Cuba, 1690. Dominican Republic, 4754. Ecuador, 1803. Honduras, 1616. Mexico, 758, 761, 1439, 2411, 4500, 4504. Venezuela, 1717, 2446.

Cold War, 2732. Jamaica, 4154.

Colha Site (Belize), 169, 252. Stone Implements, 250.

Colima, Mexico (state). Economic Conditions, 1513. Tombs, 275.

Collor de Mello, Fernando Affonso, 2121, 3702, 3717, 3765–3766, 3779.

Colombia. Ministerio de Relaciones Exteriores, 4276.

Colonial Administration. Dutch Caribbean, 3078–3079. Guadeloupe, 3081. Jamaica, 4719. Puerto Rico, 3084. Suriname, 3090, 3092.

Colonial Architecture. Mexico, 327.

Colonial Art. Ethnography, 510.

Colonial History. Bolivia, 3336. Caribbean Area, 516, 2219. Cities and Towns, 2248. Costa Rica, 468. Diseases, 2254. Haiti, 811. Land Tenure, 2398. Maps and Cartography, 2248–2249. Mexico, 2393. Transportation, 2273. Water Rights, 2393.

Colonization. Amazon Basin, 2692. Belize, 2321. Bolivia, 2522. Brazil, 2651–2652. Chile, 2604. Colombia, 2454. Dominican Republic, 493. Ecuador, 2522. Guatemala, 2344. Haiti, 493. Mexico, 2398. Paraguay, 2697. Quechua, 1061. Southern Cone, 2534. Uruguay, 2619.

Colosio Murrieta, Luis Donaldo, 2937.

Columbus, Christopher, 506, 511, 2261.
Comalcalco Site (Mexico), 231.
Comayagua, Honduras (dept.). Precolumbian Land Settlement Patterns, 246.
COMECON. *See* Council for Mutual Economic Assistance.
Comisión Económica para América Latina y el Caribe, 1116.
Comisión Sudamericana de Paz (Santiago), 4184.
Comité de Madres de Héroes y Mártires (Nicaragua), 4672.
Comité de Unidad Campesina (Guatemala), 4613.
Commerce, 1094, 1137, 1157, 1178, 1197, 1212, 1289, 3846. Argentina, 2055. Bolivia, 1984. Brazil, 2181. Caribbean Area, 1113, 1197, 1289. Chile, 1856, 1877. Cuba, 1686. Databases, 6. Ecuador, 1841, 2471. Electronic Resources, 8. Indigenous Peoples, 971. Internet, 6, 9. Jamaica, 1668. Mexico, 8, 1332, 1342, 1386, 1410, 1457, 1465, 1483, 1509, 1518–1519, 1523, 1535, 3887, 3926, 3931. North America, 1408, 1457, 1464. US, 1410, 1465, 1483, 1518–1519, 1523, 1535, 1668, 3846, 3887.
Commercial Policy. *See* Trade Policy.
Commodities, 1153, 2244. Statistics, 27.
Communal Kitchens. Peru, 4891, 4914.
Communication. Brazil, 2646. Uruguay, 5131.
Communism and Communist Parties, 2760, 2798. Brazil, 3741, 3778. Central America, 2991. Chile, 3453, 3468, 3482. Colombia, 3152. Cuba, 1682, 1685, 3102, 3107, 3109, 3111, 3121, 3134, 3139, 4130. Honduras, 2991. Panama, 3024. Peru, 4300. Uruguay, 3627.
Community Development, 53, 4430. Appropriate Technology, 4825. Argentina, 5098, 5109, 5113, 5126. Bolivia, 4954. Chile, 1869, 1911, 5046–5047. Colombia, 4825, 4919. Ecuador, 1017. Guatemala, 2347. Indigenous Peoples, 1017. Peru, 2488, 2518, 4902, 4904, 4919, 4926. Women, 4966, 5047, 5098.
Compadrazgo. Bolivia, 950. Mexico, 751.
Companhia Ford Industrial do Brasil, 2129.
Companhia Siderúrgica Nacional, 2164, 3777.
Compañía Nacional de Subsistencias Populares—Conasupo (Mexico), 1320.
Competition. Mexico, 1471.
Computers. Brazil, 2157, 2213. Mexico, 1416.

Comuneros. *See* Insurrection of the Comuneros (Colombia, 1781); Insurrection of the Comuneros (Paraguay, 1730–1735).
Concentración de Fuerzas Populares (political party, Ecuador), 3226.
Concertación de Mujeres por la Democracia (Chile), 5077.
Conchupata Site (Bolivia), 565.
Confederação Geral dos Trabalhadores (Brazil), 3695.
Confederación de Nacionalidades Indígenas del Ecuador, 1023, 3228, 3242.
Confederación de Trabajadores de México, 2919, 4471.
Confederación Nacional de Campesinos (Mexico), 4491.
Confederación Nacional de Organizaciones Populares (Mexico), 2867.
Conferência das Classes Trabalhadoras (Brazil), 3695.
Conquistadores. *See* Conquerors.
Consalvi, Simón Alberto, 4333.
Consejo Superior Universitario Centroamericano, 4624.
Conselho Monetário Nacional (Brazil), 2212.
Conservation, 2231, 2234, 2246, 2256, 2259, 2267, 2270, 2280, 2764. Amazon Basin, 2645. Bolivia, 2527, 2533. Brazil, 2643, 2647, 2677. Caribbean Area, 2280, 2302, 2307. Central America, 1540. Chile, 2595. Costa Rica, 2325, 2329, 2333. Forests and Forest Industry, 2272. Indigenous Peoples, 2233, 2267. Journalism, 2786. Mexico, 1368, 2409. Panama, 2353. Peru, 2490–2491. Saint Kitts and Nevis, 2313.
Conservation and Restoration. Archaeological Parks, 393. Peru, 709.
Conservatism. Argentina, 3526. Brazil, 3686, 3745.
Constitutional Conventions. Argentina, 3574. Brazil, 3685, 3736. Guatemala, 4618. Panama, 3023.
Constitutional History, 2770. Argentina, 3497. Brazil, 3685. Costa Rica, 2952. Curaçao, 4730. Ecuador, 3216. Honduras, 2989. Mexico, 2924. Peru, 3390.
Constitutions, 2240. Brazil, 2757, 3706, 3762, 5154. Chile, 2757, 5085. Colombia, 3158–3160, 3163, 3167, 3174–3175, 3180–3181, 3185–3186, 3196. Cuba, 4723. Dutch Caribbean, 3075. Guatemala, 4618. Haiti, 3065. Law and Legislation, 3608. Mexico,

2831, 2887, 2936. Paraguay, 3608. Uruguay, 3624. Venezuela, 3264.
Construction Industry. Brazil, 2194. Peru, 2496. Uruguay, 3658. Venezuela, 1735.
Consumption (economics). Brazil, 2179. Mexico, 1322.
Contact. *See* Cultural Contact.
Contadora. *See* Arias Peace Plan (1987); Grupo de Contadora.
Contadora Support Group. *See* Grupo de los Ocho.
Contras. *See* Counterrevolutionaries.
Contreras, Manuel, 3422.
Convención Nacional de Trabajadores (Uruguay), 3655.
Cookery. Bolivia, 4968. Caribbean Area, 818. French Guiana, 780, 806. Peru, 2490.
Cooperatives. Chile, 5042. Cuba, 4716. Mexico, 4511. Nicaragua, 4666. Peru, 4918. Uruguay, 5132, 5136.
Coordinación Nacional de Productores Agrícolas (Paraguay), 3613.
Coordinadora Regional de Pueblos Indios de Centroamérica, 4579.
Copán Site (Honduras), 67, 153, 208, 216, 257, 291, 331, 334, 400. Archaeoastronomy, 162. Architecture, 66. Cultural Collapse, 403. Demography, 172, 403. Guidebooks, 258. Precolumbian Sculpture, 296. Social Structure, 66.
COPEI (political party, Venezuela), 3247.
Copper Industry and Trade. Precolumbian Civilizations, 255.
Corn, 96, 1093, 1171. Archaeological Dating, 310. Argentina, 351. Mesoamerica, 78, 351. Mexico, 310, 1320, 1423, 2406. Peru, 2520. Political Culture, 679. Precolumbian Trade, 2406.
Corporatism, 4417. Argentina, 3505, 3523. Brazil, 3768. Mexico, 2929. Peru, 1938.
Corruption. Argentina, 3551–3552, 3567. Bolivia, 3323. Caribbean Area, 4085. Chile, 3441. Colombia, 3154. Dominican Republic, 3060. Haiti, 3066. Mexico, 2853.
Corruption in Politics. *See* Political Corruption.
Cosmology. Andean Region, 977. Aztecs, 225, 416, 451. Bororo, 887. Brazil, 870. Incas, 673. Mapuche, 975. Mayas, 404. Mesoamerica, 163, 451. Mexico, 386. Precolumbian Civilizations, 60. Quechua, 1070. The Guianas, 653.
Cost and Standard of Living, 2282. Argentina, 5114. Brazil, 2199. Caribbean Area, 1676. Chile, 2596, 3452. Colombia, 4826. Mexico, 1352, 1371, 1450. Peru, 1927. Uruguay, 2028. Venezuela, 4847.
Costa Rican Development Corporation. *See* Corporación Costarricense de Desarrollo.
Costume and Adornment. Aztecs, 227, 409. Precolumbian Civilizations, 321, 605, 696. Symbolism, 409.
Cotton Industry and Trade. Mexico, 731. Precolumbian Civilizations, 731.
Counterinsurgency. Colombia, 984. El Salvador, 2966. Guatemala, 757, 765, 4610, 4650. Peru, 3360, 3388, 3411.
Counterrevolutionaries. Dominican Republic, 3046. Nicaragua, 2997, 4034, 4680.
Counterrevolutions. Nicaragua, 4034.
Coups d'Etat, 2758. Argentina, 3515, 3521. Bolivia, 3295. Brazil, 3668. Chile, 3465, 3488, 3492. Haiti, 3064. Paraguay, 3594. Peru, 3370, 3389–3390, 3396, 3400, 3402, 3414, 4304, 4319. Suriname, 3086. Trinidad and Tobago, 781, 3100. Uruguay, 3657. Venezuela, 3253.
Courts. Colombia, 3180. Cuba, 4723. Venezuela, 3273.
Crafts. *See* Artisanry.
Credit. El Salvador, 1596.
Creole Languages. Guadeloupe, 831–833. Haiti, 830. Martinique, 833.
Crespi, Carlos, 644.
Crime and Criminals, 4438. Dominican Republic, 4745. Dutch Caribbean, 4765. Honduras, 4644. Panama, 4573. Paraguay, 4993. Uruguayans, 5139. Venezuela, 3263, 3273, 4851. Youth, 4743.
Cronistas. Dominican Republic, 509. Eurocentrism, 509. Spaniards, 509.
Cruz, Francisco da, 908.
Cuaiquer (indigenous group). Acculturation, 914. Colombia, 854. Ecuador, 854. Ethnography, 1008. Parks and Reserves, 854.
Cuba. Ministerio del Interior, 4776.
Cuban Missile Crisis (1962), 4098, 4106–4108, 4115, 4120.
Cuban Revolution (1959), 4769.
Cubans. Puerto Rico, 4726. US, 4124, 4691, 4703, 4726, 4734, 4767.
Cuenca, Ecuador (city). Cultural Identity, 1019.
Cuernavaca, Mexico (city). Domestics, 4530.
Cuervo Gómez, Manuel Antonio, 3054.
Cuexcomate Site (Mexico). Archaeological Dating, 375.
Cuiva. *See* Cuiba.
Culiacán, Mexico (city). Archaeology, 223. Excavations, 223.

Culina (indigenous group). Missionaries, 882. Sex Roles, 883. Shamanism, 883.
Cults. Argentina, 5093. Aztecs, 339. Ball Games, 462. Caribbean Area, 835. Congresses, 835. Mayas, 324. Mexico, 4566. Peru, 4910, 4929. Precolumbian Civilizations, 452. Venezuela, 4848. Warfare, 452.
Cultural Adaptation. *See* Acculturation.
Cultural Assimilation. *See* Acculturation.
Cultural Collapse. Argentina, 942. Colombia, 626. Ecuador, 626. Mayas, 151, 169. Theory, 151. Tiwanaku Culture, 566, 569.
Cultural Contact. Barbados/US, 795. Brazil, 588. Brazil/Japan, 3804. Carib/Europeans, 817. Caribbean Area, 796, 817. Dominican Republic, 509. Mayas/Europeans, 454. Mayas/Spaniards, 327, 350. Precolumbian Civilizations, 588. Precolumbian Civilizations/Africans, 54. Precolumbian Civilizations/Asians, 54. Precolumbian Civilizations/Europeans, 54, 527. Precolumbian Civilizations/Spaniards, 509, 511–512.
Cultural Destruction, 4442. Aymara, 948. Chiriguano, 930. Foreign Influences, 152. Mesoamerica, 152. Shuar, 916.
Cultural Development, 4422. Andean Region, 530. Barbados, 3074. Caribbean Area, 4697. Chile, 5016, 5021, 5030. Ecuador, 4856. Education, 4856. Mayas, 166, 205. Mesoamerica, 125, 175, 200. Opata, 214. Peru, 680. Precolumbian Civilizations, 55–56, 125, 331, 490, 541, 710, 717. Puerto Rico, 3085, 4791. Suriname, 2303. Yecuana, 895.
Cultural Geography, 2236. Andean Region, 2426. Brazil, 5186. Colombia, 4813–4814. Mexico, 271. Nicaragua, 4622. Peru, 2489.
Cultural History, 2229, 2277, 4422. Andean Region, 2425. Aymara, 962. Aztecs, 88, 128. Caribbean Area, 503–504. Chile, 5021, 5030. Coca, 3300. Colombia, 4814. Costa Rica, 467. Cuba, 799. French Guiana, 798. Goajiro, 903. Indigenous Peoples, 85. Linguistics, 59. Mayas, 71. Mesoamerica, 113. Mexico, 92. Peru, 1040, 1054. Puerto Rico, 4794. Trinidad and Tobago, 849. Uruguay, 5137. Venezuela, 4839.
Cultural Identity, 53, 779, 4442. Andean Region, 1029. Apalakiri, 865. Argentina, 5106, 5129. Aztecs, 424. Blacks, 5180. Bolivia, 952–953, 957, 3355, 4979. Brazil, 5174, 5180, 5186. Caribbean Area, 796, 824. Chile, 3434, 5007, 5052–5053. Chiriguano, 930. Colombia, 989, 1005, 3172, 4814. 4830. Cuba, 4762. Dominican Republic, 839, 4729. East Indians, 793. Ecuador, 1010, 1018–1019, 4856, 4870. Evangelicalism, 1018. Guadeloupe, 831–832. Guatemala, 769–771. Haiti, 830, 3067. Haitians, 4711. Indigenous Peoples, 735, 852, 856, 860, 968, 4856. Jamaica, 840. Mayas, 740–741, 769–771. Mexican-American Border Region, 2358. Mexico, 772, 4497. Panama, 3017, 4592. Peasants, 1064, 4945. Peru, 1030, 1039, 1045, 1054, 1057, 4895, 4930–4931, 4945. Puerto Rico, 3085, 4733, 4759. Saint Kitts and Nevis, 821. Saint Martin, 4779. Saint Vincent, 850. Trinidad and Tobago, 781, 808, 829, 844–845, 4742. Tucanoan, 899. Urban Areas, 5106. US, 795. Venezuela, 712. West Indians, 4748. Women, 5052–5053. Youth, 4895, 5129.
Cultural Policy. Barbados, 3074. Ecuador, 4290. Uruguay, 5137.
Cultural Relations. Brazil/Japan, 4377, 4391. British Caribbean/Latin America, 4157.
Culture. Aymara, 981. Caribbean Area, 810.
Cumanagoto. *See* Cumana.
Curaçaoans. The Netherlands, 4743. Youth, 4743.
Curripaco (indigenous group). Myths and Mythology, 896. Rites and Ceremonies, 896–897. Songs, 896–897.
Cusco. *See* Cuzco.
Customary Law. Antigua and Barbuda, 812. Colombia, 988.
Customs. Ecuador, 1794. Mexico, 1493. Trinidad and Tobago, 781.
Cuzco, Peru (city). City Planning, 4947. Demography, 2511. Earthquakes, 2511. Historical Geography, 2511. Population Growth, 2511.
Cuzco, Peru (dept.). Agricultural Development, 2520. Agriculture, 1034. Community Development, 2518. Corn, 2520. Development Projects, 1963. Farms, 2518. Land Reform, 2515. Peasants, 1963. Potatoes, 2520. Privatization, 2518. Small Business, 1963.
Cyclones. *See* Hurricanes.
Czamanski, Stanislaw, 1525.
Daily Report: Latin America, 50.
Dairy Industry. Peru, 1962.
Dance. Aztecs, 164. Mayas, 418. Mesoamerica, 164, 198. Peru, 1045. Precolumbian Civilizations, 164. Wamani, 1045.
Darien, Panama (prov.). Rural Development, 747.
Database Searching, 4.
Databases, 9, 26. Agricultural Geography, 43. Agriculture, 43. Archives, US, 38. Argentina, 14. Atlases, 15. Censuses, 14, 42–43.

Commerce, 6. Demography, 36, 47. Developing Countries, 48. Directories, 1–2, 11. Economic Conditions, 30, 33, 36, 39. Economic Development, 39, 44. Economic History, 37, 48. Economic Indicators, 31, 48–49. Ejidos, 43. Employment, 32. External Debt, 48. Finance, 33. International Finance, 44. International Relations, 29. International Trade, 45, 51. Law and Legislation, 29. Literature, 36. Mass Media, 34–35, 50. Mexico, 15, 30, 35–37, 40–43, 46. Military, 31. National Security, 31. Newspapers, 34–35, 50. Organisation for Economic Cooperation and Development (OECD), 39. Philosophy, 36. Political Conditions, 31, 33. Political Economy, 37. Political Science, 29. Politicians, 40. Public Opinion, 41. Reviews, 10–12. Social Indicators, 48. Social Sciences, 2–3, 28. Statistics, 39, 44, 49. Tariffs, 45. Treaties, 46. United Nations, 1. Women, 47. World Bank, 48.

Daule River Region (Ecuador). Chiefdoms, 648.

Death. Brazil, 5190.

Death Squads, 4421.

Debt. *See* External Debt; Public Debt.

Debt Conversion. *See* Debt Relief.

Debt Crisis, 1082, 1084, 1117, 1133, 1147, 1151–1152, 1167, 1172, 1236, 1246, 1256, 1258, 1263, 1271, 1281–1282, 1285, 1287, 1290, 1298, 1306. Costa Rica, 1256. Jamaica, 1673. Mexico, 1313, 1333, 1336, 1347, 1389, 1399, 1414, 1451, 1456, 1490, 1495, 1531. Peru, 1960.

Debt Relief, 1080, 1105, 1133, 1256, 1287, 1296, 3795, 3798, 3834, 3854. Argentina, 2065. Bolivia, 1996. Chile, 1867, 1896. Costa Rica, 1256. Dominican Republic, 1659. Peru, 4298.

Decentralization, 1100, 1130, 2238, 2255. Amazon Basin, 2662. Argentina, 2555, 3497, 3570. Bolivia, 2662, 3292, 3297, 3334, 3343. Brazil, 2662, 2716, 3762. Chile, 976, 2602, 5014. Cities and Towns, 2476. Colombia, 3169, 3171, 4815. Costa Rica, 2954. Ecuador, 2476. Mexico, 1321, 1538, 2400, 2824, 2849, 2915, 4469, 4495. Nicaragua, 3009. Peru, 1934, 3374, 3392. Uruguay, 2010. Venezuela, 1728, 3274.

Decolonization. Dutch Caribbean, 3078–3079. Suriname, 3092.

Decorative Arts. Suriname, 828.

Defectors. Cuba, 3144.

Defense Budgets, 1267, 2727. Peru, 4311–4312.

Defense Industries. Argentina, 1215, 4192. Brazil, 1215, 3692, 4379. Caribbean Area, 4102. Chile, 1215. Ecuador, 1786. Military, 3692. Peru, 4311. US, 4102.

Deforestation, 2276. Amazon Basin, 2191, 2233, 2475, 2673, 2681, 2692, 2698, 2721. Brazil, 2119, 2191, 2661, 2667, 2680–2681, 2722. Central America, 1552, 2317. Chile, 2579. Colombia, 614. Ecuador, 2475, 2482. Mexico, 727, 753, 2317. Paleoecology, 614. Peru, 2486, 2490, 2495–2496. Suriname, 2287, 2292.

Deities. Aztecs, 141, 154, 416, 431, 436. Mesoamerica, 316, 422, 431. Precolumbian Civilizations, 422. Precolumbian Sculpture, 316.

Democracy, 1134, 1237, 2730, 2745–2746, 2750, 2754, 2759–2760, 2767, 2775, 2792–2793, 2796, 2799, 2806, 3827, 4406. Andean Region, 2777. Argentina, 3501. Bolivia, 3286, 3297–3298, 3313–3315, 3317, 3322, 3324, 3326–3327, 3351, 3355, 4979. Brazil, 2136, 3738, 3770. British Caribbean, 3073. Caribbean Area, 3033. Central America, 2942, 4043. Chile, 1849, 1885, 1919, 3418, 3421, 3429, 3447–3449, 3469, 3479, 3483–3484, 5020, 5029. Colombia, 3166, 3171, 3196, 3211. Costa Rica, 2954, 2956. Ecuador, 3216, 3238, 3241. El Salvador, 2964, 4022. Guatemala, 2973, 2975, 2982. Honduras, 2987. Mexico, 1458, 2849, 2860, 2886. Nicaragua, 3013. Panama, 3018, 3023, 3027. Paraguay, 3615. Peru, 1959, 3368, 3370, 3378, 3389–3390, 3392, 3398, 3402–3403, 3406, 3414. South America, 2763. Southern Cone, 2535. Uruguay, 3625, 3635, 3652. Venezuela, 3248, 3253, 3257–3258, 3272.

Democratization, 1263, 2733–2735, 2746, 2749, 2765, 2769, 2772, 2778, 2781–2782, 2791, 2795, 4408, 4844. Argentina, 2736, 3505, 3558, 3576, 4386. Bolivia, 2772, 3219, 3283, 3313, 3316, 3325–3327, 3342, 3346. Brazil, 2736, 2757, 3672, 3712, 3715, 3719, 3749, 3772, 3774, 4386, 5152, 5196. Catholic Church, 5075. Central America, 4060, 4679. Chile, 970, 1885, 1892, 2757, 3421, 3434, 3437, 3447–3449, 3454, 3458, 3463, 3469, 3471, 3477–3479, 3486, 3491–3493, 4246, 4998, 5004, 5016, 5021–5022, 5026, 5031, 5048, 5060, 5075. Colombia, 993. Ecuador, 3212, 3219, 3230–3231. El Salvador, 2972, 4022. Guatemala, 4618. Haiti, 3065. Honduras, 2992, 4654, 4660. Industrialists, 3749. Mexico, 2821, 2838, 2843, 2858,

2876, 2890, 2905, 2913, 2915, 3896. Military, 2743, 3653, 3657, 5048. Nicaragua, 3004, 3011, 4598. Panama, 3980, 4002, 4007. Paraguay, 3591–3592, 3594, 3599, 3605, 3611, 4293, 4295, 4987. Peru, 1934, 3219, 3362, 3378, 3395, 3406. Political Parties, 2743. Public Opinion, 2743. Sociology, 4435. South America, 2763, 2788. Southern Cone, 2752. Suriname, 3093. Uruguay, 3626, 3653, 3656–3657, 4327. Venezuela, 3247, 3261, 3264, 3267, 3274, 4838.

Demography, 1254, 1274, 2243, 4431. Amazon Basin, 857, 861, 2692. Argentina, 5097, 5125, 5127. Bolivia, 3339, 4965. Brazil, 2184, 2201, 2696, 5155–5156, 5190, 5199. Caribbean Area, 774, 4751. Central America, 1565, 4400. Chile, 2584–2585, 2587, 5012–5013, 5028, 5084. Colombia, 4811. Cuba, 4717, 4722. Databases, 36, 47. Dominican Republic, 4727, 4773. Ecuador, 4872. French Caribbean, 4708. French Guiana, 4708. Guatemala, 2342, 4031. Indigenous Peoples, 869. Jamaica, 4741. Mayas, 172, 268, 403. Mesoamerica, 147, 172, 268, 302. Mexico, 23–24, 1352, 1386, 1398, 1434, 1476, 2364, 2389, 2405, 2407, 3900, 4470, 4474, 4516, 4519, 4534, 4537, 4539, 4543. Peru, 2508, 2511, 4912, 4925, 4944. Precolumbian Civilizations, 147, 302, 381. Refugees, 4444. Sipibo, 917. Theory, 268. Venezuela, 2445. Women, 4811. Yanomamo, 869.

Dependency, 2223, 2244, 2730, 2773, 3835, 3848, 3858. Bolivia, 3295. Brazil, 4384. Indigenous Peoples, 905. Mexico, 4501, 4541. Puerto Rico, 4733. South America, 4181.

Depressions (economic), 2263. Brazil, 2153.

Desana (indigenous group). Horticulture, 885.

Desaparecidos. *See* Disappeared Persons; Missing Persons.

Description and Travel. Belize, 104. Costa Rica, 2337. Cuba, 2304. Ecuador, 1021. French Guiana, 2309. Honduras, 2349. Martinique, 2290. Peru, 2512. Suriname, 825, 3090, 4696.

Desertification. Argentina, 2550.

Deserts. Chile, 607. Mexico, 2374, 2382.

Despradel, Fidelio, 3050.

Devaluation of Currency. *See* Exchange Rates.

Developing Countries. Databases, 48. Revolutions and Revolutionary Movements, 2739.

Development, 4408. Amazon Basin, 861, 3689. Argentina, 3516. Bolivia, 4230. Brazil, 3711, 5166. Economic Theory, 3689. Peru, 4903. Trinidad and Tobago, 4749. Women, 4433, 4749.

Development Projects, 1114, 1187, 2259, 4410. Agriculture, 1220. Amazon Basin, 875. Andean Region, 1955. Argentina, 5126. Bolivia, 3303, 4231. Brazil, 881, 2175, 2207, 2651. Chile, 1854. Costa Rica, 1587. Guatemala, 4602. Honduras, 4611. Mexico, 2412, 2859. Nongovernmental Organizations, 1955. Peasants, 4938. Peru, 1933, 1963, 4909, 4938–4939. Women, 4909.

Dictators. Chile, 3420, 3431, 3476. Nicaragua, 3015. Paraguay, 3596.

Dictatorships, 2770, 2800. Chile, 3428, 3460. Haiti, 3063, 3068.

Dictionaries. Djuka, 780. Peru, 1053.

Dieffenthaller, Ray Edwin, 3099.

Diglossia. *See* Code Switching.

Diplomatic History. Argentina, 4198, 4205, 4216, 4224. Bolivia, 4227–4228, 4234. Brazil, 4353, 4357–4358, 4363, 4367, 4383, 4385–4386, 4388. Caribbean Area, 4147, 4151. Chile, 4254, 4261. Colombia, 4103, 4276. Costa Rica, 4071. Cuba, 4103–4104, 4119. Dominican Republic, 4134–4135, 4142. Great Britain, 4332. Guyana, 4159. Haiti, 4142. Honduras, 4044. Mexico, 3902, 3950. Peru, 4318. Sweden, 3868. Uruguay, 4328. US, 3782, 3989, 4037, 4107, 4130, 4134, 4332. USSR, 4107. Venezuela, 4104, 4337.

Diplomats. Bolivia, 4236. Brazil, 4353. Chile, 4248. Colombia, 4270. Peru, 4318. USSR, 4115.

Directories. Databases, 1–2, 11. Social Sciences, 2. United Nations, 1.

Dirty War (Argentina, 1976–1983), 3496, 3499, 3514, 3541, 3583.

Disappeared Persons. Argentina, 3503, 3508, 3510. Central America, 2941. Chile, 3430. Ecuador, 3220. Honduras, 2988.

Disarmament. *See* Arms Control.

Disaster Relief. Jamaica, 2291. Peru, 2514.

Discourse Analysis. Andean Region, 1031. Argentina, 5123. Brazil, 874. Curripaco, 896. Peru, 1057. Venezuela, 891. Xavante, 874.

Discovery and Exploration. Brazil, 588. Precolumbian Civilizations, 54.

Discrimination. Chile, 4997, 5068. Labor Market, 4997. Peru, 4907.

Diseases, 2281. Brazil, 5188. Colombia, 4807. Colonial History, 2254. Mexico, 2411. Precolumbian Civilizations, 634–635. Viceroyalty of New Spain, 2415.

Disinvestment. Jamaica, 1678.
Displaced Persons. Central America, 4542. Colombia, 3204. El Salvador, 4664. Peru, 3375.
Dissenters. Cuba, 4740.
Dissertations and Theses. Ecuador, 1825.
Distribution of Wealth. *See* Income Distribution.
Distrito Federal, Mexico. Agricultural Systems, 203. Excavations, 204. Households, 4507. Poor, 4507. Precolumbian Architecture, 204. Salvage Archaeology, 204. Urban Sociology, 4507.
Djuka (Surinamese people). Dictionaries, 780. Social Life and Customs, 805.
Documentation Centers. *See* Libraries.
Dollarization, 1185, 1266, 1985. Bolivia, 1974.
Domestic Animals. Precolumbian Civilizations, 526.
Domestic Violence. *See* Family Violence.
Domestics. Bolivia, 952, 4959. Colombia, 4817. Mexico, 4530. Paraguay, 4986. Peru, 4915. Social Movements, 4915.
Dominicans (people). Puerto Rico, 4726. US, 4726, 4729.
Dos Pilas Site (Guatemala), 398–399, 413, 425.
Dreams. Aguaruna, 911.
Dresden Codex. *See* Codices. Dresden.
Droughts. Precolumbian Civilizations, 569.
Drug Abuse, 3830. Argentina, 5110. Dominican Republic, 4796. Jamaica, 4790. Uruguay, 5130. US, 4161.
Drug Enforcement, 2753, 3296, 3784, 3790, 3829–3830, 3842. Andean Region, 4161, 4182, 4186. Bolivia, 3277, 3279–3280, 3283, 3285, 3288–3289, 3293–3294, 3303, 3306–3307, 3309, 3311–3312, 3314, 3321, 3338, 3352, 3842, 4229, 4231, 4237, 4972. Caribbean Area, 4146. Colombia, 1760, 3156, 3198, 3200, 4266–4267. Costa Rica, 2953. Cuba, 3115. Ecuador, 3214. Mexico, 3907. Military, 3819. Paraguay, 3614. South America, 4161. Treaties, 3293. US, 3214, 3293, 3296, 3311–3312, 3314, 3321, 3790, 3842, 3880, 3907, 4146, 4182, 4186, 4266, 4271, 4972.
Drug Traffic, 2753, 3784, 3790, 3829–3830. Andean Region, 4161, 4452. Argentina, 4212. Bolivia, 958, 966, 1992, 3277, 3283, 3285, 3288, 3293–3294, 3306–3307, 3309–3311, 3314, 3321, 3338, 3345, 3352, 3355, 3842, 4229. Capital, 3826. Caribbean Area, 4095, 4146, 4148. Colombia, 987, 1760, 3156, 3172, 3185, 3197–3200, 3210, 4266–4267, 4271–4272, 4823. Costa Rica, 2953. Dominican Republic, 3060. Ecuador, 3214. Jamaica, 4790. Mexico, 3896, 3907, 3924–3925, 3943–3945. Panama, 4024. Paraguay, 3614. Peru, 1942, 1958, 3397, 4316. Statistics, 3199. US, 3214, 3306, 3875, 3907, 3924–3925, 3943, 3945, 4146, 4271.
Drugs and Drug Trade. *See* Drug Abuse; Drug Enforcement; Drug Traffic; Drug Utilization; Pharmaceutical Industry.
Durán, Diego, 411.
Durán-Ballén, Sixto, 3241.
Dutch Caribbean. Archaeological Surveys, 521. Archaeology, 650. Autonomy, 3077. Colonial Administration, 3078–3079. Constitutions, 3075. Crime and Criminals, 4765. Decolonization, 3078–3079. Elections, 3080. Political Culture, 3078. Political Development, 3076, 3078–3079. Political History, 3080. Politicians, 3080. Public Administration, 3075, 3077. Rastafarian Movement, 4732.
Dutra, Eurico Gaspar, 4375.
Duvalier, François, 3063, 3068.
Duvalier, Jean-Claude, 3062–3063, 3068.
Dzibilnocac Site (Mexico), 437.
Earthquakes. Peru, 2511, 2514, 4902.
East Indians. Congresses, 791. Guadeloupe, 831–832. Guyana, 842. Migration, 791. Sex Roles, 4780. Social Conditions, 791. Trinidad and Tobago, 793, 804, 808, 829, 834, 845, 4712, 4780. US, 791.
Easter Island. Maps and Cartography, 2601.
ECLAC. *See* Comisión Económica para América Latina y el Caribe.
Ecological Crisis. *See* Environmental Protection.
Ecology, 2232, 2268, 2281. Amazon Basin, 4311, 5164, 5181. Andean Region, 2418, 2433. Brazil, 4381. Central America, 4609. Colombia, 2452. Costa Rica, 2330, 2334, 2340. Economic Development, 5164. Mexico, 1367, 1530. Panama, 4010. South America, 2431. Tourism, 2330. Venezuela, 2448.
Economic Anthropology, 932. Mesoamerica, 405. Peru, 1066. Teotihuacán, 405.
Economic Assistance, 3837.
Economic Assistance, European. Central America, 1561.
Economic Assistance, US, 1227, 1299, 3815. Caribbean Area, 1638, 4079, 4102. Central

America, 3971, 4068. Costa Rica, 1586, 4671. El Salvador, 1603, 4075. Guatemala, 4001. Nicaragua, 1624.

Economic Commission for Latin America and the Caribbean. *See* Comisión Económica para América Latina y el Caribe.

Economic Conditions, 1079, 1089, 1110, 1125, 1131, 1134–1135, 1142, 1145, 1147, 1152, 1155, 1176, 1187, 1202, 1209, 1213–1214, 1238, 1258, 1272, 1283, 1289, 1297, 1336, 1346, 2250, 2252, 2787, 3811, 3826, 3836. Amazon Basin, 861, 2632. Andean Region, 1048, 1937. Argentina, 2034, 2037–2038, 2040, 2050, 2057, 2062, 2074–2075, 2079, 2091. Barbados, 2286. Bolivia, 1975, 1981, 1986, 2001, 3281. Brazil, 2102, 2109, 2118, 2121, 2127, 2132–2133, 2148, 2182, 2187, 2189–2190, 2201, 3761, 4355, 4384. British Caribbean, 3035. Caribbean Area, 1113, 1214, 1289, 1640–1641, 1651, 1665, 1674, 1705, 2296, 2804, 4093–4094. Central America, 1547, 1641, 2804, 2938, 3973, 4009, 4068, 4074. Chile, 1847, 1856, 1860, 1863, 1869, 1872–1877, 1883, 1910, 1914, 3424, 3460–3461, 3480, 4258. Chiriguano, 928. Colombia, 1761–1763, 1766–1767, 1778, 3156, 3189, 3194, 3198, 3203. Costa Rica, 1572–1573, 1575, 1580, 2332. Cuba, 1683, 1685, 1694–1695, 1699, 1702, 1705–1706, 1713, 1715, 2315, 3104, 3106, 3108, 3118–3120, 3126, 3129–3130, 3137, 3141, 3143, 3150–3151, 4126, 4782. Databases, 30, 33, 36, 39. Dominican Republic, 3059. East Asia, 1187. Ecuador, 1787, 1804, 1806, 1815, 1825, 1839, 3223, 3244–3245. El Salvador, 1603. Electronic Resources, 12. Guyana, 1671. Haiti, 4139. Jamaica, 1668. Japan, 1504, 2062. Mexican-American Border Region, 1317, 1360, 2367, 3898. Mexico, 19, 22, 30, 1308, 1311, 1314, 1316, 1324, 1330, 1336, 1344, 1372–1373, 1386, 1392, 1394, 1398, 1406, 1414, 1422, 1435, 1451–1452, 1454–1456, 1470, 1479, 1482, 1490, 1494, 1498, 1504, 1508–1509, 1513, 1515, 1525, 2356, 2376, 2402, 2829, 2862, 2873, 2888, 2934. Nicaragua, 1623–1624, 2352, 3004. Panama, 3016. Paraguay, 2006. Peru, 1034, 1043, 1940, 1944, 1948, 1950, 1952, 1957, 1959, 1968, 3367, 3386, 3398, 3401. Precolumbian Civilizations, 382. Puerto Rico, 797, 4724. Statistics, 1235. Suriname, 2303, 2314. Trinidad and Tobago, 848. Uruguay, 1572, 3640. Venezuela, 1727, 1732, 1738, 1740, 1748, 3276, 4834. West Indies, 2296.

Economic Destabilization. *See* Economic Stabilization.

Economic Development, 1091–1092, 1102, 1106, 1116, 1124, 1134, 1143, 1155, 1158, 1162, 1182, 1192, 1205, 1236, 1248, 1254–1255, 1263, 1265, 1276, 1280, 1282–1283, 1288, 1300, 1307, 2076, 2237, 2244, 2270, 2277, 2282, 2792, 3811, 3825, 3837, 3843, 3848, 3872, 4167. Amazon Basin, 2656, 2692, 5164. Andean Pact, 1937. Andean Region, 2515. Argentina, 2092, 3587. Asia, 1205. Barbados, 2286. Bolivia, 958, 1977, 1988–1989, 3303, 4302. Brazil, 2109, 2126, 2131, 2136, 2138, 2140–2142, 2147, 2161, 2168, 2172, 2192–2193, 2200, 2212, 4368, 4372, 5171. Caribbean Area, 1638–1639, 1642–1643, 1645, 1648, 1650, 1652–1653, 1665, 1676, 1679, 2237, 3042, 4079, 4082, 4088, 4102, 4752. Central America, 1553, 4009, 4585. Chile, 1853, 1868, 1892, 1900, 1907, 1917, 1919, 2580, 2593, 2602, 2615, 3421, 3432, 3435–3436, 3480. Colombia, 3186. Costa Rica, 2338. Cuba, 1698, 1704, 3122, 3140. Databases, 39, 44. Dominican Republic, 1658–1659. Ecology, 5164. Ecuador, 1792, 1796, 1832, 1835, 3227. El Salvador, 1591, 1593. Environmental Protection, 3629. Guatemala, 3977. Guyana, 1666. Haiti, 2295. Honduras, 2351. Indigenous Peoples, 881. Labor Market, 4432. Mexico, 1314, 1355, 1405, 1418, 1432, 1481, 1515, 1534, 2397, 2858–2859, 2880, 2884, 2902, 2908, 3963, 4548, 4557, 4567. Nicaragua, 1627. Peru, 1945, 1952, 1958, 1965, 3397, 4302. Uruguay, 2014, 2018, 2029, 3629. Women, 933, 4806.

Economic Development Projects. *See* Development Projects.

Economic Forecasting, 1091, 1198, 1207, 1284. Argentina, 2034. Brazil, 2146, 2176, 2186, 2188. Caribbean Area, 1647, 4092. Cuba, 1682, 1706–1707, 3109, 3132, 3143, 3148. Ecuador, 1801, 1833. Mexico, 1388, 1468, 1534, 2908, 4519.

Economic Geography, 1274. Honduras, 2350. Mexico, 2376. Nicaragua, 2352. Paraguay, 2004.

Economic Growth. *See* Economic Development.

Economic History, 1124, 1162, 1209, 1223, 1230, 1262, 1280, 1299, 1306–1307, 1549, 2229, 3869. Argentina, 2089, 3516, 3575, 4206, 4213, 4217, 5112. Bibliography, 1174. Bolivia, 1973, 3336. Brazil, 2106–2107,

2124, 2131, 2138, 2144, 2153, 3693, 4355. Caribbean Area, 4158. Central America, 4585, 4599, 4689. Chile, 1866, 1907, 1910, 2615, 3480, 5031. Colombia, 1767, 1776, 3203. Costa Rica, 4639. Cuba, 1712, 4717. Databases, 37, 48. Dominica, 777. Dominican Republic, 3059. Ecuador, 1796, 1805, 1811, 1813, 1824–1825, 1830, 1836–1837. Haiti, 4738. Honduras, 2351. Jamaica, 4741. Mexico, 23, 30, 37, 1483, 1494, 4548. North America, 1488. Peru, 1945, 3404, 4903. Trinidad and Tobago, 3099. Uruguay, 2014, 2018. Venezuela, 1720, 1731, 1752, 4850.

Economic Indicators. Cuba, 1698. Databases, 31, 48–49. Mexico, 17–19, 22–23, 1468.

Economic Integration, 1099, 1132, 1137, 1141, 1148, 1157, 1184, 1189, 1196, 1231, 1279, 1302, 1417, 2024, 2110, 3781, 3792, 3803, 3821, 3832, 3834–3835, 3841, 3845, 3849, 3860, 4212, 4359. Amazon Basin, 2692. Andean Region, 1937, 4170, 4235, 4269. Argentina/Latin America, 4190. Brazil, 4378. British Caribbean, 3035, 3038. Caribbean Area, 1113, 1184, 1190, 1644–1645, 1648–1649, 1653, 1655, 4080, 4082, 4155–4156, 4158. Central America, 1190, 1543, 1547, 1556, 1558, 1566, 3974, 3979, 4061. Central America/Mexico, 3967. Colombia/Venezuela, 4273. Costa Rica, 1583. Cuba, 4114, 4125–4126. Mexico, 767, 1372, 3903. North America, 1361, 1382, 1387, 1448, 1457, 1488, 1523, 1537, 3894, 3896, 3913, 3915, 3921, 3928, 3946, 3961. South America, 1216, 2011, 2024, 2096, 2110, 4166, 4176–4177. Southern Cone, 1210, 2139, 3781, 4176, 4187, 4321. Western Hemisphere, 3827.

Economic Liberalization, 1084, 1109, 1117, 1128–1129, 1178, 1229, 1263, 1272, 1293, 2796, 2799, 3874. Argentina, 1229, 2050, 3500, 3506, 3523, 3571, 3578–3579. Bolivia, 3219. Brazil, 2105, 3745, 3778. Caribbean Area, 4089. Central America, 1563–1564, 1568. Chile, 1873–1874, 1877, 1896, 1901, 1903, 1922, 3432, 3435, 3485, 5017–5018, 5056, 5080. Colombia, 1771. Costa Rica, 1589, 2946, 2948. Ecuador, 1803, 3219, 3241. Mexico, 1309, 1425, 1481, 1495, 1519, 1527–1528, 2837, 2895, 2907, 2928, 2930, 3946, 3960, 4554. Panama, 1636. Peru, 3219, 3414, 4301. Uruguay, 2021–2022, 2030, 3630, 3641.

Economic Models, 1071, 1077–1078, 1080–1081, 1120, 1126, 1147, 1179, 1185, 1230, 1236, 1263, 1306, 3821, 4844. Argentina, 2054, 2064, 2068, 2092, 4206. Belize, 1570. Bolivia, 1976–1977, 1985. Brazil, 2143, 2158, 2179, 2191, 2195, 2208–2209, 3711, 3745. Canada, 1492. Caribbean Area, 1652, 3042. Central America, 1544–1545, 1555. Chile, 1863, 1922, 5017. Colombia, 1769, 1776–1777. Costa Rica, 1555, 1574, 1577, 1581, 1584, 1588–1589. Cuba, 1703, 1713, 3138. East Asia, 1200. Ecuador, 1788, 1809, 1812. El Salvador, 1555, 1595, 1599, 1601–1602. Guatemala, 1555, 1606, 1608, 1611, 1613. Guyana, 1671. Mayas, 133. Mexico, 1322, 1335, 1475, 1484, 1492, 1507, 1525, 1537. Peru, 4903. Precolumbian Civilizations, 177. Uruguay, 2007, 2017, 4321. Venezuela, 4846.

Economic Planning. *See* Economic Policy.

Economic Policy, 1078, 1084, 1088–1089, 1100, 1103, 1115, 1125, 1131, 1133, 1135, 1138, 1143–1146, 1155, 1159, 1165, 1168, 1172, 1178, 1187, 1191, 1196, 1200, 1209, 1219, 1221, 1230, 1233–1234, 1238, 1241, 1247–1248, 1257, 1271–1272, 1281–1284, 1290, 1293, 1299, 1304, 1417, 2046, 2232, 2727, 2767, 2771, 3806, 3818, 3848, 3858, 3874. Amazon Basin, 2493. Andean Region, 1173. Argentina, 2037–2040, 2051, 2057, 2059, 2062, 2068, 2074, 2077, 2087, 3500, 3504, 3506, 3512, 3523, 3546, 3552, 3555–3556, 3568, 3578–3579, 3587, 4206. Bolivia, 1969, 1972–1973, 1979, 1981, 1983, 1986–1989, 1991, 1993, 1996–2000, 3284, 3287, 3291, 3332, 4235. Brazil, 1188, 2097, 2102, 2105, 2117, 2121, 2126, 2131–2133, 2141, 2143, 2148, 2150–2151, 2165, 2179, 2182, 2190, 2192, 2207, 2209, 2212, 2215, 2636, 3684, 3751, 3757, 4378. Caribbean Area, 1165, 1284, 1638, 1640–1641, 1649, 1651, 1653, 1675, 1679, 4078–4079, 4088–4090, 4092. Central America, 1539–1541, 1546–1547, 1553–1554, 1641, 3979, 4619. Chile, 1854, 1858, 1870, 1873–1876, 1880, 1890, 1892, 1897, 1904, 1910, 1918, 3420, 3435, 3441, 3443, 3485–3486, 5017, 5031, 5041, 5056, 5080. Colombia, 1761–1763, 1766, 1781, 1783. Costa Rica, 1542, 1573, 1578, 1584, 1586, 2954, 4589, 4606, 4626, 4639, 4645, 4671. Cuba, 1685, 1687, 1689, 1693–1695, 1697, 1699, 1702, 1709, 1712–1715, 3140, 3149, 4125. Developing Countries, 2739. Dominican Republic, 1662–1663, 2294. East Asia, 1187, 1413. Ecuador, 1788–1789, 1799, 1804, 1812, 1824, 1833, 1836,

1844, 3241. El Salvador, 1542, 1591, 1593, 1600–1601. Guadeloupe, 1680. Guatemala, 1542, 4670. Haiti, 1681. Honduras, 1542, 1618. Indigenous Peoples, 4670. Jamaica, 1668, 1673. Japan, 2062. Mexico, 19, 1310, 1312, 1333, 1335, 1338, 1344, 1346–1347, 1350, 1355, 1358–1359, 1363, 1372, 1385, 1389, 1395, 1401, 1411–1413, 1422, 1431–1432, 1435, 1437, 1442, 1449, 1451, 1454–1456, 1458, 1467, 1482, 1490, 1527–1528, 2815, 2827, 2884, 2895, 2923, 2927, 3963–3964, 4469, 4519, 4548, 4550, 4554, 4557, 4565. Nicaragua, 1623, 1625, 1627, 1632–1634, 3003. Panama, 1635–1637. Paraguay, 2002. Peru, 1858, 1924, 1928, 1931–1932, 1940, 1944–1946, 1950, 1952, 1954, 1956, 1993, 2493, 3401, 3568, 4298, 4301, 4308, 4903. Puerto Rico, 4718. South America, 1118. Southeast Asia, 1304. Uruguay, 2009, 2021, 2023, 2027, 2029, 3641, 3663. US, 1119, 1299, 1458, 1603, 3843. Venezuela, 1717, 1719, 1723, 1733, 1737, 1739–1740, 1743, 1748, 3252.

Economic Reform, 1227, 1240, 3825. Argentina, 2038, 2040, 2049. Chile, 1907. Colombia, 1761–1762, 1767. Cuba, 3140, 3149. Dominican Republic, 1663. Mexico, 1326, 1344. Peru, 1931, 1934.

Economic Sanctions. Haiti, 4738.

Economic Sanctions, US. Panama, 4020.

Economic Stabilization, 1084, 1133, 1147, 1168, 1172, 1180, 1207, 1227, 1247, 2046. Argentina, 1140, 2037, 2045, 2049, 2056, 2058, 2060, 2066, 2073, 2137, 2165, 3504. Asia, 1180. Bolivia, 1982–1983, 1987, 1991, 1993, 3284, 3322, 3324. Brazil, 1140, 2098, 2137, 2143, 2150, 2165, 2209, 3757. Caribbean Area, 1642. Central America, 1548. Chile, 1873, 1897, 3452. Costa Rica, 1575, 4606. Cuba, 3151. Ecuador, 1792, 1799, 1808–1809, 1844. Honduras, 1620. Mexico, 1312, 1324, 1326, 1333, 1451, 1481. Nicaragua, 1631–1633. Panama, 1636. Peru, 1942, 1946, 1993, 2137, 4308. South America, 1118, 1164. Spain, 2066. Venezuela, 1722.

Economic Theory, 1136, 1166, 1208, 1211, 1230, 1248, 1294–1295, 1298, 3432, 3806, 3848. Brazil, 2107, 4352. Chile, 3435. Costa Rica, 1571, 1579. Development, 3689. El Salvador, 1599. Guatemala, 1606. Mexico, 1390, 1395, 1531. Peru, 1929.

Economists. Chile, 1907, 3440. Nicaragua, 3003.

Education. Bolivia, 4960. Brazil, 2108. Cultural Development, 4856. Nicaragua, 4575. Social Structure, 4866.

Education and State. *See* Educational Policy.

Educational Models. Social Work, 4405.

Educational Planning. Caribbean Area, 2307.

Educational Policy. Argentina, 3524. Caribbean Area, 2307. Costa Rica, 2956. Mexico, 2847. Nicaragua, 4575. Venezuela, 1755.

Educational Reform. Bolivia, 4960.

Educational Research. Virgin Islands, 2288.

Educational Sociology. Argentina, 5086. Caribbean Area, 4720. Ecuador, 4866.

Eisenhower, Dwight David, 4134.

Ejército Guerrillero de los Pobres (Guatemala), 765.

Ejército Popular de Liberación (Colombia), 3188.

Ejército Revolucionario del Pueblo (Argentina), 3540, 3557.

Ejército Zapatista de Liberación Nacional (Mexico), 2826, 2899.

Ejidos, 4462. Databases, 43. Henequen Industry and Trade, 4462. Mexico, 1315, 1530, 2387–2388, 2845.

Ejutla Site (Mexico). Artisanry, 262. Shells and Shell Middens, 264–265.

El Alto, Bolivia (city). Community Development, 4954, 4966. Guidebooks, 4951. Medical Care, 4958. Urban Sociology, 4954. Women, 4966.

El Chorrillo (neighborhood, Panama City). Social Conditions, 4596.

El Dorado, 527.

El Mirador Site. *See* Mirador Site (Guatemala).

El Niño Current, 2266, 2548. Paleoclimatology, 584. Peru, 2498, 2514.

El Salvador. Asamblea Legislativa, 2969.

El Salvadorans. *See* Salvadorans.

Elderly Persons. *See* Aged.

Election Fraud. Dominican Republic, 3056. Mexico, 2897, 4525.

Elections, 2735, 2759, 2792, 3864. Andean Region, 2777. Argentina, 3498, 3525, 3572, 3586, 3588. Belize, 2944. Bolivia, 3315, 3317, 3324, 3326–3327, 3329, 3331, 3342, 3347, 3349. Brazil, 3674, 3683–3684, 3700, 3717, 3722, 3729, 3731–3733, 3735, 3743, 3752, 3754, 3757–3758, 3761, 3765, 3768, 3771, 3773, 3776, 5163. Central America, 4678. Chile, 3427, 3463, 3483, 3489, 3877, 5038. Colombia, 3169, 3196, 3211. Costa Rica, 2958. Dominican Republic, 3048, 3053, 3056, 3058. Dutch Caribbean, 3080.

Ecuador, 3213, 3222, 3226, 3232, 3234, 3238. El Salvador, 2961, 2969. Guatemala, 2982, 2985, 4618. Honduras, 2987, 2992–2993. Mass Media, 5163. Mexico, 2399, 2812–2813, 2816–2817, 2820–2821, 2825, 2830, 2839, 2843, 2852–2853, 2857, 2860, 2869, 2873, 2875–2878, 2886, 2891–2892, 2894, 2897–2898, 2900–2902, 2906, 2918, 2921, 2923, 2928, 2931–2932, 2935, 4484, 4525, 4564. Montserrat, 3071. Municipal Government, 3349. Newspapers, 3773. Nicaragua, 3002, 3013–3014, 3877, 3984, 3997, 4050, 4057, 4685. Panama, 3019, 3027, 3877, 4007, 4047. Paraguay, 3591–3592, 3611. Peru, 3377, 3382, 3389, 3393, 3395, 3400, 3408. Puerto Rico, 4724. Southern Cone, 2535. Television, 2803, 3588, 3765, 3771, 3776, 5038, 5163. Trinidad and Tobago, 3096. Uruguay, 3621, 3637, 3642, 3664. Venezuela, 3247, 3255, 3261, 3265–3268, 3270, 3272, 4844.

Electric Industries. Brazil, 2167–2168.

Electricity, 2218. Brazil, 2156. Southern Cone, 2537.

Electronic Industries. Argentina, 3502. Mexico, 2379.

Electronic Resources, 26.

Elementary Education. Bolivia, 4976.

Elías Calles, Plutarco, 2813.

Elites. Argentina, 3529, 3551, 3566, 3580, 3582, 5122. Bolivia, 3291. Brazil, 3749, 3754, 3759, 3770, 3772. Central America, 1554. Colombia, 3162, 3183, 4805. Cuba, 3146, 4119. Mayas, 135, 327. Mesoamerica, 135. Mexico, 2813, 2843, 2926, 4555. Panama, 3017. Peru, 1929, 3358, 3386, 3406, 4883. Political Thought, 5122. Precolumbian Civilizations, 135, 301, 716. Uruguay, 3635, 3663. Venezuela, 3183.

Embera (indigenous group). Ethnography, 747. Shamanism, 747.

Emigrant Remittances. El Salvador, 1597.

Emigration and Immigration. *See* Migration.

Employment, 1122, 1194, 1286. Argentina, 2053. Bolivia, 4964. Brazil, 2120, 2171, 2185, 2196, 5169, 5182, 5207. Central America, 1559. Children, 4641, 4879, 4937, 4964, 5169. Chile, 5039. Cuba, 1692. Databases, 32. Dominican Republic, 4694. Ecuador, 1816. Guatemala, 1609. Jamaica, 4693. Maquiladoras, 1460. Mexico, 1308, 1313, 1352, 1362, 1445, 1454, 1491, 1497, 1501, 1522, 2388, 2402, 3956, 4480. Nicaragua, 1626. Peru, 1941. Puerto Rico, 4774. Racism, 5207. Sex Roles, 5039. Statistics, 4428.

Uruguay, 2015. US, 3956. Venezuela, 1729. Women, 1559, 1609, 1941, 4428, 4434, 4516, 4523, 4553, 4694, 4774, 4967, 5182, 5203. Youth, 2015, 4864, 4892.

Encomiendas. Bolivia, 1970. Guatemala, 2344. Hispaniola, 512.

Encuentro Nacional de Mujeres del Movimiento Urbano Popular, *1st, México, 1983*, 4524.

Encuesta Nacional sobre Fecundidad y Salud de México, 1987, 4531.

Endangered Species. Caribbean Area, 2302.

Endara Galimany, Guillermo, 3025.

Energy Consumption. Mexico, 1444.

Energy Policy. Ecuador, 1807.

Energy Sources. Brazil, 2156.

Energy Supply. Argentina, 2054. Brazil, 2156. Ecuador, 1807.

Engineers. Brazil, 2168.

English Language. Mexico, 2383.

Enríquez Frödden, Edgardo, 3439.

Enterprise for the Americas Initiative, 1154, 1196, 1269, 3823, 3846.

Entrepreneurs. Argentina, 2558. Chile, 1856, 5041. Lebanese, 4698. Syrians, 4698. Trinidad and Tobago, 1677.

Environmental Degradation. Amazon Basin, 2665. Brazil, 2119, 2669. Costa Rica, 2323. Ecuador, 1838. Honduras, 4674. Mexico, 2377. Poverty, 4674.

Environmental Policy, 1098, 1123, 1138, 1270, 1278, 1834, 2237, 2257, 2267, 2269–2270, 2272, 3789, 4396. Amazon Basin, 2463–2464, 2493, 2656, 3780, 5213. Argentina, 2551–2552, 2562, 4191. Bibliography, 2232. Brazil, 2136, 2140, 2636, 2661, 2665, 2683, 4381, 4999, 5146. Caribbean Area, 1639, 2237, 2296–2297, 4088. Central America, 1541, 2943. Chile, 1862, 4999, 5057. Colombia, 1780, 4272. Ecuador, 1838, 2464. El Salvador, 1600. Italy, 2404. Martinique, 2308. Mexican-American Border Region, 1489. Mexico, 1368, 1384, 1429, 1433, 1444, 1473, 2369, 2375, 2377, 2390, 2397, 2404, 2895, 3933, 3948. Military Government, 3780, 4999, 5213. North America, 1381. Peru, 1958, 2493. Poverty, 4401. Saint Kitts and Nevis, 2313. South America, 4163. Suriname, 2292, 2314. Uruguay, 3629. US, 3933.

Environmental Pollution, 1098, 1232, 2218, 2257. Agricultural Policy, 4609. Bolivia, 4981. Brazil, 2658–2659, 2674, 2682, 2702, 2710, 2720. Central America, 4609. Chile, 1857, 2595, 5057. Dominican Republic, 2289, 2298. Honduras, 766. Mexico, 1433, 2369, 2375, 2386, 2409, 2412, 4533. Paraguay, 4985. Population Growth, 4427. Poverty, 766. Technological Development, 4985. Venezuela, 2437. Women, 4427.

Environmental Protection, 1834, 2229, 2231, 2238, 2256, 2278, 2280, 2282, 2764, 3789. Amazon Basin, 2233, 2678, 5160, 5164. Argentina, 2562, 2564, 4191. Bolivia, 2527. Brazil, 2172, 2621, 2650, 2678, 3709. Caribbean Area, 2280, 2302, 2307, 4088. Catholic Church, 3709. Central America, 2316. Chile, 2582. Colombia, 2453. Costa Rica, 2325, 2330, 2333. Economic Development, 3629. Ethics, 2316. Journalism, 2786. Martinique, 2308. Mexico, 1368, 3917. Panama, 4010. Peru, 2490–2491, 2510, 2516, 4316. Saint Kitts and Nevis, 2313. South America, 4163. Suriname, 2287, 2314. Uruguay, 3629. US, 3917.

Environmental Protection Groups. Amazon Basin, 3688. Brazil, 5146. Ecuador, 1820.

Epigraphy. Aztecs, 89, 197, 446. Mayas, 87, 90, 136, 138, 412–413, 420–421, 423, 425, 433, 460. Olmecs, 426–427.

Episcopal Church. Puerto Rico, 4690.

Escobar Cerda, Luis, 3440.

Escuintla, Guatemala (dept.). Rites and Ceremonies, 324.

Esequiba Region. *See* Essequibo Region (Guyana and Venezuela).

Esmeraldas, Ecuador (prov.). Popular Culture, 1011.

Espírito Santo, Brazil (state). Agriculture, 2095. Coffee Industry and Trade, 2095. Paleoecology, 587. Population Growth, 2714. Precolumbian Civilizations, 587. Private Enterprise, 2714. Urban Areas, 2714.

Esquipulas II. *See* Arias Peace Plan (1987).

Essequibo Region (Guyana and Venezuela), 4344. Boundary Disputes, 4332. Physical Geography, 4343.

Estructura social de la Argentina, 5095.

Ethics. Argentina, 3580, 3585. Chile, 3473. Environmental Protection, 2316.

Ethnic Groups and Ethnicity, 779. Amazon Basin, 868. Andean Region, 936. Belize, 4633. Caribbean Area, 524, 778, 792, 794, 796, 824, 837, 846. Chile, 5068. Colombia, 989, 994, 1004, 4830. Costa Rica, 813. Dominican Republic, 789–790, 839. Ecuador, 1020, 1023–1024, 4870. Guadeloupe, 832. Guatemala, 765, 2973, 2981. Guyana, 787. Haiti, 830. Honduras, 4617. Mexico, 740–741, 763. Nicaragua, 735, 750. Otomi, 773.

Panama, 4592. Peru, 1057, 1067, 3408. Precolumbian Civilizations, 538, 612. Saint Vincent, 850. Social Classes, 62. Suriname, 782, 2301, 3088. Trinidad and Tobago, 787, 793–794, 804, 809, 829, 834, 844–845, 848–849, 4793.

Ethnoarchaeology. Aztecs, 374. Brazil, 575, 594–595. Guatemala, 261. Mayas, 261, 404. Mesoamerica, 120, 146, 189–190, 192, 318, 374. Mexico, 107, 198, 207. Research, 99.

Ethnobotany, 2225. Chamí, 1007. Mesoamerica, 415. Secoya, 913.

Ethnoecology. Amazon Basin, 872. Brazil, 872. Mexico, 2363.

Ethnographers. Argentina, 939.

Ethnography, 2228. Bibliography, 949. Caribbean Area, 794, 824. Colombia, 998, 1003. Colonial Art, 510. Cuba, 776. Ecuador, 997. Pasto, 999. Peru, 1055. Quechua, 1061.

Ethnohistory. Brazil, 592. Central America, 435. Colombia, 999. Dominica, 777. Guatemala, 261. Honduras, 306. Mexico, 435, 2362. Research, 525. Sources, 592. Zapotec, 459.

Ethnologists. Biography, 894.

Ethnology. Caribbean Area, 846. Haiti, 811, 823.

Ethnomusicology. Peru, 1068.

Etymology. Argentina, 2549. Chile, 2574.

Eurocentrism. Cronistas, 509.

European Influences, 3865. Cuba, 4762.

European Union, 1198, 1563, 2110, 3792, 3812, 3817, 3820, 4082, 4164, 4215.

Evangelicalism. Ecuador, 1018. Central America, 4635–4636. Chile, 5025. Cultural Identity, 1018. Ecuador, 1018. Guatemala, 4597. Nicaragua, 2998. Public Opinion, 5025.

Excavations. Argentina, 546–547, 549, 551, 553. Aztecs, 204. Barbados, 495. Brazil, 573–574, 577–579. Chile, 534, 598–599, 602, 611. Colombia, 616, 618, 623, 627, 629. Costa Rica, 475. Dominican Republic, 486. Dutch Caribbean, 485, 650. Ecuador, 646, 649. El Salvador, 368–369. Guatemala, 205, 287, 313, 344, 363. Honduras, 202, 290, 331, 333–334. Mayas, 74, 400. Mesoamerica, 74, 291. Mexico, 182, 188, 204, 206, 210–211, 213, 221–223, 249, 256, 259, 262, 265, 272, 277, 311, 315, 329, 337, 353–354, 357, 362, 373, 375, 395, 405–407, 432. Mixtec, 179. Peru, 654, 664, 685. Research, 74. Suriname, 650. Uruguay, 711. Venezuela, 484.

Exchange Rates, 1075, 1108, 1126, 1136,

1185, 1208, 1231, 1261, 1266, 1291, 1549. Argentina, 2063. Bolivia, 1980, 1985. Brazil, 2158. Central America, 1555. Chile, 1852, 1889. Colombia, 1769. Costa Rica, 1555, 1576, 1589. El Salvador, 1555, 1595, 1599, 1605. Guatemala, 1555, 1605, 1611. Mexico, 1484, 1531, 1533. Paraguay, 2003. Peru, 1961.

Executive Power, 2750, 2770. Central America, 2984. Chile, 3423. Colombia, 3175. Ecuador, 3218. Guatemala, 2984. Mexico, 2924.

Exhibitions. Aztecs, 68. Canelo, 927. Precolumbian Art, 326. Precolumbian Sculpture, 308.

Exiles. Psychology, 4451. Research, 4767.

Exiles, Chilean, 3493, 5083.

Exiles, Cuban, 3141, 4767. US, 3150, 4124, 4767.

Exiles, German, 4395.

Exiles, Haitian, 3067.

Exiles, Jewish, 4395.

Expeditions. Suriname, 825.

Explorers. Antartica, 2573.

Export Promotion, 1085, 1092, 1094–1095, 1102, 1106–1107, 1233, 1253, 1261, 1293, 2244. Argentina, 2087. Brazil, 2112, 2169, 2642. Caribbean Area, 4083, 4092. Central America, 1567. Chile, 1900–1901, 2580, 2593. Colombia, 1765. Costa Rica, 1571, 1576, 1581, 1588. Cuba, 1693. El Salvador, 1601. Guatemala, 1614. Mexico, 1438, 1503. Nicaragua, 1629. Southeast Asia, 1180. Uruguay, 2018. US, 1518.

Export Trading Companies. Guatemala, 725. Mexico, 725.

Exports, 1153. Brazil, 2118, 2152, 2181, 2213. Caribbean Area, 1642. Central America, 1560. Chile, 5035. Southern Cone, 2152.

External Debt, 1086, 1105, 1119, 1125, 1144, 1146–1147, 1158, 1199, 1227, 1232, 1245, 1284, 1295, 1346–1347, 2076, 2250, 2256, 3795, 3798, 3818, 3823, 3834–3835, 3854, 3856, 3872, 3951. Argentina, 2043, 2058, 2060, 2065, 3538, 3567, 3584, 3847, 4195. Bolivia, 1972–1973, 3281. Brazil, 2158, 2176, 3725–3726, 3847. Caribbean Area, 1647. Central America, 1569. Chile, 1852, 1896, 1908, 1915, 3491. Colombia, 1773. Databases, 48. Mexico, 1313, 1346, 3847, 3951, 3964. Negotiation, 3725–3726. Nicaragua, 1625. Panama, 1635. Peru, 4298, 4301. Uruguay, 3618.

Falkland Islands, 4193. History, 4202, 4204, 4214. Natural Resources, 4210.

Falkland/Malvinas War (1982), 4198–4199, 4208–4210, 4216, 4220.

Family and Family Relations, 4453. Brazil, 5189. Caribbean Area, 4714. Chile, 1905. Colombia, 4803. Cuba, 4772. Ecuador, 4861. Guatemala, 4652. Guyana, 4715. Jamaica, 788, 838. Law and Legislation, 4852. Mexico, 4507. Netherlands Antilles, 2285. Nicaragua, 4612, 4651. Peasants, 5189. Peru, 4907, 4921, 4927. Trinidad and Tobago, 4749. Venezuela, 4849, 4852.

Family Planning. See Birth Control.

Family Violence. Argentina, 5096, 5099. Brazil, 5198. Chile, 5024. El Salvador, 4641. Guyana, 4715. Peru, 4927.

The far journey of Oudin, 842.

Farms. Andean Region, 1803. Brazil, 2649, 2651, 2669, 2677, 2686–2687, 2699. Dominica, 815. Dominican Republic, 1610. Ecuador, 1816. Guatemala, 1607, 1610, 1614. Mexico, 1349, 1403, 2364. North America, 1381. Paraguay, 2697. Peru, 2497, 2518. Saint Lucia, 814. Venezuela, 2446, 2448.

Fascism. Argentina, 5121. Chile, 4253. Paraguay, 3598. Uruguay, 4325.

Favelas. See Slums; Squatter Settlements.

Febres Cordero, León, 3229.

FEDECAMARAS. See Federación Venezolana de Cámaras y Asociaciones de Comercio y Producción.

Federación de Mujeres Cubanas, 3123.

Federación Popular de Mujeres de Villa El Salvador, 4890.

Federación Sindical Unica de Trabajadores Campesinos de Cochabamba, 3304.

Federal-State Relations. Brazil, 2166. Mexico, 2400, 2816, 2824.

Federalism. Brazil, 2166.

Feminism, 4406–4407, 4448. Argentina, 5116. Chile, 5077. Honduras, 2991. Paraguay, 4990. Puerto Rico, 4704, 4760.

Fernández Rojas, Max, 3320.

Fertility. See Human Fertility.

Fertilizers and Fertilizer Industry. Venezuela, 2448.

Festivals. Ecuador, 997. Indigenous Peoples, 1056. Rites and Ceremonies, 1056.

Fieldwork. Anthropologists, 1025. Barbados, 795. Saint Lucia, 814.

Figueira, Ricardo Rezende, 3699, 5167.

Finance, 1199, 1273. Argentina, 1195, 2042, 2058, 2066, 2088. Brazil, 1195, 2165, 2179, 2198, 2215. Chile, 1882, 1893. Colombia, 1759. Databases, 33. Ecuador, 1806. El Sal-

vador, 1596. Mexico, 1195, 1372, 1393, 2843. Nicaragua, 2999. Peru, 1925, 1933, 3401. Spain, 1759, 2066. Venezuela, 1195.
Financial Institutions. See Banking and Financial Institutions.
Financial Markets, 1150.
Firmenich, Mario Eduardo, 3541.
Fiscal Policy. Argentina, 2041, 2052, 2056, 2063, 2068. Bolivia, 1980. Brazil, 2150. Caribbean Area, 1654. Chile, 1889, 1898. Colombia, 1783. Costa Rica, 1576. Ecuador, 1809. Honduras, 1617. Mexico, 1335, 1441. Peru, 1954. Uruguay, 2020. Venezuela, 1728, 1737, 1753.
Fish and Fishing, 2234. Antigua, 4789. Caribbean Area, 2305, 4789. Chile, 2616. Cuba, 1690. Dominican Republic, 4789. Falkland Islands, 2539. Mexico, 4456, 4533. Puerto Rico, 797. Uruguay, 4324, 5135. Women, 5135.
Fisheries, 2234. Caribbean Area, 4789. Chile, 1886, 2616. Puerto Rico, 797. Technology Transfer, 1886.
Floods. Argentina, 2548, 2556. Bolivia, 2525. Peru, 2514.
Florida, US (state). Cocaine, 4271. Drug Traffic, 4271. Guatemalans, 722. Kanjobal, 722.
Florida Valley (Honduras). Archaeological Surveys, 333–334. Center/Periphery Relations, 331. Stone Implements, 186.
FMLN. See Frente Farabundo Martí para la Liberación Nacional.
Folk Art. Ecuador, 644, 1012–1013. Peru, 695, 3375.
Folk Medicine. See Traditional Medicine.
Folk Music. Peru, 1068.
Folk Religion. Argentina, 5093. Brazil, 5179. Cuba, 4703, 4795. Trinidad and Tobago, 4742. Venezuela, 4848, 4853.
Folklore. Andean Region, 934. Colombia, 996. Indigenous Peoples, 945. Peru, 1036, 1041–1042, 1059.
Folktales. See Folk Literature; Legends.
Food. Bolivia, 4961. Caribbean Area, 818. Ecuador, 1812. French Guiana, 780. Indigenous Peoples, 2235. Peasants, 2235. Precolumbian Civilizations, 385. Trinidad and Tobago, 808.
Food Industry and Trade. Bolivia, 1982. Mexico, 1337.
Food Supply, 1251, 2259, 4443. Andean Region, 2418, 2478. Bolivia, 1994. Brazil, 2197, 2674. Caribbean Area, 1251. Central America, 3971. Chile, 1871. Ecuador, 2478. French Guiana, 806. Guatemala, 1607,

1614. Hiwi, 898. Mexico, 1320, 1337, 1340, 1404, 2406. Nicaragua, 1622, 1628, 1630. Paleo-Indians, 563. Peru, 1034, 2487, 4894. Precolumbian Civilizations, 64, 156, 318, 508, 2406. Venezuela, 1716.
Forced Labor. Dominican Republic, 1657. Indigenous Peoples, 2413.
Forced Migration. Indigenous Peoples, 719. Mexico, 719, 737.
Ford Foundation, 4373.
Foreign Aid. See Economic Assistance.
Foreign Debt. See External Debt.
Foreign Exchange. See Exchange Rates.
Foreign Influences, 3815. Argentina, 3584. Cultural Destruction, 152. Mexico, 2383.
Foreign Influences, US. Central America, 3972.
Foreign Intervention, British. Argentina, 4214. Uruguay, 4214.
Foreign Intervention, Cuban. Angola, 4123. Dominican Republic, 3050, 3055.
Foreign Intervention, US, 3782. Argentina, 4214. Caribbean Area, 4037. Central America, 3999, 4017, 4037, 4068. Cuba, 4119. Guatemala, 2978. Latin America, 3801. Nicaragua, 3997, 4025, 4057. Panama, 3029, 3970, 3975, 3980, 4003, 4007, 4011–4012, 4023, 4028–4029, 4032, 4038–4039, 4041–4042, 4047, 4058, 4064, 4070, 4073, 4596.
Foreign Investment, 1086, 1169, 1226, 1234, 1239, 1255, 1306. Argentina, 2169. Brazil, 2138, 2169, 2215. Chile, 1848, 1854–1855, 1887, 1915, 1920, 3484, 4245. Cuba, 1704, 1714. Mexico, 1343, 1350, 1362, 1380, 1416, 1465, 1528, 3908. Uruguay, 2031.
Foreign Investment, Japanese. Mexico, 1510.
Foreign Investment, US, 3823. Argentina, 3813. Chile, 3813.
Foreign Policy, 3791, 3796, 3803, 3807, 3870, 3879, 3881–3882. Argentina, 4188, 4190, 4194, 4211–4212, 4217, 4219, 4225, 4386. Belize, 3968. Bolivia, 3285, 3354, 4227–4229, 4238–4239. Brazil, 3761, 4349–4350, 4357, 4359–4362, 4364–4365, 4370–4371, 4375–4376, 4382–4387, 4390, 4392. Canada, 3808, 3810, 3855. Caribbean Area, 4004, 4094, 4151, 4153, 4159. Central America, 3985, 3993, 4004, 4035, 4044, 4065. Chile, 4246, 4248–4249. China, 3800. Colombia, 4265, 4267, 4270, 4272, 4275, 4278–4279. Costa Rica, 2950, 4016, 4019. Cuba, 3116–3117, 3126, 3131, 3881, 4112–4113, 4116–4118, 4120, 4123, 4128. Databases, 38. Dominican Republic, 4132. Ecua-

dor, 3214–3215, 4285. Europe, 4153. Great Britain, 4109, 4264. Guatemala, 4049. Guyana, 4159. Haiti, 4138. Honduras, 4044. Jamaica, 4153. Mexico, 1335, 3883, 3885–3886, 3896, 3902, 3910, 3924, 3930, 3939, 3949, 4016. Panama, 4011, 4028, 4064. Paraguay, 4293–4296. Peru, 3363, 3401. Puerto Rico, 4136. Russia, 3788, 3809, 3856. South America, 4174. Theory, 3831. Uruguay, 4320, 4323, 4325, 4327. US, 38, 3130, 3150, 3314, 3484, 3786, 3794, 3801, 3805, 3809, 3815, 3822–3823, 3829, 3839–3840, 3850, 3857, 3861–3862, 3867, 3870, 3878, 3881, 3924, 3943, 3989, 3996, 4012, 4016, 4019, 4041, 4067–4068, 4070, 4073, 4076, 4087, 4121, 4143, 4319, 4420. USSR, 4076. Venezuela, 4330–4331, 4333, 4335, 4337–4338, 4342, 4346.

Foreign Trade. See International Trade.

Forests and Forest Industry, 2226, 2233–2234, 2274. Amazon Basin, 1936, 2623, 2627, 2660, 2675, 2698, 2718. Argentina, 2090, 2558. Bibliography, 2318. Bolivia, 4977. Brazil, 2639, 2693. Caribbean Area, 2293. Central America, 2318–2319. Chile, 2579, 2593, 2606. Conservation, 2272. Costa Rica, 2339. Ecuador, 2484. Mexico, 2356, 2370, 2384. Panama, 2353. Paraguay, 2090. Peru, 1936, 2490, 2510, 2517. Suriname, 2287, 2292. Venezuela, 2450. Yucatán Peninsula, 2370.

Fortún Suárez, Guillermo, 3278.

Fórum Nacional (Brazil), 3751.

Franciscans. Peru, 3373.

Franco, Francisco, 4222.

Franco, Itamar, 3702.

Franco, Rafael, 3597.

Free Trade, 1128–1129, 1154, 1156, 1269, 1279, 1302, 1462, 2799, 3783, 3823–3824, 3827, 3846, 3860. Argentina, 4166. Brazil, 4166. Canada, 1459. Caribbean Area, 1642, 1648, 4081. Central America, 3967. Mexican-American Border Region, 3962. Mexico, 1309, 1312, 1332, 1337, 1342, 1405, 1409–1410, 1423, 1448, 1457, 1459, 1462, 1469, 1505, 1508, 1537, 2880, 2910, 3887–3888, 3897, 3901, 3904, 3917–3918, 3928, 3935–3937, 3952–3954, 3961, 3963, 3967. North America, 1327, 1342, 1409, 1458–1459, 1464, 1523, 1535, 3888, 3901, 3915, 3937, 3946. Peru, 1930. South America, 1216. Southern Cone, 2048. US, 1156, 1410, 1448, 1459, 1469, 2048, 3846, 3887–3888, 3901, 3917–3918, 3928, 3935–3936, 3953, 3961.

Freedom. Argentina, 3507.

Frei Montalva, Eduardo, 3444.

French Caribbean. Demography, 4708. World War II (1939–1945), 3082.

French Influences. Brazil, 5153.

Frente Amplio (Uruguay), 3648, 3651, 3656, 3659, 3662.

Frente Democrático Nacional (Mexico), 2891, 2900.

Frente Democrático Nacional (Peru), 4301.

Frente Farabundo Martí para la Liberación Nacional (El Salvador), 2965, 2967, 4052.

Frente Revolucionario de Unidad Campesina (Bolivia), 3305.

Frente Sandinista de Liberación Nacional. See Sandinistas (Nicaragua).

Fresco, Manuel A., 5121.

Friendship. See Interpersonal Relationships.

Frondizi, Arturo, 3764.

Frontier and Pioneer Life. Belize, 2322. Brazil, 2652. Colombia, 1002, 4819.

Fruit Trade. Amazon Basin, 2645. Argentina, 2071. Belize, 1570, 4634, 4637. Brazil, 2206, 2642, 2668, 2674, 2706, 5158. Caribbean Area, 2219. Chile, 1903–1904, 5035.

FUCVAM (organization, Uruguay), 5136.

Fuerzas Armadas Revolucionarias de Colombia, 3152–3153, 3191.

Fujimori, Alberto, 1956, 3362, 3389, 3395–3396, 3408, 4304.

Fundación Eva Perón, 4417.

Galápagos Islands, Ecuador. Agricultural Geography, 2479. Agricultural Productivity, 2479. Climatology, 2479. Science, 2469. Scientists, 2469.

Gallegans. See Galicians.

Gambini, Jake, 3155.

Gangs. Colombia, 1001, 3195.

Garbage. See Hazardous Wastes; Organic Wastes.

García, Graciela A., 2991.

García Márquez, Gabriel, 3135.

García Villegas, René, 3446.

Gardens. Precolumbian Civilizations, 276.

Garifuna. See Black Carib.

GATT. See General Agreement on Tariffs and Trade.

Gaviria Trujillo, César, 3171, 3206, 4272.

Genealogy. Kings and Rulers, 411. Mayas, 153. Mesoamerica, 153. Mixtec, 447. Zapotec, 459.

General Agreement on Tariffs and Trade (GATT), 1089, 1310, 1314, 1433, 1436–1437, 1583, 1585, 2077, 2096, 2119, 2152, 3781, 3913, 4014, 4354.

Generals. Chile, 3465. Cuba, 3144, 3146.
Genetics. *See* Human Genetics.
Geographers, 2228. Peru, 2504. Venezuela, 2440.
Geographical History, 2227–2228, 2230. Andean Region, 2565–2566. Brazil, 2628, 2634, 2638, 2641. Chile, 2568. Guatemala, 2344, 2346. Maps and Cartography, 2249. Mexico, 2401. Peru, 2499, 2504. Uruguay, 2617.
Geographical Names. Argentina, 2549. Belize, 2321. Chile, 2574. Colombia, 620. Mayas, 449. Mesoamerica, 449.
Geography, 2262. Amazon Basin, 2644. Andean Region, 2422. Argentina, 2538, 2547. Bibliography, 2264. Bolivia, 2528. Brazil, 2663, 2688, 2719. Costa Rica, 2327. Mexico, 25. Venezuela, 2434, 2439, 2443–2444.
Geology. Andean Region, 2480. Ecuador, 2480.
Geomorphology, 2221–2222. Andean Region, 2565–2567. Argentina, 2554. Chile, 2611.
Geophysics. Mexico, 317.
Geopolítica, 4244.
Geopolitics, 3785, 3787–3789, 3793, 4244. Argentina, 2597, 4174. Bibliography, 3793. Bolivia, 2570, 2572, 3334, 4174, 4226. Brazil, 4174, 4355, 4369, 4380, 4392. Caribbean Area, 4091, 4094. Central America, 4009. Chile, 2569–2572, 2585, 2591, 2597, 2612, 2614. Ecuador, 3225. Panama Canal, 4000. Paraguay, 4294. South America, 4165, 4167, 4171, 4174, 4178–4179, 4184–4185, 4260. Southern Cone, 2569–2570.
Georgetown University. Hemispheric Migration Project, 4570.
German Influences, 2222. Chile, 4253. Guatemala, 4663. Indigenous Peoples, 4663. South America, 2427.
Germani, Gino, 5095.
Germans. Argentina, 5128. Chile, 2604. Guatemala, 4663. Mexico, 3926.
Girón de León, Andrés de Jesús, 4588.
Glaciers. Argentina, 2546. Chile, 2546, 2611. Venezuela, 2449.
Global Warming. Mexico, 2390.
Globalization. Chile, 3443. Cuba, 1702. Sociology, 4394.
Goajiro (indigenous group). Cultural History, 903. Myths and Mythology, 904. Origins, 903.
Gold. Brazil, 873. Costa Rica, 2324.
Golden Rock Site (Saint Eustatius), 485, 650.
Goldwork. Ecuador, 647. Precolumbian Civilizations, 647, 686.
Gómez, Alejandro, 3535.
Gómez de la Torre, Luisa, 3240.
González de Aedo, Felipe, 2601.
Good, Kenneth, 894.
Good, Yarima, 894.
Government, Resistance to. Argentina, 3541. Bolivia, 3319. Chile, 3456. Colombia, 3152, 3173. Cuba, 4740. Guatemala, 765. Panama, 4038. Paraguay, 3606. Venezuela, 3253, 3269.
Government Publications. Mexico, 2868.
Governors. Brazil, 3734. Mexico, 2822, 2866, 2931. Venezuela, 3270.
Grassroots Movements. *See* Social Movements.
Grazing. Argentina, 2562.
Great Savannah. *See* Gran Sabana (Venezuela).
Green Movement, 4396.
Grenadines. *See* Windward Islands.
Gross Domestic Product, 1260. Costa Rica, 1577.
Grupo de Contadora, 3882, 4055, 4270, 4333.
Grupo de Estudios sobre la Condición de la Mujer en el Uruguay, 1206.
Grupo de los Ocho, 3882.
Grupo de los Tres, 3860, 4268.
Guadalajara, Mexico (city). Americans (US), 4532. Child Abuse, 4562. City Planning, 2391. Crime and Criminals, 4562. Demography, 1398. Economic Conditions, 1398. Elections, 2820. History, 1398. Industry and Industrialization, 1398. Migration, 1398. Rape, 4562. Social Conditions, 1398, 4532. Urbanization, 2391.
Guadalupe, Costa Rica (city). Urbanization, 2328.
Guajira, Colombia. *See* La Guajira, Colombia (dept.).
Guajiro (indigenous group). *See* Goajiro.
Gualjoquito Site (Honduras). Excavations, 202.
Guanajuato, Mexico (state). Precolumbian Architecture, 240, 328.
Guanano (indigenous group). History, 867. Social Structure, 867.
Guardia Navarro, Ernesto de la, 3019.
Guatemala, Guatemala (city). Households, 4653. Indigenous Peoples, 4653. Informal Sector, 4652. Labor and Laboring Classes, 4653.
Guatemalans. Mexico, 3905, 4456, 4473, 4520, 4630. Refugees, 722. US, 722, 4687.
Guayaquil, Ecuador (city). Labor and Laboring Classes, 1810, 2474. Modern Architecture,

2466. Recessions, 1810, 2474. Social Conditions, 1793.
Guaykuru. *See* Guaycuru.
Guerrero, Mexico (state). Archaeological Surveys, 362. Artifacts, 245. Excavations, 362. Inscriptions, 457. Precolumbian Civilizations, 245, 458. Precolumbian Land Settlement Patterns, 362.
Guerrillas, 2732, 2756, 2768, 2807–2808, 3864, 3881. Argentina, 3540, 3557. Brazil, 5162. Colombia, 3152–3153, 3155, 3157, 3170, 3182, 3187, 3189–3191, 3201, 3209. Dominican Republic, 3055. El Salvador, 2967. Mexico, 2826. Nicaragua, 759, 3007. Peru, 1060, 3383, 3393, 3398, 4316. Suriname, 782. Uruguay, 3660.
Guevara, Ernesto "Che," 3142.
Guidebooks. Chile, 2575. Costa Rica, 2337. French Guiana, 2309. Mayas, 123, 258. Mexico, 123.
Guido, José María, 3547.
Gulf of Venezuela. *See* Maracaibo Gulf (Colombia and Venezuela).
Guzmán Reynoso, Abimael, 1959, 3379, 4310.
Habana. *See* La Habana, Cuba (city).
Haciendas. Bolivia, 1970. Mexico, 2398. Peru, 1935.
Haiku, Brazilian, 4391.
Haitian Revolution (1787–1794), 4746.
Haitians. Cuba, 4735. Dominican Republic, 839, 4744, 4754. Repatriation, 4744. Social Life and Customs, 4735. US, 4711.
Hallucinogenic Drugs. Peru, 921. Precolumbian Civilizations, 561.
Harakmbet. *See* Mashco.
Harris, Wilson, 842.
Harvard Chiapas Project, 768.
Havana. *See* La Habana, Cuba (city).
Health Care. *See* Medical Care.
Henequen Industry and Trade. Ejidos, 4462. Labor Movement, 4561. Mexico, 756, 4462, 4561.
Herders and Herding. Peru, 1039.
Hermann, Hamlet, 3055.
Hidalgo, Mexico (state). Political Conditions, 4498. Political Culture, 4498. Social Conditions, 4498.
Hieroglyphics. *See* Epigraphy; Inscriptions; Writing.
Higher Education, 4402. Argentina, 5124. Central America, 4624. El Salvador, 4576. Mexico, 1487. Women, 4624.
Higueras, Mexico. *See* Las Higueras, Mexico (town).

Hinduism. Trinidad and Tobago, 845.
Hirschman, Albert O., 1283, 2740.
Historia de las Indias de Nueva-España y Islas de Tierra Firme, 411.
Historic Sites, 2246. Martinique, 2290. Mayas, 741.
Historical Demography, 2254. Argentina, 2557. Bahamas, 506. Bolivia, 2523. Chile, 2588. Guatemala, 2343, 2345. Indigenous Peoples, 2415. Mayas, 2345. Mexico, 2395, 2413. Peru, 2501. Uruguay, 2619. Viceroyalty of New Spain, 2415.
Historical Geography, 2216, 2236, 2246–2247, 2261, 2264, 2268. Andean Region, 2418, 2424, 2426. Bolivia, 2529. Caribbean Area, 4094. Guatemala, 2343–2344, 2346. Mesoamerica, 83. Mexico, 2362, 2392, 2401, 2414. Peru, 2485, 2500, 2509, 2511. South America, 2423. Suriname, 2301. Uruguay, 2619. Venezuela, 4850.
Historiography, 3786. Caribbean Area, 481, 516, 525. Mayas, 425. Nicaragua, 472. Precolumbian Civilizations, 435, 525.
Hiwi (indigenous group). Food Supply, 898. Sex Roles, 898.
Ho, Stanley, 3328.
Homosexuality. Brazil, 3728. Puerto Rico, 4690.
Horticulture. Brazil, 885. Mexico, 1520. US, 1520.
Hospitals. Dominican Republic, 1619. Honduras, 1619. Jamaica, 1619.
Hostage Negotiations. Colombia, 3155.
Hostages. Colombia, 3157.
Household Surveys. Bolivia, 4951. Colombia, 4811. Ecuador, 1791. Uruguay, 2016.
Households, 1082. Argentina, 2094, 5100. Bolivia, 4954. Brazil, 2094, 2108, 2184, 5177, 5189. Caribbean Area, 774. Chile, 1865, 1905, 1916, 5065. Colombia, 1774. Costa Rica, 1582. Cuba, 4772. Ecuador, 1812, 1816, 4864. Guatemala, 1607, 4688. Honduras, 4675. Jamaica, 788. Mayas, 285, 368. Medical Care, 4869. Mexico, 1532, 4502, 4538, 4563. Peru, 1935, 4899, 4936. Precolumbian Civilizations, 146. Social Policy, 4453. Uruguay, 2010. Women, 4453.
Housing, 1259. Argentina, 5094, 5107. Bolivia, 4963. Brazil, 2650. British Caribbean, 498. Caribbean Area, 2306. Chile, 1878, 5015, 5047, 5063–5064, 5067. Colombia, 4919. Mexico, 1501, 4457, 4563. Neighborhood Associations, 4904. Nicaragua, 4632. Peru, 4878, 4904. Poor, 4424, 5063–5064.

Uruguay, 2010, 5136, 5138. Venezuela, 1735, 1747, 4847.
Housing Policy, 1259. Chile, 5000, 5058, 5067. Ecuador, 1831. Mexico, 1357, 2381, 4457. Nicaragua, 4632. Peru, 2494, 4878. Sandinistas, 4632. Urban Areas, 4424. Uruguay, 5138.
Hoy (newspaper, Ecuador), 3224.
Huaca Rajada Site (Sipán, Peru), 685, 701.
Huacaloma Site (Peru), 696.
Huandacareo Site (Mexico). Artifacts, 315. Excavations, 315.
Huari (indigenous group). Precolumbian Sculpture, 668.
Huari Site (Peru). Pottery, 535.
Huastec (indigenous group). Artifacts, 336. Social Conditions, 4551.
Huasteca Region (Mexico). Ethnic Groups and Ethnicity, 763. Land Use, 4551. Landowners, 4551. Rural Sociology, 4551. Social Conflict, 763. Violence, 4551.
Huila, Colombia (dept.). Excavations, 623.
Huilliche (indigenous group). History, 979.
Huitoto. *See* Witoto.
Humahuaca Site (Argentina), 562.
Human Adaptation. Amazon Basin, 573. Brazil, 593. Colombia, 614. Paleo-Indians, 542, 699. Peru, 672. Precolumbian Civilizations, 531, 534, 593, 600. Uruguay, 710. Venezuela, 717.
Human Ecology, 2233, 2236, 2278, 3852. Amazon Basin, 886, 893, 2464, 2690, 5181. Andean Region, 1035. Bolivia, 2531, 4982. Brazil, 580, 591, 885, 2136, 2643, 2658–2659, 2681, 2690, 5185. Caribbean Area, 4697. Central America, 2316, 2943. Chile, 607. Colombia, 630. Costa Rica, 2331. Ecuador, 2464. Honduras, 766, 4674. Mesoamerica, 152. Mexico, 753, 767, 2363, 2366, 2397, 2409. Paleo-Indians, 580. Peru, 2516. Precolumbian Civilizations, 200, 217, 302, 591. Secoya, 926. Sioni, 926. South America, 4163. Uruguay, 710. Venezuela, 717. Yuqui, 2531.
Human Fertility, 1254. Bolivia, 4967. Brazil, 5156, 5168. Caribbean Area, 800. Chile, 2588. Congresses, 4409. Cuba, 4722. Dominican Republic, 4709, 4773. Ecuador, 4866. Mexico, 4516, 4537, 4543, 4545. Paraguay, 4986. Peru, 4925, 4944. Yanomamo, 869.
Human Geography, 2243, 2258, 2265, 2281. Amazon Basin, 857–858. Andean Region, 933, 2418, 2422, 2425, 2428. Belize, 2321. Bolivia, 977. Brazil, 871, 2720, 3746. Chile, 977. Cuba, 2304, 2315. Honduras, 2350. Mesoamerica, 300. Mexico, 2358, 2414. Peru, 2489, 2500, 2516. Precolumbian Civilizations, 69. Río de la Plata, 4183. Venezuela, 4850. Women, 2245.
Human Physiology. Colombia, 996. Jamaica, 838.
Human Remains. Bolivia, 565. Brazil, 585. Caribbean Area, 520. Mexico, 366. Nazca, 660. Suriname, 651. Uruguay, 711.
Human Rights, 3876, 4110, 4411, 4420. Argentina, 3496, 3499, 3508, 3510, 3532, 3559, 3568. Bolivia, 3283, 3302. Brazil, 3753, 3770, 5162, 5172. Central America, 2941. Chile, 3422, 3426, 3430, 3446, 3451, 3476, 3487, 3493, 3622, 5081. Colombia, 984, 3158–3160, 3165, 3174, 3176, 3193, 3203–3204, 3210, 4270, 4272. Cuba, 3127, 3133, 3145, 4740, 4784. Dominican Republic, 1657, 3052. Ecuador, 3217, 3220, 4868. El Salvador, 2960, 2962–2964, 2968. Guatemala, 757, 2979–2981, 4650. Haiti, 4139. Honduras, 2986. Indigenous Peoples, 4868. Mayas, 2981. Mexican-American Border Region, 3889. Mexico, 2830, 2873, 2879, 2889, 3911, 3922, 4461. Nicaragua, 2997, 3006. Organization of American States (OAS), 3863. Panama, 3021. Paraguay, 3589, 3605, 3617. Peru, 3360, 3398, 3413, 3416, 3568, 4319. Prisons, 2762. Southern Cone, 2752. Uruguay, 3622, 5136. Venezuela, 3272.
Humboldt, Alexander von, 2427, 2433, 2440.
Humor. Mexico, 2922.
Hunting. Bolivia, 2530. Dominica, 847. Faroe Islands, 847. French Guiana, 806.
Hurricanes. Jamaica, 2291.
Hurtado, Osvaldo, 3229.
Hydration Rind Dating. *See* Archaeological Dating.
Hydroelectric Power. Amazon Basin, 875. Brazil, 2156, 2167, 2712, 2721. Chile, 973. Ecuador, 2472.
Hydrology, 2253. Caribbean Area, 2253, 2310. Colombia, 2458, 2461.
Ibáñez de Campo, Carlos, 4253.
Ica River Region (Peru). Paleo-Indians, 674.
Iconography. Aztecs, 154. Colombia, 637. Mayas, 75, 136, 138, 148, 155, 208, 270, 367, 387, 418, 425, 444. Mesoamerica, 303, 344, 401, 462. Mexico, 390. Peru, 681. Precolumbian Civilizations, 533, 535. Teotihuacán, 430, 432.
Illegal Aliens. Central America, 4623. US, 3807, 3957, 4062, 4529.

Illiteracy. *See* Literacy and Illiteracy.
Illness. *See* Diseases.
Ilo, Peru (city). Customs, 4309. Ports, 4309.
Imbabura, Ecuador (prov.). Cultural Identity, 1010.
Immigration. *See* Migration.
Imperialism, 3794. Aztecs, 374. US, 3801.
Import Substitution, 1090–1091, 1106, 1120, 1293. Brazil, 2105. North America, 1382.
Imports and Exports. *See* International Trade.
Inca Influences, 540. Ecuador, 639. Patagonia, 534. Precolumbian Civilizations, 562, 683, 694. Precolumbian Trade, 639. South America, 537.
Incas. Abstract Art, 691. Agriculture, 678. Cosmology, 673. Legends, 527. Peru, 2519. Political Culture, 639, 694. Political Institutions, 683. Precolumbian Architecture, 673, 688, 691, 693. Precolumbian Art, 682, 691. Precolumbian Pottery, 540. Precolumbian Sculpture, 691. Rites and Ceremonies, 656. Roads, 656. Sacred Space, 673. Stonemasonry, 693. Symbolism (art), 691.
Income, 1182. Brazil, 2669.
Income Distribution, 1092, 1096, 1167, 1181, 1241, 1257, 1265, 1300, 2737, 2799, 4436. Amazon Basin, 5143. Argentina, 2044, 3571, 5114. Brazil, 2097, 2116, 2122, 2143, 2185, 2189, 2214. Central America, 1546, 4585. Chile, 1882, 1902, 1914. Colombia, 1774, 1776, 3168. Costa Rica, 1572. Ecuador, 1791. El Salvador, 1604. Jamaica, 4786. Mexico, 1371, 1395, 1406, 1424, 1431, 1454, 1512, 1521–1522. Nicaragua, 1630. Uruguay, 1572, 2016. Venezuela, 1729, 1737, 1749.
Independence Movements. Puerto Rico, 3084, 4706.
Indians. *See* East Indians; Indigenous Peoples; West Indians.
Indigenismo and Indianidad, 936. Bolivia, 4980. Brazil, 864. Central America, 4579. Colombia, 995, 3167. Ecuador, 3228, 3240. Guadeloupe, 831–832. Guatemala, 769–771. Mayas, 769–771. Mexico, 772, 2871. Peru, 1054. South America, 860.
Indigenous Architecture. *See* Vernacular Architecture.
Indigenous Art, 927. Ecuador, 1014.
Indigenous Influences. Agricultural Technology, 4928. Aruba, 521. Netherlands Antilles, 816.
Indigenous Languages. Argentina, 2549. Aztecs, 154, 438. Ecuador, 1015. Mayas, 90, 420, 453. Olmecs, 426–427. Otomi, 773. Quichua, 1015. Zoque, 426–427.
Indigenous Music. Peru, 1029.
Indigenous/Non-Indigenous Relations, 856. Amazon Basin, 889, 893. Argentina, 941. Aymara, 948. Belize, 350. Brazil, 864, 892. Colombia, 905. Cuaiquer, 914. Ecuador, 1023. Guatemala, 765. Hispaniola, 493, 512. Mexico, 727. Peru, 912. Piaroa, 902. Shuar, 916. Venezuela, 892.
Indigenous Peoples, 1127, 1243. Agricultural Labor, 4867. Agricultural Systems, 2363, 4970. Agriculture, 938. Amazon Basin, 855, 868, 875, 882, 886, 893, 900, 907. Andean Region, 935–936, 947–948, 954–955, 968, 971, 977. Anthropologists, 61. Argentina, 940–941, 945. Bolivia, 928–930, 946, 961–963, 965, 967, 4956, 4980, 4982. Brazil, 580, 864–867, 871, 873–874, 876, 878–881, 883–885, 887–888, 892, 908. Caribbean Area, 483, 511, 810, 817. Catholicism, 4509. Central America, 511. Chile, 943, 969–970, 973, 979, 981–982. Civil Rights, 995. Colombia, 899, 901, 906, 986, 994, 998, 1003–1004, 3167. Community Development, 1017. Conservation, 2233, 2267. Cultural History, 85. Cultural Identity, 735, 852, 856, 860, 880, 968, 4856. Demography, 869. Dependency, 905. Economic Development, 881. Economic Policy, 4670. Ecuador, 913–914, 916, 920, 922, 925–927, 997, 1008, 1012–1014, 1016–1017, 1020–1024, 3228, 3240, 3242, 4868, 4870. Festivals, 1056. Food, 2235. Forced Labor, 2413. Forced Migration, 719. German Influences, 4663. Government Relations, 864. Guatemala, 2346–2347, 2981. Hispaniola, 501. Historical Demography, 2415. Historiography, 61. Honduras, 279, 306, 2348. Horticulture, 885. Human Ecology, 580. Human Rights, 4868. Influences on, 856, 4884. Insurrections, 57. Kinship, 915. Land Settlement, 2347. Land Tenure, 4977. Land Use, 864, 2267. Law and Legislation, 988, 993, 1032, 4880. Marriage, 915. Mexico, 85, 718–719, 721, 723, 726, 728, 732–734, 736, 744, 748, 751, 753–755, 758, 760–762, 764, 767, 772, 2356, 2398, 2415, 2870, 2899. Migration, 4856. Millennialism, 852. Missionaries, 882. Modernization, 4504. Money, 726. Myths and Mythology, 859. Natural Resources, 2278. Nicaragua, 472, 735, 759, 3007. Nutrition, 4845. Panama, 747. Paraguay, 931. Peru, 908–912, 915, 917–919, 921, 923–924, 1055–1056, 4880. Photography, 1055. Political Participation, 1024. Production (economics), 721. Protests,

4868. Relations with Spaniards, 493, 512. Religious Life and Customs, 862. Rites and Ceremonies, 1056. Sex Roles, 1022. Shamanism, 853, 862. Slaves and Slavery, 905. Social Change, 855. Social Movements, 3167, 3228, 3237, 3240, 3242. Structural Adjustment, 932. Subsistence Economy, 872. Suriname, 782, 890, 4763. Urban Areas, 4653. Venezuela, 891–892, 894–898, 902–903, 4845. Views of, 510, 1021.

Indigenous Policy. Andean Region, 936. Argentina, 860. Bolivia, 4956. Brazil, 860. Chile, 860, 976, 5044. Colombia, 854, 988, 993, 2453. Cuaiquer, 854. Ecuador, 854, 1023, 3237, 3242, 4868. Guatemala, 4579, 4620. Mexico, 724, 737, 4486. Military Government, 5044. Nicaragua, 4579. Peru, 4880. Sandinistas, 4579.

Industrial Development Projects. *See* Development Projects.

Industrial Policy, 1137, 1177. Argentina, 2204, 2555, 3502, 5111–5112. Belize, 4637. Brazil, 2111, 2113, 2139, 2145, 2167, 2204, 2213, 3694. Central America, 1567. Colombia, 4818. Ecuador, 1832. El Salvador, 1591, 1593. Mexico, 1528. Nicaragua, 1623. Peru, 1949, 4903. Southern Cone, 2139.

Industrial Productivity, 1163. Brazil, 2116. Chile, 5045. Mexico, 1377, 1407, 1477, 1485. Technology Transfer, 5045. Uruguay, 2012.

Industrial Promotion. Ecuador, 1785. Peru, 1967.

Industrial Relations, 4445. Belize, 4634. Bolivia, 3340. Brazil, 2149, 2200, 5195. Chile, 3437, 3445, 5019. Colombia, 4818, 4828. Cuba, 4725. Mexico, 1485, 2896, 4480, 4518, 4546. South America, 4445. Southern Cone, 2101, 2790. Technological Innovations, 4828.

Industrialists. Argentina, 3551, 5112. Brazil, 2162, 3749. Democratization, 3749. Peronism, 5112. Trinidad and Tobago, 3099.

Industry and Industrialization, 1094, 1159, 1239, 2244, 4434. Argentina, 2089, 2163. Brazil, 2162–2163, 2194, 2196, 2640, 2700, 2705, 2713, 5196. Caribbean Area, 774, 4736. Chile, 5023. Colombia, 4828. Dominican Republic, 2289. Ecuador, 1785, 1792, 1804, 3246. Mexico, 1398, 1400–1401, 1414, 1521, 1524, 1528, 2412, 4488, 4491. Peru, 4903. Puerto Rico, 4774. Venezuela, 1725, 1733, 1752, 4846.

Infant Mortality. Argentina, 944. Brazil, 5161. Central America, 4643. Colombia, 4820. Dominican Republic, 4773. Mexico, 4503. Statistics, 4419.

Inflation, 1185, 1219, 1376, 2046. Argentina, 1140, 1195, 2033, 2035–2036, 2043–2045, 2051, 2058, 2063–2064, 2073, 2075, 2088, 2137. Bolivia, 1982. Brazil, 1140, 1195, 2102, 2122, 2137, 2179, 2185, 2208, 5183. Chile, 1884, 1889. Colombia, 1769. Costa Rica, 1577, 1590. Dominican Republic, 1660. Ecuador, 1806, 1808–1809, 1836. El Salvador, 1602. Guatemala, 1606. Honduras, 1617. Mexico, 1195, 1335, 1372, 1395. Nicaragua, 1631, 1633. Peru, 1933, 1961, 2137. Social Conditions, 5183. South America, 1164. Uruguay, 2019, 2026. Venezuela, 1195.

Informal Labor. *See* Informal Sector.

Informal Sector, 1076, 1097, 1194, 1242, 1263, 1268, 1276, 1292. Argentina, 2069, 2080, 5107. Bibliography, 1823. Bolivia, 1998. Brazil, 2183, 2209. Caribbean Area, 4752. Central America, 1559. Chile, 1891. Colombia, 4826. Dominican Republic, 4694. East Asia, 1388. Ecuador, 1822–1823, 2483, 4874. El Salvador, 1596. Guatemala, 1612, 4652. Mexico, 1076, 1366, 1374, 1388, 1422, 1446, 1491, 1529, 4505, 4515. Nicaragua, 4586. Paraguay, 4992. Peru, 1930, 3377, 4908, 4929. Puerto Rico, 797. Uruguay, 2618. Venezuela, 1736. Women, 1222, 4874.

Inheritance and Succession. Brazil, 5189.

Innovative Telematics (online service), 6.

Inscriptions. Aztecs, 312. Mayas, 87, 410, 412, 420–421, 425, 437, 444, 449, 460. Mesoamerica, 87. Stelae, 441. Zapotec, 457.

Instituto de Aposentadoria e Pensões dos Industriários (Brazil), 3710.

Instituto Liberal (Brazil), 3745.

Instituto Mexicano del Café, 4504.

Instituto Nacional de Geografía y Estadística (Mexico), 2401.

Instituto Politécnico Nacional (Mexico), 1409.

Insurance. Brazil, 2178.

Insurrections. Andean Region, 57. Argentina, 3515, 3539. Bolivia, 965. Central America, 4033. Colombia, 3170, 3185, 3205. El Salvador, 3991–3992. Guatemala, 3991–3992. Indigenous Peoples, 57. Mexico, 2828, 2899. Peru, 1051, 1064, 3366. Suriname, 803.

Intellectual Cooperation. Brazil/Japan, 4374.

Intellectual Freedom. *See* Academic Freedom.

Intellectuals. Argentina, 3582, 5123. Chile,

3478. Developing Countries, 2739. Mexico, 4567. Peru, 1054, 3406. Political Ideology, 4567. Political Participation, 5123. Trinidad and Tobago, 3098.
Intelligence Service. Brazil, 3677.
Inter-American Development Bank, 1191–1192, 1220, 3825.
Inter-Tribal Relations. Caribbean Area, 483. Chanca/Inca, 677. Macú/Tucanoan, 886. Precolumbian Civilizations, 102, 150, 246, 263, 301, 303, 462, 539, 557.
Interest Groups. *See* Pressure Groups.
Interest Rates. Argentina, 2064. Bolivia, 1980. Colombia, 1769. Costa Rica, 1589.
Internal Migration. *See* Migration.
Internal Stability. *See* Political Stability.
International Bank for Reconstruction and Development. *See* World Bank.
International-Communist League, 3741.
International Conference on Central American Refugees, 4570.
International Court of Justice, 4025.
International Economic Relations, 3845, 3869, 3872. Argentina, 3567, 3584, 4206. Argentina/Brazil, 4177. Argentina/Chile, 4247. Argentina/Europe, 4215. Argentina/Germany, 4213. Argentina/Great Britain, 4223. Argentina/Uruguay, 2011. Bolivia, 4235, 4239. Brazil, 2148. Brazil/Chile, 4389. Brazil/Italy, 4358. Brazil/Japan, 4377. Brazil/Uruguay, 2011. Brazil/US, 4351. Caribbean Area/Pacific Area, 4077. Central America/Mexico, 3967. Central America/Spain, 4056. Chile, 4249. Chile/China, 4254. Chile/Pacific Area, 4243, 4245. Chile/Russia, 1922, 4258. Colombia/Ecuador, 4269. Colombia/Venezuela, 4269, 4273–4274. Congresses, 2141. Cuba, 4113. Ecuador/Europe, 4288. Guadeloupe/Europe, 3081. Latin America/Canada, 3859. Latin America/China, 3800, 3838. Latin America/Europe, 3792, 3820, 4215. Latin America/Japan, 3828, 4305. Latin America/Pacific Area, 1318. Latin America/Russia, 3856. Latin America/US, 3811, 3828, 3839, 4351. Mexico/South America, 3941. Mexico/US, 3897–3898, 3914, 3916, 3928, 3938, 4467, 4541. Nicaragua/US, 4054. Pacific Area, 3886. Panama/US, 4020. Peru, 4301, 4308, 4314. Peru/Pacific Area, 4317. Peru/US, 3412. South America/Europe, 4164. Southern Cone, 4187. Suriname, 3094. Uruguay, 4320.
International Finance, 1376, 1759, 1773, 3847. Argentina, 4195. Brazil, 2144, 2151, 3725–3726. Central America, 1569. Chile, 4245. Databases, 44. Mexico, 2884. Peru, 4298, 4301, 4308.
International Migration. *See* Migration.
International Monetary Fund (IMF), 1294, 1338, 1346–1347, 1431, 1638, 1647, 1931, 3568, 3663, 3847, 4093.
International Relations, 3796, 3833, 3835, 4065. Andean Region/US, 3296, 4182, 4186. Angola/South Africa, 4127. Antigua and Barbuda/US, 4144. Argentina, 4190, 4203, 4217, 4219, 4225. Argentina/Bolivia, 3321. Argentina/Brazil, 4162, 4359, 4361, 4366. Argentina/Chile, 2545, 2597, 4201, 4247, 4250, 4252, 4263. Argentina/Europe, 4188, 4211. Argentina/Great Britain, 4193, 4198–4199, 4202, 4204, 4209, 4214, 4216, 4220, 4223. Argentina/Spain, 4222. Argentina/US, 3499, 3530, 3533, 4188, 4194, 4200, 4204–4205, 4207, 4209, 4211, 4214, 4221, 4224. Argentina/USSR, 4188, 4190. Belize/Caribbean Area, 3974. Belize/Central America, 3974. Belize/Guatemala, 2944. Belize/Mexico, 2378, 3968. Bolivia, 4227, 4238–4239, 4302. Bolivia/Brazil, 2662, 4240. Bolivia/Chile, 2569–2570, 2572, 4232–4234, 4241–4242, 4259. Bolivia/Latin America, 4236. Bolivia/Peru, 4232, 4234, 4241, 4309, 4312. Bolivia/US, 3283, 3285, 3288, 3293, 3296, 3306, 3309, 3311–3312, 3321, 3354, 3842, 4231, 4972. Brazil, 4382. Brazil/Africa, 4383. Brazil/China, 4362. Brazil/Cuba, 4388. Brazil/Guyana, 4364. Brazil/Italy, 4358. Brazil/Japan, 3804, 4377. Brazil/Latin America, 4362. Brazil/Peru, 4311, 4313. Brazil/US, 4362–4363, 4373, 4375–4376, 4392. Brazil/USSR, 4362. British Caribbean/Latin America, 4157. British Caribbean/Venezuela, 4342. Caribbean Area, 3032. Caribbean Area/France, 4152. Caribbean Area/Great Britain, 4100, 4147. Caribbean Area/Japan, 4084. Caribbean Area/Pacific Area, 4077. Caribbean Area/US, 3850, 4037, 4086, 4090, 4098, 4101, 4137, 4143, 4146–4147. Caribbean Area/USSR, 2804, 3965, 4101. Central America/Colombia, 4055. Central America/Cuba, 4066. Central America/Europe, 4043. Central America/Mexico, 3930, 4074. Central America/Spain, 4056, 4060. Central America/US, 3972–3973, 3978, 3988, 3998–3999, 4013, 4021, 4045, 4048, 4055, 4059. Central America/USSR, 2804, 3965. Chile, 4246, 4248–4249, 4251. Chile/China, 4254.

Chile/Germany, 4253. Chile/Great Britain, 4264. Chile/Japan, 2616. Chile/Pacific Area, 4243. Chile/Peru, 4299, 4312. Chile/Poland, 4261. Chile/Russia, 4258. Chile/US, 3484, 4255–4257, 4262. Colombia, 4270, 4272, 4275–4276, 4278–4279. Colombia/Cuba, 4103. Colombia/Europe, 4277. Colombia/Peru, 4280, 4311, 4313. Colombia/US, 1760, 3156, 4265–4267. Colombia/Venezuela, 2457, 4270, 4273–4274, 4334, 4336, 4339, 4348. Costa Rica/Nicaragua, 2950, 4069. Costa Rica/US, 2951, 4013. Cuba, 1708, 4111. Cuba/Angola, 4118, 4127. Cuba/Great Britain, 4109. Cuba/Grenada, 4145. Cuba/Russia, 3788, 3856. Cuba/South Africa, 4127. Cuba/US, 1682, 3109, 4106, 4112, 4119–4122, 4124, 4129, 4131, 4691. Cuba/USSR, 3107, 3114, 4105–4106. Cuba/Venezuela, 4104. Curaçao/The Netherlands, 4730. Databases, 29. Dominican Republic/Haiti, 3059, 4132–4133, 4142, 4802. Dominican Republic/US, 4132, 4134–4135, 4699. Ecuador, 3225, 4285, 4290–4291. Ecuador/Europe, 4288. Ecuador/Peru, 4281–4284, 4287, 4289, 4311. Ecuador/US, 3214–3215. Ecuador/USSR, 4286. El Salvador/Honduras, 3994, 4030. El Salvador/US, 1603. Great Britain/US, 4204. Guadeloupe/Europe, 3081. Guatemala/Austria, 3977. Guatemala/Mexico, 2378, 3947, 3959, 3982. Guatemala/US, 2978. Guyana/Venezuela, 4343–4344. Haiti, 3064. Haiti/France, 4138. Haiti/US, 4139. Haiti/Venezuela, 4138. Honduras/Mexico, 3969. Honduras/Nicaragua, 4043. Honduras/US, 2995, 3969, 4063. Latin America/Canada, 3808, 3810, 3855. Latin America/China, 3800. Latin America/Europe, 3812, 3817, 3844, 3851, 3865, 4277. Latin America/Japan, 4305. Latin America/Pacific Area, 4160. Latin America/Portugal, 3803, 3853. Latin America/Russia, 3788, 3809, 3836. Latin America/Spain, 3803, 3816, 3844, 3853. Latin America/Sweden, 3868. Latin America/US, 1299, 2787, 3782, 3797, 3801, 3805, 3809, 3814–3815, 3822–3824, 3829, 3835, 3839–3840, 3849–3850, 3852, 3861–3862, 3867, 3870, 3876–3878, 4420. Latin America/USSR, 3857. Latin America/Vatican, 3799. Mexico, 3885, 3932. Mexico/Canada, 3895. Mexico/Europe, 3908. Mexico/France, 3929. Mexico/Pacific Area, 20. Mexico/Russia, 3893. Mexico/US, 1467, 2880, 2888, 3884, 3906, 3909, 3914, 3916, 3927, 3934, 3940, 3942, 3944, 3948, 4467. Mexico/USSR, 3893. Nicaragua, 4015. Nicaragua/US, 3987, 3989–3990, 3996–3997, 4034, 4036, 4057, 4067. Panama/Japan, 4000. Panama/US, 3024, 3970, 3980, 3983, 4000, 4002, 4006, 4012, 4023–4024, 4026, 4032, 4042, 4046, 4070, 4072–4073. Paraguay, 4293–4296. Paraguay/US, 4292. Peru, 3363, 4297, 4302, 4306, 4315–4316, 4318. Peru/Europe, 4314. Peru/Japan, 4305. Peru/US, 4304, 4310, 4319. Peru/USSR, 4300. Puerto Rico/US, 4137. Russia/US, 3809. South America, 4184. South America/Africa, 4171. South America/Europe, 4169. Southern Cone/US, 3813. Suriname, 2303, 3094. Suriname/The Netherlands, 2314, 3090. Theory, 3801. Uruguay, 4320, 4323, 4327. Uruguay/Germany, 4325. Uruguay/Great Britain, 4214. Uruguay/Israel, 4326. Uruguay/USSR, 4324, 4328. Venezuela, 4330–4331, 4335, 4337–4338, 4346. Venezuela/Great Britain, 4332. Venezuela/Spain, 4831. Venezuela/US, 4332–4333, 4345, 4347. Venezuela/USSR, 4341.

International Trade, 27, 1081, 1157, 1162, 1179, 2046, 3834. Argentina, 2163. Brazil, 2093, 2111, 2141, 2144, 2146, 2163. Caribbean Area, 1665. Colombia, 1772. Costa Rica, 1571, 1576. Cuba, 1688, 1698, 1711. Databases, 45, 51. Ecuador, 1794. Mexico, 1308–1309, 1327, 1436, 1469. Southern Cone, 2101, 2160. Statistics, 51. Uruguay, 2011.

International Trade Relations, 1275. Argentina/Brazil, 2114. Argentina/Germany, 4213. Argentina/Iraq, 4175. Brazil/Chile, 4389. Brazil/Mexico, 2169. Brazil/US, 2169. Caribbean Area/US, 1190, 4757. Central America/Japan, 1550. Chile/Japan, 2579. Chile/Pacific Area, 4245. Chile/US, 5035. Cuba/US, 1684. Latin America/Europe, 2110, 3812. Latin America/Southeast Asia, 1304. Latin America/US, 1091. Mexico/Canada, 1372, 1382, 1420, 1436, 1457, 1493. Mexico/Germany, 3926. Mexico/Japan, 1318, 1402, 1504, 1510. Mexico/Pacific Area, 3886, 3923. Mexico/US, 1309–1310, 1317, 1327, 1359, 1372, 1382, 1402, 1418, 1420, 1423–1424, 1436, 1453, 1456, 1469, 1493, 1511, 1518–1519, 1535, 2169, 2880, 2910, 3892, 3894, 3914, 3921, 3935, 3946, 3953. Peru/Pacific Area, 4317. South America/Europe, 4164. Southern Cone, 4187. Southern Cone/US, 2048. Venezuela/US, 4329, 4347.

Internet, 4–5, 9, 13. Commerce, 6, 9. Mexico, 7.
Interpersonal Relationships. Mexico, 4549. Saint Martin, 4779.
Intervention. *See* Foreign Intervention.
Intoxicants. *See* Stimulants.
Investments, 1237, 1271, 3834. Brazil, 2195, 2198.
Iquitos, Peru (city). Popular Religion, 4910. Religious Life and Customs, 4910.
Irrigation, 2217. Andean Region, 1035, 2513. Brazil, 2642, 2669. Caribbean Area, 2217. Mayas, 98. Mesoamerica, 98. Mexico, 340. Peru, 1947, 2513. Precolumbian Civilizations, 229, 372.
Isabela, Dominican Republic. *See* La Isabela, Dominican Republic (settlement).
Iskanwaya Site (Bolivia), 564.
Itaipú Dam, 4356.
Italian Influences. Argentina, 5121.
Italians. Peru, 4916.
Izabal, Guatemala (dept.). Excavations, 363.
Izquierda Unida (Peru), 3364.
Jade. Amazon Basin, 851. Artifacts, 321. Caribbean Area, 851. Mesoamerica, 321. Mexico, 112. Precolumbian Civilizations, 145.
Jalisco, Mexico (state). Agribusiness, 4540. Agricultural Industries, 1470. Agriculture, 1470, 4496. Archaeological Surveys, 345. Clothing Industry, 4540. Economic Conditions, 4529. Economic History, 4540. Economic Policy, 1411. Environmental Pollution, 4533. Fish and Fishing, 4533. Human Rights, 3911. Labor and Laboring Classes, 3911. Migration, 3911, 4529. Popular Culture, 4540. Social Conditions, 4540. Social History, 4540. Tombs, 272. Violence, 3911. Women, 4523.
Jama River Valley (Ecuador). Archaeological Surveys, 646. Paleobotany, 645.
Jamaat al Muslimeen (Trinidad and Tobago), 3100.
Jamaica. Legislature. House of Representatives, 3070.
Japanese, 4305. Brazil, 2625. Peru, 4920. Uruguay, 5133.
Japanese Influences, 4305. Brazil, 2138, 2155. Precolumbian Pottery, 643.
Jehovah's Witnesses. Guatemala, 4603.
Jereissati, Tasso, 3690.
Jesuits. El Salvador, 2960, 2962–2963, 4576.
Jesús María Site (Costa Rica), 477.
Jews. Peru, 4943. Venezuela, 4836.
Jivaro. *See* Shuar.
John Paul II, *Pope*, 3134.
Jōmon Culture (Japan), 643.
Jorge Blanco, Salvador, 3054.
Josselin de Jong, J.P.B. de, 650.
Journalism. Argentina, 3518. Bolivia, 3337, 3346. Brazil, 3757. Colombia, 3164. Congresses, 3346. Environmental Protection, 2786. Mexico, 2811. Nicaragua, 3001. Uruguay, 3619. US, 3996. Venezuela, 3249.
Juárez, Mexico. *See* Ciudad Juárez, Mexico.
Juchitán de Zaragoza, Mexico (town). Indigenismo and Indianidad, 772.
Judges. Chile, 3446. Mexico, 21.
Judicial Power. Chile, 3433. Colombia, 3176. Mexico, 2924. Panama, 3021.
Junín, Peru (dept.). Incas, 688.
Juruna. *See* Yuruna.
Juvenile Delinquents, 4438. Colombia, 1001, 3195.
Juvenile Literature. *See* Children's Literature.
Ka'apor. *See* Urubu.
Kagaba (indigenous group). Ethnography, 998. Myths and Mythology, 998.
Kagwahiv. *See* Parintintin.
Kalapalo. *See* Apalakiri.
Kalinya. *See* Carib.
Kamaiurá (indigenous group). Myths and Mythology, 888.
Kaminaljuyu Site (Guatemala), 287.
Kanjobal (indigenous group). Refugees, 722. US, 722.
Karajá. *See* Caraja.
Kayapó. *See* Cayapo.
Kekchi (indigenous group). Land Settlement, 4649. Land Tenure, 4649. Migration, 4649.
Kennedy, John Fitzgerald, 4135.
Kidnapping. Colombia, 3155. Mexico, 3884, 3922.
Kings and Rulers. Aztecs, 110, 281, 312, 339, 409. Genealogy, 411. Mayas, 153, 257, 277, 399, 425, 434. Mesoamerica, 153, 448. Mixtec, 447. Precolumbian Civilizations, 135, 380. Zapotec, 459.
Kinship. Antigua and Barbuda, 812. Caribbean Area, 801. Cashinawa, 880. Dominica, 815. Ecuador, 4861. Indigenous Peoples, 915. Jamaica, 838. Mapuche, 982. Mayas, 133. Mexico, 2926. Orejón, 909. Peasants, 4861. Peru, 915, 1067. Quechua, 1046.
Kogi. *See* Kagaba.
Koop, Gerhard S., 2322.
Krahó. *See* Craho.
Krannert Art Museum, 927.
Kubitschek, Juscelino, 3670, 3764, 4372.
Kuczynski, Pedro-Pablo, 1418.

Kuikuru (indigenous group). Myths and Mythology, 888.
Kuna. *See* Cuna.
Kwaiker. *See* Cuaiquer.
La Blanca Site (Guatemala). Excavations, 313. Precolumbian Pottery, 314.
La Ceiba, Honduras (city). Description and Travel, 2349.
La Ciénaga Site. *See* Ciénaga Site (El Salvador).
La Entrada, Honduras (region). Archaeological Surveys, 333. Precolumbian Trade, 187. Stone Implements, 185, 187.
La Esmeralda Site. *See* Esmeralda Site (Peru).
La Habana, Cuba (city). History, 4717.
La Milpa Site. *See* Milpa Site (Belize).
La Nación (newspaper, Buenos Aires), 5122.
La Paz, Bolivia (city). Cultural Identity, 953. Domestics, 952, 4959. Housing, 4963. Land Use, 4963. Newspapers, 3337. Religious Life and Customs, 951. Social Classes, 4959. Social Services, 4963.
La Paz, Bolivia (dept.). Indigenous Peoples, 4982.
La Rinconada Site. *See* Rinconada Site (Argentina).
La Tablada, Argentina (military base). Insurrections, 3527, 3539.
La Violencia. *See* Violencia, La (Colombia).
Labor and Laboring Classes. Argentina, 3565, 5095, 5103, 5107. Bolivia, 964. Brazil, 2100, 2110, 2135, 2149, 2155, 2173, 2196, 3710, 3760, 5141, 5157, 5195. Caribbean Area, 3031. Chile, 1872, 5019, 5023. Cuba, 4130, 4725. Databases, 32. Ecuador, 1810, 2474, 3239. El Salvador, 1591. Guatemala, 1613, 4653. Jamaica, 3069. Mexico, 1354, 1374, 2744, 2848, 2919, 3911, 4460, 4464, 4469, 4480, 4483, 4518, 4535, 4537–4538, 4546, 4565. Panama, 4581. Peru, 4885, 4946. Political Culture, 2744. Puerto Rico, 797. Uruguay, 2027. Venezuela, 2744.
Labor Market, 1082, 1107, 1194, 1268, 2046. Brazil, 2108, 2114, 2130, 2135, 2155, 2173, 2183, 2199–2200. Chile, 1851, 4997. Costa Rica, 1571, 1574. Cuba, 3120. Discrimination, 4997. Economic Development, 4432. Ecuador, 1817, 1826. Jamaica, 4693. Mexico, 1312, 1374, 1491, 1514, 1532. Nicaragua, 1626. Panama, 1637. Southern Cone, 2536. Uruguay, 2008, 2013, 2015, 2022, 2028, 2618. US, 1514. Venezuela, 1724, 1736, 1750, 1757. Women, 4432.
Labor Movement, 2741, 2747, 2766, 2776, 3348, 4445. Agricultural Labor, 5073. Argentina, 3503, 3528, 3558, 5103, 5121. Belize, 4634. Bolivia, 3318, 3325, 3340, 3348–3349. Brazil, 2098, 3691, 3695, 3697, 3733, 3760, 3763, 3777, 5144, 5195–5196, 5203. Caribbean Area, 3031. Central America, 4702. Chile, 3437, 3445, 4994, 5073. Cuba, 4725. Dominican Republic, 3057. Ecuador, 1790, 3235, 3246. El Salvador, 2969. Henequen Industry and Trade, 4561. Jamaica, 3069. Mexico, 2896, 2912, 2927, 2929, 4471, 4479, 4483, 4491, 4518, 4546, 4565. Nicaragua, 3008. Panama, 4581, 4681. Peru, 4885, 4946. Repression, 3777. Rubber Industry and Trade, 5160. Southern Cone, 2790. Uruguay, 3655, 3658, 3662. Venezuela, 3271. White Collar Workers, 3760. Women, 4994, 5203.
Labor Policy. Argentina, 5121. Colombia, 4818. Dominican Republic, 4133. El Salvador, 2969. Haiti, 4133. Mexico, 4479, 4546, 4565. Southern Cone, 2101, 2790. Venezuela, 1750.
Labor Productivity, 1218. Mexico, 1477.
Labor Supply. Argentina, 1083. Brazil, 1083, 2130, 2159, 2199, 2668. Chile, 1083, 1879, 1912, 1916. Colombia, 1784. Jamaica, 4693. Mexican-American Border Region, 1365. Mexico, 1486, 4535, 4538. Peru, 1966. Uruguay, 2017, 2022, 2027. Venezuela, 1750. Women, 1879, 1916, 2017. Youth, 2017.
Labor Supply, Brazilian. Japan, 3804.
Lacalle Herrera, Luis Alberto, 3618, 3639, 3665.
Lacam-Tún Site (Mexico), 357.
Lacandon (indigenous group). Archaeological Surveys, 357. Artifacts, 357. Human Ecology, 753. Indigenous/Non-Indigenous Relations, 727. Mortuary Customs, 357.
Laíno, Domingo, 3607.
Lake Chapala (Mexico), 4533.
Lake Pátzcuaro (Mexico). Climatology, 2394.
Lake Titicaca (Peru and Bolivia). Archaeological Surveys, 706. Economic Development, 4302. Underwater Archaeology, 572.
Lambayeque, Peru (dept.). Festivals, 1056. Precolumbian Land Settlement Patterns, 702.
Lambityeco Site (Mexico), 365. Tombs, 406.
Lame Chantre, Manuel Quintín, 995.
Land Invasions. Argentina, 5113. Brazil, 873. Costa Rica, 1587. Dominican Republic, 4756. Ecuador, 925, 4858. Guatemala, 4577. Mexico, 763. Puerto Rico, 4713. Violence, 4858.
Land Reform. Amazon Basin, 2698, 5143. An-

dean Region, 2515. Argentina, 3566. Bolivia, 1984, 2123, 3305, 4977. Brazil, 2104, 2123, 2177, 3699, 3703, 3740, 3746. Catholic Church, 4588. Central America, 2939, 4605. Chile, 3436, 5018. Colombia, 1188. Costa Rica, 4658. Ecuador, 4876. El Salvador, 1604, 4605. Guatemala, 2978, 4588, 4620, 4656. Honduras, 1615, 1618. Mexico, 1315, 1329, 1341, 1396, 1403, 1425, 2123, 2387, 2845, 4462, 4552. Nicaragua, 1621, 1629, 4605, 4608, 4668. Peru, 2123, 2515, 3404, 4934–4935. Sandinistas, 4608, 4668. Venezuela, 1726.

Land Settlement, 1123, 2268, 2271, 2276. Amazon Basin, 861, 2626, 2665, 2678, 5143. Belize, 349. Bolivia, 2522–2523, 4230, 4981. Brazil, 2622, 2649, 2661, 2678, 2693, 2695, 2707, 5143. Central America, 1552, 4400. Colombia, 2454, 3207, 4808. Cuba, 776. Ecuador, 2522. Guatemala, 2344, 2347, 4602. Honduras, 279, 2348. Mexico, 727, 737, 2387, 2398. Peru, 2492, 2510. Uruguay, 2619.

Land Tenure, 2420. Andean Region, 1038. Argentina, 5089. Aymara, 971. Bolivia, 2521, 3339, 4977. Brazil, 2177, 2648, 2679, 4380, 5189. Chamí, 1007. Chile, 943, 969, 2578. Colombia, 1000, 1006. Colonial History, 2398. Cuba, 1690. Dominican Republic, 4727, 4781. Ecuador, 2482, 4858, 4876. Guatemala, 4649, 4656. Honduras, 4611. Indigenous Peoples, 4977. Mapuche, 943. Mexico, 763, 1315, 1378, 1396, 2364, 2398. Nicaragua, 1621. Paraguay, 4989. Pehuenche, 969. Peru, 1066. Saint Lucia, 786, 814. Sumo, 2348. Viceroyalty of New Spain, 2417.

Land Use, 2267, 2420, 4443. Amazon Basin, 855, 2627, 2715. Andean Region, 2421, 2428. Argentina, 2563, 5089. Barbados, 2286. Bolivia, 2521, 4977. Brazil, 2177, 2639, 2668, 2677, 2694, 3746. Central America, 1552, 2317, 2943. Chile, 5084. Costa Rica, 730. Ecuador, 2472. Haiti, 4787. Honduras, 4611. Indigenous Peoples, 2267. Jamaica, 3069. Mesoamerica, 276. Mexico, 1348, 2317, 2356, 2396, 2416, 4551. Nicaragua, 4640. Peru, 2492, 2516, 4902. Precolumbian Civilizations, 2416. Venezuela, 2435, 2442.

Landowners. Argentina, 3553, 3566, 5089. Peru, 4916. Saint Lucia, 786.

Landslides. Chile, 2599, 2608. Peru, 4902.

Language and Languages, 59. Aztecs, 154, 438. Guadeloupe, 833. Jamaica, 785. Martinique, 833. Mayas, 90, 420, 453. Mexico, 2383. Olmecs, 426–427. Puerto Rico, 4753, 4759. Zoque, 426–427.

Laredo, Mexico (city). City Planning, 2371.

Latifundios. Brazil, 2648, 2695. Costa Rica, 730. Paraguay, 4989.

Latin American Area Studies, 3030. Caribbean Area, 4746. Congresses, 58. USSR, 3774.

Latin American Economic System. *See* Sistema Económico Latinoamericano.

Latin American Newsletters, 34.

Latin American Studies Association (LASA), 3611.

Latin Americanists, 3786. Austria, 2252. Congresses, 58. Germany, 2252.

Law and Legislation, 3782. Anthropology, 988. Antigua and Barbuda, 812. Argentina, 3563, 5118. Bolivia, 3279–3280, 3289. Brazil, 3731. Colombia, 1000, 3161. Constitutions, 3608. Cuba, 4723, 4784. Databases, 29. Ecuador, 4857. El Salvador, 2964. Family and Family Relations, 4852. Honduras, 2348. Indigenous Peoples, 988, 993, 1032, 4880. Mass Media, 2774. Mexico, 2898. Nicaragua, 4642. Paraguay, 3608. Peasants, 1058. Peru, 1032, 1058, 3374, 4880. Poor, 4843. Quechua, 1058. Venezuela, 3270, 4843, 4852. Women, 4642, 4852, 4857.

Lawyers. Cuba, 4723.

League of Nations, 4065.

Lebanese. Mexico, 4482. Trinidad and Tobago, 4698.

Legends. Ecuador, 1019. Incas, 527. Mayas, 740. Precolumbian Civilizations, 461.

Legislative Bodies. Argentina, 3562–3563, 3573. El Salvador, 2969. Mexico, 2912.

Legislative Power. Argentina, 3399. Bolivia, 3345. Central America, 2984. Chile, 3423. Ecuador, 3218. Guatemala, 2984. Mexico, 2924. Peru, 3399.

Legislators. Argentina, 3562–3563, 3573. Brazil, 3685, 3768. Jamaica, 3070. Mexico, 21, 2912.

Legislatures. *See* Legislative Bodies.

León, Mexico (city). Employment, 1497. Population Growth, 1497.

Letelier, Orlando, 4257.

Leticia Dispute (1932–1934), 4280.

Lewis, Oscar, 5034.

Lewis, William Arthur, 3042.

Lexicons. *See* Dictionaries.

Liberalism. Argentina, 3507.

Liberation Theology, 4423. Brazil, 5149.

Chile, 5009. El Salvador, 4576. Haiti, 3061. Social Change, 5009. Southern Cone, 2797.
Libertad, Peru. See La Libertad, Peru (dept.).
Library Catalogs. See Online Catalogs.
Liga Comunista Internacionalista (Brazil), 3741.
Ligua, Chile. See La Ligua, Chile (city).
Lima, Peru (city). Children, 4879. Community Development, 4902. Cultural Adaptation, 1045. Dance, 1045. Democracy, 3406. Economic Conditions, 3367. Ethnic Groups and Ethnicity, 1067. Family Violence, 4927. Housing Policy, 2494. Intellectuals, 3406. Jews, 4943. Land Use, 4913. Migration, 1026, 1067. Municipal Services, 4940. Natural Disasters, 4902. Neighborhood Associations, 4890, 4913, 4940. Political Left, 3364. Quechua, 1061. Refugees, 4911. Slums, 2494, 4927. Social Life and Customs, 4913. Social Movements, 3364, 4887, 4908. Social Services, 4940. Squatter Settlements, 2494, 4913, 4940. Street Children, 4446. Urban Sociology, 4908. Violence, 3367. Youth, 4895.
Lima, Peru (dept.). Agricultural Systems, 1043. Social Structure, 1043.
Lima Netto, Roberto Procópio de, 2164.
Limón, Costa Rica (prov.). Race and Race Relations, 826.
Linguistics. Cultural History, 59. Haiti, 830.
Literacy and Illiteracy. Andean Region, 960, 1063. Colombia, 990. Guadeloupe, 833. Martinique, 833. Mayas, 453. Precolumbian Civilizations, 453.
Literary Criticism. Guyana, 842.
Literature. Databases, 36.
Lithics. See Stone Implements.
Livestock. Brazil, 2118. Mexico, 1530.
Lleras Restrepo, Carlos, 3181.
Lobbyists. See Pressure Groups.
Local Elections. Argentina, 3525. Belize, 4637. Bolivia, 3329. Colombia, 3169. Ecuador, 3232. Mexico, 2869, 2933. Paraguay, 3600.
Local Government. See Municipal Government.
Local History. Costa Rica, 2338.
Loja, Ecuador (prov.). Water Supply, 4287.
Loltún Cave (Mexico). Bats, 201.
Lomas Entierros Site (Costa Rica), 478.
Lomé Convention, 4082.
Lopes, Lucas, 2168.
López, Vicente Fidel, 3507.
Loret de Mola, Rafael, 2878.

Los Angeles, California (city). Trinidadians, 801.
Los Roques, Venezuela (islands). Precolumbian Land Settlement Patterns, 484.
Low Intensity Conflicts. Belize, 3350. Central America, 3972, 4593.
Lula, 3682, 3717, 3754.
Lupaqa (indigenous group). Cosmology, 977.
Lusinchi, Jaime, 4331, 4342.
Luxuries. Aztecs, 209.
M-19 (Colombian guerrilla group), 3157, 3190, 3209, 3211.
Machaka, Laureano, 961.
Machiganga. See Machiguenga.
Machiguenga (indigenous group). Manioc, 918. Myths and Mythology, 923. Tools, 918.
Machismo. See Sex Roles.
Machu Picchu Site (Peru), 2510. Conservation and Restoration, 709. Excavations, 709.
Macorix (indigenous group), 501.
Macú (indigenous group). Amazon Basin, 886. Social Life and Customs, 886. Trade, 886.
Madrazo, Carlos A., 2866.
Madre de Dios, Peru (region). Bibliography, 863.
Madres de Plaza de Mayo (Argentina), 3508, 3514, 3541.
Madrid Hurtado, Miguel de la, 2868, 2877, 2927, 4565.
Magallanes, Chile (prov.). Archaeology, 598.
Magdalena, Colombia (dept.). Land Settlement, 3207. Social Conflict, 3207. Violence, 3207.
Magdalena River Valley (Colombia). Precolumbian Pottery, 613, 628.
Magic. Venezuela, 4853.
Mai Huna. See Orejón.
Maids. See Domestics.
Maisabel Site (Puerto Rico). Artifacts, 515.
Maize. See Corn.
Mak'a. See Macá.
Makú. See Macú.
Makuna. See Macuna.
Malaria. Brazil, 2708.
Malinalco Site (Mexico). Conservation and Restoration, 393.
Malla Site (Costa Rica). Excavations, 475.
Malmok Site (Aruba), 520.
Malnutrition. See Nutrition.
Maluf, Paulo, 3752.
Malvinas, Islas. See Falkland Islands.
Manabí, Ecuador (prov.). Precolumbian Civilizations, 645. Rural Sociology, 4861.
Managua, Nicaragua (city). Land Use, 4640.
Mangrove Plants. Guadeloupe, 2299.

Mangrove Swamp Ecology. Colombia, 2462. Guadeloupe, 2299.
Manioc. Machiguenga, 918.
Manley, Michael, 4750.
Manufactures. Brazil, 2113. Uruguay, 2012.
Manuscripts. Aztecs, 89. Mixtec, 85, 447.
Maps and Cartography, 2216, 2230, 2261. Chile, 2591, 2601. Colonial History, 2248–2249. Mexico, 18, 25, 2362. Peru, 2507, 2509. Puerto Rico, 2284, 2311–2312. South America, 2419, 2423.
Mapuche (indigenous group), 536. Autonomy, 976. Cosmology, 975. Discrimination, 5068. Infant Mortality, 944. Kinship, 982. Land Tenure, 943. Myths and Mythology, 978. Oral History, 945. Photography, 940. Silverwork, 980.
Maquiladoras. Employment, 1460. Guatemala, 4629. Mexican-American Border Region, 1460. Mexico, 1317–1318, 1328, 1352, 1354, 1362, 1383, 1387, 1390, 1437–1438, 1445, 1466, 1510, 1513, 1515, 1524, 2379, 4464, 4472, 4478, 4536, 4544. Puerto Rico, 4718. Technological Innovations, 1460. Trade Unions, 4544.
Maracaibo Gulf (Colombia and Venezuela), 4336. Boundary Disputes, 4334.
Marajó Island (Brazil). Precolumbian Land Settlement Patterns, 583.
Maranhão, Brazil (state). Christian Base Communities, 3667.
Marie Jean Robert, Joseph, 3082.
Mariel, Cuba (city). Exiles, 4691, 4734.
Marine Resources. Argentina, 2553. Caribbean Area, 2296–2297. Mesoamerica, 111. Mexico, 2409. Precolumbian Civilizations, 111. West Indies, 2296.
Maritime History. Peru, 4307.
Maritime Law. Bibliography, 4232. Bolivia, 4232.
Maritime Policy. Argentina, 4180. Chile, 4180. Peru, 4307. South America, 4174, 4185.
Market Surveys. Peru, 3382.
Marketing. Antigua and Barbuda, 1670.
Markets. Argentina, 4992. Bolivia, 1984. Cuba, 1689. Mexico, 4515. Paraguay, 4992. Peru, 1057.
Maroons. Archaeological Surveys, 480. Caribbean Area, 805. Jamaica, 479–480. Suriname, 803, 825.
Marriage. Caribbean Area, 774. Cuba, 4772. Guyana, 4715. Indigenous Peoples, 915. Mexico, 4543. Paraguay, 4986. Peru, 915. Quechua, 1062. Sipibo, 917. Trinidad and Tobago, 4712.

Martial Law. Colombia, 3174. Nicaragua, 3001.
Martínez Cuenca, Alejandro, 3003.
Marulanda Vélez, Manuel, 3152.
Marxism, 2760, 4395, 4406, 4460. Chile, 3453. Colombia, 3162. Jamaica, 4154. Mexico, 1471.
Mashco (indigenous group). Rites and Ceremonies, 919. Sex Roles, 919.
Masons. *See* Freemasonry.
Mass Media, 2774. Barbados, 3074. Bibliography, 4492, 4700–4701. Bolivia, 3346, 3353, 4952. Brazil, 3757, 3765, 3773, 3776, 5192. Caribbean Area, 4700. Chile, 3427, 5038, 5061. Colombia, 3164. Cuba, 4701. Databases, 34–35, 50. Elections, 2803, 5163. Environmental Protection, 2786. Mexico, 35, 2811–2812, 2820, 2885–2886, 2899, 4492, 4513, 4559. Political Culture, 4559. Political Ideology, 4513. Public Opinion, 5130. Puerto Rico, 4778, 4791. Suriname, 4792. Uruguay, 5130. US, 3011, 3861, 3996, 4050. Venezuela, 3249.
Massacres. Guatemala, 4610.
Matacapan Site (Mexico). Precolumbian Pottery, 190–191.
Matagalpa, Nicaragua (dept.). Counterrevolutions, 4672.
Matamoros, Mexico (town). Maquiladoras, 4478.
Material Culture. Aymara, 974. Aztecs, 209, 243. Brazil, 574, 576, 579, 581, 590. Caribbean Area, 805. Chile, 600, 609. Colombia, 616, 622, 625, 629, 631, 633. Mayas, 76, 205, 332. Mesoamerica, 332. Suriname, 652, 805, 825.
Mathematics. Precolumbian Civilizations, 697.
Matlatzinca (indigenous group). Artifacts, 377. Precolumbian Pottery, 377.
Mato Grosso, Brazil (state). Archaeology, 595. Economic History, 2118. Elites, 3742. Political Parties, 3742. Political Sociology, 3742.
Matos González, Ramiro, 3055.
Matrícula de Tributos, 124, 442.
Matsigenka. *See* Machiguenga.
Matthei, Evelyn, 3425.
Mayan Influences. Precolumbian Sculpture, 280.
Mayan Language. *See* Indigenous Languages.
Mayas, 165. Acculturation, 327. Agricultural Systems, 254. Agriculture, 162. Archaeoastronomy, 404. Archaeological Surveys, 71, 91, 155, 394. Artifacts, 305. Asian Influences, 410. Astronomy, 433. Ball Games,

270, 364. Belize, 104. Calendrics, 87, 421, 423, 2392. Carnival, 738. Caves, 212. Chiefdoms, 257. City Planning, 139. Codices, 423. Congresses, 138. Cosmology, 404. Cults, 324. Cultural Collapse, 151, 169. Cultural Development, 166. Cultural History, 71. Cultural Identity, 740–741, 769–771. Dance, 418. Demography, 172, 268, 403. Economic Models, 133. Elites, 135, 327. Epigraphy, 87, 90, 136, 138, 412–413, 420–421, 423, 425, 433, 460. Ethnoarchaeology, 261, 404. Excavations, 74, 400. Genealogy, 153. Geographical Names, 449. Government Relations, 757, 4650. Guatemala, 2345. Guidebooks, 123, 258. Historic Sites, 741. Historical Demography, 2345. Historiography, 425, 454. Households, 285, 368. Human Rights, 2981. Iconography, 75, 136, 138, 148, 155, 208, 270, 367, 387, 418, 425, 444. Indigenismo and Indianidad, 769–771. Inscriptions, 87, 410, 412, 420–421, 425, 437, 444, 449, 460. Irrigation, 98. Kings and Rulers, 153, 257, 277, 399, 425, 434. Kinship, 133. Language and Languages, 90, 420, 453. Legends, 740. Literacy and Illiteracy, 453. Material Culture, 76, 205, 332. Midwives, 752. Monuments, 394, 444. Myths and Mythology, 103. Navigation, 359. Obsidian, 252. Philosophy, 103. Political Development, 130, 257. Political Institutions, 155. Political Systems, 97. Population Growth, 2345. Precolumbian Architecture, 66, 208, 387, 398. Precolumbian Art, 130, 160. Precolumbian Land Settlement Patterns, 130, 267–268, 285, 332, 398. Precolumbian Pottery, 148, 269. Precolumbian Sculpture, 208, 273, 277, 367, 387, 418, 425. Precolumbian Trade, 133, 283. Relations with Spaniards, 350. Religion and Politics, 120. Religious Life and Customs, 108, 136, 138, 356, 433, 752, 4616. Research, 93, 151, 166, 171. Rites and Ceremonies, 108, 324, 356, 370–371, 418, 445. Sacred Space, 103. Shamanism, 108, 742. Social Life and Customs, 368. Social Mobility, 4616. Social Structure, 105, 130, 376, 745. Sports, 364. Stelae, 413, 434, 445, 460. Stone Implements, 181, 370–371. Symbolism, 87, 103, 136. Symbolism (art), 160. Syncretism, 410. Terracing, 254. Tombs, 413. Traditional Medicine, 742, 752. Treatment of, 757. Urbanization, 139. Warfare, 399. Water Resources Development, 98. Women, 752. Writing, 433, 435, 449, 453, 460.
Mayors. Colombia, 3169.
Mbaya. *See* Guaycuru.

MCCA. *See* Mercado Común Centroamericano.
McNamara, Robert, 4106.
Meat Industry. Mexico, 1325.
Medellín, Colombia (city). Juvenile Delinquents, 3195. Violence, 1001, 3195.
Mediation, International. *See* International Arbitration.
Medical Anthropology. Aztecs, 142. Brazil, 585. Colombia, 634–635, 991. Ecuador, 1009, 4873. Mexico, 142, 320, 4517. Venezuela, 4833, 4837.
Medical Care, 1104, 1225, 2731. Bolivia, 4958, 4969. Brazil, 2670, 2710. Chile, 2610. Cuba, 4739. Dominican Republic, 1592, 1619, 4796. Ecuador, 4869, 4873. El Salvador, 1592. Haiti, 4739. Honduras, 1619. Households, 4869. Jamaica, 838, 1619. Mexico, 4517. Nicaragua, 4615. Peru, 4896. Poor, 4430, 4873. Venezuela, 1754.
Medical Education. Nicaragua, 4615.
Medicinal Plants, 2281. Sioni, 920.
Medicine. Cuba, 3124. Venezuela, 4833, 4837.
Mella Latorre, Alejandro, 3602.
Mendes, Chico, 3688, 5160.
Mendoza, Argentina (prov.). Geography, 2547.
Menem, Carlos Saúl, 3500, 3506, 3513, 3530, 3533, 3536, 3543, 3548, 3555, 3559, 3561, 3579, 3585–3586, 4194, 4211–4212.
Mennonites. Belize, 2322.
Mental Health, 4411.
Mercado Común Centroamericano, 1588, 3973.
Merchant Marine. Peru, 4307.
Merchants. Mexico, 4515. Paraguay, 4992. Women, 4992.
Mercosul. *See* Mercosur.
Mercosur, 1143, 1148, 1210, 1216, 2003, 2006, 2008–2009, 2011–2012, 2023–2024, 2048, 2096, 2101, 2106, 2110, 2139, 2152, 2160, 2163, 2188, 2536, 3781, 4164, 4166, 4212, 4320–4321, 4378.
Merengue. Dominican Republic, 789–790.
Mérida, Mexico (city). Elections, 2878. Politicians, 2878.
Mérida, Venezuela (city). Cities and Towns, 2434. Geography, 2434, 2439. Urbanization, 2434.
Messianism. Brazil, 908. Peru, 908.
Mestizos and Mestizaje, 59, 62. Andean Region, 937. Bolivia, 3355, 4968, 4979. Brazil, 5185, 5202. Chile, 5052. Ecuador, 4870. Peru, 1028, 4895. Trinidad and Tobago, 809. US, 5202.
Metallurgy. Mesoamerica, 297. Precolumbian Civilizations, 556, 559.

Mexica. *See* Aztecs.
Mexican-American Border Region, 1432, 3890–3891, 3920, 3928. Cultural Identity, 2358. Economic Conditions, 1317, 1331, 1356, 1360, 1461, 1523, 1535, 2367, 3898. Environmental Policy, 1489. Free Trade, 3962. Human Rights, 3889. Labor and Laboring Classes, 4538. Maquiladoras, 1460. Migration, 1535, 2359, 3889. Public Finance, 1538. Social Conditions, 1331, 1360, 1461, 2367. Water Supply, 3955.
Mexican Revolution (1910–1920), 3908.
Mexicans. US, 2359, 2408, 3891, 3912, 4062, 4467, 4475–4477, 4494, 4512.
Mexico. Secretaría de Relaciones Exteriores, 3950.
México, Mexico (city). Archaeology, 322. Christian Base Communities, 4528. City Planning, 2355, 2372–2373. Climatology, 2385. Colonial History, 322. Commerce, 4505. Elections, 2820, 2902. Environmental Degradation, 2377. Environmental Pollution, 2369, 2386. Excavations, 395. Households, 4563. Housing, 4563. Informal Sector, 4505. Land Use, 4505. Merchants, 4505. Political Bosses, 2874. Population Growth, 2372–2373. Rural-Urban Migration, 744. Social Conflict, 4505. Social Movements, 4528. Urban History, 322. Urbanization, 2372–2373.
México, Mexico (state). Archaeological Parks, 393. Commerce, 1408. Demography, 2407, 4526. Economic Conditions, 1498. Elections, 2918. Industry and Industrialization, 1408. Population Growth, 2407, 4526. Public Finance, 2881. Public Opinion, 2918. Rural-Urban Migration, 2407. Social Conditions, 1498.
Mexico. Boundaries, 3899.
Mexquitic, Mexico (village). History, 4465.
Michoacán, Mexico (state). Archaeological Geography, 119. Archaeological Surveys, 224. Cattle Raising and Trade, 4463. Cities and Towns, 1521. Ecology, 119. Election Fraud, 4525. Elections, 4525. Ethnography, 754. Income Distribution, 1521. Industry and Industrialization, 1521. Land Use, 2396. Migration, 2388, 4455. Political Conditions, 2870. Precolumbian Pottery, 224. Tombs, 119. Urbanization, 1521.
Microelectronics Industry. Brazil, 3694.
Middle Classes. Brazil, 5171. Jamaica, 1672. Mexico, 4508. Peru, 3371. Political Ideology, 5171.
Midwives. Mayas, 752.

Migrant Labor. Brazil, 2701. Mexico, 4456, 4473, 4520.
Migration, 2263, 3807, 4444. Agricultural Labor, 5170. Argentina, 2557, 2561. Aztecs, 129. Bolivia, 950, 2523, 4949. Brazil, 2180, 2654, 2686, 2697, 2708, 5147, 5155, 5170. Caribbean Area, 778, 791, 801, 4714, 4736, 4800. Central America, 1551, 4400, 4542, 4591, 4623. Chile, 2589. Colombia, 3204, 4826. Costa Rica, 813. Cuba, 4122, 4691, 4726. Dominican Republic, 839, 3059, 4726. Ecuador, 2471, 4856. El Salvador, 1597. France, 2557. Guatemala, 1608, 2347. Honduras, 4617. Indigenous Peoples, 4856. Jamaica, 840, 4693. Mesoamerica, 129. Mexican-American Border Region, 1535, 2359, 3889. Mexico, 1352, 1365, 1398, 1513–1515, 2322, 2378, 2388, 2395, 2408, 2413, 3911, 4461, 4474, 4482, 4537. Nicaragua, 1626. Paraguay, 2697, 4986. Peru, 1067, 2508, 4881, 4911, 4920, 4932. Quechua, 1027. Research, 4537. Saint Kitts and Nevis, 4761. Spain, 2557, 4831. Uruguay, 5140. US, 1514, 2388, 2408, 3807, 3957, 4122. Venezuela, 2446, 4831.
Migration, Brazilian. Japan, 3804.
Migration, Caribbean. Central America, 4702. Great Britain, 4748. US, 3957, 4087, 4757, 4800.
Migration, Central American. Mexico, 3981, 4542, 4619. US, 4600, 4619.
Migration, Cuban. US, 4691, 4734.
Migration, Curaçaoan. The Netherlands, 4743.
Migration, Dominican. US, 4729.
Migration, East Indian, 791.
Migration, Guatemalan. Mexico, 4031, 4456, 4473, 4520. US, 4687.
Migration, Haitian. Cuba, 4735. Dominican Republic, 4754. US, 4711.
Migration, Japanese, 4305. Amazon Basin, 2625. Brazil, 2625. Dominican Republic, 3052. Peru, 4305.
Migration, Mexican. US, 3900–3901, 3912, 3956–3957, 4455, 4467–4468, 4472, 4475–4477, 4494, 4512, 4529, 4535, 4556–4557. Women, 4485.
Migration, Peruvian. US, 4882.
Migration, Puerto Rican. US, 4771.
Migration, Russian. Paraguay, 2769.
Migration, Women. Mexico, 4485.
Migration Policy. Mexico, 3981. US, 3900, 3912, 3956.
Milagro, Ecuador (region). Peasant Movements, 4876.

Milan, Italy (city). Uruguayans, 5139.
Militarism. Aztecs, 339. Bolivia, 3277, 3311. Peru, 3370. Precolumbian Civilizations, 117. Suriname, 3091.
Military, 2727, 2770, 2780, 2794, 2800, 3785, 3815. Argentina, 3499, 3515, 3520, 3527, 3547, 3577, 4200, 4208, 5091. Brazil, 3680, 3692, 3705, 3712, 3739, 3767, 3769, 3780, 4379, 5213. Caribbean Area, 4149. Chile, 3465, 3492, 4253, 5040, 5048. Colombia, 3179, 3209. Cuba, 3144, 4119. Databases, 31. Defense Industries, 3692. Democratization, 2743, 3653, 3657, 5048. Drug Enforcement, 3819. Ecuador, 3225, 3227, 3231, 4282. Guatemala, 2980, 2983, 4049, 4571. Honduras, 2994. Mexico, 2810, 3934. Nicaragua, 3000. Nuclear Energy, 3705. Panama, 3029. Paraguay, 3610, 3616. Peru, 3370, 3376, 3384. Political Ideology, 3680, 3739. Public Administration, 5091. Public Opinion, 5040. South America, 2788, 4173. Suriname, 3087, 3093. Uruguay, 3618, 3636, 3653, 3656–3657. US, 3815, 4026. Venezuela, 3260, 3269, 4339.
Military Assistance, Argentine. Iraq, 4175.
Military Assistance, Brazilian. Iraq, 4175.
Military Assistance, Chilean. Iraq, 4175.
Military Assistance, German. Chile, 4253.
Military Assistance, US, 3815. Central America, 4593. Chile, 4255, 4262. Panama, 4026.
Military Government, 2763, 2770, 3881. Argentina, 3549, 3583, 4225, 5112. Bolivia, 3283, 3295, 3301, 3308, 3319. Brazil, 3705, 3748, 3780, 4999, 5213. Catholic Church, 3467, 5051. Chile, 3428, 3430–3431, 3433, 3444, 3454, 3456, 3458, 3460, 3464–3465, 3467, 3475–3476, 3479, 3481, 3486–3487, 3492, 3494–3495, 4257, 4999, 5044, 5051, 5079. Dominican Republic, 4699. Ecuador, 3230. Environmental Policy, 3780, 4999, 5213. Guatemala, 4571. Haiti, 3068. Indigenous Policy, 5044. Panama, 3016–3017, 3024. Paraguay, 4292, 4294, 4296. Peru, 3385, 3388, 3412. Southern Cone, 2751–2752. Suriname, 3086. Uruguay, 3620, 3633, 3636, 3661.
Military History. Argentina, 3549. Brazil, 3739, 3767, 3769. Chile, 3443, 3488. Colombia, 3154, 3157, 4280. El Salvador, 2972. Panama, 3028. Paraguay, 3597, 3606, 3610, 3616. Peru, 3384, 3388, 4282. Suriname, 3087. US, 4003, 4017, 4037.
Military Occupation, US. Dominican Republic, 4699.

Military Policy, 2780, 3791. Argentina, 3527, 3530, 3541, 3577, 3583, 4200, 4218. Brazil, 3712, 4175. Chile, 4251, 4260, 4262. Colombia, 3177. Panama, 3020. South America, 4173, 4179, 4181, 4184. US, 4034, 4042.
Millas, Orlando, 3468.
Millennialism. Amazon Basin, 852. Indigenous Peoples, 852, 889.
Millor, Pablo, 3650.
Milpa Site (Belize). Archaeological Surveys, 394. Ball Games, 364.
Minas Gerais, Brazil (state). Agricultural Development, 2171. Agricultural Labor, 5170. Archaeology, 574, 578. Economic Conditions, 2201. Elites, 3759. Employment, 2171. Migration, 5170. Political Ideology, 3759. Social Policy, 2201.
Minerals and Mining Industry, 2260. Bolivia, 964, 1975. Brazil, 2631, 2685. Colombia, 4827. Costa Rica, 2324. Mexico, 1474, 2368, 2402, 2414, 2930. Peru, 1951. Precolumbian Civilizations, 199. Venezuela, 2438, 2441.
Mir, Pedro, 4746.
Miskito. *See* Mosquito.
Miskito Coast. *See* Mosquitia (Nicaragua and Honduras).
Missionaries. Amazon Basin, 882. Andean Region, 955.
Missions. Ashaninca, 3373. Aymara, 955. Colombia, 905. Peru, 3373.
Mitla, Mexico (village). Precolumbian Architecture, 196. Precolumbian Art, 196. Precolumbian Civilizations, 196.
Mitre, Bartolomé, 3507.
Mixtec (indigenous group), 85. Artifacts, 391. Cultural History, 113. Excavations, 179. Genealogy, 447. Human Rights, 4461. Kings and Rulers, 447–448. Manuscripts, 447. Migration, 4461. Political Development, 448. Precolumbian Pottery, 179. Writing, 435.
Mixtequilla Site (Mexico). Precolumbian Pottery, 241.
Moche. *See* Mochica.
Mochica (indigenous group). Archaeological Surveys, 666. Artifacts, 671. Bibliography, 666. Cultural Collapse, 702. Cultural Development, 702. Cultural History, 666. Ethnography, 1055. Kings and Rulers, 685. Political Culture, 655. Precolumbian Art, 671. Religious Life and Customs, 701. Symbolism, 655. Tombs, 685, 701.
Mocovi. *See* Mocobi.

Modern Architecture. Ecuador, 2466.
Modernization, 1250. Agriculture, 1073. Brazil, 2141, 5209. Chile, 973, 3471, 5023, 5030. Dominican Republic, 4797. Guatemala, 4663. Indigenous Peoples, 4504. Mexico, 4504. Peru, 1039.
Moffitt, Ronni Karpen, 4257.
Moguex (indigenous group). Ethnography, 1003.
Mojarra Site (Mexico), 426–427. Stelae, 441, 450, 460.
Monarchs. *See* Kings and Rulers.
Monetary Policy, 1161, 1376, 1894. Argentina, 2035–2036, 2068, 2088, 2137. Bolivia, 1974, 1985. Brazil, 2137, 2150, 2165, 2212, 3757. Canada, 1492. Chile, 1873, 1889, 1894–1895, 1913. Costa Rica, 1577. Dominican Republic, 1660. Ecuador, 1801, 1811. Guatemala, 1611. Honduras, 1617. Mexico, 1484, 1492. Peru, 1954, 1961, 2137. Trinidad and Tobago, 1667. Venezuela, 1717.
Monetary Unions. Caribbean Area, 1656.
Money. Indigenous Peoples, 726. Trinidad and Tobago, 1664.
Money Laundering. Dominican Republic, 3060.
Money Supply. Argentina, 2064. Brazil, 2208. Costa Rica, 1577, 1589. Ecuador, 1805. El Salvador, 1595, 1605. Guatemala, 1605.
Monopolies. Venezuela, 1756.
Monte Albán Site (Mexico), 196, 263, 320, 323, 329, 407. Architecture, 236, 256. Astronomy, 440.
Monte Negro Site (Mexico), 179.
Monte Verde Site (Chile), 543, 603.
Monterrey, Mexico (city). Cults, 4566. Religious Life and Customs, 4566. Unemployment, 1480.
Montevideo, Uruguay (city). Informal Sector, 2618. Labor Market, 2618. Municipal Government, 3648, 3651, 3662. Neighborhood Associations, 3648. Urban Policy, 3620.
Montezuma II, *Emperor of Mexico*, 110.
Montoro, André Franco, 3738.
Monuments. Aztecs, 281, 312. Guatemala, 324. Honduras, 306. Mayas, 394, 444. Mesoamerica, 280. Olmecs, 282.
Moquegua, Peru (dept.). Archaeological Surveys, 707.
Morals. *See* Ethics.
Morelos, Mexico (state). Archaeological Dating, 375. Aztec Influences, 373. Excavations, 373. Precolumbian Architecture, 239, 373.

Moreno Ocampo, Luis Gabriel, 5115.
Morley, Sylvanus Griswold, 93.
Mortality. Brazil, 5190. Mexico, 4470, 4499. Precolumbian Civilizations, 381. Venezuela, 1754. Viceroyalty of New Spain, 2415. Yanomamo, 869.
Mortuary Customs. Bolivia, 565, 571. Bororo, 887. Brazil, 582, 887. Caribbean Area, 520. Chile, 612. Colombia, 613, 625. Costa Rica, 469. Mesoamerica, 355. Mexico, 182, 223, 249, 274, 278, 319, 323, 366, 751. Precolumbian Civilizations, 272, 338, 390, 469, 571. Quechua, 959. Seri, 249. The Guianas, 653.
Mosquitia (Nicaragua and Honduras). Autonomy, 3012. Law and Legislation, 2348.
Mosquito (indigenous group). Biography, 3007. Government Relations, 759, 3007. Guerrillas, 759. Repatriation, 4648.
Motherhood. Chile, 5052. Mexico, 1532.
Movimento dos Trabalhadores sem Terra (Brazil), 3703.
Movimiento Bolivia Libre, 3331.
Movimiento Campesino Paraguayo, 3613.
Movimiento de Adolescentes y Niños Trabajadores Hijos de Obreros Cristianos (Peru), 4937.
Movimiento de Izquierda Revolucionario (Chile), 3439.
Movimiento de Liberación Nacional (Uruguay), 3442, 3645.
Movimiento Izquierda Socialista (Peru), 4298.
Movimiento Nacional Revolucionario (El Salvador), 2971.
Movimiento Nacionalista Revolucionario (Bolivia), 3284, 3292, 3330.
Movimiento Revolucionario Túpac Amaru (Peru), 3359–3360, 3365–3366.
Movimiento Todos por la Patria (Argentina), 3539, 3542.
Moxo. *See* Mojo.
Mucajaí River Valley (Brazil). Demography, 869.
Muisca. *See* Chibcha.
Multinational Corporations. Costa Rica, 2338, 2959. Mexico, 4518.
Mundurucu (indigenous group). Economic Conditions, 866. Social Conditions, 866.
Municipal Government. Argentina, 2552, 3531, 3537, 3560. Bolivia, 3329, 3342, 3349. Brazil, 3676, 3698, 3713, 3718, 3762. Chile, 3461. Colombia, 4815. Elections, 3349. Honduras, 2993, 4587. Mexico, 2399–2400, 2869–2870, 4484. Nicaragua, 2999, 3009. Peru, 3392, 4887, 4924. Sandinistas, 2999. Social Policy, 4412. Uruguay, 3648, 3651,

3662. Venezuela, 3255, 3265. Women, 4412, 4441.

Municipal Services. Italy, 2404. Mexico, 2404. Peru, 4940.

Mural Painting. Precolumbian Civilizations, 700.

Murato. *See* Candoshi.

Murder. Guatemala, 2980.

Murra, John V., 483.

Museo Casa de Colón (Canary Islands), 662.

Museo de América (Madrid), 658.

Museo Nacional de Antropología (Mexico), 244–245, 275, 377–378.

Museo Regional de Oaxaca (Mexico), 391.

Museums. Honduras, 463. Suriname, 825.

Music. Shuar, 922.

Musical Instruments. Tiwanaku Culture, 567.

Muslims. Trinidad and Tobago, 3100.

Mutilation. Zapotec, 320.

Myths and Mythology. Amazon Basin, 851, 859. Andean Region, 859, 1041. Ashaninca, 924. Aztecs, 77, 128, 141, 197, 416. Brazil, 873, 888. Canelo, 927. Caribbean Area, 817, 851. Caves, 317. Curripaco, 896. El Dorado, 2216. Goajiro, 904. Haiti, 802. Indigenous Peoples, 888. Kagaba, 998. Machiguenga, 923. Mapuche, 975, 978. Mayas, 103. Nahuas, 132. Netherlands Antilles, 816. Peru, 4884. Tarasco, 143. Yecuana, 895.

Nahuas (indigenous group). Compadrazgo, 751. Cultural History, 128. Government Relations, 763. Human Ecology, 718. Literature, 132. Mortuary Customs, 751. Myths and Mythology, 132, 718. Rites and Ceremonies, 718, 751. Witchcraft, 755.

Nahuatl (language). *See* Indigenous Languages.

Narcotic Laws. Belize, 3350. Bolivia, 3279–3280, 3289, 3345.

Narcotics Policy, 3875.

Nariño, Colombia (dept.). Precolumbian Civilizations, 620. Salvage Archaeology, 615.

Nasca. *See* Nazca.

National Autonomy. *See* Autonomy.

National Characteristics. Barbados, 3074. Brazil, 5145, 5174, 5186. Mexico, 4458–4459. Panama, 3017. Peru, 1054. Saint Vincent, 850. Trinidad and Tobago, 781.

National Defense. *See* National Security.

National Identity. *See* National Characteristics.

National Income. Brazil, 2128. Cuba, 1692.

National Parks and Reserves. *See* Parks and Reserves.

National Security. Argentina, 3520, 3527, 4197, 4200. Bolivia, 3350. Brazil, 3680, 4359. Caribbean Area, 4076, 4085, 4091, 4097, 4099–4100, 4148–4149. Chile, 3492, 4260, 4262. Colombia, 3179. Cuba, 4121. Databases, 31. Ecuador, 3225. Honduras, 2994, 4660. Mexico, 3909, 3934. Nicaragua, 3004. Pacific Area, 4097. Peru, 3363, 3385, 4303, 4307, 4312–4313, 4315. Political Ideology, 3520. South America, 4167, 4169, 4173, 4179, 4184, 4303. Uruguay, 3653.

National Security Archive, 38.

Nationalism. Argentina, 3524, 3549, 3564. Bolivia, 3301. Caribbean Area, 794, 4147, 4746, 4755. Cuba, 1696, 3150. Dominican Republic, 4802. Mexico, 1358. Panama, 3017, 3026, 3028, 4046, 4592. Race and Race Relations, 4755. Suriname, 3090. Venezuela, 4340.

Nationalization. Mexico, 2884. Peru, 3412.

Natural Disasters, 1077. Peru, 2505, 2514, 4902.

Natural Gas. Argentina, 2114. Brazil, 2684. Venezuela, 2684.

Natural History, 2220, 2228. Bolivia, 2526. Colombia, 2452. Costa Rica, 2340.

Natural Resources, 1107, 2226, 2234, 2237, 2267, 2280. Accounting, 2128. Amazon Basin, 857, 872. Andean Region, 2430. Argentina, 2551. Bolivia, 2527. Brazil, 872, 2128, 2634, 2667, 2682, 2704. Caribbean Area, 2237, 2288, 2307. Central America, 1540, 2320. Chile, 2579, 2616. Colombia, 905. Costa Rica, 1571, 2323, 2325, 2329, 2333. Cuba, 2315. Ecology, 2278. Ecuador, 1838, 2482. Falkland Islands, 4210. Mexico, 1512, 2375, 2934. Panama, 2353. Peru, 2491, 2504, 2507. Saint Kitts and Nevis, 2300. Southern Cone, 2534.

Navies. Argentina, 4203, 4218. Chile, 4255. South America, 4165, 4178, 4185.

Navigation. Mayas, 359. Precolumbian Civilizations, 487.

Nazca (indigenous group). Chiefdoms, 704. Human Remains, 660. Pilgrims and Pilgrimages, 704. Political Culture, 704. Precolumbian Art, 658. Precolumbian Land Settlement Patterns, 705. Precolumbian Pottery, 658. Social Structure, 704–705.

Nazca Culture, 704. Catalogs, 658.

Nazca Lines Site (Peru), 697–698.

Nazca Valley Region (Peru). Artifacts, 657. Precolumbian Architecture, 665. Precolumbian Sculpture, 657. Rock Art, 665.

Nazi Influences. Uruguay, 4325.

Negotiation. Bolivia, 4234. Brazil, 4381. Central America, 3986, 3991–3992, 4035. Colombia, 3170, 3188, 3201. Cuba, 4127. El Salvador, 3188, 4052–4053. External Debt, 3725–3726. Mexico, 3888, 3913, 3917–3918, 3921, 3949. North America, 3888, 3915, 3917, 3921. US, 3888, 3917.

Neighborhood Associations. Chile, 5042, 5066. Ecuador, 4863. Honduras, 4587. Housing, 4904. Nicaragua, 4651. Panama, 4583. Peru, 4878, 4890, 4904, 4913, 4921–4922, 4940. Political Participation, 4841. Poor, 5066. Social Change, 4863. Squatter Settlements, 5066. Uruguay, 3648, 5134. Venezuela, 4841–4842. Women, 4890.

Neoliberalism, 1079, 1131, 1143, 1196, 1280–1281, 1294. Chile, 1907. Colombia, 1762. Mexico, 1358, 1385, 1412.

Neuquén, Argentina (prov.). Folklore, 945. Infant Mortality, 944. Mapuche, 945. Paleo-Indians, 550. Photography, 940. Precolumbian Civilizations, 550. Precolumbian Land Settlement Patterns, 552.

Neutrality. Costa Rica, 2950, 4033.

New Left, 2732, 2768. Argentina, 3534, 3582.

New World Archaeological Foundation, 388.

New World Order, 3787, 3805–3806, 3809, 3811, 3814, 3833, 3840, 3845, 3849, 3857, 3862, 3869–3871. Argentina, 4219. Brazil, 4355, 4365, 4368–4370, 4382, 4390. Caribbean Area, 4076, 4085, 4090, 4101, 4155. Colombia, 4279. Cuba, 4111. Jamaica, 4154. Mexico, 3909, 3939. Panama, 4070. Peru, 4297, 4315. Uruguay, 4320.

New York, New York (city). Brazilians, 5178. Dominicans (people), 4729. Haitians, 4711.

Newspapers. Argentina, 3518, 5122. Bolivia, 3337. Brazil, 3773. Colombia, 3164. Databases, 34–35, 50. Ecuador, 3224. Elections, 3773. Mexico, 35. Uruguay, 3619.

Nicaragua. Ejército Popular Sandinista, 3000.

Nicaragua. Guardia Nacional, 3006.

Nicaraguan Revolution (1979), 3005. Women, 4672.

Niño Current. See El Niño Current.

Nitrate Industry. Chile, 1888.

Nivaklé. See Ashluslay.

Non-Aligned Nations. Mexico, 3902.

Nonformal Education. Nicaragua, 4575. Peru, 4919.

Nongovernmental Organizations, 1114, 1128, 1201, 1272, 1292, 2256, 2269, 2278. Argentina, 5126. Central America, 3966. Chile, 1885, 3461, 3478, 5036. Colombia, 1758.

Development Projects, 1955. Dominican Republic, 1821. Ecuador, 1820–1821. Mexico, 2856. Paraguay, 4984. Peasant Movements, 4984. Peru, 3404, 3409, 4889, 4905, 4917, 4926. Public Health, 4905. Women, 5036.

Nordeste, Brazil (region). Agricultural Labor, 5158. Economic Development, 2192. Households, 5177. Land Tenure, 2177. Peasants, 5177. Political History, 3734. Rural Sociology, 5177. Social Conditions, 2192.

Noriega, Manuel Antonio, 3016, 4023–4024, 4038.

North American Free Trade Agreement (NAFTA), 1076, 1143, 1190, 1242, 1269, 1279, 1327, 1330, 1359, 1361, 1381–1382, 1397, 1405, 1408, 1418, 1420, 1424, 1458–1459, 1464, 1469, 1483, 1488, 1493, 1515, 1537, 1563–1564, 1583, 1648, 2169, 2910, 3888, 3892, 3894, 3897, 3901, 3913–3915, 3918, 3921, 3928, 3933, 3935–3936, 3938, 3946, 3949, 3952–3954, 3961–3962, 4082, 4546.

Nuclear Energy, 1277. Argentina, 3511. Brazil, 3705, 3747. Military, 3705.

Nuclear Policy, 1277. Argentina, 2204, 3511, 4162, 4190, 4196, 4366. Brazil, 2156, 2204, 3747, 4162, 4366.

Nuclear Weapons. Argentina, 4196.

Nueva Esparta, Venezuela (state). Geography, 2444.

Nuevo León, Mexico (state). Maquiladoras, 1383.

Nuevo Orden Económico Internacional. See New International Economic Order.

Nutrition. Aztecs, 142. Bolivia, 4975. Brazil, 2197. Indigenous Peoples, 4845. Peru, 4894. Venezuela, 4845.

OAS. See Organization of American States.

Oaxaca, Mexico (city). Rural-Urban Migration, 744.

Oaxaca, Mexico (state). Archaeological Dating, 196. Archaeological Surveys, 72, 113. Archaeology, 127, 342. Artifacts, 391. Ball Games, 462. Cults, 462. Cultural Development, 125. Elections, 4484. Excavations, 179, 262, 265, 329. Forced Migration, 719. Indigenous Music, 760. Indigenous Peoples, 196, 719, 4486, 4509. Inscriptions, 457. Marine Resources, 111. Municipal Government, 4484. Nutrition, 196. Political Conditions, 4484. Precolumbian Architecture, 236, 256. Precolumbian Civilizations, 125, 127, 301. Precolumbian Land Settlement Patterns, 101, 300. Precolumbian Pottery,

266, 365. Shells and Shell Middens, 264. Sugar Industry and Trade, 720.
Obando y Bravo, Miguel, 3005.
Obraje de San Joseph de Peguchi, 1830.
Obsidian. Aztecs, 227. Belize, 252. Guatemala, 215. Honduras, 186–187, 216. Mayas, 252. Mesoamerica, 230, 252, 288, 396. Mexico, 227, 242. Precolumbian Trade, 288, 292, 396.
Occupational Training. Brazil, 2211. Mexico, 1486.
Ochoa Sánchez, Arnaldo, 3110, 3115.
O'Donnell, Guillermo A., 2740.
Oidores. *See* Judges.
Oliveira, Euclides Quandt de, 2203.
Oliveira, Lima, 4367.
Olivo, Francisco, 3271.
Ollantaytambo Site (Peru), 693.
Olmec Influences. Mesoamerica, 114. Precolumbian Sculpture, 280.
Olmecs (indigenous group). Calendrics, 426–427, 2392. Epigraphy, 426–427. Inter-Tribal Relations, 114. Monuments, 282. Precolumbian Pottery, 266. Precolumbian Sculpture, 112, 232. Research, 194.
Omagua (indigenous group, Colombia). *See* Carijona.
Ona (indigenous group), 536. Cultural Collapse, 942. Cultural History, 941.
Online Bibliographic Searching. *See* Database Searching.
Opata (indigenous group). Cultural Development, 214.
Operettas. *See* Musicals; Tonadillas; Zarzuelas.
Opposition Groups, 2800. Chile, 3474, 3478, 3482, 5027, 5049. Colombia, 3162. Mexico, 2821, 2850, 2857, 2869, 2914, 2931, 2933, 4522. Trinidad and Tobago, 3097.
Oral History. Arawak, 907. Aymara, 946. Bolivia, 946. Brazil, 865. Colombia, 1003. Cuba, 3128. Ecuador, 1014.
Oral Tradition. Colombia, 985, 990. Peru, 1059. Tunebo, 985.
Orejón (indigenous group). Kinship, 909. Sex Roles, 909. Social Life and Customs, 909.
Organisation for Economic Co-operation and Development (OECD), 39.
Organization of American States (OAS), 3863, 3866, 4116, 4278.
Organization of Eastern Caribbean States (OECS), 1645, 1674, 3034.
Oro, Ecuador. *See* El Oro, Ecuador (prov.).
Ortellado Jiménez, Hilario, 3606.
Oruro, Bolivia (dept.). Aymara, 948.
Otavalo, Ecuador (town). Social Life and Customs, 1010.
Otavalo (indigenous group). Ethnography, 1008.
Otomi (indigenous group). Acculturation, 773. Discourse Analysis, 764. Ethnic Groups and Ethnicity, 773. Ethnography, 764. Households, 732. Kinship, 732. Language and Languages, 773. Religious Life and Customs, 734. Rites and Ceremonies, 734. Social Conditions, 773. Social Life and Customs, 732. Social Structure, 764. Sociolinguistics, 764.
Otumba Site (Mexico). Obsidian, 227. Political Development, 228.
Oxkintok Site (Mexico), 356.
Oyampi (indigenous group). Myths and Mythology, 873.
Oyana (indigenous group). Social Change, 890. Suriname, 890.
Pacaritanpu Site (Peru), 673.
Pachitea, Peru (prov.). Textiles and Textile Industry, 1040.
Pacific Area. International Economic Relations, 1318, 3886. International Relations, 4160.
Pacific Coast (Colombia). Archaeological Dating, 624. Blacks, 1006. Land Tenure, 1006. Precolumbian Civilizations, 626. Race and Race Relations, 1005. Social Life and Customs, 4814.
Pacific Coast (Ecuador). Archaeological Dating, 624. Precolumbian Civilizations, 626.
Paez (indigenous group). Social Change, 992.
Paiján Culture, 659, 669, 687, 692.
Painting. Peru, 1044. Quechua, 1044.
Palenque, Carlos, 4952.
Palenque Site (Mexico), 277, 346.
Paleo-Indians, 55. Andean Region, 530. Argentina, 550. Brazil, 574–575, 580, 590. Chile, 543, 596, 599, 602–603, 606–609, 611. Colombia, 543, 615–616. Food Supply, 563. History, 544. Human Adaptation, 542, 699. Mexico, 106. Peru, 669, 674, 676, 680, 687, 699. South America, 537, 542. Stone Implements, 692. Tierra del Fuego, 563. Venezuela, 715.
Paleobotany. Caribbean Area, 518. Ecuador, 645–646. Mesoamerica, 351, 385. Mexico, 310.
Paleoclimatology. Amazon Basin, 584. Argentina, 552. Bolivia, 569. Brazil, 584. Chile, 604. Colombia, 614, 621. Peru, 672.
Paleoecology. Brazil, 587, 593–594. Chile,

599, 604, 606. Colombia, 531, 543, 614, 621, 626, 630. Deforestation, 614. Ecuador, 531, 626. Mexico, 217, 253, 358. Peru, 672, 674. Uruguay, 710. Venezuela, 717.

Paleogeography. Mexico, 217, 253.

Paleoindians. *See* Paleo-Indians.

Palés Matos, Luis, 4746.

Palestinians. Honduras, 4617.

Palms. Amazon Basin, 2675. Brazil, 2099. Mesoamerica, 307. Precolumbian Civilizations, 307.

Palynology. Mexico, 217.

Pampa, Argentina. *See* La Pampa, Argentina (prov.).

Pampa Grande Site (Peru), 702.

Pan-Americanism, 3866. Brazil, 4372.

Panama Canal, 1203, 3970, 4006, 4010. Geopolitics, 4000.

Panday, Basdeo, 3097.

Pané, Ramón, 509.

Pará, Brazil (state). Aluminum Industry and Trade, 2193. Indigenous Peoples, 864. Land Reform, 3699. Peasants, 5167, 5185. Rubber Industry and Trade, 2129. Social Conflict, 5167. Violence, 3699, 5167.

Paracas Site (Peru), 690.

Paraíso Site. *See* El Paraíso Site (Peru).

Paramaribo, Suriname (city). Indigenous Peoples, 4763. Youth, 4783.

Paramilitary Forces, 4421. Colombia, 984, 3184. Peru, 3387.

Paramos (geographical feature). Costa Rica, 2331.

Paraná, Brazil (state). Agricultural Development, 2677. Coffee Industry and Trade, 2679. Farms, 2677. Land Reform, 3740. Land Tenure, 2679. Land Use, 2677. Rural Sociology, 3740. Social Movements, 3740. Urban Areas, 2677.

Parenthood. Peru, 4899. Venezuela, 4849.

Parinacota, Chile (prov.). Land Tenure, 971. Water Supply, 972.

Parintintin (indigenous group). Shamanism, 878.

Parks and Reserves. Argentina, 2564. Caribbean Area, 2239. Central America, 2320. Colombia, 854. Costa Rica, 2334–2335, 2337, 2340. Cuaiquer, 854. Ecuador, 854. Peru, 2516–2517. Venezuela, 2435.

Parliamentary Systems. *See* Political Systems.

Parque Nacional Corcovado (Costa Rica). Guidebooks, 2337.

Parque Nacional do Xingu (Brazil), 888.

Parra Pérez, Caracciolo, 4337.

Partido Acción Nacional (Mexico), 2817, 2823, 2839, 2847, 2861, 2864, 2877, 2891, 2903, 2911, 2931, 2935–2936.

Partido Aprista Peruano. *See* APRA (Peru).

Partido Blanco. *See* Partido Nacional (Uruguay).

Partido Colorado (Paraguay), 3600.

Partido Comunista Brasileiro, 3741, 3778.

Partido Comunista de Cuba, 3102, 4725.

Partido Comunista de Cuba. Congreso, *4th, Santiago de Cuba, 1991*, 3102.

Partido Comunista de Venezuela, 4341.

Partido Comunista del Perú, 4300.

Partido Comunista del Uruguay, 3627.

Partido Comunista Dominicano, 3057.

Partido da Social Democracia Brasileira, 3690.

Partido de la Liberación Dominicana, 3044–3045.

Partido de la Revolución Democrática (Mexico), 2819, 2841, 2865, 2882, 4525, 4567.

Partido de la Revolución Mexicana, 2904.

Partido del Progreso (Costa Rica), 2958.

Partido Demócrata-Cristiano (Chile), 3472, 3488.

Partido Demócrata Cristiano (Uruguay), 3659.

Partido dos Trabalhadores (Brazil), 3673, 3682–3683, 3691, 3704, 3708, 3713, 3716, 3718, 3733, 3744, 3752, 3754, 3760.

Partido Liberación Nacional (Costa Rica), 2946, 2948, 2957.

Partido Liberal (Paraguay), 3590, 3610.

Partido Liberal Radical Auténtico (Paraguay), 3607.

Partido Nacional Revolucionario (Mexico), 2904.

Partido Obrero Revolucionario (Bolivia), 3335.

Partido Reformista Social Cristiano (Dominican Republic), 3047.

Partido Revolucionario de los Trabajadores (Argentina), 3540, 3557.

Partido Revolucionario Dominicano, 3044.

Partido Revolucionario Institucional (Mexico), 2821, 2832, 2839, 2850, 2853, 2866–2867, 2875, 2883, 2885, 2891, 2893, 2897, 2901, 2904, 2906–2907, 2909, 2912, 2926, 2931, 4498, 4510, 4525, 4550.

Partido Trabalhista Brasileiro, 3675.

Passarinho, Jarbas Gonçalves, 3748.

Pasto (indigenous group). Ethnography, 999. Precolumbian Land Settlement Patterns, 620.

Pastures. Argentina, 2550. Mexico, 2366.

Patagonia (region). Archaeological Surveys, 548, 560. Coastal Areas, 2553. Human Adaptation, 534. Marine Resources, 2553. Pre-

columbian Civilizations, 534, 536, 544, 554. Precolumbian Land Settlement Patterns, 548. Rock Art, 536.
Paz, La. *See* La Paz, Bolivia.
Paz Estenssoro, Víctor, 3322, 4228, 4238.
Paz Zamora, Jaime, 3315, 3332, 4227–4228, 4237.
Peabody Museum of Natural History (Yale University), 514.
Peace. Bolivia, 3355. Caribbean Area, 4096. Central America, 2947, 3984–3986, 3994, 4013, 4021, 4045. Colombia, 3153, 3170–3171, 3187–3188, 3201, 3208, 3210–3211. El Salvador, 2969, 2972, 3188, 3976, 3991–3992, 4018, 4052–4053. Guatemala, 2973, 3991–3992, 4040. Peru, 3403, 3417.
Peasant Movements, 2807. Bolivia, 3304. Brazil, 5189. Catholic Church, 3613. Costa Rica, 4572, 4607, 4639. Ecuador, 3242, 4859, 4876. Guatemala, 4613. Jamaica, 3069. Mexico, 746, 2871. Nicaragua, 4580, 4627, 4666, 4668. Nongovernmental Organizations, 4984. Paraguay, 3613, 4983–4984. Peru, 3380, 4917, 4924, 4931, 4934. Women, 4893, 4931, 4983.
Peasant Uprisings. Brazil, 3703, 3746. Costa Rica, 4607. Ecuador, 1016, 1023, 4876. El Salvador, 4638. Mexico, 763, 2871. Peru, 3369.
Peasants. Andean Region, 1038, 1050, 1069, 2421, 2424, 2515, 2519. Argentina, 5101. Bolivia, 956, 966, 1979, 4962, 4972, 4981. Brazil, 2635, 2642, 2649, 2655, 2695, 5177, 5185, 5189. Caribbean Area, 810. Central America, 729. Chile, 1912, 3458, 5005, 5010, 5065. Colombia, 1758, 1764, 1768, 4809. Costa Rica, 4658. Cuba, 3149, 4716. Cultural Identity, 1064, 4945. Development Projects, 4938. Dominica, 777, 815. Dominican Republic, 4727, 4781. Ecuador, 4859, 4861, 4865, 4875. Family and Family Relations, 5189. Food, 2235. Guatemala, 4613. Honduras, 4580, 4611, 4675. Kinship, 4861. Law and Legislation, 1058. Mexico, 756, 4506, 4511. Natural Resources, 2278. Nicaragua, 1629, 4608, 4612, 4622, 4627, 4673, 4680, 4682. Paraguay, 4989. Peru, 1032, 1048, 1064, 1066, 1935, 1943, 1963, 3375, 3411, 4893, 4900–4901, 4918, 4932, 4935–4936, 4938, 4941, 4945. Political Ideology, 3069. Production (economics), 720. Sex Roles, 4932. Social Structure, 4962. Sugar Industry and Trade, 5101. Venezuela, 2446. Women, 4875, 5005, 5010.
Peddlers and Peddling. *See* Informal Sector.

Pehuenche (indigenous group). Land Tenure, 969. Modernization, 973.
Pentecostalism. Brazil, 3732, 5179. Central America, 4636. Chile, 5070. Costa Rica, 4683. Guatemala, 4590, 4603. History, 5070.
People's National Party (Jamaica), 4750.
Pérez, Carlos Andrés, 3259, 4338.
Pernambuco, Brazil (state). Abandoned Children, 5208. Economic Development, 2109. Governors, 3734. Political History, 3734. Teenagers, 5208.
Perón, Eva, 4224.
Perón, Juan Domingo, 3520, 4207, 4222, 4224.
Peronism, 3509, 3512, 3517, 3519, 3534, 3543, 3546, 3558, 3578, 3586, 5121. Industrialists, 5112.
Persian Gulf War (1991). Argentina, 3530, 4203.
Personal Narrative. *See* Autobiography; Oral Tradition.
Peruvians. Japan, 4881. US, 1027, 4882.
Pesquisa Nacional por Amostra de Domicílios (Brazil), 2130, 2184, 2189.
Pesticides. Brazil, 2624.
Petén, Guatemala (dept.). Archaeological Surveys, 399. Precolumbian Land Settlement Patterns, 398.
Petrochemical Industry. Argentina, 2114. Brazil, 2114.
Petroglyphs. *See* Rock Art.
Petróleos Mexicanos, 1369.
Petroleum Industry and Trade, 1217, 1834, 2260. Argentina, 3535. Brazil, 2704. Colombia, 1775, 1780. Ecuador, 1789, 1815, 3246. Japan, 1504. Mexico, 1334, 1351, 1369, 1418, 1447, 1504, 2397. Venezuela, 1740, 1744–1745, 2438, 4834, 4850.
Pharmaceutical Industry, 2078. Brazil, 2115. Cuba, 3124.
Philippi, Rudolph A., 2604.
Phillips, Fred, 3037.
Philosophy. Databases, 36.
Philosophy of Liberation. *See* Liberation Theology.
Photography. Argentina, 940. Costa Rica, 2327. Indigenous Peoples, 1055. Mexico, 2826. Peru, 1055, 2492, 2495, 3361.
Physical Anthropology. Aruba, 520. Caribbean Area, 520. Costa Rica, 477. Dutch Caribbean, 485.
Physical Geography, 2229, 2260, 2456. Colombia, 2456. Mexico, 2382. Venezuela, 4343.
Phytogeography. *See* Biogeography.

Piaroa (indigenous group). Acculturation, 902. Indigenous/Non-Indigenous Relations, 902.
Piauí, Brazil (state). Migration, 2180.
Pictorial Works. Costa Rica, 2327. Peru, 1055, 1059.
Pilcaya Region (Mexico). Archaeological Surveys, 188.
Pilgrims and Pilgrimages. Mexico, 2410. Nazca, 704.
Piñera, Sebastián, 3425.
Pino, Rafael del, 3144.
Pinochet Ugarte, Augusto, 1922, 3428, 3431–3432, 3454, 3460, 3464, 3475–3476, 3481, 3485, 3487, 4244.
Pisco River Region (Peru). Paleo-Indians, 674.
Piura, Peru (dept.). El Niño Current, 2514. Land Reform, 4935. Peasants, 4935.
Pizarro León-Gómez, Carlos, 3190.
Plantations. Amazon Basin, 2657. British Caribbean, 498. Caribbean Area, 784, 4707. Suriname, 2301.
Plants, 2225, 2242. Bolivia, 4961. Caribbean Area, 2302. Costa Rica, 2331, 2336. French Guiana, 806. Mesoamerica, 134, 439. Panama, 2354. Precolumbian Art, 439.
Plata, Argentina. See La Plata, Argentina (city).
Plebiscites. See Referendums.
Pluralism. Caribbean Area, 792, 794, 827. El Salvador, 2961. Mexico, 2833.
Pochota Site (Costa Rica). Precolumbian Pottery, 473.
Polanyi, Karl, 483.
Poles. Chile, 4261.
Police, 4421. Brazil, 3753, 5198. El Salvador, 2970. Mexico, 4461. Peru, 3416. Repression, 3753. US Influences, 4420. Venezuela, 3263, 3269, 3273, 4851. Violence, 4420, 4438.
Political Anthropology. Andean Region, 1065. Ecuador, 1008. Latin America, 86. Mesoamerica, 150. Peru, 1051. Precolumbian Civilizations, 100.
Political Bosses. Mexico, 2874.
Political Boundaries. See Boundaries.
Political Campaigns. See Elections.
Political Candidates. See Politicians.
Political Conditions, 2726, 2755, 2769, 3836, 3864. Bibliography, 2729. Bolivia, 3344. Brazil, 3684, 3761. British Caribbean, 3035, 3072. Caribbean Area, 2804. Central America, 2804, 2938, 4008. Colombia, 3194. Cuba, 1685, 3101, 3106, 3143, 3151. Databases, 31, 33. Dominican Republic, 3059. Ecuador, 3223–3224, 3234, 3244–3245. Electronic Resources, 12. Mexico, 40, 2814–2815, 2823, 2827, 2829, 2834–2835, 2856, 2888, 2908, 2910, 2927, 2934. Nicaragua, 3004. Panama, 3025–3026. Peru, 3368, 3378, 3386, 3393, 3396–3397. Uruguay, 3621, 3626, 3640.
Political Corruption, 3192. Antigua and Barbados, 4144. Argentina, 3513, 3580, 3585, 3587, 5115. Bolivia, 3309, 3323, 3328, 3356. Brazil, 3687, 3702, 3762, 3772, 3779. Chile, 3425. Colombia, 3178, 3189, 3192. Dominican Republic, 3049. Haiti, 3066, 3068. Mexico, 2825, 2860, 2879, 2893, 2897, 2921, 4510. Panama, 3025. Paraguay, 3596, 3598. Spain, 3192. Venezuela, 3248.
Political Crimes, 4411. Argentina, 3496, 3499, 3532, 3583, 5115. Chile, 3622. Colombia, 3185. Dominican Republic, 3054. Suriname, 3086. Uruguay, 3622.
Political Culture, 2795, 2802, 4413. Advertising, 5117. Argentina, 2736, 3501, 3524, 3545, 3564, 5117–5118. Aztecs, 176. Bibliography, 3727. Brazil, 2736, 3669, 3674, 3678–3679, 3701, 3727, 3729–3731, 3735, 5174. British Caribbean, 3073. Caribbean Area, 3043. Central America, 4679. Chile, 3434, 3462, 3494, 5007, 5016, 5030, 5061. Costa Rica, 4607. Dutch Caribbean, 3078. Ecuador, 3221. Guatemala, 2977, 4670. Haiti, 3062–3063, 3066–3068. Honduras, 4654. Jamaica, 4750. Labor and Laboring Classes, 2744. Mass Media, 4559. Mexico, 1495, 2744, 2814, 2848, 2886, 2902, 2922, 2928, 4510, 4513, 4559. Montserrat, 3071. Nicaragua, 4598. Paraguay, 3603, 3612, 3615, 4987. Precolumbian Civilizations, 161, 435, 461–462. Puerto Rico, 3085, 4733, 4759. Suriname, 3089, 3093. Uruguay, 3635–3636, 3643, 3647. Venezuela, 2744, 4393.
Political Development, 2277, 2741, 2745, 2766, 2773, 2800–2801. Argentina, 3507, 3764. Aruba, 3076. Aztecs, 168, 228, 293. Brazil, 3764. Caribbean Area, 3036. Central America, 4679. Chile, 5026. Colombia, 3166, 3211. Costa Rica, 4684. Cuba, 3122, 3148. Dutch Caribbean, 3076, 3078–3079. Guadeloupe, 3081. Haiti, 3065, 3068. Honduras, 4654. Mayas, 257. Mexico, 2837, 2854, 2883, 2908, 2923. Mixtec, 448. Montserrat, 3071. Peru, 3371, 3397. Precolumbian Civilizations, 131, 137, 247–248, 300–301, 331, 382. Puerto Rico, 3083, 3085. Suriname, 3091–3093. Uruguay, 3654. Venezuela, 3265. Youth, 5026.

Political Economy, 2792, 3818, 4403. Argentina, 3505, 3523, 3555–3556, 3575, 3584. Bolivia, 3351, 3357. Brazil, 2145–2146. Caribbean Area, 3032, 4078, 4083. Chile, 3480. Christian Demcracy, 2771. Colombia, 3198. Databases, 37. Ecuador, 3212. Mexico, 37, 2920, 4460. Precolumbian Civilizations, 80. Suriname, 3094. Venezuela, 3252.

Political History, 2747–2748, 2768, 2775–2776, 2789, 2798. Argentina, 2724, 3509, 3516, 3543, 3547–3548, 3575, 3581, 4190, 4214, 4217, 4219, 5103. Bolivia, 3284, 3322, 3336, 3340–3341, 3357. Brazil, 2724, 3668, 3672, 3702, 3714–3715, 3748, 3766, 4371. Caribbean Area, 3036, 4086, 4151. Central America, 3995, 4008, 4599, 4689. Chile, 2724, 3420, 3444, 3457, 3470, 3483, 3488–3489, 3494, 4416, 5004, 5007, 5032, 5062. Colombia, 3166, 3183, 4265, 4816, 4821. Costa Rica, 2955, 4684. Cuba, 4110, 4769. Curaçao, 4730. Dominican Republic, 3046, 4699. Dutch Caribbean, 3080. Ecuador, 3213, 3236, 3238, 3243. El Salvador, 2963, 2966, 2968, 2971. Guatemala, 757, 2974, 2982, 4650. Haiti, 3063. Honduras, 2987, 2990, 2992. Jamaica, 4154, 4750. Mexico, 2813, 2841–2842, 2850, 2854, 2866, 2882, 2894, 2904, 2916, 2926, 3910, 3927. Nicaragua, 3005, 3015, 3987, 3997, 4015, 4686. Panama, 3022, 3025–3026, 4029, 4047. Paraguay, 3593, 3603, 3610, 3612. Peru, 1945, 3358, 3366, 3371, 3384, 3404, 3412. Puerto Rico, 3083. Río de la Plata, 4183. South America, 2724. Suriname, 3087, 3089–3092, 4696. Trinidad and Tobago, 3095–3096. Uruguay, 3621, 3625, 3628, 3633, 3643, 3652, 3654–3655. US, 3989. Venezuela, 3183, 3254.

Political Ideology, 2732, 2738, 2798. Argentina, 3498, 3520, 3549, 3564, 3582. Artists, 4705. Bolivia, 3341. Brazil, 3680, 3739, 3759, 5171. Chile, 3450, 3459, 3462, 3466, 3470. Cuba, 3147, 4110, 4130, 4705. Intellectuals, 4567. Jamaica, 3069. Mass Media, 4513. Mexico, 2818, 2861, 2903, 4567. Middle Classes, 5171. Military, 3680, 3739. National Security, 3520. Nicaragua, 3002. Paraguay, 3607. Peasants, 3069. Peru, 1051, 1060. Puerto Rico, 4706, 4759. Social Structure, 5105. Sociology, 4418. Suriname, 3088. Uruguay, 3643, 3645, 3647, 3652. US, 4130. Venezuela, 3251, 3256, 3262, 3266, 4340.

Political Institutions, 2240. Argentina, 3497, 3554. Brazil, 3669, 3671, 3714, 3720, 3730, 3775, 5209. Chile, 3479, 5003. Costa Rica, 4645. Guadeloupe, 3081. Incas, 683. Mexico, 2835, 2843, 2904. Uruguay, 3625, 3628, 3634.

Political Integration, 3845. Caribbean Area, 3034, 4150. Central America, 3995, 4008, 4027. Windward Islands, 3039.

Political Left, 2732, 2738, 2749, 2768, 2784. Argentina, 3534, 3539, 3542, 3582. Bolivia, 3281, 3298. Brazil, 3778. Chile, 3442–3443, 3463, 3468, 5062. Colombia, 3162. Dominican Republic, 3051. Mexico, 2819. Nicaragua, 3011. Paraguay, 3601. Peru, 3364. Puerto Rico, 3085. Uruguay, 3655, 5136.

Political Opposition. *See* Opposition Groups.

Political Participation. Argentina, 3537, 5123. Blacks, 3681. Bolivia, 3282, 3290, 3325, 3351. Brazil, 3681, 3707, 3713, 3717, 3728. Central America, 4678. Chile, 3455, 3462, 3482, 5005, 5072, 5085. Colombia, 3166, 3169, 3189, 4829. Costa Rica, 2952. Dominican Republic, 3053. Ecuador, 1024, 1813, 3244. El Salvador, 4676. Guatemala, 2974. Indigenous Peoples, 1024. Intellectuals, 5123. Mexico, 2846, 2848, 2855, 2862, 2864–2865, 2870, 2913, 4564. Neighborhood Associations, 4841. Nicaragua, 4665. Peru, 1037, 1050, 3386. Puerto Rico, 4760. Uruguay, 3620, 3648. Venezuela, 3261–3262, 3268. Women, 3282, 4429, 4676, 4760, 4829, 5005, 5072.

Political Parties, 2723, 2735, 2746, 2759, 2771, 2775, 2785, 2793, 2802, 4437. Andean Region, 2777. Argentina, 2061, 3512, 3522, 3525–3526, 3544, 3554. Belize, 2944, 4637. Bolivia, 3278, 3282, 3284, 3290–3291, 3305, 3308, 3317, 3320, 3323, 3329–3331, 3335, 3341–3342, 3349. Brazil, 2098, 3673–3675, 3678, 3683, 3704, 3708, 3716, 3722, 3729, 3733, 3737, 3742, 3744, 3775, 4349, 4371. Central America, 2940, 2942, 2984. Chile, 3438, 3443–3444, 3457, 3459–3460, 3466, 3470, 3472, 3483, 3488, 3490–3491, 5062. Colombia, 3162, 3175, 3183. Costa Rica, 2946, 2948–2949, 2955, 2957–2958, 4684. Cuba, 3132. Democratization, 2743. Dominican Republic, 3044, 3047, 3051, 3053. Ecuador, 3213, 3218, 3238. El Salvador, 2961. Guatemala, 2984–2985. Honduras, 2993, 2995, 4654, 4684. Mexico, 2817–2818, 2823, 2832, 2838–2839, 2841, 2850, 2852, 2861, 2867, 2869, 2876–2878, 2882–2883, 2885, 2892, 2900–2901, 2903–2906, 2909, 2923, 2931–2932, 2935. Nicaragua,

3002, 3014. Panama, 3022, 3027. Paraguay, 3590, 3592, 3600–3601, 3604, 3607, 3611. Peru, 3365, 3378, 3400, 3402, 3404, 3408, 3415, 4908. Puerto Rico, 3083, 4724. Trinidad and Tobago, 3098. Uruguay, 3621, 3623, 3628, 3633, 3635, 3637, 3642, 3647, 3649, 3656, 3659. Venezuela, 3183, 3247, 3252, 3254–3256, 3258, 3275.

Political Persecution. Argentina, 3499, 3508, 3568. Chile, 3422, 3430, 3439, 3446, 3451. Colombia, 3205. Cuba, 1696, 4784. El Salvador, 2960. Guatemala, 2980, 4610. Paraguay, 3589, 3617. Peru, 3568.

Political Philosophy. Bolivia, 3295.

Political Platforms. Bolivia, 3305, 3331, 3335, 3341.

Political Prisoners. Chile, 3446, 3451. Cuba, 4740. Ecuador, 3220. Nicaragua, 3006. Paraguay, 3589, 3602, 3606. Peru, 3359. Uruguay, 3660.

Political Psychology. Argentina, 3524. Brazil, 3701.

Political Reform, 2734, 2754, 2763, 2765, 2783, 2785, 2796, 3192, 3825, 4415–4416. Argentina, 3497, 3570, 3572, 3574, 3576. Bolivia, 3317–3318, 3342. Brazil, 3672, 3714, 3720, 3762, 3770, 3775. Chile, 5032. Colombia, 3163, 3178, 3181, 3186, 3192. Costa Rica, 2954, 2958, 4645. Cuba, 3102, 3104, 3121, 3139. Dominican Republic, 3048. Ecuador, 3218. El Salvador, 2970. Mexico, 2811, 2825, 2830, 2833, 2837, 2839, 2866–2867, 2890, 2893–2894, 2904, 2907, 2915, 2917, 2920, 2936, 3952, 3960, 4547, 4550, 4564. Panama, 3027. Paraguay, 3592, 3605. Uruguay, 3624, 3631, 3646. Venezuela, 3250, 3261, 3265, 3274, 4838.

Political Science. Bibliography, 2809. Databases, 29.

Political Sociology, 2741, 4403, 4416, 4435, 4437. Argentina, 5118. Bolivia, 3291, 3353, 3356. Brazil, 3681, 3707, 3742, 5152, 5209. Chile, 3427, 3450, 3462, 4998, 5041. Colombia, 4805, 4815. Dominican Republic, 3051. Guatemala, 2977. Jamaica, 4750. Mexico, 4489, 4510, 4522, 4527, 4559. Precolumbian Civilizations, 86. Teotihuacán, 126. Uruguay, 3623, 3642. Venezuela, 4838, 4842, 4850.

Political Stability, 2240, 2750, 2758, 2779, 2793. Argentina, 3521, 3554, 3569, 3572. Chile, 3421, 3493. Cuba, 3129–3130. Guatemala, 2974, 2983. Peru, 3362. Uruguay, 3625.

Political Systems, 2735, 2750, 2775, 2779. Argentina, 3399, 3497, 3574. Bibliography, 2729. Bolivia, 3308, 3317, 3326. Brazil, 3668, 3671, 3721, 3730, 3743, 3768. Chile, 3421, 3423, 3433, 3457, 3490, 3494, 5085. Colombia, 3158–3160, 3175, 3178, 3181, 3211, 4822. Cuba, 3132, 4111. Eastern Europe, 4111. Ecuador, 3218, 3222. El Salvador, 2961. Guatemala, 2985. Honduras, 2995. Jamaica, 3070. Mayas, 97. Mexico, 2838, 2844, 2852, 2854, 2863, 2872, 2876, 2905, 2917, 2924, 2928, 4527. Paraguay, 3615. Peru, 3390, 3399. Uruguay, 3624, 3631, 3634, 3644, 3646, 3661. Venezuela, 3257, 3260, 3266, 3268, 3272.

Political Theory, 2737, 2740, 2754, 2765, 2773, 2776, 2793, 2806–2807, 3806, 4394. Argentina, 3569, 4223. Brazil, 4352, 4360. Chile, 3459, 3466, 3479, 3490. Costa Rica, 4019. Cuba, 4118. El Salvador, 2961. Foreign Policy, 3831. Mexico, 2833, 2840, 2844. Populism, 2805. Revolutions and Revolutionary Movements, 2798. Uruguay, 3644, 3661.

Political Thought, 3793, 4172, 4406. Argentina, 3497, 3507, 5122–5123. Brazil, 3720–3722, 3727, 3755. Dominican Republic, 3045–3046. Elites, 5122. Haiti, 3062. Mexico, 2913, 3927, 4460. South America, 4167. Uruguay, 3624, 3631, 3645, 3654. Venezuela, 4340.

Politicians, 2754, 2792. Bolivia, 3333, 3344. Brazil, 3668, 3682, 3768. British Caribbean, 3040. Chile, 3421, 3425, 3455, 3463, 3488, 3493. Colombia, 3183, 3190. Databases, 40. Dominican Republic, 3058. Dutch Caribbean, 3080. Ecuador, 3218, 3222, 3224, 3226, 3237. Haiti, 3061. Mexico, 21, 2822, 2836, 2841, 2852, 2866, 2878, 2922, 2935, 2937. Panama, 3027. Paraguay, 3591, 3607, 3611. Saint Kitts and Nevis, 3037. Suicide, 2748. Trinidad and Tobago, 3096–3097. Uruguay, 3650, 3663. Venezuela, 3183, 3258–3259, 3275. Women, 3455.

Pollution. *See* Environmental Pollution.

Poor, 1127, 1165, 1204, 1213, 1257, 1259, 1768, 2218, 2271. Argentina, 2542, 5092, 5107. Bolivia, 1983. Brazil, 2147, 5179. Caribbean Area, 1165. Central America, 1540. Chile, 1911, 3452, 5019–5020, 5034, 5059, 5065–5066, 5069, 5076. Colombia, 1774, 3202. Costa Rica, 1572, 1582. Dominican Republic, 1592, 4710. Ecuador, 1791, 1802, 1813. El Salvador, 1592. Guatemala, 4577. Housing, 4424, 5063–5064. Law and Legislation, 4843. Medical Care, 4430, 4873.

Mexico, 1345, 1427, 1517, 2862. Peru, 3394. Popular Religion, 5179. Psychology, 5034. Uruguay, 1572, 2010. Venezuela, 1721, 1749, 4843. Youth, 5193.

Popul Vuh, 103.

Popular Art. Chile, 5074. Suriname, 805, 828.

Popular Culture, 4422, 4426. Andean Region, 1042. Brazil, 4391. Central America, 729. Chile, 5043, 5074. Colombia, 4810. Cuba, 4747. Ecuador, 1011–1013. Jamaica, 785. Mexico, 4497. Peru, 4897. Puerto Rico, 4778. Squatter Settlements, 5043. Suriname, 4792, 4799.

Popular Education. *See* Nonformal Education.

Popular Movements. *See* Social Movements.

Popular Music. African Influences, 4414. Curaçao, 775. Dominican Republic, 789–790. Jamaica, 785. Trinidad and Tobago, 781.

Popular Religion, 4423, 4439. African Influences, 841. Argentina, 5093. Brazil, 3667, 3719, 5173, 5179, 5206. Caribbean Area, 835, 841, 843. Central America, 4578, 4631, 4635–4636. Chile, 5001. Costa Rica, 4647. Cuba, 4747. Guatemala, 4661. Jamaica, 820. Peru, 4906, 4910, 4929. Poor, 5179. Research, 5001.

Population Genetics. *See* Human Genetics.

Population Growth, 4431, 4440. Birth Control, 4454. Brazil, 2650, 2653, 2668, 2687, 2696, 2714, 5168. Chile, 2578, 2587–2588. Congresses, 4409. Costa Rica, 2323. Ecuador, 2470. Environmental Pollution, 4427. Haiti, 2295. Mayas, 2345. Mexico, 1395, 1497, 1522, 2372–2373, 2395, 2407, 4519. Peru, 2506, 2508, 2511, 4944. Uruguay, 2620. Venezuela, 2442.

Population Policy, 2275. Argentina, 2085. Central America, 1565. Chile, 5028. Mexico, 1450, 1476, 2405, 4519, 4537, 4545. Peru, 4912, 4944.

Population Studies. *See* Demography.

Populism, 1241, 1257, 2730, 2733, 2805. Argentina, 3564. Bolivia, 3329. Ecuador, 3216, 3226, 3236, 3243. Panama, 3016.

Porto Alegre, Manuel de Araújo, 4367.

Porto Alegre, Brazil (city). Peasant Uprisings, 3703. Protests, 3703.

Ports, 1253. Brazil, 2181. Mexico, 2412.

Postal Cards. *See* Postcards.

Potatoes. Andean Region, 2429. Bolivia, 1977. Ecuador, 1803. History, 2429. Peru, 2520, 4888, 4928. Spain, 2429.

Potosí, Bolivia (dept.). Literacy and Illiteracy, 960. Religious Life and Customs, 959. Shamanism, 960.

Poverty, 1076, 1096, 1213, 1243, 1263, 1292, 2238, 3827, 3872, 4436, 4438. Aged, 5120. Amazon Basin, 2678. Argentina, 5108, 5114, 5120. Brazil, 2108, 2653, 5167, 5191, 5193. Central America, 1565. Chile, 1865, 1898, 3452, 3460, 5008, 5034, 5060. Colombia, 3168. Costa Rica, 1582. Dominican Republic, 4710, 4727. Ecuador, 1791. El Salvador, 1594. Environmental Degradation, 4674. Environmental Policy, 4401. Environmental Pollution, 766. Honduras, 766, 1620, 4674. Mexico, 1076, 1345, 1406, 1426–1427, 1434, 1511–1512, 1517, 1529, 2381. Panama, 4625. Peru, 1026, 1048. South America, 1101. Urban Areas, 4450. Venezuela, 1721, 1737.

Prebisch, Raúl, 1158, 1211, 1248.

Precolumbian Architecture. Aztecs, 220, 225, 341, 419. Ball Games, 289. Belize, 289, 394. Calendrics, 456. Costa Rica, 477. Honduras, 178, 247. Incas, 673, 688, 691, 693. Mayas, 66, 208, 387, 398. Mesoamerica, 121, 146, 199, 234–236, 238–240, 247, 341. Mexico, 182, 196, 204, 232–233, 256, 327–328, 346, 361, 373, 380, 390. Peru, 663, 665, 675, 688, 690, 693, 700. South America, 538.

Precolumbian Art. Argentina, 557. Aztecs, 159, 419. Brazil, 586. Caves, 330. Central America, 149. Colombia, 635–637. Dominican Republic, 487. Exhibitions, 326, 636. Goldwork, 686. Honduras, 330. Incas, 682, 691. Mayas, 148, 160. Mesoamerica, 148–149, 159, 344, 384. Mexico, 196, 326, 389, 429. Peru, 681–682, 684, 686, 690, 695, 700, 703. Plants, 439. Shells and Shell Middens, 384. Symbolism, 402, 443, 637. Teotihuacán, 443. Theory, 429. Venezuela, 715.

Precolumbian Civilizations. Acculturation, 536. Agricultural Development, 645. Agricultural Geography, 173, 302. Agricultural Productivity, 101. Agricultural Systems, 140, 229, 276, 340, 372, 390, 566, 652, 716. Agricultural Technology, 78, 2283. Agriculture, 64, 106, 109, 134, 500, 2416. Amazon Basin, 584. Andean Region, 530, 539, 544, 661, 689. Argentina, 546, 549–551, 553–554, 556–559, 561–562. Artifacts, 245. Artisanry, 262–263, 265, 405. Artisans, 686. Aruba, 521. Asian Influences, 533. Bahamas, 506. Ball Games, 122. Belize, 104, 283, 298, 364, 471. Bibliography, 157. Bolivia, 565–571, 962. Brazil, 544, 576–577, 579–580, 582–585, 587–588, 591–595. Bronze, 297. Caribbean Area, 490–491, 497, 502,

507–508, 511, 519, 526. Center/Periphery Relations, 331, 407, 448. Central America, 149, 476, 511. Chicha, 679. Chiefdoms, 86, 683. Chile, 599–602, 605, 607, 612. Classification, 352. Clothing and Dress, 612. Colombia, 613, 616, 619–621, 625, 629–631, 633–634. Communication, 600. Copper Industry and Trade, 255. Corn, 679. Cosmology, 60. Costa Rica, 467–469, 475, 477–478. Costume and Adornment, 321, 605, 696. Cotton Industry and Trade, 731. Cuba, 494, 500. Cults, 452. Cultural Contact, 588. Cultural Development, 55–56, 125, 175, 331, 490, 532, 539, 541, 616, 710, 717. Dance, 164, 198. Deities, 422. Demography, 147, 302, 381, 583. Discovery and Exploration, 54. Diseases, 634–635. Domestic Animals, 526. Dominican Republic, 487. Droughts, 569, 584. Dwellings, 554, 633. Economic Conditions, 382. Economic Models, 177. Economics, 405. Ecuador, 638, 641, 645, 648–649. El Salvador, 474. Elites, 135, 301, 716. Ethnic Groups and Ethnicity, 538, 612. Food, 385. Food Supply, 64, 156, 318, 508, 2406. Gardens, 276. Goldwork, 647, 686. Guatemala, 363. Hallucinogenic Drugs, 561. Hispaniola, 501. Historiography, 435, 525. Honduras, 290, 292, 303–304, 306. Households, 146. Human Adaptation, 531, 534, 593, 600, 699. Human Ecology, 200, 217, 302, 580, 591, 678. Human Geography, 69, 300. Iconography, 401, 533, 535. Inca Influences, 562, 683, 694. Inter-Tribal Relations, 102, 114, 150, 246, 263, 301, 303, 462, 539, 557. Jade, 145. Kings and Rulers, 135, 380, 648. Legends, 461. Literacy and Illiteracy, 453. Livestock, 566. Marine Resources, 111. Mathematics, 697. Merchants, 428. Mesoamerica, 79, 92, 105, 149, 199, 303. Metallurgy, 556, 559. Mexico, 68, 73, 83, 85, 106, 127, 222, 301, 731, 2416. Militarism, 117. Minerals and Mining Industry, 199. Mortality, 381. Mortuary Customs, 249, 272, 338, 366, 390, 469, 571. Mural Painting, 700. Musical Instruments, 567. Mutilation, 320. Names, 502. Navigation, 487. Nicaragua, 472. Pacific Coast (Colombia), 626. Pacific Coast (Ecuador), 626. Palms, 307. Patagonia, 536, 544, 554. Peru, 659, 675–676, 678–680, 685, 699. Plants, 518. Political Anthropology, 100. Political Culture, 56, 161, 435, 461–462, 675, 678–679, 708. Political Development, 131, 247–248, 300–301. Political Sociology, 86, 126. Political Systems, 117. Religion, 120, 137, 556. Religious Life and Customs, 389. Research, 157. Rites and Ceremonies, 319, 570, 689. Rock Art, 379. Sacrifice, 84. Salt and Salt Industry, 170, 335. Sex Roles, 115. Shamanism, 533. Shells and Shell Middens, 384. Social Change, 300. Social Structure, 56, 66, 80, 86, 100, 105, 137, 300, 358, 382, 537, 541, 605, 649, 678, 690, 703. Social Values, 126. South America, 532, 537. Subsistence Economy, 292, 699. Suriname, 651–652. Taxation, 458. Technology, 471. Textiles and Textile Industry, 681. The Guianas, 653. Theory, 52. Tobacco Use, 558. Tombs, 685. Tools, 553. Transportation, 94, 156. Urban History, 235, 238. Urbanization, 52, 137. Uruguay, 710–711. Venezuela, 712–713, 716–717. Warfare, 117. Water Resources Development, 98. Weights and Measures, 392. Women, 81. Writing, 435, 450.

Precolumbian Land Settlement Patterns, 52, 69. Amazon Basin, 573, 583. Andean Region, 539. Argentina, 548, 552, 554. Aztecs, 203. Belize, 267. Brazil, 573, 580, 592. Caribbean Area, 517. Chile, 604. Colombia, 531, 619, 632. Costa Rica, 477–478. Dominican Republic, 486, 517. Ecuador, 531, 640. Guatemala, 268, 313, 398. Honduras, 246, 248, 286, 290, 333–334. Mayas, 267–268, 285, 332, 398. Mesoamerica, 97, 131, 173, 200, 253, 276, 285, 299, 343. Mexico, 101, 218, 271, 300, 358, 362. Mochica, 702. Nazca, 705. Patagonia, 548. Peru, 663, 678, 702–703, 708. South America, 542. Venezuela, 484, 713–714.

Precolumbian Pottery. Argentina, 546–547, 550. Aztecs, 293–295, 378. Bolivia, 568, 571. Colombia, 613, 622, 625, 627–628, 632. Costa Rica, 470, 473. Ecuador, 641–643. Guatemala, 269, 287, 314. Honduras, 352. Incas, 540. Japanese Influences, 643. Martinique, 523. Matlatzinca, 377. Mayas, 148, 269. Mesoamerica, 190–191, 200, 226, 241, 244, 284, 304, 352. Mexico, 107, 198, 206, 213, 218, 224, 251, 266, 275, 309, 365. Mixtec, 179. Nazca, 658. Peru, 662, 664, 684, 706. South America, 535, 545. Teotihuacán Influences, 244. Tiwanaku Culture, 706. Tombs, 275. Totonac, 309.

Precolumbian Sculpture. Aztecs, 110, 159, 225, 312, 378, 436. Chinese Influences, 112. Deities, 316. Ecuador, 644. Exhibitions, 308. Guatemala, 280, 324. Huari, 668. Incas, 691. Mayas, 208, 273, 277, 367, 387,

418, 425. Mesoamerica, 112, 280, 296, 308, 316, 344, 462. Mexico, 232, 273, 346, 361, 365, 380. Olmec Influences, 280. Peru, 657, 670.
Precolumbian Textiles. Peru, 690.
Precolumbian Trade, 69. Amazon Basin, 851. Aztecs, 209, 293. Belize, 283. Caribbean Area, 851. Corn, 2406. Costa Rica, 466, 470, 478. Honduras, 246, 334. Inca Influences, 639. Martinique, 523. Mayas, 133, 283. Mesoamerica, 70, 94, 102, 116, 150, 156, 177, 187, 191, 255, 264, 283, 288, 290, 292, 298, 303, 396, 405, 428. Mexico, 219, 226, 251, 448. Nazca, 705. Obsidian, 288, 292, 396. Turquoise, 116.
Prehistory. *See* Archaeology.
Presidente Prudente, Brazil (city). Elections, 3717.
Presidential Systems. *See* Political Systems.
Presidents, 2750. Andean Region, 2725. Argentina, 2724, 3513, 3523, 3536, 3561, 3581, 3586, 4194. Bolivia, 3278, 4227–4228, 4238. Brazil, 2724, 3670, 3702, 3779. Chile, 1876, 2724, 3420, 3424, 3438, 3444, 3475–3476. Colombia, 3181, 3206. Costa Rica, 2947. Ecuador, 3218, 3222, 3229, 3243. Haiti, 3064. Honduras, 2989. Mexico, 21, 2868, 2873, 2894, 2916, 2922, 2924, 2937. Panama, 3019, 3025, 4041, 4047. Paraguay, 3597. South America, 2724–2725. Uruguay, 2026, 3639. Venezuela, 3251, 3259, 3276, 4338.
Press. *See* Mass Media.
Pressure Groups. Agriculture, 3553. Argentina, 3553–3554. Brazil, 3707. Mexico, 2840, 2864, 2872, 2924. Peru, 1929, 3371, 3402. US, 4345.
Preuss, Konrad Theodor, 998.
Price-Mars, Louis, 823.
Prices, 1122, 1208. Argentina, 2052, 2075. Bolivia, 1991. Brazil, 2158. Ecuador, 1812. Nicaragua, 1634. North America, 1382. Peru, 1945.
Primary Education. *See* Elementary Education.
Printers. *See* Printing Industry.
Printing Industry. Argentina, 5107.
Prisons. Bolivia, 3302. Human Rights, 2762. Panama, 3021. Peru, 3359.
Private Enterprise. Bolivia, 3290. Brazil, 2696, 2714, 2716. Chile, 5022, 5041. Costa Rica, 2951. Ecuador, 1845. Mexico, 2864. Peru, 1928. Trinidad and Tobago, 3099. Venezuela, 1718.
Privatization, 1090, 1109, 1112, 1117, 1121, 1139, 1160, 1175, 1218, 1246, 1253, 1255, 1301. Argentina, 1121, 2033, 2061, 2067, 2071, 2081, 3546, 3555, 3587. Brazil, 2103, 2164, 2203. Chile, 1121, 1870, 1901. Costa Rica, 4671. Ecuador, 1786. Jamaica, 1672.
Medical Care, 2731. Mexico, 1121, 1326, 1369, 1499, 2403, 2930. Paraguay, 2002. Peru, 1932, 2518. Uruguay, 3638–3639, 3641, 3665. Venezuela, 1730, 1741.
Procedure to Establish Steady and Long-Standing Peace in Central America. *See* Arias Peace Plan (1987).
PRODERM (project), 1927, 1963.
Production (economics). Aztecs, 228. Brazil, 2205.
Programa de Desarrollo Rural Integrado (Colombia), 1768.
Programa Nacional do Alcool (Brazil), 2170.
Property. Mexico, 1530.
Prostitution. Amazon Basin, 5165. Brazil, 5165, 5193. Children, 4991, 5165. Cuba, 4782. Paraguay, 4991. Uruguayans, 5139.
Protectionism, 1120. Mexico, 3904.
Protestant Churches. Nicaragua, 4601.
Protestantism. Cakchikel, 728. Central America, 4655, 4657. Dominican Republic, 4798. Ecuador, 1018. El Salvador, 4569. Guatemala, 4590, 4597. Homosexuality, 4690. Mexico, 728. Nicaragua, 2998, 4601. Social History, 4399.
Protestants. Amazon Basin, 2475. Central America, 4655. Ecuador, 2475. El Salvador, 4569. Historiography, 4399.
Protests. Argentina, 3508, 5088, 5104. Brazil, 3703. Chile, 5049, 5069, 5082. Dominican Republic, 4768. Indigenous Peoples, 4868. Squatter Settlements, 5069. Youth, 5082.
Proyecto Arqueológico Pacífico Central (Costa Rica), 465.
Przeworski, Adam, 2740.
Psychiatry. Suriname, 4801. Traditional Medicine, 4801.
Psychology. Exiles, 4451. Poor, 5034.
Public Administration, 3876. Argentina, 3538, 3561, 3587, 5091, 5111. Chile, 5002. Colombia, 3171. Congresses, 3632. Costa Rica, 4671. Cuba, 4776. Dutch Caribbean, 3075, 3077. Jamaica, 3070. Mexico, 2400, 2863, 2914. Military, 5091. Nicaragua, 4665. Paraguay, 3596. Peru, 3386. Suriname, 3094. Uruguay, 3632.
Public Debt, 4110. Argentina, 2045, 2064, 2084, 2137, 3567. Brazil, 2137. Colombia, 1769. Cuba, 4126. Mexico, 1330, 1379. Peru, 2137. Venezuela, 1753.

Public Education. Bolivia, 4976. Mexico, 2847.
Public Enterprises. Jamaica, 1672. Mexico, 2412. Venezuela, 1718.
Public Finance, 1244. Argentina, 2038, 2059, 2067, 2084. Brazil, 2102, 3750. Caribbean Area, 4093. Costa Rica, 2949. Mexico, 1323, 1509, 1538, 2827, 2881. Peru, 4301. Uruguay, 3618, 3630. Venezuela, 1728, 1731.
Public Health. Argentina, 5094, 5109–5110. Bolivia, 4960, 4969, 4975. Brazil, 2115, 2197, 5161, 5188. Caribbean Area, 4751. Chile, 5033, 5057. Colombia, 4807. Cuba, 4739. Dominican Republic, 4773. Ecuador, 4873. Haiti, 4739. Mexico, 1463, 4495, 4503, 4517. Nicaragua, 4615, 4669. Nongovernmental Organizations, 4905. Panama, 4568. Peru, 4896, 4905. Poor, 4430. Puerto Rico, 4785. Tobacco Use, 4447. Venezuela, 4833.
Public Investments, 1103, 1181, 1305. Argentina, 2085. Bolivia, 2524. Brazil, 2124. Central America, 1557. Chile, 1868. Costa Rica, 1557. Guatemala, 1557. Nicaragua, 1628. Panama, 1557.
Public Opinion, 2795, 3852. Argentina, 3528, 3588, 5117, 5124. Blacks, 5142. Brazil, 3735, 3757. Chile, 3427, 3431, 3447–3448, 3456, 3459, 3466, 3483, 4998, 5022, 5025, 5029, 5040, 5059, 5061. Databases, 41. Democratization, 2743. Ecuador, 3224. Evangelicalism, 5025. France, 4140. Mass Media, 5130. Mexico, 41, 2918, 4458–4459. Military, 5040. Panama, 3022. Paraguay, 3604. Peru, 3362, 3377, 3400, 3415. Spain, 3816. Students, 5124. Uruguay, 3637, 5130. US, 3852, 3861, 4140, 5142. Youth, 5059.
Public Policy, 1267. Argentina, 3512, 3529, 3560, 5115. Brazil, 3711. Chile, 5060. Colombia, 4812, 4815. Guatemala, 4618. Mexico, 1528, 4489. Nicaragua, 4685. Panama, 4568. Peru, 4316. Sandinistas, 4685. Women, 4812.
Public Service Employment. Costa Rica, 4614. Nicaragua, 4646. Women, 4425, 4614, 4646.
Public Transportation. *See* Local Transit; Transportation.
Public Works. Argentina, 5087. Mexico, 5087.
Puebla, Mexico (city). Ejidos, 2387. Land Reform, 2387. Land Settlement, 2387.
Puebla, Mexico (state). Archaeology, 342–343. Cooperatives, 4511. Historical Geography, 83. Peasant Movements, 746. Precolumbian Civilizations, 83.
Puebla Valley (Mexico). Historical Demography, 2395.
Puerto Ricans. Return Migration, 4721, 4771. US, 4695, 4703, 4721, 4771. Women, 4695.
Pulque. Aztecs, 451.
Pumpu Site (Peru), 688.
Puno, Peru (city). Political Participation, 1037. Youth, 1037.
Puno, Peru (dept.). Agriculture, 4941. Food Supply, 4941. Guerrillas, 1052. Land Reform, 3404. Nutrition, 4941. Peasants, 4941. Political History, 3404. Revolutions and Revolutionary Movements, 3407. Rural Development, 4948. Violence, 1052, 3407. Women, 1050.
Putumayo, Colombia (intendency). Human Rights, 3193.
Quadros, Jânio, 3668.
Quality of Life, 1207. Central America, 1553. Chile, 2592. Mexico, 2865.
Quechua (indigenous group). Colonization, 1061. Cosmology, 1070. Cultural Identity, 4898. Discourse Analysis, 1031. Dwellings, 4898. Ethnography, 1061. Human Ecology, 4982. Kinship, 1046. Law and Legislation, 1058. Marriage, 1062. Mortuary Customs, 959. Myths and Mythology, 4884. Painting, 1044. Pictorial Works, 1044. Religious Life and Customs, 959, 1070. Rites and Ceremonies, 4898. Social Life and Customs, 1028, 1062. Social Structure, 1046. Symbolism, 1061. Textiles and Textile Industry, 1040. US, 1027. Wit and Humor, 935. Youth, 1031, 1037.
Quemada Site (Mexico), 199, 338.
Quereo River Valley, Chile (region). Paleo-Indians, 608.
Querétaro, Mexico (state). Archaeological Surveys, 348. Precolumbian Civilizations, 348.
Quetzalcoatl (Aztec deity), 197.
Quiché (indigenous group). Rites and Ceremonies, 324.
Quillacinga (indigenous group). Artifacts, 615. Precolumbian Land Settlement Patterns, 620.
Quincentennial. *See* Columbus Quincentenary (1992).
Quinquén Valley (Chile). Land Tenure, 969.
Quintana Roo, Mexico (state). Archaeological Surveys, 271. Cultural Geography, 271. Economic Conditions, 1479. Environmental Conditions, 1479. Precolumbian Land Settlement Patterns, 271. Social Conditions, 1479.
Quiriguá Site (Guatemala), 363, 367.

Quiroga Santa Cruz, Marcelo, 3295.
Quito, Ecuador (city). City Planning, 2467. Clientelism, 4854. History, 2477. Informal Sector, 2483. Labor and Laboring Classes, 1810, 2474. Municipal Government, 4854. Neighborhood Associations, 4855, 4863. Recessions, 1810, 2474. Social Structure, 4854. Squatter Settlements, 4855. Transportation, 1818. Urban Areas, 2467. Urban Sociology, 4855.
Race and Race Relations, 3043. Barbados, 795. Brazil, 5142, 5151, 5200, 5202, 5207, 5211. Caribbean Area, 792, 796, 807, 827, 837, 4755. Colombia, 1000, 1005, 4830. Costa Rica, 826, 4659. Guyana, 787. Jamaica, 840. Nationalism, 4755. Nicaragua, 750. Panama, 4659. Saint Kitts and Nevis, 821. Trinidad and Tobago, 787, 809, 834, 3100, 4712. US, 5142, 5202, 5211. West Indians, 4748.
Racism, 4442. Brazil, 5200–5202, 5207. Colombia, 1005. Costa Rica, 4659. Dominican Republic, 4744. Employment, 5207. Nicaragua, 750. Panama, 4659. Peru, 4930. Television, 5201. US, 5202.
Radicalism, 3881.
Radio. Barbados, 3074. Mexico, 2885, 4513. Paraguay, 3609. Suriname, 4792.
Radio Ñandutí (Paraguay), 3609.
Railroads. Argentina, 2545, 2559. Chile, 2545, 2586.
Raimondi, Antonio, 2512.
Rain and Rainfall. Andean Region, 2567. Chile, 2599. Ecuador, 2465. Mexico, 2382. Peru, 2498. Venezuela, 2447. Virgin Islands, 2288.
Rainforests, 2233, 2274, 2276, 2420, 2764. Amazon Basin, 2681. Brazil, 2676, 2680–2681, 4352. Colombia, 2453. Costa Rica, 2336. Suriname, 2287, 2292.
Rama (indigenous group). Acculturation, 3010. Cultural Identity, 3010. Government Relations, 3010.
Rape. Argentina, 5096. El Salvador, 4662. Guatemala, 4662. Honduras, 4644. Mexico, 4562. Panama, 4573. Paraguay, 4993.
Rastafarian Movement. Dutch Caribbean, 4732. Jamaica, 4719, 4728, 4758. Saint Eustatius, 4732.
Ratzel, Friedrich, 2612.
Rebellions. *See* Insurrections.
Recessions. Argentina, 2044. Costa Rica, 1574. Ecuador, 1810, 2474. Mexico, 2381.
Recife, Brazil (city). Food Supply, 2197. Nutrition, 2197.
Reclamation of Land. *See* Land Invasions; Land Reform; Land Tenure.

Rede Globo (Brazil), 3765.
Redemocratization. *See* Democratization.
Redistribution of Wealth. *See* Income Distribution.
Referendums. Chile, 3427–3428, 3454, 3481, 5061. Cuba, 3145. Uruguay, 3636, 3638, 3641, 3665. Venezuela, 3272.
Reforestation. Haiti, 4764. Paraguay, 4985.
Refugees. Belize, 4633. Central America, 3966, 4542, 4570, 4591, 4623. Chile, 3442. Colombia, 3165. Demography, 4444. El Salvador, 4664. Guatemala, 2977. Mexico, 4630. Nicaragua, 4648. Peru, 4911. Social Movements, 4630. Women, 4444.
Refugees, Central American. Mexico, 4542. US, 4600.
Refugees, German, 4395.
Refugees, Guatemalan. Mexico, 2378, 3905, 3947, 4040. US, 722.
Refugees, Haitian. Bahamas, 4141. Dominican Republic, 4141. US, 4141.
Refugees, Jewish, 4395.
Refugees, Uruguayan. Chile, 3442.
Reggae Music. Jamaica, 4728.
Regional Development. Bolivia, 2524, 3343, 4302, 4309. Capital, 5006. Chile, 5006. Colombia, 4269, 4819. Costa Rica, 2332. Haiti, 2295. Paraguay, 2004. Peru, 1957, 4302, 4309, 4933. Venezuela, 4846.
Regional Government. Amazon Basin, 2633. Brazil, 2633. Mexico, 2816, 2822, 2829, 2857, 2872, 2874, 2931. Venezuela, 3255, 3265, 3270.
Regional Integration, 1137, 1148, 1157, 1212, 1269, 1302, 3781, 3783, 3821, 3832, 3841, 3873, 4110, 4116, 4242, 4268, 4359. Andean Region, 1792, 1797, 4170, 4235, 4269. Asia, 3821. Brazil, 4351, 4372. Caribbean Area, 3034, 4080, 4082, 4100–4101, 4156. Central America, 1543, 1556, 1563–1564, 1566, 4043. Colombia, 4268. Costa Rica, 1583. Ecuador, 1841. Europe, 3820. Mexico, 767. North America, 1372. Paraguay, 2003. Río de la Plata, 4183. South America, 2006, 4166, 4172, 4183, 4378. Southern Cone, 2030, 2106, 2160, 4187. Uruguay, 2008, 2021, 2023, 4320.
Regional Planning. Brazil, 2652, 2703. Chile, 2576, 2605. Colombia, 2459. Mexican-American Border Region, 1356. Mexico, 1364, 2405. Peru, 3374, 4316. Venezuela, 4846.
Regionalism. Argentina, 3522. Bolivia, 3342–3343. Brazil, 2648. Colombia, 3173. Ecuador, 3221. Mexico, 2360. Peru, 4933.
Reiche, María, 697–698.

Religion, 3799. African Influences, 820. Amazon Basin, 882. Aymara, 955. Brazil, 5149. Chiriguano, 929. Cuba, 4770. Mexico, 2851. Puerto Rico, 4690. Social Movements, 5149.

Religion and Politics. Brazil, 3706, 3732, 3759, 5148. Central America, 4578, 4631, 4657. Chile, 3429, 3467, 5051, 5075. Costa Rica, 4647, 4683. Cuba, 4784. El Salvador, 4569. Guatemala, 4590, 4597. Haiti, 3061. Mayas, 120. Mesoamerica, 120. Mexico, 2830, 2911. Nicaragua, 2996, 2998. Paraguay, 3594–3595. Peru, 3410. Uruguay, 4322.

Religious Life and Customs, 4423, 4439. Amazon Basin, 853. Aztecs, 88. Blacks, 820. Bolivia, 951. Bororo, 870. Caribbean Area, 841, 843. Central America, 4631, 4655, 4657. Chile, 5053. Costa Rica, 4683. Cuba, 499, 4770. Dominican Republic, 4798. Ecuador, 1018. Guatemala, 4603, 4616, 4661. Indigenous Peoples, 862. Mayas, 108, 136, 138, 356, 433, 752, 4616. Mochica, 701. Peru, 4906, 4910. Precolumbian Civilizations, 389, 556. Quechua, 959, 1070. Trinidad and Tobago, 804, 4742. Venezuela, 4853.

Remittances. *See* Emigrant Remittances.

Remote Sensing. Brazil, 2717.

Renovación Nacional (political party, Chile), 3425.

Repatriation. El Salvador, 4664. Haiti, 4744. Nicaragua, 4648.

Repression, 2770. Argentina, 3568. Bolivia, 3302, 3355, 4979. Brazil, 3677, 3753, 3777, 5162. Chile, 3433, 3475–3476, 5027. Cuba, 4740. Guatemala, 769, 2977, 4571, 4579, 4602, 4610, 4620, 4670. Labor Movement, 3777. Paraguay, 3589, 3609, 3617. Peru, 3568. Police, 3753.

Rescue Archaeology. *See* Salvage Archaeology.

Research. Anthropology, 768. Bolivia, 4974. Brazil, 2191, 2213. Caribbean Area, 801, 810, 822. Jamaica, 788. Mexico, 4493. Popular Religion, 5001. Rural Development, 5018. Social Movements, 4413, 5150. Social Sciences, 4402. Women, 1206, 4974.

Research Institutes. Social Sciences, 4402.

Return Migration. Central America, 4574. Dominican Republic, 4574. El Salvador, 4664. Haiti, 4744. Japan, 3052. Mexico, 4474, 4501. Peru, 4881. Puerto Rico, 4695, 4721, 4771.

Reviews. Archaeology, 74.

Revolution of the Comuneros. *See* Insurrection of the Comuneros (Paraguay, 1730–1735).

Revolutionaries. Bolivia, 3292. Central America, 2991. Colombia, 3182. El Salvador, 2971. Honduras, 2991. Peru, 2500, 3379. Uruguay, 3660. Women, 3405.

Revolutions and Revolutionary Movements, 2732, 2737, 2742, 2756, 2783, 2800, 2807–2808. Argentina, 3540, 3542, 3557. Bolivia, 3319, 3330, 3335. Brazil, 5162. Central America, 2995, 4585. Christianity, 2996. Cuba, 3131–3132, 3881, 4109. Developing Countries, 2739. Dominican Republic, 3050. El Salvador, 2966–2967, 2971, 4638. Mexico, 2871, 4552. Nicaragua, 2996, 3005–3006. Peru, 1060, 2515, 3359–3360, 3365–3366, 3369, 3372, 3380–3381, 3388, 3391, 3397, 3405, 3407, 3411, 4310. Political Theory, 2798. Trinidad and Tobago, 3100. Uruguay, 3645.

Reyes, Reynaldo, 759, 3007.

Rice and Rice Trade. Cuba, 1690. Ecuador, 1803.

Rico, Aldo, 3515.

Rímac River Valley, Peru. Precolumbian Civilizations, 703.

Río Azul Site (Guatemala), 180.

Rio de Janeiro, Brazil (city). Environmental Pollution, 2710. Medical Care, 2710. Political Parties, 3675. Sewage Disposal, 2710. Water Conservation, 2710. Water Supply, 2710.

Rio de Janeiro, Brazil (state). Archaeology, 581–582. Industry and Industrialization, 2700. Natural Resources, 2704. Petroleum Industry and Trade, 2704. Shells and Shell Middens, 581–582. Textiles and Textile Industry, 2700.

Río de la Plata (region). Commerce, 4187. Floods, 2548. Human Geography, 4183. Political History, 4183. Regional Integration, 4183. Satellites, 2548.

Rio Grande do Norte, Brazil (state). Income Distribution, 2116. Industrial Productivity, 2116.

Rio Grande do Sul, Brazil (state). Archaeology, 591. Cities and Towns, 2666. Civilization, 2648. Cultural Identity, 5186. Economic Conditions, 2127. Geography, 2719. Human Adaptation, 580. Industrialists, 2162. Industry and Industrialization, 2162. Latifundios, 2648. Paleo-Indians, 580. Regionalism, 2648. Social Change, 2205. Urban Policy, 2666. Urbanization, 2666.

Rio Maria, Brazil (municipality). Social Conflict, 5167. Violence, 3699.
Río Napo. See Napo River Basin.
Rioja, Argentina. See La Rioja, Argentina (prov.).
Rionegro, Santander, Colombia (city). Clientelism, 3178.
Ríos Montt, Efraín, 4571.
Rites and Ceremonies. Amazon Basin, 851. Andean Region, 689. Arhuaco, 877. Aztecs, 88, 167, 260, 311, 417, 428, 739. Caribbean Area, 851. Caves, 212. Colombia, 990. Curripaco, 896–897. Festivals, 1056. Guyana, 842. Indigenous Peoples, 1056. Mashco, 919. Mayas, 108, 324, 356, 370–371, 418, 445. Mexico, 751. Peru, 921, 1028, 1030, 1056, 4898. Precolumbian Civilizations, 319, 570, 689. Secoya, 913. Shuar, 922. Stelae, 445. Tucano, 877. Tucanoan, 901. Venezuela, 891.
Roads, 1186, 2273. Brazil, 2664. Costa Rica, 2331. Incas, 656. Mexico, 2365.
Robert, Georges, 3082.
Robinson, A.N.R., 3100.
Rocha, Uruguay (dept.). Excavations, 711.
Rock Art. Aruba, 522. Aztecs, 312. Bibliography, 496. Bolivia, 529. Brazil, 578, 586, 590, 594. Caribbean Area, 496. Chile, 529. El Salvador, 474. Honduras, 330, 379. Mexico, 221, 328. Patagonia, 536. Peru, 665. Precolumbian Civilizations, 379. Research, 529. South America, 496. Venezuela, 715.
Rock Paintings. See Rock Art.
Rockshelters. See Caves.
Rodríguez, Andrés, 3605, 4295.
Rohr, João Alfredo, 579.
Rojas, Juan, 964.
Rojas Puyo, Alberto, 3153.
Romero, Oscar Arnulfo, 2968.
Rondônia, Brazil (state). Agricultural Colonization, 5205. Boundary Disputes, 2689. Colonization, 2652. Environmental Policy, 2661. Households, 5205. Land Settlement, 2661. Regional Planning, 2652. Roads, 2664. Rural Development, 2652. Social Mobility, 5205.
Roraima, Brazil (state). Cultural History, 876. Indigenous Peoples, 876. Social Life and Customs, 876.
Rubber Industry and Trade. Amazon Basin, 2657, 2715, 5160. Bolivia, 4971. Brazil, 2129, 2639. Colombia, 905. Labor Movement, 5160. Venezuela, 2445.
Ruffo Appel, Ernesto, 2931.
Ruiz Cortines, Adolfo, 2916.

Runa. See Canelo.
Rural Conditions. Chile, 3436, 3458. Dominica, 815. Peru, 1048.
Rural Development, 1278, 1768, 4398. Amazon Basin, 1936. Bolivia, 3299, 4978. Brazil, 2652, 2699, 2705. Central America, 1540. Chile, 2576, 5018. Colombia, 1758, 1768, 4808–4809. Congresses, 4859. Costa Rica, 1587. Ecuador, 4859. History, 5018. Mexico, 1319, 1337, 2396, 2859, 4504, 4539. Panama, 747. Paraguay, 4984. Peru, 1936, 3409, 4909, 4917–4918, 4924, 4936, 4938–4939, 4948. Research, 5018. Uruguay, 2619.
Rural Development Projects. See Development Projects.
Rural Health. Ecuador, 1009. Venezuela, 4833.
Rural Sociology. Argentina, 5095, 5100. Belize, 4637. Bolivia, 956, 966, 4962. Brazil, 3740, 3746, 5177, 5185. Central America, 729. Chile, 5005, 5010, 5018, 5065. Christian Base Communities, 3719. Colombia, 4808–4809. Costa Rica, 4639, 4658. Cuba, 776. Dominican Republic, 4727, 4781. Ecuador, 4859, 4861–4862, 4865, 4875. Guatemala, 765, 4656, 4688. Haiti, 4787. Honduras, 4675. Mexico, 758, 763, 767, 4465, 4502, 4539, 4551. Nicaragua, 4612, 4622, 4666, 4668, 4673, 4680. Paraguay, 4983, 4987. Peru, 4893, 4900–4901, 4924, 4932, 4936, 4941, 4945. Venezuela, 4833.
Rural-Urban Migration, 2218. Bolivia, 4949. Brazil, 2654. Ecuador, 1019. Guatemala, 4688. Honduras, 4675. Mexico, 744, 2388, 2407, 2859, 4506, 4538. Nicaragua, 4632. Peru, 1026, 1039, 1045, 1061, 1067–1068, 4911. Quechua, 1061. Women, 4506. Youth, 4688.
Russians. Paraguay, 2769.
Saad, Pedro, 4286.
Sacred Space. Architecture, 341. Aztecs, 167, 220. Incas, 673. Mayas, 103. Mesoamerica, 121, 341. Mexico, 361. Precolumbian Architecture, 361. Precolumbian Civilizations, 389. Teotihuacán, 432.
Sacrifice. Aztecs, 77, 88, 220. Mesoamerica, 84. Precolumbian Civilizations, 84.
Safety in the Workplace. See Industrial Safety.
Sahagún, Bernardino de, 424, 438.
Saint Christopher-Nevis. See Saint Kitts and Nevis.
Saint Eustatius (island). Excavations, 485. Physical Anthropology, 485. Rastafarian Movement, 4732.

Saint Kitts and Nevis. History, 3037.
Saint Martin (island). Cultural Identity, 4779. Interpersonal Relationships, 4779. Social Life and Customs, 4779.
Saint-Pierre, Martinique (town). Description and Travel, 2290. Historic Sites, 2290.
Saint Vincent (island). Cultural Identity, 850. Social Life and Customs, 850.
Salazar Paredes, Fernando, 4239.
Salesians. Haiti, 3061.
Salinas de Gortari, Carlos, 40, 1350, 2813, 2830–2831, 2851, 2876, 2879, 2887, 2893, 2895, 2905, 2909, 2927, 2937.
Salitrón Viejo Site (Honduras). Precolumbian Pottery, 304.
Salt and Salt Industry. Chile, 2609. Guatemala, 335. Mesoamerica, 170. Precolumbian Civilizations, 170, 335.
Salta, Argentina (city). Geomorphology, 2554.
Salvador, Brazil (city). Carnival, 5180.
Salvage Archaeology. Colombia, 615. Guatemala, 335. Honduras, 286, 292. Mexico, 204, 206, 221, 223, 278, 345.
San Agustín Archaeological Park (Colombia), 618, 637.
San Agustín Culture, 621, 623. Precolumbian Art, 637.
San Antonio, Chile (prov.). Demography, 5028.
San José, Costa Rica (city). Urbanization, 2326, 2328.
San Juan, Puerto Rico (city). City Planning, 2311–2312. Maps and Cartography, 2311–2312. Urbanization, 2311–2312.
San Lorenzo Site (Mexico). Obsidian, 230. Precolumbian Sculpture, 232.
San Luis Potosí, Mexico (state). Excavations, 213.
San Martín, Peru (dept.). Development, 4933. Earthquakes, 2514. Economic History, 4933.
San Pedro de Atacama, Chile (city). History, 607. Precolumbian Civilizations, 607.
San Pedro Sula, Honduras (city). History, 2351. Palestinians, 4617.
Sánchez Carrión, Peru (prov.). Peasants, 4900. Rural Sociology, 4900.
Sandinistas (Nicaragua), 1625, 1628–1629, 1633–1634, 2996–2997, 3000–3001, 3005, 3008, 3984, 3990, 4036, 4594, 4627, 4665. Agricultural Policy, 4604, 4673, 4682. Americans (US), 3011. Democracy, 3013. Economic Policy, 3003. Education, 4575. Finance, 2999. Housing Policy, 4632. Indigenous/Non-Indigenous Relations, 3010. Indigenous Policy, 4579. Land Reform, 4608, 4668. Municipal Government, 2999. Public Health, 4615. Public Policy, 4685. Social Welfare, 4669. Statistics, 2999. Urban Policy, 4640. Women, 4672.
Sandoval Morón, Luis, 3292.
Sanema. See Yanomamo.
Sangre Grande, Trinidad and Tobago (town). Development, 4749.
Sanguinetti, Julio María, 3620.
Santa Catarina, Brazil (state). Archaeological Surveys, 577. Archaeology, 579. Excavations, 577.
Santa Clara Site (Dominican Republic). Excavations, 486.
Santa Cruz, Argentina (prov.). Desertification, 2550. Pastures, 2550. Sheep, 2550.
Santa Cruz, Bolivia (city). Social Conditions, 4957. Women, 4957.
Santa Cruz, Bolivia (dept.). Encomiendas, 1970. Haciendas, 1970. Land Tenure, 2521. Land Use, 2521.
Santa Monica de la Barca, Mexico (town). See La Barca, Mexico (town).
Santa Rita Corozal Site (Belize), 250.
Santander, Colombia (dept.). Archaeological Surveys, 622.
Santería (cult), 783, 4703, 4795. Caribbean Area, 843. Cuba, 783, 820, 4795.
Santiago, Chile (city), 2583. City Planning, 2577, 4996, 5058. Conservation, 2595. Cooperatives, 5042. Cost and Standard of Living, 2596. Economic Conditions, 1883, 3461. Economic Development, 2602. Environmental Pollution, 1857, 2595, 5057. Geopolitics, 2612. Housing, 1878, 5063–5064, 5067. Landslides, 2599. Municipal Government, 3461. Neighborhood Associations, 5042. Nongovernmental Organizations, 3461. Poor, 5063–5064. Public Health, 5057. Quality of Life, 2592. Rain and Rainfall, 2599. Slums, 2592. Social Conditions, 1883, 3461. Squatter Settlements, 5000. Urban Policy, 2577, 5054. Urban Sociology, 4996, 5054, 5058. Urbanization, 2577.
Santiago, Cuba (city). History, 4717.
Santiago del Estero, Argentina (prov.). Households, 5100.
Santo Domingo, Dominican Republic (city). Environmental Pollution, 4427. Women, 4427.
Santos, Panama. See Los Santos, Panama (prov.).
São Paulo, Brazil (city). Christian Base Com-

munities, 3696, 3719, 5173, 5176. Decentralization, 2716. Elections, 3683, 3752. Labor Market, 2173. Labor Movement, 3763, 5144. Municipal Government, 3713, 3718. Political Parties, 3752. Private Enterprise, 2716. Social Movements, 3666, 5159, 5176. Street Children, 4446. Urban Sociology, 3666, 4824.

São Paulo, Brazil (state). Agricultural Industries, 2706. Agricultural Labor, 5197. Economic Development, 2161. Elections, 3717. Environmental Pollution, 2674. Food Supply, 2674. Fruit Trade, 2674, 2706. Labor Market, 2130. Labor Supply, 2130. Sugar Industry and Trade, 2674. Sustainable Development, 2128. Technological Development, 2674.

Saramacca (Surinamese people). Social Life and Customs, 805.

Sarmiento, Domingo Faustino, 3507.

Sarney, José, 3687, 3706, 3757.

Satellites. Argentina, 2548. Physical Geography, 2242.

Saving and Investment, 1111, 1233, 1236, 1298, 1301. Argentina, 2052, 2058. Belize, 1570. Bolivia, 1971, 1999. Brazil, 2124. Central America, 1544, 1567. Chile, 1859, 1868, 1901. Ecuador, 1787. Mexico, 1308, 1322, 1393, 1430, 1475. Venezuela, 1749, 1751.

Sayil Site (Mexico), 376.

Science, 1835. Chile, 2600. Ecuador, 2469. Peru, 1923, 2504, 2600.

Scientists. Brazil, 4374. Chile, 2600. Ecuador, 2469. Japan, 4374. Peru, 2600.

Sechura Region (Peru). Excavations, 664.

Secondary Education. Catholic Church, 5037. Chile, 5037.

Secoya (indigenous group). Ethnobotany, 913. Human Ecology, 926. Land Invasions, 925. Rites and Ceremonies, 913. Shamanism, 913.

Seibal Site (Guatemala), 291.

Seineldín, Mohamed Alí, 3515.

SELA. *See* Sistema Económico Latinoamericano.

Self-Determination. *See* Autonomy; Sovereignty.

Selknam. *See* Ona.

Semiotics. Uruguay, 5131.

Sendero Luminoso (Peru), 1049, 1051–1052, 1060, 1942, 2500, 3360, 3366, 3369, 3372–3373, 3375–3376, 3379–3381, 3383, 3388, 3391, 3396–3397, 3403–3405, 3407, 3411, 3413, 4310.

Sendic, Raúl, 3660.

Separation. *See* Divorce.

Serena, Chile. *See* La Serena, Chile (city).

Seri (indigenous group). Mortuary Customs, 249.

Serra Puche, Jaime, 1397.

Servicio Exterior Mexicano, 3950.

Sewage Disposal. Brazil, 2710.

Sex and Sexual Relations. Catholic Church, 3473. Chile, 3473. Costa Rica, 4628. Cuba, 4737. Honduras, 4644. Paraguay, 4993. Suriname, 4799.

Sex Roles. Andean Region, 933. Argentina, 5102. Ashaninca, 924. Aztecs, 82. Brazil, 5141. Caribbean Area, 774, 800, 827. Chile, 5005, 5039. Colombia, 1002, 4812. East Indians, 4780. Ecuador, 1022. Employment, 5039. Hiwi, 898. Indigenous Peoples, 1022. Mashco, 919. Orejón, 909. Peasants, 4932. Peru, 4907, 4932. Precolumbian Civilizations, 115. Technological Innovations, 5141. Trinidad and Tobago, 829, 4780. Tucanoan, 900–901. Venezuela, 891, 4849.

Sexism, 1303, 4852. Argentina, 5116. Chile, 3455. Peru, 4907.

Shamanism. Amazon Basin, 853. Aymara, 954. Bolivia, 960. Brazil, 878, 883. Chamí, 1007. Ecuador, 1008. Indigenous Peoples, 853, 862. Mayas, 108, 742. Mesoamerica, 749. Mexico, 4517. Panama, 747. Peru, 921. Precolumbian Civilizations, 533. Secoya, 913. Sioni, 920. South America, 862. The Guianas, 653.

Shantytowns. *See* Squatter Settlements.

Shavante. *See* Xavante.

Sheep. Argentina, 2540, 2550.

Shells and Shell Middens. Brazil, 581–582. Mesoamerica, 264, 384. Mexico, 265, 397. Precolumbian Art, 384.

Shining Path. *See* Sendero Luminoso.

Shipbuilding. Chile, 2598.

Shipibo. *See* Sipibo.

Shipping, 1197. Caribbean Area, 1197. South America, 4178, 4185.

Shoe Industry, 1087.

Shrimp Industry. Colombia, 2455.

Shuar (indigenous group). Cultural Destruction, 916. Music, 922. Rites and Ceremonies, 922.

Sicuane (indigenous group). Music, 1029.

Silva, José Antonio Alencastro e, 2203.

Silva, Luís Inácio da. *See* Lula.

Silva, Ozires, 2203.

Silva, Vicente Paulo da, 3691.

Silverwork. Chile, 980.

Sinaloa, Mexico (state). Agriculture, 1396. City Planning, 1367. Ecology, 1367. Land Reform, 1396. Land Tenure, 1396. Urbanization, 1367.

Sindicato Unico de la Construcción y Afines (Uruguay), 3658.

Single Mothers. Peru, 4942.

Sinú River Valley (Colombia). Precolumbian Pottery, 613.

Sioni (indigenous group). Human Ecology, 926. Land Invasions, 925. Medicinal Plants, 920. Shamanism, 920.

Sipán Site. *See* Huaca Rajada Site (Sipán, Peru).

Sipibo (indigenous group). Acculturation, 917. Demography, 917. Marriage, 917. Subsistence Economy, 910.

Slaves and Slavery, 4414. Caribbean Area, 479, 784, 807, 818, 836. Indigenous Peoples, 905. Jamaica, 3069. Saint Kitts and Nevis, 821. Suriname, 803.

Slums, 2271. Brazil, 2147. Chile, 2592. Colombia, 3202. Mexico, 2355. Peru, 2494, 4927. Venezuela, 4847.

Sluyter, Andrew, 94.

Small Business. Chile, 5023. Ecuador, 1798, 1822, 1842. El Salvador, 1591. Guatemala, 1612. Mexico, 1366, 1375, 1496, 4515. Peru, 1963, 1966–1967, 4926.

Smith, Oscar, 3503.

Soap Operas, 4426. Brazil, 5187.

Social Change, 2789, 3802. Amazon Basin, 855. Anthropologists, 1065. Anthropology, 992. Bolivia, 950, 966, 1984, 4972. Brazil, 2205, 5171, 5175, 5195, 5199. Cakchikel, 728. Catholic Church, 4582. Chile, 3436, 5009, 5030. Colombia, 992. Costa Rica, 4645. Eastern Europe, 3802. Ecuador, 1016, 4863. Guatemala, 757, 4650. Honduras, 4582. Indigenous Peoples, 855. Liberation Theology, 5009. Mexico, 728, 2833, 2925, 4527. Neighborhood Associations, 4863. Peru, 1068. Precolumbian Civilizations, 300. Technological Innovations, 4449. Trinidad and Tobago, 4793. Venezuela, 4834, 4837. Women, 4731.

Social Classes, 2730, 3818, 4397. Argentina, 5086. Barbados, 795. Belize, 2944, 4637. Bolivia, 952–953, 956, 3353, 4959. Brazil, 2147, 2212, 2651, 3740, 3758, 5157. Caribbean Area, 807, 819, 827. Central America, 729. Chile, 2577. Ecuador, 3216, 3223. Ethnic Groups and Ethnicity, 62. Jamaica, 840, 3069, 4750. Mexico, 2833, 2902, 4508. Nicaragua, 472. Peru, 3358, 3408. Puerto Rico, 4706. Trinidad and Tobago, 848. Uruguay, 3623. Venezuela, 3262, 4832, 4835. Voting, 3758.

Social Conditions, 1142, 1209, 2252, 2270, 2726, 2787, 2789, 3811, 4436. Ashaninca, 3373. Bolivia, 1986, 3281. Brazil, 2192–2193, 5183, 5191. British Caribbean, 3035. Caribbean Area, 2804, 4094. Central America, 2804, 2938, 4074. Chile, 1847, 1869, 1883, 1910, 3460–3461, 5014. Colombia, 1767, 3189, 3194, 3203. Costa Rica, 826. Cuba, 3104, 3106, 3108, 3112, 3116, 3118–3120, 3123, 3128, 3135–3136, 3141. Ecuador, 1793, 1829, 3224, 3245. Haiti, 4140. Inflation, 5183. Mexican-American Border Region, 1360, 2367. Mexico, 764, 1344, 1392, 1394, 1398, 1443, 1479, 1498, 1509, 2381, 2829, 2856, 2865, 2875, 2888, 4459. Mundurucu, 866. Nicaragua, 3004, 4586. Otomi, 773. Peru, 1948, 3361, 3386, 3397, 4897. Puerto Rico, 4724. Suriname, 2303. Trinidad and Tobago, 844–845. Uruguay, 3640. Venezuela, 3267, 3276, 4851.

Social Conflict, 3864, 4438. Argentina, 3577, 5088. Belize, 4633. Chile, 5019–5020. Colombia, 987, 3184, 3187, 3189, 3205, 3207, 4816, 4819, 4827. Dominican Republic, 4768. Guatemala, 757, 4650. Honduras, 4684. Jamaica, 3069, 4719. Mexico, 724, 763, 4481. Peru, 3367, 3417. Suriname, 3087–3088. Technological Development, 4985. Venezuela, 3262–3263, 4837. Virgin Islands, 2288.

Social Customs. *See* Social Life and Customs.

Social Development. Mexico, 3963.

Social History, 2229. Argentina, 5119. Bolivia, 3336. Brazil, 5147. Caribbean Area, 819. Colombia, 4816. Cuba, 4110, 4770. Dominica, 777. Dominican Republic, 3059. Haiti, 4738. Honduras, 2351. Mexico, 23, 4508, 4548, 4555. Nicaragua, 472, 4686. Peru, 1945. Precolumbian Civilizations, 80. Protestantism, 4399. Suriname, 4696. Venezuela, 4832.

Social Indicators. Argentina, 5095, 5097, 5125. Bolivia, 4965. Brazil, 5166, 5175, 5199. Databases, 48. Honduras, 4621. Mexico, 23–24, 4503. Women, 4621.

Social Justice, 2767, 3318. Chile, 3433, 3446. Colombia, 3176.

Social Life and Customs, 4397. Aguaruna, 911. Andean Region, 999, 1047, 1063. Argentina, 5118. Aymara, 947. Aztecs, 88, 118, 168, 225. Bolivia, 4962. Bororo, 870. Brazil, 5145, 5174. Camsa, 986. Caribbean

Area, 2305. Colombia, 1004. Costa Rica, 730, 813. Dominica, 847. Ecuador, 997, 1010. Faroe Islands, 847. Guatemala, 4652. Haiti, 811, 3066. Indigenous Peoples, 931. Mayas, 368. Mesoamerica, 105. Mexico, 721, 723. Orejón, 909. Pasto, 999. Peru, 4913. Precolumbian Civilizations, 389. Quechua, 1028, 1061–1062. Saint Martin, 4779. Trinidad and Tobago, 4712. Uru, 967. Yanomamo, 894.

Social Marginality, 2756, 4438. Bolivia, 928, 3355, 4979. Chile, 5020. Chiriguano, 928. Colombia, 3203. Peru, 4929.

Social Mobility, 1286. Brazil, 5157. Guatemala, 4616, 4663. Mayas, 4616.

Social Movements, 53, 1128, 2742, 2761, 2769, 2784, 4397, 4404. Argentina, 3508, 3537, 5098, 5104, 5126. Aymara, 963. Bibliography, 4521. Bolivia, 961, 965, 3290, 3313, 3316. Brazil, 3666, 3676, 3728, 3740, 5149–5150, 5152, 5159, 5176, 5203, 5212. Catholic Church, 4797. Central America, 4678. Chile, 3462, 3474, 3477, 3482, 5047, 5055, 5069, 5076–5078. Colombia, 1006, 3167, 4919. Costa Rica, 4645. Domestics, 4915. Dominican Republic, 4710, 4768, 4797. Ecuador, 1024, 3228, 3233, 3237, 3239–3240, 3242, 3246. El Salvador, 4662. Guatemala, 2974, 4577, 4662. Honduras, 4587. Indigenous Peoples, 3167, 3228, 3237, 3240, 3242. Mexico, 2814, 2828, 2846, 2855, 2865, 2925, 4466, 4490, 4521–4522, 4524, 4528. Nicaragua, 4598, 4651, 4665. Panama, 4583, 4592. Peru, 3364, 3409, 4889–4890, 4894, 4899, 4904–4905, 4907–4908, 4915, 4919, 4921–4924, 4926. Puerto Rico, 4704, 4733. Refugees, 4630. Religion, 5149. Research, 4413, 5150. Squatter Settlements, 5076, 5078. Uruguay, 5134, 5136. US, 4600. Venezuela, 3249, 4841–4842. Women, 3233, 3240, 4407, 4429, 4448, 4595, 4630, 4651, 4662, 4704, 4887, 4899, 4907, 4921, 4923, 5055, 5078, 5159, 5176, 5212.

Social Organization. *See* Social Structure.

Social Policy, 1115, 1183, 1234, 1292, 2767, 4417. Amazon Basin, 2493. Argentina, 5092, 5099, 5108, 5119. Bolivia, 3332. Brazil, 2126, 2187, 2201, 3684, 3751, 3756, 5182, 5194. Central America, 1553–1554. Children, 5108. Chile, 1882, 1890, 1902, 5008, 5014, 5036, 5072, 5077, 5080. Costa Rica, 4589. Ecuador, 1795, 1800, 1831, 4860. Guatemala, 4577, 4584. Honduras, 4621. Households, 4453. Mexico, 1344, 1412, 1449, 1467, 1502, 1517, 2815, 2827, 4457, 4489, 4495. Municipal Government, 4412. Panama, 3016, 4568. Peru, 1934, 2493, 4887, 4912. Southern Cone, 2751. Urban Areas, 4450. Uruguay, 3630. Venezuela, 3252, 4841. Women, 4441, 4584, 5077.

Social Prediction. Mexico, 2908.

Social Relations. *See* Social Life and Customs.

Social Sciences, 4402, 4404. Bibliography, 28. Brazil, 3418, 5151. Chile, 5030. Databases, 2–3, 28. Directories, 2. Mexico, 4493. Research, 4402. Research Institutes, 4402.

Social Security, 1104, 1221–1224, 2731. Argentina, 2086. Bolivia, 1981. Brazil, 3710, 5154. Chile, 1850, 1861, 1881, 1906, 1909, 1917. Colombia, 1782. Ecuador, 1819, 1840. Mexico, 1446, 1463, 1502. Uruguay, 2016, 2025. Venezuela, 1736–1737.

Social Services. Argentina, 5090, 5108–5109. Bolivia, 4963. Brazil, 2187, 3756, 5198. Jamaica, 4786. Panama, 4568. Peru, 3394, 4940.

Social Structure, 1076. Andean Region, 1047. Antigua and Barbuda, 812. Argentina, 5086, 5095, 5125. Aztecs, 118. Bolivia, 956, 4962. Brazil, 5145, 5191. Caribbean Area, 819. Cashinawa, 880. Cayapo, 879. Central America, 4619. Dominica, 815. Ecuador, 4854, 4866. Education, 4866. Guanano, 867. Guyana, 787. Haiti, 802. Honduras, 4684. Mayas, 105, 376, 745. Mesoamerica, 80, 100, 105, 117. Mexico, 358, 1076, 4538, 4555. Nazca, 704–705. Netherlands Antilles, 2285. Nicaragua, 4686. Peasants, 4962. Peru, 1043, 1046, 1068, 4883. Political Ideology, 5105. Precolumbian Civilizations, 56, 66, 80, 86, 100, 105, 117, 137, 366, 382, 537, 541, 605, 690. Puerto Rico, 797. Quechua, 1046. Saint Kitts and Nevis, 821. Suriname, 805, 2314. Taino, 505. Trinidad and Tobago, 787, 793, 808–809, 834, 4780. Venezuela, 4835, 4850. Yucuna, 906.

Social Values. Precolumbian Civilizations, 126.

Social Welfare, 2727. Argentina, 5090, 5092, 5119. Brazil, 5154, 5166. Catholic Church, 5119. Chile, 5008, 5011. Jamaica, 4786. Mexico, 4495. Nicaragua, 4669. Sandinistas, 4669. Venezuela, 4843.

Social Work. Argentina, 5105. Educational Models, 4405.

Social Workers, 4405. Argentina, 5105. Brazil, 3756.

Socialism and Socialist Parties, 2728, 2738,

2749, 2760, 2787, 2798, 2801, 3865. Bolivia, 3295, 3298, 3305, 3335. Central America, 4060. Chile, 3418–3419, 3443, 3454, 3468, 3472, 3493. Cuba, 1701, 1704, 3108, 3125, 3131–3132, 3137–3139, 3147. Developing Countries, 2739. El Salvador, 2965. Europe, 3865. Jamaica, 4750. Nicaragua, 3013.
Sociolinguistics. Bolivia, 3333.
Sociology. Democratization, 4435. Globalization, 4394. Political Ideology, 4418.
Soil Erosion. Argentina, 5087. Bolivia, 2533. Central America, 2317. Ecuador, 2472. Mesoamerica, 140. Mexico, 2317. Precolumbian Civilizations, 2283.
Soils. Bolivia, 2525. Brazil, 885. Mexico, 182. Peru, 2503.
Soldiers. Biography, 3597. Paraguay, 3597.
Somoza, Anastasio, 3015, 3602, 3987, 4036.
Songs. Andean Region, 1029. Curripaco, 896–897.
Sonora, Mexico (state). Agricultural Development, 4541. Agriculture, 2414. Archaeological Geography, 214. Automobile Industry and Trade, 1524. Cattle Raising and Trade, 4541. Dependency, 4541. Economic Conditions, 1373, 1508, 4541. Elections, 2399, 2933. Excavations, 249. Historical Geography, 2414. Human Geography, 2414. Industrial Relations, 4480. Industry and Industrialization, 1524. Maquiladoras, 1524. Minerals and Mining Industry, 2414. Municipal Government, 2399. Opposition Groups, 2933. Technological Development, 1524.
Sonqo, Peru. *See* Sonco, Peru.
Sorcery. *See* Witchcraft.
Souza, Luiza Erundina de, 3713, 3718.
Sovereignty. British Caribbean, 3072–3073. Ecuador, 4287. Uruguay, 4323.
Soybeans. Argentina, 2558. Brazil, 2205.
Spaniards. Cronistas, 509. Relations with Mayas, 350.
Spanish-American War (1898), 4769.
Spanish Civil War (1936–1939), 4831.
Spanish Conquest, 2221, 2236. Andean Region, 1030. Argentina, 2540. Aztecs, 168, 174. Bahamas, 506. Belize, 350. Guatemala, 2344–2346. Hispaniola, 493. Historiography, 454. Mesoamerica, 174. Mexico, 110, 327, 2398, 2413.
Spanish Influences, 2263. Chile, 2600. Peru, 2600.
Spanish Language. African Influences, 1053. Bolivia, 3333. Peru, 1053. Puerto Rico, 4759.
Spence, Michael, 397.

Spiritualism. Caribbean Area, 843.
Sports. Mayas, 364. Mesoamerica, 122.
Squatter Settlements, 2271. Argentina, 1083, 5106, 5113. Brazil, 1083. Chile, 1083, 3482, 5000, 5015, 5020, 5043, 5059, 5066, 5069, 5076, 5078, 5082. Colombia, 3202. Costa Rica, 1587. Ecuador, 4855. Guatemala, 4577. Mexico, 2355. Neighborhood Associations, 5066. Peru, 2488, 2494, 4878, 4913, 4940. Popular Culture, 5043. Protests, 5069. Social Movements, 5076, 5078. Youth, 5000, 5082.
St. Kitts and Nevis. *See* Saint Kitts and Nevis.
Standard of Living. *See* Cost and Standard of Living.
State, The, 1104, 1144, 1191, 1211, 1234, 1280, 1294, 1307, 2755, 2778, 2781–2783, 2796, 4403, 4415–4417, 4421. Argentina, 3570, 3575, 3579, 3764, 5091. Bolivia, 3323–3325, 3342, 3356. Brazil, 3714, 3720, 3764. Caribbean Area, 1652, 4707, 4755. Central America, 4599. Chile, 2613, 3485, 5002, 5056, 5072. Colombia, 3156, 3164, 4808, 4822. Congresses, 5002. Costa Rica, 1586, 2954, 2959, 4589. Cuba, 3127, 3149. Dominican Republic, 4781. Ecuador, 3227, 4863. El Salvador, 1601. Environmental Policy, 4396. Guatemala, 2975, 4620. Haiti, 4738. Medical Care, 2731. Mexico, 1408, 1435, 2412, 2815, 2835, 2838, 2850–2851, 2863–2864, 2888, 2920, 2924, 2927, 3919, 4527, 4547, 4550. Nicaragua, 4598. Panama, 3018, 3026. Peru, 3358, 3386, 3392, 3410, 4926. Social Change, 4863. Suriname, 3093. Uruguay, 2029, 3639. Venezuela, 1742, 3250, 4832, 4838, 4846. Violence, 4411.
State Enterprises. *See* Public Enterprises.
State Reform, 1088, 1090, 1160, 1234, 1253, 1280, 1294. Colombia, 1761. Mexico, 1326.
States of Emergency. *See* Martial Law.
Statesmen. Argentina, 3507. Mexico, 2836.
Statistics. Argentina, 14, 16, 3525. Brazil, 2173, 2184. Colombia, 3199. Commodities, 27. Databases, 39, 44, 49. Drug Traffic, 3199. Economic Conditions, 1235. Ecuador, 4872. Infant Mortality, 4419. International Trade, 51. Mexico, 18, 20, 4534. Nicaragua, 2999. Uruguay, 3664.
Steel Industry and Trade. Argentina, 1121. Brazil, 2134, 2164. Mexico, 1505.
Stelae. Inscriptions, 441. Mayas, 413, 434, 445, 460. Rites and Ceremonies, 445.
Stock Exchanges, 1255. Chile, 1866, 1884. Mexico, 1430.
Stone Implements. Antigua, 492. Argentina,

551. Belize, 250. Brazil, 576. Caribbean Area, 492. Chile, 609. Classification, 360. Colombia, 622, 632. Honduras, 185–187. Mayas, 181, 370–371. Mexico, 213, 360. Peru, 692.

Stonemasonry. Incas, 693.

Street Art. Argentina, 2736. Brazil, 2736.

Street Children, 4438, 4446. Assassinations, 5210. Bolivia, 4964. Brazil, 5210. Chile, 5011. Colombia, 983. Dominican Republic, 4775.

Strikes and Lockouts. Brazil, 2164, 3777, 5195. Mexico, 2896. Uruguay, 3638.

Stroessner, Alfredo, 2002, 3593, 3595–3596, 3598, 3603, 4296.

Structural Adjustment, 1115, 1221, 1247. Bolivia, 1986, 1993, 1997, 4960. Caribbean Area, 1639, 1651, 1675, 4093. Central America, 1539, 1548. Chile, 1897–1898. Colombia, 1783. Costa Rica, 1574, 1578, 4606–4607. Ecuador, 1799, 3212. El Salvador, 1593, 1597, 1600. Guyana, 1666. Honduras, 1618, 1620. Indigenous Peoples, 932. Jamaica, 1668, 1673, 4786. Mexico, 1338, 1424, 1437. Nicaragua, 1631, 4586. Panama, 4625. Peru, 1924, 1931, 1948, 1993. Venezuela, 1733.

Student Movements. El Salvador, 2971.

Students. Argentina, 5124. Chile, 5037. Public Opinion, 5124.

Subsistence Economy. Brazil, 872, 5177. Indigenous Peoples, 872. Mexico, 4502. Peru, 910, 1048, 1066. Precolumbian Civilizations, 292. Sipibo, 910.

Subversive Activities. Colombia, 3184–3185. Peru, 3360, 3391.

Sugar Industry and Trade, 1646. Argentina, 5101. Brazil, 2674, 2695. Caribbean Area, 1646, 4087, 4757. Cuba, 1690. Dominican Republic, 1646, 1657. Ecuador, 4867. El Salvador, 1598. Mexico, 720, 1407, 2403, 4491. Peasants, 5101. Peru, 1953.

Suicide. Politicians, 2748.

Sula Valley (Honduras). Archaeological Surveys, 290.

Sumo (indigenous group). Land Tenure, 2348.

Sumu. See Sumo.

Sustainable Development, 1100, 2172, 2270, 3843. Accounting, 2128. Amazon Basin, 2207, 5164. Bibliography, 2270. Brazil, 2128, 2136, 2172, 2207, 2665. Caribbean Area, 1639, 2288, 2307, 4088. Congresses, 1541. Ecuador, 1838. Food Supply, 4443. Mexico, 1444, 2859. Suriname, 2292.

Suyu, Peru (town). Religious Life and Customs, 1070.

Symbolism. Aztecs, 82. Mayas, 87, 103, 136. Mochica, 655. Precolumbian Art, 402, 443, 637. Quechua, 1061.

Symbolism (art). Incas, 691. Mayas, 160. Precolumbian Civilizations, 613. Toltecs, 402.

Syncretism. Catholicism, 4509. Cuba, 499, 4762. Ecuador, 1010. Mayas, 410. Mexico, 4509.

Syrians. Trinidad and Tobago, 4698.

Tabasco, Mexico (state). Archaeological Geography, 231. Archaeological Surveys, 231. Commerce, 1509. Economic Conditions, 1509. Precolumbian Land Settlement Patterns, 299. Public Finance, 1509. Social Conditions, 1509. Urbanization, 1509.

Táchira, Venezuela (state). Minerals and Mining Industry, 2438. Petroleum Industry and Trade, 2438.

Taino (indigenous group), 501. Social Structure, 505.

Tajín Site (Mexico), 198, 309, 353.

Tajumbina Site (Colombia). Artifacts, 615.

Tamaulipas, Mexico (state). Artifacts, 383. Elections, 2816. Material Culture, 336.

Tapia House Movement (Trinidad and Tobago), 3098.

Tarahumara (indigenous group). Agriculture, 743. Land Use, 743.

Tarahumara Mountains Region (Mexico). Economic Geography, 2376. Social Conditions, 2376.

Tarasco (indigenous group), 119. Archaeological Surveys, 144. Ethnography, 754. Ethnohistory, 144. Myths and Mythology, 143.

Tariffs. Brazil, 2166. Central America, 1567. Colombia, 1772. Costa Rica, 1585. Databases, 45. Ecuador, 1828. North America, 1459. Uruguay, 2032.

Tax Evasion. Brazil, 3750. South America, 1164.

Taxation, 1136, 1181, 2047. Argentina, 2041. Aztecs, 118, 124, 209, 442. Brazil, 2166, 3750. Caribbean Area, 1654. Central America, 1558. Colombia, 1779. Mexico, 219, 1323. Nicaragua, 2999. Precolumbian Civilizations, 458. Uruguay, 2020.

Teachers. Bolivia, 4976. Mexico, 2855.

Technological Development, 1081, 3781, 4449. Argentina, 3502. Brazil, 2105, 2139, 2141, 2157, 2167, 2175, 2196, 2670, 2674. Caribbean Area, 1650, 4078. Chile, 5023. Environmental Pollution, 4985. Mexico, 1314, 1328, 1390, 1503, 1511, 1524. Paraguay, 4985. Social Conflict, 4985. Southern Cone, 2139.

Technological Innovations, 1095, 1132, 1137,

1250, 4449. Argentina, 3502. Bolivia, 2533. Brazil, 2125, 5141. Caribbean Area, 1650. Colombia, 1784, 4828. Ecuador, 1802. Industrial Relations, 4828. Maquiladoras, 1460. Mexico, 1313, 1387, 1415, 1419. Peru, 1965. Sex Roles, 5141. Social Change, 4449. Uruguay, 2009.

Technology, 1835. Peru, 1923.

Technology Transfer, 1086, 1201, 1251. Caribbean Area, 1251. Central America, 1566. Chile, 1886, 5045. Dominican Republic, 1658. Fisheries, 1886. Industrial Productivity, 5045. Mexico, 1415, 1536.

Teenagers. Brazil, 5208. Trinidad and Tobago, 829.

Tegucigalpa, Honduras (city). Municipal Government, 4587. Neighborhood Associations, 4587. Poor, 4587. Social Movements, 4587. Urban Sociology, 4667.

Telecommunication. Argentina, 1121. Brazil, 2203. Chile, 1864, 2613. Mexico, 1391. Venezuela, 1730.

Television, 4426. Argentina, 3588. Bolivia, 3344. Brazil, 3765, 3771, 3776, 5163, 5187, 5192, 5201, 5204. Chile, 5038, 5061. Congresses, 2803, 5192. Elections, 2803, 3588, 3765, 3771, 3776, 5038, 5163. Mexico, 2885, 4513. Racism, 5201.

Templo Mayor (Mexico City), 132, 243, 311, 395, 436. Animal Remains, 260. Excavations, 154.

Tenam Rosario Site (Mexico). Ball Games, 270.

Tenentismo (Brazil), 3739.

Tenochtitlán Site (Mexico), 132, 139, 144, 159, 197. Urban History, 322. Urbanization, 238.

Teotihuacán Influences. Precolumbian Pottery, 244.

Teotihuacán Site (Mexico), 73, 84, 120, 126, 139, 195, 220, 317, 319, 340, 354–355, 361, 365, 382, 389–390, 432, 439, 443, 452. Animal Remains, 397. Architecture, 233. Cosmology, 386. Demography, 381. Economic Anthropology, 405. Excavations, 182. Human Remains, 366. Mural Paintings, 429. Pottery, 244. Precolumbian Land Settlement Patterns, 69. Tombs, 366. Urbanization, 235. Writing, 430.

Tepoztlán, Mexico (town). Rites and Ceremonies, 417, 739.

Teques, Venezuela. See Los Teques, Venezuela (city).

Tereno. See Terena.

Terracing. Guatemala, 254. Mayas, 254. Mesoamerica, 254. Mexico, 372.

Territorial Waters. Colombia, 4334, 4336, 4339. Venezuela, 4334, 4336, 4339.

Terrorism, 2783, 4411. Argentina, 3539. Chile, 3426. Colombia, 3157, 3184–3185, 3209. Peru, 1049, 1051, 1060, 3379–3381, 3383, 3413, 3417.

Texas, United States. Mexicans, 2359.

Textbooks. Argentina, 3524.

Textiles and Textile Industry, 1087. Argentina, 2094. Brazil, 2094, 2153, 2186, 2700. Colombia, 4803. Ecuador, 1830. Mexico, 2370, 4483. Peru, 1040. Precolumbian Civilizations, 681. Quechua, 1040. Women, 4803. Yucatán Peninsula, 2370.

Thompson, Edward H., 76.

Tiahuanaco Site. See Tiwanaku Site.

Ticuna. See Tucuna.

Tierra del Fuego (region), 2597. Archaeological Surveys, 560. Birds, 2544. Cultural Collapse, 942. Indigenous Peoples, 941–942. Paleo-Indians, 563.

Tierradentro Site (Colombia). Archaeological Dating, 617.

Tijuana, Mexico (city). Maquiladoras, 4544. Social Movements, 4560.

Tikal Site (Guatemala), 180.

Time. Brazil, 2671.

Tin Industry. Bolivia, 964.

Tinaquillo, Venezuela (town). Social Conditions, 4834.

Tingiholo Site (Suriname). Human Remains, 651.

Tipu Site (Belize). Artifacts, 305.

Tiraque, Bolivia (canton). Social Structure, 956, 4962.

Tiriyo. See Trio.

Tirofijo. See Marulanda Vélez, Manuel.

Tiwanaku Culture, 535, 538, 570. Ceramics, 571. Chile, 612. Cultural Collapse, 566, 569. Musical Instruments, 567. Peru, 675. Precolumbian Pottery, 706. Tombs, 571.

Tiwanaku Site (Bolivia). Pottery, 535.

Tlapacoya Sites (Mexico), 337.

Tlatelolco, Mexico (zone), 419. Excavations, 206. Precolumbian Pottery, 206.

Tlatilco Site (Mexico). Tombs, 274.

Tlaxcala, Mexico (state). Archaeology, 342. Ethnoarchaeology, 207. Indigenous Peoples, 755.

Tobacco Industry and Trade. Brazil, 2624. Caribbean Area, 4766. Mexico, 4514.

Tobacco Use. Brazil, 5188. Caribbean Area, 4766. Mexico, 4514. Precolumbian Civilizations, 558. Public Health, 4447.

Tocumbo, Mexico (region). Cattle Raising and Trade, 4463. Social History, 4463.

Todman, Terence A., 3533.
Tolima, Colombia (dept.). Precolumbian Land Settlement Patterns, 632.
Toltecs (indigenous group), 121. Calendrics, 2392. Symbolism (art), 402.
Toluca Valley (Mexico). Cities and Towns, 2417. Historical Demography, 2395. Paleobotany, 385.
Tombs. Aztecs, 220. Brazil, 577. Colombia, 615, 618. Mayas, 406, 413. Mesoamerica, 180. Mexico, 272, 274–275, 323. Mochica, 701. Peru, 685. Precolumbian Civilizations, 338. Precolumbian Pottery, 275.
Tomic, Radomiro, 3488.
Tonatico Region (Mexico). Archaeological Surveys, 188.
Tools. Machiguenga, 918. Precolumbian Civilizations, 553.
Torreón, Mexico (city). Employment, 1497. Population Growth, 1497.
Torres, Camilo, 3182.
Torrijos Herrera, Omar, 4568.
Tortuguero, Costa Rica (town). Social Life and Customs, 813.
Torture. Chile, 3422, 3430, 3446. Mexico, 2889. Paraguay, 3589, 3602.
Totonac (indigenous group). Agriculture, 761. Coffee Industry and Trade, 761. Cultural Development, 748. Housing, 736. Kinship, 736. Precolumbian Pottery, 309. Social Change, 748. Social History, 748.
Touraine, Alain, 5020.
Tourism, 2258. Caribbean Area, 1669, 2293. Chile, 2575, 2603, 2609. Costa Rica, 2330. Cuba, 1692, 2315, 4782. Ecology, 2330. Ecuador, 2484. Mesoamerica, 725. Mexico, 1513, 2383, 2410, 2412.
Trade. *See* Commerce.
Trade Policy, 1075, 1098, 1129, 1177, 1229, 1260, 1279, 1462, 3783, 3858. Argentina, 1229, 2050, 2055, 2066, 2087. Brazil, 2093, 2110, 2112, 2117, 2119, 2145–2146, 2152, 2181, 4354. Canada, 1457. Caribbean Area, 4083. Central America, 1563, 1568, 4014, 4061. Chile, 1877, 1901, 1903. Colombia, 1771–1772. Costa Rica, 1583, 1585. Cuba, 1688, 1710. Ecuador, 1797, 1824, 1827. Indonesia, 2119. Japan, 1510. Mexico, 1309–1310, 1332, 1372, 1409, 1419–1420, 1433, 1436, 1453, 1456, 1462, 1481, 1483, 1493, 1500, 1507, 1510, 1514, 1518–1519, 1523, 1527, 1533, 3883, 3892, 3897, 3913, 3931, 3937, 3954. Nicaragua, 1635. North America, 1154, 1464, 1488, 3937. Peru, 1960. South America, 2096. Southern Cone, 2152. Spain, 2066. Uruguay, 2014, 2018, 2021, 2024, 2031–2032. US, 1453, 1456, 1483, 1514, 1518–1519, 1523, 1642, 1684. Venezuela, 1722.
Trade Unions, 2784, 4445. Argentina, 3503, 3517, 3528, 3548, 3558, 3568, 5095. Belize, 4634. Bolivia, 3304, 3325, 3348. Brazil, 2149, 2155, 3695, 3697, 3760, 3763, 3777, 5196. Chile, 3437, 3445. Congresses, 3629. Cuba, 4130, 4725. Dominican Republic, 3057. Ecuador, 3239, 3246. Maquiladoras, 4544. Mexico, 1477, 2855, 2896, 2912, 2919, 2929, 4464, 4471, 4491, 4518, 4544. Peru, 3568, 4885, 4946. Southern Cone, 2790. Uruguay, 2022, 3629, 3658, 3662. US, 4062. Venezuela, 1750, 3271. Virgin Islands, 2288.
Traditional Farming, 2233.
Traditional Medicine, 862. Aymara, 954. Aztec Influences, 142. Bolivia, 4958. Colombia, 991. Ecuador, 1009. Jamaica, 838. Mayas, 742, 752. Mexico, 142, 4517. Psychiatry, 4801. South America, 862. Suriname, 4801. Venezuela, 4837.
Transnational Corporations. *See* Multinational Corporations.
Transportation, 1203, 2268. Colonial History, 2273. Ecuador, 1818. Mexico, 1310, 2365. Precolumbian Civilizations, 94, 156.
Travel. French Guiana, 2309.
Treaties. Belize, 3350. Bolivia, 3312. Bolivia/US, 3293. Central America, 4013. Databases, 46. Mexico, 46.
Treaty on the Non-Proliferation of Nuclear Arms, 4196.
Trees, 2272. Central America, 2319. Colombia, 2462. Ecuador, 2473. Peru, 2499. Venezuela, 2445.
Trials. Argentina, 3583. Cuba, 3115, 3146. Dominican Republic, 3054.
Trías, Vivián, 3663.
Trinidad and Tobago. Entrepreneurs, 4698.
Trinidadians. US, 801.
Trique (indigenous group), 4486.
Trondora Site (Costa Rica). Precolumbian Pottery, 470.
Trujillo Molina, Rafael Leónidas, 4132, 4134–4135.
Tucano (indigenous group). Rites and Ceremonies, 877.
Tucanoan (indigenous group). Acculturation, 899. Cultural Identity, 899. Millennialism, 889. Rites and Ceremonies, 901. Sex Roles, 900–901. Trade, 886.
Tucma. *See* Diaguita.
Tucumán, Argentina (prov.). Etymology, 2549. Geographical Names, 2549. Indige-

nous Languages, 2549. Political History, 3550.
Tukano. See Tucano.
Tula Site (Mexico), 121, 343, 428. Mortuary Customs, 278. Obsidian, 288.
Tumaco-La Tolita Sites (Colombia and Ecuador), 531, 626, 635.
Tunebo (indigenous group). Ethnography, 985.
Tungurahua, Ecuador (prov.). Popular Culture, 1013.
Tupac Katari. See Insurrection of Tupac Katari.
Tupamaros. See Movimiento de Liberación Nacional (Uruguay).
Tupi (indigenous group). Messianism, 908.
Tupi-Guarani (indigenous group). Ethnohistory, 592.
Tupinamba (indigenous group). Migration, 871. Social Structure, 871.
Tupinikin. See Tupiniquin.
Turquoise. Mesoamerica, 116. Precolumbian Trade, 116.
Tuxtlas Region (Mexico). Ceramics, 190. Ethnoarchaeology, 192, 198. Precolumbian Pottery, 191, 198.
Tzotzil (indigenous group), 768. Agrarian Reform, 758. Agriculture, 721. Carnival, 738. Festivals, 721, 738. Political Institutions, 721. Religious Life and Customs, 721. Social Conflict, 724.
Uanano. See Guanano.
Uaxactún Site (Guatemala), 238.
Ubico, Jorge, 4663.
Umbanda (cult). Venezuela, 4848.
UN. See United Nations.
UNAM. See Universidad Nacional Autónoma de México.
Underdevelopment, 2223. Mexico, 2356.
Underwater Archaeology. Artifacts, 572. Bolivia, 572. Mesoamerica, 76. Peru, 572.
Unemployment, 1082, 1194, 2082. Argentina, 3565. Brazil, 2122, 2183, 2635, 2687. Mexico, 1335, 1480, 1487. Uruguay, 2027. Venezuela, 1727.
Unidad Cívica Solidaridad (Bolivia), 3320.
Unidad Popular (Chile), 3419, 5003, 5007, 5032, 5062.
Unión Camilista Ejército de Liberación Nacional (Colombia), 3182.
Unión Campesina Autónoma (Dominican Republic), 4756.
Unión Cívica Radical (Argentina), 3498, 3504, 3512.
Unión Nacional de Agricultores y Ganaderos (Nicaragua), 4580, 4627.

Unión Republicana Democrática (Venezuela), 3258.
United Brands Company. See United Fruit Company.
United Fruit Company, 4702.
United Nations. Databases, 1. Directories, 1. El Salvador, 3976.
United States. Agency for International Development, 1586, 1603, 4001, 4611, 4671.
United States. Central Intelligence Agency, 3060.
United States. Drug Enforcement Administration, 3321.
United States. Foreign Broadcast Information Service, 50.
United States Virgin Islands. See Virgin Islands.
Universidad Centroamericana José Simeón Cañas (El Salvador), 4576.
Universidad de Buenos Aires, 5124.
Universidad Nacional Autónoma de México. Facultad de Economía, 1316.
Universidade de São Paulo. Faculdade de Filosofia, Letras e Ciências Humanas, 5151.
Universities. Central America, 4624. El Salvador, 4576.
University Reform. Chile, 3439.
UNO. See Unión Nacional Opositora (Nicaragua).
Urban Areas, 2246, 4434. Argentina, 2561, 5127. Bolivia, 4973. Brazil, 2629, 2638, 2677, 2711, 2713–2714, 3698. Costa Rica, 2328. Cuba, 4717. Cultural Identity, 5106. Demography, 381. Ecuador, 2467, 2483. Housing Policy, 4424. Indigenous Peoples, 4653. Jamaica, 4741. Mexico, 17, 2407, 2914, 4466, 4481. Poverty, 4450. Precolumbian Civilizations, 381. Puerto Rico, 4713. Social Policy, 4450.
Urban History. El Salvador, 2341. Guatemala, 238, 2342. Mexico, 235, 238, 322, 2389. Precolumbian Civilizations, 235, 238.
Urban Planning. See City Planning.
Urban Policy. Brazil, 2666, 2691, 2709. Chile, 2577, 4996, 5054. El Salvador, 1594. Honduras, 4667. Mexico, 1443, 1506, 2405, 4481. Nicaragua, 4640. Peru, 2488. Puerto Rico, 4713. Sandinistas, 4640. Uruguay, 3620.
Urban Sociology, 2224, 4393, 4450. Argentina, 3537, 5106–5107, 5113. Bolivia, 4954, 4967. Brazil, 2147, 3666, 3676, 4824. Chile, 4996, 5054, 5058, 5071. Colombia, 4813, 4824. Ecuador, 4855. Honduras, 4667. Mexico, 1529, 4466, 4481, 4507, 4524, 4528. Panama, 4583, 4596. Peru, 4904, 4908,

4913, 4921, 4947. Puerto Rico, 4713. Uruguay, 3620. Venezuela, 4840–4841, 4847. Women, 4524, 5071.

Urban Warfare. Peru, 3391.

Urbanization, 1117, 2218, 2221, 2238, 2241, 2247, 2251, 2271, 2281, 4434, 4440. Andean Region, 2477. Argentina, 2561, 5127. Bolivia, 4965, 4973. Brazil, 2638, 2653, 2659, 2666, 2686, 2703, 2709, 2711, 3698. Central America, 4400. Chile, 1883, 2577, 2605, 2615. Colombia, 2451, 2460. Costa Rica, 2326. Ecuador, 2470, 2477. El Salvador, 2341. Guatemala, 2342. Jamaica, 4741. Mesoamerica, 139. Mexico, 1348, 1367, 1395, 1497, 1506, 1509, 1513, 1515, 1521, 1525–1526, 2355, 2358, 2371–2373, 2380, 2389, 2391, 2902, 4560. Nicaragua, 735. Peru, 1045, 2501, 2506. Precolumbian Civilizations, 52, 137. Puerto Rico, 2311–2312. Uruguay, 2620. US, 2371. Venezuela, 2434, 2442.

Uru (indigenous group). Bibliography, 949. Ethnography, 967. Social Life and Customs, 967. Witchcraft, 967.

Uruguayans. Argentina, 5140. Brazil, 5140. Chile, 3442. Prostitution, 5139.

US Influences, 3841. Archaeology, 667. Argentina, 3533. Brazil, 4373. Chile, 3484. Cities and Towns, 2221. Mexico, 3958, 4494. Police, 4420.

Utopias. Brazil, 871. Chile, 3420. Peru, 1029, 3380.

Uxmal Site (Mexico). Deities, 316. Precolumbian Sculpture, 316.

Valdés Palacio, Arturo, 3412.

Valdez Pérez del Castillo, Eduardo, 4318.

Valdivia, Chile (prov.). Etymology, 2574. Geographical Names, 2574. Geomorphology, 2611. Glaciers, 2611.

Valenzuela Morales, Andrés, 3451.

Valle Central, Costa Rica (region). Environmental Degradation, 2323.

Valle de México, Mexico (region). Archaeological Surveys, 72. Archaeology, 347. Climatology, 2386. Environmental Degradation, 2377. Environmental Pollution, 2369. Historical Demography, 2395. Mortality, 2415.

Valle del Cauca, Colombia (dept.). Archaeology, 630–631. Excavations, 616.

Valparaíso, Chile (city). Community Development, 5046. Demography, 5046. Social Indicators, 5046.

Vargas, Getúlio, 3710, 4363.

Vázquez Cobo, Alfredo, 4280.

Velasco Alvarado, Juan, 3385.

Velasco Ibarra, José María, 3216, 3236, 3243.

Velásquez, Andrés, 3275.

Velásquez, Ramón J., 3276.

Velázquez, Fidel, 2919.

Venezuela. Comisión Presidencial para la Reforma del Estado, 4838.

Venezuela. Ministerio de Relaciones Exteriores, 4337.

Venta Valley (Honduras). Archaeological Surveys, 333–334. Center/Periphery Relations, 331. Stone Implements, 186.

Veracruz, Mexico (state). Agricultural Colonization, 2364. Archaeological Surveys, 218. Ceramics, 189. Coffee Industry and Trade, 4500. Conservation, 1368. Cultural Identity, 4497. Environmental Policy, 1368. Environmental Protection, 1368. Excavations, 353. Forests and Forest Industry, 2384. Indigenous Peoples, 718, 736. Land Tenure, 2364. Natural Resources, 2934. Petroleum Industry and Trade, 1351. Popular Culture, 4497. Precolumbian Pottery, 241. Social Conditions, 4497. Social Life and Customs, 4497. Terracing, 372.

Verduga, César, 3245.

Vicaría de la Solidaridad (Chile), 5027, 5081.

Viceroyalty of New Spain (1540–1821). Diseases, 2415. Historical Demography, 2415. Land Tenure, 2417. Maps and Cartography, 2362.

Vicuña. Bolivia, 2532.

Villa de Reyes Site (Mexico), 213.

Villa El Salvador, Peru (Lima neighborhood). Neighborhood Associations, 4890. Social Movements, 4890. Squatter Settlements, 2488. Urban Policy, 2488.

Villages. Production (economics), 720.

Villalba, Jóvito, 3258.

Villas, Cuba. See Las Villas, Cuba (prov.).

Violence, 2737, 2783, 4421, 4438. Andean Region, 1033, 4452. Argentina, 3541. Bolivia, 3355. Brazil, 3699, 3753, 5172, 5193. Caribbean Area, 4085. Central America, 2939. Colombia, 984, 987, 1001, 3152, 3155–3157, 3165, 3170, 3172–3174, 3176, 3179, 3184–3185, 3187, 3189, 3195, 3203, 3207–3208, 3210–3211, 4808, 4813, 4819, 4827. Dominican Republic, 4745. Ecuador, 4858. Guatemala, 769, 2974, 2980. Indigenous Peoples, 912. Land Invasions, 4858. Mexico, 3911, 4499, 4551. Nicaragua, 2997. Peru, 1049, 1052, 1060, 3360–3361, 3366–3367, 3369, 3372, 3387, 3398, 3403, 3407, 3409, 3413, 3417. Police, 4420, 4438. State, The,

4411. Suriname, 3087. Venezuela, 3253, 3263.

Violencia, La (Colombia), 3152, 3191, 4816, 4821, 4827.

Virginity. *See* Sex and Sexual Relations.

Vogt, Evon Zartman, 768.

Volador Site (Mexico), 378.

Volcanoes. Andean Region, 2480. Ecuador, 2468, 2480.

Volkswagen de México, 4518.

Voodooism. Haiti, 820.

Voting. Brazil, 3693, 3735, 3758. Chile, 3459. Dominican Republic, 3048, 3053. Mexico, 2399, 2848, 2852–2853, 2860, 2875, 2891, 2918, 2921, 2932, 4564. Peru, 3377. Social Classes, 3758. Uruguay, 2026, 3637, 3664. Venezuela, 4844.

Wages, 1071, 1082, 1122, 1261, 2046, 2082. Argentina, 2044, 2075. Bolivia, 1988. Brazil, 2098, 2183, 2199, 2210–2211. Chile, 1851, 1889. Costa Rica, 1575. Guatemala, 1613. Mexico, 1366, 1491, 1514. Panama, 1637. Uruguay, 2007, 2019, 2028. US, 1514. Venezuela, 1724, 1729, 1736, 1757. Women, 1575.

Wakuénai. *See* Curripaco.

Wamani (indigenous group). Dance, 1045.

Wambisa. *See* Huambisa.

Wanka. *See* Huanca.

Waorani. *See* Huao.

War of the Triple Alliance. *See* Paraguayan War (1865–1870).

Warao (indigenous group). Cosmology, 653. Mortuary Customs, 653. Rites and Ceremonies, 891. Sex Roles, 891.

Warfare. Amazon Basin, 857–858. Aztecs, 197, 451. Colombia, 3205. Cults, 452. Indigenous Peoples, 892. Mayas, 399. Mesoamerica, 84. Precolumbian Civilizations, 117.

Wari. *See* Huari.

Warrau. *See* Warao.

Wars of Independence. South America, 2423.

Washington, D.C. (city). Assassinations, 4257.

Water Conservation, 2218. Brazil, 2702, 2710. Chile, 2607. Dominican Republic, 2294. Mexico, 2357, 2367. Venezuela, 2437.

Water Distribution. Andean Region, 1035. Mexico, 2393.

Water Resources Development, 1072. Andean Region, 1173. Brazil, 2699. Caribbean Area, 2310. Dominican Republic, 2294. Ecuador, 4283, 4287. Mayas, 98. Mexico, 2357. Peru, 1947, 4283, 4287. Precolumbian Civilizations, 98.

Water Rights. Andean Region, 1035. Argentina, 4356. Brazil, 4356. Colonial History, 2393. Mexico, 2393. Paraguay, 4356.

Water Supply, 1072, 1204, 2217, 2253. Andean Region, 972, 2513. Brazil, 2699, 2702, 2710. Caribbean Area, 2217, 2253, 2310. Chile, 2607. Colombia, 2458. Ecuador, 4283, 4287. Mesoamerica, 98. Mexican-American Border Region, 3955. Mexico, 2367, 2374, 2390. Peru, 2513, 4283, 4287. Southern Cone, 2537. US, 2367.

Wayana. *See* Oyana.

Wayãpi. *See* Oyampi.

Wayu. *See* Goajiro.

Wealth. Central America, 1546. Chile, 1914.

Web Sites, 9.

Weights and Measures. Mesoamerica, 392. Precolumbian Civilizations, 392.

West Indians. Central America, 4702. Costa Rica, 826. Cultural Identity, 4748. Great Britain, 4748. Race and Race Relations, 4748.

Wheat, 1171.

White Collar Workers. Brazil, 3760. Labor Movement, 3760.

Wildlife Refuges. *See* Parks and Reserves.

Windward Islands. Political Integration, 3039.

Wit and Humor. Bolivia, 4968. Quechua, 935.

Witchcraft. Bolivia, 967. Mexico, 755, 762.

Women, 1082, 1214, 1254, 1303, 5203. Aged, 4871. Agricultural Development, 4978. Agricultural Labor, 5050, 5197. Andean Region, 1050. Argentina, 3508, 5098, 5102, 5116. Authoritarianism, 5079. Aymara, 970. Aztecs, 81–82. Bibliography, 4553, 4777, 4953. Bolivia, 952–953, 1995, 3282, 4953, 4957, 4966–4968, 4974, 4978. Brazil, 5159, 5175–5176, 5182, 5194, 5197–5198, 5203, 5212. Business Administration, 4804. Caribbean Area, 800–801, 1214, 4425, 4731, 4777. Catholic Church, 3696. Central America, 1559, 4624. Chile, 970, 1879, 1905, 1916, 3455, 3477, 4994, 5010, 5024, 5036, 5039, 5047, 5050, 5052–5053, 5055, 5071, 5077–5079. Christian Base Communities, 3696. Civil Rights, 5198. Civil Service, 4425. Colombia, 1002, 4803–4804, 4806, 4811–4812, 4817, 4829. Community Development, 4966, 5047, 5098. Costa Rica, 1575, 4614. Cuba, 3123. Cultural Identity, 5052–5053. Databases, 47. Demography, 4811. Development, 4433, 4749.

Development Projects, 4909. Dominican Republic, 4694, 4709, 4729, 4745. Economic Development, 933, 4806. Ecuador, 1022, 1816–1817, 1826, 3233, 3240, 4857, 4860, 4871, 4874–4875. El Salvador, 4662, 4676. Employment, 1559, 1609, 1941, 4428, 4434, 4516, 4523, 4553, 4694, 4774, 4967, 5182. Environmental Pollution, 4427. Fish and Fishing, 5135. Guatemala, 1609, 4584, 4662. Higher Education, 4624. Honduras, 4621. Households, 4453. Human Geography, 2245. Informal Sector, 1222, 4874. Jamaica, 788, 4786. Labor Market, 4432. Labor Movement, 4994, 5203. Labor Supply, 1879, 1916. Law and Legislation, 4642, 4852, 4857. Mayas, 752. Merchants, 4992. Mexico, 81, 1532, 2846, 4465, 4506, 4523–4524, 4553, 4558. Municipal Government, 4412, 4441. Neighborhood Associations, 4890. Nicaragua, 4586, 4594–4595, 4612, 4642, 4646, 4651, 4672. Nongovernmental Organizations, 5036. Panama, 4625. Paraguay, 4983, 4990. Peasant Movements, 4893, 4931, 4983. Peasants, 4875, 5010. Peru, 1057, 1941, 3394, 3405, 4890, 4893, 4899, 4907, 4909, 4921, 4923, 4931. Political Participation, 3282, 4429, 4676, 4760, 4829, 5005, 5072. Politicians, 3455. Precolumbian Civilizations, 81. Psychology, 5116. Public Policy, 4812. Public Service Employment, 4425, 4614, 4646. Puerto Rico, 4695, 4704, 4760, 4774. Refugees, 4444. Research, 1206, 4974. Revolutionaries, 3405. Rural-Urban Migration, 4506. Sandinistas, 4672. Social Change, 4731. Social Indicators, 4621. Social Movements, 3233, 3240, 4407, 4429, 4448, 4595, 4630, 4651, 4662, 4704, 4887, 4899, 4907, 4921, 4923, 5055, 5078, 5159, 5176, 5212. Social Policy, 4441, 4584, 5077. Statistics, 4428. Suriname, 4799. Textiles and Textile Industry, 4803. Trinidad and Tobago, 4749. Urban Sociology, 4524, 5071. Uruguay, 2017, 5135. Venezuela, 1724, 1729, 1757. Wages, 1575.

Women's Rights. Congresses, 4409. Ecuador, 4857.

World Bank, 48, 1127, 1578, 1638, 1647, 1754, 2151, 4093.

World War I (1914–1918), 3908.

World War II (1939–1945), 2263. Argentina, 5128. Brazil, 4376. French Caribbean, 3082. Mexico, 1432, 3910. North America, 1382. Uruguay, 4325. Venezuela, 4337.

Writing. Aztecs, 442, 446. Mayas, 449, 453, 460. Mesoamerica, 450. Precolumbian Civilizations, 435, 450. Teotihuacán, 430. Zapotec, 457.

Xaltocán Site (Mexico). Precolumbian Trade, 219. Taxation, 219.

Xavante (indigenous group). Discourse Analysis, 874. Social Conditions, 874.

Xenophobia. Argentina, 3524.

Xingu River Valley (Brazil). Hydroelectric Power, 875. Indigenous Peoples, 875.

Xochicalco Site (Mexico). Precolumbian Architecture, 239.

Xochipala Valley (Mexico). Archaeology, 362.

Xoconochoco. *See* Soconusco.

Xoconusco. *See* Soconusco.

Xuxa, 5201.

Yalcomes Culture, 621.

Yaminahua. *See* Jaminaua.

Yanaconas (indigenous group). Ethnography, 994.

Yanoama. *See* Yanomamo.

Yanomami. *See* Yanomamo.

Yanomamo (indigenous group). Brazil, 869. Demography, 869. Kinship, 884. Rites and Ceremonies, 884. Social Life and Customs, 884, 894. Warfare, 892.

Yarumela Site (Honduras). Architecture, 247. Political Development, 248.

Yaxchilán Site (Mexico), 380, 387.

Yecuana (indigenous group). Human Ecology, 893. Indigenous/Non-Indigenous Relations, 893. Myths and Mythology, 895.

Ye'kwana. *See* Yecuana.

Youth. Andean Region, 1031. Argentina, 5129. Brazil, 2108, 5193. Chile, 5000, 5026, 5043, 5059, 5082. Crime and Criminals, 4743. Cuba, 3132, 4737, 4782. Cultural Identity, 4895, 5129. Curaçaoans, 4743. Ecuador, 4864. Employment, 2015, 4864, 4892. Guatemala, 4688. Paraguay, 4988. Peru, 1037, 4892, 4895. Political Development, 5026. Poor, 5193. Protests, 5082. Public Opinion, 5059. Quechua, 1037. Rural-Urban Migration, 4688. Squatter Settlements, 5000, 5082. Suriname, 4783. Trinidad and Tobago, 4780. Uruguay, 2017.

Yrigoyen, Hipólito, 5103.

Yucatán, Mexico (state). Ejidos, 4462. Employment, 1445. Henequen Industry and Trade, 756, 4462, 4561. Lebanese, 4482. Maquiladoras, 1445. Paleoecology, 253. Popular Culture, 4487. Precolumbian Land Settlement Patterns, 253. Social Condi-

tions, 4487. Soils, 253. Labor Movement, 4561.
Yucatán Peninsula. Archaeological Dating, 183. Archaeological Geography, 123. Archaeological Surveys, 183, 376. Bats, 201. Colonial Architecture, 327. Description and Travel, 123. Forests and Forest Industry, 2370. Guidebooks, 123. Maritime History, 359. Mayas, 740–742, 745. Precolumbian Architecture, 232, 327. Precolumbian Civilizations, 97. Precolumbian Sculpture, 273. Precolumbian Trade, 283. Textiles and Textile Industry, 2370.
Yucuna (indigenous group). Social Structure, 906.
Yuma, Dominican Republic (prov.). Precolumbian Land Settlement Patterns, 517.
Yuqui (indigenous group). Human Ecology, 2531.
Yura. *See* Yuracare.
Yuracare (indigenous group). Ethnic Groups and Ethnicity, 968.
Yuruna (indigenous group). Myths and Mythology, 888.
Zacatecas, Mexico (state). Economic Development, 4556. Migration, 2408, 4556.

Zamora, Mexico (city). Economic Conditions, 1526. Political Participation, 2865. Quality of Life, 2865.
Zapotec (indigenous group). Archaeological Dating, 329. Artifacts, 391. Child Development, 733. Cultural History, 113. Cultural Identity, 772. Dance, 723, 760. Ethnohistory, 459. Genealogy, 459. Government Relations, 772. Human Ecology, 767. Indigenismo and Indianidad, 772. Inscriptions, 457. Kings and Rulers, 459. Mexico, 723. Music, 760. Mutilation, 320. Popular Culture, 723. Rural Sociology, 767. Rural-Urban Migration, 744. Social Life and Customs, 723, 733, 762. Symbolism, 723. Violence, 733. Writing, 435, 457.
Zapotitán Valley (El Salvador). Excavations, 368.
Zavaleta Mercado, René, 4950.
Zedillo Ponce de León, Ernesto, 1451.
Zoology, 2220. Chile, 597. Costa Rica, 2334.
Zulia, Venezuela (state). Minerals and Mining Industry, 2438. Petroleum Industry and Trade, 2438.
Zurite, Peru (village). Political Development, 3410. Religion and Politics, 3410.

AUTHOR INDEX

A 50 años del protocolo de Rio de Janeiro: opiniones de actualidad, 4281
Aa, A. J. van der, 2301
Abarca Begazo, Noel, 4878
Abbaszadeh, Babak, 3666
Abdala, Manuel Angel, 2033
Abe, Masae, 178
Abele, Gerhard, 2565–2568
Abella, Gloria, 3883
Abetti, Pier A., 1215
Aboites, Jaime, 1308
Abraham, Eva, 774
Abramo, Fulvio, 3741
Abrams, Elliot Marc, 66
Abreu, Alice R. de Paiva, 5141
Abreu, Marcelo de Paiva, 2093
Abreu, Sergio, 4320
Abreu Olivo, Edgar, 1716
Abreu Páez, Víctor M., 3049
Abril-Ojeda, Galo, 1785
Ab'Saber, Aziz, 2621
Abugattás A., Juan, 3358, 4297
Abuhadba, Mario, 1071, 2007
Academia Mexicana de Derechos Humanos, 2812, 2885
Academia Nacional de Economía (Uruguay), 2013, 2024
Acción Democrática (Venezuela), 3271
Acción Nacional (Mexico), 2903
Acero, Liliana, 2094
Acevedo Marrero, Carlos Aníbal, 4690
Acha-2 y los orígenes del poblamiento humano en Arica, 596
Ackroyd, William S., 2810
Acosta, Jorge R., 179
Acosta, Maruja, 4393
Acosta, Mónica, 1827
Acosta Espinosa, Alberto, 1786, 3212
Acosta P., Orlando, 2592
Acselrad, Henri, 2136
El Acuerdo de Libre Comercio México–Estados Unidos, 1309
La acumulación del capital y los problemas de la macroeconomía ecuatoriana en el período de postguerra, 1787

Acuña, Carlos H., 3496
Acuña, Edith, 3589
Acuña, Eduardo, 4994
Acuña R., Rodrigo, 1850
Adames Mayorga, Enoch, 3016, 4568
Adams, Jan S., 3965
Adams, Richard E.W., 180
La adhesión de México al GATT: repercusiones internas e impacto sobre las relaciones México-Estados Unidos, 1310
Adler, Ilya, 2811
The administration of water resources in Latin America and the Caribbean, 1072
Adriance, Madeleine, 3667, 4995, 5167
Adriani, Alberto, 1717
Adrianzén, Alberto, 2723
Adrogué, Gerardo, 3497
Afanador H., Claudia, 997
Affonso, Almino, 3668
African-American reflections on Brazil's racial paradise, 5142
Agar, Lorenzo, 4996
Agencia Española de Cooperación Internacional, 2237
Aggarwal, Raj, 1311
Aggio, Alberto, 3418
Agorsah, Emanuel Kofi, 479–481
Agrarnyĭ kapitalizm v Latinskoĭ Amerike: tendenstsii 60–80-kh godov [Agrarian capitalism in Latin America: the tendencies of the 1960s–1980s.], 1073
La agricultura chilena durante el gobierno de las Fuerzas Armadas y de Orden: base del futuro desarrollo, 1846
La agricultura latinoamericana: crisis, transformaciones y perspectivas, 1074
Aguayo, Sergio, 2812, 3909, 3934, 3966
Aguero, Oscar Alfredo, 908
Aguerrevere, Felipe, 1730
Aguila, Juan M. del, 3101–3102
Aguila M., Marcos Tonatiuh, 2813
Aguilar, Adrián Guillermo, 2355
Aguilar, Edwin Eloy, 4569
Aguilar, Manuel, 2349
Aguilar, Pedro Pablo, 3247–3248

Aguilar, Renato, 4997
Aguilar F., Juan Manuel, 463
Aguilar Gómez, Aníbal, 3277
Aguilar P., José Alejandro, 1758
Aguilar Zinser, Adolfo, 3968–3969, 3982, 4074, 4570
Aguilera Peralta, Gabriel Edgardo, 2973, 3974, 4571
Aguirre, Benigno E., 4691
Aguirre C., Jaime, 2452
Agulla, Juan Carlos, 2543, 5086
Agurcia Fasquelle, Ricardo, 67, 258
Ahumada, Jaime, 1800
Ahumada, Jorge, 1847
Aja Espil, Jorge A., 4188
Ajara, Cesar, 2622
Ajia Keizai Kenkyūjo (Japan), 1200, 1504
Ajuria, Peru, 4831
Ajuste estructural, mercados laborales y TLC, 1312
Ajuste estructural y progreso social: la experiencia centroamericana, 1539
Alam, Asad, 1075
Alaminos, Antonio, 4998
Alape, Arturo, 3152
Alarco, Germán, 1923
Alarcón, Jorge A., 1607
Alarcón, Rafael, 4455
Alarcón Glasinovich, Walter, 4879
Alarcón Olguín, Víctor, 2838
Alario, Margarita, 4999
Alas de Franco, Carolina, 3967
Alba, Francisco, 1476
Alba Vega, Carlos, 1076
Albala-Bertrand, J. M., 1077
Alber, Erdmute, 4880
Albó, Xavier, 928, 962, 3355, 4979
Albornoz, Orlando, 4394
Albuquerque, Roberto C., 2194
Albuquerque Lloréns, Francisco, 1924
Alcántara, Elsa, 4925
Alcayá Moya, Graciela, 4456
Alcina Franch, José, 52, 68
Alcocer, Jorge, 3887
Alcorta, Ludovico, 1925
Aldenderfer, Mark S., 181, 538
Aldrete-Haas, José Antonio, 4457
Aldunate, Rafael, 1848
Alduncin Abitia, Enrique, 4458–4459
Alegría, Claribel, 3359
Aleksandrowicz, Sergio, 557
Alemán Santillán, Trinidad, 2356
Alemancia, Jesús, 3017
Alemann, Roberto T., 2034
Alencastro, Luiz Felipe de, 3669, 3681
Alés, Catherine, 2216

Alexander, Robert Jackson, 2724–2725, 3670
Alexandrenkov, Eduardo, 482
Alfaro C., Javier, 1598
Alfaro Salazar, Luis, 712
Alfie, Isaac, 2008
Alfonso, Pablo M., 4692
Algaze, Guillermo, 69
Algunos enfoques sobre la restructuración económica en México, 1313
Alianza Cívica/Observación 94 (Mexico), 2812, 2885
Alimonda, Héctor, 4321
Allegretti, Mary Helena, 2623
Allegri, Eduardo, 3577
Allen, Chris, 1078
Allen, Rose Mary, 775
Allende Gossens, Salvador, 3419
Allub, Leopoldo, 5087
Allyn, Bruce J., 4106
Almada, Vilma Paraíso Ferreira de, 2095
Almada Mireles, Hugo, 2921
Almansi, Aquiles A., 2035–2036
Almeida, Anna Luiza Ozorio de, 5143
Almeida, Joaquim Anécio, 2624
Almeida, Maria Hermínia Tavares de, 3671–3672
Almeida, Napoleón, 1012
Almeida, Paulo Roberto de, 2096, 3781, 4349–4350
Almeida, Pedro Fernando de, 2127
Almeida, Roberto Schmidt de, 2640, 2705
Alomar, Gabriel, 2241
Alonso, Jorge, 1467, 2814–2815, 2869, 4527, 4540
Alonso-Concheiro, Antonio, 2908
Alonso Gamboa, Octavio, 3
Alsina, Andres, 3618–3619
Altamirano, Teófilo, 1026–1027, 4881–4882
Altamirano Escobar, Hernán Alonso, 4282
Alternative Lateinamerika: das deutsche Exil in der Zeit des Nationalsozialismus, 4395
Altieri, Miguel, 2259
Alvarado, Arturo, 2816
Alvarado, Javier, 1926
Alvarado Ramos, Juan Antonio, 776
Alvarez, Alberto, 4103
Alvarez, Jorge, 5000
Alvarez, José E., 3782
Alvarez, Luis H., 2817–2818
Alvarez, Ticul, 201
Alvarez Bernal, María Elena, 2911
Alvarez D., Angel E., 3249
Alvarez Guerrero, Osvaldo, 3498
Alvarez Icaza, Pablo, 3968
Alvarez Martínez, Gustavo, 2988
Alvarez Mosso, Lucía, 4460

Alvarez Santiago, Héctor, 718
Alvarez Soberanis, Jaime, 1314, 3884
Alves, Maria Helena Moreira, 3673, 5144
Alzugaray Treto, Carlos, 4076
"Ama sua, ama llulla, ama quella:" entrevista con el General Hugo Banzer Suárez, líder de la ADN y ex-Presidente de Bolivia, y Guillermo Fortún Suárez, principal ideólogo del partido, 3278
Amadeo, Edward J., 2097–2098
Amador A., Freddy, 1621
Amaral Filho, Jair do, 2099
Amaringo, Pablo, 921
Amaro, Nelson, 2974
Amaya, Carlos Andrés, 2434
Amazon: Nihonjin ni yoru 60nen no Ijūshi [Amazon: 60 years of Japanese immigration], 2625
Amazônia: a fronteira agrícola 20 anos depois, 2626
Amazonía nuestra: una visión alternativa, 2463
Amazonía presente y—?, 2464
Ambiente, Estado y sociedad: crisis y conflictos socio-ambientales en América Latina y Venezuela, 4396
Ambrose, Stanley H., 464
Amend, Stephan, 2435
América Latina: crítica del neoliberalismo, 1079
América Latina en la Cuenca del Pacífico: perspectivas y dimensiones de la cooperación, 4160
América Latina y el Caribe: el manejo de la escasez de agua, 2217
American Anthropological Association, *86th, Chicago, 1987*, 105
American Chamber of Commerce of the Dominican Republic, 3058
The American southwest and Mesoamerica: systems of prehistoric exchange, 70
Americas: an anthology, 2726
Americas Watch Committee (US), 1657, 2980
Ames, Barry, 3674
Amnesty International, 2889, 2979, 2986
Amodio, Emanuele, 483
Análisis crítico de la nueva reforma agraria, 1315
Análisis de la realidad migratoria en Bolivia, 4949
Anatomía de un conjunto residencial teotihuacano en Oztoyahualco: las excavaciones, 182
Anawalt, Patricia Rieff, 89, 408–409
Anaya Franco, Eduardo, 4883

Andean Oral History Workshop, 946
Andere, Eduardo, 1457
Anders, Martha B., 654
Andersen, Martin Edwin, 3499
Anderson, Anthony B., 2627
Anderson, Jeanine, 4397
Anderson, Julie, 2100
Anderson, Leslie, 2945, 4572
Anderson, Patricia Y., 4693
Andrade, Everaldo Gaspar Lopes de, 2101
Andrade, Lúcia Mendonça Morato, 875
Andrade, Manuel Correia de Oliveira, 2628
Andrade Terán, Ramiro, 3103
Andrea, Luciano d', 2404
Andreas, Peter R., 4161
Andrews, Anthony P., 71, 183, 271
Andrews, E. Wyllys, 58
Añez Ribera, Lucio, 3279
Angell, Alan, 3420–3421, 3460
Angles Vargas, Víctor, 527
Anguiano, Arturo, 2819
Anguiano, María Eugenia, 4461
Angulo Rivas, Alfredo, 4104
Anidjar, Leonardo, 2037
Aninat U., Eduardo, 1849
Anjos, Rafael Sanzio Araújo dos, 2629
The Annals of the American Academy of Political and Social Science, 3783
Annis, Sheldon, 1540
Ansión, Juan, 4884
Antártica al iniciarse la década de 1990: contribución al 30 aniversario de la entrada en vigencia del Tratado Antártico, 4189
Antczak, Andrzej, 484
Antczak, María Magdalena, 484
Antecedentes de la reforma del Estado, 3250
Antezana Ergueta, Luis, 3330
Antezana J., Luis H., 4950
Antezana Malpartida, Oscar R., 1969
Antezana Villegas, Mauricio, 4951
Anthony, Michael, 3095
Antología del pensamiento económico de la Facultad de Economía, 1929–1989, 1316
Antropología política en el Ecuador: perspectivas desde las culturas indígenas, 1008
Antropológicas, 184
Anuario Estadístico de la República Argentina, 14
Anuario Mexicano de Relaciones Internacionales, 3885
Aoyama, Kazuo, 185–187
Aparicio, Susana, 5101
La apertura comercial y la frontera México-Texas, 1317
La apertura de México al Pacífico, 3886
La apertura económica de México y la Cuenca

del Pacífico: perspectivas de intercambio y cooperación, 1318
Apertura económica y sistema financiero, 1759
Aponte, Alejandro David, 3185
Apostolakis, Bobby E., 2727
Apoyo a la Política de Desarrollo Regional (program), 1936
Apoyo S.A. (firm), 1938
Appendini, Kirsten A. de, 1319–1320
Aprile Gniset, Jacques, 2451
Aproximación diagnóstica a las violaciones de mujeres en los distritos de Panamá y San Miguelito, 4573
Aptekar, Lewis, 983
Aquino Huerta, Armando, 3280
Aragón, Luis E., 861, 2656
Aramouni, Alberto, 2728, 3500
Arana, Mariano, 3620
Arana A., Raúl Martín, 188
Arancibia Córdova, Juan, 2987, 3969
Aranda, Misael, 1003
Aranda, Sergio, 4832
Arango, Luz Gabriela, 4803–4804
Araníbar, Antonio, 3281
Araújo, Adauto José Gonçalves de, 585
Araújo, Ana María, 4451
Araújo, Maria Celina Soares d', 3675, 3773
Araúz, Virgilio, 3970
Aravena Ricardi, Nancy, 2569–2570, 4241–4242
Araya Monge, Rolando, 2946
Arbo, Higinio, 3590
Arce, Luz, 3422
The archaeology of regions: a case for full-coverage survey, 72
The archaeology of St. Eustatius: the Golden Rock Site, 485
Archivo Histórico Nacional (Spain). Sección de Ultramar, 2284
Archondo, Rafael, 4952
Arcia, Gustavo, 2323
Arcila N., Oscar, 2460
Ardaya Salinas, Gloria, 3354
Ardila Ardila, Martha, 4265
Arditi, Benjamín, 3591–3592
Ardittis, Solon, 4574
Area promoción de la mujer: partidos políticos; ¿espacios de participación democrática para la mujer?, 3282
Arenas, Jacobo, 3153
Arenas, Nelly M., 4329
Arenas, Patricia, 939
Arenas, Reinaldo, 3145
Argenti, Gisela, 2009

Argentina. Comisión Nacional de Política Ambiental. Secretaría General, 2551
Argentina. Congreso de la Nación. Dirección de Información Parlamentaria, 4201
Argentina. Instituto Nacional de Estadística y Censos, 5094
Argentina. Presidencia de la Nación, 2551
Argentina en el mundo: 1973–1987, 4190
Argentina: from insolvency to growth, 2038
La Argentina: geografía general y los marcos regionales, 2538
Argentina: hacia una economía de mercado, 2039
Argentina: la reforma económica, 1989–1991; balance y perspectivas, 2040
Argentina: tax policy for stabilization and economic recovery, 2041
Arguedas, José María, 1028
Arguedes, Sol, 3104
Argüelles, Antonio, 1321
Arias, Arturo, 2975
Arias, Luis, 4132
Arias Chávez, José, 2357
Arias Peñate, Salvador, 1546, 3971
Arias Sánchez, Oscar, 1540, 2947
Arismendi, Rodney, 3627
Aristide, Jean-Bertrand, 3061, 3064
Armanet Armanet, Pilar, 4160, 4243
Armas Barea, Calixto A., 4189
Armendariz de Aghion, Beatriz, 1080
Armstrong, C.J., 2
Arnade, C.A., 1081
Arnáiz Burne, Stella Maris, 1479, 2378
Arnaiz Quintana, Angel, 2996
Arnaud, Expedito, 864
Arnaud, Vicente Guillermo, 4191
Arnaudo, Aldo A., 2042
Arnauld, Marie Charlotte, 261
Arnello Romo, Mario, 2571
Arnold, Denise, 947
Arnold, Marcelo, 5001
Arnold, Philip J., 189–192
Arnove, Robert F., 4575
Arnson, Cynthia, 984, 2960
Aroca, Santiago, 3105
Arocena, José, 2010
Aronovich, Selmo, 2102
Arqueología, 193–194
Arqueología Contemporánea, 528
Arqueología del Ambato, 546
Arqueología Mexicana, 195–199
Arqueologia nos empreendimentos hidrelétricos da Eletronorte: resultados preliminares = Archeology in the hydroelectric

projects of Eletronorte: preliminary results, 573
Arquivos do Museu de História Natural, 574
Arranz Recio, María José, 2284
Arrau, Patricio, 1322
Arredondo Ramírez, Pablo, 2820
Arreola, Daniel David, 2358–2359
Arriaga Conchas, Enrique, 1323
Arriagada, Irma, 1082
Arriagada Herrera, Genaro, 3423
Arriazu, Ricardo H., 1084
Arrieta, Carlos Gustavo, 1760
Arrieta Abdalla, Mario, 1970, 3299, 4977
Arrigone, Jorge L., 1083
Arriola, Carlos, 1324
Arriola Palomares, Joaquín, 1591
Arrizau, Ricardo H., 1084
Arroyo, Bárbara, 200
Arroyo Alejandre, Jesús, 1398, 1411, 2821
Arroyo-Cabrales, Joaquín, 201
Arroyo Lasa, Jesús, 4576
Arruda, Marcos, 2103
Arrúe, Wille, 5109
Arsenault, Daniel, 655
Art, ideology, and the city of Teotihuacán: a symposium at Dumbarton Oaks, 8th and 9th October 1988, 73
Artana, Daniel, 2043, 2056
Arte de la tierra: Sinú y Río Magdalena, 613
El arte y los símbolos como fuente de información arqueológica: simposio, 529
Arteaga Sáenz, Juan José, 4322
Artiles, Leopoldo, 4768
Arvide, Isabel, 2822
Aryeetey-Attoh, Samuel, 2630
Arze Vargas, Carlos, 1971
Ascencio Franco, Gabriel, 1325
Asentamientos precarios y pobladores en Guatemala, 4577
Ashmore, Wendy, 74, 202
Asiaín, Aurelio, 3106
Asociación Acción y Pensamiento Democrático (Lima), 4945
Asociación de Administradoras de Fondos de Pensiones (Santiago), 1850
Asociación de Bancos de la República Argentina, 1199, 2070
Asociación de Investigación y Estudios Sociales (Guatemala), 2984–2985
Asociación de Publicaciones Educativas TAREA (Peru), 3415
Asociación Demográfica Costarricense, 2323, 4628
Asociación Latinoamericana de Instituciones Financieras de Desarrollo, 1085
Asociación Nacional de Jóvenes Empresarios (Dominican Republic), 1663
Asociación Nicaragüense Pro-Derechos Humanos, 2997
Asociación para el Avance de las Ciencias Sociales en Guatemala, 4049
Asociación Peruana de Estudios e Investigaciones para la Paz, 3172
Aspe Armella, Pedro, 1326
Assentamentos: a resposta econômica da reforma agrária, 2104
Assessments of the North American Free Trade Agreement, 1327
Assies, Willem, 3676
Assis, Valéria Soares de, 587
Associação Brasileira das Entidades de Meio Ambiente, 2636
Association des métiers du bois et de la forêt (Basse-Terre, Guadeloupe), 2293
Association Mi Wani Sabi (French Guiana), 828
Atehortúa Cruz, Adolfo L., 3209
Atencio Bello, Heraclio, 1086
Atilio Torres, Fernando, 1379
Atlas de México, 15
Aufgang, Lidia G., 5088
Aumond, Juarês José, 2631
Automatización flexible en la industria: difusión y producción de máquinas-herramienta de control numérico en América Latina, 1087
Auty, Richard M., 1972
Avendaño Flores, Isabel, 2328
Avila López, Raúl, 203–204
Avila P., Diana, 3360
Ayala Espino, José, 1088
Ayala Mora, Enrique, 3213
Ayisi, Eric O., 2285
Ayllón, Virginia, 4953, 4973
Ayllu Sartañäni, 948
Aylwin Azócar, Arturo, 5002
Aylwin Azócar, Patricio, 3424
Azambuja, Marcos Castrioto de, 4162, 4351
Azeredo, Sandra, 4409
Aziz Nassif, Alberto, 1467, 2815, 2853, 4527
Aznar, Luis, 3501
Azpiazu, Daniel, 3502
Azpúrua, Pedro Pablo, 1718
Babiel Guitiérrez, Nelsy, 2423
Bacellar, Olavo Ivanhoé de B., 2180
Bacha, Edmar Lisboa, 1089
Bachmann, Sybille, 4578
Badilla Portuguez, Marcos, 1571
Baer, Gerhard, 862
Baer, M. Delal, 3960

Baer, Werner, 1090–1091
Báez, Clara, 4694
Báez, Francisco, 1454
Báez, René, 1788
Baéz Evertsz, Franc, 4754
Baffa, Ayrton, 3677
Bagley, Bruce Michael, 3784, 3909, 3934, 4266
Bahamonde, Marcelo Alejandro, 2572
Bahamonde Bachet, Ramón, 4311
Bahamondes, Miguel, 971, 2578
Bähr, Jürgen, 2218
Baied, Carlos A., 2418
Bailey, Adrian J., 4695
Baizán, Mario, 3503
Bakan, Abigail Bess, 3069
Baker, Mary, 75
Baker, Patrick L., 777
Bakker, Eveline, 4696
Balaguer, Alejandro, 3361
Balance y perspectivas de los estudios regionales en México, 2360
Balbachevsky, Elizabeth, 3678
The Balberta Project: the terminal formative–early classic transition on the Pacific Coast of Guatemala = El Proyecto Balberta: la transición entre el formativo terminal y el clásico temprano en la Costa Pacífica de Guatemala, 205
Balbi, Carmen Rosa, 3362, 4885
Balderrama G., Adalid, 4226, 4244
Bali, Jaime, 110
Ball, Joseph W., 76
Ballester, Horacio P., 3785
Balmaceda M., Felipe, 1851
Baloyra, Enrique A., 3116
Balze, Felipe A. M. de la, 2055
Bañas Llanos, Belén, 2219
Banck, Geert Arent, 2205, 3679
Banco Central de Chile, 1852, 1887
Banco Central del Ecuador, 1789
Banco de México, 1371
Banco de Reservas de la República Dominicana, 1661
Banco do Brasil, 2154
Baño, Rodrigo, 5003–5004
Baños Ramírez, Othón, 4462
Baños Ramos, Eneida, 206
Bansart, Andrés, 4697
Baptista Gumucio, Fernando, 1973
Baquedano, Elizabeth, 77
Baquedano Muñoz, Manuel, 4163
Barabas, Alicia Mabel, 719
Barahona Montero, Manuel, 1580
Barajas Escamilla, Rocío, 1328, 1437

Baraona Urzúa, Pablo, 1853
Barba Pingarrón, Luis Alberto, 207, 325, 362
Barbados. Town and Country Development Planning Office, 2286
Barbieri, José Carlos, 2105
Barbosa, Altair Sales, 575
Barbosa, Lívia, 5145
Barbosa, Luiz C., 4352, 5146
Barbosa, Rubens Antonio, 2106
Barboza, Mario Gibson, 4353
Barcellos, Tanya Maria Macedo de, 2666
Barceló R., Víctor Manuel, 1329
Barcelona, Eduardo, 4192
Bárcena, J. Roberto, 547, 536
Barclay, Frederica, 2485
Barclay, Lou Anne, 1677, 4698
Bareiro, Line, 4983
Barham, Bradford L., 1092, 1570, 1615
Barkey, Henri J., 3504
Barkin, David, 1093
Barnechea, Alfredo, 3401
Barragán López, Esteban, 4463
Barral, Henri, 2361
Barrantes Lingán, Alfonso, 4298
Barraza, Leticia, 2823
Barre, Marie-Chantal, 4579
Barreda, Adriana, 940
Barrera, Cristina, 4269
Barrera Bassols, Dalia, 4464
Barrera Restrepo, Efrén, 4805
Barrera Rubio, Alfredo, 93
Barrera Zapata, Rolando, 2824
Barreto, Kátia Marly Mendonça, 3680
Barrig, Maruja, 4886–4887
Barrios, Armando, 1728
Barrios, Harald, 2729
Barrios Morón, Raúl, 3283, 3355, 4227–4228, 4979
Barros, Alexandre Rands, 2107
Barros, Ricardo Paes de, 2108
Barros, Souza, 2109
Barros Filho, Clóvis de, 3736
Barros Horcasitas, José Luis, 2928, 4048
Barry, Deborah, 4593
Barry, Roger G., 2447
Barry, Tom, 2938, 3972
Barsky, Osvaldo, 4398
Barthel, Thomas S., 410
Barthélémy, Gérard, 3062
Bartholomew, Joy A., 1541
Bartolomé, Miguel Alberto, 719
Basay Vega, Rosario, 1948
Basch, Linda G., 778
Basch, Reva, 4
Bashilov, Vladimir Aleksandrovich, 633

Basso, Ellen B., 865
Bastian, Jean-Pierre, 4399
Bastida, Ricardo O., 2539
Bastos, Santiago, 4653
Basualdo, Eduardo M., 3502, 5089
Bates, Mary Ellen, 4
Batista, Ana E., 2540
Batista Júnior, Paulo Nogueira, 4354
Batt, Rosemary L., 1093
Battaglini, Elena, 2110
Baud, Michiel, 779
Baudez, Claude-François, 208
Baudoin, Mario, 2526
Bauer, Arnold J., 78
Bauer, Brian S., 656
Bauer, Mariano, 3933
Baugh, Timothy G., 70
Baumann, Renato, 1094, 2111–2112
Baumeister, Eduardo, 1616, 3973, 4580
Bayce, Rafael, 5130
Bazdresch Parada, Carlos, 1451
Beach, Timothy, 254
Beauclair, Geraldo de, 2113
Beaudry-Corbett, Marilyn, 352
Bebbington, Anthony, 4888
Beber, José Antônio Costa, 2188
Beccaria, Luis A., 2044
Becco, Horacio Jorge, 2220
Becerra, Longino, 2988
Becerra Chávez, Pablo Javier, 2825
Becerra Maldonado, Felipe, 1316
Becker, Bertha K., 2632–2633
Beckerman, Paul Ely, 2045
Bedoya Garland, Eduardo, 2486
Bedregal Gutiérrez, Guillermo, 3284–3287, 3330, 4229
Behar, Ruth, 4465
Behrens, Roberto, 1854
Beijnum, Paul van, 4954
Béjar, Ana María, 1029
Béjar, Héctor, 4889
Bekarevich, Anatoliĭ Danilovich, 2804
Bekerman, Marta, 2114
Bélanger, Louis, 3841
Belice y Centroamérica: una nueva etapa, 3974
Bellegarde-Smith, Patrick, 843
Bellier, Irene, 909
Bellinghausen, Hermann, 2826
Bello, Mónica T., 578
Beltrame, Angela da Veiga, 2634
Beltramino, Juan Carlos M., 4189
Beltz Peralta, Hernán, 1761
Beluche, Olmedo, 3975, 4581
Bénassy-Berling, Marie-Cécile, 59
Benavente, José Miguel, 1095
Benavente, María Antonieta, 597
Benavides, Marisela, 1034, 5005
Benavides Correa, Alfonso, 4299
Benbunan, Raquel, 1730
Bendaña, Alejandro, 4680
Bendel, Petra, 2729
Bender, Stephen A., 4741
Bendesky, León, 3887
Bendix, Jörg, 2465
Beneker, Tine, 4400
Benemelis, Juan F., 3107
Benencia, Roberto, 5100
Bengoa, José, 969
Bengoa y Lecanda, José María, 4833
Bengolea, Manuel, 1855
Bengtsson, Lisbeth, 657
Benítez, Andrés, 1856
Benítez Manaut, Raúl, 3930, 3976, 4599
Bennett, Vivienne, 4466
Bensabat, Remonda, 2778
Bensión, Alberto, 1572
Benson, Larry, 280
Bensusan Areous, Graciela Irma, 2827, 2896, 4546
Bentué, Antonio, 3473
Beozzo, José Oscar, 5147
Berberián, Eduardo E., 530
Berdan, Frances, 89, 209, 374, 408
Berg, Hans van den, 949
Berger, Mark T., 3786
Bergman, Roland W., 910
Bergsten, Fred, 4355
Berguño, Jorge, 2573
Berichá, 985
Berlinski, Julio, 2046
Berlo, Janet Catherine, 73
Bermejillo, Eugenio, 2828
Bermudez, Jorge, 2115
Bernal C., Fernando, 4809
Bernal Rodríguez, Francisco A., 3955
Bernales B., Enrique, 3363
Bernales Lillo, Mario, 2574
Berquó, Elza, 3681
Berrangé, Jevan P., 2324
Berretta, Nora, 2011
Berrin, Kathleen, 389
Berrios, Rubén, 4300
Berry, Albert, 1096, 1574
Berthélemy, Jean-Claude, 1330
Bertín, Hugo D., 3505
Bertrand, María, 2577
Best, Lloyd, 3098
Betances, Emelio, 4699
Betancourt, Rómulo, 3251–3252

Betto, *Frei*, 3682
Betts, Dianne C., 1331
Between St. Eustatius and The Guianas: contributions to Caribbean archaeology, 650
Beutelspacher, Ludwig, 204
Bey, George J., 226
Beyer, Harold, 5025
Bezerra, Jocildo Fernandes, 2116
Bibliografía comentada sobre el sector informal urbano en América Latina, 1975–1987, 1097
Bibliographic guide to Caribbean mass communication, 4700
Bibliography of Cuban mass communications, 4701
Biblioteca de las entidades federativas, 2829
Bicalho, Ana Maria de Souza Mello, 2635
Bifani, Patricia, 4401
Bilby, Kenneth, 780, 798
Bilyk, Azucena, 5110
Binder, Wolfgang, 836
Binford, Leigh, 720
Biondi-Morra, Brizio N., 1622
Biota y ecosistemas de Gorgona, 2452
Birch, Melissa, 1090
Birdsall, Nancy, 1098
Birth, Kevin K., 781
Bitar, Sergio, 1099
Bitrán, Eduardo, 2616
Bitran, Ricardo A., 1592
Bizberg, Ilán, 2823
Bizzozero, Lincoln J., 4164, 4323–4324
Blackman, Courtney N., 1638
Blair Trujillo, Elsa, 3154
Blancarte, Roberto, 2830–2831
Blanch, José M., 3589
Blanco, Gustavo, 4582
Blanco, Juan Antonio, 3108
Blanco Celaya, Francisco, 1480
Blanco Mendoza, Herminio, 1332
Bland, François, 2299
Bland, Gary, 1959, 4022
Blanes J., José, 4230
Blanton, Richard E., 79–80
Blasco Bosqued, María Concepción, 658
Blasier, Cole, 3787–3788, 4105, 4300
Blight, James G., 4106
Block, Alan A., 3875
Blomström, Magnus, 1288
Blondet M., Cecilia, 4890
Boada Rebata, Hugo, 1927
Bobadilla, José Luis, 4503
Bock, Marie S., 2466
Boege, Eckart, 1368

Boer, Cornelis Wilhelmus Maria den, 2205
Boerboom, Harmen, 3086
Boero Rojo, Hugo, 564
Boersner, Demetrio, 2730
Bofill, Cristián, 3425
Bogarín Navarro, Rodrigo, 5
Boggio, Ana, 4891
Boĭko, Pavel Nikolaevich, 4258
Boisier, Sergio, 1100, 1857, 5006
Bojanic, Alan, 2521
Bolaffi, Gabriel, 3683
Bolaños, Vicky, 4583
Bolaños de Aguilera, Aura Azucena, 4584
Bolaños Herrera, Andrés, 3028
Boletín del Consejo de Arqueología, 210–211
Boletín del Museo del Hombre Dominicano, 486–487
Bolivia. Congreso Nacional. Cámara de Diputados, 3288–3289
Bolivia. Congreso Nacional. Cámara de Diputados. Comisión de Derechos Humanos, 3302
Bolivia. Ministerio de Relaciones Exteriores y Culto, 4238
Bologna, Alfredo Bruno, 4193–4194
Boloña & Büchi, estrategas del cambio: reflexiones para el desarrollo, 1858
Boloña Behr, Carlos, 1858, 1928
Bolsi, Alfredo S.C., 2540
Boltvinik, Julio, 1101
Bonavia, Duccio, 654, 659
Bonelli, Regis, 2117
Bonilla, Adrián, 3214
Bonilla, Alejandro, 1881
Bonilla, Alexander, 2334
Bonilla Castro, Elssy, 4806–4807, 4811
Bonn, Marco Antonio, 2684
Bonturi, Marcos, 1102
Boomert, Arie, 851
Boomgaard, Peter, 2287
Boon, Gerard Karel, 1087
Booth, John A., 4585
Borda, Dionisio, 2002, 4984
Borelli, Silvia Helena Simões, 5187
Borge, Tomás, 3112
Borges, Fernando Tadeu de Miranda, 2118
Borland, Mark D., 340
Borner, Silvio, 1103
Borón, Atilio, 3506, 4220
Boronat, J. Yolanda, 5138
Borrego Estrada, Genaro, 2832
Borrero, Luis Alberto, 528, 548, 598, 941
Borsdorf, Axel, 2221–2224, 2252, 2534, 2575–2576
Borzutzky, Silvia, 1104, 2731

Author Index / 865

Bos, Jeanette, 4743
Bosch, Juan, 3044–3045
Bosworth, Barry, 1464
Botana, Natalio R., 3507
Botello, Alfonso Vázquez, 2409
Botero, Ana Mercedes, 4267
Botero, Libardo, 1762
Botero, Pedro, 2461
Botero Herrera, Fernando, 4808
Bothwell, Reece B., 3083
Bottome, Robert, 1719
Bouchard, Jean François, 531
Boulet, Michel, 5131
Bour, Juan Luis, 2047
Bourgois, Philippe, 4702
Bouvard, Marguerite Guzman, 3508
Bouvier, Virginia M., 3593
Bouysse-Cassagne, Thérèse, 962
Bouzas, Roberto, 1089, 1105, 2048–2049, 4195
Bove, Frederick Joseph, 205
Boven, Karen, 890
Bovo, Denise Aparencida, 2646
Boza, María Eugenia, 1730
Bracco Boksar, Roberto, 710
Brachet-Márquez, Viviane, 2833
Brackelaire, Vincent, 4231
Bradford, Colin I., 1106, 1226, 1279
Bradley, C. Paul, 3621
Brady, James E., 212
Braga, Carlos Alberto Primo, 2119
Braga, Ricardo Forin Lisboa, 2667
Brahim, A.J., 3094
Brailovsky, Vladimir, 1333
Brambila Paz, Rosa, 348
Brana-Shute, Gary, 782, 3087
Brandon, George, 783, 4703
Braniff C., Beatriz, 213–214
Brasil '92: perfil ambiental e estratégias, 2636
Braswell, Geoffrey E., 215–216
Braun, Herbert, 3155
Braveboy-Wagner, Jacqueline Anne, 4077
Bravo Barja, Rosa, 1879
Bravomalo de Espinosa, Aurelia, 638
Brawer, Moshe, 2419
Bray, Tamara, 639
Bray, Warwick, 614
Brazil. Ministério da Previdência e Assistência Social, 2187
Brazil. Secretaria do Desenvolvimento Regional, 2207
Brazil. Superintendência do Desenvolvimento da Amazônia, 2207
Brazil. Superintendência do Desenvolvimento do Nordeste, 2701

Brazil, the challenges of the 1990s, 3684
Brea, Jorge A., 1790
Brenes, Esteban R., 1598
Brenes Peña, Ada Julia, 4586
Brenner, Philip, 4107
Bressan, Delmar Antonio, 2637
Breton, Alain, 261
Briceño-León, Roberto, 4393, 4834–4835
Briceño Monzillo, Bernardo, 2436
Brigagão, Clóvis, 3789
Briggs, Charles L., 891
Briones, Carlos, 1593–1594
Brisseau-Loaiza, Jeanine, 2487
Brito, Alexandra Barahona de, 3622
Brito, Paulo Carlos de, 2126
Brock, Philip L., 1859
Brockett, Charles D., 2939
Brockmann, Andreas, 721
Browder, John O., 2638
Brown, Cynthia G., 3426
Brown, Deryck R., 1664
Brown, George E., Jr., 3888
Brown, I. Foster, 2639
Brown, Jonathan Charles, 1334, 1447
Brown, Michael F., 852, 911–912
Brown, Roy B., 217
Browne, David M., 660
Brücher, Heinz, 2225
Brüggemann, Jürgen K., 218, 353
Brugioni, Dino A., 4108
Bruhns, Karen Olsen, 532
Bruleaux, Anne Marie, 806
Brumfiel, Elizabeth M., 81, 100, 219
Bruneau, Thomas C., 5148
Brunetti, Aymo, 1103
Bruniard, Enrique D., 2541
Brunner, José Joaquín, 5007
Brunner, Markus, 3977
Brunnschweiler, Tamara, 2264
Brunstein, Fernando, 2542
Brutality unchecked: human rights abuses along the U.S. border with Mexico, 3889
Bryan, Alan L., 576–577
Bryan, Anthony T., 4091, 4157
Brysk, Alison, 932
Bubberman, F.C., 652
Büchi Buc, Hernán, 1858, 1860
Buchrucker, Cristián, 3509
Budinich, Ema, 1882
Bueno, Clodoaldo, 4357
Buenrostro Ceballos, Alfredo Félix, 3899
Buffie, Edward F., 1335
Buitelaar, Rudolf, 1107
Bulavin, V.I., 2226
Bulgheroni, Raúl, 2543

Bulletin du Bureau national d'ethnologie, 488
Bulmer-Thomas, V., 3978-3979
Burdick, John, 5149-5150
Bureau de Investigaciones Empresariales (Buenos Aires), 2081
Burela Rueda, Gilberto, 1348
Burga, Jorge, 2488
Burga, Manuel, 1030
Burger, Richard L., 661
Burgess, Katrina, 3891
Burgoa, Ignacio, 2834
Burgos Ortiz, Nilsa M., 4704
Burgueño Lomelí, Fausto, 1336
Burgwal, Gerrit, 4854-4855
Burkhalter, Brian S., 866
Burkhart, Louise M., 82
Burney, David A., 489
Burney, Lida Pigott, 489
Burns, Allan F., 722
Bursztyn, Marcel, 2172
Burt, Timothy P., 140
Busquets, José Miguel, 3630
Bussel, Gerard W. van, 308
Bustamante, Fernando, 3215, 3790
Bustamante, Jorge A., 3890, 3956, 4467-4468
Bustamante, Rosario, 4921
Bustamante Belaunde, Luis, 4301
Bustamante Institute of Public and International Affairs (Jamaica), 4101
Bustos Castillo, Raúl, 1861
Butelmann, Andrea, 1637
Butler, Edgar W., 4534-4535
Buttari, Juan J., 1542
Butzer, Karl W., 2227-2228, 2362-2363
Buxedas, Martín, 2012
Byrne, Bryan, 505
Caballero, Carlos, 1763
Caballero Martín, Víctor, 4948
Caballero Zeitun, Elsa Lily, 4587
Caballeros Otero, Rómulo, 1543
A cabeça do Congresso: quem é quem na revisão constitucional, 3685
Cabello Carro, Paz, 662
Cabeza, Angel, 599
Cabieses, Fernando, 707
Cabrera, Yolanda, 1791
Cabrera Acevedo, Gustavo, 1476
Cabrera Castro, Rubén, 220, 390
Cabrera Guerrero, Martha Eugenia, 221
Cabrera Vargas, María del Refugio, 83
Cabrero García, María Teresa, 222-223
Cacciamali, Maria Cristina, 2120
Cáceres, Luis René, 1544-1545, 1595, 1605, 1792

Cáceres Lara, Víctor, 2989
Cadavid, Gilberto, 615
Cadena Roa, Jorge, 2829, 2835
Caetano, Gerardo, 3623
Cagnoni, José Aníbal, 3624
La caída de la democracia: las bases del deterioro institucional, 1966-1973, 3625
Caipora (organization), 5212
Cajías, Lupe, 4232
Caldera T., Hilda, 2940
Calderón, Ernesto, 2577
Calderón Cockburn, Julio, 3364
Calderón G., Fernando, 1793, 3290, 4402-4404
Calderón Rodríguez, José María, 4469
Calderón Salazar, Jorge A., 1337
Calderón Viteri, Roberto, 1794
The California-Mexico connection, 3891
Calima: diez mil años de historia en el suroccidente de Colombia, 616
Cálix Suazo, Miguel, 1617
Calla Ortega, Ricardo, 3291
Caller Iberico, Clorinda, 698
Calva, José Luis, 1319, 1338-1342
Calvache U., Gladys, 1795
Calvo, Félix, 3051
Calvo, Guillermo A., 1108
Calzada Falcón, Fernando, 3892
Camacho, Jorge, 3145
Camacho, Patricio León, 1796
Camacho Guizado, Alvaro, 3156
Camacho Romero, Juan Carlos, 3292
Câmara, Nelly Lamarão, 2710
Câmara de la Producción y del Comercio de Concepción (Chile), 1849
Câmara Neto, Alcino Ferreira, 2121
Camarena L., Margarita, 1343
Camargo, José Márcio, 2097
Cambiaso, Jorge E., 1344
Cambranes, J.C., 4588, 4656
Cambrezy, Luc, 2364
Camejo, Mary Jane, 1657
Cámere, Gladys, 4921
El camino de la revolución peruana: documentos del II C.C. del MRTA., 3365
Camnitzer, Luis, 4705
Camou, María Magdalena, 4325
Camou Healy, Ernesto, 4541
Camp, Roderic Ai, 2836-2837
La campaña del NO vista por sus creadores, 3427
Campaña por los Derechos del Niño (Paraguay), 4988
Campbell, Howard, 772
Campbell, R. Keith, 4165

El campesino contemporáneo: cambios recientes en los países andinos, 4809
Campodónico, Humberto, 1929
Campos, Emma, 1345
Campos, Roberto de Oliveira, 3686
Campos Alvarez Tostado, Ricardo, 1346–1347
Campos Mitjans, Gertrudis, 799
Campos Ribeiro, Miguel Ângelo, 2640
Campos Ruiz Díaz, Daniel, 4984
Campos V., Mariana, 4589
Camposeco, Jerónimo, 722
Camposortega Cruz, Sergio, 4470
Campuzano Montoya, Irma, 4471
Camus, Manuela, 4653
Canabal Cristiani, Beatriz, 1348
Cancino, César, 2838
Cândido, Antônio, 5151
Cané, Ralph E., 533
Cánepa, María Angela, 1031
Canitrot, Adolfo, 2050
Cañón Ortegón, Leonardo, 4812
Cantera, Jaime R., 2462
Canto Chac, Manuel, 2827, 4490
Cantón Delgado, Manuela, 4590
Canziani Amico, José, 663
Capital humano y mercado de trabajo en el Uruguay, 2013
Capriles Ayala, Carlos, 3253
Capriles Méndez, Ruth, 3253, 3276
Carasales, Julio César, 4196–4197
Caravedo Molinari, Baltazar, 1962
Carbonell, Jorge, 2311
Carciente, Jacob, 4836
Cárdenas, Héctor, 3893
Cárdenas, Víctor Hugo, 962
Cárdenas Elizalde, Rosario, 4419
Cárdenas García, Efraín, 224
Cárdenas Martín, Mercedes, 664
Cárdenas Reyes, María Cristina, 3216
Cárdenas Santa-María, Mauricio, 1778
Cardich, Augusto, 549
Cardona, Diego, 4268, 4279
Cardoso, Eliana A., 1109–1112, 1682, 2122–2123, 3109
Cardoso, Maria Francisca Thereza, 2641
Cardoso, Oscar Raúl, 4198
Cardoso, Ruth Corrêa Leite, 5152
Cardozo de Da Silva, Elsa, 4330–4331
Carelli, Mario, 5153
Caribbean Conservation Association, 1639, 2296–2297, 2313
Caribbean ecology and economics, 1639
The Caribbean in the global political economy, 1113, 4078

Caribbean Perspectives, 2288
Caribbean Rights (Organization), 1657
Caribbean Studies Association, 3030
Caribbean Studies Association. Conference. *14th, Barbados, 1989*, 3042
Caribbean visions: ten presidential addresses of ten presidents of the Caribbean Studies Association, 3030
Caribbean Workers Conference, *5th, St. John's, Antigua and Barbuda, 1985*, 3031
Caribbean Workers Council, 3031
Carlessi, Carolina, 4905
Carlos, Newton, 4175
Carlson, John B., 84
Carlson, Marifran, 3510
Carneiro, Dionísio, 2124
Carnevali, Alberto, 3254
Carney, Judith Ann, 1349
Caro, Isaac, 3791
Carotenuto, Gennaro, 4109
Carothers, Thomas, 3980
Carozzi, Federico, 1658
Carpio, Jorge, 5108
Carr, Barry, 2768
Carrado-Bravo, Francisco, 1311
Carranza Valdés, Julio, 1683, 4050
Carrasco, Ana María, 970
Carrasco, David, 167, 225
Carrasco Reyes, Ella, 4405
Carrasquero, José Vicente, 3255
Carreón Blaine, Emilie A., 319
Carrera Damas, Germán, 4834
Carrera de la Torre, Luis, 4283
Carriazo Moreno, George, 1684
Carrigan, Ana, 3157
Carrillo, Arturo J., 1350
Carrillo, Mario Alejandro, 2839
Carrillo Batalla, Tomás Enrique, 1731
Carrillo Dewar, Ivonne, 1351
Carrillo Huerta, Mario M., 1352, 4472
Carrillo Viveros, Jorge, 1353–1354, 1362, 1466
Carrión, Fernando M., 2467
Carrión, Juan Manuel, 4706
Carrión, Julio, 4892
Carroll, Thomas F., 1114
Carruthers, David V., 2859
Carta-tema: a assistência social no Brasil, 1983–1990, 5154
Cartas, José María, 1115, 1355
Cartay Angulo, Rafael, 1720
Cartaya, Vanessa, 1721
Carter, Miguel, 3594–3595
Carter, William, 962

Carvalho, José Alberto Magno de, 5155–5156
Carvallo, Gastón, 4850
Casillas Hernández, Roberto, 2840
Casillas Ramírez, Rodolfo, 3981, 4542
Casimir, Jean, 4707
Caso, Alfonso, 85
Cason, Jeffrey, 3724
Casos Huamán, Victoria, 4893
Cassá, Roberto, 3046
Castañeda, Jorge G., 2732
Castañeda, Tarsicio, 5008
Castañeda Sandoval, Gilberto, 3982
Castellanos, Miguel, 2967
Castellanos de Ponciano, Eugenia, 1609
Castello, José Carlos Bruzzi, 3687
Castells, Julia, 5090
Castelnuovo, Allan, 4856
Casteñeda, Jorge E., 3894
Castiglione, Marta, 5091
Castilhos, Clarisse Chiappini, 2125
Castillero Pimentel, Ernesto, 3983
Castillo, Eduardo del, 2841
Castillo, Fernando, 5009
Castillo, Jorge, 4041
Castillo, Laureano del, 1032
Castillo, Víctor M., 1356
Castillo G., Manuel Angel, 4473, 4591
Castilo, O.A., 3075
Castro, Carlos, 4592
Castro, Fidel, 3110–3113, 4110, 4128
Castro, Milka, 971–972, 2578
Castro, Mônica Mata Machado de, 3758
Castro, Nadya, 5157
Castro, Paulo Rabello de, 2126
Castro Contreras, Jaime, 3366
Castro H., Guillermo, 3018
Castro Madero, Carlos, 3511
Castro Martignoni, Jorge, 4474
Castro Orellana, José Rodolfo, 4593
Castro Ramírez, Bernardo, 1869
Castro Rea, Julián, 2733, 3895
Castro Ruz, Raúl, 3114
Castro Suárez, Pedro, 1116
Castro y Castro, Fernando, 2842
Catalá, José Agustín, 3260
Catalán Valdés, Rafael, 1357
Catholic Institute for International Relations (London), 3984
Cattaneo, Carlos A., 1125
Caubet, Christian, 4356
Causa 1/89: fin de la conexión cubana, 3115
Cavalcante, Antonio Ricardo F., 4360
Cavallo, Ascanio, 3428
Cavallo, Domingo, 2051

Cavanagh, John, 3888
Cavarozzi, Marcelo, 2734, 3512
Caviedes, César N., 2229, 2535, 2642, 4199
Cazenave, J., 4708
Ceballos, Zenón, 4709
Cela, Jorge, 4710
CELADE (organization), 4643, 5013
Célimène, Fred, 1640
Cénatus, Bérard, 4138
Censo nacional de población y vivienda, 16
Censos económicos 1994, 17
Centeno, Miguel Angel, 1358, 2843
Center for International Private Enterprise, 1849
Centrais Elétricas do Norte do Brasil (Eletronorte), 573
Central America: braving the new world, 3984
Central Latinoamericana de Trabajadores, 3031
Central Unica dos Trabalhadores (Brazil), 2110
Centre on Transnational Corporations (United Nations), 1287, 1380, 1516
Centro Agronómico Tropical de Investigación y Enseñanza (Costa Rica), 2318–2319, 2339
Centro Andino de Acción Popular (Quito), 3217
Centro Cultural de la Villa de Madrid, 682
Centro de Altos Estudios Militares (Peru), 3363, 4313
Centro de Defensa del Menor (Paraguay), 4988, 4991
Centro de Documentación, Información y Biblioteca (Bolivia), 3293, 3296
Centro de Ecodesarrollo (Mexico), 1461
Centro de Economia Mundial (Brazil), 4355
Centro de Estudios Cristianos y Capacitación Popular (Peru), 4894
Centro de Estudios de la Mujer (Chile), 5010
Centro de Estudios de la Realidad Colombiana, 4809
Centro de Estudios de la Realidad Puertorriqueña (CEREP), 4794
Centro de Estudios de Población (Buenos Aires), 944
Centro de Estudios de Población y Paternidad Responsable (Ecuador), 4866, 4872
Centro de Estudios del Desarrollo (Chile), 2577
Centro de Estudios Internacionales Foro Interamericano (Bogotá), 2745
Centro de Estudios Minería y Desarrollo (La Paz), 1978

Centro de Estudios Monetarios Latinoamericanos (Mexico), 1161
Centro de Estudios Monetarios y Bancarios (Santo Domingo), 1660
Centro de Estudios Nueva Economía y Sociedad-CENES (Peru), 1933
Centro de Estudios para el Desarrollo Regional (Arequipa, Peru), 4878
Centro de Estudios para la Justicia Social con Libertad (Costa Rica), 2949
Centro de Estudios Paraguayos Antonio Guash (CEPAG), 3589
Centro de Estudios Públicos (Chile), 1874
Centro de Estudios Regionales Andinos Bartolomé de Las Casas (Cusco, Peru), 863
Centro de Estudios Regionales y Comunicación Alternativa (Ciudad Juárez, Mexico), 2921
Centro de Estudios Rurales Andinos Bartolomé de las Casas (Cusco, Peru), 1042
Centro de Estudios sobre América (Cuba), 3119
Centro de Estudios Sociales Solidaridad (Peru), 4934, 4938
Centro de Estudios y Datos (Ecuador), 4871
Centro de Estudios y Prevención de Desastres (Lima), 4902
Centro de Investigación de la Realidad del Norte (Iquique, Chile), 981
Centro de Investigación, Educación y Desarrollo (Lima), 2488
Centro de Investigación para el Desarrollo (Mexico), 1309
Centro de Investigación y Docencia Económicas (Mexico), 1453
Centro de Investigación y Educación Popular (Bogotá), 3172, 3210
Centro de Investigaciones de Quintana Roo (Chetumal, Mexico), 2378, 4088
Centro de Investigaciones Económicas (Guatemala), 1612
Centro de Investigaciones Económicas (Uruguay), 3640
Centro de Investigaciones Psicológicas y Sociológicas (Cuba), 4772
Centro de Investigaciones sobre Pobreza y Políticas Sociales en la Argentina, 5092
Centro de Investigaciones Tecnológicas y Científicas (San Salvador), 2969
Centro de Pesquisa e Documentação de História Contemporânea do Brasil. *See* Fundação Getúlio Vargas. Centro de Pesquisa e Documentação de História Contemporânea do Brasil

Centro de Planificación y Estudios Sociales (Ecuador), 4860, 4869
Centro de Promoción de la Mujer Gregoria Apaza (Bolivia), 4966
Centro de Servicios Legales para la Mujer (Dominican Republic), 4745
Centro di ricerca e documentazione Febbraio '74 (Rome), 2404
Centro Documental de la Mujer Adela Zamudio (Bolivia), 4953, 4974
Centro Ecuatoriano de Desarrollo de la Comunidad, 4870
Centro Interamericano de Artesanías y Artes Populares (Ecuador), 1011–1013
Centro Internacional de Estudios Superiores de Comunicación para América Latina (Ecuador), 2786
Centro Itata (Concepción, Chile), 1869
Centro Latinoamericana para el Análisis de la Democracia—CLADE (Buenos Aires), 2061
Centro Latinoamericano de Administración para el Desarrollo (Caracas), 1252
Centro Latinoamericano de Trabajo Social (Lima), 4891
Centro Nacional de Desarrollo Municipal (Mexico), 1461
Centro para el Estudio de las Relaciones Internacionales y el Desarrollo (La Paz), 3826
Centro para la Promoción, Investigación y el Desarrollo Rural y Social (Nicaragua), 4612
Centro Paraguayo para la Promoción de la Libertad Económica y de la Justicia Social, 2003, 2005–2006
Centro Peruano de Estudios Internacionales, 1958, 3363, 3402, 4302–4303, 4305, 4307, 4311–4315
Centro Uruguay Independiente, 3629
Centroamérica, escenario de "desaparición forzada.", 2941
CEPAL. *See* Comisión Económica para América Latina y el Caribe
Cepeda, Manuel José, 3158–3161
CEPROLAI (organization), 1995
Ceramic production and distribution: an integrated approach, 226
Cerdas, Rodolfo, 2735
Cereceda, Verónica, 962
Cerezo Martínez, Ricardo, 2230
Cerrado: caracterização, ocupação e perspectivas, 2643
Cerruti, Gabriela, 3513
Cervantes Carson, Alejandro, 4531
Cervo, Amado Luiz, 4357–4358
César Dachary, Alfredo A., 1479, 2378

Chabat, Jorge, 1359, 3896
Chacón, Isabel María, 2333
Chaffee, Lyman G., 2736
Chaffee, Wilber A., 2737, 3118
The challenge of integration: Europe and the Americas, 3792
Chambre de commerce et d'industrie de Cayenne, 798
Chamorro, Amalia, 4594
Chamorro C., Claudio, 1850
Changmarín, Carlos Francisco, 3024
Chanlatte Baik, Luis A., 490
Chanoff, David, 894
Chantada, Amparo, 2289
Chaparro N., Patricio, 3429
Chapman, Anne MacKaye, 942
Chapp, María Ester, 5093
Charles, Carolle, 4711
Charlton, Cynthia L. Otis, 227–228
Charlton, Thomas H., 228
Chase, Arlen Frank, 135
Chase, Diane Z., 135
Chauchat, Claude, 659, 692
Chaumeil, Jean-Pierre, 853
Chaves Mendoza, Alvaro, 617
Chávez, Eliana, 1930
Chávez, Fernando, 3887, 4489
Chavez, Gonzalo, 1991
Chávez, Leo R., 4475
Chea, José Luis, 2976
Checa, Susana, 5108
Chedid, Saad, 4224
Cheek, Charles D., 279
Chejter, Silvia, 5096
Chelala, César A., 3514
Cheng, Lu-Lin, 1388
Chernela, Janet Marion, 854, 867
Chernick, Marc W., 3162
Chias Becerril, Luis, 2365
Chibnik, Michael, 868
Chico Mendes: the defence of life, 3688
Chiefdoms: power, economy, and ideology, 86
Chiesa, Blanca, 3655
Chilcote, Ronald H., 2738
Child, Jack, 3793, 3985–3986
Childress, Malcolm, 1615
Chile. Comisión Nacional de Verdad y Reconciliación, 3430
Chile. Comisión Nacional del Medio Ambiente, 1862
Chile. Ministerio de Agricultura. División de Estudios y Presupuesto, 1846
Chile. Servicio Nacional de Menores, 5011
Chile: informe nacional a la Conferencia de las Naciones Unidas sobre Medio Ambiente y Desarrollo, Río de Janeiro, Brasil, junio de 1992, 1862
Chile: política exterior para la democracia, 4246
Chile: proyecciones y estimaciones de población, por sexo y edad; comunas 1980–1995, 5012
Chile: proyecciones y estimaciones de población, por sexo y edad; total país y regiones, 1980–2000; urbano-rural, 5013
Chile en la Cuenca del Pacífico: experiencias y perspectivas comerciales en Asia y Oceanía, 4245
Chile y Argentina: nuevos enfoques para una relación constructiva, 4247
Chinampas prehispánicas, 229
Chinchilla, Norma Stoltz, 4406, 4595, 4619
Chiriboga Zambrano, Galo A., 1841, 4288
Chisari, Omar, 2052
Chomereau-Lamotte, Marie, 2290
Chomsky, Noam, 3794
El Chorrillo: situación y alternativas, 4596
Chossudovsky, Michel, 1931
Christopher-Nunley, Mary, 126
Chudnovsky, Daniel, 4166
Chumbita, Hugo, 3515
Ciancaglini, Sergio, 3513
100 chilenos y Pinochet, 3431
Cifuentes, Mauro, 1009
Cifuentes, Miguel, 2320
CIMA, 18
Cipolletti, María Susana, 913
Circumpacifica: Festschrift für Thomas S. Barthel, 87
Ciria, Alberto, 3516–3517
Las ciudades latinoamericanas en la crisis: problemas y desafíos, 1117
Civera Cerecedo, Magalí, 362
Civilisations précolombiennnes [i.e., précolombiennes] de la Caraïbe: actes du colloque du Marin, août 1989, 491
Clarac de Briceño, Jacqueline, 4837
Clark, Paul Coe, 3987
Clark, Ricardo, 2544
Clarke, Amanda, 394
Clarke, Colin G., 3043, 4712
Clarkson, Persis, 665
Claves del periodismo argentino actual, 3518
Clavijo, Sergio, 1118
Clawson, David L., 2231
Clay, Jason W., 2420
Cleary, David, 2644, 3689
Cleary, Edward L., 3799, 4597

Clement, Charles R., 2645
Clement, David B., 2291
Clément, Jean-Pierre, 59
Clements, Benedict, 1974
Clendinnen, Inga, 88
Clichevsky, Nora, 2542
Cline, William R., 3795
Coatsworth, John H., 3988
Coba Andrade, Carlos Alberto G., 1010
Cobean, Robert H., 230
Cobos Morán, Jorge A., 2353
Cobos Palma, Rafael, 271
Coca-Cola Foods, 1570
Coca-cronología: 100 documentos sobre la problemática de la coca y la lucha contra las drogas, Bolivia, 1986–1992, 3293
Cocchi, Angel M., 3626
Cochet, Hubert, 2366
Cochran, Augustus B., III, 4598
Cochrane, James D., 3796–3797
Codevilla, Angelo, 3432
The *Codex Mendoza*, 89
Coe, Michael D., 90–91
Coeymans, Juan Eduardo, 1863
Coggins, Clemency C., 76
Cohen, Alvin, 1766
Cohen, Benjamin J., 1119
Cohen, Jeffrey, 723
Cohen, Noemí, 2053
Colantoni, Roberto, 715
Colburn, Forrest D., 2739
Colchester, Marcus, 2292
Cole, Daniel H., 3798
Cole, Julio Harold, 1120, 1606
COLEF (conference), *1st, Tijuana, Mexico, 1990*, 1360, 2367
Colegio de Geógrafos del Ecuador, 2480
Colegio de Ingenieros Agrónomos y Azucareros, 1690
Colegio de la Frontera Norte (Tijuana, Mexico), 1312, 1360, 1362, 1437, 1466, 2367
Colegio de México. Centro de Estudios Demográficos y de Desarrollo Urbano, 1476, 2404
Colegio de México. Centro de Estudios Económicos, 1320
Colegio de México. Centro de Estudios Internacionales, 3916
Colegio de México. Centro de Estudios Sociológicos, 1312
Colegio de México. Progama de Energéticos, 1504
Colegio de México. Programa Interdisciplinario de Estudios de la Mujer, 4524, 4558
Colegio de Michoacán. Centro de Estudios Antropológicos, 92, 2870
Colegio de Sociólogos de Chile, 5014
Colegio de Sonora (Mexico), 1485, 1508
Colegio Nacional de Economistas (Mexico), 1455
Coleman, Jonathan Roger, 1722
Coll-Hurtado, Atlántida, 2368
Collado, Faustino, 3047–3048
Collier, David, 2740–2741
Collier, George A., 724
Collier, Ruth Berins, 2741
Collins, Charles O., 2369
Collins, Jane L., 5158
Coloma, Germán, 2054, 2067
Coloma C., Fernando, 1864
Colombia. Departamento Administrativo Nacional de Estadística, 4804
Colombia. Departamento Nacional de Planeación, 1764–1765
Colombia. Ministerio de Agricultura, 1764–1765
Colombia. Ministerio de Gobierno, 2453
Colombia. Misión de Estudios del Sector Agropecuario, 1764–1765
The Colombian economy: issues of trade and development, 1766
Colombo, Ariel H., 3500
Colonización del bosque húmedo tropical, 2454
Coloquio en torno a la obra de un Mayista, *Mérida, Mexico, 1983*, 93
Coloquio Internacional sobre Estrategias de Desarrollo de los Países Latinoamericanos, *México, 1988*, 1125
Coloquio sobre la Cultura Moche, *1st, Trujillo, Peru, 1993*, 666
Columbia Journal of World Business, 1121
Comalcalco, 231
Combelas, Ricardo, 3256
Comercio Exterior: 1973–1993, 19
El comercio exterior argentino en la década de 1990: agenda y cursos de acción, septiembre de 1991, 2055
Comin, Alvaro A., 3690–3691
Comisión Andina de Juristas. Seccional Colombiana, 3174, 3193
Comisión Centroamericana de Ambiente y Desarrollo, 2318
Comisión de Reforma del Estado Costarricense, 2954
Comisión de Superación de la Violencia (Colombia), 3187
Comisión Económica para América Latina

y el Caribe, 1072, 1122–1123, 1155, 1193, 1197, 1213–1214, 1235, 1244, 1248–1249, 1253, 1273, 1284–1286, 1440, 1551, 1865, 1915, 2015, 2059, 2187, 2217, 2232, 2253, 2270, 4408, 4436, 4444, 4453, 5097, 5140

Comisión Habitat (Peru), 4904

Comisión Nacional del Quinto Centenario del Descubrimiento de América (Spain), 68, 682

Comisión para la Defensa de los Derechos Humanos en Centro América, 2941

Comisión por la Defensa de los Derechos Humanos (Ecuador), 4868

Comisión Sudamericana de Paz. Sesión Plenaria, *2nd, Montevideo, 1988,* 4167

Comisión Sudamericana de Paz (Santiago), 4163, 4169, 4184

Comissão Organizadora do Centenário do Tratado de Amizade Japão-Brasil, 4377

Comissão Pró-Indio/SP (São Paulo), 875

Comité de Inversiones Extranjeras (Chile), 1887

Comité Hondureño de Mujeres por la Paz "Visitación Padilla," 4644

Comité por el Plebiscito en Cuba, 3145

Comité Pro Justicia y Paz de Guatemala, 2981

Commission of the European Communities, 1561

Compañía Nacional de Subsistencias Populares—Conasupo (Mexico), 1320

Comparative perspectives on slavery in New World plantation societies, 784

Comparato, Frank E., 154

Compendio histórico del Perú, 667

Comprendre l'agriculture paysanne dans les Andes centrales: Pérou et Bolivie, 2421

Compton, John G.M., 4143

Comunidad Camëntšá (Colombia), 986

Conaghan, Catherine M., 3218–3219

Conca, Ken, 3692

Conceição, Octavio Augusto C., 2127

Concertación de Mujeres por la Democracia (Santiago), 5077

Conchello, José Angel, 3897

Conchupata: un panteón formativo temprano en el Valle de Mizque, Cochabamba, Bolivia, 565

Conde Hernández, Raúl, 1417

Confederación de Nacionalidades Indígenas del Ecuador, 1014

Confederazione Generale Italiana del Lavoro, 2110

Conference on East Indians, *4th, New York, 1988,* 791

Conference on Regional Impacts of Mexico-U.S. Economic Relations, *1st, Guanajuato, Mexico, 1991,* 3898

Conferencia do Rio de Saúde, Meio Ambiente e Desenvolvimento, *Rio de Janeiro, 1992,* 4381

Conferencia Episcopal Boliviana, 4949

Conferencia Permanente de Partidos Políticos de América Latina, 1196

Conflict and change in Cuba, 3116

Conflict and competition: the Latin American church in a changing environment, 3799

Confronting the crisis in Latin America: women organizing for change, 4407

Congrès des forestiers de la Caraïbe, *3rd, Gosier, Guadeloupe, 1986,* 2293

Congreso Ambiental de Costa Rica, *1st, San José, 1985,* 2325

Congreso Chileno de Sociología, *4th, Santiago? 1992,* 5014

Congreso de Antropología en Colombia, *6th, Bogotá, 1992,* 987

Congreso Internacional sobre Fronteras en Iberoamérica Ayer y Hoy, *1st, Tijuana, Mexico, 1989,* 3899

Congreso Nacional de Antropología, *6th, Bogotá, 1992,* 988–989

Congreso Nacional de Arqueología Argentina, *11th, Buenos Aires, 1988,* 540

Congreso Nacional de Arqueología Chilena, *11th, Santiago, 1988,* 600–601

Congreso Nacional de Arqueología Chilena, *12th, Temuco, Chile, 1991,* 534, 602

Congreso Nacional de Economía, *10th, México, 1993,* 1455

Congreso Nacional de Gerencia, *8th, Lima, 1992,* 1932

Congresos y documentos del Partido Comunista de Uruguay, 3627

Connelly, Marisela, 3800

Connolly, Michael, 1136

Conrad, Geoffrey W., 56, 120

Conroy, Michael E., 1361

Consalvi, Simón Alberto, 4332–4333

Conseil régional de la Guyanne. Bureau du patrimoine ethnologique, 828

Consejo Argentino para las Relaciones Internacionales, 2055, 4189

Consejo de Ciencia y Tecnología de Sonora (Mexico), 1485

Consejo Empresario Argentino, 2040

Consejo Federal de Inversiones (Argentina), 5097

Consejo Latinoamericano de Ciencias Sociales (CLACSO), 1152

Consejo Latinoamericano de Ciencias Sociales (CLACSO). Comisión de Estudios Rurales, 1074
Consejo Latinoamericano de Ciencias Sociales (CLACSO). Comisión de Movimientos Laborales, 2790
Consejo Latinoamericano de Ciencias Sociales (CLACSO). Grupo de Trabajo de Partidos Políticos, 2723, 2785
Consejo Nacional de Desarrollo (Ecuador), 1787
Consejo Nacional de Población (Mexico), 2405
Consejo Nacional de Universidades y Escuelas Politécnicas (Ecuador), 3239
Consejo Superior Universitario Centroamericano, 4021, 4655
Conservation of neotropical forests: working from traditional resource use, 2233
Constance, Paul, 6
La Constitución de 1991: ¿un pacto político viable?, 3163
Constructores de ciudad: nueve historias del Primer Concurso "Historia de las Poblaciones.", 5015
Construir la paz: la violencia en la vida de los pobladores del Cono Norte de Lima, 3367
Consuelo Benavides: las razones y los sueños; biografía, testimonios, documentos, 3220
Consultoría para los Derechos Humanos y el Desplazamiento (Bogotá), 3165
Contabilização econômica do meio ambiente: elementos metodológicos e ensaio de aplicação no Estado de São Paulo, 2128
Contreras, Carlos, 1032
Contreras, Héctor, 3451
Contreras, Mario, 3693
Contreras, Rafael, 2462
Contreras C., Manuel E., 1975
Contreras C., Miguel, 2602
Contreras Montellano, Oscar F., 1362
Contreras Quina, Carlos, 4169
Contreras Sánchez, Alicia del C., 2370
Contribuciones a la arqueología regional venezolana, 713
Conurbano bonaerense: aproximación a la determinación de hogares y población en riesgo sanitario a través de la Encuesta Permanente de Hogares, 5094
Convención de Bancos Privados Nacionales, 8th, Buenos Aires, 1992, 2056
Convergencia de Organismos Civiles por la Democracia (México), 2897
Conzuelo Ferreyra, María del Pilar, 2824
Cook, Anita Gwynn, 535, 668

Cook de Leonard, Carmen, 417, 739
Cooper, Carolyn, 785
Cooper, Marc, 2579
Coordinadora de la Mujer (La Paz), 1995
Coordinadora Regional de Investigaciones Económicas y Sociales (Nicaragua), 4053
Coppedge, Michael, 3257
Coram, Robert, 4144
Corbacho, Alejandro, 3562
Corbo, Vittorio, 1124
Corcoran-Nantes, Yvonne, 5159
Cordeiro, Helena Kohn, 2646
Cordera Campos, Rolando, 1363
Córdova Aguilar, Hildegardo, 2489–2490
Córdova Macías, Ricardo, 2961, 3930, 4599
Córdova Sáez, Karenia, 2684
Cornejo, Boris, 1797
Cornejo Bustamante, Romer, 3800
Cornelissen, Wilhelmus Jan, 1933
Cornelius, Wayne A., 2844–2845, 3900–3901, 4476–4477
Corona Rentería, Alfonso, 1364
Coronel, Gustavo, 1723
Corporación Araracuara (Bogotá), 2454
Corporación de Estudios Liberales (Santiago), 1880
Corporación de Estudios para el Desarrollo (Ecuador), 1793, 4284
Corporación de Investigaciones Económicas para Latinoamérica—CIEPLAN (Santiago), 1288
Corporación de Promoción Universitaria (Santiago), 3435
Corrales Quesada, Jorge, 1579
Corrales Ulloa, Francisco, 465–466
Correa, I., 2455
Correa, Jorge, 3433
Correa, Paulo Guilherme, 4389
Correa, Raquel, 3476
Correa Freitas, Ruben, 3632
Correa León, Sandra, 4857
Correa R., François, 1004
La corrupción gubernamental, 3049
Cortés, Fernando, 4478
Cortés Hernández, Jaime, 218
Cortéz Hurtado, Roger, 3294
Cortez Romero, Enrique, 3295
Cosamalón, Ana Lucía, 4895
Cosmology, values, and inter-ethnic contact in South America, 855
Costa, Francisco de Assis, 2129
Costa, Letícia B., 2130
Costa, Nadja Maria Castilho da, 2647
Costa, Rogério Haesbaert da, 2648
Costa, Thomaz Guedes da, 4359

Costa Filho, Sydney M. da, 2131
Costa Rica. Ministerio de Cultura, Juventud y Deportes, 2338
Costa Rica. Ministerio de Información y Comunicación, 4033
Costa Sobrinho, Pedro Vicente, 5160
Cotler, Julio, 3368
Cotman, John Walton, 4145
Coto Molina, Walter, 2948
Cottam, Martha L., 3801, 3989
Cotto, Liliana, 4713
Couch, N.C. Christopher, 411
Couffignal, Georges, 4479
Couriel, Alberto, 3663
Coutin, Susan Bibler, 4600
Coutto, Pedro de, 3711
Couyoumdjian, Juan Ricardo, 1866
Covarrubias V., Alejandro, 1485, 1508, 4480
Cowgill, George L., 94, 220
Cox, Donald, 1724
Coy, Martin, 2649–2652
Crafts in the world market: the impact of global exchange on Middle American artisans, 725
Crahan, Margaret E., 2742
Craig, Alan K., 2234
Craig, Ann L., 2844
Crandon-Malamud, Libbet, 950
Craske, Nikki, 2846
Creamer, Germán, 4856
Crespo, Ana María, 348
Crespo Martínez, Ismael, 2743, 3628
Crespo Oviedo, Luis Felipe, 4483
Cressa, Claudia, 2437
Criales, Lucila, 4955
Crichlow, Michaeline A., 786
Crisis, conflicto y sobrevivencia: estudios sobre la sociedad urbana en México, 4481
Crisis económica en Centroamérica y el Caribe, 1641
Crisis y crecimiento en América Latina: material para un diagnóstico, 1125
Crisis y fronteras: relaciones fronterizas binacionales de Colombia con Venezuela y Ecuador, 4269
Crivelli Montero, E.A., 550
Croan, Melvin, 3802
Croce, Arturo, 3258
Crónica de una guerra no imaginaria: cronología de las relaciones Estados Unidos–Nicaragua, 1979–1984, 3990
Cronologías de los procesos de paz: Guatemala y El Salvador, 3991–3992
Cross, Malcolm, 787
Cruxent, José María, 493

Cruz, Rafael de la, 1728
Cruz, Robert, 2214
Cruz, Roberto D., 1744
Cruz Junior, Ademar Seabra, 4360
Cruz Piñeiro, Rodolfo, 1365
CSIS-CAUSA Working Group on U.S.-Venezuelan Relations, 4347
Cuadernos de Arquitectura Mesoamericana, 232–240
40 años de desarrollo: su impacto social, 1767
Cuba 1990: realidad y futuro, 3117
Cuba, a different America, 3118
Cuba after the Cold War, 4111
Cuba at a crossroads: politics and economics after the Fourth Party Congress, 1685
Cuba García, Herberth, 4896
Cuba, handbook of trade statistics, 1686
Cuba in transition, 1687
Cuba y Estados Unidos: dos enfoques, 4112
The Cuban Revolution into the 1990s: Cuban perspectives, 3119
Cuba's ties to a changing world, 1688, 4113
Cubillos, Julio César, 618
Cuéllar, Alfredo, 2847
Cuenca del Pacífico, 20
Cuenya, Beatriz, 2542
La cuestión regional y el poder, 3221
Cueva, Agustín, 3222–3223
Cueva, Juan Martín, 1798
Cuevas, Carlos E., 1596
Cuevas Díaz, Jesús Aurelio, 4522
Cuevas Seba, Teresa, 4482
Cukierman, Alex, 1126
Cultura, autoritarismo y redemocratización en Chile, 3434, 5016
La cultura popular en el Caribe, 4810
La cultura popular en el Ecuador, 1011–1013
Cultural expression and grassroots development: cases from Latin America and the Caribbean, 53
Culturas indígenas de la Patagonia, 536
Cumberbatch, Janice A., 4146
Cumbre Iberoamericana, *1st, Guadalajara, Mexico, 1991*, 3803, 4408
Cummins, Alissandra, 504
Cunill Grau, Nuria, 1252
Cunningham, Ineke, 4785
Curet, Antonio, 241
Currie, David, 1078
Curtis, Cynthia, 4075
Curtis, James R., 2358, 2371
Curtis, Siân L., 5161
Curzio, D.E., 550
Dabaguián, Emil, 3259

Dabène, Olivier, 3993
Dagnino Pastore, José María, 2057
Daher, Antonio, 1867–1868, 2580
Daltrini, Vera Cristina, 2706
D'Amico, Deborah, 788
Damill, Mario, 2058
Dammert, Manuel, 1934
Dammert Bellido, José, 1059
Daniels Hernández, Elías Rafael, 3260
Danien, Elin C., 138
Dann, Graham M.S., 4714
Dannemann, Manuel, 973
Danns, George K., 4715
Dantas, Vera, 3694
Darcissac, Philippe, 828
Darío Correa, Hernán, 3187
Darling, J. Andrew, 242, 338
Darling, Jonathan, 3994
Daskam, Thomas, 2581
Dassin, Joan, 5162
Daudelin, Jean, 4601
Dávalos, Lorenzo, 1195
Davidovich, Fany, 2653
Davidson, William V., 2235
Dávila, Enrique, 1366
Dávila Flores, Alejandro, 1317, 1515
Dávila-Flores, José, 2491
Dávila L., Andrés, 3178
Dávila Pérez, Consuelo, 3902–3903
Davis, Charles L., 2744, 2848
Davis, Dave D., 492
Davis, Diane E., 3904
Davis, Martha Ellen, 789
Davis, Shelton, 1127
Davoust, Michel, 412
Dawsey, Cyrus B., III., 2654
Day, Jane Stevenson, 243
Daza Molina, Romelio Elías, 3164
De Cartagena a San Antonio, Texas: entre el desarrollo y la interdicción; a propósito de la Segunda Cumbre Presidencial Antidrogas, 26–27 de febrero de 1992, 3296
De Franco, Mario A., 1976–1977, 1982
De Guttry, Andrea, 3995
De Janvry, Alain, 1128, 1768, 1799
De León Arias, Adrián, 3911
De Lima, Venicio A., 5163
De Melo, Jaime, 1129
De Onis, Juan, 5164
Deagan, Kathleen, 493
Debate Regional, *5th, Santa Cruz, Bolivia, 1991*, 4956
Debate Regional, *6th, Santa Cruz, Bolivia, 1992*, 4957
Una década de opinión en el Ecuador: la página editorial de *Hoy* de 1982 a 1992, 3224
Una década de refugio en México: los refugiados guatemaltecos y los derechos humanos, 3905
Decentralization of agricultural planning systems in Latin America, 1130
Deere, Carmen Diana, 1642, 1689, 1935, 4716
Defence for Children International. Sección Bolivia, 4964
Una definición de Mesoamérica, 95
Defrance, Véronique, 806
Degregori, Carlos Iván, 3369–3370, 4897
Dehouve, Danièle, 726
Deiner, John T., 3519
Delazaro, Walter, 2105
Del Aguila, Juan M. *See* Aguila, Juan M. del
Delfim Netto, Antônio, 2132
Delgado, Gustavo, 5137
Delgado, Javier, 2372–2373
Delgado, Lucília de Almeida Neves, 3695
Delgado Fiallos, Aníbal, 2990
Delgado Súmar, Hugo E., 4898
Delhoume, Jean-Pierre, 2374
Delli Sante, Angela, 2977
Delorme, Hugues, 828
Delpar, Helen, 3906
Delpech, Claire, 2488
Delpino, Nena, 4899
Demange, Nilson Joseph, 3804
DeMar, Margaretta, 4079
Demarest, Arthur Andrew, 56, 120, 413
Demas, William G., 1643, 1665
Democracia 2000: los grandes desafíos en América Latina, 2745
Democracia sin pobreza: alternativa de desarrollo para el Istmo Centroamericano, 1546
Democracia y descentralización en Bolivia, 3297
Democracia y Socialismo (Lima), 3367
Denevan, William M., 2236, 2492, 2655
Déniz Espinós, José A., 5017
Dennis, Philip A., 4602
Denver Museum of Natural History, 243
Departamento Intersindical de Assessoria Parlamentar (Brazil), 3685
Departamento Intersindical de Estatística e Estudos Sócio-Econômicos (Brazil), 2173
D'Ercole, Robert, 2468
Los derechos humanos en El Salvador en 1989, 2962
El desafío del desarrollo centroamericano, 1547
Los desafíos del Estado en los años 90, 3435
Desafíos para la izquierda, 3298

Desarme y desarrollo: condiciones internacionales y perspectivas, 4168
Desarrollo agrícola de Cuba: ciclo de estudios para el desarrollo agrícola de Cuba, 1690
Desarrollo alternativo: utopías y realidades, 3299
El desarrollo desde dentro: un enfoque neoestructuralista para la América Latina, 1131
Desarrollo industrial y cambio tecnológico: políticas para América Latina y el Caribe en los noventa, 1132
El desarrollo regional desde el mundo social, 1869
Desarrollo sostenido de la selva: un manual técnico para promotores y extensionistas, 1936
Desarrollo urbano en Sinaloa, 1987–1992, 1367
Desarrollo y medio ambiente en América Latina y el Caribe: una visión evolutiva, 2237
Desarrollo y medio ambiente en México: diagnóstico, 1990, 2375
Desarrollo y medio ambiente en Veracruz, 1368
La desarticulación del pacto fiscal: una interpretación sobre la evolución del sector público argentino en las dos últimas décadas, 2059
Descentralización y democracia en México, 2849
Desch, Michael Charles, 3805
Desempeño y colapso de la minería nacionalizada en Bolivia: estudio técnico, económico, social y organizacional de la Corporación Minera de Bolivia, 1978
A desordem ecológica na Amazônia, 2656
Desplazamiento, derechos humanos y conflicto armado, 3165
Despradel, Fidelio, 3050
Después de Germani: exploraciones sobre *Estructura social de la Argentina*, 5095
Después de la Guerra Fría: los desafíos a la seguridad de América del Sur, 4169
Después de las privatizaciones: hacia el Estado regulador, 1870
Deuda externa, renegociación y ajuste en la América Latina, 1133
Deustua Caravedo, Alejandro, 3363, 4170, 4302–4303
Development Alternatives with Women for a New Era—DAWN (project), 4407
Development and democracy: aid policies in Latin America, 1134
Development and social change in the Chilean countryside: from the pre-land reform period to the democratic transition, 3436, 5018
Development from within: toward a neostructuralist approach for Latin America, 1135, 3806
Devereux, John, 1136
Devés V., Eduardo, 3431
Devlin, Robert, 1137
Dew, Edward M., 3032, 3088–3089
DeWalt, Billie R., 1093
Dewees, Anthony, 4575
Dewey, Kathryn G., 4502
Deza Rivasplata, Jaime, 664, 669
Dhar, Sumana, 1129
Diálogo con nuestro futuro común: perspectivas latinoamericanas del Informe Brundtland, 1138
Diamint, Rut Clara, 4200
Diamond, Ian, 5161
Dias, Ondemar, Jr, 578
Dias, Sérgio da Fonseca, 2657
Los días eran nuestros: vida y trabajo entre los obreros textiles de Atlixco, 4483
Díaz, María del Carmen, 2850
Díaz, Ramón, 2014
Díaz-Briquets, Sergio, 1691, 3120, 4717
Díaz Corvalán, Eugenio, 3437
Díaz de Guerra, María A., 2617
Díaz Montes, Fausto, 4484
Díaz Oyarzábal, Clara Luz, 244–245
Díaz S., Miguel, 5059
Díaz Saenger, Jorge, 3431
Díaz Serrano, Jorge, 1369
Díaz Vilches, Patricia, 4114
Dibbits, Ineke, 4958
Dibble, Charles E., 414
Diccionario biográfico del gobierno mexicano, 21
Dickson, Sandra H., 3996
Dictadura, corrupción y transición: compilación preliminar de los delitos económicos en el sector público durante los últimos años en el Paraguay, 3596
Diechtl, Sigrid, 727
Diessel, Wilhelm G., 670
Dietrich, Wolfgang, 3977
Dietz, James L., 4718
Diez de Medina, Rafael, 2015–2017
Díez Duarte, Raúl, 2851
Different places, different voices: gender and development in Africa, Asia, and Latin America, 933
Digges, Diana, 990

Dijk, Frank Jan van, 4719
Dijk, Meine Pieter van, 1666
Dijkstra, Geske, 1623
Dik, Evgeni, 3893
Dilla Alfonso, Haroldo, 3051
Dillehay, Tom D., 537, 603
Dimenstein, Gilberto, 5165
Diocese of Rio Branco (Brazil), 4971
Diocese of Roraima (Brazil), 876
Diplomacia de la hoja de coca: documentos de información, 3300
Directory of United Nations Information Sources, 1
Direitos reprodutivos, 4409
Dirié, Cristina, 2536
Disciplina social para la sociedad guatemalteca desde tres ópticas pentecostales, 4603
Discourses and the expression of personhood in South American inter-ethnic relations, 856
La disputa por los mercados: TLC y sector agropecuario, 1370
La distribución del ingreso en México: encuesta sobre los ingresos y gastos de las familias, 1968, 1371
Dix, Robert H., 2746
Dixon, Boyd W., 246–248
Dixon, Keith A., 249
Dobrynin, Anatoly, 4115
Dockall, John E., 250
Dockendorff, Eduardo, 2577
La doctrina de la seguridad nacional, 1958–1983, 3520
Documentos sobre el conflicto argentino-chileno en la zona austral, 4201
Documentos sobre privatización con énfasis en América Latina, 1139
Dodd, Thomas J., 3997
Doehne, E., 647
Doershuk, John F., 375
Doggenweiler S., René, 5019
Dollfus, Olivier, 2422
Domestic architecture, ethnicity, and complementarity in the south-central Andes, 538
Domínguez, Jorge I., 1145, 2747, 3033, 3121, 3807, 3835
Domínguez, Lourdes, 494
Domínguez, Roberto F., 3521
Domínguez Díaz, Marta, 2582
Dominican Republic. Consejo Nacional de Población y Familia, 4709
Dominican Republic. Oficina Nacional de Planificación, 2294, 4773

Dominika imin wa kimin datta: sengo nikkei imin no kiseki [Abandoned people: Japanese immigrants in the Dominican Republic], 3052
Donato, Katherine M., 4485, 4512
Donnan, Christopher B., 671
Donoso Pacheco, Jorge, 3488
Dore, Christopher D., 376
Dore, Elizabeth, 4604–4605
Dorn, Ronald I., 665
Dornbusch, Rudiger, 1140, 2060
Dorweiler, Jane, 96
Dosman, Edgar J., 3808, 3810
Douany, Jorge, 790
Douglas, John E., 251
Dourojeanni, Marc J., 2493
Downes, Richard, 3857
Doyle, Kate, 3907
Drago, Tito, 3438, 3998
Drake, Paul W., 1227
Dreiss, Meredith L., 252
Drewitt, Peter, 495
Driant, Jean-Claude, 2494
Drogus, Carol Ann, 3696
Droogers, André, 4439
Du Plessis, Anton, 4171
Duarte, Ozeas, 3697
Dubelaar, C.N., 496
Dubet, François, 5020
Dubly, Alain, 4858
Ducatenzeiler, Graciela, 2733
Dugard, Jane, 2469
Dugas, John, 3163
Dugini de De Candido, María Inés, 2545
Dumbarton Oaks (Washington), 73, 130
Duncan, Neville C., 4146
Duncan, W. Raymond, 3809
Dunia A., Jonny E., 1725
Dunkerley, James, 2748
Dunning, Nicholas P., 97, 253–254
Duque Corredor, Román José, 1726
Durán, Esperanza, 3908
Durán, Esteban, 1871
Durán, Reina, 714
Durán, Roberto, 4248
Durán, Victor, 551
Durand, Carlos A., 4486
Durand, Francisco, 3371
Durand Ponte, Víctor Manuel, 2827
Dux, César, 3301
Dyckerhoff, Ursula, 147
A dynamic partnership: Canada's changing role in the Hemisphere, 3810
Earle, Duncan, 728

Earle, Timothy K., 86
Early, John D., 869
The East Indian odyssey: dilemmas of a migrant people, 791
Easterly, William Russell, 1769
Eastern Caribbean Natural Area Management Program, 2296–2297
Eastwood, David A., 2522
Ebanks, G. Edward, 2238
Echavarría, Fernando R., 2495
Echegaray, Fabián, 3522
Echenique, Marcial, 2583
Echeverri Correa, Fabio, 1770
Echeverri Perico, Rafael, 4811
ECLAC. *See* Comisión Económica para América Latina y el Caribe
Ecología, medio ambiente y sindicatos, 3629
A economia brasileira em preto e branco, 2133
La economía campesina y el cultivo de la coca, 1979
Economía de la integración latinoamericana: lecturas seleccionadas, 1141
La economía mexicana actual: pobreza y desarrollo incierto, 1372
La economía mexicana en cifras, 22
Economía sonorense más allá de los valles, 1373
Economía y Trabajo en Chile, 1872
Las economías andinas: evolución y perspectivas, 1937
Economic and social progress in Latin America; annual report, 1142
Economic aspects of water management in the prehispanic New World, 98
Economic Commission for Latin America and the Caribbean. *See* Comisión Económica para América Latina y el Caribe
Economic development and social change: United States-Latin American relations in the 1990s, 3811
Economic development under democratic regimes: neo-liberalism in Latin America, 1143
Economic maladjustment in Central America, 1548
Economic reforms in Latin America: symposium held in November 1992 at the Georg-August-Universität Göttingen, 1144
Economic strategies and policies in Latin America, 1145
The economy of Latin America and the Caribbean: guidelines for a comprehensive approach to the foreign debt problem and possible action by the inter-American system; informative document, 1146

Ecuador. Ministerio de Defensa Nacional, 3225
Ecuador. Ministerio de Energía y Minas, 1834
El Ecuador frente al siglo XXI: seguridad y geopolítica, 3225
Ecuador y Perú, vecinos distantes: seminario organizado por la Corporación de Estudios para el Desarrollo (CORDES), del 7 al 10 de diciembre de 1992, Quito, Ecuador, 4284
La edad de plomo del desarrollo latinoamericano, 1147
Edelman, Marc, 729–730, 4606–4607
Education and society in the Commonwealth Caribbean, 4720
Edwards, Alejandra Cox, 1873
Edwards, Clinton R., 2261
Edwards, Sebastian, 1148, 1873, 1980
EE.UU. y el régimen militar paraguayo, 1954–1958: documentos de fuentes norteamericanas, 4292
Eguizábal, Cristina, 3999
Ehrenreich, Jeffrey, 914, 1008
Einzmann, Harald, 1012
El-Erian, Mohamed A., 1149–1150
"El Ladrillo:" bases de la política económica del gobierno militar chileno, 1874
El-Saeed, Hala Helmy, 1151
Elbow, Gary S., 497, 2326, 2341–2342, 4602
Las elecciones federales de 1988 en México, 2852
Las elecciones federales de 1991, 2853
Elera, Carlos, 672
Eletrobrás, 2167
Elía, Yolanda D', 1721
Elizalde, Antonio, 4410
Elizondo, Néstor, 1374
Ellacuría, Ignacio, 2963
Ellenberg, Ludwig, 2327
Ellis, Mark, 4695
Ellner, Steve, 2749, 2768, 3261–3262
Elorrieta Salazar, Edgar, 673
Elorrieta Salazar, Fernando E., 673
Elton, Charlotte, 4000
Eluther, Jean-Paul, 1680
Embassy of Chile (Washington), 1920
Emperaire, L., 2496
Empresarios y estado en América Latina: crisis y transformaciones, 1152
Las empresas del Perú ante el mundo, 1938
Empresas públicas y política de privatizaciones (22/11/88): ciclo de reuniones; debate sobre la coyuntura política argentina, 2061
Empresas Siderúrgicas do Brasil, 2134
En busca de la seguridad perdida: aproxima-

ciones a la seguridad nacional mexicana, 3909
Enchautegui, María E., 4721
Encinas R., Alejandro, 1370
Encontro Nacional de Estudos do Trabalho, *3rd, Rio de Janeiro, 1993,* 2135
Encontro Nacional de Estudos sobre Meio Ambiente, *2nd, Florianópolis, Brazil, 1989,* 2658
Encontro Nacional de Estudos sobre Meio Ambiente, *3rd, Londrina, Brazil, 1991,* 2659
Encuentro de Historia y Realidad Económica y Social del Ecuador, *6th, Cuenca, Ecuador, 1989,* 4859
Encuentro Latinoamericano por la Democracia y la Integración, *Bogotá, 1990,* 4172
Encuentro Nacional de Centros de Prevención de la Violencia Doméstica y Asistencia a la Mujer Golpeada, *1st, Chapadmalal, Argentina, 1988,* 5096
Encuentro Nacional de la Empresa, *10th, Santiago?, 1988,* 1875
Encuentro Nacional de la Empresa, *11th, Santiago?, 1989,* 1876
Encuentro Nacional por la Paz, *Ibagué, Colombia, 1989,* 3201
Encuentro-Panel La Corrupción Gubernamental: Posibilidades de Control, *Santo Domingo, 1986,* 3049
Encuentro sobre Investigaciones en Ciencias Sociales en Yucatán, *2nd, Mérida, Mexico, 1988,* 4487
Encuentro sobre Unidad Iberoamericana, *Huelva, Spain, 1992,* 3844
Encuentro Tres Fronteras, un Destino, *Chetumal, Mexico, 1991,* 2378
Encuesta nacional de fecundidad, 1987, Cuba, 4722
Engel, Eduardo, 1153
Engel, Frédéric André, 674
England, Suzannah, 498
The enigma of ethnicity: an analysis of race in the Caribbean and the world, 792
Enríquez, Laura J., 4608
Enríquez Frödden, Edgardo, 3439
Enríquez Hernández, Jorge, 2376
Enríquez Solano, Francisco J., 2328
Ensayos críticos para el estudio de las organizaciones en México, 4488
Ensignia, Jaime, 2790, 3469, 4445
The Enterprise for the Americas Initiative: issues and prospects for a free trade agreement in the Western Hemisphere, 1154
Entre la agresión y la cooperación: la economía nicaragüense y la cooperación externa en el período 1979-1989, 1624
Entre la guerra y la estabilidad política: el México de los 40, 3910
Entre los límites y las rupturas: las mujeres ecuatorianas en la década del 80, 4860
Environment and democracy, 2136
Epstein, Jeremiah, 255
Equidad y transformación productiva: un enfoque integrado, 1155
Equipo Técnico Campesino (Bolivia), 3299
Era de nieblas: derechos humanos, terrorismo de Estado y salud psicosocial en América Latina, 4411
Ericson, Jonathon E., 70
Eriksen, Thomas Hylland, 793-794
Erisman, H. Michael, 3122, 4080
Ernesto de la Guardia Navarro: el hombre, el político y el estadista: foro de conferencias presentadas en el acto celebrado el día 12 de enero de 1989, 3019
Errandonea, Alfredo, 5132
Errázuriz, Margarita M., 4412
Errázuriz K., Ana María, 2584-2585, 2587
Erro, Davide G., 3523
Erzan, Refik, 1156
Escamilo Cárdenas, Simón, 4900
Escavações arqueológicas do Pe. João Alfredo Rohr, S.J.: o sítio da Praia das Laranjeiras II; uma aldeia da tradição ceramista Itararé, 579
Escobar, Arturo, 4413
Escobar, Santiago, 1800, 1814, 1831
Escobar Cerda, Luis, 3440
Escobar N., Marcos E., 2438
Escobedo A., Héctor L., 155
Escoto, Jorge, 4001
Escudé, Carlos, 3524
Escuela Politécnica Nacional (Quito), 1835
Espejo, Justo, 1991
Espejo Ortega, Alberto Octavio, 2137
Espejos: el cementerio de los rehabilitados, 3302
Espín de Castro, Vilma, 3123
Espina Pérez, Darío, 1690
Espinal, Rosario, 3053
Espinar, Juan Carlos, 3029
Espino, María Dolores, 1692
Espinosa, Cristina, 4901
Espinosa, Malva, 2790, 4416, 4445, 5021-5023
Espinosa, Roque, 4861
Espinosa Goitizolo, Reinaldo, 2423
Espinosa Villarreal, Oscar, 1375
Espírito Santo, Brazil (state), 2636
Espizúa, Lydia E., 2546
Essers, J.J.A., 4765

Estadísticas de intercambio comercial de los países latinoamericanos en la década de los ochenta, 1157
Estadísticas históricas de México, 23
El estado democrático de derecho en El Salvador, 2964
El Estado y la financiación de los partidos políticos en Costa Rica, 2949
Estados Unidos y el occidente de México: estudios sobre su interacción, 3911
Estancamiento económico y crisis social en México, 1983–1988, 4489
Estay Reyno, Jaime, 1158, 3812
Estrada Lugo, Erin Ingrid Jane, 415
Estragó, Gloria, 3596
Estrategia, desarrollo y política económica, 1376
Estrategia industrial e inserción internacional, 1771
Estrategia nacional de desarrollo alternativo 1990, 3303
Estrategias para el desarrollo de la investigación agropecuaria en la costa central y sur del Perú: seminario taller; Arequipa, 25–26 de mayo 1989, 1939
Estructura social de la Argentina: indicadores de la estratificación social y de las condiciones de vida de la población en base al *Censo de población y vivenda de 1980*, 5097
Estructura y dinámica poblacional, 24
El estudio de los movimientos sociales: teoría y método, 4490
Estudio sobre el desarrollo económico de la República Argentina: informe final, 2062
Estudio sobre violencia doméstica en mujeres pobladoras chilenas, 5024
Etnoarqueología: coloquio Bosch-Gimpera, 99
Etude des relations entre la population et le développement régional en Haïti, 2295
Evaluación Económica, 1981
Los evangélicos son cada vez más políticos, 2998
Evans, Brian, 2523
Evenson, Debra, 4723
Evolución de la productividad total de los factores en la economía mexicana, 1970–1989, 1377
The evolution of the Mexican political system, 2854
Eweg, Erlijn M., 731
El "éxito" al renegociar la deuda: ¿cadena para el sandinismo?, 1625
Eyre, L. Alan, 2239
Eyzaguirre, Graciela, 1033, 3372

Ezcurra, Exequiel, 2377
Faber, Daniel, 4609
Fabian, Horst, 3138
Fabian, Stephen Michael, 870
Fackler, James, 1588
Factional competition and political development in the New World, 100
Facultad Latinoamericana de Ciencias Sociales, 1559, 3297, 4021
Facultad Latinoamericana de Ciencias Sociales. Sede Académica de México, 4048
Fähmel Beyer, Bernd Walter Federico, 256
The failure of presidential democracy, 2750
Fajnzylber, Fernando, 1159
Falchetti, Ana María, 2461
Falcoff, Mark, 4002
Falk, Pamela S., 1310
Falla, Ricardo, 4610
Fals Borda, Orlando, 3166
Fandino, Mario, 4611
Fano, Hugo, 1034
Fargeix, André, 1799
Faria, Vilmar, 5166
Fariñas Gutiérrez, María Daisy, 499
Farjado Cortez, Víctor, 1727
Faro, Clóvis de, 2150
Farrell, Terrence W., 1644, 1664, 1667
Farreras Sanz, Leonor, 3441
Farruco, 3275
Fash, Barbara, 257
Fash, William L., 258
Faucher, Philippe, 2733
Fauconnier, Françoise, 259
La fauna en el Templo Mayor, 260
Faúndes Merino, Juan Jorge, 4175
Faúndez, Julio, 4220
Fauné, Angélica, 4612
Fauriol, Georges A., 4139
Faust, Franz X., 991
Fauvet-Berthelot, Marie France, 261
Favaro, Edgardo, 2018
Favret Tondato, Rita, 1378
Faya, Ana Julia, 4116
Fearnside, Philip Martin, 2660–2661
FEDEPRICAP (organization), 1547
Federación Sindical Unica de Trabajadores Campesinos de Cochabamba. Congreso Ordinario, 6th, *Aiquile, Bolivia, 1993*, 3304
Federalismo fiscal: el costo de la descentralización en Venezuela, 1728
Fedick, Scott L., 265, 267
Feijoó, María del Carmen, 5098
Feinman, Gary M., 80, 101–102, 262–265
Feinsilver, Julie M., 3124
Feldman, Eduardo, 3573

Feldweg, Helmut, 147
Félix, David, 1160
Feliz, Raúl Aníbal, 1379
Fenske, Kurt, 1985
Fenton, R.R., 2
Fergus, Howard A., 3071
Ferguson, James, 3063
Ferguson, R. Brian, 857–858, 892
Fermandois Huerta, Joaquín, 4249
Fernandes, Edésio, 3698
Fernandes, Florestan, 871
Fernández, Ana María, 5116
Fernández, Damián J., 4117–4118
Fernández, Eduardo, 912
Fernández, Francisco, 713
Fernández, Jorge, 552
Fernández, María Augusta, 2470
Fernández, Roque B., 1161, 2063–2064
Fernández Baeza, Mario, 2751
Fernández Collado, Carlos, 4549
Fernández Dávila, Enrique, 111, 278
Fernández de la Gala, Angel, 4934
Fernández de Soto, Guillermo, 4270
Fernández Distel, Alicia A., 553
Fernández Espinosa, Iván, 3226
Fernández Estigarribia, José Félix, 4293
Fernández Faingold, Hugo, 3630
Fernández Fernández, José Manuel, 4613
Fernández Huidobro, Eleuterio, 3442
Fernández Jilberto, Alex E., 3443
Fernández Ríos, Olga, 3125
Ferradas, Pedro, 4902
Ferrán, Lourdes de, 1729
Ferrari, César A., 1937, 1940
Ferraz, João Carlos, 2196
Ferreira, Graciela S., 5099
Ferreira, Luiz Fernando, 585
Ferreira, Muniz G., 2138
Ferrer, Aldo, 2139
Ferrer Oropeza, Carlos, 2439
Ferrer Vieyra, Enrique, 4202
Ferrero Costa, Eduardo, 1958, 3402, 4304, 4312–4315
Ferro, Elia G. de, 3026
Ferrocarril de Arica a La Paz. Sección Chilena, 2586
Ffrench-Davis, Ricardo, 1105, 1162, 1877
Fichet, Gérard, 1163
Field, Leonard, 4862
Field, Les W., 992
Fifer, J. Valerie, 3813
Figueira, Ricardo Rezende, 3699, 5167
Figueirêdo, Ney Lima, 3700
Figueiredo Júnior, José Rubens de Lima, 3700
Figueras, Miguel Alejandro, 1693
Figueroa B., Eugenio, 1878
Figueroa Perea, Juan Guillermo, 4531
Filgueira, Carlos H., 5134
El fin del fantasma: las relaciones interamericanas después de la Guerra Fría, 3814
Findji, María Teresa, 3167
Fine, Kathleen Sue, 4863
Fingerhut, Eugene R., 54
Fiorin, José Luiz, 3701
Fischman, Gustavo, 943
Fish, Suzanne K., 72
Fishlow, Albert, 1164
Fitch, J. Samuel, 3815
Flakoll, Darwin J., 3359
Flanagan, Edward M., 4003
Flannery, Kent V., 266
Fleissig, Adrian, 1549
Fletcher, Richard D., 4089–4090
Flisfisch, Angel, 5003
Flor Belaunde, Pablo de la, 4305
Flores de la Vega, Margarita, 4504
Flores Lúa, Graciela, 4491
Florescano, Enrique, 103, 119
Flórez Nieto, Carmen Elisa, 4811
Flujos geográficos en el Ecuador: intercambios de bienes, personas e información, 2471
Flynn, Peter, 3702
Fogel, Jean-François, 3126
Fogel, Ramón B., 4985–4986
Uma foice longe da terra: a repressão aos sem-terra nas ruas de Porto Alegre, 3703
Folan, William J., 434
Folgado, Arístides, 482
Fondo Cultural Cafetero (Bogotá), 1781
Fondo de Cultura Económica (Mexico), 1371, 1502, 3803, 3939
Fondo Fiduciario Manuel Pérez-Guerrero para la Cooperación Económica y Técnica entre Países en Desarrollo, 3826
Fondos y programas de compensación social: experiencias en América Latina y el Caribe, 1165
Fonseca, Isaque, 3777
Fonseca Júnior, Gelson, 4387
Fonseca Z., Oscar M., 467
Fontaine Talavera, Arturo, 5025
Foo, Herminia C., 4418
Food and Agriculture Organization of the United Nations, 1130
Ford, Anabel, 104, 267–268
Ford, Robert E., 2321
Ford Hurtado, Alberto, 2265
Forde, Penelope, 1664
Foreign direct investment and industrial restructuring in Mexico: government policy,

corporate strategies and regional integration, 1380
The foreign policies of Caribbean and Central American countries: selections from PROSPEL'S 1986 Anuario *Las políticas exteriores de América Latina y el Caribe—continuidad en la crisis*, edited by Heraldo Muñoz, 4004
Forero de Saade, María Teresa, 4812
Formación cívico-política de la juventud: desafío para la democracia, 5026
La formación de la imagen de América Latina en España, 1898–1989, 3816
Forman, Johanna Mendelson, 2942
Formas del inicio: la pintura rupestre en Venezuela, 715
The formation of complex society in southeastern Mesoamerica, 105
Forni, Floreal H., 5100
Forno, Eduardo, 2526
Foro 90 (Chile). Seminario, *Jahuel, Chile, 1990*, 3435
Foro de Arqueología de Chiapas, *1st, Tuxtla Gutiérrez, Mexico, 1990*, 106
Foro Interamericano, 3814
Foro Nacional Para, Con, Por, Sobre, De Cultura, *Bogotá, 1990*, 4813
Foro Productividad, Competitividad y Relaciones Laborales en la Industria del Norte de México, *Hermosillo, Mexico, 1993*, 1485
Foro Sonora ante el TLC, *Hermosillo, Mexico, 1991*, 1508
Forsyth, Donald W., 269
Fortes, Edson, 2631
Forteza, Alvaro, 2019
Fortún, Julia Elena, 951
Forum & Forum (firm), 1938
Fórum Nacional (Brazil), 2140, 3751, 4351, 5191
Fórum Nacional Como Evitar uma Nova Década Perdida, *Rio de Janeiro, 1991*, 2140
Fórum Nacional Idéias para a Modernização do Brasil, *Rio de Janeiro, 1988*, 2141–2142
Fournier, Patricia, 107
Fournier Origgi, Luis A., 2329
Foweraker, Joe, 2855
Fowler, William R., 105
Fox, David J., 2662
Fox, John G., 270
Fox, John W., 100
Fox, Jonathan, 1381, 2856
Fox, M. Louise, 2143
Fraga, Rosendo, 3525–3530
Frambes-Buxeda, Aline, 4724, 4791
France. Office de la recherche scientifique et technique d'outre-mer (O.R.S.T.O.M.), 2626
France. Office de la recherche scientifique et technique d'outre-mer (O.R.S.T.O.M.). Centre de Cayenne, 780
France. Office national des forêts en Guadeloupe, 2299
Francés, Antonio, 1195, 1730
Franchini, Teresa, 2556
Francis, Anselm, 4005
Franco, Eliana, 1826
Franco, Gustavo Henrique Barroso, 2117, 2144–2145
Franco, Rafael, 3597
Franco, Rolando, 3632
Franco Pellotier, Víctor Manuel, 732
Frank, André Gunder, 1166
Frank, Erwin H., 3237
Frankman, Myron J., 1167, 1382
Fraser, Cary, 4147
Frayde, Martha, 3127
Frechione, John, 872, 893
Frediani, Ramón O., 1168
Freeman, Andrea K.L., 107
Fregoso Peralta, Gilberto, 2820
Frei Montalva, Eduardo, 3444
Freidel, David A., 108
French, Joan, 1668
French Guiana. Archives départementales, 806
French Guiana. Musée départemental, 806
Frente Revolucionario de Unidad Campesina (Bolivia), 3305
Freres, Christian L., 3817
Fresco, Manuel A., 5121
Fresneda, Oscar, 3168
Freter, AnnCorinne, 403
Frey, Antonio, 5023
Freyermuth Enciso, Graciela, 3905
Frias, Luiz Armando de Medeiros, 5168
Frías F., Patricio, 3445
Friedemann, Nina S. de, 1000, 4814
Frieden, Jeffry A., 3818
Friedman, Jorge, 1164
Friedrich-Ebert-Stiftung, 1437, 1444, 1485, 1508, 2061, 2375
Friedrich-Naumann-Stiftung, 4314
Fritsch, Winston, 2117, 2145–2146
Fritz, Gayle J., 109
From Tortola to Kingstown: a report on eastern Caribbean political union, 3034
Frontera norte: una década de política electoral, 2857
Frontera sur: historia y perspectiva, 2378
Frota, Luciara Silveira de Aragão e, 4361
Fruhling, Hugo, 5027
Fry, Douglas P., 733
Fuchs, Mariana, 2065

Füchtner, Hans, 2147
Fuensalida, Carlos, 2577
Fuente, Juan de la, 1370
Fuentes, Juan Alberto, 1107, 1169
Fuentes Aguilar, Luis, 2379
Fuentes Navarro, Raúl, 4492
Fujii, Gerardo, 1170
Fuller, Linda, 4725
Fullerton, Thomas M., 1801
Fundação Alexandre de Gusmão. Instituto de Pesquisa de Relações Internacionais, 4387
Fundação Carlos Chagas. Concurso de Pesquisa sobre Direitos Reprodutivos (PRODIR), 4409
Fundação de Economia e Estatística (Brazil), 2127, 2666
Fundação FIDES., 2149
Fundação Getúlio Vargas. Centro de Pesquisa e Documentação de História Contemporânea do Brasil, 2168, 4363, 4375
Fundação Getúlio Vargas. Comité de Cooperação Empresarial, 4355
Fundação Instituto Brasileiro de Geografia e Estatística, 2664
Fundação Instituto Brasileiro de Geografia e Estatística. Departamento de Emprego e Rendimento, 2184
Fundação Instituto Brasileiro de Geografia e Estatistica. Diretoria de Geosciências, 2663
Fundação Instituto de Planejamento do Ceará (Brazil), 2192
Fundación Arturo Illia para la Democracia y la Paz (Argentina), 4168
Fundación Banco Provincia del Neuquén (Argentina), 945
Fundación Cruzada Patagónica (Buenos Aires), 944
Fundación de Educación Ambiental (Costa Rica), 2325
Fundación de Investigaciones Económicas Latinoamericanas (Buenos Aires), 2039–2040, 2047, 2056
Fundación de Parques Nacionales (Costa Rica), 2325
Fundación Friedrich Ebert de Colombia, 1773
Fundación John Boulton (Caracas), 1732
Fundación Luis Carlos Galán Sarmiento (Bogotá), 4172
Fundación para el Desarrollo del Agro (Peru), 1939
Fundación para la Educación Superior y el Desarrollo (Bogotá), 1761, 1767, 1781
Fundación Polar (Caracas), 1716
Fundación Raúl Prebisch (Buenos Aires), 1125
Fundación Salvadoreña para el Desarrollo Económico y Social, 1598
Fundación Universo Veintiuno (Mexico), 2375
Funkhouser, Edward, 1597, 1626, 4726
Furlani de Civit, María E., 2547
Furtado, Celso, 2148
The future development of maize and wheat in the Third World, 1171
The future of Amazonia: destruction or sustainable development?, 2665
O futuro do sindicalismo no Brasil: o diálogo social, 2149
Gaceta informativa de la industria maquiladora, 1383
Gade, Daniel W., 2424–2425
Gadotti, Moacir, 3704
Gagnon, Mariano, 3373
Gaitán de Pombo, Pilar, 3169
Gaitán Pavía, Pilar, 4815
Galeano, Luis, 4987
Galería de Arte Nacional (Venezuela), 715
Galetovic P., Alexander, 1889
Galindo Filho, Osmil Torres, 2177
Galinier, Jacques, 734
Gallareta Negrón, Tomás, 271
Gallois, Dominique T., 873
Gallón Giraldo, Gustavo, 3174
Galvan, Cesare Giuseppe, 3705
Galván Villegas, Luis Javier, 272
Gálvez Carvallo, Eduardo, 4245
Gálvez Pérez, Thelma, 1879
Gamarra, Eduardo, 3306
Gamarra Zorrilla, José, 3307
Gamba-Stonehouse, Virginia, 4173–4174
Gándara Gallegos, Mauricio, 4006
Gandásegui, Marco A., 3020, 4007
Gangas Geisse, Mónica, 2584, 2587–2590, 5028
Gans, Paul, 2618
Ganster, Paul, 1448, 1523, 1535, 3928
Garay, Alfredo, 3531
Garay, Luis J., 1771–1773, 1778
García, Alicia S., 3520
García, Graciela A., 2991
García, Jaime E., 2333
García, Jesús Alberto, 4414
García, Mauricio, 4864
García, Miguel V., 4203
García, Rolando Víctor, 1384
García, Rubén, 660
García-Bárcena, Joaquín, 360
García Bedoy, Humberto, 1385
García Bresó, Javier, 735
García Cruz, Florentino, 273
García de León P., Guadalupe, 1524
García de Quevedo, Juan, 4540
García Durán, Mauricio, 3170
Garcia G., Rigoberto, 3978

García Gallegos, Bertha, 3227
García Guadilla, María-Pilar, 4396
García Luis, Julio, 4128
García Lupo, Rogelio, 3598, 4175
García Moll, Roberto, 110, 274
García Montaño, Jorge, 1386
García Oropeza, Guillermo, 275
García-Passalacqua, Juan M., 4136
García Pérez, Alan, 3374
García Quesada, Ana Isabel, 4614
García Ruiz, José Luis, 2066
García Tamayo, Eduardo, 4727
García V., Carlos, 1477
García Valencia, Enrique Hugo, 736
García Villegas, René, 3446
García y Griego, Manuel, 3912
Gardens of prehistory: the archaeology of settlement agriculture in Greater Mesoamerica, 276
Garfield, Richard S., 4615
Garnier, Leonardo, 1573
Garramón, C.J., 1172
Garretón, Oscar Guillermo, 1880
Garretón Merino, Manuel Antonio, 2723, 2752, 2785, 3434, 3447–3450, 4415–4416, 4437, 5016, 5029–5032
Garrido de Colasante, Henderglaist, 4728
Garrido N., Celso, 1152
Garrocho Rangel, Carlos, 2380
Garza, Enrique de la, 1477
Garza, Mercedes de la, 277
Garzón Valdés, Ernesto, 2240, 3532
Gasson, Rafael, 713
Gatto, Hebert, 3631
Gaupp, Peter, 2330, 2753, 3375
Gautier, Gérard, 4708
Gavagnin Taffarel, Osvaldo, 4903
Gayle, Dennis J., 1669
Gedda O., Juan Carlos, 980
Geddes, Barbara, 2754–2755
Geisse, Guillermo, 2241
Geldof, Lynn, 3128
Gelpi, Adriana, 2666
GEMA: geomodelos de altimetría del territorio nacional, 25
Genta Dorado, Gustavo, 5133
Gentile Lafaille, Margarita E., 934
Gentleman, Judith, 2858
Geoghegan, Tighe, 2296–2297, 2307
Georges, Eugenia, 4729
Georgetown University. Hemispheric Migration Project, 4570, 4623, 4648, 4754
Gephardt, Richard Andrew, 1465
Gerchunoff, Pablo, 2067
Gereffi, Gary, 1387–1388

German, Cristiano, 3706
Germani, Gino, 5095
Gestión del estado y desburocratización, 3632
Gestión para el desarrollo de cuencas de alta montaña en la zona andina, 1173
Gestión popular del hábitat: 7 experiencias en el Perú, 4904
Gestión popular en salud: organizaciones no gubernamentales de desarrollo y políticas sociales; algunas experiencias peruanas, 4905
Ghiringhelli, Robertino, 4493
Giacalone, Rita, 3076, 4159, 4730–4731
Giaconi, Juan, 5033
Giambiagi, Fabio, 2097
Giarracca, Norma, 5101
Gibaja, Regina Elena, 5102
Gibaja Oviedo, Arminda, 709
Gibson, Bill, 1627
Giddings, Lorrain, 2242, 2548
Gierhake, Klaus, 2524
Giesing, Cornelia, 416
Gilbert, Alan G., 2381
Gilbert, Isidoro, 3533
Gilbert, Jorge, 3439
Gill, Lesley, 952–953, 4959
Gillespie, Charles Guy, 3633
Gillespie, Richard, 2756
Gillion, Colin, 1881
Gindling, T.H., 1574–1575
Giordano, Fernando, 3620
Girón, Alicia, 1389
Girvan, Norman, 1639
Gissi Bustos, Jorge, 5034
Gitli, Eduardo, 3973
Gjerset, Heidi, 4732
Glade, William P., 1174–1175
Glascock, Michael D., 215
Glasmeier, Amy K., 1361
Glazier, Stephen D., 841
Gleijeses, Piero, 2978
Glinkin, Anatoliĭ Nikolaevich, 3836
Glover, David, 1176–1177
Gmelch, George, 795
Gnecco, Cristóbal, 625
Gobernabilidad en Colombia: retos y desafíos, 3171
Godfrey, Brian J., 2638
Godínez Plascencia, Alberto, 1390
Godio, Julio, 5103
Godoy, Ricardo, 1628, 1976–1977, 1982
Godoy Arcaya, Oscar, 3423
Góes-Filho, Luiz, 2667
Goland, Carol, 2497

Goldar, Ernesto, 3534
Goldfrank, Walter L., 5035
Goldin, Ian, 1178
Goldin, Liliana R., 4616
Goldrich, Daniel, 2859
Goldring, Luin, 4494
Goldstein, Paul, 675
Gomáriz, Enrique, 4428–4429
Gomes, Angela Maria de Castro, 4417
Gomes, Cândido Alberto da Costa, 5169
Gómez, Alejandro, 3535
Gómez, Flor Angela, 1759
Gómez, José Antonio, 1321
Gómez, Miguel, 4988
Gómez Calcaño, Luis, 4838
Gómez Carpinteiro, Francisco Javier, 4483
Gómez Chiñas, Carlos, 3913
Gómez E., Nelson, 4281
Gómez G., Rosario, 1941
Gómez Mont, Carmen, 1391
Gómez Otero, Julieta, 554
Gómez Sahagún, Lucila, 1392
Gómez Serafín, Susana, 111, 278
Gómez Solórzano, Marco A., 1393
Gómez Tagle, Silvia, 2869
Gomezjara, Francisco A., 4418
Gonçalves, Antonio Carlos Porto, 2150
Gondrie, Peter, 4960
Gonzáles, Raúl, 5036
Gonzales de Olarte, Efraín, 1924, 1942–1944
Gonzales Leiva, José I., 2591
Gonzales-Zúñiga Guzmán, Alberto, 4934
González, Alberto Rex, 555–556
González, Alfonso, 2243
González, Carlos Javier, 229
González, Chacho, 3281
González, Fernán E., 3172
González, Geraldino, 2298
González, Jason J., 364
González, Jorge, 3536
González, José E., 2354
González, José Luis, 4906
González, Luis Eduardo, 3634–3636
González, Luis J., 4294
Gonzalez, Manuel José Forero, 2151
González, Mariana, 5134
González, Mónica, 3451
González, Nancie L., 279, 796, 4617
González, Norberto, 1248
González, Raúl, 3376
González, Rodolfo, 3637
González Abuter, Tulio, 4250
González Aguirre, Iván, 1843, 4859
González-Aréchiga, Bernardo, 1437
González Arias, José Jairo, 3173

González Block, Miguel Angel, 4495
González Bombal, Inés, 5104
González C., Aurora, 2356
González Calderón, O.L., 112
González-Camino, Fernando, 4008
González Canahuate, L. Almanzor, 3054
González Carré, Enrique, 676–677
González Casanova, Pablo, 2829, 2860
González Cervera, Alfonso S., 4419
González Cháves, Alfredo, 468
González Chávez, Humberto, 4496
González Cortez, Héctor, 974, 981
González Davison, Fernando, 4618
González Encinar, José Juan, 2757
González García, Lydia Milagros, 4733
González Hinojosa, Manuel, 2861
González Kirby, Diana, 4734
González L., Ernesto, 320
González Licón, Ernesto, 113
González Marín, María Luisa, 4460
González Mejía, Hernán, 4009
González Ordosgoitti, Enrique Alí, 4839–4840
González Ramírez, Raúl S., 4535
González Sánchez, Jorge A., 4497
González-Souza, Luis, 3914
González Tiburcio, Enrique, 1363, 1454
González Vásquez, Fernando, 468
González Vela, Gabriel, 1394
Good, Kenneth, 894
Goodman, David, 2665, 4443
Goodman, Louis Wolf, 2942, 3819
Goold, J. William, 3888
Gootenberg, Paul, 1945
Gordillo, Inés, 561
Gordon C., Maribel, 1550
Goretti, Matteo, 3562–3563
Görgen, Sérgio Antônio, 2104, 3703
Gorojovsky, Néstor, 5127
Gosine, Mahine, 791
Gosner, Kevin, 57
Gossen, Gary H., 737–738
Gottret, M.V., 1179
Gougeon, Olivier, 2396
Gough, Barry M., 4204
Goulet, Denis, 1395
Gouvea Neto, Raúl de, 1180
Government spending and income distribution in Latin America, 1181
Gow, Peter, 915
Goza, Franklin, 5170
Grabendorff, Wolf, 3820
Grabois, José, 2668–2669
Grace, Alexander M., 2758
Gracia Bondía, José, 2325

Graffam, Gray, 566
Graham, Carol, 1983, 3452
Graham, Douglas H., 1596
Graham, Elizabeth, 349
Graham, John A., 280
Graham, Laura, 874
Graham, Mark Miller, 149, 476
Grammont, Hubert Carton de, 1396
Granda, Alicia, 4858
Granda, Juan Velit, 4306
Grandón, Alicia, 4907
Graneiro, Walter J.B., 2178
Granovsky, Martín, 4205
Granville, Jean-Jacques de, 2675
Grassi, Estela, 5105, 5107
Grau Guardarrama, Guillermo, 2423
Graulich, Michel, 77, 281
Gravano, Ariel, 5106
Green, Duncan, 2244
Green, Roy E., 1154
Greene, John Edward, 827, 4091
Greenow, Linda, 2245
Greenwald, Joseph A., 1459
Gregorio, José de, 1182
Grennes, Thomas, 1549
Griffin, Clifford E., 3129
Griffith, David, 797
Griffith, Ivelaw L., 4148–4149
Griffith, Kathleen Ann, 1397
Griffith, Winston H., 4081
Grigsby, Thomas L., 417, 739
Grillo, Oscar Jorge, 3537
Grimberg, Mabel, 5107
Grindle, Merilee S., 2862
Gritti, Roberto, 2759
Grompone, Romeo, 3377–3378, 4908
Groot, Jan P. de, 4668
Groot, Silvia W. de., 798
Groot de Mahecha, Ana María, 619–620
Gros, Christian, 993
Grosh, Margaret E., 1183
Grosjean, Martin, 604
Gross F., Patricio, 2592
Grosse, Robert E., 3854, 4271
Grossi, Maria, 3512
Groupe de recherches sur l'Amérique latine Toulouse-Perpignan, 936
Grove, David C., 114, 282
Grube, Nikolai, 418
Gruhn, Ruth, 576
Grün, Roberto, 5171
Grunsven, Marie-Annet van, 2303
Grupo de Análisis y Desarrollo Institucional y Social (Argentina), 5126
Grupo de Estudios para el Desarrollo (Lima), 4314

Grupo de Trabajo CLACSO Empresarios y Estado en América Latina. Seminario, *2nd, Pachuca, Mexico, 1987*, 1152
Grupo de Trabajo sobre la Cuenca del Canal de Panamá, 4010
Grupo de Trabajo sobre Reestructuración de la Industria Automotriz Mundial y Perspectivas para América Latina. Reunión Regional, *Bogotá, 1985*, 1249
Grupo Ecológico Tierra Viva (Quito), 2464
Grupos de pressão, 3707
Guadagni, Alieto Aldo, 4206
Guadalajara en el umbral del siglo XXI, 1398
Guadeloupe. Conseil régional, 3081
Guanche Pérez, Jesús, 799, 4735
Guanes de Laíno, Rafaela, 4989
Guarch Delmonte, José M., 500
Guarderas, Andrés, 1812
Guastavino, Luis, 3453
Guatemala: Amnesty International's current human rights concerns, 2979
Guatemala, getting away with murder: an Americas Watch and Physicians for Human Rights report, 2980
Gúber, Rosana, 5106
Guderjan, Thomas H., 283
Guelar, Diego R., 3538
Guengant, Jean-Pierre, 4736
Guerberoff, Simón L., 2068
Guerguil, Martine, 1137
Guerra-Borges, Alfredo, 1184
Guerra Bravo, Samuel, 3224
Guerra T., Helena, 4841
Guerra y constituyente, 3174
Guerrero, Andrés, 3228, 4865
Guerrero, José, 501
Guerrero, Natividad, 4737
Guerrero, Omar, 2863
Guerrero Barón, Javier, 4816
Guerrero Carrión, Trosky, 1802
Guerrero J., Bernardo, 981
Guerrero M., Juan V., 469
Guerrier, L. Arnault, 4738
Guevara Arze, Walter, 3308, 4233
Guhl, Ernesto, 2456
Guibbert, Jean-Jacques, 4825
Guidos Béjar, Rafael, 2965
Guidotti, Pablo E., 1185
Guier Serrano, Estrella, 2333
Guiliani Cury, Hugo, 1659
Guillén, Ann Cyphers, 115, 284
Guillén López, Tonatiuh, 2857
Guillén Velarde, Rosa, 4909
Guillet, David, 1035
Guil'liem Arroyo, Salvador, 419
Guimarães, Antônio Sérgio Alfredo, 5157

Guimarães, Eduardo Augusto, 2152
Guimarães, Raúl Borges, 2670
Guinea Bueno, Mercedes, 640
Gundermann Kroll, Hans, 974
Gunn, Gillian, 3130
Gunter, Frank R., 1766
Gurdián Guerra, Reymundo, 4011
Gurgel, Cláudio, 3708
Gurría, José A., 1399
Gurriarán, José Antonio, 3454
Guss, David, 895
Gussinyer i Alfonso, Jordi, 285
Gustafson, Lowell S., 1143
Gutiérrez, Irma Eugenia, 4498
Gutiérrez, Marcos, 3638–3639
Gutiérrez Bermedo, Hernán, 3821
Gutiérrez Colombres, Benjamín, 2549
Gutiérrez-Cuevas, Carlos, 3181
Gutiérrez de Machón, M.J., 2547
Gutiérrez Estévez, Manuel, 740–741
Gutiérrez Grageda, Blanca Estela, 1513
Gutiérrez Gutiérrez, Carlos José, 2950
Gutiérrez Lara, Abelardo Aníbal, 3892
Gutiérrez Mayorga, Alejandro, 2999
Gutiérrez Neyra, Javier, 4910
Gutiérrez Santos, Luis E., 1186
Gutiérrez Saxe, Miguel, 2951
Gutman, Margarita, 2246
Guy, James J., 3131
Guzmán, Abimael, 3379
Guzmán, Luis Humberto, 3000
Guzmán, Rolando M., 2191
Guzmán, Virginia, 4433
Guzmán Polanco, Manuel de, 4285
Gwynne, Robert N., 2593
Haar, Jerry, 3810
Haase, R., 2525
Haber, Lawrence J., 1245
Haber, Stephen H., 1400, 2153
Hacia el siglo XXI: la proyección estratégica de Chile, 4251
Hacia la estabilización y el crecimiento, 1946
Hacia una nueva política industrial, 1401
Hacía una política marítima nacional, 4307
Hacia una presencia diferente: mujeres, organización y feminismo, 4990
Haciendo la prostitución: menores prostituídos en la terminal de ómnibus de Asunción, 4991
Haedo, Javier de, 2020
Haffer, Jürgen, 2671
Haigh, Robert W., 1402
Haindl Rondanelli, Erik, 1882
Haiti. Bureau national d'ethnologie, 488
The Haitian challenge: U.S. policy considerations, 4139
Haitian Institute of Ethnology, 823
Hakim, Peter, 3822–3824
Hall, Anthony L., 2665
Hamilton, Eve, 1278
Hamilton, Nora, 4619
Hamm, Lyta, 4739
Hammen, Thomas van der, 2461
Hammond, Norman, 364, 394
Handbook of Latin American Studies CD-ROM: HLAS/CD., 26
Handwerker, W. Penn, 800
Handy, Jim, 4620
Hanks, William, 742
Hanratty, Dennis M., 3035
Harberger, Arnold C., 1187
Harbottle, Garman, 116
Hard, Robert J., 743
Harden, Carol P., 2472
Hardman, Martha, 962
Hardoy, Jorge Enrique, 2246–2248
Hardy Raskovan, Clarisa, 1883
Hargreaves, Clare, 3309
Harley, J. Brian, 2249
Harnecker, Marta, 3132
Harriet, María Inés, 5037
Harris, Allan, 3098
Harris, John Ferguson, 420–421
Harris, Olivia, 962
Harris, Richard Legé, 2760
Hartlyn, Jonathan, 3175, 3870
Harvey, Neil, 2888
Hasel, Ulrike, 422
Hasemann, George, 286, 306
Hasenbalg, Carlos Alfredo, 5200
Hasselkus, Hans, 423
Hassig, Ross, 117
Hastorf, Christine Ann, 678–679, 2499
Hatch, Marion Popenoe, 287
Hausmann, Ricardo, 1181
Healan, Dan M., 288
Healy, Paul F., 289
Heath, John Richard, 1403–1404
Hecht, Laurence, 1405
Hecht, Susanna B., 2672
Hedström, Ingemar, 2316
Heine, Jorge, 4136
Heinz, W.S., 5172
Held, Günther, 1273
Heller, Peter L., 4602
Hellman, Judith Adler, 2761
Hellwig, David J., 5142
Helwege, Ann, 1110, 1682, 2123, 3109
Henao, Marta Luz, 1774
Henderson, John S., 130, 290, 352
Hendon, Julia A., 291

Hendricks, Janet, 916
Hendriks, Jan, 1947
Henkin, Louis, 4012
Heras y Martínez, César M., 641
Heredia, Blanca, 2864
Heredia, Carlos, 3894
Heritage Foundation (Washington), 1410
Herlihy, Peter H., 2348
Hermann, Hamlet, 3055
Hern, Warren M., 917
Hernández, Carlos, 4669
Hernández, Carmen, 1803, 4864
Hernández, Isabel, 943
Hernández, José M., 4119
Hernández, Luis, 2856
Hernández, Pablo José, 3539
Hernández, Rafael, 4120–4122
Hernández, Tosca, 3263
Hernández Bringas, Héctor H., 4499
Hernández Castillo, Rosalva Aída, 3905
Hernández Cerdá, María Engracia, 2382
Hernández Laos, Enrique, 1406–1407
Hernández López, Nitza M., 4791
Hernández Madrid, Miguel J., 2865
Hernández Milián, Jairo, 4013
Hernández Ortiz, Evelyn, 4014
Hernández Rodríguez, Rogelio, 2866–2867
Hernández-Sampieri, Roberto, 4549
Hernández T., Leonardo, 1884
Hernández Valle, Rubén, 2952
Herrera, René, 4015
Herrera Falcone, Stella, 636
Herrera Núñez, José Manuel, 3892
Herrera Toledano, Salvador, 1408
Herrera Torres, Hugo, 377
Herrera Villalobos, Anayensy, 478
Herzig, Mónica, 2409
Hewitt, W.E., 3709, 5148, 5173
Hey, Jeanne A.K., 4016
Hicks, Frederic, 118
Hidalgo, Ariel, 4740
Hidalgo Saa, Susana, 3224
Hidrobo Estrada, Jorge, 1804
Hilker, Töns Henrich, 1437
Hill, Jonathan David, 856, 896–897
Hill, Kim R., 898, 918
Hill, Nancy P., 4741
Hillcoat, Guillermo, 4082–4083
Hilt, Eric, 1237
Hinojosa, Iván, 3380–3381
Hinojosa Ojeda, Raúl A., 3956
Hintze, Susana, 5107
Hinz, Eike, 424
Hirabayashi, Lane Ryo, 744
Hirmes, María Eugenia, 5038

Hiraoka, Mário, 2673
Hirschman, Albert O., 1188
Hirst, Mônica, 4176–4177, 4362–4363
Hirth, Kenneth G., 146, 292
Historia de la cerámica en el Ecuador, 642
Historia de las finanzas públicas en Venezuela, 1731
Historia del P.R.T.: 25 años en la vida política argentina, 3540
História do Banco do Brasil, 2154
Historia general de Michoacán, 119
Historia general del Perú, 680
Historia natural de un valle en los Andes, La Paz, 2526
Historia y porvenir de México ante el Tratado de Libre Comercio, 1409
History of humanity, 55
Ho, Christine G.T., 801
Hochman, Gilberto, 3710
Hockenstein, Jeremy, 1628
Hocquenghem, Anne-Marie, 539, 2498
Hodge, Mary G., 293–295
Hodges, Donald Clark, 3541–3542
Hoefle, Scott William, 2635
Hoffer, Marilyn Mona, 3373
Hoffer, William, 3373
Hoffmaister, Alexander, 1576–1577
Hoffman, Peter R., 2383
Hoffmann, Léon-François, 802
Hoffmann, Odile, 2384, 4500
Hohmann, Hasso, 296
Hojman, David E., 1885
Hola A., Eugenia, 3455, 5039
Holden, Robert H., 4017
Holiday, David, 2960, 4018
Hollihan, Mike, 1805
Holmes, Nat, 4040
Hombres de páramo y montaña: los Yanaconas del Macizo colombiano, 994
Honduras. Secretaría de Planificación, Coordinación y Presupuesto, 4621
Honduras, el ajuste estructural y la reforma agraria, 1618
Hoogbergen, Wim S.M., 798, 803, 2301
Hoopes, John W., 470
Hooykaas, Eva María, 620
Hope, Dianne L., 1670
Hoppin, Polly, 1610
Horn, Sally P., 2331
Horowicz, Alejandro, 3543
Horst, Oscar H., 2343
Hosler, Dorothy, 297, 383
Hosten-Craig, Jennifer, 1645
Hostettler, Ueli, 745
Houaiss, Antônio, 3711

Houk, James, 804, 4742
Houston, Stephen D., 425, 449
Houtart, François, 4622
Hove, Okke ten, 2301
Hualde, Alfredo, 1353
Huamantinco Cisneros, Francisco, 4911
Huanca L., Tomás, 954
Huapaya Manco, Cirilo, 664
Huchim, Eduardo, 3915
Hudgins, Edward Lee, 1410
Hufbauer, Gary Clyde, 1189
Huggins, Martha K., 4420–4421
Huguet Santos, Montserrat, 3816
Huibers, Bart, 1984
Hulme, Peter, 502, 524
Hulst, Hans van, 4743
Human rights in post-invasion Panama: justice delayed is justice denied, 3021
Human Rights Watch, 984, 2762, 2960
The Human Rights Watch global report on prisons, 2762
Humphrey, John, 2155
Huneeus, Carlos, 3456–3457, 5040
Hunter, Wendy, 3712
Huntley, Earl, 4150
Hurault, Jean, 805
Hurbon, Laënnec, 4744
Hurkmans, Dennis, 2332
Hurtado, A. Magdalena, 898, 918
Hurtado, Abelino Dagua, 1003
Hurtado, Flor de María, 2868
Hurtado, Javier, 2928
Hurtado, Osvaldo, 3229
Hurtado Salazar, Samuel, 4842
Hurwitz, Jon, 4019
Huss, Torben, 1886
Hutchinson, Gladstone A., 1190
Hydroelectric dams on Brazil's Xingu River and indigenous peoples, 875
Iákovlev, Petr, 3599
Ianni, Octávio, 5174
Ibarra Colado, Eduardo, 4488
ICARE (Chile), 1875–1876
Icaza, Ana María, 5058
Iconografía de Cajamarca, 681
Identidad y prestigio en los Andes: gorros, turbantes y diademas; exposición, noviembre 1993 a junio 1994, 605
Ideology and pre-Columbian civilizations, 56, 120
Iglesias, Alicia N., 2550
Iglesias, Enrique V., 1191–1192, 3825
Iglesias Palau, Augusto, 1850
Iguíñiz, Javier, 1948–1950
Illius, Bruno, 87

Images à croquer: l'alimentation guyanaise à travers l'iconographie ancienne, 806
Imano, Toshihiko, 3052
Imashi! Imashi!: adivinanzas poéticas de los campesinos indígenas de la sierra andina ecuatoriana/peruana, 935
Imbert, Daniel, 2299
Imbiriba, Maria de Nazaré Oliveira, 861
Imbusch, Peter, 5041
Imminck, Maarten D.C., 1607, 1614
Impacto de la agresión del gobierno de los Estados Unidos de América sobre la economía y la sociedad panameña, 4020
El impacto del capital financiero del narcotráfico en el desarrollo de América Latina: simposio internacional, 3826
El impacto económico y social de las migraciones en Centroamérica, 1551
Impactos de grandes projetos hidroelétricos e nucleares: aspectos econômicos e tecnológicos, sociais e ambientais, 2156
Impactos regionales de la apertura comercial: perspectivas del tratado de libre comercio en Jalisco, 1411
El Imperio inka: actualización y perspectivas por registros arqueológicos y etnohistóricos, 540
Impex: Foreign Trade Statistics by Commodities = Statistiques de commerce extérieur produits, 27
La importancia económica de la minería en el Perú, 1951
INADE (organization, Peru), 1936
Los incas y el antiguo Perú: 3000 años de historia, 682
Incriminación a la violencia contra la mujer, 4745
Indacochea Queirolo, Alberto, 4307
Indianidad, etnocidio, indigenismo en América Latina, 936
Indicadores de la modernización mexicana, 1412
Indigenous revolts in Chiapas and the Andean highlands, 57
Indios de Roraima: Makuxí, Taurepang, Ingarikó, Wapixana, 876
Indios, tierra y utopía: los mejores trabajos del 4o concurso de testimonios y 1o de dibujo y pintura indígena "A 500 años de resistencia, nuestros mayores cuentan su vida.", 1014
Industria, comercio y Estado: algunas experiencias en la Cuenca del Pacífico, 1413
La industria de bienes de capital en América

Latina y el Caribe: su desarrollo en un marco de cooperación regional, 1193
A indústria de informática: tendências e oportunidades, 2157
Industria y trabajo en México, 1414
Infancia y pobreza en la Argentina, 5108
Infante, Ricardo, 1194
Infante T.P., Sebastián, 1806
Inflación: economía, empresa y sociedad, 1195
Informe blanco sobre los avances logrados en el proceso de cumplimiento del acuerdo de paz para Centroamérica "Esquipulas II" a los noventa días de haberse firmado, 4021
Informe nacional a la Conferencia sobre Medio Ambiente y Desarrollo de las Naciones Unidas, 2551
INFORPRESS Centroamericana, 4051
INSOTEC (Quito), 1798
Institut haïtien de statistique et d'informatique. Division d'analyse et de recherches démographiques, 2295
Institut Latinskoĭ Ameriki (Akademiĭa nauk SSSR), 1073, 1203, 2804
Institut Latinskoĭ Ameriki (Rossiĭskaĭa Akademiĭa Nauk), 2226, 3836, 3856, 4258
Institute for Latin American Integration, 1141, 1157, 1523
Institutional analysis—4 SKN: institutional analysis in the area of natural resource management; the case of St. Kitts-Nevis, 2300
Instituto Andino de Estudios en Población y Desarrollo (Lima), 4944
Instituto Brasileiro de Administração Municipal. Núcleo de Estudos Mulher e Políticas Públicas, 5182
Instituto Brasileiro de Análises Sociais e Econômicas, 2136, 5210
Instituto Brasileiro de Siderurgia, 2134
Instituto Colombiano de Cultura, 4813
Instituto de Administración Pública del Estado de México, 2881
Instituto de Altos Estudios Nacionales (Ecuador), 1833
Instituto de Banca y Finanzas (Mexico). Centro de Investigación para el Desarrollo, 1401, 1415
Instituto de Cultura Puertorriqueña, 4794
Instituto de Defensa Legal (Lima), 3360
Instituto de Ecología (Mexico), 1368
Instituto de Estudios Contemporáneos (Buenos Aires), 2069
Instituto de Estudios de Población y Desarrollo (Dominican Republic), 4773
Instituto de Estudios e Investigaciones sobre el Medio Ambiente (Argentina), 2552
Instituto de Estudios Económicos Mineros (Peru), 1951
Instituto de Estudios Histórico-Marítimos del Perú, 4307
Instituto de Estudios Jurídicos de El Salvador, 2964
Instituto de Estudos Políticos e Sociais (Brazil), 3755
Instituto de Formación e Investigación Sindical (Uruguay), 3629
Instituto de Investigación, Capacitación y Asesoría Económica (Nicaragua), 1624
Instituto de Investigación y Desarrollo (Uruguay), 3629
Instituto de Investigaciones Cambio y Desarrollo (Peru), 1196
Instituto de Investigaciones Estadísticas (Cuba), 4722
Instituto de Pesquisa Econômica Aplicada (Brazil), 2257
Instituto de Planejamento Econômico e Social (Brazil), 2664
Instituto de Relaciones Internacionales y de Investigaciones para la Paz (Guatemala), 3991–3992
Instituto de Seguridad y Servicios Sociales de los Trabajadores del Estado (Mexico), 1502
Instituto del Sur para la Cooperación Democrática (Peru), 3401
Instituto do Desenvolvimento Econômico-Social do Pará (Brazil), 2193
Instituto dos Economistas do Rio de Janeiro, 2097
Instituto Ecuatoriano de Investigaciones y Capacitación de la Mujer, 4871, 4874
Instituto Ecuatoriano de Seguridad Social, 1840
Instituto Ecuatoriano para el Desarrollo Social, 3239
Instituto Indigenista Peruano, 863
Instituto Interamericano de Derechos Humanos (Costa Rica), 3204, 4623
Instituto Latinoamericano de Estudios Avanzados (Panama), 3019
Instituto Latinoamericano de Estudios Transnacionales, 2790
Instituto Latinoamericano de Investigaciones Sociales, 1833, 1979, 1997–1998, 2464, 3212, 3293, 3297–3299, 3348, 4288, 4956–4957, 4969
Instituto Latinoamericano y del Caribe de Planificación Económica y Social (ILPES), 3632

Instituto Matías Romero de Estudios Diplomáticos (Mexico), 3886, 3950
Instituto Mexicano de Ejecutivos de Finanzas, 1438
Instituto Mexicano del Seguro Social, 1502
Instituto Michoacano de Cultura (Mexico), 119
Instituto Nacional de Administración Pública (Mexico), 2881
Instituto Nacional de Alimentación y Nutrición (Bolivia), 4975
Instituto Nacional de Antropología e Historia (Mexico), 68
Instituto Nacional de Energía (Ecuador), 1807
Instituto Nacional de Estadísticas (Chile). Subdivisión de Estadísticas Demográficas, 5012–5013
Instituto Nicaragüense de Investigaciones Económicas y Sociales, 3990
Instituto para el Desarrollo Económico y Social de El Salvador, 4053
Instituto Peruano de Administración de Empresas, 1932
Instituto Peruano de Relaciones Internacionales, 1964
Instituto Politécnico Nacional (Mexico), 1409
Instituto Sociedade, População e Natureza (Brazil), 2686–2687, 2691–2693, 2708, 2715
Instituto Venezolano de Estudios Sociales y Políticos, 4088
Instrucción de la mujer y fecundidad, 4866
Insulza, José Miguel, 4184
Insurgencia democrática: las elecciones locales, 2869
Integración, deuda externa y relaciones económicas internacionales, 1196
Inter-American Commission on Human Rights, 3133
Inter-American Development Bank, 1209, 1226, 1283, 1303
Inter-American Dialogue, 3827
Inter-American Institute for Cooperation on Agriculture, 1172, 1251
INTERCAMPUS. Meeting, *27th, Lima, 1989*, 1946
INTERCAMPUS. Meeting, *29th, Lima, 1990*, 4912
Interdependencia: ¿un enfoque útil para el análisis de las relaciones México-Estados Unidos?, 3916
Intermediación social y procesos políticos en Michoacán, 2870
International Bank for Reconstruction and Development. *See* World Bank

International Bibliography of the Social Sciences, 28
International Center for Economic Growth (San Francisco), 1553
The international common-carrier transportation industry and the competitiveness of the foreign trade of the countries of Latin America and the Caribbean, 1197
International Conference on Nutrition, *Rome, 1992*, 4975
International Congress for Caribbean Archaeology, *12th, Cayenne, 1987*, 503
International Congress for Caribbean Archaeology, *14th, Barbados, 1991*, 504
International Congress of Americanists, *46th, Amsterdam, 1988*, 1641, 2205
International Congress of Americanists, *47th, New Orleans, 1991*, 58
International Development Research Centre (Canada), 944
International Maize and Wheat Improvement Center, 1171, 1349
International Organization for Migration, 5140
International Political Science Abstracts, 29
Intipampa, Carlos, 955
Introducción a la problemática ambiental costarricense: principios básicos y posibles soluciones, 2333
Introducción al Uruguay de los 90, 3640
Inversión extranjera directa en México en la industria informática y automotriz, 1416
Investigación Económica, 30
Invirtiendo en Chile, 1887
Irarrázaval Llona, Ignacio, 1882
Irigoyen, Soraya, 4930
Irish, J.A. George, 4746
Irrigation at high altitudes: the social organization of water control systems in the Andes, 1035
Irusta Medrano, Gerardo, 3310
Is there a transition to democracy in El Salvador?, 4022
Isaac, Barry L., 98
Isaacs, Anita, 3230–3231
Isaza, José Fernando, 1775
Isenburg, Teresa, 2674
Isis International, 4407
Island Resources Foundation (US Virgin Islands), 2313
Islands of the Commonwealth Caribbean: a regional study, 3035
Issler, João Victor, 2158
Issues in democratic consolidation: the new South American democracies in comparative perspective, 2763

Ito, Akihito, 3804
Iturralde, Marta, 3029
Izam, Miguel, 1198
Izard, Miguel, 1732
Jackisch, Carlota, 3544
Jackson, Jean E., 899–901
Jackson, Lawrence J., 298
Jacobi, Pedro, 3713
Jacobus, André Luiz, 580
Jácome, Luis Ignacio, 1808
Jaguar no sokuseki: Andesu, Amezon no shūkyo to girei [Jaguar's trace: religion and ritual in the Andes and the Amazon], 859
Jaguaribe, Hélio, 3714, 3755
Jaguaribe, Roberto, 4366
Jaime, Vicente, 2872
Jaime Barreto, Wilson, 3382
Jamaica. Committee Appointed to Advise the Jamaican Government on the Performance, Accountability, and Responsibilities of Elected Parliamentarians, 3070
Jamard, Jean-Luc, 807
James, Joel, 4747
James, Winston, 4748
Jane's Sentinel, 31
Jansana, Loreto, 5042
Jansen van Galen, John, 3090
Janssen, René, 2301
Janusek, John W., 567
Japan, the United States, and Latin America: toward a trilateral relationship in the Western Hemisphere, 3828
Jaramillo Buendía, Fidel, 1809
Jaramillo Cisneros, Hernán, 1010
Jaramillo Correa, Luis Fernando, 4272
Jarpa, Sergio G., 4178
Jarrín R., Oswaldo, 3225
Jarvis, Leslie, 1671
Jatobá, Jorge, 2159
Jauberth Rojas, Rodrigo, 3968–3969, 3982, 4074
Jáuregui O., Ernesto, 2385–2386
Jean-Louis, Marie-Paule, 828
Jeannot, Fernando, 1417
Jeftanovic, Pedro, 1888
Jereissati, Tasso, 3690
Jerez, Ariel, 3715
Jerozolimski, José, 4326
Jesus, Avelino de, 2160
Jesús D., Sara de, 4938
Jiménez, Dolores, 4084
Jiménez, Félix, 1952
Jiménez, Helga, 4624
Jiménez, Michael F., 3162
Jiménez A., Arnoldo, 1598

Jiménez Aruquipa, Domingo, 947
Jiménez Huerta, Fernando, 746
Jiménez Mejía, Lilia, 3964
Jiménez-Milón, Percy, 2491
Jiménez Ruvalcaba, María del Carmen, 3001
Jiménez Valdez, Gloria Martha, 299
Joffre Lazarini, Ruth, 4530
Jofré, Manuel Alcides, 5043
Jofré Barroso, Haydée M., 3545
Johannessen, Sissel, 679, 2499
John, George, 3096
Johns, Christina Jacqueline, 4023
Johnson, Jeffrey C., 797
Johnson, P. Ward, 4023
Johnson, Timothy Hugh, 2302
Jonas, Susanne, 2767
Jones, Claudio, 1517
Jones, E., 1672
Jones, Gareth, 2387
Jones, Grant D., 349
Jones, Jeffrey R., 1552
Jones, Lindsay, 121
Jones, Richard C., 4501
Jones-Hendrickson, Simon B., 3030
Jongkind, Fred, 1733
Joosten, Elisabeth M., 731
Jordan, Ekkehard, 2473
Jordán Pozo, Rolando, 1978
Jorge, Antonio, 3811, 3834
Jorge, Graciela, 3442
Jornadas Amazónicas Internacionales, *1st, Quito, 1990*, 2463
Jornadas Bancarias de la República Argentina, *1st?, Buenos Aires?, 1991*, 2070
Jornadas Bancarias de la República Argentina, *2nd, Buenos Aires?, 1991?*, 1199
Jornadas Territoriales, *2nd, Santiago, 1987?*, 2594
Jorrat, Jorge Raúl, 5095
Jouravlev, Andrei, 1204
Jované, Juan, 1635
Joyce, Arthur A., 300–302
Joyce, Rosemary A., 303, 444
El juego de pelota en Mesoamérica: raíces y supervivencia, 122
Julien, Daniel G., 683
Junco, Silvia, 2050
Junguito Bonnet, Roberto, 1781
Junquera, Carlos, 919
Juri, María E., 2071
Justeson, John S., 426
Justicia, derechos humanos e impunidad, 3176
Justicia y pobreza en Venezuela, 4843
Juteram, Dolores, 4749

Kagami, Mitsuhiro, 1200
Kahn, Francis, 2675
Kaimowitz, David, 1201
Kaiser, Lucia L., 4502
Kaláshnikov, Nikolái, 2764
Kalinsky, Beatriz, 5109
Kaller, Martina, 2871, 3977
Kalter, Eliot, 1456
Kane, Stephanie C., 747
Kapfhammer, Wolfgang, 877
Kaplan, Marcos, 3829–3830
Kaplowitz, Donna Rich, 1688, 4113
Kapsoli, Wilfredo, 1036
Kapstein Lomboy, Glenda, 2603
Karepovs, Dainis, 3741
Karl, Terry Lynn, 2765
Karlsson, Weine, 3868
Kasburg, Carola, 748
Katz, Jorge M., 2072
Katz, Ricardo, 2595
Kaufer, Erich, 2250
Kauffmann Doig, Federico, 684
Kaufman, Terrence, 426
Kaufmann, Jorge, 1854
Kaulicke, Peter, 680
Kawamura, Lili Katsuco, 3804
Kay, Cristóbal, 3436, 3458, 5018
Keegan, William F., 505–507
Keifman, Saúl, 4195
Keith, Nelson W., 4750
Keith, Novella Zett, 4750
Keller, Tom, 1418
Kelley, David H., 427
Kellogg International Fellowship Program in Food Systems, 1172
Kelly, Joyce, 123
Kelly, Mary E., 3917
Kelly, Thomas C., 471
Kennedy, Nedenia, 304
Kent, Robert B., 2500–2501
Kern, Arno Alvarez, 580
Kessel, Georgina, 1419, 1457
Kessel, Juan van, 5044
Khan, Aisha, 808–809
Khavisse, Miguel, 5089
Khazanov, Anatoly, 4123
Khemradj, Roy, 4792
Kice, David A., 338
Kietz, Renate, 4961
Kiguel, Miguel A., 1126, 2073
Killion, Thomas W., 276
Kim, Kwan S., 1395
Kim, Sŭng-ho, 4045
Kimber, Clarissa, 810
Kincaid, A. Douglas, 2726

King, Jonathan, 1420
King, Philippa, 504
Kinnaird, Vivian, 933
Kinzo, Maria D'Alva Gil, 3684, 3716–3717
Kirk, John M., 3134
Kirk, Robin, 984
Kirkpatrick, Sidney, 685
Kirschbaum, Ricardo, 4198
Kirton, Mark, 4364
Kisic, Drago, 4308
Klaiber, Jeffrey, 3383
Klak, Thomas, 1810, 2251, 2474
Klaveren, Alberto van, 3817, 3831
Klein, Harald, 4314
Kleinpenning, Jan M.G., 2619
Kleinpenning, Johan, 2090
Kleymeyer, Charles D., 53, 935
Klich, Ignacio, 4207
Klinken, G.J. van, 508
Klochkovskiĭ, Lev L'vovich, 4258
Klugmann, Mark, 3459
Knapp, Gregory, 2426
Kneip, Lina Maria, 581–582
Knight, Alan, 1447
Kobayashi, Munehiro, 124
Köhler, Ulrich, 749
Kohlhepp, Gerd, 2676–2680
Kohut, Karl, 4395
Kokusai Kyōryoku Jigyōdan, 2062
Kolata, Alan L., 569
Kon, Anita, 2161
Kondracke, Morton, 3918
Koonings, Kees, 2162
Koop, Gerhard S., 2322
Kooy, Eduardo van der, 4198
Köppen, Elke, 4521
Kornblit, Ana Lía, 5110
Kornblith, Miriam, 3264–3265
Korzeniewicz, Roberto P., 2766
Kosacoff, Bernardo P., 2072
Kóssov, Igor, 3546
Kowalewski, Stephen A., 72, 80, 125
Kowarick, Lúcio, 3718
Kowii, Ariruma, 1015
Kracke, Waud, 878
Kramer, Wendy, 2344
Kranenburg, Ronald H., 4954
Krapovickas, Pedro, 557
Kras, Eva Simonsen, 1421
Krauthausen, Ciro, 3197, 4823
Krestianinov, Vladimir, 4286
Krischke, Paulo J., 3719
Kristan-Graham, Cynthia, 428
Kruijt, Dirk, 1076, 3384–3385
Kuant, Elia María, 3002

Kulakova, N.N., 811
Kuperman, Nelson, 2170
Kurtz, Donald V., 126
Kuzma, Lynn M., 4016
Kuznar, Lawrence A., 2502
Kvaternik, Eugenio, 3547
Kvist, Lars Peter, 2681
Kwant, Verónica de, 4909
Kwassteniet, Peter de, 1984
La Gamma, Alisa, 429
Laboratorio de Investigación y Desarrollo Regional (Mexico), 1439
Laborde, Miguel, 975
LABORDOC, 32
Labra, Armando, 1422
Lacalle Herrera, Luis Alberto, 3639
Laens, Silvia, 2021
Laet, Sigfried J. de, 55
Lafer, Celso, 4365
Lage, Carlos, 1694
Lagiglia, Humberto A., 558
Lagos, Maria L., 956–957, 4962
Lagos, Marta, 3447–3448, 5029
Lagos M., Luis Felipe, 1889
Lamaziere, Georges, 4366
Lambert, Joseph B., 305
Lamboglia, Ramón, 4024
Lame Chantre, Manuel Quintín, 995
Lameiras, Brigitte Boehm de, 92
Lamounier, Bolívar, 3693, 3720–3723
Lanata, José Luis, 528
Lancaster, Roger N., 750
Lancini V., Abdem Ramón, 2427, 2440
Landaburu, Carlos Augusto, 4208
Landazábal Reyes, Fernando, 3177
Langdon, E. Jean Matteson, 862, 920
Lange, Frederick W., 145
Langebaek, Carl Henrik, 541
Langer, Ana, 4503
Langley, James C., 430
Langues et cultures en Amérique espagnole coloniale: colloque international, Université de la Sorbonne nouvelle-Paris III, 22–23 novembre 1991, 59
Lanza, Gregorio, 3311–3312
Laporte, Juan Pedro, 155
Lara Enríquez, Blanca, 1485
Lara Peña, Pedro José, 4334
Lara Pinto, Gloria, 306
Laredo, Iris Mabel, 3832
Lares Monserratte, Eloy, 1718
Larios, F., 1734
Laroche, Rose Claire, 2682
Larraín, Felipe, 1295
Larraín Arroyo, Luis, 1890

Larraín Navarro, Patricio, 2596
Larrañaga, Osvaldo, 1902
Larrea, José Enrique, 4913
Larrea Maldonado, Carlos, 1176
Larrea Stacey, Eduardo, 1811
Larroulet Vignau, Cristián, 1890
Larson, Donald F., 1722
Laserna, Roberto, 3313–3316
Lasserre, Guy, 4151
Lateinamerika, Krise ohne Ende?: Beiträge zu einer Ringvorlesung im Wintersemester 1993/94 an der Leopold-Franzens-Universität Innsbruck, 2252
Latin America and the Caribbean: inventory of water resources and their use, 2253
Latin America faces the twenty-first century: reconstructing a social justice agenda, 2767
Latin America in a New World, 3833
Latin America in graphs, 1202
Latin American and Caribbean Institute for Economic and Social Planning (Santiago), 1252
Latin American Business Intelligence, 33
Latin American Centre for Economic and Social Documentation (Santiago), 1139
The Latin American debt, 3834
Latin American Economic System, 1132
Latin American horizons: a symposium at Dumbarton Oaks, 11th and 12th October 1986, 60
The Latin American left: from the fall of Allende to Perestroika, 2768
Latin American Newletters, 34
Latin American News, 35
Latin American Studies Association (LASA), 3611
Latin America's international relations and their domestic consequences: war and peace, dependency and autonomy, integration and disintegration, 3835
Latinamerika-institutet i Stockholm, 3625, 3868, 3978
Latino 3, 36
Latinskaia Amerika: problemy i tendentsii razvitiia transporta [Latin America: problems and tendencies in the growth of transportation], 1203
Latinskaia Amerika: demokratiia i vlast' [Latin America: democracy and power, 2769
Latinskaia Amerika: sobytiia i liudi; analiticheskiĭ obzor [Latin America: events and people; an analytical survey], 3836
Latorre, Adolfo Paul, 2597, 4252
Latorre, Eduardo, 1646

Lattuada, Mario J., 3566, 5111
Lau, Rubén, 2872
Laubscher, Matthias Samuel, 87
Laudy, Marion, 4025
Lauer, Wilhelm, 2428, 2465
Laurell, Ana Cristina, 2873
Lavados, Iván, 3837
Lavagna, Roberto, 2139, 2163
Lavalle, José Antonio, 686
Lawless, Robert, 4140
Lawrence, Robert Z., 1464
Lazarte Rojas, Jorge, 3317–3318
Lazarus-Black, Mindie, 812
Le Goff, Jean-Luis, 5045
Le Marin, Martinique. Office municipal de la culture, 491
Lea, Vanessa, 879
Leake, Andrew P., 2348
Leal, Juan Felipe, 2852
Leal Buitrago, Francisco, 3178–3179
Leante, César, 3135
Leão, Valdemar Carneiro, 4387
Lechuga Montenegro, Jesús, 4489
Lecturas históricas del estado de Oaxaca, 127
Lee, David, 1812
Lee, Naeyoung, 3724
Lee, Susana, 1695
Lee, Terence, 1204
Lefever, Harry G., 813
Leff, Nathaniel H., 1205
LeFranc, Elsie, 814–815, 4751
The legacy of dictatorship: political, economic and social change in Pinochet's Chile, 3460
The legacy of the disinherited: popular culture in Latin America; modernity, globalization, hybridity and authenticity, 4422
Legassa, Victoria, 1883, 3461
Legorreta, Jorge, 2874
Lehman, Howard P., 3725–3726
Leiderman, Leonardo, 1108
Leis, Héctor, 3789
Leis Romero, Raúl Alberto, 4026
Leiva, Alicia Ximena, 1891
Leiva, Fernando Ignacio, 5060
Leiva Lavalle, Patricio, 1877
Lemercinier, Geneviève, 4622
Lémuz Aguirre, Carlos, 568
Léna, Philippe, 2626
Lent, John A., 4700–4701
Lenten, Roelie, 4914
Lentz, Carola, 4867
Lentz, David, 307
LeoGrande, William M., 2942
León, Arturo, 4504

León, Eduardo de, 3641
León, Juan, 2471
León, Ninfa, 3245
León, Ramón, 4348
León, Samuel, 2896
León de Bernal, Aracelly de, 4625
León de Leal, Magdalena, 1206, 4817
León Portilla, Miguel, 68, 128
León Trujillo, Jorge, 1016
León Velasco, Juan Bernardo, 3232
Leons, Madeline Barbara, 958
Lerner, Susana, 4539
Leslie, Valeria, 3527–3528
L'Etang, Thierry, 816
Letelier L., Marta, 5046
El levantamiento indígena y la cuestión nacional, 4868
Lever D., George, 1878
Levi de Lopez, Silvana, 2388
Levin, Sander, 1465
Levine, Barry B., 3874
Levine, Daniel H., 4423
Levine, Jonathan D., 1705
Levine, Ruth E., 1532
Levitt, Kari, 1647, 1673
Levy, Santiago, 1423–1427, 1599
Lewis, Daniel K., 3548
Lewis, David E., 1648, 4752
Lewis, Gordon, 3072
Lewis, Maureen A., 1619
Lewis, Paul, 3549
Lewis, Vaughan A., 1649
Leyenarr, Ted J.J., 308
Lezama, José Luis, 2389, 4505
Lezcano, Carlos María, 3601
Li, He, 3838
Li Kam, Sui Moy, 4626
Liang, Zai, 4512
Libby, Justin H., 4137
Libonatti, Oscar, 2043
El libro blanco de la tutela: Vicepresidencia de la República, 3180
Lidzinski, Silvia, 431
Lifschitz, Edgardo, 1428
Liguori, Ana Luisa, 4506
Lima, Gerson Portela, 2180
Lima, Marcelo Coutinho de, 3752
Lima Barrios, Francisca G., 4507
Lima Júnior, Olavo Brasil de, 3727
Lima Netto, Roberto Procópio de, 2164
Linares Zapata, Luis, 2875
Lind, Amy, 3233
Lindenberg, Marc, 1553
Lindert, Paul van, 4424, 4963
Lindner, Bernardo, 1037

Lindo, Ricardo, 474
Linz, Juan J., 2750
Lipietz, Alain, 2165
Lira López, Yamile, 309
Lisboa, Marcos, 2186
Little, Marilyn, 1207
Litvak King, Jaime, 95
Liu, Peter C., 1208
Liverman, Diana, 2390
Liviatan, Nissan, 1126
Lizama, Jaime, 3462
Lizano Fait, Eduardo, 1554, 1578
Llanes, Marlen I., 4608
Llanos Vargas, Héctor, 621
Lleras Restrepo, Carlos, 3181
Llorente Sardi, María Victoria, 4273
Loaeza, Soledad, 2876–2877, 3919
Loch, Carlos, 2631
Logan, Kathleen, 2726, 4425
Lok, Rossana, 751
Loker, W.M., 2503
Londoño de la Cuesta, Juan Luis, 1776
Londoño Paredes, Julio, 2457
Long, Austin, 310
Long, Brian, 2475
Long-term trends in Latin American economic development, 1209
Longhi Zunino, Augusto, 2022
Longo, Carlos Alberto, 2166
Lopes, Francisco Lafaiete, 2167
Lopes, Lucas, 2168
López, Carlos, 1344
López, José Roberto, 1555–1556
López, Luis Ignacio, 3463
López, Luz, 5135
López, Sinesio, 3386
López Austin, Alfredo, 129, 432
López Cámara, Francisco, 2923, 4508
López Castaño, Carlos Eduardo, 622
López Castro, Gustavo, 1521
López de Alba, Federico, 1429
López Duarte, José Luis, 1503
López Echagüe, Hernán, 3550
López Esparza, Víctor Manuel, 1430
López G., Julio, 1431
López Gómez, Emilio, 1315
López Laguerre, María M., 4753
López Linage, Javier, 2429
López Luján, Leonardo, 311–312, 432
López Maya, Margarita, 3274, 4838
López Mazz, José M., 711
López Moreno R., Eduardo, 2391
López Murphy, Ricardo, 2056
López-Ocón Cabrera, Leoncio, 2504
López Oliver, Alberto, 3272

López Valadez, Gerardo, 1440
López Vigil, María, 3182
Lora, Eduardo, 1761, 1763
Lora, Guillermo, 3319–3320
Lorenzo, Fernando, 2011, 2021
Loret, John, 2553
Loret de Mola, Rafael, 2878
Lorey, David E., 1411, 1432, 3920
Loser, Claudio M., 1456
Loser, Eva, 3980
Love, Bruce, 433
Love, Michael W., 313–314
Lovell, Peggy A., 5175
Lovell, W. George, 2254, 2344–2347
Loveman, Brian, 2770
Lovera, Alberto, 1735
Low, Patrick, 1433
Lowder, Stella, 2255, 2476
Lowenthal, Abraham F., 3833, 3839–3840, 3891
Lowland Maya civilization in the eighth century A.D.: a symposium at Dumbarton Oaks, 7th and 8th October 1989, 130
Loyola Díaz, Rafael, 3910
Loza, Martha, 4915
Lozano, Rafael, 4503
Lozano, Wilfredo, 3056, 4754
Lozano Ascencio, Fernando, 2408, 4556–4557
Lozano Castro, Alfredo, 2477
Lucchini, Cristina, 5112
Lucena, Tibisay, 3273, 4851
Luciak, Ilya A., 4627
Lücker, Reinhold, 2649, 2651
Lucrèce, André, 491
Lugo, Félix, 3596
Luiselli, Cassio, 1453
Luján, Carlos, 4327
Luna, Luis Eduardo, 921
Lundahl, Mats, 1641, 1681
Lungo Uclés, Mario, 2966, 4027, 4640
Lustig, Nora, 1434–1436, 1464, 3921
Lutz, Christopher H., 2344–2345
Lutz, Ellen L., 2879, 3922
Luyando, Elda, 2386
Lynch, Edward A., 2771
Lyndon B. Johnson School of Public Affairs. U.S.-Mexican Policy Studies Program, 1489
Lyra, Heitor, 4367
Mabileau, Albert, 4151
Mac Gregor, Felipe E., 4452
MacDonald, Joan, 5047
Mace, Gordon, 3841
Macedo Martínez, Javier, 1408
MacEwan, Arthur, 2880
Machado, Claudio Caetano, 5155

Machado, Eduardo Paes, 2699
Machado, João Bosco M., 1210, 2169, 4389
Machado, Leda Maria Vieira, 5176
Machado, Lilia Cheuiche, 582
Machado, Roberto Crivano, 2170
Machado C., Absalón, 1777
Machicado, Fernando, 4953
Machicote, Eduardo, 4224
Machovec, Frank M., 1211
Macias Goytia, Angelina, 315
Mack, Carlos, 4209
Mackinlay, Horacio, 1370
Maclachlan, Morgan, 505
MacRae, Edward, 3728
Macroeconomía de los flujos de capital en Colombia y América Latina, 1778
Madariaga, Marta, 2540
Mader, Ron, 7
Madrid, Roberto, 1877
Madrid. Concejalía de Cultura, 682
Madrigal Carmona, Alfredo, 4536
Madrigal Pana, Johnny, 4628
Magalhães, João Paulo de Almeida, 2170
Magallanes, Manuel Vicente, 3247, 3266, 3272
Magaña, Edmundo, 817
Maggiora, Ferruccio, 1212
Magnitud de la pobreza en América Latina en los años ochenta, 1213
La magnitud y la integración de la administración pública en el Estado de México: ángulo de interpretación cuantificable, 2881
Magnusson, Åke, 3868
Maia, Sylvia M. dos Reis, 5177
Maiguashca, Juan, 1813
Maihold, Günther, 1138
Maimon, Dália, 2683
Maingón, Thais, 4844
Maingot, Anthony P., 4085–4086, 4755
Maino, Valeria, 2598
Mainwaring, Scott, 2763, 3729–3731
Major changes and crisis: the impact on women in Latin America and the Caribbean, 1214
Majul, Luis, 3551–3552
Malacrida, María Gabriela, 3526
Malamud Goti, Jaime E., 3321, 3842
Malaret, Antonio Emilio, 2539
Maldifassi, José O., 1215
Maldonado, Olga, 2349
Maldonado Bautista, Samuel, 2882
Maldonado C., Rubén, 316
Maldonado Prieto, Carlos, 4253
Malengreau, Jacques, 1038
Mallat, Gustavo, 1600

Mallet-Guy Guerra, Sinclair, 2684
Malloy, James M., 3219, 3322
Malmström, Vincent H., 2392
Malpass, Michael A., 687, 694
Malta, I.M., 574
Mamalakis, Markos J., 1892
Mamani, Mauricio, 962
Mañana Plasencio, Miguel, 4482
Mandle, Jay R., 1650
Manitzas, Elena S., 3387
Manning, Roswitha, 752
Manrique, Nelson, 1039
Mansilla, H.C.F., 2772, 3323
Mansutti Rodríguez, Alexander, 902
Mantilla, Lucía, 4523
Manual de planeamiento andino comunitario: el PAC en el Ecuador, 1017
Manzanilla, Linda, 182, 317–319
Manzetti, Luigi, 1216, 3553–3555
Manzo, Kate, 2773
La maquila en Guatemala, 4629
Las maquiladoras: ajuste estructural y desarrollo regional, 1437
Maquiladoras, su estructura y operación, 1438
Marcadent, Philippe, 1439
Marchi, Jane G., 4004
Marcus, Joyce, 131, 266, 434–435
Los mares nos unen: México y la Cuenca del Pacífico, 3923
Margain, Hugo B., 3924
Margheritis, Ana, 4210
Margolis, Maxine L., 5178
Margulis, Sérgio, 2257
María y Campos, Mauricio de, 1440
Mariangel Candia, Walter B., 2599
Mariano, Ricardo, 3732
Marie Jean Robert, Joseph, 3082
Marimán, José, 976
Marín, Gustavo, 1914
Marín, Raúl, 3136
Marín Ramírez, Rodrigo, 2458
Marini, Onildo J., 2685
Mariño Ferro, Xosé Ramón, 959
Marinović Pino, Milan, 5048
Marion Singer, Marie-Odile, 753
Mariz, Cecília Loreto, 5179
Marks, Siegfried, 1217
Marques, Maria Inez Medeiros, 2669
Márquez, Gustavo, 1736–1737, 1814
Márquez de la Plata Yrarrázaval, Alfonso, 3464
Márquez de Pérez, Amelia, 4573
Márquez Morfín, Lourdes, 320
Marras, Sergio, 3465

Marray, Michael, 2256
Marrero, Leví, 1696
Marrero, Rosita, 3064
Marroquín, Enrique, 4509
Marroquín, Manfredo, 4001
Marshall R., Enrique, 1893
Marshall Silva, Jorge, 1815, 1894–1895
Martel, Jorge Hugo I., 2707
Martin, JoAnn, 4510
Martín, Juan, 1252
Martin, Philip L., 3901
Martín del Campo, Antonio, 1218
Martine, George, 2686–2687, 2693, 2715
Martínez, Carolina, 4539
Martínez, Cruz, 662
Martínez, Daniel, 1557
Martínez, Elena, 2554
Martínez, Gabriel, 977
Martínez, Guillermo E., 1880, 3433, 5002
Martínez, Javier, 5049
Martínez, José, 2909
Martínez, José de Jesús, 4028
Martínez, José María, 1474
Martínez, Judith, 1791
Martínez, Milton, 4029
Martínez, Philip R., 1629
Martínez, Regino, 4756
Martínez Almazán, Raúl, 1441
Martínez Assad, Carlos R., 2360
Martínez Blanco, Gerardo, 4030
Martínez Borrego, Estela, 4511
Martínez Borrero, Juan, 1011–1013
Martínez Cuenca, Alejandro, 3003
Martínez de Hoz, José Alfredo, 2074
Martínez Heredia, Fernando, 3137
Martínez Legorreta, Omar, 1413
Martínez Moya, Arturo, 1660
Martínez Nogueira, Roberto, 5126
Martínez Olavarría, Leopoldo, 3274
Martínez Olivé, Alba, 4553
Martínez Portilla, Isabel María, 4630
Martínez Rocha, Abelino, 4631
Martínez Rodríguez, Antonia, 2883
Martínez Valle, Luciano, 1816–1817
Martínez Velasco, Germán, 4031
Martini, Carlos, 3600–3601
Martinic B., Mateo, 2600
Martinique. Conseil régional, 2308
Martins, Luciano, 4368
Martner, Gonzalo, 1099, 3419
Martos López, Luis Alberto, 321
Martz, John D., 3183
Marúm Espinosa, Elia, 1442
Marzal, Manuel María, 61
Maskrey, Andrew, 2505, 4933

Massad, Carlos, 1084, 1273
Massenmedien in Lateinamerika, 2774
Massey, Douglas S., 4512
Massolo, Alejandra, 1443, 4524
Masson Meiss, Luis, 2429
Massone M., Mauricio, 606
Massuh, Héctor Daniel, 3556
Mastro V., Marco del, 4916
Masuda, Yoshio, 2430
Mata García, Bernardino, 1315
Mathéy, Kosta, 4632
Matos González, Ramiro, 3055
Matos Mendieta, Ramiro, 688
Matos Moctezuma, Eduardo, 68, 132, 225, 322, 395, 436
Matos Moquete, Manuel, 509
Matsumoto, Ryozo, 859
Matta, Javier Eduardo, 4254
Matta, Paulina, 5015
Matta, Roberto da, 5145, 5183
Máttar, Jorge, 1568
Mattini, Luis, 3557
Mattos, Carlos de Meira, 4369
Mattos Cintrón, Wilfredo, 3084
Mauceri, Philip, 3388
Mauro, Amalia, 4869
Maxfield, Sylvia, 2843, 2884
The Mayan People and Human Rights, 2981
Maybury-Lewis, David, 860
Mayer, Karl Herbert, 437
Maynard, Kent, 1018
Máynez, Pilar, 438
Mayor Mora, Alberto, 4818
Mayorga, René Antonio, 3324–3327
Maza Zavala, Domingo Felipe, 1738
Mc Phail, Elsie, 4558
McAfee, Kathy, 1651
McAnany, Patricia A., 133
McCafferty, Geoffrey G., 323
McCafferty, Sharisse D., 323
McCalla, Jocelyn, 4141
McCallum, Cecilia, 880
McCallum, Malcolm E., 2441
McCaughan, Ed, 2767
McCleary, William A., 1779
McClintock, Cynthia, 3389–3390
McClung de Tapia, Emily, 134, 385, 439
McCommon, Carolyn S., 4633
McCormick, Gordon H., 3391
McCort, Robert F., 4108
McCoy, Alfred W., 3875
McCoy, Jennifer L., 3725
McCoy, Terry L., 4087, 4757
McDonald, John W., 5161
McDougall, Derek, 4152

McGuire, James W., 3558
McGurk, Russell, 3843
McInnes, D. Keith, 1592
McIntyre, Arnold M., 1674
McLeod, Darryl, 1453
McMahon, Gary, 1219
McNelis, Paul D., 1896
McPherson, Everton S.P., 4758
Meassick, Mark A, 4646
Medel R., Julia, 5050
The media and the 1994 federal elections in Mexico: a content analysis of television news coverage of the political parties and presidential candidates, 2885
Medina, Abraham, 1661
Medina, Juvenal, 2514
Medina, Laurie Kroshus, 4634
Medina, Manuel, 3844
Medina Cano, Federico, 4426
Medina Gallego, Carlos, 3184
Medio ambiente, seguridad y cooperación regional en el Caribe, 4088
Medios, democracia y fines, 2886
Meditz, Sandra W., 3035
Medrano, Sonia, 324
Meel, Peter, 3091–3092
Mege R., Pedro, 978
Meggers, Betty J., 583–584, 643
Meinardus, Marc, 1841
Meio ambiente: aspectos técnicos e econômicos, 2257
Meissner, Frank, 1220
Mejía, Cristina, 4960
Mejía, José Manuel, 1953
Mejía Barquera, Fernando, 4513
Mejía Navarrete, Julio Víctor, 3392
Mejía Pérez Campos, Elizabeth, 231, 325
Meléndez, Edwin, 1642
Meléndez, Guillermo, 4635–4636
Melgar, Ricardo, 3393
Melhado, Oscar, 1558
Meliá, Bartomeu, 929
Mella Latorre, Alejandro, 3602
Mellén Blanco, Francisco, 2601
Meller, Patricio, 1288, 1897–1898
Mello, Celso Albuquerque, 4370
Mello, João Baptista Ferreira de, 2688
Mello, José Octávio de Arruda, 4371
Mello, Mauro Pereira de, 2689
Mello e Silva, Alexandra de, 4372
Melo M., Cándida, 4427
Méluzin, Sylvia, 440–441
Melvin, Michael, 1985
Memoria: energía, medio ambiente y desarrollo sustentable, 1444

Mendelson, Johanna S.R., 3819
Mendes, Ana Gláucia, 2171
Mendes, Armando Dias, 2172
Mendes, Chico, 5160
Mendes Diz, Ana M., 5110
Méndez, José Luis, 4759
Méndez, Roberto N., 1636, 3447–3448, 3466, 5029
Méndez Asensio, Luis, 3925
Méndez Dávila, Lionel, 4032
Méndez de Herrera, Genoveva, 1818
Méndez de Pérez, Betty, 4845
Méndez Valenzuela, Juan, 3328
Mendiburu Mendocilla, Armando, 4942
Mendizábal Losack, Emilio, 1040
Mendizábal P., Ana Beatriz, 1560
Mendonça, Rosane Silva Pinto de, 2108
Mendoza, Iván, 4917
Mendoza B., Rodolfo E., 2354
Mendoza Fernández, María Teresa, 1445
Mendoza Fletes, Orlando, 1630
Mendoza Morales, Alberto, 2459
Meneguello, Rachel, 3723, 3733
Menem, Carlos Saúl, 4211–4212
Menéndez-Carrión, Amparo, 3212
Menéndez D'Avila, Lionel, 3022
Meneses C., Aldo, 3467, 5051
Meneses Ciuffardi, Emilio, 4255
Menjívar, Rafael, 1559
El menor trabajador asalariado en la ciudad de Cochabamba, 4964
Mensbrugghe, Dominique van der, 1178
Mentz, Brígida von, 3926
Mercado, Silvia, 3503
Mercado de trabalho na Grande São Paulo, 2173
El Mercado de Valores: la política económica de Mexico, 1946–1994, 37
Mercado García, Alfonso, 1087
Mercado Jarrín, Edgardo, 4179
Mercau, Raúl, 2071
Mercosur: claroscuro de una integración; ciclo de conferencias realizadas en la Facultad de Ciencias Económicas y Administración, 2023
Mergal, Margarita, 4760
Merin, B.M., 2769
Merino Huerta, Mauricio, 2883
Merklen, Denis, 5113
Merrill, Tim, 3004
Merrill, William L., 743
Mertins, Günter, 2218
Mesa Gisbert, Carlos D., 3329, 3344
Mesa-Lago, Carmelo, 1221–1224, 1446, 1697–1700, 1819, 3138–3140, 4111

Mesa Redonda Hacía una Política Marítima Nacional, *Lima, 1990*, 4307
Mesoamerican elites: an archaeological assessment, 135
Messina, Ernesto G., 3927
Messner, Dirk, 1899
Metropolitan Museum of Art (New York), 326
Metz, Allan, 2887
Meurs, Mieke, 1689, 1701, 4716
The Mexican petroleum industry in the twentieth century, 1447
The Mexican-U.S. border region and the Free Trade Agreement, 1448, 3928
Mexico. Dirección General de Empleo, 1362
Mexico. Poder Ejecutivo Federal, 1449
Mexico. Secretaría de Comercio y Fomento Industrial, 3931
Mexico. Secretaría de Energía, Minas e Industria Paraestatal, 1417
Mexico. Secretaría de Relaciones Exteriores, 3929, 3932, 3950
Mexico. Secretaría del Trabajo y Previsión Social. Subsecretaría B., 1377
Mexico, 1450, 4514
México: auge, crisis y ajuste, 1451
Mexico: dilemmas of transition, 2888
México en Centroamérica: expediente de documentos fundamentales, 1979–1986, 3930
México en el comercio internacional, 3931
México en la década de los ochenta: la modernización en cifras, 1452
México en las Naciones Unidas, 3932
México-Estados Unidos: la interacción macroeconómica, 1453
México-Estados Unidos: energía y medio ambiente, 3933
Mexico: in search of security, 3934
México: informe sobre la crisis, 1982–1986, 1454
México: perspectivas de una economía abierta, 1455
Mexico: splendors of thirty centuries, 326
Mexico: the strategy to achieve sustained economic growth, 1456
Mexico: torture with impunity, 2889
México y el Tratado Trilateral de Libre Comercio: impacto sectorial, 1457
Meyer, Carrie A., 1820–1821
Meyer, Henry O.A., 2441
Meyer, Lorenzo, 2890, 3935
Meyer-Arendt, Klaus J., 2258
Meyers de Ortiz, Carol, 4515
Meza Ocampo, Tobías, 2334–2335
Miceli, Sergio, 4373

Michalopoulos, Constantine, 1563
Michel López, Marcos, 568
Michel Vega, Javier, 1398
Michelet, Dominique, 2396
Micheli T., Jordy, 3887
Midaglia, Carmen, 5136
Middleton, Alan, 1822
Middleton, William D., 262
Mielnik, Otávio, 2156
Mier y Terán, Marta, 4516
Mieres, Pablo, 3628, 3642
Mignone, Emilio Fermín, 3559
Miguens, José Enrique, 3509
Milbrath, Susan, 510
Miles, Ann, 1019
Milhou, Alain, 59
Milla Batres, Carlos, 667
Millar, René, 1866
Millas, Orlando, 3468
Miller, Errol, 4720
Miller, Eurico T., 573
Miller, Mary Ellen, 136
Millet Cámara, Luis, 327
Millett, Richard, 4002
Millones, Luis, 689
Millor, Pablo, 3650
Mills, Don, 4153
Mills, Frank L., 4761
Minà, Gianni, 3110
Minella, Ary Cesar, 2174
Minello, Nelson, 2816
Mingo M., Orlando, 2602
Minian, Isaac, 1416
Minoliti, Claudia, 3560
Mintz, Sidney W., 818
Minujin Z., Alberto, 5108, 5114
Mirambell Silva, Lorena, 179, 231
Miranda, Anibal, 4292
Miranda, Carlos R., 3603
Miranda, Edson, 3734
Miranda Francisco, Olivia, 4762
Miranda R., Ernesto, 1225
Miras, Claude de, 1823
Mires, Alfredo, 681, 1059
Misetic Yurac, Vladimir, 2603
Misión de Cooperación Técnica Holandesa (La Paz), 1995
Mitchell, Christopher, 3807
Mitchell, William P., 1035
Miyar Bolio, María Teresa, 4124
Mizala, Alejandra, 1900
Moberg, Mark, 4637
Mobilising international investment for Latin America, 1226
Moctezuma Barragán, Javier, 1502

Modellfall Chile?: ein Jahr nach dem demokratischen Neuanfang, 3469
Módena, María Eugenia, 4517
Moguel Cos, María Antonieta, 328
Mohar Betancourt, Luz María, 442
Moïse, Claude, 3065
Moisés, José Alvaro, 3735, 4382
Molano, Alfredo, 4819
Moldau, Juan Hersztajn, 2175
Molendijk, Mathilde, 4763
Molina, Manuel Jesús, 644
Molina Chocano, Guillermo, 2992
Molina Otárola, Raúl, 979
Molina Piñeiro, Luis, 2834
Molina Vega, José Enrique, 3267–3268
Molinar Horcasitas, Juan, 2891–2892
Molinié Fioravanti, Antoinette, 937, 1041
Moltedo Castaña, Cecilia, 5024
Momsen, Janet Henshall, 933
Monasterio, M., 2448
Moncada Sánchez, José, 1824
Moncayo, Juan Fernando, 2468
Monclaire, Stéphane, 3736
Moneta, Carlos Juan, 3869
Money doctors, foreign debts, and economic reforms in Latin America from the 1890s to the present, 1227
Monge, Luis Alberto, 4033
Monge Oviedo, Rodolfo, 4606
Monreal González, Pedro M., 4125–4126
Montañez G., Gustavo, 2460
Montaño García, Jaime, 4965
Montaño Hirose, Luis, 4488
Monte Albán: estudios recientes, 329
Montecino Aguirre, Sonia, 5052–5053, 5077
Montecinos, Camila, 2259
Monteiro, Brandão, 3737
Montenegro, Carlos, 3330
Montenegro, Walter, 4234
Montes, Noemí, 1950
Montes, Segundo, 4638
Montesino, Estervino, 4116
Montiel H., Yolanda, 4518
Montiel Masís, Nancy, 1571
Montilla, M., 2448
Montis, Malena de, 4646
Montoro, André Franco, 3738
Montuschi, Luisa, 2075
Mora, Hernán, 4918
Mora, Jorge A., 4639
Mora Lomelí, Raúl H., 1412
Moraes, João Quartim de, 3739
Moral, Octavio del, 3023
Morales, Anamaria, 5180
Morales, Cecile B. de, 2527
Morales, David, 378
Morales, Eduardo, 5054
Morales, Josefina, 1393
Morales, Juan Antonio, 1986–1991
Morales, Róger, 2320
Morales, Rolando, 1791
Morales Carazo, Jaime, 4034
Morales Chocano, Daniel, 667
Morales Gómez, Jorge, 996
Morales Mena, Faustino, 2442
Morales Ortega, Ninette, 4640
Morales Paúl, Isidro, 3247, 4335
Morales Solá, Joaquín, 3561
Morales Yordán, Jorge, 3041
Moran, Emilio F., 2690, 5181
Morán de Ferrer, Rhina, 4641
Morandé Lavín, José, 4256
Moreano, Alejandro, 3234
Moreira, Marcílio Marques, 2176
Morelli Pando, Jorge, 4309
Morelos García, Noel, 137, 312, 390
Moreno, Dario, 4035
Moreno, Dennis, 4735
Moreno, Lorenzo, 4519
Moreno, Rafael, 3845
Moreno González, Leonardo, 623
Moreno Ocampo, Luis Gabriel, 5115
Moreno Ospina, Carlos, 4815
Moreno V., José Luis, 1508
Moreno Yáñez, Segundo, 642, 1020, 4870
Morgan, María de la Luz, 4919
Morgan, Martha I., 4642
Morici, Peter, 1405, 1458, 3846, 3936
Morimoto, Amelia, 4920
Morínigo, José Nicolás, 3604
Morley, Morris H., 2787, 4036
Morley, Samuel A., 2143
Morlino, Leonardo, 2775
Mörner, Magnus, 819
Moro, Celia O., 2541
Morote Best, Efraín, 1042
Morris, Arthur, 2555
Morris, James A., 3116
Morris, Michael A., 4180
Morris, Stephen D., 2821, 2893–2894
Morris von Bennewitz, Raúl, 980
Morrison, Andrew R., 1608
Morrison, Nancy, 1220
La mortalidad en la niñez: Centroamérica, Panamá, y Belice, 4643
Morton, Colleen Shores, 1459
Moscoso, Francisco, 472, 592
Moseley, Michael Edward, 707
Mosquera Aguilar, Antonio, 4520
Moss, Ambler H., 1327

Mossbrucker, Harald, 1043
Mostajo de Muente, Patricia, 4944
Mothes, Patricia, 2480
Motoyama, Shozo, 4374
Motta, Raúl Domingo, 2265
Moulian, Tomás, 3470, 5032
Moura, Alexandrina Saldanha Sobreira de, 2177
Moura, Gerson, 4375–4376
Moura, Hélio Augusto, 2180
Moura, José Tupy Caldas de, 2178
Mourik, Jan van, 2314
Movimento Nacional de Meninos e Meninas de Rua (Brazil), 5210
Movimentos sociais no campo, 3740
Movimiento Bolivia Libre. Congreso Nacional, *1st, La Paz, 1991*, 3331
Movimiento de Adolescentes y Niños Trabajadores Hijos de Obreros Cristianos (Peru), 4937
Movimiento de la Izquierda Revolucionaria (Bolivia). Secretaría Nacional Obrera, 3332
El movimiento popular de mujeres como respuesta a la crisis, 4921
Movimiento Revolucionario Túpac Amaru (Peru). Comité Central, 3365
Movimientos sociales: elementos para una relectura, 4922
Movimientos sociales en México, 1968–1987, 4521
Movimientos sociales en México durante la década de los 80, 4522
Mower, Roland D., 2260
Moya, Alba, 2471
Moyano Berrios, Eduardo, 1901
Mozzillo, Elizabeth Oster, 58
Mueller, Charles Curt, 2691–2693
Mueller, Raymond G., 302
La mujer de la tercera edad en el Ecuador, 4871
Mujer en el desarrollo: balance y propuestas, 4923
Mujer, género y desarrollo local urbano, 4966
La mujer jalisciense: clase, género y generación, 4523
Mujer, trabajo y reproducción humana en tres contextos urbanos de Bolivia, 1986–1987, 4967
La mujer y la ciudad: propuestas de políticas a los gobiernos locales, 3394
Las mujeres en la imaginación colectiva: una historia de discriminación y resistencias, 5116
Mujeres latinoamericanas en cifras: avances de investigación, 4428–4429

Mujeres y ciudades: participación social, vivienda y vida cotidiana, 4524
Mujeres y empleo en Ciudad de Guatemala, 1609
Mujica, María Eugenia, 1954
Mujica, Patricio, 1902
Mujica, Rodrigo, 1903
Mujica B., Elias, 666
Mujica Miranda, Estela, 1228
Mulher e políticas públicas, 5182
Muller, Frits, 4430
Muller, Keith D., 2642, 2694–2695
Muller-Karger, Frank, 2443
Müller Rojas, Alberto A., 3269
Mumme, Stephen P., 2895
Mummert, Gail, 4538
Munck, Gerardo L., 2776, 3471
Mundigo, Axel, 4431
Mundlak, Yair, 1863
El mundo ceremonial andino, 689
Mungaray Lagarda, Alejandro, 1460
Los municipios de las fronteras de México, 1461
Munneke, Harold F., 3077
Muñoz, Heraldo, 3814, 4246
Muñoz, Ismael, 1949
Muñoz, Oscar, 1904
Muñoz Dálbora, Adriana, 4432, 5055
Muñoz García, Humberto, 4537
Muñoz Gomá, Oscar, 1870, 5056
Muñoz L., Carlos A., 4846
Muñoz Reyes, Jorge, 2528
Muñoz Salazar, Patricia, 5063
Muñoz V., Cecilia, 4820
Muñoz V., Mario, 5057
Muñoz Vicuña, Elías, 3235
Munroe, Trevor, 4154
Murakami, Yusuke, 3395
Muraro, Heriberto, 5117
Muratorio, Blanca, 1021
Murcio, F. Javier, 1462
Murillo Castaño, Gabriel, 2777, 4273
Murillo Lozoya, Irma Laura, 1480
Muro, Víctor Gabriel, 4490
Murphy, Joseph M., 820
Murphy, Robert F., 866
Murphy, Vincent, 330
Murra, John, 962
Murray, Douglas L., 1610
Murray, Gerald F., 4764
Musacchio, Andrés, 4213
Muscar Benasayag, Eduardo F., 2556
Muschietti, Ulises M., 4214
Museo Arqueológico Nacional (Madrid), 68
Museo Casa de Colón (Canary Islands), 662

Museo Chileno de Arte Precolombino, 605, 974
Museo de América (Madrid), 658
Museo del Oro (Colombia), 636
Museo di Sant'Agostino (Genova, Italy), 636
Museo Nacional de Antropología (Mexico), 244–245, 377–378
Museo Nacional de Arqueología y Etnología (Guatemala), 155
Museo Regional de Oaxaca (Mexico), 391
Museu Paraense Emílio Goeldi (Brazil), 2626
Musicant, Ivan, 4037
Musset, Alain, 2393
Mustapic, Ana M., 3562–3563
Mychaszula, Sonia M., 944
Na contracorrente da história: documentos da Liga Comunista Internacionalista, 1930–1933, 3741
Na corda bamba: doze estudos sobre a cultura da inflação, 5183
Nabel Pérez, Blas, 511
Nabuco, Maria Regina, 2171
Nachtigall, Horst, 754
Naím, Moisés, 1739
Nakamura, Seiichi, 331–334
Nance, C. Roger, 335
Napolitano, Emanuela, 922
Naranco, Rafael del, 3253
Naranjo, Marcelo Fernando, 1011, 1013
Nárez, Jesús, 336–337
Narganes Storde, Yvonne M., 490
Narro R., José, 1463, 1502
Nash, June C., 725, 964
Nasi L., Carlo, 4274
National Coalition for Haitian Refugees, 1657
National Geographic Society (US), 165
National Security Archive index, 38
Nations, James D., 2317
Nava Hernández, Eduardo, 4525
Navarrete, Emma Liliana, 4526
Navarrete Talavera, Ela, 4038
Navarro, Jorge, 1620
Navarro, Juan Carlos, 3273, 4851
Navarro L., Daniel, 1479
Necochea, Andrés, 5058
Nef, Jorge, 2778
Negociación y conflicto laboral en México, 2896
Neild, Rachel, 3605
Neilson, James, 3564
Neiman, Guillermo, 5100
Neira Cuadra, Oscar, 1631
Nelen, J.M., 4765
Nelson, Ben A., 338

Nentwig Silva, Bárbara-Christine, 2696
Neri, Anita Liberalesso, 5184
Neri, Marcelo Cortes, 2179
Netherlands. Ministerie van Buitenlandse Zaken. Voorlichtingsdienst Ontwikkelingssamenwerking, 1292
Neto, Abdon Portela N., 2180
Neupert, Ricardo F., 2697
Neurath, Johannes, 339
Neves, Maria Manuela Renha de Novis, 3742
New theories on the ancient Maya, 138
Newton, Adrian C., 2272
Ni héroes ni villanas: género e informalidad urbana en Centroamérica, 1559
Nicaragua: a country study, 3004
Nichaus, Bernd, 2953
Nicholas, Linda M., 101–102, 262–265
Nichols, Deborah L., 228, 340
Nicholson, Henry B., 341
Nicolau, Jairo César Marconi, 3727, 3743
Nicolini, José Luis, 2076
Niekerk, N.C.M. van, 3340, 4960
Niekerk, Nicolaas G.W., 1955
Niemeyer Fernández, Hans, 534, 600–602
Nierenberg, Claudia, 3843
Nieto, Elba María, 4644
Nieto C., Carlos, 2478
Nieto Calleja, Rosalba, 346
Nieuwolt, Simon, 2479
Nino, Carlos Santiago, 5118
Niño Rodríguez, Antonio, 3816
Nippon Brazil Kōryūshi = História das relações nipo-Brasileiras, 4377
Njaim, Humberto, 3270
Nochteff, Hugo, 3502
Noé Pino, Hugo, 1618
Nogales Taborga, Ivonne, 4976
Nogués, Julio, 1229, 2077–2078
Nohlen, Dieter, 2779, 3497
Nolasco Armas, Margarita, 1461
Nolte, Detlef, 3469
Nolte Maldonado, Rosa María Josefa, 1044
Noordegraaf, Wim, 2303
Norden, Deborah L., 2740
Norr, Lynette, 464
North American free trade: assessing the impact, 1464
North American Free Trade Agreement: U.S.-Mexican trade and investment data: report to the Honorable Richard A. Gephardt, Majority Leader, and to the Honorable Sander Levin, House of Representatives, 1465
North-South Institute (Canada), 1910
Norvell, Scott, 8–9
Notario Castro, Nelson, 4039

Notas Mesoamericanas, 342–343
Novaes, Adauto, 5192
Novaes, Carlos Alberto Marques, 3690, 3744
Novo, Evlyn M.L.M., 2702
Novoa Goicochea, Daniel I., 2506
Novoa Magallanes, César, 139
Noyola Vázquez, Juan F., 1230
Nuestra palabra: el fraude electoral de 1991 y la participación ciudadana en la lucha por la democracia, 2897
La nueva era de la industria automotriz en México: cambio tecnológico, organizacional y en las estructuras de control, 1466
La nueva etapa de la integración regional, 1231
Una nueva etapa en el proceso económico latinoamericano, 2024
Una nueva lectura: género en el desarrollo, 4433
El nuevo Estado mexicano, 1467, 4527
El nuevo rostro de Costa Rica, 4645
Nugent, Stephen L., 2698, 5185
Nun, José, 3565–3566
Nunes, Maria Cavaliari, 3776
Núñez, Arturo, 2898
Núñez Atencio, Lautaro, 542, 604, 607–608
Núñez de la Peña, Francisco J., 1468
Núñez González, Oscar, 4528
Núñez Jiménez, Antonio, 2304
Núñez Pinto, Jorge, 2604
Núñez Rebaza, Lucy, 1045
Núñez Regueiro, Víctor A., 559
Núñez-Sandoval, Oscar A., 1595, 1605
Nunn, Frederick M., 2780
Nunn, George E., 2261
Nutini, Hugo G., 755
Nweihed, Kaldone G., 2305
Nylen, William R., 3745
OAS. *See* Organization of American States
Obando Arbulú, Enrique, 4181, 4310
Obando y Bravo, Miguel, 3005
Oberstar, James L., 4139
Obregón T., Liliana, 4274
O'Brien, Philip J., 1232
Ocampo, José Antonio, 1632
Ocampos, Lorraine, 2003
Ochoa García, Carlos, 4040
Odio Orozco, Eduardo, 473
O'Donnell, Guillermo A., 2763, 2781–2782
O'Donnell, Mario, 4215
OECD Statistical Compendium, 39
O'Hara, Sarah L., 140, 2394
Ohio University. Center for International Studies, 4045

Ojeda, Mario, 3937
Ojeda M., Heber, 327
Ojeda Segovia, Lautaro, 1786
Oka, Christine K., 10
O'Kane, Rosemary H.T., 2783
O'Kane, Trish, 3002
Olavarría, Jorge, 4336
O'Leary, Mick, 11
Oleas Montalvo, Julio, 1825
Olinto Camacho, Oscar, 4847
Olivares, Juan C., 2609
Oliveira, Adélia Engrácia de, 2626
Oliveira, Ariovaldo Umbelino de, 3746
Oliveira, Carlos Alberto Pereira, 3737
Oliveira, Carlos Tavares de, 2181
Oliveira, Elizabeth Homem, 2699
Oliveira, Fabrício Augusto de, 2133
Oliveira, Juarez de Castro, 5168
Oliveira, Márcio de, 2700
Oliveira, Naia, 2666
Oliveira, Odete Maria de, 3747
Oliveira, Omar Souki, 5204
Oliveira, Orlandina de, 4434
Oliven, Ruben George, 5186
Oliver, José R., 903
Olivera, Mercedes, 4646
Olivera Cárdenas, Luis, 4924
Oliveri, Ernest J., 3847
Oliveri López, Angel M., 4216
Olivo, Francisco, 3271
Olivo Chacín, Beatriz, 2444
Olivos M., Soledad, 5050
Olmos, Alejandro, 3567
Olsen, Organ, 3776
Olvera L., Guillermo, 2355
Olwig, Karen Fog, 821–822
O'Mack, Scott, 141
Ondarts, Guillermo, 1233
Ondarza Linares, Franz, 4235
Onufriev, IU. G., 1073
Ōnuki, Yoshio, 689
Oostindie, Gert, 3078–3079
Opazo Bernales, Andrés, 4647, 5059
Opiniones y Análisis, 4236
OPTIM Corporation, 51
Oranje, Joost, 3086
Ordóñez, Hernán, 615
Orellana, Mario, 609–610
Organización de Estados Iberoamericanos para la Educación, la Ciencia y la Cultura, 4810
Organization for Economic Cooperation and Development, 27, 1080, 1134, 1226, 1425
Organization for Tropical Studies (Costa Rica), 2339

Organization of American States (OAS), 1146, 2294, 2300, 2338, 3133
Organization of Eastern Caribbean States. Natural Resource Management Unit, 2300
Oriol, Jacques, 823
Orlove, Benjamin, 2507
Orme, William A., Jr, 1469
Oropeza, Luis José, 3272
Orozco, Concepción, 4041
Orozco, Juan Luis, 4529
Orozco, Román, 3141
Orozco, Víctor, 2872
Orozco Abad, Iván, 3185
Orozco Alvarado, Javier, 1470
Orozco Vílchez, Lorena, 2336
Orrego Corzo, Miguel, 344
Orrego Vicuña, Francisco, 4247, 4257
Ortega, Hugo, 1904
Ortega, Marvin, 4648
Ortega Benayas, María Angeles, 2284
Ortega Frei, Eugenio, 3472
Ortellado Jiménez, Hilario, 3606
Ortiz, Agustín, 207
Ortiz, Jorge, 4925
Ortiz, Renato, 5187
Ortiz-Colón, Reinaldo, 4785
Ortiz Crespo, Gonzalo, 3226
Ortiz Cruz, Etelberto, 1471
Ortiz de Montellano, Bernard, 142
Ortiz de Zevallos M., Felipe, 1956
Ortiz Martínez, Guillermo, 1472
Ortiz Pérez, Irene, 4530
Ortiz Sarmiento, Carlos Miguel, 4821
Ortiz-Troncoso, Omar R., 560
Ortloff, Charles R., 569
Ortuño Cos, Francisco, 345
Orvig, Helen, 4921
Osella, Vilma, 3536
Osimani, Rosa, 2021
Osobennosti razvitiia chiliĭskoĭ ėkonomiki i perspektivy rossiĭsko-chiliĭskogo delovogo sotrudnichestva. [Features of the development of the Chilean economy and perspectives on Russian-Chilean economic relations.], 4258
Osorio, Jaime, 4435
Osorio, Marta, 4277
Ossio A., Juan M., 1046
Ostiguy, Pierre, 2079
Ostrowitz, Judith, 443
Otero, Hernán, 2557
Ouweneel, Arij, 57, 2395, 2899
Ovalles, Eduardo, 3530
Pablo, Juan Carlos de, 2060
Pabón Tarantino, Elvyra Elena, 3186

Pacheco, Juan Carlos, 2460
Pacheco Méndez, Guadalupe, 2900–2902
Pacheco Méndez, Teresa, 1473
Pachón C., Ximena, 4820
Pacificar la paz: lo que no se ha negociado en los acuerdos de paz, 3187
Packenham, Robert A., 3848
Packer, George, 3066
Padoch, Christine, 2233
Paerregaard, Karsten, 1047
Pagell, Ruth A., 12
Paillés Hernández, María de la Cruz, 346
Painter, Michael, 1048
El paisaje volcánico de la sierra ecuatoriana: geomorfología, fenómenos volcánicos y recursos asociados, 2480
Paisajes Geográficos, 2262
Paisajes rurales en el norte de Michoacán, 2396
Palacios, Félix, 962
Palacios, Sergio, 2349
Palacios Lara, Juan José, 1318
Palau, Tomás, 2004, 3596
Palazón Ferrando, Salvador, 2263
Paleopatologia e paleoepidemiologia: estudos multidisciplinares, 585
Palma, Eduardo, 1800
Palma, Norman, 1234
Palma C., Pedro A., 1740
Palma Cabrera, Yolanda, 4531
Palma Mora, María Dolores Mónica, 4532
Palmer, David Scott, 1060, 3396–3398
Palmer-Moloney, Jean, 2350
Palmie, Stephan, 824
Palza Medina, Javier, 1992
Pan American Health Organization, 1165, 4447, 4514, 4766, 5188
Panaia, Marta, 2080, 5095
Panama. Dirección de Planificación Económica y Social, 4020
Panama. Ministerio de Planificación y Política Económica, 1637
Panamá: autodeterminación contra intervención de Estados Unidos, 4042
Panarello, Héctor O., 552
Panday, Basdeo, 3097
Panizza, Francisco E., 3643
Panorama Económico de América Latina, 1235
Panorama social de América Latina, 4436
Pantelides, Edith A., 4986
Pantoja Andrade, Willy, 570
Pantojas-García, Emilio, 4718
Paoli Bolio, Francisco José, 2829
Paolino, Carlos, 2011

Papadópulos, Jorge, 2025
Papma, Frans, 5189
Paracas art & architecture: object and context in south coastal Peru, 690
Paradiso, José, 4217
Pardo, Rodrigo, 4275
Pardo V., Lucía, 1905
Paré, Luisa, 4491, 4533
Paredes Candia, Antonio, 4968
Pareja, Carlos, 3644
Parker, Dick, 2784
Parker, Joy, 108
Parker G., Cristián, 5026
Parmenter, Ross, 447
Parra Pérez, Caracciolo, 4337
Parra Rizo, Jaime Hernando, 997
Parra Vázquez, Manuel Roberto, 2356
Parrini Roces, Vicente, 3473
Parrochia Beguin, Juan, 2605
Parsons, James J., 2264
Parsons, Jeffrey R., 72, 347
Partido Acción Nacional (Mexico), 2861, 2903, 2936
Partido Comunista del Uruguay, 3627
Partido Demócrata Cristiano (Uruguay), 3659
El partido en el poder: seis ensayos, 2904
Partido Liberal Radical Auténtico (Paraguay), 3607
Partido Revolucionario de los Trabajadores (Argentina), 3540
Partido Revolucionario Institucional (Mexico). Instituto de Estudios Políticos, Económicos y Sociales, 2904
Partidos = pleitos = politiquería, 2993
Los partidos y la transformación política de América Latina, 2785, 4437
Partridge, William, 1127
A party politics for Trinidad and Tobago: the flowering of an idea, 3098
Pásara, Luis, 3399, 4926
Passarinho, Jarbas Gonçalves, 3748
Pastor, Aníbal, 3474
Pastor, Manuel, 1236–1237, 1993
Pastor, Robert A., 3849–3850, 3938, 4089–4090
Pastor Fasquelle, Rodolfo, 2351
Pastori Fillol, Alejandro, 4320
Pastrana, Alejandro, 348
Pasztory, Esther, 389
Paternosto, César, 691
Patiño, Diógenes, 624–626
Patiño, Ninfa, 3237
Pau Trayner, María, 4672
Paul, Anne, 690
Paúl Fresno, Luis Hernán, 1855
Paula, Sergio Goes de, 5190

Paula de Teresa, Ana, 756
Paunero, Rafael S., 549
Pavez Reyes, María Isabel, 2605
Paxton, Julia A., 1596
Payne, Anthony, 3036, 4097, 4100, 4155
Payne, Leigh A., 3749
Payne, William O., 266
La paz: más allá de la guerra, 3188
Paz Barnica, Edgardo, 4043–4044
La paz de Cuito Cuanavale: documentos de un proceso, 4127
Paz y Miño C., Juan J., 4281
Paz Zamora, Jaime, 3332, 4237
Pazos, Felipe, 1238
Pazos, Luis, 1742
Peabody Museum of Archaeology and Ethnology (Harvard University), 76
Peace, development, and security in the Caribbean: perspectives to the year 2000, 4091
Pearce, Jenny, 3189
Pearsall, Deborah M., 351, 645–646
Pedersen, Art, 2337
Pedone, Luiz, 3771, 4360
Pedraza, Silvia, 4767
Pedraza Gallardo, Consuelo, 4276
Pedrazzini, Yves, 4438
Pedroni, Guillermo, 4649
Peffley, Mark, 4019
Peixoto, João Paulo Machado, 3707
Pelegrin, Jacques, 692
Pellizzari, Deoni, 3750
Peltre, Pierre, 2471, 2481
Pelupessy, Wim, 1548, 1601, 1641
Peña Cazas, Waldo, 3333
Peña Hasbún, Paula, 3645
Peña León, Germán Alberto, 627
Peña Marazuela, María Teresa de la, 2284
Peña N., Jaime, 642
Peña Sánchez, Julián, 3057
Peñaflor G., Giovanna, 3400
Peñaherrera Padilla, Blasco, 3236
Peñaranda, Ricardo, 3191
Pendergast, David M., 349–350
Pensamiento Iberoamericano, 3851
Pensar al Ché, 3142
Peralta Ramírez, Orem, 1474
Pereira, Edithe, 586
Pereira, Luiz Carlos Bresser, 1241, 2182, 4378
Pereira, Otaviano, 3704
Perera, Miguel A., 2445
Perera, Victor, 757, 4650
Peres, Wilson, 1239
Pereyra Quinto, Armando, 218
Pérez, Carlos Andrés, 4338
Pérez, César, 4768
Pérez, Fernando, 3024

Pérez, Niurka, 4716
Pérez Alemán, Paola, 4651
Pérez Antón, Romeo, 3628, 3646–3647
Pérez Brignoli, Héctor, 4677
Pérez C., Edelmira, 1758
Pérez Castro, Ana Bella, 758
Pérez Fernández del Castillo, Germán, 2928
Pérez Gollán, José Antonio, 561
Pérez Herrero, Pedro, 3816
Pérez López, Emma Paulina, 1474
Pérez-López, Jorge F., 1268, 1685, 1698, 1702–1704, 1713, 3120, 3143
Pérez Luciani, Ramiro, 4339
Pérez Miranda, Rafael, 4493
Pérez Perdomo, Rogelio, 3273, 4843, 4851
Pérez Piera, Adolfo, 3648
Pérez Sáinz, Juan Pablo, 1559, 1609, 4652–4653
Pérez Santarcieri, María Emilia, 3649
Pérez-Stable, Marifeli, 4769
Perfil socio-demográfico provincial, 4872
Perina, Rubén M., 4190
Periodismo y medio ambiente: memoria del seminario realizado en Quito, entre el 28 de noviembre y el 1 de diciembre de 1990, 2786
Periodistas Asociados Televisión (La Paz), 3344
Perl, Raphael F., 4182
Pero, Valéria Lúcia, 2183
Perota, Celso, 587
Perrin, Michel, 904
Perry, Guillermo, 1780
Perspectivas de desarrollo de la Región Inka, 1957
As perspectivas do Brasil e o novo governo, 3751
Perspectives on war and peace in Central America, 4045
Pertusio, Roberto L., 4218
Peru. Congreso. Senado. Comisión Especial sobre las Causas de la Violencia y Alternativas de Pacificación Nacional, 3417
El Perú de los 90, 3401
El Perú, el medio ambiente y el desarrollo, 1958
Peru in crisis: dictatorship or democracy?, 1959
Perzabal, Carlos, 1475
Peschard, Jacqueline, 2852–2853, 2905
Pesquisa: trabalho temporário e migrações na agricultura baiana, 2701
Pesquisa nacional por amostra de domicílios, 2184
Pessoa, Dirceu Murilo, 2177
Pessoa, Ricardo, 3650

Peters, John F., 869
Petras, James F., 2787, 5060
Petriella, Angel, 2265
PETROECUADOR. Asociación de Economistas, 1834
Petróleo y ecodesarrollo en el sureste de México, 2397
Philander, S. George, 2266
Phillips, Fred, 3037
Physicians for Human Rights (US), 2980
Piazza, María del Carmen, 4924
Picchi, Debra, 881
Pick, James B., 4534–4535
Pierucci, Antônio Flávio, 3732, 3752
Pifarré, Francisco, 930
Pike, Fredrick B., 3852
Pimentel González, Nuri, 2906
Pimentel Sevilla, Carmen, 4927
Piñeda Bañuelos, Gilberto J., 4536
Pineda C., Roberto, 905
Pineda Duque, Javier Armando, 4812
Pinedo, Teócrito, 4933
Pinedo Antezana, Alvaro, 3328
Piñera Echenique, José, 1906
Pinheiro, Paulo Sérgio, 3753
Pinilla, José, 672
Pino, Rafael del, 3144
Pino Navarro, Arturo, 1690
Pinochet, Fernando, 2606
Pinochet de la Barra, Oscar, 3444, 4259
Pinochet Ugarte, Augusto, 3475–3477
Pinto, Aníbal, 1907
Pinto, Maria Novaes, 2643
Pinto, Nuno Renan de Figueiredo, 2185
Piñuel Raigada, J.L., 5061
Pinzón de Lewin, Patricia, 3196
Pion-Berlin, David, 2788, 3568
Piperno, Dolores R., 351
Pires, J.S.R., 2702
Pírez, Pedro, 3560
Pischedda, Gabriela, 3455
Piscitelli, Alejandro, 4404
Pitt-Rivers, Julian, 62
Pizano Salazar, Diego, 1781
Pizarro, Juan Antonio, 3190
Pizarro, Roberto, 3003
Pizarro Leongómez, Eduardo, 3191
Pizarro Tapia, Roberto, 2607
Pizzurno Gelós, Patricia, 4046–4047
Placencia, María Mercedes, 1826
Plan de acción forestal tropical para América Central: bibliografía, 2318
Planning Institute of Jamaica, 4786
Plataforma de la Mujer (La Paz), 1995
Platt, Tristan, 960, 962
Plazas, Clemencia, 2461

Un plebiscito a Fidel Castro, 3145
Ploeg, Jan Douwe van der, 4928
La población en el desarrollo contemporáneo de México, 1476
Población y sociedad en México, 4537
Población y trabajo en contextos regionales, 4538
Poblamiento: desarrollo agrícola y regional, 4539
Poblete, Patricio, 2609
Pobreza urbana: interrelaciones económicas y marginalidad religiosa, 4929
Pobreza y salud en Bolivia, 4969
Podestà, Bruno, 4314
Polaco, Oscar J., 260
La política exterior argentina en el nuevo orden mundial, 4219
Política exterior de Bolivia, 4238
La política exterior de México en el nuevo orden mundial: antología de principios y tesis, 3939
La política exterior norteamericana hacia Centroamérica: reflexiones y perspectivas, 4048
Política exterior y estabilidad estatal, 4049
Política y región: Los Altos de Jalisco, 4540
Political and economic liberalization in Mexico: at a critical juncture?, 2907
The political economy of policy reform, 1240
Political parties and democracy in Central America, 2942
Los políticos y los indígenas: diez entrevistas a candidatos presidenciales y máximos representantes de partidos políticos del Ecuador sobre la cuestión indígena, 3237
Politics and social change in Latin America: still a distinct tradition?, 2789
Pollack, Benny, 3460
Pollak-Eltz, Angelina, 4848
Pollard, Helen Perlstein, 143–144
Pollitzer, Germán, 944
Pollock, Donald, 882–883
Pomar, Wladimir, 3754
Ponce G., Dolores, 2908
Ponce Sanginés, Carlos, 571–572
Pontificia Universidad Católica del Ecuador, 3239
Pontificia Universidad Católica del Perú. Centro de Investigaciones Sociales, Económicas, Políticas y Antropológicas, 3415
Pool, Christopher A., 226
Poole, Deborah A., 1049
Poole, Peter, 2267
Populações humanas e desenvolvimento amazônico, 861

The popular use of popular religion in Latin America, 4439
Populismo económico: ortodoxia, desenvolvimentismo e populismo na América Latina, 1241
Portals of power: shamanism in South America, 862
Portes, Alejandro, 1242, 4440
Portillo, Alvaro, 3651
Porto, Walter Costa, 3707
Portocarrero, Patricia, 4433, 4923
Portocarrero Maisch, Gonzalo, 4930
Portugal M., Pedro, 961
Posas, Mario, 4654
Posey, Darrell A., 872, 2672
Potreros, vegas y mahuechis: sociedad y ganadería en la sierra sonorense, 4541
Potter, Robert B., 2306
Pottery of prehistoric Honduras: regional classification and analysis, 352
Pozo, José del, 5062
Pozzo Medina, Julio, 3334
Prado, María Teresa, 5001
Prahl, Henry von, 2462
A pré-história no século do descobrimento, 588
Prebisch, Raúl, 1248
Precolumbian jade: new geological and cultural interpretations, 145
Preeg, Ernest H., 1705
Prehispanic domestic units in western Mesoamerica: studies of the household, compound, and residence, 146
Prem, Hanns J., 147, 2398
Premdas, Ralph R., 792, 3042
Preston, Julia, 3146
Preuss, Konrad Theodor, 998
Prevost, Gary, 3013, 3118
Price, Marie, 2446
Price, Richard, 825
Price, Sally, 825
Price-Mars, Jean, 4142
Prieto, Alfredo, 611
Prieto, Jaire Brito, 4379
Prieto, Justo José, 3608
Prieto González, Alfredo, 4050
Prieto Vial, Daniel, 4260
Primera cumbre iberoamericana: memoria, 3853
Los primeros americanos, 543
El principio del pez gordo: estrategias para combatir la corrupción, 3192
Private sector solutions to the Latin American debt problem, 3854
Privatizaciones en Argentina = Privatizations

in Argentina: conferencia, Bs. As., Argentina, Septiembre de 1991, 2081
Proceso, 40
Proceso de paz en Centro América: compendio, 4051
El proceso de paz en El Salvador, 4052
Proceso de paz en El Salvador: la solución política negociada, 4053
Proceso de retorno a la institucionalidad democrática en el Perú, 3402
Los procesos migratorios centroamericanos y sus efectos regionales, 4542
O processo de urbanização no Oeste Baiano, 2703
Prochnick, Víctor, 2186
PRODERM (project), 1933
Producción de café en Colombia, 1781
Productividad: distintas experiencias, 1477
Programa de Economía del Trabajo (Santiago), 1872
Programa de Estudios Centroamericanos del Centro de Investigación y Docencia Económicas (Mexico), 4042
Programa de Estudios Conjuntos sobre las Relaciones Internacionales de América Latina, 1089
Programa de Seguimiento de las Políticas Exteriores Latinoamericanas (PROSPEL), 4004
Programa del Partido Obrero Revolucionario: aprobado en el 32 Congreso, 3335
Programa Manejo Integrado de Recursos Naturales (Costa Rica), 2320
Projeto: a política social em tempo de crise; articulação institucional e descentralização, 2187
Projeto de Proteção do Meio Ambiente e das Comunidades Indígenas (Brazil), 2664
¿Promesa o espejismo?: exportaciones agrícolas no tradicionales; su análisis y evaluación en el Istmo Centroamericano, 1560
A proposta social-democrata: a social-democracia na atualidade européia, hispano-americana e brasileira, 3755
Proskouriakoff, Tatiana, 444
Prospects for democracy & peace in Peru, 3403
Protestantismos y procesos sociales en Centroamérica, 4655
Protzen, Jean-Pierre, 693
Prous, André, 574, 589
Provincial Inca: archaeological and ethnohistorical assessment of the impact of the Inca state, 694
Provisions on conversion of external debt =
Disposiciones sobre conversión de deuda externa, 1908
Provoste Fernández, Patricia, 4402
Proyecto Andino de Tecnologías Campesinas (Peru), 938, 4939
Proyecto Nacional Abaj Takalik (Guatemala), 344
Proyecto Piloto de Ecosistemas Andinos (Peru), 938
Proyecto Tajín, 353
Prudencio B., Julio, 1994
Psacharopoulos, George, 1243, 1724
Public finances in Latin America in the 1980s, 1244
Puente, Patricio de la, 5063
Puerto Restrepo, Mauricio, 617
Puga, Cristina, 1478
Puig, Max, 4133
Puig, Juan Carlos, 4220
Pulido E., Roberto, 3425
Pulido Méndez, Salvador, 345
Pulwarty, Roger S., 2447
Purcell, Trevor W., 826
Puryear, Jeffrey, 3478
Putumayo, 3193
Puyau, Hermes A., 3569
Queiroz, Emanuel T., 2685
Queiser Morales, Waltrud, 3336
Queisser, Monika, 1909
Quenan, Carlos, 4082–4083
Quesada, Gustavo Martín, 2188
Quesada Camacho, Juan Rafael, 2338
A questão social no Brasil, 5191
Quick, Stephen A., 4092
Quijano, Carlos, 4054
Quijano, Edgardo, 474
Quilodrán, Julieta, 4543
500 años de lucha por la tierra: estudios sobre propiedad rural y refoma [i.e., reforma] agraria en Guatemala, 4656
Quintana Condarco, Raúl de la, 3337
Quintana Roo: los retos del fin de siglo, 1479
Quintanilla Jiménez, Ifigenia, 466, 475
Quintanilla P., Víctor, 2608
Quintero, Rafael, 3221
Quintero Ramírez, Cirila, 4544
Quintero Rivera, Angel G., 3085
Quiroga, Hugo, 3570
Quiroga Garza, Julián, 1480
Quiroga T., José Antonio, 3338
Quiroz, Daniel, 2609
Raat, William Dirk, 3940
Rabkin, Rhoda, 3147, 3479–3480
Race, class & gender in the future of the Caribbean, 827

Raczynski, Dagmar, 4441
Radcliffe, Sarah A., 1050, 2508, 4931–4932
Radio Ñandutí (Paraguay), 3609
Raffino, Rodolfo A., 530, 562
Ragster, LaVerne E., 2307
Rahnama-Moghadam, Mashaalah, 1245
Raíces de América: el mundo Aymara, 962
Raíces institucionales de la política económica costarricense, 1579
Raíces y bosques: San Martín, modelo para armar, 4933
Raichelis, Raquel, 3756
Raiol, Osvaldino da Silva, 4380
Rajapatirana, Sarath, 1075
Rama, Claudio, 5137
Rama, Germán W., 3652
Ramamurti, Ravi, 1246
Ramdas, Anil, 4442
Ramires, Julio Cesar de Lima, 2704
Ramírez, Carlos, 2909
Ramírez, Gabriel, 3653
Ramírez, Miguel Angel, 1353–1354
Ramírez, Miguel D., 1247, 1481
Ramírez, Sergio, 3003
Ramírez Avendaño, Victoria, 2338
Ramírez Brun, J. Ricardo, 1482
Ramírez Calzadilla, Jorge, 4770
Ramírez García, Agustín, 1483
Ramírez León, José Luis, 4277
Ramírez López, Berenice Patricia, 3941
Ramírez Ocampo, Augusto, 4055, 4278
Ramírez Ramírez, José Luis, 179
Ramírez Rodríguez, Edwin, 1580
Ramírez Rodríguez, Juan Carlos, 4545
Ramón, Armando de, 5064
Ramos, Alcida Rita, 884
Ramos, Alejandro, 2909
Ramos, Fernando A., 4726, 4771
Ramos, Hugo H., 1827
Ramos, José María, 3942–3943
Ramos, José Mário Ortiz, 5187
Ramos, Lauro Roberto Albrecht, 2189
Ramos, Roberto, 3757
Ramos-Bellido, Carlos Gil, 4785
Ramos Gómez, Luis Javier, 512, 658
Ramos Sánchez, Pablo, 3281
Ramos Tercero, Raúl M., 1484
Ramphal, Shridath, 4156
Ramsaran, Ramesh, 1675–1676, 4093
Rangel, Inácio, 2190
Rangel, Susana Regina Salum, 2719
Rangel Ch., Orlando, 2452
Rapoport, Mario, 4221
Rappaport, Joanne, 990, 999
Raqaypampa: los complejos caminos de una comunidad andina; estrategias campesinas, mercado, revolución verde, 4970
Rattner, Henrique, 2175
Rattray, Evelyn Childs, 354–355
Raúl Prebisch: un aporte al estudio de su pensamiento; las cinco etapas de su pensamiento sobre el desarrollo, su última intervención pública, bibliografía de su obra entre 1920 y 1986, 1248
Ravines, Rogger, 695–696
Ravines, Tristán, 2509
Reaching for the future: a timely trilogy, 3038
Real de Azúa, Carlos, 3654
Realidade dos seringueiros brasileiros na Bolívia: pesquisa, 4971
Reboratti, Carlos, 2558
Reca Moreira, Inés, 4772
Recalde, Héctor, 5119
Recio Adrados, Juan Luis, 4657
Recio Pinto, Alejandro, 1741
The reconstruction of Central America: the role of the European Community, 1561
Recursos naturales andinos, 2430
Redclift, Michael, 4443
Rede imaginária: televisão e democracia, 5192
Redford, Kent Hubbard, 2233, 2530
Reding, Andrew, 2910
Redmond, Elsa M., 716
Redondo, Nélida, 5120
Reents-Budet, Dorie, 148
Rees, Peter W., 2268
Reestructuración de la industria automotriz mundial y perspectivas para América Latina, 1249
Reestructuración industrial y cambio tecnológico: consecuencias para América Latina, 1250
La reforma agraria peruana, 20 años después, 4934
Reforma del Estado en Costa Rica, 2954
Reforma estructural al sistema arancelario, 1828
Refugee and displaced women in Latin America and the Caribbean, 4444
Regalsky, Pablo, 4970
Regards sur l'art boni aujourd'hui: Bureau du patrimoine ethnologique, Association Mi Wani Sabi, 22 avril-13 mai 1989, 828
La région et l'environnement, 2308
Regional archaeology in Northern Manabí, Ecuador = Arqueología regional del Norte de Manabí, Ecuador, 646
Regional Constituent Assembly of the Wind-

ward Islands. Meeting, *1st, Kingstown, Saint Vincent, and the Grenadines, 1991,* 3039
Regional Employment Program for Latin America and the Caribbean, 1097
Regional overview of food security in Latin America and the Caribbean with a focus on agricultural research, technology transfer and application, 1251
Regional Workshop on Water Resources Management, *Saint Augustine, Trinidad and Tobago, 1990,* 2310
El regreso de Fidel a Caracas, 1989, 4128
Reiche, María, 697–698
Reiche C., Carlos, 2319
Reichel-Dolmatoff, Alicia, 628
Reichel-Dolmatoff, Gerardo, 628
Reid, Basil, 513
Reilly, Charles A., 2269
Reimann, Elisabeth, 3006
Rein, Raanan, 4222
Reinders, Alex, 3080
Reinert, Kenneth A., 1581
Reinhart, Carmen M., 1108
Reinterpreting prehistory of Central America, 149, 476
Reis, Eustáquio J., 2191
Reis, Fábio Wanderley, 3758
Reisen, Helmut, 1080
Reitano, Emir, 5121
Relación gobierno central-empresas públicas en América Latina, 1252
Relaciones del Perú con Brasil, Colombia y Ecuador, 4311
Relaciones del Perú con Chile y Bolivia, 4312
Relaciones del Perú con los países vecinos, 4313
Relaciones económicas del Perú con la Comunidad Europea, 4314
Las relaciones entre España y América Central, 1976–1989, 4056
Relaciones Iglesia-Estado: cambios necesarios; tésis del Partido Acción Nacional, 2911
Relaciones industriales y productividad en el norte de México: tendencias y problemas, 1485
Las relaciones laborales y el Tratado de Libre Comercio, 4546
Relaciones laborales y modelos de acción sindical: experiencias europeas y latinoamericanas, 2790, 4445
Relacje Polska-Chile: doświadczenia i stan obecny. [Polish-Chilean relations: past and present.], 4261

Relatório Norte-Sul: o Brasil e o Nordeste, 2192
Remmer, Karen L., 2791–2793
Renard-Casevitz, France Marie, 923
Rengel, Jorge Hugo, 4287
Rénique, Gerardo, 1049
Rénique, José Luis, 1051, 3404
Rens, Marjan, 1022
Renzi, María Roza, 3003
Repercussões sócio-econômicas do complexo industrial ALBRAS-ALUNORTE em sua área de influência imediata, 2193
Repetto, Fabián, 3571
Repetto Tío, Beatriz, 316
República Dominicana: encuesta demográfica y de salud, 1991, 4773
Reseñas de documentos sobre desarrollo ambientalmente sustentable, 2270
Resende, Maria Efigênia Lage de, 3759
Resnick, Rosalind, 13
Resources, power, and interregional interaction, 150
Restrepo Moreno, Luis Alberto, 3194
The restructuring of public-sector enterprises: the case of Latin American and Caribbean ports, 1253
Reuter, Peter, 3944–3945
Revenga, Ana, 1486–1487
Revesz, Bruno, 4935
Revilla C., Esther, 4941
Revilla C., Víctor, 1960
Revista de Arqueologia, 590
Revista de la Integración y Desarrollo de Centroamérica, 1562
Revista Este País, 41
Reyes, Ana María, 3655
Reyes, Paulina de los, 5065
Reyes, Reynaldo, 759, 3007
Reyes del Campillo, Juan, 2912
Reyes García, Cayetano, 2396
Reyes Heroles, Federico, 2818, 2913
Reyes Heroles González Garza, Jesús, 1414, 4547
Reyes Posada, Alejandro, 3187
Reyna, Carlos, 3405
Reyna, José Luis, 4404, 4548
Reyna de Roche, Carmen Luisa, 4849
Reyna Espinosa, Rafael, 3
Reynolds, Clark Winton, 1488, 1937, 3956
Reynoso, Víctor Manuel, 2399
Rial Roade, Juan, 2735, 2794, 3656–3657
Riaño, Yvonne, 2271
Riart, Gustavo Adolfo, 3610
Ribadeneira, Juan Carlos, 4873
Ribbink, Gerardo, 1621

Ribeiro, Berta G., 885
Ribeiro, Gilberto de Assis, 2707
Ribeiro, Miguel Angelo Campos, 2705
Ribeiro, Pedro Augusto Mentz, 591
Riboud, Michelle, 1486–1487
Ricalde, David G., 2510
Ricciardi, Joseph, 1633
Rice, Don Stephen, 60
Rich, Jan Gilbreath, 1489
Richards, Mike, 3074
Richardson, Bonham C., 4094
Richardson, Francis B., 154
Richardson, James B., 699
Richman, Karen, 3067
Ricupero, Rubens, 4381–4382
Riehl, Herbert, 2447
Rigobón, Roberto, 1181
Rilla, José P., 3623
Rinaldi, Milagro, 716
Ríos, Palmira N., 4774
Ríos, Roberto José, 1254
Ríos de Hernández, Josefina, 4850
Rioseco Hormazábal, Reinaldo, 2610
Riquelme, Marcial A., 3611
Riquelme G., Verónica, 5050
Riquelme U., Horacio, 4411
Risso, Marta R., 5138
Ritter, Archibald R.M., 1706–1707, 1910, 3148
Rius, Andrés, 2026
Rivas, Carlos, 2043
Rivas Aguilar, Ramón, 3254
Rivas Sánchez, Fernando, 3006
Rivera, Carlos, 3370
Rivera, Carlos Alá Santiago, 3008
Rivera, Concepción, 2852
Rivera, Jorge B., 3518
Rivera, Ramiro, 3238
Rivera Agüero, Rigoberto, 1911, 4936
Rivera Campos, Roberto, 1602
Rivera Cusicanqui, Silvia, 963, 3355, 4979
Rivera Dorado, Miguel, 356
Rivera Escobar, Sergio, 629
Rivera Estrada, Araceli, 336
Rivera Pizarro, Alberto, 3339
Rivera Ríos, Miguel Angel, 1490
Rivero Torres, Sonia E., 357–358
Riveros, Luis A., 2082
Riz, Liliana de, 3497, 3572–3574
Rizzini, Irene, 5193
Robalino Bolle, Isabel, 3239
Roberts, Bryan R., 1491, 4434
Roberts, John M., 755
Robichaux, Hubert R., 180
Robinson, Linda Sickler, 503

Robinson, William I., 4057
Robledo, Marcos, 4262
Robles Parra, Jesús, 1510
Robles Piquer, Carlos, 3117
Roca, Sergio G., 1708
Roccatagliata, Juan A., 2538, 2559
Rocha, Sonia, 2194
Rocha V., Alberto, 3406
Rochlin, James, 3855
Rodan, Bruce D., 2272
Rodas, Raquel, 3240
Rodas, Sonia, 1829
Roddick, Jacqueline, 3340
Rodman, Amy Oakland, 612
Rodrigues, Gilda de Castro, 5194
Rodrigues, José Carlos Valim, 2710
Rodrigues, Leôncio Martins, 3760
Rodríguez, Adrián G., 1582
Rodríguez, Alfredo, 5015
Rodríguez, Allen M., 1255
Rodríguez, Carlos Alfredo, 1185, 2036, 2083–2084
Rodríguez, Carlos Armando, 630–631, 633
Rodríguez, Daniel, 1912
Rodríguez, Ennio, 1547, 1583
Rodríguez, Gustavo, 1823
Rodríguez, Hipólito, 1368
Rodríguez, Jaime Arocha, 1000
Rodríguez, José Luis, 1709–1711
Rodríguez, José Ramón, 4727
Rodríguez, Juan Manuel, 3662
Rodríguez, Mario Augusto, 4058
Rodríguez, Marta, 3237
Rodríguez, Miguel, 514
Rodríguez, Miguel Angel, 1539
Rodríguez, Oscar, 1782
Rodríguez, Víctor Manuel, 4775
Rodríguez, Victoria E., 2400, 2914–2915
Rodríguez, Yolanda, 1052, 3407
Rodríguez Amurrio, María, 4978
Rodríguez Benitez, Leonel, 2401
Rodríguez Castro, Ignacio, 3954
Rodríguez Garcia, Ignacio, 390
Rodríguez Gil, Adolfo, 3009
Rodríguez Lapuente, Manuel, 2814
Rodríguez López, Jorge, 3658
Rodríguez-Menier, Juan Antonio, 4776
Rodríguez O., Jaime E., 2854
Rodríguez Prats, Juan José, 2916
Rodríguez Ramírez, Camilo, 632
Rodríguez Sehk, Penélope, 4806
Rodríguez Solera, Carlos Rafael, 4658
Rodríguez T., Lindaura, 4938
Rodríguez Villouta, Mili, 5083
Rodríguez Wong, Laura, 5156
Roett, Riordan, 2907, 3612, 3761

Rogers, John H., 1492
Roggenbuck, Stefan, 4446
Roggiero, Roberto, 1823
Rojas, Jorge Enrique, 3165
Rojas, Josefa, 4933
Rojas, Juan, 964
Rojas Aravena, Francisco, 4059
Rojas Díaz Durán, Alfredo, 3964
Rojas H., Fernando, 4822
Rojas Hoppe, Carlos, 2611
Rojas Martínez, José Luis, 336
Rojas Ramírez, Policarpio, 965
Rojas-Suárez, Liliana, 1961
Rojas U., Javier, 2967
Rojas Zolezzi, Enrique Carlos, 924
Rolfes, Irene, 4777
Rolón Anaya, Mario, 3341
Romaguera, Pilar, 1071
Román, Alicia J., 547
Román Calleros, Jesús, 3955
Romano, Eduardo, 3518
Romão, Maurício Costa, 2116
Romero, Aníbal, 3251
Romero, Carlos A., 4340–4341
Romero, Fernando, 1053
Romero, José, 1493
Romero, María Teresa, 3251, 4342
Romero, Oscar Arnulfo, 2968
Romero, Rocío, 2514
Romero Frizzi, María de los Angeles, 127
Romero Molina, Javier, 179
Romero Pérez, Jorge Enrique, 1256
Romero Pittari, Salvador, 3342–3343
Romero Rivera, María Eugenia, 359
Ronci, Márcio Valério, 2195
Rondón, Melania, 4745
Ronfeldt, David F., 3944–3945
Roniger, Luis, 2795
Ropp, Steve C., 3025
Ros, Jaime, 1147, 3946
Rosa, Alberto de Sousa Amorim, 2177
Rosa, Herman, 1603
Rosa, Luiz Pinguelli, 2156
Rosada-Granados, Héctor, 2982
Rosales, Roberto, 1599
Rosario Quiles, Luis Antonio, 4778
Rosasco de Chacón, Carlota, 1042
Rosenberg, Jonathan, 1712, 3149
Rosenberg, Mark B., 2726
Rosenberg, Robin L., 4060
Rosenberg, Tina, 2796
Rosende R., Francisco, 1913
Rosenfeld, Alex, 5015
Rosenthal, Bertrand, 3126
Rosenzweig, Fernando, 1494
Roslund, Curt, 657

Rospigliosi, Fernando, 3408
Ross A., Alicia, 2602
Rossetti, Josefina, 5077
Rossi, Diana, 5109
Rossiia i Latinskaia Amerika: k novomu partnerstvu [Russia and Latin America: to a new partnership], 3856
Rossum, Peter van, 172
Rostas, Susanna, 4439
Rottenberg, Simon, 1572
Rottmann, Jürgen, 2581
Roulet, Elva, 2560
Roy, Joaquín, 1561
Royce, Anya Peterson, 760
Rozas, Patricio, 1914–1915
Rubalcava, Rosa María, 4478
Rubén, Raúl, 1604, 4668
Rubin, Rebecca B., 4549
Rubin, Vera, 784
Rubín de Celis T., Emma, 4430
Rubinstein, Eugenia Muchnik de, 1916
Rubinstein, Juan Carlos, 2730
Rubio, Luis, 1495, 2917, 4550
Rubio, Mónica, 1948
Rue, David, 403
Rueda, Guillermina G. de, 3026
Rueda Castillo, J. Francisco, 2906
Rueda Novoa, Rocío, 1830
Rueda Peña, Mario, 3344
Ruff, Bernard, 2309
Ruiz, Arminda C., 521
Ruiz, Carmen Beatriz, 4966
Ruiz, María Teresa, 4659
Ruiz Durán, Clemente, 1496
Ruiz-Giménez, Guadalupe, 3817
Ruiz Lombardo, Andrés, 761
Ruiz M., Lucy, 2463
Ruiz Vásquez, Juan Carlos, 2777
Rummens, Joanna W.A., 4779
La ruptura del Frente Amplio: antecedentes y documentos, 3659
Rürup, Luise, 2790, 4445
Rush, Howard, 2196
Russell, Roberto, 4190, 4219
Russian views of Russian-Latin American relations in the post-Cold War world, 3857
Russier, François, 2299
Ruvalcaba Mercado, Jesús, 4551
Ryan, Selwyn D., 1677, 3099–3100
Saad, Pedro, 4286
Saavedra Lucero, José Luis, 2607
Las sabanas americanas: aspecto de su biogeografía, ecología y utilización, 2431
Sabatini, Francisco, 5066–5067
Sabino, Carlos, 1742
Sabloff, Jeremy A., 130, 151

Saboia, João, 2135
Saborío Alvarado, Sylvia, 1563–1564, 3846, 4061
Sabrovsky J., Eduardo, 4449
Sachs, Jeffrey D., 1257–1258, 1743, 1987
Sacks, Richard Scott, 3612
Sadek, Maria Tereza Aina, 3762
Sadoulet, Elisabeth, 1128, 1799
Sáenz, O., 1584
Sáenz Carrete, Erasmo, 3947
Saenz Porras, Verónica, 3361
Sagastizábal, Leandro de, 3575
Saidel, Vivian, 5001
Saiz, José L., 5068
Salas, Carmen, 1923
Salas Serrano, Julián, 1259
Salazar-Carrillo, Jorge, 1260, 1744, 3834
Salazar J., Alonso, 1001, 3195
Salazar Manrique, Jorge Arturo, 3164
Salazar Mora, Jorge Mario, 2955
Salazar Mora, Orlando, 2955
Salazar Paredes, Fernando, 3345, 4239
Salazar Salvo, Manuel, 3428
Salazar Sánchez, Héctor, 1497
Salcedo S., Luis Eduardo, 1775
Salgado, Germánico, 1141
Salgado López, Héctor, 633
Salgado Vega, Jesús, 1498
Salinas, Maximiliano, 2797
Salisbury, Lutishoor, 2310
Salles, Vania, 4558
Salman, Ton, 4422, 5069
Salomon, Jean-Noël, 2537
Salomón, Leticia, 2994, 4660
Salvat, Pablo, 5026
Salyano Rodríguez, Raúl, 2918
Samamé, Lilian, 1944
Samán Mancero, Alfredo, 1831
Samandú, Luis, 4655, 4661
Samanez Argumedo, Roberto, 2511
Samaniego, Ricardo, 1419
Samaniego Román, Lorenzo, 700
Samavati, Hedayeh, 1245
Samayoa Urrea, Otto Arturo, 1611
Samoilovich, Daniel, 3576
Sampaio, Sílvia Selingardl, 2706
Sampaio, Yoni, 1261, 2197
Sampath, Niels M., 829, 4780
Samper K., Mario, 4677
San Antonio Summit, *San Antonio, Texas, 1992*, 3296
San Diego State University. Institute for Regional Studies of the Californias, 1523
San Miguel, Pedro L., 4781
San Sebastián, Koldo, 4831
Sanabria, Harry, 966, 4972
Sancen Contreras, Fernando, 1417
Sánchez, Antulio, 4782
Sánchez, Carlos E., 2085
Sánchez, Magaly, 4438
Sánchez, Manuel, 1499
Sánchez, Mónica, 1798
Sánchez Almanza, Adolfo, 1522
Sánchez Botero, Esther, 988
Sánchez Correa, Sergio Arturo, 328
Sánchez-Crispín, Alvaro, 2402
Sánchez David, Rubén, 3196
Sánchez-Gijón, Antonio, 4183
Sánchez Gleason, Juan, 1364
Sánchez González, Agustín, 2919
Sánchez Guerrero, Gustavo, 1745
Sánchez Montañés, Emma, 445
Sánchez Parga, José, 3241
Sánchez Roa, Adriano, 1662
Sánchez Rodríguez, Roberto, 3948
Sánchez-Salazar, María Teresa, 2368, 2403
Sánchez Susarrey, Jaime, 2920
Sánchez Torres, Fabio, 1262
Sanders, Ronald, 4095
Sanders, William T., 152, 172
Sanderson, Steven E., 3858
Sandoval, Salvador A.M., 5195
Sandoval Z., Godofredo, 4973
Sangmeister, Hartmut, 1263
Sansone, Livio, 4783
Santacruz Guzmán, Fabián, 1832
Santamaría, Marco, 1917
Santamaría Estévez, Diana, 360
Santamaría Gómez, Arturo, 4062
Santana, Roberto, 1023–1024
Santiago Quijada, Guadalupe, 2921
Santibáñez, Abraham, 3481
Santillana Cantella, Tomás Guillermo, 2512
Santis Arenas, Hernán, 2612–2614, 4263
Santley, Robert S., 146
Santos, Eduardo A., 1264
Santos, Guarino Fernandes dos, 3763
Santos, Leinad Ayer O., 875
Santos, Mario R. dos, 4403
Santos, Mílton, 2165, 2703
Santos M., Benjamín, 2940
São Paulo, Brazil (state). Fundação Sistema Estadual de Análise de Dados, 2173
São Paulo, Brazil (state). Secretaria de Economia e Planejamento, 2173
São Paulo, Brazil (state). Secretaria do Meio Ambiente, 2128, 2636
Sapelli, Claudio, 2018, 2020, 2027–2029
Saragossi, Maggy, 3859
Saragoussi, Muriel, 2707

Saraiva, José Flávio Sombra, 4383
Sargent, Charles S., 2561, 2620
Sarkisyanz, Manuel, 1054
Sarmento, Walney Moraes, 4384
Sarmiento, Guillermo, 2431
Sarmiento, L., 2448
Sarmiento, Luis Fernando, 3197, 4823
Sarmiento Palacio, Eduardo, 1265, 1783
Sarro, Patricia Joan, 361
Sato, Kazuo, 1205
Saul, Nestor, 2198
Sault, Nicole L., 762
Saurwein, Anton, 446
Sautter, Hermann, 1144
Sautu, Ruth, 5095
Savastano, Miguel A., 1266
Savedoff, William D., 2199
Sawyer, Donald Rolfe, 2708
Sayago, José Manuel, 2562
Scarborough, Vernon L., 98
Scarpaci, Joseph L., 2615
Scatamacchia, Maria Cristina Mineiro, 592
Schackt, Jon, 906
Schaedel, Richard P., 1055
Schatan, Claudia, 1453, 1500
Schauffler, Richard, 1242
Scheetz, Thomas, 1267
Schele, Linda, 108, 153
Schenone, Osvaldo H., 2086
Schiavini, Adrián, 563
Schiavoni, Lidia, 4992
Schibotto, Giangi, 4937
Schieffelin, Bambi B., 830
Schifter, Jacobo, 4628
Schiller, Nina Glick, 778
Schinkel, Kees, 485
Schirmer, Jennifer, 2983, 4662
Schkolnik, Mariana, 1918
Schmelz, Bernd, 1056
Schmidt, Samuel, 2922
Schmidt-Hebbel, Klaus, 1896
Schmidt Schoenberg, Paul, 362
Schmink, Marianne, 3780, 5213
Schmitt, Rogério Augusto, 3727
Schmitter, Philippe C., 2765
Schmitz, Pedro Ignacio, 593
Schmölz-Häberlein, Michaela, 4663
Schneider, Cathy, 3482
Schneider, Pablo R., 1612
Schneider, Robin, 3010
Schnepel, Ellen M., 831–833
Schoepfle, Gregory K., 1268
Schortman, Edward M., 150, 363
Schott, Jeffrey J., 1189, 1269
Schoultz, Lars, 3870

Schrading, Roger, 4664
Schreiner, Jorge, 2005
Schryer, Frans J., 763
Schteingart, Martha, 1117, 1501, 2404
Schubert, Carlos, 2449
Schuldt, Jurgen, 1829
Schultz, Kevan C., 364
Schulz, Deborah Sundloff, 4063
Schulz, Donald E., 2995, 3150, 4063, 4129
Schumacher, Ute, 1190
Schuster, Angela M.H., 701
Schwab y Etchebarne, Martín, 2087
Schwartz, Gerd, 1974
Schwartz, Stephen, 3011
Schweigert, Thomas E., 1613
SCINCE., 42
Scott, Catherine V., 4598
Scott, David A., 647
Scott, Gregory, 1812
Scott, Steven L., 2369
Scott, Sue A., 365
Scranton, Margaret E., 4064
Scrinivasan, T.G., 1078
Scully, Timothy R., 3483
Scurrah, Martín J., 1962
Sealy, Theodore, 3040
Sector agropecuario, 43
Seda, Paulo, 578
Seddon, Matthew T., 706
Sedoc-Dahlberg, Betty, 3093
Segal, Aaron, 3068
Segal, Daniel A., 834
Segond, Cláudia Rodrigues, 2647
Segre, Magdalena, 4362
Segura, Jorge Rhenán, 4065
Segura Altamirano, José, 2513, 4938
Segura Bonilla, Olman, 1541, 1585
Segura Covalón, Jerónimo F., 3613
La seguridad del Perú frente al nuevo contexto internacional, 4315
Seguridad democrática regional: una concepción alternativa, 4184
Seguridad personal: un asalto al tema, 3273, 4851
La seguridad social y el estado moderno, 1502
Seidman, Gay W., 5196
Sejas, Lidia, 2563
Seki, Yūji, 545
Selbin, Eric, 2798
Seler, Eduard, 154
Seligmann, Linda J., 1057–1058
Seligson, Mitchell A., 4019, 4066
Seltzer, Geoffrey O., 2432

Selverston, Melina, 3242
Seminario Alternativas Económicas para América Latina, *México, 1990,* 1079
Seminario Análisis Medio Ambiental de la Región Amazónica, *Quito? 1987,* 2464
Seminario ANJE, *12th, Santo Domingo?, 1990,* 1663
Seminario de Periodismo, *1st, Cochabamba, Bolivia, 1990,* 3346
Seminario Democracia y Descentralización en Bolivia, *La Paz?, 1988,* 3297
Seminario Desarrollo Andino y Cultura Aymara en el Norte de Chile, *Iquique, Chile, 1989,* 981
Seminário Desenvolvimento Econômico, Investimento, Mercado de Trabalho e Distribuição da Renda, *Rio de Janeiro, 1992,* 2200
Seminario Ecología, Medio Ambiente y Sindicatos, *Juan Lacaze, Uruguay, 1988,* 3629
Seminario El Ecuador del Siglo XXI, *Quito, 1991,* 1833
Seminario El Movimiento Sindical en la Asamblea Legislativa, *El Salvador, 1991,* 2969
Seminario Hacia una Democracia Moderna: la Opción Parlamentaria, *Santiago, 1990,* 3423
Seminario Industria, Comercio y Estado, *México, 1990,* 1413
Seminario Internacional de Economía y Energía 1990–2000, *Quito, 1990,* 1834
Seminário Internacional Populações Humanas e Desenvolvimento Amazônico, *Florencia, Colombia, 1988,* 861
Seminario Internacional Relaciones Económicas del Perú con la Comunidad Europea en Perspectiva, *Lima, 1990,* 4314
Seminario Internacional Relaciones Ecuador-Comunidad Europea 1992, *Guayaquil, Ecuador, 1990,* 4288
Seminário Internacional sobre a Social-Democracia, *Rio de Janeiro, 1987,* 3755
Seminario Internacional sobre Economía y Derecho, *Santiago, 1989,* 1880
Seminario La Campaña del NO, Análisis y Perspectivas, *Santiago, 1988,* 3427
Seminario La Política Exterior Argentina en el Orden Mundial de la Pos-Guerra Fría: Supuestos Teóricos y Alternativas de Inserción Externa, *Buenos Aires, 1992,* 4219
Seminario La Proyección Estratégica de Chile Hacia el Siglo XXI, *Santiago, 1988,* 4251
Seminario Latinoamericano sobre Medio Ambiente y Desarrollo, *San Carlos de Bariloche, Argentina, 1990,* 1270
Seminario Los Retos Tecnológicos del Ecodesarrollo, *Quito, 1991,* 1835
Seminario México-Estados Unidos: Energía, Medio Ambiente y el Tratado de Libre Comercio, *México, 1992,* 3933
Seminario Nacional: Mujer, Género y Desarrollo, *1st, La Paz, 1992,* 1995
Seminario Regional Plan de Desarrollo del Trópico, *1st, Cochabamba, Bolivia, 1992,* 3299
Seminario Relaciones España-América Central, *Barcelona, 1989,* 4056
Seminário sobre a Economia Mineira, *6th, Diamantina, Brazil, 1992,* 2201
Seminario sobre Administración de las Pesquerías Chilenas, *Santiago, 1989,* 2616
Seminario sobre Defensa Nacional y Relaciones Internacionales, *1st, Lima?, 1987,* 4313
Seminario sobre Defensa Nacional y Relaciones Internacionales, *2nd, Lima?, 1988,* 4312
Seminario sobre Defensa Nacional y Relaciones Internacionales, *3rd, Lima, 1989,* 4311
Seminario sobre Defensa Nacional y Relaciones Internacionales, *4th, Lima, 1990,* 4315
Seminario sobre Defensa Nacional y Relaciones Internacionales, *5th, Lima?, 1991,* 3363
Seminario sobre el Acuerdo de Libre Comercio y su Impacto en la Agricultura, *Culiacán, Mexico, 1991,* 1503
Seminario sobre el Rol de los Partidos Políticos, *4th, Guatemala?, 1988,* 2984
Seminario sobre el Rol de los Partidos Políticos, *7th, Guatemala, 1991,* 2985
Seminario sobre el Sistema Político Mexicano, *Cuernavaca, Mexico, 1988,* 2923
Seminario sobre Gestión del Estado y Desburocratización, *Montevideo, 1988,* 3632
Seminario sobre la Relación Gobierno Central-Empresas Públicas en América Latina, *Montevideo, 1986,* 1252
Seminario-Taller sobre Agricultura Andina y Proyecto Campesino, *1st, Ayacucho, Peru, 1987,* 4939
Seminario-Taller Tecnología Apropiada para la Mitigación de Desastres, *Moyobamba, Peru, 1990,* 2514
Sempowski, Martha Lou, 366
Sendic, Raúl, 3660
Sepúlveda, Narciso, 5070
Sepúlveda Pacheco, Oscar, 3428

Sepúlveda-Rivera, Aníbal, 2311–2312
Sequera Tamayo, Isbelia, 4343
Serbín, Andrés, 3860, 4088, 4096, 4157
Sereno, Jorge, 2047
Los seres del más acá: muestras sobrenaturales en la tradición oral cajamarquina, 1059
Serra, Geraldo, 2709
Serra, Luis H., 4665–4666
Serra P., Mari Carmen, 99
Serra Puche, Jaime, 3937, 3949
Serrano, Claudia, 4441, 5071–5072
Serrato, Marcela, 1504
Serrera Contreras, Ramón María, 2273
Serven, Luis, 1271
El Servicio Exterior Mexicano, 3950
Los servicios del Estado en los pueblos jóvenes y la participación ciudadana: situación actual y perspectivas, 4940
Servicios urbanos, gestión local y medio ambiente, 2404
El servico civil en el Caribe = The civil service in the Caribbean, 3041
Setti, Eduardo Pablo, 2088
Severin, K.P., 515
Sgambatti, Sonia, 4852
Sguiglia, Eduardo, 2089
Shafer, Harry J., 250
Shao, Lixin, 4552
Shapiro, Helen, 2202
Sharbach, Sarah E., 3861
Sharer, Robert J., 138, 367
Shaw, Timothy M., 4091
Sheets, Payson D., 368–369
Sheinin, David, 3347, 3862
Shelton, Dinah L., 3863
Sheremet'ev, Igor' Konstantinovich, 1073, 3546
Sherman, Amy L., 1272
Shimada, Izumi, 702
The Shining Path of Peru, 1060
Shinohara, Miyohei, 1504
Shiw Parsad, Basmat, 4715
Short, Margaret I., 4784
Shugart, Matthew Soberg, 3864
El SIDA en Puerto Rico: acercamientos multidisciplinarios, 4785
Sidicaro, Ricardo, 5122
Siebers, Hans, 4661
Siegel, Peter E., 515
Siemen, Alfred H., 372
Sierra, María Teresa, 764
Sierra, Oliva, 1774
Sierra, Oscar, 4661
Sierra Mejía, Marcio E., 4667

Sierra Valenzuela, Vitalia, 1836
Sievert, April Kay, 370–371
Sigal, Silvia, 5123
Sigaud, Lygia, 2156
Sigmund, Paul E., 3484
Sikkink, Kathryn, 3764
Silva, Carlos Eduardo Lins da, 3765–3766
Silva, Carlos Rafael, 1746
Silva, Ciléa Souza de, 2710
Silva, Eduardo, 3485–3486
Silva, Hélio, 3767
Silva, José Afonso da, 3768
Silva, José Luiz Foresti Werneck da, 4385
Silva, José Wilson da, 3769
Silva, Luiz Francelino da, 872
Silva, Maria A. Moraes, 5197
Silva, Marlise Vinagre, 5198
Silva, Mauro José da, 2668–2669
Silva, Nelson do Valle, 5199–5200
Silva, Patricio, 3436, 4386, 5018, 5073
Silva, Paulo Napoleão Nogueira da, 3770
Silva, Sylvio Bandeira de Mello e, 2711
Silva, Verónica, 5006
Silva, Vicente Paulo da, 3691
Silva Castillo, Jorge, 3929
Silva Hernández, Ana Margarita, 2956
Silva-Herzog Flores, Jesús, 3951
Silva S., Jorge E., 703
Silva Téllez, Armando, 4824
Silveira, M.J., 550
Silverman, Helaine, 660, 704–705
Silvero, Ilde, 3604
Silverwood-Cope, Peter L., 886
Simeoni, Héctor Rubén, 3577
Simmons, David Alan, 1639
Simón G., José Luis, 3614, 3617, 4295
Simposio de Investigaciones Arqueológicas en Guatemala, *5th, Guatemala, 1991*, 155
Simposio Internacional Cultos Religiosos a los Antepasados en el Caribe, *Río Piedras, Puerto Rico, 1990*, 835
Simpósio Internacional sobre o Futuro do Sindicalismo no Brasil, *1st, São Paulo, 1990*, 2149
Simposio Recursos Documentales sobre la Mujer en Bolivia, *La Paz, 1990*, 4974
Simposio sobre la Reforma Agraria Nicaragüense, *Amsterdam, 1988*, 4668
Simposio sobre Mercado de Capitales, *12th, Cali, Colombia, 1990*, 1759
Simposio TLC: Impactos en la Frontera Norte, *1st, Saltillo, Mexico, 1992*, 1515
Simpson, Amelia S., 5201
Sims, Harold Dana, 4130

Sinaloa, Mexico (state). Secretaría de Planeación y Desarrollo, 1367
Sindicalismo latinoamericano: el desafío del cambio, 3348
Sindicato dos Jornalistas Profissionais do Rio Grande do Sul (Brazil), 3703
Singer, André, 3718
Singer, Paul Israel, 3693
Sips, P., 2274
Siqueira, Ethevaldo, 2203
Sir Arthur Lewis: an economic and political portrait, 3042
Sistema BNDES (Brazil), 2200
Sistema de ciudades y distribución espacial de la población en México, 2405
Sistema Económico Latinoamericano (SELA), 1231
Sistema financiero y asignación de recursos—experiencias latinoamericanas y del Caribe: Colombia, Costa Rica, Chile, República Dominicana, Venezuela, 1273
El sistema presidencial mexicano: algunas reflexiones, 2924
Sistema regional de áreas silvestres protegidas en América Central: plan de acción, 1989–2000, 2320
Sistemas agroforestales: principios y aplicaciones en los trópicos, 2339
Sistemas eleitorais e processos políticos comparados: a promessa de democracia na América Latina e Caribe, 3771
Situación alimentaria y nutricional de Bolivia, 1992, 4975
La situación de la mujer en la economía informal: caso ecuatoriano, 4874
Situación de la niñez nicaragüense, 4669
Situación habitacional en Venezuela, 1747
Situation analysis of the status of children and women in Jamaica, 4786
Siu, Ivonne, 4669
Sives, Amanda, 3772
Size and survival: the politics of security in the Caribbean and the Pacific, 4097
Skar, Sarah Lund, 1061–1063
Skidmore, Thomas E., 2803, 5202
Sklar, Holly, 4067
Skoczek, M., 1274
Slavery in the Americas, 836
Sloan, Tod Stratton, 759, 3007
Slottje, Daniel Jonathan, 1331
Sluyter, Andrew, 156, 372, 2406
Small, Michael, 2721
Smith, Carol A., 4670
Smith, Gavin, 1064
Smith, M.G., 792, 837

Smith, Mary Elizabeth, 447
Smith, Michael Ernest, 63, 157–158, 373–375
Smith, Michael L., 3409
Smith, Peter H., 2742, 3792, 3952
Smith, Robert Freeman, 4098
Smith, Sherry, 5074
Smith, Stephen M., 1582
Smith, William C., 2799, 3578–3579
Smith W., David A., 3027
Smoking and health in the Americas: a 1992 report of the Surgeon General, in collaboration with the Pan American Health Organization, 4447
Smulovitz, Catalina, 3496
Smyth, Michael P., 376
Snodgrass, B. Warren, 1505
Snyder, Richard, 2800
Soares, Dino Magalhães, 2624
Soares, Gláucio Ary Dillon, 3773
Soberanes Reyes, José Luis, 1506
Sobo, Elisa Janine, 838
Sobrino, Raúl Augusto, 3580
Soche, Gladys, 2280
Social democracy in Latin America: prospects for change, 2801, 3865
Sociedad Antropológica de Colombia, 988
Sociedad Chilena de Arqueología, 602
Sociedad Mexicana de Antropología. Mesa Redonda, *21st, Mérida, 1989*, 194
Sociedad Mexicana de Demografía, 4539
Sociedad Mexicana de Geografía y Estadística, 1509
Sociedade de Arqueologia Brasileira, *6th, Rio de Janeiro, 1991*, 594
Las sociedades americanas del postpleistoceno temprano, 544
Las sociedades americanas y los orígenes de la producción de alimentos, 64
Society and politics in the Caribbean, 3043
Society for American Archaeology. Meeting. *52nd, Toronto, 1987*, 276
Sodi Miranda, Federica, 377
Sojo, Ana, 2275, 4068
Sojo, Carlos, 1586, 4069, 4671
Solà, Roser, 2352, 4672
Solano B., Federico, 469
Solano de la Sala Veintimilla, Germán, 1837
Solares Serrano, Humberto, 2529
Soler, Giancarlo, 3028, 4070
Solimano, Andrés, 1271
Solingen, Etel, 2204
Solís, Manuel Antonio, 2957
Solís Alpízar, Olman E., 477
Solís de Alba, Ana Alicia, 4553
Solís del Vecchio, Felipe, 478

Solís Olguín, Felipe R., 110, 159, 378
Solís Rivera, Luis Guillermo, 4059
Solís Soberón, Fernando, 1507
Sollis, Peter, 3012
Solorio P., Fortunata, 4941
Somavía, Juan, 4184
Somoza, Jorge L., 944
Sondrol, Paul C., 3615–3616, 3661
Sonntag, Heinz R., 4844
Sonora ante el Tratado de Libre Comercio, 1508
Sorensen, Ninna Nyberg, 839
Soria Galvarro, Carlos, 1996
Soruco, Juan Cristóbal, 3349
Sosa, Erasmo, 379
Sosa Abascal, Arturo, 3252
Sosa Rodríguez, Raúl, 1275
Sosnovskiĭ, Anatoliĭ Aleksandrovich, 3774
Sosnowski, Saúl, 3434, 5016
Sotelo Santos, Laura Elena, 160, 380
Soto, Hernando de, 1276
Soto, Willy, 4071
Soto-Heim, Patricia, 982
Soto Sánchez, Oscar David, 1726
Sotomayor, Héctor, 3812
Sotomayor Tribín, Hugo Armando, 634–635
Sottoli, Susana, 4988
Souffrant, Claude, 4787
Sousa Filho, Francisco Romualdo de, 2188
Southgate, Douglas DeWitt, 1838, 2276, 2482
Souza, Amaury de, 3775
Souza, Ayda Connia de, 2802
Souza, Célia Ferraz de, 2713
Souza, Jaimeval Caetano de, 2711
Souza, Nali de Jesus de, 2185
Souza-Lobo, Elisabeth, 5203
Soviet Union. Ministerstvo inostrannykh del, 4328
Sowing the whirlwind: soya expansion and social change in southern Brazil, 2205
Spain. Secretaría General de Medio Ambiente, 2237
Spalding, Hobart, 1206
Speer, John G., 2744
Spence, Michael W., 340, 366
Spencer, Charles S., 161, 716
Spencer-Strachan, Louise, 840
Spier, Fred, 3410
Spiguel, Claudio, 4221
Spinanger, Dean, 1637
Spiritual Baptists, shango, and others: African derived religions in the Caribbean, 841
Spooner, Mary Helen, 3487
Spores, Ronald, 65, 448
Sposati, Aldaíza de Oliveira, 5154

Šprajc, Ivan, 162–163
Spreafico, Alberto, 2775
St. Cyr, Eric, 3042
St. Kitts and Nevis country environmental profile, 2313
Stadel, Christoph, 2433
Stahler-Sholk, Richard, 1634, 4673
Stallings, Barbara, 3828
Stanish, Charles, 706
Stanley, William, 2970, 4018
Starn, Orin, 1065, 3411
Statistical Masterfile, 44
Statistics Canada. International Trade Division, 51
Stavenhagen, Rodolfo, 4554
Steadman, Lee, 706
Stearman, Allyn MacLean, 2530–2531
Stearns, Stephen K., 421
Stédile, João Pedro, 2104
Steen, L.J. van der, 651
Steger, Hanns-Albert, 2277
Stein, Eduardo, 1546
Stein, William W., 1066
Steinhauf, Andreas, 1067
Stemper, David Michael, 630, 648
Sten, María, 164
Stephanides, Stephanos, 842
Stephens, Evelyne Huber, 1652
Stephens, John D., 1652
Sterling Arango, Rolando, 3029
Sternbach, Nancy Saporta, 4448
Sternberg, Rolf, 2712
Sterner, Thomas, 1277
Stetson, Jeane H., 4185
Stewart, Thelma, 4788
Stewart-Gambino, Hannah W., 3799, 5075
Stoetzer, O. Carlos, 3866
Stoffle, Richard W., 4789
Stolcke, Verena, 4409
Stølen, Kristi Anne, 4875
Stoll, David, 765
Stolovich, Luis, 2022, 2030, 3662
Stone, Carl, 1653, 1678, 3070, 4790
Stone, Roger D., 1278
Stonich, Susan C., 766, 4674–4675
Storey, Rebecca, 366, 381–382
Stothert, Karen E., 649
Strategic options for Latin America in the 1990s, 1279
Straubhaar, Joseph, 3776
Street, Susan, 2925
Street-Perrott, Alayne, 140
Stresser-Péan, Guy, 297, 383
Strohaecker, Tânia Marques, 2713
Strubbia, Mario, 3581

Struck, Ernst, 2714
Stuart, David, 449
Stuart, Gene S., 165
Stuart, George E., 165, 450
Stunnenberg, Peter, 2090
Suárez, Carlos O., 3867
Suárez A., Vicente, 327
Suárez Diez, Lourdes, 384
Suárez Farías, Francisco José, 2926, 4555
Suasnávar, José, 398
Subercaseaux, Bernardo, 3434, 5016
Subercaseaux, Elizabeth, 3476
Subervi-Velez, Federico A., 4791, 5204
Subirats Ferreres, José, 4976
Subramaniam, V., 1672
Suchlicki, Jaime, 4131
Suecia-Latinoamérica: relaciones y cooperación, 3868
Sued, Ronaldo, 2206
Sued-Badillo, Jalil, 516
Sugiura Y., Yoko, 99, 385
Sugiyama, Saburo, 220, 386, 432
Sunkel, Osvaldo, 1131, 1135, 1280–1281, 1919, 3806
Superintendência Baiana para o Trabalho (Brazil), 2701
Supervielle, Marcos, 5132
Sureda Delgado, Rafael Angel, 4344
Suriname: twintig jaar benarde onafhankelijkheid [Suriname: 20 years of perilous independence], 2314
Suriname in het jaar 2000 [Suriname in the year 2000], 3094
Suriname jaarboek 1995: radioprogramma Zorg en Hoop [Suriname yearbook 1995: radio program Zorg en Hoop], 4792
Susmel, Nuria, 2047
Sustainable development of the Amazon: development strategy and investment alternatives, 2207
Sutton, Paul K., 3036, 3073, 4097, 4099–4100, 4155
Svejnar, Jan, 1713
Sweden. Beredningen för u-landsforskning, 1206
Swezey, William R., 2347
Sydenstricker, John M., 5205
Szanton Blanc, Cristina, 778
Szasz Pianta, Ivonne, 2407
Székely, Gabriel, 3828
Tabasco: realidad y perspectivas, 1509
Tacoma, J., 520, 651
Taddei Bringas, Cristina, 1510
Tagle D., Matías, 3435

Takacs, Esteban A., 3511
Taller de Investigaciones Socio-Económicas, *12th, La Paz, 1991*, 1997
Taller de Política Social, *4th, La Paz?, 1991*, 1998
Taller Internacional sobre la Transformación del Medio Geográfico en Cuba, *1st, La Habana, 1988*, 2315
Taller sobre Tenencia y Uso de la Tierra en Santa Cruz, *4th, Santa Cruz, Bolivia, 1987*, 2521
Tamayo, Jaime, 1467, 2815, 4527
Tamayo, Jesús, 2408, 4556–4557
Tamez, Silvia, 1313
Tan, Hong W., 1486
Taniguchi Kōgyō Shōreikai. Division of Ethnology. International Symposium, *16th, Osaka and Ōtsu-shi, Japan, 1992*, 689
Tapia Montaño, Rafael, 3350
Tapia Santamaría, Jesús, 2870
Tappatá, Ricardo, 2006
Tarasov, Konstantin Sergeevich, 1203, 2226
Tarhan, Ariana, 4847
Tarn, Nathaniel, 772
Tate, Carolyn Elaine, 387
Taube, Karl A., 136, 451–452
Tavares, Froilán J.R., 3058
Tavares, Maria da Conceição, 2121
Tax reform in the Caribbean, 1654
Taylor, Patrick, 4793
Tecnología y modernidad en Latinoamérica: ética, política y cultura, 4449
Tecnologías urbanas socialmente apropiadas: experiencias colombianas: Red Colombiana de Tecnología Apropiada, 4825
Tedesco, Laura, 4210
Tedlock, Barbara, 166
Tedlock, Dennis, 453–454
Teichman, Judith, 2927
Teitel, Simón, 1283
Teixeira, Faustino Luiz Couto, 5206
Teixeira, Nelson Gomes, 2149
Tejada Bouscayrol, Mario, 388
Television, politics, and the transition to democracy in Latin America, 2803
Telles, Edward E., 5207
Téllez Ardila, Mireya, 3184
Téllez Kuenzler, Luis, 1511
Tello, Carlos, 1454, 1512
Teltscher, Suzanne, 2483
Temas de política externa brasileira, 4387
Tena, Rafael, 455
Tenencia actual de la tierra en Bolivia, 4977

Teotihuacán: art from the City of the Gods, 389
Teotihuacán 1980–1982: nuevas interpretaciones, 390
Terán, Oscar, 3582
La tercera raíz: presencia africana en Puerto Rico, 4794
Terríquez, Ernesto, 1513
Teslenko, A. I͡U., 1203
I tesori delle città perdute: oro della Colombia, 636
Tesoros del Museo Regional de Oaxaca =Oaxaca Regional Museum treasures, 391
Testimonio de la represión política en Paraguay, 1954–1974, 3617
Testimonios mapuches en Neuquén, 945
Textos y pre-textos: once estudios sobre la mujer, 4558
The Netherlands. Ministry of Foreign Affairs. Development Cooperation Information Department, 4450
Theodore, Karl, 1654
Thérien, Jean Philippe, 3841
Thijsen, Theo, 1963
Thompson, Andrew, 4223
Thompson, Gary D., 1514, 1530
Thompson, John Eric Sidney, 154
Thomson, Marilyn, 4676
Thornberry, Guillermo, 4316
Thorp, Rosemary, 1839
Thorpe, Andy, 1618
Thorstensen, Vera, 4378
Thorup, Cathryn L., 3953
Thoumi, Francisco E., 3198–3200
Tichit, Muriel, 2532
Tichy, Franz, 392, 456
Tierra, café y sociedad: ensayos sobre la historia agraria centroamericana, 4677
Tinoco, Elizabeth, 3251
Tironi Barrios, Eugenio, 5020, 5076
TLC: impactos en la frontera norte, 1515
To change place: Aztec ceremonial landscapes, 167
Tobón, Evamaría Uribe, 1779
Tocón Armas, Carmen, 4942
Tocornal, Josefina, 1866
Todaro C., Rosalba, 5039
Toer, Mario, 5124
Tokatlian, Juan, 4112, 4275, 4279
Toledo, Alejandro, 2397, 2409
Toledo, Víctor M., 2278
Toledo Rivera, Héctor, 2596
El Tolima: una respuesta pacífica, 3201
Tollefson, Scott D., 3761
Tolosa, Hamilton C., 2194
Tomassini, Luciano, 1282, 3869
Tomic, Radomiro, 3488
Tomoeda, Hiroyasu, 859
Toranzo Roca, Carlos F., 1997–1999, 3297–3298, 3351
Torello, Mariella, 2031
Toro Hardy, Alfredo, 4345–4346
Toro Hardy, José, 1748
Torrado, Susana, 5125
Torre, Carlos de la, 3243
Torre, Mario de la, 159
Torres, Haroldo, 2693, 2715, 5205
Torres, Victor, 4317
Torres Abrego, José Eulogio, 3028
Torres C., Alfonso, 3202
Torres C., Víctor, 1964
Torres-Lima, Pablo Alberto, 1348
Torres Montes, Luis, 393
Torres Ramírez, Blanca, 1310, 2849, 3916
Torres-Rivas, Edelberto, 4678–4679
Torres Rodríguez, Luis, 1840
Torres Rojas, Emilio, 5063
Torrico Flores, Gonzalo, 3352
Torrico V., Erick R., 3353
Torrijos Herrera, Omar, 4072
Tourtellot, Gair, 394
Toussaint Ribot, Mónica, 2944
Tovar Piérola, Raúl F., 2000
Towards a new development strategy for Latin America: pathways from Hirschman's thought, 1283
Towards sustained development in Latin America and the Caribbean: restrictions and requisites, 1284
Townroe, Peter, 2716
Townsend, Janet G., 1002
Townsend, Richard F., 168
Trabajos arqueológicos en el centro de la Ciudad de México, 395
Trabajos arqueológicos en Moquegua, Perú, 707
Trabas no arancelarias al comercio andino en el Ecuador, 1841
Trade and investment relations between Chile and the United States during 1989, 1920
Trade policies and practices by sector, 2032
Traditional spirituality in the African diaspora, 843
Una tragedia campesina: testimonios de la resistencia, 4680
Trahtemberg Siederer, León, 4943
TRAINS, 45

La trama solidaria: pobreza y microproyectos de desarrollo social, 5126
Tramas para un nuevo destino: propuestas de la Concertación de Mujeres por la Democracia, 5077
Transborder data flows and Mexico: a technical paper, 1516
La transferencia de recursos externos de América Latina en la posguerra, 1285
Transformación ocupacional y crisis social en América Latina, 1286
Transición a la democracia y reforma del Estado en México, 2928
Transnational banks and the international debt crisis, 1287
El tratado de libre comercio—y usted, 3954
Tratados internacionales celebrados por México, 46
Trava Manzanilla, José Luis, 3955
Trayectorias divergentes: comparación de un siglo de desarrollo económico latinoamericano y escandinavo, 1288
Trejo, Guillermo, 1517
Trejo Delarbre, Raúl, 2820, 2929, 4559
Trejos, Gerardo, 2958
Treverton, Gregory F., 3833
Trías, Vivián, 3663
Tricart, Jean L.F., 2717
Triches, Divanildo, 2208
Trinidad ethnicity, 844
Trobo, Claudio, 3658
Trombold, Charles D., 396
Troncoso Muñoz, Oscar, 1871
Trueba Lara, José Luis, 2930
Trueblood, Beatrice, 3923
TSentral'naĭa Amerika i Kariby: nachalo 90-kh godov [Central America and the Caribbean: the beginning of the nineties], 2804
Tuden, Arthur, 784
Tuijtelaars de Quitón, Christiane, 4978
Tulchin, Joseph S., 1959, 4022
Tulet, Jean-Christian, 2279
Tuller, Lawrence W., 1289
Turino, Thomas, 1068
Turizo Callejas, Alfredo, 4430
Turner, Billie Lee, 173, 2416
Turner, Frederick C., 3509
Turner, Jorge, 4681
Turner, Terence, 855
Twomey, Michael J., 1290–1291
Tyrakowski, Konrad, 2410
Tyrtania, Leonardo, 767
U.S. Council of the Mexico-U.S. Business Committee, 1459

U.S. exports to Mexico: a state-by-state overview, 1987–1991, 1518
U.S.-Mexico relations: labor market interdependence, 3956
U.S.-Mexico trade: impact of liberalization in the agricultural sector: report to the Chairman, Committee on Agriculture, House of Representatives, 1519
U.S.-Mexico trade: extent to which Mexican horticultural exports complement U.S. production; briefing report to the Chairman, Committee on Agriculture, House of Representatives, 1520
Uceda, Santiago, 666
Uggen, John F., 1601, 4876
UN. See United Nations
UNAM. See Universidad Nacional Autónoma de México
Ungar Bleier, Elisabeth, 3171, 4826
UNICEF, 4786
United Nations Centre for Human Settlements, 2286
United Nations Conference on Environment and Development, Rio de Janeiro, 1992, 1862, 2551
United Nations Conference on Trade and Development, 1877
United Nations Development Programme, 1273, 1728, 1964, 2010, 2200, 2257
United Nations Environment Programme, 2237
United Nations Industrial Development Organization, 1249
United Nations Research Institute for Social Development, 1384
United States. Central Intelligence Agency. Directorate of Intelligence, 1686
United States. Commission for the Study of International Migration and Cooperative Economic Development, 3957
United States. Congress. House. Committee on Agriculture, 1519–1520
United States. Congress. House. Committee on Government Operations, 4186
United States. Department of Health and Human Services, 4447
United States. General Accounting Office, 1519–1520
United States. International Trade Administration, 1518
United States. Office of the Under Secretary of Defense for Policy, 3945
United States. Library of Congress. Federal Research Division, 3035

The United States and Latin America in the 1990s: beyond the Cold War, 3870

The United States and Venezuela: new opportunities in an established relationship; the report of the CSIS-CAUSA Working Group on U.S.-Venezuelan Relations, 4347

Universidad Academia de Humanismo Cristiano (Santiago). Grupo de Investigaciones Agrarias, 1074

Universidad Autónoma Chapingo (Mexico), 2356

Universidad Autónoma de Baja California (Mexico), 3899

Universidad Autónoma de Ciudad Juárez (Mexico), 1360, 2367

Universidad Autónoma de Nuevo León (Mexico), 1480

Universidad Autónoma Metropolitana (Mexico). Unidad Iztapalapa, 1372, 1376

Universidad Centroamericana José Simeón Cañas (El Salvador). Instituto de Derechos Humanos, 2962

Universidad de Chile. Departamento de Ingeniería Industrial, 2616

Universidad de Costa Rica, 2325

Universidad de Guadalajara (Mexico). Centro de Investigación Educativa, 4523

Universidad de la Amazonía (Colombia), 861

Universidad de la Rábida (Spain), 3844

Universidad de la República (Uruguay). Facultad de Ciencias Económicas y de Administración, 2023

Universidad de los Andes (Bogotá). Departamento de Ciencia Política, 3163, 3171

Universidad de los Andes (Mérida, Venezuela), 1716, 2431

Universidad de Santiago de Chile. Instituto de Investigaciones del Patrimonio Territorial de Chile, 2594

Universidad de Sonora (Mexico), 1373

Universidad Juárez Autónoma de Tabasco (Mexico), 1509

Universidad Mayor de San Andrés (La Paz). Instituto de Ecología, 2526

Universidad Mayor de San Simón (Cochabamba, Bolivia), 3299

Universidad Nacional (Heredia, Costa Rica), 1541

Universidad Nacional Andrés Bello (Santiago), 1860

Universidad Nacional Autónoma de Honduras, 1618

Universidad Nacional Autónoma de México. Centro de Investigaciones Interdisciplinarias en Humanidades, 1454, 2360, 2829, 4521

Universidad Nacional Autónoma de México. Centro de Investigaciones sobre Estados Unidos de América, 3933, 4048

Universidad Nacional Autónoma de México. Escuela Nacional de Estudios Profesionales Acatlán, 1444

Universidad Nacional Autónoma de México. Facultad de Economía, 1316

Universidad Nacional Autónoma de México. Instituto de Investigaciones Jurídicas, 2924

Universidad Nacional Autónoma de México. Instituto de Investigaciones Sociales, 3910

Universidad Nacional Autónoma de México. Programa Universitario de Alimentos, 1319

Universidad Nacional Autónoma de México. Programa Universitario de Energía, 3933

Universidad Nacional Autónoma de México. Seminario de Estudios Prehispánicos para la Descolonización de México, 460

Universidad Nacional de Córdoba (Argentina). Facultad de Filosofía y Humanidades, 546

Universidad Nacional Mayor de San Marcos (Lima). Seminario de Historia Rural Andina, 1070

Universidad Pontificia de México, 2851

Universidad Tecnológica de la Mixteca, 2357

Universidade de São Paulo. Núcleo de Estudos da Violência, 5210

Universidade Federal de Minas Gerais (Brazil). Centro de Desenvolvimento e Planejamento Regional, 2201

Universidade Federal de Santa Catarina (Brazil), 2658

Universidade Federal do Pará (Brazil), 861

Universidade Federal do Rio de Janeiro. Coordenação dos Programas de Pós-Graduação de Engenharia. Area Interdisciplinar de Energia, 2156

Universität Innsbruck. Institut für Geographie, 2252

University for Peace, 4021

University of Guyana. Women's Studies Unit, 4715

University of Illinois-Urbana. Krannert Art Museum, 927

University of London. Institute of Latin American Studies, 3684

University of Miami. European Community Research Institute, 1561

University of Miami. Iberian Studies Institute, 1561

University of Miami. Institute of Interamerican Studies, 4004
University of Miami. North-South Center, 3811, 3854, 3934
University of Pennsylvania. University Museum of Archaeology and Anthropology, 138
University of the West Indies (Mona, Jamaica). Institute of Social and Economic Research, 815, 1639, 4751
University of the West Indies (Saint Augustine, Trinidad and Tobago), 1677, 2310
Uniwersytet Warszawski (Poland). Centrum Studiów Latynoamerykańskich, 4261
Uprimny, Rodrigo, 3203
Urani, André, 2209
Urban, Patricia A., 150
Urban poverty alleviation in Latin America, 1292, 4450
Urbanización y desarrollo en Michoacán, 1521
Urbina Fuentes, Manuel, 1522
Urcid, Javier, 457
Urdaneta, Alberto, 3274, 4348
Uriarte, María Teresa, 122
Uribe, Alvaro, 3022
Uribe, Carlos A., 989
Uribe, Maruja, 2280
Uribe Alarcón, María Victoria, 4827
Urquidi, Víctor L., 1125, 1138, 1293
Urrea, Fernando, 4828
Urriola, Rafael, 1785, 1803, 1841, 4288
Urrutia, Jaime, 1069
Urrutia, Miguel, 1209, 1767
Urruzola, María, 5139
Uruguay. Ministerio de Economía y Finanzas, 2015
Uruguay. Ministerio de Relaciones Exteriores, 4328
Uruguay. Ministerio de Trabajo y Seguridad Social, 2010
Uruguay. Oficina Nacional del Servicio Civil, 3632
Uruguay. Presidencia de la República, 4326
Uruguay-URSS: 60 años de relaciones diplomáticas, 1926–1986—documentos y materiales, 4328
Uruguayos en Argentina y Brasil: movimientos de población entre los países del Plata, 5140
Urzúa Valenzuela, Germán, 3489
USA, USSR, and the Caribbean, 4101
Useche Aldaña, Helena, 4826
Useche K., Aurelio, 1718
Ushino, Tsuyoshi, 545

Uthoff B., Andras, 1565
Utting, Peter, 4682
Vachon, Michael, 2411
Vail, Gabriela, 183
Vain, Leonor, 5096
Vainsencher, Semira Adler, 5208
Valadez Azúa, Raúl, 397
Valcárcel C., Marcel, 4929
Valdeavellano, Rocío, 3364
Valderrábano, Azucena, 2931
Valdés, Juan Antonio, 398–399
Valdés, Teresa, 4428–4429, 5078
Valdés, Yrmino, 4795
Valdés Palacio, Arturo, 3412
Valdés Paz, Juan, 3871
Valdés Pizzini, Manuel, 797
Valdés Zurita, Leonardo, 2932
Valdez, Fred, 169
Valdez, José, 4796
Valdez, Lidio M., 708
Valdez Pérez del Castillo, Eduardo, 4318
Valdivieso, Gabriel, 4994, 5037
Valdivieso Eguiguren, Sergio, 4245
Valecillos Toro, Héctor, 1749–1751
Valencia Cárdenas, Alberto, 3413
Valencia Espinoza, Abraham, 1070
Valencia García, Guadalupe, 2829
Valencia Rodríguez, Luis, 4289
Valencia Villa, Hernando, 3204–3205
Valencia Zegarra, Alfredo, 709
Valenciano, Eugenio O., 1448, 1523, 1535, 3928
Valencio, Susana A., 552
Valentín, Isidro, 4930
Valenzuela, Arturo, 2750, 3490–3491
Valenzuela, Julio Samuel, 2763, 3483
Valenzuela, María Helena, 5079
Valenzuela Arce, José Manuel, 4560
Valenzuela Feijóo, José, 1294
Vallance, Michel, 2293
Valle, Carlos del, 3872
Valle, Víctor Manuel, 2971
Valverde, Jaime, 4582, 4683
Valverde, María del Carmen, 160
Van Der Karr, Jane, 4224
Vanden, Harry E., 3013
Vapnarsky, César A., 5127
Varas, Augusto, 3869–3870, 3873, 4251
Varela, Ramón J., 2443
Vargas, César, 4853
Vargas, Mauricio, 3206
Vargas, Oscar-René, 3014
Vargas, Rosío, 3933
Vargas, Virginia, 4433
Vargas Cullell, Jorge, 2951

Vargas-Garcia, Jesus, 1331
Vargas González, Pablo E., 2933
Vargas Llosa, Alvaro, 3414
Vargas Llosa, Mario, 3874
Vargas Ramírez, Antonio J., 712
Vargas Velásquez, Alejo, 3207
Várguez Pasos, Luis A., 4487, 4561
Varillas Montenegro, Alberto, 4944
Vasco Uribe, Luis Guillermo, 1003–1004
Vasconcelos, Jorge, 2006
Vasconcelos, Luiz L., 4388
Vásquez, Ana, 4451
Vásquez, Víctor, 672
Vásquez de Labandera, Edwin, 1842
Vázquez, Josefina Zoraida, 3958
Vázquez, Paciente, 1843
Vázquez Cobo, Alfredo, 4280
Vázquez Colmenares, Pedro, 3959
Vázquez L., Ricardo, 469
Vázquez Machicado, Humberto, 4240
Vázquez Ocampo, José María, 4225
Vázquez-Presedo, Vicente, 2091
Vázquez Ruiz, Miguel Angel, 1373, 1524
¿Vecinos indiferentes?: el Caribe de habla inglesa y América Latina, 4157
Veeris, Milena, 5128
Vega, Abel U. de la, 3583
Vega, Bernardo, 3059, 4134–4135
Vega, Celsa, 4983
Vega, Luis, 3492
Vega Carballo, José Luis, 4684
Vega Centeno, Máximo, 1965
Vega-Centeno B., Imelda, 3415
Vega Sosa, Constanza, 458
Vega Vega, Juan, 1714
Vegas Torres, José Martín, 3416
Veiga, Pedro da Motta, 4389
Veiga, Sandra Mayrink, 3777
Veillon, Jean-Pierre, 2450
25 años, encuesta industrial, 1752
Velandia Jagua, César Augusto, 637
Velarde, Julio, 4945
Velarde, Patricio, 3244
Velasco, Andrés, 1295
Velasco L., Mónica, 1994
Velásquez, Andrés, 3275
Velásquez, Ramón J., 3276
Velázquez, Efraín J., 1753
Velázquez Gutiérrez, Luis Arturo, 1398
Velázquez Hernández, Emilia, 2934
Velázquez-Mainardi, Miguel Angel, 3060
Velde, P. van de, 520
Vélez, Félix, 1345
Vélez de Piedrahita, Rocío, 3208
Vélez R., Humberto, 3209

Velho, Otávio, 5209
Velis Meza, Héctor, 1921
Véliz, Vito, 400
Vellinga, Menno, 2801, 3865
Velloso, João Paulo dos Reis, 2140–2142, 3751, 4351, 5191
Veloz Maggiolo, Marcio, 501, 517–519
Veltmeyer, Henry, 5060
Venegas, Sylvia, 1912
Venezuela. Comisión Presidencial para la Reforma del Estado, 1728, 3250
Venezuela. Ministerio de Justicia, 4843
Venezuela. Ministerio de Relaciones Exteriores, 4337
Venezuela. Oficina Central de Estadística e Informática, 1747, 1752
Venezuela: health sector review, 1754
Venezuela 2000: education for growth and social equity, 1755
Veni, George, 212
Venturini, Angel R., 3664
Vera Bolaños, Marta G., 4526
Verbitsky, Horacio, 3584–3585
La verdad del '93: paz, derechos humanos y violencia, 3210
Verdesoto, Luis, 3234, 3354
Verduga, César, 3245
Verduzco Chávez, Basilio, 1525
Verduzco Igartúa, Gustavo, 1526
Verea, Mónica, 3912, 4048
Vergara, Dulce Helena, 2210
Vergara, Pilar, 5080
Vergara, Rafael, 3211
Vergara, Ricardo, 4924
Verhine, Robert Evan, 2211
Veríssimo, Adalberto, 2272, 2718
Verkoren, Otto, 4424
Verlin, Evgueni, 1922
Verna, Maria Alejandra, 931
Verón, Carlos, 2004
Verschoor, Gerard, 2959
Versteeg, Aad H., 485, 520–521, 650–652
Vertovec, Steven, 845
Vetencourt Guerra, Lola, 1756
Vial, Joaquín, 1109
Vianna, Luiz Werneck, 3778
Vianna, Maria Lúcia Teixeira Werneck, 2212
Vicaría de la Solidaridad: historia de su trabajo social, 5081
Vicariato de Pando (Bolivia), 4971
Vicencio Acevedo, Gustavo, 2935
Vicencio Tovar, Abel, 2936
Vickers, William T., 925–926
Victoria, José Guadalupe, 378
Vicuña Izquierdo, Leonardo, 1844

Vidales, Carlos, 3868
Vidas em risco: assassinatos de crianças e adolescentes no Brasil, 5210
Videla, Pedro, 1637
Vidigal, Armando Amorim Ferreira, 4390
Vidrio, Martha, 4562
Vieira, Eurípedes Falcão, 2719
Vieira, José Ribas, 5183
Vieira, Luiz, 2213
Vieira, Paulo Freire, 2720
Vieira, Renata, 2213
Viera-Gallo, José Antonio, 3493–3494
Viertler, Renate Brigitte, 887
Vigorización de la chacra andina, 938
Vilas, Carlos M., 2805, 4685–4686
Vildoso Chirinos, Carmen, 4946
Villacrés Moscoso, Jorge W., 4290–4291
Villagra, María Susana, 3604
Villagrán de Brady, Sandra, 155
Villalba, Ricardo, 3024
Villalonga, Julio, 4192
Villamán P., Marcos, 4797–4798
Villamonte Blas, Ricardo N., 1296
Villanueva, Javier, 1125
Villar, Luis del, 2937
Villarán de la Puente, Fernando, 1966–1967
Villarreal Méndez, Norma, 4829
Villarreal P., Alonso, 4073
Villars, Rina, 2991
Villas Bôas, Claudio, 888
Villas Bôas, Orlando, 888
Villasuso Estomba, Juan Manuel, 1566, 4645
Villavicencio, Judith, 4563
Villers Ruiz, Lourdes, 2412
Vindel de Cálix, Zonia, 1617
Vines, David, 1078
Vinholes, Luiz Carlos, 4391
Vinhosa, Maria Celina Arraes, 4187
Vinocur, Pablo, 5108
La violación sexual en el Paraguay: aspectos psicológico, social y jurídico, 4993
Violence in the Andean region, 4452
Violencia y pacificación en 1991, 3417
Violencias encubiertas en Bolivia, 3355, 4979
Viramontes, Carlos, 170
Viscarra Pando, Ruddy, 3285, 4229
Visintini, Alfredo, 2092
Vivanco, Cirilo, 708
Vlach, Norita, 4687
Vogt, Evon Zartman, 768
Vol'skiĭ, Viktor Vatslavovich, 3856
Von Braun, Joachim, 1614
Von Winning, Hasso, 401–402
Vonós a la capital: estudio sobre la emigración rural reciente en Guatemala, 4688

Vourc'h, Ann, 1330
Vries, Jaap de, 4947
Vries, Peter de, 1587
La vulnerabilidad de los hogares con jefatura femenina: preguntas y opciones de política para América Latina y el Caribe, 4453
Vuskovic, Pedro, 1297
Vusković Céspedes, Pedro, 4074
Wachtel, Nathan, 967–968
Wade, Peter, 1005–1006, 4830
Wagenaar Hummelinck, Pieter, 522
Wagner, Erika, 717
Wahl Kleiser, Lissie, 863
Waksman, Guillermo, 3665
Walker, Thomas W., 4045
Wallace, Elizabeth, 4158
Walter, Knut, 2972, 3015
Walter, Véronique, 523
War on drugs: studies in the failure of U.S. narcotics policy, 3875
Ward, Peter M., 2914
Warner, Andrew M., 1298
Warren, Kay B., 769–770
Washington Office on Latin America, 2812, 2885, 3593, 3605, 3876–3877, 4319
Watanabe, Luis K., 707
Watson, Hilbourne A., 1113, 4078, 4102
Watson, Patrick, 1640
Watson, Rodney, 2413
Watters, R.F., 2515
Weaver, Frederick Stirton, 4689
Weaver, Muriel Porter, 171
Webb, Michael, 1588
Weber, Gesine, 308
Weber, Jutta, 4980
Webster, David, 172, 403
Weder, Beatrice, 1103
Weeks, John, 1548, 4605
Weigand, Phil C., 92, 116
Weil, Connie, 2281, 4981
Weil, Jim, 4981
Weinberg, Bill, 2943
Weinberger, Mary Beth, 4454
Weinberger V., Karen, 1941
Weinstein, Brian, 3068
Weinstein, José, 5082
Weinstein C., Marisa, 5078
Weintraub, Sidney, 1299, 3960–3961
Weis, W. Michael, 4392
Weiss, John, 1527
Weiss, Wendy A., 1025
Weisskoff, Richard, 1300
Wekker, Gloria, 4799
Welch, David A., 4106
Weldon, Jeffrey, 2891

Weller, Jürgen, 1560
Welsch, Friedrich, 3255
Werneck, Rogério L.F., 2124
Wertime, Richard A., 404
Wesche, Rolf, 2271, 2484, 2721
West, Peter J., 1301
West, Robert Cooper, 2414
West Indian Commission, 1649, 1655, 3038
Wettmann, Reinhart W., 1937
Weyland, Kurt, 3779
Whalley, John, 1302
Wheeler, David, 1098
Wheeler, Jane C., 2418
Whitaker, Morris D., 1838, 2482
Whitecotton, Joseph W., 459
Whiteford, Scott, 2282
Whitehead, Laurence, 2806
Whitehead, Neil L., 524, 846
Whiting, Van R., 1528
Whitmore, Thomas M., 173, 2415–2416
Whitten, Dorothea S., 927
Whitten, Norman E., 927, 4877
Wiarda, Howard J., 2789, 3878
Wickham-Crowley, Timothy P., 2807–2808
Widmer, Randolph J., 405
Wijnbergen, Sweder van, 1322, 1423–1425
Wild Majesty: encounters with Caribs from Columbus to the present day; an anthology, 524
Wiley, David, 2282
Wilhelmy von Wolff, Manfred, 3821, 3879
Wilke, Jürgen, 2774
Wilkerson, S. Jeffrey K., 174
Wilkie, James Wallace, 1414
Wilkinson, Michael D., 4264
Wille Trejos, Alvaro, 2340
Willey, Gordon, 175
Williams, Barbara J., 2362
Williams, Denis, 653
Williams, Glen, 4615
Williams, Lynden S., 2283
Williams, Philip J., 2972
Williamson, John, 1240
Willis, Edwin O., 2722
Willmore, Larry, 1567–1568
Willumsen, Maria José F., 2214
Wilson, Fiona, 1529
Wilson, Judith Kay, 759, 3007
Wilson, Paul N., 1530
Wilson, Richard, 771
Wilson, Samuel M., 525
Wilson, Suzanne, 3880
Wiltshire, Rosina, 4800
Winant, Howard, 5211
Winfield Capitaine, Fernando, 460

Wing, Elizabeth S., 526
Winkler, Donald R., 1218
Winograd, Alejandro, 2564
Winograd, Carlos D., 2209
Winter, Carolyn, 1757
Winter, Marcus, 127, 329, 391
Winters, Cecilia Ann, 1531
Wionczek, Miguel S., 1504
Wise, Carol, 932
Wistat: women's indicators and statistics database, 47
Witte, Lothar, 1224
Woldenberg, José, 4564
Wolves from the sea: readings in the anthropology of the native Caribbean, 846
Women in Brazil, 5212
Women in the Americas: bridging the gender gap, 1303
Won Choi, Dae, 1304
Wong, Rebecca, 1532
Wong-González, Pablo, 1533
Wood, Charles H., 3780, 5213
Wood, Christopher, 1534
Wood, Stephanie, 2417
Wooding, Charles J., 4801
Workshop El Acuerdo de Libre Comercio México-Estados Unidos y Repercusiones en la Frontera, *Tijuana, Mexico, 1991*, 1535
World Bank, 48, 1486, 1592, 1769, 2038, 2041
World Data, 48
World databases in social sciences, 2
World Health Organization, 1165
World marketing data and statistics, 49
World News Connection, 50
World Trade Database = Base de donées sur le commerce mondial, 51
Woroniuk, Beth, 1303
Worrell, DeLisle, 1656, 1679
Wortman, Ana, 5129
Wright, Robin R., 889
Wright, Thomas C., 3881
Wurgaft, José, 1305
Wust, Irmhild, 595
Wylie, Jonathan, 847
Wylie, Kenneth, 2282
Wynia, Gary W., 2809, 3586
Ya nunca me verás como me vieras, 5083
Yamaguchi, Yutaka, 1306
Yanes, Hugo, 5023
Yáñez Rojas, Eugenio, 3495
Yapita, Juan de Dios, 947
Yariv, Danielle, 4075
Ycaza Cortez, Patricio, 3246
Yearwood, Gladstone Lloyd, 3074
Yeats, Alexander J., 1156

Yelvington, Kevin A., 844, 848–849
Yepes, Ernesto, 1968
Yong Chacón, Marlon, 1566
Yopo H., Boris, 3882
Yopo H., Mladen, 4296
Yore, Fátima Myriam, 4987
Yoshio, Onuki, 545
Young, Frank, 5084
Young, Kenneth R., 2516–2517
Young, Linda Wilcox, 1536
Young, Virginia Heyer, 850
Younger, Stephen D., 1845
Youngers, Coletta, 4319
Yukihiro, Takahashi, 3052
Yúnez-Naude, Antonio, 1537
Zaglul, Jesús M., 4802
Zahler, Roberto, 1084
Zaĭt͡sev, Nikolaĭ Grigor'evich, 4258
Zalles Cueto, Alberto Augusto, 4982
Zambrano, Marta, 3880
Zamora, Gerardo, 4565
Zantwijk, Rudolf A.M. van, 176, 461
Zapata Novoa, Juan, 4566
Zapata Pericón, Hugo, 3356
Zapotec struggles: histories, politics, and representations from Juchitán, Oaxaca, 772
Zárate Márquez, Eduardo, 1503
Zárate Morán, Roberto, 406
Zavala, Juan Ovidio, 3587
Zavaleta Mercado, René, 3357, 4950
Zeballos Hurtado, Hernán, 2001
Zeidler, James A., 646
Zeitlin, Judith Francis, 462
Zeitlin, Robert N., 177, 407
Zelaya, Ricardo, 3311
Zeller, Ludwig, 391
Zemelman, Hugo, 5085
Zepeda Miramontes, Eduardo, 1538, 3962
Zerda Sarmiento, Alvaro, 1784
Zermeño, Sergio, 3963, 4522, 4567
Zimbalist, Andrew, 1715, 3151
Zimmerer, Karl S., 2518–2520, 2533
Zimmermann, Klaus, 773
Zockun, Maria Helena Garcia Pallares, 2215
Zoghbi Pérez, Jorge Alberto, 3964
Zovatto, Daniel, 2735
Zubek, Voytek, 2858
Zubirán Schadtler, Carlos, 1496
Zucchi M., Alberta, 907
Zulawska, Ursula, 1307
Zuleta-Puceiro, Enrique, 3588
Zuluaga Gómez, Víctor, 1007
Zúñiga F., Norberto, 1589–1590
Zur Mühlen, Patrik von, 4395
Zurita, Dante, 4948
Zuvekas, Clarence, 1569